TABLE OF CONTENTS

1. A Brief History of Microbiology 1
2. The Chemistry of Microbiology 27
3. Cell Structure and Function 58
4. Microscopy, Staining, and Classification 96
5. Microbial Metabolism 124
6. Microbial Nutrition and Growth 161
7. Microbial Genetics 193
8. Recombinant DNA Technology 237
9. Controlling Microbial Growth in the Environment 262
10. Controlling Microbial Growth in the Body: Antimicrobial Drugs 288
11. Characterizing and Classifying Prokaryotes 324
12. Characterizing and Classifying Eukaryotes 353
13. Characterizing and Classifying Viruses, Viroids, and Prions 384
14. Infection, Infectious Diseases, and Epidemiology 413
15. Innate Immunity 446
16. Adaptive Immunity 472
17. Immunization and Immunoassays 503
18. Immune Disorders 525
19. Pathogenic Gram-Positive Bacteria 546
20. Pathogenic Gram-Negative Cocci and Bacilli 586
21. Rickettsias, Chlamydias, Spirochetes, and Vibrios 620
22. Pathogenic Fungi 647
23. Parasitic Protozoa, Helminths, and Arthropod Vectors 675
24. Pathogenic DNA Viruses 709
25. Pathogenic RNA Viruses 734
26. Applied and Industrial Microbiology 778
27. Microbial Ecology and Microbiomes 802

VIDEO TUTORS

1. The Scientific Method
2. The Structure of Nucleotides
3. Bacterial Cell Walls
4. The Light Microscope
5. Glycolysis, Electron Transport Chains
6. Bacterial Growth Media
7. Initiation of Translation, Elongation in Translation
8. Action of Restriction Enzymes
9. Principles of Autoclaving
10. Actions of Some Drugs That Inhibit Prokaryotic Protein Synthesis
11. Arrangements of Prokaryotic Cells
12. Principles of Sexual Reproduction in Fungi
13. The Lytic Cycle of Viral Replication
14. Some Virulence Factors
15. Inflammation
16. Clonal Deletion, Processing of Antigen
17. ELISA
18. Hemolytic Disease of the Newborn
19. Diseases in Depth: Necrotizing Fasciitis, Listeriosis, Tuberculosis
20. Disease in Depth: Bacterial Urinary Tract Infections
21. Disease in Depth: Rocky Mountain Spotted Fever
22. Disease in Depth: Candidiasis
23. Diseases in Depth: Giardiasis, Malaria
24. Disease in Depth: Papillomas
25. Diseases in Depth: Ebola, Influenza

Explore the Invisible World of Microbiology and Why it Matters to Human Life

***Microbiology With Diseases by Taxonomy*, Sixth Edition** encourages a deep understanding of why microbiology matters to health as well as disease. Just-in-time remediation and checkpoints throughout the eText help students to progress in their understanding of course concepts. Students expand their learning in Mastering Microbiology with Interactive Microbiology case studies which give students the opportunity to practice making predictions and observing outcomes. The **Sixth Edition** also includes discussions about emerging research on recombinant DNA and CRISPR.

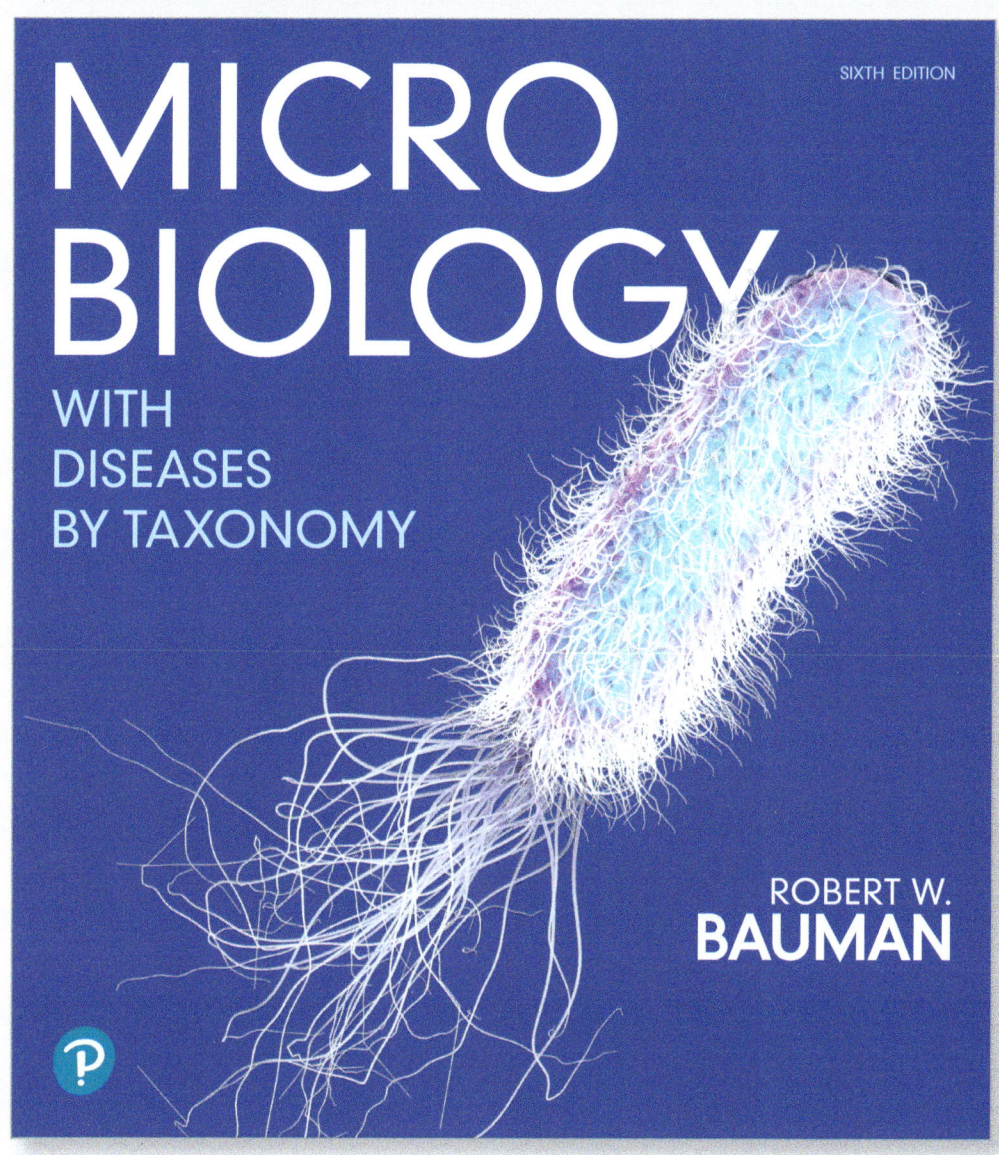

Maximize Learning with Effective Study Tools

Before You Begin
1. Which cellular structure is present in eukaryotes and not in prokaryotes?
2. Which of the following organisms are classified as eukaryotes: archaea, bacteria, fungi, helminths, and protozoa?
3. What are the differences between the structure of DNA in eukaryotic nuclei and that in bacteria?
4. What organelle generates most of the energy in eukaryotic cells?
5. What is the difference between eukaryotic and prokaryotic flagella?

NEW! Before You Begin Questions help students revisit critical chemistry and biology knowledge for their success in this course. Discover them in Chapter Openers and in Mastering Microbiology where they are interactive providing feedback with embedded hints. This just-in-time remediation allows students to fill their own skill gaps before reading the chapter.

NEW! Micro Checks allow students to assess their understanding as new concepts are presented. These formative assessments are also assignable in Mastering Microbiology and include animations and videos to reinforce concepts and fill in skill gaps.

MICRO CHECK
1. During which reproductive process are gametes formed?
2. What is the difference between the results of mitosis and meiosis?
3. How many daughter nuclei are produced by mitosis in contrast to the number produced by meiosis?
4. How are coenocytes formed?

Most favored by students! 60% of students use Dynamic Study Modules . . . even when not assigned for test preparation. Adaptive learning technology from the most recent cognitive science allow students to practice their knowledge of course topics in a fun and positive format, building confidence. Topics progress in difficulty based on Bloom's Taxonomy, and students are provided just-in-time learning resources to review where needed.

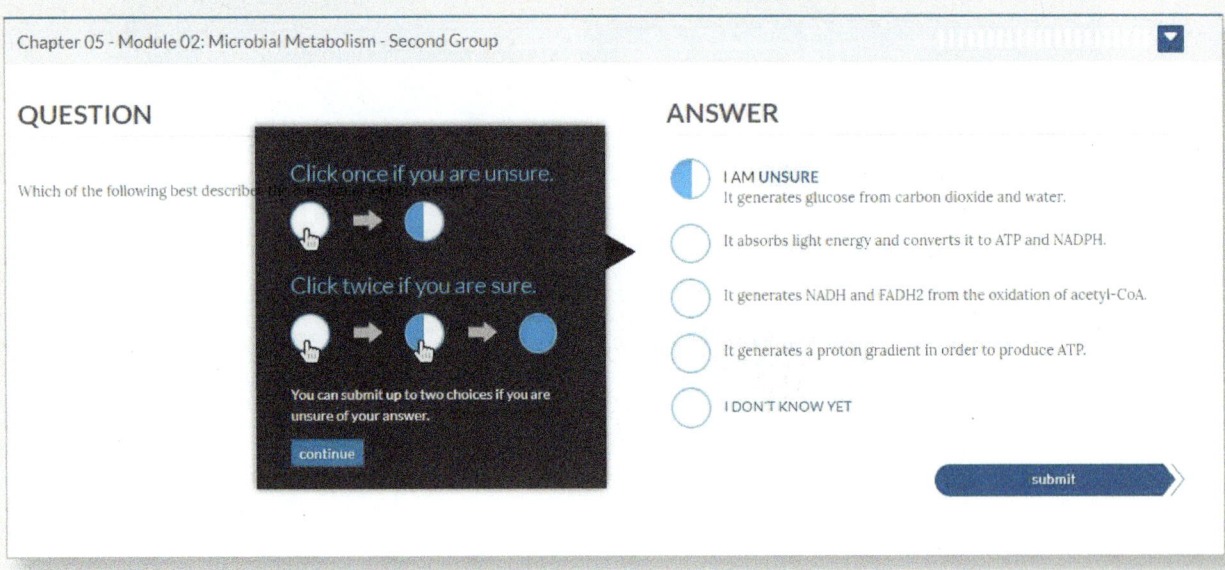

Connect Lecture and Lab

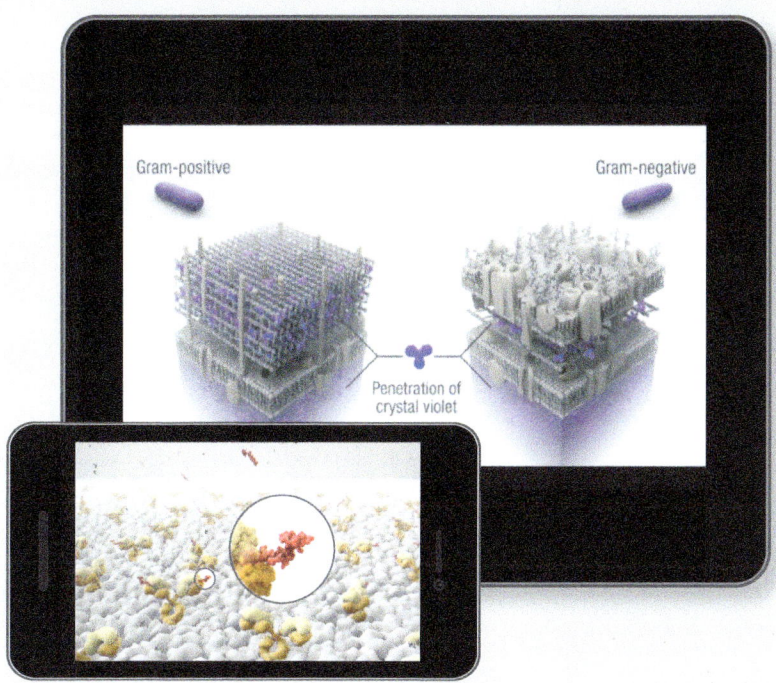

Micro Lab Tutor Videos and Coaching Activities help instructors and students get the most out of lab time. Students can practice their lab skills virtually, reviewing proper lab techniques with real-world applications. Live action video combined with molecular animation with assessment and feedback coach students how to interpret and analyze different lab results.

Lab Technique Videos give students an opportunity to see techniques performed correctly and quiz themselves on lab procedures both before and after lab time, improving confidence and proficiency. Assign as pre-lab quizzes in Mastering Microbiology and include coaching and feedback on theses lab techniques:

- NEW! The Scientific Method
- NEW! How to Write a Lab Report
- Acid-Fast Staining
- Amylase Production

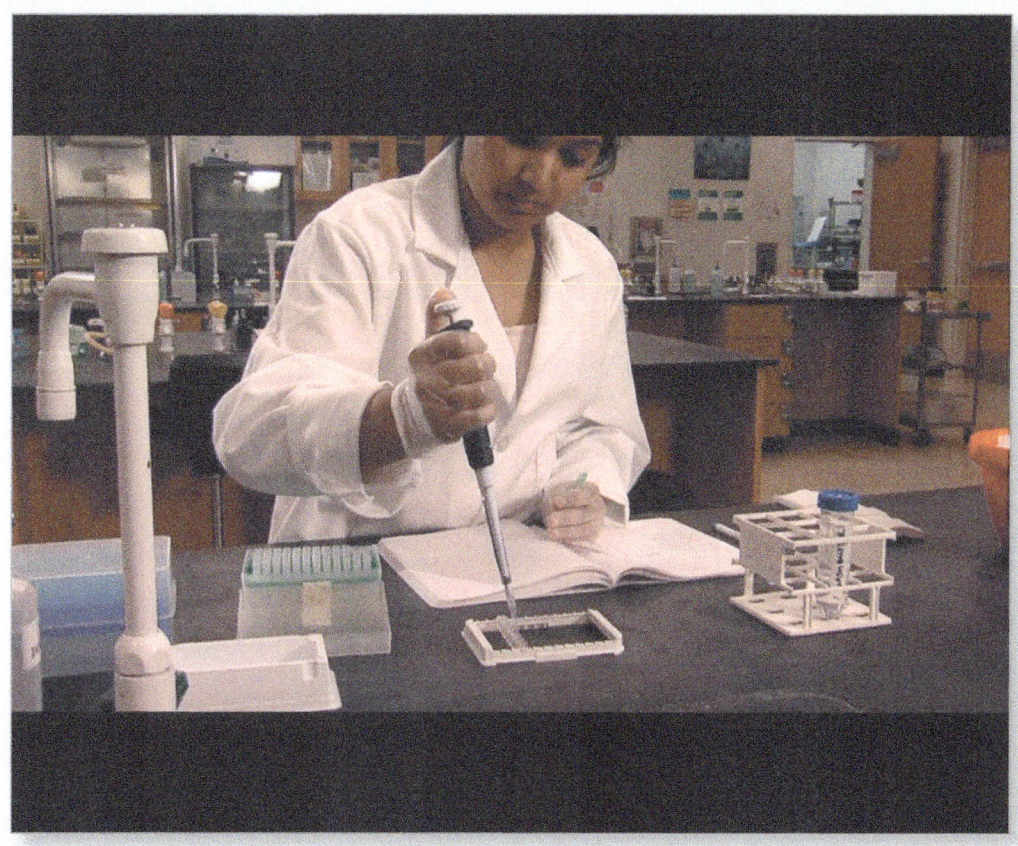

Why Microbiology Matters in Health Care

MICRO IN THE CLINIC
Another Day Care Cold?

CLAIRE IS A YOUNG WOMAN TRYING to get used to life as a single parent. Since her husband died, she's had to find a new place to live and a new day care for her three-year-old daughter, Hailey. It's been a hard transition, but Claire is starting to feel like things are back on track.

Early one afternoon, Claire receives a call from the day care—Hailey has a fever and a severe cough that almost seems like she cannot breathe. Claire rushes to the day care to pick her up as soon as possible. When she arrives, one of the caregivers tells Claire that Hailey had been a bit fussy and cranky in the morning before she worsened. She also said that a couple of other kids had been sick earlier in the week with a similar cough. Claire takes Hailey home.

After a few days, Claire is more concerned—Hailey has a fever, has started coughing more frequently, and seems to be

Micro In the Clinic proposes real clinical scenarios at the beginning of each chapter. Students remain curious and engaged throughout the chapter as they learn the microbiology behind the case, and the chapter closing **Micro In the Clinic Follow-Up** provides the case result.

MICRO IN THE CLINIC FOLLOW-UP
Another Day Care Cold?

Dr. Fischer completes a preliminary examination of Hailey. She learns that Claire and her husband had not felt childhood immunizations were necessary for Hailey. Based on Hailey's symptoms, in particular her acute cough that has been occurring for about a week with the distinctive coughing sound upon inspiration, she believes that Hailey has pertussis, otherwise known as whooping cough. Dr. Fischer collects a nasopharyngeal swab to send for laboratory confirmation.

Dr. Fischer tells Claire that Hailey most likely became infected at day care. It is a highly infectious disease that is transmitted via droplets from infected persons. Because Hailey isn't immunized against the bacterium that causes the disease, *Bordetella pertussis*, she was susceptible to infection.

The lab analysis confirms Dr. Fischer's diagnosis. Because Hailey's pertussis was diagnosed relatively early, Dr. Fischer prescribes a course of azithromycin: 500 mg the first day, then 250 mg for four days. After completing five days of antibiotics, Hailey should be able to safely return to day care without spreading the disease further.

Pertussis is a mandatory reportable disease to the state and local health departments, so Dr. Fischer reports the case to her local health department. The other sick kids at the day care also had whooping cough—none of them had been immunized against pertussis.

1. Explain how *B. pertussis* affects the trachea and how this damage leads to the characteristic cough associated with pertussis.
2. Hailey's infection could have been prevented with the DTaP vaccine. Hailey's parents' argument against the vaccine was that immunization was a personal choice that affected only Hailey. As a future health care professional, do you agree with Claire's argument? Why or why not?

Check your answers to Micro in the Clinic Follow-Up questions in the Mastering Microbiology Study Area.

SOLVE THE PROBLEM

Rights versus Responsibilities

In the early 20th century, almost all children contracted measles, with 4 million new infections per year, resulting in approximately 50,000 hospitalizations and 6,000 deaths. But in 2000, the Centers for Disease Control and Prevention (CDC) declared measles eliminated (defined by lack of disease transmission for a year) in the United States. This medical advance capped a campaign of immunization that began in 1963. However, multiple measles outbreaks, beginning in 2014, resulted in the greatest number of cases of measles in the United States since 1995. How did measles reemerge? Several factors contributed.

Measles is still common in other parts of the world, and travelers can bring the measles virus here. A measles outbreak occurs when there are significant numbers of unimmunized people in a community. When the immunization rate in a community is above 95 percent, even unimmunized individuals are protected because susceptible people are too rarely in contact with infected individuals to allow disease transmission. This effect is called herd immunity. The currently recommended vaccine against measles is MMR, which protects against measles, mumps, and rubella. One dose is to be given at age 12–15 months, followed by a booster at 4–6 years, and a second booster between 19 and 40.

There is no controversy in the scientific community on MMR efficacy: the vaccine is highly effective. However, there is an ethical dilemma between the individual **rights** of people to refuse immunization as compared to their **responsibility** toward public health, since lack of immunization can allow the virus to spread, resulting in outbreaks.

If you were on a panel advising the CDC, how would you answer these questions?

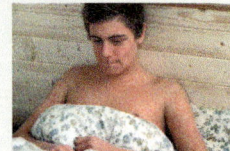

1. Should we force people to be immunized?
2. What procedures should be used to ensure the highest immunization rates?
3. What other facts would help make decisions related to immunization?

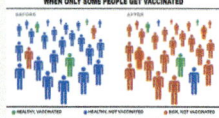

Solve the Problem boxes are problem-based learning explorations of current microbiological challenges. Paired active learning instructor activities and Mastering Microbiology assessments encourage critical thinking.

Sweat the Small Stuff

CLINICAL CASE STUDY

Evaluating an Abnormal CBC

CBC Profile			
Name: Brown, Roger	Age/Sex: 61/M		Attend Dr: Kevin, Larry
Acct#: 04797747	Status: ADM IN		
Reg: 11/27/12			
SPEC #: 0303:AS:H00102T	COLL: 12/03/12-0620	STATUS: COMP	
	RECD: 12/03/12-0647	SUBM DR: Kevin, Larry	
ENTERED: 12/03/12-0002		OTHER DR: NONE, PER PT	
ORDERED: CBC W/ MAN DIFF		REQ #: 01797367	

Test	Result	Normal range
CBC		
WBC (white blood cells)	0.8	4.8–10.8 K/mm3
RBC (red blood cells)	3.09	4.20–5.40 M/mm3
HGB (hemoglobin)	9.6	12.0–16.0 g/dL
HCT (hematocrit)	28.2	37.0–47.0 %
MCV	91.3	81.0–99.0 fL
MCH	31.1	27.0–31.0 pg
MCHC	34.1	32.0–36.0 g/dl
RDW	17.1	11.5–14.5 %
PLT (platelets)	21	150–450 K/mm3
MPV	8.7	7.4–10.4 fL
DIFF		
CELLS COUNTED	100	#CELLS
SEGS	39	
BAND	4	
LYMPH (lymphocytes)	41	
MONO (monocytes)	15	
EOS (eosinophils)	1	
NEUT# (# neutrophils)	0.3	1.9–8.0 K/mm3
LYMPH#	0.3	0.9–5.2 K/mm3
MONO#	0.1	0.1–1.2 K/mm3
EOS#	0.0	0–0.8 K/mm3
PLATELET EST	DECREASED	

Roger Brown, an African American cancer patient, received a chemotherapeutic agent as a treatment for his disease. The drug used to destroy the cancer also produced an undesirable condition known as *bone marrow suppression*. The complete blood count (CBC) profile shown here indicates that this patient is in trouble. Review the lab values, and answer the following questions.

1. Note that the platelet count is very low. How does this

Clinical Case and Emerging Disease Case Studies discuss a patient's experience with a microbial disease and conclude with questions for the student to consider.

EMERGING DISEASE CASE STUDY

Chikungunya

An old man arrived at the doctor's office in Ravenna, Italy, with a combination of signs and symptoms the physician had never heard of: a widespread, itchy rash covering both arms and trunk; difficulty in breathing; high fever; nausea; and extreme joint pain. Chikungunya (chik-en-gun'ya) had arrived in Europe. Though scientists had known of chikungunya virus, which is related to equine encephalitis viruses, for over 50 years, most considered the tropical disease benign—a limited, mild irritation, not a catastrophe. Therefore, few researchers studied chikungunya virus or its disease. Now, they know better.

Over the past decade, chikungunya virus has spread throughout the nations of the Indian Ocean and across Africa. In 2006, officials on the French-owned island of La Réunion in the Indian Ocean reported 47,000 cases of chikungunya in a single week! That same year chikungunya reemerged in India for the first time in four decades with more than 1.5 million reported cases, and in 2010 it emerged in China. Why?

Aedes albopictus (Asian tiger mosquito), one species of *Aedes* that carries the virus, has moved into temperate climates, including Europe and the United States, as the climate has warmed. With the mosquito comes the possibility of viral proliferation—the insects have spread the tropical disease as far north in Europe as France.

And our Italian patient? His crippling pain lasted 10 days, but he survived. Now that he knows about mosquito-borne chikungunya, he insists that his family and friends use mosquito repellent liberally. Officials in the rest of Europe and in the United States join in his concern: With the coming of *Ae. albopictus*, is incurable chikungunya far behind?

Crystal of chikungunya viruses.

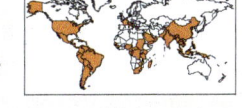

1. Besides using repellents, how can people protect themselves from mosquito-borne pathogens?
2. Why is *Aedes* commonly known as the tiger mosquito?
3. Why aren't antibiotics such as penicillin, erythromycin, and ciprofloxacin effective in preventing and treating chikungunya?

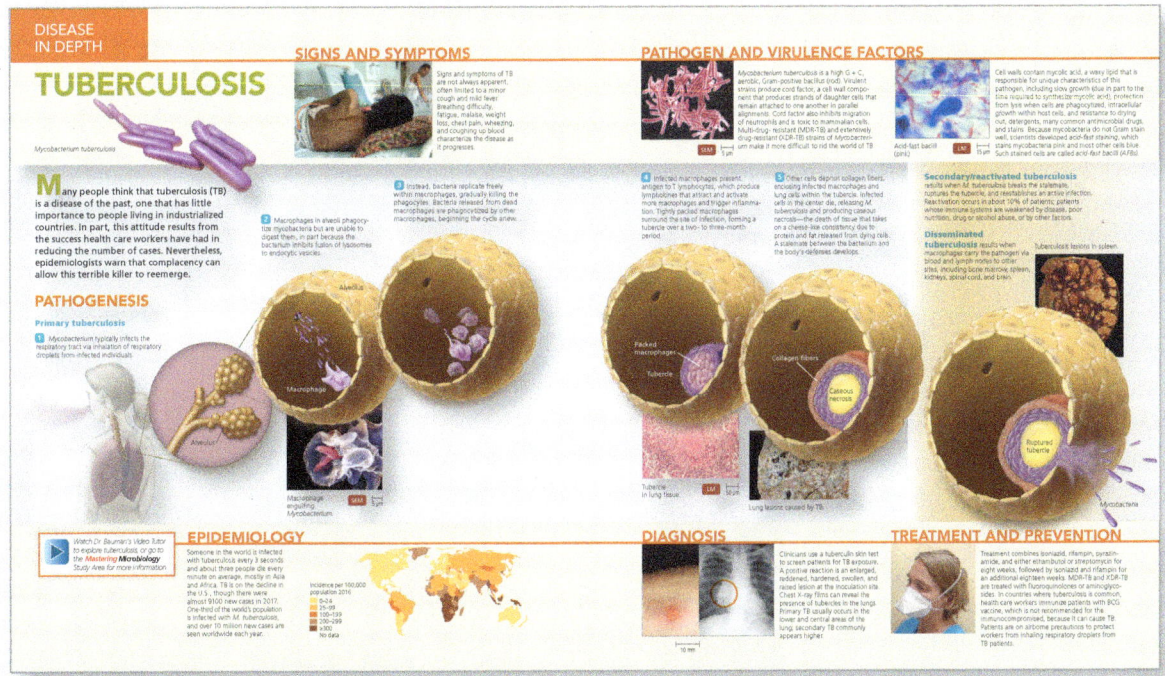

Disease in Depth Eleven complex diseases are presented in a visual layer highlighting the entire cycle of the disease encouraging students to think critically. Accompanying Video Tutors are provided in Mastering Microbiology.

Expand and Apply Learning in Mastering Microbiology™

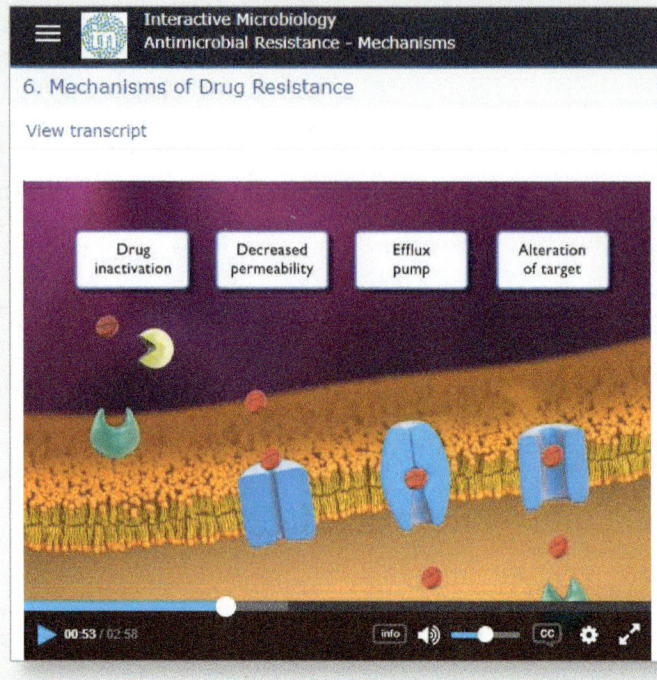

NEW! Interactive Microbiology is a dynamic suite of interactive tutorials and animations that present concepts within a real helathcare scenario emphasizng problem solving. Topics include Antimicrobial Resistance, Human Microbiota, Operons, Biofilms and Quorum Sensing, Complement, and more.

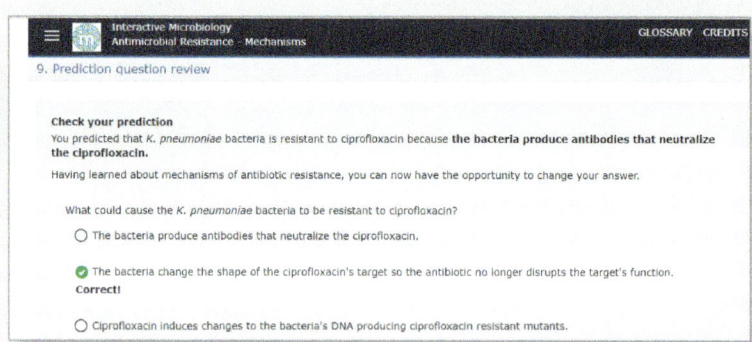

Make and Check Predictions. Formative and summative assessments encourage students to think critically as they make and check predictions.

Predict Outcomes in a fun and engaging format! Students are presented with a clinical scenario that allows them to manipulate variables that change outcomes in a fun and engaging learning activity.

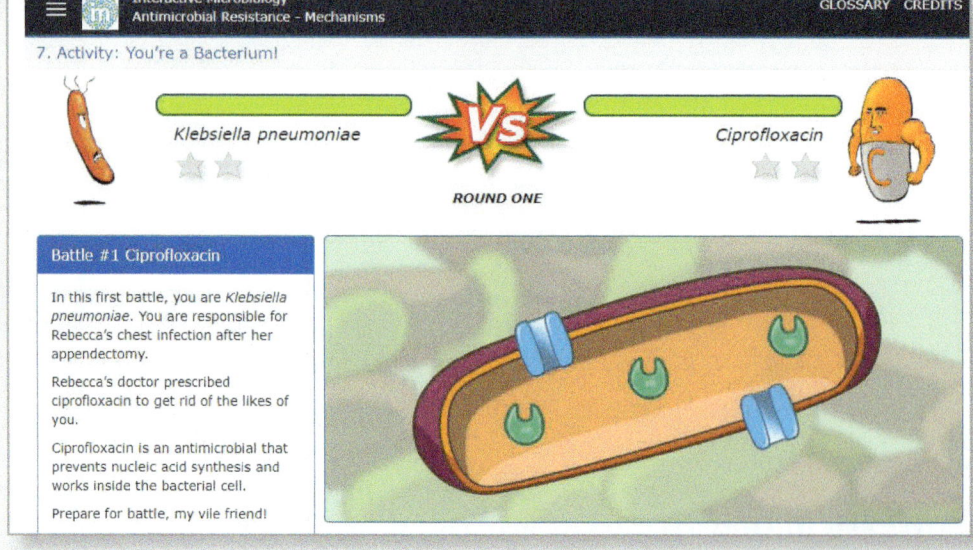

Give Students Anytime, Anywhere Access with the Pearson eText App

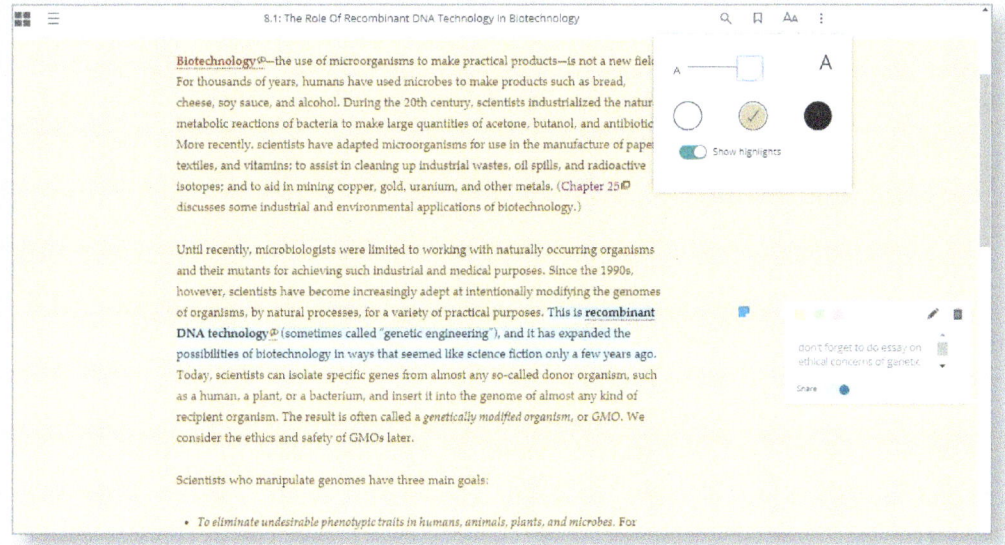

Pearson eText is a simple-to-use, mobile-optimized, personalized reading experience available within Mastering. It allows students to easily highlight, take notes, and review key vocabulary all in one place—even when offline. Seamlessly integrated videos and other rich media engage students and give them access to the help they need, when they need it. Pearson eText is available within Mastering when packaged with a new book; students can also purchase *Mastering with Pearson eText* during the student registration process.

Instructors can rearrange the table of contents at the chapter and section level to match the way they teach. It's easy for instructors to share their notes and highlights with students to add their personal teaching style to important topics, call out need-to-know information, or clarify difficult concepts. Further, the dashboard lets educators see how students are working in their eTextbook so that they can plan more effective instruction in and out of class.

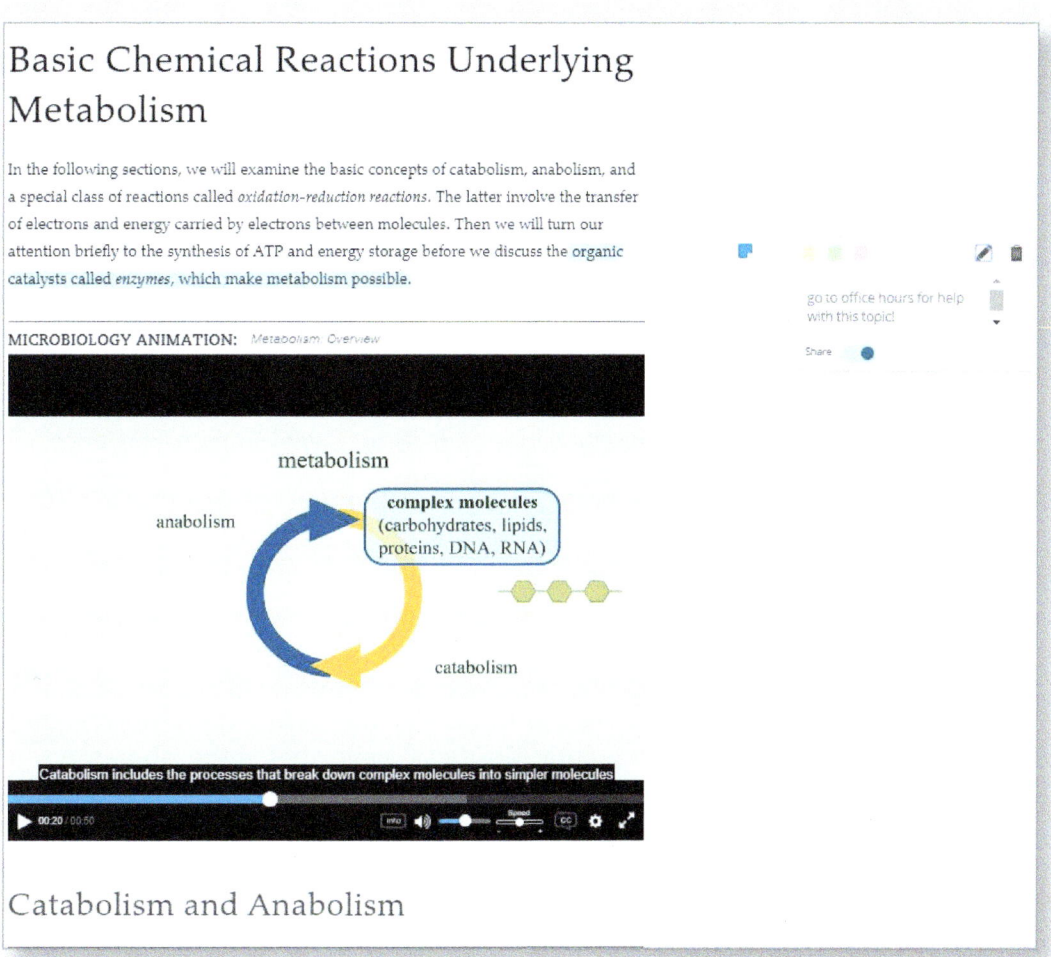

The *Very Best* Instructor Resources and Support

The following resources can be found within a Mastering Microbiology instructor account and are organized by chapter:

- Learning Catalytics is a "bring your own device" student engagement, assessment, and classroom intelligence system. With Learning Catalytics, instructors can assess student understanding in real time using open-ended tasks to probe student understanding. Learning Catalytics is included in a Mastering with eText student subscription.
- All fgures, photos, and tables from the textbook in both labeled and unlabeled formats
- Test Gen test bank®
- MicroFlix Animations™, Video Tutors, Lab Technique videos and more
- Instructor's Guide
- Step-edit PowerPoint® presentations that present multi-step process figures step by step

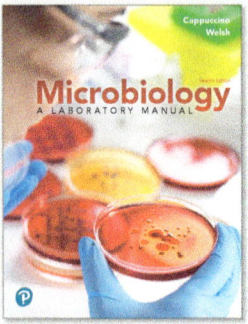

Cappuccino/Welsh's *Microbiology: A Laboratory Manual, 12e* is easy to adapt for almost any microbiology lab course. This versatile, comprehensive, and clearly written manual is competitively priced and can be paired with any undergraduate microbiology text. Known for its thorough coverage, straightforward procedures, and minimal equipment requirements, this lab manual incorporates current safety protocols from governing bodies such as the EPA, ASM, and AOAC.

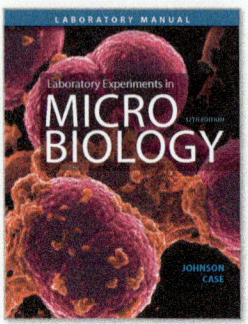

Johnson/Case's *Laboratory Experiments in Microbiology, 12e* features 57 thoroughly class-tested and easily customizable exercises that teach basic microbiology techniques and applications. The manual provides comprehensive coverage of every area of microbiology for undergraduate students across diverse disciplines, including the biological sciences, allied health sciences, agriculture, environmental science, nutrition, pharmacy, and various pre-professional programs.

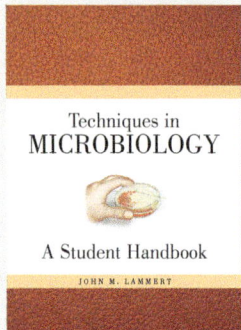

Lammert's *Techniques in Microbiology* is visual and incorporates "voice balloons" that keep the student focused on the process described. The techniques are those that will be used frequently for studying microbes in the laboratory, and include those identified by the American Society for Microbiology in its recommendations for the Microbiology Laboratory Core Curriculum (recommendations in which the author participated).

SIXTH EDITION

MICRO BIOLOGY

WITH DISEASES BY TAXONOMY

ROBERT W. BAUMAN, PH.D.
Amarillo College

Contributions By: Todd P. Primm, Ph.D.
Sam Houston State University

Asha Brunings, Ph.D.
Santa Fe College

Terry Austin
Temple College

Elizabeth Machunis-Masuoka, Ph.D.
University of Virginia

Amy M. Siegesmund, Ph.D.
Pacific Lutheran University

Clinical Consultants: Jordan A. Roeder MS, AGACNP-C, CCRN, CVRN-BC

Cecily D. Cosby, Ph.D., FNP-C, PA-C

Courseware Portfolio Manager: *Jennifer McGill Walker*
Director of Portfolio Management: *Serina Beauparlant*
Content Producer: *Cheryll Linthicum*
Managing Producer: *Nancy Tabor*
Courseware Director, Content Development: *Barbara Yien*
Development Editor: *Marie Beaugureau and Laura Cheu*
Courseware Analyst, Art: *Jay McElroy*
Courseware Editorial Assistants: *Katrina Taylor and Jordan Barron Goldman*
Rich Media Content Producer: *Lucinda Bingham and Tod Regan*
Full-Service Vendor: *Pearson CSC*
Copyeditor: *Anita Hueftle*
Art Coordinator: *Morgan Ewald, Lachina Creative*

Design Manager: *Mark Stuart Ong, Side By Side Studios*
Interior Designer: *Preston Thomas, Cadence Design Studio*
Cover Designer: *Preston Thomas, Cadence Design Studio*
Illustrators: *Lachina Creative*
Rights & Permissions Project Manager: *Matthew Perry, SPi Global US, Inc.*
Rights & Permissions Management: *Ben Ferrini*
Photo Researcher: *Maureen Spuhler*
Manufacturing Buyer: *Stacey Weinberger, LSC Communications*
Product Marketer: *Wendy Mears*
Director of Product Marketing: *Allison Rona*
Field Marketer: *Kelly Galli*
Director of Field Marketing: *Tim Galligan*

Cover Photo Credit: *Kim Kwangshin / Getty Images*

Copyright © 2020, 2017, 2014 Pearson Education, Inc. 221 River Street, Hoboken, NJ 07030. All Rights Reserved. Printed in the United States of America. This publication is protected by copyright, and permission should be obtained from the publisher prior to any prohibited reproduction, storage in a retrieval system, or transmission in any form or by any means, electronic, mechanical, photocopying, recording, or otherwise. For information regarding permissions, request forms and the appropriate contacts within the Pearson Education Global Rights & Permissions department.

Acknowledgments of third party content appear on page C-1, which constitutes an extension of this copyright page.

PEARSON, ALWAYS LEARNING Mastering Microbiology®, MicroFlix™, and Interactive Microbiology™ are exclusive trademarks in the U.S. and/or other countries, owned by Pearson Education, Inc. or its affiliates.

Unless otherwise indicated herein, any third-party trademarks that may appear in this work are the property of their respective owners and any references to third-party trademarks, logos or other trade dress are for demonstrative or descriptive purposes only. Such references are not intended to imply any sponsorship, endorsement, authorization, or promotion of Pearson's products by the owners of such marks, or any relationship between the owner and Pearson Education, Inc. or its affiliates, authors, licensees or distributors.

Library of Congress Cataloging-in-Publication Data
Names: Bauman, Robert W., author. | Primm, Todd P., contributor. |
  Brunings, Asha, contributor. | Austin, Terry (Terry A.), contributor. |
  Elizabeth Machunis-Masuoka, contributor. | Siegesmund, Amy M., contributor.
Title: Microbiology. With diseases by taxonomy / Robert W. Bauman ;
  contributions by Todd P. Primm, Asha Brunings, Terry Austin,
  Elizabeth Machunis-Masuoka, Amy M. Siegesmund; clinical consultants,
  Jordan A. Roeder, Cecily B. Cosby.
Other titles: With diseases by taxonomy
Description: Sixth edition. | San Francisco : Pearson, [2018] | Includes
  bibliographical references and index.
Identifiers: LCCN 2018051922 | ISBN 9780134832302 | ISBN 0134832302
Subjects: | MESH: Microbiological Phenomena | Microbiological
Techniques
Classification: LCC QR41.2 | NLM QW 4 | DDC 616.9/041--dc23 LC record available at https://lccn.loc.gov/2018051922

ISBN 10: 0-134-83230-2 ISBN 13: 978-0-134-83230-2 (Student Edition)
ISBN 10: 0-135-17470-8 ISBN 13: 978-0-135-17470-8 (Looseleaf Edition)
ISBN 10: 0-135-20398-8 ISBN 13: 978-0-135-20398-9 (Instructor's Review Copy)

 www.pearsonhighered.com

14 2024

Michelle, you are my best friend, my closest confidant, my cheerleader, my partner, my love for over thirty-five years. Thank you for working alongside me to bring this book to fruition. I love you more now than then.

—*Robert*

ABOUT THE AUTHOR

ROBERT W. BAUMAN

is a professor of biology and past chairman of the Department of Biological Sciences at Amarillo College in Amarillo, Texas. He has taught microbiology, human anatomy and physiology, and botany for over 30 years. In 2004, the students of Amarillo College selected Dr. Bauman as the recipient of the John F. Mead Faculty Excellence Award. He is gratified to have been nominated by the students every year since then, though College policy currently allows a teacher to win only once. He received an M.A. degree in botany and oceanography from the University of Texas at Austin and a Ph.D. in biology from Stanford University. His research interests have included the morphology, ecology, and culturing of freshwater algae and photosynthetic protozoa, the cell biology of marine algae (particularly the deposition of cell walls and intercellular communication), environmentally triggered chromogenesis in butterflies, and remediation of terrestrial oil pollution by naturally occurring bacteria. He is a member of the American Society of Microbiology (ASM) where he has held national offices; Texas Community College Teachers Association (TCCTA), where he serves in a statewide position of leadership; American Association for the Advancement of Science (AAAS); Human Anatomy and Physiology Society (HAPS); and the Lepidopterists' Society. When he is not writing books, he enjoys spending time with his family: gardening, hiking, camping, rock climbing, backpacking, cycling, skiing, and reading by a crackling fire in the winter and in a gently swaying hammock in the summer.

TODD P. PRIMM (Contributor)

is a professor of pre-nursing and general microbiology at Sam Houston State University and is Director of the Professional and Academic Center for Excellence, which focuses on improving pedagogy. In 2010, he was Distinguished Alumnus of the Graduate School of Biomedical Sciences of Baylor College of Medicine, where he earned his Ph.D. in biochemistry. He is active in ASM and received the Texas Branch 2015 Faculty Teaching Award. He was chair of the 2013 ASM Conference for Undergraduate Educators and currently serves on the editorial board for the Journal of Microbiology & Biology Education. He is also an affiliate staff member with the international organization Cru. He loves mentoring students and spending time with his wife of 25 years and their five children.

ASHA M. BRUNINGS (Contributor)

is an associate professor of microbiology at Santa Fe College in Gainesville, Florida. In 2013, she was a recipient of the John & Suanne Roueche Excellence Awards of the League for Innovation in the Community College. Born in the Netherlands, raised in Suriname, she received a B.S. in Agronomy (Anton de Kom University of Suriname), an M.S. in Plant Molecular Cellular Biology and a Ph.D. in Plant Pathology (University of Florida). Her research interests include the host-pathogen interactions of citrus canker, the benefits of silicon on plant disease resistance, and strawberry genetics. She is a member of ASM and the National Science Teachers Association, and coaches Brain Bowl, an academic quiz tournament. She enjoys swimming, dancing, sci-fi, and spending time with her family.

TERRY A. AUSTIN (Contributor)

is a professor of biology and former chairman of the Department of Biology at Temple College in Temple, Texas. He has taught microbiology, human anatomy and physiology, and other biological science courses for over twenty-five years. He received a B.S. in organismal biology, a M.S. in biology specializing in mammalogy at Midwestern State University, and completed doctoral coursework in neuroscience at the University of North Texas. Terry is a member of and regular presenter at HAPS.

About the Clinical Consultants

JORDAN A. ROEDER

is nationally certified as an Adult Geriatric-Acute Care Nurse Practitioner through the American Association of Critical Care Nurses. He received his associate degree in nursing from Amarillo College in 2012, a BSN from Texas Tech University in 2014, and a MSN from the University of Colorado-Anschutz Medical Campus in 2018. He has over six years of hospital nursing experience including three years ICU nursing at Northwest Texas Hospital in Amarillo Texas where he was the recipient of the Daisy Award for extraordinary nursing. From 2018 to 2019 he served as a Critical Care Fellow at OhioHealth Riverside Methodist Hospital in Columbus, Ohio.

CECILY D. COSBY

is nationally certified as both a family nurse practitioner and physician assistant. She is a professor of nursing at Samuel Merritt University in Oakland, California and has been in clinical practice since 1979. She received her Ph.D. and M.S. from the University of California, San Francisco; her BSN from California State University, Long Beach; and her P.A. certificate from the Stanford Primary Care program.

PREFACE

The re-emergence of measles, an historical epidemic of Ebola hemorrhagic fever, and the emergence of Zika virus infections, spotted fever rickettsioses, Middle East respiratory syndrome, and other diseases; cases of strep throat, MRSA, and tuberculosis; the progress of research into more effective and new vaccines, advance in microbial genetics, including genetic engineering and gene therapy; the challenge of increasingly drug-resistant pathogens; the continual discovery of microorganisms previously unknown—these are just a few examples of why exploring microbiology has never been more exciting, or more important. Welcome!

I have explored microbiology with undergraduates for over 30 years and witnessed firsthand how students struggle with the same topics and concepts year after year. My students begged me to write this book to address these challenging topics. My aim was to help students see complex topics of microbiology—especially metabolism, genetics, and immunology—in a way that they can understand, while at the same time presenting a thorough and accurate overview of microbiology. I also wished to highlight the many positive effects of microorganisms on our lives, along with the medically important microorganisms that cause disease.

New to This Edition

In approaching the sixth edition, my goal was to build upon the strengths and success of the previous editions by updating it with the latest scientific and educational research and data available and by incorporating many terrific suggestions received from colleagues and students alike. The feedback from instructors who adopted previous editions has been immensely gratifying and is much appreciated. Seven new Solve the Problem! features use problem-based learning, encouraging students to put knowledge into practice. New case studies were developed to open each chapter to engage students with real world, practical problems in clinical microbiology. Further, responding to recent pedagogical research that shows that learning occurs more readily with frequent opportunities to review and apply information, Micro Check questions were added throughout every chapter. Micro Checks are short questions that encourage students to focus attention on the Learning Outcomes, increase their learning, and enhance metacognition—awareness of student's own learning. Diligent application of Micro Checks will increase student success.

In this new edition:

- NEW **Before You Begin questions** refresh students' critical chemistry and general biology knowledge from pre-requisite courses and previous chapters for success in the course. Before You Begin questions appear in the chapter openers and can be assigned in Mastering Microbiology. In Mastering, the Before You Begin questions are interactive and provide wrong and correct answer feedback with embedded hints that provide just-in-time remediation via tutorials that fill in students' individual skill gaps.

- NEW **Solve the Problem** features carry education to a new level with problem-based learning exercises that excite, inspire, and stimulate students to apply critical thinking skills to current microbiological quandaries. Each of the seven Solve the Problem features challenges students to work together to devise and articulate possible resolutions to a timely, real world problem. Solve the Problem exercises can stand alone or be expanded with ambitious extensions and resources available in Mastering Microbiology®.

- NEW **MicroChecks** provide an intuitive check in for students to ensure mastery of the new chapter content. These formative assessments appear in the text and can be assigned in Mastering Microbiology. In Mastering, the interactive questions provide immediate feedback for just-in-time remediation, directing students to videos and animations that reinforce the concepts.

- **Micro Matters** features tie together subjects from different chapters to encourage students to apply and synthesize new knowledge as they explore medical cases and answer pertinent questions. Each of the five Micro Matters video tutorials is paired with assessments in Mastering Microbiology®.

- The genetics and immunology chapters (Chapters 7, 8, 15, and 16) have been reviewed and revised by specialists. These now reflect the most current understanding of these rapidly evolving fields, including new discussions of next-generation DNA sequencing and of CRISPR.

- Nearly 300 NEW and revised micrographs, photos, and figures enhance student understanding of the text and boxed features.

  - NEW Chapter 27 is dedicated to microbial ecology and microbiomes.

- NEW AND EXPANDED MasteringMicrobiology® includes:

  - NEW **Concept Maps** give students the opportunity to construct their knowledge using a list of key terms as higher level Bloom's activities in a scaffolded, drag and drop format. The interactive end-of-chapter activities can be assigned in Mastering Microbiology, providing students with immediate feedback for just-in-time remediation that directs them to specific content to reinforce the concepts.
  - NEW Interactive Microbiology Modules, a dynamic suite of interactive tutorials and animations that teach key concepts in the context of a clinical setting. Students actively engage with each topic and learn from manipulating variables, predicting outcomes, and answering formative

and summative assessments. New modules include Antimicrobial Resistance, Aerobic Resistance, and the Human Microbiome.

- Micro Matters case tutorials and assessments connect chapter concepts and coach students through applying and synthesizing new knowledge.
- MicroBoosters pair video tutorials and assessments covering key concepts that students often need to review, including Basic Chemistry, Cell Biology, Biology and more!
- The Microbiology Lab resources include MicroLab Tutors, which use lab technique videos, 3-D molecular animations, and step-by-step tutorials to help students make connections between lecture and lab.
- Lab Technique Videos and pre-lab quizzes ensure students come prepared for lab time.
- Lab Practical and post-lab quizzes reinforce what students have learned.

Mastering Microbiology® also provides access to Dynamic Study Modules to help students acquire, retain, and recall information faster and more efficiently than ever before, with textbook-specific explanations and art. Dynamic Study Modules are available for use as a self-study tool or as assignments. Additionally, Mastering Microbiology® includes Learning Catalytics—a "bring your own device" student engagement, assessment, and classroom intelligence system. With Learning Catalytics, instructors can assess students in real time using open-ended tasks to probe student understanding using Pearson's library of questions or designing their own.

The following section provides a detailed outline of this edition's chapter-by-chapter revisions.

CHAPTER-BY-CHAPTER REVISIONS

- More closely aligned assessment questions with Learning Outcomes
- Every chapter now begins and ends with **Micro in the Clinic**—a clinical case study covering a disease relevant to the chapter
- Every chapter opener contains **Before You Begin** questions to spur students to test their foundational knowledge before beginning the chapter
- New **Micro Check** questions throughout every chapter challenge students to use newly acquired knowledge as they proceed
- Clinical consultants evaluated, rewrote, and improved each case study to be more medically relevant and accurate

1 A Brief History of Microbiology

- New case study opens the chapter
- Eight other new photos (1.2, 1.3, 1.11, 1.15, 1.16, 1.17, 1.20, 1.22)
- Two figures revised for better pedagogy, clarity, and accuracy (1.10, 1.21)
- Updated map showing countries having transmission of variant Creutzfeldt-Jakob disease (vJCD)
- New **Solve the Problem: Smallpox: To Be or Not to Be?** problem-based learning exercise concerning complete smallpox virus destruction
- Simplified the language and sentence structure
- Added discussion of the contribution of Fanny Hesse to the success of Koch's laboratory with her introduction of agar
- Increased coverage of Florence Nightingale and her accomplishments
- Added the work of Lady Mary Worley Montague to the discussion of the development of vaccination
- Introduced the success of gene therapy to treat several inherited immune deficiencies
- Added to list of current problems in microbiology: Ebola control and the problems associated with emerging diseases
- Added a Tell Me Why concerning the accomplishments of Nightingale leading to consideration of her as Mother of Medical Microbiology
- Added a critical thinking question concerning the "fatherhood" of microbiology (Leeuwenhoek vs. Pasteur)
- Deleted Learning Outcome 1.21 concerning fastest-growing fields within microbiology; this topic is better covered by an instructor in class rather than in a textbook

2 The Chemistry of Microbiology

- New case study opens the chapter
- Representation of ether bond in Table 2.3 clarified for better pedagogy
- Clarified that most organisms code for 21 amino acids, though 20 are more common
- New Learning Outcome regarding *electron shells* and *valence electrons*
- Added Learning Outcomes 2.3 regarding the chemistry of carbohydrates and Learning Outcome 2.25 regarding the general structure of amino acids

3 Cell Structure and Function

- New case study opens the chapter
- Seven other new/upgraded photos (3.1b, 3.7b & c, 3.11, 3.24, 3.25b, 3.37)
- Revised and enhanced artwork in eleven figures for enhanced pedagogy (3.2, 3.3, 3.15, 3.16, 3.21, 3.22, 3.23, 3.32, 3.33, 3.37, 3.39)
- Enhanced discussion of chemistry and function of lipids in archaeal cytoplasmic membranes
- Clarified that *endotoxin* refers to lipopolysaccharide (LPS), which contains the toxic molecule lipid A
- Recognized Lynn Margulis as a proponent of the theory of endosymbiosis
- Added **Clinical Case Study: The Big Game** about strep throat

4 Microscopy, Staining, and Classification

- New case study opens the chapter
- Five other new/upgraded photos (4.10c, 4.11c, 4.12, 4.18, 4.24)
- Revised and enhanced artwork in Figure 4.4 for enhanced pedagogy
- Expanded discussion of resolution; immersion oil; the appearance of scanning electron, scanning tunneling, and atomic force micrographs; mordants; definitions of microbial species and serotypes; and the role of George Fox in the discovery of the archaea and three domains of life
- Added contribution of Rebecca Lancefield to serological identification of streptococci
- Added discussion of matrix-assisted laser desorption/ionization time-of-flight (MALDI/TOF) mass spectrometry as a method of identification of microbial species and diagnoses
- At request of reviewers and instructors, reduced complexity and chapter length by removing detailed figures for dark field, phase, and scanning electron microscopy

5 Microbial Metabolism

- New case study opens the chapter
- One new photo (5.23b)
- Revised fifteen figures for greater clarity and better pedagogy (5.3, 5.6, 5.9, 5.10, 5.11, 5.12, 5.14, 5.15, 5.17, 5.18, 5.21, 5.22, 5.25, 5.31, Concept map)
- Revised textual explanation of glycolysis to make it more clear and pedagogically efficient
- Fully adopted current name *citric acid cycle* instead of eponymous *Krebs cycle*

6 Microbial Nutrition and Growth

- New case study opens the chapter
- New **Solve the Problem: The Microbes Ate My Homework** problem-based learning exercise concerning using engineered microbes to reduce cardboard waste
- Revised five figures for greater clarity and better pedagogy (6.2, 6.6, 6.10, 6.21, 6.23)
- One new figure on quorum sensing (6.7)
- Two new photos (6.10b, 6.24)
- Revised Learning Outcome 6.8 to be more specific (Describe methods for collecting clinical specimens from the skin, blood, cerebrospinal fluid, stomach, urine, lungs, and diseased tissue.)
- Expanded discussion of singlet oxygen and superoxide radicals as oxidizing agents, the nature of extracellular matrix in biofilms, and quorum sensing
- Clarified the method of counting microbes using a cell counter

CHAPTER-BY-CHAPTER REVISIONS IX

7 Microbial Genetics

- New case study opens the chapter
- Upgraded twenty-three figures for greater clarity, accuracy, ease of reading, and better pedagogy (7.1, 7.3, 7.5, 7.6, 7.8, 7.9, 7.10, 7.11, 7.12, 7.13, 7.15, 7.16, 7.17, 7.18, 7.19, 7.20, 7.21, 7.25, 7.26, 7.27, 7.28, 7.29, 7.33)
- One new figure concerning termination of translation (7.20)
- Added new Learning Outcome (7.7) concerning the concept that nucleotide base complementarity ensures precise and accurate DNA replication
- Added critical thinking question to Figure 7.13 concerning central dogma of genetics
- Noted some contributions of Esther Lederberg to microbial genetics (development of replica plating and the discovery of the F plasmid)
- Clarified the structure of eukaryotic chromosomes
- Added term *constitutive genes* to contrast with operon genes
- Moved **Beneficial Microbes: Life in a Hot Tub (Taq polymerase)** to chapter 8 where its direct application for PCR is more apparent
- Moved **Emerging Disease Case Study: *Vibrio vulnificus* Infection** to chapter 6: the microbe's use of quorum sensing increases its virulence

8 Recombinant DNA Technology

- New case study opens the chapter
- Modified two figures for better pedagogy (8.7, 8.8)
- One new figure illustrating a form of CRISPR (8.4)
- Added seven NEW Learning Outcomes concerning CRISPR, uses of synthetic nucleic acids, PCR, fluorescent *in situ* hybridization (FISH), functional genomics, Sanger sequencing, and next generation sequencing
- Deleted figures for Southern blots and Sanger automated DNA sequencing as these techniques are more historical than current
- Enhanced or added discussion of restriction enzymes *Hae*III and *Msp*I; CRISPR-Cas; real-time PCR (RT-PCR); Sanger sequencing methods; next generation DNA sequencing (NGS), including pyrosequencing and fluorescent methods; functional genomics; microbiomes; transcriptomics, metabolomics, biomedical animal models; heat shocking for the uptake of DNA; and successful gene therapies
- Moved **Beneficial Microbes: Life in a Hot Tub (Taq polymerase)** from chapter 7 to chapter 8 where its direct application for PCR is more apparent
- Fully adopted current name *recombinant DNA technology* instead of the sometimes controversial or problematic *genetic engineering*
- Added two end of chapter questions over CRISPR

9 Controlling Microbial Growth in the Environment

- New case study opens the chapter
- Revised one figure for better accuracy, currency, and pedagogy (9.11)
- Two new photos (9.4, 9.11)
- New **Solve the Problem: How Clean is Too Clean** problem-based learning investigation concerning the potential overuse of household and industrial disinfectants
- Revised definition of heavy-metal ions
- Updated techniques for deactivation of prions, coverage of thimerosal in vaccines, and activity of AOAC International in developing disinfection standards

10 Controlling Microbial Growth in the Body: Antimicrobial Drugs

- New case study opens the chapter
- Updated and revised tables of antimicrobials to include all new antimicrobials mentioned in disease chapters, including antibacterial capreomycin, antihelmintic bithionol, anti-influenzavirus peramivir; updated sources of drugs, modes of action, clinical considerations, and methods of resistance
- Eight revised figures for greater clarity, accuracy, ease of reading, and better pedagogy (10.2, 10.3, 10.4, 10.5, 10.6, 10.7, 10.16; map of worldwide, community-associated MRSA
- One new figure showing Therapeutic index (10.15)
- Two new photos (10.10, 10.14)
- Added discussion of: 1) drugs that inhibit protein synthesis by interfering with the charging of tRNA molecules, the new antibiotic teixobactin, 2) the importance of persister cells for antimicrobial resistance, 3) the CDC's threat levels for microbial drug resistance
- Enhanced and clarified discussion of therapeutic index, therapeutic window, and adverse effects of gramicidin
- Added Learning Outcome over persister cells
- Increased discussion of women's contributions to development of antimicrobial drugs: developing the first effective antiviral agent (acyclovir), the first anti-HIV agent (azidothymidine), and an antifungal agent (nystatin); elucidation of the 3-D structure of penicillin, which allowed the synthesis of synthetic penicillins; and as leaders of medical organizations, director-general of WHO and director of the CDC
- Removed amantadine as a treatment for influenza A
- Added new Learning Outcome on action of mupirocin, which interferes with charging of tRNA^{Ile}

11 Characterizing and Classifying Prokaryotes

- New case study opens the chapter
- Added four **Tell Me Why** critical thinking questions to text
- Three revised figures for better pedagogy (11.6, 11.10, 11.24)
- Two new photos (11.14b, 11.18)
- Clarified binary fission of cocci to form tetrads; clarified taxonomic groupings among the bacteria, expanded on the properties mycoplasmas use to protect themselves from osmosis; distinguished between true reproductive spores and endospores; expanded coverage of acid-fast staining and acid-fast bacilli (AFBs); expanded on the zetaproteobacteria
- New Learning Outcomes concerning: classes of proteobacteria, the use of *Agrobacterium* in genetic modification of plants
- New end-of-chapter Short Answer Question concerning the special features of snapping division

12 Characterizing and Classifying Eukaryotes

- New case study opens the chapter
- Eight new photos (12.7, 12.13, 12.16b, 12.20b, 12.21a(2), 12.25, **Emerging Disease Case Study: Chronic Aspergillosis**)
- Four revised artwork for better pedagogy (12.1, 12.2, 12.8, 12.18)
- Updated, revised, or clarified discussion of cytokinesis; asexual spores of fungi; truffles; expanded discussion and location of algal flagella
- New Learning Outcomes concerning fungal nutrition and fungal morphology

13 Characterizing and Classifying Viruses, Viroids, and Prions

- New case study opens the chapter
- New **Solve the Problem: Microengineer a Better Mosquito?** problem-based learning exercise concerning using genetically engineered *Wolbachia* bacteria to reduce mosquito populations
- Updated viral nomenclature to correspond to changes approved by the International Committee on Taxonomy of Viruses (ICTV) (2016)

- Six new photos (13.3b, 13.5a, 13.7b, **Beneficial Microbes: Prescription Viruses?**, 13.19, 13.21)
- Upgraded three figures for better pedagogy and currency (13.22, 13.23, 13.24)
- Added discussion of the work of Esther Lederberg in the discovery of lambda phage and lysogeny
- Revised discussion of prions; expanded discussion of bacteriophages for treatment of human diseases and synthesis strategies of animal viruses

14 Infection, Infectious Diseases, and Epidemiology

- New case study opens the chapter
- New **Solve the Problem: Microbes in the Produce Aisle** problem-based learning exercise concerning epidemiological case study concerning *Legionella* outbreak
- Epidemiology charts, tables, graphs, and maps updated
- Updated list of nationally notifiable infectious conditions (changed AIDS to HIV Stage III; added cancer, carbapenems-resistant infections of Enterobacteriaceae; Zika virus disease and infections; changed SARS to severe acute respiratory syndrome-associated coronavirus disease)
- Changed four figures for better pedagogy, timeliness, or clarity (14.7, 14.14, 14.16, 14.18)
- One new photo (14.12)
- Changed the term *normal microbiota* to *resident microbiome* (*resident microbiota*) to reflect current vocabulary
- Clarified the fact that microbiota normal for one person may be different in another
- New question for **Clinical Case Study: A Deadly Carrier** concerning control of an epidemic when antimicrobial drugs are unavailable
- Clarified and expanded discussion of parenteral infections and the list of arthropod vectors for microbial diseases

15 Innate Immunity

- New case study opens the chapter
- Added new **Tell Me Why** critical thinking questions to **The Body's First Line of Defense** concerning lysozyme
- Expanded coverage of the action of antimicrobial peptides (defensins), probiotics, defensive action of resident microbiome, and phagocytosis
- Clarified the processes involved in the first line of defense
- Added nature and action of dermicidins, fatty acids, and mucins
- Modified one figure for enhanced clarity and better pedagogy (15.8)
- Three new photos [15.5 (2), 15.10]

16 Specific Defense: Adaptive Immunity

- New case study opens the chapter
- Five revised figures for greater accuracy and pedagogical efficiency (16.3, 16.4, 16.7, 16.9, 16.12, 16.14)
- One new photo (16.5)
- Revised and added to Learning Outcomes (16.27), concerning the events in antibody immune responses
- New table (16.1) summarizing differences between innate and adaptive immunity
- Revised and clarified terminology for CD8+ and CD4+ cells, genetic basis for creation of BCR and TCR diversity, binding capability of MHC

17 Immunization and Immune Testing

- New case study opens the chapter
- New **Solve the Problem: Rights versus Responsibilities** problem-based learning exercise concerning balancing the benefits of immunization versus the rights of families to direct their own health care
- Added a **Tell Me Why** critical thinking question to text
- Updated to newly revised CDC 2018 vaccination schedule for children, adolescents, and adults
- Updated table of vaccine preventable diseases in USA
- Three revised figures for better pedagogy (17.1, 17.2, 17.3)
- Two new photos (17.10, 17.11b)
- Updated, clarified, and in some cases, increased coverage of safety and efficacy of immunizations, treatment with passive immunotherapy using monoclonal antibodies, determination of blood type by agglutination, use of indirect fluorescent immunoassays, antibody sandwich ELISA, and immunochromatographic assays
- Standardized usage of *immunization* versus *vaccination*, which specifically describes immunization against smallpox
- Reworded labeled antibody tests as labeled immunoassays

18 Hypersensitivities, Autoimmune Diseases, and Immune Deficiencies

- New case study opens the chapter
- Three revised figures for better pedagogy, accuracy, and clarity (18.1, 18.5, 18.6)
- Updated and clarified the role of IgE, basophils, and mast cells in type I hypersensitivity

19 Pathogenic Gram-Positive Bacteria

- New case study opens the chapter
- Updated diagnoses, incidence data, treatments, and prevention, particularly for necrotizing fasciitis, streptococcal pneumonia, tetanus, tuberculosis
- Seven new photos (19.2, 19.7, 19.11, 19.18, 19.19, 19.21, **Microbe at a Glance:** *Streptococcus* pneumoniae
- Six revisions to figures for consistency, currency, accuracy, and better pedagogy [19.5, 19.13, **Disease in Depth: Tuberculosis** (2), **Listeriosis** (2)]
- Added discussion of cellulitis, HA-MRSA, CA-MRSA, use of term *GAS* for group A *Streptococcus* (*S. pyogenes*), red squirrels as host for *Mycobacterium leprae*
- Updated and clarified discussion of erysipelas, rheumatic fever, puerperal fever, necrotizing fasciitis (it is polymicrobial rather than monomicrobial), Lancefield classification of *Enterococcus*, atypical (walking) pneumonia
- Updated immunization schedule for diphtheria
- Enhanced Concept Mapping to include new treatment regimen for tuberculosis
- New **Clinical Case Study** over MRSA cellulitis

20 Pathogenic Gram-Negative Cocci and Bacilli

- New case study opens the chapter
- New Learning Outcome regarding antibody immune responses
- Updated all diagnoses, treatments, and incidence data
- Twelve new photos (20.2, 20.4a, 20.7b, 20.12, **Disease in Depth: Urinary Tract Infections** (2), 20.13, 20.20, 20.21, 20.25, 20.28, 20.29)
- Fourteen revised figures for better pedagogy (20.3, 20.7a, 20.8, 20.9, 20.10, **Disease in Depth: Urinary Tract Infections** (3); 20.15, 20.16, 20.17, 20.22, 20.23, 20.25)
- Added discussion of strains of *E. coli*, carbapenem-resistant enterobacteriaceae (CREs); Margaret Pittman's role in elucidating the strains of *H. influenzae*; emergence of extremely drug resistant (XDR) *Salmonella typhi* in Pakistan in 2018
- Updated and clarified discussion of breakdown of maltose by *Neisseria*

CHAPTER-BY-CHAPTER REVISIONS   XI

21 Rickettsias, Chlamydias, Spirochetes, and Vibrios

- New case study opens the chapter
- Updated all diagnoses, treatments, and incidence data
- Four new photos [21.5, 21.9(a-c)]
- Eight revised figures for better pedagogy (21.1, 21.2, **Disease in Depth: Rocky Mountain Spotted Fever** (1), 21.3, **Microbe at a Glance:** *Treponema*, 21.8, 21.12, 21.13)
- Added discussion of potential vaccine against *Chlamydia trachomatis*, manifestations of *Chlamydia trachomatis*, and a new vaccine against Lyme disease

22 Pathogenic Fungi

- New case study opens the chapter
- New Learning Outcome concerning microsporidia and microsporidioses
- Updated all diagnoses, treatments, and incidence data, particularly histoplasmosis, blastomycosis, and coccidioidomycosis
- Updated name of *Penicillium marneffei* to *Talaromyces marneffei* (renamed in 2015)
- Modified six Questions for Review at the end of the chapter (MC: 4, 13; MTF: 3, 10; FIB: 7; SA: 4)
- Modified Concept Mapping for the up to date treatment for aspergillosis (vericonazole)
- Modified five figures for enhanced pedagogy and lucidity (22.2, 22.3, 22.5, 22.7, **Disease in Depth: Candidiasis**)
- One new figure concerning microsporidial life cycle (22.14)
- Four new photos for enhanced pedagogy (22.4b, 22.9, 22.11, 22.20)
- Updated and enhanced discussion of potential vaccines against fungi
- Added discussion of emerging pathogens—microsporidia

23 Parasitic Protozoa, Helminths, and Arthropod Vectors

- New case study opens the chapter
- Updated all diagnoses, incidence data, and treatments, particularly giardiasis, schistosomiasis, and babesiosis
- Ten new, more engaging photos (23.2, 23.6, 23.8, **Disease in Depth: Giardiasis** (1), 23.10, 23.12, 23.15, 23.18b, 23.19b and c)
- Fourteen revised, updated, enhanced, and pedagogically more effective figures [23.1, 23.3, 23.5, 23.6, **Disease in Depth: Giardiasis** (3), **Disease in Depth: Malaria** (4), 23.9, 23.14, **Emerging Disease Case Study: Babesiosis**]
- Clarified features of the life cycles of trypanosomes, nematodes
- Redrew Concept Mapping to more accurately reflect parasitic protozoa

24 Pathogenic DNA Viruses

- New case study opens the chapter
- Updated all diagnoses, incidence data, and treatments, adenoviral diseases, hepatitis B, and newly-approved treatment for smallpox
- Three new photos (24.3, 24.7, 24.15)
- Eight revised, updated, enhanced, and pedagogically more effective figures [**Microbe at a Glance:** *Orthopoxvirus variola* (Smallpox Virus (1)), 24.12, **Disease in Depth: Papillomas** (2), 24.16, **Microbe at a Glance:** *Adenovirus*, 24.18, 24.20, 24.21]
- Updated and enhanced discussion of adenoviral diseases, the role of papillomaviruses in the development of oral and throat cancers, and sign of erythema infectiosum

25 Pathogenic RNA Viruses

- New case study opens the chapter
- Updated all diagnoses, treatments, incidence data, and case studies, particularly hepatitis C, West Nile virus encephalitis, dengue and dengue hemorrhagic fever, SARS and MERS, HIV/AIDS, measles, Ebola hemorrhagic fever, rotavirus diarrhea
- Added new diseases: Zika fever and Zika congenital syndrome
- New Learning Outcome concerning Zika virus and its diseases
- Replaced **Emerging Disease Case Study: H5N1 Influenza** with new **Emerging Disease Case Study: A Deadly Mosquito Bite?**, concerning Zika virus, Zika fever, and Zika congenital syndrome
- Twenty figures revised, updated, or enhanced for better pedagogy [25.2, 25.9, 25.10, 25.11, 25.12, 25.14, 25.18, 25.20, 25.22, 25.24, 25.27, **Microbe at a Glance:** *Morbillivirus measles virus* (1), 25.29, 25.30, 25.32, **Disease in Depth: Ebola** (3), 25.37]; Concept Mapping: Hepatitis
- Twelve new photos [25.6, 25.7, 25.9, **Beneficial Microbe: Eliminating** *Aedes*-**Borne Diseases?** (1), **Emerging Disease Case Study: A Deadly Mosquito Bite?** (1), 25.16, **Microbe at a Glance:** *Morbillivirus measles virus* (1), 25.33, **Disease in Depth: Ebola** (3), **Clinical Case Study: A Case of AIDS** (1)]
- Increased discussion of encephalitis arboviruses, coxsackievirus B, hepatitis E, dengue fever and dengue hemorrhagic fever, rubella, measles, hemorrhagic fevers, and Zika virus disease

26 Applied and Industrial Microbiology

- Split the chapter into two so as to give more attention to applied microbiology
- New chapter opener case study and photo
- One new photo (26.14)
- Five figures revised, updated, or enhanced for better pedagogy (26.1, **Emerging Disease: Primary Amebic Meningoencephalitis**, 26.6, 26.8, 26.9, 26.10)
- Revised and enhanced discussion of pickling, "live" yogurt, rennin, cheese making, butanol as an alternative fuel, seaweed as biosensors, drinking water treatment, wastewater treatment
- Added four fill-in-the-blank questions and two new Visualize It! questions to the Questions for Review at the end of chapter

27 Microbial Ecology and Microbiomes

- Split the chapter from Chapter 26 so as give more attention to the worldwide microbiome as well as ecology of the human body and its microbiome
- New case study opens the chapter
- New **Solve the Problem: Fecal Microbiome Transfer: Medicine of Magic?** problem-based learning exercise on ecology of human microbiome and the use of fecal transplantation to treat chronic gastrointestinal diseases
- Four figures revised, updated, or enhanced for better pedagogy (27.1, 27.4, 27.5, 27.6)
- One other new photo (27.7)
- New **Clinical Case Study: Bioterrorism in the Mail** concerning anthrax
- Added coverage of the World Microbiome Project
- One new Multiple Choice question, one new Modified True/False question, one new Fill in the Blanks question, and three new Critical Thinking questions added to the end of chapter Questions for Review

REVIEWERS FOR THE SIXTH EDITION

"Thank you!" to the hundreds of instructors and students who participated in reviews, class tests, and focus groups for earlier editions of the textbook. Your comments have informed this book from beginning to end, and I am deeply grateful. For the sixth edition, I especially extend my deepest appreciation to the following reviewers.

Book Reviewers

Dena Berg
Tarrant County College

Carroll Bottoms
Collin College

Nick Butkevitch
Schoolcraft College

Kari Cargill
Montana State University

Richard J. Cristiano
Houston Community College Northwest—Spring Branch campus

Ann Evancoe
Hudson Valley Community College

Tod Fairbanks
Palm Beach State College

Teresa G. Fischer
Indian River State College

Sandra M. Fox-Moon
Anne Arundel Community College and *University of Maryland*

Eric Gillock
Fort Hays State University

Raymond Harris
Prince George's Community College

Jennifer Hatchel
College of Coastal Georgia

Barbara R. Heard
Atlantic Cape Community College

Nazanin Hebel
Houston Community College—Northwest

Amy Helms
Collin College

David T. Jenkins
University of Alabama at Birmingham

Denice D. King
Cleveland State Community College

Todd Martin
Metropolitan Community College, Blue River

Jennifer Metzler
Ball State University

Mary Miller
Baton Rouge Community College

Alicia Musser
Lansing Community College

Gregory Nasello
Lewis and Clark Community College

Dana Newton
College of the Albemarle

Johanna Porter-Kelley
Winston-Salem State University

Jennifer Reaves
Jackson State Community College

Jackie Reynolds
Richland College

Steven Scott
Merritt College

Amy Siegesmund
Pacific Lutheran University

Tony A. Slieman
University of South Dakota

Lori Smith
American River College

Vetaley Stashenko
Palm Beach State College, Belle Glade

Jennifer Swartz
Pikes Peak Community College

Christopher Thompson
Loyola University, Maryland

Marie N. Yearling
Laramie County Community College

Video Tutor Reviewers

Jason Adams
College of Dupage

Abiodun Adibi
Hampton University

Melody J. Bernot
Ball State University

Denise Foley
Santiago Canyon College

Emily Getty
Ivy Tech Community College

Mary Ann Arnold Hedrick
Wytheville Community College

Cristi Hunnes
Rocky Mountain College

Sudeep Majumdar
Temple College

Bhavya Mathur
Chattahoochee Technical College

Daniel Brian Nichols
Seton Hall University

Kevin Sorensen
Snow College

Sandra L. Specht
Sinclair Community College

ACKNOWLEDGMENTS

As has been the case with all previous editions, I am ever more cognizant that this book is a team effort. I am grateful to Jennifer McGill Walker of Pearson Science and to the team she gathered to produce this sixth edition. Jen and extremely dedicated project manager Cheryll Linthicum helped develop the vision for this edition, generating ideas to make it more effective and compelling. Thanks also to Marie Beaugureau and Laura Cheu, who were invaluable in developmental editing. Thank you to Barbara Yien, project editor of the first two editions, for years of support and for introducing me to chocolate truffles. I am thrilled to watch your family grow; thank you for letting me practice grandparenting. Thanks as well to editorial assistants Katrina Taylor and Jordan Barron Goldman for their work securing reviewer feedback and on many other "behind-the-scenes" tasks.

The incredible Anita Wagner Hueftle copyedited the manuscript thoroughly and meticulously, suggesting important changes for clarity, accuracy, and consistency. My thanks to art development editor Jay McElroy for his assistance with new figures in this edition, as well as Morgan Ewald at Lachina Creative for rendering the art. Rose Kernan and SPi Global expertly guided the project through production. Maureen "Mo" Spuhler remains the most amazing photo researcher. I am in your debt, "Molybdenum," for superb photos, excellent suggestions, and as the model for several patients and nurses. I'll never forget the poison ivy blisters on your arm. Rich Robison and Brent Selinger supplied many of the text's wonderful and unique micrographs, and Matt Perry provided much needed help in updating the photo collection. Preston Thomas created the beautiful interior design and the stunning cover; it's beautiful—thank you Preston!.

Thanks to Lucinda Bingham and Nicole Constantine for their work on the media supplements for this edition and for their management of the extraordinary array of media resources for students and instructors, especially Mastering Microbiology®.

I am grateful to Wendy Mears, Allison Rona, and Kelly Galli in Marketing. They have led the amazing Pearson sales representatives to do a terrific job of keeping in touch with the professors and students who provide so many wonderful suggestions for this textbook. As always, I am humbled, inspired, and encouraged by all the sales representatives; your role on the team deserves more gratitude than I can express here or with grapefruit.

I am grateful to Phil Mixter of Washington State University, Mary Jane Niles of the University of San Francisco, Bronwen Steele of Estrella Mountain Community College, Jan Miller of American River College, and Tammy Tollison of Wake Technical Community College for their expertise and advice. I also thank Asha Brunings, Jordan Roeder, Todd Primm, and Terry Austin for their contributions to this edition.

I am indebted to my colleagues, Sam Schwarzlose, Nichol Dolby, and Brandon Moore for their expertise and suggestions.

On the home front: Thank you, Jennie and Nick Knapp, Elizabeth Bauman, Seth Daniel, Jeremy Bauman, Larry Latham, Josh Wood, and Mike Isley. You keep me even-keeled. My wife Michelle deserves more recognition than I can possibly express.

Robert W. Bauman
Amarillo, Texas

CONTENTS

1 A Brief History of Microbiology 1

The Early Years of Microbiology 2
    What Does Life Really Look Like? 2
    How Can Microbes Be Classified? 4
The Golden Age of Microbiology 7
    Does Microbial Life Spontaneously Generate? 7
    What Causes Fermentation? 10
    What Causes Disease? 12
    How Can We Prevent Infection and Disease? 15
The Modern Age of Microbiology 18
    What Are the Basic Chemical Reactions of Life? 19
    How Do Genes Work? 19
    What Roles Do Microorganisms Play in the Environment? 20
    How Do We Defend Against Disease? 20
    What Will the Future Hold? 22

CHAPTER SUMMARY 23 • QUESTIONS FOR REVIEW 23
CRITICAL THINKING 25 • CONCEPT MAPPING 26

2 The Chemistry of Microbiology 27

Atoms 28
    Atomic Structure 28
    Isotopes 28
    Electron Configurations 29
Chemical Bonds 31
    Nonpolar Covalent Bonds 31
    Polar Covalent Bonds 33
    Ionic Bonds 33
    Hydrogen Bonds 34
Chemical Reactions 35
    Synthesis Reactions 35
    Decomposition Reactions 36
    Exchange Reactions 36
Water, Acids, Bases, and Salts 37
    Water 37
    Acids and Bases 38
    Salts 39
Organic Macromolecules 40
    Functional Groups 40
    Lipids 40
    Carbohydrates 44
    Proteins 46
    Nucleotides and Nucleic Acids 50

CHAPTER SUMMARY 53 • QUESTIONS FOR REVIEW 55
CRITICAL THINKING 56 • CONCEPT MAPPING 57

3 Cell Structure and Function 58

Processes of Life 59
Prokaryotic and Eukaryotic Cells: An Overview 59
External Structures of Bacterial Cells 62
    Glycocalyces 62
    Flagella 62
    Fimbriae and Pili 65
Bacterial Cell Walls 66
    Gram-Positive Bacterial Cell Walls 67
    Gram-Negative Bacterial Cell Walls 67
    Bacteria Without Cell Walls 69
Bacterial Cytoplasmic Membranes 69
    Structure 69
    Function 70
Cytoplasm of Bacteria 74
    Cytosol 74
    Inclusions 74
    Endospores 75
    Nonmembranous Organelles 76
External Structures of Archaea 77
    Glycocalyces 77
    Flagella 77
    Fimbriae and Hami 78
Archaeal Cell Walls and Cytoplasmic Membranes 78
Cytoplasm of Archaea 79
External Structure of Eukaryotic Cells 80
    Glycocalyces 80
Eukaryotic Cell Walls and Cytoplasmic Membranes 80
Cytoplasm of Eukaryotes 82
    Flagella 82
    Cilia 82
    Other Nonmembranous Organelles 82
    Membranous Organelles 85
    Endosymbiotic Theory 88

CHAPTER SUMMARY 89 • QUESTIONS FOR REVIEW 92
CRITICAL THINKING 94 • CONCEPT MAPPING 95

4 Microscopy, Staining, and Classification 96

Units of Measurement 97
Microscopy 97
    General Principles of Microscopy 98
    Light Microscopy 100
    Electron Microscopy 104
    Probe Microscopy 105

Staining 106
    Preparing Specimens for Staining 106
    Principles of Staining 106
    Simple Stains 108
    Differential Stains 108
    Special Stains 110
    Staining for Electron Microscopy 111
Classification and Identification of Microorganisms 112
    Linnaeus and Taxonomic Categories 112
    Domains 115
    Taxonomic and Identifying Characteristics 115
    Taxonomic Keys 119
MICRO MATTERS 120
CHAPTER SUMMARY 120 • QUESTIONS FOR REVIEW 121
CRITICAL THINKING 123 • CONCEPT MAPPING 123

5 Microbial Metabolism 124

Basic Chemical Reactions Underlying Metabolism 125
    Catabolism and Anabolism 125
    Oxidation and Reduction Reactions 126
    ATP Production and Energy Storage 126
    The Roles of Enzymes in Metabolism 127
Carbohydrate Catabolism 133
    Glycolysis 133
    Cellular Respiration 135
    Metabolic Diversity 140
    Fermentation 140
Other Catabolic Pathways 142
    Lipid Catabolism 143
    Protein Catabolism 143
Photosynthesis 144
    Chemicals and Structures 144
    Light-Dependent Reactions 145
    Light-Independent Reactions 146
Other Anabolic Pathways 149
    Carbohydrate Biosynthesis 149
    Lipid Biosynthesis 150
    Amino Acid Biosynthesis 150
    Nucleotide Biosynthesis 151
Integration and Regulation of Metabolic Functions 151
CHAPTER SUMMARY 154 • QUESTIONS FOR REVIEW 156
CRITICAL THINKING 159 • CONCEPT MAPPING 160

6 Microbial Nutrition and Growth 161

Growth Requirements 162
    Nutrients: Chemical and Energy Requirements 162
    Physical Requirements 165
    Associations and Biofilms 168
Culturing Microorganisms 171
    Clinical Sampling 172
    Obtaining Pure Cultures 172
    Culture Media 174
    Special Culture Techniques 177
    Preserving Cultures 178
Growth of Microbial Populations 179
    Generation Time 180
    Mathematical Considerations in Population Growth 180
    Phases of Microbial Population Growth 180
    Continuous Culture in a Chemostat 182
    Measuring Microbial Reproduction 183
MICRO MATTERS 188
CHAPTER SUMMARY 188 • QUESTIONS FOR REVIEW 189
CRITICAL THINKING 191 • CONCEPT MAPPING 192

7 Microbial Genetics 193

The Structure and Replication of Genomes 194
    The Structure of Nucleic Acids 194
    The Structure of Prokaryotic Genomes 194
    The Structure of Eukaryotic Genomes 196
    DNA Replication 198
Gene Function 202
    The Relationship Between Genotype and Phenotype 202
    The Transfer of Genetic Information 203
    The Events in Transcription 203
    Translation 207
    Regulation of Genetic Expression 212
Mutations of Genes 216
    Types of Mutations 216
    Effects of Point Mutations 216
    Mutagens 217
    Frequency of Mutation 219
    DNA Repair 219
    Identifying Mutants, Mutagens, and Carcinogens 221
Genetic Recombination and Transfer 223
    Horizontal Gene Transfer Among Prokaryotes 223
    Transposons and Transposition 228
CHAPTER SUMMARY 231 • QUESTIONS FOR REVIEW 233
CRITICAL THINKING 235 • CONCEPT MAPPING 236

8 Recombinant DNA Technology 237

The Role of Recombinant DNA Technology in Biotechnology 238
Tools of Recombinant DNA Technology 238
    Mutagens 238
    The Use of Reverse Transcriptase to Synthesize cDNA 239
    Synthetic Nucleic Acids 239
    Restriction Enzymes 240
    Vectors 241
    CRISPR 242
    Gene Libraries 243
Techniques of Recombinant DNA Technology 243
    Multiplying DNA In Vitro: The Polymerase Chain Reaction 244
    Selecting a Clone of Recombinant Cells 246

Separating DNA Molecules: Gel Electrophoresis and the Southern Blot 246
DNA Microarrays 246
Inserting DNA into Cells 248

Applications of Recombinant DNA Technology 248
Genetic Mapping 248
Microbial Community Studies 252
Pharmaceutical and Therapeutic Applications 252
Agricultural Applications 254

The Ethics and Safety of Recombinant DNA Technology 256

MICRO MATTERS 257

CHAPTER SUMMARY 257 • QUESTIONS FOR REVIEW 258
CRITICAL THINKING 260 • CONCEPT MAPPING 261

9 Controlling Microbial Growth in the Environment 262

Basic Principles of Microbial Control 263
Terminology of Microbial Control 263
Microbial Death Rates 264
Action of Antimicrobial Agents 264

The Selection of Microbial Control Methods 266
Factors Affecting the Efficacy of Antimicrobial Methods 266
Biosafety Levels 268

Physical Methods of Microbial Control 268
Heat-Related Methods 268
Refrigeration and Freezing 271
Desiccation and Lyophilization 272
Filtration 272
Osmotic Pressure 273
Radiation 273

Chemical Methods of Microbial Control 276
Phenol and Phenolics 276
Alcohols 276
Halogens 277
Oxidizing Agents 278
Surfactants 278
Heavy Metals 279
Aldehydes 279
Gaseous Agents 280
Enzymes 280
Antimicrobial Drugs 280
Methods for Evaluating Disinfectants and Antiseptics 280
Development of Resistant Microbes 282

CHAPTER SUMMARY 283 • QUESTIONS FOR REVIEW 284
CRITICAL THINKING 286 • CONCEPT MAPPING 287

10 Controlling Microbial Growth in the Body: Antimicrobial Drugs 288

The History of Antimicrobial Agents 289
Mechanisms of Antimicrobial Action 289
Inhibition of Cell Wall Synthesis 290
Inhibition of Protein Synthesis 293
Disruption of Cytoplasmic Membranes 293
Inhibition of Metabolic Pathways 295
Inhibition of Nucleic Acid Synthesis 295
Prevention of Virus Attachment, Entry, or Uncoating 298

Clinical Considerations in Prescribing Antimicrobial Drugs 298
Spectrum of Action 298
Effectiveness 298
Routes of Administration 300
Safety and Side Effects 302

Resistance to Antimicrobial Drugs 303
The Development of Resistance and Persister Cells 303
Mechanisms of Resistance 305
Multiple Resistance and Cross Resistance 306
Retarding Resistance 307

MICRO MATTERS 319

CHAPTER SUMMARY 319 • QUESTIONS FOR REVIEW 320
CRITICAL THINKING 322 • CONCEPT MAPPING 323

11 Characterizing and Classifying Prokaryotes 324

General Characteristics of Prokaryotic Organisms 325
Morphology of Prokaryotic Cells 325
Endospores 325
Reproduction of Prokaryotic Cells 326
Arrangements of Prokaryotic Cells 327

Modern Prokaryotic Classification 329

Survey of Archaea 329
Extremophiles 329
Methanogens 332

Survey of Bacteria 332
Deeply Branching and Phototrophic Bacteria 332
Low G + C Gram-Positive Bacteria 334
High G + C Gram-Positive Bacteria 336
Gram-Negative Proteobacteria 338
Other Gram-Negative Bacteria 347

CHAPTER SUMMARY 348 • QUESTIONS FOR REVIEW 350
CRITICAL THINKING 351 • CONCEPT MAPPING 352

12 Characterizing and Classifying Eukaryotes 353

General Characteristics of Eukaryotic Organisms 354
Reproduction of Eukaryotes 354
Classification of Eukaryotic Organisms 357

Protozoa 358
Distribution of Protozoa 358
Morphology of Protozoa 359
Nutrition of Protozoa 359
Reproduction of Protozoa 360
Classification of Protozoa 360

Fungi 364
The Significance of Fungi 365

Morphology of Fungi   365
Nutrition of Fungi   366
Reproduction of Fungi   367
Classification of Fungi   368
Lichens   371

Algae   373
Distribution of Algae   373
Morphology of Algae   373
Reproduction of Algae   373
Classification of Algae   374

Water Molds   376

Other Eukaryotes of Microbiological Interest: Parasitic Helminths and Vectors   377
Arachnids   377
Insects   378

CHAPTER SUMMARY   379 • QUESTIONS FOR REVIEW   381
CRITICAL THINKING   382 • CONCEPT MAPPING   383

13 Characterizing and Classifying Viruses, Viroids, and Prions   384

Characteristics of Viruses   385
Genetic Material of Viruses   386
Hosts of Viruses   386
Sizes of Viruses   388
Capsid Morphology   389
Viral Shapes   389
The Viral Envelope   389

Classification of Viruses   390

Viral Replication   392
Lytic Replication of Bacteriophages   392
Lysogenic Replication of Bacteriophages   395
Replication of Animal Viruses   396

The Role of Viruses in Cancer   401

Culturing Viruses in the Laboratory   402
Culturing Viruses in Mature Organisms   402
Culturing Viruses in Embryonated Chicken Eggs   403
Culturing Viruses in Cell (Tissue) Culture   403

Are Viruses Alive?   404

Other Parasitic Particles: Viroids and Prions   404
Characteristics of Viroids   404
Characteristics of Prions   405

MICRO MATTERS   408

CHAPTER SUMMARY   408 • QUESTIONS FOR REVIEW   410
CRITICAL THINKING   411 • CONCEPT MAPPING   412

14 Infection, Infectious Diseases, and Epidemiology   413

Symbiotic Relationships Between Microbes and Their Hosts   414
Types of Symbiosis   414

Microbiome of Humans   416
How Normal Microbiota Become Opportunistic Pathogens   417

Reservoirs of Infectious Diseases of Humans   418
Animal Reservoirs   418
Human Carriers   418
Nonliving Reservoirs   419

The Invasion and Establishment of Microbes in Hosts: Infection   420
Exposure to Microbes: Contamination and Infection   420
Portals of Entry   420
The Role of Adhesion in Infection   421

The Nature of Infectious Disease   422
Manifestations of Disease: Symptoms, Signs, and Syndromes   423
Causation of Disease: Etiology   423
Virulence Factors of Infectious Agents   426
The Stages of Infectious Diseases   429

The Movement of Pathogens Out of Hosts: Portals of Exit   430

Modes of Infectious Disease Transmission   430
Contact Transmission   431
Vehicle Transmission   431
Vector Transmission   432

Classification of Infectious Diseases   432

Epidemiology of Infectious Diseases   434
Frequency of Disease   434
Epidemiological Studies   436
Hospital Epidemiology: Healthcare-Associated (Nosocomial) Infections   437
Epidemiology and Public Health   440

CHAPTER SUMMARY   441 • QUESTIONS FOR REVIEW   443
CRITICAL THINKING   445 • CONCEPT MAPPING   445

15 Innate Immunity   446

An Overview of the Body's Defenses   447

The Body's First Line of Defense   447
The Role of Skin in Innate Immunity   447
The Role of Mucous Membranes in Innate Immunity   448
The Role of the Lacrimal Apparatus in Innate Immunity   449
The Role of the Microbiome in Innate Immunity   449
Other First-Line Defenses   450

The Body's Second Line of Defense   451
Defense Components of Blood   451
Phagocytosis   454
Nonphagocytic Killing   456
Nonspecific Chemical Defenses Against Pathogens   457
Inflammation   462
Fever   464

CHAPTER SUMMARY   467 • QUESTIONS FOR REVIEW   468
CRITICAL THINKING   470 • CONCEPT MAPPING   471

XVIII TABLE OF CONTENTS

16 Adaptive Immunity 472

Overview of Adaptive Immunity 473
Elements of Adaptive Immunity 474
The Tissues and Organs of the Lymphatic System 474
Antigens 476
Preparation for an Adaptive Immune Response 478
B Lymphocytes (B Cells) and Antibodies 483
Immune Response Cytokines 488
Cell-Mediated Immune Responses 489
Activation of Cytotoxic T Cell Clones and Their Functions 489
The Perforin-Granzyme Cytotoxic Pathway 491
The CD95 Cytotoxic Pathway 491
Memory T Cells 491
T Cell Regulation 492
Antibody Immune Responses 492
Inducement of T-Dependent Antibody Immunity with Clonal Selection 492
Memory Cells and the Establishment of Immunological Memory 494
Types of Acquired Immunity 495
Naturally Acquired Active Immunity 495
Naturally Acquired Passive Immunity 495
Artificially Acquired Active Immunity 496
Artificially Acquired Passive Immunotherapy 496

CHAPTER SUMMARY 498 • QUESTIONS FOR REVIEW 499
CRITICAL THINKING 501 • CONCEPT MAPPING 502

17 Immunization and Immunoassays 503

Immunization 504
Brief History of Immunization 504
Active Immunization 505
Passive Immunotherapy 510
Serological Tests That Use Antigens and Corresponding Antibodies 512
Precipitation Tests 512
Turbidimetric and Nephelometric Tests 513
Agglutination Tests 513
Neutralization Tests 515
The Complement Fixation Test 515
Labeled Immunoassays 515
Point-of-Care Testing 518

CHAPTER SUMMARY 521 • QUESTIONS FOR REVIEW 522
CRITICAL THINKING 524 • CONCEPT MAPPING 524

18 Immune Disorders 525

Hypersensitivities 526
Type I (Immediate) Hypersensitivity 526
Type II (Cytotoxic) Hypersensitivity 530
Type III (Immune Complex–Mediated) Hypersensitivity 532
Type IV (Delayed or Cell-Mediated) Hypersensitivity 534

Autoimmune Diseases 538
Causes of Autoimmune Diseases 538
Examples of Autoimmune Diseases 539
Immunodeficiency Diseases 540
Primary Immunodeficiency Diseases 540
Acquired Immunodeficiency Diseases 540

MICRO MATTERS 541

CHAPTER SUMMARY 541 • QUESTIONS FOR REVIEW 543
CRITICAL THINKING 544 • CONCEPT MAPPING 545

19 Pathogenic Gram-Positive Bacteria 546

Staphylococcus 547
Structure and Physiology 547
Pathogenicity 547
Epidemiology 548
Staphylococcal Diseases 549
Diagnosis, Treatment, and Prevention 550
Streptococcus 552
Group A *Streptococcus*: *Streptococcus pyogenes* 552
Group B *Streptococcus*: *Streptococcus agalactiae* 556
Other Beta-Hemolytic Streptococci 557
Alpha-Hemolytic Streptococci: The Viridans Group 557
Streptococcus pneumoniae 557
Enterococcus 559
Structure and Physiology 560
Pathogenesis, Epidemiology, and Diseases 560
Diagnosis, Treatment, and Prevention 560
Bacillus 560
Structure, Physiology, and Pathogenicity 560
Epidemiology 561
Disease 561
Diagnosis, Treatment, and Prevention 562
Clostridium 562
Clostridium difficile 562
Clostridium botulinum 563
Clostridium tetani 565
Clostridium perfringens 567
Listeria 568
Mycoplasmas 568
Mycoplasma pneumoniae 569
Other Mycoplasmas 572
Corynebacterium 572
Pathogenesis, Epidemiology, and Disease 572
Diagnosis, Treatment, and Prevention 573
Mycobacterium 573
Tuberculosis 573
Leprosy 574
Other Mycobacterial Infections 575
Propionibacterium 578
Nocardia and *Actinomyces* 579
Nocardia asteroides 579
Actinomyces 580

CHAPTER SUMMARY 581 • QUESTIONS FOR REVIEW 583
CRITICAL THINKING 584 • CONCEPT MAPPING 585

20 Pathogenic Gram-Negative Cocci and Bacilli 586

Pathogenic Gram-Negative Cocci: *Neisseria* 587
- Structure and Physiology of *Neisseria* 587
- The Gonococcus: *Neisseria gonorrhoeae* 588
- The Meningococcus: *Neisseria Meningitidis* 589

Pathogenic, Gram-Negative, Facultatively Anaerobic Bacilli 591
- The Enterobacteriaceae: An Overview 591
- Coliform Opportunistic Enterobacteriaceae 595
- Noncoliform Opportunistic Enterobacteriaceae 597
- Truly Pathogenic Enterobacteriaceae 597
- The Pasteurellaceae 603

Pathogenic, Gram-Negative, Aerobic Bacilli 605
- *Bartonella* 605
- *Brucella* 606
- *Bordetella* 606
- *Burkholderia* 608
- Pseudomonads 608
- *Francisella* 611
- *Legionella* 611
- *Coxiella* 613

Pathogenic, Gram-Negative, Anaerobic Bacilli 614
- *Bacteroides* 614
- *Prevotella* 614

CHAPTER SUMMARY 615 • QUESTIONS FOR REVIEW 617
CRITICAL THINKING 618 • CONCEPT MAPPING 619

21 Rickettsias, Chlamydias, Spirochetes, and Vibrios 620

Rickettsias 621
- *Rickettsia* 621
- *Orientia tsutsugamushi* 622
- *Ehrlichia* and *Anaplasma* 623

Chlamydias 626
- *Chlamydia trachomatis* 626
- *Chlamydophila pneumoniae* 629
- *Chlamydophila psittaci* 629

Spirochetes 629
- *Treponema* 630
- *Borrelia* 633
- *Leptospira* 636

Pathogenic Gram-Negative Vibrios 637
- *Vibrio* 637
- *Campylobacter jejuni* 639
- *Helicobacter pylori* 640

CHAPTER SUMMARY 642 • QUESTIONS FOR REVIEW 643
CRITICAL THINKING 645 • CONCEPT MAPPING 646

22 Pathogenic Fungi 647

An Overview of Medical Mycology 648
- The Epidemiology of Mycoses 648
- Categories of Fungal Agents: True Fungal Pathogens and Opportunistic Fungi 648
- Clinical Manifestations of Fungal Diseases 649
- The Diagnosis of Fungal Infections 649
- Antifungal Therapies 650
- Antifungal Vaccines 650

Systemic Mycoses Caused by Pathogenic Fungi 651
- Histoplasmosis 651
- Blastomycosis 653
- Coccidioidomycosis 654
- Paracoccidioidomycosis 656

Systemic Mycoses Caused by Opportunistic Fungi 656
- *Pneumocystis* Pneumonia 656
- Candidiasis 657
- Aspergillosis 657
- Cryptococcosis 659
- Zygomycoses and Microsporidiosis 662
- The Emergence of Fungal Opportunists in AIDS Patients 664

Superficial, Cutaneous, and Subcutaneous Mycoses 664
- Superficial Mycoses 664
- Cutaneous and Subcutaneous Mycoses 666

Fungal Intoxications and Allergies 668
- Mycotoxicoses 668
- Mushroom Poisoning (Mycetismus) 668
- Allergies to Fungi 669

CHAPTER SUMMARY 670 • QUESTIONS FOR REVIEW 671
CRITICAL THINKING 674 • CONCEPT MAPPING 674

23 Parasitic Protozoa, Helminths, and Arthropod Vectors 675

Parasitology 676
Arthropod Vectors 676
Protozoan Parasites of Humans 677
- Ciliates 677
- Amoebas 678
- Flagellates 679
- Apicomplexans 686

Helminthic Parasites of Humans 693
- Cestodes 693
- Trematodes 695
- Nematodes 698

CHAPTER SUMMARY 704 • QUESTIONS FOR REVIEW 705
CRITICAL THINKING 707 • CONCEPT MAPPING 708

24 Pathogenic DNA Viruses 709

Poxviridae 710
   Smallpox 710
   Molluscum Contagiosum 712
   Other Poxvirus Infections 712
Herpesviridae 713
   Infections of *Human Herpesvirus 1* and *2* 714
   Human Herpesvirus 3 (Varicella-Zoster Virus) Infections 716
   Human Herpesvirus 4 (Epstein-Barr Virus) Infections 718
   Human Herpesvirus 5 (Cytomegalovirus) Infections 720
   Other Herpesvirus Infections 721
Papillomaviridae and *Polyomaviridae* 722
   Papillomavirus Infections 722
   Polyomavirus Infections 722
Adenoviridae 722
Hepadnaviridae 725
   Hepatitis B Infections 726
   The Role of Hepatitis B Virus in Hepatic Cancer 728
Parvoviridae 728

CHAPTER SUMMARY 729 • QUESTIONS FOR REVIEW 730
CRITICAL THINKING 732 • CONCEPT MAPPING 733

25 Pathogenic RNA Viruses 734

Naked, Positive ssRNA Viruses: *Picornaviridae, Caliciviridae, Astroviridae,* and *Hepeviridae* 735
   Common Colds Caused by Rhinoviruses 735
   Diseases of Enteroviruses 736
   Hepatitis A 739
   Acute Gastroenteritis 739
   Hepatitis E 740
Enveloped, Positive ssRNA Viruses: *Togaviridae, Flaviviridae,* and *Coronaviridae* 741
   Diseases of +RNA Arboviruses 741
   Other Diseases of Enveloped +ssRNA Viruses 747
Enveloped, Positive ssRNA Viruses with Reverse Transcriptase: *Retroviridae* 750
   Oncogenic Retroviruses (*Deltaretrovirus*) 750
   Immunosuppressive Retroviruses (*Lentivirus*) and Acquired Immunodeficiency Syndrome 751
Enveloped, Unsegmented, Negative ssRNA Viruses: *Paramyxoviridae, Rhabdoviridae,* and *Filoviridae* 758
   Measles 758
   Diseases of Parainfluenza Virus 759
   Mumps 760
   Disease of Respiratory Syncytial Virus 760
   Rabies 761
   Filoviruses (Hemorrhagic Fever Viruses) 763
Enveloped, Segmented, Negative ssRNA Viruses: *Orthomyxoviridae, Bunyaviridae,* and *Arenaviridae* 763
   Influenza 763
   Diseases of Bunyaviruses 766
   Diseases of Arenaviruses 767
Naked, Segmented dsRNA Viruses: *Reoviridae* 770
   Rotavirus Infections 770
   Coltivirus Infections 770

CHAPTER SUMMARY 772 • QUESTIONS FOR REVIEW 774
CRITICAL THINKING 776 • CONCEPT MAPPING 777

26 Applied and Industrial Microbiology 778

Food Microbiology 779
   The Roles of Microorganisms in Food Production 779
   The Causes and Prevention of Food Spoilage 782
   Foodborne Illnesses 786
Industrial Microbiology 786
   The Roles of Microbes in Industrial Fermentations 786
   Industrial Products of Microorganisms 787
   Water Treatment 789

CHAPTER SUMMARY 796 • QUESTIONS FOR REVIEW 797
CRITICAL THINKING 800 • CONCEPT MAPPING 801

27 Microbial Ecology and Microbiomes 802

Environmental Microbiology 803
   Microbial Ecology 803
   Bioremediation 805
   The Problem of Acid Mine Drainage 805
   The Roles of Microorganisms in Biogeochemical Cycles 806
   Soil Microbiology 809
   Aquatic Microbiology 811
Biological Warfare and Bioterrorism 812
   Assessing Microorganisms as Potential Agents of Warfare or Terror 812
   Known Microbial Threats 813
   Defense Against Bioterrorism 814
   Roles of Recombinant Genetic Technology in Bioterrorism 815

CHAPTER SUMMARY 816 • QUESTIONS FOR REVIEW 817
CRITICAL THINKING 819 • CONCEPT MAPPING 819

Answers to Questions for Review AN-1
Appendix A MINAH Standards A-1
Appendix B Pronunciation Guide A-4
Glossary G-1
Credits C-1
Subject Index I-1

FEATURE BOXES

BENEFICIAL MICROBES

Bread, Wine, and Beer 10
Architecture-Preserving Bacteria 39
Plastics Made Perfect? 75
Glowing Viruses 113
A Nuclear Waste-Eating Microbe? 169
Life in a Hot Tub 246
Our Other "Organ" 253
Controlling Bacteria with Viruses 277
Probiotics: Using Live Microorganisms to Treat or Prevent Disease 304
A Microtube of Superglue 341
Fungi for $100,000 a Pound 371
Good Viruses? Who Knew? 390
Prescription Viruses? 395
A Bioterrorist Worm 415
Cowpox: To Vaccinate or Not to Vaccinate? 510
Eliminating *Aedes*-Borne Diseases? 746

EMERGING DISEASE CASE STUDY

Variant Creutzfeldt-Jakob Disease 21
Necrotizing Fasciitis 118
Vibrio vulnificus Infection 170
Acanthamoeba Keratitis 267
Community-Associated MRSA 305
Pertussis 345
Chronic Aspergillosis 370
Chikungunya 401
Microsporidiosis 496
Buruli Ulcer 578
Melioidosis 610
A New Cause of Spots 622
Pulmonary Blastomycosis 654
Babesiosis 691
Snail Fever Reemerges in China 699
Monkeypox 713
Norovirus in the Dorm 740
Tick-Borne Encephalitis 743
A Deadly Mosquito Bite? 746
Attack in the Lake 790

CLINICAL CASE STUDY

Remedy for Fever or Prescription for Death? 16
Raw Oysters and Antacids: A Deadly Mix? 40
The Big Game 70
Cavities Gone Wild 171
Boils in the Locker Room 182
Deadly Horizontal Gene Transfer 230
Antibiotic Overkill 299
Battling the Enemy 301
Treating TB 307
Invasion from Within or Without? 407
A Deadly Carrier 419
TB in the Nursery 432
Evaluating an Abnormal CBC 454
The Stealth Invader 460
The First Time's Not the Problem 535
A Fatal Case of Methicillin-Resistant *Staphylococcus aureus* (MRSA) 551
This Cough Can Kill 574
A Painful Problem 594
A Heart-Rending Experience 595
A Sick Camper 604
When "Health Food" Isn't 609
Nightmare on The Island 612
The Case of the Lactovegetarians 641
What's Ailing the Bird Enthusiast? 652
Disease from a Cave 663
A Protozoan Mystery 682
A Sick Soldier 682
A Fluke Disease? 697
Grandfather's Shingles 719
A Child with Warts 724
The Eyes Have It 727
A Threat from the Wild 762
A Case of AIDS 766
Bioterrorism in the Mail? 816

MICROBE AT A GLANCE

Streptococcus pneumoniae 558
Clostridium botulinum 565
Neisseria gonorrhoeae 589
Treponema pallidum 631
Helicobacter pylori 640
Histoplasma capsulatum 653
Aspergillus 659
Orthopoxvirus variola (*Smallpox Virus*) 712
Adenovirus 725
Lentivirus human immunodeficiency virus (HIV) 754
Morbillivirus measles virus 759

DISEASE IN DEPTH

Necrotizing Fasciitis 554
Listeriosis 570
Tuberculosis 576
Bacterial Urinary Tract Infections 598
Rocky Mountain Spotted Fever 624
Candidiasis 660
Giardiasis 683
Malaria 688
Papillomas 723
Ebola Viral Disease 764
Influenza 768

SOLVE THE PROBLEM

Smallpox: To Be or Not To Be? 2
The Microbes Ate My Homework 162
How Clean Is Too Clean? 263
Microengineer a Better Mosquito? 385
Microbes in the Produce Aisle 414
Rights Versus Responsibilities 504
Fecal Microbiome Transfer: Medicine or Magic? 803

1 A Brief History of Microbiology

Before You Begin

1. What does the science of microbiology study?
2. Are most microorganisms harmful or harmless to people?

MICRO IN THE CLINIC

Too Much Cake, or Something Worse?

PATTY IS A MOTHER TO 14-YEAR-OLD twins and works full time. Between her own job, driving her kids to practices and events, and spending time with her husband, Patty is constantly on the go. This past weekend was no exception. On Friday night, her office group went out for happy hour to celebrate a colleague's promotion. They had a great time—eating sushi, drinking wine, and relaxing. Saturday morning, her daughter's soccer team had a brunch, and in the afternoon her son's Little League team had an end-of-the-season barbecue. Saturday night she felt a little bit bloated but thought it was from the all the food she had eaten at the brunch—she was so full from the brunch that she had eaten only fruit salad at the barbecue.

Late Sunday afternoon, Patty and her husband returned home from his sister's birthday party. As they began to prepare dinner, Patty started to have a stomachache and felt a bit nauseated. She suspected it was from eating too much birthday cake; however, when she woke up in the middle of the night with diarrhea, she thought it might be something more than the cake. Monday morning Patty was unable to go to work—she had a terrible headache, and the diarrhea had lasted all night long. When she starts vomiting early Monday afternoon, she decides that she needs to go to the doctor.

1. Is it just a case of too much cake?
2. What else could be causing Patty's symptoms?

Turn to the end of the chapter (p. 22) to find out.

Our microbial world can appear beautiful, as shown in this artist's rendition of *Norovirus*.

SOLVE THE PROBLEM

Smallpox: To Be or Not To Be?

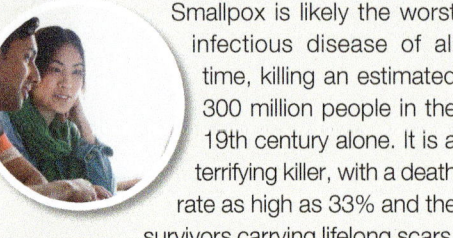

Smallpox is likely the worst infectious disease of all time, killing an estimated 300 million people in the 19th century alone. It is a terrifying killer, with a death rate as high as 33% and the survivors carrying lifelong scars.

British medical doctor Edward Jenner is credited with inventing smallpox vaccination—the world's first immunization. On May 14, 1796, Jenner collected secretions from a cowpox sore on the hand of a milkmaid and rubbed them into scratches he made on the skin of an eight-year-old boy. Then, about a month later, he injected the boy with secretions from a lesion on a smallpox patient. The child did not get smallpox; he was immune. Jenner termed his technique *vaccination*, which comes from the Latin term for cow, *vacca*.

Medical doctors began vaccinating people with special two-pronged needles, and eventually smallpox was eradicated worldwide. The last case was documented on October 26, 1977.

Eradication represents one of the great triumphs of modern medicine, but smallpox virus itself still exists. Stocks are kept frozen in secure laboratories at the Centers for Disease Control and Prevention (CDC) in Atlanta, Georgia, and in the State Research Center of Virology and Biotechnology in Koltsovo, Russia.

Imagine you are assigned to be part of a team tasked to determine what to do with the world's remaining stores of smallpox virus.

- Should governments and laboratories keep them?
- Or should they be destroyed? In other words, should we intentionally make a species extinct forever?

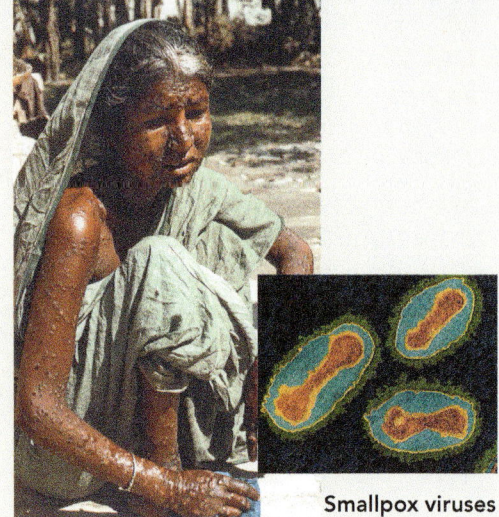

Smallpox viruses

- What facts do you need to make an informed decision?
- If the decision were to be made today, how would you vote?

Science is the study of nature that proceeds by posing questions about observations. Why are there seasons? What is the function of the nodules at the base of this plant? Why does this bread taste sour? What does plaque from between teeth look like when magnified? What causes the spread of diseases?

Many early written records show that people have always asked questions like these. For example, the Greek physician Hippocrates (ca. 460–ca. 377 B.C.) wondered whether there is a link between environment and disease, and the Greek historian Thucydides (ca. 460–ca. 404 B.C) questioned why he and other survivors of the plague could have close contact with victims and not fall ill again. For many centuries, the answers to these and other fundamental questions about the nature of life remained largely unanswered. But about 350 years ago, the invention of the microscope began to provide some clues.

In this chapter, we'll see how one man's determination to answer a fundamental question about the nature of life—What does life really look like?—led to the birth of a new science called *microbiology*. We'll then see how the search for answers to other questions—such as those concerning spontaneous generation, the reason fermentation occurs, and the cause of disease—prompted advances in this new science. Finally, we'll look briefly at some of the key questions microbiologists are asking today.

The Early Years of Microbiology

The early years of microbiology brought the first observations of microbial life and the initial efforts to organize them into logical classifications.

What Does Life Really Look Like?

LEARNING OUTCOMES

1.1 Describe the world-changing scientific contributions of van Leeuwenhoek.
1.2 Define microbes in the words of van Leeuwenhoek and as we know them today.

A few people have changed the world of science forever. We've all heard of Galileo, Newton, and Einstein, but the list also includes Antoni van Leeuwenhoek (lā'věn-huk; 1632–1723), a Dutch clothier, merchant, and lens grinder, and the man who first discovered the bacterial world (FIGURE 1.1).

Van Leeuwenhoek was born in Delft, the Netherlands, and lived most of his 90 years in the city of his birth. What set van Leeuwenhoek apart from many other men of his generation was an unending curiosity coupled with an almost stubborn desire

▲ FIGURE 1.1 **Antoni van Leeuwenhoek.** Van Leeuwenhoek reported the existence of protozoa in 1674 and of bacteria in 1676. *Why did van Leeuwenhoek discover protozoa before bacteria?*

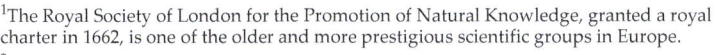

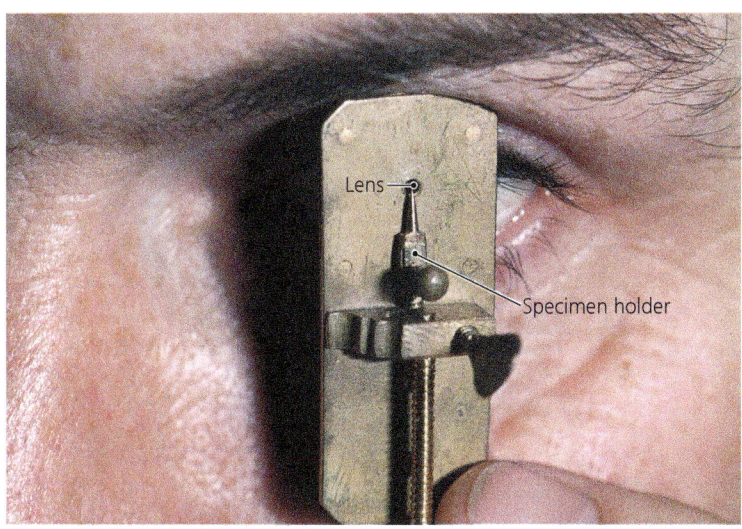

▲ FIGURE 1.2 **Reproduction of van Leeuwenhoek's microscope.** This simple device is little more than a magnifying glass with screws for manipulating the specimen, yet with it, van Leeuwenhoek changed the way we see our world. The lens, which is convex on both sides, is about the size of a pinhead. The object to be viewed was mounted either directly on the specimen holder or inside a small glass tube, which was then mounted on the specimen holder.

to do everything for himself. His journey to fame began simply enough, when as a cloth merchant he needed to examine the quality of cloth. Rather than merely buying a magnifying lens, he learned to make glass lenses of his own (**FIGURE 1.2**). Soon he began asking, "What does it really look like?" of everything in his world: the stinger of a bee, the brain of a fly, the leg of a louse, a drop of blood, flakes of his own skin. To find answers, he spent hours examining, reexamining, and recording every detail of each object he observed.

Making and looking through his simple microscopes, really no more than magnifying glasses, became the overwhelming passion of his life. He often made a new microscope for each specimen, which remained mounted so that he could view it again and again. Then one day, he turned a lens onto a drop of water. We don't know what he expected to see, but certainly he saw more than he had anticipated. As he reported to the Royal Society of London[1] in 1674, he was surprised and delighted by

> some green streaks, spirally wound serpent-wise, and orderly arranged.... Among these there were, besides, very many little animalcules, some were round, while others a bit bigger consisted of an oval. On these last, I saw two little legs near the head, and two little fins at the hind most end of the body.... And the motion of most of these animalcules in the water was so swift, and so various, upwards, downwards, and round about, that 'twas wonderful to see.[2]

Van Leeuwenhoek had discovered the microbial world, which today we know to be populated with tiny animals, fungi, algae, and single-celled protozoa (**FIGURE 1.3**). In a later report to the Royal Society, he noted that

> the number of these animals in the plaque of a man's teeth, are so many that I believe they exceed the number of men in a kingdom.... in a quantity of matter no bigger than the 1/100 part of a [grain of] sand.

▲ FIGURE 1.3 **The microbial world.** Van Leeuwenhoek reported seeing a scene very much like this, full of numerous fantastic, cavorting creatures.

---

[1] The Royal Society of London for the Promotion of Natural Knowledge, granted a royal charter in 1662, is one of the older and more prestigious scientific groups in Europe.

[2] Antoni van Leeuwenhoek, in a letter to the Royal Society of London for the Promotion of Natural Knowledge.

From the figure accompanying his report and the precise description of the size of these organisms from between his teeth, we know that van Leeuwenhoek was reporting the existence of bacteria. By the end of the 19th century, van Leeuwenhoek's "beasties," as he sometimes dubbed them, were called **microorganisms**, and today we also know them as **microbes**. Both terms include all organisms that are too small to be seen without a microscope.

Because of the quality of his microscopes, his profound observational skills, his detailed reports over a 50 year period, and his report of the discovery of many types of microorganisms, the Dutchman Antoni van Leeuwenhoek was elected to the Royal Society of London in 1680. He was one of the more famous scientists of his time.

How Can Microbes Be Classified?

LEARNING OUTCOMES

1.3 List six groups of microorganisms.
1.4 Explain why protozoa, algae, and nonmicrobial parasitic worms are studied in microbiology.
1.5 Differentiate prokaryotic from eukaryotic organisms.

Shortly after van Leeuwenhoek made his discoveries, the Swedish botanist Carolus Linnaeus (1707–1778) developed a **taxonomic system**—a system for naming plants and animals and grouping similar organisms together. For instance, Linnaeus and other scientists of the period grouped all organisms into either the animal kingdom or the plant kingdom. Today, biologists still use this basic system, but they have modified Linnaeus's scheme by adding categories that more realistically reflect the relationships among organisms. For example, scientists no longer classify yeasts, molds, and mushrooms as plants but instead as fungi. (We examine taxonomic schemes in more detail in Chapter 4.)

The microorganisms that van Leeuwenhoek described can be grouped into six basic categories: bacteria, archaea, fungi, protozoa, algae, and small multicellular animals. The only type of microbes not described by van Leeuwenhoek is *viruses*,[3] which are too small to be seen without an electron microscope. We briefly consider organisms in the first five categories in the following sections.

Bacteria and Archaea

Bacteria and **archaea** are **prokaryotic**,[4] meaning that their cells lack nuclei; that is, their genes are not surrounded by a membrane. Bacterial cell walls are composed of a polysaccharide called *peptidoglycan*, though some bacteria lack cell walls. The cell walls of archaea lack peptidoglycan and instead are composed of other chemicals. Members of both groups reproduce asexually. (Chapters 3, 4, and 11 examine other differences between bacteria and archaea, and Chapters 19–21 discuss pathogenic [disease-causing] bacteria.)

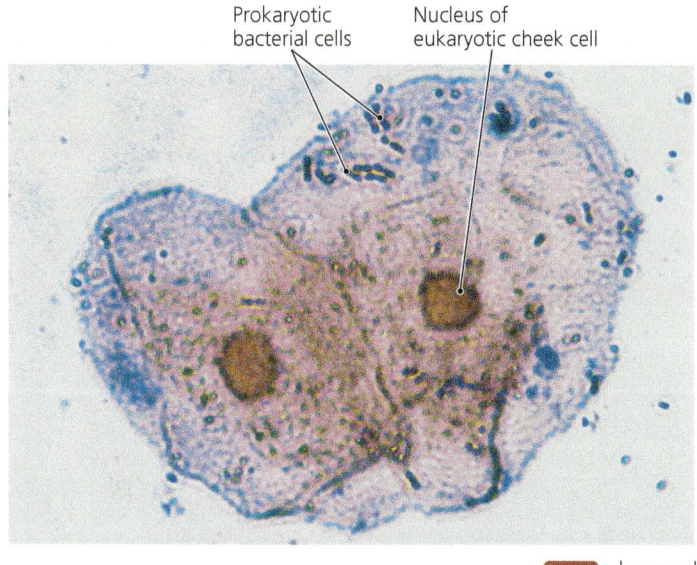

▲ FIGURE 1.4 **Cells of the bacterium *Streptococcus* (dark blue) and two human cheek cells.** Notice the size difference.

Most archaea and bacteria are much smaller than eukaryotic cells (**FIGURE 1.4**). They live singly or in pairs, chains, or clusters in almost every habitat containing sufficient moisture. Scientists first discovered archaea in extreme environments, such as the highly saline and arsenic-rich Mono Lake in California, acidic hot springs in Yellowstone National Park, and oxygen-depleted mud at the bottom of swamps. No archaea are known to cause diseases in humans.

Though bacteria may have a poor reputation, the great majority do not cause disease in animals, humans, or crops. Indeed, bacteria are beneficial to us in many ways. For example, without beneficial bacteria, our bodies would be much more susceptible to disease. Also, bacteria (and fungi) degrade dead plants and animals to release phosphorus, sulfur, nitrogen, and carbon back into the air, soil, and water to be used by new generations of organisms. Without microbial recyclers, the world would be buried under the corpses of uncountable dead organisms.

Fungi

Fungi (fŭn'jī)[5] are **eukaryotic**;[6] that is, each of their cells contains a nucleus composed of genetic material surrounded by a distinct membrane. Fungal cells differ from plant cells in that fungi obtain their food from other organisms (rather than making it for themselves). Fungal cells differ from animal cells by having cell walls.

Microscopic fungi include some molds and yeasts. **Molds** are typically multicellular organisms that grow as long filaments that intertwine to make up the body of the mold. Molds reproduce by sexual and asexual spores, which are cells that produce a new individual without fusing with another cell (**FIGURE 1.5a**). The cottony growths on cheese, bread, and jams are molds. *Penicillium chrysogenum* (pen-i-sil'ē-ŭm krī-so'jĕn-ŭm) is a mold that produces penicillin.

---

[3]Technically, viruses are not "organisms," because they neither replicate themselves nor carry on the chemical reactions of living things.

[4]From Greek *pro*, meaning "before," and *karyon*, meaning "kernel" (which, in this case, refers to the nucleus of a cell).

[5]Plural of the Latin *fungus*, meaning "mushroom."

[6]From Greek *eu*, meaning "true," and *karyon*, meaning "kernel."

▲ FIGURE 1.5 Fungi. (a) The mold *Penicillium chrysogenum*, which produces penicillin, has long filaments that intertwine to form its body (not shown). It reproduces by spores. (b) The yeast *Saccharomyces cerevisiae*. Yeasts are round to oval and typically reproduce by budding.

Yeasts are unicellular and typically oval to round. They reproduce asexually by *budding*, a process in which a daughter cell grows off the mother cell. Some yeasts also produce sexual spores. An example of a useful yeast is *Saccharomyces cerevisiae* (sak-ă-rō-mī′sēz se-ri-vis′ē-ī; FIGURE 1.5b), which causes bread to rise and produces alcohol from sugar (see **Beneficial Microbes: Bread, Wine, and Beer** on p. 10). Another example of a yeast is *Candida albicans* (kan′did-ă al′bi-kanz), which causes most cases of yeast infections in women. (Chapters 12, 22, and 26 discuss fungi and their significance in the environment, as agents of human disease, and in food production.)

Protozoa

Protozoa are single-celled eukaryotes that are similar to animals in their nutritional needs and cellular structure. In fact, *protozoa* is Greek for "first animals," though scientists today classify them in their own groups rather than as animals. Most protozoa are capable of locomotion, and one way scientists categorize protozoa is according to their locomotive structures: *pseudopods*,[7] *cilia*,[8] or *flagella*.[9] Pseudopods are extensions of a cell that flow in the direction of travel (FIGURE 1.6a). Cilia are numerous short protrusions of a cell that beat rhythmically to propel a protozoan through its environment (FIGURE 1.6b). Flagella are also extensions of a cell but are fewer, longer, and more whiplike than cilia (FIGURE 1.6c). Some protozoa, such as the malaria-causing *Plasmodium* (plaz-mō′dē-ŭm), are nonmotile in their mature forms.

Many protozoa live freely in water, but some live inside animal hosts, where they can cause disease. Most protozoa reproduce

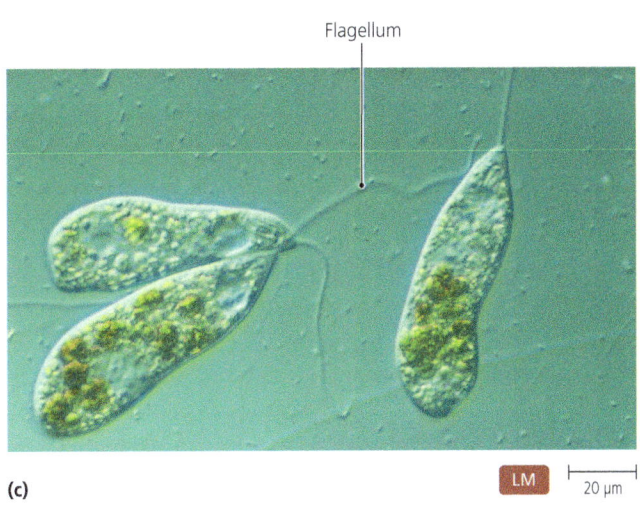

▲ FIGURE 1.6 Locomotive structures of protozoa. (a) Pseudopods are cellular extensions used for locomotion and feeding, as seen in *Amoeba proteus*. (b) *Blepharisma americana* moves by means of cilia. (c) Flagella are whiplike extensions that are less numerous and longer than cilia, as seen in three individuals of *Peranema*. *How do cilia and flagella differ?*

Figure 1.6 Cilia are short and numerous and often cover the cell, whereas flagella are long and relatively few in number.

---

[7]Plural Greek *pseudes*, meaning "false," and *podos*, meaning "foot."
[8]Plural of the Latin *cilium*, meaning "eyelid."
[9]Plural of the Latin *flagellum*, meaning "whip."

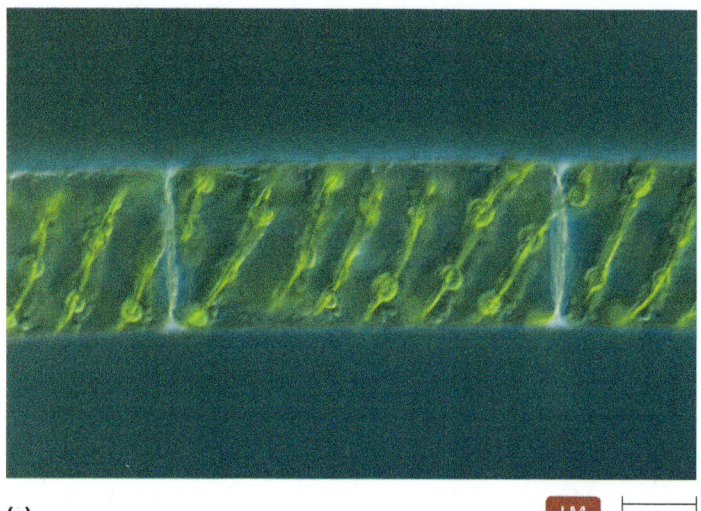

▲ FIGURE 1.7 **Algae.** (a) *Spirogyra*. These microscopic algae grow as chains of cells containing helical photosynthetic structures. (b) Diatoms. These beautiful algae have glasslike cell walls.

asexually, though some are sexual as well. (Chapters 12 and 23 further examine protozoa and some diseases they cause.)

Algae

Algae[10] are unicellular or multicellular *photosynthetic* eukaryotes; that is, like plants, they make their own food from carbon dioxide and water using energy from sunlight. They differ from plants in the relative simplicity of their reproductive structures. Algae are categorized on the basis of their pigmentation and the composition of their cell walls.

Large algae, commonly called seaweeds and kelps, are common in the world's oceans. Manufacturers use gelatinous chemicals from the cell walls of some large algae as thickeners and emulsifiers in many foods and cosmetics, while health professionals use them as lubricants. Scientists use the algae-derived chemical called *agar* to solidify laboratory media.

Unicellular algae (FIGURE 1.7) are common in freshwater ponds, streams, and lakes and in the oceans as well. They are the major food of small aquatic and marine animals and provide much of the world's oxygen as a by-product of photosynthesis. The glasslike cell walls of diatoms provide grit for many polishing compounds. (Chapter 12 discusses other aspects of the biology of algae.)

Other Organisms of Importance to Microbiologists

Microbiologists also study parasitic worms, which range in size from microscopic forms (FIGURE 1.8) to adult tapeworms over 10 meters (approximately 33 feet) in length. Even though most parasitic worms are not microscopic as adults, many of them cause diseases that were studied by early microbiologists, so microbiology books and classes often discuss parasitic worms. Further, laboratory scientists diagnose infections of parasitic worms by finding microscopic eggs and immature stages in blood, fecal, urine, and lymph specimens. (Chapter 23 discusses parasitic worms.)

The only type of microbe that remained hidden from van Leeuwenhoek and other early microbiologists was the virus, which is typically much smaller than the smallest prokaryote and is not usually visible by light microscopy (FIGURE 1.9). Viruses were not seen until the electron microscope was invented in 1932. All viruses are acellular (not composed of cells) obligatory parasites composed of small amounts of genetic material (either DNA or RNA) surrounded by a protein coat. (Chapter 13 examines the general characteristics of viruses, and Chapters 24 and 25 discuss specific viral pathogens.)

Van Leeuwenhoek first reported the existence of most types of microorganisms in the late 1600s, but microbiology did not develop significantly as a field of study for almost two centuries. There were a number of reasons for this delay. First, van Leeuwenhoek was a suspicious and secretive man. Though he

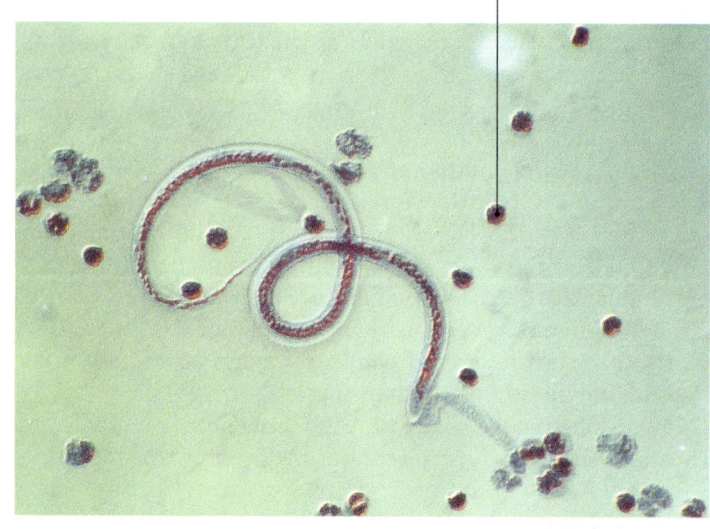

▲ FIGURE 1.8 An immature stage of a parasitic worm in blood.

---

[10]Plural of the Latin *alga*, meaning "seaweed."

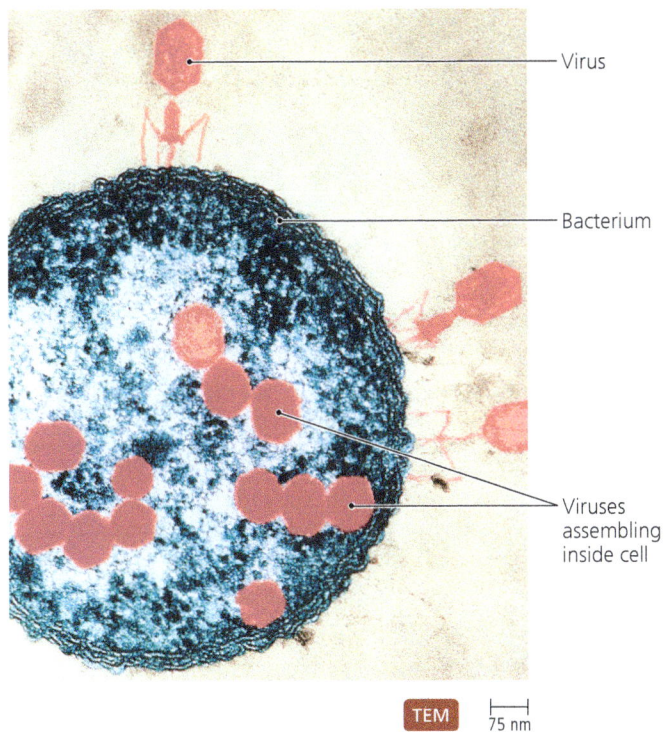

▲ FIGURE 1.9 **A colorized electron microscope image of viruses infecting a bacterium.** Viruses, which are acellular obligatory parasites, are generally too small to be seen with a light microscope. Notice how small the viruses are compared to the bacterium.

built over 400 microscopes, he never trained an apprentice, and he never sold or gave away a microscope. In fact, he never let *anyone*—not his family or such distinguished visitors as the czar of Russia—so much as peek through his very best instruments. When van Leeuwenhoek died, the secret of creating superior microscopes was lost. It took almost 100 years for scientists to make microscopes of equivalent quality.

Another reason that microbiology was slow to develop as a science is that scientists in the 1700s considered microbes to be curiosities of nature and insignificant to human affairs. But in the late 1800s, scientists began to adopt a new philosophy, one that demanded experimental evidence rather than mere acceptance of traditional knowledge. This fresh philosophical foundation, accompanied by improved microscopes, new laboratory techniques, and a drive to answer a series of pivotal questions, propelled microbiology to the forefront as a scientific discipline.

MICRO CHECK

1. What scientific device did van Leeuwenhoek create?
2. What is the modern name for organisms that are too small to be seen without a microscope?
3. Van Leeuwenhoek described bacteria, archaea, fungi, algae, small multicellular animals, and one other type of microorganism. What other type of microorganism did he describe?
4. All eukaryotic cells contain most of their genetic material inside what structure?

TELL ME WHY

Some people consider van Leeuwenhoek the "Father of Microbiology." Explain why this moniker makes sense.

The Golden Age of Microbiology

LEARNING OUTCOMES

1.6 List four questions that propelled research in what is called the "Golden Age of Microbiology."
1.7 Explain why some consider Pasteur to be the "Father of Microbiology."

For about 50 years, during what is sometimes called the "Golden Age of Microbiology," scientists and the blossoming field of microbiology were driven by the search for answers to the following four questions:

- Is spontaneous generation of microbial life possible?
- What causes fermentation?
- What causes disease?
- How can we prevent infection and disease?

Competition among scientists who were striving to be the first to answer these questions drove exploration and discovery in microbiology during the late 1800s and early 1900s. These scientists' discoveries and the fields of study they initiated continue to shape the course of microbiological research today.

In the next sections, we consider these questions and how the women and men of the era accumulated the experimental evidence that answered them.

Does Microbial Life Spontaneously Generate?

LEARNING OUTCOMES

1.8 Identify the scientists who argued in favor of spontaneous generation.
1.9 Compare and contrast the investigations of Redi, Needham, Spallanzani, and Pasteur concerning spontaneous generation.
1.10 List four steps in the scientific method of investigation.

A dry lake bed has lain under the relentless North African desert sun for eight long months. The cracks in the baked, parched mud are wider than a man's hand. There is no sign of life anywhere in the scorched terrain. With the abruptness characteristic of desert storms, rain falls in a torrent, and a raging flood of water and mud crashes down the dry streambed and fills the lake. Within hours, what had been a lifeless, dry mudflat becomes a pool of water teeming with billions of shrimp; by the next day it is home to hundreds of toads. Where did these animals come from?

Many philosophers and scientists of past ages thought that living things arose via three processes: through asexual reproduction, through sexual reproduction, or from nonliving matter. The appearance of shrimp and toads in the mud of what so recently was a dry lake bed was seen as an example of the third process, which came to be known as *abiogenesis*,[11] or **spontaneous generation**. The theory of spontaneous generation as promulgated by Aristotle (384–322 B.C.) was widely accepted for over 2000 years because it seemed to explain a variety of commonly observed phenomena, such as the appearance of maggots on spoiling meat. However, the validity of the theory came under challenge in the 17th century.

Redi's Experiments

In the late 1600s, the Italian physician Francesco Redi (1626–1697) demonstrated by a series of experiments that when decaying meat was kept isolated from flies, maggots never developed, whereas meat exposed to flies was soon infested with maggots (**FIGURE 1.10**). As a result of experiments such as these, scientists began to doubt Aristotle's view and adopt the idea that animals come only from other animals.

Needham's Experiments

The debate over spontaneous generation was rekindled when van Leeuwenhoek discovered microbes and showed that they appeared after a few days in freshly collected rainwater. Though scientists agreed that larger animals could not arise spontaneously, they disagreed about van Leeuwenhoek's "wee animalcules"; surely they did not have parents, did they? They must arise spontaneously.

The proponents of spontaneous generation pointed to the careful demonstrations of British investigator John T. Needham (1713–1781). He boiled beef gravy and infusions[12] of plant material in vials, which he then tightly sealed with corks. Some days later, Needham observed that the vials were cloudy, and examination revealed an abundance of "microscopical animals of most dimensions." As he explained it, there must be a "life force" that causes inanimate matter to spontaneously come to life because he had heated the vials sufficiently to kill everything. Needham's experiments so impressed the Royal Society in Britain that they elected him a member.

Spallanzani's Experiments

Then, in 1799, the Italian Catholic priest and scientist Lazzaro Spallanzani (1729–1799) reported results that contradicted Needham's findings. Spallanzani boiled infusions for almost an hour and sealed the vials by melting their slender necks closed. His infusions remained clear unless he broke the seal and exposed the infusion to air, after which they became cloudy with microorganisms. He concluded three things:

- Needham either had failed to heat his vials sufficiently to kill all microbes or had not sealed them tightly enough.
- Microorganisms exist in the air and can contaminate experiments.
- Spontaneous generation of microorganisms does not occur; all living things arise from other living things.

Although Spallanzani's experiments would appear to have settled the controversy once and for all, it proved difficult to dethrone a theory that had held sway for centuries, especially when so notable

[11]From Greek *a*, meaning "not"; *bios*, meaning "life"; and *genein*, meaning "to produce."
[12]Infusions are broths made by heating water containing plant or animal material.

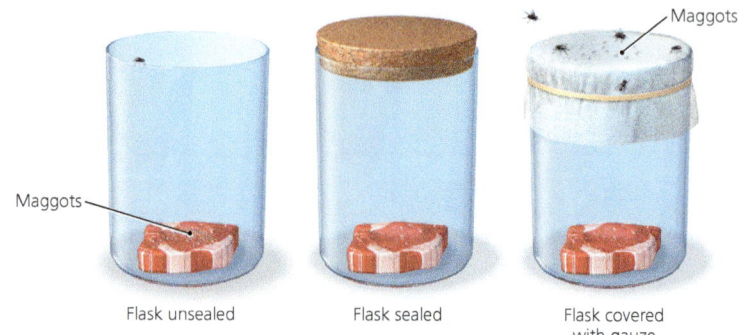

▲ FIGURE 1.10 **Redi's experiments.** When the flask remained unsealed, maggots covered the meat within a few days. When the flask was sealed, flies were kept away, and no maggots appeared on the meat. When the flask opening was covered with gauze, flies were kept away from the meat, and no maggots appeared there, although a few appeared on top of the gauze.

a man as Aristotle had propounded it. One of the criticisms of Spallanzani's work was that his sealed vials did not allow enough air for organisms to thrive; another objection was that his prolonged heating destroyed the "life force." The debate continued until the French chemist Louis Pasteur (**FIGURE 1.11**) conducted experiments that finally laid the theory of spontaneous generation to rest.

Pasteur's Experiments

Louis Pasteur (1822–1895) was a tireless worker who pushed himself as hard as he pushed others. As he wrote his sisters, "To *will* is a great thing dear sisters, for Action and Work usually follow Will, and almost always Work is accompanied by Success. These three things, Work, Will, Success, fill human existence. Will opens the door to success both brilliant and happy; Work passes these doors, and at the end of the journey Success comes to crown one's efforts." When his wife complained about his long hours in the laboratory, he replied, "I will lead you to fame."

▲ FIGURE 1.11 **Louis Pasteur.** Often called the "Father of Microbiology", he disproved spontaneous generation. In this depiction, Pasteur examines some bacterial cultures.

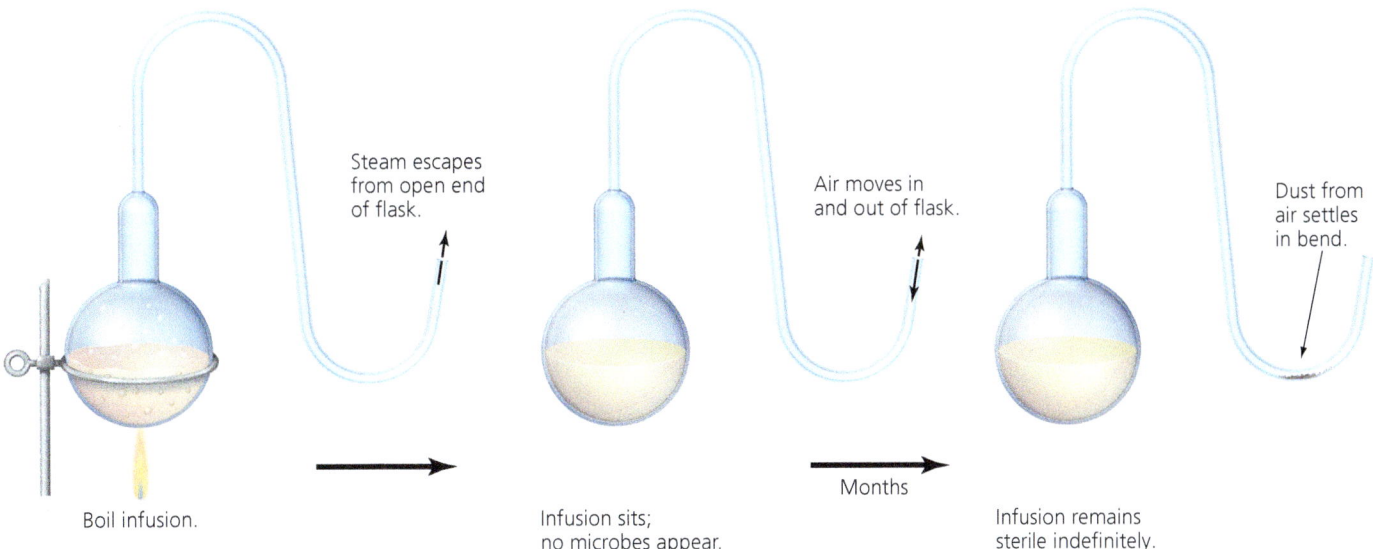

▲ FIGURE 1.12 Pasteur's experiments with "swan-necked flasks." As long as the flask remained upright, no microbial growth appeared in the infusion.

Pasteur's determination and hard work are apparent in his investigations of spontaneous generation. Like Spallanzani, he boiled infusions long enough to kill everything. But instead of sealing the flasks, he bent their necks into an S-shape, which allowed air to enter while preventing the introduction of dust and microbes into the broth (FIGURE 1.12).

Crowded for space and lacking funds, he improvised an incubator in the opening under a staircase. Day after day, he crawled on hands and knees into this small space and examined his flasks for the cloudiness that would indicate the presence of living organisms. In 1861, he reported that his "swan-necked flasks" remained free of microbes even 18 months later. Because the flasks contained all the nutrients (including air) known to be required by living things, he concluded, "Never will spontaneous generation recover from the mortal blow of this simple experiment."

Pasteur followed this experiment with demonstrations that microbes in the air were the "parents" of Needham's microorganisms. He broke the necks off some flasks, exposing the liquid in them directly to the air, and he carefully tilted others so that the liquid touched the dust that had accumulated in their necks. The next day, all of these flasks were cloudy with microbes. He concluded that the microbes in the liquid were the progeny of microbes that had been on the dust particles in the air.

The Scientific Method

The debate over spontaneous generation led in part to the development of a generalized **scientific method** by which questions are answered through observations of the outcomes of carefully controlled experiments instead of by conjecture or according to the opinions of any authority figure. The scientific method, which provides a framework for conducting an investigation rather than a rigid set of specific "rules," consists of four basic steps (FIGURE 1.13):

1. A group of observations leads a scientist to ask a question about some phenomenon.
2. The scientist generates a hypothesis—that is, a potential answer to the question.
3. The scientist designs and conducts an experiment to test the hypothesis.
4. Based on the observed results of the experiment, the scientist either accepts, rejects, or modifies the hypothesis.

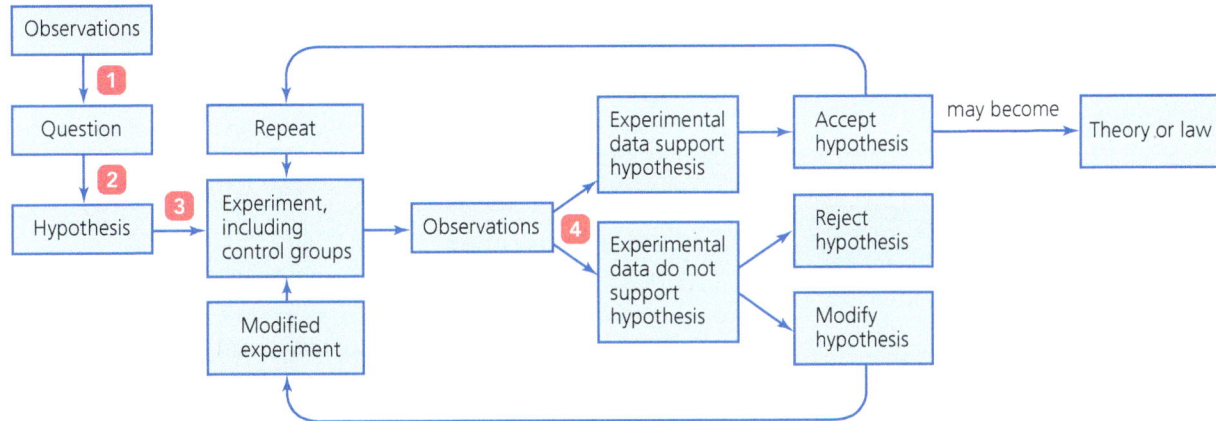

▲ FIGURE 1.13 The scientific method, which forms a framework for scientific research.

BENEFICIAL MICROBES

Bread, Wine, and Beer

Microorganisms play important roles in people's lives; for example, pathogens have undeniably altered the course of history. However, what may be the most important microbiological event—one that has had a greater impact on culture and society than that of any disease or epidemic—was the domestication of the yeast used by bakers and brewers. Its scientific name, *Saccharomyces cerevisiae*, translates from Latin as "sugar fungus [that makes] beer."

The earliest record of the use of yeast comes from Persia (modern Iran), where archaeologists have found the remains of grapes and wine preservatives in pottery vessels more than 7000 years old. Brewing of beer likely started even earlier, its beginnings undocumented. The earliest examples of leavened bread are from Egypt and show that bread making was routine about 6000 years ago. Before that time, bread was unleavened and flat.

It is likely that making wine and brewing beer occurred earlier than the use of leavened bread because *Saccharomyces* is naturally found on grapes, which can begin to ferment while still on the vine.

Historians hypothesize that early bakers may have exposed bread dough to circulating air, hoping that the invisible and inexplicable "fermentation principle" would inoculate the bread. Another hypothesis is that bakers learned to add small amounts of beer or wine to the bread, intentionally inoculating the dough with yeast. Of course, all those millennia before van Leeuwenhoek and Pasteur, no one knew that the fermenting ingredient of wine was a living organism.

Besides its role in baking and in making alcoholic beverages, *S. cerevisiae* is an important tool for the study of cells. Scientists use yeast to delve into the mysteries of cellular function, organization, and genetics, making *Saccharomyces* the most intensely studied eukaryote. In fact, molecular biologists published the complete sequence of the genes of *S. cerevisiae* in 1996—the first complete sequence published for any eukaryotic cell.

Today, scientists are working toward using *S. cerevisiae* in novel ways. For example, some nutritionists and gastroenterologists are examining the use of *Saccharomyces* as a *probiotic*, that is, a microorganism intentionally taken to ward off disease and promote good health. Research suggests that the yeast helps treat diarrhea and colitis and may even help prevent these and other gastrointestinal diseases.

The scientist then returns to earlier steps in the method, either modifying hypotheses and then testing them or repeatedly testing accepted hypotheses until the evidence for a hypothesis is convincing. Accepted hypotheses that explain many observations and are repeatedly verified by numerous scientists over many years are called *theories* or *laws*.

Note that for the scientific community to accept experiments (and their results) as valid, they must include appropriate *control groups*—groups that are treated exactly the same as the other groups in the experiment except for the one variable that the experiment is designed to test. In Pasteur's experiments on spontaneous generation, for example, his "control flasks" were the sterile infusion composed of all the nutrients living things need as well as air made available through the flasks' "swan necks." His "experimental flasks" for testing his hypothesis were exposed to exactly the same conditions *plus* contact with the dust in the bend in the neck. Because exposure to the dust was the *only* difference between the control and experimental groups, Pasteur was able to conclude that the microbes growing in the infusion arrived on the dust particles.

MICRO CHECK

5. What term describes the idea that living organisms can come from nonliving substances?
6. The investigations of which researcher finally disproved that living organisms can come from nonliving substances?

What Causes Fermentation?

LEARNING OUTCOMES

1.11 Discuss the significance of Pasteur's fermentation experiments to our world today.
1.12 Identify the scientist whose experiments led to the field of biochemistry and the study of metabolism.

The controversy over spontaneous generation was largely a philosophical exercise among people who conducted research to gain basic scientific knowledge and not to apply the knowledge they gained. However, the second question that moved microbial studies forward in the 1800s had tremendous practical applications.

Our story resumes in 19th-century France, where spoiled, acidic wine was threatening the livelihood of many grape growers. This led to a fundamental question, "What causes the fermentation of grape juice into wine?" This question was so important to wine producers that they funded research concerning fermentation, hoping scientists could develop methods to promote the production of alcohol and prevent spoilage by acid during fermentation.

Pasteur's Experiments

Scientists of the 1800s used the word *fermentation* to mean not only the formation of alcohol from sugar but also other chemical reactions, such as the formation of lactic acid, the putrefaction of meat, and the decomposition of waste. Many scientists asserted that air caused fermentation reactions; others insisted that living organisms were responsible.

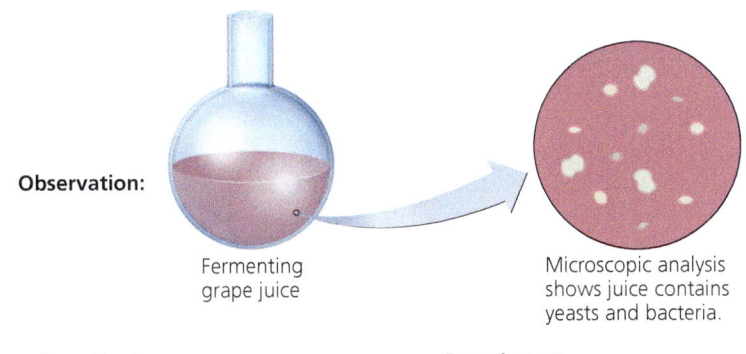
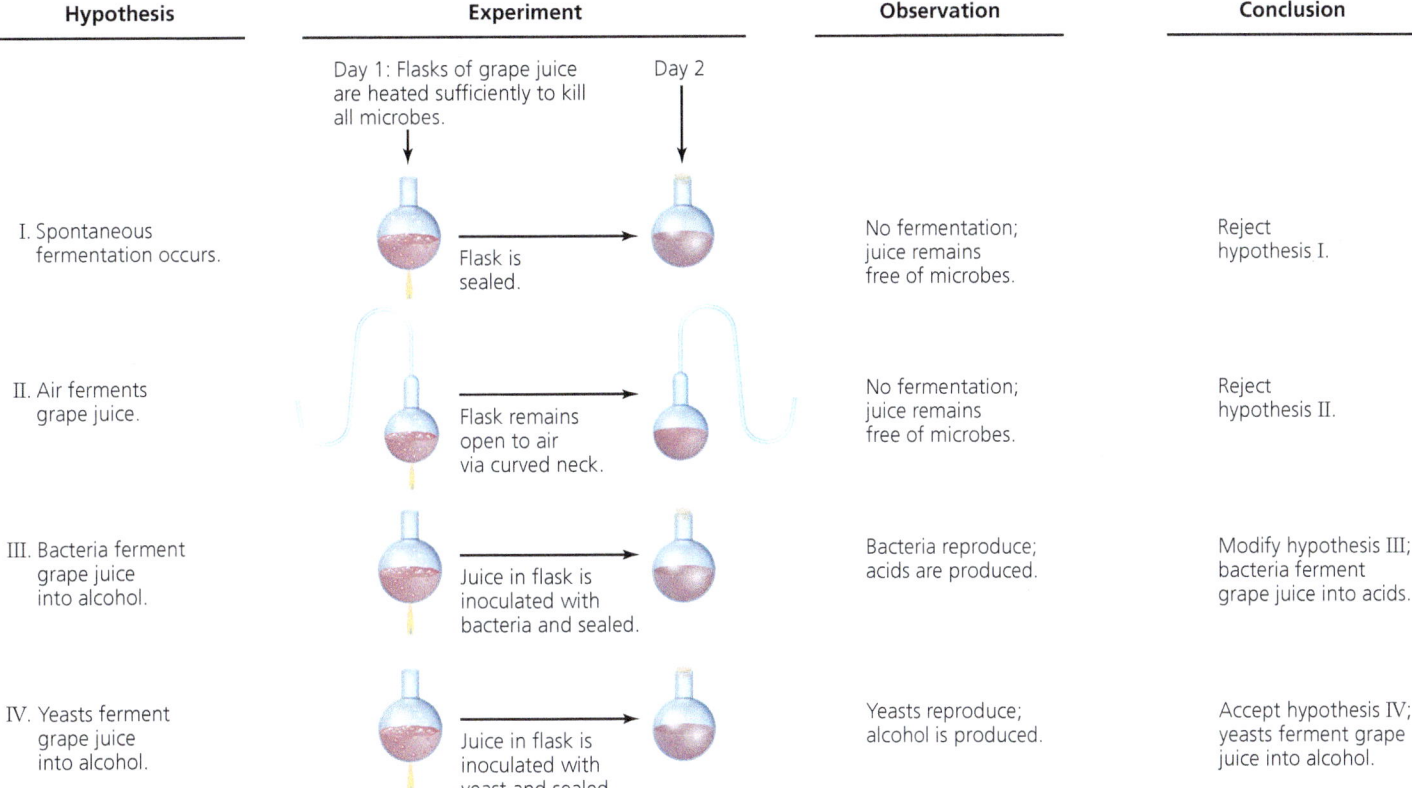

▲ FIGURE 1.14 **How Pasteur applied the scientific method in investigating the nature of fermentation.** After observing that fermenting grape juice contained both yeasts and bacteria, Pasteur hypothesized that these organisms cause fermentation. On eliminating the possibility that fermentation could occur spontaneously or be caused by air (hypotheses I and II), he concluded that fermentation requires the presence of living cells. The results of additional experiments (those testing hypotheses III and IV) indicated that bacteria ferment grape juice to produce acids and that yeasts ferment grape juice to produce alcohol. *Which of Pasteur's flasks was the control?*

Figure 1.14 The sealed flask that remained free of microorganisms served as the control.

The debate over the cause of fermentation reactions was linked to the debate over spontaneous generation. Some scientists proposed that the yeasts observed in fermenting juices were nonliving globules of chemicals and gases. Others thought that yeasts were alive and were spontaneously generated during fermentation. Still others asserted that yeasts not only were living organisms but also caused fermentation.

Pasteur conducted a series of careful observations and experiments that answered the question, "What causes fermentation?" First, he observed yeast cells growing and budding in grape juice and conducted experiments showing that they arise only from other yeast cells. Then, by sealing some sterile flasks containing grape juice and yeast and by leaving others open to the air, he demonstrated that yeast could grow with or without oxygen; that is, he discovered that yeasts are *facultative anaerobes*[13]—organisms that can live with or without oxygen. Finally, by introducing bacteria and yeast cells into different flasks of sterile grape juice, he proved that bacteria ferment grape juice to produce acids and that yeast cells ferment grape juice to produce alcohol (**FIGURE 1.14**).

---

[13]From Greek *an*, meaning "not"; *aer*, meaning "air" (i.e., oxygen); and *bios*, meaning "life."

TABLE 1.1 Some Industrial Uses of Microbes

Product or Process	Contribution of Microorganism
Foods and Beverages	
Cheese	Flavoring and ripening produced by bacteria and fungi; flavors dependent on the source of milk and the type of microorganism
Alcoholic beverages	Alcohol produced by bacteria or yeast by fermentation of sugars in fruit juice or grain
Soy sauce	Produced by fungal fermentation of soybeans
Vinegar	Produced by bacterial fermentation of sugar
Yogurt	Produced by certain bacteria growing in milk
Sour cream	Produced by bacteria growing in cream
Artificial sweetener	Amino acids synthesized by bacteria from sugar
Bread	Rising of dough produced by action of yeast; sourdough results from bacteria-produced acids
Other Products	
Antibiotics	Produced by bacteria and fungi
Human growth hormone, human insulin	Produced by genetically engineered bacteria
Laundry enzymes	Isolated from bacteria
Vitamins	Isolated from bacteria
Diatomaceous earth (in polishes and buffing compounds)	Composed of cell walls of microscopic algae
Pest control chemicals	Insect pests killed or inhibited by insect-destroying bacteria
Drain opener	Protein-digesting and fat-digesting enzymes produced by bacteria

Pasteur's discovery that **anaerobic** bacteria fermented grape juice into acids suggested a method for preventing the spoilage of wine. His name became a household word when he developed *pasteurization*, a process of heating the grape juice just enough to kill most contaminating bacteria without changing the juice's basic qualities. After pasteurization, wine makers added yeast to ensure that alcohol fermentation occurred. Pasteur thus began the field of **industrial microbiology** (or **biotechnology**), in which microbes are intentionally used to manufacture products (TABLE 1.1; see also Chapter 26). Today, pasteurization is used routinely on milk to eliminate pathogens that cause such diseases as tuberculosis; it is also used to eliminate pathogens in juices and other beverages.

These are just a few of the many experiments Pasteur conducted with microbes. Although a few of Pasteur's successes can be attributed to the superior microscopes available in the late 1800s, his genius is clearly evident in his carefully designed and straightforward experiments. Because of his many, varied, and significant accomplishments in working with microbes, Pasteur rivals van Leeuwenhoek as the Father of Microbiology.

Buchner's Experiments

Studies on fermentation began with the idea that fermentation reactions were strictly chemical and did not involve living organisms. This idea was supplanted by Pasteur's work showing that fermentation proceeded only when living cells were present and that different types of microorganisms growing under varied conditions produced different end products.

In 1897, the German scientist Eduard Buchner (1860–1917) resurrected the chemical explanation by showing that fermentation does not require living cells. Buchner's experiments demonstrated the presence of *enzymes*, which are cell-produced proteins that promote chemical reactions. Buchner's work began the field of **biochemistry** and the study of **metabolism**, a term that refers to the sum of all chemical reactions within an organism.

MICRO CHECK

7. Today we understand that yeasts and bacteria can cause fermentation. That understanding comes to us from the work of which noted historical scientist?
8. What industry has the work of Pasteur most influenced?

What Causes Disease?

LEARNING OUTCOMES

1.13 List at least seven contributions made by Koch and his colleagues to the field of microbiology.
1.14 List the four steps that must be taken to prove the cause of an infectious disease.
1.15 Describe the contribution of Gram to the field of microbiology.

You are a physician in London, and it is August 1854. It is past midnight, and you have been visiting patients since before dawn. As you enter the room of your next patient, you observe with frustration and despair that this case is like hundreds of others you and your colleagues have attended in the neighborhood over the past month.

A five-year-old boy with a vacant stare lies in bed listlessly. As you watch, he is suddenly gripped by severe abdominal cramps, and his gastrointestinal tract empties in an explosion of watery diarrhea. The voided fluid is clear, colorless, odorless, and streaked with thin flecks of white mucus, reminiscent of water poured off a pot of cooking rice. His anxious mother changes his bedclothes as his father gives him a sip of water, but it is of little use. With a heavy heart, you confirm the parents' fear—their child has cholera, and there is nothing you can do. He will likely die before morning. As you despondently turn to go, the question that has haunted you for two months is foremost in your mind: What causes such a disease?

The third question that propelled the advance of microbiology concerned disease, defined generally as any abnormal condition in the body. Prior to the 1800s, disease was attributed to various factors, including evil spirits, astrological signs, imbalances in body fluids, and foul vapors. Although the

▲ FIGURE 1.15 **Robert Koch.** Koch was instrumental in modifying the scientific method to prove that a given pathogen caused a specific disease.

Italian philosopher Girolamo Fracastoro (1478–1553) conjectured as early as 1546 that "germs[14] of contagion" cause disease, the idea that germs might be living organisms awaited investigations in the 1800s.

Pasteur's discovery that bacteria are responsible for spoiling wine led naturally to his hypothesis in 1857 that microorganisms are also responsible for diseases. This idea came to be known as the **germ theory of disease**. Because a particular disease is typically accompanied by the same symptoms in all affected individuals, early investigators suspected that diseases such as cholera, tuberculosis, and anthrax are each caused by a specific germ, called a **pathogen**.[15] Today, we know that some diseases are genetic and that allergic reactions and environmental toxins cause others, so the germ theory applies only to *infectious*[16] *diseases*.

Just as Pasteur was the chief investigator in disproving spontaneous generation and determining the cause of fermentation, so Robert Koch (1843–1910) dominated **etiology**[17] (disease causation) (FIGURE 1.15).

Koch's Experiments

Koch was a country doctor in Germany when he began a race with Pasteur to discover the cause of anthrax, which is a potentially fatal disease, primarily of animals, in which toxins produce ulceration of the skin. Anthrax, which can spread to humans, caused untold financial losses to farmers and ranchers in the 1800s.

Koch carefully examined the blood of infected animals, and in every case he identified rod-shaped bacteria[18] in chains. He observed the formation of resting stages (*endospores*) within the bacterial cells and showed that the endospores always produced anthrax when they were injected into mice. This was the first time that a bacterium was proven to cause a disease. Koch published his findings in 1876. As a result of his successful work on anthrax, Koch moved to Berlin and was given facilities and funding to continue his research.

Heartened by his success, Koch turned his attention to other diseases. He had been fortunate when he chose anthrax for his initial investigations, because anthrax bacteria are quite large and easily identified with the microscopes of that time. However, most bacteria are very small, and different types exhibit few or no visible differences. Koch was puzzled regarding how he was to distinguish among these bacteria.

He attempted to solve the problem by taking specimens (e.g., blood, pus, or sputum) from disease victims and then smearing the specimens onto a solid surface such as a slice of potato or a gelatin medium. Unfortunately, many bacteria ate the potatoes and the gelatin, resulting in a slimy mess.

An illustrator working in the Koch laboratory, Fanny Hesse (1850–1934), solved the problem when she introduced Koch to *agar*, a gel derived from red algae (FIGURE 1.16). She and her husband, who also worked with Koch, streaked specimens onto the surface of agar. Bacteria in the specimens multiplied and formed distinct colonies (FIGURE 1.17).

Koch hypothesized that each colony consisted of the progeny of a single cell. He inoculated samples from each colony into laboratory animals to see which caused disease. These scientists' method of isolating bacteria is a standard technique in microbiological and medical labs to this day.

▲ FIGURE 1.16 **Fanny Hesse.** Hesse was instrumental in developing the use of agar as a solidifying agent in microbiological media.

---

[14]From Latin *germen*, meaning "sprout."
[15]From Greek *pathos*, meaning "disease," and *genein*, meaning "to produce."
[16]From Latin *inficere*, meaning "to taint" (i.e., with a pathogen).
[17]From Greek *aitia*, meaning "cause," and *logos*, meaning "word" or "study."
[18]Now known as *Bacillus anthracis*—Latin for "the rod of anthrax."

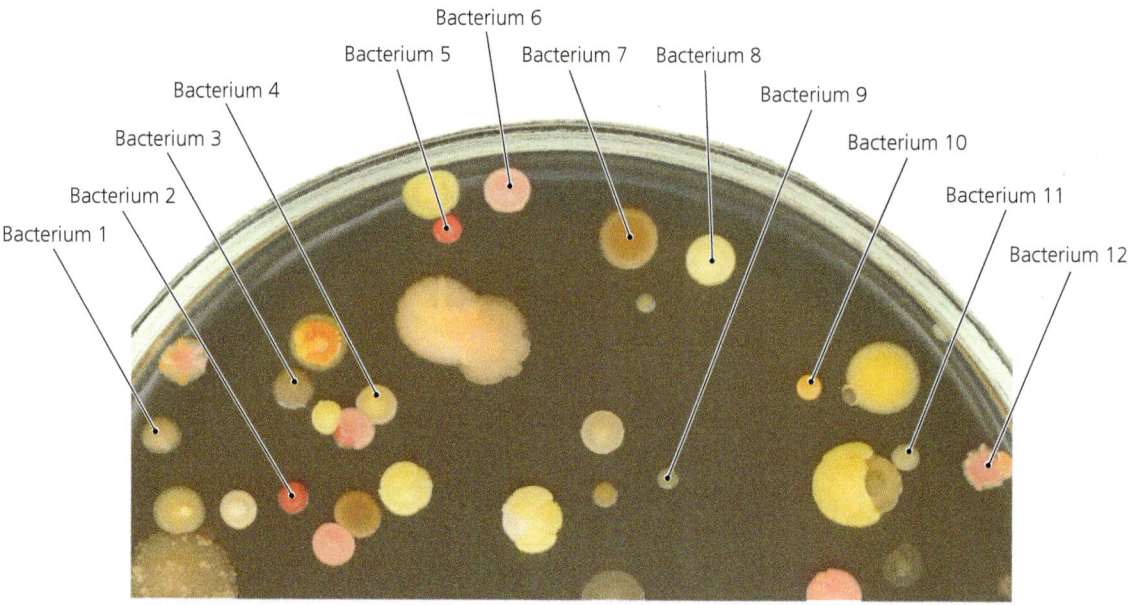

▲ FIGURE 1.17 **Bacterial colonies on a solid surface (agar).** Differences in colony shape, size, and color indicate the presence of different species (only twelve are labeled). Such differences allowed Koch to isolate specific types of bacteria that could be tested for their ability to cause diseases.

Koch and his colleagues are also responsible for many other advances in laboratory microbiology, including the following:

- Simple staining techniques for bacterial cells and flagella
- The first photomicrograph of bacteria
- The first photograph of bacteria in diseased tissue
- Techniques for estimating the number of bacteria in a solution based on the number of colonies that form after inoculation onto a solid surface
- The use of steam to sterilize growth media
- The use of Petri[19] dishes to hold solid growth media
- Laboratory techniques such as transferring bacteria between media using a metal wire that has been heat-sterilized in a flame
- Elucidation of bacteria as distinct species

Koch's Postulates

After discovering the anthrax bacterium, Koch continued to search for disease agents. In two pivotal scientific publications in 1882 and 1884, he announced that the cause of tuberculosis was a rod-shaped bacterium, *Mycobacterium tuberculosis* (mī′kō-bak-tēr′ē-ŭm too-ber-kyū-lō′sis). In 1905, he received the Nobel Prize in Physiology or Medicine for this work.

In his publications on tuberculosis, Koch elucidated a series of steps that must be taken to prove the cause of any infectious disease. These steps, now known as **Koch's postulates**, are one of his important contributions to microbiology. His postulates (which we discuss in more detail in Chapter 14) are the following:

1. The suspected causative agent must be found in every case of the disease and be absent from healthy hosts.
2. The agent must be isolated and grown outside the host.
3. When the agent is introduced to a healthy, susceptible host, the host should get the disease.
4. The same agent must be found in the diseased experimental host.

We use the term *suspected causative agent* because it is merely "suspected" until the postulates have been fulfilled, and "agent" can refer to any fungus, protozoan, bacterium, virus, or other pathogen. There are practical and ethical limits in the application of Koch's postulates, but in almost every case they must be satisfied before the cause of an infectious disease is proven.

During microbiology's Golden Age, other scientists used Koch's postulates and laboratory techniques introduced in Koch's and Pasteur's laboratories to discover the causes of most protozoan and bacterial diseases. Dmitri Ivanovsky (1864–1920) and Martinus Beijerinck (1851–1931) discovered that a certain disease in tobacco plants is caused by a pathogen that passes through filters with such extremely small pores that bacteria cannot pass through. Beijerinck, recognizing that the pathogen was not bacterial, called it a *filterable virus*. Now such pathogens are simply called *viruses*. As previously noted, viruses could not be seen until electron microscopes were invented in 1932. The American physician Walter Reed (1851–1902) proved in 1900 that viruses can cause such diseases as yellow fever in humans. (Chapter 13 deals with virology, and Chapters 24 and 25 deal with viral diseases.)

A partial list of scientists and the pathogens they discovered is provided in **TABLE 1.2**.

---
[19]Named for Richard Petri, Koch's coworker, who invented them in 1887.

TABLE 1.2 Other Notable Scientists of the Golden Age of Microbiology and the Agents of Disease They Discovered

Scientist	Year	Disease	Agent
Edwin Klebs	1883	Diphtheria	*Corynebacterium diphtheriae* (bacterium)
Theodor Escherich	1884	Traveler's diarrhea; bladder infection	*Escherichia coli* (bacterium)
Albert Fraenkel	1884	Pneumonia	*Streptococcus pneumoniae* (bacterium)
David Bruce	1887	Undulant fever (brucellosis)	*Brucella melitensis* (bacterium)
Anton Weichselbaum	1887	Meningococcal meningitis	*Neisseria meningitidis* (bacterium)
A. A. Gartner	1888	Salmonellosis (form of food poisoning)	*Salmonella* species (bacterium)
Shibasaburo Kitasato	1889	Tetanus	*Clostridium tetani* (bacterium)
Dmitri Ivanovsky and Martinus Beijerinck	1892 1898	Tobacco mosaic disease	*Tobamovirus tobacco mosaic virus*
William Welch and George Nuttall	1892	Gas gangrene	*Clostridium perfringens* (bacterium)
Alexandre Yersin and Shibasaburo Kitasato	1894	Bubonic plague	*Yersinia pestis* (bacterium)
Kiyoshi Shiga	1898	Shigellosis (a type of severe diarrhea)	*Shigella dysenteriae* (bacterium)
Walter Reed	1900	Yellow fever	*Flavivirus yellow fever virus*
Robert Forde and Joseph Dutton	1902	African sleeping sickness	*Trypanosoma brucei gambiense* (protozoan)

Gram's Stain

The first of Koch's postulates demands that the suspected agent be found in every case of a given disease, which presupposes that minute microbes can be seen and identified. However, because most microbes are colorless and difficult to see, scientists began to use dyes to stain them and make them more visible under the microscope.

Though Koch reported a simple staining technique in 1877, the Danish scientist Hans Christian Gram (1853–1938) developed a more important staining technique in 1884. His procedure, which involves the application of a series of dyes, leaves some microbes purple and others pink. We now label the purple cells as *Gram positive* and the pink ones as *Gram negative*, and we use the Gram procedure to separate bacteria into these two large groups (**FIGURE 1.18**).

The **Gram stain** is still the most widely used staining technique in microbiology. It is one of the first steps carried out when bacteria are being identified, and it is one of the procedures you will learn in microbiology lab. (Chapter 4 discusses the full procedure.)

MICRO CHECK

9. Which researcher ultimately gave us a method for proving that a particular microbe caused a particular disease?
10. Which researcher developed the staining technique most widely used in microbiology to visualize and to begin identifying microbes under the microscope?

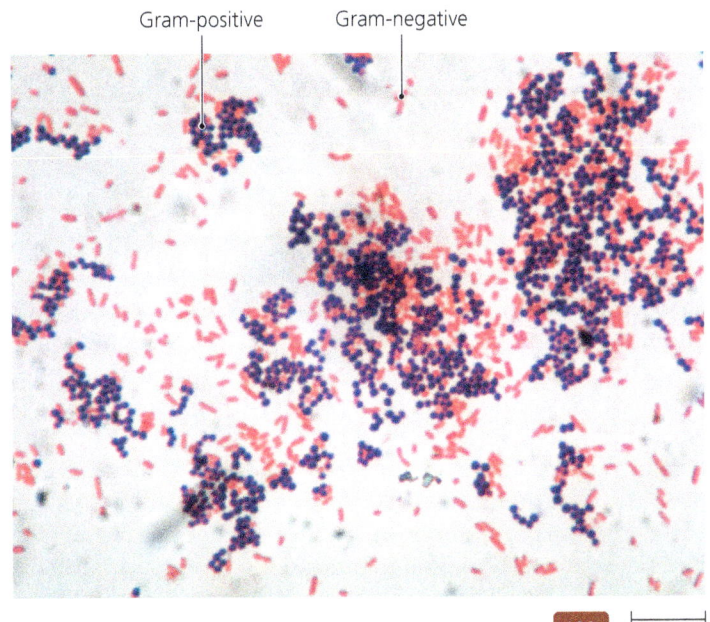

▲ **FIGURE 1.18 Results of Gram staining.** Gram-positive cells (in this case, *Staphylococcus aureus*) are purple; Gram-negative cells (in this case, *Escherichia coli*) are pink.

How Can We Prevent Infection and Disease?

LEARNING OUTCOMES

1.16 Identify six health care practitioners who did pioneering research in the areas of public health microbiology and epidemiology.
1.17 Name three individuals whose work began the field of immunology.
1.18 Describe the quest for a "magic bullet."

Another great question that drove microbiological research during the Golden Age was how to prevent infectious diseases. Though some methods of preventing or limiting disease were discovered even before it was understood that microorganisms

CLINICAL CASE STUDY

Remedy for Fever or Prescription for Death

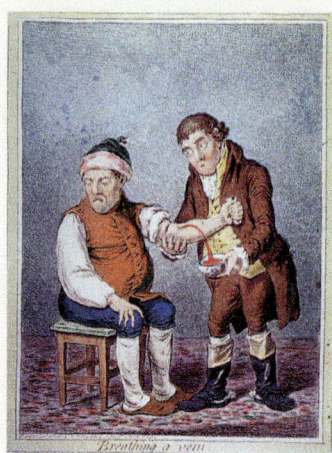

In the late 18th century, Philadelphia was one of the larger and wealthier cities in the United States and served as the capital. That changed in 1793. The city had an unusually wet spring, which left behind stagnant pools that became breeding grounds for mosquitoes. At about the same time, refugees from the slave revolution in Haiti fled to Philadelphia, carrying the yellow fever virus. In late August 1793, a female *Aedes aegypti* mosquito bit an infected refugee and then bit a healthy Philadelphian. This began a yellow fever epidemic that killed 10% of the city's population within three months and led another 30% to flee for their lives. Victims suffered from high fever, nausea, skin eruptions, black vomit, and jaundice.

The treatment for yellow fever in the 18th century was often worse than the disease: physicians administered potions to purge the victims' intestines and drained up to four-fifths of their patients' blood in the mistaken belief the bloodletting would stem fever. These attempted remedies often left patients tired, weak, and unable to fight disease. Without effective treatments, the epidemic stopped only when the first frost arrived.

1. People who left the city seemed to have milder cases of yellow fever or avoided the infection altogether. Explain why.
2. The story mentions that the coming of the first frost brought an end to the epidemic. Discuss the possible reasons why this would provide at least temporary relief from the epidemic.

---

caused contagious diseases, great advances occurred after Pasteur and Koch showed that life comes from life and that microorganisms can cause diseases.

In the mid-1800s, modern principles of hygiene, such as those involving sewage and water treatment, personal cleanliness, and pest control, were not widely practiced. Typically, medical personnel and health care facilities lacked adequate cleanliness. **Healthcare associated infections (HAI)**, formerly called nosocomial[20] infections, were rampant. For example, surgical patients frequently succumbed to gangrene acquired while under their doctor's care, and many women who gave birth in hospitals died from puerperal[21] fever. Seven health care practitioners who were especially instrumental in changing health care delivery methods were Ignaz Semmelweis, Joseph Lister, Florence Nightingale, John Snow, Mary Wortley Montagu, Edward Jenner, and Paul Ehrlich.

Semmelweis and Hand Washing

Ignaz Semmelweis (1818–1865) was a physician on the obstetric ward of a teaching hospital in Vienna. In about 1848, he observed that women giving birth in the wing where medical students were trained died from puerperal fever at a rate 20 times higher than the mortality rates of either women attended by midwives in an adjoining wing or women who gave birth at home.

Though Pasteur had not yet elaborated his germ theory of disease, Semmelweis hypothesized that medical students carried "cadaver particles" from their autopsy studies into the delivery rooms and that these "particles" resulted in puerperal fever. Today, we know that the primary cause of puerperal fever is a bacterium in the genus *Streptococcus* (strep-tō-kok'ŭs; see Figure 1.4), which is usually harmless on the skin or in the mouth but causes severe complications when it enters the blood.

Semmelweis began requiring medical students to wash their hands with chlorinated lime water, a substance long used to eliminate the smell of cadavers. Mortality in the subsequent year dropped from 18.3% to 1.3%. Despite his success, Semmelweis was ridiculed by the director of the hospital and eventually was forced to leave.

Though his impressive record made it easier for later doctors to institute changes, Semmelweis was unsuccessful in gaining support for his method from most European doctors. He became severely depressed and was committed to a mental hospital, where he died from an infection of *Streptococcus*, the very organism he had fought for so long.

Lister's Antiseptic Technique

Shortly after Semmelweis was rejected in Vienna, the English physician Joseph Lister (1827–1912) modified and advanced the idea of *antisepsis*[22] in health care settings. As a surgeon, Lister was aware of the dreadful consequences that resulted from the infection of wounds. Therefore, he began spraying wounds, surgical incisions, and dressings with carbolic acid (phenol), a chemical that had previously proven effective in reducing odor and decay in sewage. Like Semmelweis, he initially met with some resistance, but when he showed that it reduced deaths among his patients by two-thirds, his method was accepted into common practice. In this manner, Lister vindicated Semmelweis, became the founder of antiseptic surgery, and opened new fields of research into antisepsis and disinfection.

---

[20]From Greek *nosos*, meaning "disease," and *komein*, meaning "to care for" (relating to a hospital).
[21]From Latin *puerperus*, meaning "childbirth."
[22]From Greek *anti*, meaning "against," and *sepein*, meaning "putrefaction."

▲ FIGURE 1.19 **Florence Nightingale.** The founder of modern nursing, she was influential in introducing antiseptic technique into nursing practice.

Nightingale and Nursing

Florence Nightingale (1820–1910) (FIGURE 1.19) was a dedicated English nurse who introduced cleanliness and other antiseptic techniques into nursing practice. She was instrumental in setting standards of hygiene that saved innumerable lives during the Crimean War of 1854–1856. One of her first requisitions in the military hospital was for 200 scrubbing brushes, which she and her assistants used diligently in the squalid wards. She next arranged for each patient's filthy clothes and dressings to be replaced or cleaned at a different location, thus removing many sources of infection. As a gifted mathematician, she thoroughly documented statistical comparisons using a type of pie chart graph that she developed to show that poor food and unsanitary conditions in the hospitals were responsible for the deaths of many soldiers.

After the war, Nightingale returned to England, where she actively exerted political pressure to reform hospitals and implement public health policies in India and in England. As a direct result of her recommendations, mortality among English soldiers in India dropped 74% and the average life expectancy in Britain increased by 20 years.

Perhaps her greatest achievements were in nursing. Nightingale founded modern nursing with her development of methods and standards for patient care, her groundbreaking statistical analyses, and her dedicated mentoring of nurses around the world. She founded the Nightingale School for Nurses—the first of its kind in the world. The school, now called the Florence Nightingale Faculty of Nursing and Midwifery, is ranked third among all nursing schools in the world.

Snow and Epidemiology

Another English physician, John Snow (1813–1858), also played a key role in setting standards for good public hygiene to prevent the spread of infectious diseases. Snow had been studying the propagation of cholera and suspected that the disease was spread by a contaminating agent in water. In 1854, he mapped the occurrence of cholera cases during an epidemic in London and showed that they centered around a public water supply on Broad Street.

Though Snow did not know the cause of cholera, his careful documentation of the epidemic highlighted the critical need for adequate sewage treatment and a pure water supply. His study was the foundation for two branches of microbiology—**infection control** and **epidemiology**,[23] which is the study of the occurrence, distribution, and spread of disease in humans.

Jenner's Vaccine

Lady Mary Wortley Montagu (1689–1762) introduced inoculation against smallpox to Europe (FIGURE 1.20). She had learned during her travels in the Ottoman Empire that it was possible to provide some protection against the deadly disease by inoculating children with pus taken from a patient suffering with a mild case of the disease. Upon returning to England, she had her daughter inoculated in 1721—the first time the procedure was performed in England. Lady Montagu successfully promoted the procedure, called *variolation*, because it reduced the chance that variolized children might die from smallpox.

▲ FIGURE 1.20 **Lady Mary Wortley Montagu.** Montagu introduced England to variolization—inoculating patients with pus from a mild smallpox case to protect them against a more severe case.

In 1796, the English physician Edward Jenner (1749–1823) expanded on Lady Montagu's work, testing his hypothesis that inoculation with infectious material from a milder disease, called cowpox, would provide even safer protection against smallpox. After he intentionally inoculated a boy with pus collected from a milkmaid's cowpox lesion, the boy developed cowpox and survived. When Jenner then infected the boy with smallpox pus, he

---
[23]From Greek *epi*, meaning "upon"; *demos*, meaning "people"; and *logos*, meaning "word" or "study."

found that the boy had become immune[24] to smallpox. (Today, experiments that intentionally expose human subjects to deadly pathogens are unethical.) In 1798, Jenner reported similar results from additional experiments, demonstrating the validity of the procedure he named *vaccination* after *Vaccinia virus*,[25] the virus that causes cowpox. Because vaccination stimulates a long-lasting response by the body's protective immune system, the term *immunization* is often used synonymously today. Jenner began the field of **immunology**—the study of the body's defenses against pathogens. (Chapters 15–18 discuss immunology.)

Pasteur, almost a century later, capitalized on Jenner's work by producing weakened strains of various pathogens for use in preventing the serious diseases they cause. In honor of Jenner's work with cowpox, Pasteur used the term *vaccine* to refer to all weakened, protective strains of pathogens. He developed successful vaccines against chicken cholera, anthrax, and rabies.

Ehrlich's "Magic Bullets"

Gram's discovery that stained bacteria could be differentiated into two types by color suggested to the German microbiologist Paul Ehrlich (1854–1915) that chemicals could be used to kill microorganisms differentially. To investigate this idea, Ehrlich undertook an exhaustive survey of chemicals to find a "magic bullet" that would destroy pathogens while remaining nontoxic to humans. By 1908, he had discovered a chemical active against the causative agent of syphilis, though the arsenic-based drug can have serious side effects in humans. His discoveries began the branch of medical microbiology known as **chemotherapy**.

In summary, the Golden Age of Microbiology was a time when researchers proved that living things come from other living things, that microorganisms can cause fermentation and disease, and that certain procedures and chemicals can limit, prevent, and cure infectious diseases. These discoveries were made by scientists who applied the scientific method to biological investigation, and they led to an explosion of knowledge in a number of scientific disciplines (FIGURE 1.21).

MICRO CHECK

11. Which health care practitioner improved our understanding of the importance of hand washing?
12. The use of antiseptic chemicals during surgical techniques was the direct result of which surgeon's innovations?
13. The work of Paul Ehrlich brought us chemicals that destroy pathogens while remaining nontoxic to humans. What is the modern term for this concept?

TELL ME WHY

Why might Nightingale be considered the Mother of Medical Microbiology?

---

[24]From Latin *immunis*, meaning "free."
[25]From Latin *vacca*, meaning "cow."

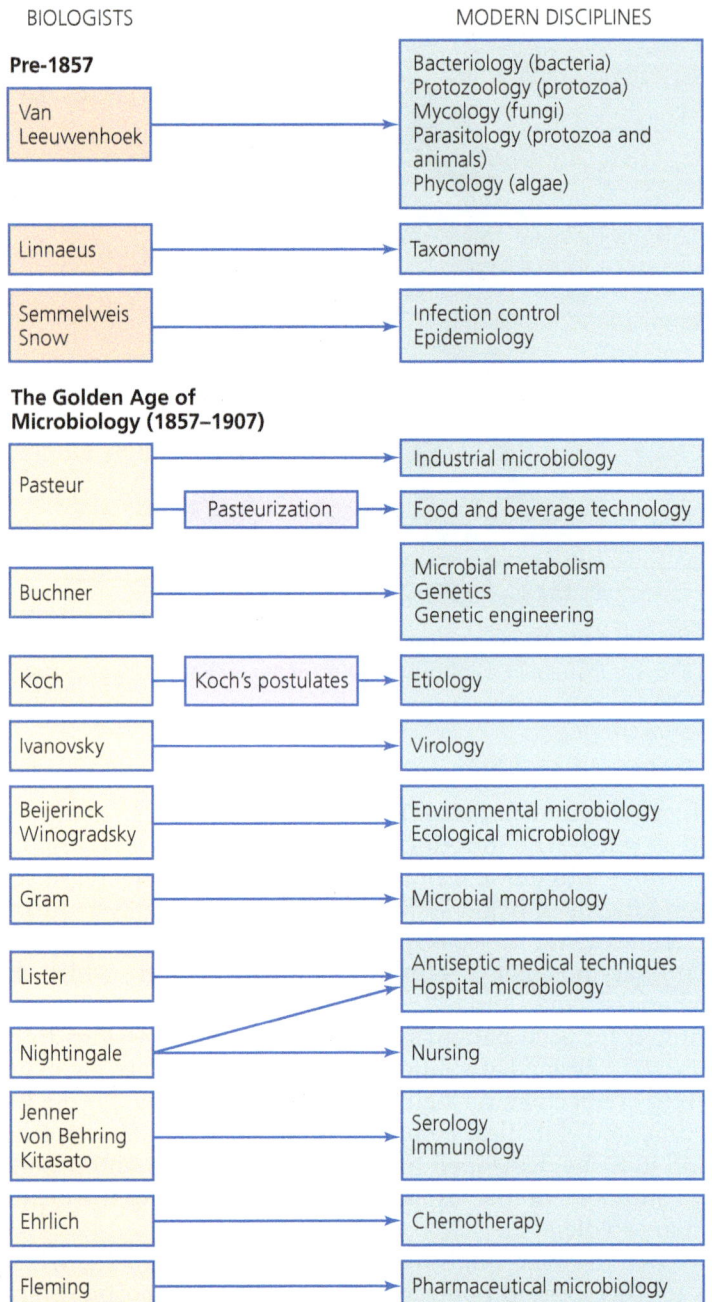

▲ FIGURE 1.21 Some of the many scientific disciplines and applications that arose from the pioneering work of scientists just before and around the time of the Golden Age of Microbiology.

The Modern Age of Microbiology

LEARNING OUTCOME

1.19 List four major questions that drive microbiological investigations today.

The vast increase in the number of microbiological investigations and in scientific knowledge during the 1800s opened new fields of science, including disciplines called environmental microbiology, immunology, epidemiology, chemotherapy, and genetic engineering (TABLE 1.3). Microorganisms played a significant

TABLE 1.3 Fields of Microbiology

Disciplines	Subject(s) of Study
Basic Research	
Microbe Centered	
Bacteriology	Bacteria and archaea
Phycology	Algae
Mycology	Fungi
Protozoology	Protozoa
Parasitology	Parasitic protozoa and parasitic animals
Virology	Viruses
Process Centered	
Microbial metabolism	Biochemistry: chemical reactions within cells
Microbial genetics	Functions of DNA and RNA
Environmental microbiology	Relationships between microbes and among microbes, other organisms, and their environment
Applied Microbiology	
Medical Microbiology	
Serology	Antibodies in blood serum, particularly as an indicator of infection
Immunology	Body's defenses against specific diseases
Epidemiology	Frequency, distribution, and spread of disease
Etiology	Causes of disease
Infection control	Hygiene in health care settings and control of healthcare-associated infections
Chemotherapy	Development and use of drugs to treat infectious diseases
Applied Environmental Microbiology	
Bioremediation	Use of microbes to remove pollutants
Public health microbiology	Sewage treatment, water purification, and control of insects that spread disease
Agricultural microbiology	Use of microbes to control insect pests
Industrial Microbiology (Biotechnology)	
Food and beverage technology	Reduction or elimination of harmful microbes in food and drink
Pharmaceutical microbiology	Manufacture of vaccines and antibiotics
Recombinant DNA technology (genetic engineering)	Alteration of microbial genes to synthesize useful products

role in the development of these disciplines because microorganisms are relatively easy to grow, take up little space, and are available by the trillions. Much of what has been learned about microbes also applies to other organisms, including humans. In the rest of this book we examine advances made in these branches of microbiology, though it would require thousands of books this size to deal with all that is known.

Once the developing science of microbiology had successfully answered questions about spontaneous generation, fermentation, and disease, additional questions arose in each branch of the new science. In this section, we briefly consider some of the early 21st century's overarching questions in both basic and applied research. The chapter concludes with a look at some of the questions that might propel microbiological research for the next 50 years.

What Are the Basic Chemical Reactions of Life?

Biochemistry is the study of metabolism—that is, the chemical reactions that occur in living organisms. Biochemistry began with Pasteur's work on fermentation by yeast and bacteria and with Buchner's discovery of enzymes in yeast extract, but in the early 1900s, many scientists still thought that the metabolic reactions of microbes had little to do with the metabolism of plants and animals.

In contrast, microbiologists Albert Kluyver (1888–1956) and his student C. B. van Niel (1897–1985) proposed that basic biochemical reactions are shared by all living things, that these reactions are relatively few in number, and that their primary feature is the transfer of electrons and hydrogen ions. In adopting this view, scientists could use microbes as model systems to answer questions about metabolism in all organisms. Research during the 20th century validated this approach to understanding basic metabolic processes, but scientists have also documented an amazing metabolic diversity. (Chapter 5 discusses basic metabolic processes, and Chapter 6 considers metabolic diversity.)

Basic biochemical research has many practical applications, including the following:

- The design of herbicides and pesticides that are specific in their action and have no long-term adverse effects on the environment.
- The diagnosis of illnesses and the monitoring of a patient's responses to treatment. For example, physicians routinely monitor liver disease by measuring blood levels of certain enzymes and products of liver metabolism.
- The treatment of metabolic diseases. One example is treating phenylketonuria, a disease resulting from the inability to properly metabolize the amino acid phenylalanine, by eliminating foods containing phenylalanine from the diet.
- The design of drugs to treat leukemia, gout, bacterial infections, malaria, herpes, AIDS, asthma, and heart attacks.

How Do Genes Work?

Genetics, the scientific study of inheritance, started in the mid-1800s as an offshoot of botany, but scientists studying microbes made most of the great advances in this discipline.

Microbial Genetics

While working with the bacterium *Streptococcus pneumoniae* (strep-tō-kok'ŭs nū-mō'nē-ī), Oswald Avery (1877–1955), Colin MacLeod (1909–1972), and Maclyn McCarty (1911–2005) determined that genes are contained in molecules of DNA. In 1958, George Beadle (1903–1989) and Edward Tatum (1909–1975), working with the bread mold *Neurospora crassa* (noo-ros'pōr-ă kras'ă), established that a gene's activity is related to the function of the specific protein coded by that gene. Other researchers, such as Esther Lederberg (1922–2006), also working with microbes, determined the exact way in which genetic information is translated into a protein, the rates and mechanisms of genetic mutation, and the methods by which cells and viruses control genetic expression. (Chapter 7 examines all these aspects of microbial genetics.)

Over the past 40 years, advances in microbial genetics developed into several new disciplines that are among the faster-growing areas of scientific research today, including *molecular biology*, *recombinant DNA technology*, and *gene therapy*.

Molecular Biology

Molecular biology combines aspects of biochemistry, cell biology, and genetics to explain cell function at the molecular level. Molecular biologists are particularly concerned with *genome*[26] *sequencing*. Using techniques perfected on microorganisms, molecular biologists have sequenced the genomes of many organisms, including humans and many of their pathogens. It is hoped that a fuller understanding of the genomes of organisms will result in practical ways to limit disease, repair genetic defects, and enhance agricultural yield.

The American Nobel Prize winner Linus Pauling (1901–1994) proposed in 1965 that gene sequences could provide a means of understanding evolutionary relationships and processes, establishing taxonomic categories that more closely reflect these relationships, and identifying the existence of microbes that have never been cultured in a laboratory. Two examples illustrate such uses of gene sequencing data:

- In the 1970s, Carl Woese (1928–2012) and George Fox (1945–) discovered that significant differences in nucleic acid sequences among organisms clearly reveal that cells belong to one of *three* major groups—bacteria, archaea, and eukaryotes—and not merely two groups (prokaryotes and eukaryotes), as previously thought.
- Scientists showed in 1990 that cat scratch disease is caused by a bacterium that had not been cultured. The bacterium was discovered by comparing the sequence of a portion of its ribonucleic acid with ribonucleic acid sequences from all other known bacteria. This is a standard technique in modern bacterial identification.
- Microbiologists such as Jill Banfield (1959–) use gene sequencing data to analyze and understand microbes that have not been grown in laboratories.

Recombinant DNA Technology

Molecular biology is applied in **recombinant DNA technology**,[27] commonly called *genetic engineering*, which was first developed using microbial models. Geneticists manipulate genes in microbes, plants, and animals for practical applications. For instance, once scientists have inserted the gene for a human blood-clotting factor into the bacterium *Escherichia coli* (esh-ĕ-rik′ē-ă kō′lī), the bacterium produces the factor in a pure form. This technology is a benefit to hemophiliacs, who previously depended on clotting factor derived from donated blood, which was possibly contaminated by life-threatening viral pathogens.

Gene Therapy

An exciting new area of study is the use of recombinant DNA technology for **gene therapy**, a process that involves inserting a missing gene or repairing a defective one in human cells. In such procedures, researchers insert a desired gene into host cells, where it is incorporated into a chromosome and begins to function normally. Doctors have successfully treated several inherited immune deficiencies with gene therapy. (Chapter 8 examines recombinant DNA technology and gene therapy in more detail.)

What Roles Do Microorganisms Play in the Environment?

LEARNING OUTCOME

1.20 Identify the field of microbiology that studies the role of microorganisms in the environment.

Ever since Koch and Pasteur, most research in microbiology has focused on pure cultures of individual species; however, microorganisms are not alone in the "real world." Instead, they live in natural microbial communities in the soil, water, the human body, and other habitats. Bonnie Bassler (1962–), Barbara Iglewski (1938–), and other microbiologists have shown that microbes in communities transmit information among themselves via a process called *quorum sensing*. These communities play critical roles in such processes as causing human and animal diseases, the production of vitamins, and *bioremediation*—the use of living bacteria, fungi, and algae to detoxify polluted environments.

Microbial communities also play an essential role in the decay of dead organisms and the recycling of chemicals such as carbon, nitrogen, and sulfur. Martinus Beijerinck discovered bacteria capable of converting nitrogen gas (N_2) from the air into nitrate (NO_3), the form of nitrogen used by plants, and the Russian microbiologist Sergei Winogradsky (1856–1953) elucidated the role of microorganisms in the recycling of sulfur. Together these two microbiologists developed laboratory techniques for several important aspects of **environmental microbiology**.

Another role of microbes in the environment is the causation of disease. Although most microorganisms are not pathogenic, in this text (particularly in Chapters 18–25), we focus on pathogenic microbes because of the threat they pose to human health. We examine their characteristics and the diseases they cause as well as the steps we can take to limit their abundance and control their spread in the environment, such as sewage treatment, water purification, disinfection, pasteurization, and sterilization.

How Do We Defend Against Disease?

Why do some people get sick during the flu season while their close friends and family remain well? The germ theory of disease

---

[26] A genome is the total genetic information of an organism.
[27] Recombinant DNA is DNA composed of genes from more than one organism.

EMERGING DISEASE CASE STUDY

Variant Creutzfeldt-Jakob Disease

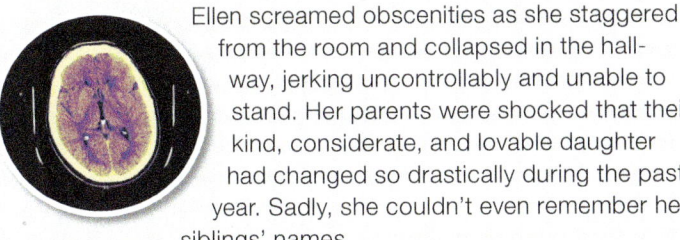

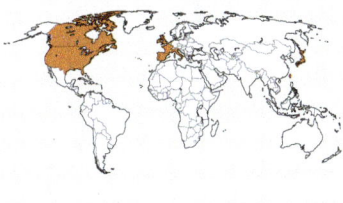

Ellen screamed obscenities as she staggered from the room and collapsed in the hallway, jerking uncontrollably and unable to stand. Her parents were shocked that their kind, considerate, and lovable daughter had changed so drastically during the past year. Sadly, she couldn't even remember her siblings' names.

Ellen had joined the nearly 200 people afflicted with variant Creutzfeldt-Jakob disease (vCJD; what the media call "mad cow disease" because most humans with the condition acquired the pathogen from eating infected beef). Because vCJD affects the brain by slowly eroding nervous tissue and leaving the brain full of spongelike holes, the signs and symptoms of vCJD are neurological. Ellen's disease started with insomnia, depression, and confusion, but eventually it led to uncontrollable emotional and verbal outbursts, inability to coordinate movements, coma, and death. Typically, the disease lasts about a year, and there is no treatment.

Variant CJD resembles the rare genetic disorder Creutzfeldt-Jakob disease (named for its discoverers), which is caused by a mutation and occurs in the elderly. The difference is that the variant form of CJD results from an acquired infection and often strikes and kills college-aged people, like Ellen in our story. (For more about vCJD, see pp. 405–407.)

1. The vCJD pathogen is primarily transmitted when a person or animal consumes nervous tissue (brains). How could cattle become infected?
2. Why is vCJD called *variant*?
3. What effect does this pathogen have on cattle?

---

showed not only that microorganisms can cause diseases, but also that the body can defend itself—otherwise, everyone would be sick most of the time.

The work of Montagu, Jenner, and Pasteur on immunization showed that the body can protect itself from repeated diseases by the same organism. The German bacteriologist Emil von Behring (1854–1917) and the Japanese microbiologist Shibasaburo Kitasato (1852–1931), working in Koch's laboratory, reported the existence in the blood of chemicals and cells that fight infection. Their studies developed into the fields of *serology*, the study of blood serum[28]—specifically, the chemicals in the liquid portion of blood that fight disease—and *immunology*, the study of the body's defense against specific pathogens. (Chapters 15–18 cover these aspects of microbiology, which are of utmost importance to physicians, nurses, and other health care practitioners.)

Ehrlich introduced the idea of a "magic bullet" that would kill pathogens, but it was not until Alexander Fleming (1881–1955) discovered penicillin (**FIGURE 1.22**) in 1929 and Gerhard Domagk (1895–1964) discovered sulfa drugs in 1935 that medical personnel finally had drugs effective against a wide range of bacteria. (We study chemotherapy and some physical and chemical agents used to control microorganisms in the environment in Chapters 9 and 10.)

Health care workers are concerned about emerging diseases—ones that have been diagnosed in a population for the first time or are rapidly increasing in incidence or geographic range. Notable examples include Middle East respiratory syndrome (MERS), Zika fever, chikungunya, and Ebola and Marburg hemorrhagic fevers. The Centers for Disease Control and Prevention (CDC) in Atlanta, Georgia, have called these diseases "the new normal" in health care. Further, diseases once thought to be near eradication, such as measles, whooping cough, and tuberculosis, have reemerged in troubling outbreaks. Other near-vanquished pathogens

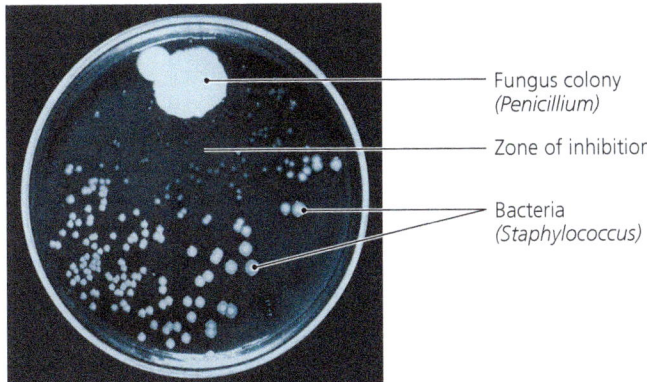

▲ **FIGURE 1.22 The effects of penicillin on a bacterial "lawn" in a Petri dish.** The clear area (zone of inhibition) surrounding the fungus colony, which is producing the antibiotic, is where the penicillin prevented bacterial growth.

---

[28] Latin, meaning "whey." Serum is the liquid that remains after blood coagulates.

such as smallpox or anthrax could become potential weapons in bioterrorist attacks.

How do emerging and reemerging diseases arise? Some are introduced to humans as we move into remote jungles and contact infected animals, some are carried by insects whose range is spreading, and some previously harmless microbes acquire new genes that allow them to be infective and cause disease. Some emerging and reemerging pathogens spread with the speed of jet planes carrying infected people around the globe, and still others arise when previously treatable microbes develop resistance to our antibiotics. Throughout this text, you will encounter many boxed discussions of such emerging and reemerging diseases.

What Will the Future Hold?

Science is built on asking and answering questions. What began with the questioning curiosity of a dedicated lens grinder in the Netherlands has come far in the past 350 years, expanding into disciplines as diverse as immunology, recombinant DNA technology, and bioremediation. However, the adage remains true: *The more questions we answer, the more questions we have.*

What will microbiologists discover next? Among the questions for the next 50 years are the following:

- How can we develop successful programs to control or eradicate diseases such as tuberculosis, malaria, AIDS, and Ebola?
- How can we surveil, treat, and prevent emerging diseases such as Zika fever, Marburg hemorrhagic fever, and MERS?
- What is it about the physiology of life forms known only by their nucleic acid sequences that has prevented researchers from growing them in the laboratory?
- Can bacteria and archaea be used in ultraminiature technologies, such as living computer circuit boards?
- How can an understanding of microbial *biofilms*—aggregates of microbial cells growing together on a surface—help us understand aspects of microbial action in preventing and curing diseases, recycling nutrients, degrading pollutants, and moderating climate change?
- How can we reduce the threat from microbes resistant to antimicrobial drugs, particularly so-called *persistent* cells that resist antimicrobial treatment without acquiring genetic changes?
- Can we develop inexpensive, rapid, accurate, and simple tests for infections?
- Should we use recombinant DNA technology to modify pathogens or the insects that carry them in order to reduce the spread of diseases?
- Can microorganisms that grow normally in the body bolster our ability to fight disease, lose weight, and maintain good mental health?
- Can microorganisms or their genes be used to develop sustainable fuels or bioremediate synthetic chemicals?

TELL ME WHY

Why are so many modern questions in microbiology related to genetics?

MICRO IN THE CLINIC FOLLOW-UP
Too Much Cake or Something Worse?

Dr. Andrews has a lot of questions for Patty. When did she start feeling poorly? When did the diarrhea start? How long after the diarrhea began did she start vomiting? Patty feels terrible and is somewhat annoyed with the doctor—she already told the nurse most of this information, so why does he need to ask again? Dr. Andrews tells Patty that he suspects she may have food poisoning, and for public health reasons it's necessary to try to track down the source. Dr. Andrews gets a complete list of the foods that Patty consumed over the last 48 hours, including where they were consumed. The doctor collects a stool (feces) sample from Patty and sends it to the lab for analysis. The results come back positive for norovirus.

Dr. Andrews asks Patty whether anyone else at her weekend events has gotten sick or whether anyone at the parties had been sick in the not-too-distant past. Patty replies that some of the other people at the barbecue who had eaten meat were feeling poorly; she thinks that since she ate only fruit salad at the barbecue, she couldn't have the same thing they had. She also knew that the family that hosted the barbecue had a child home sick from day care the week before. Dr. Andrews tells Patty that it's likely that the fruit salad became contaminated with noroviruses and that this is where Patty got the infection. Since norovirus is viral, typical antibacterial drugs such as penicillin cannot be used as treatment. Dr. Andrews tells Patty to rest as much as she can and drink a lot of water to prevent dehydration; she should start feeling better in a day or two.

1. **How does Patty's diagnostic conversation with Dr. Andrews parallel the scientific method?**
2. **Based on what you know about Patty's case, can you form a hypothesis about how the fruit salad became contaminated?**

Check your answers to the Micro in the Clinic Follow-Up questions in the Mastering Microbiology Study Area.

CHAPTER SUMMARY

Go to Mastering Microbiology for Interactive Microbiology, Dr. Bauman's Video Tutors, Micro Matters, MicroFlix, MicroBoosters, practice quizzes, and more.

The Early Years of Microbiology (pp. 2–7)

1. Van Leeuwenhoek's observations of **microbes** introduced most types of **microorganisms** to the world. His discoveries were named and classified by Linnaeus in his **taxonomic system**.
2. Small **prokaryotes**—**bacteria** and **archaea**—live in a variety of communities and in most habitats. Even though some cause disease, most are beneficial.
3. Relatively large microscopic **eukaryotic fungi** include **molds** and **yeasts**.
4. Animal-like **protozoa** are single-celled eukaryotes. Some cause disease.
5. Plantlike eukaryotic **algae** are important providers of oxygen, serve as food for many marine animals, and make chemicals used in microbiological growth media.
6. Parasitic worms, the largest organisms studied by microbiologists, are often visible without a microscope, although their immature stages are microscopic.
7. Viruses, the smallest microbes, are so small they can be seen only with an electron microscope.

The Golden Age of Microbiology (pp. 7–18)

1. The study of the Golden Age of Microbiology includes a look at the men who proposed or refuted the theory of **spontaneous generation**: Aristotle, Redi, Needham, Spallanzani, and Pasteur (the Father of Microbiology). The **scientific method** that emerged then remains the accepted sequence of study today.
2. The study of fermentation by Pasteur and Buchner led to the discovery of **anaerobic** bacteria, to the fields of **industrial microbiology (biotechnology)** and **biochemistry**, and to the study of **metabolism**.
3. Koch, Pasteur, and others proved that **pathogens** cause infectious diseases, an idea that is known as the **germ theory of disease**. **Etiology** is the study of the causation of diseases.
4. Koch initiated careful microbiological laboratory techniques in his search for disease agents. **Koch's postulates**, the logical steps he followed to prove the cause of an infectious disease, remain an important part of microbiology today.
5. The procedure for the **Gram stain** was developed in the 1880s and is still used to differentiate bacteria into two categories: Gram positive and Gram negative.
6. The investigations of Semmelweis, Lister, Nightingale, and Snow are the foundations on which **infection control**, including control of **healthcare-associated infections (HAI)**, and **epidemiology** are built.
7. Jenner's use of a cowpox-based vaccine for preventing smallpox began the field of **immunology**. Pasteur significantly advanced the field.
8. Ehrlich's search for "magic bullets"—chemicals that differentially kill microorganisms—laid the foundations for the field of **chemotherapy**.

The Modern Age of Microbiology (pp. 18–22)

1. Microbiology in the modern age has focused on answering questions regarding **biochemistry**, which is the study of metabolism; microbial genetics, which is the study of inheritance in microorganisms; and **molecular biology**, which involves investigations of cell function at the molecular level.
2. Scientists have applied knowledge from basic research to answer questions in **recombinant DNA technology** and **gene therapy**.
3. The study of microorganisms in their natural environment is **environmental microbiology**.
4. The discovery of chemicals in the blood that are active against specific pathogens advanced immunology and began the field of serology.
5. Advancements in chemotherapy were made in the 1900s with the discovery of numerous substances, such as penicillin and sulfa drugs, that inhibit pathogens.

QUESTIONS FOR REVIEW

Answers to the Questions for Review (except Short Answer questions) begin on p. A-1.

Multiple Choice

1. Which of the following microorganisms are *not* eukaryotic?
   a. bacteria
   b. yeasts
   c. molds
   d. protozoa

2. Which microorganisms are used to make microbiological growth media?
   a. bacteria
   b. fungi
   c. algae
   d. protozoa

3. In which habitat would you most likely find archaea?
   a. acidic hot springs
   b. swamp mud
   c. Great Salt Lake
   d. all of the above

4. Of the following scientists, who first promulgated the theory of abiogenesis?
   a. Aristotle
   b. Pasteur
   c. Needham
   d. Spallanzani

5. Which of the following scientists hypothesized that a bacterial colony arises from a single bacterial cell?
   a. Antoni van Leeuwenhoek
   b. Louis Pasteur
   c. Robert Koch
   d. Richard Petri

6. Which scientist first hypothesized that medical personnel can infect patients with pathogens?
   a. Florence Nightingale
   b. Joseph Lister
   c. John Snow
   d. Ignaz Semmelweis

7. Van Leeuwenhoek described microorganisms as _____.
   a. animalcules
   b. prokaryotes
   c. eukaryotes
   d. protozoa

8. Which of the following favored the theory of spontaneous generation?
   a. Spallanzani
   b. Needham
   c. Pasteur
   d. Koch

9. A scientist who studies the role of microorganisms in the environment is a(n) _____.
   a. genetic technologist
   b. earth microbiologist
   c. epidemiologist
   d. environmental microbiologist

10. The laboratory of Robert Koch contributed which of the following to the field of microbiology?
    a. simple staining technique
    b. use of Petri dishes
    c. first photomicrograph of bacteria
    d. all of the above

Fill in the Blanks

Fill in the blanks with the name(s) of the scientist(s) whose investigations led to the following fields of study in microbiology.

1. Environmental microbiology _____ and _____
2. Biochemistry _____ and _____
3. Chemotherapy _____
4. Immunology _____
5. Infection control _____
6. Etiology _____
7. Epidemiology _____
8. Biotechnology _____
9. Food microbiology _____

Short Answer

1. Why was the theory of spontaneous generation a hindrance to the development of the field of microbiology?
2. Discuss the significant difference between the flasks used by Pasteur and Spallanzani. How did Pasteur's investigation settle the dispute about spontaneous generation?
3. List six types of microorganisms.
4. Defend this statement: "The investigations of Antoni van Leeuwenhoek changed the world forever."
5. Why would a *macro*scopic tapeworm be studied in *micro*biology?
6. Describe what has been called the "Golden Age of Microbiology" with reference to four major questions that propelled scientists during that period.
7. List four major questions that drive microbiological investigations today.
8. Refer to the four steps in the scientific method in describing Pasteur's fermentation experiments.
9. List Koch's postulates, and explain why they are significant.
10. What does the term HAI (*nosocomial infection*) have to do with patient care?

Matching

Match each of the following descriptions with the person it best describes more than once or not at all. An answer may be used more than once.

1. _____ Developed smallpox immunization
2. _____ First photomicrograph of bacteria
3. _____ Used mathematical data to improve nursing
4. _____ Germs cause disease
5. _____ Sought a "magic bullet" to destroy pathogens
6. _____ Early epidemiologist
7. _____ Father of Microbiology
8. _____ Classification system
9. _____ Discoverer of bacteria
10. _____ Discoverer of protozoa
11. _____ Founder of antiseptic surgery
12. _____ Developed the most widely used bacterial staining technique

A. John Snow
B. Paul Ehrlich
C. Louis Pasteur
D. Antoni van Leeuwenhoek
E. Carolus Linnaeus
F. John Needham
G. Eduard Buchner
H. Robert Koch
I. Joseph Lister
J. Edward Jenner
K. Girolamo Fracastoro
L. Hans Christian Gram
M. Florence Nightingale

VISUALIZE IT!

1. On the following photos, label *cilium, flagellum, nucleus,* and *pseudopod.*

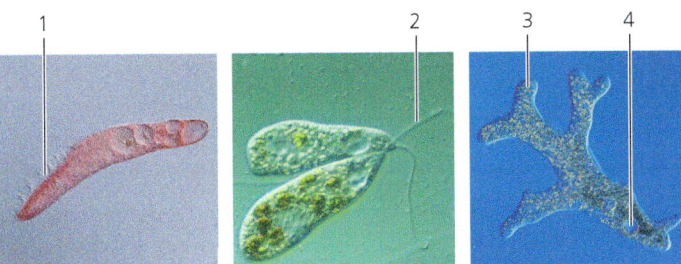

2. Show where microbes ended up in Pasteur's experiment.

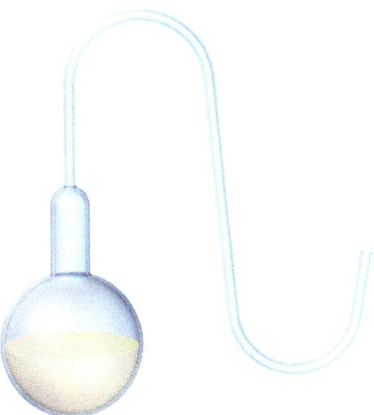

CRITICAL THINKING

1. If Robert Koch had become interested in a viral disease, such as influenza, instead of anthrax (caused by a bacterium), how might his list of lifetime accomplishments be different? Why?

2. In 1911, the Polish scientist Casimir Funk proposed that a limited diet of polished white rice (rice without the husks) caused beriberi, a disease of the central nervous system. Even though history has proven him correct—beriberi is caused by a thiamine deficiency, which in his day resulted from unsophisticated milling techniques that removed the thiamine-rich husks—Funk was criticized by his contemporaries, who told him to find the microbe that caused beriberi. Explain how the prevailing scientific philosophy of the day shaped the point of view of Funk's detractors.

3. *Haemophilus influenzae* does not cause flu, but it received its name because it was once thought to be the cause. Explain how a proper application of Koch's postulates would have prevented this error in nomenclature.

4. Just before winter break in early December, your roommate stocks the refrigerator with a gallon of milk, but both of you leave before opening it. When you return in January, the milk has soured. Your roommate is annoyed because the milk was pasteurized and thus should not have spoiled. Explain why your roommate's position is unreasonable.

5. Design an experiment to prove that microbes do not spontaneously generate in milk.

6. The British General Board of Health concluded in 1855 that the Broad Street cholera epidemic discussed in the chapter (see p. 17) resulted from fermentation of "nocturnal clouds of vapor" from the polluted Thames River. How could an epidemiologist prove or disprove this claim?

7. Compare and contrast the investigations of Redi, Needham, Spallanzani, and Pasteur in relation to the idea of spontaneous generation.

8. If you were a career counselor directing a student in the field of applied microbiology, describe three possible disciplines you could suggest.

9. A few bacteria produce disease because they derive nutrition from human cells and produce toxic wastes. Algae do not typically cause disease. Why not?

10. How might the debate over spontaneous generation have been different if Buchner had conducted his experiments in 1857 instead of 1897?

11. French microbiologists, led by Pasteur, tried to isolate a single bacterium by diluting liquid media until only a single type of bacterium could be microscopically observed in a sample of the diluted medium. What advantages does Koch's method have over the French method?

12. Why aren't Koch's postulates always useful in proving the cause of a given disease? Consider a variety of diseases, such as cholera, pneumonia, Alzheimer's, AIDS, Down syndrome, and lung cancer.

13. Albert Kluyver said, "From elephant to . . . bacterium—it is all the same!" What did he mean?

14. The ability of farmers around the world to produce crops such as corn, wheat, and rice is often limited by the lack of nitrogen-based fertilizer. How might scientists use Beijerinck's discovery to increase world supplies of grain?

15. Two students are arguing about who is the "father" of microbiology. One claims that van Leeuwenhoek deserves the title, while the other insists that Pasteur should be so honored. What facts can you bring to the argument to support each of their positions?

CONCEPT MAPPING

Using the terms provided, fill in this concept map that describes what microbiologists study. You can also complete this and other concept maps online by going to the Mastering Microbiology Study Area.

Acellular
Algae
Archaea
Bacteria
Eukaryotes
Molds
Multicellular
Obligate intracellular parasites
Protozoa
Unicellular (X2)
Yeasts

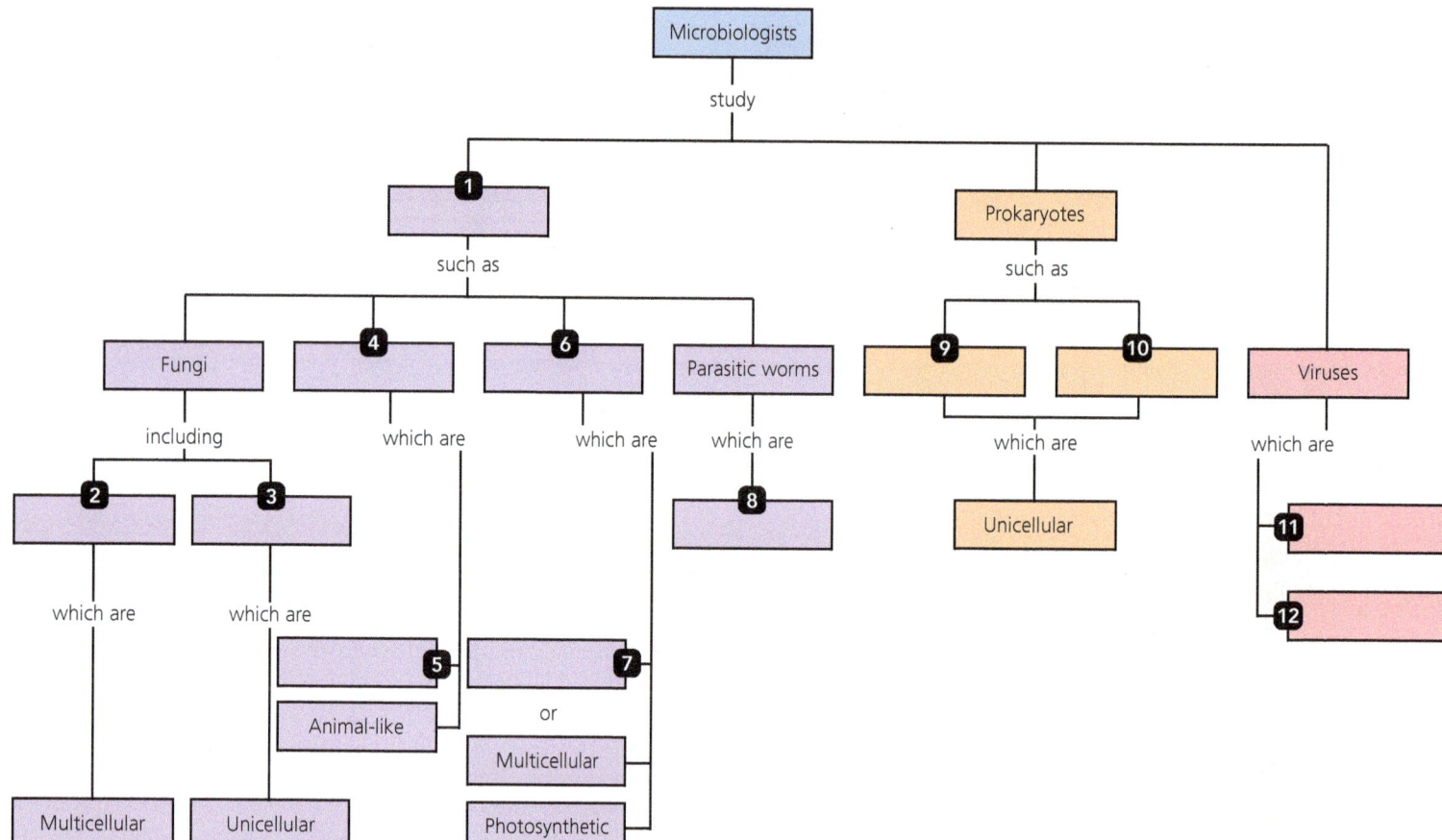

2 The Chemistry of Microbiology

Before You Begin

1. What is the nucleus of an atom composed of?
2. What substance makes up most of the mass in living organisms?

MICRO IN THE CLINIC

Can Spicy Food Cause Ulcers?

RAMONA IS A YOUNG MOM who works during the day and takes pre-nursing classes at night. Juggling the needs of her family and her studies means a hectic schedule, late nights, very little sleep, and eating on the run. Ramona particularly loves spicy food, and she eats a lot of it. She adds hot sauce to nearly every meal, which tends to be Mexican fast food. She also likes to drink wine with dinner on the weekends and sneaks an occasional cigarette when her children aren't watching.

One night, Ramona notices a burning pain in her upper abdomen. It disappears after a few minutes but then comes back a couple of nights later. Pretty soon she is feeling the pain every night—sometimes accompanied by nausea. She mentions her symptoms to her friend Brian, who suggests that she might have an ulcer. Brian knows about Ramona's love of spicy food and advises her to cut back on the hot sauce to see whether that improves her symptoms. Ramona takes his advice, but the pain and nausea continue. Sometimes the symptoms occur during the day, but mostly they flare up during the night.

1. What do you think?
2. Has eating too much spicy food given Ramona an ulcer?

Turn to the end of the chapter (p. 53) to find out.

The bacterium *Helicobacter pylori* can cause gastric ulcers.

Learning or reviewing some basic concepts of chemistry will enable you to understand more fully the variety of interactions between microorganisms and their environment—which includes you. If you plan a career in health care, you will find microbial chemistry involved in the diagnosis of disease, the response of the immune system, the growth and identification of pathogenic microorganisms in the laboratory, and the function and selection of antimicrobial drugs. Understanding the fundamentals of chemistry will even help you preserve your own health.

In this chapter, we study atoms, which are the basic units of chemistry, and we consider how atoms react with one another to form chemical bonds and molecules. Then we examine the three major categories of chemical reactions. The chapter concludes with a look at the molecules of greatest importance to life: water, acids, bases, lipids, carbohydrates, proteins, nucleic acids, and ATP.

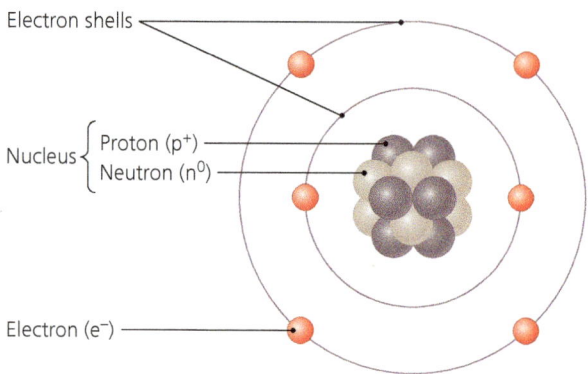

▲ **FIGURE 2.1 An example of a Bohr model of atomic structure.** This drawing is not to scale; for the electrons to be shown in scale with the greatly magnified nucleus, the electrons would have to occupy orbits located many miles from the nucleus. Put another way, the volume of an entire atom is about 100 trillion times the volume of its nucleus.

Atoms

LEARNING OUTCOME

2.1 Define *matter* and *atom*, and explain how these terms relate to one another.

Matter is defined as anything that takes up space and has mass.[1] The smallest chemical units of matter are **atoms**. Atoms are extremely small, and only the very largest of them can be seen using the most powerful microscopes. Therefore, scientists have developed various models to conceptualize and illustrate the structure of atoms.

Atomic Structure

LEARNING OUTCOMES

2.2 Draw and label an atom, showing the parts of the nucleus and orbiting electrons.
2.3 Define *element, atomic number, atomic mass,* and *dalton,* and explain how these terms relate to one another.

In 1913, the Danish physicist Niels H. D. Bohr (1885–1962) proposed a simple model in which negatively charged subatomic particles called **electrons** orbit a centrally located nucleus like planets in a miniature solar system (**FIGURE 2.1**). A nucleus is composed of both uncharged **neutrons** and positively charged **protons**. (The only exception is the nucleus of a normal hydrogen atom, which is composed of only a single proton and no neutrons.)

An **element** is matter that is composed of a single type of atom. For example, gold is an element because it consists of only gold atoms. In contrast, the ink in your pen is not an element because it is composed of many different kinds of atoms.

Elements differ from one another in their **atomic number**, which is the number of protons in their nuclei. For example, the atomic numbers of hydrogen, carbon, and oxygen are 1, 6, and 8, respectively, because all hydrogen nuclei contain a single proton, all carbon nuclei have six protons, and all oxygen nuclei have eight protons.

The **atomic mass** of an atom (sometimes called its *atomic weight*) is the sum of the masses of its protons, neutrons, and electrons. Protons and neutrons each have a mass of approximately 1 *atomic mass unit*,[2] which is also called a *dalton*.[3] An electron is much less massive, with a mass of about 0.00054 dalton. Electrons are often ignored in discussions of atomic mass because their contribution to the overall mass is negligible. Therefore, the sum of the number of protons and neutrons approximates the atomic mass of an atom.

There are 93 naturally occurring elements known;[4] however, organisms typically utilize only about 20 elements, each of which has its own symbol that is derived from its English or Latin name (**TABLE 2.1**).

Isotopes

LEARNING OUTCOME

2.4 List at least four ways that radioactive isotopes are useful.

Every atom of an element has the same number of protons, but atoms of a given element can differ in the number of neutrons in their nuclei. Atoms that differ in this way are called **isotopes**. For example, there are three naturally occurring isotopes of carbon, each having six protons and six electrons (**FIGURE 2.2**). Over 95% of carbon atoms also have six neutrons. Because these carbon atoms have six protons and six neutrons, the atomic mass of this isotope is about 12 daltons, and it is known as carbon-12, symbolized as ^{12}C. Atoms of carbon-13 (^{13}C) have seven neutrons per nucleus, and ^{14}C atoms each have eight neutrons.

---

[1]Mass and weight are sometimes confused. Mass is the quantity of material in something, whereas weight is the effect of gravity on mass. Even though an astronaut is weightless in space, his mass is the same in space as on Earth.

[2]An atomic mass unit (dalton) is 1/597,728,630,000,000,000,000,000, or 1.673×10^{-24}, grams.

[3]Named for John Dalton, the British chemist who helped develop atomic theory around 1800.

[4]For many years, scientists thought that there were only 92 naturally occurring elements, but natural plutonium was discovered in Africa in 1997.

TABLE 2.1 Common Elements of Life

Element	Symbol	Atomic Number	Atomic Mass[a] (daltons)	Biological Significance
Hydrogen	H	1	1	Component of organic molecules and water; H^+ released by acids
Boron	B	5	11	Essential for plant growth
Carbon	C	6	12	Backbone of organic molecules
Nitrogen	N	7	14	Component of amino acids, proteins, and nucleic acids
Oxygen	O	8	16	Component of many organic molecules and water; OH^- released by bases; necessary for aerobic metabolism
Sodium (Natrium)	Na	11	23	Principal cation outside cells
Magnesium	Mg	12	24	Component of many energy-transferring enzymes
Silicon	Si	14	28	Component of cell wall of diatoms
Phosphorus	P	15	31	Component of nucleic acids and ATP
Sulfur	S	16	32	Component of proteins
Chlorine	Cl	17	35	Principal anion outside cells
Potassium (Kalium)	K	19	39	Principal cation inside cells; essential for nerve impulses
Calcium	Ca	20	40	Utilized in many intercellular signaling processes; essential for muscular contraction
Manganese	Mn	25	54	Component of some enzymes; acts as intracellular antioxidant; used in photosynthesis
Iron (Ferrum)	Fe	26	56	Component of energy-transferring proteins; transports oxygen in the blood of many animals
Cobalt	Co	27	59	Component of vitamin B_{12}
Copper (Cuprum)	Cu	29	64	Component of some enzymes; used in photosynthesis
Zinc	Zn	30	65	Component of some enzymes
Molybdenum	Mo	42	96	Component of some enzymes
Iodine	I	53	127	Component of many brown and red algae

[a]Rounded to nearest whole number.

Unlike carbon-12 and carbon-13, the nucleus of carbon-14 is unstable because of the ratio of its protons and neutrons. Unstable atomic nuclei release energy and subatomic particles such as neutrons, protons, and electrons in a process called *radioactive decay*. Atoms that undergo radioactive decay are *radioactive isotopes*. Radioactive decay and radioactive isotopes play important roles in microbiological research, medical diagnosis, the treatment of disease, and the complete destruction of contaminating microbes (sterilization) of medical equipment and chemicals.

MICRO CHECK

1. What is the smallest chemical unit of matter?
2. What atomic particles orbit around the nucleus?
3. The number of protons identifies an element. What characteristic tells us how many protons an atom has?
4. Isotopes are forms of atoms that differ from one another by having varying numbers of what particle?

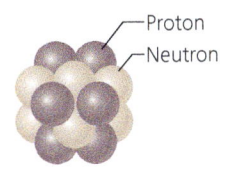
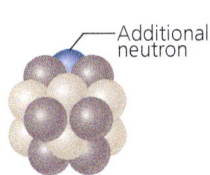

(a) Carbon-12
6 Protons
6 Neutrons

(b) Carbon-13
6 Protons
7 Neutrons

(c) Carbon-14
6 Protons
8 Neutrons

▲ **FIGURE 2.2 Nuclei of the three naturally occurring isotopes of carbon.** Each isotope also has six electrons, which are not shown. *What are the atomic number and atomic mass of each of these isotopes?*

Figure 2.2 *The atomic number of all three is 6; their atomic masses are 12, 13, and 14, respectively.*

Electron Configurations

LEARNING OUTCOME

2.5 Define *electron shell* and *valence electron*, and list how many electrons each of the first three shells can contain.

Although the nuclei of atoms determine their identities, it is electrons that determine an atom's *chemical behavior*. Nuclei of different atoms almost never come close enough together to interact.[5] Typically, only the electrons of atoms interact. Thus, because all of the isotopes of carbon (for example) have the same number of electrons, all these isotopes behave the same way in chemical reactions, even though their nuclei are different.

---
[5]Except during nuclear reactions, such as occur in nuclear power plants.

Scientists know that electrons do not really orbit the nucleus in a two-dimensional circle, as indicated by a Bohr model; instead, they speed around a nucleus 100 quadrillion times per second in three-dimensional *electron shells* or *clouds* that assume unique shapes dependent on the energy of the electrons (FIGURE 2.3a). More accurately put, an electron shell depicts the *probable* locations of electrons at a given time; nevertheless, it is simpler and more convenient to draw electron shells as circles (FIGURE 2.3b).

Each electron shell can hold only a certain maximum number of electrons. For example, the first shell (the one nearest the nucleus) can accommodate a maximum of two electrons, and the second shell can hold no more than eight electrons. Atoms of hydrogen and helium have one and two electrons, respectively; thus, these two elements have only a single electron shell. A lithium atom, which has three electrons, has two shells.

Atoms with more than 10 electrons require more shells. The third shell holds up to eight electrons when it is the outermost shell, though its capacity increases to 18 when the fourth shell contains two electrons. Heavier atoms have even more shells, but these atoms do not play significant roles in the processes of life.

Electrons in the outermost shell of atoms are called *valence electrons*. FIGURE 2.4 depicts the electron configurations of atoms of some elements important to microbial life. Notice that except for helium, atoms of all elements in a given column of the periodic table of elements have the same number of valence electrons. Helium is placed in the far right-hand column with the

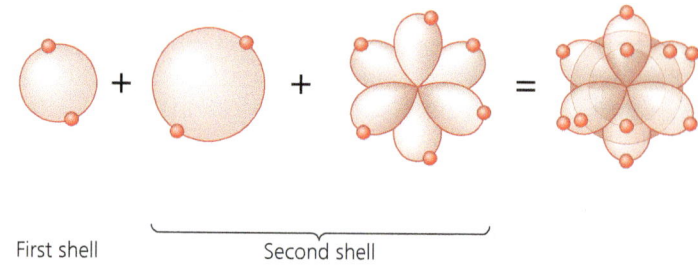

First shell   Second shell

(a) Electron shells of neon: three-dimensional view

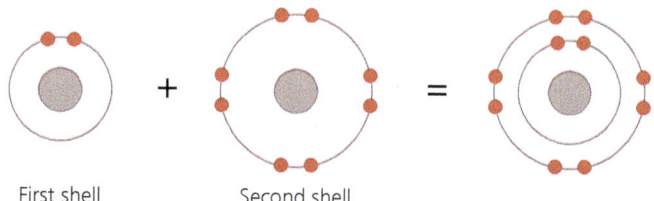

First shell   Second shell

(b) Electron shells of neon: two-dimensional view

▲ **FIGURE 2.3 Electron configurations.** (a) Three-dimensional model of the electron shells of neon. In this model, the first shell is a small sphere, whereas the second shell consists of a larger sphere plus three pairs of ellipses that extend from the nucleus at right angles. Larger shells (not shown) are even more complex. (b) Two-dimensional model (Bohr diagram) of the electron shells of neon. *How many shells does a sodium atom need to hold its 11 electrons?*

Figure 2.3 *Three shells: two electrons in the first shell, eight in the second, and a single electron in the third.*

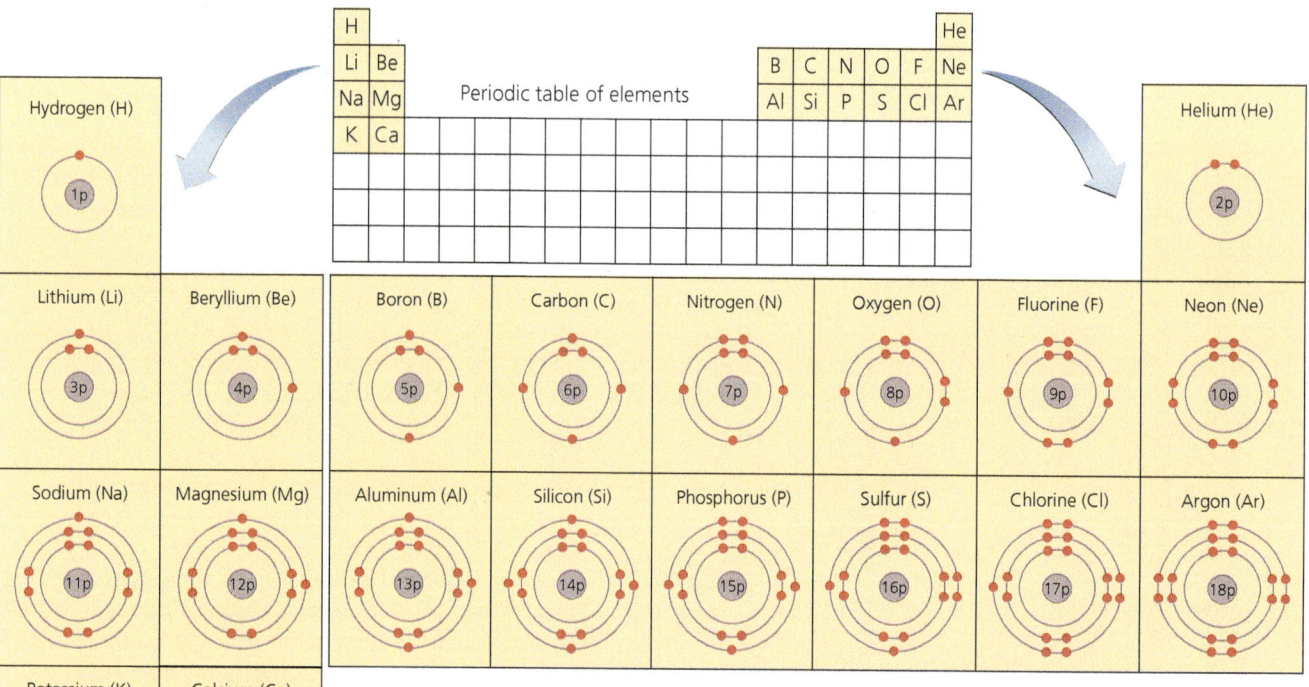

▲ **FIGURE 2.4 Bohr diagrams of the first 20 elements and their places within the chart known as the periodic table of the elements.** Note that the number of valence electrons increases from left to right in each row and that every element in a column has the same number of valence electrons (with the exception of helium). Heavier atoms have been omitted because most heavy elements are less important to living organisms. Neutrons are not shown because they have little effect on chemistry.

other inert gases because its outer shell is full, though it has two rather than eight valence electrons. Valence electrons are critical for interactions between atoms. Next, we consider these interactions, which are called chemical bonds.

TELL ME WHY

Electrons zip around the nucleus at about 5 million miles per hour. Why don't they fly off?

Chemical Bonds

LEARNING OUTCOMES

2.6 Describe the configuration of electrons in a stable atom, and explain how valence electrons form chemical bonds.
2.7 Contrast molecules and compounds.

Outer electron shells are stable when they contain eight electrons (except for the first electron shell, which is stable with only two electrons, because that is its maximum number). When atoms' outer shells are not filled with eight electrons, they either have room for more electrons or have "extra" electrons, depending on whether it is easier for them to gain electrons or lose electrons. For example, an oxygen atom, with six electrons in its outer shell, has two "unfilled spaces" (see Figure 2.4) because it requires less energy for the oxygen atom to gain two electrons than to lose six electrons. A calcium atom, by contrast, has two "extra" electrons in its outer (fourth) shell because it requires less energy to lose these two electrons than to gain six new ones. When a calcium atom loses two electrons, its third shell, which is then its outer shell, is full and stable with eight electrons.

As previously noted, an atom's outermost electrons are called valence electrons, and thus the outermost shell of an atom is the *valence shell*. An atom's **valence**,[6] defined as its combining capacity, is considered to be positive if its valence shell has extra electrons to give up and to be negative if its valence shell has spaces to fill. Thus, a calcium atom, with two electrons in its valence shell, has a valence of +2, whereas an oxygen atom, with two spaces to fill in its valence shell, has a valence of −2.

Atoms combine with one another by either sharing or transferring valence electrons in such a way as to fill their valence shells. Such interactions between atoms are called **chemical bonds**. Two or more atoms held together by chemical bonds form a **molecule**. A molecule that contains atoms of more than one element is a **compound**. Two hydrogen atoms bonded together form a hydrogen molecule, which is not a compound because only one element is involved. However, two hydrogen atoms bonded to an oxygen atom form a molecule of water (H_2O), which is a compound.

In this section, we discuss the three principal types of chemical bonds: nonpolar covalent bonds, polar covalent bonds, and ionic bonds. We also consider hydrogen bonds, which are weak forces that act with polar covalent bonds to give certain large chemicals their characteristic three-dimensional shapes.

MICRO CHECK

5. An atom with a total of six electrons would have how many electrons in its inner shell?
6. An atom with a total of six electrons would have how many valence electrons?
7. What do you call a molecule that contains atoms of two or more different elements?

Nonpolar Covalent Bonds

LEARNING OUTCOMES

2.8 Contrast nonpolar covalent, polar covalent, and ionic bonds.
2.9 Define *organic compound*.

A **covalent**[7] **bond** is the sharing of a pair of electrons between two atoms. Consider, for example, what happens when two hydrogen atoms approach one another. Each hydrogen atom consists of a single proton orbited by a single electron. Because the valence shell of each hydrogen atom requires two electrons in order to be full, each atom shares its single electron with the other, forming a hydrogen molecule in which both atoms have full shells (**FIGURE 2.5a**). Similarly, two oxygen atoms can share electrons, but they must share *two* pairs of electrons for their valence shells to be full (**FIGURE 2.5b**). Because two pairs of electrons are involved, oxygen atoms form two covalent bonds, or a *double covalent bond*, with one another.

The attraction of an atom for electrons is called its **electronegativity**. The more electronegative an atom, the greater the pull its nucleus exerts on electrons. Note in **FIGURE 2.6**, which displays the electronegativities of atoms of several elements, that electronegativities tend to increase from left to right in the chart. The reason is that elements toward the right of the chart have more protons and thus exert a greater pull on electrons. Electronegativities of elements decrease from top to bottom in the chart because the distance between the nucleus and the valence shell increases as elements get larger.

Atoms with equal or nearly equal electronegativities, such as two hydrogen atoms or a hydrogen and a carbon, share electrons equally or nearly equally. In chemistry and physics, "poles" are opposed forces, such as north and south magnetic poles or positive and negative terminals of a battery. In the case of atoms with similar electronegativities, the shared electrons tend to spend an equal amount of time around each nucleus of the pair, and no poles exist; therefore, the bond between them is a **nonpolar covalent bond**. (The covalent bonds illustrated in Figure 2.5a–c are nonpolar; formaldehyde, shown in Figure 2.5d, is polar.)

A hydrogen molecule can be symbolized a number of ways:

$$H-H \quad H{:}H \quad H_2$$

In the first symbol, the dash represents the chemical bond between the atoms. In the second symbol, the dots represent the electron pair of the covalent bond. These two symbols are known

---

[6] From Latin *valentia*, meaning "strength."

[7] From Latin *co*, meaning "with" or "together," and *valentia*, meaning "strength."

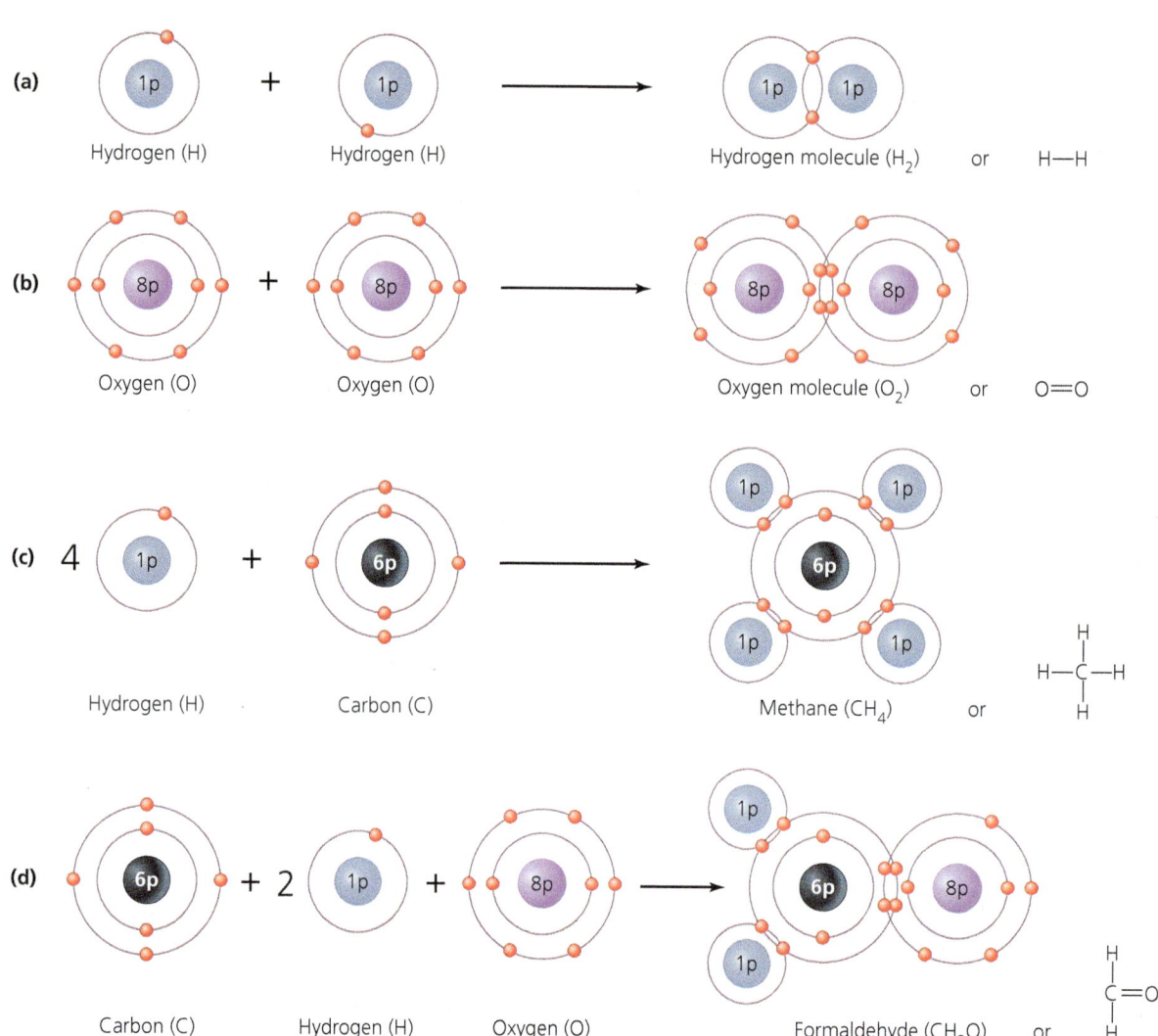

▲ **FIGURE 2.5 Four molecules formed by covalent bonds.** (a) Hydrogen. Each hydrogen atom needs another electron to have a full valence shell. The two atoms share their electrons, forming a covalent bond. (b) Oxygen. Oxygen atoms have six electrons in their valence shells; thus, they need two electrons each. When they share with each other, two covalent bonds are formed. Note that the valence electrons of oxygen atoms are in the second shell. (c) A methane molecule, which has four single covalent bonds. (d) Formaldehyde. The carbon atom forms a double bond with the oxygen atom and single bonds with two hydrogen atoms. *Which of these molecules are also compounds? Why?*

Figure 2.5 Methane and formaldehyde molecules are also compounds because they are composed of more than one element.

as *structural formulas*. In the third symbol, known as a molecular formula, the subscript "2" indicates the number of hydrogen atoms that are bonded, not the number of shared electrons. Each of these symbols indicates the same thing—two hydrogen atoms are sharing a pair of electrons.

Many atoms need more than one electron to fill their valence shell. For instance, a carbon atom has four valence electrons and needs to gain four more if it is to have eight in its valence shell. **FIGURE 2.5c** illustrates a carbon atom sharing electrons with four hydrogen atoms. As before, a line in the structural formula represents a covalent bond formed from the sharing of two electrons. Two covalent bonds are formed between an oxygen atom and a carbon atom in formaldehyde (**FIGURE 2.5d**). This fact is represented by a double line, which indicates that the carbon atom shares four electrons with the oxygen atom.

Carbon atoms are critical to life. Because a carbon atom has four electrons in its valence shell, it has equal tendencies to either lose four electrons or gain four electrons. Either event produces a full outer shell. The result is that carbon atoms tend to share electrons and form four covalent bonds with one another and with many other types of atoms. Each carbon atom, in effect, acts as a four-way intersection where different components of a molecule can attach. One result of this feature is that carbon atoms can form very large chains that constitute the "backbone" of many biologically important molecules. Carbon chains can be branched or unbranched, and some even close back on

I	II	III	IV	V	VI	VII	Inert gases
H 2.1							He 0.0
Li 1.0	Be 1.5	B 2.0	C 2.5	N 3.0	O 3.5	F 4.0	Ne 0.0
Na 0.9	Mg 1.2	Al 1.5	Si 1.8	P 2.1	S 2.5	Cl 3.0	Ar 0.0
K 0.8	Ca 1.0	Ga 1.6	Ge 1.8	As 2.0	Se 2.4	Br 2.8	Kr 0.0

▲ FIGURE 2.6 **Electronegativity values of selected elements.** The values are expressed according to the Pauling scale, named for the Nobel Prize-winning chemist Linus Pauling, who based the scale on bond energies. Pauling chose to compare the electronegativity of each element to that of fluorine, to which he assigned a value of 4.0.

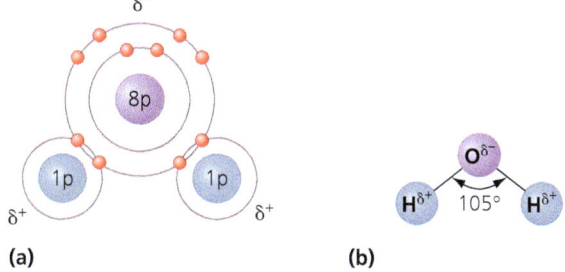

▲ FIGURE 2.7 **Polar covalent bonding in a water molecule.** **(a)** A Bohr model of a water molecule, which has two polar covalent bonds. When the electronegativities of two atoms are significantly different, the shared electrons of covalent bonds spend more time around the more electronegative atom, giving it a transient negative charge (δ^-). Its partner has a transient positive charge (δ^+). **(b)** The bond angle in a water molecule. Atoms maximize the distances between electron orbitals in polar and nonpolar covalent bonds.

themselves to form rings. Compounds that contain carbon and hydrogen atoms are called **organic compounds**. Among the many biologically important organic compounds are proteins and carbohydrates, which are discussed later in the chapter.

MICRO CHECK

8. Organic compounds always contain which elements?

Polar Covalent Bonds

LEARNING OUTCOME

2.10 Explain the relationship between electronegativity and the polarity of a covalent bond.

If two covalently bound atoms have significantly different electronegativities, their electrons will not be shared equally. Instead, the electron pair will spend more time orbiting the nucleus of the atom with greater electronegativity. This type of bond, in which there is unequal sharing of electrons, is a polar covalent bond. An example of a molecule with **polar covalent bonds** is water (FIGURE 2.7a).

Because oxygen is more electronegative than hydrogen, the electrons spend more time near the oxygen nucleus than near the hydrogen nuclei, and thus the oxygen atom acquires a transient (partial) negative charge (symbolized as δ^-). Each of the hydrogen nuclei has a corresponding transient positive charge (δ^+). The covalent bond between an oxygen atom and a hydrogen atom is called polar because the atoms have opposite partial electrical charges.

Polar covalent bonds can form between many different elements. Generally, molecules with polar covalent bonds are water soluble, and nonpolar molecules are not. The most important polar covalent bonds for life are those that involve hydrogen because they allow hydrogen bonding, which we discuss shortly.

Both nonpolar and polar covalent bonds form angles between atoms such that the distances between electron orbits are maximized. The bond angle for water is shown in FIGURE 2.7b. However, it is more convenient to simply draw molecules as though all the atoms were in one plane; for example, H—O—H.

Ionic Bonds

LEARNING OUTCOMES

2.11 Define *ionization* using the terms *cation* and *anion*.
2.12 Contrast the formation of an ionic bond with that of a covalent bond.

Consider what happens when two atoms with vastly different electronegativities—for example, sodium, with one electron in its valence shell and an electronegativity of 0.9, and chlorine, with seven electrons in its valence shell and an electronegativity of 3.0—come together (FIGURE 2.8 on the next page). Chlorine has so much higher electronegativity that it very strongly attracts sodium's single valence electron, and the result is that the sodium loses that electron to chlorine 1.

Now that the chlorine atom has one more electron than it has protons, it has a full negative charge, and the sodium atom, which has lost an electron, now has a full positive charge 2. An atom or group of atoms that has either a full negative charge or a full positive charge is called an *ion*. Positively charged ions are called **cations**, whereas negatively charged ions are called **anions**.

Because of their opposite charges, cations and anions attract each other and form what is termed an **ionic bond** 3. They form crystalline compounds composed of metallic and nonmetallic ions. These compounds are known as **salts**. Examples include sodium chloride (NaCl), also known as table salt, and potassium chloride (KCl, sodium-free table salt). Ionic bonds differ from covalent bonds in that ions do not share electrons. Instead, the bond is formed from the attraction of opposite electrical charges.

34 CHAPTER 2 The Chemistry of Microbiology

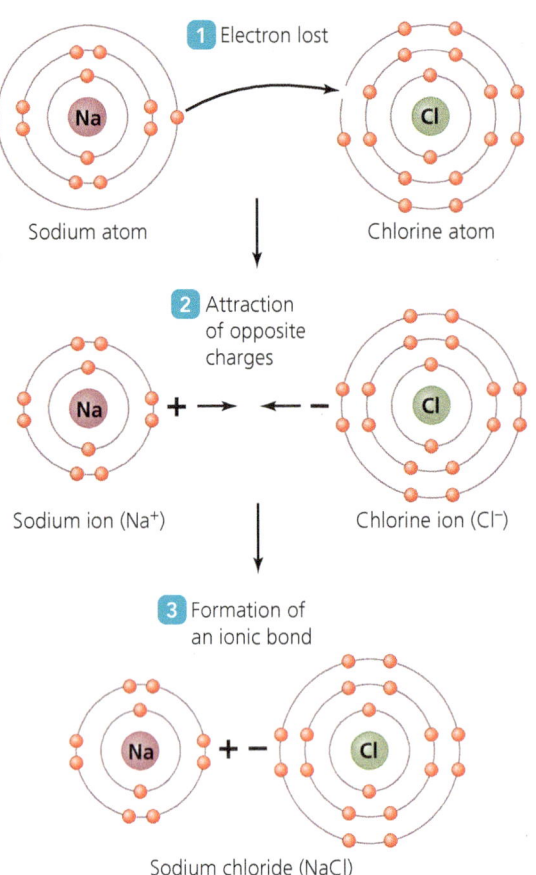

▲ FIGURE 2.8 The interaction of sodium and chlorine to form an ionic bond.

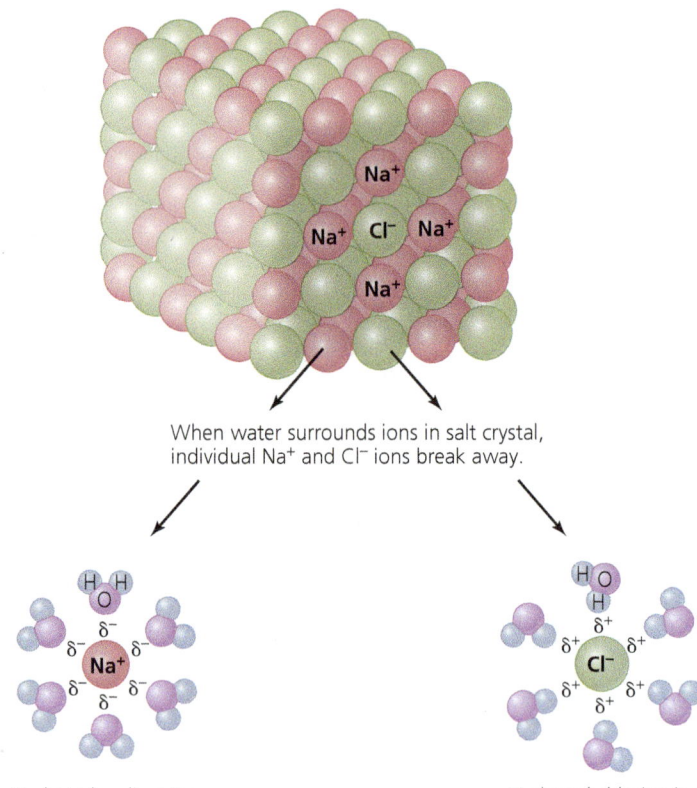

▲ FIGURE 2.9 **Dissociation of NaCl in water.** When water surrounds the ions in a NaCl crystal, the partial charges on water molecules are attracted to charged ions, and the water molecules hydrate the ions by surrounding them. The partial negative charges (δ^-) on oxygen atoms are attracted to cations (in this case, the sodium ions), and the partial positive charges (δ^+) on hydrogen atoms are attracted to anions (the chlorine ions). Because the ions no longer attract one another, the salt crystal dissolves. Hydrated ions are called electrolytes.

The polar bonds of water molecules interfere with the ionic bonds of salts, causing *dissociation* (also called *ionization*) (**FIGURE 2.9**). This occurs as the partial negative charge on the oxygen atom of water attracts cations, and the partial positive charge on hydrogen atoms attracts anions. The presence of polar bonds interferes with the attraction between the cation and anion.

When cations and anions dissociate from one another and become surrounded by water molecules (are hydrated), they are called **electrolytes** because they can conduct electricity through the solution. Electrolytes are critical for life because they stabilize a variety of compounds, act as electron carriers, and allow electrical gradients to exist within cells. (We examine these functions of electrolytes in later chapters.)

In nature, chemical bonds range from nonpolar bonds to polar bonds to ionic bonds. The important thing to remember is that electrons are shared between atoms in covalent bonds and transferred from one atom to another in ionic bonds.

MICRO CHECK

9. What is the name for an atom that has a full positive or negative charge?
10. Ionic bonds involve electrons being donated from one atom to another. How are the electrons treated in covalent bonds?

Hydrogen Bonds

LEARNING OUTCOME

2.13 Describe hydrogen bonds, and discuss their importance in living organisms.

As we have seen, hydrogen atoms bind to oxygen atoms by means of polar covalent bonds, resulting in transient positive charges on the hydrogen atoms. Hydrogen atoms form polar covalent bonds with atoms of other elements as well.

The electrical attraction between a partially charged hydrogen atom and a full or partial negative charge on either a different region of the same molecule or another molecule is called a **hydrogen bond** (**FIGURE 2.10**). Hydrogen bonds can be likened to weak ionic bonds in that they arise from the attraction of positive and negative charges. Notice also that although they are a consequence of polar covalent bonds between hydrogen atoms and other, more electronegative atoms, hydrogen bonds themselves are not covalent bonds—they do not involve the sharing of electrons.

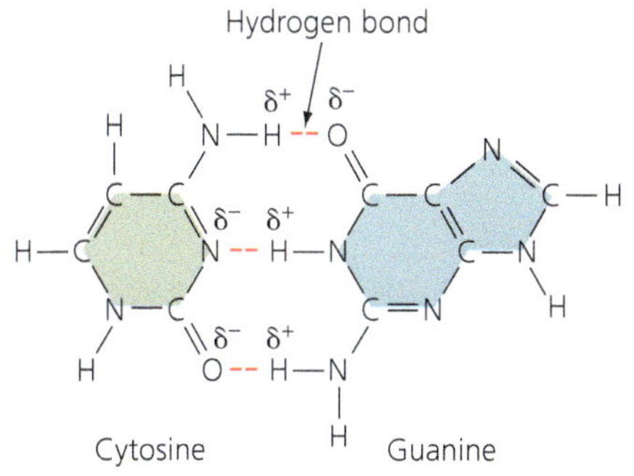

FIGURE 2.10 Hydrogen bonds. The transient positive charge (symbolized δ^+) on a hydrogen atom is attracted to a transient negative charge (δ^-) on another atom. Such attraction is a hydrogen bond. Hydrogen bonds can hold together portions of the same molecule or hold two different molecules together. In this case, three hydrogen bonds are holding molecules of cytosine and guanine together.

As we have seen, covalent bonds are essential for life because they strongly link atoms together to form molecules. Hydrogen bonds, though weaker than covalent bonds, are also essential. The cumulative effect of numerous hydrogen bonds is to stabilize the three-dimensional shapes of large molecules. For example, the familiar double-helix shape of DNA is due in part to the stabilizing effects of thousands of hydrogen bonds holding the molecule together. Hydrogen bonds play a role in the recognition of target cells by pathogens and also play a role in maintaining the exact shape of enzymes, antibodies, and intercellular chemical messengers, which is critical for their correct function. Further, because hydrogen bonds are weak, they can be overcome when necessary. For example, the two complementary halves of a DNA molecule are held together primarily by hydrogen bonds, and they can be separated for DNA replication and other processes (see Chapter 7, Figure 7.6).

TABLE 2.2 summarizes some characteristics of chemical bonds.

TELL ME WHY

Chlorine and potassium atoms form ionic bonds, carbon atoms form nonpolar covalent bonds with nitrogen atoms, and oxygen forms polar covalent bonds with phosphorus. Explain why these bonds are the types they are.

Chemical Reactions

LEARNING OUTCOME

2.14 Describe three general types of chemical reactions found in living things.

You are already familiar with many consequences of chemical reactions: you add yeast to bread dough, and it rises; enzymes in your laundry detergent remove grass stains; and gasoline burned in your car releases energy to speed you on your way. What exactly is happening in these reactions? What is the precise definition of *chemical reaction*?

We have discussed how bonds are formed via the sharing of electrons or the attraction of positive and negative charges. Scientists define **chemical reactions** as the making or breaking of such chemical bonds. All chemical reactions begin with **reactants**—the atoms, ions, or molecules that exist at the beginning of a reaction. Similarly, all chemical reactions result in **products**—the atoms, ions, or molecules left after the reaction is complete. *Biochemistry* involves the chemical reactions of living things.

Reactants and products may have very different physical and chemical characteristics. For example, hydrogen and oxygen are gases and have very different properties from water, which is composed of hydrogen and oxygen atoms. However, the numbers and types of atoms never change in a chemical reaction; atoms are neither destroyed nor created, only rearranged.

Now let's turn our attention to three general categories of biochemical reactions (reactions that occur in organisms): *synthesis*, *decomposition*, and *exchange reactions*.

Synthesis Reactions

LEARNING OUTCOMES

2.15 Give an example of a synthesis reaction that involves the formation of a water molecule.
2.16 Contrast endothermic and exothermic chemical reactions.

Synthesis reactions involve the formation of larger, more complex molecules. Synthesis reactions can be expressed symbolically as

$$\text{Reactant} + \text{Reactant} \rightarrow \text{Product(s)}$$

The arrow indicates the direction of the reaction and the formation of new chemical bonds. For example, algae make their own glucose (sugar) using the following reaction:

$$6\,H_2O + 6\,CO_2 \rightarrow C_6H_{12}O_6 + 6\,O_2$$

TABLE 2.2 Characteristics of Chemical Bonds

Type of Bond	Description	Relative Strength
Nonpolar covalent bond	Pair of electrons is nearly equally shared between two atoms	Strong
Polar covalent bond	Electrons spend more time around the more electronegative of two atoms	Strong
Ionic bond	Electrons are stripped from a cation by an anion	Weaker than covalent in aqueous environments
Hydrogen bond	Partial positive charges on hydrogen atoms are attracted to full and partial negative charges on other molecules or other regions of the same molecule	Weaker than ionic

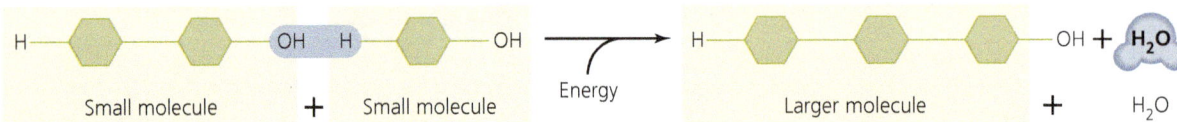

(a) **Dehydration synthesis**

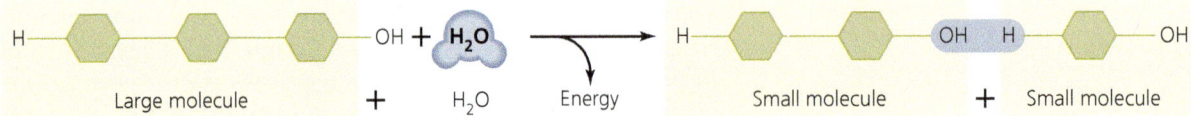

(b) **Hydrolysis**

▲ **FIGURE 2.11 Two types of chemical reactions in living things. (a)** Dehydration synthesis. In this energy-requiring reaction, a hydroxyl ion (OH⁻) removed from one reactant and a hydrogen ion (H⁺) removed from another reactant combine to form hydrogen hydroxide (HOH), which is water. **(b)** Hydrolysis, an energy-yielding reaction that is the reverse of a dehydration synthesis reaction. *What are the scientific words meaning "energy-requiring" and "energy-releasing"?*

Figure 2.11 Endothermic means "energy-requiring," and exothermic means "energy-releasing."

The reaction is read, "Six molecules of water plus six molecules of carbon dioxide yield one molecule of glucose and six molecules of oxygen." Notice that the total number and kind of atoms are the same on both sides of the reaction.

A common synthesis reaction in biochemistry is a **dehydration synthesis**, in which two smaller molecules are joined together by a covalent bond and a water molecule is also formed (**FIGURE 2.11a**). The word *dehydration* in the name of this type of reaction refers to the fact that one of the products is a water molecule formed when a hydrogen ion (H⁺) from one reactant combines with a hydroxyl ion (OH⁻) from another reactant.

Synthesis reactions require energy to break bonds in the reactants and to form new bonds to make products. Reactions that require energy are said to be **endothermic**[8] **reactions** because they trap energy within new molecular bonds. An energy supply for fueling synthesis reactions is one common requirement of all living things (Chapter 6).

Taken together, all of the synthesis reactions in an organism are called **anabolism**.

Decomposition Reactions

LEARNING OUTCOME

2.17 Give an example of a decomposition reaction that involves breaking the bonds of a water molecule.

Decomposition reactions are the reverse of synthesis reactions in that they break bonds within larger molecules to form smaller atoms, ions, and molecules. These reactions release energy and are therefore **exothermic**[9] **reactions**. In general, decomposition reactions can be represented by the following formula:

$$\text{Reactant} + \text{Product} \rightarrow \text{Product(s)}$$

An example of a biologically important decomposition reaction is the aerobic decomposition of glucose to form carbon dioxide and water:

$$C_6H_{12}O_6 + 6\,O_2 \rightarrow 6\,H_2O + 6\,CO_2$$

Note that this reaction is exactly the reverse of the synthesis reaction in algae that we examined previously. Synthesis and decomposition reactions are often reversible in living things.

A common type of decomposition reaction in biochemistry is **hydrolysis**,[10] the reverse of dehydration synthesis (**FIGURE 2.11b**). In hydrolytic reactions, a covalent bond in a large molecule is broken, and the ionic components of water (H⁺ and OH⁻) are added to the products.

Collectively, all of the decomposition reactions in an organism are called **catabolism**.

Exchange Reactions

LEARNING OUTCOME

2.18 Compare exchange reactions to synthesis and decomposition reactions.

Exchange reactions (also called *transfer reactions*) have features similar to both synthesis and decomposition reactions. For instance, they involve breaking and forming covalent bonds, and they involve both endothermic and exothermic steps. As the name suggests, atoms are moved from one molecule to another. In general, these reactions can be represented as either

$$A + BC \rightarrow AB + C$$

or

$$AB + CD \rightarrow AD + BC$$

---

[8]From Greek *endon*, meaning "within," and *thermos*, meaning "heat" (energy).
[9]From Greek *exo*, meaning "outside," and *thermos*, meaning "heat" (energy).
[10]From Greek *hydor*, meaning "water," and *lysis*, meaning "loosening."

An important exchange reaction within organisms is the phosphorylation of glucose:

$$C_6H_{12}O_6 + A-P-P-P \longrightarrow C_6H_{11}O_6-P + A-P-P + H^+$$

Glucose — Adenosine triphosphate — Glucose phosphate — Adenosine diphosphate

The sum of all of the chemical reactions in an organism, including catabolic, anabolic, and exchange reactions, is called **metabolism**. (Chapter 5 examines metabolism in more detail.)

MICRO CHECK

11. Dehydration synthesis is a chemical reaction that joins two molecules through the removal of a hydrogen atom from one and hydroxide from the other, forming a new bond. What molecule is always released during this kind of reaction?
12. Some chemical reactions require energy in order to occur. What do you call those reactions?
13. Exchange reactions contain both endothermic and exothermic steps. What other type of chemical reaction in living things is endothermic?

TELL ME WHY

Why are decomposition reactions exothermic, that is, energy releasing?

Water, Acids, Bases, and Salts

As previously noted, living things depend on organic compounds, those that contain carbon and hydrogen atoms. Living things also require a variety of **inorganic chemicals**, which typically lack carbon. Such inorganic substances include water, oxygen molecules, metal ions, and many acids, bases, and salts. In this section, we examine the characteristics of some of these inorganic substances.

Water

LEARNING OUTCOME

2.19 Describe five qualities of water that make it vital to life.

Water is the most abundant substance in organisms, constituting 50% to 99% of their mass. Most of the special characteristics that make water vital result from the fact that a water molecule has two polar covalent bonds, which allow hydrogen bonding between water molecules and their neighbors. Among the special properties of water are the following:

- Water molecules are cohesive; that is, they tend to stick to one another through hydrogen bonding (**FIGURE 2.12**). This property generates many special characteristics of water, including *surface tension*, which allows water to form a thin layer on the surface of cells. This aqueous layer is necessary for the transport of dissolved materials into and out of a cell.
- Water is an excellent *solvent*; that is, it dissolves salts and other electrically charged molecules because it is attracted to both positive and negative charges (see Figure 2.9).
- Water remains a liquid across a wider range of temperatures than other molecules of its size. This is critical because living things require water in liquid form.
- Water can absorb significant amounts of heat energy without itself changing temperature. Further, when heated water molecules eventually evaporate, they take much of this absorbed energy with them. These properties moderate temperature fluctuations that would otherwise damage organisms.
- Water molecules participate in many chemical reactions within cells both as reactants in hydrolysis and as products of dehydration synthesis.

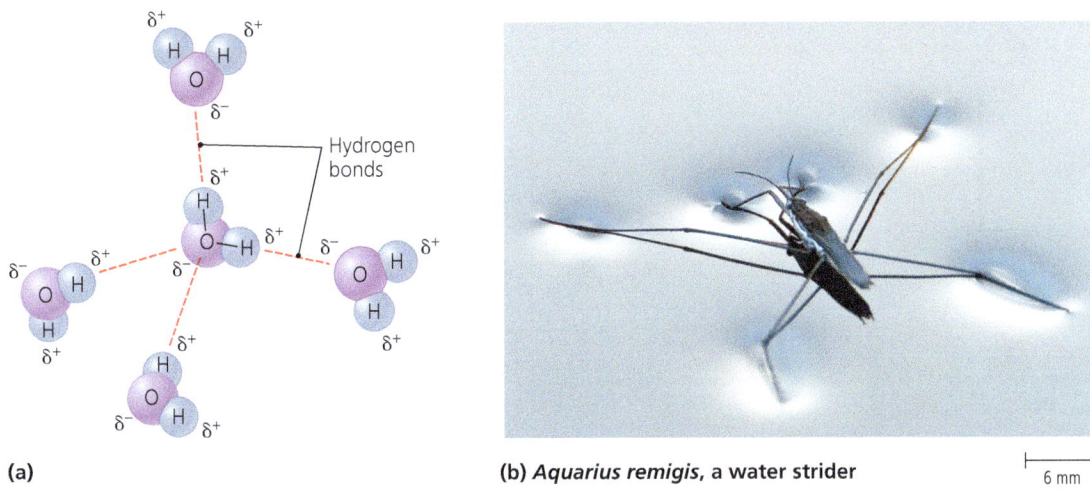

(b) *Aquarius remigis*, a water strider

▲ FIGURE 2.12 **The cohesiveness of liquid water.** (a) Water molecules are cohesive because hydrogen bonds cause them to stick to one another. (b) One result of cohesiveness in water is surface tension, which can be strong enough to support the weight of insects known as water striders.

Acids and Bases

LEARNING OUTCOME

2.20 Contrast acids, bases, and salts, and explain the role of buffers.

As we have seen, the polar bonds of water molecules dissociate salts into their component cations and anions. A similar process occurs with substances known as acids and bases.

An **acid** is a substance that dissociates into one or more hydrogen ions (H^+) and one or more anions (**FIGURE 2.13a**). Acids can be inorganic molecules, such as hydrochloric acid (HCl) and sulfuric acid (H_2SO_4), or organic molecules, such as amino acids and nucleic acids. Familiar organic acids are found in lemon juice, black coffee, and tea. Of course, the anions of organic acids contain carbon, whereas those of inorganic acids do not.

A **base** is a molecule that binds with H^+ when dissolved in water. Some bases dissociate into cations and *hydroxyl ions* (OH^-) (**FIGURE 2.13b**), which then combine with hydrogen ions to form water molecules:

$$H^+ + OH^- \rightarrow H_2O$$

Other bases, such as household ammonia (NH_3), directly accept hydrogen ions and become compound ions such as NH_4^+ (ammonium). Another common household base is baking soda (sodium bicarbonate, $NaHCO_3$).

Metabolism requires a relatively constant balance of acids and bases because hydrogen ions and hydroxyl ions are involved in many chemical reactions. Further, many complex molecules such as proteins lose their functional shapes when acidity changes. If the concentration of either hydrogen ions or hydroxyl ions deviates too far from normal, metabolism ceases.

The concentration of hydrogen ions in a solution is expressed using a logarithmic **pH scale** (**FIGURE 2.14**). The term *pH* comes from *potential hydrogen*, which is the negative of the logarithm of the concentration of hydrogen ions. In this logarithmic scale, it is important to notice that acidity increases as pH values decrease and that each decrease by a whole number in pH indicates a 10-fold increase in acidity (hydrogen ion concentration). For example, a glass of grapefruit juice, which has a pH of 3.0, contains 10 times as many hydrogen ions as the same volume of tomato juice, which has a pH of 4.0. Similarly, tomato juice is 1000 times more acidic than pure water, which has a pH of 7.0 (neutral). Water is neutral because it dissociates into one hydrogen cation and one hydroxyl anion:

$$H_2O \rightarrow H^- + OH^-$$

Alkaline (basic) substances have pH values greater than 7.0. They reduce the number of free hydrogen ions by combining with them. For bases that produce hydroxyl ions, the concentration of hydroxyl ions is inversely related to the concentration of hydrogen ions.

Organisms can tolerate only a certain, relatively narrow pH range. Fluctuations outside an organism's preferred range inhibit its metabolism and may even be fatal. Most organisms contain natural **buffers**—substances, such as proteins, that prevent drastic changes in internal pH by removing excess hydrogen and hydroxyl ions. In a laboratory culture, the metabolic activity of

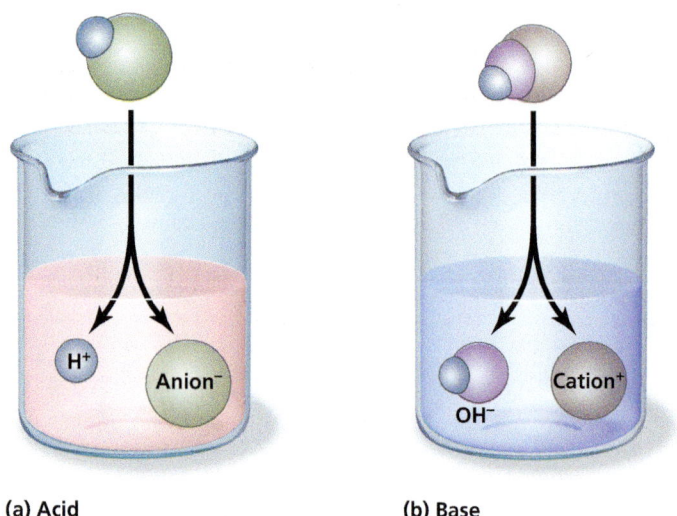

▲ **FIGURE 2.13 Acids and bases.** (a) Acids dissociate in water into hydrogen ions and anions. (b) Many bases dissociate into hydroxyl ions and cations.

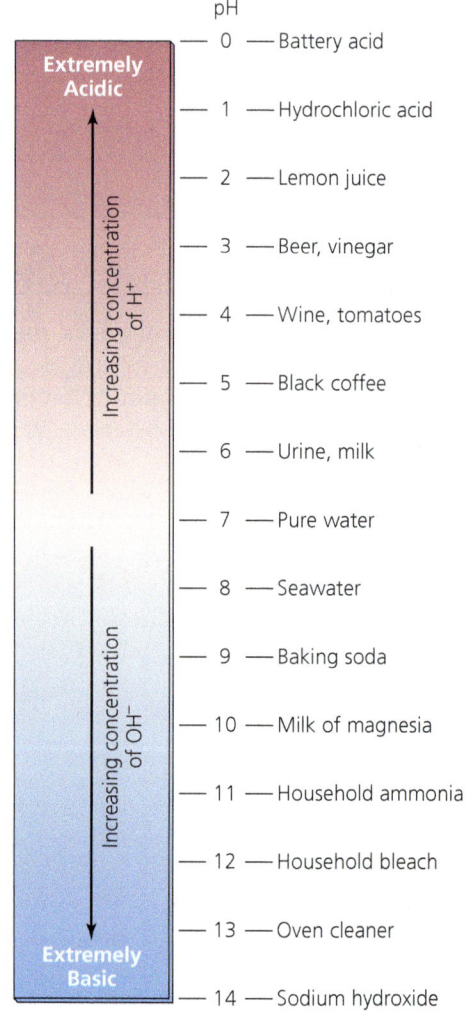

▲ **FIGURE 2.14 The pH scale.** Values below 7 are acidic; values above 7 are basic.

BENEFICIAL MICROBES

Architecture-Preserving Bacteria

The Alhambra, a Moorish palace constructed of limestone and marble beginning in the 9th century, was built to last. But not even stone lasts forever. Wind and rain wear away the surface. Acid rain reacts with the calcite crystals in limestone and marble. As years pass, stone slowly crumbles.

Those who would preserve the Alhambra and other historic structures face a dilemma. The microscopic pores that riddle limestone and marble make these materials particularly susceptible to weathering and decay. Sealing the stone's pores can reduce weathering but can also lock in moisture that speeds the stone's decay.

With the help of *Myxococcus xanthus*, a bacterium commonly found in soil, a team of researchers led by mineralogist Carlos Rodríguez-Navarro of the University of Granada may have found a way to protect the stone of structures like the Alhambra.

In many natural environments, bacteria instigate the formation of calcite crystals like the ones in limestone. In tests conducted using samples of the limestone commonly used in historic Spanish buildings, *M. xanthus* formed calcite crystals that lined the stone's pores rather than plugging them. The crystals formed by the bacteria are even more durable than the original stone, offering the potential for long-term protection.

---

microorganisms can change the pH of microbial growth solutions as nutrients are taken up and wastes are released; therefore, pH buffers are often added to them. One common buffer used in microbiological media is KH_2PO_4 (potassium dihydrogen phosphate), which exists as either a weak acid or a weak base, depending on the pH of its environment. Under acidic conditions, KH_2PO_4 is a base that combines with H^+, neutralizing the acidic environment; in alkaline conditions, however, KH_2PO_4 acts as an acid, releasing hydrogen ions.

Microorganisms differ in their ability to tolerate various ranges of pH. Many grow best when the pH is between 6.5 and 8.5. Photosynthetic bacteria known as *cyanobacteria* grow well in more basic solutions. Fungi generally tolerate acidic environments better than most prokaryotes, though acid-loving prokaryotes, called *acidophiles*, require acidic conditions. Some bacteria are tolerant of acid. One such bacterium is *Propionibacterium acnes* (prō-pē-on-i-bak-tēr´ē-ŭm ak´nēz). This bacterium can cause acne in the skin, which normally has a pH of about 4.0. Another is *Helicobacter pylori*[11] (hel´ĭ-kō-bak´ter pī´lō-rē), a curved bacterium that has been shown to cause ulcers in the stomach, where pH can fall as low as 1.5 when acid is being actively secreted. **Clinical Case Study: Raw Oysters and Antacids: A Deadly Mix?** on p. 40 focuses on how the use of antacids may increase the survival rates of certain disease-causing bacteria in the stomach.

Microorganisms can change the pH of their environment by utilizing acids and bases and by producing acidic or basic wastes. For example, fermentative microorganisms form organic acids from the decomposition of sugar, and the bacterium *Thiobacillus* (thī-ō-bă-sil´ŭs) can reduce the pH of its environment to 0.0. Some mining companies rely on acid produced by this bacterium to dissolve uranium and copper from low-grade ore.

Scientists measure pH with a pH meter or with test papers impregnated with chemicals (such as litmus or phenol red) that change color in response to pH. In a microbiological laboratory, changes in color of such pH indicators incorporated into microbial growth media are commonly used to distinguish among bacterial genera.

Salts

As we have seen, a salt is a compound that dissociates in water into cations and anions other than H^+ and OH^-. Acids and hydroxyl-yielding bases neutralize each other during exchange reactions that produce water and salt. For instance, milk of magnesia (magnesium hydroxide) is an antacid used to neutralize excess stomach acid. The chemical reaction is

$$\underset{\text{Magnesium hydroxide}}{Mg(OH)_2} + \underset{\text{Hydrochloric acid}}{2\,HCl} \longrightarrow \underset{\text{Magnesium chloride (salt)}}{MgCl_2} + \underset{\text{Water}}{2H_2O}$$

Cations and anions of salts are electrolytes. A cell uses electrolytes to create electrical differences between its inside and outside, to transfer electrons from one location to another, and as important components of many enzymes. Certain organisms also use salts such as calcium carbonate ($CaCO_3$) to provide structure and support for their cells.

MICRO CHECK

14. What is the most abundant substance in organisms, constituting between 50% and 99% of the total mass of living cells?
15. Water dissolves a wide variety of molecules. What is the term for a substance that can dissolve molecules?
16. What is the name for a substance that dissociates into one or more hydrogen ions and one or more anions?
17. What do you call a substance that helps remove excess hydrogen or hydroxyl ions from a solution, helping it maintain a stable internal pH?

TELL ME WHY

Why does the neutralization of an acid by a base often produce water?

---

[11] The name *pylori* refers to the pylorus, a region of the stomach.

CLINICAL CASE STUDY

Raw Oysters and Antacids: A Deadly Mix?

The highly acidic environment of the stomach kills most bacteria before they cause disease. One bacterium that can slightly tolerate conditions as it passes through the stomach is *Vibrio vulnificus*—a bacterium commonly ingested by eating raw tainted oysters. The bacterium cannot be seen, tasted, or smelled in food or water.

V. vulnificus is an emerging pathogen and a growing cause of food poisoning in the United States: it triggers vomiting, diarrhea, and abdominal pain. The pathogen can also infect the bloodstream, causing life-threatening illness characterized by fever, chills, skin lesions, and deadly loss of blood pressure. About 50% of patients with bloodstream infections die. *V. vulnificus* especially affects the immunocompromised and people with long-term liver disease.

Researchers have discovered that taking antacids may make people more susceptible to becoming ill from *V. vulnificus*. They found that antacids in a simulated gastric environment significantly increased the survival rate of *V. vulnificus*.

1. Why are patients who take antacids at greater risk for infections with *V. vulnificus*?
2. Will antacids raise or lower the pH of the stomach?
3. Other than refraining from antacids, what can people do to reduce their risk of infection?

Organic Macromolecules

Inorganic molecules play important roles in an organism's metabolism; however, water excluded, they compose only about 1.5% of its mass. Inorganic molecules are typically too small and too simple to constitute an organism's basic structures or to perform the complicated chemical reactions required of life. These functions are fulfilled by organic molecules, which are generally larger and much more complex.

Functional Groups

LEARNING OUTCOME

2.21 Define *functional group* as it relates to organic chemistry.

As we have seen, organic molecules contain carbon and hydrogen atoms, and each carbon atom can form four covalent bonds with other atoms (see Figure 2.5c and d). Carbon atoms that are linked together in branched chains, unbranched chains, and rings provide the basic frameworks of organic molecules.

Atoms of other elements are bound to these carbon frameworks to form an unlimited number of compounds. Besides carbon and hydrogen, the most common elements in organic compounds are oxygen, nitrogen, phosphorus, and sulfur. Other elements, such as iron, copper, molybdenum, manganese, zinc, and iodine, are important in some proteins.

Atoms often appear in certain common arrangements called **functional groups** that are common in organic molecules. For example, $-NH_2$, the amino functional group, is found in all amino acids, and $-OH$, the hydroxyl functional group,[12] is common to all alcohols. When a class of organic molecules is discussed, the letter **R** (for *residue*) designates atoms in the compound that vary from one molecule to another. The symbol R—OH, therefore, represents the general formula for an alcohol. **TABLE 2.3** describes some common functional groups of organic molecules.

There is a great variety of organic compounds, but all organisms use certain basic types. These molecules—known as *macromolecules* because they are very large—are lipids, carbohydrates, proteins, and nucleic acids.

Lipids

LEARNING OUTCOMES

2.22 Describe the structure of a molecule of fat (triglyceride), and compare it to that of phospholipids, waxes, and steroids.

2.23 Distinguish among saturated, unsaturated, and polyunsaturated fatty acids.

Lipids are a diverse group of organic macromolecules not composed of regular subunits. They have one common trait—they are **hydrophobic**;[13] that is, they are insoluble in water. Lipids have little or no affinity for water because they are composed almost entirely of carbon and hydrogen atoms linked by nonpolar covalent bonds. Because these bonds are nonpolar, they have no attraction to the polar bonds of water molecules. To look at it another way, the polar water molecules are attracted to each other and exclude the nonpolar lipid molecules. There are four major groups of lipids in cells: fats (triglycerides), phospholipids, waxes, and steroids.

---

[12]Note that the hydroxyl functional group is not the same thing as the hydroxyl ion, because the former is covalently bonded to a carbon atom and the latter is unbound.

[13]From Greek *hydor*, meaning "water," and *phobos*, meaning "fear."

TABLE 2.3 Functional Groups of Organic Molecules and Some Classes of Compounds in Which They Are Found

Structure	Name	Class of Compounds
—OH	Hydroxyl	Alcohol Monosaccharide Amino acid
R—CH$_2$—O—CH$_2$—R'	Ether	Disaccharide Polysaccharide
R—C(=O)—R'	Internal carbonyl—a carbon atom (in R group) on each side	Ketone Carbohydrate
R—C(=O)—H	Terminal carbonyl—a carbon atom (in R group) on only one side	Aldehyde
R—C(=O)—O—H	Carboxyl	Amino acid Protein Fatty acid
R—CH(H)(H)—NH$_2$	Amino	Amino acid Protein
R—C(=O)—O—R'	Ester	Fat Wax
R—CH$_2$—SH	Sulfhydryl	Amino acid Protein
R—CH$_2$—O—P(=O)(OH)(OH)	Organic phosphate	Phospholipid Nucleotide ATP

Fats

Organisms make **fats** via dehydration synthesis reactions that form esters between an alcohol named glycerol and three fatty acids, which are long hydrocarbon chains capped by a carboxyl functional group (**FIGURE 2.15a**). Fats are also called *triglycerides* because they contain three fatty acid molecules linked to glycerol.

The three fatty acids in a fat molecule may be identical or different from one another, but each usually has 12–20 carbon atoms. An important difference among fatty acids is the presence and location of double bonds between carbon atoms. When carbon atoms are linked solely by single bonds, every carbon atom, with the exception of the terminal ones, is covalently linked to two hydrogen atoms. Such a fatty acid is *saturated* with hydrogen and is thus termed a **saturated fatty acid** (**FIGURE 2.15b**). In contrast, **unsaturated fatty acids** contain one double bond between adjacent carbon atoms (*monounsaturated fatty acid*) or more than one double bond between carbon atoms (*polyunsaturated fatty acid*). Triglycerides with at least one polyunsaturated fatty acid are polyunsaturated fats.

Saturated fats (composed solely of saturated fatty acids), like those found in animals, are usually solid at room temperature because their fatty acids can be packed closely together. Unsaturated fatty acids, by contrast, are bent at every double bond, and so unsaturated fats cannot be packed tightly; they remain liquid at room temperature. Most fats in plants are unsaturated

(a)

(b)

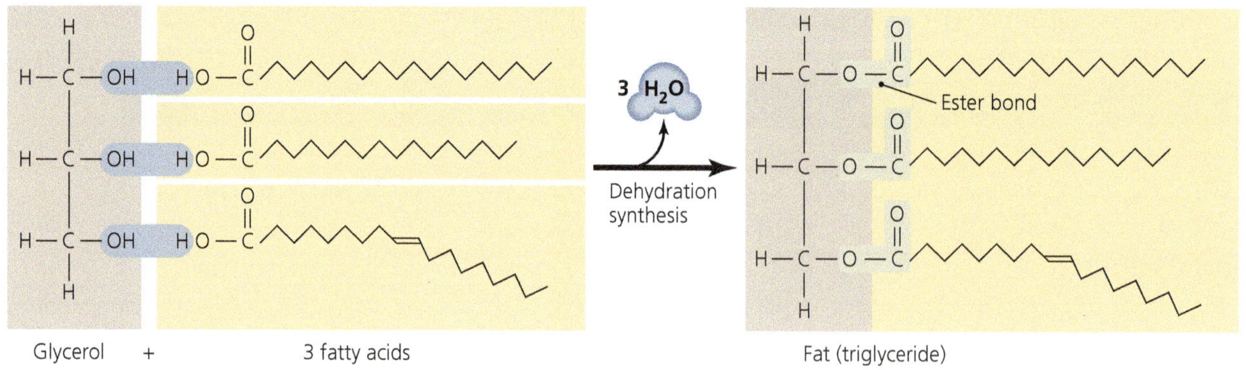

▲ **FIGURE 2.15 Fats (triglycerides). (a)** Fats are made in dehydration synthesis reactions that form ester bonds between a glycerol molecule and three fatty acids. **(b)** Saturated fatty acids have only single bonds between their carbon atoms, whereas unsaturated fatty acids have double bonds between carbon atoms. Scientists often use abbreviated diagrams of fatty acids in which each angle represents a carbon atom and most hydrogen atoms are not shown, as seen in part (a) of this figure. *According to Table 2.4, which fatty acids are shown in Figure 2.15a?*

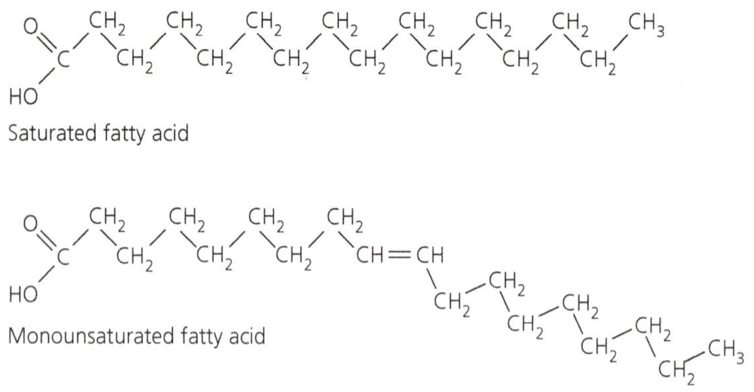

Figure 2.15 *Stearic acid, palmitic acid, and oleic acid.*

or polyunsaturated. TABLE 2.4 (on p. 43) compares the structures and melting points of four common fatty acids.

Fats contain an abundance of energy stored in their carbon-carbon covalent bonds. Indeed, a major role of fats in organisms is to store energy. Fats can be catabolized to provide energy for movement, synthesis, and transport (Chapter 5).

Phospholipids

Phospholipids are similar to fats, but they contain only two fatty acid chains instead of three. In phospholipids, the third carbon atom of glycerol is linked to a phosphate (PO_4) functional group instead of a fatty acid (FIGURE 2.16a). Like fats, different phospholipids contain different fatty acids. Small organic groups

(a) Phospholipid

(b) Symbol **(c) Phospholipid bilayer**

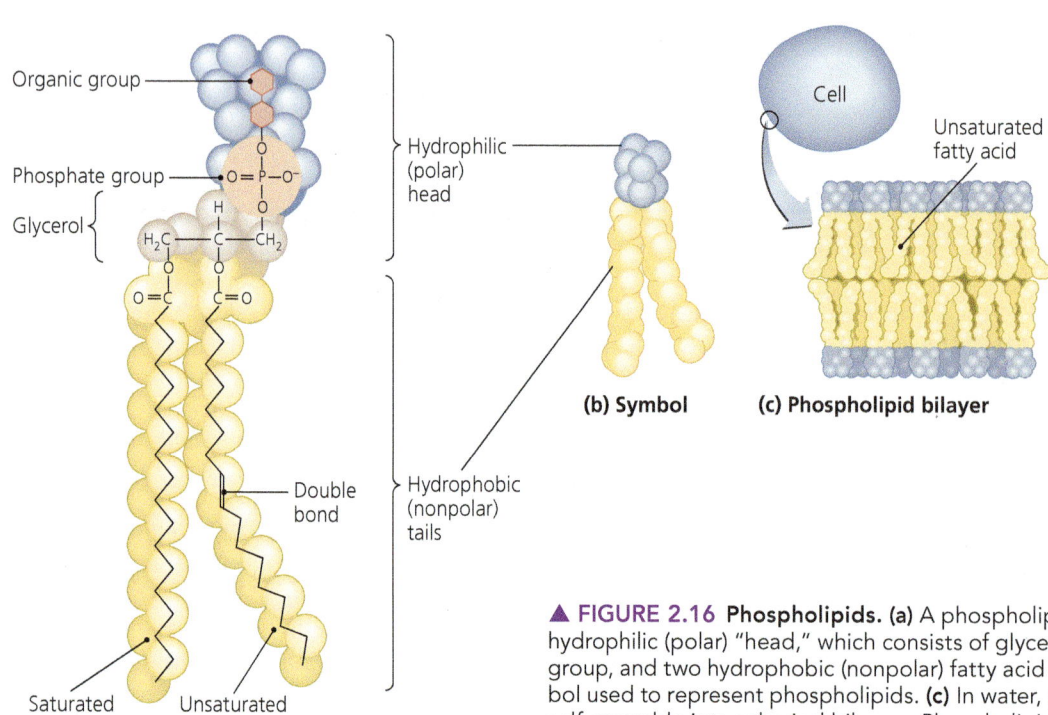

▲ **FIGURE 2.16 Phospholipids. (a)** A phospholipid is composed of a hydrophilic (polar) "head," which consists of glycerol and a phosphate group, and two hydrophobic (nonpolar) fatty acid "tails." **(b)** The symbol used to represent phospholipids. **(c)** In water, phospholipids can self-assemble into spherical bilayers. Phospholipids containing unsaturated fatty acids do not pack together as tightly as those containing saturated fatty acids.

TABLE 2.4 Common Fatty Acids in Fats and Cell Membranes

Numbers of Carbon Atoms: Double Bonds	Type of Fatty Acid	Structure and Formula	Common Name	Melting Point
16:0	Saturated	$CH_3(CH_2)_{14}COOH$	Palmitic acid	63°C
18:0	Saturated	$CH_3(CH_2)_{16}COOH$	Stearic acid	70°C
18:1	Monounsaturated	$CH_3(CH_2)_7CH=CH(CH_2)_7COOH$	Oleic acid	16°C
18:2	Polyunsaturated	$CH_3(CH_2)_4(CH=CHCH_2)_2(CH_2)_6COOH$	Linoleic acid	−5°C

linked to the phosphate group provide additional variety among phospholipid molecules.

The fatty acid "tail" portion of a phospholipid molecule is nonpolar and thus hydrophobic, whereas the phospholipid "head" is polar and thus **hydrophilic**.[14] As a result, phospholipids placed in a watery environment will always self-assemble into forms that keep the fatty acid tails away from water. One way they do this is to form a phospholipid bilayer (**FIGURE 2.16c**). The fatty acid tails, which are hydrophobic, congregate in the water-free interior of bilayers. The polar phosphate heads, which are hydrophilic, orient toward the water on either side of the bilayer. Phospholipid bilayers make up the membranes surrounding cells as well as the internal membranes of plant, fungal, and animal cells.

Waxes

Waxes contain one long-chain fatty acid linked covalently to a long-chain alcohol by an ester bond. Waxes do not have hydrophilic heads; thus, they are completely water insoluble. Certain microorganisms, such as *Mycobacterium tuberculosis* (mī′kō-bak-tēr′ē-ŭm too-ber-kyū-lō′sis), are surrounded by a waxy wall, making them resistant to drying. Some marine microbes use waxes instead of fats as energy storage molecules.

Steroids

A final group of lipids are **steroids**. Steroids consist of four rings (each containing five or six carbon atoms) that are fused to one another and attached to various side chains and functional groups (**FIGURE 2.17a**). Steroids play many roles in human metabolism. Some act as hormones; another steroid, *cholesterol*, is perhaps familiar to you as a less desirable component of food. However, cholesterol is also an essential part of the phospholipid bilayer membrane surrounding an animal cell. Cells of fungi, plants, and one group of bacteria (mycoplasmas) have similar sterol molecules in their membranes. Sterols, which are steroids with an —OH functional group, interfere with the tight packing of the fatty acid chains of phospholipids (**FIGURE 2.17b**). This keeps the membranes fluid and

---
[14]From Greek *philos*, meaning "love."

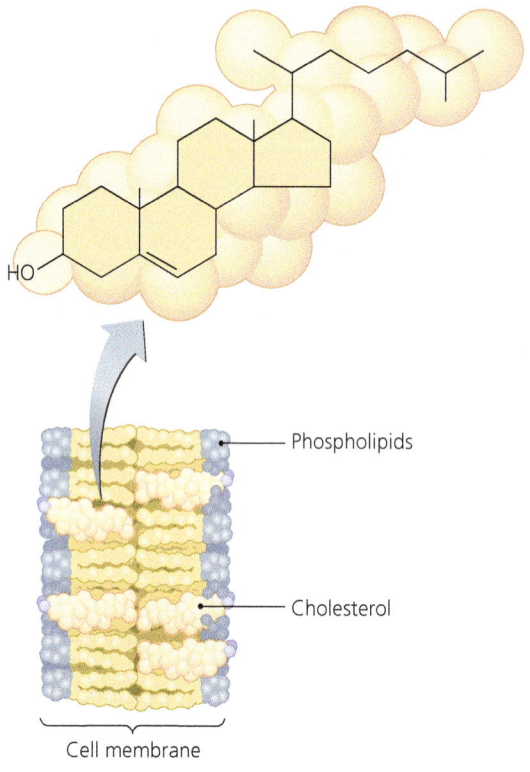

▲ FIGURE 2.17 **Steroids.** (a) Steroids are lipids characterized by four "fused" rings. (b) The steroid cholesterol, which has a short chain of atoms extending from the fourth ring, functions in animal and protozoan cell membranes to prevent packing of phospholipids, thereby keeping the membranes fluid at low temperatures.

flexible at low temperatures. Without steroids such as cholesterol, the membranes of cells would become stiff and inflexible in the cold.

Carbohydrates, proteins, and nucleic acid macromolecules are composed of simpler subunits known as **monomers**,[15] which are basic building blocks. The monomers of these macromolecules are joined together to form chains of monomers called **polymers**.[16] Some macromolecular polymers are composed of hundreds of thousands of monomers.

MICRO CHECK

18. What lipid macromolecule forms most membranes of cells?

---

[15]From Greek *mono*, meaning "one," and *meris*, meaning "part."
[16]From Greek *poly*, meaning "many," and *meris*, meaning "part."

Carbohydrates

LEARNING OUTCOME
2.24 Discuss the chemical makeup of carbohydrates and the roles carbohydrates play in living systems.

Carbohydrates are organic molecules composed solely of atoms of carbon, hydrogen, and oxygen. Most carbohydrate compounds contain an equal number of oxygen and carbon atoms and twice as many hydrogen atoms as carbon atoms, so the general formula for a carbohydrate is $(CH_2O)_n$, where n indicates the number of CH_2O units.

Carbohydrates play many important roles in organisms. Large carbohydrates, such as starch and glycogen, are used for the long-term storage of chemical energy, and smaller carbohydrate molecules serve as a ready energy source, form part of the backbones of DNA and RNA, and may be converted into amino acids. Additionally, polymers of carbohydrate form the cell walls of most fungi, plants, algae, and prokaryotes and are involved in intercellular interactions between animal cells.

Monosaccharides

The simplest carbohydrates are **monosaccharides**[17]—simple sugars (**FIGURE 2.18**). The general names for the classes of monosaccharides are formed from a prefix indicating the number of carbon atoms and from the suffix *-ose*. For example, pentoses are sugars with five carbon atoms, and *hexoses* are sugars with six carbon atoms. Pentoses and hexoses are particularly important in cellular metabolism. For example, deoxyribose, which is the sugar component of DNA, is a pentose. Glucose is a hexose and the primary energy molecule of cells, and fructose is a hexose found in fruit. Chemists assign numbers to the carbon atoms.

Monosaccharides may exist as linear molecules, but because of energy dynamics, each usually takes a cyclic (ring) form. In some cases, more than one cyclic structure may exist. For example, glucose can assume an alpha (α) configuration or a beta (β) configuration (see **FIGURE 2.18a**). As we will see, these configurations play important roles in the formation of different polymers.

Disaccharides

When two monosaccharide molecules are linked together via dehydration synthesis, the result is a **disaccharide**. For example, the linkage of two hexoses, glucose and fructose, forms sucrose (table sugar) in a dehydration reaction (**FIGURE 2.19a**). Other disaccharides include maltose (malt sugar) and lactose (milk sugar). Disaccharides can be broken down via hydrolysis into their constituent monosaccharides (**FIGURE 2.19b**).

---

[17]From Greek *sakcharon*, meaning "sugar."

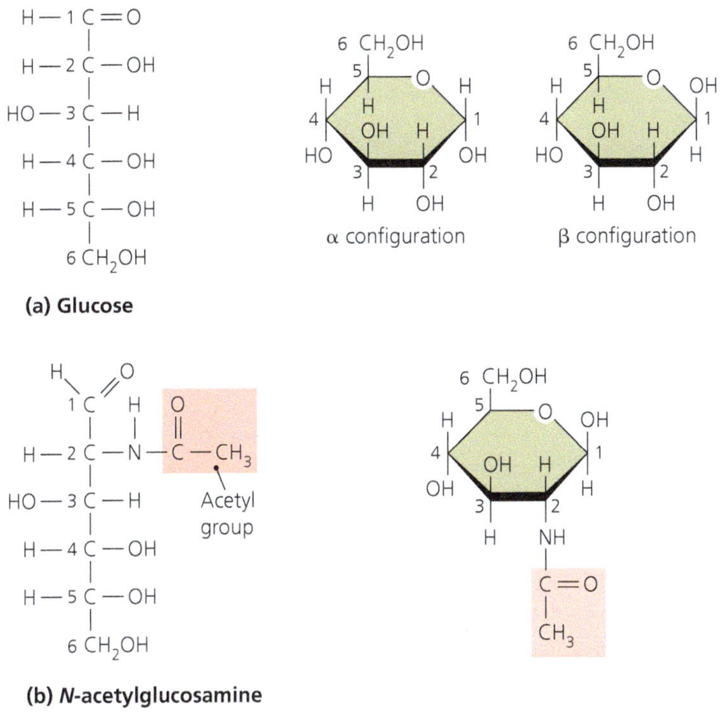

Organic Macromolecules

(a) Glucose — α configuration, β configuration

(b) N-acetylglucosamine

▲ **FIGURE 2.18 Monosaccharides (simple sugars).** Although simple sugars may exist as either linear molecules (at left) or rings (at right), energy dynamics in the watery cytoplasm of cells generally favor ring forms. **(a)** Glucose, a hexose, is the primary energy source for cellular metabolism and an important monomer in many larger carbohydrates. Chemists number the carbon atoms as shown. Alpha and beta ring configurations differ in the location of oxygen bound to carbon 1. **(b)** N-acetylglucosamine (NAG), a monomer in bacterial cell walls.

Polysaccharides

Polysaccharides are polymers composed of tens, hundreds, or thousands of monosaccharides that have been covalently linked in dehydration synthesis reactions. Even polysaccharides that contain only glucose monomers can be quite diverse because they can differ according to their monosaccharide monomer configurations (either alpha or beta) and their shapes (either branched or unbranched). Cellulose, the main constituent of the cell walls of plants and some green algae, is a long unbranched molecule that contains only β-monomers of glucose linked between carbons 1 and 4 of alternating monomers; such bonds are termed β-1,4 bonds (**FIGURE 2.20a**). Amylose, a starch storage compound in plants, has only α-1,4 bonds and is unbranched (**FIGURE 2.20b**); glycogen, a storage molecule formed in the liver and muscle cells of animals, is a highly branched molecule with both α-1,4 and α-1,6 bonds (**FIGURE 2.20c**).

The cell walls of bacteria are composed of *peptidoglycan*, which is made of polysaccharides and amino acids (see Figure 3.15). Polysaccharides may also be linked to lipids to form glycolipids, which can form cell markers such as those involved in the ABO blood typing system in humans.

MICRO CHECK

19. What is the simplest type of carbohydrate?

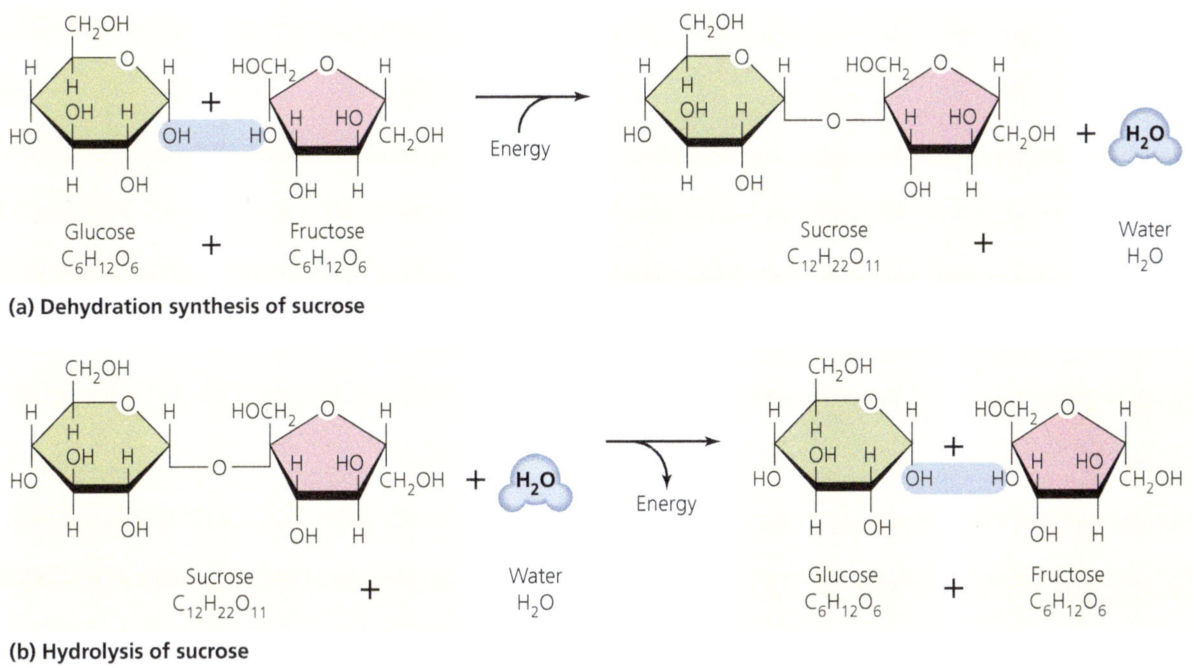

(a) Dehydration synthesis of sucrose

Glucose $C_6H_{12}O_6$ + Fructose $C_6H_{12}O_6$ → Sucrose $C_{12}H_{22}O_{11}$ + Water H_2O

(b) Hydrolysis of sucrose

Sucrose $C_{12}H_{22}O_{11}$ + Water H_2O → Glucose $C_6H_{12}O_6$ + Fructose $C_6H_{12}O_6$

▲ **FIGURE 2.19 Disaccharides. (a)** Formation of the disaccharide sucrose via dehydration synthesis. **(b)** Breakdown of sucrose via hydrolysis.

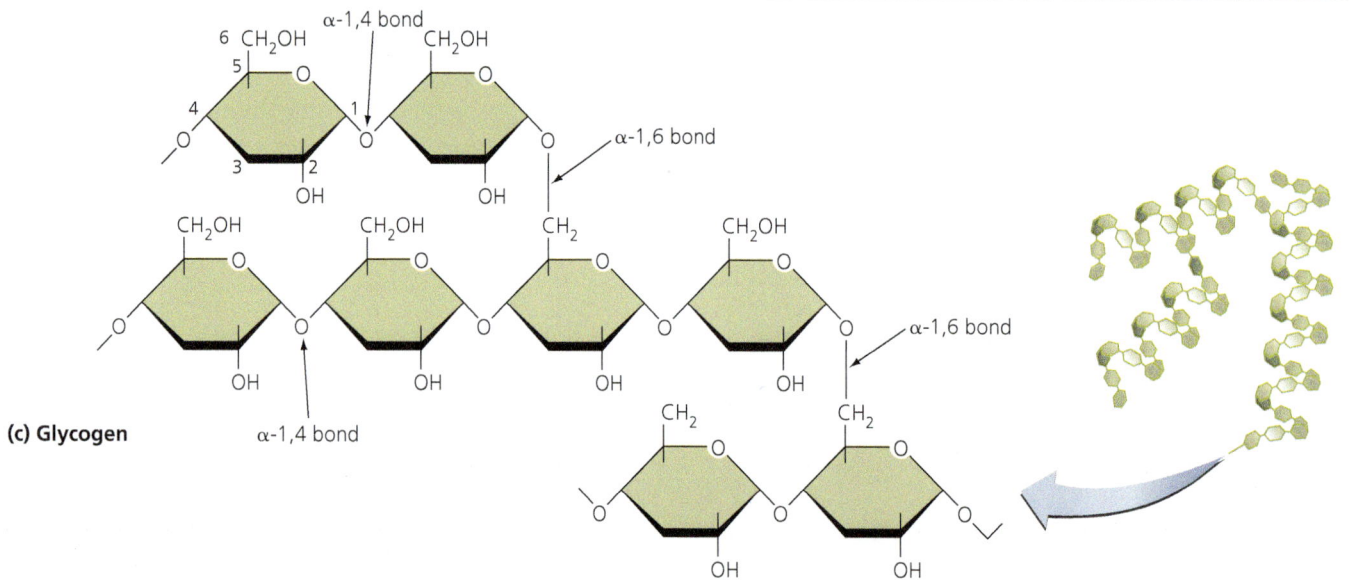

▲ FIGURE 2.20 **Polysaccharides.** All three polysaccharides shown here are composed solely of glucose but differ in the configuration of the glucose monomers and the amount of branching. **(a)** Cellulose, the major structural material in plants, is unbranched and contains only β-1,4 bonds. **(b)** Amylose is an unbranched plant starch with only α-1,4 bonds. **(c)** Glycogen, a highly branched storage molecule in animals, is composed of glucose monomers linked by α-1,4 or α-1,6 bonds.

Proteins

LEARNING OUTCOMES

2.25 Describe five general functions of proteins in organisms.
2.26 Describe the general structure of an amino acid and sketch and label four levels of protein structure.

The most complex organic compounds are **proteins**, which are composed mostly of carbon, hydrogen, oxygen, nitrogen, and sulfur. Proteins perform many functions in cells, including the following:

- *Structure.* Proteins are structural components found in cell walls, in membranes, and within cells themselves. Proteins are also the primary structural material of hair, nails, the

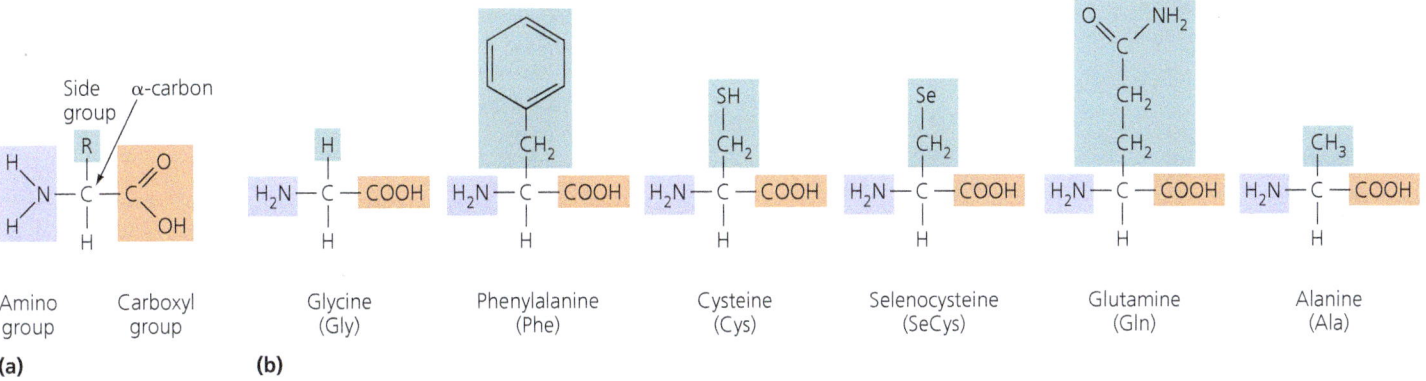

▲ **FIGURE 2.21 Amino acids.** (a) The basic structure of an amino acid. The central α-carbon is attached to an amino group, a hydrogen atom, a carboxyl group, and a side group (—R group) that varies among amino acids. (b) Some selected amino acids, with their side groups highlighted. Note that each amino acid has a distinctive abbreviation.

outer cells of skin, muscle, and flagella and cilia (the last two act to move microorganisms through their environment).

- *Enzymatic catalysis.* Catalysts are chemicals that enhance the speed or likelihood of a chemical reaction. Protein catalysts in cells are called *enzymes.*
- *Regulation.* Some proteins regulate cell function by stimulating or hindering either the action of other proteins or the expression of genes. Hormones are examples of regulatory proteins.
- *Transportation.* Certain proteins act as channels and "pumps" that move substances into or out of cells.
- *Defense and offense. Antibodies* and *complement* are examples of proteins that defend your body against pathogens.

A protein's function is dependent on its shape, which is determined by the molecular structures of its constituent parts and by the bonds within the molecule.

Amino Acids

Proteins are polymers composed of monomers called **amino acids**. Amino acids contain a basic amino group (—NH_2), a hydrogen atom, and an acidic carboxyl group (—COOH). All attach to the same carbon atom, which is known as the α-carbon (**FIGURE 2.21**). A fourth bond attaches the α-carbon to a side group (—R) that varies among different amino acids. The side group may be a single hydrogen atom, various chains, or various complex ring structures. Hundreds of amino acids are possible, but most organisms use only 21 amino acids in synthesizing proteins.[18] The different side groups affect the way amino acids interact with one another within a given protein as well as how a protein interacts with other molecules. A change in an amino acid's side group may seriously interfere with a protein's normal function.

Because amino acids contain both an acidic carboxyl group and a basic amino group, they have both positive and negative charges and are easily soluble in water. Aqueous solutions of organic molecules such as amino acids and simple sugars bend light rays passing through the solution. Molecules known as D *forms*[19] bend light rays clockwise; other molecules bend light rays counterclockwise and are known as L *forms.*[20]

Many organic molecules exist as both D and L forms that are *stereoisomers* of one another; that is, they have the same atoms and functional groups but are mirror images of each other (**FIGURE 2.22** on the next page). Amino acids in proteins are almost always L forms—except for glycine, which does not have a stereoisomer. Interestingly, organisms almost always use D sugars in metabolism and polysaccharides. Rare stereoisomers—D amino acids and L sugars—do exist in some bacterial cell walls and in some antibiotics.

Peptide Bonds

Cells link amino acids together in chains that somewhat resemble beads on a necklace. By a dehydration synthesis reaction, a covalent bond is formed between the carbon of the carboxyl group of one amino acid and the nitrogen of the amino group of the next amino acid in the chain (**FIGURE 2.23** on the next page). Cells follow the organism's genetic instructions to link amino acids together in precise sequences. (Chapter 7 examines this process in more detail.)

Scientists refer to covalent bonds between amino acids by a special name: **peptide**[21] **bonds**. A molecule composed of two amino acids linked together by a single peptide bond is called a dipeptide; longer chains of amino acids are called polypeptides.

---

[18]Although 20 amino acids are commonly listed, the genes of almost all organisms code for 21 amino acids; the 21st is selenocysteine.

[19]From Latin *dexter*, meaning "on the right."

[20]From Latin *laevus*, meaning "on the left."

[21]From *peptone*, the name given to short chains of amino acids resulting from the partial digestion of protein.

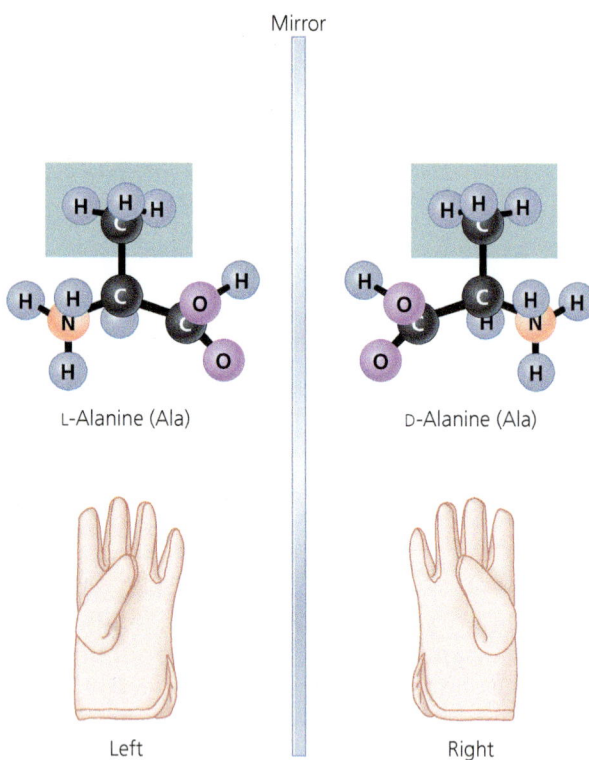

▲ FIGURE 2.22 **Stereoisomers, molecules that are mirror images of one another.** When dissolved in water, D isomers bend light clockwise, and L forms bend light counterclockwise. Just as a right-handed glove does not fit a left hand, so a D stereoisomer cannot be substituted for an L stereoisomer in metabolic reactions.

Protein Structure

Proteins are unbranched polypeptides composed of hundreds to thousands of amino acids linked together in specific patterns as determined by genes. The structure of a protein molecule is directly related to its function; therefore, understanding protein structure is critical to understanding certain specific chemical reactions, the action of antibiotics, and specific defenses against pathogens. Every protein has at least three levels of structure, and some proteins have four levels.

- *Primary structure.* The primary structure of a protein is its sequence of amino acids (FIGURE 2.24a). Cells use many different types of amino acids in proteins, though not every protein contains all types. The primary structures of proteins vary widely in length and amino acid sequence.

A change in a single amino acid in the sequence can drastically affect a protein's overall structure and function, though this is not always the case. For instance, a change in a particular amino acid in the primary structure of a sheep brain protein, called cellular prion (prē'on) protein, may result in a disease called *scrapie*. The altered protein has spread into cows, causing *mad cow disease*, and spread from cows into humans, causing *variant Creutzfeldt-Jakob* (kroytsfelt-yah-kŭb) *disease.*[22]

- *Secondary structure.* Ionic bonds, hydrogen bonds, and hydrophobic and hydrophilic characteristics cause many polypeptide chains to fold into either coils called α-*helices* or accordion-like structures called β-*pleated sheets* (FIGURE 2.24b). Proteins are typically composed of both α-helices and β-pleated sheets linked by short sequences of amino acids that do not show such secondary structure. Because of its primary structure, the protein that causes variant Creutzfeldt-Jakob disease has β-pleated sheets in locations where the normal protein has α-helices (see Figure 13.22).

- *Tertiary structure.* Polypeptides further fold into complex three-dimensional shapes that are not repetitive like α-helices and β-pleated sheets (FIGURE 2.24c) but are uniquely designed to accomplish the function of the protein. Scientists are only beginning to understand the interactions that determine tertiary structure, but it is clear that covalent bonds between —R groups of amino acids, hydrogen bonds, ionic bonds, and other molecular interactions are important. For instance, nonpolar (hydrophobic) side chains fold into the interior of molecules, away from the presence of water.

- Some proteins form strong covalent bonds between sulfur atoms of cysteine amino acids that are brought into proximity by the folding of the polypeptide. These disulfide bridges are critical in maintaining tertiary structure of many proteins.

- *Quaternary structure.* Some proteins are composed of two or more polypeptide chains linked together by disulfide bridges or other bonds. The overall shape of such a protein may be fibrous (threadlike) or more globular (FIGURE 2.24d).

---

[22]Named for the two German neurobiologists who first described the disease.

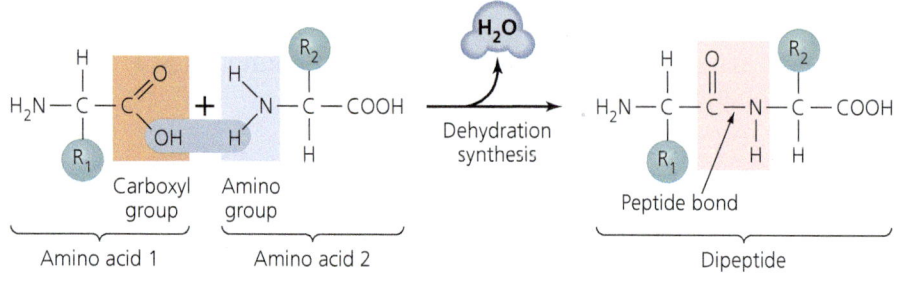

◀ FIGURE 2.23 **The linkage of amino acids by peptide bonds via a dehydration reaction.** In this reaction, removing a hydroxyl group from amino acid 1 and a hydrogen atom from amino acid 2 produces a dipeptide—which is two amino acids linked by a single peptide bond—and a molecule of water.

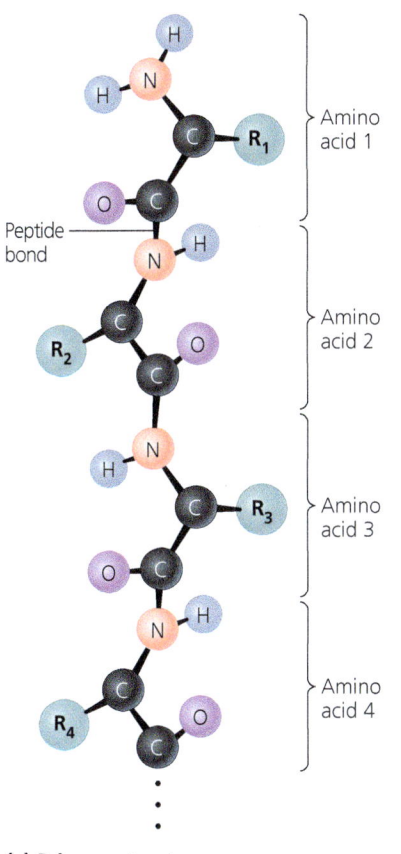
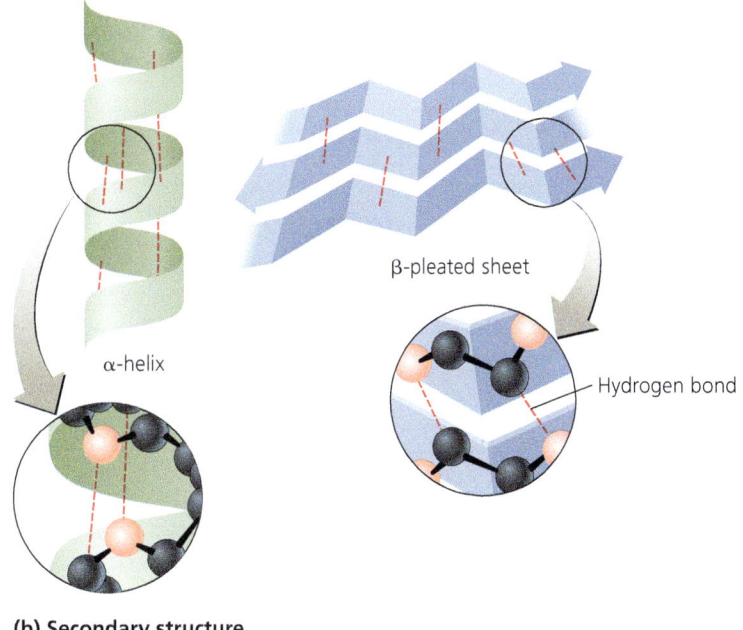
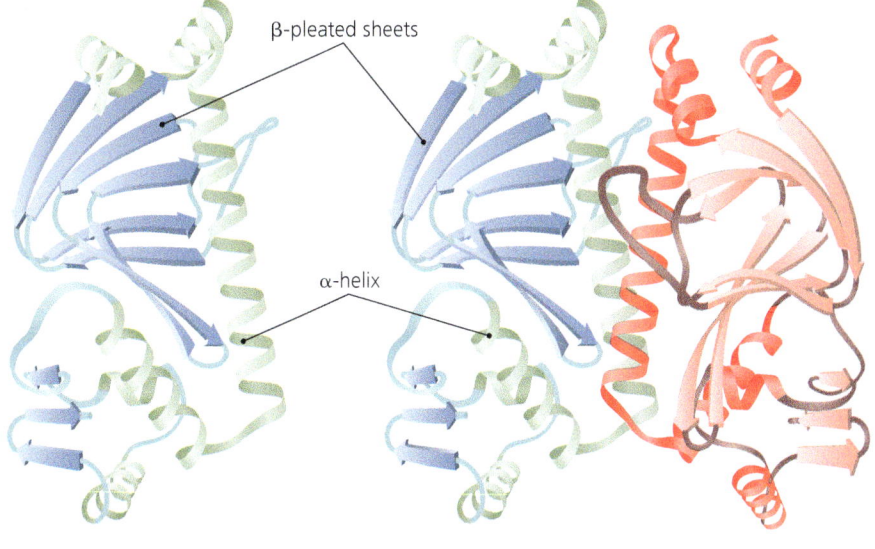

▲ FIGURE 2.24 **Levels of protein structure.**
(a) A protein's primary structure is the sequence of amino acids in a polypeptide. **(b)** Secondary structure arises as a result of interactions, such as hydrogen bonding, between regions of the polypeptide. Secondary structure takes two basic shapes: α-helices and β-pleated sheets. **(c)** The more complex tertiary structure is a three-dimensional shape defined by further hydrogen bonding as well as disulfide bridges between neighboring cysteine amino acid molecules. **(d)** Proteins that are composed of more than one polypeptide chain have a quaternary structure.

(a) Primary structure

(b) Secondary structure

(c) Tertiary structure

(d) Quaternary structure: Two or more polypeptides act together as a single protein.

Organisms may further modify proteins by combining them with other organic or inorganic molecules. For instance, *glycoproteins* are proteins covalently bound with carbohydrates, lipoproteins are proteins bonded with lipids, and *nucleoproteins* are proteins bonded with nucleic acids.

Because protein shape determines protein function, anything that severely interrupts shape also disrupts function. As we have seen, amino acid substitution can alter shape and function. Additionally, physical and chemical factors, such as heat, changes in pH, and salt concentration, can interfere with hydrogen and ionic bonding between parts within a protein. This in turn can disrupt the three-dimensional structure. This process is called **denaturation**. Denaturation can be temporary (if the denatured protein is able to return to its original shape) or permanent.

MICRO CHECK

20. What are two examples of proteins that protect your body from pathogens?
21. What type of bond joins amino acids together to form proteins?

Nucleotides and Nucleic Acids

LEARNING OUTCOMES

2.27 Describe the basic structure of a nucleotide.
2.28 Compare and contrast the structure and functions of DNA and RNA.
2.29 Contrast the structures of ATP, ADP, and AMP.

The nucleic acids **deoxyribonucleic acid (DNA)** and **ribonucleic acid (RNA)** are the vital genetic material of cells and viruses. Moreover, RNA, acting as an enzyme, binds amino acids together to form polypeptides. Both DNA and RNA are unbranched macromolecular polymers that differ primarily in the structures of their monomers, which we discuss next.

Nucleotides and Nucleosides

Each monomer of nucleic acids is a **nucleotide** and consists of three parts (FIGURE 2.25a):

- Phosphate (PO_4^{3-})
- A pentose sugar, either deoxyribose or ribose (FIGURE 2.25b)
- One of five cyclic (ring-shaped) nitrogenous bases: **adenine (A)**, **guanine (G)**, **cytosine (C)**, **thymine (T)**, or **uracil (U)** (FIGURE 2.25c)

Adenine and guanine are double-ringed molecules of a class called *purines*, whereas cytosine, thymine, and uracil have single rings and are *pyrimidines*. DNA contains A, G, C, and T bases, whereas RNA contains A, G, C, and U bases. As their names suggest, DNA nucleotides contain deoxyribose, and RNA nucleotides contain ribose. The similarly named **nucleosides** are nucleotides lacking phosphate; that is, a nucleoside is one of the nitrogenous bases attached only to a sugar.

Each nucleotide or nucleoside is also named for the base it contains. Thus, a nucleotide made with ribose, uracil, and phosphate is a uracil RNA nucleotide, which is also called a uracil *ribonucleotide*. Likewise, a nucleoside composed of adenine and deoxyribose is an adenine DNA nucleoside (or adenine *deoxyribonucleoside*).

Nucleic Acid Structure

Nucleic acids, like polysaccharides and proteins, are polymers. They are composed of nucleotides linked by covalent bonds

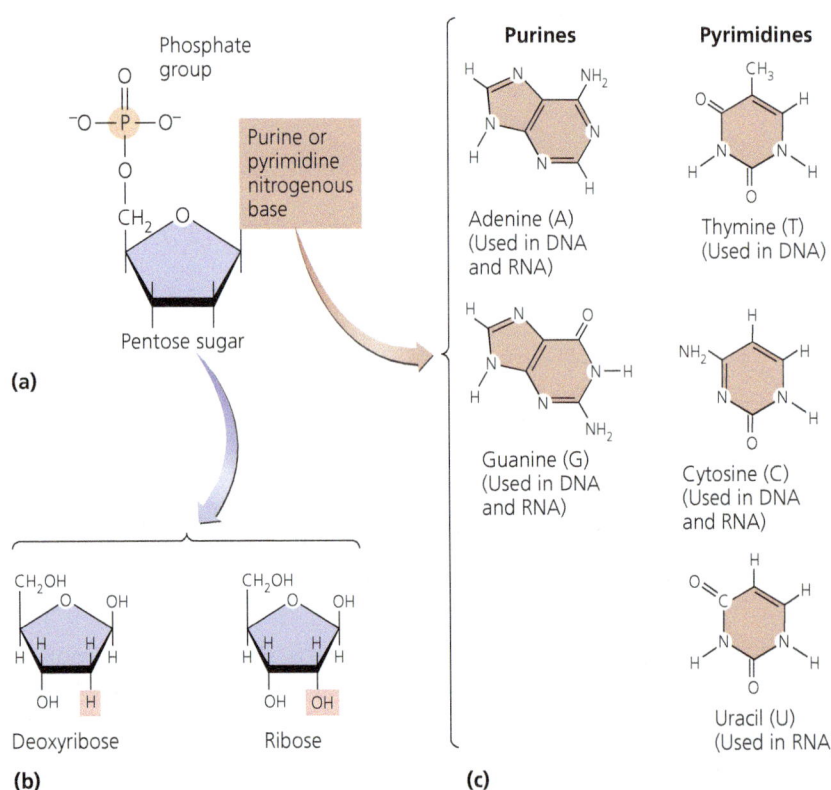

▲ **FIGURE 2.25 Nucleotides. (a)** The basic structure of nucleotides, each of which is composed of a phosphate, a pentose sugar, and a nitrogenous base. **(b)** The pentose sugars deoxyribose, which is found in deoxyribonucleic acid (DNA), and ribose, which is found in ribonucleic acid (RNA). **(c)** The nitrogenous bases, which are either the double-ringed purines adenine or guanine, or the single-ringed pyrimidines thymine, cytosine, or uracil. *How does a nucleoside differ from a nucleotide?*

Figure 2.25 A nucleoside is composed only of a nitrogenous base and a sugar, whereas a nucleotide has a base, sugar, and phosphate.

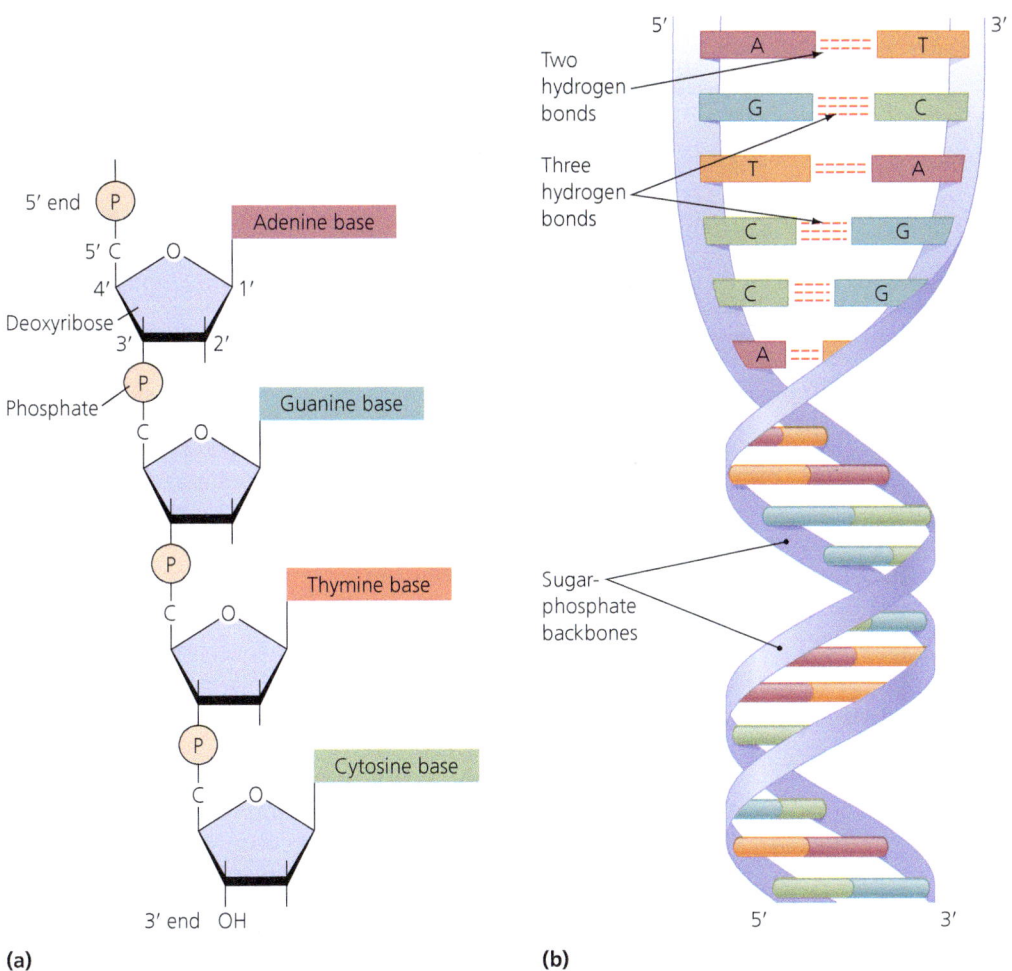

▲ **FIGURE 2.26 General nucleic acid structure. (a)** Nucleotides are polymerized to form chains in which the nitrogenous bases extend from a sugar-phosphate backbone like the teeth of a comb. **(b)** Specific pairs of nitrogenous bases form hydrogen bonds between adjacent nucleotide chains to form the familiar DNA double helix. *How can you determine that the molecule in (a) is DNA and not RNA?*

Figure 2.26 It is DNA because its nucleotides have deoxyribose sugar and because some of them have thymine bases (not uracil, as in RNA).

between the phosphate of one nucleotide and the sugar of the next. Polymerization results in a linear spine composed of alternating sugars and phosphates, with bases extending from it rather like the teeth of a comb (**FIGURE 2.26a**). The two ends of a chain of nucleotides are different. At one end, called the 5′ end[23] (five prime end), carbon 5′ of the sugar is attached to a phosphate group. At the other end (the 3′ end), carbon 3′ of the sugar is attached to a hydroxyl group.

The atoms of the bases in nucleotides are arranged in such a manner that hydrogen bonds readily form between specific bases of two adjacent nucleic acid chains. Three hydrogen bonds form between an adjacent pair composed of cytosine (C) and guanine (G), whereas two hydrogen bonds form between an adjacent pair composed of adenine (A) and thymine (T) in DNA (**FIGURE 2.26b**) or between an adjacent pair composed of adenine (A) and uracil (U) in RNA. Hydrogen bonds do not readily form between other combinations of nucleotide bases; for example, adenine does not readily pair with cytosine, guanine, or another adenine nucleotide.

In cells, DNA molecules are double stranded. Most DNA viruses (viruses that use DNA as a genome) also use double-stranded DNA, though some single-stranded DNA viruses are known. The two strands of double-stranded DNA are complementary to one another; that is, the specificity of nucleotide base pairing ensures that opposite strands are composed of complementary nucleotides. For instance, if one strand has the sequence AATGCT, then its complement has TTACGA.

The two strands are also *antiparallel*; that is, they run in opposite directions. One strand runs from the 3′ end to the 5′ end,

---

[23]Carbon atoms in organic molecules are commonly identified by numbers. In a nucleotide, carbon atoms 1, 2, 3, and so on, belong to the base, and carbon atoms 1′, 2′, 3′, and so on, belong to the sugar.

TABLE 2.5 Comparison of Nucleic Acids

Characteristic	DNA	RNA
Sugar	Deoxyribose	Ribose
Purine nucleotides	A and G	A and G
Pyrimidine nucleotides	T and C	U and C
Number of strands	Double stranded in cells and in most DNA viruses; single stranded in parvoviruses	Single stranded in cells and in most RNA viruses; double stranded in reoviruses
Function	Genetic material of all cells and DNA viruses	Protein synthesis in all cells; genetic material of RNA viruses

whereas its complement runs in the opposite direction, from its 5' end to its 3' end. Though hydrogen bonds are relatively weak bonds, thousands of them exist at normal temperatures, forming a stable, double-stranded DNA molecule that looks much like a ladder: the two deoxyribose-phosphate chains are the side rails, and base pairs form the rungs. Hydrogen bonding also twists the phosphate-deoxyribose backbones into a helix. Thus, typical DNA is a double helix.

Nucleic Acid Function

DNA is the genetic material of all organisms and of DNA viruses; it carries instructions for the synthesis of RNA molecules and proteins. By controlling the synthesis of enzymes and regulatory proteins, DNA controls the synthesis of all other molecules in an organism. Genetic instructions are carried in the sequence of nucleotides that make up the nucleic acid. Even though only four kinds of bases are found in DNA (A, T, G, and C), they can be sequenced in distinctive patterns that create genetic diversity and code for an infinite number of proteins, just as an alphabet of only four letters could spell a very large number of words. Cells replicate their DNA molecules and pass copies to their descendants, ensuring that each has the instructions necessary for life.

Several kinds of ribonucleic acids, such as messenger RNA, transfer RNA, and ribosomal RNA, play roles in the formation of proteins, including catalyzing the synthesis of proteins. RNA molecules also function in place of DNA as the genome of RNA viruses.

TABLE 2.5 compares and contrasts RNA and DNA. (We examine the synthesis and function of DNA and RNA in detail in Chapter 7.)

ATP (Adenosine Triphosphate)

Phosphate in nucleotides and other molecules is a highly reactive functional group and can form covalent bonds with other phosphate groups to make diphosphate and triphosphate molecules. Such molecules made from ribose nucleotides are important in many metabolic reactions. The names of these molecules indicate the nucleotide base and the number of phosphate groups they contain. Thus, cells make adenosine monophosphate (AMP) from the nitrogenous base adenine, ribose sugar, and one phosphate group; adenosine diphosphate (ADP), which has two phosphate groups; and **adenosine triphosphate** (ă-den'ō-sēn trī-fos'fāt) or **ATP**, which has three phosphate groups (FIGURE 2.27).

ATP is the principal, short-term, recyclable energy supply for cells. When the phosphate bonds of ATP are broken, a significant amount of energy is released; in fact, more energy is released from phosphate bonds than is released from most other covalent bonds. For this reason, the phosphate-phosphate bonds of ATP are known as *high-energy bonds*, and to show these specialized bonds, ATP can be symbolized as

$$A-\text{P} \sim \text{P} \sim \text{P}$$

Energy is released when ATP is converted to ADP and when phosphate is removed from ADP to form AMP, though the latter reaction is not as common in cells. Energy released from the phosphate bonds of ATP is used for important life-sustaining activities, such as synthesis reactions, locomotion, and transportation of substances into and out of cells.

Cells also use ATP as a structural molecule in the formation of *coenzymes*. Coenzymes such as *flavin adenine dinucleotide*, *nicotinamide nucleotide*, and *coenzyme A* function in many metabolic reactions (as discussed in Chapter 5).

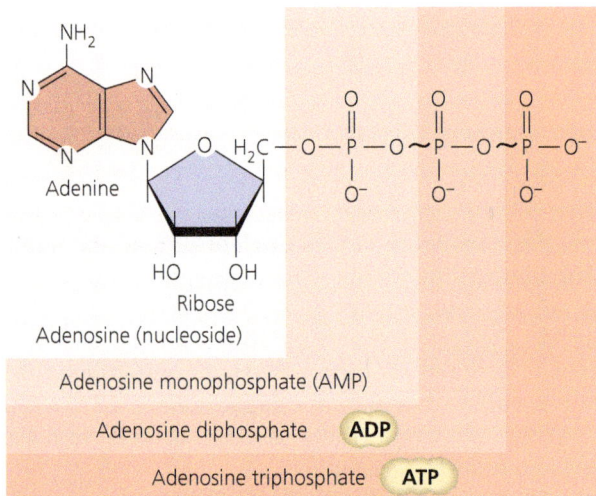

▲ **FIGURE 2.27 ATP.** Adenosine triphosphate (ATP), the main short-term, recyclable energy supply for cells. Energy is stored in high-energy bonds between the phosphate groups. *What is the relationship between AMP and adenine ribonucleotide?*

Figure 2.27 *AMP and adenine ribonucleotide are two names for the same thing.*

MICRO CHECK

22. DNA contains the sugar deoxyribose. What sugar does RNA contain?
23. Which nucleic acid controls the synthesis of enzymes and regulatory proteins?
24. Breaking the bond attaching the terminal phosphate group of which molecule produces the main short-term energy supply for cells?

A cell's supply of ATP is limited; therefore, an important part of cellular metabolism is to replenish ATP stores. (Chapter 5 discusses the important ATP-generating reactions.)

TELL ME WHY

Why do the cell membranes of microbes living in Arctic water likely contain more unsaturated fatty acids than do membranes of microbes living in hot springs?

MICRO IN THE CLINIC FOLLOW-UP
Can Spicy Food Cause Ulcers?

The pain and nausea continue for a month before Ramona finally decides to see a doctor. It turns out that she does, indeed, have an ulcer—but eating spicy food had nothing to do with it. Her doctor explains that most ulcers are caused by bacteria called *Helicobacter pylori*, which thrive in low-pH, highly acidic environments, such as the stomach. *H. pylori* usually lives in the stomach without causing any problems, but occasionally it causes inflammation in the stomach's mucous lining, resulting in an ulcer. The doctor notes that it's a common misconception that spicy food causes ulcers. He adds, however, that alcohol, tobacco, and stress can aggravate ulcers and slow the healing process. He prescribes antibiotics to Ramona to kill the bacteria. Within a short time, Ramona is feeling better—and back to enjoying her spicy food.

1. **Antacids are sometimes taken to treat ulcers. Given what you've learned about the pH environment that *H. pylori* prefers, explain how antacids might help relieve the symptoms of an ulcer.**
2. **What do you think would happen if Ramona consumed a large amount of milk? Would you expect her to feel temporary relief from her ulcer, or would you expect her to feel worse?**

Check your answers to the Micro in the Clinic Follow-Up questions in the Mastering Microbiology Study Area.

CHAPTER SUMMARY

Go to Mastering Microbiology for Interactive Microbiology, Dr. Bauman's Video Tutors, Micro Matters, MicroFlix, MicroBoosters, practice quizzes, and more.

Atoms (pp. 28–31)

1. **Matter** is anything that takes up space and has mass. Its smallest chemical units, **atoms**, contain negatively charged **electrons** orbiting a nucleus composed of uncharged **neutrons** and positively charged **protons**.
2. An **element** is matter composed of a single type of atom.
3. The number of protons in the nucleus of an atom is its **atomic number**. The sum of the masses of its protons, neutrons, and electrons is an atom's **atomic mass**, which is estimated by adding the number of neutrons and protons (because electrons have little mass).
4. **Isotopes** are atoms of an element that differ only in the numbers of neutrons they contain.

Chemical Bonds (pp. 31–35)

1. The region of space occupied by electrons is an electron shell. The number of electrons in the outermost shell, or **valence** shell, of an atom determines the atom's reactivity. Most valence shells hold a maximum of eight electrons. Sharing or transferring valence electrons to fill a valence shell results in **chemical bonds**.
2. A chemical bond results when two atoms share a pair of electrons. The **electronegativities** of each of the atoms, which is the strength of their attraction for electrons, determine whether the bond between them will be a **nonpolar covalent bond** (equal sharing of electrons), **a polar covalent bond** (unequal sharing of electrons), or an **ionic bond** (giving up of electrons from one atom to another).
3. A **molecule** that contains atoms of more than one element is a **compound**. **Organic compounds** are those that contain carbon and hydrogen atoms.
4. An **anion** is an atom with an extra electron and thus a negative charge. A **cation** has lost an electron and thus has a positive charge. Ionic bonding between the two types of ions makes **salt**. When salts dissolve in water, their ions are called **electrolytes**.
5. **Hydrogen bonds** are relatively weak but important chemical bonds. They hold molecules in specific shapes and confer unique properties to water molecules.

Chemical Reactions (pp. 35–37)

1. **Chemical reactions** result from the making or breaking of chemical bonds in a process in which **reactants** are changed into **products**. Biochemistry involves chemical reactions of life.
2. **Synthesis reactions** form larger, more complex molecules. In **dehydration synthesis**, a molecule of water is removed from the reactants as the larger molecule is formed. **Endothermic reactions** require energy. **Anabolism** is the sum of all synthesis reactions in an organism.
3. **Decomposition reactions** break larger molecules into smaller molecules and are **exothermic reactions** because they release energy. **Hydrolysis** is a decomposition reaction that uses water as one of the reactants. The sum of all decomposition reactions in an organism is called **catabolism**.
4. **Exchange reactions** involve exchanging atoms between reactants.
5. **Metabolism** is the sum of all anabolic, catabolic, and exchange chemical reactions in an organism.

Water, Acids, Bases, and Salts (pp. 37–40)

1. **Inorganic chemicals** typically lack carbon.
2. Water is a vital inorganic compound because of its properties as a solvent, its liquidity, its great capacity to absorb heat, and its participation in chemical reactions.
3. **Acids** release hydrogen ions. **Bases** release hydroxyl anions. The relative strength of each is assessed on a logarithmic **pH scale**, which measures the hydrogen ion concentration in a substance.
4. **Buffers** are substances that prevent drastic changes in pH.

Organic Macromolecules (pp. 40–53)

1. Certain groups of atoms in common arrangements, called **functional groups**, are found in organic macromolecules. **Monomers** are simple subunits that can be covalently linked to form chainlike **polymers**.
2. **Lipids**, which include fats, phospholipids, waxes, and steroids, are **hydrophobic** (insoluble in water) macromolecules.
3. **Fat** (triglyceride) molecules are formed from a glycerol and three chainlike fatty acids. **Saturated fatty acids** contain more hydrogen in their structural formulas than **unsaturated fatty acids**, which contain double bonds between some carbon atoms. If several double bonds exist in the fatty acids of a molecule of fat, it is a polyunsaturated fat.
4. **Phospholipids** contain two fatty acid chains and a phosphate functional group. The phospholipid head is **hydrophilic**, whereas the fatty acid portion of the molecule is hydrophobic.
5. **Waxes** contain a long-chain fatty acid covalently linked to a long-chain alcohol. Waxes, which are water insoluble, are components of cell walls and are sometimes used as energy storage molecules.
6. **Steroid** lipids such as cholesterol help maintain the structural integrity of membranes as temperature fluctuates.
7. **Carbohydrates** such as **monosaccharides**, **disaccharides**, and **polysaccharides** serve as energy sources, structural molecules, and recognition sites during intercellular interactions.
8. **Proteins** are structural components of cells, enzymatic catalysts, regulators of various activities, molecules involved in the transportation of substances, and defensive molecules. They are composed of **amino acids** linked by **peptide bonds**, and they possess primary, secondary, tertiary, and (sometimes) quaternary structures that affect their function. **Denaturation** of a protein disrupts its structure and consequently its function.
9. **Deoxyribonucleic acid (DNA)** and **ribonucleic acid (RNA)** are unbranched macromolecular polymers of **nucleotides**, each composed of either deoxyribose or ribose sugar, ionized phosphate, and a nitrogenous base. Five different bases exist: **adenine**, **guanine**, **cytosine**, **thymine**, and **uracil**. DNA contains A, G, C, and T nucleotides. RNA uses U nucleotides instead of T nucleotides. **Nucleosides** are nucleotides lacking phosphate.
10. The structure of nucleic acids allows for genetic diversity, the correct copying of genes for their passage on to the next generation, and the accurate synthesis of proteins.
11. **Adenosine triphosphate (ATP)**, which is related to adenine nucleotide, is the most important short-term energy storage molecule in cells. It is also incorporated into the structure of many coenzymes.

QUESTIONS FOR REVIEW

Answers to the Questions for Review (except Short Answer questions) begin on p. A-1.

Multiple Choice

1. Which of the following structures have *no* electrical charge?
   a. protons
   b. electrons
   c. neutrons
   d. ions

2. The atomic mass of an atom most closely approximates the sum of the masses of all its _____.
   a. protons
   b. isotopes
   c. electrons
   d. protons and neutrons

3. One isotope of iodine differs from another in _____.
   a. the number of protons
   b. the number of electrons
   c. the number of neutrons
   d. atomic number

4. Which of the following is *not* an organic compound?
   a. monosaccharide
   b. formaldehyde
   c. water
   d. steroid

5. Which of the following terms most correctly describes the bonds in a molecule of water?
   a. nonpolar covalent bond
   b. polar covalent bond
   c. ionic bond
   d. hydrogen bond

6. In water, cations and anions of salts dissociate from one another and become surrounded by water molecules. In this state, the ions are also called _____.
   a. electrically negative
   b. ionically bonded
   c. electrolytes
   d. hydrogen bonds

7. Which of the following can be most accurately described as a decomposition reaction?
   a. $C_6H_{12}O_6 + 6\ O_2 \rightarrow 6\ H_2O + 6\ CO_2$
   b. glucose + ATP $\rightarrow$ glucose phosphate + ADP
   c. $6\ H_2O + 6\ CO_2 \rightarrow C_6H_{12}O_6 + 6\ O_2$
   d. $A + BC \rightarrow AB + C$

8. Which of the following statements about a carbonated cola beverage with a pH of 2.9 is true?
   a. It has a relatively high concentration of hydrogen ions.
   b. It has a relatively low concentration of hydrogen ions.
   c. It has equal amounts of hydroxyl and hydrogen ions.
   d. Cola is a buffered solution.

9. Proteins are polymers of _____.
   a. amino acids
   b. fatty acids
   c. nucleic acids
   d. monosaccharides

10. Which of the following are hydrophobic organic molecules?
    a. proteins
    b. carbohydrates
    c. lipids
    d. nucleic acids

Fill in the Blanks

1. The outermost electron shell of an atom is known as the _____ shell.

2. The type of chemical bond between atoms with nearly equal electronegativities is called a(n) _____ bond.

3. The principal short-term energy storage molecule in cells is _____.

4. Common long-term energy storage molecules are _____, _____, _____, and _____.

5. Groups of atoms such as NH_2 or OH that appear in certain common arrangements are called _____.

6. The reverse of dehydration synthesis is _____.

7. Reactions that release energy are called _____ reactions.

8. All chemical reactions begin with reactants and result in new molecules called _____.

9. The _____ scale is a measure of the concentration of hydrogen ions in a solution.

10. A nucleic acid containing the base uracil would also contain _____ sugar.

VISUALIZE IT!

1. In the following molecule, label a portion that shows only primary structure; label two types of secondary structure; circle the tertiary structure.

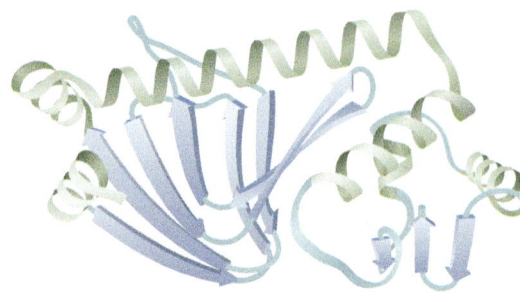

2. Shown is the amino acid tryptophan. Put the letter "C" at the site of every carbon atom. Label the amino group, the carboxyl group, and the side group.

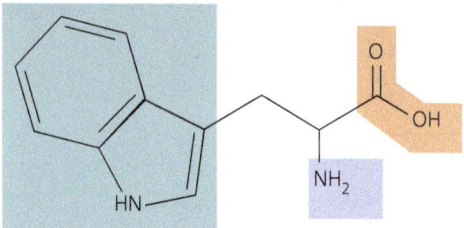

Short Answer

1. List three main types of chemical bonds, and give an example of each.
2. Name five properties of water that are vital to life.
3. Describe the difference(s) among saturated fatty acids, unsaturated fatty acids, and polyunsaturated fatty acids.
4. What is the difference between atomic oxygen and molecular oxygen?
5. Explain how the polarity of water molecules makes water an excellent solvent.

CRITICAL THINKING

1. Anthrax is caused by a bacterium, *Bacillus anthracis*, that avoids the body's defenses against disease by synthesizing an outer glycoprotein covering made from D-glutamic acid. This covering is not digestible by white blood cells that normally engulf bacteria. Why is the covering indigestible?
2. Dehydrogenation is a chemical reaction in which a saturated fat is converted to an unsaturated fat. Explain why the name for this reaction is an appropriate one.
3. Two freshmen disagree about an aspect of chemistry. The nursing major insists that H^+ is the symbol for a hydrogen ion. The physics major insists that H^+ is the symbol for a proton. How can you help them resolve their disagreement?
4. When an egg white is heated, it changes from liquid to solid. When gelatin is cooled, it changes from liquid to solid. Both gelatin and egg white are proteins. From what you have learned about proteins, why can the gelatin be changed back to liquid but the cooked egg cannot?
5. When amino acids are synthesized in a test tube, D and L forms occur in equal amounts. However, cells use only L forms in their proteins. Occasionally, meteorites are found to contain amino acids. Based on these facts, how could NASA scientists determine whether the amino acids recovered from space are evidence of Earth-like extraterrestrial life rather than the result of nonmetabolic processes?
6. The poison glands of many bees and wasps contain acidic compounds. What common household chemical could be used to neutralize this poison?
7. Neon (atomic mass 10) and argon (atomic mass 18) are inert elements, which means that they very rarely form chemical bonds. Give the electron configuration of their atoms, and explain why these elements are inert.
8. An article in the local newspaper about gangrene states that the tissue-destroying toxin involved, lecithinase, is an "organic compound." But many people consider "organic" chemicals to mean something is good. Explain the apparent contradiction.
9. The deadly poison hydrogen cyanide has the chemical formula $H-C\equiv N$. Describe the bonds between carbon and hydrogen and between carbon and nitrogen in terms of the number of electrons involved.
10. Triple covalent bonds are stronger and more difficult to break than single covalent bonds. Explain why by referring to the stability of a valence shell that contains eight electrons.
11. How can hydrogen bonding between water molecules help explain water's ability to absorb large amounts of energy before evaporating?
12. How can a single molecule of magnesium hydroxide neutralize two molecules of hydrochloric acid?

13. We have seen that it is important that biological membranes remain flexible. Most bacteria lack sterols in their membranes and instead incorporate unsaturated phospholipids in the membranes to resist tight packing and solidification. Examine Table 2.4 (on p. 43). Which fatty acid might best protect the membranes of an ice-dwelling bacterium?

14. Why is there no stereoisomer of glycine?

15. A textbook states that only five nucleotide bases are found in cells, but a laboratory worker reports that she has isolated eight different nucleotides. Explain why both are correct.

CONCEPT MAPPING

Using the terms provided, fill in this concept map that describes nucleic acids. You can also complete this and other concept maps online by going to the Mastering Microbiology Study Area.

Adenine
Cytosine
Deoxyribonucleotides
Deoxyribose
DNA
Double-stranded
Guanine
Messenger RNA
Nitrogenous bases (2)
Phosphate
Ribonucleotides
Ribose
RNA
Ribosomal RNA
Transfer RNA

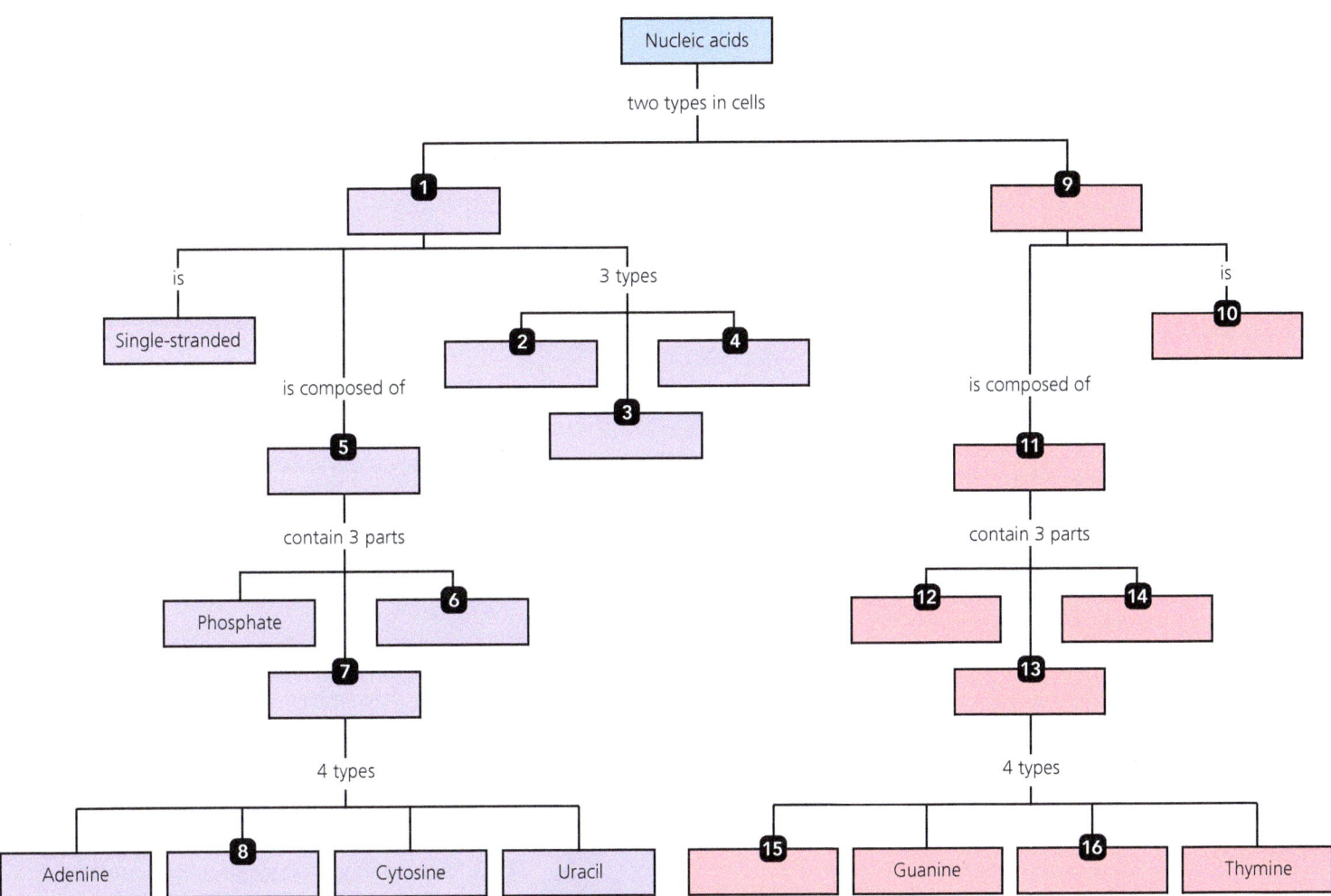

3 Cell Structure and Function

Before You Begin

1. What is the most abundant substance in organisms?
2. What is the characteristic of phospholipids that allows them to form double-layered membranes spontaneously?
3. Which macromolecules catalyze (speed up) chemical reactions inside cells?
4. Cells have buffers. What is the function of buffers inside cells?

MICRO IN THE CLINIC

Kidney Infection Leading to Mental Confusion?

PETE IS A 68-YEAR-OLD retired engineer. He remains mentally and physically active and is generally in good health. That's part of the reason he was so surprised a couple of weeks ago when he found blood in his urine. Pete went to his doctor and was diagnosed with a kidney infection; the doctor prescribed a 14-day course of antibiotics. After the first week or so, Pete started to feel better, so he stopped taking the antibiotics. About two days ago, he noticed that the pain in his back had returned. He also felt a little dizzy and seemed to get winded more quickly than usual. Pete thinks he should go back to the doctor but wants to wait until after his granddaughter, Sara, visits. She doesn't get the chance to visit very often, and they have a great day planned—a picnic in the park and a nice long walk along the lakeshore.

Pete is thrilled when Sara arrives—but Sara looks worried when she sees her grandpa. He's quite flushed, is a little bit sweaty, and seems to be a little bit confused or disoriented. Although Pete insists that he's fine, Sara really wants to take him to the emergency room (ER). Pete continues to argue, but Sara guarantees that she won't have a bit of fun if she's worried about her grandpa all day long. Finally, Pete agrees to go to the ER.

1. Is Pete's earlier kidney infection connected to his current symptoms?

Turn to the end of the chapter (p. 89) to find out.

Macrophage (blue) devouring *Escherichia coli* (red).

Processes of Life

LEARNING OUTCOME

3.1 Describe four major processes of living cells.

Microbiology is the study of particularly small living things. That raises a question: What does *living* mean; how do we define *life*? Scientists once thought that living things were composed of special organic chemicals, such as glucose and amino acids, that carried a "life force" found only in living organisms. These organic chemicals were thought to be formed only by living things and to be very different from the inorganic chemicals of nonliving things.

The idea that organic chemicals could come only from living organisms had to be abandoned in 1828, when Friedrich Wöhler (1800–1882) synthesized an organic molecule, urea, using only inorganic reactants in his laboratory. Today, we know that all living things contain both organic and inorganic chemicals and that many organic chemicals can be made from inorganic chemicals by laboratory processes. If organic chemicals can be made even in the absence of life, what is the difference between a living thing and a nonliving thing? What is life?

At first, this may seem a simple question. After all, you can usually tell when something is alive. However, defining "life" itself can prove troublesome, so biologists generally avoid setting a definition, preferring instead to describe characteristics common to all living things. Biologists agree that all living things share at least four processes of life: growth, reproduction, responsiveness, and metabolism.

- **Growth.** Living things can grow; that is, they can increase in size.
- **Reproduction.** Organisms normally have the ability to reproduce themselves. *Reproduction* means that they increase in number, producing more organisms organized like themselves. Reproduction may be accomplished asexually (alone) or sexually with gametes (sex cells). Note that reproduction is an increase in number, whereas growth is an increase in size. Growth and reproduction often occur simultaneously. (We consider several methods of reproduction when we examine microbes in detail in Chapters 11–13.)
- **Responsiveness.** All living things respond to their environment. They have the ability to change themselves in reaction to changing conditions around or within them. Many organisms also have the ability to move toward or away from environmental stimuli—a response called *taxis*.
- **Metabolism.** *Metabolism* can be defined as the ability of organisms to take in nutrients from outside themselves and use the nutrients in a series of controlled chemical reactions to provide the energy and structures needed to grow, reproduce, and be responsive. Metabolism is a unique process of living things; nonliving things cannot metabolize. Cells store metabolic energy in the chemical bonds of *adenosine triphosphate* (ă-den'ō-sēn trī-fos'fāt), or *ATP*. (Major processes of microbial metabolism, including the generation of ATP, are discussed in Chapters 5–7.)

TABLE 3.1 shows how these characteristics, along with cell structure, relate to various kinds of microbes.

Organisms may not exhibit these four processes at all times. For instance, in some organisms reproduction may be postponed or curtailed by age or disease or, in humans at least, by choice. Likewise, the rate of metabolism may be reduced, as occurs in a seed, a hibernating animal, or a bacterial endospore,[1] and growth often stops when an animal reaches a certain size. However, microorganisms typically grow, reproduce, respond, and metabolize as long as conditions are suitable. (Chapter 6 discusses the proper conditions for the metabolism and growth of various types of microorganisms.)

TELL ME WHY

The smallest free-living microbe—the bacterium *Mycoplasma*—is nonmotile. Why is it alive, even though it cannot move?

Prokaryotic and Eukaryotic Cells: An Overview

LEARNING OUTCOME

3.2 Compare and contrast prokaryotic and eukaryotic cells.

In the 1800s, two German biologists, Theodor Schwann (1810–1882) and Matthias Schleiden (1804–1881), developed the theory that all living things are composed of cells. *Cells* are living entities, surrounded by a membrane, that are capable of growing,

TABLE 3.1 Characteristics of Life and Their Distribution in Microbes

Characteristic	Bacteria, Archaea, Eukaryotes	Viruses
Growth: increase in size	Occurs in all	Growth does not occur
Reproduction: increase in number	Occurs in all	Host cell replicates the virus
Responsiveness: ability to react to environmental stimuli	Occurs in all	Reaction to host cells seen in some viruses
Metabolism: controlled chemical reactions of organisms	Occurs in all	Viruses use host cell's metabolism
Cellular structure: membrane-bound structure capable of all of the above functions	Present in all	Viruses lack cytoplasmic membrane or cellular structure

---

[1]Endospores are resting stages, produced by some bacteria, that are tolerant of environmental extremes.

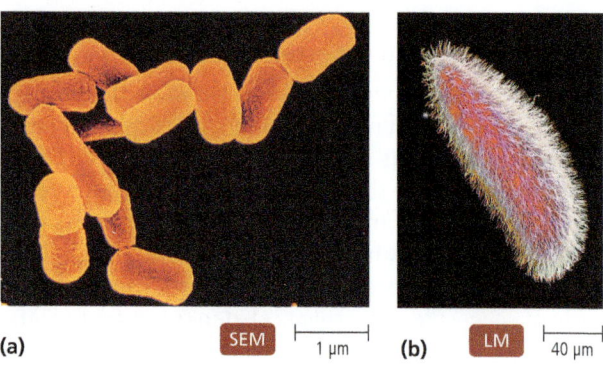

▲ FIGURE 3.1 **Examples of types of cells.** (a) *Escherichia coli* bacterial cells. (b) *Paramecium*, a single-celled eukaryote. Note the differences in magnification.

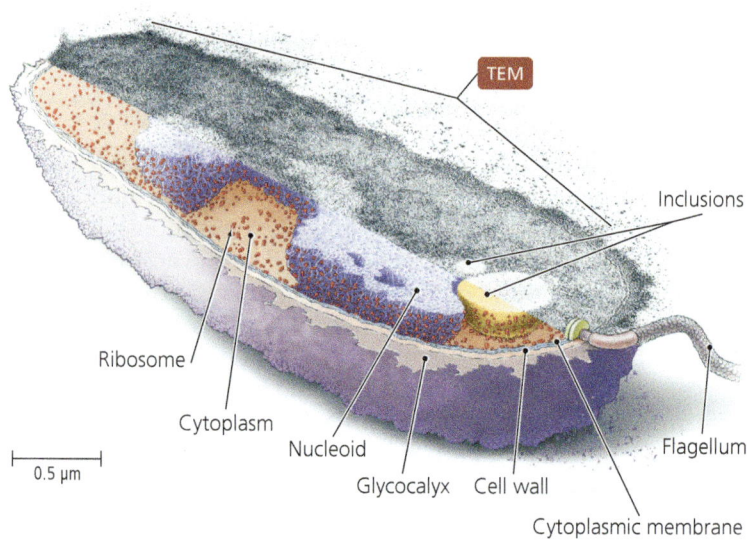

▲ FIGURE 3.2 **Typical prokaryotic cell.** Prokaryotes include archaea and bacteria. The artist has extended an electron micrograph to show three dimensions. Not all prokaryotic cells contain all these features.

reproducing, responding, and metabolizing. The smallest living things are single-celled microorganisms.

There are many different kinds of cells (**FIGURE 3.1**). Some cells are free-living, independent organisms; others live together in colonies or form the bodies of multicellular organisms. Cells also exist in various sizes, from the smallest bacteria to bird eggs, which are the largest of cells. All cells may be described as either *prokaryotes* (prō-kar′ē-ōts) or *eukaryotes* (yū-kar′ē-ōts).

Scientists categorize organisms based on shared characteristics into groups called *taxa*. Two taxa—*domain Archaea* and *domain Bacteria*—consist of organisms that are prokaryotic, but "prokaryote" is not itself a taxon. The distinctive feature of **prokaryotes** is that they can read their DNA genetic code and simultaneously make proteins—a typical prokaryote does not have a membrane surrounding its genetic material. In other words, a typical prokaryote does not have a nucleus (**FIGURE 3.2**). (Researchers have discovered a few prokaryotes with internal membranes that look like nuclei, but further investigation is needed to determine what these structures are.) The word *prokaryote* comes from Greek words meaning "before nucleus." Moreover, electron microscopy has revealed that prokaryotes typically lack various types of internal structures bound with membranes that are present in eukaryotic cells.

Bacteria and archaea differ fundamentally in such ways as the type of lipids in their cytoplasmic membranes and in the chemistry of their cell walls. In many ways, archaea are more like eukaryotes than they are like bacteria. (Chapter 11 discusses archaea and bacteria in more detail.)

Eukaryotes have a membrane called a nuclear envelope surrounding their DNA, forming a nucleus (**FIGURE 3.3**), which sets eukaryotes in *domain Eukarya*. Indeed, the term *eukaryote* comes from Greek words meaning "true nucleus." Besides the nuclear membrane, eukaryotes have numerous other internal membranes that compartmentalize cellular functions. These compartments are membrane-bound organelles—specialized structures that act like tiny organs to carry on the various functions of the cell. **Organelles** and their functions are discussed later in this chapter. The cells of algae, protozoa, fungi, animals, and plants are eukaryotic. Eukaryotes are usually larger, with cells 10–100 μm in diameter,[2] and more complex than prokaryotes, which are often about 1.0 μm in diameter or smaller (**FIGURE 3.4**).

[2]Micrometer (μm), which is one-millionth of a meter.

Although there are many kinds of cells, they all share the characteristic processes of life as previously described, as well as certain physical features. In this chapter, we will distinguish among bacterial, archaeal, and eukaryotic "versions" of physical features common to cells, including (1) external structures, (2) the cell wall, (3) the cytoplasmic membrane, and (4) the cytoplasm. We will also discuss features unique to each type. (Chapters 11, 12, and 19–23 examine further details of prokaryotic and eukaryotic organisms, their classification, and their ability to cause disease.)

MICRO CHECK

1. All living cells take in nutrients, which provide building materials and energy for a complex set of chemical reactions inside the cell. What are these reactions called?
2. Through what process do living cells increase in number?
3. When comparing size, how would you describe prokaryotes versus eukaryotes?
4. What feature is missing in prokaryotes but commonly found in eukaryotes, as hinted at by their names?

TELL ME WHY

In 1985, an Israeli scientist discovered the single-celled microbe *Epulopiscium fishelsoni*. This organism is visible with the naked eye. Why did the scientist initially think *Epulopiscium* was eukaryotic?

What discovery revealed that the microbe is really a giant bacterium?

Next, we explore characteristics of bacterial cells, beginning with external features and working into the cell.

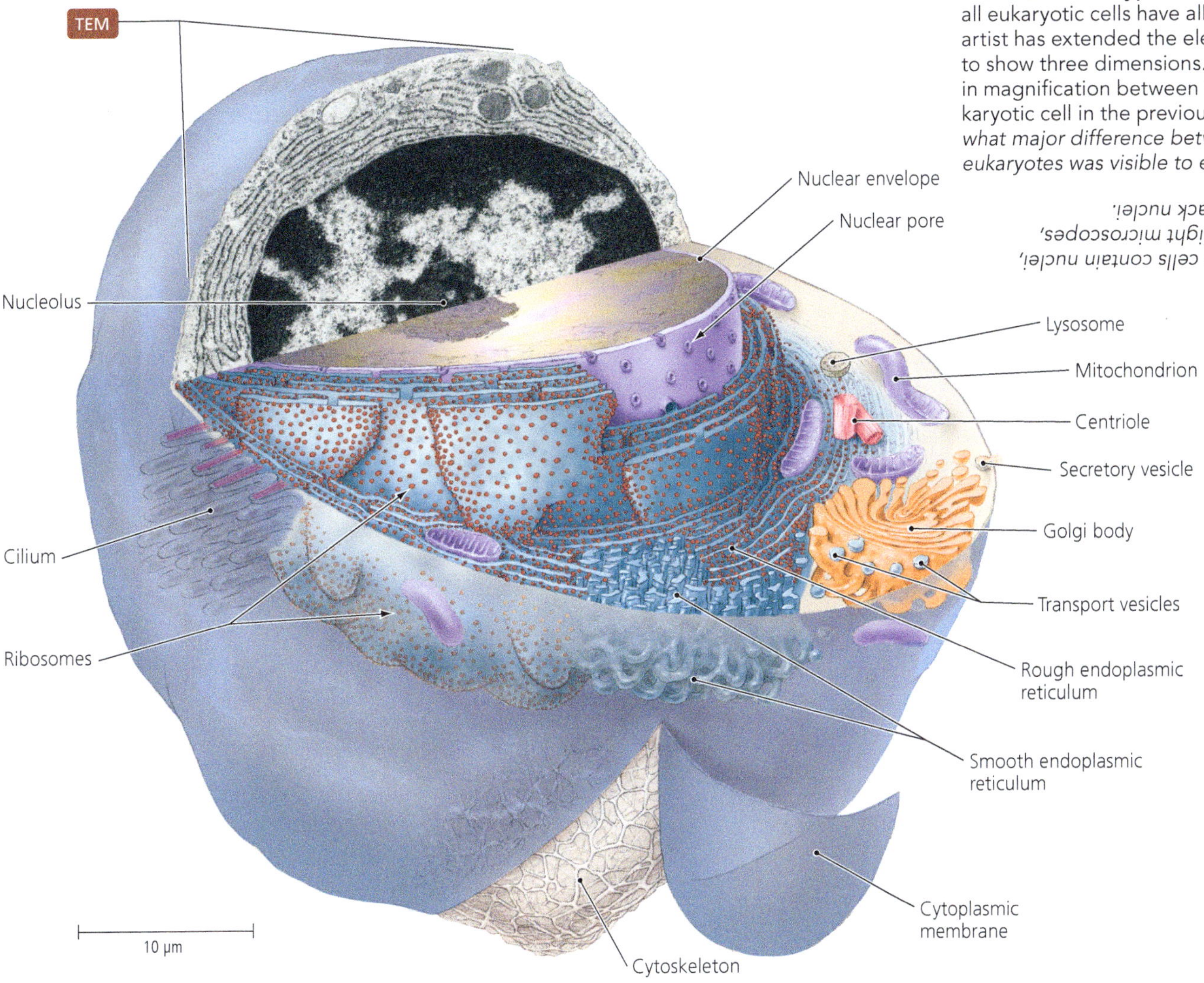

▼ **FIGURE 3.3 Typical eukaryotic cell.** Not all eukaryotic cells have all these features. The artist has extended the electron micrograph to show three dimensions. Note the difference in magnification between this cell and the prokaryotic cell in the previous figure. *Besides size, what major difference between prokaryotes and eukaryotes was visible to early microscopists?*

Figure 3.3 Eukaryotic cells contain nuclei, which are visible with light microscopes, whereas prokaryotes lack nuclei.

▼ **FIGURE 3.4 Approximate size of various types of cells.** Birds' eggs are the largest cells. Note that *Staphylococcus*, a bacterium, is smaller than *Giardia*, a unicellular eukaryote. A smallpox virus (*Orthopoxvirus*) is shown only for comparison; viruses are not cellular.

External Structures of Bacterial Cells

Many cells have special external features that enable them to respond to other cells and their environment. In bacteria, these features include glycocalyces, flagella, fimbriae, and pili.

Glycocalyces

LEARNING OUTCOMES

3.3 Describe the composition, function, and relevance to human health of glycocalyces.
3.4 Distinguish capsules from slime layers.

Some cells have a gelatinous, sticky substance that surrounds the outside of the cell. This substance is known as a **glycocalyx** (plural: *glycocalyces*), which literally means "sweet cup." The glycocalyx may be composed of polysaccharides, polypeptides, or both. These chemicals are produced inside the cell and are extruded onto the cell's surface.

When the glycocalyx of a bacterium is composed of organized repeating units of organic chemicals firmly attached to the cell's surface, the glycocalyx is called a **capsule** (FIGURE 3.5a). In contrast, a loose, water-soluble glycocalyx is called a **slime layer** (FIGURE 3.5b).

Glycocalyces protect cells from drying (desiccation) and can also play a role in the ability of pathogens to survive and cause disease. For example, slime layers are often sticky and provide one means for bacteria to attach to surfaces as *biofilms*, which are aggregates of many bacteria living together on a surface. Oral bacteria colonize the teeth as a biofilm called dental plaque. Bacteria in a dental biofilm can produce acid and cause *dental caries* (cavities).

The chemicals in many bacterial capsules can be similar to chemicals normally found in the body, preventing bacteria from being recognized or devoured by defensive cells of the host. For example, the capsules of *Streptococcus pneumoniae* (strep-tō-kok′ŭs nū-mō′nē-ī) and *Klebsiella pneumoniae* (kleb-sē-el′ă nū-mō′nē-ī) enable these prokaryotes to avoid destruction by defensive cells in the respiratory tract and to cause pneumonia. Unencapsulated strains of these same bacterial species do not cause disease because the body's defensive cells destroy them.

Flagella

LEARNING OUTCOMES

3.5 Discuss the structure and function of bacterial flagella.
3.6 List and describe four bacterial flagellar arrangements.

A cell's motility may enable it to flee from a harmful environment or move toward a favorable environment, such as one where food or light is available. The most notable structures responsible for such bacterial movement are flagella. Bacterial **flagella** (singular: *flagellum*) are long structures that extend beyond the surface of a cell and its glycocalyx and propel the cell through its environment. Not all bacteria have flagella, but for those that do, their flagella are very similar in composition, structure, and development.

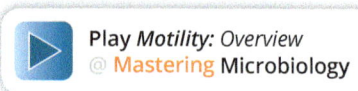
Play *Motility: Overview*
@ Mastering Microbiology

Structure

Bacterial flagella are composed of three parts: a *filament*, a *hook*, and a *basal body* (FIGURE 3.6). A filament is a long hollow shaft, about 20 nm in diameter,[3] that extends out of the cell into its environment. No membrane covers a filament.

A bacterial flagellum is composed of many identical globular molecules of a protein called *flagellin*. A flagellum lengthens by growing at its tip as the cell secretes molecules of flagellin through the hollow core of the flagellum, to be deposited in a clockwise helix at the tip of the filament. Bacterial flagella react to external wetness, inhibiting their own growth in dry habitats.

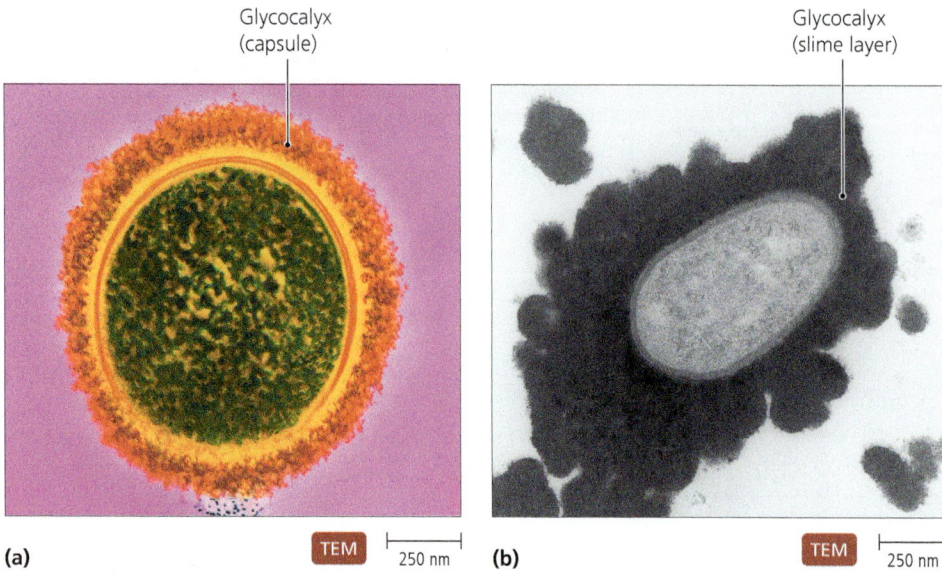

◀ FIGURE 3.5 **Glycocalyces.** (a) Micrograph of a single cell of the bacterium *Streptococcus*, showing a prominent capsule. (b) An unidentified skin bacterium has a slime layer surrounding the cell. *What advantage does a glycocalyx provide a cell?*

Figure 3.5 A glycocalyx provides protection from drying and from being devoured; it may also help attach cells to one another and to surfaces in the environment.

---

[3]Nanometer (nm), which is a billionth of a meter.

▲ FIGURE 3.6 **Proximal structure of bacterial flagella. (a)** Detail of flagellar structure of a Gram-positive cell. **(b)** Detail of the flagellum of a Gram-negative bacterium. *How do flagella of Gram-positive bacteria differ from those of Gram-negative bacteria?*

Figure 3.6 *Flagella of Gram-positive cells have a single pair of rings in the basal body that function to attach the flagellum to the cytoplasmic membrane. The flagella of Gram-negative cells have two pairs of rings; one pair anchors the flagellum to the cytoplasmic membrane, the other pair to the cell wall.*

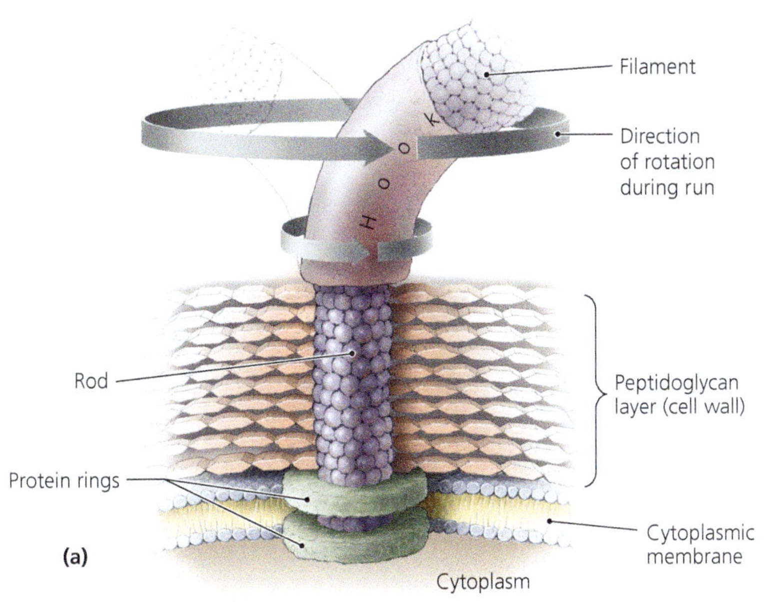

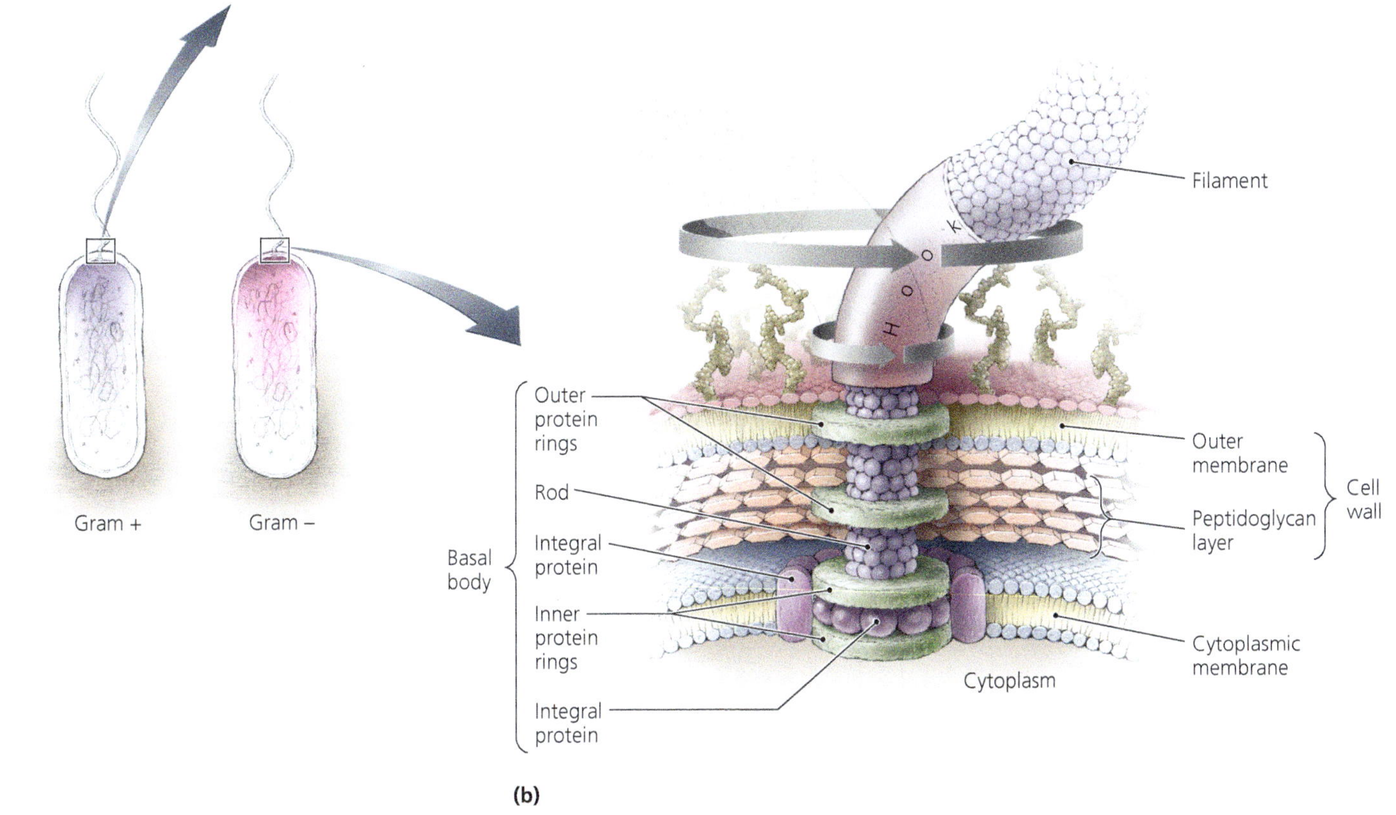

At its base, a filament inserts into a curved structure, the hook, which is composed of a different protein. The basal body, which is composed of still different proteins, anchors the filament and hook to the cell wall and cytoplasmic membrane by means of a rod and a series of either two or four rings of proteins. Together, the hook, rod, and rings allow the filament to rotate 360°. Differences in the proteins associated with bacterial flagella vary enough to allow classification of species into strains called *serovars*.

Play *Flagella: Structure*
@ Mastering Microbiology

Arrangement

Bacteria may have one of several flagellar arrangements (**FIGURE 3.7**). For example, flagella that cover the surface of the cell are termed **peritrichous**;[4] in contrast, **polar** flagella are only at the ends. Other bacteria have tufts of polar flagella.

Some spiral-shaped bacteria, called *spirochetes* (spī'rō-kēts),[5] have flagella at both ends that spiral tightly around the cell instead of protruding into the surrounding medium. These

[4]From Greek *peri*, meaning "around," and *trichos*, meaning "hair."
[5]From Greek *speira*, meaning "coil," and *chaeta*, meaning "hair."

(a) Peritrichous flagella — SEM 0.5 μm

(b) Single polar flagellum — SEM 0.5 μm

(c) Tuft of polar flagella — TEM 0.25 μm

▲ FIGURE 3.7 Micrographs of basic arrangements of bacterial flagella.

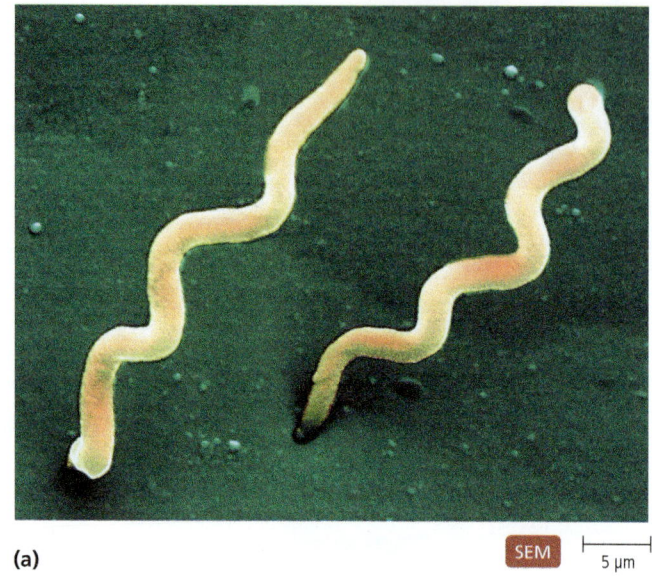

(a) SEM 5 μm

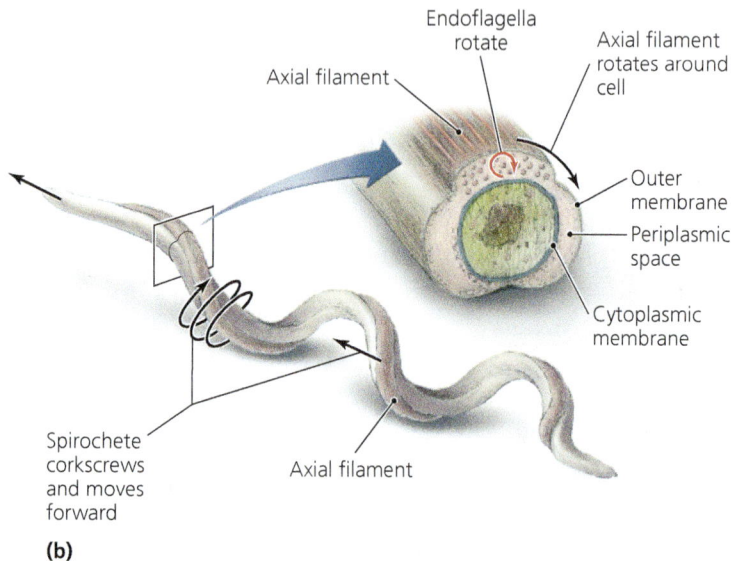

(b)

▲ FIGURE 3.8 Axial filament. (a) Scanning electron micrograph of two spirochete cells of *Borrelia burgdorferi*, which causes Lyme disease. (b) Diagram of axial filament wrapped around a spirochete. Cross section reveals that an axial filament is composed of endoflagella.

flagella, called **endoflagella**, form an **axial filament** that wraps around the cell between its cytoplasmic membrane and an outer membrane (**FIGURE 3.8**). Rotation of endoflagella evidently causes the axial filament to rotate around the cell, causing the spirochete to "corkscrew" through its medium. *Treponema pallidum* (trep-ō-nē′mă pal′li-dŭm), the agent of syphilis, and *Borrelia burgdorferi* (bō-rē′lē-ă burg-dōr′fer-ē), the cause of Lyme disease, are notable spirochetes. Some scientists think that the corkscrew motility of these pathogens allows them to invade human tissues.

▶ Play *Flagella: Arrangement; Spirochetes* @ Mastering Microbiology

Function

Although the precise mechanism by which bacterial flagella operate is not completely understood, we do know that they rotate 360° like boat or airplane propellers rather than whipping from side to side. The flow of ions (electrically charged atoms) through the cytoplasmic membrane near the basal body powers the rotation, propelling the bacterium through the environment at about 60 cell lengths per second—equivalent to a car traveling at 670 miles per hour! Flagella can rotate at more than 100,000 rpm and can change direction from counterclockwise to clockwise.

Bacteria move with a series of "runs" interrupted by "tumbles." Counterclockwise flagellar rotation produces movements of a cell in a single direction for some time; this is called a *run*. If

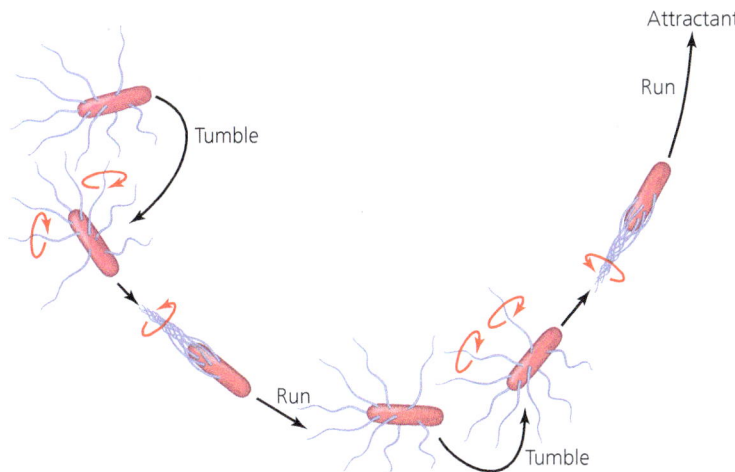

▲ FIGURE 3.9 **Motion of a peritrichous bacterium.** In peritrichous bacteria, runs occur when all of the flagella rotate counterclockwise and become bundled. Tumbles occur when the flagella rotate clockwise, become unbundled, and the cell spins randomly. In positive chemotaxis (shown), runs last longer than tumbles, resulting in motion toward the chemical attractant. *What triggers a bacterial flagellum to rotate counterclockwise, producing a run?*

Figure 3.9 *Favorable environmental conditions induce runs.*

more than one flagellum is present, the flagella align and rotate together as a bundle. Runs are interrupted by brief, abrupt, random changes in direction called *tumbles*. Tumbles result from clockwise flagellar rotation where each flagellum rotates independently.

Receptors for light or chemicals on the surface of the cell send signals to the flagella, which then adjust their speed and direction of rotation. If, during its random movements, a cell finds itself approaching a favorable stimulus, it increases the duration of the runs, which tends to move the cell toward an attractant—the bacterium moves itself into a more favorable environment (**FIGURE 3.9**). Unfavorable stimuli increase the duration of runs that are in a direction away from a repellent. This tends to move the cell toward a more favorable environment.

Movement in response to a stimulus is termed **taxis**. Stimuli are often either light (**phototaxis**) or a chemical (**chemotaxis**). Movement toward a favorable stimulus is *positive taxis*, whereas movement away from an unfavorable stimulus is *negative taxis*. For example, movement toward a nutrient would be *positive chemotaxis*.

Play *Flagella: Movement*
Mastering Microbiology

MICRO CHECK

5. What are the three parts of bacterial flagella?
6. What term describes the flagellar arrangement where flagella cover the entire surface of a bacterial cell?

Fimbriae and Pili

LEARNING OUTCOME

3.7 Compare and contrast the structures and functions of fimbriae, pili, and flagella.

Many bacteria have rodlike proteinaceous extensions called **fimbriae** (fim′brē-ī; singular: *fimbria*). These sticky, bristlelike projections adhere to one another and to substances in the environment. There may be hundreds of fimbriae per cell, and they are usually shorter than flagella (**FIGURE 3.10**). An example of a bacterium with fimbriae is *Neisseria gonorrhoeae* (nī-se′rē-ă go-nor-rē′ī), which causes gonorrhea. Pathogens must be able to adhere to their hosts if they are to survive and cause disease. This bacterium is able to colonize the mucous membrane of the reproductive tract by attaching with fimbriae. *Neisseria* cells that lack fimbriae are nonpathogenic. Some fimbriae carry enzymes that render soluble, toxic metal ions into insoluble, nontoxic forms.

Bacteria may use fimbriae to move across a surface via a process similar to pulling an object with a rope. The bacterium extends a fimbria, which attaches at its tip to the surface; then the bacterium retracts the fimbria, pulling itself toward the attachment point.

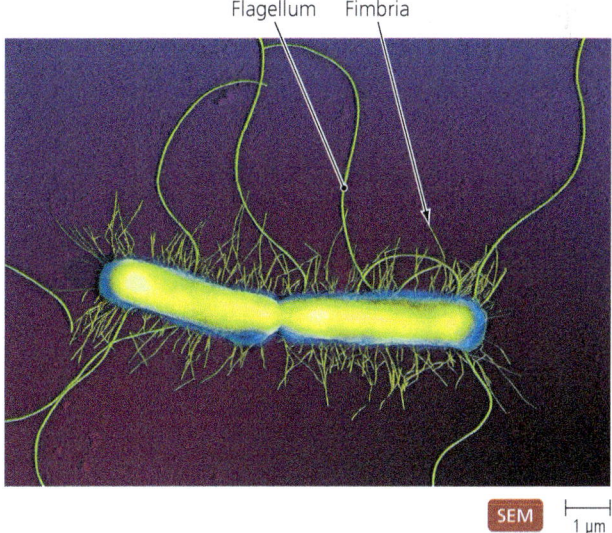

▲ FIGURE 3.10 **Fimbriae.** *Proteus vulgaris* has flagella and fimbriae.

Fimbriae also serve an important function in **biofilms**, three-dimensional slimy masses of microbes adhering to a substrate and to one another by means of fimbriae and glycocalyces (**FIGURE 3.11**). It has been estimated that at least 99%

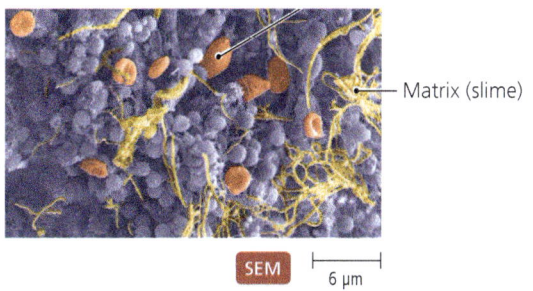

▲ FIGURE 3.11 **Biofilms.** Biofilms are organized layered systems of microbes attached to surfaces—here, the inside surface of medical tubing.

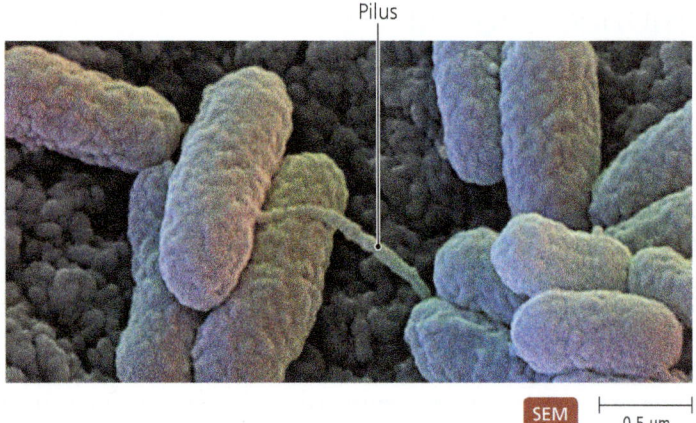

▲ FIGURE 3.12 **Pili.** Two *Escherichia coli* cells are connected by a pilus. *How are pili different from bacterial flagella?*

Figure 3.12 Bacterial flagella are flexible structures that rotate to propel the cell; pili are hollow tubes used to transfer DNA from one cell to another.

of bacteria in nature exist in biofilms. Some fimbriae act as electrical wires, conducting electrical signals among cells in a biofilm.

Researchers are interested in biofilms because of the roles they play in human diseases and in industry. For example, biofilms form plaque on teeth, form slime inside medical catheters, are involved in about two-thirds of bacterial human infections, clog drains, and corrode industrial pipes. Further, the architecture of a biofilm provides protection that free-floating bacteria lack. For example, the lower concentrations of oxygen found in the interior of biofilms thwart the effectiveness of some antibiotics.

A special type of fimbria is a **pilus** (pī′lus; plural: *pili*, pī′lī), also called a *conjugation pilus*. Pili are usually longer than other fimbriae but shorter than flagella. Typically, only one to a few pili are present per cell in bacteria that have them. Cells use pili to transfer DNA from one cell to the other via a process termed *conjugation* (FIGURE 3.12). (Chapter 7 deals with conjugation in more detail.)

TELL ME WHY

Why is a pilus a type of fimbria, but a flagellum is not a type of fimbria?

Bacterial Cell Walls

LEARNING OUTCOMES

3.8 Describe common shapes and arrangements of bacterial cells.
3.9 Describe the sugar and peptide portions of peptidoglycan.
3.10 Compare and contrast the cell walls of Gram-positive and Gram-negative bacteria in terms of structure and Gram staining.

The cells of most prokaryotes are surrounded by a **cell wall** that provides structure and shape to the cell and protects it from osmotic forces (described shortly). In addition, a cell wall assists some cells in attaching to other cells or in resisting antimicrobial drugs. Note that animal cells do not have cell walls, a difference that plays a key role in treatment of many bacterial diseases with certain types of antibiotics. For example, penicillin attacks the cell wall of bacteria but is harmless to human cells, because the latter lack walls.

Cell walls give bacterial cells characteristic shapes. Spherical cells, called *cocci* (kok′sī), may appear in various arrangements, including singly or in chains (streptococci), clusters (staphylococci), or cuboidal packets (sarcinae, sar′si-nī) (FIGURE 3.13), depending on the planes of cell division. Rod-shaped cells, called *bacilli* (bă-sil′ī), typically appear singly or in chains.

Bacterial cell walls are composed of **peptidoglycan** (pep′ti-dō-glī′kan), a meshlike complex polysaccharide. Peptidoglycan in turn is composed of two types of regularly alternating sugar molecules, called N-*acetylglucosamine* (NAG) and N-*acetylmuramic acid* (NAM), which are structurally similar to glucose (FIGURE 3.14). Millions of NAG and NAM molecules are covalently linked in chains in which NAG and NAM alternate. These chains are the "glycan" portions of peptidoglycan.

Chains of NAG and NAM are attached to other NAG-NAM chains by crossbridges of four amino acids (tetrapeptides)

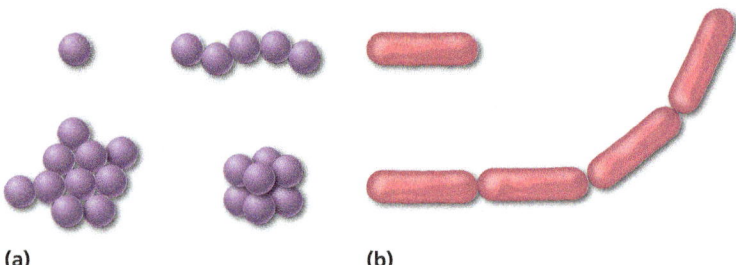

▲ FIGURE 3.13 **Bacterial shapes and arrangements. (a)** Spherical cocci may be single or in arrangements such as chains (streptococci), clusters (staphylococci), or cuboidal packets (sarcinae). **(b)** Rod-shaped bacilli may also be single or in arrangements such as chains.

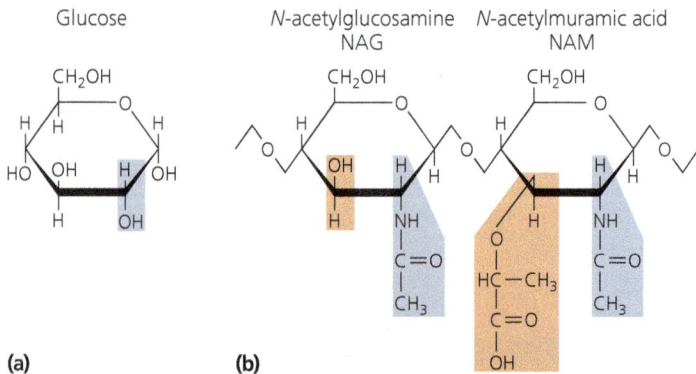

▲ FIGURE 3.14 **Comparison of the structures of glucose, NAG, and NAM. (a)** Glucose. **(b)** N-acetylglucosamine (NAG) and N-acetylmuramic acid (NAM) molecules linked as in peptidoglycan. Blue shading indicates the differences between glucose and the other two sugars. Orange boxes highlight the difference between NAG and NAM.

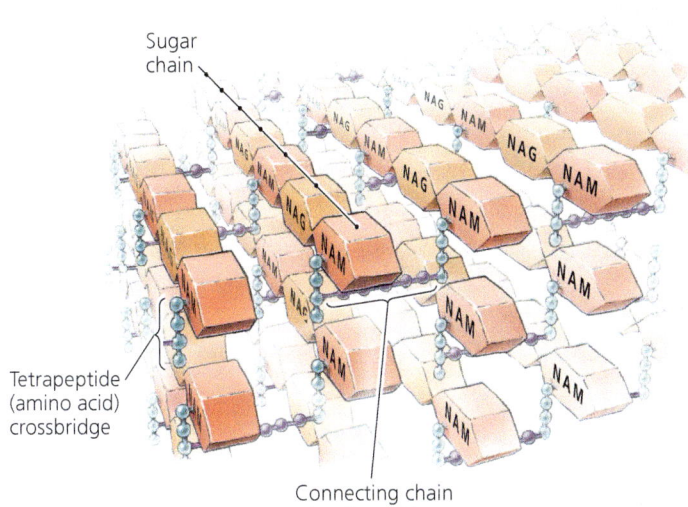

▲ **FIGURE 3.15 Possible structure of peptidoglycan.** Peptidoglycan is composed of chains of NAG and NAM linked by tetrapeptide crossbridges and, in some cases, as shown here, connecting chains of amino acids to form a tough yet flexible structure. The amino acids of the crossbridges differ among bacterial species.

between neighboring NAMs. **FIGURE 3.15** illustrates one possible configuration. Such peptide crossbridges are the "peptido" portion of peptidoglycan. Depending on the bacterium, tetrapeptide bridges are either bonded to one another or held together by *short connecting chains* of other amino acids, as shown in Figure 3.15. Peptidoglycan covers the entire surface of a cell, which must insert millions of new NAG and NAM subunits if it is to grow and divide.

Scientists describe two basic types of bacterial cell walls as *Gram-positive* cell walls or *Gram-negative* cell walls. They distinguish Gram-positive and Gram-negative cells by the use of the Gram staining procedure (described in Chapter 4), which was invented long before the structure and chemical nature of bacterial cell walls were known.

Gram-Positive Bacterial Cell Walls

LEARNING OUTCOME

3.11 Compare and contrast the cell walls of acid-fast bacteria with typical Gram-positive cell walls.

Gram-positive bacterial cell walls have a relatively thick layer of peptidoglycan that also contains unique chemicals called *teichoic*[6] *acids* (tī-kō′ik). Some teichoic acids are covalently linked to lipids, forming *lipoteichoic acids* that anchor the peptidoglycan to the *cytoplasmic membrane* of the cell (**FIGURE 3.16a**). Teichoic acids have negative electrical charges, which help give the surface of a Gram-positive bacterium a negative charge and may play a role in the passage of ions through the wall. The thick cell wall of a Gram-positive bacterium retains the crystal violet dye used in the Gram staining procedure, so the stained cells appear purple under magnification.

Some additional chemicals are associated with the walls of some Gram-positive bacteria. For example, species of *Mycobacterium* (mī′kō-bak-tēr′ē-ŭm), which include the causative agents of tuberculosis and leprosy, have walls with up to 60% mycolic acid, a waxy lipid. Mycolic acid helps these cells survive desiccation (drying out) and makes them difficult to stain with regular water-based dyes. Researchers have developed a special staining procedure called the *acid-fast stain* to stain these Gram-positive cells that contain large amounts of waxy lipids. Such cells are called *acid-fast bacteria* (see Chapter 4).

Gram-Negative Bacterial Cell Walls

LEARNING OUTCOME

3.12 Describe the clinical implications of the structure of the Gram-negative cell wall.

Gram-negative cell walls have only a thin layer of peptidoglycan outside the cytoplasmic membrane (**FIGURE 3.16b**), but outside the peptidoglycan there is another, *outer membrane*. This outer membrane is a bilayer; that is, it is composed of two different layers, or leaflets. The inner leaflet of the outer membrane is composed of phospholipids and proteins, but the outer leaflet is made of **lipopolysaccharide (LPS)**. Integral proteins called *porins* form channels through both leaflets of the outer membrane, allowing midsized molecules such as glucose to move freely across the outer membrane.

LPS is a union of lipid with sugar. The lipid portion of LPS, known as **lipid A**, can be toxic. The erroneous idea that lipid A is inside Gram-negative cells led to the use of the term **endotoxin**[7] for LPS. Dead Gram-negative cells release lipid A when the outer membrane disintegrates. This is medically important because lipid A may trigger fever, vasodilation, inflammation, shock, and blood clotting in humans. Killing large numbers of Gram-negative bacteria with antimicrobial drugs within a short time period releases large amounts of lipid A, which might threaten the patient more than the live bacteria; thus, any internal infection by Gram-negative bacteria is cause for concern.

The outer membrane of a Gram-negative cell can also be an impediment to the treatment of disease. For example, the outer membrane may prevent the movement of penicillin to the underlying peptidoglycan, thus rendering this drug ineffectual against many Gram-negative pathogens.

Between the cytoplasmic membrane and the outer membrane of Gram-negative bacteria is a **periplasmic space** (see Figure 3.16b). The periplasmic space contains the peptidoglycan and *periplasm*, the name given to the gel between the membranes. Periplasm contains water, nutrients, and substances secreted by

---

[6]From Greek *teichos*, meaning "wall."

[7]From Greek *endo*, meaning "inside," and *toxikon*, meaning "poison."

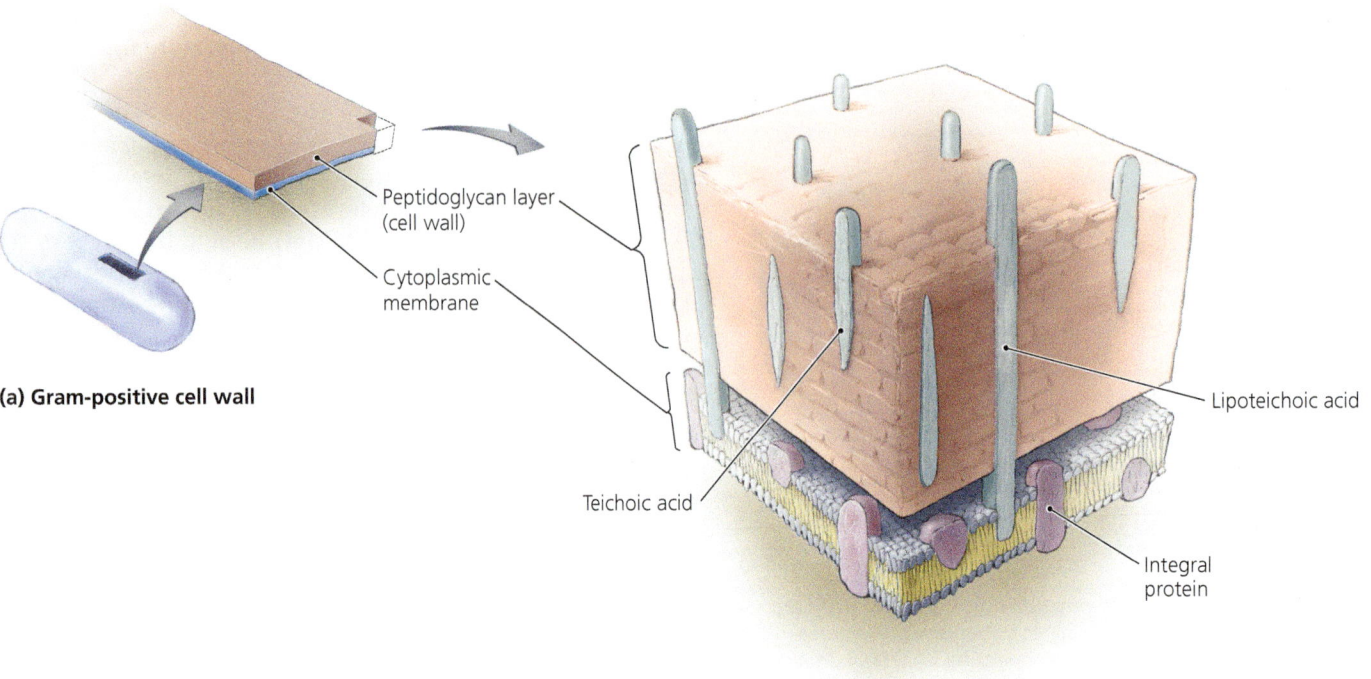
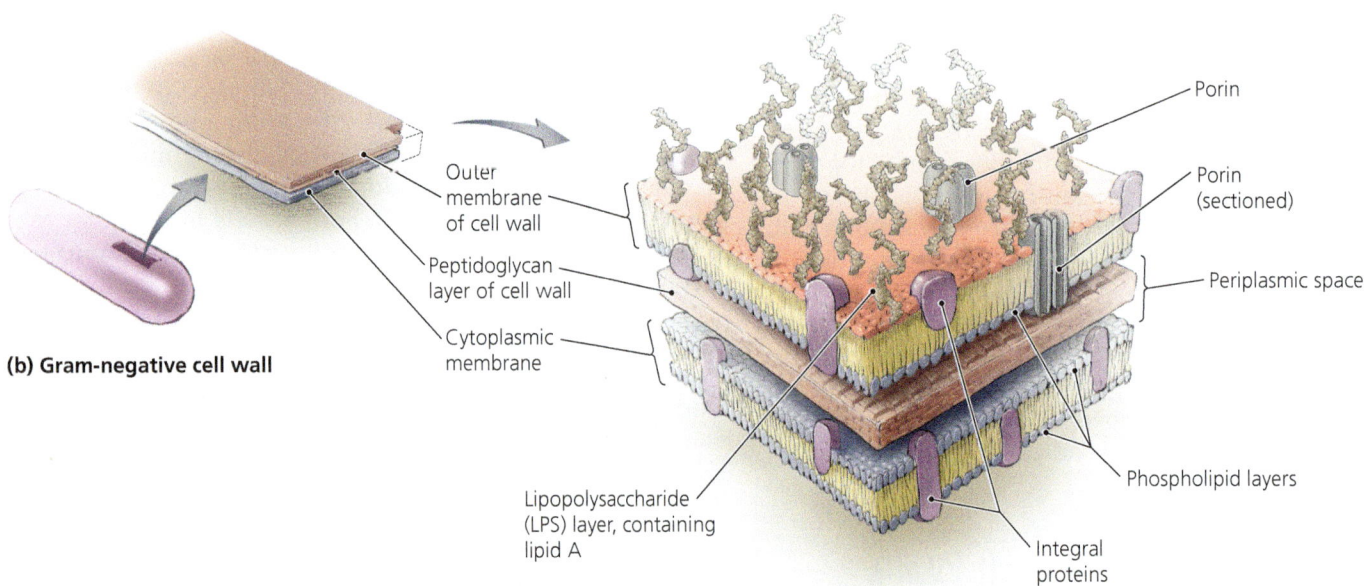

▲ FIGURE 3.16 Comparison of cell walls of Gram-positive and Gram-negative bacteria. (a) The Gram-positive cell wall has a thick layer of peptidoglycan and lipoteichoic acids that anchor the wall to the cytoplasmic membrane. (b) The Gram-negative cell wall has a thin layer of peptidoglycan and an outer membrane composed of lipopolysaccharide (LPS), phospholipids, and proteins. *What effects can lipid A have on human physiology?*

Figure 3.16 *Lipid A can cause shock, blood clotting, and fever in humans.*

the cell, such as digestive enzymes and proteins involved in specific transport. The enzymes function to catabolize large nutrient molecules into smaller molecules that can be absorbed or transported into the cell.

Because the cell walls of Gram-positive and Gram-negative bacteria differ, the Gram stain is an important diagnostic tool. After the Gram staining procedure, Gram-negative cells appear pink, and Gram-positive cells appear purple.

Bacteria Without Cell Walls

A few bacteria, such as *Mycoplasma pneumoniae* (mī'kō-plaz-mă nū-mō'nē-ī), lack cell walls entirely. In the past, these bacteria were often mistaken for viruses because of their small size and lack of walls. However, they do have other features of prokaryotic cells, such as prokaryotic ribosomes (discussed later in the chapter).

MICRO CHECK

7. Two of the most common shapes for bacteria are spherical and rod shaped. What is the proper term for spherical cells?
8. Bacteria may live as single cells or may cluster together in specific arrangements. What prefix is used for organisms that grow in chains?
9. One distinction between Gram-positive and Gram-negative bacteria is the relative thickness of peptidoglycan. Which group has a thicker layer of peptidoglycan?
10. What chemical gives the surfaces of Gram-positive and acid-fast bacteria their negative charges?

TELL ME WHY

Why is the microbe illustrated in Figure 3.2 more likely a Gram-positive bacterium than a Gram-negative one?

Bacterial Cytoplasmic Membranes

Beneath the glycocalyx and the cell wall is a **cytoplasmic membrane**. The cytoplasmic membrane may also be referred to as the *cell membrane* or a *plasma membrane*.

Structure

LEARNING OUTCOMES

3.13 Diagram a phospholipid bilayer, and explain its significance in reference to a cytoplasmic membrane.
3.14 Explain the fluid mosaic model of membrane structure.

Cytoplasmic membranes are about 8 nm thick and are composed of phospholipids (see Chapter 2; Figure 2.16) and associated proteins. Some bacterial membranes also contain sterol-like molecules, called *hopanoids*, that help stabilize the membrane.

The structure of a cytoplasmic membrane is referred to as a **phospholipid bilayer** (FIGURE 3.17). A phospholipid molecule is bipolar; that is, the two ends of the molecule are different. The phosphate-containing heads of each phospholipid molecule are *hydrophilic*,[8] meaning that they are attracted to water at the two surfaces of the membrane. The hydrocarbon tails of each phospholipid molecule are *hydrophobic*[9] and huddle together with other tails in the interior of the membrane, away from water. Phospholipids placed in a watery environment naturally form a bilayer because of their bipolar nature.

About half of a bacterial cytoplasmic membrane is composed of *integral proteins* inserted amidst the phospholipids. Some integral proteins penetrate the entire bilayer; others are found in only half the bilayer. In contrast, *peripheral proteins* are loosely attached to the membrane on one side or the other. Proteins of cell membranes may act as recognition proteins, enzymes, receptors, carriers, or channels.

The **fluid mosaic model** describes our current understanding of membrane structure. The term *mosaic* indicates that the membrane proteins are arranged in a way that resembles the tiles in a mosaic, and *fluid* indicates that the proteins and lipids are free to move laterally within a membrane.

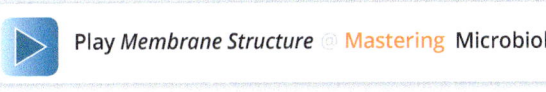
Play *Membrane Structure* Mastering Microbiology

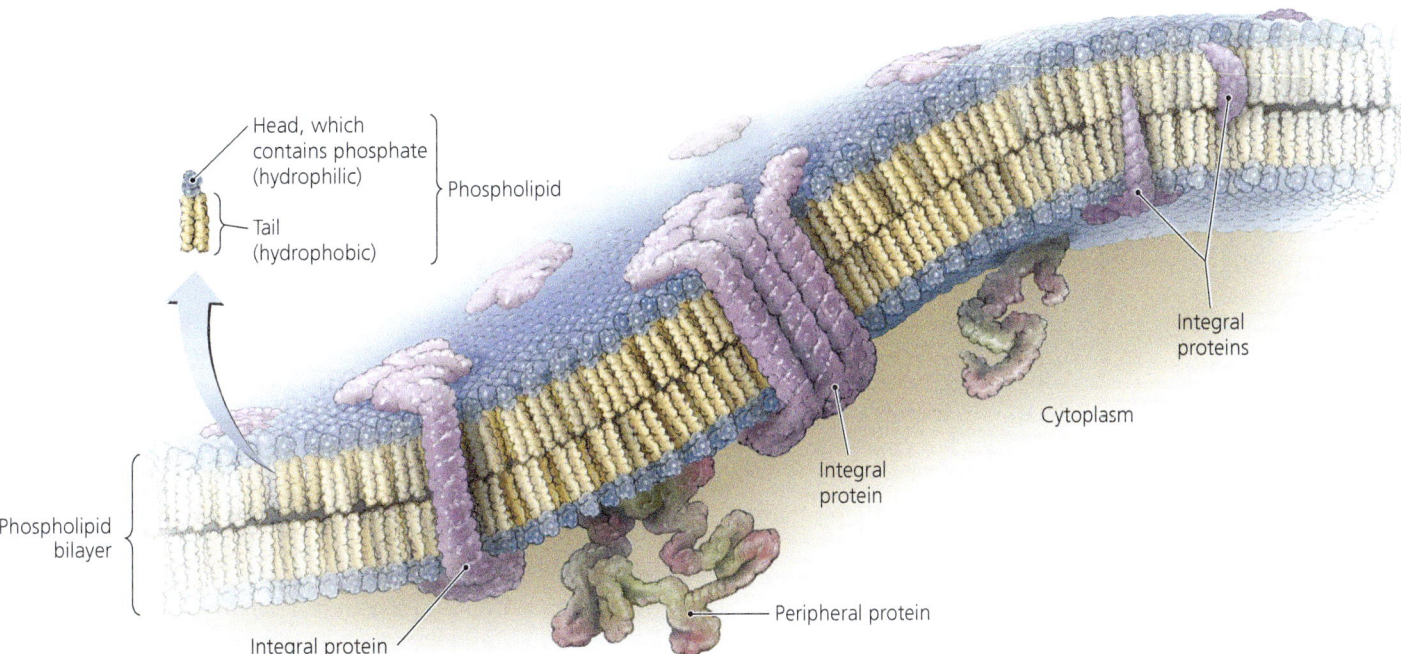

▲ FIGURE 3.17 Structure of a prokaryotic cytoplasmic membrane: a phospholipid bilayer.

---

[8]From Greek *hydor*, meaning "water," and *philos*, meaning "love."
[9]From Greek *hydor*, meaning "water," and *phobos*, meaning "fear."

MICRO CHECK

11. Phospholipid molecules have heads that are described as "loving water." What is the proper term for this behavior?
12. Which end of a phospholipid orients itself toward the interior of a phospholipid bilayer?

Function

LEARNING OUTCOMES

3.15 Describe the functions of a cytoplasmic membrane as they relate to permeability.
3.16 Compare and contrast the passive and active processes by which materials cross a cytoplasmic membrane.
3.17 Define *osmosis*, and distinguish among isotonic, hypertonic, and hypotonic solutions.

A cytoplasmic membrane does more than separate the contents of a cell from its outside environment. The cytoplasmic membrane also controls the passage of substances into and out of the cell. Nutrients are brought into the cell, and wastes are removed. The membrane also functions for harvesting light energy in photosynthetic bacteria producing molecules of ATP. (Chapter 5 discusses photosynthesis and ATP synthesis.)

In its function of controlling the contents of a cell, the cytoplasmic membrane is **selectively permeable**, meaning that it allows some substances to cross while preventing the crossing of others. How does a membrane exert control over substances that move across it and the contents of the cell?

Play *Membrane Permeability* @ Mastering Microbiology

A phospholipid bilayer is naturally impermeable to most substances. Large molecules cannot cross between the packed phospholipids; ions and molecules with an electrical charge are repelled by it; and hydrophilic substances cannot easily cross its hydrophobic interior. However, cytoplasmic membranes contain proteins, and some of these proteins function as pores, channels, or carriers that allow substances to cross the membrane.

Movement across a cytoplasmic membrane occurs by either passive or active processes. Passive processes do not require the expenditure of a cell's metabolic energy store (usually ATP), whereas active processes do require the expenditure of cellular energy, either directly or indirectly. Active and passive processes will be discussed shortly, but first you must understand another feature of selectively permeable cytoplasmic membranes: their ability to maintain a *concentration gradient*.

Membranes enable a cell to concentrate chemicals on one side of the membrane or the other. The difference in concentration of a chemical on the two sides of a membrane is the chemical's **concentration gradient** (also known as a *chemical gradient*).

Because many of the substances that have concentration gradients across cell membranes are electrically charged chemicals, a corresponding **electrical gradient**, or voltage, also exists across the membrane (FIGURE 3.18). For example, a greater concentration of negatively charged proteins exists inside the membrane, and positively charged sodium ions are more concentrated outside the membrane. One result of the segregation of electrical charges by a membrane is that the interior of a cell is usually electrically negative compared to the exterior. This tends to repel negatively charged chemicals and attract positively charged substances into cells.

Play *Passive Transport: Principles of Diffusion* @ Mastering Microbiology

CLINICAL CASE STUDY

The Big Game

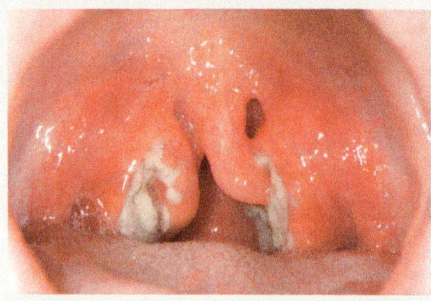

College sophomore Nadia is a star point guard for her school's basketball team. She is excited about the divisional finals Friday night—she's even heard rumors that a professional scout will be in the stands. On Thursday morning, she wakes up with a sore throat. Her forehead doesn't feel warm, so she forces herself to attend her Thursday classes; but when she wakes up on Friday morning, her throat is noticeably worse. Still, she forces herself to attend Friday morning class but feels tired and much worse by noon. It is downright painful to swallow, and she skips lunch.

Nearly crying, she heads back to the dorm and checks her temperature—101°F. Desperate, she walks to the student health center, where a nurse practitioner notices white patches on the back of Nadia's throat and on her tonsils. The divisional basketball game starts in six hours, but it only takes a few minutes for the nurse practitioner to perform a rapid streptococcal antigen test and determine that Nadia has streptococcal, also known as group A Streptococcus (GAS), pharyngitis—strep throat. She will miss the big game.

Strep throat is caused by an encapsulated, Gram-positive bacterium, *Streptococcus pyogenes*. The only good news is that by taking the prescribed penicillin, Nadia should be ready for her next big game—hopefully, the quarterfinals.

1. How does the capsule of *Streptococcus* contribute to the bacterium's ability to cause disease?
2. What bacterial structures, besides the capsule, may be allowing *Streptococcus* to infect Nadia's throat?
3. Penicillin works by interrupting the formation of peptidoglycan. What bacterial structure contains peptidoglycan?
4. In a Gram-positive organism such as *Streptococcus*, is peptidoglycan typically thicker or thinner than it would be in a Gram-negative bacterium?

Passive processes

In passive processes, electrochemical gradients provide energy to move substances across the membrane; the cell does not expend its ATP. Passive processes include diffusion, facilitated diffusion, and osmosis.

Diffusion Diffusion is the net movement of a chemical down its concentration gradient—that is, from an area of higher concentration to an area of lower concentration. It requires no energy output by the cell, a common feature of all passive processes. In fact, diffusion occurs even in the absence of cells or their membranes. In the case of diffusion into or out of cells, only chemicals that are small or lipid soluble can diffuse through the lipid portion of the membrane (FIGURE 3.19a). For example, oxygen, carbon dioxide, alcohol, and fatty acids can freely diffuse through the cytoplasmic membrane, but molecules such as glucose and proteins cannot.

Facilitated Diffusion The phospholipid bilayer blocks the movement of large or electrically charged molecules, so they do not cross the membrane unless there is a pathway for diffusion. As we have seen, cytoplasmic membranes contain integral proteins. Some of these proteins act as channels or carriers to allow certain molecules to diffuse down their concentration gradients into or out of the cell. This process is called **facilitated diffusion** because the proteins facilitate the process by providing a pathway for diffusion. The cell expends no energy in facilitated diffusion; electrochemical gradients provide all of the energy necessary.

Some channel proteins allow the passage of a range of chemicals that have the right size or electrical charge (FIGURE 3.19b). Other channel proteins, known as *permeases*, are more specific, carrying only certain substrates (FIGURE 3.19c). A permease has a binding site that is selective for one substance.

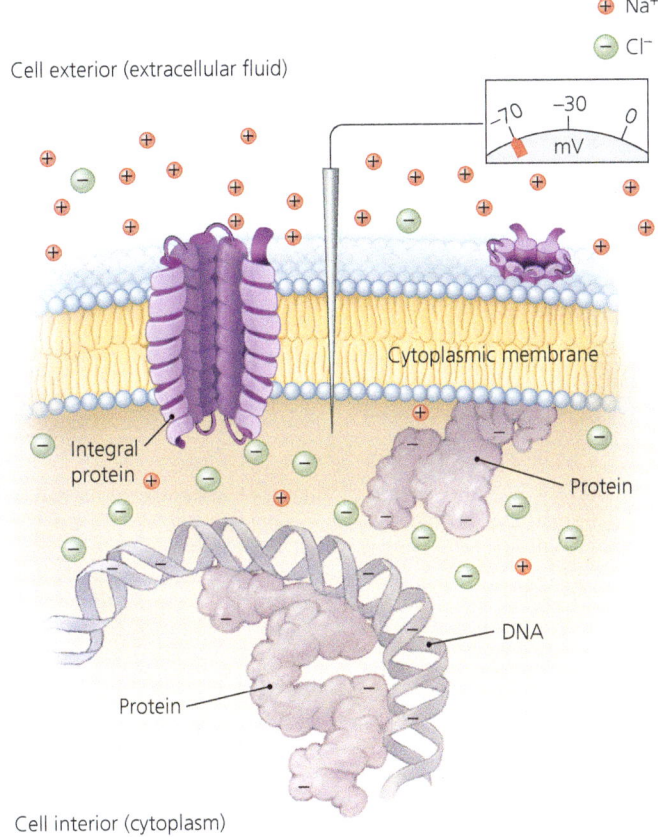

▲ FIGURE 3.18 **Electrical potential of a cytoplasmic membrane.** The electrical potential, in this case −70 mV, exists across a membrane because there are more negative charges inside the cell than outside it.

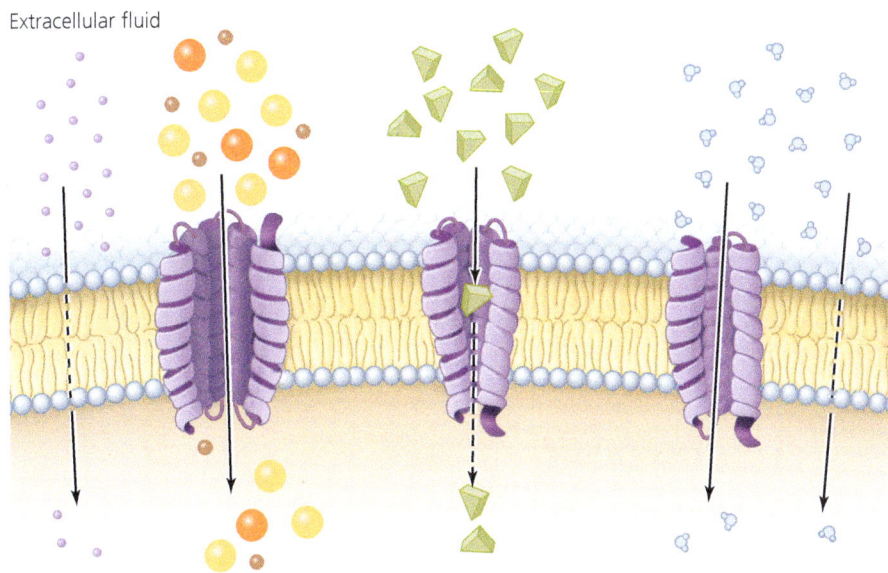

◀ FIGURE 3.19 **Passive processes of movement across a cytoplasmic membrane.** Passive processes always involve movement down an electrochemical gradient.

(a) Diffusion of small or lipid-soluble chemicals through the membrane

(b) Facilitated diffusion of several types of chemicals through a nonspecific channel

(c) Facilitated diffusion through a permease of a specific chemical

(d) Osmosis (net movement of water through a specific channel or through the membrane)

Osmosis When discussing simple and facilitated diffusion, we considered a solution in terms of the *solutes* (dissolved chemicals) it contains because it is those solutes that move into and out of the cell. In contrast, with osmosis it is useful to consider the concentration of the solvent, which in organisms is always water. **Osmosis** is the special name given to the diffusion of water across a semipermeable membrane—that is, across a membrane that is permeable to water molecules but not to most solutes that are present, such as proteins, amino acids, salts, or glucose (FIGURE 3.19d). Because these solutes cannot freely penetrate the membrane, they cannot diffuse across the membrane, no matter how unequal their concentrations on either side may be. Instead, the water diffuses. Water molecules cross from the side of the membrane that contains a higher concentration of water (lower concentration of solute) to the side that contains a lower concentration of water (higher concentration of solute). In osmosis, water moves across the membrane until equilibrium is reached, or until the pressure of water is equal to the force of osmosis (FIGURE 3.20).

We commonly compare solutions according to their concentrations of solutes. When solutions on either side of a selectively permeable membrane have the same concentration of solutes, the two solutions are said to be **isotonic**.[10] In an isotonic situation, neither side of a selectively permeable membrane will experience a net loss or gain of water (FIGURE 3.21a).

When the concentrations of solutions are unequal, the solution with the higher concentration of solutes is said to be **hypertonic**[11] to the other. The solution with a lower concentration of solutes is **hypotonic**[12] in comparison. Note that the terms *hypertonic* and *hypotonic* refer to the concentration of solute, even though osmosis refers to the movement of water. The terms *isotonic*, *hypertonic*, and *hypotonic* are relative. For example, a glass of tap water is isotonic to another glass of the same water, but it is hypertonic compared to distilled water, and hypotonic when compared to seawater. In biology, the three terms are traditionally used relative

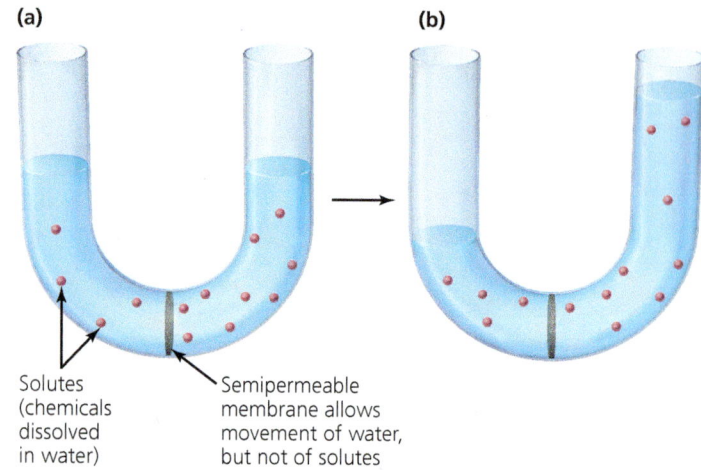

▲ FIGURE 3.20 **Osmosis, the diffusion of water across a semipermeable membrane.** (a) A membrane separates two solutions of different concentrations in a U-shaped tube. The membrane is permeable to water, but not to the solute. (b) After time has passed, water has moved down its concentration gradient until water pressure prevented the osmosis of any additional water. *Which side of the tube more closely represents a living cell?*

Figure 3.20 *The right-hand side represents the cell, because cells are typically hypertonic to their environment.*

to the interior of cells. Most cells are hypertonic to their environments.

A hypertonic solution, with its higher concentration of solutes, necessarily means a lower concentration of water; that is, a hypertonic solution has a lower concentration of water than does a hypotonic solution. Like other chemicals, water moves down its concentration gradient from a hypotonic solution into a hypertonic solution. A cell placed in a hypertonic solution will therefore lose water and shrivel (FIGURE 3.21b).

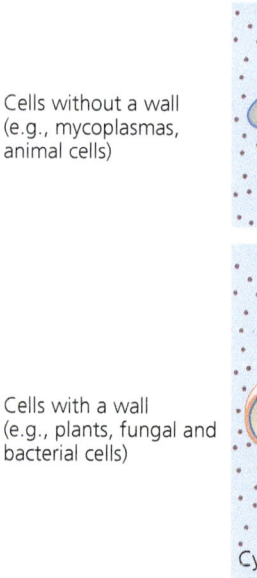 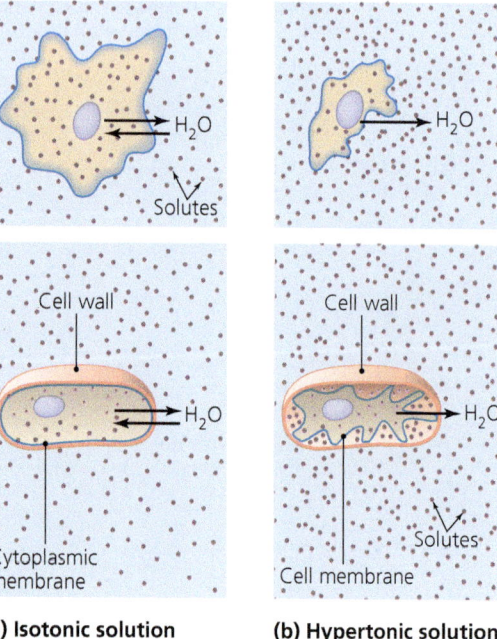 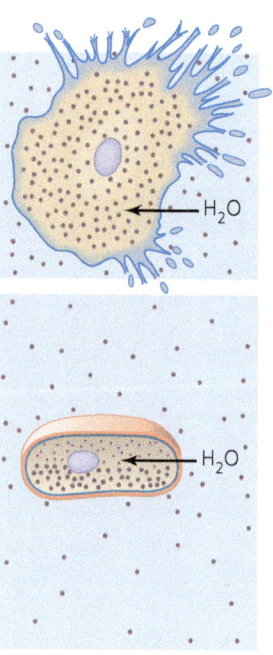

◀ FIGURE 3.21 **Effects of isotonic, hypertonic, and hypotonic solutions on cells.** (a) Cells in isotonic solutions experience no net movement of water. (b) Cells in hypertonic solutions shrink because of the net movement of water out of the cell. (c) Cells in hypotonic solutions undergo a net gain of water. Animal cells burst because they lack a cell wall; in cells with a cell wall, the pressure of water pushing against the interior of the wall eventually stops the movement of water into the cell.

(a) Isotonic solution   **(b) Hypertonic solution**   **(c) Hypotonic solution**

---

[10] From Greek *isos*, meaning "equal," and *tonos*, meaning "tone."
[11] From Greek *hyper*, meaning "more" or "over."
[12] From Greek *hypo*, meaning "less" or "under."

In contrast, water will diffuse into a cell placed in a hypotonic solution because the cell has a higher solutes-to-water concentration. As water moves into the cell, water pressure against its cytoplasmic membrane increases, and the cell expands (FIGURE 3.21c). One function of a cell wall, such as the peptidoglycan of bacteria, is to resist further osmosis and prevent cells from bursting.

It is useful to compare solutions to the concentration of solutes in a patient's blood cells. Isotonic saline solutions administered to a patient have the same percent dissolved solute (in this case, salt) as do the patient's cells. Thus, the patient's intracellular and extracellular environments remain in equilibrium when an isotonic saline solution is administered. However, if the patient is infused with a hypertonic solution, water will move out of the patient's cells, and the cells will shrivel, a condition called *crenation*. Conversely, if a patient is infused with a hypotonic solution, water will move into the patient's cells, which will swell and possibly burst.

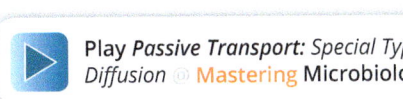
Play *Passive Transport: Special Types of Diffusion* @ Mastering Microbiology

Active Processes

As stated previously, active processes expend energy (often stored in ATP) to move material against its electrochemical gradient across a membrane. As we will see, ATP may be utilized directly during transport or indirectly at some other site or at some other time. Active processes in bacteria include active transport by means of carrier proteins and a special process termed *group translocation*.

Play *Active Transport: Overview* @ Mastering Microbiology

Active Transport Like facilitated diffusion, **active transport** utilizes transmembrane permease proteins; however, the functioning of active transport proteins requires the cell to expend ATP to move molecules across the membrane. Some such proteins are referred to as *gated channels* or *ports* because they are controlled. When the cell is in need of a substance, the protein becomes functional (the gate "opens"). At other times, the gate is "closed."

If only one substance is transported at a time, the permease is called a *uniport* (FIGURE 3.22a). In contrast, *antiports* simultaneously transport two chemicals, but in opposite directions; that is, one substance is transported into the cell at the same time that a second substance is transferred out of the cell (FIGURE 3.22b). In other types of active transport, two substances move together in the same direction across the membrane by means of a single carrier protein. Such proteins are known as *symports* (FIGURE 3.22c).

In all cases, active transport moves substances against their electrochemical gradients. Typically, the transport protein acts as an ATPase—an enzyme that breaks down ATP into ADP during transport, releasing energy that is used to move the substance across the membrane against its electrochemical gradient.

With symports and antiports, one chemical's electrochemical gradient may provide the energy needed to transport a second chemical, a mechanism called *coupled transport* (Figure 3.22c). For example, a uniport can pump H^+ out of a cell using ATP to establish an electrochemical gradient of hydrogen ions. Then H^+ moving back into the cell down its electrochemical gradient provides energy to bring a second chemical with it through a symport. At first glance it may appear that the second chemical is moving passively against its electrochemical gradient—no ATP is directly involved in movement through the symport. However, active transport utilizing ATP is involved in pumping hydrogen ions out of the cell, and this is coupled with the movement of the second chemical into the cell.

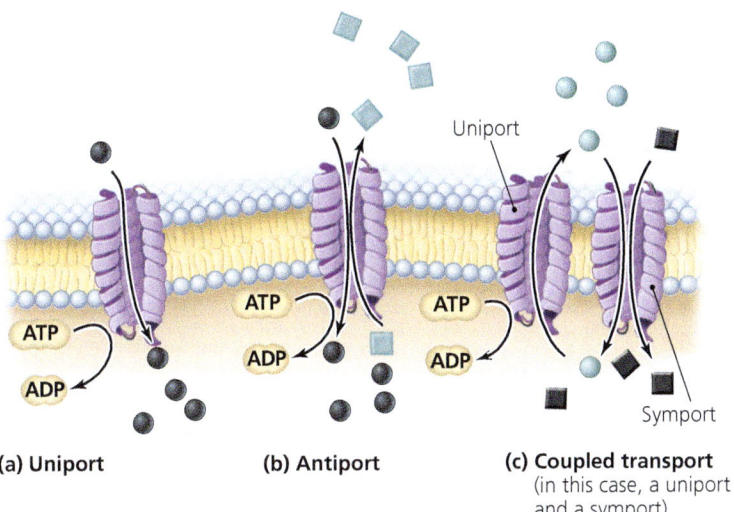

▲ **FIGURE 3.22 Mechanisms of active transport.** Cellular energy is used to transport the chemical against its electrochemical gradient. **(a)** Via a uniport. **(b)** Via an antiport. **(c)** Via a uniport coupled with a symport. In this example, the membrane uses ATP energy to pump one substance out through a uniport. As this substance flows back into the cell, it brings another substance with it through the symport. *What is the usual source of energy for active transport?*

Figure 3.22 *ATP is the usual source of energy for active transport processes.*

Play *Active Transport: Types* @ Mastering Microbiology

Group Translocation Many bacteria use an active process called **group translocation**. In group translocation, the substance transported across the membrane is chemically changed during transport (FIGURE 3.23). The membrane is impermeable to the altered

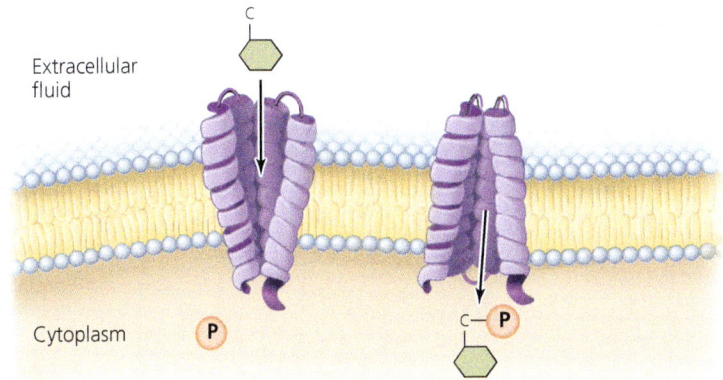

▲ **FIGURE 3.23 Group translocation.** This process involves a chemical change to the substance being transported, in this case, the addition of phosphate.

TABLE 3.2 Transport Processes Across Bacterial Cytoplasmic Membranes

	Description	Examples of Transported Substances
Passive Transport Processes	Processes require no use of energy by the cell; the electrochemical gradient provides energy.	
Diffusion	Molecules move down their electrochemical gradient through the phospholipid bilayer of the membrane.	Oxygen, carbon dioxide, lipid-soluble chemicals
Facilitated diffusion	Molecules move down their electrochemical gradient through channels or carrier proteins.	Glucose, fructose, urea, some vitamins
Osmosis	Water molecules move down their concentration gradient across a selectively permeable membrane.	Water
Active Transport Processes	Cell expends energy in the form of ATP to move a substance against its electrochemical gradient.	
Active transport	ATP-dependent carrier proteins bring substances into cell.	Na^+, K^+, Ca^{2+}, H^+, Cl^-
Group translocation	The substance is chemically altered during transport.	Glucose, mannose, fructose

substance, trapping it inside the cell. Group translocation is very efficient at bringing substances into a cell. It can operate effectively even if the external concentration of the chemical being transported is as low as 1 part per million (ppm).

One well-studied example of group translocation is the accumulation of glucose inside a bacterial cell. As glucose is transported across the membrane, the glucose is phosphorylated; that is, a phosphate group is added to change it into glucose 6-phosphate, a sugar that can be used in the ATP-producing metabolism of the cell. Other carbohydrates, fatty acids, purines, and pyrimidines are also brought into bacterial cells by group translocation. A summary of bacterial transport processes is shown in TABLE 3.2.

MICRO CHECK

13. The cell membrane determines what is permitted to enter or exit a cell. What is the term used to describe this property?
14. What form of diffusion is used by large molecules that must use a passageway provided by integral proteins?
15. What term describes an aqueous solution in which animal cells swell and burst?

TELL ME WHY

When the bacterium *Escherichia coli* is grown in a hypertonic solution, it turns on a gene to synthesize a protein that transports potassium into the cell. Why?

Cytoplasm of Bacteria

LEARNING OUTCOMES

3.18 Describe bacterial cytoplasm and its basic contents.
3.19 Define *inclusion*, and give two examples.
3.20 Describe the formation and function of endospores.

Cytoplasm is the general term used to describe the gelatinous material inside a cell. Cytoplasm is semitransparent, fluid, elastic, and aqueous. It is composed of cytosol, inclusions, ribosomes, and, in many cells, a cytoskeleton. Some bacterial cells produce internal, resistant, dormant forms called *endospores*.

Cytosol

The liquid portion of the cytoplasm is called **cytosol**. It is mostly water, but it also contains dissolved and suspended substances, including ions, carbohydrates, proteins (mostly enzymes), lipids, and wastes. The cytosol of prokaryotes also contains the cell's DNA in a region called the **nucleoid**. Recall that a distinctive feature of prokaryotes is lack of a membrane surrounding this DNA.

Most bacteria have a single, circular DNA molecule organized as a chromosome. Some bacteria, such as *Vibrio cholerae* (vib′rē-ō kol′er-ī), the bacterium that causes cholera, have two chromosomes.

The cytosol is the site of some chemical reactions. For example, enzymes within the cytosol function to produce amino acids and degrade sugar.

Inclusions

Deposits, called **inclusions**, are often found within bacterial cytosol. Rarely, a cell surrounds its inclusions with a polypeptide membrane. Inclusions may include reserve deposits of lipids, starch, or compounds containing nitrogen, phosphate, or sulfur. Such chemicals may be taken in and stored in the cytosol when nutrients are in abundance and then utilized when nutrients are scarce. The presence of specific inclusions is diagnostic for several pathogenic bacteria.

Some bacteria store carbon and energy in molecules of a lipid polymer called *polyhydroxybutyrate (PHB)* (FIGURE 3.24). Long chains of PHB accumulate as inclusion granules in the cytoplasm. Slight chemical modification of PHB produces a plastic that can be used for packaging and other applications (see **Beneficial Microbes: Plastics Made Perfect?**). PHB plastics are biodegradable, breaking down in a landfill in a few weeks rather than persisting for years as petroleum-based plastics do.

Many aquatic cyanobacteria (blue-green photosynthetic bacteria) contain inclusions called *gas vesicles* that store gases in protein sacs. The gases buoy the cells to the surface and into the light needed for photosynthesis.

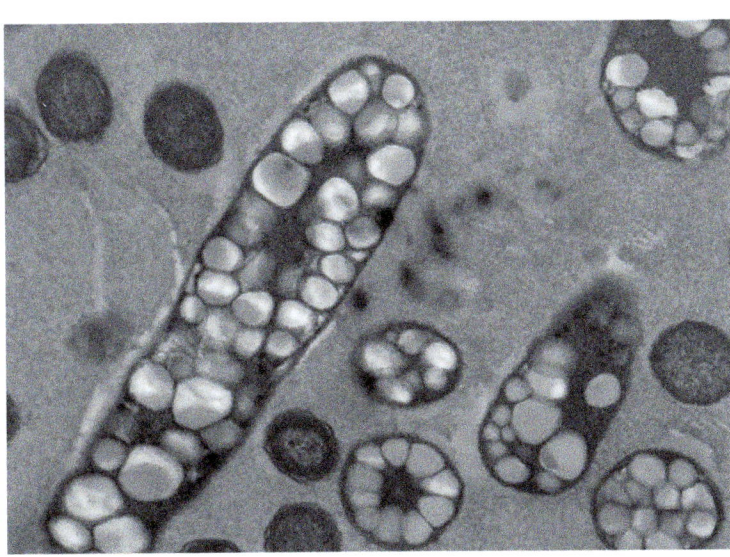

▲ FIGURE 3.24 Granules of PHB in a bacterium.

Endospores

Some bacteria, notably *Bacillus* (ba-sil'ŭs) and *Clostridium* (klos-trid'ē-ŭm), are characterized by the ability to produce unique dormant entities called **endospores**. Endospores are important for several reasons, including their durability and potential pathogenicity. Though some people refer to endospores simply as "spores," endospores should not be confused with the reproductive spores of actinobacteria, algae, and fungi. A bacterial cell, called a *vegetative cell* to distinguish it from an endospore, transforms into only one endospore, which later germinates to grow into only one vegetative cell; therefore, endospores are not reproductive structures. Instead, endospores constitute a defensive strategy against hostile or unfavorable conditions.

A vegetative cell normally transforms itself into an endospore only when one or more nutrients (such as carbon or nitrogen) are in limited supply. The process of endospore formation, called *sporulation*, requires 8–10 hours (**FIGURE 3.25**). During the process, two membranes, a thick layer of peptidoglycan, and a spore coat form around a copy of the cell's DNA and a small portion of cytoplasm. The cell deposits large quantities of dipicolinic acid, calcium, and DNA-binding proteins within the endospore while removing most of the water. The remains of the vegetative cell die, and the endospore is released. Depending on the species, a cell forms an endospore either *centrally*, *subterminally* (near one end), or *terminally* (at one end). Sometimes an endospore is so large it swells its vegetative cell.

Endospores are extremely resistant to drying, heat, radiation, and lethal chemicals. For example, they remain alive in boiling water for several hours; are unharmed by alcohol, peroxide, bleach, and other toxic chemicals; and can tolerate over 400 rad of radiation, which is more than five times the dose that is lethal to most humans. Endospores are stable resting stages that barely metabolize—they are essentially in a state of suspended animation—and they germinate only when conditions improve. Scientists do not know how endospores are able to resist harsh conditions, but it appears that the double membrane, spore coats, dipicolinic acid, calcium, and DNA-binding proteins serve to stabilize DNA and enzymes, protecting them from adverse conditions.

The ability to survive harsh conditions makes endospores the most resistant and enduring cells. In one case, scientists were

BENEFICIAL MICROBES

Plastics Made Perfect?

PHB bottle caps. One partially biodegraded in 60 days.

Petroleum-based plastics play a considerable role in modern life, appearing in packaging, bottles, appliances, furniture, automobiles, disposable diapers, and many other synthetic goods. Despite the good that plastic brings to our lives, this artificial polymer also brings problems.

Manufacturers make plastic from oil, exacerbating U.S. dependence on foreign supplies of crude oil. Consumers discard plastic, filling landfills with more than 15 million tons of plastic every year in the United States. Because plastic is artificial, microorganisms do not break it down effectively, and discarded plastic might remain in landfills for decades or centuries. What is needed is a functional "green" plastic—a plastic that is strong and light and that can be shaped and colored as needed yet is more easily biodegradable.

Enter the bacteria. Many bacterial cells, particularly Gram-negative bacteria, use polyhydroxybutyrate (PHB) as a storage molecule and energy source, much as humans use fat. PHB and similar storage molecules turn out to be rather versatile plastics that are produced when bacteria metabolizing certain types of sugar are simultaneously deprived of an essential element, such as nitrogen, phosphorus, or potassium. The bacteria, faced with such a nutritionally stressed environment, convert the sugar to PHB, which they store as intracellular inclusions. Scientists harvest these biologically created molecules by breaking the cells open and treating the cytoplasm with chemicals to isolate the plastics and remove dangerous endotoxin.

Purified PHB possesses many of the properties of petrochemically derived plastic—its melting point, crystal structure, molecular weight, and strength are very similar. Further, PHB has a singular, overwhelming advantage compared to artificial plastic: PHB is naturally and completely biodegradable—bacteria catabolize PHB into carbon dioxide and water. The positive effect that increasing PHB use would have on our overtaxed landfills would be tremendous, and replacing just half of the oil-based plastic used in the United States with PHB could reduce oil usage by more than 250 million barrels per year.

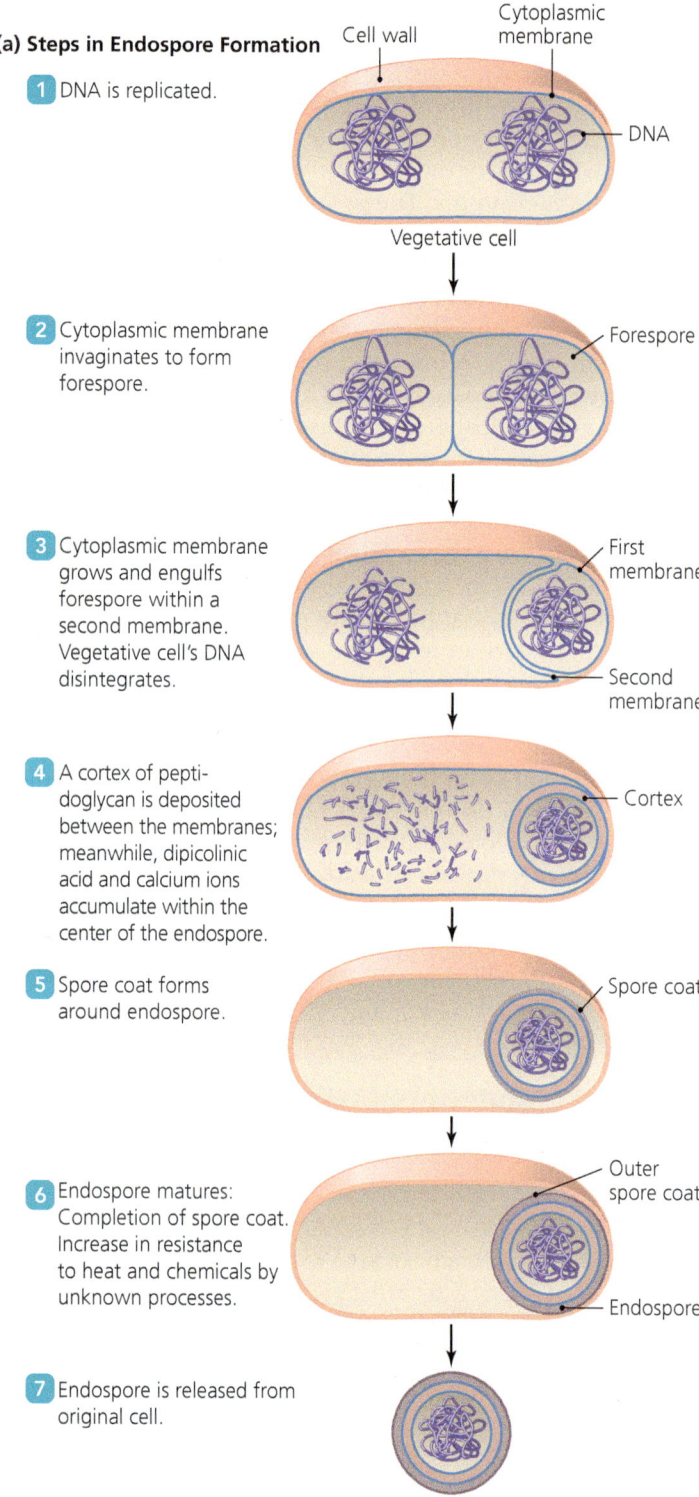

(a) Steps in Endospore Formation

1. DNA is replicated.
2. Cytoplasmic membrane invaginates to form forespore.
3. Cytoplasmic membrane grows and engulfs forespore within a second membrane. Vegetative cell's DNA disintegrates.
4. A cortex of peptidoglycan is deposited between the membranes; meanwhile, dipicolinic acid and calcium ions accumulate within the center of the endospore.
5. Spore coat forms around endospore.
6. Endospore matures: Completion of spore coat. Increase in resistance to heat and chemicals by unknown processes.
7. Endospore is released from original cell.

(b) TEM of an endospore in *Bacillus* or *Clodsridium*

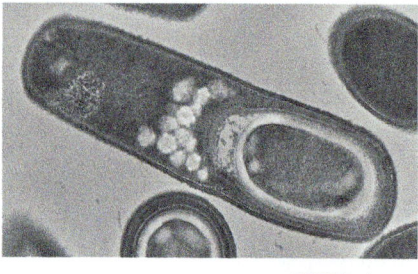

▲ FIGURE 3.25 Endospores. (a) Formation of an endospore occurs over a period of 8–10 hours. (b) An endospore inside a vegetative cell of *Bacillus*.

able to revive endospores of *Clostridium* that had been sealed in a test tube for 34 years. This record pales, however, beside other researchers' claim to have revived *Bacillus* endospores from inside 250-million-year-old salt crystals retrieved from an underground site near Carlsbad, New Mexico. Some scientists question this claim, suggesting that the bacteria might be recent contaminants. In any case, there is little doubt that endospores can remain viable for a minimum of tens, if not thousands, of years.

Endospore formation is a serious concern to food processors, health care professionals, and governments because endospores are resistant to treatments that inhibit other microbes, and because endospore-forming bacteria produce deadly toxins that cause such fatal diseases as anthrax, tetanus, and gangrene. (Chapter 9 considers techniques for controlling endospore formers.)

Nonmembranous Organelles

LEARNING OUTCOME

3.21 Describe the structure and function of ribosomes and the cytoskeleton.

As previously noted, prokaryotes do not usually have membranes surrounding their organelles. However, two types of *nonmembranous organelles* are found in direct contact with the cytosol in bacterial cytoplasm: ribosomes and the cytoskeleton. Some investigators do not consider them to be true organelles because they lack a membrane, but other scientists consider them nonmembranous organelles.

Ribosomes

Ribosomes synthesize proteins. Bacterial cells have thousands of ribosomes in their cytoplasm, which gives cytoplasm a grainy appearance (see Figure 3.2). The approximate size of ribosomes—and indeed other cellular structures—is expressed in Svedbergs (S)[13] and is determined by their sedimentation rate: the rate at which they move to the bottom of a test tube during centrifugation. As you might expect, large, compact, heavy particles sediment faster than small, loosely packed, or light ones and are assigned a higher number. Prokaryotic ribosomes are 70S; in contrast, the larger ribosomes of eukaryotes are 80S.

All ribosomes are composed of two subunits, each of which is composed of polypeptides and molecules of RNA called **ribosomal RNA (rRNA)**. The subunits of prokaryotic 70S ribosomes are a smaller 30S subunit and a larger 50S subunit; the 30S subunit contains polypeptides and a single rRNA molecule, whereas the 50S subunit has polypeptides and two rRNA molecules. Because sedimentation rates depend not only on mass and size but also on shape, the sedimentation rates of subunits do not add up to the sedimentation rate of a whole ribosome.

Many antibacterial drugs act on bacterial 70S ribosomes or their subunits without deleterious effects on the larger 80S ribosomes of eukaryotic cells (see Chapter 10). This is why such

---

[13]Svedberg units are named for Theodor Svedberg, a Nobel Prize winner and the inventor of the ultracentrifuge.

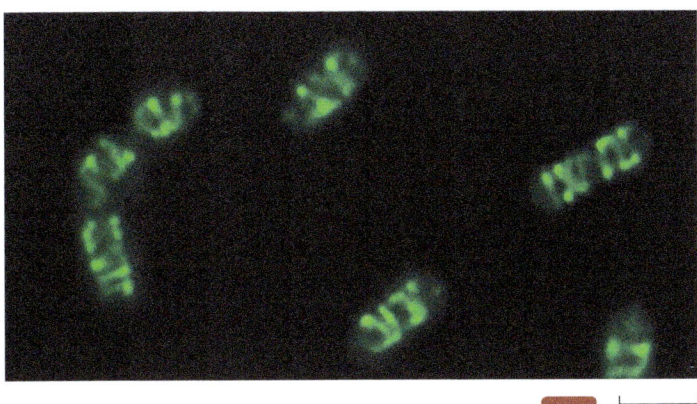

▲ FIGURE 3.26 **A simple helical cytoskeleton.** The rod-shaped bacterium *Bacillus subtilis* has a helical cytoskeleton composed of only a single protein, which has been stained with a fluorescent dye.

drugs can stop protein synthesis in bacteria without affecting protein synthesis in a patient.

Cytoskeleton

Cells have an internal scaffolding called a **cytoskeleton**, which is composed of three or four types of protein fibers. Bacterial cytoskeletons play a variety of roles in the cell. For example, one type of cytoskeleton fiber wraps around the equator of a cell and constricts, dividing the cell into two. Another type of fiber forms a helix down the length of some cells (**FIGURE 3.26**). Such helical fibers appear to play a role in the orientation and deposition of strands of NAG and NAM sugars in the peptidoglycan wall, thereby determining the shape of the cell. Other fibers help keep DNA molecules segregated to certain areas within bacterial cells. An unusual motile bacterium, *Spiroplasma* (spī′rō-plaz-mă), which lacks flagella, uses contractile elements of its cytoskeleton to swim through its environment.

MICRO CHECK

16. What is the term for the region inside a bacterial cell where the DNA can be found?
17. Some bacteria store energy reserves and other useful molecules. What are these stored deposits called?
18. Why do some bacteria form endospores?
19. What organelle is found in both prokaryotes and eukaryotes and produces new proteins?

TELL ME WHY

The 2001 bioterrorist anthrax attacks in the United States involved *Bacillus anthracis*. Why is *B. anthracis* able to survive in mail?

We have considered bacterial cells. Next we turn our attention to archaea—the other prokaryotic cells—and compare them to bacterial cells.

External Structures of Archaea

Archaeal cells have external structures similar to those seen in bacteria. These include glycocalyces, flagella, and fimbriae. Some archaea have another kind of proteinaceous appendage called a *hamus*. We consider each of these in order beginning with the outermost structures—glycocalyces.

Glycocalyces

LEARNING OUTCOME

3.22 Compare the structure and chemistry of archaeal and bacterial glycocalyces.

Like those of bacteria, archaeal **glycocalyces** are gelatinous, sticky, extracellular structures composed of polysaccharides, polypeptides, or both. Scientists have not studied archaeal glycocalyces as much as those of bacteria, but archaeal glycocalyces function at a minimum in the formation of **biofilms**—adhering cells to one another, to other types of cells, and to nonliving surfaces in the environment. Organized glycocalyces (capsules) of bacteria and bacterial biofilms are often associated with disease. Some research has demonstrated the presence of archaea in some biofilms associated with oral gum disease; however, no archaeon has been shown conclusively to be pathogenic.

Flagella

LEARNING OUTCOMES

3.23 Describe the structure and formation of archaeal flagella.
3.24 Compare and contrast archaeal flagella with bacterial flagella.

Archaea use **flagella** to move through their environments, though at a slower speed than bacteria. An archaeal flagellum is superficially similar to a bacterial flagellum: it consists of a basal body, hook, and filament, each composed of protein. The basal body anchors the flagellum in the cell wall and cytoplasmic membrane. As with bacterial flagella, archaeal flagella extend outside the cell, are not covered by a membrane, and rotate like propellers; however, scientists have discovered many differences between archaeal and bacterial flagella:

- Archaeal flagella are 10–14 nm in diameter, which is about half the thickness of bacterial flagella.
- Archaeal flagella are not hollow.
- Archaeal flagella, lacking a central channel, grow with the addition of subunits at the base of the filament rather than at the tip.
- The proteins making up archaeal flagella share common amino acid sequences across archaeal species. These sequences are very different from the amino acid sequences common to bacterial flagella.
- Sugar molecules are attached to the filaments of many archaeal flagella, a condition that is rare in bacteria.

- Archaeal flagella are powered with energy stored in molecules of ATP, whereas the flow of hydrogen ions across the membrane powers bacterial flagella.
- Archaeal flagella rotate together as a bundle both when they rotate clockwise and when they rotate counterclockwise. In contrast, bacterial flagella operate independently when rotating clockwise.

These differences indicate that archaeal flagella arose independently of bacterial flagella; they are analogous structures—having similar structure without having a common ancestor.

Fimbriae and Hami

LEARNING OUTCOMES

3.25 Compare the structure and function of archaeal and bacterial fimbriae.
3.26 Describe the structure and function of hami.

Many archaea have **fimbriae**—nonmotile, rodlike, sticky projections. As with bacteria, archaeal fimbriae are composed of protein and anchor the cells to one another and to environmental surfaces.

Some archaea make unique proteinaceous, fimbriae-like structures called **hami**[14] (singular: *hamus*). More than 100 hami may radiate from the surface of a single archaeon (FIGURE 3.27).

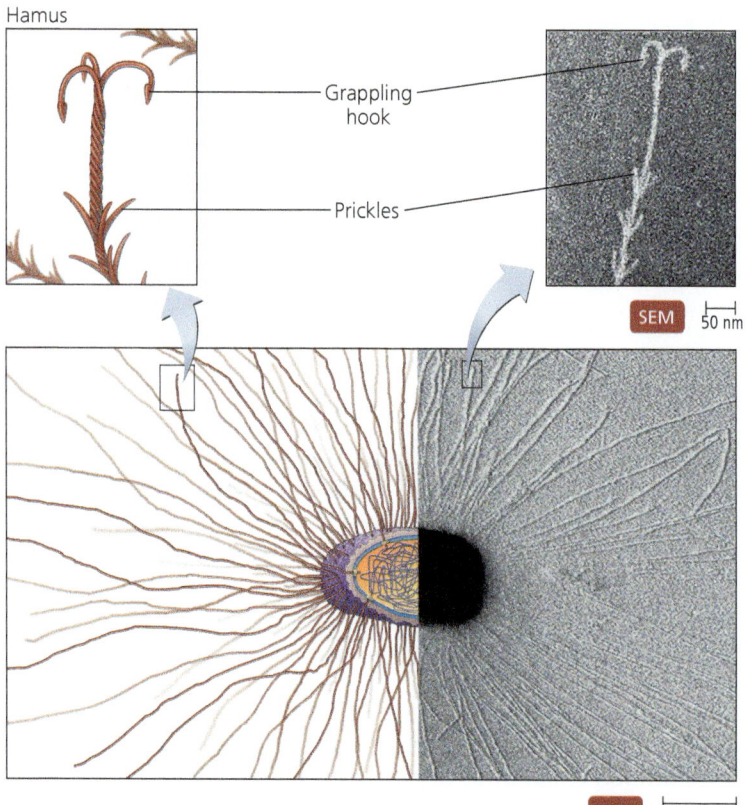

▲ **FIGURE 3.27 Archaeal hami.** Archaea use hami, which are shaped like grappling hooks on barbed wire, to attach themselves to structures in the environment.

---
[14]From Latin *hamus*, meaning "prickle," "claw," "hook," or "barb."

Each hamus is a helical filament with tiny prickles sticking out at regular intervals, much like barbed wire. The end of the hamus is frayed into three distinct arms, each of which has a thickened end and bends back toward the cell to make the entire structure look like a grappling hook. Indeed, hami function to securely attach archaea to surfaces.

TELL ME WHY

Why do scientists consider bacterial and archaeal flagella to be analogous rather than evolutionary relations?

Archaeal Cell Walls and Cytoplasmic Membranes

LEARNING OUTCOMES

3.27 Contrast types of archaeal cell walls with each other and with bacterial cell walls.
3.28 Contrast the archaeal cytoplasmic membrane with that of bacteria.

Most archaea, like most bacteria, have cell walls. All archaea have cytoplasmic membranes. However, there are distinct differences between archaeal and bacterial walls and membranes, further emphasizing the uniqueness of archaea.

Archaeal cell walls are composed of specialized proteins or polysaccharides. In some species, the outermost protein molecules form an array that coats the cell like chain mail. Archaeal walls lack peptidoglycan, which is common to all bacterial cell walls. Gram-negative archaeal cells, which appear pink when Gram stained, have an outer layer of protein rather than an outer lipid bilayer as seen in Gram-negative bacteria. Gram-positive archaea have a thick cell wall and Gram stain purple, like Gram-positive bacteria.

Archaeal cells are typically spherical or rod shaped, though irregularly shaped, needle-like, rectangular, and square archaea exist (FIGURE 3.28).

The cytoplasmic membranes of archaea often contain lipids that have branched hydrocarbons linked to glycerol by ether linkages rather than by ester linkages as in bacterial membranes (see Table 2.3 on p. 41). Ether linkages are stronger in many ways than ester linkages. They are mechanically stronger, more stable at high temperatures, and more salt tolerant, allowing archaea to live in extreme environments such as near-boiling water and hypersaline lakes.

An archaeal cytoplasmic membrane maintains electrical and chemical gradients in the cell. It also functions to control the import and export of substances from the cell using membrane proteins as ports and pumps, just as proteins are used in bacterial cytoplasmic membranes.

TELL ME WHY

Why did scientists in the 19th and early 20th centuries think that archaea were bacteria?

 (a) SEM 0.5 μm
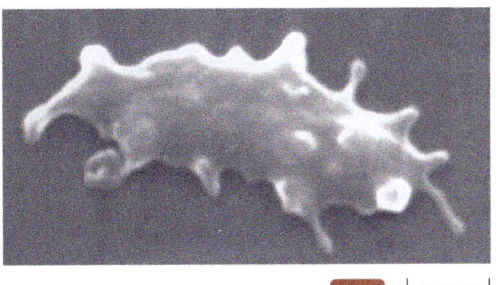 (b) SEM 1 μm
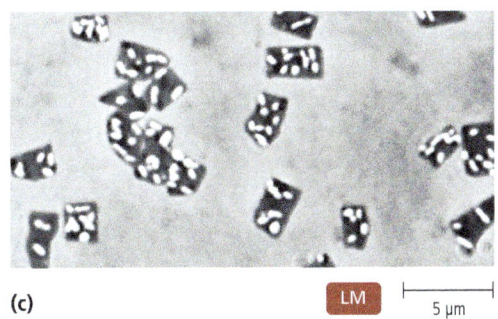 (c) LM 5 μm

▲ FIGURE 3.28 **Representative shapes of archaea. (a)** Cocci, *Pyrococcus furiosus*, attached to rod-shaped *Methanopyrus kandleri*. **(b)** Irregularly shaped archaeon, *Thermoplasma acidophilum*. **(c)** Square archaeon, *Haloquadratum walsbyi*. What are the stringlike extensions of Pyrococcus?

Figure 3.28 *The extensions are fimbriae or hami.*

Cytoplasm of Archaea

LEARNING OUTCOME

3.29 Compare and contrast the cytoplasm of archaea with that of bacteria.

Cytoplasm is the gel-like substance found in all cells, including archaea. Like bacteria, archaeal cells have 70S **ribosomes**, a fibrous cytoskeleton, and circular DNA suspended in a liquid cytosol. Also like bacteria, they lack membranous organelles.

However, archaeal cytoplasm differs from that of bacteria in several ways. For example, the protein molecules in archaeal ribosomes differ from those in the ribosomes of bacteria; indeed, archaeal ribosomal proteins are more similar to eukaryotic ribosomal proteins. Scientists further distinguish archaea from bacteria in that archaea use different metabolic enzymes to make RNA and use a genetic code more similar to the code used by eukaryotes. (Chapter 7 discusses these genetic differences in more detail.) TABLE 3.3 contrasts features of archaea and bacteria.

To this point, we have discussed basic features of bacterial and archaeal prokaryotic cells. (Chapter 11 discusses the classification of prokaryotic organisms in more detail.) Next we turn our attention to eukaryotic cells.

MICRO CHECK

20. What are archaeal glycocalyces composed of?
21. What are the grappling-hook-like structures used by archaea to attach to surfaces in their environment?
22. What substance do bacterial cell walls contain that archaeal cell walls lack?
23. What are the benefits of the ether linkages in archaeal cytoplasmic membranes, compared to the ester linkages in bacterial cytoplasmic membranes?
24. How do archaeal ribosomes differ from bacterial ribosomes?

TABLE 3.3 Some Structural Characteristics of Prokaryotes

Feature	Archaea	Bacteria
Glycocalyx	Polypeptide or polysaccharide	Polypeptide or polysaccharide
Flagella	Present in some; 10–14 nm in diameter; grow at base; rotate both counterclockwise and clockwise as bundles	Present in some; about 20 nm in diameter; grow at the tip; rotate counterclockwise in bundles to cause runs; rotate independently clockwise to cause tumbles
Fimbriae	Proteinaceous; used for attachment and in formation of biofilms	Proteinaceous; used for attachment, gliding motility, and in formation of biofilms
Pili	None discovered	Present in some; proteinaceous; used in bacterial exchange of DNA
Hami	Present in some; used for attachment	Absent
Cell Walls	Present in most; composed of polysaccharides (not peptidoglycan) or proteins	Present in most; composed of peptidoglycan, a polysaccharide
Cytoplasmic Membrane	Present in all; membrane lipids are made with ether linkages; some have single lipid layer	Present in all; phospholipids are made with ester linkages in bilayer
Cytoplasm	Cytosol contains circular DNA molecule and 70S ribosomes; ribosomal proteins are similar to eukaryotic ribosomal proteins	Cytosol contains a circular DNA molecule and 70S ribosomes with bacterial proteins

TELL ME WHY

Why do some scientists consider archaea, which are prokaryotic, more closely related to eukaryotes than they are to the other prokaryotes—bacteria?

External Structure of Eukaryotic Cells

Some eukaryotic cells have glycocalyces, which are similar to those of prokaryotes.

Glycocalyces

LEARNING OUTCOME

3.30 Describe the composition, function, and importance of eukaryotic glycocalyces.

Animal and most protozoan cells lack cell walls, but such a cell may have a sticky **glycocalyx**[15] that is anchored to the cell's cytoplasmic membrane via covalent bonds to membrane proteins and lipids. Eukaryotic glycocalyces are not as structurally organized as prokaryotic capsules. Animal cell glycocalyces help to anchor cells to each other, strengthen the cell surface, provide some protection against dehydration, and function in cell-to-cell recognition and communication. Glycocalyces are absent in eukaryotes that have cell walls, such as plants and fungi.

TELL ME WHY

Why are eukaryotic glycocalyces covalently bound to cytoplasmic membranes, and why don't eukaryotes with cell walls have glycocalyces?

Eukaryotic Cell Walls and Cytoplasmic Membranes

LEARNING OUTCOMES

3.31 Compare and contrast prokaryotic and eukaryotic cell walls and cytoplasmic membranes.
3.32 Contrast exocytosis and endocytosis.
3.33 Describe the role of pseudopods in eukaryotic cells.

The eukaryotic cells of fungi, algae, plants, and some protozoa have cell walls. Recall that glycocalyces are absent from eukaryotes with cell walls; instead, the cell wall takes on some of the functions of a glycocalyx by providing protection from the environment and anchoring neighboring cells to one another. The wall also provides shape and support against osmotic pressure. Most eukaryotic cell walls are composed of various polysaccharides, but not the peptidoglycan seen in the walls of bacteria.

The walls of plant cells are composed of *cellulose*, a polysaccharide that is familiar to you as paper and dietary fiber. Fungi also have walls of polysaccharides, including cellulose, *chitin*, or *glucomannan*. The walls of algae (FIGURE 3.29) are composed of a variety of polysaccharides or other chemicals, depending on the type of alga. These chemicals include cellulose, proteins, *agar, carrageenan, silicates, algin, calcium carbonate,* or a combination of these substances. (Chapter 12 discusses fungi and algae in more detail.)

All eukaryotic cells have cytoplasmic membranes (FIGURE 3.30). A eukaryotic cytoplasmic membrane, like those of bacteria, is a fluid mosaic of phospholipids and proteins, which act as recognition molecules, enzymes, receptors, carriers, or channels. Channel proteins for facilitated diffusion are

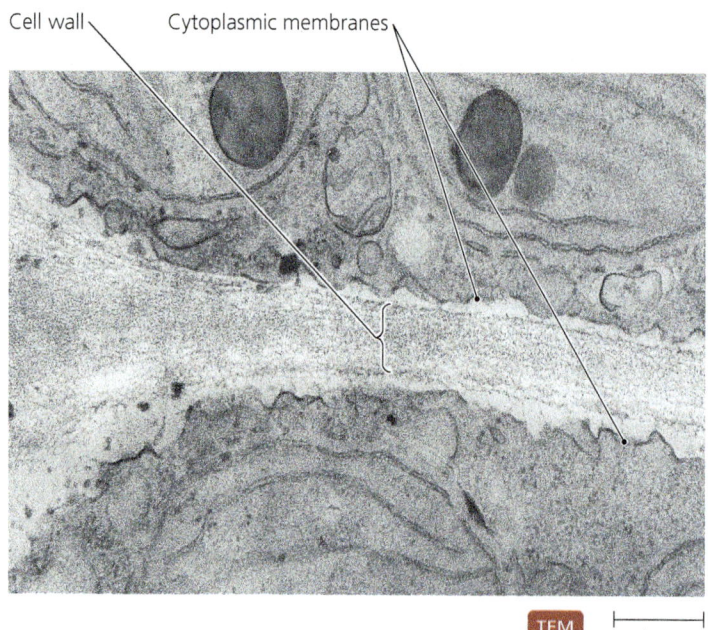

▲ FIGURE 3.29 **A eukaryotic cell wall.** The cell wall of the red alga *Gelidium* is composed of layers of the polysaccharide called agar. *What is the function of a cell wall?*

Figure 3.29 *The cell wall provides support, protection, and resistance to osmotic forces.*

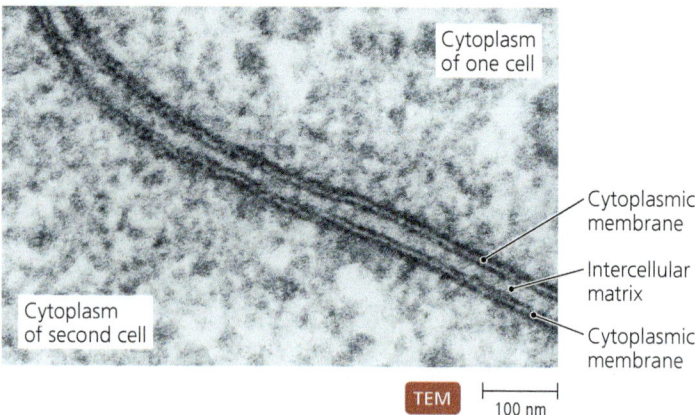

▲ FIGURE 3.30 **Eukaryotic cytoplasmic membrane.** Note that this micrograph depicts the cytoplasmic membranes of two adjoining cells.

---

[15]From Greek *glykys*, meaning "sweet," and *kalyx*, meaning "cup" or "husk."

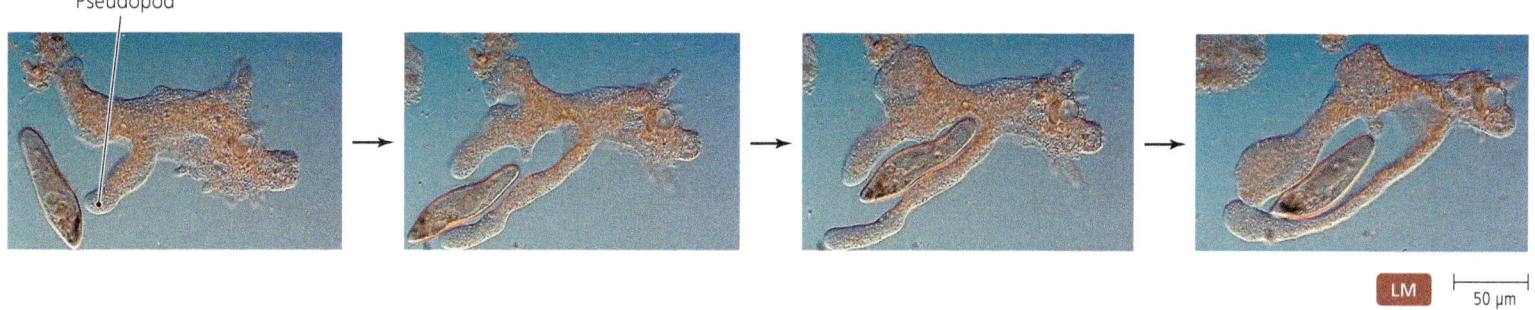

▲ FIGURE 3.31 **Endocytosis.** Pseudopods extend to surround solid and/or liquid nutrients, which become incorporated into a food vesicle inside the cytoplasm. *What is the difference between phagocytosis and pinocytosis?*

Figure 3.31 *Phagocytosis is endocytosis of a solid; pinocytosis is endocytosis of a liquid.*

more common in eukaryotes than in prokaryotes. Additionally, within multicellular organisms some membrane proteins serve to anchor cells to each other.

Eukaryotic cytoplasmic membranes may differ from prokaryotic membranes in several ways. Eukaryotic membranes contain steroid lipids (sterols), such as cholesterol in animal cells, that help maintain membrane fluidity. Paradoxically, at high temperatures sterols stabilize a phospholipid bilayer by making it less fluid, but at low temperatures sterols have the opposite effect—they prevent phospholipid packing, making the membrane more fluid.

Eukaryotic cytoplasmic membranes may contain small, distinctive assemblages of lipids and proteins that remain together in the membrane as a functional group and do not "float" independently amidst other membrane components. Such distinct regions are called **membrane rafts**. Eukaryotic cells use membrane rafts to localize cellular processes, including signaling the inside of the cell, protein sorting, and some kinds of cell movement. Some viruses, including those of AIDS, Ebola, measles, and flu, use membrane rafts to enter human cells or during viral replication. Researchers hope that blocking molecules in membrane rafts will provide a way to limit the spread of these viruses.

Eukaryotic cells frequently attach chains of sugar molecules to the outer surfaces of lipids and proteins in their cytoplasmic membranes; prokaryotes rarely do this. Sugar molecules may act in intercellular signaling, in cellular attachment, and in other roles.

Like its prokaryotic counterpart, a eukaryotic cytoplasmic membrane controls the movement of materials into and out of a cell. Eukaryotic cytoplasmic membranes use both passive processes (diffusion, facilitated diffusion, and osmosis; see Figure 3.19) and active transport (see Figure 3.22). Eukaryotic membranes do not perform group translocation, which occurs only in prokaryotes, but many perform another type of active transport—**endocytosis** (FIGURE 3.31), which involves physical manipulation of the cytoplasmic membrane around the cytoskeleton. Endocytosis occurs when the membrane distends to form **pseudopods**[16] (soo′dō-podz) that surround a substance, bringing it into the cell. Endocytosis is termed **phagocytosis** if a solid is brought into the cell and **pinocytosis** if only liquid is brought into the cell. Nutrients brought into a cell by endocytosis are then enclosed in a *food vesicle*. Vesicles and digestion of the nutrients they contain are discussed in more detail shortly. (Chapter 15 considers the process of phagocytosis as it relates to the defense of the body against disease.)

Some eukaryotes also use pseudopods as a means of locomotion. The cell extends a pseudopod, and then the cytoplasm streams into it, a process called *amoeboid action*.

Exocytosis, another solely eukaryotic process, is the reverse of endocytosis in that it enables substances to be exported from the cell. Not all eukaryotic cells can perform endocytosis or exocytosis.

TABLE 3.4 lists some of the features of endocytosis and exocytosis.

TABLE 3.4 Active Transport Processes Found Only in Eukaryotes: Endocytosis and Exocytosis

	Description	Examples of Transported Substances
Endocytosis: Phagocytosis and Pinocytosis	Substances are surrounded by pseudopods and brought into the cell. Phagocytosis involves solid substances; pinocytosis involves liquids.	Bacteria, viruses, aged and dead cells; liquid nutrients in extracellular solutions
Exocytosis	Vesicles containing substances are fused with cytoplasmic membrane, dumping their contents to the outside.	Wastes, secretions

[16]From Greek *pseudes*, meaning "false," and *podos*, meaning "foot."

> **MICRO CHECK**
>
> 25. Eukaryotic cells contain steroid-based lipids in their cytoplasmic membranes that help maintain membrane fluidity. What example of this is found in animal cells?
> 26. Some eukaryotic cells bring solid materials inside the cell by extending portions of the cell membrane. What is the term for those elongated extensions?

TELL ME WHY

Many antimicrobial drugs target bacterial cell walls. Why aren't there many drugs that act against bacterial cytoplasmic membranes?

Cytoplasm of Eukaryotes

LEARNING OUTCOMES

3.34 Compare and contrast the cytoplasm of prokaryotes and eukaryotes.
3.35 Identify nonmembranous and membranous organelles.

The gel-like **cytoplasm** of eukaryotic cells is more complex than that of either bacteria or archaea. The most distinctive difference is the presence of numerous membranous organelles in eukaryotes. However, before we discuss these membranous organelles, we will consider organelles of locomotion and other nonmembranous organelles in eukaryotes.

Flagella

LEARNING OUTCOME

3.36 Compare and contrast the structure and function of prokaryotic and eukaryotic flagella.

Structure and Arrangement

Some eukaryotic cells have whiplike extensions called **flagella**. Flagella of eukaryotes (FIGURE 3.32a) differ structurally and functionally from flagella of prokaryotes. First, eukaryotic flagella are inside the cell; the cytoplasmic membrane completely surrounds a eukaryotic flagellum. Second, the shaft of a eukaryotic flagellum is composed of molecules of a globular protein called *tubulin* arranged in chains to form hollow *microtubules*. Nine pairs of microtubules surround two microtubules in the center (FIGURE 3.32c). This "9 + 2" arrangement of microtubules is common to all flagellated eukaryotic cells, whether they are found in protozoa, algae, animals, or plants. The filaments of eukaryotic flagella are anchored in the cytoplasm by a basal body, but no hook connects the two parts as it does in prokaryotes. The basal body has *triplets* of microtubules instead of pairs, and there are no microtubules in the center, so scientists say it has a "9 + 0" arrangement of microtubules. Eukaryotic flagella may be single or multiple and are generally found at one pole of the cell.

Function

The flagella of eukaryotes also move differently from those of prokaryotes. Rather than rotating, the flagella of eukaryotes undulate rhythmically (FIGURE 3.33a). Some eukaryotic flagella push the cell through the medium (as occurs in animal sperm), whereas others pull the cell through the medium (as occurs in many protozoa). Positive and negative phototaxis and chemotaxis are seen in eukaryotic cells, but such cells do not move in runs and tumbles.

Cilia

LEARNING OUTCOMES

3.37 Describe the structure and function of cilia.
3.38 Compare and contrast eukaryotic cilia and flagella.

Other eukaryotic cells move by means of motile, hairlike structures called **cilia**, which extend the surface of the cell and are shorter and more numerous than flagella (FIGURE 3.32b). No prokaryotic cells have cilia. Like flagella, cilia are surrounded by the cytoplasmic membrane (they are inside the cell) and are composed of tubulin microtubules arranged in a "9 + 2" arrangement of pairs in their shafts and a "9 + 0" arrangement of triplets in their basal bodies.

A single cell may have hundreds or even thousands of motile cilia. Such cilia beat rhythmically, much like a swimmer doing a butterfly stroke (FIGURE 3.33b). Coordinated beating of cilia propels single-celled eukaryotes through their environment. Cilia are also used within some multicellular eukaryotes to move substances in the local environment past the surface of the cell. For example, such movement of cilia helps cleanse the human respiratory tract of dust and microorganisms.

Other Nonmembranous Organelles

LEARNING OUTCOMES

3.39 Describe the structure and function of ribosomes, cytoskeletons, and centrioles.
3.40 Compare and contrast the ribosomes of prokaryotes and eukaryotes.
3.41 List and describe the three filaments of a eukaryotic cytoskeleton.

Here we discuss three nonmembranous organelles found in eukaryotes: ribosomes and cytoskeleton (both of which are also present in prokaryotes) and centrioles (which are present only in certain kinds of eukaryotic cells).

Ribosomes

The cytosol of eukaryotes, like that of prokaryotes, is a semitransparent fluid composed primarily of water containing dissolved and suspended proteins, ions, carbohydrates, lipids, and wastes. Within the cytosol of eukaryotic cells are protein-synthesizing **ribosomes** that are larger than prokaryotic ribosomes; instead of 70S ribosomes, eukaryotic ribosomes are 80S and are composed of 60S and 40S subunits. In addition to the 80S

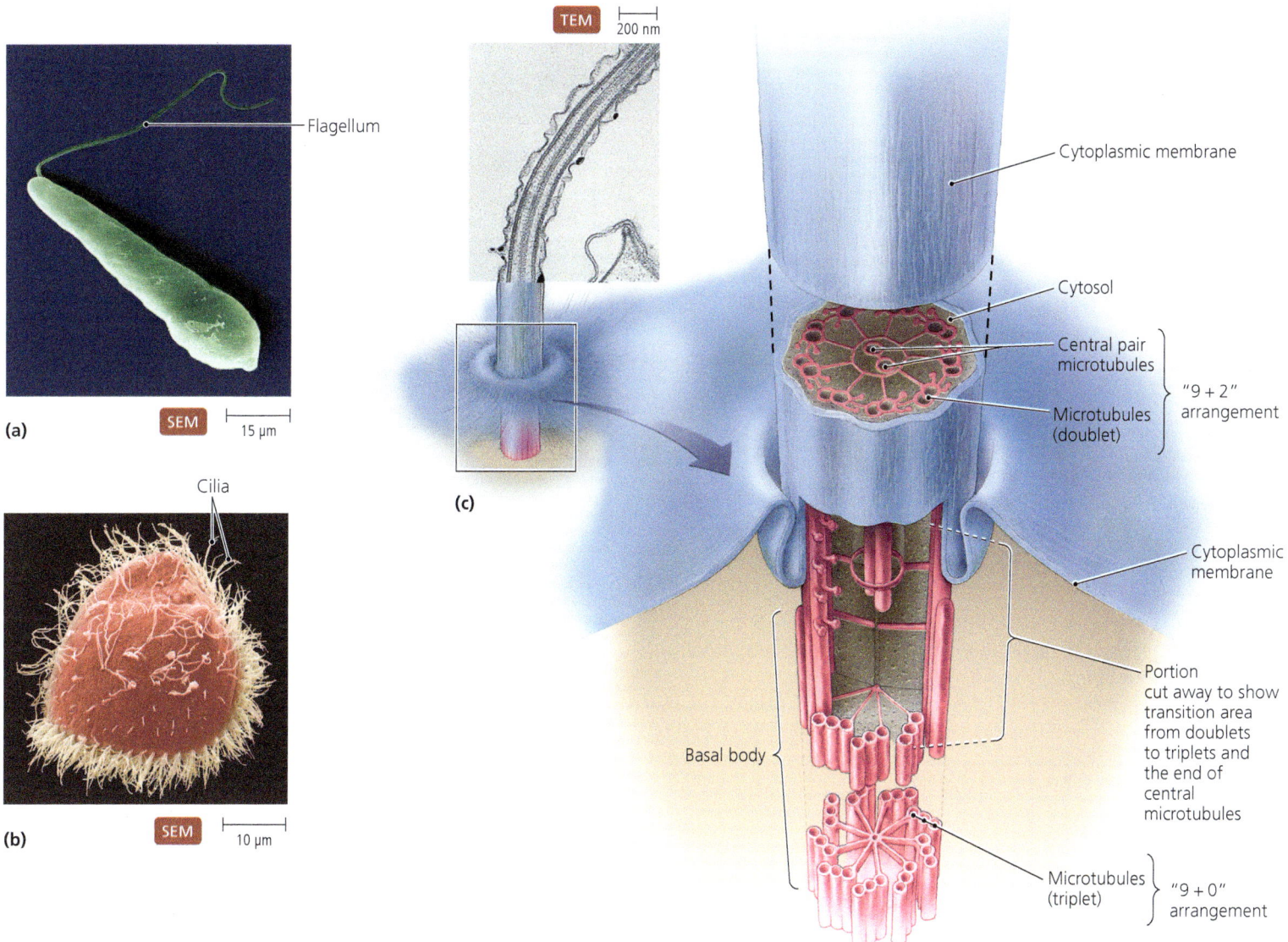

▲ **FIGURE 3.32 Eukaryotic flagella and cilia. (a)** Micrograph of *Euglena*, which possesses a single flagellum. **(b)** Scanning electron micrograph of a protozoan, *Blepharisma*, which has numerous cilia. **(c)** Details of the arrangement of microtubules of eukaryotic flagella and cilia. Both flagella and cilia have the same internal structure. *How do eukaryotic cilia differ from flagella?*

Figure 3.32 *Flagella are longer and less numerous than cilia.*

ribosomes found within the cytosol, many eukaryotic ribosomes are attached to the membranes of endoplasmic reticulum (discussed shortly).

Cytoskeleton

Eukaryotic cells contain an extensive **cytoskeleton** composed of an internal scaffolding of fibers and tubules that help maintain the basic shape of the cell. Eukaryotic cytoskeletons also act to anchor organelles, function in cytoplasmic streaming, and can move organelles within the cytosol. Cytoskeletons in some cells enable the cells to move their cytoplasmic membranes for endocytosis and amoeboid action.

Eukaryotic cytoskeletons are made up of *tubulin microtubules* (also found in flagella and cilia), thinner *microfilaments* composed of *actin,* and *intermediate filaments* composed of various proteins (**FIGURE 3.34**).

Centrioles and Centrosome

Animal cells and some fungal cells contain two **centrioles**, which lie at right angles to each other near the nucleus, in a region of the cytoplasm called the **centrosome** (**FIGURE 3.35**). Plants, algae, and most fungi lack centrioles but usually have a region of cytoplasm corresponding to a centrosome. Centrioles are composed of nine *triplets* of tubulin microtubules arranged in a way that

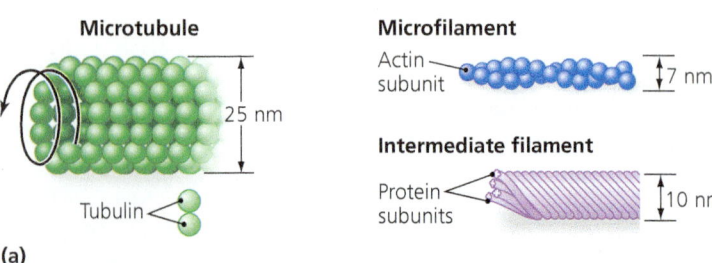

(a) Flagella

(b) Cilia

▲ **FIGURE 3.33 Movement of eukaryotic flagella and cilia. (a)** Eukaryotic flagella undulate in waves that begin at one end and traverse the length of the flagellum. **(b)** Cilia move with a power stroke followed by a return stroke. In the power stroke, a cilium is stiff; it relaxes during the return stroke. *How is the movement of eukaryotic flagella different from that of prokaryotic flagella?*

Figure 3.33 *Eukaryotic flagella undulate in a wave that moves down the flagellum; the flagella of prokaryotes rotate about the basal body.*

resembles the "9 + 0" arrangement seen at the base of eukaryotic flagella and cilia.

Centrosomes play a role in *mitosis* (nuclear division), *cytokinesis* (cell division), and the formation of flagella and cilia. However, because many eukaryotic cells that lack centrioles, such as

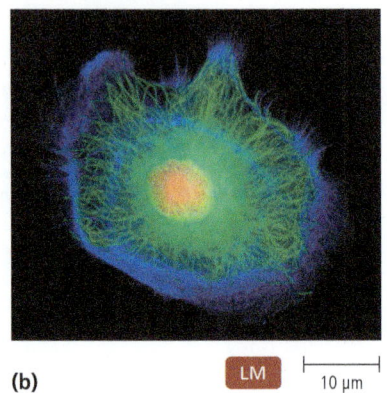

▲ **FIGURE 3.34 Eukaryotic cytoskeleton.** The cytoskeleton of eukaryotic cells serves to anchor organelles, provides a "track" for the movement of organelles throughout the cell, and provides shape to animal cells. Eukaryotic cytoskeletons are composed of microtubules, microfilaments, and intermediate filaments. **(a)** Artist's rendition of cytoskeleton filaments. **(b)** Various elements of the cytoskeleton shown here have been stained with different fluorescent dyes. DNA in the nucleus is stained yellow-orange.

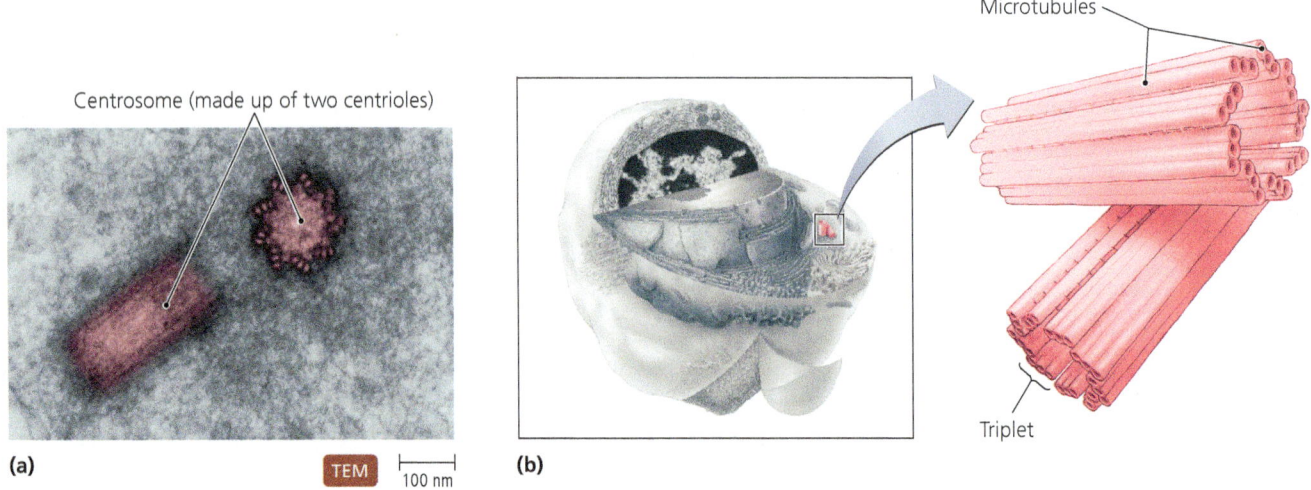

▲ **FIGURE 3.35 Centrosome.** A centrosome is a region of cytoplasm that in animal cells contains two centrioles at right angles to one another; each centriole has nine triplets of microtubules. **(a)** Transmission electron micrograph of centrosome and centrioles. **(b)** Artist's rendition of a centrosome. *How do centrioles compare with the basal body and shafts of eukaryotic flagella and cilia (see Figure 3.32c)?*

Figure 3.35 *Centrioles have the same "9 + 0" arrangement of microtubules that is found in the basal bodies of eukaryotic cilia and flagella.*

TABLE 3.5 Nonmembranous and Membranous Organelles of Cells

	General Function	Prokaryotes	Eukaryotes
Nonmembranous Organelles			
Ribosomes	Protein synthesis	Present in all	Present in all
Cytoskeleton	Shape in prokaryotes; support, cytoplasmic streaming, and endocytosis in eukaryotes	Present in some	Present in all
Centrosome	Appears to play a role in mitosis, cytokinesis, and formation of flagella and cilia in animal cells	Absent in all	Present in animals
Membranous Organelles	Sequester chemical reactions within the cell		
Nucleus	"Control center" of the cell	Absent in all	Present in all
Endoplasmic reticulum	Transport within the cell; lipid synthesis	Absent in all	Present in all
Golgi bodies	Exocytosis; secretion	Absent in all	Present in some
Lysosomes	Breakdown of nutrients; self-destruction of damaged or aged cells	Absent in all	Present in some
Peroxisomes	Neutralization of toxins	Absent in all	Present in some
Vacuoles	Storage	Absent in all	Present in some
Vesicles	Storage, digestion, transport	Absent in all	Present in all
Mitochondria	Aerobic ATP production	Absent in all	Present in most
Chloroplasts	Photosynthesis	Absent in all, though infoldings of cytoplasmic membrane called *photosynthetic lamellae* have same function in photosynthetic prokaryotes	Present in plants and algae

brown algal sperm and numerous one-celled algae, are still able to form flagella and undergo mitosis and cytokinesis, scientists are still researching the function of centrioles.

TABLE 3.5 summarizes characteristics of nonmembranous organelles of cells and contrasts them with characteristics of membranous organelles, which we consider next.

MICRO CHECK

27. The ribosomes found in bacteria are significantly different from those in eukaryotes. What is the Svedberg value for uniquely eukaryotic ribosomes?
28. The names of both the largest and smallest cytoskeletal fibers in eukaryotic cells begin with the same prefix, which (perhaps confusingly) means "small." What is the name of the largest cytoskeletal fibers?

Membranous Organelles

LEARNING OUTCOMES

3.42 Discuss the function of each of the following membranous organelles: nucleus, endoplasmic reticulum, Golgi body, lysosome, peroxisome, vesicle, vacuole, mitochondrion, and chloroplast.
3.43 Label the structures associated with each of the membranous organelles.

Eukaryotic cells contain a variety of organelles that are surrounded by phospholipid bilayer membranes similar to the cytoplasmic membrane. These membranous organelles include the nucleus, endoplasmic reticulum, Golgi body, lysosomes, peroxisomes, vacuoles, vesicles, mitochondria, and chloroplasts. Prokaryotic cells lack these structures.

Nucleus

The **nucleus** is usually spherical to ovoid and is often the largest organelle in a cell[17] (**FIGURE 3.36**). Some eukaryotic cells have a single nucleus; others are multinucleate, while still others lose their nuclei. The nucleus is often referred to as "the control center of the cell" because it contains most of the cell's genetic instructions in the form of DNA. Cells that lose their nuclei, such as mammalian red blood cells, can survive for only a few months.

Just as the semiliquid portion of the cell is called cytoplasm, the semiliquid matrix of the nucleus is called **nucleoplasm**. Within the nucleoplasm may be one or more **nucleoli** (noo-klē′ō-lī; singular: *nucleolus*), which are specialized regions where RNA is synthesized. The nucleoplasm also contains **chromatin**, which is a threadlike mass of DNA associated with special proteins called *histones* that play a role in packaging nuclear DNA. During mitosis (nuclear division), chromatin becomes visible as *chromosomes*. (Chapter 12 discusses mitosis in more detail.)

Surrounding the nucleus is a double membrane called the **nuclear envelope**, which is composed of two phospholipid bilayers, for a total of four phospholipid layers. The nuclear envelope contains **nuclear pores**, which are protein-lined channels that pierce the two membranes. Nuclear pores control the import and export of substances through the envelope.

---

[17]Historically, the nucleus was not considered an organelle because it is large and not considered part of the cytoplasm.

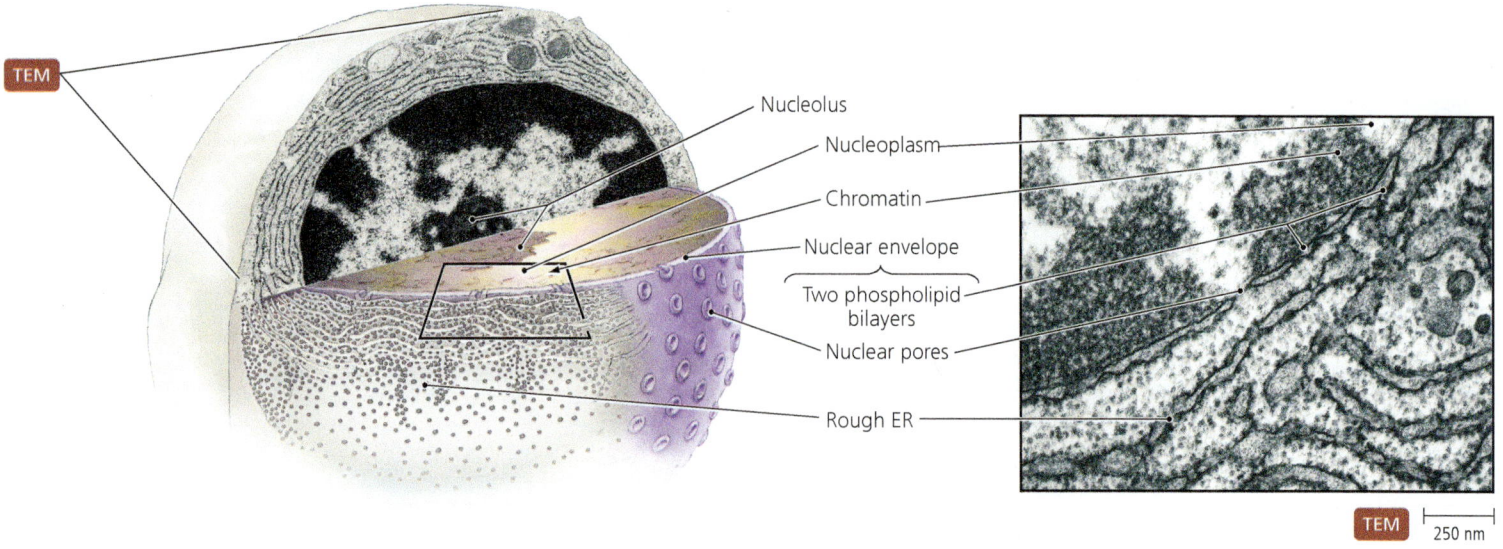

▲ FIGURE 3.36 **Eukaryotic nucleus.** Micrograph and artist's conception of a nucleus showing chromatin, a nucleolus, and the nuclear envelope. Nuclear pores punctuate the two membranes of the nuclear envelope.

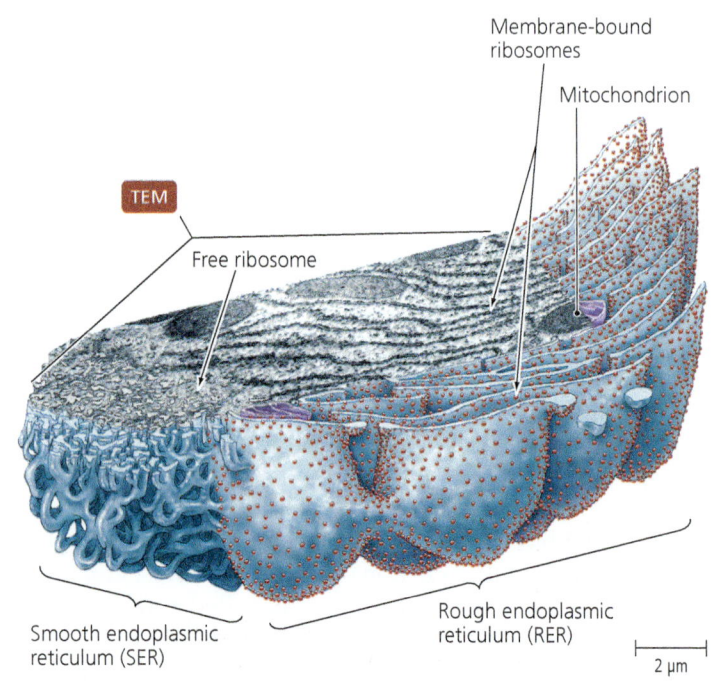

▲ FIGURE 3.37 **Endoplasmic reticulum.** ER functions in transport throughout the cell. Ribosomes are on the surface of rough ER; smooth ER lacks ribosomes.

▲ FIGURE 3.38 **Golgi body.** A Golgi body is composed of flattened sacs. Proteins synthesized by ribosomes on RER are transported via vesicles to a Golgi body. The Golgi body then modifies the proteins and sends them via secretory vesicles to the cytoplasmic membrane, where they can be secreted from the cell by exocytosis.

Endoplasmic Reticulum

Continuous with the outer membrane of the nuclear envelope is a netlike arrangement of flattened hollow tubules called **endoplasmic reticulum (ER)** (FIGURE 3.37). The ER traverses the cytoplasm of eukaryotic cells and functions as a transport system. There are two forms of ER: **smooth endoplasmic reticulum (SER)** and **rough endoplasmic reticulum (RER)**. SER plays a role in lipid synthesis as well as transport. Rough endoplasmic reticulum is rough because ribosomes adhere to its outer surface. Proteins produced by ribosomes on the RER are inserted into the lumen (central canal) of the RER and transported throughout the cell.

Golgi Body

A **Golgi body**[18] is like the "shipping department" of a cell: it receives, processes, and packages large molecules for export from the cell (FIGURE 3.38). The Golgi body packages secretions in sacs

---
[18]Camillo Golgi was an Italian histologist who first described the organelle in 1898. This organelle is also known as a *Golgi complex* or *Golgi apparatus,* and in plants and algae as a *dictyosome.*

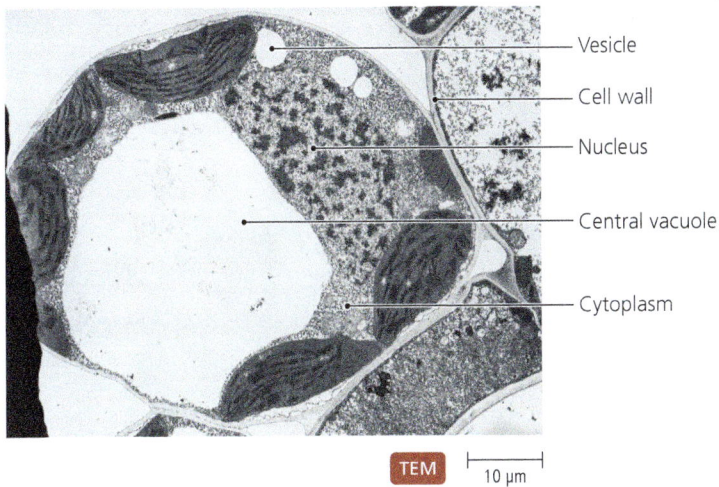

▲ FIGURE 3.39 **Vacuole.** The large central vacuole of a plant cell, which constitutes a storehouse for the cell, presses the cytoplasm against the cell wall.

called **secretory vesicles**, which then fuse with the cytoplasmic membrane before dumping their contents outside the cell via exocytosis. Golgi bodies are composed of a series of flattened hollow sacs that are circumscribed by a phospholipid bilayer. Not all eukaryotic cells contain Golgi bodies.

Lysosomes, Peroxisomes, Vacuoles, and Vesicles

Lysosomes, peroxisomes, vacuoles, and vesicles are membranous sacs that function to store and transfer chemicals within eukaryotic cells. Both **vesicle** and **vacuole** are general terms for such sacs. Large vacuoles are found in plant and algal cells that store starch, lipids, and other substances in the center of the cell. Often, this central vacuole is so large that the rest of the cytoplasm is pressed against the cell wall in a thin layer (FIGURE 3.39).

Lysosomes, which are found in animal cells, contain catabolic enzymes that damage the cell if they are released from their packaging into the cytosol. The enzymes are used during the self-destruction of old, damaged, and diseased cells and to digest nutrients that have been phagocytized. For example, white blood cells utilize the digestive enzymes in lysosomes to destroy phagocytized pathogens (FIGURE 3.40).

Peroxisomes are vesicles derived from ER. They contain *oxidase* and *catalase*, which are enzymes that degrade poisonous metabolic wastes (such as free radicals and hydrogen peroxide) resulting from some oxygen-dependent reactions. Peroxisomes are found in all types of eukaryotic cells but are especially prominent in the kidney and liver cells of mammals.

Mitochondria

Mitochondria are spherical to elongated structures found in most eukaryotic cells (FIGURE 3.41). Like nuclei, they have two membranes, each composed of a phospholipid bilayer, making four layers in total. The inner bilayer forms numerous folds called *cristae* that increase the inner membrane's surface area. Mitochondria are often called the "powerhouses of the cell" because their cristae produce most of the ATP in many eukaryotic cells. (Chapter 5 discusses the chemical reactions that produce ATP in more detail.)

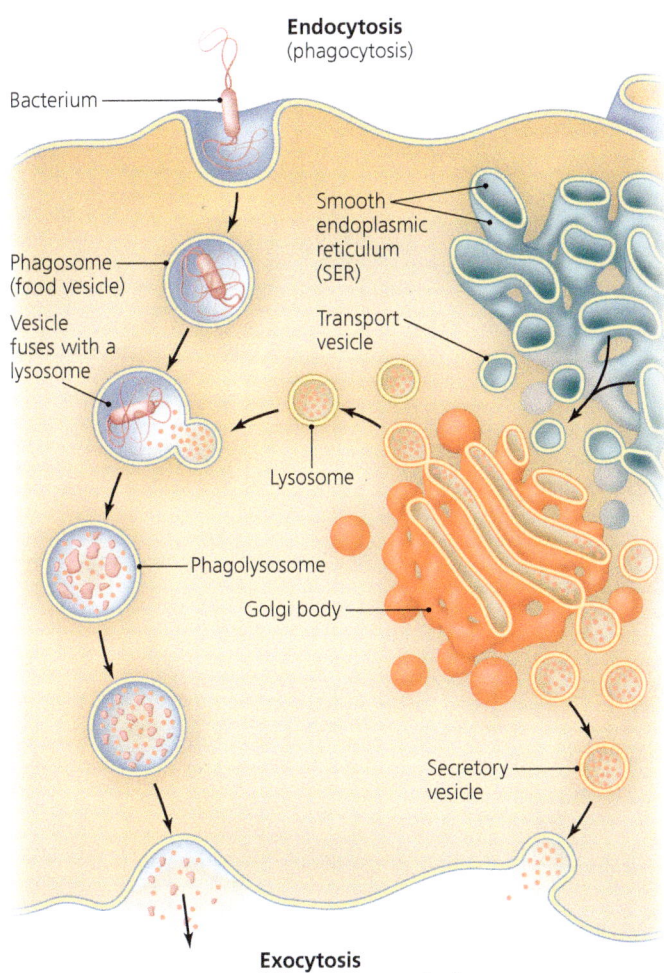

▲ FIGURE 3.40 **Roles of vesicles in endocytosis and exocytosis.** Before endocytosis, vesicles from the endoplasmic reticulum deliver digestive enzymes to a Golgi body, which then packages them into lysosomes. During endocytosis, phagocytized particles (in this case, a bacterium) are enclosed within a vesicle called a phagosome (food vesicle), which then fuses with a lysosome to form a phagolysosome vesicle. Once digestion within the phagolysosome is complete, the resulting wastes can be expelled from the cell via exocytosis. A Golgi body can also form secretory vesicles that deliver secretions outside the cell.

The interior matrix of a mitochondrion contains ("prokaryotic") 70S ribosomes and a circular molecule of DNA. This DNA contains genes for some RNA molecules and for a few mitochondrial polypeptides that are manufactured by mitochondrial ribosomes; however, most mitochondrial proteins are coded by nuclear DNA and synthesized by cytoplasmic ribosomes.

Chloroplasts

Chloroplasts are light-harvesting structures found in photosynthetic eukaryotes (FIGURE 3.42). Like mitochondria and the nucleus, chloroplasts have two phospholipid bilayer membranes and DNA. Scientists long thought chloroplast DNA was circular, but they now know that it is linear—like nuclear DNA. Like mitochondria, chloroplasts synthesize a few polypeptides with their own 70S ribosomes. The pigments of chloroplasts gather light energy to produce ATP and form sugar from carbon dioxide. Numerous membranous sacs called *thylakoids* form an extensive

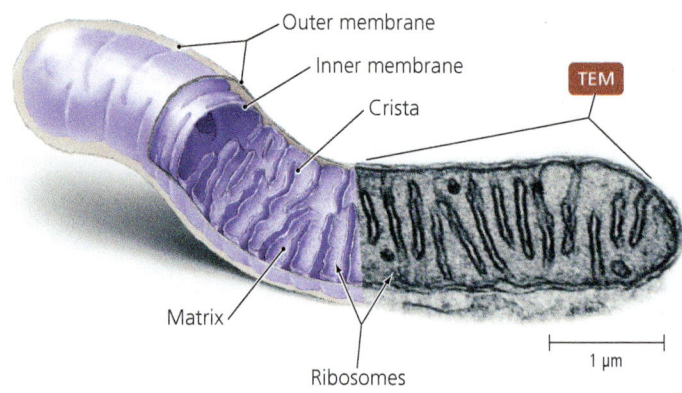

▲ FIGURE 3.41 **Mitochondrion.** Note the double membrane. The inner membrane is folded into cristae that increase its surface area. *What is the importance of the increased surface area of the inner membrane that cristae make possible?*

Figure 3.41 The chemicals involved in aerobic ATP production are located on the inner membranes of mitochondria. Increased surface area provides more space for more chemicals.

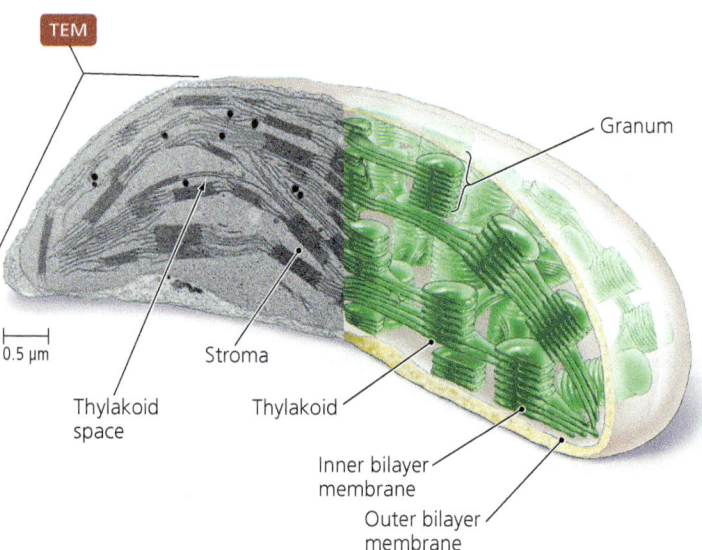

▲ FIGURE 3.42 **Chloroplast.** Chloroplasts have an ornate internal structure designed to harvest light energy for photosynthesis.

surface area with which chloroplasts photosynthesize, that is, use light energy to make sugar molecules. (Chapter 5 discusses the details of photosynthesis.)

The functions of the nonmembranous and membranous organelles, and their distribution among prokaryotic and eukaryotic cells, are summarized in Table 3.5 (on p. 85).

MICRO CHECK

29. Several organelles inside eukaryotic cells have double membranes. Which of these organelles is the "energy powerhouse"?
30. Which organelle receives, processes, and packages large molecules for export outside the cell?
31. Which animal cell organelle, other than the nucleus, contains DNA?

Endosymbiotic Theory

LEARNING OUTCOMES

3.44 Describe the endosymbiotic theory of the origin of mitochondria, chloroplasts, and eukaryotic cells.
3.45 List evidence for the endosymbiotic theory.

Mitochondria and chloroplasts are semiautonomous; that is, they divide independently of the cell but remain dependent on the cell for most of their proteins. As we have seen, both mitochondria and chloroplasts contain some DNA and 70S ribosomes, and each can produce a few polypeptides with its own ribosomes. The presence of DNA, 70S ribosomes, and two bilipid membranes in these semiautonomous organelles led scientists, such as Lynn Margulis (1938–2011), to the **endosymbiotic**[19] **theory** for the formation of eukaryotic cells. This theory suggests that eukaryotes formed when a larger anaerobic prokaryote surrounded and formed a union with a much smaller aerobic[20] prokaryote. The smaller prokaryote somehow became an internal parasite that remained surrounded by a vesicular membrane of the host.

According to the theory, the parasite eventually lost the ability to exist independently, though it retained a portion of its DNA, some ribosomes, and its cytoplasmic membrane. During the same time, the larger cell became dependent on the parasite for much of its ATP production. According to the theory, the internal, aerobic prokaryote eventually evolved into mitochondria, and its cytoplasmic membrane became cristae. A similar scenario explains the origin of chloroplasts from engulfed photosynthetic prokaryotes called *cyanobacteria*. The theory provides an explanation for the presence of 70S ribosomes and circular DNA within mitochondria and of ribosomes within chloroplasts, and it accounts for the presence of two membranes in each of the organelles.

The endosymbiotic theory does not explain the two membranes of the nuclear envelope, nor does it explain why most of the organelles' proteins come from nuclear DNA and cytoplasmic ribosomes.

TABLE 3.6 summarizes features of prokaryotic and eukaryotic cells.

MICRO CHECK

32. The endosymbiotic theory is based on the somewhat independent nature of two organelles within eukaryotic cells. What are those organelles?
33. The endosymbiotic theory explains the presence of what two things within mitochondria?

TELL ME WHY

Colchicine is a drug that inhibits microtubule formation. Why does colchicine inhibit phagocytosis, movement of organelles within the cell, and formation of flagella and cilia?

---

[19]From Greek *endo*, meaning "inside," and *symbiosis*, meaning "to live with."

[20]*Aerobic* means "requiring oxygen;" *anaerobic* is the opposite.

TABLE 3.6 Comparison of Archaeal, Bacterial, and Eukaryotic Cells

Characteristic	Archaea	Bacteria	Eukaryotes
Nucleus	Absent	Absent	Present
Free organelles bound with phospholipid membranes	Absent in all	Present in few	Various types present in all; include ER, Golgi bodies, lysosomes, mitochondria, and chloroplasts
Glycocalyx	Present	Present as organized capsule or unorganized slime layer	Present in some, e.g., surrounding some animal cells
Motility	Present in some	Present in some	Present in some; some have flagella, cilia, or pseudopods
Flagella	Some have flagella, each composed of basal body, hook, and filament; flagella rotate	Some have flagella, each composed of basal body, hook, and filament; flagella rotate	Some have flagella or cilia composed of a "9 + 2" arrangement of microtubules; flagella and cilia undulate
Cilia	Absent in all	Absent in all	Present in some
Fimbriae or pili	Present in some	Present in some	Absent in all
Hami	Present in some	Absent in all	Absent in all
Cell wall	Present in most; lack peptidoglycan	Present in most; composed of peptidoglycan	Present in plants, algae, and fungi
Cytoplasmic membrane	Present in all	Present in all	Present in all
Cytosol	Present in all	Present in all	Present in all
Inclusions	Present in most	Present in most	Present in some
Endospores	Absent in all	Present in some	Absent in all
Ribosomes	Small (70S)	Small (70S)	Large (80S) in cytosol and on ER, smaller (70S) in mitochondria and chloroplasts
Chromosomes	Commonly single and circular	Commonly single and circular	Linear and more than one chromosome per cell

MICRO IN THE CLINIC FOLLOW-UP

Kidney Infection Leading to Mental Confusion?

At the ER, Pete's vital signs are taken, his mental state is assessed, and samples of blood and urine are collected. Pete has elevated heart and respiratory rates, and his blood pressure is low. Blood work reveals elevated white blood cells (a sign of infection) and presence of *E. coli*, which is also detected in the urine sample. Pete is admitted to the hospital and started on intravenous antibiotics. The doctor explains that Pete stopped taking his antibiotics before the original infection was fully under control; this allowed *E. coli* to continue growing and move out of his urinary system (bladder, kidneys) into his bloodstream. He explains to Sara that her grandpa is so sick because he is having a systemic inflammatory response to the *E. coli*. The doctor also notes that it was good she insisted on bringing Pete into the hospital—one day longer, and Pete could have been suffering from organ failure. Pete stays in the hospital for two weeks; when he is discharged, he is still a bit weak but recovering. The doctor continues Pete on oral antibiotics for another two weeks; ultimately Pete makes a full recovery.

1. ***E. coli*** **is a Gram-negative bacterium. What cellular component is most likely to be causing Pete's systemic inflammation?**

2. **Pete's current situation developed from his original kidney infection, which probably started as a urinary tract infection. What cellular structures would help *E. coli* establish an infection in the urethra and then move into the bladder, kidneys, and bloodstream?**

Check your answers to Micro in the Clinic Follow-Up questions in the Mastering Microbiology Study Area.

Go to Mastering Microbiology for Interactive Microbiology, Dr. Bauman's Video Tutors, Micro Matters, MicroFlix, MicroBoosters, practice quizzes, and more.

CHAPTER SUMMARY

Processes of Life (p. 59)

1. All living things have some common features, including **growth**, an increase in size; **reproduction**, an increase in number; **responsiveness**, reactions to environmental stimuli; **metabolism**, controlled chemical reactions in an organism; and cellular structure.

2. Viruses are acellular and do not grow, self-reproduce, or metabolize.

Prokaryotic and Eukaryotic Cells: An Overview (pp. 59–61)

1. All cells are either **prokaryotes** or **eukaryotes**. These descriptive terms help scientists categorize organisms in groups called taxa. Generally, prokaryotic cells simultaneously read the genetic code and make proteins, and they lack a nucleus and organelles surrounded by phospholipid membranes. Domain Bacteria and domain Archaea are prokaryotic taxa.

2. Eukaryotes (domain Eukarya) have internal, membrane-bound **organelles**, including nuclei. Animals, plants, algae, fungi, and protozoa are eukaryotic.

3. Cells share common structural features. These include external structures, cell walls, cytoplasmic membranes, and cytoplasm.

External Structures of Bacterial Cells (pp. 62–66)

1. The external structures of bacterial cells include glycocalyces, flagella, fimbriae, and pili.

2. **Glycocalyces** are sticky external sheaths of cells. They may be loosely attached **slime layers** or firmly attached **capsules**. Glycocalyces prevent cells from drying out. Capsules protect cells from phagocytosis by other cells, and slime layers enable cells to stick to each other and to surfaces in their environment.

3. A prokaryotic **flagellum** is a long, whiplike protrusion of some cells that is composed of a basal body, hook, and filament. Flagella allow cells to move toward favorable conditions such as nutrients or light or to move away from unfavorable stimuli such as poisons. Play *Motility: Overview; Flagella: Structure* @ Mastering Microbiology

4. Bacterial flagella may be **polar** (single or tufts) or cover the cell (**peritrichous**). **Endoflagella**, which are special flagella of a spirochete, form an **axial filament**, located in the periplasmic space. 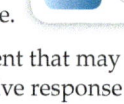 Play *Flagella: Arrangement; Spirochetes* @ Mastering Microbiology

5. **Taxis** is movement that may be either a positive response or a negative response to light (**phototaxis**) or chemicals (**chemotaxis**).  Play *Flagella: Movement* @ Mastering Microbiology

6. **Fimbriae** are extensions of some bacterial cells that function along with glycocalyces to adhere cells to one another and to environmental surfaces. A mass of such bacteria on a surface is termed a **biofilm**. Cells may also use fimbriae to pull themselves across a surface or to conduct signals to neighboring cells.

7. **Pili**, also known as *conjugation pili*, are hollow, nonmotile tubes of protein that allow bacteria to pull themselves forward and mediate the movement of DNA from one cell to another. Not all bacteria have fimbriae or pili.

Bacterial Cell Walls (pp. 66–69)

1. Most prokaryotic cells have **cell walls** that provide shape and support against osmotic pressure. Cell walls are composed primarily of polysaccharide chains.

2. Cell walls of bacteria are composed of a large, interconnected molecule of **peptidoglycan**. Peptidoglycan is composed of alternating sugar molecules called *N*-acetylglucosamine (NAG) and *N*-acetylmuramic acid (NAM).

3. A **Gram-positive** bacterial cell has a thick layer of peptidoglycan.

4. A **Gram-negative** bacterial cell has a thin layer of peptidoglycan and an external wall membrane with a **periplasmic space** between. The outer layer of this outer membrane is **lipopolysaccharide (LPS)**, which contains **lipid A**. LPS is also known as **endotoxin**. During an infection with Gram-negative bacteria, endotoxins can accumulate in the blood, causing shock, fever, and blood clotting.

5. Acid-fast bacteria have waxy lipids in their cell walls.

Bacterial Cytoplasmic Membranes (pp. 69–74)

1. A **cytoplasmic membrane** is typically composed of phospholipid molecules arranged in a double-layer configuration called a **phospholipid bilayer**. Proteins associated with the membrane vary in location and function and are able to flow laterally within the membrane. The **fluid mosaic model** is descriptive of the current understanding of membrane structure.  Play *Membrane Structure* @ Mastering Microbiology

2. The **selectively permeable** cytoplasmic membrane prevents the passage of some substances while allowing other substances to pass through protein pores or channels, sometimes requiring carrier molecules.  Play *Membrane Permeability* @ Mastering Microbiology

3. The relative concentrations of a chemical inside and outside the cell create a **concentration gradient**. Differences of electrical charges on the two sides of a membrane create an **electrical gradient** across the membrane. The gradients have a predictable effect on the passage of substances through the membrane.  Play *Passive Transport: Principles of Diffusion* @ Mastering Microbiology

4. Passive processes that move chemicals across the cytoplasmic membrane require no energy expenditure by the cell. Molecular size and concentration gradients determine the rate of simple **diffusion**. **Facilitated diffusion** depends on the electrochemical gradient and carriers within the membrane that allow certain substances to pass through the membrane. **Osmosis** specifically refers to the diffusion of water molecules across a selectively permeable membrane.

5. The concentrations of solutions can be compared. **Hypertonic** solutions have a higher concentration of solutes than **hypotonic** solutions, which have a lower concentration of solutes. Two **isotonic** solutions have the same concentrations of solutes. In biology, comparisons are usually made with the cytoplasm of cells. Play *Passive Transport: Special Types of Diffusion* @ Mastering Microbiology

6. Active transport processes require cell energy from ATP. **Active transport** moves a substance against its electrochemical gradient via carrier proteins. These carriers may move a single substance (uniport), two substances in the same direction at once (symports), or substances in opposite directions (antiports). Carrier proteins may work together in coupled transport.

7. **Group translocation** is a form of active transport in prokaryotes, during which the substance being transported is chemically altered in transit.  Play *Active Transport: Overview, Types* @ Mastering Microbiology

Cytoplasm of Bacteria (pp. 74–77)

1. **Cytoplasm** is composed of the liquid **cytosol** inside a cell plus nonmembranous organelles and inclusions. Inclusions in the cytosol are deposits of various substances.

2. Both prokaryotic and eukaryotic cells contain nonmembranous organelles.

3. The **nucleoid** is the nuclear region in prokaryotic cytosol. It has no membrane and usually contains a single circular molecule of DNA.

4. **Inclusions** include reserve deposits of lipids, starch, or compounds containing nitrogen, phosphate, or sulfur. Inclusions called gas vesicles store gases.

5. Some bacteria produce dormant, resistant **endospores** within vegetative cells.

6. **Ribosomes**, composed of protein and **ribosomal RNA (rRNA)**, are nonmembranous organelles, found in both prokaryotes and eukaryotes, that function to make proteins. The 70S ribosomes of prokaryotes are smaller than the 80S ribosomes of eukaryotes.

7. A **cytoskeleton** is a network of fibers that appears to help maintain the basic shape of cells.

External Structures of Archaea (pp. 77–78)

1. Archaea form polysaccharide and polypeptide **glycocalyces** that function in attachment and **biofilm** formation but are evidently not associated with diseases.

2. Archaeal **flagella** differ from bacterial flagella. For example, archaeal flagella are thinner than bacterial flagella.

3. Archaeal flagella rotate together as a bundle in both directions and are powered by molecules of ATP.

4. Archaea may have **fimbriae** and grappling-hook-like **hami** that serve to anchor the cells to environmental surfaces.

Archaeal Cell Walls and Cytoplasmic Membranes (p. 78-79)

1. Archaeal cell walls are composed of protein or polysaccharides but not peptidoglycan.

2. Phospholipids in archaeal cytoplasmic membranes are built with ether linkages rather than ester linkages, which occur in bacterial membranes.

Cytoplasm of Archaea (pp. 79–80)

1. Gel-like archaeal **cytoplasm** is similar to the cytoplasm of bacteria, having DNA, ribosomes, and a fibrous cytoskeleton, all suspended in the liquid cytosol.

2. 70S **ribosomes** of archaea have proteins more similar to those of eukaryotic ribosomes than to bacterial ribosomes.

External Structure of Eukaryotic Cells (p. 80)

1. Eukaryotic animal and some protozoan cells lack cell walls but have **glycocalyces** that prevent desiccation, provide support, and enable cells to stick together.

2. Eukaryotic cells that have walls lack glycocalyces.

Eukaryotic Cell Walls and Cytoplasmic Membranes (pp. 80–82)

1. Fungal, plant, algal, and some protozoan cells have cell walls composed of polysaccharides or other chemicals. Cell walls provide support, shape, and protection from osmotic forces.

2. Fungal cell walls are composed of chitin or other polysaccharides. Plant cell walls are composed of cellulose. Algal cell walls contain agar, carrageenan, algin, cellulose, or other chemicals.

3. Eukaryotic cytoplasmic membranes contain sterols such as cholesterol, which act to strengthen and solidify the membranes when temperatures rise and provide fluidity when temperatures fall.

4. **Membrane rafts** are distinct assemblages of certain lipids and proteins that remain together in the cytoplasmic membrane. Some viruses use membrane rafts during their infections of cells.

5. Some eukaryotic cells transport substances into the cytoplasm via **endocytosis**, which is an active process requiring the expenditure of energy from ATP. In endocytosis, **pseudopods**—movable extensions of the cytoplasm and membrane of the cell—surround a substance and move it into the cell. When solids are brought into the cell, endocytosis is called **phagocytosis**; the incorporation of liquids by endocytosis is called **pinocytosis**.

6. **Exocytosis** is an active export of substances out of a cell; in this process, vesicles in the cytoplasm fuse with the cytoplasmic membrane to release vesicular contents outside the cell.

Cytoplasm of Eukaryotes (pp. 82–89)

1. Eukaryotic **cytoplasm** is characterized by membranous organelles, particularly a nucleus. It also contains nonmembranous organelles and cytosol.

2. Some eukaryotic cells have long, whiplike **flagella** that differ from the flagella of prokaryotes. They have no hook, and the basal bodies and shafts are arrangements of microtubules. Further, eukaryotic flagella are internal to the cytoplasmic membrane and undulate rather than rotate.

3. Some eukaryotic cells have **cilia**, which have the same structure as eukaryotic flagella but are much shorter and more numerous. Cilia are internal to the cytoplasmic membrane.

4. The 80S **ribosomes** of eukaryotic cells are composed of 60S and 40S subunits. They are found free in the cytosol and attached to endoplasmic reticulum. The ribosomes within mitochondria and chloroplasts are 70S.

5. The eukaryotic **cytoskeleton** is composed of microtubules, intermediate filaments, and microfilaments. It provides an infrastructure and aids in movement of cytoplasm and organelles.

6. **Centrioles**, which are nonmembranous organelles in animal and some fungal cells only, are found in a region of the cytoplasm called the **centrosome** and are composed of triplets of microtubules in a "9 + 0" arrangement. Centrosomes function in the formation of flagella and cilia and in cell division.

7. The **nucleus**, a membranous structure in eukaryotic cells, contains **nucleoplasm**, in which are found one or more **nucleoli** and **chromatin**. Chromatin consists of the multiple strands of DNA and associated histone proteins that become obvious as chromosomes during mitosis. **Nuclear pores** penetrate the four phospholipid layers of the **nuclear envelope** (membrane).

8. The **endoplasmic reticulum (ER)** functions as a transport system. It can be **rough ER (RER)**, which has ribosomes on its surface, or **smooth ER (SER)**, which lacks ribosomes.

9. A **Golgi body** is a series of flattened hollow sacs surrounded by phospholipid bilayers. It packages large molecules destined for export from the cell in **secretory vesicles**, which release these molecules from the cell via exocytosis.

10. **Vesicles** and **vacuoles** are general terms for membranous sacs that store or carry substances. More specifically, **lysosomes** of animal cells contain digestive enzymes, and **peroxisomes** contain enzymes that neutralize poisonous free radicals and hydrogen peroxide.

11. Two phospholipid bilayers (four layers in total) surround **mitochondria**, site of production of ATP in a eukaryotic cell. The inner bilayer is folded into cristae, which greatly increase the surface area available for chemicals that generate ATP.

12. Photosynthetic eukaryotes possess **chloroplasts**, which are organelles containing membranous thylakoids that provide increased surface area for photosynthetic reactions.

13. The **endosymbiotic theory** explains why mitochondria and chloroplasts have 70S ribosomes and two membranes, and why mitochondria have circular DNA. The theory states that the ancestors of these organelles were prokaryotic cells that were internalized by other prokaryotes and then lost the ability to exist outside their host—thus forming early eukaryotes.

QUESTIONS FOR REVIEW

Answers to the Questions for Review (except Short Answer questions) begin on p. A-1.

Multiple Choice

1. A cell may allow a large or charged chemical to move across the cytoplasmic membrane, down the chemical's electrical and chemical gradients, in a process called _____.
   a. active transport
   b. facilitated diffusion
   c. endocytosis
   d. pinocytosis

2. Which of the following statements concerning growth and reproduction is *false*?
   a. Growth and reproduction may occur simultaneously in living organisms.
   b. A living organism must reproduce to be considered alive.
   c. Living things may stop growing and reproducing yet still be alive.
   d. Normally, living organisms have the ability to grow and reproduce themselves.

3. A "9 + 2" arrangement of microtubules is seen in _____.
   a. archaeal flagella
   b. bacterial flagella
   c. eukaryotic flagella
   d. all prokaryotic flagella

4. Which of the following is most associated with diffusion?
   a. symports
   b. antiports
   c. carrier proteins
   d. endocytosis

5. Which of the following is *not* associated with prokaryotic organisms?
   a. nucleoid
   b. glycocalyx
   c. cilia
   d. circular DNA

6. Which of the following is true of Svedbergs?
   a. They are not exact but are useful for comparisons.
   b. They are abbreviated "sv."
   c. They are prokaryotic in nature but exhibit some eukaryotic characteristics.
   d. They are an expression of sedimentation rate during high-speed centrifugation.

7. Which of the following statements is true?
   a. The cell walls of bacteria are composed of peptidoglycan.
   b. Peptidoglycan is a fatty acid.
   c. Gram-positive bacterial walls have a relatively thin layer of peptidoglycan anchored to the cytoplasmic membrane by teichoic acids.
   d. Peptidoglycan is found mainly in the cell walls of fungi, algae, and plants.

8. Which of the following is *not* a function of a glycocalyx?
   a. It forms pseudopods for faster mobility of an organism.
   b. It can protect a bacterial cell from drying out.
   c. It hides a bacterial cell from other cells.
   d. It allows a bacterium to stick to a host.

9. Bacterial flagella are _____.
   a. anchored to the cell by a basal body
   b. composed of hami
   c. surrounded by an extension of the cytoplasmic membrane
   d. composed of tubulin in hollow microtubules in a "9 + 2" arrangement

10. Which cellular structure is important in classifying a bacterial species as Gram positive or Gram negative?
    a. flagella
    b. cell wall
    c. cilia
    d. glycocalyx

11. A Gram-negative cell is moving uric acid across the cytoplasmic membrane against its chemical gradient. Which of the following statements is true?
    a. The exterior of the cell is probably electrically negative compared to the interior of the cell.
    b. The acid probably moves by a passive means such as facilitated diffusion.
    c. The acid moves by an active process such as active transport.
    d. The movement of the acid requires phagocytosis.

12. Gram-positive bacteria _____.
    a. have a thick cell wall, which retains crystal violet dye
    b. contain teichoic acids in their cell walls
    c. appear purple after Gram staining
    d. all of the above

13. Endospores _____.
    a. are reproductive structures of some bacteria
    b. occur in some archaea
    c. can cause shock, fever, and inflammation
    d. are dormant, resistant cells

14. Inclusions have been found to contain _____.
    a. DNA
    b. sulfur globules
    c. dipicolinic acid
    d. tubulin

15. Dipicolinic acid is an important component of _____.
    a. Gram-positive archaeal walls
    b. cytoplasmic membranes in eukaryotes
    c. endospores
    d. Golgi bodies

Questions for Review

Matching

1. Match the structures with their descriptions following. A letter may be used more than once or not at all, and more than one letter may be correct for each blank.

   _____ Glycocalyx
   _____ Flagella
   _____ Axial filaments
   _____ Cilia
   _____ Fimbriae
   _____ Pili
   _____ Hami

   A. Bristlelike projections found in quantities of 100 or more
   B. Long whip
   C. Responsible for conjugation
   D. "Sweet cup" composed of polysaccharides and/or polypeptides
   E. Numerous "grappling-hook" projections
   F. Responsible for motility of spirochetes
   G. Extensions not used for cell motility
   H. Made of tubulin in eukaryotes
   I. Made of flagellin in bacteria

2. Match the terms with their descriptions following. Only one description is intended for each term.

   _____ Ribosome
   _____ Cytoskeleton
   _____ Centriole
   _____ Nucleus
   _____ Mitochondrion
   _____ Chloroplast
   _____ ER
   _____ Golgi body
   _____ Peroxisome

   A. Site of protein synthesis
   B. Contains enzymes to neutralize hydrogen peroxide
   C. Functions as the transport system within a eukaryotic cell
   D. Allows contraction of the cell
   E. Site of most DNA in eukaryotes
   F. Contains microtubules in "9 + 0" arrangement
   G. Light-harvesting organelle
   H. Packages large molecules for export from a cell
   I. Its internal membranes are sites for ATP production

VISUALIZE IT!

1. Label the structures of the following prokaryotic and eukaryotic cells. With a single word or short phrase, explain the function of each structure.

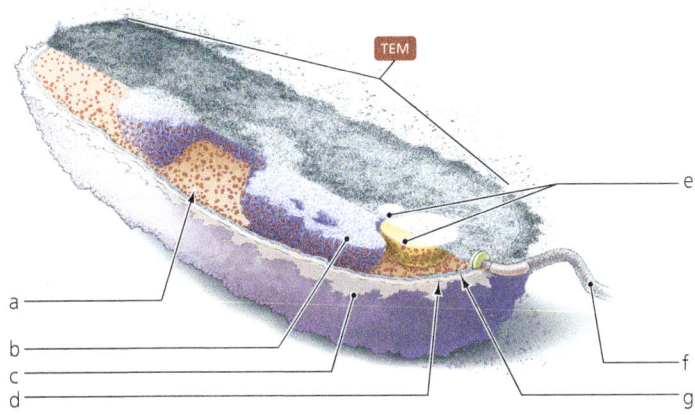
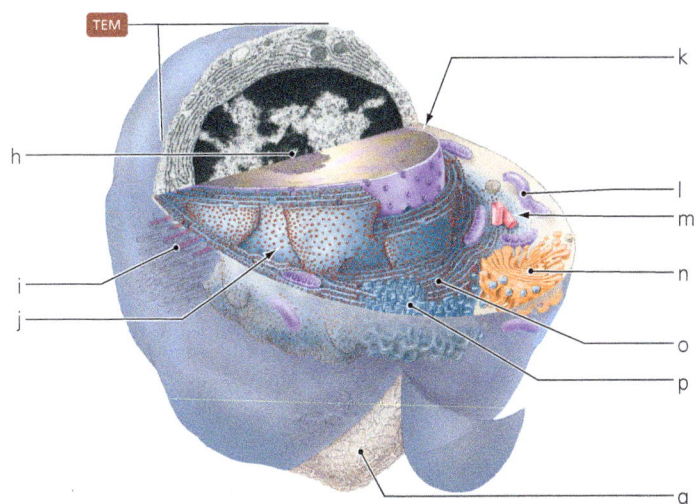

2. Label each type of flagellar arrangement.

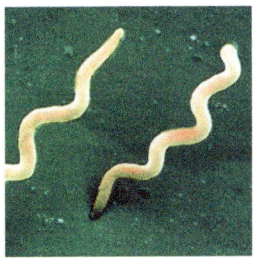
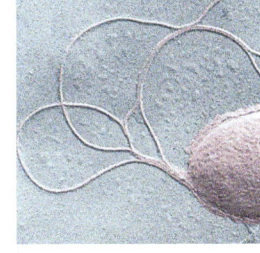
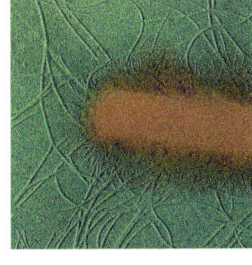
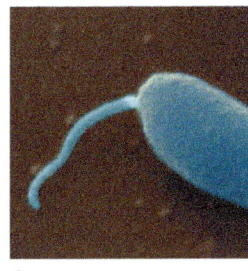

a. _____ b. _____ c. _____ d. _____

3. A scientist who is studying passive movement of chemicals across the cytoplasmic membrane of *Salmonella enterica* serotype Typhi measures the rate at which two chemicals diffuse into a cell as a function of external concentration. The results are shown in the following figure. Chemical A diffuses into the cell more rapidly than does B at lower external concentrations, but the rate levels off as the external concentration increases. The rate of diffusion of chemical B continues to increase as the external concentration increases.

   a. How can you explain the differences in the diffusion rates of chemicals A and B?
   b. Why does the diffusion rate of chemical A taper off?
   c. How could the cell increase the diffusion rate of chemical A?
   d. How could the cell increase the diffusion rate of chemical B?

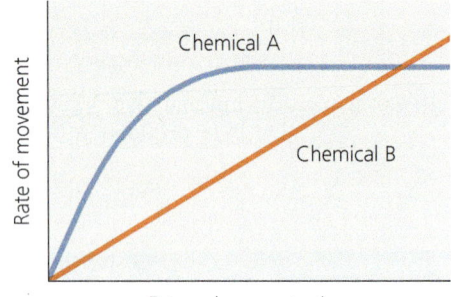

Short Answer

1. Describe (or draw) an example of diffusion down a concentration gradient.
2. Sketch, name, and describe three flagellar arrangements in bacteria.
3. Define *cytosol*.
4. The term *fluid mosaic* has been used in describing the cytoplasmic membrane. How does each word of that phrase accurately describe our current understanding of a cell membrane?
5. A local newspaper writer has contacted you, an educated microbiology student from a respected college. He wants to obtain scientific information for an article he is writing about "life" and poses the following query: "What is the difference between a living thing and a nonliving thing?" Knowing that he will edit your material to fit the article, give an intelligent, scientific response.
6. What is the difference between growth and reproduction?
7. Compare bacterial cells and algal cells, giving at least four similarities and four differences.
8. Contrast a cell of *Streptococcus pyogenes* (a bacterium) with the unicellular protozoan *Entamoeba histolytica*, listing at least eight differences.
9. Differentiate among pili, fimbriae, and cilia, using sketches and descriptive labels.
10. Can nonliving things metabolize? Explain your answer.
11. How do archaeal flagella differ from bacterial flagella and eukaryotic flagella?
12. Contrast bacterial and eukaryotic cells by filling in the following table.

Characteristic	Bacteria	Eukaryotes
Size		
Presence of nucleus		
Presence of membrane-bound organelles		
Structure of flagella		
Chemicals in cell walls		
Type of ribosomes		
Structure of chromosomes		

13. What is the function of glycocalyces and fimbriae in forming a biofilm?
14. What factors may prevent a molecule from moving across a cell membrane?
15. Compare and contrast three types of passive transport across a cell membrane.
16. Contrast the following active processes for transporting materials into or out of a cell: active transport, group translocation, endocytosis, exocytosis.
17. Contrast symports and antiports.
18. Describe the endosymbiotic theory. What evidence supports the theory? Which features of eukaryotic cells are *not* explained by the theory?

CRITICAL THINKING

1. A scientist develops a chemical that prevents Golgi bodies from functioning. Contrast the specific effects the chemical would have on human cells versus bacterial cells.
2. Methylene blue binds to DNA. What structures in a yeast cell would be stained by this dye?
3. A new chemotherapeutic drug kills bacteria but not humans. Discuss the possible ways the drug may act selectively on bacterial cells.
4. Some bacterial toxins cause cells lining the digestive tract to secrete ions, making the contents of the tract hypertonic. What effect does this have on a patient's water balance?
5. A researcher carefully inserts an electrode into an algal cell. He determines that the electrical charge across the cytoplasmic membrane is −70 millivolts. Then he slips the electrode deeper into the cell across another membrane and measures an electrical charge of −90 millivolts compared to the outside. What large organelle does the second membrane surround? Explain your answer.
6. The smallest single-celled, free-living eukaryote known is a green alga, *Ostreococcus luuris*. What membranous organelles must this photosynthetic cell have?
7. An electron micrograph of a newly discovered cell shows long projections with a basal body in the cell wall. What kind of projections are these? Is the cell prokaryotic or eukaryotic?

How will this cell behave in its environment because of these projections?

8. An entry in a recent scientific journal reports positive phototaxis in a newly described species. What condition could you create in the lab to encourage the growth of these organisms?

9. A medical microbiological lab report indicates that a sample contained a biofilm and that one species in the biofilm was identified as *Neisseria gonorrhoeae*. Is this strain of *Neisseria* likely to be pathogenic? Why or why not?

10. A researcher treats a cell to block the function of SER only. Describe the initial effects this would have on the cell.

11. After a man infected with the bacterium *Escherichia coli* was treated with the correct antibiotic for this pathogen, the bacterium was no longer found in the man's blood, but his symptoms of fever and inflammation worsened. What caused the man's response to the treatment? Why was his condition worsened by the treatment?

12. Solutions hypertonic to bacteria and fungi are used for food preservation. For instance, jams and jellies are hypertonic with sugar, and pickles are hypertonic with salt. How do hypertonic solutions kill bacteria and fungi that would otherwise spoil these foods?

13. Following the bioterrorist anthrax attacks in the fall of 2001, a news commentator suggested that people steam their mail for 30 seconds before opening it. Would the technique protect people from anthrax infections? Why or why not?

14. Eukaryotic cells are almost always larger than prokaryotic cells. What structures might allow for their larger size?

CONCEPT MAPPING

Using the terms provided, fill in this concept map that describes nucleic acids. You can also complete this and other concept maps online by going to the Mastering Microbiology Study Area.

Glycan chains
Gram-negative cell wall
Gram-positive cell wall
Lipid A

Lipopolysaccharide (LPS)
N-acetylglucosamine
Peptidoglycan
Periplasm

Porins
Teichoic acids

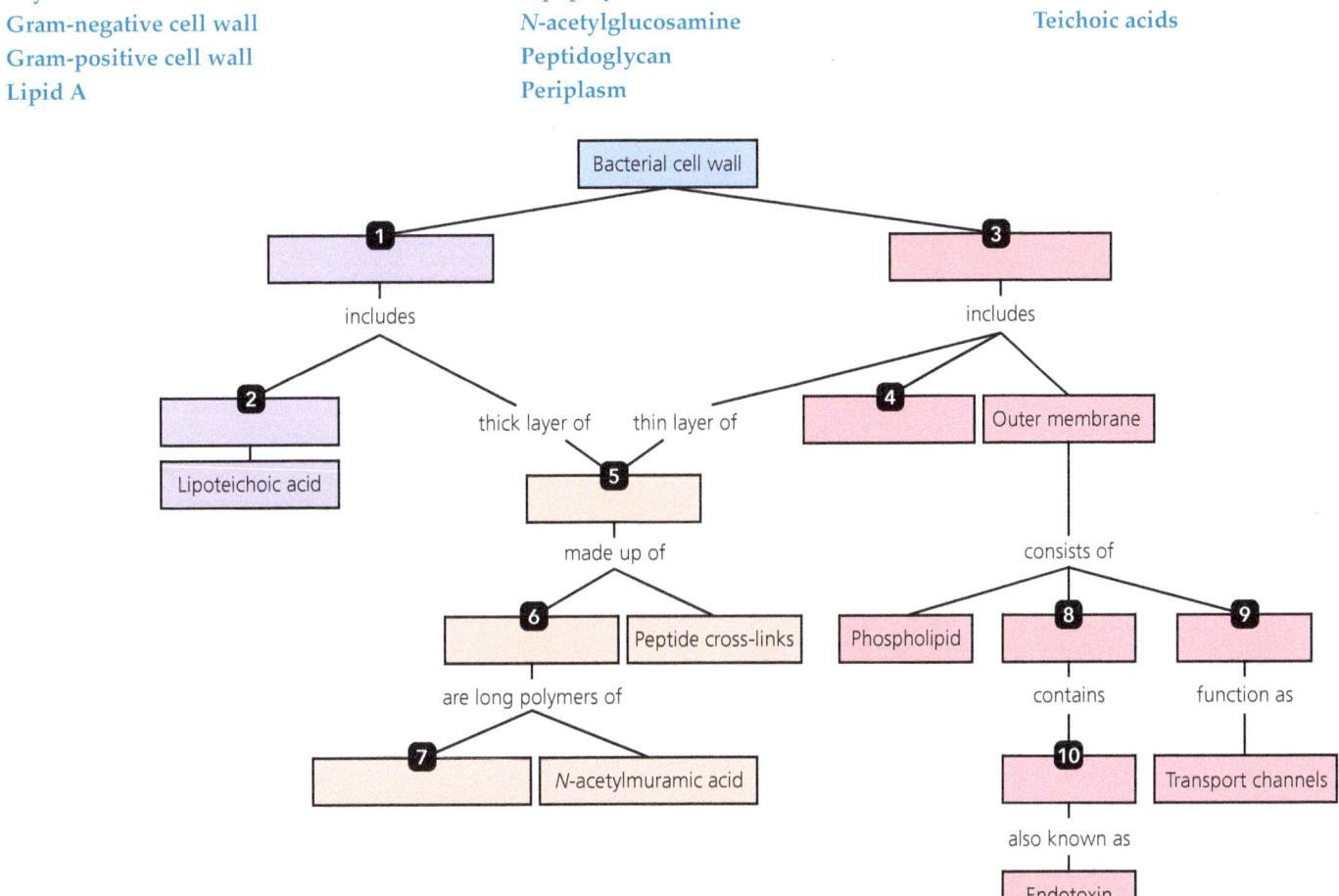

4 Microscopy, Staining, and Classification

Before You Begin

1. Consider bacteria, fungi, protozoa, and viruses. Which of these is generally smallest in size?
2. What is the main difference between the cell walls of Gram-positive and Gram-negative bacteria?
3. Among endospores, fimbriae, flagella, and inclusions, which allow bacteria to survive very harsh conditions?

Scanning electron micrograph of two strains of bacteria, isolated from a housefly's feet, growing together.

MICRO IN THE CLINIC

End of the Camping Trip

MARISOL IS A SOPHOMORE studying biology at the University of Minnesota. This has been a hard year—starting up level courses, volunteering off campus, and playing on women's softball team. While she loves all of it, Marisol really looking forward to the two-week camping trip th she and her girlfriends are taking right after finals week Finals come and go, and Marisol and her friends head the Boundary Waters Wilderness area in northern Minneso The first week of the trip is amazing—great weather, can ing, hiking, hanging out with friends. At the start of the second wee Marisol wakes up feeling really achy and tired. They've been paddli canoes for the last seven days, so she assumes that it's just mus fatigue. However, the next day she still feels even more tired and ac and she thinks that she has a slight fever. The group decides to take day off from canoeing and just hang out at their camp, giving Maris chance to rest. One of Marisol's friends notices something on the ba of Marisol's right leg. It looks like a bruise—it has a blue center, a the edges are a little red. Since they've been hiking, camping, haul gear, and climbing in and out of canoes, her friend disregards it.
 The morning after the rest day, Marisol says that she feels little bit better, but she still has the fever and aches. Her frien think they should paddle out and cut the trip short; Maris thinks she's OK to continue on the trip.

1. What do you think?
2. Are Marisol's symptoms severe enough to warrant an early end to the trip?

Turn to the end of the chapter (p. 12 to find out.

Scientists can see the marvelous microbial world through advances in microscopy. With the invention of new laboratory techniques and the construction of new instruments, biologists are still discovering wonders in the microbial world. In this chapter, we will examine some of the techniques microbiologists use to enter that world. We begin with a discussion of metric units as they relate to measuring the size of microbes. We then examine the instruments and staining techniques used in microbiology. Finally, we consider the classification schemes used to categorize the inhabitants of the microbial wonderland.

Play *Microscopy and Staining: Overview*
@ Mastering Microbiology

Units of Measurement

LEARNING OUTCOMES

4.1 Identify the two primary metric units used to measure the diameters of microbes.

4.2 List the metric units of length in order, from meter to nanometer.

Microorganisms are small. This may seem an obvious statement, but it is one that should not be taken for granted. Exactly how small are they? How can we measure the width and length of microbes?

Typically, a unit of measurement is smaller than the object being measured. For example, we measure a person's height in feet or inches, not in miles. Likewise, the diameter of a dime is measured in fractions of an inch, not in feet. So, measuring the size of a microbe requires units that are smaller than even the smallest interval on a ruler marked with English units (typically 1/16 inch). Even smaller units, such as 1/64 inch or 1/128 inch, become quite cumbersome and very difficult to use when we are dealing with microorganisms.

So that they can work with units that are simpler and in standard use the world over, scientists use metric units of measurement. Unlike the English system, the metric system is a decimal system, so each unit is one-tenth the size of the next largest unit. Even extremely small metric units are much easier to use than the fractions involved in the English system.

The unit of length in the metric system is the *meter (m)*, which is slightly longer than a yard. One-tenth of a meter is a *decimeter (dm)*, and one-hundredth of a meter is a *centimeter (cm)*, which is equivalent to about a third of an inch. One-tenth of a centimeter is a *millimeter (mm)*, which is the thickness of a dime. A millimeter is still too large to measure the size of most microorganisms, but in the metric system we continue to divide by multiples of 10 until we have a unit appropriate for use. Thus, one-thousandth of a millimeter is a *micrometer (μm)*, which is small enough to be useful in measuring the size of cells. One-thousandth of a micrometer is a *nanometer (nm)*, a unit used to measure the smallest cellular organelles and viruses. A nanometer is one-billionth of a meter.

TABLE 4.1 presents these metric units and some English equivalents. (Refer to Chapter 3, Figure 3.4, for a visual size comparison of a typical eukaryotic cell, prokaryotic cell, and virus.)

TELL ME WHY

Why do scientists use metric rather than English units?

Microscopy

LEARNING OUTCOME

4.3 Define *microscopy*.

Microscopy[1] refers to the use of light or electrons to magnify objects. The science of microbiology began when Antoni van Leeuwenhoek (lā'ven-hŭk, 1632–1723) used primitive microscopes to observe and report the existence of microorganisms. Since that time, scientists and engineers have developed a variety of light and electron microscopes.

TABLE 4.1 Metric Units of Length

Metric Unit (abbreviation)	Meaning of Prefix	Metric Equivalent	U.S. Equivalent	Representative Microbiological Application of the Unit
Meter (m)	—[a]	1 m	39.37 in (about a yard)	Length of pork tapeworm, *Taenia solium* (e.g., 1.8–8.0 m)
Decimeter (dm)	1/10	0.1 m = 10^{-1} m	3.94 in	—[b]
Centimeter (cm)	1/100	0.01 m = 10^{-2} m	0.39 in; 1 in = 2.54 cm	Diameter of a mushroom cap (e.g., 12 cm)
Millimeter (mm)	1/1000	0.001 m = 10^{-3} m	—	Diameter of a bacterial colony (e.g., 2.3 mm); length of a tick (e.g., 5.7 mm)
Micrometer (μm)	1/1,000,000	0.000001 m = 10^{-6} m	—	Diameter of white blood cells (e.g., 5–25 μm)
Nanometer (nm)	1/1,000,000,000	0.000000001 m = 10^{-9} m	—	Diameter of a poliovirus (e.g., 25 nm)

[a]The meter is the standard metric unit of length.
[b]Decimeters are rarely used.

[1]From Greek *micro*, meaning "small," and *skopein*, meaning "to view."

MICRO CHECK

1. What metric unit is one-thousandth of a millimeter and is a useful unit to use when measuring whole cells?
2. Other than light, what is used to magnify objects in microscopy?

General Principles of Microscopy

LEARNING OUTCOMES

4.4 Explain the relevance of electromagnetic radiation to microscopy.
4.5 Define *magnification*.
4.6 List and explain two factors that determine resolving power.
4.7 Discuss the relationship between contrast and staining in microscopy.

General principles involved in both light and electron microscopy include the wavelength of radiation, the magnification of an image, the resolving power of the instrument, and contrast in the specimen.

Wavelength of Radiation

Visible light is one part of a spectrum of electromagnetic radiation that includes X rays, microwaves, and radio waves (FIGURE 4.1). Note that beams of radiation may be referred to as either *rays* or *waves*. These various forms of radiation differ in **wavelength**— the distance between two corresponding parts of a wave. The human eye discriminates among different wavelengths of visible light and sends patterns of nerve impulses to the brain, which interprets the impulses as different colors. For example, we see wavelengths of 400 nm as violet and of 650 nm as red. White light, composed of many colors (wavelengths), has an average wavelength of 550 nm.

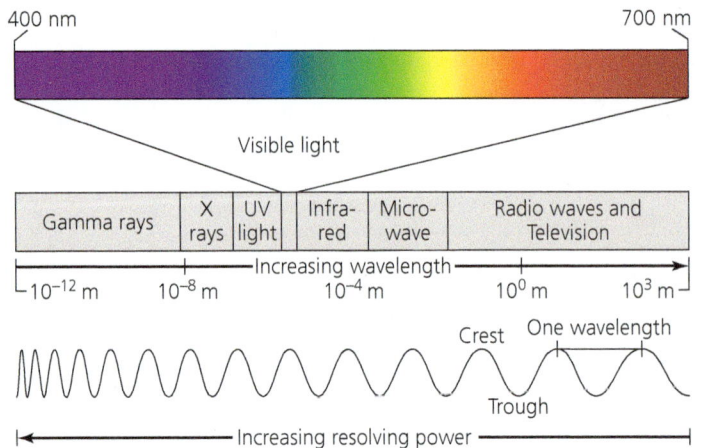

▲ **FIGURE 4.1 The electromagnetic spectrum.** Visible light is made up of a narrow band of wavelengths of radiation. Visible light and ultraviolet (UV) light are used in microscopy.

Electrons are negatively charged particles that orbit the nuclei of atoms. Besides being particulate, moving electrons also act as waves, with wavelengths dependent on the voltage of an electron beam. For example, the wavelength of electrons at 10,000 volts (V) is 0.01 nm; that of electrons at 1,000,000 V is 0.001 nm. As we will see, using radiation of smaller wavelengths results in enhanced microscopy.

Magnification

Magnification is an apparent increase in the size of an object. It is indicated by a number and ×, which is read "times." For example, 16,000× is 16,000 times. Magnification results when a beam of radiation *refracts* (bends) as it passes through a lens. Curved glass lenses refract light, and magnetic fields act as lenses to refract electron beams. Let's consider the magnifying power of a glass lens that is convex on both sides.

A lens refracts light because the lens is *optically dense* compared to the surrounding medium (such as air); that is, light travels more slowly through the lens than through air. Think of a car moving at an angle from a paved road onto a dirt shoulder. As the right front tire leaves the pavement, it has less traction and slows down. Since the other wheels continue at their original speed, the car veers toward the dirt, and the line of travel bends to the right. Likewise, the leading edge of a light beam slows as it enters glass, and the beam bends (FIGURE 4.2a). Light also bends as it leaves the glass and reenters the air.

Because of its curvature, a lens refracts light rays that pass through its periphery more than light rays that pass through its center, so that the lens focuses light rays on a *focal point*. Importantly for the purpose of microscopy, light rays spread

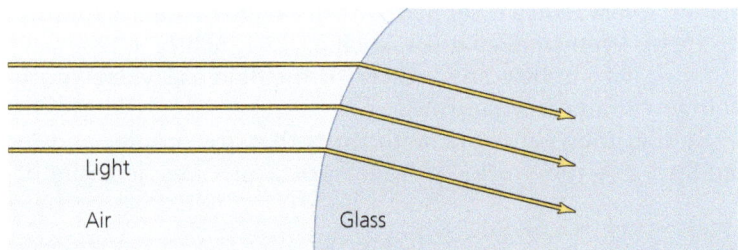

(a)

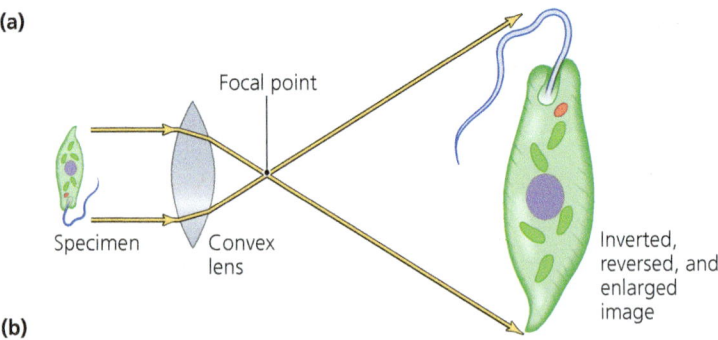

(b)

▲ **FIGURE 4.2 Light refraction and image magnification by a convex glass lens. (a)** Light passing through a lens refracts (bends) because light rays slow down as they enter the glass. Light at the leading edge of a beam that strikes the glass at an angle slows first. **(b)** A convex lens focuses light on a focal point. The image is enlarged and inverted as light rays pass the focal point and spread apart.

apart as they travel past the focal point and produce an enlarged, inverted image (FIGURE 4.2b). The degree to which the image is enlarged depends on the thickness of the lens, its curvature, and the speed of light through its substance.

Microscopists could combine lenses to obtain an image magnified millions of times, but the image would be so faint and blurry that it would be useless. Such magnification is said to be *empty magnification*. The properties that determine the clarity of an image, which in turn determines the useful magnification of a microscope, are *resolution* and *contrast*.

Resolution

Resolution, also called *resolving power*, is the ability to distinguish two points that are close together. An optometrist's eye chart is a test of resolution at a distance of 20 feet (6.1 m). Van Leeuwenhoek's microscopes had a resolving power of about 1 μm; that is, he could distinguish objects if they were more than about 1 μm apart, whereas objects closer together than 1 μm appeared as a single object. The better the resolution, the better the ability to distinguish two objects that are close to one another. Modern microscopes have fivefold better resolution than van Leeuwenhoek's; they can distinguish objects as close together as 0.2 μm. FIGURE 4.3 illustrates the size of various objects that can be resolved by the unaided human eye and by various types of microscopes.

Why do modern microscopes have better resolution than van Leeuwenhoek's microscopes? A principle of microscopy is that resolution is dependent on (1) the wavelength of the electromagnetic radiation and (2) the **numerical aperture** of the lens, which refers to the ability of a lens to gather light.

Resolution may be calculated using the following formula:

$$\text{resolution} = \frac{0.61 \times \text{wavelength}}{\text{numerical aperture}}$$

The resolution of today's microscopes is greater than that of van Leeuwenhoek's microscopes because modern microscopes use shorter wavelength radiation, such as blue light or electron beams, and because they have lenses with larger numerical apertures.

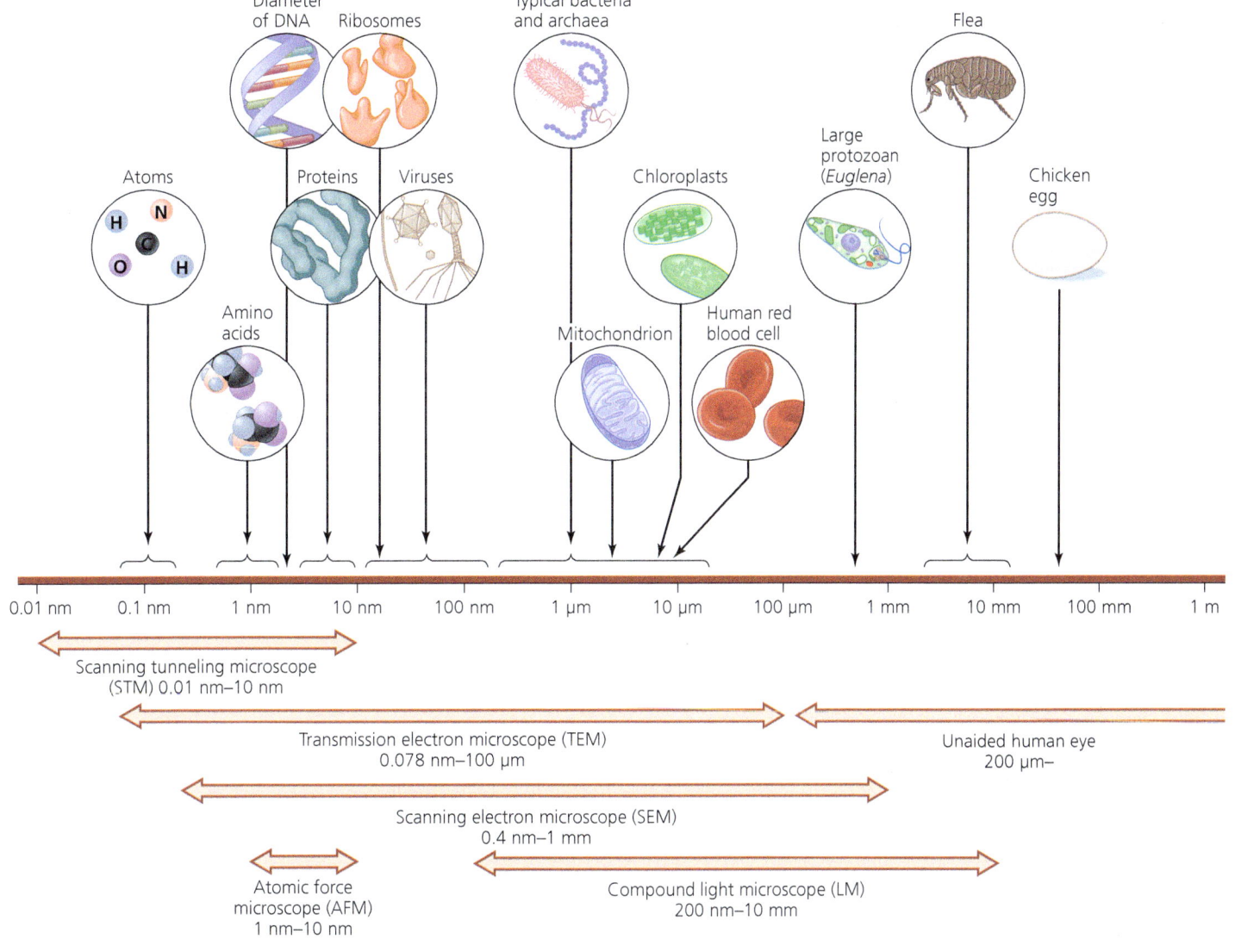

▲ FIGURE 4.3 The limits of resolution (and some representative objects within those ranges) of the human eye and of various types of microscopes.

Contrast

Contrast refers to differences in intensity between two objects or between an object and its background. Contrast is important in determining resolution. For example, although you can easily distinguish two golf balls lying side by side on a putting green 15 m away, at that distance it is much more difficult to distinguish them if they are lying on a white towel.

Most microorganisms are colorless and have very little contrast whether one uses light or electrons. One way to increase the contrast between microorganisms and their background is to stain them. Stains and staining techniques are covered later in the chapter. As we will see, the use of light that is in *phase*—that is, in which all of the waves' crests and troughs are aligned—can also enhance contrast.

> **MICRO CHECK**
>
> 3. What term best describes the ability to distinguish two separate points that are close together when observing objects through a microscope?

Light Microscopy

LEARNING OUTCOMES

4.8 Describe the difference between simple and compound microscopes.

4.9 Compare and contrast bright-field microscopy, dark-field microscopy, and phase microscopy.

4.10 Compare and contrast fluorescence and confocal microscopes.

Several classes of microscopes use various types of light to examine microscopic specimens. The most common microscopes are *bright-field microscopes*, in which the background (or *field*) is illuminated. In *dark-field microscopes*, the specimen is made to appear light against a dark background. *Phase microscopes* use the alignment or misalignment of light waves to achieve the desired contrast between a living specimen and its background. *Fluorescence microscopes* use invisible ultraviolet light to cause specimens to radiate visible light, a phenomenon called *fluorescence*. Microscopes that use lasers to illuminate fluorescent chemicals in a thin plane of a specimen are called *confocal microscopes*. Next, we examine each of these kinds of light microscope in turn.

Bright-Field Microscopes

There are two basic types of bright-field microscopes: *simple microscopes* and *compound microscopes*.

Simple Microscopes Van Leeuwenhoek first reported his observations of microorganisms using a simple microscope in 1674. A **simple microscope**, which contains a single magnifying lens, is more similar to a magnifying glass than to a modern microscope (see Figure 1.2). Though van Leeuwenhoek did not invent the microscope, he was the finest lens maker of his day and produced microscopes of exceptional quality. They were capable of approximately 300× magnification and achieved excellent clarity, far surpassing other microscopes of his time.

Compound Microscopes Simple microscopes have been replaced in modern laboratories by compound microscopes. A **compound microscope** uses a series of lenses for magnification (FIGURE 4.4a). Many scientists, including Galileo Galilei (1564–1642), made compound microscopes as early as 1590, but it was not until about 1830 that scientists developed compound microscopes that exceeded the clarity and magnification of van Leeuwenhoek's simple microscope.

In a basic compound microscope, magnification is achieved as light rays pass through a specimen and into an **objective lens**, which is the lens immediately above the object being magnified (FIGURE 4.4b). An objective lens is really a series of lenses that not only create a magnified image but also are engineered to reduce aberrations in the shape and color of the image. Most light microscopes used in biology have three or four objective lenses mounted on a **revolving nosepiece**. The objective lenses on a typical microscope are *scanning objective lens* (4×), *low-power objective lens* (10×), *high-power lens* or *high dry objective lens* (40×), and *oil immersion objective lens* (100×).

An oil immersion lens increases not only magnification but also resolution. As we have seen, light refracts as it travels from air into glass and also from glass into air; therefore, some of the light passing out of a glass slide is bent so much that it bypasses the lens (FIGURE 4.5a). Placing *immersion oil* (historically, cedarwood oil; today, more commonly a synthetic oil) between the slide and an *oil immersion objective lens* enables the lens to capture this light because light travels through immersion oil at the same speed as through glass. Because light is traveling at a uniform speed through the slide, the immersion oil, and the glass lens, it does not refract (FIGURE 4.5b). Immersion oil increases the numerical aperture, which increases resolution because more light rays are gathered into the lens to produce the image. Obviously, the space between the slide and the lens can be filled with oil only if the distance between the lens and the specimen, called the *working distance*, is small.

An objective lens bends the light rays, which then pass up through one or two **ocular lenses**, which are the lenses closest to the eyes. Microscopes with a single ocular lens are *monocular*, and those with two are *binocular*. Ocular lenses magnify the image created by the objective lens, typically another 10×.

The **total magnification** of a compound microscope is determined by multiplying the magnification of the objective lens by the magnification of the ocular lens. Thus, total magnification using a 10× ocular lens and a 10× low-power objective lens is 100×. Using the same ocular and a 100× oil immersion objective produces 1000× magnification. Some light microscopes, using higher-magnification oil immersion objective lenses and ocular lenses, can achieve 2000× magnification, but this is the limit of useful magnification for light microscopes because their resolution is restricted by the wavelength of visible light.

Modern compound microscopes also have a **condenser lens** (or lenses), which directs light through the specimen, as well as one or more mirrors or prisms that deflect the path of the light rays from an objective lens to the ocular lens (see Figure 4.4b). Some microscopes have mirrors or prisms that direct light to a

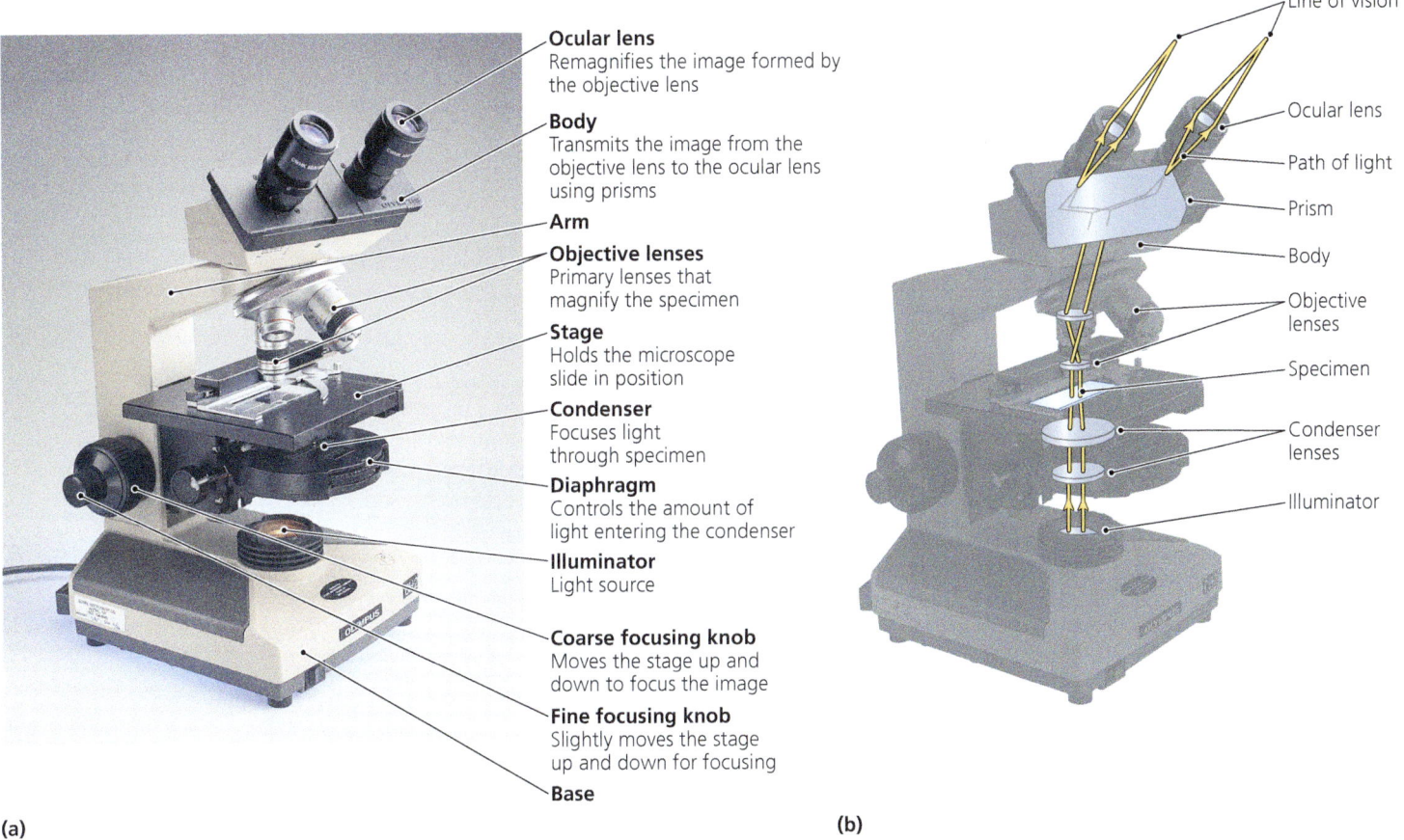

▲ FIGURE 4.4 **A bright-field compound light microscope.** (a) The parts of a compound microscope, which uses a series of lenses to produce an image at up to 2000× magnification. (b) The path of light in a compound microscope; light travels from bottom to top. *Why can't light microscopes produce clear images that are magnified 10,000×?*

Figure 4.4 Although it is possible for a light microscope to produce an image that is magnified 10,000×, magnification above 2000× is empty magnification because pairs of objects in the specimen are too close together to resolve even with the shortest-wavelength (blue) light.

camera through a special tube. A photograph of such a microscopic image is called a *light micrograph (LM)*; **micrograph** refers to any microscopic image.

Dark-Field Microscopes

Pale objects are often better observed with **dark-field microscopes**. These microscopes prevent light from directly entering the objective lens. Instead, light rays are reflected inside the condenser, so that they pass into the slide at such an oblique angle that they completely miss the objective lens. Only light rays scattered by a specimen can enter the objective lens and be seen, so the specimen appears light against a dark background—the field. Dark-field microscopy increases contrast and enables observation of more details than are visible in bright-field microscopy. Dark-field microscopes are especially useful for examining small or colorless cells.

Phase Microscopes

Scientists use **phase microscopes** to examine living microorganisms or specimens that would be damaged or altered by attaching them to slides or staining them. Basically, phase microscopes treat one set of light rays differently from another set of light rays.

When a phase microscope's lenses bring the two sets of rays together, contrast is created. There are two types of phase microscopes: phase-contrast and differential interference contrast microscopes.

Phase-Contrast Microscopes The simplest phase microscopes, **phase-contrast microscopes**, produce sharply defined images in which fine structures can be seen in living cells. These microscopes are particularly useful for observing cilia and flagella.

Differential Interference Contrast Microscopes Differential interference contrast microscopes (also called Nomarski[2] microscopes) significantly increase contrast and give an image a dramatic three-dimensional and shadowed appearance, almost as though light were striking the specimen from one side. This technique can also produce unnatural colors, which enhance contrast.

---
[2]After the French physicist Georges Nomarski, who invented the differential interference contrast microscope.

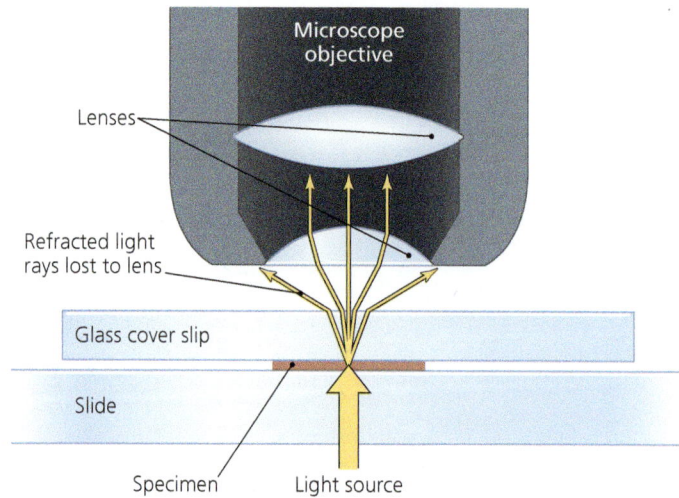

(a) Without immersion oil

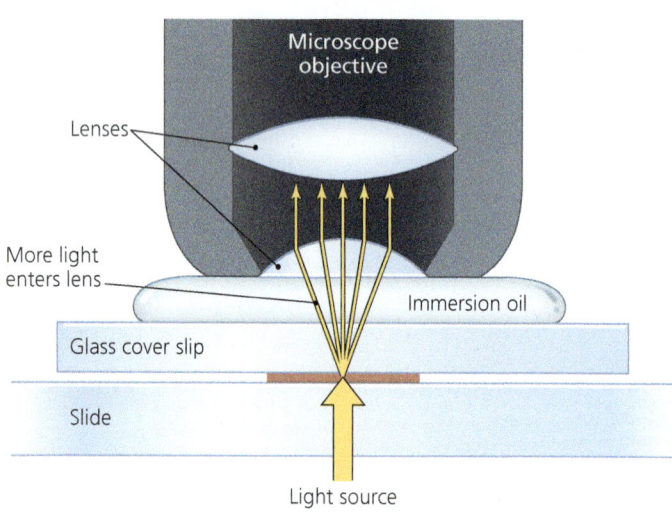

(b) With immersion oil

▲ FIGURE 4.5 **The effect of immersion oil on resolution.** (a) Without immersion oil, light is refracted as it moves from the glass cover slip into the air. Part of the scattered light misses the objective lens. (b) With immersion oil. Because light travels through the oil at the same speed as it does through glass, no light is refracted as it leaves the specimen and more light enters the lens, which increases resolution.

FIGURE 4.6 illustrates the differences that can be observed in a single specimen when viewed using four different types of light microscopy.

Fluorescence Microscopes

Molecules that absorb energy from invisible radiation (such as ultraviolet light) and then radiate the energy back as a longer, visible wavelength are said to be *fluorescent*. **Fluorescence microscopes** use an ultraviolet (UV) light source to fluoresce objects. UV light increases resolution because it has a shorter wavelength than visible light, and contrast is improved because fluorescing structures are visible against a nonfluorescing, black background.

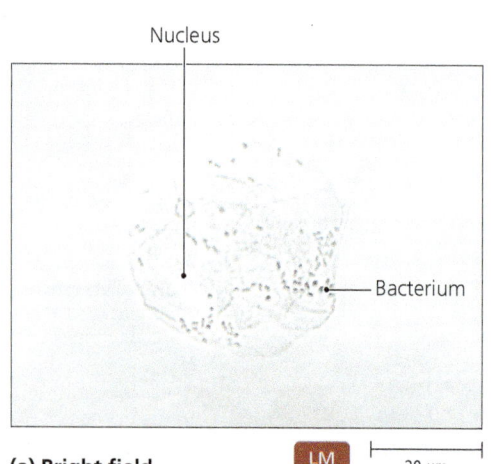

(a) Bright field

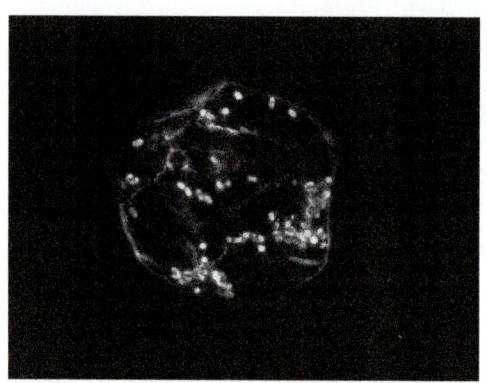

(b) Dark field

(c) Phase contrast

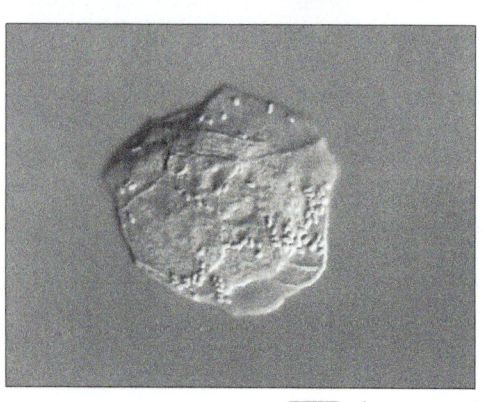

(d) Differential interference contrast (Nomarski)

◀ FIGURE 4.6 **Four kinds of light microscopy.** All four photos show the same human cheek cell and bacteria. (a) Bright-field microscopy reveals some internal structures. (b) Dark-field microscopy increases contrast between some internal structures and between the edges of the cell and the surrounding medium. (c) Phase-contrast microscopy provides greater resolution of internal structures. (d) Differential interference contrast (Nomarski) microscopy produces a three-dimensional effect.

Some cells—for example, the pathogen *Pseudomonas aeruginosa* (soo-dō-mō'nas ā-roo-ji-nō'să)—and some cellular molecules (such as chlorophyll in photosynthetic organisms) are naturally fluorescent. Other cells and cellular structures can be stained with fluorescent dyes. When these dyes are bombarded with ultraviolet light, they emit visible light and show up as bright orange, green, yellow, or other colors (see Figure 3.34b).

Some fluorescent dyes are specific for certain cells. For example, the dye fluorescein isothiocyanate attaches to cells of *Bacillus anthracis* (ba-sil'ŭs an-thrā'sis), the causative agent of anthrax, and appears apple green when viewed in a fluorescence microscope. Another fluorescent dye, auramine O, stains *Mycobacterium tuberculosis* (mī'kō-bak-tēr'ē-ŭm too-ber-kyū-lō'sis) (FIGURE 4.7).

Fluorescence microscopy is also used in a process called *immunofluorescence*. First, fluorescent dyes are chemically linked to Y-shaped immune system proteins called *antibodies* (FIGURE 4.8a). Antibodies will bind specifically to complementary-shaped *antigens*, which are portions of molecules that are present, for example, on the surface of microbial cells. When viewed under UV light, a microbial specimen that has bound dye-tagged antibodies becomes visible (FIGURE 4.8b). In addition to identifying pathogens, including those that cause syphilis, rabies, and Lyme disease, scientists use immunofluorescence to locate and make visible a variety of proteins of interest.

Confocal Microscopes

Confocal[3] microscopes also use fluorescent dyes or fluorescent antibodies, but these microscopes use ultraviolet lasers to illuminate the fluorescent chemicals in only a single plane that is no thicker than 1.0 μm; the rest of the specimen remains dark and out of focus. Visible light emitted by the dyes passes through a pinhole aperture that helps eliminate blurring that can occur with other types of microscopes and increases

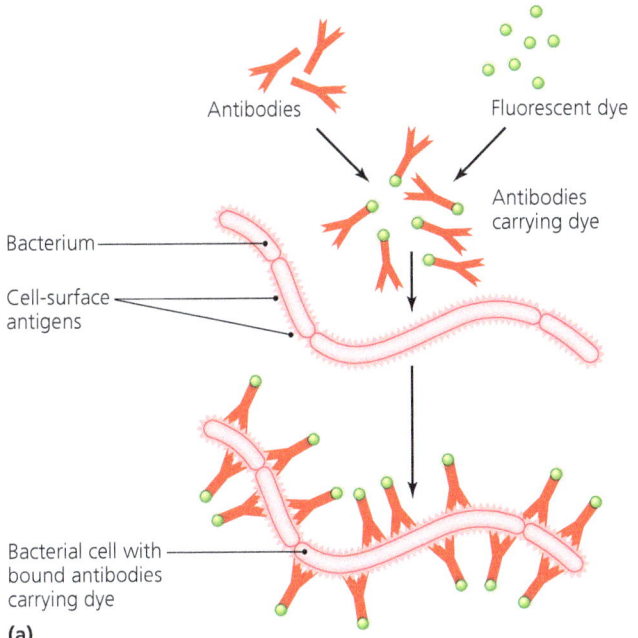

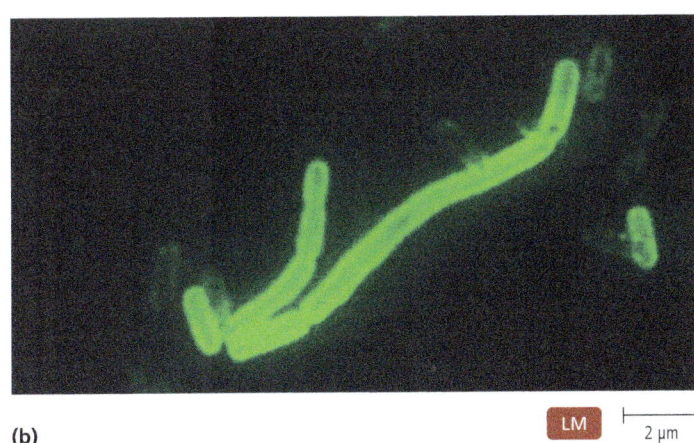

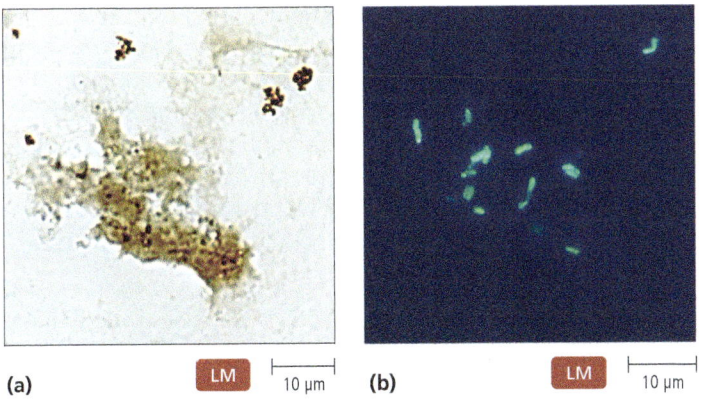

▲ **FIGURE 4.7 Fluorescence microscopy.** Fluorescent chemicals absorb invisible short-wavelength radiation and emit visible (longer-wavelength) radiation. **(a)** When viewed under normal illumination, *Mycobacterium tuberculosis* cells stained with the fluorescent dye auramine O are invisible amid the mucus and debris in a sputum smear. **(b)** When the same smear is viewed under UV light, the bacteria fluoresce and are clearly visible.

▲ **FIGURE 4.8 Immunofluorescence. (a)** After a fluorescent dye is covalently linked to an antibody, the dye-antibody combination binds to the antibody's target, making the target visible under fluorescent microscopy. **(b)** Immunofluorescent staining of *Yersinia pestis*, the causative agent of bubonic plague. The bacteria are brightly colored against a dark background.

resolution by up to 40%. Each image from a confocal microscope is thus an "optical slice" through the specimen, as if it had been thinly cut. Once individual images are digitized, a computer is used to construct a three-dimensional representation, which can be rotated and viewed from any direction (FIGURE 4.9 on the next page). Confocal microscopes have been particularly useful for examining the relationships among various organisms within complex microbial communities called *biofilms*. Regular light microscopy cannot produce clear images of structures within a living biofilm, and removing surface layers from a biofilm would change the dynamics of a biofilm community.

---

[3]From *coinciding focal* points of (laser) light.

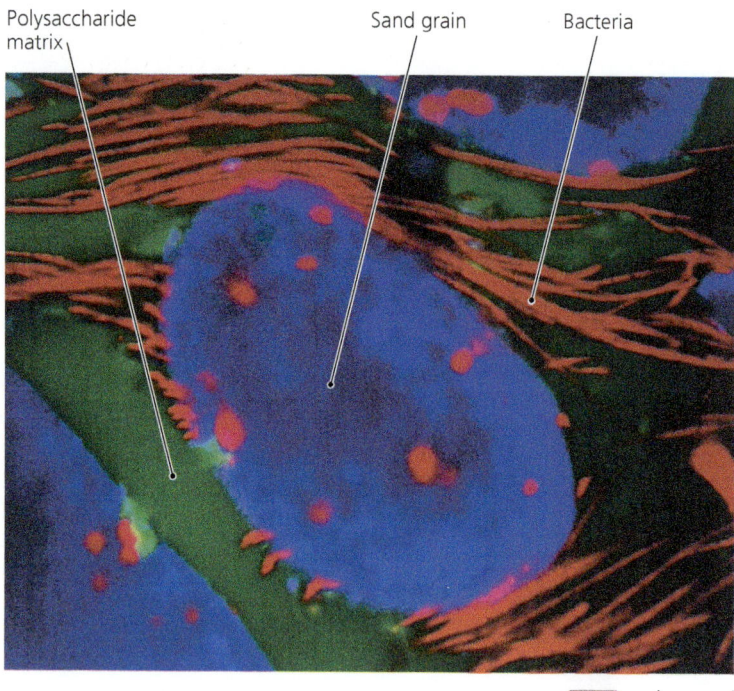

▲ FIGURE 4.9 **Confocal microscopy.** A laser is used to stimulate fluorescent dyes to produce single "optical slices." A computer then constructs a three-dimensional representation—here, a marine biofilm.

MICRO CHECK

4. What type of microscope is essentially a magnifying glass?
5. What type of light microscope uses a series of lenses that together contribute to the total magnification of a specimen?
6. Which style of microscope, used in virtually every school science lab, uses a white light source below the specimen to illuminate the specimen?

Electron Microscopy

LEARNING OUTCOME

4.11 Contrast transmission electron microscopes with scanning electron microscopes in terms of how they work, the images they produce, and the advantages of each.

Even with the most expensive phase microscope using the best oil immersion lens with the highest numerical aperture, resolution is still limited by the wavelength of visible light. Because the shortest visible radiation (violet) has a wavelength of about 400 nm, structures closer together than about 200 nm cannot be distinguished using even the best light microscope. By contrast, electrons traveling as waves can have wavelengths of 0.001 nm, which is one hundred-thousandth the wavelength of visible light. The resolving power of electron microscopes is therefore much greater than that of light microscopes, and with greater resolving power comes the possibility of greater magnification.

Generally, researchers use electron microscopes to magnify objects 10,000× to 100,000×, although millions of times magnification with good resolution is possible. Electron microscopes provide detailed views of the smallest bacteria, viruses, internal cellular structures, and even molecules and large atoms. Cellular structures that can be seen only by using electron microscopy are referred to as a cell's *ultrastructure*. Ultrastructural details cannot be made visible by light microscopy because they are too small to be resolved.

There are two general types of electron microscopes: *transmission electron microscopes* and *scanning electron microscopes*.

Transmission Electron Microscopes

A **transmission electron microscope (TEM)** generates a beam of electrons that ultimately produces an image on a fluorescent screen (FIGURE 4.10a). The path of electrons is similar to the path of light in a light microscope. From their source, the electrons pass through the specimen, through magnetic fields (instead of glass lenses) that manipulate and focus the beam, and then onto a fluorescent screen that changes some of their energy into visible light (FIGURE 4.10b). Dense areas of the specimen block electrons, resulting in a dark area on the screen. In regions where the specimen is less dense, the screen fluoresces more brightly. As with light microscopy, contrast and resolution can be enhanced through the use of electron-dense stains, which are discussed later. The brightness of each region of the screen corresponds to the number of electrons striking it. Therefore, the image on the screen is composed of light and dark areas.

The screen can be folded out of the way to enable the electrons to strike a photographic film, located in the base of the microscope. Prints made from the film are called *transmission electron micrographs* or *TEM images* (FIGURE 4.10c). Photographers often colorize such images to emphasize certain features.

Matter, including air, absorbs electrons, so the specimen must be very thin, and the column of a transmission electron microscope must be a vacuum. When dealing with thicker specimens, technicians dehydrate them, embed them in plastic, and cut them to a thickness of about 100 nm with a diamond or glass knife mounted in a slicing machine called an *ultramicrotome* so that the electron beam can pass through them. Because the vacuum and slicing of the specimens are required, transmission electron microscopes cannot be used to study living organisms.

Scanning Electron Microscopes

A **scanning electron microscope (SEM)** also uses magnetic fields within a vacuum tube to manipulate a beam of electrons; however, rather than passing electrons through a specimen, a SEM rapidly focuses the electrons back and forth across a specimen's surface, which has previously been coated with a metal such as platinum or gold. The electrons knock other electrons off the surface of the coated specimen, and these scattered electrons pass through a detector and a photomultiplier, producing an amplified signal that is displayed on a monitor. Typically, scanning microscopes are used to magnify up to 10,000× with a resolution of about 20 nm.

One advantage of scanning microscopy over transmission microscopy is that whole specimens can be observed because sectioning is not required. Scanning electron micrographs can be beautifully realistic and appear three-dimensional (FIGURE 4.11).

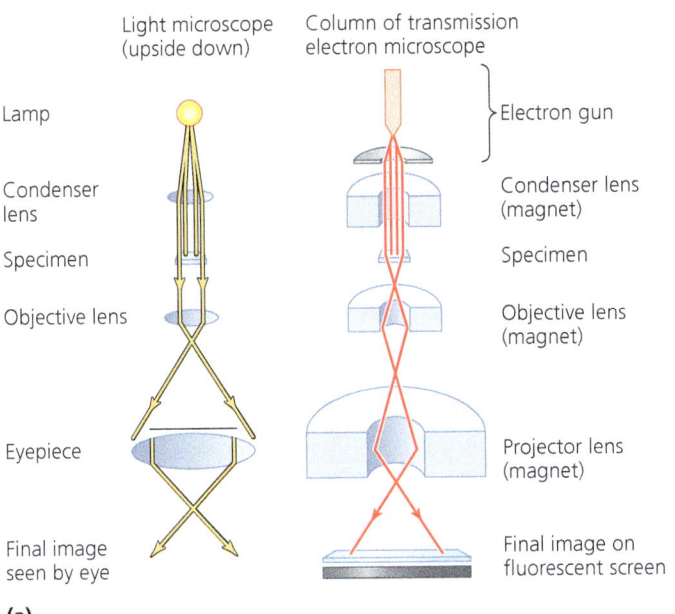
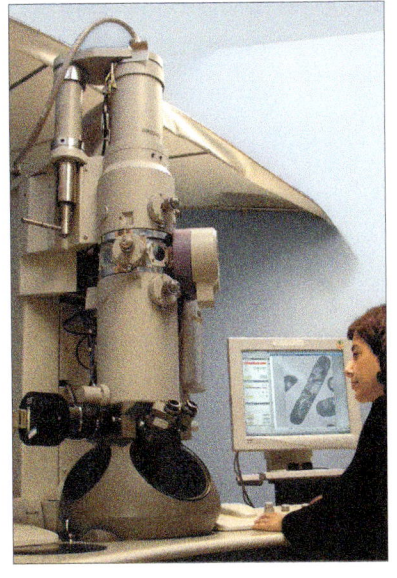
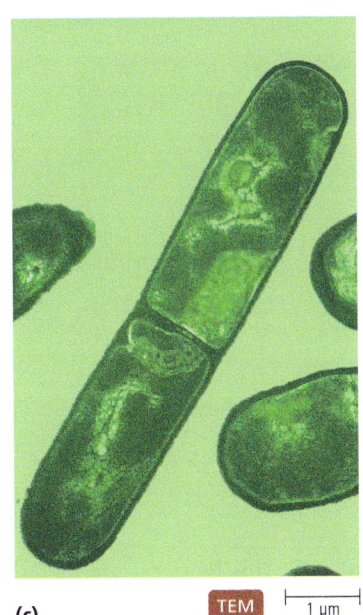

▲ **FIGURE 4.10 A transmission electron microscope (TEM).** (a) The path of electrons through a TEM, as compared to the path of light through a light microscope (at left, drawn upside down to facilitate the comparison). (b) TEMs are much larger than light microscopes. (c) Transmission electron image of a bacterium, *Bacillus subtilis*. A transmission electron micrograph reveals much internal detail not visible by light microscopy. *Why must air be evacuated from the column of an electron microscope?*

Figure 4.10 Air would absorb electrons, so there would be no radiation to produce an image.

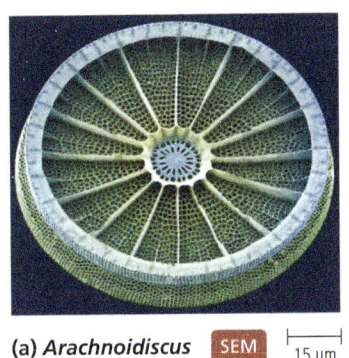

(a) *Arachnoidiscus* SEM 15 µm

(b) *Aspergillus* SEM 50 µm

(c) *Paramecium*, SEM 20 µm
a unicellular
protozoan on top of
filamentous bacteria.

(d) *Streptococcus* SEM 2 µm

▲ **FIGURE 4.11 Scanning electron microscope (SEM) images.**
(a) *Arachnoidiscus*, a marine diatom (alga). (b) *Aspergillus*, a fungus.
(c) *Paramecium*, a unicellular "animal" on top of rod-shaped bacteria.
(d) *Streptococcus*, a bacterium.

Two disadvantages of a scanning electron microscope are that it magnifies only the external surface of a specimen and that, like TEM, it requires a vacuum and thus can examine only dead organisms.

Play *Electron Microscopy*
@ Mastering Microbiology

Probe Microscopy

LEARNING OUTCOME

4.12 Describe two variations of probe microscopes.

A recent advance in microscopy utilizes minuscule, pointed electronic probes to magnify more than 100,000,000×. There are two variations of probe microscopes: *scanning tunneling microscopes* and *atomic force microscopes*.

Scanning Tunneling Microscopes

A **scanning tunneling microscope (STM)** passes a metallic probe, sharpened to end in a single atom, back and forth across and slightly above the surface of a specimen. Rather than scattering a beam of electrons into a detector, as in scanning electron microscopy, a scanning tunneling microscope measures the flow of electrons to and from the probe and the specimen's surface. The amount of electron flow, called a *tunneling current*, is directly proportional to the distance from the probe to the specimen's surface. A scanning tunneling microscope can measure distances as small as 0.01 nm and reveal three-dimensional details on the surface of a specimen at the atomic level (**FIGURE 4.12a**). A requirement for scanning tunneling microscopy is that the specimen be electrically conductive.

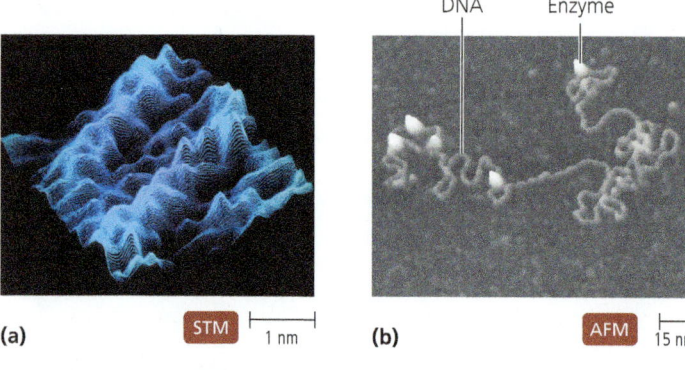

▲ FIGURE 4.12 Probe microscopy. (a) Scanning tunneling microscopes reveal surface detail; in this case, three turns of a DNA double helix. (b) Plasmid DNA being digested by an enzyme, viewed through atomic force microscopy.

Atomic Force Microscopes

An **atomic force microscope (AFM)** also uses a pointed probe, but it traverses the tip of the probe lightly on the surface of the specimen rather than at a distance. This might be likened to the way a person reads Braille. Deflection of a laser beam aimed at the probe's tip measures vertical movements, which when translated by a computer reveals the three-dimensional atomic topography.

Unlike tunneling microscopes, atomic force microscopes can magnify specimens that do not conduct electrons. They can also magnify living specimens because neither an electron beam nor a vacuum is required (FIGURE 4.12b). Researchers have used atomic force microscopes to magnify the surfaces of bacteria, viruses, proteins, and amino acids. Recent studies using them have examined single living bacteria in three dimensions as they grow and divide.

TABLE 4.2 on p. 107 summarizes the features of the various types of microscopes.

> **MICRO CHECK**
> 7. What sort of electron microscope is used in order to see the internal details of cells?

> **TELL ME WHY**
> Why is magnification high and color absent in an unretouched electron micrograph?

Staining

Earlier we discussed the difficulty of resolving two distant white golf balls viewed against a white background. If the balls were painted black, they could be distinguished more readily from the background and from one another. This illustrates why staining increases contrast and resolution.

Most microorganisms are colorless and difficult to view with bright-field microscopes. Microscopists use stains to make microorganisms and their parts more visible because stains increase contrast between structures and between a specimen and its background. Electron microscopy requires that specimens be treated with stains or coatings to enhance contrast.

In this section, we examine how scientists prepare specimens for staining and how stains work, and we consider seven kinds of stains used for light microscopy. We conclude with a look at staining for electron microscopy.

Preparing Specimens for Staining

LEARNING OUTCOME

4.13 Explain the purposes of a smear, heat fixation, and chemical fixation in the preparation of a specimen for microscopic viewing.

Many investigations of microorganisms, especially those seeking to identify pathogens, begin with light microscopic observation of stained specimens. **Staining** simply means coloring specimens with stains, which are also called dyes.

Before microbiologists stain microorganisms, they must place them on and then firmly attach them to a microscope slide. Typically, this involves making a *smear* and *fixing* it to the slide (FIGURE 4.13). If the organisms are growing in a liquid, the microscopist spreads a small drop of the broth across the surface of the slide. If the organisms are growing on a solid surface, such as an agar plate, they are mixed into a small drop of water on the slide. Either way, the thin film of organisms on the slide is called a **smear**.

The smear is air-dried completely and then attached or fixed to the surface of the slide. In **heat fixation**, developed more than a hundred years ago by Robert Koch, the slide is gently heated by passing the slide, smear up, through a flame. Alternatively, **chemical fixation** involves applying a chemical such as methyl alcohol to the smear for one minute. Desiccation (drying) and fixation kill the microorganisms, attach them firmly to the slide, and generally preserve their shape and size. It is important to smear and fix specimens properly so that they are not lost during staining.

Specimens prepared for electron microscopy must be dry because water vapor from a wet specimen would stop an electron beam. As we have seen, transmission electron microscopy requires that the desiccated sample also be sliced very thin, generally before staining. Specimens for scanning electron microscopy are coated, not stained.

> **MICRO CHECK**
> 8. What is the process by which microbes are air dried and "attached" to a microscope slide using flame?

Principles of Staining

LEARNING OUTCOME

4.14 Describe the uses of acidic and basic dyes, mentioning ionic bonding and pH.

Dyes used as microbiological stains for light microscopy are usually salts. A salt is composed of a positively charged *cation* and a negatively charged *anion*. At least one of the two ions in

TABLE 4.2 Comparison of Types of Microscopes

Type of Microscope	Typical Image	Description of Image	Special Features	Typical Uses
Light Microscopes		Useful magnification 1× to 2000×; resolution to 200 nm	Use visible light; shorter, blue wavelengths provide better resolution	
Bright-field		Colored or clear specimen against bright background	Simple to use; relatively inexpensive; stained specimens often required	To observe killed stained specimens and naturally colored live ones; also used to count microorganisms
Dark-field		Bright specimen against dark background	Uses a special filter in the condenser that prevents light from directly passing through a specimen; only light scattered by the specimen is visible	To observe living, colorless, unstained organisms
Phase-contrast		Specimen has light and dark areas	Uses a special condenser that splits a polarized light beam into two beams, one of which passes through the specimen, and one of which bypasses the specimen; the beams are then rejoined before entering the oculars; contrast in the image results from the interactions of the two beams	To observe internal structures of living microbes
Differential interference contrast (Nomarski)		Image appears three-dimensional	Uses two separate beams instead of a split beam; false color and a three-dimensional effect result from interactions of light beams and lenses; no staining required	To observe internal structures of living microbes
Fluorescence		Brightly colored fluorescent structures against dark background	An ultraviolet light source causes fluorescent natural chemicals or dyes to emit visible light	To localize specific chemicals or structures; used as an accurate and quick diagnostic tool for detection of pathogens
Confocal		Single plane of structures or cells that have been specifically stained with fluorescent dyes	Uses a laser to fluoresce only one plane of the specimen at a time	Detailed observation of structures of cells within communities
Electron Microscopes		Typical magnification 1000× to 100,000×; resolution to 0.001 nm	Use electrons traveling as waves with short wavelengths; require specimens to be in a vacuum, so cannot be used to examine living microbes	
Transmission		Monotone, two-dimensional, highly magnified images; may be color enhanced	Produces two-dimensional image of ultrastructure of cells	To observe internal ultrastructural detail of cells and observation of viruses and small bacteria
Scanning		Monotone, three-dimensional, surface images; may be color enhanced	Produces three-dimensional view of the surface of microbes and cellular structures	To observe the surface details of structures
Probe Microscopes		Magnification greater than 100,000,000× with resolving power greater than that of electron microscopes	Uses microscopic probes that move over the surface of a specimen	
Scanning tunneling		Individual molecules and atoms visible	Measures the flow of electrical current between the tip of a probe and the specimen to produce an image of the surface at atomic level	To observe the surface of objects; provide extremely fine detail, high magnification, and great resolution
Atomic force		Individual molecules and atoms visible	Measures the deflection of a laser beam aimed at the tip of a probe that travels across the surface of the specimen	To observe living specimens at the molecular and atomic levels

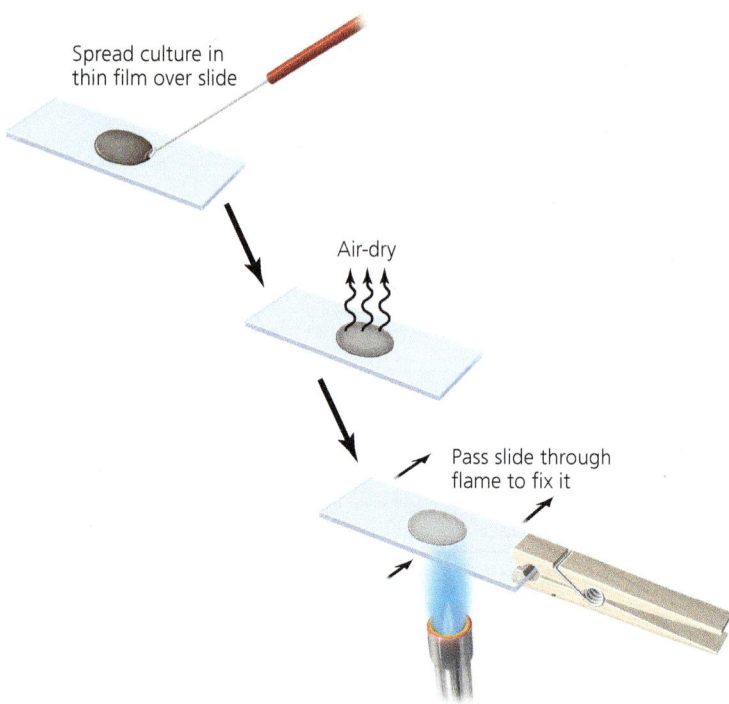

▲ FIGURE 4.13 **Preparing a specimen for staining.** Microorganisms are spread in liquid across the surface of a slide using a circular motion. After drying in the air, the smear is passed through the flame of a Bunsen burner to fix the cells to the glass. Alternatively, chemical fixation can be used. *Why must a smear be fixed to the slide?*

Figure 4.13 *Fixation causes the specimen to adhere to the glass so that it does not easily wash off during staining.*

the molecular makeup of dyes is colored; this colored portion of a dye is known as the *chromophore*. Chromophores bind to chemicals via covalent, ionic, or hydrogen bonds. For example, methylene blue chloride is composed of a cationic chromophore, methylene blue, and a chloride anion. Because methylene blue is positively charged, it ionically bonds to negatively charged molecules in cells, including DNA and many proteins. In contrast, anionic dyes, for example, eosin, bind to positively charged molecules, such as some amino acids.

Anionic chromophores are also called **acidic dyes** because they stain alkaline structures and work best in acidic (low pH) environments. Positively charged, cationic chromophores are called **basic dyes** because they combine with and stain acidic structures; further, they work best under basic (higher pH) conditions. In microbiology, basic dyes are used more commonly than acidic dyes because most cells are negatively charged. Acidic dyes are used in negative staining, which is discussed shortly.

Some stains do not form bonds with cellular chemicals but rather function because of their solubility characteristics. For example, Sudan black selectively stains membranes because it is lipid soluble and accumulates in phospholipid bilayers.

Simple Stains

LEARNING OUTCOME

4.15 Describe simple stains, four kinds of differential stains, and two kinds of special stains.

Simple stains are composed of a single basic dye, such as crystal violet, safranin, or methylene blue. They are "simple" because they involve no more than soaking the smear in the dye for 30–60 seconds and then rinsing off the slide with water. (A properly fixed specimen will remain attached to the slide despite this treatment.) After carefully blotting the slide dry, the microbiologist observes the smear under the microscope. Simple stains are used to determine size, shape, and arrangement of cells (**FIGURE 4.14**).

Differential Stains

Most stains used in microbiology are **differential stains**, which use more than one dye so that different cells, chemicals, or structures can be distinguished when microscopically examined. Common differential stains are the *Gram stain*, the *acid-fast stain*, the *endospore stain, Gomori methenamine silver stain*, and *hematoxylin and eosin stain*.

Gram Stain

In 1884, the Danish scientist Hans Christian Gram (1853–1938) developed the most frequently used differential stain, which now bears his name. The **Gram stain** differentiates between two large groups of microorganisms: purple-staining Gram-positive cells and pink-staining Gram-negative cells. These cells differ significantly in the chemical and physical structures of their cell walls (see Figure 3.16). Typically, a Gram stain is the first step a medical laboratory scientist performs to identify bacterial pathogens.

Let's examine the Gram staining procedure as it was originally developed and as it is typically performed today. For the purposes of our discussion, we'll assume that a smear has been made on a slide and heat fixed and that the smear contains both Gram-positive and Gram-negative colorless bacteria.

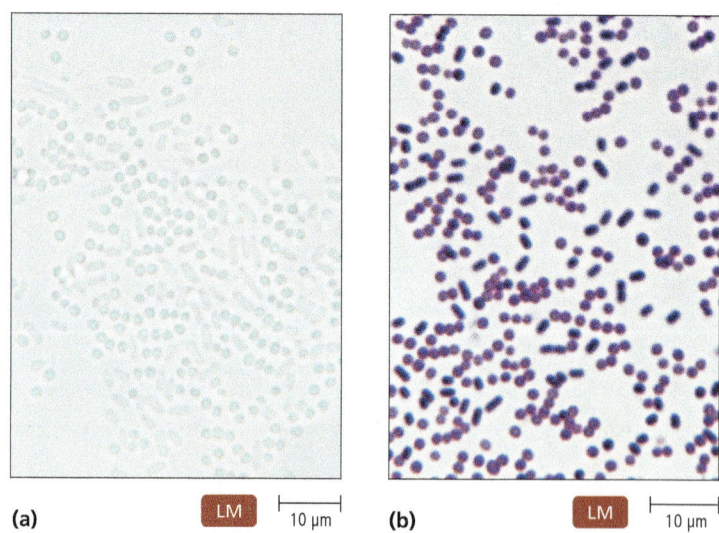

▲ FIGURE 4.14 **Simple stains.** Simply and quickly performed, simple stains increase contrast and allow determination of size, shape, and arrangement of cells. **(a)** Unstained *Escherichia coli* and *Staphylococcus aureus*. **(b)** Same mixture stained with crystal violet. Note that all cells, no matter their type, stain almost the same color with a simple stain because only one dye is used.

The classical Gram staining procedure involves the following four steps (FIGURE 4.15):

1. Flood the smear with the basic dye crystal violet for 1 minute, and then rinse with water. Crystal violet, which is called the **primary stain**, colors all cells.

2. Flood the smear with an iodine solution for 1 minute, and then rinse with water. Iodine is a **mordant**, a substance that binds to a dye and makes it less soluble. After this step, all cells remain purple.

3. Rinse the smear with a solution of ethanol and acetone for 10–30 seconds, and then rinse with water. This solution, which acts as a **decolorizing agent**, breaks down the thin cell wall of Gram-negative cells, allowing the stain and mordant to be washed away; these Gram-negative cells are now colorless. Gram-positive cells, with their thicker cell walls, remain purple.

4. Flood the smear with safranin for 1 minute, and then rinse with water. This red **counterstain** provides a contrasting color to the primary stain. Although all types of cells may absorb safranin, the resulting pink color is masked by the darker purple dye already in Gram-positive cells. After this step, Gram-negative cells now appear pink, whereas Gram-positive cells remain purple.

After the final step, the slide is blotted dry in preparation for light microscopy.

The Gram procedure works best with young cells. Older Gram-positive cells bleach more easily than younger cells and can therefore appear pink, which makes them appear to be Gram-negative cells. Therefore, smears for Gram staining should come from freshly grown bacteria.

Microscopists have developed minor variations on Gram's original procedure. For example, 95% ethanol may be used to decolorize instead of Gram's ethanol-acetone mixture. In a three-step variation, safranin dissolved in ethanol simultaneously decolorizes and counterstains.

Acid-Fast Stain

The **acid-fast stain** is another important differential stain because it stains cells of the genera *Mycobacterium* and *Nocardia* (nō-kar′dē-ă), which cause many human diseases, including tuberculosis, leprosy, and other lung and skin infections. Cells of these bacteria have large amounts of waxy lipid in their cell walls, so they do not readily stain with the water-soluble dyes used in Gram staining.

Microbiological laboratories can use a variation of the acid-fast stain developed by Franz Ziehl (1857–1926) and Friedrich Neelsen (1854–1894) in 1883. Their procedure is as follows:

1. Cover the smear with a small piece of tissue paper to retain the dye during the procedure.

2. Flood the slide with the red primary stain, carbolfuchsin, for several minutes while warming it over steaming water. In this procedure, heat is used to drive the stain through the waxy wall and into the cell, where it remains trapped.

3. Remove the tissue paper, cool the slide, and then decolorize the smear by rinsing it with a solution of hydrochloric acid (pH < 1.0) and alcohol. The bleaching action of

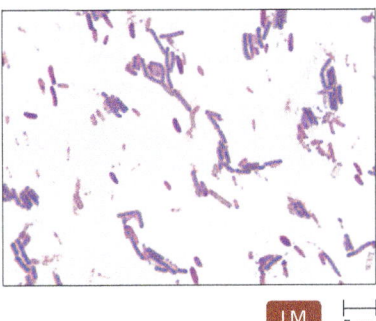

1. Slide is flooded with crystal violet for 1 min, then rinsed with water.

   Result: All cells are stained purple.

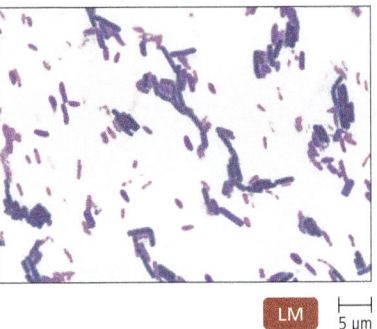

2. Slide is flooded with iodine for 1 min, then rinsed with water.

   Result: Iodine acts as a mordant; all cells remain purple.

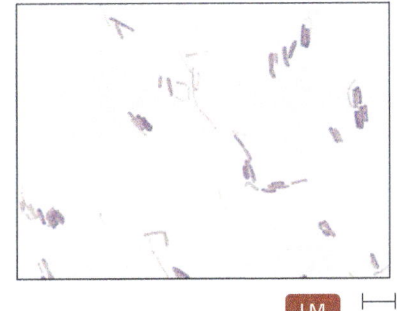

3. Slide is rinsed with solution of ethanol and acetone for 10–30 sec, then rinsed with water.

   Result: Smear is decolorized; Gram-positive cells remain purple, but Gram-negative cells are now colorless.

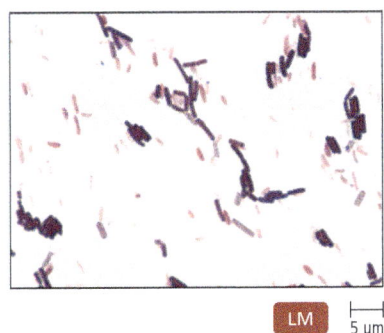

4. Slide is flooded with safranin for 1 min, then rinsed with water and blotted dry.

   Result: Gram-positive cells remain purple, Gram-negative cells are pink.

▲ **FIGURE 4.15 The Gram staining procedure.** A specimen is smeared and fixed to a slide. The classical procedure consists of four steps. Gram-positive cells (in this case, *Bacillus cereus*) remain purple throughout the procedure; Gram-negative cells (here, *Escherichia coli*) end up pink.

acid-alcohol removes color from both non-acid-fast cells and the background. Acid-fast cells retain their red color because the acid cannot penetrate the waxy wall. The name of the procedure is derived from this step; that is, the cells are colorfast in acid.

4. Counterstain with methylene blue, which stains only bleached, non-acid-fast cells.

The Ziehl-Neelsen acid-fast staining procedure results in pink acid-fast cells, which can be differentiated from blue non-acid-fast cells, including human cells (FIGURE 4.16). The presence of *acid-fast bacilli* (AFBs) in sputum is indicative of mycobacterial infection.

Endospore Stain

Some bacteria—notably those of the genera *Bacillus* and *Clostridium* (klos-trid′ē-ŭm), which contain species that cause such diseases as anthrax, gangrene, and tetanus—produce **endospores**. These dormant, highly resistant cells form inside the cytoplasm of the bacteria and can survive environmental extremes such as desiccation, heat, and harmful chemicals. Endospores cannot be stained by normal staining procedures because their walls are practically impermeable at room temperature. The **Schaeffer-Fulton endospore stain** uses heat to drive the primary stain, *malachite green*, into the endospore. After cooling, the slide is decolorized with water and counterstained with safranin. This staining procedure results in green-stained endospores and red-colored vegetative cells (FIGURE 4.17).

Histological Stains

Laboratory technicians use two popular stains to stain histological specimens, that is, tissue samples. *Gomori*[4] *methenamine silver (GMS) stain* is commonly used to screen for the presence of fungi and the locations of carbohydrates in tissues. *Hematoxylin and eosin (HE) stain*, which involves applying the basic dye hematoxylin and the acidic dye eosin, is used to delineate many features of histological specimens, such as the presence of cancer cells.

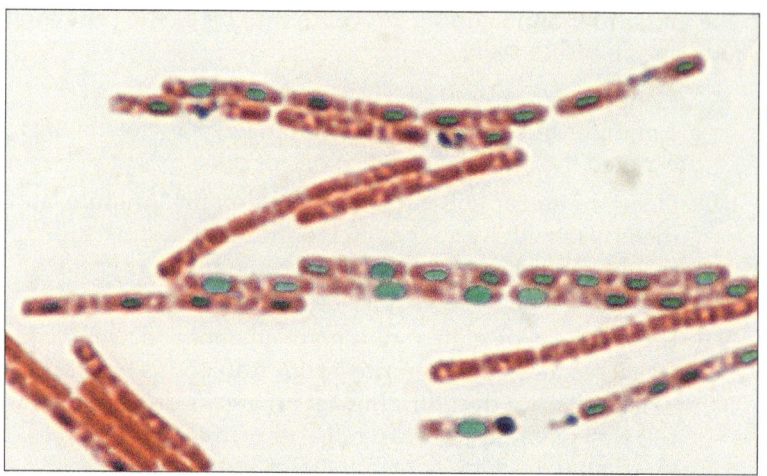

▲ **FIGURE 4.17 Schaeffer-Fulton endospore stain of *Bacillus anthracis*.** The nearly impermeable spore wall retains the green dye during decolorization. Vegetative cells, which lack spores, pick up the counterstain and appear red. *Why don't the spores stain red as well?*

Figure 4.17 *Heat from steam is used to drive the green primary stain into the endospores. Counterstaining is performed at room temperature, and the thick, impermeable walls of the endospores resist the counterstain.*

Special Stains

Special stains are simple stains designed to reveal specific microbial structures. Three types of special stains are *negative stains*, *flagellar stains*, and *fluorescent stains* (which we already discussed in the section on fluorescence microscopy).

Negative (Capsule) Stain

Most dyes used to stain bacterial cells, such as crystal violet, methylene blue, malachite green, and safranin, are basic dyes. These dyes stain cells by attaching to negatively charged molecules within them.

Acidic dyes, by contrast, are repulsed by the negative charges on the surface of cells and, therefore, do not stain them. Such stains are called **negative stains** because they stain the background and leave cells colorless. Eosin and nigrosin are examples of acidic dyes used for negative staining.

Negative stains are used primarily to reveal the presence of negatively charged bacterial capsules. Therefore, they are also called **capsule stains**. Encapsulated cells appear to have a halo surrounding them (FIGURE 4.18).

Flagellar Stain

Bacterial flagella are extremely thin and thus normally invisible with light microscopy, but their presence, number, and arrangement are important in identifying some species, including some pathogens. Flagellar stains, such as pararosaniline and carbolfuchsin, in combination with *mordants*—chemicals that combine with a dye and make it less soluble and therefore affix it in a material—are applied in a series of steps. Flagellar stains bind to flagella, increasing their diameter and colorizing them, which increases contrast and makes the flagella visible (FIGURE 4.19).

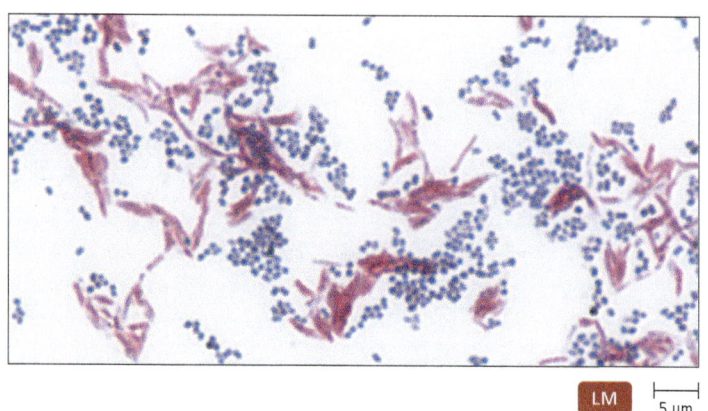

▲ **FIGURE 4.16 Ziehl-Neelsen acid-fast stain.** Acid-fast cells such as these rod-shaped *Mycobacterium bovis* cells stain pink or red. Non-acid-fast cells—in this case, *Staphylococcus*—stain blue. *Why isn't the Gram stain utilized to stain* Mycobacterium?

Figure 4.16 *Cell walls of Mycobacterium are composed of waxy materials that repel the water-based dyes of the Gram stain.*

---

[4]Named for George Gomori, a noted Hungarian American histologist.

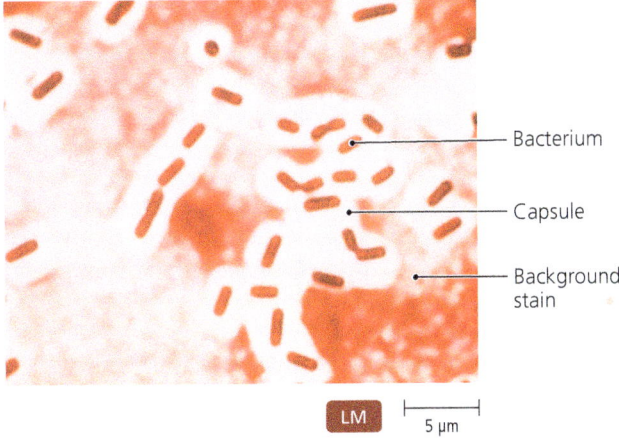

▲ FIGURE 4.18 **Negative (capsule) stain of *Klebsiella pneumoniae*.** Notice that the acidic dye stains the background and does not penetrate the capsule.

Stains used for light microscopy are summarized in TABLE 4.3. **Beneficial Microbes: Glowing Viruses** (on p. 113) illustrates a unique kind of stain that uses fluorescent dye attached to viruses to stain specific strains of bacteria.

Staining for Electron Microscopy

LEARNING OUTCOME

4.16 Explain how stains used for electron microscopy differ from those used for light microscopy.

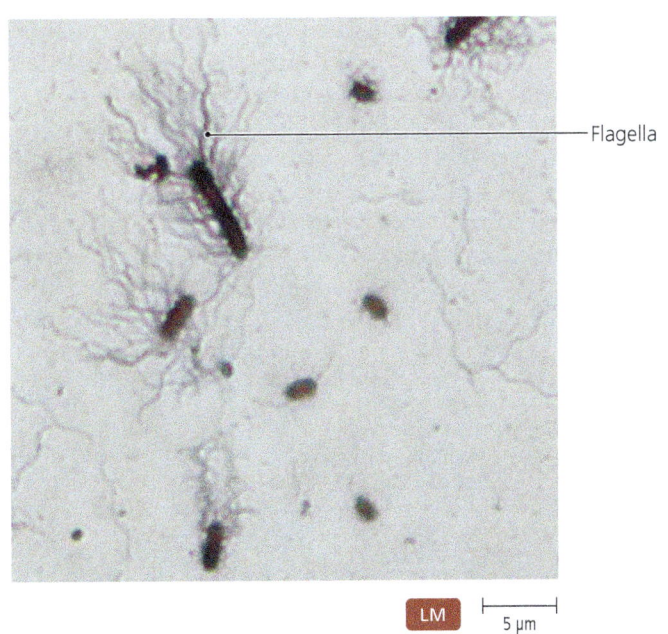

▲ FIGURE 4.19 **Flagellar stain of *Proteus vulgaris*.** Various bacteria have different numbers and arrangements of flagella, features that might be important in identifying some species. *How can the flagellar arrangement shown here be described?*

Figure 4.19 Peritrichous

Laboratory technicians increase contrast and resolution for transmission electron microscopy by using stains, just as they do for light microscopy. However, stains used for transmission electron microscopy are not colored dyes but instead chemicals

TABLE 4.3 Some Stains Used for Light Microscopy

Type of Stain	Examples	Results	Typical Image	Representative Uses
Simple Stains (use a single dye)	Crystal violet	Uniform purple stain		Reveals size, morphology, and arrangement of cells
	Methylene blue	Uniform blue stain		
Differential Stains (use two or more dyes to differentiate between cells or structures)	Gram stain	Gram-positive cells are purple; Gram-negative cells are pink		Differentiates Gram-positive and Gram-negative bacteria, which is typically the first step in their identification
	Ziehl-Neelsen acid-fast stain	Pink to red acid-fast cells and blue non-acid-fast cells		Distinguishes the genera *Mycobacterium* and *Nocardia* from other bacteria
	Schaeffer-Fulton endospore stain	Green endospores and pink to red vegetative cells		Highlights the presence of endospores produced by species in the genera *Bacillus* and *Clostridium*
Special Stains	Negative stain for capsules	Background is dark; cells unstained or stained with simple stain		Reveals bacterial capsules
	Flagellar stain	Bacterial flagella become visible		Allows determination of number and location of bacterial flagella

containing atoms of heavy metals, such as lead, osmium, tungsten, and uranium, which absorb electrons. Electron-dense stains may bind to molecules within specimens, or they may stain the background. The latter type of negative staining is used to provide contrast for extremely small specimens, such as viruses and molecules.

Stains for electron microscopy can be general in that they stain most objects to some degree, or they may be highly specific. For example, osmium tetroxide (OsO_4) has an affinity for lipids and is thus used to enhance the contrast of membranes. Electron-dense stains can also be linked to antibodies to provide an even greater degree of staining specificity because antibodies bind only to their specific target molecules.

MICRO CHECK

9. What is the name of the process that uses a single dye such as crystal violet or methylene blue to color a microscope specimen?
10. Which stain is the most frequently used differential stain in modern microbiology labs, differentiating microorganisms into two distinct groups?
11. What type of chemical used in the staining process does not add color itself but binds to a dye, making the dye less soluble and more likely to be attached to cellular materials?
12. Atoms of what class of elements are used to absorb electrons in electron-dense stains?

TELL ME WHY

Why is a Gram-negative bacterium colorless but a Gram-positive bacterium purple after it is rinsed with decolorizer?

Classification and Identification of Microorganisms

LEARNING OUTCOME

4.17 Discuss the purposes of taxonomy.

Biologists classify organisms for several reasons: to bring a sense of order and organization to the variety and diversity of living things, to enhance communication, to make predictions about the structure and function of similar organisms, and to uncover and understand potential evolutionary connections. They sort organisms on the basis of mutual similarities into nonoverlapping groups called **taxa**.[5] **Taxonomy**[6] is the science of classifying and naming organisms. Taxonomy consists of *classification*, which is the assigning of organisms to taxa based on similarities; *nomenclature*, which is concerned with the rules of naming organisms; and *identification*, which is the practical science of determining that an isolated individual or population belongs to a particular taxon. In this text, we concentrate on classification and identification.

Because all members of any given taxon share certain common features, taxonomy enables scientists both to organize large amounts of information about organisms and to make predictions based on knowledge of similar organisms. For example, if one member of a taxon is important in recycling nitrogen in the environment, it is possible that others in the group will play a similar ecological role. Similarly, a clinician might suggest a treatment against one pathogen based on what has been effective against other pathogens in the same taxon.

Identification of organisms is an essential part of taxonomy because it enables scientists to communicate effectively and be confident that they are discussing the same organism. Further, identification is often essential for treating groups of diseases, such as meningitis and pneumonia, which can be caused by pathogens as different as fungi, bacteria, and viruses.

In this section, we examine the historical basis of taxonomy, consider modern advances in this field, and briefly consider various taxonomic methods. This chapter also presents a general overview of the taxonomy of prokaryotes (which are considered in greater detail in Chapter 11); of animals, protozoa, fungi, and algae (Chapter 12); and of viruses, viroids, and prions (Chapter 13).

Linnaeus and Taxonomic Categories

LEARNING OUTCOMES

4.18 Discuss the difficulties in defining species of microorganisms.
4.19 List the hierarchy of taxa from general to specific.
4.20 Define *binomial nomenclature*.
4.21 Describe a few modifications of the Linnaean system of taxonomy.

Our current system of taxonomy began in 1753 with the publication of *Species Plantarum* by the Swedish botanist Carolus Linnaeus (1707–1778). Until his time, the names of organisms were often strings of descriptive terms that varied from country to country and from one scientist to another. Linnaeus provided a system that standardized the naming and classification of organisms based on characteristics they have in common. He grouped similar organisms that can successfully interbreed into categories called **species**.

The definition of *species* as "a group of organisms that interbreed to produce viable offspring" works relatively well for sexually reproducing organisms, but it is not satisfactory for asexual organisms such as most microorganisms. As a result, some scientists define a microbial species as a collection of *strains* or *serotypes*—populations of cells that arose from a single cell—that share many stable properties, differ from other strains, and evolve as a group. Alternatively, biologists define a microbial species as cells that share at least 97% common genetic sequences. Not surprisingly, these definitions sometimes result in disagreements and inconsistencies in the classification of microbial life. Some researchers question whether unique microbial species exist at all.

---

[5]From Greek *taxis*, meaning "order."
[6]From *taxis* and Greek *nomos*, meaning "rule."

BENEFICIAL MICROBES

Glowing Viruses

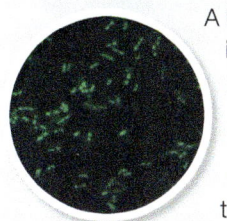

Fluorescent phages light up bacteria.

A bacteriophage is a virus that inserts its DNA into a bacterium. Commonly called a *phage*, it adheres only to a select bacterial strain for which each phage type has a specific adhesion factor. Many phages are so specialized for their particular bacterial strain that scientists have used phages to identify and classify bacteria. Such identification is called *phage typing*.

Scientists at San Diego State University have taken phage specificity a step further. They successfully linked a fluorescent dye to the DNA of phages of the bacterium *Salmonella* and used the phages to detect and identify *Salmonella* species. Such fluorescent phages rapidly and accurately detect specific strains of *Salmonella* in mixed bacterial cultures.

Fluorescent phages have advantages over fluorescent antibodies: Unlike antibodies, phages are not metabolized by bacteria. Phages are also more stable over time and are not as sensitive to vagaries in temperature, pH, and ionic strength. Further, fluorescent dyes within phages have a long shelf life; the phage coat protects the dye, which is attached to the DNA within the phage.

There are numerous uses for test kits using fluorescent phages. Environmental scientists could use them to detect bacterial contamination of streams and lakes, food processors could identify potentially fatal *Escherichia coli* strain O157:H7 in meat and vegetables, or homeland security agents could positively establish the presence or absence of bacteria used for biological warfare. Antibody-based kits frequently failed to accurately detect *Bacillus anthracis* used in the 2001 terrorist attacks. Fluorescent phage field kits should be much more robust and precise.

In Linnaeus's system, which forms the basis of modern taxonomy, similar species are grouped into **genera**[7] and similar genera into still larger taxonomic categories. That is, genera sharing common features are grouped together to form **families**; similar families are grouped into **orders**; orders are grouped into **classes**; classes into **phyla**;[8] and phyla into **kingdoms** (FIGURE 4.20).

All these categories, except strains (serotypes), and including species and genera, are taxa, which are hierarchical; that is, each successive taxon has a broader description than the preceding one, and each taxon includes all the taxa beneath it. For all groups except species of viruses, the rules of nomenclature require that every taxon have a Latin or Latinized name, in part because the language of science during Linnaeus's time was Latin. The name *Chondrus crispus* (kon'drŭs krisp'ŭs) describes the exact same algal species all over the world despite the fact that in England its common name is Irish moss, in Ireland it is carragheen, in North America it is curly moss, and, in reality, it isn't a moss at all.

When new microscopic, genetic, or biochemical techniques identify new or more detailed characteristics of organisms, a taxon may be split into two or more taxa. Alternatively, several taxa may be lumped together into a single taxon. For example, the genus "*Diplococcus*" (dip'lō-kok'ŭs) has been united (synonymized) with the genus *Streptococcus* (strep-tō-kok'ŭs), so the name *Diplococcus pneumoniae* (nū-mō'nē-ī) is now *Streptococcus pneumoniae* to reflect this synonymy.

Linnaeus assigned each species a descriptive name consisting of its genus name and a **specific epithet**. The genus name is always a noun, and it is written first and capitalized. The specific epithet contains only lowercase letters and is usually an adjective. Both names, together called a *binomial*, are either printed in italics or underlined. Because the Linnaean system assigns two names to every organism, it is said to use **binomial**[9] **nomenclature**.

Consider the following examples of binomial names of species: *Enterococcus faecalis* (en'ter-ō-kok'ŭs fē-kă'lis)[10] is a fecal bacterium, whereas *Enterococcus faecium* (fē-sē'ŭm) in the same genus is a different species because of certain differing characteristics. Therefore, it has a different specific epithet.

In some cases, binomials honor people. Examples include *Pasteurella haemolytica* (pas-ter-el'ă hē-mō-lit'i-kă), a bacterium named after the microbiologist Louis Pasteur; *Escherichia coli* (esh-ĕ-rik'ē-ă kō'lī), a bacterium named after the physician Theodor Escherich (1857–1911); and *Izziella abbottiae* (iz-ē-el'lă ab'ot-tē-ī), a marine alga named after the phycologist and taxonomist Isabella Abbott (1919–2010). Even though binomials are often descriptive of an organism, sometimes they can be misleading. For instance, *Haemophilus influenzae* (hē-mof'i-lŭs in-flu-en'zī) does not cause influenza.

Most scientists still use the Linnaean system today, though significant modifications have been adopted. For example, scientists sometimes use additional categories, such as *tribes, sections, subfamilies*, and *serotypes*.[11] Further, Linnaeus divided all organisms into only two kingdoms (Plantae and Animalia), and he did not know of the existence of viruses. As scientists learned more about organisms, they adopted taxonomic schemes to reflect advances in knowledge. For example, a widely accepted taxonomic approach of the last century was based on five kingdoms: Animalia, Plantae, Fungi, Protista, and Prokaryotae. Scientists created the kingdom Protista for eukaryotic organisms that did not fit neatly with true plants, animals, or fungi. This

---

[7]Plural of Latin *genus*, meaning "race" or "birth."
[8]Plural of *phylum*, from Greek *phyllon*, meaning "tribe." The term *phylum* is used for animals and bacteria; the corresponding taxon in mycology and botany is called a *division*.
[9]From Greek *bi*, meaning "two," and *nomos*, meaning "rule."
[10]From Greek *enteron*, meaning "intestine;" *kokkos*, meaning "berry"; and Latin *faeces*.
[11]Serotypes, which are also called *varieties, strains*, or *subspecies*, differ only slightly from each other.

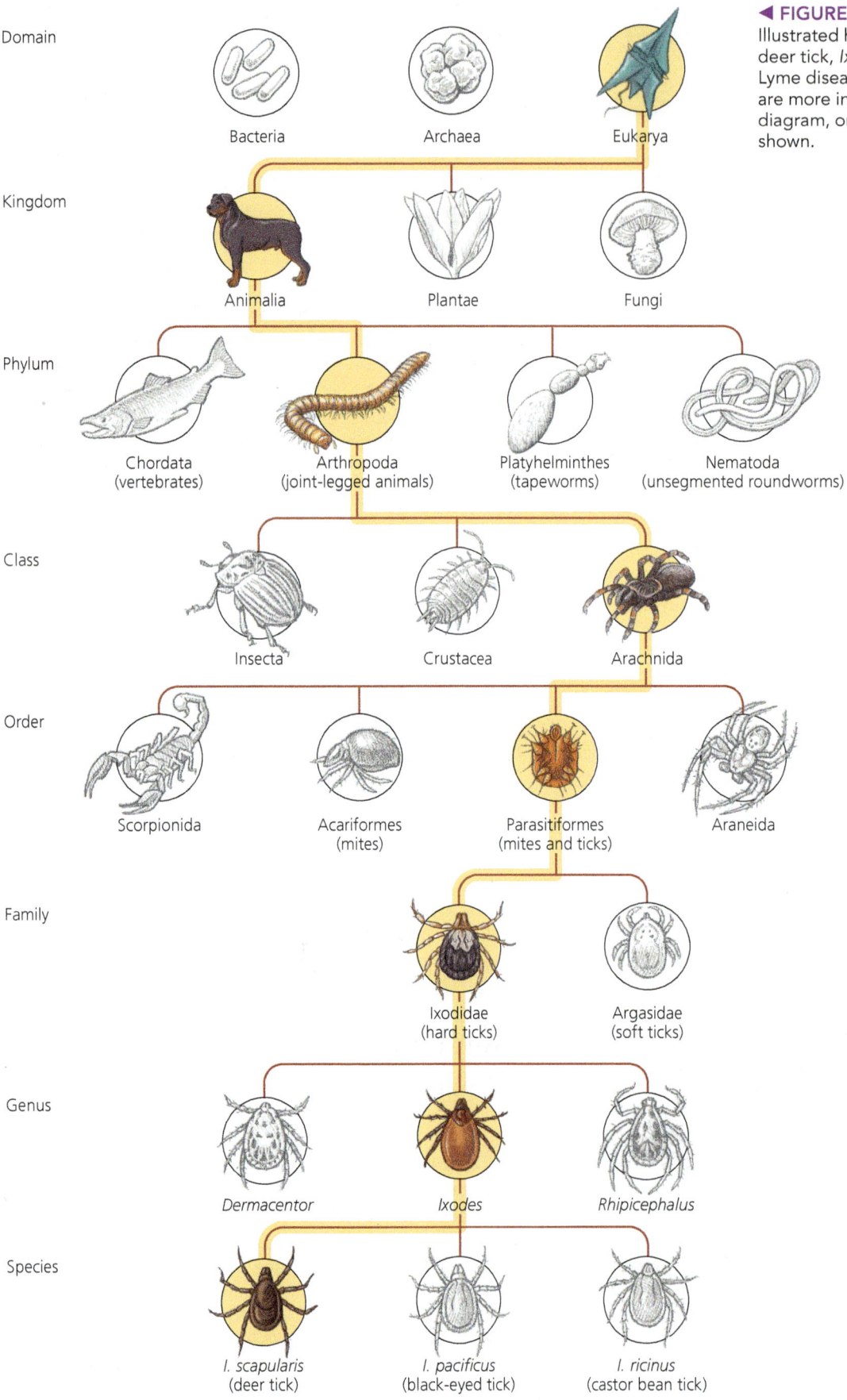

◀ **FIGURE 4.20 Levels in a taxonomic scheme.** Illustrated here are the taxonomic categories for the deer tick, *Ixodes scapularis*, the primary vector of Lyme disease. Notice that higher taxonomic categories are more inclusive than lower ones. To simplify the diagram, only selected classification possibilities are shown.

scheme grouped together such obviously different eukaryotes as massive brown seaweeds (kelps) and unicellular microbes, such as *Euglena* (yū-glēn'ă). Because this five-kingdom approach does not fully address the taxonomy of prokaryotes or the many differences among protists, scientists have proposed other taxonomic schemes that have from 5 to more than 50 kingdoms.

Sometimes students are upset that taxonomists do not agree about all the taxonomic categories or the species they contain, but it must be realized that classification of organisms reflects the state of our current knowledge and theories. Taxonomists change their schemes to accommodate new information, and not every expert agrees with every proposed modification.

Another significant development in taxonomy is a shift in its basic goal—from Linnaeus's goal of classifying and naming organisms as a means of cataloging them to the more modern goal of understanding the relationships among groups of organisms. Linnaeus based his taxonomic scheme primarily on organisms' structural similarities, and whereas such terms as *family* may suggest the existence of some common lineage, he and his contemporaries thought of species as divinely created. However, when Charles Darwin (1809–1882) propounded his theory of the evolution of species by natural selection (a century after Linnaeus published his pivotal work on taxonomy), taxonomists came to consider that common ancestry explains the similarities among organisms in the various taxa. Today, most taxonomists agree that a major goal of modern taxonomy is to reflect a *phylogenetic*[12] *hierarchy*—that is, that the ways in which organisms are grouped should reflect their evolution from common ancestors.

Taxonomists' efforts to classify organisms according to their ancestry have resulted in reduced emphasis on comparisons of physical and chemical traits, and in greater emphasis on comparisons of their genetic material. Such work led to the addition of a new, most inclusive taxon: the *domain*.

Domains

LEARNING OUTCOME

4.22 List and describe the three domains proposed by Woese and Fox.

Carl Woese (1928–2012) and George Fox (1945–) labored for years to understand the taxonomic relationships among prokaryotes. Morphology (shape) and biochemical tests did not provide enough information to classify organisms fully, so for over a decade Woese and Fox painstakingly sequenced the DNA nucleotides coding for the smaller subunits of ribosomal RNA (rRNA) in an effort to unravel the relationships among these organisms. Because rRNA molecules are present in all cells and are crucial to protein synthesis, changes in the nucleotide sequences coding for them are presumably very rare.

In 1976, they sequenced DNA from an odd group of prokaryotes that produce methane gas as a metabolic waste. Woese and Fox were surprised that the DNA sequences coding for the rRNA of these cells did not contain nucleotide sequences characteristic of bacteria. Repeated testing showed that methanogens, as they are called, were not like other prokaryotic or eukaryotic organisms. They were something new to science, a third branch of life. Woese and Fox had discovered that there are three basic types of ribosome, leading them to propose a new classification scheme in which a new taxon, called a **domain**, contains the Linnaean taxa of kingdoms. The three domains identified by Woese and Fox are **Eukarya**, **Bacteria**, and **Archaea**.

Domain Eukarya includes all eukaryotic cells, all of which contain eukaryotic rRNA sequences. Domains Bacteria and Archaea include all prokaryotic cells. They contain bacterial and archaeal rRNA sequences, respectively, which differ significantly from one another and from those in eukaryotic cells. In addition to differences in rRNA sequences, cells of the three domains differ in many other characteristics, including the lipids in their cell membranes, transfer RNA (tRNA) molecules, and sensitivity to antibiotics. (Chapters 11 and 12, which cover prokaryotes and eukaryotes respectively, discuss the taxonomy of organisms within the three domains.)

Ribosomal nucleotide sequences further suggest that there may be more than 50 kingdoms of Bacteria and five or more kingdoms of Archaea. Further, scientists examining substances such as human saliva, water, soil, and rock regularly discover novel nucleotide sequences that have been released from cells. When these sequences are compared to known sequences stored in a computer database, they cannot be associated with any previously identified organism. This suggests that many curious new forms of microbial life have never been grown in a laboratory and still await discovery.

Ribosomal nucleotide sequences have also given microbiologists a new way to define prokaryotic species. Some scientists propose that prokaryotes whose rRNA sequence differs from that of other prokaryotes by more than 3% be classified as a distinct species. Although this definition has the advantage of being precise, not all taxonomists agree with it.

MICRO CHECK

13. What taxonomic group appears between kingdom and class?
14. What historical figure developed our current system of taxonomy, by which we assign names to organisms?
15. What were Woese and Fox studying that led them to propose the new taxon *domain*?

Taxonomic and Identifying Characteristics

LEARNING OUTCOME

4.23 Describe six procedures taxonomists use to identify and classify microorganisms.

Criteria and laboratory techniques used for classifying and identifying microorganisms are quite numerous and include macroscopic and microscopic examination of physical characteristics,

---
[12]From Greek *phyllon*, meaning "tribe," and Latin *genus*, meaning "birth" (i.e., origin of a group).

differential staining characteristics, growth characteristics, microorganisms' interactions with antibodies, microorganisms' susceptibilities to viruses, nucleic acid analysis, biochemical tests, and organisms' environmental requirements, including the temperature and pH ranges of their various types of habitats. Clearly, then, microbial taxonomy is too broad a subject to cover in one chapter, and thus the details of the criteria for the classification of major groups are provided in subsequent chapters.

It is important to note that even though scientists may use a given technique to either classify or identify microorganisms, the criteria used to identify a particular organism are not always the same as those that were used to classify it. For example, even though medical laboratory scientists distinguish the genus *Escherichia* from other bacterial genera by its inability to utilize citric acid (citrate) as a sole carbon source, this characteristic is not vital in the classification of *Escherichia*.

Bergey's Manual of Determinative Bacteriology, first published in 1923 and now in its ninth edition (1994), contains information used for the laboratory identification of prokaryotes. *Bergey's Manual of Systematic Bacteriology* (second edition, 2001, 2005, 2009, 2010) is a similar reference work that is used for classification based on ribosomal RNA sequences, which taxonomists use to describe relationships among organisms. Each of these manuals is known as "Bergey's Manual."

Linnaeus did not know of the existence of viruses and thus did not include them in his original taxonomic hierarchy, nor are viruses assigned to any of the five kingdoms or Woese and Fox's three domains because viruses are acellular and generally lack rRNA. Virologists do classify viruses into families and genera, but higher taxa are poorly defined for viruses. (Chapter 13 further discusses viral taxonomy.)

With this background, let's turn now to some brief discussions of five types of information that microbiologists commonly use to distinguish among microorganisms: physical characteristics, biochemical tests, serological tests, phage typing, and analysis of nucleic acids. In addition, MALDI-TOF mass spectrometry is used in medical laboratories.

Physical Characteristics

Many physical characteristics are used to identify microorganisms. Scientists can usually identify protozoa, fungi, algae, and parasitic worms based solely on their *morphology* (shape). Medical laboratory scientists can also use the physical appearance of a bacterial colony[13] to help identify microorganisms. As we have discussed, stains are used to view the size and shape of individual bacterial cells and to show the presence or absence of identifying features such as endospores and flagella.

Linnaeus categorized prokaryotic cells into two genera based on two prevalent shapes. He classified spherical prokaryotes in the genus "*Coccus*,"[14] and he placed rod-shaped cells in the genus *Bacillus*.[15] However, subsequent studies have revealed vast differences among many of the thousands of spherical and rod-shaped prokaryotes, and thus visible characteristics alone are not sufficient to classify prokaryotes. Instead, taxonomists rely primarily on genetic differences as revealed by metabolic dissimilarities and, more and more frequently, on DNA sequences that code for subunits of rRNA.

Biochemical Tests

Microbiologists distinguish many prokaryotes on the basis of differences in their ability to utilize or produce certain chemicals. Biochemical tests include procedures that determine an organism's ability to ferment various carbohydrates; utilize various substrates, such as specific amino acids, starch, citrate, and gelatin; or produce waste products, such as hydrogen sulfide (H_2S) gas (**FIGURE 4.21**). Differences in fatty acid composition of bacteria are also used to distinguish among bacteria. Obviously, biochemical tests can be used to identify only those microbes that can be grown under laboratory conditions.

Laboratory scientists utilize biochemical tests to identify pathogens, allowing physicians to prescribe appropriate treatments. Many tests require that the microorganisms be *cultured* (grown) for 12–24 hours, though this time can be greatly reduced by the use of rapid identification tools. Such tools exist for many groups of medically important pathogens, such as Gram-negative bacteria in the family Enterobacteriaceae, Gram-positive bacteria, yeasts, and filamentous fungi. Automated systems for identifying pathogens use the results of a whole battery of biochemical tests performed in a plastic plate containing numerous small wells (**FIGURE 4.22**). A color change in a well indicates

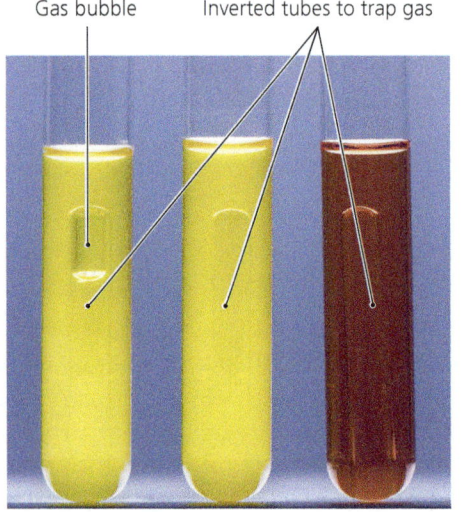

(a)   (b)

▲ **FIGURE 4.21 Two biochemical tests for identifying bacteria.** (a) A carbohydrate utilization test. At left is a tube in which the bacteria have metabolized a particular carbohydrate to produce acid (which changes the color of a pH indicator, phenol red, to yellow) and gas, as indicated by the bubble. At center is a tube with another bacterium that metabolized the carbohydrate to produce acid but no gas. At right is a tube inoculated with bacteria that are "inert" with respect to this test. (b) A hydrogen sulfide (H_2S) test. Bacteria that produce H_2S are identified by the black precipitate formed by the reaction of the H_2S with iron present in the medium.

---

[13]A group of bacteria that has arisen from a single cell grown on a solid laboratory medium.

[14]From Greek *kokkos*, meaning "berry." This genus name has been supplanted by many genera, including *Staphylococcus*, *Micrococcus*, and *Streptococcus*.

[15]From Latin *bacillum*.

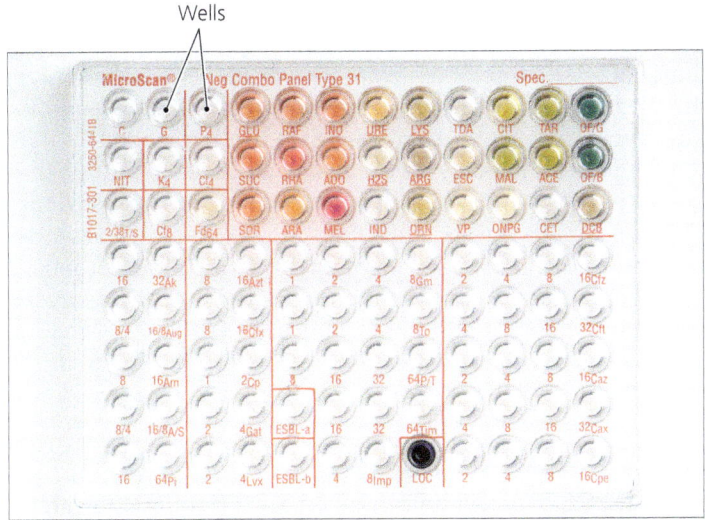

▲ FIGURE 4.22 One tool for the rapid identification of bacteria, the automated MicroScan system. A MicroScan panel is a plate containing numerous wells, each the site of a particular biochemical test. The instrument ascertains the identity of the organism by reading the pattern of colors in the wells after the biochemical tests have been performed.

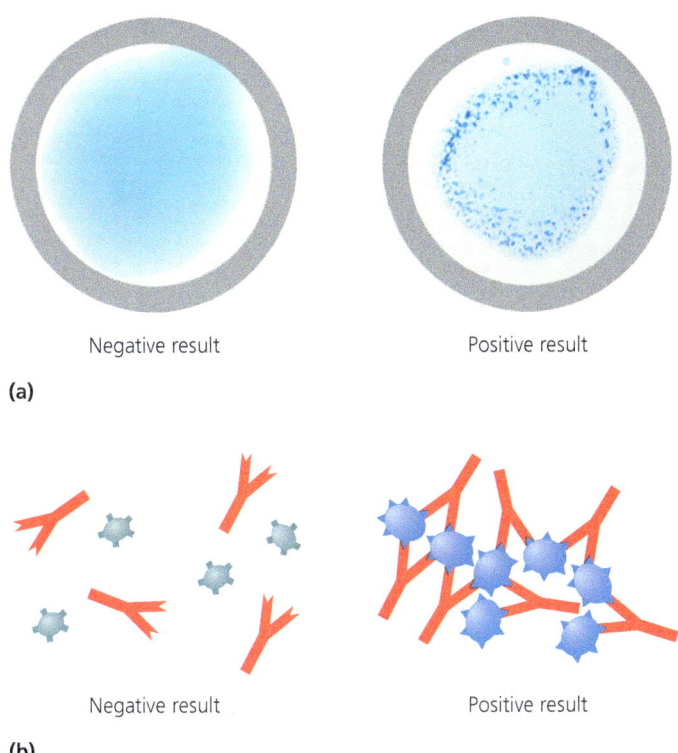

▲ FIGURE 4.23 An agglutination test, one type of serological test. (a) In a positive agglutination test, visible clumps are formed by the binding of antibodies to their target antigens present on cells. (b) The processes involved in agglutination tests. In a negative result, antibody binding cannot occur because its specific target is not present; in a positive result, specific binding does occur. Note that agglutination occurs because each antibody molecule can bind simultaneously to two antigen molecules.

the presence of a particular metabolic reaction, and the machine reads the pattern of colors in the plate to ascertain the identity of the pathogen.

Serological Tests

In the narrowest sense, serology is the study of *serum*, the liquid portion of blood after the clotting factors have been removed and an important site of antibodies. In its most practical application, serology is the study of antigen-antibody reactions in laboratory settings. Antibodies are immune system proteins that bind very specifically to target antigens (Chapter 16). In this section, we briefly consider the use of serological testing to identify microorganisms.

Many microorganisms are *antigenic*; that is, within a host organism they trigger an immune response that results in the production of antibodies. Suppose, for example, that a scientist injects a sample of *Borrelia burgdorferi* (bō-rē′lē-ă burg-dōr′fer-ē), the bacterium that causes Lyme disease, into a rabbit. The bacterium has many surface proteins and carbohydrates that are antigenic because they are foreign to the rabbit. The rabbit responds to these foreign antigens by producing antibodies against them. These antibodies can be isolated from the rabbit's serum and concentrated into a solution known as an **antiserum** (plural: *antisera*). Antisera bind to the antigens that triggered their production.

In a procedure called an **agglutination test**, antiserum is mixed with a sample that potentially contains its target cells. If the antigenic cells are present, antibodies in the antiserum will clump (*agglutinate*) the antigen (**FIGURE 4.23**). Other antigens, and therefore other organisms, remain unaffected because antibodies are highly specific for their targets.

Scientists and laboratory technicians use antisera to distinguish among species and even among serotypes (strains) of the same species. For example, Rebecca Lancefield (1895–1981) increased our ability to distinguish different streptococcal bacteria and diagnose their diseases using a serological classification system that she devised. Her work is memorialized as what are now known as *Lancefield groupings*, a common way to identify streptococci. (Chapter 17 examines other serological tests, such as *enzyme-linked immunosorbent assay*, or *ELISA*, and *immunoblotting*.)

Phage Typing

Bacteriophages (or simply **phages**) are viruses that infect and usually destroy bacterial cells. Just as antibodies are specific for their target antigens, phages are specific for the hosts they can infect. **Phage typing**, like serological testing, works because of such specificity. One bacterial strain may be susceptible to a particular phage while a related strain is not.

In phage typing, a technician spreads a solution containing the bacterium to be identified across a solid surface of growth medium and then adds small drops of solutions containing different types of bacteriophage. Wherever a specific phage is able to infect and kill bacteria, the resulting lack of bacterial growth produces within the bacterial lawn a clear area called a **plaque** (**FIGURE 4.24**). A microbiologist can identify an unknown bacterium by comparing the phages that form plaques with known phage-bacteria interactions.

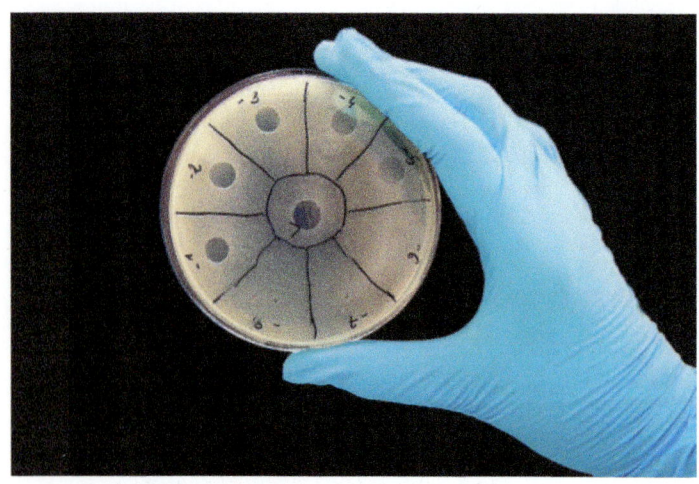

▲ FIGURE 4.24 **Phage typing.** Drops containing bacteriophages were added to this plate after its entire surface was inoculated with a bacterium. After 12 hours of bacterial growth, clear zones, called plaques, developed where the phages killed bacteria. Given the great specificity of phages for infecting and killing its host, the strain of bacterium can be identified.

MALDI-TOF Mass Spectrometry

Medical laboratory scientists can use *matrix-assisted laser desorption/ionization time-of-flight mass spectrometry*, more simply known as **MALDI-TOF** (mal'de-toff) to identify microbial species based upon unique proteins that the microbes synthesize. The technique uses a matrix to stabilize ions of large organic molecules, such as proteins, which have been ionized with laser energy. The organic molecules remain stable and retain their structure so that scientists can sort them by mass spectrometry—a process that sorts chemicals based on their mass-to-charge ratios. Though the details of MALDI-TOF and mass spectrometry are beyond the scope of this text, scientists use MALDI-TOF to sort and identify proteins unique to specific microbes and thereby show that particular microbes are present in a sample. Scientists and physicians use MALDI-TOF to identify species, to predict resistance to particular antibiotics, and to rapidly diagnose diseases.

Analysis of Nucleic Acids

As we have discussed, the sequence of nucleotides in nucleic acid molecules (either DNA or RNA) provides a powerful tool for classifying and identifying microbes. In many cases, nucleic acid analysis has confirmed classical taxonomic hierarchies. In other cases, as in Woese and Fox's discovery of archaea, curious new organisms and relationships not obvious from classical methodologies have come to light. Techniques of nucleotide sequencing and comparison, such as *polymerase chain reaction (PCR)* (Chapter 8), are best understood after we have discussed microbial genetics (Chapter 7).

Determining the percentage of a cell's DNA that is guanine and cytosine, a quantity referred to as the cell's *G + C content*

EMERGING DISEASE CASE STUDY

Necrotizing Fasciitis

Fever, chills, nausea, weakness, and general yuckiness. Carlos thought he was getting the flu. Further, he had pulled a cactus thorn from his arm the day before, and the tiny wound had swollen to a centimeter in diameter. It was red, extremely hot, and much more painful than such a puncture had a right to be. Everything was against him. He couldn't afford to miss days at work, but he had no choice.

He shivered in bed with fever for the next two days and suffered more pain than he had ever experienced, certainly more than the time he broke his leg. Even more than passing a kidney stone. The red, purple, and black inflammation on his arm had grown to the size of a baseball. It was hard to the touch and excruciatingly painful. He decided it was time to call his brother to take him to the doctor. That decision saved his life.

Carlos's blood pressure dropped severely, and he was unconscious by the time they arrived. The physician immediately admitted Carlos to the hospital, where the medical team raced to treat necrotizing fasciitis, commonly called "flesh-eating" disease. This reemerging disease is caused by group A *Streptococcus*, a serotype of Gram-positive bacteria also known as *Streptococcus pyogenes*. Group A strep invades through a break in the skin and travels along the fascia—the protective covering of muscles—producing toxins that destroy human tissues, affecting about 750 people each year in the United States.

By cutting away all the infected tissue; using high-pressure, pure oxygen to inhibit bacterial growth; and applying antimicrobial drugs to kill the bacterium, the doctors stabilized Carlos. After months of skin grafts and rehabilitation, he returned to work, grateful to be alive. (For more about necrotizing fasciitis, see pp. 554–555.)

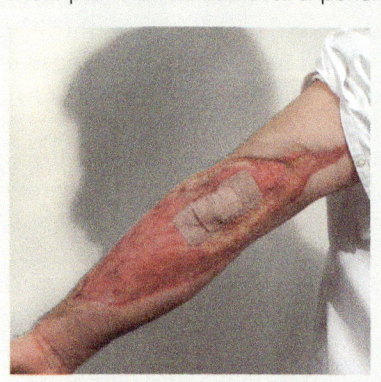
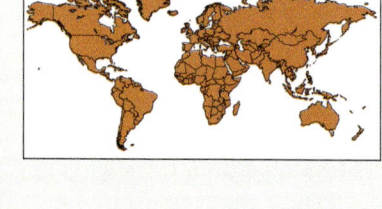

1. What color do cells of *S. pyogenes* appear after the Gram staining procedure?

(or *G + C percentage*), has also become a part of prokaryotic taxonomy. Scientists express the content as follows:

$$\frac{G + C}{A + T + G + C} \times 100$$

G + C content varies from 20% to 80% among prokaryotes. Often (but not always), organisms that share characteristics have similar G + C content. Organisms that were once thought to be closely related but have widely different G + C percentages are invariably not as closely related as had been thought.

Taxonomic Keys

As we have seen, taxonomists, medical clinicians, and researchers can use a wide variety of information—including morphology, chemical characteristics, nucleotide sequencing, and results from biochemical, serological, and phage typing tests—in their efforts to identify microorganisms, including pathogens. But how can all these characteristics and results be organized so that they can be used efficiently to identify an unknown organism? All this information can be arranged in **dichotomous keys**, which contain a series of paired statements worded so that only one of two choices applies to any particular organism (FIGURE 4.25). Depending on which of the two statements applies, the key either directs the user to another pair of statements or provides the name of the organism in question. Note that more than one key can be created to enable the identification of a given set of organisms, but all such keys involve mutually exclusive, "either/or" choices that send the user along a path that leads to the identity of the unknown organism.

 Play *Dichotomous Keys: Overview, Sample with Flowchart, Practice* @ Mastering Microbiology

MICRO CHECK

16. What type of testing allows lab technicians to test pathogenic bacterial specimens for reaction with known antibodies?

TELL ME WHY

Why didn't Linnaeus create taxonomic groups for viruses?

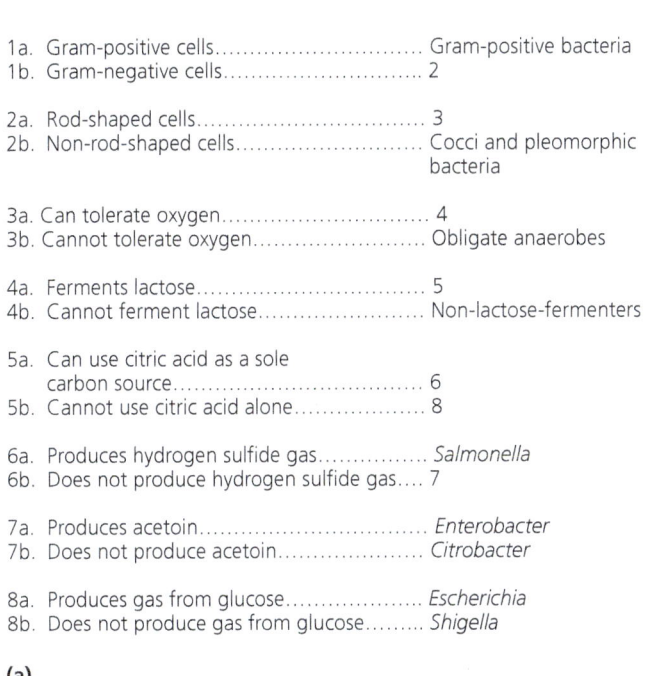

1a. Gram-positive cells............................ Gram-positive bacteria
1b. Gram-negative cells........................... 2

2a. Rod-shaped cells................................ 3
2b. Non-rod-shaped cells........................ Cocci and pleomorphic bacteria

3a. Can tolerate oxygen........................... 4
3b. Cannot tolerate oxygen..................... Obligate anaerobes

4a. Ferments lactose................................ 5
4b. Cannot ferment lactose..................... Non-lactose-fermenters

5a. Can use citric acid as a sole
     carbon source................................... 6
5b. Cannot use citric acid alone.............. 8

6a. Produces hydrogen sulfide gas......... *Salmonella*
6b. Does not produce hydrogen sulfide gas.... 7

7a. Produces acetoin................................ *Enterobacter*
7b. Does not produce acetoin.................. *Citrobacter*

8a. Produces gas from glucose................ *Escherichia*
8b. Does not produce gas from glucose........ *Shigella*

(a)

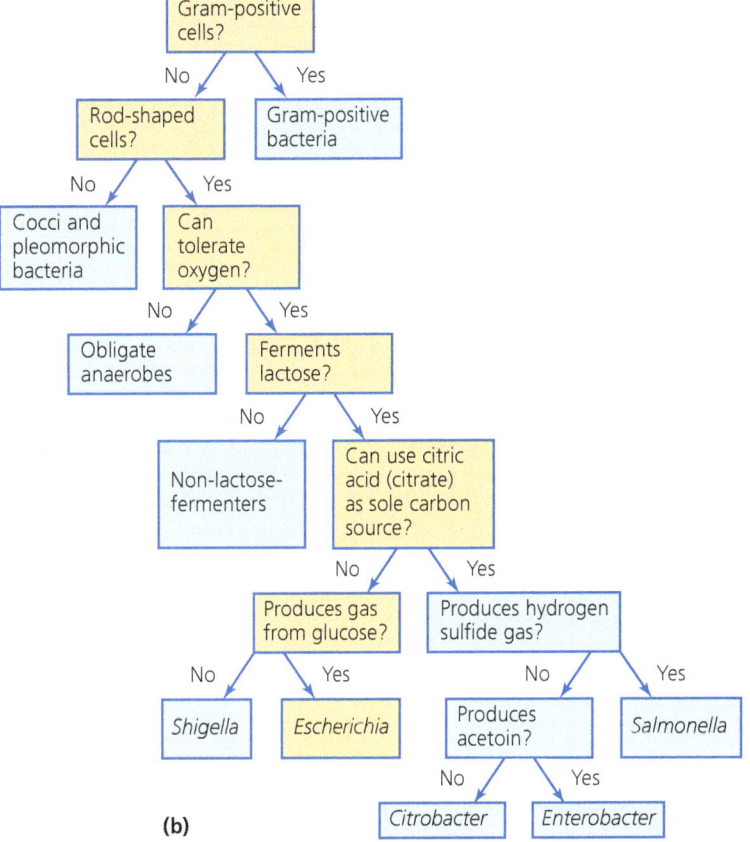

(b)

▲ **FIGURE 4.25 Use of a dichotomous taxonomic key.** The example presented here involves identifying the genera of potentially pathogenic intestinal bacteria. **(a)** A sample key. To use it, choose the one statement in a pair that applies to the organism to be identified, and then either refer to another key (as indicated by words) or go to the appropriate place within this key (as indicated by a number). **(b)** A flowchart that shows the various paths that might be followed in using the key presented in part (a). Highlighted is the path taken when the bacterium in question is *Escherichia*.

MICRO IN THE CLINIC FOLLOW-UP

End of the Camping Trip?

Upon the insistence of all her friends, Marisol agrees to end the trip early. When she gets back into town, she makes an appointment to see her doctor. Unlike Marisol's friend, the doctor is concerned about the discolored area on the back of Marisol's leg. The doctor tells Marisol that the area isn't a bruise; it's actually a rash. The presence of the rash, in combination with her other symptoms, makes him suspect that she may have Lyme disease. He draws blood for laboratory tests and to confirm the diagnosis. Marisol wants to know how her blood will be used to diagnose infection. The doctor explains that they will be using tests to detect the bacterium *Borrelia burgdorferi*. Marisol is surprised; she remembers learning about Lyme disease in her biology class but thought that a "bull's-eye" rash was the telltale sign. The doctor explains that although the bull's-eye rash is the best-known and perhaps the most distinctive sign, there are several rashes that are indicative of Lyme disease.

The blood tests indicate that Marisol indeed has Lyme disease. Her doctor prescribes a course of the antimicrobial drug doxycycline and explains that Marisol should start feeling better in the next day or so. Within a couple of weeks she should be back to normal. He also reminds her that it's important to use insect repellents containing the chemical DEET and thoroughly check for ticks when outdoors in areas where ticks may be present.

1. **One of the tests used to positively diagnose Lyme disease is immunofluorescence microscopy. How can this method be used to identify a pathogen such as *B. burgdorferi* in a blood sample?**

2. **Identify and describe a serological test that could be used to detect *B. burgdorferi* in blood.**

Check your answers to Micro in the Clinic Follow-Up questions in the Mastering Microbiology Study Area.

Make the connection among Chapters 2, 3, and 4. Watch MICRO MATTERS videos in the Mastering Microbiology Study Area.

CHAPTER SUMMARY

Play *Microscopy and Staining: Overview*
@ Mastering Microbiology

Units of Measurement (p. 97)

1. The metric system is a decimal system in which each unit is one-tenth the size of the next largest unit.
2. The basic unit of length in the metric system is the meter.

Microscopy (pp. 97–106)

1. **Microscopy** refers to the passage of light or electrons of various **wavelengths** through lenses to **magnify** objects and provide **resolution** and contrast so that those objects can be viewed and studied.
2. Immersion oil is used in light microscopy to fill the space between the specimen and a lens to reduce light refraction and thus increase the **numerical aperture** and resolution.
3. Staining techniques and polarized light may be used to enhance **contrast** between an object and its background.
4. **Simple microscopes** contain a single magnifying lens, whereas **compound microscopes** use a series of lenses for magnification.
5. The lens closest to the object being magnified is the **objective lens**, several of which are mounted on a **revolving nosepiece**. The lenses closest to the eyes are **ocular lenses**. **Condenser lenses** lie beneath the stage and direct light through the slide.
6. The magnifications of the objective lens and the ocular lens are multiplied together to give **total magnification**.
7. A photograph of a microscopic image is a **micrograph**.
8. **Dark-field microscopes** provide a dark background for small or colorless specimens.
9. **Phase microscopes**, such as **phase-contrast microscopes** and **differential interference contrast microscopes** (Nomarski microscopes), cause light rays that pass through a specimen to be out of phase with light rays that pass through the field, producing contrast.
10. **Fluorescence microscopes** use ultraviolet light and fluorescent dyes to fluoresce specimens and enhance contrast.
11. A **confocal microscope** uses fluorescent dyes in conjunction with computers to provide three-dimensional images of a specimen.

    Play *Light Microscopy* @ Mastering Microbiology

12. A **transmission electron microscope (TEM)** provides an image produced by the transmission of electrons through a thinly sliced, dehydrated specimen.
13. A **scanning electron microscope (SEM)** provides a three-dimensional image by scattering electrons from the metal-coated surface of a specimen.
14. Minuscule electronic probes are used in **scanning tunneling microscopes (STM)** and in **atomic force microscopes (AFM)** to reveal details at the atomic level.

    Play *Electron Microscopy* @ Mastering Microbiology

Staining (pp. 106–112)

1. Preparing to **stain** organisms with dyes for light microscopy involves making a **smear**, or thin film, of the specimens on a slide and then either passing the slide through a flame (**heat fixation**) or applying a chemical (**chemical fixation**) to attach the specimens to the slide. **Acidic dyes** or **basic dyes** are used to stain different portions of an organism to aid viewing and identification.

2. **Simple stains** involve the simple process of soaking the smear with one dye and then rinsing with water. **Differential stains** such as the **Gram stain**, **acid-fast stain**, endospore stain, Gomori methenamine silver (GMS) stain, and the hematoxylin and eosin (HE) stain use more than one dye to differentiate different cells, chemicals, or structures.

3. The Gram stain procedure includes use of a **primary stain**, a **mordant**, a **decolorizing agent**, and a **counterstain** that results in either purple (Gram-positive) or pink (Gram-negative) organisms, depending on the chemical structures of their cell walls.

4. The acid-fast stain is used to differentiate cells with waxy cell walls. **Endospores** are stained by the **Schaeffer-Fulton endospore stain** procedure.

5. Dyes that stain the background and leave the cells colorless are called **negative stains** (or **capsule stains**).

6. Stains for electron microscopy use atoms of heavy metals, which absorb electrons.

Play *Staining*
@ Mastering Microbiology

Classification and Identification of Microorganisms (pp. 112–119)

1. **Taxa** are nonoverlapping groups of organisms that are studied and named in **taxonomy**. Carolus Linnaeus invented a system of taxonomy, grouping similar interbreeding organisms into **species**, species into **genera**, genera into **families**, families into **orders**, orders into **classes**, classes into **phyla**, and phyla into **kingdoms**.

2. Linnaeus gave each species a descriptive name consisting of a genus name and **specific epithet**. This practice of naming organisms with two names is called **binomial nomenclature**.

3. Carl Woese and George Fox proposed the existence of three taxonomic **domains** based on three cell types revealed by rRNA sequencing: **Eukarya**, **Bacteria**, and **Archaea**.

4. Taxonomists rely primarily on genetic differences revealed by morphological and metabolic dissimilarities to classify organisms. Species or strains within species may be distinguished by using **antisera**; **agglutination tests**; nucleic acid analysis, particularly G + C content; or **phage typing** with **bacteriophages** (or **phages**), in which unknown bacteria are identified by observing **plaques** (regions of a bacterial lawn where the phage has killed bacterial cells).

5. Medical laboratory scientists can use **MALDI-TOF** (matrix-assisted laser desorption/ionization time-of-flight mass spectrometry) to identify species, to predict resistance to particular antibiotics, and to rapidly diagnose diseases.

6. Microbiologists use **dichotomous keys**, which involve stepwise choices between paired characteristics, to help them identify microbes.

Play *Dichotomous Keys:* Overview, Sample with Flowchart, Practice @ Mastering Microbiology

QUESTIONS FOR REVIEW

Answers to the Questions for Review (except Short Answer questions) begin on p. A-1.

Multiple Choice

1. Which of the following is smallest?
   a. decimeter
   b. millimeter
   c. nanometer
   d. micrometer

2. A nanometer is _____ than a micrometer.
   a. 10 times larger
   b. 10 times smaller
   c. 1000 times larger
   d. 1000 times smaller

3. Resolution is best described as the _____.
   a. ability to view something that is small
   b. ability to magnify a specimen
   c. ability to distinguish between two adjacent objects
   d. difference between two waves of electromagnetic radiation

4. Curved glass lenses _____ light.
   a. refract
   b. bend
   c. magnify
   d. both a and b

5. Which of the following factors is important in making an image appear larger?
   a. thickness of the lens
   b. curvature of the lens
   c. speed of the light passing through the lens
   d. all of the above

6. Which of the following is different between light microscopy and transmission electron microscopy?
   a. magnification
   b. resolution
   c. wavelengths
   d. all of the above

7. Which of the following types of microscopes produces a three-dimensional image with a shadowed appearance?
   a. simple microscope
   b. differential interference contrast microscope
   c. fluorescence microscope
   d. transmission electron microscope

8. Which of the following microscopes combines the greatest magnification with the best resolution?
   a. confocal microscope
   b. phase-contrast microscope
   c. dark-field microscope
   d. bright-field microscope

9. Negative stains such as eosin are also called _____.
   a. capsule stains
   b. endospore stains
   c. simple stains
   d. acid-fast stains

10. In the binomial system of nomenclature, which term is always written in lowercase letters?
    a. kingdom
    b. domain
    c. genus
    d. specific epithet

Fill in the Blanks

1. If an objective magnifies 40× and each binocular lens magnifies 15×, the total magnification of the object being viewed is _____.

2. The type of fixation developed by Koch for bacteria is _____.

3. Immersion oil _____ (increases/decreases) the numerical aperture, which _____ (increases/decreases) resolution because _____ (more/fewer) light rays are involved.

4. _____ refers to differences in intensity between two objects.

5. Cationic chromophores such as methylene blue ionically bond to _____ (positively/negatively) charged chemicals such as DNA and proteins.

VISUALIZE IT!

1. Label each photograph with the type of microscope used to acquire the image.

a. _____

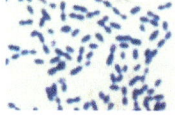
b. _____

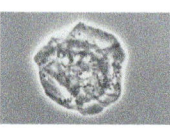
c. _____

d. _____

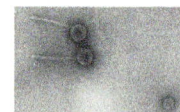
e. _____

f. _____

2. Label the microscope.

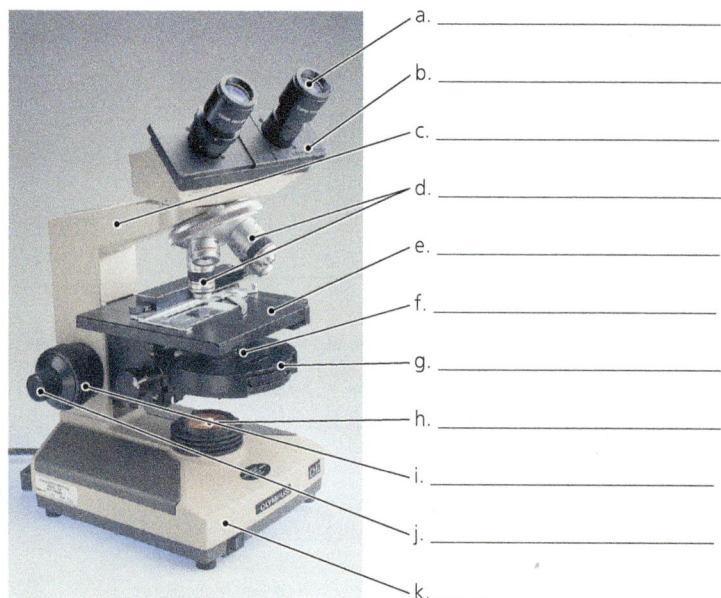

a. _____
b. _____
c. _____
d. _____
e. _____
f. _____
g. _____
h. _____
i. _____
j. _____
k. _____

Short Answer

1. Explain how the principle, "electrons travel as waves," applies to microscopy.

2. Critique the following definition of *magnification* given by a student on a microbiology test: "Magnification makes things bigger."

3. Why can electron microscopes magnify only dead organisms?

4. Put the following substances in the order they are used in a Gram stain: counterstain, decolorizing agent, mordant, primary stain.

5. Why is Latin used in taxonomic nomenclature?

6. Give three characteristics of a "specific epithet."

7. How does the study of the nucleotide sequences of ribosomal RNA fit into a discussion of taxonomy?

8. An atomic force microscope can magnify a living cell, whereas electron microscopes and scanning tunneling microscopes cannot. What requirement of scanning tunneling microscopes precludes the imaging of living specimens?

CRITICAL THINKING

1. Miki came home from microbiology lab with green fingers and a bad grade. When asked about this, she replied that she was doing a Gram stain but that it never worked the way the book said it should. Christina overheard the conversation and said that she must have used the wrong chemicals. What dye was she probably using, and what structure does that chemical normally stain?

2. Why is the definition of *species* as "successfully interbreeding organisms" not satisfactory for most microorganisms?

3. With the exception of the discovery of new organisms, is it logical to assume that taxonomy as we currently know it will stay the same? Why or why not?

4. A novice microbiology student incorrectly explains that immersion oil increases the magnification of his microscope. What is the function of immersion oil?

5. A light microscope has 10× oculars and 0.3-μm resolution. Using the oil immersion lens (100×), will you be able to resolve two objects 400 nm apart? Will you be able to resolve two objects 40 nm apart?

6. In what ways are the Gram stain and the acid-fast staining procedures similar? In the acid-fast procedure, what takes the place of Gram's iodine mordant?

7. Microbiologists have announced the discovery of over 30 new species of bacteria that thrive between the teeth and gums of humans. The bacteria could not be grown in the researchers' laboratories, nor were any of them ever observed via any kind of microscopy.

    If they couldn't culture them or see them, how could the researchers know they had discovered new species? If they couldn't examine the cells for the presence of a nucleus, how did they determine that the organisms were prokaryotes and not eukaryotes?

8. Why is the genus name *"Coccus"* placed within quotation marks but not the genus name *Bacillus*?

9. A clinician obtains a specimen of urine from a patient suspected to have a bladder infection. From the specimen she cultures a Gram-negative, rod-shaped bacterium that ferments lactose in the presence of oxygen, utilizes citrate, and produces acetoin but not hydrogen sulfide. Using the key presented in Figure 4.25, identify the genus of the infective bacterium.

CONCEPT MAPPING

Using the terms provided, fill in this concept map that describes Gram stain. You can also complete this and other concept maps online by going to the Mastering Microbiology Study Area.

- Decolorizer
- Ethanol and acetone
- Gram-negative bacteria
- Gram-positive bacteria
- Iodine
- Primary stain
- Purple
- Safranin
- Thin peptidoglycan layer

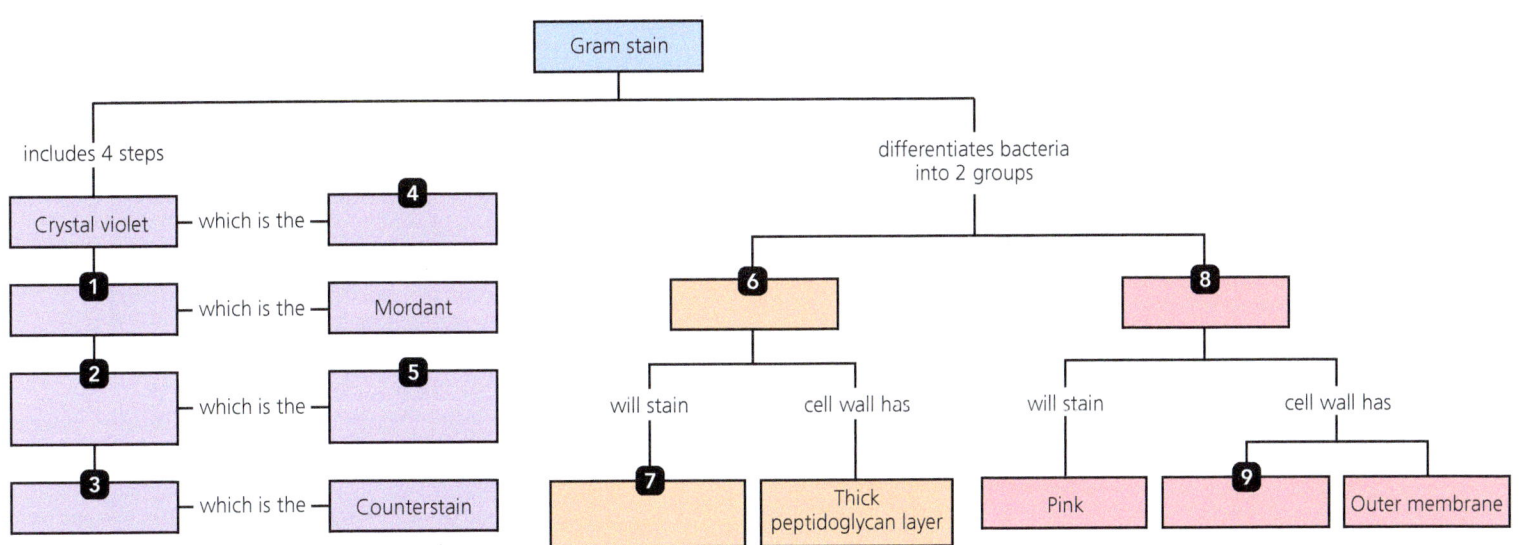

5 Microbial Metabolism

Before You Begin

1. What class of biochemical reactions generally release energy?
2. Why is it difficult for protons (H^+) to move across phospholipid bilayer membranes?
3. Large carbohydrates, such as starch and glycogen, are polymers. What monomers compose large carbohydrates?

The spherical cells of *Streptococcus pyogenes* grow together in chains.

MICRO IN THE CLINIC

Scraped Knee and Loss of a Leg?

ARJUN IS AN ACTIVE 15-YEAR-OLD high school student; he plays soccer, runs track, and volunteers at his local food bank every week. A couple of days ago, while skateboarding, he scraped his knee pretty badly. He didn't tell his mom because he knew that she would lecture him about being careful, and he was worried that she wouldn't let him go back to the skate park. He washed his knee well with soap and water when he got home and didn't think more about it.

The next morning, Arjun's knee is stiff, swollen, and quite red. He washes it again when he takes a shower, and then he heads to school. Later in the afternoon, Arjun realizes that his knee is really sore, and he feels hot. He shows up for soccer practice, but his coach tells him that because he has a fever, he should go home and have his mom take him to the doctor to look at his knee.

The doctor examines Arjun's knee and tells him that the wound is infected. The doctor is concerned by the rapid progression of the infection and Arjun's fever. He admits Arjun to the hospital and orders an intravenous course of antibiotics. After two days of treatment, his knee actually looks worse; the infection is spreading to surrounding tissues. Arjun's mom is very worried and anxious for the doctor to visit in the afternoon.

1. **What do you think? Is there cause for Arjun's mom to be worried?**

    Turn to the end of the chapter (p. 154) to find out.

How do pathogens acquire energy and nutrients at the expense of a patient's health? How does grape juice turn into wine, and how does yeast cause bread to rise? When laboratory personnel perform biochemical tests to identify unknown microorganisms and help diagnose disease, what exactly are they doing?

The answers to all of these questions require an understanding of microbial **metabolism**,[1] the collection of controlled biochemical reactions that takes place within a microbe. Although it is true that metabolism in its entirety is complex, consisting of thousands of chemical reactions and control mechanisms, the reactions are nevertheless elegantly logical and can be understood in a simplified form. In this chapter, we concern ourselves only with central metabolic pathways and energy metabolism.

Your study of metabolism will be manageable if you keep in mind that the ultimate function of an organism's metabolism is to reproduce the organism and that metabolic processes are guided by the following eight elementary statements:

- Every cell acquires *nutrients*, which are the chemicals necessary as building blocks and energy sources for metabolism.
- Metabolism requires energy from light or from the *catabolism* (kă-tab´ō-lizm; breakdown) of acquired nutrients.
- Energy is often stored in the chemical bonds of *adenosine triphosphate (ATP)*.
- Using *enzymes*, cells catabolize nutrient molecules to form elementary building blocks called *precursor metabolites*.
- Using precursor metabolites, other enzymes, and energy from ATP, cells construct larger building blocks in *anabolic* (an-ă-bol´ik; biosynthetic) reactions.
- Cells use enzymes and additional energy from ATP to anabolically link building blocks together to form macromolecules in *polymerization* reactions.
- Cells grow by assembling macromolecules into cellular structures such as ribosomes, membranes, and cell walls.
- Cells typically divide into two once they have doubled in size.

We will discuss each aspect of metabolism in the chapters that most directly apply. (For instance, Chapter 3 discussed the first step of metabolism—the active and passive transport of nutrients into cells.) The current chapter examines the importance of enzymes in catabolic and anabolic reactions, studies three ways that ATP molecules are synthesized, and shows that catabolic and anabolic reactions are linked. We also examine the catabolism of nutrient molecules; the anabolic reactions involved in the synthesis of carbohydrates, lipids, amino acids, and nucleotides; and a few ways that cells control their metabolic activities. (Chapter 7 discusses genetic control of metabolism and the polymerization of DNA, RNA, and proteins, and Chapters 11 and 12 cover specifics of cell division.)

[1]From Greek *metabole*, meaning "change."

Basic Chemical Reactions Underlying Metabolism

In the following sections, we will examine the basic concepts of catabolism, anabolism, and a special class of reactions called *oxidation-reduction reactions*. The latter involve the transfer of electrons and energy carried by electrons between molecules. Then we will turn our attention briefly to the synthesis of ATP and energy storage before we discuss the organic catalysts called *enzymes*, which make metabolism possible.

Play *Metabolism:* Overview
@ Mastering Microbiology

Catabolism and Anabolism

LEARNING OUTCOME

5.1 Distinguish among metabolism, anabolism, and catabolism.

Metabolism, which is all of the chemical reactions in an organism, can be divided into two major classes of reactions: **catabolism** and **anabolism** (FIGURE 5.1). A series of such reactions is called a *pathway*. Cells have *catabolic pathways*, which break larger molecules into smaller products, and *anabolic pathways*, which synthesize large molecules from smaller ones. Even though catabolic and anabolic pathways are intimately linked in cells, it is often useful to study the two types of pathways as though they were separate.

When catabolic pathways break down large molecules, they release energy; that is, catabolic pathways are *exergonic* (ek-ser-gon´ik). Cells store some of this released energy in the bonds of ATP, though much of the energy is lost as heat. Another result

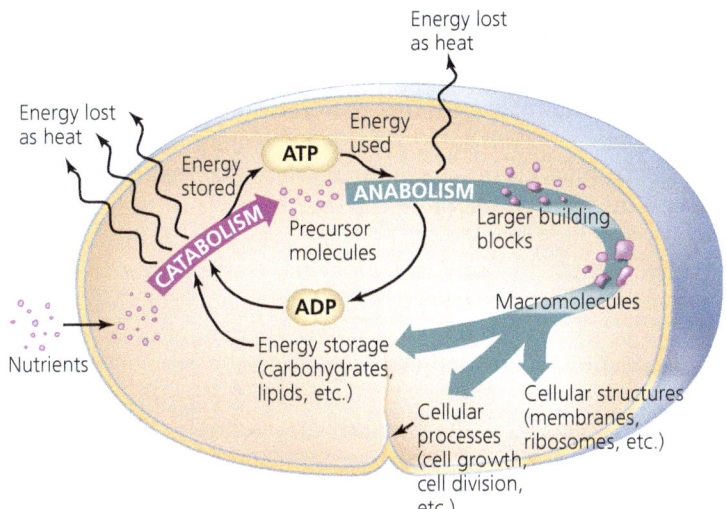

▲ FIGURE 5.1 **Metabolism is composed of catabolic and anabolic reactions.** Catabolic reactions are exergonic—they release energy, some of which is stored in ATP molecules, though most of the energy is lost as heat. Anabolic reactions are endergonic—they require energy, typically provided by ATP. There is some heat loss in anabolism as well. The products of catabolism provide many of the building blocks (precursor metabolites) for anabolic reactions. These reactions produce macromolecules and cellular structures, leading to cell growth and division.

of the breakdown of large molecules by catabolic pathways is the production of numerous smaller molecules, some of which are **precursor metabolites** of anabolism. Some organisms, such as *Escherichia coli* (esh-ĕ-rik´ē-ă kō´lī), can synthesize everything in their cells from precursor metabolites alone; other organisms must acquire at least some of their anabolic building blocks from outside themselves as nutrients. Catabolic *pathways*, but not necessarily *individual catabolic reactions*, produce ATP, or metabolites, or both. An example of a catabolic pathway is the breakdown of lipids into glycerol and fatty acids.

Anabolic pathways are functionally the opposite of catabolic pathways in that anabolic pathways synthesize macromolecules and cellular structures. Because building anything requires energy, anabolic pathways are *endergonic* (en-der-gon´ik); that is, they require more energy than they release. The energy required for anabolic pathways usually comes from ATP molecules produced during catabolism. An example of an anabolic pathway is the synthesis of cell membrane lipids from glycerol and fatty acids.

To summarize, a cell's metabolism involves both catabolic pathways that break down macromolecules to supply molecular building blocks and energy in the form of ATP, and anabolic pathways that use the building blocks and ATP to synthesize macromolecules needed for growth and reproduction.

Oxidation and Reduction Reactions

LEARNING OUTCOME

5.2 Contrast oxidation and reduction reactions.

Many metabolic reactions involve the transfer of electrons, and keeping track of these electrons is important for your understanding of metabolism. Why is it important? Electrons carry energy, and you need to understand how a cell uses that energy. Electrons are transferred from an *electron donor* (a molecule that donates an electron) to an *electron acceptor* (a molecule that accepts an electron). Such electron transfers are called **oxidation-reduction reactions**, or **redox reactions** (FIGURE 5.2). An electron acceptor is said to be *reduced*. This may seem backward, but electron acceptors are reduced because their gain in electrons reduces their overall electrical charge (i.e., they are more negatively charged). Molecules that lose electrons are said to be *oxidized* because in many reactions their electrons are donated to oxygen atoms. An acronym to help you remember these concepts is OIL RIG: oxidation involves loss; reduction involves gain.

Reduction and oxidation reactions always happen simultaneously because every electron donated by one chemical is accepted by another chemical. A chemical may be reduced by gaining either a simple electron or an electron that is part of a hydrogen atom—which is composed of one proton and one electron.

In contrast, a molecule may be oxidized in one of three ways: by losing a simple electron, by losing a hydrogen atom, or by gaining an oxygen atom. Biological oxidations often involve the loss of hydrogen atoms; such reactions are also called *dehydrogenation reactions* (dē-hī´drō-jen-ā´shŭn).

Electrons rarely exist freely in cytoplasm; instead, they orbit atomic nuclei. Therefore, cells use electron carrier molecules to carry electrons (often in hydrogen atoms) from one location in a

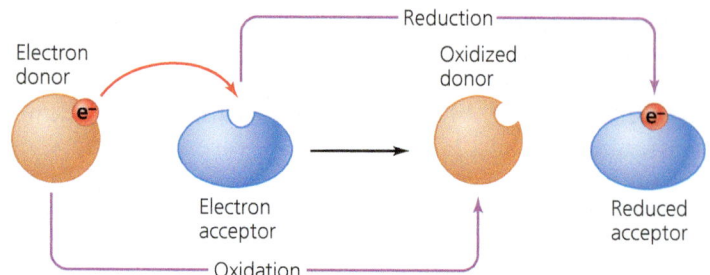

▲ **FIGURE 5.2 Oxidation-reduction, or redox, reactions.** When electrons are transferred from donor molecules to acceptor molecules, donors become oxidized, and acceptors become reduced. *Why are acceptor molecules said to be reduced when they are gaining electrons?*

Figure 5.2 Reduction refers to the overall electrical charge on a molecule. Because electrons have a negative charge, the gain of an electron reduces the molecule's overall charge.

cell to another. Three important electron carrier molecules are **nicotinamide adenine dinucleotide (NAD$^+$)**, **nicotinamide adenine dinucleotide phosphate (NADP$^+$)**, and **flavin adenine dinucleotide (FAD)**. These molecules are derived from vitamins. Cells use each of these electron carriers in specific metabolic pathways to carry pairs of electrons. One of the electrons carried by either NAD$^+$ or NADP$^+$ is part of a hydrogen atom, forming NADH or NADPH. FAD carries two electrons as hydrogen atoms (FADH$_2$). Many metabolic pathways, including those that synthesize ATP, require electron carrier molecules.

Play *Oxidation-Reduction Reactions*
@ Mastering Microbiology

MICRO CHECK

1. Which type of chemical reactions in cells release energy?
2. Which type of metabolic reaction synthesizes large molecules from smaller ones?
3. What term is used to describe the gain of one or more electrons by a molecule?
4. What is the defining characteristic of a redox reaction?

ATP Production and Energy Storage

LEARNING OUTCOME

5.3 Compare and contrast the three types of ATP phosphorylation.

Nutrients contain energy, but that energy is spread throughout their chemical bonds, and cells cannot use the energy in its this form to do work or perform anabolic reactions. During catabolism, the cells release the energy and concentrate it in high-energy phosphate bonds of molecules such as ATP. ATP and similar molecules serve as the "energy currency" of a cell. High-energy phosphate bonds form by a general process called *phosphorylation* (fos´fōr-i-lā´shŭn), in which inorganic phosphate (PO_4^{3-}) is added to a substrate. For example, cells phosphorylate adenosine diphosphate (ADP), which has two phosphate groups, to form adenosine triphosphate (ATP), which has three phosphate groups (see Chapter 2, Figure 2.27). Why is ATP well suited to serve as the primary short-term energy carrier in metabolic pathways? First, it is multifunctional as a ribonucleotide for use in synthesizing RNA. Second, it is highly water soluble and can accumulate to high concentrations in cells with no ill effects. Third, it has two different levels of energy donation, ATP to ADP and ATP to AMP, depending on what is needed for a reaction. Fourth, ATP can also serve as a phosphate donor.

As we will examine in the following sections, cells phosphorylate ADP to form ATP in three specific ways:

- *Substrate-level phosphorylation* (on p. 135), which involves the transfer of phosphate to ADP from another phosphorylated organic compound
- *Oxidative phosphorylation* (on pp. 139–140), in which energy from redox reactions of respiration (described shortly) is used to attach inorganic phosphate to ADP
- *Photophosphorylation* (on pp. 145–146), in which light energy is used to phosphorylate ADP with inorganic phosphate

We will investigate each of these in more detail as we proceed through the chapter.

Anabolic pathways use some energy of ATP by breaking one phosphate bond of ATP, so it changes back to ADP. ATP molecules last for a very short time, because their energy is used for anabolism and cellular activity such as active transport (see Figure 3.22). Thus, the cyclical interconversion of ADP and ATP functions somewhat like rechargeable batteries: ATP molecules store energy from catabolic reactions (and from light in photosynthetic organisms) and then release stored energy to drive cellular processes (including anabolic reactions, active transport, and movement). ADP molecules can be "recharged" to ATP again and again. Cells use ATP as a very short-term energy storage molecule; the average half-life of a molecule of ATP is measured in seconds.

The Roles of Enzymes in Metabolism

LEARNING OUTCOMES

5.4 Make a table listing the six basic types of enzymes, their activities, and an example of each.
5.5 Describe the components of a holoenzyme, and contrast protein and RNA enzymes.
5.6 Define *activation energy*, *enzyme*, *apoenzyme*, *cofactor*, *coenzyme*, *active site*, and *substrate*, and describe their roles in enzyme activity.
5.7 Describe how temperature, pH, substrate concentration, and competitive and noncompetitive inhibition affect enzyme activity.

Chemical reactions occur when bonds are broken or formed between atoms. In catabolic reactions, a bond must be destabilized before it will break, whereas in anabolic reactions, reactants collide with sufficient energy for bonds to form between them. In anabolism, increasing either the concentrations of reactants or ambient temperatures increases the number of collisions and produces more chemical reactions; however, in living organisms, neither reactant concentration nor temperature is usually high enough to ensure that bonds will form. Therefore, the chemical reactions of life depend on *catalysts*, which are chemicals that increase the likelihood of a reaction but are not permanently changed in the process. Organic catalysts are known as **enzymes**.

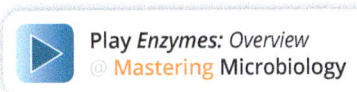
Play *Enzymes: Overview*
@ Mastering Microbiology

Naming and Classifying Enzymes

The names of enzymes usually end with the suffix *-ase*, and the name of each enzyme often incorporates the name of that enzyme's **substrate**, which is the molecule the enzyme acts on. Based on their mode of action, enzymes can be grouped into six basic categories:

- *Hydrolases* catabolize molecules by adding water in a decomposition process known as *hydrolysis*. Hydrolases are used primarily in depolymerizing catabolism of macromolecules.
- *Isomerases*[2] rearrange atoms within a molecule but do not add or remove anything (so they are neither catabolic nor anabolic).
- *Ligases, or polymerases*, join two molecules together (and are thus anabolic). They often use energy supplied by ATP.
- *Lyases* split large molecules (and are thus catabolic) without using water in the process.
- *Oxidoreductases* remove electrons from (oxidize) or add electrons to (reduce) various substrates. They are used in both catabolic and anabolic pathways.
- *Transferases* transfer functional groups, such as an amino group (NH_2), a phosphate group (PO_4^{3-}), or a two-carbon (acetyl) group, between molecules. Transferases can be anabolic.

TABLE 5.1 (on p. 128) summarizes these types of enzymes and gives examples of each.

The Makeup of Enzymes

Many protein enzymes are complete and functional by themselves. Others are combinations of both protein and nonprotein portions. The proteins in these combinations are called **apoenzymes** (ap´ō-en-zīms). Apoenzymes are inactive until they are bound to one or more nonprotein substances called **cofactors**. Cofactors are either inorganic ions (such as iron, magnesium, zinc, or copper ions) or certain organic molecules called **coenzymes** (ko-en´zīms). All coenzymes are either vitamins or contain vitamins, which are organic molecules that are required for metabolism but cannot be synthesized by certain organisms (particularly mammals). Some apoenzymes bind with inorganic cofactors, some bind with coenzymes, and some bind with both. The binding of an apoenzyme and its cofactor(s) forms an active enzyme called a **holoenzyme** (hol-ō-en´zīm; FIGURE 5.3).

---

[2]An isomer is a compound with the same molecular formula as another molecule, but with a different arrangement of atoms.

TABLE 5.1 Enzyme Classification Based on Reaction Types

Class	Type of Reaction Catalyzed	Example
Hydrolase	Hydrolysis (catabolic)	Lipase—breaks down lipid molecules
Isomerase	Rearrangement of atoms within a molecule (neither catabolic nor anabolic)	Phosphoglucoisomerase—converts glucose 6-phosphate into fructose 6-phosphate during glycolysis
Ligase or Polymerase	Joining two or more chemicals together (anabolic)	Acetyl-CoA synthetase—combines acetate and coenzyme A to form acetyl-CoA for the citric acid cycle
Lyase	Splitting a chemical into smaller parts without using water (catabolic)	Fructose-1,6-bisphosphate aldolase—splits fructose 1,6-bisphosphate into G3P and DHAP
Oxidoreductase	Transfer of electrons or hydrogen atoms from one molecule to another	Lactic acid dehydrogenase—oxidizes lactic acid to form pyruvic acid during fermentation
Transferase	Moving a functional group from one molecule to another (may be anabolic)	Hexokinase—transfers phosphate from ATP to glucose in the first step of glycolysis

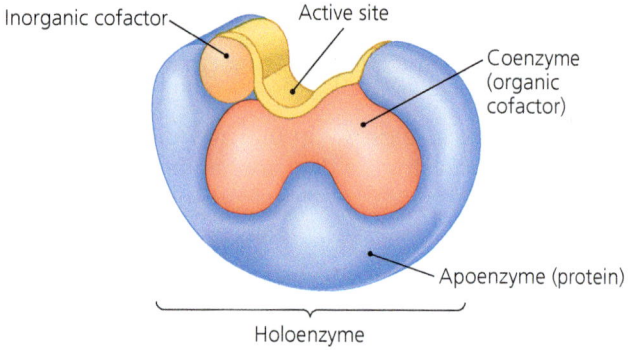

▲ **FIGURE 5.3 Makeup of a protein holoenzyme.** The combination of a proteinaceous apoenzyme with one or more cofactors forms a holoenzyme, which is the active form. A cofactor is either an inorganic ion or a coenzyme, which is an organic cofactor derived from a vitamin. An apoenzyme is inactive unless it is bound to its cofactors. *Name four metal ions that can act as cofactors.*

Figure 5.3 *Iron, magnesium, zinc, and copper ions can act as cofactors.*

TABLE 5.2 lists several examples of inorganic cofactors and organic cofactors (coenzymes). Note that three important coenzymes are the electron carriers NAD^+, $NADP^+$, and FAD, which, as we have seen, carry electrons as part of hydrogen atoms from place to place within cells. We will examine more closely the roles of these coenzymes in the generation of ATP later in the chapter.

Not all enzymes are proteins; some are RNA enzymes called **ribozymes**. In eukaryotes, ribozymes process other RNA molecules by removing sections of RNA and splicing the remaining pieces together. Recently, researchers have discovered that a ribozyme makes up the functional core of a ribosome, which is a cell's protein factory; in other words, all protein enzymes are made by ribozymes.

Enzyme Activity

Within cells, enzymes catalyze reactions by lowering the **activation energy**, which is the amount of energy needed to trigger a chemical reaction (**FIGURE 5.4**). Heat can provide

TABLE 5.2 Representative Cofactors of Enzymes

Cofactors	Examples of Use in Enzymatic Activity	Substance Transferred in Enzymatic Activity	Vitamin Source (of Coenzyme)
Inorganic (Metal Ion)			
Magnesium (Mg^{2+})	Forms bond with ADP during phosphorylation	Phosphate	None
Organic (Coenzymes)			
Nicotinamide adenine dinucleotide (NAD^+)	Carrier of reducing power	Two electrons and a hydrogen ion	Niacin (B_3)
Nicotinamide adenine dinucleotide phosphate ($NADP^+$)	Carrier of reducing power	Two electrons and a hydrogen ion	Niacin (B_3)
Flavin adenine dinucleotide (FAD)	Carrier of reducing power	Two hydrogen atoms	Riboflavin (B_2)
Tetrahydrofolate	Used in synthesis of nucleotides and some amino acids	One-carbon molecule	Folic acid (B_9)
Coenzyme A	Formation of acetyl-CoA in citric acid cycle and beta-oxidation	Two-carbon molecule	Pantothenic acid (B_5)
Pyridoxal phosphate	Transaminations in the synthesis of amino acids	Amine group	Pyridoxine (B_6)
Thiamine pyrophosphate	Decarboxylation of pyruvic acid	Aldehyde group (—CHO)	Thiamine (B_1)

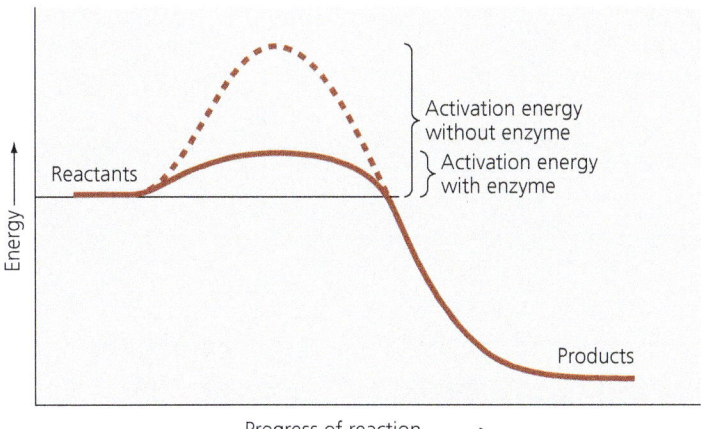

▲ **FIGURE 5.4 The effect of enzymes on chemical reactions.** Enzymes catalyze reactions by lowering the activation energy—that is, the energy needed to trigger the reaction. *Does this graph represent a catabolic or an anabolic reaction?*

Figure 5.4 *Energy of the products is lower than energy of the reactants. Thus, the graph represents catabolism.*

energy to trigger reactions, but for most metabolic reactions the temperatures needed to reach activation energy are often too high to allow cells to survive. So, for metabolism to occur, enzymes are needed. This is true regardless of whether the enzyme is a protein or RNA or whether the chemical reaction is anabolic or catabolic.

The activity of enzymes depends on the closeness of fit between the functional sites of an enzyme and its substrate. The shape of an enzyme's functional site, called its **active site**, is complementary to the shape of the enzyme's substrate. Generally, the shapes and locations of only a few amino acids in a protein enzyme or nucleotides in a ribozyme determine the shape of that enzyme's active site. A change in a single component—for instance, through mutation—can render the enzyme less effective or even completely nonfunctional.

Enzyme-substrate specificity, which is critical to enzyme activity, has been likened to the fit between a lock and key. This analogy is not completely apt because enzymes' active sites change shape slightly when they bind to their substrate, almost as if a lock could grasp its key once it had been inserted. This latter description of enhanced enzyme-substrate specificity is called the **induced-fit model** (FIGURE 5.5).

In some cases, several different enzymes possess active sites that are complementary to various portions of a single substrate molecule. For example, an important precursor metabolite called phosphoenolpyruvic acid (PEP) is the substrate for at least five enzymes. Depending on the enzyme involved, various products are produced from PEP. In one catabolic pathway, PEP is converted to pyruvic acid, whereas in a particular anabolic pathway, PEP is converted to the amino acid phenylalanine.

Although the exact ways that enzymes lower activation energy are not known, it appears that several mechanisms are involved. Some enzymes appear to bring reactants into sufficiently close proximity to enable a bond to form, whereas other enzymes change the shape of a reactant, inducing a bond to be broken. In any case, enzymes increase the likelihood that bonds will form or break.

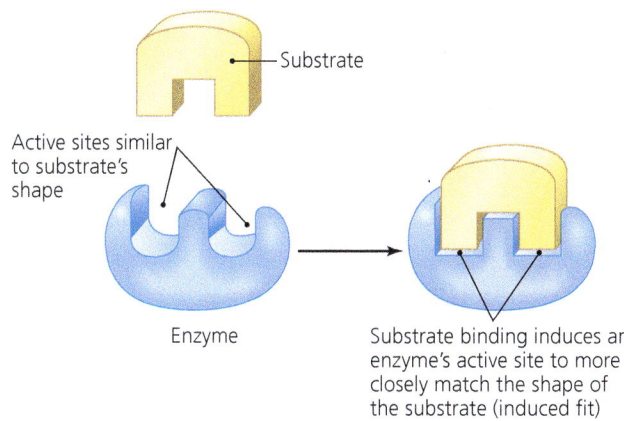

▲ **FIGURE 5.5 Enzymes fitted to substrates.** The induced-fit model of enzyme-substrate interaction. An enzyme's active site is generally complementary to the shape of its substrate, but a perfect fit between them does not occur until the substrate and enzyme bind to form a complex, at which time the enzyme's active site changes to fit the substrate more closely.

The activity of enzymes is believed to follow the process illustrated in **FIGURE 5.6**, which depicts the catabolic lysis of a molecule called fructose 1,6-bisphosphate:

1. An enzyme associates with its specific substrate molecule, which has a shape complementary to that enzyme's active site.

2. The enzyme and its substrate bind to form a temporary intermediate compound called an *enzyme-substrate complex*. The binding of the substrate induces the enzyme to fit the shape of the substrate even more closely—the induced-fit model.

3. Bonds within the substrate are broken, forming two or more products in catabolic reactions. (In anabolic reactions, reactants are linked together to form products.)

4. The enzyme dissociates from the newly formed products, which diffuse away from the site of the reaction.

5. The enzyme resumes its original configuration and is ready to associate with another substrate molecule.

Many factors influence the rate of enzymatic reactions, including temperature, pH, enzyme and substrate concentrations, and the presence of inhibitors.

Play *Enzymes: Steps in a Reaction*
@ Mastering Microbiology

Temperature As mentioned, higher temperatures tend to increase the rate of most chemical reactions because molecules are moving faster and collide more frequently, encouraging bonds to form or break. However, this is not entirely true of enzymatic reactions because as the temperature rises above the stability point for the enzyme, the enzyme loses structure and the active site is lost.

Each enzyme has an optimal temperature for its activity (**FIGURE 5.7a**). The optimum temperature for the enzymes in the human body is 37°C, which is normal body temperature. Part of the reason certain pathogens can cause disease in humans is that the optimal temperature for the enzymes in those microorganisms is also 37°C. The enzymes of some other microorganisms,

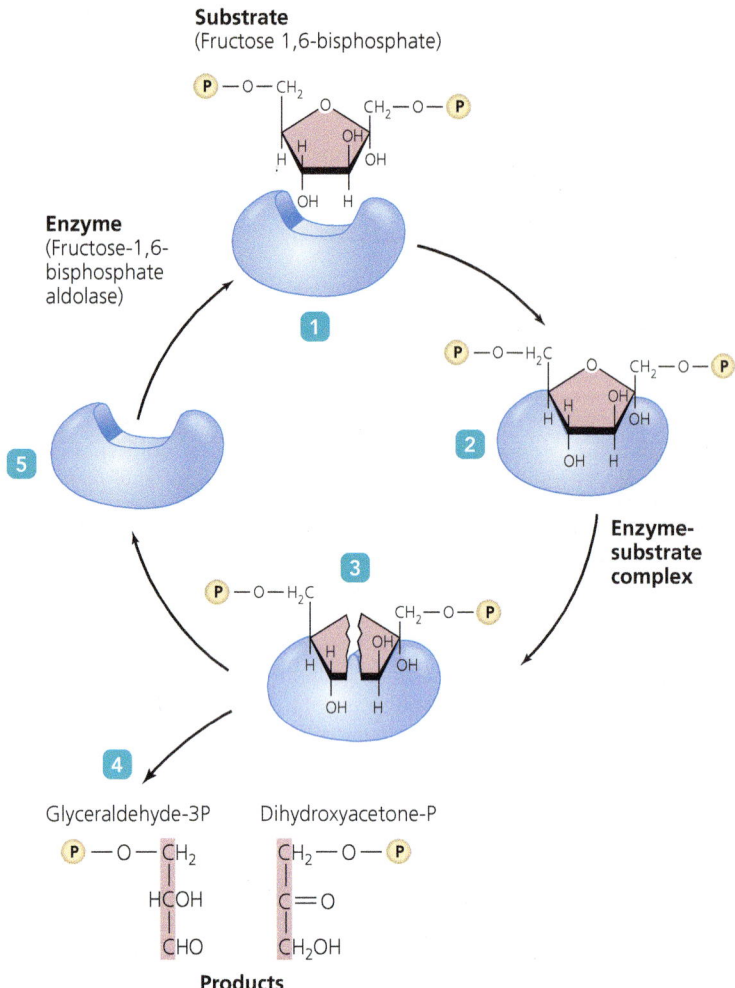

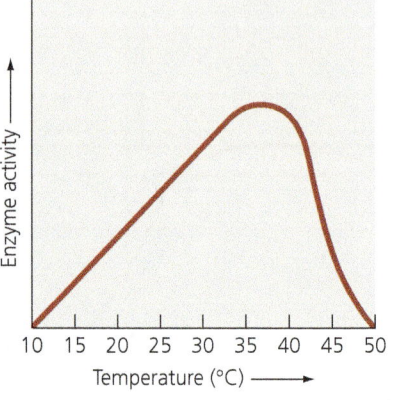

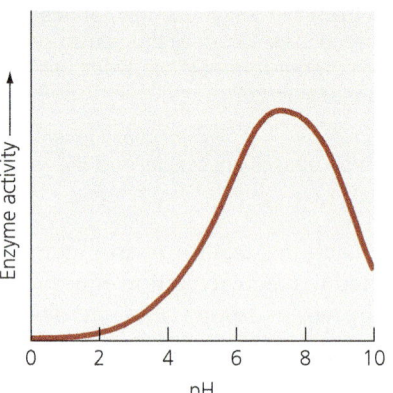

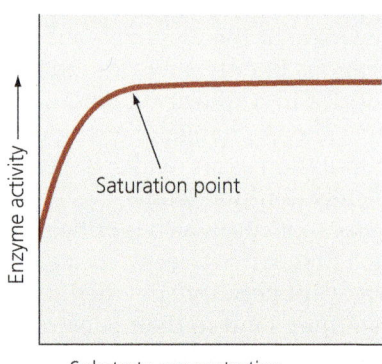

▲ **FIGURE 5.6 The process of enzymatic activity.** Shown here is the lysis of fructose 1,6-bisphosphate by the enzyme fructose-1,6-bisphosphate aldolase (a catabolic reaction). After the enzyme associates with the substrate ①, the two molecules bind to form an enzyme-substrate complex ②. As a result of binding, the enzyme's active site is induced to fit the substrate even more closely; then bonds within the substrate are broken ③, after which the enzyme dissociates from the two products ④. The enzyme resumes its initial shape and is then ready to associate with another substrate molecule ⑤. This entire process occurs 14 times per second at 37°C (body temperature) for a single enzyme.

▲ **FIGURE 5.7 Representative effects of temperature, pH, and substrate concentration on enzyme activity.** Effects on each enzyme will vary. **(a)** Rising temperature enhances enzymatic activity to a point, but above some optimal temperature an enzyme denatures and loses function. **(b)** Enzymes typically have some optimal pH, at which point enzymatic activity reaches a maximum. **(c)** At lower substrate concentrations, enzyme activity increases as the substrate concentration increases and as more and more active sites are utilized. At the substrate concentration at which all active sites are utilized, termed the *saturation point*, enzymatic activity reaches a maximum, and any additional increase in substrate concentration has no effect on enzyme activity. *What is the optimal pH of the enzyme shown in part (b)?*

Figure 5.7 The enzyme's optimal pH is approximately 7.2.

however, function best at different temperatures; this is the case for *hyperthermophiles*, organisms that grow best at temperatures above 80°C.

If temperature rises beyond a certain critical point, the noncovalent bonds within an enzyme (such as the hydrogen bonds between amino acids) will break, and the enzyme will **denature** (FIGURE 5.8). Denatured enzymes lose the specific three-dimensional structure of the active site, so they are no longer functional. Denaturation is said to be *permanent* when an enzyme cannot regain its original structure once conditions return to normal, much like the irreversible solidification of the protein albumin when egg whites are cooked and then cooled. In other cases, denaturation is *reversible*—the denatured enzyme's noncovalent bonds re-form on the return of normal conditions.

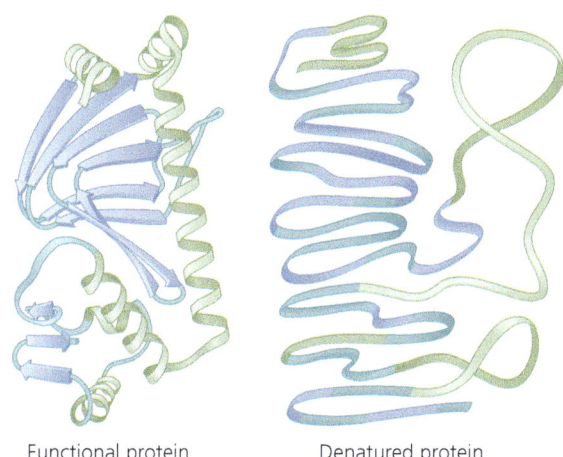

Functional protein   Denatured protein

▲ **FIGURE 5.8 Denaturation of protein enzymes.** Breakage of noncovalent bonds (such as hydrogen bonds) causes the protein to lose its secondary and tertiary structure and become denatured; as a result, the enzyme is no longer functional.

pH Extremes of pH also denature enzymes when ions released from acids and bases interfere with hydrogen bonding and distort and disrupt an enzyme's secondary and tertiary structures. Therefore, each enzyme has an optimal pH (**FIGURE 5.7b**).

Changing the pH provides a way to control the growth of unwanted microorganisms by denaturing their proteins. For example, vinegar (acetic acid, pH 3.0) acts as a preservative in dill pickles, and ammonia (pH 11.5) can be used as a disinfectant.

Enzyme and Substrate Concentration Another factor that determines the rate of enzymatic activity within cells is the concentration of substrate present (**FIGURE 5.7c**). As substrate concentration increases, enzymatic activity increases as more and more enzyme active sites bind more and more substrate molecules. Eventually, when all enzyme active sites have bound substrate, the enzymes have reached their saturation point, and the addition of more substrate will not increase the rate of enzymatic activity.

Obviously, the rate of enzymatic activity is also affected by the concentration of enzyme within cells. In fact, one way that organisms regulate their metabolism is by controlling the quantity and timing of enzyme synthesis. In other words, many enzymes are produced in the amounts and at the times they are needed to maintain metabolic activity. (Chapter 7 discusses the role of genetic mechanisms in regulating enzyme synthesis.) Additionally, eukaryotic cells control some enzymatic activities by compartmentalizing enzymes inside membranes so that certain metabolic reactions proceed physically separated from the rest of the cell. For example, white blood cells catabolize phagocytized pathogens using enzymes packaged within lysosomes.

Control of Enzymatic Activity Organisms can influence enzymatic activity to control metabolic activity. Here, we consider both positive and negative control.

An enzyme can be activated by the binding of a cofactor to the enzyme at a site located away from the active site—a site called an *allosteric site* (al-ō-stār´ik). In *allosteric activation*, the binding of an activator, such as a heavy-metal ion or other

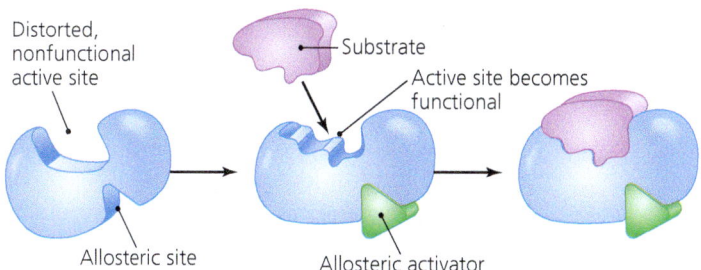

▲ **FIGURE 5.9 Allosteric activation.** Allosteric activation results when the binding of an activator molecule to an allosteric site causes a change in the active site that makes it capable of binding substrate.

cofactor, to an allosteric site causes the enzyme's active site to change shape, which activates the enzyme (**FIGURE 5.9**).

Substances that block enzyme activity are called *inhibitors*. Enzymatic inhibitors may be either competitive or noncompetitive. **Competitive inhibitors** are shaped such that they fit into an enzyme's active site and thus prevent the normal substrate from binding (**FIGURE 5.10a**). Such inhibitors do not undergo a chemical reaction to form products. Competitive inhibitors can bind permanently or reversibly to an active site. Permanent binding results in permanent loss of enzymatic activity; reversible competition can be overcome by an increase in the concentration of substrate molecules, increasing the likelihood that active sites will be filled with substrate instead of inhibitor (**FIGURE 5.10b**).

Play *Enzymes: Competitive Inhibition*
@ Mastering Microbiology

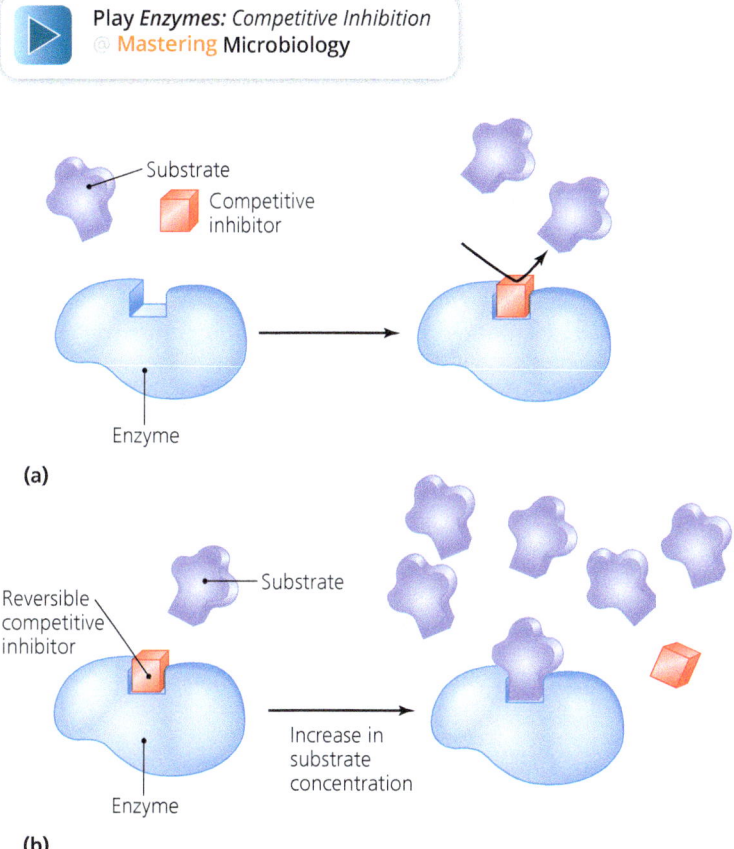

▲ **FIGURE 5.10 Competitive inhibition of enzyme activity.** (a) Inhibitory molecules, which are similar in shape to substrate molecules, compete for and block active sites. (b) Reversible inhibition can be overcome by an increase in substrate concentration.

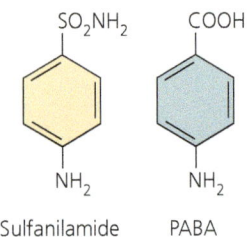

An example of competitive inhibition is the action of the antibacterial drug sulfanilamide, which has a shape similar to that of a metabolite called para-aminobenzoic acid (PABA). Sulfanilamide can fill the active site of an enzyme required for the conversion of PABA into folic acid. Folic acid is essential for RNA and DNA synthesis. Once sulfanilamide is bound to the enzyme, it stays bound and inhibits synthesis of folic acid and thus prevents synthesis of RNA and DNA. Thus, sulfanilamide effectively inhibits microbes that make their own folic acid for synthesizing nucleic acids. Humans do not synthesize folic acid—we must acquire it as a vitamin in our diets—so sulfanilamide does not affect us in this way.

Noncompetitive inhibitors do not attach to the active site but instead bind to an allosteric site on the enzyme. This binding alters the shape of the active site so that enzymatic activity is reduced or blocked completely (**FIGURE 5.11**). Some enzymes have more than one allosteric site, some inhibitory sites and some activation sites (described earlier), allowing the cell to regulate enzyme function precisely.

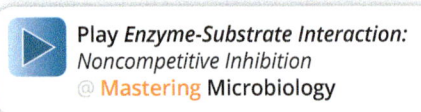
Play *Enzyme-Substrate Interaction: Noncompetitive Inhibition*
@ Mastering Microbiology

Cells often control the action of enzymes through **feedback inhibition** (also called *negative feedback* or *end-product inhibition*). Allosteric feedback inhibition functions in much the way a thermostat controls a heater. As the room gets warmer, a sensor inside the thermostat changes shape and sends a signal that turns off the heater. Similarly, in metabolic feedback inhibition, the end-product of a series of reactions is an allosteric inhibitor of an enzyme in an earlier part of the pathway (**FIGURE 5.12**). Because the product of each reaction in the pathway is the substrate for the next reaction, inhibition of the first enzyme in the series inhibits the entire pathway, thereby saving the cell energy. For example, in *Escherichia coli*, the presence of the amino acid isoleucine allosterically inhibits the first enzyme in the pathway that produces isoleucine. In this manner, the bacterium prevents the synthesis of isoleucine when this amino acid is already available. When isoleucine is depleted, the pathway is no longer inhibited, and isoleucine production resumes.

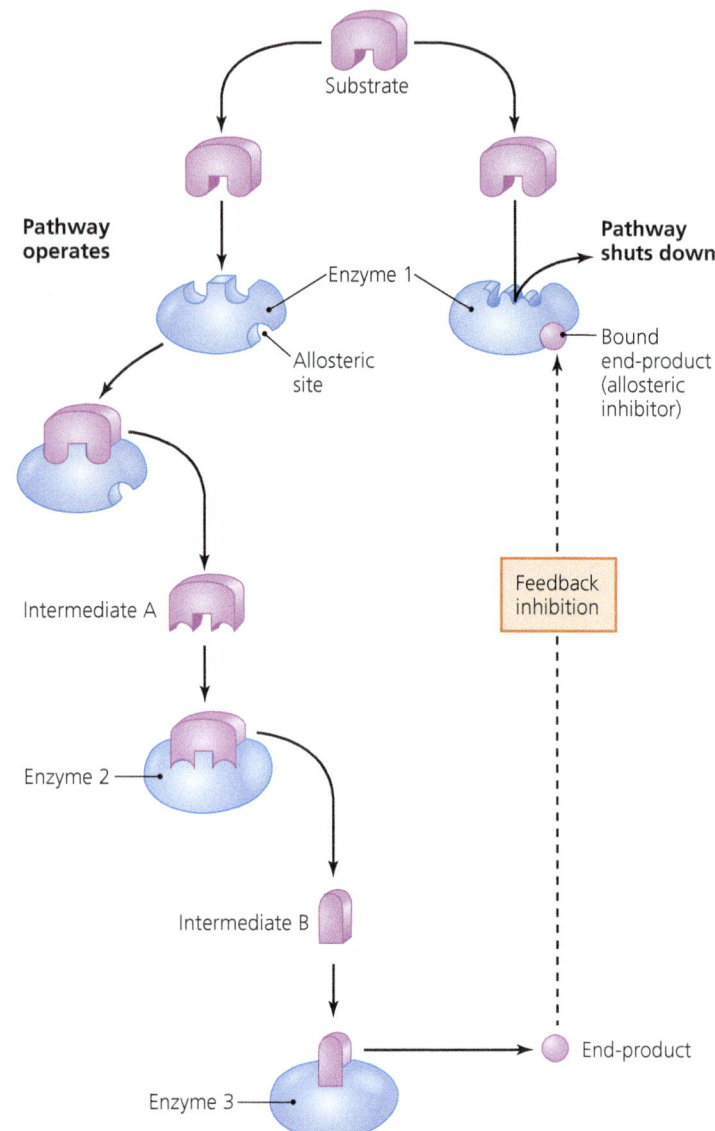

▲ **FIGURE 5.12 Feedback inhibition.** The end-product of a metabolic pathway allosterically inhibits the initial step, shutting down the pathway.

To this point, we have viewed the concept of metabolism as a collection of chemical reactions (pathways) that can be categorized as either catabolic (breaking down) or anabolic (building up). Because enzymes are required to lower the activation energy of these reactions, we examined them in some detail.

Energy is also critical to metabolism, so we examined redox reactions as a means of transferring energy within cells. We saw, for example, that redox reactions and carrier molecules are used to transfer energy from catabolic pathways to ATP, a molecule that stores energy in cells.

MICRO CHECK

5. What is a holoenzyme?
6. What type of enzyme is not made of protein?
7. Why are denatured enzymes no longer functional?
8. How does a noncompetitive inhibitor limit an enzyme's activity?

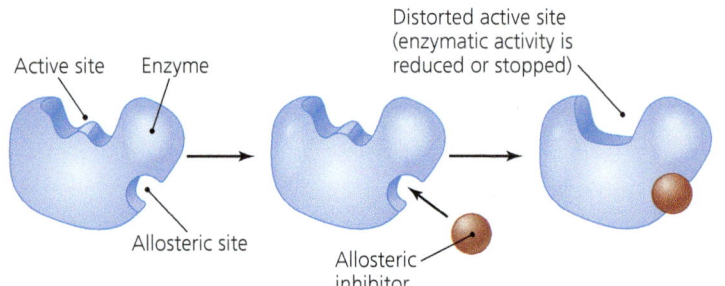

▲ **FIGURE 5.11 Noncompetitive inhibition at an allosteric site.** Inhibition results from a change in the shape of the active site when an inhibitor binds to an allosteric site.

> **TELL ME WHY**
>
> How can oxidation take place in an anaerobic environment, that is, without oxygen?

We will now consider how cells acquire and utilize metabolites, which are used to synthesize the macromolecules necessary for growth and, eventually, reproduction—the ultimate goal of metabolism. We will also consider in more detail the phosphorylation of ADP to make ATP.

Carbohydrate Catabolism

LEARNING OUTCOME

5.8 In general terms, describe the three stages of aerobic glucose catabolism (glycolysis, the citric acid cycle, and an electron transport chain), including their substrates, products, and net energy production.

Many organisms oxidize carbohydrates as their primary energy source for anabolic reactions. Glucose is used most commonly, though other sugars, amino acids, and fats are also utilized, often by first being converted into glucose. Glucose is catabolized via one of two processes: either via *cellular respiration*[3]—a process that results in the complete breakdown of glucose to carbon dioxide and water—or via *fermentation*, which results in organic waste products.

As shown in **FIGURE 5.13**, both cellular respiration and fermentation begin with *glycolysis* (glī-kol´i-sis), a process that catabolizes a single molecule of glucose to two molecules of pyruvic acid (also called *pyruvate*) and results in a small amount of ATP production. Respiration then continues via the *citric acid cycle* and an *electron transport chain*, which results in a significant amount of ATP production. Fermentation involves the conversion of pyruvic acid into other organic compounds. Because it lacks the citric acid cycle and an electron transport chain, fermentation results in the production of much less ATP than does respiration.

The following is a simplified discussion of glucose catabolism. To help understand the basic reactions in each of the pathways of glucose catabolism, pay special attention to three things: the number of carbon atoms in each of the intermediate products, the relative numbers of ATP molecules produced in each pathway, and the changes in the coenzymes NAD^+ and FAD as they are reduced and then oxidized back to their original forms.

Glycolysis

Glycolysis[4] is commonly the first step in the catabolism of glucose via both respiration and fermentation. Glycolysis in most organisms follows the *Embden-Meyerhof-Parnas (EMP) pathway*, which is named after the scientists who discovered it.

In general, as its name implies, glycolysis involves the splitting of a six-carbon glucose molecule into two three-carbon sugar molecules. When these three-carbon molecules are oxidized to pyruvic acid, some of the energy released is stored in molecules

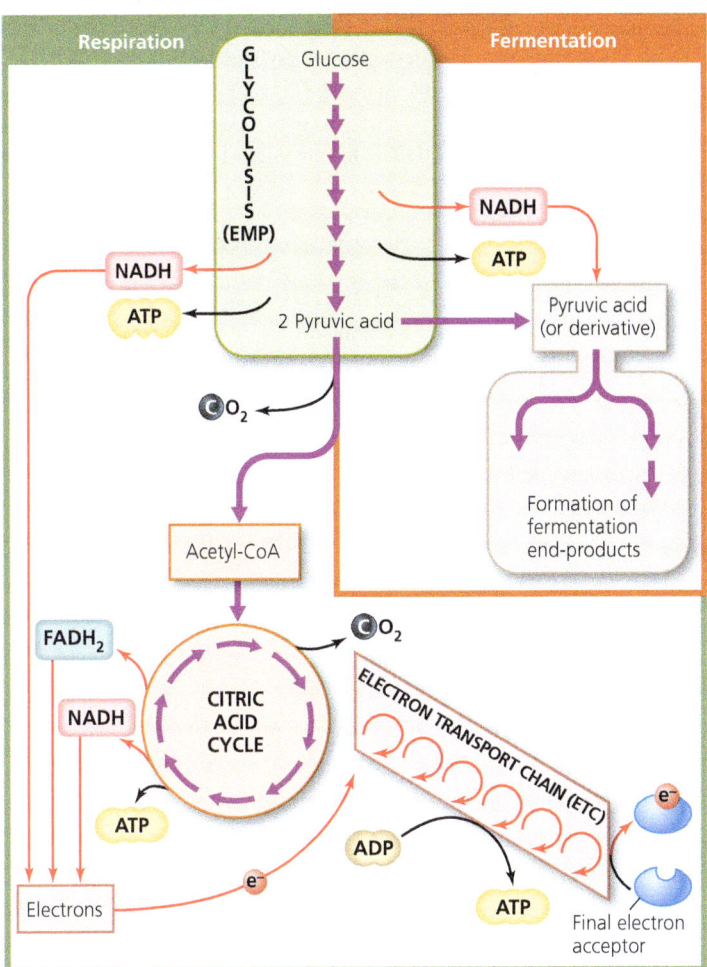

▲ **FIGURE 5.13 Summary of glucose catabolism.** Glucose catabolism begins with glycolysis, which forms pyruvic acid and two molecules of both ATP and NADH. Two pathways branch from pyruvic acid: respiration and fermentation. In aerobic respiration (shown here), the citric acid cycle and the electron transport chain completely oxidize pyruvic acid to CO_2 and H_2O, in the process synthesizing many molecules of ATP. Fermentation results in the incomplete oxidation of pyruvic acid to form organic fermentation products. *Which of the blue ovals represents a reduced chemical?*

Figure 5.13 The upper one, with an additional electron, is reduced.

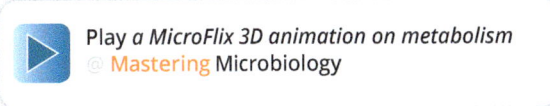

Play *a MicroFlix 3D animation on metabolism*
@ **Mastering** Microbiology

of ATP and NADH. Cells can use many of the intermediate molecules in glycolysis as precursor metabolites.

Play *Glycolysis: Overview*
@ **Mastering** Microbiology

The EMP pathway of glycolysis occurs in the cytosol and can be divided into three stages involving a total of 10 steps (**FIGURE 5.14**). Each step is catalyzed by its own enzyme:

1. *Energy-investment stage* (steps **1**–**3**). In finance, one must invest before a profit can be made. In glycolysis, a cell "invests" the energy in two molecules of ATP by phosphorylating a six-carbon glucose molecule and rearranging its atoms to form fructose 1,6-bisphosphate.

---
[3]Cellular respiration is often referred to simply as *respiration*, which should not be confused with breathing, also called respiration.
[4]From Greek *gleukos*, meaning "sweet wine," and *lysis*, meaning "to loosen."

2. *Lysis stage* (steps 4 and 5). The cell then lyses fructose 1,6-bisphosphate into glyceraldehyde 3-phosphate (G3P)[5] and dihydroxyacetone phosphate (DHAP). Each of these compounds contains three carbon atoms and is freely convertible into the other. DHAP is converted to G3P before the next stage, so the cell has two molecules of G3P at the end of the lysis stage.

3. *Energy-conserving stage* (steps 6–10). The cell oxidizes the two molecules of G3P to pyruvic acid, yielding two ATP molecules from each oxidation. So, in this stage, the cell has conserved

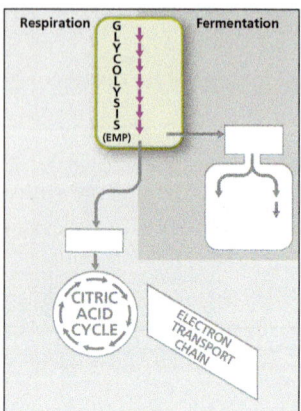

ENERGY-INVESTMENT STAGE

Step 1. Glucose is phosphorylated by ATP to form glucose 6-phosphate.

Steps 2 and 3. The atoms of glucose 6-phosphate are rearranged to form fructose 6-phosphate. Fructose 6-phosphate is phosphorylated by ATP to form fructose 1,6-bisphosphate.

LYSIS STAGE

Step 4. Fructose 1,6-bisphosphate is cleaved to form glyceraldehyde 3-phosphate (G3P) and dihydroxyacetone phosphate (DHAP).

Step 5. DHAP is rearranged to form another G3P.

ENERGY-CONSERVING STAGE

Step 6. Inorganic phosphates are added to the two G3P, and two NAD^+ are reduced.

Step 7. Two ADP are phosphorylated by substrate-level phosphorylation to form two ATP.

Steps 8 and 9. The remaining phosphates are moved to the middle carbons. A water molecule is removed from each substrate.

Step 10. Two ADP are phosphorylated by substrate-level phosphorylation to form two ATP. Two pyruvic acid are formed.

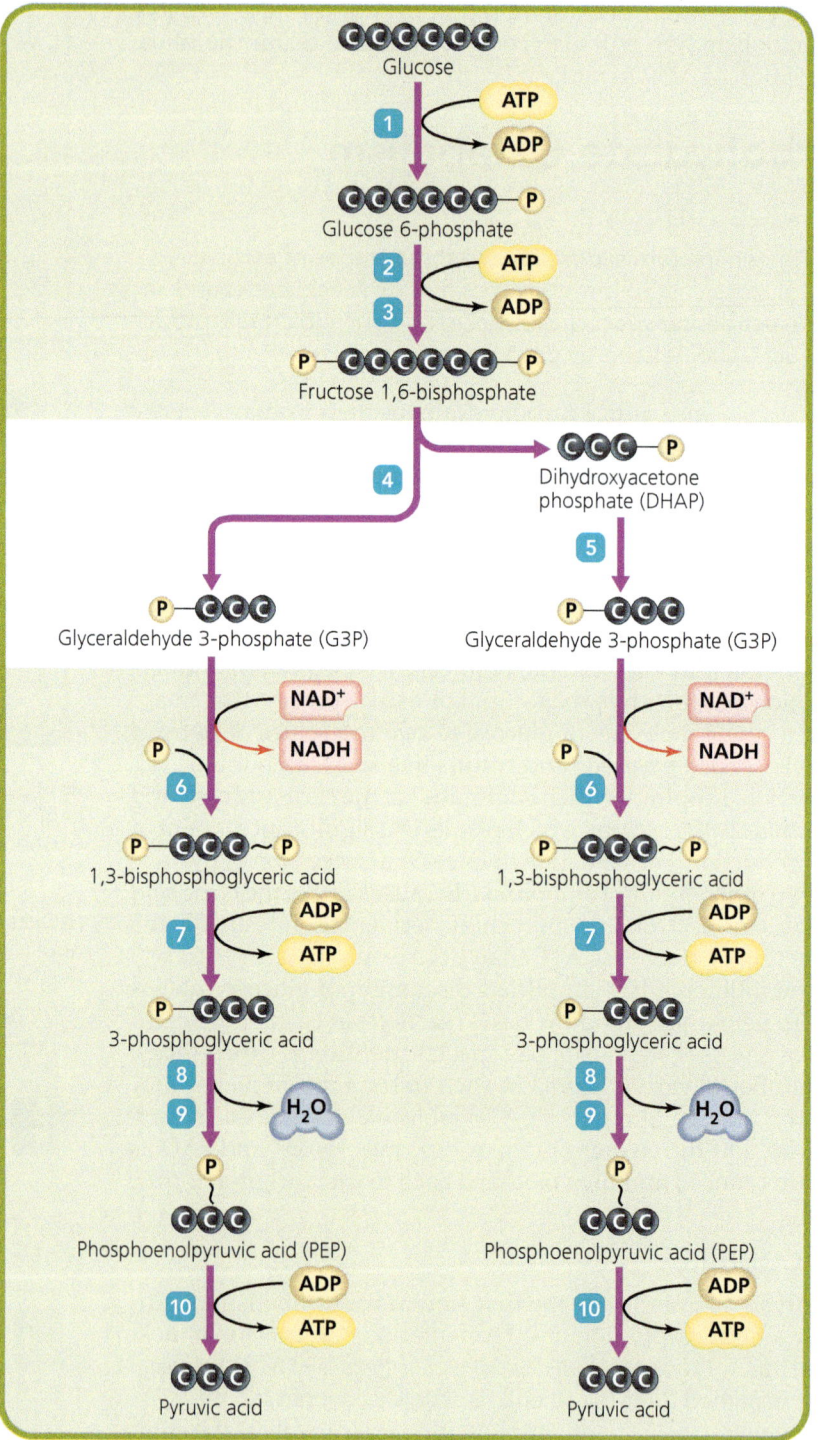

▲ **FIGURE 5.14 Glycolysis by the EMP pathway.** Glucose is cleaved and ultimately transformed into two molecules of pyruvic acid in this process. Four ATPs are formed and two ATPs are used, so the result is a net gain of two ATPs. Two molecules of NAD^+ are reduced to NADH. (For simplification, only carbon atoms and phosphate are shown.)

---

[5] G3P is also known as phosphoglyceraldehyde, or PGAL.

and stored energy from glucose into four molecules of ATP. (Of course, two molecules of ATP were used in the energy-investment stage, so the net gain is two molecules of ATP.)

Play *Glycolysis: Steps*
Mastering Microbiology

Our study of glycolysis provides our first opportunity to study *substrate-level phosphorylation* (see steps 1, 3, 7, and 10). Let's examine this important process more closely by considering the 10th and final step of glycolysis.

Each of the two phosphoenolpyruvic acid (PEP) molecules produced in step 9 of glycolysis is a three-carbon compound containing a high-energy phosphate bond. In the presence of a specific holoenzyme (which requires a Mg^{2+} cofactor), the high-energy phosphate in PEP (one substrate) is transferred to an ADP molecule (a second substrate) to form ATP (step 10 and FIGURE 5.15); the direct transfer of the phosphate between the two substrates is the reason the process is called **substrate-level phosphorylation**. A variety of substrate-level phosphorylations occur in metabolism. As you might expect, each type has its own enzyme that recognizes both ADP and the enzyme's specific substrate molecule.

In glycolysis, two ATP molecules are invested by substrate-level phosphorylation to prime glucose for lysis, and four molecules of ATP are produced by substrate-level phosphorylation. Therefore, a net gain of two ATP molecules occurs for each molecule of glucose that is oxidized to pyruvic acid. Glycolysis also yields two molecules of NADH.

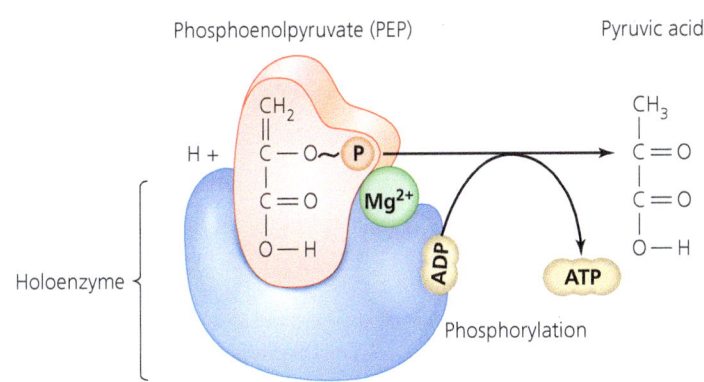

▲ **FIGURE 5.15 Example of substrate-level phosphorylation.** High-energy phosphate bonds are transferred from phosphoenolpyruvate (PEP) to ATP in this process. *What role does Mg^{2+} play in this reaction?*

Figure 5.15 Mg^{2+} is a cofactor of the enzyme.

(as′e-til kō-ā; FIGURE 5.16). Enzymes in the pyruvate dehydrogenase complex carry this out, first by removing one carbon from pyruvic acid as CO_2—a reaction called a *decarboxylation* 1. An enzyme then joins the remaining two-carbon molecule (called acetate) to coenzyme A with a high-energy bond to form acetyl-CoA 2. This reaction also produces one molecule of NADH 3.

Cellular Respiration

Play an Interactive Microbiology video
Mastering Microbiology for more information about Aerobic Respiration in Prokaryotes.

LEARNING OUTCOMES

5.9 Discuss the roles of acetyl-CoA, the citric acid cycle, and electron transport in carbohydrate catabolism.
5.10 Contrast electron transport in aerobic and anaerobic respiration.
5.11 Identify four classes of carriers in electron transport chains.
5.12 Describe the role of chemiosmosis in oxidative phosphorylation of ATP.

After glucose has been oxidized via glycolysis or one of the alternative pathways considered shortly, a cell uses the resultant pyruvic acid molecules to complete either cellular respiration or fermentation (which we will discuss in a later section). Our topic here—**cellular respiration**—is a metabolic process that involves the complete oxidation of substrate molecules and then production of ATP by a series of redox reactions. The three stages of cellular respiration are (1) synthesis of acetyl-CoA, (2) the citric acid cycle, and (3) a final series of redox reactions, called an electron transport chain, that passes electrons to a chemical not derived from cellular metabolism.

Synthesis of Acetyl-CoA

Before pyruvic acid can enter the citric acid cycle for respiration, it must first be converted to *acetyl-coenzyme A*, or **acetyl-CoA**

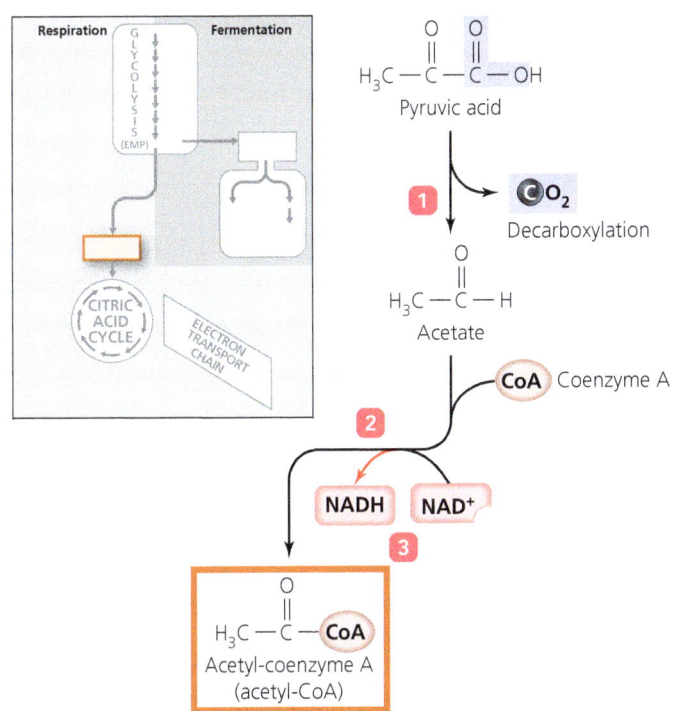

▲ **FIGURE 5.16 The pyruvate dehydrogenase complex.** The enzyme complex acts in a stepwise manner to 1 remove CO_2 from pyruvic acid, 2 attach the remaining two-carbon acetate to coenzyme A, and 3 simultaneously reduce a molecule of NAD^+ to NADH. Because glycolysis produces two molecules of pyruvic acid, two molecules of acetyl-CoA and NADH are produced (only one of each is shown here).

Recall that glycolysis produced two molecules of pyruvic acid from each molecule of glucose. Therefore, this stage produces two molecules of acetyl-CoA, two molecules of CO_2, and two molecules of NADH.

The Citric Acid (Krebs) Cycle

At this point in the catabolism of a molecule of glucose, a great amount of energy remains in the bonds of acetyl-CoA. The **citric acid cycle**[6] is a "circular" series of eight enzymatically catalyzed reactions that transfer much of this stored energy via electrons to the coenzymes NAD^+ and FAD (the cycle is diagrammed in **FIGURE 5.17**). The two carbon atoms in acetate are oxidized, and the coenzymes are reduced. The citric acid cycle occurs in the cytosol of prokaryotes and in the matrix of mitochondria in eukaryotes. It is also known as the *tricarboxylic acid (TCA) cycle* because many of its compounds have three carboxyl (—COOH) groups. Still another name for the citric acid cycle is the **Krebs cycle**, for the scientist who discovered details of the reactions. As we will see, cells also use the citric acid cycle for catabolism of lipids and proteins.

There are six types of reactions in the citric acid cycle:

- Anabolism (of citric acid; step **1**)
- Isomerization (step **2**)
- Redox reactions (steps **3**, **4**, **6**, and **8**)
- Decarboxylations (steps **3** and **4**)
- Substrate-level phosphorylation (step **5**)
- Hydration (step **7**)

In the first step of the citric acid cycle, splitting of the high-energy bond between acetate and coenzyme A releases enough energy to enable the binding of the freed two-carbon acetate to a four-carbon compound called oxaloacetic acid, forming the six-carbon compound citric acid (step **1**).

As you study Figure 5.17, notice that after isomerization (step **2**), the decarboxylations of the citric acid cycle release two molecules of CO_2 for each acetyl-CoA that enters (steps **3** and **4**). Thus, for every pair of carbon atoms that enter the cycle, two are lost to the environment. At this juncture, all six carbon atoms of a molecule of glucose have been lost to the environment: two as CO_2 produced during decarboxylation of two molecules of pyruvic acid to form acetyl-CoA, and four as CO_2 molecules released in decarboxylations during *two* turns through the citric acid cycle. (One molecule of acetyl-CoA enters the cycle at a time.)

For every two molecules of acetyl-CoA that pass through the citric acid cycle, two molecules of ATP are generated by substrate-level phosphorylation (step **5**). A molecule of guanosine triphosphate (GTP), which is similar to ATP, can serve as an intermediary in this process. The two molecules of ATP produced by the citric acid cycle are a small amount compared to the energy available. Most of the energy contained in the original glucose molecule is still in electrons, which are carried away by NADH and $FADH_2$.

Redox reactions reduce FAD to $FADH_2$ (step **6**) and NAD^+ to NADH (steps **3**, **4**, and **8**); so, for the two molecules of acetyl-CoA derived from the original glucose molecule, six molecules of NADH and two of $FADH_2$ are formed. These coenzymes are the most important molecules of respiration because they carry a large amount of energy that is subsequently used to phosphorylate ADP to ATP. Many of the intermediates of the citric acid cycle are also precursor metabolites; for example, cells can use oxaloacetic acid to make several kinds of amino acids (see Figure 5.29).

Play *Krebs Cycle: Overview, Steps* @ Mastering Microbiology

Electron Transport

Scientists estimate that each day an average human synthesizes his or her own weight in ATP molecules and uses them for metabolism, responsiveness, growth, and cell reproduction. ATP turnover in microbes is relatively as copious. In organisms that use respiration, the most significant production of ATP does not occur through glycolysis or the citric acid cycle, but rather through the stepwise release of energy from a series of redox reactions known as an **electron transport chain** (**FIGURE 5.18a**).

Play *Electron Transport Chain: Overview* @ Mastering Microbiology

An electron transport chain consists of a series of membrane-bound *carrier molecules* that pass electrons from one to another and ultimately to a *final electron acceptor*. Typically, as we have seen, electrons come from the catabolism of an organic molecule such as glucose; however, microorganisms called *lithotrophs* (lith´ō-trōfs) acquire electrons from inorganic sources such as H_2, NO_2^-, or Fe^{2+}. (Chapter 6 discusses lithotrophs further.) In any case, molecules in the chain pass electrons down the chain to the final acceptor like firefighters of old, who passed buckets of water from one to another until the last one threw the water on a fire. In electron transport, NAD^+ is an "empty-handed firefighter"; NADH is a "firefighter with a bucket." As with a bucket brigade, the final step of electron transport is irreversible. Energy from the electrons is used to actively transport (pump) protons (H^+) across the membrane, establishing a *proton gradient* that generates ATP via a process called *chemiosmosis*, which we will discuss shortly.

To avoid getting lost in the details of electron transport, keep the following critical concepts in mind:

- Electrons pass sequentially from one membrane-bound carrier molecule to another, each time losing some energy. Eventually, they pass to a final acceptor molecule.
- The electrons' energy is used to pump protons across the membrane.

Electron transport chains are located in the cytoplasmic membranes of prokaryotes (and in the inner mitochondrial membranes of eukaryotes) (**FIGURE 5.18b**). Though NADH and $FADH_2$ donate electrons as hydrogen atoms (electrons and protons), many carriers pass only the electrons down the chain. There are four categories of carriers in the transport chains:

- *Flavoproteins* are integral membrane proteins, many of which contain flavin, a coenzyme derived from riboflavin (vitamin B_2). One form of flavin is flavin mononucleotide (FMN), which is the initial carrier molecule of electron transport chains of mitochondria. The familiar FAD is a coenzyme for other flavoproteins. Like all carrier molecules in an electron transport chain, flavoproteins alternate between reduced and oxidized states.

---
[6]Named for the first product of the cycle (citric acid). Biochemist Hans Krebs elucidated its reactions in the 1940s.

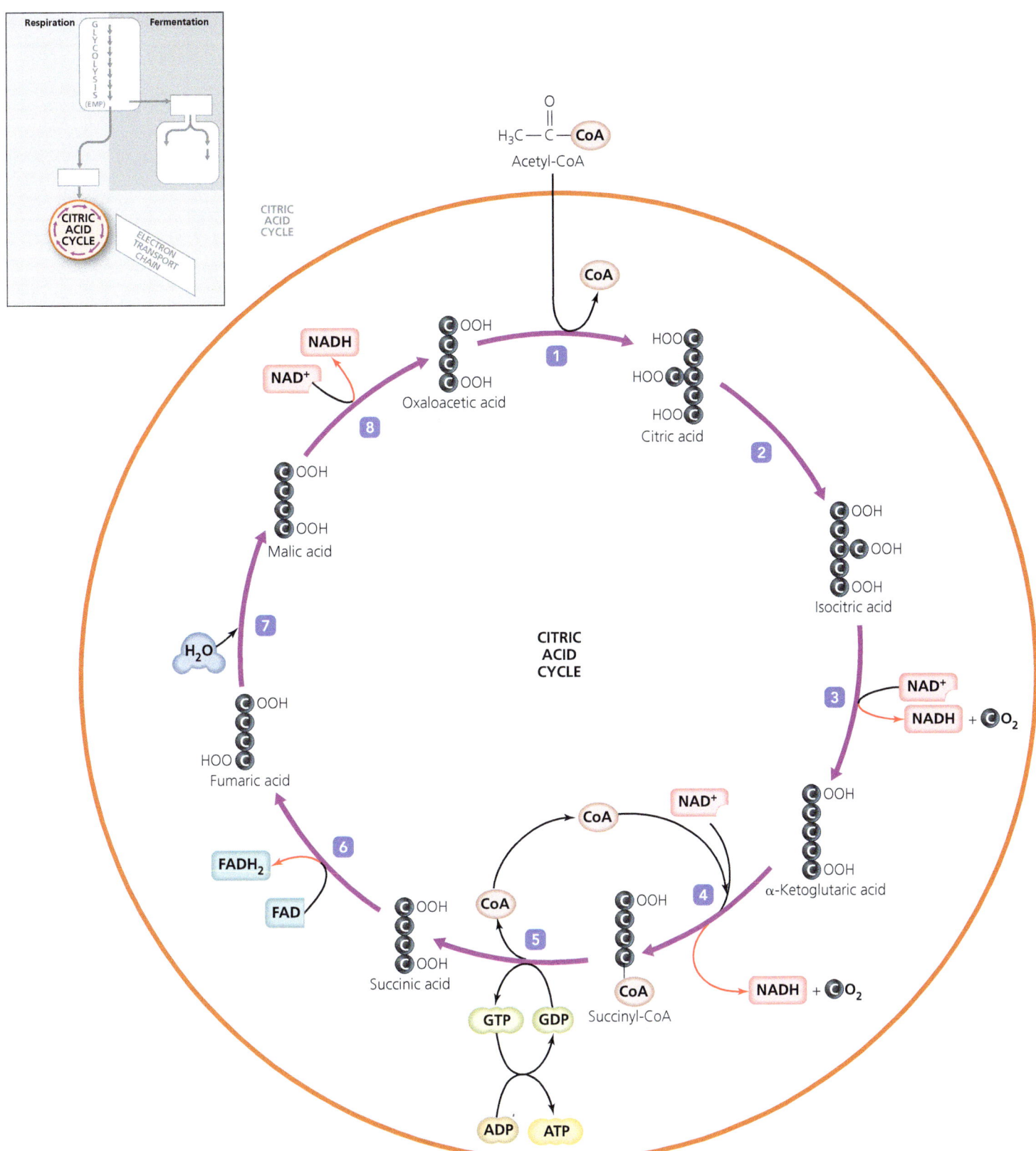

▲ **FIGURE 5.17 The citric acid cycle.** ① Acetyl-CoA enters the citric acid cycle by joining with oxaloacetic acid to form citric acid, releasing coenzyme A. ②–④ Two oxidations and two decarboxylations and the addition of coenzyme A yield succinyl-CoA. ⑤ Substrate-level phosphorylation produces ATP and again releases coenzyme A. ⑥–⑧ Further oxidations and rearrangements regenerate oxaloacetic acid, and the cycle can begin anew. Two molecules of acetyl-CoA enter the cycle for each molecule of glucose undergoing glycolysis.

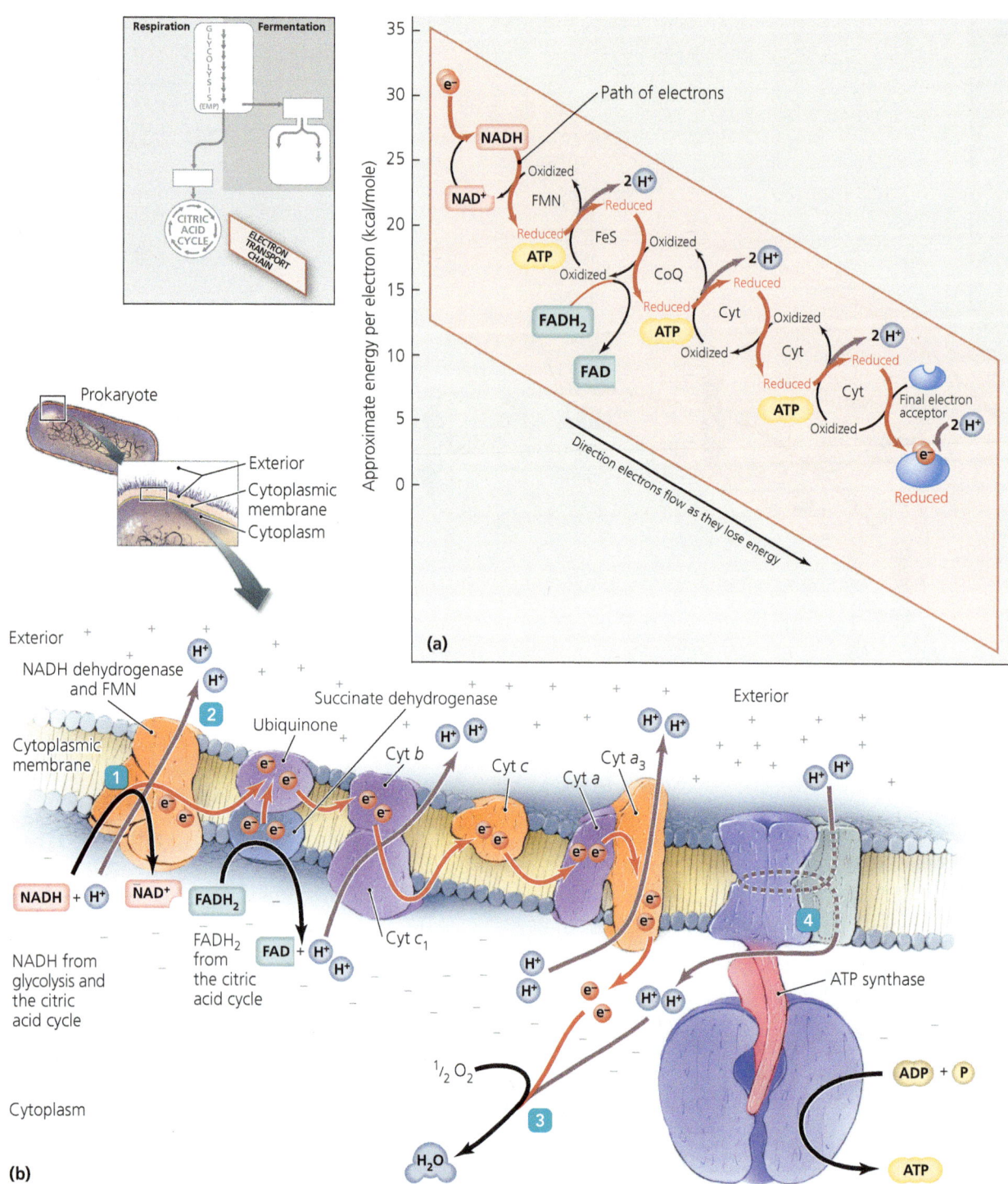

▲ FIGURE 5.18 **One possible arrangement of electron transport chain molecules.** The exact types and sequences of carrier molecules in electron transport chains vary among organisms. **(a)** Diagrammatic view of electron and energy transfer. As electrons move down the chain from molecule to molecule (red arrows), energy is released. This energy can be used to synthesize ATP. The ATP icons indicate the approximate point in the chain where enough energy has been released to make an ATP molecule, but the molecules of the chain do not actually synthesize ATP. **(b)** Electron transport chains are located in the cytoplasmic membranes of prokaryotes (as illustrated) and in the inner membranes of mitochondria in eukaryotes. Energy of electrons moving down the chain (red arrows) **1**, is used to pump protons (H^+) across the membrane **2**. Eventually, the electrons pass to a final acceptor, in this case, oxygen **3**. Protons then flow through ATP synthase (ATPase), which phosphorylates ADP to make ATP **4**. Approximately one molecule of ATP is generated for every three protons that cross the membrane. *Why is it essential that electron carriers of an electron transport chain be membrane-bound?*

Figure 5.18 The carrier molecules of electron transport chains are membrane-bound for at least two reasons: (1) Carrier molecules need to be in fairly close proximity so they can pass electrons between each other. (2) The goal of the chain is the pumping of H^+ across a membrane to create a proton gradient; without a membrane, there could be no gradient.

- *Ubiquinones* (yū-bik′wi-nōns) are lipid-soluble, nonprotein carriers that are so named because they are ubiquitous (present universally) in cells. Ubiquinones are derived from vitamin K. In mitochondria, the ubiquinone is called *coenzyme Q*.

- *Metal-containing proteins* are a mixed group of integral proteins with a wide-ranging number of iron, sulfur, or copper atoms that can alternate between reduced and oxidized states. Iron-sulfur proteins occur in various places in electron transport chains of many organisms. Copper proteins are found only in electron transport chains involved in photosynthesis (discussed shortly).

- *Cytochromes* (sī′tō-krōms) are integral proteins associated with *heme*, which is the same iron-containing, nonprotein, pigmented molecule found in the hemoglobin of blood. Iron can alternate between a reduced (Fe^{2+}) state and an oxidized (Fe^{3+}) state. Researchers named cytochromes with letters and numbers based on the order in which they identified them, so the sequence of carrier molecules can seem out of order; for example, cytochrome *b* may occur before cytochrome *a* in an electron transport chain.

Play *Electron Transport Chain: The Process* @ Mastering Microbiology

The carrier molecules in electron transport chains are diverse—bacteria typically have different carrier molecules arranged in different sequences from those of archaea or the mitochondria of eukaryotes. Even among bacteria, the makeup of carrier molecules can vary. For example, the pathogens *Neisseria* (nī-se′rē-ă) and *Pseudomonas* (soo-dō-mō′nas) contain two cytochromes, *a* and *a*₃ —together called *cytochrome oxidase*—that oxidize cytochrome *c*; such bacteria are said to be *oxidase positive*. In contrast, other bacterial pathogens, such as *Escherichia*, *Salmonella* (sal′mŏ-nel′ă), and *Proteus* (prō′tē-ŭs), lack cytochrome oxidase and are thus considered to be *oxidase negative*.

Electrons carried by NADH enter a transport chain earlier than electrons carried by $FADH_2$, which are passed to a ubiquinone. Thus, more energy can be harvested from electrons carried by NADH than from those carried by $FADH_2$ and more molecules of ATP can be generated from NADH than from $FADH_2$.

The order and types of electron carriers in electron transport chains vary among organisms. Some bacteria, including *E. coli*, can even change their carrier molecules to accommodate different environmental conditions. Researchers are not agreed on which carrier molecules are the actual proton pumps or on the number of protons that are pumped as energy from electrons is harvested. Figure 5.18b shows one possibility.

In some organisms, the final electron acceptors are oxygen atoms, which, with the addition of hydrogen ions, generate H_2O. Because they use oxygen as the final electron acceptor, these organisms are called *aerobes* and conduct **aerobic**[7] **respiration**.

Other organisms, called *anaerobes*,[8] perform **anaerobic respiration** by using inorganic chemicals other than oxygen (or rarely an organic molecule) as the final electron acceptor. The anaerobic bacterium *Desulfovibrio* (dē′sul-fō-vib′rē-ō), for example, reduces sulfate (SO_4^{2-}) to hydrogen sulfide gas (H_2S), whereas anaerobes in the genera *Bacillus* (ba-sil′ŭs) and *Pseudomonas* utilize nitrate (NO_3^-) to produce nitrite ions (NO_2^-), nitrous oxide (N_2O), or nitrogen gas (N_2). Some prokaryotes—particularly archaea called methanogens—reduce carbonate (CO_3^{2-}) to methane gas (CH_4). Medical laboratory scientists can test for products of anaerobic respiration, such as nitrite, to help identify some species of microbes. (Chapter 27 discusses anaerobic respiration as a critical process in recycling nitrogen and sulfur in nature.)

Play *Electron Transport Chain: Factors Affecting ATP Yield* @ Mastering Microbiology

In summary, a number of redox reactions in glycolysis and in the citric acid cycle strip electrons, which carry energy, from glucose molecules and transfer them to molecules of NADH and $FADH_2$. In turn, NADH and $FADH_2$ pass the electrons to electron transport chains. As the electrons move down an electron transport chain, proton pumps use the electrons' energy to actively transport protons (H^+) across the membrane, creating a proton concentration gradient.

Recall, however, that the significance of electron transport is not merely pumping protons across a membrane but the synthesis of ATP. We turn now to the process by which cells synthesize ATP using the proton gradient.

Chemiosmosis

Chemiosmosis is a general term for the use of ion gradients to generate ATP; that is, ATP is synthesized utilizing energy released by the flow of ions down their electrochemical gradient across a membrane. The term should not be confused with osmosis of water. To understand chemiosmosis, we need to review several concepts concerning diffusion and phospholipid membranes (see pp. 70–71).

Recall that chemicals diffuse from areas of high concentration to areas of low concentration and toward an electrical charge opposite their own. We call the composite of differences in concentration and charge an *electrochemical gradient*. Chemicals diffuse down their electrochemical gradients. Recall as well that membranes of cells and organelles are impermeable to most chemicals unless a specific protein channel allows their passage across the membrane. A membrane maintains an electrochemical gradient by keeping one or more chemicals in a higher concentration on one side. The membrane's blockage of diffusion creates potential energy, like water behind a dam.

Chemiosmosis uses the potential energy of electrochemical gradients to phosphorylate ADP into ATP. Even though chemiosmosis is a general principle relevant to both *oxidative phosphorylation* and *photophosphorylation*, here we consider it as it relates to oxidative phosphorylation.

As we have seen, cells use the energy released during the redox reactions of electron transport chains to actively transport protons (H^+) across a membrane. Theoretically, an electron transport chain pumps three pairs of protons for each pair of electrons contributed by NADH, and it pumps two pairs of protons for each electron pair delivered by $FADH_2$. This difference

---

[7] From Greek *aer*, meaning "air" (i.e., oxygen), and *bios*, meaning "life."

[8] The Greek prefix *an* means "not."

results from the fact that NADH delivers electrons farther up the chain than does FADH$_2$; therefore, energy carried by FADH$_2$ is used to transport one-third fewer protons (see Figure 5.18a). Because lipid bilayers are impermeable to protons, the transport of protons to one side of the membrane creates an electrochemical gradient known as a **proton gradient**, which has potential energy known as a *proton motive force*.

Protons, propelled by the proton motive force, flow down their electrochemical gradient through protein channels, called **ATP synthases (ATPases)**, that phosphorylate molecules of ADP to ATP (see Figure 5.18b ④). Such phosphorylation is called **oxidative phosphorylation** because the proton gradient is created by the oxidation of carriers in an electron transport chain.

In the past, scientists attempted to calculate the exact number of ATP molecules synthesized per pair of electrons that travel down an electron transport chain. However, it is now apparent that phosphorylation and oxidation are not directly coupled. In other words, chemiosmosis does not require exact constant relationships among the number of molecules of NADH and FADH$_2$ reduced, the number of electrons that move down an electron transport chain, and the number of molecules of ATP that are synthesized. Additionally, cells use proton gradients for other cellular processes, including active transport and bacterial flagellar motion, so not every transported electron results in ATP production.

Nevertheless, about 34 molecules of ADP per molecule of glucose are oxidatively phosphorylated to ATP via chemiosmosis: three from each of the 10 molecules of NADH generated from glycolysis, the synthesis of acetyl-CoA, and the citric acid cycle; and two from each of the two molecules of FADH$_2$ generated in the citric acid cycle. Given that glycolysis produces a net two molecules of ATP by substrate-level phosphorylation and that the citric acid cycle produces two more, the complete aerobic oxidation of one molecule of glucose by a prokaryote can theoretically yield a net total of 38 molecules of ATP **(TABLE 5.3)**. The theoretical net maximum for eukaryotic cells is generally given as 36 molecules of ATP because energy from two ATP molecules is required to transport NADH (generated by glycolysis) from the cytoplasm into the mitochondria.

MICRO CHECK

9. What role do NADH and FADH$_2$ serve in catabolism?

Metabolic Diversity

LEARNING OUTCOMES

5.13 Compare and contrast the ED and pentose phosphate pathway with EMP glycolysis in terms of energy production and products.

5.14 Describe several examples of the vast metabolic diversity in bacteria.

Some bacteria substitute the *Entner-Doudoroff (ED) pathway* for the EMP pathway. Bacteria using the ED pathway instead of EMP include potentially pathogenic *Pseudomonas aeruginosa* (ā-roo-ji-nōsă) and *Enterococcus faecalis* (en-ter-ō-kok´ŭs fē-kă´lis). Some bacteria, such as *E. coli*, have both ED and EMP pathways. To date, ED has been discovered only in prokaryotes. Whereas EMP generates two ATP, two pyruvate, and two NADH molecules per glucose oxidized, ED produces one ATP, one NADH, and one NADPH molecule. NADPH and NADH both carry high-energy electrons, but NADH is devoted to catabolic pathways and NADPH to anabolic pathways.

The **pentose phosphate pathway** is an alternative to glycolysis for the breakdown of glucose and is named for the phosphorylated pentose (five-carbon) sugars (ribulose, xylulose, and ribose) that it produces. The pathway also yields molecules of nicotinamide adenine dinucleotide phosphate (NADPH) that carry electrons used in anabolic reactions such as *photosynthesis* (discussed in a later section).

Breakdown of glucose by the pentose phosphate pathway nets a single molecule of ATP from each molecule of glucose, so the pathway is half as energy efficient as glycolysis. However, the pathway is essential because the precursor metabolites and NADPH synthesized in the pathway are required to make many important molecules, including DNA nucleotides, steroids, and fatty acids.

Fermentation

LEARNING OUTCOMES

5.15 Describe fermentation, and contrast it with respiration.

5.16 List three useful end-products of fermentation, and explain how fermentation reactions are used to identify bacteria.

5.17 Discuss how biochemical tests for metabolic enzymes and products are used in the identification of bacteria.

Sometimes cells cannot completely oxidize glucose by cellular respiration. For instance, they may lack sufficient final electron acceptors, as in the case, for example, of a bacterium that lacks

TABLE 5.3 Summary of Ideal Prokaryotic Aerobic Respiration of One Molecule of Glucose

Pathway	ATP Produced	ATP Used	NADH Produced	FADH$_2$ Produced
Glycolysis	4	2	2	0
Synthesis of acetyl-CoA and the citric acid cycle	2	0	8	2
Electron transport chain	34	0	0	0
Total	40	2		
Net Total	**38**			

oxygen in the anaerobic environment of the colon. Electrons cannot flow down an electron transport chain unless oxidized carrier molecules are available to receive them at the end. Our bucket brigade analogy can help clarify this point.

Suppose the last person in the brigade did not throw the water but instead held onto two full buckets. What would happen? The entire brigade would soon consist of firefighters holding full buckets of water. An analogous situation occurs in an electron transport chain: All the carrier molecules are forced to remain in their reduced states when there is no final electron acceptor. Without the continual movement of electrons down the chain, protons cannot be transported, the proton motive force is lost, and oxidative phosphorylation of ADP to ATP ceases. Without sufficient ATP, a cell is unable to anabolize, grow, or divide.

ATP can be synthesized in glycolysis and the citric acid cycle by substrate-level phosphorylation; after all, together these pathways produce four molecules of ATP per molecule of glucose. However, careful consideration reveals that in respiration, the processes of glycolysis, formation of acetyl-CoA, and the citric acid cycle require a continual supply of oxidized NAD^+ molecules (see Figures 5.14, 5.16, and 5.17). Electron transport yields the required NAD^+, but without a final electron acceptor, this source of NAD^+ ceases to be available. A cell in such a predicament must use an alternative source of NAD^+ provided by alternative pathways, called *fermentation pathways*.

In everyday language, *fermentation* refers to the production of alcohol from sugar, but in microbiology, *fermentation* has an expanded meaning. **Fermentation** is the partial oxidation of sugar (or other metabolites) to release energy using an organic molecule from within the cell as the final electron acceptor. In other words, fermentation pathways are metabolic reactions that oxidize NADH to NAD^+ while reducing cellular organic molecules. Although energetically less efficient than respiration, fermentation has no need for electron acceptors from the environment. **FIGURE 5.19** illustrates two common fermentation pathways that reduce pyruvic acid to lactic acid or ethanol, oxidizing NADH in the process. **TABLE 5.4** (on p. 142) compares fermentation to aerobic and anaerobic respiration with respect to four crucial aspects of these processes.

Microorganisms produce a variety of fermentation products depending on the enzymes and substrates available to each. Though fermentation products are wastes to the cells that make them, many are useful to humans, including ethanol (drinking alcohol) and lactic acid (used in the production of cheese, sauerkraut, and pickles) (**FIGURE 5.20**).

Other fermentation products are harmful to human health and industry. For example, fermentation products of the bacterium *Clostridium perfringens* (klos-trid´ē-ŭm per-frin´jens) are involved in the necrosis (death) of muscle tissue associated with gangrene. Pasteur discovered that bacterial contaminants in grape juice fermented the sugar into unwanted products such as acetic acid and lactic acid, which spoiled the wine (see Chapter 1).

Fermentation reactions can be used to identify microbes. For example, *Proteus* ferments glucose but not lactose, whereas *Escherichia* and *Enterobacter* (en´ter-ō-bak´ter) ferment both.

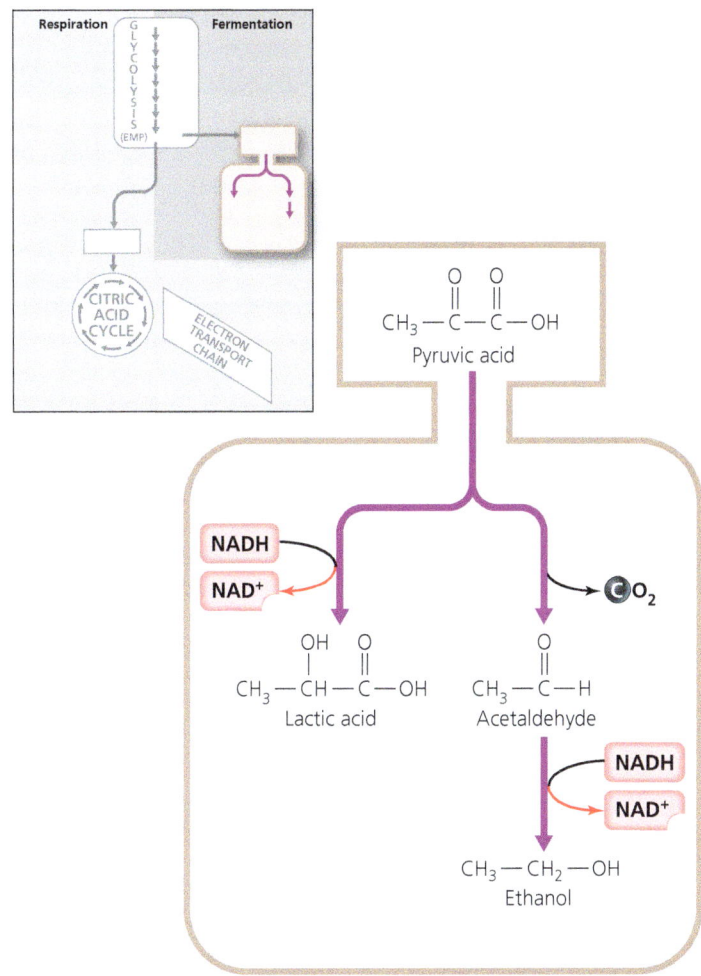

▲ **FIGURE 5.19 Examples of fermentation.** In the simplest fermentation reaction, NADH reduces pyruvic acid to form lactic acid. Another simple fermentation pathway involves a decarboxylation reaction and reduction to form ethanol.

Further, glucose fermentation by *Escherichia* produces mixed acids (acetic, lactic, succinic, and formic), whereas *Enterobacter* produces 2,3-butanediol. Common fermentation tests contain a carbohydrate and a pH indicator, which is a molecule that changes color as the pH changes. An organism that utilizes the carbohydrate causes a change in pH, causing the pH indicator to change color.

Play *Fermentation*
Mastering Microbiology

In this section on carbohydrate catabolism, we have spent some time examining glycolysis, alternatives to glycolysis, the citric acid cycle, and electron transport because these pathways are central to metabolism. They generate all of the precursor metabolites and most of the ATP needed for anabolism. We have seen that some ATP is generated in respiration by substrate-level phosphorylation (in both glycolysis and the citric acid cycle) but that most ATP is generated by oxidative phosphorylation via chemiosmosis utilizing the reducing power of NADH and $FADH_2$. We also saw that some microorganisms use fermentation to provide an alternative source of NAD^+.

142 CHAPTER 5 Microbial Metabolism

TABLE 5.4 Comparison of Aerobic Respiration, Anaerobic Respiration, and Fermentation

	Aerobic Respiration	Anaerobic Respiration	Fermentation
Oxygen Required	Yes	No	No
Type of Phosphorylation	Substrate-level and oxidative	Substrate-level and oxidative	Substrate level
Final Electron (Hydrogen) Acceptor	Oxygen	NO_3^-, SO_4^{2-}, CO_3^{2-}, or externally acquired organic molecules	Cellular organic molecules
Potential Molecules of ATP Produced per Molecule of Glucose	38 in prokaryotes, 36 in eukaryotes	4–36	2

TELL ME WHY

Why do electrons carried by NADH allow for production of 50% more ATP molecules than do electrons carried by $FADH_2$?

Thus far, we have concentrated on the catabolism of glucose as a representative carbohydrate, but microorganisms can also use other molecules as energy sources. In the next section, we will examine catabolic pathways that utilize lipids and proteins.

Other Catabolic Pathways

Lipid and protein molecules contain abundant energy in their chemical bonds (for example, one C_{16} palmitate fatty acid molecule can generate 129 ATP!) and can also be converted into precursor metabolites. These molecules are first catabolized to produce their constituent monomers, which serve as substrates in glycolysis and the citric acid cycle.

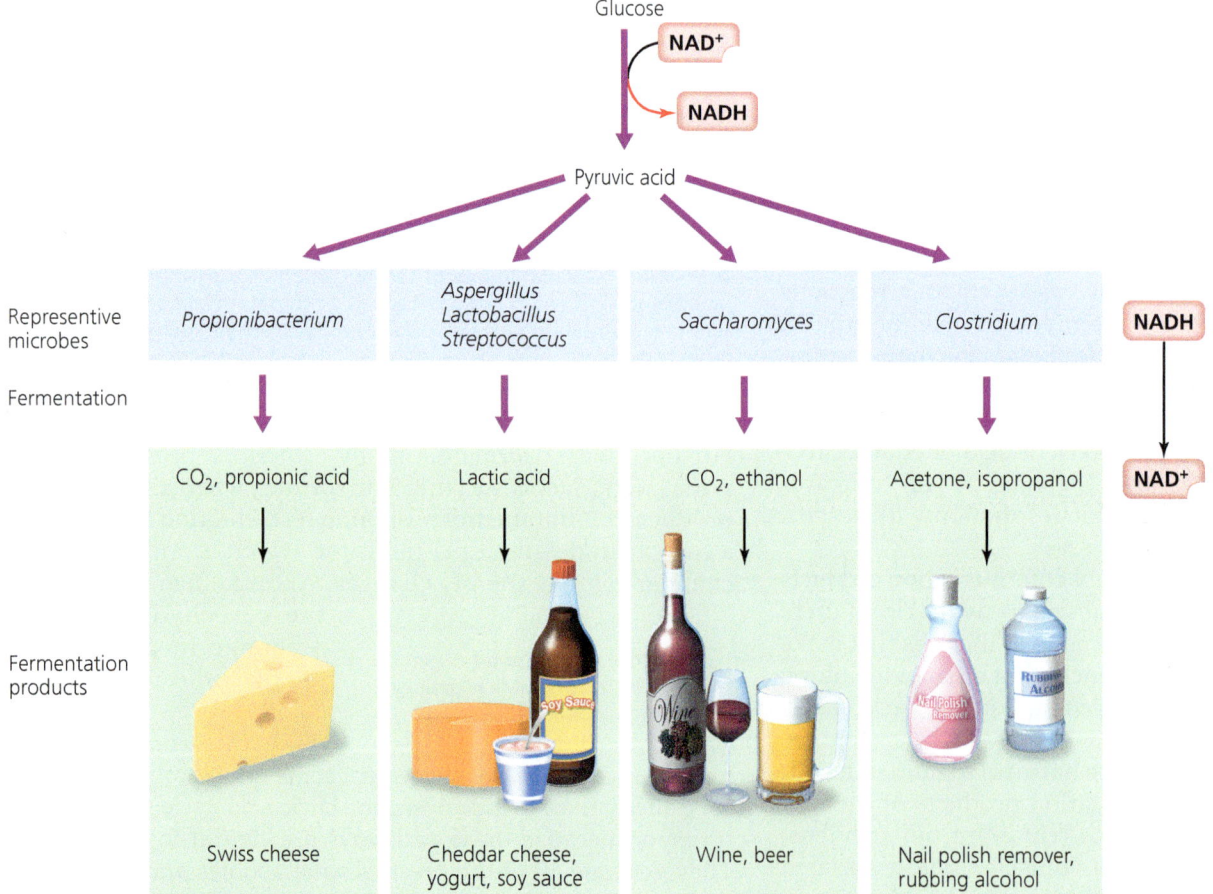

▲ **FIGURE 5.20 Representative fermentation products and the organisms that produce them.** All of the organisms are bacteria except *Saccharomyces* and *Aspergillus*, which are fungi. *Why does Swiss cheese have holes, but cheddar does not?*

Figure 5.20 *Swiss cheese is a product of Propionibacterium, which ferments pyruvic acid to produce CO_2; bubbles of this gas cause the holes in the cheese. Cheddar is a product of Lactobacillus, which does not produce gas.*

Lipid Catabolism

LEARNING OUTCOME

5.18 Explain how lipids are catabolized for energy and metabolite production.

The most common lipids involved in ATP and metabolite production are fats, which typically consist of glycerol and fatty acids (as we saw in Chapter 2). In the first step of triglyceride catabolism, enzymes called *lipases* hydrolyze the bonds attaching glycerol to the fatty acid chains (**FIGURE 5.21a**).

Subsequent reactions further catabolize the glycerol and fatty acid molecules. Glycerol is converted to DHAP, which, as one of the substrates of glycolysis, is oxidized to pyruvic acid (see Figure 5.14 5–10). The fatty acids are degraded in a catabolic process known as **beta-oxidation**[9] (**FIGURE 5.21b**). In this process, enzymes repeatedly split off pairs of the hydrogenated carbon atoms that make up a fatty acid and join each pair to coenzyme A to form acetyl-CoA; this continues until the entire fatty acid has been converted to molecules of acetyl-CoA, which can then feed into the citric acid cycle (see Figure 5.17). Beta-oxidation also generates NADH and FADH$_2$, which supply electrons to electron transport chains, ultimately resulting in ATP production. As is the case for the citric acid cycle, the enzymes involved in beta-oxidation are located in the cytosol of prokaryotes and in the mitochondria of eukaryotes.

MICRO CHECK

10. What part of a lipid can be converted to DHAP during lipid catabolism and then enter the glycolysis pathway?

Protein Catabolism

LEARNING OUTCOME

5.19 Explain how proteins are catabolized for energy and metabolite production.

Some microorganisms—notably food-spoilage bacteria, pathogenic bacteria, and fungi—normally catabolize proteins as an important source of energy and metabolites. Most cells catabolize proteins and their constituent amino acids only when carbon sources such as glucose and fat are not available.

Generally, proteins are too large to cross cytoplasmic membranes, so prokaryotes typically conduct the first step in the process of protein catabolism outside the cell by secreting **proteases** (prō′tē-ās-ez)—enzymes that split proteins into their constituent amino acids (**FIGURE 5.22**). Once released by the action of proteases, amino acids are transported into the cell, where special enzymes split off amino groups in a reaction called **deamination**. The resulting altered molecules enter the citric acid cycle, and the amino groups are either recycled to synthesize other amino acids or excreted as nitrogenous wastes such as ammonia (NH_3) or ammonium ion (NH_4^+).

---

[9]*Beta* is part of the name of this process because enzymes break the bond at the second carbon atom from the end of a fatty acid, and beta is the second letter in the Greek alphabet.

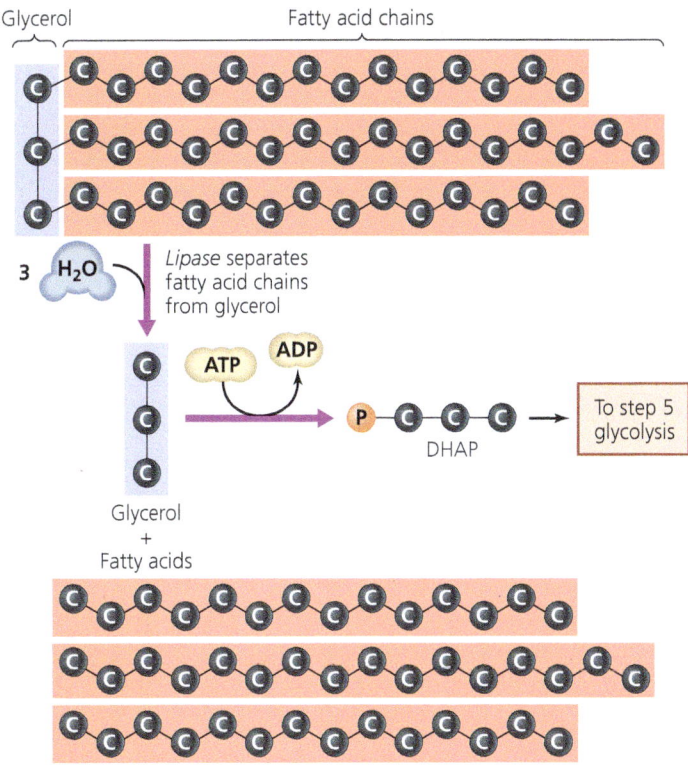

(a) Hydrolysis

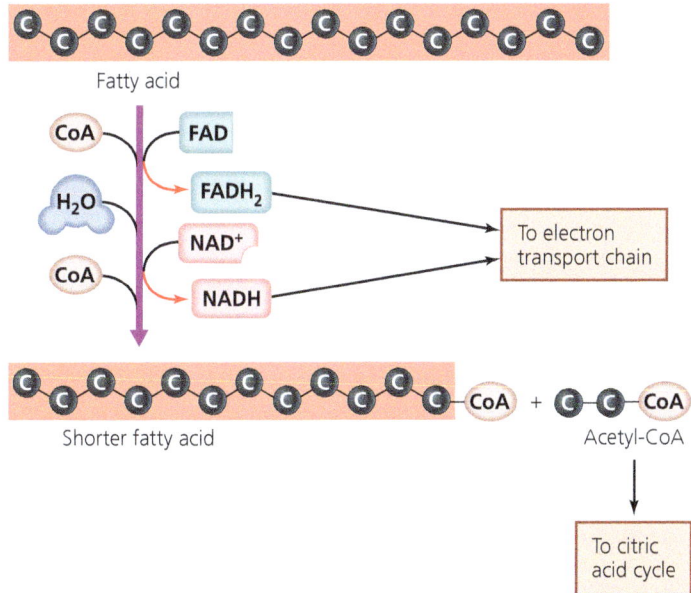

(b) Beta-oxidation

▲ **FIGURE 5.21 Catabolism of a triglyceride molecule. (a)** Lipase breaks triglycerides into glycerol and three fatty acids by hydrolysis. Glycerol is converted to DHAP, which can be catabolized via glycolysis and the citric acid cycle. **(b)** Beta-oxidation reactions catabolize fatty acid chains, reducing them by two carbon atoms each time, to produce molecules of acetyl-CoA and reduced coenzymes (NADH and FADH$_2$).

MICRO CHECK

11. What is the first step in protein catabolism for prokaryotes?

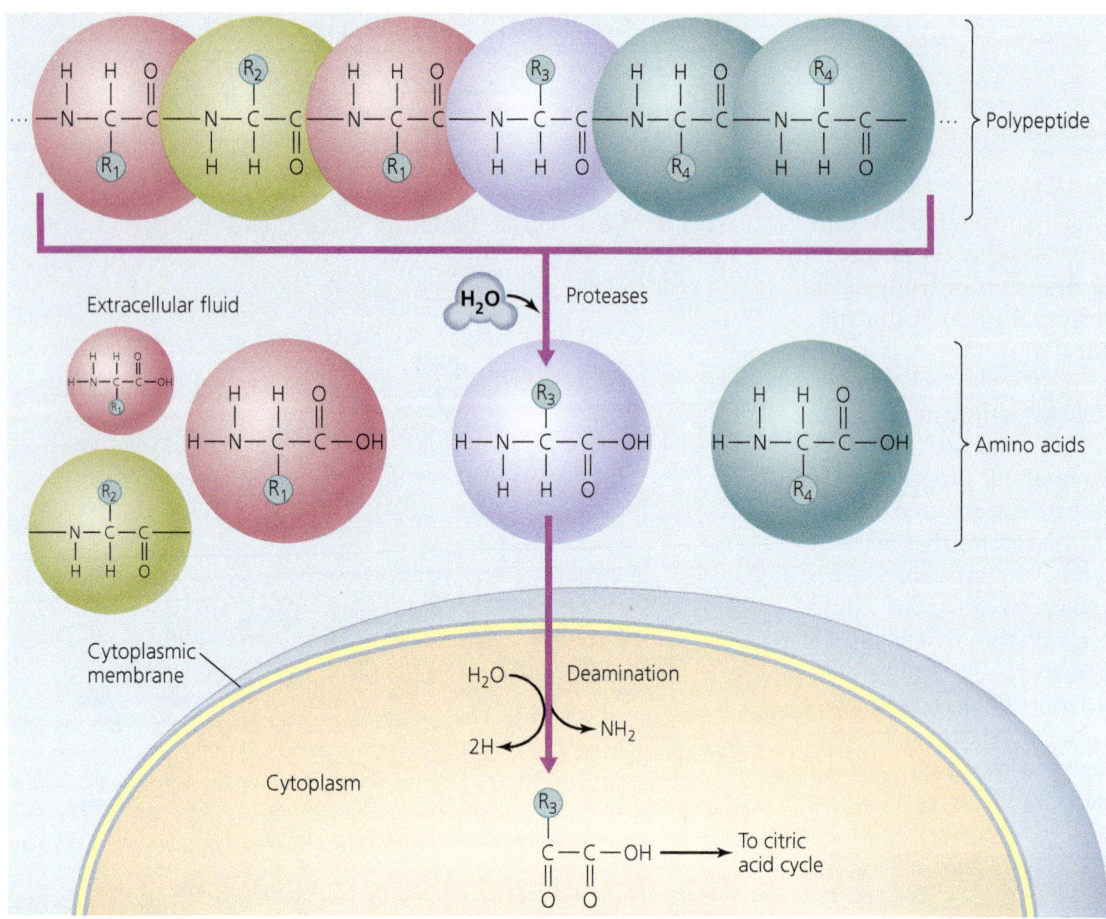

◀ **FIGURE 5.22 Protein catabolism.** Secreted proteases hydrolyze proteins, releasing amino acids, which are deaminated after uptake to produce molecules used as substrates in the citric acid cycle. R indicates the side group, which varies among amino acids.

TELL ME WHY
Why does catabolism of amino acids for energy result in ammonia and other nitrogenous wastes?

Thus far, we have examined the catabolism of carbohydrates and briefly considered the breakdown of lipids and proteins. Now we turn our attention to the synthesis of these molecules, beginning with the anabolic reactions of photosynthesis.

Photosynthesis

LEARNING OUTCOME
5.20 Define *photosynthesis*.

Many organisms use only organic molecules as a source of energy and metabolites, but where do they acquire organic molecules? Ultimately, every food chain begins with anabolic pathways in organisms that synthesize their own organic molecules from inorganic carbon dioxide. Most of these organisms capture light energy from the sun and use it to drive the synthesis of carbohydrates from CO_2 and H_2O by a process called **photosynthesis**. Cyanobacteria, purple sulfur bacteria, green sulfur bacteria, green nonsulfur bacteria, purple nonsulfur bacteria, algae, green plants, and a few protozoa are photosynthetic.

Play *Photosynthesis: Overview*
@ **Mastering** Microbiology

Chemicals and Structures

LEARNING OUTCOME
5.21 Compare and contrast the basic chemicals and structures involved in photosynthesis in prokaryotes and eukaryotes.

Photosynthetic organisms capture light energy with pigment molecules, the most important of which are **chlorophylls**. Chlorophyll molecules are composed of a hydrocarbon tail attached to a light-absorbing *active site* centered around a magnesium ion (Mg^{2+}) (**FIGURE 5.23a**). The active sites of chlorophyll molecules are structurally similar to the cytochrome molecules found in electron transport chains, except that chlorophylls use Mg^{2+} rather than Fe^{2+}. Chlorophylls, typically designated with letters—for example, chlorophyll *a*, chlorophyll *b*, and bacteriochlorophyll *a*—vary slightly in the lengths and structures of their hydrocarbon tails and in the atoms that extend from their active sites. Green plants, algae, photosynthetic protozoa, and cyanobacteria principally use chlorophyll *a*, whereas green and purple bacteria use bacteriochlorophylls.

The slight structural differences among chlorophylls cause them to absorb light of different wavelengths. For example, chlorophyll *a* from algae best absorbs light with wavelengths of about 425 nm and 660 nm (violet and red), whereas bacteriochlorophyll *a* from purple bacteria best absorbs light with wavelengths of about 350 nm and 880 nm (ultraviolet and infrared). Because they best use light with differing wavelengths, algae and purple bacteria successfully occupy different ecological niches.

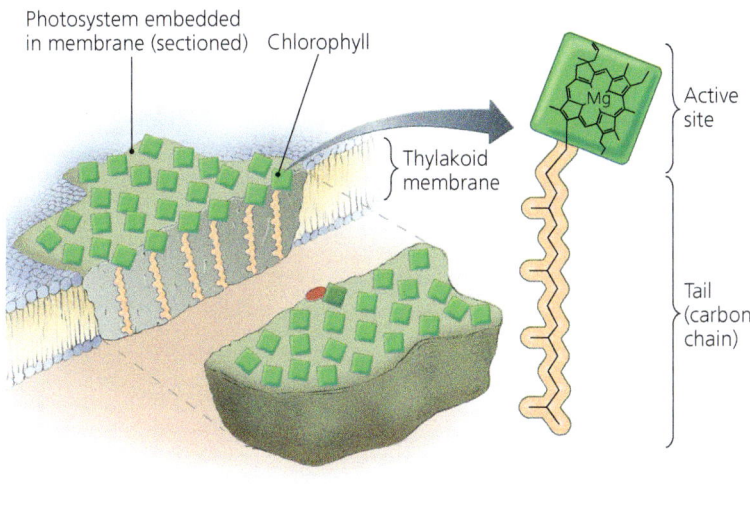
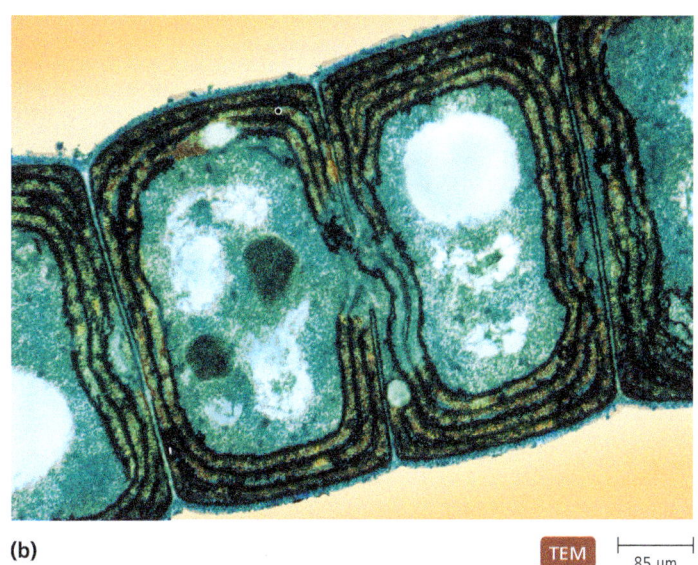

▲ **FIGURE 5.23 Photosynthetic structures in a prokaryote. (a)** Chlorophyll molecules are grouped together in thylakoid membranes to form photosystems. Chlorophyll contains a light-absorbing active site that is connected to a long hydrocarbon tail. The chlorophyll shown here is bacteriochlorophyll *a*; other chlorophyll types differ in the side chains protruding from the central ring. **(b)** Prokaryotic thylakoids are infoldings of the cytoplasmic membrane, here in a cyanobacterium.

Cells arrange numerous molecules of chlorophyll and other pigments within a protein matrix to form light-harvesting matrices called **photosystems** that are embedded in cellular membranes called **thylakoids**. Thylakoids of photosynthetic prokaryotes are invaginations of their cytoplasmic membranes (FIGURE 5.23b). The thylakoids of eukaryotes appear to be formed from infoldings of the inner membranes of chloroplasts, though thylakoid membranes and the inner membranes are not connected in mature chloroplasts (see Figure 3.42). Thylakoids of chloroplasts are arranged in stacks called *grana*. An outer chloroplast membrane surrounds the grana, and the space between the outer membrane and the thylakoid membrane is known as the *stroma*. The thylakoids enclose a narrow, convoluted cavity called the *thylakoid space*.

There are two types of photosystems, named photosystem I (PS I) and photosystem II (PS II), in the order of their discovery. Photosystems absorb light energy and use redox reactions to store this energy in molecules of ATP and nicotinamide adenine dinucleotide phosphate (NADPH). Because they depend on light energy, these reactions of photosynthesis are classified as **light-dependent reactions**. Photosynthesis also involves **light-independent reactions** that actually synthesize glucose from carbon dioxide and water. Historically, these reactions were called *light* and *dark reactions*, respectively, but this older terminology implies that light-independent reactions occur only in the dark, which is not the case. We first consider the light-dependent reactions of the two photosystems.

Light-Dependent Reactions

LEARNING OUTCOMES

5.22 Describe the components and function of the two photosystems, PS I and PS II.
5.23 Contrast cyclic and noncyclic photophosphorylation.

The pigments of photosystem I absorb light energy and transfer it to a neighboring molecule within the photosystem until the energy eventually arrives at a special chlorophyll molecule called the **reaction center chlorophyll** (FIGURE 5.24). Light energy from hundreds of such transfers excites electrons in the reaction center chlorophyll, which passes its excited electrons to the reaction center's electron acceptor, which is the initial carrier of an electron

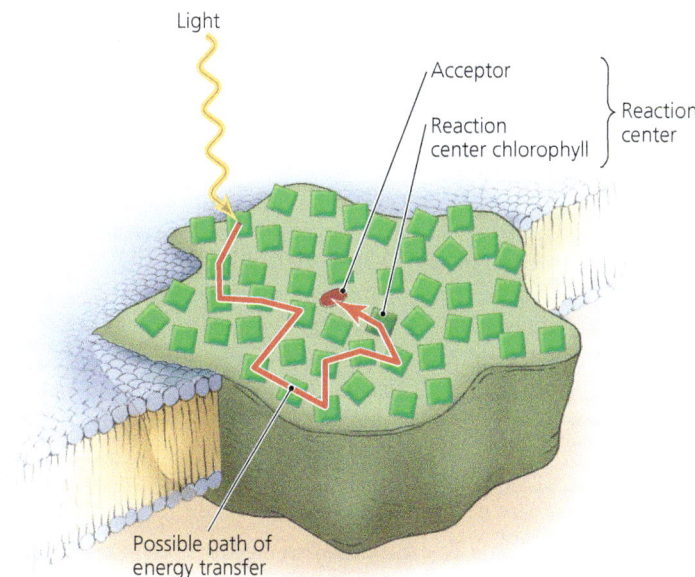

▲ **FIGURE 5.24 Reaction center of a photosystem.** Light energy absorbed by pigments anywhere in the photosystem is transferred to the reaction center chlorophyll, where it is used to excite electrons for delivery to the reaction center's electron acceptor.

transport chain. As electrons move down this electron transport chain, their energy is used to pump protons across the membrane, creating a proton motive force. In prokaryotes, protons are pumped out of the cell; in eukaryotes, they are pumped from the stroma into the interior of the thylakoids—the thylakoid space.

The proton motive force is used in chemiosmosis, in which, you should recall, protons flow down their electrochemical gradient through ATPases, which generate ATP. In photosynthesis, this process is called *photophosphorylation*, which can be either cyclic or noncyclic.

Cyclic Photophosphorylation

Electrons moving from one carrier molecule to another in a thylakoid must eventually pass to a final electron acceptor. In **cyclic photophosphorylation**, which occurs in all photosynthetic organisms, the final electron acceptor is the original reaction center chlorophyll that donated the electrons (**FIGURE 5.25a**). In other words, when light energy excites electrons in PS I, they pass down an electron transport chain and return to PS I. The energy from the electrons is used to establish a proton gradient that drives the phosphorylation of ADP to ATP by chemiosmosis.

Play *Photosynthesis: Cyclic Photophosphorylation*
@ Mastering Microbiology

Noncyclic Photophosphorylation

Some photosynthetic bacteria and all plants, algae, and photosynthetic protozoa utilize **noncyclic photophosphorylation** as well. (Exceptions are green and purple sulfur bacteria.) Noncyclic photophosphorylation, which requires both PS I and PS II, generates molecules of ATP and also reduces molecules of coenzyme $NADP^+$ to NADPH (**FIGURE 5.25b**).

When light energy excites electrons of PS II, they are passed to PS I through an electron transport chain. (Note that photosystem II occurs first in the pathway and photosystem I second; as mentioned earlier, the photosystems were named in the order in which they were discovered, not the order in which they operate.) PS I further energizes the electrons with additional light energy and transfers them through an electron transport chain to $NADP^+$, which is thereby reduced to NADPH. The hydrogen ions added to NADPH come from the stroma in chloroplasts or from the cytosol in photosynthetic bacteria. NADPH subsequently participates in the synthesis of glucose in the light-independent reactions, which we will examine in the next section.

In noncyclic photophosphorylation, a cell must constantly replenish electrons to the reaction center of photosystem II. *Oxygenic* (oxygen-producing) organisms, such as algae, green plants, and cyanobacteria, derive electrons from the dissociation of H_2O (see Figure 5.25b). In these organisms, two molecules of water give up their electrons, producing molecular oxygen (O_2) as a waste product of photosynthesis. *Anoxygenic* photosynthetic bacteria get electrons from inorganic compounds such as H_2S, resulting in a nonoxygen waste, such as sulfur.

TABLE 5.5 (on p. 148) compares photophosphorylation to substrate-level and oxidative phosphorylation.

To this point, we have examined the use of photosynthetic pigments and thylakoid structure to harvest light energy to produce both ATP and reducing power in the form of NADPH. Next, we examine the light-independent reactions of photosynthesis.

Play *Photosynthesis: Noncyclic Photophosphorylation*
@ Mastering Microbiology

Light-Independent Reactions

LEARNING OUTCOMES
5.24 Contrast the light-dependent and light-independent reactions of photosynthesis.
5.25 Describe the reactants and products of the Calvin-Benson cycle.

Light-independent reactions of photosynthesis do not require light directly; instead, they use large quantities of ATP and NADPH generated by the light-dependent reactions. The key reaction of the light-independent pathway of photosynthesis is **carbon fixation** by the **Calvin-Benson cycle**,[10] which involves the attachment of molecules of CO_2 to molecules of a five-carbon organic compound called *ribulose 1,5-bisphosphate (RuBP)*. RuBP is derived initially from phosphorylation of a precursor metabolite produced by the pentose phosphate pathway. The Calvin-Benson cycle is very endergonic—it requires a great deal of energy. Life on Earth is dependent on carbon fixation by this cycle. This cycle occurs in the cytoplasm of photosynthetic bacteria or the interior stroma of chloroplasts in eukaryotes (see Figure 3.42).

The Calvin-Benson cycle is reminiscent of the citric acid cycle in that the substrates of the cycle are regenerated. It is helpful to notice the number of carbon atoms during each part of the three steps of the Calvin-Benson cycle (**FIGURE 5.26**):

1. **Fixation of CO_2.** An enzyme attaches three molecules of carbon dioxide (three carbon atoms) to three molecules of RuBP (15 carbon atoms), which are then split to form six molecules of 3-phosphoglyceric acid (18 carbon atoms). The rubisco enzyme that catalyzes this reaction may be the most abundant protein on the planet.

2. **Reduction.** Molecules of NADPH reduce the six molecules of 3-phosphoglyceric acid to form six molecules of glyceraldehyde 3-phosphate (G3P; 18 carbon atoms). These reactions require six molecules of ATP and six molecules of NADPH, which are generated in the light-dependent reactions.

3. **Regeneration of RuBP.** The cell regenerates three molecules of RuBP (15 carbon atoms) from five molecules of G3P (15 carbon atoms). It uses the remaining molecule of glyceraldehyde 3-phosphate to synthesize glucose by reversing the reactions of glycolysis.

In summary, ATP and NADPH from the light-dependent reactions drive the synthesis of glucose from CO_2 in the light-independent reactions of the Calvin-Benson cycle. For every three molecules of CO_2 that enter the Calvin-Benson cycle, a molecule of glyceraldehyde 3-phosphate (G3P) leaves. Glycolysis is

---
[10]Named for the men who elucidated its pathways.

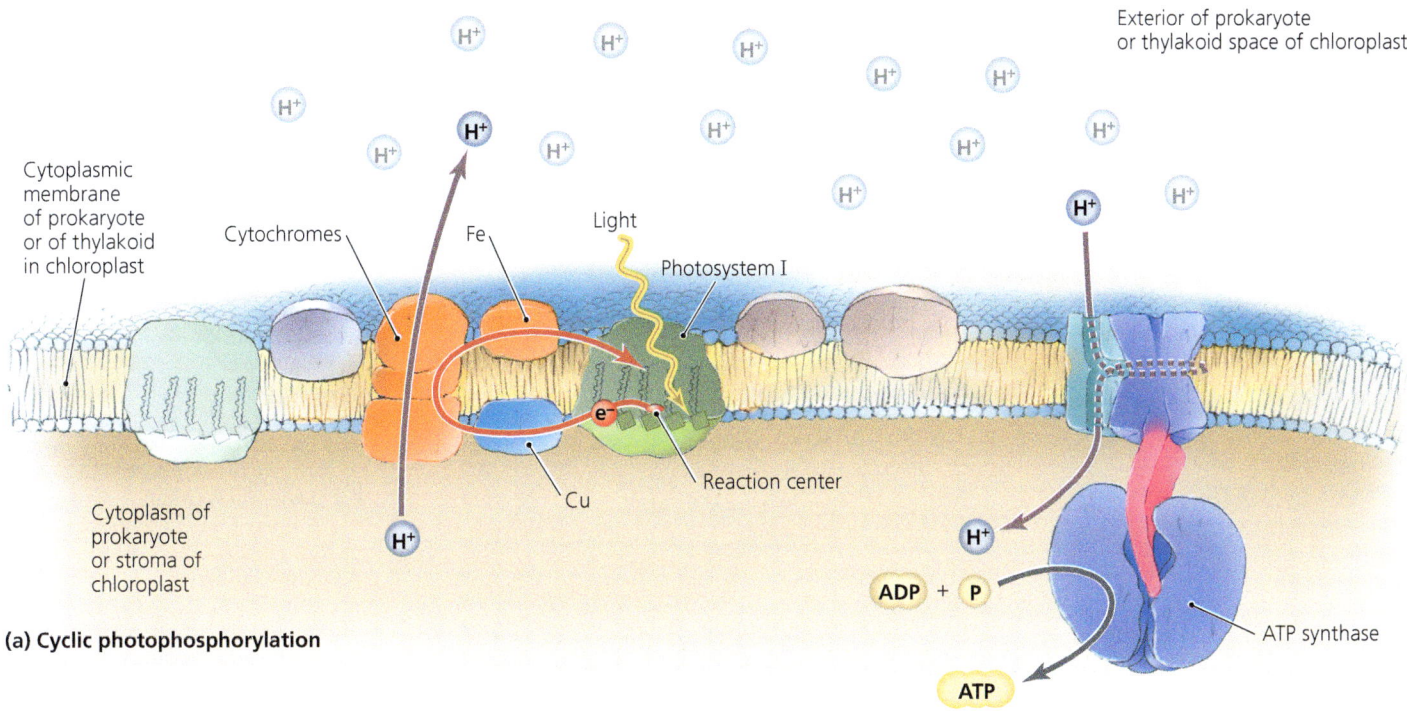

(a) Cyclic photophosphorylation

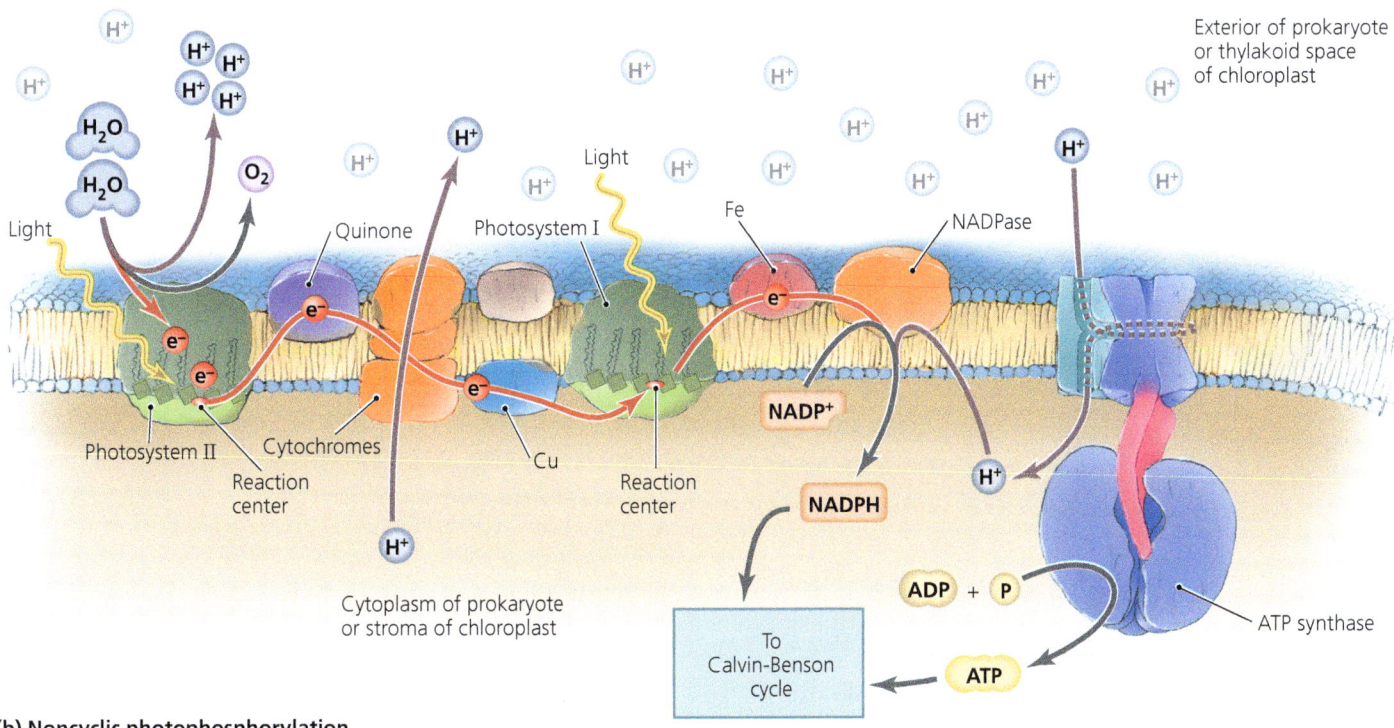

(b) Noncyclic photophosphorylation

▲ FIGURE 5.25 **The light-dependent reactions of photosynthesis: cyclic and noncyclic photophosphorylation.** (a) In cyclic photophosphorylation, electrons excited by light striking photosystem I travel down an electron transport chain (red arrows) and then return to the photosystem. (b) In noncyclic photophosphorylation, light striking photosystem II excites electrons that are passed to photosystem I, simultaneously establishing a proton gradient. Light energy collected by photosystem I further excites the electrons, which are used to reduce NADP⁺ to NADPH via an electron transport chain. In oxygenic organisms, new electrons are provided by the lysis of H_2O (left). In both types of photophosphorylation, the proton gradient drives the phosphorylation of ADP by ATP synthase (chemiosmosis).

TABLE 5.5 A Comparison of the Three Types of Phosphorylation

	Source of Phosphate	Source of Energy	Location in Eukaryotic Cell	Location in Prokaryotic Cell
Substrate-Level Phosphorylation	Organic molecule	High-energy phosphate bond of donor	Cytosol and mitochondrial matrix	Cytosol
Oxidative Phosphorylation	Inorganic phosphate (PO_4^{3-})	Proton motive force	Inner membrane of mitochondrion	Cytoplasmic membrane
Photophosphorylation	Inorganic phosphate (PO_4^{3-})	Proton motive force	Thylakoid of chloroplast	Thylakoid of cytoplasmic membrane

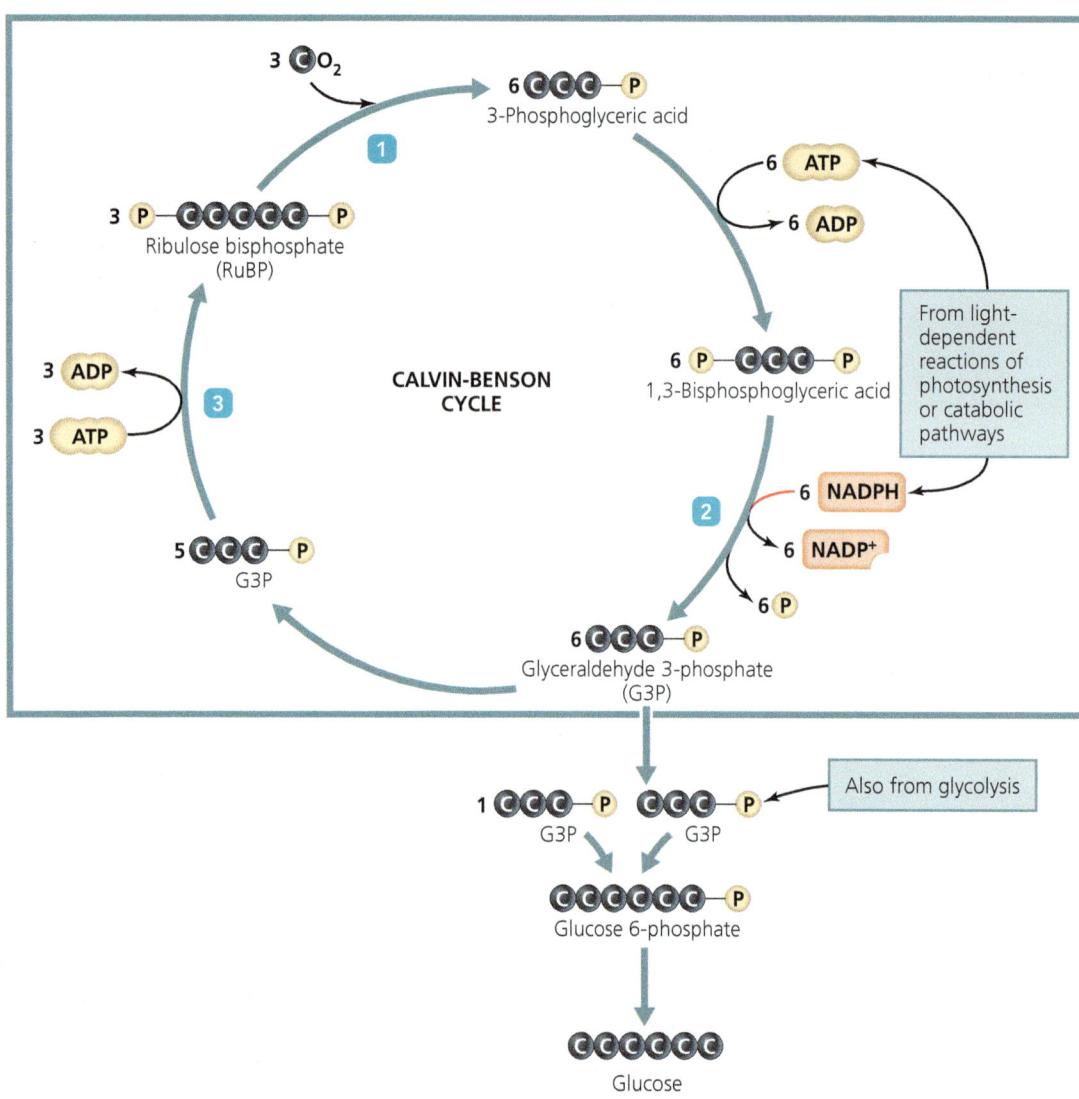

◀ **FIGURE 5.26 Simplified diagram of the Calvin-Benson cycle.**
1 Fixation of CO_2. Three molecules of RuBP combine with three molecules of CO_2. **2** Reduction. The resulting molecules are reduced to form six molecules of G3P. **3** Generation of RuBP. Five molecules of G3P are converted to three molecules of RuBP, which completes the cycle. Two turns of the cycle also yield two molecules of G3P, which are polymerized to synthesize glucose 6-phosphate.

subsequently reversed to anabolically combine two molecules of G3P to synthesize glucose 6-phosphate.

Play *Photosynthesis: Light-Independent Reaction*
@ Mastering Microbiology

The processes of oxygenic photosynthesis and aerobic respiration complement one another to complete both a carbon cycle and an oxygen cycle. During the synthesis of glucose in oxygenic photosynthesis, water and carbon dioxide are used, and oxygen is released as a waste product; in aerobic respiration, oxygen serves as the final electron acceptor in the oxidation of glucose to produce carbon dioxide and water.

In this section, we examined an essential anabolic pathway for life on Earth—photosynthesis, which produces glucose. Next, we will consider anabolic pathways involved in the synthesis of other organic molecules.

MICRO CHECK

12. What is the main product of the light-dependent reactions of photosynthesis?
13. How is ribulose 1,5-bisphospate (RuBP) used in the first step of the Calvin-Benson cycle?

> **TELL ME WHY**
>
> An uninformed student describes the Calvin-Benson cycle as "cellular respiration in reverse." Why is the student incorrect?

Other Anabolic Pathways

LEARNING OUTCOME

5.26 Define *amphibolic reaction*.

Anabolic reactions are synthesis reactions. As such, they require energy and a source of precursor metabolites. ATP generated in the catabolic reactions of aerobic respiration, anaerobic respiration, and fermentation and in the initial redox reactions of photosynthesis provides energy for anabolism. Glycolysis, the citric acid cycle, and the pentose phosphate pathway provide 12 basic precursor metabolites from which all macromolecules and cellular structures can be made **(TABLE 5.6)**. Some microorganisms, such as *E. coli*, can synthesize all 12 precursors, whereas other organisms, such as humans, must acquire some precursors in their diets.

Many anabolic pathways are the reversal of the catabolic pathways we have discussed; therefore, much of the material that follows has *in a sense* been discussed in the sections on catabolism. Reactions that can proceed in either direction—toward catabolism or toward anabolism—are said to be **amphibolic reactions**. The following sections discuss the synthesis of carbohydrates, lipids, amino acids, and nucleotides. (Chapter 7 covers anabolic reactions that are closely linked to genetics—the polymerizations of amino acids into proteins and of nucleotides into RNA and DNA.)

Carbohydrate Biosynthesis

LEARNING OUTCOME

5.27 Describe the biosynthesis of carbohydrates.

As we have seen, anabolism begins in photosynthetic organisms with carbon fixation by the enzymes of the Calvin-Benson cycle to form molecules of G3P. Enzymes use G3P as the starting point for synthesizing sugars, complex polysaccharides such as starch, cellulose for cell walls in algae, and peptidoglycan for cell walls of bacteria. Animals and protozoa synthesize the storage molecule glycogen.

Some cells are able to synthesize sugars from noncarbohydrate precursors, such as amino acids, glycerol, and fatty acids, by pathways collectively called *gluconeogenesis*[11] (glū´kō-nē-ō-jen´ĕ-sis; **FIGURE 5.27**). Most of the reactions of gluconeogenesis are amphibolic, using enzymes of glycolysis in reverse, but four of the reactions require unique enzymes. Gluconeogenesis is highly endergonic; that is, it requires of energy. The reactions of gluconeogenesis can proceed only when there is an adequate supply of energy.

> **MICRO CHECK**
>
> 14. What anabolic reaction is the opposite of what happens during protein catabolism?
> 15. What carbohydrate storage molecule do animals and protozoa synthesize?

TABLE 5.6 The 12 Precursor Metabolites

	Pathway That Generates the Metabolite	Examples of Macromolecule Synthesized from Metabolite[a]	Examples of Functional Use
Glucose 6-Phosphate	Glycolysis	Lipopolysaccharide	Outer membrane of cell wall
Fructose 6-Phosphate	Glycolysis	Peptidoglycan	Cell wall
Glyceraldehyde 3-Phosphate (G3P)	Glycolysis	Glycerol portion of lipids	Fats—energy storage
Phosphoglyceric Acid	Glycolysis	Amino acids: cysteine, selenocysteine, glycine, and serine	Enzymes
Phosphoenolpyruvic Acid (PEP)	Glycolysis	Amino acids: phenylalanine, tryptophan, and tyrosine	Enzymes
Pyruvic Acid	Glycolysis	Amino acids: alanine, leucine, and valine	Enzymes
Ribose 5-Phosphate	Pentose phosphate pathway	DNA, RNA, amino acid, and histidine	Genome, enzymes
Erythrose 4-Phosphate	Pentose phosphate pathway	Amino acids: phenylalanine, tryptophan, and tyrosine	Enzymes
Acetyl-CoA	Citric acid cycle	Fatty acid portion of lipids	Cytoplasmic membrane
α-Ketoglutaric Acid	Citric acid cycle	Amino acids: arginine, glutamic acid, glutamine, and proline	Enzymes
Succinyl-CoA	Citric acid cycle	Heme	Cytochrome electron carrier
Oxaloacetate	Citric acid cycle	Amino acids: aspartic acid, asparagine, isoleucine, lysine, methionine, and threonine	Enzymes

[a] Examples given apply to the bacterium *E. coli*.

[11] From Greek *glukus*, meaning "sweet;" *neo*, meaning "new;" and *genesis*, meaning "generate."

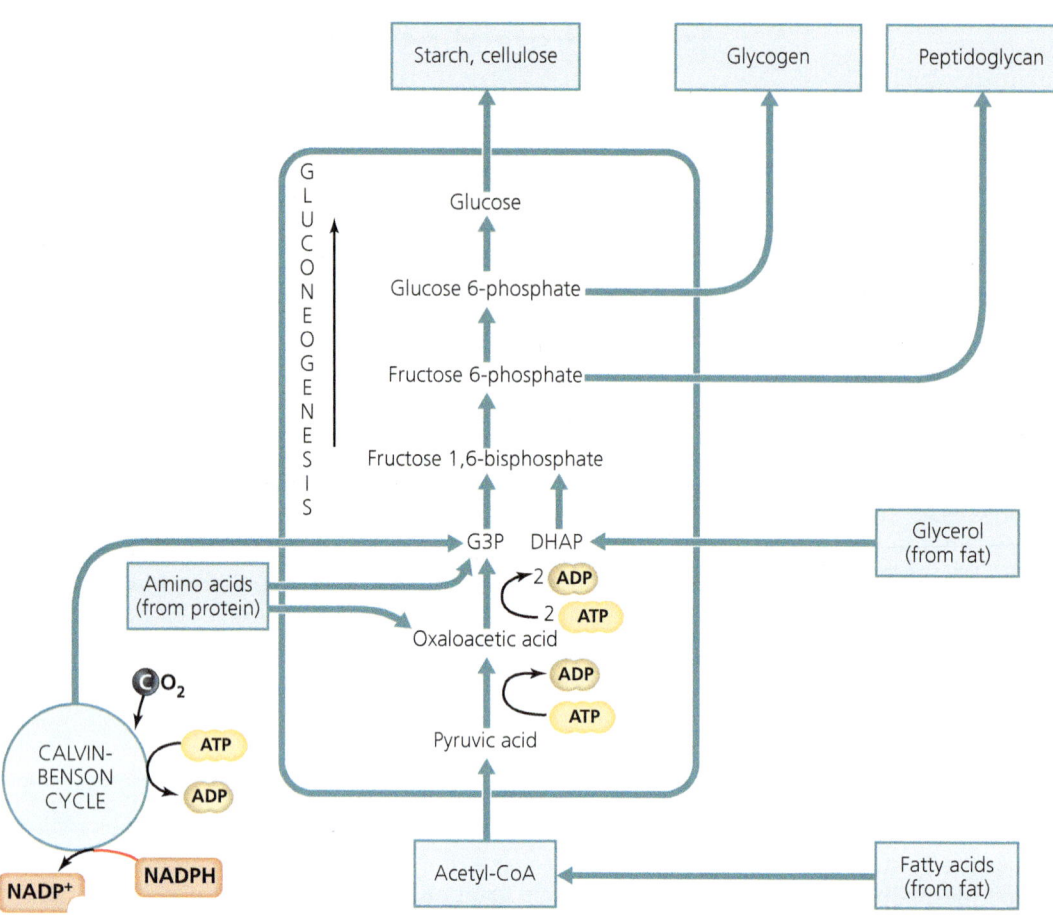

▲ **FIGURE 5.27 The role of gluconeogenesis in the biosynthesis of complex carbohydrates.** Complex carbohydrates are synthesized from simple sugar molecules such as glucose, glucose 6-phosphate, and fructose 6-phosphate. Starch and cellulose are found in algae; glycogen is found in animals, protozoa, and fungi; and peptidoglycan is found in bacteria.

Lipid Biosynthesis

LEARNING OUTCOME

5.28 Describe the biosynthesis of lipids.

Lipids are a diverse group of organic molecules that function as energy storage compounds and as components of membranes (as we saw in Chapter 2). The most energy-efficient energy storage form in cells is triglycerides. The most common form of fat synthesized in cells is phospholipids, to generate membranes. *Carotenoids*, which are reddish pigments found in many bacterial and plant photosystems, are also lipids.

Because of their variety, it is not surprising that lipids are synthesized by a variety of routes. For example, fats are typically synthesized in anabolic reactions that are the reverse of their catabolism; in the case of triglycerides, cells polymerize glycerol and three fatty acids (**FIGURE 5.28**). Glycerol is derived from G3P generated by the Calvin-Benson cycle and glycolysis; the fatty acids are produced by the linkage of two-carbon acetyl-CoA molecules to one another by a sequence of endergonic reactions that effectively reverse the catabolic reactions of beta-oxidation. Other lipids, such as steroids, are synthesized in complex pathways involving polymerizations and isomerizations of sugar and amino acid metabolites.

Amino Acid Biosynthesis

LEARNING OUTCOME

5.29 Describe the biosynthesis of amino acids.

Cells synthesize amino acids from precursor metabolites derived from glycolysis, the citric acid cycle, and the pentose phosphate pathway. Additionally, cells can synthesize many amino acids from other amino acids. Some organisms (such as *E. coli* and most plants and algae) synthesize all their amino acids from precursor metabolites. Other organisms, including humans, cannot synthesize certain amino acids, called *essential amino acids*; these must be acquired in the diet. One extreme example is *Lactobacillus* (lak´tō-bă-sil´ŭs), a bacterium that ferments milk and produces some cheeses. This microorganism cannot synthesize any amino acids; it acquires all of them by catabolizing proteins in its environment.

Precursor metabolites are converted to amino acids by the addition of an amine group. This process is called **amination** when the amine group comes from ammonia (NH_3); an example is the formation of aspartic acid from amination of oxaloacetic acid, which is an intermediate in the citric acid cycle (**FIGURE 5.29a**). Amination reactions are the reverse of the catabolic deamination reactions we discussed previously.

More commonly, however, an amine group is moved from one amino acid to a metabolite, producing a different amino acid. This process is called **transamination** because an amine group is transferred from one amino acid to form another amino acid (**FIGURE 5.29b**). All transamination enzymes use a coenzyme, *pyridoxal phosphate*, which is derived from vitamin B_6.

Ribozymes of ribosomes polymerize amino acids into proteins. (Chapter 7 examines this energy-demanding process because it is intimately linked with genetics.)

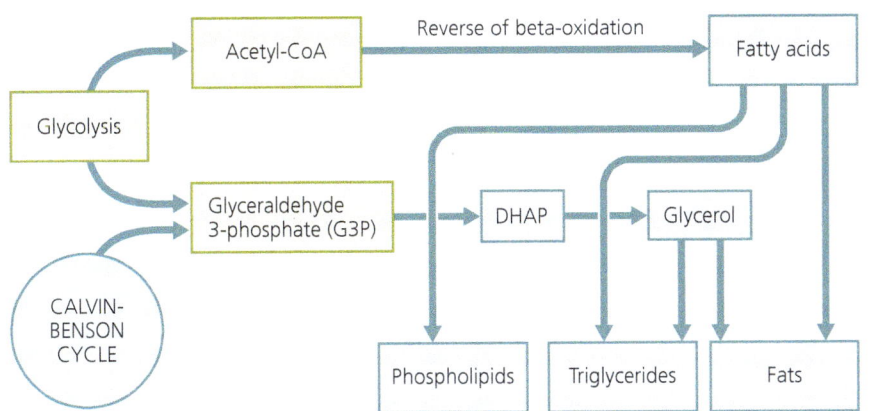

◀ **FIGURE 5.28 Biosynthesis of a triglyceride fat, a lipid.** A triglyceride molecule is synthesized from glycerol and three molecules of fatty acid, the precursors of which are produced in glycolysis.

Nucleotide Biosynthesis

LEARNING OUTCOME

5.30 Describe the biosynthesis of nucleotides.

The building blocks of nucleic acids are nucleotides, each of which consists of a five-carbon sugar, a phosphate group, and a purine or pyrimidine base (see Figure 2.25a). Nucleotides are produced from precursor metabolites of glycolysis and the citric acid cycle (**FIGURE 5.30**):

- The five-carbon sugars—ribose in RNA and deoxyribose in DNA—are derived from ribose 5-phosphate from the pentose phosphate pathway.
- The phosphate group is derived ultimately from ATP.

Purines and pyrimidines are synthesized in a series of ATP-requiring reactions from ribose 5-phosphate, folic acid, and the amino acids glutamine and aspartic acid, which are derived from intermediates of the citric acid cycle. *Folic acid* is synthesized by many bacteria and protozoa but is a vitamin for humans. (Chapter 7 examines the anabolic reactions by which polymerases polymerize nucleotides to form DNA and RNA.)

TELL ME WHY

Why is nitrogen required for the production of amino acids by amination?

Integration and Regulation of Metabolic Functions

LEARNING OUTCOMES

5.31 Describe interrelationships between catabolism and anabolism in terms of ATP and substrates.
5.32 Discuss regulation of metabolic activity.

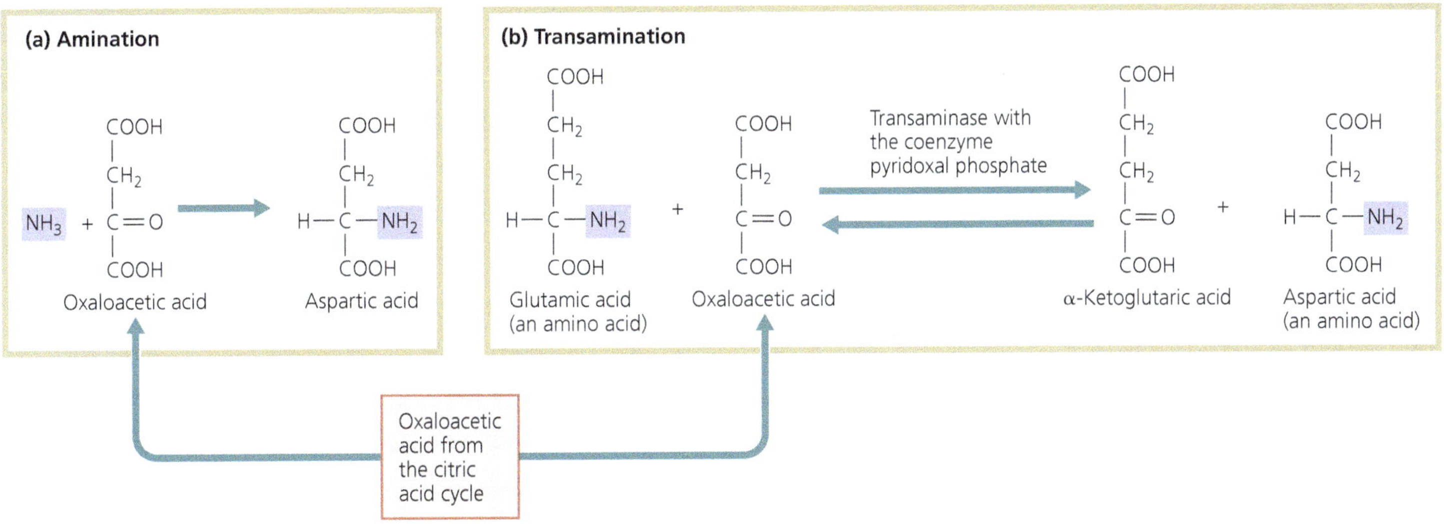

▲ **FIGURE 5.29 Examples of the synthesis of amino acids via amination and transamination.** (a) In amination, an amine group from ammonia is added to a precursor metabolite. In this example, oxaloacetic acid is converted into the amino acid aspartic acid. (b) In transamination, the amine group is derived from an existing amino acid. In the example shown here, the amino acid glutamic acid donates its amine group to oxaloacetic acid, which is converted into aspartic acid. Note that transamination is a reversible reaction.

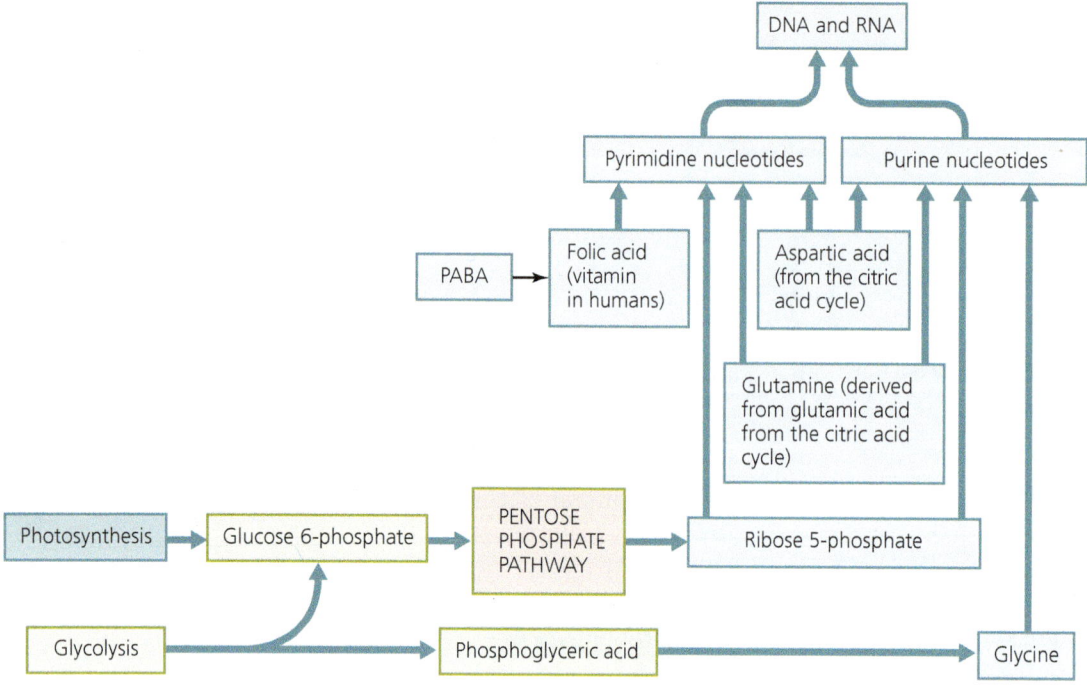

► FIGURE 5.30 The biosynthesis of nucleotides.

As we have seen, catabolic and anabolic reactions interact with one another in several ways. First, ATP molecules produced by catabolism are used to drive anabolic reactions. Second, catabolic pathways produce precursor metabolites to use as substrates for anabolic reactions. Additionally, most metabolic pathways are amphibolic; they function as part of either catabolism or anabolism as needed.

Cells regulate metabolism in a variety of ways to maximize efficiency in growth and reproductive rate. Among the mechanisms involved are the following:

- Cells synthesize or degrade channel and transport proteins to increase or decrease the concentration of chemicals in the cytosol or organelles.
- Cells often synthesize the enzymes needed to catabolize a particular substrate only when that substrate is available. For instance, the enzymes of beta-oxidation are not produced when there are no fatty acids to catabolize.
- If two energy sources are available, cells catabolize the more energy efficient of the two. For example, a bacterium growing in the presence of both glucose and lactose will produce enzymes only for the transport and catabolism of glucose. Once the supply of glucose is depleted, lactose-utilizing proteins are produced.
- Cells synthesize the metabolites they need, but they typically cease synthesis if a metabolite is available as a nutrient. For instance, bacteria grown in an environment containing an excess of aspartic acid will cease the amination of oxaloacetic acid (see Figure 5.29).
- Eukaryotic cells keep metabolic processes from interfering with each other by isolating particular enzymes within membrane-bound organelles. For example, proteases sequestered within lysosomes digest phagocytized proteins without destroying vital proteins in the cytosol.

- Cells use inhibitory and excitatory allosteric sites on enzymes to control the activity of enzymes (see Figures 5.9 and 5.11).
- Feedback inhibition slows or stops anabolic pathways when the product is in abundance (see Figure 5.12).
- Cells regulate catabolic and anabolic pathways that use the same substrate molecules by requiring different coenzymes for each. For instance, NADH is used almost exclusively with catabolic enzymes, whereas NADPH is typically used for anabolism.

Note that these regulatory mechanisms are generally of two types: *control of gene expression*, in which cells control the amount and timing of protein (enzyme) production, and *control of metabolic expression*, in which cells control the activity of proteins (enzymes) once they have been produced.

FIGURE 5.31 diagrams some of the numerous interrelationships among the metabolic pathways discussed in this chapter.

Play *Metabolism: The Big Picture*
@ Mastering Microbiology

MICRO CHECK

16. Which type of metabolic reaction (catabolic or anabolic) requires ATP?

TELL ME WHY

Why is feedback inhibition necessary for controlling anabolic pathways?

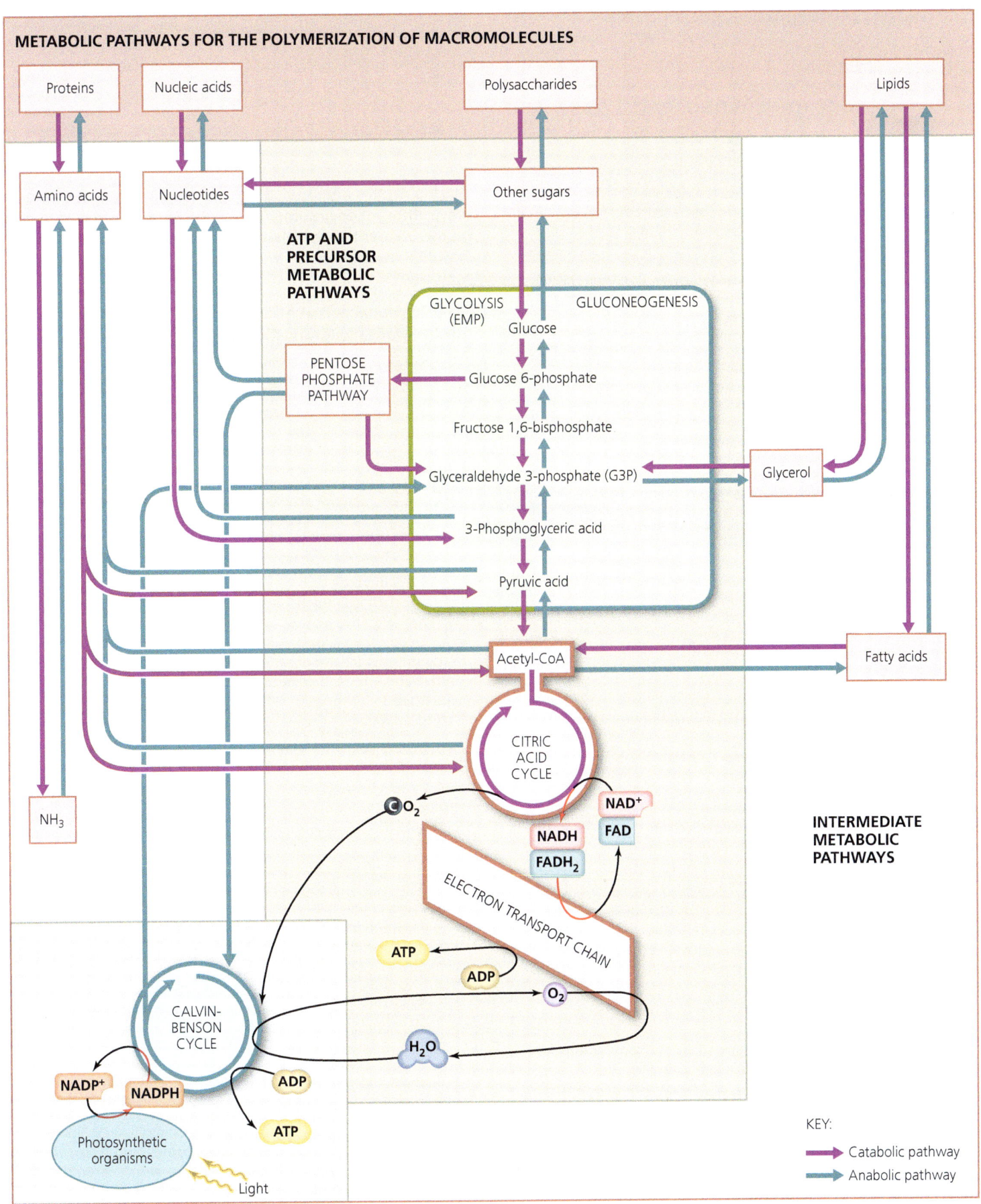

▲ **FIGURE 5.31 Integration of cellular metabolism (shown in an aerobic organism).** Cells possess three major categories of metabolic pathways: pathways for the polymerization of macromolecules (proteins, nucleic acids, polysaccharides, and lipids), intermediate pathways, and ATP and precursor pathways (glycolysis, citric acid cycle, the pentose phosphate pathway, and the Entner-Doudoroff pathway [not shown]). Cells of photosynthetic organisms also have the Calvin-Benson cycle.

MICRO IN THE CLINIC FOLLOW-UP

Scraped Knee and Loss of a Leg?

The doctor confirms that the infection has spread to deeper tissues, and he notes areas of necrosis (dead tissue). He tells Arjun and his mom that the fact that the infection hasn't responded to antibiotics and has spread to deeper tissues is quite serious. The necrotic tissue will need to be surgically removed as soon as possible.

During surgery, the dead tissue is removed, and samples are sent to the lab to identify the causative organism. The lab analysis identifies *Streptococcus pyogenes* in the samples. Intravenous antibiotics are continued post-surgery, and Arjun undergoes a second surgery to remove dead tissue. He remains in the hospital for an additional two weeks and continues oral antibiotics for five weeks after leaving the hospital. With the exception of a few scars, Arjun's knee ultimately regains full function, and he makes a complete recovery.

1. During Arjun's infection, *S. pyogenes* was growing in two very different environments—on the surface of his skin and in the deeper tissue. What types of metabolic pathways would *S. pyogenes* use in each of these environments, and why?

2. Why do you think that Arjun must continue to take antibiotics for five weeks following his release from the hospital?

Check your answers to Micro in the Clinic Follow-Up questions in the Mastering Microbiology Study Area.

CHAPTER SUMMARY

Go to Mastering Microbiology for Interactive Microbiology, Dr. Bauman's Videos Tutors, Micro Matters, MicroFlix, Micro-Boosters, practice quizzes, and more.

This chapter has MicroFlix. Go to the Mastering Microbiology Study Area to view movie-quality animations for metabolism.

Basic Chemical Reactions Underlying Metabolism (pp. 125–133)

1. **Metabolism** is the sum of biochemical reactions within the cells of an organism, including **catabolism**, which breaks down molecules and releases energy, and **anabolism**, which synthesizes molecules and uses energy.

   Play *Metabolism: Overview* @ Mastering Microbiology

2. **Precursor metabolites**, often produced in catabolic reactions, are used to synthesize all other organic compounds.

3. Reduction reactions are those in which electrons are added to a chemical. A  Play *Oxidation-Reduction Reactions* @ Mastering Microbiology
molecule that donates an electron is oxidized. If the electron is part of a hydrogen atom, an oxidation reaction is also called dehydrogenation. Oxidation and reduction reactions always occur in pairs called **oxidation-reduction (redox) reactions**.

4. Three important electron carrier molecules are **nicotinamide adenine dinucleotide (NAD$^+$)**, **nicotinamide adenine dinucleotide phosphate (NADP$^+$)**, and **flavin adenine dinucleotide (FAD)**.

5. Phosphorylation is the addition of phosphate to a molecule. Three types of phosphorylation form ATP: **Substrate-level phosphorylation** involves the transfer of phosphate from a phosphorylated organic compound to ADP. In **oxidative phosphorylation**, energy from redox reactions of respiration is used to attach inorganic phosphate (PO_4^{3-}) to ADP. **Photophosphorylation** is the phosphorylation of ADP with inorganic phosphate using energy from light.

6. Catalysts increase the rates of chemical reactions and are not permanently changed in the process.

    Play *Enzymes: Overview* @ Mastering Microbiology

   Enzymes, which are organic catalysts, are often named for their **substrates**—the molecules on which they act. Enzymes can be classified as hydrolases, isomerases, ligases (polymerases), lyases, oxidoreductases, or transferases, reflecting their mode of action.

7. **Apoenzymes** are the portions of enzymes that may require one or more **cofactors** such as inorganic ions or organic cofactors (also called **coenzymes**). The combination of an apoenzyme and its cofactors is a **holoenzyme**. RNA molecules functioning as enzymes are called **ribozymes**.

8. **Activation energy** is the amount of energy required to initiate a chemical reaction.

9. Substrates fit into the specifically shaped **active sites** of the enzymes that catalyze  Play *Enzymes: Steps in a Reaction* @ Mastering Microbiology
their reactions. Active sites change shape slightly to better fit their substrates—a process described as the **induced-fit model**.

10. Enzymes may be **denatured** by physical and chemical factors such as heat and pH. Denaturation may be reversible or permanent.

11. Enzyme activity proceeds at a rate proportional to the concentration of substrate molecules until all the active sites are filled.

12. **Competitive inhibitors** block active sites and thereby block enzyme activity. **Noncompetitive inhibitors** bind to an allosteric site on an enzyme. This binding alters the active site so that it is no longer functional.

 Play *Enzymes: Competitive Inhibition; Enzyme-Substrate Interaction: Noncompetitive Inhibition* @ Mastering Microbiology

Carbohydrate Catabolism (pp. 133–142)

1. Typical **glycolysis** involves splitting a glucose molecule in a three-stage, 10-step process that ultimately results in two molecules of pyruvic acid and a net gain of two ATP and two NADH molecules.

   Play *Glycolysis: Overview, Steps*
   @ Mastering Microbiology

2. **Cellular respiration** is a metabolic process that involves the complete oxidation of substrate molecules and the production of ATP following a series of redox reactions.

3. Two carbons from pyruvic acid join coenzyme A to form acetyl-coenzyme A (**acetyl-CoA**), which then enters the **citric acid cycle (Krebs cycle)**, a series of eight enzymatic steps that transfer electrons from acetyl-CoA to coenzymes NAD^+ and FAD.

   Play *Krebs Cycle: Overview, Steps*
   @ Mastering Microbiology

4. An **electron transport chain** is a series of redox reactions that pass electrons from one membrane-bound carrier to another and then to a final electron acceptor. The energy from these electrons is used to pump protons across the membrane.

   Play *Electron Transport Chain: Overview* @ Mastering Microbiology

5. The four classes of carrier molecules in electron transport systems are flavoproteins, ubiquinones, metal-containing proteins, and cytochromes.

   Play *Electron Transport Chain: The Process* @ Mastering Microbiology

6. Aerobes use oxygen atoms as final electron acceptors in their electron transport chains in a process known as **aerobic respiration**, whereas anaerobes use other inorganic molecules (such as NO_3^-, SO_4^{2-}, and CO_3^{2-}, or rarely, an externally acquired organic molecule as the final electron acceptor in **anaerobic respiration**.

   Play *Electron Transport Chain: Factors Affecting ATP Yield* @ Mastering Microbiology

7. In **chemiosmosis**, ions flow down their electrochemical gradient across a membrane through **ATP synthase (ATPase)** to synthesize ATP.

8. A **proton gradient** is an electrochemical gradient of hydrogen ions across a membrane. It has potential energy known as a proton motive force.

9. Oxidative phosphorylation and photophosphorylation use chemiosmosis.

10. The **pentose phosphate pathway** is an alternative means for the catabolism of glucose that yields fewer ATP molecules than does glycolysis. The pathway produces precursor metabolites and NADPH, which are necessary for synthesis of many cellular chemicals.

11. **Fermentation** is the partial oxidation of sugar to release energy using a cellular organic molecule rather than an electron transport chain as the final electron acceptor. End products of fermentation, which are often useful to humans and aid in laboratory identification of microbes, include acids, alcohols, and gases.

    Play *Fermentation*
    @ Mastering Microbiology

Other Catabolic Pathways (pp. 142–144)

1. Lipids and proteins can be catabolized into smaller molecules, which can be used as substrates for glycolysis and the citric acid cycle.

2. **Beta-oxidation** is a catabolic process in which enzymes split pairs of hydrogenated carbon atoms from a fatty acid and join them to coenzyme A to form acetyl-CoA.

3. **Proteases** secreted by microorganisms digest proteins outside the microbes' cell walls. The resulting amino acids are moved into the cell and used in anabolism or are **deaminated** and catabolized for energy.

Photosynthesis (pp. 144–149)

1. **Photosynthesis** is a process in which light energy is captured by pigment molecules called **chlorophylls** (bacteriochlorophylls in some bacteria) and transferred to ATP and metabolites. **Photosystems** are networks of light-absorbing chlorophyll molecules and other pigments held within a protein matrix on membranes called **thylakoids**.

   Play *Photosynthesis: Overview*
   @ Mastering Microbiology

2. The redox reactions of photosynthesis are classified as **light-dependent reactions** and **light-independent reactions**.

3. A **reaction center chlorophyll** is a special chlorophyll molecule in a photosystem in which electrons are excited by energy harvested from light by many other chlorophyll molecules. The excited electrons of the reaction center are passed to a carrier molecule in an electron transport chain.

4. In **cyclic photophosphorylation**, the electrons return to the original reaction center after passing down the electron transport chain.

   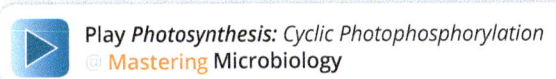
   Play *Photosynthesis: Cyclic Photophosphorylation*
   @ Mastering Microbiology

5. In **noncyclic photophosphorylation**, photosystem II works with photosystem I, and the electrons are used to reduce $NADP^+$ to NADPH. In oxygenic photosynthesis, cyanobacteria, algae, and green plants replenish electrons to the reaction center by dissociation of H_2O molecules, resulting in the release of O_2 molecules. Anoxygenic bacteria derive electrons from inorganic compounds such as H_2S, producing waste such as sulfur.

   Play *Photosynthesis: Noncyclic Photophosphorylation*
   @ Mastering Microbiology

6. In the light-independent pathway of photosynthesis, **carbon fixation** occurs in the **Calvin-Benson cycle**, in which CO_2 is reduced to produce glucose.

   Play *Photosynthesis: Light-Independent Reaction* @ Mastering Microbiology

Other Anabolic Pathways (pp. 149–151)

1. **Amphibolic reactions** are metabolic reactions that are reversible—they can operate catabolically or anabolically.
2. Some cells are able to synthesize glucose from amino acids, glycerol, and fatty acids via a process called gluconeogenesis.
3. **Amination** reactions involve adding an amine group from ammonia to a metabolite to make an amino acid. **Transamination** occurs when an amine group is transferred from one amino acid to another.
4. Nucleotides are synthesized from precursor metabolites produced by glycolysis, the citric acid cycle, and the pentose phosphate pathway.

Integration and Regulation of Metabolic Functions (pp. 151–153)

1. Cells regulate metabolism by control of gene expression or metabolic expression. They control the latter in a variety of ways, including synthesizing or degrading channel proteins and enzymes, sequestering reactions in membrane-bound organelles (seen in eukaryotes), and feedback inhibition.

Play *Metabolism: The Big Picture*
@ Mastering Microbiology

QUESTIONS FOR REVIEW

Answers to the Questions for Review (except Short Answer questions) begin on p. A-1.

Multiple Choice

For each of the phrases in questions 1–7, indicate the type of metabolism referred to, using the following choices:
  a. anabolism only
  b. both anabolism and catabolism (amphibolic)
  c. catabolism only

1. Breaks a large molecule into smaller ones
2. Includes dehydration synthesis reactions
3. Is exergonic
4. Is endergonic
5. Involves the production of cell membrane constituents
6. Includes hydrolytic reactions
7. Includes metabolism

8. Redox reactions___.
   a. transfer energy
   b. transfer electrons
   c. involve oxidation and reduction
   d. are involved in all of the above

9. A reduced molecule___.
   a. has gained electrons
   b. has become more positive in charge
   c. has lost electrons
   d. is an electron donor

10. Activation energy___.
    a. is the amount of energy required during an activity such as flagellar motion
    b. requires the addition of nutrients in the presence of water
    c. is lowered by the action of organic catalysts
    d. results from the movement of molecules

11. Coenzymes are___.
    a. types of apoenzymes
    b. proteins
    c. inorganic cofactors
    d. organic cofactors

12. Which of the following statements best describes ribozymes?
    a. Ribozymes are proteins that aid in the production of ribosomes.
    b. Ribozymes are nucleic acids that produce ribose sugars.
    c. Ribozymes store enzymes in ribosomes.
    d. Ribozymes process RNA molecules in eukaryotes.

13. Which of the following does *not* affect the function of enzymes?
    a. ubiquinone
    b. substrate concentration
    c. temperature
    d. competitive inhibitors

14. Most oxidation reactions in bacteria involve the___.
    a. removal of hydrogen ions and electrons
    b. removal of oxygen
    c. addition of hydrogen ions and electrons
    d. addition of hydrogen ions

15. Under ideal conditions, the fermentation of one glucose molecule by a bacterium allows a net gain of how many ATP molecules?
    a. 2
    b. 4
    c. 38
    d. 0

16. Under ideal conditions, the complete aerobic oxidation of one molecule of glucose by a bacterium allows a net gain of how many ATP molecules?
    a. 2
    b. 4
    c. 38
    d. 0

17. Which of the following statements about the Entner-Doudoroff pathway is *false*?
    a. It is a series of reactions that synthesizes glucose.
    b. Its products are sometimes used to determine the presence of *Pseudomonas*.
    c. It is a pathway of chemical reactions that catabolizes glucose.
    d. It is an alternative pathway to glycolysis.

18. Reactions involved in the light-independent reactions of photosynthesis constitute the___.
    a. citric acid cycle
    b. Entner-Doudoroff pathway
    c. Calvin-Benson cycle
    d. pentose phosphate pathway

19. The glycolysis pathway is basically___.
    a. catabolic
    b. amphibolic
    c. anabolic
    d. cyclical

20. A major difference between anaerobic respiration and anaerobic fermentation is___.
    a. in the use of oxygen
    b. that the former requires breathing
    c. that the latter uses organic molecules within the cell as final electron acceptors
    d. that fermentation only produces alcohol

Matching

Match the descriptions below with their corresponding terms.

1. ___ Occurs when energy from a compound containing phosphate reacts with ADP to form ATP
2. ___ Involves formation of ATP via reduction of coenzymes in the electron transport chain
3. ___ Begins with glycolysis
4. ___ Occurs when all active sites on substrate molecules are filled

A. Saturation
B. Oxidative phosphorylation
C. Substrate-level phosphorylation
D. Photophosphorylation
E. Carbohydrate catabolism

Fill in the Blanks

1. The final electron acceptor in cyclic photophosphorylation is _____.
2. Two ATP molecules are used to initiate glycolysis. Enzymes generate molecules of ATP for each molecule of glucose that undergoes glycolysis. Thus, a net gain of _____ molecules of ATP is produced in glycolysis.
3. The initial catabolism of glucose occurs by glycolysis and/or the _____ and _____ pathways.
4. _____ is a cyclic series of eight reactions involved in the catabolism of acetyl-CoA that yields eight molecules of NADH and two molecules of $FADH_2$.
5. The final electron acceptor in aerobic respiration is_____.
6. Three common inorganic electron acceptors in anaerobic respiration are _____, _____, and _____.
7. Anaerobic respiration typically uses (organic/inorganic) _____ molecules as final electron acceptors.
8. Complete the following chart:

Category of Enzymes	Description
	Catabolizes substrate by adding water
Isomerase	
Ligase/polymerase	
	Moves functional groups such as an acetyl group
	Adds or removes electrons
Lyase	

9. The use of a proton motive force to generate ATP is _____.
10. The main coenzymes that carry electrons in catabolic pathways are _____ and _____.

VISUALIZE IT!

1. Label the mitochondrion to indicate the location of glycolysis, the citric acid cycle, and electron transport chains.

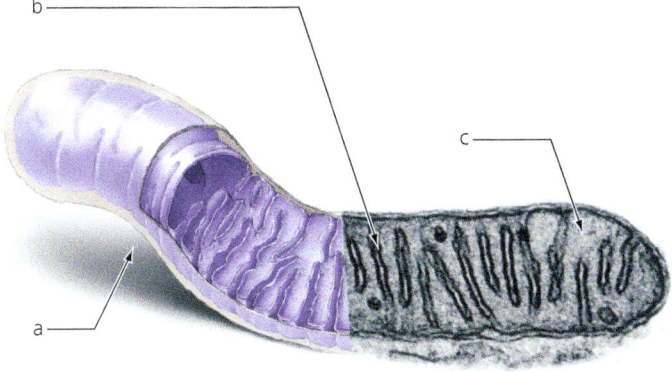

2. Label the diagram below to indicate acetyl-CoA, electron transport chain, $FADH_2$, fermentation, glycolysis, citric acid cycle, NADH, and respiration. Indicate the net number of molecules of ATP that could be synthesized at each stage during bacterial respiration of one molecule of glucose.

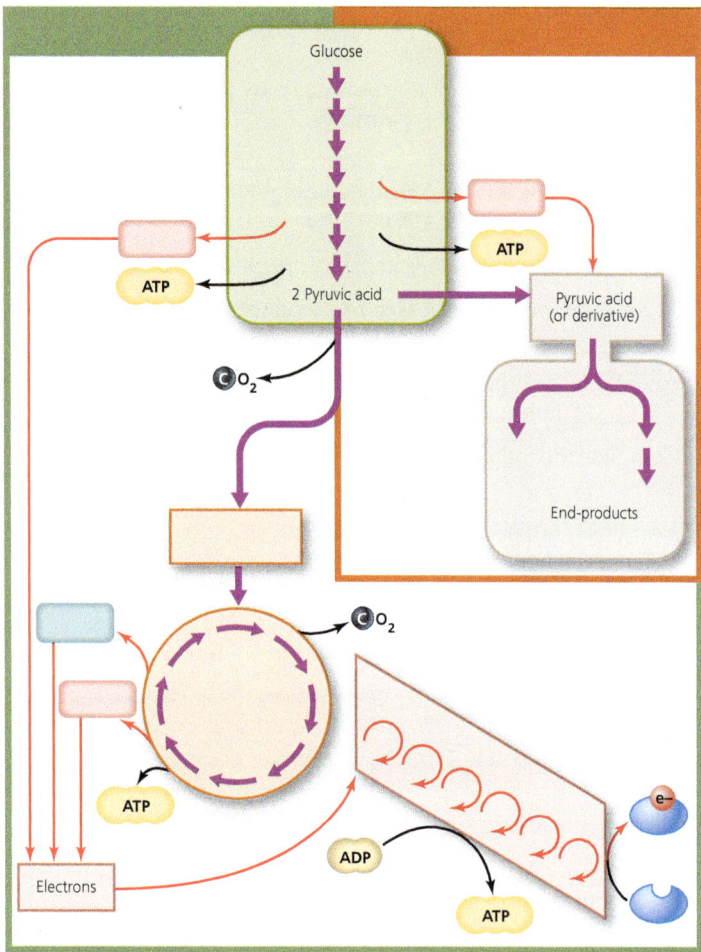

Short Answer

1. How does amination differ from transamination?
2. Why are enzymes necessary for anabolic reactions to occur in living organisms?
3. How do organisms control the rate of metabolic activities in their cells?
4. How does a noncompetitive inhibitor at a single allosteric site affect a whole pathway of enzymatic reactions?
5. Explain the mechanism of negative feedback with respect to enzyme action.
6. Facultative anaerobes can live under either aerobic or anaerobic conditions. What metabolic pathways allow these organisms to continue to harvest energy from sugar molecules in the absence of oxygen?
7. How does oxidation of a molecule occur without oxygen?
8. List at least four groups of microorganisms that are photosynthetic.
9. Why do we breathe oxygen and give off carbon dioxide?
10. Why do cyanobacteria and algae take in carbon dioxide and give off oxygen?
11. What happens to the carbon atoms in sugar catabolized by *Escherichia coli*?
12. How do yeast cells make alcohol and cause bread to rise?
13. Where specifically does the most significant production of ATP occur in prokaryotic and eukaryotic cells?
14. Why are vitamins essential metabolic factors for microbial metabolism?
15. A laboratory scientist notices that a certain bacterium does not utilize lactose when glucose is available in its environment. Describe a cellular regulatory mechanism that would explain this observation.

3. Examine the biosynthetic pathway for the production of the amino acids tryptophan, tyrosine, and phenylalanine in the figure. Where do the initial reactants (erythrose 4-phosphate and PEP) originate?

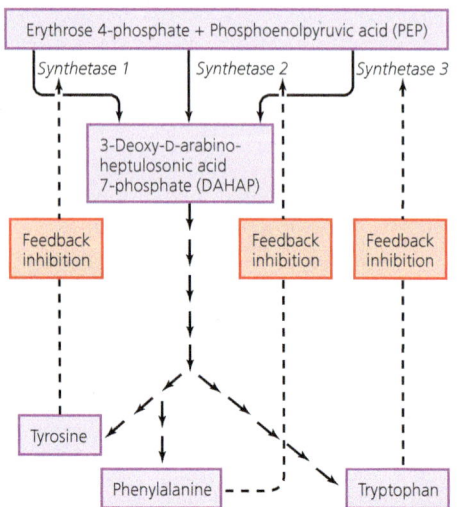

CRITICAL THINKING

1. Arsenic is a poison that exists in two states in the environment—arsenite ($H_2AsO_3^-$) and arsenate ($H_2AsO_4^-$). Arsenite dissolves in water, making the water dangerous to drink. Arsenate is less soluble and binds to minerals, making this form of arsenic less toxic in the environment. Some strains of the bacterium *Thermus* oxidize arsenic in an aerobic environment but reduce arsenic under anaerobic conditions. In case of arsenic contamination of water, how could scientists use *Thermus* to remediate the problem?

2. Explain why an excess of all three of the amino acids mentioned in Visualize It! question 3 (above) is required to inhibit production of DAHAP.

3. Why might an organism that uses glycolysis and the citric acid cycle also need the pentose phosphate pathway?

4. Describe how bacterial fermentation causes milk to sour.

5. *Giardia intestinalis* and *Entamoeba histolytica* are protozoa that live in the colons of mammals and can cause life-threatening diarrhea. Interestingly, these microbes lack mitochondria. What kind of pathway must they have for carbohydrate catabolism?

6. Two cultures of a facultative anaerobe are grown in the same type of medium, but one is exposed to air, and the other is maintained under anaerobic conditions. Which of the two cultures will contain more cells at the end of a week? Why?

7. What is the maximum number of molecules of ATP that can be generated by a bacterium after the complete aerobic oxidation of a fat molecule containing three 12-carbon chains? (Assume that all the available energy released during catabolism goes to ATP production.)

8. In terms of its effects on human metabolism, why is a fever over 40°C often life threatening?

9. Cyanide is a potent poison because it irreversibly blocks cytochrome a_3. What effect would its action have on the rest of the electron transport chain? What would be the redox state (reduced or oxidized) of ubiquinone in the presence of cyanide?

10. How are photophosphorylation and oxidative phosphorylation similar? How are they different?

11. Members of the pathogenic bacterial genus *Haemophilus* require NAD^+ and heme from their environment. For what purpose does *Haemophilus* use these growth factors?

12. Compare and contrast aerobic respiration, anaerobic respiration, and fermentation.

13. Scientists estimate that up to one-third of Earth's biomass is composed of methanogenic prokaryotes in ocean sediments (*Science* 295: 2067–2070). Describe the metabolism of these organisms.

14. A young student was troubled by the idea that a bacterium is able to control its diverse and complex metabolic activities even though it lacks a brain. How would you explain its metabolic control?

15. If a bacterium uses beta-oxidation to catabolize a molecule of the fatty acid arachidic acid, which contains 20 carbon atoms, how many acetyl-CoA molecules will be generated?

16. Some desert rodents rarely have water to drink. How do they get enough water for their cells without drinking it?

17. Why do fatty acids typically contain an even number of carbon atoms?

18. We have examined the total ATP, NADH, and $FADH_2$ production in the citric acid cycle for each molecule of glucose coming through Embden-Meyerhof-Parnas glycolysis. How many of each of these molecules would be produced if the Entner-Doudoroff pathway were used instead of EMP glycolysis?

19. Explain why hyperthermophiles do not cause disease in humans.

20. In addition to extremes in temperature and pH, other chemical and physical agents denature proteins. These agents include ionizing radiation, alcohol, enzymes, and heavy-metal ions. For example, the first antimicrobial drug, arsphenamine, contained the heavy metal arsenic and was used to inhibit the enzymes of the bacterium *Treponema pallidum*, the causative agent of syphilis.

    Given that both human and bacterial enzymes are denatured by heavy metals, how was arsphenamine used to treat syphilis without poisoning the patient? Why is syphilis no longer treated with arsenic-containing compounds?

21. Figure 5.18b illustrates events in aerobic respiration where oxygen acts as the final electron acceptor to yield water. How would the figure be changed to reflect anaerobic respiration?

22. Suppose you could insert a tiny pH probe into the space between mitochondrial membranes. Would the pH be above or below 7.0? Why?

23. Even though *Pseudomonas aeruginosa* and *Enterococcus faecalis* usually grow harmlessly, they can cause disease. Because these bacteria use the Entner-Doudoroff pathway instead of glycolysis to catabolize glucose, investigators can use clinical tests that provide evidence of the Entner-Doudoroff pathway to identify the presence of these potential pathogens.

    Suppose you were able to identify the presence of any specific organic compound. Name a substrate molecule you would find in *Pseudomonas* and *Enterococcus* cells but not in human cells.

24. Photosynthetic organisms are rarely pathogenic. Why?

25. We have seen that of the two ways ATP is generated via chemiosmosis—photophosphorylation and oxidative phosphorylation—the former can be cyclical, but the latter is never cyclical. Why can't oxidative phosphorylation be cyclical; that is, why aren't electrons passed back to the molecules that donated them?

26. A scientist moves a green plant grown in sunlight to a room with 24 hours of artificial green light. Will this increase or decrease the plant's rate of photosynthesis? Why?

27. What class of enzyme is involved in amination reactions? What class of enzyme catalyzes transaminations?

CONCEPT MAPPING

Using the terms provided, fill in the concept map that describes aerobic respiration. You can also complete this and other concept maps online by going to the **Mastering Microbiology** Study Area.

2 ATP (X2)
34 ATP
Chemiosmosis

Electron transport chain
Glycolysis
Citric acid cycle

Oxidative phosphorylation
Substrate-level phosphorylation (2)
Synthesis of acetyl-CoA

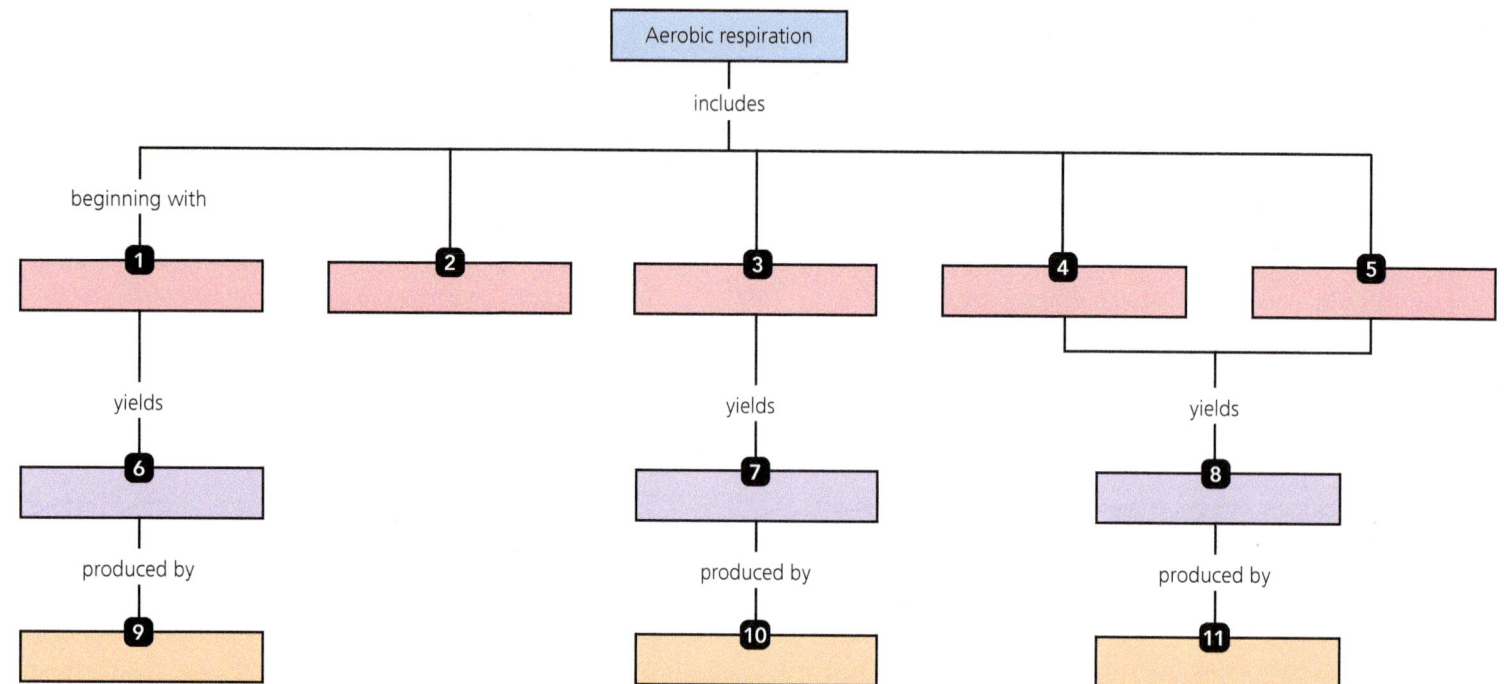

6 Microbial Nutrition and Growth

Before You Begin

1. List these catabolic pathways—fermentation, anaerobic respiration, and aerobic respiration—in order of the most energy produced in the form of ATP to the least.
2. Of the catabolic pathways: glycolysis, fermentation, anaerobic respiration, and aerobic respiration, which can proceed without molecular oxygen (O_2) present?
3. Which of the major macromolecules—lipids, carbohydrates, proteins, and nucleic acids—contain nitrogen atoms (N)?
4. If an environment is acidic, is its pH lower than, equal to, or greater than 7?

MICRO IN THE CLINIC

Just a Sore Throat and Cough?

JENNIFER IS A COLLEGE STUDENT who hasn't been feeling like herself lately. Usually she's full of energy, but in the last couple of weeks she's been more tired than usual and has had a sore throat and cough. Neither has gotten worse, so she assumes she's just a little run down. After all, midterm week has just passed, and she was up late every night studying. As a member of the university volleyball team, Jennifer traveled to three matches in ten days. Since she doesn't have anything major planned for the upcoming weekend, she's focused on getting a lot of sleep and feeling better.

Monday morning, Jennifer doesn't feel quite as tired and her throat doesn't hurt anymore, but her cough is worse. She uses throat lozenges and cough syrup to decrease the coughing, but over the next week it worsens—getting so bad that she can't sleep through the night. Her roommate is annoyed; Jennifer's coughing has been keeping her up, too. While running drills at volleyball practice, Jennifer gets winded more quickly than usual; when she can't catch her breath, the coach decides that it's time for a trip to the health clinic.

1. Why can't Jennifer catch her breath?
2. Why is her cough getting worse when her other symptoms (fatigue, sore throat) have gotten better?

Turn to the end of the chapter (p. 187) to find out.

The cells of *Mycoplasma* assume many different shapes.

SOLVE THE PROBLEM

The Microbes Ate My Homework

Paper in the form of cardboard and shipping containers is one of the largest contributors to landfills. What if we could turn this waste into fuel? Paper is mainly composed of cellulose, a branched polymer of glucose (see Figure 2.20). Cellulose makes up the majority of the cell wall material of plants. A number of species of bacteria, including members of the phyla Actinobacteria and Firmicutes, can degrade cellulose by releasing free sugars from the polymer using enzymes called cellulases. Other bacteria can ferment these sugars into fuels such as ethanol or butanol, which are called biofuels when they are derived by metabolic activity. An exciting idea is to genetically engineer a single bacterial species to secrete many different kinds of cellulases and also ferment the released sugars into biofuel. This new organism could be added to cardboard waste dissolved in water and would turn the waste into fuel, solving two problems at once. Adversaries of this idea are opposed to genetically modifying bacteria in this manner. What if such a microbe escaped into the wild? Could it become a new plant pathogen, attacking the very fabric of plants' cell walls?

Imagine you are on a scientific team at the Environmental Protection Agency charged with reviewing an application from a company that wants to use engineered bacteria in cellulosic bioreactors.

- How can these organisms be used safely?
- Would you approve this application?
- What additional information is required to make an informed decision?

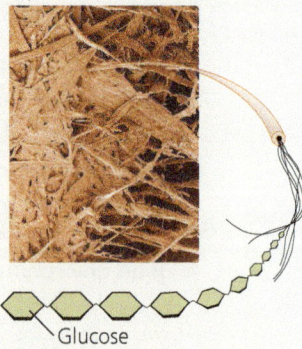

Cellulose fibers

Tree decaying through the actions of microbes

Metabolism—the set of controlled chemical reactions within cells—is an essential characteristic of all living things. The ultimate outcome of metabolic activity is reproduction, an increase in the number of individual cells or organisms. When speaking of the reproductive activities of microbes in general and of bacteria in particular, microbiologists typically use the term *growth*, referring to an increase in the size of a population of microbes rather than to an increase in size of an individual. The result of such microbial growth is either a discrete *colony*, which is an aggregation of cells arising from a single parent cell, or a *biofilm*, which is a collection of microbes living on a surface in a complex community. Put another way, the reproduction of individual microorganisms results in the growth of a colony or biofilm. Further, common expressions such as "The microorganisms *grow* in salt-containing media" are widely understood to mean that the organisms metabolize and reproduce rather than that they increase in size.

In this chapter, we consider the characteristics of microbial growth from two different but related perspectives: We examine the requirements of microbes in natural settings, including their chemical, physical, and energy requirements, and we explore how microbiologists try to create similar conditions to grow microorganisms in the laboratory so that they can be transported, identified, and studied. We conclude by examining laboratory analysis of bacterial population dynamics and some techniques for measuring bacterial population growth.

Growth Requirements

Organisms use a variety of chemicals—called **nutrients**—to meet their energy needs and to build organic molecules and cellular structures. The most common of these nutrients are compounds containing necessary elements such as carbon, oxygen, nitrogen, and hydrogen. Like all organisms, microbes obtain nutrients from a variety of sources in their environment, and they must bring nutrients into their cells by passive and active transport processes (see Chapter 3). When they acquire their nutrients by living in or on another organism, they may cause disease as they interfere with their hosts' metabolism and nutrition.

Nutrients: Chemical and Energy Requirements

LEARNING OUTCOMES

6.1 Describe the roles of carbon, hydrogen, oxygen, nitrogen, trace elements, and vitamins in microbial growth and reproduction.

6.2 Compare four basic categories of organisms based on their carbon and energy sources.

6.3 Distinguish among anaerobes, aerobes, aerotolerant anaerobes, facultative anaerobes, and microaerophiles.

6.4 Explain how oxygen can be fatal to organisms by discussing singlet oxygen, superoxide radical, peroxide anion, and hydroxyl radical; describe how organisms protect themselves from toxic forms of oxygen.

6.5 Define *nitrogen fixation*, and explain its importance.

We begin our examination of microbial growth requirements by considering three things all cells need for metabolism: a source of carbon, a source of energy, and a source of electrons or hydrogen atoms.

Sources of Carbon, Energy, and Electrons

Organisms can be categorized into two broad groups based on their source of carbon. Organisms that utilize an inorganic source of carbon (i.e., carbon dioxide) as their sole source of carbon are called *autotrophs*[1] (aw´tō-trōfs), so named because they "feed themselves." More precisely, autotrophs make organic compounds from CO_2 and thus need not acquire carbon in organic compounds from other organisms. In contrast, organisms called *heterotrophs*[2] (het´er-ō-trōfs) catabolize organic molecules (such as proteins, carbohydrates, amino acids, and fatty acids) they acquire from other organisms.

Organisms can also be categorized according to whether they use chemicals or light as an energy source for such cellular processes as anabolism, intracellular transport, and motility. Organisms that acquire energy from redox reactions involving inorganic and organic chemicals are called *chemotrophs*[3] (kēm´ō-trōfs). These reactions are either aerobic respiration, anaerobic respiration, or fermentation, depending on the final electron acceptor (see Chapter 5). Organisms that use light as an energy source are called *phototrophs*[4] (fō´tō-trōfs).

Thus, we see that organisms can be categorized on the basis of their carbon and energy sources into one of four basic groups: **photoautotrophs**, **chemoautotrophs**, **photoheterotrophs**, and **chemoheterotrophs** (FIGURE 6.1). Plants, some protozoa, and algae are photoautotrophs, whereas animals, fungi, and other protozoa are chemoheterotrophs. Bacteria and archaea exhibit greater metabolic diversity, uniquely having members in all four groups.

In addition to carbon and energy sources, cells require electrons or hydrogen atoms (which are composed of an electron orbiting a simple nucleus) for redox reactions. Hydrogen is the most common chemical element in cells, and it is so common in organic molecules and water that it is never a *limiting nutrient*; that is, metabolism is never interrupted by a lack of hydrogen. Hydrogen is essential for hydrogen bonding and in electron transfer. Heterotrophs acquire electrons (typically as part of hydrogen atoms) from the same organic molecules that provide them carbon and are called **organotrophs** (ōr´gān-ō-trōfs); alternatively, autotrophic organisms acquire electrons or hydrogen atoms from inorganic molecules (such as H_2, NO_2^-, H_2S, and Fe^{2+}) and are called **lithotrophs**[5] (lith´ō-trōfs).

Oxygen Requirements

Oxygen is essential for **obligate aerobes** because it serves as the final electron acceptor of electron transport chains, which produce most of the ATP in these organisms. By contrast, oxygen is a deadly poison for **obligate anaerobes**. How can oxygen be essential for one group of organisms and yet be a fatal toxin for others?

The key to understanding this apparent incongruity is understanding that neither atmospheric oxygen (O_2) nor covalently bound oxygen in compounds such as carbohydrates and water

◀ **FIGURE 6.1 Four basic groups of organisms based on their carbon and energy sources.** Additionally, organotrophs utilize electrons from organic molecules, and lithotrophs utilize electrons from inorganic molecules.

Carbon source		Energy source	
		Light (photo-)	**Chemical compounds (chemo-)**
Carbon dioxide (auto-)		*Photoautotrophs* • Plants, algae, and cyanobacteria use H_2O as an electron source to reduce CO_2, producing O_2 as a by-product • Green sulfur bacteria and purple sulfur bacteria use H_2S as an electron source; they do not produce O_2	*Chemoautotrophs* • Hydrogen, sulfur, and nitrifying bacteria, some archaea
Organic compounds (hetero-)		*Photoheterotrophs* • Green nonsulfur bacteria and purple nonsulfur bacteria, some archaea	*Chemoheterotrophs* • Aerobic respiration: most animals, fungi, and protozoa, and many bacteria • Anaerobic respiration: some animals, protozoa, bacteria, and archaea • Fermentation: some bacteria, yeasts, and archaea

---

[1] From Greek *auto*, meaning "self," and *trophe*, meaning "nutrition."
[2] From Greek *hetero*, meaning "other," and *trophe*, meaning "nutrition."
[3] From Gr. *chemeia*, meaning "alchemy," and *trophy*.
[4] From Greek *photos*, meaning "light," and *trophe*, meaning "nutrition."
[5] From Greek *lithos*, meaning "rock," and *trophe*, meaning "nutrition."

is poisonous. Rather, the toxic forms of oxygen are those that are highly reactive. They are toxic for the same reason that oxygen is the final electron acceptor for aerobes: They are excellent oxidizing agents, so they steal electrons from other compounds, which in turn steal electrons from still other compounds. The resulting chain of vigorous oxidations causes irreparable damage to cells by oxidizing important compounds, including proteins and lipids. The cellular components most sensitive to oxidative damage are proteins with active sites involved in redox reactions, especially iron-sulfur clusters.

There are four toxic forms of oxygen:

- **Singlet oxygen ($^1O_2$).** Singlet oxygen is molecular oxygen with electrons that have been boosted to a higher energy state, typically during aerobic metabolism. Singlet oxygen is a very reactive oxidizing agent. Phagocytic cells, such as certain human white blood cells, use it to oxidize pathogens. Singlet oxygen is photochemically produced by the reaction of light with oxygen in the presence of certain light-absorbing pigments, such as chlorophyll. To protect themselves from singlet oxygen, phototrophic microorganisms often contain pigments called **carotenoids** (ka-rot′e-noyds) that prevent toxicity by removing the excess energy of singlet oxygen.

- **Superoxide radical (O_2^-).** A few superoxide radicals form during the incomplete reduction of O_2 during electron transport in aerobes and during metabolism by anaerobes in the presence of oxygen. Superoxide radicals are so reactive and toxic that aerobic organisms must produce enzymes called *superoxide dismutases* (dis′myu-tās-es) to detoxify them. These enzymes, which have active sites that contain metal ions—Zn^{2+}, Mn^{2+}, Fe^{2+}, Ni^{2+}, or Cu^{2+}, depending on the organism—combine two superoxide radicals and two protons to form hydrogen peroxide (H_2O_2) and molecular oxygen (O_2):

$$2\, O_2^- + 2\, H^+ \rightarrow H_2O_2 + O_2$$

One reason that anaerobes are susceptible to oxygen is that they lack superoxide dismutase; the oxidizing reactions of superoxide radicals that have been formed in the presence of oxygen cause their death.

- **Peroxide anion (O_2^{2-}).** Hydrogen peroxide formed during reactions catalyzed by superoxide dismutase (and during other metabolic reactions) contains peroxide anion, another highly reactive oxidant. It is peroxide anion that makes hydrogen peroxide an antimicrobial agent. Aerobes contain catalase or peroxidase enzymes that detoxify peroxide anion.

Catalase converts hydrogen peroxide to water and molecular oxygen:

$$2\, H_2O_2 \rightarrow 2\, H_2O + O_2$$

A simple test for catalase involves adding a sample from a bacterial colony to a drop of hydrogen peroxide. The production of bubbles of oxygen indicates the presence of catalase.

Peroxidase breaks down hydrogen peroxide without forming oxygen, using a reducing agent such as the coenzyme NADH:

$$H_2O_2 + NADH + H^+ \rightarrow 2\, H_2O + NAD^+$$

Obligate anaerobes either lack both catalase and peroxidase or have only a small amount of them, so they are susceptible to the toxic action of hydrogen peroxide.

- **Hydroxyl radical (OH·).** Hydroxyl radicals result from ionizing radiation and from the incomplete reduction of hydrogen peroxide:

$$H_2O_2 + e^- + H^+ \rightarrow H_2O + OH\cdot$$

Hydroxyl radicals are the most reactive of the four toxic forms of oxygen, but hydrogen peroxide does not accumulate in aerobic cells (because of the action of catalase and peroxidase); so in aerobic cells, the threat of hydroxyl radical is virtually eliminated.

Besides the enzymes superoxide dismutase, catalase, and peroxidase, aerobes use other antioxidants, such as vitamins C and E, to protect themselves from toxic oxygen products. These antioxidants provide electrons that reduce toxic forms of oxygen.

Not all organisms are either strict **aerobes** or **anaerobes**; many organisms can live in various oxygen concentrations between these two extremes. For example, some aerobic organisms can maintain life via fermentation or anaerobic respiration, though their metabolic efficiency is often reduced in the absence of oxygen. Such organisms are called **facultative anaerobes**. *Escherichia coli* (esh-ĕ-rik′ē-ă kō′lī) is an example of a facultatively anaerobic bacterium. Note that although the term *facultative aerobe* could possibly have the same meaning, for consistency scientists use only the term *facultative anaerobe*.

Aerotolerant anaerobes do not use aerobic metabolism, but they tolerate oxygen by having some of the enzymes that detoxify oxygen's poisonous forms. The lactobacilli that transform cucumbers into pickles and milk into cheese are aerotolerant. These organisms can be kept in a laboratory without the special conditions required by obligate anaerobes.

Microaerophiles, such as the ulcer-causing pathogen *Helicobacter pylori*[6] (hel′ĭ-kō-bak′ter pī′lō-rē), require oxygen levels of 2% to 10%. This concentration of oxygen is found in the stomach. The 21% concentration of oxygen in the atmosphere damages microaerophiles, presumably because they have limited ability to detoxify hydrogen peroxide and superoxide radicals.

Microbial groups contain members with each of the five types of oxygen requirement. Algae, most fungi and protozoa, and many prokaryotes are obligate aerobes. A few yeasts and numerous prokaryotes are facultative anaerobes. Many prokaryotes and a few protozoa are aerotolerant, microaerophilic, or obligate anaerobes. The oxygen requirement of an organism can be identified by growing it in a medium that contains an oxygen gradient from top to bottom (FIGURE 6.2).

Nitrogen Requirements

Another essential element is nitrogen, which is an important element in many organic compounds, including the amine group of amino acids and as part of nucleotide bases. Nitrogen makes up about 14% of the dry weight of microbial cells.

---

[6]From the semihelical shape of the cell and from Greek *pyle*, meaning "gate," in reference to the pylorus, the distal portion of the stomach, which is the gate to the small intestine.

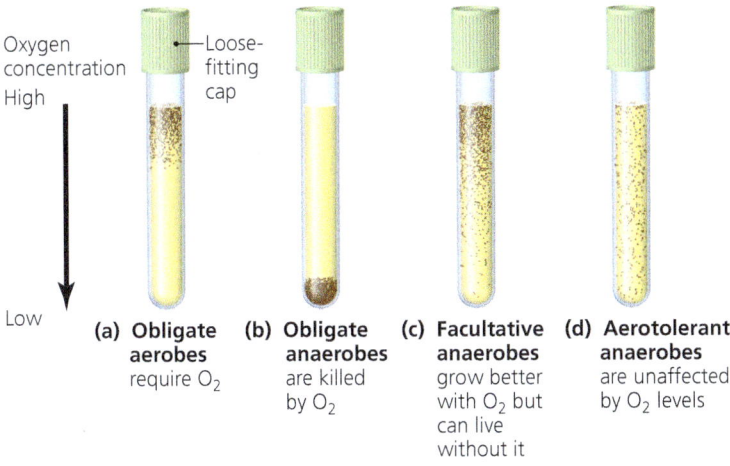

▲ FIGURE 6.2 Using a liquid thioglycolate growth medium to identify the oxygen requirements of organisms. Thioglycolate absorbs oxygen, providing a range of oxygen concentrations from aerobic, at the top, to anaerobic at the bottom. (a) Obligate aerobes cannot survive below the depth to which oxygen penetrates the medium. (b) Obligate anaerobes cannot tolerate any oxygen. (c) Facultative anaerobes can grow with or without oxygen, but their ability to use aerobic respiration pathways enhances their growth near the surface. (d) Aerotolerant anaerobes can grow equally well with or without oxygen; their growth is relatively evenly distributed throughout the medium. *Where in such a test tube would the growth zone be for a microaerophilic aerobe?*

Figure 6.2 *Microaerophiles would be found slightly below the surface but neither directly at the surface nor in the depths of the tube.*

Nitrogen is often a growth-limiting nutrient for many organisms; that is, their anabolism ceases because they do not have sufficient nitrogen to build proteins and nucleotides. Organisms acquire nitrogen from organic and inorganic nutrients. For example, most photosynthetic organisms can reduce nitrate (NO_3^-) to ammonium (NH_4^+), which can then be used for biosynthesis. In addition, all cells recycle nitrogen from their amino acids and nucleotides.

Though nitrogen constitutes about 79% of the atmosphere, relatively few organisms can utilize nitrogen gas (N_2), because it is a triple-bonded, highly stable gas. A few bacteria, notably many cyanobacteria and *Rhizobium* (rī-zō′bē-ŭm), reduce nitrogen gas to ammonia (NH_3) via a process called **nitrogen fixation**. Nitrogen fixation is essential for life on Earth because nitrogen fixers provide nitrogen in a form usable by other organisms. (Chapters 11 and 27 discuss nitrogen-fixing prokaryotes as well as nitrifying prokaryotes—those that oxidize nitrogenous compounds to acquire electrons for electron transport.)

Other Chemical Requirements

Together, carbon, hydrogen, oxygen, and nitrogen make up more than 95% of the dry weight of cells; phosphorus, sulfur, calcium, manganese, magnesium, copper, iron, and a few other elements constitute the rest. Phosphorus is a component of phospholipid membranes, DNA, RNA, ATP, and some proteins. Sulfur is a component of sulfur-containing amino acids, which bind to one another via disulfide bonds that are critical to the tertiary structure of proteins, and in vitamins such as thiamine (B_1) and biotin.

TABLE 6.1 Some Growth Factors of Microorganisms and Their Functions

Growth Factor	Function
Amino acids	Components of proteins
Cholesterol	Used by mycoplasmas (bacteria) for cell membranes
Heme	Functional portion of cytochromes in electron transport system
NADH	Electron carrier
Niacin (nicotinic acid, vitamin B_3)	Precursor of NAD^+ and $NADP^+$
Para-aminobenzoic acid (PABA)	Precursor of folic acid, which is involved in metabolism of one-carbon compounds and nucleic acid synthesis

Other elements are called **trace elements** because they are required in very small ("trace") amounts. For example, a few atoms of selenium dissolved out of the walls of glass test tubes provide the total requirement for the growth of green algae in a laboratory. Other trace elements are usually found in sufficient quantities dissolved in water. For this reason, tap water can sometimes be used instead of distilled or deionized water to grow microorganisms in the laboratory.

Some microorganisms—for example, algae and photosynthetic bacteria—are lithotrophic photoautotrophs; that is, they can synthesize all of their metabolic and structural needs from inorganic nutrients. They have every enzyme and cofactor they need to produce all their cellular components. Most organisms, however, require small amounts of certain organic chemicals that they cannot synthesize in addition to those that provide carbon and energy. These necessary organic chemicals are called **growth factors** (TABLE 6.1). For example, vitamins are growth factors for some microorganisms. Recall that vitamins constitute all or part of many coenzymes. (Note that vitamins are not growth factors for microorganisms that can manufacture them, such as *E. coli*.) Growth factors for various microbes include some amino acids, purines, pyrimidines, cholesterol, NADH, and heme.

MICRO CHECK

1. What category of organisms acquire both energy and carbon from complex organic compounds?
2. A bacterial culture that grows only near the surface of a thioglycolate broth is an example of what type of organism?

Physical Requirements

LEARNING OUTCOME

6.6 Explain how extremes of temperature, pH, and osmotic and hydrostatic pressure limit microbial growth.

166 CHAPTER 6 Microbial Nutrition and Growth

In addition to chemical nutrients, organisms have physical requirements for growth, including specific conditions of temperature, pH, osmolarity, and pressure.

Temperature

Temperature plays an important role in microbial life through its effects on the three-dimensional configurations of biological molecules. Recall that to function properly, proteins require a specific three-dimensional shape that is determined in part by temperature-sensitive hydrogen bonds, which are more likely to form at lower temperatures and more likely to break at higher temperatures. Proteins denature and lose their shape and function when hydrogen bonds break. Additionally, lipids, such as those that are components of the membranes of cells and organelles, are temperature sensitive. If the temperature is too low, membranes become rigid and fragile; if the temperature is too high, lipids can become too fluid, and the membrane cannot contain the cell or organelle.

Because temperature plays an important role in the three-dimensional structure of many types of biological molecules, different temperatures have different effects on the survival and growth of microbes (**FIGURE 6.3**). The lowest temperature at which an organism is able to conduct metabolism is called the *minimum growth temperature*. Note, however, that many microbes, particularly bacteria, survive (although they do not thrive) at temperatures far below their minimum growth temperature despite the fact that cell membranes are less fluid and transport processes are too slow to support metabolic activity. The highest temperature at which an organism continues to metabolize is called its *maximum growth temperature*; when the temperature exceeds this value, the organism's proteins are permanently denatured, and it dies. The temperature at which an organism's metabolic activities produce the highest growth rate is its **optimum growth temperature**. Each organism thus survives over a *temperature range* within which its growth and metabolism are supported.

Based on their preferred temperature ranges—the temperatures within which their metabolic activity and growth are best supported—microbes can be categorized into five overlapping groups (**FIGURE 6.4**). **Psychrophiles**[7] (sī'krō-fīls) grow best at temperatures below about 15°C. The current record holder for growth at subzero temperatures is a bacterium—*Psychrobacter arcticus* (sī'krō-bak-ter ark'ti-kus)—which lives inside ice at −10°C. Most die at temperatures above 20°C. In nature, psychrophilic algae, fungi, archaea, and bacteria live in snowfields, ice, and cold water (**FIGURE 6.5**). They do not cause disease in humans because they cannot thrive at body temperature, though some do spoil food in refrigerators. **Psychrotolerant** organisms (also called *psychrotrophs*) tolerate but don't grow best in the cold. They may infect warm-blooded animals, including humans; for example, *Listeria monocytogenes* (lis-tēr'ē-ă mo-nō-sī-tah'jē-nēz) is a pathogen that can grow in refrigerated foods. **Mesophiles**[8] (mez'ō-fīls) are organisms that grow best in temperatures ranging from 20°C to about 40°C, though they can

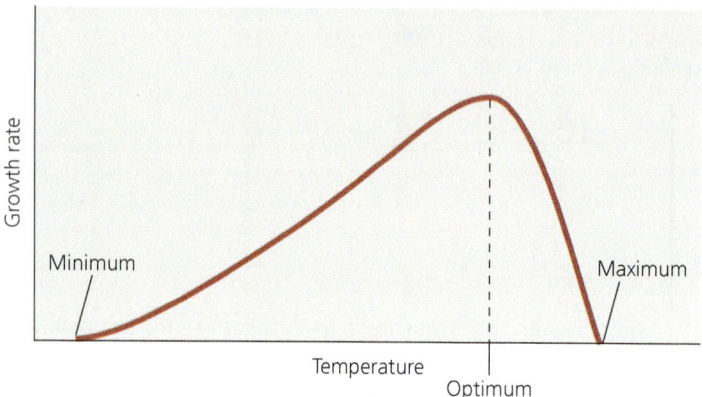

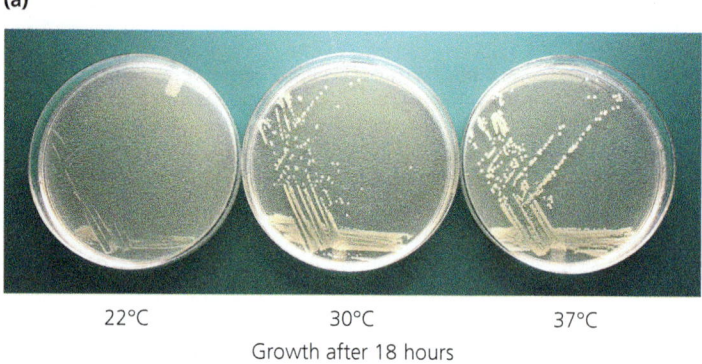

▲ **FIGURE 6.3 The effects of temperature on microbial growth.** (a) Minimum, optimum, and maximum growth temperatures are revealed by growth rate plotted against temperature. (b) Growth of *Escherichia coli* on nutrient agar after 18 hours of incubation at three different temperatures. *If microorganisms can survive at temperatures lower than their minimum growth temperature, then why is it called "minimum"?*

Figure 6.3 *The minimum growth temperature is defined as the lowest temperature that supports metabolism. Though organisms might survive at lower temperatures, they do not actively metabolize, grow, or reproduce.*

survive at higher and lower temperatures. Because normal body temperature is approximately 37°C, most human pathogens are mesophiles. *Thermoduric*[9] organisms are mesophiles that can survive brief periods at higher temperatures. Inadequate heating during pasteurization or canning can result in food spoilage by thermoduric mesophiles.

Thermophiles[10] (ther'mō-fīls) grow at temperatures above 45°C in habitats such as compost piles and hot springs. Some bacteria and archaea, called **hyperthermophiles**, grow best at temperatures above 80°C; others live in water heated above its boiling point (100°C).[11] The current record holder is an archaeon, *Geogemma barossii* (jē'ō-jem-ă ba-rōs'ē-ē). *Geogemma* grows and reproduces near submarine hot springs at temperatures between 85°C and 121°C and can survive for at least two hours at 130°C!

---

[7] From Greek *psuchros*, meaning "cold," and *philos*, meaning "love."
[8] From Greek *mesos*, meaning "middle," and *philos*, meaning "love."
[9] From Greek *therme*, meaning "hot." The organisms are so named because of their ability to endure or tolerate heat.
[10] From Greek *therme*, meaning "hot," and *philos*, meaning "love."
[11] Water can remain a liquid above 100°C if it has a high salt content or is under pressure, such as occurs in geysers or deep ocean troughs.

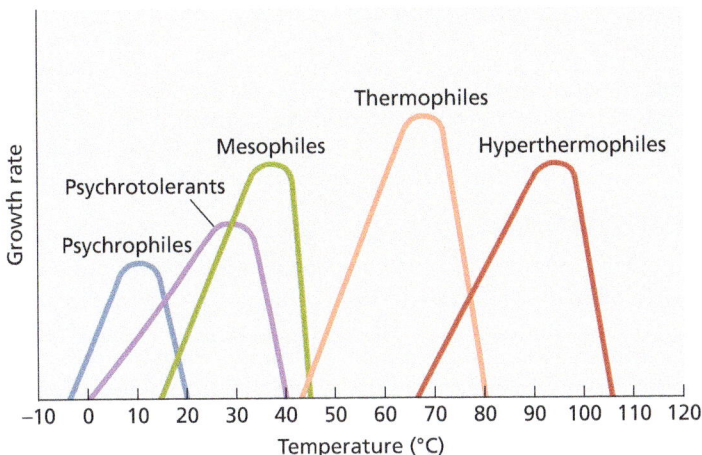

▲ FIGURE 6.4 Five categories of microbes based on temperature ranges for growth. *Categorize the bacterium* Thermus aquaticus, *which has an optimum growth temperature of 70°C.*

Figure 6.4 Thermus aquaticus is a thermophile.

Thermophiles and hyperthermophiles stabilize their proteins with extra hydrogen and covalent bonds between amino acids. Heat-stable enzymes are useful in industrial, engineering, and research applications. Heat-loving organisms do not cause disease because they "freeze" at body temperature. **Beneficial Microbes: A Nuclear Waste-Eating Microbe?** (p. 169) highlights an unusual thermophile that can also withstand radiation.

pH

pH is a measure of the concentration of hydrogen ions in a solution; that is, it is a measure of the acidity or alkalinity of a substance. A pH below 7.0 is acidic; the lower the pH value, the more acidic a substance is. Alkaline (basic) pH values are higher than 7.0. Organisms are sensitive to changes in acidity because hydrogen ions interfere with hydrogen bonding within proteins and nucleic acids; as a result, organisms have ranges of acidity that they prefer and can tolerate.

Most bacteria and protozoa, including most pathogens, are **neutrophiles** (nū′trō-fīls), because they grow best in a narrow range around a neutral pH—that is, between pH 6.5 and pH 7.5, which is also the pH range of most tissues and organs in the human body. By contrast, other bacteria and many fungi are **acidophiles** (ā-sīd′ō-phīls), organisms that grow best in acidic habitats. One example of acidophilic microbes is the chemoautotrophic prokaryotes that live in mines and in water that runs through mine tailings (waste rock), habitats that have pH values as low as 0.0. These prokaryotes oxidize sulfur to sulfuric acid, further lowering the pH of their environment. Whereas *obligate acidophiles* require an acidic environment and die if the pH approaches 7.0, *acid-tolerant microbes* merely survive in acid without preferring it.

Many organisms produce acidic waste products that accumulate in their environment until eventually they inhibit further growth. For example, many cheeses are acidic because of lactic acid produced by fermenting bacteria and fungi. The low pH of these cheeses then acts as a preservative by preventing any further microbial growth. Other acidic foods, such as sauerkraut and dill pickles, are also kept from spoiling because most organisms cannot tolerate the foods' low pH.

The normal acidity of certain regions of the body also inhibits microbial growth and retards many kinds of infection. At one site, the vaginas of adult women, acidity results from the fermentation of carbohydrates by normal resident bacteria. If the growth of these normal residents is disrupted—for instance, by antibiotic therapy—the resulting higher pH may allow yeasts to grow and lead to a yeast infection. Another site, the stomach, is inhospitable to most microbes because of the normal production of stomach acid. However, the acid-tolerant bacterium *Helicobacter pylori* neutralizes stomach acid by secreting bicarbonate and urease, an enzyme that converts urea to alkaline ammonia. The growth of *Helicobacter* is the cause of most gastric ulcers.

Alkaline conditions also inhibit the growth of most microbes, but **alkalinophiles** live in alkaline soils and water up to pH 11.5. For example, *Vibrio cholerae* (vib′rē-ō kol′er-ī), the causative agent of cholera, grows best outside the body in water at pH 9.0.

Physical Effects of Water

Microorganisms require water; they must be in a moist environment if they are to be metabolically active. Water is needed to dissolve enzymes and nutrients; also, it is an important reactant in many metabolic reactions. Even though most cells die without water, some microorganisms—for example, the bacterium *Mycobacterium tuberculosis* (mī′kō-bak-tēr′ē-ŭm too-ber-kyū-lō′-sis)—have cell walls that retain water, allowing them to survive for months under dry conditions. Additionally, some microbes have endospores and cysts that cease most metabolic activity in a dry environment and might survive for years; these cells are in essence in a state of suspended animation because they neither grow nor reproduce in their dry condition.

(a)

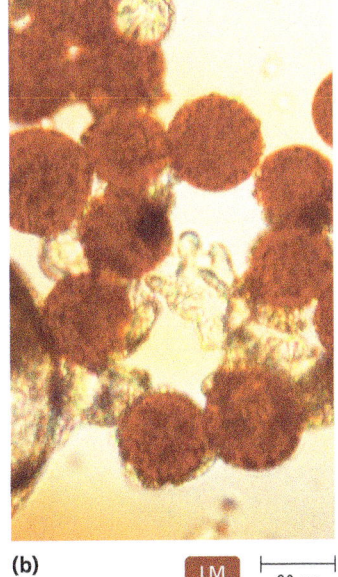
(b) LM 80 μm

▲ FIGURE 6.5 An example of a psychrophile. (a) The alga *Chlamydomonas nivalis* colors this summertime snowbank on Cuverville Island, Antarctica. (b) Microscopic view of the red-pigmented spores of *C. nivalis.*

We now consider the physical effects of water on microbes by examining two topics: osmotic pressure and hydrostatic pressure.

Osmotic Pressure *Osmosis* is the diffusion of water across a membrane and is driven by unequal solute concentrations on the two sides of such a membrane. The *osmotic pressure* of a solution is the pressure exerted on a membrane by a solution containing solutes (dissolved material) that cannot freely cross the membrane. Osmotic pressure is related to the concentration of dissolved molecules and ions in a solution. Solutions with greater concentrations of such solutes are **hypertonic** relative to those with a lower solute concentration, which are **hypotonic**.

Osmotic pressure can have dire effects on cells. For example, a cell placed in freshwater (a hypotonic solution relative to the cell's cytoplasm) gains water from its environment and swells to the limit of its cell wall. Cells that lack a cell wall—animal cells and some bacterial, fungal, and protozoan cells—will swell until they burst in hypotonic solutions. By contrast, a cell placed in seawater, which is a solution containing about 3.5% solutes and thus hypertonic to most cells, loses water into the surrounding saltwater. Such a cell can die from **crenation**, or shriveling of its cytoplasm. Osmotic pressure accounts for the preserving action of salt in jerky and salted fish and of sugar in jellies, preserves, and honey. In those foods, the salt and sugar are solutes that draw water out of any microbial cells that are present, preventing growth and reproduction.

Osmotic pressure restricts organisms to certain environments. Some microbes, called **obligate halophiles**,[12] are adapted to growth under high osmotic pressure such as exists in the Great Salt Lake and smaller salt ponds. They may grow in environments with up to 30% salt and will burst if placed in freshwater. Other microbes are *facultative halophiles*; that is, although they do not require high salt concentrations, they can tolerate them. One potential bacterial pathogen, *Staphylococcus aureus* (staf´i-lō-kok´ŭs o´rē-ŭs), can tolerate up to 20% salt, which allows it to colonize the surface of the skin—an environment that is too salty for most microbes. *S. aureus* causes a number of different skin and mucous membrane diseases ranging from pimples, sties, and boils to life-threatening scalded skin syndrome and toxic shock syndrome. (Chapter 19 covers these diseases more fully.)

Hydrostatic Pressure Water exerts pressure in proportion to its depth. For every additional 10 m of depth, water pressure increases 1 atmosphere (atm). Therefore, the pressure at 100 m below the surface is 10 atm—10 times greater than at the surface. Obviously, the pressure in deep ocean basins and trenches, which are thousands of meters below the surface, is tremendous. Organisms that live under such extreme pressure are called **barophiles**[13] (bar´ō-fīls). Their membranes and enzymes do not merely tolerate pressure; they also depend on pressure to maintain their three-dimensional, functional shapes. Thus, barophiles brought to the surface die quickly because their proteins denature. Obviously, barophiles cannot cause diseases in humans, plants, or animals that do not live at great depths.

---
[12]From Greek *halos*, meaning "salt," and *philos*, meaning "love."
[13]From Greek *baros*, meaning "weight," and *philos*, meaning "love."

Associations and Biofilms

Play an Interactive Microbiology video @ **Mastering** Microbiology for more information about biofilms and quorum sensing.

LEARNING OUTCOME

6.7 Describe how quorum sensing can lead to formation of a biofilm.

Organisms in solitary culture in a laboratory are living very differently from organisms in nature, which live in association with other individuals of their own and different species. The relationships between organisms can be viewed as falling along a continuum stretching from causing harm to providing benefits.

Relationships in which a microbe harms or even kills another organism are considered *antagonistic relationships*. Viruses are especially clear examples of antagonistic microbes; they require cells in which to replicate themselves and almost always kill their cellular hosts.

Beneficial relationships take at least two forms: synergistic relationships and symbiotic relationships. In *synergistic relationships*, the individual members of an association cooperate such that each receives benefits that exceed those that would result if each lived by itself, even though each member could live separately. In *symbiotic relationships*, organisms live in such close nutritional or physical contact that they become interdependent; the members rarely (if ever) live outside the relationship. (Chapter 14 discusses symbiotic relationships in greater detail, particularly as they relate to the production of disease.)

Biofilms are examples of complex, synergistic relationships among numerous microorganisms, often different species, attached to surfaces such as teeth (dental plaque), rocks in streams, shower curtains ("soap scum" is really a biofilm), implanted medical devices (e.g., catheters), and mucous membranes of the digestive system. Biofilms are the primary residence of microorganisms in nature. For example, one study showed that more than 10 billion bacteria per square centimeter form the slippery biofilm on rocks in a streambed. The Centers for Disease Control and Prevention estimates that biofilms cause up to 70% of bacterial diseases in industrialized countries, including kidney infections, tooth and gum decay, infections that occur with cystic fibrosis, and healthcare-associated infections, such as might occur with implantation of medical devices. Cells within biofilms communicate and coordinate with one another, acting similar to a tissue in a multicellular organism.

Biofilm development can involve at least six steps (**FIGURE 6.6**). Free-living cells settle on a surface and attach 2. They develop a gooey, *extracellular polymeric substance (EPS)*

MICRO CHECK

3. What type of organisms could spoil macaroni and cheese stored in a closed container in the refrigerator, but are unlikely to infect humans?
4. What do you call a solution that contains higher levels of dissolved solids than does a cell's cytoplasm?

BENEFICIAL MICROBES

A Nuclear Waste-Eating Microbe?

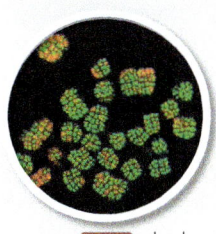

Kineococcus radiotolerans

Gamma rays emitted by radioactive decay are usually deadly. However, some fungi and bacteria can survive being bombarded with radiation at levels higher than thousands of times the level that would kill a person. Currently, researchers are studying these organisms in the hope that we can use their biochemistry to develop ways to protect our cells from radiation and use the microbes to clean up nuclear wastes.

Microbes that can survive in extremely hostile environments are called *extremophiles*. Scientists have discovered amazing species of fungi in the genera *Wangiella* and *Cladosporium* inhabiting the walls of the damaged nuclear reactor in Chernobyl, Ukraine, which is considered the site of the worst nuclear accident in history. The fungi absorb radiation with the help of a pigment called *melanin*, the same pigment that colors human skin and hair. Incredibly, the fungi harness radioactive energy so as to grow faster—the first time we have discovered organisms that can use an energy source other than light or chemicals. Since radiation is available in space, perhaps future astronauts will be able to grow fungi as a continual food supply for long space voyages.

The bacterium *Kineococcus radiotolerans* is notable among radiation-tolerant extremophiles because it can also break down herbicides, chlorinated compounds, and other toxic substances. Researchers at the U.S. Department of Energy would like to shape *K. radiotolerans* into a biological tool that can clean up environments contaminated with radioactive wastes. Using microbes to break down toxic chemicals in the environment, a process known as *bioremediation*, is often cheaper, quicker, and more effective than conventional methods. *K. radiotolerans* could potentially slash the cost of nuclear cleanup.

---

or *matrix*, often composed of DNA, proteins, and polysaccharides ③. This slimy matrix adheres cells to one another, sticks the cells to their substrate, forms microenvironments within the biofilm, sequesters nutrients, forms water channels between groups of cells, and may protect individuals in a biofilm from environmental stresses, including ultraviolet radiation, antimicrobial drugs, and changes in pH, temperature, and humidity.

Biofilms form as a result of a process called **quorum sensing**, in which microorganisms respond to the density of nearby microorganisms (FIGURE 6.7). The microbes secrete *quorum-sensing molecules* that act to communicate number and types of cells among members of the biofilm ④. Many cells possess receptors for these signal molecules. When the density of microorganisms increases, the concentration of quorum-sensing molecules also increases such that more and more receptors bind them. Once the binding exceeds a certain threshold amount, just like the quorum required for Congress to vote, the cells activate previously suppressed genes. The microbes develop new characteristics, such as the production of enzymes, changes in cell shape, the formation of mating types, and the ability to form and

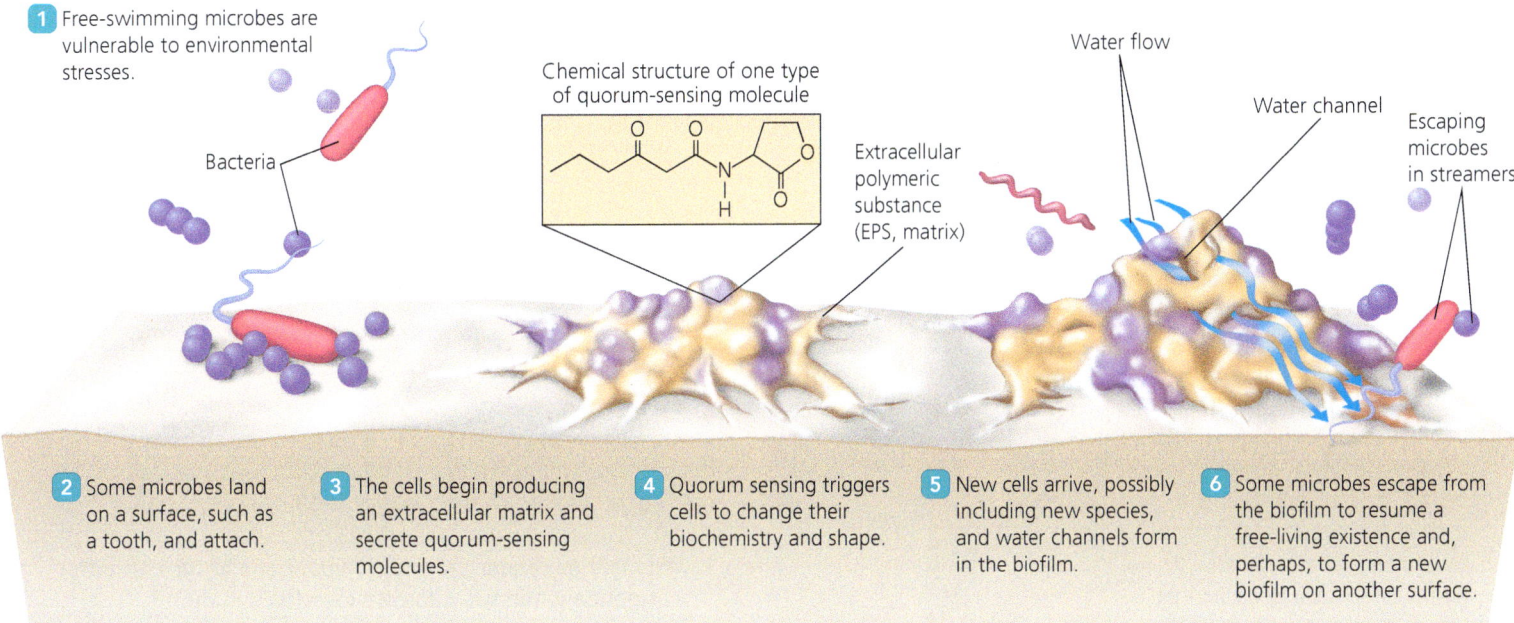

▲ **FIGURE 6.6 Biofilm development.** Quorum sensing allows microbes to change their behavior in the presence of other microbes to form microbial communities called biofilms.

maintain biofilms. Some researchers estimate quorum sensing may regulate 10% of the genes in a cell. An EPS not only attaches a biofilm and positions the cells but also may allow members of the biofilm to concentrate and conserve digestive enzymes, directing them against the underlying structure rather than having them diffuse away in the surrounding medium. New microbes can arrive, and the synergistic relationships allowed in a biofilm continue to organize the biofilm community, so that individual members display metabolic and structural traits different from those expressed by the same cells living individually 5 . Members assume different roles in different areas of a biofilm, much as cells and tissues of multicellular organisms have different functions in different parts of the body. Some individual cells or groups of cells in *streamers* may leave the biofilm 6 .

Given that many microorganisms become more harmful when they are part of a biofilm, scientists are seeking ways to prevent biofilms from forming in the first place. Researchers have learned that drugs that block microbial cell receptors, thereby disrupting communication among cells, can inhibit biofilm formation and thus prevent disease. Such receptor-blocking drugs have successfully blocked biofilm formation and prevented disease in mice; they are being considered for use in humans. Scientists are investigating another potential new tool against biofilm infections in the lungs—enzymes that degrade the EPS material to enhance penetration of antimicrobial drugs.

Another possible approach to preventing biofilm formation involves artificially amplifying quorum sensing while bacteria are still relatively few in number. Some pathogens hide within capsules and in blood clots, and only after they have multiplied significantly and formed a biofilm do they emerge and become "visible" to the immune system. With amplified quorum sensing, bacteria might "reveal" themselves sooner to the immune system, which may then eliminate them before they can cause disease.

Dental plaque is a common biofilm that can lead to dental caries—cavities. Plaque formation usually begins with colonization of the teeth by *Streptococcus mutans* (strep-tō-kok´ŭs mū´tanz). This bacterium breaks down carbohydrates, particularly the disaccharide sucrose (table sugar), to provide itself with nutrition and a glycocalyx. One of its enzymes catabolizes sucrose into its component monosaccharides—glucose and fructose—which the cells use as energy sources. A second enzyme polymerizes glucose into long, insoluble polysaccharide strands called *glucan* molecules, which form a sticky matrix around the bacteria. Glucan adheres *S. mutans* to the tooth, provides a home for other species of oral bacteria, and traps food particles. A biofilm has formed.

Bacteria in the biofilm digest nutrients and release acid, which is held against the teeth by the biofilm's matrix. The acid gradually eats away the minerals that compose the tooth, resulting in dental caries and eventually total loss of the teeth. **Clinical Case Study: Cavities Gone Wild** deals with an especially severe case of a biofilm running amok in a small boy's mouth.

EMERGING DISEASE CASE STUDY

Vibrio vulnificus Infection

Greg enjoyed Florida's beaches; swimming in the warm water was his favorite pastime. Of course, the saltwater did sting his leg where he had cut himself on some coral, but it didn't sting enough to stop Greg from enjoying the beach. He spent the afternoon jogging in the pure sand, throwing a disc, watching people, drinking a few beers, and of course spending more time in the water.

That evening, he felt chilled, a condition he associated with having gotten too much sun during the day, but by midnight he thought he must have caught a rare summertime flu. He was definitely feverish, extremely weak, and tired. His leg felt strangely tight, as though the underlying muscles were trying to burst through his skin.

The next morning, he felt better, except for his leg. It was swollen, dark red, tremendously painful, and covered with fluid-filled blisters. The ugly sight motivated him to head straight for the hospital, a decision that likely saved his life.

Greg was the victim of an emerging pathogen, *Vibrio vulnificus*—a slightly curved,

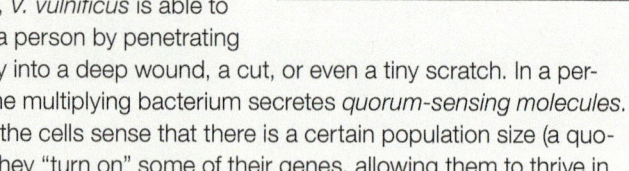

Gram-negative bacterium with DNA similar to that of *V. cholerae* (cholera bacterium). *V. vulnificus* lives in salty, warm water around the globe. Unlike the cholera bacterium, *V. vulnificus* is able to infect a person by penetrating directly into a deep wound, a cut, or even a tiny scratch. In a person, the multiplying bacterium secretes *quorum-sensing molecules*. When the cells sense that there is a certain population size (a quorum), they "turn on" some of their genes, allowing them to thrive in the body and cause disease.

Greg's doctor cut away the dead tissue and prescribed doxycycline and cephalosporin for two weeks. Greg survived and kept his leg. Half of the victims of *V. vulnificus* are not so lucky; they lose a limb or die. Who knew that a beach could be so dangerous? (For more about quorum sensing, see pp. 169–171.)

1. Why is it necessary for *Vibrio vulnificus* to turn on different genes when the microbe invades a human?
2. Why do you think the related microbe *V. cholerae* is unable to infect through the skin?

CLINICAL CASE STUDY

Cavities Gone Wild

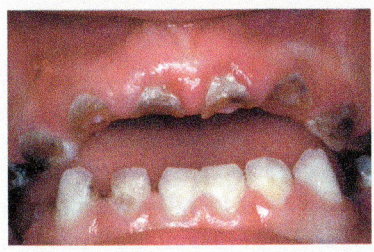

Five-year-old Daniel appears to be shy. He always looks at the floor, has no friends, never plays with the other children, and will rarely speak to adults. When he does speak, it is difficult to understand his broken enunciation. His skinny frame makes him appear malnourished. Daniel cries frequently and misses many days of school.

A speech specialist at school finds that only two of Daniel's teeth are healthy; all the others have rotted away to the gum line. The little guy is in constant pain, has a fever and it hurts to chew. A doctor later determines that bacteria from the cavities in his mouth have entered his bloodstream and infected his heart, causing an irregular heartbeat and poor blood circulation.

1. How does knowledge of biofilms help explain the bulk of Daniel's problems?
2. What can Daniel, his parents, and health care professionals do to cure his diseases?
3. What is the scientific name of the nutrient that Daniel's parents should eliminate from his diet to help prevent a repeat of this condition?
4. A recent study has shown that brushing does more than decrease plaque from the teeth; it also disrupts associations between oral bacteria. How does this simple act help prevent the formation of biofilms?

▲ FIGURE 6.7 **Quorum sensing.** To measure their population density, bacteria synthesize and secrete a small signaling molecule. A protein receptor can detect this signal. At low density, the signal diffuses away from cells. At high density, the signal reaches the receptor.

MICRO CHECK

5. Biofilms consist of several different organisms working in complex relationships. These organisms communicate with each other using what sort of chemicals?

TELL ME WHY

Why should cardiac nurses and respiratory therapists care about biofilms?

Culturing Microorganisms

One of Koch's postulates for demonstrating that a certain agent causes a specific disease requires that microorganisms be isolated and cultivated (Chapter 1). Medical laboratory personnel must also grow pathogens as a step in the diagnosis of many diseases. To cultivate or *culture* microorganisms, a sample called an **inoculum** (plural: *inocula*) is introduced into a collection of nutrients called a **medium**. Microorganisms that grow from an inoculum are also called a *culture*; thus, **culture** can refer to the act of cultivating microorganisms or to the microorganisms that are cultivated.

Cultures can be grown in liquid media called **broths** or on the surface of solid media. Cultures that are visible on the surface of solid media are called **colonies**. Bacterial and fungal colonies often have distinctive characteristics—including color, size, shape, elevation, texture, and appearance of the colony's margin (edge)—that taken together help identify the microbial species that formed the colony (**FIGURE 6.8**).

Microbiologists obtain inocula from a variety of sources. *Environmental specimens* are taken from such sources as ponds, streams, soil, and air. *Clinical specimens* are taken from patients and handled in ways that facilitate the examination of microorganisms or testing for their presence. Another source of inocula is a culture originally grown from an environmental or clinical specimen and maintained in storage in a laboratory. Next, we briefly examine clinical sampling.

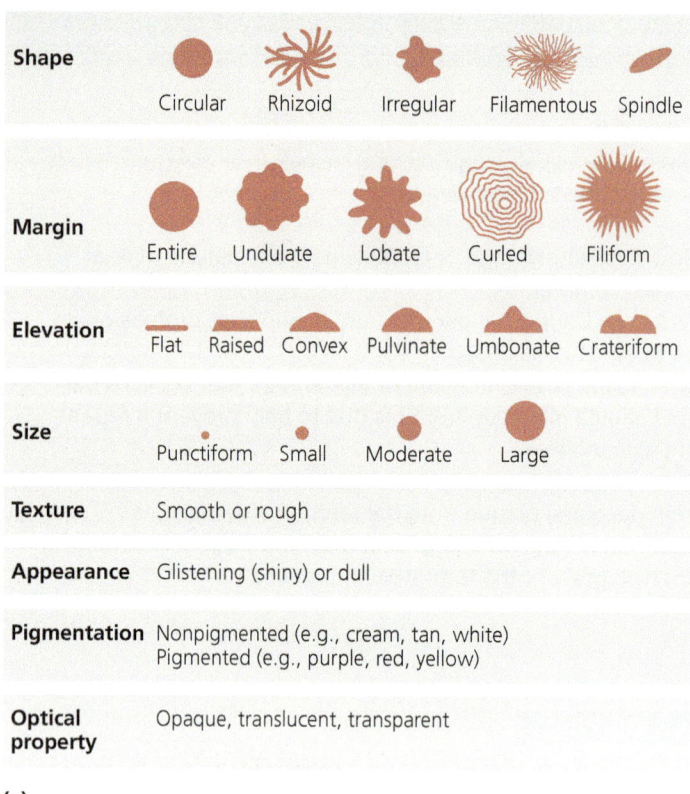

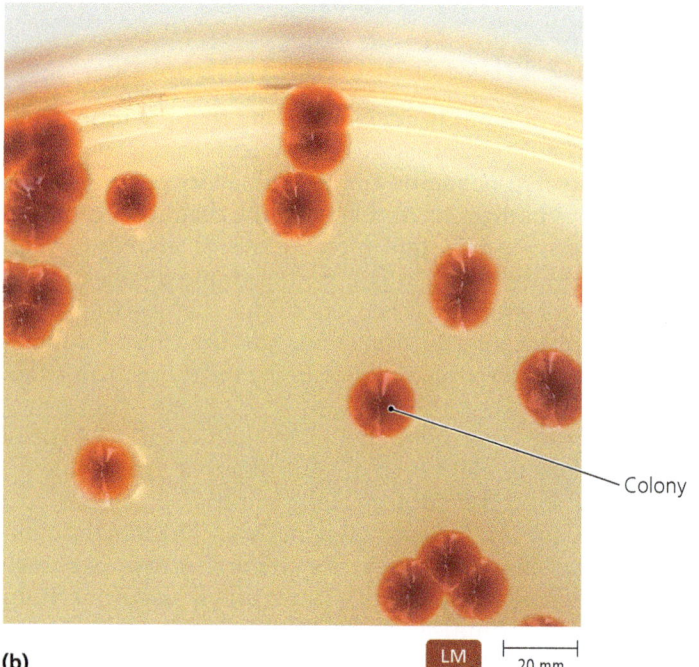

▲ FIGURE 6.8 **Characteristics of bacterial colonies.** (a) Shape, margin, elevation (side view), size, texture, appearance, pigmentation (color), and optical properties are described by a variety of terms. (b) *Serratia marcescens* growing on an agar surface. These colonies are circular, entire, convex, large, smooth, shiny, red, and opaque.

Clinical Sampling

LEARNING OUTCOME

6.8 Describe methods for collecting clinical specimens from the skin, blood, cerebrospinal fluid, stomach, urine, lungs, and diseased tissue.

Diagnosing and treating disease often depend on isolating and correctly identifying pathogens. Health care professionals must properly obtain samples from their patients and must then transport them quickly and correctly to a microbiology laboratory for culture and identification. They must take care to prevent contaminating samples with microorganisms from the environment or other regions of the patient's body, and they must prevent infecting themselves with pathogens while sampling. In this regard, the Centers for Disease Control and Prevention has established a set of guidelines, called *standard precautions*, to protect health care professionals from contamination by pathogens.

In clinical microbiology, a **clinical specimen** is a sample of human material, such as feces, saliva, cerebrospinal fluid, or blood, that is examined or tested for the presence of microorganisms. As summarized in TABLE 6.2, health care professionals collect clinical specimens using a variety of techniques and equipment. Specimens must be properly labeled and promptly transported to a microbiological laboratory to avoid death of the pathogens and to minimize the growth of microbes normally found at the collection site. Clinical specimens are often transported in special *transport media* that are chemically formulated to maintain the relative abundance of different microbial species or to maintain an anaerobic environment.

MICRO CHECK

6. What type of clinical specimen is collected through the "clean catch" method?

Obtaining Pure Cultures

LEARNING OUTCOME

6.9 Describe the two most common methods by which microorganisms can be isolated for culture.

Clinical specimens are collected in order to identify a suspected pathogen, but they also contain *normal microbiota*, which are microorganisms associated with a certain area of the body without causing diseases (see Table 14.2 on p. 417). As a result, the suspected pathogen in a specimen must be isolated from the normal microbiota in culture. Scientists use several techniques to isolate organisms in **pure cultures**, that is, cultures composed of cells arising from a single progenitor. The word *axenic*[14] (ā-zen′ik) is also used to refer to a pure culture. The progenitor from which

---

[14]From Greek *a*, meaning "no," and *xenos*, meaning "stranger."

TABLE 6.2 Clinical Specimens and the Methods Used to Collect Them

Type or Location of Specimen	Collection Method
Skin, accessible membrane (including eye, outer ear, nose, throat, vagina, cervix, urethra), or open wounds	Sterile swab brushed across the surface; care should be taken not to contact neighboring tissues
Blood	Needle aspiration from vein; anticoagulants are included in the specimen transfer tube
Cerebrospinal fluid	Needle aspiration from subarachnoid space of spinal column
Stomach	Intubation, which involves inserting a tube into the stomach, often via a nostril
Urine	In aseptic collection, a catheter is inserted into the bladder through the urethra; in the "clean catch" method, initial urination washes the urethra, and the specimen is midstream urine
Lungs	Collection of sputum either dislodged by coughing or acquired via a catheter
Diseased tissue	Surgical removal (biopsy)

a particular pure culture is derived may be either a single cell or a group of related cells; therefore, the progenitor is termed a **colony-forming unit (CFU)**.

In all microbiological procedures, care must be taken to reduce the chance of contamination, which occurs, for example, when instruments or air currents carry foreign microbes into culture vessels. All media, vessels, and instruments must be **sterile**—that is, free of any microbial contaminants. Sterilization and *aseptic techniques* are designed to limit contamination (discussed in Chapter 9). Now we examine two common isolation techniques: streak plates and pour plates. We consider another method, *serial dilution*, in a later section.

Streak Plates

The most commonly used isolation technique in microbiological laboratories is the **streak-plate** method. In this technique, a sterile inoculating loop (or sometimes a needle) is used to spread an inoculum across the surface of a solid medium in *Petri*[15] *dishes*, which are clear, flat culture dishes with loose-fitting lids. The loop is used to lightly streak a set pattern that gradually dilutes the sample to a point that CFUs are isolated from one another (FIGURE 6.9a). After an appropriate period of time called *incubation*, colonies develop from each isolate (FIGURE 6.9b). The various types of organisms present are distinguished from one another by differences in colonial characteristics (Figure 6.8). Samples from each variety can then be inoculated in new media to establish axenic cultures.

Pour Plates

In the **pour-plate** technique, CFUs are separated from one another using a series of dilutions. There are various ways to perform pour-plate isolations. In one method, an initial 1-milliliter sample is mixed into 9 ml of sterile medium in a test tube. After mixing, a new sample from this mixture is then used to inoculate a second tube of liquid medium. The process is repeated to establish a series of dilutions (FIGURE 6.10a). Samples from the more diluted media are mixed in Petri dishes with sterile, warm medium containing *agar*—a gelling agent derived from the cell walls of red algae. After the agar cools and solidifies, the filled dishes are called **Petri plates**. Isolated colonies—colonies that are separate and distinct from all others—form in the plates from CFUs that have been separated via the dilution series (FIGURE 6.10b). One difference between this method and the streak-plate technique is that colonies form both at and below the surface of the medium. As before, pure cultures can be established from distinct colonies.

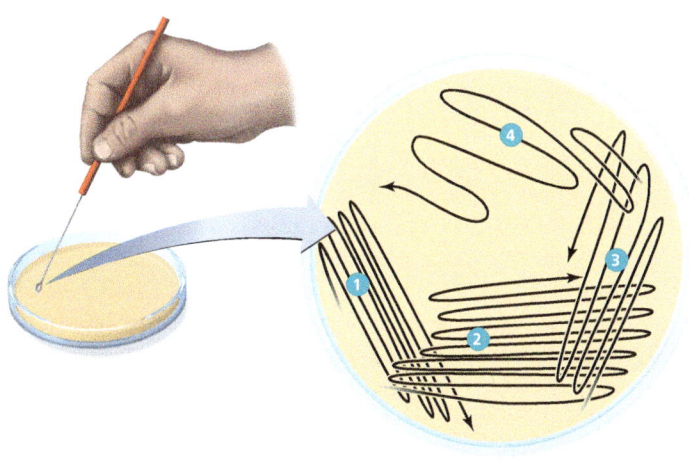

(a)

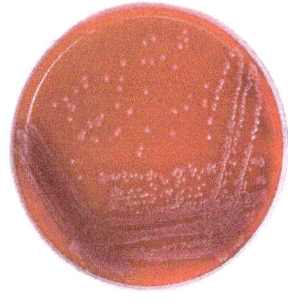

(b)

◀ **FIGURE 6.9 The streak-plate method of isolation. (a)** An inoculum is spread across the surface of an agar plate in a sequential pattern of streaks (as indicated by the numbers and arrows). The loop is sterilized between streaks. In streaks 2, 3, and 4, bacteria are picked up from a previous streak; thus the number of cells is diluted each time. **(b)** A streak plate showing colonies of *Escherichia coli* on blood agar.

---

[15]Named for Richard Petri, Robert Koch's assistant, who invented them in 1887.

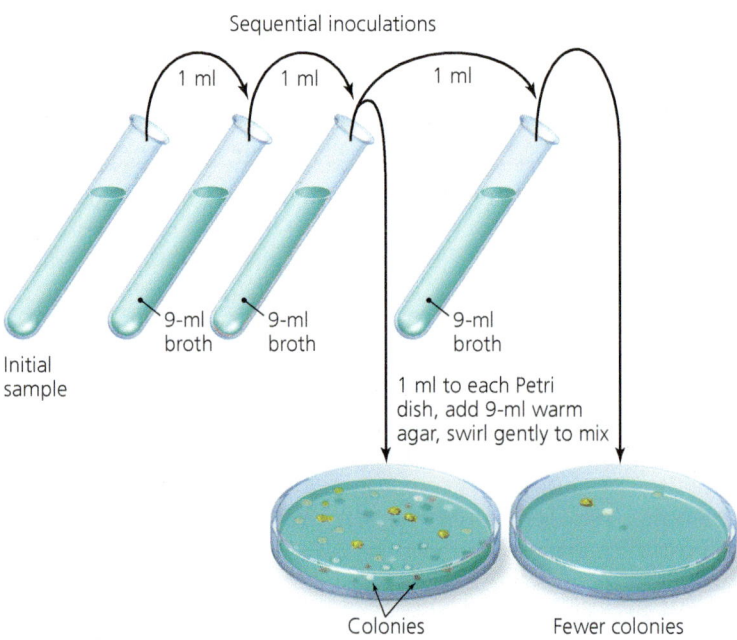
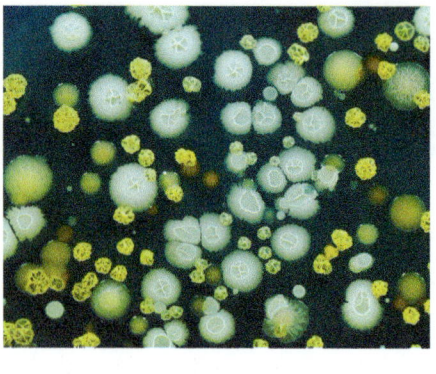

▲ FIGURE 6.10 **The pour-plate method of isolation.** (a) After an initial sample is diluted through a series of transfers, the final dilutions are mixed with warm agar in Petri plates. Individual CFUs form colonies in and on the agar. (b) A portion of a plate showing the results of isolation.

Isolation techniques work well only if a relatively large number of CFUs of the organism of interest are present in the initial sample and if the medium supports the growth of that microbe. As discussed later, special media and enriching techniques can be used to increase the likelihood of success.

Other Isolation Techniques

Streak plates and pour plates are used primarily to establish pure cultures of bacteria, but they can also be used for some fungi, particularly yeasts. Protozoa and motile unicellular algae are not usually cultured on solid media because they do not remain in one location to form colonies. Instead, they are isolated through a series of dilutions but remain in broth culture media. In cases of fairly large microorganisms, such as the protists *Euglena* (yū-glēn´ă) and *Amoeba* (am-ē´bă), hollow tubes with small diameters called *micropipettes* can be used to pick up single cells that are then used to establish axenic cultures.

MICRO CHECK

7. What method of isolation involves a series of liquid dilutions?

Culture Media

LEARNING OUTCOMES

6.10 Describe six types of general culture media available for bacterial culture.
6.11 Describe enrichment culture as a means of enhancing the growth of less abundant microbes.

Culturing microorganisms can be an exacting science. Although some microbes, such as *E. coli*, are not particular about their nutritional needs and can be grown in a variety of media, some other bacteria, such as *Neisseria gonorrhoeae* (nī-se´rē-ă go-nor-rē´ī) and *Haemophilus influenzae* (hē-mof´i-lŭs in-flu-en´zī), and many archaea require specific nutrients, including specific growth factors. The majority of microorganisms have never been successfully grown in any culture medium in part because scientists have concentrated their efforts on culturing commercially important species and pathogens. However, some pathogens, such as the syphilis bacterium *Treponema pallidum* (trep-ō-nē´mă pal´li-dŭm), have never been cultured in any laboratory medium despite over a century of effort.

A variety of media are available for microbiological cultures, and more are developed each year, to support the needs of food, water, industrial, and clinical microbiologists. Most media are available from commercial sources and come in powdered forms that require only the addition of water to make broths. A common medium, for example, is *nutrient broth*, which contains powdered beef extract and peptones (short chains of amino acids produced by enzymatic digestion of protein) dissolved in water. For some purposes broths are adequate, but if solid media are needed, dissolving about 1.5% agar into hot broth, pouring the liquid mixture into an appropriate vessel, and allowing it to cool provides a solid surface to support colonial growth. Media made solid by the addition of agar to a broth have the word *agar* in their names; thus, *nutrient agar* is nutrient broth to which 1.5% agar has been added.

Agar, a complex polysaccharide derived from the cell walls of certain red algae, is a useful compound in microbiology for several reasons:

- Most microbes cannot digest agar; therefore, agar media remain solid even when bacteria and fungi are growing on them.

- Powdered agar dissolves in water at 100°C, a temperature at which most nutrients remain undamaged.
- Agar solidifies at temperatures below 40°C, so temperature-sensitive, sterile nutrients such as vitamins and blood can be added without detriment to cooling agar before it solidifies. Further, cooling liquid agar can be poured over most bacterial cells without harming them. The latter technique plays a role in the pour-plate isolation technique.
- Solid agar does not melt below 100°C; thus, it can even be used to culture some hyperthermophiles.

Still-warm liquid agar media can be poured into Petri dishes to make Petri plates. When warm agar media are poured into test tubes that are then placed at an angle and left to cool until the agar solidifies, the result is **slant tubes**, or **slants** (FIGURE 6.11). The slanted surface provides a larger surface area for aerobic microbial growth while the butt of the tube remains almost anaerobic.

Next we examine six types of general culture media: defined media, complex media, selective media (including enrichment culture), differential media, anaerobic media, and transport media. It is important to note that these types of media are not mutually exclusive categories; that is, in some cases a given medium can belong to more than one category.

Defined Media

If microorganisms are to grow and multiply in culture, the medium must provide essential nutrients (including an appropriate energy source for chemotrophs), water, an appropriate oxygen level, and the required physical conditions (such as the correct pH and suitable osmotic pressure and temperature). A **defined medium** (also called a **synthetic medium**) is one in which the exact chemical composition is known. TABLE 6.3 gives a recipe for one example. Relatively simple defined media containing inorganic salts and a source of CO_2 (such as sodium bicarbonate) are available for autotrophs, particularly cyanobacteria and algae. Defined media for chemoheterotrophs can contain necessary organic molecules, such as glucose, amino acids, or vitamins, which supply carbon and energy or are vital growth factors.

Organisms that require a relatively large number of growth factors are termed *fastidious*. Such organisms may be used as living assays for the presence of growth factors. For example, a scientist wanting to know whether a particular environment contains vitamin B_{12} could inoculate a sample from the environment with *Euglena granulata* (gran-yū-lă'tă), an organism that requires the vitamin. If the microbe grows, the vitamin is present in the sample and therefore in the environment. The amount of growth provides an estimate of the amount of vitamin present, scant growth indicating a small amount.

TABLE 6.3 Ingredients of a Representative Defined (Synthetic) Medium for Culturing *E. coli*

Ingredient	Amount
Glucose	1.0 g
Na_2HPO_4	16.4 g
KH_2PO_4	1.5 g
$(NH_4)_3PO_4$	2.0 g
$MgSO_4 \cdot 7H_2O$	0.2 g
$CaCl_2$	0.01 g
$FeSO_4 \cdot 7H_2O$	0.005 g
Distilled or deionized water	Enough to bring volume to 1 L

Complex Media

For most clinical cultures, defined media are unnecessarily troublesome to prepare. Most chemoheterotrophs, including pathogens, are routinely grown on **complex media** that contain nutrients released by the partial chemical breakdown of yeast, beef, soy, or proteins, such as casein from milk. The exact chemical composition of a complex medium is unknown because partial degradation releases many different chemicals in a variety of concentrations each time it takes place.

Complex media have advantages over defined media. Because a complex medium contains a variety of nutrients, including growth factors, it can support a wider variety of different microorganisms. Complex media are also used to culture organisms whose exact nutritional needs are unknown. Nutrient broth, Trypticase soy agar, and MacConkey agar are some common complex media. Blood is often added to complex media to provide additional growth factors, such as NADH and heme. Such a fortified medium is said to be *enriched* and can support the growth of many fastidious microorganisms.

Selective Media

Selective media typically contain substances that either favor the growth of particular microorganisms or inhibit the growth of unwanted ones. Eosin, methylene blue, and crystal violet dyes as well as bile salts are included in media to inhibit the growth of Gram-positive bacteria without adversely affecting most Gram-negatives. A high concentration of NaCl (table salt) in a medium selects for halophiles and for salt-tolerant bacteria, such as the pathogen *Staphylococcus aureus*. Sabouraud dextrose agar

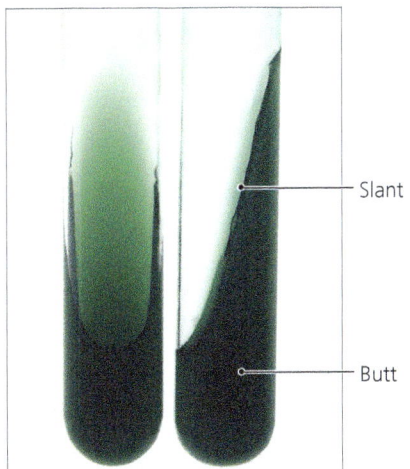

▲ FIGURE 6.11 **Slant tubes containing solid media.** In this case, citrate agar is the medium.

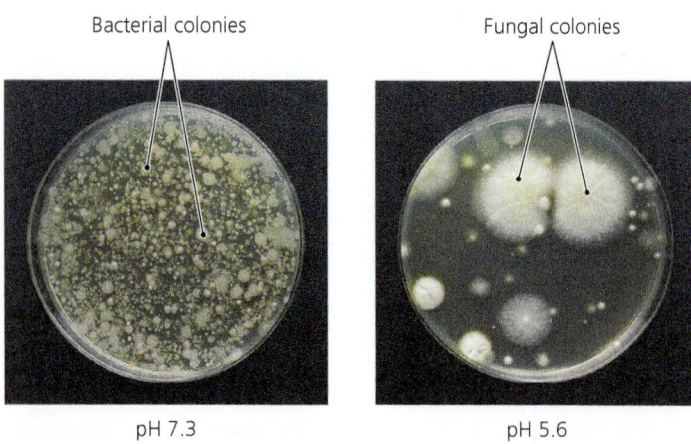

▲ FIGURE 6.12 **An example of the use of a selective medium.** After the medium is inoculated with a diluted soil sample, acidic pH in Sabouraud dextrose agar (right) makes the medium selective for fungi by inhibiting the growth of bacteria. At left, for comparison, is a nutrient agar plate inoculated with an identical sample.

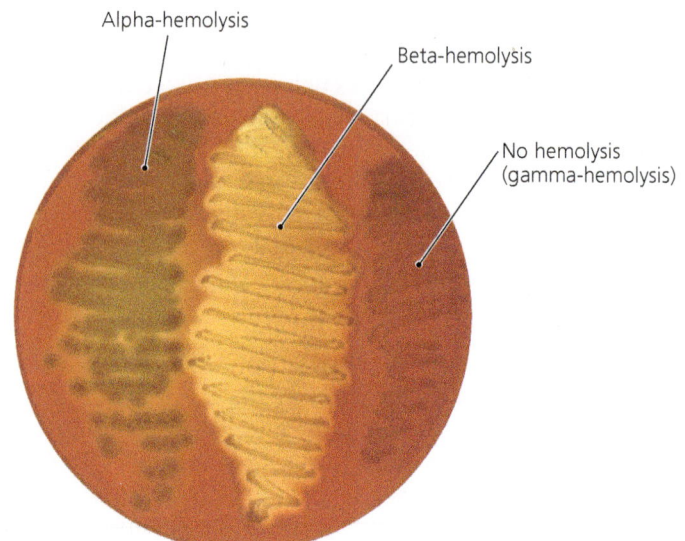

▲ FIGURE 6.13 **The use of blood agar as a differential medium.** *Streptococcus pneumoniae* (left) partially uses red blood cells, producing a discoloration termed *alpha-hemolysis*. *Streptococcus pyogenes* (middle) completely uses red blood cells, producing a clear zone termed *beta-hemolysis*. *Enterococcus faecalis* (right) does not use red blood cells; the lack of any change in the medium around colonies is termed *gamma-hemolysis*, even though no red cells are hemolyzed.

has a slightly low pH, which inhibits the growth of bacteria and thereby is selective for fungi (FIGURE 6.12).

A medium can also become a selective medium when a single crucial nutrient is left out of it. For example, leaving glucose out of Trypticase soy agar makes the resulting medium selective for organisms that can meet all their carbon requirements by catabolizing amino acids.

Enrichment Culture Bacteria that are present in small numbers may be overlooked on a streak plate or overwhelmed by faster-growing, more abundant strains. This is especially true of organisms in soil and fecal samples that contain a wide variety of microbial species. To isolate potentially important microbes that might otherwise be overlooked, microbiologists enhance the growth of less abundant organisms by a variety of techniques.

In the late 1800s, the Dutch microbiologist Martinus Beijerinck (1851–1931) introduced the most common of these methods, called simply **enrichment culture**. Enrichment cultures use a selective medium and are designed to increase very small numbers of a chosen microbe to observable levels. For example, suppose a microbiologist specializing in environmental cleanup wanted to isolate an organism capable of digesting crude oil on an oil-soaked beach. Even though a sample of the beach sand might contain a few such organisms, it would also likely contain many millions of unwanted common bacteria. To isolate oil-utilizing microbes, the scientist would inoculate a sample of the sand into a tube of selective medium containing oil as the sole carbon source and then incubate it. Then a small amount of the culture would be transferred into a new tube of the same medium to be incubated again. After a series of such enrichment transfers, any remaining bacteria will be oil-utilizing organisms. Different species could be isolated by either streak-plate or pour-plate methods.

Cold enrichment is another technique used to enrich a culture with cold-tolerant species, such as *Vibrio cholerae*, the bacterium that causes cholera. Stool specimens or water samples suspected of containing the bacterium are incubated in a refrigerator. Cold enrichment works because *Vibrio* cells are much less sensitive to cold than are more common fecal bacteria, such as *E. coli*; therefore, *Vibrio* continues to grow in the cold while the other species are inhibited. The result of cold enrichment is a culture with a greater percentage of *Vibrio* cells than the original sample; the *Vibrio* cells can then be isolated by other methods.

Differential Media

Differential media are formulated such that either the presence of visible changes in the medium or differences in the appearance of colonies help microbiologists differentiate among the kinds of bacteria growing on the medium. Such media take advantage of the fact that different bacteria utilize the ingredients of any given medium in different ways. One example of the use of a differential medium involves the differences in organisms' utilization of red blood cells in blood agar (FIGURE 6.13). *Streptococcus pneumoniae* (nū-mō′nē-ī) partially digests (lyses) red blood cells, producing around its colonies a greenish brown discoloration denoted *alpha-hemolysis*. By contrast, *Streptococcus pyogenes* (pī-oj′en-ēz) completely digests red blood cells, producing around its colonies clear zones termed *beta-hemolysis*. *Enterococcus faecalis* (en′ter-ō-kok′ŭs fē-kă′lis) does not digest red blood cells, so the agar appears little changed, a reaction called *gamma-hemolysis* even though no lysis occurs. In some differential media, such as carbohydrate utilization broth tubes, a pH-sensitive dye changes color when bacteria metabolizing sugars produce acid waste products (FIGURE 6.14). Some common differential complex media are described in TABLE 6.4.

Many media are both selective and differential; that is, they enhance the growth of certain species that can then be distinguished from other species by variations in appearance. For example, bile salts and crystal violet in MacConkey agar inhibit the growth of Gram-positive bacteria and also differentiate between

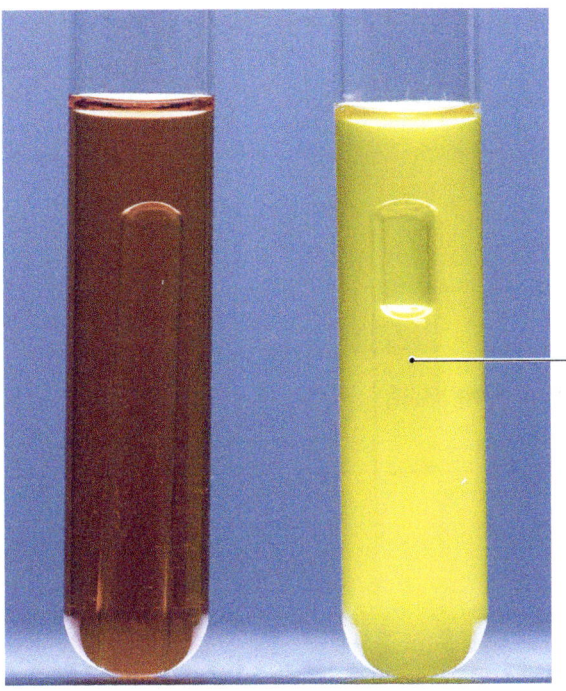

No fermentation | Acid fermentation with gas

▲ **FIGURE 6.14 The use of carbohydrate utilization tubes as differential media.** Each tube contains a single kind of simple carbohydrate (a sugar) as a carbon source and the dye phenol red as a pH indicator. *Alcaligenes faecalis* in the tube on the left did not ferment this carbohydrate; because no acid was produced, the medium did not turn yellow. *Escherichia coli* in the tube on the right fermented the sugar, producing acid and lowering the pH enough to cause the phenol red to turn yellow. This bacterium also produced gas, which is visible as a small bubble in the Durham tube.

TABLE 6.4 Representative Differential Complex Media

Medium and Ingredients	Use and Interpretation of Results
MacConkey Medium	
Peptone (20.0 g)	For the culture and differentiation of enteric bacteria based on the ability to ferment lactose
Agar (12.0 g)	
Lactose (10.0 g)	
Bile salts (5.0 g)	Lactose-fermenters produce red-to-pink colonies; non-lactose-fermenters form colorless or transparent colonies
NaCl (5.0 g)	
Neutral red (0.075 g)	
Crystal violet (0.001 g)	
Water to bring volume to 1 L	
Blood Agar	
Agar (15.0 g)	For culture of fastidious microorganisms and differentiation of hemolytic microorganisms
Pancreatic digest of casein (15.0 g)	
Papaic digest of soybean meal (5.0 g)	
NaCl (5.0 g)	Partial digestion of blood: alpha-hemolysis; complete digestion of blood: beta-hemolysis; no digestion of blood: gamma-hemolysis
Water to bring volume to 950.0 ml	
Sterile blood (50.0 ml, added to medium after autoclaving and cooling)	

lactose-fermenting and non-lactose-fermenting Gram-negative bacteria (**FIGURE 6.15**).

Anaerobic Media

Obligate anaerobes require special culture conditions to protect their cells from free oxygen. Anaerobes can be introduced with a straight inoculating wire into the anoxic (oxygen-free) depths of solid media to form a *stab culture*, but special media called **reducing media** provide better anaerobic culturing conditions. These media contain compounds, such as sodium thioglycolate, that chemically combine with free oxygen and remove it from the medium. Heat is used to drive absorbed oxygen from thioglycolate immediately before such a medium is inoculated.

The use of Petri plates presents special problems for the culture of anaerobes because each dish has a loose-fitting lid that allows the entry of air. For the culture of anaerobes, inoculated Petri plates are placed in sealable containers containing reducing chemicals (**FIGURE 6.16**). Of course, the airtight lids of anaerobic culture vessels must be sealed so that oxygen cannot enter. Only anaerobes that can tolerate brief exposure to oxygen can be cultured by this method because inoculation and transfer occur outside the anaerobic environment. Laboratories that routinely study strict anaerobes have large anaerobic glove boxes, which are transparent, airtight chambers with special airtight rubber gloves, chemicals that remove oxygen, and air locks. These chambers allow scientists to manipulate equipment and anaerobic cultures in an oxygen-free environment.

Transport Media

Health care personnel use special **transport media** to carry clinical specimens of feces, urine, saliva, sputum, blood, and other bodily fluids in such a way as to ensure that people are not infected and that the specimens are not contaminated. Speed in transporting clinical specimens to the laboratory is extremely important because pathogens often do not survive long outside the body. Specimens are transported in buffered media designed to maintain the ratio and life of different microbes. Anaerobic specimens may occasionally be transported for less than an hour inside a syringe, but longer times require the use of anaerobic transport media.

MICRO CHECK

8. What sort of media can be used to encourage growth of some organisms, while discouraging the growth of others?

Special Culture Techniques

LEARNING OUTCOME

6.12 Discuss the use of animal and cell culture and low-oxygen culture.

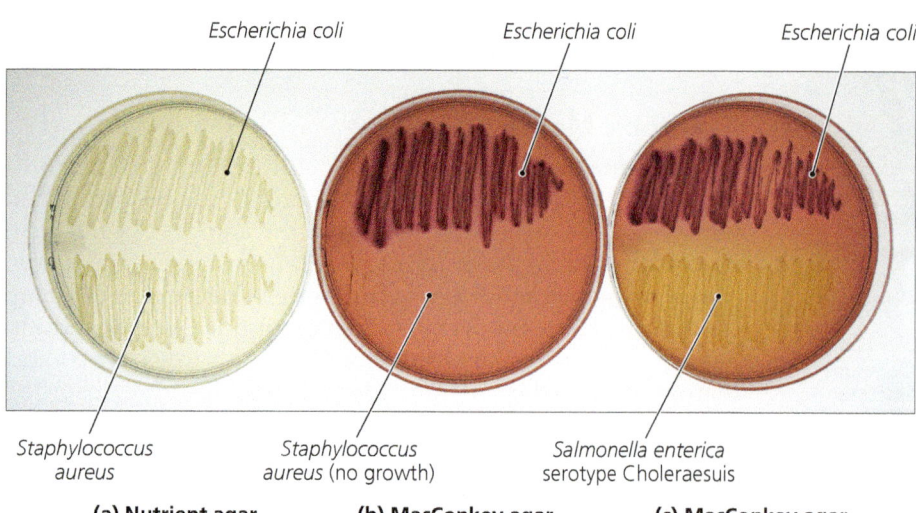

▲ FIGURE 6.15 **The use of MacConkey agar as a selective and differential medium.** (a) Whereas both the Gram-positive *Staphylococcus aureus* and the Gram-negative *Escherichia coli* grow on nutrient agar, MacConkey agar (b) selects for Gram-negative bacteria and inhibits Gram-positive bacteria. (c) MacConkey agar also differentiates Gram-negative bacteria based on their ability to ferment lactose. The colonies of the lactose-fermenting *E. coli* are easily distinguished from those of the non-lactose-fermenting *Salmonella enterica* serotype Choleraesuis.

Not all organisms can be grown under the culture conditions we have discussed. Scientists have developed other techniques to culture many of these organisms.

Animal and Cell Culture

Microbiologists have developed animal and cell culture techniques for growing some microbes for which artificial media are inadequate. For example, the causative agents of leprosy and syphilis are grown in animals because all attempts to grow these bacteria using standard culture techniques have been unsuccessful. *Mycobacterium leprae* (lep′rī) is cultured in armadillos, whose internal conditions (including a relatively low body temperature) provide the conditions this microbe prefers. Rabbits meet the culture needs for *Treponema pallidum*, the bacterium that causes syphilis. Because viruses and small bacteria called rickettsias and chlamydias are obligate intracellular parasites—that is, they grow and reproduce only within living cells—bird eggs and cultures of living cells are used to culture these microbes.

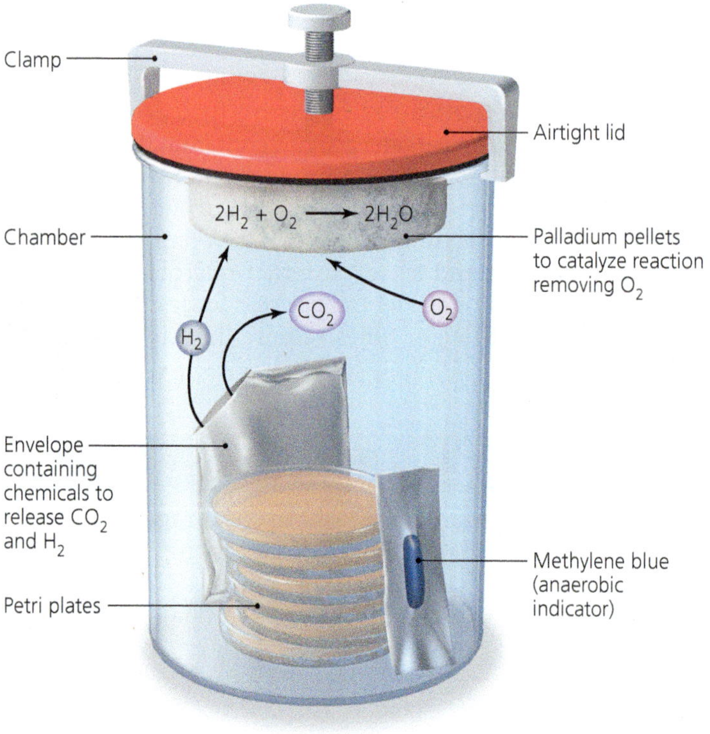

Low-Oxygen Culture

As we have discussed, many types of organisms prefer oxygen conditions that are intermediate between strictly aerobic and anaerobic environments. *Carbon dioxide incubators*, machines that electronically monitor and control CO_2 levels, provide atmospheres that mimic the environments of the intestinal tract, the respiratory tract, and other body tissues and thus are useful for culturing these kinds of organisms. Smaller and much less expensive alternatives to CO_2 incubators are *candle jars*. In these simple but effective devices, culture plates are sealed in a jar along with a lit candle; the flame consumes much of the O_2, replacing it with CO_2. The candle eventually extinguishes itself, creating an environment that is ideal for many aerotolerant anaerobes, microaerophiles, and **capnophiles**, which are organisms such as *Neisseria gonorrhoeae* that grow best with a relatively high concentration of carbon dioxide (3–10%) in addition to low oxygen levels. Remaining oxygen in the jar prevents the growth of strict anaerobes. The use of packets of chemicals that remove most of the oxygen from the jar has replaced candles in modern microbiology labs.

▲ FIGURE 6.16 **An anaerobic culture system.** The system utilizes chemicals to create an anaerobic environment inside a sealable, airtight jar. Methylene blue, which turns colorless in the absence of oxygen, indicates when the environment within the jar is anaerobic.

Preserving Cultures

LEARNING OUTCOME

6.13 Contrast refrigeration, deep-freezing, and lyophilization as methods for preserving cultures of microbes.

To store living cells, a scientist slows the cells' metabolism to prevent the excessive accumulation of waste products and the exhaustion of all nutrients in a medium. Refrigeration is often the best technique for storing bacterial cultures for short periods of time.

Deep-freezing and lyophilization are used for long-term storage of bacterial cultures. **Deep-freezing** involves freezing the cells at temperatures from −50°C to −95°C. Deep-frozen cultures can be restored years later by thawing them and placing a sample in an appropriate medium.

Lyophilization (lī-of´i-li-zā´shŭn; freeze-drying) involves removing water from a frozen culture using an intense vacuum. Under these conditions, ice sublimates (directly becomes a gas) and is removed from cells without permanently damaging cellular structures and chemicals. Lyophilized cultures can last for decades and are revived by adding lyophilized cells to liquid culture media.

> **TELL ME WHY**
>
> Why do clinical laboratory scientists keep many different kinds of culture media on hand?

Growth of Microbial Populations

LEARNING OUTCOME

6.14 Describe binary fission as a means of reproduction.

Most unicellular microorganisms reproduce by *binary fission*, a process in which a cell grows to twice its normal size and divides in half to produce two daughter cells of equal size. (Chapters 11 and 12 discuss other reproductive strategies of prokaryotes and eukaryotes.)

Binary fission generally involves five steps, as illustrated in **FIGURE 6.17**, for a prokaryotic cell.

Play *Bacterial Growth: Overview*
@ Mastering Microbiology

1. The cell replicates its chromosome (DNA molecule). The duplicated chromosomes are attached to the cytoplasmic membrane. (In eukaryotic cells, chromosomes are attached to microtubules.)
2. The cell elongates, and growth between attachment sites pushes the chromosomes apart. (Eukaryotic cells segregate their chromosomes by the process of *mitosis*, described in Chapter 12.)
3. The cell forms a new cytoplasmic membrane and wall (septum) across the midline.
4. When the septum is completed, the daughter cells may remain attached, as shown in **FIGURE 6.17**, or they may separate completely. When the cells remain attached, further binary fission in parallel planes produces a chain.

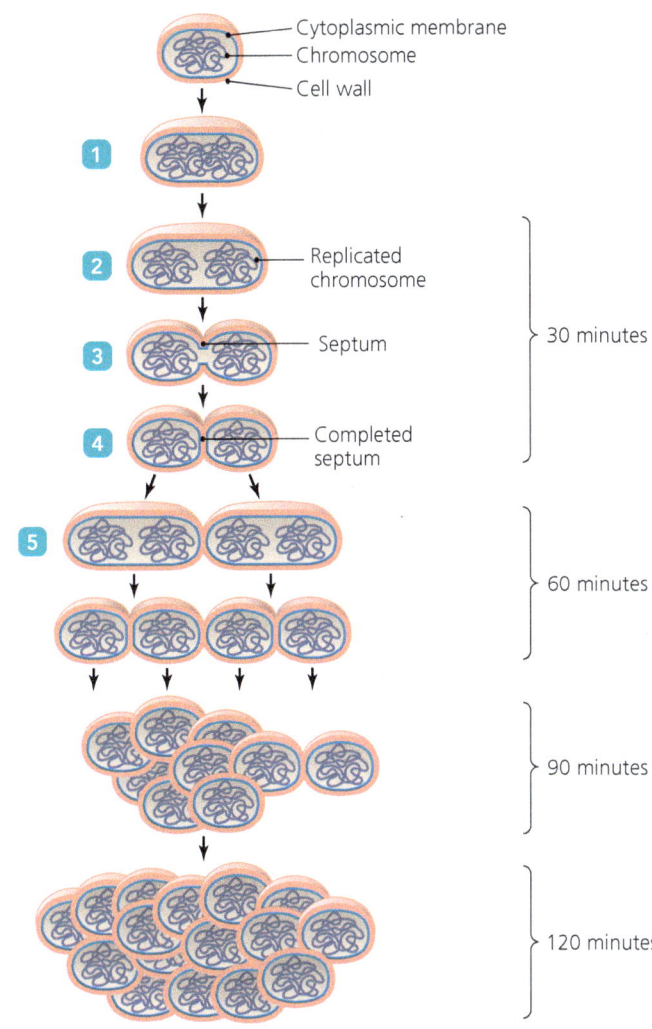

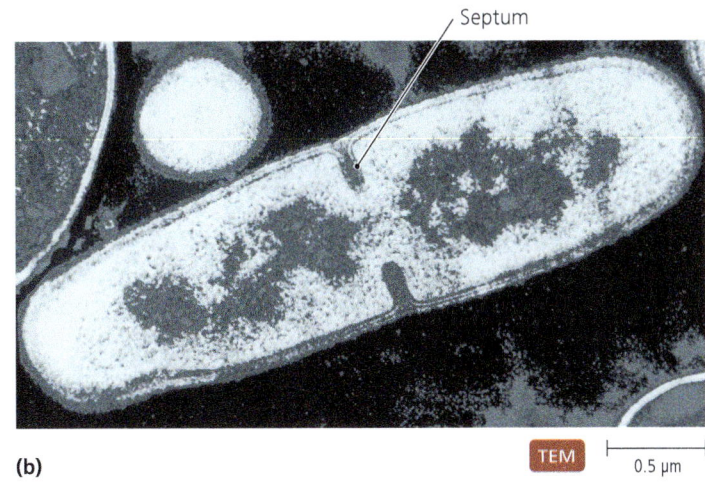

▲ **FIGURE 6.17 Binary fission. (a)** The events in binary fission. All the cells may divide in parallel planes and remain attached to form a chain, or they may divide in different planes to form a cluster (as shown at 120 minutes). **(b)** Transmission electron micrograph of *Bacillus licheniformis* undergoing binary fission.

Clusters form when attached cells continue to divide in different planes (as shown in Figure 6.17).

5 The process repeats.

Here we consider the growth of populations by binary fission, using bacterial cultures as examples. We begin with a brief discussion of the mathematics of population growth.

Generation Time

LEARNING OUTCOME

6.15 Explain what is meant by the *generation time* of bacteria.

The time required for a bacterial cell to grow and divide is its **generation time**. Viewed another way, generation time is also the time required for a population of cells to double in number. Generation times vary among populations and are dependent on chemical and physical conditions. Under optimal conditions, some bacteria (such as *E. coli* and *S. aureus*) have a generation time of 20 minutes or less. For this reason, food contaminated with only a few of these organisms can cause food poisoning if not properly refrigerated and cooked. Most bacteria have a generation time of 1 to 3 hours, though some slow-growing species such as *Mycobacterium leprae* require more than 10 days before they double.

Mathematical Considerations in Population Growth

LEARNING OUTCOME

6.16 Describe logarithmic growth.

With binary fission, any given cell divides to form two cells; then each of these new cells divides in two to make four, and then four become eight, and so on. This type of growth, called **logarithmic growth** or **exponential growth**, produces results very different from simple addition, known as *arithmetic growth*. We can compare these two types of growth by considering what would happen over time to two hypothetical populations, as shown in **FIGURE 6.18**. In this case, we assume that a hypothetical population of a pink species increases by adding one new cell every 20 minutes, whereas the cells of a purple species divide by binary fission every 20 minutes. After the initial 20 minutes, each population, which started with a single cell, would have two cells; after 40 minutes, the pink species would have three cells, whereas the purple species would have four cells. At this point, there is little difference in the growth of the two populations, but after 2 hours, the arithmetically growing pink species would have only seven cells, whereas the logarithmically growing purple species would have 64 cells. Clearly, logarithmic growth can increase a population's size dramatically—after only 7 hours, the purple species will have over 2 million cells!

The number of cells arising from a single cell by binary fission is calculated as 2^n, where n is the number of generations;

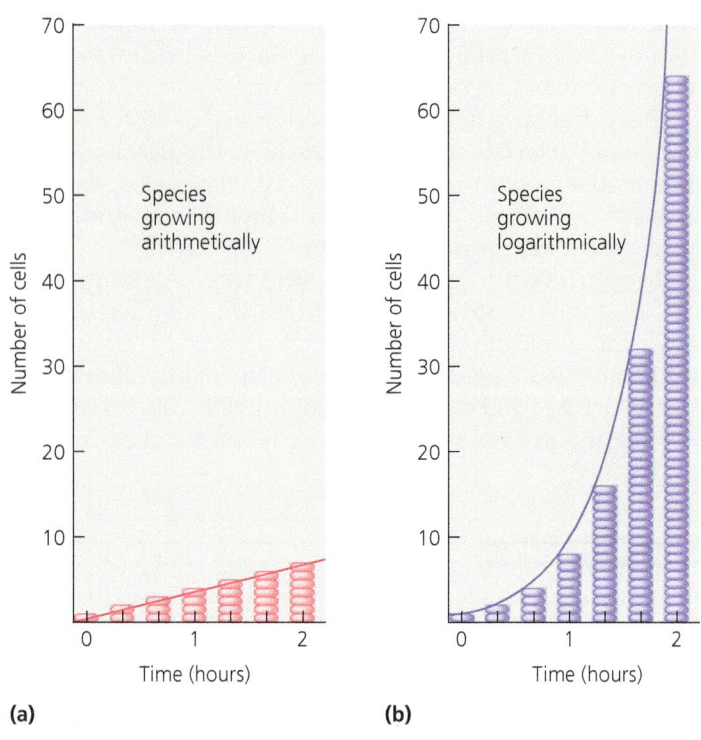

▲ **FIGURE 6.18 A comparison of arithmetic and logarithmic growth.** Given two hypothetical initial populations consisting of a single cell each with a generation time of 20 minutes, **(a)** after 2 hours an arithmetically growing species will have 7 cells, whereas **(b)** a logarithmically growing species will have 64 cells.

in other words, multiply 2 times itself n number of times. To calculate the total number of cells in a population, we multiply the original number of cells by 2^n. If, for example, we had begun with three purple cells instead of one, then after 2 hours there would be 192 cells ($3 \times 2^6 = 3 \times 64 = 192$).

A visible culture of bacteria may consist of trillions of cells, so microbiologists use scientific notation to deal with the huge numbers involved. One advantage of scientific notation is that large numbers are expressed as powers of 10, making them easier to read and write. For example, consider our culture of the purple species. After 10 hours (30 generations), there would be 1,073,741,824 (2^{30}) cells. This large number can be rounded off and expressed more succinctly in scientific notation as 1.07×10^9. After 30 more generations, scientific notation would be the only practical way to express the huge number of cells in the culture, which would be 1.15×10^{18} (2^{60}). Such a number, if written out (the digits 115 followed by 16 zeros), would be impractically large.

Phases of Microbial Population Growth

LEARNING OUTCOMES

6.17 Draw and label a bacterial growth curve.
6.18 Describe what occurs at each phase of a population's growth.

A graph that plots the numbers of organisms in a growing population over time is known as a **growth curve**. When drawn using

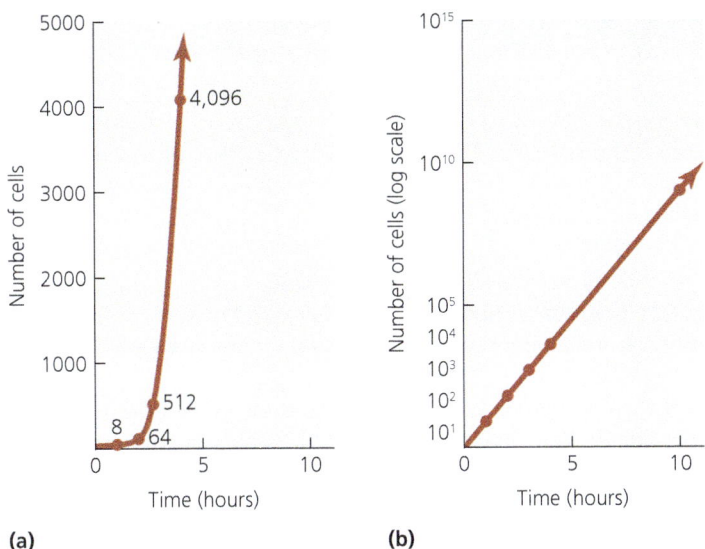

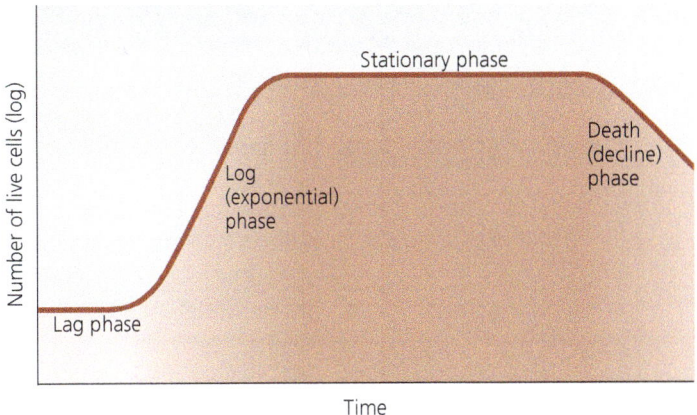

▲ FIGURE 6.20 **A typical population growth curve.** The curve shows four phases of population growth. *Why do cells trail behind their optimum reproductive potential during the lag phase?*

Figure 6.20 During lag phase, cells are synthesizing the metabolic machinery and chemicals required for optimal reproduction.

▲ FIGURE 6.19 **Two growth curves of logarithmic growth.** The generation time for this *E. coli* population is 20 minutes. **(a)** An arithmetic graph. Using an arithmetic scale for the *y*-axis makes it difficult to ascertain actual numbers of cells near the beginning and impossible to plot points after only a short time. **(b)** A semilogarithmic graph. Using a logarithmic scale for the *y*-axis solves both of these problems. Note that a plot of logarithmic population growth using a logarithmic scale produces a straight line.

an arithmetic scale on the *y*-axis, a plot of exponential growth presents two problems (FIGURE 6.19a): it is difficult or impossible to distinguish numbers in early generations from the baseline, and as the population grows it becomes impossible to accommodate the graph on a single page.

The solution to these problems is to replace the arithmetic scale on the *y*-axis with a logarithmic (log) scale (FIGURE 6.19b). Such a log scale, in which each division is 10 times larger than the preceding one, can accommodate small numbers at the lower end of the graph and very large numbers at the upper end. This kind of graph is semilogarithmic because only one axis uses a log scale.

When bacteria are inoculated into a closed vessel of liquid medium, the population's growth curve has four distinct phases: the lag, log, stationary, and death phases (FIGURE 6.20).

Play *Bacterial Growth Curve*
@ Mastering Microbiology

Lag Phase

During a **lag phase**, the cells are adjusting to their new environment; most cells do not reproduce immediately but instead actively synthesize enzymes to utilize novel nutrients in the medium. For example, bacteria inoculated from a medium containing glucose as a carbon source into a medium containing lactose must synthesize two new types of proteins: membrane proteins to transport lactose into the cell and the enzyme lactase to catabolize lactose. The lag phase can last less than an hour or can last for days, depending on the species and the chemical and physical conditions of the medium.

Log Phase

Eventually, the bacteria synthesize the necessary chemicals for conducting metabolism in their new environment and then enter a phase of rapid chromosome replication, growth, and reproduction. This is the **log phase**, so named because the population increases logarithmically, and the reproductive rate reaches a constant as DNA and protein syntheses are maximized.

Researchers are interested in the log phase for many reasons. Populations in log phase are more susceptible to antimicrobial drugs that interfere with metabolism, such as erythromycin, and to drugs that interfere with the formation of cell structures, such as the inhibition of cell wall synthesis by penicillin. Populations in log phase are preferred for Gram staining because most cells' walls are intact—an important characteristic for correct staining. Further, because the metabolic rate of individual cells is at a maximum during log phase, this phase may be preferred for industrial and laboratory purposes.

Stationary Phase

If bacterial growth continued at the exponential rate of the log phase, bacteria would soon overwhelm the Earth. This does not occur because as nutrients are depleted and wastes accumulate, the rate of reproduction decreases. Eventually, the number of dying cells equals the number of cells being produced, and the size of the population remains constant—hence the name **stationary phase**. During this phase, the metabolic rate of surviving cells declines. Cells in this phase are stress tolerant because they are adapted for challenging conditions, such as heat, high salt concentrations, and the presence of antibiotics.

Death Phase

If nutrients are not added and wastes are not removed, a population reaches a point at which cells die at a faster rate than they are produced. Such a culture has entered the **death phase** (or *decline phase*). Bear in mind that during the death phase, some

CLINICAL CASE STUDY

Boils in the Locker Room

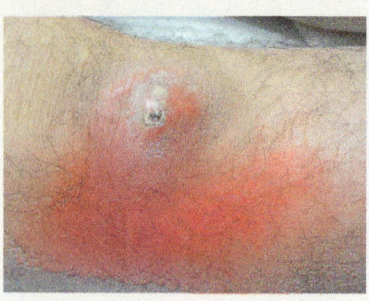

For several weeks, faculty, students, and staff at Rayburn High School have been dealing with an epidemic of unsightly, painful boils and skin infections. Over 40 athletes have been too sore to play sports, and 13 students have been hospitalized. Nurses take clinical samples of the drainage of the infections from hospitalized students, and medical laboratory scientists culture the samples. Isolated bacterial colonies are circular, convex, and golden. The bacterium is Gram-positive, mesophilic, and facultatively halophilic and can grow with or without oxygen. The cells are spherical and remain attached in clusters.

1. What color are the Gram-stained cells?
2. What does the term *facultatively halophilic* mean?
3. What is the scientific description of the bacterium's oxygen requirement?
4. If the bacterium divides every 30 minutes under laboratory conditions, how many cells would there be in a colony after 24 hours?

---

cells remain alive and continue metabolizing and reproducing, but the number of dying cells exceeds the number of new cells produced so that eventually the population decreases to a fraction of its previous abundance. In some cases, all the cells die, whereas in others, a few survivors may remain indefinitely. The latter case is especially true for cultures of bacteria that can develop resting structures called *endospores* (see Chapter 3).

MICRO CHECK

9. What type of microorganisms reproduce by binary fission?
10. What influences bacterial generation times?
11. During which phase of a microbial growth curve is the population growing fastest?

Continuous Culture in a Chemostat

LEARNING OUTCOME

6.19 Explain how a chemostat can maintain a microbial culture in a continuous phase.

Researchers and industrialists can continuously maintain a particular phase of microbial population growth by using a special culture device called a **chemostat** (FIGURE 6.21). A chemostat is an *open system*; that is, fresh medium is continuously supplied while an equal amount of old medium (containing microbes) is removed. As nutrients enter the culture vessel of a chemostat, the cells can metabolize, grow, and reproduce—but only to the extent that a limiting nutrient (e.g., a particular amino acid) is available. By controlling the amount of limiting nutrient entering the culture vessel, a chemostat maintains a culture in a particular phase, typically log phase. This is impossible in a *closed system* because as nutrients are depleted and wastes accumulate, the culture enters the stationary phase and then the death phase.

Chemostats make possible the study of microbial population growth at steady but low nutrient levels, such as might be found in biofilms. Such studies are essential to understand interactions of microbial species in nature and can reveal aspects of microbial interactions that are not apparent in a closed system. Scientists also use chemostats to maintain log phase population growth for experimental inquiries into aspects of microbial metabolism, such as enzyme activities. Food and industrial microbiologists use chemostats to maintain constant production of useful microbial products.

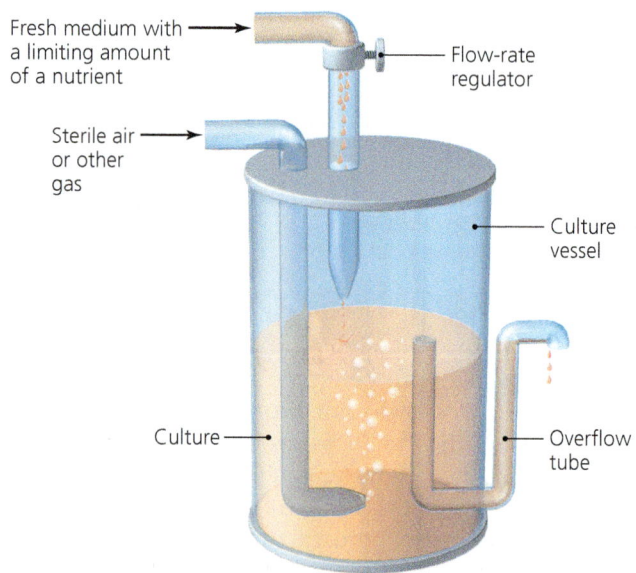

▲ FIGURE 6.21 **Schematic of chemostat.** A microbial culture is continuously maintained in a predetermined phase, typically log phase, by controlling the amount and type of fresh medium added while removing an equal amount of old culture.

Measuring Microbial Reproduction

LEARNING OUTCOME

6.20 Contrast direct and indirect methods of measuring bacterial reproduction.

We have discussed the concepts of population growth and have seen that large numbers result from logarithmic growth, but we have not discussed practical methods of determining the size of a microbial population. Because of each cell's small size and incredible rate of reproduction, it is not possible to actually count every one in a population. For one thing, they grow so rapidly that their number changes during the count. Therefore, laboratory personnel must estimate the number of cells in a population by counting the number in a small, representative sample and then multiplying to estimate the number in the whole specimen. For example, if there are 25 cells in a microliter (μl) sample of urine, then there are approximately 25 million cells in a liter of urine.

Estimating the number of microorganisms in a sample is useful for determining such things as the severity of urinary tract infections, the effectiveness of pasteurization and other methods of food preservation, the degree of fecal contamination of water supplies, and the usefulness of particular disinfectants and antibiotics.

Microbiologists use either direct or indirect methods to estimate the number of cells. We begin with direct methods of measuring bacterial reproduction.

Direct Methods Not Requiring Incubation

It is possible to directly count cells without having to incubate cultures. Here we consider two such direct methods.

Microscopic Counts Microbiologists can count microorganisms directly through a microscope rather than inoculating them onto the surface of a solid medium. In this method, particularly suitable for stained prokaryotes and relatively large eukaryotes, a sample is placed on a *cell counter* (also called a *Petroff-Hausser counting chamber*), which is a special glass slide composed of an etched grid positioned beneath a glass cover slip (**FIGURE 6.22**). A cell counter is designed to precisely hold the cover slip 0.02 mm above the grid; therefore, the volume of a suspension of microbes in water over a 1 mm^2 portion of the grid is 1 mm $\times$ 1 mm $\times$ 0.02 mm = 0.02 mm^3. Each 1 mm^2 grid contains 25 large squares, so a microbiologist can count the number of bacteria in several of the large squares and then calculate the mean number of bacteria per square. The number of bacteria per milliliter (cm^3) can be calculated as follows:

mean number of bacteria per square $\times$ 25 squares
= number of bacteria per 0.02 mm^3
no. of bacteria per 0.02 mm^3 $\times$ 50 = no. of bacteria per mm^3
no. of bacteria per mm^3 $\times$ 1000 = no. of bacteria per cm^3 (ml)

This means that one needs only to multiply the mean number of bacteria per square by 1,250,000 (25 $\times$ 50 $\times$ 1000) to calculate the number of bacteria per milliliter of bacterial suspension.

Direct microscopic counts are advantageous when there are more than 10 million cells per milliliter or when a speedy

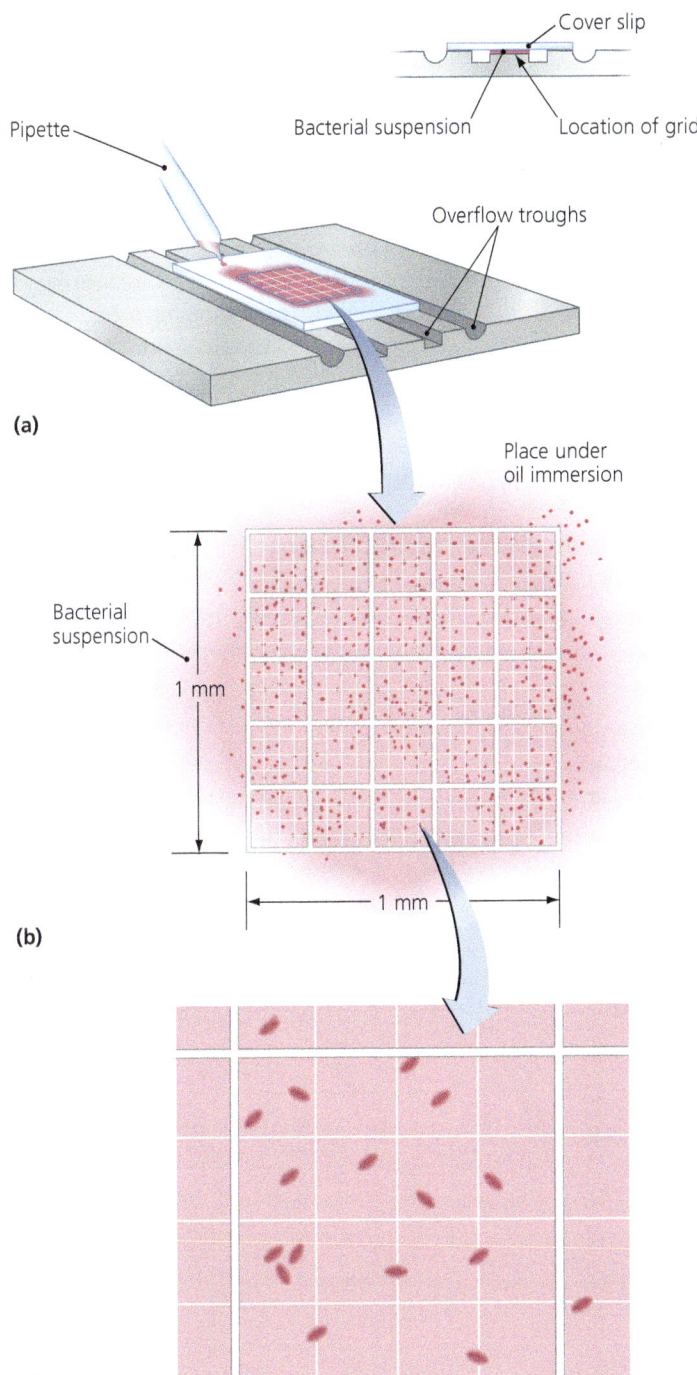

▲ **FIGURE 6.22 The use of a cell counter for estimating microbial numbers.** (a) The counter is a glass slide with an etched grid that is exactly 0.02 mm lower than the bottom of the cover slip. A bacterial suspension placed next to the cover slip through a pipette moves under the cover slip and over the grid by capillary action. (b) View of a 1 mm^2 portion of the grid through the microscope. Each square millimeter of the grid has 25 large squares, each of which is divided into 16 small squares. (c) Enlarged view of one large square containing 15 cells. The number of bacteria in several large squares is counted and averaged. The calculations involved in estimating the number of bacteria per milliliter (cm^3) of suspension are described in the text.

estimate of population size is required. However, direct counts can be problematic because it is often difficult to differentiate between living and dead cells and nearly impossible to count rapidly moving microbes.

Electronic Counters A *Coulter*[16] *counter* is a device that directly counts cells as they interrupt an electrical current flowing across a narrow tube held in front of an electronic detector. This device is useful for counting the larger cells of yeasts, unicellular algae, and protozoa; it is less useful for bacterial counts because of debris in the media and the presence of filaments and clumps of cells.

Flow cytometry is a variation of counting with a Coulter counter. A cytometer uses a light-sensitive detector to record changes in light transmission through the tube as cells pass. Scientists use this technique to distinguish among cells that have been differentially stained with fluorescent dyes or tagged with fluorescent antibodies. They can count bacteria in a solution and even count host cells that contain fluorescently stained intracellular parasites.

We have considered direct methods not requiring incubation. Now we consider direct methods with incubation.

Direct Methods Requiring Incubation

Among the many direct techniques are techniques requiring incubation—viable plate counts following dilution, membrane filtration, and the most probable number method.

Serial Dilution and Viable Plate Counts What if the number of cells in even a very small sample is still too great to count? If, for example, a 1-ml sample of milk containing 20,000 bacterial cells per milliliter were plated on a Petri plate, there would be too many colonies to count. In such cases, microbiologists make a **serial dilution**, which is the stepwise dilution of a liquid culture in which the dilution factor at each step is constant. The scientists plate a set amount of each dilution onto an agar surface and count the number of colonies resulting on a plate from each dilution. They count the colonies on plates with 25 to 250 colonies and multiply the number on each plate by the reciprocal of the dilution used to inoculate that plate to estimate the number of bacteria per milliliter of the original culture. This method is called a **viable plate count** (FIGURE 6.23). When a plate has fewer than 25 colonies, it is not used to estimate the number of bacteria in the original sample because the chance of underestimating the population increases when the number of colonies is small.

Recall that the number of colonies on a plate indicates the number of CFUs that were inoculated onto the plate. This number differs from the actual number of cells when the CFUs are composed of more than one cell. In cases where a colony-forming unit consists of more than one cell, a viable plate count underestimates the number of cells present in the sample.

The accuracy of a viable plate count also depends on the homogeneity of the dilutions, the ability of the bacteria to grow on the medium used, the number of cell deaths, and the growth phase of the sample population. Thoroughly mixing each dilution, inoculating multiple plates per dilution, and using log-phase cultures minimize errors.

Membrane Filtration Viable plate counts allow scientists to estimate the number of microorganisms when the population is very large, but if the population density is very small—as is the case, for example, for fecal bacteria in a stream or lake—microbes are more accurately counted by **membrane filtration** (FIGURE 6.24).

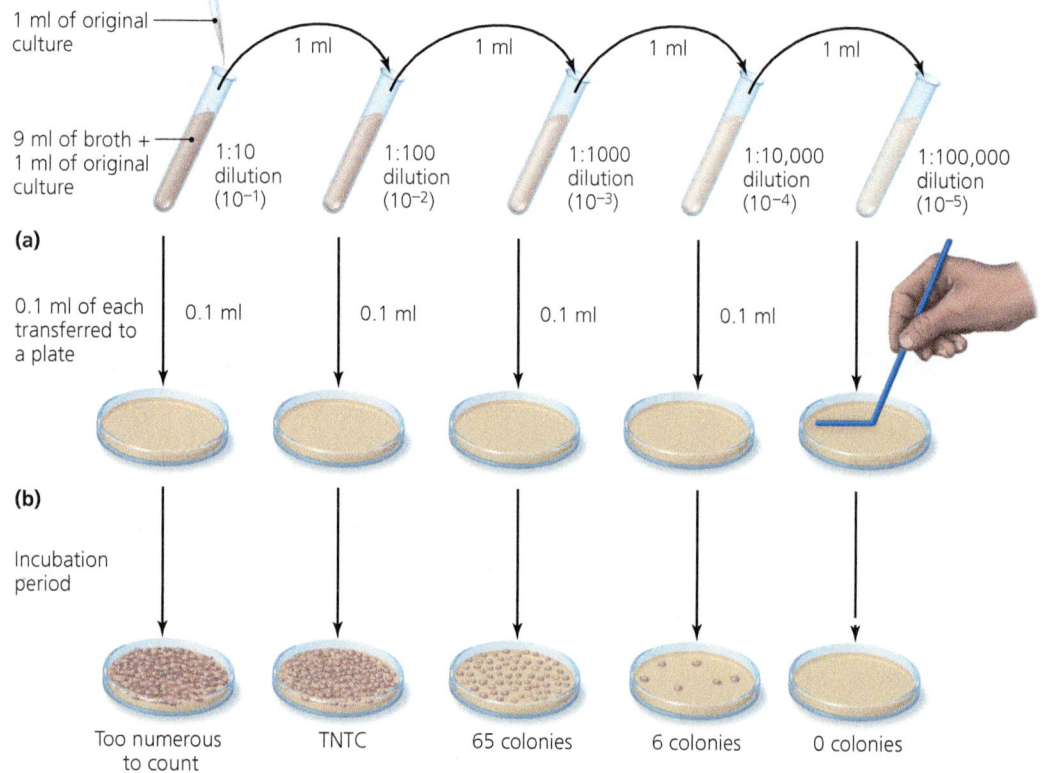

◀ **FIGURE 6.23 A serial dilution and viable plate count for estimating microbial population size.** (a) Serial dilutions. A series of 10-fold dilutions is made. (b) Plating. A 0.1-ml sample from each dilution is poured onto a plate and spread with a sterile spreader. Alternatively, 0.1 ml of each dilution can be mixed with melted agar medium and poured into plates. (c) Counting. Plates are examined after incubation. Some plates may contain so many colonies that they are too numerous to count (TNTC). The number of colonies is multiplied by 10 (because 0.1 ml was plated instead of 1 ml) and then by the reciprocal of the dilution to estimate the concentration of bacteria in the original culture—in this case, 65 colonies × 10 × 1000 = 650,000 bacteria/ml.

---

[16]Named for Wallace Coulter, American inventor.

Most Probable Number

The **most probable number (MPN) method** is a statistical estimation technique based on the fact that the more bacteria are in a sample, the more dilutions are required to reduce their number to zero.

Let's consider an example of the use of the MPN method to estimate the number of fecal bacteria contaminating a stream. A researcher inoculates a set of test tubes of a broth medium with a sample of stream water. The more tubes that are used, the more accurate is the MPN method; even so, accuracy must be balanced against the time and cost involved in inoculating and incubating numerous tubes. Typically, a set of five tubes is inoculated.

The researcher also inoculates a set of five tubes with a 1:10 dilution and another set of five tubes with a 1:100 dilution of stream water. Thus, there are 15 test tubes—the first set of five tubes inoculated with undiluted sample, the second set with a 1:10 dilution, and the third set with a 1:100 dilution (**FIGURE 6.25**).

After incubation for 48 hours, the researcher counts the number of test tubes in each set that show growth. This generates three numbers in this example—growth occurs in four of the undiluted broth tubes, two of the 1:10 tubes, and in only one of the 1:100 tubes (4, 2, 1). The numbers are compared to the numbers in an MPN table (**TABLE 6.5** on p. 186). How statisticians develop MPN tables is beyond the scope of our discussion, but they accurately estimate the number of cells in a solution. In this case, the MPN table estimates that there were 26 bacteria per 100 ml of stream water.

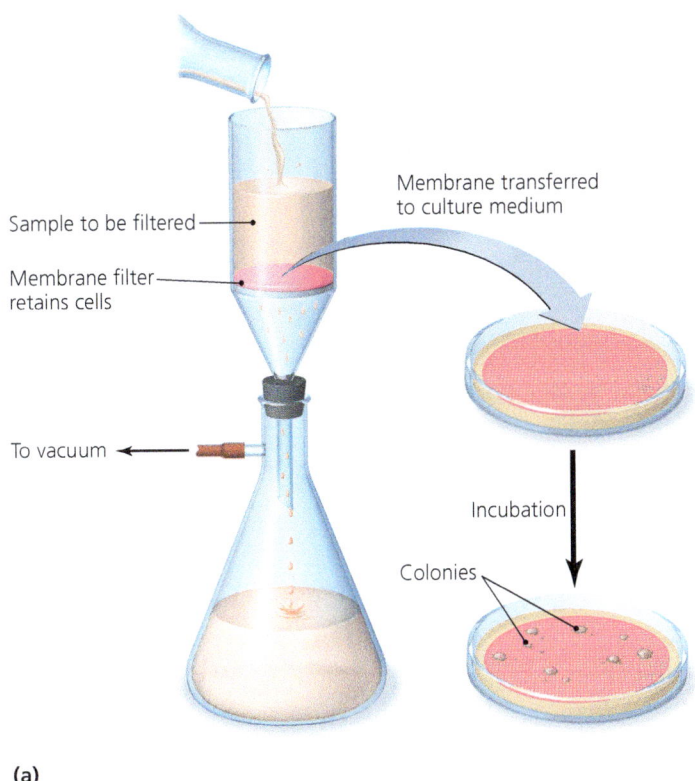

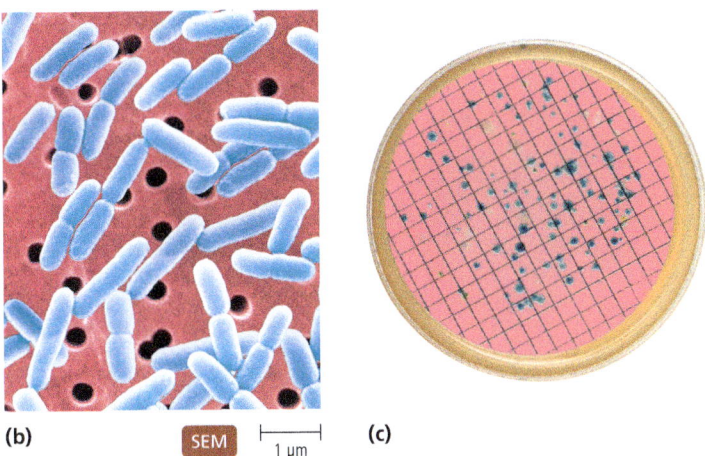

▲ **FIGURE 6.24 The use of membrane filtration to estimate microbial population size.** (a) After all the bacteria in a given volume of sample are trapped on a membrane filter, the filter is transferred onto an appropriate medium and incubated. The microbial population is estimated by multiplying the number of colonies counted by the volume of sample filtered. (b) Bacteria trapped on the surface of a membrane filter. (c) Colonies growing on a solid medium after being transferred from a membrane filter. Scientists use a superimposed grid to help them count the colonies. *If the colonies in (c) resulted from filtering 2.5 liters of stream water, what is the minimum number of bacteria per liter in the stream?*

Figure 6.24 83 colonies/2.5 L = 33.2 colonies/L.

In this method, a large sample (perhaps as large as several liters) is poured (or drawn under a vacuum) through a membrane filter with pores small enough to trap the cells. The membrane is then transferred onto a solid medium, and the colonies present after incubation are counted. In this case, the number of colonies is equal to the number of CFUs in the original large sample.

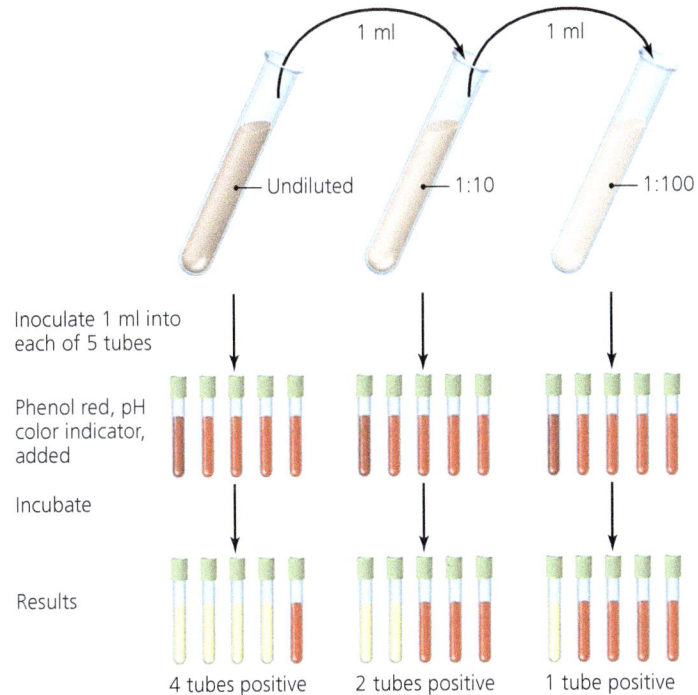

▲ **FIGURE 6.25 The most probable number (MPN) method for estimating microbial numbers.** Typically, sets of five test tubes are used for each of three dilutions. After incubation, the number of tubes showing growth in each set is used to enter an MPN table (see Table 6.5), which provides an estimate of the number of cells per 100 ml of liquid. *If the results were 5, 3, 1, what would be the most probable number of microorganisms in the original broth?*

Figure 6.25 The MPN is 110/100 ml.

TABLE 6.5 Most Probable Number Table (Partial)

Number Out of Five Tubes Giving Positive Results in Three Dilutions			Most Probable Number of Bacteria per 100 ml
4	0	0	13
4	0	1	17
4	1	0	17
4	1	1	21
4	1	2	26
4	2	0	22
4	2	1	26
4	3	0	27
4	3	1	33
4	4	0	34
5	0	0	23
5	0	1	30
5	0	2	40
5	1	0	30
5	1	1	50
5	1	2	60
5	2	0	50
5	2	1	70
5	2	2	90
5	3	0	80
5	3	1	110
5	3	2	140
5	3	3	170
5	4	0	130
5	4	1	170
5	4	2	220
5	4	3	280
5	4	4	350

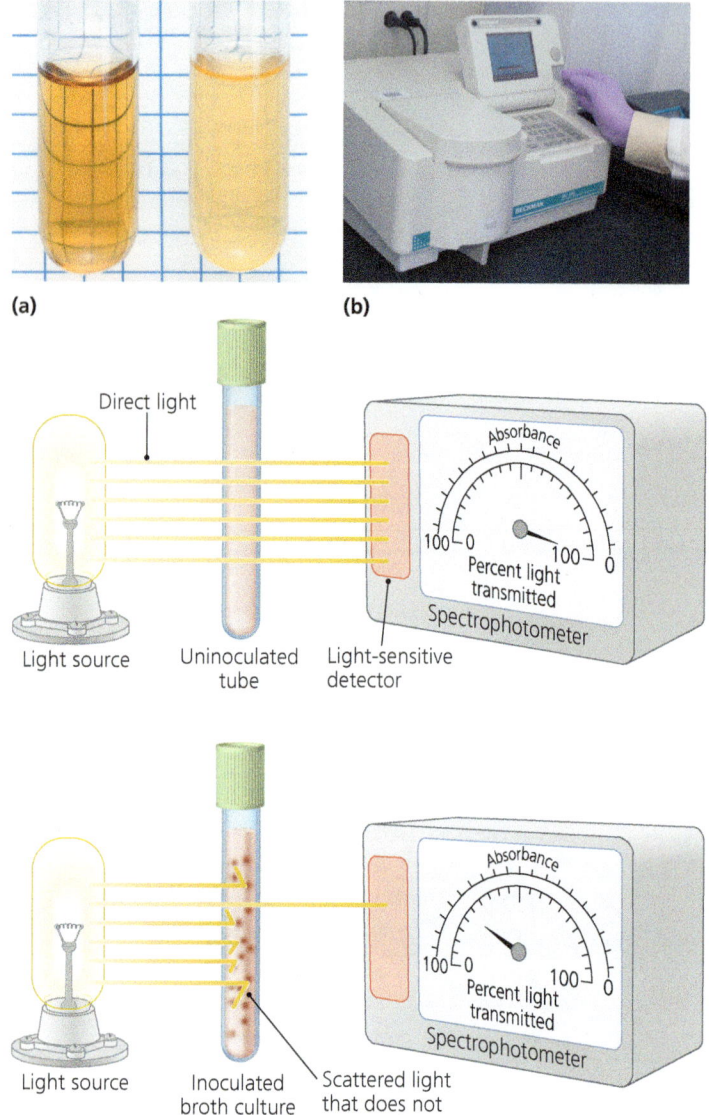

▲ **FIGURE 6.26 Turbidity and the use of spectrophotometry in indirect measurement of population size.** (a) Turbidity (right), an increased optical density or cloudiness of a solution. (b) A spectrophotometer. (c) The principle of spectrophotometry. After a light beam is passed through an uninoculated sample of the culture medium, the scale is set at 100% transmission. In an inoculated sample, the microbial cells absorb and scatter light, reducing the amount reaching the detector. The percentage of light transmitted is inversely proportional to population density.

The most probable number method is useful for counting microorganisms that do not grow on solid media, when bacterial counts are required routinely, and when samples of wastewater, drinking water, and food samples contain too few organisms to use a viable plate count. The MPN method is also used to count algal cells because algae seldom form distinct colonies on solid media.

Indirect Methods

It is not always necessary to count microorganisms to estimate population size or density. Industrial and research microbiologists use indirect methods that measure such variables as turbidity, metabolic activity, and dry weight instead of counting microorganisms, colonies, or MPN tubes. Scientists can also estimate population size and diversity by analyzing the unique sequences of DNA present in a sample.

Turbidity As bacteria reproduce in a broth culture, the broth often becomes *turbid* (cloudy) (FIGURE 6.26a). Generally, the greater the bacterial population, the more turbid a broth will be.

An indirect method for estimating the growth of a microbial population involves measuring changes in turbidity using a device called a *spectrophotometer* (FIGURE 6.26b). Researchers most often use this method.

A spectrophotometer measures the amount of light transmitted through a culture under standardized conditions (FIGURE 6.26c). The greater the concentration of bacteria within a broth, the more light will be absorbed and scattered, and the less light will pass through and strike a light-sensitive detector. Generally, transmission is inversely proportional to the population size; that is, the larger the population grows, the less light will reach the detector.

Scales on the gauge of a spectrophotometer report *percentage of transmission* and *absorbance*. These are two ways of looking at the same things; for example, 25% transmission is the same thing as 75% absorbance. Direct counts must be calibrated with transmission and absorbance readings to provide estimates of population size. Once these values are determined, spectrophotometry provides estimates of population size more quickly than any direct method.

The benefits of measuring turbidity to estimate population growth include ease of use and speed. However, the technique is useful only if the concentration of cells exceeds 1 million per milliliter; densities below this value generally do not produce turbidity. Further, the technique is accurate only if the cells are suspended uniformly in the medium. If they form either a *pellicle* (a film of cells at the surface) or a *sediment* (an accumulation of cells at the bottom), their number will be underestimated. Further, spectrophotometry does not distinguish between living and dead cells.

Metabolic Activity Under standard temperature conditions, the rate at which a population of cells utilizes nutrients and produces wastes depends on their number. Once they establish the metabolic rate of a microorganism, scientists can indirectly estimate the number of cells in a culture by measuring changes in such things as nutrient utilization, waste production, or pH. Scientists studying environmental water samples often use this method.

Dry Weight The abundance of some microorganisms, particularly filamentous microorganisms, is difficult to measure by direct methods. Instead, these organisms are filtered from their culture medium, dried, and weighed. The *dry weight method* is suitable for broth cultures, but growth cannot be followed over time because the organisms are killed during the process.

Molecular Methods The majority of bacteria and archaea have not been grown in the laboratory, and representatives of most species are too few in number to study by direct observation. How do scientists estimate the number of such uncultured microbes?

Scientists can isolate unique DNA sequences representing uncultured prokaryotic species using genetic techniques such as *polymerase chain reaction (PCR)* and *hybridization* of DNA that codes for ribosomal RNA. For example, one study estimated that a single gram of garden soil contains more than 100 billion bacteria and archaea, representing more than 10 million different species. In the clinical environment, molecular methods have three major advantages. First, PCR and hybridization probes are highly specific, much more than selective media. Second, they are faster because no culture is required. For example, probe tests for *Chlamydia* can give a result in 2 hours. Third, molecular methods are based on DNA and thus are generic, avoiding the necessity of using a different media type for each microbe. (Chapter 8 discusses molecular methods in more detail.)

MICRO CHECK

12. Which form of direct measurement uses statistical estimation to estimate the number of microbes in a sample?

TELL ME WHY

Some students transfer some "gunk" from a 2-week-old bacterial culture into new media. Why shouldn't they be surprised when this "death-phase" sample grows?

MICRO IN THE CLINIC FOLLOW-UP

Just a Sore Throat and Cough?

At the clinic, Dr. Wyatt gets a medical history, listens to Jennifer's lungs, and uses a pulse oximeter to measure her pulse and the level of oxygen in her blood. He tells her that he hears a crackling sound in her lungs and that her oxygen levels are slightly lower than normal. Dr. Wyatt orders a chest X-ray exam and blood work, and he obtains a sputum (mixture of saliva and mucus) sample from Jennifer by having her cough deeply. As they view the chest image, Dr. Wyatt points out an area of cloudiness in Jennifer's right lung and explains that this indicates inflammation in the lung, most likely due to an infection. This also explains the crackling in her lungs, the pain in her chest, her difficulty in breathing, and the lower oxygen saturation. He adds that, based on her medical history, he thinks her sore throat and cough have progressed to pneumonia. Although they need to wait for the blood and sputum test results to come back from the lab in order to make a definitive diagnosis, Dr. Wyatt believes that it's best to start Jennifer on antibiotics right away. He also prescribes a steroid inhaler to help open up her airways and make breathing easier.

Shortly after her visit, Dr. Wyatt phones Jennifer to follow up with her lab results. Molecular analysis (using polymerase chain reaction, or PCR) of the sputum test identified *Mycoplasma pneumoniae*, a bacterium commonly associated with pneumonia. He tells her to continue taking the antibiotics and to schedule a follow-up appointment. After just a few days on the antibiotics, Jennifer feels a lot better. Her cough tapers off and eventually disappears, and by the time of her follow-up appointment, her right lung is totally clear, and her lung sounds and oxygen levels have returned to normal.

1. Sputum samples are typically plated on blood agar, chocolate agar (blood agar containing lysed red blood cells), and MacConkey agar. Explain why each of these media may be helpful in identifying pathogens in lower respiratory infections.

2. The *M. pneumoniae* in Jennifer's sputum was identified using PCR, not the culture plates. What are some possible reasons that *M. pneumoniae* was not found on the plates?

Check your answers to Micro in the Clinic Follow-Up questions in the Mastering Microbiology Study Area.

CHAPTER SUMMARY

 Make the connection between Chapters 5 and 6. Watch MICRO MATTERS videos in the Mastering Microbiology Study Area.

Growth Requirements (pp. 162–171)

1. A **colony**, which is a visible population of microorganisms arising from a single cell or colony-forming unit living in one place, grows in size as the number of cells increases. Most microbes live as **biofilms**, that is, in association with one another on surfaces.

2. Chemical **nutrients** such as carbon, hydrogen, oxygen, and nitrogen are required for the growth of microbial populations.

3. **Photoautotrophs** use carbon dioxide as a carbon source and light energy to make their own food; **chemoautotrophs** use carbon dioxide as a carbon source but catabolize organic or inorganic molecules for energy. **Photoheterotrophs** are photosynthetic organisms that acquire energy from light and acquire nutrients via catabolism of organic compounds; **chemoheterotrophs** use organic compounds for both energy and carbon. **Organotrophs** acquire electrons for redox reactions from organic sources, whereas **lithotrophs** acquire electrons from inorganic sources.

4. **Obligate aerobes** require oxygen molecules as the final electron acceptor of their electron transport chains, whereas **obligate anaerobes** cannot tolerate oxygen and must use an electron acceptor other than oxygen.

5. The four toxic forms of oxygen are **singlet oxygen** ($^1O_2$), which is neutralized by pigments called **carotenoids**; **superoxide radicals** (O_2^-), which are detoxified by superoxide dismutase; **peroxide anion** (O_2^{2-}), which is detoxified by catalase or peroxidase; and **hydroxyl radicals** ($OH \cdot$), the most reactive of the toxic forms of oxygen.

6. Microbes are described in terms of their oxygen requirements and limitations as strict **aerobes**, which require oxygen; as strict **anaerobes**, which cannot tolerate oxygen; as **facultative anaerobes**, which can live with or without oxygen; as **aerotolerant anaerobes**, which prefer anaerobic conditions but can tolerate exposure to low levels of oxygen; or as **microaerophiles**, which require low levels of oxygen.

7. Nitrogen, acquired from organic or inorganic sources, is an essential element for microorganisms. Some bacteria can reduce nitrogen gas into a more usable form via a process called **nitrogen fixation**.

8. In addition to the main elements found in microbes, very small amounts of **trace elements** are required. Vitamins are among the **growth factors**, which are organic chemicals required in small amounts for metabolism.

9. Though microbes survive within the limits imposed by a minimum growth temperature and a maximum growth temperature, an organism's metabolic activities produce the highest growth rate at the **optimum growth temperature**.

10. Microbes are described in terms of their temperature requirements as (from coldest to warmest) **psychrophiles**, **mesophiles**, **thermophiles**, or **hyperthermophiles**. **Psychrotolerant** microbes do not normally live in the cold but can tolerate cold temperatures such as in refrigeration.

11. **Neutrophiles** grow best at neutral pH, **acidophiles** grow best in acidic surroundings, and **alkalinophiles** live in alkaline habitats.

12. Osmotic pressure can cause cells to die from either swelling and bursting in **hypotonic** solutions or from **crenation** (shriveling) in **hypertonic** solutions. The cell walls of some microorganisms protect them from osmotic shock. **Obligate halophiles** require high osmotic pressure, whereas facultative halophiles do not require but can tolerate such conditions.

13. **Barophiles**, organisms that normally live under the extreme hydrostatic pressure at great depth below the surface of a body of water, often cannot live at the pressure found at the surface.

14. **Quorum sensing** is the process by which bacteria respond to changes in microbial density by utilizing signal and receptor molecules. **Biofilms**, which are communities of cells attached to surfaces, use quorum sensing.

Culturing Microorganisms (pp. 171–179)

1. Microbiologists culture microorganisms by transferring an **inoculum** from a clinical or environmental specimen into a **medium** such as **broth** or solid media. The microorganisms grow into a **culture**. On solid surfaces, cultures are seen as **colonies**.

2. A **clinical specimen** is a sample of human material. Standard precautions are the guidelines to protect health care professionals from infection.

3. **Pure cultures** (axenic cultures) contain cells of only one species and are derived from a **colony-forming unit (CFU)** composed of a single cell or group of related cells. To obtain pure cultures, **sterile** equipment and use of aseptic techniques are critical.

4. The **streak-plate** method allows CFUs to be isolated by streaking. The **pour-plate** technique isolates CFUs via a series of dilutions.

5. Petri dishes that are filled with solid media are called **Petri plates**. **Slant tubes (slants)** are test tubes containing agar media that solidified while the tube was resting at an angle.

6. A **defined medium** (also known as **synthetic medium**) provides exact known amounts of nutrients for the growth of a particular microbe. **Complex media** contain a variety of growth factors. **Selective media** either inhibit the growth of unwanted microorganisms or favor the growth of particular microbes. Microbiologists use **differential media** to distinguish among groups of bacteria. **Reducing media** provide conditions conducive to culturing anaerobes. **Transport media** are designed to move specimens safely from one location to another while maintaining the relative abundance of organisms and preventing contamination of the specimen or environment.

7. Special culture techniques include the use of animal and cell cultures, low-oxygen cultures, **enrichment cultures**, and **cold enrichment** cultures. A **capnophile** grows best with high CO_2 levels in addition to low oxygen levels.

8. Cultures can be preserved in the short term by *refrigeration* and in the long term by **deep-freezing** and **lyophilization**.

Growth of Microbial Populations (pp. 179–187)

1. A population of microorganisms doubles during its **generation time**—the time also required for a single cell to grow and divide.

2. Bacteria grow by **logarithmic growth**, or **exponential growth**.

Play *Bacterial Growth: Overview; Binary Fission*
@ Mastering Microbiology

3. A graph that plots the number of organisms growing in a population over time is called a **growth curve**. When organisms are grown in a broth and the growth curve is plotted on a semilogarithmic scale, the population's growth curve has four phases. In the **lag phase**, the organisms are adjusting to their environment. In the **log phase**, the population is most actively growing. In the **stationary phase**, new organisms are being produced at the same rate at which they are dying. In the **death phase**, the organisms are dying more quickly than they can be replaced by new organisms.

Play *Bacterial Growth Curve*
@ Mastering Microbiology

4. A **chemostat** is a continuous culture device that maintains a desired phase of microbial population growth by adding limited amounts and kinds of nutrients while removing an equal amount of old medium.

5. Direct methods for estimating population size that do not require incubation are microscopic counts and electronic counters, including flow cytometry. Direct methods requiring incubation include **serial dilution** with **viable plate counts**, **membrane filtration**, and the **most probable number (MPN) method**.

6. Indirect methods include measurements of turbidity, metabolic activity, and dry weight and analysis of numbers and kinds of unique genetic sequences.

QUESTIONS FOR REVIEW

Answers to the Questions for Review (except Short Answer questions) begin on p. A-1.

Multiple Choice

1. Which of the following can grow in a Petri plate on a laboratory table?
   a. an anaerobic bacterium
   b. an aerobic bacterium
   c. viruses on an agar surface
   d. all of the above

2. "In the laboratory, a sterile inoculating loop is moved across the agar surface in a culture dish, thinning a sample and isolating individuals." This statement describes which of the following?
   a. broth culture
   b. pour plate
   c. streak plate
   d. dilution plate

3. Superoxide dismutase _____.
   a. causes hydrogen peroxide to become toxic
   b. detoxifies superoxide radicals
   c. neutralizes singlet oxygen
   d. is missing in aerobes

4. The most reactive of the four toxic forms of oxygen is _____.
   a. the hydroxyl radical
   b. the peroxide anion
   c. the superoxide radical
   d. singlet oxygen

5. Microaerophiles that grow best with a high concentration of carbon dioxide in addition to a low level of oxygen are called _____.
   a. aerotolerant
   b. capnophiles
   c. facultative anaerobes
   d. fastidious

6. Which of the following is *not* a growth factor for various microbes?
   a. cholesterol
   b. water
   c. vitamins
   d. heme

7. Organisms that preferentially may thrive in icy waters are described as _____.
   a. barophiles
   b. thermophiles
   c. mesophiles
   d. psychrophiles

8. Barophiles _____.
   a. cannot cause diseases in humans
   b. live at normal barometric pressure
   c. die if put under high pressure
   d. thrive in warm air

9. Which of the following terms best describes an organism that *cannot* exist in the presence of oxygen?
   a. obligate aerobe
   b. facultative aerobe
   c. obligate anaerobe
   d. facultative anaerobe

10. In a defined medium, _____.
    a. the exact chemical composition of the medium is known
    b. agar is available for microbial nutrition
    c. blood may be included
    d. organic chemicals are excluded

11. Which of the following is most useful in representing population growth on a graph?
    a. logarithmic reproduction of the growth curve
    b. a semilogarithmic graph using a log scale on the *y*-axis
    c. an arithmetic graph of the lag phase followed by a logarithmic section for the log, stationary, and death phases
    d. None of the above would best represent a population growth curve.

12. Which of the following methods is best for counting fecal bacteria from a stream to determine the safety of the water for drinking?
    a. dry weight
    b. turbidity
    c. viable plate counts
    d. membrane filtration

13. A Coulter counter is a(n) _____.
    a. statistical estimation using 15 dilution tubes and a table of numbers to estimate the number of bacteria per milliliter
    b. indirect method of counting microorganisms
    c. device that directly counts microbes as they pass through a tube in front of an electronic detector
    d. device that directly counts microbes that are differentially stained with fluorescent dyes

14. Lyophilization can be described as _____.
    a. freeze-drying
    b. deep-freezing
    c. refrigeration
    d. pickling

15. Quorum sensing is _____.
    a. the ability to respond to changes in population density
    b. a characteristic allowing secretion of a matrix
    c. dependent on direct contact among cells
    d. associated with colonies in broth culture

Fill in the Blanks

1. All cells require a source of _____ for redox reactions.

2. A toxic form of oxygen, _____ oxygen, is molecular oxygen with electrons that have been boosted to a higher energy state.

3. All cells recycle the essential element _____ from amino acids and nucleotides.

4. _____ are small organic molecules that are required in minute amounts for metabolism.

5. The lowest temperature at which a microbe continues to metabolize is called its _____.

6. Cells that shrink in hypertonic solutions such as saltwater are responding to _____ pressure.

7. Obligate _____ exist in salt ponds because of their ability to withstand high osmotic pressure.

8. _____ pigments protect many phototrophic organisms from photochemically produced singlet oxygen.

9. Microbes that reduce N_2 to NH_3 engage in nitrogen _____.

10. A student observes a researcher streaking a plate numerous times, flaming the loop between streaks. The researcher is likely using the _____ method to isolate microorganisms.

11. Chemolithotrophs *acquire* electrons from (organic/inorganic) _____ compounds.

VISUALIZE IT!

1. Label each of these thioglycolate tubes to indicate the oxygen requirements of the microbes growing in them.

(a)　(b)　(c)　(d)

2. Describe the type of hemolysis shown by the pathogen *Staphylococcus aureus* pictured here.

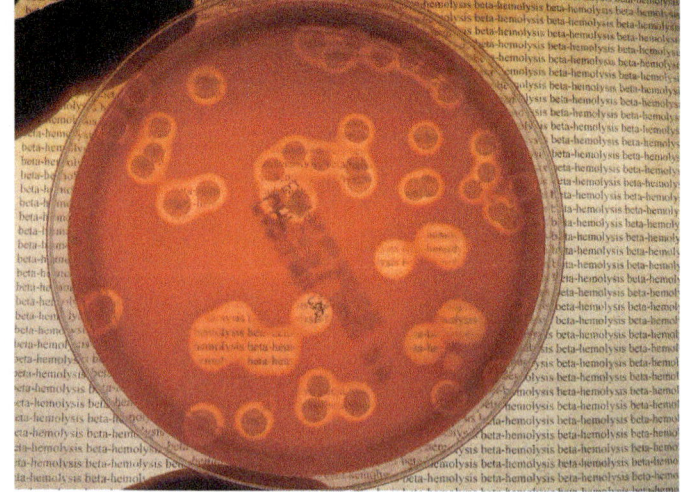

Short Answer

1. High temperature affects the shape of particular molecules. How does this affect the life of a microbe?
2. Support or refute the following statement: Microbes cannot tolerate the low pH of the human stomach.
3. Explain quorum sensing, and describe how it is related to biofilm formation.
4. Why must media, vessels, and instruments be sterilized before they are used for microbiological procedures?
5. Why is agar used in microbiology?
6. What is the difference between complex media and defined media?
7. Draw and label the four distinct phases of a bacterial growth curve. Describe what is happening within the culture as it passes through the phases.
8. If there are 47 cells in 1 µl of sewage, how many cells are there in a liter?
9. List three indirect methods of counting microbes.
10. List five direct methods of counting microbes.
11. Explain the differences among photoautotrophs, chemoautotrophs, photoheterotrophs, chemoheterotrophs, organotrophs, and lithotrophs.
12. Contrast the media described in Tables 6.3 and 6.4 (on pp. 175 and 177). Why is *E. coli* medium described as defined, whereas MacConkey medium and blood agar are defined as complex?
13. How does a chemostat maintain a constant population size?

CRITICAL THINKING

1. A scientist describes an organism as a chemoheterotrophic, aerotolerant, mesophilic, facultatively halophilic coccus. Describe the cell's metabolic and structural features in plain English.
2. Pasteurization is a technique that uses temperatures of about 72°C to neutralize potential pathogens in foods. What effect does this temperature have on the enzymes and cellular metabolism of pathogens? Why does the heat of pasteurization kill some microorganisms yet fail to affect thermophiles?
3. Two cultures of a facultative anaerobe are grown under identical conditions, except that one was exposed to oxygen and the other was completely deprived of oxygen. What differences would you expect to see between the dry weights of the cultures? Why?
4. Some organisms require riboflavin (vitamin B_2) to make FAD. For what purpose do they use FAD?
5. A scientist inoculates a bacterium into a complex nutrient slant tube. The bacterium forms only a few colonies on the slanted surface but grows prolifically in the depth of the agar. Describe the oxygen requirements of the bacterium.
6. How can regions within biofilms differ in their chemical content?
7. A scientific article describes a bacterium as an obligate microaerophilic chemoorganoheterotroph. Describe the oxygen and nutritional characteristics of the bacterium in everyday language.
8. Microorganisms require phosphorus, sulfur, iron, and magnesium for metabolism. What specifically are these elements used for in microbial metabolism? (Review Chapters 2 and 5.)
9. The bacterium *Desulforudis audaxviator* lives almost two miles underground, deriving energy from sulfate, acquiring electrons from hydrogen, and building organic molecules from inorganic carbon found in surrounding rocks. Describe the nutritional classifications of *D. audaxviator*.
10. Starting with 10 bacterial cells per milliliter in a sufficient amount of complete culture medium with a 1-hour lag phase and a 30-minute generation time, how many cells will there be in a liter of medium at the end of 2 hours? At the end of 7 hours?
11. Suppose you perform a serial dilution of 0.1-ml sample from a liter of culture medium as illustrated in Figure 6.23. The 10^{-3} plate gives 440 colonies, and the 10^{-4} plate gives 45 colonies. Calculate the approximate number of bacteria in the original liter.
12. How might the study of biofilms benefit humans?
13. The filamentous bacterium *Beggiatoa* gets its carbon from carbon dioxide and its electrons and energy from hydrogen sulfide. What is its nutritional classification?

    Not all organisms are easy to classify. For instance, the single-celled eukaryote *Euglena granulata* typically uses light energy and gets its carbon from carbon dioxide. However, when it is cultured on a suitable medium in the dark, this microbe utilizes energy and carbon solely from organic compounds. What is the nutritional classification of *Euglena*?
14. Given that *Haemophilus ducreyi* is a chemoheterotrophic pathogen that requires heme as a growth factor, deduce how this bacterium phosphorylates most of its ADP to form ATP. Defend your answer.
15. Examine the graph in Figure 6.3. Note that the growth rate increases slowly until the optimum is reached and then declines steeply at higher temperatures. In other words, organisms tolerate a wider range of temperatures below their optimal temperature than they do above the optimum. Explain this observation.
16. Over 100 years ago, doctors infected syphilis victims with malaria parasites to induce a high fever. Surprisingly, such treatment often cured the syphilis infection. Explain how this could occur.
17. Using the terms in Figure 6.8a, describe the shape, margin, pigmentation, and optical properties of two bacterial colonies seen in Figure 6.10b.
18. Why have scientists been unable to axenically culture *Treponema pallidum* in a laboratory medium?
19. Examine the ingredients of MacConkey agar as listed in Table 6.4 (on p. 177). Does this medium select for Gram-positive or Gram-negative bacteria? Explain your reasoning.

20. The sole carbon source in citrate medium is citric acid (citrate). Why might a laboratory microbiologist use this medium?

21. Using as many of the following terms as apply—selective, differential, broth, solid, defined, and complex—categorize each of the media listed in Table 6.3 (on p. 175) and Table 6.4 (on p. 177).

22. Beijerinck used the concept of enrichment culture to isolate aerobic and anaerobic nitrogen-fixing bacteria, sulfate-reducing bacteria, and sulfur-oxidizing bacteria. What kind of selective media could he have used for isolating each of these four types of microbes?

23. Viable plate counts are used to estimate population size when the density of microorganisms is high, whereas membrane filtration is used when the density is low. Why is a viable plate count appropriate when the density is high but not when the density is low?

CONCEPT MAPPING

Using the terms provided, fill in the concept map that describes culture media. You can also complete this and other concept maps online by going to the **Mastering** Microbiology Study Area.

Differential
Enriched
Fastidious microorganisms
Fermentation broths
General purpose
Selective
Selective and differential
Trypticase soy agar
Unwanted microorganisms

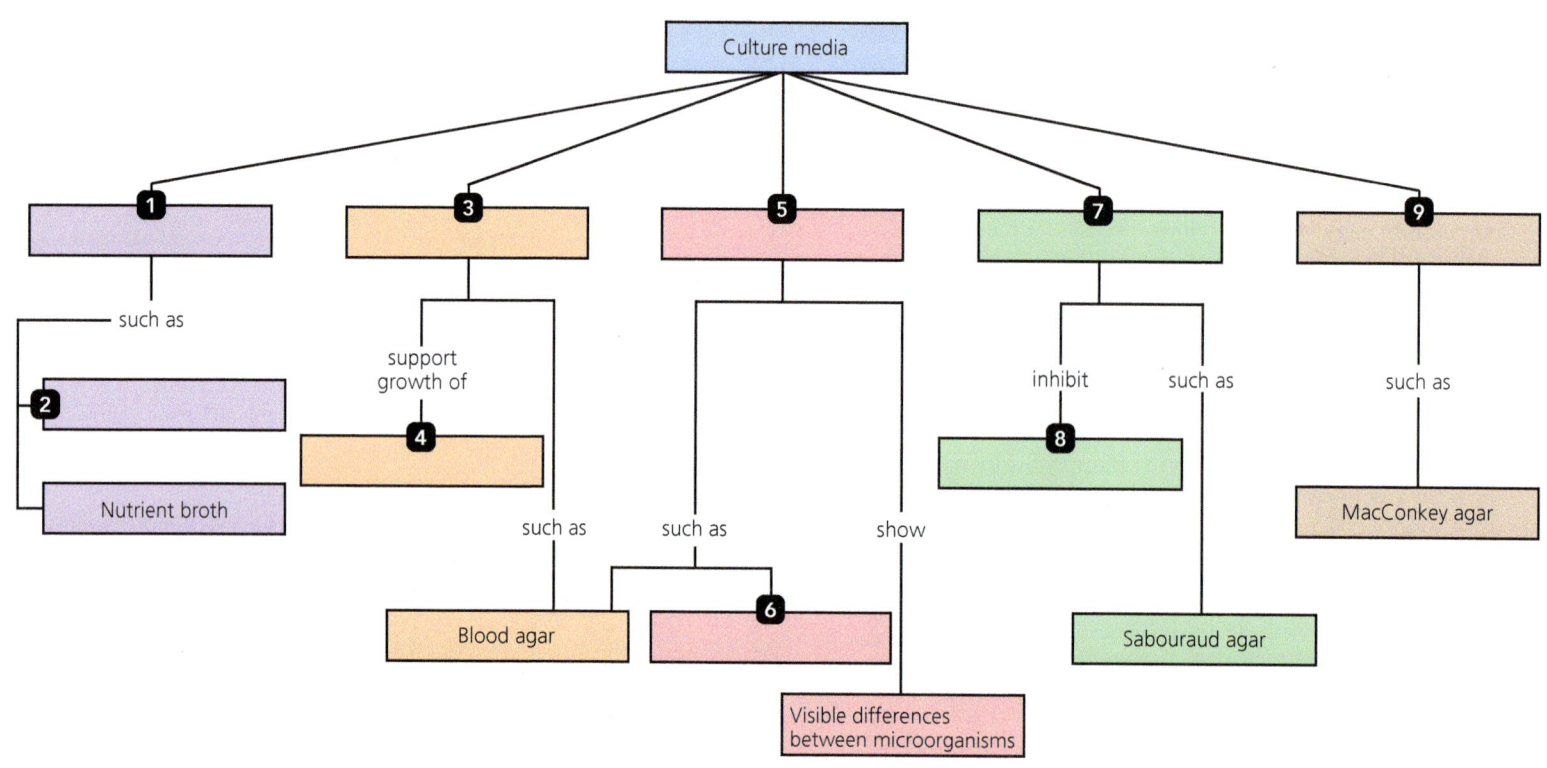

7 Microbial Genetics

Before You Begin

1. Deoxyribonucleic acid (DNA) is the genetic material in cells. Which of the following nucleotide bases are found in DNA: adenine (A), cytosine (C), guanine (G), thymine (T), or uracil (U)?
2. DNA is a double-stranded molecule, with the two strands running in opposite directions. Hydrogen bonds between nucleotides on opposite strands hold the two strands together. If there is a cytosine nucleotide on one strand of the DNA molecule, which nucleotide is found opposite the cytosine?
3. Which of the following bacterial structures transfers DNA horizontally by conjugation: cell wall, flagellum, glycocalyx, or pilus?
4. In this chapter you will learn about the biochemical reactions that synthesize DNA, RNA, and proteins in cells. Are these reactions anabolic or catabolic, and do they require energy or do they produce energy?
5. Do cells need to replicate their genomes and synthesize proteins continuously throughout their lives, or only just before cell division?

MICRO IN THE CLINIC

Another Stroke?

BEN SWANSON, A 68-YEAR-OLD MAN, is returning to his nursing home after spending two weeks in the hospital recovering from a stroke. His doctor is impressed with his progress and thinks that by continuing to work with a physical therapist and rehabilitation nurse, Ben will be able to function as independently as he did before the stroke. Ben is excited to get back to the home—he misses his friends, especially his roommate, Chester. But he remembers the doctor's orders—he's supposed to take it easy and rest, particularly on his first few days back at home.

Ben follows the doctor's instructions—his physical therapist and rehab nurse visit him every day, and he and Chester spend time just chatting or playing cards and dominoes. On Ben's third day home, Chester decides to surprise him by bringing some ice cream from the dining room back to their room as a treat. Chester is shocked when he opens the door to their room and finds Ben lying on the floor. Ben doesn't respond when Chester calls his name; Chester quickly hits the emergency call button located in every room. The nurses and doctor arrive. Chester is certain that Ben has suffered another stroke.

1. Do you think Ben has suffered another stroke?
2. Could there be another explanation for his condition?

Turn to the end of the chapter (p. 230) to find out.

The familiar double helix of DNA provides the genes for all cells and many viruses.

Genetics is the study of inheritance and inheritable traits as expressed in an organism's genetic material. Geneticists study many aspects of inheritance, including the physical structure and function of genetic material, mutations, and the transfer of genetic material among organisms. In this chapter, we will examine these topics as they apply to microorganisms, the study of which has formed much of the basis of our understanding of human, animal, and plant genetics.

The Structure and Replication of Genomes

LEARNING OUTCOME

7.1 Compare and contrast the genomes of prokaryotes and eukaryotes.

The **genome** (jē'nōm) of a cell or virus is its entire genetic complement, including both its **genes**—specific sequences of nucleotides that code for RNA or polypeptide molecules—and nucleotide sequences that connect genes to one another. The genomes of cells and DNA viruses are composed solely of molecules of deoxyribonucleic acid (DNA), whereas RNA viruses use ribonucleic acid instead. (We will examine the genomes of viruses in more detail in Chapter 13.) The remainder of this chapter focuses on bacterial genomes—their structure, replication, function, mutation, and repair and how they compare and contrast with eukaryotic genomes and with the genomes of archaea. We begin by examining the structure of nucleic acids.

The Structure of Nucleic Acids

LEARNING OUTCOME

7.2 Describe the structure of DNA, and discuss how it facilitates the ability of DNA to act as genetic material.

Nucleic acids are polymers of basic building blocks called **nucleotides**. Each nucleotide is made up of phosphate attached to a *nucleoside*. A nucleoside is in turn made up of a pentose sugar (ribose in RNA and deoxyribose in DNA) attached to one of five nitrogenous bases: guanine (G), cytosine (C), thymine (T), adenine (A), or uracil (U) (**FIGURE 7.1**).

The bases of nucleotides hydrogen-bond to one another in specific ways called complementary **base pairs (bp)**: In DNA, the complementary bases thymine and adenine bond to one another with two hydrogen bonds (**FIGURE 7.1a**), whereas in RNA, uracil, not thymine, forms two hydrogen bonds with adenine (**FIGURE 7.1b**). In both DNA and RNA, the complementary bases guanine and cytosine bond to one another with three hydrogen bonds (**FIGURE 7.1c**). The critical function of DNA is to carry information in sequences of nucleotide bases. Just as computer code is a series of zeros and ones with rules to transmit information, so DNA and RNA use specific base sequences to transmit information. They are the "language of life."

Deoxyribonucleotides are linked through their sugars and phosphates to form the two backbones of the familiar, helical, double-stranded DNA (dsDNA) molecule (**FIGURE 7.1d**). The carbon atoms of deoxyribose are numbered 1' (pronounced "one prime") through 5' ("five prime"). One end of a DNA strand is called the 5' end because it terminates in a phosphate group attached to a 5' carbon. The opposite end of that DNA strand terminates with a hydroxyl group bound to a 3' carbon of deoxyribose, so that end is called the 3' end. The two strands are constructed similarly but are oriented in opposite directions to each other; one strand runs in a 5' to 3' direction, while the other runs 3' to 5'. Scientists say the two strands are *antiparallel*. The nucleotide base pairs extend into the middle of the molecule in a way reminiscent of the steps of a spiral staircase. The hydrophobic bases are tucked inside the molecule—away from water—while the hydrophilic sugars and phosphates are outside. This provides an explanation for the double-helical structure of cellular DNA.

The lengths of DNA molecules are not usually given in metric units; instead, the length of a DNA molecule is expressed in base pairs. For example, the genome of an insect-dwelling bacterium tentatively named *Nasuia deltocephalinicola* (nă-soo'ē-ă del-to-sef'ă-lĭn-ĭ-kō-lă) is 112,091 bp long, making it the smallest known cellular genome discovered to date. Most known bacteria have genomes of 1.5 to 6 million base pairs.

The structure of DNA helps explain its ability to act as genetic material. First, the linear sequence of nucleotides carries the instructions for the synthesis of polypeptides and RNA molecules—in much the way a sequence of letters carries information used to form words and sentences. Second, the complementary structure of the two strands allows a cell to make exact copies to pass to its progeny. We will examine the genetic code and DNA replication shortly.

The amount of DNA in a genome can be extraordinary, as some examples will illustrate. A cell of the bacterium *Escherichia coli* (esh-ĕ-rik'ē-ă kō'lī) is approximately 2 μm long, but its genome consists primarily of a 4.6×10^6 bp DNA molecule that is about 1600 μm long—800 times longer than the cell. The human genome has about 6 billion bp in 46 nuclear DNA molecules and numerous copies of a unique mitochondrial DNA molecule, and the entire genome would be about 3 meters (3 million μm) long if all DNA molecules from a single cell were laid end to end. Most of a human cellular genome is packed into a nucleus that is typically only 5 μm in diameter. This is like packing 45 miles of thread into a golf ball and still being able to access any particular section of the thread. To understand how cells package such prodigious amounts of DNA into such small spaces, we must first understand that bacteria, archaea, and eukaryotes package DNA in different ways. We begin by examining the structure of prokaryotic genomes.

The Structure of Prokaryotic Genomes

LEARNING OUTCOMES

7.3 Describe the structure, shape, and number of chromosomes in typical prokaryotic cells.
7.4 Describe the structure and function of plasmids.

The DNA of prokaryotic genomes is found in two structures: chromosomes and plasmids.

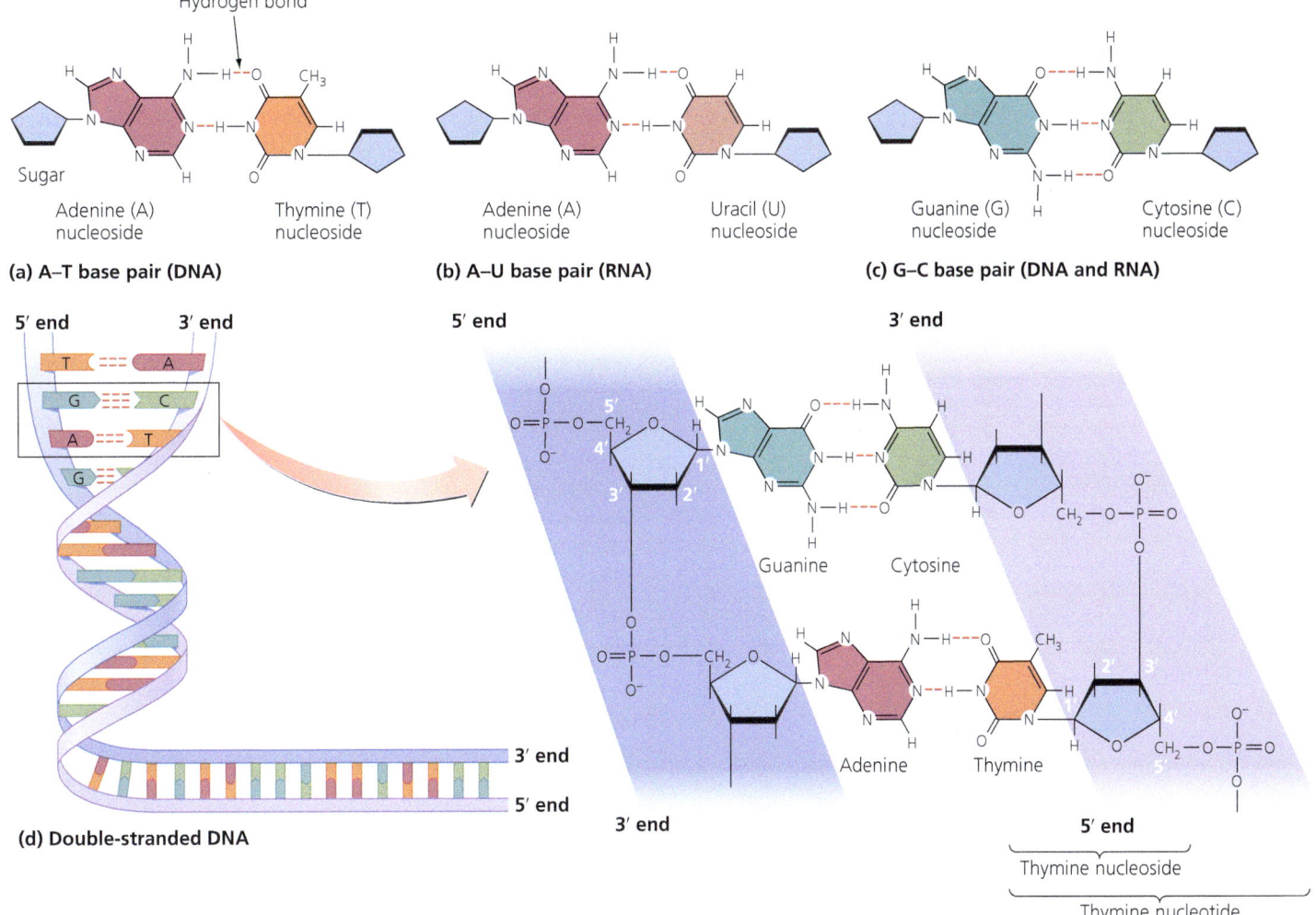

▲ **FIGURE 7.1 The structure of nucleic acids.** Nucleic acids are polymers of nucleotides consisting of nucleosides (a pentose sugar and a nitrogenous base) bound to a phosphate. **(a)** Base pairing between the complementary bases adenine (A) and thymine (T) formed by two hydrogen bonds, found in DNA only. **(b)** Base pairing between adenine and uracil (U), found in RNA only. Notice the structural similarities between thymine and uracil. **(c)** Base pairing between the complementary bases guanine (G) and cytosine (C) formed by three hydrogen bonds, found in both DNA and RNA. **(d)** Double-stranded DNA, which consists of antiparallel strands of nucleotides held to one another by the hydrogen bonding between complementary bases. Note that scientists number the carbon atoms in a nucleotide's sugar component as 1 prime through 5 prime. *What structures do DNA nucleotides and RNA nucleotides have in common?*

Figure 7.1 DNA and RNA nucleotides are each composed of a pentose sugar, a phosphate, and a nitrogenous base.

Prokaryotic Chromosomes

Bacterial cells package the main portion of their DNA, along with associated molecules of protein and RNA, as distinct **chromosomes**.[1] Each cell has a single copy of the genome and is therefore called a *haploid* cell. A typical chromosome consists of a circular molecule of DNA localized in a region of the cytoplasm called the **nucleoid** (FIGURE 7.2a). With few exceptions, no membrane surrounds a nucleoid, though the chromosome is packed in such a way that a distinct boundary is visible between a nucleoid and the rest of the cytoplasm. Chromosomal DNA is folded into loops that are 50,000 to 100,000 bp long (FIGURE 7.2b) held in place by molecules of protein and RNA. Archaeal DNA is also haploid, located in a nucleoid, and typically circular; however, in an archaeal chromosome, DNA is wrapped around globular proteins called **histones**, which makes it different from a bacterial chromosome.

For many years, scientists thought that each prokaryote had only a single circular chromosome, but we now know that there are many exceptions. For example, *Epulopiscium* (ep'yoo-lō-pis'sē-ŭm), a giant bacterium, has as many as hundreds or thousands of identical chromosomes. Other bacterial species

---

[1]From Greek *chroma*, meaning "color" (because they typically stain dark in eukaryotes, where they were first discovered), and *soma*, meaning "body."

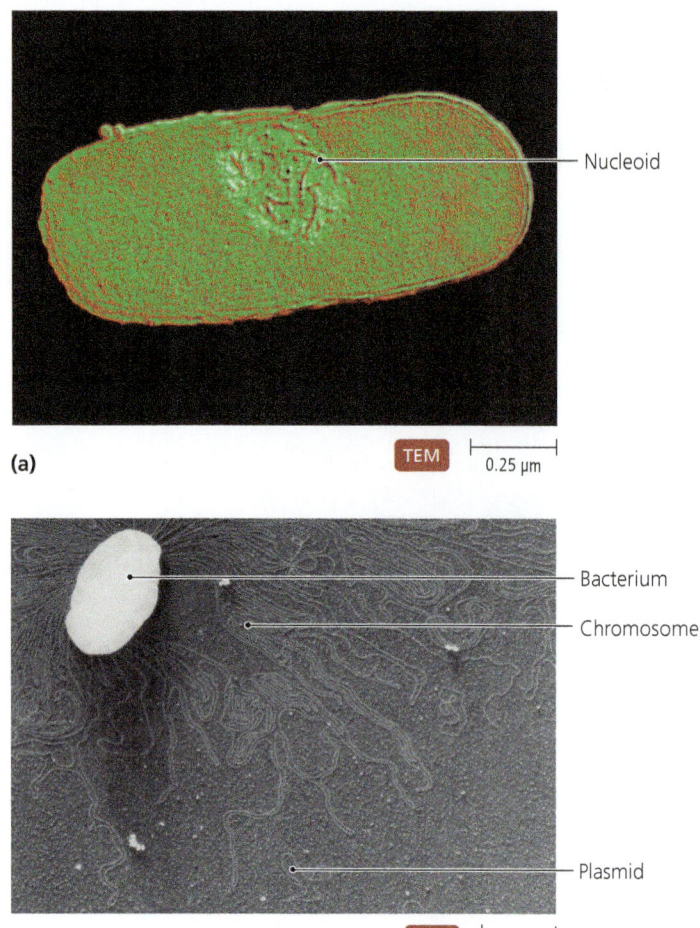

▲ FIGURE 7.2 **Bacterial genome.** (a) Bacterial chromosomes are packaged in a region of the cytosol called the nucleoid, which is not surrounded by a membrane. (b) The packing of a circular bacterial chromosome into loops, as seen after the cell was gently broken open to release the chromosome. Extrachromosomal DNA in the form of plasmids is also visible.

have linear chromosomes. For example, *Agrobacterium tumefaciens* (ag′rō-bak-tēr′ē-um tū′me-fāsh-enz), a bacterium used to transfer genes into plants, has two chromosomes, one circular and one linear.

Plasmids

In addition to chromosomes, many prokaryotic cells contain one or more **plasmids**, which are small molecules of DNA that replicate independently of the chromosome. Plasmids are usually circular and 1% to 20% of the size of a prokaryotic chromosome (see Figure 7.2b), ranging in size from a few thousand base pairs to a few million base pairs. Each plasmid carries information required for its own replication and often for one or more cellular traits. Typically, genes carried on plasmids are not essential for normal metabolism, for growth, or for cellular reproduction but can confer advantages to the cells that carry them.

Researchers have identified many types of plasmids (sometimes called *factors*), including the following:

- *Fertility (F) plasmids* carry instructions for *conjugation*, a process by which some bacterial cells transfer DNA to other bacterial cells. We will consider conjugation in more detail near the end of this chapter.

- *Resistance (R) plasmids* carry genes for resistance to one or more antimicrobial drugs or heavy metals. By processes we will discuss shortly, certain cells can transfer resistance plasmids to other cells, which then acquire resistance to the same antimicrobial chemicals. One example of the effects of an R plasmid involves strains of *E. coli* that have acquired resistance to the antimicrobials ampicillin, tetracycline, and kanamycin from a strain of bacteria in the genus *Pseudomonas* (soo-dō-mō′nas).

- *Bacteriocin plasmids* (bak-tēr′ē-ō-sin) carry genes for proteinaceous toxins called *bacteriocins*, which kill bacterial cells of the same or similar species that lack the plasmid. In this way, a bacterium containing this plasmid can kill its competitors.

- *Virulence plasmids* carry instructions for structures, enzymes, or toxins that enable a bacterium to become pathogenic. For example, *E. coli*, a normal resident of the human gastrointestinal tract, can cause diarrhea when it carries plasmids that code for certain toxins.

Now that we have examined the structure of prokaryotic genomes, we turn to the structure of eukaryotic genomes.

The Structure of Eukaryotic Genomes

LEARNING OUTCOME

7.5 Compare and contrast prokaryotic and eukaryotic chromosomes.

Eukaryotic genomes consist of both nuclear and extranuclear chromosomes. Some eukaryotic cells contain plasmids.

Nuclear Chromosomes

Eukaryotic cells have more than one nuclear chromosome in their genomes, though some eukaryotic cells, such as mammalian red blood cells, lose their chromosomes as they mature. Eukaryotic cells are often *diploid*; that is, they have two copies of each chromosome.

Eukaryotic nuclear chromosomes differ from typical prokaryotic chromosomes in that nuclear chromosomes are all linear (recall that many prokaryotic chromosomes are circular) and are sequestered within a nucleus. A nucleus is an organelle surrounded by two membranes, which together are called the *nuclear envelope*. Given that a typical eukaryotic cell must package substantially more DNA than its prokaryotic counterpart, it is not surprising that its nuclear chromosomes are more elaborate than those of prokaryotes.

Most eukaryotic chromosomes have globular eukaryotic histones,[2] which are similar to archaeal histones. DNA, which has an overall negative electrical charge, wraps around the positively charged histone proteins to form 10-nm-diameter beads called **nucleosomes** (FIGURE 7.3a). Nucleosomes wind into a helix with other proteins to form **chromatin fibers** that are about 30 nm in

---

[2]Dinoflagellates, a group of single-celled aquatic microorganisms, are the only eukaryotes without histones.

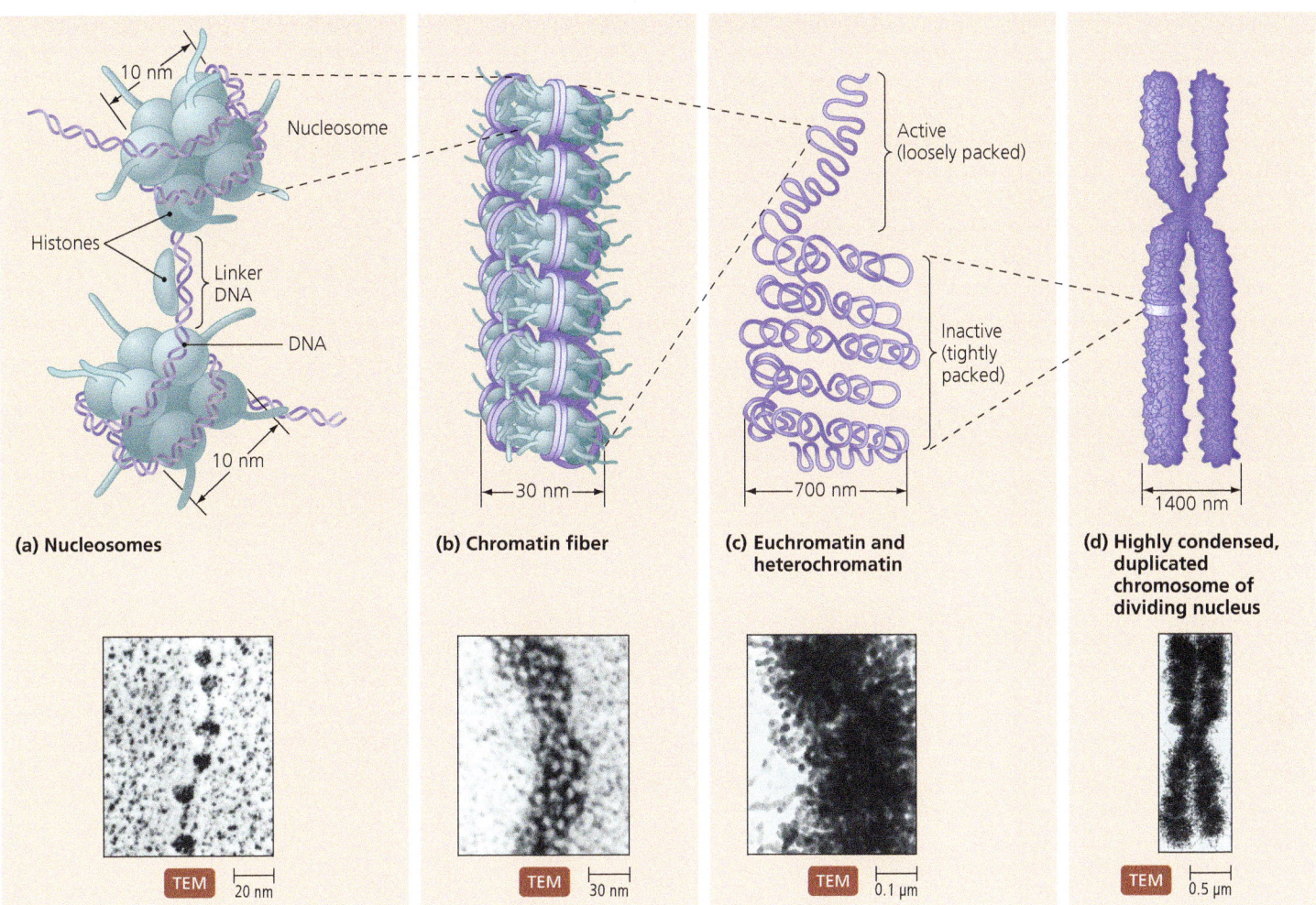

▲ FIGURE 7.3 **Eukaryotic nuclear chromosomal packaging. (a)** Histones stabilize and package DNA to form nucleosomes connected by linker DNA. **(b)** Nucleosomes organize to form chromatin fibers. **(c)** Chromatin fibers fold and are organized into active euchromatin and inactive heterochromatin. **(d)** During nuclear division (mitosis), duplicated chromatin fully condenses into a mitotic chromosome that is visible by light microscopy. If the nucleosomes were actually the size shown in the artist's illustration (a), the chromosome in (d) would be 20 m (about 65 feet) long.

diameter (**FIGURE 7.3b**). Except during *mitosis* (nuclear division), chromatin fibers are dispersed throughout the nucleus and are too thin to be resolved without the extremely high magnification of electron microscopes. In regions of the chromosome where genes are active, chromatin fibers are loosely packed to form *euchromatin* (yū-krō′mă-tin); inactive DNA is more tightly packed and is called *heterochromatin* (het′er-ō-krō′mă-tin) (**FIGURE 7.3c**).

Prior to mitosis, a cell replicates its chromosomes and then condenses them into pairs of chromosomes visible by light microscopy (**FIGURE 7.3d**). One molecule of each pair is destined for each daughter nucleus. (Chapter 12 discusses mitosis in more detail.) The net result is that each DNA molecule is packaged as a mitotic chromosome that is 50,000 times shorter than its extended length.

Extranuclear Chromosomes in Eukaryotes

Not all of the DNA of a eukaryotic genome is contained in its nuclear chromosomes: most eukaryotic cells also have mitochondria; and plant, algal, and some protozoan cells have chloroplasts.

Both mitochondria and chloroplasts contain chromosomes. Chromosomes of mitochondria are circular, and though researchers had thought the chromosomes of chloroplasts were also circular, recent research has shown that chloroplast chromosomes are linear. The extranuclear chromosomes of both mitochondria and chloroplasts resemble the chromosomes of prokaryotes.

Genes located on extranuclear chromosomes code for about 5% of the RNA and polypeptides required for each organelle's replication and function; nuclear DNA codes for the remaining 95% of the organelle's RNA molecules and polypeptides. Some proteins have a quaternary structure formed from the association of individual polypeptides (see Chapter 2, Figure 2.24). Interestingly, mitochondrial or chloroplast chromosomes do not alone code for functional proteins. Rather, mitochondria-coded and chloroplast-coded polypeptides become functional only when associated with polypeptides coded by nuclear chromosomes.

In addition to the extranuclear DNA in their mitochondria and chloroplasts, some fungi, algae, and protozoa carry plasmids. For instance, most strains of brewer's yeast (*Saccharomyces cerevisiae*, sak-ă-rō-mī′sēz se-ri-vis′ē-ī) contain about 70 copies of a plasmid known as a *2-μm circle*. Each 2-μm circle is about 6300 bp long and has four protein-encoding genes that are involved solely in replicating the plasmid and confer no other traits to the cell.

In summary, the genome of a prokaryotic cell consists of haploid chromosomal DNA and all extrachromosomal DNA in the form of plasmids that are present. In contrast, a eukaryotic genome consists of nuclear chromosomal DNA in one or more linear chromosomes, plus all the extranuclear DNA in mitochondria, chloroplasts, and any plasmids that are present. The genomes of prokaryotes and eukaryotes are compared and contrasted in TABLE 7.1.

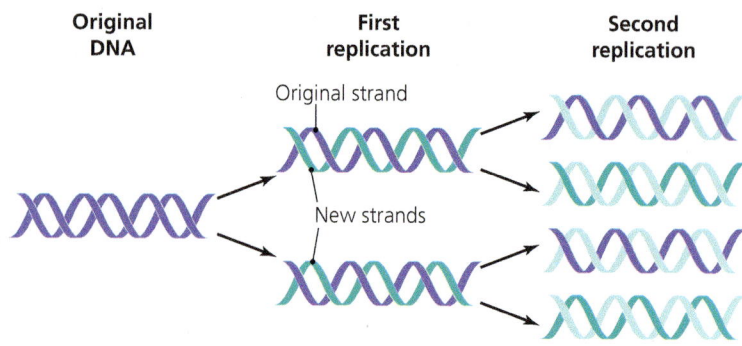

▲ FIGURE 7.4 **Semiconservative model of DNA replication.** Each of the two strands of the original molecule (shown dark) serves as a template for the synthesis of a new, complementary strand (shown lighter).

MICRO CHECK

1. Many bacteria have additional small molecules of DNA in addition to their main genome. What are these small molecules called?
2. What type of cells "package" DNA with proteins called histones?

DNA Replication

LEARNING OUTCOMES

7.6 Describe the replication of DNA as a semiconservative process.
7.7 Explain how the complementarity of nucleotides in opposite DNA strands ensures that DNA replication is precise and accurate.
7.8 Compare and contrast the synthesis of leading and lagging strands in DNA replication.
7.9 Contrast bacterial DNA replication with that of eukaryotes.

DNA replication is an anabolic polymerization process that allows a cell to make copies of its genome. Though bacterial, archaeal, and eukaryotic cells package DNA differently, all three types employ similar mechanisms for DNA replication.

Play *DNA Replication: Overview*
@ Mastering Microbiology

The key to DNA replication is the complementary structure of the two strands: Adenine and guanine in one strand bond with thymine and cytosine, respectively, in the other. DNA replication is a simple concept—a cell separates the two original strands and uses each as a template for the synthesis of a new complementary strand. Biologists say that DNA replication is *semiconservative* because each daughter DNA molecule is composed of one original strand and one new strand (FIGURE 7.4). When replication occurs accurately, the nucleotide sequence of each new daughter strand is precisely the same as the sequence of an original strand.

All polymerization processes require monomers (building blocks) and energy. The monomers of DNA are four *triphosphate deoxyribonucleotides* composed of a base, deoxyribose sugar, and three phosphate groups. DNA nucleotides differ from one another only in the kind of base present. Guanosine triphosphate deoxyribonucleotide (dGTP) has a guanine base (FIGURE 7.5a). Cytidine triphosphate deoxyribonucleotide (dCTP) has a cytosine base. Thymine is the base in thymidine triphosphate deoxyribonucleotide (dTTP), and adenine is the base in adenosine triphosphate deoxyribonucleotide (dATP). The structure of dATP is similar to that of the energy storage molecule ATP, except that ATP uses ribose sugar; it is a ribonucleotide rather than a deoxyribonucleotide (see Figure 2.25b).

The three phosphate groups of a nucleotide are linked together by high-energy bonds. During DNA synthesis, energy is released when two of the phosphate groups are removed

TABLE 7.1 Characteristics of Microbial Genomes

	Bacteria	Archaea	Eukarya
Number of Chromosomes	Single (haploid) copies of one or more	One (haploid)	With one exception, two or more, typically diploid
Plasmids Present?	In some cells; frequently more than one per cell	In some cells	In some fungi, algae, and protozoa
Type of Nucleic Acid	Circular or linear dsDNA	Circular dsDNA	Linear dsDNA in nucleus and chloroplasts; circular dsDNA in mitochondria and plasmids
Location of DNA	In nucleoid of cytoplasm and in plasmids	In nucleoid of cytoplasm and in plasmids	In nucleus and in mitochondria, chloroplasts, and plasmids in cytosol
Histones Present?	No, though chromosome is associated with a small amount of nonhistone protein	Yes	Yes, in nuclear chromosomes; not in extranuclear chromosomes

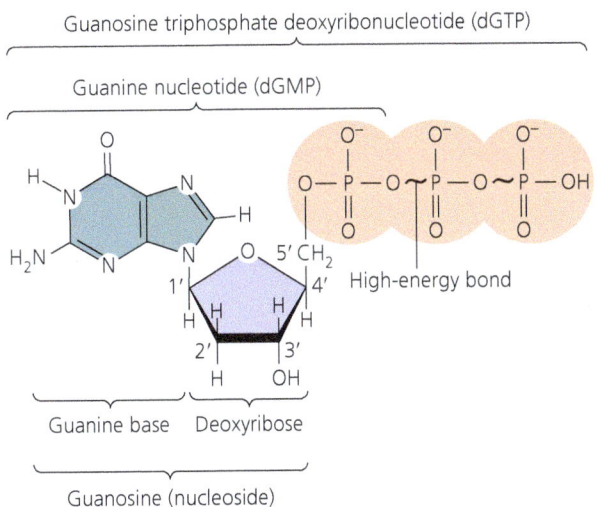

◀ **FIGURE 7.5 The dual role of triphosphate deoxyribonucleotides as building blocks and energy sources in DNA synthesis. (a)** Guanosine triphosphate deoxyribonucleotide (dGTP), like all the triphosphate monomers of DNA, is a nucleotide to which two additional phosphate groups are attached. **(b)** The energy required for DNA polymerization (the addition of nucleotide building blocks to a DNA strand) is carried by each triphosphate nucleotide in the high-energy bonds between phosphate groups. *What is the difference between dGTP and guanosine triphosphate ribonucleotide (rGTP)?*

Figure 7.5 *This molecule (dGTP) contains deoxyribose; rGTP contains ribose (see Figure 2.25b).*

from a nucleotide as a *pyrophosphate* (also called *diphosphate*) (**FIGURE 7.5b**). Thus, a nucleotide is both a building block and an energy source.

The following sections focus on bacterial DNA replication and then consider small differences in the process in eukaryotes. Archaeal processes are not as well characterized and are not examined here.

Initial Processes in Bacterial DNA Replication

DNA replication begins at a specific sequence of nucleotides called an *origin* and follows a series of steps (Figure 7.6a):

1. An enzyme called *DNA helicase* locally "unzips" the DNA molecule by breaking the hydrogen bonds between complementary nucleotide bases, which exposes the nucleotide bases in a *replication fork* (**FIGURE 7.6a**).

2. Other protein molecules stabilize the separated single strands so that they do not rejoin while replication proceeds.

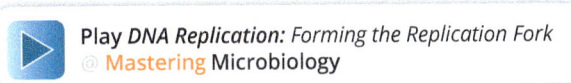
Play *DNA Replication: Forming the Replication Fork*
@ Mastering Microbiology

3. After helicase untwists and separates the strands, *DNA polymerase* (po-lim′er-ās) binds to each strand. Scientists have identified several kinds of bacterial DNA polymerase. These vary in their specific functions, but all of them share one important feature—they synthesize DNA by adding nucleotides only to the hydroxyl group at the 3′ end of a nucleic acid. All DNA polymerases replicate DNA by adding nucleotides in only one direction—5′ to 3′—like a jeweler stringing pearls to make a necklace, adding them one at a time, always moving from one end of the string to the other. DNA polymerase III is the usual enzyme of DNA replication in bacteria.

Because the two original strands are antiparallel, cells synthesize new strands in two different ways. One new strand, called the **leading strand**, is synthesized continuously—5′ to 3′—as a single long chain of nucleotides. The other new strand, called the **lagging strand**, is also synthesized 5′ to 3′, but in short segments that are later joined. We will consider synthesis of the leading strand before examining replication of the lagging strand, but you should understand that the two processes occur simultaneously.

Play *DNA Replication: Replication Proteins*
@ Mastering Microbiology

Continual Synthesis of the Leading Strand

A cell synthesizes a leading strand toward the replication fork in the following series of five steps, the first three of which are shown in **FIGURE 7.6b**:

4. An enzyme called *primase* synthesizes a short RNA molecule (4–15 nucleotides long) that is complementary to the DNA strand being copied. This *RNA primer* provides a 3′ hydroxyl group, which is required by DNA polymerase III.

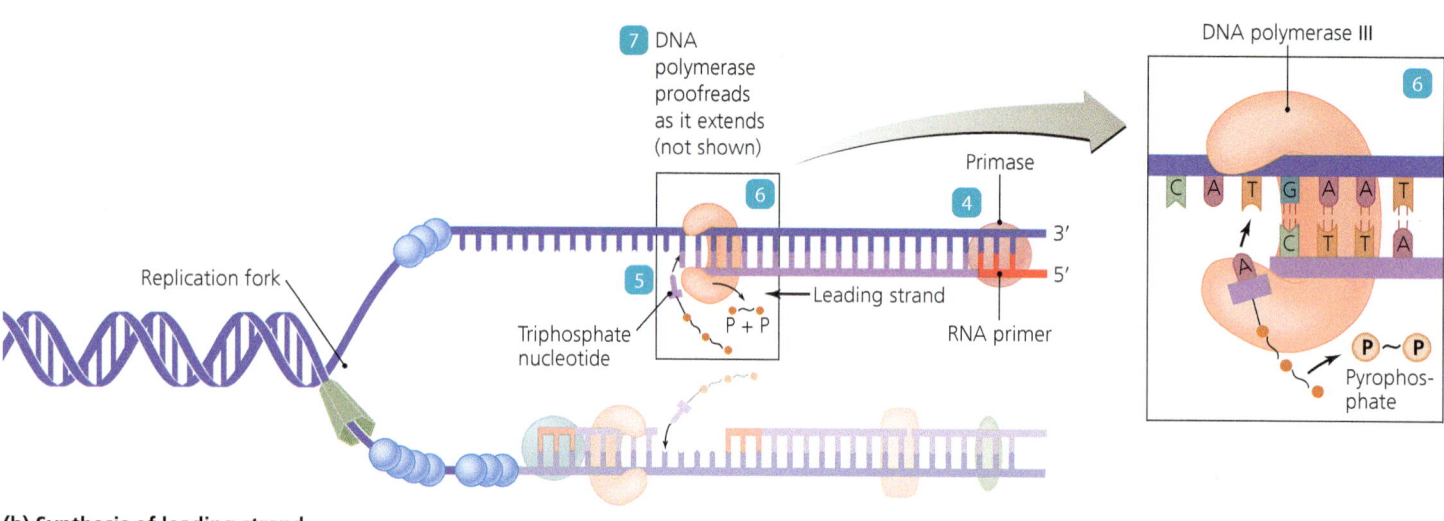

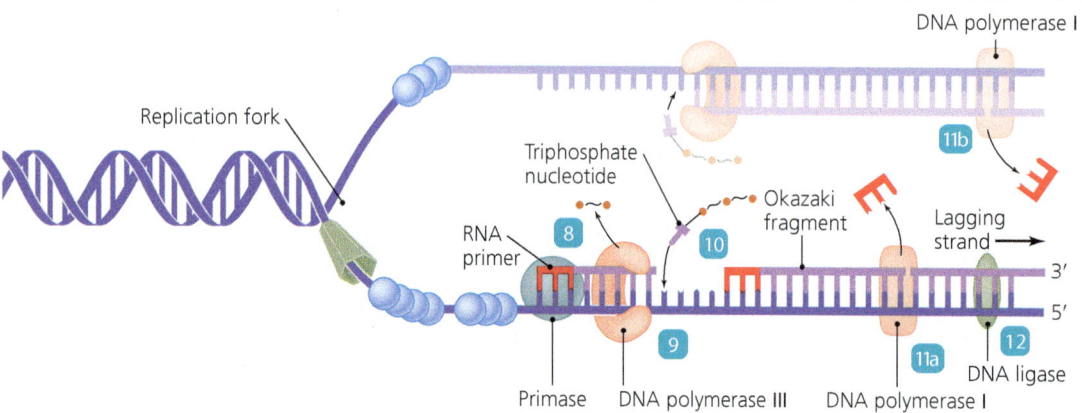

▲ FIGURE 7.6 **DNA replication.** (a) Initial processes. The cell removes proteins (called *histones* in eukaryotes and archaea) from the DNA molecule. Helicase unzips the double helix—breaking hydrogen bonds between complementary base pairs—to form a replication fork. (b) Continuous synthesis of the leading strand. DNA synthesis always moves in the 5′ to 3′ direction, so the leading strand is synthesized toward the replication fork. The numbers refer to the steps in the process, which are described in the text. The proofreading function of DNA polymerase III occurs at step 7, though proofreading is not detailed in the artwork. Replacement of the leading strand's RNA primer with DNA is shown in step 11b. (c) Discontinuous synthesis of the lagging strand, which proceeds moving away from the replication fork. Actual Okazaki fragments are about 1000 nucleotides long. *Why is DNA replication termed "semiconservative"?*

Figure 7.6 *"Semiconservative"* refers to the fact that each of the daughter molecules retains one parental strand and has one new strand; in other words, each is half new and half old.

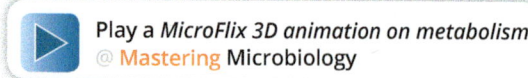

5. Triphosphate deoxyribonucleotides form hydrogen bonds with their complements in the parental strand. Adenine nucleotides bind to thymine nucleotides, and guanine nucleotides bind to cytosine nucleotides.

6. Using the energy in the high-energy bonds of the triphosphate deoxyribonucleotides, DNA polymerase III covalently joins them one at a time to the leading strand. DNA polymerase III can add about 500 to 1000 nucleotides per second to a new strand.

7. DNA polymerase III also performs a proofreading function. About one out of every 100,000 nucleotides is mismatched with its template; for instance, a guanine might become incorrectly paired with a thymine. DNA polymerase III recognizes most of these errors and removes the incorrect nucleotides before proceeding with synthesis. This role, known as the *proofreading exonuclease function*, acts like the backspace key on a keyboard, removing the most recent error. Because of this proofreading exonuclease function and other repair strategies beyond the scope of this discussion, only about one error remains for every 10 billion (10^{10}) bp replicated, so cells make an almost exact copy of their genome.

As replication continues, another DNA polymerase—DNA polymerase I—replaces the RNA primer with DNA nucleotides (as shown in FIGURE 7.6c, steps 11a and 11b). Note that researchers named DNA polymerase enzymes in the order of their discovery, not the order of their actions.

Fragmented Synthesis of the Lagging Strand

Because DNA polymerase III adds nucleotides only to the 3' end of the new strand, the enzyme moves away from the replication fork as it synthesizes a lagging strand. As a result, the lagging strand is synthesized discontinuously and always lags behind the process occurring in the leading strand. The steps in the synthesis of a lagging strand are as follows (FIGURE 7.6c):

8. Primase synthesizes RNA primers, but in contrast to its action on the leading strand, primase synthesizes multiple primers—one every 1000 to 2000 DNA bases of the template strand.

9. DNA polymerase III joins neighboring nucleotides and proofreads. In contrast to synthesis of the leading strand, however, the lagging strand is synthesized in discontinuous segments called *Okazaki fragments*, named for the Japanese scientist Reiji Okazaki (1930–1975), who first identified them. Each Okazaki fragment uses one of the new RNA primers, so each fragment consists of 1000 to 2000 nucleotides.

10. Nucleotides pair up with their complements in the template—adenine with thymine, and cytosine with guanine.

11. DNA polymerase I replaces the RNA primers of Okazaki fragments as well as the RNA primer in the leading strand with DNA and proofreads the short DNA segment it has just synthesized.

12. *DNA ligase* seals the nicks between adjacent Okazaki fragments to form a continuous DNA strand.

In summary, synthesis of the leading strand proceeds continuously toward the replication fork from a single RNA primer at the origin, following helicase and the replication fork down the DNA. The lagging strand is synthesized away from the replication fork discontinuously as a series of Okazaki fragments, each of which begins with its own RNA primer. All the primers are eventually replaced with DNA nucleotides, and ligase joins the Okazaki fragments.

As noted earlier, DNA replication is semiconservative; each daughter molecule is composed of one parental strand and one daughter strand. The replication process produces double-stranded daughter molecules with a nucleotide sequence virtually identical to that in the original double helix, ensuring that the integrity of an organism's genome is maintained each time it is copied.

Play *DNA Replication:* Synthesis
@ Mastering Microbiology

Other Characteristics of Bacterial DNA Replication

DNA replication is *bidirectional*; that is, DNA synthesis proceeds in both directions from the origin. In bacteria, the process of replication typically proceeds from a single origin, so it involves two sets of enzymes and two replication forks, each having a leading strand and a lagging strand (FIGURE 7.7). One replication fork moves clockwise and the other moves counterclockwise until replication ends about 180° from the origin at termination sequences in the DNA molecule.

The unzipping and unwinding action of helicase introduces supercoils into the DNA molecule ahead of the replication forks. Excessive supercoiling creates tension on the DNA molecule—like your grandmother's overwound phone cord—and would stop DNA replication. The enzymes *gyrase* and *topoisomerase* remove such supercoils by cutting the DNA, rotating the cut ends in the direction opposite the supercoiling, and then rejoining the cut ends.

DNA replication is further complicated by **methylation** of the daughter strands, in which an enzyme adds a methyl group to one or two bases that are part of specific nucleotide sequences. Bacteria typically methylate adenine bases and only rarely a cytosine base.

Methylation plays a role in a variety of cellular processes, including the following:

- *Control of genetic expression.* In some cases, genes that are methylated are "turned off" and are not transcribed, whereas in other cases methylated genes are "turned on" and are transcribed.
- *Initiation of DNA replication.* In many bacteria, methylated nucleotide sequences play a role in initiating DNA replication.
- *Protection against viral infection.* Methylation at specific sites in a nucleotide sequence enables cells to distinguish their DNA from viral DNA, which lacks methylation. The cells can then selectively degrade viral DNA.
- *Repair of DNA.* The role of methylation in some DNA repair mechanisms is discussed later in the chapter (p. 221).

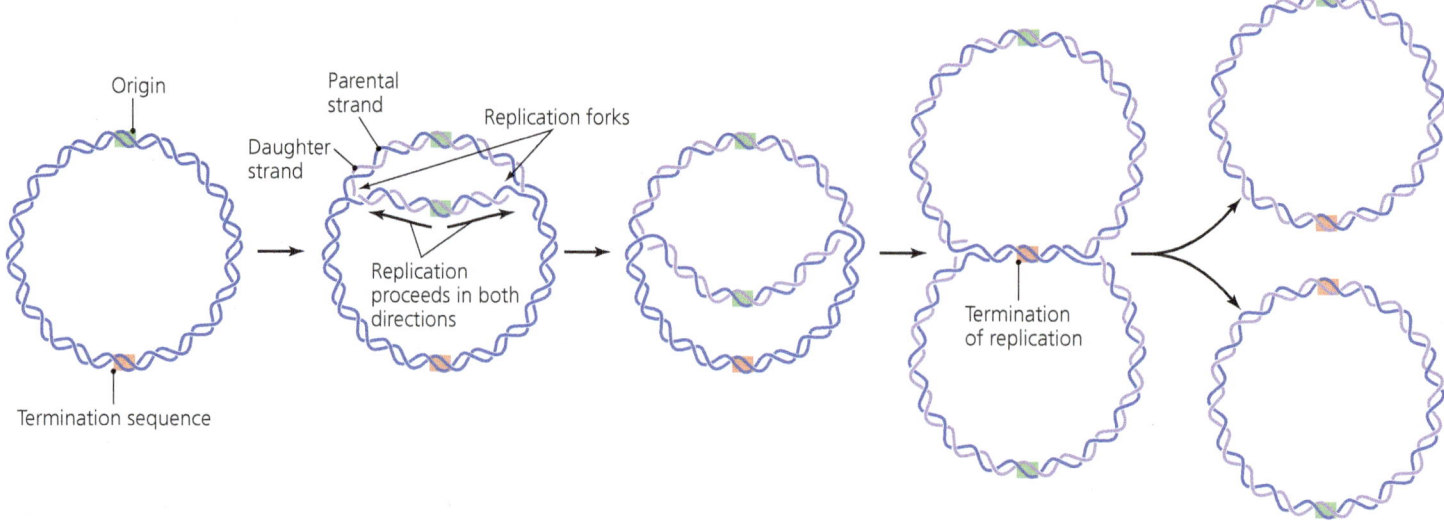

▲ FIGURE 7.7 **The bidirectionality of DNA replication in prokaryotes.** Replication begins at an origin and proceeds in both directions to a termination sequence. The origin in *E. coli* is 80 bp in size. Bacterial chromosomes (shown here) have a single origin, but eukaryotic chromosomes have thousands of origins.

Replication of Eukaryotic DNA

Eukaryotes replicate DNA in much the same way as do bacteria; helicases and topoisomerases unwind DNA, protein molecules stabilize single-stranded DNA, and molecules of DNA polymerase synthesize leading and lagging strands simultaneously. However, eukaryotic replication differs from prokaryotic replication in some significant ways:

- Eukaryotic cells use different DNA polymerases to replicate DNA. There are four basic types: One initiates replication, including synthesis of a primer—the function performed by primase in bacteria. A second type of eukaryotic DNA polymerase elongates the leading strand, and a third appears to be responsible for replicating the lagging strand. A fourth type replicates mitochondrial DNA.
- The large size of eukaryotic chromosomes necessitates thousands of origins per molecule, each generating two replication forks; otherwise, the replication of eukaryotic genomes would take days instead of hours.
- Eukaryotic Okazaki fragments are shorter than those of bacteria—100 to 400 nucleotides long.
- Plant and animal cells methylate cytosine bases exclusively.
- DNA replication in eukaryotic cells precedes the process called mitosis, which includes the critical step of separation of chromosome pairs by the mitotic spindle apparatus.

MICRO CHECK

3. How does a cell ensure accurate DNA replication?
4. Which enzyme is responsible for replacing RNA primers on newly synthesized DNA strands and is needed much more frequently on the lagging strand than on the leading strand?
5. During DNA synthesis, what enzyme works more or less alone to synthesize the leading strand?

TELL ME WHY

DNA replication requires a large amount of energy, yet none of a cell's ATP energy supply is used. Why isn't it?

We have examined the physical structure of cellular genes—the specific sequences of DNA nucleotides—and the way cells replicate their genes. Now we will consider how genes function and how cells control genetic expression.

Gene Function

The first topic we must consider if we are to understand gene function is the relationship between an organism's genotype and its phenotype.

The Relationship Between Genotype and Phenotype

LEARNING OUTCOME

7.10 Explain how the genotype of an organism determines its phenotype.

The **genotype**[3] (jēn′ō-tīp) of an organism is the actual set of genes in its genome. A genotype differs from a genome in that a genome also includes nucleotides that are not part of genes, such as the nucleotide sequences that link one gene to another. At the molecular level, the genotype consists of all the series of DNA nucleotides that carry instructions for an organism's life.

Phenotype[4] (fē′nō-tīp) refers to the physical features and functional traits of an organism, including characteristics such

---
[3] From Greek *genos*, meaning "race," and *typos*, meaning "type."
[4] From Greek *phainein*, meaning "to show."

as structures, morphology, and metabolism. For example, the shape of a cell, the presence and location of flagella, the enzymes and cytochromes of electron transport chains, and membrane receptors that trigger chemotaxis are all phenotypic traits.

Genotype determines phenotype by specifying what kinds of RNA and which structural, enzymatic, and regulatory protein molecules are produced. Though genes do not code *directly* for such molecules as phospholipids or for behaviors such as chemotaxis, ultimately phenotypic traits result from the actions of RNA and protein molecules that are themselves coded by DNA. Not all genes are active at all times; that is, the information of a genotype is not always expressed as a phenotype. For example, *E. coli* activates genes for lactose catabolism only when it detects lactose in its environment.

The Transfer of Genetic Information

LEARNING OUTCOME

7.11 State the central dogma of genetics, and explain the roles of DNA and RNA in polypeptide synthesis.

Cells must continually synthesize proteins required for growth, reproduction, metabolism, and regulation. This synthesis requires that they accurately transfer the genetic information contained in DNA nucleotide sequences to the amino acid sequences of proteins. Cells do not convert the information coded in DNA directly into proteins but first make an RNA copy of the gene. In this copying process, called **transcription**,[5] the information in DNA is copied as RNA nucleotide sequences; RNA molecules in ribosomes then synthesize polypeptides in a process called **translation**.[6] These processes make up the **central dogma** of genetics: DNA is transcribed to RNA, which is translated to form polypeptides (**FIGURE 7.8**).

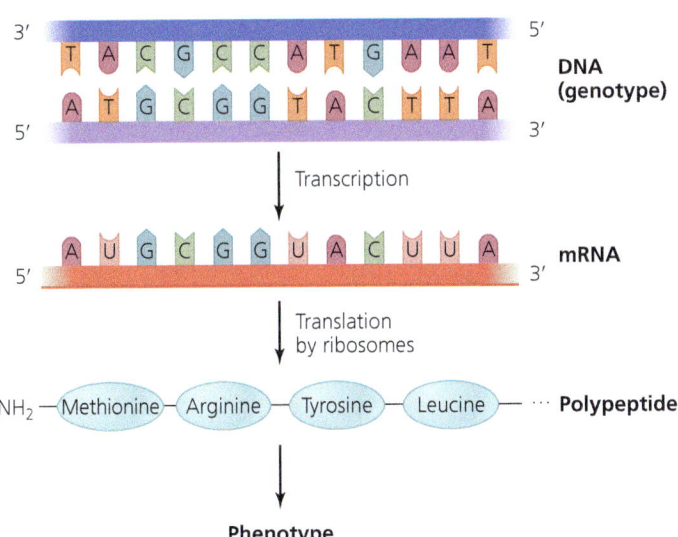

▲ **FIGURE 7.8 The central dogma of genetics.** A cell transcribes RNA from a DNA gene and then synthesizes polypeptides by translating the code carried by the RNA molecules. Polypeptides determine phenotype by acting as structural, enzymatic, and regulatory proteins.

---
[5]From Latin *trans*, meaning "across," and *scribere*, meaning "to write"—that is, to transfer in writing.

[6]From Latin *translatus*, meaning "transferred."

An analogy serves to illustrate the central dogma. Suppose you were trying to understand the following message (which is a portion of the oath of Hippocrates, written in the Greek alphabet):

ΔΙΑΙΤΗΜΑΣΙΤΕΧΡΗΣΟΜΑΙΕΠΩΦΕΛΕΙΝ
ΚΑΜΝΟΝΤΩΝΚΑΤΑΔΥΝΑΜΙΝΚΑΙΚΡΙΣΙΝΕΜΗΝ
ΕΠΙΔΗΛΗΣΕΙΔΕΚΑΙΑΔΙΚΙΗ

If the Greek alphabet is foreign to you, you might have the Greek characters *transcribed* into the familiar English alphabet as a first step in understanding the message:

Diaiteimasi te chreisomai ep ophelein
kamnonton kata dunamin kai krisin emein
epi deileisei de kai adikiei eirzein

Then you could begin the process of having the Greek words, now expressed in English letters, *translated* into English words:

I will prescribe treatment to the best of my ability and judgment to help the sick and never for a harmful or illicit purpose

To a ribosome, DNA is like a foreign language written in a foreign alphabet. Thus, a cell must use processes analogous to those just described: It must first *transcribe* the "foreign alphabet" of DNA nucleotides (genes) into the more "familiar alphabet" of RNA nucleotides; then it must *translate* the message formed by these "letters" into the "words" (amino acids) that make up the "message" (a polypeptide). In this way, a genotype can be expressed as a phenotype. There are a few exceptions to the central dogma. For example, some RNA viruses transcribe DNA from an RNA template—a process that is the reverse of cellular transcription.

In the following sections, we will examine the processes of transcription and translation.

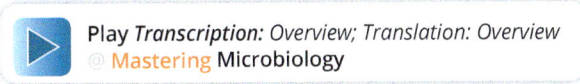

The Events in Transcription

LEARNING OUTCOMES

7.12 Describe three steps in RNA transcription, mentioning the following: DNA, RNA polymerase, promoter, initiation, 5′ to 3′ direction, terminator, and Rho.
7.13 Contrast bacterial transcription with that of eukaryotes.

Cells transcribe six main types of RNA from DNA:

- **RNA primer** molecules for DNA polymerase to use during DNA replication
- **Messenger RNA (mRNA)** molecules, which carry genetic information from chromosomes to ribosomes
- **Ribosomal RNA (rRNA)** molecules, which combine with ribosomal polypeptides to form ribosomes—the organelles that synthesize polypeptides
- **Transfer RNA (tRNA)** molecules, which deliver the correct amino acids to ribosomes based on the sequence of nucleotides in mRNA

- **Regulatory RNA** molecules, which interact with DNA to control gene expression
- **Ribozymes**, which are RNA molecules that function as metabolic enzymes in a cell

We have already considered the role of RNA primer in DNA replication and will more closely examine the functions of the other types of RNA shortly. Next we examine transcription in bacteria and contrast it with eukaryotic transcription; archaeal processes are not as well known.

Transcription occurs in the nucleoid region of the cytoplasm in bacteria. The three steps of RNA transcription are ① *initiation of transcription*, ② *elongation of the RNA transcript*, and ③ *termination of transcription*. FIGURE 7.9 depicts the events in transcription.

Play *Transcription: The Process* @ Mastering Microbiology

Initiation of Transcription

RNA polymerases—the enzymes that synthesize RNA—bind to specific DNA nucleotide sequences called **promoters**, each of which is located near the beginning of a gene and serves to initiate transcription (①a in FIGURE 7.9a). In bacteria, a polypeptide subunit of RNA polymerase called a *sigma factor* is necessary for recognition of a promoter. Once it adheres to a promoter sequence, an RNA polymerase unzips and unwinds the DNA molecule in the promoter region and then travels along the DNA, unzipping the double helix to form a "bubble" as it moves ①b. Sigma factor often drops off the RNA polymerase as transcription proceeds, though it remains attached in some bacteria.

Cells may use different sigma factors and different promoter sequences to provide some control over transcription. RNA polymerases using different sigma factors do not adhere equally strongly to all promoters; there is about a millionfold difference between the strongest attraction and the weakest one. The greater the attraction between a particular sigma factor and a promoter, the more likely that a particular gene will be transcribed; thus, variations in sigma factors and promoters affect the amounts and kinds of polypeptides produced.

Elongation of the RNA Transcript

RNA transcription does not actually begin in the promoter region but, rather, at a spot 10 nucleotides away. There, triphosphate ribonucleotides (ATP, UTP, GTP, and CTP) align opposite their complements in the open DNA "bubble." RNA polymerase links together two adjacent ribonucleotide molecules using energy from the phosphate bonds of the first ribonucleotide (② in FIGURE 7.9b). The enzyme then moves down the DNA strand, elongating RNA by repeating the process. Only one of the separated DNA strands is transcribed.

In prokaryotes, many molecules of RNA polymerase may concurrently transcribe the same gene (FIGURE 7.10). In this way, a cell simultaneously produces numerous identical copies of RNA from a single gene—much as many identical prints can be made from a single photographic negative.

Like DNA polymerase, RNA polymerase links nucleotides only to the 3' end of the growing molecule; that is, transcription is in the 5' to 3' direction. However, RNA polymerase differs from DNA polymerase in the following ways:

- RNA polymerase unwinds and opens DNA by itself; helicase is not required.
- RNA polymerase does not need a primer.
- RNA polymerase copies only one of the DNA strands.
- RNA polymerase is slower than DNA polymerase III, proceeding at a rate of about 50 nucleotides per second.
- RNA polymerase incorporates ribonucleotides instead of deoxyribonucleotides.
- Uracil nucleotides are incorporated instead of thymine nucleotides.
- The proofreading function of RNA polymerase is less efficient, leaving a base-pair error about every 10,000 nucleotides.

Termination of Transcription

Transcription terminates when RNA polymerase and the transcribed RNA are released from DNA (FIGURE 7.9c). RNA polymerase is tightly associated with the DNA molecule and cannot be removed easily; therefore, the termination of transcription is complicated. Scientists have elucidated two types of termination processes in bacteria—those that are self-terminating and those that depend on the action of an additional protein called *Rho*. These processes of transcription termination should not be confused with termination of translation, which we examine in a later section.

Self-Termination Self-termination occurs when RNA polymerase transcribes a **terminator** sequence of DNA composed of two regions: one that is symmetrical in guanine and cytosine bases, followed by a region rich in adenine bases (see ③a in Figure 7.9c). RNA polymerase slows down during transcription of the GC-rich portion of the terminator because the three hydrogen bonds between each guanine and cytosine base pair make unwinding the DNA helix more difficult. This pause in transcription, which lasts about 60 seconds, provides enough time for the RNA molecule to form hydrogen bonds between its own symmetrical sequences, forming a hairpin loop structure that puts tension on the union of RNA polymerase and the DNA. When RNA polymerase transcribes the adenine-rich portion of the terminator, the relatively few hydrogen bonds between the adenine bases of DNA and the uracil bases of RNA cannot withstand the tension, and the RNA transcript breaks away from the DNA, releasing RNA polymerase.

Rho-Dependent Termination The second type of termination depends on the termination protein called *Rho*. Rho binds to a specific RNA sequence near the end of an RNA transcript. Rho moves toward RNA polymerase at the 3' end of the growing RNA molecule, pushing between RNA polymerase and the DNA strand and forcing them apart; this releases RNA polymerase, the RNA transcript, and Rho (see ③b in Figure 7.9c).

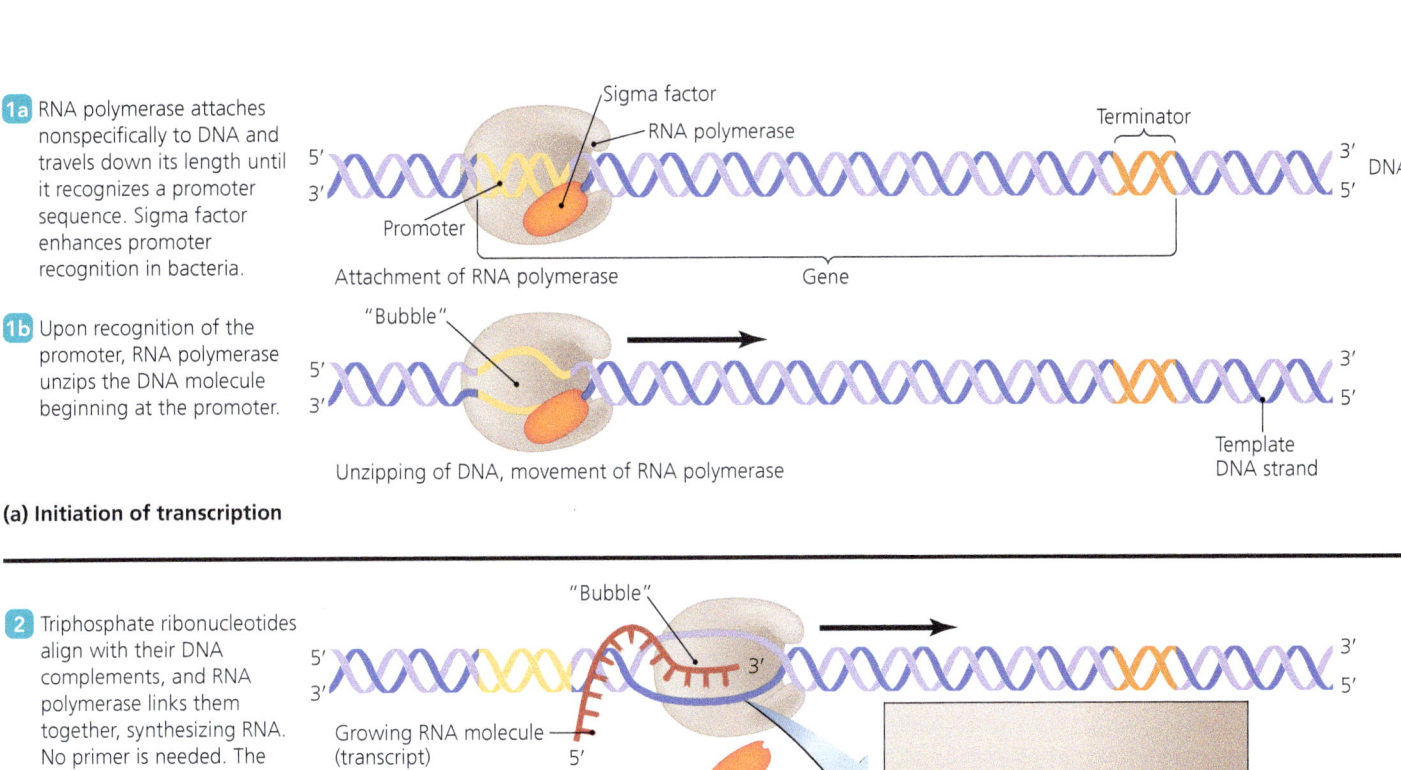

1a RNA polymerase attaches nonspecifically to DNA and travels down its length until it recognizes a promoter sequence. Sigma factor enhances promoter recognition in bacteria.

1b Upon recognition of the promoter, RNA polymerase unzips the DNA molecule beginning at the promoter.

(a) Initiation of transcription

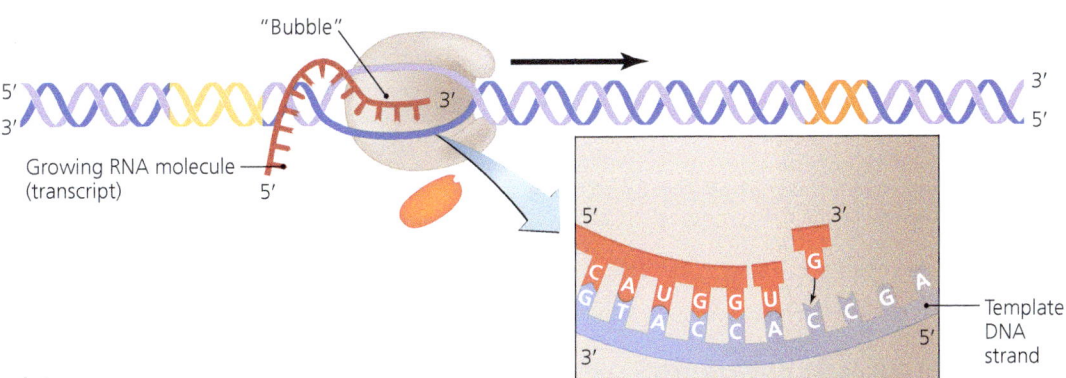

2 Triphosphate ribonucleotides align with their DNA complements, and RNA polymerase links them together, synthesizing RNA. No primer is needed. The triphosphate ribonucleotides also provide the energy required for RNA synthesis.

(b) Elongation of the RNA transcript

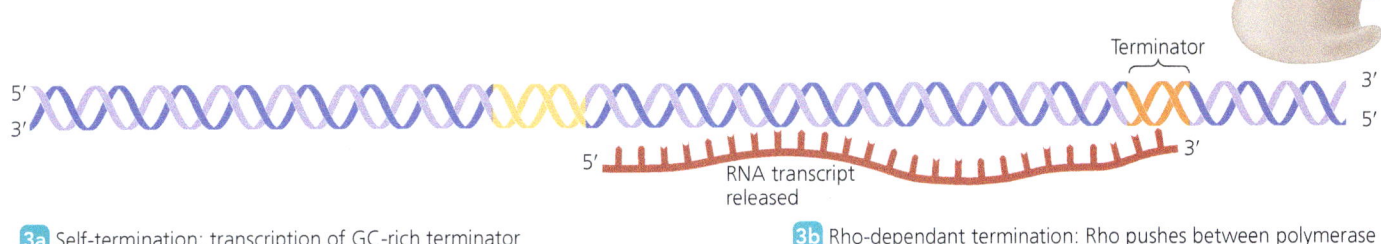

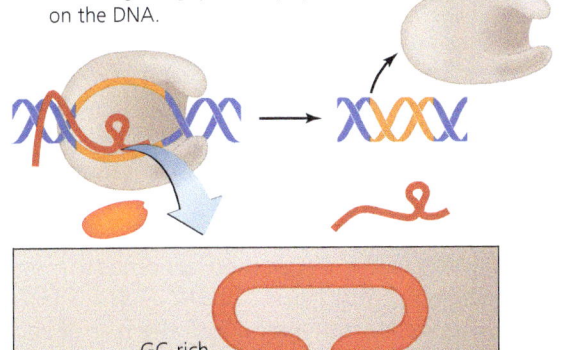

3a Self-termination: transcription of GC-rich terminator region produces a hairpin loop, which creates tension, loosening the grip of the polymerase on the DNA.

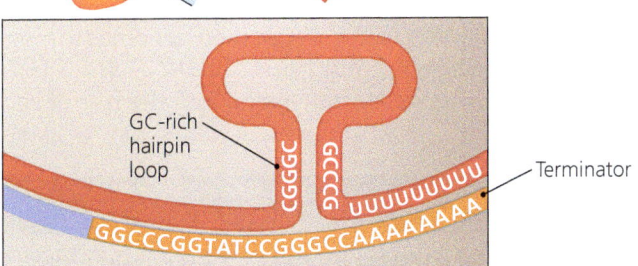

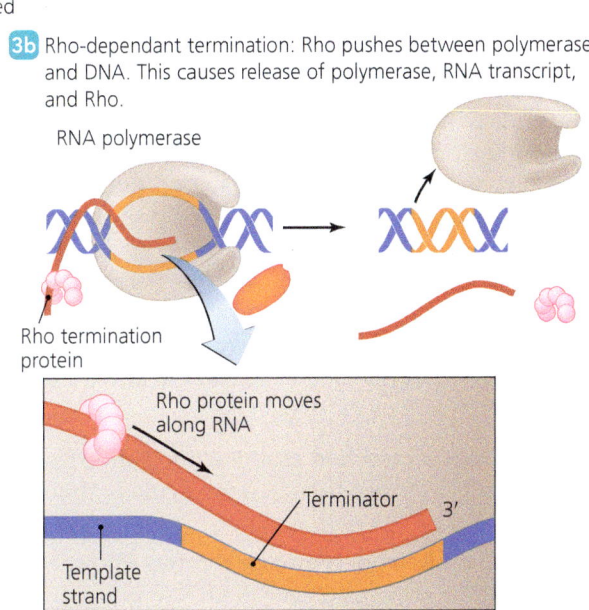

3b Rho-dependant termination: Rho pushes between polymerase and DNA. This causes release of polymerase, RNA transcript, and Rho.

(c) Termination of transcription: release of RNA polymerase

▲ **FIGURE 7.9 The events in the transcription of RNA in prokaryotes.** (a) Initiation of transcription. (b) Elongation of the RNA transcript. (c) Termination of transcription: release of RNA polymerase by one of two methods. *What is the difference between a promoter sequence and an origin?*

Figure 7.9 A promoter is a DNA sequence that initiates transcription; an origin is a point where DNA replication begins.

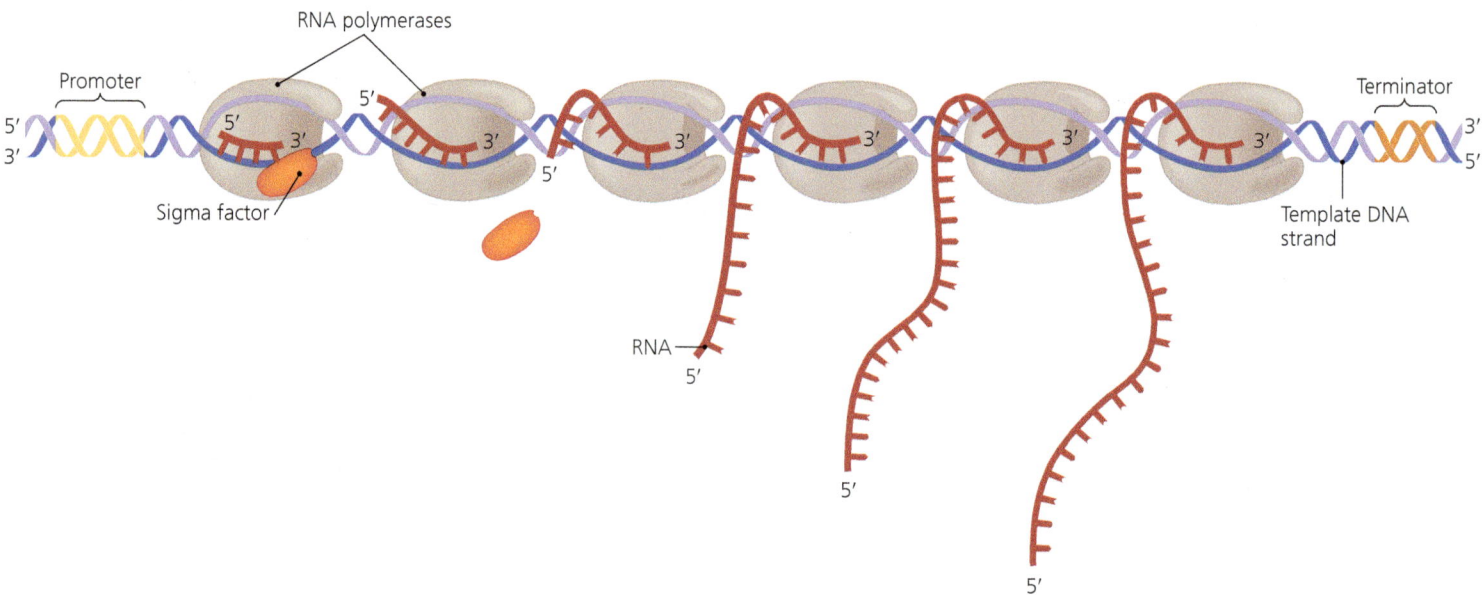

▲ **FIGURE 7.10 Concurrent RNA transcription.** Once an RNA polymerase molecule has cleared the promoter, another molecule can recognize the promoter and initiate transcription. In this manner, multiple copies of RNA are transcribed simultaneously. The art shows six molecules of RNA being transcribed at the same time along a gene.

Transcriptional Differences in Eukaryotes

Eukaryotic transcription differs from bacterial transcription in several ways. First, a eukaryotic cell transcribes RNA inside its nucleus, primarily in the region of the nucleus called the *nucleolus*, as well as inside any mitochondria and chloroplasts that are present. In contrast, transcription in prokaryotes occurs in the cytosol.

Another difference between eukaryotes and bacteria is that eukaryotes have three types of nuclear RNA polymerase—one for transcribing mRNA, one for transcribing the major rRNA gene, and one for transcribing tRNA and smaller rRNA molecules. Mitochondria use a fourth type of RNA polymerase. Further, several separate protein *transcription factors* at a time assist in binding eukaryotic RNA polymerase to promoter sequences, in contrast to a single sigma factor in bacteria. After initiating transcription, eukaryotic RNA polymerases shed most of the transcription factors and recruit another set of polypeptides called *elongation factors*.

Finally, eukaryotic cells must process mRNA before beginning polypeptide translation (**FIGURE 7.11**). In general, the functions of RNA processing are to aid in export of RNA from the nucleus, to stabilize mRNA in the cytoplasm, and to aid in translation. RNA processing involves three events:

1. *Capping.* The cell adds a modified guanine nucleotide to the 5′ end of the mRNA when the RNA molecule is about 30 nucleotides long.
2. *Polyadenylation.* When RNA polymerase reaches the end of a gene, termination proteins cleave the RNA molecule and add 100 to 250 adenine nucleotides, depending on the organism, to the 3′ end. Polyadenylation occurs without a DNA template.
3. *Splicing.* Newly capped and polyadenylated mRNA molecules are called *pre-messenger RNA (pre-mRNA)* because they contain **introns**, which are noncoding sequences that may be thousands of nucleotides long. (Few prokaryotic mRNA molecules contain introns.) A cell removes introns to make functional mRNA containing only coding regions called **exons**, each of which is about 150 nucleotides long. The "in" in *intron* refers to *intervening* sequences (i.e., they lie between coding regions), whereas the "ex" in *exon* refers to the fact that these regions are expressed. Five small RNA molecules associate with about 300 polypeptides to form a *spliceosome* that acts as a ribozyme (ribosomal enzyme) to splice pre-mRNA into mRNA—it removes introns and splices the exons to produce a functional mRNA molecule that exits the nucleus.

A major advantage of splicing is the ability to form multiple proteins from one gene, called *alternative splicing*. For example, Figure 7.11 shows the joining of exons 1, 2, 3, and 4. However, a different protein could be made from the same gene by splicing together 1, 2, and 4 (skipping exon 3). This is how human cells generate over 150,000 proteins from only 25,000 genes.

Now that we have discussed how cells use DNA as the genetic material, maintain the integrity of their genomes through semiconservative replication, and transcribe RNA from DNA genes, we turn to the process of translation and the role of each type of RNA.

MICRO CHECK

6. Which step in the pathway from DNA to polypeptides produces mRNA?
7. Where does transcription occur in prokaryotes?

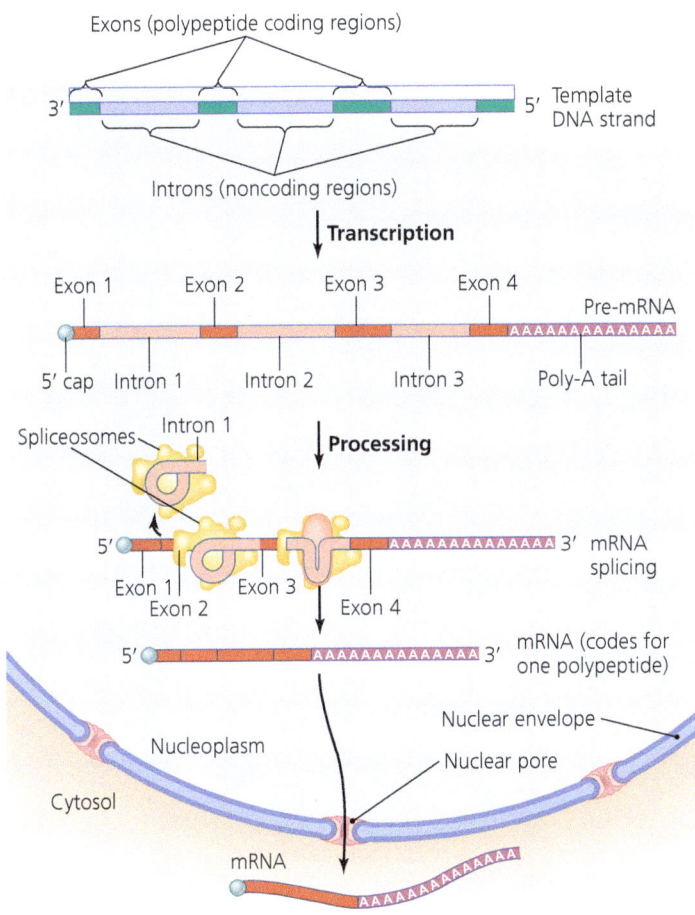

▲ FIGURE 7.11 **Processing eukaryotic mRNA.** Within the nucleus, transcription produces pre-mRNA, which contains coding exons and noncoding introns. Enzymes cap the 5' end with a modified guanine nucleotide and add hundreds of adenine nucleotides to the 3' end, a process known as polyadenylation. Ribozymes further process pre-mRNA by removing introns and splicing together exons to form a molecule that codes for a single polypeptide. Eukaryotic mRNA then moves from the nucleus to the cytoplasm.

Translation

LEARNING OUTCOMES

7.14 Describe the genetic code in general, and identify the relationship between codons and amino acids.
7.15 Describe the synthesis of polypeptides, identifying the roles of three types of RNA.
7.16 Contrast translation in bacteria from that in eukaryotes.

Translation is the process whereby ribosomes use the genetic information of nucleotide sequences to synthesize polypeptides composed of specific amino acid sequences. Some proteins are single polypeptides, whereas other proteins are composed of several polypeptides bound together in a quaternary structure (see Figure 2.24).

Ribosomes can be thought of as "polypeptide factories," so consider the following analogy between translation and a hypothetical automobile factory. Trucks deliver auto parts to the factory at the correct times and in the correct order to manufacture one of a large variety of automobile models, depending on instructions from corporate headquarters delivered by special courier. Similarly, molecules of tRNA (the trucks) deliver preformed amino acids (the parts) to a ribosome (the factory), which can manufacture an infinite variety of polypeptides (the car models) by assembling amino acids in the correct order according to the instructions from DNA (corporate headquarters) delivered via mRNA (the special courier).

How do ribosomes interpret the nucleotide sequence of mRNA to determine the correct order in which to assemble amino acids? To answer this question, we will consider the genetic code, examine in more detail the RNA molecules that participate in translation, and describe the specific steps of translation.

The Genetic Code

When geneticists in the early 20th century began to consider that DNA might be the genetic molecule, they were confronted with a problem: How can four kinds of DNA nucleotide bases—adenine, thymine, guanine, and cytosine (abbreviated A, T, G, and C)—specify the 20 common amino acids found in proteins? If each nucleotide coded for a single amino acid, only four amino acids could be specified. Eventually scientists showed that genes are composed of sequences of three nucleotides where each triplet specifies a particular amino acid. For example, the DNA nucleotide sequence AAA specifies the amino acid phenylalanine, and GTA codes for histidine. There are 64 possible arrangements of the four nucleotides in triplets (4^3)—more than enough to code all 20 amino acids.

These examples—AAA and GTA—are DNA triplets, but ribosomes do not directly access genetic information on a DNA molecule. Instead, molecules of mRNA carry the code to the ribosomes; therefore, scientists define the genetic code (FIGURE 7.12) as triplets of mRNA nucleotides called **codons** (kō'donz), which code for specific amino acids. UUU is a codon for phenylalanine, and CAU is a codon for histidine.

In most cases, 61 codons specify amino acids, and three codons—UAA, UAG, and UGA—instruct ribosomes to stop translating; however, under some conditions, UGA codes for a 21st amino acid, selenocysteine. Codon AUG also has a dual function, acting both as a start signal and as the codon for the amino acid methionine. In prokaryotes, mitochondria, and chloroplasts, AUG as a start codon codes for *N*-formylmethionine (fMet), a modified amino acid:

N-formylmethionine (fMet)   Methionine (Met)

First nucleotide base (5' position)	Second nucleotide base				Third nucleotide base (3' position)
	U	**C**	**A**	**G**	
U	UUU Phenylalanine (Phe)	UCU Serine (Ser)	UAU Tyrosine (Tyr)	UGU Cysteine (Cys)	U
	UUC (Phe)	UCC (Ser)	UAC (Tyr)	UGC (Cys)	C
	UUA Leucine (Leu)	UCA (Ser)	UAA STOP	UGA STOP & Selenocysteine (SeCys)	A
	UUG (Leu)	UCG (Ser)	UAG STOP*	UGG Tryptophan (Trp)	G
C	CUU (Leu)	CCU Proline (Pro)	CAU Histidine (His)	CGU Arginine (Arg)	U
	CUC (Leu)	CCC (Pro)	CAC (His)	CGC (Arg)	C
	CUA (Leu)	CCA (Pro)	CAA Glutamine (Gln)	CGA (Arg)	A
	CUG (Leu)	CCG (Pro)	CAG (Gln)	CGG (Arg)	G
A	AUU Isoleucine (Ile)	ACU Threonine (Thr)	AAU Asparagine (Asn)	AGU Serine (Ser)	U
	AUC (Ile)	ACC (Thr)	AAC (Asn)	AGC (Ser)	C
	AUA (Ile)	ACA (Thr)	AAA Lysine (Lys)	AGA Arginine (Arg)	A
	AUG START & Methionine (Met)	ACG (Thr)	AAG (Lys)	AGG (Arg)	G
G	GUU Valine (Val)	GCU Alanine (Ala)	GAU Aspartic acid (Asp)	GGU Glycine (Gly)	U
	GUC (Val)	GCC (Ala)	GAC (Asp)	GGC (Gly)	C
	GUA (Val)	GCA (Ala)	GAA Glutamic acid (Glu)	GGA (Gly)	A
	GUG (Val)	GCG (Ala)	GAG (Glu)	GGG (Gly)	G

*Also codes for a 22nd amino acid, pyrrolysine, in some prokaryotes.

▲ **FIGURE 7.12 The genetic code.** The table shows the set of mRNA codons and the amino acids for which they code. AUG not only is the start codon but also specifies methionine (Met) in eukaryotes and N-formylmethionine (fMet) in prokaryotes, mitochondria, and chloroplasts. Two codons (UAA and UAG) are stop codons that do not typically specify amino acids. UGA functions as a stop codon and also specifies selenocysteine.

As you examine the genetic code, notice that it is redundant; that is, more than one codon is associated with every amino acid except methionine and tryptophan. With most redundant codons, the first two nucleotides determine the amino acid, and the third nucleotide is inconsequential. For example, the codons GUU, GUC, GUA, and GUG all specify the amino acid valine.

Interestingly, the genetic code is nearly universal; that is, with few exceptions, ribosomes in archaeal, bacterial, plant, fungal, protozoan, and animal cells use the same genetic code.

Play *Translation: Genetic Code*
@ Mastering Microbiology

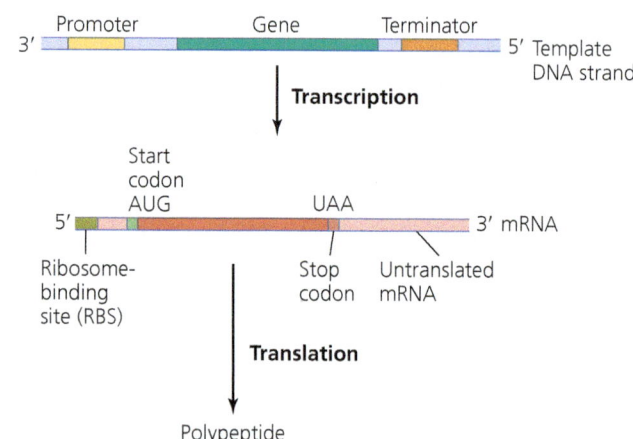

▲ **FIGURE 7.13 Prokaryotic mRNA molecules typically lack introns and exons.** Almost all prokaryotic genes, and therefore prokaryotic mRNA molecules, lack intervening sequences. Each gene codes for a single polypeptide. *Compare and contrast Figure 7.13 and Figure 7.8. What is each showing about the central dogma?*

Figure 7.13 Figure 7.8 focuses on information transfer, whereas Figure 7.13 focuses on gene structure.

Participants in Translation

As we have discussed, transcription produces messenger RNA, transfer RNA, and ribosomal RNA—each of which is involved in translation. We now discuss each kind of RNA in turn.

Messenger RNA Messenger RNA molecules carry genetic information from chromosomes to ribosomes as codons (triplet sequences of RNA nucleotides) that encode the order of amino acid sequences in a polypeptide. In prokaryotes, a basic mRNA molecule contains codons that are recognized by ribosomes: an AUG start codon, sequential codons for other amino acids in the polypeptide, and one of the three stop codons (FIGURE 7.13).

A single molecule of prokaryotic mRNA often contains start codons and instructions for more than one polypeptide arranged in series. Because both transcription and the subsequent events of translation occur in the cytosol of prokaryotes, prokaryotic ribosomes can begin translation before transcription is finished.

which varies with the tRNA. In other words, each transfer RNA carries a specific amino acid and recognizes mRNA codons only for that amino acid. A tRNA molecule is designated by a superscript abbreviation of its amino acid. For example, tRNA^{Phe} carries phenylalanine, and tRNA^{Ser} carries serine.

The fact that 61 codons specify the 20 amino acids commonly used by cells does not mean that there must be 61 different types of tRNA, because many tRNA molecules recognize more than one codon. *E. coli*, for example, has only about 40 different tRNAs. The variability in codon recognition by tRNA is due to "wobble" of the anticodon's third nucleotide. Wobble, which is a change of angle from the normal axis of the molecule, allows the third nucleotide to hydrogen bond to a nucleotide other than its usual complement. For example, a guanine nucleotide in the third position normally bonds to cytosine, but it can wobble and also pair with uracil; therefore, whether a codon has cytosine or uracil in the third position makes no difference because the same tRNA recognizes either nucleotide in the third position. For example, the codons UUU and UUC specify the amino acid phenylalanine because the anticodon AAG recognizes both of them. Similarly, UCU and UCC code for serine, and UAU and UAC code for tyrosine. This redundancy in the genetic code helps protect cells against the effects of errors in replication and transcription.

Ribosomes and Ribosomal RNA Prokaryotic ribosomes, which are also called *70S ribosomes* based on their sedimentation rate in an ultracentrifuge, are extremely complex associations of ribosomal RNAs and polypeptides. Each ribosome is composed of two subunits: 50S and 30S (**FIGURE 7.15a**). The 50S subunit is in turn composed of two rRNA molecules (23S and 5S) and about 34 different polypeptides, whereas the 30S subunit consists of

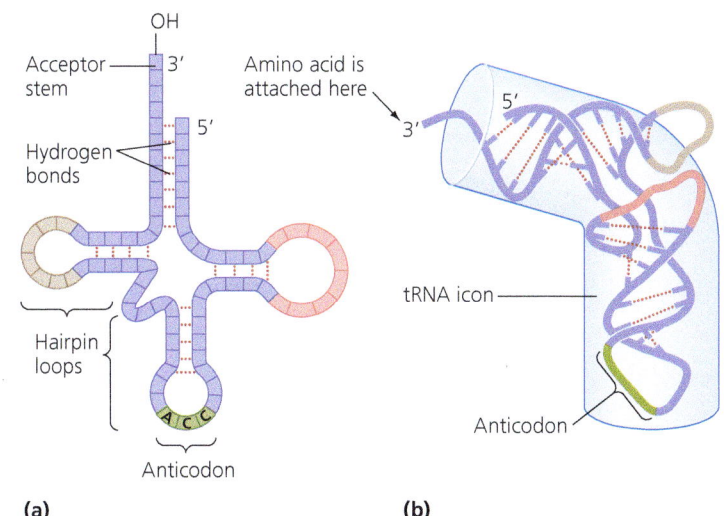

▲ **FIGURE 7.14 Transfer RNA. (a)** A two-dimensional "cloverleaf" representation of tRNA showing three hairpin loops held in place by intramolecular hydrogen bonding. **(b)** A three-dimensional drawing of the same tRNA. A specific amino acid attaches to the 3′ acceptor stem; the anticodon is a nucleotide triplet that is complementary to the mRNA codon for that amino acid. *Which amino acid would be attached to the acceptor stem of this tRNA?*

Figure 7.14 UGG is the codon for tryptophan, which is complementary to the anticodon ACC.

Eukaryotic mRNA differs from prokaryotic mRNA in several ways:

- As we have seen, eukaryotic cells extensively process pre-mRNA to make mRNA (see Figure 7.11).
- A processed molecule of eukaryotic mRNA contains instructions for only one polypeptide.
- Eukaryotic mRNA is not translated until it is fully transcribed and processed and has left the nucleus. In other words, transcription and translation of a molecule of eukaryotic mRNA cannot occur simultaneously because transcription occurs in the nucleus, and translation by eukaryotic ribosomes occurs in the cytoplasm.

Transfer RNA A tRNA molecule is a sequence of about 75 ribonucleotides that curves back on itself to form three main hairpin loops held in place by hydrogen bonding between complementary nucleotides (**FIGURE 7.14a**). Although transfer RNA molecules can be modeled simplistically by a cloverleaf structure, their three-dimensional shape is more complex (**FIGURE 7.14b**). For simplicity, tRNA will be represented in subsequent figures by an icon shaped like the curved cylinder in Figure 7.14b.

Amino acids have no direct affinity for or recognition of mRNA codons; a molecule of tRNA carries the correct amino acid to a ribosome during polypeptide synthesis. To this end, tRNA has an **anticodon** (an-tē-kō′don) triplet in its bottom loop that is complemetary to a codon and an *acceptor stem* for the corresponding amino acid at its 3′ end. Specific enzymes in the cytoplasm, called *aminoacyl tRNA synthetases*, each "charge" a specific tRNA molecule; that is, they attach the appropriate amino acid to the acceptor stem.

Anticodons are complementary to mRNA codons, and each acceptor stem is designed to carry one particular amino acid,

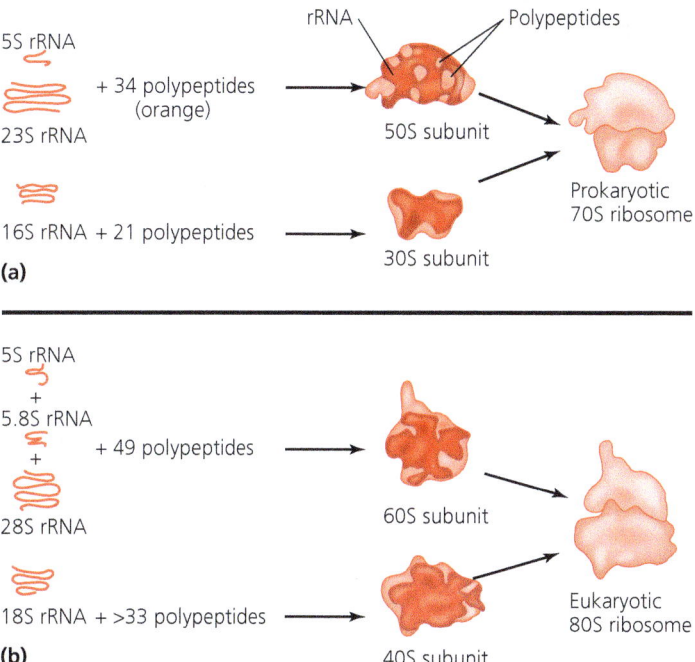

▲ **FIGURE 7.15 Ribosomal structures.** Polypeptides shown in beige and RNA molecules shown in red. **(a)** The 70S prokaryotic ribosome, which is composed of polypeptides and three rRNA molecules arranged in 50S and 30S subunits. **(b)** The 80S eukaryotic ribosome, which is composed of molecules of rRNA and polypeptides, arranged in 60S and 40S subunits.

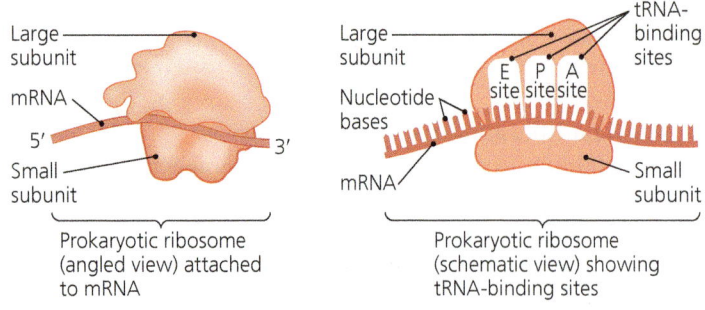

▲ **FIGURE 7.16** Assembled ribosome and its tRNA-binding sites. The A site accepts the tRNA carrying the next amino acid to be added to the growing polypeptide, whereas the P site holds the tRNA carrying the polypeptide. Empty tRNA molecules exit from the E site.

one molecule of 16S rRNA and about 21 ribosomal polypeptides. The ribosomes of mitochondria and chloroplasts are composed of similar subunits and polypeptides.

In contrast, both the cytosol and the rough endoplasmic reticulum (RER) of eukaryotic cells have 80S ribosomes composed of 60S and 40S subunits (FIGURE 7.15b). These subunits contain larger molecules of rRNA and more polypeptides than the corresponding prokaryotic subunits, though researchers do not agree on the exact number of polypeptides. The term *eukaryotic ribosome* is understood to mean only the 80S ribosomes of the cytosol and RER. Because the ribosomes of mitochondria and chloroplasts are 70S, they are "prokaryotic" ribosomes even though they are in eukaryotic cells.

The structural differences between prokaryotic and eukaryotic ribosomes play a crucial role in the efficacy and safety of antimicrobial drugs. Because erythromycin, for example, binds only to the 23S rRNA found in prokaryotic ribosomes, it has no effect on eukaryotic 80S ribosomes and thus little deleterious effect on a patient. (Chapter 10 discusses antimicrobial drugs in more detail.)

The smaller subunit of a ribosome is shaped to accommodate three codons at one time—that is, nine nucleotide bases of a molecule of mRNA. Each ribosome also has three tRNA-binding sites that are named for their function (FIGURE 7.16):

- The **A site** accommodates a tRNA delivering an *amino acid*.
- The **P site** holds a tRNA and the growing *polypeptide*.
- Discharged tRNAs *exit* from the **E site**.

We will now examine translation, the process whereby ribosomes actually synthesize polypeptides using amino acids delivered by tRNAs in the sequence dictated by the order of codons in mRNA.

Events in Translation

Molecular biologists divide translation into three stages: *initiation*, *elongation*, and *termination*. All three stages require additional protein factors that assist the ribosomes. Initiation and elongation also require energy provided by molecules of the ribonucleotide GTP, which are free in the cytosol (i.e., they are not part of an RNA molecule). Here, we consider bacterial translation.

Initiation During initiation, the two ribosomal subunits, mRNA, several protein factors, and tRNA^{fMet} form an *initiation complex*. Initiation in prokaryotes may occur while the cell is still transcribing mRNA from DNA. The events of initiation in a bacterium are as follows (FIGURE 7.17):

1. The smaller ribosomal subunit attaches to mRNA at a ribosome-binding site (also known as a *Shine-Dalgarno sequence* after its discoverers) so as to position a start codon (AUG) at the ribosomal subunit's P site.

2. tRNA^{fMet} (whose anticodon, UAC, is complementary to the start codon, AUG) attaches at the ribosome's P site. This is the only tRNA that enters the ribosome at the P site. It can do so because the two subunits have not yet joined to form a complete ribosome. All other tRNAs enter at the A site.

3. The larger ribosomal subunit then attaches to form a complete initiation complex.

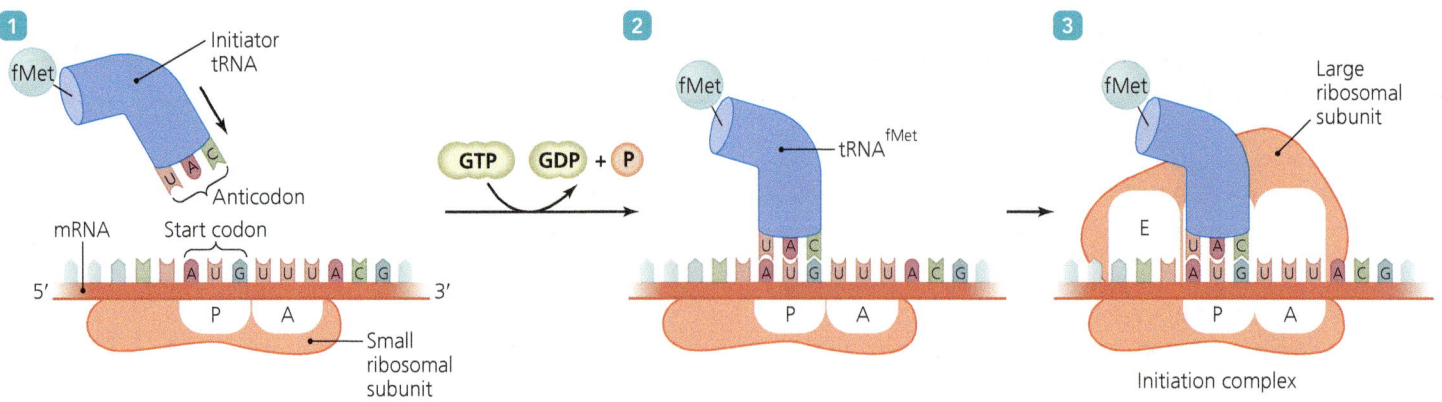

▲ **FIGURE 7.17 The initiation of translation in prokaryotes.** 1 The smaller ribosomal subunit attaches to mRNA at a ribosome-binding site near a start codon (AUG). 2 The anticodon of tRNA^{fMet} aligns with the start codon on the mRNA; energy from GTP is used to bind the tRNA in place. 3 The larger ribosomal subunit attaches to form an initiation complex—a complete ribosome attached to mRNA.

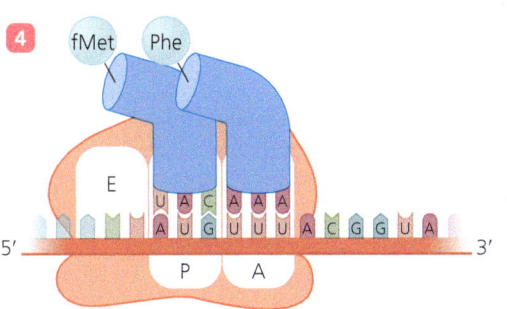

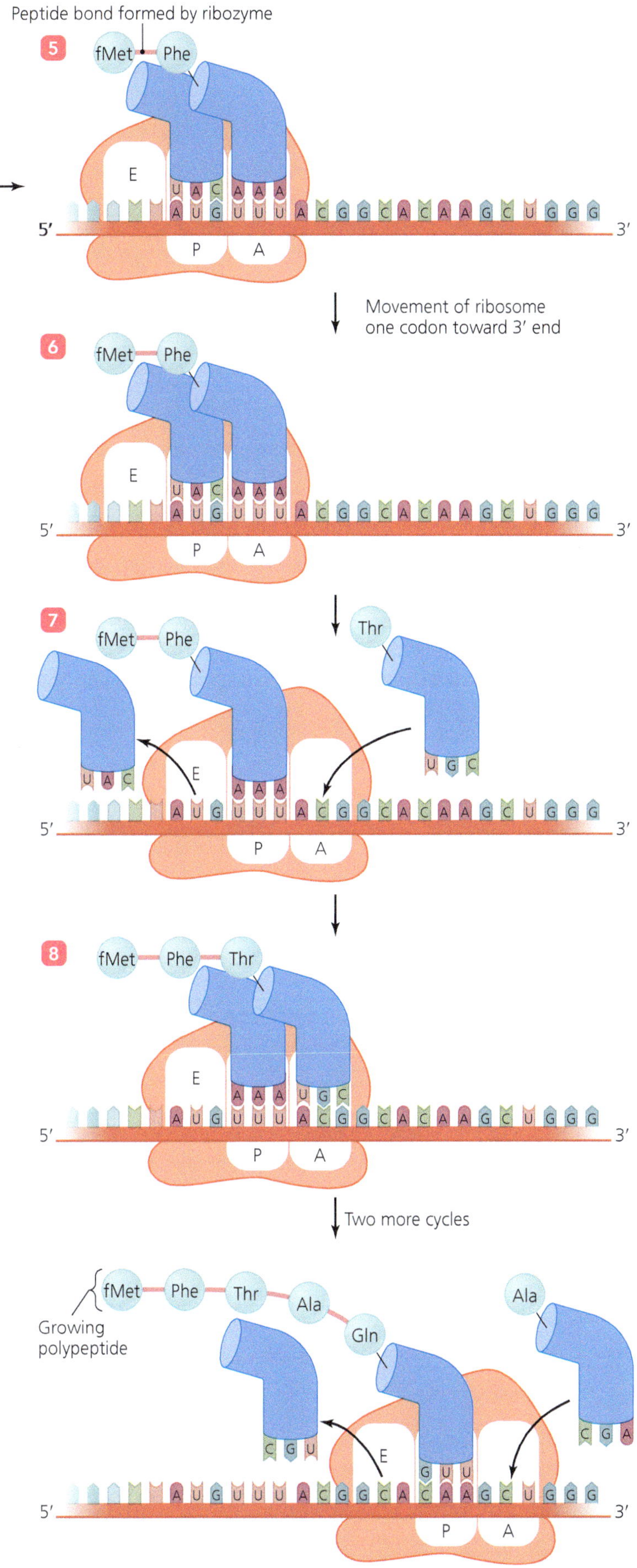

▶ **FIGURE 7.18 The elongation stage of translation.** Transfer RNAs sequentially deliver amino acids as directed by the codons of the mRNA. Ribosomal RNA in the large ribosomal subunit catalyzes a peptide bond between the amino acid at the A site and the growing polypeptide at the P site. The steps in the process are described in the text. *How would eukaryotic translation differ?*

Figure 7.18 *The initial amino acid in eukaryotic polypeptide is methionine.*

Elongation Elongation of a polypeptide is a cyclical process that involves the sequential addition of amino acids to a polypeptide chain growing at the P site. **FIGURE 7.18** illustrates several cycles of the process. The steps of each cycle occur as follows:

4. The tRNA whose anticodon is complementary to the next codon—in this example, AAA complementary to the codon UUU—delivers its amino acid, in this case, phenylalanine (Phe), to the A site. Proteins called *elongation factors* escort the tRNA along with a molecule of GTP (not shown). Energy from GTP is used to stabilize each tRNA as it binds at the A site. All incoming tRNAs enter the A site, as the P site is now buried.

5. An enzymatic RNA molecule—a ribozyme—in the larger ribosomal subunit forms a peptide bond between the terminal amino acid of the growing polypeptide chain (in this case, *N*-formylmethionine) and the newly introduced amino acid. The polypeptide is now attached to the tRNA occupying the A site.

6. Using energy supplied by more GTP, the ribosome moves one codon down the mRNA. This transfers each tRNA to the adjacent binding site; that is, the first tRNA moves from the P site to the E site, and the second tRNA (with the attached polypeptide) moves to the vacated P site.

7. The ribosome releases the "empty" tRNA from the E site. In the cytosol, the appropriate enzyme recharges the empty tRNA with another molecule of the type of amino acid carried by that tRNA.

8. The cycle repeats, each time adding another amino acid, at a rate of about 15 amino acids per second (in this example, threonine, then alanine, and then glutamine).

As elongation proceeds, ribosomal movement exposes the start codon, allowing another ribosome to attach behind the first one. In this way, one ribosome after another attaches at the start codon and begins to translate identical polypeptide molecules from the same message. Such a group of ribosomes, called a *polyribosome*, resembles beads on a string (**FIGURE 7.19**).

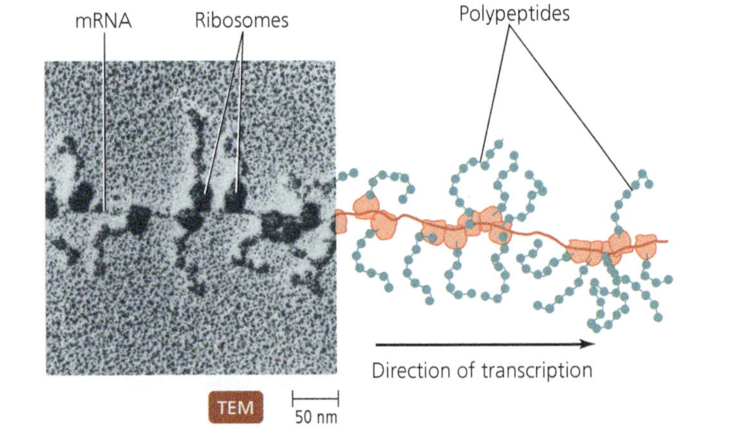

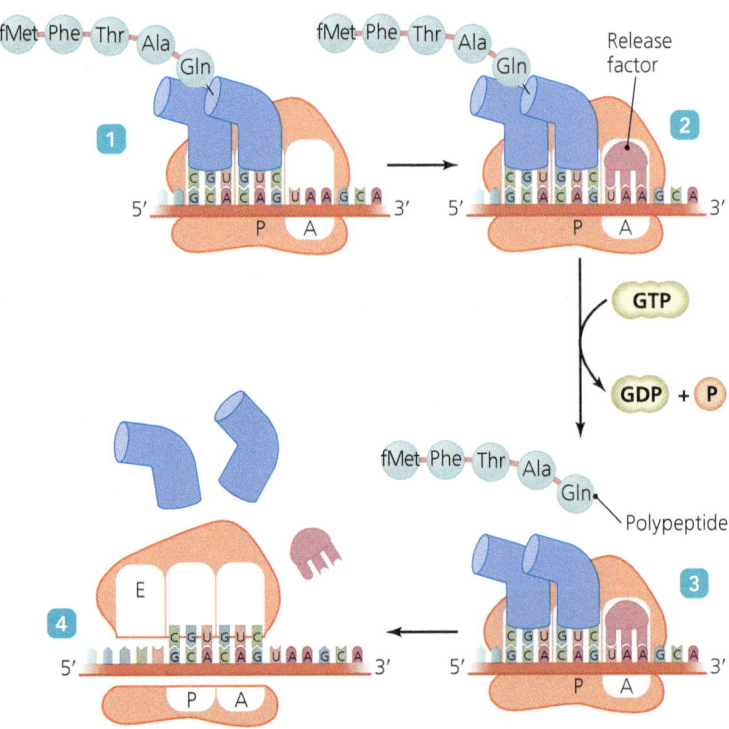

▲ **FIGURE 7.19 Polyribosome in prokaryotes—one mRNA and many ribosomes and polypeptides.** As each ribosome moves down the mRNA, the start codon (AUG) becomes available to another ribosome. In this manner, numerous identical polypeptides are translated simultaneously from a single mRNA molecule. The artist has extended an electron micrograph of a polyribosome in a prokaryotic cell with a schematic drawing.

Termination Termination occurs at one of the three stop codons (**FIGURE 7.20**). ❶ Termination does not involve tRNA; instead, proteins called *release factors* halt elongation. It appears that release factors somehow recognize stop codons ❷ and modify the larger ribosomal subunit in such a way as to activate another of its ribozymes that severs the polypeptide from the final tRNA

▲ **FIGURE 7.20 Termination of translation.** ❶ Because a tRNA under normal circumstances does not recognize a stop codon (here UAA), the A site is empty. ❷ A protein release factor enters the A site and ❸ modifies the large subunit to sever the polypeptide from the tRNA at the P site using energy from GTP. ❹ The ribosome dissociates; the subunits can now bind to another start codon to begin another round of translation.

(resident at the P site) ❸. The ribosome then dissociates into its subunits ❹. Termination of translation should not be confused with termination of transcription, covered in a previous section. The polypeptides released from a ribosome at termination may function alone as proteins, or they may function with other polypeptides in quaternary protein structures.

Translational Differences in Eukaryotes

Eukaryotic translation is similar to that of bacteria, with some notable differences, including the following:

- Initiation of translation in eukaryotes occurs when the small ribosomal subunit binds to the 5' guanine cap rather than a specific nucleotide sequence.
- The first amino acid in eukaryotic polypeptides is methionine rather than formylmethionine.
- Ribosomes attached to membranes of endoplasmic reticulum (ER), forming rough ER (RER), can synthesize polypeptides into the cavity of the RER.

Archaeal translation is more similar to that of eukaryotes than to that of bacteria; however, archaea lack ER.

The processes we have examined thus far—how a cell replicates DNA, transcribes RNA, and translates RNA into polypeptides—are summarized in **TABLE 7.2**. Next, we examine the way cells control the process of transcription.

MICRO CHECK

8. What is the term for three RNA nucleotides that code for a specific amino acid during polypeptide synthesis?
9. Which codon is considered the "start codon" and codes for methionine (in eukaryotes) and *N*-formylmethionine (in prokaryotes, mitochondria, and chloroplasts)?
10. Which molecules carry new amino acids to the location where they are assembled into polypeptides?

Regulation of Genetic Expression

LEARNING OUTCOMES

7.17 Explain the operon model of transcriptional control in prokaryotes.
7.18 Contrast the regulation of an inducible operon with that of a repressible operon, and give an example of each.
7.19 Describe the use of microRNA, small interfering RNA, and riboswitches in genetic control.

Many of a bacterium's genes are expressed at all times; that is, they are constantly transcribed and translated and play a persistent role in the phenotype. Such genes are called *constitutive genes*. Constitutive genes code for RNAs and polypeptides that are needed continuously by the cell—for example, integral proteins of the cytoplasmic membrane, structural proteins of ribosomes, and enzymes of glycolysis.

Other genes are regulated so that the polypeptides they encode are synthesized only in response to a change in the

TABLE 7.2 Comparison of Genetic Processes

	Replication	Transcription	Translation
Enzyme	DNA polymerases	RNA polymerases	Ribosomes
Template	Both parental strands of DNA	One strand of DNA	mRNA
Start Site	Origin of replication	Promoter	AUG start codon
Fidelity Mechanism	Polymerase proofreading, followed by mismatch repair systems	None	Specificity of enzymes that charge tRNAs
Termination	Termination sequences	Terminator	UAA, UAG, or UGA stop codons
Location	Prokaryotes: cytosol Eukaryotes: nucleus	Prokaryotes: cytosol; can be simultaneous with replication Eukaryotes: Nucleolus in nucleus	Prokaryotes: cytosol; can be simultaneous with transcription Eukaryotes: cytosol or on RER
Product	Two daughter DNA strands, each paired with one original strand (semiconservative replication)	RNA, single stranded	Polypeptides; some function alone as proteins, and others work together as a single protein
Energy Source for Process	Deoxyribonucleotides	Ribonucleotides	GTP; also ATP for charging tRNAs
Direction of Polymerization	5' to 3'	5' to 3'	N terminus (end with amino group) to C terminus (end with carboxyl group)

environment; after all, protein synthesis requires a large amount of energy, which can be conserved if a cell forgoes production of unneeded polypeptides. For example, *Pseudomonas aeruginosa* (ā-roo-ji-nō'să), a potential pathogen of cystic fibrosis patients, can synthesize harmful proteins. Early in an infection, *Pseudomonas* does not produce the proteins: The body would respond defensively and eliminate the bacterium. Instead, the pathogen uses **quorum sensing**—a process whereby cells secrete quorum-sensing molecules into their environment and other cells detect these signals so as to measure their density. The result is that *Pseudomonas* cells synthesize harmful proteins only after there are numerous bacterial cells, overwhelming the body's defenses.

Cells regulate polypeptide synthesis in many ways. They may initiate or stop transcription of mRNA or may stop translation directly. Much of our knowledge of regulation has come from the study of microorganisms such as *E. coli*. We will examine two types of regulation of transcription in this bacterium—*induction* and *repression*. But before we examine these processes, we must first consider *operons*—special arrangements of prokaryotic genes that play roles in gene regulation.

The Nature of Prokaryotic Operons

As originally described, a prokaryotic **operon** consists of a promoter, a series of genes that code for enzymes and structures (such as channel proteins), and an adjacent regulatory element called an **operator** (FIGURE 7.21), which controls movement of RNA polymerase. Operons can be *polycistronic*; that is, they may code for more than one polypeptide.

Inducible operons are usually inactive and must be induced, or "turned on," by *inducers*. Such operons often code for proteins that transport and catabolize nutrients that are only sometimes available to a cell.

Repressible operons operate in reverse fashion—they are typically continually active until *repressors* deactivate them. Repressible operons often produce proteins that are involved in anabolic (synthesis) pathways. To clarify these concepts, we next examine an inducible operon and a repressible operon found in *E. coli*.

The Lactose Operon, an Inducible Operon

The *lactose (lac) operon* of *E. coli* is an inducible operon and the first operon whose structure and action were elucidated. The *lac* operon includes a promoter, an operator, and three genes that encode proteins involved in the transport and catabolism of lactose, a disaccharide sugar. *E. coli* can catabolize lactose to release glucose and galactose, which can then be further catabolized as energy sources. Two events induce, or activate, the *lac* operon: (1) positive regulation by a protein called CAP and (2) deactivation of a repressor molecule. Let's examine these two events.

Play *Operons:* Overview
@ Mastering Microbiology

Positive Regulation by CAP *E. coli* uses glucose more efficiently than other sugar molecules, so the bacterium avoids using lactose when glucose is available. When glucose levels are low,

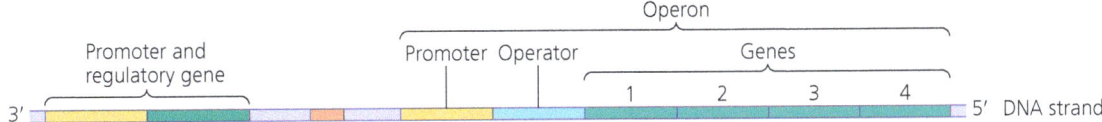

▲ FIGURE 7.21 **An operon.** An operon consists of genes, their promoter, and an operator. The genes code for enzymes and structures such as channel and carrier proteins. A separate regulatory gene that is not part of the operon codes for a protein that controls the operon.

▲ FIGURE 7.22 **CAP-cAMP enhances lac transcription.** Cyclic adenosine monophosphate (cAMP), accumulating in *E. coli* when glucose is absent, binds to catabolite activator protein (CAP). CAP-cAMP binds to a CAP-binding site of DNA, allowing RNA polymerase to effectively bind to the *lac* promoter and begin transcription.

a small molecule called *cyclic adenosine monophosphate (cAMP)* accumulates in the cell. Buildup of cAMP acts as a glucose starvation signal.

Cyclic AMP binds to a regulatory protein called *catabolite activator protein (CAP)*. CAP then attaches to a CAP-binding site, also called an *enhancer*, on DNA near operons controlling sugar catabolism, including the *lac* operon. With CAP attached to the nearby CAP-binding site, RNA polymerase is more attracted to the *lac* operon's promoter (**FIGURE 7.22**). Thus, CAP positively enhances *lac* transcription in the absence of glucose.

Repression and Induction The *lac* operon is controlled by a regulatory gene located outside the operon (**FIGURE 7.23a**).

Usually, the regulatory gene is constantly transcribed and translated to produce a repressor ❶ protein that attaches to DNA at the *lac* operator. The presence of this repressor prevents RNA polymerase from binding to the promoter. Thus, the *lac* operon is usually inactive even if CAP is present ❷.

When lactose is available, *E. coli* converts it to allolactose. Allolactose induces the *lac* operon by inactivating the repressor so that it no longer attaches to DNA (**FIGURE 7.23b**) ❸. With the operator unblocked, RNA polymerase binds to the promoter and transcribes the three *lac* genes—the operon has been induced ❹.

Ribosomes then translate the newly synthesized mRNA to produce enzymes that catabolize lactose (not shown). Once the lactose supply has been depleted, there is no more allolactose inducer, so the repressor once again becomes active, binds to the operator, and suppresses the *lac* operon. In this manner, *E. coli* cells synthesize enzymes for the catabolism of lactose only when two conditions are met: glucose is absent (CAP is bound), and lactose is present.

Inducible operons are often involved in controlling catabolic pathways whose polypeptides are not needed unless a particular nutrient is available. A different situation occurs with anabolic pathways, such as those that synthesize amino acids.

Play *Operons: Induction* @ Mastering Microbiology

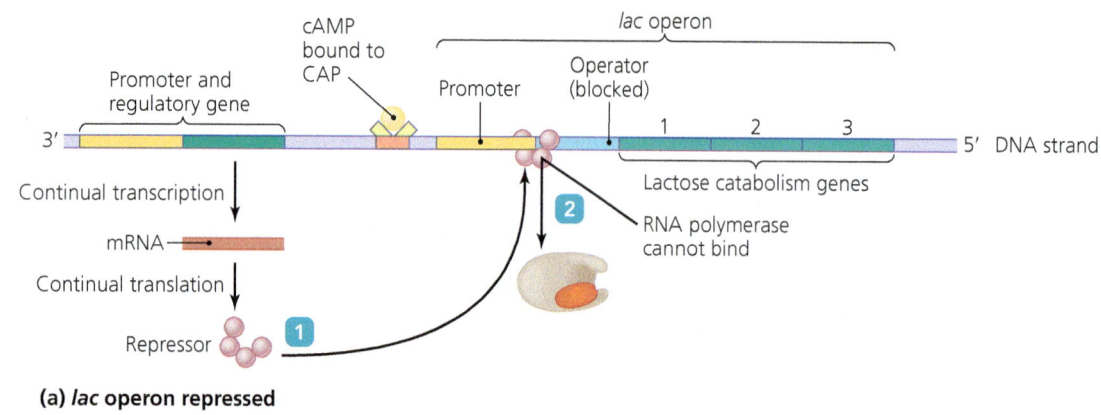

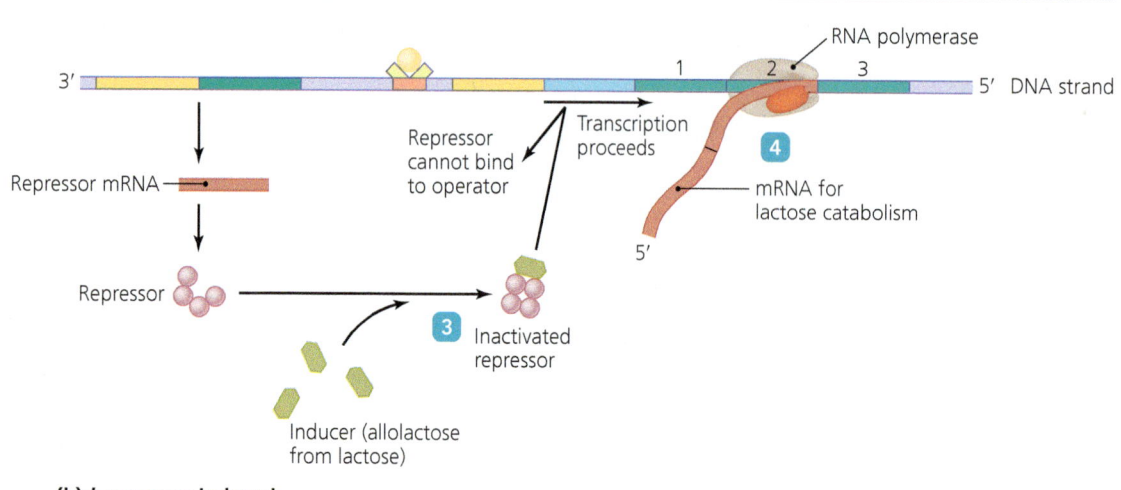

◀ FIGURE 7.23 **The *lac* operon, an example of an inducible operon.** The repressor, a protein encoded by a regulatory gene, is constantly synthesized. **(a)** When lactose is absent from the cell's environment, the repressor binds to the operator, blocking the movement of RNA polymerase and halting transcription. **(b)** When lactose is present in the cell's environment, its derivative, allolactose, acts as an inducer by inactivating the repressor so that the repressor cannot bind to the operator, allowing transcription to proceed.

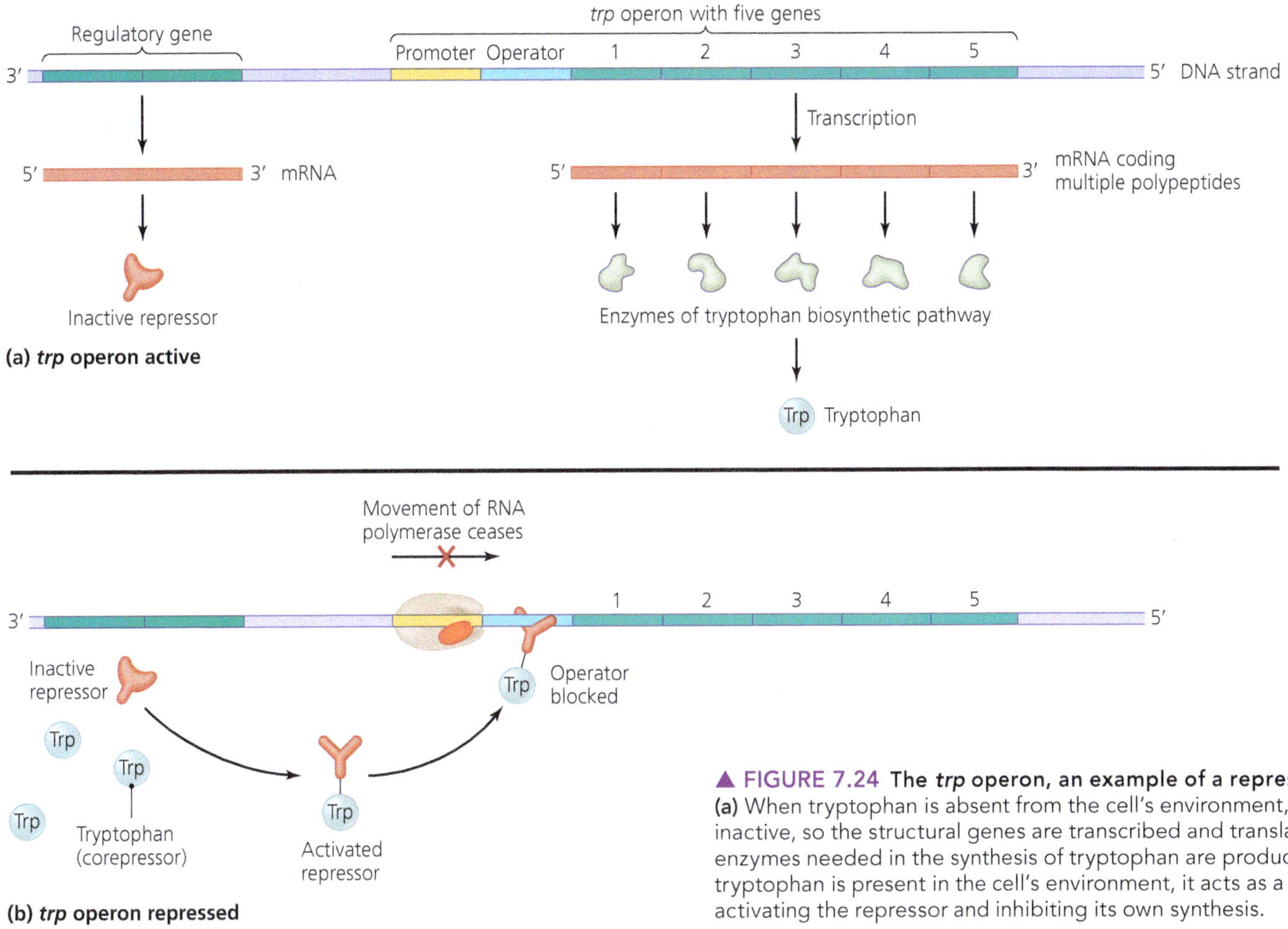

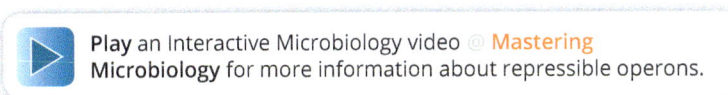

▲ **FIGURE 7.24** The *trp* operon, an example of a repressible operon. (a) When tryptophan is absent from the cell's environment, the repressor is inactive, so the structural genes are transcribed and translated, and the five enzymes needed in the synthesis of tryptophan are produced. (b) When tryptophan is present in the cell's environment, it acts as a corepressor, activating the repressor and inhibiting its own synthesis.

The Tryptophan Operon, a Repressible Operon

E. coli can synthesize all the amino acids it needs for polypeptide synthesis; however, it can save energy by using amino acids available in its environment. In such cases, *E. coli* represses the genes for a given amino acid's synthetic pathway.

The *tryptophan (trp) operon* is a repressible polycistronic operon consisting of a promoter, an operator, and five genes that code for the enzymes involved in the synthesis of tryptophan. Just as with the *lac* operon, a regulatory gene codes for a repressor molecule that is constantly synthesized. In contrast to an inducible operon's repressor, this repressor is normally inactive. Thus, the *trp* operon is usually active: the cell transcribes mRNA, translates the enzymes for tryptophan synthesis, and synthesizes tryptophan (**FIGURE 7.24a**).

However, when tryptophan is available, it binds to the repressor, thus activating it. The activated repressor then binds to the operator, halting the movement of RNA polymerase and transcription (**FIGURE 7.24b**). In other words, tryptophan stops its own synthesis.

TABLE 7.3 summarizes how the basic characteristics of inducible and repressible operons relate to the regulation of transcription.

> Play an Interactive Microbiology video @ Mastering Microbiology for more information about repressible operons.

> Play *Operons:* Repression @ **Mastering** Microbiology

TABLE 7.3 Basic Roles of Operons in Regulating Transcription

Type of Regulation	Type of Metabolic Pathway Regulated	Regulating Condition
Inducible Operons	Catabolic pathways	Presence of substrate of pathway
Repressible Operons	Anabolic pathways	Presence of product of pathway

RNA Molecules Can Control Translation

Besides using inducible and repressible operons, cells can also use molecules of RNA to regulate translation of polypeptides. Regulatory RNA molecules include *microRNA, small interfering RNA,* and *riboswitches.*

Eukaryotic cells transcribe single-stranded RNA molecules about 22 nucleotides long called **microRNAs (miRNAs)**, which are not translated by ribosomes. Instead, miRNA joins with regulatory proteins to form a *miRNA-induced silencing complex (miRISC).*

miRISC binds to messenger RNA that is complementary to the microRNA within the miRISC. Once bound, miRISC performs one of two functions: In some cases, miRISC cuts the messenger RNA molecule, rendering it useless. In other cases, miRISC remains bound to messenger RNA, effectively hiding the mRNA molecule from ribosomes. In both cases, no polypeptide is formed from the messenger RNA molecule; thus, miRNAs (associated with RISC) regulate gene expression by blocking translation. Eukaryotic cells use miRISC to regulate a number of processes, including embryogenesis, cell division, apoptosis (programmed cell death), blood cell formation, and development of cancer.

Another method of regulation involving RNA uses **small interfering RNA (siRNA)**. siRNAs are about the same length as miRNAs but differ from miRNAs in that siRNAs are double stranded. Further, strands of siRNAs may be complementary to mRNA, tRNA, or DNA. siRNAs unwind and join RISC proteins to form siRISC, which appears to always bind to and cut the target nucleic acid. This result is called *gene silencing*. Scientists can create siRNA molecules so as to artificially regulate gene expression in laboratory studies.

A **riboswitch** is another RNA molecule that helps regulate translation. Riboswitches change shape in response to environmental conditions such as changes in temperature or shifts in the concentration of specific nutrients, including vitamins, nucleotide bases, or amino acids.

Some messenger RNA molecules themselves act as riboswitches. When conditions warrant, riboswitch mRNA folds to either favor or block translation, depending on the need by the cell for the polypeptide it encodes. For example, mRNA for the virulence regulator of the plague bacterium *Yersinia pestis* (yer-sin′ē-ă pes′tis) folds in such a way as to prevent translation when the temperature is below human body temperature (37°C). When the bacterium enters a human, the mRNA refolds into a shape that allows translation, and virulence regulator is synthesized; plague ensues.

MICRO CHECK

11. What kind of gene sequence is typically active and producing proteins but can be shut down?

TELL ME WHY

In bacteria, polypeptide translation can begin even before mRNA transcription is complete. Why can't this happen in eukaryotes?

Mutations of Genes

LEARNING OUTCOME

7.20 Define *mutation*.

The life of a cell is dependent on both the integrity and accurate control of the cell's genes; however, the nucleotide sequences of genes are not always accurately maintained. A **mutation** is a heritable change in the nucleotide base sequence of a genome. Not all mutations occur in genes; some mutations occur in DNA between genes. Mutations of genes are almost always deleterious, though a few make no difference to an organism. Even more rarely, a mutation leads to a novel property that might improve the ability of the organism and its descendants to survive and reproduce. This is evolution in action. Mutations in unicellular organisms are passed on to the organism's progeny, but mutations in multicellular organisms typically are passed to offspring only if a mutation occurs in gametes (sex cells) or gamete-producing cells.

Types of Mutations

LEARNING OUTCOME

7.21 Define *point mutation*, and describe three types.

Mutations range from large changes in an organism's genome, such as the loss or gain of an entire chromosome, to **point mutations**, in which just a single nucleotide base pair is affected. Point mutations include *substitutions* and *frameshift mutations*, which result from either *insertions* or *deletions*. The following analogy illustrates some types of point mutations. Suppose that the DNA code was represented by the letters THECATATEELK. Grouping the letters into triplets (like codons) yields THE CAT ATE ELK. The substitution of a single letter at one point could either change the meaning of the sentence, as in THE RAT ATE ELK, or result in a meaningless phrase, such as THE CAT RTE ELK. Insertions or deletions cause frameshift mutations because nucleotide triplets following the mutation are displaced, creating new sequences of codons that result in vastly altered polypeptide sequences. Such point mutations produce more serious changes, such as TRH ECA TAT EEL K (insertion) or TEC ATA TEE LK (deletion). Frameshift mutations can affect a cell much more seriously than mere substitutions because a frameshift affects all codons subsequent to the mutation.

Mutations can also involve *inversion* (THE ACT ATE KLE), *duplication* (THE CAT CAT ATE ELK ELK), or *transposition* (THE ELK ATE CAT). Such mutations and even larger deletions and insertions are **gross mutations**.

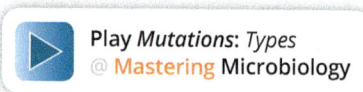

Play *Mutations: Types*
@ Mastering Microbiology

Effects of Point Mutations

LEARNING OUTCOME

7.22 List three effects of point mutations.

Some base-pair substitutions produce **silent mutations** because redundancy in the genetic code prevents the substitution from altering the amino acid sequence (compare FIGURE 7.25a and b). For example, when the DNA triplet AAA is changed to AAG, the mRNA codon will be changed from UUU to UUC; however, because both codons specify phenylalanine, there is no change in the phenotype—the mutation is silent because it affects the genotype only.

Of greater concern are substitutions that change a codon for one amino acid into a codon for a different amino acid. A change that specifies a different amino acid is called a **missense mutation** (FIGURE 7.25c); what gets transcribed and translated makes sense,

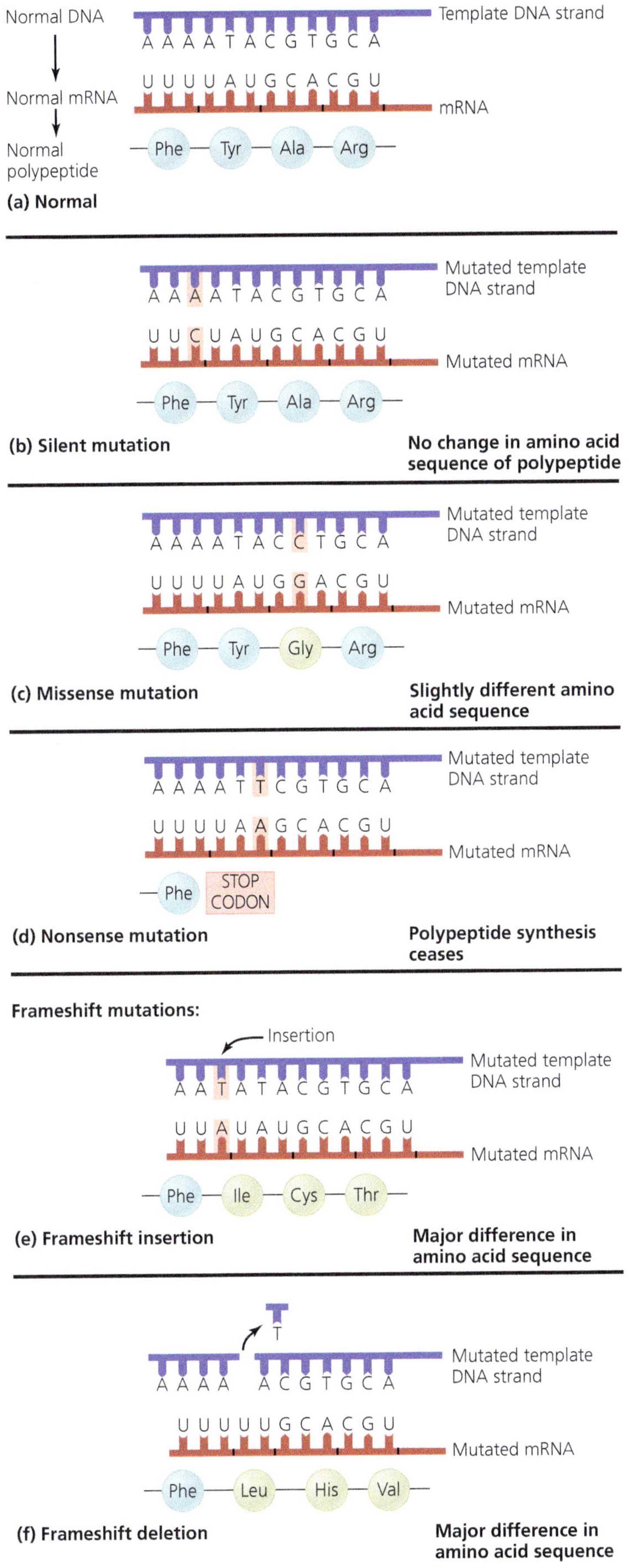

Figure 7.25

(a) Normal
- Normal DNA: A A A A T A C G T G C A (Template DNA strand)
- Normal mRNA: U U U U A U G C A C G U (mRNA)
- Normal polypeptide: Phe — Tyr — Ala — Arg

(b) Silent mutation — No change in amino acid sequence of polypeptide
- Mutated template DNA strand: A A A A T A C G T G C A
- Mutated mRNA: U U C U A U G C A C G U
- Phe — Tyr — Ala — Arg

(c) Missense mutation — Slightly different amino acid sequence
- Mutated template DNA strand: A A A A T A C C T G C A
- Mutated mRNA: U U U U A U G G A C G U
- Phe — Tyr — Gly — Arg

(d) Nonsense mutation — Polypeptide synthesis ceases
- Mutated template DNA strand: A A A A T T C G T G C A
- Mutated mRNA: U U U U A A G C A C G U
- Phe — STOP CODON

Frameshift mutations:

(e) Frameshift insertion — Major difference in amino acid sequence
- Insertion
- Mutated template DNA strand: A A T A T A C G T G C A
- Mutated mRNA: U U A U A U G C A C G U
- Phe — Ile — Cys — Thr

(f) Frameshift deletion — Major difference in amino acid sequence
- Mutated template DNA strand: A A A A   A C G T G C A (T deleted)
- Mutated mRNA: U U U U U G C A C G U
- Phe — Leu — His — Val

◀ **FIGURE 7.25 The effects of the various types of point mutations.** (a) Normal gene. Base-pair substitutions can result in (b) silent mutations, (c) missense mutations, or (d) nonsense mutations. (e) Frameshift insertions and (f) frameshift deletions usually result in severe missense or nonsense mutations because all codons downstream from the mutation are altered.

but not the right sense. The effect of missense mutations depends on where in the protein the changed amino acid occurs. When the different amino acid is in a critical region of a protein, the protein becomes nonfunctional; however, when the different amino acid is in a less important region, the mutation may have no adverse effect.

A third type of mutation occurs when a base-pair substitution changes an amino acid codon into a stop codon. This is called a **nonsense mutation** (FIGURE 7.25d). Nearly all nonsense mutations result in nonfunctional proteins.

Frameshift mutations (i.e., insertions or deletions) typically result in drastic missense and nonsense mutations (FIGURE 7.25e and f) except when the insertion or deletion is very close to the end of a gene.

TABLE 7.4 (on p. 218) summarizes the types and effects of point mutations.

Mutagens

LEARNING OUTCOMES

7.23 Discuss how different types of radiation cause mutations in a genome.
7.24 Describe three kinds of chemical mutagens and their effects.

Mutations occur naturally during the life of an organism. Such *spontaneous mutations* result from errors in replication and repair as well as from *recombination* in which relatively long stretches of DNA move among chromosomes, plasmids, and viruses, introducing frameshift mutations. For example, we have seen that mismatched base pairing during DNA replication results in one error in every 10 billion (10^{10}) base pairs. Since an average gene has 10^3 base pairs, about one of every 10^7 (10 million) genes contains an error. Further, though cells have repair mechanisms to reduce the effect of mutations, the repair process itself can introduce additional errors. Physical or chemical agents called **mutagens** (myū′tă-jenz), which include radiation and several types of DNA-altering chemicals, induce mutations.

Play *Mutagens*
Mastering Microbiology

Radiation

In the 1920s, Hermann Muller (1890–1967) discovered that X rays increased phenotypic variability in fruit flies by causing mutations. *Gamma rays* also damage DNA. X rays and gamma rays are *ionizing radiation*; that is, they energize electrons in atoms, causing some of the electrons to escape from their atoms (see Chapter 9). These free electrons strike other atoms, producing ions that can react with the structure of DNA, thereby creating mutations. More seriously, electrons and ions can break the covalent bonds between

TABLE 7.4 The Types of Point Mutations and Their Effects

Type of Point Mutation	Description	Effects
Substitution	Mismatching of nucleotides or replacement of one base pair by another	*Silent mutation* if change results in redundant codon, as amino acid sequence in polypeptide is not changed. *Missense mutation* if change results in codon for a different amino acid; effect depends on location of different amino acid in polypeptide. *Nonsense mutation* if codon for an amino acid is changed to a stop codon.
Frameshift (insertion)	Addition of one or a few nucleotide pairs creates new sequence of codons	Missense and nonsense mutations
Frameshift (deletion)	Removal of one or a few nucleotide pairs creates new sequence of codons	Missense and nonsense mutations

the sugars and phosphates of a DNA backbone, causing physical breaks in chromosomes and complete loss of cellular control.

Nonionizing radiation in the form of *ultraviolet (UV) light* is also mutagenic because it causes adjacent pyrimidine bases to covalently bond to one another, forming **pyrimidine dimers** (FIGURE 7.26). The presence of dimers prevents hydrogen bonding with nucleotides in the complementary strand, distorts the sugar-phosphate backbone, and prevents proper replication and transcription. Cells have several methods of repairing dimers (discussed shortly).

Chemical Mutagens

Here we consider three of the many basic types of mutagenic chemicals.

Nucleotide Analogs Compounds that are structurally similar to normal nucleotides are called **nucleotide analogs** (FIGURE 7.27a). When nucleotide analogs are available to replicating cells, they may be incorporated into DNA in place of normal nucleotides, where their structural differences either inhibit DNA replication or result in mismatched base pairing. For example, when thymine is replaced by the analog 5′-bromouracil, a wrong complement can form during subsequent replications—5′-bromouracil can pair with guanine rather than with adenine, resulting in a point mutation (FIGURE 7.27b). (Figure 10.7 illustrates other nucleotide analogs.)

Nucleotide (or nucleoside) analogs make potent antiviral and anticancer drugs. Because viruses and cancer cells typically

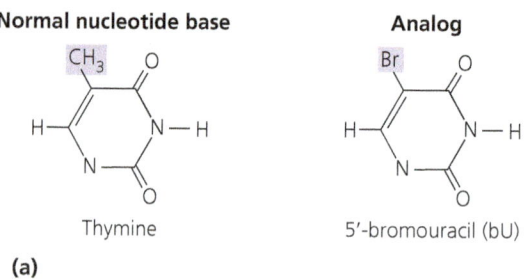

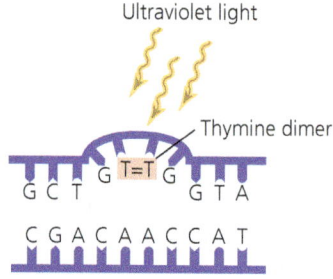

▲ FIGURE 7.26 **A pyrimidine (in this case, thymine) dimer.** Ultraviolet light causes adjacent pyrimidine bases to covalently bond to each other, preventing hydrogen bonding with bases in the complementary strand. The resulting distortion of the sugar-phosphate backbone prevents proper replication and transcription.

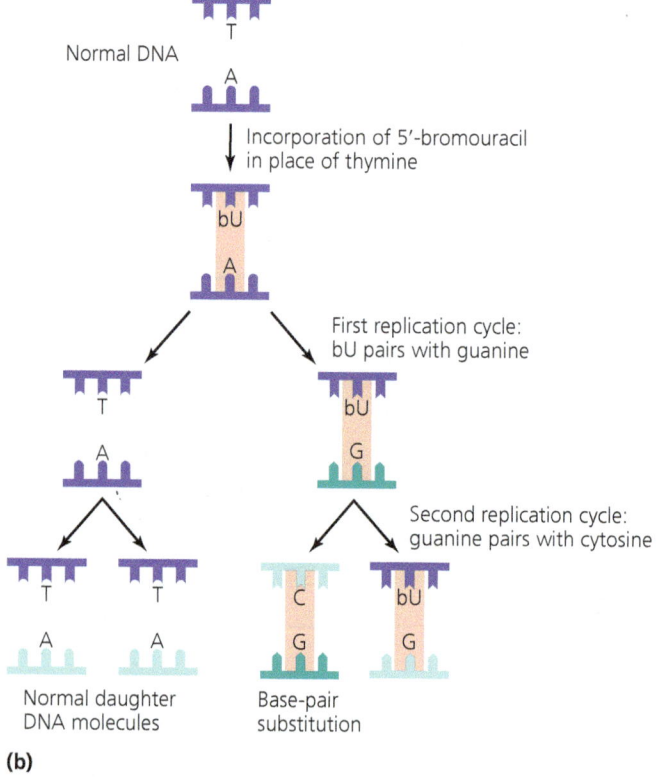

▲ FIGURE 7.27 **The structure and effects of a nucleotide analog.** (a) The structures of thymine and one of its nucleotide analogs, 5′-bromouracil (bU). (b) When 5′-bromouracil is incorporated into DNA, the result is a point mutation: a single replication cycle results in one normal DNA molecule and one mutated molecule in which bU pairs with guanine. After a second replication cycle, a complete base-pair substitution has occurred in one of the four DNA molecules. CG has been substituted for TA.

replicate faster than normal cells, they more quickly incorporate nucleotide or nucleoside analogs into their DNA and RNA. The presence of such analogs is deleterious; therefore, nucleotide analogs kill viruses and cancer cells more quickly than they do normal cells. (Chapter 10 considers the use of nucleotide analogs as antimicrobial agents.)

Nucleotide-Altering Chemicals Some chemical mutagens alter the structure of nucleotides. For example, *Aspergillus* (as-per-jil'ŭs) molds growing on grains and nuts produce nucleotide-altering chemicals called *aflatoxins*. When the liver catabolizes aflatoxins, a by-product can convert guanine nucleotides into thymine nucleotides so that a GC base pair is converted to a TA base pair, resulting in missense mutations and possibly cancer. The Food and Drug Administration prohibits excessive amounts of aflatoxins in human and animal food. Another example of a nucleotide-altering chemical is *nitrous acid* (HNO_2), which converts adenine into a guanine analog by removing the amine group from adenine. When a cell replicates DNA containing this guanine analog, an AT base pair is changed to a GC base pair in one daughter molecule—a base-pair-substitution mutation.

Frameshift Mutagens Still other mutagenic chemical agents insert or delete nucleotide base pairs, resulting in frameshift mutations. Examples of frameshift mutagens are *benzopyrene*, which is found in smoke; *ethidium bromide*, which is used to stain DNA; and *acridine*, one of a class of dyes commonly used as mutagens in genetic research. These chemicals are exactly the right size to slip between adjoining nucleotides in DNA, producing a bulge in the molecule (**FIGURE 7.28**). When DNA polymerase copies the misshapen strands, one or more base pairs may be inserted or deleted in the daughter strands.

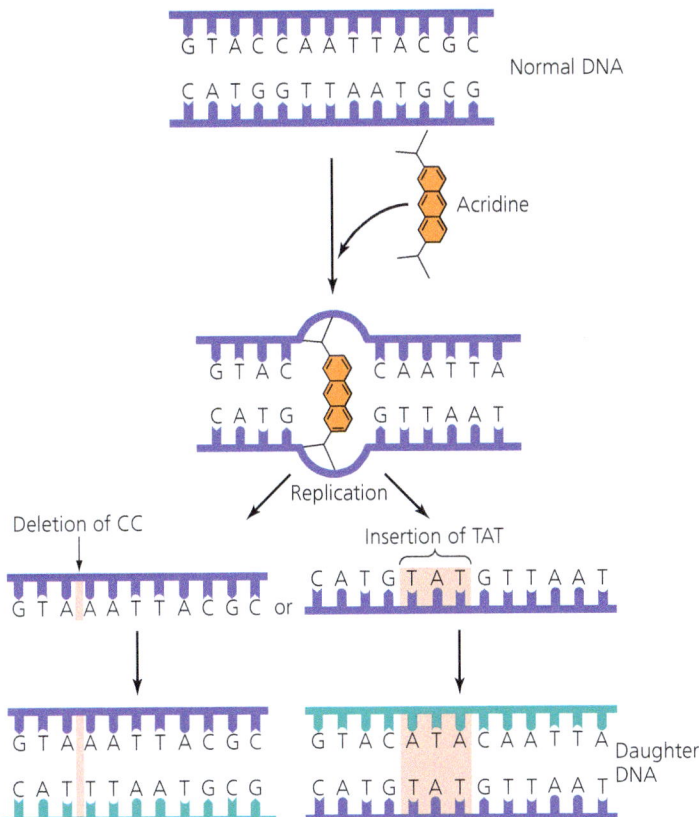

▲ **FIGURE 7.28 The action of a frameshift mutagen.** When DNA polymerase III passes the bulge caused by the insertion of acridine between the two DNA strands, it incorrectly synthesizes a daughter strand that contains either a deletion or an insertion mutation.

Frequency of Mutation

LEARNING OUTCOME

7.25 Discuss the relative frequency of deleterious and useful mutations.

Mutations are rare events. If they occurred often, organisms could not live or effectively reproduce themselves. As we have seen, about 1 of every 10 million (10^7) genes contains an error. Mutagens typically increase the mutation rate by a factor of 10 to 1000 times; that is, mutagens induce an error in one of every 10^6 to 10^4 genes.

Many mutations are deleterious because they code for nonfunctional proteins or stop transcription entirely. Cells without functional proteins cannot metabolize; therefore, deleterious mutations are removed from the population when the cells die. Rarely, however, a cell acquires a beneficial mutation that allows it to survive, reproduce, and pass the mutation to its descendants. Evolution involves a change in gene frequency in a population. For example, the tuberculosis bacterium has acquired a mutation that confers resistance to the antimicrobial drug rifampin. In patients taking rifampin, mutated bacterial cells survive and reproduce, increasing the frequency of the mutated gene in the population because cells without the mutation die. As long as rifampin is present, cells with such a mutation have an advantage over cells without the mutation, and the population evolves resistance to rifampin.

MICRO CHECK

12. What kind of mutation is characterized by having no effect on the final protein?
13. What is the effect of ultraviolet light on DNA?
14. Is a cell more likely to pass on a harmful or a beneficial mutation to its descendants?

DNA Repair

LEARNING OUTCOME

7.26 Describe base-excision repair, light repair of pyrimidine dimers, nucleotide-excision repair, mismatch repair, and error-prone repair, such as the SOS response.

We have seen that mutations rarely convey an advantage; most mutations are deleterious. To respond to the dangers mutations pose, cells have numerous methods for repairing damaged DNA. Here we consider three categories of repair: direct repair, single-strand repair, and repair of double-strand damage. The latter is highly error prone.

Play *Mutations: Repair*
@ Mastering Microbiology

Direct Repair

Direct repair systems are highly accurate because they correct damage to nucleotides in cases where an error exists in only one strand of DNA, so the correct base can be ascertained from the complementary strand. Direct repair systems can correct only one specific type of mutation.

One example of direct repair is **base-excision repair** in which one of several enzymes recognizes a specific incorrect base and removes and replaces only that base. For example, an enzyme replaces hypoxanthine (abbreviated as H), which is an altered adenine base, with a normal adenine base (**FIGURE 7.29a**).

Another form of direct repair is the repair of pyrimidine dimers by so-called **light repair**. Pyrimidine dimers form when adjoining pyrimidines (thymine or cytosine nucleotides) absorb ultraviolet light (UV) and covalently bond to one another, which prevents them from hydrogen bonding to their complementary purines. Many cells contain *photolyase*, an enzyme that is activated by visible light to break pyrimidine dimers, reversing the mutation and restoring the original DNA sequence (**FIGURE 7.29b**). However, each system can only fix one type of damage.

Single-Strand Repair

When damage cannot be repaired directly, then a cell cuts away a section of a single strand containing the damage and uses the complementary strand as a template to repair the gap. An example is **nucleotide-excision repair** (sometimes called *dark repair* because it can correct a pyrimidine dimer without using light). In this repair system, an enzyme removes a short section of the strand (about a dozen nucleotides) containing the dimer

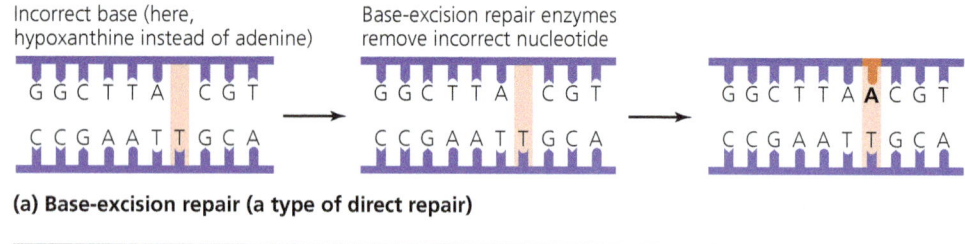
(a) Base-excision repair (a type of direct repair)

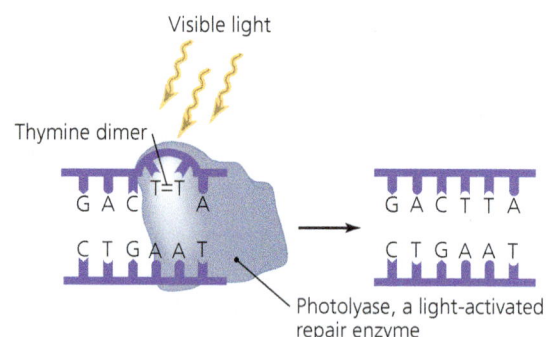
(b) Light repair (a type of direct repair)

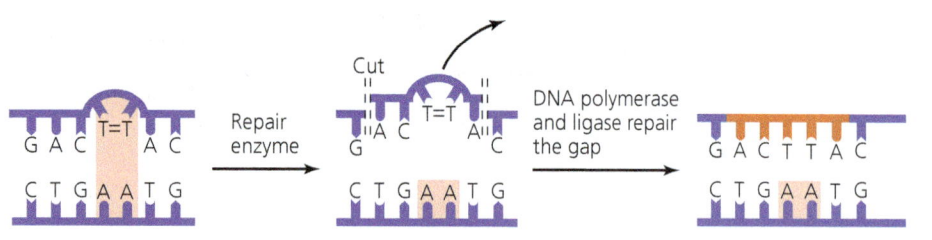
(c) Nucleotide-excision repair (a type of single-strand repair)

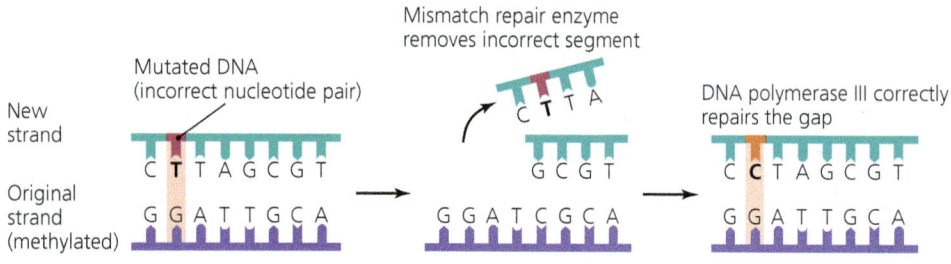
(d) Mismatch repair (a type of single-strand repair)

◀ **FIGURE 7.29 DNA repair mechanisms.** (a) Base-excision repair (type of direct repair), in which enzymes remove incorrect bases. Here, hypoxanthine (H) is corrected with adenine. (b) In another type of direct repair, called light repair, a light-activated enzyme, photolyase, breaks pyrimidine-to-pyrimidine bonds. (c) Nucleotide-excision repair is a type of single-strand repair in which a repair enzyme removes the entire damaged section from one strand of DNA, and DNA polymerase I and DNA ligase repair the breach. (d) Another type of single-strand repair is mismatch repair, which involves excision of a segment of a single strand. The repair enzyme distinguishes the original strand from the new strand because the original strand is methylated.

and then DNA polymerase I and DNA ligase repair the breach using the complementary, undamaged strand as a template (FIGURE 7.29c).

Another type of single-strand repair is mismatch repair, which is a repair mechanism that corrects errors produced during replication. **Mismatch repair** enzymes scan newly synthesized DNA looking for mismatched bases, which they remove and replace (FIGURE 7.29d). How does the mismatch repair system determine which of the two strands is in error? If it chose randomly, 50% of the time it would choose the wrong strand and introduce mutations. Mismatch repair enzymes, however, do not choose randomly. They distinguish a new DNA strand from an old strand because old strands are methylated. Recognition of an error as far as 1000 bp away from a methylated portion of DNA triggers mismatch repair enzymes. Once a new DNA strand is methylated, mismatch repair enzymes cannot correct any errors that remain, so mismatch repair must immediately follow DNA replication.

Error-Prone Repair

Sometimes damage to DNA is so extensive, perhaps even corrupting both strands of DNA, that regular repair mechanisms cannot cope with the damage. The repair system of last resort is *error-prone repair* (not shown in Figure 7.29). Geneticists best understand error-prone repair in *E. coli*, where it is called an **SOS response**. This form of repair is a last-ditch effort by the cell to salvage enough DNA following severe damage to allow the cell to remain alive. SOS repair involves a variety of processes, such as the production of novel DNA polymerases (called IV and V) capable of copying less-than-perfect DNA. These polymerases replicate DNA with little regard to the base sequence of the strand being repaired. Of course, this introduces many new and potentially fatal mutations, but presumably SOS repair allows a few offspring of these bacteria to survive.

Identifying Mutants, Mutagens, and Carcinogens

LEARNING OUTCOMES

7.27 Contrast positive and negative selection techniques for isolating mutants.

7.28 Describe the Ames test, and discuss its use in discovering carcinogens.

If a cell does not successfully repair a mutation, it and its descendants are called **mutants**. In contrast, cells normally found in nature (in the wild) are called **wild-type cells**. Scientists distinguish mutants from wild-type cells by observing or testing for altered phenotypes. Because mutations are rare and nonfatal mutations are even rarer, mutants can easily be "lost in the crowd." Therefore, researchers have developed methods to recognize mutants amidst their wild-type neighbors.

Positive Selection

Positive selection involves selecting a mutant by eliminating wild-type phenotypes. Assume, for example, that researchers

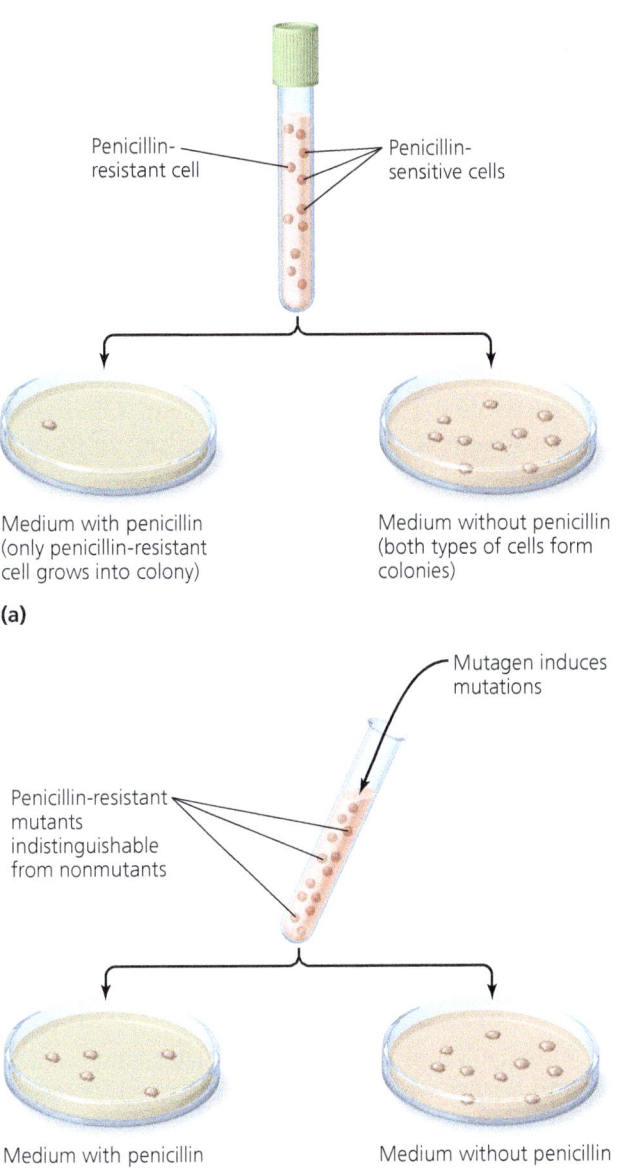

▲ FIGURE 7.30 **Positive selection of mutants.** Only mutants that are resistant to penicillin can survive on the plate containing this antibiotic. (a) Normally, a population includes very few mutants. (b) Introduction of a mutagen increases the number of mutants and thus the number of colonies that grow in the presence of penicillin. *What is the rate of mutation induced by the mutagen in (b)?*

Figure 7.30 $\frac{1}{5-1} \times 100\% = 400\%$

want to isolate penicillin-resistant bacterial mutants from a liquid culture. To do so, they spread the liquid medium, which contains mostly penicillin-sensitive cells but also the few penicillin-resistant mutants, onto a medium that includes penicillin. Only the penicillin-resistant mutants multiply on this medium and produce colonies (FIGURE 7.30a).

When a mutagenic agent is added to a liquid culture, it increases the number of mutants (FIGURE 7.30b). While the researchers are isolating mutants, they can also determine the rate of mutation by comparing the number of mutant colonies formed after use of the mutagen with the number formed before treatment. The rate of mutation can be calculated as follows:

$$\frac{\text{number of colonies seen} \atop \text{with use of mutagen} - \text{number of colonies seen} \atop \text{without use of mutagen}}{\text{number of colonies seen without the use of mutagen}} \times 100\%$$

Negative (Indirect) Selection

An organism that requires different nutrients than its wild-type phenotype requires is known as an *auxotroph*[7] (awk′sō-trōf). For example, a mutant bacterium that has lost the ability to synthesize tryptophan is auxotrophic for this amino acid—it must acquire tryptophan from the environment. Obviously, if a researcher attempts to grow tryptophan auxotrophs on media lacking tryptophan, it will not be possible; the bacteria will die because tryptophan is a necessary amino acid for proper synthesis of proteins. Therefore, to isolate such auxotrophs, scientists can use a technique called **negative (indirect) selection**, which was developed by microbiologist Esther Lederberg (1922–2006). The process by which a researcher uses negative selection to culture a tryptophan auxotroph is as follows (FIGURE 7.31).

1. The researcher inoculates a sample of a bacterial suspension containing potential mutants onto a plate containing complete media (including tryptophan). The sample is diluted such that the plate receives only about 100 cells.
2. Both auxotrophs and wild-type cells reproduce and form colonies on the plate, though the colonies are indistinguishable.
3. The researcher picks up cells from all the colonies on the plate simultaneously with a sterile velvet pad by pressing the pad onto the plate.
4. The researcher inoculates two new plates—one containing tryptophan, the other lacking tryptophan—by pressing the pad onto each of them. This technique is called *replica plating*.
5. After the plates have incubated for several hours, the researcher compares the two replica plates. Tryptophan auxotrophs growing on medium containing tryptophan are revealed by the absence of a corresponding colony on the plate lacking tryptophan.
6. The researcher takes cells of the auxotroph colony from the replica plate and inoculates them into a complete medium. The auxotroph is now isolated.

The Ames Test for Identifying Mutagens

Numerous chemicals in food, the workplace, and the environment in general have been suspected of being carcinogenic (kar′si-nō-jen′ik) mutagens; that is, of causing mutations that result in cancer. Because animal tests to prove that they are indeed **carcinogenic** are expensive and time consuming, researchers have used a fast and inexpensive method of screening for mutagens called an **Ames test**, which is named for its inventor, Bruce Ames (1928–).

---

[7]From Greek *auxein*, meaning "to increase," and *trophe*, meaning "nutrition."

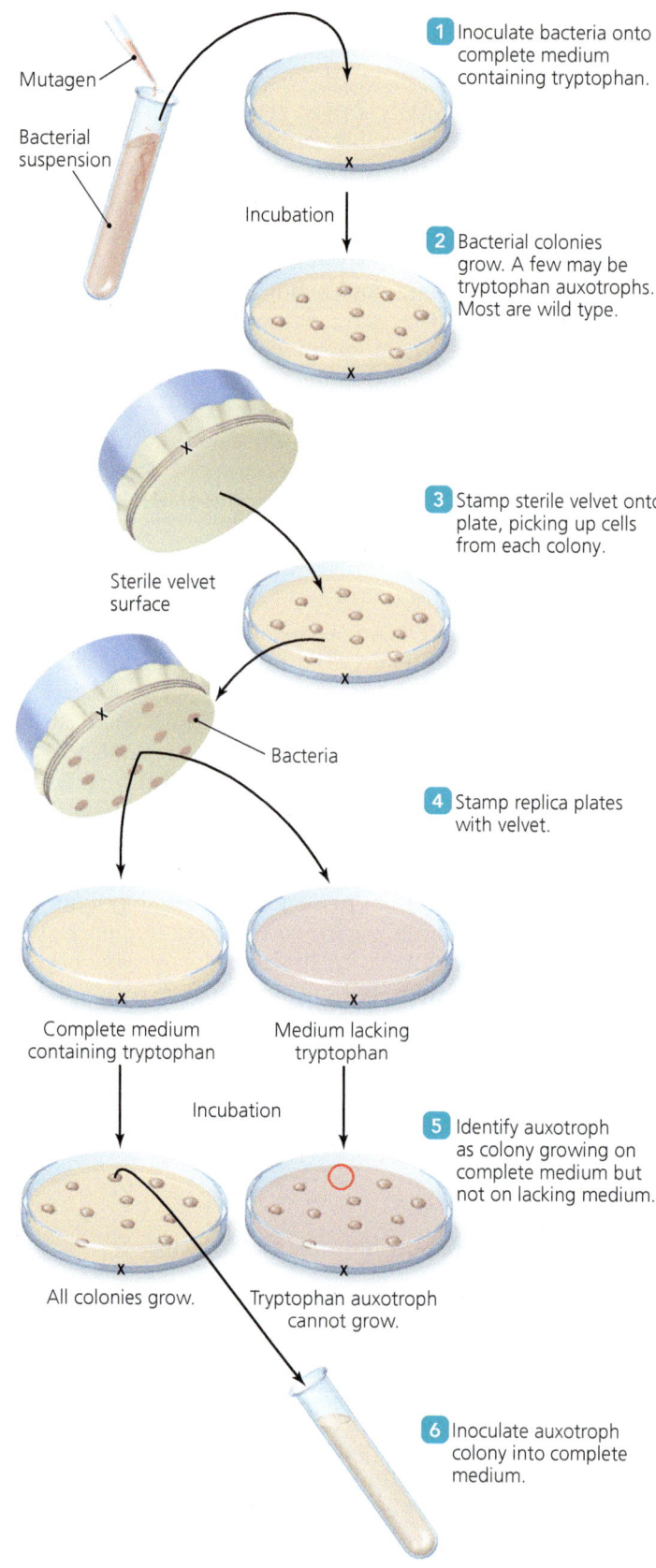

▲ FIGURE 7.31 **The use of negative (indirect) selection to isolate a tryptophan auxotroph.** The plates are marked (in this case, with an X) so that their orientation can be maintained throughout the procedure. Researchers may have to inoculate hundreds of such plates to identify a single mutant.

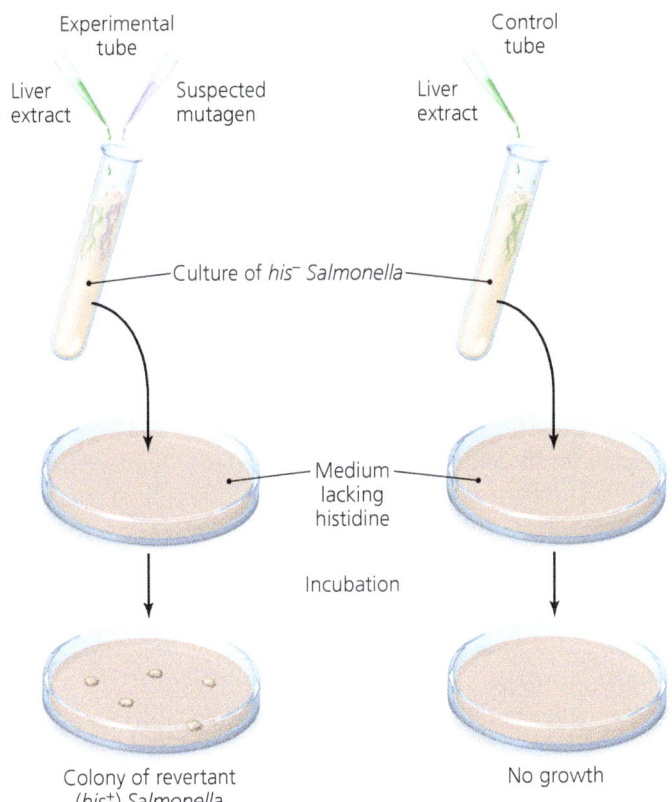

▲ FIGURE 7.32 **The Ames test.** A mixture containing *his⁻ Salmonella* mutants, rat liver extract, and the suspected mutagen is inoculated onto a plate lacking histidine. Colonies will form only if a mutagen reverses the *his⁻* mutation, producing revertant *his⁺* organisms with the ability to synthesize histidine. A control tube that lacks the suspected mutagen demonstrates that reversion did not occur in the absence of the mutagen. *What is the purpose of liver extract in an Ames test?*

Figure 7.32 *Liver extract simulates conditions in the body by providing enzymes that may degrade harmless substances into mutagens.*

An Ames test uses mutant *Salmonella* (sal'mŏ-nel'ă) bacteria possessing a point mutation that prevents the synthesis of the amino acid histidine; in other words, they are histidine auxotrophs, indicated by the abbreviation *his⁻*. To perform the test, an investigator mixes *his⁻* mutants with the substance suspected to be a mutagen and with liver extract (FIGURE 7.32). The presence of liver extract simulates the conditions in the body under which liver enzymes can turn harmless chemicals into mutagens. The researcher then spreads the mixture on a solid medium lacking histidine. If the suspected substance does in fact cause mutations, some of the mutations will likely reverse the effect of the original point mutation, producing *revertant cells* (designated *his⁺*), which are cells that have regained the wild-type phenotype; in this case, they have regained the ability to synthesize histidine and thus can survive on a medium lacking histidine. Thus, the presence of colonies during an Ames test reveals that the suspected substance is mutagenic in *Salmonella*.

Given that DNA in all cells is very similar, the ability of a chemical to cause mutations in *Salmonella* indicates that it is likely to cause mutations in humans as well. Some mutations cause cancers, so the mutagenic chemical may also be carcinogenic. To prove that the substance can in fact cause cancer, scientists must test it in laboratory animals, a process that usually is warranted only if a chemical is mutagenic in *Salmonella*. Ames testing reduces the cost and time required to assay chemicals for carcinogenicity in animals.

MICRO CHECK

15. A researcher testing for an antibiotic resistance mutation places a culture on a plate containing the antibiotic. What type of selection technique is she most likely using?

TELL ME WHY

Changes in RNA sequences resulting from poor transcription are not as deleterious to an organism as changes to DNA resulting from mutations. Why is this the case?

Genetic Recombination and Transfer

LEARNING OUTCOME
7.29 Define *genetic recombination*.

Genetic recombination refers to the exchange of nucleotide sequences between two DNA molecules and often involves segments that are composed of identical or nearly identical nucleotide sequences called *homologous sequences*. Scientists have discovered a number of molecular mechanisms for genetic recombination, one of which is illustrated simply in FIGURE 7.33. In this type of recombination, enzymes nick one strand of DNA at the homologous sequence, and another enzyme inserts the nicked strand into the second DNA molecule. Ligase then reconnects the strands in new combinations, and the molecules resolve themselves into novel molecules. DNA molecules that contain such new arrangements of nucleotide sequences are called *recombinant DNA*. An example of genetic recombination is crossing over during gamete formation, part of sexual reproduction in eukaryotes. Cells that contain recombinant DNA are called *recombinants*. (Chapter 12 discusses crossing over in more detail.)

MICRO CHECK

16. What is genetic recombination?

Horizontal Gene Transfer Among Prokaryotes

LEARNING OUTCOMES
7.30 Contrast vertical gene transfer with horizontal gene transfer.
7.31 Explain the roles of an F factor, F⁺ cells, and Hfr cells in bacterial conjugation.
7.32 Compare and contrast crossing over, transformation, transduction, and conjugation.

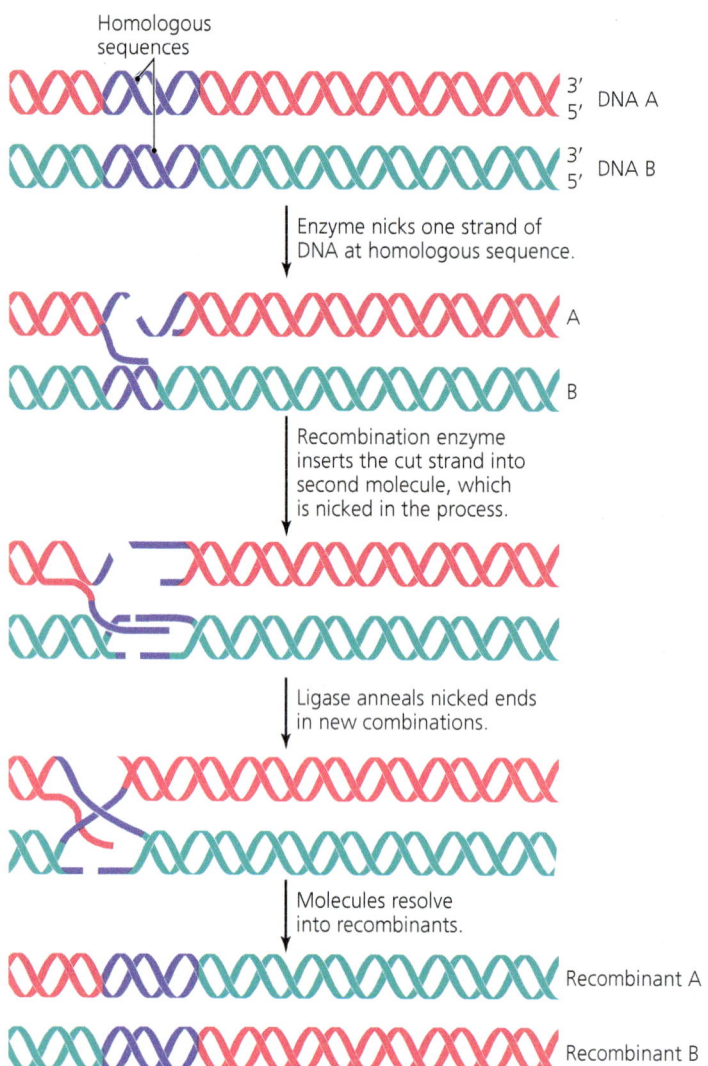

▲ FIGURE 7.33 **One type of genetic recombination.** Simplified depiction of one type of recombination between two DNA molecules. After an enzyme nicks one strand (here, in DNA), a recombination enzyme rearranges the strands, and ligase seals the gaps to form recombinant molecules. *What is the normal function of ligase (during DNA replication)?*

Figure 7.33 *Ligase functions to anneal Okazaki fragments during replication of the lagging strand.*

As we have discussed, both prokaryotes and eukaryotes replicate their genomes and supply copies to their descendants. This is known as *vertical gene transfer*—the passing of genes to the next generation. Many prokaryotes can acquire genes from other microbes of the same generation—a process termed **horizontal (lateral) gene transfer**. In horizontal gene transfer, a donor cell contributes part of its genome to a **recipient cell**, which may be of a different species from the donor. Typically, a recipient cell inserts part of a donor's DNA into its own chromosome, becoming a **recombinant cell**. Cellular enzymes then usually degrade remaining unincorporated donor DNA. Horizontal gene transfer is a rare event, typically occurring in less than 1% of a population of prokaryotes.

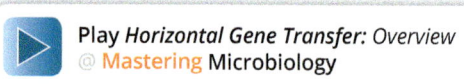

Here, we consider the three types of horizontal gene transfer: *transformation*, *transduction*, and *bacterial conjugation*.

Transformation

In **transformation**, a recipient cell takes up DNA from the environment, such as DNA that might be released by dead organisms. Frederick Griffith (1879–1941) discovered this process in 1928 while studying pneumonia caused by *Streptococcus pneumoniae* (strep-tō-kok'ŭs nū-mō'nē-ī). Griffith worked with two strains of *Streptococcus*. Cells of the first strain have protective capsules that enable them to escape a body's defensive white blood cells; thus, these encapsulated cells cause deadly septicemia when injected into mice (FIGURE 7.34a). They are called *strain S* because they form smooth colonies on an agar surface. The application of heat kills such encapsulated cells and renders them harmless when injected into mice (FIGURE 7.34b). In contrast, cells of the second strain (called *strain R* because they form rough colonies) are mutants that cannot make capsules. The unencapsulated cells of strain R are much less virulent, only infrequently causing disease (FIGURE 7.34c) because a mouse's defensive white blood cells quickly devour them.

Griffith discovered that when he injected both heat-killed strain S and living strain R into a mouse, the mouse died even though neither of the injected strains was harmful when administered alone (FIGURE 7.34d). Further, and most significantly, Griffith isolated numerous living bacterial cells from the dead mouse, cells that had capsules. He realized that harmless, unencapsulated strain R bacteria had been transformed into deadly, encapsulated strain S bacteria. Subsequent investigations showed that transformation, by which cells take up DNA from the environment, also occurs *in vitro*[8] (FIGURE 7.34e).

The fact that the living encapsulated cells retrieved at the end of this experiment outnumbered the dead encapsulated cells injected at the beginning indicated that strain R cells were not merely appropriating capsules released from dead strain S cells. Instead, strain R cells had acquired the capability of producing their own capsules by assimilating the capsule-coding genes of strain S cells. In 1944, Oswald Avery (1877–1955), Colin MacLeod (1909–1972), and Maclyn McCarty (1911–2005) extracted various chemicals from S cells and determined that the transforming agent was DNA. This discovery was one of the conclusive pieces of evidence that DNA is the genetic material of cells.

Cells that have the ability to take up DNA from their environment are said to be **competent**. Competence results from expression of a transporter protein that crosses the cytoplasmic membrane and brings DNA into the cell. Scientists have observed natural competence in a few types of bacteria, including pathogens belonging to the genera *Streptococcus*, *Haemophilus* (hē-mof'i-lŭs), *Neisseria* (nī-se'rē-ă), *Bacillus* (ba-sil'ŭs), *Staphylococcus* (staf'i-lō-kok'ŭs), and *Pseudomonas*. Scientists can also generate competence artificially in *Escherichia* and other bacteria by manipulating the temperature and salt content of

---
[8]Latin, meaning "within glassware."

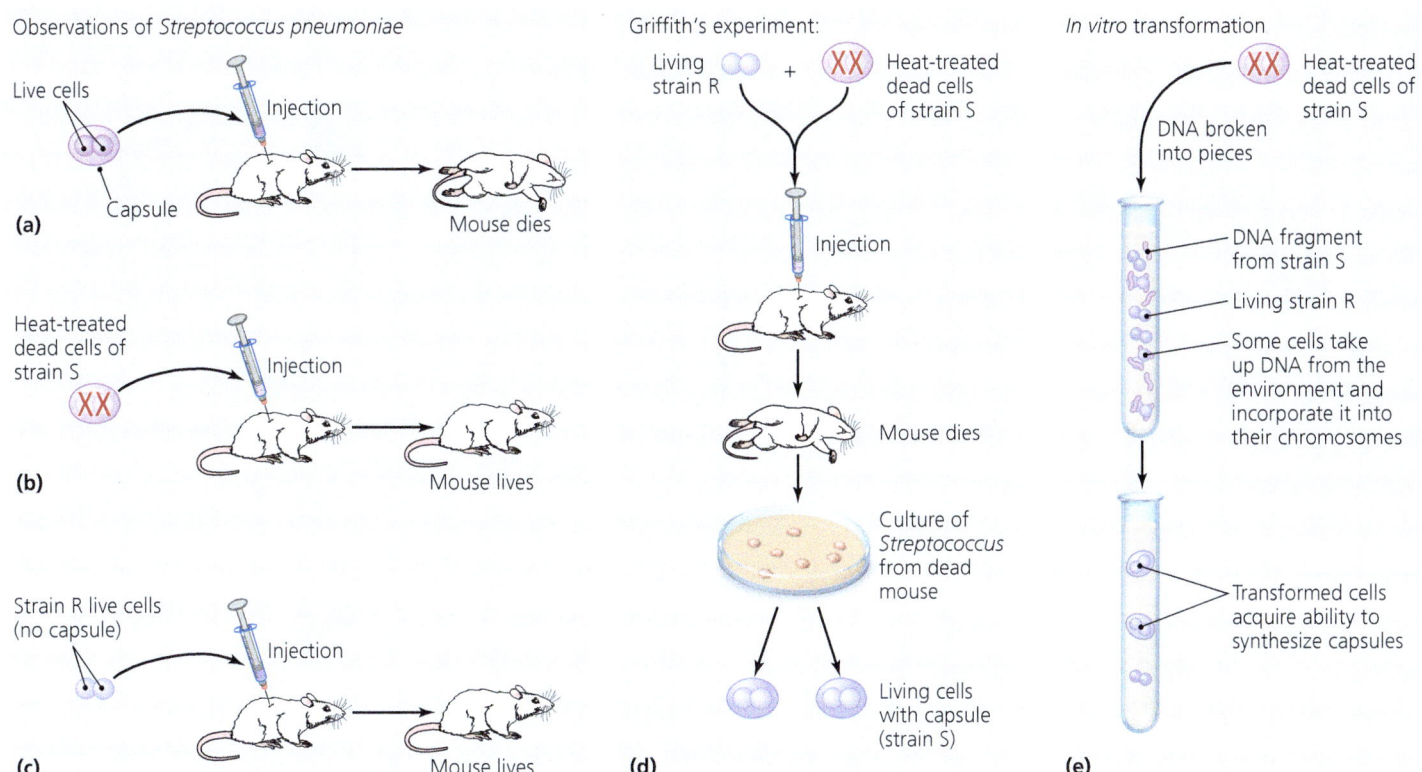

▲ **FIGURE 7.34 Transformation of *Streptococcus pneumoniae*.** Griffith's observations revealed that **(a)** encapsulated strain S killed mice, **(b)** heating renders strain S harmless to mice, and **(c)** unencapsulated strain R did not harm mice. **(d)** In Griffith's experiment, a mouse injected concurrently with killed strain S and live strain R (each harmless) died and was found to contain numerous living, encapsulated bacteria. **(e)** A demonstration that transformation of R cells to S cells also occurs *in vitro*.

the medium. Because competent cells take up DNA from any donor genome, competence and transformation are important tools in recombinant DNA technology, commonly called *genetic engineering*.

Transduction

A second method of horizontal gene transfer, called **transduction**, involves the transfer of DNA from one cell to another via a replicating virus. Transduction can occur either between prokaryotic cells or between eukaryotic cells; it is limited only by the availability of a virus that can infect both donor and recipient cells. Here, we will consider transduction in bacteria.

A virus that infects bacteria is called a **bacteriophage** or simply a **phage** (fāj).[9] The process by which a phage participates in transduction is depicted in **FIGURE 7.35**. To replicate, a bacteriophage attaches to a bacterial host cell and injects its genome into the cell ❶. Phage enzymes (enzymes coded by the phage's genome even though translated by bacterial ribosomes) degrade the cell's DNA ❷. The phage genome now controls the cell's functions and directs the cell to synthesize new phage DNA and phage proteins. Normally, phage proteins assemble around phage DNA to form new daughter phages, but some phages mistakenly incorporate remaining fragments of bacterial DNA. This forms **transducing phages** ❸. Eventually, the host cell lyses, releasing daughter and transducing phages. Transduction occurs when a transducing phage injects donor DNA into a new host cell (the recipient) ❹. The recipient host cell incorporates the donated DNA into its chromosome by recombination ❺.

In *generalized transduction*, the transducing phage carries a random DNA segment from a donor host cell's chromosome or plasmids to a recipient host cell. Generalized transduction is not limited to a particular DNA sequence. In *specialized transduction*, only certain host sequences are transferred. In nature, specialized transduction is important in transferring genes encoding certain bacterial toxins—including those responsible for diphtheria, scarlet fever, and the bloody, life-threatening diarrhea caused by *E. coli* O157:H7—into cells that would otherwise be harmless. (Chapter 8 discusses the use of specialized transduction to intentionally insert genes into cells.)

---

[9]From Greek *phagein*, meaning "to eat."

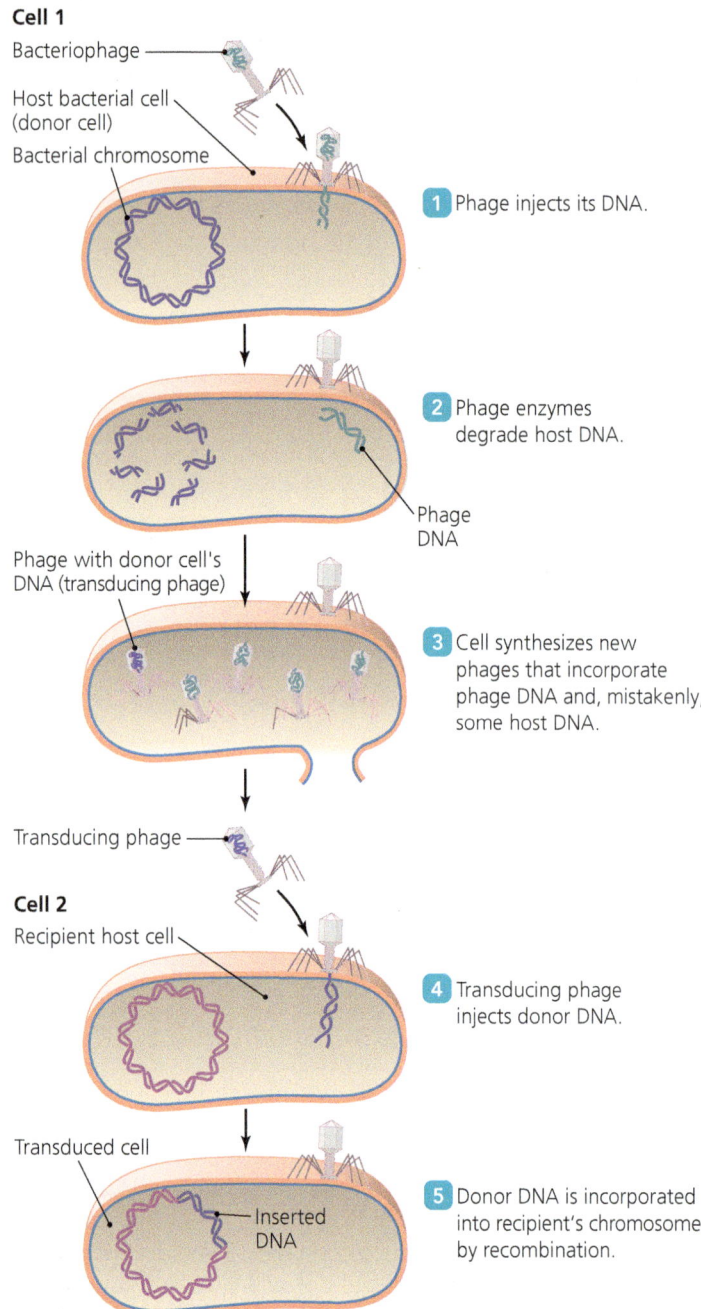

▲ FIGURE 7.35 **Transduction.** After a virus called a bacteriophage (phage) attaches to a host bacterial cell, it injects its genome into the cell and directs the cell to synthesize new phages. During assembly of new phages, some host DNA may be incorporated, forming transducing phages, which subsequently carry donor DNA to a recipient host cell.

Bacterial Conjugation

A third method of genetic transfer in bacteria is **conjugation**.[10] Unlike the typical donor cells in transformation and transduction, a donor cell in conjugation remains alive. Further, conjugation requires physical contact between donor and recipient cells.

Scientists discovered conjugation between cells of *E. coli*, and it is best understood in this species. Thus, the remainder of our discussion will focus on conjugation in this bacterium.

Conjugation is mediated by **pili** (pī′lī; singular: *pilus*, pī′lus), also called *conjugation* or *sex pili*,[11] which are thin, proteinaceous tubes extending from the surface of a cell. The gene coding for conjugation pili is located on a plasmid called an **F (fertility) plasmid**, which was discovered by Esther Lederberg. (Recall that a plasmid is a small, circular, extrachromosomal molecule of DNA; see Figure 7.2b.) Cells that contain an F plasmid are called F^+ cells ("ef plus"), and they serve as donors during conjugation. Recipient cells are F^-; that is, they lack an F plasmid and therefore have no pili.

Despite the historical use of the terms *sex* and *fertility*, conjugation is completely unrelated to sexual reproduction, which involves mixing DNA from two organisms. Conjugation is a unidirectional method of horizontal gene transfer from a donor to a recipient.

The process of bacterial conjugation is illustrated in **FIGURE 7.36**. First, a sex pilus connects a donor cell (F^+) to a recipient cell (F^-) 1. The pilus may draw the cells together 2, though scientists have filmed DNA transfer between cells 12 μm apart. An enzyme, coded by the plasmid, nicks one strand of the plasmid DNA, and this single strand transfers into the recipient, beginning with a section called the *origin of transfer* 3. Both cells synthesize complementary strands of F plasmid DNA to form complete double-stranded DNA circles. The F^- recipient is now an F^+ cell 4.

In some bacterial cells, the F plasmid does not remain independent in the cytosol but instead integrates into the cellular chromosome. Such cells, which are called **Hfr (high frequency of recombination) cells**, can conjugate with F^- cells (**FIGURE 7.37**). After the F plasmid has integrated 1 and the Hfr and F^- cells join via a sex pilus 2, DNA transfer begins at the F plasmid's origin of transfer and begins to carry a copy of the donor's chromosome into the recipient 3. In most cases, movement of the cells breaks the intercellular connection before the entire donor chromosome is transferred 4. Since the recipient receives only a portion of the F plasmid, it remains an F^- cell; however, it acquires some chromosomal genes from the donor. Recombination can integrate the donor's DNA into the recipient's chromosome 5. The recipient is now a recombinant cell because it contains some donor genes.

Because an F plasmid can integrate into a chromosome at only a few locations, the order in which donor genes are transferred into the recipient cell is consistent. Scientists produced the first gene maps of bacterial chromosomes by noting the time, in

---

[10]From Latin *conjugatus*, meaning "yoked together."

[11]Bacterial conjugation may resemble intercourse, but because no gametes are involved, it is not true sex.

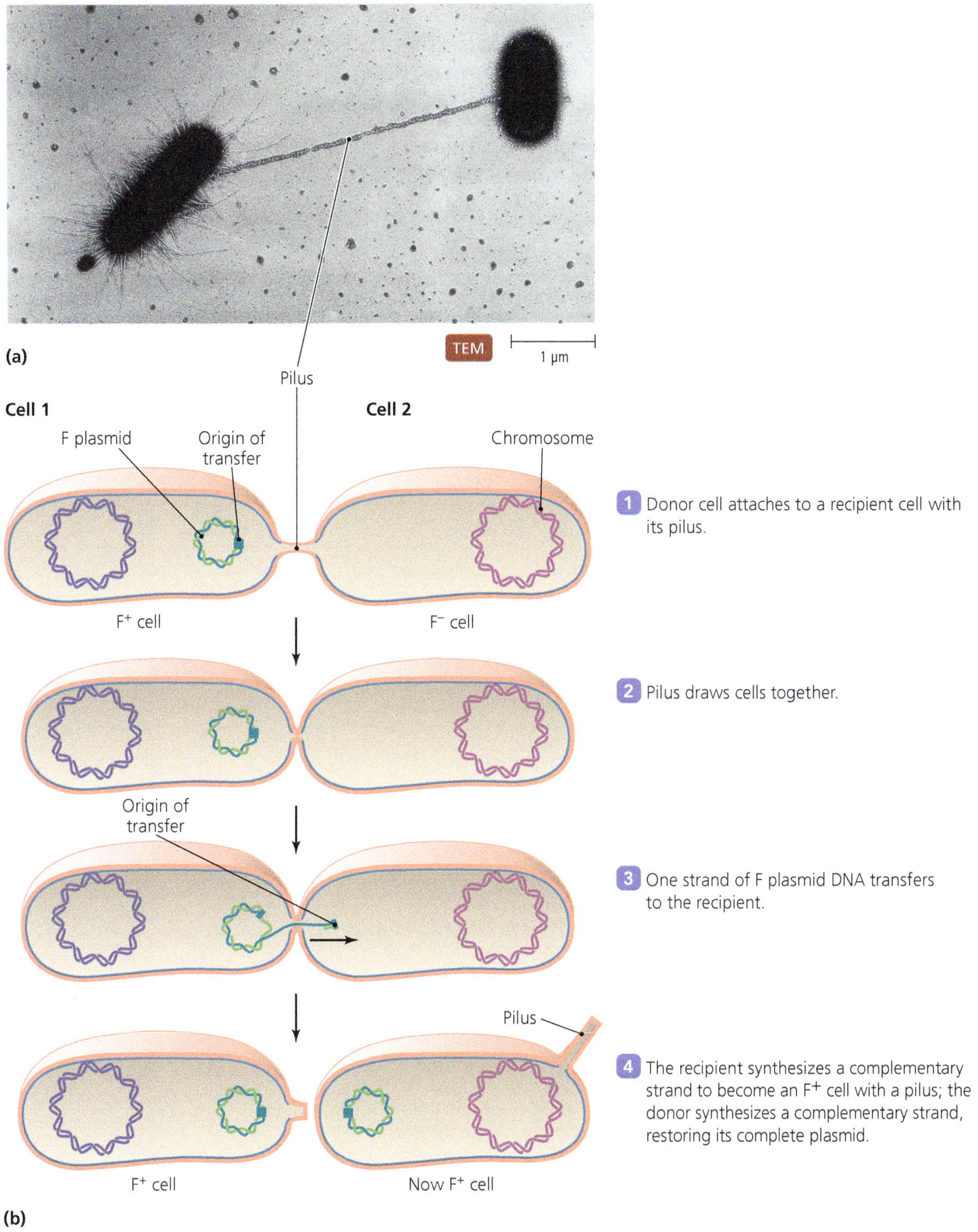

FIGURE 7.36 Bacterial conjugation.
A conjugation pilus connecting two cells mediates the transfer of DNA between the cells. (a) Two *E. coli* cells are connected by a pilus. (b) Artist's rendition of the process.

1. Donor cell attaches to a recipient cell with its pilus.
2. Pilus draws cells together.
3. One strand of F plasmid DNA transfers to the recipient.
4. The recipient synthesizes a complementary strand to become an F⁺ cell with a pilus; the donor synthesizes a complementary strand, restoring its complete plasmid.

minutes, required for a particular gene to transfer from donor cell to recipient cell.

Play *Conjugation: Chromosome Mapping*
@ Mastering Microbiology

Conjugation occurs in several species of bacteria, which can be quite promiscuous; that is, conjugation can occur among bacteria of widely varying kinds and even between a bacterium and a yeast cell or between a bacterium and a plant cell. For example, the crown gall bacterium, *Agrobacterium*, transfers some of its genes into the chromosome of its host plant.

The natural transfer of genes by conjugation among diverse organisms heightens some scientists' concerns about the spread of resistance (R) plasmids among pathogens. These scientists note that antibiotic resistance developed by one pathogen can spread to several other pathogens. This is of particular concern in health care settings.

TABLE 7.5 (on p. 228) summarizes the mechanisms of horizontal gene transfer in bacteria.

228 CHAPTER 7 Microbial Genetics

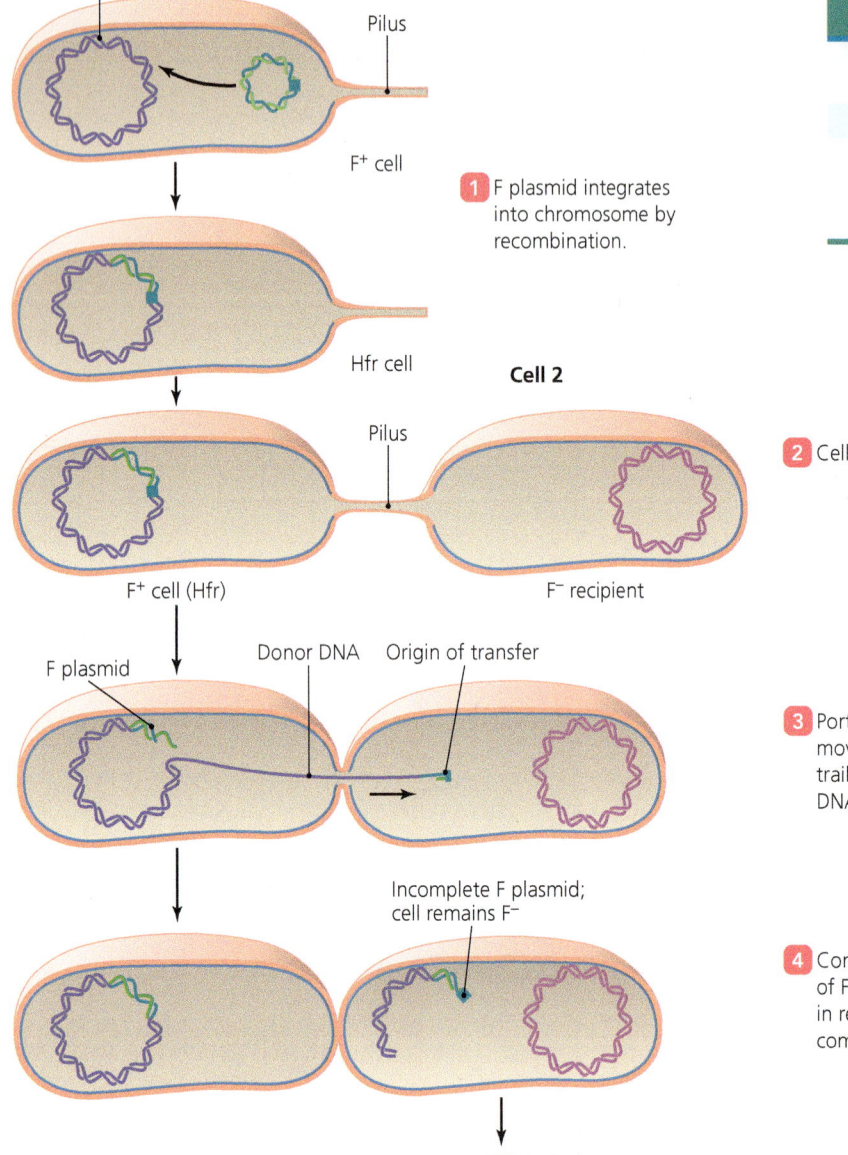

▲ **FIGURE 7.37 Conjugation involving an Hfr cell.** An Hfr cell is formed when an F⁺ cell integrates its F plasmid into its chromosome. Hfr cells donate a partial copy of their DNA and a portion of the F plasmid to a recipient, which is rendered a recombinant cell but remains F⁻.

TABLE 7.5 Natural Mechanisms of Horizontal Genetic Transfer in Bacteria

Mechanism	Requirements
Transformation	Free DNA in the environment and a competent recipient
Transduction	Bacteriophage
Conjugation	Cell-to-cell contact and F plasmid, which is either in cytosol or incorporated into chromosome of donor (Hfr) cell

Transposons and Transposition

LEARNING OUTCOMES

7.33 Describe the structures and actions of simple and complex transposons.
7.34 Describe transposition and its effects.

Transposons[12] are segments of DNA, 700 to 40,000 bp in length, that transpose (move) themselves from one location in a DNA molecule to another location in the same or a different molecule.

---
[12]From *transposable* elements.

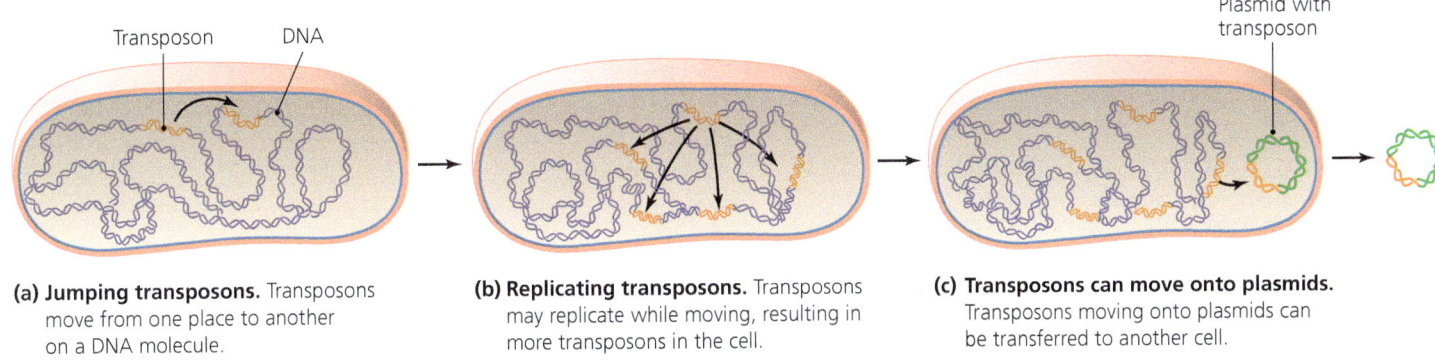

(a) **Jumping transposons.** Transposons move from one place to another on a DNA molecule.

(b) **Replicating transposons.** Transposons may replicate while moving, resulting in more transposons in the cell.

(c) **Transposons can move onto plasmids.** Transposons moving onto plasmids can be transferred to another cell.

▲ **FIGURE 7.38 Transposition.** Transposons move from place to place within, among, and between chromosomes and plasmids. Plasmids can carry transposons to and from cells.

American geneticist Barbara McClintock (1902–1992) discovered these "jumping genes" through a painstaking analysis of the colors of the kernels of corn. She discovered that the genes for kernel color were turned on and off by the insertion of transposons. Subsequent research has shown that transposons are found in many, if not all, prokaryotes and eukaryotes and in many viruses.

The result of the action of a transposon is termed **transposition**; in effect, it is a kind of frameshift insertion (see Figure 7.25e.). This "illegitimate" recombination does not need a region of homology, unlike other recombination events. Transposition occurs between plasmids and chromosomes and within and among chromosomes (FIGURE 7.38). Further, plasmids can carry transposons from one cell to another. Fortunately, although transposons are common, transposition and the frameshift mutations it causes are relatively rare occurrences.

Play *Transposons: Overview*
Mastering Microbiology

Transposons vary in their nucleotide sequences, but all of them contain palindromic sequences at each end. A *palindrome*[13] is a word, phrase, or sentence that has the same sequence of letters when read backward or forward—for example, "Madam, I'm Adam." In genetics, a palindrome is a region of DNA in which the sequence of nucleotides is identical to an inverted sequence in the complementary strand. For example, GAATTC is the palindrome of CTTAAG. Such a palindromic sequence is also known as an **inverted repeat (IR)**.

The simplest transposons, called **insertion sequences (IS)**, consist of no more than two inverted repeats and a gene that encodes the enzyme *transposase* (FIGURE 7.39a). Transposase recognizes its own inverted repeat in a target site, cuts the DNA at that site, and inserts the transposon (or a copy of it) into the DNA molecule at that site (FIGURE 7.39b). Such transposition also produces a duplicate copy of the target site.

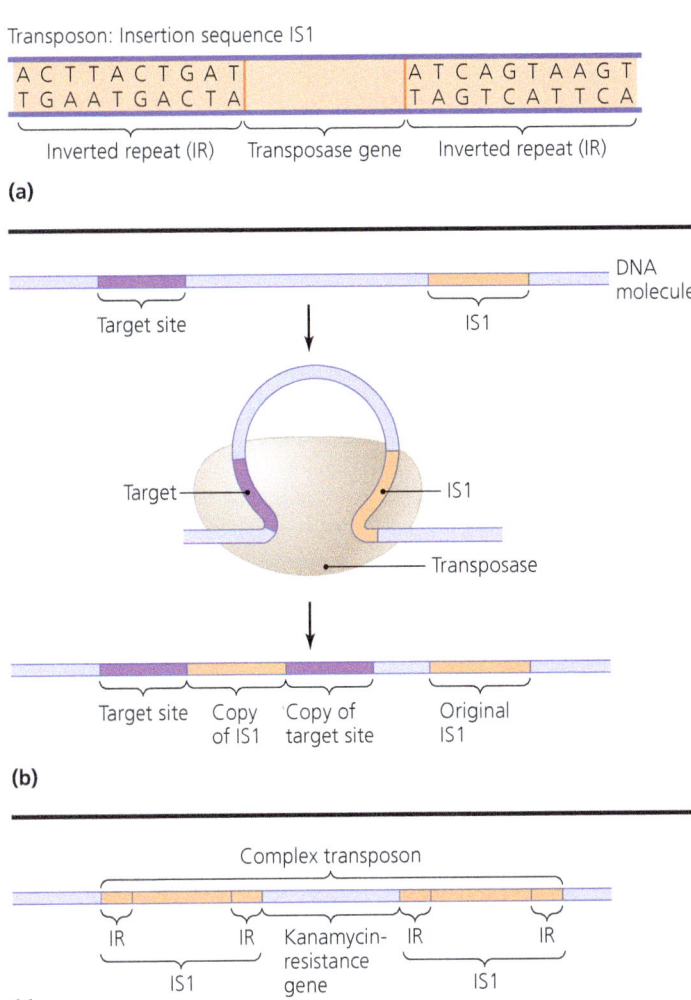

▲ **FIGURE 7.39 Transposons. (a)** A simple transposon, or insertion sequence. Shown here is insertion sequence I (IS1), which consists of a gene for the enzyme transposase bounded by identical (though inverted) repeats of nucleotides. **(b)** Transposase recognizes its target site (elsewhere in the same DNA molecule, as shown here, or in a different DNA molecule); then it moves the transposon (or, as in the case of IS1, a copy of the transposon) to its new site. The target site is duplicated in the process. **(c)** A complex transposon, which contains genes not related to transposition. Shown here is Tn5, which consists of a gene for kanamycin resistance between two IS1 transposons.

[13] From Greek *palin*, meaning "again," and *dramein*, meaning "to run."

Complex transposons contain one or more genes not connected with transposition, such as genes for antibiotic resistance (FIGURE 7.39c). *R plasmids* often contain transposons. R factors are of great clinical concern because they spread antibiotic resistance among pathogens.

 Play *Transposons: Insertion Sequences, Complex Transposons* @ Mastering Microbiology

(Chapter 8 discusses how scientists have adapted the natural processes of transformation, transduction, conjugation, and transposition to manipulate the genes of organisms.)

> **MICRO CHECK**
> 17. What kind of gene transfer between bacterial cells in the same generation utilizes the help of a virus?
> 18. How does transposition differ from other recombination events?

> **TELL ME WHY**
> Why is the genetic ancestry of microbes much more difficult to ascertain than the ancestry of animals?

CLINICAL CASE STUDY

Deadly Horizontal Gene Transfer

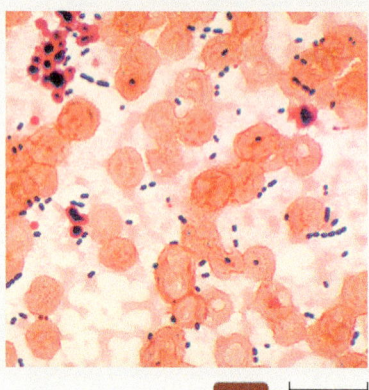

Sarah might have been 85 years old, but her mind was still sharp; it was her body that was failing. After she broke her hip by falling in the bathroom, she had succumbed to a series of complications that ultimately required admittance to a large university-associated hospital in New York City. Treatment necessitated a urinary catheter, a feeding tube, and intensive antibiotic therapy. While hospitalized, she acquired a life-threatening urinary tract infection with multi-drug-resistant (MDR) *Enterococcus faecium*.

Enterococci are normal members of the microbiota of the colon, yet these Gram-positive bacteria have a propensity to acquire genes that convey resistance to antimicrobial drugs. As a result of such horizontal gene transfer, *E. faecium* has become an MDR bacterium—that is, a strain resistant to many different kinds of antimicrobials. Drug-resistant enterococci are a leading cause of healthcare-associated infections (nosocomial infections).

Physicians treated Sarah with a series of antimicrobial drugs, including metronidazole, penicillin, erythromycin, ciprofloxacin, and vancomycin, but to no avail. After 90 days of hospitalization, she died.

1. Define *healthcare-associated infection*.
2. What is the likely source of Sarah's infection?
3. List three ways by which *E. faecium* might have acquired genes for drug resistance.
4. How can hospital personnel prevent the spread of resistant *E. faecium* throughout the hospital?

MICRO IN THE CLINIC FOLLOW-UP

Another Stroke?

The doctor and nurses are able to stabilize Ben, and he is transferred to the hospital. Upon arrival, Ben has a fever of 103.1°F, and his white blood cell (WBC) count is elevated. Blood and urine cultures are collected and sent for lab analysis. The elevated WBC count suggests an infection, so the doctor starts Ben on meropenem, an ultra-broad-spectrum antibiotic.

The next day, Ben does not appear to be responding to the meropenem treatment. The lab results indicate multi-drug-resistant *Klebsiella pneumoniae* present in Ben's blood sample. This diagnosis is cause for concern because this organism is likely a carbapenem-resistant member of the Enterobacteriaceae, a family of bacteria that have high levels of resistance to multiple antibiotics. The doctor puts Ben on a course of high-dose tigecycline, an antibiotic that has shown effectiveness against carbapenem-resistant Enterobacteriaceae (CRE). The next few days bring little improvement to Ben's condition. His condition continues to decline, and he ultimately suffers multiple organ failure and dies.

1. The prevalence of CRE infections in health care settings has increased significantly in recent years, in part because these organisms are able to share plasmid-encoded resistance genes. Which method of horizontal gene transfer is most likely responsible for transfer of the plasmid between bacteria? Explain your choice.
2. Ben most likely acquired the infection while he was in the hospital recovering from his stroke. What are some possible sources of infection in the hospital that might have led to Ben's infection?

Check your answers to Micro in the Clinic Follow-Up questions in the Mastering Microbiology Study Area.

CHAPTER SUMMARY

Go to Mastering Microbiology for Interactive Microbiology, Dr. Bauman's Video Tutors, Micro Matters, MicroFlix, Micro-Boosters, practice quizzes, and more.

The Structure and Replication of Genomes (pp. 194–202)

1. **Genetics** is the study of inheritance and inheritable traits. **Genes** are composed of specific sequences of nucleotides that code for polypeptides or RNA molecules. A **genome** is the sum of all the genes and linking nucleotide sequences in a cell or virus. Prokaryotic and eukaryotic cells use DNA as their genetic material; some viruses use DNA, and other viruses use RNA.

2. The two strands of DNA are held together by hydrogen bonds between complementary **base pairs (bp)** of **nucleotides**. Adenine bonds with thymine, and guanine bonds with cytosine.

3. Bacterial and archaeal genomes consist of one (or, rarely, two or more) **chromosomes**, which are typically circular molecules of DNA associated with protein and RNA molecules, localized in a region of the cytosol called the **nucleoid**. Archaeal DNA organizes around globular proteins called **histones**. Prokaryotic cells may also contain one or more extrachromosomal DNA molecules called **plasmids**, which contain genes that regulate nonessential life functions, such as bacterial conjugation and resistance to antibiotics.

4. In addition to DNA, eukaryotic chromosomes contain eukaryotic histones and are arranged as **nucleosomes** (beads of DNA) that clump to form **chromatin fibers**. Eukaryotic cells may also contain extranuclear DNA in mitochondria, chloroplasts, and plasmids.

5. DNA replication is semiconservative; that is, each newly synthesized strand of DNA remains associated with one of the parental strands. After helicase unwinds and unzips the original molecule, synthesis of each of the two daughter strands—called the **leading strand** and the **lagging strand**—occurs from 5′ to 3′. Synthesis is mediated by enzymes that prime, join, and proofread the pairing of new nucleotides.

Play *DNA Replication: Overview, Forming the Replication Fork, Replication Proteins, Synthesis* @ Mastering Microbiology

6. After DNA replication, **methylation** occurs. Methylation plays several roles, including the control of gene expression, the initiation of DNA replication, recognition of a cell's own DNA, and repair.

Gene Function (pp. 202–216)

1. The **genotype** of an organism is the actual set of genes in its genome, whereas its **phenotype** refers to the physical and functional traits expressed by those genes.

2. RNA has several forms. These include **RNA primer**; **messenger RNA (mRNA)**, which carries genetic information from DNA to a ribosome; **transfer RNA (tRNA)**, which carries amino acids to the ribosome; **ribosomal RNA (rRNA)**, which, together with polypeptides, makes up the structure of ribosomes; **regulatory RNA**, which interacts with DNA to control gene expression; and **ribozymes**, which are RNA molecules that function as metabolic enzymes.

3. The **central dogma** of genetics states that genetic information is transferred from DNA to RNA to polypeptides, which function alone or in conjunction as proteins.

4. The transfer of genetic information begins with **transcription** of the genetic code from DNA to RNA, in which **RNA polymerase** links RNA nucleotides that are complementary to genetic sequences in DNA. Transcription begins at a region of DNA called a **promoter** (recognized by RNA polymerase) and ends with a sequence called a **terminator**. In bacteria, Rho protein may assist in termination, or termination may depend solely on the nucleotide sequence of the transcribed RNA.

Play *Transcription: Overview, The Process* @ Mastering Microbiology

5. Eukaryotic mRNA is synthesized as pre-messenger RNA. Before translation can occur, a spliceosome removes noncoding **introns** from pre-mRNA and splices together the **exons**, which are the coding sections.

6. In **translation**, the sequence of genetic information carried by mRNA is used by ribosomes to construct polypeptides with specific amino acid sequences. Initiation of translation involves three steps.

Play *Translation: Overview, Genetic Code, The Process* @ Mastering Microbiology

7. The genetic code consists of triplets of mRNA nucleotides, called **codons**. During translation, these bind with complementary **anticodons** on transfer RNAs (tRNAs), which are molecules that carry specific amino acids. A ribozyme (rRNA), which is part of a ribosome, catalyzes the bonding of one amino acid to another to form a polypeptide. A sequence of nucleotides thus codes for a sequence of amino acids.

8. A ribosome contains three tRNA-binding sites: an **A site** (associated with incoming amino acids), a **P site** (associated with elongation of the polypeptide), and an **E site** from which tRNA exits the ribosome.

9. Prokaryotes measure their cellular density by a process called **quorum sensing** in which cells secrete quorum-sensing molecules that neighboring cells detect.

10. An **operon** consists of a series of prokaryotic genes, a promoter, and an **operator** sequence, all controlled by one regulatory gene. The operon model explains gene regulation in prokaryotes.

Play *Operons: Overview* @ Mastering Microbiology

11. **Inducible operons** are normally "turned off" and are activated when the repressor no longer binds to the operator site, whereas **repressible operons** are normally "on" and are deactivated when the repressor binds to the operator site.

Play *Operons: Induction, Repression* @ Mastering Microbiology

12. Cells and some viruses use short **microRNAs (miRNAs)** in conjunction with proteins to form RNA silencing complexes (RISC) that attach to mRNA sequences to inhibit polypeptide translation.

13. Genetic expression can also be controlled with **riboswitches** or with **small interfering RNA (siRNA)** associated with protein in RISC.

Mutations of Genes (pp. 216–223)

1. A **mutation** is a change in the nucleotide base sequence of a genome. **Point mutations** involve a change in a single nucleotide base pair and include substitutions and two types of frameshift mutations: insertions and deletions.

2. **Gross mutations** include major changes to the DNA sequence, such as inversions, duplications, transpositions, and large deletions or insertions.

3. Mutations can be categorized by their effects as **silent mutations**, **missense mutations**, or **nonsense mutations**. Most mutations are spontaneous.

    Play *Mutations: Types* @ Mastering Microbiology

4. Physical or chemical agents called **mutagens** can increase the normal rate of mutation. Physical mutagens include ionizing radiation, such as X rays and gamma rays, and nonionizing ultraviolet light. Ultraviolet light causes adjacent pyrimidine bases to bond to one another to form **pyrimidine dimers**. Mutagenic chemicals include **nucleotide analogs**, chemicals that are structurally similar to nucleotides and can result in mismatched base pairing, and chemicals that insert or delete nucleotide base pairs, producing frameshift mutations.

    Play *Mutagens* @ Mastering Microbiology

5. Cells repair damaged DNA via direct repair, including **base-excision repair** and **light repair** of pyrimidine dimers; single-strand repair, such as **nucleotide-excision repair** (dark repair) and **mismatch repair**. Error-prone repair is adopted when damage is extensive; the **SOS response** of *E. coli* is an example.

    Play *Mutations: Repair* @ Mastering Microbiology

6. Researchers have developed methods to distinguish **mutants**, which carry mutations, from normal **wild-type cells**. These methods include **positive selection**, **negative (indirect) selection**, and the **Ames test**, which is used to identify mutagens, which may be potential **carcinogens**.

Genetic Recombination and Transfer (pp. 223–230)

1. Organisms acquire new genes through **genetic recombination**, which is the exchange of segments of DNA. Crossing over occurs during gamete formation, part of sexual reproduction in eukaryotes.

2. Vertical gene transfer is the transmission of genes from parents to offspring. In **horizontal (lateral) gene transfer**, DNA from a donor cell is transmitted to a **recipient cell**. A **recombinant** cell results from genetic recombination between donated and recipient DNA. Transformation, transduction, and bacterial conjugation are types of horizontal gene transfer.

    Play *Horizontal Gene Transfer: Overview* @ Mastering Microbiology

3. In **transformation**, a **competent** recipient prokaryote takes up DNA from its environment. Competence is found naturally or can be created artificially in some cells.

    Play *Transformation* @ Mastering Microbiology

4. In **transduction**, a virus such as a **bacteriophage**, or **phage**, carries DNA from a donor cell to a recipient cell. Donor DNA is accidentally incorporated in such **transducing phages**.

    Play *Transduction: Generalized Transduction, Specialized Transduction* @ Mastering Microbiology

5. In **conjugation**, an F⁺ bacterium—that is, one containing an **F (fertility) plasmid**—forms a **pilus** (conjugation pilus) that attaches to an F⁻ recipient bacterium. Plasmid genes are transferred to the recipient, which becomes F⁺ as a result.

    Play *Conjugation: Overview, F Factor, Hfr Conjugation, Chromosome Mapping* @ Mastering Microbiology

6. **Hfr (high frequency of recombination) cells** result when an F plasmid integrates into a prokaryotic chromosome. Hfr cells form conjugation pili and transfer cellular genes more frequently than regular F⁺ cells do.

7. **Transposons** are DNA segments that code for the enzyme transposase and have palindromic sequences known as **inverted repeats (IR)** at each end. Transposons move among locations in chromosomes in eukaryotes and prokaryotes—a process called **transposition**. The simplest transposons, known as **insertion sequences (IS)**, consist only of inverted repeats and transposase. **Complex transposons** contain other genes as well.

    Play *Transposons: Overview, Insertion Sequences, Complex Transposons* @ Mastering Microbiology

QUESTIONS FOR REVIEW

Answers to the Questions for Review (except Short Answer Questions) begin on p. A-I.

Multiple Choice

1. Which of the following is most likely the number of base pairs in a bacterial chromosome?
   a. 4,000,000
   b. 4000
   c. 400
   d. 40

2. Which of the following is a true statement concerning prokaryotic chromosomes?
   a. They typically have two or three origins of replication.
   b. They contain single-stranded DNA.
   c. They are located in the cytosol.
   d. They are associated in linear pairs.

3. A plasmid is _____.
   a. a molecule of RNA found in bacterial cells
   b. distinguished from a chromosome by being circular
   c. a structure in bacterial cells formed from plasma membrane
   d. extrachromosomal DNA

4. Which of the following forms ionic bonds with eukaryotic DNA and stabilizes it?
   a. chromatin
   b. bacteriocin
   c. histone
   d. nucleoid

5. Nucleotides used in the replication of DNA _____.
   a. carry energy
   b. are found in four forms, each with a deoxyribose sugar, a phosphate, and a base
   c. are present in cells as triphosphate nucleotides
   d. All of the above are correct.

6. Which of the following molecules functions as a "proofreader" for a newly replicated strand of DNA?
   a. DNA polymerase III
   b. primase
   c. helicase
   d. ligase

7. The addition of $-CH_3$ to a cytosine nucleotide after DNA replication is called _____.
   a. methylation
   b. restriction
   c. transcription
   d. transversion

8. In translation, the site through which tRNA molecules leave a ribosome is called the _____.
   a. A site
   b. X site
   c. P site
   d. E site

9. The Ames test _____.
   a. uses auxotrophs and liver extract to reveal mutagens
   b. is time intensive and costly
   c. involves the isolation of a mutant by eliminating wild-type phenotypes with specific media
   d. proves that suspected chemicals are carcinogenic

10. Which of the following methods of DNA repair involves enzymes that recognize and correct nucleotide errors in unmethylated strands of DNA?
    a. light repair of T dimers
    b. dark repair of P dimers
    c. mismatch repair
    d. SOS response

11. Which of the following is *not* a mechanism of genetic transfer between cells?
    a. transduction
    b. transformation
    c. transcription
    d. conjugation

12. Cells that have the ability to take up DNA from their environment are said to be _____.
    a. Hfr cells
    b. transposing
    c. genomic
    d. competent

13. Which of the following statements is true?
    a. Conjugation requires a sex pilus extending from the surface of a cell.
    b. Conjugation involves a C factor.
    c. Conjugation is an artificial genetic engineering technique.
    d. Conjugation involves DNA that has been released into the environment.

14. Which of the following are called "jumping genes"?
    a. Hfr cells
    b. transducing phages
    c. palindromic sequences
    d. transposons

15. Although two cells are totally unrelated, one cell receives DNA from the other cell and incorporates this new DNA into its chromosome. This process is _____.
    a. crossing over of DNA from the two cells
    b. vertical gene transfer
    c. horizontal gene transfer
    d. transposition

16. Transcription produces _____.
    a. DNA molecules
    b. RNA molecules
    c. polypeptides
    d. palindromes

17. A nucleotide is made of _____.
    a. a five-carbon sugar
    b. phosphate
    c. a nitrogenous base
    d. all of the above

18. In DNA, adenine forms _____ hydrogen bonds with _____.
    a. three / uracil
    b. two / uracil
    c. two / thymine
    d. three / thymine

19. A sequence of nucleotides formed during replication of the lagging DNA strand is a(n) _____.
    a. palindrome
    b. Okazaki fragment
    c. coding strand
    d. operon

20. Which of the following is *not* part of an operon?
    a. operator
    b. promoter
    c. origin
    d. gene

21. Repressible operons are important in regulating prokaryotic _____.
    a. DNA replication
    b. RNA transcription
    c. rRNA processing
    d. sugar catabolism

22. Which of the following is part of each molecule of mRNA?
    a. palindrome
    b. codon
    c. anticodon
    d. base pair

23. Ligase plays a major role in _____.
    a. replication of lagging strands
    b. mRNA processing in eukaryotes
    c. polypeptide synthesis by ribosomes
    d. RNA transcription

24. Before mutations can affect a population permanently, they must be _____.
    a. lasting
    b. inheritable
    c. beneficial
    d. all of the above

25. The *trp* operon is repressible. This means it is usually _____ and is directly controlled by a(n) _____.
    a. active / inducer
    b. active / repressor
    c. inactive / inducer
    d. inactive / repressor

Fill in the Blanks

1. The three steps in RNA transcription are _____, _____, and _____.

2. A triplet of mRNA nucleotides that specifies a particular amino acid is called a _____.

3. Three effects of point mutations are _____, _____, and _____.

4. Insertions and deletions in the genetic code are also called _____ mutations.

5. An operon consists of _____, _____, and _____, and is associated with a regulatory gene.

6. In general, _____ operons are inactive until the substrate of their genes' polypeptides is present.

7. A daughter DNA molecule is composed of one original strand and one new strand because DNA replication is _____.

8. A gene for antibiotic resistance can move horizontally among bacterial cells by _____, _____, and _____.

9. _____ are nucleotide sequences containing palindromes and genes for proteins that cut DNA strands.

10. _____ _____ is a recombination event that occurs during gamete formation in eukaryotes.

11. _____ RNA carries amino acids.

12. _____ RNA and _____ RNA are antisense; that is, they are complementary to another nucleic acid molecule.

Short Answer

1. How does the genotype of a bacterium determine its phenotype? Use the terms *gene*, *mRNA*, *ribosome*, and *polypeptide* in your answer.

2. List several ways in which eukaryotic messenger RNA differs from prokaryotic mRNA.

3. Compare and contrast introns and exons.

4. Polypeptide synthesis requires large amounts of energy. How do cells regulate synthesis to conserve energy? Describe one specific example.

5. Describe the operon model of gene regulation.

6. Compare and contrast the structure and components of DNA and RNA in prokaryotes.

7. Besides the fact that it synthesizes RNA, how does RNA polymerase differ in function from DNA polymerase?

8. Describe the formation and function of mRNA, rRNA, and tRNA in prokaryotes and eukaryotes.

9. Describe how DNA is packaged in both prokaryotes and eukaryotes.

10. Explain the central dogma of genetics.

11. Compare and contrast the processes of transformation, transduction, and conjugation.

12. Fill in the following table:

Process	Purpose	Beginning Point	Ending Point
Replication			Origin or end of molecule
Transcription		Promoter	
Translation	Synthesis of polypeptides		

VISUALIZE IT!

1. On the accompanying figure, label DNA polymerase I, DNA polymerase III, helicase, lagging strand, leading strand, ligase, nucleotide (triphosphate), Okazaki fragment, primase, replication fork, RNA primer, and stabilizing proteins.

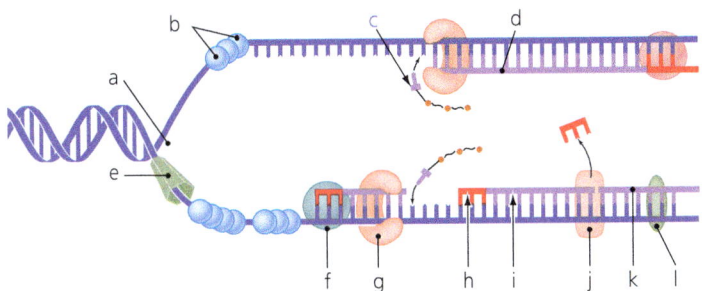

2. This bacteriophage DNA molecule has been warmed. Label the portions that likely have a higher ratio of GC base pairs and the portions that have a higher ratio of AT base pairs.

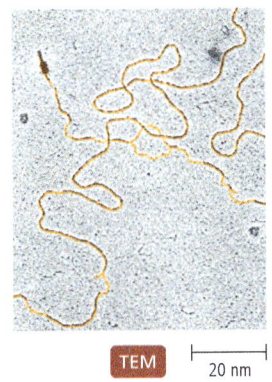

3. The drugs ddC and AZT are used to treat AIDS.

ddC (2′,3′-dideoxycytidine)   AZT (3′-azido-2′,3′-dideoxythymidine)

Based on their chemical structures, what is their mode of action?

CRITICAL THINKING

1. If molecules of mRNA have the following nucleotide base sequences, what will be the sequence of amino acids in polypeptides synthesized by eukaryotic ribosomes? Remember, transcription starts with an AUG codon.
   a. 5′–AUGGGGAUACGCUACCCC–3′
   b. 5′–CCGUACAUGCUAAUCCCU–3′
   c. 5′–CCGAUGUAACCUCGAUCC–3′
   d. 5′–AUGCGGUCAGCCCCGUGA–3′

2. A scientist uses a molecule of DNA composed of nucleotides containing radioactive deoxyribose as a template for replication and transcription in a nonradioactive environment. What percentage of DNA strands will be radioactive after three DNA replication cycles? What percentage of RNA molecules will be radioactive?

3. Explain why an insertion of three nucleotides is less likely to result in a deleterious effect than an insertion of a single nucleotide.

4. How could scientists use siRNA to turn off a cancer-inducing gene?

5. The chromosome of *Mycobacterium tuberculosis* is 4,411,529 bp long. A scientist who isolates and counts the number of nucleotides in its DNA molecule discovers that there are 2,893,963 molecules of guanine. How many molecules of the other three nucleotides are in the original DNA?

6. Suppose that the *E. coli* gene for the *lac* operon repressor has a mutation that makes the repressor ineffective. What effect would this have on the bacterium's ability to catabolize lactose? Would the mutant strain have an advantage over wild-type cells? Explain your answer.

7. A student claims that nucleotide analogs can be dangerously carcinogenic. Another student in the study group disagrees, insisting that nucleotide analogs are used to treat cancer. Explain why both students are correct.

8. Why is DNA polymerase so named?

9. *Corynebacterium diphtheriae*, the causative agent of diphtheria, secretes a toxin that enzymatically inactivates all molecules of elongation factor in a eukaryotic cell. What immediate and long-term effects does this have on cellular metabolism?

10. How can knowledge of nucleotide analogs be useful to a cancer researcher?

11. The endosymbiotic theory states that mitochondria and chloroplasts evolved from prokaryotes living within other prokaryotes (see p. 88). What aspects of the eukaryotic genome support this theory? What aspects do *not* support the theory?

12. Hydrogen bonds between complementary nucleotides are crucial to the structure of dsDNA because they hold the two strands together. Why couldn't the two strands be effectively linked by covalent bonds?

13. On average, RNA polymerase makes one error for every 10,000 nucleotides it incorporates in RNA. By contrast, only one base-pair error remains for every 10 billion bp during DNA replication. Explain why the accuracy of RNA transcription is not as critical as the accuracy of DNA replication.

14. We have seen that wobble makes the genetic code redundant in the third position for C and U. After reexamining the genetic code in Figure 7.12, state what other nucleotides in the third position appear to accommodate anticodon wobbling.

15. If a scientist synthesizes a DNA molecule with the nucleotide base sequence 3'–TACGGGGGAGGGGGAGGGGGA–5' and then uses it for transcription and translation, what would be the amino acid sequence of the product?

16. What DNA nucleotide triplet codes for codon UGU? Identify a base-pair substitution that would produce a silent mutation at this codon. Identify a base-pair substitution that would result in a missense mutation at this codon. Identify a base-pair substitution that would produce a nonsense mutation at this codon.

17. Suppose you want to insert into your dog a gene that encodes a protein that protects dogs from heartworms. A dog's cells are not competent, so they cannot take up the gene from the environment; but you have a plasmid, a competent bacterium, and a related (though incompetent) F$^+$ bacterium that lives as an intracellular parasite in dogs. Describe a possible scenario by which you could use natural processes to genetically alter your dog to be heartworm resistant.

CONCEPT MAPPING

Fill in the concept map that describes point mutations using the following terms. You can also complete this and other concept maps online by going to the **Mastering Microbiology** Study Area.

Change in DNA sequence
Deleted
Effect on amino acid sequence
Incorrect amino acid substituted
Inserted
Insertion mutation
Nonsense mutation
One or few nucleotide base pairs
Silent mutation
Substituted
Substitution mutation

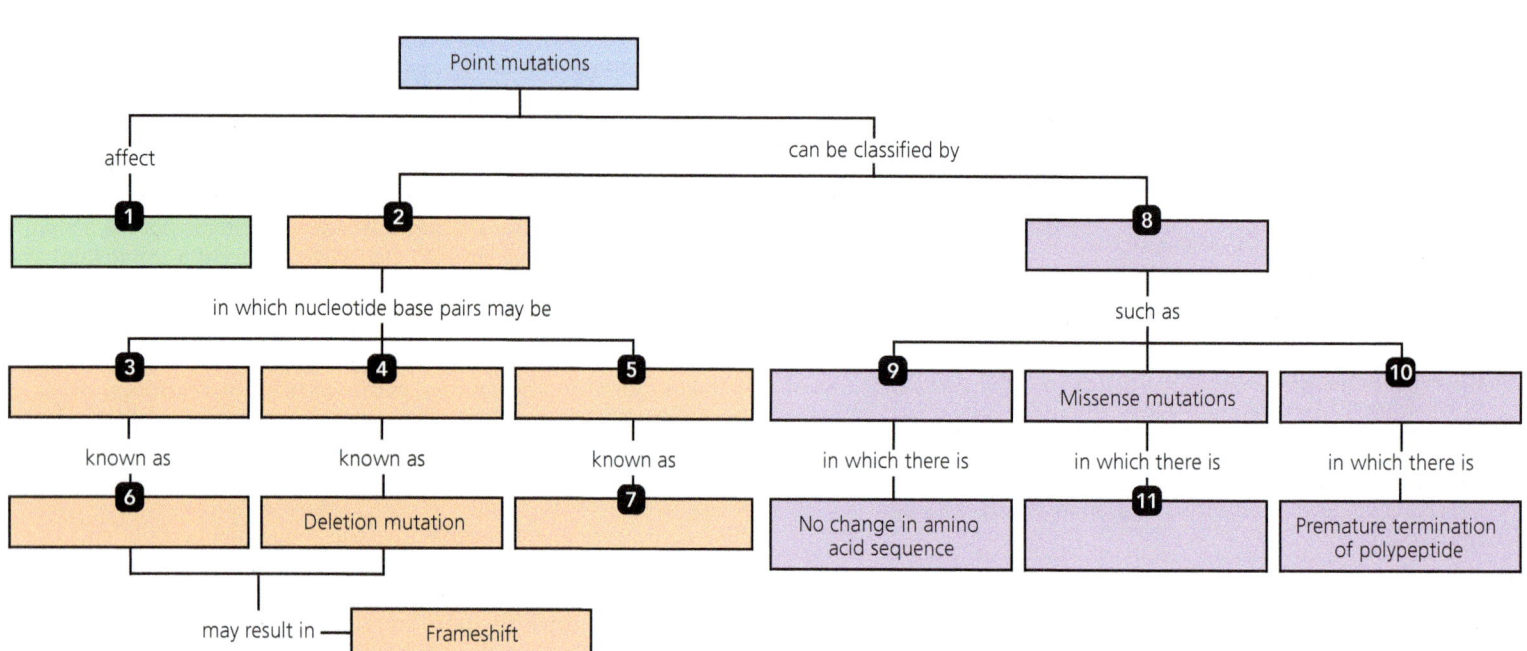

8 Recombinant DNA Technology

Before You Begin

1. What is the purpose of a primer during DNA replication?
2. Which type of molecule serves as a template during transcription, and which type of molecule is synthesized?
3. What is the difference between a missense mutation and a nonsense mutation?
4. What are plasmids?

DNA sequences called CRISPR, in conjunction with Cas enzymes, allow scientists to edit DNA almost at will.

MICRO IN THE CLINIC

Cause for Concern?

SINCE SHE WAS A YOUNG CHILD, Caroline and her family have traveled to Brazil every year to visit her grandparents. She loves being able to spend time with family, but she also loves being able to go hiking in the mountains, swimming in the ocean, and wandering through the little town near her grandparents' home. It's always a fun but busy trip, and Caroline usually returns home tired.

After being home a couple of days from her latest trip, Caroline starts having a headache and low-grade fever. It isn't too bad at first, but it continues over the next couple of days and gets quite severe. Her headache and fever (39.4°C) are worsening, she develops a sharp pain behind her eyes (retro-orbital pain), and she feels achy all over—her muscles and joints really hurt. Caroline stays home from work and focuses on getting plenty of rest and drinking water. She takes acetaminophen to help with the headache and body aches.

Caroline's grandmother calls to see how she's settling in after her trip. Caroline tells her grandmother how she's been feeling, and her grandmother insists that she go to the doctor immediately.

1. Why is Caroline's grandmother so concerned?
2. Do you think that Caroline's symptoms warrant an immediate trip to the doctor?

Turn to the end of the chapter (p. 257) to find out.

This chapter examines how genetic researchers have adopted the natural enzymes and processes of DNA recombination, replication, transcription, transformation, transduction, and conjugation to manipulate genes for industrial, medical, and agricultural purposes. Together these techniques are termed *recombinant DNA technology*. We end the chapter with a discussion of the ethics and safety of these techniques.

The Role of Recombinant DNA Technology in Biotechnology

LEARNING OUTCOMES

- **8.1** Define *biotechnology* and *recombinant DNA technology*.
- **8.2** List several examples of useful products made possible by biotechnology.
- **8.3** Identify the three main goals of recombinant DNA technology.

Biotechnology—the use of microorganisms to make practical products—is not a new field. For thousands of years, humans have used microbes to make products such as bread, cheese, soy sauce, and alcohol. During the 20th century, scientists industrialized the natural metabolic reactions of bacteria to make large quantities of acetone, butanol, and antibiotics. More recently, scientists have adapted microorganisms for use in the manufacture of paper, textiles, and vitamins; to assist in cleaning up industrial wastes, oil spills, and radioactive isotopes; and to aid in mining copper, gold, uranium, and other metals. (Chapter 26 discusses some industrial and environmental applications of biotechnology.)

Until recently, microbiologists were limited to working with naturally occurring organisms and their mutants for achieving such industrial and medical purposes. Since the 1990s, however, scientists have become increasingly adept at intentionally modifying the genomes of organisms, by natural processes, for a variety of practical purposes. This is **recombinant DNA technology** (sometimes called "genetic engineering"), and it has expanded the possibilities of biotechnology in ways that seemed like science fiction only a few years ago. Today, scientists can isolate specific genes from almost any so-called donor organism, such as a human, a plant, or a bacterium, and insert it into the genome of almost any kind of recipient organism. The result is often called a *genetically modified organism*, or *GMO*. We consider the ethics and safety of GMOs later.

Scientists who manipulate genomes have three main goals:

- *To eliminate undesirable phenotypic traits in humans, animals, plants, and microbes.* For example, scientists have inserted genes from microbes into plants to make them resistant to pests or freezing, and they have used gene therapy to treat children born with previously untreatable genetic disorders.
- *To combine beneficial traits of two or more organisms to create valuable new organisms,* such as laboratory animals that mimic human susceptibility to HIV.
- *To create organisms that synthesize products that humans need,* such as paint solvents, vaccines, antibiotics, enzymes, and hormones. For instance, geneticists have successfully inserted the human gene for insulin into the bacterium *Escherichia coli* (esh-ĕ-rik'ē-ă kō'lī) as well as into the yeast *Saccharomyces cerevisiae* (sak-ă-rō-mī'sēz se-ri-vis'ē-ī) so that they synthesize human insulin, which is cheaper and safer than insulin derived from animals.

Recombinant DNA technology is not a single procedure or technique but rather a collection of tools and techniques scientists use to manipulate the genomes of organisms. In general, they isolate a gene from a cell, manipulate it *in vitro*,[1] and insert it into another organism. **FIGURE 8.1** illustrates the basic processes involved in recombinant DNA technology.

MICRO CHECK

1. Baking bread, creating a genetically modified vegetable, using a microbe to make human insulin . . . Which of these is NOT considered "biotechnology"?

TELL ME WHY

Why aren't the terms *recombinant DNA technology (genetic engineering)* and *biotechnology* synonymous?

Tools of Recombinant DNA Technology

Scientists use a variety of physical agents, naturally occurring enzymes, and synthetic molecules to manipulate genes and genomes. These tools of recombinant DNA technology include *mutagens, reverse transcriptase, synthetic nucleic acids, restriction enzymes, vectors,* and *CRISPR*. Scientists use these molecular tools to create *gene libraries*, which are a time-saving tool for genetic researchers.

Mutagens

LEARNING OUTCOME

- **8.4** Describe how gene researchers use mutagens.

Mutagens are physical and chemical agents that produce mutations (heritable changes in the nucleotide sequence of a genome). Scientists deliberately use mutagens to create changes in microbes' genomes so that the microbes' phenotypes are changed. They then select for and culture cells with characteristics considered beneficial for a given biotechnological application. For example, scientists exposed the fungus *Penicillium* (pen-i-sil'ē-ŭm) to mutagenic agents and then selected strains that produce greater amounts of penicillin. In this manner, they developed a strain of *Penicillium* that secretes over 25 times as much penicillin as did the strain originally isolated by Alexander Fleming (1881–1955). Today, with recombinant DNA techniques (discussed shortly), researchers can isolate mutated genes rather than dealing with entire organisms.

---
[1] Latin, meaning "within glassware;" that is, in a laboratory.

The Use of Reverse Transcriptase to Synthesize cDNA

LEARNING OUTCOME

8.5 Explain the function and use of reverse transcriptase in synthesizing cDNA.

Transcription involves the transmission of genetic information from molecules of DNA to molecules of RNA (see Chapter 7, Figure 7.9). The discovery of retroviruses, which have genomes consisting of RNA instead of DNA, led to the discovery of the enzyme **reverse transcriptase**. Reverse transcriptase creates a flow of genetic information in the opposite direction from the flow in conventional transcription: it uses an RNA template to transcribe a molecule of DNA, which is called **complementary DNA (cDNA)** because it is complementary to an RNA template.

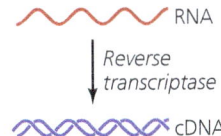

Because hundreds to millions of copies of mRNA exist for every active gene, it is frequently easier to produce a desired gene by first isolating the mRNA molecules that code for a particular polypeptide and then use reverse transcription to synthesize a cDNA gene from the mRNA template. Further, eukaryotic DNA is not normally expressible by prokaryotic cells, which cannot remove the introns (noncoding sequences) present in eukaryotic pre-mRNA. However, since eukaryotic mRNA has already been processed to remove introns, cDNA produced from processed mRNA lacks noncoding sequences. Therefore, scientists can successfully insert cDNA into prokaryotic cells, making it possible for the prokaryotes to produce eukaryotic proteins such as human growth factor, insulin, or blood-clotting factors.

Synthetic Nucleic Acids

LEARNING OUTCOMES

8.6 Explain how gene researchers synthesize nucleic acids.
8.7 Describe three uses of synthetic nucleic acids.

The enzymes of DNA replication and RNA transcription function not only *in vivo*[2] but also *in vitro*, making it possible for scientists to produce molecules of DNA and RNA in cell-free solutions for genetic research. In fact, scientists have so mechanized the processes of nucleic acid replication and transcription that they can produce molecules of DNA and RNA with any nucleotide sequence; all they must do is enter the desired sequence into a synthesis machine's four-letter keyboard. A computer controls the actual synthesis, using a supply of nucleotides and other required reagents. Nucleic acid synthesis machines synthesize molecules over 100 nucleotides long in a few hours, and scientists can join two or more of these molecules end to end with ligase to create even longer synthetic molecules.

[2]Latin, meaning "in life" (i.e., within a cell).

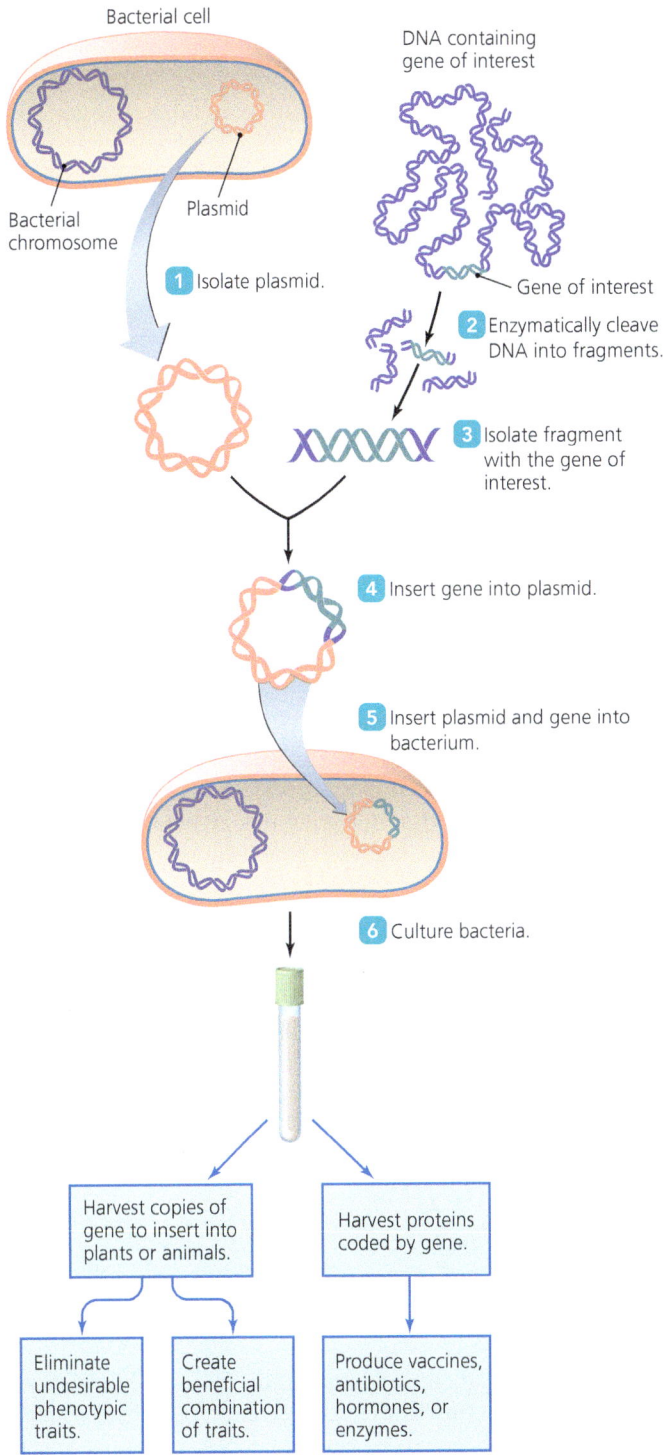

▲ FIGURE 8.1 Overview of a type of recombinant DNA technology.

Researchers have used synthetic nucleic acids in many ways, including the following:

- **Elucidating the genetic code.** Using synthetic molecules of varying nucleotide sequences and observing the amino acids in the resulting polypeptides, scientists elucidated the genetic code. For example, synthetic DNA consisting only of adenine nucleotides yields a polypeptide consisting solely of the amino acid phenylalanine. Therefore, the mRNA codon UUU (transcribed from the DNA triplet AAA) must code for phenylalanine (see Figure 7.12).

- **Creating genes for specific proteins.** Once they know the genetic code and the amino acid sequence of a protein, scientists can create a gene for that protein. In this manner, scientists synthesized a gene for human insulin. The genetic code is universal and most amino acids have more than one codon—a condition called *redundancy* (see Figure 7.12). Organisms preferentially use some redundant codons as opposed to others, so scientists designing a gene will likely use the preferred codon.

- **Synthesizing DNA and RNA probes to locate specific sequences of nucleotides.** *Probes* are nucleic acid molecules with a specific nucleotide sequence that have been labeled with radioactive or fluorescent chemicals so that their locations can be detected. The use of probes to locate specific sequences of nucleotides is based on the fact that any given nucleotide sequence will preferentially bond to its complementary sequence. Thus, a probe constructed with the nucleotide sequence 5′-ATGCT-3′ will bond to a DNA strand with the sequence 3′-TACGA-5′, and the probe's label allows researchers to then detect the complementary site. Scientists design probes to have a nucleotide sequence complementary to a target sequence so that the probe binds to the target via hydrogen bonds. Researchers can use such probes in many ways, such as to monitor which cells are making a particular protein by locating cells that have mRNA for that protein. It may then be inferred that the protein is present in those cells.

- **Synthesizing antisense nucleic acid molecules.** Antisense nucleic acid molecules have nucleotide sequences that bind to and interfere with genes and mRNA molecules. Scientists are developing antisense molecules to control genetic diseases.

- **Synthesizing PCR primers.** The most common usage for synthetic nucleic acid molecules is as single-stranded primers for a *polymerase chain reaction* (PCR) (see Figure 8.6). Primers are typically 20–30 nucleotides long.

Restriction Enzymes

LEARNING OUTCOMES

8.8 Explain the source and names of restriction enzymes.
8.9 Describe the importance and action of restriction enzymes.

An important development in recombinant DNA technology was the discovery of **restriction enzymes** (also called *restriction endonucleases*) in bacterial cells. Such enzymes cut DNA molecules and are restricted in their action—they cut DNA only at locations called *restriction sites*. Restriction sites are specific nucleotide sequences, which are usually *palindromes*[3]—they have the same sequence when read forward or backward. In nature, bacterial cells use restriction enzymes to protect themselves from phages by cutting phage DNA into nonfunctional pieces. The bacterial cells protect their own DNA by methylation[4] of some of their nucleotides, in effect hiding their DNA from the restriction enzymes (see Chapter 7).

Researchers name restriction enzymes with three letters (denoting the genus and specific epithet of the source bacterium) and Roman numerals (to indicate the order in which enzymes from the same bacterium were discovered). In some cases, a fourth letter denotes the strain of the bacterium. Thus, *E. coli* strain R produces the restriction enzymes *Eco*RI and *Eco*RII. *Hind*III is the third restriction enzyme isolated from *Haemophilus influenzae* (hē-mof′i-lŭs in-flu-en′zī) strain Rd.

Scientists have discovered hundreds of restriction enzymes and categorize them in two groups on the basis of types of cuts they make. The first type, as exemplified by *Eco*RI, makes staggered cuts of the two strands of DNA, producing fragments that terminate in mortise-like *sticky ends*. Each sticky end is composed of up to four nucleotides that form hydrogen bonds with its complementary sticky end (**FIGURE 8.2a**). Scientists can use these bits of single-stranded DNA to combine pieces of DNA from different organisms into a single recombinant DNA molecule (the enzyme ligase unites the sugar-phosphate backbones of the pieces) (**FIGURE 8.2b**). Other restriction enzymes, such as *Hind*II and *Sma*I (the latter from *Serratia marcescens,* ser-rat′ē-a mar-ses′enz), cut both strands of DNA at the same point, resulting in *blunt ends* (**FIGURE 8.2c**). It is more difficult to make recombinant DNA from blunt-ended fragments because they are not sticky, but they have a potential advantage—blunt ends are nonspecific. This enables any two blunt-ended fragments, even those produced by different restriction enzymes, to be combined (**FIGURE 8.2d**). In contrast, sticky-ended fragments bind only to complementary sticky-ended fragments.

TABLE 8.1 identifies several restriction enzymes and their target DNA sequences.

Play *Recombinant DNA Technology*
@ Mastering Microbiology

TABLE 8.1 Properties of Some Restriction Enzymes

Enzyme	Bacterial Source	Restriction Site[a]
BamHI	*Bacillus amyloliquefaciens* H	G↓GATCC
		CCTAG↑G
EcoRI	*Escherichia coli* RY13	G↓AATTC
		CTTAA↑G
HaeIII	*Haemophilus aegyptius*	GG↓CC
		CC↑GG
HindII	*H. influenzae* Rd	GTPy↓PuAC
		CAPu↑PyTG
HindIII	*H. influenzae* Rd	A↓AGCTT
		TTCGA↑A
HinfI	*H. influenzae* Rf	G↓ANTC
		CTNA↑G
HpaI	*H. parainfluenzae*	GTT↓AAC
		CAA↑TTG
MspI	*Moraxella* sp.	C↓CGG
		GGC↑C
SmaI	*Serratia marcescens*	CCC↓GGG
		GGG↑CCC

[a]Arrows indicate sites of cleavage. Py = pyrimidine (either thymine [T] or cytosine [C]); Pu = purine (either adenine [A] or guanine [G]); N = any nucleotide (A, T, G, or C).

---
[3]From Greek *palin*, meaning "again," and *dramein*, meaning "to run."
[4]Adding a methyl group, —CH$_3$, to a chemical.

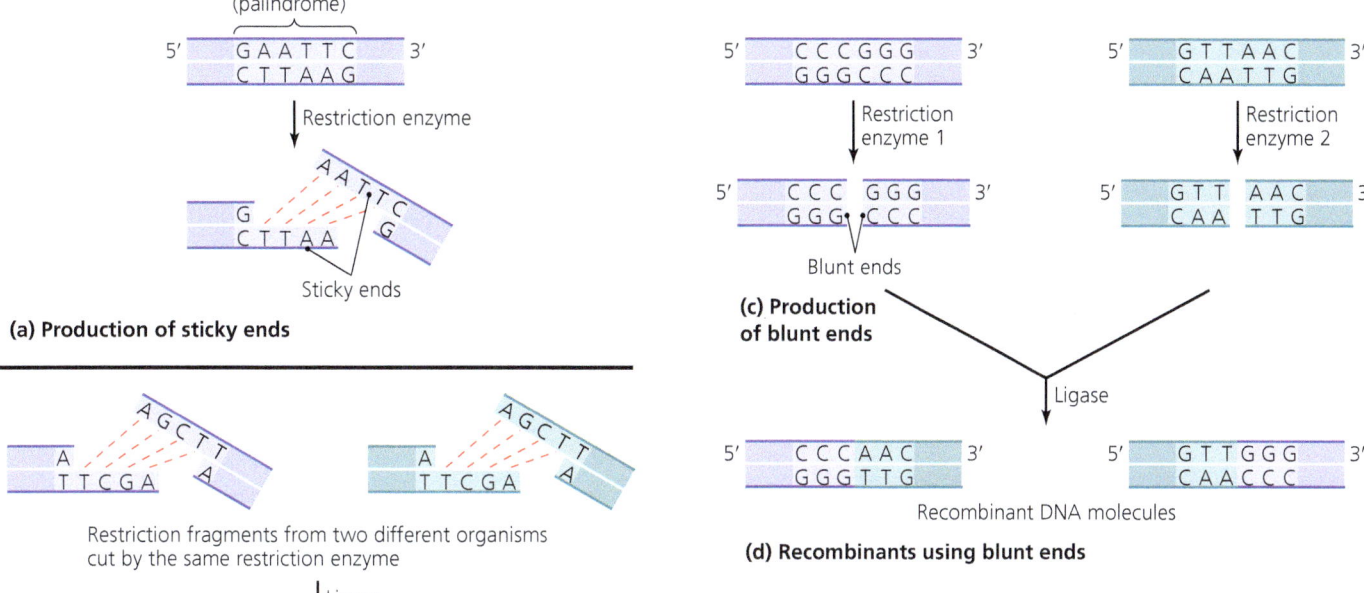

▲ FIGURE 8.2 **Actions of representative restriction enzymes.** Restriction enzymes recognize and cut both strands of a DNA molecule at a specific (usually palindromic) restriction site. **(a)** Certain restriction enzymes produce staggered cuts with complementary "sticky ends." **(b)** When two complementary sticky-ended fragments come from different organisms, their bonding (catalyzed by ligase) produces recombinant DNA. **(c)** Other restriction enzymes produce blunt-ended fragments. **(d)** A lack of specificity enables blunt-ended fragments produced by different restriction enzymes to be combined easily into recombinant DNA. *Which restriction enzymes act at the restriction sites shown?*

Figure 8.2 (a) EcoRI, (b) HindIII, and (c) SmaI and HpaI.

Vectors

LEARNING OUTCOME

8.10 Define *vector* as the term applies to genetic manipulation.

One goal of recombinant DNA technology is to insert a useful gene into a cell so that the cell has a new phenotype—for example, the ability to synthesize a novel protein. Researchers use **vectors**, which are nucleic acid molecules, such as viral genomes, transposons, and plasmids, that can deliver genes into a cell.

Genetic vectors share several useful properties:

- **Vectors are small enough to manipulate in a laboratory.** Large DNA molecules the size of entire chromosomes are generally too fragile to serve as vectors.
- **Plasmid vectors are stable inside cells.** Plasmids, which are circular DNA, make good vectors because cellular enzymes do not degrade circular DNA as quickly as they degrade linear fragments of DNA. However, some linear vectors, such as transposons and certain viruses, can insert themselves rapidly into a host's chromosome before they can be degraded.
- **Vectors contain a recognizable genetic marker** so that researchers can identify the cells that have received the vector and thereby the specific gene of interest. Genetic markers can be either phenotypic markers, such as those that confer antibiotic resistance or code for enzymes that metabolize a unique nutrient, or radioactive or fluorescent labels.
- **Vectors can ensure genetic expression** by providing required genetic elements, such as promoters.

An example of a process used to produce a vector containing a specific gene is depicted in **FIGURE 8.3**. After a given restriction enzyme cuts both the DNA molecule containing the gene of interest (in this example, the gene for human growth hormone [HGH]) and the vector DNA (here, a plasmid containing a gene for antibiotic resistance as a marker) into fragments with sticky ends ❶, ligase anneals the fragments to produce a recombinant plasmid ❷. After the recombinant plasmid has been inserted into a bacterial cell ❸, the bacteria are grown on a medium containing the antibiotic ❹; only those cells that contain the recombinant plasmid (and thus the human growth hormone gene as well) can grow on the medium.

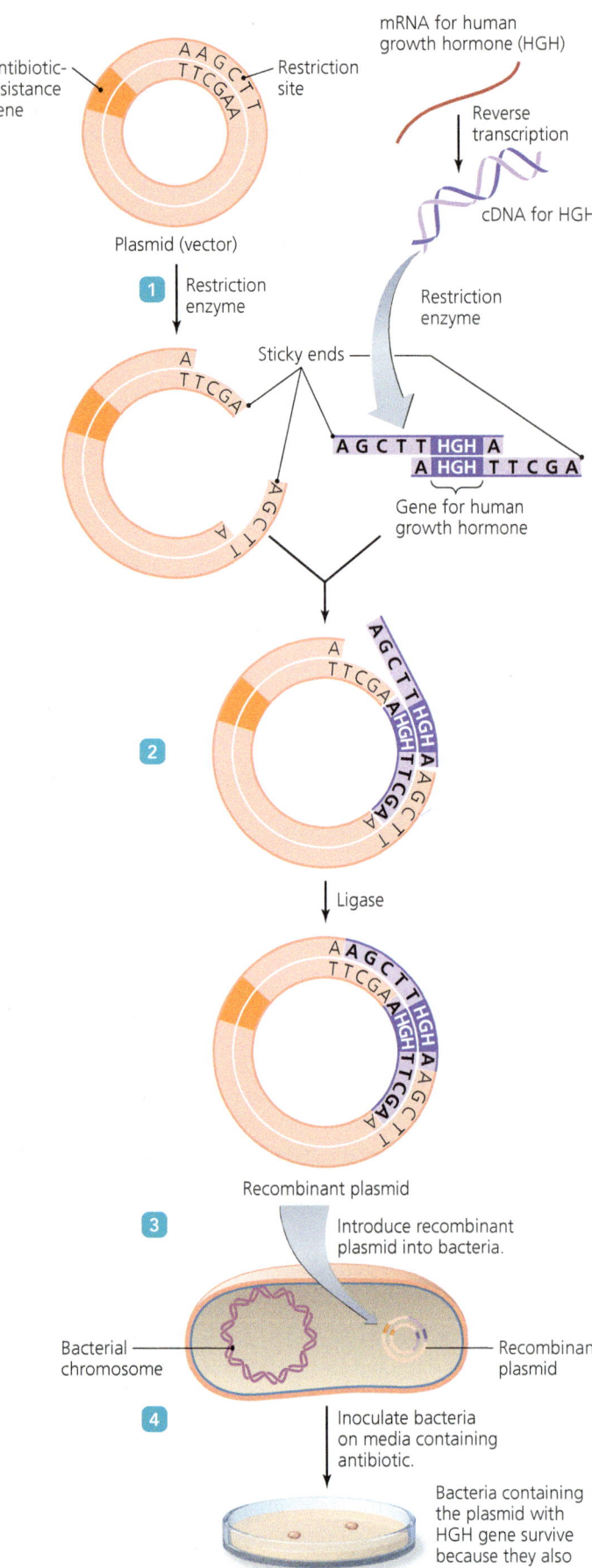

◀ **FIGURE 8.3 An example of a process for producing a recombinant vector.** In this case, the vector is a plasmid. *Which restriction enzyme was used in this example?*

Figure 8.3 HindIII.

Generally, viruses and transposons are able to carry larger genes than can plasmids. Researchers are developing vectors from adenoviruses, poxviruses, and a genetically modified form of the human immunodeficiency virus (HIV). HIV in particular might make an excellent vector because HIV inserts itself directly into human chromosomes; however, scientists must ensure that viral vectors do not insert DNA into the middle of a necessary gene, leading to mutation and possibly causing cancer.

CRISPR

LEARNING OUTCOME

8.11 Describe the CRISPR system.

In the 1980s, scientists reported the existence in many prokaryotes of sequences of DNA called **CRISPR** (*c*lustered, *r*egularly *i*nterspaced, *s*hort *p*alindromic *r*epeats, pronounced "crisper") regions, which act like a primitive immune system, protecting the prokaryotes from viruses. Let's consider the components of CRISPR, how CRISPR works, and how scientists can exploit this natural prokaryotic defense mechanism as a tool in recombinant DNA technology.

CRISPR consists of *repeats*—stretches of DNA 25–50 bp long—interspersed with *spacers*, which are 20–80 bp long (**FIGURE 8.4**). Spacer regions contain sequences derived from viruses that previously invaded the prokaryote or its ancestors. When a virus infects the prokaryote ❶, a piece of that virus's DNA is added as a new spacer to the CRISPR region of the prokaryotic chromosome ❷. The cell transcribes the repeat and spacer sequences of CRISPR to make CRISPR-RNA (crRNA, ❸). Subsequently, when a virus with the same DNA sequence attempts to infect the prokaryote ❹, crRNA complementary to the phage DNA guides *CRISPR-associated enzymes (Cas)* to the phage DNA, and Cas disables the virus by cutting its DNA ❺. In this way, CRISPR-Cas protects the prokaryote from future infections by this virus. Since CRISPR with all its associated spacers is passed to all the prokaryote's descendants, CRISPR-Cas provides protection to them all.

There are different types of CRISPR-Cas systems in prokaryotes. Many researchers use *CRISPR-Cas9*. CRISPR-Cas9 is revolutionary in the speed, precision, and efficiency with which it can target genes in living cells. Even though scientists derived CRISPR-Cas9 from bacteria, they can use it to manipulate DNA in any organism.

Scientists have found numerous ways to utilize the editing capabilities of CRISPR-Cas9 systems to alter DNA by deleting or adding nucleotide sequences. For example, they can replace CRISPR spacer sequences with sequences complementary to those in one or more target genes and thereby inactivate the target genes. In another use, scientists can replace a segment of DNA with a different sequence by providing the new desired sequence at the same time that they provide CRISPR-Cas9.

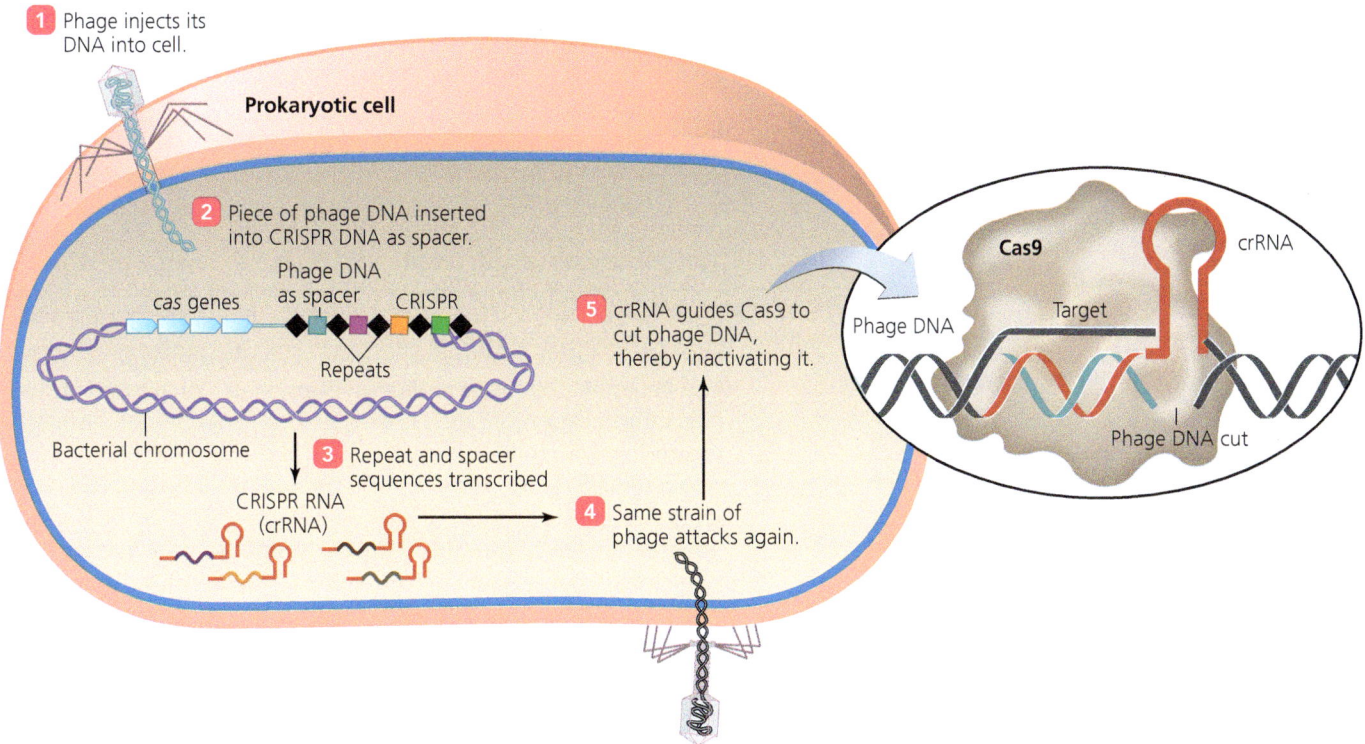

▲ FIGURE 8.4 **CRISPR-Cas protects prokaryotes from viral infection.** Researchers can use CRISPR-Cas to edit the genomes of living cells with high efficiency and precision.

In such cases, after a target sequence is cut, it is replaced by homologous recombination with the provided sequence. An exciting potential use of CRISPR-Cas9 is the treatment of serious, even fatal, genetic diseases in humans. We consider the ethics of such manipulation at the end of the chapter.

Gene Libraries

LEARNING OUTCOME

8.12 Explain the significance of gene libraries.

Suppose you were a scientist investigating the effects of the genes for 24 different kinds of interleukins (proteins that mediate certain aspects of immunity). Having to isolate the specific genes for each type of interleukin would require much time, labor, and expense. Your task would be made much easier if you could obtain the genes you need from a **gene library**, a collection of bacterial or phage clones—identical descendants—each of which contains a portion of the genetic material of interest. In effect, each clone is like one book in a library in that it contains one fragment (typically a single gene) of an organism's entire genome. Alternatively, a gene library may contain clones with all the genes of a single chromosome or the full set of cDNA that is complementary to an organism's mRNA.

As depicted in FIGURE 8.5, genetic researchers can create each of the clones in a gene library by using restriction enzymes to generate fragments of the DNA of interest and then using ligase to synthesize recombinant vectors. They insert the vectors into bacterial cells, which are then grown on culture media.

Once a scientist isolates a recombinant clone and places it in a gene library, the gene that the clone carries becomes available to other investigators, saving them the time and effort required to isolate that gene. Many gene libraries are now commercially available. For example, one library contains every gene from *E. coli* strain K-12 controlled by inducible promoters, so scientists can control gene expression at will.

MICRO CHECK

2. Which viral enzyme is used to copy nucleotide sequences to make cDNA?
3. What function do restriction enzymes perform?
4. What is the name of genetic tools that can deliver genes into a cell?

TELL ME WHY

Why did the discovery and development of restriction enzymes speed up the study of recombinant DNA technology?

Techniques of Recombinant DNA Technology

Scientists use the tools of recombinant DNA technology in a number of basic techniques to multiply, identify, manipulate, isolate, map, and sequence the nucleotides of genes.

Multiplying DNA *In Vitro*: The Polymerase Chain Reaction

LEARNING OUTCOME

8.13 Describe the purpose and application of the polymerase chain reaction.

The **polymerase chain reaction (PCR)** is a technique by which scientists produce a large number of identical molecules of a specific segment of DNA *in vitro*. Using PCR, researchers start with a small amount of DNA (theoretically, even a single molecule is enough) and generate billions of replicas within hours. Such rapid and specific amplification of DNA is critical in a variety of situations. For example, epidemiologists used PCR to amplify the genomes of Ebola viruses from both western and central Africa in 2014. The large number of identical DNA molecules produced by PCR allowed scientists to determine and compare the nucleotide sequences of the two viruses. Because the two viruses differed significantly, epidemiologists were able to determine that the western epidemic had not escaped their quarantine procedures and spread to central Africa; instead, the central African outbreak was a separate epidemic.

▶ Play *Polymerase Chain Reaction (PCR): Overview, Components* @ Mastering Microbiology

PCR is a repetitive process that alternately separates and replicates the two strands of DNA. Each cycle of PCR consists of the following three steps (FIGURE 8.6a):

1 **Denaturation.** Exposure to heat (about 94°C) separates the two strands of the target DNA by breaking the hydrogen bonds between base pairs but otherwise leaves the two strands unaltered.

2 **Priming.** A mixture containing an excess of DNA primers (synthesized such that they are complementary to nucleotide sequences at the ends of the target DNA), DNA polymerase, and an abundance of the four deoxyribonucleotide triphosphates (A, T, G, and C) is added to the target DNA. Because there is an excess of primers, single strands are more likely to bind to a primer than to one another. The primers provide DNA polymerase with the 3′ hydroxyl group it requires for DNA synthesis and bind only to specific sequences of 20–30 nucleotides.

3 **Extension.** Raising the temperature to about 72°C increases the rate at which DNA polymerase replicates each strand to produce more DNA.

▶ Play: *PCR: The Process* @ Mastering Microbiology

These steps are repeated over and over **4**, so the number of DNA molecules increases exponentially (FIGURE 8.6b). After 30 cycles—which require only a few hours to complete—PCR produces over 1 billion identical copies of the original DNA molecule.

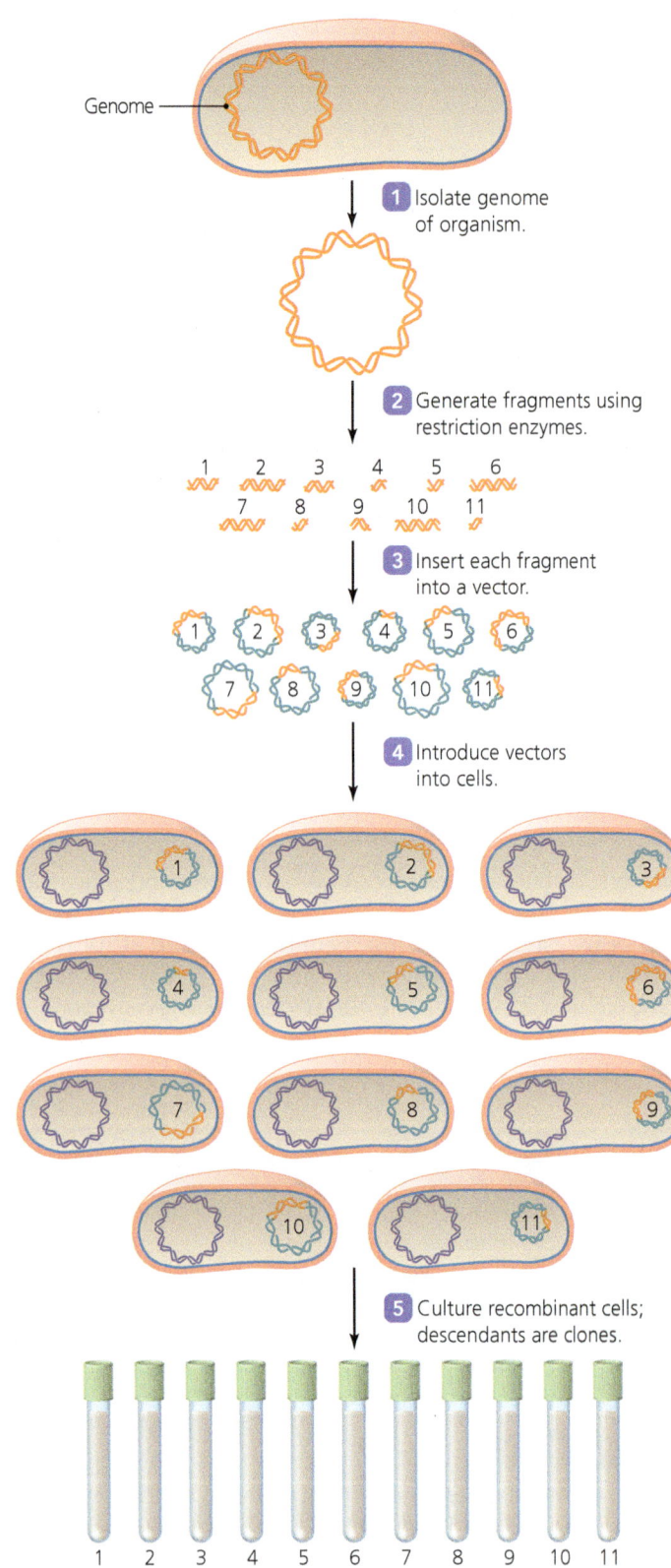

▲ **FIGURE 8.5 Production of a gene library.** A gene library is the population of all cells or phages that together contain all of the genetic material of interest. In this figure, each clone of cells carries a portion of a bacterium's genome.

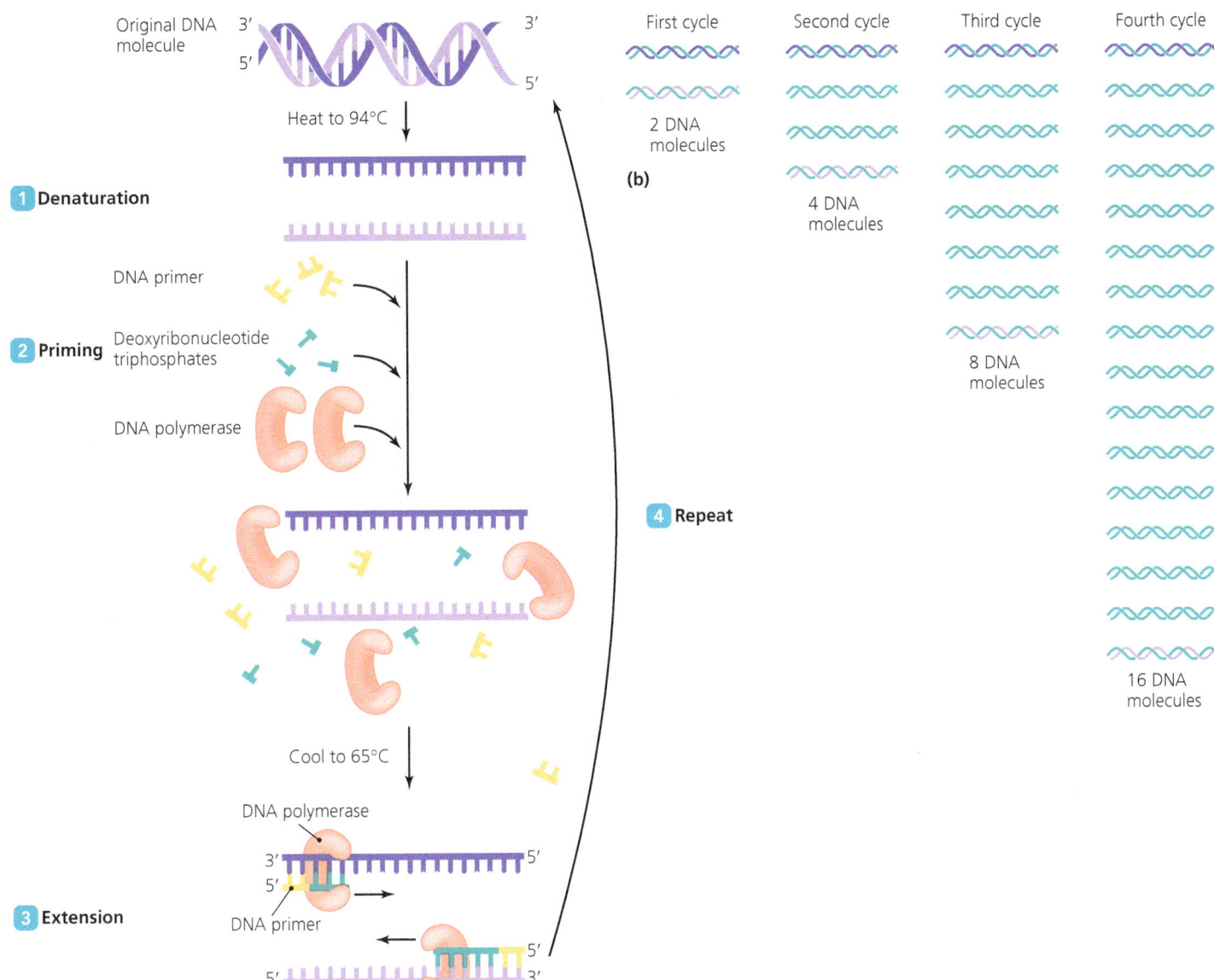

▲ FIGURE 8.6 **Use of the polymerase chain reaction (PCR) to replicate DNA.** The artwork is not rendered to scale; for example, primers are typically 20–30 nucleotides long. **(a)** Each cycle of PCR consists of three steps: ❶ During *denaturation* at 94°C, the hydrogen bonds between DNA strands are broken, and the strands separate. ❷ Cooling to 65°C in the presence of deoxyribonucleotide triphosphates, primers, and DNA polymerase allows *priming* to occur. ❸ During *extension* at 72°C, warming speeds the action of DNA polymerase in replicating strands to produce more DNA. **(b)** Each cycle of PCR doubles the amount of DNA ❹; over 1 billion copies of the original DNA molecule are produced by 30 cycles of PCR.

The process can be automated using a *thermocycler,* a device that automatically performs PCR by continuously cycling all the necessary reagents—DNA, DNA polymerase, primers, and triphosphate deoxynucleotides—through the three temperature regimes. A thermocycler uses DNA polymerase derived from hyperthermophilic archaea or bacteria, such as *Thermus aquaticus* (ther'mŭs a-kwa'ti-kŭs; see **Beneficial Microbes: Life in a Hot Tub** on p. 246). This enzyme, called *Taq DNA polymerase* or simply *Taq,* is not denatured at 94°C, so the machine need not be replenished with DNA polymerase after each cycle.

Investigators use a variation of PCR—called **real-time PCR** or *quantitative PCR (qPCR)*—to accurately measure the number of DNA or RNA molecules in the sequence of interest present in a sample. For example, health care providers can use real-time PCR to determine the number of HIV genome copies in a patient's blood sample. This is important in monitoring the progression of disease and effectiveness of treatment of diseases such as AIDS.

Investigators perform real-time PCR by adding a dye that fluoresces when bound to double-stranded DNA (dsDNA). A photometer measures the amount of fluorescence—an amount that corresponds to the number of dsDNA molecules. As the amount of fluorescence increases with each cycle of PCR, the number of new molecules of dsDNA can be ascertained. The number of original molecules can then be calculated.

BENEFICIAL MICROBES

Life in a Hot Tub

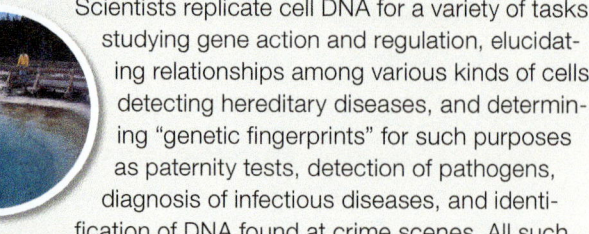

Scientists replicate cell DNA for a variety of tasks: studying gene action and regulation, elucidating relationships among various kinds of cells, detecting hereditary diseases, and determining "genetic fingerprints" for such purposes as paternity tests, detection of pathogens, diagnosis of infectious diseases, and identification of DNA found at crime scenes. All such studies use millions or billions of identical copies of DNA that are produced using a process called polymerase chain reaction (PCR). PCR replicates DNA enzymatically without using living cells.

The concept of PCR is relatively simple. As its inventor wrote in *Scientific American*, "Beginning with a single molecule of the genetic material DNA, the PCR can generate 100 billion similar molecules in an afternoon. The reaction is easy to execute. It requires no more than a test tube, a few simple reagents, and a source of heat." In the latter, however, lies a problem.

The temperature required to perform PCR is about 94°C. This temperature, which is almost that of boiling water, is the temperature required to break the hydrogen bonds of DNA and unzip the double helix, but it also permanently denatures most DNA polymerase enzymes.

Enter *Thermus aquaticus*, a bacterium that thrives in hot springs such as those of Yellowstone National Park. Since this bacterium loves hot water, it is not surprising that its enzymes are heat stable. Its DNA polymerase—called Taq polymerase or Taq—was the first polymerase used for PCR replication of DNA. Though *Science* magazine declared Taq "Molecule of the Year" in 1989, scientists now have many other heat-stable polymerases available for PCR from bacterial and archaeal hyperthermophiles.

Combined with a step of reverse transcription, researchers can use real-time PCR to accurately determine the number of copies of a specific mRNA in a cell; with this information, they can determine the expression level of the corresponding gene.

Selecting a Clone of Recombinant Cells

LEARNING OUTCOME

8.14 Explain how researchers use DNA probes to identify recombinant cells.

Before recombinant DNA technology can have practical applications, a scientist must be able to select and isolate recombinant cells that contain particular genes of interest. For example, once researchers have created a gene library, they must find the clone that contains the gene of interest from among thousands of clones in the library. They can identify the appropriate clone by performing PCR (only clones from which a PCR product is obtained contain the desired gene) or by using a fluorescent or radioactive probe—which, as explained earlier, binds specifically and exclusively to complementary nucleotide sequences. A researcher then isolates and cultures cells that have the gene of interest.

Separating DNA Molecules: Gel Electrophoresis and the Southern Blot

LEARNING OUTCOME

8.15 Describe the process and use of gel electrophoresis, particularly as it is used in a Southern blot.

Electrophoresis (ē-lek-trō-fōr-ē′sis) is a technique that involves separating molecules based on their electrical charge, size, and shape. Scientists can use **gel electrophoresis** to isolate fragments of DNA molecules that can then be inserted into vectors, multiplied by PCR, or preserved in a gene library.

In gel electrophoresis, DNA molecules, which have an overall negative charge, are drawn through a semisolid gel by an electric current toward the positive electrode within an electrophoresis chamber (**FIGURE 8.7**). The gel is typically composed of a purified sugar component of agar, called *agarose*, which acts as a molecular sieve that retards the movement of DNA fragments and separates the fragments by size. Smaller DNA fragments move faster and farther than larger ones. Scientists can determine the size of a fragment by comparing the distance it travels to the distances traveled by standard DNA fragments of known sizes.

In 1975, Ed Southern (1938–) devised a method called the **Southern blot** to transfer DNA from electrophoresis gels onto membranes, which are less delicate than gels. A researcher adds probes to reveal the presence of DNA of interest. A northern blot is a similar technique used to detect specific RNA molecules.

DNA Microarrays

LEARNING OUTCOME

8.16 Describe the manufacture and use of DNA microarrays.

Another tool of biotechnology is a **DNA microarray**. An array consists of molecules of single-stranded DNA (ssDNA), either genomic DNA or cDNA, immobilized on glass slides, silicon chips, or nylon membranes. Robots, similar to those that construct computer chips, deposit PCR-derived copies of hundreds or thousands of different DNA sequences in precise locations on an array (**FIGURE 8.8**). An array may consist of DNA from a single species (e.g., DNA microarrays containing sequences from all the genes of *E. coli* are available commercially), or an array may contain DNA sequences from numerous species. In any case, single

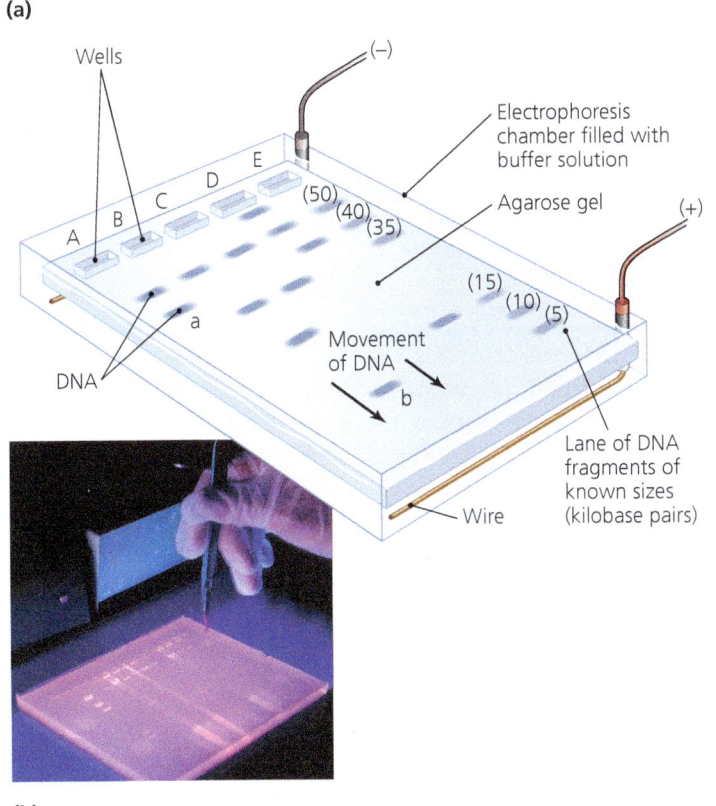

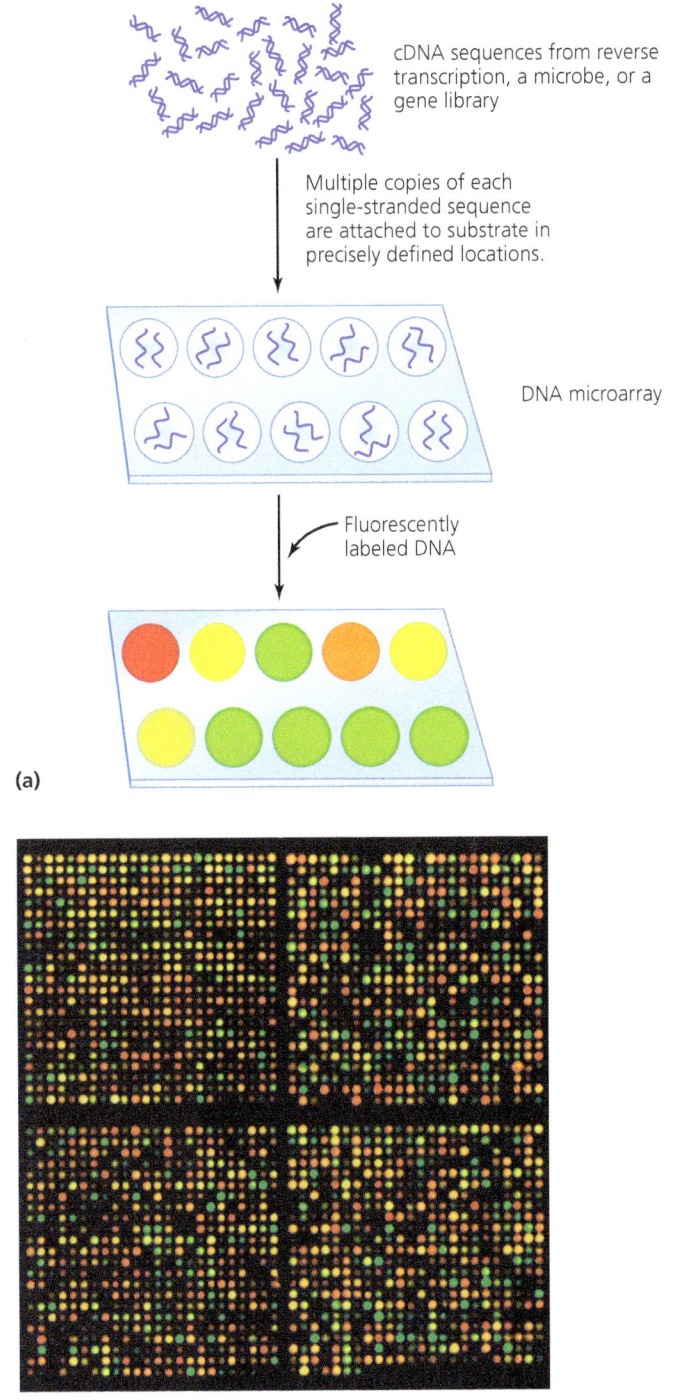

Researchers using DNA microarrays can simultaneously monitor the expression of thousands of genes and compare and contrast genetic expression under different conditions. Different colors of fluorescent dye can be used to label DNA from microbes grown in each condition.

▲ FIGURE 8.7 **Gel electrophoresis.** (a) After DNA is cleaved into fragments by restriction enzymes, it is loaded into wells, which are small holes cut into the agarose gel. DNA fragments of known sizes, typically in thousands (kilo) of nucleotide base pairs, are often loaded into one well (in this case, E) to serve as standards. After the DNA fragments are drawn toward the positive electrode by an electric current, they are stained with a dye. (b) Ethidium bromide dye fluoresces under ultraviolet illumination to reveal the locations of DNA within a gel. *Compare the positions of the fragments in lanes A and B of the diagram to the positions of the fragments of known sizes. What sizes are the fragments labeled a and b?*

Figure 8.7 a: 40 kilobase pairs; b: 10 kilobase pairs.

strands of fluorescently labeled DNA in a sample washed over an array adhere only to locations on the array where there are complementary DNA sequences.

Scientists use DNA microarrays in a number of ways, including the following:

- **Monitoring gene expression.** One way organisms control metabolism is by controlling RNA transcription. Scientists use DNA microarrays to monitor which genes a cell is transcribing at a particular time by making fluorescently labeled, single-stranded cDNA from mRNA in the cell. These DNA strands bind to complementary DNA sequences on an array, and the location of fluorescence on the array at specific sites reveals which genes the cell was transcribing at the time. For example, activity of normal cells can be compared to activity in cancerous cells to determine how gene expression is related to oncogenesis.

▲ FIGURE 8.8 **DNA microarray.** (a) Construction and use of a microarray. Multiple copies of single-stranded DNA with known sequences are affixed in precise locations on a glass slide, silicon chip, nylon membrane, or other substrate. Fluorescently labeled DNA washed over the microarray binds to complementary strands. (b) Photograph of a DNA microarray showing locations of differently labeled cDNA molecules.

- **Diagnosing infection.** DNA microarrays made with DNA sequences of numerous pathogens reveal the presence of those pathogens in medical samples.
- **Identifying organisms in an environmental sample.** Microbial ecologists monitor the presence or absence of microbes in an environment by using microarrays of DNA from the organisms to simultaneously detect and identify numerous bacterial and fungal species in environmental samples.

Inserting DNA into Cells

LEARNING OUTCOME

8.17 List and explain four artificial techniques for introducing DNA into cells.

A goal of recombinant DNA technology is the insertion of a gene into a cell. In addition to using vectors and the natural methods of transformation of competent cells, transduction, and conjugation, scientists have developed several artificial methods to introduce DNA into cells, including the following:

- **Electroporation (FIGURE 8.9a).** Electroporation involves using an electrical current to puncture microscopic holes through a cell's membrane so that DNA can enter the cell from the environment. Electroporation can be used on all types of cells, though the thick-walled cells of fungi and algae must first be converted to *protoplasts*, which are cells whose cell walls have been enzymatically removed. Cells treated by electroporation repair their membranes and cell walls after a time.
- **Protoplast fusion (FIGURE 8.9b).** When protoplasts encounter one another, their cytoplasmic membranes may fuse to form a single cell that contains the genomes of both "parent" cells. Exposure to polyethylene glycol increases the rate of fusion. The DNA from the two fused cells recombines to form a recombinant molecule. Scientists often use protoplast fusion for the genetic modification of plants.
- **Injection.** Two types of injection are used with larger eukaryotic cells. Researchers use a *gene gun* powered by a blank .22-caliber cartridge or compressed gas to fire tiny tungsten or gold beads coated with DNA into a target cell (FIGURE 8.9c). The cell eventually eliminates the inert metal beads. In *microinjection*, a geneticist inserts DNA into a target cell with a glass micropipette having a tip diameter smaller than that of the cell or nucleus (FIGURE 8.9d). Unlike electroporation and protoplast fusion, injection can be used on intact tissues, such as in plant seeds.
- **Heat shock.** With this technique, scientists shock bacterial cells into becoming **competent**—that is, the cells gain the ability to take up DNA from their environment. To do this, researchers mix the cells with liquid medium containing the DNA of interest and add calcium chloride or rubidium chloride. They then shock the cells by placing them in a warm water bath at 42°C for 30–60 seconds, quickly chill them with an ice bath, mix them with liquid medium, and incubate them at the preferred temperature for that particular bacterium. The cells can take up the DNA of interest during the temperature shock treatment.

In every case, foreign DNA that enters a cell remains in a cell's progeny only if the DNA is self-replicating, as in the case of plasmid and viral vectors, or if the DNA integrates into a cellular chromosome by recombination.

MICRO CHECK

5. Where could a research scientist find copies of all the genes of a microbe without having to isolate them from the microbe?
6. What process is used to "amplify" DNA, making large numbers of copies from a small sample?
7. What characteristic(s) of gel electrophoresis make it an effective technique in genetic research?

TELL ME WHY

Why wasn't polymerase chain reaction (PCR) practical before the discovery of hyperthermophilic bacteria?

Applications of Recombinant DNA Technology

The importance of recombinant DNA technology lies not in the novelty, cleverness, or elegance of its procedures but in its wide range of applications. In this section we consider how recombinant DNA technology is used to solve various problems and create research, medical, and agricultural products.

Genetic Mapping

LEARNING OUTCOMES

8.18 Describe genetic mapping and fluorescent *in situ* hybridization, and explain their usefulness.
8.19 Define *genomics*, *transcriptomics*, *metabolomics*, and *functional genomics*.
8.20 Contrast Sanger sequencing and next-generation sequencing.

One application of the tools and techniques of recombinant DNA technology is **genetic mapping**, which involves locating genes on a nucleic acid molecule. Genetic maps provide scientists with useful facts, including information concerning an organism's metabolism and growth characteristics, as well as its potential relatedness to other microbes. For example, scientists have discovered a virus with a genetic map similar to those of certain hepatitis viruses. They named the new discovery *hepatitis G virus* because it presumably causes hepatitis, though it has not been demonstrated that the virus actually causes the disease.

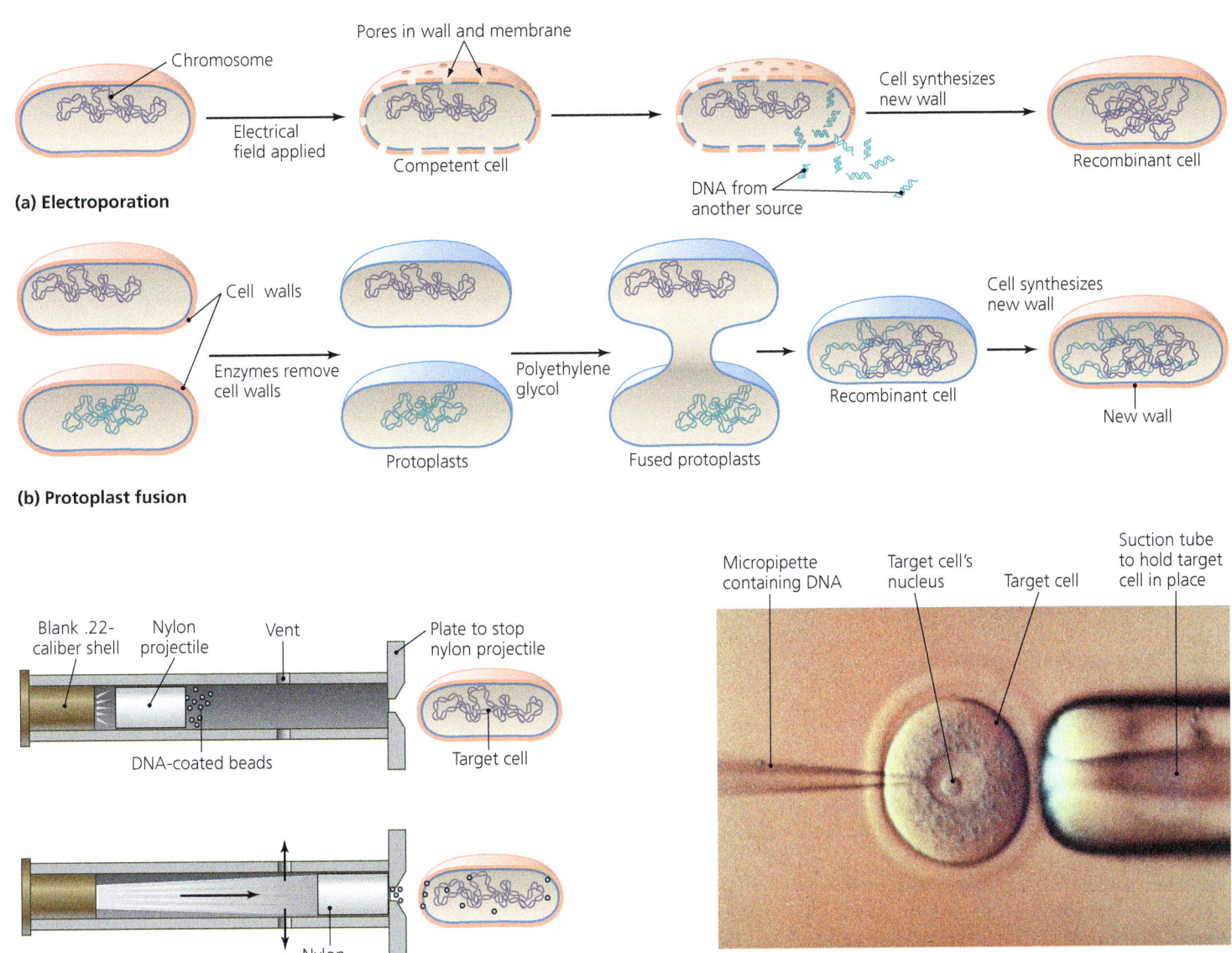

▲ FIGURE 8.9 **Some methods of inserting DNA into cells.** (a) Electroporation, in which an electrical current applied to a cell makes it competent to take up DNA. (b) Protoplast fusion, in which enzymes digest cell walls to create protoplasts that fuse at a high rate when treated with polyethylene glycol. (c) A gene gun, which fires DNA-coated beads into a cell. (d) Microinjection, in which a solution of DNA is introduced into a cell through a micropipette.

Locating Genes

Until about 1970, scientists identified the specific location of genes on chromosomes by cumbersome, time-consuming, labor-intensive methods. Recombinant DNA techniques provide simpler and universal methods for genetic mapping.

One technique for locating genes, called *restriction fragmentation*, was one of the earliest applications of restriction enzymes. In this technique, which is used for mapping the relative locations of genes in plasmids and viruses, researchers compare DNA fragments resulting from cleavages by several restriction enzymes to determine each fragment's location relative to the others. If the researchers know the locations of specific genes on specific fragments, then elucidation of the correct arrangement of the fragments will reveal the relative locations of the genes on the entire DNA molecule.

Using this method, scientists first completed the entire gene map of a cellular microbe—the bacterium *H. influenzae*—in 1995. Since then, geneticists have elucidated complete gene maps of numerous viruses and prokaryotic and eukaryotic organisms.

Often a scientist wants to know where in the environment, clinical sample, or biofilm a particular microbial species is located. When researchers know of a particular gene exclusive to that organism, they can locate the gene and thereby the microbe using *fluorescent in situ hybridization (FISH)*.

In this method, scientists use fluorescent DNA probes that hybridize with their complementary target. Scientists using fluorescent microscopes to view such probes can determine where the gene and its organism are located (**FIGURE 8.10** on the next page). Using a number of different colors of fluorescent probes, researchers can locate numerous genes and the microbes that

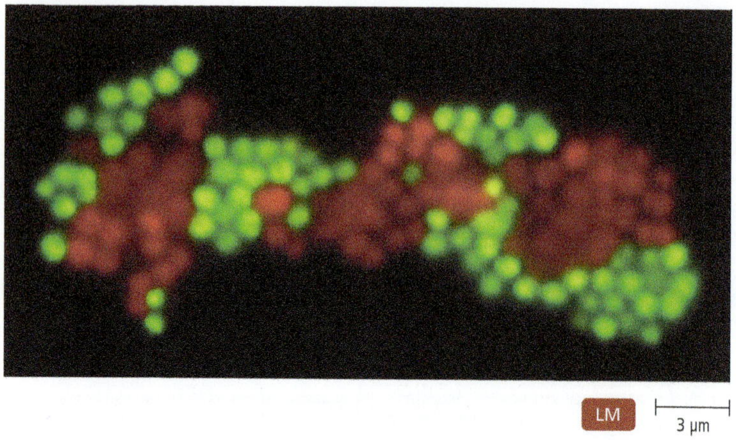

▲ FIGURE 8.10 Fluorescent *in situ* hybridization (FISH). Green-fluorescing cells are *Staphylococcus aureus*, whereas red-fluorescing cells are another species of *Staphylococcus*. The absence of fluorescence in some cells indicates that at least three species are present in this blood sample.

carry them simultaneously. FISH is used for a variety of purposes, including diagnosing disease, identifying microbes in environmental samples, and analyzing biofilms.

Genomics and Nucleotide Sequencing

Genomics is the sequencing and analysis of the nucleotide bases of genomes of organisms, viruses, and eukaryotic mitochondria and chloroplasts. Genomics is related to transcriptomics and metabolomics. *Transcriptomics* is the study of *transcriptomes*—all the mRNAs that are made in a cell under a given set of circumstances—and their functions. Researchers might be interested in transcriptomics for a variety of reasons. For example, they might elucidate the metabolic pathways that a soil bacterium utilizes at human body temperature by contrasting the transcriptome at 37°C with that at 25°C. *Metabolomics* involves the study of all the chemicals present in a particular cell. This includes metabolic intermediates and the end products of all the biochemical reactions of the cell. Together, genomics, transcriptomics, and metabolomics improve our understanding of cellular processes and how cells respond to environmental stimuli.

At first, scientists sequenced DNA molecules by selectively cleaving DNA at A, T, G, or C bases, separating the fragments by gel electrophoresis, and mapping the order in which overlapping fragments occur in a complete DNA molecule. Such cumbersome sequencing was limited to short DNA molecules, such as those of plasmids.

Modern DNA sequencing began when Frederick Sanger (1918–2013) began using modified versions of each of the four DNA nucleotides—versions that terminate DNA replication. By using various amounts of these modified nucleotides and then examining numerous DNA fragments produced by termination, the genetic sequence could be pieced together.

Scientists improved the Sanger method by incorporating four different fluorescent dyes—a different color for each nucleotide base—and then using an automated DNA sequencer to determine the sequence of colors emitted by the dyes. Scientists using the Sanger method, with machines running 24 hours a day for months, sequenced the entire genomes of numerous viruses, bacteria, and eukaryotic organisms. The Human Genome Project reached a milestone in 2001 by using automated Sanger techniques to sequence approximately 3 billion nucleotide base pairs that constitute a human genome.

Sanger sequencing methods are considered first-generation DNA sequencing. Researchers now use **next-generation sequencing (NGS)**. Among the variety of NGS methods, a popular one is *four-color reversible termination sequencing*. In this method, DNA synthesis is temporarily stopped after the addition of a single modified nucleotide. Each of the four kinds of nucleotide fluoresces a color specific to its base (A, T, G, or C): one color for each of the four bases. After a record is made of the location of each dye, the dyes are removed and a new nucleotide is added as the process repeats.

This process is illustrated in FIGURE 8.11: Researchers attach identical ssDNA fragments (produced by PCR) at known locations on a slide **1**. The great advantage of NGS is to sequence not just one piece of DNA at a time, but hundreds of millions; it is massively parallel. Once the fragments are attached to the slide, an automated sequencer passes modified molecules of all four types of DNA nucleotides over the slide at once **2**. Each nucleotide carries a dye that can fluoresce a specific color—one color for each of the four nucleotide bases—and a modification to the nucleotide that stops DNA elongation. Each time a nucleotide is added to a growing DNA strand, synthesis of that strand ceases. The sequencer shines a laser across the slide, causing each fluorescent dye to flash its color, each color corresponding to its particular nucleotide **3**. A light detector records the event **4**. The sequencer then chemically removes the fluorescent dye and the chemical group that stopped elongation **5**. DNA synthesis begins again but stops as soon as modified nucleotides are added. The laser stimulates the dyes, and another image is captured before the dyes and stop groups are removed **6**. The process repeats until a complementary DNA strand has been synthesized for each fragment on the slide **7**. The sequence of colors recorded by the light detector at each location corresponds to the sequence of nucleotides in the ssDNA molecules at that site **8**.

NGS, four-color reversible termination sequencing instruments can sequence an astounding 80 billion nucleotide bases in a few days. In contrast, the automated Sanger method would require almost seven years of nonstop sequencing to process the same number.

With various sequencing methods, particularly NGS, researchers have ascertained the complete genomes of over 20,000 bacteria, about a thousand archaea, and of numerous eukaryotes. We are well into the genomic age.

Elucidation of the gene sequences of pathogens, particularly those affecting hundreds of millions of people and those with potential bioterrorist uses, is a current priority of researchers. Scientists hope to use the information to develop novel drugs and more effective therapies and vaccines.

Another use for genomics is to relate DNA sequence data to protein function. For instance, scientists are investigating the genes and proteins of *Deinococcus radiodurans* (dī-nō-kok'ŭs rā-dē-ō-dur'anz), a microorganism that is remarkably resistant to damage of its DNA by radiation. Such studies may lead to

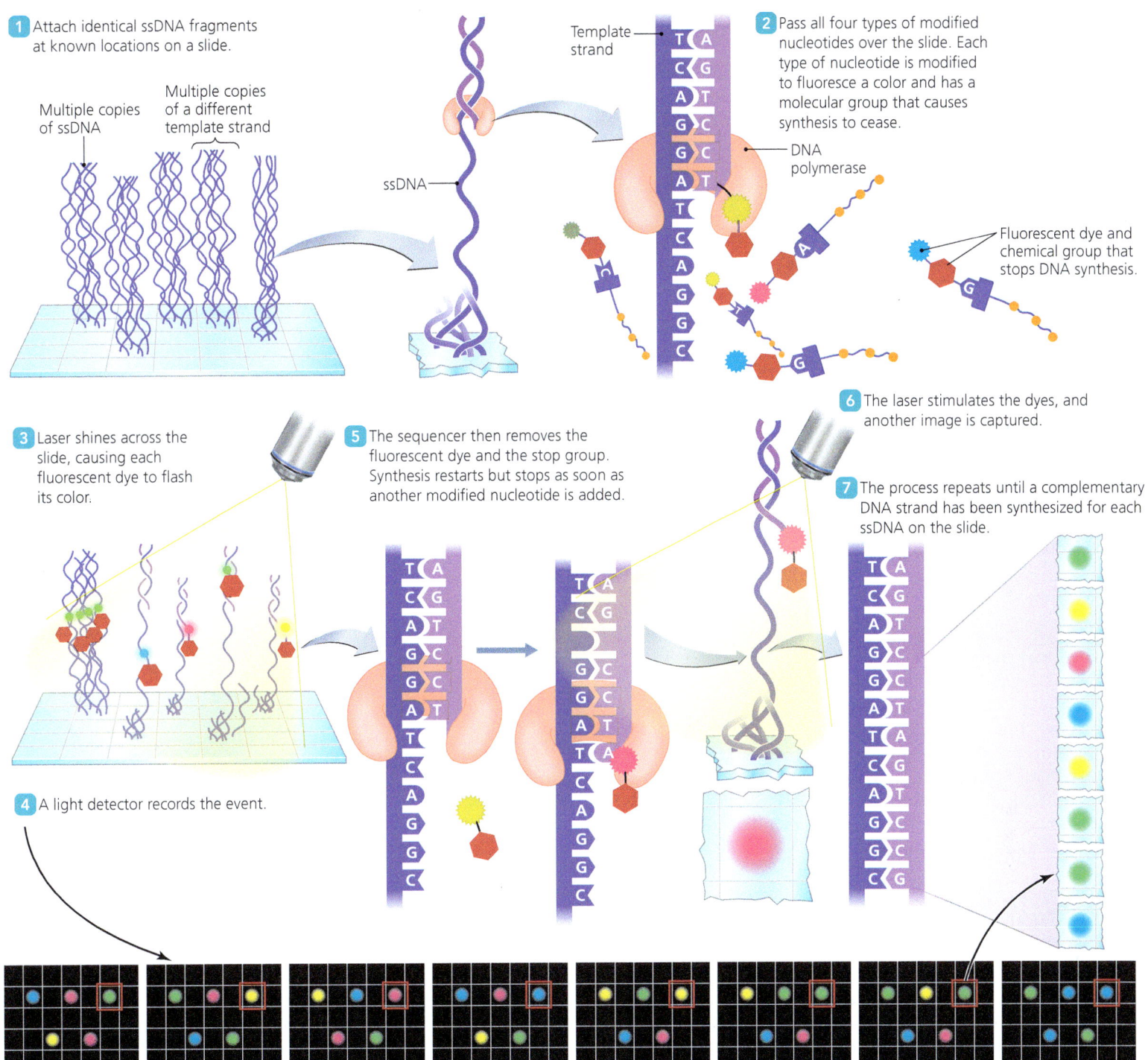

▲ **FIGURE 8.11 Next-generation DNA sequencing (NGS).** Shown here is four-color reversible termination sequencing.

methods of reversing genetic damage in cancer patients undergoing radiation therapy. Researchers are also investigating the genetic basis of the enzymes of psychrophiles, which are microorganisms that thrive at temperatures below 20°C. Such enzymes have potential applications in food processing and in the manufacture of drugs.

TABLE 8.2 on p. 252 summarizes some tools and techniques of recombinant DNA technology.

Functional Genomics

Currently, a major emphasis in genomics is to determine what genes' products do—this is the field of *functional genomics*. A classic method to determine the function of a gene is a *gene knockout*, which involves removing a single gene from a bacterial genome while leaving the rest of the genome intact. Phenotypic analysis of the knockout strain may then reveal the gene's function, since the bacteria will lack this function. The *E. coli* Keio Knockout

TABLE 8.2 Some Tools and Techniques of Recombinant DNA Technology

Tool or Technique	Description	Potential Application
Mutagen	Chemical or physical agent that creates mutations	Creating novel genotypes and phenotypes
Reverse transcriptase	Enzyme from RNA retrovirus that synthesizes cDNA from an RNA template	Synthesizing a gene using an mRNA template
Synthetic nucleic acid	DNA molecule prepared *in vitro*	Creating DNA probes to localize genes within a genome
Restriction enzyme	Bacterial enzyme that cleaves DNA at specific sites	Creating recombinant DNA by joining fragments
Vector	Transposon, plasmid, or virus that carries DNA into cells	Altering the genome of a cell
Gene library	Collection of cells or viruses, each of which carries a portion of a given organism's genome	Providing a ready source of genetic material
Polymerase chain reaction (PCR)	Produces multiple copies of a DNA molecule	Multiplying DNA for various applications
Gel electrophoresis	Uses electrical charge to separate molecules according to their size	Separating DNA fragments by size
Electroporation	Uses electrical current to make cells competent	Inserting a novel gene into a cell
Protoplast fusion	Fuses two cells to create recombinants	Inserting a novel gene into a cell
Gene gun	Blasts genes into target cells	Inserting a novel gene into a cell
Microinjection	Uses micropipette to inject genes into cells	Inserting a novel gene into a cell
Southern blot	Localizes specific DNA sequences on a stable membrane	Identifying a strain of pathogen
Nucleic acid probes	RNA or DNA molecules labeled with radioactive or fluorescent tags	Localizing specific genes in a Southern blot
Genetic mapping	Uses restriction enzymes to locate relative positions of restriction sites	Locating genes in an organism's genome
DNA sequencing	Determines the sequence of nucleotide bases in DNA	Comparing genomes of organisms
DNA microarray	Reveals presence of specific DNA or RNA molecules in a sample	Diagnosing infection

collection (from Keio University in Japan) is a powerful resource for functional analysis: it contains 3985 strains, each having one nonessential gene replaced by a kanamycin-resistance gene.

Another basic method in functional genomics is to overexpress a particular gene by enhancing its transcription or translation, resulting in an abundance of that gene's product in the cell. Scientists can then analyze the phenotype or purify the individual protein for further analysis. There are gene expression libraries that contain every gene in a plasmid under control of a strong promoter for this purpose.

Microbial Community Studies

It is estimated that more than 99% of microorganisms have never been grown in a laboratory; indeed, scientists know them only by unique DNA sequences, sometimes called *signatures* or *DNA fingerprints*. For example, based on such unique DNA sequences, scientists have detected DNA from over 500 species of bacteria from human mouths; however, they have been able to positively identify only about 150 species. An understanding of the biology of the other 350 species may lead to a better understanding of tooth and gum decay, diagnosis of disease, and advances in oral health care.

Because next-generation DNA sequencing is relatively inexpensive and rapid, investigators can now determine complete genetic sequences for and relative abundance of all the members making up a **microbiome**, which is all the microorganisms in a particular environment. In the developing field of metagenomics, all the genetic material from an environment is sequenced, allowing insight into the microbiome regardless of whether the organisms can be cultured. Metagenomics is particularly suitable for studying microbial communities and diversity (see Chapter 27).

Beneficial Microbes: Our Other "Organ" (on p. 253) examines the microbiome of humans and the effects it may have on our health.

Pharmaceutical and Therapeutic Applications

LEARNING OUTCOMES
8.21 Describe seven potential medical applications of recombinant DNA technology.
8.22 Describe the steps and uses of genetic screening.
8.23 Define *gene therapy*.

Researchers now supplement traditional biotechnology with recombinant DNA technology to produce a variety of pharmaceutical and therapeutic substances and to perform a host of medically important tasks. Here, we explore the use of recombinant DNA technology to synthesize medically important proteins, produce vaccines, screen for genetic diseases, treat genetic illnesses, diagnose diseases by matching DNA specimens to the organisms from which they came, aid in organ transplantation, and make animals that respond more similarly to humans when treated with various pathogens or experimental treatments.

Protein Synthesis
In 1978, researchers, including lead author Lydia Villa-Komaroff (1947–), first reported their successful insertion of a gene for insulin into *E. coli*, allowing the bacterium to synthesize insulin for diabetics less expensively and more safely than insulin derived from animals. Scientists have since inserted genes for

BENEFICIAL MICROBES

Our Other "Organ"

The number of cells in an adult human body is an astounding 10–40 trillion. However, scientists estimate that perhaps ten times as many bacteria, fungi, and other microorganisms, plus innumerable viruses, cover the surfaces of our bodies. Four hundred trillion "guests"! You support this gigantic, symbiotic ecosystem—a microbiome, which can be defined as the community of all microbes in a particular environment. A major health benefit of the microbiome is that it prevents invading pathogens from attaching or persisting on your body surfaces. What if we could identify and categorize all of the different species of microorganisms that live in and on a body? This was the ambitious goal of the Human Microbiome Project (HMP).

To begin, several hundred scientists from 80 institutions collected samples from 242 healthy volunteers. Eighteen different body sites were sampled, including nine different oral locations, nostrils, ears, and the gut, which was represented by fecal samples. Researchers utilized direct DNA extraction and next-generation sequencing to discover DNA from almost every microbe in each sample. Because most microorganisms have not been cultured in the lab, the majority of the members of the microbiome as determined by DNA sequences had never before been observed in humans.

As expected, the researchers didn't find obligate pathogens in the healthy humans, though they commonly found opportunistic pathogens. A surprise to many researchers was that the fecal bacterium *Escherichia coli*, though present in the gut of most people, is actually rare, present at a level of less than 1 in 1000 bacteria.

The diversity of microbes among people was much greater than expected. For example, before the HMP, microbiologists thought that anaerobic bacteria in the phylum *Bacteroidetes* were the most common bacteria in everyone's gut. Now, we know this is true for only some people; the actual percentage varies among these healthy individuals from 20% to 80% of the total number of bacteria. Obviously, the normal and healthy number of bacteroids for one person can be very different from the number in another.

Another surprising finding was that although the taxonomic diversity at a particular body site varied widely among individuals, the abundance of bacterial genes representing certain metabolic pathways was highly consistent, suggesting that metabolism controls what organisms are there. This could possibly be medically significant: future therapies may focus on restoring certain biochemical activities to a patient's gut microbiome, rather than simply adding specific bacteria.

---

interferon (a natural antiviral chemical), human growth hormone, blood clotting factor VIII, and other proteins into bacteria and yeast cells so that the microbes synthesize these proteins in vast quantities. In the past, such proteins were isolated from cadavers, donated blood, or animals—labor-intensive processes that carry the risk of inducing allergies or of transferring pathogens, such as hepatitis B and HIV. "Genetically engineered" proteins are safer and less expensive than their naturally occurring counterparts.

Vaccine Production

Vaccines contain *antigens*—foreign substances such as weakened bacteria, viruses, and toxins that stimulate the body's immune system to respond to and subsequently remember these foreign materials. In effect, a vaccine primes the immune system to respond quickly and effectively when confronted with pathogens and their toxins. However, the use of some vaccines entails a risk—they may cause the disease they are designed to prevent.

Scientists now use recombinant DNA technology to produce safer vaccines. Once they have inserted the gene that codes for a pathogen's antigens into a vector, they can inject the recombinant vector or the proteins it produces into a patient. Thus, the patient's immune system is exposed to a subunit of the pathogen—one of the pathogen's antigens—but not to the pathogen itself. Such **subunit vaccines** are especially useful in safely protecting against pathogens that either cannot be cultured or cause incurable fatal diseases. Hepatitis B vaccine is an example of a successful subunit vaccine. Scientists are also pursuing subunit vaccines against HIV.

A promising future approach to vaccination involves introducing genes coding for antigenic proteins of pathogens into common fruits or vegetables, such as bananas or beans. The immune systems of people or animals eating such altered produce would be exposed to the pathogen's antigens and theoretically would develop immunological memory against the pathogen. Such a vaccine would have the advantages of being painless and easy to administer, and vaccination would not require a visit to a health care provider.

Another type of vaccination involves producing a recombinant plasmid carrying a gene from a pathogen and injecting the plasmid into a human whose body then synthesizes polypeptides characteristic of the pathogen. The polypeptides stimulate immunological memory, readying the body to mount a vigorous immune response and prevent infection should it be exposed to the real pathogen. These *DNA vaccines* have great potential because the production and delivery method remains the same—the pathogen providing the antigen is the only difference—which should reduce costs.

Genetic Screening

Genetic mutations cause some diseases, such as inherited forms of breast cancer and Huntington's disease. Laboratory technicians use DNA microarrays to screen patients, prospective parents, and fetuses for such mutant genes. This procedure, called **genetic screening**, can also identify viral DNA sequences in a patient's blood or other tissues. For instance, genetic screening can identify HIV in a patient's cells even before the patient shows any other sign of infection.

Gene Therapy

Gene therapy is the use of recombinant DNA technology to introduce DNA into patients' cells to treat or cure a disease that is caused by missing or defective genes. Gene therapies have tremendous potential because they could cure diseases at their root. There are two main approaches to gene therapy: scientists clone DNA and deliver it directly into a patient's cells to introduce missing DNA or supplement the patient's DNA, or they isolate and culture some of a patient's cells, edit the cells' genomes, and then put the cells back into the patient.

There are potential problems when introducing genes directly into a patient:

- The patient's immune system may neutralize the vector, so that the therapy is ineffective.
- The gene may not be delivered to the correct cell type.
- It might not be possible to introduce DNA into enough cells to have a therapeutic effect.
- The DNA may insert in a part of the patient's genome that normally is not transcribed, so the desired mRNA and protein would not be synthesized.
- The DNA may insert into another gene, disrupting its function.
- The insertion may cause unintended problems, such as triggering cancer.
- The inserted gene may become active in cells where it normally is not active, causing unforeseen side effects.

Editing a patient's cells and returning them to the patient addresses some of these concerns. For example, once a cellular genome is edited, the insertion location can be verified by sequencing, though inadvertent genome changes in locations other than the preferred site may not be noticed.

For the above reasons, gene therapy is still largely an experimental approach and only considered a viable option in cases of severe diseases. Doctors have successfully treated patients for metachromatic leukodystrophy, a hereditary neurological disorder; Wiskott-Aldrich syndrome, an immune system disorder; and choroidoeremia and retinitis pigmentosa, types of hereditary blindness. Other conditions that may respond well to gene therapy are cystic fibrosis, sickle-cell anemia, some types of hemophilia, muscular dystrophy, and diabetes.

Medical Diagnosis

Clinical microbiologists use PCR, fluorescent genetic probes, and DNA microarrays in diagnostic applications. They examine specimens from patients for the presence of gene sequences unique to certain pathogens, such as particular hepatitis viruses, cytomegalovirus, human immunodeficiency virus, or the bacterial pathogens of gonorrhea, tuberculosis, trachoma, and ulcers. These tests are rapid and highly specific. For almost all viruses and for many bacteria, molecular detection has replaced culturing and identification by biochemical testing in clinical labs.

Xenotransplantation

Xenotransplants[5] are animal cells, tissues, or organs introduced into the human body. For years physicians have performed xenotransplants, for instance, using valves from pig hearts to repair severely damaged human hearts. However, recombinant DNA technology may expand the possibilities. It is theoretically feasible to insert functional human genes into animals to direct them to produce organs and tissues for transplantation into humans. For example, scientists could induce pigs to produce humanlike cytoplasmic membrane proteins so that entire organs from pigs would not be rejected as foreign tissue by a transplant recipient.

Biomedical Animal Models

Researchers use common animals such as mice or zebra fish in biomedical research to understand mechanisms of cancer, genetic disorders, infectious disease, and microbiome interactions. The animals allow scientists to develop new diagnostic and therapeutic procedures for human and veterinary medicine. Although the most common biomedical model, the mouse, is very similar to humans, it does not always accurately predict the responses of a human body. Using recombinant DNA technology, mouse physiology and tissues can be made "more human," thus making the genetically altered animal more useful. For example, scientists have replaced mouse genes that code for antibody proteins with their human equivalents. The genetically altered mice make human antibodies when exposed to pathogens, making them useful models for research of infectious disease and vaccines.

Agricultural Applications

LEARNING OUTCOME

8.24 Identify six agricultural applications of recombinant DNA technology.

Recombinant DNA technology has been applied to the realm of agriculture to produce **transgenic organisms**—recombinant plants and animals that have been altered for specific purposes by the addition of genes from other organisms. The purposes for which transgenic organisms, also called *genetically modified organisms* (*GMOs*), have been produced are many and varied and include herbicide resistance, tolerance to salty soils, resistance to freezing and pests, and improvements in nutritional value and yield. More than 15 million farmers in 29 countries grow transgenic crops on over 2.5 billion acres worldwide—more land than in the entire United States.

Herbicide Tolerance

The biodegradable herbicide *glyphosate* (Roundup) normally kills all plants—weeds and crops alike—by blocking an enzyme that is essential for plants to synthesize several amino acids. After scientists discovered and isolated a gene from *Agrobacterium tumefaciens* (ag'rō-bak-tēr'ē-um tū'me-fāsh-enz) that conveys tolerance to glyphosate, they produced transgenic crop

---
[5]From Greek *xenos*, meaning "stranger."

plants containing the gene. As a result of this application of recombinant DNA technology, farmers can now spray glyphosate, which has a relatively low toxicity, on a field of transgenic plants to kill weeds without damaging the crop. An added benefit is that farmers do not need to till the soil to suppress weeds during the growing season, reducing soil erosion by 80%. Most of the soybeans, corn, and cotton grown in the United States are genetically modified in this manner to be "Roundup ready." Glyphosate-tolerant rice, wheat, sugar beets, and alfalfa strains are also available, as are soybeans and corn tolerant of other herbicides.

Pest Resistance

Strains of the bacterium *Bacillus thuringiensis* (ba-sil'ŭs thur-in-jē-en'sis) produce a protein that, when modified by enzymes in the intestinal tracts of insects, becomes **Bt toxin (Bt)**. Bt binds to receptors lining the insect's digestive tract and causes the tissue to dissolve. Unlike some insecticides, Bt is naturally occurring, harmful only to insects, and biodegradable. Farmers, particularly organic growers, have used Bt for over 30 years to reduce insect damage to their crops.

Now, genes for Bt toxin have been inserted into a variety of crop plants, including potatoes, cotton, rice, and corn, so that they produce Bt for themselves. Insects feeding on such plants are killed, while humans and other animals that eat them are unharmed. By inserting genes for more than one type of Bt, scientists hope to forestall evolution of resistant insects.

Scientists have also developed crops that are resistant to microbial diseases. Two cases illustrate.

The water mold *Phytophthora infestans* (fī-tof'tho-ră in-fes'tanz) is the most devastating potato pathogen; it caused the great Irish potato famine of the 19th century, resulting in the deaths of at least a million people. The mold still causes many billions of dollars of crop damage each year. Scientists have now cloned genes from potato species resistant to *Phytophthora*; the cloned genes, multiplied by PCR, can be inserted into potato crops to reduce losses to farmers and increase available food for a growing world population.

Papaya ringspot virus causes devastating harm to papayas, mottling and malforming leaves, streaking fruit, and eventually killing the entire plant. Researchers centered at the University of Hawaii saved that state's papaya industry by inserting a gene for ringspot virus capsid into the papaya's genome. When the plant produces the viral protein, it triggers protection by an unknown mechanism against infection by the ringspot virus. Genetically modified plants grow normally (FIGURE 8.12).

Salt Tolerance

Years of irrigation have resulted in excessive salt buildup in farmland throughout the world, rendering the land useless for farming. Though salt-tolerant plants can grow under these conditions, they are not edible. Scientists have now successfully removed the gene for salt tolerance from these plants and inserted it into tomato and canola plants to create food crops that can grow in soil so salty that it would poison normal crops. Not only do such transgenic plants survive and produce fruit, but they also remove salt from the soil, restoring the soil and making it suitable to grow

▲ FIGURE 8.12 **Genetically modified papaya plants.** Plants on the left are dying from ringspot virus infection. Those on the right are modified to produce viral protein, which prevents infection by the virus.

unmodified crops as well. Researchers are now attempting to insert the gene for salt tolerance into rice, cotton, wheat, and corn.

Freeze Resistance

Ice crystals form more readily when bacterial proteins (from natural bacteria present in a field) are available as crystallization nuclei. Scientists have modified strains of the bacterium *Pseudomonas* (soo-dō-mō'nas) with a gene for a polypeptide that prevents ice crystals from forming. Crops sprayed with genetically modified bacteria can tolerate mild freezes, so the farmers no longer lose their crops to unseasonable cold snaps.

Improvements in Nutritional Value and Yield

Genetic researchers have increased crop and animal yields in several ways. For example, MacGregor tomatoes remain firm after harvest because the gene for the enzyme that breaks down pectin has been suppressed. This allows farmers to let the tomatoes ripen on the vine before harvesting and increases the tomatoes' shelf life. Scientists suppressed the gene indirectly by inserting a promoter in the noncoding DNA strand that allows transcription of antisense RNA. The antisense RNA then binds to the gene's mRNA, forming double-stranded RNA that makes it impossible to translate the pectin-catabolizing enzyme.

Another example of agricultural improvement involves bovine growth hormone (BGH), which when injected into cattle enables them to more rapidly gain weight and produce 10% more milk. Also, use of BGH reduces the fat content in beef. Though BGH can be derived from animal tissue, it is more economical to insert the BGH gene into bacteria so that they produce the hormone, which is then purified and injected into farm and ranch animals.

In yet another application, scientists have improved the nutritional value of rice by adding a gene for beta-carotene, which is a precursor to vitamin A. Vitamin A is required for human embryonic development and for vision in adults, and it is an important antioxidant that plays a role in ameliorating cancer and atherosclerosis (hardening of arteries).

Recombinant DNA technology has progressed to the point that scientists are now considering transplanting genes coding for entire metabolic pathways rather than merely genes encoding single proteins. For instance, researchers are attempting to transfer into corn and rice all the genes that bacteria use to convert atmospheric nitrogen into nitrogenous fertilizer, in effect allowing the recombinant plants to produce their own fertilizer.

Recombinant DNA tools and techniques allow scientists to examine, compare, and manipulate the genomes of microorganisms, plants, animals, and humans for a variety of purposes, including gene mapping and forensic, medical, and agricultural applications. However, as with many scientific advances, concerns arise about the ethics and safety of genetic manipulations. The next section deals with these issues.

MICRO CHECK

8. Which technique would be most useful in diagnosing an infection?
9. Which technique would be useful for locating a particular gene in living cells?
10. Which of the following would not be a potential medical application of recombinant DNA technology: gene library, genetic mapping, gene therapy, xenotransplantaton?

TELL ME WHY

Why don't doctors routinely insert genes into their patients to cure the common cold, flu, or tuberculosis?

The Ethics and Safety of Recombinant DNA Technology

LEARNING OUTCOME

8.25 Discuss the pros and cons concerning the safety and ethics of recombinant DNA technology.

Recombinant DNA technology provides the opportunity to transfer genes among unrelated organisms, even among organisms in different kingdoms, but how safe and ethical is it? No other change in agricultural practices has generated so much controversy. "Biological Russian roulette" and "Frankenfood" are some of the terms opponents use to denigrate gene therapy and genetically modified organisms. Critics of GMOs correctly state that the long-term effects of transgenic manipulations are unknown and that unanticipated problems arise from every new technology and procedure. Recombinant DNA technology may burden society with complex and unforeseen regulatory, administrative, financial, legal, social, medical, and environmental problems.

Critics also argue that natural genetic transfer through sexual reproduction and processes such as transformation and transduction could deliver genes from transgenic plants and animals into other organisms. For example, if herbicide-resistant plants cross-pollinate with related weed species, we might be cursed with weeds that are more difficult to kill. Opponents further express concern that transgenic organisms might trigger allergies or cause harmless organisms to become pathogenic. Some opponents of recombinant DNA technology argue that we should ban all genetically modified organisms and all products derived from them.

The U.S. National Academy of Sciences, the U.S. National Research Council, the European Food Safety Authority (EFSA), and 81 research projects conducted between 1985 and 2012 by the European Union have not revealed any risks to human health or the environment from genetically modified agricultural products beyond those found with conventional plant breeding. In fact, the European Union concluded in 2001 that "the use of more precise technology and the greater regulatory scrutiny probably make them [GMOs] even safer than conventional plants and foods." Additionally, studies have shown that Bt crops spare harmless and beneficial insects that would be killed if pesticides were used. With fewer harmful insects to transmit fungi, Bt corn crops contain less cancer-causing fungal toxin, a potential benefit to human health.

As the debate continues, governments impose additional standards on laboratories involved in recombinant DNA technology. These are intended to prevent the accidental release of altered organisms or exposure of laboratory workers to potential dangers. Additionally, genetic researchers often design organisms to lack a vital gene so that they could not survive for long outside a laboratory.

Security concerns arise because terrorists could apply the procedures used to create beneficial crops and animals to engineer biological weapons that are more infective or more resistant to treatment than their natural counterparts are. Though international treaties prohibit the development of biological weapons, *Bacillus anthracis* (an-thrā'sis) spores were used in bioterrorist attacks in the United States in 2001. Thankfully, the strain utilized was not genetically altered to realize its deadliest potential.

Until recently, gene therapy was technically difficult and inefficient, and therefore it was only an option for very serious diseases caused by changes in a single gene. However, technological advances, in particular CRISPR-Cas9, have increased the efficiency and precision with which DNA can be introduced into human cells. The relative ease with which gene editing can be performed raises serious ethical questions, for example:

- Should people be routinely screened for diseases?
- Should editing of sperm or egg cells be permitted? All the patient's future children and their descendants would have altered genomes. Further, there could be unintended effects on fetal development.
- What types of alterations to the human genome should be considered? For example, is a change that would allow a child afflicted with a fatal genetic disease to grow into adulthood a good candidate? How about a genetic change that augments intelligence or "beauty"? Further, in some cases there is no straightforward definition of what is "normal" or what is a "disorder."
- Genetic diseases or disorders are relatively rare and gene therapy is expensive. Can treatment be made available to all regardless of income, or will only wealthy patients be able afford to genome editing? Who should pay for these procedures: individuals, employers, prospective employers, insurance companies, health maintenance organizations (HMOs), or government agencies?

- What rights do individuals have to genetic privacy? If entities other than individuals pay the costs involved in genetic screening, should those entities have access to *all* the genetic information that results?
- Should businesses be allowed to have patents on and make profits from living organisms they have genetically altered?
- Should governments be allowed to require genetic screening and then force genetic manipulations on individuals to correct perceived genetic abnormalities that some claim are the bases of criminality, manic depression, risk-taking behavior, and alcoholism?
- Should HMOs, physicians, or the government demand genetic screening and then refuse to provide services related to the birth or care of supposedly "defective" children?

We as a society will have to confront these and other ethical considerations as the genomic revolution continues to affect people's lives in many unpredictable ways.

TELL ME WHY

Why don't scientists who work with recombinant DNA know all the long-term effects of their work?

MICRO IN THE CLINIC FOLLOW-UP

Cause for Concern?

Caroline's grandmother suspected that she has either dengue fever or Zika virus. Both are transmitted via mosquito bites, and their clinical manifestations are similar. Dr. Watson does an initial physical exam, gets a medical history (including recent travel), and draws blood for lab analysis. In addition, Dr. Watson notices some bleeding along Caroline's gum line and small, localized rashes (petechiae) on her arms that Caroline says are new since her trip to Brazil. The physician rules out Zika virus since Caroline does not have conjunctivitis. Based on Caroline's symptoms and her recent travel to Brazil, Dr. Watson agrees with the grandmother—Caroline may have dengue fever—and admits her to the hospital, where she can be closely monitored. Dr. Watson orders a reverse transcriptase polymerase chain reaction (RT-PCR) as well as an immunoglobulin M (IgM) antibody test.

Caroline's blood work confirms the diagnosis of dengue fever. There is no antiviral therapy or cure for dengue fever—management is supportive with rest, intravenous or oral fluid replacement (to prevent dengue shock syndrome), pain relievers (avoiding aspirins and anticoagulants), and time for the immune system to conquer the infection. Caroline remains in the hospital, and within a week, she has recovered enough to be discharged. Dr. Watson tells Caroline to thank her grandmother for pushing her to go to the doctor—the relatively early diagnosis allowed for Caroline's quick and complete recovery from dengue.

1. **A sample of Caroline's blood was sent to the Centers for Disease Control and Prevention (CDC) for confirmation of dengue. The CDC uses real-time polymerase chain reaction (PCR) to confirm the presence of dengue virus in a blood sample. Explain how real-time PCR can be used to identify the presence of a specific pathogen in a blood sample.**

2. **The CDC also determines the particular strain of dengue in each sample using genetic analysis. How do they do this?**

3. **Similarities in DNA sequences indicate relatedness of different viruses. What method could the CDC use to determine the DNA sequence of the dengue virus isolated from Caroline's blood sample? Provide a rationale for your choice.**

Check your answers to Micro in the Clinic Follow-Up questions in the Mastering Microbiology Study Area.

Make the connection between Chapters 7 and 8. Watch Micro Matters videos in the Mastering Microbiology Study Area.

CHAPTER SUMMARY

The Role of Recombinant DNA Technology in Biotechnology (p. 238)

1. **Biotechnology** is the use of microorganisms to make useful products. Historically these include bread, cheese, soy sauce, and alcohol.
2. **Recombinant DNA technology**, also known as "genetic engineering," is a new type of biotechnology in which scientists change the genotypes and phenotypes of organisms to benefit humans.

Tools of Recombinant DNA Technology (p. 238–243)

1. The tools of recombinant DNA technology include mutagens, reverse transcriptase, synthetic nucleic acids, restriction enzymes, vectors, CRISPR, and gene libraries.
2. **Mutagens** are chemical and physical agents used to create changes in a microbe's genome to effect desired changes in the microbe's phenotype.
3. The enzyme **reverse transcriptase** transcribes DNA from an RNA template; genetic researchers use reverse transcriptase to make **complementary DNA (cDNA)**.
4. Scientists used synthetic nucleic acids to elucidate the genetic code, and they now use them to create genes for specific proteins and to synthesize DNA and RNA **probes** labeled with radioactive or fluorescent markers.
5. **Restriction enzymes** cut DNA at specific (usually palindromic) nucleotide sequences and are used to produce recombinant DNA molecules.

6. In recombinant DNA technology, a **vector** is a small DNA molecule (such as a viral genome, transposon, or plasmid) that carries a particular gene and a recognizable genetic marker into a cell.

7. Bacteria use **CRISPR**—clustered, regularly interspaced, short palindromic repeats—as a tool to target and cut specific sequences of DNA from bacteriophages. Scientists can use CRISPR to manipulate DNA in other cells.

8. A **gene library** is a collection of bacterial or phage clones, each of which carries a fragment (typically a single gene) of an organism's genome.

Techniques of Recombinant DNA Technology (p. 243–248)

1. The **polymerase chain reaction (PCR)** allows researchers to replicate molecules of DNA rapidly. **Real-time PCR** can be used to measure the number of specific DNA or mRNA sequences in a sample.

2. **Gel electrophoresis** is a technique for separating molecules (including fragments of nucleic acids) by size, shape, and electrical charge.

3. The **Southern blot** technique allows researchers to stabilize DNA sequences from an electrophoresis gel and then localize them using DNA dyes or probes.

4. **DNA microarrays**, containing nucleotide sequences of thousands of genes, are used to monitor gene activity and the presence of microbes in patients and the environment.

5. Geneticists artificially insert DNA into cells by electroporation, protoplast fusion, or injection. They use heat shock to make cells **competent**—able to take up DNA from the environment.

Applications of Recombinant DNA Technology (pp. 248–256)

1. **Genomics** is the sequencing (**genetic mapping**), analysis, and comparison of genomes.

2. Fluorescent *in situ* hybridization (FISH) uses fluorescent nucleic acid probes to localize specific genetic sequences.

3. **Next-generation sequencing (NGS)** has sped up determination of nucleotide sequences. Unique DNA sequences reveal the presence of microbes that have never been cultured in a laboratory.

4. Functional genomics involves determination of the function of a gene. This may be through the use of knockouts, which remove a single gene from a genome, or by overexpressing a gene so that its product builds up in a cell, allowing scientists to analyze the effect.

5. Determination of DNA sequences isolated in a particular environment allows scientists to deduce the presence of microbes that have been cultured or isolated by other methods. Complete sequences and relative abundance can be determined for the members of a **microbiome**, all the microorganisms in a specific environment.

6. Scientists synthesize **subunit vaccines** by introducing genes for a pathogen's polypeptides into cells or viruses. When the cells, the viruses, or the polypeptides they produce are injected into a human, the body's immune system is exposed to and reacts against relatively harmless antigens instead of the potentially harmful pathogen.

7. **Genetic screening** can detect infections and inherited diseases before a patient shows any sign of disease.

8. **Gene therapy** cures various diseases by replacing defective genes with normal genes.

9. Clinicians can use PCR, genetic probes, and DNA microarrays to rapidly diagnose diseases.

10. In **xenotransplants** involving recombinant DNA technology, human genes would be inserted into animals to produce cells, tissues, or organs for introduction into the human body.

11. **Transgenic organisms**, also called genetically modified organisms (GMOs), are plants and animals that have been genetically altered by the inclusion of genes from other organisms.

12. Agricultural uses of recombinant DNA technology include advances in herbicide tolerance; salt tolerance; freeze resistance; pest resistance, such as insertion of the gene for **Bt toxin (Bt)** into plants; and improvements in nutritional value, yield, and shelf life.

The Ethics and Safety of Recombinant DNA Technology (pp. 256–257)

1. Among the ethical and safety issues surrounding recombinant DNA technology are concerns over the accidental release of altered organisms into the environment and the potential for creating genetically modified biological weapons.

QUESTIONS FOR REVIEW

Answers to the Questions for Review (except Short Answer questions) begin on p. A-1.

Multiple Choice

1. Which of the following statements is true concerning recombinant DNA technology?
   a. It will replace biotechnology in the future.
   b. It is a single technique for genetic manipulation.
   c. It is useful in manipulating genotypes but not phenotypes.
   d. It involves modification of an organism's genome.

2. A DNA gene synthesized from an RNA template is _____.
   a. reverse transcriptase
   b. complementary DNA
   c. recombinant DNA
   d. probe DNA

3. After scientists exposed cultures of *Penicillium* to agents X, Y, and Z, they examined the type and amount of penicillin produced by the altered fungi to find the one that is most effective. Agents X, Y, and Z were probably _____.
   a. recombinant cells
   b. competent
   c. mutagens
   d. phages

4. Which of the following is *false* concerning vectors in recombinant DNA technology?
   a. Vectors are small enough to manipulate outside a cell.
   b. Vectors contain a recognizable genetic marker.
   c. Vectors survive inside cells.
   d. Vectors must contain genes for self-replication.

5. Which recombinant DNA technique is used to replicate copies of a DNA molecule?
   a. PCR
   b. gel electrophoresis
   c. electroporation
   d. reverse transcription

6. Which of the following would be most useful in following gene expression in a yeast cell?
   a. Southern blot
   b. PCR
   c. DNA microarray
   d. restriction enzymes

7. Which of the following techniques is used regularly in the study of genomics?
   a. Clones are selected using a vector with two genetic markers.
   b. Genes are inserted to produce an antigenic protein from a pathogen.
   c. Fluorescent nucleotide bases are sequenced.
   d. Defective organs are replaced with those made in animal hosts.

8. The CRISPR-associated enzyme Cas9 _____.
   a. cuts RNA at a specific nucleotide sequence
   b. cuts DNA at a specific nucleotide sequence
   c. is likely derived from a bacteriophage
   d. All of the above are correct.

9. Which application of recombinant DNA technology involves replacing a nonfunctional, mutated gene with a functional gene?
   a. gene therapy
   b. functional genomics
   c. genetic screening
   d. protein synthesis

10. A DNA microarray consists of _____.
    a. a series of clones containing the entire genome of a microbe
    b. recombinant microbial cells
    c. restriction enzyme fragments of DNA molecules
    d. single-stranded DNA localized on a substrate

Modified True/False

Indicate which of the following are true and which are false. Rewrite any false statements to make them true by changing the underlined words.

1. _____ Restriction enzymes <u>inhibit the movement of</u> DNA at specific sites.

2. _____ Restriction enzymes act at <u>specific</u> nucleotide sequences within a double-stranded DNA molecule.

3. _____ <u>A thermocycler</u> separates molecules based on their size, shape, and electrical charge.

4. _____ Protoplast fusion is often used in the genetic modification of <u>plants</u>.

5. _____ Gel electrophoresis is used in <u>DNA microarrays</u>.

Short Answer

1. Describe three artificial methods of introducing DNA into cells.
2. Why is cloning a practical technique for medical researchers?
3. Describe three ways scientists use synthetic nucleic acids.
4. Describe a CRISPR system and its usefulness.
5. List three potential problems of recombinant DNA technology.

VISUALIZE IT!

1. Label the reagents and steps of PCR on the following figure. Indicate the temperature of the reaction at each numbered step.

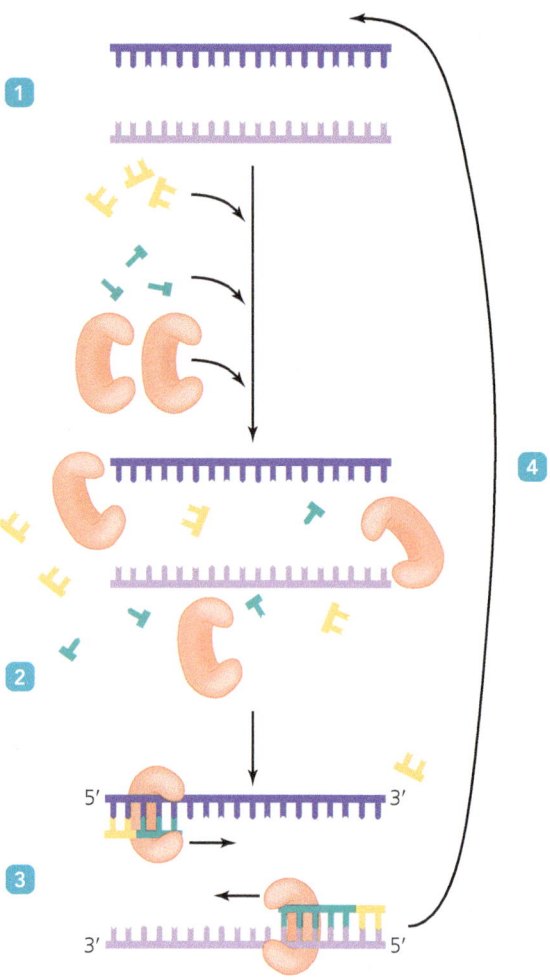

CRITICAL THINKING

1. Examine the restriction sites listed in Table 8.1 (on p. 240). Which restriction enzymes produce restriction fragments with sticky ends? Which produce fragments with blunt ends?

2. A cancer-inducing virus, HTLV-1, inserts itself into a human chromosome, where it remains. How can a laboratory technician prove that a patient is infected with HTLV-1 even when there is no sign of cancer?

3. A thermocycler uses DNA polymerase from hyperthermophilic prokaryotes, but it cannot use DNA polymerase derived from *E. coli*. Why not?

4. How is the result of a Southern blot similar to the result of surveillance using a DNA microarray?

5. *Hha*I recognizes and cuts this DNA sequence at the sites indicated:

   GCG↓C

   C↑GCG

   Describe the fragments resulting from the use of this enzyme.

6. PCR replication of DNA is similar to bacterial population growth. If a scientist starts PCR with 15 DNA helices and runs the reaction for 15 cycles, how many DNA molecules will be present at the end? Show your calculations.

7. If a gene contains the sequence 3'TACAATCGCATTGAA5', what antisense RNA could be used to stop translation directly?

8. Suppose researchers learn that a particular congenital disease is caused by synthesis of a protein coded by a mutated gene. Describe a way in which recombinant DNA technology might be used to prevent translation of the protein.

9. Even though some students correctly synthesize a fluorescent cDNA probe complementary to mRNA for a particular yeast protein, they find that the probe does not attach to any portion of the yeast's genome. Explain why the students' probe does not work.

CONCEPT MAPPING

Using the terms provided, fill in the following concept map that describes recombinant DNA technology. You can also complete this and other concept maps online by going to the **Mastering Microbiology** Study Area.

Blunt ends
DNA probes
Electroporation
Fluorescent
Injection

Mutagens
Plasmids
Protoplast fusion
Radioactive
Restriction enzymes

Reverse transcriptase
Southern blot
Transposons

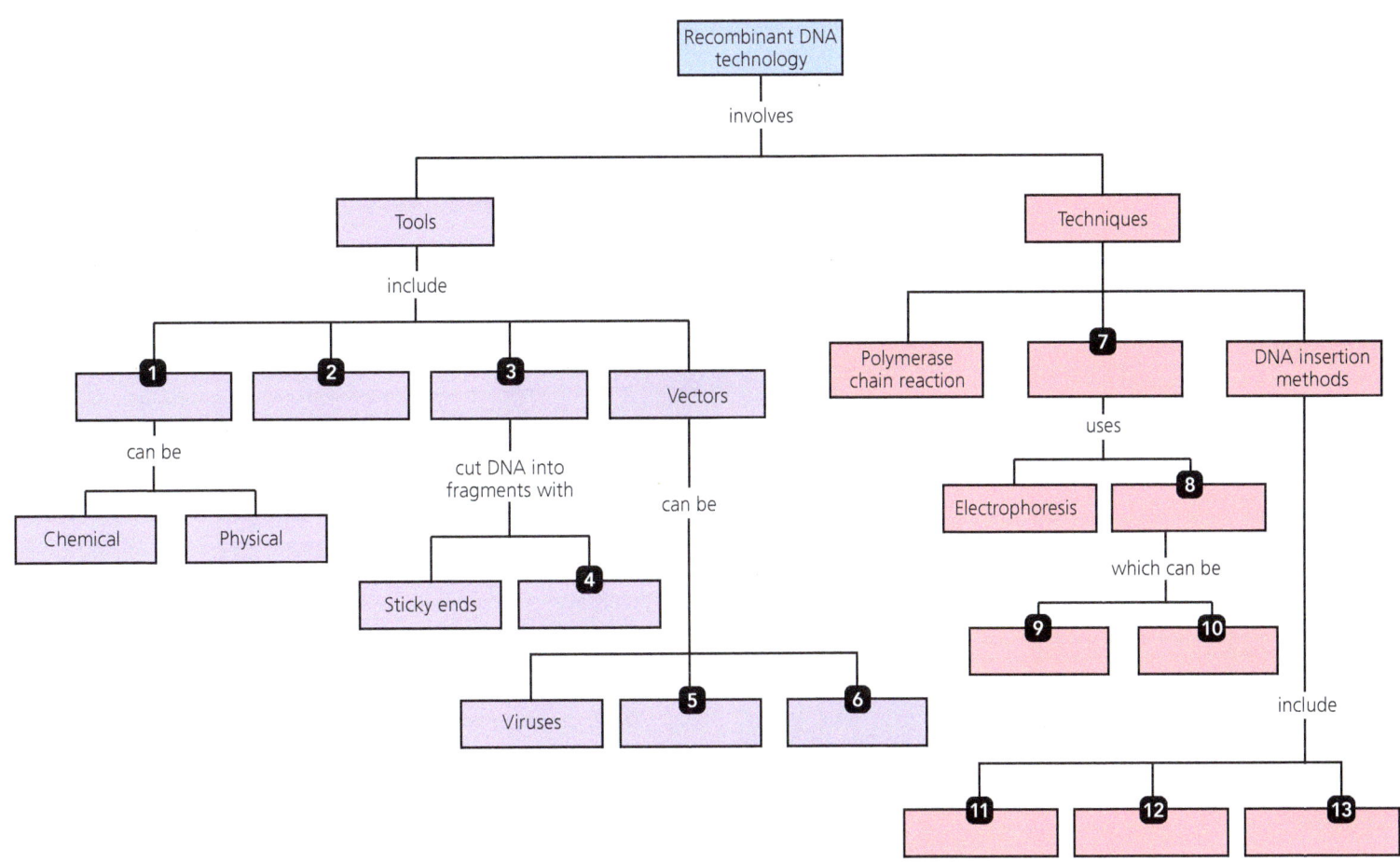

9 Controlling Microbial Growth in the Environment

Before You Begin

1. Of the microbes classified as aerotolerant anaerobes, facultative anaerobes, microaerophiles, obligate aerobes, and obligate anaerobes, which would be harmed by a high concentration of oxygen in their environment?
2. Consider a mesophilic bacterium. What would happen if this bacterium were exposed to a temperature much higher or lower than its optimal temperature?
3. Some microbes are halophiles and can withstand a high concentration of salt in their environment. What happens to nonhalophiles under such a condition?
4. Why do endospores form?

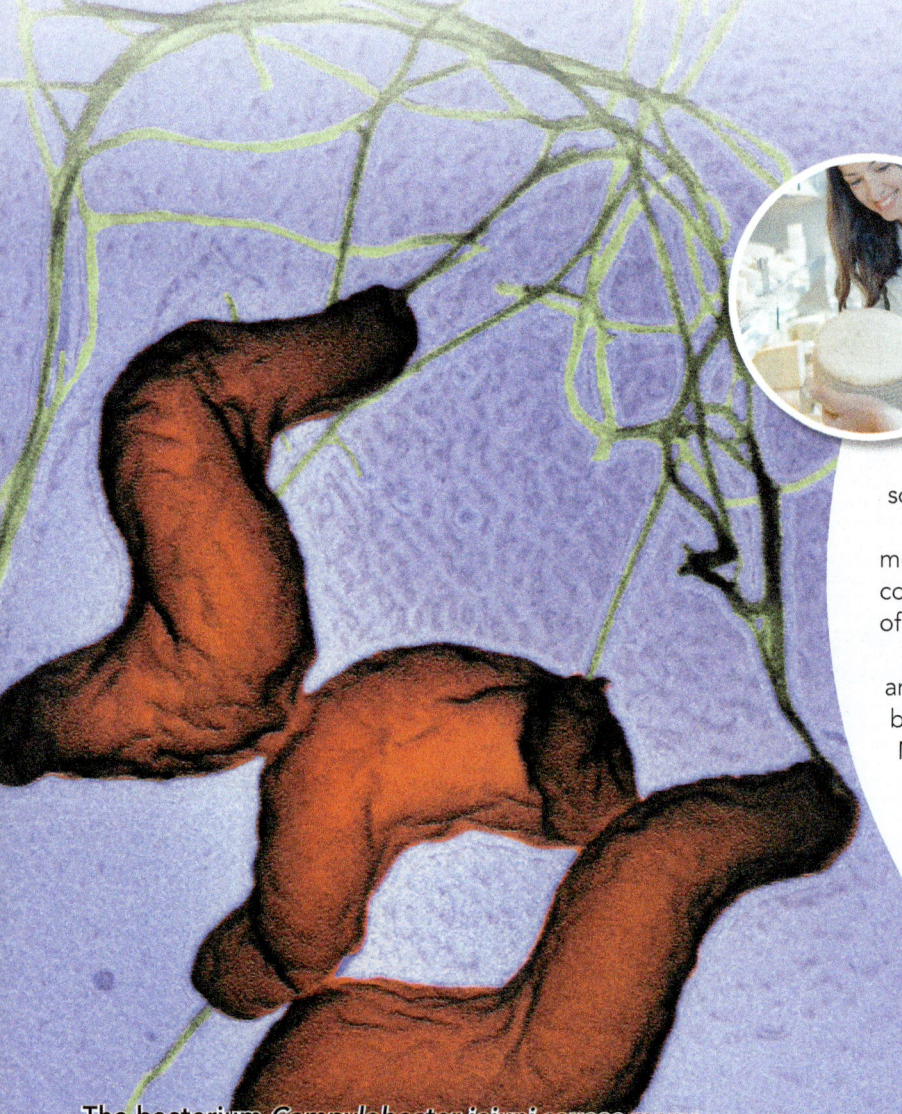

The bacterium *Campylobacter jejuni* causes many cases of food poisoning.

MICRO IN THE CLINIC

An Uninvited Dinner Guest?

MARCELLA IS A 43-YEAR-OLD woman who has been working hard toward getting healthy and fit. Over the last year, she has been eating more fresh fruits and vegetables and has been working out regularly six days a week. She has also committed herself to buying her fruits and vegetables from her local farmer's market. While she likes the fresh produce at the market, her favorite stall is Johnson's Dairy, a local farm that sells raw milk and cheeses. Marcella is having her friend Sarah over for dinner, so she decides to splurge and buy a variety of cheeses for appetizers.

Marcella and Sarah have a great dinner—cheese and cured meat appetizer, grilled fresh fish, vegetables, salad, and a peach cobbler for dessert. They decide to get together again in a couple of days to go for a long bike ride.

The morning of the bike ride, Marcella wakes up with a low fever and abdominal cramps. She's really looking forward to the bike ride, but when she starts having diarrhea, she calls Sarah and cancels. Marcella feels pretty awful for the rest of the day—the cramping and the diarrhea continue. That evening, Sarah calls Marcella and tells her that she has started having cramps and diarrhea as well. They decide that it must have been something they ate at dinner earlier in the week. Both decide that it will probably "run its course" and that they'll be fine the next day.

1. Do you think that the dinner is to blame for Marcella and Sarah's sickness?
2. What part of dinner could be the culprit?

Turn to the end of the chapter (p. 282) to find out.

SOLVE THE PROBLEM

How Clean Is Too Clean?

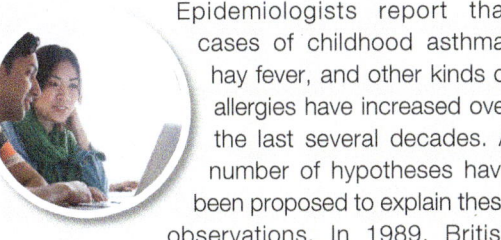

Epidemiologists report that cases of childhood asthma, hay fever, and other kinds of allergies have increased over the last several decades. A number of hypotheses have been proposed to explain these observations. In 1989, British epidemiologist David Strachan suggested that the increase in immune disorders has arisen because children are less frequently exposed to childhood illnesses. This decreased exposure is due to a number of factors. Today children have fewer "opportunities" to be exposed to communicable diseases. Moreover, improvements in sanitation and advances in medicine have significantly reduced the incidence of numerous infectious diseases in the developed world.

Strachan's so-called "hygiene hypothesis" has been expanded to consider children's decreased contact not only with disease-causing microbes but also with everyday microbes in homes and schools. The idea is that fewer childhood illnesses and higher degrees of personal hygiene and cleanliness have so reduced children's exposure to microbes that when the body does encounter common microbes, the immune system overreacts—causing allergies. The hypothesis has also been applied to a wide range of other disorders that may involve the immune system, such as multiple sclerosis, type 1 diabetes, and inflammatory bowel diseases.

An idea as sweeping as the hygiene hypothesis is almost impossible to experimentally demonstrate. Decreased rates of infectious diseases and better hygiene are not the only changes that our modern world has undergone. Many of today's children live in an environment that differs in other ways from the one their ancestors grew up in. For example, their diet tends to include more processed foods, and they may have greater exposure to chemicals through a variety of personal and household products. Lifestyle is also different; today's children typically spend more time indoors than their great-grandparents did. These and other factors may also play a role.

Your group is writing a guide for first-time parents.

- What will you advise a young couple concerning standards of cleanliness in their home?
- Should their child sleep with the family pet?

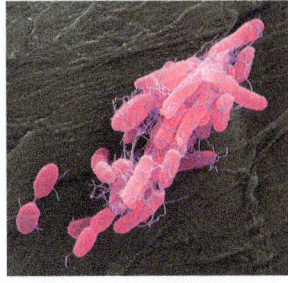

Basic Principles of Microbial Control

The control of microbes in health care facilities, in laboratories, and at home is a significant and practical aspect of microbiology. In this chapter, we study the terminology and principles of microbial control, the factors affecting the efficacy of microbial control, and various chemical and physical means to control microorganisms and viruses. (Chapter 10 considers the use of antimicrobial drugs in patients to assist the body's defenses against pathogens.)

Scientists, health care professionals, researchers, and government workers should use precise terminology when referring to microbial control in the environment. In the following sections we consider the terminology of microbial control, the concept of microbial death rates, and the action of antimicrobial agents.

Terminology of Microbial Control

LEARNING OUTCOMES

9.1 Contrast sterilization, disinfection, and antisepsis, and describe their practical uses.
9.2 Contrast the terms *degerming*, *sanitization*, and *pasteurization*.
9.3 Compare the effects of *-static* versus *-cidal* control agents on microbial growth.

It is important for microbiologists, health care workers, and others to use correct terminology for describing microbial control. Although many of these terms are familiar to the general public, they are often misused.

In its strictest sense, **sterilization** refers to the removal or destruction of *all* microbes, including viruses and bacterial endospores, in or on an object. (The term does not apply to infectious proteins called *prions*, because standard sterilizing techniques do not destroy them.)

In practical terms, *sterilization* indicates only the eradication of harmful microorganisms and viruses; some innocuous microbes may still be present and viable in an environment that is considered sterile. For instance, *commercial sterilization* of canned food does not kill all hyperthermophilic microbes; because they do not cause disease and cannot grow and spoil food at ambient temperatures, they are of no practical concern. Likewise, some hyperthermophiles may survive sterilization by laboratory methods (discussed shortly), but they are of no practical concern to technicians because they cannot grow or reproduce under normal laboratory conditions.

The term **aseptic**[1] (ā-sep'tik) describes an environment or procedure that is free of contamination by *pathogens*. For example, vegetables and fruit juices are available in aseptic packaging, and surgeons and laboratory technicians use aseptic techniques to avoid contaminating a surgical field or laboratory equipment.

Disinfection[2] refers to the use of physical or chemical agents known as **disinfectants**, including ultraviolet light, heat, alcohol, and bleach, to inhibit or destroy microorganisms, especially pathogens. Unlike sterilization, disinfection does not guarantee that all pathogens are eliminated; indeed, disinfectants alone cannot inhibit endospores or some viruses. Further, the term *disinfection* is used only in reference to treatment of inanimate objects. When a chemical is used on skin or other tissue, the process is called **antisepsis**[3] (an-tē-sep'sis), and the chemical is called an **antiseptic**. Antiseptics and disinfectants often have the same components, but disinfectants are more concentrated or can be left on a surface for longer periods of time. Of course, some disinfectants, such as steam or concentrated bleach, are not suitable for use as antiseptics.

Degerming is the removal of microbes from a surface by scrubbing, such as when you wash your hands or a nurse prepares an area of skin for an injection. Though chemicals such as soap or alcohol are commonly used during degerming, the action of thoroughly scrubbing the surface may be more important than the chemical in removing microbes.

Sanitization[4] is the process of disinfecting places and utensils used by the public to reduce the number of pathogenic microbes to meet accepted public health standards. For example, steam, high-pressure hot water, and scrubbing are used to sanitize restaurant utensils and dishes, and chemicals are used to sanitize public toilets. Thus, the difference between *disinfecting* dishes at home and *sanitizing* dishes in a restaurant is the arena—private versus public—in which the activity takes place. **Pasteurization**[5] is the use of heat to kill pathogens and reduce the number of spoilage microorganisms in food and beverages. Milk, fruit juices, wine, and beer are commonly pasteurized.

So far, we have seen that there are two major types of microbial control—sterilization, which is the elimination of all microbes, and antisepsis or disinfection, which each denote the destruction of vegetative (nonspore) cells and many viruses. Modifications of disinfection include degerming, sanitization, and pasteurization. Some scientists and clinicians apply these terms only to pathogenic microorganisms.

Additionally, scientists and health care professionals use the suffixes -*stasis* or -*static*[6] to indicate that a chemical or physical agent inhibits microbial metabolism and growth but does not necessarily kill microbes. Thus, refrigeration is bacteriostatic for most bacterial species; it inhibits their growth, but they can resume metabolism when the optimal temperature is restored. By contrast, words ending in -*cide* or -*cidal*[7] refer to agents that destroy or permanently inactivate a particular type of microbe; *virucides* inactivate viruses, *bactericides* kill bacteria, and *fungicides* kill fungal hyphae, spores, and yeasts. *Germicides* are chemical agents that destroy pathogenic microorganisms in general.

TABLE 9.1 summarizes the terminology used to describe the control of microbial growth.

Microbial Death Rates

LEARNING OUTCOME

9.4 Define *microbial death rate*, and describe its significance in microbial control.

Scientists define **microbial death** as the permanent loss of reproductive ability under ideal environmental conditions. One technique for evaluating the efficacy of an antimicrobial agent is to calculate the **microbial death rate**, which is usually found to be constant over time for any particular microorganism under a particular set of conditions (**FIGURE 9.1**). Suppose, for example, that a scientist treats a broth containing 1 billion (10^9) microbes with an agent that kills 90% of them in 1 minute. The most susceptible cells die first, leaving 100 million (10^8) hardier cells after the first minute. After another minute of treatment, another 90% die, leaving 10 million (10^7) cells that have even greater resistance to and require longer exposure to the agent before they die. Notice that in this case, each full minute decreases the number of living cells 10-fold. The broth will be sterile when all the cells are dead. When these results are plotted on a semilogarithmic graph—in which the *y*-axis is logarithmic and the *x*-axis is arithmetic—the plot of microbial death rate is a straight line; that is, the microbial death rate is constant.

Action of Antimicrobial Agents

LEARNING OUTCOME

9.5 Describe how antimicrobial agents act against cell walls, cytoplasmic membranes, proteins, and nucleic acids.

There are many types of chemical and physical microbial controls, but their modes of action fall into two basic categories: those that disrupt the integrity of cells by adversely altering their

---

[1] From Greek *a*, meaning "not," and *sepsis*, meaning "decay."
[2] From Latin *dis*, meaning "reversal," and *inficere*, meaning "to corrupt."
[3] From Greek *anti*, meaning "against," and *sepsis*, meaning "decay."
[4] From Latin *sanitas*, meaning "healthy."
[5] Named for Louis Pasteur, inventor of the process.
[6] Greek, meaning "to stand"—that is, to remain relatively unchanged.
[7] From Latin *cidium*, meaning "a slaying."

TABLE 9.1 Terminology of Microbial Control

Term	Definition	Examples	Comments
Antisepsis	Reduction in the number of microorganisms and viruses, particularly potential pathogens, on living tissue	Use of iodine or alcohol to prepare skin for an injection	Antiseptics are frequently disinfectants whose strength has been reduced to make them safe for living tissues.
Aseptic	Refers to an environment or procedure free of pathogenic contaminants	Preparation of surgical field; hand washing; flame sterilization of laboratory equipment	Scientists, laboratory technicians, and health care workers routinely follow standardized aseptic techniques.
-cide -cidal	Suffixes indicating destruction of a type of microbe	Bactericide; fungicide; germicide; virucide	Germicides include ethylene oxide, propylene oxide, and aldehydes.
Degerming	Removal of microbes by mechanical means	Hand washing; alcohol swabbing at site of injection	Chemicals play a secondary role to the mechanical removal of microbes.
Disinfection	Destruction of most microorganisms and viruses on nonliving tissue	Use of phenolics, alcohols, aldehydes, or soaps on equipment or surfaces	The term is used primarily in relation to pathogens.
Pasteurization	Use of heat to destroy pathogens and reduce the number of spoilage microorganisms in foods and beverages	Pasteurized milk and fruit juices	Heat treatment is brief to minimize alteration of taste and nutrients; microbes remain and eventually cause spoilage.
Sanitization	Removal of pathogens from objects to meet public health standards	Washing tableware in scalding water in restaurants	Standards of sanitization vary among governmental jurisdictions.
-stasis -static	Suffixes indicating inhibition but not complete destruction of a type of microbe	Bacteriostatic; fungistatic; virustatic	Germistatic agents include some chemicals, refrigeration, and freezing.
Sterilization	Destruction of all microorganisms and viruses in or on an object	Preparation of microbiological culture media and canned food	Typically achieved by steam under pressure, by incineration, or by ethylene oxide gas.

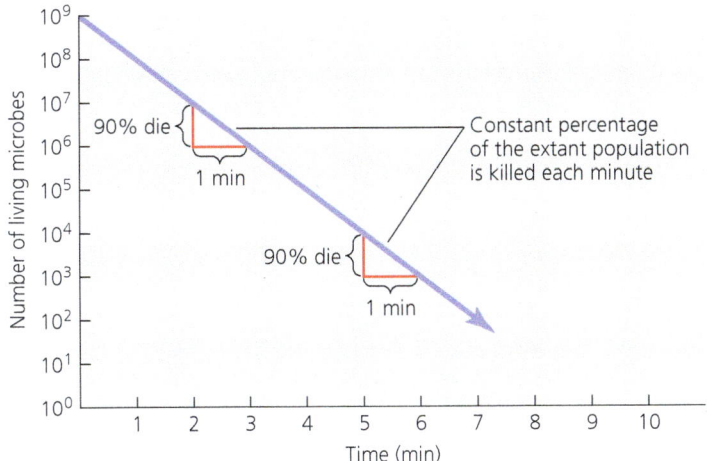

▲ FIGURE 9.1 **A plot of microbial death rate.** Microbicidal agents do not simultaneously kill all cells. Rather, they kill a constant percentage of cells over time—in this case, 90% per minute. On this semilogarithmic graph, a straight line indicates a constant death rate. *How many minutes are required for sterilization in this case?*

Figure 9.1 *For this microbe under these conditions, this microbicidal agent requires 9 minutes to achieve sterilization.*

cell walls or cytoplasmic membranes, and those that interrupt cellular metabolism and reproduction by interfering with the structures of proteins and nucleic acids.

Alteration of Cell Walls and Membranes

A cell wall maintains cellular integrity by counteracting the effects of osmosis when the cell is in a hypotonic solution. If the wall is disrupted by physical or chemical agents, it no longer prevents the cell from bursting as water moves into the cell by osmosis (see Chapter 3, Figure 3.21c).

Beneath a cell wall, the cytoplasmic membrane essentially acts as a bag that contains the cytoplasm and controls the passage of chemicals into and out of the cell. Extensive damage to a membrane's proteins or phospholipids by any physical or chemical agent allows the cellular contents to leak out—damage that, if not immediately repaired, causes death.

In enveloped viruses, the envelope is a membrane composed of proteins and phospholipids that is responsible for the attachment of the virus to its target cell. Damage to the envelope by physical or chemical agents prevents viral replication. The lack of an envelope in nonenveloped viruses accounts for their greater tolerance of harsh environmental conditions, including antimicrobial agents.

Damage to Proteins and Nucleic Acids

Proteins regulate cellular metabolism, function as enzymes in most metabolic reactions, and form structural components in membranes and cytoplasm. A protein's function depends on its exact three-dimensional shape, which is maintained by hydrogen and disulfide bonds between amino acids. When these bonds are broken by extreme heat or certain chemicals, the protein's shape changes (see Figure 5.8). Such *denatured* proteins cease to function, bringing about cellular death.

Chemicals, radiation, and heat can also alter and even destroy nucleic acids. Given that the genes of a cell or virus are composed of nucleic acids, disruption of these molecules can produce fatal mutations. Additionally, that portion of a ribosome that actually catalyzes the synthesis of proteins is a *ribozyme*—that is, an enzymatic RNA molecule. For this reason, physical or chemical agents that interfere with nucleic acids also stop protein synthesis.

Scientists and health care workers have at their disposal many chemical and physical agents to control microbial growth and activity. In the next section, we consider the factors and conditions that should be considered in choosing a particular control method, as well as some ways to evaluate a method's effectiveness.

MICRO CHECK

1. What term describes a procedure that destroys all microbes?

TELL ME WHY

Why does milk eventually go "bad" despite being pasteurized?

The Selection of Microbial Control Methods

Ideally, agents used for the control of microbes should be inexpensive, fast acting, and stable during storage. Further, a perfect agent would control the growth and reproduction of every type of microbe while being harmless to humans, animals, and objects. Unfortunately, such ideal products and procedures do not exist—every agent has limitations and disadvantages. In the next section, we consider the factors that affect the efficacy of antimicrobial methods.

Factors Affecting the Efficacy of Antimicrobial Methods

LEARNING OUTCOMES

9.6 List factors to consider in selecting a microbial control method.
9.7 Identify the three most resistant groups of microbes, and explain why they are resistant to many antimicrobial agents.
9.8 Discuss environmental conditions that can influence the effectiveness of antimicrobial agents.

In each situation, microbiologists, laboratory personnel, and medical staff must consider at least three factors: the nature of the sites to be treated, the degree of susceptibility of the microbes involved, and the environmental conditions that pertain.

Site to Be Treated

In many cases, the choice of an antimicrobial method depends on the nature of the site to be treated. For example, harsh chemicals and extreme heat cannot be used on humans, animals, and fragile objects, such as artificial heart valves and plastic utensils.

Moreover, when performing medical procedures, medical personnel must choose a method and level of microbial control based on the site of the procedure because the site greatly affects the potential for subsequent infection. For example, the use of medical instruments that penetrate the outer defenses of the body, such as needles and scalpels, carries a greater potential for infection, so they must be sterilized; however, disinfection may be adequate for items that contact only the surface of a mucous membrane or the skin. In the latter case, sterilization is usually required only if the patient is immunocompromised.

Relative Susceptibility of Microorganisms

Though microbial death rate is usually constant for a particular agent acting against a single microbe, death rates do vary—sometimes dramatically—among microorganisms and viruses. Microbes fall along a continuum from most susceptible to most resistant to antimicrobial agents. For example, *enveloped* viruses, such as HIV, are more susceptible to antimicrobial agents and heat than are *nonenveloped viruses*, such as poliovirus, because viral envelopes are more easily disrupted than the protein coats of nonenveloped viruses. The relative susceptibility of microbes to antimicrobial agents is illustrated in FIGURE 9.2.

Often, scientists and medical personnel select a method to kill the hardiest microorganisms present, assuming that such a treatment will kill more fragile microbes as well. The most resistant microbes include the following:

- *Bacterial endospores.* The endospores of *Bacillus* (ba-sil'ŭs) and *Clostridium* (klos-trid'ē-ŭm) are the most resilient forms of life. They can survive environmental extremes of temperature, acidity, and dryness and can withstand many

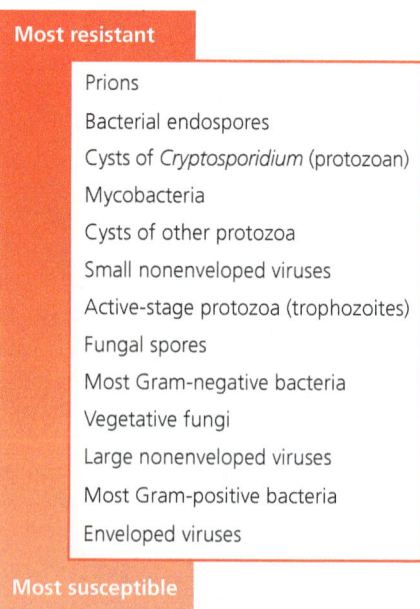

▲ FIGURE 9.2 Relative susceptibilities of microbes to antimicrobial agents. *Why are nonenveloped viruses more resistant than enveloped viruses?*

Figure 9.2 *A protein coat is less fragile than a phospholipid membrane (envelope).*

chemical disinfectants. For example, endospores have survived more than 20 years in 70% alcohol, and scientists have recovered viable endospores that were embalmed with Egyptian mummies thousands of years ago.

- *Species of mycobacteria.* The cell walls of members of the genus *Mycobacterium* (mī-kō-bak-tēr′ē-ŭm), such as *Mycobacterium tuberculosis* (too-ber-kyū-lō′sis), contain a large amount of a waxy lipid. The wax allows these bacteria to survive drying and protects them from most water-based chemicals; therefore, medical personnel must use strong disinfectants or heat to treat whatever comes into contact with tuberculosis patients, including utensils, equipment, and patients' rooms.
- *Cysts of protozoa.* A protozoan cyst's wall prevents entry of most disinfectants, protects against drying, and shields against radiation and heat.

Prions, which are infectious proteins that cause degenerative diseases of the brain, are more resistant than any cell or virus; it takes heating to 482°C for 4 hours to denature them.

The effectiveness of germicides can be classified as high, intermediate, or low, depending on their proficiency in inactivating or destroying microorganisms on medical instruments that cannot be sterilized with heat. *High-level germicides* kill all pathogens, including bacterial endospores. Health care professionals use them to sterilize invasive instruments such as catheters, implants, and parts of heart-lung machines. *Intermediate-level germicides* kill fungal spores, protozoan cysts, viruses, and pathogenic bacteria, but not bacterial endospores. They are used to disinfect instruments that come in contact with mucous membranes but are noninvasive, such as respiratory equipment and endoscopes. *Low-level germicides*

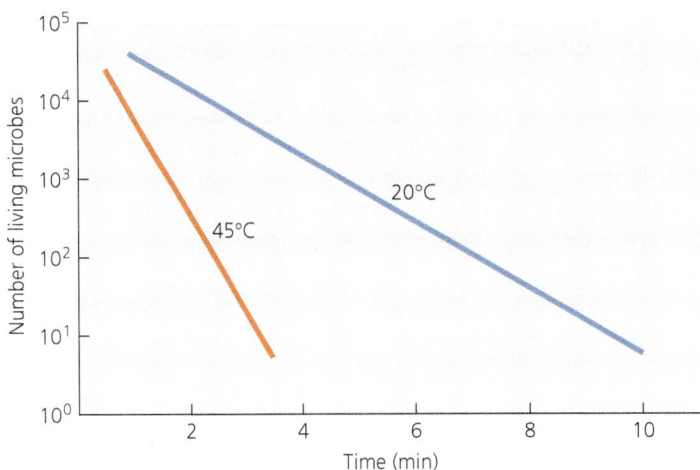

▲ **FIGURE 9.3 Effect of temperature on the efficacy of an antimicrobial chemical.** This semilogarithmic graph shows that the microbial death rate is higher at higher temperatures; to kill the same number of microbes, this disinfectant required only 2 minutes at 45°C but 7 minutes at 20°C.

eliminate vegetative bacteria, fungi, protozoa, and some viruses; they are used to disinfect items that contact only the skin of patients, such as furniture and electrodes.

Environmental Conditions

Temperature and pH affect microbial death rates and the efficacy of antimicrobial methods. Warm disinfectants, for example, generally work better than cool ones because chemicals react faster at higher temperatures (**FIGURE 9.3**). Acidic conditions enhance

EMERGING DISEASE CASE STUDY

Acanthamoeba Keratitis

Tim liked the lake; in fact, his girlfriend suggested he was more fish than man. They spent all of their free time swimming, waterskiing, diving, and sunbathing, until Tim met a single-celled amoeba, *Acanthamoeba*.

Tim noticed something was wrong when his left eye began to hurt, turned red, and became so sensitive to light that he could not stand to be outside. Two days later, the pain was excruciating, like nothing Tim had ever experienced. It felt as if someone were pounding pieces of broken glass into the front of his eye and at the same time quickly inserting a thousand tiny needles into the back of the eye. With his eye swollen shut and tears flowing down his face, Tim sought medical aid.

The doctor diagnosed *Acanthamoeba* keratitis, which is inflammation of the covering of the eye (the cornea) caused by the amoeba. This eukaryotic microbe commonly lives in water, including rivers, hot springs, and lakes. When trapped under a contact lens, the amoeba can penetrate the eye to cause keratitis. Very occasionally, *Acanthamoeba* may also enter the body through the nasal mucous membrane or through a cut in the skin. It has become an emerging menace in our modern society because it can live in hot tubs, pools, showerheads, and sink taps.

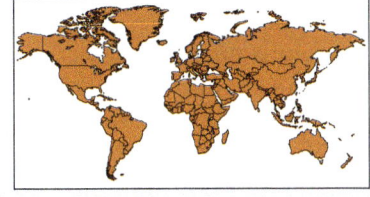

The physician prescribed a solution of antiseptic agent that Tim had to drop into his eyes every 30 minutes, day and night, for three weeks. The treatment was painful and time consuming, but at least Tim retained his sight without needing a corneal transplant. He also learned to remove his contact lenses at the lake!

1. Why are there many more cases of *Acanthamoeba* keratitis than there were 100 years ago?

the antimicrobial effect of heat. Some chemical disinfectants, such as household chlorine bleach, are more effective at low pH.

Organic materials, such as fat, feces, vomit, blood, and the intercellular matrix of biofilms, interfere with the penetration of heat, chemicals, and some forms of radiation, and in some cases these materials inactivate chemical disinfectants. For this reason, it is important to clean objects before sterilization or disinfection so that antimicrobial agents can thoroughly contact all the object's surfaces.

Biosafety Levels

LEARNING OUTCOME

9.9 Describe four levels of biosafety, and give examples of microbes handled at each level.

The Centers for Disease Control and Prevention (CDC) has established guidelines for four levels of safety in microbiological laboratories dealing with pathogens. Each level raises personnel and environmental safety by specifying increasingly strict laboratory techniques, use of safety equipment, and design of facilities.

Biosafety Level 1 (BSL-1) is suitable for handling microbes, such as *Escherichia coli* (esh-ĕ-rik′ē-ă kō′lī), not known to cause disease in healthy humans. Precautions in BSL-1 are minimal and include hand washing with antibacterial soap and disinfecting surfaces.

BSL-2 facilities are similar to those of BSL-1 but are designed for handling moderately hazardous agents, such as hepatitis and influenza viruses and methicillin-resistant *Staphylococcus aureus* (staf′i-lō-kok′ŭs o′rē-ŭs) (MRSA). Access to BSL-2 labs is limited when work is being conducted, extreme precautions are taken with contaminated sharp objects, and procedures that might produce aerosols are conducted within safety cabinets (see Figure 9.11).

BSL-3 is stricter, requiring that all manipulations be done within safety cabinets containing *high-efficiency particulate air (HEPA)* filters. BSL-3 also specifies special design features for the laboratory. These include entry through double sets of doors and lower air pressure in the laboratory, such that an open door only allows air to move into, not out of, the room. Air leaving the room is HEPA-filtered before being discharged. BSL-3 is designed for experimentation on microbes such as bacteria of tuberculosis and anthrax and viruses of yellow fever and Rocky Mountain spotted fever.

The most secure laboratories are *BSL-4* facilities, designated for working with dangerous or exotic microbes that cause severe or fatal diseases in humans, such as Ebola, smallpox, and Lassa fever viruses. BSL-4 labs are either separate buildings or completely isolated from all other areas of their buildings. Entry and exit are strictly controlled through electronically sealed airlocks with multiple showers, a vacuum room, an ultraviolet light room, and other safety precautions designed to destroy all traces of the biohazard. All air and water entering and leaving the facility are filtered to prevent accidental release. Personnel wear "space suits" supplied with air hoses (**FIGURE 9.4**). Suits and the laboratory itself are pressurized such that microbes are swept away from workers.

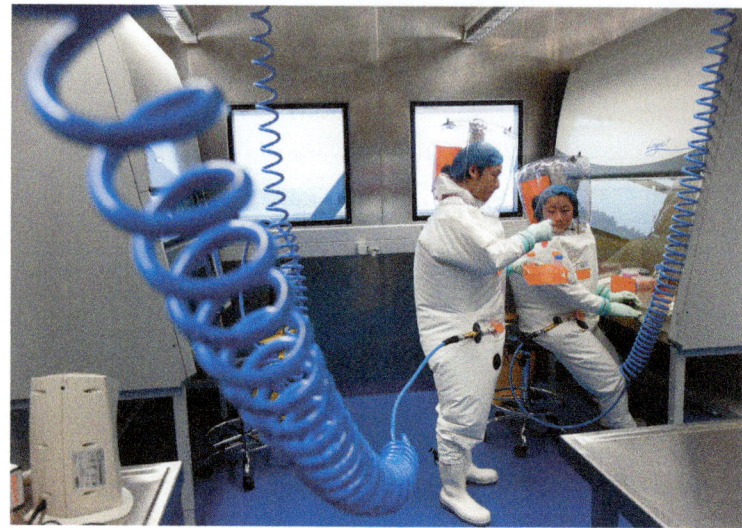

▲ **FIGURE 9.4** BSL-4 scientists working with level 4 biohazards.

MICRO CHECK

2. What microbial structures are the most resilient forms of life?
3. What is the most likely biosafety level of your introductory microbiology lab classroom?

TELL ME WHY

Why are BSL-4 suits pressurized? Why not just wear tough regular suits?

Now that we have studied the terminology and general principles of microbial control and biosafety levels, we turn our attention to the actual physical and chemical agents available to scientists, medical personnel, and the general public to control microbial growth.

Physical Methods of Microbial Control

LEARNING OUTCOME

9.10 Describe five types of physical methods of microbial control.

Physical methods of microbial control include exposure of the microbes to extremes of heat and cold, desiccation, filtration, osmotic pressure, and radiation.

Heat-Related Methods

LEARNING OUTCOMES

9.11 Discuss the advantages and disadvantages of using moist heat in an autoclave and dry heat in an oven for sterilization.

9.12 Define and contrast *thermal death point*, *thermal death time*, and *decimal reduction time (D)*.
9.13 Explain the use of *Bacillus* endospores in sterilization techniques.
9.14 Explain the importance of pasteurization, and describe three different pasteurization methods.

Heat is one of the older and more common means of microbial control. High temperatures denature proteins, interfere with the integrity of cytoplasmic membranes and cell walls, and disrupt the function and structure of nucleic acids. Heat can be used for sterilization, in which case all cells and viruses are deactivated, or for commercial preparation of canned goods. In so-called commercial sterilization, hyperthermophilic prokaryotes remain viable but are harmless because they cannot grow at the normal (room) temperatures in which canned foods are stored.

Though microorganisms vary in their susceptibility to heat, it can be an important agent of microbial control. As a result, scientists have developed concepts and terminology to convey these differences in susceptibility. **Thermal death point** is the lowest temperature that kills all cells in a broth in 10 minutes, and **thermal death time** is the time it takes to sterilize a particular volume of liquid at a set temperature.

As we have discussed, cell death occurs logarithmically. When measuring the effectiveness of heat sterilization, researchers calculate the **decimal reduction time (D)**, which is the time required to destroy 90% of the microbes in a sample (**FIGURE 9.5**). This concept is especially useful to food processors because they must heat foods to eliminate the endospores of anaerobic *Clostridium botulinum* (bo-tū-lī′num), which could germinate and produce life-threatening botulism toxin inside sealed cans. The standard in food processing is to apply heat such that a population of 10^{12} *C. botulinum* endospores is reduced to 10^{0} (i.e., 1) endospore (a 12-fold reduction), which leaves only a very small chance that any particular can of food contains a single endospore. Researchers have calculated that the D value for *C. botulinum* endospores at 121°C is 0.204 minute, so it takes 2.5 minutes (0.204 × 12) to reduce 10^{12} endospores to 1 endospore.

Moist Heat

Moist heat (which is commonly used to disinfect, sanitize, sterilize, and pasteurize) kills cells by denaturing proteins and destroying cytoplasmic membranes. Moist heat is more effective in microbial control than dry heat because water is a better conductor of heat than air. An example from your kitchen readily demonstrates this: you can safely stick your hand into an oven at 350°F for a few moments, but putting a hand into boiling water at the lower temperature of 212°F would burn you severely.

The first method we consider for controlling microbes using moist heat is boiling.

Boiling Boiling kills the vegetative cells of bacteria and fungi, the trophozoites of protozoa, and most viruses within 10 minutes at sea level. Contrary to popular belief, water at a rapid boil is no hotter than water at a slow boil; boiling water at normal atmospheric pressure cannot exceed boiling temperature (100°C at sea level), because escaping steam carries excess heat away. It is impossible to boil something more quickly simply by applying more heat; the added heat is carried away by the escaping steam. Boiling *time* is the critical factor. Further, it is important to realize that at higher elevations, water boils at lower temperatures because atmospheric pressure is lower; thus, a longer boiling time is required in Denver than in Los Angeles to get the same antimicrobial effect.

Bacterial endospores, protozoan cysts, and some viruses (such as hepatitis viruses) can survive boiling at sea level for many minutes or even hours. In fact, because bacterial endospores can withstand boiling for more than 20 hours, boiling is not recommended when true sterilization is required. Boiling is effective for sanitizing restaurant tableware or disinfecting baby bottles.

Autoclaving Practically speaking, true sterilization using heat requires higher temperatures than that of boiling water. To achieve the required temperature, pressure is applied to boiling water to prevent the escape of heat in steam. The reason that applying pressure succeeds in achieving sterilization is that the temperature at which water boils (and steam is formed) increases as pressure increases (**FIGURE 9.6**). Scientists and medical personnel routinely use a piece of equipment called an *autoclave* to sterilize chemicals and objects that can tolerate moist heat. Alternative techniques (discussed shortly) must be used for items that are damaged by heat or water, such as some plastics and vitamins.

An **autoclave** consists of a pressure chamber, pipes to introduce and evacuate steam, valves to remove air and control pressure, and pressure and temperature gauges to monitor the procedure (**FIGURE 9.7**). As steam enters an autoclave chamber, it forces air out, raises the temperature of the contents, and increases the pressure until a set temperature and pressure are reached.

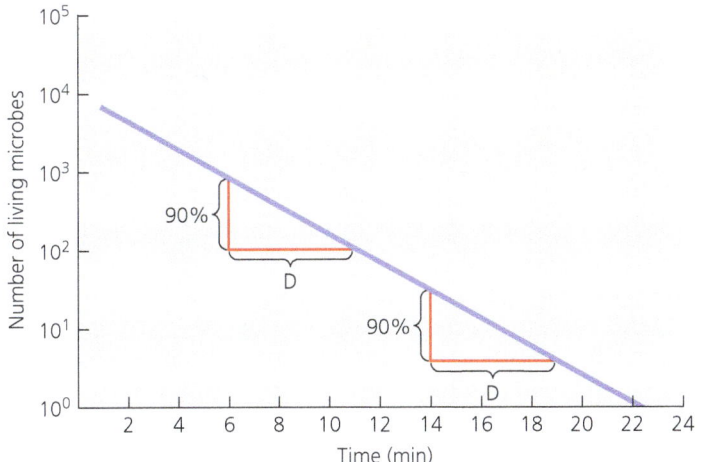

▲ **FIGURE 9.5 Decimal reduction time (D) as a measure of microbial death rate.** D is defined as the time it takes to kill 90% of a microbial population. Note that D is a constant that is independent of the initial density of the population. *What is the decimal reduction time of this heat treatment against this organism? What is the thermal death time?*

Figure 9.5 D = 5 minutes; thermal death time = 22.5 minutes.

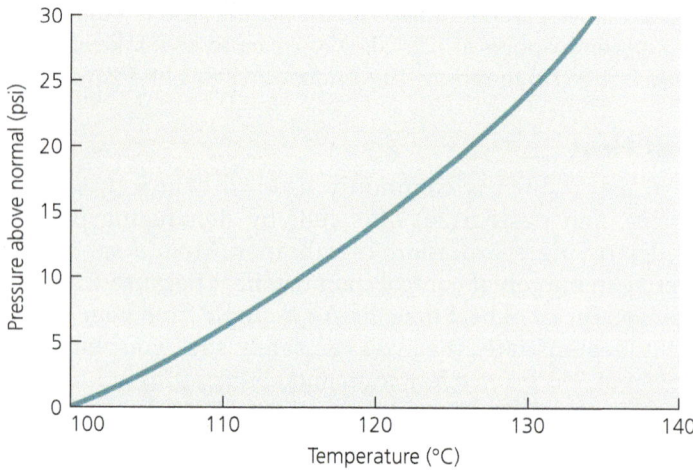

▲ **FIGURE 9.6 The relationship between temperature and pressure.** Note that higher temperatures—and consequently, greater antimicrobial action—are associated with higher pressures. *Ultra-high-temperature pasteurization of milk requires a temperature of 134°C; what pressure must be applied to the milk to achieve this temperature?*

Figure 9.6 29 psi.

Scientists have determined that a temperature of 121°C, which requires the addition of 15 pounds per square inch (psi)[8] of pressure above that of normal air pressure (see Figure 9.6), destroys microbes generally encountered in a laboratory or medical setting in a small volume in about 15 minutes. This treatment sterilizes the medium for all practical purposes, though some extremely hyperthermophilic microbes may survive autoclaving for 15 minutes. Sterilizing large volumes of liquids or solids slows the process because they require more time for heat to penetrate. Thus, it requires more time to sterilize 1 liter of fluid in a flask than the same volume of fluid distributed into smaller tubes. Autoclaving solid substances, such as meat, also requires extra time because it takes longer for heat to penetrate to their centers.

Sterilization in an autoclave requires that steam be able to contact all liquids and surfaces that might be contaminated with microbes; therefore, solid objects must be wrapped in porous cloth or paper, not sealed in plastic or aluminum foil, both of which are impermeable to steam. Containers of liquids must be sealed loosely enough to allow steam to circulate freely, and all air must be forced out by steam. Since steam is lighter than air, it cannot force air from the bottom of an empty vessel; therefore, empty containers must be tipped so that air can flow out of them.

Scientists use several means to ensure that an autoclave has sterilized its contents. A common one is a chemical that changes color when the proper combination of temperature and time has been reached. Often such a color indicator is impressed in a pattern on tape or paper so that the word *sterile* or a pattern or design appears. Another technique uses plastic beads that melt when proper conditions are met.

A biological indicator of sterility uses endospores of bacteria such as *Bacillus* impregnated into tape. After autoclaving, the tape is aseptically inoculated into sterile broth. If no bacterial growth

---
[8]The Standard International (SI) equivalent of 1 psi is 6.9×10^3 pascals.

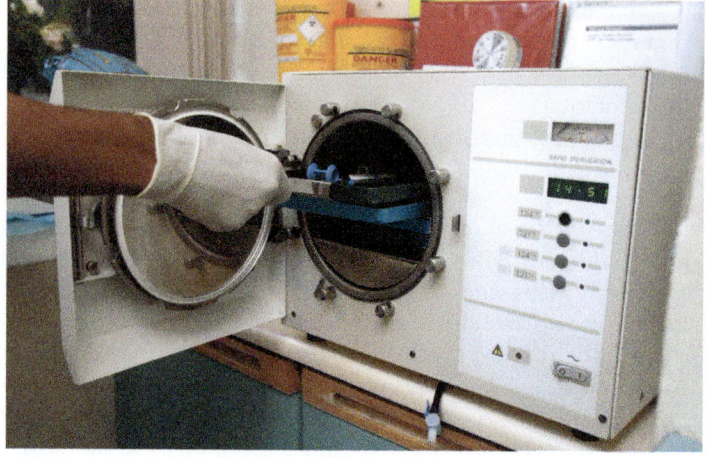

(a)

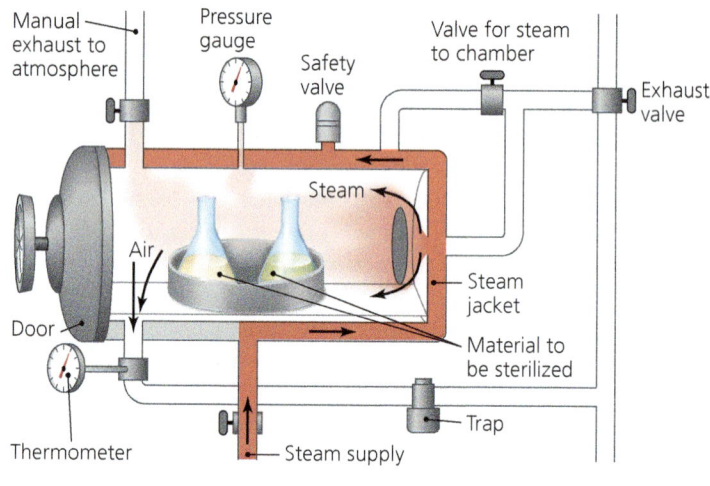

(b)

▲ **FIGURE 9.7 An autoclave.** (a) A photo of a laboratory autoclave. (b) A schematic of an autoclave, showing how it functions.

appears, the original material is considered sterile. In a variation on this technique, the endospores are on a strip in one compartment of a vial that also includes, in a separate compartment, a growth medium containing a pH color indicator. After autoclaving, a barrier between the two compartments is broken, putting the endospores into contact with the medium (**FIGURE 9.8**). In this case, the absence of a color change after incubation indicates sterility.

Pasteurization Louis Pasteur developed a method of heating beer and wine just enough to destroy the microorganisms that cause spoilage without ruining the taste. Today, pasteurization is also used to kill pathogens in milk, ice cream, yogurt, and fruit juices. *Brucella melitensis* (broo-sel'lă me-li-ten'sis), *Mycobacterium bovis* (bō'vis), and *Escherichia coli,* the causative agents of undulant fever, bovine tuberculosis, and one kind of diarrhea, respectively, are controlled in this manner.

Pasteurization is not sterilization. *Thermoduric* and *thermophilic*—heat-tolerant and heat-loving—prokaryotes survive pasteurization, but they do not cause spoilage over the relatively short times during which properly refrigerated and

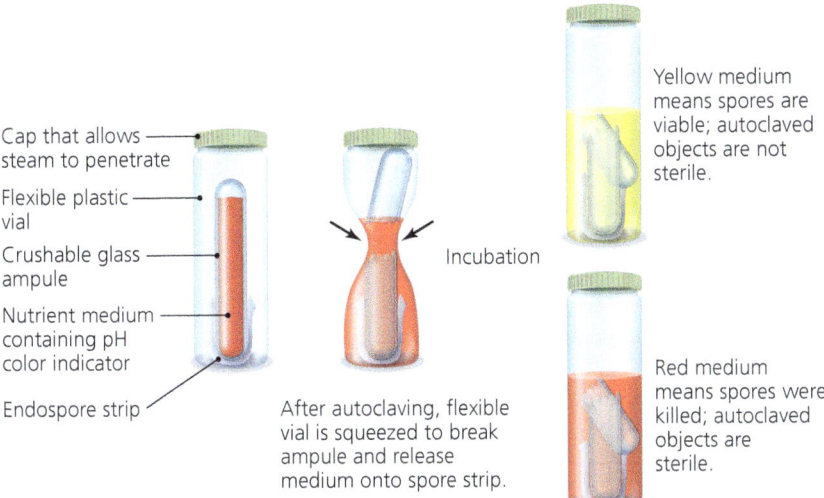

◀ **FIGURE 9.8 Sterility indicators.** A commercial endospore-test ampule, which is included among objects to be sterilized. After autoclaving is complete, the ampule is broken, and the medium, which contains a pH color indicator, is released onto the endospore strip. If the endospores are still alive, their metabolic wastes lower the pH, changing the color of the medium.

pasteurized foods are stored before consumption. In addition, such prokaryotes are generally not pathogenic.

The combination of time and temperature required for effective pasteurization varies with the product. Because milk is the most familiar pasteurized product, we consider the pasteurization of milk in some detail. Historically, milk was pasteurized by the *batch method* for 30 minutes at 63°C, but most milk processors today use *high-temperature, short-time pasteurization (HTST)*, which is also known as *flash pasteurization*, in which milk flows through heated tubes that raise its temperature to 72°C for only 15 seconds. This treatment effectively destroys all pathogens. *Ultra-high-temperature pasteurization* heats the milk to at least 135°C for only 1 second, but some consumers claim that it adversely affects the taste.

Ultra-High-Temperature Sterilization The dairy industry and other food processors can also use *ultra-high-temperature sterilization*, which involves flash heating milk or other liquids to rid them of all living microbes. The process involves passing the liquid through superheated steam at about 140°C for 1 to 3 seconds and then cooling it rapidly. Treated liquids can be stored indefinitely at room temperature without microbial spoilage, though chemical degradation after months of storage results in flavor changes. Small packages of dairy creamer served in restaurants are often sterilized by the ultra-high-temperature method. **TABLE 9.2** summarizes the dairy industry's use of moist heat for controlling microbes in milk.

Dry Heat

Some substances, such as powders and oils, cannot be sterilized by boiling or with steam. Others, such as some metal objects, can be damaged by repeated exposure to steam. For such materials, sterilization can be achieved by the use of dry heat, as occurs in an oven.

Hot air is an effective sterilizing agent because it denatures proteins and fosters the oxidation of metabolic and structural chemicals; however, in order to sterilize, dry heat requires higher temperatures for longer times than moist heat because dry heat penetrates more slowly. For instance, whereas an autoclave needs about 15 minutes to sterilize an object at 121°C, an oven at the same temperature requires at least 16 hours to achieve sterility. Scientists typically use higher temperatures—171°C for 1 hour or 160°C for 2 hours—to sterilize objects in an oven, but objects made of rubber, paper, and many types of plastic oxidize rapidly (combust) under these conditions.

Complete incineration is the ultimate means of sterilization. As part of standard aseptic technique in microbiological laboratories, inoculating loops are sterilized by heating them in the flame of a Bunsen burner or with an electric heating coil until they glow red (about 1500°C). Health care workers incinerate contaminated dressings, bags, and paper cups, and field epidemiologists incinerate the carcasses of animals that have diseases such as anthrax or bovine spongiform encephalopathy (mad cow disease).

Refrigeration and Freezing

LEARNING OUTCOME

9.15 Describe the use and importance of refrigeration and freezing in limiting microbial growth.

In many situations, particularly in food preparation and storage, the most convenient method of microbial control is either refrigeration (temperatures between 0°C and 7°C) or freezing

TABLE 9.2 Moist Heat Treatments of Milk

Process	Treatment
Historical (batch) pasteurization	63°C for 30 minutes
Flash pasteurization	72°C for 15 seconds
Ultra-high-temperature pasteurization	135°C for 1 second
Ultra-high-temperature sterilization	140°C for 1–3 seconds

(temperatures below 0°C). These processes decrease microbial metabolism, growth, and reproduction because chemical reactions occur more slowly at low temperatures and because liquid water is not available at subzero temperatures. Note, however, that psychrophilic (cold-loving) microbes can still multiply in refrigerated food and spoil its taste and suitability for consumption.

Refrigeration halts the growth of most pathogens, which are predominantly mesophiles. Notable exceptions are the bacteria *Listeria* (lis-tēr'ē-ă), which can reproduce to dangerous levels in refrigerated food, and *Yersinia* (yer-sin'ē-ă), which can multiply in refrigerated blood products and be passed on to blood recipients. (Chapters 19 and 20 discuss these pathogens in more detail.)

Slow freezing, during which ice crystals have time to form and puncture cell membranes, is more effective than quick freezing in inhibiting microbial metabolism, though microorganisms also vary in their susceptibility to freezing. Whereas the cysts of tapeworms perish after several days in frozen meat, many vegetative bacterial cells, bacterial endospores, and viruses can survive subfreezing temperatures for years. In fact, scientists store many bacteria and viruses in low-temperature freezers at $-30°C$ to $-80°C$ and are able to reconstitute the microbes into viable populations by warming them in media containing proper nutrients. Therefore, we must take care in thawing and cooking frozen food because it can still contain pathogenic microbes.

Desiccation and Lyophilization

LEARNING OUTCOME

9.16 Compare and contrast desiccation and lyophilization.

Desiccation, or drying, has been used for thousands of years to preserve such foods as fruits, peas, beans, grain, nuts, and yeast (**FIGURE 9.9**). Desiccation inhibits microbial growth because metabolism requires liquid water. Drying inhibits the spread of most pathogens, including the bacteria that cause syphilis, gonorrhea, and the more common forms of bacterial pneumonia and diarrhea. However, many molds can grow on dried raisins and apricots, which have as little as 16% water content.

Scientists use **lyophilization** (lī-of'i-li-zā'shŭn), a technique that combines freezing and drying, to preserve microbes and other cells for many years. In this process, scientists instantly freeze a culture in liquid nitrogen or frozen carbon dioxide (dry ice); then they subject it to a vacuum that removes frozen water through a process called *sublimation*, in which the water is transformed directly from a solid to a gas. Lyophilization prevents the formation of large, damaging ice crystals. Although not all cells survive, enough are viable to enable the culture to be reconstituted many years later.

Filtration

LEARNING OUTCOME

9.17 Describe the use of filters for disinfection and sterilization.

Filtration is the passage of a fluid (either a liquid or a gas) through a sieve designed to trap particles—in this case, cells or viruses—and separate them from the fluid. Researchers often use a vacuum to assist the movement of fluid through the filter (**FIGURE 9.10a**). Filtration traps microbes larger than the pore size, allowing smaller microbes to pass through. In the late 1800s, filters were able to trap cells, but their pores were too large to trap the pathogens of such diseases as rabies and measles. These pathogens were thus named *filterable viruses*, which today has been shortened to *viruses*.[9] Now, filters with pores small enough to trap even viruses are available, so filtration can be used to sterilize such heat-sensitive materials as ophthalmic solutions, antibiotics, vaccines, liquid vitamins, enzymes, and culture media.

Over the years, filters have been constructed from porcelain, glass, cotton, asbestos, and diatomaceous earth, a substance composed of the innumerable glasslike cell walls of single-celled algae called diatoms. Scientists today typically use thin (only 0.1 mm thick), circular **membrane filters** manufactured of nitrocellulose or plastic and containing specific pore sizes ranging from 25 μm to less than 0.01 μm in diameter (**FIGURE 9.10b**). The pores of the latter filters are small enough to trap small viruses and even some large protein molecules. Microbiologists also use filtration to estimate the number of microbes in a fluid by counting the number deposited on the filter after passing a given volume through the filter (see Figure 6.24). **TABLE 9.3** lists some pore sizes of membrane filters and the microbes they do not allow through.

Health care and laboratory workers routinely use filtration to prevent airborne contamination by microbes. Medical personnel wear surgical masks to prevent exhaled microbes from contaminating the environment, and cotton plugs are placed in culture vessels to prevent contamination by airborne microbes. Additionally, high-efficiency particulate air (HEPA) filters are crucial parts of biological safety cabinets (**FIGURE 9.11**). HEPA

▲ **FIGURE 9.9 The use of desiccation as a means of preserving apricots in Pakistan.** In this time-honored practice, drying inhibits microbial growth by removing the water that microbes need for metabolism.

---
[9]Latin, meaning "poisons."

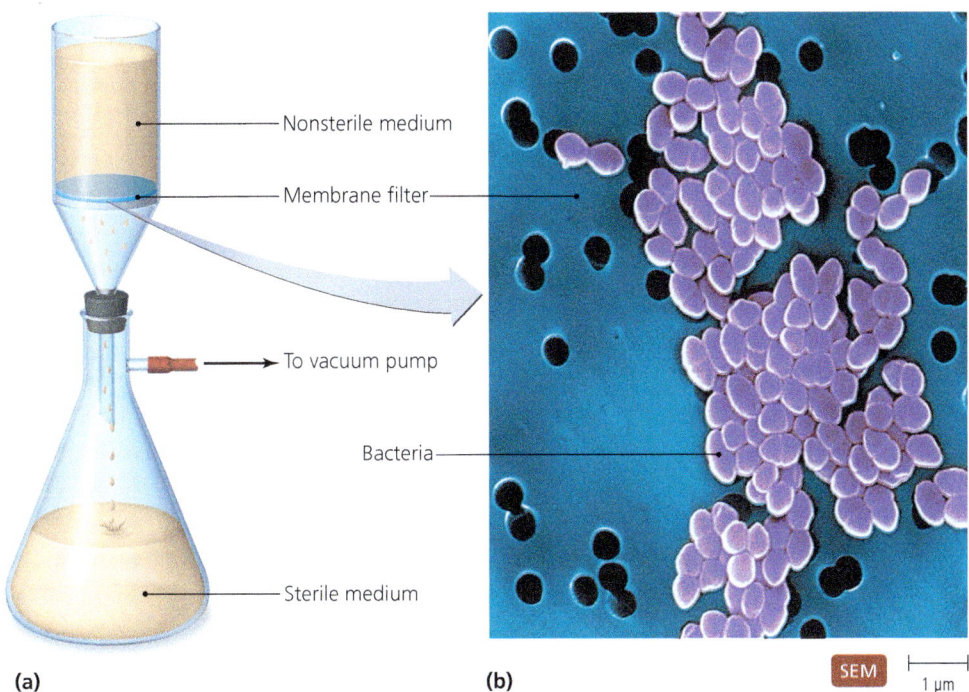

◀ **FIGURE 9.10** Filtration equipment used for microbial control. **(a)** Assembly for sterilization by vacuum filtration. **(b)** Membrane filters composed of various substances and with pores of various sizes can be used to trap diverse microbes. Here, a bacterium known as vancomycin-resistant *Enterococcus* (VRE) is trapped.

TABLE 9.3 Membrane Filters

Pore Size (μm)	Smallest Microbes That Are Trapped
5	Multicellular algae, animals, and fungi
3	Yeasts and larger unicellular algae
1.2	Protozoa and small unicellular algae
0.45	Largest bacteria
0.22	Largest viruses and most bacteria
0.025	Larger viruses and pliable bacteria (mycoplasmas, rickettsias, chlamydias, and some spirochetes)
0.01	Smallest viruses

filters are mounted in the air ducts of some operating rooms, rooms occupied by patients with airborne diseases such as tuberculosis, and rooms of immunocompromised patients, such as burn victims and AIDS patients.

Osmotic Pressure

LEARNING OUTCOME

9.18 Discuss the use of hypertonic solutions in microbial control.

Another ancient method of microbial control is the use of high concentrations of salt or sugar in foods to inhibit microbial growth by **osmotic pressure**. Osmosis is the net movement of water across a semipermeable membrane (such as a cytoplasmic membrane) from an area of higher water concentration to an area of lower water concentration. Cells in a hypertonic solution of salt or sugar lose water, and the cell shrinks (see Figure 3.21b). The removal of water inhibits cellular metabolism because enzymes are fully functional only in aqueous environments. Thus, osmosis preserves honey, jerky, jams, jellies, salted fish, and some types of pickles from most microbial attacks.

Fungi have a greater ability than bacteria to tolerate hypertonic environments with little moisture, which explains why jelly in your refrigerator may grow a colony of *Penicillium* (pen-i-sil'ē-ŭm) mold but is not likely to grow the bacterium *Salmonella* (sal'mŏ-nel'ă).

Radiation

LEARNING OUTCOME

9.19 Differentiate ionizing radiation from nonionizing radiation as they relate to microbial control.

Another physical method of microbial control is the use of **radiation**. There are two types of radiation: particulate radiation and electromagnetic radiation. Particulate radiation consists of high-speed subatomic particles, such as protons, that have been freed from their atoms. *Electromagnetic radiation* can be defined as energy without mass traveling in waves at the speed of light (3×10^5 km/sec). Electromagnetic energy is released from atoms that have undergone internal changes. The *wavelength* of electromagnetic radiation, defined as the distance between two crests of a wave, ranges from very short gamma rays; to X rays, ultraviolet light, and visible light; to long infrared rays; and, finally, to very long radio waves (see Figure 4.1). Though they are particles, electrons also have a wave nature, with wavelengths that are even shorter than gamma rays.

The shorter the wavelength of an electromagnetic wave, the more energy it carries; therefore, shorter-wavelength radiation is more suitable for microbial control than longer-wavelength radiation, which carries less energy and is less penetrating. Scientists describe all types of radiation as either *ionizing* or *nonionizing* according to its effects on the chemicals within cells.

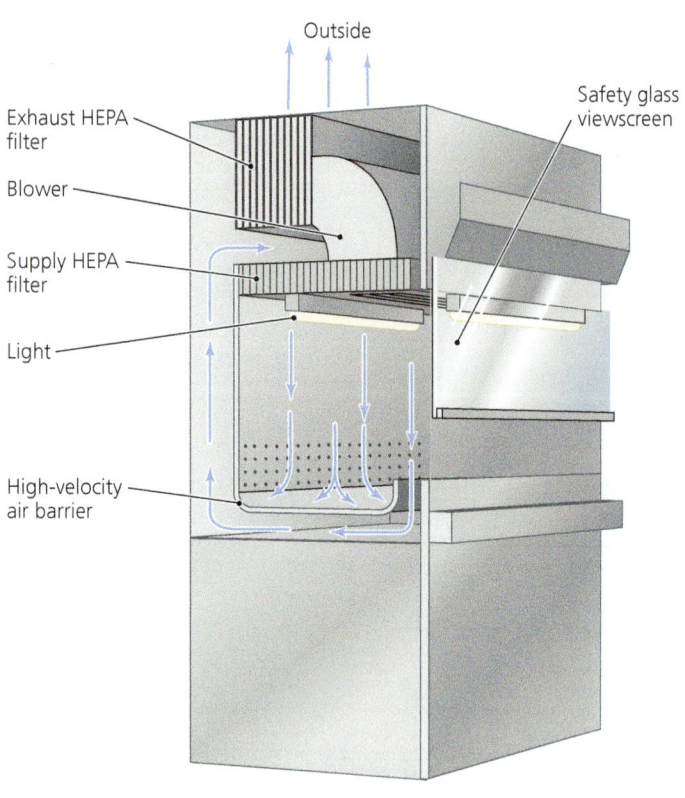
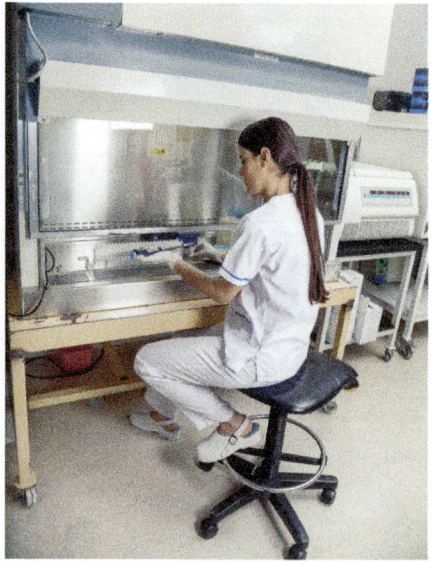

▶ FIGURE 9.11 **The roles of high-efficiency particulate air (HEPA) filters in biological safety cabinets.** HEPA filters protect workers from exposure to microbes (by maintaining a barrier of filtered air across the opening of the cabinet). Hospitals also use HEPA filters in air ducts of operating rooms and of the rooms of highly contagious or immunocompromised patients.

Ionizing Radiation

Electron beams, gamma rays, and some X rays, all of which have wavelengths shorter than 1 nm, are **ionizing radiation** because when they strike molecules, they have sufficient energy to eject electrons from atoms, creating ions. Such ions disrupt hydrogen bonding, oxidize double covalent bonds, and create highly reactive hydroxyl radicals (see Chapter 6). These ions in turn denature other molecules, particularly DNA, causing fatal mutations and cell death.

Electron beams are produced by *cathode ray machines*. Electron beams are highly energetic and therefore very effective in killing microbes in just a few seconds, but they cannot sterilize thick objects or objects coated with large amounts of organic matter. They are used to sterilize spices, meats, microbiological plastic ware, and dental and medical supplies, such as gloves, syringes, and suturing material.

Gamma rays, which are emitted by some radioactive elements, such as radioactive cobalt, penetrate much farther than electron beams but require hours to kill microbes. The U.S. Food and Drug Administration (FDA) has approved the use of gamma irradiation for microbial control in meats, spices, and fresh fruits and vegetables (**FIGURE 9.12**). Irradiation with gamma rays kills not only microbes but also the larvae and eggs of insects; it also kills the cells of fruits and vegetables, preventing both microbial spoilage and overripening.

Consumers have been reluctant to accept irradiated food. A number of reasons have been cited, including fear that radiation makes food radioactive and claims that it changes the taste and nutritive value of foods or produces potentially carcinogenic (cancer-causing) chemicals. Supporters of irradiation reply that gamma radiation passes through food and cannot make it radioactive any more than a dental X ray produces radioactive teeth, and they cite numerous studies that conclude that irradiated foods are tasty, nutritious, and safe.

▲ FIGURE 9.12 **A demonstration of the increased shelf life of food achieved by ionizing radiation.** Both sets of strawberries are the same age and type. The circular radura symbol is used internationally to label irradiated foods.

X rays travel the farthest through matter, but they have less energy than gamma rays and require a prohibitive amount of time to make them practical for microbial control.

Nonionizing Radiation

Electromagnetic radiation with a wavelength greater than 1 nm does not have enough energy to force electrons out of orbit, so it is **nonionizing radiation**. However, such radiation does contain enough energy to excite electrons and cause them to make new covalent bonds, which can affect the three-dimensional structure of proteins and nucleic acids.

Ultraviolet (UV) light, visible light, infrared radiation, and radio waves are nonionizing radiation. Of these, only UV light has sufficient energy to be a practical antimicrobial agent. Visible light and microwaves (radio waves of extremely short wavelength) have little value in microbial control, though microwaves heat food, inhibiting microbial growth and reproduction if the food gets hot enough. The more energetic microwaves produced by some commercial microwave ovens can kill fungal spores, preventing treated food from molding.

UV light with a wavelength of 260 nm is specifically absorbed by adjacent pyrimidine nucleotide bases in DNA, causing them to form covalent bonds with each other rather than forming hydrogen bonds with bases in the complementary DNA strand (see Figure 7.26). Such *pyrimidine dimers* distort the shape of DNA, making it impossible for the cell to accurately transcribe or replicate its genetic material. If dimers remain uncorrected, an affected cell may die.

The effectiveness of UV irradiation is tempered by the fact that UV light does not penetrate well. UV light is therefore suitable primarily for disinfecting air, transparent fluids, and the surfaces of objects, such as barber's shears and operating tables. Some cities use UV irradiation in sewage treatment. By passing wastewater past banks of UV lights, they reduce the number of bacteria without using chlorine, which might damage the environment.

TABLE 9.4 summarizes the physical methods of microbial control discussed in the previous pages.

TABLE 9.4 Physical Methods of Microbial Control

Method	Conditions	Action	Representative Use(s)
Moist Heat			
Boiling	10 min at 100°C (at sea level)	Denatures proteins and destroys membranes	Disinfection of baby bottles and sanitization of restaurant cookware and tableware
Autoclaving (pressure cooking)	15 min at 121°C	Denatures proteins and destroys membranes	Autoclave: sterilization of medical and laboratory supplies that can tolerate heat and moisture; pressure cooker: sterilization of canned food
Pasteurization	30 minutes at 63°C; 15 sec at 72°C	Denatures proteins and destroys membranes	Destruction of all pathogens and most spoilage microbes in dairy products, fruit juices, beer, and wine
Ultra-high-temperature sterilization	1–3 sec at 140°C	Denatures proteins and destroys membranes	Sterilization of dairy products
Dry Heat			
Hot air	2 h at 160°C or 1 h at 171°C	Denatures proteins, destroys membranes, oxidizes metabolic compounds	Sterilization of water-sensitive materials, such as powders, oils, and metals
Incineration	1 sec at approximately 1500°C	Oxidizes everything completely	Sterilization of inoculating loops, flammable contaminated medical waste, and diseased carcasses
Refrigeration	0–7°C	Inhibits metabolism	Preservation of food
Freezing	Below 0°C	Inhibits metabolism	Long-term preservation of foods, drugs, and cultures
Desiccation (drying)	Varies with amount of water to be removed	Inhibits metabolism	Preservation of food
Lyophilization (freeze-drying)	−196°C for a few minutes while drying	Inhibits metabolism	Long-term storage of bacterial cultures
Filtration	Filter retains microbes	Physically separates microbes from air and liquids	Sterilization of air and of heat-sensitive ophthalmic and enzymatic solutions, vaccines, and antibiotics
Osmotic Pressure	Exposure to hypertonic solutions	Inhibits metabolism	Preservation of food
Ionizing Radiation (electron beams, gamma rays, X rays)	Seconds to hours of exposure (depending on wavelength of radiation)	Destroys DNA	Sterilization of medical and laboratory equipment and preservation of food
Nonionizing Radiation (ultraviolet light)	Irradiation with 260-nm-wavelength radiation	Formation of thymine dimers inhibits DNA transcription and replication	Disinfection and sterilization of surfaces and of transparent fluids and gases

MICRO CHECK

4. Which heat-related method of microbial control most effectively destroys bacterial endospores?
5. Which drying method of microbial control can also be used to preserve microbial cultures?
6. When might a researcher more likely use filtration as a sterilization method?
7. Which form of radiation is most likely to control microbes in a barbershop or hair salon?

TELL ME WHY

Why are *Bacillus* endospores used as sterility indicators? (See Figure 9.8.)

Chemical Methods of Microbial Control

LEARNING OUTCOME

9.20 Compare and contrast nine major types of antimicrobial chemicals, and discuss the positive and negative aspects of each.

Although physical agents are sometimes used for disinfection, antisepsis, and preservation, more often chemical agents are used for these purposes. As we have seen, chemical agents act to adversely affect microbes' cell walls, cytoplasmic membranes, proteins, or DNA. As with physical agents, the effect of a chemical agent varies with temperature, length of exposure, and the amount of contaminating organic matter in the environment. The effect also varies with pH, concentration, and freshness of the chemical. Chemical agents tend to destroy or inhibit the growth of enveloped viruses and the vegetative cells of bacteria, fungi, and protozoa more than fungal spores, protozoan cysts, or bacterial endospores. The latter are particularly resistant to chemical agents.

In the following sections, we discuss nine major categories of antimicrobial chemicals used as antiseptics and disinfectants: *phenols, alcohols, halogens, oxidizing agents, surfactants, heavy metals, aldehydes, gaseous agents,* and *enzymes*. Some chemical agents combine one or more of these. Additionally, researchers and food processors sometimes use antimicrobial drugs—substances normally used to treat diseases—as disinfectants. (Chapter 10 discusses antimicrobial drugs in more detail.)

Phenol and Phenolics

LEARNING OUTCOME

9.21 Distinguish between phenol and the types of phenolics, and discuss their action as antimicrobial agents.

In 1867, Dr. Joseph Lister (1827–1912) began using phenol (also known as *carbolic acid*) (FIGURE 9.13a) to reduce infection during surgery. **Phenolics** are compounds derived from phenol

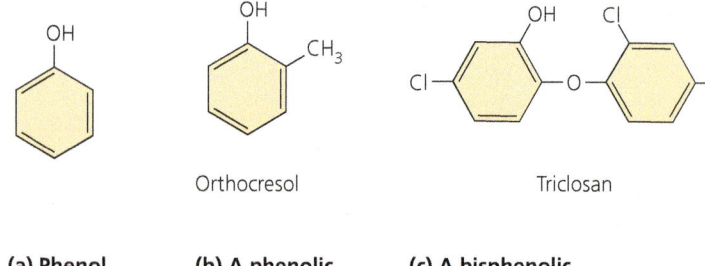

▲ FIGURE 9.13 **Phenol and phenolics. (a)** Phenol, a naturally occurring molecule that is also called *carbolic acid*. **(b)** Phenolics, which are compounds synthesized from phenol, have greater antimicrobial efficacy with fewer side effects. **(c)** Bisphenolics are paired, covalently linked phenolics.

molecules that have been chemically modified by the addition of organic functional groups or reactive atoms (FIGURE 9.13b). For instance, chlorinated phenolics contain one or more atoms of chlorine and have enhanced antimicrobial action and a less annoying odor than phenol. Natural oils, such as pine and clove oils, are also phenolics and can be used as antiseptics.

Bisphenolics are composed of two covalently linked phenolics. An example is *triclosan*, which is incorporated into numerous consumer products, including garbage bags, diapers, and cutting boards (FIGURE 9.13c).

Phenol and phenolics denature proteins and disrupt cell membranes in a wide variety of pathogens. They are effective even in the presence of contaminating organic material, such as vomit, pus, saliva, and feces, and they remain active on surfaces for a prolonged time. For these reasons, phenolics are commonly used in health care settings, laboratories, and households.

Negative aspects of phenolics include their disagreeable odor and possible side effects; for example, phenolics irritate the skin of some individuals.

Alcohols

LEARNING OUTCOME

9.22 Discuss the action of alcohols as antimicrobial agents, and explain why solutions of 70% to 90% alcohol are more effective than pure alcohols.

Alcohols are bactericidal, fungicidal, and virucidal against enveloped viruses; however, they are not effective against fungal spores or bacterial endospores. Alcohols are considered intermediate-level disinfectants. Commonly used alcohols include rubbing alcohol (isopropanol) and drinking alcohol (ethanol):

$$CH_3-\underset{\underset{\text{Isopropanol}}{}}{\overset{\overset{OH}{|}}{CH}}-CH_3 \qquad \underset{\text{Ethanol}}{CH_3-CH_2OH}$$

Isopropanol is slightly superior to ethanol as a disinfectant and antiseptic. *Tinctures* (tingk'chŭrs) of antimicrobial chemicals, which are solutions of the chemicals in alcohol, are often more effective than the same chemicals dissolved in water.

Alcohols denature proteins and disrupt cytoplasmic membranes. Surprisingly, pure alcohol is not an effective antimicrobial agent because the denaturation of proteins requires water; therefore, solutions of 70% to 90% alcohol are typically used to control microbes. Alcohols evaporate rapidly, which is advantageous in that they leave no residue but disadvantageous in that they may not contact microbes long enough to be effective. Alcohol-based antiseptics are more effective than soap in removing bacteria from hands but not effective against some viruses, such as diarrhea-causing noroviruses. Swabbing the skin with alcohol prior to an injection removes more microbes by physical action (degerming) than by chemical action.

Halogens

LEARNING OUTCOME

9.23 Discuss the types and uses of halogen-containing antimicrobial agents.

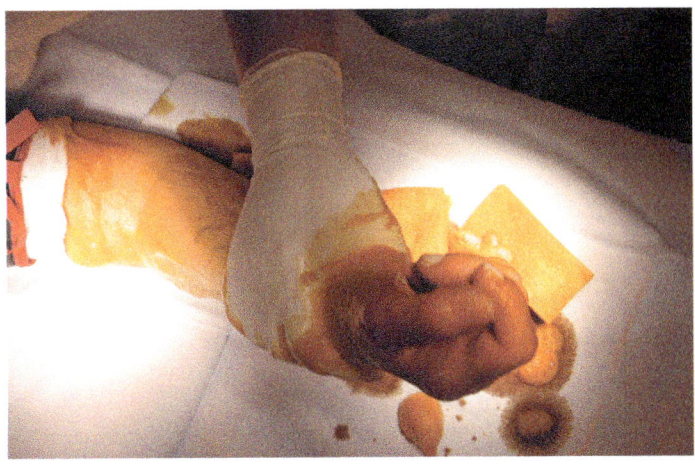

▲ FIGURE 9.14 **Degerming in preparation for surgery on a hand.** Betadine, an iodophor, is the antiseptic used here.

Halogens are the four very reactive, nonmetallic chemical elements: iodine, chlorine, bromine, and fluorine. Halogens are intermediate-level antimicrobial chemicals that are effective against vegetative bacterial and fungal cells, fungal spores, some bacterial endospores and protozoan cysts, and many viruses. Halogens are used both alone and combined with other elements in organic and inorganic compounds. Halogens exert their antimicrobial effect by unfolding and thereby denaturing essential proteins, including enzymes.

Iodine is a well-known antiseptic. In the past, backpackers and campers disinfected water with iodine tablets, but experience has shown that some protozoan cysts can survive iodine treatment unless the iodine concentration is so great that the water is undrinkable.

Medically, iodine is used either as a tincture or as an *iodophor*, which is an iodine-containing organic compound that slowly releases iodine. Iodophors have the advantage of being long lasting and nonirritating to the skin. Betadine is an example of an iodophor used in medical institutions to prepare skin for surgery (**FIGURE 9.14**) and injections and to treat burns.

Municipalities commonly use *chlorine* in its elemental form (Cl_2) to disinfect drinking water, swimming pools, and wastewater from sewage treatment plants. Compounds containing chlorine are also effective disinfectants. Examples include *sodium hypochlorite* (NaOCl), which is household chlorine bleach, and *calcium hypochlorite* ($Ca(OCl)_2$). The dairy industry and restaurants use these compounds to disinfect utensils, and the medical field uses them to disinfect hemodialysis systems. Household bleach diluted by adding two drops to a liter of water can be used in an emergency to make water safer to drink, but it does not kill all protozoan cysts, bacterial endospores, or viruses. *Chlorine dioxide* (ClO_2) is a gas that can be used to disinfect large spaces; for example, officials used it to sterilize federal buildings contaminated with anthrax spores following the 2001 bioterrorism

BENEFICIAL MICROBES

Controlling Bacteria with Viruses

Controlling bacteria in the environment is becoming more difficult because they are developing resistance to common disinfectants and antiseptics. So, scientists are turning to natural parasites of bacteria—bacteriophages, also simply called *phages*—to control bacterial contamination. *Bacteriophages*, which literally means "bacteria eaters," is the term for viruses that specifically attack particular strains of bacteria. Phages are like "smart bombs;" unlike disinfectants and most antimicrobial drugs, phages attack specific bacterial strains and leave neighboring bacteria unharmed. A phage injects its genetic material into a bacterial cell, causing the bacterial cell to produce hundreds of new phages that burst out of the bacterium and kill it. Researchers are developing phage solutions to control bacteria in medical settings, in food, and in patients.

In 2006, the U.S. Food and Drug Administration (FDA) approved the nonmedical use of a phage that specifically kills *Listeria monocytogenes*, which is frequently a bacterial contaminant of cheese and lunch meat. *Listeria* kills about 20% of people who get sick from ingesting it. The approved anti-*Listeria* phage is available in a solution that food processors, delicatessen owners, and consumers can spray on food to reduce the number of *Listeria* cells. Some people find the idea hard to swallow—deliberately contaminating food and equipment with viruses sounds like poor hygiene.

Proponents of using phages point out that an individual consumes millions of phages daily in water and food without ill effect. Indeed, phages in restaurants are more common than mustard and mayonnaise. Medical professionals in the former Soviet Union, particularly the country of Georgia, have used phages to successfully treat disease for over six decades without deleterious side effects.

attack. Chloramines—chemical combinations of chlorine and ammonia—are used in wound dressings, as skin antiseptics, and in some municipal water supplies. Chloramines are less effective antimicrobial agents than other forms of chlorine, but they release chlorine slowly and are thus longer lasting.

Bromine is an effective disinfectant in hot tubs because it evaporates more slowly than chlorine at high temperatures. Bromine is also used as an alternative to chlorine in the disinfection of swimming pools, cooling towers, and other water containers.

Fluorine in the form of fluoride is antibacterial in drinking water and toothpastes and can help reduce the incidence of dental caries (cavities). Fluorine works in part by disrupting metabolism in the biofilm of dental plaque.

Oxidizing Agents

LEARNING OUTCOME

9.24 Describe the use and action of oxidizing agents in microbial control.

Peroxides, *ozone*, and *peracetic acid* kill microbes by oxidizing their enzymes, thereby preventing metabolism. **Oxidizing agents** are high-level disinfectants and antiseptics that work by releasing oxygen radicals, which are particularly effective against anaerobic microorganisms. Health care workers use oxidizing agents to kill anaerobes in deep puncture wounds.

Hydrogen peroxide (H_2O_2) is a common household chemical that can disinfect and even sterilize the surfaces of inanimate objects such as contact lenses, but it is often mistakenly used to treat open wounds. Hydrogen peroxide does not make a good antiseptic for open wounds because *catalase*—an enzyme released from damaged human cells—quickly neutralizes hydrogen peroxide by breaking it down into water and oxygen gas, which can be seen as escaping bubbles. Though aerobes and facultative anaerobes on inanimate surfaces also contain catalase, the volume of peroxide used as a disinfectant overwhelms the enzyme, making hydrogen peroxide a useful disinfectant. Food processors use hot hydrogen peroxide to sterilize packages such as juice boxes.

Ozone (O_3) is a reactive form of oxygen that is generated when molecular oxygen (O_2) is subjected to electrical discharge. Ozone gives air its "fresh smell" after a thunderstorm. Some Canadian and European municipalities treat their drinking water with ozone rather than chlorine. Ozone is a more effective antimicrobial agent than chlorine, but it is more expensive, and it is more difficult to maintain an effective concentration of ozone in water.

Peracetic acid is an extremely effective sporicide that can be used to sterilize surfaces. Food processors and medical personnel use peracetic acid to sterilize equipment because it is not adversely affected by organic contaminants, and it leaves no toxic residue.

Surfactants

LEARNING OUTCOME

9.25 Define *surfactants*, and describe their antimicrobial action.

Surfactants are "surface active" chemicals. One of the ways surfactants act is to reduce the surface tension of solvents such as water by decreasing the attraction among molecules. One result of this reduction in surface tension is that the solvent becomes more effective at dissolving solute molecules.

Two common surfactants involved in microbial control are soaps and detergents. One end of a soap molecule is hydrophobic because it is composed of fatty acids, and the other end is hydrophilic and negatively charged. When soap is used to wash skin, for instance, the hydrophobic ends of soap molecules are effective at breaking oily deposits into tiny droplets, and the hydrophilic ends attract water molecules; the result is that the tiny droplets of oily material—and any bacteria they harbor—are more easily dissolved in and washed away by water. Thus, soaps by themselves are good degerming agents though poor antimicrobial agents; when household soaps are antiseptic, it is largely because they contain antimicrobial chemicals.

Synthetic **detergents** are positively charged organic surfactants that are more soluble in water than soaps. The most popular detergents for microbial control are **quaternary ammonium compounds (quats)**, which are composed of an ammonium cation (NH_4^+) in which the hydrogen atoms are replaced by other functional groups or hydrocarbon chains (FIGURE 9.15). Quats are not only antimicrobial but also colorless, tasteless, and harmless to humans (except at high concentrations), making them ideal for many industrial and medical applications. If your mouthwash foams, it probably contains a quaternary ammonium compound.

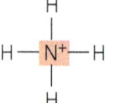

◀ **FIGURE 9.15 Quaternary ammonium compounds (quats).** Quats are surfactants in which the hydrogen atoms of an ammonium ion **(a)** are replaced by other functional groups **(b)**.

(a) Ammonium ion

Benzalkonium

Cetylpyridinium

(b) Quaternary ammonium ions (quats)

Examples of quats are benzalkonium chloride (Zephiran®) and cetylpyridinium (used in Cepacol® mouthwash).

Quats function by disrupting cellular membranes so that affected cells lose essential internal ions, such as potassium ions (K^+). Quats are bactericidal (particularly against Gram-positive bacteria), fungicidal, and virucidal against enveloped viruses, but they are not effective against nonenveloped viruses, mycobacteria, or endospores. The action of quaternary ammonium compounds is retarded by organic contaminants, and they are deactivated by soaps. Some pathogens, such as *Pseudomonas aeruginosa* (soo-dō-mō'nas ā-roo-ji-nō'să), actually thrive in quats; therefore, quats are classified as low-level disinfectants.

Heavy Metals

LEARNING OUTCOME

9.26 Define *heavy-metal ions*, give several examples, and describe their use in microbial control.

Heavy-metal ions are ions of relatively high-density metals, such as arsenic, zinc, mercury, silver, and copper, that are toxic at low concentrations. Heavy-metal ions are antimicrobial because they combine with sulfur atoms in molecules of cysteine, an amino acid. Such bonding denatures proteins, inhibiting or eliminating their function. Heavy-metal ions are low-level bacteriostatic and fungistatic agents, and with few exceptions their use has been superseded by more effective antimicrobial agents. **FIGURE 9.16** illustrates the effectiveness of heavy metals in inhibiting bacterial reproduction on a Petri plate.

At one time, many states required that the eyes of newborns be treated with 1% *silver nitrate* ($AgNO_3$) to prevent blindness caused by *Neisseria gonorrhoeae* (nī-se'rē-ă go-nor-rē'ī), which can enter babies' eyes while they pass through an infected birth canal. Today, hospitals have replaced silver nitrate with less irritating antimicrobial ointments that are also effective against other pathogens. Silver still plays an antimicrobial role in some surgical dressings, burn creams, and catheters.

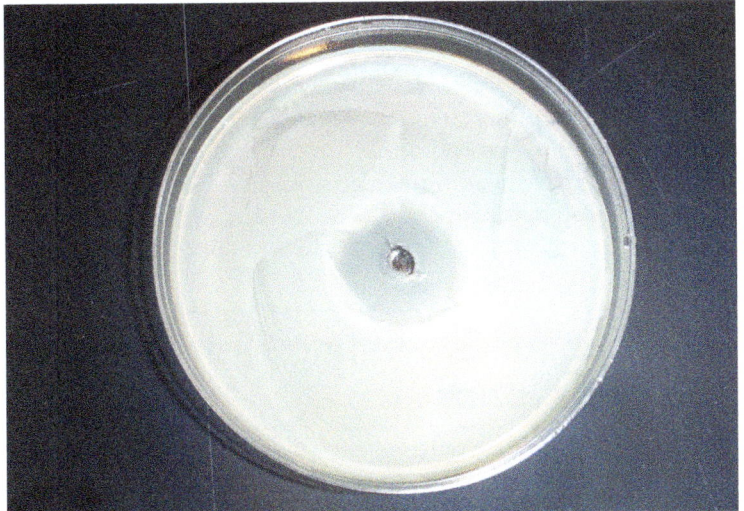

▲ **FIGURE 9.16 The effect of heavy-metal ions on bacterial growth.** Zones of inhibition can form because ions of heavy metals, such as dental amalgam used in fillings and shown in the center of the plate, inhibit bacterial reproduction through their effects on protein function.

For over 70 years, drug companies used *thimerosal*, a mercury-containing compound, to preserve vaccines. In 1999, the U.S. Public Health Service recommended that alternatives be used because mercury is a metabolic poison, though the very small amount of mercury in vaccines is considered safe. Today, the recommended children's vaccines are thimerosal-free, except for influenza (flu) vaccines distributed in multi-dose vials. A few adult vaccines also contain thimerosal, including the whole-cell pertussis vaccine and some vaccines against tetanus, flu, and meningococcal meningitis.

Copper, which interferes with chlorophyll, is used to control algal growth in reservoirs, fish tanks, swimming pools, and water storage tanks. In the absence of organic contaminants, copper is an effective algicide in concentrations as low as 1 ppm (part per million). Copper, zinc, or mercury is used to control mildew in some paint.

Aldehydes

LEARNING OUTCOME

9.27 Compare and contrast formaldehyde and glutaraldehyde as antimicrobial agents.

Aldehydes are compounds containing terminal —CHO groups (Table 2.3 on p. 41). *Glutaraldehyde*, which is a liquid, and *formaldehyde*, which is a gas, are highly reactive chemicals with the following structural formulas:

$$\underset{\text{Glutaraldehyde}}{\overset{O}{\overset{\|}{C}}H-CH_2-CH_2-CH_2-\overset{O}{\overset{\|}{C}}H} \qquad \underset{\text{Formaldehyde}}{\overset{O}{\overset{\|}{H}}CH}$$

Aldehydes function in microbial control by cross-linking organic functional groups, including amino, hydroxyl, sulfhydryl, and carboxyl (Table 2.3 on p. 41), thereby denaturing proteins and inactivating nucleic acids.

Hospital personnel and scientists can use 2% solutions of glutaraldehyde to kill bacteria, viruses, and fungi; a 10-minute treatment effectively disinfects most objects. When the time of exposure is increased to 10 hours, glutaraldehyde sterilizes. Glutaraldehyde is less irritating and more effective than formaldehyde, but it is more expensive.

Morticians and health care workers use formaldehyde dissolved in water to make a 37% solution called *formalin*. They use formalin for embalming and to disinfect hospital rooms, instruments, and machines. Formaldehyde must be handled with care because it irritates mucous membranes and is carcinogenic (cancer causing).

MICRO CHECK

8. Which halogen is commonly used to degerm surgical sites?
9. Which oxidizing agent is sometimes used to treat drinking water?
10. How do aldehydes denature proteins to control microbes?

Gaseous Agents

LEARNING OUTCOME

9.28 Describe the advantages and disadvantages of gaseous agents of microbial control.

Many items, such as heart-lung machine components, sutures, plastic laboratory ware, mattresses, pillows, artificial heart valves, catheters, electronic equipment, and dried or powdered foods, cannot be sterilized easily with heat or water-soluble chemicals, nor is irradiation always practical for large or bulky items. However, they can be sterilized within a closed chamber containing highly reactive microbicidal and sporicidal gases such as *ethylene oxide*, *propylene oxide*, and *beta-propiolactone*:

Ethylene oxide   Propylene oxide   Beta-propiolactone

These gases rapidly penetrate paper and plastic wraps and diffuse into every crack. Over time (usually 4–18 hours), they denature proteins and DNA by cross-linking organic functional groups, thereby killing everything they contact without harming inanimate objects.

Ethylene oxide is frequently used as a gaseous sterilizing agent in hospitals and dental offices, and NASA uses the gas to sterilize spacecraft designed to land on other worlds lest they accidentally export earthly microbes. Large hospitals often use ethylene oxide chambers, which are similar in appearance to autoclaves, to sterilize instruments and equipment sensitive to heat.

Despite their advantages, gaseous agents are far from perfect: They are often highly explosive, and they are extremely poisonous, so workers must extensively flush sterilized objects with air to remove every trace of the gas (which adds to the time required to use them). Finally, gaseous agents, especially beta-propiolactone, are potentially carcinogenic.

Enzymes

LEARNING OUTCOME

9.29 Describe the use of an enzyme to remove most bacteria from food and to remove prions from medical instruments.

Many organisms produce chemicals that inhibit or destroy a variety of fungi, bacteria, or viruses. Among these are **antimicrobial enzymes**, which are enzymes that act against microorganisms. For example, human tears contain the enzyme *lysozyme*, which is a protein that digests the peptidoglycan cell walls of bacteria, causing the bacteria to rupture because of osmotic pressure and thus protecting the eye from most bacterial infections.

Scientists, food processors, and medical personnel are researching ways to use natural and chemically modified antimicrobial enzymes to control microbes in the environment, inhibit microbial decay of foods and beverages, and reduce the number and kinds of microbes on medical equipment. For example, food processors use lysozyme to reduce the number of bacteria in cheese, and some vintners use lysozyme instead of poisonous sulfur dioxide (SO_2) to remove bacteria that would spoil wine.

One exciting development is the use of an enzyme to eliminate the prion that causes variant Creutzfeldt-Jakob disease (vCJD), also called *mad cow disease*. The brain, spinal cord, placenta, eye, liver, kidney, pituitary gland, spleen, lung, and lymph nodes, as well as cerebrospinal fluid, can harbor prions. Medical instruments contaminated by these highly infectious and deadly proteins may remain infectious even after normal autoclaving; boiling; exposure to formaldehyde, glutaraldehyde, or ethylene oxide; or 24 hours of dry heat at 160°C. Until recently, harsh methods, such as autoclaving in sodium hydroxide for 30 minutes or complete incineration, were required to eliminate prions. In 2006, the European Union approved the use of the enzyme *Prionzyme* to safely and completely remove prions on medical instruments. Prionzyme is the first certified, noncaustic chemical to target prions.

Antimicrobial Drugs

LEARNING OUTCOME

9.30 Describe three general types of antimicrobial drugs and their use in environmental control of microorganisms.

Antimicrobial drugs include antibiotics, semisynthetics, and synthetics. Specifically, *antibiotics* are antimicrobial chemicals produced naturally by microorganisms. When scientists chemically modify an antibiotic, the agent is called a *semisynthetic*. Scientists have also developed wholly *synthetic* antimicrobial drugs. The main difference between these antimicrobials and the chemical agents we have discussed in this chapter is that antimicrobial drugs are typically used for treatment of disease and not for environmental control of microbes. Nevertheless, some antimicrobial drugs are used for control outside the body. For example, *nisin* and *natamycin* are used to reduce the growth of bacteria and fungi, respectively, in cheese. (Chapter 10 discusses in more detail the nature and use of antimicrobial drugs to treat infectious diseases.)

TABLE 9.5 summarizes the chemical methods of microbial control discussed in this chapter.

> ### MICRO CHECK
>
> 11. Which enzyme is used to reduce the number of bacteria in some cheeses and wine?
> 12. Technically speaking, which antimicrobial drugs are produced naturally?

Methods for Evaluating Disinfectants and Antiseptics

LEARNING OUTCOME

9.31 Compare and contrast four methods used to measure the effectiveness of disinfectants and antiseptics.

With few exceptions, higher concentrations and fresher solutions of a disinfectant are more effective than more dilute, older

TABLE 9.5 Chemical Methods of Microbial Control

Method	Action(s)	Level of Activity	Some Uses
Phenol (carbolic acid)	Denatures proteins and disrupts cell membranes	Intermediate to low	Original surgical antiseptic; now replaced by less odorous and injurious phenolics
Phenolics (chemically altered phenol; bisphenolics are composed of a pair of linked phenolics)	Denature proteins and disrupt cell membranes	Intermediate to low	Disinfectants and antiseptics
Alcohols	Denature proteins and disrupt cell membranes	Intermediate	Disinfectants, antiseptics, and as a solvent in tinctures
Halogens (iodine, chlorine, bromine, and fluorine)	Denature proteins	Intermediate	Disinfectants, antiseptics, and water purification
Oxidizing Agents (peroxides, ozone, and peracetic acid)	Denature proteins by oxidation	High	Disinfectants, antiseptics for deep wounds, water purification, and sterilization of food processing and medical equipment
Surfactants (soaps and detergents)	Decrease surface tension of water and disrupt cell membranes	Low	Soaps: degerming; detergents: antiseptic
Heavy Metals (arsenic, zinc, mercury, silver, copper, etc.)	Denature proteins	Low	Fungistats in paints; silver nitrate cream: surgical dressings, burn creams, and catheters; copper: algicide in water reservoirs, swimming pools, and aquariums
Aldehydes (glutaraldehyde and formaldehyde)	Denature proteins	High	Disinfectant and embalming fluid
Gaseous Agents (ethylene oxide, propylene oxide, and beta-propiolactone)	Denature proteins	High	Sterilization of heat- and water-sensitive objects
Enzymes	Denature proteins	High against target substrate	Removal of prions on medical instruments
Antimicrobials (antibiotics, semisynthetics, and synthetics)	Act against cell walls, cell membranes, protein synthesis, and DNA transcription and replication	Intermediate to low	Disinfectants and treatment of infectious diseases

solutions. We have also seen that longer exposure times ensure the deaths of more microorganisms. However, anyone using disinfectants must consider whether higher concentrations and longer exposures may damage an object or injure animals or people.

Scientists have developed several methods to measure the efficacy of antimicrobial agents. These include the *phenol coefficient*, the *use-dilution test*, the *Kelsey-Sykes capacity test*, and the *in-use test*.

Phenol Coefficient

Joseph Lister introduced the widespread use of phenol as an antiseptic during surgery. Since then, researchers have evaluated the efficacy of various disinfectants and antiseptics by calculating a ratio that compares a given agent's ability to control microbes to that of phenol under standardized conditions. This ratio is referred to as the **phenol coefficient**. A phenol coefficient greater than 1.0 indicates that an agent is more effective than phenol, and the larger the ratio, the greater the effectiveness. For example, *chloramine,* a mixture of chlorine and ammonia, has a phenol coefficient of 133.0 when used against the bacterium *Staphylococcus aureus* and a phenol coefficient of 100.0 when used against *Salmonella enterica* (en-ter'i-kă). This indicates that chloramine is at least 133 times more effective than phenol against *Staphylococcus* but only 100 times more effective against *Salmonella*. Measurement of an agent's phenol coefficient has been replaced by newer methods because scientists have developed disinfectants and antiseptics much more effective than phenol.

Use-Dilution Test

Another method for measuring the efficacy of disinfectants and antiseptics against specific microbes is the **use-dilution test**. In this test, a researcher dips several metal cylinders into broth cultures of bacteria and briefly dries them at 37°C. The bacteria used in the standard test are *Pseudomonas aeruginosa*, *Salmonella enterica* serotype Choleraesuis (kol-er-a-su'is), and *Staphylococcus aureus*. The researcher then immerses each contaminated cylinder into a different dilution of the disinfectants being evaluated. After 10 minutes, each cylinder is removed, rinsed with water to remove excess disinfectant, and placed into a fresh tube of sterile medium for 48 hours of incubation. The most effective agent is the one that entirely prevents microbial growth at the highest dilution.

The use-dilution test is the current standard test in the United States, though it was developed several decades ago before the appearance of many of today's more troubling pathogens, including hepatitis C virus, HIV, and antibiotic-resistant bacteria and protozoa. Moreover, the disinfectants in use at the time the test was developed were far less powerful than many used today. Some government agencies have expressed concern that the test is neither accurate, reliable, nor relevant;

Kelsey-Sykes Capacity Test

The **Kelsey-Sykes capacity test** is the standard alternative assessment approved by the European Union to determine the capacity of a given chemical to inhibit bacterial growth. In this test, researchers add a suspension of a bacterium such as *Psuedomonas aeruginosa* or *Staphylococcus aureus* to a suitable concentration of the chemical being tested. Then at predetermined times, they move samples of the mixture into growth medium containing a disinfectant deactivator. After incubation for 48 hours, turbidity in the medium indicates that bacteria survived treatment. Lack of turbidity, indicating lack of bacterial reproduction, reveals the minimum time required for the disinfectant to be effective.

In-Use Test

Though phenol coefficient, use-dilution, and Kelsey-Sykes capacity tests can be beneficial for initial screening of disinfectants, they can also be misleading. These types of evaluation are measures of effectiveness under controlled conditions against one or, at most, a few species of microbes at a time, but disinfectants are generally used in various environments against a diverse population of organisms that are often associated with one another in complex biofilms affording mutual protection.

A more realistic (though more time-consuming) method for determining the efficacy of a chemical is called an **in-use test**. In this procedure, swabs are taken from actual objects, such as operating room equipment, both before and after the application of a disinfectant or an antiseptic. The swabs are then inoculated into appropriate growth media that, after incubation, are examined for microbial growth. The in-use test allows a more accurate determination of the proper strength and application procedure of a given disinfection agent for each specific situation.

Development of Resistant Microbes

Many scientists are concerned that Americans have become overly preoccupied with antisepsis and disinfection, as evidenced by the proliferation of products containing antiseptic and disinfecting chemicals. For example, you can now buy hand soap, shampoo, toothpaste, hand lotion, foot pads for shoes, deodorants, and bath sponges that contain antiseptics, as well as kitty litter, cutting boards, scrubbing pads, garbage bags, children's toys, and laundry detergents that contain disinfectants. There is little evidence that the extensive use of such products adds to human or animal health, but it does promote the development of strains of microbes resistant to antimicrobial chemicals: While susceptible cells die, resistant cells remain to proliferate. Scientists have already isolated strains of pathogenic bacteria, including *M. tuberculosis*, *P. aeruginosa*, *E. coli*, and *S. aureus*, that are less susceptible to common disinfectants and antiseptics.

MICRO CHECK

13. Which method would be most accurate for evaluating disinfectants or antiseptics by an American industrialist for any specific situation?

TELL ME WHY

Many chemical disinfectants and antiseptics act by denaturing proteins. Why does denaturation kill cells?

MICRO IN THE CLINIC FOLLOW-UP

An Uninvited Dinner Guest?

Rather than feeling fine the next morning, Marcella feels far worse. She also notices blood in her diarrhea; this scares her, so she makes an appointment to see her doctor as soon as possible.

Dr. Walker examines Marcella and collects a stool (feces) sample. She tells Marcella that, based on her symptoms, it's likely that Marcella has food poisoning. They will know for certain after the lab analysis of the stool, but for now the doctor prescribes azithromycin, fluids, and rest. Dr. Walker asks Marcella to try to remember all the food she consumed in the last few days and where she got the food; it's possible that Marcella and Sarah aren't the only ones affected, and public health authorities may need to investigate.

A few days later, Marcella is feeling much better. Dr. Walker calls and tells Marcella that *Campylobacter jejuni* was identified in her stool sample and that she suspects the raw cheese was the cause of the infection. Public health officials contact Johnson Dairy and learn that a number of customers have reported getting sick within the last week after drinking their milk or eating their cheese. The health department requests that Johnson Dairy cease selling their products until the source of the contamination can be determined.

1. Why do you think that Dr. Walker suspected the milk as the source of infection?
2. Imagine that you are working for Johnson Dairy. What kinds of practices (both during and after production) would help limit the chance of contamination in raw milk?

Check your answers to Micro in the Clinic Follow-Up questions in the **Mastering Microbiology** Study Area.

CHAPTER SUMMARY

Go to Mastering Microbiology for Interactive Microbiology, Dr. Bauman's Video Tutors, Micro Matters, MicroFlix, Micro-Boosters, practice quizzes, and more.

Basic Principles of Microbial Control (pp. 263–266)

1. **Sterilization** is the eradication of microorganisms and viruses; the term is not usually applied to the destruction of prions.
2. An **aseptic** environment or procedure is free of contamination by pathogens.
3. **Antisepsis** is the inhibition/killing of microorganisms (particularly pathogens) on skin or tissue by the use of a chemical **antiseptic**, whereas **disinfection** refers to the use of agents (called **disinfectants**) to inhibit microbes on inanimate objects.
4. **Degerming** refers to the removal of microbes from a surface by scrubbing.
5. **Sanitization** is the reduction of a prescribed number of pathogens from surfaces and utensils in public settings.
6. **Pasteurization** is a process using heat to kill pathogens and control microbes that cause spoilage of food and beverages.
7. The suffixes *-stasis* and *-static* indicate that an antimicrobial agent inhibits microbes, whereas the suffixes *-cide* and *-cidal* indicate that the agent kills or permanently inactivates a particular type of microbe.
8. **Microbial death** is the permanent loss of reproductive capacity. **Microbial death rate** measures the efficacy of an antimicrobial agent.
9. Antimicrobial agents destroy microbes either by altering their cell walls and membranes or by interrupting their metabolism and reproduction via interference with proteins and nucleic acids.

The Selection of Microbial Control Methods (pp. 266–268)

1. Factors affecting the efficacy of antimicrobial methods include the site to be treated, the relative susceptibility of microorganisms, and environmental conditions.
2. The CDC has established four biosafety levels (BSL) for microbiological laboratories. BSL-1 is minimal; BSL-4 requires special suits, rooms, and other precautions.

Physical Methods of Microbial Control (pp. 268–276)

1. **Thermal death point** is the lowest temperature that kills all cells in a broth in 10 minutes, whereas **thermal death time** is the time it takes to completely sterilize a particular volume of liquid at a set temperature. **Decimal reduction time (D)** is the time required to destroy 90% of the microbes in a sample.
2. An **autoclave** uses steam heat under pressure to sterilize chemicals and objects that can tolerate moist heat.
3. Pasteurization, a method of heating foods to kill pathogens and control spoilage organisms without altering the quality of the food, can be achieved by several methods: the historical (batch) method, flash pasteurization, and ultra-high-temperature pasteurization. The methods differ in their combinations of temperature and time of exposure.
4. Under certain circumstances, microbes can be controlled using ultra-high-temperature sterilization, dry-heat sterilization, incineration, refrigeration, or freezing.
5. Antimicrobial methods involving drying are **desiccation**, used to preserve food, and **lyophilization** (freeze-drying), used for the long-term preservation of cells or microbes.
6. When used as a microbial control method, **filtration** is the passage of air or a liquid through a material that traps and removes microbes. Some **membrane filters** have pores small enough to trap the smallest viruses. HEPA (high-efficiency particulate air) filters remove microbes and particles from air.
7. The high **osmotic pressure** exerted by hypertonic solutions of salt or sugar can preserve foods such as jerky and jams by removing from microbes the water that they need to carry out their metabolic functions.
8. **Radiation** includes high-speed subatomic particles and even more energetic electromagnetic waves released from atoms. **Ionizing radiation** (wavelengths shorter than 1 nm) produces ions that denature important molecules and kill cells. **Nonionizing radiation** (wavelengths longer than 1 nm) is less effective in microbial control, although UV light causes pyrimidine dimers, which can kill affected cells.

Chemical Methods of Microbial Control (pp. 276–282)

1. **Phenolics**, which are chemically modified phenol molecules, are intermediate- to low-level disinfectants that denature proteins and disrupt cell membranes in a wide variety of pathogens.
2. **Alcohols** are intermediate-level disinfectants that denature proteins and disrupt cell membranes; they are used either as 70% to 90% aqueous solutions or in a tincture, which is a combination of an alcohol and another antimicrobial chemical.
3. **Halogens** (iodine, chlorine, bromine, and fluorine) are used as intermediate-level disinfectants and antiseptics to kill microbes by protein denaturation in water or on medical instruments or skin.
4. **Oxidizing agents** such as hydrogen peroxide, ozone, and peracetic acid are high-level disinfectants and antiseptics that release oxygen radicals, which are toxic to many microbes, especially anaerobes.
5. **Surfactants** include soaps, which act primarily to break up oils during degerming, and **detergents**, such as **quaternary ammonium compounds (quats)**, which are low-level disinfectants.
6. **Heavy-metal ions**, such as ions of arsenic, silver, mercury, copper, and zinc, are low-level disinfectants that denature proteins. For most applications they have been superseded by less toxic alternatives.
7. **Aldehydes** are high-level disinfectants that cross-link organic functional groups in proteins and nucleic acids. A 2% solution of glutaraldehyde or a 37% aqueous solution of formaldehyde (called formalin) is used to disinfect or sterilize medical or dental equipment and in embalming fluid.

8. Gaseous agents of microbial control, which include ethylene oxide, propylene oxide, and beta-propiolactone, are high-level disinfecting agents used to sterilize heat-sensitive equipment and large objects. These gases are explosive and potentially carcinogenic.

9. Many organisms use **antimicrobial enzymes** to combat microbes. Humans use them commercially in food preservation and as a noncaustic, nondestructive way to eliminate prions on medical instruments.

10. **Antimicrobial drugs**, which include antibiotics, semisynthetics, and synthetics, are compounds that are typically used to treat diseases but can also function as intermediate-level disinfectants.

11. Four methods for evaluating the effectiveness of a disinfectant or antiseptic are the **phenol coefficient**, the **use-dilution test**, the **Kelsey-Sykes capacity test**, and the **in-use test**. The in-use test provides a more accurate determination of efficacy under real-life conditions.

QUESTIONS FOR REVIEW

Answers to the Questions for Review (except Short Answer questions) begin on p. A-1.

Multiple Choice

1. In practical terms in everyday use, which of the following statements provides the definition of *sterilization*?
    a. Sterilization eliminates all organisms and viruses.
    b. Sterilization eliminates harmful microorganisms and viruses.
    c. Sterilization eliminates prions.
    d. Sterilization eliminates hyperthermophiles.

2. Which of the following substances or processes kills microorganisms on laboratory surfaces?
    a. antiseptics
    b. disinfectants
    c. degermers
    d. pasteurization

3. Which of the following terms best describes the disinfecting of cafeteria plates?
    a. pasteurization
    b. antisepsis
    c. sterilization
    d. sanitization

4. The microbial death rate is used to measure the effectiveness of _____.
    a. a detergent
    b. an antiseptic
    c. sanitization techniques
    d. all of the above

5. Which of the following statements is true concerning the selection of an antimicrobial agent?
    a. An ideal antimicrobial agent is stable during storage.
    b. An ideal antimicrobial agent is fast acting.
    c. Ideal microbial agents do not exist.
    d. All of the above are correct.

6. The endospores of which organism can be used as a biological indicator of sterilization?
    a. *Bacillus stearothermophilus*
    b. *Salmonella enterica*
    c. *Mycobacterium tuberculosis*
    d. *Staphylococcus aureus*

7. A company that manufactures an antimicrobial cleaner for kitchen counters claims that its product is effective when used in a 50% water solution. By what means might scientists best verify this statement?
    a. disk-diffusion test
    b. phenol coefficient
    c. filter paper test
    d. in-use test

8. Which of the following items functions most like an autoclave?
    a. boiling pan
    b. incinerator
    c. microwave oven
    d. pressure cooker

9. The preservation of beef jerky from microbial growth relies on which method of microbial control?
    a. filtration
    b. lyophilization
    c. desiccation
    d. radiation

10. Which of the following types of radiation is more widely used as an antimicrobial technique?
    a. electron beams
    b. visible light waves
    c. radio waves
    d. microwaves

11. Which of the following substances would most effectively inhibit anaerobes?
    a. phenol
    b. silver
    c. ethanol
    d. hydrogen peroxide

12. Which of the following adjectives best describes a surgical procedure that is free of microbial contaminants?
    a. disinfected
    b. sanitized
    c. degermed
    d. aseptic

13. Biosafety Level 3 includes _____.
    a. double sets of entry doors
    b. pressurized suits
    c. showers in entryways
    d. all of the above

14. A sample of *E. coli* has been subjected to heat for a specified time, and 90% of the cells have been destroyed. Which of the following terms best describes this event?
    a. thermal death point
    b. thermal death time
    c. decimal reduction time
    d. none of the above

15. Which of the following substances is least toxic to humans?
    a. carbolic acid
    b. glutaraldehyde
    c. hydrogen peroxide
    d. formalin

16. Which of the following chemicals is active against bacterial endospores?
    a. copper ions
    b. ethylene oxide
    c. ethanol
    d. triclosan

17. Which of the following disinfectants acts against cell membranes?
    a. phenol
    b. peracetic acid
    c. silver nitrate
    d. glutaraldehyde

18. Which of the following disinfectants contains alcohol?
    a. iodophor
    b. quat
    c. formalin
    d. tincture of bromine

19. Which antimicrobial chemical has been used to sterilize spacecraft?
    a. phenol
    b. alcohol
    c. heavy metal
    d. ethylene oxide

20. Which class of surfactant is most soluble in water?
    a. quaternary ammonium compounds
    b. alcohols
    c. soaps
    d. peracetic acids

Short Answer

1. Describe three types of microbes that are extremely resistant to antimicrobial treatment, and explain why they are resistant.
2. Compare and contrast four tests that have been developed to measure the effectiveness of disinfectants.
3. Why is it necessary to use strong disinfectants in areas exposed to tuberculosis patients?
4. Why do warm disinfectant chemicals generally work better than cool ones?
5. Why are Gram-negative bacteria more susceptible to heat than Gram-positive bacteria?
6. Describe five physical methods of microbial control.
7. What is the difference between thermal death point and thermal death time?
8. Defend the following statement: "Pasteurization is not sterilization."
9. Compare and contrast desiccation and lyophilization.
10. Compare and contrast the action of alcohols, halogens, and oxidizing agents in controlling microbial growth.
11. Hyperthermophilic prokaryotes may remain viable in canned goods after commercial sterilization. Why is this situation not dangerous to consumers?
12. Why are alcohols more effective in a 70% solution than in a 100% solution?
13. Contrast the structures and actions of soaps and quats.
14. What are some advantages and disadvantages of using ionizing radiation to sterilize food?
15. How can campers effectively treat stream water to remove pathogenic protozoa, bacteria, and viruses?

VISUALIZE IT!

1. Calculate the decimal reduction time (D) for the two temperatures in the following graph.

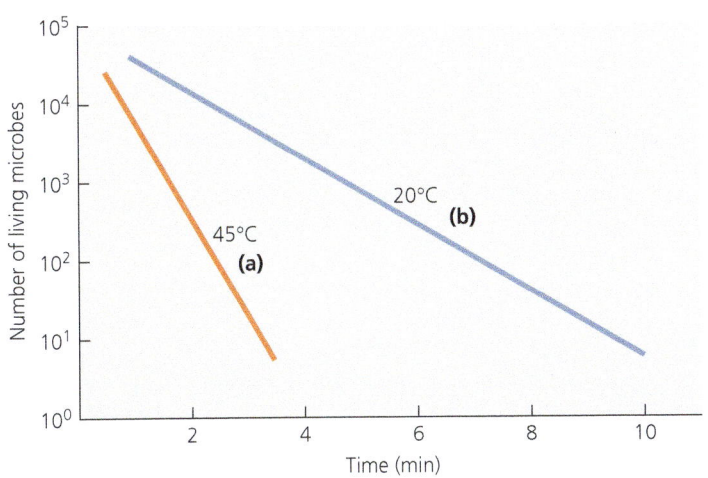

2. Indian tradition holds that storing water in brass pitchers prevents disease. Scientists have discovered that there is some truth in the tradition. The researchers collected river water samples and found fecal bacterial counts as high as 1 million bacteria per milliliter. However, they could detect no bacteria in the water after it had been stored for two days in traditional brass pitchers. Bacterial levels in plastic or earthenware containers remained high over the same period. How can brass, which is an alloy of copper mixed with zinc, make water safer to drink?

CRITICAL THINKING

1. In 2004, a casino paid $28,000 for a grilled cheese sandwich that was purported to have an image of the Virgin Mary on it. The seller had stored the sandwich in a less-than-airtight box for 10 years without decay or the growth of mold. What antimicrobial chemical and physical agents might account for the longevity of the sandwich?

2. Is desiccation the only antimicrobial effect operating when grapes are dried in the sun to make raisins? Explain.

3. How long would it take to reduce a population of 100 trillion (10^{14}) bacteria to 10 viable cells if the D value of the treatment is 3 minutes?

4. Some potentially pathogenic bacteria and fungi, including strains of *Enterococcus*, *Staphylococcus*, *Candida*, and *Aspergillus*, can survive for one to three months on a variety of materials found in hospitals, including scrub suits, lab coats, and plastic aprons and computer keyboards. What can hospital personnel do to reduce the spread of these pathogens?

5. Over 1000 people developed severe diarrhea, and at least four died, from a strain of *Salmonella enterica* in the fall of 2012. Epidemiologists determined that infection resulted from the consumption of contaminated smoked salmon. Based on this chapter, what methods to control microbial growth are available to fishermen, packers, and stores that might have prevented such an outbreak? What other precautions could consumers have taken?

6. An over-the-counter medicated foot powder contains camphor, eucalyptus oil, lemon oil, and zinc oxide. Only one of the ingredients is a proven antimicrobial. Which one? How does it act against fungi?

7. Tsunamis and hurricanes severely contaminate water wells and disrupt water supply lines. What immediate steps should people take to lessen the spread of waterborne illnesses such as cholera?

8. In what ways might it be argued that the widespread commercial use of antiseptics and disinfectants has hurt rather than helped American health?

9. Explain why quaternary ammonium compounds are not very effective against mycobacteria such as *Mycobacterium tuberculosis*.

10. A student inoculates *Escherichia coli* into two test tubes containing the same sterile liquid medium, except that the first tube also contains a drop of a chemical with an antimicrobial effect. After 24 hours of incubation, the first tube remains clear, whereas the second tube has become cloudy with bacteria. Design an experiment to determine whether this amount of the antimicrobial chemical is *bacteriostatic* or *bactericidal* against *E. coli*.

11. Would you expect Gram-negative bacteria or Gram-positive bacteria to be more susceptible to antimicrobial chemicals that act against cell walls? Explain your answer, which you should base solely on the nature of the cells' walls (see Figure 3.16).

12. Where should you place a sterilization indicator within an autoclave? Explain your reasoning.

13. Why is liquid water necessary for microbial metabolism?

14. A virologist needs to remove all bacteria from a solution containing viruses without removing the viruses. What size membrane filter should the scientist use?

15. During the fall 2001 bioterrorist attack in which anthrax endospores were sent through the mail, one news commentator suggested that people should iron all their incoming mail with a regular household iron as a means of destroying endospores. Would you agree that this is a good way to disinfect mail? Explain your answer. Which disinfectant methods would be both more effective and more practical?

16. What common household antiseptic contains a heavy metal as its active ingredient?

17. What is the phenol coefficient of phenol when used against *Staphylococcus*?

CONCEPT MAPPING

Using the terms provided, fill in this concept map that describes moist heat applications to control microorganisms. You can also complete this and other concept maps online by going to the Mastering Microbiology Study Area.

140°C, 1–3 sec
63°C, 30 min
Autoclave
Boiling water
Equivalent treatments

Flash pasteurization
Fungi
Most viruses
Pasteurization
Protozoan trophozoites

Sterilization technique
Thermoduric microorganisms (2)
Ultra-high-temperature pasteurization
Ultra-high-temperature sterilization
Vegetative bacterial cells

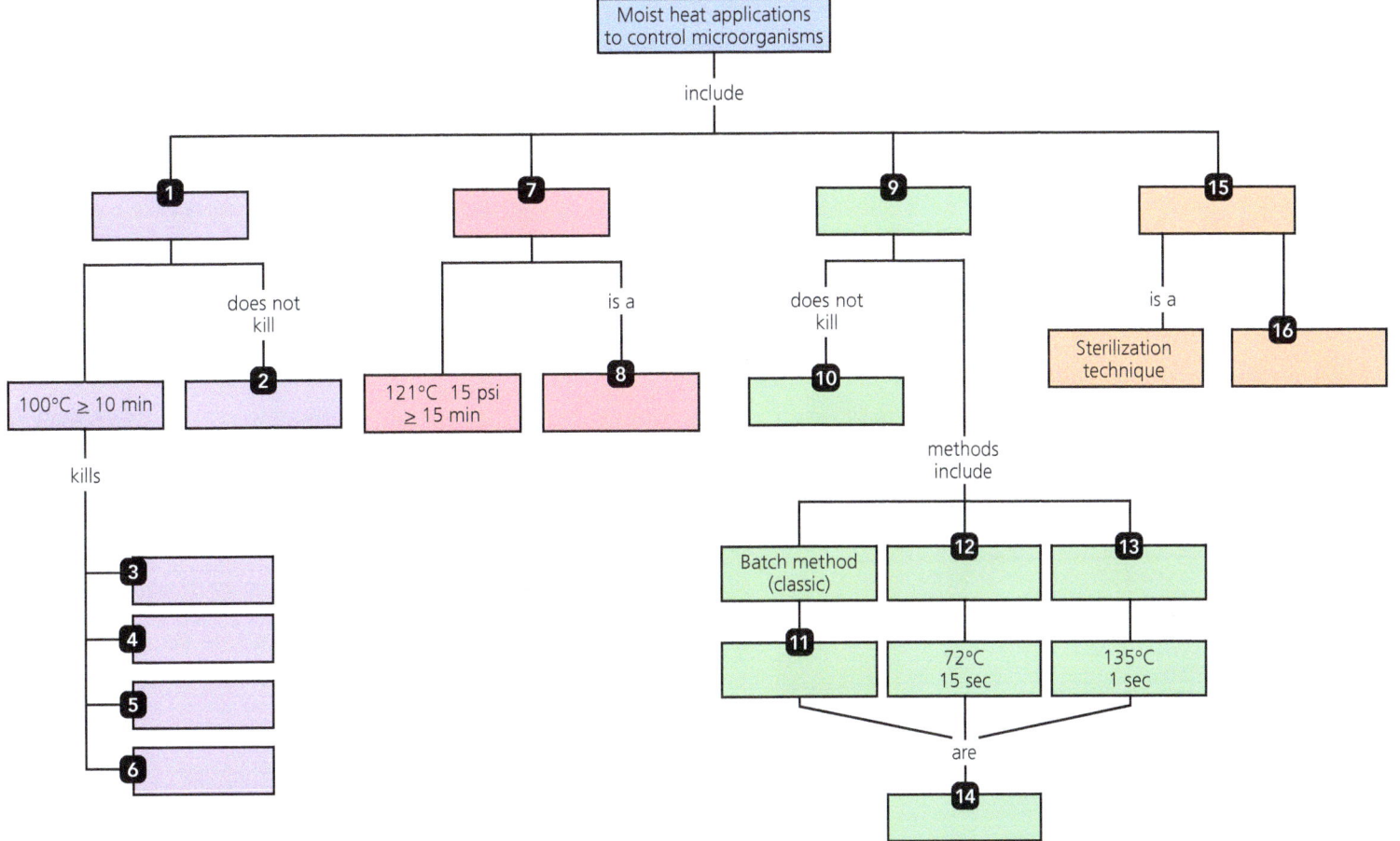

10 Controlling Microbial Growth in the Body: Antimicrobial Drugs

Before You Begin

1. The microbial growth curve has four phases: lag, log, stationary, and death. If a chemical could prevent microbial cell division, in which phase of growth would it be most effective?
2. Considering the cell structure of mycoplasmas and other bacteria that stain either Gram negative, Gram positive, or with acid-fast stains, which group should you expect relies most on the presence of peptidoglycan for strength?
3. What is the function of tRNA during the process of translation?
4. What is the difference between prokaryotic and eukaryotic ribosomes?

MICRO IN THE CLINIC

A Cure Leading to Disease?

THOMAS IS A 63-YEAR-OLD MAN who has had a persistent cough for the past three weeks. Thomas's symptoms started as a sore throat but then moved into his chest. When he coughs, there is a rattling sound in his chest, and the cough is dry and nonproductive. His doctor prescribes a short course of azithromycin for his assumed bronchitis. He returns to the doctor a few days after completing the azithromycin—his symptoms have only worsened. The doctor orders a chest X ray that shows consolidation (liquid instead of air in a lung); he suspects mild pneumonia. Dr. Warner prescribes a five-day course of levofloxacin (an antimicrobial drug) and tells Thomas he should start feeling better in a couple of days.

A week later, the cough is essentially gone, and Thomas feels a lot better—at least his respiratory system feels better. The previous day, Thomas started feeling some cramping in his lower abdomen and began having diarrhea. Today, the abdominal cramping is more intense, and he has more frequent bouts of diarrhea that are quite smelly. He takes an antidiarrheal medication he got at the drugstore and spends the rest of the day in bed.

The next morning, Thomas feels terrible—he has a fever, the cramping is severe, and the diarrhea hasn't stopped. He's worried that he's gotten food poisoning, so he calls and makes an appointment to see his doctor that afternoon.

1. Are Thomas's symptoms due to food poisoning?
2. Are there other possible explanations for his illness?

Turn to the end of the chapter (p. 319) to find out.

Thousands of different antimicrobial drugs are available for treating a variety of infectious diseases.

Chemicals that affect physiology in any manner, such as caffeine, alcohol, and tobacco, are called *drugs*. Drugs that act against diseases are called *chemotherapeutic agents*. Examples include insulin, anticancer drugs, and drugs for treating infections—called **antimicrobial agents** or **antimicrobials**, the subject of this chapter.

In the pages that follow, we'll examine the mechanisms by which antimicrobial agents act, the factors that must be considered in the use of antimicrobials, and several issues surrounding resistance to antimicrobial agents among microorganisms. First, however, we begin with a brief history of antimicrobial chemotherapy.

The History of Antimicrobial Agents

LEARNING OUTCOMES

10.1 Describe the contributions of Paul Ehrlich, Alexander Fleming, and Gerhard Domagk in the development of antimicrobials.

10.2 Explain how semisynthetic and synthetic antimicrobials differ from antibiotics.

The little girl lay struggling to breathe as her parents stood mutely by, willing the doctor to do something—anything—to relieve the symptoms that had so quickly consumed their four-year-old daughter's vitality. Sadly, there was little the doctor could do. The thick "pseudomembrane" of diphtheria, composed of bacteria, mucus, blood-clotting factors, and white blood cells, adhered tenaciously to her pharynx, tonsils, and vocal cords. He knew that trying to remove it could rip open the underlying mucous membrane, resulting in bleeding, possibly additional infections, and death. In 1902, there was little medical science could offer for the treatment of diphtheria; all physicians could do was wait and hope.

At the beginning of the 20th century, much of medicine involved diagnosing illness, describing its expected course, and telling family members either how long a patient might be sick or when they might expect her to die. Even though physicians and scientists had recently accepted the germ theory of disease and knew the causes of many diseases, very little could be done to inhibit pathogens, including *Corynebacterium diphtheriae* (kŏ-rī'nē-bak-tēr'ē-ŭm dif-thi'rē-ī), and alter the course of infections. In fact, one-third of children born in the early 1900s died from infectious diseases before the age of five.

It was at this time that Paul Ehrlich (1854–1915), a visionary German scientist, proposed the term *chemotherapy* to describe the use of chemicals that would selectively kill pathogens while having little or no effect on a patient. He wrote of "magic bullets" that would bind to receptors on germs to bring about their death while ignoring host cells, which lacked the receptor molecules.

Ehrlich's search for antimicrobial agents resulted in the discovery of arsenic compounds, one of which worked against the bacterial agent of syphilis. A few years later, in 1928, the British bacteriologist Alexander Fleming (1881–1955) reported the antibacterial action of penicillin released from *Penicillium* (pen-i-sil'ē-ŭm) mold, which creates a zone where bacteria don't grow (**FIGURE 10.1**).

Though arsenic compounds and penicillin were discovered first, they were not the first antimicrobials in widespread

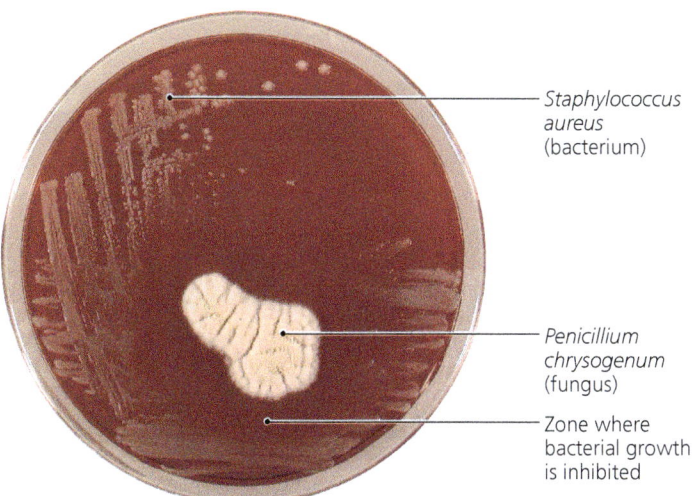

▲ **FIGURE 10.1 Antibiotic effect of the mold *Penicillium chrysogenum*.** Alexander Fleming observed that this mold secretes penicillin, which inhibits the growth of bacteria, as is apparent with *Staphylococcus aureus* growing on this blood agar plate.

use: Ehrlich's arsenic compounds are toxic to humans, and penicillin was not available in large enough quantities to be useful until the late 1940s. Instead, *sulfanilamide*, discovered in 1932 by the German chemist Gerhard Domagk (1895–1964), was the first practical antimicrobial agent efficacious in treating a wide array of bacterial infections.

Selman Waksman (1888–1973) discovered that other microorganisms are sources of useful antimicrobials, most notably species of soil-dwelling bacteria in the genus *Streptomyces* (strep-tō-mī'sēz). Waksman coined the term **antibiotics** to describe antimicrobial agents that are produced naturally by an organism. In today's common usage, *antibiotic* denotes an antibacterial agent, including synthetic compounds and excluding agents with antiviral and antifungal activity.

Other scientists produced **semisynthetic antimicrobials**—chemically altered antibiotics—that are more effective, longer lasting, or easier to administer than naturally occurring antibiotics. Antimicrobials that are completely synthesized in a laboratory are called **synthetic drugs**. Most commonly used antimicrobials are either natural or semisynthetic.

TABLE 10.1 (p. 290) provides a list of some common antibiotics and semisynthetics and their sources.

TELL ME WHY

Why aren't antibiotics effective against the common cold?

Mechanisms of Antimicrobial Action

LEARNING OUTCOMES

10.3 Explain the principle of selective toxicity.

10.4 List six mechanisms by which antimicrobial drugs affect pathogens.

TABLE 10.1 Sources of Some Common Antibiotics and Semisynthetics

Microorganism	Antimicrobial
Fungi	
Penicillium chrysogenum	Penicillin G
Penicillium griseofulvum	Griseofulvin
Acremonium[a] spp.[b]	Cephalothin
Bacteria	
Amycolatopsis orientalis	Vancomycin
Amycolatopsis rifamycinica	Rifampin
Bacillus licheniformis	Bacitracin
Bacillus polymyxa	Polymyxin
Micromonospora purpurea	Gentamicin
Pseudomonas fluorescens	Mupirocin
Saccharopolyspora erythraea	Erythromycin
Streptomyces griseus	Streptomycin
Streptomyces fradiae	Neomycin
Streptomyces aureofaciens	Tetracycline
Streptomyces venezuelae	Chloramphenicol
Streptomyces nodosus	Amphotericin B
Streptomyces avermitilis	Ivermectin
Streptomyces noursei	Nystatin

[a]This genus was formerly called *Cephalosporium*.
[b]spp. is the abbreviation for multiple species of a genus.

As Ehrlich foresaw, the key to successful chemotherapy against microbes is **selective toxicity**; that is, an effective antimicrobial agent must be more toxic to a pathogen than to the pathogen's host. Selective toxicity is possible because of differences in structure or metabolism between the pathogen and its host. Typically, the more differences, the easier it is to discover or create an effective antimicrobial agent.

Because there are many differences between the structure and metabolism of pathogenic bacteria and their eukaryotic hosts, antibacterial drugs constitute the greatest number and diversity of antimicrobial agents. Fewer antifungal, antiprotozoan, and anthelmintic drugs are available because fungi, protozoa, and helminths—like their animal and human hosts—are eukaryotic chemoheterotrophs (see Chapter 6) and thus share many common features. The number of effective antiviral drugs is also limited, despite major differences in structure, because viruses utilize their host cells' enzymes and ribosomes to metabolize and replicate. Therefore, drugs that are effective against viral replication are likely toxic to the host as well.

Although they can have a variety of effects on pathogens, antimicrobial drugs can be categorized into several general groups according to their mechanisms of action (FIGURE 10.2):

- Drugs that inhibit cell wall synthesis. These drugs are selectively toxic to certain fungal or bacterial cells, which have cell walls, but not to animals, which lack cell walls.
- Drugs that inhibit protein synthesis (translation) by targeting the differences between prokaryotic and eukaryotic ribosomes.
- Drugs that disrupt unique components of a cytoplasmic membrane.
- Drugs that inhibit general metabolic pathways not used by humans.
- Drugs that inhibit nucleic acid synthesis.
- Drugs that block a pathogen's recognition of or attachment to its host.

In the following sections, we examine these mechanisms in turn.

Play *Chemotherapeutic Agents: Modes of Action*
@ Mastering Microbiology

MICRO CHECK

1. Which type of antimicrobial drug is produced entirely in a lab?
2. What modern term best describes the property of an antimicrobial compound that lets it target bacteria while having little effect on humans?

Inhibition of Cell Wall Synthesis

LEARNING OUTCOME

10.5 Describe the actions and give examples of drugs that affect the cell walls of bacteria or fungi.

A cell wall protects a cell from the effects of osmotic pressure. Both pathogenic bacteria and fungi have cell walls, which animals and humans lack. First, we examine drugs that act against bacterial cell walls.

Inhibition of Synthesis of Bacterial Walls

The major structural component of a bacterial cell wall is its peptidoglycan layer. Peptidoglycan is a huge macromolecule composed of polysaccharide chains of alternating *N*-acetylglucosamine (NAG) and *N*-acetylmuramic acid (NAM) molecules that are cross-linked by short peptide chains extending between NAM subunits (see Chapter 3, Figure 3.15). To enlarge or divide, a cell synthesizes more peptidoglycan by adding new NAG and NAM subunits to existing NAG-NAM chains, and the new NAM subunits must then be bonded to other new, nearby NAM subunits (FIGURE 10.3a,b).

Many common antibacterial agents act by preventing the cross-linkage of NAM subunits. Most prominent among these drugs are **beta-lactams** (such as penicillins, cephalosporins, and carbapenems), which are antimicrobials whose functional portions are called *beta-lactam* (β-*lactam*) rings (FIGURE 10.3c). Chemist Dorothy Hodgkin (1910–1994) developed protein X-ray crystallography and used her technique to elucidate the structure of penicillin's beta-lactam ring, which inhibits peptidoglycan formation by irreversibly binding to

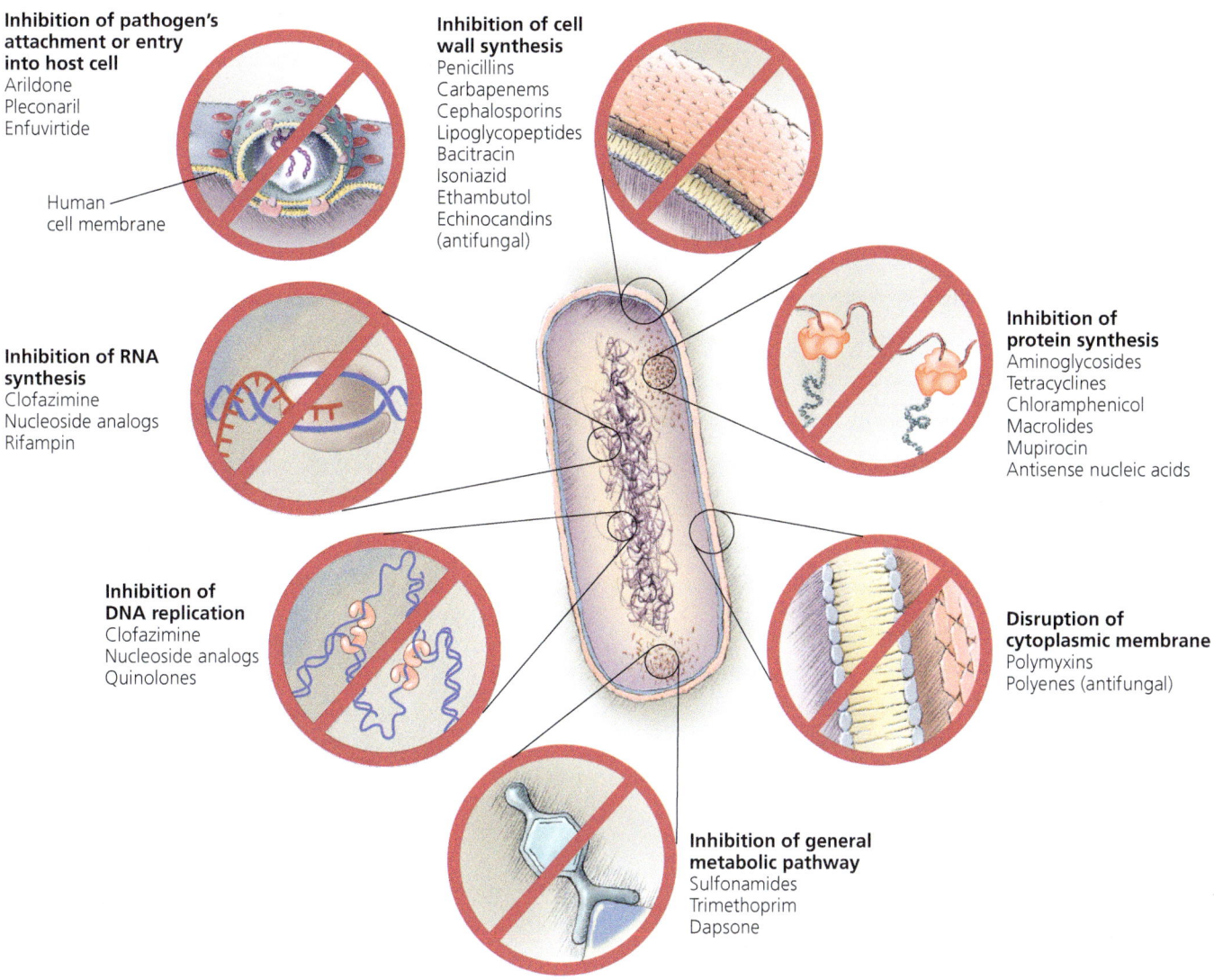

▲ **FIGURE 10.2 Mechanisms of action of microbial drugs.** Also listed are representative drugs for each type of action.

the enzymes that cross-link NAM subunits (**FIGURE 10.3d**). In the absence of correctly formed peptidoglycan, growing bacterial cells have weakened cell walls that are less resistant to the effects of osmotic pressure. The underlying cytoplasmic membrane bulges through the weakened portions of the cell wall as water moves into the cell, and eventually the cell lyses (**FIGURE 10.3e**).

Chemists have made alterations to natural beta-lactams, such as penicillin G, to create semisynthetic derivatives such as methicillin and imipenem (see Figure 10.3c), which are more stable in the acidic environment of the stomach, more readily absorbed in the intestinal tract, less susceptible to deactivation by bacterial enzymes, or more active against more types of bacteria.

Other antimicrobials, such as *lipoglycopeptides*—for example, *vancomycin* (van-kō-mī'sin)—and *cycloserine*, a semisynthetic, disrupt cell wall formation in a different manner. They directly interfere with particular alanine-alanine bridges that link the NAM subunits in many Gram-positive bacteria. Those bacteria that lack this type of alanine-alanine crossbridge are naturally resistant to these drugs. Still another drug that prevents cell wall formation, **bacitracin** (bas-i-trā'sin), blocks the transport of NAG and NAM across the cytoplasmic membrane to the wall. Like beta-lactams, lipoglycopeptides, cycloserine, and bacitracin result in cell lysis due to the effects of osmotic pressure.

Since all these drugs prevent bacteria from *increasing* the amount of cell wall material but have no effect on existing peptidoglycan, they are effective only on bacterial cells that are growing or reproducing; dormant cells are unaffected.

Bacteria of the genus *Mycobacterium* (mī'kō-bak-tēr'ē-ŭm), notably the agents of leprosy and tuberculosis, are characterized by unique, complex cell walls that contain waxy lipids called *mycolic acids* in addition to the usual peptidoglycan of bacterial cells. **Isoniazid** (ī-sō-nī'ă-zid), or **INH**,[1] and **ethambutol**

---

[1]From *isonicotinic acid hydrazide*, the correct chemical name for isoniazid.

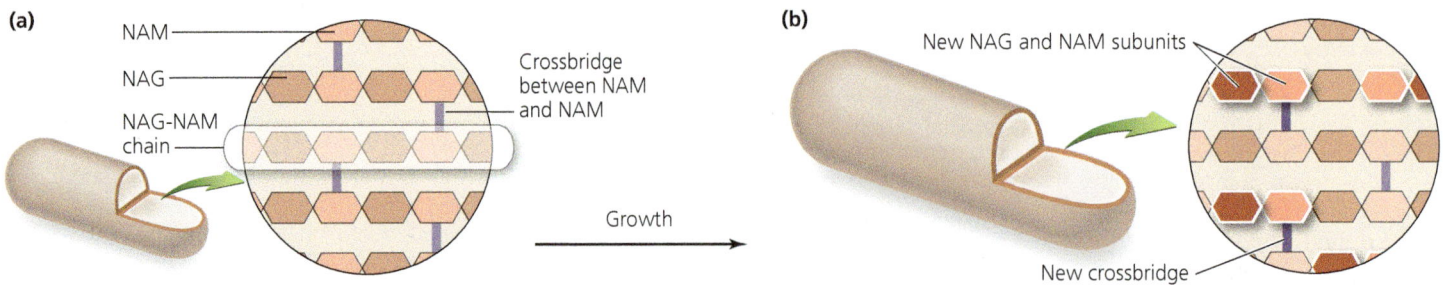

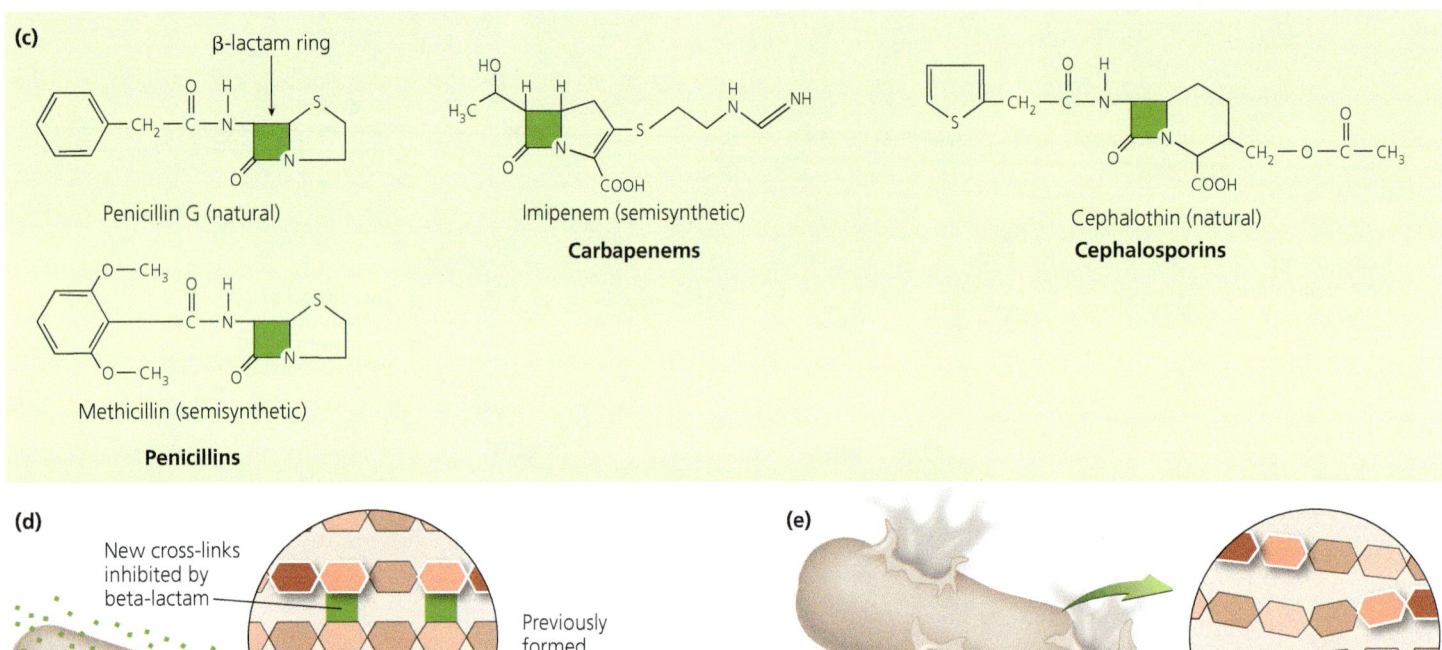

▲ FIGURE 10.3 **Bacterial cell wall synthesis and the inhibitory effects of beta-lactams on it.** (a) A schematic depiction of a normal peptidoglycan cell wall showing NAG-NAM chains and cross-linked NAM subunits. (b) A bacterium grows by adding new NAG and NAM subunits and linking new NAM subunits to older ones. (c) Structural formulas of some beta-lactam drugs. Their functional portion is the beta-lactam ring. (d) A schematic depiction of the effect of penicillin on peptidoglycan in preventing NAM-NAM cross-links. (e) Bacterial lysis due to the effects of a beta-lactam drug. *After a beta-lactam drug weakens a peptidoglycan molecule by preventing NAM-NAM cross-linkages, what force actually kills an affected bacterial cell?*

Figure 10.3 *Osmotic movement of water into the cell causes cells to lyse.*

(eth-am'boo-tol) disrupt the formation of mycolic acid. Mycobacteria typically reproduce only every 12 to 24 hours, in part because of the complexity of their cell walls, so antimicrobial agents that act against mycobacteria must be administered for months or even years to be effective. It is often difficult to ensure that patients continue such a long regimen of treatment.

Inhibition of Synthesis of Fungal Walls

Fungal cell walls are composed of various polysaccharides containing a sugar—1,3-D-glucan—that is not found in mammalian cells. A new class of antifungal drugs called **echinocandins**, among them *caspofungin*, inhibit the enzyme that synthesizes glucan; without glucan, fungal cells cannot make cell walls, leading to osmotic rupture.

Inhibition of Protein Synthesis

LEARNING OUTCOMES

10.6 Describe the actions and give examples of six antimicrobial drugs that interfere with ribosomes.

10.7 Describe how mupirocin interferes with protein synthesis.

Cells use proteins for structure and regulation, as enzymes in metabolism, and as channels and pumps to move materials across cell membranes. Thus, a consistent supply of proteins is vital for the active life of a cell. Given that all cells, including human cells, use ribosomes to translate proteins using information from messenger RNA templates, it is not immediately obvious that drugs could selectively target differences related to protein synthesis. However, there are differences between prokaryotic and eukaryotic ribosomes as well as differences in the enzymes that "load" amino acids onto tRNA molecules.

Interference with Prokaryotic Ribosomes

Prokaryotic ribosomes differ from eukaryotic ribosomes in structure and size: Prokaryotic ribosomes are 70S and composed of 30S and 50S subunits, whereas eukaryotic ribosomes are 80S with 60S and 40S subunits (see Figure 7.15).

Many antimicrobial agents take advantage of the differences between ribosomes to selectively target bacterial protein translation without significantly affecting eukaryotes. Note, however, that because some of these drugs affect eukaryotic mitochondria, which also contain 70S ribosomes like those of prokaryotes, such drugs may be harmful to animals and humans, especially very active cells in the liver and bone marrow.

Understanding the actions of antimicrobials that inhibit protein synthesis requires an understanding of the process of translation because various parts of ribosomes are the targets of antimicrobial drugs. Both the 30S and 50S subunits of a prokaryotic ribosome play a role in the initiation of protein synthesis, in codon recognition, and in the docking of tRNA–amino acid complexes. Additionally, the 50S subunit contains the enzymatic portion that actually forms peptide bonds (see Figure 7.18).

Among the antimicrobials that target the 30S ribosomal subunit are aminoglycosides and tetracyclines. **Aminoglycosides** (am'i-nō-glī'kō-sīds), such as *streptomycin* and *gentamicin*, change the shape of the 30S subunit, making it impossible for the ribosome to read the codons of mRNA correctly (FIGURE 10.4a). Other aminoglycosides and **tetracyclines** (tet-ră-sī'klēns) block the tRNA docking site (A site), which then prevents the incorporation of additional amino acids into a growing polypeptide (FIGURE 10.4b).

Some antimicrobials interfere with the function of the 50S subunit. **Chloramphenicol** and similar drugs block the enzymatic site of the 50S subunit (FIGURE 10.4c), which prevents translation. **Lincosamides**, **streptogramins**, and **macrolides** (mak'rō-līds, such as *erythromycin*) bind to a different portion of the 50S subunit, preventing movement of the ribosome from one codon to the next (FIGURE 10.4d); as a result, translation is frozen, and protein synthesis is halted.

Other drugs that block protein synthesis are **antisense nucleic acids** (FIGURE 10.4e). These RNA or single-stranded DNA molecules are designed to be complementary to specific mRNA molecules of pathogens. They block ribosomal subunits from attaching to that mRNA with no effect on human mRNA. *Fomivirsen* is the first of this class of drugs to be approved. It inactivates cytomegalovirus and is used to treat eye infections.

Oxazolidinones (oks-ă'zō-lĭ-dĭ-nōnz) are antimicrobial drugs that work to stop protein synthesis by blocking initiation of translation (FIGURE 10.4f). Oxazolidinones are used as a last resort in treating infections of Gram-positive bacteria resistant to other antimicrobials, including vancomycin- and methicillin-resistant *Staphylococcus aureus* (staf'i-lō-kok'ŭs o'rē-ŭs).

Interference with Charging of Transfer RNA Molecules

At the base of each transfer RNA molecule there is an anticodon composed of three nucleotides complementary to the nucleotides of a codon that codes for a specific amino acid (see Figure 7.14). Enzymes called *aminoacyl-tRNA synthetases* "charge" or "load" amino acids onto tRNA molecules. There is one type of aminoacyl-tRNA synthetase for each amino acid. Specificity of the enzymes ensures that the amino acid loaded at the top of a tRNA molecule (on the acceptor stem) is appropriate and correct for the anticodon at the base of each particular tRNA.

A topical drug, **mupirocin**, selectively binds to isoleucyl-tRNA synthetase, the enzyme that loads isoleucine in Gram-positive bacteria, particularly *Staphylococcus* and *Streptococcus* (strep-tō-kok'ŭs). The drug is selectively toxic because it does not bind to any eukaryotic tRNA molecules. In the presence of the drug, Gram-positive cells cannot load isoleucine, so protein synthesis cannot proceed past isoleucine codons; thus, mupirocin cripples a bacterium's protein production. The World Health Organization (WHO) lists mupirocin as an essential medicine needed in a basic health system in order to treat bacterial skin infections.

Disruption of Cytoplasmic Membranes

LEARNING OUTCOME

10.8 Describe the action of antimicrobial drugs that interfere with cytoplasmic membranes.

Some antibacterial drugs, such as the short polypeptide *gramicidin*, disrupt the cytoplasmic membrane of a targeted cell, often by forming a channel through the membrane, damaging its integrity. This is also the mechanism of action of a group of antifungal and antiprotozoan drugs called **polyenes** (pol-ē-ēns'). The polyenes *nystatin*[2] and *amphotericin B* (FIGURE 10.5a on the next page) are fungicidal because they attach to *ergosterol*, a lipid constituent of fungal membranes (FIGURE 10.5b), in the process disrupting the membrane and causing lysis of the cell. The cytoplasmic membranes of humans are somewhat susceptible to amphotericin B because they contain cholesterol, which is similar to ergosterol, though cholesterol does not bind amphotericin B as well as does ergosterol.

---

[2]From New York State Department of Health, where Rachel Brown and Elizabeth Hazen discovered the drug.

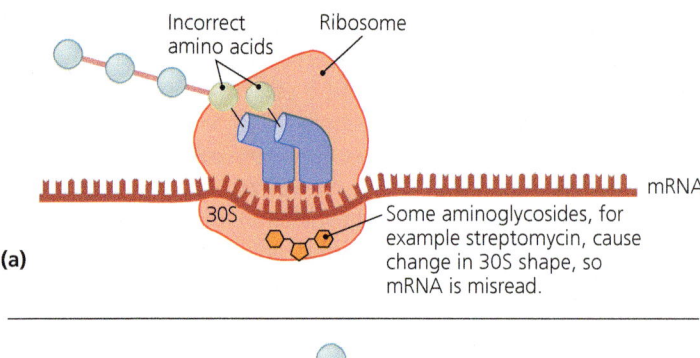
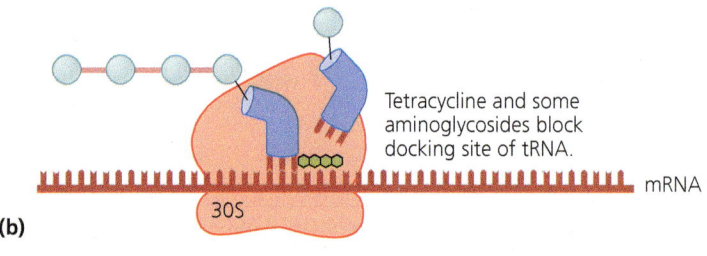
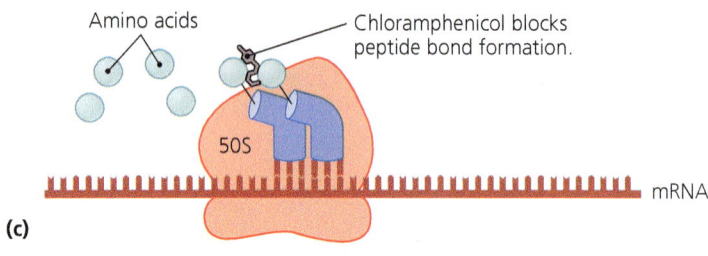
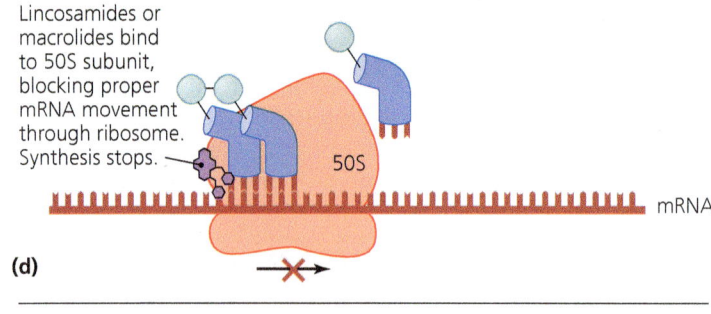
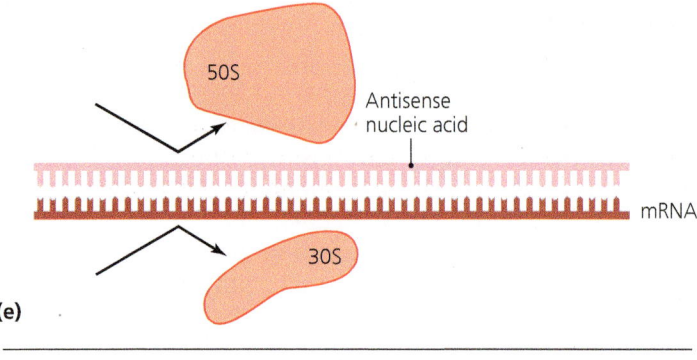
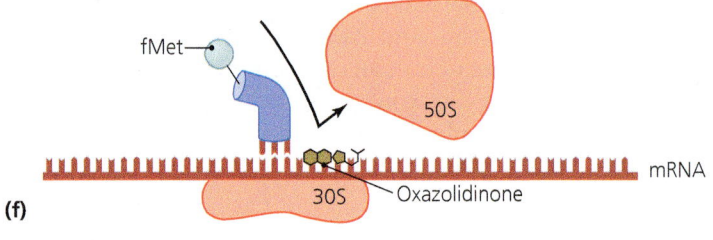

◀ **FIGURE 10.4 Some mechanisms by which antimicrobials target prokaryotic ribosomes to inhibit protein synthesis.**
(a) Aminoglycosides change the shape of the 30S subunit, causing incorrect pairing of tRNA anticodons with mRNA codons.
(b) Tetracyclines block the tRNA docking site (A site) on the 30S subunit, preventing protein elongation. **(c)** Chloramphenicol blocks enzymatic activity of the 50S subunit, preventing the formation of peptide bonds between amino acids. **(d)** Lincosamides or macrolides bind to the 50S subunit, preventing movement of the ribosome along the mRNA.
(e) Antisense nucleic acids bind to mRNA, blocking ribosomal subunits.
(f) Oxazolidinones inhibit initiation of translation. *Which tRNA anticodon should align with the codon CUG?*

Figure 10.4 Anticodon GAC is the correct complement for the codon CUG.

Azoles, such as *fluconazole*, and **allylamines**, such as *terbinafine*, are two other classes of antifungal and antiprotozoan drugs that disrupt cytoplasmic membranes. They act by inhibiting the synthesis of ergosterol; without ergosterol, a fungal cell's cytoplasmic membrane does not remain intact, and the cell dies. Azoles and allylamines are generally harmless to humans because human cells do not manufacture ergosterol.

Most bacterial membranes lack sterols, so these bacteria are naturally resistant to polyenes, azoles, and allylamines; however, there are other agents that disrupt bacterial membranes. An example of these antibacterial agents is *polymyxin*, produced by *Bacillus polymyxa* (ba-sil'ŭs po-lē-miks'ă). Polymyxin is effective against Gram-negative bacteria, particularly *Pseudomonas* (soo-dō-mō'nas), but because it is toxic to human kidneys, it is usually reserved for use against external pathogens that are resistant to other antibacterial drugs.

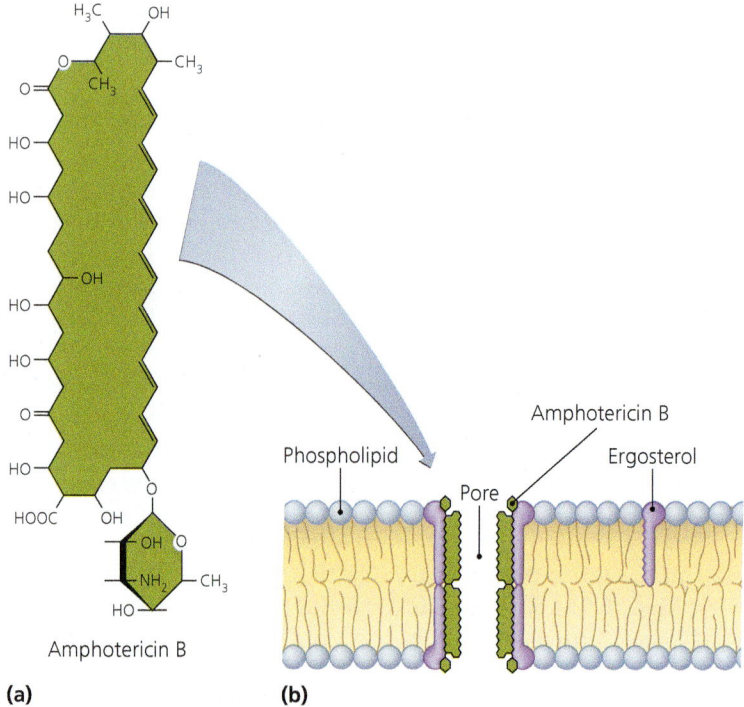

▲ **FIGURE 10.5 Disruption of the cytoplasmic membrane by the antifungal amphotericin B. (a)** The structure of amphotericin B. **(b)** The proposed action of amphotericin B. The drug binds to molecules of ergosterol, which then congregate, forming a pore.

Pyrazinamide disrupts transport across the cytoplasmic membrane of *Mycobacterium tuberculosis* (too-ber-kyū-lō′sis). The pathogen uniquely activates and accumulates the drug. Unlike many other antimicrobials, pyrazinamide is most effective against intracellular, nonreplicating bacterial cells.

Some antiparasitic drugs also act against cytoplasmic membranes. For example, *praziquantel* and *ivermectin* change the permeability of cell membranes of several types of parasitic worms.

MICRO CHECK

3. What major cellular feature that might be targeted by antimicrobial drugs is found in both bacteria and fungi but not in human cells?
4. What effect does penicillin have on susceptible bacterial cells?
5. The ribosomes of bacteria differ significantly from our own, making them a prime target for antimicrobial drugs. Which commonly prescribed antimicrobial drug for eye infections blocks the formation of peptide bonds in prokaryotes?
6. How does mupirocin interfere with protein synthesis in Gram-positive bacteria such as *Staphylococcus* and *Streptococcus*?
7. Which group of drugs includes nystatin and amphotericin B and cause cell lysis by forming channels in the cytoplasmic membranes of fungal cells?

Inhibition of Metabolic Pathways

LEARNING OUTCOMES

10.9 Explain the action of antimicrobials that disrupt synthesis of folic acid.
10.10 Define the term *analog* as it relates to antimicrobial drugs.
10.11 Describe the action of antiviral drugs that interfere with metabolism.

Metabolism can be defined simply as the sum of all chemical reactions that take place within an organism. Whereas most living things share certain metabolic reactions—for example, glycolysis—other chemical reactions are unique to certain organisms. Whenever differences exist between the metabolic processes of a pathogen and its host, *antimetabolic agents* can be effective.

Various kinds of antimetabolic agents are available, including:

- *Atovaquone*, which interferes with electron transport in protozoa and fungi
- Heavy metals (such as arsenic, mercury, and antimony), which inactivate enzymes
- Agents that disrupt tubulin polymerization and glucose uptake by many protozoa and parasitic worms
- Drugs that block the activation of viruses
- Metabolic antagonists, such as sulfanilamide, the first commercially available antimicrobial agent

Sulfanilamide and similar compounds, collectively called **sulfonamides** or *sulfa drugs*, act as antimetabolic drugs because they are structural **analogs** of—that is, are chemically very similar to—*para-aminobenzoic acid (PABA)* (FIGURE 10.6a,b). PABA is crucial in the synthesis of nucleotides required for DNA and RNA synthesis. Many organisms, including some pathogens, enzymatically convert PABA into dihydrofolic acid and then dihydrofolic acid into tetrahydrofolic acid (THF), a form of folic acid that is used as a coenzyme in the synthesis of purine and pyrimidine nucleotides (FIGURE 10.6c). As analogs of PABA, sulfonamides compete with PABA molecules for the active site of the enzyme involved in the production of dihydrofolic acid (FIGURE 10.6d). This competition leads to a decrease in the production of THF and thus of DNA and RNA. The end result of sulfonamide competition with PABA is the cessation of cell metabolism, which leads to cell death.

Note that humans do not synthesize THF from PABA; instead, we take simple folic acids found in our diets and convert them into THF. As a result, human metabolism is unaffected by sulfonamides.

Another antimetabolic agent, *trimethoprim*, also interferes with nucleic acid synthesis. However, instead of binding to the enzyme that converts PABA to dihydrofolic acid, trimethoprim binds to the enzyme involved in the conversion of dihydrofolic acid to THF, the second step in this metabolic pathway.

Some antiviral agents target the unique aspects of the metabolism of viruses. After attachment to a host cell, viruses must penetrate the cell's membrane and be uncoated to release viral genetic instructions and assume control of the cell's metabolic machinery. Some viruses of eukaryotes are uncoated as a result of the acidic environment within phagolysosomes (vesicles derived after a lysosome fuses with a phagosome in phagocytosis). *Amantadine*, *rimantadine*, and weak organic bases can neutralize the acid of phagolysosomes and thereby prevent viral uncoating; thus, these are antiviral drugs. The Centers for Disease Control and Prevention (CDC) showed in 2008 that amantadine is no longer effective against influenza type A virus, so physicians currently do not prescribe this drug for use against type A influenzaviruses.

Protease inhibitors interfere with the action of protease—an enzyme that HIV needs near the end of its replication cycle. These drugs, when used as part of a "cocktail" of drugs including reverse transcriptase inhibitors (discussed shortly), have revolutionized treatment of AIDS patients in industrialized countries. Researchers have reduced the number of pills in the daily cocktail to just a few that contain all the drugs formerly found in 35 or more daily pills.

Inhibition of Nucleic Acid Synthesis

LEARNING OUTCOME

10.12 Describe the antimicrobial action of nucleotide and nucleoside analogs, quinolones, drugs that bind to RNA or DNA, and reverse transcriptase inhibitors.

The nucleic acids DNA and RNA are built from purine and pyrimidine nucleotides and are critical to the survival of cells. Several drugs function by blocking either the replication of DNA or its transcription into RNA.

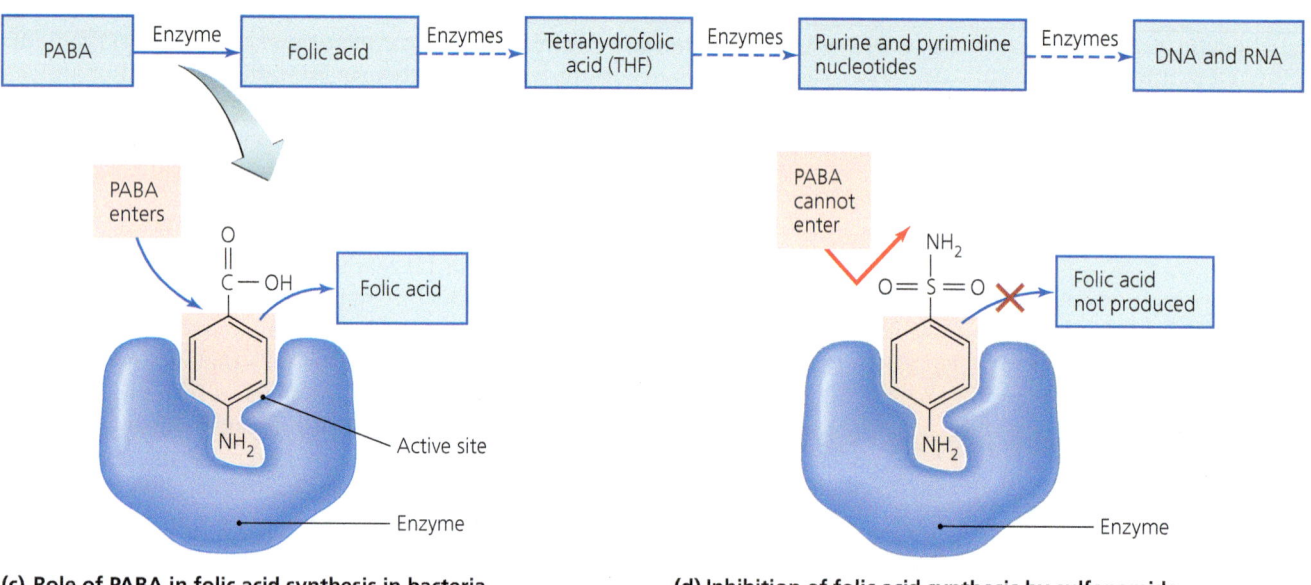

▲ FIGURE 10.6 **The antimetabolic action of sulfonamides in inhibiting nucleic acid synthesis.** (a) Para-aminobenzoic acid (PABA) and representative members of its structural analogs, the sulfonamides. The analogous portions of the compounds are shaded. (b) The metabolic pathway in some bacteria and protozoa by which DNA and RNA are synthesized from PABA. (c) The inhibition of folic acid synthesis by the presence of a sulfonamide, which deactivates the enzyme by binding irreversibly to the enzyme's active site.

Because only slight differences exist between the DNA of prokaryotes and eukaryotes, drugs that affect DNA replication often act against both types of cells. For example, *actinomycin* binds to DNA and effectively blocks DNA synthesis and RNA transcription not only in bacterial pathogens but in their hosts as well. Generally, drugs of this kind are not used to treat infections, though they are used in research of DNA replication and may be used carefully to slow replication of cancer cells.

One exception is the synthetic drugs called **quinolones**, commonly *fluoroquinolones*, which contain fluorine. Quinolones inhibit *DNA gyrase*, the bacterial enzyme necessary for correct coiling and uncoiling of replicating bacterial DNA; they typically have little effect on eukaryotes or viruses, though they may act against replication of mitochondrial DNA in some eukaryotes. Recall that mitochondrial DNA is similar to bacterial DNA.

Other compounds that can act as antimicrobials by interfering with the function of nucleic acids are **nucleotide**[3] **analogs** or **nucleoside analogs**, which are molecules with structural similarities to the normal nucleotide building blocks of nucleic acids (FIGURE 10.7). Gertrude Elion (1918–1999) received the Nobel Prize in Physiology or Medicine in 1988 in part for her discovery of the first effective antiviral agent, the nucleoside analog *acyclovir*, and the first drug against HIV, the nucleoside analog *azidothymidine*

---

[3]Nucleotides are composed of a pentose sugar, phosphate, and a nitrogenous base; nucleosides lack the phosphate.

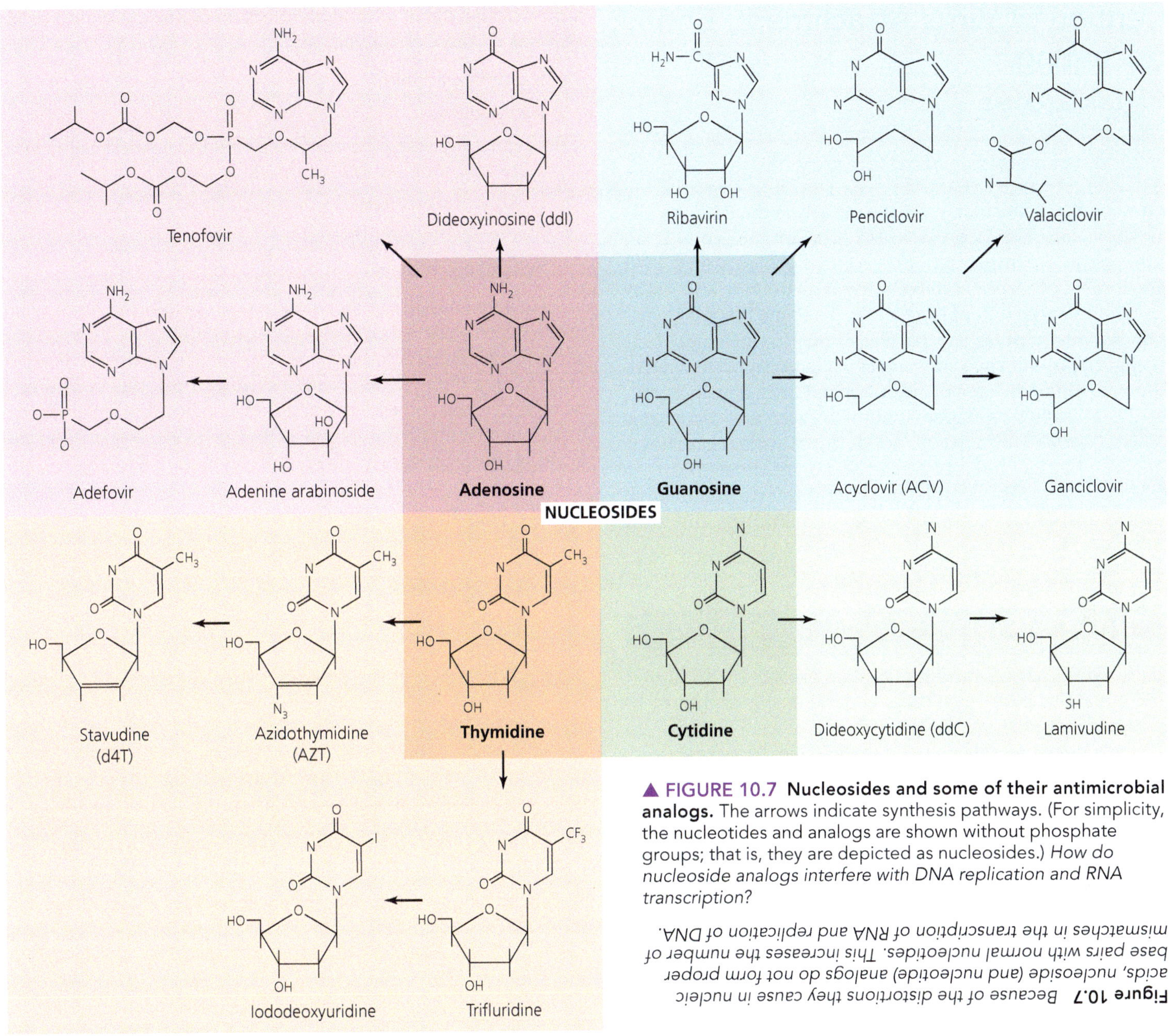

▲ **FIGURE 10.7 Nucleosides and some of their antimicrobial analogs.** The arrows indicate synthesis pathways. (For simplicity, the nucleotides and analogs are shown without phosphate groups; that is, they are depicted as nucleosides.) *How do nucleoside analogs interfere with DNA replication and RNA transcription?*

Figure 10.7 Because of the distortions they cause in nucleic acids, nucleoside (and nucleotide) analogs do not form proper base pairs with normal nucleotides. This increases the number of mismatches in the transcription of RNA and replication of DNA.

(AZT). Nucleotide analogs are often incorporated into the DNA or RNA of pathogens, where they distort the shapes of the nucleic acid molecules and prevent further replication, transcription, or translation. Nucleotide and nucleoside analogs are most often used against viruses because viral DNA polymerases are tens to hundreds of times more likely to incorporate these nonfunctional nucleotides into nucleic acids than is human DNA polymerase. Additionally, complete viral nucleic acid synthesis is more rapid than cellular nucleic acid synthesis. These characteristics make viruses more susceptible to nucleotide analogs than their hosts are, though nucleotide analogs can also be effective against rapidly dividing cancer cells.

Other antimicrobial agents function by binding to and inhibiting the action of RNA polymerases during the synthesis of RNA from a DNA template. Several drugs, including *rifampin* (rif′am-pin), bind more readily to prokaryotic RNA polymerase than to eukaryotic RNA polymerase; as a result, rifampin is more toxic to prokaryotes than to eukaryotes. *Clofazimine* binds to bacterial DNA and evidently prevents normal transcription (and DNA replication). Rifampin and clofazimine are used primarily against species of mycobacteria such as *M. tuberculosis* and *M. leprae* (lep′rī), the causative agents of tuberculosis and leprosy, respectively.

Pentamidine and *propamidine isethionate* bind to protozoan DNA, inhibiting DNA replication and RNA transcription. The result is that the protozoa cannot reproduce or metabolize.

Reverse transcriptase inhibitors, such as emtricitabine—a part of AIDS cocktails—act against reverse transcriptase, which is an enzyme HIV uses early in its replication cycle to make DNA copies of its RNA genome (reverse transcription). Since people lack reverse transcriptase, the inhibitor does not harm patients.

Prevention of Virus Attachment, Entry, or Uncoating

LEARNING OUTCOME

10.13 Describe the action of antimicrobial attachment antagonists and of drugs that inhibit viral entry.

Many pathogens, particularly viruses, attach to their host's cells via the chemical interaction between attachment proteins on the pathogen and complementary receptor proteins on a host cell. Attachment of viruses can be blocked by peptide and sugar analogs of either attachment or receptor proteins. When analogs block these sites, viruses can neither attach to nor enter their host's cells. The use of such substances, called *attachment antagonists*, is an exciting new area of antimicrobial drug development. *Pleconaril* is a synthetic antagonist of the receptors of picornaviruses such as some cold viruses, polioviruses, and coxsackievirus. The drug blocks viral attachment and deters infections.

Arildone is a synthetic antiviral that prevents removal of poliovirus capsids (uncoating), thereby interrupting the viral replication cycle.

MICRO CHECK

8. Sulfonamide drugs inhibit microbial growth by disrupting metabolic pathways. What general mechanism do these drugs use?
9. What is the general term used to describe drugs that are chemically similar enough to a metabolite (such as PABA or a nucleotide) that they can substitute for that metabolite in a chemical reaction, possibly disrupting the chemical reaction sufficiently to halt the process?
10. How do the antiviral drugs amantadine and rimantadine work?
11. A class of antiviral drugs known as attachment antagonists help deter viral infections. What mechanism do they use?

TELL ME WHY

Some antimicrobial drugs are harmful to humans. Why can physicians safely use such drugs despite the potential danger?

Clinical Considerations in Prescribing Antimicrobial Drugs

LEARNING OUTCOME

10.14 List six characteristics of an ideal, hypothetical, antimicrobial drug.

Even though many fungi and bacteria produce antibiotics, most of these chemicals are not effective for treating diseases because they are toxic to humans and animals, are too expensive, are produced in minute quantities, or lack adequate potency. The ideal antimicrobial agent to treat an infection or disease would be one that has all these characteristics:

- Readily available
- Inexpensive
- Chemically stable (so that it can be transported easily and stored for long periods of time)
- Easily administered
- Nontoxic and nonallergenic
- Selectively toxic against a wide range of pathogens

No agent has all of these qualities, so doctors and medical laboratory technicians must evaluate antimicrobials with respect to several characteristics: the types of pathogens against which they are effective; their effectiveness, including the dosages required to be effective; the routes by which they can be administered; their overall safety; and any side effects they produce. We consider each of these characteristics of antimicrobials in the following sections.

Spectrum of Action

LEARNING OUTCOME

10.15 Distinguish narrow-spectrum drugs from broad-spectrum drugs in terms of their targets and side effects.

The number of different kinds of pathogens a drug acts against is known as its **spectrum of action** (FIGURE 10.8); drugs that work against only a few kinds of pathogens are **narrow-spectrum drugs**, whereas those that are effective against many different kinds of pathogens are **broad-spectrum drugs**. For instance, because tetracycline acts against many different kinds of bacteria, including Gram-negatives, Gram-positives, chlamydias, and rickettsias, it is considered a broad-spectrum antibiotic. In contrast, penicillin cannot easily penetrate the outer membrane of a Gram-negative bacterium to reach and prevent the formation of peptidoglycan, so its efficacy is limited largely to Gram-positive bacteria. Thus, penicillin has a narrower spectrum of action than tetracycline.

The use of broad-spectrum antimicrobials is not always as desirable as it might seem. Broad-spectrum antimicrobials can also open the door to serious secondary infections by transient pathogens or *superinfections* by members of the normal microbiota unaffected by the antimicrobial. This results because the killing of normal microbiota reduces *microbial antagonism*, the competition between normal microbes and pathogens for nutrients and space. Microbial antagonism reinforces the body's defense by limiting the ability of pathogens to colonize the skin and mucous membranes. Thus, a woman using erythromycin to treat strep throat (a bacterial disease) could develop vaginitis resulting from the excessive growth of *Candida albicans* (kan'did-ă al'bi-kanz), a yeast that is unaffected by erythromycin and is freed from microbial antagonism when the antibiotic kills normal bacteria in the vagina.

Effectiveness

LEARNING OUTCOME

10.16 Compare and contrast diffusion susceptibility tests, Etest, MIC test, and MBC test.

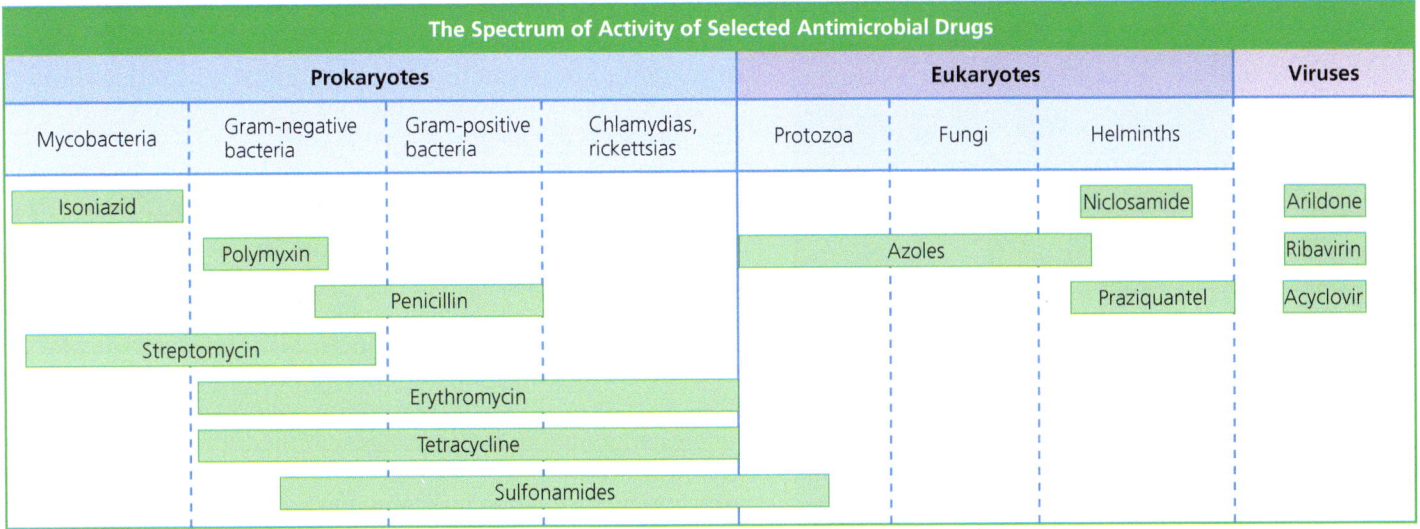

▲ FIGURE 10.8 **Spectrum of action for selected antimicrobial agents.** The more kinds of pathogens a drug affects, the broader its spectrum of action.

To effectively treat infectious diseases, physicians must know which antimicrobial agent is most effective against a particular pathogen. To ascertain the efficacy of antimicrobials, microbiologists conduct a variety of tests, including *diffusion susceptibility tests*, the *minimum inhibitory concentration test*, and the *minimum bactericidal concentration test*.

Diffusion Susceptibility Test

Diffusion susceptibility tests, also known as *Kirby-Bauer tests*, involve uniformly inoculating a Petri plate with a standardized amount of the pathogen in question. Then small disks of paper containing standard concentrations of the drugs to be tested are firmly arranged on the surface of the plate. The plate is incubated, and the bacteria grow and reproduce to form a "lawn" everywhere except the areas where effective antimicrobial drugs diffuse through the agar. After incubation, the plates are examined for the presence of a **zone of inhibition**—that is, a clear area where bacteria do not grow (FIGURE 10.9). A zone of inhibition is measured as the diameter (to the closest millimeter) of the clear region.

If all antimicrobials diffused at the same rate, then a larger zone of inhibition would indicate a more effective drug. However, drugs with low molecular weights generally diffuse more quickly and so might have a larger zone of inhibition than a more effective but larger molecule. The size of a zone of inhibition must be compared to a standard table for that particular drug before accurate comparisons can be made. Diffusion susceptibility tests enable scientists to classify pathogens as *susceptible*, *intermediate*, or *resistant* to each drug.

Minimum Inhibitory Concentration (MIC) Test

Once scientists identify an effective antimicrobial agent, they quantitatively express its potency as a **minimum inhibitory concentration (MIC)**, often using the unit μg/ml. As the name suggests, the MIC is the smallest amount of the drug that will

CLINICAL CASE STUDY

Antibiotic Overkill

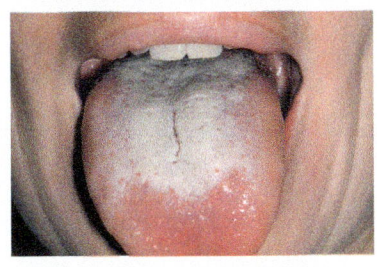

A young woman was taking antibiotic pills for a urinary infection. A few days after finishing her course of medication, she began to experience peculiar symptoms. At first, they were hardly noticeable. Very quickly, however, they worsened and became embarrassing and unbearable.

She noticed a white coating on her tongue, bad breath, and an awful taste in her mouth. Despite persistent brushing and mouthwash applications, she was unable to completely remove the film. Furthermore, she had excessive vaginal discharges consisting of a cheeselike white substance. When she began to have vaginal itching and painful, burning urination, she finally decided it was time to seek help.

Reluctantly, she revisited her personal physician and described the symptoms. Her doctor explained the symptoms and provided additional prescriptions to alleviate her distress.

1. What happened to the young woman in this situation?
2. How had her body's defenses been violated?
3. How can she avoid a repeat of this situation?

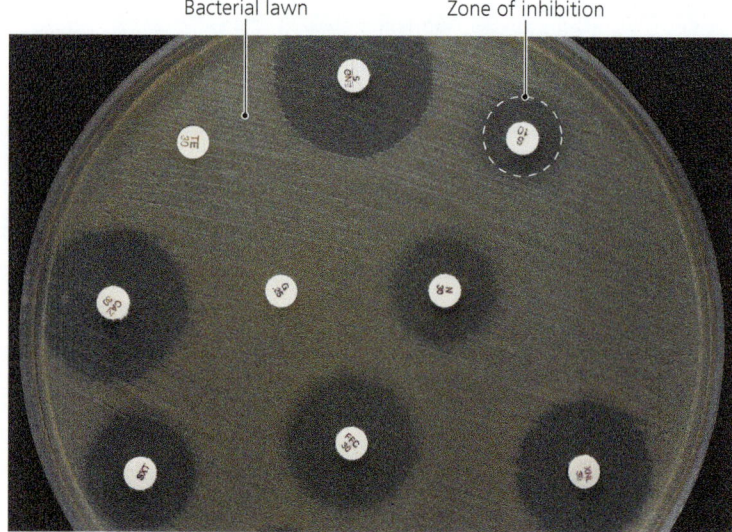

▲ FIGURE 10.9 **Zones of inhibition in a diffusion susceptibility (Kirby-Bauer) test.** Disks are impregnated with an antimicrobial agent. In general, the larger the zone of inhibition around the disks, the more effective that antimicrobial is against the organism growing on the plate. The organism is classified as either susceptible, intermediate, or resistant to the antimicrobials tested based on the sizes of the zones of inhibition. *If all of these antimicrobial agents diffuse at the same rate and are equally safe and easily administered, which one would be the drug of choice for killing this pathogen?*

Figure 10.9 *The drug in the uppermost disk.*

inhibit growth and reproduction of the pathogen. The MIC can be determined via a **broth dilution test**, in which a standardized amount of bacteria is added to serial dilutions of antimicrobial agents in tubes or wells containing broth. After incubation, turbidity (cloudiness) indicates bacterial growth; lack of turbidity indicates that the bacteria were either inhibited or killed by the antimicrobial agent (FIGURE 10.10). Dilution tests can be conducted simultaneously in wells, and the entire process can be automated, with turbidity measured by special scanners connected to computers.

Another test that determines minimum inhibitory concentration combines aspects of an MIC test and a diffusion susceptibility test. This test, called an **Etest**,[4] involves placing a plastic strip containing a gradient of the antimicrobial agent being tested on a plate uniformly inoculated with the organism of interest (FIGURE 10.11). After incubation, an elliptical zone of inhibition indicates antimicrobial activity, and the minimum inhibitory concentration can be noted where the zone of inhibition intersects a scale printed on the strip.

Minimum Bactericidal Concentration (MBC) Test

Similar to the MIC test is a **minimum bactericidal concentration (MBC) test**, though an MBC test determines the amount of drug required to kill the microbe rather than just the amount to inhibit it, as the MIC does. In an MBC test, samples taken from clear MIC tubes (or, alternatively, from zones of inhibition from a series of diffusion susceptibility tests) are transferred to plates containing a drug-free growth medium (FIGURE 10.12). The appearance of bacterial growth in these subcultures after appropriate incubation indicates that at least some bacterial cells survived that concentration of the antimicrobial drug and were able to grow and multiply once placed in a drug-free medium. Any drug concentration at which growth occurs in subculture is *bacteriostatic*, not *bactericidal*, for that bacterium. The lowest concentration of drug for which no growth occurs in the subcultures is the minimum bactericidal concentration (MBC).

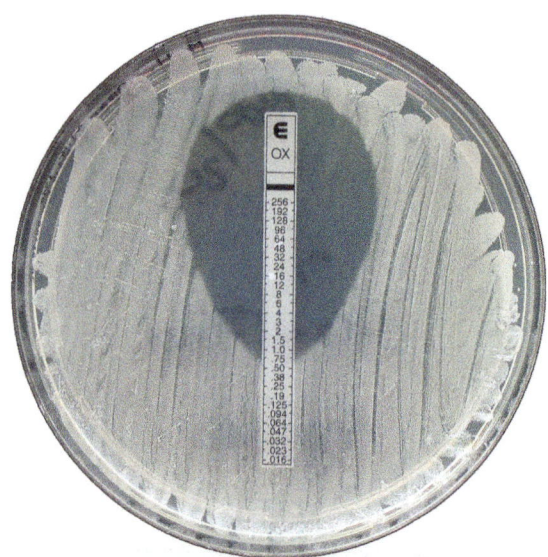

▲ FIGURE 10.11 **An Etest, which combines aspects of Kirby-Bauer and MIC tests.** The plastic strip contains a gradient of the antimicrobial agent of interest. The MIC is estimated to be the concentration printed on the strip where the zone of inhibition intersects the strip. In this example, the MIC is 0.75 μg/ml.

Routes of Administration

LEARNING OUTCOME

10.17 Discuss the advantages and disadvantages of the different routes of administration of antimicrobial drugs.

For an antimicrobial agent to be effective, an adequate amount of it must reach a site of infection. For external infections such as athlete's foot, drugs can be applied directly. This is known as *topical* or *local* administration. For internal infections, drugs can

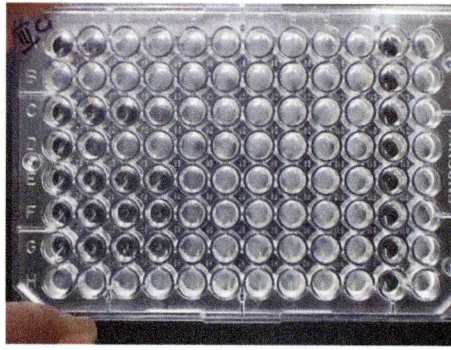

▲ FIGURE 10.10 **Minimum inhibitory concentration (MIC) test in wells.** Wells with bacterial growth are darker.

[4]The name *Etest* has no specific origin.

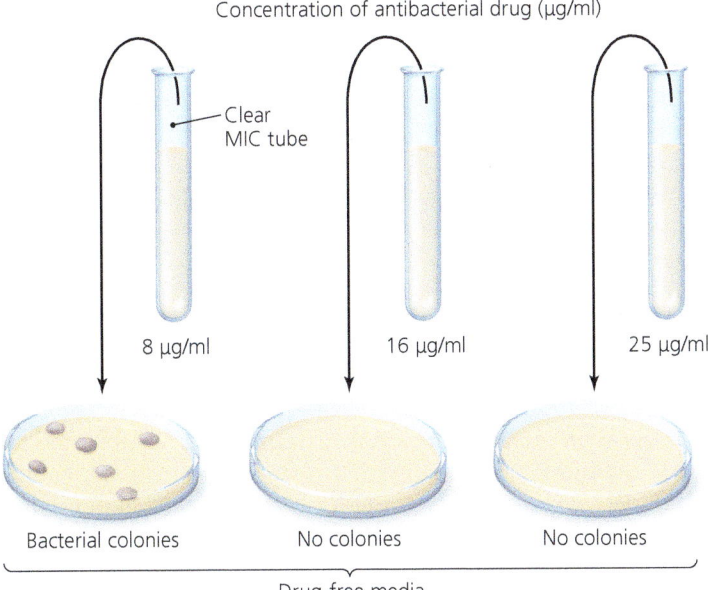

▲ **FIGURE 10.12 A minimum bactericidal concentration (MBC) test.** In this test, plates containing a drug-free growth medium are inoculated with samples taken from zones of inhibition or from clear MIC tubes. After incubation, growth of bacterial colonies on a plate indicates that the concentration of antimicrobial drug (in this case, 8 μg/ml) is bacteriostatic. The lowest concentration for which no bacterial growth occurs on the plate is the minimum bactericidal concentration; in this case, the MBC is 16 μg/ml.

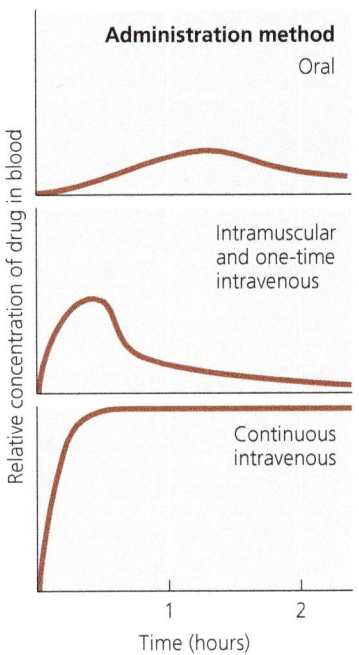

▲ **FIGURE 10.13 The effect of route of administration on blood levels of a chemotherapeutic agent.** Although intravenous (IV) and intramuscular (IM) administration achieve higher drug concentrations in the blood, oral administration has the advantage of simplicity.

be administered *orally*, *intramuscularly (IM)*, or *intravenously (IV)*. Each route has advantages and disadvantages.

Even though the oral route is simplest (it requires no needles and is self-administered), the drug concentrations achieved in the body are lower than occur via other routes of administration (**FIGURE 10.13**). Further, because patients administer the drug themselves, they do not always follow prescribed timetables.

IM administration via a hypodermic needle allows a drug to diffuse slowly into the many blood vessels within muscle tissue, but the concentration of the drug in the blood is never as high as that achieved by IV administration, which delivers the drug directly into the bloodstream through either a needle or a catheter (a plastic or rubber tube). The amount of a drug in the blood is initially very high for the IV route, but unless the drug is continuously administered, the concentration can rapidly diminish as the liver and kidneys remove the drug from circulation. Physicians can administer non-antimicrobial chemicals that prolong an antimicrobial's life span in the body; for example, cilastatin inhibits a kidney enzyme that would destroy imipenem—a type of beta-lactam.

CLINICAL CASE STUDY

Battling an Enemy

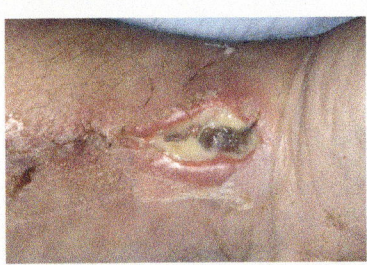

Inflammation and tissue damage resulting from a wound infection

Like his father and grandfather before him, Ben is a Marine sergeant and proud to serve his country. He was nearing the end of his second tour of duty in the Middle East and was looking forward to returning home to his wife and two sons. However, two weeks before he was scheduled to leave, Ben was seriously wounded in a suicide bomber's attack and almost lost his hand. He initially responded well to treatment, but three days into his recovery, his wound was obviously infected and getting worse. His doctors needed to act fast. They prescribed ceftriaxone, which is a broad-spectrum, semisynthetic cephalosporin. The nurse swabbed the wound and sent a sample to the lab for a diffusion susceptibility test.

1. What does the term *semisynthetic* mean in reference to an antimicrobial drug?
2. What is the name of the active part of the ceftriaxone molecule?
3. If the bacterium proves to be resistant to ceftriaxone, resistance is likely due to what enzyme?
4. What is another name for a diffusion susceptibility test?

In addition to considering the route of administration, physicians must consider how the blood distributes the antimicrobial agents to infected tissues. For example, an agent removed rapidly from the blood by the kidneys might be the drug of choice for a bladder infection but would not be chosen to treat an infection of the heart. Finally, given that blood vessels in the brain, spinal cord, and eye are almost impermeable to many antimicrobial agents (because the tight structure of capillary walls in these structures creates the blood-brain barrier), infections there are often difficult to treat.

Safety and Side Effects

LEARNING OUTCOMES

10.18 Identify three main categories of side effects of antimicrobial therapy.
10.19 Define *therapeutic index* and *therapeutic window*.

Another aspect of chemotherapy that physicians must consider is the possibility of adverse side effects. These fall into three main categories—toxicity, allergies, and disruption of normal microbiota.

Toxicity

Though antimicrobial drugs are ideally selectively toxic against microbes and harmless to humans, many antimicrobials in fact have toxic side effects. The exact cause of many adverse reactions is poorly understood, but drugs may be toxic to the kidneys, the liver, or nerves. For example, polymyxin and aminoglycosides can have fatally toxic effects on kidneys. Not all toxic side effects are so serious. *Metronidazole (Flagyl)*, a drug effective against a variety of anaerobic protozoa and bacteria, may cause a harmless temporary condition called "black hairy tongue," which results when the breakdown products of hemoglobin accumulate in the papillae of the tongue (**FIGURE 10.14a**).

Doctors must be especially careful when prescribing drugs for pregnant women because many drugs that are safe for adults can have adverse affects when absorbed by a fetus. For instance, tetracyclines combine with calcium that is incorporated into bones and developing teeth, causing malformation of the skull and stained, weakened tooth enamel (**FIGURE 10.14b**).

Researchers are able to estimate the safety of an antimicrobial drug by calculating the drug's **therapeutic index (TI)**, which is essentially the margin of safety of the drug—a ratio of the dose of a drug that is toxic to 50% of a patient population to the dose of the drug that is therapeutic in 50% of patients (**FIGURE 10.15**). The higher the TI, the safer the drug. Caregivers must be careful when administering drugs with a low TI and must monitor the patient closely for signs of drug toxicity. Clinicians also refer to a drug's **therapeutic window**, which is the range of concentrations of the drug that are effective without being excessively toxic.

Allergies

In addition to toxicity, some drugs trigger allergic immune responses in sensitive patients. Although relatively rare, such reactions may be life threatening, especially an immediate, violent reaction called *anaphylactic shock*. For example, about 0.1% of Americans have an anaphylactic reaction to penicillin, resulting in approximately 300 deaths per year. However, not every allergy to an antimicrobial agent is so serious. Recent studies indicate that patients with mild allergies to penicillin frequently lose their sensitivity to it over time. Thus, an initial mild reaction to penicillin need not preclude its use in treating future infections. (Chapter 18 discusses allergies in more detail.)

Disruption of Normal Microbiota

Drugs that disrupt the normal microbiota and their microbial antagonism may allow opportunistic pathogens to proliferate as secondary infections. In some cases, when a member of the normal microbiota is not affected by a drug, it can become an opportunistic pathogen and can overgrow, causing a disease. For example, long-term use of broad-spectrum antibacterials often results in explosions in the growth rate of *Candida albicans* in the vagina (vaginitis) or mouth (thrush) and the multiplication of *Clostridium difficile* (klos-trid′ē-ŭm di′fe-sēl) in the colon, causing a potentially fatal condition called *pseudomembranous colitis*. Such opportunistic pathogens are of great concern for hospitalized patients, who are often not only debilitated and susceptible to opportunistic infections but are also more likely to be exposed to other pathogens with resistance to antimicrobial drugs—the topic of the next section.

Beneficial Microbes: Probiotics: Using Live Microorganisms to Treat or Prevent Disease (p. 304) focuses on how normal microbiota may help keep pathogens in check.

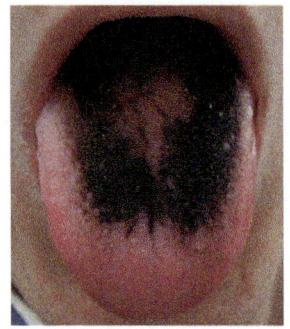
(a)

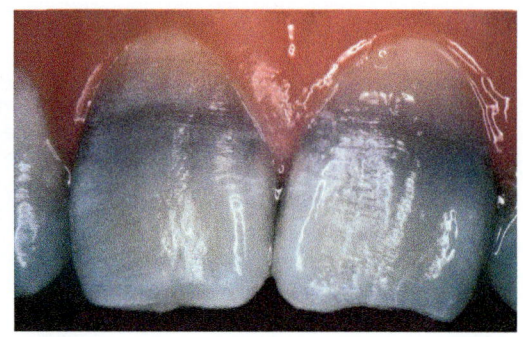
(b)

◀ **FIGURE 10.14 Some side effects resulting from toxicity of antimicrobial agents.** (a) "Black hairy tongue," caused by the antiprotozoan drug metronidazole (Flagyl). (b) Discoloration and damage to tooth enamel caused by tetracycline. *Who should avoid taking tetracycline?*

Figure 10.14 Pregnant women and children should not use tetracycline.

MICRO CHECK

12. Using a broad-spectrum antibiotic to treat an infection might seem like a good idea at first, but its use can allow unaffected members of the microbiome to cause what sort of infection?
13. In order to determine which antibiotic might treat an infection most effectively, a plate is inoculated with a pathogen. A variety of paper disks, each containing a known amount of antibiotic, are placed on the surface of the plate. After incubation, the plates are examined for zones of inhibition. Which type of test is this?
14. There are several factors to consider when choosing a route of administration for antimicrobial drugs. Which would likely be the most effective route of administration when a patient has an infection that could result in death if not treated promptly?
15. Many antimicrobials have side effects. Which side effect can be circumvented, at least in part, by the use of probiotics?

TELL ME WHY

Why don't physicians invariably prescribe the antimicrobial with the largest zone of inhibition?

Resistance to Antimicrobial Drugs

Among the major challenges facing microbiologists today are the problems presented by pathogens that are resistant to antimicrobial agents (see Emerging Disease Case Study: Community-Associated MRSA on p. 305). In the sections that follow, we examine the development 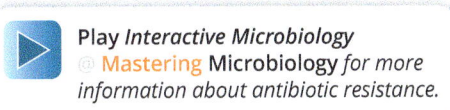 of resistant populations of pathogens, the mechanisms by which pathogens are resistant to antimicrobials, and some ways that resistance can be retarded.

The Development of Resistance and Persister Cells

LEARNING OUTCOMES
10.20 Describe how populations of resistant microbes can arise.
10.21 Describe the relationship between R plasmids and resistant cells.
10.22 Define and describe *persister cells*.

Antimicrobial resistance is a global crisis—a slow moving tsunami.

—Margaret Chan, former director-general of the World Health Organization

Antibiotic resistance has the potential to impact all Americans at every stage of life.

—CDC director Brenda Fitzgerald

Not all pathogens are equally sensitive to a given therapeutic agent; a population may contain a few individuals that are naturally either partially or completely resistant. Among bacteria, individual cells can acquire such resistance in two ways: through new mutations of chromosomal genes or by acquiring resistance genes on extrachromosomal pieces of DNA called **R plasmids** (or *R factors*) via the processes of horizontal gene transfer—transformation, transduction, or conjugation. We focus here on resistance in populations of bacteria, but resistance is known to occur among other microbes as well, including protozoa, fungi, and viruses.

A process by which a bacterial strain develops resistance is depicted in **FIGURE 10.16** on the next page. In the absence of an antimicrobial drug, resistant cells are usually less efficient than their normal neighbors because they must expend extra energy to maintain resistance genes and proteins. Under these circumstances, resistant individuals remain the minority in a population because they reproduce more slowly. However, when an antimicrobial agent is present, the majority of microbes, which are sensitive to the antimicrobial, are inhibited or die while the resistant microbes continue to grow and multiply, often more rapidly because they then face less competition. The result is that resistant organisms soon replace the sensitive ones as the majority in the population. The microbe has evolved resistance. It should be noted that the presence of an antimicrobial agent does not *produce* resistance but instead selects for the replication of resistant microbes that were already present in the population.

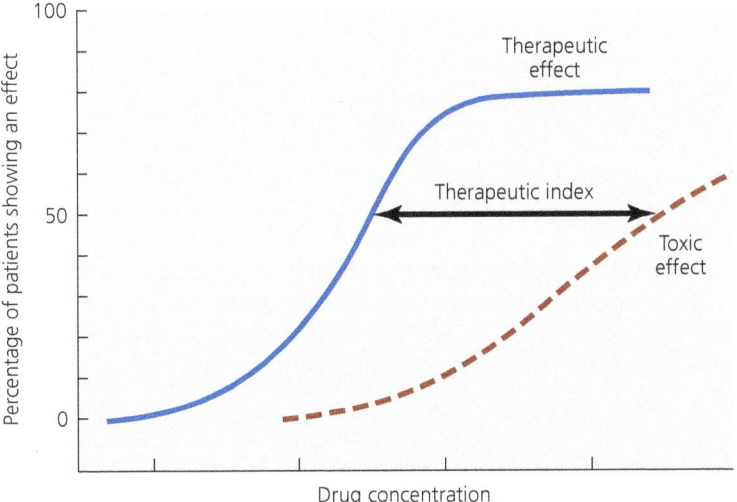

▲ **FIGURE 10.15 Therapeutic index (TI).** TI is calculated by dividing the dose that is toxic to 50% of patients by the minimum dose that is effective in 50% of patients. The higher the TI, the safer the drug.

Another form of antimicrobial resistance arises from bacterial **persister cells**—cells of a clonal population that

BENEFICIAL MICROBES

Probiotics: Using Live Microorganisms to Treat or Prevent Disease

Antibiotics are powerful tools in modern medicine that kill or inhibit the growth of microbes. The general concept behind antibiotics is to remove pathogenic microbes from a patient's body, or at least inhibit them so that the immune system can do its job. As scientists have learned more about the microbial ecology of the human body, they have discovered that some diseases arise from an imbalance between "good" microbes and "bad" ones. *Probiotics* is the concept that it is sometimes better to introduce beneficial microbes to resolve or prevent a disease rather than kill pathogens after they attack. One advantage of this proactive approach is that it mitigates the development of antibiotic resistance.

Lactobacillus is a well-studied bacterium that can act as a probiotic. Lactobacilli are the predominant organisms in yogurt and also live naturally in the human gut and vagina. Researchers have known for a while that *Lactobacillus* can help reduce the frequency and duration of diarrhea, relieve milk allergies, and possibly treat inflammatory bowel diseases. Now, they are beginning to understand how this probiotic bacterium might accomplish its beneficial tasks.

Some strains of *Lactobacillus* convert the amino acid histidine into histamine, which can then bind to histamine receptors on the surface epithelial cells that line the human gut. This dampens inflammation and may be one of the major ways that our normal bacteria protect themselves from a strong immune response so as to persist in the gut. Further, some lactobacilli can synthesize and secrete an antimicrobial compound called *reuterin,* which kills many pathogens, including *E. coli* O157:H7 and *Listeria monocytogenes*, but which doesn't harm most bacteria that normally live in humans, including *Lactobacillus*. Thus, reuterin allows lactobacilli to "protect their own turf," which is our gut. We benefit from this effect. Continuing research may result in wider use of probiotics to promote human health.

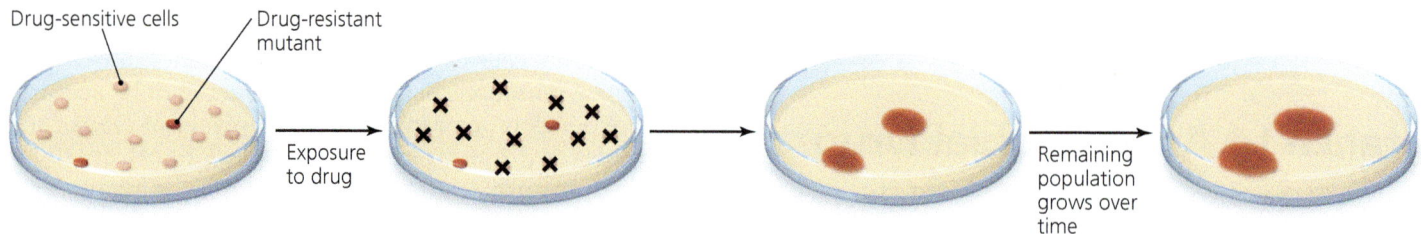

▲ **FIGURE 10.16** The development of a resistant strain of bacteria. **(a)** A bacterial population contains both drug-sensitive and drug-resistant cells, although sensitive cells constitute the vast majority of the population. **(b)** Exposure to an antimicrobial drug inhibits the sensitive cells; as long as the drug is present, reduced competition from sensitive cells facilitates the multiplication of resistant cells. **(c)** Eventually, resistant cells constitute the majority of the population. *Why do resistant strains of bacteria more often develop in hospitals and nursing homes than in college dormitories?*

Figure 10.16 *Resistant strains are more likely to develop in hospitals and other health care facilities because the extensive use of antimicrobial agents in those places inhibits the growth of sensitive strains and selects for the growth of resistant strains.*

resist the effects of antimicrobial drugs, even though they are genetically identical to susceptible cells and are living in the same environmental conditions. Persister cells can tolerate a wide range of stresses, in addition to antimicrobial drugs. Unlike resistant cells that bestow resistance to their offspring, persisters do not preferentially pass on resistance. The percentage of descendants of persister cells susceptible to antimicrobial drugs is the same percentage as in the original population.

Scientists do not know how persistence develops nor if there is a single mechanism; however, it appears that many if not all persister cells are metabolically inactive and slow- or non-growing. Persister cells are often associated with chronic infections.

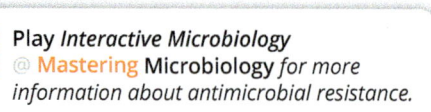

Play *Interactive Microbiology*
@ **Mastering** Microbiology *for more information about antimicrobial resistance.*

EMERGING DISEASE CASE STUDY

Community-Associated MRSA

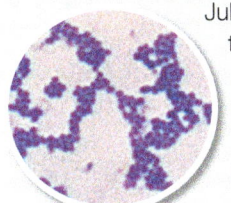

Julie was proud and excited about her new tattoo; it was cool! Jason, her boyfriend, had bought it as a gift for her 18th birthday, and she loved the way it covered her arm. A week later something else covered her arm . . . and his leg.

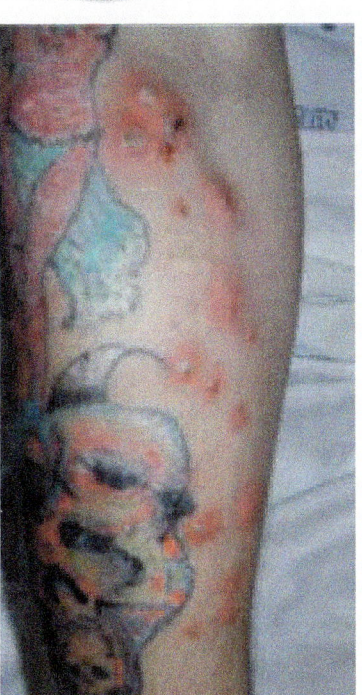

Julie and Jason had fallen victim to an emerging disease—community-associated, methicillin-resistant *Staphylococcus aureus* (CA-MRSA) infection. Julie's arm was covered with red, swollen, painful, pus-containing lesions, and she had a fever.

Jason thought he had a spider bite on Wednesday, but by Thursday morning a bright red line extended from his ankle to his groin. He could barely walk, and then his condition worsened. By the weekend, he was hospitalized, and his fever hit 103°F (39.2°C) during his 10 days there. MRSA had entered his blood (bacteremia) and infected his bones (osteomyelitis). Jason's pain was so severe that he wondered only when he might die to be free of the agony.

For years, health care workers have battled healthcare-associated MRSA (HA-MRSA), which commonly afflicts patients in hospitals. Now, researchers are concerned that victims who have never been in a hospital are succumbing; MRSA has escaped hospitals and travels communities worldwide, including among athletes, students in middle schools, and customers of unsafe tattooists. The bacterium can be spread between individuals who share fomites—towels, razors, clothing, or sheets.

Physicians drained Julie's lesions and prescribed oral antimicrobials, including trimethoprim-sulfamethoxazole, doxycycline, and clindamycin. Jason received intravenous vancomycin. Both eventually recovered, but her tattoo is now a reminder of a terrible experience and not a happy birthday. (For more about staphylococcal infections, see Chapter 19.)

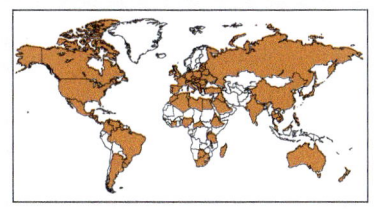

1. How were Julie and then Jason likely infected?
2. Why is CA-MRSA considered an "emerging disease," given that MRSA has been around for years?
3. Why doesn't the map show more countries in Africa and South America with cases of CA-MRSA?

Mechanisms of Resistance

 Play an *Interactive Microbiology* video @ Mastering Microbiology for more information about antimicrobial resistance.

LEARNING OUTCOMES

10.23 List seven ways by which microorganisms can be resistant to antimicrobial drugs.

10.24 List two ways that genes for drug resistance are spread between bacteria.

The problem of resistance to antimicrobial drugs is a major health threat to our world. An "alphabet soup" of resistant pathogens and diseases plague health care professionals: MRSA,[5] VRSA,[6] VISA,[7] VRE,[8] MDR-TB,[9] and XDR-TB[10] are just some of these.

---

[5]Methicillin-resistant *S. aureus*.
[6]Vancomycin-resistant *S. aureus*.
[7]*S. aureus* with intermediate level of resistance to vancomycin.
[8]Vancomycin-resistant enterococci.
[9]Multi-drug-resistant tuberculosis.
[10]Extensively drug-resistant tuberculosis.

The CDC categorizes the threat of microbial resistance in three levels:

- *Urgent:* High consequences of resistance because the microbes cause life-threatening diseases that may become widespread, for example, *Clostridium difficile*.

- *Serious:* The threat is not as severe as the urgent level because alternative antimicrobials are available or the incidence of disease is low or declining, for example, methicillin-resistant *Staphylococcus aureus* (MRSA).

- *Concerning:* The threat of resistance is low or a variety of other antimicrobials are available, though the diseases themselves can be life threatening, for example, erythromycin-resistant group A *Streptococcus*.

How do microbes gain resistance to antimicrobial drugs? Consider the path a typical antimicrobial drug must take to affect a microbe: The drug must cross the cell's wall, then cross the cytoplasmic membrane to enter the cell; there the antimicrobial binds to its target (receptor) molecule. Only then can it inhibit or kill the microbe. Microbes gain resistance by blocking some

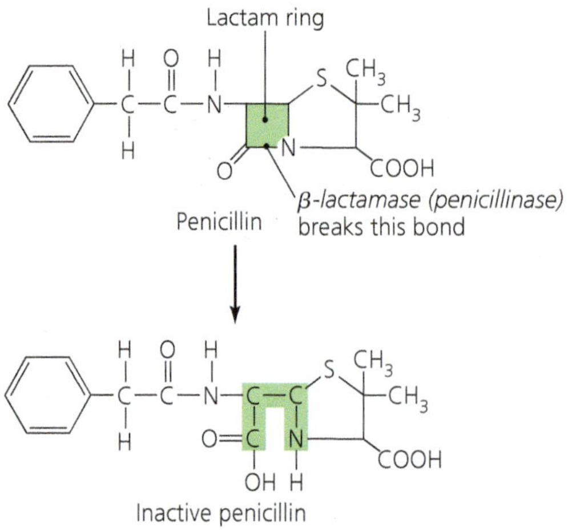

▲ FIGURE 10.17 **How beta-lactamase (penicillinase) renders penicillin inactive.** The enzyme acts by breaking a bond in the lactam ring, the functional portion of the drug.

point in this pathway. There are at least seven mechanisms of resistance:

- Resistant cells may produce an enzyme that destroys or deactivates the drug. This common mode of resistance is exemplified by **beta (β)-lactamases** (penicillinases), which are enzymes that break the beta-lactam rings of penicillin and similar molecules, rendering them inactive (FIGURE 10.17). Many MRSA strains have evolved resistance to penicillin-derived antimicrobials in this manner. Over 200 different lactamases have been identified. Frequently their genes are located on R plasmids.
- Resistant microbes may slow or prevent the entry of the drug into the cell. This mechanism typically involves changes in the structure or electrical charge of the cytoplasmic membrane proteins that constitute channels or pores. Such proteins in the outer membranes of Gram-negative bacteria are called *porins* (see Figure 3.16b). Altered pore proteins result from mutations in chromosomal genes. Resistance against tetracycline and penicillin is known to occur via this mechanism.
- Resistant cells may alter the target of the drug so that the drug either cannot attach to it or binds it less effectively. This form of resistance is often seen against antimetabolites (such as sulfonamides) and against drugs that thwart protein translation (such as erythromycin).
- Resistant cells may alter their metabolic chemistry, or they may abandon the sensitive metabolic step altogether. For example, a cell may become resistant to a drug by producing more enzyme molecules for the affected metabolic pathway, effectively reducing the power of the drug. Alternatively, cells become resistant to sulfonamides by abandoning the synthesis of folic acid, absorbing it from the environment instead.
- Resistant cells may pump the antimicrobial out of the cell before the drug can act. So-called **efflux pumps**, which are typically powered by ATP, are often able to pump more

than one type of antimicrobial from a cell. Some microbes become multi-drug resistant (perhaps to as many as 10 or more drugs) by utilizing resistance pumps.
- Bacteria within biofilms resist antimicrobials more effectively than free-living cells. Biofilms retard diffusion of the drugs and often slow metabolic rates of species making up the biofilm. Lower metabolic rates reduce the effectiveness of antimetabolic drugs.
- Some resistant strains of the bacterium *Mycobacterium tuberculosis* have a novel method of resistance against fluoroquinolone drugs that bind to DNA gyrase. These strains synthesize an unusual protein that forms a negatively charged, rodlike helix about the width of a DNA molecule. This protein, called *MfpA*, binds to DNA gyrase in place of DNA, depriving fluoroquinolone of its target site. DNA competes with and replaces MfpA in binding to DNA gyrase, allowing gyrase to function normally. This is the first known method of antibiotic resistance that involves a decoy to protect the target of an antimicrobial drug rather than, say, changing the target or deactivating the drug. MfpA protein probably slows down cellular division of *M. tuberculosis*, but that is better for the bacterium than being killed.

Play *Antibiotic Resistance: Forms of Resistance*
@ Mastering Microbiology

Despite decades of research, scientists still do not fully comprehend what makes certain species, such as *Staphylococcus aureus*, more prone to developing resistance; how and with what frequency resistance to antimicrobials spreads through a population; and what methods are most effective to limit proliferation of resistance. It has been shown that horizontal gene transfer, most often by conjugation and less often by transformation, accounts for the spread of antimicrobial resistance in biofilms. Often, several resistance genes travel together between cells.

Multiple Resistance and Cross Resistance

LEARNING OUTCOME

10.25 Define *cross resistance*, and distinguish it from *multiple resistance*.

A given pathogen can acquire resistance to more than one drug at a time, especially when resistance is conferred by R plasmids, which are exchanged readily among bacterial cells. Such *multi-resistant* strains of bacteria frequently develop in hospitals and nursing homes, where the constant use of many kinds of antimicrobial agents eliminates sensitive cells and encourages the development of resistant strains.

Multiple-drug-resistant pathogens (erroneously called *superbugs* in the popular press) are resistant to three or more different types of antimicrobial agents. Multiple-drug-resistant strains of *Staphylococcus*, *Streptococcus*, *Enterococcus* (en'ter-ō-kok'ŭs), *Pseudomonas*, *Mycobacterium tuberculosis*, and *Plasmodium* (plaz-mō'dē-ŭm) pose unique problems. Caregivers must treat infected patients without effective antimicrobials while taking care to protect themselves and others from infection.

Resistance to one antimicrobial agent may confer resistance to similar drugs, a phenomenon called **cross resistance**. Cross resistance typically occurs when drugs are similar in structure. For example, resistance to one aminoglycoside drug, such as streptomycin, may confer resistance to similar aminoglycoside drugs.

Retarding Resistance

LEARNING OUTCOME

10.26 Describe four ways to retard development of resistance.

The development of resistant populations of pathogens can be averted in at least four ways. The first way is to maintain sufficiently high concentrations of the drug in a patient's body for a long enough time to inhibit the pathogens, allowing the body's defenses to defeat them. Discontinuing a drug too early may promote the development of resistant strains. For this reason, physicians advise that patients finish their entire antimicrobial prescription and resist the temptation to "save some for another day." Research continues to ascertain if this advice is correct in all cases and with all antimicrobials.

A second way to avert resistance is to use antimicrobial agents in combination so that pathogens resistant to one drug will be killed by other drugs, and vice versa. Additionally, one drug sometimes enhances the effect of a second drug in a process called **synergism** (sin′er-jizm) (**FIGURE 10.18**). For example, the inhibition of cell wall formation by penicillin makes it easier for streptomycin molecules to enter bacteria and interfere with protein synthesis. Synergism can also result from combining an antimicrobial drug and a non-antimicrobial chemical, as occurs when *clavulanic acid* enhances the effect of penicillin by deactivating β−lactamase. (Not all drugs act synergistically; some combinations of drugs can be *antagonistic*—interfering with each other. For example, drugs that slow bacterial growth are antagonistic to the action of penicillin, which acts only against growing and dividing cells.)

A third way to reduce the development of resistance is to limit the use of antimicrobials to necessary cases. Unfortunately, many antimicrobial agents are used indiscriminately, both in developed countries and in less developed regions, where many agents are available without a physician's prescription. For example, antibacterial drugs have no effect on cold and flu viruses; therefore, 100% of antibacterial prescriptions for treating

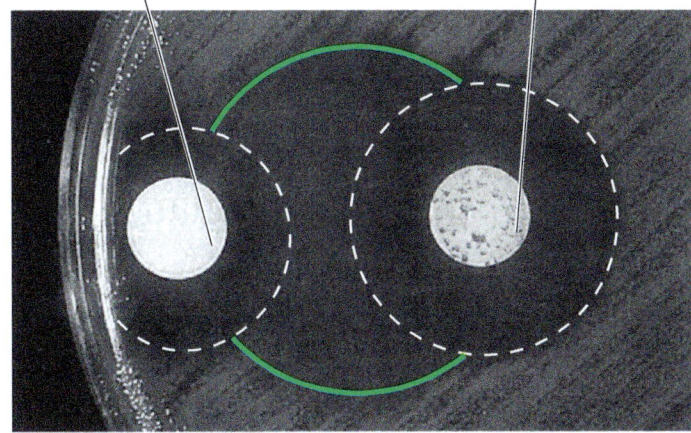

▲ **FIGURE 10.18 An example of synergism between two antimicrobial agents.** The portion of the zone of inhibition outlined in green represents the synergistic enhancement of antimicrobial activity beyond the activities of the individual drugs (outlined in white). *What does clavulanic acid do?*

Figure 10.18 *Clavulanic acid deactivates β-lactamase, allowing penicillins to work.*

CLINICAL CASE STUDY

Treating TB

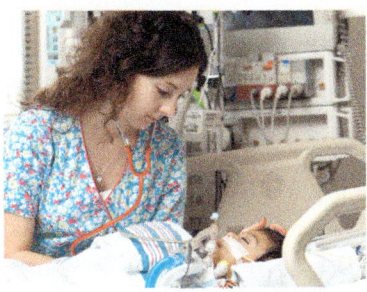

A young Hispanic mother brought her frail infant son to a southern Texas emergency room. While she waited, her baby began to have seizures. The staff stabilized the child while the mother explained that he had been ill for almost two weeks with a fever and cough that came and went. Her son had lost weight because his excessive coughing impaired his ability to breast-feed. Other family members were ill with bloody coughs.

The doctor ordered a chest X ray and acid-fast bacilli (AFB) gastric cultures. The child was suspected to have pulmonary tuberculosis. He was immediately hospitalized and isolated with airborne precautions. Health department officials discovered that two individuals had exposed the baby to TB: an uncle and his father. The strain infecting the uncle and father was a multi-drug-resistant strain (MDR-TB). The doctors ordered a test for rapid detection of TB resistant to antitubercular drugs. It would take about 24 hours to determine if they have resistant TB.

Doctors had to make a tough decision: Should they empirically treat the infant for MDR-TB, which involves using five different drugs with multiple and painful side effects for many months, or should they treat him for the more normal form of TB, using a more typical and less stressful drug regimen, but which increases the risk of death if it turns out to be MDR-TB?

1. What issues must be considered to determine the drug therapy for the baby?
2. What treatment would you guess was used for the infant?

these diseases are superfluous. The use of antimicrobial agents encourages the reproduction of resistant bacteria by limiting the growth of sensitive cells; therefore, inappropriate use of such drugs increases the likelihood that resistant strains of bacteria will multiply.

Finally, scientists can combat resistant strains by developing semisynthetic drugs, perhaps by adding different side chains to the original molecule. In this way, scientists develop *second-generation* drugs. If resistance develops to these drugs, *third-generation* drugs may be developed to replace them.

Alternatively, scientists search for new antimicrobials. Researchers explore diverse habitats, such as peat bogs, ocean sediments, soil, and people's mouths, for organisms that produce novel antibiotics. Researchers see potential in two new drugs: bacteriocins and teixobactin. Bacteria synthesize polypeptides called *bacteriocins* that inhibit other bacterial strains; now we can do the same. *Teixobactin* (tiks´ō-bak-tin)—a natural product from a harmless soil bacterium, *Eleftheria terrae* (el-ef-ther-ē´ă ter´ī)—inhibits the formation of cell walls by Gram-positive bacteria by binding to precursors of both peptidoglycan and teichoic acid. Since it binds to two different precursor molecules, it is unlikely that resistance could easily develop. Researchers are currently studying the effectiveness and safety of these antimicrobials in clinical settings. Tinkering with these and other antibiotics is yielding promising new semisynthetics, though the cost of bringing a new antimicrobial into production can be staggering—the cost to prove a drug's effectiveness and safety can be almost $2 billion.

With the advent of genome sequencing and enhanced understanding of protein folding, some scientists predict that we are moving into a new era of antimicrobial drug discovery. They point out that researchers who know the exact shapes of microbial proteins should be able to design drugs complementary to those shapes—drugs that will inhibit microbial proteins without affecting humans.

Currently, scientists are examining ways to inhibit microbes by inhibiting the following: secretion systems, which allow bacteria to attach to and inject proteins into a cell; attachment molecules and their receptors; biofilm signaling molecules; and bacterial RNA polymerase.

Despite researchers' best efforts, other scientists wonder whether drug developers can stay ahead of the development of resistance by pathogens.

Selected antimicrobial agents, their modes of action, clinical considerations, and other features are summarized in **TABLES 10.2** through **10.6**. Particular use of antimicrobial drugs against specific pathogens is covered in the relevant chapters of this text.

MICRO CHECK

16. What benefit would the presence of an R plasmid provide to a microbe that has this extra piece of DNA?
17. By what mechanism do bacteria develop resistance to tetracycline?
18. Which method of drug resistance renders bacteria resistant to penicillin and similar drugs?
19. What term *best* describes an organism that is resistant to three or more different antimicrobial agents?
20. Some antimicrobial drugs can work in tandem with one another to produce an enhanced antimicrobial effect. What is this process called?

TELL ME WHY

Why is it incorrect to say that an individual bacterium develops resistance in response to an antibiotic?

TABLE 10.2 Antibacterial Drugs

Drug	Description and Mode of Action	Clinical Considerations	Method of Resistance
Antibacterial Drugs That Inhibit Cell Wall Synthesis			

Bacitracin	Isolated from *Bacillus licheniformis* growing on a patient named Tracy; appears to have three modes of action:
	
	    Interference with the movement of peptidoglycan precursors through the bacterial cell membrane to the cell wall
	
	    Inhibition of RNA transcription
	
	    Damage to the bacterial cytoplasmic membrane
	
	The latter two modes of action have not been proven definitely
		Spectrum of action: Gram-positive (G+) bacteria
		Route of administration: Topical
		Adverse effects: Toxic to kidneys
			Resistance most often involves changes in bacterial cell membranes that prevent bacitracin from entering the cell

Beta-Lactams
Representative natural penicillins:
  Penicillin G
  Penicillin V
Representative semisynthetic penicillins:
  Amoxicillin
  Ampicillin
  Cloxacillin
  Dicloxacillin
  Methicillin
  Nafcillin
  Oxacillin
  Piperacillin
Representative semisynthetic carbapenems:
  Imipenem
  Meropenem
Representative natural cephalosporin:
  Cephalothin
Representative semisynthetic cephalosporins:
  Cefotaxime
  Ceftriaxone
  Cefuroxime
  Cephalexin
  Cephamycin
Representative semisynthetic monobactam:
  Aztreonam
	Large number of natural and semisynthetic derivatives from the fungi *Penicillium* (penicillins) and *Acremonium* or *Streptomyces* (cephalosporins); bind to and deactivate the enzyme that cross-links the NAM subunits of peptidoglycan
	
	Monobactams have only a single ring instead of the two rings seen in other beta-lactams
		Spectrum of action: Natural drugs have limited action against most Gram-negative (G−) bacteria because they do not readily cross the outer membrane; semisynthetics have broader spectra of action
		
		Monobactams have a limited spectrum of action, affecting only aerobic, G− bacteria
		
		Route of administration: Penicillin V, a few cephalosporins (e.g., cephalexin), and monobactams: oral; penicillin G and many semisynthetics (e.g., methicillin, ampicillin, carbenicillin, cephalothin): IM or IV
		
		Adverse effects: Allergic reactions against beta-lactams in some adults; monobactams are least allergenic
			Develops in three ways in G− bacteria:
			
			  Change their outer membrane structure to prevent entrance of the drug
			
			  Modify the enzyme so that the drug no longer binds
			
			  Synthesize beta-lactamases that cleave the functional lactam ring of the drug; genes for lactamases are often carried on R plasmids

Cycloserine	Analog of alanine that interferes with the formation of alanine-alanine bridges between NAM subunits	**Spectrum of action:** Some G+ bacteria, mycobacteria **Route of administration:** Oral **Adverse effects:** Toxic to nervous system, producing depression, aggression, confusion, and headache	Some G+ bacteria enzymatically deactivate the drug

Continued—

TABLE 10.2 Antibacterial Drugs—Continued

Drug	Description and Mode of Action	Clinical Considerations	Method of Resistance
Antibacterial Drugs That Inhibit Cell Wall Synthesis—Continued			
Ethambutol	Prevents the formation of mycolic acid; used in combination with other antimycobacterial drugs	**Spectrum of action:** Mycobacteria, including *M. tuberculosis* and *M. leprae* **Route of administration:** Oral **Adverse effects:** None	Resistance is due to random mutations of bacterial chromosomes that result in alteration of target site
Isoniazid (isonicotinic acid hydrazide, INH)	Analog of the vitamins nicotinamide and pyridoxine; blocks the gene for an enzyme that forms mycolic acid	**Spectrum of action:** Mycobacteria, including *M. tuberculosis* and *M. leprae* **Route of administration:** Oral **Adverse effects:** May be toxic to liver	Resistance is due to random mutations of bacterial chromosomes that result in alteration of target site or overproduction of target molecules
Lipoglycopeptide Representatives: Dalbavancin Oritavancin Vancomycin	Directly interfere with the formation of certain alanine-alanine bridges between NAM subunits	**Spectrum of action:** Effective against most G+ bacteria but often reserved for use against strains resistant to other drugs, such as methicillin-resistant *Staphylococcus aureus* (MRSA) **Route of administration:** IV **Adverse effects:** Damage to ears and kidneys, allergic reactions, may cause depression	G− bacteria are naturally resistant because the drug is too large to pass through the outer membrane; some G+ bacteria (e.g., *Lactobacillus*) are naturally resistant because they do not form alanine-alanine bonds between NAM subunits
Antibacterial Drugs That Inhibit Protein Synthesis			
Aminoglycosides Representatives: Amikacin (semisynthetic) Capreomycin Gentamicin Kanamycin Neomycin Paromomycin Spiramycin Streptomycin Tobramycin	Compounds in which two or more amino sugars are linked with glycosidic bonds; were originally isolated from species of the bacterial genera *Streptomyces* and *Micromonospora*. Inhibit protein synthesis by irreversibly binding to the 30S subunit of prokaryotic ribosomes, which either causes the ribosome to mistranslate mRNA, producing aberrant proteins, or causes premature release of the ribosome from mRNA, stopping synthesis. Also bactericidal by destroying outer membranes of G− bacteria	**Spectrum of action:** Broad: effective against most G− bacteria and some protozoa **Route of administration:** Oral, IV, or IM, depending on drug; do not traverse blood-brain barrier **Adverse effects:** Often toxic to kidneys and to organs of inner ear or auditory nerves, causing deafness	Uptake of these drugs is energy dependent, so anaerobes (with less ATP available) are less susceptible; aerobic bacteria alter membrane pores to prevent uptake or synthesize enzymes that alter or degrade the drug once it enters; rarely, bacteria alter the binding site on the ribosome; some bacteria make biofilms when exposed to the drugs
Chloramphenicol	Rarely used drug produced by *Streptomyces venezuelae* that binds to the 50S subunits of prokaryotic ribosomes, preventing them from moving along mRNA	**Spectrum of action:** Broad but rarely used except in treatment of typhoid fever **Route of administration:** Oral; traverses blood-brain barrier **Adverse effects:** In 1 of 24,000 patients, causes aplastic anemia, a potentially fatal condition in which blood cells fail to form; can also cause neurological damage	Develops via gene carried on an R plasmid that codes for an enzyme that deactivates drug
Lincosamides Representative semisynthetic: Clindamycin	Produced by *Streptomyces lincolnensis*; binds to 50S ribosomal subunit and stops protein elongation	**Spectrum of action:** Effective against G+ and anaerobic G− bacteria and some protozoa **Route of administration:** Oral or IV; does not traverse blood-brain barrier **Adverse effects:** Gastrointestinal distress, including nausea, diarrhea, vomiting, and pain	Develops via changes in ribosomal structure that prevent drug from binding; resistance genes are same as those of aminoglycosides and macrolides

TABLE 10.2 Antibacterial Drugs—Continued

Drug	Description and Mode of Action	Clinical Considerations	Method of Resistance
Antibacterial Drugs That Inhibit Protein Synthesis—Continued			
Macrolides Representatives: Azithromycin Clarithromycin Erythromycin Natamycin Telithromycin	Group of antimicrobials typified by a macrocyclic lactone ring; the most prescribed is erythromycin, which is produced by *Saccharopolyspora erythraea*; act by binding to the 50S subunit of prokaryotic ribosomes and preventing the elongation of the nascent protein	**Spectrum of action:** Effective against G+ and a few G− bacteria and fungi (natamycin) **Route of administration:** Oral; do not traverse blood-brain barrier **Adverse effects:** Nausea, mild gastrointestinal pain, vomiting; associated with narrowing of the stomach in children under six months old; erythromycin increases risk of cardiac arrest	Develops via changes in ribosomal RNA that prevent drugs from binding, or via R plasmid genes coding for the production of macrolide-digesting enzymes; resistance genes are same as those of lincosamides and aminoglycosides
Mupirocin	Produced by *Pseudomonas fluorescens*; binds to bacterial tRNA^{Ile} synthetase, which prevents charging of isoleucine onto tRNA, blocking polypeptide synthesis at isoleucine codons	**Spectrum of action:** Effective primarily against G+ bacteria **Route of administration:** Topical cream **Adverse effects:** None reported	Resistance develops from mutations that change the shape of tRNA^{Ile}
Oxazolidinones Representatives: Linezolid Tedizolid	Synthetic; inhibits initiation of polypeptide synthesis	**Spectrum of action:** G+ bacteria **Route of administration:** Oral, IV **Adverse effects:** Rash, diarrhea, loss of appetite, constipation, fever	Method of resistance not known
Streptogramins Representatives: Quinupristin Dalfopristin	Produced by species of *Streptomyces*; binds to 50S ribosomal subunit, stops protein synthesis; the synergistic drugs quinupristin and dalfopristin are taken together	**Spectrum of action:** Broad but reserved for use against multiple-drug-resistant strains **Route of administration:** IV **Adverse effects:** Muscle and joint pain	Not known
Tetracyclines Representatives: Doxycycline Minocycline Tetracycline Tigecycline	Produced by species of *Streptomyces*, composed of four hexagonal rings with various side groups; prevent tRNA molecules, which carry amino acids, from binding to ribosomes at the 30S subunit's docking site	**Spectrum of action:** Most are broad: effective against many G+ and G− bacteria as well as against bacteria that lack cell walls, such as *Mycoplasma* **Route of administration:** Oral; crosses poorly into brain **Adverse effects:** Nausea, diarrhea, sensitivity to light; forms complexes with calcium, which stains developing teeth and adversely affects the strength and shape of bones	Develops in three ways; bacteria may: Alter gene for pores in outer membrane such that new pore prevents drug from entering cell Alter binding site on the ribosome to allow tRNA to bind even in presence of drug Actively pump drug from cell
Antibacterial Drugs That Alter Cytoplasmic Membranes			
Gramicidin	Short polypeptide produced by *Bacillus brevis* that forms pore across cytoplasmic membrane, allowing single-charged cations to cross freely	**Spectrum of action:** G+ bacteria **Route of administration:** Topical **Adverse effects:** In allergic individuals, itching, swelling, and local irritation	Not known
Nisin	A bacteriocin naturally produced by *Lactococcus lactis* that disrupts a bacterial cytoplasmic membrane	**Spectrum of action:** G+ and G− bacteria in food **Route of administration:** Used as a preservative in food **Adverse effect:** None reported	Not known

Continued—

TABLE 10.2 Antibacterial Drugs—Continued

Drug	Description and Mode of Action	Clinical Considerations	Method of Resistance
Antibacterial Drugs That Alter Cytoplasmic Membranes—Continued			
Polymyxin	Produced by *Bacillus polymyxa*; destroys cytoplasmic membranes of susceptible cells	**Spectrum of action:** Effective against G− bacteria, particularly *Pseudomonas* (and some amoebas) **Route of administration:** Topical **Adverse effect:** Toxic to kidneys	Results from changes in cell membrane that prohibit entrance of the drug
Pyrazinamide	Disrupts membrane transport and prevents *Mycobacterium* from repairing damaged proteins	**Spectrum of action:** *Mycobacterium tuberculosis* **Route of administration:** Oral **Adverse effects:** Malaise, nausea, diarrhea	Results from point mutations in bacterial gene for enzyme necessary to activate drug
Antibacterial Drugs That Are Antimetabolites			
Bedaquiline	Synthetic drug that inhibits an ATP synthesis enzyme unique to mycobacteria	**Spectrum of action:** Mycobacteria; reserved for use for multi-drug-resistant tuberculosis **Route of administration:** Oral **Adverse effects:** Disturbance in heart's electrical activity, nausea, headache, joint pain	Not known
Dapsone	Synthetic analog of para-aminobenzoic acid; interferes with synthesis of folic acid	**Spectrum of action:** *M. leprae*, *M. tuberculosis* **Route of administration:** Oral **Adverse effects:** Insomnia, headache, nausea, vomiting, increased heart rate	Not known
Sulfonamides Representatives: 　Sulfadiazine 　Sulfadoxine 　Sulfamethoxazole 　Sulfanilamide	Synthetic drugs; first produced as a dye; analogs of PABA that bind irreversibly to enzyme that produces dihydrofolic acid; synergistic with trimethoprim	**Spectrum of action:** Broad: effective against G+ and G− bacteria and some protozoa and fungi; however, resistance is widespread **Route of administration:** Oral **Adverse effects:** Rare: allergic reactions, anemia, jaundice, mental retardation of fetus if administered in last trimester of pregnancy	*Pseudomonas* is naturally resistant because of permeability barriers; organisms that use folic acid from the diet are also naturally resistant; chromosomal mutations result in lowered affinity for the drugs
Thiazolides Nitazoxanide	Interferes with electron transport chains	**Spectrum of action:** Broad: effective against helminths, protozoa, influenzaviruses, and RNA and DNA hepatitis viruses **Route of administration:** Oral **Adverse effects:** Allergic reactions in some patients	Not known
Trimethoprim	Blocks second metabolic step in the formation of folic acid from PABA; synergistic with sulfonamides, such as sulfamethoxazole	**Spectrum of action:** Broad: effective against G+ and G− bacteria and some protozoa and fungi; however, resistance is widespread **Route of administration:** Oral **Adverse effects:** Allergic reactions or liver damage in some patients	*Pseudomonas* is naturally resistant because of permeability barriers; organisms that use folic acid from the diet are also naturally resistant; chromosomal mutations result in lowered affinity for the drug
Antibacterial Drugs That Inhibit Nucleic Acid Synthesis			
Clofazimine	Binds to DNA, preventing replication and transcription	**Spectrum of action:** Mycobacteria, especially *M. tuberculosis*, *M. leprae*, and *M. ulcerans* **Route of administration:** Oral **Adverse effects:** Diarrhea, discoloration of skin and eyes	Not known

TABLE 10.2 Antibacterial Drugs—Continued

Drug	Description and Mode of Action	Clinical Considerations	Method of Resistance
Antibacterial Drugs That Inhibit Nucleic Acid Synthesis—*Continued*			
Quinolones Representatives: Ciprofloxacin, Levofloxacin, Moxifloxacin, Nalidixic acid, Ofloxacin, Tedizolid	Synthetic agents that inhibit DNA gyrase, which is needed to correctly replicate bacterial DNA; penetrate cytoplasm of cells	**Spectrum of action:** Broad: G+ and G− bacteria are affected **Route of administration:** Oral **Adverse effects:** Tendonitis, tendon rupture; tedizolid not approved for children	Results from chromosomal mutations that lower affinity for drug, reduce its uptake, or protect gyrase from drug
Nitroimidazoles Representatives: Benznidazole, Metronidazole, Tinidazole	Anaerobic conditions reduce these synthetic molecules, which then damage DNA, preventing correct DNA replication	**Spectrum of action:** Obligate anaerobic bacteria and protozoa	Not known
Rifamycins Representatives: Rifampin, Rifaximin	Natural and semisynthetic derivatives from *Amycolatopsis rifamycinica* that bind to bacterial RNA polymerase, preventing transcription of RNA; used with other antimicrobial bacterial drugs	**Spectrum of action:** Bacteriostatic against aerobic G+ bacteria; bactericidal against mycobacteria **Route of administration:** Oral **Adverse effects:** None of major significance	Results from chromosomal mutation that alters binding site on enzyme; G− bacteria are naturally resistant because of poor uptake

TABLE 10.3 Antiviral Drugs

Drug	Description and Mode of Action	Clinical Considerations	Method of Resistance
Attachment Antagonists			
Pleconaril	Synthetic molecule that blocks attachment molecule on host cell or pathogen	**Spectrum of action:** Picornaviruses (e.g., poliovirus, some cold viruses, and, coxsackieviruses) **Route of administration:** Oral **Adverse effects:** None	Not known
Neuraminidase Inhibitors Representatives: Oseltamivir, Peramivir, Zanamivir	Synthetic drugs that prevent influenzaviruses from exiting from cells	**Spectrum of action:** Influenzavirus **Route of administration:** Oral (oseltamivir), IV (peramivir), or aerosol (zanamivir) **Adverse effects:** None	Not known
Antiviral Drugs That Inhibit Viral Entry			
Enfuvirtide	Synthetic drug that binds viral protein gp41, preventing fusion of viral envelope with host cell membrane, thereby blocking viral entry	**Spectrum of action:** HIV-1 **Route of administration:** IV **Adverse effects:** Tingling, allergic reaction at injection site	Not known
Antiviral Drugs That Inhibit Viral Uncoating			
Amantadine	Synthetic drug that neutralizes acid environment within phagolysosomes that is necessary for viral uncoating	**Spectrum of action:** Influenza A virus **Route of administration:** Oral **Adverse effects:** Toxic to central nervous system; results in nervousness, irritability, insomnia, and blurred vision	Mutation resulting in a single amino acid change in a membrane ion channel leads to viral resistance
Arildone	Synthetic antiviral that prevents removal of poliovirus capsids (uncoating)	**Spectrum of action:** Polioviruses **Route of administration:** Topical **Adverse effects:** None reported	Not known

Continued—

TABLE 10.3 Antiviral Drugs—Continued

Drug	Description and Mode of Action	Clinical Considerations	Method of Resistance
Antiviral Drugs That Inhibit Viral Uncoating—*Continued*			
Rimantadine	Synthetic drug that neutralizes phagolysosomal acid, preventing viral uncoating	**Spectrum of action:** Influenza A virus **Route of administration:** Oral, adults only **Adverse effects:** Toxic to central nervous system; results in nervousness, irritability, insomnia, and blurred vision	Mutation resulting in a single amino acid change in a membrane ion channel leads to viral resistance
Antiviral Drugs That Inhibit Nucleic Acid Synthesis			
Acyclovir (ACV) Ganciclovir Penciclovir	Phosphorylation by virally coded kinase enzyme activates these synthetic nucleotide analogs; inhibit DNA and RNA synthesis	**Spectrum of action:** Viruses that code for kinase enzymes: human herpesvirus, Epstein-Barr virus, cytomegalovirus, and varicella virus **Route of administration:** Oral **Adverse effects:** None	Mutations in genes for kinase enzymes may render the drugs ineffective at drug activation
Adenine Arabinoside	Phosphorylation by cell-coded kinase enzyme activates this synthetic drug; inhibits DNA synthesis; viral DNA polymerase more likely to incorporate the drugs than human DNA polymerase	**Spectrum of action:** Herpesvirus **Route of administration:** IV **Adverse effects:** Fatal to host cells that incorporate the drug into cellular DNA; anemia	Results from mutation of viral DNA polymerase
Sofosbuvir	Cell-coded kinase phosphorylates this nucleotide analog, which inhibits DNA replication	**Spectrum of action:** Hepatitis C virus **Route of administration:** Oral **Adverse effects:** Fatigue, headache, nausea, insomnia, rash, diarrhea	Not known

Other Nucleotide Analogs
Representatives (see also Figure 10.7):
  Adefovir
  Azidothymidine (AZT)
  Brincidofovir
  Cidifovir
  Emtricitabine
  Entecavir
  Lamivudine
  Ribavarin
  Tenofovir
  Valaciclovir
	Synthetic molecules phosphorylated by cell-coded kinase enzyme that activates these drugs; inhibits DNA synthesis; HIV and hepatitis B virus reverse transcriptases more likely to incorporate some of these drugs; used in conjunction with protease inhibitor to treat HIV	**Spectrum of action:** Human immunodeficiency, hepatitis B and C, respiratory syncytial, influenza A, measles, and some adeno- and hemorrhagic fever viruses
		Route of administration: Oral, aerosol, or IV, depending on drug
		Adverse effects: Vary with drug but include nausea, bone marrow toxicity, and fetal harm
			Results from mutation of HIV reverse transcriptase

Antiviral Drugs That Inhibit Protein Synthesis			
Antisense Nucleic Acids Representative: Fomivirsen	Synthetic nucleic acid complementary to specific mRNA; binding prevents protein synthesis by blocking ribosomes	**Spectrum of action:** Specific to species with complementary mRNA; fomivirsen specific against cytomegalovirus **Route of administration:** Fomivirsen injected weekly into eyes **Adverse effects:** Possible glaucoma	Not known
Antiviral Drugs That Inhibit Viral Proteins			

Protease Inhibitors
Representatives:
  Boceprevir
  Darunavir
  Fosamprenavir
  Ledipasvir
  Telaprevir
	Computer-assisted modeling of protease enzymes, which are unique to HIV and hepatitis C virus, allowed the creation of drugs that block the enzymes' active sites, often used in conjunction with drugs active against nucleic acid synthesis	**Spectrum of action:** HIV and hepatitis C
		Route of administration: Oral
		Adverse effects: None
			Results from mutation in protease gene

Envelope Protein Inhibitors			
Tecovirimat	Synthetic drug that deactivates an envelope protein necessary for viral release from a cell.	**Spectrum of action:** Orthopoxviruses **Route of administration:** Oral **Adverse effects:** No serious events reported	Not known

TABLE 10.4 Antimicrobials Against Eukaryotes: Antifungal Drugs

Drug	Description and Mode of Action	Clinical Considerations	Method of Resistance
Antifungal Drugs that Inhibit Cell Wall Synthesis			
Echinocandins Representative: Caspofungin	Natural product of *Aspergillus* and semisynthetic derivatives that inhibit synthesis of glucan subunits in some fungal cell walls	**Spectrum of action:** *Candida, Aspergillus* **Route of administration:** IV **Adverse effects:** Rash, facial swelling, respiratory spasms, gastrointestinal distress, toxicity to human embryos	Results from mutation in glucan synthase gene
Antifungal Drugs That Alter Cell Membranes			
Allylamines Representative: Terbinafine	Antifungal action due to inhibition of ergosterol synthesis, which is essential for fungal cytoplasmic membranes	**Spectrum of action:** Fungi **Route of administration:** Oral, IV **Adverse effects:** Headache, nausea, vomiting, diarrhea, liver damage, rash	Not known
Azoles Representatives: Fluconazole, Itraconazole, Ketoconazole, Voriconazole	Synthetic antifungal and antiprotozoan drug that inhibits synthesis of membrane sterols	**Spectrum of action:** Fungi and protozoa **Route of administration:** Topical, IV **Adverse effects:** Possibly causes cancer in humans	Mutation in gene for target enzyme
Polyenes Representatives: Amphotericin B, Nystatin	Associate with molecules of ergosterol, forming a pore through the fungal membrane, which leads to leakage of essential ions from the cell; amphotericin B is produced by *Streptomyces nodosus*	**Spectrum of action:** Fungi, some amoebas **Route of administration:** Amphotericin B: IV; nystatin: topical **Adverse effects:** Chills, vomiting, fever	Rare; decrease in amount or change in chemistry of ergosterol
Antifungal Drugs that are Antimetabolites			
Atovaquone	Synthetic analog of coenzyme Q of *Pneumocystis* and of several protozoa; interrupts electron transport	**Spectrum of action:** *Pneumocystis*, protozoa (especially *Plasmodium*) **Route of administration:** Oral **Adverse effects:** Possible rash, diarrhea, headache	Cells modify the structure of their electron transport chain proteins
Ciclopirox	Synthetic antimetabolite that appears to disrupt DNA repair, inhibit mitosis, and interfere with some aspects of intracellular transport	**Spectrum of action:** Fungi, especially *Trichophyton* **Route of administration:** Topical **Adverse effects:** Possible burning sensation at application site	Not known
Griseofulvin	Isolated from *Penicillium griseofulvum*; deactivates tubulin, preventing cytokinesis and segregation of chromosomes during mitosis (see Chapter 12)	**Spectrum of action:** Fungal skin infection **Route of administration:** Topical, oral **Adverse effects:** None	Not known
Antifungal Drugs that Inhibit Nucleic Acid Synthesis			
5-Fluorocytosine	Fungi, but not mammals, have an enzyme that converts this synthetic drug into 5-fluorouracil, an analog of uracil that inhibits RNA function	**Spectrum of action:** *Candida, Cryptococcus, Aspergillus* **Route of administration:** Oral **Adverse effects:** None	Develops from mutations in the genes for enzymes necessary for utilization of uracil

TABLE 10.5 Antimicrobials Against Eukaryotes: Anthelmintic Drugs

Drug	Description and Mode of Action	Clinical Considerations	Method of Resistance
Anthelmintic Drugs That Are Antimetabolites			
Bithionol	Interferes with neuromuscular physiology of helminths; also possibly uncouples electron transport from phosphorylation (ATP production)	**Spectrum of action:** Helminths **Route of administration:** Oral **Adverse effects:** Diarrhea, abdominal cramping, nausea	Not known
Benzimidazole Derivatives Representatives: Albendazole Mebendazole Thiabendazole Triclabendazole	Inhibit microtubule formation and glucose uptake	**Spectrum of action:** Helminths, protozoa **Route of administration:** Oral **Adverse effects:** Possible diarrhea	Not known
Ivermectin Metrifonate	Semisynthetic drugs modified from a natural product of *Streptomyces*; produce flaccid paralysis by blocking neurotransmitters and changing membrane permeability	**Spectrum of action:** Helminths **Route of administration:** Oral **Adverse effects:** Allergic reactions may result from antigens of dead helminths	Not known
Niclosamide	Inhibits oxidative phosphorylation of ATP by mitochondria	**Spectrum of action:** Cestodes **Route of administration:** Oral **Adverse effects:** Abdominal pain, nausea, diarrhea	Not known
Praziquantel	Synthetic drug that changes membrane permeability to calcium ions, which are required for muscular contraction; induces complete muscular contraction in helminths	**Spectrum of action:** Cestodes, trematodes **Route of administration:** Oral **Adverse effects:** None	Not known
Pyrantel Pamoate Diethylcarbamazine	Synthetic drugs that bind to neurotransmitter receptors, causing complete muscular contraction of helminths	**Spectrum of** action: Nematodes **Route of administration:** Oral **Adverse effects:** None	Not known
Anthelmintic Drugs That Inhibit Nucleic Acid Synthesis			
Niridazole	When partially catabolized by schistosome enzymes, binds to DNA, preventing replication	**Spectrum of action:** *Schistosoma* **Route of administration:** Oral **Adverse effects:** Central nervous system toxicity, hallucinations	Not known
Oltipraz	Synthetic drug that possibly acts by reducing the supply of deoxyribonucleotides	**Spectrum of action:** *Schistosoma* **Route of administration:** Oral **Adverse effects:** Central nervous system and gastrointestinal toxicity	Not known
Oxamniquine	Schistosome enzyme activates semisynthetic drug, which then inhibits DNA synthesis	**Spectrum of action:** *Schistosoma mansoni* **Route of administration:** Oral **Adverse effects:** Central nervous system toxicity, hallucinations	Not known

TABLE 10.6   Antimicrobials Against Eukaryotes: Antiprotozoan Drugs

Drug	Description and Mode of Action	Clinical Considerations	Method of Resistance
Antiprotozoan Drugs That Inhibit Protein Synthesis			
Lincosamides Representative: Clindamycin	Produced by *Streptomyces lincolnensis*; binds to 50S ribosomal subunit and stops protein elongation	**Spectrum of action:** Effective against some protozoa and G+ and anaerobic G− bacteria **Route of administration:** Oral or IV; does not traverse blood-brain barrier **Adverse effects:** Gastrointestinal distress, including nausea, diarrhea, vomiting, and pain	Develops via changes in ribosomal structure that prevent drug from binding; resistance genes are same as those for resistance to aminoglycosides and macrolides
Paromomycin	Aminoglycoside produced by *Streptomyces rimosus* that interferes with bacterial 30S ribosomal subunits; its method of action against protozoa is not known	**Spectrum of action:** Effective against some protozoa, e.g., *Leishmania* and *Entamoeba*, and most G− and G+ bacteria **Route of administration:** Oral **Adverse effects:** Allergic reactions in some patients; intestinal blockage	Decreased drug uptake due to altered cytoplasmic membrane
Antiprotozoan Drugs That Are Antimetabolites			
Artemisinins Representative: Artesunate	Derived from Chinese wormwood shrub; interferes with heme detoxification and with Ca^{2+} transport	**Spectrum of action:** *Plasmodium* **Route of administration:** Oral **Adverse effects:** Nausea, vomiting, itching, dizziness	Mutation in Ca^{2+} transporter gene
Atovaquone	Synthetic analog of coenzyme Q of several protozoa and of *Pneumocystis*; interrupts electron transport	**Spectrum of action:** Protozoa (especially *Plasmodium*), fungus—*Pneumocystis* **Route of administration:** Oral **Adverse effects:** Possible rash, diarrhea, headache	Cells modify the structure of their electron transport chain proteins
Azoles Representatives: Albendazole Ketoconazole Mebendazole	Synthetic antiprotozoan and anthelmintic drugs that inhibit microtubule formation and glucose uptake	**Spectrum of action:** Helminths, protozoa **Route of administration:** Oral **Adverse effects:** Possible diarrhea	Not known
Heavy Metals (e.g., Hg, As, Cr, Sb) Representatives: Meglumine antimonate Melarsoprol (contains As) Salvarsan (contains As) Sodium stibogluconate (contains Sb)	Deactivate enzymes by breaking hydrogen bonds necessary for effective tertiary structure; drugs containing arsenic were the first recognized selectively toxic chemotherapeutic agents	**Spectrum of action:** Metabolically active cells **Route of administration:** Topical, oral **Adverse effects:** Toxic to active cells, such as those of the brain, kidney, liver, and bone marrow	Not known
Iodoquinol	Mode of action unknown	**Spectrum of action:** Intestinal amoebas **Route of administration:** Oral **Adverse effects:** Fever, chills, rash	Not known
Lumefantrine	Synthetic drug that prevents detoxification of heme released from hemoglobin of damaged red blood cells	**Spectrum of action:** *Plasmodium* **Route of administration:** Oral **Adverse effects:** Headache, dizziness, weakness	Not known
Nifurtimox	Synthetic drug that interferes with electron transport	**Spectrum of action:** *Trypanosoma cruzi* **Route of administration:** Oral **Adverse effects:** Abdominal pain, nausea	Not known
Nitazoxanide	Synthetic drug believed to interfere with anaerobic electron transport chain	**Spectrum of action:** Protozoa **Route of administration:** Oral **Adverse effects:** Gastrointestinal distress	Not known

Continued—

TABLE 10.6 Antimicrobials Against Eukaryotes: Antiprotozoan Drugs—Continued

Drug	Description and Mode of Action	Clinical Considerations	Method of Resistance
Antiprotozoan Drugs That Are Antimetabolites—*Continued*			
Proguanil **Pyrimethamine**	Synthetic drugs that block second metabolic step in the formation of folic acid from PABA; synergistic with sulfonamides	**Spectrum of action:** Broad: effective against G+ and G− bacteria and some protozoa and fungi; however, resistance is widespread **Route of administration:** Oral **Adverse effects:** Allergic reactions in some patients	Cells that require folic acid as a vitamin are also naturally resistant; chromosomal mutations result in lowered affinity for the drugs

Sulfonamides
 Representatives:
  Sulfadiazine
  Sulfadoxine
  Sulfanilamide
	Synthetic drugs; first produced as a dye; analogs of PABA that bind irreversibly to enzyme that produces dihydrofolic acid; synergistic with trimethoprim	**Spectrum of action:** Broad: effective against G+ and G− bacteria and some protozoa and fungi; however, resistance is widespread
		 Route of administration: Oral
		 Adverse effects: Rare: Allergic reactions, anemia, jaundice, mental retardation of fetus if administered in last trimester of pregnancy
			Organisms that do not synthesize folic acid are naturally resistant; chromosomal mutations result in lowered affinity for the drugs

Suramin	Synthetic drug that inhibits specific enzymes in some protozoa	**Spectrum of action:** *Trypanosoma brucei* (East African variant) **Route of administration:** Oral **Adverse effects:** None	Not known
Trimethoprim	Synthetic drug that blocks second metabolic step in the formation of folic acid from PABA; synergistic with sulfonamides	**Spectrum of action:** Broad: effective against some protozoa, fungi, and some G+ and G− bacteria; however, resistance is widespread **Route of administration:** Oral **Adverse effects:** Allergic reactions or liver damage in some patients	Organisms that cannot synthesize folic acid are naturally resistant; chromosomal mutations result in lowered affinity for the drugs
Antiprotozoan Drugs That Inhibit DNA Synthesis			
Eflornithine	Synthetic drug that inhibits synthesis of precursors of nucleic acids	**Spectrum of action:** *Trypanosoma brucei gambiense* (West African variant) **Route of administration:** Oral **Adverse effects:** Anemia, inhibition of blood clotting, nausea, vomiting	Not known

Nitroimidazoles
 Representatives:
  Benznidazole
  Metronidazole
  Tinidazole
	Anaerobic conditions reduce these synthetic drugs, which then appear to damage DNA, preventing correct replication and transcription	**Spectrum of action:** Obligate anaerobic protozoa and bacteria
		 Route of administration: Oral
		 Adverse effects: Metronidazole and tinidazole cause cancer in laboratory rodents
			Not known

Pentamidine	Synthetic drug that binds to nucleic acids, inhibiting replication, transcription, and translation	**Spectrum of action:** Protozoa, including *Trypanosoma brucei* (West African variant), and fungi (e.g., *Pneumocystis*) **Route of administration:** IM, IV **Adverse effects:** Rash, low blood pressure, irregular heartbeat, kidney and liver failure	Not known

Quinolones
 Representatives:
 Natural quinine
 Semisynthetic quinines:
  Chloroquine
  Mefloquine
  Piperaquine
  Primaquine
	Natural and semisynthetic drugs derived from the bark of cinchona tree; though considered antibacterial, they inhibit metabolism of malaria parasites by one or more unknown methods	**Spectrum of action:** *Plasmodium*
		 Route of administration: Oral
		 Adverse effects: Allergic reactions, visual disturbances
			Results from the presence of quinoline pumps that remove the drugs from parasite's cells

MICRO IN THE CLINIC FOLLOW-UP
A Cure Leading to Disease?

Dr. Warner tells Thomas that his symptoms indicate he most likely has antimicrobial-associated diarrhea (AAD) caused by *Clostridium difficile*, which is a common side effect of taking antibiotics, especially fluoroquinolones, which include levofloxacin. The doctor tells Thomas that antibiotics increase the risk for AAD up to 7–10 times, but assures him the condition can be cured. Dr. Warner collects a stool (fecal) sample for a PCR test for *C. difficile* and tells Thomas to stop taking antidiarrheal medication immediately. He also instructs Thomas to drink more water, to eat saltine crackers and bananas to help restore electrolytes, and to wash his hands very carefully, since *C. diff.* endospores are easily transmitted via the fecal-oral route.

The results of the stool analysis confirm that Thomas has a *Clostridium difficile* infection. Dr. Warner prescribes metronidazole (another antimicrobial that inhibits DNA synthesis by *C. difficile*).

Within a few days, Thomas's cramping and diarrhea are gone. He finishes his course of metronidazole, and at his follow-up visit with Dr. Warner he is told that he has made a full recovery, though the doctor informs him that he is at a higher risk (15–35%) for having a recurrent *C. diff.* infection because of his age (over 60) and already having been infected.

1. Thomas's pneumonia was treated with levofloxacin, a fluoroquinolone. What is the mechanism of action for this class of antibiotics?
2. How could antibiotic treatment, which is meant to cure an infection, lead to an infection with another organism?

Check your answers to Micro in the Clinic Follow-Up questions in the Mastering Microbiology Study Area.

CHAPTER SUMMARY

 Go to Mastering Microbiology for Interactive Microbiology, Dr. Bauman's Video Tutors, Micro Matters, MicroFlix, Micro-Boosters, practice quizzes, and more.

 Make the connection between Chapters 9 and 10. Watch *Micro Matters* videos in the Mastering Microbiology Study Area.

The History of Antimicrobial Agents (pp. 289)

1. Chemotherapeutic agents are chemicals used to treat diseases. Among them are **antimicrobial agents**, or **antimicrobials**, which include **antibiotics** (biologically produced agents), **semisynthetic antimicrobials** (chemically modified antibiotics), and **synthetic drugs**.

Mechanisms of Antimicrobial Action (pp. 289–298)

1. Successful chemotherapy against microbes is based on **selective toxicity**, that is, using antimicrobial agents that are more toxic to pathogens than to the patient.
2. Antimicrobial drugs affect pathogens by inhibiting cell wall synthesis, inhibiting the translation of proteins, disrupting cytoplasmic membranes, inhibiting general metabolic pathways, inhibiting nucleic acid synthesis, blocking the attachment of viruses to their hosts, or blocking a pathogen's recognition of its host.

 Play *Chemotherapeutic Agents: Modes of Action* @ Mastering Microbiology

3. **Beta-lactams**—penicillins, carbapenems, cephalosporins—have a functional β-lactam ring. They prevent bacteria from cross-linking NAM subunits of peptidoglycan in the bacterial cell wall during growth. Lipoglycopeptides (for example, vancomycin) and cycloserine also disrupt cell wall formation in many Gram-positive bacteria. **Bacitracin** blocks NAG and NAM transport from the cytoplasm. **Isoniazid (INH)** and **ethambutol** block mycolic acid synthesis in the walls of mycobacteria. **Echinocandins** block synthesis of fungal cell walls.

4. Antimicrobial agents that inhibit protein synthesis by interfering with 70S ribosomes include **aminoglycosides** and **tetracyclines**, which inhibit functions of the 30S ribosomal subunit, and **chloramphenicol**, **lincosamides**, **streptogramins**, and **macrolides**, which inhibit 50S subunits. **Oxazolidinones** block initiation of translation. **Antisense nucleic acid** molecules inhibit protein synthesis by preventing ribosomes from assembling on mRNA.

5. **Mupirocin** binds to isoleucine aminoacyl-tRNA synthetase—the enzyme that loads isoleucine onto the tRNA.

6. **Polyenes**, **azoles**, and **allylamines** disrupt the cytoplasmic membranes of fungi. Polymyxin acts against the membranes of Gram-negative bacteria.

7. **Sulfonamides** are structural **analogs** of para-aminobenzoic acid (PABA), a chemical needed by some microorganisms but not by humans. The substitution of sulfonamides in the metabolic pathway leading to nucleic acid synthesis kills those organisms. Trimethoprim also blocks this pathway.

8. Drugs that inhibit nucleic acid replication in pathogens include actinomycin, **nucleotide analogs** and **nucleoside analogs**, **quinolones**, and rifampin.

Clinical Considerations in Prescribing Antimicrobial Drugs (pp. 298–303)

1. Chemotherapeutic agents have a **spectrum of action** and may be classed as either **narrow-spectrum drugs** or **broad-spectrum drugs**, depending on how many kinds of pathogens they affect.

2. **Diffusion susceptibility tests**, or Kirby-Bauer tests, reveal which drug is most effective against a particular pathogen; in general,

the larger the **zone of inhibition** around a drug-soaked disk on a Petri plate, the more effective the drug.

3. The **minimum inhibitory concentration (MIC)**, usually determined by either a **broth dilution test** or an **Etest**, is the smallest amount of a drug that will inhibit a pathogen.

4. A **minimum bactericidal concentration (MBC) test** ascertains whether a drug is bacteriostatic and the lowest concentration of a drug that is bactericidal.

5. In choosing antimicrobials, physicians must consider effectiveness, how a drug is best administered—orally, intramuscularly, or intravenously—and possible side effects, including toxicity and allergic responses.

6. A drug's **therapeutic index (TI)** is essentially a ratio of the drug's tolerated dose to its effective dose. The higher the TI, the safer the drug.

7. Clinicians use the term **therapeutic window** to indicate the range of concentrations of a drug that are effective without being excessively toxic.

Resistance to Antimicrobial Drugs (pp. 303–308)

1. Some members of a pathogenic population may develop resistance to a drug because of extra DNA pieces called **R plasmids** or the mutation of genes. Microorganisms may resist a drug by producing enzymes such as **beta (β)-lactamase** that deactivate the drug, by inducing changes in the cell membrane that prevent entry of the drug, by altering the drug's target to prevent its binding, by altering the cell's metabolic pathways, by removing the drug from the cell with **efflux pumps**, or by protecting the drug's target by binding another molecule to it.

Play *Antibiotic Resistance: Origins of Resistance, Forms of Resistance* @ Mastering Microbiology

2. **Persister cells** are bacterial cells that resist the action of antimicrobial drugs even though they are genetically identical to susceptible cells. Persistence may be involved in chronic infections and in the development of resistance. The mechanisms of persistence are not known.

3. **Cross resistance** occurs when resistance to one chemotherapeutic agent confers resistance to similar drugs. **Multiple-drug-resistant pathogens** are resistant to three or more types of antimicrobial drugs.

4. **Synergism** describes the interplay between drugs that results in efficacy that exceeds the efficacy of either drug alone. Some drug combinations are antagonistic.

QUESTIONS FOR REVIEW

Answers to the Questions for Review (except Short Answer questions) begin on p. A-1.

Multiple Choice

1. Diffusion and dilution tests that expose pathogens to antimicrobials are designed to determine _____.
   a. the spectrum of action of a drug
   b. which drug is most effective against a particular pathogen
   c. the amount of a drug to use against a particular pathogen
   d. both b and c

2. In a Kirby-Bauer susceptibility test, the presence of a zone of inhibition around disks containing antimicrobial agents indicates _____.
   a. that the microbe does not grow in the presence of the agents
   b. that the microbe grows well in the presence of the agents
   c. the smallest amount of the agent that will inhibit the growth of the microbe
   d. the minimum amount of an agent that kills the microbe in question

3. The key to successful chemotherapy is _____.
   a. selective toxicity
   b. a diffusion test
   c. the minimum inhibitory concentration test
   d. the spectrum of action

4. Which of the following statements is relevant in explaining why sulfonamides are effective?
   a. Sulfonamides attach to sterol lipids in the pathogen, disrupt the membranes, and lyse the cells.
   b. Sulfonamides prevent the incorporation of amino acids into polypeptide chains.
   c. Humans and microbes use PABA differently in their metabolism.
   d. Sulfonamides inhibit DNA replication in both pathogens and human cells.

5. Cross resistance is _____.
   a. the deactivation of an antimicrobial agent by a bacterial enzyme
   b. alteration of the resistant cells so that an antimicrobial agent cannot attach
   c. the mutation of genes that affect the cytoplasmic membrane channels so that antimicrobial agents cannot cross into the cell's interior
   d. resistance to one antimicrobial agent because of its similarity to another antimicrobial agent

6. Multiple-drug-resistant microbes _____.
   a. are resistant to all antimicrobial agents
   b. respond to new antimicrobials by developing resistance
   c. frequently develop in hospitals
   d. all of the above

7. Which of the following is most closely associated with a beta-lactam ring?
   a. penicillin
   b. vancomycin
   c. bacitracin
   d. isoniazid

8. Drugs that act against protein synthesis include _____.
   a. beta-lactams
   b. trimethoprim
   c. polymyxin
   d. aminoglycosides

9. Which of the following statements is *false* concerning antiviral drugs?
   a. Macrolide drugs block attachment sites on the host cell wall and prevent viruses from entering.
   b. Drugs that neutralize the acidity of phagolysosomes prevent viral uncoating.
   c. Nucleotide analogs can be used to stop microbial replication.
   d. Drugs containing protease inhibitors retard viral growth by blocking the production of essential viral proteins.

10. PABA is _____.
    a. a substrate used in the production of penicillin
    b. a type of β-lactamase
    c. molecularly similar to cephalosporins
    d. used to synthesize folic acid

VISUALIZE IT!

1. Label each of the accompanying figures to indicate the class of drug that is stopping polypeptide translation.

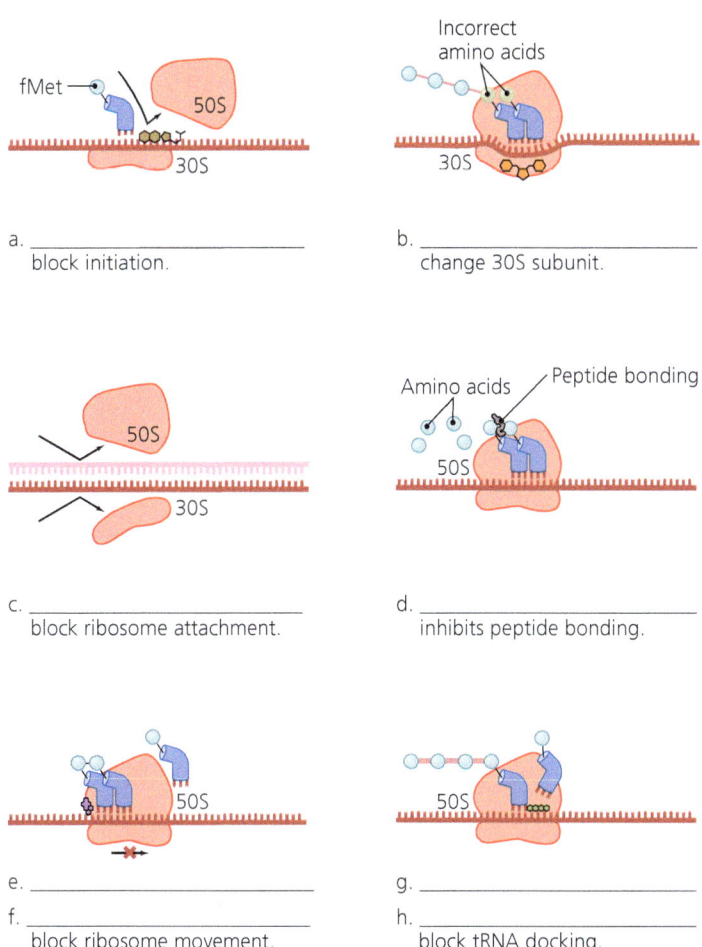

a. _____ block initiation.

b. _____ change 30S subunit.

c. _____ block ribosome attachment.

d. _____ inhibits peptide bonding.

e. _____
f. _____ block ribosome movement.

g. _____
h. _____ block tRNA docking.

Short Answer

1. What characteristics would an ideal chemotherapeutic agent have? Which drug has these qualities?

2. Contrast narrow-spectrum and broad-spectrum drugs. Which are more effective?

3. Why is the fact that drug Z destroys the NAM portions of a cell's wall structure an important factor in considering the drug for chemotherapy?

4. Given that both human cells and pathogens synthesize proteins at ribosomal sites, how can antimicrobial agents that target this process be safe to use in humans?

5. Support or refute the following statement: Antimicrobial agents make cells become resistant.

6. Given that resistant strains of pathogens are a concern to the general health of a population, what can be done to prevent their development?

7. Why are antiviral drugs difficult to develop?

8. A man has been given a broad-spectrum antibiotic for his stomach ulcer. What unintended consequences could arise from this therapy?

9. Compare and contrast the actions of polyenes, azoles, allylamines, and polymyxin.

10. What is the difference in drug action of synergists contrasted with that of antagonists?

2. What specific test for antimicrobial efficacy is shown? What does this test measure? Draw an oval to predict the size and shape of the zone of inhibition if the drug concentration on the strip were increased twofold.

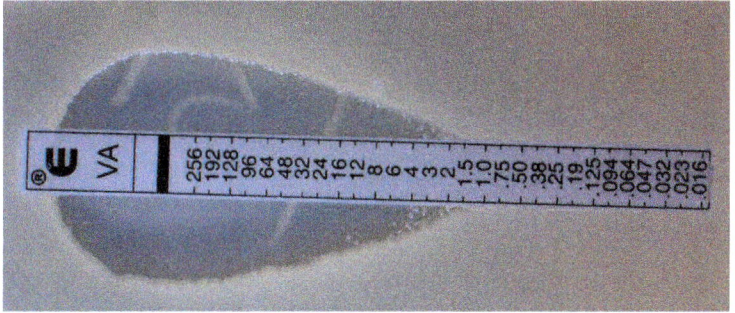

CRITICAL THINKING

1. AIDS is treated with a "cocktail" of several antiviral agents at once. Why is the cocktail more effective than a single agent? What is a physician trying to prevent by prescribing several drugs at once?

2. How does *Penicillium* escape the effects of the penicillin it secretes?

3. How might a colony of *Bacillus licheniformis* escape the effects of its own bacitracin?

4. Fewer than 1% of known antibiotics have any practical value in treatment of disease. Why is this so?

5. In an issue of *News of the Lepidopterists' Society*, a recommendation was made to moth and butterfly collectors to use antimicrobials to combat disease in the young of these insects. What are the possible ramifications for human health of such usage of antimicrobials?

6. Even though aminoglycosides such as gentamicin can cause deafness, there are still times when they are the best choice for treating some infections. What laboratory test would a clinical scientist use to show that gentamicin is the best choice to treat a particular *Pseudomonas* infection?

7. Your pregnant neighbor has a sore throat and tells you that she is taking some tetracycline she had left over from a previous infection. Give two reasons why her decision is a poor one.

8. Acyclovir has replaced adenine arabinoside as treatment for herpes infections. Compare the ways these drugs are activated (see Table 10.3 on pp. 313–314). Why is acyclovir a better choice?

9. Why might amphotericin B affect the kidneys more than other human organs?

10. Antiparasitic drugs in the benzimidazole family inhibit the polymerization of tubulin. What effect might these drugs have on mitosis and flagella?

11. Your cousin reads in a blog that the U.S. Food and Drug Administration (FDA) has approved an antimicrobial called tigecycline. The blog says that the drug is an analog of tetracycline. She asks you what that means and how the drug works. What should you tell your cousin?

12. Scientists have cultured bacteria isolated from within frozen mammoths, which are thousands of years old. Why would it not be surprising if these microbes were to show some resistance to modern antimicrobials that didn't exist when the mammoths died?

13. It would be impractical and expensive for every American to take oseltamivir during the entire flu season to prevent influenza infections. For what group of people might oseltamivir prophylaxis be cost effective?

14. Sometimes it is not possible to conduct a susceptibility test because of either a lack of time or an inability to access the bacteria (for example, from an inner-ear infection). How could a physician select an appropriate therapeutic agent in such cases?

15. *Enterococcus faecium* is frequently resistant to vancomycin. Why might this be of concern in a hospital setting in terms of developing resistant strains of *other* genera of bacteria?

CONCEPT MAPPING

Using the terms provided, fill in the concept map that describes antimicrobial resistance. You can also complete this and other concept maps online by going to the Mastering Microbiology Study Area.

Altered targets
Beta-lactamase
Cell division
Conjugation
Efflux pumps
Entry of antimicrobials into cell
Mutation
Pathogen's enzymes
Penicillin
Transduction
Transformation

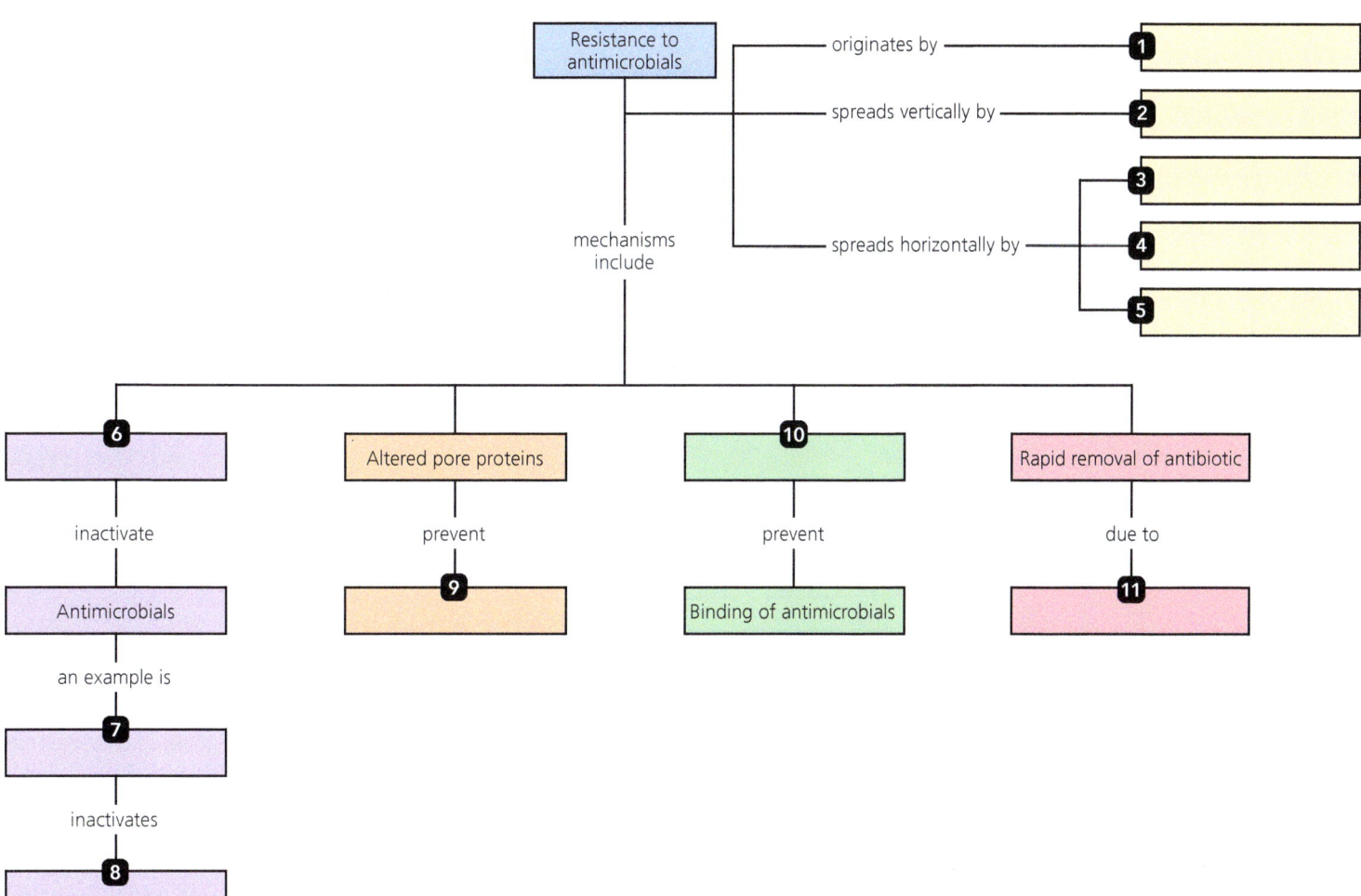

11 Characterizing and Classifying Prokaryotes

Before You Begin

1. What type of microbes have a cell wall that consists mostly of peptidoglycan?
2. During DNA replication, what is the difference between synthesis of the leading and of the lagging strand?
3. The pentose phosphate pathway is an alternative to which more common metabolic pathway?
4. How much ATP can be produced per molecule of glucose by sulfur reducers, in contrast to fermenters and aerobic respirators?

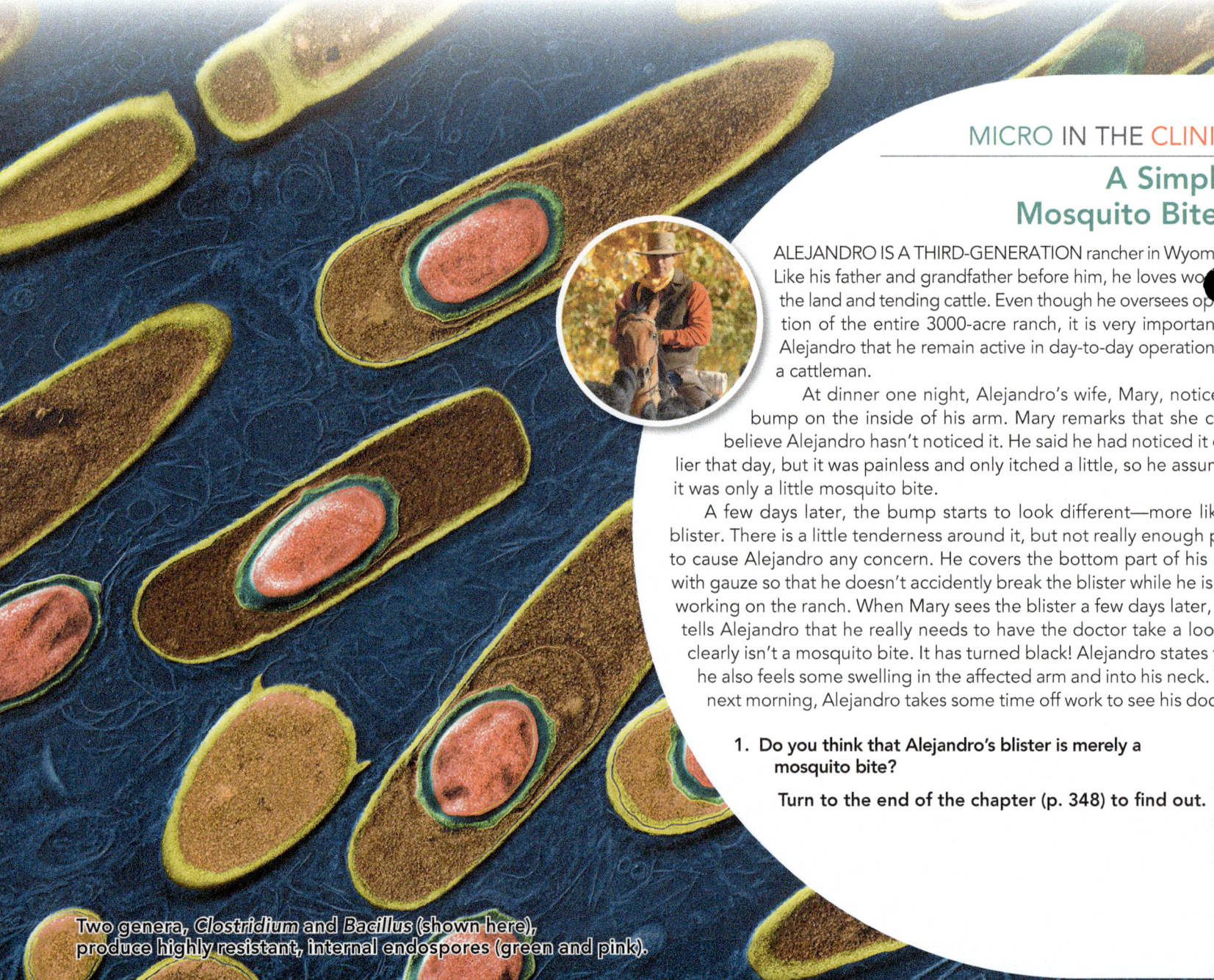

MICRO IN THE CLINI[C]

A Simpl[e] Mosquito Bite

ALEJANDRO IS A THIRD-GENERATION rancher in Wyom[ing]. Like his father and grandfather before him, he loves wo[rking] the land and tending cattle. Even though he oversees op[era]tion of the entire 3000-acre ranch, it is very importan[t to] Alejandro that he remain active in day-to-day operation[s as] a cattleman.

At dinner one night, Alejandro's wife, Mary, notice[s a] bump on the inside of his arm. Mary remarks that she c[an't] believe Alejandro hasn't noticed it. He said he had noticed it e[ar]lier that day, but it was painless and only itched a little, so he assum[ed] it was only a little mosquito bite.

A few days later, the bump starts to look different—more lik[e a] blister. There is a little tenderness around it, but not really enough p[ain] to cause Alejandro any concern. He covers the bottom part of his [arm] with gauze so that he doesn't accidentally break the blister while he is working on the ranch. When Mary sees the blister a few days later, [she] tells Alejandro that he really needs to have the doctor take a loo[k. It] clearly isn't a mosquito bite. It has turned black! Alejandro states t[hat] he also feels some swelling in the affected arm and into his neck. [The] next morning, Alejandro takes some time off work to see his doc[tor.]

1. Do you think that Alejandro's blister is merely a mosquito bite?

Turn to the end of the chapter (p. 348) to find out.

Two genera, *Clostridium* and *Bacillus* (shown here), produce highly resistant, internal endospores (green and pink).

Prokaryotes are by far the most numerous and diverse group of cellular microbes. Scientists estimate there are more than 5×10^{30} prokaryotes on Earth. If laid end to end, they would more than encircle the entire Milky Way galaxy. They thrive in every habitat: from Antarctic glaciers to thermal hot springs to the dry sands of deserts, from the colons of animals to the cytoplasm of other prokaryotes, from distilled water to supersaturated brine, and from disinfectant solutions to basalt rocks thousands of meters below the Earth's surface. Among this great diversity of prokaryotes, a very few have virulence factors (enzymes, toxins, or cellular structures) that enable them to colonize humans or cause diseases. We begin this chapter by examining general prokaryotic characteristics and conclude with a survey of specific prokaryotic taxa, briefly mentioning human pathogens throughout the chapter. (More detailed discussions of disease-causing prokaryotes appear in Chapters 19–21.)

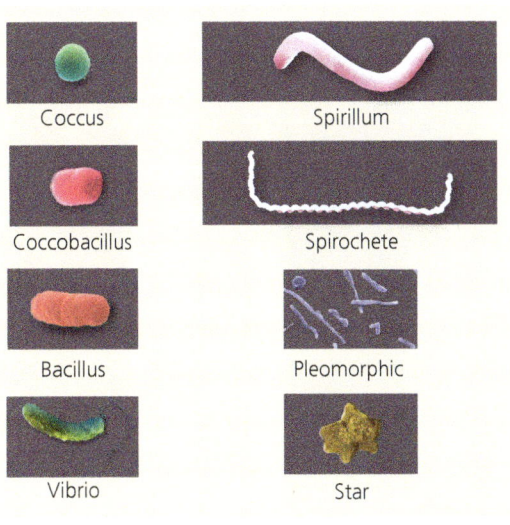

▲ **FIGURE 11.1 Typical prokaryotic morphologies.** *What is one difference between a spirillum and a spirochete?*

Figure 11.1 Generally, spirilla are stiff, whereas spirochetes are flexible.

General Characteristics of Prokaryotic Organisms

In this section, we consider shapes of prokaryotic cells, the ability of some prokaryotes to form endospores that allow them to survive unfavorable conditions, some reproduction strategies, and spatial arrangements of prokaryotic cells. (Chapters 3 and 5–7 consider other general characteristics of prokaryotic cells, including cellular structure, metabolism, growth, and genetics.)

Morphology of Prokaryotic Cells

LEARNING OUTCOME

11.1 Identify six basic shapes of prokaryotic cells.

Prokaryotic cells exist in a variety of shapes, or morphologies (**FIGURE 11.1**). The three basic shapes are **coccus** (kok'ŭs; roughly spherical; plural: *cocci*, kok'sī), **bacillus** (ba-sil'ŭs; rod shaped; plural: *bacilli*, bă-sil'ī), and **spiral**. Cocci are not all perfectly spherical; for example, there are pointed, kidney-shaped, and oval cocci. Similarly, bacilli vary in shape; for example, some bacilli are pointed, others are spindle-shaped, and still others are threadlike (filamentous). Spiral-shaped prokaryotes are either **spirilla**, which are stiff, or **spirochetes** (spī'rō-kētz), which are flexible. Curved rods are **vibrios**. The term **coccobacillus** is used to describe cells that are intermediate in shape between cocci and bacilli; that is, it is difficult to ascertain whether a coccobacillus is an elongated coccus or a short bacillus. In addition to these basic shapes, there are star-shaped, triangular, and rectangular prokaryotes as well as prokaryotes that are **pleomorphic**[1] (plē-ō-mōr'fik), that is, vary in shape and size.

Endospores

LEARNING OUTCOME

11.2 Describe the formation and function of bacterial endospores.

The Gram-positive bacteria *Bacillus* (ba-sil'ŭs) and *Clostridium* (klos-trid'ē-ŭm) produce **endospores**, which are important for several reasons, including their durability and potential pathogenicity. Endospores constitute a defensive strategy: They are stable resting stages that remain dormant and viable in hostile or unfavorable conditions, and they germinate when conditions improve.

Though some people refer to endospores as "spores," endospores should not be confused with reproductive spores, such as those of algae and fungi. A single bacterial cell, called a *vegetative cell* to distinguish it from an endospore, transforms into only one endospore, which reactivates to grow into a single vegetative cell; therefore, endospores are not reproductive structures. No new cells are produced from endospores.

The process of endospore formation, called *sporulation*, requires 8 to 10 hours and proceeds in seven steps (see Chapter 3, Figure 3.25). Depending on the species, a vegetative cell forms an endospore either *centrally*, *subterminally* (near one end), or *terminally* (at one end) (**FIGURE 11.2**).

Endospore formation is a concern to food processors, health care professionals, and governments. Endospores are difficult to kill; moreover, many endospore-forming bacteria produce deadly toxins that cause fatal diseases, such as anthrax, tetanus, and gangrene.

---

[1] From Greek *pleon*, meaning "more," and *morphe*, meaning "form."

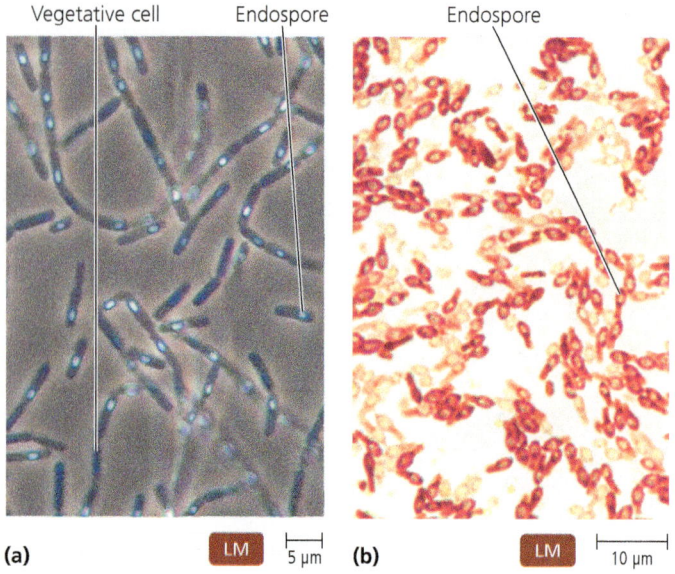

▲ FIGURE 11.2 Locations of endospores. (a) Central endospores of *Bacillus*. (b) Subterminal endospores of *Clostridium botulinum*. The enlarged endospores have swollen the vegetative cells that produced them.

Reproduction of Prokaryotic Cells

LEARNING OUTCOMES

11.3 List three common types of reproduction in prokaryotes.
11.4 Describe snapping division as a type of binary fission.

All prokaryotes reproduce asexually; none reproduce sexually. The most common method of asexual reproduction is **binary fission**, which proceeds as follows (FIGURE 11.3): ① The cell replicates its DNA by *semiconservative replication*, meaning that each molecule of DNA consists of one strand of original DNA and one strand of new DNA (see Chapter 7). Each DNA molecule remains attached to the cytoplasmic membrane. ② The cell grows, and as the cytoplasmic membrane elongates, it moves the daughter molecules of DNA apart. ③ The cell forms a cross wall, invaginating the cytoplasmic membrane. ④ The cross wall completely divides daughter cells. ⑤ The daughter cells may or may not separate.

Play *Bacterial Growth: Overview*
@ Mastering Microbiology

The parental cell disappears with the formation of progeny.

A variation of binary fission called **snapping division** occurs in some Gram-positive bacilli (FIGURE 11.4). In snapping division, only the inner portion of a cell wall is deposited across the dividing cell. As the cells grow and the new wall thickens, they put tension on the outer layer of the old cell wall, which still holds the two cells together. Eventually, as the tension increases, the outer wall breaks at its weakest point with a snapping movement that tears it most of the way around. The daughter cells then remain hanging together almost side by side, held at an angle by a small remnant of the original outer wall that acts like a hinge.

Some prokaryotes have other methods of reproduction in which a parental cell retains its identity. For example, the *actinomycetes* (ak′ti-nō-mī-sētz) produce reproductive cells called

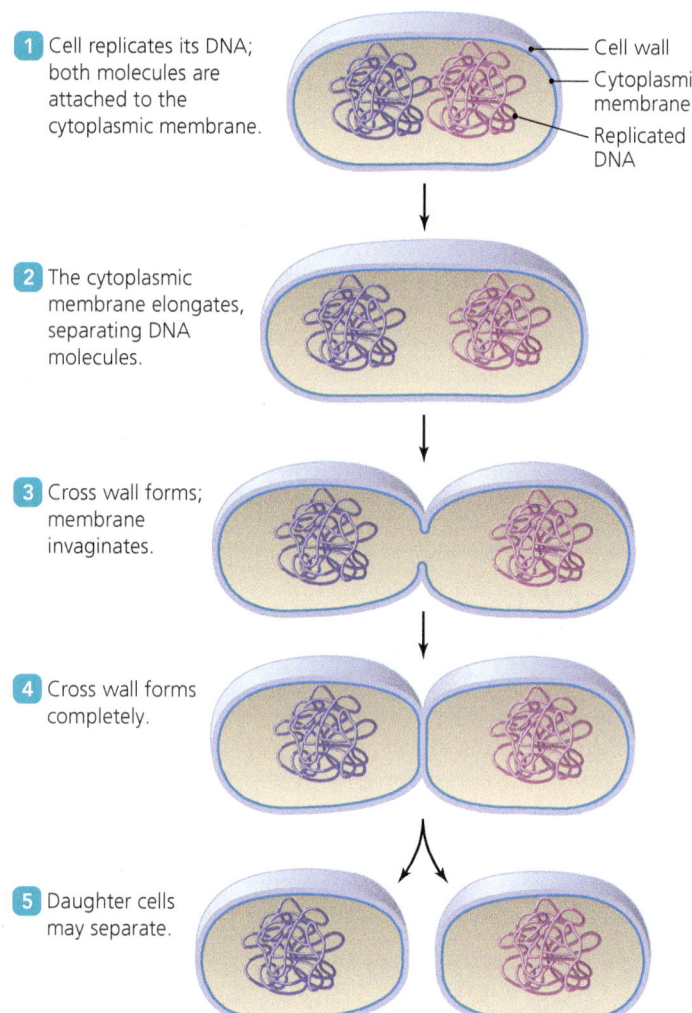

▲ FIGURE 11.3 Binary fission.

spores[2] at the ends of filamentous cells (FIGURE 11.5). These are true spores and should not be confused with endospores. Each actinomycete spore contains complete genetic information and can develop into a clone of the original organism.

Other prokaryotes reproduce by fragmentation into small motile filaments with a full complement of DNA that glide away from the parental strand. Still others, such as the bacterium *Planctomyces* (plank-tō-mī′sēz), reproduce by **budding**, in which an outgrowth of the original cell (a bud) receives a copy of DNA and enlarges. Eventually the bud is cut off from the parental cell, typically while it is still quite small (FIGURE 11.6).

Epulopiscium (ep′yoo-lō-pis′sē-ŭm), a giant bacterium that lives inside surgeonfish, and many of its relatives have a truly unique method of reproduction among prokaryotes: They give "birth" to as many as 12 live offspring that emerge from the body of their dead mother cell (FIGURE 11.7). The production of live offspring within a mother is called *viviparity*, and this is the first documented case of viviparous behavior in the prokaryotic world. In these bacteria, formation of internal offspring proceeds in a manner similar to the early stages of endospore formation.

---
[2]These true spores, which are reproductive, should not be confused with endospores, which are resting stages only.

General Characteristics of Prokaryotic Organisms 327

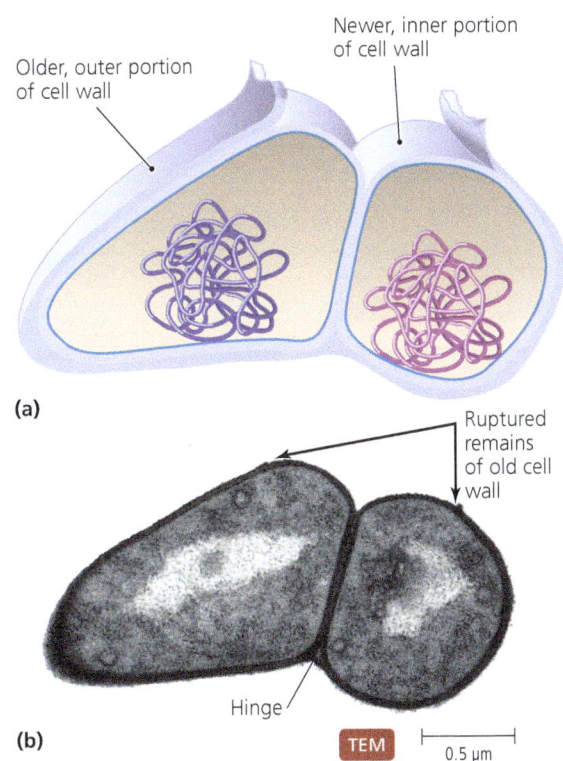

▲ FIGURE 11.4 Snapping division, a variation of binary fission. (a) Only the inner portion of the cell wall forms a cross wall. (b) As the daughter cells grow, tension snaps the outer portion of the cell wall, leaving the daughter cells connected by a hinge of old cell wall material.

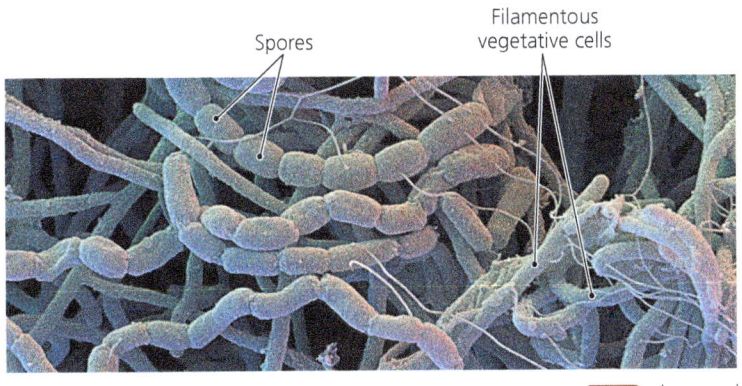

▲ FIGURE 11.5 Spores of actinomycetes. Filamentous vegetative cells produce long, filamentous cells and chains of spores, shown here in *Streptomyces*.

Arrangements of Prokaryotic Cells

LEARNING OUTCOME

11.5 Draw and label nine arrangements of prokaryotes.

The arrangements of prokaryotic cells result from two aspects of division during binary fission: the planes in which cells divide and whether or not daughter cells remain attached to each other. Cocci that remain attached in pairs are **diplococci** (FIGURE 11.8a),

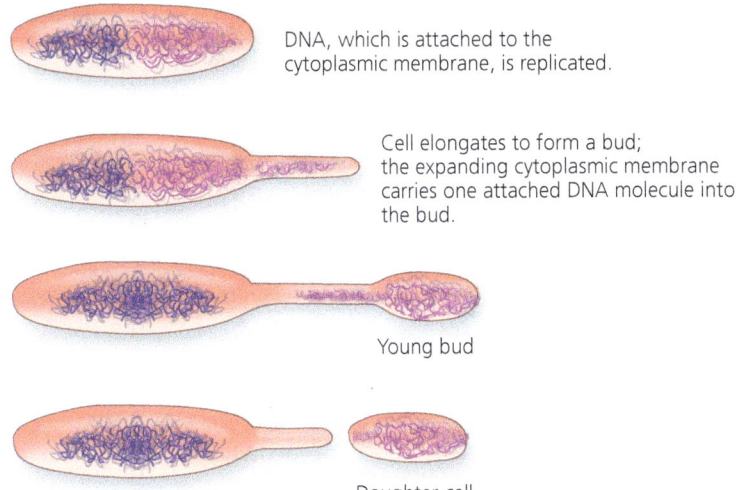

▲ FIGURE 11.6 Budding. How does budding differ from binary fission?

Figure 11.6 In binary fission, the parent cell disappears with the formation of two equal-sized offspring; a bud, in contrast, is often much smaller than its parent, and the parent remains to produce more buds.

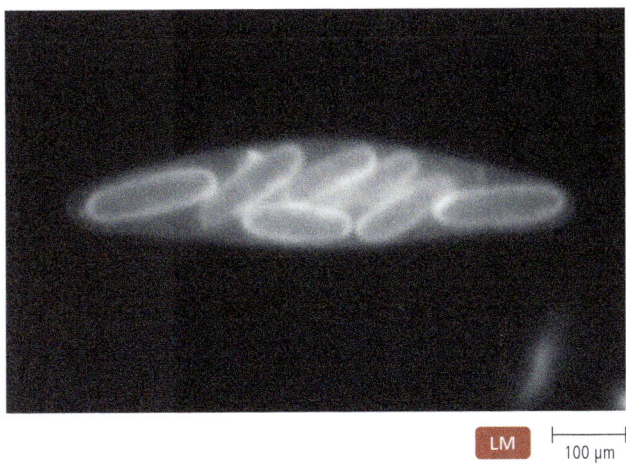

▲ FIGURE 11.7 Viviparity in *Epulopiscium*. Multiple offspring simultaneously develop inside a mother cell.

and long chains of cocci are called **streptococci**[3] (FIGURE 11.8b). Some cocci divide in two perpendicular planes and remain attached to form **tetrads** (FIGURE 11.8c); others divide in three planes to form cuboidal packets called **sarcinae**[4] (sar'si-nī) (FIGURE 11.8d). Clusters called **staphylococci**[5] (staf'i-lo-kok-sī), which look like bunches of grapes, form when the planes of cell division are random (FIGURE 11.8e).

Bacilli are less varied in their arrangements than cocci because bacilli divide transversely—that is, across their long axis. Daughter bacilli may separate to become single cells or stay attached as either pairs or chains (FIGURE 11.9a–c). Because the

---
[3]From Greek *streptos*, meaning "twisted," because long chains tend to twist.
[4]Latin for "bundles."
[5]From Greek *staphyle*, meaning "bunch of grapes."

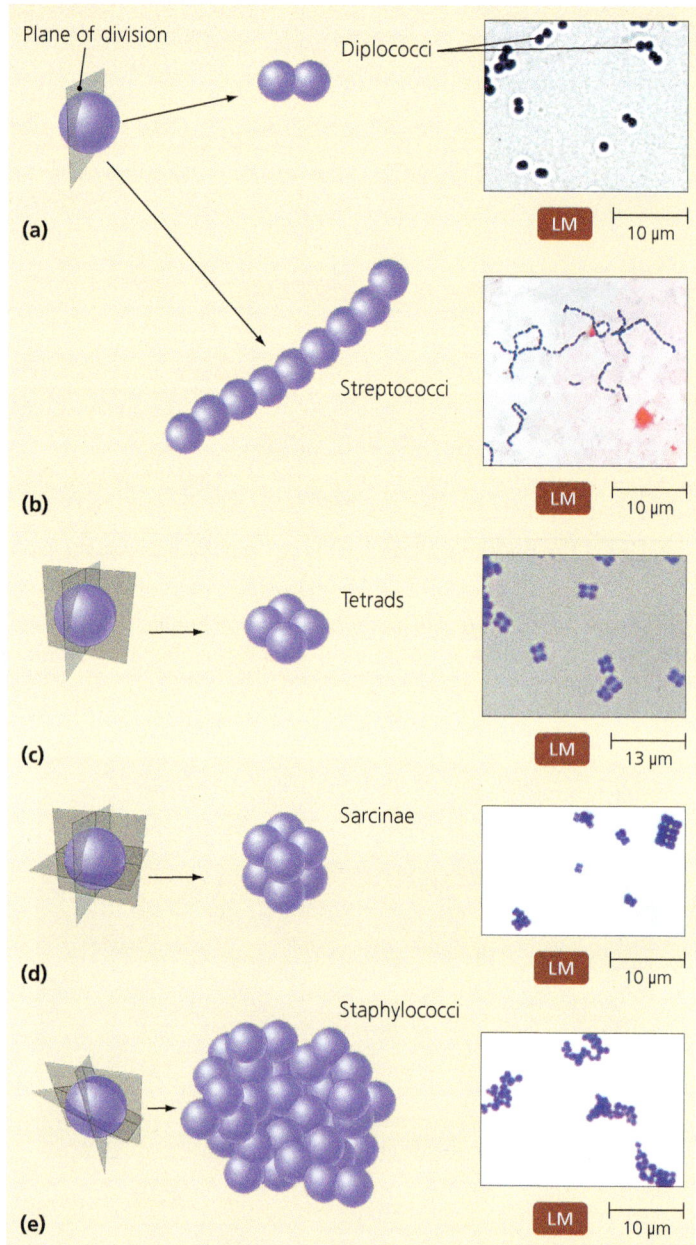

▲ FIGURE 11.8 Arrangements of cocci. (a) The diplococci of *Streptococcus pneumoniae*. (b) The streptococci of *Streptococcus pyogenes*. (c) Tetrads, in this case of *Micrococcus luteus*. (d) The genus *Sarcina* is characterized by sarcinae. (e) The staphylococci of *Staphylococcus aureus*.

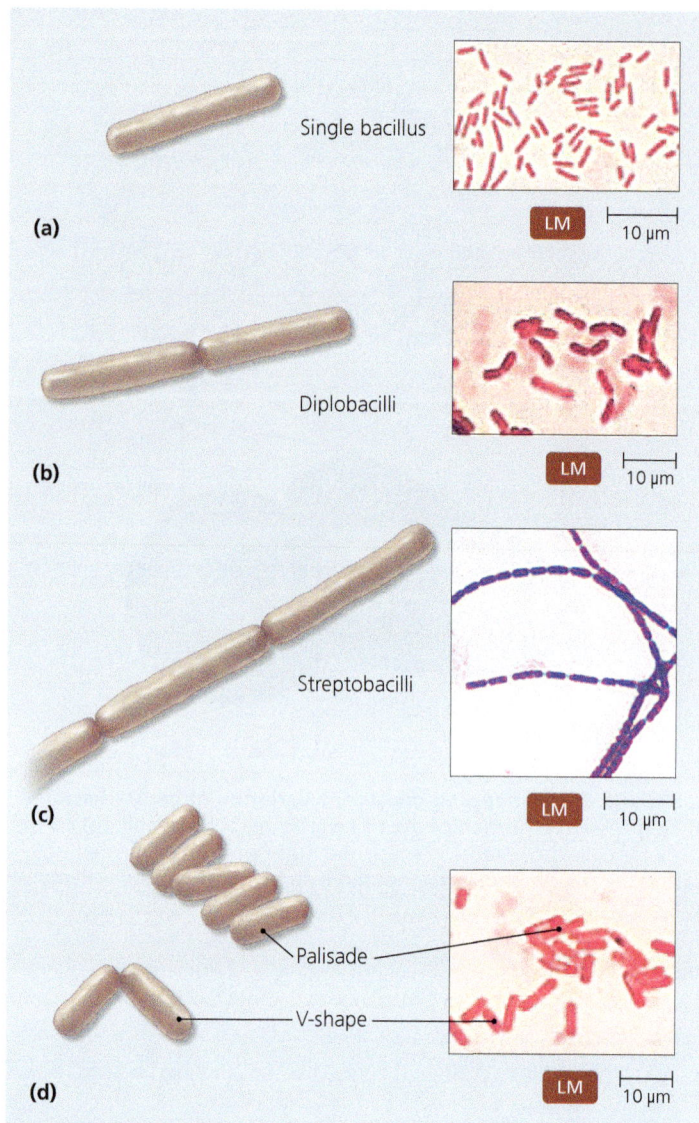

▲ FIGURE 11.9 Arrangements of bacilli. (a) A single bacillus of *Escherichia coli*. (b) Diplobacilli in a young culture of *Bacillus cereus*. (c) Streptobacilli in an older culture of *Bacillus cereus*. (d) V-shapes and palisades of *Corynebacterium diphtheriae*.

cells of *Corynebacterium diphtheriae* (kŏ-rī′nē-bak-tēr′ē-ŭm dif-thi′rē-ī), the causative agent of diphtheria, divide by snapping division, the daughter cells remain attached to form V-shapes and a side-by-side arrangement called a **palisade**[6] (FIGURE 11.9d).

Descriptive words can be used to refer either to a general shape and/or arrangement or to a specific genus. Thus, the characteristic shape of the genus *Bacillus* is a rod-shaped bacterium, and the characteristic arrangement of bacteria in the genus *Sarcina* (sar′si-nă) is cuboidal. In such potentially confusing cases, the meaning can be distinguished because genus names are always capitalized and italicized. In other cases, the arrangement uses the plural form, whereas a genus name is singular; thus, streptococci—spherical cells arranged in a chain—are characteristic of the genus *Streptococcus* (strep-tō-kok′ŭs).

Spirilla, spirochetes, and star-shaped, triangular, and pleomorphic prokaryotes are typically singular. Rectangular and square prokaryotes often form sheets of cells.

[6]From Latin *palus*, meaning "stake," referring to a fence made of adjoining stakes.

MICRO CHECK

1. What are flexible, spiral-shaped bacteria called?
2. What function do bacterial endospores perform?
3. What type of arrangement forms when multiple planes of division occur randomly in cocci?
4. What type of arrangement occurs when bacilli divide by snapping division?

> **TELL ME WHY**
>
> Why does binary fission produce chains of cocci in some species but clusters of cocci in other species?

Modern Prokaryotic Classification

LEARNING OUTCOMES

11.6 Explain the general purpose of *Bergey's Manual of Systematic Bacteriology*.
11.7 Discuss the veracity and limitations of any taxonomic scheme.

Scientists called *taxonomists* group similar organisms into categories called *taxa*. At one time, the least inclusive taxa of prokaryotes (i.e., genera and species) were based solely on growth habits and the characteristics we considered in the previous section, especially morphology and arrangement. More recently, the classification of living things has been based more on genetic similarities. Accordingly, modern taxonomists place all organisms into three *domains*—Archaea (ar-kē'ǎ), Bacteria, and Eukarya—which are the largest, most inclusive taxa. Scientists recognize bacterial and archaeal (both prokaryotic) taxa primarily on the basis of similarities in RNA, DNA, and protein sequences.

The vast majority of prokaryotes—perhaps as many as 99.5%, and probably millions of species—have never been isolated or cultured and are known, if at all, only from their ribosomal (rRNA) "fingerprints;" that is, they are known only from sequences of rRNA that do not match any known rRNA sequences. In light of this information, taxonomists now construct modern classification schemes of prokaryotes based primarily on the relative similarities of rRNA sequences found in various prokaryotic taxa (**FIGURE 11.10**).

An authoritative reference in modern prokaryotic systematics is *Bergey's Manual of Systematic Bacteriology*, which classifies prokaryotes into 29 phyla—5 in Archaea and 24 in Bacteria. The five volumes of the second edition of *Bergey's Manual* discuss the great diversity of prokaryotes based in large part (but not exclusively) on their possible evolutionary relationships, as reflected in their rRNA sequences.

Our examination of prokaryotic diversity in this text is for the most part organized to reflect the taxonomic scheme that appears in *Bergey's Manual*, but it is important to note that as authoritative as *Bergey's Manual* is, it is not an "official" list of prokaryotic taxa. The reason is that taxonomy is partly a matter of opinion and judgment, and not all taxonomists agree. Legitimately differing views often change as more information is uncovered and examined, so *Bergey's Manual* is merely a consensus of experts at a given time. More information about *Bergey's Manual* can be found on Mastering Microbiology.

In the following sections, we examine representative prokaryotes of major phyla. We begin our exploration of prokaryotic diversity with a survey of archaea.

> **TELL ME WHY**
>
> Why are taxonomic names and categories in our current taxonomic scheme different from those in 1990?

Survey of Archaea

LEARNING OUTCOME

11.8 Identify the common features of microbes in the domain Archaea.

Scientists originally identified archaea as a distinct type of prokaryote on the basis of unique rRNA sequences. Archaea also share other common features that distinguish them from bacteria:

- Archaea lack true peptidoglycan in their cell walls.
- Their cytoplasmic membrane lipids have branched or ring-form hydrocarbon chains, whereas bacterial membrane lipids have straight chains.
- The initial amino acid in their polypeptide chains, coded by the AUG start codon, is methionine (as in eukaryotes and in contrast to the *N*-formylmethionine used by bacteria).

Archaea are currently classified in two major phyla—Crenarchaeota (kren-ar'kē-ō-ta) and Euryarchaeota (yŭr-ē-ar'kē-ō-ta)—and three minor phyla, which each contain only one or a few described species. Researchers discovered these minor phyla by analyzing rRNA sequences. They have cultured few of the archaeal species from the minor phyla.

Archaea reproduce by binary fission, budding, or fragmentation. Known archaeal cells are cocci, bacilli, spirals, or pleomorphs (**FIGURE 11.11**). Archaeal cell walls vary among taxa and are composed of a variety of compounds, including proteins, glycoproteins, lipoproteins, and polysaccharides; all lack peptidoglycan. Another surprising feature for an entire domain of life is that to date microbiologists have not discovered any archaeon that causes disease.

The first archaea identified were found in extreme environments considerably hostile to most life; these archaea are *extremophiles*, which we discuss next. Other archaea are *methanogens* (discussed shortly). We now know that most archaea live in moderate environmental conditions, including the mouths of people. It should be noted that some bacteria are also extremophiles.

> **MICRO CHECK**
>
> 5. What is the main basis for classifying prokaryotes in *Bergey's Manual of Systematic Bacteriology*?
> 6. In which of these ways do archaea differ from bacteria?

Extremophiles

LEARNING OUTCOME

11.9 Compare and contrast the two kinds of extremophiles discussed in this section.

Extremophiles are microbes that require what humans consider to be extreme conditions of temperature, pH, pressure, and/or salinity to survive. There are extremophilic bacteria (mentioned later) as well as archaea. Prominent among extremophiles are *thermophiles* and *halophiles*.

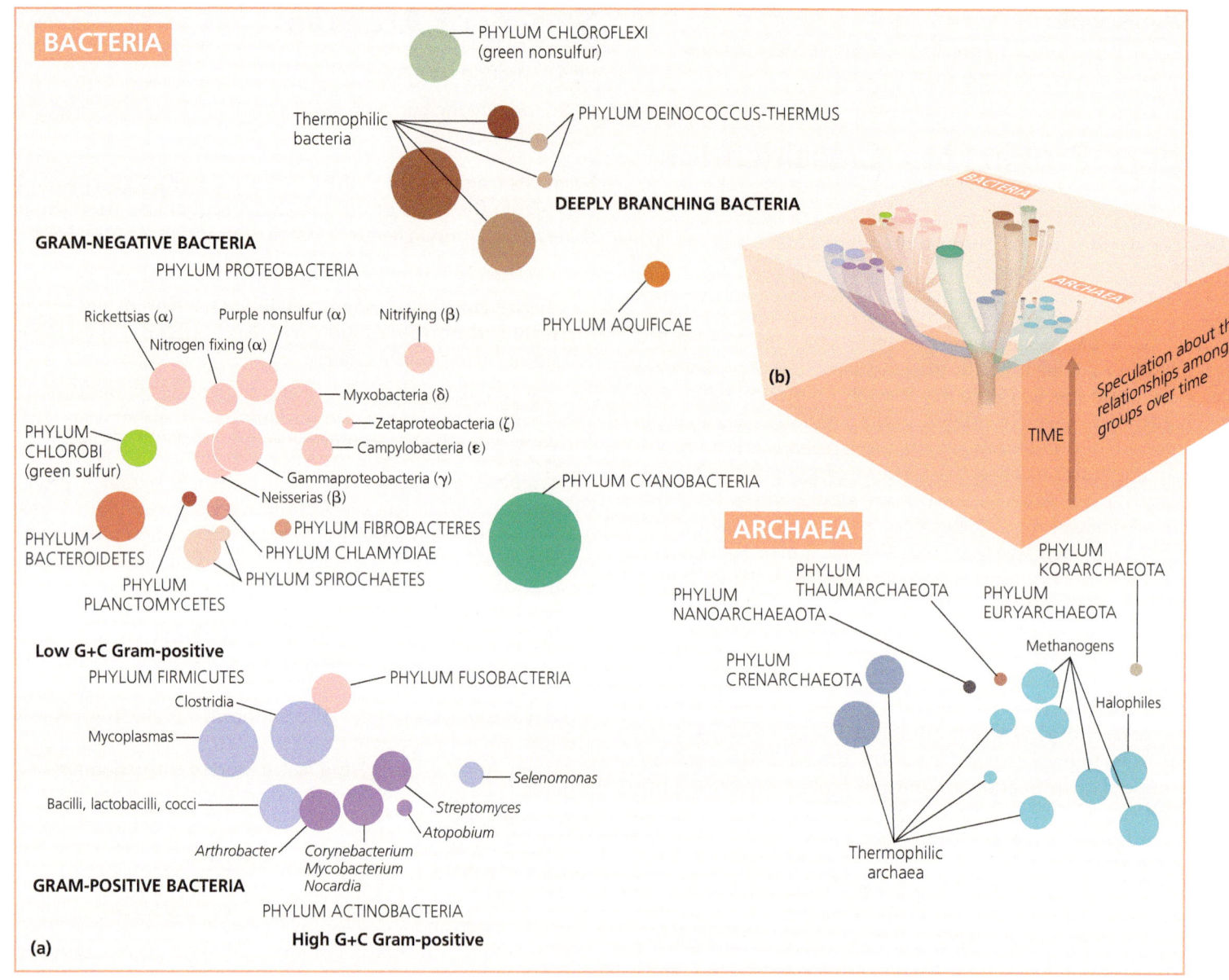

▲ FIGURE 11.10 **Prokaryotic taxonomy.** (a) This scheme is based on relatedness according to rRNA sequences. The closer together the disks, the more similar are the rRNA sequences of the species within the taxon. The sizes of disks are proportional to the number of species known for that group. Note that archaea are distinctly separate from bacteria. The discussion in this chapter is based largely on the scheme depicted in this figure. Not all phyla are shown. (In phylum Proteobacteria, classes are indicated by a Greek letter: α = alpha, β = beta, γ = gamma, δ = delta, ε = epsilon, and ζ = zeta.) (b) The exact relationships between and among groups over time are speculative. (Adapted from *Road Map to Bergey's*, 2002, Bergey's Manual Trust.)

Thermophiles

Thermophiles[7] are prokaryotes whose DNA, RNA, cytoplasmic membranes, and proteins do not function properly at temperatures lower than about 45°C. Prokaryotes that require temperatures over 80°C are called **hyperthermophiles**. Most thermophilic archaea are in the phylum Crenarchaeota, though some are also found in the phylum Euryarchaeota.

Two representative genera of thermophilic archaea are *Thermococcus* (ther-mō-kok'us) and *Pyrodictium* (pī-rō-dik'tē-um; see Figure 11.11). These microorganisms live in acidic hot springs such as those found in deep ocean rifts and similar terrestrial volcanic habitats (**FIGURE 11.12**). *Geogemma* (jē'ō-jem-ă) is the current record holder for surviving high temperatures—it can survive two hours at 130°C! The cells of *Pyrodictium*, which live in deep-sea hydrothermal vents, are irregular disks with elongated protein tubules that attach them to grains of sulfur that they use as final electron acceptors in respiration.

---
[7]From Greek *thermos*, meaning "heat," and *philos*, meaning "love."

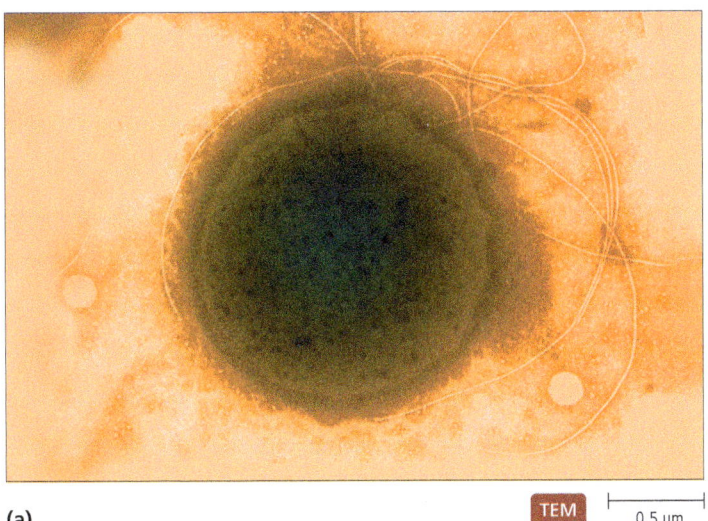

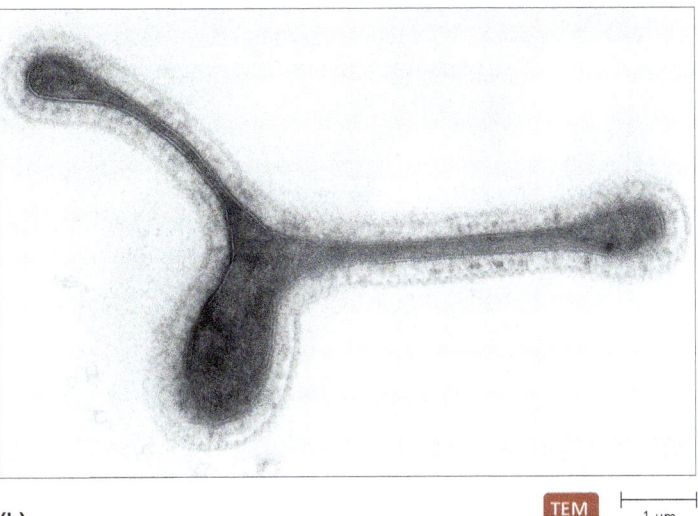

▲ **FIGURE 11.11 Archaea. (a)** *Thermococcus*, which has a tuft of flagella. **(b)** *Pyrodictium*, which has disk-shaped cells with filamentous extensions.

▲ **FIGURE 11.12 Some hyperthermophilic archaea live in hot springs.** Orange archaea thrive along the edge of this steaming pool in Yellowstone National Park.

▲ **FIGURE 11.13 The habitat of halophiles: highly saline water.** These are solar evaporation ponds near San Francisco. Halophiles often contain red-to-orange pigments, possibly to protect them from intense solar energy. Different colors in the ponds indicate the presence of different species and numbers of halophiles.

Scientists use thermophiles and their enzymes in recombinant DNA technology applications because thermophiles' cellular structure and enzymes are stable and functional at temperatures that denature most proteins and nucleic acids and kill other cells. DNA polymerase from hyperthermophilic archaea makes possible the automated amplification of DNA in thermocyclers (see Chapter 8). Heat-stable enzymes are also ideal for many industrial applications, including their use as additives in laundry detergents.

Halophiles

Halophiles[8] are organisms that inhabit extremely saline habitats, such as the Dead Sea, the Great Salt Lake, and solar evaporation ponds used to concentrate salt for use in seasoning and for the production of fertilizer (**FIGURE 11.13**). Halophiles can also colonize and spoil such foods as salted fish, sausages, and pork. Halophilic archaea are classified in the phylum Euryarchaeota.

[8] From Greek *halos*, meaning "salt."

The distinctive characteristic of halophiles is their absolute dependence on a salt concentration greater than 9% to maintain the integrity of their cell walls. Most halophiles grow and reproduce within an optimum range of 17% to 23% salt, and many species can survive in a saturated saline solution (35% sodium chloride). Many halophiles contain red to orange pigments that probably play a role in protecting them from intense sunlight.

A well-studied halophile is *Halobacterium salinarum* (hā'lō-bak-tēr'ē-ŭm sal-ē-nar'ŭm), which is an archaeon despite its name. It is a photoheterotroph, using light energy to drive the synthesis of ATP while deriving carbon from organic compounds. *Halobacterium* lacks photosynthetic pigments—chlorophylls and bacteriochlorophylls. Instead, it synthesizes purple proteins called **bacteriorhodopsins** (bak-tēr'ē-ō-rō-dop'sinz) that absorb light energy to pump protons across the cytoplasmic membrane to establish a proton gradient. Cells use the energy of proton gradients to produce ATP (see Chapter 5). *Halobacterium* also rotates its flagella with energy from the proton gradient so as to position itself at the proper water depth for maximum light absorption.

Methanogens

LEARNING OUTCOME

11.10 List at least four significant roles that methanogens play in the environment.

Methanogens are obligate anaerobes that convert CO_2, H_2, and organic acids into methane gas (CH_4), which is the major component of natural gas. These microbes constitute the largest known group of archaea in the phylum Euryarchaeota. A few thermophilic methanogens are known. For example, *Methanopyrus*[9] (meth'a-nō-pī'rŭs) has an optimum growth temperature of 98°C and grows well in 110°C seawater around submarine hydrothermal vents. Scientists have also discovered halophilic methanogens.

Methanogens play significant roles in the environment by converting organic wastes in pond, lake, and ocean sediments into methane. Other methanogens living in the colons of animals are one of the primary sources of environmental methane. Methanogens dwelling in the intestinal tract of a cow, for example, can produce 400 liters of methane a day. Sometimes the production of methane in swamps and bogs is so great that bubbles rise to the surface as "swamp gas."

Methane is a so-called *greenhouse gas*; that is, methane in the atmosphere traps heat, which adds to global warming. It is about 25 times more potent as a greenhouse gas than carbon dioxide. Methanogens have produced about 10 trillion tons of methane—twice the known amount of oil, natural gas, and coal combined—that lies buried in mud on the ocean floor. If all the methane trapped in ocean sediments were released, it would wreak havoc with the world's climate.

Methanogens also have useful industrial applications. An important step in sewage treatment is the digestion of sludge by methanogens, and some sewage treatment plants burn methane to heat buildings and generate electricity.

Though we have concentrated our discussion on extremophiles and methanogens, most archaea live in more moderate habitats and don't produce methane. For example, archaea make up about a third of the prokaryotic biomass in coastal Antarctic water, providing food for marine animals.

> **TELL ME WHY**
> Why is it that archaea don't seem to cause disease of humans, plants, or animals?

Survey of Bacteria

As we noted previously, our survey of prokaryotes in this chapter reflects the classification scheme that is featured in the second edition of *Bergey's Manual*. Whereas the classification scheme for bacteria in the first edition of the *Manual* emphasized morphology, Gram reaction, and biochemical characteristics, the second edition bases its classification of bacteria largely on differences in 16S rRNA sequences. We begin our survey of bacteria by considering the deeply branching and phototrophic bacteria.

Deeply Branching and Phototrophic Bacteria

LEARNING OUTCOMES

11.11 Provide a rationale for the name *deeply branching bacteria*.

11.12 Explain the function of heterocysts in terms of both photosynthesis and nitrogen fixation.

Deeply Branching Bacteria

The **deeply branching bacteria** are so named because their rRNA sequences and growth characteristics lead scientists to conclude that these organisms are similar to the earliest bacteria; that is, they appear to have branched off the "tree of life" at an early stage. For example, the deeply branching bacteria are autotrophic, and early organisms must have been autotrophs because heterotrophs by definition must derive their carbon from autotrophs. Further, many of the deeply branching bacteria live in extreme habitats similar to those some scientists think existed on the early Earth—hot, acidic, and exposed to intense ultraviolet radiation from the sun.

One representative of these microbes—the Gram-negative *Aquifex* (ăk'wē-feks), a bacterium in the phylum Aquificae—is considered to represent the earliest branch of bacteria. It is chemoautotrophic, hyperthermophilic, and microaerophilic, deriving energy and carbon from inorganic sources in very hot habitats containing little oxygen.

Another representative of deeply branching bacteria is *Deinococcus* (dī-nō-kok'ŭs), in phylum Deinococcus-Thermus, which grows as tetrads of Gram-positive cocci. Interestingly, the cell wall of *Deinococcus* has an outer membrane similar to that of Gram-negative bacteria, but the cells stain purple like typical Gram-positive microbes. *Deinococcus* is extremely resistant to radiation because of the way it packages its DNA, the presence of radiation-absorbing pigments, unique lipids within its membranes, and high cytoplasmic levels of manganese that protect its DNA repair proteins from radiation damage. Even when exposed to 5 million rad of radiation, which shatters its chromosome into hundreds of fragments, *Deinococcus* has enzymes that repair the damage. Not surprisingly, researchers have isolated *Deinococcus* from sites severely contaminated with radioactive wastes.

Phototrophic Bacteria

Phototrophic bacteria acquire the energy needed for anabolism by absorbing light with pigments located in thylakoids called *photosynthetic lamellae*. They lack the membrane-bound thylakoids seen in eukaryotic chloroplasts. Most phototrophic bacteria are also autotrophic—they produce organic compounds from carbon dioxide.

Phototrophs are a diverse group of microbes that are taxonomically confusing. On the basis of their pigments and their

---
[9]From Greek *pyrus*, meaning "fire."

source of electrons for photosynthesis, phototrophic bacteria can be divided into five groups (though they are classified in four phyla based on 16S rRNA sequencing).

- Blue-green bacteria (phylum Cyanobacteria)
- Green sulfur bacteria (phylum Chlorobi)
- Green nonsulfur bacteria (phylum Chloroflexi)
- Purple sulfur bacteria (phylum Proteobacteria)
- Purple nonsulfur bacteria (phylum Proteobacteria)

We consider them together in the following sections because of their common phototrophic metabolism.

Cyanobacteria Cyanobacteria are Gram-negative phototrophs that vary greatly in shape, size, and method of reproduction. They range in size from 1 µm to 10 µm in diameter and are typically coccoid or disk-shaped. Coccal forms can be single or arranged in pairs, tetrads, chains, or sheets (FIGURE 11.14a,b). Disk-shaped forms are often tightly appressed end to end to form filaments that can be either straight, branched, or helical and are frequently contained in a gelatinous glycocalyx called a *sheath* (FIGURE 11.14c). Some filamentous cyanobacteria are motile, moving along surfaces by *gliding*. Cyanobacteria generally reproduce by binary fission, with some species also reproducing by motile fragments or by thick-walled spores called *akinetes* (ā-kin-ēts'; see Figure 11.14a).

Like plants and algae, cyanobacteria utilize chlorophyll *a* and are oxygenic (generate oxygen) during photosynthesis:

$$12 H_2O + 6 CO_2 \xrightarrow{light} C_6H_{12}O_6 + 6 H_2O + 6 O_2 \qquad (1)$$

For this reason, cyanobacteria were formerly called *blue-green algae*; however, the name *cyanobacteria* properly emphasizes their true bacterial nature: They are prokaryotic and have peptidoglycan cell walls. Additionally, they lack membranous cellular organelles such as nuclei, mitochondria, and chloroplasts.

Photosynthesis by cyanobacteria is thought to have transformed the anaerobic atmosphere of the early Earth into our oxygen-containing one, and according to the endosymbiotic theory, chloroplasts developed from cyanobacteria. Indeed, chloroplasts and cyanobacteria have similar rRNA and structures, such as 70S ribosomes and photosynthetic membranes.

Nitrogen is an essential element in proteins and nucleic acids. Though nitrogen constitutes about 79% of the atmosphere, relatively few organisms can utilize the element in gaseous form. A few species of filamentous cyanobacteria as well as some proteobacteria (discussed later in the chapter) reduce nitrogen gas (N_2) to ammonia (NH_3) via a process called **nitrogen fixation**. Nitrogen fixation is essential for life on Earth because nitrogen fixers not only enrich their own growth but also provide nitrogen in a usable form to other organisms.

Because the enzyme responsible for nitrogen fixation is inhibited by oxygen, nitrogen-fixing cyanobacteria are faced with a problem—how to segregate nitrogen fixation, which is inhibited by oxygen, from oxygenic photosynthesis, which produces oxygen. Nitrogen-fixing cyanobacteria solve this problem in one of two ways. Most cyanobacteria isolate the enzymes of nitrogen fixation in specialized, thick-walled, nonphotosynthetic cells called **heterocysts** (see Figure 11.14a). Heterocysts transport reduced nitrogen to neighboring cells in exchange for glucose. A few types of cyanobacteria photosynthesize during daylight hours and fix nitrogen at night, thereby separating nitrogen fixation from photosynthesis in time rather than in space.

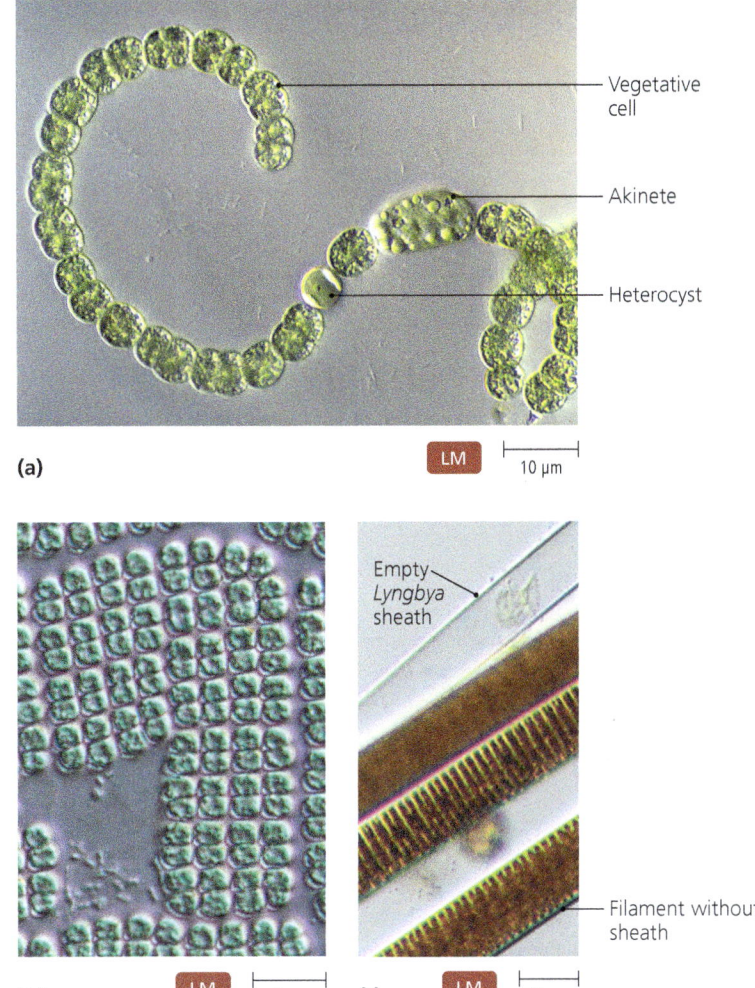

▲ FIGURE 11.14 Examples of cyanobacteria with different growth habits. (a) *Anabaena*, which grows as a filament of cocci with differentiated cells. Heterocysts fix nitrogen; akinetes are reproductive cells. (b) *Merismopedia*, which grows as a flat sheet of cocci surrounded by a gelatinous glycocalyx. (c) *Lyngbya*, which forms a filament of tightly appressed disk-shaped cells within a sheath.

Green and Purple Phototrophic Bacteria Green and purple bacteria differ from plants, algae, and cyanobacteria in two ways: They use *bacteriochlorophylls* for photosynthesis instead of chlorophyll *a*, and they are *anoxygenic*; that is, they do not generate oxygen during photosynthesis. Green and purple phototrophic bacteria commonly inhabit anaerobic muds rich in hydrogen sulfide at the bottoms of ponds and lakes. These microbes are not necessarily green and purple in color; rather, the terms refer to pigments in some of the better-known members of the groups.

As previously indicated, the green and purple phototrophic bacteria include both sulfur and nonsulfur forms. Whereas nonsulfur bacteria derive electrons for the reduction of CO_2 from organic compounds such as carbohydrates and organic acids, sulfur bacteria derive electrons from the oxidation of hydrogen sulfide to sulfur, as follows:

$$12 H_2S + 6 CO_2 \xrightarrow{light} C_6H_{12}O_6 + 6 H_2O + 12 S \qquad (2)$$

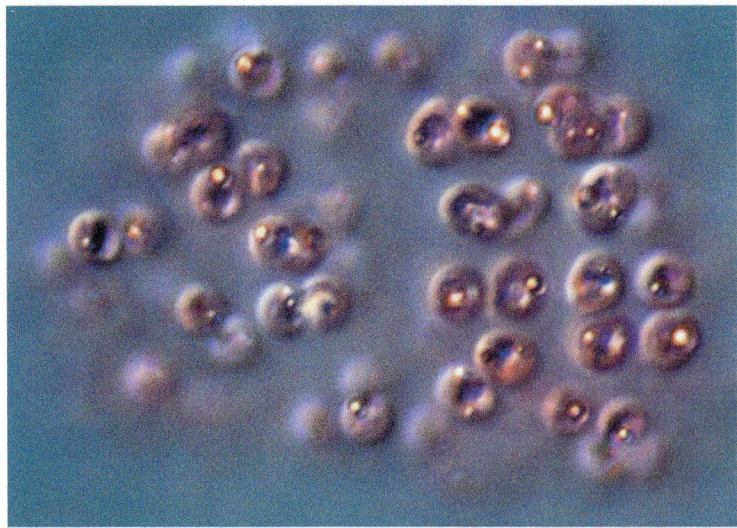

▲ **FIGURE 11.15 Deposits of sulfur within purple sulfur bacteria.** These bacteria oxidize H_2S to produce the granules of elemental sulfur evident inside the cells. *Where do green sulfur bacteria deposit sulfur grains?*

Figure 11.15 Green sulfur bacteria deposit sulfur grains externally.

Green sulfur bacteria deposit the resultant sulfur outside their cells, whereas purple sulfur bacteria deposit sulfur within their cells (FIGURE 11.15).

At the beginning of the 20th century, a prominent question in biology concerned the origin of the oxygen released by photosynthetic plants. It was initially thought that oxygen was derived from carbon dioxide, but a comparison of photosynthesis in cyanobacteria (equation 1) with that in sulfur bacteria (equation 2) provided evidence that free oxygen is derived from water.

Whereas green sulfur bacteria are placed in phylum Chlorobi, green nonsulfur bacteria are members of phylum Chloroflexi. The purple bacteria (both sulfur and nonsulfur) are placed in three classes of phylum Proteobacteria, which is composed of Gram-negative bacteria and is discussed shortly. TABLE 11.1 summarizes the characteristics of phototrophic bacteria.

Play *Photosynthesis: Comparing Prokaryotes and Eukaryotes* @ **Mastering Microbiology**

MICRO CHECK

7. What is the difference between halophiles and thermophiles?
8. What is a potential role of methanogens in global warming?
9. Deeply branching bacteria are thought to be similar to some of Earth's earliest life forms. What is a characteristic that the earliest life forms would likely have had?
10. What is the function of heterocysts?

Low G + C Gram-Positive Bacteria

LEARNING OUTCOMES

11.13 Discuss the lack of cell walls in mycoplasmas.
11.14 Identify significant beneficial or detrimental effects of the genera *Clostridium*, *Bacillus*, *Listeria*, *Lactobacillus*, *Streptococcus*, and *Staphylococcus*.
11.15 Describe the difference between low and high G + C content.

Now we turn our attention to various taxa of Gram-positive bacteria and to a different characteristic of microbes that is used in the classification of Gram-positive bacteria—*G + C content*. This is the percentage of all base pairs in a genome that are guanine-cytosine base pairs, and it is a useful criterion in classifying

TABLE 11.1 Characteristics of the Major Groups of Phototrophic Bacteria

	Phylum				
	Cyanobacteria	Chlorobi	Chloroflexi	Proteobacteria	Proteobacteria
Class	Cyanobacteria	Chlorobia	Chloroflexi	Gammaproteobacteria	Alphaproteobacteria and one genus in Betaproteobacteria
Common Name(s)	Blue-green bacteria ("blue-green algae")	Green sulfur bacteria	Green nonsulfur bacteria	Purple sulfur bacteria	Purple nonsulfur bacteria
Major Photosynthetic Pigments	Chlorophyll *a*	Bacteriochlorophyll *a* plus *c*, *d*, or *e*	Bacteriochlorophylls *a* and *c*	Bacteriochlorophyll *a* or *b*	Bacteriochlorophyll *a* or *b*
Types of Photosynthesis	Oxygenic	Anoxygenic	Anoxygenic	Anoxygenic	Anoxygenic
Electron Donor in Photosynthesis	H_2O	H_2, H_2S, or S	Organic compounds	H_2, H_2S, or S	Organic compounds
Sulfur Deposition	None	Outside cell	None	Inside cell	None
Nitrogen Fixation	Some species	None	None	None	None
Motility	Nonmotile or gliding	Nonmotile	Gliding	Motile with polar or peritrichous flagella	Nonmotile or motile with polar flagella

Gram-positive bacteria. Those with G + C content below 50% are considered "low G + C bacteria;" the remainder are considered "high G + C bacteria." Because taxonomists have discovered that Gram-positive bacteria with low G + C content have similar sequences in their 16S rRNA and that those with high G + C content also have rRNA sequences in common, they have assigned low G + C bacteria and high G + C bacteria to different phyla. We discuss the low G + C bacteria first.

The low G + C Gram-positive bacteria are classified within phylum Firmicutes (fer-mik'ū-tēz), which includes three groups: clostridia, mycoplasmas, and other low G + C Gram-positive bacilli and cocci. Next, we consider these three groups in turn.

Clostridia

Clostridia are rod-shaped, obligate anaerobes, many of which form endospores. The group is named for the genus *Clostridium*,[10] which is important both in medicine—in large part because its members produce potent toxins, which cause a variety of diseases in humans—and in industry because their endospores enable them to survive harsh conditions, including many types of disinfection and antisepsis. Examples of clostridia include *C. tetani* (te'tan-ē, which causes tetanus), *C. perfringens* (per- frin'jens; gangrene), *C. botulinum* (bo-tū-lī'num; botulism), and *C. difficile* (di'fe-sēl, severe diarrhea). (Chapter 19 examines these pathogens and the diseases they cause in greater detail.)

Microbes related to *Clostridium* include *Epulopiscium*, a giant bacterium that "gives birth" (discussed on pp. 326); sulfate-reducing microbes, which produce H_2S from elemental sulfur during anaerobic respiration; and *Selenomonas* (sĕ-lē'nō-mō'nas), a genus of vibrio-shaped bacteria that includes members that live as part of the biofilm (plaque) that forms on the teeth of warm-blooded animals. *Selenomonas* is unusual because even though it has a typical Gram-positive rRNA sequence, it has a negative Gram reaction—it stains pink. Researchers have linked one species of *Selenomonas* to obesity.

Mycoplasmas

A second group of low G + C bacteria are the **mycoplasmas**[11] (mī'kō-plaz'mas). These facultative or obligate anaerobes lack cell walls, meaning that they stain pink when Gram-stained. Indeed, until their nucleic acid sequences proved their similarity to Gram-positive organisms, mycoplasmas were classified as Gram-negative microbes. Now, of course, they are classified in phylum Firmicutes with other low G + C Gram-positive bacteria.

Mycoplasmas are able to survive without cell walls in part because they colonize osmotically protected habitats in and on animals and humans, mostly inside cells. Further, mycoplasmas have cytoplasmic membranes that are stabilized by lipids called *sterols* that give the membranes rigidity.

Because mycoplasmas lack cell walls, they are pleomorphic. They were named "mycoplasmas" because their filamentous forms resemble the filaments of fungi. Mycoplasmas have diameters ranging from 0.2 μm to 0.8 μm, making them the smallest free-living cells, and many mycoplasmas have a terminal structure that is used for attachment to eukaryotic cells and that gives the bacterium a pearlike shape. They require organic growth factors, such as cholesterol, fatty acids, vitamins, amino acids, and nucleotides, which they acquire from their host or which must be added to laboratory media. When growing on solid media, most species form a distinctive "fried egg" appearance because cells in the center of the colony grow into the agar while those around the perimeter only spread across the surface (FIGURE 11.16).

In animals, mycoplasmas colonize mucous membranes of the respiratory and urinary tracts and are associated with pneumonia and urinary tract infections. (Chapter 19 discusses pathogenic mycoplasmas and the diseases they cause more fully.)

Other Low G + C Gram-Positive Bacilli and Cocci

A third group of low G + C Gram-positive organisms is composed of bacilli and cocci that are significant in environmental, industrial, and health care settings. Among the genera in this group are *Bacillus*, *Listeria* (lis-tēr'ē-ă), *Lactobacillus* (lak'tō-bă-sil'ŭs), *Streptococcus*, *Enterococcus* (en'ter-ō-kok'ŭs), and *Staphylococcus* (staf'i-lō-kok'ŭs). (Chapter 19 examines the pathogens in these genera in greater detail.)

Bacillus The genus *Bacillus* includes endospore-forming aerobes and facultative anaerobes that typically move by means of peritrichous flagella. The genus name *Bacillus* should not be confused with the general term *bacillus*. The latter refers to any rod-shaped cell of any genus.

Numerous species of *Bacillus* are common in soil. *Bacillus thuringiensis* (thur-in-jē-en'sis) is beneficial to farmers and gardeners. During sporulation, this bacterium produces a crystalline protein that is toxic to caterpillars that ingest it (FIGURE 11.17). Gardeners spray *Bt toxin*, as preparations of the bacterium and toxin are known, on plants to protect them from caterpillars. Scientists have achieved the same effect, without the need of spraying, by

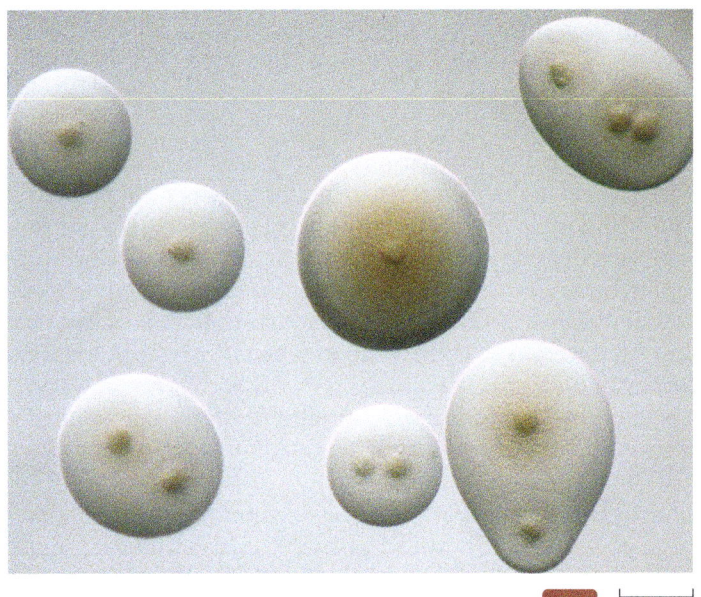

▲ FIGURE 11.16 The distinctive "fried egg" appearance of *Mycoplasma* colonies. This visual feature is unique to some species of this genus of bacteria when growing on an agar surface.

---

[10]From Greek *kloster*, meaning "spindle."

[11]From Greek *mycos*, meaning "fungus," and *plassein*, meaning "to mold."

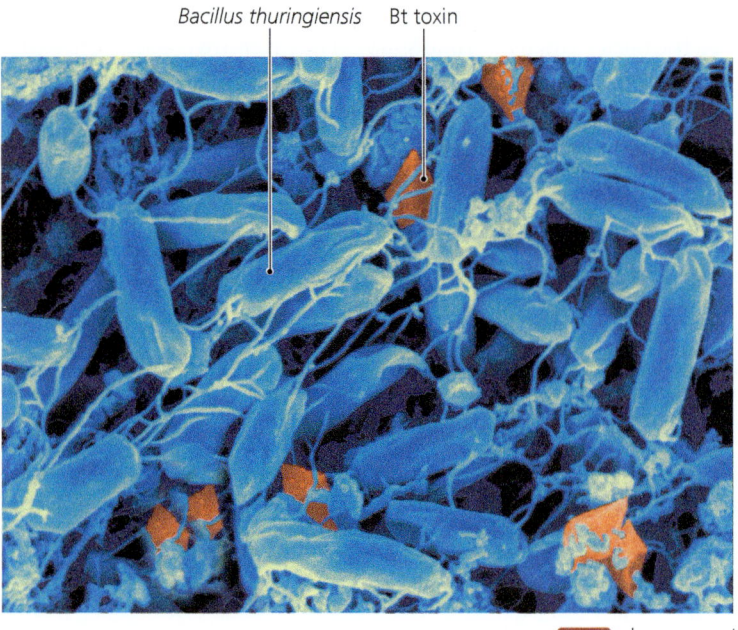

▲ FIGURE 11.17 **Crystals of Bt toxin, produced by the endospore-forming *Bacillus thuringiensis*.** The crystalline protein kills caterpillars that ingest it.

introducing the gene for Bt toxin into plants' chromosomes. Other beneficial species of *Bacillus* include *B. polymyxa* (po-lē-miks'ă) and *B. licheniformis* (lī-ken-i-for'mis), which synthesize the antibiotics polymyxin and bacitracin, respectively. (Chapter 10 discusses the production and effects of antibiotics in more detail.)

Bacillus anthracis (an-thrā'sis), which causes anthrax, gained notoriety in 2001 as an agent of bioterrorism. Its endospores either can be inhaled or can enter the body through breaks in the skin. Each endospore germinates and becomes a vegetative cell producing toxins that kill surrounding tissues. Untreated *cutaneous anthrax* is fatal in 20% of patients; untreated *inhalational anthrax* is generally 100% fatal without prompt, aggressive treatment. Refrigeration of food prevents excessive growth of this and many other bacterial contaminants (see Chapter 9).

Listeria Another pathogenic low G + C Gram-positive rod is *Listeria monocytogenes* (mo-nō-sī-tah'je-nēz), which can contaminate milk and meat products. This microbe, which does not produce endospores, is notable because it continues to reproduce under refrigeration and because it can survive inside phagocytic white blood cells. *Listeria* rarely causes disease in adults, but it can kill a fetus in an infected woman when it crosses the placental barrier. It also causes meningitis[12] and bacteremia[13] when it infects immunocompromised patients, such as the aged and patients with AIDS, cancer, or diabetes.

Lactobacillus Organisms in the genus *Lactobacillus* are non-spore-forming rods normally found growing in the human mouth, stomach, intestinal tract, and vagina. These organisms rarely cause disease; instead, they protect the body by inhibiting the growth of pathogens—a situation called *microbial antagonism*. Lactobacilli are used in industry in the production of yogurt, buttermilk, pickles, and sauerkraut. (The Beneficial Microbes: Probiotics box on p. 304 discusses the potential role of *Lactobacillus* in promoting human health.)

Streptococcus and *Enterococcus* The genera *Streptococcus* and *Enterococcus* are diverse taxa of Gram-positive cocci associated in pairs and chains (see Figure 11.8a and b). They cause numerous human diseases, including pharyngitis, scarlet fever, impetigo, fetal meningitis, wound infections, pneumonia, and diseases of the inner ear, skin, blood, and kidneys. In recent years, health care providers have become concerned over strains of multi-drug-resistant enterococci and streptococci. Of particular concern are so-called *flesh-eating streptococci* (see Disease in Depth on pp. 554–555), which produce toxins that destroy muscle and fat tissue; and enterococci, which are leading causes of healthcare-associated (nosocomial) infections.

Staphylococcus Among the common inhabitants of humans is *Staphylococcus aureus*[14] (o'rē-ŭs), which is typically found growing harmlessly in clusters in the nasal passages.

A variety of toxins and enzymes allow some strains of *S. aureus* to invade the body and cause such diseases as bacteremia, pneumonia, wound infections, food poisoning, toxic shock syndrome, and diseases of the joints, bones, heart, and blood.

Horizontal gene transfer, in which genes are shared between and among neighboring cells (discussed in Chapter 7), results in many different strains of *S. aureus*. Strains vary widely in their capacity to make toxins and harmful enzymes, some making one or a few and others generating dozens. Further, strains of *S. aureus* can become resistant to multiple antibiotics. This is especially true of the infamous methicillin-resistant *S. aureus* (MRSA), which is resistant to penicillin and cephalosporin antimicrobial drugs.

The characteristics of the low G + C Gram-positive bacteria are summarized in the first part of **TABLE 11.2**. Next, we consider Gram-positive bacteria that have high G + C ratios.

High G + C Gram-Positive Bacteria

LEARNING OUTCOMES

11.16 Explain the slow growth of *Mycobacterium*.
11.17 Identify significant beneficial or detrimental properties of the genera *Corynebacterium*, *Mycobacterium*, *Actinomyces*, *Nocardia*, and *Streptomyces*.

Taxonomists classify Gram-positive bacteria with a G + C percentage greater than 50% in the phylum Actinobacteria. The phylum includes species with rod-shaped cells (many of which are significant human pathogens) and filamentous bacteria, which resemble fungi in their growth habit and in the production of reproductive spores. Here, we examine briefly some prominent high G + C Gram-positive bacteria. (Chapter 19 discusses the numerous pathogens in this group more fully.)

---

[12]Inflammation of the membranes covering the brain and spinal cord; from Greek *meninx*, meaning "membrane."
[13]The presence of bacteria, particularly those that produce disease symptoms, in the blood.

[14]From Latin *aurum*, meaning "gold," because it produces yellow pigments.

TABLE 11.2 Characteristics of Selected Gram-Positive Bacteria

Phylum/Class	G + C Percentage	Representative Genera	Special Characteristics	Diseases
Firmicutes				
Clostridia	Low (less than 50%)	Clostridium	Obligate anaerobic rod; endospore former	Tetanus, botulism, gangrene, severe diarrhea
		Epulopiscium	Giant rod	—
		Selenomonas	Part of oral biofilm on human teeth; stains like Gram-negative bacteria (pink)	Dental caries
Mollicutes	Low (less than 50%)	Mycoplasma	Lacks cell walls; pleomorphic; smallest free-living cell; stains like Gram-negative bacteria (pink)	Pneumonia, urinary tract infections
Bacilli	Low (less than 50%)	Bacillus	Facultative anaerobic rod; endospore former	Anthrax
		Listeria	Contaminates dairy products	Listeriosis
		Lactobacillus	Produces yogurt, buttermilk, pickles, sauerkraut	Rare blood infections
		Streptococcus	Cocci in chains	Strep throat, scarlet fever, and others
		Staphylococcus	Cocci in clusters	Bacteremia, food poisoning, and others
Actinobacteria				
Actinobacteria	High (greater than 50%)	Corynebacterium	Snapping division; metachromatic granules in cytoplasm	Diphtheria
		Mycobacterium	Waxy cell walls (mycolic acid)	Tuberculosis and leprosy
		Actinomyces	Filaments	Actinomycosis
		Nocardia	Filaments; degrade pollutants	Lesions
		Streptomyces	Produce antibiotics	Rare sinus infections

Corynebacterium

Members of the genus *Corynebacterium* are pleomorphic—though generally rod-shaped—aerobes and facultative anaerobes. They reproduce by snapping division, which often causes the cells to form V-shapes and palisades (see Figures 11.4 and 11.9d). Corynebacteria are also characterized by their stores of phosphate within inclusions called **metachromatic granules**, which stain differently from the rest of the cytoplasm when the cells are stained with methylene blue or toluidine blue. The best-known species is *C. diphtheriae*, which causes diphtheria.

Mycobacterium

The genus *Mycobacterium* (mī′kō-bak-tēr′ē-ŭm) is composed of aerobic species that are slightly curved or straight rods that sometimes form filaments. Mycobacteria grow very slowly, often requiring a month or more to form a visible colony on an agar surface. Their slow growth is partly due to the time and energy required to enrich their cell walls with high concentrations of long carbon-chain waxes called **mycolic acids**, which make the cells resistant to desiccation and to staining with water-based stains, such as the Gram stain. Microbiologists developed the *acid-fast stains* such as the Ziehl-Neelsen stain[15] (see Chapter 4). This stain results in mycobacteria appearing red against a blue background; for this reason they are also called *acid-fast bacilli* (AFB) (see **FIGURE 11.18**).

Though some mycobacteria are free-living, the most prominent species are pathogens of animals and humans, including *Mycobacterium tuberculosis* (too-ber-kyū-lō′sis) and *Mycobacterium leprae* (lep′rī), which cause tuberculosis and leprosy, respectively. Mycobacteria should not be confused with the low G + C mycoplasmas discussed earlier.

Actinomycetes

Actinomycetes (ak′ti-nō-mī-sētz) are high G + C Gram-positive bacteria that form branching filaments resembling fungi (**FIGURE 11.19**). Of course, in contrast to fungi, the filaments of

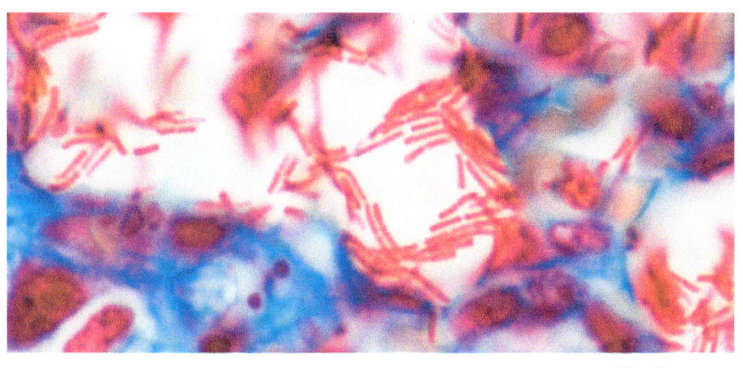

▲ **FIGURE 11.18 Acid-fast bacilli (AFB).** Acid-fast staining, as with the Ziehl-Neelsen stain of *Mycobacterium intracellulare* (shown here), which often infects AIDS patients. Such acid-fast staining results in pink or red acid-fast cells and blue non-acid-fast cells.

---
[15]Named for Franz Ziehl and Friedrich Neelsen in the 1800s.

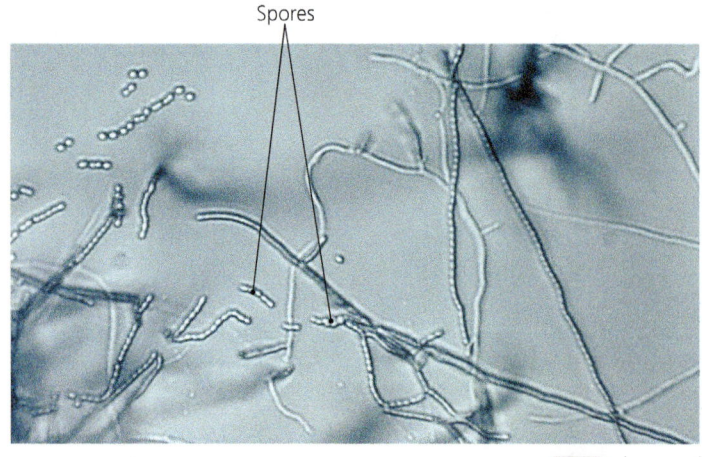

▲ FIGURE 11.19 **The branching filaments of actinomycetes.** This photograph shows filaments of a colony of *Streptomyces* sp. growing on agar. *How do the filaments of actinomycetes compare to the filaments of fungi?*

Figure 11.19 Filaments of actinomycetes are thinner than those of fungi, and they are composed of prokaryotic cells.

actinomycetes are composed of prokaryotic cells. As we have seen, some actinomycetes also resemble fungi in the production of chains of reproductive spores at the ends of their filaments. These spores should not be confused with endospores, which are resting stages and not reproductive cells. Actinomycetes may cause disease, particularly in immunocompromised patients. Among the important actinomycete genera are *Actinomyces* (which gives this group its name), *Nocardia*, and *Streptomyces*.

Actinomyces Species of *Actinomyces* (ak'ti-nō-mī'sēz) are facultative capneic[16] filaments that are normal inhabitants of the mucous membranes lining the oral cavity and throats of humans. *Actinomyces israelii* (is-rā'el-ē-ē) growing as an opportunistic pathogen in humans destroys tissue to form abscesses and can spread throughout the abdomen, consuming every vital organ.

Nocardia Species of *Nocardia* (nō-kar'dē-ă) are soil- and water-dwelling aerobes that typically form aerial and subterranean filaments that make them resemble fungi. *Nocardia* is notable because it can degrade many pollutants of landfills, lakes, and streams, including waxes, petroleum hydrocarbons, detergents, benzene, polychlorinated biphenyls (PCBs), pesticides, and rubber. Some species of *Nocardia* cause lesions in humans, particularly in the brain and skin, though they can occur in other organs and tissues as well.

Streptomyces Bacteria in the genus *Streptomyces* (strep-tō-mī'sēz) are important in several realms. Ecologically, they recycle nutrients in the soil by degrading a number of carbohydrates, including cellulose, lignin (the woody part of plants), and chitin (the outer skeletal material of insects and crustaceans). They can also degrade latex, aromatic chemicals (organic compounds containing a benzene ring), and keratin (the protein that forms hair, nails, and horns). The metabolic by-products of *Streptomyces* give soil

[16]Meaning that they grow best with a relatively high concentration of carbon dioxide.

its musty smell. Medically, *Streptomyces* species produce most of the important antibiotics, including chloramphenicol, erythromycin, and tetracycline (discussed in Chapter 10).

Table 11.2 on p. 337 includes a summary of the characteristics of the genera of high G + C Gram-positive bacteria (phylum Actinobacteria) discussed in this chapter.

To this point, we have discussed archaea and deeply branching, phototrophic, and Gram-positive bacteria. Now we turn our attention to the Gram-negative bacteria that are grouped together within the phylum Proteobacteria.

MICRO CHECK

11. Mycoplasmas are pleomorphic bacteria. How do the unique structural characteristics of mycoplasmas relate to their bacterial morphology?
12. Which of the low G + C Gram-positive bacterial genera (*Bacillus*, *Clostridium*, *Lactobacillus*, *Listeria*, *Staphylococcus*, and *Streptococcus*) can produce endospores?
13. Which genus of high G + C Gram-positive bacteria synthesize antimicrobials such as chloramphenicol, erythromycin, and tetracycline?

Gram-Negative Proteobacteria

LEARNING OUTCOME

11.18 Name six distinct classes of proteobacteria.

Phylum **Proteobacteria**[17] constitutes the largest and most diverse taxon of bacteria. Scientists have identified over 2000 species of proteobacteria. Though they have a variety of shapes, reproductive strategies, and nutritional types, they are all Gram-negative and share common 16S rRNA nucleotide sequences. The G + C percentage of Gram-negative species is not critical in delineating taxa of most Gram-negative organisms, so we will not consider this characteristic in our discussion of these bacteria.

There are six distinct classes of proteobacteria, designated by the first six letters of the Greek alphabet—alpha, beta, gamma, delta, epsilon, and zeta. These classes are distinguished by minor differences in their rRNA sequences. Here, we will focus our attention on species with novel characteristics as well as species with practical importance.

Class Alphaproteobacteria

LEARNING OUTCOMES

11.19 Describe the appearance and function of prosthecae in alphaproteobacteria.
11.20 Describe three genera of nitrogen-fixing alphaproteobacteria and their association with crops or biofuels.
11.21 Describe nitrification and name a nitrifying bacterium.
11.22 Name two diseases caused by alphaproteobacteria.
11.23 Explain how scientists can use *Agrobacterium* for the genetic manipulation of plants.

[17]Named for the Greek god Proteus, who could assume many shapes.

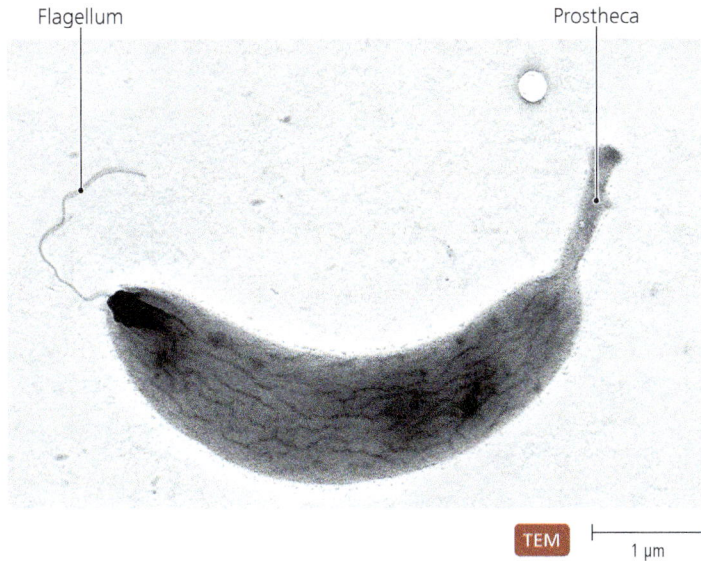

▲ FIGURE 11.20 **A prostheca.** This extension of an alphaproteobacterial cell increases surface area for absorbing nutrients and serves as an organ of attachment.

▲ FIGURE 11.21 **Nodules on pea plant roots.** *Rhizobium*, a nitrogen-fixing alphaproteobacterium, stimulates the growth of such nodules.

Alphaproteobacteria are typically aerobes capable of growing at very low nutrient levels. Many have unusual methods of metabolism, as we will see shortly. They may be rods, curved rods, spirals, coccobacilli, or pleomorphic. Many species have unusual extensions called *prosthecae* (pros-thē′kē), which are composed of cytoplasm surrounded by the cytoplasmic membrane and cell wall (**FIGURE 11.20**). They use prosthecae for attachment and to increase surface area for nutrient absorption. Some prosthecate species produce buds at the ends of the extensions.

Nitrogen Fixers Two genera of nitrogen fixers in the class Alphaproteobacteria—*Azospirillum* (ā-zō-spī′ril-ŭm) and *Rhizobium* (rī-zō′bē-ŭm)—are important in agriculture. They grow in association with the roots of plants, where they make atmospheric nitrogen available to the plants as ammonia (NH_3), which is one kind of fixed nitrogen:

$$N_2 + 8\,H^+ + 8\,e^- \longrightarrow 2\,NH_3 + H_2 \tag{3}$$

Azospirillum associates with the outer surfaces of roots of tropical grasses, such as sugarcane. In addition to supplying nitrogen to the grass, this bacterium also releases chemicals that stimulate the plant to produce numerous root hairs, increasing a root's surface area and thus its uptake of nutrients.

Rhizobium grows within the roots of leguminous plants, such as peas, beans, and clover, stimulating the formation of nodules on their roots (**FIGURE 11.21**). *Rhizobium* cells within the nodules make ammonia available to the plant, encouraging growth. Scientists are actively seeking ways to successfully insert the genes of nitrogen fixation into plants such as corn, which requires large amounts of nitrogen.

Rhodopseudomonas palustris (rō-dō′soo-dō-mō′ nas pal-us′tris) is another nitrogen-fixing alphaproteobacterium. Scientists are actively studying this bacterium because it can also reduce hydrogen, forming hydrogen gas (H_2), which can be used as a clean-burning fuel. Biofuel producers such as *R. palustris* may help relieve the world's dependence on oil and gas.

Nitrifying Bacteria Organisms need a source of electrons for redox reactions of metabolism (see Chapters 5 and 6). Bacteria that derive electrons from the oxidation of nitrogenous compounds are called **nitrifying bacteria**. These microbes are important in the environment and in agriculture because they convert reduced nitrogenous compounds, such as ammonia (NH_3), into nitrate (NO_3^-)—a two-step process called **nitrification**. Nitrate moves more easily through soil and is therefore more available to plants.

The first step in nitrification is the oxidation of ammonia into nitrite (NO_2^-); the second step is further oxidation of nitrite into nitrate (NO_3^-):

$$2\,NH_3 + 3\,O_2 \rightarrow 2\,NO_2^- + 2\,H_2O + 2\,H^+ \tag{4a}$$

$$2\,NO_2^- + O_2 \longrightarrow 2\,NO_3^- \tag{4b}$$

No microbe can perform both reactions. Nitrifying alphaproteobacteria include species of *Nitrobacter* (nī-trō-bak′ter), which perform the second step of nitrification. The first step of nitrification is performed either by archaea or by betaproteobacteria.

Purple Nonsulfur Phototrophs With one exception (the betaproteobacterium *Rhodocyclus*), purple nonsulfur phototrophs are classified as alphaproteobacteria. Purple nonsulfur bacteria grow in the upper layer of mud at the bottoms of lakes and ponds. As we discussed earlier, they harvest light as an energy source by using bacteriochlorophylls, and they do not generate oxygen during photosynthesis. Morphologically, they may be rods, curved rods, or spirals; some species are prosthecate. Refer to Table 11.1 (p. 334) to review the characteristics of all phototrophic bacteria.

Pathogenic Alphaproteobacteria Notable pathogens among the alphaproteobacteria include *Rickettsia* (ri-ket′sē-ă) and *Brucella* (broo-sel′lă).

Rickettsia is a genus of small, Gram-negative, aerobic rods that live and reproduce inside mammalian cells. Outside a host cell, rickettsias are unstable and quickly die; therefore, they require a vector for transmission from host to host. They are transmitted through the bites of arthropods (fleas, lice, ticks, and mites). Rickettsias cause a number of human diseases, including typhus and Rocky Mountain spotted fever.

Brucella is a coccobacillus that causes *brucellosis*, a disease of mammals characterized by spontaneous abortions and sterility in animals. In contrast, infected humans suffer chills, sweating, fatigue, and fever. *Brucella* is notable because it survives phagocytosis by white blood cells—normally an important step in the body's defense against disease.

Other Alphaproteobacteria Other alphaproteobacteria are important in industry and the environment. For example, *Acetobacter* (a-sē'tō-bak-ter) and *Gluconobacter* (gloo-kon'ō-bak-ter) are used to synthesize acetic acid in the production of vinegar. *Caulobacter* (kaw'lō-bak-ter) is a common prosthecate rod-shaped microbe that inhabits nutrient-poor seawater and freshwater; it can also be found in laboratory water baths.

Caulobacter has a unique reproductive strategy (FIGURE 11.22). A cell that has attached to a substrate with its prostheca ❶ grows ❷ until it has doubled in size, at which time it produces a flagellum at its apex; it then divides by asymmetric binary fission ❸. The flagellated daughter cell, which is called a *swarmer cell*, then swims away. The swarmer cell can either attach to a substrate with a new prostheca that replaces the flagellum ❹ₐ or attach to other swarmer cells to form a rosette of cells ❹ᵦ. The process of reproduction repeats about every two hours. **Beneficial Microbes: A Microtube of Superglue** (p. 341) examines an amazing property of *Caulobacter* prosthecae.

Scientists are interested in the usefulness of another alphaproteobacterium, *Agrobacterium* (ag'rō-bak-tēr'ē-um), which infects plants to form tumors called *galls* (FIGURE 11.23). The bacterium inserts a plasmid (an extrachromosomal DNA molecule) that carries a gene for a plant growth hormone through a pilus-like hollow filament into a plant cell. The plant cell then makes extra growth hormone and the plant's cells proliferate into a gall that provides nutrients for the bacterium. Scientists have discovered that they can insert almost any DNA sequence into the plasmid, making it an ideal vector for genetic manipulation of plants.

Characteristics of selected members of the alphaproteobacteria are listed in Table 11.4 (p. 344).

Class Betaproteobacteria

LEARNING OUTCOME

11.24 Name three pathogenic, three useful, and one problematic betaproteobacteria.

Betaproteobacteria are another diverse taxon of Gram-negative bacteria that thrive in habitats with low levels of nutrients. They differ from alphaproteobacteria in their rRNA sequences, though metabolically the two taxa overlap. One example of such metabolic overlap is seen with *Nitrosomonas* (nī-trō-sō-mō'nas), an important nitrifying soil bacterium that performs the first

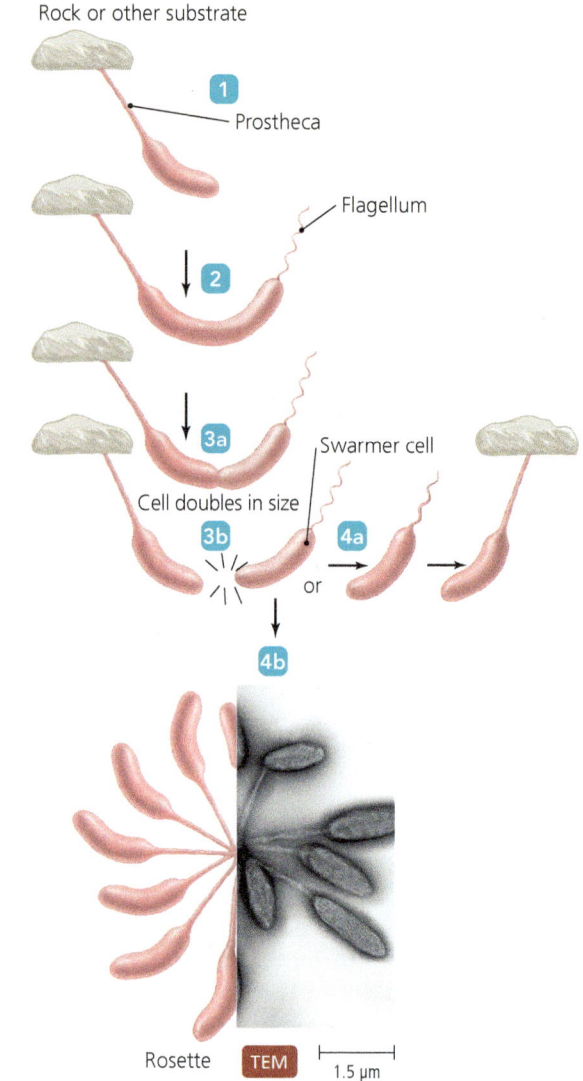

▲ **FIGURE 11.22 Growth and reproduction of *Caulobacter*.** ❶ A cell attached to a substrate by its prostheca. ❷ Growth and production of an apical flagellum. ❸ Division by asymmetrical binary fission to produce a flagellated daughter cell called a swarmer cell. ❹ₐ Attachment of a swarmer cell to a substrate by a prostheca that replaces the flagellum. ❹ᵦ Swarmer cells can attach to one another to form rosettes.

reaction of nitrification—the conversion of ammonia into nitrite (equation 4a on p. 339). In this section, we will discuss a few other interesting betaproteobacteria.

Pathogenic Betaproteobacteria Species of *Neisseria* (nī-se'rē-ă) are Gram-negative diplococci that inhabit the mucous membranes of mammals and cause such diseases as gonorrhea, meningitis, pelvic inflammatory disease, and inflammation of the cervix, pharynx, and external lining of the eye.

Other pathogenic betaproteobacteria include *Bordetella* (bōr-dĕ-tel'ă), which is the cause of pertussis (whooping cough; see **Emerging Disease Case Study: Pertussis** on p. 345), and *Burkholderia* (burk-hol-der'ē-ă), which can colonize moist environmental surfaces (including laboratory and medical equipment) and the respiratory passages of patients with cystic fibrosis. *Burkholderia* can be beneficial as when it recycles organic

▲ FIGURE 11.23 **A plant gall.** Infecting cells of *Agrobacterium* have inserted into a plant chromosome a plasmid carrying a plant growth hormone gene. The hormone causes the proliferation of undifferentiated plant cells. These gall cells synthesize nutrients for the bacteria.

compounds contaminating soil, such as crude oil, herbicides, and gasoline additives. (Chapters 20 and 21 discuss pathogenic betaproteobacteria in more detail.)

Other Betaproteobacteria Members of the genus *Thiobacillus* (thī-ō-bă-sil′ŭs) are colorless sulfur bacteria that are important in recycling sulfur in the environment by oxidizing hydrogen sulfide (H_2S) or elemental sulfur (S^0) to sulfate (SO_4^{2-}). Miners use *Thiobacillus* to leach metals from low-grade ore, though the bacterium does cause extensive pollution when it releases metals and acid from mine wastes. (Chapter 27 discusses the sulfur cycle.)

Some sewage treatment supervisors are interested in *Zoogloea* (zō′ō-glē-ă) and *Sphaerotilus* (sfēr-ō′til-us), two genera that form *flocs*—slimy, tangled masses of bacteria and organic matter in sewage (FIGURE 11.24). *Zoogloea* forms compact flocs that settle to the bottom of treatment tanks and assist in the purification process (see Figure 11.24). *Sphaerotilus*, in contrast, forms loose flocs that float, impede the proper flow of water, and can overflow onto walkways, creating a hazard to staff and interfering with proper function of the treatment plant.

The characteristics of the genera of betaproteobacteria discussed in this section are summarized in Table 11.4 (p. 344).

MICRO CHECK

14. Many species of alphaproteobacteria have prosthecae. What function do prosthecae serve?
15. How do the nitrogen-fixing genera *Azospirillum* and *Rhizobium* associate with plant roots?
16. What is nitrification?
17. Of the genera *Agrobacterium*, *Brucella*, *Gluconobacter*, and *Rickettsia*, which causes human disease and is transmitted by arthropod vectors?

Class Gammaproteobacteria

LEARNING OUTCOMES

11.25 Describe the gammaproteobacteria.
11.26 Describe the metabolism of the largest taxon of gammaproteobacteria.
11.27 Contrast gammaproteobacterial nitrogen fixers with alphaproteobacterial nitrogen fixers.

The **gammaproteobacteria** make up the largest and most diverse class of proteobacteria; almost every shape, arrangement of cells, metabolic type, and reproductive strategy are represented

BENEFICIAL MICROBES

A Microtube of Superglue

Caulobacter crescentus

A swarmer cell of the Gram-negative alphaproteobacterium *Caulobacter crescentus* attaches itself to an environmental substrate by secreting an organic adhesive from its prostheca as if it were a tube of glue. This polysaccharide-based bonding agent is the strongest known glue of biological origin, beating out such contenders as barnacle glue, mussel glue, and the adhesion of gecko lizard bristles.

One way scientists gauge adhesive strength is to measure the force required to break apart two glued objects. Commercial superglues typically lose their grip when confronted with a shear force of 18 to 28 newtons (N) per square millimeter.[a] Dental cements bond with strengths up to 30 N/mm², but *Caulobacter* glue is more than twice as adhesive. It maintains its grip up to 68 N/mm²! That is equivalent to being able to hang an adult female elephant on a wall with a spot of glue the size of a U.S. quarter. And remember, this bacterium lives in water, so its glue works even when submerged.

Scientists are researching the biophysical and chemical mechanisms that give this biological glue such incredible gripping power. One critical component of the glue is *N*-acetylglucosamine, one of the sugar subunits of peptidoglycan found in bacterial cell walls. Scientists are struggling to characterize the other molecules that make up the glue. The problem? They cannot pry the glue free to analyze it.

Some potential applications of such a bacterial superglue include use as a biodegradable suture in surgery, as a more durable dental adhesive, or to stick anti-biofilm disinfectants onto surfaces such as medical devices and ships' hulls.

[a] 1 newton is the amount of force needed to accelerate a 1-kg mass 1 meter per second.

▲ FIGURE 11.24 Flocs—slimy, tangled masses of bacteria and organic matter. *Sphaerotilus* forms flocs that float and can clog sewage treatment plant pipes. *Zoogloea* forms "good" flocs that settle, as in the beaker on the right.

in this group. Ribosomal studies indicate that gammaproteobacteria can be divided into several subgroups:

- Purple sulfur bacteria
- Intracellular pathogens
- Methane oxidizers
- Facultative anaerobes that utilize Embden-Meyerhof-Parnas glycolysis and the pentose phosphate pathway
- Pseudomonads, which are aerobes that catabolize carbohydrates by the Entner-Doudoroff and pentose phosphate pathways

Here we will examine some representatives of these groups.

Purple Sulfur Bacteria Whereas the purple nonsulfur bacteria are distributed among the alpha- and betaproteobacteria, **purple sulfur bacteria** are all gammaproteobacteria (FIGURE 11.25a). Purple sulfur bacteria are obligate anaerobes that oxidize hydrogen sulfide to sulfur, which they deposit as internal granules. They are found in sulfur-rich zones in lakes, bogs, and oceans. Some species form intimate relationships with marine worms, covering the body of a worm like strands of a rope (FIGURE 11.25b).

Intracellular Pathogens Organisms in the genera *Legionella* (lē-jŭ-nel'lă) and *Coxiella* (kok-sē-el'ă) are pathogens of humans that avoid digestion by white blood cells, which are normally part of a body's defense; in fact, they thrive inside these defensive cells. *Legionella* derives energy from the metabolism of amino acids, which are more prevalent inside cells than outside. *Coxiella* grows best at low pH, like that found in the phagolysosomes of white blood cells. The bacteria in these genera cause Legionnaires' disease and Q fever, respectively.

Methane Oxidizers Gram-negative bacteria that utilize methane as a carbon source and as an energy source are called **methane oxidizers**. Like archaeal methanogens, bacterial methane oxidizers inhabit anaerobic environments worldwide,

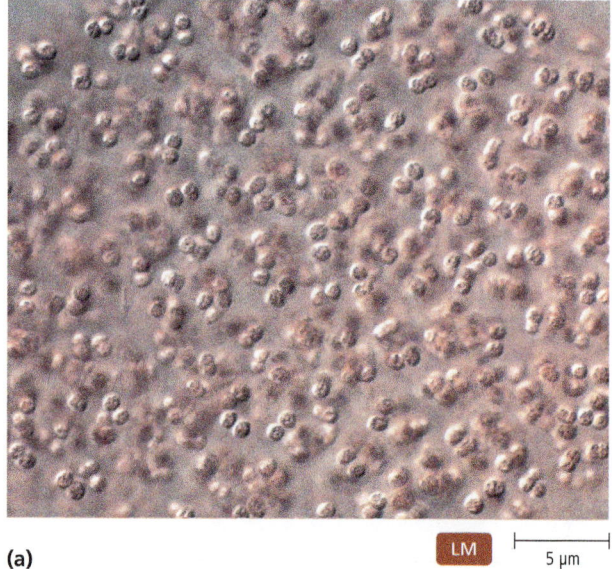

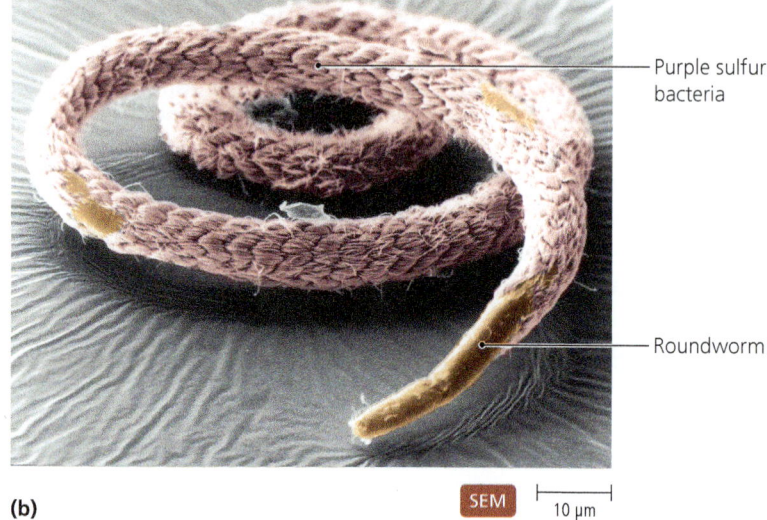

▲ FIGURE 11.25 Purple sulfur bacteria. (a) These bacteria deposit sulfur granules internally. (b) Numerous bacteria on the surface of a nematode (roundworm, colored yellow) give the appearance of a rope.

growing just above the anaerobic layers that contain methanogens, which generate methane. Methane is one of the so-called greenhouse gases that retains heat in the atmosphere, but methane oxidizers can digest methane in their local environment, ideally before it can adversely affect the world's climate.

Glycolytic Facultative Anaerobes The largest group of gammaproteobacteria is composed of Gram-negative, facultatively anaerobic rods that catabolize carbohydrates by glycolysis and the pentose phosphate pathway. This group, which is divided into three families (TABLE 11.3, p. 343), contains numerous human pathogens.

Members of the family Enterobacteriaceae live in the guts of animals and can be harmless, beneficial, or pathogenic. Because of their effect on human health, scientists have studied enterobacteria, particularly *Escherichia coli* (esh-ĕ-rik'ē-ă ko'lī), more than any other bacteria. Researchers frequently use enterobacteria for laboratory studies of metabolism, genetics, and recombinant DNA technology. (Chapter 20 examines several pathogenic gammaproteobacteria in detail.)

TABLE 11.3 Representative Glycolytic Facultative Anaerobes of the Class Gammaproteobacteria

Family	Special Characteristics	Representative Genera	Typical Human Diseases
Enterobacteriaceae	Straight rods; oxidase negative; peritrichous flagella or nonmotile	Escherichia	Gastroenteritis
		Enterobacter	(Rarely pathogenic)
		Serratia	(Rarely pathogenic)
		Salmonella	Enteritis
		Proteus	Urinary tract infection
		Shigella	Shigellosis
		Yersinia	Plague
		Klebsiella	Pneumonia
Vibrionaceae	Vibrios; oxidase positive; polar flagella	Vibrio	Cholera
Pasteurellaceae	Cocci or straight rods; oxidase positive; nonmotile	Haemophilus	Meningitis in children, middle ear infections, pneumonia

Pseudomonads Bacteria called **pseudomonads** are Gram-negative, aerobic, flagellated, straight or slightly curved rods that catabolize carbohydrates by the Entner-Doudoroff and pentose phosphate pathways. These organisms are noted for their ability to break down numerous organic compounds; some strains can catabolize more than 80 different organic compounds, including oil, rubber, and plastic. Many pseudomonads are important pathogens of humans and animals and are involved in the spoilage of refrigerated milk, eggs, and meat because they can grow and catabolize proteins and lipids at 4°C. The pseudomonad group is named for its most important genus, *Pseudomonas* (soo-dō-mō′nas) (FIGURE 11.26), which causes diseases such as urinary tract infections, external otitis (swimmer's ear), and lung infections in cystic fibrosis patients. Other pseudomonads, such as *Azotobacter* (ā-zō-tō-bak′ter) and *Azomonas* (ā-zō-mō′nas), are soil-dwelling, nonpathogenic nitrogen fixers; however, in contrast to nitrogen-fixing alphaproteobacteria, these gammaproteobacteria do not associate with the roots of plants.

The characteristics of the members of the gammaproteobacteria discussed in this section are summarized in TABLE 11.4 (p. 344).

Class Deltaproteobacteria

LEARNING OUTCOME

11.28 List several members of the deltaproteobacteria.

The **deltaproteobacteria** are not a large assemblage, but, like other proteobacteria, they include a wide variety of metabolic types. *Desulfovibrio* (dē′sul-fō-vib′rē-ō) is a sulfate-reducing microbe that is important in recycling sulfur in the environment. It is also an important member of bacterial communities living in the sediments of polluted streams and sewage treatment lagoons, where its presence is often apparent by the odor of hydrogen sulfide that it releases during anaerobic respiration. Hydrogen sulfide reacts with iron to form iron sulfide, so sulfate-reducing bacteria, such as *Desulfovibrio*, play a primary role in the corrosion of iron pipes in heating systems, sewer lines, and other structures.

Bdellovibrio (del-lō-vib′rē-ō) is another deltaproteobacterium. It attacks and destroys other Gram-negative bacteria in a complex and unusual way (FIGURE 11.27a):

1 A free *Bdellovibrio* swims rapidly through the medium until it attaches via fimbriae to a Gram-negative bacterium.

2 It rapidly drills through the cell wall of its prey by secreting hydrolytic enzymes and rotating in excess of 100 revolutions per second (FIGURE 11.27b) on p. 345.

3 Once inside, *Bdellovibrio* lives in the periplasmic space—the space between the cytoplasmic membrane and the outer membrane of the cell wall. It kills its host by disrupting the host's cytoplasmic membrane and inhibiting DNA, RNA, and protein synthesis.

4 The invading bacterium uses the nutrients released from its dying prey and grows into a long filament.

5 Eventually, the filament divides into as many as nine smaller cells at once, each of which, when released from the dead cell, produces a flagellum and swims off to repeat the process. Undergoing multiple fissions to produce many offspring is a rare form of reproduction.

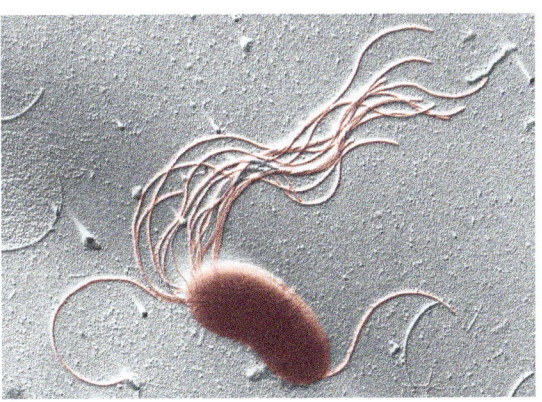

▲ FIGURE 11.26 *Pseudomonas* has notable polar flagella.

TABLE 11.4 Characteristics of Selected Gram-Negative Bacteria

Phylum/Class	Representative Members	Special Characteristics	Diseases
Proteobacteria			
Alphaproteobacteria	*Azospirillum*	Nitrogen fixer	
	Rhizobium	Nitrogen fixer	
	Nitrobacter	Nitrifying bacterium	
	Purple nonsulfur bacteria	Anoxygenic phototrophs	
	Rickettsia	Intracellular pathogen	Typhus and Rocky Mountain spotted fever
	Brucella	Coccobacillus	Brucellosis
	Acetobacter, Gluconobacter	Synthesize acetic acid	
	Caulobacter	Prosthecate bacterium	
	Agrobacterium	Causes galls in plants; vector for gene transfer in plants	
Betaproteobacteria	*Nitrosomonas*	Nitrifying bacterium	
	Neisseria	Diplococcus	Gonorrhea and meningitis
	Bordetella		Pertussis
	Burkholderia		Lung infection of cystic fibrosis patients
	Thiobacillus	Colorless sulfur bacterium	
	Zoogloea	Forms beneficial flocs in sewage treatment	
	Sphaerotilus	Floating flocs can block sewage treatment pipes	
Gammaproteobacteria	Purple sulfur bacteria	Anoxygenic phototrophs	
	Legionella	Intracellular pathogen	Legionnaires' disease
	Coxiella	Intracellular pathogen	Q fever
	Methylococcus	Oxidizes methane	
	Glycolytic facultative anaerobes: *Escherichia, Enterobacter, Serratia, Salmonella, Proteus, Shigella, Yersinia, Klebsiella, Vibrio, Haemophilus*	Facultative anaerobes that catabolize carbohydrates via glycolysis and the pentose phosphate pathway	See Table 11.3 (p. 343)
	Pseudomonas	Aerobe that catabolizes carbohydrates via Entner-Doudoroff and pentose phosphate pathways	Urinary tract infections, external otitis
	Azotobacter, Azomonas	Nitrogen fixers not associated with plant roots	
Deltaproteobacteria	*Desulfovibrio*	Sulfate reducer	
	Bdellovibrio	Pathogen of Gram-negative bacteria	
	Myxobacteria	Reproduces by forming differentiated fruiting bodies	
Epsilonproteobacteria	*Campylobacter*	Curved rod	Gastroenteritis
	Helicobacter	Spiral	Gastric ulcers
Zetaproteobacteria	*Mariprofundus*	Oxidizes iron	
Chlamydiae			
Chlamydiae	*Chlamydia*	Intracellular pathogen; lacks peptidoglycan	Neonatal blindness and lymphogranuloma venereum
Spirochaetes			
Spirochaetes	*Treponema*	Motile by axial filaments	Syphilis
	Borrelia	Motile by axial filaments	Lyme disease
Bacteroidetes			
"Bacteroidia"[a]	*Bacteroides*	Anaerobe that lives in animal colons	Abdominal infections
"Sphingobacteria"	*Cytophaga*	Digests complex polysaccharides	

[a]The names of taxa within quotation marks are not officially recognized.

EMERGING DISEASE CASE STUDY

Pertussis

Jeeyun was coughing again. She had been coughing off and on for two weeks. If she had thought of it, she might have noticed that the coughing spells began soon after her return from New York, where she had visited her grandmother. Within a week of arriving home in California, she had developed coldlike signs—runny nose, sneezing, and a slight fever—but she didn't associate these things with the cough. What she did know was that the coughing was worse; her chest hurt from the constant hacking. Then the coughing broke two ribs.

The surprising pain caused Jeeyun to involuntarily urinate. When the coughing stopped, she vomited and fainted in a heap on the floor. This was no ordinary cough! Jeeyun's roommate called 911. Later, emergency room staff at the hospital bandaged her chest to stabilize the broken ribs and diagnosed the cough as pertussis.

A bacterium, *Bordetella pertussis*, causes pertussis, commonly known as whooping cough. The disease is highly contagious and usually considered a childhood ailment, though it can affect adults. Older patients seldom develop the severe symptoms or characteristic "whooping" sound associated with gasping inhalation, but in Jeeyun's case, the severity and length of the coughing led to a speedy diagnosis.

Immunization is the only way to control pertussis, but adults have been lax in vaccinating children and receiving boosters for themselves. As a result, whooping cough is reemerging as a major problem in the industrialized world. Infected people spread *Bordetella* in respiratory droplets to their neighbors. (For more about pertussis, see pp. 606–608).

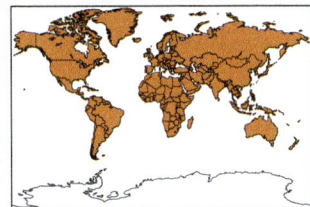

1. Why might adults neglect to get a pertussis booster shot for themselves?
2. Why might parents leave their children unimmunized against pertussis?
3. Why is immunization a societal and not merely a personal decision?
4. How might Jeeyun's friends and family protect themselves from an infection arising from the bacteria spread from her coughing?

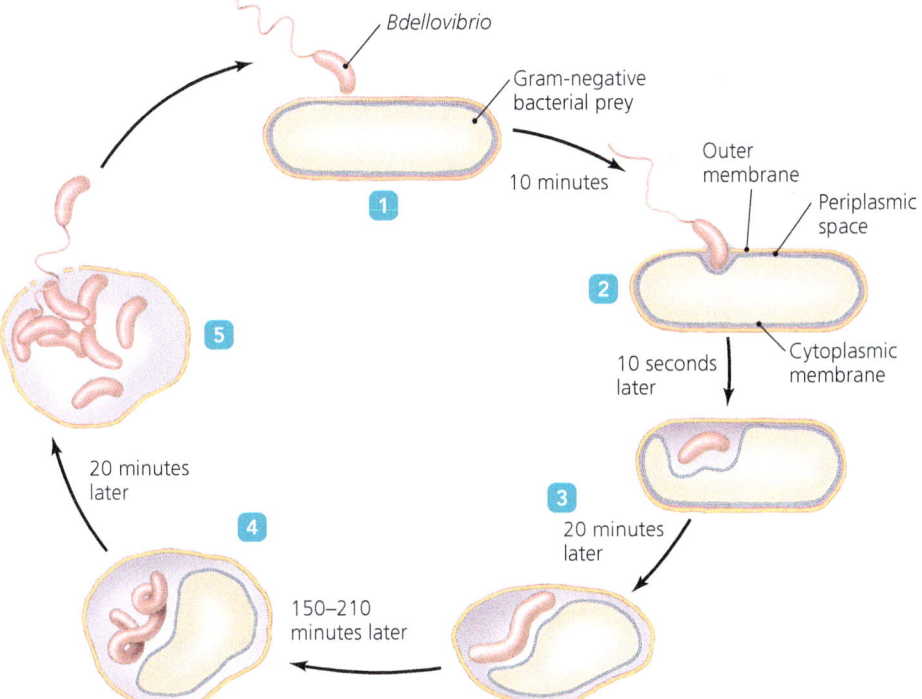
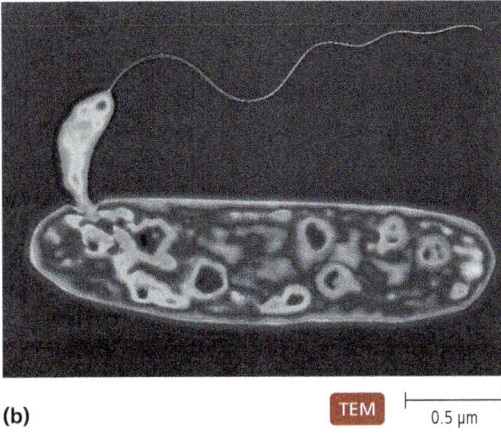

▲ **FIGURE 11.27** *Bdellovibrio*, a Gram-negative pathogen of other Gram-negative bacteria. **(a)** Life cycle, including the elapsed times between events. **(b)** *Bdellovibrio* invading the periplasmic space of its host.

Myxobacteria are Gram-negative, aerobic, soil-dwelling bacteria with a unique life cycle for prokaryotes in that individuals cooperate to produce differentiated reproductive structures. In fact, myxobacteria demonstrate many features that scientists previously thought were only found in eukaryotes, such as moving together as a group, cell differentiation, and cell-to-cell communication via hormones. Not surprisingly, the genome of a myxobacterium can be over 10 million bp in size. The life cycle of myxobacteria can be summarized as follows (FIGURE 11.28a):

1. Vegetative myxobacteria glide on slime trails through their environment, digesting yeasts and other bacteria or scavenging nutrients released from dead cells. When nutrients and cells are plentiful, the myxobacteria divide by binary fission; when nutrients are depleted, however, they aggregate by gliding into a mound of cells.
2. Myxobacteria within the mound differentiate to form a macroscopic *fruiting body* ranging in height from 50 μm to 700 μm
3. Some cells within the fruiting body develop into dormant *myxospores* that are enclosed within walled structures called *sporangia* (singular: *sporangium*) (FIGURE 11.28b).
4. The sporangia release the myxospores, which can resist desiccation and nutrient deprivation for a decade or more.
5. When nutrients are again plentiful, the myxospores germinate and become vegetative cells.

Myxobacteria live worldwide in soils that have decaying plant material or animal dung. Though certain species live in the Arctic and others in the tropics, most myxobacteria live in temperate regions.

Class Epsilonproteobacteria

LEARNING OUTCOME

11.29 List two epsilonproteobacteria that cause human diseases.

The class **Epsilonproteobacteria** includes Gram-negative rods, vibrios, or spirals. Important genera are *Campylobacter* (kam′pi-lō-bak′ter), which causes blood poisoning and inflammation of the intestinal tract, and *Helicobacter* (hel′ĭ-kō-bak′ter), which causes ulcers. (Chapter 21 examines these pathogens more fully.)

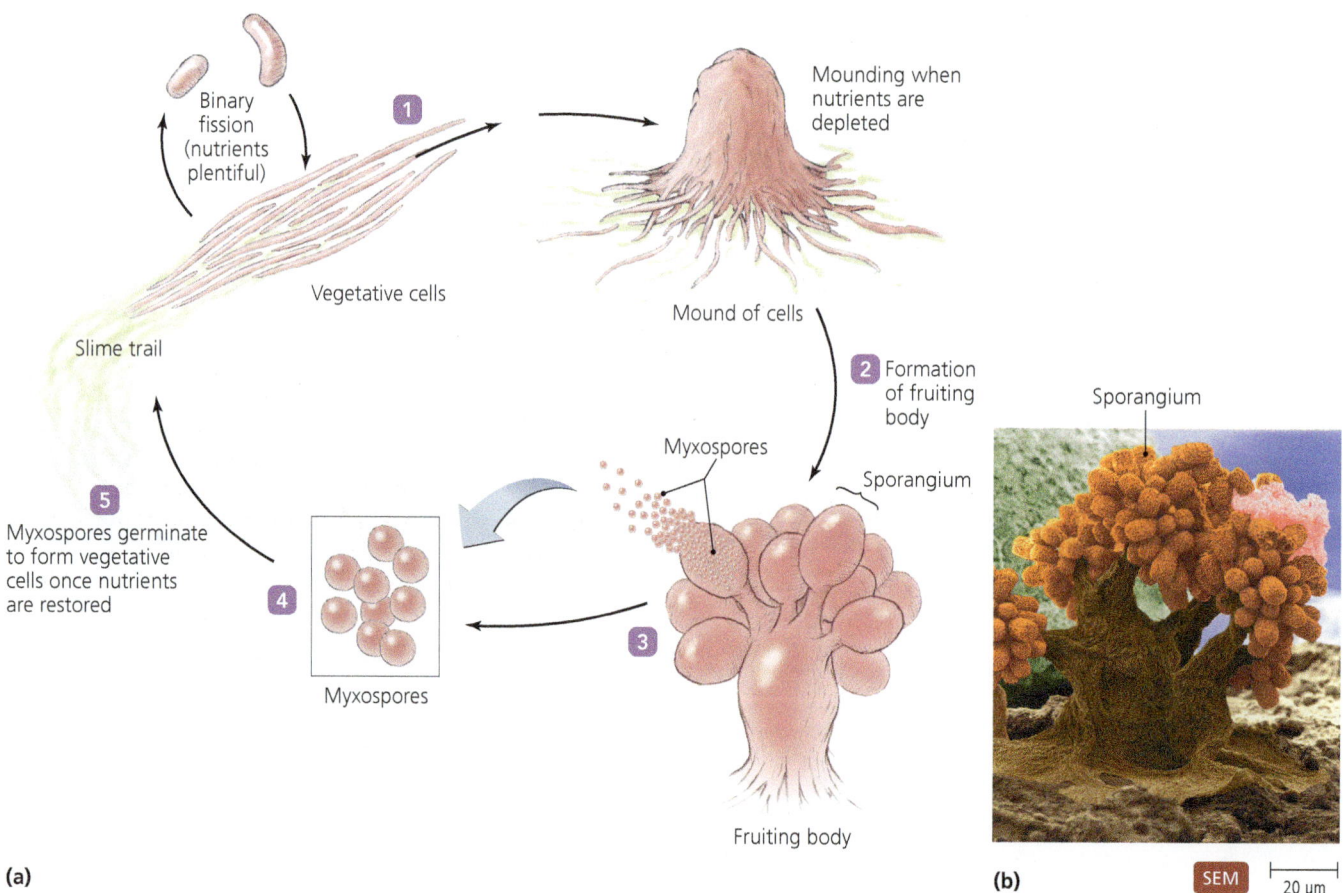

▲ FIGURE 11.28 **Life cycle of myxobacteria.** (a) 1 Vegetative myxobacteria divide by binary fission when nutrients are plentiful. They aggregate in mounds when nutrients are depleted. 2 The mound forms fruiting bodies. 3 Some cells in a fruiting body develop into dormant myxospores within a sporangium. The sporangia release the myxospores 4, which remain dormant until nutrients are again plentiful, at which time they germinate and become vegetative cells 5. (b) Fruiting body.

Class Zetaproteobacteria

LEARNING OUTCOME

11.30 Describe both the discovery of zetaproteobacteria and the type of metabolism most zetaproteobacteria use.

Researchers discovered the new class **Zetaproteobacteria** in the mid-2000s based on DNA sequences. Discoveries of never-before-identified microbes are becoming common occurrences as modern scientists use *metagenomic sequencing*—sequencing of all the DNA from a site to discover all of the microbes present, not just the 1–2% that can currently be grown in the lab. While scientists have identified genetic sequences of over 20 different zetaproteobacteria, only a few have been cultured, including one species formally named *Mariprofundus ferrooxydans* (mĕ-rē-prō-fun'dus fĕ-rō-oks'ē-danz). The name reflects the fact that these organisms oxidize iron as an energy source and use oxygen as the final electron acceptor. Based on their DNA sequences, most zetaproteobacteria appear to be autotrophic, using carbon dioxide as a carbon source. Scientists have discovered DNA from zetaproteobacteria in oceans around the world, so it is likely these bacteria play an important role in Earth's iron cycle.

The characteristics of the delta-, epsilon-, and zetaproteobacteria are summarized in Table 11.4 (p. 344)

MICRO CHECK

18. The largest group of Gram-negative gammaproteobacteria contains many human pathogens. What metabolic characteristics do they share?
19. Which deltaproteobacterium invades and replicates inside other Gram-negative bacteria?

Other Gram-Negative Bacteria

LEARNING OUTCOMES

11.31 Describe the unique features of chlamydias and spirochetes.
11.32 Describe the ecological importance of bacteroids.

In the final section of this chapter we consider an assortment of Gram-negative bacteria that are classified in the second edition of *Bergey's Manual* into nine phyla that are grouped together for convenience rather than because of genetic relatedness. Species in six of the nine phyla are of relatively minor importance. Here, we discuss representatives from the three phyla that are either of particular ecological concern or significantly affect human health: the chlamydias (phylum Chlamydiae), the spirochetes (phylum Spirochaetes), and the bacteroids (phylum Bacteroidetes).

Chlamydias

Microorganisms called **chlamydias** (kla-mid'ē-ăz) are small, Gram-negative cocci that grow and reproduce only within the cells of mammals, birds, and a few invertebrates. The smallest chlamydias—0.2 μm in diameter—are smaller than the largest viruses; however, in contrast to viruses, chlamydias have both DNA and RNA, cytoplasmic membranes, functioning ribosomes, reproduction by binary fission, and metabolic pathways. Like other Gram-negative prokaryotes, chlamydias have two membranes, but unlike most prokaryotes, they lack peptidoglycan.

Because chlamydias and rickettsias share an obvious characteristic—both require an intracellular lifestyle—these two types of organisms were classified together in a single taxon in the first edition of *Bergey's Manual*. Now, however, the rickettsias are classified with the alphaproteobacteria, and the chlamydias are in their own phylum. Chlamydias cause neonatal blindness, pneumonia, and a sexually transmitted disease called *lymphogranuloma venereum*; in fact, chlamydias are the most common sexually transmitted bacterium in the United States. (Chapter 21 examines chlamydial diseases. Figure 21.3 illustrates the chlamydial life cycle.)

Spirochetes

Spirochetes (also spelled *spirochaetes*) are unique helical bacteria that are motile by means of axial filaments composed of flagella that lie within the periplasmic space (see Figure 3.8). When an axial filament rotates, the entire cell corkscrews through the medium.

Spirochetes have a variety of types of metabolism and live in diverse habitats. They are frequently isolated from the human mouth, marine environments, moist soil, and the surfaces of protozoa that live in termites' guts. In the latter case, they may coat the protozoan so thickly that they look and act like cilia. The spirochetes *Treponema* (trep-ō-ne'mă) and *Borrelia* (bō-rē'lē-ă) cause syphilis and Lyme disease, respectively, in humans.

Bacteroids

Bacteroids are yet another diverse taxon of Gram-negative microbes that are grouped together on the basis of similarities in their rRNA nucleotide sequences. The group is named for *Bacteroides* (bak-ter-oy'dēz), a genus of obligately anaerobic rods that normally inhabit the digestive tracts of humans and animals. Bacteroids assist in digestion by catabolizing substances such as cellulose and other complex carbohydrates that are indigestible by mammals. Between 5% and 50% of the bacteria isolated from various humans' feces are *Bacteroides*. Some species cause abdominal, pelvic, blood, and other infections. They are the most common anaerobic human pathogen, causing diarrhea, fever, foul-smelling lesions, gas, and pain.

Bacteroids in the genus *Cytophaga* (sī-tof'ă-gă) are aquatic, gliding, rod-shaped aerobes with pointed ends. These bacteria degrade complex polysaccharides, such as agar, pectin, chitin, and cellulose, so they cause damage to wooden boats and piers; they also play an important role in the degradation of raw sewage. Their ability to glide allows these bacteria to position themselves at sites with optimum nutrients, pH, temperature, and oxygen levels. Organisms in *Cytophaga* differ from other gliding bacteria, such as cyanobacteria and myxobacteria, in that they are nonphotosynthetic and do not form fruiting bodies.

The characteristics of these groups of Gram-negative bacteria are listed in Table 11.4 on p. 344.

MICRO CHECK

20. Which group of bacteria do the pathogenic species of *Treponema* and *Borrelia* belong to?

TELL ME WHY

Why are bacteria all classified in the same domain (Bacteria) despite their widely divergent oxygen tolerance, sizes, shapes, and nutritional requirements?

MICRO IN THE CLINIC FOLLOW-UP
A Simple Mosquito Bite?

When Dr. Emerson removes the gauze from Alejandro's arm, she doesn't find a blister, but she does see an inflamed vesicle with dead tissue in the middle. As part of her initial examination, she swabs the lesion and sends it to the lab for Gram staining, culturing, and polymerase chain reaction (PCR). Since the lesion isn't inflamed and Alejandro doesn't have any other obvious signs of infection, Dr. Emerson instructs him to treat the lesion with a topical antibiotic.

Three days later, Dr. Emerson calls Alejandro to come to the hospital as soon as possible. The urgency of Dr. Emerson's request worries Alejandro and Mary—what could be so bad that she couldn't tell them the results over the phone?

Dr. Emerson meets them at the hospital. She explains that the lab results indicate that Alejandro's lesion is due to *Bacillus anthracis*—he has anthrax! Dr. Emerson explains that Alejandro has a type called cutaneous anthrax—the most common and least severe form. Further good news is that cutaneous anthrax is rarely contagious; however, since the swelling is worsening and may be spreading, she starts Alejandro on intravenous (IV) antibiotics: ciprofloxacin and clindamycin. Alejandro should make a full recovery.

Dr. Emerson also notifies public health authorities of the anthrax infection. The cattle at Alejandro's ranch and neighboring ranches are tested for anthrax. Four other ranch workers are found with lesions on their hands, arms, neck, or face; all recover with antibiotic treatment. Ultimately, 63 cattle die on two of the ranches. Diseased cattle are confirmed as the source of infection for the ranchers.

1. The cattle likely became infected after ingesting *B. anthracis* endospores in the soil. How could ingestion of endospores by the cattle lead to an infection with vegetative cells of *B. anthracis* in the cattle?

2. What is one hypothesis to explain how the ranchers became infected from the cattle?

Check your answers to Micro in the Clinic Follow-Up questions in the Mastering Microbiology Study Area.

Go to Mastering Microbiology for Interactive Microbiology, Dr. Bauman's Video Tutors, Micro Matters, MicroFlix, MicroBoosters, practice quizzes, and more.

CHAPTER SUMMARY

General Characteristics of Prokaryotic Organisms (pp. 325–329)

1. Three basic shapes of prokaryotic cells are spherical **cocci**, rod-shaped **bacilli**, and **spirals**. Spirals may be stiff (**spirilla**) or flexible (**spirochetes**).

2. Other variations in shapes include **vibrios** (slightly curved rods), **coccobacilli** (intermediate to cocci and bacilli), and **pleomorphic** (variable shape and size).

3. Environmentally resistant **endospores** are produced within vegetative cells of the Gram-positive genera *Bacillus* and *Clostridium*. Depending on the species in which they are formed, the endospores may be terminal, subterminal, or centrally located.

4. Prokaryotes reproduce asexually by **binary fission**, **snapping division** (a type of binary fission), **spore** formation, and **budding**.

   Play *Bacterial Growth: Overview* @ Mastering Microbiology

5. Cocci may typically be found in groups, including long chains (**streptococci**), pairs (**diplococci**), foursomes (**tetrads**), cuboidal packets (**sarcinae**), and clusters (**staphylococci**).

6. Bacilli are found singly, in pairs, in chains, or in a **palisade** arrangement.

Modern Prokaryotic Classification (p. 329)

1. Living things are now classified into three domains—Archaea, Bacteria, and Eukarya—based largely on genetic similarities.

2. The most authoritative reference in modern prokaryotic systematics is *Bergey's Manual of Systematic Bacteriology*, second edition. Currently, there are five known phyla of Archaea and 24 phyla of Bacteria. The organization of this text's survey of prokaryotes largely follows Bergey's classification scheme.

Survey of Archaea (pp. 329–332)

1. The domain Archaea includes **extremophiles**, microbes that require extreme conditions of temperature, pH, pressure, and/or salinity to survive.

2. **Thermophiles** and **hyperthermophiles** (in the phyla Crenarchaeota and Euryarchaeota) live at temperatures above 45°C and 80°C, respectively, because their DNA, membranes, and proteins do not function properly at lower temperatures.

3. **Halophiles** (phylum Euryarchaeota) depend on high concentrations of salt to keep their cell walls intact. Halophiles such as *Halobacterium salinarum* synthesize purple proteins called **bacteriorhodopsins** that harvest light energy to synthesize ATP.

4. **Methanogens** (phylum Euryarchaeota) are obligate anaerobes that produce methane gas and are useful in sewage treatment.

Survey of Bacteria (pp. 332–348)

1. **Deeply branching bacteria** have rRNA sequences thought to be similar to those of earliest bacteria. They are autotrophic and live in hot, acidic environments, often with intense exposure to sun.

2. Phototrophic bacteria trap light energy with photosynthetic lamellae. The five groups of phototrophic bacteria are cyanobacteria, green sulfur bacteria, green nonsulfur bacteria, purple sulfur bacteria, and purple nonsulfur bacteria.

3. Many **cyanobacteria** reduce atmospheric N_2 to NH_3 via a process called **nitrogen fixation**. Cyanobacteria must separate (in either time or space) the metabolic pathways of nitrogen fixation from those of oxygenic photosynthesis because nitrogen fixation is inhibited by the oxygen generated during photosynthesis. Many cyanobacteria fix nitrogen in thick-walled cells called **heterocysts**.

4. Green and purple bacteria use bacteriochlorophylls for anoxygenic photosynthesis. Nonsulfur forms derive electrons from organic compounds; sulfur forms derive electrons from H_2S.

 Play *Photosynthesis: Comparing Prokaryotes and Eukaryotes* @ Mastering Microbiology

5. The phylum Firmicutes contains bacteria with a G + C content (the percentage of all base pairs that are guanine-cytosine base pairs) of less than 50%. Firmicutes includes clostridia, mycoplasmas, and other low G + C cocci and bacilli.

6. Clostridia include the genus *Clostridium* (pathogenic bacteria that cause gangrene, tetanus, botulism, and diarrhea), *Epulopiscium* (which is large enough to be seen without a microscope), and *Selenomonas* (often found in dental plaque).

7. **Mycoplasmas** are Gram-positive, pleomorphic, facultative anaerobes and obligate anaerobes that lack cell walls and therefore stain pink with Gram stain. They are frequently associated with pneumonia and urinary tract infections.

8. Low G + C Gram-positive bacilli and cocci important to human health and industry include *Bacillus* (which contains species that cause anthrax and food poisoning and includes beneficial bacteria that produce Bt toxin), *Listeria* (which causes bacteremia and meningitis), *Lactobacillus* (used to produce yogurt and pickles), *Streptococcus*, *Enterococcus*, and *Staphylococcus*, each of which causes a number of human diseases.

9. High G + C Gram-positive bacteria (*Corynebacterium*, *Mycobacterium*, and actinomycetes) are classified in phylum Actinobacteria.

10. Bacteria in *Corynebacterium* store phosphates in **metachromatic granules**; *C. diphtheriae* causes diphtheria.

11. Members of the genus *Mycobacterium*, including species that cause tuberculosis and leprosy, grow slowly and have unique, resistant cell walls containing waxy **mycolic acids**.

12. **Actinomycetes** resemble fungi in that they produce spores and form filaments; this group includes *Actinomyces* (normally found in human mouths), *Nocardia* (useful in degradation of pollutants), and *Streptomyces* (produces important antibiotics).

13. Phylum **Proteobacteria** is a very large taxon of Gram-negative bacteria divided into six classes—the alpha-, beta-, gamma-, delta-, epsilon-, and zetaproteobacteria.

14. The **alphaproteobacteria** include a variety of aerobes, many of which have unusual cellular extensions called prosthecae. *Azospirillum* and *Rhizobium* are nitrogen fixers that are important in agriculture.

15. Some members of the alphaproteobacteria are **nitrifying bacteria**, which oxidize NH_3 to NO_3 via a process called **nitrification**. Nitrifying alphaproteobacteria are in the genus *Nitrobacter*.

16. Most purple nonsulfur phototrophs are alphaproteobacteria.

17. Pathogenic alphaproteobacteria include *Rickettsia* (typhus and Rocky Mountain spotted fever) and *Brucella* (brucellosis).

18. There are many beneficial alphaproteobacteria, including *Acetobacter* and *Gluconobacter*, both of which are used to synthesize acetic acid. *Caulobacter* is of interest in reproductive studies, and *Agrobacterium* is used in genetic recombination in plants.

19. The **betaproteobacteria** include the nitrifying *Nitrosomonas* and pathogenic species such as *Neisseria* (gonorrhea), *Bordetella* (whooping cough), and *Burkholderia* (which colonizes the lungs of cystic fibrosis patients).

20. Other betaproteobacteria include *Thiobacillus* (ecologically important), *Zoogloea* (useful in sewage treatment), and *Sphaerotilus* (hampers sewage treatment).

21. The **gammaproteobacteria** constitute the largest class of proteobacteria; they include **purple sulfur bacteria**, intracellular pathogens, methane oxidizers, facultative anaerobes that utilize glycolysis and the pentose phosphate pathway, and pseudomonads.

22. Both *Legionella* and *Coxiella* are intracellular, pathogenic gammaproteobacteria.

23. **Methane oxidizers** are anaerobic bacteria that use methane for both carbon and energy.

24. Numerous human pathogens are facultatively anaerobic gammaproteobacteria that catabolize carbohydrates by glycolysis.

25. **Pseudomonads**, including pathogenic *Pseudomonas* and nitrogen-fixing *Azotobacter* and *Azomonas*, utilize the Entner-Doudoroff and pentose phosphate pathways for catabolism of glucose.

26. The **deltaproteobacteria** include *Desulfovibrio* (important in the sulfur cycle and in corrosion of pipes), *Bdellovibrio* (pathogenic to bacteria), and **myxobacteria**. The latter form stalked fruiting bodies containing resistant, dormant myxospores.

27. The **epsilonproteobacteria** include some important human pathogens, including *Campylobacter* and *Helicobacter*.

28. Scientists discovered **zetaproteobacteria** using metagenomic sequencing, which entails sequencing all the DNA from a site so as to ascertain all microbes present. Zetaproteobacteria oxidize iron in marine environments.

29. **Chlamydias** are Gram-negative cocci typified by the genus *Chlamydia*; they cause neonatal blindness, pneumonia, and a sexually transmitted disease.

30. **Spirochetes** are flexible, helical bacteria that live in diverse environments. The pathogens *Treponema* (syphilis) and *Borrelia* (Lyme disease) are important spirochetes.

31. **Bacteroids** include *Bacteroides*, an obligate anaerobic rod that inhabits the digestive tract, and *Cytophaga*, an aerobic rod that degrades wood and raw sewage.

QUESTIONS FOR REVIEW

Answers to the Questions for Review (except Short Answer questions) begin on p. A-I.

Modified True/False

For each of the following statements that is true, write "true" in the blank. For each statement that is false, write the word(s) that should be substituted for the underlined word(s) to make the statement correct.

1. _____ All prokaryotes reproduce <u>sexually</u>.
2. _____ A <u>bacillus</u> is a bacterium with a slightly curved rod shape.
3. _____ If you were to view staphylococci, you should expect to see <u>clusters</u> of cells.
4. _____ Chlamydias <u>have</u> peptidoglycan cell walls.
5. _____ Archaea are classified into phyla based primarily on <u>tRNA</u> sequences.
6. _____ <u>Halophiles</u> inhabit extremely saline habitats, such as the Great Salt Lake.
7. _____ Pigments located in <u>thylakoids</u> in phototrophic bacteria trap light energy for metabolic processes.
8. _____ Most cyanobacteria form <u>heterocysts</u> in which nitrogen fixation occurs.
9. _____ A giant bacterium that is large enough to be seen without a microscope is <u>Selenomonas</u>.
10. _____ When environmental nutrients are depleted, <u>myxobacteria</u> aggregate in mounds to form fruiting bodies.

Multiple Choice

1. The type of reproduction in prokaryotes that results in a palisade arrangement of cells is called _____.
   a. pleomorphic division
   b. endospore formation
   c. snapping division
   d. binary fission

2. The thick-walled reproductive spores produced in the middle of cyanobacterial filaments are called _____.
   a. akinetes
   b. terminal endospores
   c. metachromatic granules
   d. heterocysts

3. Which of the following terms best describes stiff, spiral-shaped prokaryotic cells?
   a. cocci
   b. bacilli
   c. spirilla
   d. spirochetes

4. Endospores _____.
   a. remain alive for decades
   b. can remain alive in boiling water
   c. exist in a state of suspended animation
   d. All of the above are correct.

5. How is *Halobacterium salinarum* distinctive?
   a. It is absolutely dependent on high salt concentrations to maintain its cell wall.
   b. It is found in terrestrial volcanic habitats.
   c. It photosynthesizes without chlorophyll.
   d. It can survive 5 million rad of radiation.

6. Photosynthetic bacteria that also fix nitrogen are _____.
   a. mycoplasmas
   b. spirilla
   c. bacteroids
   d. cyanobacteria

7. Which genus is the most common anaerobic human pathogen?
   a. *Bacteroides*
   b. *Spirochetes*
   c. *Chlamydia*
   d. *Methanopyrus*

8. Flexible spiral-shaped prokaryotes are _____.
   a. spirilla
   b. spirochetes
   c. vibrios
   d. rickettsias

9. Bacteria that convert nitrogen gas into ammonia are _____.
   a. nitrifying bacteria
   b. nitrogenous
   c. nitrogen fixers
   d. nitrification bacteria

10. The presence of mycolic acid in the cell wall characterizes _____.
    a. *Corynebacterium*
    b. *Listeria*
    c. *Nocardia*
    d. *Mycobacterium*

Matching

Match the bacteria in the numbered list with the terms in the lettered list.

1. ___ *Bacillus anthracis*
2. ___ *Selenomonas*
3. ___ *Clostridium perfringens*
4. ___ *Clostridium botulinum*
5. ___ *Bacillus licheniformis*
6. ___ *Streptococcus*
7. ___ *Streptomyces*
8. ___ *Corynebacterium*
9. ___ *Gluconobacter*
10. ___ *Bordetella*
11. ___ *Zoogloea*
12. ___ *Rhizobium*
13. ___ *Desulfovibrio*
14. ___ *Chlamydia*
15. ___ *Cytophaga*

A. wood damage
B. dental biofilm (plaque)
C. gangrene
D. botulism
E. anthrax
F. lymphogranuloma venereum
G. leprosy
H. tetracycline
I. vinegar
J. yogurt
K. impetigo
L. bacitracin
M. iron pipe corrosion
N. pertussis
O. nitrogen fixation
P. floc formation
Q. diphtheria

Short Answer

1. Whereas the first edition of *Bergey's Manual* relied on morphological and biochemical characteristics to classify microbes, the new edition focuses on ribosomal RNA sequences. List several other criteria for grouping and classifying bacteria.

2. What are extremophiles? Describe two kinds, and give examples.

3. Name and describe three types of bacteria mentioned in this chapter that "glide."

4. Name three groups of low G + C Gram-positive bacteria.

5. Compare and contrast bacterial and archaeal cells.

6. A student was memorizing the arrangements of bacteria and noticed that there are more arrangements for cocci than for bacilli. Why might this be so?

7. How is *Agrobacterium* used in recombinant DNA technology?

8. Name and describe six distinct classes of phylum Proteobacteria.

9. Describe the special features of snapping division that distinguish it from regular binary fission.

VISUALIZE IT!

1. Label the shapes of these prokaryotic cells.

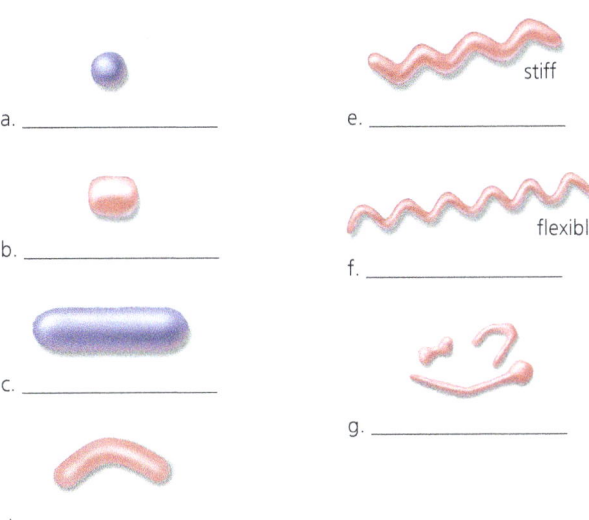

2. Describe the location of these endospores within their cells.

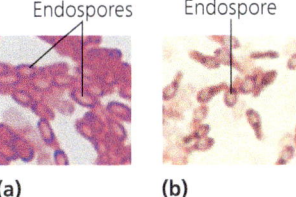

CRITICAL THINKING

1. A microbiology student described "deeply branching bacteria" as having a branched filamentous growth habit akin to *Streptomyces*. Do you agree with this description? Why or why not?

2. Iron oxide (rust) forms when iron is exposed to oxygen, particularly in the presence of water. Nevertheless, iron pipes typically corrode more quickly when they are buried in moist *anaerobic* soil than when they are buried in soil containing oxygen. Explain why this is the case.

3. Why is it that Gram-positive species don't have axial filaments?

4. Even though *Clostridium* is strictly an anaerobic bacterium, it can be isolated easily from the exposed surface of your skin. Explain how this can be.

5. Louis Pasteur said, "The role of the infinitely small in nature is infinitely large." Explain what he meant, using examples of the roles of microorganisms in health, industry, and the environment.

6. How are bacterial endospores different from the spores of actinomycetes?

7. A scientist who discovers a prokaryote living in a hot spring at 100°C suspects that it belongs to the archaea. Why does she think it might be archaeal? How could she prove that it is not bacterial?

8. Contrast the processes of nitrogen fixation and nitrification.

9. What do the names *Desulfovibrio* and *Bdellovibrio* tell you about the shape of these deltaproteobacteria?

10. How do scientists know there are millions of species of bacteria and of archaea, if most have never been cultured in a lab?

11. Why might soil-dwelling organisms such as *Pseudomonas* have large genomes and be highly metabolically flexible?

CONCEPT MAPPING

Using the terms provided, fill in the concept map that describes archaeal extremophiles. You can also complete this and other concept maps online by going to the **Mastering** Microbiology Study Area.

>80°C
Acidophiles
Animal colons
Disease
Extremophiles

Great Salt Lake
Halophiles
Hydrothermal vents
Methane
Peptidoglycan

Prokaryotes
Sewage treatment
Thermophiles

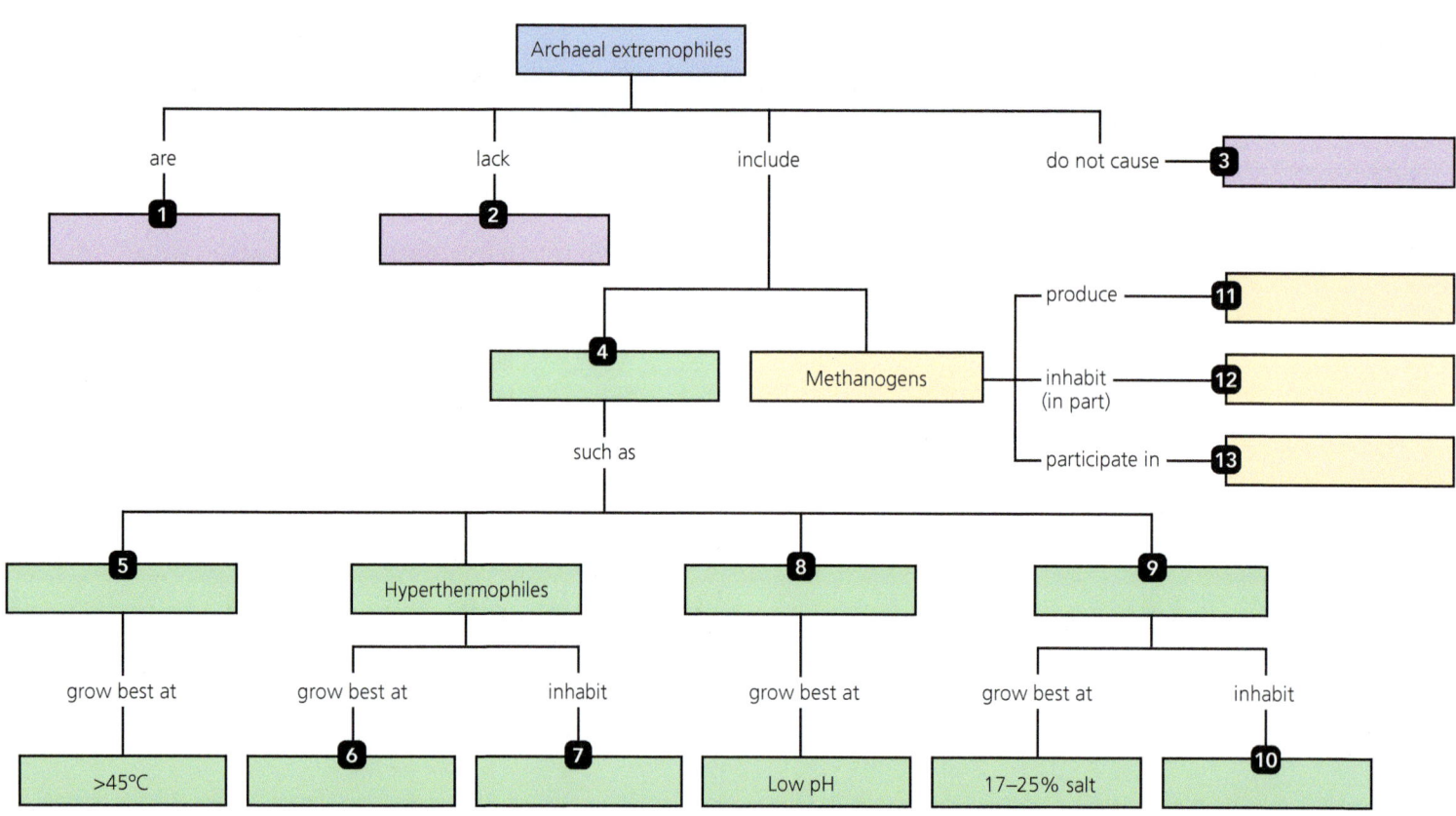

12 Characterizing and Classifying Eukaryotes

Before You Begin

1. Which major cellular structure is present in eukaryotes and not in prokaryotes?
2. Which of the following organisms are classified as eukaryotes: archaea, bacteria, fungi, helminths, or protozoa?
3. What are the differences between the structure of chromosomes in eukaryotic nuclei and that in bacteria?
4. What organelle generates most of the energy in eukaryotic cells?
5. What is are the differences between eukaryotic and prokaryotic flagella?

MICRO IN THE CLINIC
Shellfish Allergy?

BECKY IS A 53-YEAR-OLD WOMAN LIVING IN SEATTLE, Washington, with her husband Tim and their dog Tipper. Even though they both have full-time jobs, Becky and Tim try to get out every weekend and do something active—kayaking, hiking, or cycling. While Becky loves the physical activities, her favorite is the two-week vacation that she and Tim take every year. They rent a house on the southeast coast of Alaska and spend their time gathering crabs, harvesting clams, and relaxing. On the last weekend of their vacation, Becky and Tim always host a big dinner party featuring fresh crabs and clams.

This year the dinner party is a huge success—everyone has a great time, and the saltwater mussels are particularly delicious. After two hours of consuming the delicious bivalve mollusks, Becky begins to experience what appear to be severe allergic symptoms. She has tingling around her mouth and starts having difficulty swallowing. Within 30 minutes, her symptoms rapidly worsen, and her legs tingle and move involuntarily. Tim becomes very concerned when Becky can hardly breathe, and he immediately takes her to the nearest emergency room.

1. Is Becky's condition a medical emergency?
2. Could her symptoms result from eating seafood at the party, even though she has never had trouble with seafood before?

Turn to the end of the chapter (p. 379) to find out.

Diatoms are eukaryotic microbes with intricate "shells" that inhabit streams, lakes, ponds, and oceans worldwide.

Eukaryotic microbes include a fascinating and almost bewilderingly diverse assemblage. They include unicellular and multicellular protozoa,[1] fungi,[2] algae,[3] slime molds, and water molds. Additionally, microbiologists study parasitic helminths[4] because they have microscopic stages, and they study arthropod vectors because they are intimately involved in the transmission of microbial pathogens. Eukaryotes include both human pathogens and organisms that are vital for human life. For example, one group of marine algae called *diatoms*, together with a set of protozoa called *dinoflagellates* (dī'nō-flaj'ĕ-lātz), provide the basis for the oceans' food chains and produce most of the world's oxygen. Eukaryotic fungi produce penicillin, and tiny baker's and brewer's yeasts are essential for making bread and alcoholic beverages.

Among the 20 most frequent microbial causes of death worldwide, six are eukaryotic, including the agents of malaria, African sleeping sickness, and amebic dysentery. *Pneumocystis* pneumonia, toxoplasmosis, and cryptosporidiosis—common afflictions of AIDS patients—are all caused by eukaryotic pathogens.

In previous chapters, we discussed characteristics of *cells*—their metabolism, growth, and genetics. In this chapter, we discuss eukaryotic *organisms* of interest to microbiologists—protozoa (single-celled "animals"), fungi, algae, slime molds, and water molds—and conclude with a brief discussion of the relationship of parasitic helminths and vectors to microbiology.

We begin by discussing general features of eukaryotic reproduction and classification; the following sections survey some representative members of microbiologically important eukaryotic groups, focusing on beneficial, environmentally significant, and unusual species. (Chapters 22 and 23 discuss fungal and parasitic agents and vectors of human disease in more detail.)

General Characteristics of Eukaryotic Organisms

Our discussion of the general characteristics of eukaryotes begins with a survey of the events in eukaryotic reproduction; then we consider some aspects of the complexity of classifying the great variety of eukaryotic organisms.

Reproduction of Eukaryotes

LEARNING OUTCOMES

12.1 State four reasons why eukaryotic reproduction is more complex than prokaryotic reproduction.
12.2 Describe the phases of mitosis, mentioning chromosomes, chromatids, centromeres, and spindle.
12.3 Contrast meiosis with mitosis, mentioning homologous chromosomes, tetrads, and crossing over.
12.4 Distinguish among nuclear division, cytokinesis, and schizogony.

A unique characteristic of living things is the ability to reproduce themselves. Prokaryotic reproduction typically involves replication of DNA and binary fission of the cytoplasm to produce two identical offspring. Reproduction of eukaryotes is more complicated and varied than reproduction in prokaryotes for a number of reasons:

- Most of the DNA in eukaryotes is packaged with histone proteins as *chromosomes* in the form of *chromatin fibers* (krō'ma-tin) located within nuclei. The remaining DNA in eukaryotic cells is found in mitochondria and chloroplasts, organelles that reproduce by binary fission in a manner similar to prokaryotic reproduction. In this chapter we will discuss only the nuclear portion of eukaryotic genomes.
- Eukaryotes have a variety of methods of asexual reproduction, including binary fission, budding, fragmentation, spore formation, and *schizogony* (ski-zog'ō-nē) (discussed later).
- Many eukaryotes reproduce sexually—that is, via a process that involves the formation of sexual cells called *gametes*, and the subsequent fusion of two gametes to form a cell called a *zygote*.
- Additionally, algae, fungi, and some protozoa reproduce both sexually and asexually. (Animals generally reproduce only one way or the other.)

Eukaryotic reproduction involves two types of division: nuclear division and cytoplasmic division (also called *cytokinesis*). After we discuss the various aspects of these two types of division, we will consider schizogony.

Nuclear Division

Typically, a eukaryotic nucleus has either one or two complete copies of the chromosomal portion of a cell's genome. A nucleus with a single copy of each chromosome is called a **haploid**,[5] or 1*n*, nucleus, and one with two sets of chromosomes is a **diploid**,[6] or 2*n*, nucleus. Generally, each organism has a consistent number of chromosomes. For example, each haploid cell of the brewer's yeast *Saccharomyces cerevisiae*[7] (sak-ă-rō-mī'sēz se-ri-vis'ē-ī) has 16 chromosomes.

The cells of most fungi, many algae, and some protozoa are haploid, and the cells of most plants and animals and the remaining fungi, algae, and protozoa are diploid. Typically, gametes are haploid, and a zygote (formed from the union of gametes) is diploid.

A cell divides its nucleus so as to pass a copy of its chromosomal DNA to each of its descendants; in this way, each new generation has the necessary genetic instructions to carry on life. There are two types of nuclear division—*mitosis* (mī-tō'sis) and *meiosis* (mī-ō'sis).

Mitosis Eukaryotic cells have two main stages in their life cycle: a stage called *interphase*,[8] during which the cells grow and

---

[1] From Greek *protos*, meaning "first," and *zoion*, meaning "animal."
[2] Plural of Latin *fungus*, meaning "mushroom."
[3] Plural of Latin *alga*, meaning "seaweed."
[4] From Greek *helmins*, meaning "worm."
[5] From Greek *haploos*, meaning "single."
[6] From Greek *diploos*, meaning "double."
[7] From Greek *sakcharon*, meaning "sugar," and *mykes*, meaning "fungus," and Latin *cerevisiae*, meaning "beer."
[8] Latin, meaning "between phases."

replicate their DNA, and a stage during which the nucleus divides. In the type of nuclear division called **mitosis**,[9] which begins after the cell has duplicated its DNA such that there are two exact DNA copies (see Chapter 7, Figure 7.6), the cell partitions its replicated DNA equally between two nuclei. Thus, mitosis maintains the ploidy of the parent nucleus; that is, a haploid nucleus that undergoes mitosis forms two haploid nuclei, and a diploid nucleus that undergoes mitosis produces two diploid nuclei.

Scientists recognize four phases in mitosis: *prophase*,[10] *metaphase*,[11] *anaphase*,[12] and *telophase*.[13] The events of mitosis proceed as follows (**FIGURE 12.1**):

1. **Prophase**. The cell condenses its DNA molecules into visible threads called *chromatids* (krō′mă-tidz). Two identical chromatids, sister DNA molecules, are joined together in a region called a *centromere* to form one chromosome. Also during prophase, a set of microtubules is constructed in the cytosol to form a *spindle*. In most cells, the nuclear envelope disintegrates during prophase so that mitosis occurs freely in the cytosol; however, many fungi and some unicellular microbes (e.g., diatoms and dinoflagellates) maintain their nuclear envelopes so that mitosis occurs completely inside their nuclei.

2. **Metaphase**. The chromosomes line up on a plane in the middle of the cell and attach near their centromeres to microtubules of the spindle.

3. **Anaphase**. Sister chromatids separate and crawl along the microtubules toward opposite poles of the spindle. Each chromatid is now called a *chromosome*.

4. **Telophase**. The cell restores its chromosomes to their less compact, nonmitotic state, and nuclear envelopes form around the daughter nuclei. A cell might divide during telophase, but mitosis is nuclear division, not cell division. Cell division is called *cytokinesis*.

Though certain specific events distinguish each of the four phases of mitosis, the phases are not discrete steps; that is, mitosis is a continuous process, and there are no clear boundaries between succeeding phases—one phase leads seamlessly to the next. For example, late anaphase and early telophase are indistinguishable.

Students sometimes confuse the terms *chromosome* and *chromatid*, in part because early microscopists used the word *chromosome* for two different things. During prophase and metaphase, a chromosome consists of two chromatids (DNA molecules) joined at a centromere. However, during anaphase and telophase, the chromatids separate, and each chromatid is then called a *chromosome*. In other words, a prophase or metaphase "chromosome" is composed of two chromatids, whereas *chromatid* and *chromosome* are synonymous terms during anaphase and telophase.

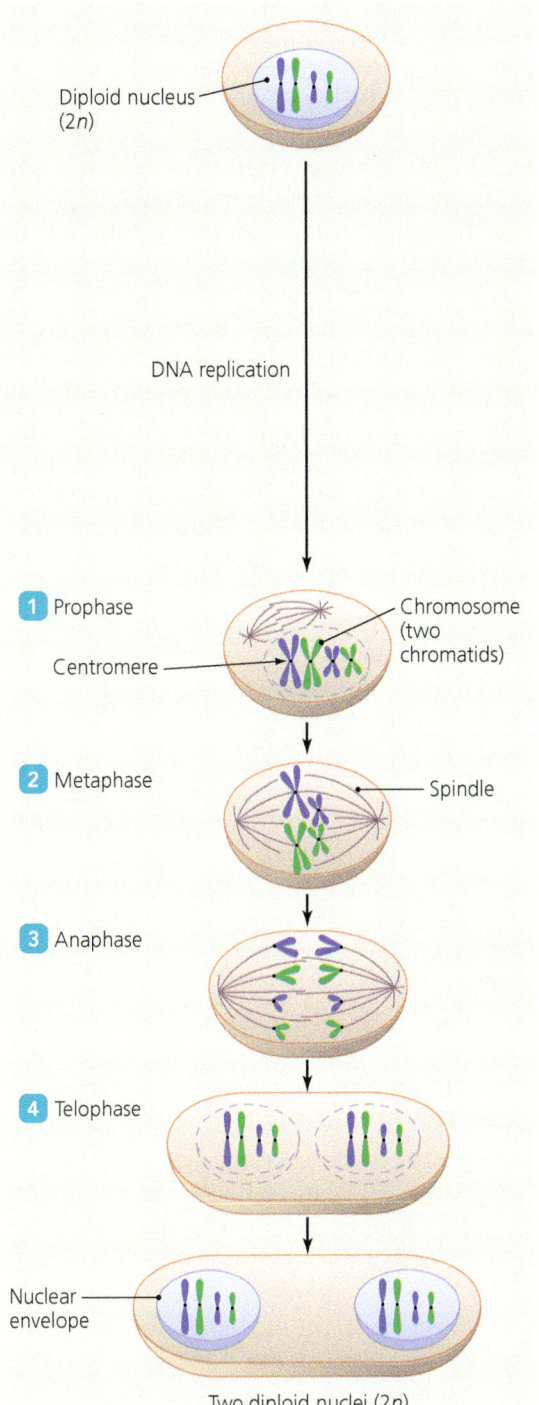

▲ **FIGURE 12.1 Mitosis.** The number of chromosomes (ploidy) in the daughter nuclei is the same as in the parent nucleus. Cell division (cytokinesis) may occur simultaneously with mitosis, but mitosis is not cell division. The text describes the events occurring during each of the numbered phases.

---

[9]From Greek *mitos*, meaning "thread," after the threadlike appearance of chromosomes during nuclear division.
[10]From Greek *pro*, meaning "before," and *phasis*, meaning "appearance."
[11]From Greek *meta*, meaning "in the middle."
[12]From Greek *ana*, meaning "back."
[13]From Greek *telos*, meaning "end."

Meiosis In contrast to mitosis, **meiosis**[14] is nuclear division that involves the partitioning of chromatids into four nuclei such that each nucleus receives only half the original amount of DNA. Thus, diploid nuclei use meiosis to produce haploid daughter nuclei. Meiosis is a necessary condition for sexual reproduction (in which nuclei from two different cells fuse to form a single nucleus) because if cells lacked meiosis, each nuclear fusion to form a zygote would cause the number of chromosomes to double, and their number would soon become unmanageable.

Meiosis occurs in two stages known as *meiosis I* and *meiosis II* (FIGURE 12.2). As in mitosis, each stage has four phases, named prophase, metaphase, anaphase, and telophase. The events in meiosis as they occur in a diploid nucleus proceed as follows:

1. **Early prophase I** (prophase of meiosis I). As with mitosis, DNA replication during interphase results in pairs of identical chromatids joined at centromeres, forming chromosomes. But in meiosis I, additional pairing occurs: *homologous chromosomes*—that is, chromosomes carrying similar or identical genetic sequences—line up side by side. Because these are prophase chromosomes, each of them consists of two identical chromatids; therefore, four DNA molecules are involved in this pairing. An aligned pair of homologous chromosomes is known as a *tetrad*.

2. **Late prophase I.** Once tetrads have formed, the homologous chromosomes exchange sections of DNA in a random fashion via a process called *crossing over*. This results in recombinations of their DNA. It is because of meiotic crossing over that the offspring produced by sexual reproduction have different genetic makeups from their siblings. Prophase I can last for days or longer.

3. **Metaphase I.** Tetrads align on a plane in the center of the cell and attach to spindle microtubules. Metaphase I differs from metaphase of mitosis in that homologous chromosomes remain as tetrads.

4. **Anaphase I.** Homologous chromosomes of the tetrads, carrying crossed-over DNA, move apart from one another. In contrast to mitotic anaphase, sister chromatids remain attached to one another.

5. **Telophase I.** The first stage of meiosis is completed as the spindle disintegrates. Typically, the cell divides at this phase to form two cells. Nuclear envelopes may form, though prophase II may begin immediately without a nuclear envelope forming. Each daughter nucleus is now haploid, though each meiotic haploid chromosome at this stage consists of two chromatids.

6. **Prophase II.** Nuclear envelopes disintegrate, if they formed, and new spindles form.

7. **Metaphase II.** The chromosomes align in the middle of each cell and attach to microtubules of the spindles.

8. **Anaphase II.** Sister chromatids separate as in mitosis.

9. **Telophase II.** Daughter nuclei form. The cells divide, yielding four haploid cells.

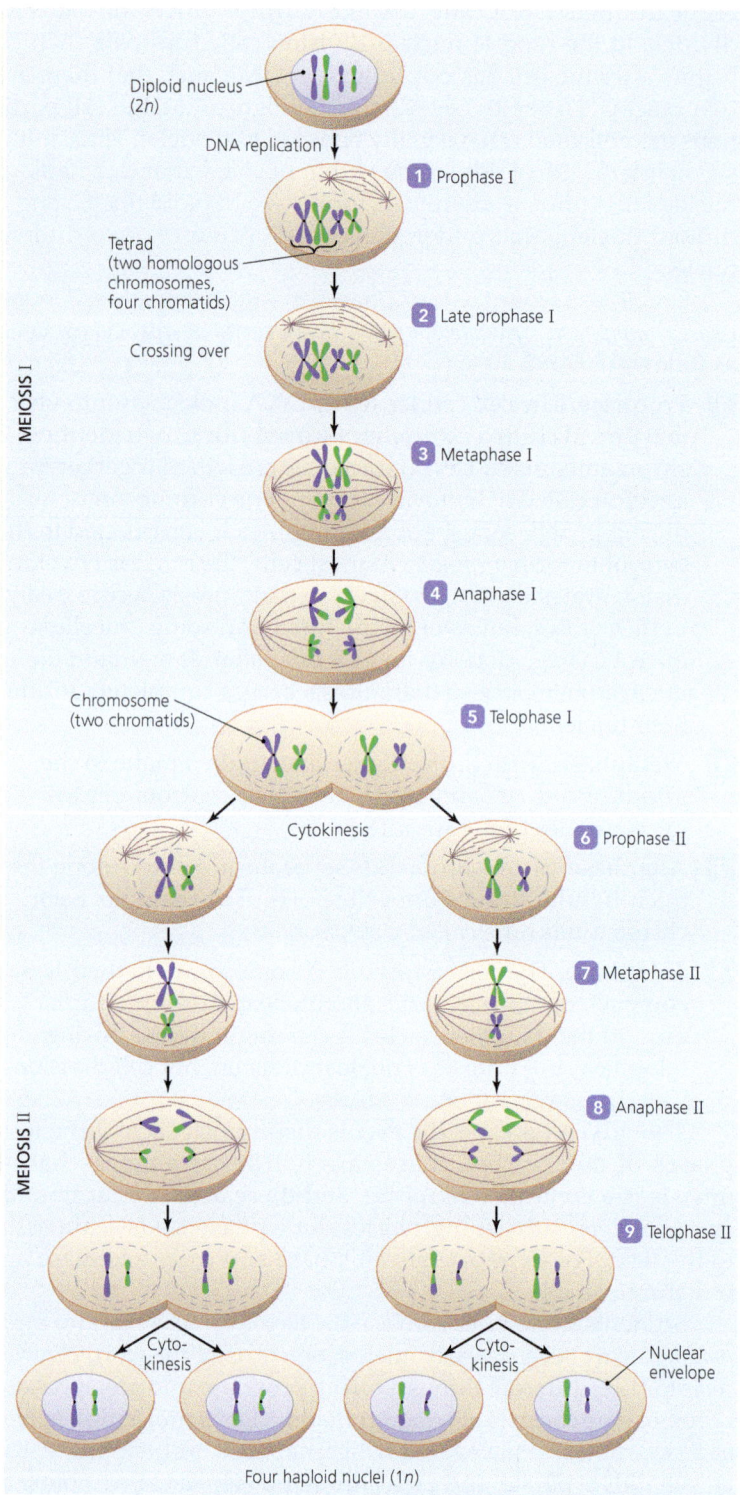

▲ **FIGURE 12.2 Meiosis results in four haploid nuclei.** Each nucleus has half the number of chromosomes of the parent nucleus. The text describes the events occurring during each of the numbered phases.

In summary, meiosis produces four haploid nuclei from a single diploid nucleus. Meiosis can be considered back-to-back mitoses without the DNA replication of interphase between them, though the four phases of meiosis I differ from those of

---
[14]From Greek *meioun*, meaning "to make smaller."

mitosis. The phases of meiosis II are equivalent to those in mitosis. Additionally, crossing over during meiosis I produces genetic recombinations, ensuring that the chromosomes resulting from meiosis are different from the parental chromosomes. This provides genetic variety in the next generation. TABLE 12.1 compares and contrasts mitotic and meiotic nuclear divisions.

Cytokinesis (Cytoplasmic Division)

Cytoplasmic division—also called **cytokinesis** (sī'tō-ki-nē'sis)—typically occurs simultaneously with telophase of mitosis, though in some algae and fungi it may be postponed or may not occur at all. In these cases, mitosis produces multinucleate cells called **coenocytes** (sē'nō-sītz) in which all the nuclei are identical.

In plant and algal cells, cytokinesis occurs as vesicles deposit wall material at the equatorial plane between nuclei to form a *cell plate*, which eventually becomes a transverse wall between daughter cells (FIGURE 12.3a). Cytokinesis of protozoan, animal, and some fungal cells occurs when an equatorial ring of actin microfilaments contracts just below the cytoplasmic membrane, pinching the cell in two (FIGURE 12.3b). Single-celled fungi called *yeasts* form a bud, which receives one of the daughter nuclei and pinches off from the parent cell (FIGURE 12.3c).

Schizogony

Some protozoa, such as *Plasmodium* (plaz-mō'dē-ŭm)—the cause of malaria—reproduce asexually via a special type of reproduction called **schizogony** (ski-zog'ō-nē; FIGURE 12.4). In schizogony, multiple mitoses form a multinucleate cell called a **schizont** (skiz'ont); only then does cytokinesis occur, simultaneously releasing numerous uninucleate daughter cells.

Classification of Eukaryotic Organisms

LEARNING OUTCOMES

12.5 Briefly describe the major groups of eukaryotes as they were first classified in the late 18th century and as they were classified in the late 20th century.

12.6 List some of the problems involved in the classification of protists in particular.

Historically, the classification of many eukaryotic microbes has been fraught with difficulty and characterized by change. From the late 18th century, when Carolus Linnaeus (1707–1778) began modern taxonomy, until near the end of the 20th century taxonomists grouped organisms together largely according to readily observable structural traits. Linnaeus classified unicellular algae and fungi as plants, and he classified protozoa—as the name suggests—as animals (FIGURE 12.5a). By the late 20th century, taxonomists had placed fungi in their own kingdom and grouped protozoa and algae together within the kingdom Protista (FIGURE 12.5b), though some taxonomists kept the green algae in the kingdom Plantae.

This scheme was troublesome, in part because the Protista included large, photosynthetic, multicellular kelps; nonphotosynthetic unicellular protozoa; and organisms that have both animal-like characteristics and characteristics more typical of plants. An example of the latter is *Euglena* (yŭ-glēn'ă; see Figure 3.32a), which swims with a flagellum, can move its cytoplasmic membrane, and devours food—all animal-like qualities—but also has a wall-like structure and can photosynthesize sugar in chloroplasts, like plants. Adding to the confusion is the fact that taxonomists who classify plants use the term *divisions* to refer to the same taxonomic level that zoologists call *phyla*.

More recently, many taxonomists have abandoned classification schemes that are so strongly grounded in large-scale structural similarities in favor of schemes based on similarities in nucleotide sequences and cellular ultrastructure as revealed by electron microscopy. One of the most evident results of such taxonomic studies is that modern schemes no longer include the taxa "Protozoa" or "Protista"; instead, such eukaryotic microbes belong in several kingdoms.

Though no one classification scheme has garnered universal support, and more thorough understanding based on new information will almost certainly dictate changes, many taxonomists favor a scheme similar to the one shown in FIGURE 12.5c. In this scheme, on which the discussions of eukaryotic microbes in this chapter are largely based, the organisms commonly referred to as protozoa are classified in six kingdoms: Parabasala, Diplomonadida, Euglenozoa, Alveolata, Rhizaria, and Amoebozoa; fungi are in the kingdom Fungi; slime molds are in the kingdom Amoebozoa; water molds are in the kingdom Stramenopila; and algae are distributed among

TABLE 12.1 Characteristics of the Two Types of Nuclear Division

	Mitosis	Meiosis
DNA Replication	During interphase, before nuclear division	During interphase, before meiosis I begins
Phases	Prophase, metaphase, anaphase, telophase	Meiosis I—prophase I, metaphase I, anaphase I, telophase I; Meiosis II—prophase II, metaphase II, anaphase II, telophase II
Formation of Tetrads (alignment of homologous chromosomes)	Does not occur	Early in prophase I
Crossing Over	Does not occur	Following formation of tetrads during prophase I
Number of Accompanying Cytoplasmic Divisions That May Occur	One	Two
Resulting Nuclei	Two nuclei with same ploidy as the original	Four nuclei with half the ploidy of the original

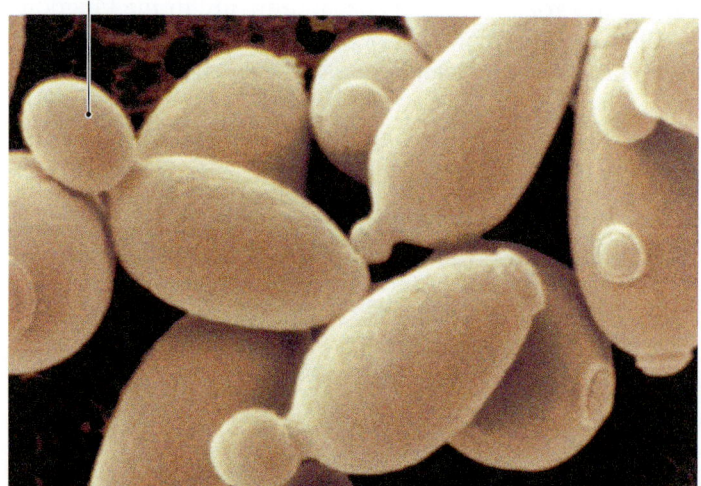

◀ FIGURE 12.3 **Different types of cytoplasmic division.**
(a) Cytokinesis in a plant cell, in which vesicles form a cell plate.
(b) Cytokinesis as it occurs in animals, protozoa, and some fungi.
(c) Budding in yeast cells.

the kingdoms Stramenopila, Rhodophyta, and Plantae. As we study the eukaryotic microbes discussed in this chapter, bear in mind that because the relationships among eukaryotic microbes are not fully understood, not all taxonomists agree with this scheme, and new information will shape future alterations in our understanding of the taxonomy of eukaryotic microbes.

We begin our survey of eukaryotic microbes with the group of organisms commonly known as protozoa.

MICRO CHECK

1. During which reproductive process are gametes formed?
2. What is the difference between the results of mitosis and meiosis?
3. How many daughter nuclei are produced by mitosis in contrast to the number produced by meiosis?
4. How are coenocytes formed?

TELL ME WHY

Why is it incorrect to call mitosis "cell division"?

Protozoa

LEARNING OUTCOME

12.7 List three characteristics shared by all protozoa.

The microorganisms called **protozoa** (prō-tō-zō′ă) are a diverse assemblage defined by three characteristics: they are eukaryotic, unicellular, and lack a cell wall. Note that "protozoa" is not a currently accepted taxon. With the exception of one subgroup (called apicomplexans), protozoa are motile by means of cilia, flagella, and/or pseudopods. The scientific study of protozoa is *protozoology*, and scientists who study these microbes are *protozoologists*.

In the following sections, we discuss the distribution, morphology, nutrition, reproduction, and classification of various groups of protozoa.

Distribution of Protozoa

Protozoa require moist environments; most species live worldwide in ponds, streams, lakes, and oceans, where they are critical members of the *plankton*—free-living, drifting organisms that form the basis of aquatic food chains. Other protozoa live in moist soil, beach sand, and decaying organic matter, and a very few are pathogens—that is, disease-causing microbes—of animals and humans.

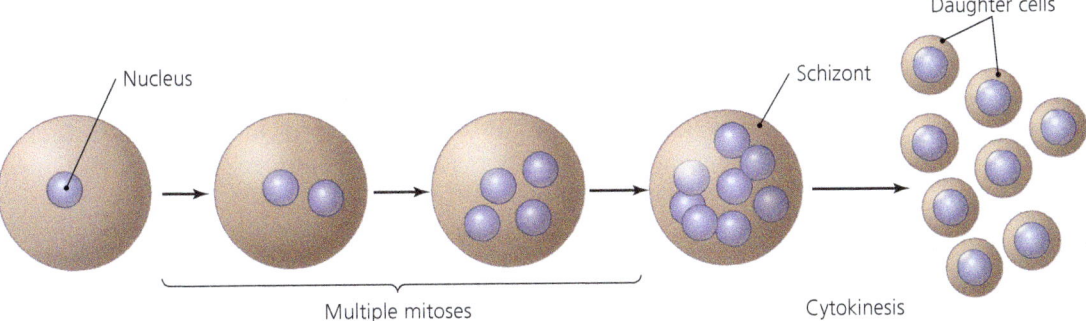

◀ FIGURE 12.4 Schizogony. Sequential mitoses without intervening cytokineses produce a multinucleate schizont, which later undergoes cytokinesis to produce many daughter cells.

Morphology of Protozoa

Though protozoa have most of the features of eukaryotic cells (discussed in Chapter 3, pp. 59–61, and illustrated in Figure 3.3), this group of eukaryotic microbes is characterized by great morphological diversity. Some ciliates have two nuclei: a larger *macronucleus*, which contains many copies of the genome (often more than $50n$) and controls metabolism, growth, and sexual reproduction, and a smaller *micronucleus*, which is involved in genetic recombination, sexual reproduction, and regeneration of macronuclei.

Protozoa also show variety in the number and kind of mitochondria they contain. Several groups lack mitochondria, whereas all the others have mitochondria with discoid or tubular cristae rather than the platelike cristae seen in animals, plants, fungi, and many algae. Additionally, some protozoa have *contractile vacuoles* that actively pump water from the cells, protecting them from osmotic lysis (FIGURE 12.6).

All free-living aquatic and pathogenic protozoa exist as a motile feeding stage called a **trophozoite** (trof-ō-zō′īt), and many have a hardy resting stage called a **cyst**, which is characterized by a thick capsule and a low metabolic rate. Cysts of protozoa are not reproductive structures: one trophozoite forms one cyst, which later becomes one trophozoite. Such cysts allow intestinal protozoa to pass from one host to another and to survive harsh environmental conditions such as desiccation, nutrient deficiency, extremes of pH and temperature, and lack of oxygen.

Nutrition of Protozoa

Most protozoa are chemoheterotrophic; that is, they obtain nutrients by phagocytizing bacteria, decayed organic matter, other protozoa, or the tissues of a host; a few protozoa absorb nutrients from the surrounding water. Because the protozoa called dinoflagellates and euglenids are photoautotrophic, botanists historically classified them as algal plants rather than as protozoa.

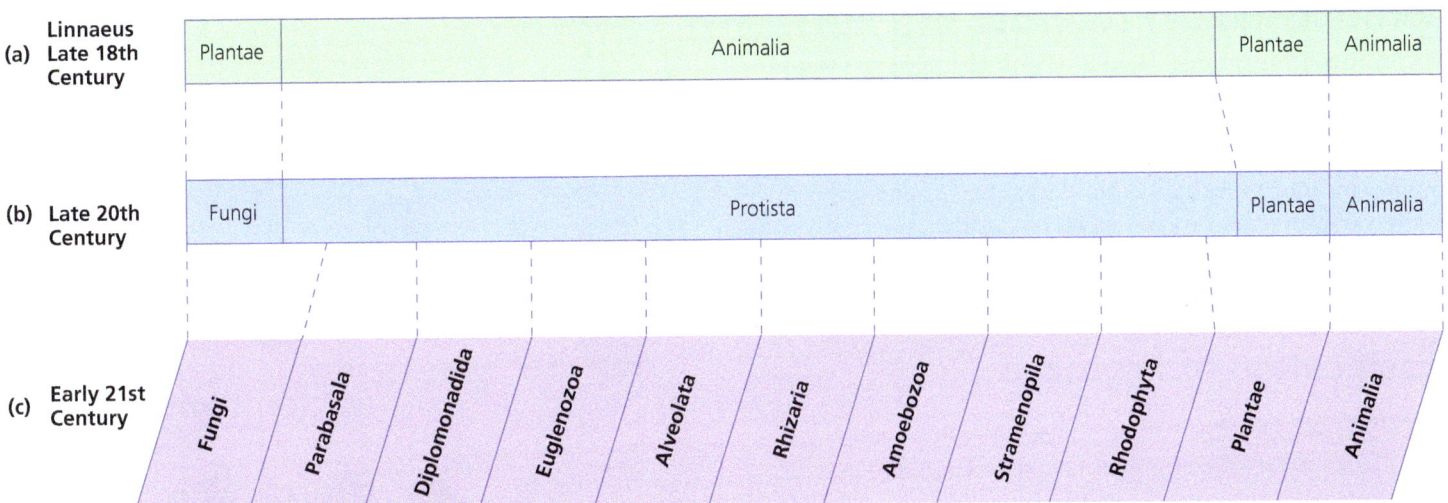

▲ FIGURE 12.5 The changing classification of eukaryotes over the centuries. (a) In the late 18th century, Linnaeus classified all organisms as either plants or animals. (b) In the late 20th century, taxonomists placed fungi in their own group and recognized a new kingdom, Protista. (c) Today, microbial eukaryotes are classified into numerous kingdoms based largely on their genetic relatedness. Though not all taxonomists would agree about every detail of this scheme, it forms the basis for the discussion of eukaryotic organisms in this chapter.

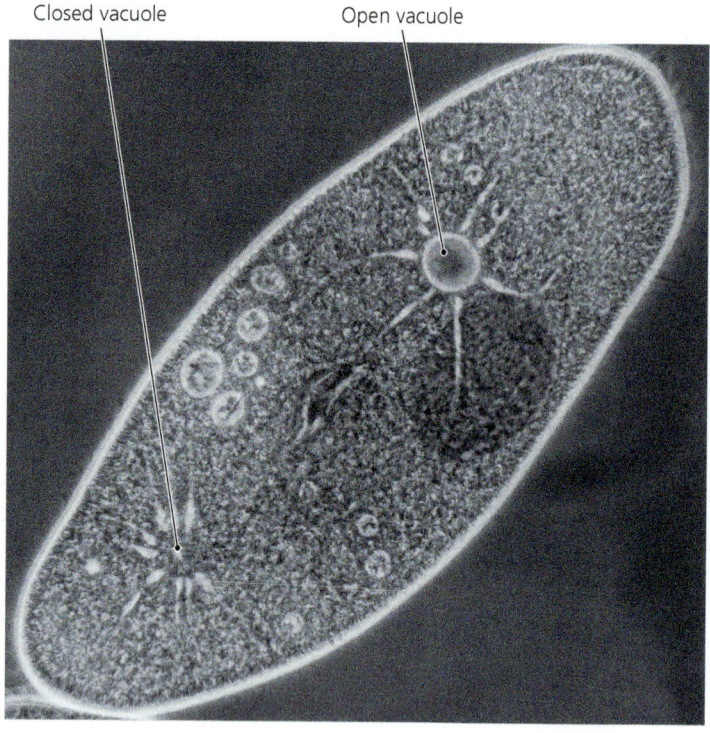

▲ FIGURE 12.6 **Contractile vacuoles.** Many protozoa, such as *Paramecium*, have this prominent feature. Open vacuoles fill with water that entered the cell via osmosis; closed vacuoles are contracted, pumping the water out of the cell. *Is the environment in this case hypertonic, hypotonic, or isotonic to the cell? Explain.*

Figure 12.6 The cell is pumping water out, so water has moved into the cell. Water moves into a cell by osmosis when the environment is hypotonic to the cell.

Reproduction of Protozoa

Most protozoa reproduce asexually only, by binary fission or schizogony; a few protozoa also carry out sexual reproduction in which two individuals exchange genetic material. Some sexually reproducing protozoa become **gametocytes** (gametes) that fuse with one another to form a diploid **zygote**. Ciliates reproduce sexually via a complex process called *conjugation*, which involves the coupling of two compatible mating cells and the exchange of nuclei.

Classification of Protozoa

LEARNING OUTCOMES

12.8 Discuss the reasons for the many different taxonomic schemes for protozoa.
12.9 Identify several features of a typical euglenid.
12.10 Compare and contrast three types of alveolates.
12.11 Compare and contrast three types of amoebas.
12.12 Describe characteristic features of parabasalids, diplomonads, rhizaria, and amoebozoa.

As we have seen, over two centuries ago Linnaeus classified protozoa as animals; later taxonomists grouped protozoa into kingdom Protista. Furthermore, some taxonomists divided the protozoa into four groups based on the organisms' mode of locomotion: Sarcodina (motile by means of pseudopods), Mastigophora (flagella), Ciliophora (cilia), and Sporozoa (nonmotile). Other taxonomists lumped the first two groups together into a single group called Sarcomastigophora. Grouping of the protozoa according to locomotory features is in common usage for many practical applications.

Taxonomists today recognize that grouping protozoa by locomotion does not reflect genetic relationships either among protozoa or between protozoa and other organisms. Accordingly, taxonomists continue to revise and refine the classification of protozoa based on nucleotide sequencing and features made visible by electron microscopy. One genetic scheme classifies protozoa into the six taxa Parabasala through Amoebozoa shown in Figure 12.5c, which various taxonomists consider kingdoms (as here), subkingdoms, phyla, or other taxa.

In the following sections, we will briefly discuss members of these six taxa of protozoa, formed largely according to similarities in nucleotide sequences and ultrastructure. We begin with parabasalids.

Parabasala

Parabasalids lack mitochondria, but each has a single nucleus and a *parabasal body*, which is a Golgi body–like structure. A well-known parabasalid is *Trichomonas* (trik-ō-mō'nas), which lives in the human vagina (**FIGURE 12.7**). When the normally acidic pH of a vagina is raised, *Trichomonas* proliferates and causes severe inflammation that can lead to sterility. Sexual intercourse spreads *Trichomonas*; trichomoniasis is usually asymptomatic in males.

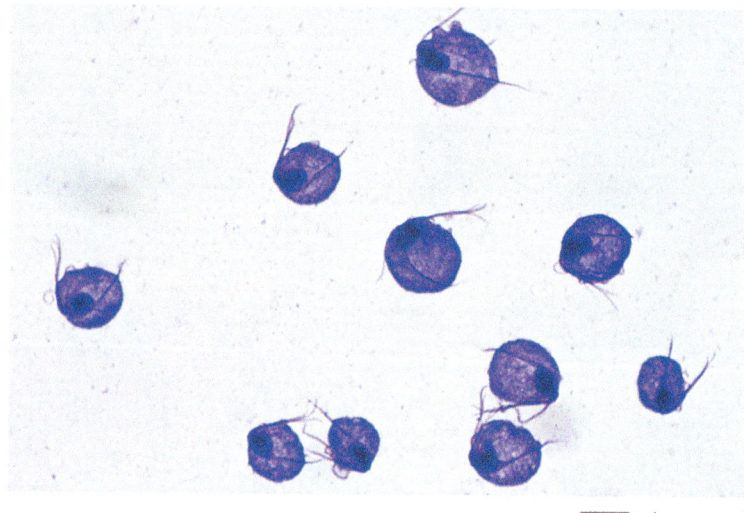

▲ FIGURE 12.7 *Trichomonas vaginalis*, a sexually transmitted parabasalid.

Diplomonadida

Because members of the group Diplomonadida[15] lack mitochondria, Golgi bodies, and peroxisomes, biologists once thought these organisms were descended from ancient eukaryotes that had not yet phagocytized the prokaryotic ancestors of mitochondria. More recently, however, geneticists have discovered rudimentary mitochondria-like structures called *mitosomes* in the cytoplasm as well as mitochondrial genes in the nuclear chromosomes, a finding that suggests that diplomonads might be descended from typical eukaryotes that somehow lost their organelles.

Diplomonads have two equal-sized nuclei and multiple flagella. A prominent example is *Giardia* (jē-ar′dē-ă), a diarrhea-causing pathogen of animals and humans that is spread to new hosts when they ingest resistant *Giardia* cysts (see Disease in Depth: Giardiasis, pp. 684–685).

Euglenozoa

Part of the reason that taxonomists established the kingdom Protista in the 1960s was to create a "dumping ground" for *euglenids*, eukaryotic microbes that share certain characteristics of both plants and animals. More recently, based on similar 18S rRNA sequences, the presence of a crystalline rod of unknown function in the flagella, and the presence of mitochondria with disk-shaped cristae, some taxonomists have created a new taxon: kingdom Euglenozoa. The euglenozoa include euglenids and some flagellated protozoa called *kinetoplastids*.

Euglenids The group of euglenozoa called **euglenids**, which are named for the genus *Euglena* (FIGURE 12.8a), are photoautotrophic, unicellular microbes with chloroplasts containing light-absorbing pigments—chlorophylls *a* and *b* and carotene. For this reason, botanists historically classified euglenids in the kingdom Plantae. However, one reason for not including euglenids with plants is that euglenids store food as a unique polysaccharide called *paramylon* instead of starch, as in true plants. Euglenids are similar to animals in that they lack cell walls, have flagella, are chemoheterotrophic phagocytes (in the dark), and move by using their flagella as well as by flowing, contracting, and expanding their cytoplasm. Such a squirming movement, which is similar to amoeboid movement but does not involve pseudopods, is called *euglenoid movement*.

A euglenid has a flexible, proteinaceous, helical *pellicle* that underlies its cytoplasmic membrane and helps maintain its shape. Typically, each euglenid also has a red "eyespot," which plays a role in positive phototaxis by casting a shadow on a photoreceptor at the flagellar base, triggering movement in that direction. Euglenids reproduce by mitosis followed by longitudinal cytokinesis. They form cysts when exposed to harsh conditions.

Kinetoplastids Euglenozoa called **kinetoplastids** (FIGURE 12.7b) each have a single large mitochondrion that contains a unique

---
[15]From Greek *diploos*, meaning "double," and *monas*, meaning "unit," referring to two nuclei.

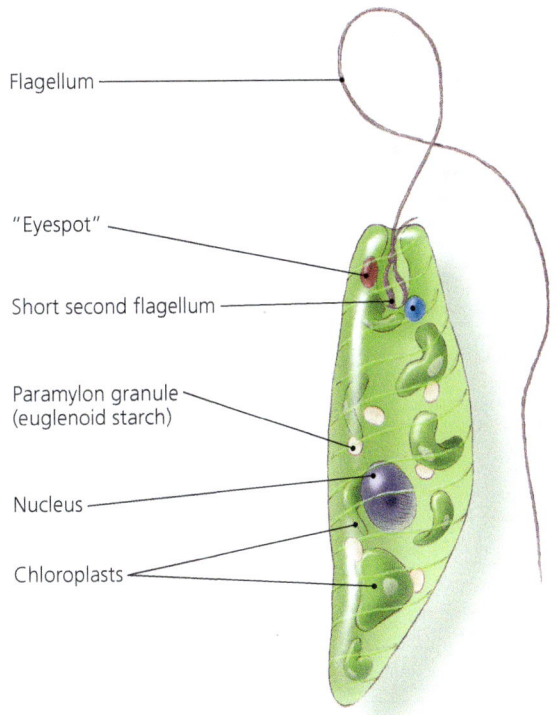

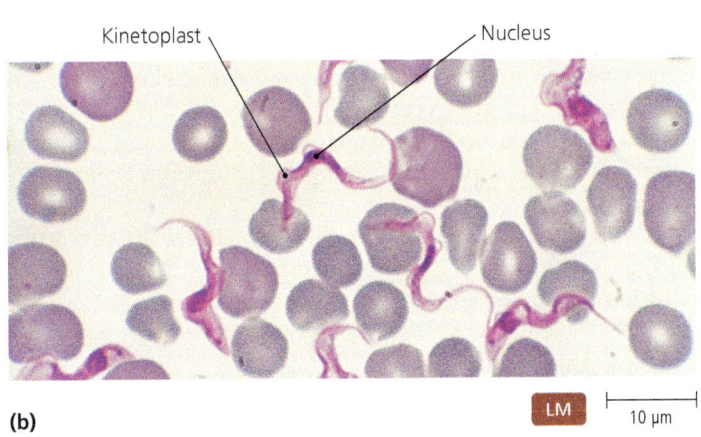

▲ FIGURE 12.8 Two representatives of the kingdom euglenozoa. (a) The euglenid *Euglena*. Euglenids have characteristics that are similar to both plants and animals. (b) Kinetoplastids called *Trypanosoma* in blood. *What is the function of the kinetoplast in* Trypanosoma?

Figure 12.8 A kinetoplast is mitochondrial DNA, which codes for some mitochondrial polypeptides.

region of mitochondrial DNA called a *kinetoplast*. As in all mitochondria, this DNA codes for some mitochondrial polypeptides, which join with polypeptides coded by nuclear DNA to make functional proteins.

Kinetoplastids live inside animals, and some are pathogenic. Among the latter are the genera *Trypanosoma* (tri-pan′ō-sō-mă) and *Leishmania* (lēsh-man′ē-ă), certain species of which cause potentially fatal diseases of mammals, including humans (see Chapter 23).

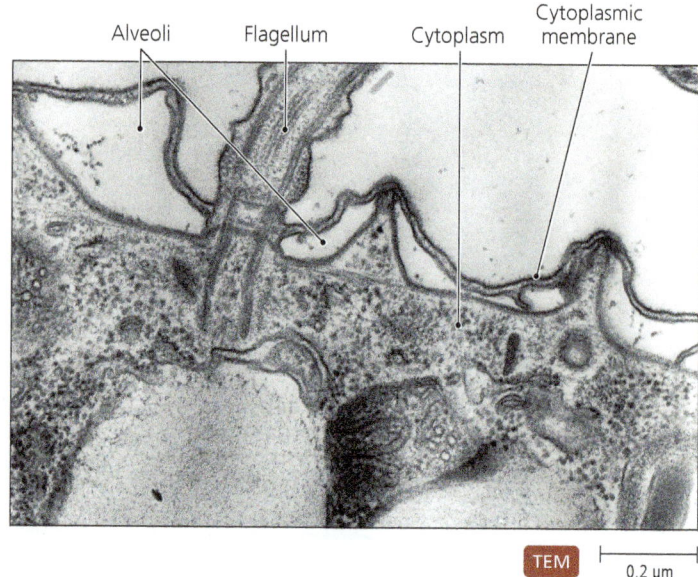

▲ FIGURE 12.9 **Membrane-bound alveoli found in some protozoa.** Even though their function is not yet known, alveoli are present in eukaryotic microbes with similar 18S rRNA sequences, indicating genetic relatedness and forming the basis of the group of eukaryotes called alveolates.

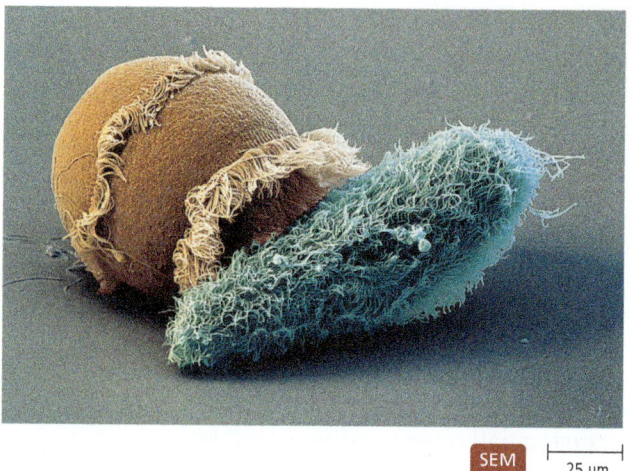

▲ FIGURE 12.10 A predatory ciliate, *Didinium* (left), devouring another ciliate, *Paramecium*.

Alveolates

Alveolates (al-vē'ō-lātz) are protozoa with alveoli[16] (al-vē'ō-lē), which are small membrane-bound cavities packed as a continuous layer beneath the cytoplasmic membrane (FIGURE 12.9). Scientists do not know the purpose of alveoli, though alveoli are essential for some parasitic alveolates' motility and invasion of host cells.

Alveolates share at least one other characteristic—tubular mitochondrial cristae. This group is further divided into three subgroups: *ciliates*, *apicomplexans*, and *dinoflagellates*.

Ciliates As their name indicates, **ciliate** (sil'ē-āt) alveolates have cilia by which they either move themselves or move water past their cell surfaces. (The structure and function of cilia are discussed on p. 82.) Some ciliates are covered with cilia, whereas others have only a few isolated tufts. All ciliates are chemoheterotrophs and have two nuclei—one macronucleus and one micronucleus. Some taxonomists consider them the sole members of phylum Ciliophora.

Notable ciliates include *Vorticella* (vōr-ti-sel'ă), whose apical cilia create a whirlpool-like current to direct food into its "mouth;" *Balantidium* (bal-an-tid'ē-ŭm), which is the only ciliate pathogenic to humans; and the carnivorous *Didinium* (dī-di'nē-ŭm), which phagocytizes other protozoa, such as the well-known pond-water ciliate *Paramecium* (par-ă-mē'sē-ŭm) (FIGURE 12.10).

Apicomplexans The alveolates called **apicomplexans** (ap-i-kom-plek'sănz) are all chemoheterotrophic pathogens of animals.

The name of this group refers to the *complex* of special intracellular organelles, located at the *apices* of the infective stages of these microbes, that enables them to penetrate host cells. Many apicomplexans reproduce using schizogony.

Examples of apicomplexans are *Plasmodium*, *Cryptosporidium* (krip-tō-spō-rid'ē-ŭm), and *Toxoplasma* (tok-sō-plaz'mă), which cause malaria, cryptosporidiosis, and toxoplasmosis, respectively. (Chapter 23 considers in more detail these apicomplexans and the diseases they cause.)

Dinoflagellates The group of alveolates called **dinoflagellates** are unicellular microbes that have photosynthetic pigments, such as carotene and chlorophylls a, c_1, and c_2. As with many plants and algae, their food reserves are starch and oil, and their cells are often strengthened by plates of cellulose within their alveoli. Even though botanists historically classified dinoflagellates as algae because dinoflagellates are photoautotrophic and share other characteristics with algae, taxonomists today note that their 18S rRNA sequences and the presence of alveoli indicate that dinoflagellates are more closely related to ciliates and apicomplexans than they are to either plants or algae. Interestingly, unlike other eukaryotic chromosomes, dinoflagellate chromosomes lack histone proteins.

Dinoflagellates make up a large proportion of freshwater and marine plankton. Motile dinoflagellates have two flagella of unequal length (FIGURE 12.11). The transverse flagellum wraps around the equator of the cell in a groove in the cell wall, and its beat causes the cell to spin; the second flagellum extends posteriorly and propels the cell forward.

Many dinoflagellates are bioluminescent—that is, able to produce light via metabolic reactions. When luminescent dinoflagellates are present in large numbers, the ocean water lights up with every crashing wave, passing ship, or jumping fish. Other dinoflagellates produce a red pigment, and their abundance in marine water is one cause of a phenomenon called a **red tide**.

Some dinoflagellates, such as *Gymnodinium* (jim-nō-din'ē-um) and *Gonyaulax* (gon-ē-aw'laks), produce *neurotoxins*—poisons

---

[16]Latin, meaning "small hollows."

that can act against the human nervous system. Humans can become exposed when they eat shellfish that have ingested planktonic dinoflagellates and concentrated their toxins.

The neurotoxin of another dinoflagellate, *Pfiesteria*[17] (fes-tēr'ē-ă), may be even more potent: It has been claimed that the toxin poisons people who merely handle infected fish or breathe air laden with the microbes, resulting in memory loss, confusion, headache, respiratory difficulties, skin rash, muscle cramps, diarrhea, nausea, and vomiting. The Centers for Disease Control and Prevention (CDC) calls such poisoning *possible estuary-associated syndrome (PEAS)*.

Rhizaria

Unicellular eukaryotes called **amoebas** (or *amoebae*)[18] are protozoa that move and feed by means of pseudopods (see Figure 3.31). Beyond this common feature and the fact that they all reproduce via binary fission, amoebas exhibit little uniformity. Some taxonomists currently classify amoebas into two kingdoms: Rhizaria and Amoebozoa. We consider them in order.

Rhizaria is a group of amoebas with threadlike pseudopods. A major taxon is composed of armored marine amoebas known as **foraminifera**. A foraminiferan has a porous shell composed of calcium carbonate arranged on an organic matrix in a snail-like manner (**FIGURE 12.12**). Pseudopods extend through holes in the shell. Commonly, foraminifera live attached to sand grains on the ocean floor. Most foraminifera are microscopic, though scientists have discovered species several centimeters in diameter.

Over 90% of known foraminifera are fossil species, some of which form layers of limestone hundreds of meters thick. The great pyramids of Giza outside Cairo, Egypt, are built of foraminiferan limestone. Geologists correlate the ages of sedimentary rocks from different parts of the world by finding identical foraminiferan fossils embedded in them.

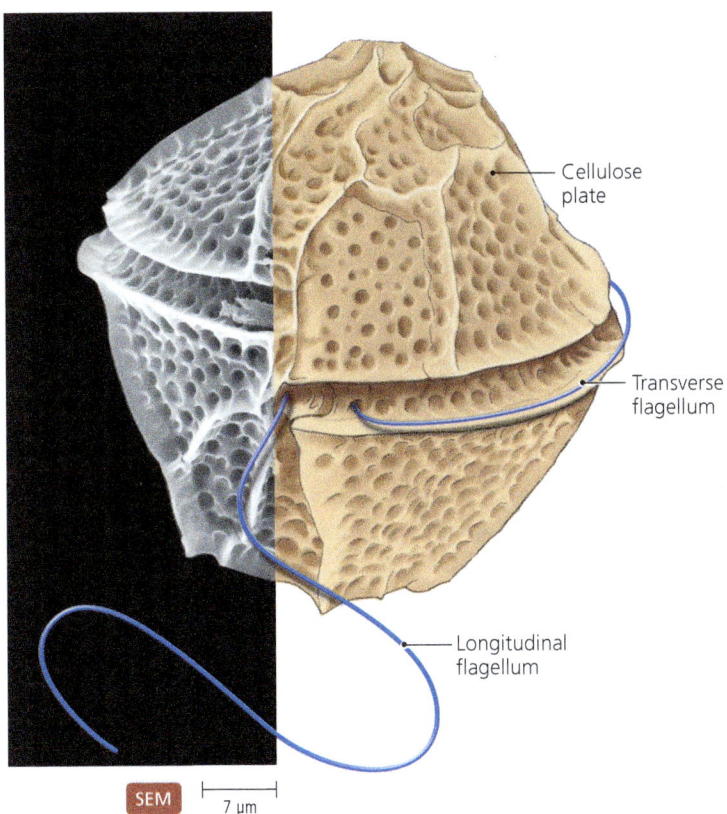

▲ **FIGURE 12.11** *Gonyaulax*, a motile armored dinoflagellate. Dinoflagellates have two flagella. The transverse flagellum spins the cell; the longitudinal flagellum propels the cell forward.

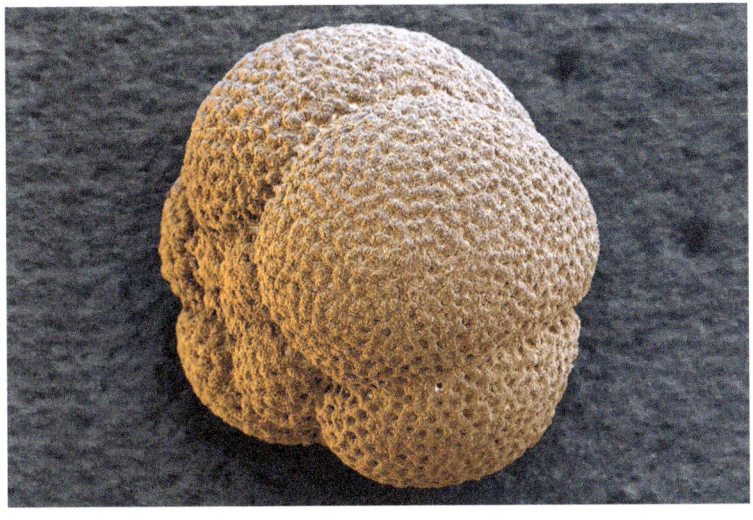
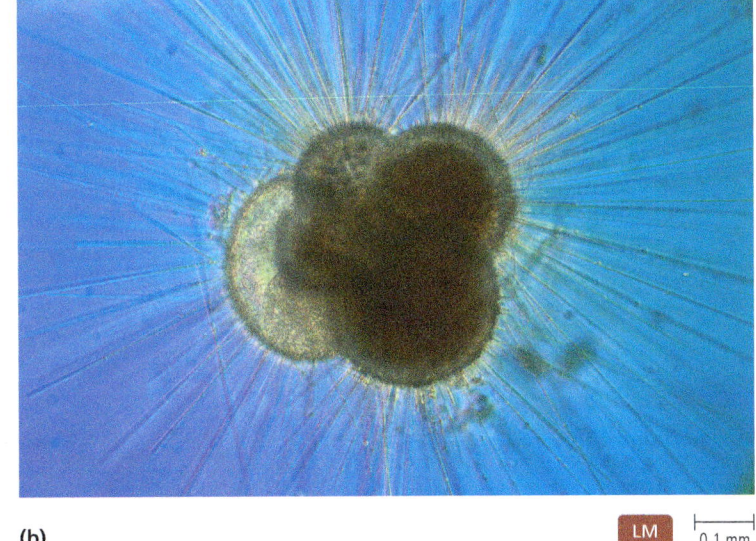

▲ **FIGURE 12.12** Rhizaria called foraminifera (here, *Globigerina*) have multichambered, snail-like shells of calcium carbonate. **(a)** Pseudopods are not present in this dead specimen. **(b)** Long, thin pseudopods of a living specimen.

---

[17] Named for dinoflagellate biologist Lois Pfiester.
[18] From Greek *ameibein*, meaning "to change."

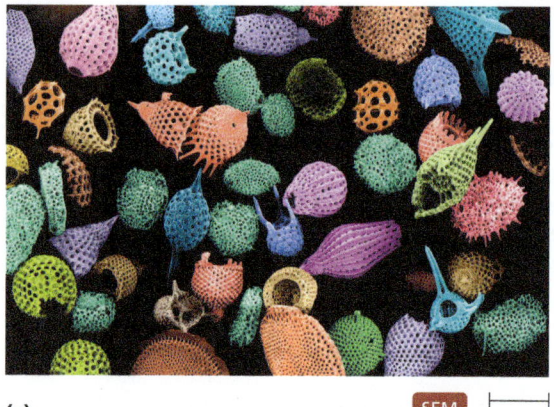
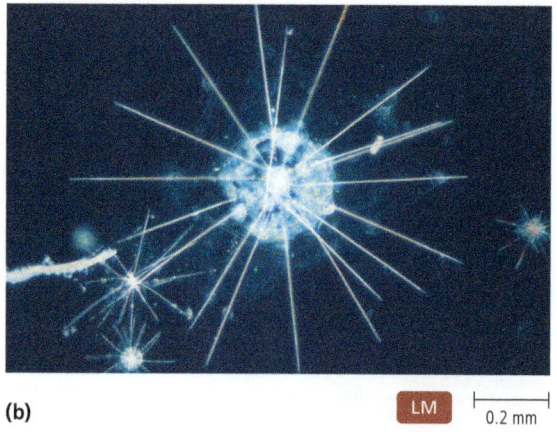

◀ **FIGURE 12.13**
Radiolarians, a type of rhizaria. (a) Radiolarians have ornate shells of silica (colorized). **(b)** Thin pseudopods extend through the holes in the shells.

Amoebas called **radiolaria** make up another group of rhizaria, but they have ornate shells composed of silica (SiO_2, the mineral found in opal) (**FIGURE 12.13a**) and live unattached as part of the marine plankton. Radiolarians reinforce their pseudopods with stiff internal bundles of microtubules so that the pseudopods radiate from the central body like spokes of a spherical wheel (**FIGURE 12.13b**). The dead bodies of radiolarians settle to the bottom of the ocean, where they form ooze that is hundreds of meters thick in some locations.

Amoebozoa

Amoebozoa constitute a second kingdom of amoebas distinguished from rhizaria by having lobe-shaped pseudopods and no shells. Amoebozoa include the normally free-living amoebas *Naegleria* (nā-glē'rē-ă) and *Acanthamoeba* (ă-kan-thă-mē'bă), which can each cause diseases of the eyes or brains of humans and animals that swim in water containing them. Other amoebozoa, such as *Entamoeba* (ent-ă-mē'bă), always live inside animals, where they produce potentially fatal amebic dysentery. (Chapter 23 examines these pathogenic amoebas in more detail.)

Taxonomists formerly considered another group of amoebozoa—**slime molds**—to be fungi, but their nucleotide sequences, as well as the lobe-shaped pseudopods by which they feed and move, show that they are amoebozoa. Scientists have identified two types of slime molds: *plasmodial slime molds* and *cellular slime molds*.

Slime molds differ from true fungi in two main ways:

- They lack cell walls, more closely resembling the amoebas in this regard.
- They are phagocytic rather than absorptive in their nutrition.

Species in the two groups of slime molds differ based on their morphology, reproduction, and 18S rRNA sequences. Slime molds are important to humans primarily as excellent laboratory systems for the study of developmental and molecular biology.

In summary, protozoa are a heterogeneous collection of single-celled, mostly chemoheterotrophic organisms that lack cell walls. Some taxonomists classify them in Parabasala, Diplomonadida, Euglenozoa, Alveolata, Rhizaria, and Amoebozoa, though their relationships with one another and with other eukaryotic organisms are still unclear. **TABLE 12.2** summarizes the incredible diversity of these microbes.

We next turn our attention to another group of chemoheterotrophs, the fungi, which differ from the protozoa in that fungi have cell walls.

MICRO CHECK

5. What are three defining characteristics of protozoa?
6. *Giardia* species cause diarrhea in humans and animals. They have two nuclei and lack Golgi bodies, peroxisomes, and true mitochondria. To which group of protozoa does *Giardia* belong?

TELL ME WHY

Why did early taxonomists categorize such obviously different microorganisms as parabasalids, diplomonads, euglenozoa, alveolates, rhizaria, and amoebozoa in a single taxon, Protozoa?

Fungi

LEARNING OUTCOME

12.13 Cite at least three characteristics that distinguish fungi from other groups of eukaryotes.

Organisms in the kingdom **Fungi** (fŭn'jī), such as molds, mushrooms, and yeasts, are like most protozoa in that they are chemoheterotrophic; however, unlike protozoa, they have cell walls, which typically are composed of a strong, flexible, nitrogenous polysaccharide called **chitin**. (The chitin in fungi is chemically identical to that in the exoskeletons of insects and other arthropods, such as grasshoppers, lobsters, and crabs.) Fungi differ from plants in that they lack chlorophyll and do not perform photosynthesis; they differ from animals by having cell walls, although genetic sequencing of fungal and animal genomes has shown that fungi and animals are related. The study of fungi is *mycology*,[19] and scientists who study fungi are *mycologists*.

---
[19]From Greek *mykes*, meaning "mushroom," and *logos*, meaning "discourse."

TABLE 12.2 Characteristics of Protozoa

Category	Distinguishing Features	Representative Genera Mentioned in the Text
Parabasala	Parabasal body; single nucleus; lack mitochondria	*Trichomonas*
Diplomonadida	Two equal-sized nuclei; lack mitochondria, Golgi bodies, and peroxisomes; multiple flagella	*Giardia*
Euglenozoa	Flagella with internal crystalline rod; disk-shaped mitochondrial cristae	
Euglenids	Photosynthesis; pellicle; "eyespot"	*Euglena*
Kinetoplastids	Single mitochondrion with DNA localized in kinetoplast	*Trypanosoma, Leishmania*
Alveolates	Alveoli (membrane-bound cavities underlying the cytoplasmic membrane); tubular cristae in mitochondria	
Ciliates	Cilia	*Balantidium, Paramecium, Didinium*
Apicomplexans	Apical complex of organelles	*Plasmodium, Cryptosporidium, Toxoplasma*
Dinoflagellates	Photosynthesis; two flagella; internal cellulose plates	*Gymnodinium, Gonyaulax, Pfiesteria*
Rhizaria	Threadlike pseudopods	
Foraminifera	Shells of calcium carbonate	
Radiolarians	Shells of silica	
Amoebozoa	Lobe-shaped pseudopods; no shells	
Free-living and parasitic forms	Do not form aggregates	*Naegleria, Acanthamoeba, Entamoeba*
Plasmodial (acellular) slime molds	Multinucleate body (called *plasmodium*)	*Physarum*
Cellular slime molds	Cells aggregate but retain individual (cellular) nature	*Dictyostelium*

The Significance of Fungi

LEARNING OUTCOME

12.14 List five ways in which fungi are beneficial.

Fungi are extremely beneficial microorganisms. In nature, they decompose dead organisms (particularly plants) and recycle their nutrients. Additionally, the roots of about 90% of vascular plants form *mycorrhizae*,[20] which are beneficial associations between roots and fungi that help the plants absorb water and dissolved minerals.

Humans use fungi for food (mushrooms and truffles), in religious ceremonies (because of their hallucinogenic properties), and in the manufacture of foods and beverages, including bread, alcoholic beverages, citric acid (the basis of the soft drink industry), soy sauce, and some cheeses. Fungi also produce antibiotics, such as penicillin and cephalosporin; the immunosuppressive drug *cyclosporine*, which makes organ transplants possible; and *mevinic acids*, which are cholesterol-reducing agents.

Fungi are also important research tools in the study of metabolism, growth, and development and in genetics and biotechnology. For instance, based on their work with the fungus *Neurospora* (noo-ros'pōr-ă) in the 1950s, George Beadle (1903–1989) and Edward Tatum (1909–1975) developed their Nobel Prize-winning theory that one gene codes for one enzyme. Because of similar research, *Saccharomyces* (brewer's yeast) is the best-understood eukaryote and the first eukaryote to have its entire genome sequenced. (Chapter 26 highlights some uses of fungi in agriculture and industry.)

Not all fungi are beneficial—about 30% of known fungal species produce **mycoses** (mī-kō'sēz), which are fungal diseases of plants, animals, and humans (see Chapter 22). For example, Dutch elm disease is a mycosis of elm trees, and athlete's foot is a fungal disease of humans. Because fungi tolerate concentrations of salt, acid, and sugar that inhibit bacteria, fungi are responsible for the spoilage of fruit, pickles, jams, and jellies exposed to air.

In the following sections, we will consider the basic characteristics of fungal morphology, nutrition, and reproduction before turning to a brief survey of the major groups of fungi.

Morphology of Fungi

LEARNING OUTCOME

12.15 Describe the main two body shapes of fungi, explain what dimorphic means, and distinguish among septate hyphae, aseptate hyphae, and mycelia.

Fungi have two basic body shapes. The bodies of *molds* are relatively large and composed of long, branched, tubular filaments called **hyphae**.[21] Hyphae are either **septate** (divided into cells by cross walls called *septa*;[22] FIGURE 12.14a) or **aseptate** (not divided by septa; FIGURE 12.14b). Aseptate hyphae are multinucleate. The bodies of *yeasts* are typically small, globular, and composed of a single cell, which may produce buds (FIGURE 12.14c).

In response to environmental conditions such as temperature or carbon dioxide concentration, some fungi produce both yeastlike and moldlike shapes (FIGURE 12.14d); fungi that produce two types of body shapes are said to be **dimorphic** (which

---
[20]From Greek *rhiza*, meaning "root."
[21]From Greek *hyphe*, meaning "weaving" or "web."
[22]Latin, meaning "partitions" or "fences."

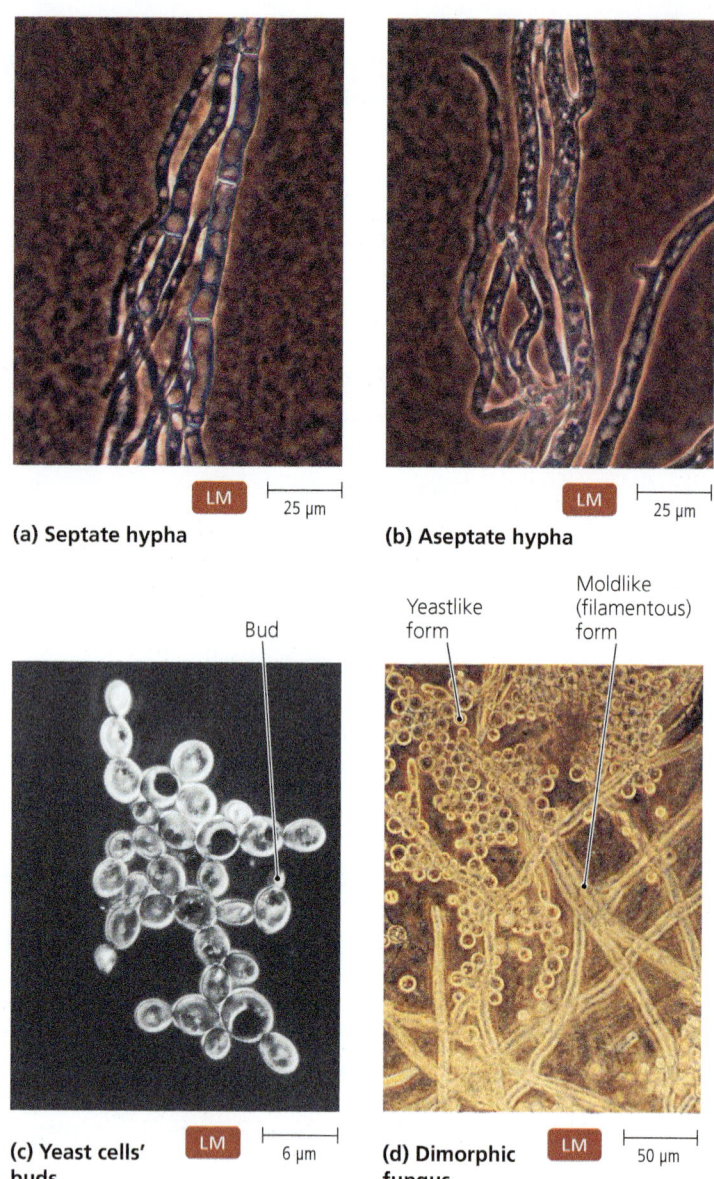

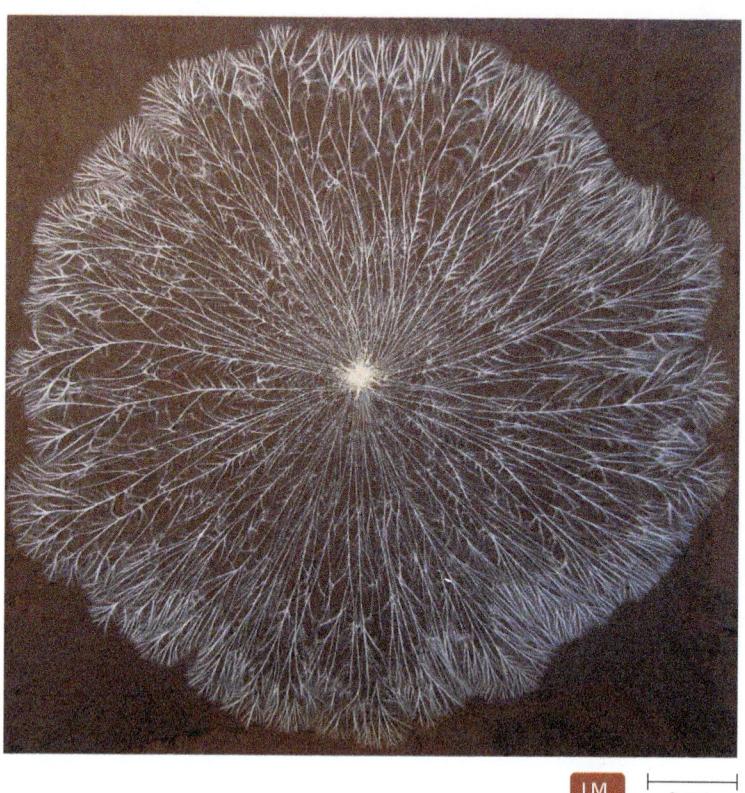

▲ FIGURE 12.15 A fungal mycelium.

▲ FIGURE 12.14 Fungal morphology. (a) Septate hyphae have cross walls; (b) aseptate hyphae do not. (c) Cells of *Saccharomyces* (baker's or brewer's yeast), which are unicellular and spherical to irregularly oval in shape. (d) The bodies of a dimorphic fungus, *Mucor rouxii*, showing both yeastlike and moldlike growth in response to environmental conditions.

means "two-shaped"). Many medically important fungi are thermally dimorphic; that is, they change growth habits in response to the temperature in their immediate vicinity. Such fungi include *Histoplasma capsulatum* (his-tō-plaz′mă kap-soo-lā′tŭm), which causes a respiratory disease called histoplasmosis, and *Coccidioides immitis* (kok-sid-ē-oy′dēz im′mi-tis), which causes a flulike disease called coccidioidomycosis (kok-sid-ē-oy′dō-mī-kō′sis). Generally, the yeast form of a dimorphic fungal pathogen causes disease, whereas the filamentous form does not.

The body of a mold is composed of hyphae intertwined to form a tangled mass called a **mycelium** (plural: *mycelia*; FIGURE 12.15). Mycelia are typically subterranean and thus usually escape our notice, though they can be very large. In fact, the largest known organisms on Earth are fungi in the genus *Armillaria*, the mycelia of which can spread through thousands of acres of forest to a depth of several feet and weigh many hundreds of tons. (In contrast, blue whales, the largest living animals, weigh only about 150 tons.) *Fruiting bodies*, such as puffballs and mushrooms, are the hyphae of the reproductive structures of molds and are only small visible extensions of vast underground mycelia.

Nutrition of Fungi

LEARNING OUTCOME

12.16 Describe two ways that fungi acquire nutrients and the way molds transport nutrients throughout their bodies.

Fungi acquire nutrients by absorption; that is, they secrete catabolic enzymes outside their bodies to break large organic molecules into smaller molecules, which they then transport throughout their bodies. Most fungi are **saprobes**[23] (sap′rōbz)—they absorb nutrients from the remnants of dead organisms—though some species trap and kill microscopic soil-dwelling nematodes (worms; FIGURE 12.16). Fungi that derive their nutrients from living plants and animals usually have modified hyphae called **haustoria**[24] (haw-stō′rē-ă), which penetrate the tissue of the host to withdraw nutrients. Absorptive nutrition is important in the role that fungi play as decomposers and recyclers of organic waste. Cytoplasmic streaming frequently transports nutrients and organelles, including nuclei, throughout a mycelium. Streaming between cells of septate mycelia occurs through pores in the septa.

---

[23]From Greek *sapros*, meaning "rotten," and *bios*, meaning "life."
[24]From Latin *haustor*, meaning "someone who draws water from a well."

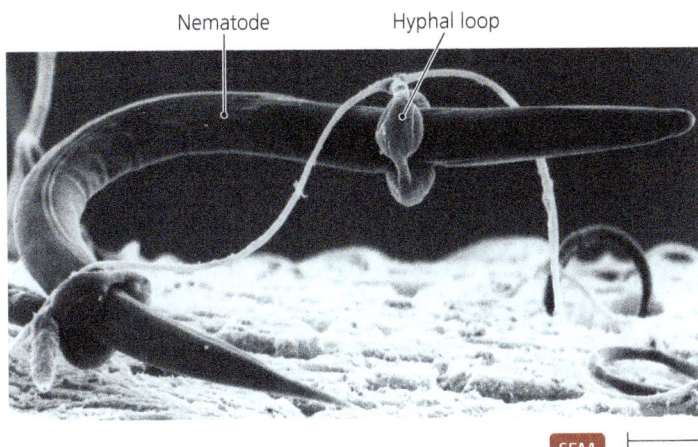

▲ **FIGURE 12.16 Predation of a nematode by the fungus *Drechslerella*.** The fungus produces special looped hyphae that constrict when the worm contacts the inside of the loop. The fungus secretes enzymes that digest the nematode and then absorbs the resulting nutrients. *What is the more typical mode of nutrition found in fungi?*

Figure 12.16 Most fungi are saprobic.

Scientists have discovered that some fungi use ionizing radiation (radioactivity) as an energy source for metabolism. The fungi absorb radiation with a black pigment, melanin, and transform absorbed gamma radiation into chemical energy, which the fungi then use to grow.

Most fungi are aerobic, though many yeasts (e.g., *Saccharomyces*) are facultative anaerobes that obtain energy from fermentation, such as occurs in the reactions that produce alcohol. Anaerobic fungi are found in the digestive systems of many herbivores, such as cattle and deer, where they assist in the catabolism of plant material.

MICRO CHECK

7. When viewed through the microscope, many fungi look very plantlike. Which characteristic makes them distinctly different from plants?
8. Which type of fungi has distinct cross walls between neighboring cells?

Reproduction of Fungi

LEARNING OUTCOMES

12.17 Describe asexual and sexual reproduction in fungi.
12.18 List three basic types of asexual spores found in molds.

Whereas all fungi have some means of asexual reproduction involving mitosis followed by cytokinesis, most fungi also reproduce sexually. In the next sections we briefly examine asexual and sexual reproduction in fungi.

Budding and Asexual Spore Formation

Yeasts typically bud in a manner similar to prokaryotic budding. Following mitosis, one daughter nucleus is sequestered in a small bleb (a blisterlike outgrowth) of cytoplasm that is isolated from the parent cell by the formation of a new wall (see Figure 12.13c). In some species, especially *Candida albicans* (kan'did-ă al'bi-kanz), which causes oral thrush and vaginal yeast infections in humans, a series of buds remain attached to one another and to the parent cell, forming a long filament called a *pseudohypha* (see Disease in Depth: Candidiasis on pp. 660–661). *Candida* invades human tissues by means of such pseudohyphae, which can penetrate intercellular cracks.

Filamentous fungi reproduce asexually by producing lightweight spores, which enable a fungus to disperse vast distances on the wind. Researchers have isolated fungal spores from wind currents many miles above the surface of the Earth.

Scientists categorize the asexual spores of molds according to their mode of development. Following are examples of some types of asexual spores:

- *Sporangiospores* form inside a sac called a *sporangium*,[25] which is often borne on a spore-bearing stalk, called a *sporangiophore*,[26] at either the tips or the sides of hyphae (**FIGURE 12.17a**).
- *Chlamydospores* form with a thickened cell wall inside hyphae (**FIGURE 12.17b**).
- *Conidiospores* (also called *conidia*) are produced at the tips or sides of hyphae, but not within a sac. There are many types of conidia, some of which develop in chains on stalks called *conidiophores* (**FIGURE 12.17c**).

Medical lab technologists use the presence and type of asexual spores in clinical samples to identify many fungal pathogens.

Sexual Spore Formation

Scientists can arbitrarily designate fungal mating types as "plus (+)" and "minus (−)" rather than as male and female, in part because their bodies are morphologically indistinguishable. The process of sexual reproduction in many fungi has four basic steps (**FIGURE 12.18**):

1. Many fungi form *dikaryons* in the course of sexual reproduction. Haploid (n) cells from a plus fungus and a minus fungus fuse to form a dikaryon, a cell containing two different nuclei, both + and − nuclei. A dikaryotic stage is neither diploid nor haploid but instead is designated ($n + n$). Dikaryotic cells can reproduce to form daughter cells, each of which is also a dikaryon.

2. After a period of time that typically ranges from hours to years but can be centuries, a pair of nuclei within a dikaryon fuse to form one diploid ($2n$) nucleus.

3. Meiosis of the diploid nucleus restores the haploid state.

4. The haploid nuclei are partitioned into + and − spores, which reestablish + and − fungi by mitoses and cell divisions.

Fungi differ in the ways they form dikaryons and in the site at which meiosis occurs.

---
[25]From Greek *spora*, meaning "seed," and *angeion*, meaning "vessel."
[26]From Greek *phoros*, meaning "bearing."

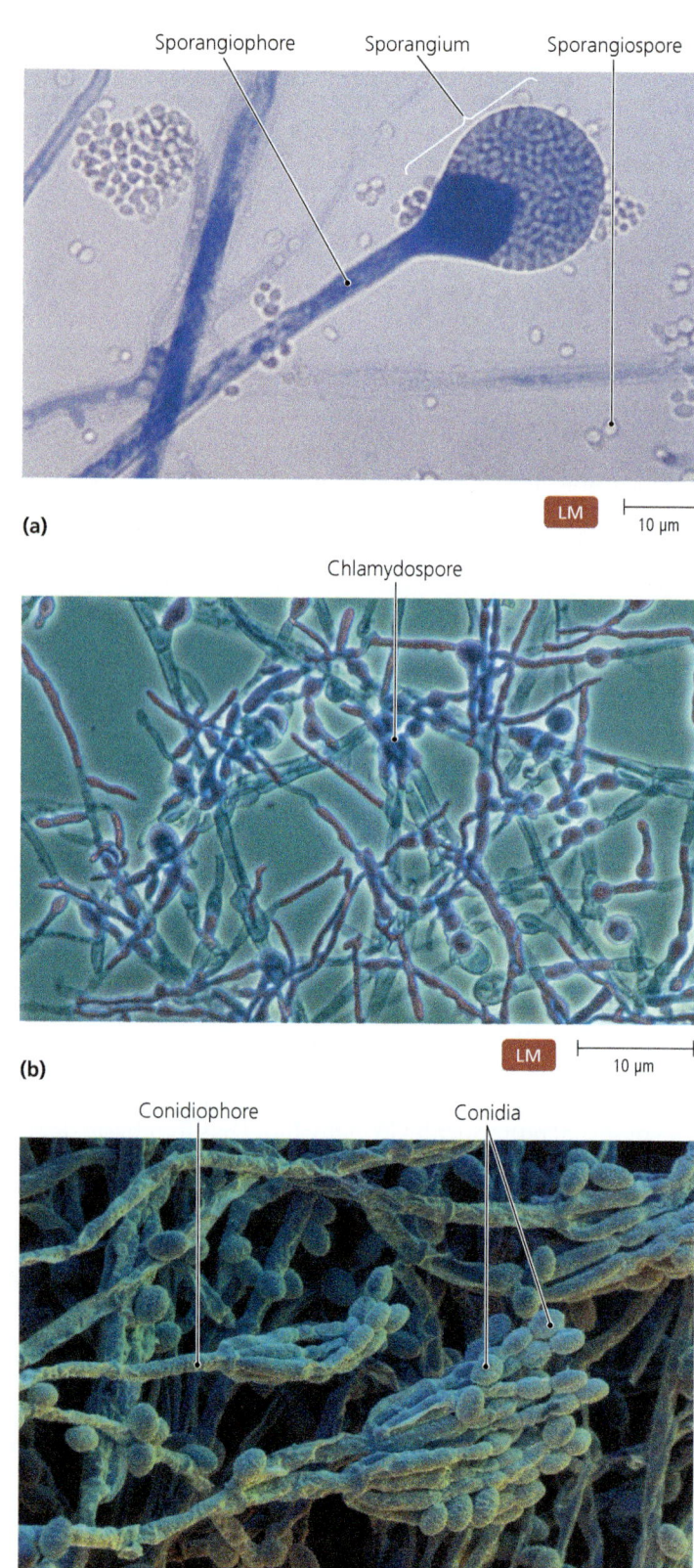

(a)

(b)

(c)

▲ FIGURE 12.17 **Representative asexual spores of molds.**
(a) Sporangiospores, which develop within a sac called a sporangium that is borne on a sporangiophore, here of *Mucor*. (b) Chlamydospores are thick-walled spores that form inside hyphae, here of *Aureobasidium*. (c) Conidiospores (conidia) develop at the ends of hyphae, here of *Penicillium*. *How do conidia differ from sporangiospores?*

Figure 12.17 Conidia are never enclosed in a sac.

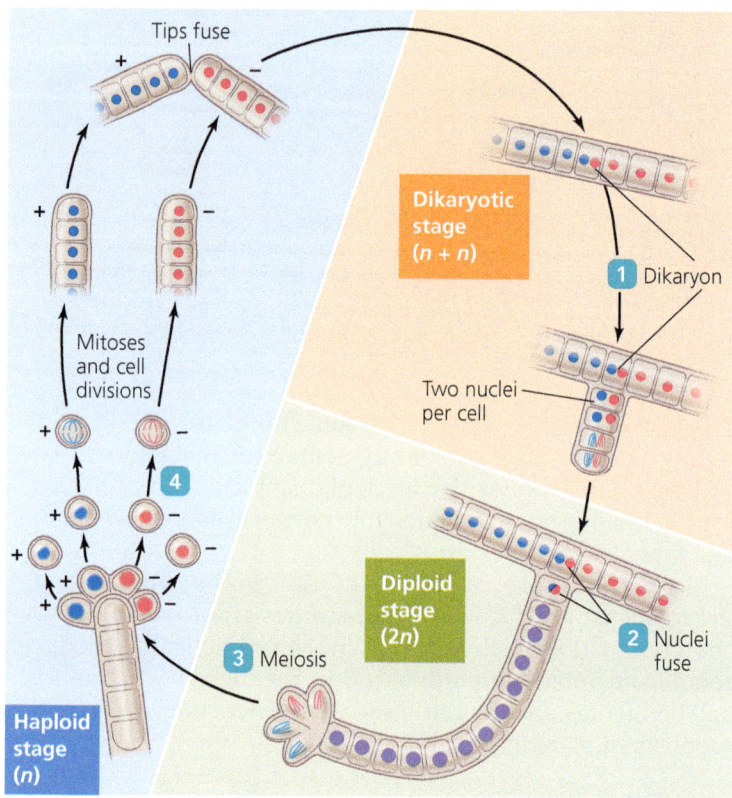

▲ FIGURE 12.18 **The process of sexual reproduction in fungi.** The steps in the process are described in the text. Haploid cells appear against a blue background, diploid cells against a green background, and dikaryotic cells against an orange background.

Classification of Fungi

LEARNING OUTCOMES

12.19 Compare and contrast the three divisions of fungi with respect to the formation of sexual spores.
12.20 Describe the deuteromycetes, and explain why this group no longer constitutes a formal taxon.

In the following sections, we will consider the four major subgroups into which taxonomists traditionally divided the kingdom Fungi. Three of these subgroups, which are taxa called *divisions* that are equivalent to phyla in other kingdoms, are based on the type of sexual spore produced (divisions Zygomycota, Ascomycota, and Basidiomycota); the fourth (the deuteromycetes) refers to an abandoned taxon that was a repository of fungi for which no sexual stage was known. We begin by considering the Zygomycota.

Division Zygomycota

Fungi in the division **Zygomycota** are multinucleate molds called *zygomycetes* (zī′gō-mī-sēts). Of the approximately 1100 species known, most are saprobes; the rest are obligate parasites of insects and other fungi.

Typical zygomycetes reproduce asexually via sporangiospores, but the distinctive feature of most zygomycetes is the formation of sexual structures called **zygosporangia** (sometimes incorrectly termed "zygospores") (FIGURE 12.19). The dikaryon of a zygomycete is typically a single cell that develops from the fusion of sexually compatible hyphal tips. The dikaryon develops into a zygosporangium where nuclei from one hypha (+) fuse with nuclei from the other hypha (−) to form many diploid nuclei. Each diploid nucleus undergoes meiosis, but only one of the four meiotic daughters of each nucleus survives. The zygosporangium then produces a haploid sporangium, which is filled with haploid spores (true zygospores). The sporangium releases these spores, each of which germinates to produce either a + or a − mycelium. This completes the life cycle.

Microsporidia Microsporidia are small organisms that are difficult to classify. Until 2003, taxonomists thought microsporidia were protozoa, but genetic analysis indicates they are more similar to zygomycetes.

They are obligatory intracellular parasites, that is, organisms that must live within their hosts' cells. Microsporidia spread from host to host as small, resistant spores. An example is *Nosema* (nō-sē′mă), which is parasitic on insects, such as silkworms and honeybees. The Environmental Protection Agency has approved one species of *Nosema* as a biological control agent for grasshoppers. Seven genera of microsporidia, including *Nosema, Encephalitozoon* (en-sef-a-lat-e′zō-an), and *Microsporidium* (mī-krō-spor-i′dē-ŭm), are known to cause diseases in immunocompromised patients (see Emerging Disease Case Study: Microsporidiosis, in Chapter 16 on p. 496).

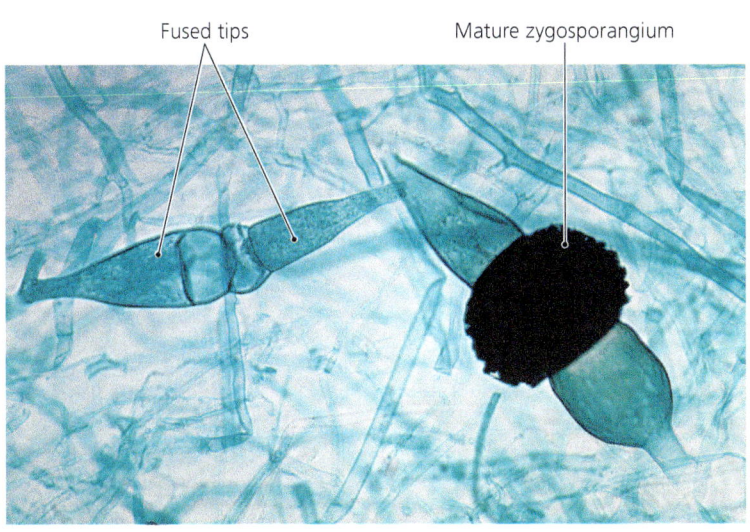

▲ FIGURE 12.19 **Zygosporangium.** The tips of + and − hyphae (here, of the bread mold *Rhizopus*) fuse and form a dark, diploid zygosporangium that matures and undergoes meiosis to produce internal haploid sporangiospores.

▲ FIGURE 12.20 Ascocarps (fruiting bodies) of the common morel, *Morchella esculenta*. (a) Though it might look like a mushroom, it isn't; *Morchella* is a delectable edible ascomycete. The pits visible in this photograph are lined with asci. (b) Ascospores align in groups of four or eight in a sac called an ascus.

Division Ascomycota

The division **Ascomycota** contains about 32,000 known species of molds and yeasts that are characterized by the formation of haploid **ascospores** within sacs called **asci** (singular: *ascus*).[27] Asci occur in fruiting bodies called *ascocarps*, which have various shapes (FIGURE 12.20). Ascomycetes (as′kō-mī-sēts), as they are called, also reproduce asexually by conidiospores.

Sexual reproduction of an ascomycete proceeds when hyphal tips of opposite mating types fuse to form a dikaryon. The dikaryon undergoes mitosis and cell division to produce a hypha composed of dikaryotic cells. Eventually, nuclei of opposite mating types within the dikaryons fuse to form diploid nuclei that then undergo meiosis and cytokinesis to form four haploid cells lined up within an ascus. The four haploid cells may each undergo mitosis and cytokinesis to form eight haploid ascospores. The asci open to release their ascospores, which germinate to produce either a + or a − hypha.

Ascomycetes are familiar and economically important fungi. For example, most of the fungi that spoil food are ascomycetes. This group also includes plant pathogens, such as the causative agents of Dutch elm disease and chestnut blight, which have almost eliminated their host trees in many parts of the United States. *Claviceps purpurea* (klav′i-seps poor-poo′rē-ă) growing on grain produces *lysergic acid*, also known as *ergot*, which causes abortions in cattle and hallucinations in humans. *Aspergillus* (ăs-per-jil′ŭs) can also infect humans (see **Emerging Disease Case Study: Aspergillosis** on p. 370).

---

[27]From Greek *askus*, meaning "wineskin."

EMERGING DISEASE CASE STUDY

Chronic Aspergillosis

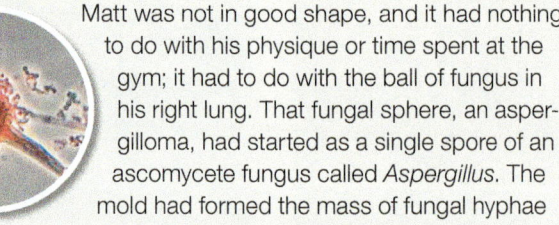

Matt was not in good shape, and it had nothing to do with his physique or time spent at the gym; it had to do with the ball of fungus in his right lung. That fungal sphere, an aspergilloma, had started as a single spore of an ascomycete fungus called *Aspergillus*. The mold had formed the mass of fungal hyphae that was invading the airways of his lung and slowly killing him. Such bronchopulmonary aspergillosis is a rare but increasingly frequent pathogen of immunocompromised people. Besides his difficulty in breathing, fever, and chest pain, Matt most hated coughing up wads of bloody mucus—as if he were expelling the very fabric of his life. In fact, he literally was coughing up pieces of his lung.

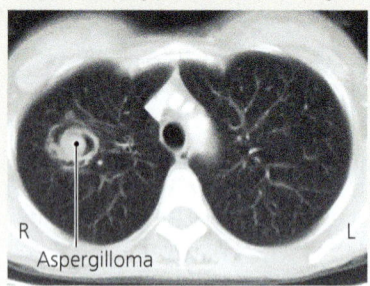

Aspergilloma

Unfortunately for Matt, the disease had not yet peaked. *Aspergillus* had invaded his blood and was even now progressing toward his brain. Soon the signs and symptoms of invasive aspergillosis would be his—extreme tiredness, excessive weakness, severe headaches, and delirium.

All would be his daily companions. He might also be paralyzed on one side of his body.

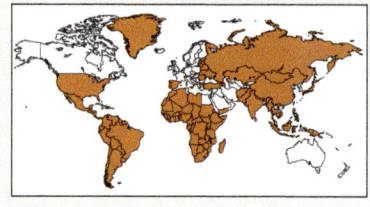

Matt spent four weeks in the hospital. The worst days were those when he was aware. The sheer terror of knowing his brain was being pierced by thin hyphal threads, digesting his personality away, was almost more than Matt could stand. He looked forward to the times he would lapse into unconsciousness, even though he knew that each period of wakefulness might be his last.

A medical miracle in the form of a new antifungal drug (voriconazole) brought Matt back to life. The invasive mold was defeated, and Matt returned home grateful, aware that life is precious and hopeful that no more spores floated his way.

1. How could a microbiologist determine that the fungus in Matt's lung was an ascomycete rather than a basidiomycete?
2. Why is aspergillosis emerging as a "new" disease?
3. Why is a powerful drug like voriconazole effective against fungi but relatively harmless to humans?

On the other hand, many ascomycetes are beneficial. For example, *Penicillium* (pen-i-sil'ē-ŭm) mold is the source of penicillin; *Saccharomyces*, which ferments sugar to produce alcohol and carbon dioxide gas, is the basis of the baking and brewing industries; and *truffles* (varieties of *Tuber*) grow as mycorrhizae in association with oak and beech trees to form culinary delights (see **Beneficial Microbes: Fungi for $100,000 a Pound**). As previously noted, another ascomycete, the pink bread mold *Neurospora*, has been an important tool in genetics and biochemistry. Many ascomycetes partner with green algae or cyanobacteria to form *lichens* (discussed in more detail shortly).

Division Basidiomycota

A walk through fields and woods in most parts of the world may reveal mushrooms, puffballs, stinkhorns, jelly fungi, bird's nest fungi, or bracket fungi, all of which are the visible fruiting bodies of the almost 22,000 known species of fungi in the division **Basidiomycota**. Poisonous mushrooms are sometimes called *toadstools*, but there is no sure way to always distinguish an edible "mushroom" from a poisonous "toadstool" except by eating them—a truly risky practice!

The major mycelium of a basidiomycete is dikaryotic; every cell is a dikaryon. The aboveground mushrooms or other fruiting bodies of *basidiomycetes* (ba-sid'ē-ō-mi-sēts) are called **basidiocarps (FIGURE 12.21)**. The entire structure of a basidiocarp consists of tightly woven dikaryotic hyphae that extend into multiple, often club-shaped projections called *basidia*, the ends of which produce sexual *basidiospores* (typically four on each basidium) as a result of meiosis and cell division. Basidiospores disperse in the wind and germinate in suitable, moist environments to form small haploid hyphae that have either + or − nuclei. The tips of opposite mating types can fuse to form dikaryons that develop into the expansive mycelia of these fungi.

Besides the edible mushrooms—most notably, cultivated *Agaricus* (a-gār'i-kus)—basidiomycetes affect humans in several ways. Most basidiomycetes are important decomposers that digest chemicals such as cellulose and lignin in dead plants and return nutrients to the soil. Many mushrooms produce toxins or hallucinatory chemicals. An example of the latter is *Psilocybe cubensis* (sil-ō-sī'bē kū-ben'sis), which produces *psilocybin*, a hallucinogen. The basidiomycete yeast *Cryptococcus neoformans* (krip-tō-kok'ŭs nē-ō-for'manz) is the leading cause of fungal meningitis. *Cryptococcus* is an unusual basidiomycete in that it doesn't form hyphae, but it does form sexual spores in a basidium—it is a true basidiomycete. Other basidiomycetes are *rusts* and *smuts*, which attack crop plants, causing millions of dollars of damage each year.

Deuteromycetes

As noted previously, the divisions Zygomycota, Ascomycota, and Basidiomycota are based on type of sexual spore produced. Because scientists have not observed sexual reproduction in all fungi, taxonomists in the middle of the 20th century created the

BENEFICIAL MICROBES

Fungi for $100,000 a Pound

Truffles—rare, intensely flavored ascocarps that grow underground—are one of the most luxurious and expensive foods on Earth, selling on average for more than $800 per pound. There are many different varieties of truffles. The most coveted include *Tuber melanosporum*, a black truffle that is also known as the "black diamond," and *Tuber magnatum*, a white truffle that can sell for $100,000 per pound!

Tuber forms symbiotic relationships with the roots of trees. Its underground habitat requires it to use animals for spore dispersal—the animals dig up the truffles, eat them, and spread the fungal spores in their feces to other locations.

The staff at the University of Truffle Hunting Dogs, founded in 1880 in Roddi, Italy, train dogs to sniff out and uncover the underground truffles. While both pigs and dogs can smell truffles growing underground, dogs are more easily trained not to devour the tasty ascocarps.

Incidentally, chocolate truffles derive only their name from their prized fungal counterparts.

---

division *Deuteromycota* (also called *imperfect fungi*) to contain the fungi whose sexual stages are unknown—either because they do not produce sexual spores or because their sexual spores have not been observed. More recently, however, the analysis of rRNA sequences has revealed that most deuteromycetes in fact belong in the division Ascomycota, and thus modern taxonomists have abandoned Deuteromycota as a formal taxon. Nevertheless, many medical laboratory technologists, health care practitioners, and scientists continue to refer to "deuteromycetes" or "imperfect fungi" because these are traditional names.

MICRO CHECK

9. Which type of asexual spores is formed at the tips or sides of hyphae, but not in sacs?
10. What is a dikaryon?
11. Fungi in which fungal taxon can form a structure with eight haploid sexual spores in a sac during sexual production?

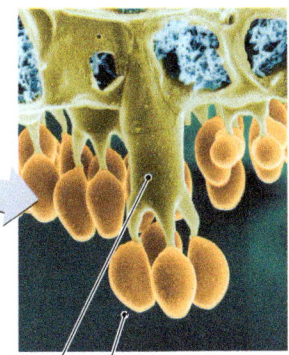

Basidium
Basidiospore   LM   8 μm

(a)

(b)

▲ FIGURE 12.21 **Basidiocarps (fruiting bodies).** The familiar shapes of mushrooms and bracket fungi are the basidiocarps of extensive basidiomycete mycelia. Shown here are **(a)** *Amanita muscaria* with basidiospores in the enlargement, and **(b)** *Laetiporus sulphureus* growing on a tree.

Lichens

LEARNING OUTCOMES

12.21 Describe a lichen's members.
12.22 List several beneficial roles or functions of lichens.

A discussion of fungi is incomplete without considering **lichens**. A lichen is a partnership between fungi and photosynthetic microbes—commonly cyanobacteria or, less frequently, green algae. For over a century, scientists thought that each lichen contained one fungal species and one photosynthetic species, but recent genetic research shows that most lichens have two fungi—a filamentous ascomycete and a basidiomycete yeast. The relationship between the fungi and their photosynthetic partner is so intimate that scientists classify lichens as distinct species, even though a lichen is composed of distinct organisms.

In a lichen, the hyphae of the ascomycete surround the yeast and photosynthetic cells (**FIGURE 12.22**) and provide them nutrients, water, and protection from desiccation and harsh light. In return, each alga or cyanobacterium provides the fungi with products of photosynthesis—carbohydrates and oxygen. In some lichens, the phototroph releases 60% of its carbohydrates to the fungi.

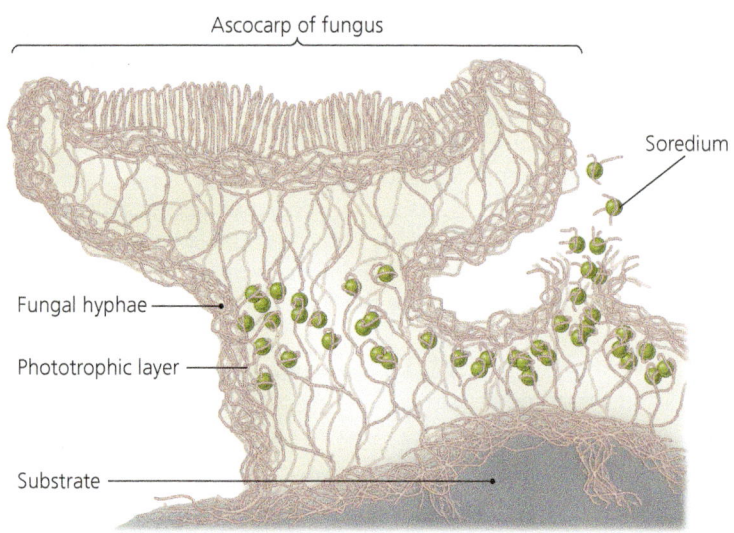

▲ FIGURE 12.22 **Makeup of a lichen.** The hyphae of a fungus (most commonly an ascomycete) constitute the major portion of the body of the lichen. A basidiomycete yeast (not shown) is also present. Cells of the photosynthetic member of the lichen are concentrated near the lichen's surface. Soredia, which are bits of fungus surrounding phototrophic cells, are the means by which some lichens propagate themselves. *Members of what three groups of photosynthetic microbes can be part of a lichen?*

Figure 12.22 Lichens are composed of an ascomycete, a basidiomycete, and a cyanobacterium or an alga.

The partnership in a lichen is not always mutually beneficial; in some lichens, the ascomycete produces haustoria that penetrate and kill the alga. Such lichens are maintained only because the alga reproduces faster than the fungus can devour it.

The ascomycete of a lichen reproduces by spores, which must germinate and develop into hyphae that capture appropriate partners. Alternatively, wind, rain, and small animals disperse bits of lichen called *soredia*, which contain both phototrophs and bits of fungi, to new locations where they can establish a new lichen if there is suitable substrate.

Scientists have identified over 14,000 species of lichens, which are abundant throughout the world, particularly in pristine unpolluted habitats, growing on soil, rocks, leaves, tree bark, other lichens, and even the backs of tortoises. Indeed, lichens grow in almost every habitat—from high-elevation alpine tundra to submerged rocks on the oceans' shores, from frozen Antarctic soil to hot desert climes. The only unpolluted places where lichens do not consistently grow are in the dark depths of the oceans and the black world of caves—after all, lichens require light.

Lichens grow slowly, but they can live for hundreds and possibly thousands of years. They occur in three basic shapes (FIGURE 12.23):

- *Foliose* lichens are leaflike, with margins that grow free from the substrate.
- *Crustose* lichens grow appressed to their substrates and may extend into the substrate for several millimeters.
- *Fruticose* lichens are either erect or are hanging cylinders.

Lichens create soil from weathered rocks, and lichens containing nitrogen-fixing cyanobacteria provide significant amounts of usable nitrogen to nutrient-poor environments. Many animals eat lichens; for example, reindeer and caribou subsist primarily on lichens throughout the winter. Birds use lichens for nesting materials, and some insects camouflage themselves with bits of living lichen. Humans also use lichens in the production

▲ FIGURE 12.23 **Gross morphology of lichens.** Crustose forms are flat and tightly joined to the substrate, here a tree branch; foliose forms are leaflike with free margins; and fruticose forms are either erect (as shown) or pendulant.

of foods, dyes, clothing, perfumes, medicines, and the litmus of indicator paper. Because lichens will not grow well in polluted environments, ecologists use them as sensitive living assays for monitoring air pollution.

In the preceding sections, we have seen that fungi are chemoheterotrophic yeasts and molds that function primarily to decompose and degrade dead organisms. Some fungi are pathogenic, and others associate with cyanobacteria or algae to form lichens. All fungi reproduce asexually either by budding or via asexual spores, and most fungi also produce sexual spores, by which taxonomists classify them. TABLE 12.3 summarizes the characteristics of fungi.

TELL ME WHY

Why isn't a fungal dikaryon—with its two haploid (*n*) nuclei—considered diploid?

Algae

LEARNING OUTCOME

12.23 Describe the distinguishing characteristic of algae.

The Romans used the word *alga* (al'ga) to refer to any simple aquatic plant, particularly one found in marine habitats. Their usage thus included the organisms we recognize as algae (al'jē), cyanobacteria, sea grasses, and other aquatic plants. Today, the word **algae** properly refers to eukaryotic photoautotrophic organisms that have sexual reproductive structures in which every cell becomes a gamete. In plants, by contrast, a portion of the reproductive structure always remains vegetative.

Algae are not a unified group; rather, they differ widely in distribution, morphology, reproduction, and biochemical traits. Moreover, the word *algae* is not synonymous with any taxon; in the taxonomic scheme shown in Figure 12.4c, algae can be found in several kingdoms, including kingdoms Stramenopila, Rhodophyta, and Plantae. The study of algae is *phycology*,[28] and the scientists who study them are called *phycologists*.

TABLE 12.3 Characteristics of Fungi

Division and Type of Sexual Spore	Distinguishing Features	Representative Genera
Zygomycota Zygospores	Multinucleate (aseptate)	*Rhizopus*
Ascomycota Ascospores	Septate; some associate with cyanobacteria or green algae to form lichens	*Claviceps, Neurospora, Penicillium, Saccharomyces, Tuber*
Basidiomycota Basidiospores	Septate	*Agaricus, Amanita, Cryptococcus*

[28]From Greek *phykos*, meaning "seaweed," and *logos*, meaning "discourse."

Distribution of Algae

Even though some algae grow in such diverse habitats as in soil and ice, in intimate association with fungi as lichens, and on plants, most algae are aquatic, living in the *photic zone* (penetrated by sunlight) of fresh, brackish, and salt bodies of water. This watery environment provides some benefits and also presents some difficulties for photosynthetic organisms. Whereas most bodies of water contain sufficient dissolved chemicals to provide nutrients for algae, water also differentially absorbs longer wavelengths of light (including red light), so only shorter (blue) wavelengths penetrate more than a meter below the surface. This is problematic for algae because their primary photosynthetic pigment—chlorophyll *a*—captures red light. Thus, to grow in deeper waters, algae must have *accessory photosynthetic pigments* that trap the energy of penetrating, short-wavelength light and pass that energy to chlorophyll *a*. Members of the group of algae known as red algae, for example, contain a red pigment that absorbs blue light, enabling red algae to inhabit even the deepest parts of the photic zone.

Morphology of Algae

Algae can be unicellular or colonial, or they can have simple multicellular bodies, which are commonly composed of branched filaments or sheets. The bodies of large marine algae, commonly called seaweeds, can be relatively complex, with branched *holdfasts* to anchor them to rocks, stemlike *stipes*, and leaflike *blades*. Many of the larger marine algae are buoyed in the water by gas-filled bulbs called *pneumocysts* (see Figure 12.26). Though the bodies of some marine algae can surpass land plants in length, they lack the well-developed transport systems common to vascular plants.

Reproduction of Algae

LEARNING OUTCOME

12.24 Describe the alternation of generations in algae.

In unicellular algae, asexual reproduction involves mitosis followed by cytokinesis. In unicellular algae that reproduce sexually, each algal cell acts as a gamete and fuses with another such gamete to form a zygote, which then undergoes meiosis to return to the haploid state.

Multicellular algae may reproduce asexually by fragmentation, in which each piece of a parent alga develops into a new individual, or by motile or nonmotile asexual spores. As noted previously, in multicellular algae that reproduce sexually, every cell in the reproductive structures of the alga becomes a gamete—a feature that distinguishes algae from all other photosynthetic eukaryotes.

Many multicellular algae reproduce sexually with an **alternation of generations** of haploid and diploid individuals (FIGURE 12.24). In such life cycles, diploid individuals undergo meiosis to produce male and female haploid spores that develop into haploid male and female bodies, which may look identical to the diploid body. Every cell of the reproductive portions of the haploid bodies becomes a gamete—the definition of an alga.

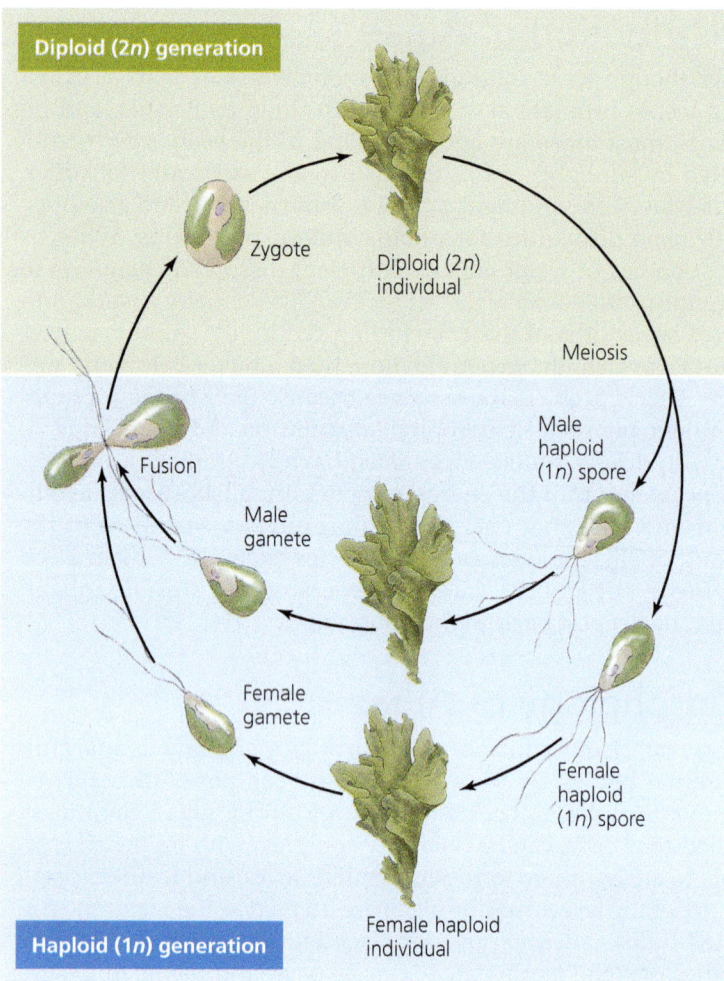

▲ FIGURE 12.24 Alternation of generations in algae, as occurs in the green alga *Ulva*. A diploid individual meiotically produces haploid spores with anterior flagella. These germinate and grow into male and female bodies. These haploid algae produce gametes, also with flagella in front, that fuse to form a diploid zygote, which grows into a new diploid individual. Each generation can also reproduce asexually by fragmentation and by spores.

Some haploid algae produce gametes that fuse to form a zygote, which grows into a new diploid alga. Both haploid and diploid algae may reproduce asexually as well.

Classification of Algae

LEARNING OUTCOMES

12.25 List four groups of algae, and describe the distinguishing characteristics of each.
12.26 List several economic benefits derived from algae.

The classification of algae is not yet settled. Historically, taxonomists have used differences in photosynthetic pigments, storage products, and cell wall composition to classify algae into several groups that are named for the colors of their photosynthetic pigments: green algae, red algae, brown algae, golden algae, and yellow-green algae. The following sections present some of these groups. We begin with the green algae of the division Chlorophyta.

Division Chlorophyta (Green Algae)

Chlorophyta[29] are green algae that share numerous characteristics with plants—they have chlorophylls *a* and *b* and use sugar and starch as food reserves. Most have cell walls composed of cellulose, though some have walls of glycoprotein or lack walls entirely. In addition, the nucleotide sequences of the 18S rRNA of green algae and plants are comparable. Because of the similarities, green algae are often considered to be the progenitors of plants, and in some taxonomic schemes the Chlorophyta are placed in the kingdom Plantae.

Most green algae are unicellular or filamentous (see Figure 1.7a) and live in freshwater ponds, lakes, and pools, where they form green to yellow scum. Some multicellular forms grow in the marine intertidal zone—that is, in the region exposed to air during low tide.

Prototheca (prō-tō-thē′kă) is an unusual green alga in that it lacks pigments, making it colorless. This chemoheterotrophic alga causes a skin rash in sensitive individuals. *Codium* (kō′dē-ŭm) is a member of a group of marine green algae that do not form cross walls after mitosis; thus, its entire body is a single, large, multinucleate cell. Some Polynesians dry and grind *Codium* for use as seasoning pepper. The green alga *Trebouxia* (tre-book′sē-a) is the most common alga found in association with fungi in lichens.

Kingdom Rhodophyta (Red Algae)

Algae of division **Rhodophyta**,[30] which had historically been placed in kingdom Plantae and then Protista, are now in their own kingdom—Rhodophyta. They are characterized by the red accessory pigment **phycoerythrin**; the storage molecule glycogen; cell walls of **agar** or **carrageenan** (kar-ă-gē′nan), sometimes supplemented with calcium carbonate; and nonmotile male gametes called *spermatia*. Phycoerythrin allows red algae to absorb short-wavelength blue light and photosynthesize at depths greater than 100 meters. Because the relative proportions of phycoerythrin and chlorophyll *a* vary, red algae range in color from green to black in the intertidal zone to red in deeper water (FIGURE 12.25). Most red algae are marine, although a few freshwater genera are known.

The gel-like polysaccharides agar and carrageenan, once they have been isolated from red algae such as *Gelidium* (jel-li′dē-ŭm) and *Chondrus* (kon′drŭs), are used as thickening agents for the production of solid microbiological media and numerous consumer products, including ice cream, toothpaste, syrup, salad dressings, and snack foods. Some studies suggest that ingested carrageenan can induce inflammation of the colon.

---

[29]From Greek *chloros*, meaning "green," and *phyton*, meaning "plant."

[30]From Greek *rhodon*, meaning "rose."

▲ FIGURE 12.25 **Palmaria palmata, a red alga.** Commonly called dulse, this alga is a traditional food in Ireland.

▲ FIGURE 12.27 **Portion of the giant kelp *Macrocystis*, a brown alga.** A kelp's blades are kept afloat by pneumocysts.

Brown algae use the polysaccharide *laminarin* and oils as food reserves and have cell walls composed of cellulose and **alginic acid** (alginate). Food processors use alginic acid as a thickening agent, and dental hygienists use it in the preparation of dental impressions.

Chrysophyta (Golden Algae, Yellow-Green Algae, and Diatoms)

Chrysophyta[33] is a group of algae that are diverse with respect to cell wall composition and pigments. They are unified in using the polysaccharide *chrysolaminarin* as a storage product. Some additionally store oils. Modern taxonomists group these algae with brown algae and water molds (discussed shortly) in the kingdom Stramenopila based on similarities in nucleotide sequences and flagellar structure. Whereas some chrysophytes lack cell walls, others have ornate external coverings, such as scales or plates. Members of one taxon of chrysophytes, Bacillariophyceae—the **diatoms** (dī′ă-tomz)—are unique in having silica in their cell walls, which consist of two halves called *frustules* that fit together like a Petri dish (FIGURE 12.28).

Most chrysophytes are unicellular or colonial. All chrysophytes contain more orange-colored *carotene* pigment than they do chlorophylls *a* and *c*, accounting for the common names of two major classes of chrysophytes—*golden algae* and *yellow-green algae*.

Diatoms are a major component of marine *phytoplankton*: free-floating photosynthetic microorganisms that form the basis of food chains in the oceans. Further, because of their enormous number, diatoms are the major source of the world's oxygen. The silica frustules of diatoms contain minute holes for the exchange of gases, nutrients, and wastes with the environment. Organic gardeners use *diatomaceous earth*, composed of innumerable frustules of dead diatoms, as a pesticide against harmful insects

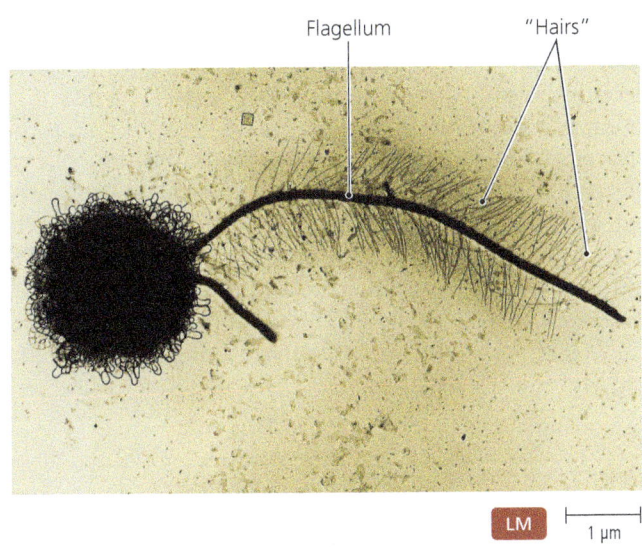

▲ FIGURE 12.26 **Hairy flagellum.** The "hairy" flagellum of gametes of brown algae and other stramenopiles is distinctive.

Phaeophyta (Brown Algae)

The **Phaeophyta**[31] are in kingdom Stramenopila[32] based in large part on their gametes being motile by means of two flagella—one whiplike and one with hollow projections giving it a "hairy" appearance (FIGURE 12.26). They have chlorophylls *a* and *c*, carotene, and brown pigments called *xanthophylls* (zan′thō-fils). Depending on the relative amounts of these pigments, brown algae may appear dark brown, tan, yellow-brown, greenish brown, or green. Most brown algae are marine organisms, and some of the giant kelps, such as *Macrocystis* (FIGURE 12.27), form "kelp forests." Kelps can surpass the tallest trees in length though not in girth.

---

[31]From Greek *phaeo*, meaning "brown."

[32]From Latin *stramen*, meaning "straw," and *pilos*, meaning "hair."

[33]From Greek *chrysos*, meaning "gold."

▲ FIGURE 12.28 *Paralia sulcata*, a diatom. Diatoms have frustules, composed of silica and cellulose, that fit together like a Petri dish.

and worms—the frustules appear to draw water from the cells of the pests. Diatomaceous earth is also used in polishing compounds, detergents, and paint removers and as a component of firebrick, soundproofing products, swimming pool filters, and reflective paints.

In summary, algae are unicellular or multicellular photoautotrophs characterized by sexual reproductive structures in which every cell becomes a gamete. The colors produced by the combination of their primary and accessory photosynthetic pigments give them their common names and provide the basis of at least one classification scheme. **TABLE 12.4** summarizes the characteristics of the major groups of algae.

TELL ME WHY

Why are there few pathogenic algae?

Water Molds

LEARNING OUTCOME

12.27 List four ways in which water molds differ from true fungi.

Scientists once classified the microbes commonly known as **water molds** as fungi because they resemble filamentous fungi in having finely branched filaments. However, water molds are not true molds—they are not fungi. Water molds differ from fungi in the following ways:

- They have tubular cristae in their mitochondria.
- They have cell walls of cellulose instead of chitin.
- Their spores have two flagella—one whiplike and one "hairy."
- They have true diploid bodies rather than haploid bodies.

Because water molds have "hairy" flagella and certain similarities in rRNA sequence to sequences of diatoms, other chrysophytes, and brown algae, taxonomists classify all these organisms in kingdom Stramenopila (see Figure 12.5c).

Water molds decompose dead animals and return nutrients to the environment (**FIGURE 12.29**). Some species are detrimental pathogens of crops, such as grapes, tobacco, and soybeans. In 1845, the water mold *Phytophthora infestans* (fī-tof'tho-ră in-fes'tanz) was accidentally introduced into Ireland and devastated the potato crop, causing the great famine that killed over 1 million people and forced a greater number to emigrate to the United States and Canada.

TABLE 12.4 Characteristics of Various Algae

Group (Common Name)	Kingdom	Pigments	Storage Product(s)	Cell Wall Component(s)	Habitat	Representative Genera
Chlorophyta (green algae)	Plantae	Chlorophylls *a* and *b*, carotene, xanthophylls	Sugar, starch	Cellulose or glycoprotein; absent in some	Freshwater, brackish water, and saltwater; terrestrial	*Spirogyra* *Prototheca* *Codium* *Trebouxia*
Rhodophyta (red algae)	Rhodophyta	Chlorophyll *a*, phycoerythrin, phycocyanin, xanthophylls	Glycogen	Agar or carrageenan, some with calcium carbonate	Mostly saltwater	*Chondrus* *Gelidium* *Antithamnion*
Phaeophyta (brown algae)	Stramenopila	Chlorophylls *a* and *c*, carotene, xanthophylls	Laminarin, oils	Cellulose and alginic acid	Brackish water and saltwater	*Macrocystis*
Chrysophyta (golden algae, yellow-green algae, diatoms)	Stramenopila	Chlorophylls *a*, c_1, and c_2; carotene; xanthophylls	Chrysolaminarin, oils	Cellulose, silica, calcium carbonate	Freshwater, brackish water, and saltwater; terrestrial; ice	*Stephanodiscus*

▲ FIGURE 12.29 **Water molds help recycle organic nutrients in aquatic habitats.** Here a water mold decomposes a dead goldfish.

MICRO CHECK

12. Lichens are composed of what types of microbes?
13. What is the one characteristic that distinguishes algae from other photoautotrophic eukaryotes?
14. What is the general taxon of algae that produce agar, which is used to solidify laboratory media to culture bacteria?
15. How are water molds different from fungi?

TELL ME WHY

How do scientists know that water molds are more closely related to brown algae than to true molds?

Other Eukaryotes of Microbiological Interest: Parasitic Helminths and Vectors

LEARNING OUTCOMES

12.28 Explain why microbiologists study large organisms such as parasitic worms.
12.29 Discuss the inclusion of vectors in a study of microbiology.

Microbiologists are also interested in two other groups of eukaryotes, although they are not microorganisms. The first group is the parasitic *helminths*, commonly called parasitic worms. Microbiologists became interested in parasitic helminths because they observed the microscopic infective and diagnostic stages of the helminths—usually eggs or larvae (immature forms)—in samples of blood, feces, and urine. Thus, microbiologists study parasitic helminths in part because they must distinguish the parasites' microscopic forms from other microbes. (Chapter 23 discusses some helminth parasites of the blood and of the digestive system.)

The second group of macroorganisms of interest to microbiologists is **arthropod vectors**[34]—animals that carry pathogens and have segmented bodies, hard external skeletons, and jointed legs. Some arthropods are *mechanical vectors*, meaning they merely carry pathogens; others are *biological vectors*, meaning they also serve as hosts for microbial pathogens. Given that arthropods are small organisms (so small that we generally don't notice them until they bite us) and given that they produce numerous offspring, controlling arthropod vectors to eliminate their role in the transmission of important human diseases is an almost insurmountable task.

Disease vectors belong to two classes of arthropods: *Arachnida* and *Insecta*. Ticks and mites are arachnoid vectors. Spiders are also arachnids, but they do not transmit diseases. Insects account for the greatest number of vectors, and within this group are fleas, lice, flies (such as tsetse flies and mosquitoes), and true bugs (such as kissing bugs). **FIGURE 12.30** shows representatives of major types of arthropod vectors. Most vectors are found on a host only when they are actively feeding. Lice are the only arthropods that may spend their entire lives in association with a single individual host. (Chapters 20, 21, 23, and 25 discuss in more detail vectors and some diseases they carry.)

Arachnids

LEARNING OUTCOMES

12.30 Describe the distinctive features of arachnids.
12.31 List five diseases vectored by ticks and two diseases vectored by mites.

All adult **arachnids** (ă-rak'nidz) have four pairs of legs. Both **ticks** and **mites** (commonly known as *chiggers*) go through a six-leg stage when they are juveniles, but they display the characteristic eight legs as adults. Ticks and mites resemble each other morphologically, having disk-shaped bodies. Ticks are roughly the size of a small rice grain, whereas mites are usually much smaller (**FIGURE 12.30a,b**).

Ticks are the most important arachnid vectors. They are distributed worldwide and serve as vectors for bacterial, viral, and protozoan diseases. Ticks are second only to mosquitoes in the number of diseases that they transmit. Hard ticks—those with a hard plate on their dorsal surfaces—are the most prominent tick vectors. They wait on stalks of grasses and brush for their hosts to come by. When a human, for example, walks past them, the ticks leap onto the person and begin searching for exposed skin. They use their mouthparts to cut holes in the skin and attach themselves with a gluelike compound to prevent being dislodged. As they feed on blood, their bodies swell to several times normal size. Some tick-borne diseases are Lyme disease, Rocky Mountain spotted fever, tularemia, relapsing fever, and tick-borne encephalitis.

---

[34]From Latin *vectus*, meaning "carried."

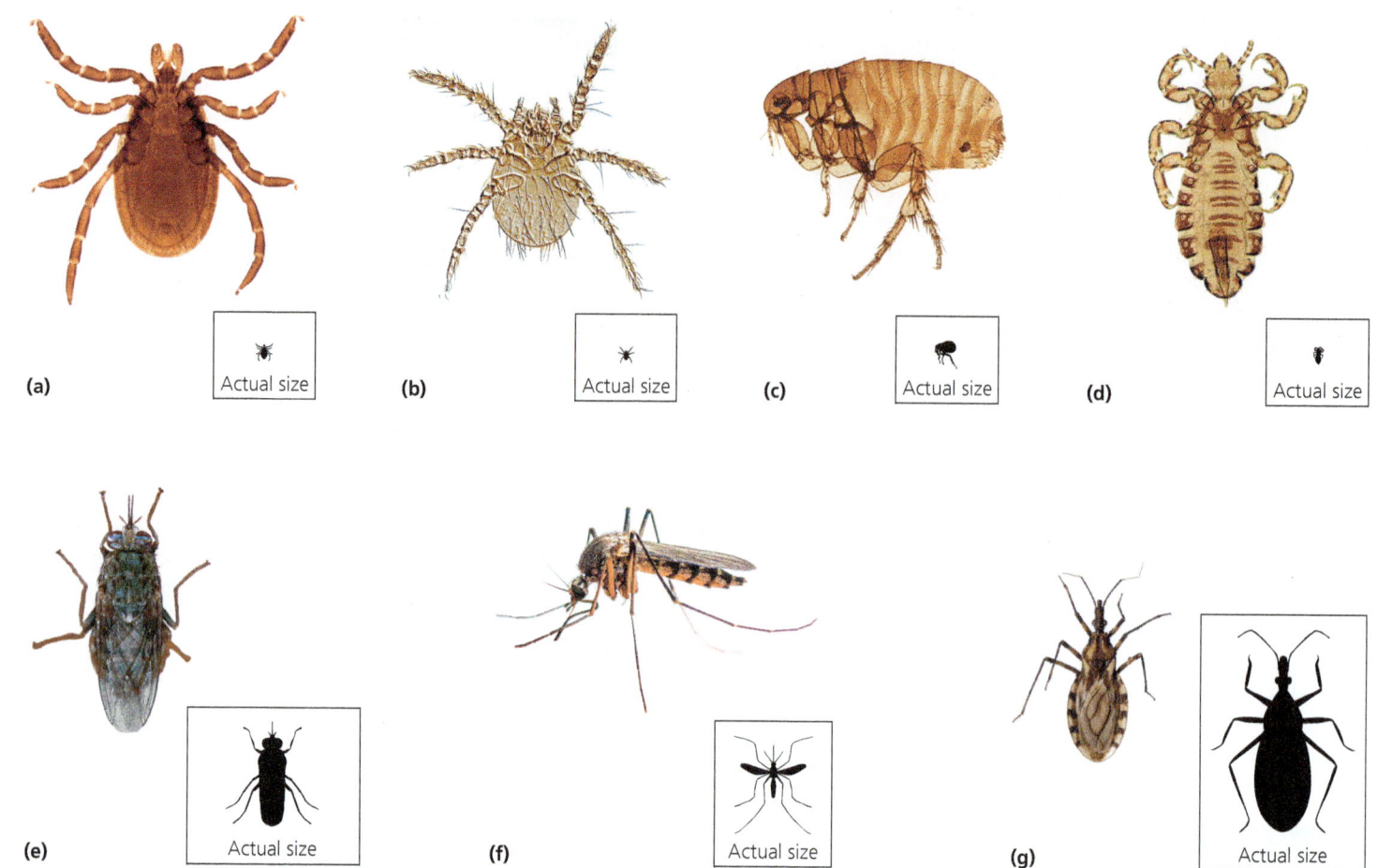

▲ FIGURE 12.30 **Representative arthropod vectors.** Arachnid vectors include **(a)** ticks and **(b)** mites; insect vectors include **(c)** fleas, **(d)** lice, **(e)** true flies, such as this tsetse fly, **(f)** mosquitoes (a type of fly), and **(g)** true bugs.

Parasitic mites of humans live around the world, wherever humans and animals coexist. A few mite species transmit rickettsial diseases (rickettsial pox and scrub typhus) among animals and humans.

Insects

LEARNING OUTCOMES

12.32 Describe the general physical features of insects and the features specific to the various groups of insect vectors.
12.33 List diseases transmitted by fleas, lice, true flies, mosquitoes, and kissing bugs.

As adults, all **insects** have three pairs of legs and three body regions—head, thorax (chest), and abdomen. Adult insects are, however, far from uniform in appearance. Some have two wings, others have four wings, and others are wingless; some have long legs, some short; and some have biting mouthparts, whereas others have sucking mouthparts. Many insects have larval stages that look very different from the adults, complicating their identification.

The fact that many insects can fly has epidemiological implications. Flying insects have broader home ranges than nonflying insects, and some migrate, making control difficult.

Fleas are small, vertically flattened, wingless insects that are found worldwide, though some species have geographically limited ranges. Most are found in association with wild rodents, bats, and birds, and are not encountered by humans. A few species, however, feed on humans. Cat and dog fleas are usually just pests (to the animal and its owner), but they can also serve as the intermediate host for a dog tapeworm, *Dipylidium* (dip-ĭ-lid′ē-ŭm). The most significant microbial disease transmitted by fleas is plague, carried by rat fleas.

Lice (singular: *louse*) are parasites that can also transmit disease. They are horizontally flat, soft-bodied, wingless insects found worldwide among humans and their habitations. They live in clothing and bedding and move onto humans to feed. Lice are most common in unsanitary conditions or overcrowded communities. Lice are the vectors involved in epidemic outbreaks of typhus in developing countries.

Flies are among the more common insects, and many different species are found around the world. Flies differ greatly in size, but all have at least two wings and fairly well-developed body segments. Not all flies transmit disease, but those that do are usually bloodsuckers. Female sand flies (*Phlebotomus*; fle-bot′ō-mŭs) transmit leishmaniasis in North Africa, the Middle East, Europe, and parts of Asia. Tsetse flies (*Glossina*; glo-sī′nă) are limited geographically to tropical Africa, where they are found in brushy areas and transmit African sleeping sickness.

Mosquitoes are a type of fly, though they are morphologically distinct from other fly species. Female mosquitoes are thin and have wings, elongated bodies, long antennae, long legs, and a long proboscis for feeding on blood. Male mosquitoes do not feed on blood. Mosquitoes are found throughout the world, but particular species are geographically limited. Mosquitoes are the most important arthropod vectors of diseases, and they carry the pathogens that cause malaria, yellow fever, dengue fever, filariasis, viral encephalitis, and Rift Valley fever.

Kissing bugs are relatively large, winged, true bugs with cone-shaped heads and wide abdomens. They are called kissing bugs because of their tendency to take blood meals near the mouths of their hosts. Both sexes feed nocturnally while their victims sleep. Kissing bugs transmit disease in Central and South America. The most important disease they transmit is Chagas' disease.

MICRO CHECK

16. What organisms transmit the bacterial pathogens that cause Lyme disease and Rocky Mountain spotted fever?

TELL ME WHY

Why are large eukaryotes such as mosquitoes and ticks considered in a microbiology class?

MICRO IN THE CLINIC FOLLOW-UP
Shellfish Allergy?

At the ER, Becky's physical exam is abnormal, and she is admitted to the hospital. She is now having fewer involuntary leg movements; instead she is unable to move her legs at all—they appear paralyzed. She also begins complaining of shortness of breath. The doctor orders blood and urine lab analyses to search for a diagnosis. Becky continues to rapidly decline and goes into respiratory failure. The staff places a plastic tube down her trachea to supply air and she is stabilized. Because her blood tests are normal, there is no skin rash, and she isn't itching, the doctor rules out an allergic reaction.

Urgently seeking a diagnosis, the doctor sits down with Tim to go over the preceding days again—where had they gone, what had they done, what had they eaten? As Tim informs the doctor of the rapid progression of symptoms, he recounts their meals. Tim said he had only a bite of mussels and clams. He preferred the crabs, but mussels and clams are Becky's favorite, and she ate more of them than anyone. The doctor suspects that Becky was poisoned by the mollusks and orders a urinary lab test for toxins. The results of the test come back positive for saxitoxins.

He diagnoses Becky with paralytic shellfish poisoning (PSP) from contaminated clams and mussels. Although there is no antidote for the saxitoxin, the toxin will eventually pass from the system. Becky receives supportive care, including mechanical pulmonary ventilation for 24 hours, and is discharged after spending only two days in the hospital. Her symptoms completely resolve on their own, and she makes a complete recovery.

1. Why did Becky's doctor provide only supportive care? Why weren't antimicrobials part of her treatment?

2. Becky's illness was caused by saxitoxins from a dinoflagellate in the genus *Alexandrium*. How are diatoms different from other eukaryotic cells? Do they have any features found in prokaryotes?

Check your answers to Micro in the Clinic Follow-Up questions in the Mastering Microbiology Study Area.

> Go to Mastering Microbiology for Interactive Microbiology, Dr. Bauman's Video Tutors, Micro Matters, MicroFlix, Micro-Boosters, practice quizzes, and more.

CHAPTER SUMMARY

General Characteristics of Eukaryotic Organisms (pp. 354–358)

1. A typical eukaryotic nucleus may be **haploid** (having a single copy of each chromosome) or **diploid** (having two copies). It divides by **mitosis** in four phases—**prophase**, **metaphase**, **anaphase**, and **telophase**—resulting in two nuclei with the same ploidy as the original.

2. **Meiosis** is nuclear division that results in four nuclei, each with half the ploidy of the original.

3. A cell's cytoplasm divides by **cytokinesis** either during or after nuclear division.

4. **Coenocytes** are multinucleate cells resulting from repeated mitoses but postponed or no cytokinesis.

5. Some microbes undergo multiple mitoses by **schizogony** to form a multinucleate **schizont**, which then undergoes cytokinesis to produce numerous daughter cells.

6. The classification of eukaryotic microbes is problematic and has changed frequently. Historical schemes based on similarities in morphology and chemistry have been replaced with schemes based on nucleotide sequences and ultrastructural features.

Protozoa (pp. 358–364)

1. **Protozoa** (studied by protozoologists) are eukaryotic, unicellular organisms that lack cell walls.

2. A motile **trophozoite** is the feeding stage of a typical protozoan. Some protozoa form a **cyst**, a resting stage that is resilient to environmental changes.

3. Most protozoa are chemoheterotrophs, getting their energy from chemicals and their carbon from other organisms, but dinoflagellates and euglenoids are photoautotrophs.

4. A few protozoa undergo sexual reproduction by forming **gametocytes** that fuse to form a **zygote**.

5. Protozoa may be classified into six groups: parabasalids, diplomonads, euglenozoa, alveolates, rhizaria, and amoebozoa.

6. Parabasalids (e.g., *Trichomonas*) are characterized by a Golgi body-like structure called a parabasal body.

7. Members of the Diplomonadida lack mitochondria, Golgi bodies, and peroxisomes.

8. Unicellular flagellated **euglenids** are euglenozoa that store food as paramylon, lack cell walls, and have eyespots used in positive phototaxis. Because they exhibit characteristics of both animals and plants, euglenids are a taxonomic problem.

9. A **kinetoplastid** is a euglenozoan with a single, large, apical mitochondrion that contains a kinetoplast, which is a region of DNA.

10. Alveolates, with cavities called alveoli beneath their cell surfaces, include **ciliate** alveolates (characterized by cilia), **apicomplexans** (all of which are pathogenic), and **dinoflagellates** (responsible for **red tides**).

11. Protozoa that move and feed with pseudopods are **amoebas**, which are classified into two kingdoms: Rhizaria and Amoebozoa.

12. Rhizaria include **foraminifera**, which have threadlike pseudopods and calcium carbonate shells, and **radiolaria**, which have threadlike pseudopods and silica shells.

13. Amoebozoa have lobe-shaped pseudopods. They include free-living amoebas, parasitic amoebas, and slime molds. **Slime molds**, despite their name, are not fungi but are a kind of amoeba. They lack cell walls and are phagocytic in their nutrition.

Fungi (pp. 364–373)

1. **Fungi** (studied by mycologists) are chemoheterotrophic eukaryotes with cell walls usually composed of **chitin**.

2. Most fungi are beneficial, but some produce **mycoses** (fungal diseases).

3. Mold bodies are composed of tubular filaments called **hyphae**. Yeasts are globular, single cells.

4. Hyphae are described as either **septate** or **aseptate**, depending on the presence of cross walls. A **mycelium** is a tangled mass of hyphae.

5. A **dimorphic** fungus is either moldlike (with hyphae) or yeastlike, depending on environmental conditions.

6. Most fungi are **saprobes**—they acquire nutrients by absorption from dead organisms. Other fungi get nutrients from living organisms using **haustoria** that penetrate host tissues.

7. Fungi reproduce asexually either by budding or via asexual spores, which are categorized according to their mode of development. Most fungi also reproduce sexually via sexual spores.

8. Most fungi in the division **Zygomycota** produce rough-walled **zygosporangia**. **Microsporidia** are intracellular parasites formerly classified as protozoa but now classed with zygomycetes based on genetic analysis.

9. Fungi in the division **Ascomycota**, a group of economically important fungi, produce **ascospores** within sacs called **asci**.

10. Fungi of the division **Basidiomycota** have fruiting bodies called **basidiocarps** that include mushrooms, puffballs, and bracket fungi. Basidiocarps produce basidiospores at the ends of basidia.

11. Deuteromycetes (imperfect fungi) are an informal grouping of fungi having no known sexual stage.

12. **Lichens** are economically and environmentally important organisms composed of fungi living in partnership with photosynthetic microbes, either green algae or cyanobacteria.

Algae (pp. 373–376)

1. **Algae** are photosynthetic organisms in which every cell of the reproductive structures becomes a gamete.

2. The study of algae is phycology, and the scientists who study them are phycologists.

3. Most algae are aquatic, but they can live in moist soil and in ice.

4. Small algae grow as single cells, in colonies, or as simple multicellular bodies. Large algae are multicellular with stemlike stipes, leaflike blades, and holdfasts that attach them to substrates.

5. Algae typically reproduce by an **alternation of generations**, in which a haploid body alternates with a diploid body.

6. Division **Chlorophyta** contains green algae, which are metabolically similar to land plants.

7. **Rhodophyta**, red algae, contain the pigment **phycoerythrin**, the storage molecule glycogen, and cell walls of **agar** or **carrageenan**, substances used as thickening agents.

8. **Phaeophyta**, brown algae, contain xanthophylls, laminarin, and oils. They have cell walls composed of cellulose and **alginic acid**, which is another thickening agent. A brown algal spore is motile by means of one "hairy" flagellum and one whiplike flagellum.

9. **Chrysophyta**—the golden algae, yellow-green algae, and **diatoms**—contain chrysolaminarin as a storage product. The silica cell walls of diatoms are arranged in nesting halves called frustules.

Water Molds (pp. 376–377)

1. **Water molds** have tubular cristae in their mitochondria, cell walls of cellulose, spores having two different flagella, and diploid bodies. They are placed in the kingdom Stramenopila along with chrysophytes and brown algae.

Other Eukaryotes of Microbiological Interest: Parasitic Helminths and Vectors (pp. 377–379)

1. Parasitic helminths are significant to microbiologists in part because their infective stages are usually microscopic.

2. Also important to microbiologists are **arthropod vectors**, animals that carry and transmit pathogens. Mechanical vectors merely carry microbes; biological vectors also serve as microbial hosts.

3. **Ticks** and **mites** (chiggers) are **arachnids**—eight-legged arthropods.

4. **Insect** vectors include **fleas**; **lice**; bloodsucking **flies**, including **mosquitoes**; and **kissing bugs**.

QUESTIONS FOR REVIEW

Answers to the Questions for Review (except Short Answer questions) begin on p. A-1.

Multiple Choice

1. Haploid nuclei ——————.
   a. contain one set of chromosomes
   b. contain two sets of chromosomes
   c. contain half a set of chromosomes
   d. are found in the cytosol of eukaryotic organisms

2. Which of the following sequences reflects the correct order of events in mitosis?
   a. telophase, anaphase, metaphase, prophase
   b. prophase, anaphase, metaphase, telophase
   c. telophase, prophase, metaphase, anaphase
   d. prophase, metaphase, anaphase, telophase

3. Which of the following statements accurately describes prophase?
   a. The cell appears to have a line of chromosomes across the midregion.
   b. The nuclear envelope becomes visible.
   c. The cell constructs microtubules to form a spindle.
   d. Chromatids separate and become known as chromosomes.

4. Multiple nuclear divisions without cytoplasmic divisions result in cells called ——————.
   a. mycoses
   b. coenocytes
   c. haustoria
   d. a pseudohypha

5. Tubular filaments with cross walls found in large fungi are ——————.
   a. septate hyphae
   b. aseptate hyphae
   c. aseptate haustoria
   d. dimorphic mycelia

6. The type of asexual fungal spore that forms within hyphae is called a ——————.
   a. sporangiospore
   b. conidiospore
   c. blastospore
   d. chlamydospore

7. A phycologist studies which of the following?
   a. classification of eukaryotes
   b. alternation of generations in algae
   c. rusts, smuts, and yeasts
   d. parasitic worms

8. The stemlike portion of a seaweed is called its ——————.
   a. trunk
   b. holdfast
   c. stipe
   d. blade

9. Carrageenan is found in the cell walls of which group of algae?
   a. red algae
   b. green algae
   c. dinoflagellates
   d. yellow-green algae

10. Chrysolaminarin is a storage product found in which group of microbes?
    a. dinoflagellates
    b. euglenids
    c. golden algae
    d. brown algae

11. Which of the following features characterizes diatoms?
    a. laminarin and oils as food reserves
    b. protective plates of cellulose within their cells
    c. chlorophylls *a* and *c* and carotene
    d. paramylon as a food storage molecule

12. Amoebas include microbes with ——————.
    a. threadlike pseudopods
    b. eyespots
    c. parabasal bodies
    d. alveoli

13. The motile feeding stage of a protozoan is called a(n) ——————.
    a. apicomplexan
    b. gametocyte
    c. cyst
    d. trophozoite

14. Which of the following is common to mitosis and meiosis?
    a. spindle
    b. crossing over
    c. tetrad of chromatids
    d. cytokinesis

15. Which taxon is characterized by "hairy" flagella?
    a. Apicomplexa
    b. Euglenozoa
    c. Alveolata
    d. Stramenopila

Matching

Match the terms below with their corresponding definitions.

1. ___ Mitosis
2. ___ Meiosis
3. ___ Homologous chromosomes
4. ___ Crossing over
5. ___ Cytokinesis

A. Cytoplasmic division
B. Diploid nuclei producing haploid nuclei
C. Results in genetic variation
D. Carry similar genes
E. Diploid nuclei producing diploid nuclei

Match the terms below with their corresponding definitions.

1. ___ Chitin
2. ___ Basidiospore
3. ___ Zygosporangium
4. ___ Hypha
5. ___ Ascospore
6. ___ Lichen

A. Fungal cell wall component
B. Fungus + alga or bacterium
C. Fungal filament
D. Fungal spore formed in a sac
E. Diploid fungal zygote with a thick wall
F. Fungal spore formed on club-shaped hypha

Match the terms below with their corresponding definitions.

1. ___ Chlorophyta
2. ___ Rhodophyta
3. ___ Chrysophyta
4. ___ Phaeophyta
5. ___ Rhizaria

A. Foraminifera
B. Yellow-green algae
C. Green algae
D. Brown algae
E. Red algae

VISUALIZE IT!

1. Label the photos below with the type of fungal spore, and indicate whether the spore is asexual or sexual.

2. Describe the features of a general fungal life cycle.

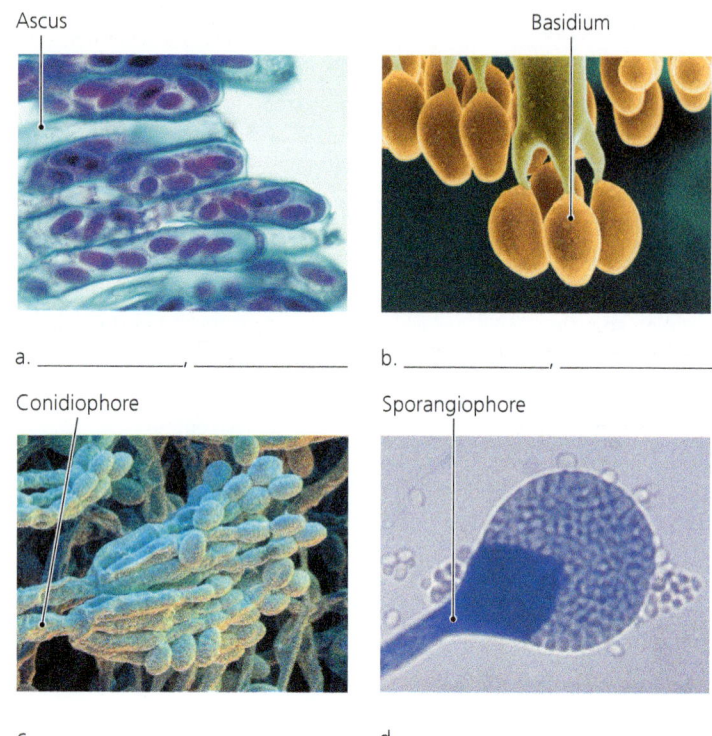

Ascus

Basidium

a. _____, _____   b. _____, _____

Conidiophore

Sporangiophore

c. _____, _____   d. _____, _____

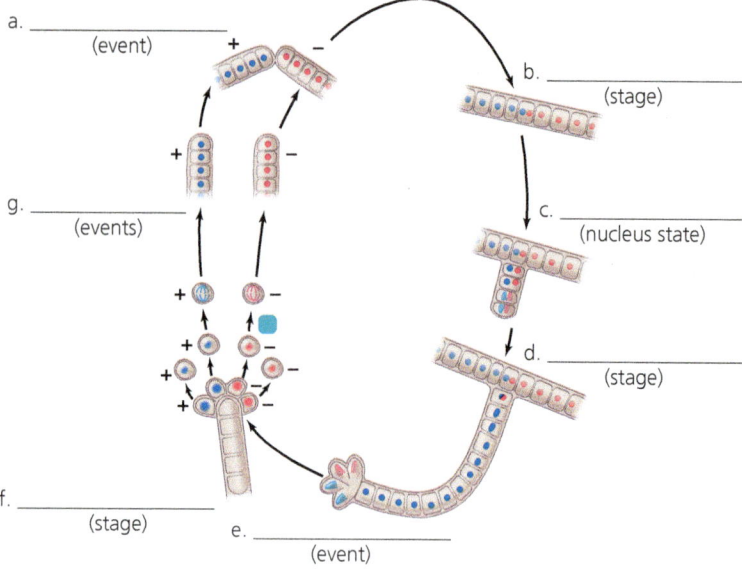

a. _____ (event)
b. _____ (stage)
c. _____ (nucleus state)
d. _____ (stage)
e. _____ (event)
f. _____ (stage)
g. _____ (events)

Short Answer

1. Compare and contrast the following closely related terms:
   a. Chromatid and chromosome
   b. Mitosis and meiosis II
   c. Hypha and mycelium
2. How do fungi acquire nutrients?
3. How are lichens useful in environmental protection studies?
4. What are the taxonomic challenges in classifying euglenids?
5. List several economic benefits of algae.
6. Why are relatively large animals such as parasitic worms studied in microbiology?
7. Name two ways that slime molds differ from true fungi.
8. What is the role of rRNA sequencing in the classification of eukaryotic microbes?
9. Describe the nuclear divisions that produce eight ascospores in an ascus.

Fill in the Blanks

1. The study of protozoa is called _____.
2. The study of fungi is called _____.
3. The study of algae is called _____.
4. Fungal diseases are called _____.
5. Amoebas with stiff pseudopods and silica shells are _____.

CRITICAL THINKING

1. How are cysts of protozoa similar to bacterial endospores? How are they different?
2. The host of a home improvement show suggests periodically emptying a package of yeast into a drain leading to a septic tank. Explain why this might be beneficial.
3. Why doesn't penicillin act against any of the pathogens discussed in this chapter?
4. How can one distinguish a filamentous fungus from a colorless alga?
5. Why do scientists as a group spend more time and money studying protozoa than they do studying algae?

6. Why are there more antibacterial drugs than antifungal drugs?
7. Which metabolic pathways are present in protozoa that lack mitochondria (amoebas, diplomonads, and parabasalids)? Which metabolic pathways are absent?
8. Without reference to genetic sequences, mycologists are certain that none of the septate, filamentous deuteromycetes will be shown to make zygospores. How can they be certain when the sexual stages of deuteromycetes are still unknown?
9. Until 1988, *Pneumocystis jirovecii*, a pathogen of immunocompromised patients that causes pneumonia, was classified as a protozoan because it is a chemoheterotroph that lacks a cell wall. However, taxonomists today classify *Pneumocystis* as a fungus. What advances in genetic sequencing caused this pathogen to be reclassified?
10. Explain why both dinoflagellates and euglenids were originally classified by zoologists as protozoa and by botanists as algae.
11. Fungi tend to reproduce sexually when nutrients are limited or other conditions are unfavorable, but they reproduce asexually when conditions are more ideal. Why is this a successful strategy?
12. Given that *Prototheca* is colorless, how do scientists know that it is really a green alga?

CONCEPT MAPPING

Using the terms provided, fill in the concept map that describes eukaryotic microorganisms. You can also complete this and other concept maps online by going to the Mastering Microbiology Study Area.

Algae
Alginic acid
Carrageenan
Cell walls
Chitin
Colonial
Cryptosporidium
Fungi
Giardia
Molds
Plasmodium
Protozoa
Unicellular (2)
Yeasts

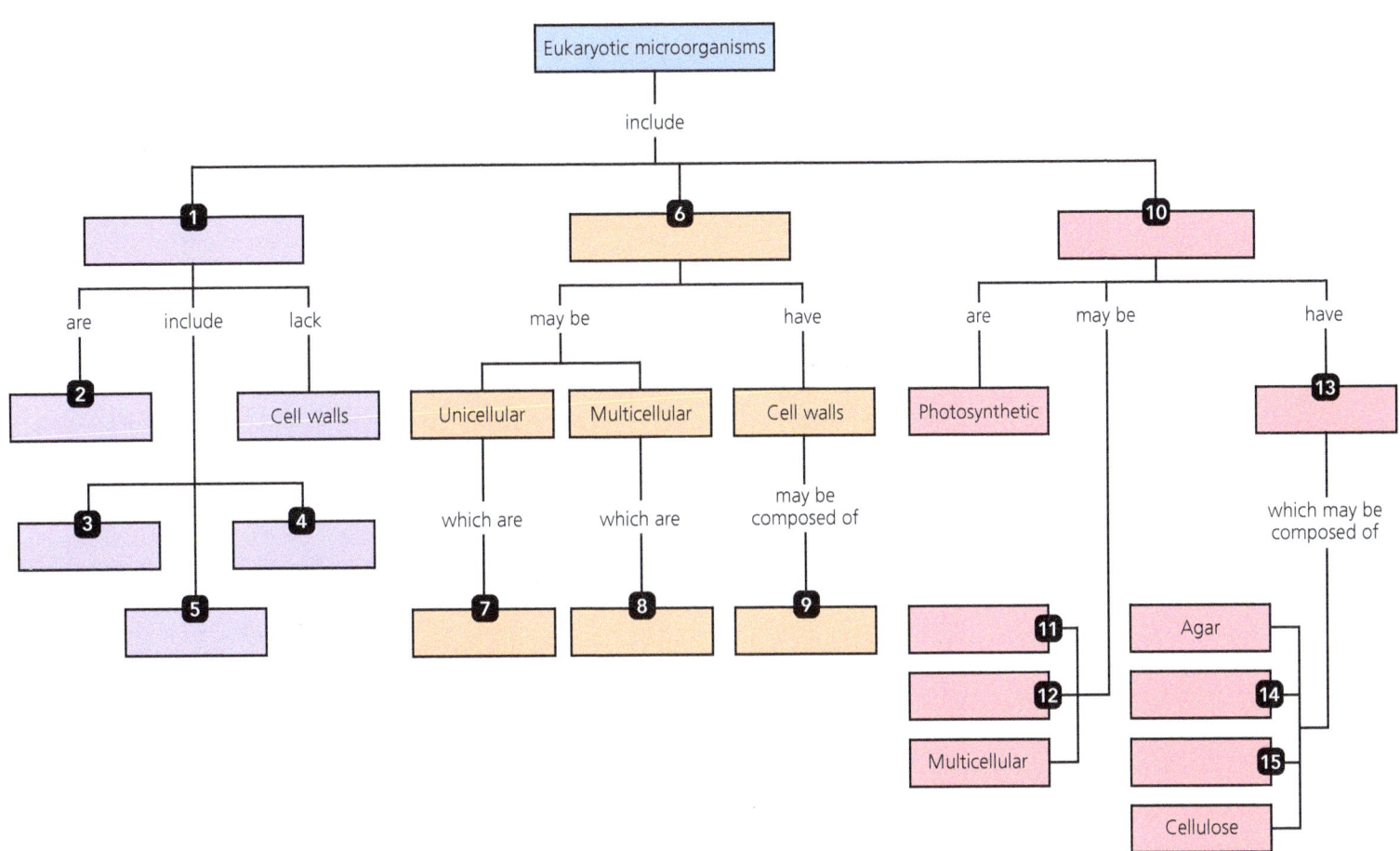

13 Characterizing and Classifying Viruses, Viroids, and Prions

Before You Begin

1. How can viruses be distinguished from eukaryotic and prokaryotic organisms?
2. What are the differences between DNA and RNA?
3. What effect does lysozyme have on bacteria?
4. Where does transcription occur in eukaryotic cells?
5. What are the template and end-product of the enzyme reverse transcriptase?

Viruses (green)—genetic material surrounded by protein and in some cases a phospholipid envelope—attach to and enter cells.

MICRO IN THE CLINIC
Working Too Hard?

DAVID IS A PARK RANGER FOR THE National Park Service in New Mexico. He's been hiking and camping all of his life, and he feels lucky to have a job where he gets to be outside protecting the land and wildlife. More important, he gets to help other people appreciate and enjoy nature. The past couple of months have been exciting: David has been part of a crew working on renovating some old cabins. Over the past four weeks he has spent more time inside than he has in years. Much of his time was spent sweeping up old nests and feces from deer mice, a prevalent local rodent. When the renovations are complete, there will be new areas for visitors to explore.

The work has been rewarding but hard. For the last two days, David's legs and back have been aching, and he has had a headache. He keeps working, but a few days later, he still has the headache and muscle aches, and now he has a fever. He's also had some dizzy spells while he's been working; he assumes it's because of the heat. David visits his doctor, and aside from a cut on his arm that has become infected, his physical exam is normal. The doctor prescribes antibiotics for the cut and tells David that headache and dizziness are likely due to dehydration—David should remain well hydrated while working in the heat and keep his arm bandaged.

About a week later, the cut on David's arm is healed, and the muscle soreness has lessened. But he still has a headache and has developed a cough and a tight feeling in his chest. When he wakes up in the middle of the night gasping for breath, he calls for an ambulance.

1. Why is David's condition worsening even though he's been taking antibiotics?

Turn to the end of the chapter (p. 408) to find out.

SOLVE THE PROBLEM

Microengineer a Better Mosquito?

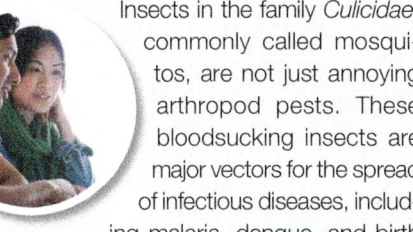

Insects in the family *Culicidae*, commonly called mosquitos, are not just annoying arthropod pests. These bloodsucking insects are major vectors for the spread of infectious diseases, including malaria, dengue, and birth defects caused by the newly emerging Zika virus. The importance of such vectors and their diseases is indicated by the fact that epidemiologists have given a special name to viruses spread by arthropods, the *arboviruses*, and their disease are arboviral diseases.

Microbiologists have discovered a genus of bacteria, *Wolbachia*, that is an obligate intracellular parasite in arthropods, including mosquitos. These bacteria can change their hosts' characteristics.

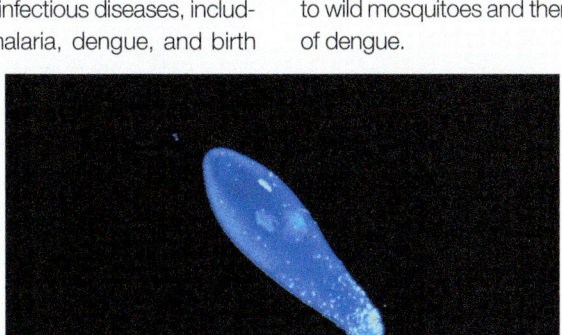

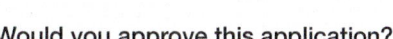
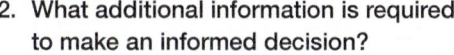

Elongated wasp egg infected with *Wolbachia*.

For example, certain strains of *Wolbachia* can reduce the ability of *Aedes aegypti* mosquitos to carry and transmit dengue virus. Since the bacteria can spread between mosquitos, some scientists have proposed releasing *Wolbachia*-infected insects into the environment to spread *Wolbachia* to wild mosquitoes and thereby reduce the spread of dengue.

Imagine you are on a scientific team at the Environmental Protection Agency (EPA) that is charged with reviewing an application to release mosquitos infected with genetically engineered *Wolbachia* to prevent the spread of Zika in the United States.

1. Would you approve this application?
2. What additional information is required to make an informed decision?

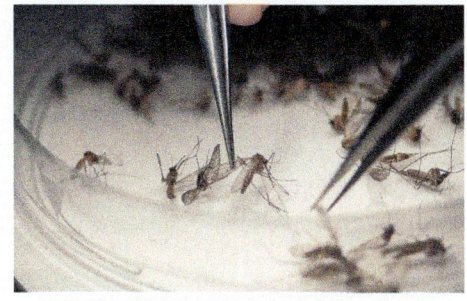

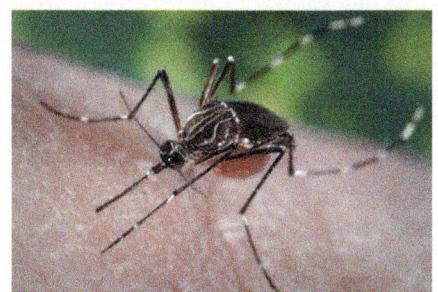

Not all pathogens are cellular. Many infections of humans, animals, and plants (and even of bacteria) are caused by **acellular** (noncellular) agents, including viruses and other pathogenic particles called *viroids* and *prions*. Although these agents are similar to eukaryotic and prokaryotic pathogens in that they cause disease when they invade susceptible cells, they are simple compared to a cell—lacking cell membranes and composed of only a few organic molecules. In addition to lacking a cellular structure, they lack most of the characteristics of life: They cannot carry out any metabolic pathway; they can neither grow nor respond to the environment; and they cannot reproduce independently but instead must utilize the chemical and structural components of the cells they infect. They must recruit the cell's metabolic pathways in order to increase their numbers.

In this chapter, we first examine a range of topics concerning viruses: what their characteristics are, how they are classified, how they replicate, what role they play in some kinds of cancers, how they are maintained in the laboratory, and whether viruses are alive. Then we consider the nature of viroids and prions.

Characteristics of Viruses

LEARNING OUTCOME

13.1 Define *virus* and *virion*, and describe the parts of a virion.

Viruses cause most of the diseases that still plague the industrialized world: common colds, influenza, herpes, and AIDS, to name a few. Although we have immunizations against many viral diseases and are adept at treating the symptoms of others, the characteristics of viruses and the means by which they attack their hosts make cures for viral diseases elusive. Throughout this section, we consider the clinical implications of viral characteristics.

We begin by looking at the characteristics viruses have in common. A **virus** is a minuscule, acellular, infectious agent having one or several pieces of nucleic acid—either DNA or RNA. The nucleic acid is the genetic material (genome) of the virus. Being acellular, viruses have no cytoplasmic membrane (though, as we will see, some viruses possess a membrane-like *envelope*). Viruses also lack cytosol and functional organelles.

They are not capable of metabolic activity on their own; instead, once viruses have invaded a cell, they take control of the cell's metabolism to produce more molecules of viral nucleic acid and viral proteins, which then assemble into new viruses via a process we will examine shortly. A biological virus is similar to a computer virus—it is basically a code that is inactive and harmless until it infects its host; then, it can be devastating.

Viruses have an extracellular and an intracellular state. Outside a cell, in the extracellular state, a virus is called a **virion** (vir′ē-on). Basically, a virion consists of a protein coat, called a **capsid**, surrounding a nucleic acid core (FIGURE 13.1a). Together, a viral nucleic acid and its capsid are also called a *nucleocapsid*, which in many cases can crystallize like crystalline chemicals (FIGURE 13.1b). Some virions have a phospholipid membrane called an **envelope** surrounding the nucleocapsid. The outermost layer of a virion (capsid or envelope) provides the virus both protection and recognition sites that bind to complementary chemicals on the surfaces of their specific host cells. Once a virus is inside, the intracellular state is initiated, and the capsid (and envelope, if there is one) is removed. A virus without its capsid exists solely as nucleic acid but is still referred to as a virus.

Now that we have examined ways in which viruses are alike, we will consider the characteristics that are used to distinguish different viral groups. Viruses differ in the type of genetic material they contain, the kinds of cells they attack, their size, the nature of their capsid coat, their shapes, and the presence or absence of an envelope.

Genetic Material of Viruses

LEARNING OUTCOME

13.2 Discuss viral genomes in terms of dsDNA, ssDNA, ssRNA, dsRNA, and the number of segments of nucleic acid.

Viruses show more variety in the nature of their genomes than do cells. Whereas the genome of every cell is double-stranded DNA, a viral genome is either DNA or RNA. The primary way in which scientists categorize and classify viruses is based on the type of genetic material that makes up the viral genome.

Some viral genomes, such as those of herpesvirus and smallpox virus, are double-stranded DNA (dsDNA), like the genomes of cells. Other viral genomes are either single-stranded RNA (ssRNA), single-stranded DNA (ssDNA), or double-stranded RNA (dsRNA). Cells never use these molecules for genomes—in fact, ssDNA and dsRNA are almost nonexistent in cells. Further, the genome of any particular virus may be either linear and composed of several molecules of nucleic acid, as in eukaryotic cells, or circular and singular, as in most prokaryotic cells. For example, the genome of an influenzavirus is composed of eight linear segments of single-stranded RNA, whereas the genome of poliovirus is one molecule of single-stranded RNA.

Viral genomes are usually smaller than the genomes of cells. For example, the genome of the smallest bacterium that infects humans (a species of *Chlamydia*) has almost 1000 protein-encoding genes, whereas the genome of a virus that infects *Escherichia coli* (esh-ĕ-rik′e-ă kō′lī), MS2, has only three. FIGURE 13.2 compares the physical size of the genome of a virus with the genome of its bacterial host, *E. coli*.

Hosts of Viruses

LEARNING OUTCOMES

13.3 Explain the mechanism by which viruses are specific for their host cells.
13.4 Compare and contrast viruses of fungi, plants, animals, and bacteria.

Most viruses infect only particular hosts' cells. This specificity is due to the precise affinity of viral attachment molecules—proteins or glycoproteins on the viral surface—for complementary proteins or glycoproteins on the surface of the host cell.

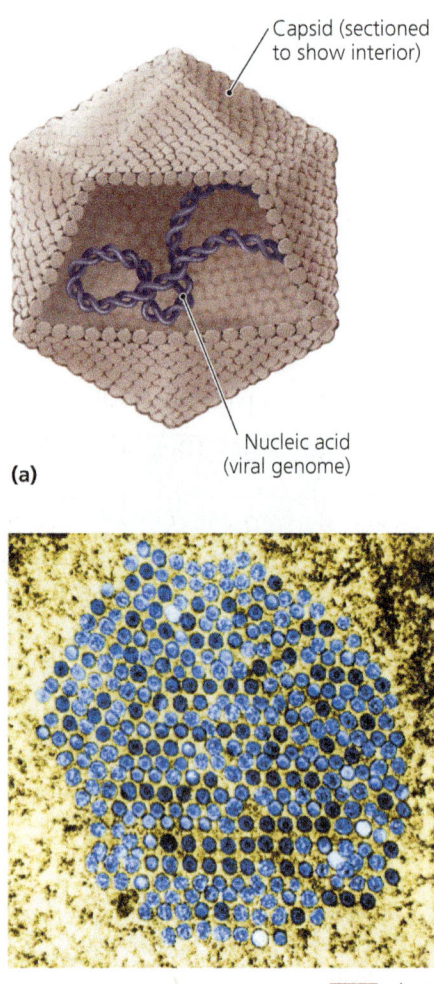

▲ FIGURE 13.1 Virions are complete virus particles and include a nucleic acid, a capsid, and in some cases an envelope. (a) A drawing of a nonenveloped polyhedral virus containing DNA. (b) A transmission electron microscope image of crystallized tobacco mosaic virus. Like many chemicals and unlike cells, some viruses can form crystals.

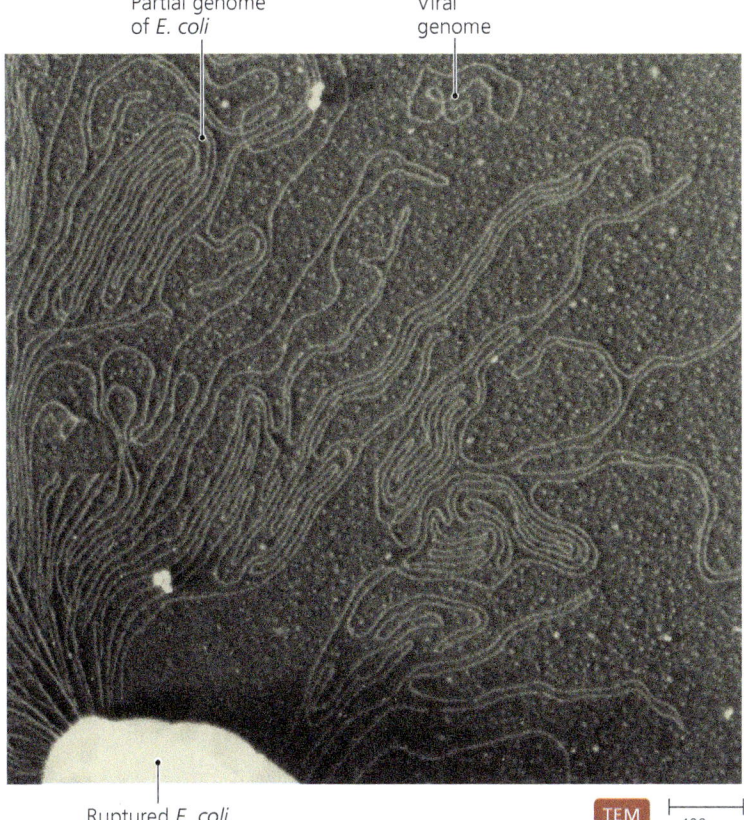

▲ FIGURE 13.2 **The relative sizes of genomes.** A cell of *Escherichia coli*, just visible at lower left, ruptured, releasing DNA.

Viruses may be so specific that they infect not only a particular host but also a particular kind of cell in that host. For example, HIV (human immunodeficiency virus, the agent that causes AIDS) specifically attacks helper T lymphocytes (a type of white blood cell) in humans and has no effect on, say, human muscle or bone cells. By contrast, some viruses are *generalists*; they infect many kinds of cells or many different hosts. An example of a generalist virus is West Nile virus, which can infect humans, most species of birds, several mammalian species, and some reptiles.

All types of organisms are susceptible to some sort of viral attack. There are viruses that infect archaeal, bacterial, plant, protozoan, fungal, and animal cells **(FIGURE 13.3)**. There are even tiny viruses that appear to attack specific larger viruses.

Most viral research and scientific study has focused on bacterial and animal viruses. A virus that infects bacteria is referred to as a **bacteriophage** (bak-tēr′ē-ō-fāj) or simply a **phage** (fāj). Scientists have determined that the number of bacteriophages is more than all bacteria, archaea, and eukaryotes put together. We will return our attention to bacteriophages and animal viruses later in this chapter.

Viruses of plants are less well known than bacterial and animal viruses, even though viruses were first identified and isolated from tobacco plants. Plant viruses infect many food crops, including corn, beans, sugarcane, tobacco, and potatoes, resulting in billions of dollars in losses each year. Viruses of plants are introduced into plant cells either through abrasions of the

(a)

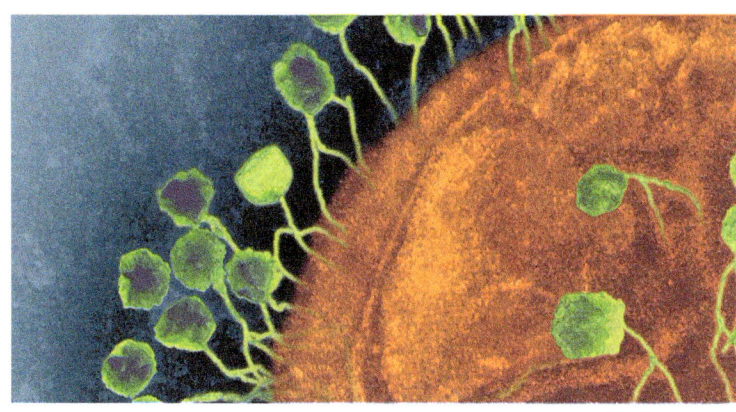

(b)

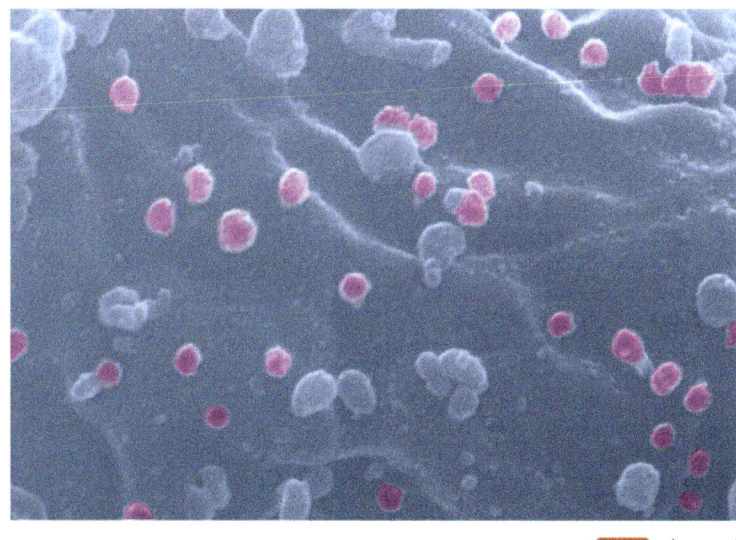

(c)

▲ FIGURE 13.3 **Some examples of plant, bacterial, and human hosts of viral infections.** (a) Tobacco mosaic virus—the first virus isolated—causes yellow discolorations of tobacco leaves. (b) A bacterial cell (colored orange) under attack by numerous bacteriophages (colored green). (c) A human white blood cell's cytoplasmic membrane, to which HIV (pink) is attached.

cell wall or by plant parasites, such as nematodes and aphids. After entry, plant viruses follow the replication cycle discussed below for animal viruses.

Fungal viruses have been little studied. We do know that fungal viruses are different from animal and bacterial viruses in that fungal viruses appear to exist only within cells; that is, they seemingly have no extracellular state. Presumably, fungal viruses cannot penetrate a thick fungal cell wall. However, because fusion of cells is typically a part of a fungal life cycle, viral infections can easily be propagated by the fusion of an infected fungal cell with an uninfected one.

Not all viruses are deleterious. The box **Beneficial Microbes: Good Viruses? Who Knew?** (p. 390) illustrates some useful aspects of viruses in the environment.

Sizes of Viruses

In the late 1800s, scientists hypothesized that the cause of many diseases, including polio and smallpox, was an agent smaller than a bacterium. They named these tiny agents "viruses," which is a Latin word for "poison." Most viruses are small; for example, 100 million polioviruses could fit side by side on the period at the end of this sentence. Even smaller viruses have diameters as small as 17 nm, while one of the larger viruses, *Megavirus*, is about 500 nm in diameter—about the diameter of many bacterial cells. FIGURE 13.4 compares the sizes of selected viruses to *E. coli* and a human red blood cell.

In 1892, Russian microbiologist Dmitri Ivanowski (1864–1920) first demonstrated that viruses are acellular with an experiment designed to elucidate the cause of tobacco mosaic disease. He filtered the sap of infected tobacco plants through a porcelain filter fine enough to trap even the smallest of cells. Viruses, however, were not trapped but instead passed through the filter with the liquid, which remained infectious to tobacco plants. This experiment proved the existence of an acellular disease-causing entity smaller than a bacterium. Tobacco mosaic virus (TMV) was isolated and characterized in 1935 by an American chemist, Wendell Stanley (1904–1971). The invention of electron microscopy allowed scientists to finally see TMV and other viruses.

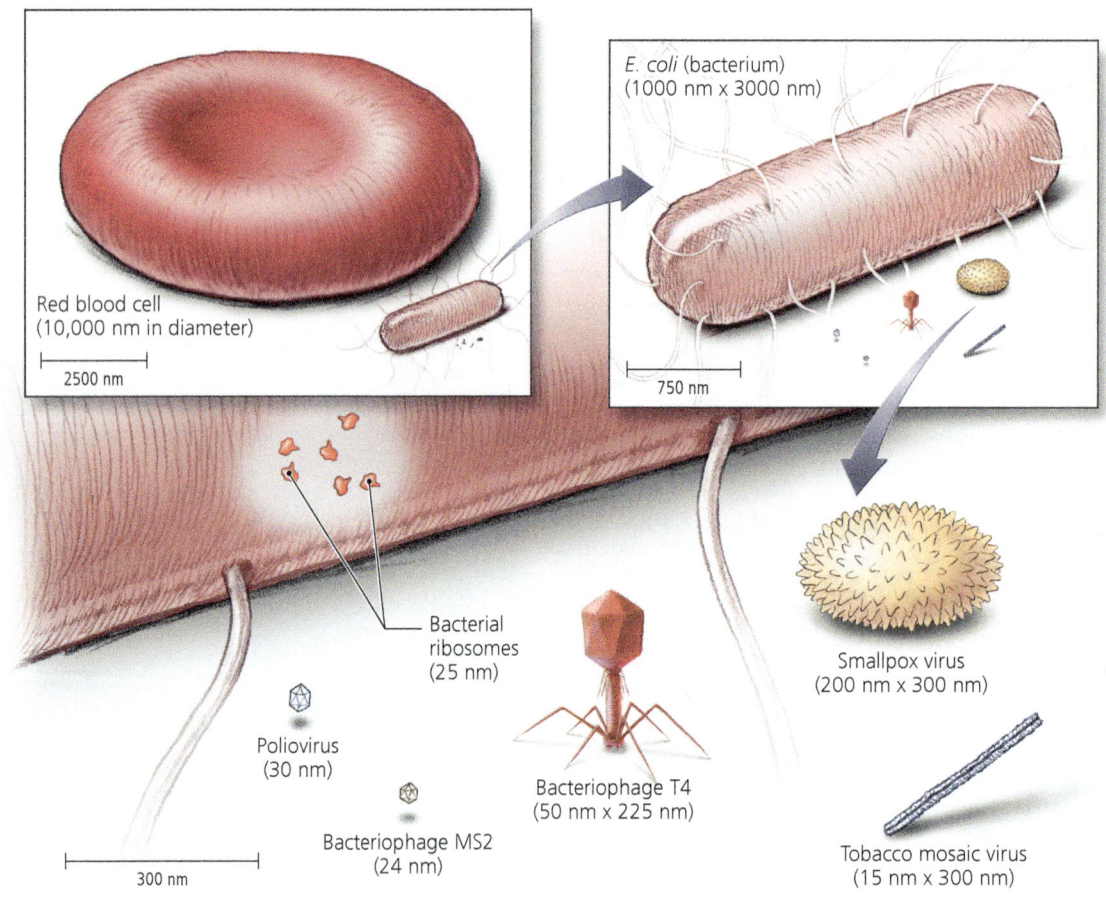

▲ FIGURE 13.4 **Sizes of selected virions.** Selected viruses are compared in size to a bacterium, *Escherichia coli*, and to a human red blood cell. *How can viruses be so small and yet still be pathogenic?*

Figure 13.4 *Viruses utilize a host cell's enzymes, organelles, and membranes to complete their replication cycle.*

Capsid Morphology

LEARNING OUTCOME

13.5 Discuss the structure and function of viral capsids and list their three basic shapes.

As we have seen, viruses in their extracellular state as virions have capsids—protein coats that provide both protection for viral nucleic acid and a means by which many viruses attach to their hosts' cells. The capsid of a virus is composed of proteinaceous subunits called **capsomeres** (or *capsomers*). Some capsomeres are composed of only a single type of protein, whereas others are composed of several different kinds of proteins. Recall that viral nucleic acid surrounded by its capsid is termed a *nucleocapsid*.

Viral Shapes

The shapes of virions are also used to classify viruses. There are three basic types of viral shapes: helical, polyhedral, and complex (**FIGURE 13.5**). The capsid of a helical virus is composed of capsomeres that bond together in a circling fashion to form a tube around the nucleic acid. The capsid of a polyhedral virus is roughly spherical, with a shape similar to a geodesic dome. The most common type of polyhedral capsid is an icosahedron, which has 20 sides.

Complex viruses have capsids of many different shapes that do not readily fit into either of the other two categories. An example of a complex virus is smallpox virus, which has several covering layers (including lipid) and no easily identifiable capsid. The complex shapes of many bacteriophages include icosahedral heads, which contain the genome, attached to helical tails with tail fibers. The complex capsids of such bacteriophages somewhat resemble NASA's lunar lander (**FIGURE 13.6**).

The Viral Envelope

LEARNING OUTCOME

13.6 Discuss the origin, structure, and function of the viral envelope.

All viruses lack cell membranes (after all, they are not cells), but some, particularly animal viruses, have an envelope similar in composition to a cell membrane surrounding their capsids. Other viral proteins called *matrix proteins* fill the region between capsid and envelope. A virion with a membrane is an *enveloped virion* (**FIGURE 13.7**); a virion without an envelope is called a *nonenveloped* or *naked virion*.

An enveloped virus acquires its envelope from its host cell during viral replication or release (discussed shortly). Indeed, the envelope of a virus is a portion of the membrane system of a host cell. Like a cytoplasmic membrane, a viral envelope is composed of a phospholipid bilayer and proteins. Some of the proteins are virally coded glycoproteins, which appear as spikes protruding outward from the envelope's surface (see Figure 13.7). Host DNA carries the genetic code required for the assembly of the phospholipids and some of the proteins in the envelope, while the viral genome specifies the other membrane proteins.

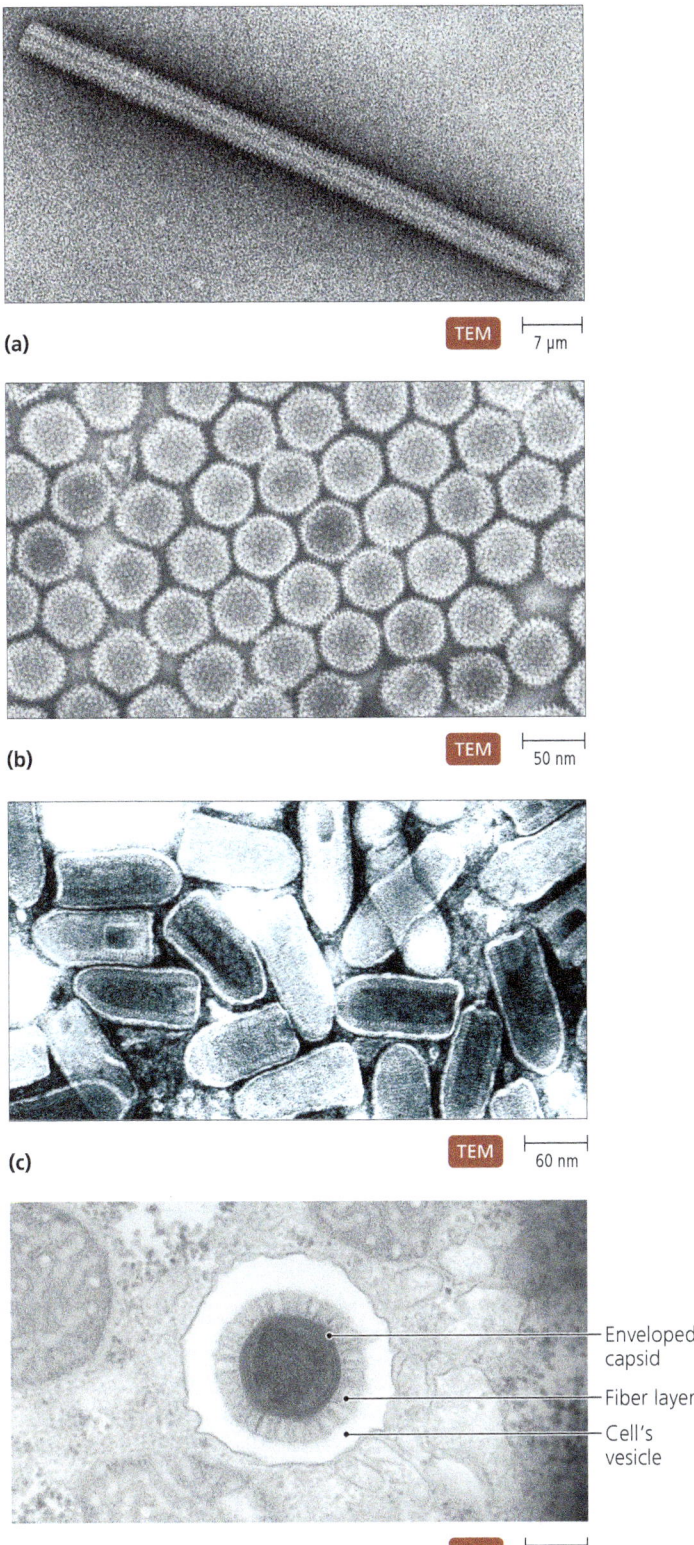

▲ **FIGURE 13.5 The shapes of virions.** (a) A helical virus, tobacco mosaic virus. The tubular shape of the capsid results from the tight arrangement of several rows of helical capsomeres. (b) Polyhedral virions of a virus that causes common colds. (c) The complex shape of rabies virus, which results from the shapes of the capsid and bullet-shaped envelope. (d) Complex shape of *Megavirus*, which is shown inside a cell's vesicle. These viruses have a layer of protein and sugar fibers outside the capsid.

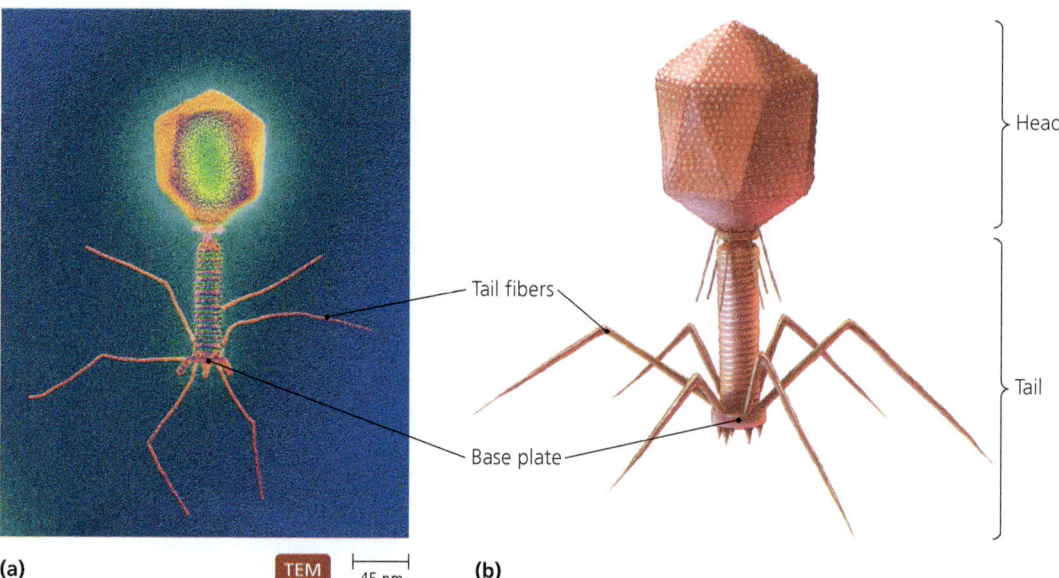

◀ **FIGURE 13.6 The complex shape of bacteriophage T4.** It includes an icosahedral head and an ornate tail that enables viral attachment and penetration.

An envelope's proteins and glycoproteins often play a role in the virion's recognition of host cells. A viral envelope does not perform other physiological roles of a cytoplasmic membrane, such as endocytosis or active transport.

An envelope provides some protection to the virus from the immune system; after all, enveloped viruses are carrying membrane from host cells and thus are more chemically similar to the host. However, membranes are more susceptible to detergents, alcohol, and drying out, so enveloped viruses are more fragile than naked ones.

Naked viruses are more stable outside a host than are enveloped viruses, but the naked capsid exposes more viral proteins to the environment and are thus more susceptible to recognition and attack by the immune system once inside a host.

TABLE 13.1 summarizes the novel properties of viruses and how those properties differ from the corresponding characteristics of cells. Next we turn our attention to the criteria by which virologists classify viruses.

TELL ME WHY

Why are naked *icosahedral* viruses able to crystallize?

Classification of Viruses

LEARNING OUTCOME

13.7 List the characteristics by which viruses are classified.

Virologists established the International Committee on Taxonomy of Viruses (ICTV) in 1966 to provide a single taxonomic scheme for viral classification and identification. They classify

BENEFICIAL MICROBES

Good Viruses? Who Knew?

An algal bloom off the coast of Seattle, Washington.

Viruses, though normally pathogenic to their host cells, do have positive effects, including what appear to be extensive roles in the environment. Some important ways viruses affect our world:

- Scientists have found a previously unknown virus that attacks a tiny marine alga that forms algal "blooms" consisting of hundreds of thousands to millions of algal cells per milliliter of water, which are visible from space (see photo). Algal blooms like these can deplete the water of oxygen at night, potentially harming fish and other marine life. The newly discovered virus stops blooms by killing the alga, a result that is good for animal life.

- When the algal cells die by this means, they release an airborne sulfate compound that acts to seed clouds. The resulting increase in cloudiness noticeably shades the ocean, measurably lowering water temperature. Thus, a marine virus both protects wildlife and helps to reduce global warming!

- Researchers have also discovered a bacteriophage of oceanic cyanobacteria that transfers genes for photosynthetic machinery into its hosts' cells so that the cells' photosynthetic rate increases. There are up to 10 million of these viruses in a single milliliter of seawater, so the scientists estimate that much of the oxygen we breathe may be attributable to the action of this virus on blue-green bacteria.

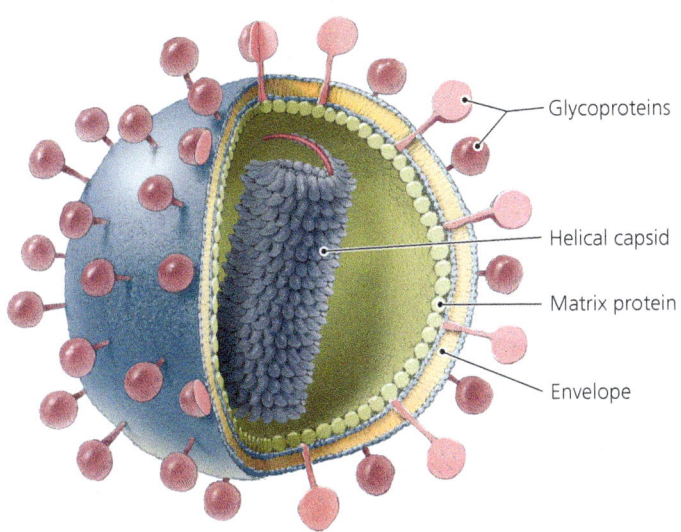
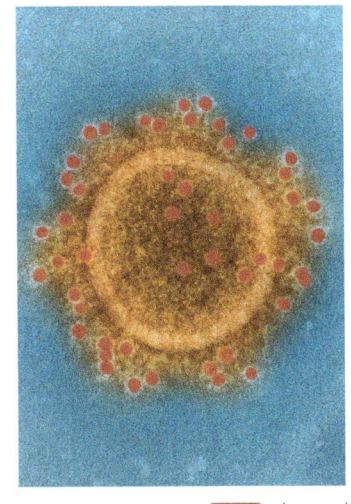

▶ **FIGURE 13.7 Enveloped virion.** Artist's rendition and electron micrograph of severe acute respiratory syndrome (SARS) virus, an enveloped virus with a helical capsid.

TABLE 13.1 The Novel Properties of Viruses

Viruses	Cells
Inert macromolecules outside a cell but become active inside a cell	Metabolize on their own
Do not divide or grow	Divide and grow
Acellular	Cellular
Obligate intracellular parasites	Most are free-living
Contain either DNA or RNA	Contain both DNA and RNA
Genome can be dsDNA, ssDNA, dsRNA, +ssRNA, or −ssRNA	Genome is dsDNA
Usually ultramicroscopic in size, ranging from 10 nm to over 500 nm	200 nm to 12 cm in diameter
Have a proteinaceous capsid around genome; some have an envelope around the capsid	Surrounded by a phospholipid membrane and often a cell wall
Replicate in an assembly-line manner using the enzymes and organelles of a host cell	Self-replicating by asexual and/or sexual means

viruses by the type of nucleic acid, presence of an envelope, shape, and size. So far, virologists have established families for all viral genera, but only nine viral orders are described. Taxonomists have not described kingdoms, divisions, and classes for viruses because the relationships among viruses are not well understood. Some researchers consider viruses to be a fourth domain of life (in addition to Bacteria, Archaea, and Eukarya).

Family names are typically derived either from special characteristics of viruses within the family or from the name of an important member of the family. For example, family *Picornaviridae* contains very small[1] RNA viruses, and *Hepadnaviridae* contains the DNA virus that causes hepatitis B. Family *Herpesviridae* is named for herpes simplex, a virus that can cause genital herpes. TABLE 13.2 (p. 392) lists the major families of human viruses, grouped according to the type of nucleic acid each contains.

Specific epithets for viruses are their common English designations written in italics. Generally, viruses are known solely by their English names. As examples, the nomenclature for two important viral pathogens, HIV and rabies virus, is as follows:

	HIV	Rabies Virus
Order	Not yet assigned	*Mononegavirales*
Family	*Retroviridae*	*Rhabdoviridae*
Genus	*Lentivirus* (len'ti-vī-rŭs)	*Lyssavirus* (lis'a-vī-rŭs)
Specific Epithet	*human immunodeficiency virus*	*rabies virus*

MICRO CHECK

1. What type of genome do viruses have?
2. What type of organisms do phages infect?
3. What is the term used to describe viral nucleic acid surrounded by a protein "shell"?
4. Where does an enveloped virus get its envelope?
5. Which characteristics are used to classify viruses?

TELL ME WHY

What characteristics of the genomes of parvoviruses and of reoviruses make them very different from cells?

---
[1]*Pico* means one-trillionth, 10^{-12}.

TABLE 13.2 Families of Human Viruses

Family	Strand Type	Representative Genera (Diseases)
DNA Viruses		
Poxviridae	Double	Orthopoxvirus (smallpox)
Herpesviridae	Double	Simplexvirus (herpes type 1: fever blisters, respiratory infections; herpes type 2: genital infections); Varicellovirus (chicken pox); Lymphocryptovirus, Epstein-Barr virus (infectious mononucleosis, Burkitt's lymphoma); Cytomegalovirus (birth defects); Roseolovirus (roseola)
Papillomaviridae	Double	Papillomavirus (benign tumors, warts, cervical and penile cancers)
Polyomaviridae	Double	Polyomavirus (progressive multifocal leukoencephalopathy)
Adenoviridae	Double	Mastadenovirus (conjunctivitis, respiratory infections)
Hepadnaviridae	Partial single and partial double	Orthohepadnavirus (hepatitis b)
Parvoviridae	Single	Erythrovirus (erythema infectiosum)
RNA Viruses		
Picornaviridae	Single, +[a]	Enterovirus (polio); Hepatovirus (hepatitis A)
Caliciviridae	Single, +	Norovirus (gastroenteritis)
Astroviridae	Single, +	Astrovirus (gastroenteritis)
Hepeviridae	Single, +	Hepevirus (hepatitis E)
Togaviridae	Single, +	Alphavirus (encephalitis); Rubivirus (rubella)
Flaviviridae	Single, +	Flavivirus (yellow fever, Japanese encephalitis); Hepacivirus (hepatitis C)
Coronaviridae	Single, +	Coronavirus (common cold, severe acute respiratory syndrome)
Retroviridae	Single, +, segmented	Deltaretrovirus (leukemia); Lentivirus (AIDS)
Paramyxoviridae	Single, −[b]	Paramyxovirus (common cold, respiratory infections); Pneumovirus (pneumonia, common cold); Morbillivirus (measles); Rubulavirus (mumps)
Rhabdoviridae	Single, −	Lyssavirus (rabies)
Filoviridae	Single, −	Filovirus (Ebola hemorrhagic fever); Marburgvirus (hemorrhagic fever)
Bunyaviridae	Single, −, segmented	Bunyavirus (California encephalitis virus); Hantavirus (pneumonia)
Orthomyxoviridae	Single, − segmented	Influenzavirus (flu)
Arenaviridae	Single, −, segmented	Lassavirus (hemorrhagic fever)
Reoviridae	Double, segmented	Orbivirus (encephalitis); Rotavirus (diarrhea); Coltivirus (Colorado tick fever)

[a] Positive-sense (+RNA) is equivalent to mRNA; that is, it instructs ribosomes in protein translations.
[b] Negative-sense (−RNA) is complementary to mRNA; it cannot be directly translated.

Viral Replication

As previously noted, viruses cannot reproduce themselves because they lack the genes for all the enzymes necessary for replication; in addition, they do not possess functional ribosomes for protein synthesis. Instead, viruses depend on their hosts' enzymes and organelles to produce new virions. Once a host cell falls under control of a viral genome, it is forced to replicate viral genetic material and translate viral proteins, including viral capsomeres and viral enzymes.

The replication cycle of a virus usually results in the death and lysis of the host cell. Because the cell undergoes lysis near the end of the cycle, this type of replication is called a **lytic replication cycle**. In general, a lytic replication cycle consists of the following five stages:

- **Attachment** of the virion to the host cell
- **Entry** of the virion or its genome into the host cell
- **Synthesis** of new nucleic acids and viral proteins by the host cell's enzymes and ribosomes
- **Assembly** of new virions within the host cell
- **Release** of the new virions from the host cell

In the following sections, we examine the events that occur in the replication of bacteriophages and animal viruses. We begin with lytic replication in bacteriophages; then we turn to *lysogenic replication* (a modification of replication); and finally, we consider the replication of animal viruses.

Play *Viral Replication*: Overview @ Mastering Microbiology

Lytic Replication of Bacteriophages

LEARNING OUTCOME

13.8 Sketch and describe the five stages of the lytic replication cycle as it typically occurs in bacteriophages.

Studies of phages revealed the basics of viral biology. Indeed, bacteriophages make excellent tools for the general study of viruses because bacteria are easier and less expensive to culture than animal or human cells. **Beneficial Microbes: Prescription Viruses?** (p.395) is an interesting side note on the use of bacteriophages as an alternative to antibiotics.

Here, we examine the replication of a much-studied dsDNA phage of *E. coli* called *type 4 (T4)*. T4 virions are complex, having the polyhedral heads and helical tails seen in many bacteriophages. We begin with attachment, the first stage of replication **(FIGURE 13.8)**.

Attachment 1

Because phages, like all virions, are nonmotile, contact with bacteria occurs by purely random collision, brought about as molecular bombardment and currents move virions through the environment. The tail fibers are responsible for the attachment of T4 to its host bacterium. Attachment proteins on the tail fibers precisely fit to complementary receptor proteins on the surface of the cell wall of *E. coli*. The specificity of the attachment proteins for the receptors ensures that the virus will attach only to *E. coli*. Various bacteriophages may attach to receptor proteins on their host cells' walls, flagella, or pili. Phage attachment can be so specific that a phage might infect only one particular strain of a bacterial species. Thus, scientists can use phage attachment for bacterial identification in the lab. (Chapter 4 discusses so-called phage typing.)

Entry 2

Once phage T4 has attached to the bacterium's cell wall, it must still overcome the formidable barrier posed by the cell wall and cytoplasmic membrane if it is to enter the cell. T4 overcomes this obstacle in an elegant way. Upon contact with *E. coli*, T4 releases *lysozyme* (li'sō-zīm), a protein enzyme carried within the capsid that weakens the peptidoglycan of the cell wall. The phage's tail sheath then contracts, forcing an internal hollow tube within the tail through the cell wall and membrane, much as a hypodermic needle penetrates the skin. The phage injects its genome through the tube and into the bacterium. The empty capsid, having performed its task, is left on the outside of the cell looking like an abandoned spacecraft.

After entry, viral enzymes (which either are carried within the capsid and delivered with the viral genome or are coded by viral genes and made by the bacterium) degrade the bacterial DNA into its constituent nucleotides.

Synthesis 3

After losing its chromosome, the bacterium stops synthesizing its own molecules and begins synthesizing only viral parts under control of the viral genome.

For dsDNA viruses like T4, protein synthesis is straightforward and similar to cellular transcription and translation, except that mRNA is transcribed from viral DNA instead of cellular DNA. Translation by the host cell's ribosomes results in viral proteins, including capsomeres, components of the tail, viral DNA polymerase (which replicates viral DNA), and lysozyme (which weakens the bacterial cell wall from within, enabling the virions to leave the cell once they have been assembled).

Assembly 4

Scientists do not understand completely how phages are assembled inside a host cell, but it appears that as capsomeres accumulate within the cell, they spontaneously attach to one another to form new capsid heads. Likewise, tails assemble and attach to heads, and tail fibers attach to tails, forming mature virions. Such capsid assembly is a spontaneous process, requiring little or no enzymatic activity. For many years, it was assumed that capsids form around a genome; however, research has shown that for some viruses, enzymes pump the genome into the assembled capsid under high pressure—five times that used in a paintball gun. This process resembles stuffing a strand of cooked spaghetti into a matchbox through a single small hole.

Sometimes a capsid assembles around leftover pieces of host DNA instead of viral DNA. A virion formed in this manner is called a *transducing virion*. It is still able to attach to a new host by means of its tail fibers, but instead of inserting phage DNA, it transfers DNA from the first host into a new host. This process is known as *transduction* (described in more detail in Chapter 7).

Release 5

Newly assembled virions are released from the cell as lysozyme completes its work on the cell wall and the bacterium disintegrates. Areas of disintegrating bacterial cells in a lawn of bacteria in a Petri plate look as if the lawn were being eaten, and it was the appearance of these *plaques* that prompted early scientists to give the name *bacteriophage*, "bacterial eater," to these viruses.

For phage T4, the process of lytic replication takes about 25 minutes and can produce as many as 100 to 200 new virions for each bacterial cell lysed. For any phage undergoing lytic replication, the period of time required to complete the entire process, from attachment to release, is called the *burst time*, and the number of new virions released from each lysed bacterial cell is called the *burst size* (FIGURE 13.9).

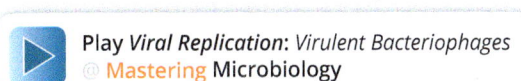

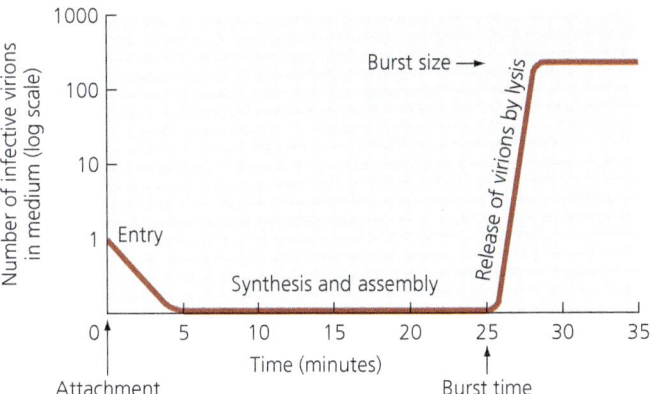

▲ FIGURE 13.9 **Pattern of virion abundance in lytic cycle.** Shown is virion abundance over time for a single lytic replication cycle. New virions are not observed in the culture medium until synthesis, assembly, and release (lysis) are complete, at which time (the burst time) the new virions are released all at once. Burst size is the number of new virions released per lysed host cell.

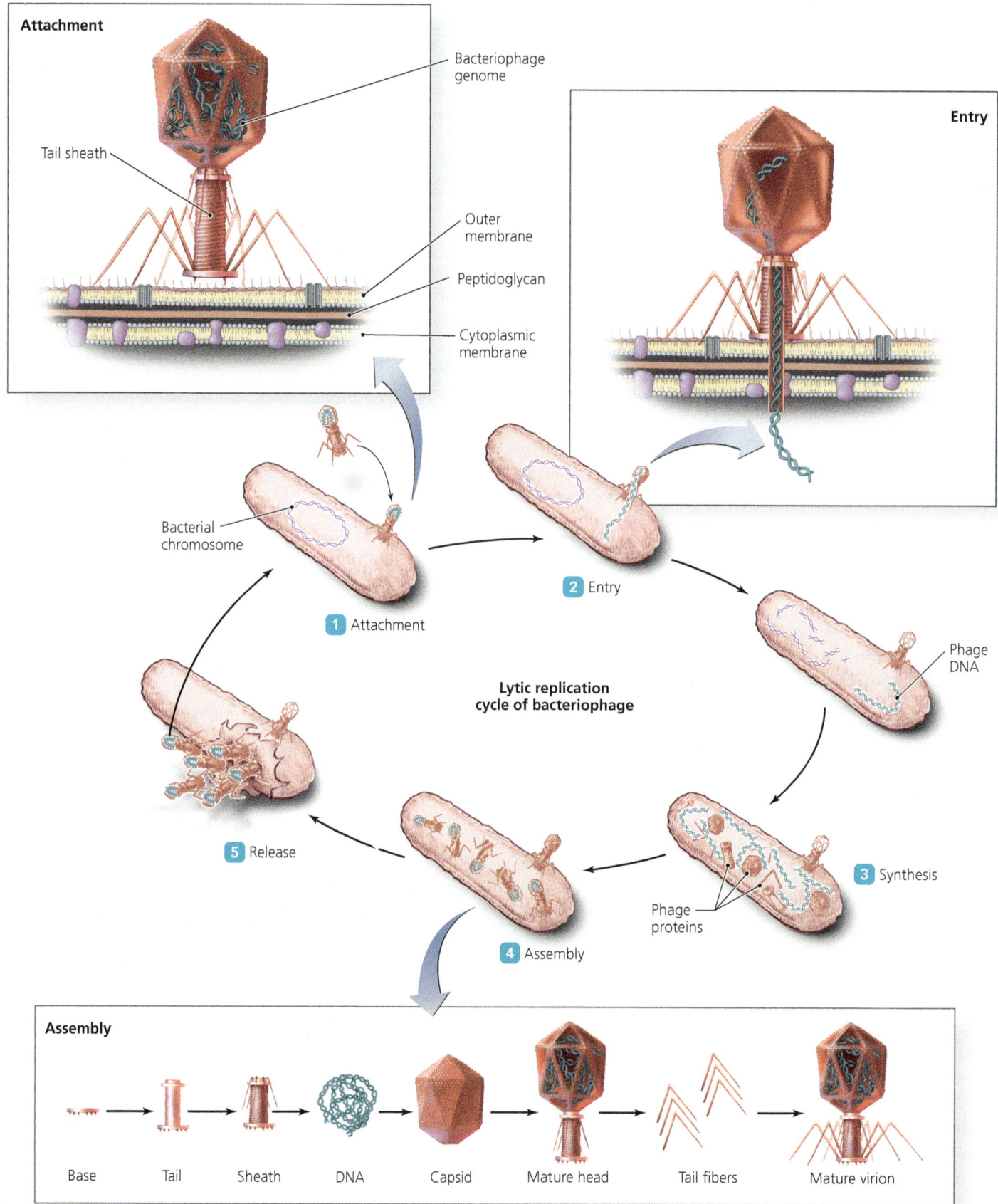

▲ FIGURE 13.8 **The lytic replication cycle in bacteriophages.** The phage shown in this illustration is T4, and the bacterium shown is E. coli. The circular bacterial chromosome is represented diagrammatically; in reality, it would be much longer.

BENEFICIAL MICROBES

Prescription Viruses?

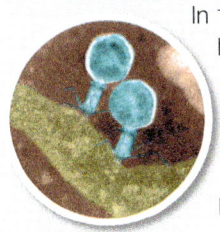

Bacteriophages (blue) attacking *E. coli* (amber).

In 1917, Canadian biologist Felix d'Herelle published a paper announcing the discovery of *bacteriophages*, viruses that prey on bacteria. In fact, half the bacteria on Earth succumb to phages every two days! D'Herelle proposed that phages could be natural weapons against bacterial pathogens.

Phage therapy was used in the early 1900s to combat dysentery, typhus, and cholera but was largely abandoned in the 1940s in the United States, eclipsed by the development of antibiotics, such as penicillin. Phage therapy continued in the Soviet Union and Eastern Europe, where research has been ongoing. In 2018, motivated by the growing problem of antibiotic-resistant bacteria, scientists and physicians in the United States established a medical research center to investigate phage therapy in a clinical setting.

A phage reproduces by inserting genetic material into a bacterium, causing the bacterium to build copies of the virus that burst out of the cell to infect other bacteria. A single phage can become 10 trillion phages within 2 hours, killing 99.9% of its host bacteria. Since each type of phage attacks a specific strain of bacteria, phage treatment is effective only if the phages are carefully matched to the disease-causing bacterium. It also means that phage treatment, unlike the use of antibiotics, can be effective without killing the body's helpful bacteria.

Lysogenic Replication of Bacteriophages

LEARNING OUTCOME

13.9 Compare and contrast the lysogenic replication cycle of viruses with the lytic cycle.

Not all viruses follow the lytic pattern of phage T4 we just examined. Some bacteriophages have a modified replication cycle in which infected host cells grow and reproduce normally for many generations before they lyse. Such a replication cycle is called a **lysogenic replication cycle** or **lysogeny** (lī-soj'ĕ-nē), and the phages are called **temperate phages** or *lysogenic phages*.

Here, we examine lysogenic replication as it occurs in a much-studied temperate phage discovered by a leading geneticist and microbiologist, Esther Lederberg (1922–2006), in 1950. This phage, *lambda phage*, is another parasite of *E. coli*. A lambda phage has a linear molecule of dsDNA in a complex capsid consisting of an icosahedral head attached to a tail that lacks tail fibers (**FIGURE 13.10**).

FIGURE 13.11 illustrates lysogeny with lambda phage. First, the virion randomly contacts an *E. coli* cell and attaches via its tail **1**. The viral DNA enters the cell, just as occurs with phage T4, but the host cell's DNA is not destroyed, and the phage's genome does not immediately assume control of the cell. Instead, the virus remains inactive. Such an inactive bacteriophage is called a **prophage** (pro'fāj) **2**. A prophage remains inactive by coding for a protein that suppresses prophage genes. A side effect of this repressor protein is that it renders the bacterium resistant to additional infection by other viruses of the same type.

Another difference between a lysogenic cycle and a lytic cycle is that the prophage inserts into the DNA of the bacterium, becoming a physical part of the bacterial chromosome **3**. For DNA viruses like lambda phage, this is a simple process of fusing two pieces of DNA: One piece of DNA, the virus, is fused to another piece of DNA, the chromosome of the cell. Every time the cell replicates its infected chromosome, the prophage is also replicated **4**. All daughter cells of a lysogenic cell are thus infected with the nearly inactive virus. A prophage and its descendants may remain a part of bacterial chromosomes for generations or even forever.

Lysogenic phages can change the phenotype of a bacterium, for example, from a harmless form into a pathogen—a process called **lysogenic conversion**. Bacteriophage genes are responsible for toxins and other disease-evoking proteins found in the bacterial agents of diphtheria, cholera, rheumatic fever, and certain severe cases of diarrhea caused by *E. coli*.

At some later time, a prophage might be excised from the chromosome by recombination or some other genetic event; it then reenters the lytic phase. The process whereby a prophage is excised from the host chromosome is called **induction** **5**. Inductive agents are typically the same physical and chemical agents that damage DNA molecules, including ultraviolet light, X rays, and carcinogenic chemicals.

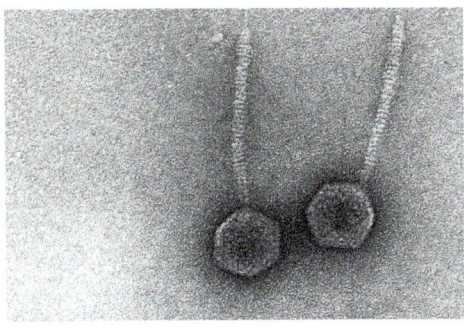

▲ **FIGURE 13.10 Bacteriophage lambda.** Note the absence of tail fibers. *Phage T4 attaches by means of molecules on its tail fibers. How does lambda phage, which lacks fibers, attach?*

Figure 13.10 *Lambda has attachment molecules at the end of its tail rather than on its tail fibers.*

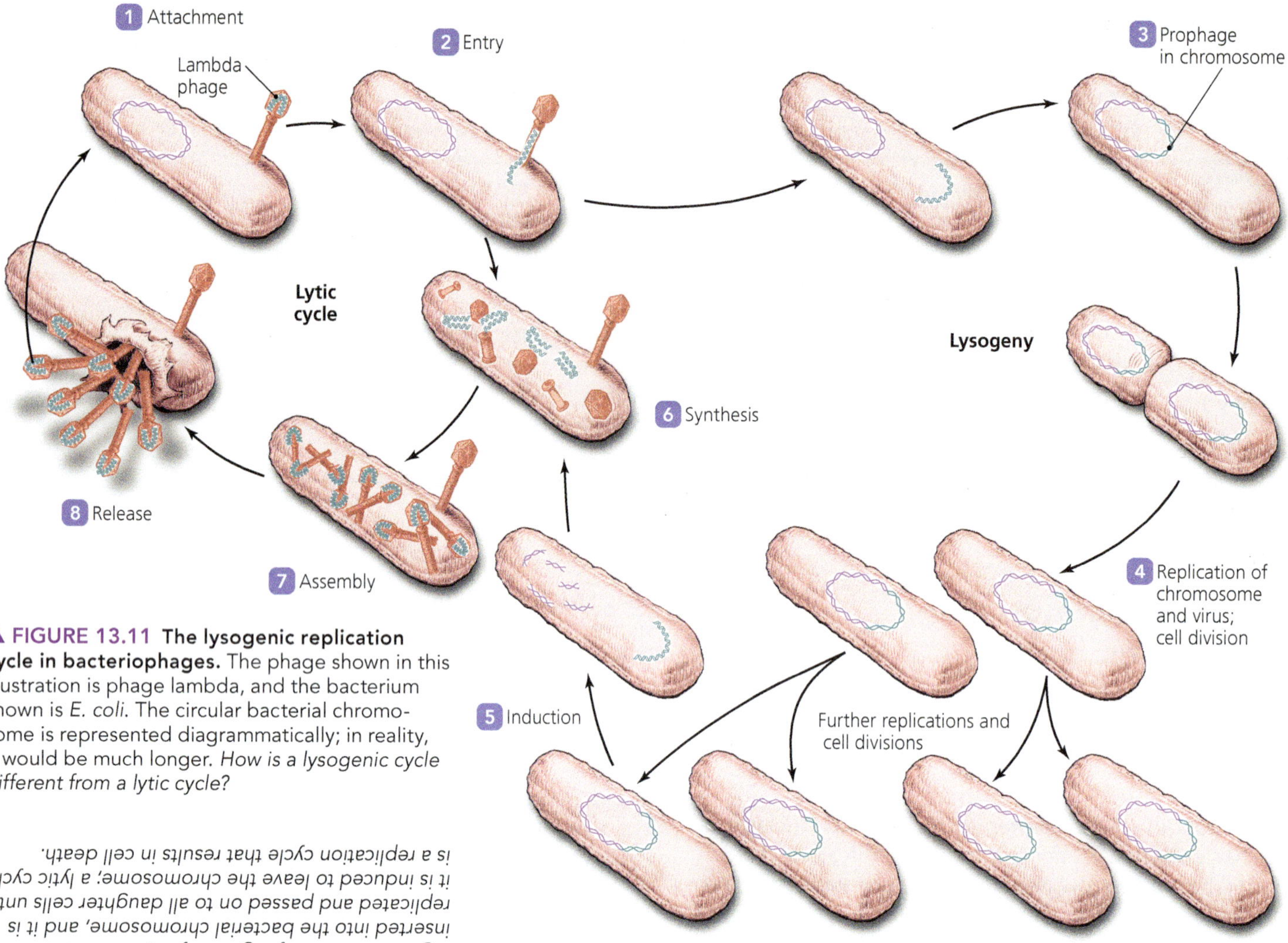

▲ **FIGURE 13.11 The lysogenic replication cycle in bacteriophages.** The phage shown in this illustration is phage lambda, and the bacterium shown is *E. coli*. The circular bacterial chromosome is represented diagrammatically; in reality, it would be much longer. *How is a lysogenic cycle different from a lytic cycle?*

Figure 13.11 In a lysogenic cycle, the virus is inserted into the bacterial chromosome, and it is replicated and passed on to all daughter cells until it is induced to leave the chromosome; a lytic cycle is a replication cycle that results in cell death.

After induction, the lytic steps of synthesis 6, assembly 7, and release 8 resume from the point at which they stopped. The cell becomes filled with virions and breaks open.

Bacteriophages T4 and lambda demonstrate two replication strategies that are typical for many DNA viruses. RNA viruses and enveloped viruses present variations on the lytic and lysogenic cycles we have examined. We will next examine some variations seen in replication of animal viruses.

Play *Viral Replication: Temperate Bacteriophages* @ **Mastering** Microbiology

MICRO CHECK

6. List the steps of lytic phage replication in order.
7. What is transduction?
8. How are lysogenic phages different from lytic phages?

Replication of Animal Viruses

LEARNING OUTCOMES

13.10 Explain the differences between bacteriophage replication and animal viral replication.

13.11 Compare and contrast the replication and synthesis of DNA, −RNA, and +RNA viruses.

13.12 Compare and contrast the release of viral particles by lysis and budding.

13.13 Compare and contrast latency in animal viruses with phage lysogeny.

Animal viruses have the same five basic steps in their replication pathways as bacteriophages—that is, attachment, entry, synthesis, assembly, and release. However, there are significant differences in the replication of animal viruses that result in part from the presence of envelopes around some of the viruses and in part from the eukaryotic nature of animal cells and the lack of an animal cell wall.

In this section, we examine the replication processes that are shared by DNA and RNA animal viruses, compare these processes with those of bacteriophages, and discuss how the synthesis of DNA and RNA viruses differ.

Attachment of Animal Viruses

As with bacteriophages, attachment of an animal virion is dependent on the chemical attraction and exact fit between attachment molecules on the virion and complementary receptors on the animal cell's surface. Unlike the bacteriophages we have examined, animal viruses lack both tails and tail fibers. Instead, animal viruses often have glycoprotein spikes on their capsids or envelopes.

Entry and Uncoating of Animal Viruses

Animal viruses enter a host cell shortly after attachment. Even though entry of animal viruses is not as well understood as entry of bacteriophages, there are at least three different mechanisms: direct penetration, membrane fusion, and endocytosis.

Some naked viruses enter their hosts' cells by *direct penetration*—a process in which the viral capsid attaches and sinks into the cytoplasmic membrane, creating a pore through which the genome alone enters the cell, leaving the empty capsid on the surface **(FIGURE 13.12a)**. Poliovirus infects host cells via direct penetration.

With other animal viruses, by contrast, the entire capsid and its contents enter a host cell by either membrane fusion or endocytosis. With viruses using *membrane fusion*, such as measles virus, the viral envelope and the host's cytoplasmic membrane fuse, releasing the capsid into the cell's cytoplasm while leaving the envelope glycoproteins as part of the cell membrane **(FIGURE 13.12b)**.

Most enveloped viruses and some naked viruses enter host cells by triggering *endocytosis*. Attachment of the virus to receptor molecules on the cell's surface stimulates the cell to endocytize the entire virus **(FIGURE 13.12c)**. For example, adenoviruses (naked) and herpesviruses (enveloped) enter human host cells via endocytosis.

For a virus that penetrates a host cell with its capsid intact, the capsid must be removed to release the genome before replication of the virus can continue. The process, called **uncoating**, is poorly understood. Uncoating occurs by different means in different viruses; some viruses are uncoated within vesicles by cellular enzymes, whereas others are uncoated by enzymes within the cell's cytosol.

Synthesis of DNA Viruses of Animals

Synthesis of animal viruses also differs from synthesis of bacteriophages. Each type of animal virus requires a different strategy for synthesis that depends on the kind of nucleic acid involved—whether it is DNA or RNA and whether it is double-stranded or single-stranded. Some DNA animal viruses enter the cell's nucleus, whereas other DNA viruses and most RNA animal viruses are replicated in the cytoplasm. Further, in synthesis of enveloped viruses, some viral proteins are inserted into cellular membranes. These membranes become the viral envelopes while the proteins become the envelope's spikes.

As we discuss the synthesis and assembly of each type of animal virus, consider the following two questions:

- How is mRNA—needed for the translation of viral proteins—synthesized?
- What molecule serves as a template for nucleic acid replication?

dsDNA Viruses Synthesis of new double-stranded DNA (dsDNA) virions is similar to the normal replication of cellular

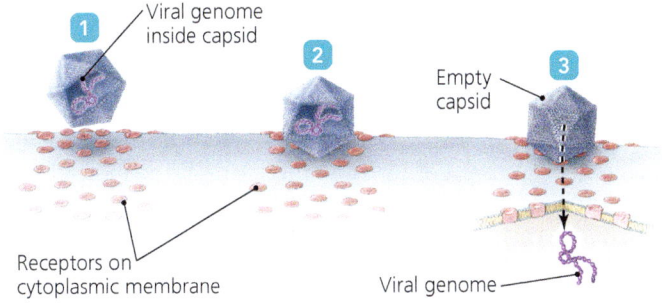

(a) Direct penetration

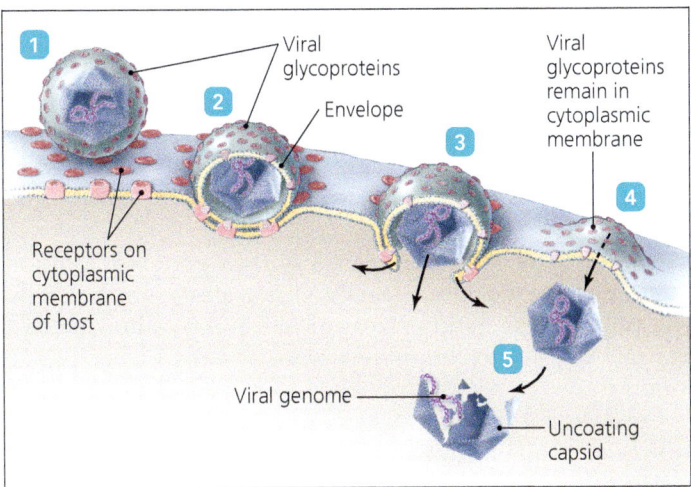

(b) Membrane fusion

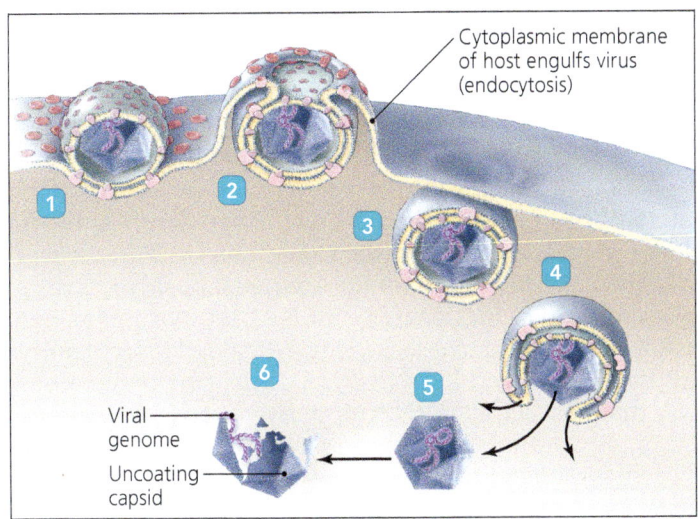

(c) Endocytosis

▲ **FIGURE 13.12 Three mechanisms of entry of animal viruses.** (a) Direct penetration, a process whereby naked virions inject their genomes into their animal cell hosts. (b) Membrane fusion, in which fusion of the viral envelope and cell membrane dumps the capsid into the cell. (c) Endocytosis, in which attachment of a naked or an enveloped virus (as shown here) stimulates the host cell to engulf the entire virus. *After penetration, many animal viruses must be uncoated, but bacteriophages need not be. Why is this so?*

Figure 13.12 Generally, bacteriophages inject their DNA during penetration, so the capsid does not enter the cell.

DNA (which is double stranded) and the normal translation of cellular proteins. The genome of most dsDNA viruses enters the nucleus of the host cell, where cellular enzymes replicate the viral genome in the same manner as they replicate the host DNA—using each strand of viral DNA as a template for its complement. Messenger RNA is transcribed from viral DNA in the nucleus and exported into the cytoplasm, where host ribosomes make capsomere proteins. The capsomeres then enter the nucleus, where new virions spontaneously assemble. This method of replication is seen with herpes and papilloma (wart) viruses.

Replication of hepatitis B virus is one exception to this scenario: The genomes of hepatitis B and similar viruses are replicated using an RNA intermediary instead of replicating DNA from a DNA template. In other words, the genome of such viruses is transcribed into RNA, which is then used as a template to make multiple copies of viral DNA genome. The latter process, which is the reverse of normal transcription, is mediated by a viral enzyme, *reverse transcriptase*.

ssDNA Viruses A human virus with a genome composed of single-stranded DNA (ssDNA) is a parvovirus. Cellular RNA polymerase synthesizes mRNA complementary to the viral ssDNA, and cellular ribosomes can then translate viral proteins.

Cells do not have ssDNA, so synthesizing ssDNA viruses requires a unique form of DNA replication. The ssDNA of parvovirus folds back on itself to form a hairpin loop that acts as double-stranded DNA, which can be replicated by cellular DNA polymerase. However, when DNA polymerase replicates this "false" dsDNA, the newly replicated strand of DNA is released as ssDNA. This single-stranded DNA is packaged into a viral capsid.

Synthesis of RNA Viruses of Animals

As previously noted, cells don't have RNA genes, so it follows that synthesis of RNA viruses must differ significantly from cellular processes and from the replication of DNA viruses as well. There are four types of RNA viruses: positive-sense, single-stranded RNA (designated +ssRNA); retroviruses (a kind of +ssRNA virus); negative-sense, single-stranded RNA (−ssRNA); and double-stranded RNA (dsRNA). Scientists designate viral RNA that can act as mRNA as positive-sense RNA; that is, ribosomes can translate polypeptides directly from +ssRNA. Ribosomes cannot directly translate negative-sense RNA. A complementary copy of negative-sense RNA—a positive-sense molecule—must first be transcribed before a ribosome can translate the message carried by −ssRNA.

The synthesis processes for the four types of RNA viruses are varied and rather complex. We start with the synthesis of +ssRNA viruses.

Positive-Sense ssRNA Viruses Single-stranded viral RNA that can act directly as mRNA is called **positive-sense single-stranded RNA (+ssRNA)**. Ribosomes can translate polypeptides using the codons of such RNA. An example of a +ssRNA virus is poliovirus. In many +ssRNA viruses, a virally coded RNA polymerase transcribes complementary negative-sense, single-stranded RNA (−ssRNA) from the +ssRNA genome; −ssRNA then serves as the template for the transcription of multiple +ssRNA genomes (FIGURE 13.13a). Such transcription of RNA from RNA is unique to viruses; no cell transcribes RNA from RNA.

Retroviruses Unlike other +ssRNA viruses, the +ssRNA viruses called **retroviruses** do not use their genome as mRNA. Instead, reverse transcriptase carried within the retrovirus's capsid transcribes a dsDNA intermediary from the +ssRNA genome. This DNA intermediary then serves as the template for the synthesis of additional +ssRNA molecules, which act both as mRNA for protein synthesis and as genomes for new virions (see Chapter 25, Figures 25.18 and 25.19). Human immunodeficiency virus (HIV) is a prominent retrovirus.

Negative-Sense ssRNA Viruses Other single-stranded RNA virions are negative sense (−ssRNA). Their **negative-sense single-stranded RNA** genomes must overcome a unique problem: In order to synthesize a protein, a cell's ribosome can use only mRNA, which is positive sense (i.e., +ssRNA). Ribosomes don't recognize −ssRNA. The virus overcomes this problem by carrying within its capsid an enzyme, *RNA-dependent RNA transcriptase*, which is released into the host cell's cytoplasm during uncoating and then transcribes +ssRNA molecules from the virus's −ssRNA genome. Translation of proteins can then occur as usual. The newly transcribed +ssRNA also serves as a template for transcription of additional copies of −ssRNA (FIGURE 13.13b). Diseases caused by −ssRNA viruses include rabies and flu.

dsRNA Viruses Viruses that have double-stranded RNA (dsRNA) use yet another method of synthesis. The positive strand of the molecule serves as mRNA for the translation of proteins, one of which is an RNA polymerase that transcribes dsRNA. Each strand of the dsRNA acts as a template for transcription of its opposite, which is reminiscent of DNA replication in cells (FIGURE 13.13c). Double-stranded RNA rotaviruses cause most cases of diarrhea in infants.

TABLE 13.3 summarizes the various strategies by which animal viruses are synthesized.

Assembly and Release of Animal Viruses

As with bacteriophages, once the components of animal viruses are synthesized they assemble into virions that are then released from the host cell. Most DNA viruses assemble in and are released from the nucleus into the cytosol, whereas most RNA viruses develop solely in the cytoplasm. The number of viruses produced and released depends on both the type of virus and the size and initial health of the host cell.

Generally, replication of animal viruses takes more time than replication of bacteriophages. Herpesviruses, for example, require almost 24 hours to replicate, as compared to 25 minutes for bacteriophage T4.

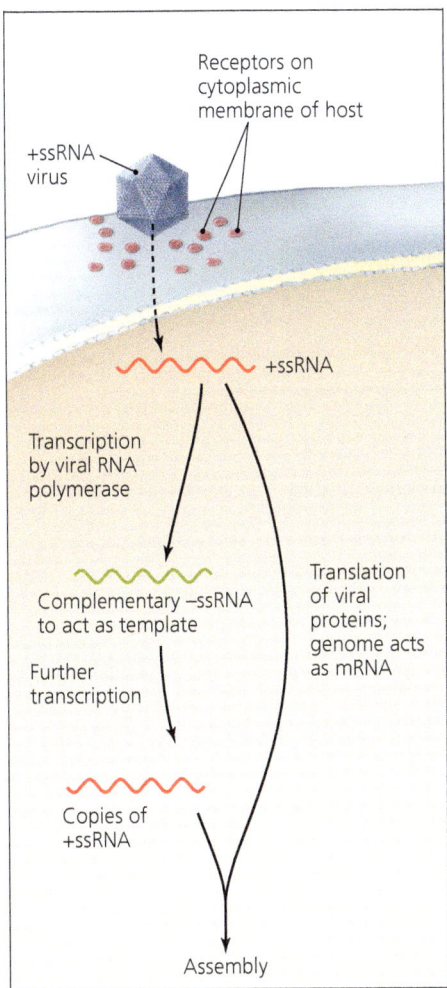
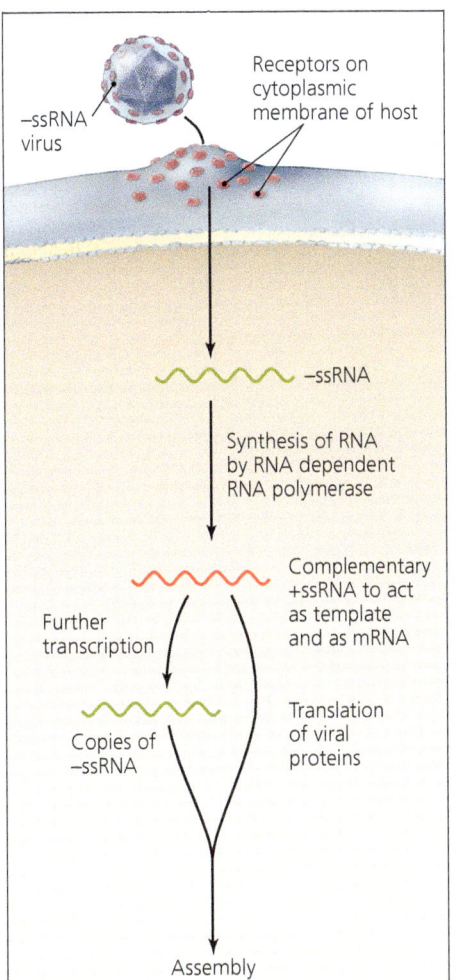
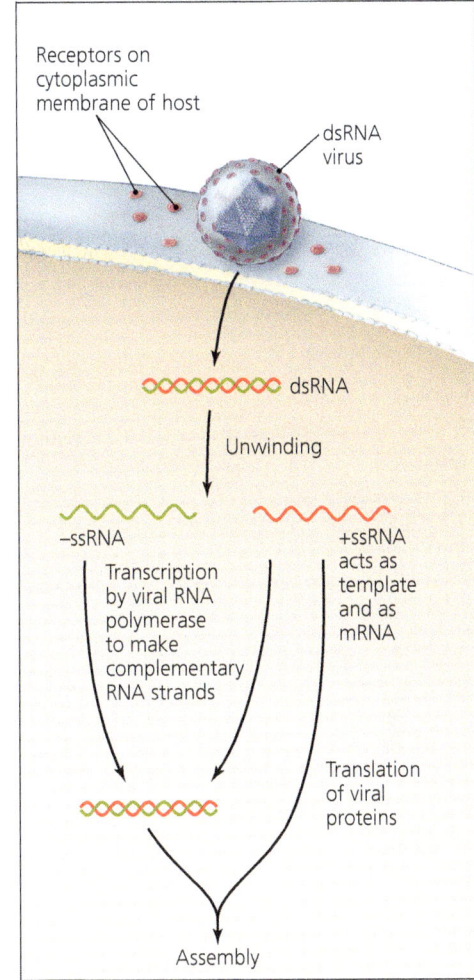

(a) Positive-sense ssRNA virus (b) Negative-sense ssRNA virus (c) Double-stranded RNA virus

▲ **FIGURE 13.13** **Synthesis of proteins and genomes in animal RNA viruses.** **(a)** Positive-sense ssRNA virus, in which +ssRNA acts as mRNA and −ssRNA is synthesized for a viral genome template. **(b)** Negative-sense ssRNA virus: transcription forms +ssRNA to serve both as mRNA and as template for viral −ssRNA genome. **(c)** dsRNA virus genome unwinds so that the positive-sense strand serves as mRNA and each strand serves as a template for its complement.

TABLE 13.3 Synthesis Strategies of Animal Viruses

Genome	How Is mRNA Synthesized?	What Molecule is the Template for Genome Replication?
dsDNA	By cellular RNA polymerase (in nucleus) or by viral RNA polymerase in cytoplasm	Each strand of DNA serves as template for its complement (except for certain viruses such as hepatitis B virus, which synthesizes RNA to act as the template for new DNA)
ssDNA	By cellular RNA polymerase (in nucleus)	Complementary strand of DNA is synthesized to act as template for synthesis in the nucleus
+ssRNA	Genome acts as mRNA	−ssRNA complementary to the genome is synthesized by viral RNA polymerase in the cytoplasm to act as template
+ssRNA (*Retroviridae*)	DNA is synthesized from viral RNA by viral reverse transcriptase carried in capsid; mRNA is transcribed from DNA by cellular RNA polymerase	DNA (synthesized from viral RNA by viral reverse transcriptase), which is integrated into cellular DNA as a provirus, acts as template
−ssRNA	By viral, RNA-dependent, RNA transcriptase in cytoplasm	+RNA (mRNA) complementary to the genome
dsRNA	Positive strand of genome acts as mRNA	Each strand of genome acts as template for synthesis of its complement by viral RNA polymerase

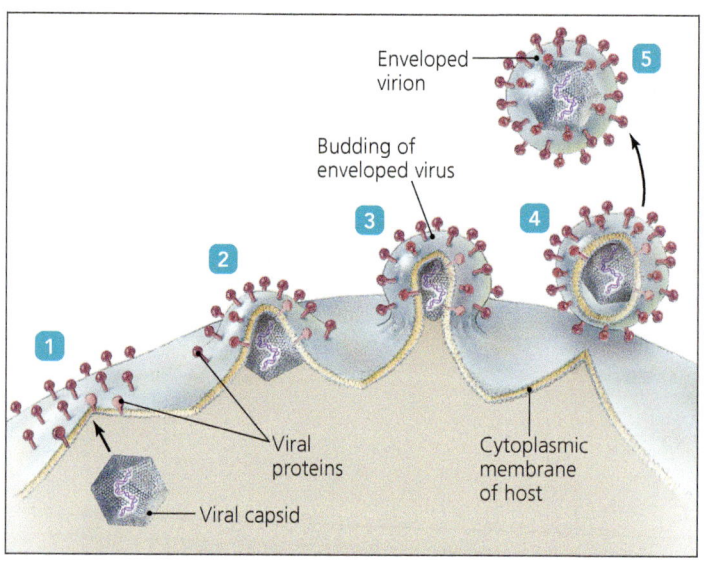

▲ FIGURE 13.14 **The process of budding in enveloped viruses.**
What term describes a nonenveloped virus?

Enveloped animal viruses are released via a process called **budding** (FIGURE 13.14). As virions are assembled, they are extruded through one of the cell's membranes—the nuclear, endoplasmic reticulum, or cytoplasmic membrane. Each enveloped virion acquires a portion of a cellular membrane, into which viral proteins were inserted during the synthesis stage. This membrane becomes the viral envelope.

Because the animal cell is not quickly lysed, as occurs in bacteriophage replication, budding allows an infected cell to remain alive for some time. Infections with enveloped viruses in which host cells shed viruses slowly and relatively steadily are called *persistent infections*; a curve showing virus abundance over time during a persistent infection lacks the burst of new virions seen in lytic replication cycles (FIGURE 13.15; compare to Figure 13.9).

Naked animal viruses may be released in one of two ways: Either they may be extruded from the cell by exocytosis, in a manner similar to budding but without the acquisition of an envelope, or they may cause lysis and death of the cell, reminiscent of bacteriophage release.

> Play *Viral Replication: Animal Viruses* @ Mastering Microbiology

Because viral replication uses cellular structures and pathways involved in the growth and maintenance of healthy cells, any strategy for the treatment of viral diseases that involves disrupting viral replication may disrupt normal cellular processes as well. This is one reason it is difficult to treat viral diseases. (The modes of action of some available antiviral drugs are discussed in Chapter 10; Chapters 15 and 16 discuss the body's naturally produced antiviral chemicals—interferons and antibodies.)

Latency of Animal Viruses

Some animal viruses, including AIDS, chicken pox, hepatitis B, and herpes viruses, may remain dormant in cells in a process known as **latency**; the viruses involved in latency are called **latent viruses** or **proviruses**. Latency may be prolonged for years with no viral activity, signs, or symptoms. Though latency is similar to lysogeny as seen with bacteriophages, there are differences. Typically, when latent animal viruses are incorporated into host DNA, the condition is permanent; induction does not occur. Thus, an incorporated, latent provirus becomes a permanent, physical part of the host's chromosome, and all descendants of the infected cell will carry the provirus.

Given that RNA cannot be incorporated directly into a chromosome molecule, how does the ssRNA of HIV become a provirus incorporated into the DNA of its host cell? HIV can become a permanent part of a host's chromosome because it, like all retroviruses, carries reverse transcriptase, which transcribes the genetic information of the +RNA molecule to a DNA molecule—which can then become incorporated into a host cell's genome. **TABLE 13.4** compares the features of the replication of bacteriophages and animal viruses. Next we turn our attention to the part viruses can play in cancer, beginning with a brief consideration of the terminology needed to understand the basic nature of cancer.

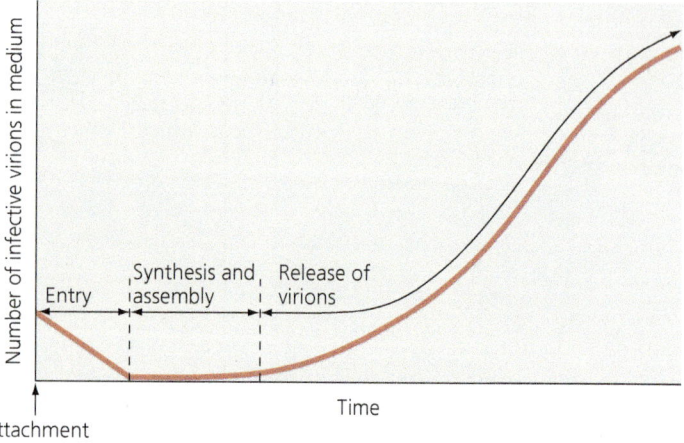

▲ FIGURE 13.15 **Pattern of virion abundance in persistent infections.** A generalized curve of virion abundance for persistent infections by budding enveloped viruses. Because the curve does not represent any actual infection, units for the graph's axes are omitted.

MICRO CHECK

9. Name a structural difference between animal viruses and bacteriophages.
10. Which part of animal viruses is involved in membrane fusion with a host cell?
11. How does the replication of animal RNA viruses differ from that of animal DNA viruses?
12. How is latency of animal viruses different from lysogeny of phages?
13. How do negative-sense single-stranded RNA (−ssRNA) viruses generate mRNA?

TELL ME WHY

Why are lysogenic and latent viral infections generally longer lasting than lytic infections?

TABLE 13.4   A Comparison of Bacteriophage and Animal Virus Replication

	Bacteriophage	Animal Virus
Attachment	Proteins on tails attach to proteins on cell wall	Spikes, capsids, or envelope proteins attach to proteins or glycoproteins on cell membrane
Penetration	Genome is injected into cell or diffuses into cell	Capsid enters cell by direct penetration, fusion, or endocytosis
Uncoating	None	Removal of capsid by cell enzymes
Site of Synthesis	In cytoplasm	RNA viruses: in cytoplasm; most DNA viruses: in nucleus
Site of Assembly	In cytoplasm	RNA viruses: in cytoplasm; most DNA viruses: in nucleus
Mechanism of Release	Lysis	Naked virions: exocytosis or lysis; enveloped virions: budding
Nature of Chronic Infection	Lysogeny, always incorporated into host chromosome, may leave host chromosome	Latency, with or without incorporation into host DNA; incorporation is permanent

The Role of Viruses in Cancer

LEARNING OUTCOMES

13.14 Define the terms *neoplasia*, *tumor*, *benign*, *malignant*, *cancer*, and *metastasis*.

13.15 Explain in simple terms how a cell may become cancerous, with special reference to the role of viruses.

Under normal conditions, the division of cells in a mature multicellular animal is under strict genetic control; that is, the animal's genes dictate that some types of cells no longer divide at all and that those that can divide are prevented from unlimited division. In this genetic control, either genes for cell division are "turned off," or genes that inhibit division are "turned on," or some combination of both these genetic events occurs. However, if something upsets the genetic control, cells begin to divide uncontrollably. This phenomenon of uncontrolled cell division in a multicellular animal is called **neoplasia**[2] (nē-ō-plā'zē-ă). Cells undergoing neoplasia are said to be *neoplastic*, and a mass of neoplastic cells is a **tumor**.

Some tumors are **benign tumors**; that is, they remain in one place and are not generally harmful, although occasionally such noninvasive tumors are painful and rob adjacent normal cells of space and nutrients. Other tumors are **malignant tumors**, invading neighboring tissues and even traveling throughout the body to invade other organs and tissues to produce new tumors—a process called **metastasis** (mĕ-tas'tă-sis). Malignant tumors are also

EMERGING DISEASE CASE STUDY

Chikungunya

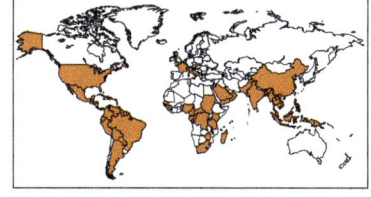

An old man arrived at the doctor's office in Ravenna, Italy, with a combination of signs and symptoms the physician had never heard of: a widespread, itchy rash covering both arms and trunk; difficulty in breathing; high fever; nausea; and extreme joint pain. Chikungunya (chik-en-gun'ya) had arrived in Europe. Though scientists had known of chikungunya virus, which is related to equine encephalitis viruses, for over 50 years, most considered the tropical disease benign—a limited, mild irritation, not a catastrophe. Therefore, few researchers studied chikungunya virus or its disease. Now, they know better.

Crystal of chikungunya viruses.

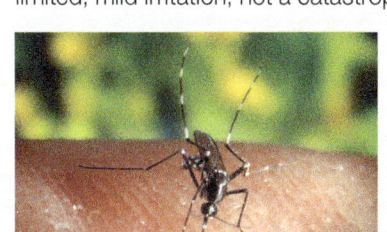

Over the past decade, chikungunya virus has spread throughout the nations of the Indian Ocean and across Africa. In 2006, officials on the French-owned island of La Réunion in the Indian Ocean reported 47,000 cases of chikungunya in a single week! That same year chikungunya reemerged in India for the first time in four decades with more than 1.5 million reported cases, and in 2010 it emerged in China. Why?

Aedes albopictus (Asian tiger mosquito), one species of *Aedes* that carries the virus, has moved into temperate climates, including Europe and the United States, as the climate has warmed. With the mosquito comes the possibility of viral proliferation—the insects have spread the tropical disease as far north in Europe as France.

And our Italian patient? His crippling pain lasted 10 days, but he survived. Now that he knows about mosquito-borne chikungunya, he insists that his family and friends use mosquito repellent liberally. Officials in the rest of Europe and in the United States join in his concern: With the coming of *Ae. albopictus*, is incurable chikungunya far behind?

1. Besides using repellents, how can people protect themselves from mosquito-borne pathogens?
2. Why is *Aedes* commonly known as the tiger mosquito?
3. Why aren't antibiotics such as penicillin, erythromycin, and ciprofloxacin effective in preventing and treating chikungunya?

---

[2]From Greek *neo*, meaning "new," and *plassein*, meaning "to mold."

called **cancers**. Cancers rob normal cells of space and nutrients and cause pain; in some kinds of cancer, malignant cells derange the function of the affected tissues, until eventually the body can no longer withstand the loss of normal function and dies.

Several theories have been proposed to explain the role viruses play in the development of cancers. These theories revolve around the presence of *proto-oncogenes* (prō-tō-ong'kō-jēnz)—genes that promote cell growth and division. As long as proto-oncogenes are appropriately expressed under controlled conditions, no cancer results. However, activity of oncogenes (their name when they are active) or inactivation of oncogene repressors can cause cancer to develop. In most cases, several genetic changes must occur before cancer develops. Put another way, "multiple hits" to the genome must occur for cancer to result (FIGURE 13.16).

A variety of environmental factors contribute to the inhibition of oncogene repressors and the activation of oncogenes. Ultraviolet light, radiation, certain chemicals called *carcinogens* (kar-si'nō-jenz), and viruses have all been implicated in the development of cancer.

Viruses may contribute to the development of 20% to 25% of human cancers, and they do so in several ways. Some viruses carry copies of oncogenes as part of their genomes, other viruses insert near to and promote oncogenes already present in the host, and still other viruses interfere with normal tumor repression when they insert (as latent proviruses) into repressor genes.

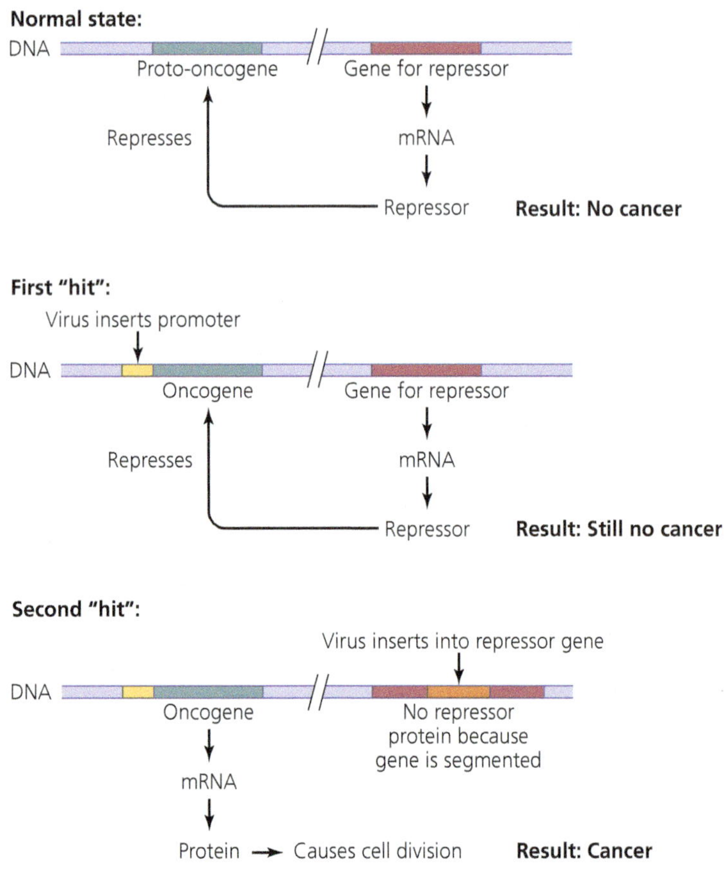

That viruses lead to some animal cancers is well established. In the first decade of the 1900s, virologist F. Peyton Rous (1879–1970) proved that viruses induce cancer in chickens. Though several DNA and RNA viruses are associated with human cancers, proving that viruses cause human cancers has been difficult. Among the human cancers proven to be virally induced are Burkitt's lymphoma, Hodgkin's disease, Kaposi's sarcoma, and cervical cancer. DNA viruses in the families *Adenoviridae*, *Herpesviridae*, *Hepadnaviridae*, *Papillomaviridae*, and *Polyomaviridae* and two RNA viruses in the family *Retroviridae* lead to these and other human cancers. (Chapters 24 and 25 discuss diseases caused by DNA and RNA viruses.)

TELL ME WHY

Why are DNA viruses more likely to cause neoplasias than are RNA viruses?

Culturing Viruses in the Laboratory

LEARNING OUTCOMES

13.16 Describe some ethical and practical difficulties to overcome in culturing viruses.

13.17 Describe three types of media used for culturing viruses.

Scientists must culture viruses in order to conduct research and develop vaccines and treatments. But because viruses cannot metabolize or replicate by themselves, they cannot be grown in standard microbiological broths or on agar plates. Instead, viruses must be cultured inside suitable host cells, a requirement that complicates the detection, identification, and characterization of viruses. Virologists have developed three types of media for culturing viruses: media consisting of mature organisms (bacteria, plants, or animals), embryonated (fertilized) eggs, and cell cultures. We begin by considering the culture of viruses in organisms.

Culturing Viruses in Mature Organisms

LEARNING OUTCOMES

13.18 Explain the use of a plaque assay in culturing viruses in bacteria.

13.19 List three problems with growing viruses in animals.

In the following sections, we consider the use of bacterial cells as a virus culture medium before considering the issues involved in growing viruses in living animals.

Culturing Viruses in Bacteria

Most of our knowledge of viral replication has been derived from research on bacteriophages, which are relatively easy to culture because some bacteria are easily grown and maintained. Phages can be grown in bacteria maintained either in liquid cultures or on agar plates. In the latter case, bacteria and phages are mixed with warm (liquid) nutrient agar and poured in a thin layer across

▲ **FIGURE 13.16 The oncogene theory of the induction of cancer in humans.** The theory suggests that more than one "hit" to the DNA (i.e., any change or mutation), whether caused by a virus (as shown here) or various physical or chemical agents, is required to induce cancer.

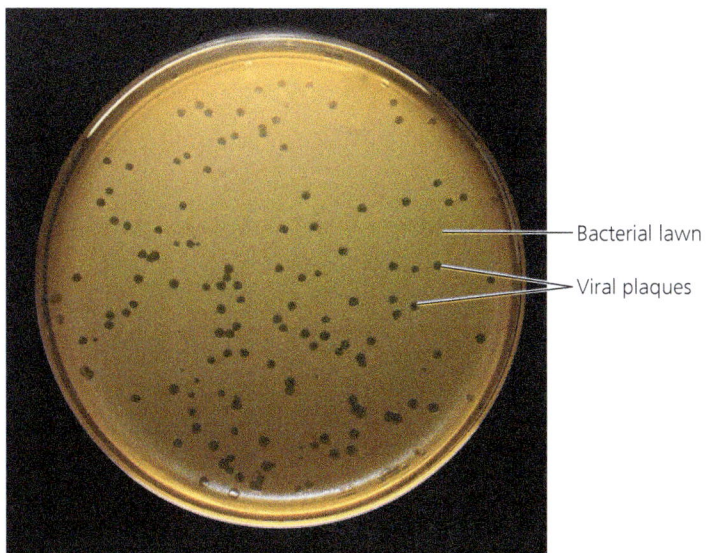

▲ FIGURE 13.17 Viral plaques in a lawn of bacterial growth on the surface of an agar plate. *What is the cause of viral plaques?*

Figure 13.17 Each plaque is an area in a bacterial lawn where bacteria have succumbed to phage infections.

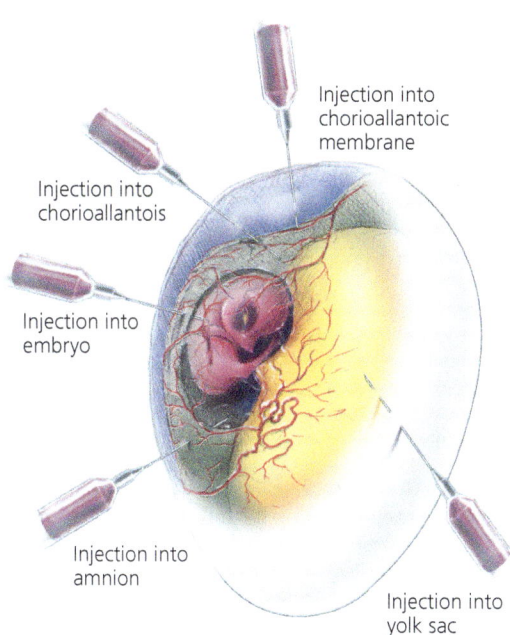

▲ FIGURE 13.18 Inoculation sites for the culture of viruses in embryonated chicken eggs. *Why are eggs often used to grow animal viruses?*

Figure 13.18 Eggs are large, sterile, self-sufficient cells that contain a number of different sites suitable for viral replication.

the surface of an agar plate. During incubation, bacteria infected by phages lyse and release new phages that infect nearby bacteria, while uninfected bacteria grow and reproduce normally. After incubation, the appearance of the plate includes a uniform bacterial lawn interrupted by clear zones called **plaques**, which are areas where phages have lysed the bacteria (FIGURE 13.17). Such plates enable virologists to estimate phage numbers via a technique called **plaque assay**, in which they assume that each plaque corresponds to a single phage spreading and multiplying through the original bacterium–virus mixture.

Culturing Viruses in Plants and Animals

Plant and animal viruses can be grown in laboratory plants and animals. Recall that the first discovery and isolation of a virus was the discovery of tobacco mosaic virus in tobacco plants. Rats, mice, guinea pigs, rabbits, pigs, and primates have been used to culture and study animal viruses. However, maintaining laboratory animals can be difficult and expensive, and this practice raises ethical issues for some people. Growing viruses that infect only humans raises additional ethical complications. Scientists have developed alternative ways of culturing animal and human viruses using fertilized chicken eggs or cell cultures.

Culturing Viruses in Embryonated Chicken Eggs

Chicken eggs are a useful culture medium for viruses because they are inexpensive, are among the largest of cells, are free of contaminating microbes, and contain a nourishing yolk (which makes them self-sufficient). Most suitable for culturing viruses are chicken eggs that have been fertilized and thus contain a developing embryo. Embryonic tissues (called *membranes*, which should not be confused with cellular membranes) provide ideal inoculation sites for growing viruses (FIGURE 13.18). Researchers inject samples of virus into embryonated eggs at the sites that are best suited for the particular virus's replication. Vaccines against some viruses can also be prepared in eggs. You may have been asked whether you are allergic to eggs before you received such a vaccine; the reason is that egg protein may remain as a contaminant in the vaccine.

Culturing Viruses in Cell (Tissue) Culture

LEARNING OUTCOME

13.20 Compare and contrast diploid cell culture and continuous cell culture.

Viruses can also be grown in **cell culture**, which consists of cells isolated from an organism and grown on the surface of a medium or in broth (FIGURE 13.19). Such cultures became practical when antibiotics provided a way to limit the growth of contaminating bacteria. Cell culture can be less expensive than maintaining research animals, plants, or eggs, and it avoids some of the moral problems associated with experiments performed on animals and humans. Cell cultures are sometimes called *tissue cultures*, but the term *cell culture* is more accurate because only a single type of cell is used in the culture. (By definition, a tissue is composed of at least two kinds of cells.)

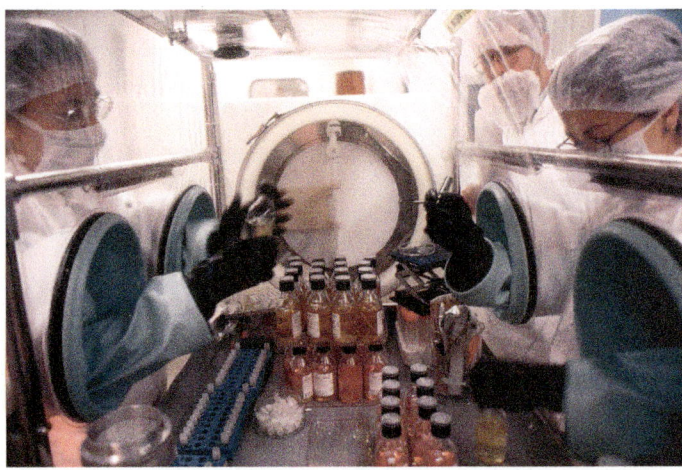

▲ **FIGURE 13.19 An example of cell culture.** The vials contain cells being cultured inside an isolation chamber. The cells, which have been infected with viruses, will produce millions of new viruses to be used as vectors in gene therapy.

Cell cultures are of two types. The first type, **diploid cell culture**, is created from embryonic animal, plant, or human cells that have been isolated and provided appropriate growth conditions. The cells in diploid cell culture generally last no more than about 100 generations (cell divisions) before they die.

The second type of culture, **continuous cell culture**, is longer lasting because it is derived from tumor cells. Recall that a characteristic of neoplastic cells is that they divide relentlessly, providing a never-ending supply of new cells. One of the more famous continuous cell cultures is of HeLa cells, derived from *He*nrietta *La*cks, who died of cervical cancer in 1951. Though she is dead, Mrs. Lacks's cells live on in laboratories throughout the world.

It is interesting that HeLa cells have lost some of their original characteristics. For example, they are no longer diploid, because they have lost many chromosomes. Thus, HeLa cells provide only a semistandard[3] human tissue culture medium for studies on cell metabolism, aging, and (of course) viral infection.

MICRO CHECK

14. What is a continuous cell culture?

TELL ME WHY

HIV replicates only in certain types of human cells, and one early problem in AIDS research was culturing those cells. Why are scientists now able to culture HIV?

---

[3]HeLa cells are "semistandard" because various strains have lost different chromosomes and acquired different mutations over the years. The result is that HeLa cells in one laboratory may be slightly different from HeLa cells in another laboratory.

Are Viruses Alive?

LEARNING OUTCOME

13.21 Discuss aspects of viral replication that are lifelike and nonlifelike.

Now that we have studied the characteristics and replication processes of viruses, let's ask a question: Are viruses alive?

To be able to wrestle with the answer, we must first recall four characteristics of life: growth, self-reproduction, responsiveness, and the ability to metabolize, all within structures called cells. According to these criteria, viruses lack the qualities of living things, prompting some scientists to consider them nothing more than complex pathogenic chemicals. For other scientists, however, at least four observations—viruses use sophisticated methods to invade cells, have the means of taking control of their host cells, possess genomes containing instructions for replicating themselves, and evolve over time—indicate that viruses are the ultimate parasites because they use cells to make more viruses. According to this viewpoint, viruses are the least complex living entities.

At the least, viruses are biological entities on the threshold of life—outside cells they do not appear to be alive, but within cells they direct the synthesis and assembly required to make copies of themselves.

TELL ME WHY

Why are viruses seemingly alive and yet not alive?

Other Parasitic Particles: Viroids and Prions

Viruses are not the only submicroscopic entities capable of causing disorders within cells. In this section, we will consider the characteristics of two molecular particles that infect cells: viroids and prions.

Characteristics of Viroids

LEARNING OUTCOMES

13.22 Define and describe viroids.
13.23 Compare and contrast viroids and viruses.

Viroids are extremely small (as few as 239 nucleotides long), circular pieces of ssRNA that are infectious and pathogenic in plants **(FIGURE 13.20)**. Viroids are similar to RNA viruses except that they lack capsids and viroid RNA does not code for proteins. Viroids appear to cause disease by adhering to complementary sequences of plant mRNA, forming double-stranded RNA, which a plant enzyme degrades as if it were a dsRNA virus. In the case of a viroid disease, the enzyme is destroying the plant's own mRNA, which results in the disease state. Several plant diseases, including some of coconut palm, chrysanthemum, cucumber, avocado, and potato, are caused by viroids, including the stunting shown in **FIGURE 13.21**.

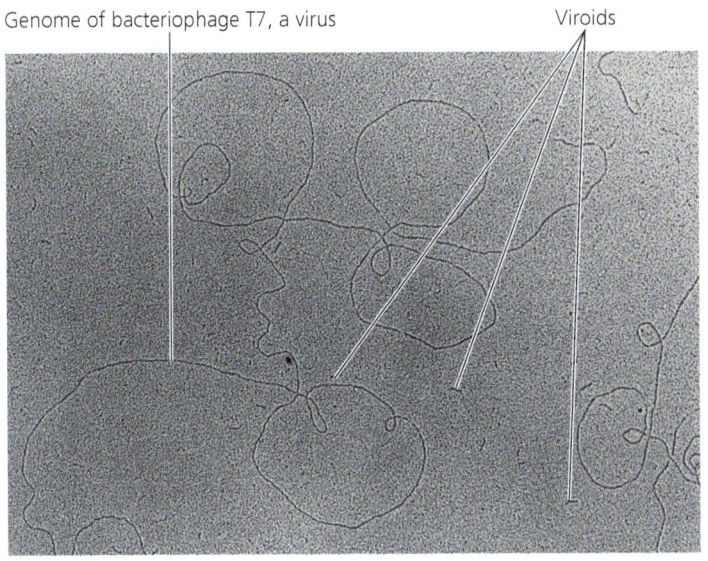

▲ FIGURE 13.20 **The RNA strand of potato viroid.** Though circular, viroids can appear linear, as here, because bonding of complementary bases on opposite sides of the circle can flatten the circle. Also shown for comparison is the longer DNA genome of bacteriophage T7. *Compare both to the size of a bacterial genome in Figure 13.2.* How are viroids similar to and different from viruses?

Figure 13.20 *Viroids are similar to certain viruses in that they are infectious and contain a single strand of RNA; they are different from viruses in that they lack a proteinaceous capsid.*

▲ FIGURE 13.21 **One effect of viroids on plants.** The potatoes at right are stunted as the result of infection with potato viroids.

Viroidlike agents—infectious, pathogenic RNA particles that lack capsids but do not infect plants—affect some fungi. (They are not called viroids because they do not infect plants.) No animal diseases are known to be caused by viroidlike molecules, though the possibility exists that infectious RNA may be responsible for some diseases in humans.

Characteristics of Prions

LEARNING OUTCOMES

13.24 Define and describe prions, including their replication process, and contrast them with viruses.
13.25 Contrast prions with viruses and viroids.
13.26 Describe methods to control and destroy prions.
13.27 List four diseases caused by prions.

In 1982, Stanley Prusiner (1942–) described infectious agents of mammals that differ from other infectious agents because they lack nucleic acid. Prusiner named such agents of disease **prions** (prē′onz), for *pro*teinaceous *in*fective particles. Before his discovery, researchers thought prion diseases were caused by what they called "slow viruses," so named because 60 years might lapse between an infection and the onset of signs and symptoms. Through experiments, Prusiner and his colleagues showed that prions are not viruses because they completely lack nucleic acid.

Some scientists resisted the concept of prions because particles that lack any nucleic acid violate the "universal" rule of protein synthesis—that proteins are translated from a molecule of mRNA. Given that prions lack RNA or DNA genes, how can they replicate themselves?

All mammals' brain cells make a cytoplasmic membrane protein called *PrP*. PrP is anchored in lipid rafts and plays a role in the normal activity of the brain, though the exact function of PrP is unknown. The amino acid sequence in PrP is such that the protein can fold into two stable tertiary structures: The normal, functional structure, called *cellular PrP (c-PrP)*, has several prominent α-helices, whereas a disease-causing form—*prion PrP (p-PrP)*—is characterized by β-pleated sheets **(FIGURE 13.22)**.

Scientists have determined that prion PrP acts like a bad influence in a crowd of teenagers, encouraging molecules of normal, cellular PrP to misbehave by refolding, thereby turning into prion PrP molecules. This process is called **templating**—prion PrP acts as a template to refold molecules of normal, cellular PrP **(FIGURE 13.23)**. As clumps of prion PrP propagate throughout the brain, neurons stop working properly and eventually die, leaving holes and a spongy appearance **(FIGURE 13.24)**. Because of this characteristic, clinicians call prion diseases of the brain *spongiform encephalopathies* (spŭn′ji-fōrm en-sef′ă-lop′ă-thēz).

Why don't prions develop in all mammals, given that all mammals have PrP? Under normal circumstances, it appears that other nearby proteins and polysaccharides in lipid rafts in the cytoplasmic membrane force PrP into the correct (cellular) shape. Further, human PrP misfolds only if it contains methionine as the 129th amino acid. Only about 40% of humans have this amino acid at this place in their cellular PrP and are thus susceptible to prion disease.

Prion disease can be inherited, sporadic, or infectious. In the inherited form, a mutation in the cellular PrP gene results in the initial formation of prion PrP. Sporadic prion disease develops following a rare random change of c-PrP in the membrane into p-PrP; note that this is a change in protein shape—the PrP gene remains

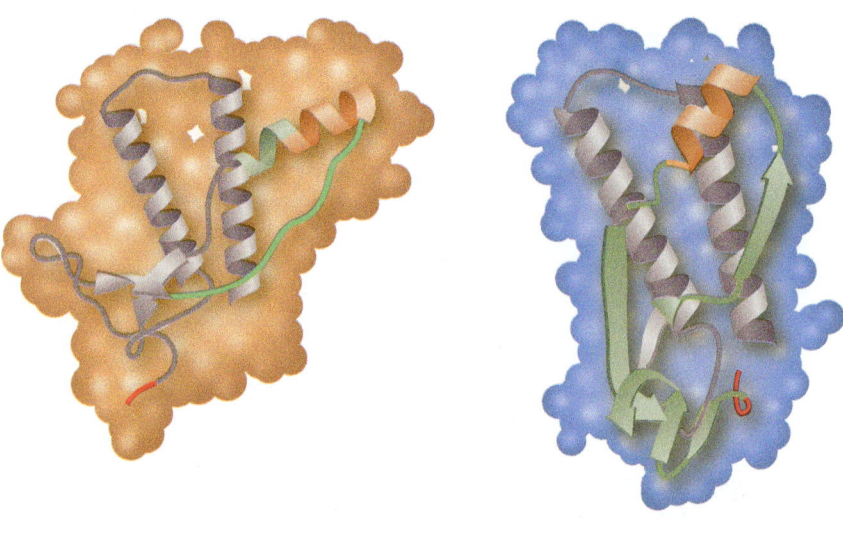

► FIGURE 13.22 The two stable, three-dimensional forms of prion protein (PrP). (a) Cellular prion protein (normal form) found in functional cells has a preponderance of alpha-helices. (b) Prion PrP (abnormal form), which has the same amino acid sequence, is folded to produce more beta-pleated sheets.

(a) Cellular PrP (c-PrP)

(b) Prion PrP (p-PrP)

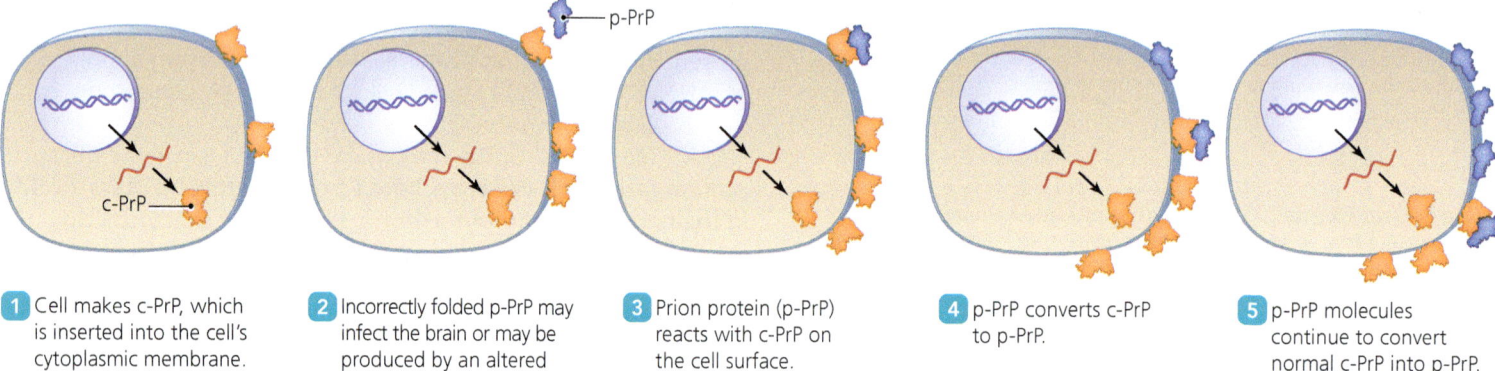

1. Cell makes c-PrP, which is inserted into the cell's cytoplasmic membrane.
2. Incorrectly folded p-PrP may infect the brain or may be produced by an altered c-PrP gene (not shown).
3. Prion protein (p-PrP) reacts with c-PrP on the cell surface.
4. p-PrP converts c-PrP to p-PrP.
5. p-PrP molecules continue to convert normal c-PrP into p-PrP.

▲ FIGURE 13.23 Templating action of prions.

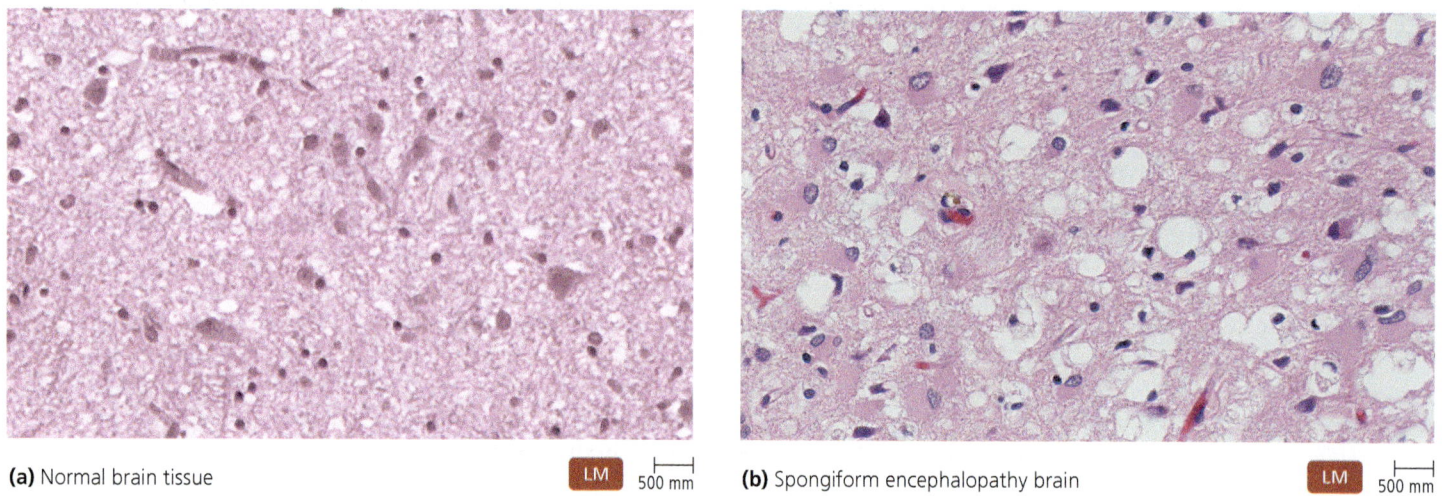

(a) Normal brain tissue

(b) Spongiform encephalopathy brain

▲ FIGURE 13.24 Brain tissue. (a) Normal human brain tissue appears nearly solid. (b) Large vacuoles and a spongy appearance are typical in prion-induced diseases. Shown here is the brain from a patient with variant Creutzfeldt-Jakob disease.

normal in these patients. Infectious prion disease follows ingestion of infected tissue, transplants of infected tissue, or contact between infected tissue and mucous membranes or skin abrasions.

Generally, prions of one host affect only related hosts, though some prions are able to "jump" species. Prions are associated with several diseases, including *bovine spongiform encephalitis (BSE)*, so-called mad cow disease, which can spread from cattle to people; *scrapie* in sheep; *kuru*, a human disease that has been eliminated; *chronic wasting disease (CWD)* in deer, moose, and elk; and *variant Creutzfeldt-Jakob disease*[4] *(vCJD)* in humans. Some scientists suggest that noninfectious prions may also be involved in Alzheimer's disease, Parkinson's disease, amyotrophic lateral sclerosis (ALS), and even some types of cancer.

Play *Prions: Diseases*
Mastering Microbiology

Cooking doesn't deactivate prions, nor do normal sterilization procedures; it takes much more severe heat treatment to destroy prions—heating to 482°C for four hours or autoclaving at 132°C in concentrated sodium hydroxide for one hour. Treatment with a 10% solution of some phenolic compounds for an hour can also be effective. The European Union recently approved the use of enzymes developed using biotechnology to remove prions from medical equipment.

Two different blood tests can detect misfolded proteins such as prion PrP in human blood samples. This is important, because healthcare providers should start treatment as soon as possible to mitigate damage to the brain. There is no standard treatment for any prion disease, though some drugs hold promise. These include the antimalarial drug quinacrine, the antipsychotic drug chlorpromazine, the antihistamine astemizole, as well as antibodies against p-PrP. These have each mitigated prion diseases in animals. Scientists must perform human trials to see whether these drugs are safe and effective in humans.

PrP proteins differ between species, and at one time it was thought unlikely that prions could move from one species to another; however, an epidemic of BSE in Great Britain in the late 1980s resulted in the spread of prions from cattle into humans who ate infected beef from cattle that had fed on nervous tissue from infected sheep. To forestall the spread of prions from one species to another, and resulting infections between and among species, most countries ban the use of animal-derived protein in animal feed. Unfortunately, this step was too late for more than 200 Europeans who developed fatal vCJD. (See **Clinical Case Study: Invasion from Within or Without?**)

MICRO CHECK

15. What are viroids?
16. How can prions be destroyed?

TELL ME WHY

Why did scientists initially resist the idea of an infectious protein?

In this chapter, we have seen that humans, animals, plants, fungi, bacteria, and archaea are susceptible to infection by acellular pathogens: viruses, viroids, and prions. **TABLE 13.5** (p. 408) summarizes the differences and similarities among these pathogenic agents and bacterial pathogens.

CLINICAL CASE STUDY

Invasion from Within or Without?

In between John's visits to the UK, his wife noticed changes in his demeanor: signs of depression, anxiety, insomnia, and social withdrawal. Later that year, John told his wife that he was having occasional numbness alternating with tingling on one side of his face, trunk, legs, and feet. John refused to go to the doctor, since these sensations returned to normal after a while. Another six months passed and his wife noticed he was becoming forgetful. Frequently, he didn't remember to pick up things from the store and even forgot that his wife had called to remind him. Later that year, he was unable to complete paperwork at his business and had difficulty performing even basic math. In England on business, he had forgotten his home phone number in the United States and couldn't remember how to spell his name for directory assistance.

His wife became very concerned with the increasingly alarming symptoms and insisted he seek medical care. All the standard blood tests came back normal. A doctor diagnosed depression, but all CT and MRI scans of his brain were normal. In a last-ditch effort, the doctor ordered a biopsy of his tonsils that revealed prion protein (p-PrP). His family was told he had about six weeks to live because there is no treatment for this disease.

1. What is the likely diagnosis?
2. John's wife wondered, "Can we catch this disease from my husband?" How would you respond?
3. Where and how was John probably infected?

---

[4]"Variant" because it is derived from BSE prions in cattle, as opposed to the regular form of CJD, which is a genetic disease.

TABLE 13.5 Comparison of Viruses, Viroids, and Prions to Bacterial Cells

	Bacteria	Viruses	Viroids	Prions
Width	200–2000 nm	10–400 nm	2 nm	5 nm
Length	200–550,000 nm	20–800 nm	40–130 nm	5 nm
Nucleic Acid	Both DNA and RNA	Either DNA or RNA, never both	RNA only	None
Protein	Present	Present	Absent	Present (PrP)
Cellular	Yes	No	No	No
Cytoplasmic Membrane	Present	Absent (though some viruses do have a membranous envelope)	Absent	Absent
Functional Ribosomes	Present	Absent	Absent	Absent
Growth	Present	Absent	Absent	Absent
Self-Replicating	Yes	No	No	No; transforms PrP protein already present in cell
Responsiveness	Present	Some bacteriophages respond to a host cell by injecting their genomes	Absent	Absent
Metabolism	Present	Absent	Absent	Absent

MICRO IN THE CLINIC FOLLOW-UP
Working Too Hard?

At the emergency room, David's white blood cell counts are normal, but he is hypotensive (low blood pressure) and bradycardic (slow heartbeat, <60 beats per minute). His respiratory rate is fast, his breathing is labored with use of accessory muscles, and oxygen levels in his blood are low. A cardiac workup reveals no abnormalities, but a chest X-ray exam indicates fluid in both lungs. The ER doctor orders labs and blood cultures to identify the cause of the infection.

Because David's condition is so severe, he is admitted to the intensive care unit (ICU). He is given frequent breathing treatments to help with his shortness of breath. Blood work results are normal except for a low platelet count and elevated white blood cell counts. Further evaluation shows that David has anti-*Hantavirus* antibodies, and David is diagnosed with *Hantavirus* cardiopulmonary syndrome.

There is no cure for *Hantavirus* infection; David remains in the ICU and is placed on mechanical ventilation for airway support. After ten days, his respiratory and cardiac condition begins to improve. His breathing tube is removed two days later. David makes a complete recovery and is able to return to his job within the next month.

1. The *Hantavirus* genome consists of three segments of negative-sense, single-stranded RNA. How are *Hantavirus* proteins synthesized in host cells?

2. *Hantavirus* is an enveloped virus. How do enveloped viruses acquire their envelopes? What role does the envelope play in viral replication?

Check your answers to Micro in the Clinic Follow-Up questions in the **Mastering Microbiology** Study Area.

> Make the connection between Chapters 11, 12, and 13. Watch MICRO MATTERS videos in the Mastering Microbiology Study Area.

CHAPTER SUMMARY

1. Viruses, viroids, and prions are **acellular** disease-causing agents that lack cell structure and cannot metabolize, grow, self-reproduce, or respond to their environment.

Characteristics of Viruses (pp. 385–390)

1. A **virus** is a tiny infectious agent with nucleic acid surrounded by proteinaceous **capsomeres** that form a coat called a **capsid**. A virus exists in an extracellular state and an intracellular state. A **virion** is a complete viral particle, including a nucleic acid and a capsid, outside a cell.

2. The genomes of viruses include either DNA or RNA. Viral genomes may be dsDNA, ssDNA, dsRNA, or ssRNA. They may exist as linear or circular and singular or multiple molecules of nucleic acid, depending on the type of virus.

3. Viruses are specific for their hosts' cells because viral attachment molecules are complementary in shape to specific receptor molecules on the host's cells.

4. All types of organisms can be infected by viruses. A **bacteriophage** (or **phage**) is a virus that infects a bacterial cell.

5. Virions can have a membranous **envelope** or be naked—that is, have no envelope.

Classification of Viruses (pp. 390–392)

1. Viruses are classified based on type of nucleic acid, presence of an envelope, shape, and size.
2. The International Committee on Taxonomy of Viruses (ICTV) has recognized viral family and genus names. With the exception of nine orders, higher taxa are not established.

Viral Replication (pp. 392–401)

1. Viruses depend on random contact with a specific host cell type for replication. Typically, a virus in a cell proceeds with a **lytic replication cycle** consisting of five stages: **attachment, entry, synthesis, assembly,** and **release.**

   Play *Viral Replication*: Overview
   @ Mastering Microbiology

2. Once the virion has attached to the host cell, the nucleic acid enters the cell. With phages, only the nucleic acid enters the host cell. With animal viruses, the entire virion often enters the cell, where the capsid is then removed in a process called **uncoating.**

   Play *Viral Replication*: Animal Viruses
   @ Mastering Microbiology

3. Within the host cell, the viral nucleic acid directs synthesis of more viruses using metabolic enzymes and ribosomes of the host cell.
4. Assembly of synthesized virions occurs in the host cell, typically as capsomeres surround replicated or transcribed nucleic acids to form new virions.
5. Virions are released from the host cell either by lysis of the host cell (seen with phages and animal viruses) or by the extrusion of enveloped virions through the host's cytoplasmic membrane (called **budding**), a process seen only with certain animal viruses. If budding continues over time, the infection is persistent. An envelope is derived from a cell membrane.

   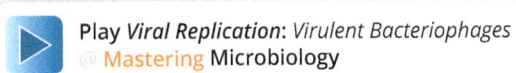
   Play *Viral Replication*: Virulent Bacteriophages
   @ Mastering Microbiology

6. **Temperate phages** (lysogenic phages) enter a bacterial cell and remain inactive in a process called **lysogeny** or **lysogenic replication cycle**. Such inactive phages are called **prophages** and are inserted into the chromosome of the cell and passed to its daughter cells. **Lysogenic conversion** results when phages carry genes that alter the phenotype of a bacterium. At some point in the generations that follow, a prophage may be excised from the chromosome in a process known as **induction**. At that point, the prophage again becomes a lytic virus.

   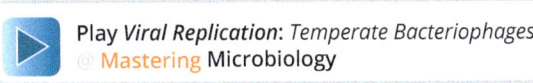
   Play *Viral Replication*: Temperate Bacteriophages
   @ Mastering Microbiology

7. With the exception of hepatitis B virus, dsDNA viruses use their DNA like cellular DNA in transcription and replication.
8. Some ssRNA viruses have positive-sense single-stranded RNA (+ssRNA), which can be directly translated by ribosomes to synthesize protein. From the **positive-sense single-stranded RNA (+ssRNA)**, complementary negative-sense single-stranded RNA (−ssRNA) is transcribed to serve as a template for more +ssRNA.
9. **Retroviruses**, such as HIV, are +ssRNA viruses that carry reverse transcriptase, which transcribes DNA from RNA. This reverse process (DNA transcribed from RNA) is reflected in the name retrovirus.
10. **Negative-sense single-stranded RNA (−ssRNA)** viruses carry an RNA-dependent RNA transcriptase for transcribing mRNA from the −ssRNA genome, and the mRNA is translated to protein. Transcription of RNA from RNA does not occur in uninfected cells.
11. In dsRNA viruses, the positive strand of RNA functions as mRNA, and each strand functions as a template for an RNA complement.
12. In **latency**, a process similar to lysogeny, an animal virus remains inactive in a cell, possibly for years, as part of a chromosome or in the cytosol. A **latent virus** is also known as a **provirus**. A provirus that has become incorporated into a host's chromosome remains there.

The Role of Viruses in Cancer (pp. 401–402)

1. **Neoplasia** is uncontrolled cellular reproduction in a multicellular animal. A mass of neoplastic cells, called a **tumor**, may be relatively harmless (**benign tumor**) or invasive (**malignant tumor**). Malignant tumors are also called **cancer**. **Metastasis** describes the spreading of malignant tumors. Environmental factors or oncogenic viruses may cause neoplasia.

Culturing Viruses in the Laboratory (pp. 402–404)

1. In the laboratory, viruses must be cultured inside mature organisms, in embryonated chicken eggs, or in cell cultures because viruses cannot metabolize or replicate alone.
2. When a mixture of bacteria and phages is grown on an agar plate, bacteria infected with phages lyse, producing clear areas called **plaques** on the bacterial lawn. A technique called **plaque assay** is used to estimate phage numbers.
3. Viruses can be grown in two types of **cell cultures**. Whereas **diploid cell cultures** last about 100 generations, **continuous cell cultures**, derived from cancer cells, last longer.

Are Viruses Alive? (p. 404)

1. Outside cells, viruses do not appear to be alive, but within cells, they exhibit lifelike qualities, such as the ability to replicate.

Other Parasitic Particles: Viroids and Prions (pp. 404–408)

1. **Viroids** are small circular pieces of RNA with no capsid that infect and cause disease in plants. Similar pathogenic RNA molecules called viroidlike agents have been found in fungi.
2. **Prions** are infectious protein particles that lack nucleic acids and replicate by inducing similar, normal proteins to misfold into new prions—a process called **templating**. Diseases caused by prions are spongiform encephalopathies, which involve fatal neurological degeneration.

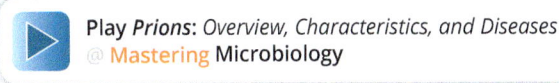
Play *Prions*: Overview, Characteristics, and Diseases
@ Mastering Microbiology

QUESTIONS FOR REVIEW

Answers to the Questions for Review (except Short Answer questions) begin on p. A-l.

Multiple Choice

1. Which of the following is *not* an acellular agent?
   a. viroid
   b. virus
   c. rickettsia
   d. prion

2. Which of the following statements is true?
   a. Viruses move toward their host cells.
   b. Viruses are capable of metabolism.
   c. Viruses lack a cytoplasmic membrane.
   d. Viruses grow in response to their environmental conditions.

3. A virus that is specific for a bacterial host is called a _____.
   a. phage
   b. prion
   c. virion
   d. viroid

4. A naked virus _____.
   a. lacks a membranous envelope
   b. has injected its DNA or RNA into a host cell
   c. is devoid of capsomeres
   d. is one that is unattached to a host cell

5. Which of the following statements is *false*?
   a. Viruses may have circular DNA.
   b. dsRNA is found in bacteria more often than in viruses.
   c. Viral DNA may be linear.
   d. Typically, viruses have DNA or RNA but not both.

6. When a eukaryotic cell is infected with an enveloped virus and sheds viruses slowly over time, this infection _____.
   a. is called a lytic infection
   b. is a prophage cycle
   c. is called a persistent infection
   d. is caused by a quiescent virus

7. Another name for a complete virus is _____.
   a. virion
   b. viroid
   c. prion
   d. capsid

8. Which of the following viruses can be latent?
   a. HIV
   b. chicken pox virus
   c. herpesviruses
   d. all of the above

9. Which of the following is *not* a criterion for specific family classification of viruses?
   a. the type of nucleic acid
   b. envelope structure
   c. capsid type present
   d. lipid composition

10. A clear zone of phage infection in a bacterial lawn is _____.
    a. a prophage
    b. a plaque
    c. naked
    d. a zone of inhibition

Matching

Match each numbered term with its following description.

1. ___ uncoating
2. ___ prophage
3. ___ retrovirus
4. ___ bacteriophage
5. ___ capsid
6. ___ envelope
7. ___ virion
8. ___ provirus
9. ___ benign tumor
10. ___ cancer

A. dormant virus in a eukaryotic cell
B. a virus that infects a bacterium
C. transcribes DNA from RNA
D. protein coat of virus
E. a membrane on the outside of a virus
F. complete viral particle
G. inactive virus within bacterial cell
H. removal of capsomeres from a virion
I. invasive neoplastic cells
J. harmless neoplastic cells

VISUALIZE IT

1. Label each step in the bacteriophage replication cycle below.

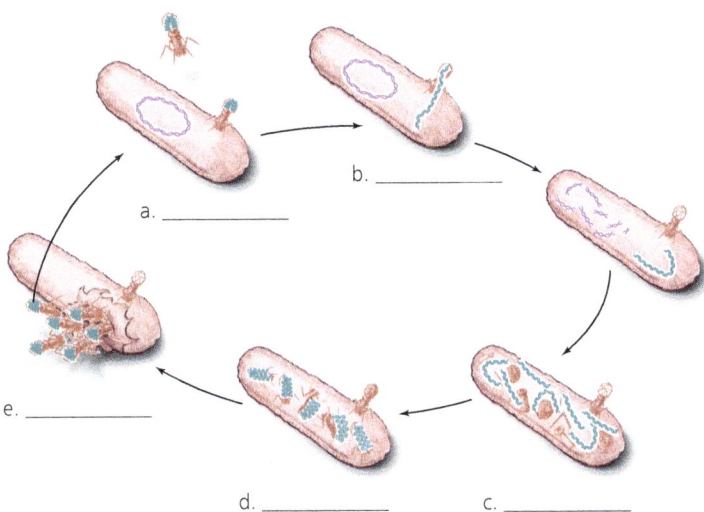

a. _____
b. _____
c. _____
d. _____
e. _____

2. Identify the viral capsid shapes.

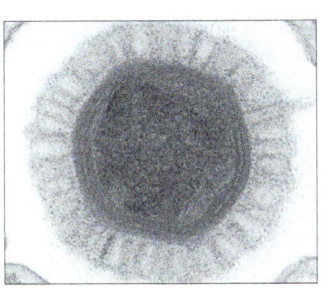

a. _____

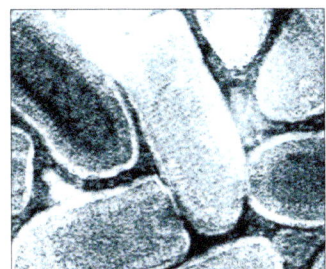

b. _____

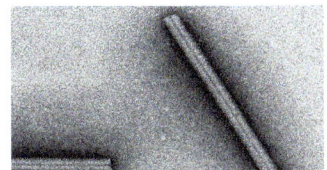

c. _____

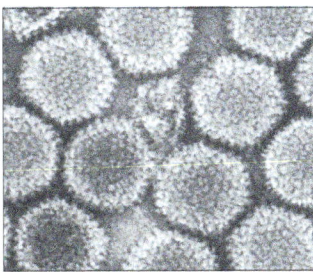

d. _____

Short Answer

1. Compare and contrast a bacterium and a virus by writing either "Present" or "Absent" for each of the following structures.

Structure	Bacterium	Virus
Cell membrane		
Functional ribosome		
Cytoplasm		
Nucleic acid		
Nuclear membrane		

2. Describe the five phases of a generalized lytic replication cycle.
3. Why is it difficult to treat viral infections?
4. Describe four different ways that viral nucleic acid can enter a host cell.
5. Contrast lysis and budding as means of release of virions from a host cell.
6. What is the difference between a virion and a virus particle?
7. How is a provirus like a prophage? How is it different?
8. Describe lysogeny.
9. How are viruses specific for their host's cells?
10. Compare and contrast diploid cell culture and continuous cell culture.

CRITICAL THINKING

1. Larger viruses usually have a double-stranded genome, whereas small viruses typically have a single-stranded genome. What reasonable explanation can you offer for this observation?
2. What are the advantages and disadvantages to bacteriophages of the lytic and lysogenic reproductive strategies?
3. How are computer viruses similar to biological viruses? Are computer viruses alive? Why or why not?
4. Compare and contrast lysogeny by a prophage and latency by a provirus.
5. An agricultural microbiologist wants to stop the spread of a viral infection of a crop. Is stopping viral attachment a viable option? Why or why not?
6. Some viral genomes, composed of single-stranded RNA, act as mRNA. What advantage might these viruses have over other kinds of viruses?
7. In some viruses, the capsomeres act enzymatically as well as structurally. What advantage might this provide the virus?
8. Why has it been difficult to develop a complete taxonomy for viruses?

9. If a colony of 1.5 billion *E. coli* cells were infected with a single phage T4 and each lytic replication cycle of the phage produced 200 new phages, how many replication cycles would it take for T4 phages to overwhelm the entire bacterial colony? (Assume for the sake of simplicity that every phage completes its replication cycle in a different cell and that the bacteria themselves do not reproduce.)

10. What differences would you expect in the replication cycles of RNA phages from those of DNA phages? (Hints: Think about the processes of transcription, translation, and replication of nucleic acids. Also, note that RNA is not normally inserted into a DNA molecule.)

11. Although many +ssRNA viruses use their genome directly as messenger RNA, +ssRNA retroviruses do not. Instead, their +RNA is transcribed into DNA by reverse transcriptase. What advantage do retroviruses gain by using reverse transcriptase?

12. If an enveloped virus were somehow released from a cell without budding, it would not have an envelope. What effect would this have on the virulence of the virus? Why?

13. A latent virus that is incorporated into a host cell's chromosome is never induced; that is, it never emerges from the host cell's chromosome to become a free virus. Given that it cannot emerge from the host cell's chromosome, can such a latent virus be considered "safe"? Why or why not?

CONCEPT MAPPING

Using the terms provided, fill in the concept map that describes the replication of animal viruses. You can also complete this and other concept maps online by going to the Mastering Microbiology Study Area.

+ssRNA
+ssRNA retrovirus
−ssRNA
Attachment
Budding
dsDNA
dsRNA
Endocytosis
Exocytosis
Fusion
Host cell
New virions
Release
ssDNA
Synthesis
Uncoating
Viral nucleic acid
Viral proteins

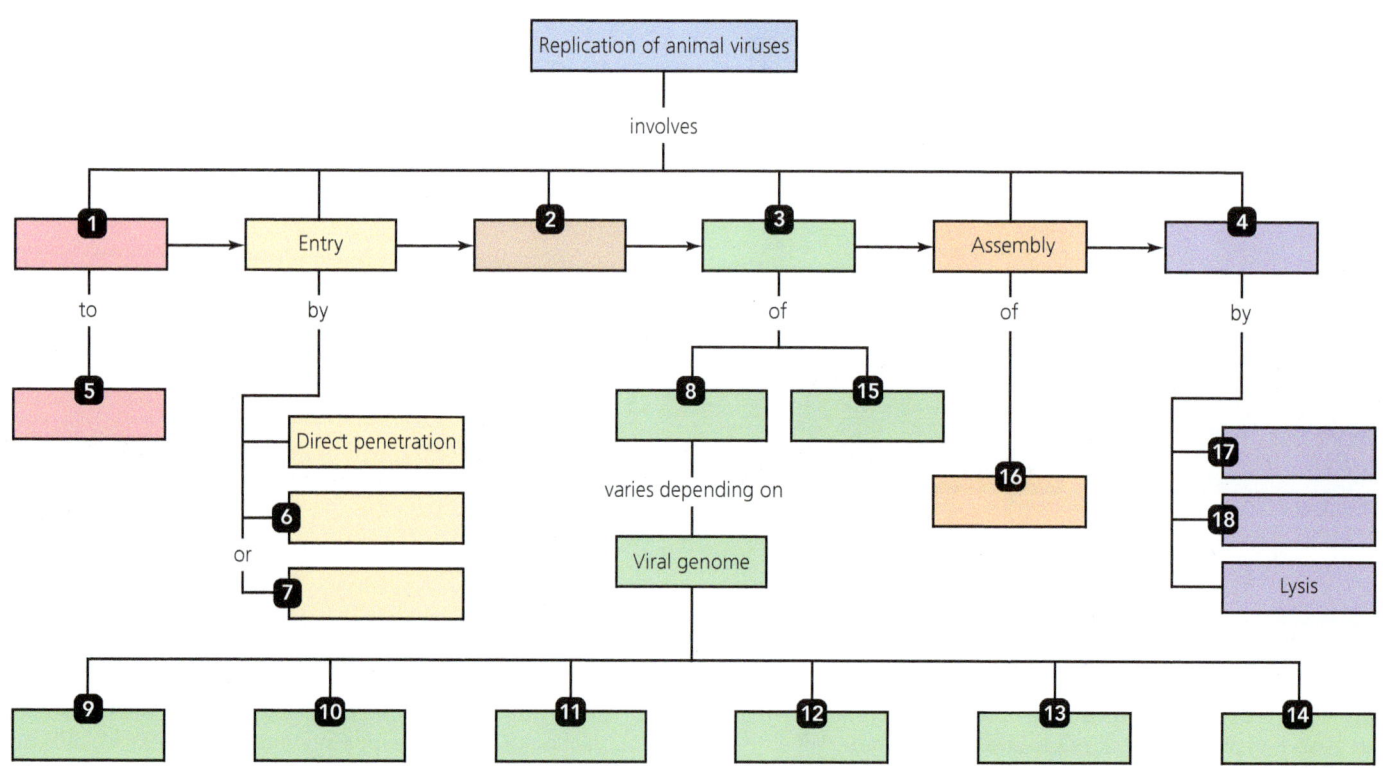

14 Infection, Infectious Diseases, and Epidemiology

Before You Begin

1. What cellular structure protects bacteria from drying out?
2. What type of microbe has lipid A?
3. Which general group of microbes cannot be isolated in pure culture?

Plasmodium, the cause of malaria, is endemic in many tropical countries; here shown replicating inside cells.

MICRO IN THE CLINIC

Just Travel Fatigue?

ELLEN RECENTLY RETURNED from one of the best trips of her life. For three weeks, she traveled through India as part of group working to bring clean water and sanitation to rural villages. It was incredibly rewarding—living and working with the villagers while helping them get access to life-changing resources.

She's been home for two months and hasn't been feeling well. She has a little bit of a sore throat, she's tired, and she has a low fever. She assumes that it's just tiredness from traveling and perhaps she caught a cold on the long plane journey home and hasn't fully recovered. After several more days, though, she isn't feeling much better, so she makes an appointment to see her doctor.

Dr. Dalton does a basic physical exam and instructs her to drink plenty of water and rest as much as possible. Ellen should start feeling better in a couple of days, the doctor tells her.

A week later, Ellen still has a sore throat, and she feels more tired than ever. She also feels extremely weak and dizzy. Worried that she has something much worse than a simple sore throat, Ellen makes an appointment to see Dr. Dalton the next day.

1. Is Ellen's illness travel fatigue or the result of something she caught on the plane?
2. Could her recent trip have anything to do with her symptoms?

Turn to the end of the chapter (p. 441) to find out.

SOLVE THE PROBLEM

Microbes in the Produce Aisle

In October, discussions in a small town in Louisiana centered on dramatic world events, especially terrorism. However, an outbreak of pneumonia in 33 people from their community within a single week changed the talk in the coffee shops and community center. Epidemiologists with the state Department of Health began examining the patients for the most common causes of pneumonia in adults: the bacterial species *Streptococcus pneumoniae*, *Mycoplasma pneumoniae*, and *Haemophilus influenzae* and influenzaviruses and adenoviruses. The patients were negative for all of these pathogens, and officials contacted the Centers for Disease Control and Prevention (CDC). Basic epidemiological strategy is to look for factors common to those infected but not common in nonvictims that match the victims in age, geographical location, and similar factors. One factor the scientists noticed was that patients were more likely to have shopped at a particular grocery store before becoming hospitalized. This is puzzling, because while one might expect foodborne infections to be related to a grocery store, what would be a link for pneumonia?

Imagine you are on the field team of the CDC sent to investigate this outbreak.

1. What first steps would you take to solve this mystery?
2. How would you isolate and identify the cause of the pneumonia?

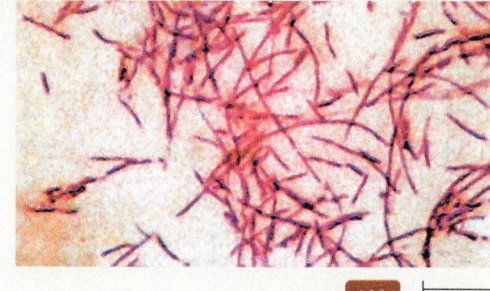

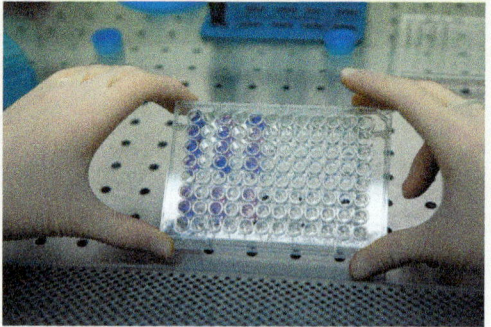

In this chapter, we examine how microbes directly affect our bodies. Bear in mind that most microorganisms are neither harmful nor expressly beneficial to humans—they live their lives, and we live ours. Relatively few microbes either directly benefit us or harm us.

We first examine the general types of relationships that microbes can have with the bodies of their hosts. Then we discuss sources of infectious diseases of humans, how microbes enter and attach to their hosts, the nature of infectious diseases, and how microbes leave hosts to become available to enter new hosts. Next, we explore the ways that infectious diseases are spread among hosts. Finally, we consider *epidemiology*, the study of the occurrence and spread of diseases within groups of humans and the methods by which we can limit the spread of pathogens within society.

Symbiotic Relationships Between Microbes and Their Hosts

Symbiosis (sim-bī-o'sis) means "living together." Each of us has symbiotic relationships with countless microorganisms, although often we are completely unaware of them. Your body's 37 trillion or so cells provide homes for trillions of bacteria, protozoa, and fungi. How many? No one knows for sure; estimates range from 100 to more than 700 trillion microbes in and on a single person; and there are even more viruses, perhaps more than 100 trillion of a single type. We begin this section by considering types of symbiotic relationships.

Types of Symbiosis

LEARNING OUTCOMES

14.1 Distinguish among four types of symbiosis.
14.2 Describe the relationships among the terms *parasite*, *host*, and *pathogen*.

Members of a symbiotic relationship are called *symbionts*. Biologists see symbiotic relationships as a continuum from cases that are beneficial to both symbionts to situations in which one symbiont lives at a damaging expense to the other. In the following sections, we examine four conditions along the continuum: mutualism, commensalism, amensalism, and parasitism.

Mutualism

In **mutualism** (mū'tū-ăl-izm), both symbionts benefit from their interaction. For example, bacteria in your colon receive a warm, moist, nutrient-rich environment in which to thrive, while you absorb vitamin precursors and other nutrients released from the bacteria. In this example, the relationship is beneficial to microbe and human alike but is not required by either. Many of the bacteria could live elsewhere, and you could get vitamins from your diet.

Some mutualistic relationships provide such important benefits that one or both of the parties cannot live without the other. For instance, termites, which cannot digest the cellulose in wood by themselves, would die without a mutualistic relationship with colonies of wood-digesting protozoa and bacteria living in their intestines **(FIGURE 14.1)**. The termites provide wood pulp and a home to the protozoa, while the protozoa break down the wood within the termite intestines using enzymes from the bacteria, which live on the protozoa. The microbes share with the termites the nutrients released. **Beneficial Microbes: A Bioterrorist Worm** describes another mutualistic relationship.

Commensalism

In **commensalism** (kŏ-men′săl-izm), a second type of symbiosis, one symbiont benefits without significantly affecting the other. For example, microscopic animals—hair follicle mites—live on your skin without causing measurable harm to you. However, an absolute example of commensalism is difficult to prove because the host may experience unobserved benefits.

Amensalism

Amensalism (ā-men′săl-izm) is a symbiotic relationship in which one symbiont is harmed by a second symbiont, while the second is neither harmed nor helped by the first. For example, the fungus *Penicillium* (pen-i-sil′ē-ŭm) produces penicillin, which inhibits nearby bacteria, but the bacteria have no effect on the fungus.

▲ **FIGURE 14.1 Mutualism.** Wood-eating termites in the genus *Reticulitermes* cannot digest the cellulose in wood, but the protozoan *Trichonympha*, which lives in their intestines, can digest cellulose with the help of bacteria. The three organisms (termite, protozoan, bacterium) maintain a mutualistic relationship that is crucial for the life of the termite. *What benefits accrue to each of the symbionts in this mutualistic relationship?*

Figure 14.1 *The termite gets digested cellulose from the protozoa, which get a constant supply of wood pulp to digest, so the bacteria get a food source as well.*

Parasitism

Of concern to health care professionals is a fourth type of symbiosis called **parasitism**[1] (par′ă-si-tizm). A **parasite** derives benefit from its **host** while harming it, though some hosts sustain only slight damage. In the most severe cases, a parasite kills its host,

BENEFICIAL MICROBES

A Bioterrorist Worm

The worm *Steinernema* (white) attacks insects by infecting them with lethal *Xenorhabdus* bacteria.

Bioterrorism has been defined as the deliberate release of viruses, bacteria, or other germs to cause illness or death. A nematode worm, *Steinernema*, is by that definition a bioterrorist, releasing a mutualistic bacterial symbiont, *Xenorhabdus*.

Nematodes are microscopic, unsegmented roundworms that live in soil worldwide. *Steinernema* preys on insects, including ants, termites, and the immature stages of various beetles, weevils, worms, fleas, ticks, and gnats. *Steinernema* crawls into an insect's mouth or anus and then crosses the intestinal walls to enter the insect's blood. The worm then releases its symbiotic *Xenorhabdus*.

This bacterium inactivates the insect's defensive systems; generates antibacterial compounds, which eliminate other bacteria; produces insecticidal toxins to kill the insect; and secretes digestive enzymes. The bacterial enzymes turn the insect's body into a slimy porridge of nutrients within 48 hours.

Meanwhile, *Steinernema* nematodes mature, mate, and reproduce within the insect's liquefying body. The nematodes' offspring feed on the gooey fluid until they are old enough to emerge from the insect's skeleton but not before taking up a supply of bacteria for their own future bioterrorist raids on new insect hosts.

This is a true mutualistic symbiosis: the nematode depends on the bacterium to kill and digest the insect host, while the bacterium depends on the nematode to deliver it to new hosts. Both benefit. Farmers, too, can benefit. They can use *Steinernema* and its bacterium to control insect pests without damaging their crops, animals, or people.

---

[1]From Greek *parasitos*, meaning "one who eats at the table of another."

in the process destroying its own home, which is not very beneficial for the parasite; therefore, parasites that allow their hosts to survive are more likely to spread. Similarly, hosts that tolerate a parasite are more likely to reproduce. The result over time will be *coevolution* toward commensalism or mutualism. Any parasite that causes disease is called a **pathogen** (path′ō-jen).

A variety of protozoa, fungi, and bacteria are microscopic parasites of humans. Larger parasites of humans include parasitic worms and biting arthropods,[2] including mites (skin-eating chiggers, not the commensal hair follicle mites mentioned previously), ticks, mosquitoes, fleas, and bloodsucking flies.

You should realize that microbes that are parasitic may become mutualistic, or vice versa; that is, the relationships between and among organisms can change over time. TABLE 14.1 summarizes the types of symbioses.

Microbiome of Humans

 Play *an Interactive Microbiology* video @ Mastering Microbiology for more information about human microbiota.

LEARNING OUTCOME

14.3 Describe the microbiome, including resident and transient members.

Even though many parts of your body are *axenic*[3] (ā-zen′ik) environments—that is, sites free of any microbes—other parts of your body shelter trillions of mutualistic and commensal symbionts. Each square centimeter of your skin, for example, contains millions of bacteria, and your large intestine contains perhaps more than 5000 different strains of bacteria alone, plus countless fungi, protozoa, and viruses. Microbes that colonize a body without normally causing disease constitute the body's **microbiome** (FIGURE 14.2), also sometimes called the *normal microbiota*, or *normal flora*,[4] or *indigenous microbiota*. The microbiome contains two main types of organisms: resident microbiota and transient microbiota.

TABLE 14.1 Types of Symbiotic Relationships

	Organism 1	Organism 2	Example
Mutualism	Benefits	Benefits	Bacteria in human colon
Commensalism	Benefits	Neither benefits nor is harmed	Mites in human hair follicles
Amensalism	Is harmed	Neither benefits nor is harmed	Fungus secreting an antibiotic, inhibiting nearby bacteria
Parasitism	Benefits	Is harmed	Tuberculosis bacteria in human lung

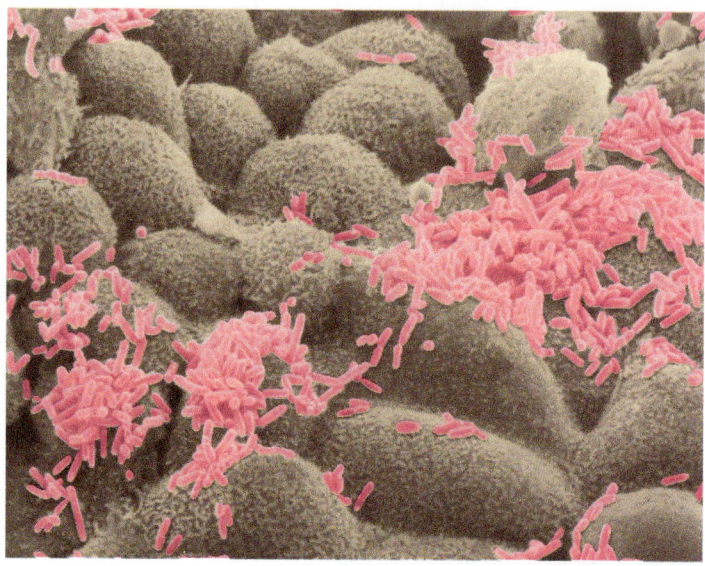

▲ **FIGURE 14.2 An example of microbiota.** Bacteria (pink) colonize the surfaces of human nasal cells covered with mucus (tan).

Resident Microbiota

Resident microbiota remain a part of the microbiome for most of a person's life. These organisms are found on the skin and on the mucous membranes of the digestive tract, upper respiratory tract, distal[5] portion of the urethra,[6] and vagina. TABLE 14.2 illustrates these environments and lists some of the resident bacteria, fungi, and protozoa. Most resident microbiota are commensal; that is, they feed on excreted cellular wastes and dead cells without causing harm to the person.

Transient Microbiota

Transient microbiota remain in the body for only a few hours, days, or months before disappearing. They are found in the same locations as resident microbiota, but transient microbes cannot persist because of competition from other microorganisms, elimination by the body's defenses, or chemical and physical changes in the body that dislodge them.

Acquisition of the Microbiome

You developed in your mother's womb without a microbiome because you had surrounded yourself with an amniotic membrane and fluid, which generally kept microorganisms at bay, and because your mother's uterus provided additional protection. Your microbiome began to develop when your protective amniotic membrane ruptured and microorganisms came in contact with you during birth. Microbes entered your mouth and nose as you passed through the birth canal, and your first breath was loaded with thousands of microbes, many of which quickly established themselves in your upper respiratory tract. Your first meals provided the progenitors of resident microbiota for

---

[2]From Greek *arthros*, meaning "jointed," and *pod*, meaning "foot."
[3]From Greek *a*, meaning "no," and *xenos*, meaning "foreigner."
[4]This usage of the term *flora*, which literally means "plants," derives from the fact that bacteria and fungi were considered plants in early taxonomic schemes. *Microbiome* is preferable because it properly applies to all microbes (whether eukaryotic, prokaryotic, or viral).

[5]*Distal* refers to the end of some structure, as opposed to *proximal*, which refers to the beginning of a structure.
[6]The tube that empties the bladder.

TABLE 14.2 Some Resident Microbiota[a]

Upper Respiratory Tract

Genera	Notes
Fusobacterium, Haemophilus, Lactobacillus, Moraxella, Staphylococcus, Streptococcus, Veillonella, Candida (fungus)	The nose is cooler than the rest of the respiratory system and has some unique microbiota. The microbiota of the trachea and bronchi are sparse compared to those of the nose and mouth. The alveoli of the lungs, which are too small to depict in this illustration, have no natural microbiota—they are axenic.

Upper Digestive Tract

Genera	Notes
Actinomyces, Bacteroides, Corynebacterium, Haemophilus, Lactobacillus, Neisseria, Staphylococcus, Treponema, Entamoeba (protozoan), *Trichomonas* (protozoan)	Microbes colonize surfaces of teeth, gingiva, lining of cheeks, and pharynx, and they are found in saliva in large numbers. Dozens of species have never been identified.

Lower Digestive Tract

Genera	Notes
Bacteroides, Bifidobacterium, Clostridium, Enterococcus, Escherichia, Fusobacterium, Lactobacillus, Proteus, Shigella, Candida (fungus), *Entamoeba* (protozoan), *Trichomonas* (protozoan)	The bacteria are mostly strict anaerobes, though some facultative anaerobes are also resident.

Female Urinary and Reproductive Systems

Genera	Notes
Bacteroides, Clostridium, Lactobacillus, Staphylococcus, Streptococcus, Candida (fungus), *Trichomonas* (protozoan)	Microbiota change as acidity in the vagina changes during menstrual cycle. The flow of urine prevents extensive colonization of the urinary bladder or urethra.

Male Urinary and Reproductive Systems

Genera	Notes
Bacteroides, Fusobacterium, Lactobacillus, Mycobacterium, Peptostreptococcus, Staphylococcus, Streptococcus	The flow of urine prevents extensive colonization of the urinary bladder or urethra.

Eyes and Skin

Genera	Notes
Skin: *Corynebacterium, Micrococcus, Propionibacterium, Staphylococcus, Candida* (fungus), *Malassezia* (fungus) Conjunctiva: *Staphylococcus*	Microbiota live on the outer, dead layers of the skin and in hair follicles and pores of glands. The deeper layers (dermis and hypodermis) are axenic. Tears wash most microbiota from the eyes, so there are few compared to the number on the skin.

[a]Genera are bacteria unless otherwise noted.

your colon, while hair follicle mites, bacteria, and other microbes transferred from the skin of both the medical staff and your parents and colonized your skin. Although you continue to add to and subtract from your transient microbiota, most of your resident microbiota was initially established during your first months of life.

How Normal Microbiota Become Opportunistic Pathogens

LEARNING OUTCOME

14.4 Describe three conditions that create opportunities for normal microbiota to cause disease.

Under ordinary circumstances, normal microbiota do not cause disease. However, these same microbes may become harmful if an opportunity to do so arises. In this case, normal microbiota (or other normally harmless transient microbes from the environment) become **opportunistic pathogens**, or *opportunists*. Conditions that create opportunities for pathogens include the following:

- *Introduction of a member of the normal microbiota into an unusual site in the body.* As we have seen, normal microbiota are present only in certain body sites, and each site has only certain species of microbiota. If a member of the normal microbiota in one site is introduced into a site it normally does not inhabit, the organism may become an opportunistic pathogen. For example, *Escherichia coli* (esh-ĕ-rik′ē-ă kō′lī) is mutualistic in the colon, but should it enter the urethra, it becomes an opportunist that produces a urinary tract infection and disease.

- *Immune suppression.* Anything that suppresses the body's immune system—including disease, malnutrition, emotional or physical stress, extremes of age (either very young or very old), the use of radiation or chemotherapy to combat cancer, or the use of immunosuppressive drugs in transplant patients—can enable opportunistic pathogens. AIDS patients often die from opportunistic infections that are typically controlled by a healthy immune system because HIV infection suppresses immune system function.

- *Changes in the normal microbiome.* Normal microbiota use nutrients, take up space, and release toxic waste products, all of which make it less likely that arriving pathogens can compete well enough to become established and produce disease. This situation is known as **microbial antagonism** or **microbial competition**. However, changes in the relative abundance of normal microbiota, for whatever reason, may allow a member of the normal microbiome to become an opportunistic pathogen and thrive. For example, when a woman must undergo long-term antimicrobial treatment for a bacterial blood infection, the antimicrobial drug may also kill normal bacterial microbiota in the vagina. In the absence of competition from bacteria, *Candida albicans* (kan'did-ă al'bi-kanz), a yeast and member of the normal vaginal microbiota, grows prolifically, producing an opportunistic vaginal yeast infection.

- *Stressful conditions.* Anything that strains a person's normal metabolism or emotional state can disrupt the normal microbiome. Such conditions include hormonal changes, unresolved emotional stress, abrupt changes in diet, or exposure to overwhelming numbers of pathogens.

Now that we have considered the relationships of the normal microbiota to their hosts, we turn our attention to the movement of microbes into new hosts by considering aspects of human diseases.

TELL ME WHY

Why is an absolute commensalism difficult to prove?

Reservoirs of Infectious Diseases of Humans

LEARNING OUTCOME

14.5 Describe three types of reservoirs of infection in humans.

Most pathogens of humans cannot survive for long in the relatively harsh conditions they encounter outside the human body. If these pathogens are to enter new people, they must survive in some site from which they can infect new hosts. Sites where pathogens are normally maintained as a source of infection are called **reservoirs of infection**. In this section, we discuss three types of reservoirs: animal reservoirs, human carriers, and nonliving reservoirs.

Animal Reservoirs

Many pathogens that normally infect either domesticated or sylvatic[7] (wild) animals can also infect humans. The more similar an animal's physiology is to human physiology, the more likely its pathogens can affect human health. Diseases that spread naturally from animal hosts to humans are called **zoonoses** (zō-ō-nō'sēz). Over 200 zoonoses have been identified throughout the world. Well-known examples include anthrax, bubonic plague, rabies, and Ebola hemorrhagic fever.

Humans may acquire zoonoses from animals via a number of routes, including various types of direct contact with animals or their wastes, by eating animals, or via bloodsucking arthropods, which act as vectors. Human infections with zoonoses are difficult to eradicate because extensive animal reservoirs are often involved. The larger the animal reservoir (i.e., the greater the number and types of infected animals) and the greater the contact between humans and the animals, the more difficult and costly it is to control the spread of the disease to humans. This is especially true when the animal reservoir consists of both sylvatic and domesticated animals. In the case of rabies, for example, the disease typically spreads from a sylvatic reservoir (often bats, foxes, or skunks) to domestic pets, from which humans may be infected. The wild animals constitute a reservoir for the rabies virus, though transmission to humans can be limited by vaccinating domestic pets.

TABLE 14.3 provides a brief view of some common zoonoses. Humans are usually dead-end hosts for zoonotic pathogens—that is, humans do not act as significant reservoirs for the reinfection of animal hosts—largely because the circumstances under which zoonoses are transmitted favor movement from animals to humans but not in the opposite direction. For example, animals do not often eat humans these days, and animals less frequently have contact with human wastes than humans have contact with animal wastes. Zoonotic diseases transmitted via the bites of bloodsucking arthropods are the most likely type to be transmitted back to animal hosts.

Human Carriers

Experience tells you that humans with active diseases are important reservoirs of infection for other humans. What may not be so obvious is that people with no obvious symptoms before or after a disease may also be infective in some cases. Further, some infected people can remain both asymptomatic and infective for years. This is true of tuberculosis, syphilis, and AIDS, for example. Whereas some of these **carriers** incubate the pathogen in their body and eventually develop the disease, others remain a continued source of infection without ever becoming sick. Presumably many such healthy carriers have defensive systems that protect them from illness. A true example of such a carrier is given in **Clinical Case Study: A Deadly Carrier**.

---

[7]From Latin *sylva*, meaning "woodland."

TABLE 14.3 Some Common Zoonoses

Disease	Causative Agent	Animal Reservoir	Mode of Transmission
Helminthic			
Tapeworm infestation	*Dipylidium caninum*	Dogs	Ingestion of larvae transmitted in dog saliva
Fasciola infestation	*Fasciola hepatica*	Sheep, cattle	Ingestion of contaminated vegetation
Protozoan			
Malaria	*Plasmodium* spp.[a]	Monkeys	Bite of *Anopheles* mosquito
Toxoplasmosis	*Toxoplasma gondii*	Cats and other animals	Ingestion of contaminated meat, inhalation of pathogen, direct contact with infected tissues
Fungal			
Ringworm	*Trichophyton* spp. *Microsporum* spp. *Epidermophyton floccosum*	Domestic animals	Direct contact
Bacterial			
Anthrax	*Bacillus anthracis*	Domestic livestock	Direct contact with infected animals, inhalation
Bubonic plague	*Yersinia pestis*	Rodents	Flea bites
Lyme disease	*Borrelia burgdorferi*	Deer	Tick bites
Salmonellosis	*Salmonella* spp.	Birds, rodents, reptiles	Ingestion of fecally contaminated water or food
Viral			
Rabies	*Lyssavirus* sp.[b]	Bats, skunks, foxes, dogs	Bite of infected animal
Hantavirus pulmonary syndrome	*Hantavirus* sp.	Deer mice	Inhalation of viruses in dried feces and urine
Yellow fever	*Flavivirus* sp.	Monkeys	Bite of *Aedes* mosquito

[a]Abbreviation for species plural; it refers to all species within a genus.
[b]Abbreviation for a single species within a genus.

CLINICAL CASE STUDY

A Deadly Carrier

In 1937, a man employed to lay water pipes was found to be the source of a severe epidemic of typhoid fever. The man, who had recovered from typhoid, was still an asymptomatic carrier of *Salmonella enterica* serotype Typhi, the bacterium that causes the disease. He habitually urinated at his job site and in the process contaminated the town's water supply with bacteria from his bladder. Over 300 cases of typhoid fever developed, and 43 people died before the man was identified as the carrier.

1. How do you think health officials were able to identify the source of this typhoid epidemic?
2. Given that antibiotics were not generally available in 1937, how could health officials end the epidemic, short of removing the man from the job site?

Nonliving Reservoirs

Inanimate things such as soil, water, and food can be **nonliving reservoirs of infection**. Soil can harbor species of *Clostridium* (klos-trid'ē-ŭm) bacteria, which cause botulism, tetanus, and other diseases. Fecally contaminated soil may serve as a reservoir for the pathogens of toxoplasmosis, hookworm infestation, or gastrointestinal bacteria and viruses. Water can be contaminated with feces and urine containing parasitic worm eggs, pathogenic protozoa, bacteria, and viruses. Meats and vegetables can also harbor pathogens.

> **MICRO CHECK**
>
> 1. What is a relationship called between two types of organisms in which one benefits while the other remains unaffected?
> 2. What is a zoonosis?

> **TELL ME WHY**
>
> Why might animal reservoirs be involved in more diseases than are human reservoirs?

The Invasion and Establishment of Microbes in Hosts: Infection

In this section, we examine events that occur when hosts are exposed to microbes from a reservoir, the sites at which microbes can gain entry into hosts, and the ways entering microbes become established in new hosts.

Exposure to Microbes: Contamination and Infection

LEARNING OUTCOME

14.6 Describe the relationship between contamination and infection.

In the context of the interaction between microbes and their hosts, **contamination** refers to the mere presence of microbes in or on the body. Some microbial contaminants reach the body in food, drink, or the air, whereas others are introduced via wounds, biting arthropods, or sexual intercourse. Several outcomes of contamination by microbes are possible. Some microbial contaminants remain where they first contacted the body without causing harm and subsequently become part of the resident microbiota; other microbial contaminants remain on the body for only a short time as part of the transient microbiota. Pathogens overcome the body's external defenses, multiply, and become established in the body; such a successful invasion of the body is called an **infection**. An infection may or may not result in disease; that is, it may not adversely affect the body.

Next, we consider in greater detail the various sites through which pathogens gain entry into the body.

Portals of Entry

LEARNING OUTCOME

14.7 Identify and describe the portals through which pathogens invade the body.

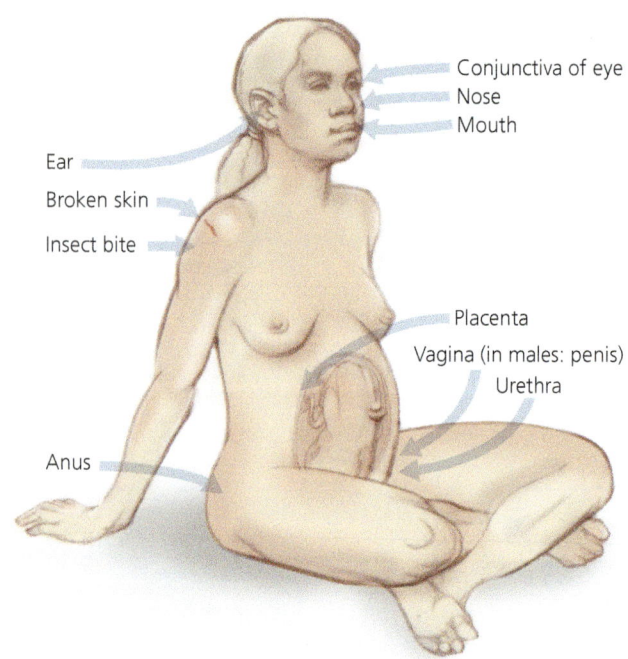

▲ **FIGURE 14.3 Routes of entry for invading pathogens.** The portals of entry include the skin, placenta, conjunctiva, and mucous membranes of the respiratory, gastrointestinal, urinary, and reproductive tracts. The parenteral route involves a puncture through the skin.

The sites through which most pathogens enter the body can be likened to the great gates, or portals, of a castle because those sites constitute the routes by which microbes gain entry. Pathogens can enter the body at several sites, called **portals of entry** (FIGURE 14.3), which are of three major types: the skin, the mucous membranes, and the placenta. A fourth entry point, the so-called *parenteral route* (pă-ren′ter-ăl), is not a portal but a way of circumventing the usual portals. Next we consider each type in turn.

Skin

Because the outer layer of skin is composed of relatively thick layers of tightly packed, dead, dry cells (FIGURE 14.4), it forms a formidable barrier to most pathogens as long as it remains intact. Still, some pathogens can enter the body through natural openings in the skin, such as hair follicles and sweat glands. Abrasions, cuts, bites, scrapes, stab wounds, and surgeries open the skin to infection by contaminants. Additionally, the larvae of some parasitic worms are capable of burrowing through the skin to reach underlying tissues, and some fungi can digest the dead outer layers of skin, thereby gaining access to deeper, moister areas within the body.

Mucous Membranes

The major portals of entry for pathogens are the *mucous membranes*, which line all the body cavities that are open to the outside world. These include the linings of the respiratory, gastrointestinal, urinary, and reproductive tracts as well as the *conjunctiva* (kon-jŭnk-tī′vă), the thin membrane covering the surface of the eyeball and the underside of each eyelid. Like the skin, mucous membranes are composed of tightly packed

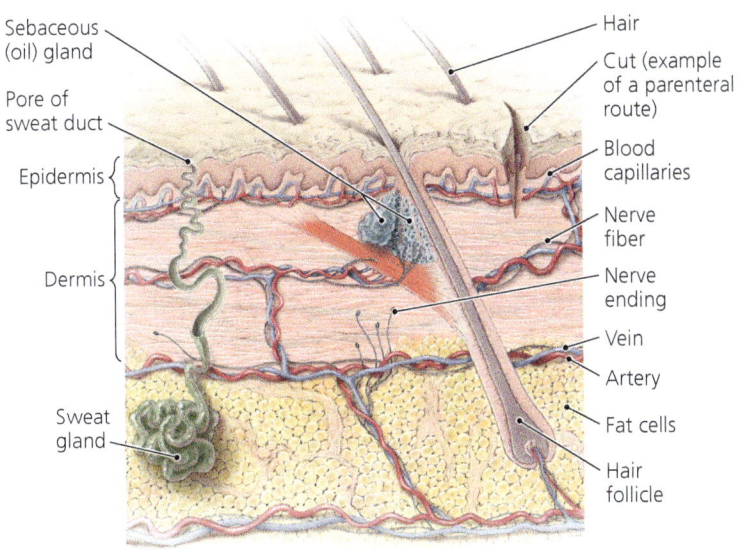

▲ FIGURE 14.4 **A cross section of skin.** The layers of cells constitute a barrier to most microbes as long as the skin remains intact. Some pathogens can enter the body through hair follicles, through the ducts of sweat glands, and parenterally.

cells, but unlike the skin, mucous membranes are relatively thin, moist, and warm, and their cells are living. Therefore, pathogens find mucous membranes more hospitable and easier portals of entry.

The respiratory tract is the most frequently used portal of entry. Pathogens enter the mouth and nose in the air, on dust particles, and in droplets of moisture. For example, the bacteria that cause whooping cough, diphtheria, pneumonia, strep throat, and meningitis; the viruses of measles, influenza, and colds; as well as some diseases caused by fungi or protozoa enter through the respiratory tract.

Surprisingly, many viruses enter the respiratory tract via the eyes. They are introduced onto the conjunctiva by contaminated fingers and are washed into the nasal cavity with tears. Cold and influenza viruses often enter the body in this manner.

Some parasitic protozoa, helminths, bacteria, and viruses infect the body through the gastrointestinal mucous membranes. These parasites are able to survive the acidic pH of the stomach and the digestive juices of the intestinal tract. Noncellular pathogens called *prions* can also enter the body through oral mucous membranes.

Placenta

A developing embryo forms an organ, called the *placenta*, through which the embryo obtains nutrients from the mother. The placenta is in such intimate contact with the wall of the mother's uterus that nutrients and wastes diffuse between the blood vessels of the developing child and of the mother; however, because the two blood supplies do not actually contact each other, the placenta typically forms an effective barrier to most pathogens.

Even so, in about 2% of pregnancies, pathogens cross the placenta and infect the embryo or fetus, sometimes causing spontaneous abortion, birth defects, or premature birth. Some pathogens that can cross the placenta are listed in TABLE 14.4.

The Parenteral Route

The **parenteral route** is not a regular portal of entry but instead a means by which the skin and gastrointestinal tract are circumvented. Generally, this means that pathogens are deposited directly into tissues beneath the skin or mucous membranes, such as occurs in punctures by a nail, thorn, or hypodermic needle. Some experts include in the parenteral route breaks in the skin by cuts, bites, deep abrasions, or surgery.

The Role of Adhesion in Infection

LEARNING OUTCOMES

14.8 List the types of adhesion factors and the roles they play in infection.

14.9 Explain how a biofilm may facilitate contamination and infection.

TABLE 14.4 Some Pathogens That Cross the Placenta

	Pathogen	Condition in the Adult	Effect on Embryo or Fetus
Protozoan	*Toxoplasma gondii*	Toxoplasmosis	Abortion, epilepsy, encephalitis, microcephaly, mental retardation, blindness, anemia, jaundice, rash, pneumonia, diarrhea, hypothermia, deafness
Bacteria	*Treponema pallidum*	Syphilis	Abortion, multiorgan birth defects, syphilis
	Listeria monocytogenes	Listeriosis	Granulomatosis infantiseptica (nodular inflammatory lesions and infant blood poisoning), death
DNA Viruses	Cytomegalovirus	Usually asymptomatic	Deafness, microcephaly, mental retardation
	Erythrovirus	Erythema infectiosum	Abortion
RNA Viruses	Lentivirus (HIV)	AIDS	Immunosuppression (AIDS)
	Rubivirus	German measles	Severe birth defects or death

After entering the body, symbionts must adhere to cells if they are to be successful in establishing colonies. The process by which microorganisms attach themselves to cells is called **adhesion** (ad-hē'zhŭn) or *attachment*. To accomplish adhesion, pathogens use **adhesion factors**, which are either specialized structures or attachment molecules. Examples of such specialized structures are adhesion disks in some protozoa and suckers and hooks in some helminths (see Chapter 23). In contrast, viruses and many bacteria have surface lipoprotein or glycoprotein molecules called *ligands* that enable them to bind to complementary receptors on host cells **(FIGURE 14.5)**. Ligands are also called *adhesins* on bacteria and *attachment proteins* on viruses. Adhesins are found on fimbriae, flagella, and glycocalyces of many pathogenic bacteria. Receptor molecules on host cells are typically glycoproteins, which are proteins bound to sugar molecules such as mannose or galactose. If ligands or their receptors can be changed or blocked, infection can often be prevented.

The specific interaction of adhesins and receptors with chemicals on host cells often determines the specificity of pathogens for particular hosts. For example, *Neisseria gonorrhoeae* (nī-se'rē-ă go-nor-rē'ī) has adhesins on its fimbriae that adhere to cells of humans; therefore, this pathogen cannot affect other hosts.

Some bacteria, including *Bordetella* (bōr-dĕ-tel'ă; the cause of whooping cough), have more than one type of adhesin. Other pathogens, such as *Plasmodium* (plaz-mō'dē-ŭm; the cause of malaria), change their adhesins over time, helping the pathogen evade the body's immune system and allowing the pathogen to attack more than one kind of cell. Bacterial cells and viruses that have lost the ability to make ligands—whether as the result of some genetic change (mutation) or exposure to certain physical or chemical agents (as occurs in the production of some vaccines)—become harmless, or **avirulent** (ā-vir'ū-lent).

Some bacterial pathogens do not attach to host cells directly but instead interact with each other to form a sticky web of bacteria and polysaccharides called a **biofilm**, which adheres to a surface within a host. A prominent example of a biofilm is dental plaque **(FIGURE 14.6)**, which contains the bacteria that cause dental caries (tooth decay).

MICRO CHECK

3. On which microbes do glycoproteins serve as adhesion factors?

TELL ME WHY

Why does every infection start with contamination but not every contamination results in an infection?

The Nature of Infectious Disease

LEARNING OUTCOME

14.10 Compare and contrast the terms *infection*, *disease*, *morbidity*, *pathogenicity*, and *virulence*.

Most infections succumb to the body's defenses (discussed in detail in Chapters 15 and 16), but some infectious agents not only evade the defenses and multiply but also affect body functions. By definition, every parasite injures its host. When the injury is significant enough to interfere with the normal functioning of the host's body, the result is termed **disease**. Thus, disease, also known as **morbidity**, is any change from a state of health.

Infection and disease are not the same. Infection is the invasion of a pathogen; disease results only if the pathogen multiplies sufficiently to adversely affect the body. Some illustrations will help to clarify this point. While caring for a patient, a nurse may become contaminated with the bacterium *Staphylococcus aureus* (staf'i-lō-kok'us o'rē-ŭs). If the pathogen is able to gain access to his body, perhaps through a break in his skin, he will become infected. Only if the multiplication of *Staphylococcus*

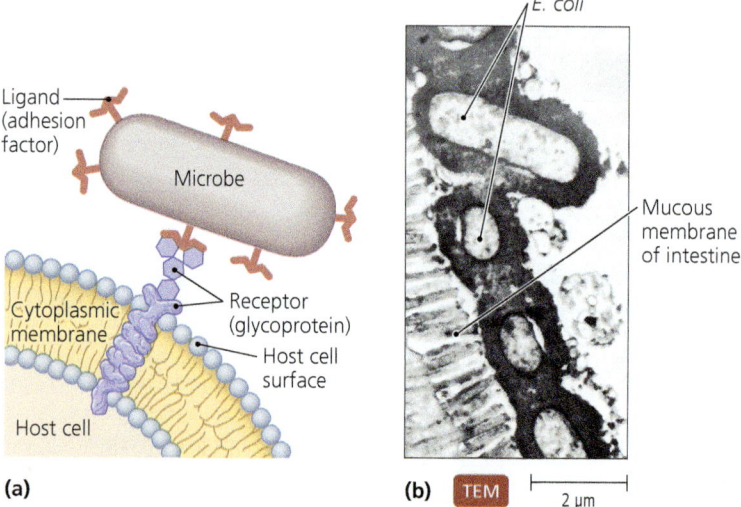

▲ **FIGURE 14.5 The adhesion of pathogens to host cells.** (a) An artist's rendition of the attachment of a microbial attachment molecule (ligand) to a complementary surface receptor on a host cell. (b) A photomicrograph of cells of a pathogenic strain of *E. coli* attached to the mucous membrane of the intestine. Although the thick glycocalyces (black) of the bacteria are visible, a bacterium's attachment molecules are too small to be seen at this magnification.

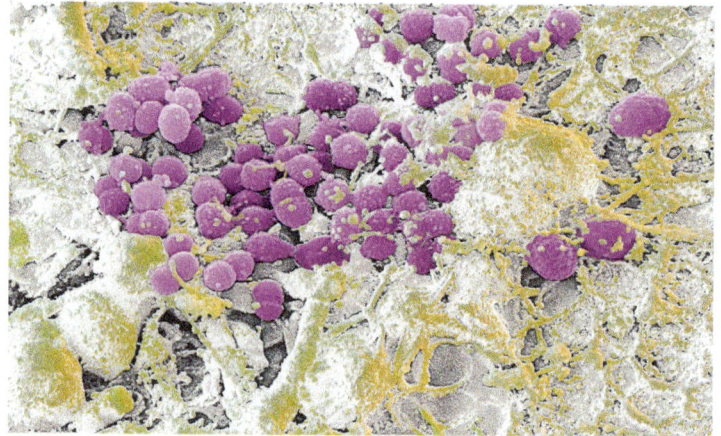

▲ **FIGURE 14.6 Dental plaque.** This biofilm consists of bacteria that adhere to one another and to teeth by means of sticky capsules; an extracellular, slimy matrix; and fimbriae.

results in adverse conditions, for example, in the production of a boil, will the nurse have a disease. Another example: A drug addict who shares needles may become contaminated and subsequently infected with HIV without experiencing visible change in body function for years. Such a person is infected but not yet diseased.

Manifestations of Disease: Symptoms, Signs, and Syndromes

LEARNING OUTCOME

14.11 Contrast symptoms, signs, and syndromes.

Diseases may become manifest in different ways. **Symptoms** are *subjective* characteristics of a disease that can be felt by the patient alone, whereas **signs** are *objective* manifestations of disease that can be observed or measured by others. Symptoms include pain, headache, dizziness, and fatigue; signs include swelling, rash, redness, and fever. Sometimes pairs of symptoms and signs reflect the same underlying cause: nausea is a symptom, and vomiting a sign; chills are a symptom, whereas shivering is a sign. Note, however, that even though signs and symptoms may have the same underlying cause, they need not occur together. Thus, for example, a viral infection of the brain may result in both a headache (a symptom) and the presence of viruses in the cerebrospinal fluid (a sign), but viruses in the cerebrospinal fluid are not invariably accompanied by headache. Symptoms and signs are used in conjunction with laboratory tests to make diagnoses. Some typical symptoms and signs are listed in TABLE 14.5.

A **syndrome** is a group of symptoms and signs that collectively characterizes a particular disease or abnormal condition. For example, acquired immunodeficiency syndrome (AIDS) is characterized by malaise, loss of certain white blood cells, diarrhea, weight loss, pneumonia, toxoplasmosis, and tuberculosis.

TABLE 14.5 Typical Manifestations of Disease

Symptoms (Sensed by the Patient)	Signs (Detected or Measured by an Observer)
Pain, nausea, headache, chills, sore throat, fatigue or lethargy (sluggishness, tiredness), malaise (discomfort), itching, abdominal cramps	Swelling, rash or redness, vomiting, diarrhea, fever, pus formation, anemia, leukocytosis/leukopenia (increase/decrease in the number of circulating white blood cells), bubo (swollen lymph node), tachycardia/bradycardia (increase/decrease in heart rate)

Some infections go unnoticed because they have no symptoms. Such cases are **asymptomatic**, or **subclinical**, infections. Note that even though asymptomatic infections by definition lack symptoms, in some cases certain signs may still be detected if the proper tests are performed. For example, leukocytosis (an excess of white blood cells) may be detected in a blood sample from an individual who feels completely healthy.

TABLE 14.6 lists some prefixes and suffixes used to describe various aspects of diseases and syndromes.

Causation of Disease: Etiology

LEARNING OUTCOMES

14.12 Define *etiology*.
14.13 List Koch's postulates, explain their function, and describe their limitations.

Even though our focus in this chapter is infectious disease, it's obvious that not all diseases result from infections. Some diseases are *hereditary*, which means they are genetically transmitted from parents to offspring. Other diseases, called *congenital diseases*, are diseases that are present at birth, regardless of the cause (whether hereditary, environmental, or infectious). Still other

TABLE 14.6 Terminology of Disease

Prefix/Suffix	Meaning	Example
carcino-	Cancer	Carcinogenic: giving rise to cancer
col-, colo-	Colon	Colitis: inflammation of the colon
dermato-	Skin	Dermatitis: inflammation of the skin
-emia	Pertaining to the blood	Viremia: viruses in the blood
endo-	Inside	Endocarditis: inflammation of lining of heart
-gen, gen-	Give rise to	Pathogen: giving rise to disease
hepat-	Liver	Hepatitis: inflammation of the liver
idio-	Unknown	Idiopathic: pertaining to a disease of unknown cause
-itis	Inflammation of a structure	Meningitis: inflammation of the meninges (covering of the brain)
-oma	Tumor or swelling	Papilloma: wart
-osis	Condition of	Toxoplasmosis: being infected with *Toxoplasma*
-patho, patho-	Abnormal	Pathology: study of disease
septi-	Literally, *rotting*; refers to presence of pathogens	Septicemia: pathogens in the blood
terato-	Defects	Teratogenic: causing birth defects
tox-	Poison	Toxin: harmful compound

TABLE 14.7　Categories of Diseases[a]

	Description	Examples
Hereditary	Caused by errors in the genetic code received from parents	Sickle-cell anemia, diabetes mellitus, Down syndrome
Congenital	Anatomical and physiological (structural and functional) defects present at birth; caused by drugs (legal and illegal), X-ray exposure, or infections	Fetal alcohol syndrome, deafness from rubella infection
Degenerative	Result from aging	Renal failure, age-related farsightedness
Nutritional	Result from lack of some essential nutrients in diet	Kwashiorkor, rickets
Endocrine (hormonal)	Due to excesses or deficiencies of hormones	Addison's disease
Mental	Emotional or psychosomatic	Skin rash, gastrointestinal distress
Immunological	Hyperactive or hypoactive immunity	Allergies, autoimmune diseases, agammaglobulinemia
Neoplastic (tumor)	Abnormal cell growth	Benign tumors, cancers
Infectious	Caused by an infectious agent	Colds, influenza, herpes infections
Iatrogenic[b]	Caused by medical treatment or procedures; are a subgroup of healthcare-associated diseases	Surgical error, yeast vaginitis resulting from antimicrobial therapy
Idiopathic[c]	Unknown cause	Alzheimer's disease, multiple sclerosis
Healthcare-Associated (nosocomial[d]**)**	Disease acquired in health care setting	*Pseudomonas* infection in burn patient

[a] Some diseases may fall in more than one category.
[b] From Greek *iatros*, meaning "physician."
[c] From Greek *idiotes*, meaning "ignorant person," and *pathos*, meaning "disease."
[d] From Greek *nosokomeion*, meaning "hospital."

diseases are classified as *degenerative, nutritional, endocrine, mental, immunological,* or *neoplastic*.[8] The various categories of diseases are described in **TABLE 14.7**. Note that some diseases can fall into more than one category. For example, liver cancer, which is usually associated with infection by hepatitis B and D viruses, can be classified as both neoplastic and infectious.

The study of the cause of a disease is called **etiology** (ē-tē-ol'ō-jē). Because our focus here (and in subsequent chapters) is on infectious diseases, we next examine how microbiologists investigate the causation of infectious diseases, beginning with an examination of the work of Robert Koch.

Using Koch's Postulates

In our modern world, we take for granted the idea that specific microbes cause specific diseases, but this has not always been the case. In the past, disease was thought to result from a variety of causes, including bad air, imbalances in body fluids, or astrological forces. In the 19th century, Louis Pasteur, Robert Koch, and other microbiologists proposed the **germ theory of disease**, which states that disease is caused by infections of pathogenic microorganisms (at the time called *germs*).

But which pathogen causes a specific disease? How can we distinguish the pathogen that causes a disease from all the other biological agents (fungi, bacteria, protozoa, and viruses) that are part of the normal microbiome and are in effect "innocent bystanders"?

Koch developed a series of essential conditions, or *postulates*, that scientists must demonstrate or satisfy to prove that a particular microbe is pathogenic and causes a particular disease. Using his postulates, Koch proved that *Bacillus anthracis* (ba-sil'ŭs an-thrā'sis) causes anthrax and that *Mycobacterium tuberculosis* (mī-kō-bak-tēr'ē-ŭm too-ber-kyū-lō'sis) causes tuberculosis.

To prove that a given infectious agent causes a given disease, a scientist must satisfy all of **Koch's postulates** (FIGURE 14.7):

1. The suspected agent (bacterium, virus, etc.) must be present in every case of the disease.
2. That agent must be isolated and grown in pure culture.
3. The cultured agent must cause the disease when it is inoculated into a healthy, susceptible experimental host.
4. The same agent should be found in the diseased experimental host.

It is critical that all the postulates be satisfied in order. The mere presence of an agent does not prove that it causes a disease. Although Koch's postulates have been used to prove the cause of many infectious diseases in humans, animals, and plants, inadequate attention to the postulates has resulted in incorrect conclusions concerning some disease causation. For instance, in the early 1900s *Haemophilus influenzae* (hē-mof'i-lŭs in-flū-en'zī) was found in the respiratory systems of flu victims and identified as the causative agent of influenza based on its presence. Later, flu victims who lacked *H. influenzae* in their lungs were discovered. This discovery violated the first postulate, so *H. influenzae* cannot be the cause of flu. Today, we know that an RNA virus causes flu and that *H. influenzae* was incidentally part of the microbiota of those early flu patients. (Later studies, correctly using Koch's postulates, showed that *H. influenzae* can cause an often-fatal meningitis in children.)

---

[8] Neoplasms are tumors, which may either remain in one place (benign tumors) or spread (cancers).

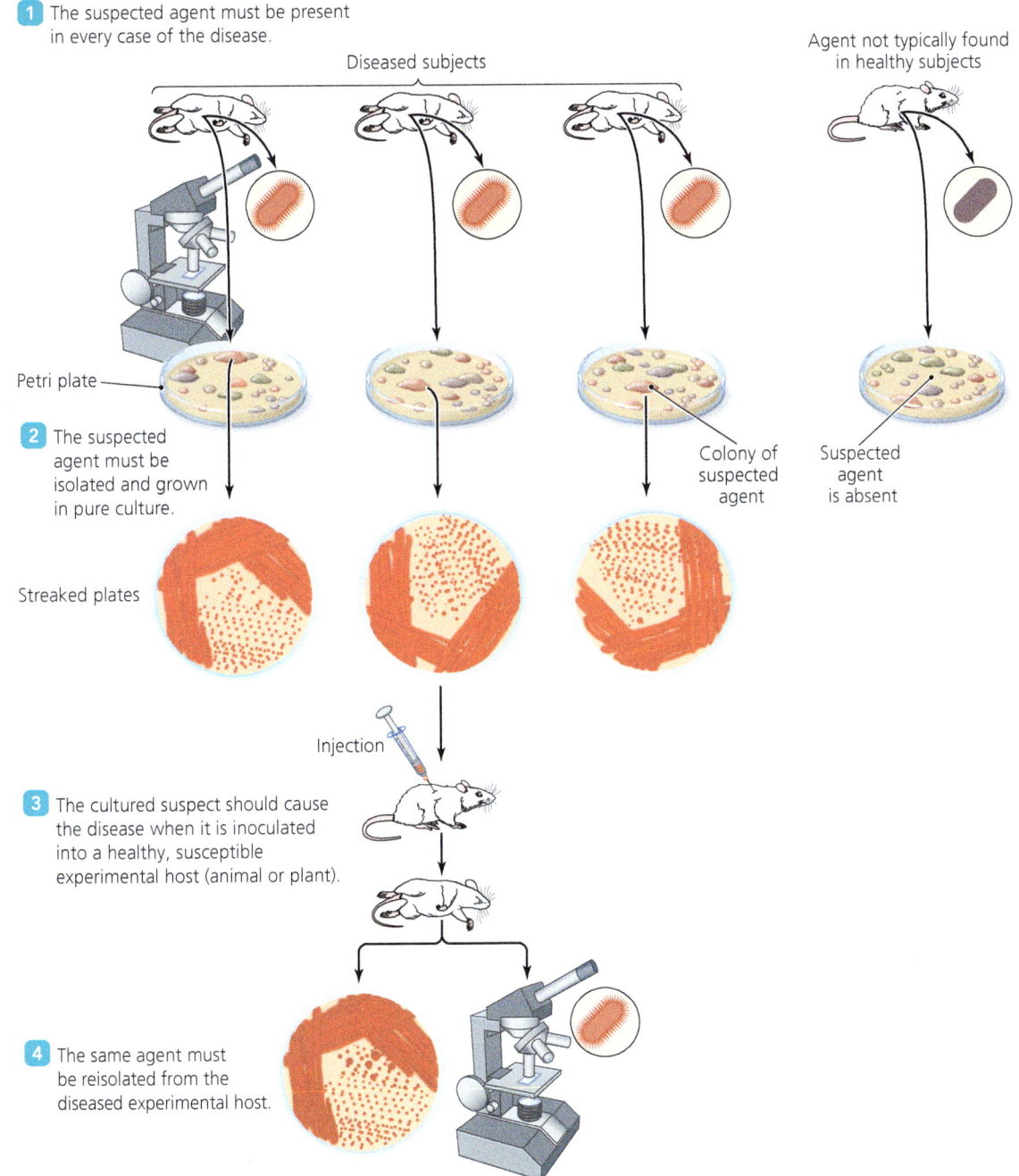

▲ FIGURE 14.7 Koch's postulates.

Exceptions to Koch's Postulates

Clearly, Koch's postulates are a cornerstone of the etiology of infectious diseases, but using them is not always feasible, for the following reasons:

- Some pathogens have not been cultured in the laboratory. For example, pathogenic strains of *Mycobacterium leprae* (lep'rī), which causes leprosy, have never been grown on laboratory media.
- Some diseases are caused by a combination of pathogens or by a combination of a pathogen and physical, environmental, or genetic cofactors. In such cases, the pathogen alone is unable to cause disease, but when accompanied by another pathogen or the appropriate cofactor, disease results.
- Ethical considerations prevent applying Koch's postulates to diseases and pathogens that occur in humans only. In such cases, the third postulate, which involves inoculation of a healthy susceptible host, cannot be satisfied within ethical boundaries. For this reason, scientists have never applied Koch's postulates to prove that HIV causes AIDS; however, observations of fetuses that were naturally exposed to HIV across their placentas from their infected mothers, as well as accidentally infected health care and laboratory workers, have in effect satisfied the third postulate.

Some additional circumstances may make satisfying Koch's postulates difficult:

- It is not possible to establish a single cause for such infectious diseases as pneumonia, meningitis, and hepatitis because the names of these diseases refer to conditions that can be caused by more than one pathogen. For these diseases, laboratory technicians must identify the etiologic agent involved in any given case.
- Some pathogens may have been ignored. For example, gastric ulcers were long thought to be caused by excessive production of stomach acid in response to stress, but the majority of such ulcers are now known to be caused by a long-overlooked bacterium, *Helicobacter pylori* (hel'ĭ-kō-bak'ter pī'lō-rē).

If Koch's postulates cannot be applied to a disease condition for whatever reason, how can we positively know the causative agent of a disease? *Epidemiological* studies, discussed later in this chapter, can give statistical support to causation theories but not absolute proof. For example, some researchers have proposed that the bacterium *Chlamydophila pneumoniae* (kla-mē-dof'ĭ-lă noo-mō'nē-ī) causes many cases of arteriosclerosis, in part on the basis of the bacterium's presence in patients with the disease. Debate among scientists about such cases fosters a continued drive for knowledge and discovery. For example, we are still searching for the causes of chronic fatigue syndrome, multiple sclerosis, and Alzheimer's disease.

Virulence Factors of Infectious Agents

LEARNING OUTCOME

14.14 Explain how microbial extracellular enzymes, toxins, adhesion factors, and antiphagocytic factors affect virulence.

Microbiologists characterize the disease-related capabilities of microbes by using two related terms. The ability of a microbe to cause disease is termed **pathogenicity** (path'ō- jĕ-nis'i-tē), and the *degree* of pathogenicity is termed **virulence**. In other words, virulence is the relative ability of a pathogen to infect a host and cause disease. Neither term addresses the severity of a disease; the pathogen causing rabbit fever is highly virulent, but the disease is relatively mild. Organisms can be placed along a virulence continuum (FIGURE 14.8); highly virulent organisms almost always cause disease, whereas less virulent organisms (including opportunistic pathogens) cause disease only in weakened hosts or when present in overwhelming numbers.

Pathogens have a variety of traits that interact with a host and enable the pathogen to enter a host, adhere to host cells, gain access to nutrients, and escape detection or removal by the immune system. These traits are collectively called **virulence factors**. Virulent pathogens have one or more virulence factors that nonvirulent microbes lack. We discussed two virulence factors—adhesion factors and biofilm formation—previously; now we examine three other virulence factors: extracellular enzymes, toxins, and antiphagocytic factors. Other virulence factors can be examined online at the Mastering Microbiology Study Area.

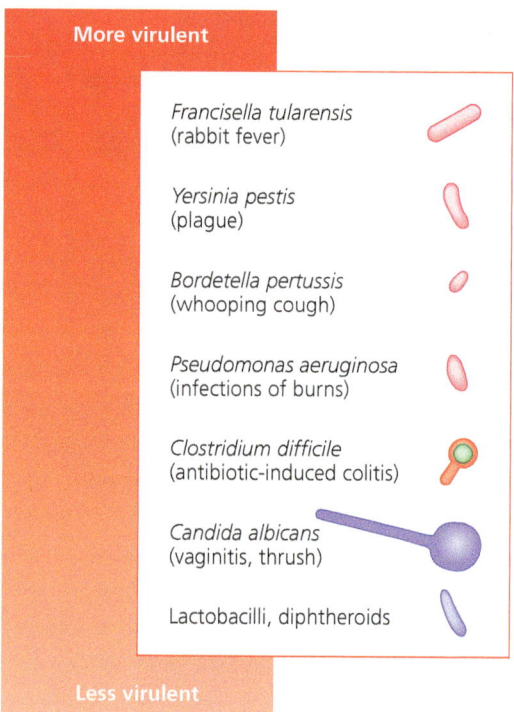

▲ **FIGURE 14.8 Relative virulence of some microbial pathogens.** Virulence involves ease of infection and the ability of a pathogen to cause disease; the term does not indicate the seriousness of a disease.

Play *Virulence Factors: Inactivating Host Defenses* @ Mastering Microbiology

Extracellular Enzymes

Many pathogens secrete enzymes that enable them to dissolve structural chemicals in the body and thereby maintain an infection, invade further, and avoid body defenses. FIGURE 14.9a illustrates the action of some extracellular enzymes of bacteria:

- Hyaluronidase and collagenase degrade specific molecules to enable bacteria to invade deeper tissues. Hyaluronidase digests hyaluronic acid—the "glue" that holds animal cells together—and collagenase, which breaks down collagen, the body's chief structural protein.
- Coagulase causes blood proteins to clot, providing a "hiding place" for bacteria within a clot.
- Kinases, such as staphylokinase and streptokinase, digest blood clots, allowing subsequent invasion of damaged tissues.

Many bacteria with these enzymes are virulent; mutant strains of the same species that have defective genes for these extracellular enzymes are usually avirulent.

Pathogenic eukaryotes also secrete enzymes that contribute to virulence. For example, fungi that cause "ringworm" produce *keratinase*, which enzymatically digests keratin—the main component of skin, hair, and nails. *Entamoeba histolytica* (ent-ă-mē'bă his-tō-li'ti-kă) secretes *mucinase* to digest the mucus

The Nature of Infectious Disease

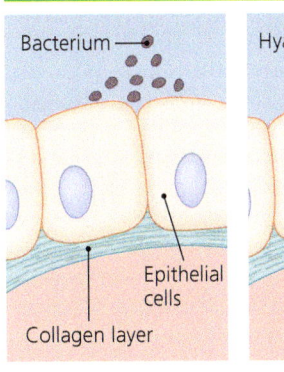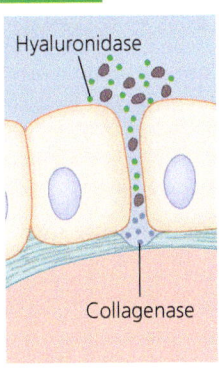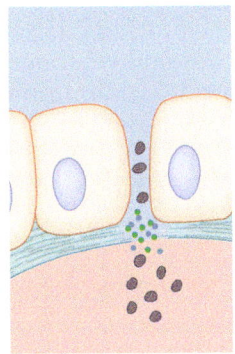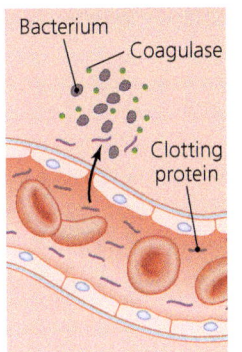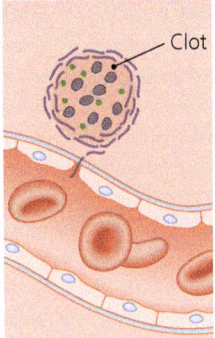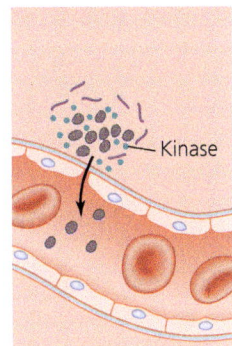

(a) Extracellular enzymes

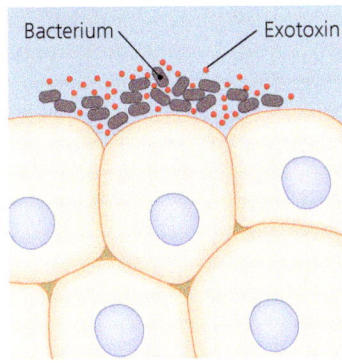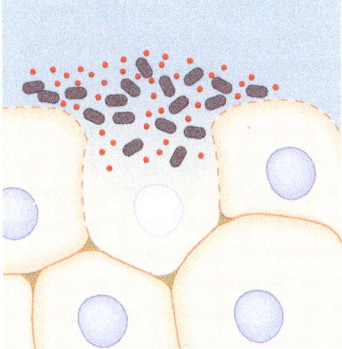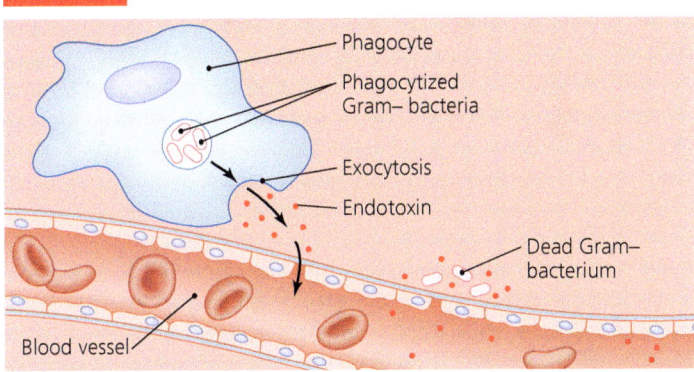

(b) Toxins

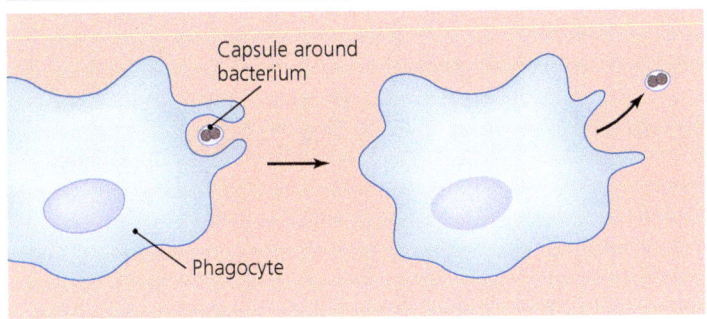

(c) Antiphagocytic factors

▲ **FIGURE 14.9 Some virulence factors.** (a) Extracellular enzymes. Hyaluronidase and collagenase digest structural materials in the body. Coagulase in effect "camouflages" bacteria inside a blood clot, whereas kinases digest clots to release bacteria. (b) Toxins. Exotoxins (including cytotoxin, shown here) are released from living pathogens and harm neighboring cells. Endotoxins are released from dead Gram-negative bacteria and can trigger widespread disruption of normal body functions. (c) Antiphagocytic factors. Capsules, one antiphagocytic factor, can prevent phagocytosis or stop digestion by a phagocyte. *How can bacteria prevent a phagocytic cell from digesting them once they have been engulfed by pseudopods?*

Figure 14.9 Some bacteria secrete a chemical that prevents the fusion of lysosomes with the phagosome containing the bacteria.

lining the intestinal tract, allowing the amoeba entry to the underlying cells, where it causes amebic dysentery.

Play *Virulence Factors:* Penetrating Host Tissues @ Mastering Microbiology

Toxins

Toxins are chemicals that either harm tissues or trigger host immune responses that cause damage. The distinction between extracellular enzymes and toxins is not always clear because many enzymes are toxic and many toxins have enzymatic action. In a condition called **toxemia** (tok-sē′mē-ă), toxins enter the bloodstream and are carried to other parts of the body, including sites that may be far removed from the site of infection. There are two types of toxin: exotoxins and endotoxins.

Exotoxins Many microorganisms secrete **exotoxins** that are central to their pathogenicity in that they destroy host cells or interfere with host metabolism. Exotoxins are of three principal types:

- *Cytotoxins*, which kill host cells in general or affect their function **(FIGURE 14.9b)**
- *Neurotoxins*, which specifically interfere with nerve cell function
- *Enterotoxins*, which affect cells lining the gastrointestinal tract

Examples of pathogenic bacteria that secrete exotoxins are the clostridia that cause gangrene, botulism, and tetanus; pathogenic strains of *S. aureus* that cause food poisoning and other ailments; and diarrhea-causing *E. coli, Salmonella enterica* (sal′mŏ-nel′ă en-ter′i-kă), and *Shigella* (shē-gel′lă) species. Some fungi and marine dinoflagellates (protozoa) also secrete exotoxins. (Specific exotoxins are discussed when we examine specific pathogens in Chapters 19–25.)

Play *Virulence Factors:* Exotoxins @ Mastering Microbiology

The body protects itself from exotoxins with protective molecules called *antibodies* that bind to specific toxins and neutralize them. Antibodies against toxins are called **antitoxins**. Health care workers stimulate the production of antitoxins by administering immunizations composed of *toxoids*, which are toxins that have been treated with heat, formaldehyde, chlorine, or other chemicals to make them nontoxic but still capable of stimulating the production of antibodies. (Chapters 16 and 17 further discuss antibodies, toxoids, and immunizations.)

Endotoxins Gram-negative bacteria have an outer (wall) membrane composed of lipopolysaccharides, phospholipids, and proteins (see Chapter 3, Figure 3.16). Lipopolysaccharide (LPS) is called **endotoxin** because early microbiologists thought this poisonous molecule came from within Gram-negative bacteria. The toxic part of LPS is the lipid, which is called **lipid A**. Lipid A is not a single molecular structure but an array of similar lipid molecules that vary between different Gram-negative bacteria and among bacterial strains.

Endotoxins can be released when Gram-negative bacteria divide, die naturally, or are digested by phagocytic cells such as macrophages (see Figure 14.9b). Lipid A stimulates the body to release chemicals that can cause fever, inflammation, diarrhea, hemorrhaging, shock, and blood coagulation. Most Gram-negative pathogens can be potentially life threatening because the release of endotoxin from dead cells can produce serious systemic effects in the host.

Play *Virulence Factors:* Endotoxins @ Mastering Microbiology

TABLE 14.8 summarizes differences between exotoxins and endotoxins.

Antiphagocytic Factors

Typically, the longer a pathogen remains in a host, the greater the damage, and the more severe the disease. To limit the extent and duration of infections, the body's phagocytic cells, such as the white blood cells called macrophages, engulf and remove

TABLE 14.8 A Comparison of Bacterial Exotoxins and Endotoxins

	Exotoxins	Endotoxins
Source	Mainly Gram-positive and Gram-negative bacteria	Gram-negative bacteria
Relation to Bacteria	Metabolic product secreted from living cell	Lipopolysaccharide (LPS) portion of the outer (cell wall) membrane released upon cell death
Chemical Nature	Protein or short peptide	Lipid portion (lipid A) of LPS
Toxicity	High	Low but may be fatal in high doses
Heat Stability	Typically unstable at temperatures above 60°C	Stable for up to 1 hour at autoclave temperature (121°C)
Effect on Host	Variable, depending on source; may be cytotoxin, neurotoxin, enterotoxin	Fever, lethargy, malaise, shock, blood coagulation
Fever Producing?	No	Yes
Antigenicity[a]	Strong: stimulates antitoxin (antibody) production	Weak
Toxoid Formation for Immunization?	By treatment with heat or formaldehyde	Not feasible
Representative Diseases	Botulism, tetanus, gas gangrene, diphtheria, cholera, plague, staphylococcal food poisoning	Typhoid fever, tularemia, endotoxic shock, urinary tract infections, meningococcal meningitis

[a] Refers to the ability of a chemical to trigger a specific immune response, particularly the formation of antibodies.

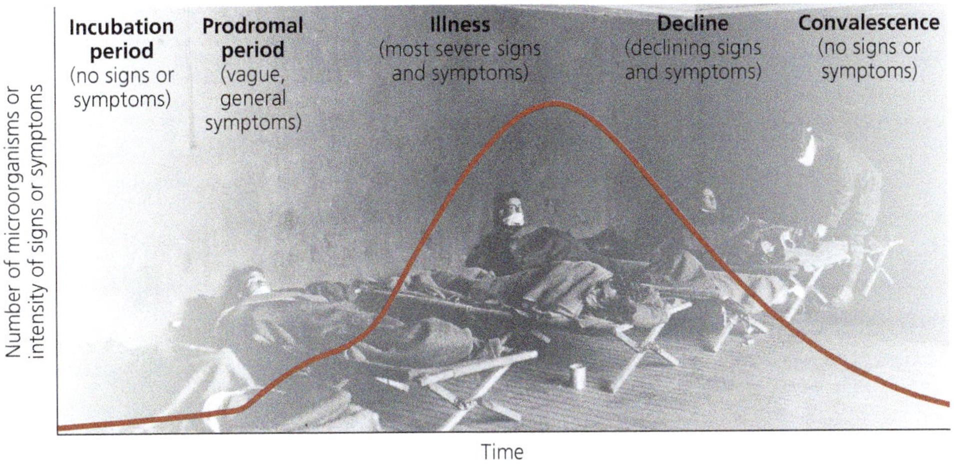

◀ **FIGURE 14.10 The stages of infectious diseases.** Not every stage occurs in every disease.

invading pathogens. Here, we consider some virulence factors related to the evasion of phagocytosis, beginning with bacterial capsules.

Capsules The capsules of many pathogenic bacteria (see Figure 3.5a) are effective virulence factors because they are composed of chemicals normally found in the body; as a result, they do not stimulate an immune response. For example, hyaluronic acid capsules in effect deceive phagocytic cells into treating the encapsulated bacteria as if they were a normal part of the body. Additionally, capsules are often "slippery," making it difficult for phagocytes to surround and phagocytize them—their pseudopods cannot grip the capsule, much as wet hands have difficulty holding a wet bar of soap **(FIGURE 14.9c)**.

Antiphagocytic Chemicals Some bacteria, including the cause of gonorrhea, produce chemicals that prevent the fusion of lysosomes with phagocytic vesicles, allowing the bacteria to survive inside phagocytes (see Figure 14.9c). *Streptococcus pyogenes* (strep-tō-kok′ŭs pī-oj′en-ēz) produces a protein on its cell wall and fimbriae, called *M protein*, that inhibits phagocytosis and thus increases virulence. Other bacteria produce *leukocidins*, which are chemicals capable of destroying phagocytic white blood cells outright.

Play *Phagocytosis: Microbes That Evade It* Mastering Microbiology

Play *Virulence Factors: Hiding from Host Defenses* Mastering Microbiology

The Stages of Infectious Diseases

LEARNING OUTCOME

14.15 List and describe the five typical stages of infectious diseases.

Following exposure and infection, a sequence of events called the **disease process** can occur. Many infectious diseases proceed through five stages following infection: an incubation period, a prodromal period, illness, decline, and convalescence **(FIGURE 14.10)**.

Incubation Period

The **incubation period** is the time between infection and occurrence of the first symptoms or signs of disease. The length of the incubation period depends on the virulence of the infective agent, the infective dose (initial number of pathogens), the state and health of the patient's immune system, the nature of the pathogen and its reproduction time, and the site of infection. Some diseases have typical incubation periods, whereas for others the incubation period varies considerably. **TABLE 14.9** lists typical incubation periods for selected diseases.

TABLE 14.9 Incubation Periods of Selected Infectious Diseases

Disease	Incubation Period
Staphylococcus foodborne infection	<1 day
Influenza	About 1 day
Cholera	2 to 3 days
Genital herpes	About 5 days
Tetanus	5 to 15 days
Syphilis	10 to 21 days
Hepatitis B	70 to 100 days
AIDS	1 to >8 years
Leprosy	10 to >30 years

MICRO CHECK

4. What is the difference between the terms *infection* and *disease*?
5. What is an example of a sign (as opposed to a symptom) of disease?
6. What is the purpose of Koch's postulates?
7. How do the enzymes hyaluronidase and collagenase increase bacterial virulence?
8. What effect do enterotoxins have?

Prodromal Period

The **prodromal period**[9] (prō-drō'măl) is a short time of generalized, mild symptoms (such as malaise and muscle aches) that precedes illness. Not all infectious diseases have a prodromal stage.

Illness

Illness is the most severe stage of an infectious disease. Signs and symptoms are most evident during this time. Typically, the patient's immune system has not yet fully responded to the pathogens, and their presence is harming the body. This stage is usually when a physician first sees the patient.

Decline

During the period of **decline**, the body gradually returns to normal as the patient's immune response and/or medical treatment vanquish the pathogens. Fever and other signs and symptoms subside. Normally the immune response and its products (such as antibodies in the blood) peak during this stage. If the disease doesn't decline, then the disease is fatal—the patient dies.

Convalescence

During **convalescence** (kon-vă-les'ens), the patient recovers from the illness; tissues are repaired and returned to normal. The length of a convalescent period depends on the amount of damage, the nature of the pathogen, the site of infection, and the overall health of the patient. Thus, whereas recovery from staphylococcal food poisoning may take less than a day, recovery from Lyme disease may take years.

A patient is likely to be infectious during every stage of disease. Even though most of us realize we are infective during the symptomatic periods, many people are unaware that infections can often be spread during incubation and convalescence as well. For example, a patient who no longer has any obvious herpes sores is always capable of transmitting herpesviruses, and a convalescent carrier of *Salmonella* may excrete the bacterium for a year or more (see **Clinical Case Study: A Deadly Carrier** on p. 419). Good aseptic technique can limit the spread of many pathogens from recovering patients.

> **TELL ME WHY**
>
> Why is mutated *Streptococcus pneumoniae*, which cannot make a capsule, unable to cause pneumonia?

The Movement of Pathogens Out of Hosts: Portals of Exit

Just as infections occur through portals of entry, so pathogens must leave infected patients through **portals of exit** in order to infect others (FIGURE 14.11). Many portals of exit are essentially identical to portals of entry. However, pathogens often exit hosts

[9]From Greek *prodromos*, meaning "forerunner."

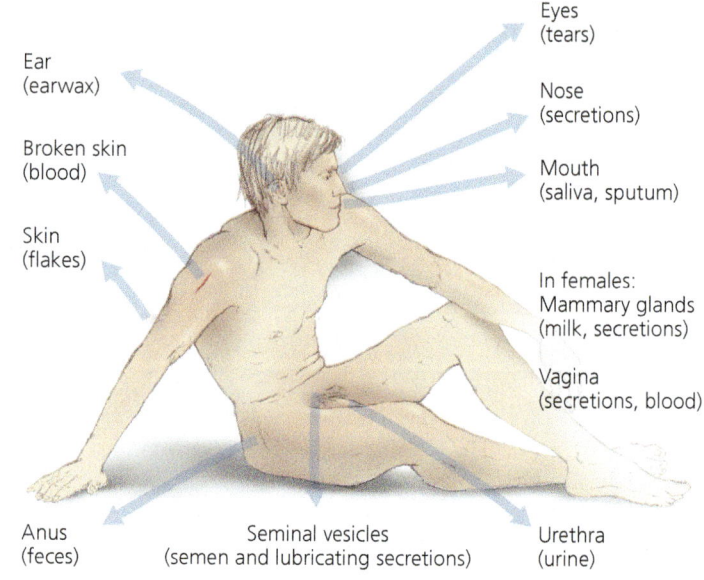

▲ **FIGURE 14.11 Portals of exit.** Many portals of exit are also portals of entry. Pathogens often leave the body via bodily secretions and excretions produced at those sites.

in materials that the body secretes or excretes. Thus, pathogens may leave hosts in secretions (earwax, tears, nasal secretions, saliva, sputum, and respiratory droplets), in blood (via arthropod bites, hypodermic needles, or wounds), in vaginal secretions or semen, in milk produced by the mammary glands, and in excreted bodily wastes (feces and urine). As we will see, health care personnel must consider the portals of entry and exit in their efforts to understand and control the spread of diseases within populations.

> **TELL ME WHY**
>
> Why is the tube emptying the bladder (the urethra) more likely to be a portal of exit than a portal of entry?

Modes of Infectious Disease Transmission

LEARNING OUTCOMES

14.16 Contrast contact, vehicle, and vector transmission of pathogens.
14.17 Contrast droplet transmission and airborne transmission.
14.18 Contrast mechanical and biological vectors.

By definition, an infectious disease agent must be transmitted from either a reservoir or a portal of exit to another host's portal of entry. Transmission can occur by numerous modes that are somewhat arbitrarily categorized into three groups: *contact transmission*, *vehicle transmission*, and *vector transmission*.

Play *Epidemiology: Transmission of Disease* @ **Mastering** Microbiology

Contact Transmission

Contact transmission is the spread of pathogens from one host to another by direct contact, indirect contact, or respiratory droplets.

Direct contact transmission, including *person-to-person spread*, typically involves body contact between hosts. Touching, kissing, and sexual intercourse are involved in the transmission of such diseases as warts, herpes, and gonorrhea. Touching, biting, or scratching can transmit zoonoses such as rabies, ringworm, and tularemia from an animal reservoir to a human. The transfer of pathogens from an infected mother to a developing baby across the placenta is another form of direct contact transmission. Direct transmission within a single individual can also occur if the person transfers pathogens from a portal of exit directly to a portal of entry—as occurs, for example, when people with poor personal hygiene unthinkingly place fingers contaminated with fecal pathogens into their mouths.

Indirect contact transmission occurs when pathogens are spread from one host to another by **fomites** (fōm′i-tēz; singular: *fomes*, fō′mēz, or *fomite*, fō′mīt), which are inanimate objects that are inadvertently used to transfer pathogens to new hosts. Fomites include needles, toothbrushes, paper tissues, toys, money, diapers, drinking glasses, bedsheets, medical equipment, and other objects that can harbor or transmit pathogens. Contaminated needles are a major source of infection of the hepatitis B and AIDS viruses.

Droplet transmission is a third type of contact transmission. Pathogens can be transmitted within *droplet nuclei* (droplets of mucus) that travel less than a meter from the body during exhaling, coughing, and sneezing **(FIGURE 14.12)**. Pathogens such as cold and flu viruses may be spread in this manner. If pathogens travel more than 1 meter in respiratory droplets, the mode is considered to be *airborne transmission* (discussed shortly) rather than contact transmission.

Vehicle Transmission

Vehicle transmission is the spread of pathogens via air, drinking water, and food, as well as bodily fluids being handled outside the body.

Airborne transmission involves the spread of pathogens farther than 1 meter to the respiratory mucous membranes of a new host via an **aerosol** (ār′ō-sol)—a cloud of small droplets and solid particles suspended in the air. Aerosols may contain pathogens, either on dust or inside droplets. (Recall that transmission via droplet nuclei that travel less than 1 meter is considered to be a form of direct contact transmission.) Aerosols can come from sneezing and coughing, or they can be generated by such means as air-conditioning systems, sweeping, mopping, changing clothes or bed linens, or even from flaming inoculating loops in microbiology labs. Dust particles can carry *Staphylococcus*, *Streptococcus*, and *Hantavirus* (han′ta-vī-rŭs), whereas measles virus and tuberculosis bacilli can be transmitted in airborne droplets. Fungal spores of *Histoplasma* (his-tō-plaz′-mă) and *Coccidioides* (kok-sid-ē-oy′-dēz) are typically inhaled.

Waterborne transmission is important in the spread of many gastrointestinal diseases, including giardiasis, amebic dysentery, and cholera. Note that water can act as a reservoir as well as a vehicle of infection. **Fecal–oral infection** is a major source of disease in the world. Some waterborne pathogens, such as *Schistosoma* (shis-tō-sō′-mă) worms (see Chapter 23) and enteroviruses (see Chapter 25), are shed in the feces, contaminate a water source, and enter through the gastrointestinal mucous membrane or skin. They subsequently can cause disease elsewhere in the body.

Foodborne transmission involves pathogens in and on foods that are inadequately processed, undercooked, or poorly refrigerated **(FIGURE 14.13)**. Foods may be contaminated with typical members of the microbiota (e.g., *E. coli* and *S. aureus*), with zoonotic pathogens such as *Mycobacterium bovis* (bō′-vis)

▲ **FIGURE 14.12 Droplet transmission.** In this case, droplets are propelled, primarily from the mouth, during a sneeze. By convention, such transmission is considered contact transmission only if droplets transmit pathogens to a new host within 1 meter of their source. *By what portal of entry does airborne transmission most likely occur?*

Figure 14.12 The most likely portal of entry of airborne pathogens is the respiratory mucous membrane.

▲ **FIGURE 14.13 Poorly stored foods can harbor pathogens and transmit diseases.**

and *Toxoplasma* (tok-sō-plaz′-mă), and with parasitic worms that alternate between human and animal hosts. Contamination of food with feces and pathogens such as hepatitis A virus is another kind of fecal–oral transmission. Because milk is particularly rich in nutrients that microorganisms use (protein, lipids, vitamins, and sugars), it can be associated with the transmission of many diseases from infected animals and milk handlers if it is not properly pasteurized.

Because blood, urine, saliva, and other bodily fluids can contain pathogens, everyone—but especially health care workers—must take precautions when handling these fluids in order to prevent **bodily fluid transmission**. Special care must be taken to prevent such fluids—all of which should be considered potentially contaminated with pathogens—from contacting the conjunctiva or any breaks in the skin or mucous membranes. Examples of diseases that can be transmitted via bodily fluids are AIDS, hepatitis, and herpes, which as we have seen can also be transmitted via direct contact.

Vector Transmission

Vectors are arthropods that transmit diseases. Vectors can be either biological or mechanical.

Biological vectors not only transmit pathogens but also serve as hosts for the multiplication of a pathogen during some stage of the pathogen's life cycle. The biological vectors of diseases affecting humans are typically biting arthropods, including mosquitoes, ticks, lice, fleas, bloodsucking flies, bloodsucking bugs, and mites (see Figure 12.29). After pathogens replicate within a biological vector, often in its gut or salivary gland, the pathogens enter a new host through a bite. The bite site becomes contaminated with the vector's feces, or the vector's bite directly introduces pathogens into the new host.

Mechanical vectors are not required as hosts by the pathogens they transmit; such vectors only passively carry pathogens to new hosts on their feet or other body parts. Mechanical vectors, such as houseflies and cockroaches, may introduce pathogens such as *Salmonella* and *Shigella* into drinking water and food or onto the skin.

TABLE 14.10 lists some arthropod vectors and the diseases they transmit. **TABLE 14.11** summarizes the modes of disease transmission.

TELL ME WHY

Why can't we correctly say that all arthropod vectors are reservoirs?

Classification of Infectious Diseases

LEARNING OUTCOMES

14.19 Describe the basis for each of the various classification schemes of infectious diseases.

14.20 Distinguish among acute, subacute, chronic, and latent diseases.

14.21 Distinguish among communicable, contagious, and noncommunicable infectious diseases.

CLINICAL CASE STUDY

TB in the Nursery

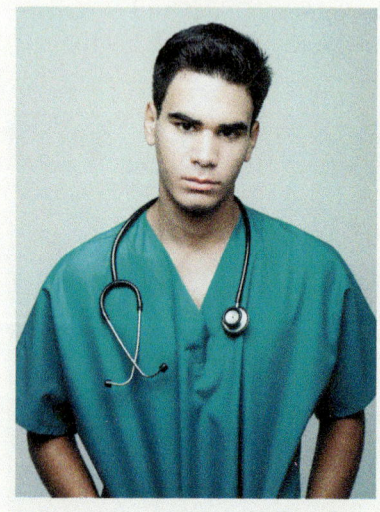

In the early fall, a neonatal nurse in a large metropolitan hospital became ill with a cough and low-grade fever. He initially thought he had a mild cold and took some time off work with oral hydration and acetaminophen for his headache. He returned to work in the hospital's nursery after two days.

Three weeks later, his condition had worsened, including elevated temperature (38.6°C), chills, night sweats, and generalized fatigue. His symptoms were complicated by shortness of breath, which was worse on inhalation, and bloody sputum. He reported to his doctor that he was in the United States on a work visa and was a native of South Africa. He had a positive skin test for tuberculosis (TB) but had always believed this was his body's natural reaction to the TB vaccine he had received as a child. His chest X-ray films in the past had always been clear of infection.

This time, however, his sputum tested positive for acid-fast bacilli (AFB) and his chest X ray showed a tubercle in the upper lobe of his right lung. He was diagnosed with active tuberculosis and immediately admitted to the hospital, where he began a standard four-drug regimen: isoniazid, rifampin, ethambutol, and pyrazinamide. He was restricted from work and placed in respiratory isolation until he had been on the antitubercular drugs for at least two weeks, his symptoms resolved, and he had three negative AFB smears. Unfortunately, during the three weeks that he had continued to work in the nursery, he had exposed over 620 newborns to TB, an airborne infectious disease.

1. How can private physicians quickly assess their clients for the possibility of an infectious disease?
2. What policies should be in place at hospitals to protect patients from exposure to infectious staff members?
3. How could doctors' offices improve public knowledge and protect the public from tuberculosis and other infectious diseases?

TABLE 14.10 Selected Arthropod Vectors

	Disease	Causative Agent (bacteria unless otherwise indicated)
Biological Vectors		
Mosquitoes		
Anopheles spp.	Malaria	*Plasmodium* spp. (protozoa)
Aedes spp.	Yellow fever	*Flavivirus* sp. (viruses)
Aedes spp., *Anopheles* spp., *Culex* spp.	Elephantiasis	*Wuchereria* sp. (nematode helminth)
Aedes sp.	Dengue	*Flavivirus* spp. (viruses)
Aedes spp., *Culiseta* sp., *Culex* spp.	Viral encephalitis	*Alphavirus*, *Flavivirus*, *Orthobunyavirus* (viruses)
Ticks		
Ixodes	Lyme disease	*Borrelia burgdorferi*
Dermacentor	Rocky Mountain spotted fever	*Rickettsia rickettsii*
Fleas		
Xenopsylla	Bubonic plague	*Yersinia pestis*
	Endemic typhus	*Rickettsia typhi*
Lice		
Pediculus	Epidemic typhus	*Rickettsia prowazekii*
Bloodsucking Flies		
Glossina	African sleeping sickness	*Trypanosoma brucei*
Simulium	River blindness	*Onchocerca volvulus* (helminth)
Bloodsucking Bugs		
Triatoma	Chagas' disease	*Trypanosoma cruzi* (protozoan)
Mites (chiggers)		
Leptotrombidium	Scrub typhus	*Orientia tsutsugamushi*
Mechanical Vectors		
Houseflies		
Musca	Foodborne infections	*Shigella* spp., *Salmonella* spp., *Escherichia coli*
Cockroaches		
Blatella, *Periplaneta*	Foodborne infections	*Shigella* spp., *Salmonella* spp., *Escherichia coli*

Infectious diseases can be classified in a number of ways. One scheme groups diseases based upon the body systems affected. A difficulty with this method of classification is that many diseases involve more than one organ system. For example, AIDS begins as a sexually transmitted infection (reproductive system), or as a parenteral infection of the blood (cardiovascular system). It then becomes an infection of the lymphatic system as viruses invade lymphocytes. Finally, the syndrome involves diseases and degeneration of the respiratory, nervous, digestive, cardiovascular, and lymphatic systems.

Another classification system deals with diseases according to taxonomic groups. (Chapters 19–25 examine diseases based on this approach.)

Every disease (not just infectious diseases) can also be classified according to its longevity and severity. If a disease develops rapidly but lasts a relatively short time, it is called an **acute disease**. An example is the common cold. In contrast, **chronic diseases** develop slowly (usually with less severe symptoms) and are continual or recurrent. Infectious mononucleosis, hepatitis C, tuberculosis, and leprosy are chronic diseases. **Subacute diseases** have durations and severities that lie somewhere between acute and chronic. Subacute bacterial endocarditis, a disease of heart valves, is one example. **Latent diseases** are those in which a pathogen remains inactive for a long period of time before becoming active. Herpes is an example of a latent disease.

When an infectious disease comes from another infected host, either directly or indirectly, it is a **communicable disease**. Influenza, herpes, and tuberculosis are examples of communicable diseases. If a communicable disease is easily transmitted between hosts, as is the case for chicken pox or measles, it is also called a **contagious disease**. **Noncommunicable diseases** arise outside hosts or from the resident microbiome. In other words, they are not spread from one host to another, and diseased patients are not a source of contamination for others. Tooth decay, acne, and tetanus are examples of noncommunicable diseases. **TABLE 14.12** (p. 434) defines these and other terms used to classify infectious diseases.

Yet another way in which all infectious diseases may be classified is by the effects they have on populations rather than on individuals. Is a certain disease consistently found in a given group of people or geographic area? Under what circumstances is it more prevalent than usual in a given geographic area? How prevalent is "normal"? How is the disease transmitted throughout a population? We next examine issues related to diseases at the population level.

TABLE 14.11 Modes of Disease Transmission

Mode of Transmission	Examples of Diseases Spread
Contact Transmission	
Direct Contact (e.g., handshaking, kissing, sexual intercourse, bites)	Cutaneous anthrax, genital warts, gonorrhea, herpes, rabies, staphylococcal infections, syphilis
Indirect Contact (e.g., drinking glasses, toothbrushes, toys, punctures)	Common cold, enterovirus infections, influenza, measles, Q fever, pneumonia, tetanus
Droplet Transmission (e.g., droplets from sneezing, within 1 meter)	Whooping cough, streptococcal pharyngitis (strep throat)
Vehicle Transmission	
Airborne (e.g., dust particles or droplets carried more than 1 meter)	Chicken pox, coccidioidomycosis, histoplasmosis, influenza, measles, pulmonary anthrax, tuberculosis
Waterborne (e.g., streams, swimming pools)	*Campylobacter* infections, cholera, *Giardia* diarrhea
Foodborne (e.g., poultry, seafood, meat)	Food poisoning (botulism, staphylococcal); hepatitis A, listeriosis, tapeworms, toxoplasmosis, typhoid fever
Vector Transmission	
Mechanical (e.g., on bodies of flies, roaches)	*E. coli* diarrhea, salmonellosis, trachoma
Biological (e.g., lice, mites, mosquitoes, ticks)	Chagas' disease, Lyme disease, malaria, plague, Rocky Mountain spotted fever, typhus fever, yellow fever

TABLE 14.12 Terms Used to Classify Infectious Diseases

Term	Definition
Acute disease	Disease in which symptoms develop rapidly and that runs its course quickly
Chronic disease	Disease with usually mild symptoms that develop slowly and last a long time
Subacute disease	Disease whose time course and symptoms range between acute and chronic
Asymptomatic disease	Disease without symptoms
Latent disease	Disease that appears a long time after infection
Communicable disease	Disease transmitted from one host to another
Contagious disease	Communicable disease that is easily spread
Noncommunicable disease	Disease not passed from person to person
Local infection	Infection confined to a small region of the body
Systemic infection	Widespread infection in many systems of the body; often travels in the blood or lymph
Focal infection	Infection site that serves as a source of pathogens for infections at other sites in the body
Primary infection	Initial infection within a given patient
Secondary infection	Infections that follow a primary infection; often by opportunistic pathogens

MICRO CHECK

9. If someone contracts food poisoning from drinking contaminated raw milk, what type of transmission has occurred?
10. The viral pathogen that causes cold sores can remain inactive in the body for a long time between disease outbreaks. How is this disease classified?

TELL ME WHY

Why is an acute disease with a high rate of mortality unlikely to be associated with a pandemic?

Epidemiology of Infectious Diseases

LEARNING OUTCOME

14.22 Define *epidemiology*.

Our discussion so far has centered on the negative impact of microorganisms on *individuals*. Now we turn our attention to the effects of pathogens on *populations*. **Epidemiology**[10] (ep-i-dē-mē-ol'ō-jē) is the study of where and when diseases occur and how they are transmitted within populations. During the 20th century, epidemiologists expanded the scope of their work beyond infectious diseases to also consider injuries and deaths related to automobile and fireworks accidents, cigarette smoking, lead poisoning, and other causes; however, we will limit our discussion primarily to the epidemiology of infectious diseases.

Play *Epidemiology: Overview*
@ Mastering Microbiology

Frequency of Disease

LEARNING OUTCOMES

14.23 Contrast incidence and prevalence.
14.24 Differentiate among the terms *endemic*, *sporadic*, *epidemic*, and *pandemic*.

Epidemiologists keep track of the occurrence of diseases by using two measures: incidence and prevalence. **Incidence** is the number of *new* cases of a disease in a given area or population during a given period of time; **prevalence** is the *total number* of cases, both new and already existing, in a given area or population during a given period of time. In other words, prevalence is a cumulative number. Thus, for example, the estimated number of new cases—the incidence—of HIV infection/AIDS in adults in the United States in 2013 was 47,352. However, the prevalence of HIV infection/AIDS in the same year was 1,194,039 because more than 1,100,000 patients who got the disease prior to 2013 still survived. **FIGURE 14.14** illustrates this relationship between incidence and prevalence.

Epidemiologists report their data in many ways, including maps, graphs, charts, and tables **(FIGURE 14.15)**. Why do they report their data in so many different ways? Using a variety of

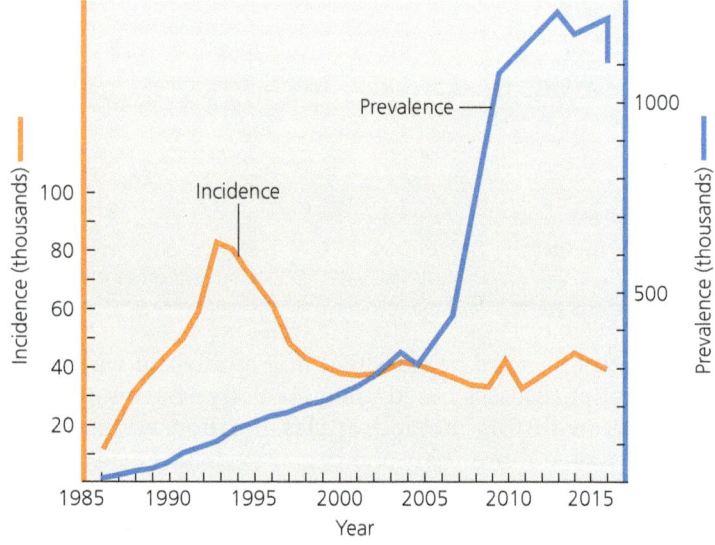

▲ **FIGURE 14.14 Curves representing the incidence and the estimated prevalence of AIDS among U.S. adults.** Note that the scales for the two curves are different. *Why can the incidence of a disease never exceed the prevalence of that disease?*

Figure 14.14 Because prevalence includes all cases, both old and new, prevalence must always be larger than incidence.

---
[10]From Greek *epidemios*, meaning "among the people," and *logos*, meaning "study of."

Epidemiology of Infectious Diseases 435

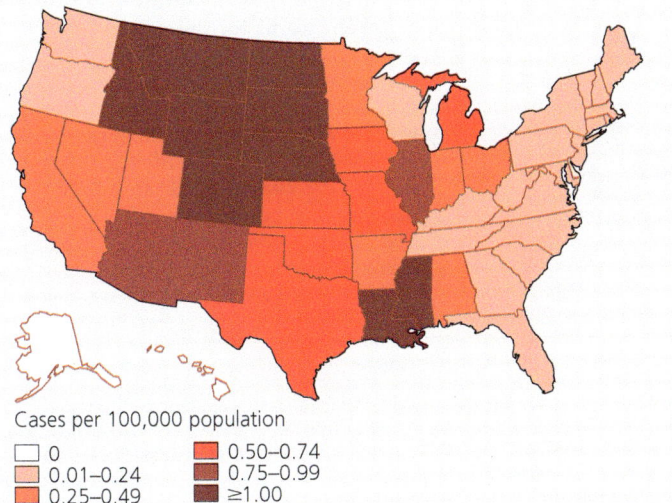

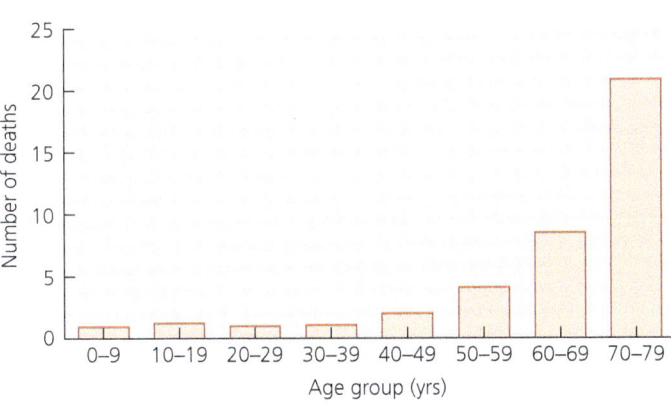

▲ FIGURE 14.15 **Epidemiologists report data in a variety of ways.** Here, incidence of cases of an emerging disease—West Nile virus disease—in which the virus invades the nervous system. Data for the decade 1999–2008 are presented. **(a)** Average annual incidence for the United States is on a map by state. **(b)** Percentage of cases by week of onset. **(c)** Percentage of deaths by age. *How do the data help support the idea that mosquitoes carry West Nile virus?*

Figure 14.15 *Most cases of the disease occur in late summer, when mosquitoes are likely biting and people are more likely to be outdoors.*

formats enables epidemiologists to observe patterns that may give clues about the causes of or ways to prevent diseases. For example, West Nile virus encephalitis occurs across the United States, but when data are reported by age group, it becomes obvious that elderly Americans are most at risk.

The occurrence of a disease can also be considered in terms of a combination of frequency and geographic distribution (FIGURE 14.16). A disease that typically occurs continually (at moderately regular intervals) at a relatively stable incidence within a given population or geographical area is said to be **endemic**[11] to that population or region. A disease is considered **sporadic** when only a few scattered cases occur within an area or population. Whenever a disease occurs at a greater frequency than is usual for an area or population, the disease is said to be **epidemic** within that area or population.

Play *Epidemiology: Occurrence of Disease* @ Mastering Microbiology

The commonly held belief that a disease must infect thousands or millions to be considered an epidemic is mistaken. The time period and the number of cases necessary for an outbreak of disease to be classified as an epidemic are not specified; the important fact is that there are more cases than expected, based on historical statistics. For example, fewer than 70 cases of an emerging disease—hemolytic uremic syndrome caused by a strain of *E. coli*—occurred in Germany in 2011, but because there are typically fewer than five cases annually, the 2011 outbreak was considered an epidemic. The same year, thousands of cases of flu occurred in Germany, but there was no flu epidemic because the number of cases observed did not exceed the number expected. FIGURE 14.17 illustrates how epidemics are defined according to the number of expected cases and not according to the absolute number of cases.

If an epidemic occurs simultaneously on more than one continent, it is referred to as a **pandemic** (see FIGURE 14.16d). H1N1 flu, so-called swine flu, became pandemic worldwide in 2009.

Obviously, for disease prevalence to be classified as either endemic, sporadic, epidemic, or pandemic, good records must be kept for each region and population. From such records, incidence and prevalence can be calculated, and then changes in these data can be noted. Health departments at the local and state levels require doctors and hospitals to report certain infectious conditions. Some are also nationally notifiable; that is, their occurrence must be reported to the Centers for Disease Control and Prevention (CDC) in Atlanta, Georgia, which is the headquarters and clearinghouse for national epidemiological research. Nationally notifiable conditions are listed in TABLE 14.13 (p. 437). Each week, the CDC reports the number of cases of most of the nationally notifiable conditions in the *Morbidity and Mortality Weekly Report (MMWR)* (FIGURE 14.18).

MICRO CHECK

11. What is an endemic disease?

---

[11]From Greek *endemos,* meaning "native."

◀ **FIGURE 14.16 Illustrations of the different terms for the occurrence of disease.** (a) An endemic disease is typically present in a region. (b) A sporadic disease occurs irregularly and infrequently. (c) An epidemic disease is present in greater frequency than is usual. (d) A pandemic disease is an epidemic disease occurring on more than one continent at a given time. *Regardless of where you live, name a disease that is endemic, one that is sporadic, and one that is epidemic in your state.*

Figure 14.16 Some possible answers: Flu is endemic in every state; tuberculosis is sporadic in most states; AIDS is epidemic in every state.

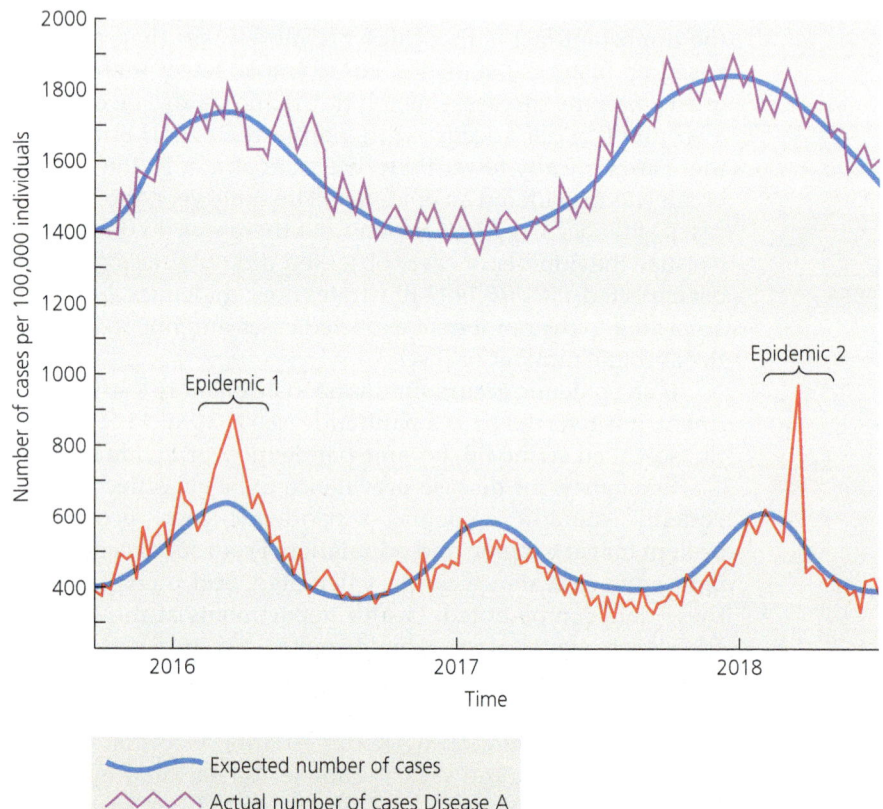

◀ **FIGURE 14.17 Epidemics may have fewer cases than nonepidemics.** These two graphs demonstrate the independence between the absolute number of cases and a disease's designation as an epidemic. Even though the number of cases of disease A always exceeds the number of cases of disease B, only disease B is considered epidemic—at those times when the number of cases observed exceeds the number of cases expected. *How can scientists know the usual prevalence of a given disease?*

Figure 14.17 Scientists record every case of the disease so that they will have a "baseline" prevalence, which then becomes the expected prevalence for each disease.

Epidemiological Studies

LEARNING OUTCOME

14.25 Explain three approaches epidemiologists use to study diseases in populations.

Epidemiologists conduct research to study the dynamics of diseases in populations by taking three different approaches, called *descriptive, analytical,* and *experimental epidemiology*.

Descriptive Epidemiology

Descriptive epidemiology involves the careful tabulation of data concerning a disease. Relevant information includes the location and time of cases of the disease as well as information about the patients, such as ages, gender, occupations, health histories, and socioeconomic groups. Because the time course and chains of transmission of a disease are an important part of descriptive epidemiology, epidemiologists strive to identify the **index case** (the first case) of the disease in a given area or population.

TABLE 14.13 Nationally Notifiable Infectious Diseases[a]

Anthrax	*Haemophilus influenzae*, invasive disease	Mumps	Streptococcal toxic-shock syndrome
Arboviral diseases (e.g., West Nile encephalitis, chikungunya, etc.)	Hansen disease (leprosy)	Novel influenza type A infections	Syphilis
Babesiosis	*Hantavirus* pulmonary syndrome	Pertussis	Tetanus
Botulism	Hemolytic uremic syndrome, postdiarrheal	Plague	Toxic shock syndrome, nonstreptococcal
Brucellosis	Hepatitis A	Poliomyelitis	Trichinellosis
Campylobacteriosis	Hepatitis B	Psittacosis	Tuberculosis
Cancer	Hepatitis C	Q fever	Tularemia
Carbapenem-resistant infections of Enterobacteriaceae	HIV infection (including stage III, which is AIDS)	Rabies, animal and human	Typhoid fever
Chancroid	Influenza-associated infant deaths	Rubella	Vancomycin-intermediate *Staphylococcus aureus*
Chlamydia trachomatis infection	Invasive pneumococcal disease	Rubella, congenital syndrome	Vancomycin-resistant *S. aureus*
Cholera	Legionellosis	Salmonellosis	Varicella (chickenpox)
Coccidioidomycosis	Leptospirosis	Severe acute respiratory syndrome–associated coronavirus disease	Vibriosis
Cryptosporidiosis	Listeriosis	Shiga-toxin-producing *Escherichia coli*	Viral hemorrhagic fevers
Cyclosporiasis	Lyme disease	Shigellosis	Yellow fever
Dengue virus infections	Malaria	Smallpox	Zika virus disease (and infection)
Diphtheria	Measles	Spotted fever rickettsiosis	
Ehrlichiosis/anaplasmosis	Meningococcal disease		
Giardiasis			
Gonorrhea			

[a]Diseases for which hospitals, physicians, and other health care workers are required to report cases to state health departments and then forward the data to the CDC.

Sometimes it is difficult or impossible to identify an index case because the patient has recovered, moved, or died.

The earliest descriptive epidemiological study was by John Snow (1813–1858), who studied a cholera outbreak in London in 1854. By carefully mapping the locations of the cholera cases in a particular part of the city, Snow found that the cases were clustered around the Broad Street water pump (FIGURE 14.19). This distribution of cases, plus the voluminous watery diarrhea of cholera patients, suggested that the disease was spread via contamination of drinking water by sewage.

Analytical Epidemiology

Analytical epidemiology investigates a disease in detail, including analysis of data acquired in descriptive epidemiological studies, to determine the probable cause, mode of transmission, and possible means of prevention of the disease. Analytical epidemiology may be used in situations where it is not ethical to apply Koch's postulates. Thus, even though Koch's third postulate has never been fulfilled in the case of AIDS (because it is unethical to intentionally inoculate a human with HIV), analytical epidemiological studies indicate that HIV causes AIDS and that it is transmitted primarily sexually.

Often analytical studies are *retrospective*; that is, they attempt to identify causation and mode of transmission after an outbreak has occurred. Epidemiologists compare a group of people who had the disease with a group who did not. The groups are carefully matched by factors such as gender, environment, and diet and then compared to determine which pathogens and factors may play a role in morbidity.

Experimental Epidemiology

Experimental epidemiology involves testing a hypothesis concerning the cause of a disease. The application of Koch's postulates to determine etiology is an example of experimental epidemiology. Experimental epidemiology also involves studies to test a hypothesis resulting from an analytical study, such as the efficacy of a preventive measure or certain treatment. For example, analytical epidemiological studies suggested that the bacterium *Chlamydophila pneumoniae* may contribute to arteriosclerosis, resulting in heart attacks. If so, antimicrobial drugs that kill *Chlamydophila* should prevent some heart attacks by killing the bacteria. However, when researchers administered these drugs to 7700 patients with a history of heart disease, infection with *C. pneumoniae* was reduced, but the number of heart attacks was not. Thus, an experimental epidemiological study disproved a hypothesis suggested by an analytical epidemiological analysis.

Hospital Epidemiology: Healthcare-Associated (Nosocomial) Infections

LEARNING OUTCOMES

14.26 Explain how healthcare-associated infections differ from other infections.
14.27 Describe the factors that influence the development of healthcare-associated infections.
14.28 Describe three types of healthcare-associated infections and how they may be prevented.

National Notifiable Infectious Diseases: Weekly Tables

TABLE 2g. Weekly cases* of selected notifiable diseases (≥ 1,000 cases reported during the preceding year), and selected low frequency diseases, United States and U.S. territories, week ending November 3, 2018 (WEEK 44)
(Accessible Version: https://wonder.cdc.gov/nndss/static/2018/44/2018-44-table2G.html)

Reporting Area	Giardiasis					Gonorrhea					Haemophilus influenzae, invasive disease (all ages, all serotypes)				
	Current week	Previous 52 weeks		Cum 2018	Cum 2017	Current week	Previous 52 weeks		Cum 2018	Cum 2017	Current week	Previous 52 weeks		Cum 2018	Cum 2017
		Med	Max				Med	Max				Med	Max		
United States	133	258	931	11,383	12,545	4,948	10,568	12,718	450,026	466,170	42	106	301	4,168	4,427
New England	11	19	39	919	934	5	217	385	8,663	11,649	2	5	20	225	238
Connecticut	1	3	11	160	194	-	0	156	U	3,335	-	1	6	47	60
Maine	2	2	13	125	113	-	12	24	548	524	-	0	2	18	27
Massachusetts	7	11	20	492	484	-	160	202	6,607	6,386	2	2	8	106	112
New Hampshire	1	2	9	97	106	5	11	24	487	444	-	0	3	21	19
Rhode Island	-	1	5	45	37	-	25	107	976	816	-	0	7	17	11
Vermont	N	0	0	N	N	-	1	14	45	144	-	0	2	16	9
Middle Atlantic	57	53	85	2,360	2,362	789	1,152	1,479	51,321	49,666	16	16	38	665	711
New Jersey	-	4	11	133	336	64	169	251	7,236	7,991	-	3	8	120	127
New York (excluding New York City)	27	15	34	754	714	204	215	452	9,283	8,707	9	5	24	198	219
New York City	18	20	32	901	830	285	482	693	21,818	20,005	2	3	8	132	102
Pennsylvania	12	11	28	572	482	236	297	360	12,984	12,963	5	5	14	215	263
East North Central	13	30	75	1,445	1,658	657	1,534	1,898	65,182	70,032	2	16	36	680	767
Illinois	N	0	0	N	N	30	420	549	15,861	20,037	-	4	11	192	191
Indiana	-	3	10	137	171	78	235	302	9,920	9,938	-	3	9	120	119
Michigan	-	7	17	345	526	119	287	409	12,640	13,331	-	3	8	129	172
Ohio	8	7	29	429	352	290	468	547	20,313	20,385	2	5	13	224	188
Wisconsin	5	10	30	534	609	140	153	208	6,448	6,341	-	0	5	15	97
West North Central	1	18	34	748	1,348	278	706	871	31,484	27,484	3	5	14	222	333
Iowa	-	4	10	202	240	6	91	115	3,838	3,067	-	0	1	-	2
Kansas	-	2	6	88	92	51	102	131	4,481	3,847	-	0	2	14	49
Minnesota	-	0	17	-	566	-	143	190	6,057	5,396	-	0	7	12	100
Missouri	-	4	17	214	211	195	288	392	12,815	11,017	1	3	10	133	116
Nebraska	1	2	6	108	117	24	47	106	2,316	2,283	2	1	3	28	36
North Dakota	-	1	6	44	30	2	20	38	966	808	-	0	2	15	13
South Dakota	-	2	7	92	92	-	27	44	1,011	1,066	-	0	3	20	17
South Atlantic	30	42	63	1,820	1,999	1,084	2,157	2,701	90,429	102,314	4	23	58	983	1,030
Delaware	-	0	2	1	23	-	30	79	1,389	1,500	-	0	2	13	20
District of Columbia	-	1	10	62	47	-	48	130	1,417	3,835	-	0	3	15	13
Florida	19	20	36	931	851	373	622	850	26,835	26,629	4	6	14	275	232
Georgia	-	5	34	265	507	155	350	478	14,155	19,548	-	3	14	176	185
Maryland	5	3	8	127	147	12	188	340	7,682	8,862	-	2	9	116	109
North Carolina	-	0	0	-	-	-	432	497	17,846	19,465	-	4	12	156	161
South Carolina	1	4	9	167	131	248	240	462	10,575	10,632	-	1	6	81	149
Virginia	5	6	31	241	207	296	215	430	9,630	10,722	-	3	21	104	125
West Virginia	-	0	5	26	86	-	21	42	900	1,121	-	1	8	47	36
East South Central	-	4	12	197	155	95	796	1,005	31,987	34,223	2	8	22	354	330
Alabama	-	4	12	197	155	-	234	386	9,864	9,892	-	2	7	77	73
Kentucky	N	0	0	N	N	-	107	200	3,396	6,261	2	2	8	88	64
Mississippi	N	0	0	N	N	13	187	306	7,808	7,514	-	1	5	64	51
Tennessee	N	0	0	N	N	82	270	322	10,919	10,556	-	3	11	125	142
West South Central	2	3	17	135	291	1,231	1,385	1,955	59,817	62,427	9	12	96	545	479
Arkansas	2	2	12	116	121	-	125	289	3,612	5,047	-	1	11	49	45
Louisiana	-	0	12	19	170	82	230	329	9,727	10,139	-	1	5	68	53

▲ **FIGURE 14.18** A page from the *MMWR*. The CDC's *Morbidity and Mortality Weekly Report (MMWR)* provides epidemiological data state by state for the current week, for the current year to date, and for the corresponding week in the previous year to date. The *MMWR* also publishes reports on epidemiological case studies. This page shows the incidence of three diseases in one week in 2018.

Epidemiology of Infectious Diseases 439

▲ FIGURE 14.19 **A map showing cholera deaths in a section of London, 1854.** From the map he compiled, Dr. John Snow showed that cholera cases centered around the Broad Street pump. Snow's work was a landmark in epidemiological research.

Of special concern to epidemiologists and health care workers are **healthcare-associated infections (HAIs)**, which were formerly called *nosocomial* (nos-ō-kō′mē-ăl) *infections*. HAIs are infections acquired by patients or health care workers while they are in health care facilities, including hospitals, dental offices, nursing homes, and doctors' waiting rooms. The CDC estimates that about 10% of American patients acquire an HAI each year. **Healthcare-associated diseases** (formerly called *nosocomial diseases*) increase the duration and cost of medical care and result in some 100,000 deaths annually in the United States.

Play *Nosocomial Infections: Overview*
@ Mastering Microbiology

Types of Healthcare-Associated Infections

When most people think of HAIs, what likely comes to mind are **exogenous HAIs** (eks-oj′ĕ-nŭs), which are caused by pathogens acquired from the health care environment. After all, hospitals are filled with sick people shedding pathogens from every type of portal of exit. However, we have seen that members of the resident microbiome can become opportunistic pathogens as a result of hospitalization or medical treatments such as chemotherapy. Such opportunists cause **endogenous HAIs** (en-doj′ĕ-nŭs); that is, they arise from normal microbiota within the patient that become pathogenic because of factors within the health care setting.

Iatrogenic infections (ī-at-rō-jen′ik; literally meaning "doctor-induced" infections) are a subset of HAIs that ironically are the direct result of modern medical procedures such as the use of catheters, invasive diagnostic procedures, and surgery.

Superinfections may result from the use of antimicrobial drugs that, by inhibiting some resident microbiota, allow others to thrive in the absence of competition. For instance, long-term antimicrobial therapy to inhibit a bacterial infection may allow *Clostridium difficile* (di′fe-sēl), a transient microbe of the colon, to grow excessively and cause a painful condition called *pseudomembranous colitis*. Such superinfections are not limited to health care settings.

HAIs most often occur in the urinary, respiratory, cardiovascular, and integumentary (skin) systems, though surgical wounds can become infected and result in HAIs in any part of the body.

Factors Influencing Healthcare-Associated Infections

HAIs arise from the interaction of several factors in the health care environment, which include the following (FIGURE 14.20):

- Exposure to numerous pathogens present in the health care setting, including many that are resistant to antimicrobial agents
- The weakened immune systems of patients who are ill, making them more susceptible to opportunistic pathogens
- Transmission of pathogens among patients and health care workers—from staff and visitors, to patients, and even from one patient to another via activities of staff members (including invasive procedures and other iatrogenic factors)

Control of Healthcare-Associated Infections

Aggressive control measures can noticeably reduce the incidence of HAIs. These include disinfection; medical asepsis, including good housekeeping, hand washing, bathing, sanitary handling of food, proper hygiene, and precautionary measures to avoid the spread of pathogens among patients; surgical asepsis and sterile procedures, including thorough cleansing of the surgical

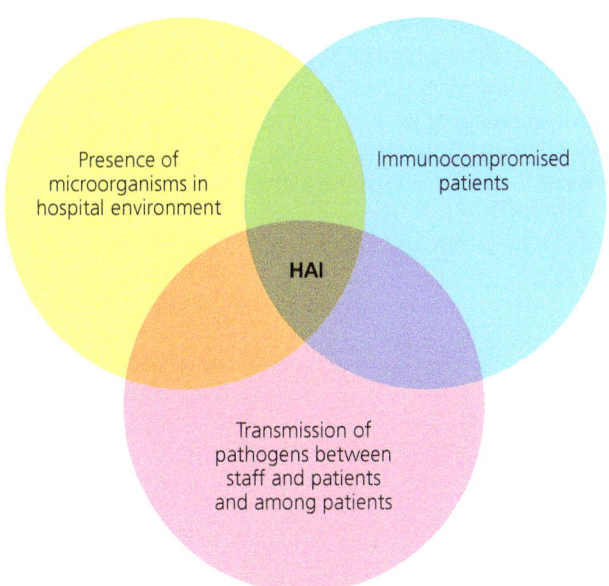

▲ FIGURE 14.20 **The interplay of factors that result in healthcare-associated infections (HAIs).** Although HAIs can result from any one of the factors shown, most are the product of the interaction of all three factors.

field, use of sterile instruments, and use of sterile gloves, gowns, caps, and masks; isolation of particularly contagious or susceptible patients; and establishment of an HAI control committee charged with surveillance of nosocomial diseases and review of control measures.

Play *Nosocomial Infections: Prevention* @ Mastering Microbiology

Numerous studies have shown that the single most effective way to reduce HAIs is effective hand washing by all medical and support staff. In one study, deaths from HAIs were reduced by over 50% when hospital personnel followed strict guidelines about washing their hands frequently.

Epidemiology and Public Health

LEARNING OUTCOME

14.29 List three ways public health agencies work to limit the spread of diseases.

As you have likely realized by now, epidemiologists gather information concerning the spread of disease within populations so that they can take steps to reduce the number of cases and improve the health of individuals within a community. In the following sections, we will examine how the various public health agencies share epidemiological data, facilitate the interruption of disease transmission, and educate the public about public health issues. Public health agencies also implement immunization programs (Chapter 17 discusses immunization).

The Sharing of Data Among Public Health Organizations

Numerous agencies at the local, state, national, and global levels work together with the entire spectrum of health care personnel to promote public health. By submitting reports on incidence and prevalence of disease to public officials, physicians can subsequently learn of current disease trends. Additionally, public health agencies often provide physicians with laboratory and diagnostic assistance.

City and county health departments report data on disease incidence to state agencies. Because state laws govern disease reporting, state agencies play a vital role in epidemiological studies. States accumulate data similar to those in the *MMWR* and assist local health departments and medical practitioners with diagnostic testing for diseases such as rabies and Lyme disease.

Data collected by the states are forwarded to the CDC, which is but one branch of the U.S. Public Health Service, the national public health agency. In addition to epidemiological studies, the CDC and other branches of the Public Health Service conduct research in disease etiology and prevention, make recommendations concerning immunization schedules, and work with public health organizations of other countries.

The World Health Organization (WHO) coordinates efforts to improve public health throughout the world, particularly in poorer countries, and the WHO has undertaken ambitious projects to eradicate such diseases as polio, measles, and mumps. Some other current campaigns involve AIDS education, malaria control, and childhood immunization programs in poor countries.

The Role of Public Health Agencies in Interrupting Disease Transmission

As we have seen, pathogens can be transmitted in air, food, and water as well as by vectors and via fomites. Public health agencies work to limit disease transmission by a number of methods:

- Enforce standards of cleanliness in water and food supplies.
- Work to reduce the number of disease vectors and reservoirs.
- Establish and enforce immunization schedules (see Figure 17.3).
- Locate and prophylactically treat individuals exposed to contagious pathogens.
- Establish isolation and quarantine measures to control the spread of pathogens.

A water supply that is **potable** (fit to drink) is vital to good health. Organisms that cause dysentery, cholera, and typhoid fever are just some of the pathogens that can be spread in water contaminated by sewage. Filtration and chlorination processes are used to reduce the number of pathogens in water supplies. Local, state, and national agencies work to ensure that water supplies remain clean and healthful by monitoring both sewage treatment facilities and the water supply.

Food can harbor infective stages of parasitic worms, protozoa, bacteria, and viruses. National and state health officials ensure the safety of the food supply by enforcing standards in the use of canning, pasteurization, irradiation, and chemical preservatives and by insisting that food preparers and handlers wash their hands and use sanitized utensils. The U.S. Department of Agriculture also provides for the inspection of meats for the presence of pathogens, such as *E. coli* and tapeworms.

Milk is an especially rich food, and the same nutrients that nourish us also facilitate the growth of many microorganisms. In the past, contaminated milk has been responsible for epidemics of tuberculosis, brucellosis, typhoid fever, scarlet fever, and diphtheria. Today, public health agencies require the pasteurization of milk, and as a result disease transmission via milk has been practically eliminated in the United States.

Individuals should assume responsibility for their own health by washing their hands before and during food preparation, using disinfectants on kitchen surfaces, and using proper refrigeration and freezing procedures. It is also important to thoroughly cook all meats.

Public health officials also work to control vectors, especially mosquitoes and rodents, by eliminating breeding grounds, such as stagnant pools of water and garbage dumps. Insecticides have been used with some success to control insects.

Public Health Education

Diseases that are transmitted sexually or through the air are particularly difficult for public health officials to control. In these cases, individuals must take responsibility for their own health, and health departments can only educate the public to make healthy choices.

Colds and flu remain the most common diseases in the United States because of the ubiquity of the viral pathogens and their mode of transmission in aerosols and via fomites. Health departments encourage afflicted people to remain at home, use disposable tissues to reduce the spread of viruses, and avoid crowds of coughing, sneezing people. As we have discussed, hand washing is also important in preventing the introduction of cold and flu viruses onto the conjunctiva.

We as a society are faced with several epidemics of sexually transmitted diseases. Syphilis, gonorrhea, genital warts, and sexually transmitted AIDS are completely preventable if the chain of transmission is interrupted by abstinence or mutually faithful monogamy. Their incidence can be reduced but not eliminated by the use of condoms. Based on the premise that "an ounce of prevention is worth a pound of cure," public health agencies expend considerable effort on public campaigns to educate people to make good choices—those that can result in healthier individuals and improved health for the public at large.

MICRO CHECK

12. What is a superinfection?
13. Why are sexually transmitted diseases particularly difficult to control through public health measures?

TELL ME WHY

Why are all iatrogenic infections healthcare associated, but not all healthcare-associated infections are iatrogenic?

MICRO IN THE CLINIC FOLLOW-UP
Just Travel Fatigue?

The doctor does another physical exam and orders blood work. Ellen has a fever (101°F), increased respiratory rate, body aches, and generalized weakness. The results of the blood work indicate that Ellen is anemic (low hemoglobin levels in the blood) and has elevated liver enzymes, indicating damage to her liver. Microscopic analysis of her blood reveals the presence of *Plasmodium*, the causative agent of malaria. Ellen is shocked—she used chloroquine as an antimalarial prophylaxis (a drug taken to prevent a disease) during her time in India—how could she have malaria? She explains that a friend had some chloroquine left over from a trip to Africa, so Ellen used it rather than pay for a prescription. Dr. Dalton explains that chloroquine-resistant *Plasmodium* is common in India, and therefore chloroquine isn't the most effective prophylactic.

The good news is that there are drugs that can effectively treat chloroquine-resistant *Plasmodium*. Dr. Dalton prescribes a seven-day course of atovaquone-proguanil (250 mg/100 mg). At a follow-up visit, Ellen has no remaining physical symptoms and no trace of *Plasmodium* in her blood.

1. **What is the difference between signs and symptoms of an infection? Review Ellen's case, and distinguish her signs from her symptoms.**

2. **Malaria is a zoonotic disease; Ellen was infected by a bite from the *Anopheles* mosquito. Identify each of the following for Ellen's case: portal of entry, mode of transmission, and reservoir.**

Check your answers to Micro in the Clinic Follow-Up questions in the Mastering Microbiology Study Area.

Go to Mastering Microbiology for Interactive Microbiology, Dr. Bauman's Video Tutors, Micro Matters, MicroFlix, MicroBoosters, practice quizzes, and more.

CHAPTER SUMMARY

Symbiotic Relationships Between Microbes and Their Hosts (pp. 414–418)

1. Microbes live with their hosts in **symbiotic** relationships, including **mutualism**, in which both members benefit; **parasitism**, in which a **parasite** benefits while the **host** is harmed; **amensalism**, in which one member is harmed while the second is neither harmed nor helped; and, more rarely, **commensalism**, in which one member benefits while the other is relatively unaffected. Any parasite that causes disease is called a **pathogen**.

2. Organisms making up the **microbiome** live in and on the body. Some of these microbes are resident microbiota, whereas others are transient microbiota.

3. **Opportunistic pathogens** cause disease when the immune system is suppressed, when normal **microbial antagonism (microbial competition)** is affected by certain changes in the body, or when a member of the normal microbiome is introduced into an area of the body unusual for that microbe.

Reservoirs of Infectious Diseases of Humans (pp. 418–420)

1. Living and nonliving continuous sources of infectious disease are called **reservoirs of infection**. Animal reservoirs harbor agents of **zoonoses**, which are diseases of animals that may be spread to humans via direct contact with the animal or its waste products or via an arthropod. Humans may be asymptomatic **carriers**.

2. **Nonliving reservoirs of infection** are inanimate objects, including soil, water, and food.

The Invasion and Establishment of Microbes in Hosts: Infection (pp. 420–422)

1. Microbial **contamination** refers to the mere presence of microbes in or on the body or object. Microbial contaminants include harmless resident and transient members of the microbiome as well as pathogens, which after a successful invasion cause an **infection**.
2. **Portals of entry** of pathogens into the body include skin, mucous membranes, and the placenta. These portals may be bypassed via the **parenteral route**, by which microbes are directly deposited into deeper tissues.
3. Pathogens attach to cells—a process called **adhesion**—via a variety of structures or attachment proteins called **adhesion factors**. Some bacteria and viruses lose the ability to make adhesion factors called adhesins and thereby become **avirulent**.
4. Some bacteria interact to produce a sticky web of cells and polysaccharides called a **biofilm** that adheres to a surface.

The Nature of Infectious Disease (pp. 422–430)

1. **Disease**, also known as **morbidity**, is a condition sufficiently adverse to interfere with normal functioning of the body.
2. **Symptoms** are subjectively felt by a patient, whereas an outside observer can observe **signs**. A **syndrome** is a group of symptoms and signs that collectively characterizes a particular abnormal condition.
3. **Asymptomatic**, or **subclinical**, infections may go unnoticed because of the absence of symptoms, even though clinical tests might reveal signs of disease.
4. **Etiology** is the study of the cause of a disease.
5. Nineteenth-century microbiologists proposed the **germ theory of disease**, and Robert Koch developed a series of essential conditions called **Koch's postulates** to prove the cause of infectious diseases. Certain circumstances can make the use of these postulates difficult or even impossible.
6. **Pathogenicity** is a microorganism's ability to cause disease; **virulence** is a measure of pathogenicity. **Virulence factors**, such as adhesion factors, extracellular enzymes, toxins, and antiphagocytic factors, affect the relative ability of a pathogen to infect and cause disease.

 Play *Virulence Factors: Hiding from Host Defenses, Inactivating Host Defenses, Penetrating Host Tissues; Phagocytosis: Microbes That Evade It* @ **Mastering** Microbiology

7. **Toxemia** is the presence in the blood of poisons called **toxins**. **Exotoxins** are secreted by pathogens into their environment. **Endotoxins** are lipopolysaccharides (LPS) released from the cell wall of dead and dying Gram-negative bacteria and can have fatal effects. The toxic portion of LPS is the lipid, called **lipid A**.

 Play *Virulence Factors: Exotoxins, Endotoxins* @ **Mastering** Microbiology

8. **Antitoxins** are antibodies the host forms against exotoxins.
9. The **disease process**—the stages of infectious diseases—typically consists of the **incubation period**, **prodromal period**, **illness**, **decline**, and **convalescence**.

The Movement of Pathogens Out of Hosts: Portals of Exit (p. 430)

1. **Portals of exit**, such as the nose, mouth, and urethra, allow pathogens to leave the body and are of interest in studying the spread of disease.

Modes of Infectious Disease Transmission (pp. 430–432)

1. **Direct contact transmission** of infectious diseases involves person-to-person spread by body contact. Transmission of pathogens via inanimate objects (called **fomites**) is called **indirect contact transmission**.

 Play *Epidemiology: Transmission of Disease* @ **Mastering** Microbiology

2. **Droplet transmission** (a third type of contact transmission) occurs when pathogens travel less than 1 meter in droplets of mucus to a new host as a result of speaking, coughing, or sneezing.
3. **Vehicle transmission** involves **airborne transmission**, **waterborne transmission**, and **foodborne transmission**. **Aerosols** are clouds of water droplets that travel more than 1 meter in airborne transmission. **Fecal–oral infection** can result from drinking sewage-contaminated water or from ingesting fecal contaminants. **Bodily fluid transmission** is the spread of pathogens via blood, urine, saliva, or other fluids.
4. **Vectors** transmit pathogens between hosts. **Biological vectors** are animals, usually biting arthropods, that serve as both host and vector of pathogens. **Mechanical vectors** are not hosts to the pathogens they carry.

Classification of Infectious Diseases (pp. 432–434)

1. There are various ways in which infectious disease may be grouped and studied. When grouped by time course and severity, disease may be described as **acute disease**, **subacute disease**, **chronic disease**, or **latent disease**.
2. When an infectious disease comes either directly or indirectly from another host, it is considered a **communicable disease**. If a communicable disease is easily transmitted from a reservoir or patient, it is called a **contagious disease**. **Noncommunicable diseases** arise either from outside of hosts or from members of the microbiome.

Epidemiology of Infectious Diseases (pp. 434–441)

1. **Epidemiology** is the study of where and when diseases occur and of how they are transmitted within populations.

Play *Epidemiology: Overview* @ **Mastering** Microbiology

2. Epidemiologists track the **incidence** (number of new cases) and **prevalence** (total number of cases) of a disease and classify disease outbreaks as **endemic** (usually present), **sporadic** (occasional), **epidemic** (more cases than usual), or **pandemic** (epidemic on more than one continent).

 Play *Epidemiology: Occurrence of Disease* @ **Mastering** Microbiology

3. **Descriptive epidemiology** is the careful recording of data concerning a disease; it often includes detection of the **index**

case—the first case of the disease in a given area or population. **Analytical epidemiology** seeks to determine the probable cause of a disease. **Experimental epidemiology** involves testing a hypothesis resulting from analytical studies.

4. **Healthcare-associated infections (HAIs)** (nosocomial infections) and **healthcare-associated diseases** (nosocomial diseases) are acquired by patients or workers in health care facilities. They may be **exogenous HAIs** (acquired from the health care environment), **endogenous HAIs** (derived from the normal microbiota that become opportunistic while in the hospital setting), or **iatrogenic infections** (induced by treatment or medical procedures).

   Play *Nosocomial Infections: Overview* @ Mastering Microbiology

5. Health care workers can help protect their patients and themselves from exposure to pathogens by hand washing and other aseptic and disinfecting techniques.

   Play *Nosocomial Infections: Prevention* @ Mastering Microbiology

6. Public health organizations, such as the World Health Organization (WHO), use epidemiological data to promulgate rules and standards for clean, **potable** water and safe food; to prevent disease by controlling vectors and animal reservoirs; and to educate people to make healthy choices concerning the prevention of disease.

QUESTIONS FOR REVIEW

Answers to the Questions for Review (except Short Answer questions) begin on p. A-I.

Multiple Choice

1. In which type of symbiosis do both members benefit from their interaction?
   a. mutualism
   b. parasitism
   c. commensalism
   d. pathogenesis

2. An axenic environment is one that _____.
   a. exists in the human mouth
   b. contains only one species
   c. exists in the human colon
   d. both a and c

3. Which of the following is *false* concerning microbial contaminants?
   a. Contaminants may become opportunistic pathogens.
   b. Most microbial contaminants will eventually cause harm.
   c. Contaminants may be a part of the transient microbiota.
   d. Contaminants may be introduced by a mosquito bite.

4. The most frequent portal of entry for pathogens is _____.
   a. the respiratory tract
   b. the skin
   c. the conjunctiva
   d. a cut or wound

5. The process by which microorganisms attach themselves to cells is _____.
   a. infection
   b. contamination
   c. disease
   d. adhesion

6. Which of the following is the correct sequence of events in infectious diseases?
   a. incubation, prodromal period, illness, decline, convalescence
   b. incubation, decline, prodromal period, illness, convalescence
   c. prodromal period, incubation, illness, decline, convalescence
   d. convalescence, prodromal period, incubation, illness, decline

7. Which of the following are most likely to cause disease?
   a. opportunistic pathogens in a weakened host
   b. pathogens lacking the enzyme kinase
   c. pathogens lacking the enzyme collagenase
   d. highly virulent organisms

8. The nature of bacterial capsules _____.
   a. causes widespread blood clotting
   b. allows phagocytes to readily engulf these bacteria
   c. affects the virulence of these bacteria
   d. has no effect on the virulence of bacteria

9. When pathogenic bacterial cells lose the ability to make adhesins, they typically _____.
   a. become avirulent
   b. produce endotoxin
   c. absorb endotoxin
   d. increase in virulence

10. A disease in which a pathogen remains inactive for a long period of time before becoming active is termed a(n) _____.
    a. subacute disease
    b. acute disease
    c. chronic disease
    d. latent disease

11. Which of the following statements is the best definition of a pandemic disease?
    a. It normally occurs in a given geographic area.
    b. It is a disease that occurs more frequently than usual for a geographical area or group of people.
    c. It occurs infrequently at no predictable time scattered over a large area or population.
    d. It is an epidemic that occurs on more than one continent at the same time.

12. Which of the following types of epidemiologists is most like a detective?
    a. descriptive epidemiologist
    b. analytical epidemiologist
    c. experimental epidemiologist
    d. reservoir epidemiologist

13. Consider the following case. An animal was infected with a virus. A mosquito bit the animal, was contaminated with the virus, and proceeded to bite and infect a person. Which was the vector?
    a. animal
    b. virus
    c. mosquito
    d. person

14. A patient contracted athlete's foot after long-term use of a medication. His physician explained that the malady was directly related to the medication. Such infections are termed _____.
    a. healthcare-associated infections
    b. exogenous infections
    c. iatrogenic infections
    d. endogenous infections

15. Which of the following phrases describes a contagious disease?
    a. a disease arising from fomites
    b. a disease that is easily passed from host to host in aerosols
    c. a disease that arises from opportunistic members of the resident microbiome
    d. both a and b

Fill in the Blanks

1. A microbe that causes disease is called a _____.
2. Infections that may go unnoticed because of the absence of symptoms are called _____ infections.
3. The study of the cause of a disease is _____.
4. The study of where and when diseases occur and how they are transmitted within populations is _____.
5. Diseases that are naturally spread from their usual animal hosts to humans are called _____.
6. Nonliving reservoirs of disease, such as a toothbrush, drinking glass, and needle, are called _____.
7. _____ infections are those acquired by patients or staff while in health care facilities.
8. The total number of cases of a disease in a given area is its _____.
9. An animal that carries a pathogen and also serves as host for the pathogen is a _____ vector.
10. Endotoxin, also known as _____, is part of the outer (wall) membrane of Gram-negative bacteria.

Short Answer

1. List four types of symbiotic relationships, and give an example of each.
2. List three conditions that create opportunities for pathogens to become harmful in a human.
3. List three portals through which pathogens enter the body.
4. List Koch's four postulates, and describe situations in which *not* all may be applicable.
5. List in the correct sequence the five stages of infectious diseases.
6. Describe three modes of disease transmission.
7. Describe the parenteral route of infection.
8. In general, contrast transient microbiota with resident microbiota.
9. Contrast the terms *infection* and *morbidity*.
10. Contrast iatrogenic and healthcare-associated diseases.

VISUALIZE IT!

1. Each of the three accompanying maps shows the locations (dots) of cases of a disease that normally occurs in the Western Hemisphere. Label each map with a correct epidemiological description of the disease's occurrence.

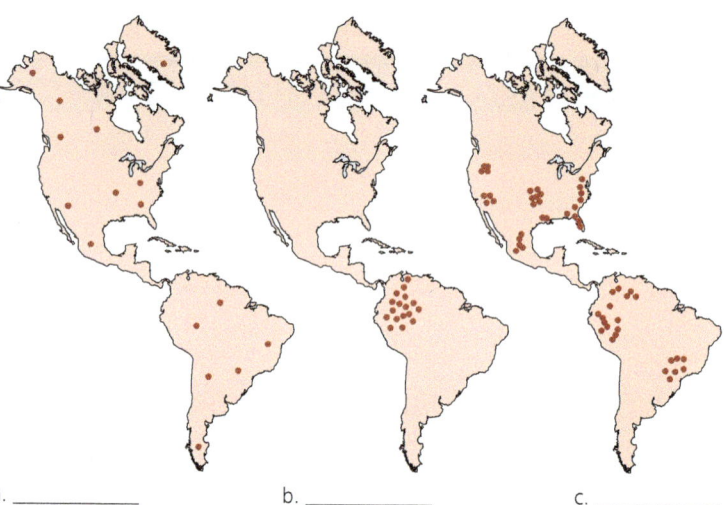

a. _____   b. _____   c. _____

2. Examine the graph of the red epidemic and the blue epidemic. Neither red disease nor blue disease is treatable. A person with either disease is ill for only one day and recovers fully. Both epidemics began at the same time. Which epidemic affected more people during the first three days? What could explain the short time course for the red epidemic? Why was the blue epidemic longer lasting?

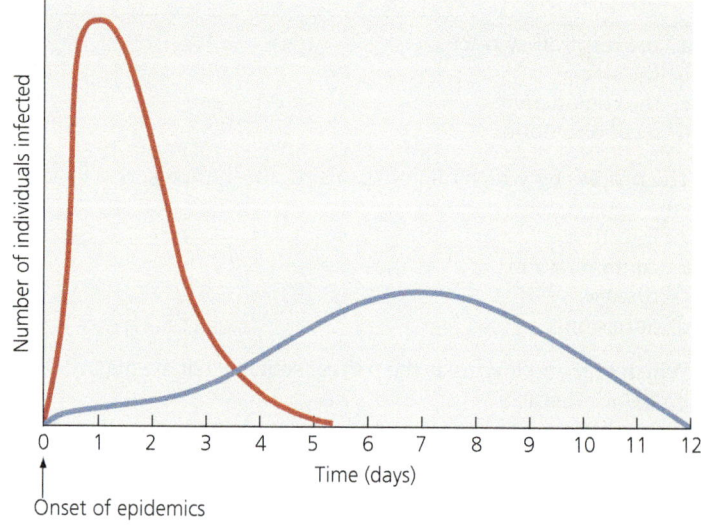

CRITICAL THINKING

1. Explain why Ellen H., a menopausal woman, may have developed gingivitis from her resident microbiome.

2. Will P. died of *E. coli* infection after an intestinal puncture. Explain why this microbe, which normally lives in the colon, could kill this patient.

3. A 27-year-old woman went to her doctor's office with a widespread rash, fever, malaise, and severe muscle pain. The symptoms had begun with a mild headache three days previously. She reported being bitten by a tick one week prior to that. The doctor correctly diagnosed Rocky Mountain spotted fever (RMSF) and prescribed tetracycline. The rash and other signs and symptoms disappeared in a couple of days, but she continued on her antibiotic therapy for two weeks. Draw a graph showing the course of disease and the relative numbers of pathogens over time. Label the stages of the disease.

4. Over 30 children younger than three years of age developed gastroenteritis after visiting a local water park. These cases represented 44% of the park visitors in this age group on the day in question. No older individuals were affected. The causative agent was determined to be a member of the bacterial genus *Shigella*. The disease resulted from oral transmission to the children. Based only on the information given, can you classify this outbreak as an epidemic? Why or why not? If you were an epidemiologist, how would you go about determining which pools in the water park were contaminated? What factors might account for the fact that no older children or adults developed disease? What steps could the park operators take to reduce the chance of future outbreaks of gastroenteritis?

5. A lichen is an intimate relationship between fungi and a photosynthetic microbe in which the fungi deliver water and minerals to the photosynthetic partner, which delivers sugar to the fungi. What type of symbiosis is involved?

6. Corals are colonial marine animals that feed by filtering small microbes from seawater in tropical oceans worldwide. Biologists have discovered that the cells of most corals are hosts to microscopic algae called *zooxanthellae*. Design an experiment to ascertain whether corals and zooxanthellae coexist in a mutual, commensal, or parasitic relationship.

7. If a mutation occurred in *Escherichia coli* that deleted the gene for an adhesin, what effect might it have on the ability of *E. coli* to cause urinary tract infections?

CONCEPT MAPPING

Using the terms provided, fill in this concept map that describes disease transmission. You can also complete this and other concept maps online by going to the Mastering Microbiology Study Area.

- Arthropods
- Biological
- Body
- Contact transmission
- Droplet transmission
- Foodborne
- Indirect contact
- Mosquito
- Vector transmission
- Vehicle transmission
- Waterborne

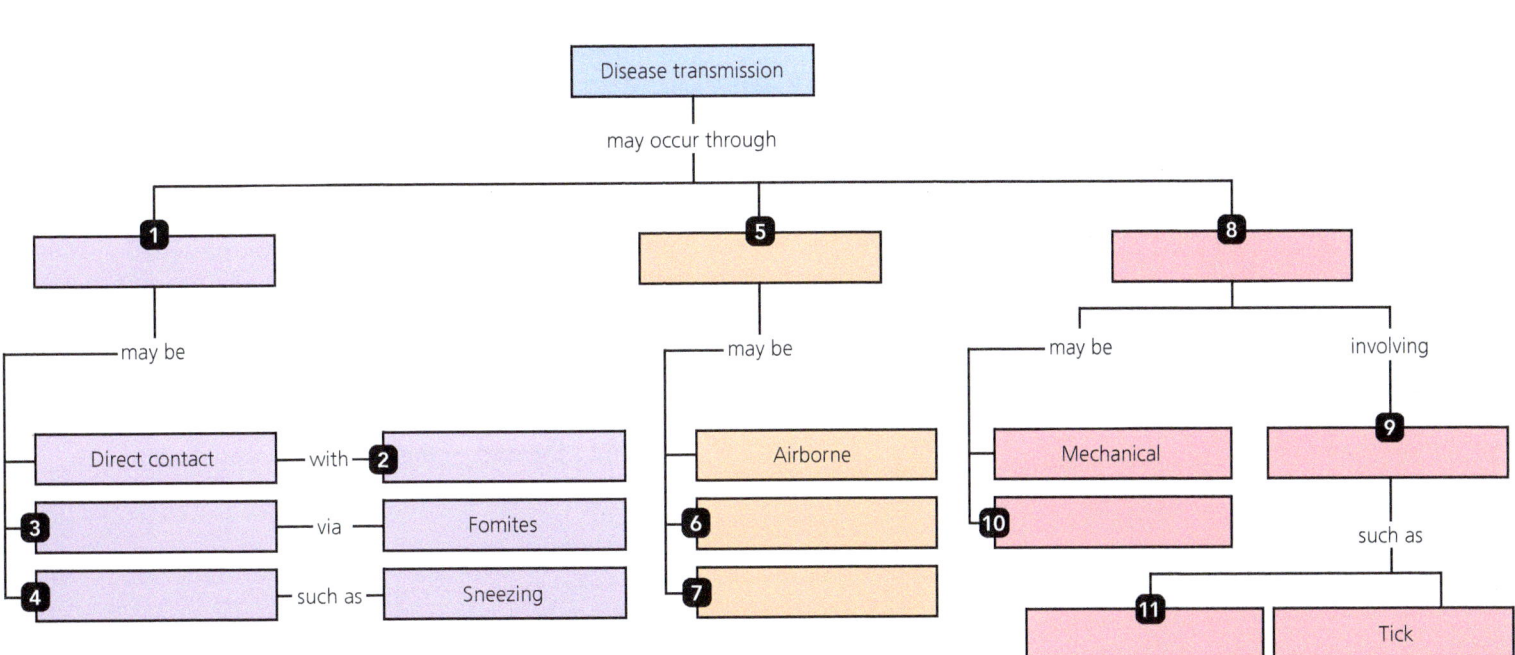

15 Innate Immunity

Before You Begin

1. Which eukaryotic organelle has catabolic enzymes that degrade old, damaged, and diseased cells?
2. What does it typically mean when a pathogen is transmitted via the *parenteral route*?
3. Which cellular structure can help bacteria avoid destruction by phagocytic cells?

Macrophages, such as this one which is phagocytizing a bacterium, are a critical part of innate immunity and can initiate adaptive immunity

MICRO IN THE CLINIC

Allergies in Bloom?

SPRING IS ALICE'S FAVORITE TIME OF THE YEAR. The days are getting longer and warmer, and flowers are in bloom. Alice is a preschool teacher, and she loves being able to take the kids outside for lessons and playtime in the park.

The only downside of the spring weather is that Alice's seasonal allergies flare up every spring. One morning she wakes up with a runny nose and notices that her right eye is very red, worse than normal for her allergy flare-ups. She goes to her local drugstore and gets some over-the-counter allergy medicine and eye drops and starts using them immediately.

A couple of days later, Alice's eye is much worse. The redness has increased, and her eye is watery, with tears welling up almost constantly. When Alice takes her class to the park, her sunglasses don't seem dark enough; her eye is very sensitive to the light. Alice decides that she should probably see her doctor to get some prescription-strength allergy medicine. She makes an appointment for the next morning.

1. Are Alice's symptoms the result of allergies?
2. Will prescription-strength allergy medicine bring Alice relief?

Turn to the end of the chapter (p. 466) to find out.

A pathogen causes a disease only if it can (1) gain access, either by penetrating the surface of the skin or by entering through some other portal of entry; (2) attach itself to host cells; and (3) evade the body's defensive mechanisms long enough to produce harmful changes. In this chapter, we examine the structures, processes, and chemicals that respond in a general way to protect the body from all types of pathogens.

An Overview of the Body's Defenses

LEARNING OUTCOMES

15.1 List and briefly describe the three lines of defense in the human body.

15.2 Explain the phrases *species resistance* and *innate immunity*.

Because the cells and certain basic physiological processes of humans are incompatible with structures and physiology of most plant and animal pathogens, humans have what is termed **species resistance** to these pathogens. In many cases, the chemical receptors that these pathogens require for attachment to a host cell do not exist in the human body; in other cases, the pH or temperature of the human body is incompatible with the conditions under which these pathogens can survive. Thus, for example, all humans have species resistance to tobacco mosaic virus, which affects only tobacco plants, and to the virus that causes feline immunodeficiency syndrome, which attaches only to cellular receptors found on cats' cells.

Nevertheless, we are confronted every day with other pathogens that can and do cause disease in humans. Bacteria, viruses, fungi, protozoa, and parasitic worms come in contact with your body in the air you breathe, in the water you drink, in the food you eat, and during the contacts you have with other people. Your body must defend itself from these potential pathogens and, in some cases, from members of your normal microbiome, which may become opportunistic pathogens.

It is convenient to cluster the structures, cells, and chemicals that act against pathogens into three so-called lines of defense, each of which overlaps and reinforces the other two. The first line of defense is composed chiefly of external physical barriers and associated chemicals and processes. The second line of defense is internal and composed of protective cells, blood-borne chemicals, and processes that inactivate or kill invaders. Together, the first two lines of defense are called **innate immunity** because they are present at birth. Innate immunity, which is also called *natural* or *native immunity*, is always active, nonspecific, and rapid, and it works against a wide variety of pathogens, including parasitic worms, protozoa, fungi, bacteria, and viruses.

Lymphocytes are the main cells of the third line of defense; unlike the participants in the first and second lines of defense, lymphocytes respond against unique species and strains of pathogens and alter the body's defenses such that they act more effectively upon subsequent infection by that same specific strain. In other words, they are specific and have memory. For this reason, scientists call the third line of defense **adaptive immunity** (see Chapter 16). Now we turn our attention to the two lines of innate immunity.

Play *Host Defenses*: Overview
@ Mastering Microbiology

TELL ME WHY

Why aren't the body's skin and mucous membrane barriers significant factors in your resistance to infection by hyperthermophiles?

The Body's First Line of Defense

The body's initial line of defense is made up of structures, chemicals, and processes (such as flow of mucus and blood coagulation) that work together to prevent pathogens from entering the body in the first place. Here, we discuss the main components of the first line of defense: the skin and the mucous membranes of the respiratory, digestive, urinary, and reproductive systems. These structures provide a formidable barrier to the entrance of microorganisms. When these barriers are pierced, broken, or otherwise damaged, they become portals of entry for pathogens. In this section, we examine aspects of the first line of defense, including the role of the microbiome, which is all the microorganisms in a given location, in this case the human body.

The Role of Skin in Innate Immunity

LEARNING OUTCOME

15.3 Identify the physical and chemical aspects of skin that enable it to prevent the entry of pathogens.

The skin—the organ of the body with the greatest surface area—is composed of two major layers: an outer **epidermis** and a deeper **dermis**, which contains hair follicles, glands, and nerve endings (see Chapter 14, Figure 14.4). Both the physical structure and the chemical components of skin enable it to act as an effective defense.

The epidermis is composed of multiple layers of tightly packed cells that constitute a physical barrier to most bacteria, fungi, and viruses. Few pathogens can penetrate the layers of epidermal cells unless the skin has been burned, broken, or cut, so wound care is a critical area for good health.

The deepest cells of the epidermis continually divide, pushing their daughter cells toward the surface. As the daughter cells are pushed toward the surface, they flatten and die and are eventually shed in flakes (FIGURE 15.1). Microorganisms that attach to the skin's surface are sloughed off with the flakes of dead cells. The skin illustrates two basic characteristics of the first line of defense in innate immunity: barriers and clearance. In this case, the tightly joined layer of epithelial cells acts as a barrier, and clearance comes as the outermost dead cells slough off.

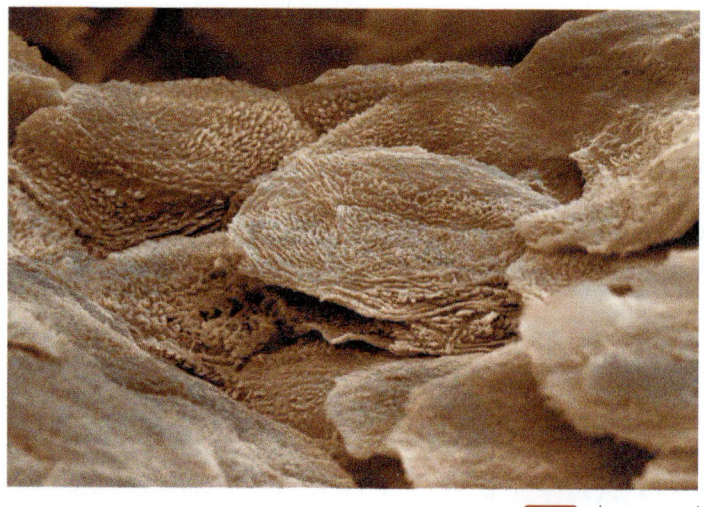

▲ FIGURE 15.1 **A scanning electron micrograph of the surface of human skin.** Epidermal cells are dead and dry and slough off, providing an effective barrier to most microorganisms.

The epidermis also contains defensive cells called **dendritic cells**.[1] The slender, fingerlike processes of dendritic cells extend among the surrounding cells, forming an almost continuous network to intercept invaders. Dendritic cells both devour pathogens nonspecifically and play a role in adaptive immunity (see Chapter 16).

The combination of the barrier function of the epidermis, its continual replacement, and the presence of phagocytic dendritic cells provides significant nonspecific defense against colonization and infection by pathogens.

The dermis also defends nonspecifically. It contains tough fibers of a protein called *collagen*, which give the skin strength and pliability to prevent many jabs and scrapes from penetrating the skin and introducing microorganisms. Blood vessels in the dermis deliver defensive cells and chemicals, which will be discussed shortly.

In addition to its physical structure, the skin has a number of chemical substances that nonspecifically defend against pathogens. Dermal cells secrete antimicrobial peptides, and sweat glands secrete perspiration, which contains salt, antimicrobial peptides, and an enzyme—lysozyme. Salt draws water osmotically from invading cells, inhibiting their growth and killing them.

Antimicrobial peptides are positively charged chains of 20 to 50 amino acids that act against microorganisms. Sweat glands secrete a class of antimicrobial peptides called *dermcidins*. Dermcidins are broad-spectrum antimicrobials that are active against many Gram-negative and Gram-positive bacteria and fungi. As expected of a peptide active on the surface of the skin, dermcidins are insensitive to low pH and salt. Dermcidins form permeable channels in bacterial cytoplasmic membranes. This causes them to leak, leading to membrane depolarization and cell death.

Lysozyme (līʹsō-zīm) is an enzyme that destroys the cell walls of bacteria by cleaving the bonds between the sugar subunits of the peptidoglycan walls. Bacteria without cell walls are more susceptible to osmotic shock and digestion by other enzymes within phagocytes (discussed later in the chapter as part of the second line of defense).

The skin also contains sebaceous (oil) glands, which secrete **sebum** (seʹbŭm). This oily substance not only helps keep the skin pliable and less sensitive to breaking or tearing but also contains fatty acids that lower the pH of the skin's surface to about pH 5, which is inhibitory to many bacteria. Free fatty acids also destabilize cells' membranes.

Although salt, antimicrobial peptides, lysozyme, and acidity make the surface of the skin an inhospitable environment for most microorganisms, some bacteria, such as *Staphylococcus epidermidis* (stafʹi-lō-kokʹŭs ep-i-der-midʹis), find the skin a suitable environment for growth and reproduction. Bacteria are particularly abundant in crevices around hairs and in the ducts of glands; usually they are nonpathogenic.

In summary, the skin is a complex barrier that limits access by microbes.

The Role of Mucous Membranes in Innate Immunity

LEARNING OUTCOMES

15.4 Identify the locations of the body's mucous membranes.
15.5 Explain how mucous membranes protect the body both physically and chemically.

Mucus-secreting (mucous) membranes, another major part of the first line of defense, cover all body cavities that are open to the outside environment. Thus, mucous membranes line the lumens[2] of the respiratory, urinary, digestive, and reproductive tracts. Like the skin, mucous membranes act nonspecifically to limit infection both physically and chemically.

Like the skin, mucous membranes have two distinct layers: an *epithelium*, in which cells form a covering that is superficial (closer to the surface, in this case, the lumen), and a deeper connective tissue layer that provides mechanical and nutritive support for the mucous membrane epithelium. As with the skin, epithelial cells of mucous membranes are packed closely together, but they form only a thin layer. Indeed, in some mucous membranes, the epithelium is only a single cell thick. Unlike surface epidermal cells, the surface cells of mucous membranes are alive and play roles in the diffusion of nutrients and oxygen (in the digestive, respiratory, and female reproductive systems) and in the elimination of wastes (in the urinary, respiratory, and female reproductive systems). The skin has a dry surface that inhibits bacterial growth, while mucous surfaces are moist; therefore the number of resident microbes is much higher on mucous membranes than on the skin.

The thin epithelium on the surface of a mucous membrane provides a less efficient barrier to the entrance of pathogens than the multiple layers of dead cells found at the skin's surface. So how are microorganisms kept from invading through thin mucous membranes? In some cases, they are not, which is why mucous membranes, especially those of the respiratory and reproductive systems, are common portals of entry for pathogens. Still, the epithelial cells of mucous membranes are tightly packed to prevent the entry of many pathogens, and the cells are continually shed

---

[1]From Greek *dendron*, meaning "tree," referring to their branched appearance.

[2]A lumen is a cavity or channel within any tubular structure or organ.

and then replaced by **stem cells**, which are generative cells capable of dividing to form daughter cells of various types. One effect of mucosal shedding is that it carries attached microorganisms away.

Phagocytic dendritic cells, like those in the epidermis, reside below the mucous epithelium to devour invaders. These cells are also able to extend cytoplasmic extensions (*pseudopods*) between epithelial cells to "sample" the contents of the lumen, helping to prepare adaptive immune responses against particular pathogens that might breach the mucosal barrier (a subject covered more fully in Chapter 16).

In addition, epithelia of some mucous membranes have still other means of removing pathogens. In the mucous membrane of the trachea, for example, stem cells produce both *goblet cells*, which secrete an extremely sticky mucus that acts as a barrier, trapping bacteria and other pathogens, and *ciliated columnar cells*, whose cilia propel the mucus and its trapped particles and pathogens up and away from the lungs (FIGURE 15.2). The effect of the action of the cilia is often likened to that of an escalator. Mucus carried into the throat is coughed up and either swallowed or expelled, again illustrating the barrier and clearance characteristics of innate immunity. Because poisons and tars in tobacco smoke damage cilia, the lungs of smokers are not properly cleared of mucus, so smokers may develop severe coughs as their respiratory tracts attempt to expel excess mucus, often leading to *COPD (chronic obstructive pulmonary disease)*. Smokers may succumb to many respiratory pathogens because they are unable to effectively clear pathogens from their lungs.

In addition to these physical actions, mucous membranes produce mucus, as mentioned. Mucus is mainly made of *mucins*—large proteins to which numerous sugar molecules are covalently bound. These *glycoproteins* are sticky and trap most microorganisms. Additionally, mucus contains chemicals that defend against pathogens. For example, mucus contains antimicrobial peptides. Nasal mucus also contains lysozyme. TABLE 15.1 compares some physical and chemical actions of the skin and mucous membranes in the body's first line of defense.

TABLE 15.1 The First Line of Defense: A Comparison of the Skin and Mucous Membranes

	Skin	Mucous Membrane
Number of Cell Layers	Many	One to a few
Cells Tightly Packed?	Yes	Yes
Cells Dead or Alive?	Outer layers: dead; inner layers: alive	Alive
Mucus Present?	No	Yes
Relative Water Content	Dry	Moist
Defensins Present?	Yes	Yes
Lysozyme Present?	Yes	With some
Sebum Present?	Yes	No
Cilia Present?	No	Trachea, uterine tubes
Constant Shedding and Replacement of Cells?	Yes	Yes

The Role of the Lacrimal Apparatus in Innate Immunity

LEARNING OUTCOME

15.6 Describe the lacrimal apparatus and the role of tears in combating infection.

The lacrimal apparatus is a group of structures that produce and drain away tears (FIGURE 15.3). Lacrimal glands, located above and to the sides of the eyes, secrete tears into lacrimal gland ducts and onto the surface of the eyes. The tears either evaporate or drain into small lacrimal canals, which carry them into nasolacrimal ducts that empty into the nose. There, the tears join the nasal mucus and flow into the pharynx, where they are swallowed. The blinking action of eyelids spreads the tears and washes the surfaces of the eyes. Normally, evaporation and flow into the nose balance the flow of tears onto the eye. However, if the eyes are irritated, increased tear production floods the eyes, carrying the irritant away. In addition to their washing action, tears contain *lysozyme*, which destroys bacteria.

The Role of the Microbiome in Innate Immunity

LEARNING OUTCOME

15.7 Define *microbiome*, and explain how it helps provide protection against disease.

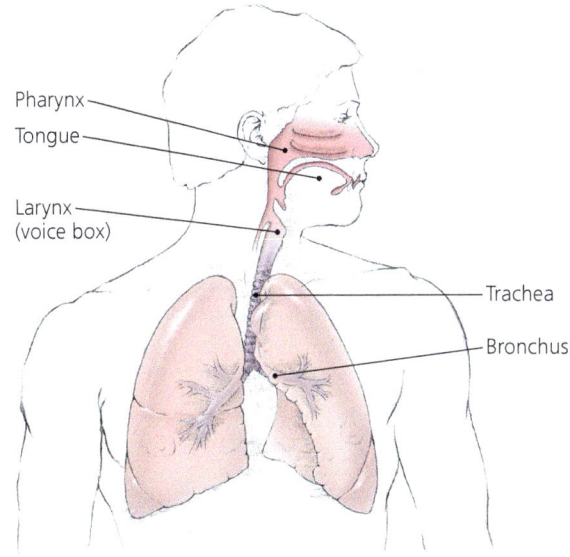

▲ **FIGURE 15.2 The structure of the respiratory system, which is lined with a mucous membrane.** The epithelium of the trachea contains mucus-secreting goblet cells and ciliated cells whose cilia propel the mucus (and the microbes trapped within it) up to the larynx for removal. *What is the function of stem cells within the respiratory epithelium?*

Figure 15.2 Stem cells in the respiratory epithelium undergo cytokinesis to form both ciliated cells and goblet cells to replace those lost during normal shedding.

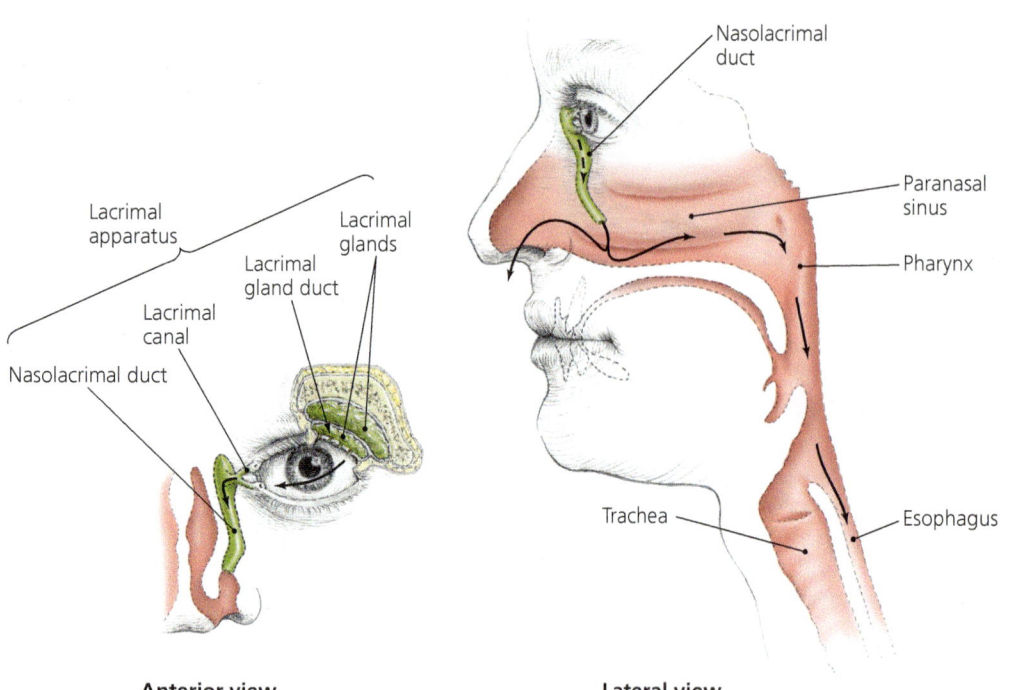

◀ **FIGURE 15.3 The lacrimal apparatus.** These structures (in green) function in the body's first line of defense by bathing the eye with tears, washing microbes away. Arrows indicate the route tears take across the eye and into the throat. *Name an antimicrobial protein found in tears.*

Figure 15.3 Tears contain lysozyme, an antimicrobial protein which acts against the peptidoglycan of bacterial cell walls.

The skin and mucous membranes of the body are home to a variety of protozoa, fungi, bacteria, and viruses. This **microbiome** (including normal members, the *normal microbiota*) plays a role in protecting the body by competing with potential pathogens in a variety of ways, a situation called **competitive inhibition** or *microbial antagonism*.

The microbiome carries out various activities that impede a pathogen's efforts to invade and produce disease:

- Members of the microbiome consume nutrients, making them unavailable to pathogens.
- The microbiome can change the pH of regions of the body, creating an environment that is favorable for themselves but unfavorable to other microorganisms, such as the acid produced in the vagina by resident *Lactobacillus*.
- The microbiome can attach to all of the sites on human cells, blocking pathogens from attaching.
- The microbiome stimulates the body's second line of defense (discussed shortly), boosting the body's production of antimicrobial substances. Researchers have observed that animals raised in an *axenic*[3] (ā-zen′ik) environment—that is, one free of all other organisms or viruses—are slower to defend themselves when exposed to a pathogen.
- Members of the microbiome can themselves generate antimicrobial compounds to defend their "turf," the human body.
- Nutritionally, the resident microbiota of the intestines improve overall health by providing several vitamins, including biotin (vitamin B_7) and pantothenic acid (vitamin B_5), which are important in glucose metabolism;

folic acid, which is essential for the production of the purine and pyrimidine bases of nucleic acids; and the precursor of vitamin K, which has an important role in blood clotting.

- Microbiota can generate compounds that modulate immunity. For example, gut bacteria ferment sugars to generate short-chain fatty acids that trigger inflammation and stimulate the gut epithelium to produce mucus and antimicrobial peptides.

Thus, the human microbiome plays many roles in protecting us from infection, from inhibiting invading pathogens and training the immune system, to strengthening immune responses. Given these facts, you can see that using antimicrobial soaps and shampoos to remove all microbes from your body is neither medically wise nor ecologically sound; in fact, it may make you more sensitive to infection. A significant component of medicine in the future may include the use of **probiotics**—giving live bacteria to patients to rebalance their microbiome to prevent or treat infection. While numerous probiotic products are available over the counter, there is currently no consistent scientific evidence to effectively guide their use.

Other First-Line Defenses

LEARNING OUTCOME

15.8 Describe antimicrobial peptides as part of the body's defenses.

Besides the physical barrier of the skin and mucous membranes, there are other hindrances to microbial invasion. Among these are additional antimicrobial peptides and other processes and chemicals.

---

[3]From Greek *a*, meaning "no," and *xenos*, meaning "foreigner."

Antimicrobial Peptides

As we saw in our examination of skin and mucous membranes, antimicrobial peptides (sometimes called *defensins*) act against microorganisms. Scientists have discovered hundreds of types of antimicrobial peptides in organisms as diverse as silkworms, frogs, and humans. Besides being secreted onto the surface of the skin, antimicrobial peptides are found in mucous membranes and in certain white blood cells called neutrophils. These peptides act against a variety of potential pathogens, being triggered by sugar and protein molecules on the external surfaces of microbes. Some antimicrobial peptides act only against Gram-positive bacteria or Gram-negative bacteria, others act against both, and still others act against protozoa, enveloped viruses, or fungi.

Researchers have elucidated several ways in which antimicrobial peptides work. Some punch holes in the cytoplasmic membranes of the pathogens, and others interrupt internal signaling or enzymatic action. Some antimicrobial peptides act as *chemotactic factors*—chemicals that recruit leukocytes to a site—while others assemble to form fibers and microscopic nets that ensnare invading bacteria.

Other Processes and Chemicals

Many other body organs contribute to the first line of defense by secreting chemicals with antimicrobial properties that are secondary to their prime function. For example, stomach acid is present primarily to aid digestion of proteins, but it also prevents the growth of many potential pathogens. The contributions of these and other processes and chemicals to the first line of defense are listed in **TABLE 15.2**.

MICRO CHECK

1. What is an example of species resistance?
2. How does lysozyme on the skin and in tears protect against bacterial pathogens?
3. How do mucous membranes protect the body?
4. What is the role of normal microbiota in preventing disease?

TELL ME WHY

Why doesn't lysozyme released in tears harm a person's eyes?

The Body's Second Line of Defense

LEARNING OUTCOME

15.9 Compare and contrast the body's first and second lines of defense against disease.

When pathogens succeed in penetrating the skin or mucous membranes, the body's second line of innate defense comes into play. Like the first line of defense, the second line operates against a wide variety of pathogens, from parasitic worms to viruses. But unlike the first line of defense, the second line includes no barriers; instead, it is composed of cells (especially phagocytes), antimicrobial chemicals (peptides, complement, interferons), and processes (inflammation, fever). Some cells and chemicals from the first line of defense play additional roles in the second line of defense. We will consider each component of the second line of defense in some detail shortly, but because many of them either are contained in or originate in the blood, we first consider the components of blood.

TABLE 15.2 Secretions and Activities That Contribute to the First Line of Defense

Secretion/Activity	Function
Digestive System	
Saliva	Washes microbes from teeth, gums, tongue, and palate; contains lysozyme, an antibacterial enzyme
Stomach acid	Digests and/or inhibits microorganisms
Gastroferritin	Sequesters iron being absorbed, making it unavailable for microbial use
Bile	Inhibitory to most microorganisms
Intestinal secretions	Digests and/or inhibits microorganisms
Peristalsis	Moves gastrointestinal (GI) contents through GI tract, constantly eliminating potential pathogens
Defecation	Eliminates microorganisms
Vomiting	Eliminates microorganisms
Urinary System	
Urine	Contains lysozyme; urine's acidity inhibits microorganisms; may wash microbes from ureters and urethra during urination
Reproductive System	
Vaginal secretions	Acidity inhibits microorganisms; contains iron-binding proteins that sequester iron, making it unavailable for microbial use
Menstrual flow	Cleanses uterus and vagina
Prostate secretion	Contains iron-binding proteins that sequester iron, making it unavailable for microbial use
Cardiovascular System	
Blood flow	Removes microorganisms from wounds
Coagulation	Prevents entrance of many pathogens

Defense Components of Blood

LEARNING OUTCOMES

15.10 Discuss the components of blood and their functions in the body's defense.
15.11 Explain how macrophages are named.

Blood is a complex liquid tissue composed of cells and portions of cells within a fluid called *plasma*. We begin our discussion of the defense functions of blood by briefly considering plasma.

Plasma

Plasma is mostly water containing electrolytes (ions), dissolved gases, nutrients, and—most relevant to the body's defenses—a variety of proteins. Some plasma proteins are

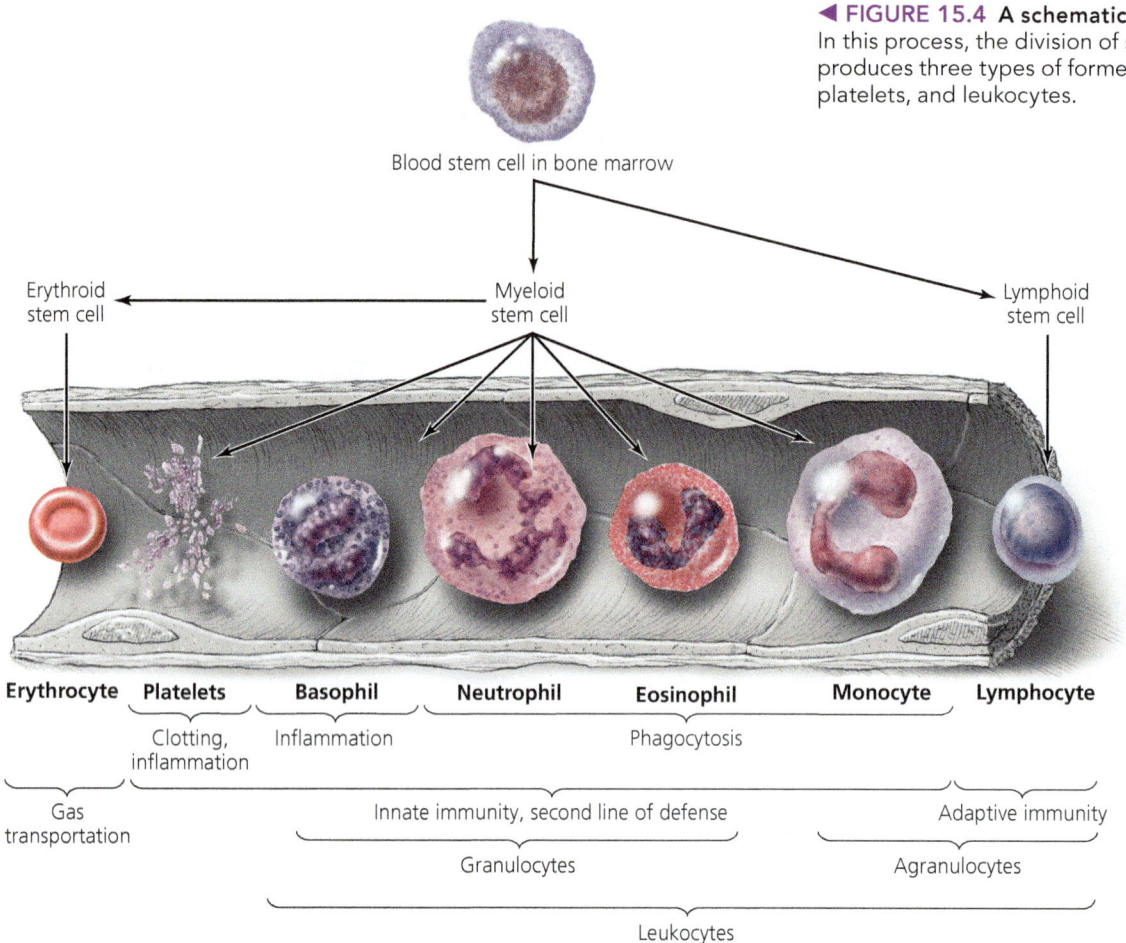

◀ **FIGURE 15.4 A schematic representation of hematopoiesis.** In this process, the division of stem cells in the bone marrow produces three types of formed (cellular) elements: erythrocytes, platelets, and leukocytes.

involved in inflammation (discussed later) and in blood clotting, a defense mechanism that reduces both blood loss and the risk of infection. When clotting factors have been removed from the plasma, as when blood clots, the remaining liquid is called *serum*.

Humans require iron for metabolism: it is a component of cytochromes of electron transport chains, functions as an enzyme cofactor, and is an essential part of hemoglobin—the oxygen-carrying protein of red blood cells. Because iron is relatively insoluble, in humans it is transported in plasma to cells by a transport protein called *transferrin*. When transferrin-iron complexes reach cells with receptors for transferrin, the binding of the protein to the receptor stimulates the cell to take up the iron via endocytosis. Excess iron is stored in the liver bound to another protein called *ferritin*. Though the main function of iron-binding proteins is transporting and storing iron, they play a secondary, defensive role—sequestering iron so that it is unavailable to microorganisms.

Some bacteria, such as *Staphylococcus aureus* (o'rē-ŭs), respond to a shortage of iron by secreting their own iron-binding compounds called *siderophores*. Because siderophores have a greater affinity for iron than does transferrin or other proteins, bacteria that produce siderophores can in effect steal iron from the body.

Some pathogens bypass this contest altogether. For example, *S. aureus* and related pathogens can secrete the protein *hemolysin*, which punches holes in the cytoplasmic membranes of red blood cells, releasing iron-containing hemoglobin. Other bacterial proteins then bind hemoglobin to the bacterial membrane and strip it of its iron. *Neisseria meningitidis* (nī-se'rē-ă me-nin-ji'ti-dis), a pathogen that causes often fatal meningitis, produces receptors for transferrin and plucks iron from the bloodstream as it flows by.

Another group of plasma proteins, called *complement proteins*, is an important part of the second line of defense and is discussed shortly. Still other plasma proteins, called *antibodies* or *immunoglobulins*, are a part of adaptive immunity, the body's third line of defense.

Defensive Blood Cells: Leukocytes

Cells and cell fragments suspended in the plasma are called **formed elements**. In a process called *hematopoiesis*,[4] blood stem cells located principally in the bone marrow within the hollow cavities of larger bones produce three types of formed elements **(FIGURE 15.4)**:

- So-called erythroid stem cells form **erythrocytes**[5] (ĕ-rith'rō-sītz; red blood cells).

---
[4] From Greek *haima*, meaning "blood," and *poiein*, meaning "to make."
[5] From Greek *erythro*, meaning "red," and *cytos*, meaning "cell."

- Myeloid stem cells produce **platelets**[6] (plāt'letz) and four kinds of **leukocytes**[7] (loo'kō-sīts; white blood cells).
- Lymphoid stem cells produce a fifth kind of leukocyte.

Erythrocytes are the most numerous of the formed elements and carry oxygen and carbon dioxide in the blood. Platelets, which are pieces of large cells called *megakaryocytes* that have split into small portions of cytoplasm surrounded by cytoplasmic membranes, are involved in blood clotting. Leukocytes, the formed elements that are directly involved in defending the body against invaders, are commonly called *white blood cells* because they form a whitish layer when the components of blood are separated within a test tube.

Based on their appearance in stained blood smears when viewed under the microscope, leukocytes are divided into two groups: *granulocytes* (gran'ū-lō-sītz) and *agranulocytes* (ā-gran'ū-lō-sītz) **(FIGURE 15.5)**.

Granulocytes have large granules in their cytoplasm that stain different colors depending on the type of granulocyte and the dyes used: **basophils** (bā'sō-fils) stain blue with the basic dye methylene blue, **eosinophils** (ē-ō-sin'ō-fils) stain red to orange with the acidic dye eosin, and **neutrophils** (noo'trō-fils), also known as *polymorphonuclear leukocytes (PMNs)*, stain lilac with a mixture of acidic and basic dyes. Both neutrophils and eosinophils phagocytize pathogens, and both can exit the blood to attack invading microbes in the tissues by squeezing between the cells lining capillaries (the smallest blood vessels). This process is called **diapedesis**[8] (dī'a-pĕ-dē'sis). As we will see later in the chapter, eosinophils are also involved in defending the body against parasitic worms and are present in large number during many allergic reactions, though their exact function in allergies is disputed. Basophils can also leave the blood by diapedesis, though they are not phagocytic; instead, they release inflammatory chemicals, an aspect of the second line of defense that will be discussed shortly.

The cytoplasm of **agranulocytes** appears uniform when viewed via light microscopy, though granules are visible with an electron microscope. Agranulocytes are of two types: **monocytes** (mon'ō-sītz), which are large agranulocytes with slightly lobed nuclei, and **lymphocytes** (lim'fō-sītz), which are the smallest leukocytes and have nuclei that nearly fill the cells. Although most lymphocytes are involved in adaptive immunity, *natural killer (NK) lymphocytes* function in innate defense and thus are discussed later in this chapter. Monocytes also leave the blood by diapedesis and mature into **macrophages** (mak'rō-fāj-ĕz), which are phagocytic cells of the second line of defense. Their initial function is to devour foreign objects, including bacteria, fungi, spores, and dust as well as dead body cells.

Macrophages are often named for their location in the body, because they specialize for each body tissue. *Wandering macrophages* leave the blood via diapedesis and perform their scavenger function while traveling throughout the body in intercellular

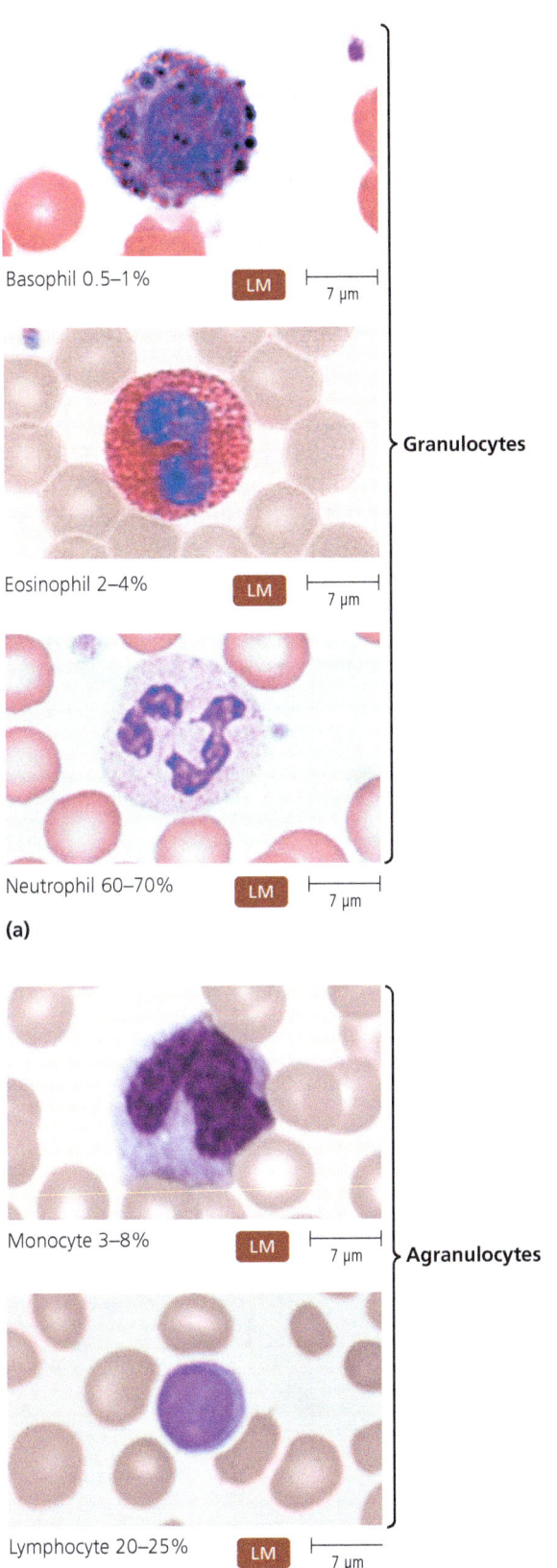

▲ **FIGURE 15.5 Leukocytes as seen in stained blood smears. (a)** Granulocytes: basophil, eosinophil, and neutrophil. **(b)** Agranulocytes: monocyte and lymphocyte. The numbers are typical percentages for each cell type among all leukocytes.

---

[6]French for "small plates." Platelets are also called *thrombocytes*, from Greek *thrombos*, meaning "lump," and *cytos*, meaning "cell," though they are technically not cells but instead pieces of cells.
[7]From Greek *leuko*, meaning "white," and *cytos*, meaning "cell."
[8]From Greek *dia*, meaning "through," and *pedan*, meaning "to leap."

spaces. Other macrophages are fixed and do not wander. These include *alveolar macrophages* (al-vē'ō-lăr)[9] of the lungs and *microglia* (mī-krog'lē-ă) of the central nervous system. Fixed macrophages generally phagocytize within specific organs, such as the heart chambers, blood vessels, and lymphatic vessels. (The lymphatic system is discussed in Chapter 16.)

Dendritic cells (mentioned previously) are a special group of phagocytes derived from myeloid stem cells. These multi-branched cells are plentiful throughout the body, particularly in the skin and mucous membranes. Dendritic cells await microbial invaders, phagocytize them, and inform cells of adaptive immunity that there is a microbial invasion.

Lab Analysis of Leukocytes Analysis of blood for diagnostic purposes, including white blood cell counts, is one task of medical lab technologists. The proportions of leukocytes, as determined in a **differential white blood cell count**, can serve as a sign of disease. For example, an increase in the percentage of eosinophils can indicate allergies or infection with parasitic worms; bacterial diseases typically result in an increase in the number of leukocytes and an increase in the percentage of neutrophils, whereas viral infections are associated with an increase in the relative number of lymphocytes. An increase in the number of a certain leukocyte over normal can indicate cancer, and a decrease in the number can reveal immunodeficiency. The ranges for the normal values for each kind of white blood cell, expressed as a percentage of the total leukocyte population, are shown in Figure 15.5.

Now that we have some background concerning the defensive properties of plasma components and leukocytes, we turn our attention to the details of the body's second line of defense: phagocytosis, nonphagocytic killing by leukocytes, nonspecific chemical defenses, inflammation, and fever.

Phagocytosis

LEARNING OUTCOME

15.12 Name and describe the six stages of phagocytosis.

Phagocytosis, which means "eating by a cell," is a way that some microbes obtain nutrients (see Chapter 3), but **phagocytes** (fag'ō-sītz)—phagocytic defense cells of the body—use phagocytosis to rid the body of pathogens that have evaded the body's first line of defense.

Play *Phagocytosis*: Overview
@ **Mastering** Microbiology

For improved understanding, we will consider the process of phagocytosis in six steps: chemotaxis, adhesion, ingestion,

---

[9]Alveoli are small pockets at the end of respiratory passages where oxygen and carbon dioxide exchange occurs between the lungs and the blood.

CLINICAL CASE STUDY

Evaluating an Abnormal CBC

CBC Profile			
Name: Brown, Roger	Age/Sex: 61/M		Attend Dr: Kevin, Larry
Acct#: 04797747	Status: ADM IN		
Reg: 11/27/12			
SPEC #: 0303:AS:H00102T	COLL: 12 / 03 / 12-0620	STATUS: COMP	
	RECD: 12 / 03 / 12-0647	SUBM DR: Kevin, Larry	
ENTERED: 12 / 03 / 12-0002		OTHER DR: NONE, PER PT	
ORDERED: CBC W/ MAN DIFF			REQ #: 01797367

Test	Result	Normal range
CBC		
WBC (white blood cells)	0.8	4.8–10.8 K/mm3
RBC (red blood cells)	3.09	4.20–5.40 M/mm3
HGB (hemoglobin)	9.6	12.0–16.0 g/dL
HCT (hematocrit)	28.2	37.0–47.0 %
MCV	91.3	81.0–99.0 fL
MCH	31.1	27.0–31.0 pg
MCHC	34.1	32.0–36.0 g/dl
RDW	17.1	11.5–14.5 %
PLT (platelets)	21	150–450 K/mm3
MPV	8.7	7.4–10.4 fL
DIFF		
CELLS COUNTED	100	#CELLS
SEGS	39	
BAND	4	
LYMPH (lymphocytes)	41	
MONO (monocytes)	15	
EOS (eosinophils)	1	
NEUT# (# neutrophils)	0.3	1.9–8.0 K/mm3
LYMPH#	0.3	0.9–5.2 K/mm3
MONO#	0.1	0.1–1.2 K/mm3
EOS#	0.0	0–0.8 K/mm3
PLATELET EST	DECREASED	

Roger Brown, an African American cancer patient, received a chemotherapeutic agent as a treatment for his disease. The drug used to destroy the cancer also produced an undesirable condition known as *bone marrow suppression*. The complete blood count (CBC) profile shown here indicates that this patient is in trouble. Review the lab values, and answer the following questions.

1. Note that the platelet count is very low. How does this affect the patient? Discuss measures to protect him.
2. Note that the white blood cell count is also abnormally low. With the second line of defense impaired, how can the first line of defense be protected?

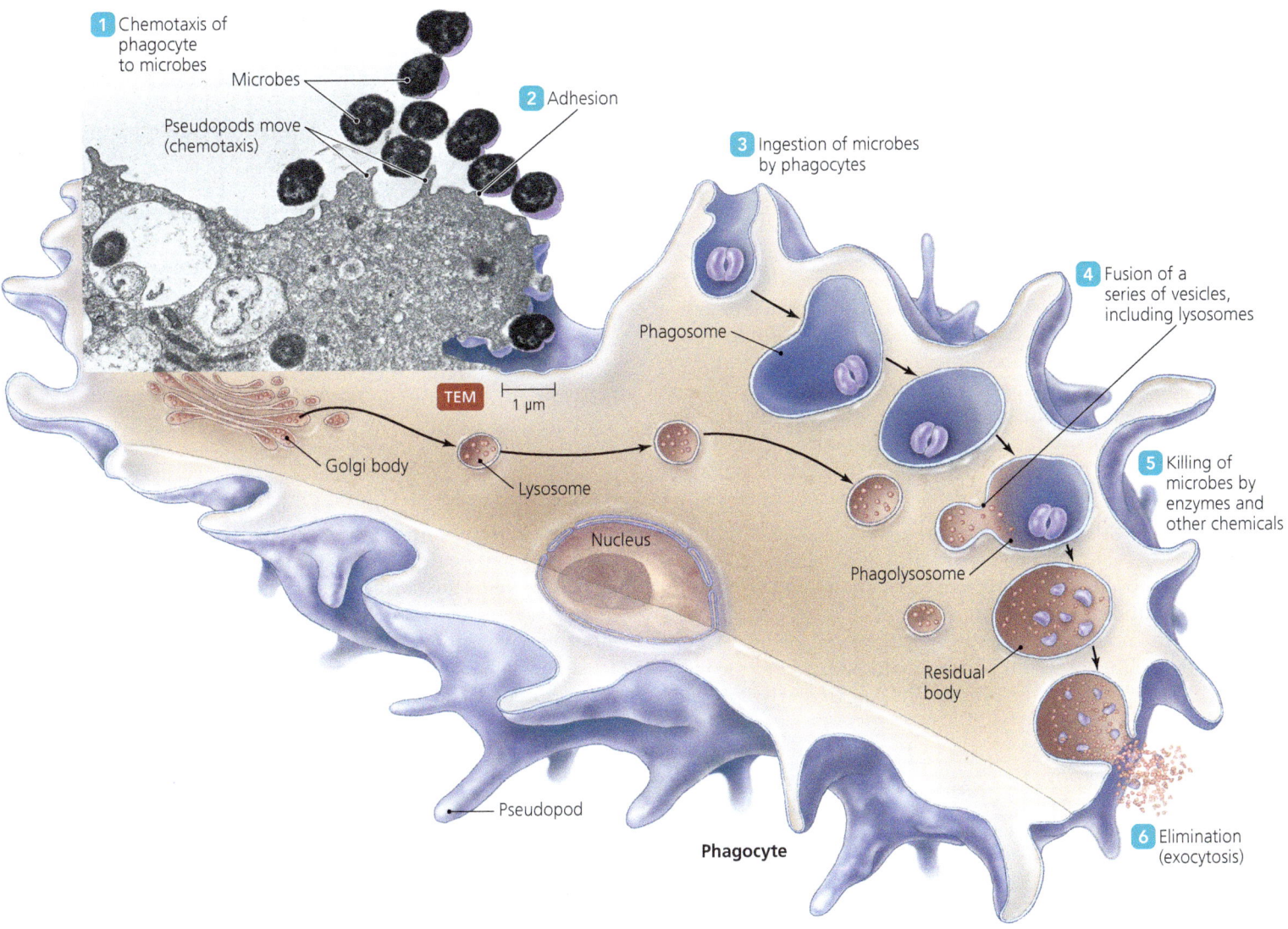

▲ **FIGURE 15.6 The events in phagocytosis.** Here, a neutrophil phagocytizes *Neisseria gonorrhoeae*.

maturation, killing, and elimination **(FIGURE 15.6)**. However, the process is actually continuous and doesn't proceed in a discrete stepwise manner.

Chemotaxis

Chemotaxis is movement of a cell either toward a chemical stimulus (positive chemotaxis) or away from a chemical stimulus (negative chemotaxis). In the case of phagocytes, positive chemotaxis involves the use of *pseudopods* (soo'dō-podz) to crawl toward microorganisms at the site of an infection **1**. Chemicals that attract phagocytic leukocytes include microbial secretions, parts of microbial cells, components of damaged tissues and white blood cells, and **chemotactic factors** (kem-ō-tak'tik). Chemotactic factors include defensins, peptides derived from complement (discussed later in this chapter), and chemicals called **chemokines** (kē'mō-kīnz), which are released by leukocytes already at the site of infection.

Adhesion

After arriving at the site of an infection, phagocytes attach to microorganisms through the binding of receptors to complementary chemicals, such as glycoproteins, found on the membranes of cells **2**. This process is called **adhesion** or *attachment*. Some bacteria have virulence factors, such as slippery capsules, which hinder adhesion by phagocytes.

All pathogens are more readily phagocytized if they are first covered with antimicrobial proteins, such as complement proteins (discussed later) or specific antimicrobial proteins of adaptive immunity called *antibodies* (discussed in Chapter 16). This coating process is called **opsonization**[10] (op'sŭ-nī-za'shun), and the proteins are called **opsonins**. Generally, opsonins increase the number and kinds of binding sites on a microbe's surface.

---

[10]From Greek *opsonein*, meaning "to supply food," and *izein*, meaning "to cause"; thus, loosely, "to prepare for dinner."

Ingestion

After phagocytes adhere to pathogens, they extend pseudopods to surround the microbe ③. The encompassed microbe is internalized as the pseudopods fuse to form a food vesicle called a **phagosome**.

Phagosome Maturation and Microbial Killing

A membranous organelle called a *lysosome* adds digestive chemicals to the maturing phagosome, which is now called a **phagolysosome** (fag-ō-lī′sō-sōm) ④. Phagolysosomes contain antimicrobial substances, such as highly reactive, toxic forms of oxygen, in an environment with a pH of about 5.5 due to active pumping of H^+ from the cytosol into the phagolysosome. These substances, along with 30 or so different enzymes, such as lipases, proteases, and nucleases, destroy the engulfed microbes ⑤.

Most pathogens are dead within 30 minutes of the phagolysosome's formation, though some bacteria contain virulence factors (such as waxy cell walls) that resist a lysosome's action. In the end, a phagolysosome is known as a *residual body*.

Elimination

Digestion is not always complete, and phagocytes eliminate remnants of microorganisms via *exocytosis*, a process that is essentially the reverse of ingestion ⑥. Some microbial components are specially processed and remain attached to the cytoplasmic membrane of some phagocytes, particularly dendritic cells, a phenomenon that plays a role in the adaptive immune response (discussed in Chapter 16).

Play *Phagocytosis: Mechanism*
@ Mastering Microbiology

How is it that phagocytes destroy invading pathogens and leave the body's own healthy cells unharmed? At least two mechanisms are responsible for this:

- Some phagocytes have cytoplasmic membrane receptors for various microbial surface components lacking on the body's cells, such as microbial cell wall components or flagellar proteins.
- Opsonins such as certain blood proteins and immune molecules called *antibodies* provide a signal to the phagocyte.

Nonphagocytic Killing

LEARNING OUTCOME

15.13 Describe the role of eosinophils, NK cells, and neutrophils in nonphagocytic killing of microorganisms and parasitic helminths.

Phagocytosis involves killing a pathogen once it has been ingested—that is, once it is inside the phagocyte. In contrast, eosinophils, natural killer cells, and neutrophils can accomplish killing without phagocytosis.

Killing by Eosinophils

As discussed earlier, eosinophils can phagocytize; however, this is not their usual mode of attack. Instead, eosinophils secrete antimicrobial chemicals against pathogens that are too large to ingest. They attack parasitic helminths (worms) by adhering to the worm's surface, where they secrete toxins onto the surface of the worms. These toxins weaken the helminth and may even kill it. **Eosinophilia** (ē-ō-sin′ō-fil-e-ă), an abnormally high number of eosinophils in the blood, is often indicative of helminth infestation, though it is also commonly seen in patients with allergies. Besides their attacks against parasitic helminths, eosinophils have recently been discovered to use a never-before-seen tactic against bacteria: Lipopolysaccharide from Gram-negative bacterial cell walls triggers eosinophils to rapidly eject their mitochondrial DNA, which combines with previously extruded eosinophil proteins to form a physical barrier. This extracellular structure binds to and then kills the bacteria. This is the first evidence that DNA can have antimicrobial activity, and scientists are investigating exactly how mitochondrial DNA might act as an antibacterial agent.

Killing by Natural Killer Lymphocytes

Natural killer lymphocytes (NK cells) are another type of defensive leukocyte of innate immunity that works by secreting toxins onto the surfaces of virally infected cells and cancer cells. NK cells identify and spare normal body cells because the latter express membrane proteins similar to those on the NK cells. The name *natural killer* comes from the fact that NK cells are programmed to kill other cells unless they receive an inhibitory signal.

Killing by Neutrophils

Neutrophils can also destroy nearby microbial cells without phagocytosis, though phagocytosis is more common. They do this in several ways:

- Enzymes in a neutrophil's cytoplasmic membrane add electrons to oxygen, creating highly reactive superoxide radical O_2^- and hydrogen peroxide (H_2O_2). Another enzyme converts these into hypochlorite, the active antimicrobial ingredient in household bleach. These chemicals can kill nearby invaders.
- Yet another enzyme in the neutrophil's membrane makes nitric oxide, which is a powerful inducer of inflammation.
- Neutrophils also generate webs of extracellular fibers nicknamed *NETs*, for *neutrophil extracellular traps*. Neutrophils synthesize NETs via a unique form of cellular suicide involving the disintegration of their nuclei. As the nuclear envelope breaks down, DNA and histones are released into the cytosol, and the mixing of nuclear components with membranes and proteins derived from cytoplasmic granules combine to form NET fibers. The neutrophil then kills itself using superoxide and peroxide. The dying cell releases the NETs as its cytoplasmic membrane ruptures. NETs trap both Gram-positive and Gram-negative bacteria, immobilizing them and sequestering them along with antimicrobial peptides, which kill the bacteria. Thus even in their dying moments, neutrophils fulfill their role as defensive cells.

MICRO CHECK

5. Which formed blood element is responsible for blood clotting?
6. What is diapedesis?
7. How do phagocytes digest foreign agents such as pathogens?
8. Which white blood cells secrete antimicrobial chemicals onto large parasites such as helminths?

Nonspecific Chemical Defenses Against Pathogens

LEARNING OUTCOMES

15.14 Define *Toll-like receptors*, and describe their action in relation to pathogen-associated molecular patterns.
15.15 Describe the location and functions of NOD proteins.
15.16 Explain the roles of interferons in innate immunity.
15.17 Describe the complement system, including its three activation pathways.

Chemical defenses augment phagocytosis in the second line of defense. The chemicals assist phagocytic cells either by enhancing other features of innate immunity or by directly attacking pathogens. Defensive chemicals include lysozyme and antimicrobial peptides (examined previously) as well as Toll-like receptors, NOD proteins, interferons, and complement.

Toll-Like Receptors

A key function of immunity is to distinguish normal chemical shapes, "self," from foreign chemical shapes, "nonself." The immune system then degrades and removes objects with nonself features, such as pathogens, while leaving self-tissues and cells alone. How does innate immunity accomplish recognition of nonself chemicals? Certain structures are present on bacteria, fungi, and other microbes that are lacking on human cells. These structures, called **pathogen-associated molecular patterns (PAMPs)**, are critical components of the microbes and thus cannot be easily altered by mutation. Some examples of PAMPs are lipopolysaccharide, peptidoglycan, and flagellin (see Chapter 3).

Pattern recognition receptors are proteins on the surface of and inside human cells that can recognize PAMPs. The best-understood receptors, **Toll-like receptors (TLRs)**,[11] are integral proteins of the cytoplasmic membranes of phagocytic cells.

Ten TLRs are known for humans. TLRs 1, 2, 4, 5, and 6 are found in cytoplasmic membranes, whereas TLRs 3, 7, 8, and 9 are added to phagosome membranes. Some TLRs act alone; others act in pairs to recognize a particular PAMP. For example, TLR3 binds to double-stranded RNA from viruses such as West Nile virus, and TLR2 and TLR6 in conjunction bind to lipoteichoic acid—a component of Gram-positive cell walls. **TABLE 15.3**

TABLE 15.3 Toll-Like Receptors and Their Natural Microbial Binding Partners

TLR	PAMP (Microbial Molecule)
In Cytoplasmic Membrane	
TLR1	Bacterial lipopeptides and certain proteins in multicellular parasites
TLR2	Peptidoglycan, lipoteichoic acid (found in Gram-positive cell wall), and cell wall of yeast
TLR4	Lipid A in LPS (found in outer membrane of Gram-negative bacteria)
TLR5	Flagellin (bacterial flagella)
TLR6	Bacterial lipopeptides, lipoteichoic acid, and cell wall of yeast
TLR10	Unknown component of influenzaviruses
In Phagosome Membrane	
TLR3	Double-stranded RNA (found only in viruses)
TLR7	Single-stranded viral RNA
TLR8	Single-stranded viral RNA
TLR9	Unmethylated cytosine-guanine pairs of viral and bacterial DNA

summarizes the membrane locations of the 10 known TLRs of humans and the PAMPs they bind.

Binding of a PAMP to a Toll-like receptor initiates a number of possible defensive responses, including secretion of inflammatory mediators or interferons (both discussed shortly), production of chemical stimulants of adaptive immune responses (discussed in Chapter 16), or apoptosis (cell suicide) of an infected cell. If TLRs fail, much of an immune response collapses, leaving the body open to attack by myriad pathogens.

Scientists are actively seeking ways to stimulate TLRs so as to enhance a body's immune response to pathogens and immunizations. In contrast, methods to inhibit TLRs may provide us with ways to counter inflammatory disorders and hyperimmune responses.

NOD Proteins

NOD proteins[12] are another set of pattern recognition receptors for microbial molecules, such as PAMPs, but NOD proteins are located in a cell's cytoplasm rather than as part of a membrane. Scientists have studied NOD proteins that bind to components of Gram-negative bacteria's cell walls and RNA of viruses, such as those that cause AIDS, hepatitis C, and mononucleosis. NOD proteins trigger inflammation, apoptosis, and other innate immune responses against bacterial pathogens, though researchers are still elucidating their exact method of action. Mutations in NOD genes are associated with several inflammatory bowel diseases, including Crohn's disease.

Interferons

So far in this chapter we have focused primarily on how a body defends itself against bacterial and eukaryotic pathogens. Now we consider how chemicals in the second line of defense act against viruses.

---

[11]*Toll* is a German word meaning "fantastic," originally referring to a gene of fruit flies, mutations of which cause the flies to look bizarre. Toll-like proteins are similar to fruit fly Toll proteins in their amino acid sequence, though not in their function.

[12]Nucleotide-binding oligomerization domains, referring to their ability to bind a region of a finite number (oligomer) of DNA nucleotides.

Viruses use their host's metabolic machinery to produce new viruses. For this reason, it is often difficult to interfere with virus replication without harming the host. **Interferons** (in-ter-fēr'onz; IFNs) are protein molecules released by host cells to nonspecifically inhibit the spread of viral infections. Their lack of specificity means that interferons produced against one viral invader protect somewhat against infection by other types of viruses as well. However, interferons also cause malaise, muscle aches, chills, headache, and fever, which are typically associated with viral infections.

Different cell types produce one of two basic types of interferon when stimulated by viral nucleic acid binding to certain Toll-like receptors (TLR3, TLR7, or TLR8). Interferons within any given type share certain physical and chemical features, though they are specific to the species that produces them. In general, type I interferons—also known as alpha and beta interferons—are present early in viral infections, whereas type II (gamma) interferon appears somewhat later in the course of infection.

Type I (Alpha and Beta) Interferons Within hours after infection, virally infected monocytes, macrophages, and some lymphocytes secrete small amounts of **alpha interferon (IFN-α)**; similarly, fibroblasts, which are undifferentiated cells in such connective tissues as cartilage, tendon, and bone, secrete small amounts of **beta interferon (IFN-β)** when they are infected by viruses. The structures of alpha and beta interferons are similar, and their actions are identical; therefore, they are classed together as *type I interferons*.

Type I interferons can act on the cells that secrete them, cells already infected with viruses, and trigger protective steps in neighboring uninfected cells. Alpha and beta interferons bind to interferon receptors on the cytoplasmic membrane. Such binding triggers the production of **antiviral proteins (AVPs)**, which remain inactive within these cells until AVPs encounter viral nucleic acids, particularly double-stranded RNA, a molecule that is common among viruses but generally absent in eukaryotic cells (FIGURE 15.7).

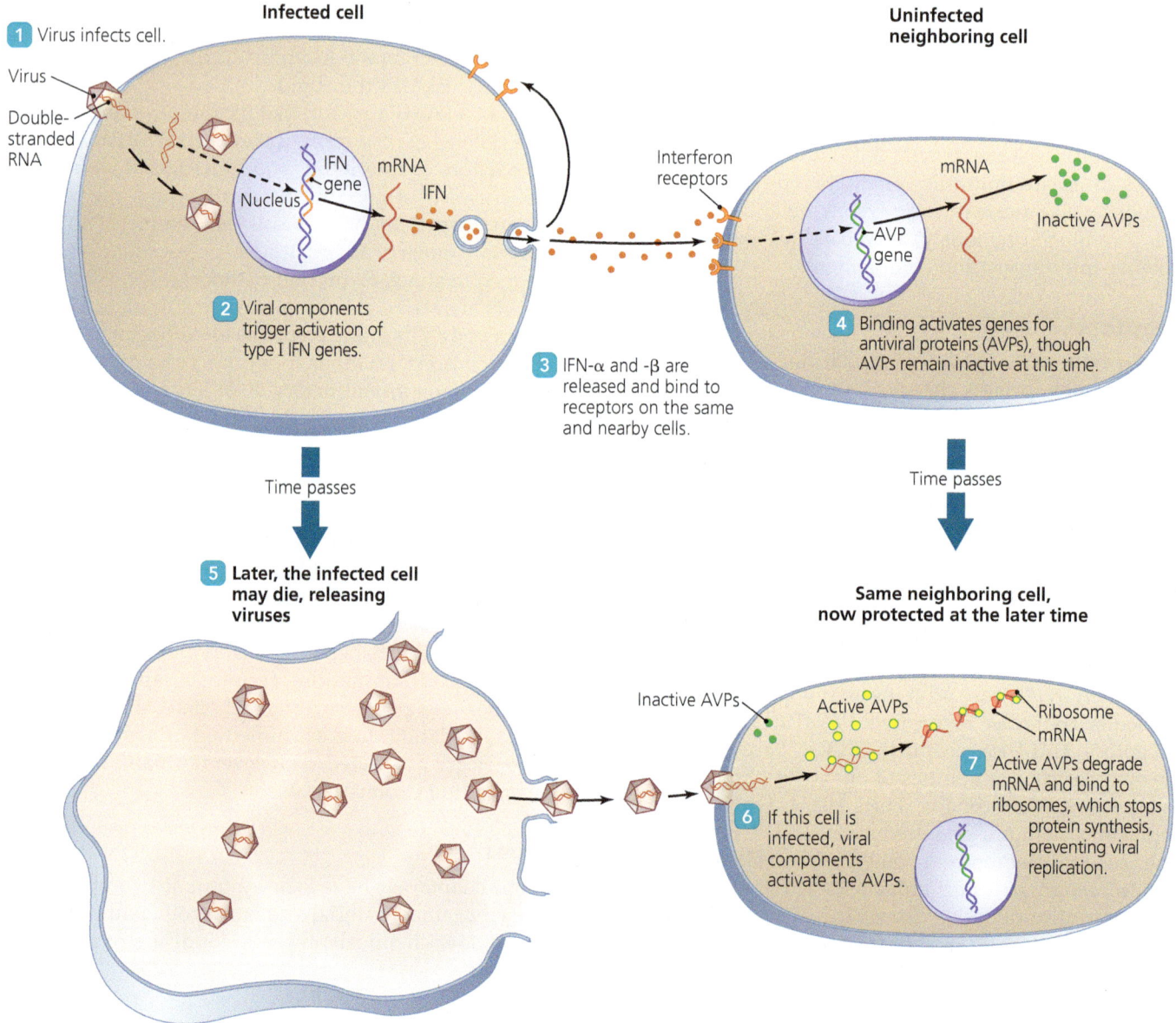

▲ **FIGURE 15.7** The actions of alpha and beta interferons.

TABLE 15.4 The Characteristics of Human Interferons

Property	Type I		Type II
	Alpha Interferon (IFN-α)	Beta Interferon (IFN-β)	Gamma Interferon (IFN-γ)
Principal source	Epithelium, leukocytes	Fibroblasts	Activated T lymphocytes and NK lymphocytes
Inducing agent	Viruses	Viruses	Adaptive immune responses
Action	Stimulates production of antiviral proteins	Stimulates production of antiviral proteins	Stimulates phagocytic activity of macrophages and neutrophils
Other names	Leukocyte-IFN	Fibroblast-IFN	Immune-IFN, macrophage activation factor

At least two types of antiviral proteins are produced: *oligoadenylate synthetase*, the action of which results in the destruction of RNA, and *protein kinase R*, which inhibits protein synthesis by ribosomes. Between them, these AVP enzymes essentially destroy the protein production system of the cell, preventing viruses from being replicated. Of course, cellular metabolism is also affected negatively. The antiviral state lasts three to four days, which may be long enough for oligoadenylate synthetase to rid the cell of viral RNA but still a short enough period for the cell to survive without protein production. Type I interferons also activate NK and other lymphocytes, which kill virally infected cells, halting virus reproduction.

Type II (Gamma) Interferon Gamma interferon (IFN-γ) is produced by activated T lymphocytes and by natural killer (NK) lymphocytes. Because T lymphocytes are usually activated as part of an adaptive immune response (see Chapter 16) days after an infection has occurred, gamma interferon appears later than either alpha or beta interferon. Its action in stimulating the activity of macrophages gives IFN-γ its other name: *macrophage activation factor*. Gamma interferon plays a small role in protecting the body against viral infections; mostly, IFN-γ regulates the immune system, as in its activation of phagocytic activity.

Physicians can administer interferons created using recombinant DNA techniques to patients with viral infections or cancers. This treatment can activate immunological responses.

TABLE 15.4 summarizes various properties of interferons in humans.

MICRO CHECK

9. What are Toll-like receptors (TLRs)?
10. What is one function of interferons?

Complement

 Play an *Interactive Microbiology* video @ Mastering Microbiology for more information about the complement system.

The **complement system**—or **complement** for short—is a set of serum proteins, many of which are designated numerically according to the order of their discovery (C1, C2, C3, etc.). These proteins initially act as opsonins and chemotactic factors and indirectly trigger inflammation and fever. The end result of full complement activation can be destruction of foreign cells. Complement needs to act quickly without destroying normal cells surrounding a pathogen or virally infected cell. This is accomplished by utilizing proteins in the blood that are not activated until needed, like horses in a starting cage, poised to run when the gate opens.

 Play *Complement*: Overview @ Mastering Microbiology

Complement Activation Complement is activated in three ways (FIGURE 15.8):

- In the *classical pathway*, antibodies, which are part of adaptive immunity (and discussed in detail in Chapter 16), bind

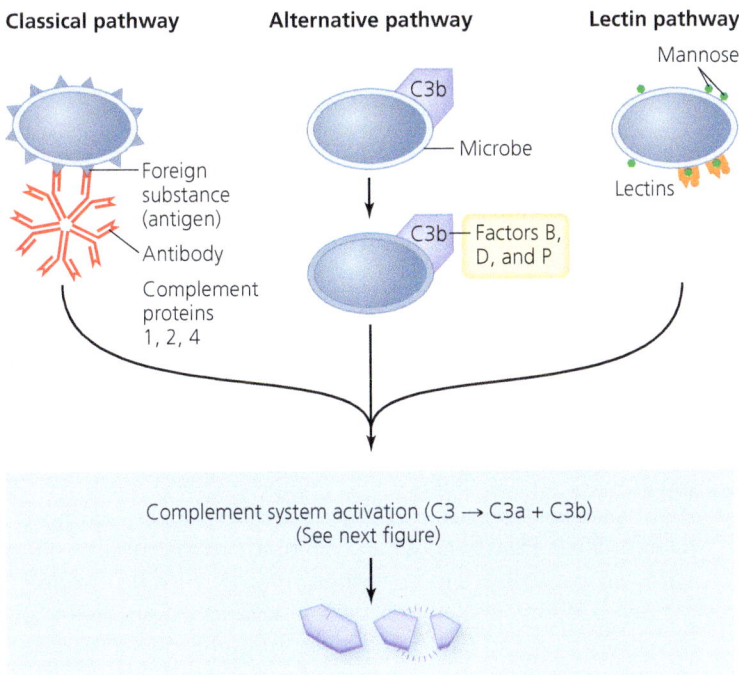

▲ FIGURE 15.8 **Pathways by which complement is activated.** In the classical pathway, the binding of antibodies to foreign substances (antigens) activates complement. In the alternative pathway, the binding of complement fragment C3b and factors B, D, and P to surfaces of microbes activates complement. In the lectin pathway, mannose, a sugar found on many fungi, bacteria, and viruses, binds to lectins, thereby activating complement. *How did complement get that name?*

Figure 15.8 Complement proteins add to—or complement—the action of antibodies.

CLINICAL CASE STUDY

The Stealth Invader

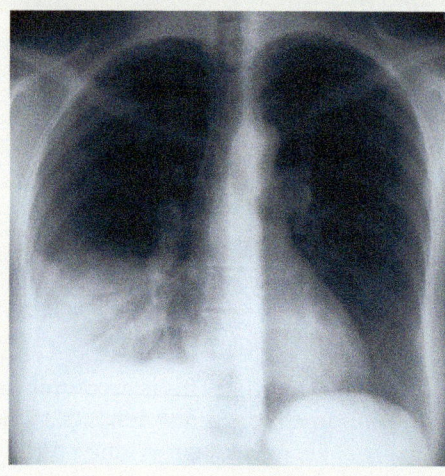

Tim is often seen walking around campus, hanging out at the coffee shop, laughing with his friends, and, as he puts it, "investing time with the ladies." Tim started smoking cigarettes in high school and has never tried to kick the habit. He jokes about the "smoker's cough" that has punctuated his conversations over the summer and start of the fall semester. Recently, the cough has been getting worse, his the throat is scratchy, his head hurts, he is tired all the time, and he might have a little fever. Tim wonders if he has some long-lasting flu, though he has had little or no fever. He decides not to waste time with a doctor's visit, though the cough starts keeping him up at night. In October, after a month of progressively worse coughing, he's had enough; he heads to the campus clinic right after his morning classes.

The clinic physician notes the persistence and worsening of Tim's cough and confirms that Tim has a low-grade fever. To rule out a bacterial infection, she orders a routine sputum culture (a test of the material coughed up from Tim's lungs), which comes back negative. Tim is sent home with the advice to quit smoking.

A week later, he's back, coughing nonstop, short of breath, sweaty, and aching. The doctor orders a chest X-ray exam and several blood tests. The X-ray film reveals fluid in Tim's lungs, and real-time PCR confirms an infection with *Mycoplasma pneumoniae*, which is an atypical Gram-positive bacterium. Ordinary sputum microscopy doesn't detect this pathogen, which lacks cell walls and doesn't stain well.

Mycoplasma pneumoniae infects and disrupts the mucous membranes of the lungs, invading and disrupting the epithelium. Pneumonia caused by *M. pneumoniae* is called "walking pneumonia" because the symptoms are often milder and come and go, making this disease different from streptococcal pneumonia. Most patients get better on their own, but not Tim. Smoke has compromised the lungs' innate immunity. A five-day course of an antimicrobial drug—azithromycin—clears Tim's lungs of the bacterium within a week. Tim decides to quit smoking.

1. Which Toll-like receptors (TLRs) might have been involved in Tim's innate immune response to *Mycoplasma*?
2. Why didn't Tim's naturally occurring interferons help clear the infection?
3. What structures and chemicals normally fend off lung infections?
4. Why didn't Tim's innate lung defenses operate properly?

---

to foreign substances (called *antigens*), and this binding begins a series of enzyme activities called a *complement cascade*. The complement system was so named because its various proteins act to "complement," or act in conjunction with, antibodies.

- In the *alternative pathway*, activation occurs independently of antibodies with the cleavage of C3 into two fragments, C3a and C3b. This naturally occurs at a slow rate in the plasma, but in the absence of infection proceeds no further because C3b is cleaved into inactive fragments almost immediately. However, when C3b binds to microbial surfaces, it stabilizes long enough to work with three other plasma proteins (called factors B, D, and P) to split more C3 into more C3a and C3b. As C3b accumulates on the surface of the microbe, it acts as an opsonin and also triggers a complement cascade. The alternative pathway is useful in the early stages of an infection, before the adaptive immune response has created the antibodies needed to activate the classical pathway.

- In the *lectin pathway*, mannose sugar, which is found on the surfaces of fungi, bacteria, and viruses, attaches to activating molecules called *lectins*.[13] This begins the series of enzyme activities as in the classical activation pathway, leading to a complement cascade.

 Play *Complement*: Activation, Results @ Mastering Microbiology

Complement Cascade Complement proteins react with one another in an amplifying sequence of chemical reactions in which the product of each reaction becomes an enzyme that catalyzes the next reaction many times over. For example, a single molecule of complement protein might activate 50 molecules of a second complement protein, each of which might in turn activate 30 molecules of a third complement protein, ending up with activation of 1500 of the third molecule. Such reactions are called *cascades* because they progress in a way that can be likened to a rock avalanche in which one rock dislodges several other rocks, each of which dislodges many others until a whole cascade of rocks is tumbling down the mountain.

As you study the depiction of the classical complement cascade in FIGURE 15.9, keep the following concepts in mind:

- Complement enzymes cleave other, inactive complement proteins, producing *fragments* that are active and are designated with lowercase letters. For example, inactive complement protein 3 (C3) is cleaved into active fragments C3a and C3b.

---
[13] From Latin *lēct*, meaning "selected;" these carbohydrate-binding proteins are highly specific for certain sugar molecules.

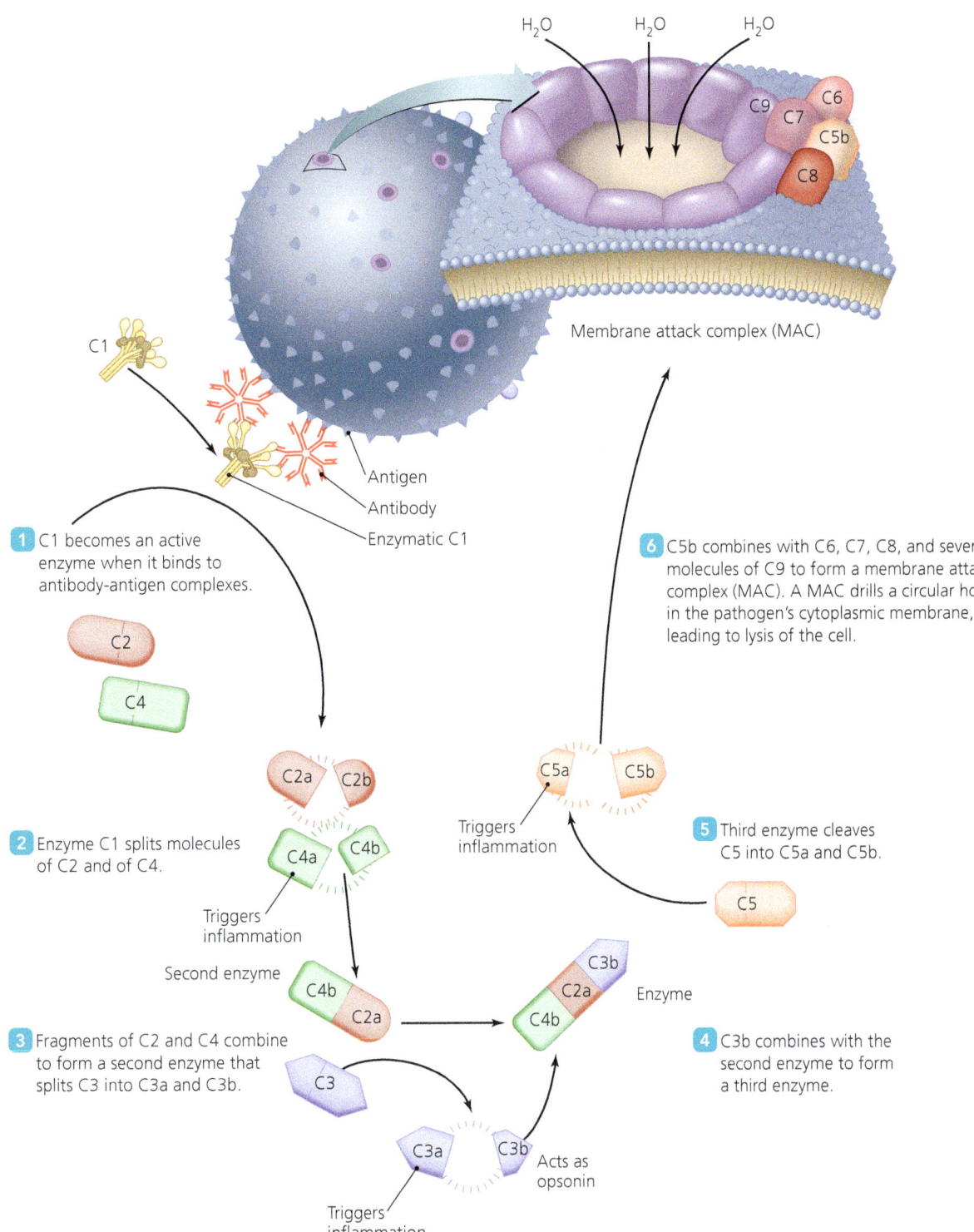

▲ FIGURE 15.9 **The classical pathway and the complement cascade.** Steps 2 to 6 are illustrated apart from the pathogen's membrane for clarity; in reality, a complement cascade occurs at the site where antibodies on the membrane activate C1. The major functions of complement are opsonization and mediation of chemotaxis and inflammation. A membrane attack complex can form in a pathogen's membrane, resulting in lysis of the pathogen. *What proteins would be involved in activating a complement cascade if this were the alternative pathway?*

Figure 15.9 Whereas the classical pathway of complement activation involves proteins C1, C2, and C4, the alternative pathway involves factors B, D, and P.

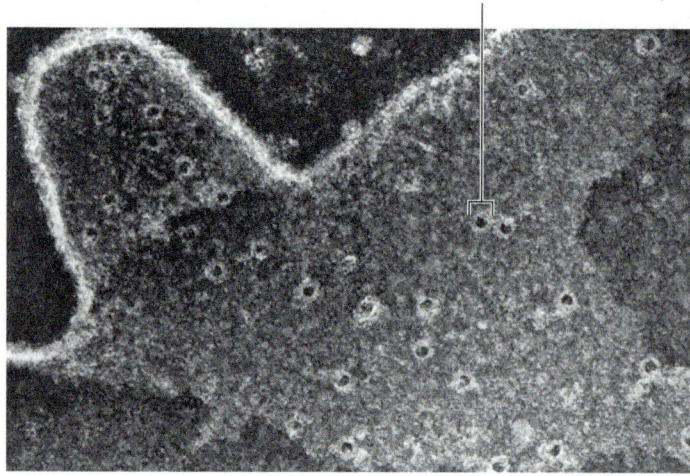

▲ FIGURE 15.10 **Membrane attack complexes.** Transmission electron micrograph of a cell damaged by multiple punctures produced by membrane attack complexes.

MICRO CHECK

11. Which of the pathways to activate complement proteins is triggered by mannose on the surface of fungi, bacteria, and viruses?
12. What is the end result of the formation of the membrane attack complex by complement proteins?

Inflammation

LEARNING OUTCOME

15.18 Discuss the process and benefits of inflammation.

Inflammation is a general, nonspecific response to tissue damage resulting from a variety of causes, including heat, chemicals, ultraviolet light (sunburn), abrasions, cuts, and pathogens. **Acute inflammation** develops quickly, is short-lived, is typically beneficial, and results in the elimination or resolution of whatever condition precipitated it. Long-lasting **chronic inflammation** causes damage (even death) to tissues, resulting in disease. Both acute and chronic inflammation exhibit similar signs and symptoms,

Play *Inflammation:* Overview
@ Mastering Microbiology

including redness in light-colored skin (rubor), localized heat (calor), edema (swelling), and pain (dolor).

It may not be obvious from this list of signs and symptoms that acute inflammation is beneficial; however, acute inflammation is an important part of the second line of defense because it results in (1) dilation and increased permeability of blood vessels, (2) migration of phagocytes, and (3) tissue repair. Although the chemical details of inflammation are beyond the scope of our study, we now consider these three aspects of acute inflammation.

Dilation and Increased Permeability of Blood Vessels

Part of the body's initial response to an injury or invasion of pathogens is localized *vasodilation*—an increase in diameter of blood vessels in the affected region **(FIGURE 15.11).** Dilation of blood vessels results in the redness and localized heat associated with inflammation.

- Most fragments have specific and important roles in achieving the functions of the complement system. Some combine to form new enzymes; some act to increase vascular permeability, which increases diapedesis; others enhance inflammation; some act as opsonins; and still others are involved as positive chemotactic factors, attracting phagocytes.

- One end product of a full cascade is a **membrane attack complex (MAC)**, which forms a circular hole in a pathogen's membrane. The production of numerous MACs **(FIGURE 15.10)** leads to lysis of the pathogen. The bacterium causing gonorrhea is particularly sensitive to MACs.

In addition to its enzymatic role, fragment C3b acts as an opsonin. Fragment C5a is chemotactic, attracting phagocytes to the site of infection. C3a, C4a, and C5a are each inflammatory agents that cause localized vascular dilation, leading to inflammation. C3a and C5a also trigger the release of other inflammatory chemicals; their role is discussed in more detail shortly.

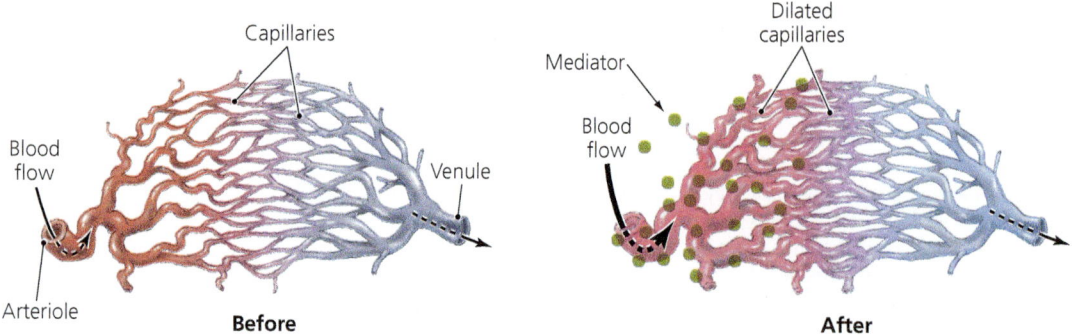

▲ FIGURE 15.11 **The dilating effect of inflammatory mediators on small blood vessels.** The release of mediators from damaged tissue causes nearby arterioles to dilate. Vasodilation causes capillaries to expand and enables more blood to be delivered to the affected site. Increased blood flow causes the reddening and heat associated with inflammation.

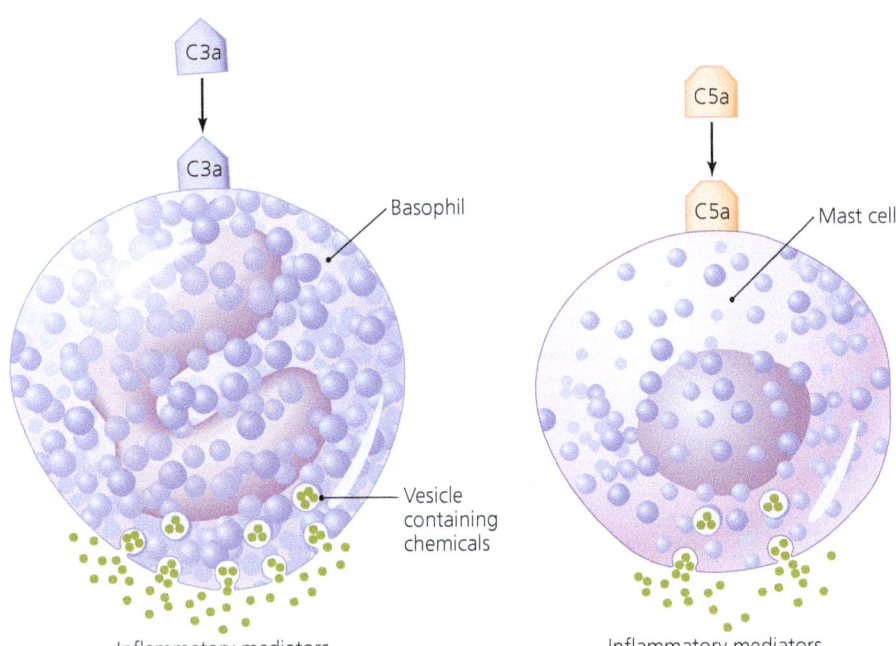

◀ **FIGURE 15.12 The stimulation of inflammation by complement.** Complement fragments C3a and C5a can each bind to platelets (not shown), basophils, or mast cells, causing them to release inflammatory mediators such as histamine, which in turn stimulates the dilation of arterioles. *What kind of white blood cell does the drawing on the left represent?*

Figure 15.12 Basophil.

Many chemicals can trigger and promote dilation. These include:

- **Bradykinin** (brad-e-kī′nin), which forms when the processes of blood clotting following damage to a blood vessel trigger conversion of a plasma protein into a peptide that is a powerful vasodilator.
- **Prostaglandins** (pros-tă-glan′dinz) released by patrolling macrophages, using Toll-like receptors and NOD proteins to identify invaders.
- **Leukotrienes** (loo-kō-trī′ēnz), which are released by the same cells that release prostaglandins.
- **Histamine** (his′tă-mēn) released by basophils, platelets, and **mast cells**—specialized cells located in connective tissue—when they are exposed to either complement fragment C3a or C5a (FIGURE 15.12). Recall that these complement fragments were cleaved from larger complement proteins during a complement cascade.

Bradykinin and histamine cause vasodilation of the body's smallest arteries (arterioles), resulting in the delivery of more blood and thus more phagocytes, oxygen, and nutrients to the site of infection. These two vasodilators, prostaglandins, and leukotrienes also make small veins (venules) more permeable—that is, they cause cells lining the vessels to contract and pull apart, leaving gaps in the walls through which phagocytes can move into the damaged tissue and fight invaders (FIGURE 15.13). Increased permeability also allows delivery

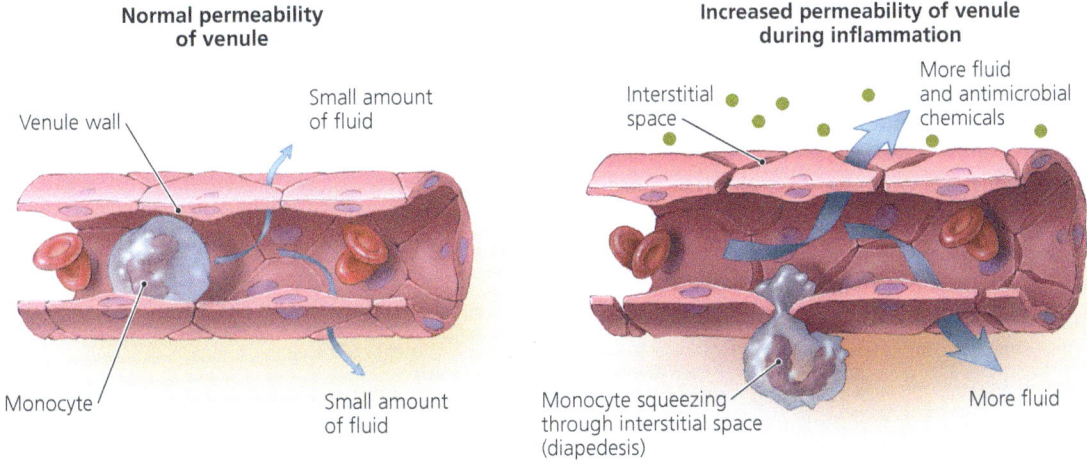

▲ **FIGURE 15.13 Increased vascular permeability during inflammation.** The presence of bradykinin, prostaglandins, leukotrienes, or histamine causes cells lining venules to pull apart, allowing phagocytes to leave the bloodstream and more easily reach a site of infection. The leakage of fluid and cells causes the edema and pain associated with inflammation.

of more bloodborne antimicrobial chemicals to the site. Fluid leaking from the more permeable blood vessels accumulates in the surrounding tissue, resulting in edema (swelling), which puts pressure on nerve endings, producing much of the pain of inflammation.

Vasodilation and increased permeability also deliver fibrinogen, the blood's clotting protein. Clots forming at the site of injury or infection wall off the area and help prevent pathogens and their toxins from spreading. One result is the formation of *pus*, a fluid containing dead tissue cells, leukocytes, and pathogens in the walled-off area. Pus may push up toward the surface and erupt, or it may remain isolated in the body, where it is slowly absorbed over a period of days. Such an isolated site of infection is called an **abscess**. Pimples, boils, and pustules are examples of abscesses.

The signs and symptoms of inflammation can be treated with antihistamines, which block histamine receptors on blood vessel walls, or with antiprostaglandins. One of the ways aspirin and ibuprofen reduce pain is by preventing synthesis of prostaglandins, which drive fever and pain. Physicians are divided on whether relief from the symptoms outweighs the loss of the positive effects of histamine and prostaglandins.

Migration of Phagocytes

Increased blood flow due to vasodilation delivers more neutrophils and monocytes to the site of infection. These leukocytes arrive at the site of dilation where they adhere to receptors lining the vessels in a process called **margination**. They then squeeze between the cells of the vessel's wall (diapedesis) and enter the site of infection, usually within an hour of tissue damage. The phagocytes can then destroy pathogens via phagocytosis.

As mentioned previously, phagocytes are attracted to the site of infection by chemotactic factors, including C5a, leukotrienes, and microbial components and toxins. The first phagocytes to arrive are often neutrophils, which are then followed by monocytes. Once monocytes leave the blood, they are called *wandering macrophages*, which are especially active phagocytic cells that devour pathogens, damaged tissue cells, and dead neutrophils. Wandering macrophages are a major component of pus.

Tissue Repair

The final stage of inflammation is tissue repair, which in part involves the delivery of extra nutrients and oxygen to the site via dilated blood vessels. Areas of the body where cells regularly divide, such as the skin and mucous membranes, are usually repaired rapidly. Some sites are not fully reparable and instead form scar tissue.

If a damaged tissue contains undifferentiated stem cells, tissues can be fully restored. For example, a minor skin cut is repaired to such an extent it is no longer visible. However, if undifferentiated connective tissue cells called *fibroblasts* are involved to a significant extent, scar tissue forms, inhibiting normal function. Some tissues, such as cardiac muscle and parts of the brain, do not replicate, and thus tissue damage cannot be repaired. As a result, these tissues remain damaged following heart attacks and strokes.

FIGURE 15.14 gives an overview of the entire inflammatory process. TABLE 15.5 summarizes the chemicals involved in inflammation.

Play *Inflammation: Steps* @ Mastering Microbiology

MICRO CHECK

13. What is the cause of swelling during acute inflammation?

Fever

LEARNING OUTCOME

15.19 Explain the benefits of fever in fighting infection.

Fever is a body temperature above normal, which averages 37°C. Fever augments the beneficial effects of inflammation, but like inflammation it also has unpleasant side effects, including malaise, body aches, and tiredness.

The hypothalamus, a portion of the brain just above the brain stem, controls the body's internal (core) temperature. Fever results when the presence of chemicals called **pyrogens**[14] (pī'rō-jenz) trigger the hypothalamic "thermostat" to reset at a higher temperature. Pyrogens include bacterial toxins, cytoplasmic contents of bacteria that are released upon lysis, antibody-antigen complexes formed in adaptive immune responses, and pyrogens released by phagocytes that have phagocytized bacteria. Although the exact mechanism of fever production is not known, the following discussion and FIGURE 15.15 present one possible explanation.

Chemicals produced by phagocytes **1** cause the hypothalamus to secrete prostaglandin, which resets the hypothalamic thermostat by an unknown mechanism **2**. The hypothalamus then communicates the new temperature setting to other parts of the brain, initiating nerve impulses that produce rapid and repetitive muscle contractions (shivering), an increase in metabolic rate, inhibition of sweating, and constriction of blood vessels of the skin **3**. These processes combine to raise the body's core temperature until it equals the prescribed temperature setting **4**. Because blood vessels in the skin constrict as fever progresses, one effect of inflammation (vasodilation) is undone. The constricted vessels carry less blood to the skin of the trunk, arms, and legs, causing them to appear less red and feel cold to the touch, even though the body's core temperature is higher. This symptom is the *chill* associated with fever.

Fever continues as long as pyrogens are present. As an infection comes under control and fewer active phagocytes are involved, the level of pyrogens decreases, the thermostat is reset

---

[14]From Greek *pyr*, meaning "fire," and *genein*, meaning "to produce."

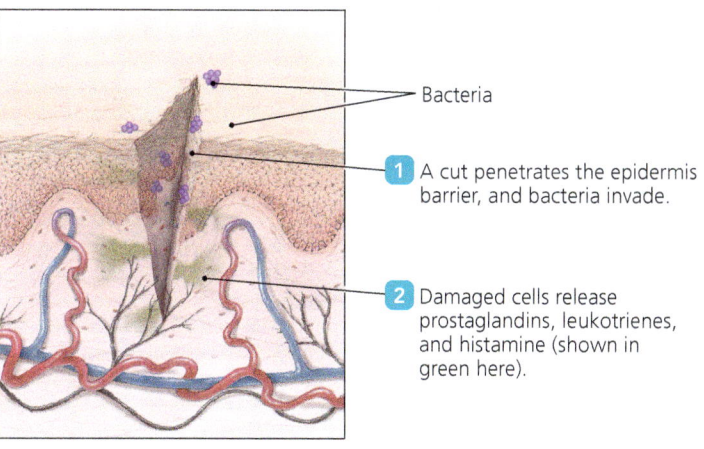
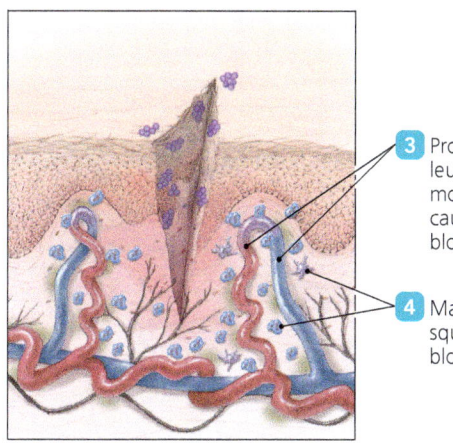
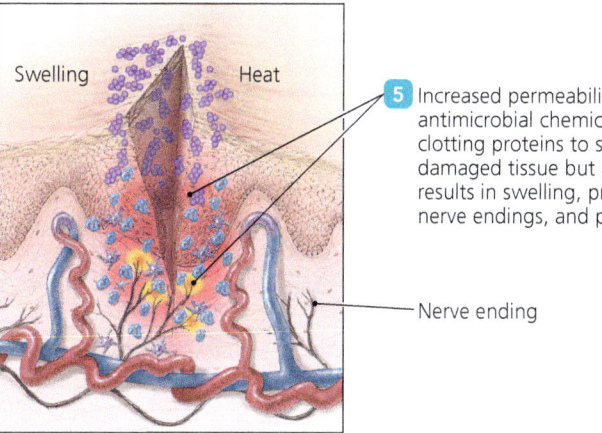
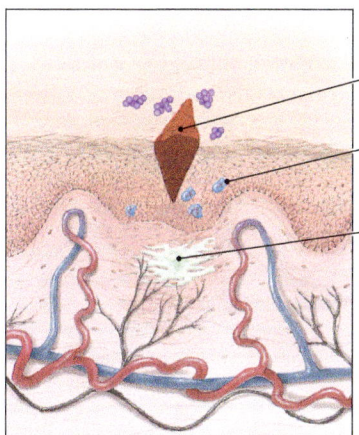

- Bacteria
- **1** A cut penetrates the epidermis barrier, and bacteria invade.
- **2** Damaged cells release prostaglandins, leukotrienes, and histamine (shown in green here).
- **3** Prostaglandins and leukotrienes make vessels more permeable. Histamine causes vasodilation, increasing blood flow to the site.
- **4** Macrophages and neutrophils squeeze through walls of blood vessels (diapedesis).
- **5** Increased permeability allows antimicrobial chemicals and clotting proteins to seep into damaged tissue but also results in swelling, pressure on nerve endings, and pain.
- Nerve ending
- **6** Blood clot forms.
- **7** More phagocytes migrate to the site and devour bacteria.
- **8** Accumulation of damaged tissue and leukocytes forms pus.
- **9** Undifferentiated stem cells repair the damaged tissue. Blood clot is absorbed or falls off as a scab.

◀ **FIGURE 15.14 An overview of the events in inflammation following a cut and infection.** The process, which is characterized by redness, swelling, heat, and pain, ends with tissue repair. *In general, what types of cells are involved in tissue repair?*

Figure 15.14 *Tissue repair is effected by cells that are capable of cytokinesis and differentiation. If fibroblasts are among them, scar tissue is laid down.*

TABLE 15.5 Chemical Mediators of Inflammation

Vasodilating Chemicals	Histamine, serotonin, bradykinin, prostaglandins
Chemotactic Factors	Fibrin, collagen, mast cell chemotactic factors, bacterial peptides
Substances with Both Vasodilating and Chemotactic Effects	Complement fragment C5a, interferons, interleukins, leukotrienes, platelet secretions

to 37°C, and the body begins to cool by perspiring, lowering the metabolic rate, and dilating blood vessels in the skin. These processes, collectively called the *crisis* of a fever, are a sign that the infection has been overcome and that body temperature is returning to normal.

The increased temperature of fever enhances the effects of interferons, inhibits the growth of some microorganisms, and is thought to enhance the performance of phagocytes, the activity

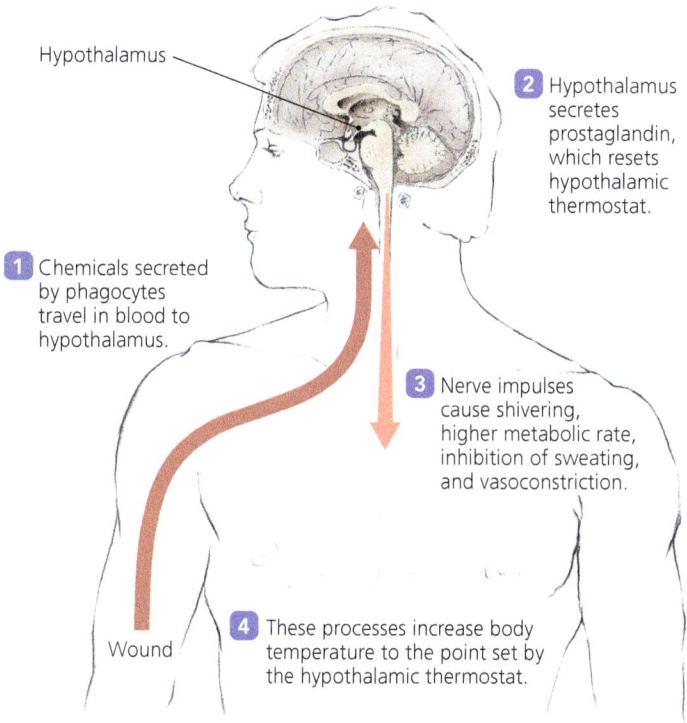

▲ **FIGURE 15.15 One theoretical explanation for the production of fever in response to infection.**

TABLE 15.6 Summary of Some Nonspecific Components of the First and Second Lines of Defense (Innate Immunity)

First Line				Second Line			
Barriers and Associated Chemicals	Phagocytes	Extracellular Killing	Complement	Interferons	Antimicrobial Peptides	Inflammation	Fever
Skin and mucous membranes prevent entry of pathogens; chemicals (e.g., sweat, acid, lysozyme, mucus) enhance the protection	Macrophages, neutrophils, and eosinophils ingest and destroy pathogens	Eosinophils and NK lymphocytes kill pathogens without phagocytizing them; neutrophils can also kill without phagocytosis	Components attract phagocytes, stimulate inflammation, and attack a pathogen's cytoplasmic membrane	Increase resistance of cells to viral infection, slow the spread of disease	Interfere with membranes, internal signaling, and metabolism; act against pathogens	Increases blood flow, capillary permeability, and migration of leukocytes into infected area; walls off infected region; increases local temperature	Mobilizes defenses, accelerates repairs, inhibits pathogens

of lymphocytes involved in adaptive immunity, and the process of tissue repair. However, if fever is too high, critical proteins are denatured; additionally, nerve impulses are inhibited, resulting in hallucinations, coma, and even death.

Because of the potential benefits of fever, many doctors recommend that patients refrain from taking fever-reducing drugs unless the fever is prolonged or extremely high. Other physicians believe that the benefits of fever are too slight to justify enduring the adverse symptoms.

TABLE 15.6 summarizes the barriers, cells, chemicals, and processes involved in the body's first two, nonspecific lines of defense.

TELL ME WHY

Why are pathogen-associated molecular patterns (PAMPs) necessary for Toll-like receptors (TLRs) to fully function?

MICRO IN THE CLINIC FOLLOW-UP

Allergies in Bloom?

When Alice wakes up the next morning, she can barely open her right eye. Her eyelid is swollen shut and crusty, with a thick white discharge. She puts a warm washcloth on her eye to help get rid of the crust and drainage, but the swelling is significant enough that she asks her roommate to give her a ride to the doctor's office.

Dr. Danby examines Alice's eye and tells her that it's not allergies—Alice has conjunctivitis, or pinkeye, that is probably caused by a bacterium. It's likely that one of the children in Alice's preschool had pinkeye and transmitted it to Alice. Pinkeye is highly contagious; to minimize the chance of transmission, Alice needs to wash her hands frequently with soap and water, avoid touching her eyes, wash all of her bedding and towels, and get rid of her makeup and brushes.

Dr. Danby prescribes antibiotic eye drops and instructs Alice to thoroughly wash her hands before and after using the drops. Dr. Danby also tells Alice to have her roommate follow all of the precautions to avoid infection, and to go to her doctor immediately if she begins to experience symptoms.

Alice alerts the preschool to her infection so other teachers and parents can watch for signs of infection in the children. Within a couple of days of using the antibiotic eye drops, Alice's symptoms begin to disappear; and within a week, her eyes are completely back to normal.

1. Alice's symptoms are a result of an infection of the conjunctiva, the lining that covers the white part of the eye and eyelid. What innate mechanisms protect the eye from infection?

2. Conjunctivitis is inflammation of the conjunctiva. What are the steps in an inflammatory response? How do these steps correspond to Alice's symptoms?

Check your answers to Micro in the Clinic Follow-Up questions in the Mastering Microbiology Study Area.

CHAPTER SUMMARY

Go to Mastering Microbiology for Interactive Microbiology, Dr. Bauman's Video Tutors, Micro Matters, MicroFlix, MicroBoosters, practice quizzes, and more.

An Overview of the Body's Defenses (p. 447)

1. Humans have **species resistance** to certain pathogens as well as three overlapping lines of defense. The first two lines of defense comprise **innate immunity**, which is generally nonspecific and protects the body against a wide variety of potential pathogens. A third line of defense is **adaptive immunity**, in which each response is specifically generated against each particular antigen.

Play *Host Defenses: Overview* @ Mastering Microbiology

The Body's First Line of Defense (pp. 447–451)

1. The first line of defense includes the skin, composed of an outer **epidermis** and a deeper **dermis**. **Dendritic cells** of the epidermis devour pathogens. Sweat glands of the skin produce salty sweat containing the enzyme called **lysozyme** and **antimicrobial peptides** (defensins), which are small peptide chains that act against a broad range of pathogens. **Sebum** is an oily substance of the skin that lowers pH, deterring the growth of many pathogens.

2. The mucous membranes, another part of the body's first line of defense, are composed of tightly packed cells that are replaced frequently by **stem cell** division and often coated with sticky mucus secreted by goblet cells.

3. Tears contain antibacterial lysozyme and also flush invaders from the eyes. Saliva similarly protects the teeth. The low pH of the stomach inhibits most microbes that are swallowed.

4. **Competitive inhibition**, the competition between the **microbiome** and potential pathogens, also contributes to the body's first line of defense. A **probiotic** is a beneficial microbe administered to improve health or prevent disease.

5. Antimicrobial peptides (defensins) act against pathogens on the skin and mucous membranes and in neutrophils.

The Body's Second Line of Defense (pp. 451–466)

1. The second line of defense includes cells (especially **phagocytes**), antimicrobial chemicals (Toll-like receptors, NOD proteins, interferons, complement, lysozyme, and antimicrobial peptides), and processes (phagocytosis, inflammation, and fever).

2. Blood is composed of **formed elements** (cells and parts of cells) within a fluid called **plasma**. Serum is that portion of plasma without clotting factors. The formed elements are **erythrocytes** (red blood cells), **leukocytes** (white blood cells), and **platelets**.

3. Based on their appearance in stained blood smears, leukocytes are grouped as either **granulocytes** (**basophils**, **eosinophils**, and **neutrophils**) or **agranulocytes** (**lymphocytes**, **monocytes**). When monocytes leave the blood, they become **macrophages**.

4. Basophils function to release histamine during inflammation, whereas eosinophils and neutrophils phagocytize pathogens. All three exit capillaries via **diapedesis**.

5. Macrophages, neutrophils, and dendritic cells are phagocytic cells of the second line of defense. Many macrophages are named for their location in the body, for example, alveolar macrophages (the lungs) and microglia (the nervous system).

6. A **differential white blood cell count** is a lab technique that indicates the relative numbers of leukocyte types; it can be helpful in diagnosing disease.

7. **Chemotactic factors**, such as chemicals called **chemokines**, attract phagocytic leukocytes to the site of damage or invasion. Phagocytes attach to pathogens via a process called **adhesion**.

Play *Phagocytosis: Overview* @ Mastering Microbiology

8. **Opsonization**, the coating of pathogens by proteins called **opsonins**, makes those pathogens more vulnerable to phagocytes. A phagocyte's pseudopods then surround the microbe to form a sac called a **phagosome**, which fuses with a lysosome to form a **phagolysosome**, in which the pathogen is killed.

Play *Phagocytosis: Mechanism* @ Mastering Microbiology

9. Leukocytes can distinguish the body's normal cells from foreign cells because leukocytes have receptor molecules for foreign cells' components or because the foreign cells are opsonized by complement or antibodies.

10. Eosinophils and **natural killer lymphocytes (NK cells)** attack nonphagocytically, especially in the case of helminth infections and cancerous cells. **Eosinophilia**—an abnormally high number of eosinophils in the blood—typically indicates such a helminth infection. Neutrophils can also kill nonphagocytically by making toxic forms of oxygen and NETs—neutrophil extracellular traps.

11. Microbial molecules called **pathogen-associated molecular patterns (PAMPs)** bind to **Toll-like receptors (TLRs)** on host cells' membranes or to **NOD proteins** inside cells, triggering innate immune responses.

12. **Interferons (IFNs)** are protein molecules that inhibit the spread of viral infections. **Alpha interferons (IFN-α)** and **beta interferons (IFN-β)**, which are released within hours of infection, trigger **antiviral proteins (AVPs)** to prevent viral reproduction in neighboring cells. **Gamma interferons (IFN-γ)**, produced days after initial infection, activate macrophages and neutrophils.

13. The **complement system**, or **complement**, is a set of proteins that act as chemotactic attractants, trigger inflammation and fever, and can effect the destruction of foreign cells via the formation of **membrane attack complexes (MACs)**, which result in multiple, fatal holes in pathogens' membranes. A complement cascade is activated by a classical pathway involving antibodies, by an alternative pathway triggered by bacterial chemicals, or by a lectin pathway triggered by mannose found on microbial surfaces.

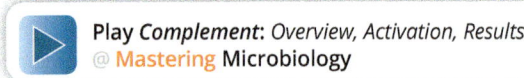
Play *Complement:* Overview, Activation, Results
@ Mastering Microbiology

14. **Acute inflammation** develops quickly and damages pathogens, whereas **chronic inflammation** develops slowly and can cause tissue damage that can lead to disease. Signs and symptoms of inflammation include redness, heat, swelling, and pain.

Play *Inflammation:* Overview, Steps
@ Mastering Microbiology

15. The process of blood clotting triggers formation of **bradykinin**—a potent mediator of inflammation.
16. Macrophages with Toll-like receptors or NOD proteins release **prostaglandins** and **leukotrienes**, which increase permeability of blood vessels. **Mast cells**, basophils, and platelets release **histamine** when exposed to peptides from the complement system. Blood clots may isolate an infected area to form an **abscess**, such as a pimple or boil.
17. When leukocytes rolling along blood vessel walls reach a site of infection, they stick to the wall in a process called **margination** and then undergo diapedesis to arrive at the site of tissue damage. The increased blood flow of inflammation also brings extra nutrients and oxygen to the infection site to aid in repair.
18. **Fever** results when chemicals called **pyrogens**, including substances released by bacteria and phagocytes, affect the hypothalamus in a way that causes it to reset body temperature to a higher level. The exact process of fever and its control are not fully understood.

QUESTIONS FOR REVIEW

Answers to the Questions for Review (except Short Answer questions) begin on p. A-1.

Multiple Choice

1. Phagocytes of the epidermis are called _____.
   a. microglia
   b. goblet cells
   c. alveolar macrophages
   d. dendritic cells

2. Mucus-secreting membranes are found in _____.
   a. the urinary system
   b. the digestive cavity
   c. the respiratory passages
   d. all of the above

3. The complement system involves _____.
   a. the production of antigens and antibodies
   b. serum proteins involved in nonspecific defense
   c. a set of genes that distinguish foreign cells from body cells
   d. the elimination of undigested remnants of microorganisms

4. The alternative complement activation pathway involves _____.
   a. factors B, D, and P
   b. the cleavage of C5 to form C9
   c. binding to mannose sugar
   d. recognition of antigens bound to specific antibodies

5. Which of the complement fragments is inflammatory?
   a. C3a
   b. C4a
   c. C5a
   d. all of the above

6. The type of interferon present late in an infection is _____.
   a. alpha interferon
   b. beta interferon
   c. gamma interferon
   d. delta interferon

7. Interferons _____.
   a. do not protect the cell that secretes them
   b. stimulate the activity of macrophages
   c. cause muscle aches, chills, and fever
   d. all of the above

8. Which of the following is *not* targeted by a Toll-like receptor?
   a. lipid A
   b. eukaryotic flagellar protein
   c. single-stranded RNA
   d. lipoteichoic acid

9. Toll-like receptors (TLRs) act to _____.
   a. bind microbial proteins and polysaccharides
   b. induce phagocytosis
   c. cause phagocytic chemotaxis
   d. destroy microbial cells

10. Which of the following binds iron?
    a. lactoferrin
    b. siderophores
    c. transferrin
    d. all of the above

Modified True/False

Indicate which statements are true. Correct all false statements by changing the underlined words.

1. _____ The surface cells of the epidermis of the skin are <u>alive</u>.
2. _____ The surface cells of mucous membranes are <u>alive</u>.
3. _____ Wandering macrophages experience <u>diapedesis</u>.
4. _____ <u>Monocytes</u> are immature macrophages.
5. _____ <u>Lymphocytes</u> are large agranulocytes.
6. _____ <u>Phagocytes</u> exhibit chemotaxis toward a pathogen.
7. _____ In phagocytosis, adhesion involves the binding between complementary chemicals on a phagocyte and on the membrane of a <u>body cell</u>.

8. _____ Opsonization occurs when a phagocyte's pseudopods surround a microbe and fuse to form a sac.
9. _____ Lysosomes fuse with phagosomes to form peroxisomes.
10. _____ A membrane attack complex drills circular holes in a macrophage.
11. _____ Rubor, calor, swelling, and dolor are associated with fever.
12. _____ Acute and chronic inflammation exhibit similar signs and symptoms.
13. _____ The hypothalamus of the brain controls body temperature.
14. _____ Defensins are phagocytic parts of the first line of defense.
15. _____ NETs are webs produced by neutrophils to trap microbes.

Matching

Which line of defense first applies to each term?

1. ___ Inflammation
2. ___ Monocytes
3. ___ Lactoferrin
4. ___ Fever
5. ___ Dendritic cells
6. ___ Alpha interferon
7. ___ Mucous membrane of the digestive tract
8. ___ Neutrophils
9. ___ Epidermis
10. ___ Lysozyme
11. ___ Goblet cells
12. ___ Phagocytes
13. ___ Sebum
14. ___ T lymphocytes
15. ___ Antimicrobial peptides

A. First line of defense
B. Second line of defense
C. Third line of defense

Write the letter of the description that applies to each of the following terms.

1. ___ Goblet cell
2. ___ Lysozyme
3. ___ Stem cell
4. ___ Dendritic cell
5. ___ Cell from sebaceous gland
6. ___ Bone marrow stem cell
7. ___ Eosinophil
8. ___ Alveolar macrophage
9. ___ Microglia
10. ___ Wandering macrophage

A. Leukocyte that primarily attacks parasitic worms
B. Phagocytic cell in lungs
C. Secretes sebum
D. Devours pathogens in epidermis
E. Breaks bonds in bacterial cell wall
F. Phagocytic cell in central nervous system
G. Generative cell with many types of offspring
H. Develops into formed elements of blood
I. Intercellular scavenger
J. Secretes mucus

Short Answer

1. In order for a pathogen to cause disease, what three things must happen?
2. How does a phagocyte "know" it is in contact with a pathogen instead of another body cell?
3. Give three characteristics of the epidermis that make it an intolerable environment for most microorganisms.
4. What is the role of Toll-like receptors in innate immune responses?
5. Describe the classical complement cascade pathway from C1 to the MAC.
6. How do NOD proteins differ from Toll-like receptors?

VISUALIZE IT!

1. Label the steps of phagocytosis.

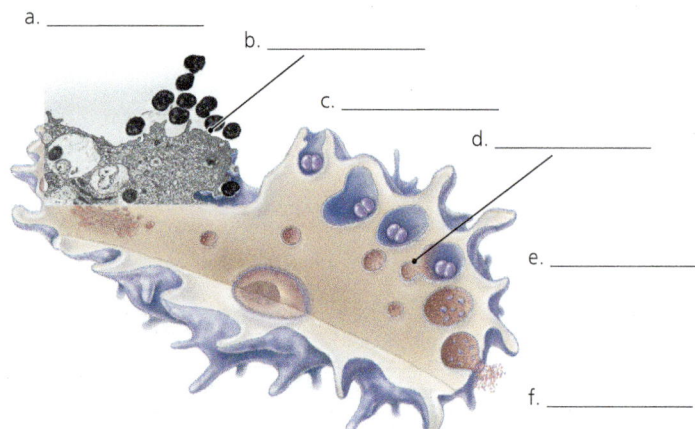

a. _____
b. _____
c. _____
d. _____
e. _____
f. _____

2. Name the cells.

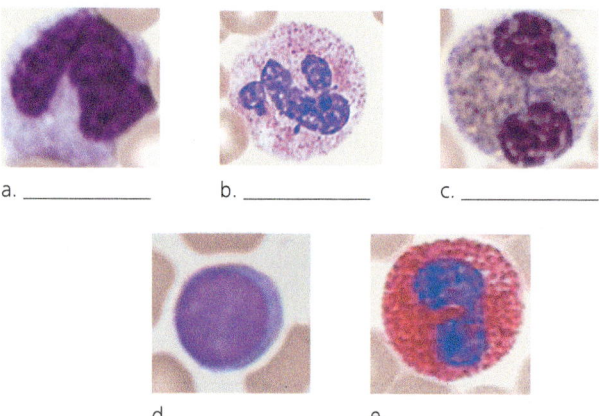

a. _____ b. _____ c. _____
d. _____ e. _____

CRITICAL THINKING

1. John received a chemical burn on his arm and was instructed by his physician to take an over-the-counter anti-inflammatory medication for the painful, red, swollen lesions. When Charles suffered pain, redness, and swelling from an infected cut on his foot, he decided to take the same anti-inflammatory drug because his symptoms matched John's symptoms. How is Charles's inflammation like that of John? How is it different? Is it appropriate for Charles to medicate his cut with the same medicine John used?

2. What might happen to someone whose body did not produce C3? C5?

3. Mary, age 65, has had diabetes for 40 years, with resulting damage to the small blood vessels in her feet and toes. Her circulation is impaired. How might this condition affect her vulnerability to infection?

4. A patient's chart shows that eosinophils make up 8% of his white blood cells. What does this lead you to suspect? Would your suspicions change if you learned that the patient had spent the previous three years as an anthropologist living in an African jungle village? What is the normal percentage of eosinophils?

5. There are two kinds of agranulocytes in the blood—monocytes and lymphocytes. Janice noted that monocytes are phagocytic and that lymphocytes are not. She wondered why two agranulocytes would be so different. What facts of hematopoiesis can help her answer her question?

6. A patient has a genetic disorder that prevents him from synthesizing C8 and C9. What effect does this have on his ability to resist bloodborne Gram-negative and Gram-positive bacteria? What would happen if C3 and C5 fragments were also inactivated?

7. Sweat glands in the armpits secrete perspiration with a pH close to neutral (7.0). How does this fact help explain body odor in this area as compared to other parts of the skin?

8. Scientists can raise "germ-free" animals in axenic environments. Would such animals be as healthy as their worldly counterparts?

9. Compare and contrast the protective structures and chemicals of the skin and mucous membranes.

10. Scientists are interested in developing antimicrobial drugs that act like the body's normal antimicrobial peptides. What advantage might such a drug have over antibiotics?

11. A medical laboratory scientist argues that granulocytes are a natural group, whereas agranulocytes are an artificial grouping. Based on Figure 15.4, do you agree or disagree with the lab scientist? What evidence can you cite to justify your conclusion?

12. A patient has a genetic disorder that makes it impossible for her to synthesize complement protein 8 (C8). Is her complement system nonfunctional? What major effects of complement could still be produced?

13. While using a microscope to examine a sample of pus from a pimple, Maria observed a large number of macrophages. Is the pus from an early or a late stage of infection? How do you know?

14. How do drugs such as aspirin and ibuprofen act to reduce fever? Should you take fever-reducing drugs or let a fever run its course?

CONCEPT MAPPING

Using the terms provided, fill in the concept map that describes phagocytosis. You can also complete this and other concept maps online by going to the Mastering Microbiology Study Area.

Adhesion
Chemotactic factors
Chemotaxis
Dendritic cells
Elimination
Eosinophils
Exocytosis
Ingestion
Killing
Lysosome
Macrophages
Neutrophils
Opsonins
Phagolysosome
Phagosome

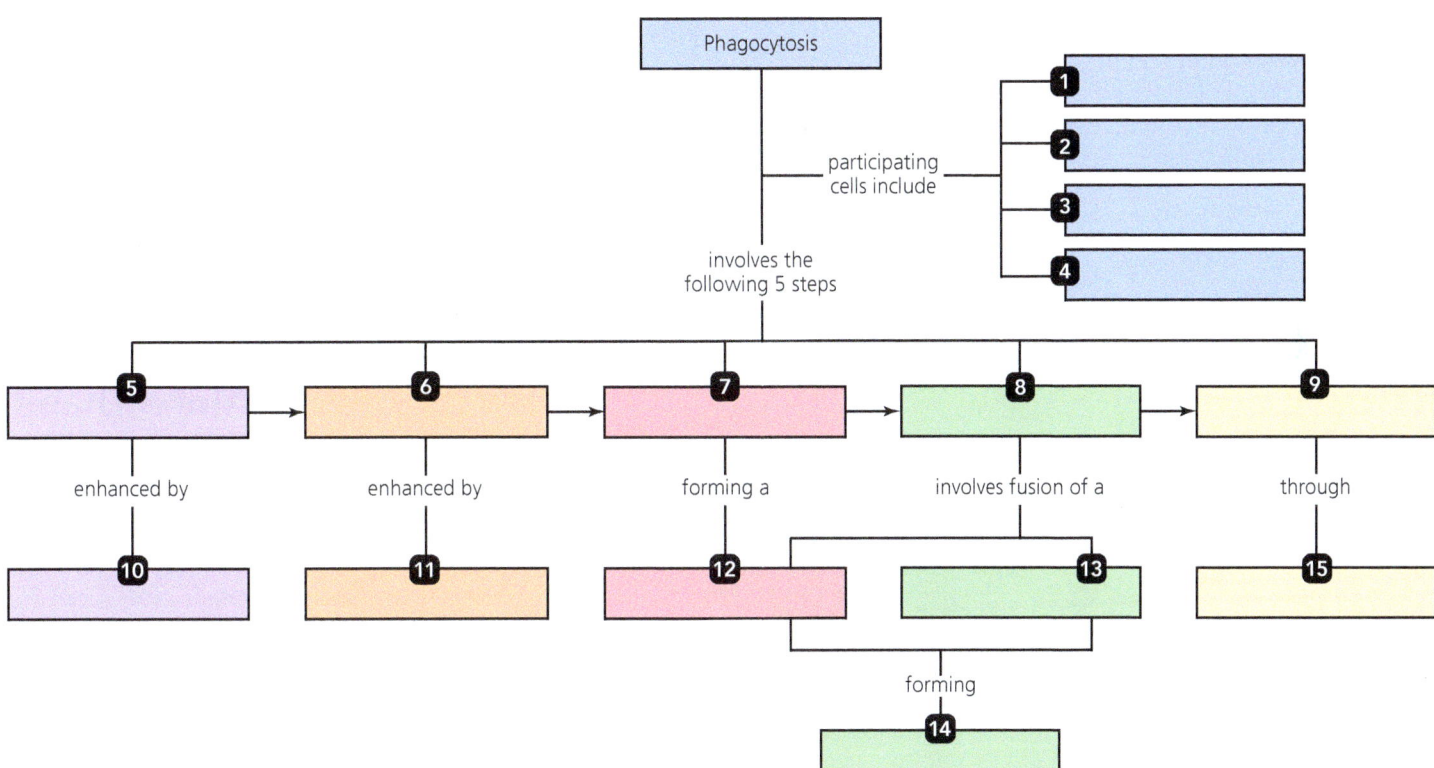

16 Adaptive Immunity

Before You Begin

1. Which type of cell forms an almost continuous network of phagocytes in the skin's epidermis?
2. How do phagocytes distinguish between invading pathogens and the body's own healthy cells?
3. Which membranous organelles package proteins for secretion to the outside of the cell?

MICRO IN THE CLINIC

More Than the Common Cold?

BILL RECENTLY GRADUATED FROM COLLEGE, and his parents surprised him with a graduation trip to Taiwan, China, and the Philippines. As a political science major with an emphasis in Asian studies, Bill couldn't have asked for a better present. The trip was amazing—Bill had a lot of fun, and he also learned much about the history and experienced the culture of the region. He can't wait to return.

On his second day back home, Bill's friend Connie asks him to join her at a local street fair. There will be food trucks, music, artists, and local breweries showcasing their beers. Never one to turn down a beer tasting, Bill gladly joins Connie at the fair. The fair has a great turnout, and Bill and Connie have a lovely evening.

When Bill wakes up the next morning, he feels like he has the flu—he has a runny nose, a cough, and a fever. A few days later, Bill notices that he has a rash on his face and neck. The rash isn't itchy, so Bill doesn't worry about it. But when the rash spreads to his chest and arms, Bill decides that he should visit his doctor.

1. Is there a connection between Bill's flulike symptoms and the rash?

Turn to the end of the chapter (p. 497) to find out.

White blood cells called lymphocytes (here colored blue) roll along the inside walls of blood vessels. Lymphocytes are the main cells of adaptive immunity.

Innate, or native, immunity includes two lines of rapid defense against microbial pathogens (see Chapter 15). The first line includes intact skin and mucous membranes, and the second line includes phagocytosis, nonspecific chemical defenses (such as complement), inflammation, and fever. Innate defenses do not always offer enough protection in defending the body. Although the mechanisms of innate immunity are readily available and fast acting, they respond in the same ways to the great variety of pathogens confronting the body—they do not adapt to make their response more effective against specific pathogens. For example, a fever triggered by a flu virus is similar to a fever triggered by Ebola virus. Vertebrates have a third line of defense that augments the mechanisms of innate immunity by destroying and targeting specific invaders while adapting to be more effective in the process. This response is called *adaptive immunity* or *acquired immunity*. TABLE 16.1 compares innate and adaptive immunity.

Overview of Adaptive Immunity

LEARNING OUTCOMES

16.1 Describe five distinctive characteristics of adaptive immunity.
16.2 List the two basic types of white blood cells involved in adaptive immunity.
16.3 List two basic divisions of adaptive immunity and describe their targets.

Adaptive immunity is a vertebrate's ability to recognize and then mount a defense against distinct invaders and their products, whether they are protozoa, fungi, bacteria, viruses, or toxins. Only vertebrates have adaptive immunity (see Table 16.1). *Immunologists*—scientists who study the cells and chemicals involved in immunity—are continually refining and revising our knowledge of adaptive immunity. In this chapter, we will examine some of what they have discovered.

Adaptive immunity has five distinctive attributes:

- **Specificity.** Any particular adaptive immune response acts against only one particular molecular shape, called an *antigen*, and not against others. Adaptive immune responses are precisely tailored reactions against specific attackers, whereas innate immunity involves more generalized responses to pathogen-associated molecular patterns (PAMPs)—molecular shapes common to many microbes. Thus, adaptive immunity can act against pathogenic *Escherichia coli* (esh-ĕ-rik'ē-ă kō'lī) strain O157:H7 without harming *E. coli* strain MG1655, which is a beneficial member of the human microbiome.

- **Inducibility.** The specific antigen-containing pathogen activates or induces cells of adaptive immunity.

- **Clonality.** Once induced, cells of adaptive immunity proliferate to form many generations of nearly identical cells, which are collectively called *clones*.

- **Unresponsiveness to self.** As a rule, adaptive immunity does not act against normal body cells; in other words, adaptive immune responses are self-tolerant. Several mechanisms help ensure that immune responses, when functioning properly, do not attack the body itself.

- **Memory.** An adaptive immune response has *immunological memory* for specific pathogens; that is, it adapts to respond faster and more effectively in subsequent encounters with a particular type of pathogen or toxin. This memory distinguishes adaptive immunity from innate immunity (see Table 16.1).

These aspects of adaptive immunity involve the activities of **lymphocytes** (lim'fō-sītz), which are a type of leukocyte (white blood cell) that acts against specific pathogens. Lymphocytes in their resting state are the smallest white blood cells, and each is characterized by a large, round, central nucleus surrounded by a thin rim of cytoplasm (FIGURE 16.1). Initially, human lymphocytes form in the *red bone marrow*, located in the ends of long bones in juveniles and in the centers of adult flat bones such as the ribs and hip bones. These sites contain *blood stem cells (hematopoietic stem cells)*, which are cells that give rise to all types of blood cells, including lymphocytes (see Chapter 15, Figure 15.4). This is why a bone marrow transplant can restore a person's entire adaptive immune system.

Although lymphocytes appear identical when viewed with a microscope, scientists make distinctions between two main types—*B lymphocytes* and *T lymphocytes*—according to surface proteins that are part of each lymphocyte's cytoplasmic membrane. These proteins allow lymphocytes to recognize specific pathogens and toxins by their molecular shapes, and the proteins play roles in intercellular communication among immune cells.

Lymphocytes must undergo a maturation process. **B lymphocytes**, which are also called **B cells**, arise and mature in the red bone marrow of adults. **T lymphocytes**, also known as **T cells**, begin in bone marrow as well but do not mature there. Instead, T cells travel to and mature in the *thymus*, located in

TABLE 16.1 A Comparison of Innate and Adaptive (Acquired) Branches of Immunity

	Innate	Adaptive
Distribution	Almost all multicellular eukaryotes	Only in vertebrates
Targets	Limited number of key structures present in many microbes (PAMPs), perhaps 20–50	Antigens, which are mostly proteins; different ones number in the billions
Immune Receptors	Pattern recognition receptors, such as Toll-like receptors (TLRs)	T cell receptors and antibodies
Cellular Presence	Almost all cells	Lymphocytes only
Discrimination	Host cells do not contain PAMPs	Tolerance for self-antigens can break down, resulting in autoimmune disease
Immunological Memory	Absent	Present

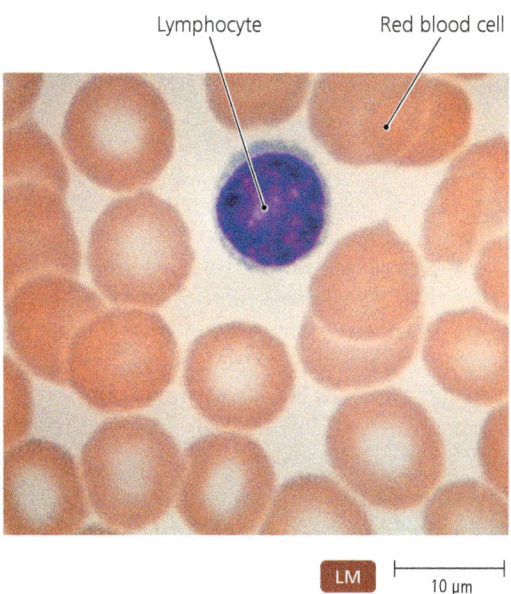

▲ **FIGURE 16.1 Lymphocytes play a central role in adaptive immunity.** An inactive lymphocyte is the smallest leukocyte—slightly larger than a red blood cell.

the chest near the heart in humans. T lymphocytes are so called because of the role the thymus plays in their maturation.

Adaptive immunity consists of many different immune responses that can be considered under two broad categories: **cell-mediated immune responses** and **antibody immune responses**. Long-lived B and T lymphocytes in each of these categories retain the ability to fight specific pathogens as long as they live—the characteristic called *immunological memory*.

Both cell-mediated and antibody immune responses stimulate and regulate innate immunity as well as act directly against the specific pathogen that initiated the adaptive response. Cell-mediated immune responses are controlled and carried out by T cells and often act against *intracellular pathogens*, such as viruses replicating inside a cell. B cells carry out antibody immune responses, though T cells also play roles in regulating and fulfilling antibody immune responses. Antibody immune responses are often directed against extracellular pathogens and toxins. An *antibody* is a protective protein secreted by descendants of a B cell that recognizes and strongly binds to the specific biochemical shape of the antigen. We will consider antibodies in more detail later. Scientists have also used another term for antibody immune responses—*humoral immune responses*[1]—because many antibody molecules circulate in the liquid portion of the blood.

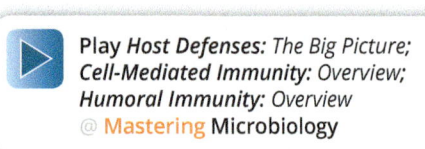

Play *Host Defenses: The Big Picture; Cell-Mediated Immunity: Overview; Humoral Immunity: Overview*
@ **Mastering** Microbiology

Both cell-mediated and antibody immune responses are powerful defensive reactions that have the potential to severely and fatally attack the body's own cells. Therefore, the body regulates adaptive immune responses to prevent damage; for example, an immune response requires multiple chemical signals before proceeding, thus reducing the possibility that an immune response will be randomly triggered against uninfected healthy tissue. Autoimmune disorders, hypersensitivities, or immunodeficiency diseases result when regulation is insufficient or overexcited. (Chapter 18 deals with such disorders.)

TELL ME WHY

Why are the activities of B and T cells called *adaptive*?

Elements of Adaptive Immunity

When you attend a theatrical performance, you are given a program that presents a synopsis of the play and introduces the actors and their roles. In that vein, the following sections present elements of adaptive immunity by describing the "stage" and introducing the "cast of characters" involved. We will examine the *lymphatic system*—the organs, tissues, and cells of adaptive immunity; then we consider the molecules, called *antigens*, that trigger adaptive immune responses; next, we take a look at *antibodies*; and finally, we examine special *chemical signals* and *mediators* involved in coordinating and controlling a specific immune response.

First, we turn our attention to the "stage"—the lymphatic system, which plays an important role in the production, maturation, and housing of the cells that function in adaptive immunity.

The Tissues and Organs of the Lymphatic System

LEARNING OUTCOMES

16.4 Compare and contrast the flow of lymph with the flow of blood.

16.5 Describe the primary and secondary organs of the lymphatic system.

16.6 Describe the importance to immunity of red bone marrow, the thymus, lymph nodes, spleen, tonsils, and mucosa-associated lymphoid tissue.

The **lymphatic system** is composed of the *lymphatic vessels*, which conduct the flow of a liquid called *lymph* (pronounced "limf"); and lymphoid cells, tissues, and organs, which are directly involved in adaptive immunity (**FIGURE 16.2a**). Taken together, the components of the lymphatic system constitute a surveillance system that screens the tissues of the body—particularly possible places of entry such as the throat and intestinal tract—for foreign molecules. We begin by examining lymphatic vessels.

The Lymphatic Vessels and the Flow of Lymph

Lymphatic vessels form a one-way system that conducts *lymph* from local tissues and returns it to the circulatory system. Most importantly for immune responses, lymph carries toxins and pathogens to areas where lymphocytes are concentrated.

---

[1]From Latin *humor*, meaning "liquid," referring to bodily fluids, such as blood.

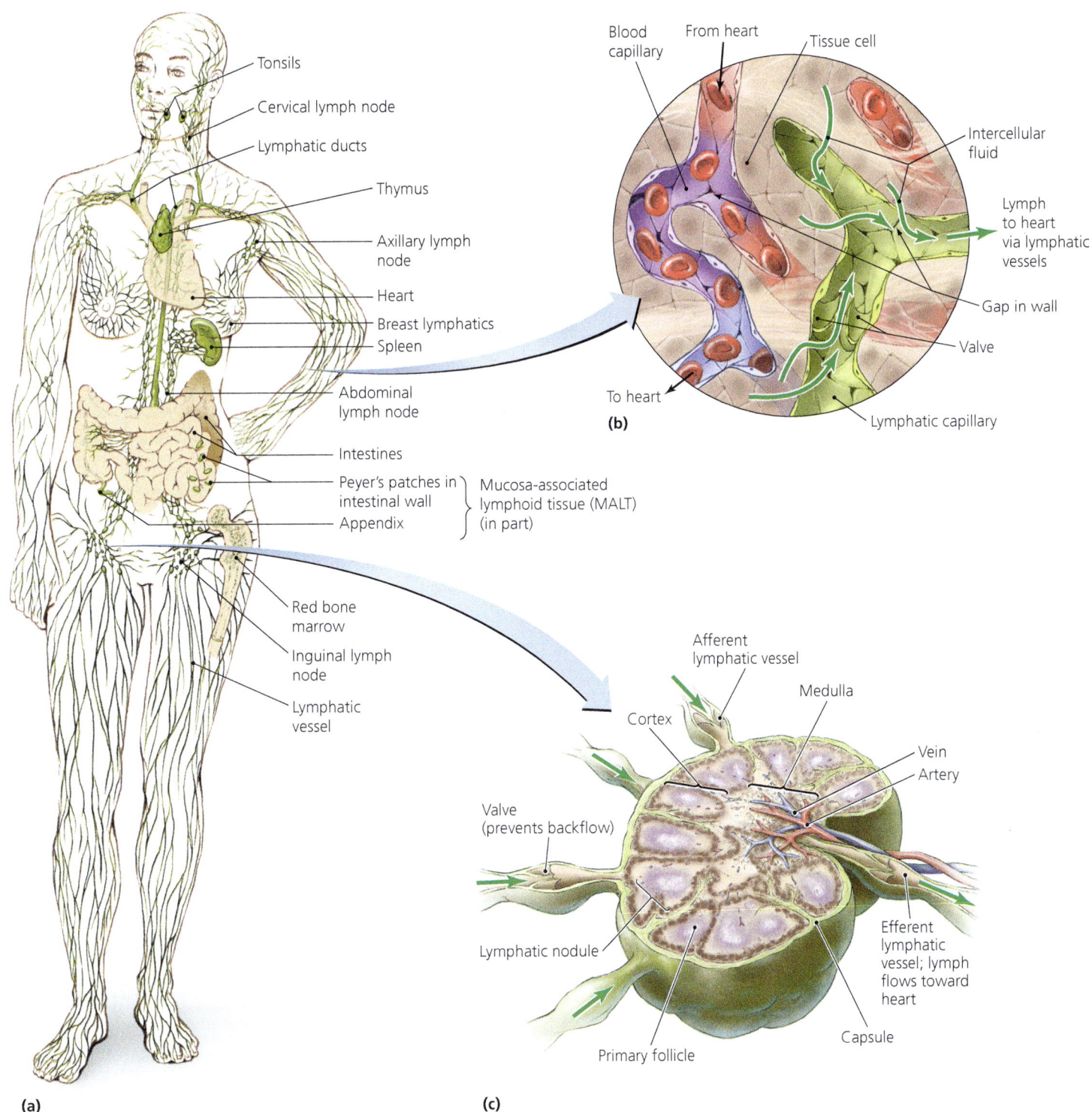

▲ **FIGURE 16.2 The lymphatic system.** (a) The system consists of primary lymphoid organs—bone marrow and thymus—and secondary lymphoid organs, including lymphatic vessels, lymph nodes, tonsils, and other lymphoid tissue. (b) Lymphatic capillaries collect lymph from intercellular spaces. (c) Afferent lymphatic vessels carry lymph into lymph nodes; efferent vessels carry it away. The cortex contains primary follicles, where B lymphocytes proliferate; the lymphocytes in the medulla encounter foreign molecules. *Why do lymph nodes enlarge during an infection?*

Figure 16.2 During an infection, lymphocytes multiply profusely in lymph nodes. This proliferation and swelling cause lymph nodes to enlarge.

Lymph is a colorless, watery liquid similar in composition to blood plasma; indeed, lymph arises from fluid that has leaked out of blood vessels into the surrounding intercellular spaces. Lymph is first collected by remarkably permeable *lymphatic capillaries* (FIGURE 16.2b), which are located in most parts of the body (exceptions include the bone marrow, brain, and spinal cord). From the lymphatic capillaries, lymph passes into increasingly larger lymphatic vessels until it finally flows via two large *lymphatic ducts* into blood vessels near the heart. Unlike the cardiovascular system, the lymphatic system has no unique pump and is not circular; that is, lymph flows in only one direction—toward the heart. One-way valves ensure the direction of lymphatic flow as contractions of skeletal muscles squeeze the lymphatic vessels. Located at various points within the system of lymphatic vessels are hundreds of *lymph nodes*, which house numerous white blood cells, including B and T lymphocytes. These lymphocytes recognize and attack foreigners present in the lymph, allowing for immune system surveillance and interactions.

Lymphoid Organs

Once lymphocytes have arisen and matured in the *primary lymphoid organs* of the red bone marrow and thymus, they migrate to *secondary lymphoid organs* and *tissues*, including lymph nodes, spleen, and other less organized accumulations of lymphoid tissue, where they in effect lie in wait for foreign microbes. **Lymph nodes** are located throughout the body but are concentrated in the cervical (neck), inguinal (groin), axillary (armpit), and abdominal regions. Each lymph node receives lymph from numerous *afferent* (inbound) *lymphatic vessels*. Just one or two *efferent* (outbound) *lymphatic vessels* drain each lymph node (FIGURE 16.2c). Essentially, lymph nodes are sites to facilitate interactions among immune cells and between immune cells and material arriving at lymph nodes from throughout the body.

A node has a *medulla*, or central region, consisting of a maze of passages that filter the lymph as it passes through. The medulla also houses numerous lymphocytes that survey the lymph for foreign molecules and mount specific immune responses against them. The cortex, or outer portion, of a lymph node consists of a tough capsule surrounding *primary follicles*, which is where clones of B cells replicate.

The lymphatic system contains additional secondary lymphoid tissues and organs, including the spleen, the tonsils, and *mucosa-associated lymphoid tissue (MALT)*. The spleen is similar in structure and function to lymph nodes, except that it filters blood instead of lymph. The spleen removes bacteria, viruses, toxins, and other foreign matter from the blood. It also cleanses the blood of old and damaged blood cells, stores blood platelets (which are required for the proper clotting of blood), and stores blood components, such as iron.

Tonsils are two mounds of tissue, one on each side of the tongue, similar to lymph nodes without afferent or efferent vessels. They sample microbes that enter the mouth or nose.

MALT includes lymphoid tissue associated with the mucous membranes of the appendix; lymphoid tissue of the respiratory tract, vagina, urinary bladder, and mammary glands; and discrete bits of lymphoid tissue called *Peyer's patches* in the wall of the small intestine. MALT contains most of the body's lymphocytes.

The tonsils and MALT lack the tough outer capsules of lymph nodes and the spleen, but they function in the same way by physically trapping foreign particles and microbes.

We are considering adaptive immunity in terms of a "play" taking place on the stage of the lymphatic system. Now, let's meet the "villains"—the foreign molecules that lymphocytes recognize.

Antigens

LEARNING OUTCOMES

16.7 Identify the characteristics of antigens that stimulate effective immune responses.

16.8 Distinguish among exogenous antigens, endogenous antigens, and autoantigens.

Adaptive immune responses are directed not against whole bacteria, fungi, protozoa, or viruses but instead against portions of cells, viruses, and even parts of single molecules that the body recognizes as foreign and worthy of attack. Immunologists call these biochemical shapes **antigens**[2] (an'ti-jenz). Lymphocytes bind to antigens and can then trigger adaptive immune responses. Antigens from pathogens and toxins are the "villains" of our story.

Properties of Antigens

Not every molecule is an effective antigen. Among the properties that make certain molecules more effective at provoking adaptive immunity are a molecule's *shape*, *size*, and *complexity*. The body recognizes antigens by the three-dimensional shapes of regions called **epitopes**, which are also known as *antigenic determinants* because they are the actual part of an antigen that determines an immune response (FIGURE 16.3a).

In general, larger molecules with molecular masses (sometimes misidentified as molecular weights) between 5000 and 100,000 daltons[3] are better antigens than smaller ones. The most effective antigens are large foreign macromolecules, such as proteins and glycoproteins, but carbohydrates and lipids can also be antigenic. Small molecules, especially those with a molecular mass under 5000 daltons, make poor antigens by themselves because they evade detection; however, they can become antigenic when bound to larger carrier molecules (often proteins). For example, the fungal product penicillin is too small by itself (molecular mass: 302 daltons) to trigger a specific immune response. However, bound to a carrier protein in the blood, penicillin can become antigenic and elicit an allergic response in some patients.

Complex molecules make better antigens than simple ones because they have more epitopes, like a gemstone with its many

---

[2]From *anti*body *gen*erator; antibodies are proteins secreted during an antibody immune response that bind to specific regions of antigens.
[3]Named for John Dalton, the British chemist who helped develop atomic theory around 1800.

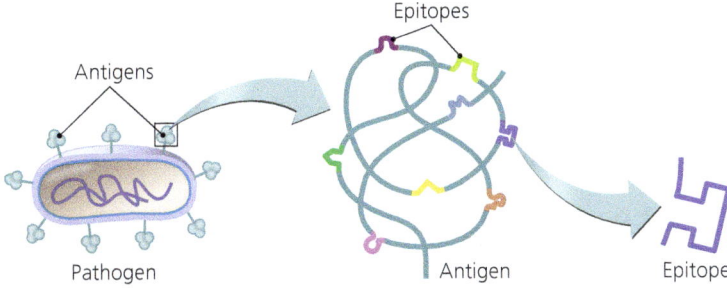
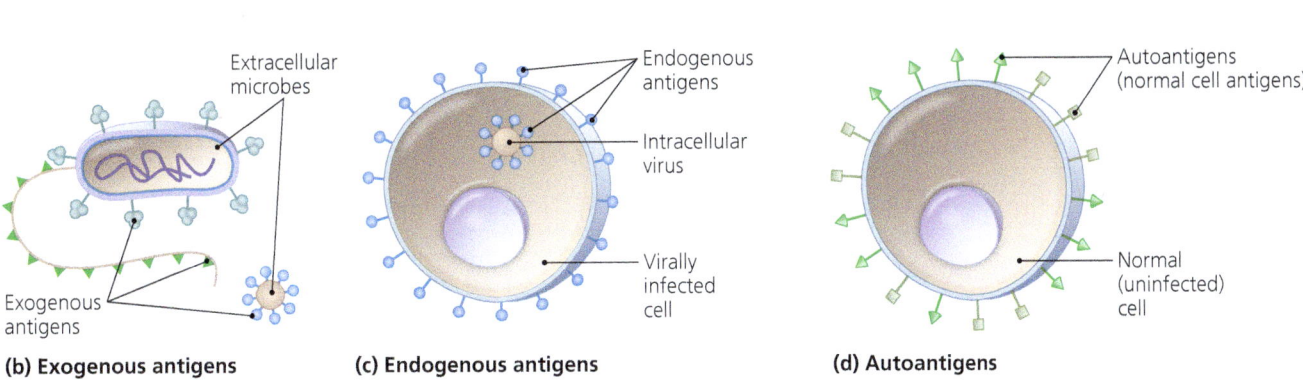

▲ FIGURE 16.3 **Antigens, molecules that provoke a specific immune response. (a)** Each pathogen has hundreds to thousands of antigens. Each antigen contains one to dozens of epitopes (antigenic determinants). Epitopes are three-dimensional regions of antigens whose shapes are recognized by immune receptors. **(b–d)** Categories of antigens are based on their relationship to the host cells. Exogenous antigens originate from microbes located outside the body's cells; endogenous antigens are produced by intracellular microbes and are typically incorporated into a host cell's cytoplasmic membrane; autoantigens are components of normal body cells.

facets. For example, starch, which is a very large polymer of repeating glucose subunits, is not a good antigen, despite its large size, because it lacks structural complexity. In contrast, complicated molecules, such as glycoproteins and phospholipids, have multiple distinctive shapes and novel combinations of subunits that cells of the immune system recognize as foreign.

Examples of antigens include components of bacterial cell walls, capsules, pili, and flagella as well as the external and internal proteins of viruses, fungi, and protozoa. Many toxins and some nucleic acid molecules are also antigenic. Invading microorganisms are not the only source of antigens; for example, food may contain antigens called *allergens* that provoke allergic reactions, and inhaled dust contains mite feces, pollen grains, dander (flakes of skin), and other antigenic and allergenic particles. (Chapter 18 covers allergies in more detail.)

Types of Antigens

Though immunologists categorize antigens in various ways, one especially important way is to group antigens according to their relationship to the body:

- **Exogenous**[4] **antigens** (FIGURE 16.3b). Exogenous (eks-oj'en-us) antigens come from outside the body's cells and include toxins and other secretions and components of microbial cell walls, membranes, flagella, and pili.
- **Endogenous**[5] **antigens** (FIGURE 16.3c). Protozoa, fungi, bacteria, and viruses that reproduce inside a body's cells produce endogenous antigens. The immune system cannot assess the internal health of the body's cells; it responds to endogenous antigens only if the body's cells incorporate such antigens into their cytoplasmic membranes, leading to their external display.
- **Autoantigens**[6] **(FIGURE 16.3d).** Antigenic molecules derived from normal cellular processes are autoantigens (or *self-antigens*). As we will discuss more fully in a later section, immune cells that treat autoantigens as if they were foreign are normally eliminated during the development of the immune system. This phenomenon, called *self-tolerance*, prevents the body from mounting an immune response against itself.

So far, we have examined two aspects of immune responses in terms of an analogy to a stage play—the lymphoid organs and tissues provide the stage, and antigens of pathogens are the villains that induce an immune response with their

---

[4]From Greek *exo*, meaning "without," and *genein*, meaning "to produce."

[5]From Greek *endon*, meaning "within."

[6]From Greek *autos*, meaning "self."

epitopes (antigenic determinants). Now, we will examine more important players in the drama—membrane proteins called *major histocompatibility complex (MHC) proteins*. The body prepares for an adaptive immune response by making MHC proteins and processing antigens. Cells of adaptive immunity only recognize epitopes carried on MHC proteins.

> ## MICRO CHECK
>
> 1. Adaptive immunity has characteristics that distinguish it from innate immunity, one of which is *inducibility*. What is inducibility?
> 2. What are the targets of the two basic divisions of adaptive immunity?
> 3. How are the lymphatic system and cardiovascular system similar or dissimilar?
> 4. Which part of a lymph node is the site of B clone replication?
> 5. What is a characteristic of a chemical that makes it a good antigen for the immune system?

Preparation for an Adaptive Immune Response

LEARNING OUTCOMES

16.9 Describe the two classes of major histocompatibility complex (MHC) proteins with regard to their location and function.

16.10 Explain the roles of antigen-presenting cells (e.g., dendritic cells and macrophages) and MHC molecules in antigen processing and presentation.

16.11 Contrast endogenous antigen processing with exogenous antigen processing.

The Roles of the Major Histocompatibility Complex and Antigen-Presenting Cells

When scientists first tried grafting skin tissue from one animal onto another to treat burn victims, they discovered that if the animals were not closely related, the recipients swiftly rejected the grafts. When they analyzed the reason for such rapid rejection, they found that a graft recipient mounted a very strong immune response against a specific protein found on the cells of unrelated grafts, resulting in rapid rejection. This helped scientists understand how the body is able to distinguish "self" from "nonself."

Immunologists named the cluster of genes that code for these proteins the **major histocompatibility complex (MHC)**. (The prefix *histo-* means "tissue," so the name indicates the importance in determining tissue compatibility during transplantation.) In humans, an MHC is located on each copy of chromosome 6. The protein product of MHC genes is *MHC protein*, though scientists often use *MHC* to refer to both the genes and the protein products of the genes. This might be confusing at first, but context should make it clear whether MHC refers to the genes or their products.

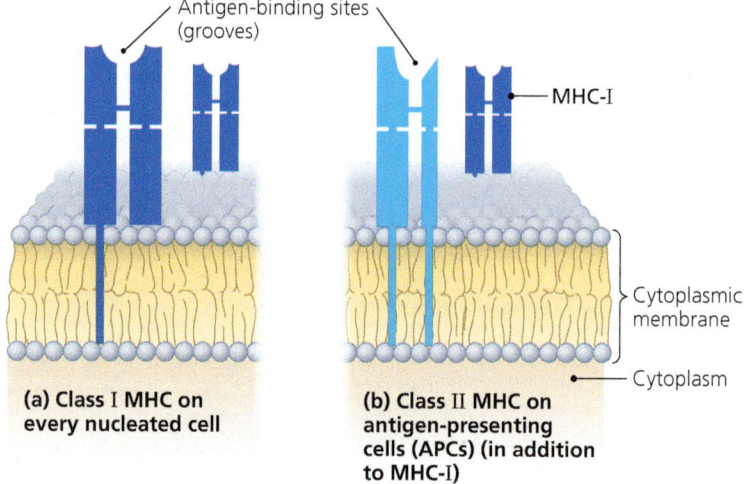

▲ **FIGURE 16.4** The two classes of major histocompatibility complex (MHC) proteins. Each is composed of two polypeptides that form an antigen-binding groove. **(a)** MHC class I glycoproteins are found on all cells except red blood cells. **(b)** In addition to MHC-I, MHC class II glycoproteins are expressed only by B cells and special antigen-presenting cells (APCs).

Because organ grafting is a modern surgical procedure with no counterpart in nature, scientists reasoned that MHC must have some natural function. Indeed, immunologists have determined that MHC proteins in cytoplasmic membranes function to hold and position epitopes for presentation to immune cells. Each MHC molecule has an *antigen-binding groove* that lies between two polypeptides that make up an MHC. Inherited variations in the amino acid sequences of the polypeptides modify the shapes of MHC binding sites and determine which epitopes can be bound and presented. Cells of adaptive immunity recognize epitopes that are bound to MHC molecules.

There are two classes of MHC proteins (**FIGURE 16.4**). Class I MHC molecules are found on the cytoplasmic membranes of all human cells except red blood cells. Special cells called **antigen-presenting cells (APCs)** also have class II MHC proteins. Professional antigen-presenting cells regularly present antigens. They include macrophages, B cells, and, most importantly, **dendritic cells**. Dendritic cells are so named because they have many long, thin cytoplasmic processes called *dendrites*[7] (**FIGURE 16.5**). Phagocytic dendritic cells are found under the surface of the skin and mucous membranes. Some dendritic cells extend dendrites between skin cells or through a mucous surface to sample antigens, much like a submarine extends a periscope to get a view of surface activity. After acquiring antigens, dendritic cells migrate to lymph nodes to interact with cells of adaptive immunity. Certain other phagocytes, such as *microglia* in the brain and *stellate macrophages* (formerly called *Kupffer cells*) in the liver, may also present antigens under certain conditions. These cells are termed *nonprofessional antigen-presenting cells*.

The cytoplasmic membrane of a professional APC has about 100,000 MHC II molecules. Both MHC I and MHC II are broad in their binding, so they can bind thousands of different epitopes (unlike enzymes, they are not specific). Their binding diversity

---

[7]From Greek *dendron*, meaning "tree," referring to long cellular extensions that look like branches of a tree.

Elements of Adaptive Immunity   479

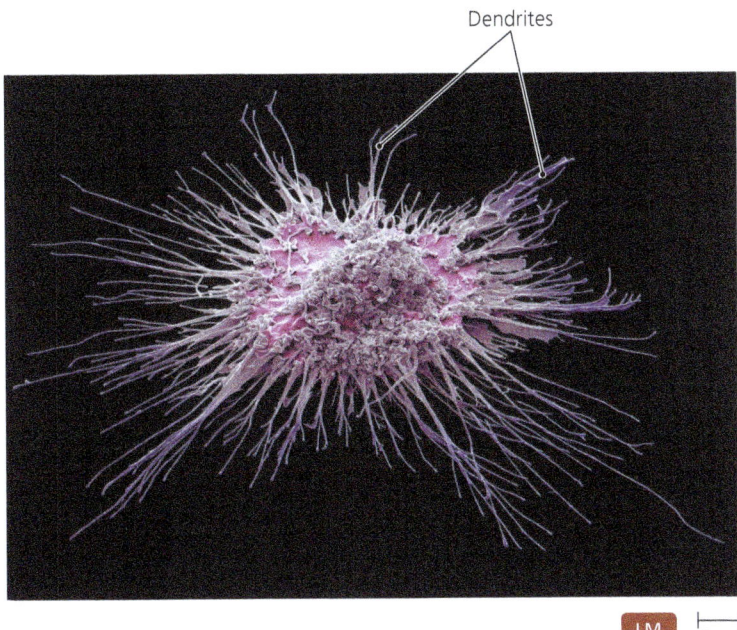

▲ FIGURE 16.5 **Dendritic cell.** These important antigen-presenting cells in the body are found in the skin and mucous membranes. When active, they migrate to lymph nodes. They have numerous thin cytoplasmic processes called dendrites.

is dependent on an individual's genes. Some epitopes won't be bound to the MHC of a particular person, and such epitopes typically don't trigger an immune response. Thus, MHC molecules determine which epitopes might initiate immune responses. This explains why some people are allergic to peanuts and others are not.

Antigen Processing

Before MHC proteins can display epitopes, antigens must be processed to isolate epitopes. Antigen processing occurs via somewhat different processes according to whether an antigen is endogenous or exogenous. Recall that endogenous antigens come from a cell's cytoplasm or from pathogens living within the cell, whereas exogenous antigens have extracellular sources, such as from pathogens in the lymph.

 Play *Antigen Processing and Presentation: Overview* @ Mastering Microbiology

Processing Endogenous Antigens FIGURE 16.6 illustrates the processing of endogenous antigens. A few molecules of each polypeptide produced within nucleated cells—including polypeptides produced by intracellular bacteria or polypeptides coded by viruses—are cut by proteases into smaller pieces containing about 8 to 12 amino acids ❶. These pieces, which will include epitopes of the polypeptides, move into the endoplasmic reticulum (ER) and bind onto complementary antigen-binding grooves of MHC class I molecules that were previously inserted into the membrane of the ER ❷. The ER membrane, now loaded with MHC class I proteins and epitopes, is packaged by a Golgi body to form vesicles ❸. Each vesicle fuses with the cytoplasmic

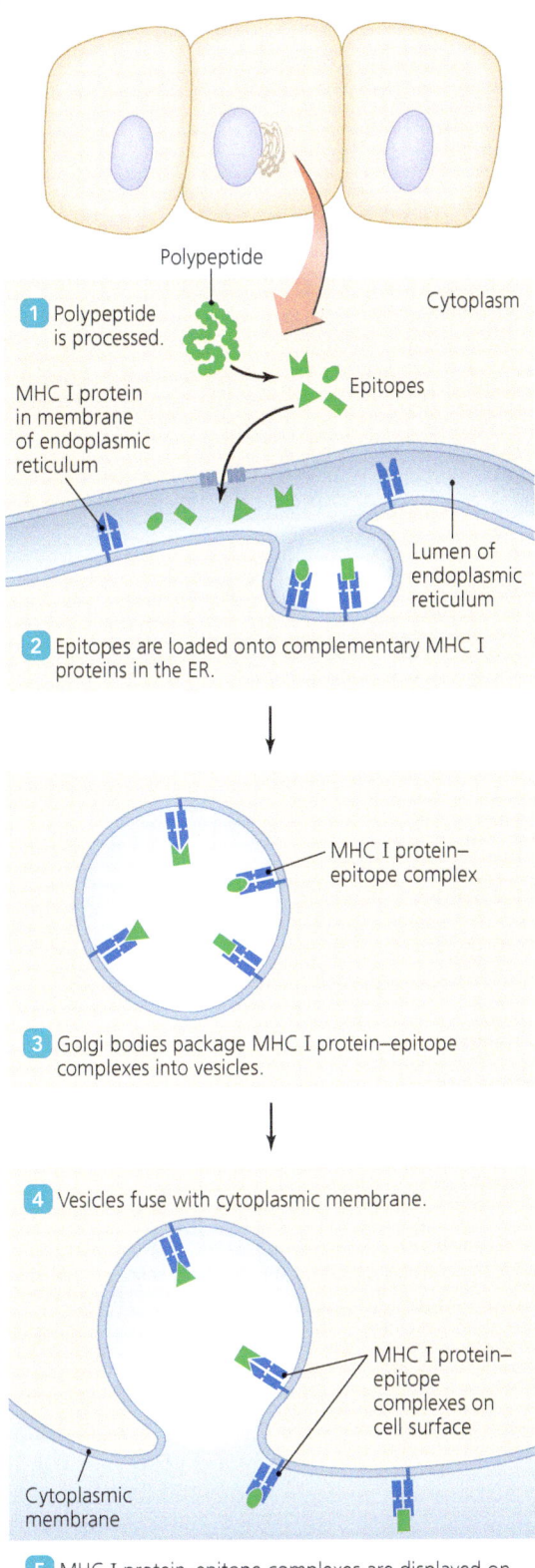

▲ FIGURE 16.6 **The processing of endogenous antigens.** Epitopes from all polypeptides synthesized within a nucleated cell load onto complementary MHC I proteins in the endoplasmic reticulum (ER), which are exported to the cytoplasmic membrane.

membrane in such a way that the vesicle's membrane becomes part of the cytoplasmic membrane **4**. The result is that the cell displays the MHC I protein–epitope complex on the cell's surface **5**. Since every nucleated cell in the body makes MHC class I, each cell displays epitopes from every endogenous antigen and every autoantigen inside that cell. This allows immune cells to detect all the antigens inside cells.

Processing Exogenous Antigens In contrast, only antigen-presenting cells (APCs)—usually dendritic cells—process exogenous antigens. Processing these antigens from outside the body's cells differs from the processing done for antigens produced within cells (FIGURE 16.7). First, a dendritic cell phagocytizes an invading pathogen, enclosing it in a phagosome **1**. Another vesicle, a lysosome already containing MHC class II molecules in its membrane, fuses with the phagosome, and lytic enzymes of the lysosome clip antigens into epitopes within this phagolysosome. MHC II molecules bind complementary epitopes **2**. The vesicle then fuses with the cytoplasmic membrane **3**, leaving MHC II–epitope complexes on the cell's surface **4**. Empty MHC II molecules (those not bound to epitopes) are not stable on a cell's surface; they degrade.

Play *Antigen Processing and Presentation: Steps, MHC* @ Mastering Microbiology

Having considered the "villains" and set the stage, it's time to introduce the "stars," or the "heroes," of the play—the T and B lymphocytes. First, we consider various T lymphocytes that function in cell-mediated immune responses.

T Lymphocytes (T Cells)

LEARNING OUTCOMES

16.12 Describe the importance of the thymus in the development of T lymphocytes.
16.13 Describe basic structural characteristics common to T lymphocytes.
16.14 Compare and contrast three types of T cells.
16.15 Describe apoptosis, and explain its role in lymphocyte editing by clonal deletion.

T lymphocytes often act against body cells that harbor intracellular pathogens, such as viruses, and against body cells that produce abnormal antigens (such as cancer cells). Because T cells act directly against antigens, T lymphocyte immune activities are called *cell-mediated immune responses*.

A human adult's red bone marrow produces T lymphocytes (T cells), which are released into the blood. Chemotactic molecules attract T lymphocytes from blood vessels to the thymus, where they adhere and mature under the influence of sticky, adhesive chemicals and molecular signals from the thymus.

▶ **FIGURE 16.7 The processing of exogenous antigens.** Antigens arising outside the body's cells are phagocytized by an APC and digested to release epitopes, which are loaded into complementary antigen-binding grooves of MHC II molecules. The MHC II–epitope complexes are then displayed on the outside of the APC's cytoplasmic membrane.

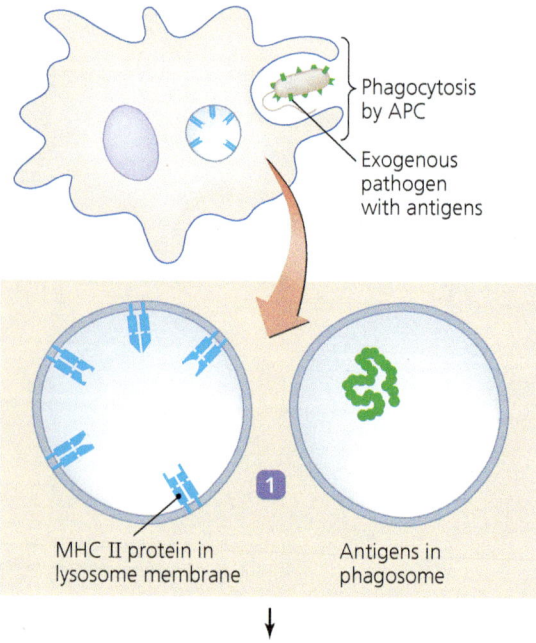

1 MHC II protein in lysosome membrane | Antigens in phagosome

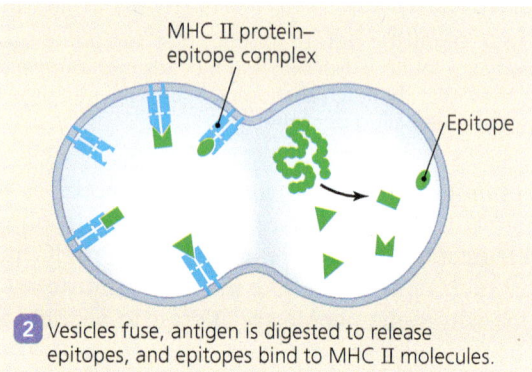

2 Vesicles fuse, antigen is digested to release epitopes, and epitopes bind to MHC II molecules.

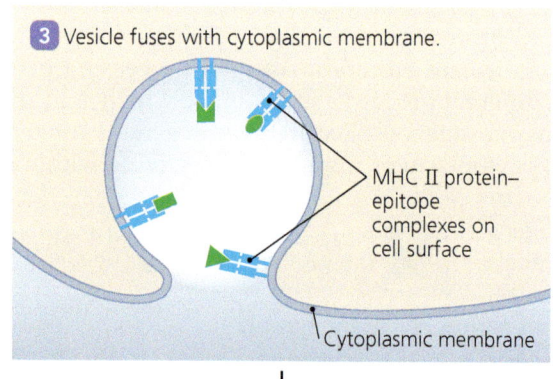

3 Vesicle fuses with cytoplasmic membrane.

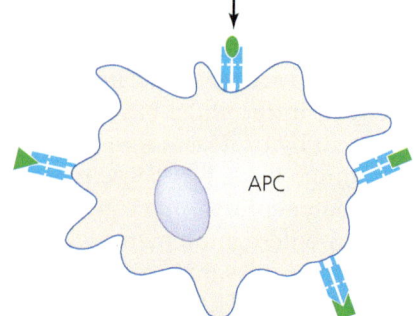

4 MHC II protein–epitope complexes are displayed on cytoplasmic membranes of antigen-presenting cell.

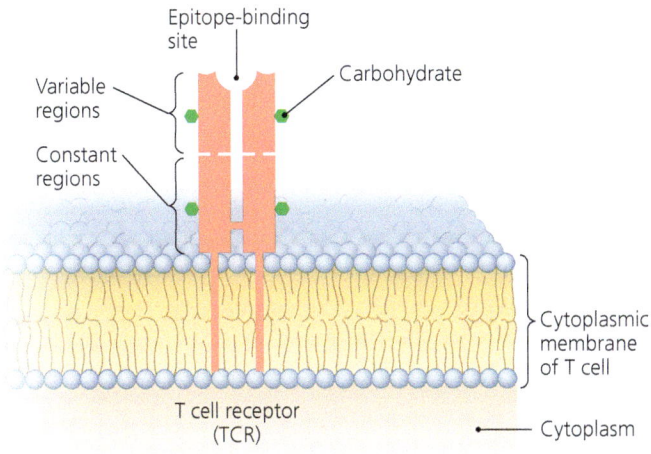

▲ FIGURE 16.8 **A T cell receptor (TCR).** A TCR is a surface molecule composed of two polypeptides containing a single epitope-binding site between them.

Following maturation, they circulate in the lymph and blood and migrate to the lymph nodes, spleen, and Peyer's patches. T cells account for about 70% to 85% of all lymphocytes in the blood.

T cell maturation involves production of about half a million copies of a protein called a **T cell receptor (TCR)** on each T cell's cytoplasmic membrane. Every T cell randomly chooses and combines segments of DNA from TCR genes to create a new genetic combination specific to that cell, which codes for the cell's unique and specific TCR.

Specificity of the T Cell Receptor (TCR)

In your body, there are at least 10^9 different TCRs—each specific TCR type on a different T cell—enough to recognize every possible epitope. Each T cell expresses only one TCR, and so it displays one specific epitope-binding site, which recognizes and binds to a complementary shape (FIGURE 16.8). So, unlike the MHC, each TCR binds only one unique epitope.

An analogy will serve to clarify this point. Imagine a locksmith who has a copy of every possible key to fit every possible lock. If a customer arrives at the shop with a lock needing a key, the locksmith can provide it (though it may take a while to find the correct key). Similarly, you have a TCR complementary to every possible epitope in the environment, though you will encounter only some of them. For example, you have lymphocytes with TCRs complementary to epitopes of stingray venom, though it is unlikely you will ever be stabbed by a stingray.

Types of T Lymphocytes

Immunologists recognize three main types of T cells based on surface molecules and characteristic functions. These are *cytotoxic T cells*, *helper T cells*, and *regulatory T cells*.

Cytotoxic T Lymphocyte Every **cytotoxic T cell (Tc cell)** is distinguished by copies of its own unique TCR as well as the presence of CD8 cell-surface glycoproteins.[8] For this reason, cytotoxic T cells are also known as *CD8 positive (CD8⁺) cells*. As the name *cytotoxic T cell* implies, these lymphocytes directly kill other cells, such as those infected with viruses or other intracellular pathogens. They also kill abnormal cells, such as cancer cells.

Helper T Lymphocyte Immunologists distinguish **helper T cells (Th cells)** by the presence of CD4 glycoproteins in the cells' cytoplasmic membranes, so Th cells are also called *CD4 positive (CD4⁺) cells*. These cells are called "helpers" because their function is to help regulate the activity of B cells and cytotoxic T cells during immune responses by providing necessary signals and growth factors. During an immune response, there are two main subpopulations of helper T cells: *type 1 helper T cells (Th1 cells)*, which assist cytotoxic T cells and stimulate and regulate innate immunity, and *type 2 helper T cells (Th2 cells)*, which function in conjunction with B cells. Immunologists distinguish Th1 from Th2 cells on the basis of their secretions and by characteristic cell-surface proteins.

Helper T cells secrete various soluble protein messengers called *cytokines* that regulate the entire immune system, both adaptive and innate portions. We will consider the types and effects of cytokines shortly.

Regulatory T Lymphocyte Immunologists recognize **regulatory T cells (Tr cells)**, previously known as *suppressor T cells*, by the presence of both CD4 and CD25 glycoproteins. Scientists have not fully characterized the manner in which Tr cells work, but it is known that they are activated by contact with other immune cells and that they secrete cytokines different from those secreted by helper T cells. Regulatory T cells generally suppress immune responses and promote tolerance of certain antigens.

TABLE 16.2 (p. 482) compares and contrasts the features of various types of T lymphocytes. Now, we turn our attention to the way in which the body eliminates T cells that recognize the body's normal autoantigens.

Clonal Deletion of T Cells

Given that T lymphocytes randomly generate the variable region shapes of their TCRs, a population of new lymphocytes will include numerous cells with receptors complementary to normal body components (the autoantigens mentioned earlier). It is vitally important that adaptive immune responses not be directed against autoantigens; in other words, the immune system must be tolerant of "self." When self-tolerance is impaired, the result is an *autoimmune disease* (see Chapter 18).

The body eliminates self-reactive lymphocytes via **clonal deletion**, so named because elimination of a self-reactive cell deletes its potential daughter cells (clones). In this process, lymphocytes are exposed to autoantigens, and those lymphocytes that react to autoantigens undergo **apoptosis**[9] (programmed cell suicide) and are thereby deleted from the repertoire (collection)

---
[8]Scientists name molecules on cells' surfaces with internationally accepted designations consisting of *CD*, which stands for *cluster of differentiation*, followed by a number that reflects the order in which the CD was discovered.

[9]Greek, meaning "falling off."

TABLE 16.2 Characteristics of T Lymphocytes

Lymphocyte	Site of Maturation	Representative Cell-Surface Glycoproteins	Notable Secretions
Helper T cell type 1 (Th1)	Thymus	CD4 and distinctive TCR	Interleukin 2, IFN-γ
Helper T cell type 2 (Th2)	Thymus	CD4 and distinctive TCR	Interleukin 4 and 5
Cytotoxic T cell (Tc)	Thymus	CD8, CD95L, and distinctive TCR	Perforin, granzyme
Regulatory T cell (Tr)	Thymus	CD4, CD25, and distinctive TCR	Interleukin 10

of lymphocytes. Apoptosis is the critical feature of clonal deletion and the development of self-tolerance. Clonal deletion ensures that surviving lymphocytes respond only to foreign antigens. In humans, clonal deletion for T lymphocytes occurs in the thymus.

As we have seen, T cells recognize epitopes only when the epitopes are bound to MHC protein. Immature T lymphocytes spend about a week in the thymus being exposed to all of the body's natural epitopes through a unique feature of thymus cells. As a group, these cells express all of the body's normal proteins, including proteins that have no function in the thymus. For example, some thymus cells synthesize lysozyme, hemoglobin, and muscle cell proteins, though these proteins are not expressed externally to the cells. Rather, thymus cells process these autoantigens so as to express their epitopes in association with an MHC protein. Since the cells collectively synthesize polypeptides from all the body's proteins, together they process and present all the body's autoantigens to immature T cells.

FIGURE 16.9 illustrates the development of T cells. Stem cells in the bone marrow produce T cells **1** that derive their unique TCRs in the thymus **2**. These T cells undergo several steps of selection:

- Those T cells that do not have a functional TCR that binds the body's MHC proteins will undergo apoptosis, in other words, clonal deletion. Because they do not bind the body's own MHC protein, they will be of no use identifying foreign epitopes carried by the body's MHC protein. T cells that do bind MHC protein receive a signal to survive **3**, a result called positive selection.

- Those T cells that subsequently recognize autoantigen in conjunction with MHC protein mostly die by apoptosis—further clonal deletion **4**. This negative selection prevents autoimmune disease and results in tolerance of autoantigens.

- A few of these "self-recognizing" T cells remain alive to become regulatory T cells, to dampen immune responses to autoantigens **5**.

- The remaining T cells are those that will bind the body's own MHC protein in conjunction with foreign epitopes and not with autoantigens. These T lymphocytes become the repertoire of protective T cells, which leave the thymus to circulate in the blood and lymph **6**.

Now that we have considered major histocompatibility proteins, antigen-presenting cells, and T lymphocytes, we consider the other star player in adaptive immune responses—B lymphocytes.

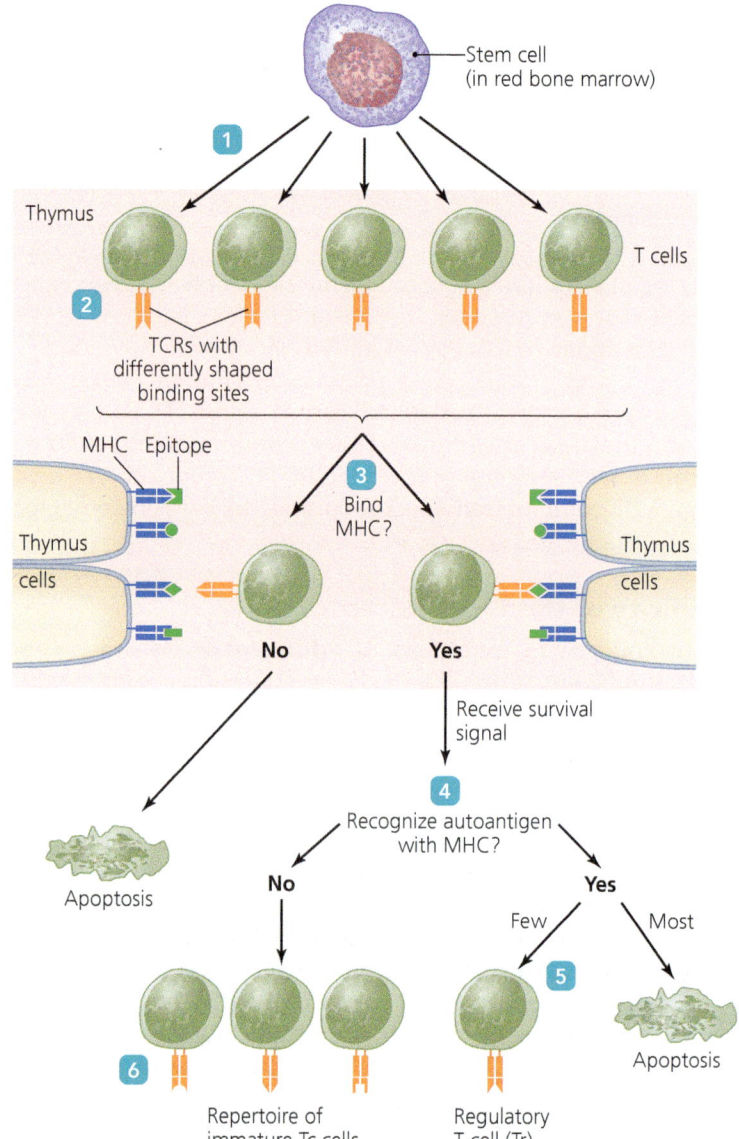

▲ **FIGURE 16.9 Development and clonal deletion of T cells.** **1** Stem cells in the red bone marrow generate a host of lymphocytes that move to the thymus. **2** In the thymus, each lymphocyte randomly generates a TCR with one particular shape. Note that each cell's TCR binding sites differ from those of other cells. **3** T cells pass through a series of "decision questions" in the thymus: Are their TCRs complementary to the body's MHC protein? If no, they undergo apoptosis—*clonal deletion*. If yes, they survive. **4** Do the surviving cells recognize MHC protein bound to any autoantigen? If yes, then most undergo apoptosis (more clonal deletion); **5** a few survive as regulatory T cells (Tr). **6** If no, they survive and become the repertoire of immature T cells.

MICRO CHECK

6. On which types of cells are the two classes of major histocompatibility complex (MHC) proteins located and what type of antigen do they display?
7. Which type of T cell positively affects both B cells and T cells?
8. T lymphocytes mature in the thymus. Which T lymphocytes are deleted by apoptosis during the maturation process?

B Lymphocytes (B Cells) and Antibodies

LEARNING OUTCOMES

16.16 Describe the characteristic of B lymphocytes that gives them specificity.
16.17 Describe the basic structure of an immunoglobulin molecule.
16.18 Contrast the structure, function, and prevalence of the five classes of immunoglobulins.
16.19 Compare and contrast clonal deletion of B cells with clonal deletion of T cells.

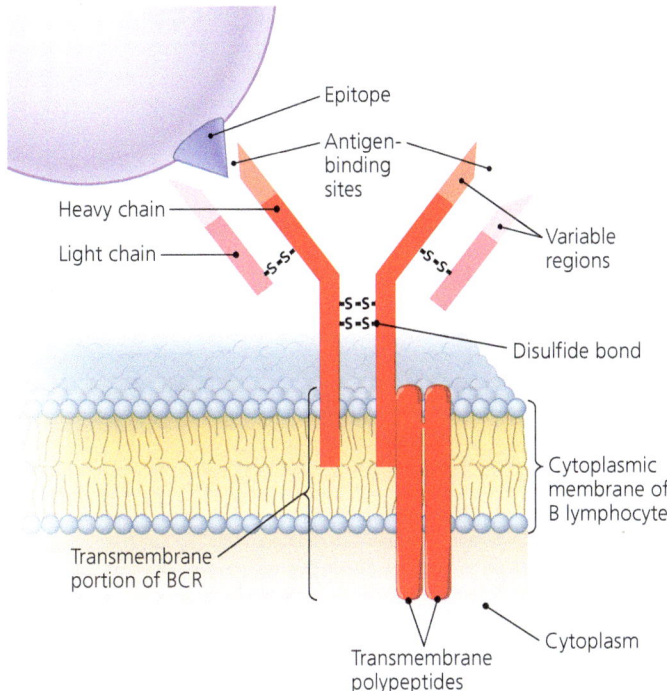

▲ **FIGURE 16.10 B cell receptor (BCR).** The B cell receptor is composed of a symmetrical, epitope-binding, Y-shaped protein composed of two heavy and two light polypeptide chains in association with two additional transmembrane polypeptides.

B lymphocytes are found in the spleen, in MALT, and in the primary follicles of lymph nodes. A small percentage of B cells circulates in the blood. The major function of B cells is the secretion of soluble antibodies, which we examine in more detail shortly. As we have established, B cells function in antibody immune responses, each of which is against only a particular epitope. As with T cells, such specificity comes from membrane proteins, in this case called *B cell receptors*.

Specificity of the B Cell Receptor (BCR)

The surface of each B lymphocyte is covered with about 500,000 identical copies of a **B cell receptor (BCR)** unique to that cell. A BCR is a type of *immunoglobulin* (im′yū-nō-glob′yū-lin; Ig) made of six polypeptide chains: Two identical longer chains called *heavy chains* and two identical shorter *light chains* are arranged in such a way that they form two *arms* that extend outside the cell and look like the letter Y (**FIGURE 16.10**). Two additional polypeptide chains attach to the stem of the "Y" and anchor the BCR into the membrane, forming a *transmembrane portion* of the BCR.

The terms *heavy* and *light* refer to the relative molecular masses of the polypeptide chains that extend beyond the surface of the cell. Disulfide bonds, which are covalent bonds between sulfur atoms in adjoining amino acids, link the heavy and light chains together. The end of each arm is a *variable region*; that is, the ends of BCRs on any one B cell differ from the ends of any other B cell's BCRs, because the ends of each heavy and each light chain vary in amino acid sequence among B cells. Together the two variable regions, one at the end of each arm of the "Y," form **antigen-binding sites** (see Figure 16.10). Antigen-binding sites are complementary in shape to the three-dimensional shape of an epitope and bind precisely to it. Exact binding between antigen-binding site and epitope accounts for the specificity of an antibody immune response.

Though all of the BCRs on a single B cell are the same, the BCRs of one cell differ from the BCRs of all other B cells, much as each snowflake is distinct from all others. Scientists estimate that each person forms no fewer than 10^9 and likely as many as 10^{13} B lymphocytes—each with its own unique BCR. Because an antigen (e.g., a bacterial protein) typically has numerous epitopes of various shapes, many different BCRs will recognize any particular antigen, though each BCR recognizes only one epitope of that antigen.

When an antigenic epitope stimulates a specific B cell via the B cell's unique BCR, the B cell responds by undergoing cell division, giving rise to nearly identical offspring that secrete immunoglobulins into the blood or lymph. The immunoglobulins act against the same epitope shape that stimulated the B cell. Activated, immunoglobulin-secreting B lymphocytes are called **plasma cells**. They no longer display BCRs and have extensive rough endoplasmic reticulum and many Golgi bodies involved in the synthesis, packaging, and secreting of the immunoglobulins (**FIGURE 16.11**).

Lymphocyte Receptor Diversity

BCR genes are randomly generated in sufficient numbers to render the entire collection of BCRs capable of recognizing almost every one of the billions of different epitopes. In other words, at least one BCR is fortuitously complementary to any given specific epitope that the body may or may not encounter. Just as T cells recombine their TCR genes to generate a repertoire of TCRs complementary to every epitope, B cells also recombine their genes to make unique B cell receptors complementary to the same epitopes.

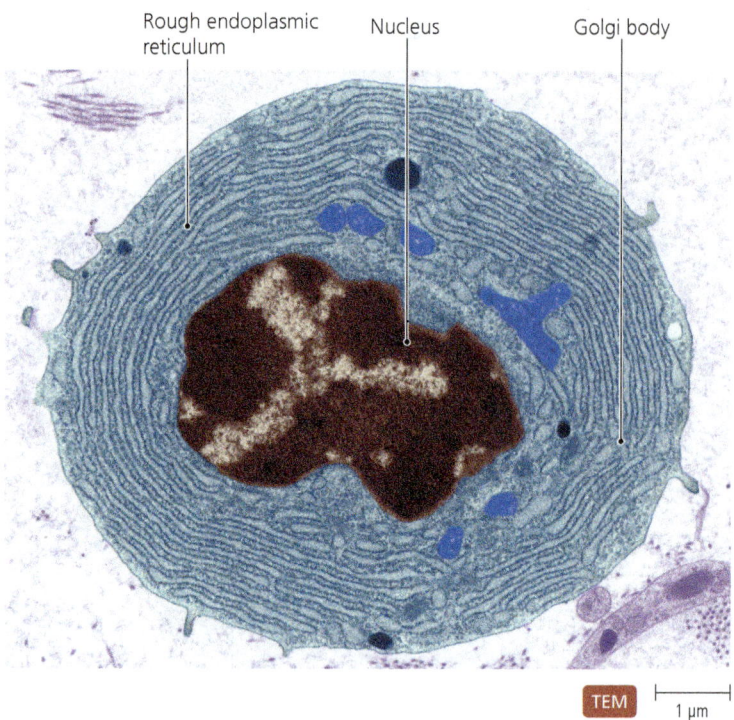

▲ FIGURE 16.11 **A plasma cell.** Plasma cells are almost twice as large as inactive B cells and are filled with rough endoplasmic reticulum and Golgi bodies for the synthesis and secretion of antibodies.

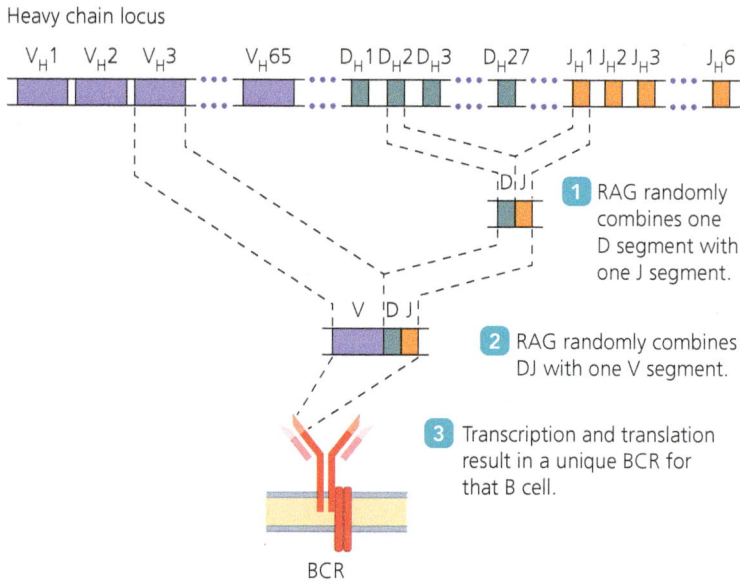

▲ FIGURE 16.12 **Immunoglobulin heavy chain loci have three genetic segments, V, D, and J.** Each B cell randomly recombines an exon from each of the three segments to generate an antigen-binding site that is unique for immunoglobulins of each B cell.

Each of us has billions of different BCR proteins, yet we have only about 25,000 genes. Obviously, a person cannot have a separate gene for each of the different BCRs, so how does such variability among B cells develop? Basically, each B cell randomly generates its own BCR genes.

Genes for the constant regions and for the variable regions of BCRs occur in discrete stretches of DNA called *loci* (singular: *locus*). The variable loci are where diversity is generated, and the variable regions code for the epitope-binding sites in the antibody proteins.

BCR variable region genes occur in three loci on each of two chromosomes—one locus for the heavy chain variable region and two loci (called *kappa* and *lambda*) for the light chain variable region. Each cell is diploid—having two of each type of chromosome—so there are six loci altogether, though an individual B cell uses only one chromosome's loci for each of its heavy and light chains. Once a locus on a particular chromosome is chosen, the other chromosome's corresponding locus is inhibited.

Each locus is divided further into distinct genetic segments coding for portions of its respective chain. The heavy chain variable region locus (FIGURE 16.12) has segments called *variable* (V_H), *diversity* (D_H), and *junction* (J_H) *segments*. In each developing B cell, there are 65 variable segments, 27 diversity segments, and 6 junction segments. Each light chain variable region locus has additional *variable segments* (V_L) and *junction segments* (J_L). For the kappa locus, there are 40 variable segments and 5 junction segments. The lambda locus is simpler, having only 30 V segments and a single J segment.

A great variety of BCRs form during B cell maturation when each B cell uses an enzyme called RAG—the *recombination activating gene* protein—to randomly combine one of each kind of the various segments to form its BCR gene. First, RAG randomly combines one D segment with one J segment **1**, then RAG adds one randomly chosen V segment **2** to create that cell's unique BCR gene. The cell transcribes and translates the BCR gene to give each B cell its own characteristic BCR protein protruding from its cytoplasmic membrane **3**.

An analogy will help you understand this concept. Imagine that you have a wardrobe consisting of 65 different pairs of shoes, 27 different shirts, and 6 different pairs of pants that can be combined to create outfits. (These correspond to the V, D, and J segments of the heavy chain locus.) You can choose any one pair of shoes, any one shirt, and any one pair of pants each day, so with these 98 pieces of clothing, you will potentially have 10,530 different outfits ($65 \times 27 \times 6$). Your roommate, who happens to be the same size as you, might have the same numbers of items but in different colors and styles. With each of you having 10,530 potential outfits, which either of you could wear, there are 21,060 potential outfits! Similarly, each developing B cell has 10,530 possible combinations of V_H, D_H, and J_H segments on each of its two chromosomes, for a total of 21,060 possible gene combinations for the heavy chain variable region.

The developing B cell also uses RAG to recombine the segments of the light chain variable loci. Because there are 200 (40×5) possible kappa genetic combinations and 30 (30×1) possible lambda combinations on each chromosome, there is a total of 460 ($230 + 230$) possible light chain genes on the two chromosomes. Therefore, each B cell can make one of a possible 9,687,600 ($21,060 \times 460$) different BCRs using only 348 genetic segments in the heavy and light chain loci on its two chromosomes.

Still, these nearly 10 million different BCRs do not account for all the variability seen among BCRs. Each B cell creates additional diversity when RAG randomly removes portions of D and J segments before joining the two together. Another enzyme randomly adds nucleotides to each heavy chain VDJ combination, and B cells in the primary follicles of lymph nodes undergo random point mutations in their V regions. The result of RAG recombinations, random deletions, random insertions, and point mutations is tremendous potential variability. Scientists estimate that you may have about 10^{23} B cell receptor possibilities. That's one hundred billion trillion—a number 10 times greater than all the stars in the known universe!

T cells generate their receptors in a similar way, so there are billions of possible TCRs as well.

Next, we consider the structure and functions of secreted immunoglobulins—called antibodies.

Specificity and Antibody Structure

Antibodies are free immunoglobulins—not attached to a membrane—and similar to BCRs in shape except they lack most of the transmembrane portions of a BCR. Thus, a basic antibody molecule is Y-shaped with two identical heavy chains and two identical light chains. The ends of the four chains form two identical antigen-binding sites (**FIGURE 16.13**). The antigen-binding sites of antibodies from a given plasma cell are identical to one another and to the antigen-binding sites of that cell's BCR; thus, antibodies carry the same specificity for an epitope as the BCR of the activated B cell.

Because the arms of an antibody molecule contain antigen-binding sites, they are also known as the F_{ab} *regions (fragment, antigen-binding region)*. The angle between the arms and the stem can change because the point at which they join is hinge-like. F_{ab} regions of all the antibodies produced by a single cell are nearly identical and unique to each B cell.

An antibody stem, which is formed of the lower portions of the two heavy chains, is also called the F_c *region* (because it forms a *fragment* that is *crystallizable*). There are five basic types of stems (F_c regions), designated by the Greek letters *mu, gamma, alpha, epsilon,* and *delta* or by the corresponding English letters, M, G, A, E, and D. A plasma cell attaches DNA coding for its heavy chain variable region to one of five DNA sequences coding for the five types of stems to form one of five classes of antibodies known as IgM (immunoglobulin class mu), IgG, IgA, IgE, or IgD. The F_c regions of all antibodies of each class in a person's body are identical; that is, the F_c regions of all IgM molecules are identical, the F_c regions of all IgA molecules are identical, and so on for each antibody class.

Antibody Function

As we have seen, antigen-binding sites of antibodies are complementary to epitopes; in fact, the shapes of the two can match so closely that most water molecules are excluded from the area of contact, producing a strong, highly specific, noncovalent interaction. Additionally, hydrogen bonds and other molecular attractions mediate antibody binding to epitope. The binding of antibody to epitope is the central functional feature of antibody adaptive immune responses. Once bound, antibodies function in several ways. These include activation of complement and inflammation, neutralization, opsonization, agglutination, and antibody-dependent cell-mediated cytotoxicity.

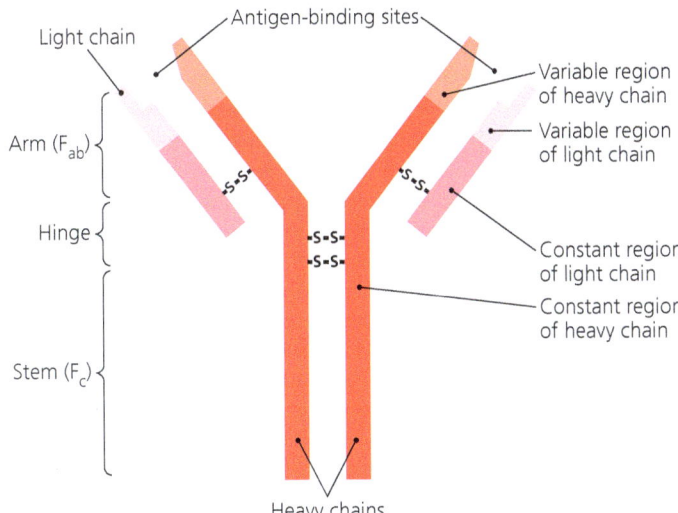

▲ **FIGURE 16.13 Basic antibody structure. (a)** Artist's rendition. Each antibody molecule, which is shaped like the letter Y, consists of two identical heavy chains and two identical light chains held together by disulfide bonds. Five different kinds of heavy chains form an antibody's stem (F_c region.) The arms (F_{ab} regions) terminate in variable regions to form two antigen-binding sites. The hinge region is flexible, allowing the arms to bend in almost any direction. **(b)** Three-dimensional shape of an antibody based on X-ray crystallography. *Why are the two antigen-binding sites on a typical antibody molecule identical?*

Figure 16.13 *The amino acid sequences of the two light chains of an antibody molecule are identical, as are the sequences of the two heavy chains, so the binding sites—each composed of the ends of one light chain and one heavy chain—must be identical.*

Activation of Complement and Inflammation Stems of two or more IgM antibodies bind to complement protein 1 (C1), activating it to become enzymatic. This begins the classical complement pathway, which releases inflammatory mediators. In addition, IgE bound to antigen attaches via its stem to mast cells, eosinophils, and basophils. Attachment triggers the release of inflammatory chemicals. This is what is seen in allergies. (Figures 15.8 and 15.13 more fully illustrate the defense reactions of complement activation and inflammation.)

Neutralization Antibodies can **neutralize** a toxin by binding to a critical portion of the toxin so that it can no longer harm the body. Similarly, antibodies can block adhesion molecules on the surface of a bacterium or virus, neutralizing the pathogen's virulence because it cannot adhere to its target cell (**FIGURE 16.14a**).

Opsonization Antibodies act as **opsonins**[10]—molecules that stimulate phagocytosis. Neutrophils and macrophages have receptors for the stems of IgG molecules; therefore, these leukocytes bind to the stems of antibodies. Once antibodies are so bound, the leukocytes phagocytize them, along with the antigens they carry, at a faster rate compared to antigens lacking bound antibody. Changing the surface of an antigen so as to enhance phagocytosis is called **opsonization** (op′sŏ-nī-zā′shŭn; **FIGURE 16.14b**).

Agglutination Because each antibody has at least two antigen-binding sites, each can attach to multiple epitopes at once. Numerous antibodies can aggregate antigens together—a state called **agglutination** (ă-glū-ti-nā′shŭn; **FIGURE 16.14c**). Agglutination of soluble molecules typically causes them to become insoluble and precipitate. Agglutination may hinder the activity of pathogenic organisms and increases the chance that they will be phagocytized or filtered out of the blood by the spleen. (Chapter 17 examines some uses scientists make of the agglutinating nature of antibodies.)

Antibody-Dependent Cell-Mediated Cytotoxicity (ADCC) Antibodies often coat a target cell by binding to epitopes all over the target's surface. The antibodies' stems can then bind to receptors on special lymphocytes called *natural killer lymphocytes (NK cells)*, which are neither B nor T cells. NK lymphocytes destroy target cells with proteins called **perforin** (per′fŏr-in) and **granzyme** (gran′zīm). Perforin molecules form into a tubular structure in the target cell's membrane, forming a channel through which granzyme enters the cell and triggers apoptosis (**FIGURE 16.14d**). **Antibody-dependent cell-mediated cytotoxicity (ADCC)** is similar to opsonization in that antibodies cover the target cell; however, with ADCC the target dies by apoptosis, whereas with opsonization the target is phagocytized.

Play *Humoral Immunity: Antibody Function*
@ Mastering Microbiology

[10]From Greek *opsonein*, meaning "to supply food," and *izein*, meaning "to cause;" thus, loosely, "to prepare for dinner."

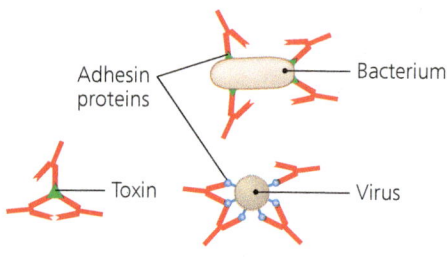

(a) Neutralization

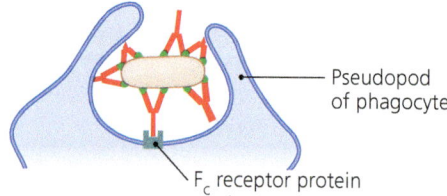

(b) Opsonization

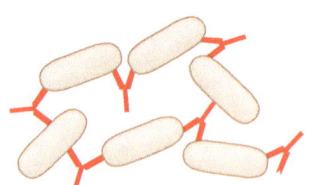

(c) Agglutination

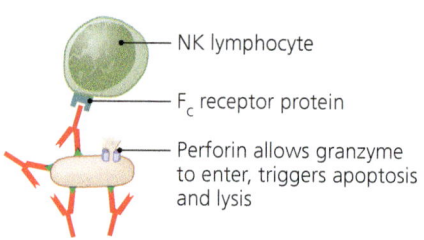

(d) Antibody-dependent cell-mediated cytotoxicity (ADCC)

▲ **FIGURE 16.14 Four functions of antibodies.** Drawings are not to scale. **(a)** Neutralization of toxins and microbes. **(b)** Opsonization. **(c)** Agglutination. **(d)** Antibody-dependent cellular cytotoxicity. Antibodies have two other functions—activation of complement and participation in inflammation (see Figure 15.9).

Classes of Antibodies

Threats confronting the body can be extremely variable, so it is not surprising that there are several classes of antibody. The class involved in any given antibody immune response depends on the type of invading foreign antigens, the portal of entry involved, and the antibody function required. Here, we consider the structure and functions of the five classes of antibodies.

Every plasma cell (activated B lymphocyte) begins by attaching its variable region gene to the gene for the mu stem and thus begins by making **immunoglobulin M (IgM)**. Most IgM is secreted during the initial stages of an immune response. A secreted IgM molecule is more than five times larger than the basic Y-shape because secreted IgM is a pentamer, consisting of five basic units linked together in a circular fashion via disulfide bonds and a short polypeptide *joining (J) chain*. Each IgM subunit has a conventional immunoglobulin structure, consisting of two light chains and two mu heavy chains. IgM is most efficient at complement activation, which also triggers inflammation, and IgM can be involved in agglutination and neutralization.

In a process called **class switching**, a plasma cell then combines its variable region gene to the gene for a different stem and begins secreting a new class of antibodies; that is, the cell switches from synthesizing IgM to secreting IgA or IgE or, most commonly, IgG. **Immunoglobulin G (IgG)** molecules are the most common and longest-lasting class of antibodies, accounting for about 80% of serum antibodies, possibly because IgG has many functions. Each molecule of IgG has the basic Y-shaped antibody structure.

IgG molecules play a major role in antibody-mediated defense mechanisms, including complement activation, opsonization, neutralization, and antibody-dependent cellular cytotoxicity (ADCC). IgG molecules can leave blood vessels to enter extracellular spaces more easily than can other types of immunoglobulins. This is especially important during inflammation because it enables IgG to bind to invading pathogens before they get into the circulatory systems. IgG molecules are also the only antibodies that cross a placenta to protect a developing child.

Immunoglobulin A (IgA), which has alpha heavy chains, is the immunoglobulin most closely associated with various body secretions. Some of the body's IgA is a monomer with the basic Y-shape that circulates in the blood, constituting about 12% of total serum antibody. However, plasma cells in the tear ducts, mammary glands, and mucous membranes synthesize **secretory IgA**, which is composed of two IgA molecules linked via a J chain and another short polypeptide (called a *secretory component*). Plasma cells add secretory component during the transport of secretory IgA across mucous membranes. Secretory component protects secretory IgA from digestion by intestinal enzymes.

Secretory IgA agglutinates and neutralizes antigens and is of critical importance in protecting the body from infections arising in the gastrointestinal, respiratory, urinary, and reproductive tracts. IgA provides nursing newborns some protection against foreign antigens because mammary glands secrete IgA into milk. Thus, nursing babies receive antibodies directed against antigens that have infected their mothers and are likely to infect them as well.

Immunoglobulin E (IgE) is a typical Y-shaped immunoglobulin with two epsilon heavy chains. Because it is found in extremely low concentrations in serum (less than 1% of total antibody), it is not critical for most antibody immune responses. Instead, IgE antibodies act as signal molecules—they attach to receptors on eosinophil cytoplasmic membranes to trigger the release of cell-damaging molecules onto the surface of parasites, particularly parasitic worms. IgE antibodies also trigger mast cells and basophils to release inflammatory chemicals, such as histamine. In more-developed countries, IgE is more likely associated with allergies than with parasitic worms.

Immunoglobulin D (IgD) is characterized by delta heavy chains. IgD molecules are not secreted but are membrane-bound antigen receptors on B cells that are often seen during the initial phases of an antibody immune response. As mentioned previously, a BCR is composed of two transmembrane subunits bound to an immunoglobulin, which is either IgM or IgD (see Figure 16.10). Not all mammals have IgD, and animals that lack IgD show no observable ill effects; therefore, scientists are uncertain of all the functions or importance of the D class of antibody.

TABLE 16.3 (p. 488) compares the different classes of immunoglobulins and adds some details of antibody structure.

Clonal Deletion of B Cells

Clonal deletion of B cells occurs in the bone marrow in a manner similar to deletion of T cells (**FIGURE 16.15**), though self-reactive B cells may become inactive or change their BCR rather than undergo apoptosis. In any case, self-reactive B cells are removed from the active B cell repertoire so that an antibody immune response does not act against autoantigens. Tolerant B cells leave the bone marrow and travel to the spleen, where they undergo further maturation before circulating in the blood and lymph.

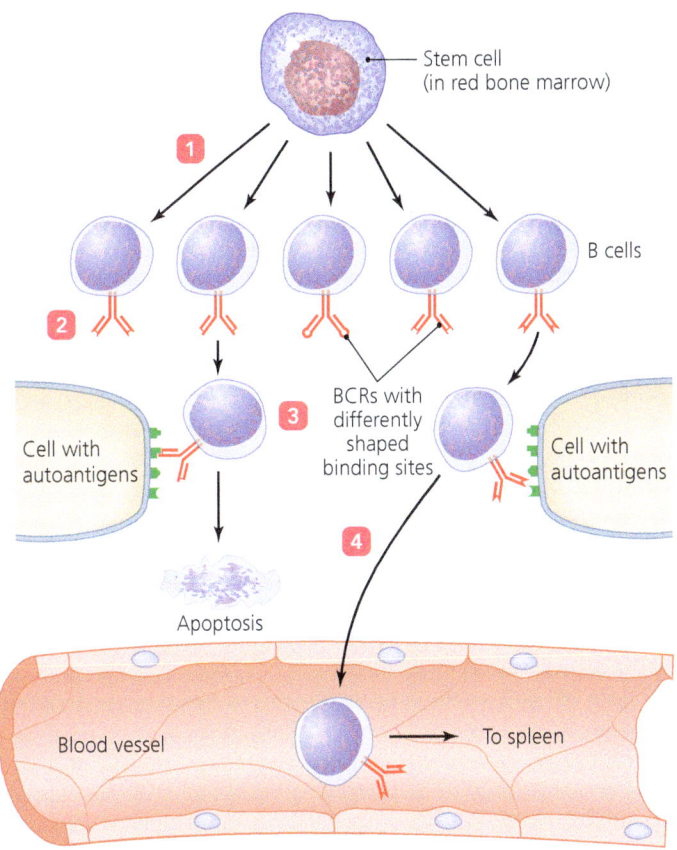

▲ **FIGURE 16.15 Clonal deletion of B cells.** **1** Stem cells in the red bone marrow generate a host of B lymphocytes. **2** Each newly formed B cell randomly generates a BCR with a particular shape. Note that each cell's BCR binding sites differ from those of other cells. **3** Cells whose BCR is complementary to some autoantigen bind with that autoantigen, stimulating the cell to undergo apoptosis. Thus, an entire set of potential daughter B cells (a clone) that are reactive with the body's own cells is eliminated—*clonal deletion*. **4** B cells with a BCR that is not complementary to any autoantigen are released from the bone marrow and into the blood. *Of the B cells shown, which is likely to undergo apoptosis?*

Figure 16.15 The second and third B lymphocytes from the left will undergo apoptosis; their active sites are complementary to the second and fourth autoantigens from the top, respectively.

TABLE 16.3 Characteristics of the Five Classes of Antibodies

	IgM	IgG	IgA	IgE	IgD
Structure, Number of Binding Sites	Monomer, 2; Pentamer, 10	Monomer, 2	Monomer, 2; Dimer, 4	Monomer, 2	Monomer, 2
Type of Heavy Chain	Mu (μ)	Gamma (γ)	Alpha (α)	Epsilon (ε)	Delta (δ)
Functions	Monomer can act as BCR; pentamer acts in complement activation, neutralization, agglutination	Complement activation, neutralization, opsonization, agglutination, and antibody-dependent cell-mediated cytotoxicity (ADCC); crosses placenta to protect fetus	Neutralization and agglutination; dimer is secretory antibody	Triggers release of antiparasitic molecules from eosinophils and of histamines from basophils and mast cells (allergic reactions)	Unknown, but perhaps acts as BCR
Locations	Monomer on B cell surface, pentamer in serum	Serum, mast cell surfaces	Monomer: serum; Dimer: mucous membrane secretions (e.g., tears, saliva, mucus); milk	Serum, mast cell surfaces	B cell surface
Approximate Half-Life[a] in Blood	5 days	23 days	6 days	2 days	3 days
Percentage of Serum Antibodies	5–10%	80%	10–15%	<1%	<0.05%
Size (mass in kilodaltons)	Monomer: 194; Dimer: 970	150	Monomer: 160 Dimer: 385	188	184

[a]Half-life = time it takes for concentration to reduce by half.

Surviving B and T lymphocytes move into the blood and lymph, where they collectively scan for antigens. They communicate among one another and with other body cells via chemical signals called *cytokines*. In terms of our analogy of a stage play, there is dialogue among the cast of characters—in the story of adaptive immunity, the dialogue consists of chemical signals.

MICRO CHECK

9. Which part(s) of B cell receptors form the antigen-binding sites?
10. Which class of antibodies functions as a dimer?
11. Which class of antibodies is the most versatile (meaning it can have the greatest variety of effects) and is longest lasting?

Immune Response Cytokines

LEARNING OUTCOME

16.20 Describe five types of cytokines.

Cytokines (sī′tō-kīnz) are soluble regulatory proteins that act as intercellular messages when released by certain body cells, including those of the kidney, skin, and immune system. Here, we are concerned with cytokines that signal among various immune leukocytes. For example, cytotoxic T cells (Tc) do not respond to antigens unless they are first signaled by cytokines.

Immune system cytokines are secreted by various leukocytes and affect diverse cells. Many cytokines are redundant; that is, their effects are almost identical to those of other cytokines. Such complexity has given rise to the concept of a *cytokine network*—a complex web of signals among all the cell types of the immune system. The nomenclature of cytokines is not based on a systematic relationship among them; instead, scientists named cytokines after their cells of origin, their function, and/or the order in which they were discovered. Cytokines of the immune system include the following substances:

- **Interleukins (ILs)**[11] (in-ter-lū′kinz). As their name suggests, ILs signal among leukocytes, though cells other than leukocytes may also use interleukins. Immunologists named interleukins sequentially as they were discovered. Scientists have identified about 37 human interleukins.
- **Interferons (IFNs)** (in-ter-fēr′onz). These proteins, which inhibit the spread of viral infections (as discussed in Chapter 15), may also act as cytokines. The most important interferon with such a dual function is gamma interferon (IFN-γ), which is a potent phagocytic activator secreted by type 1 helper T cells.
- **Growth factors**. These proteins stimulate leukocyte stem cells to divide, ensuring that the body is supplied with sufficient white blood cells of all types. The body can control the progression of an adaptive immune response by limiting the production of growth factors.
- **Tumor necrosis factor (TNF)**.[12] Macrophages and T cells secrete TNF to kill tumor cells and to regulate immune responses and inflammation.
- **Chemokines** (kē′mō-kīnz). Chemokines are chemotactic cytokines; that is, they signal leukocytes to move—for example, to rush to a site of inflammation or infection or to move within tissues.

TABLE 16.4 summarizes some properties of selected cytokines.

TELL ME WHY

Why are exogenous epitopes processed in vesicles instead of in endoplasmic reticulum, as endogenous epitopes are?

We have been considering adaptive immunity as a stage play. To this point, we have examined the "stage" (the tissues and organs of the lymphatic system) and the "cast of characters" involved in adaptive immunity. The "villainous" characters are antigens with their epitopes; the "heroes" are T cells with their TCRs, B cells with their BCRs, and plasma cells and their antibodies. The dialogue consists of chemical signals—cytokines.

We have set the stage for the immune system "play" by examining the preparatory steps of antigen processing and antigen presentation. We have seen that adaptive immunity is specific because of the precise and accurate fit of lymphocyte receptors with their complements; that adaptive immunity involves clones of T and B cells, which are unresponsive to self (because of clonal deletion); and that immune responses are inducible against foreign antigens. Now, we can examine cell-mediated and antibody immune responses in more detail. We begin with cell-mediated immunity.

Cell-Mediated Immune Responses

LEARNING OUTCOMES
16.21 Describe a cell-mediated immune response.
16.22 Compare and contrast the two pathways of cytotoxic T cell action.

The body uses cell-mediated immune responses primarily to fight intracellular pathogens and abnormal body cells. Recall that inducibility and specificity are two hallmark characteristics of adaptive immunity. The body induces cell-mediated immune responses only against specific endogenous antigens. Given that many common intracellular invaders are viruses, our examination of cell-mediated immunity will focus on these pathogens. However, cell-mediated immune responses are also mounted against cancer cells, intracellular parasitic protozoa, and intracellular bacteria, such as the tuberculosis bacterium, *Mycobacterium tuberculosis* (mī′kō-bak-tēr′ ē- ŭm too-ber-kyū-lō′sis).

Activation of Cytotoxic T Cell Clones and Their Functions

Adaptive immune responses do not initiate at the site of an infection but rather in lymphoid organs, usually lymph nodes, where antigen-presenting cells interact with lymphocytes. The initial event in cell-mediated immunity is the activation of a specific

TABLE 16.4 Selected Immune Response Cytokines

Cytokine	Representative Source	Representative Target	Representative Action
Interleukin 2 (IL-2)	Type 1 helper T (Th1) cell, cytotoxic T (Tc) cell	Tc cell	Cloning of Tc cell
Interleukin 4 (IL-4)	Type 2 helper T (Th2) cell	B cell	B cell differentiates into plasma cell
Interleukin 12 (IL-12)	Dendritic cell	Helper T (Th) cell	Th cell differentiates into Th1 cell
Gamma interferon (IFN-γ)	Th1 cell	Macrophage	Increases phagocytosis
Tumor necrosis factor (TNF)	Macrophages, T cells	Body tissues	Triggers inflammation or apoptosis

---
[11]From Latin *inter*, meaning "between," and Greek *leukos*, meaning "white."
[12]From Latin *necare*, meaning "to kill."

clone of cytotoxic T cells, as depicted for a dendritic cell infected by a virus in FIGURE 16.16:

1. **Antigen presentation.** The infected dendritic cell migrates to a nearby lymph node, where it presents virus epitopes in conjunction with its MHC I protein. Because of the vast diversity of randomly generated T cell receptors (TCRs), at least one cytotoxic T (Tc) cell will have a TCR complementary to the presented MHC I protein–epitope complex. This Tc cell binds to the dendritic cell to form a cell-cell contact site called an *immunological synapse*. CD8 glycoprotein of the Tc cell, which specifically binds to MHC I protein, stabilizes the synapse.

2. **Helper T cell differentiation.** A nearby CD4-bearing helper T (Th) lymphocyte assists by binding to MHC II on the dendritic cell via the TCR of the helper cell. (Recall that dendritic cells present the same epitopes on MHC II and MHC I.) Association with a helper T cell induces the APC to more vigorously signal the Tc cell. Viruses and some intracellular bacteria induce dendritic cells to secrete interleukin 12 (IL-12), which stimulates the helper T cells to differentiate into many clones of type 1 helper T (Th1) cells. Th1 cells in turn secrete IL-2. Lacking the assistance of Th cells, an immunological synapse between the APC and the Tc cell fails to progress. This limits improper immune responses.

3. **Clonal expansion.** The dendritic cell imparts a second signal (not shown) in the immunological synapse. This signal, in conjunction with IL-2 from a Th1 cell, activates the cytotoxic T (Tc) cell to secrete its own IL-2. Interleukin 2 triggers cell division by Tc cells. Activated Tc cells reproduce to form memory T cells (discussed shortly) and more Tc progeny—a process known as **clonal expansion**.

4. **Self-stimulation.** Daughter Tc cells activate and produce both IL-2 receptors (IL-2R) and more IL-2, thereby becoming self-stimulating; they no longer require either an APC or a helper T cell. They leave the lymph node and are now ready to attack virally infected cells.

As previously discussed, when any nucleated cell synthesizes proteins, it displays epitopes from them in the antigen-binding grooves of MHC class I molecules on its cytoplasmic membrane. Thus, when viruses are replicated inside cells, epitopes of viral proteins are displayed on the host cell's surface.

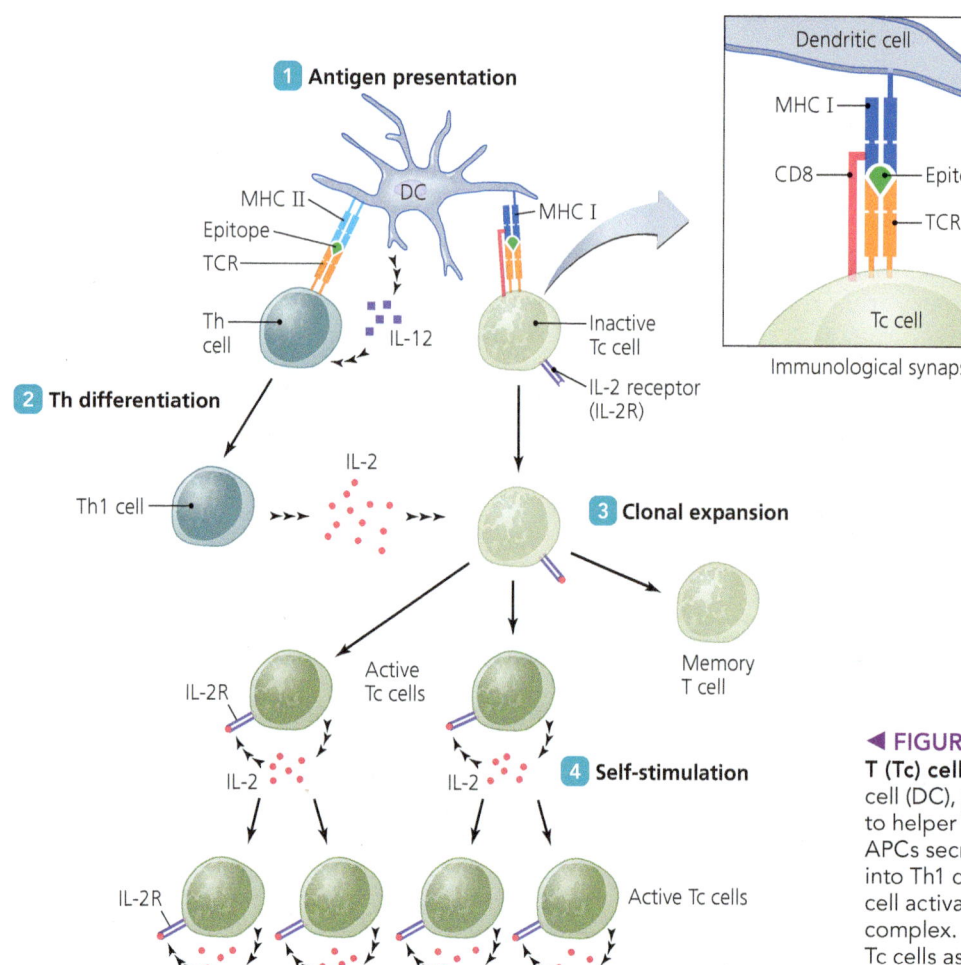

◀ FIGURE 16.16 Activation of a clone of cytotoxic T (Tc) cells. ① Antigen-presenting cells (APCs), here a dendritic cell (DC), present epitopes in conjunction with MHC II protein to helper T (Th) cells and with MHC I to Tc cells. ② Infected APCs secrete IL-12, which causes helper T cells to differentiate into Th1 cells. ③ Signaling from the APC and IL-2 from the Th1 cell activate Tc cells that recognize the MHC I protein–epitope complex. IL-2 triggers Tc cells to divide, forming a clone of active Tc cells as well as memory T cells. ④ Active Tc cells secrete IL-2, becoming self-stimulatory.

An active Tc cell binds to an infected cell via its TCR, which is complementary to the MHC I protein–epitope complex, and via its CD8 glycoprotein, which is complementary to the MHC class I protein of the infected cell (FIGURE 16.17a).

Cytotoxic T cells kill their targets through one of two pathways: the *perforin-granzyme pathway*, which involves the synthesis of special killing proteins, or the *CD95 pathway*, which is mediated through a glycoprotein found on the body's cells.

Play *Cell-Mediated Immunity: Cytotoxic T Cells* @ Mastering Microbiology

The Perforin-Granzyme Cytotoxic Pathway

The cytoplasm of cytotoxic T cells has vesicles containing two key protein cytotoxins—*perforin* and *granzyme*, which are also used by NK cells in conjunction with antibody-dependent cellular cytotoxicity (discussed previously). When a cytotoxic T cell first attaches to its target, vesicles containing the cytotoxins release their contents. Perforin molecules associate into a channel in the membrane of the infected cell through which granzyme enters, activating apoptosis in the target cell (FIGURE 16.17b). Having forced its target to commit suicide, the cytotoxic T cell disengages and moves on to another infected cell.

The CD95 Cytotoxic Pathway

The **CD95 pathway** of cell-mediated cytotoxicity involves the integral glycoprotein CD95, also called Fas, that is present in the cytoplasmic membranes of many body cells. Activated Tc cells insert CD95L—the receptor for CD95—into their cytoplasmic membranes. When an activated Tc cell comes into contact with its target via its TCR, CD95L binds to CD95 on the target, which then activates enzymes that trigger apoptosis, killing the target cells (FIGURE 16.17c).

Memory T Cells

LEARNING OUTCOME

16.23 Describe the establishment of memory T cells.

Some activated T cells become **memory T cells**, which may persist in a state of "suspended animation" for months or years in lymphoid tissues. If a memory T cell subsequently contacts an epitope–MHC I protein complex matching its TCR, it responds immediately (without a need for interaction with APCs) and produces cytotoxic T cell clones that recognize the offending epitope. These cells need fewer regulatory signals and become functional immediately. Further, because the number of memory T cells is greater than the number of T cells that recognized the antigen during the initial exposure, a subsequent cell-mediated immune response to a previously encountered antigen is much

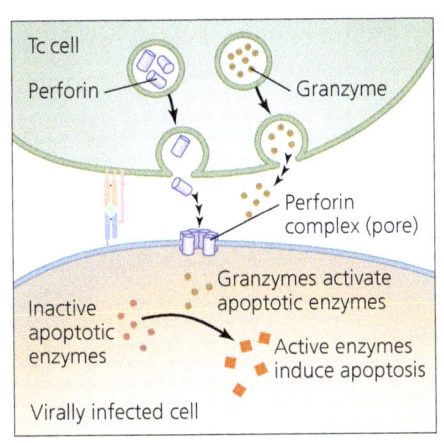

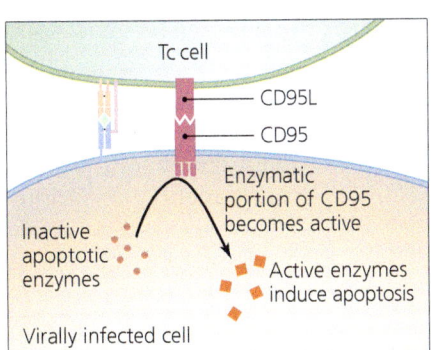

◀ FIGURE 16.17 **A cell-mediated immune response.** (a) The binding of a virus-infected cell by an active cytotoxic T (Tc) cell. (b) The perforin-granzyme cytotoxic pathway. After perforins and granzymes have been released from Tc cell vesicles, granzymes enter the infected cell through the perforin complex pore and activate the enzymes of apoptosis. (c) The CD95 cytotoxic pathway. Binding of CD95L on the Tc cell activates the enzymatic portion of the infected cell's CD95 such that apoptosis is induced.

more effective than a primary response. An enhanced cell-mediated immune response upon subsequent exposure to the same antigen is called a **memory response**.[13]

T Cell Regulation

LEARNING OUTCOME

16.24 Explain the process and significance of the regulation of cell-mediated immunity.

The body carefully regulates cell-mediated immune responses so that T cells do not respond to autoantigens. As we have seen, T cells require several signals from an antigen-presenting cell to activate. If the T cells do not receive these signals in a specific sequence—like the sequence of numbers in a combination padlock—they will not respond. Thus, when a T cell and an antigen-presenting cell interact in an immunological synapse, the two cell types have a chemical dialogue that stimulates the T cell to fully respond to the antigen. If a T cell does not receive the signals required for its activation, it will "shut down" as a precaution against autoimmune responses.

Regulatory T (Tr) cells also modulate cytotoxic T cells by mechanisms that are beyond the scope of our discussion. Suffice it to say that Tr cells provide one more level of control over potentially dangerous cell-mediated autoimmune responses.

MICRO CHECK

12. Which type of cytokine signals white blood cells to move toward a site of infection?
13. Once an inactive cytotoxic T cell recognizes and binds to an epitope that is presented by a dendritic cell, what type of cell activates the T cell, resulting in clonal expansion?
14. What process involves the production of more activated Tc cells?
15. What is the difference between memory T cells and cytotoxic T cells?

TELL ME WHY

Why did scientists give the name *perforin* to a molecule secreted by Tc cells?

Antibody Immune Responses

As we have discussed, the body induces antibody immune responses against the antigens of exogenous pathogens and toxins. Recall that inducibility is one of the main characteristics of adaptive immunity: antibody immunity activates only in response to specific pathogens. The following sections examine the activity of B lymphocytes in antibody immune responses.

---

[13]Sometimes also called *anamnestic responses*, from Greek *ana*, meaning "again," and *mimneskein* "to call to mind."

Inducement of T-Dependent Antibody Immunity with Clonal Selection

LEARNING OUTCOMES

16.25 Describe the formation and functions of plasma cells and memory B cells.
16.26 Describe the steps and effect of clonal selection.
16.27 Describe the events in antibody immune responses.

Most antibody immune responses depend on interaction with helper T cells, so they are called **T-dependent antibody immunity**. (*T-independent immune responses* act against molecules with multiple repeating epitopes, such as polysaccharides. T-independent responses are beyond the scope of this introductory explanation.)

A T-dependent response begins with the action of an antigen-presenting dendritic cell (APC). After endocytosis and processing of the antigen, the dendritic cell presents epitopes in conjunction with its MHC II proteins. This APC will induce the specific helper T (Th, $CD4^+$) lymphocyte with a TCR complementary to the MHC II–epitope complex presented by the APC. The activated Th2 cell must in turn induce the specific B cell that recognizes the same antigen. Lymph nodes facilitate and cytokines mediate interactions among the antigen-presenting cells and lymphocytes, increasing the chance that the appropriate cells find each other.

Thus, a T-dependent antibody immune response involves a series of interactions among antigen-presenting cells, helper T cells, and B cells, all of which are mediated and enhanced by cytokines. Now we will examine each step in more detail (**FIGURE 16.18**):

1 **Antigen presentation for Th activation and proliferation.** A dendritic cell, after acquiring antigens in the skin or mucous membrane, moves via the lymph to a local lymph node. The trip takes about a day. As helper T (Th, $CD4^+$) cells pass through the lymph node, they survey all the resident APCs for complementary epitopes in conjunction with MHC II proteins. Antigen presentation depends on chance encounters between Th cells and the dendritic cells, but immunologists estimate that every lymphocyte browses the dendritic cells in every lymph node every day; therefore, complementary cells eventually find each other. Once they have established an immunological synapse, CD4 molecules in membrane rafts of the Th cell's cytoplasmic membrane recognize and bind to MHC II, stabilizing the synapse.

As we saw in cell-mediated immune responses, helper T cells need further stimulation before they activate. The requirement for a second signal helps prevent accidental inducement of an immune response. As before, the APC imparts the second signal by displaying an integral membrane protein in the immunological synapse. This induces the Th cell to proliferate, producing clones.

2 **Differentiation of helper T cells into Th2 cells.** In antibody immune responses, the cytokine interleukin 4 (IL-4) acts as a signal to the Th cells to become type 2 helper

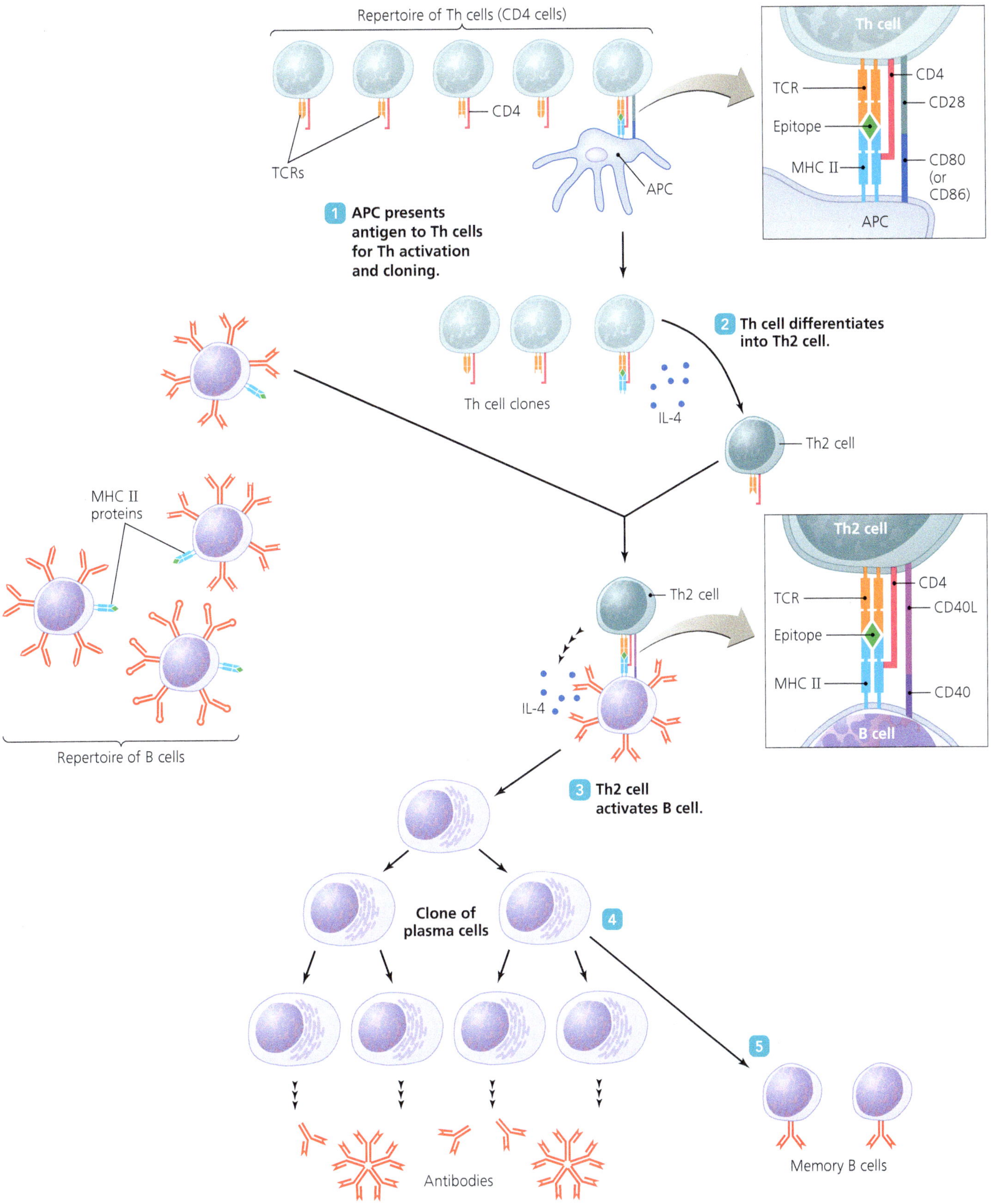

▲ FIGURE 16.18 **A T-dependent antibody immune response.** ① Antigen presentation, in which an APC, typically a dendritic cell, presents antigen to a complementary Th cell. ② Differentiation of the Th cell into a Th2 cell. ③ Activation of the B cell in response to secretion of IL-4 by the Th2 cell, which causes the B cell to differentiate into ④ antibody-secreting plasma cells and ⑤ long-lived memory cells.

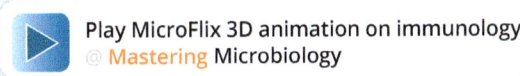

Play MicroFlix 3D animation on immunology
Mastering Microbiology

T cells (Th2 cells). Immunologists do not know the source of IL-4, but it may be secreted initially by innate cells, such as mast cells, or secreted later in a response by the Th cells themselves.

3. **Activation of B cells.** B cells and newly formed Th2 cells survey one another. A Th2 cell binds to the B cell with an MHC II protein–epitope complex that is complementary to the TCR of the Th2 cell. CD4 glycoprotein again stabilizes the immunological synapse.

   Th2 cells secrete more IL-4, which induces the selected B cell to move to the cortex of the lymph node. A Th2 cell in contact with an MHC II protein–epitope on the B cell is stimulated, expresses new gene products, and inserts a protein called CD40L into its cytoplasmic membrane. CD40L binds to CD40, which is found on B cells. This provides a second signal in the immunological synapse, triggering B cell activation. These processes are called **clonal selection** because in effect they select a particular B cell.

4. **Proliferation and differentiation of B cells.** The selected B cell proliferates rapidly to produce a population of cells (clones) that make up a primary follicle in the lymph node. The clones differentiate into two types of cells—*memory B cells* (discussed shortly) and antibody-secreting *plasma cells*.

Most differentiated B cells become **plasma cells**.

Play *Humoral Immunity: Clonal Selection and Expansion* @ Mastering Microbiology

The initial plasma cell descendants of any single activated B cell secrete antibodies with binding sites identical to one another and complementary to the specific antigen recognized by their parent cell. However, as the plasma cell clones replicate, each one slightly modifies its antigen-binding-site genes such that each descendant secretes antibodies with slightly different variable regions. Plasma cells that secrete antibodies with a higher affinity for the epitope have a selective survival advantage over plasma cells secreting antibodies with a less good fit; that is, active B cells with BCRs that bind the epitope more closely survive at a higher rate. Thus, as the antibody immune response progresses, there are more and more plasma cells, secreting antibodies whose affinity gets progressively better.

Each plasma cell produces different classes of antibodies. They begin by secreting IgM and then, through class switching, secrete IgG or IgA or IgE. Class switching is primarily controlled by Th cells and is irreversible. Plasma cells are able to secrete their own weight in IgG every day. As discussed previously, antibodies activate complement, trigger inflammation, agglutinate and neutralize antigen, act as opsonins, and induce antibody-dependent cellular cytotoxicity.

Individual plasma cells are short lived, at least in part because of their high metabolic rate; they die within a few days of activation, although their antibodies can remain in body fluids for several weeks. Providentially, their descendants persist for years to maintain long-term adaptive responses.

Memory Cells and the Establishment of Immunological Memory

LEARNING OUTCOME

16.28 Contrast primary and secondary immune responses.

A small percentage of the cells produced during B cell proliferation do not secrete antibodies but survive as **memory B cells**—that is, long-lived cells with BCRs complementary to the specific epitope that triggered their production (Figure 16.18 ⑤). In contrast to plasma cells, memory cells retain their BCRs and persist in lymphoid tissues, surviving for more than 20 years, ready to initiate antibody production if the same epitope is encountered again. Let's examine how memory cells provide the basis for immunization to prevent disease, using tetanus immunization as an example.

Because the body produces an enormous variety of T and B cells (and therefore TCRs and BCRs), a few Th cells and B cells bind to and respond to epitopes of *tetanus toxoid* (deactivated tetanus toxin), which is used for tetanus immunization. In a **primary response** (FIGURE 16.19a), relatively small amounts of antibodies are produced, and it may take days before sufficient antibodies are made to completely eliminate the toxoid from the body. Though some antibody molecules may persist for three

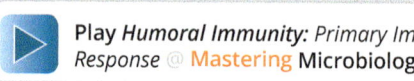
Play *Humoral Immunity: Primary Immune Response* @ Mastering Microbiology

weeks, a primary immune response basically ends when the plasma cells have lived out their normal life spans.

Memory B cells (and memory T cells), surviving in lymphoid tissue, constitute a reserve of antigen-sensitive cells that become active when there is another exposure to the antigen, in this case, toxin from infecting tetanus bacteria. Exposure may occur many years later. Thus, tetanus toxin produced during the course of a bacterial infection directly stimulates a population of memory cells, which proliferate and differentiate rapidly into plasma cells without having to be activated by antigen-presenting cells. Newly differentiated plasma cells produce large amounts of antibody within a few days (FIGURE 16.19b), and the tetanus toxin is neutralized before it can cause disease. Because many memory cells recognize and respond to the antigen, such a **secondary immune response** is much faster and more effective than the primary response.

Play *Humoral Immunity: Secondary Immune Response* @ Mastering Microbiology

As you might expect, a third exposure (whether to tetanus toxin or to toxoid in an immunization booster) results in an even more effective response. Enhanced immune responses triggered by subsequent exposure to antigens are memory responses, which are the basis of *immunization*. (Chapter 17 discusses immunization in more detail.)

In summary, the body's response to infectious agents seldom relies on one mechanism alone because this course of action would be far too risky. Therefore, the body typically uses several different mechanisms to combat infections. An initial response to intruders is inflammation (a nonspecific, innate response), but

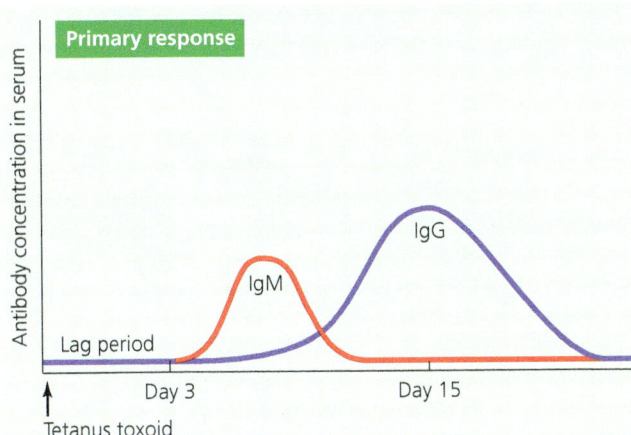

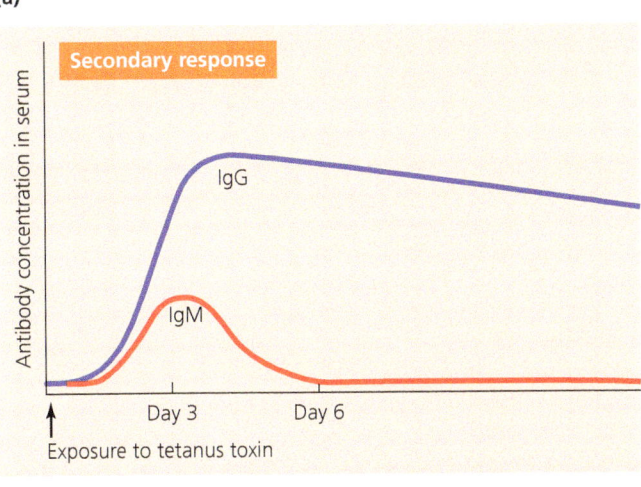

▲ FIGURE 16.19 **The production of primary and secondary antibody immune responses.** This example depicts some events following the administration of a tetanus toxoid in immunization. **(a)** Primary response. After the tetanus toxoid is introduced into the body, the body slowly removes the toxoid while producing memory B cells. **(b)** Secondary response. Upon exposure to active tetanus toxin during the course of an infection, memory B cells immediately differentiate into plasma cells and proliferate, producing a response that is faster and results in greater antibody production than occurs in the primary response.

a specific immune response against the invading microorganisms is also sometimes necessary. APCs phagocytize some of the invaders, process their epitopes, and induce clones of lymphocytes in both antibody and cell-mediated immune responses. Key to enduring protection are the facts that adaptive immunity is unresponsive to self and involves immunological memory brought about by long-lived memory B and T cells.

Cell-mediated adaptive immune responses involve the activity of cytotoxic T lymphocytes in killing cells infected with intracellular bacteria and viruses. Antibody adaptive immunity involves the secretion of specific antibodies that have a variety of functions. A T-dependent antibody immune response occurs when an APC binds to a specific Th cell and signals the Th cell to proliferate.

The relative importance of each of these pathways depends on the type of pathogen involved and on the mechanisms by which it causes disease. In any case, adaptive immune responses are specific, are inducible, involve clones, are unresponsive to self, and give the body long-term memory against their antigenic triggers.

TELL ME WHY

Plasma cells are vital for protection against infection, but memory B cells are not. Why not?

Types of Acquired Immunity

LEARNING OUTCOME

16.29 Contrast active versus passive acquired immunity and naturally acquired versus artificially acquired immunity.

As we have seen, adaptive immunity is acquired during an individual's life. Immunologists categorize immunity as either naturally or artificially acquired. **Naturally acquired immunity** occurs when the body mounts an immune response against antigens, such as influenza viruses or food antigens, encountered during the course of daily life. **Artificial immunity** is the body's response to antigens introduced in vaccines, as occurs with immunization against tetanus and flu. Immunologists further distinguish acquired immune responses as either *active* or *passive*; that is, either the immune system responds actively to antigens via antibody or cell-mediated responses, or the body passively receives antibodies from another individual. Next, we consider each of four types of acquired immunity.

Naturally Acquired Active Immunity

Naturally acquired active immunity occurs when the body responds to exposure to pathogens and environmental antigens by mounting specific immune responses. The body is naturally and actively engaged in its own protection. As we have seen, once an immune response occurs, immunological memory persists—on subsequent exposure to the same antigen, the immune response will be rapid and powerful and often provides the body complete protection.

Naturally Acquired Passive Immunity

Although newborns possess the cells and tissues needed to mount an immune response, they respond slowly to antigens. If required to protect themselves solely via naturally acquired active immunity, they might die of infectious disease before their immune systems were mature enough to respond adequately. However, they are not on their own; in the womb, IgG molecules cross the placenta from the mother's bloodstream to provide protection, and after birth, children receive secretory IgA in breast milk. Via these two processes, a mother provides her baby with antibodies that protect it during its early months. Because the baby is not actively producing its own antibodies, this type of protection is known as **naturally acquired passive immunity**.

EMERGING DISEASE CASE STUDY

Microsporidiosis

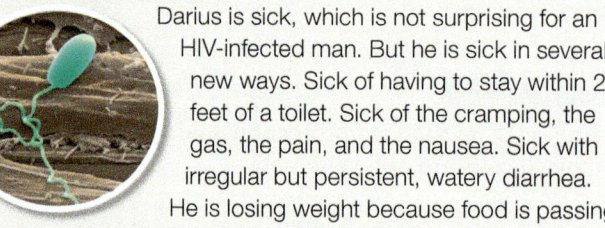

Darius is sick, which is not surprising for an HIV-infected man. But he is sick in several new ways. Sick of having to stay within 20 feet of a toilet. Sick of the cramping, the gas, the pain, and the nausea. Sick with irregular but persistent, watery diarrhea. He is losing weight because food is passing through him undigested. Most days over the past seven months have been disgusting despite his use of over-the-counter remedies, which provide a few days of intermittent relief. His belief that these normal days signaled the end of the ordeal have kept him from the doctor. But now his eyes have begun to hurt, and his vision is blurry. Whatever it is, it's attacking him at both ends.

Time to get stronger drugs from his physician.

Microscopic examination of Darius's stool sample reveals that he is being assaulted by *Encephalitozoon intestinalis*, a member of a group of opportunistic emerging pathogens called microsporidia. The single-celled pathogens are also seen on smears from Darius's nose and eyes. Microsporidia were long thought to be simple single-celled animals, but genetic analysis and comparison with other organisms reveal that they are closer to zygomycete yeasts.

Microsporidia appear to infect humans who engage in unprotected sexual activity, consume contaminated food or drink, or swim in contaminated water. People with active T cells rarely have symptoms, but people with suppressed immunity become easy targets for the fungus.

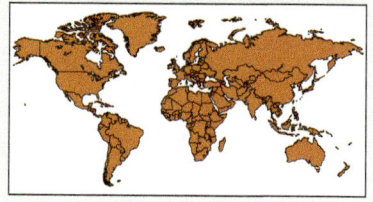

Microsporidia attack by uncoiling a flexible, hollow filament that stabs a host cell and serves as a conduit for the microsporidium's cytoplasm to invade. In this way, the pathogens become intracellular parasites. They can destroy the intestinal lining, causing diarrhea, and spread to the eyes, muscles, or lungs.

Fortunately for Darius, an antimicrobial drug, albendazole, kills the parasite, and the effects of the infection are reversed. Unfortunately for Darius, the loss of helper T cells in AIDS means that another emerging, reemerging, or opportunistic infection is likely to follow. (Chapter 22 discusses microsporidia and microspoidiosis in more detail.)

1. Why are microsporidia considered to be opportunistic pathogens?
2. How could the discovery that microsporidia are fungi rather than animals improve treatment of microsporidiosis?
3. Microsporidia are intracellular pathogens. Which immune cells likely fight off the infection in people with a normal immune system?

Artificially Acquired Active Immunity

Physicians induce immunity in their patients by introducing antigens in the form of vaccines. The patients' own immune systems then mount active responses against the foreign antigens, just as if the antigens were part of a naturally acquired pathogen. Such **artificially acquired active immunity** is the basis of immunization (see Chapter 17).

Artificially Acquired Passive Immunotherapy

Active immunity usually requires days to weeks to develop fully, and in some cases such a delay can prove detrimental or even fatal. For instance, an active immune response may be too slow to protect against infection with rabies or exposure to rattlesnake venom. Therefore, medical personnel routinely harvest antibodies specific for toxins and pathogens that are so deadly or so fast acting that an individual's active immune response is inadequate. They acquire these antibodies from the blood of immune humans or animals, typically a horse. Physicians then inject such *antisera* or *antitoxins* into infected patients to confer **artificially acquired passive immunotherapy**. (Chapter 17 discusses this type of treatment in greater detail.)

Active immune responses, whether naturally or artificially induced, are advantageous because they result in immunological memory and protection against future infections. However, they are slow acting. Passive processes, in which individuals are provided fully formed antibodies, have the advantage of speed but do not confer immunological memory because B and T lymphocytes are not activated. **TABLE 16.5** summarizes the four types of acquired immunity.

MICRO CHECK

16. How is the result of clonal expansion of activated B cells different than or similar to the clonal expansion of T cells?
17. If someone gets measles, and is immune to measles as a result, what type of immunity is this?

TELL ME WHY

Why is passive immunity effective more quickly than active immunity?

TABLE 16.5 A Comparison of the Types of Acquired Immunity

	Active	Passive
Naturally Acquired	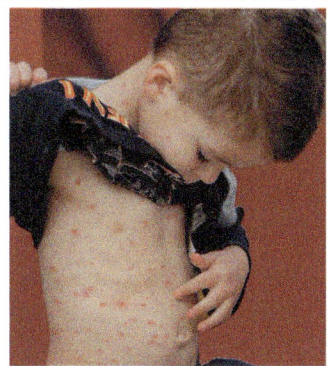 The body responds to antigens that enter naturally, such as during infections.	 Antibodies are transferred from mother to offspring, either across the placenta (IgG) or in breast milk (secretory IgA).
Artificially Acquired	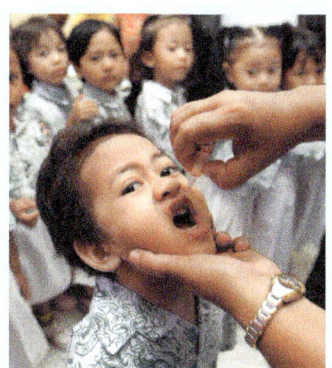 Health care workers introduce antigens in vaccines; the body responds with antibody or cell-mediated immune responses, including the production of memory cells.	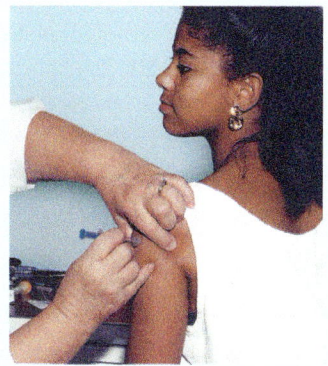 Health care workers give patients antisera or antitoxins, which are preformed antibodies obtained from immune individuals or animals.

MICRO IN THE CLINIC FOLLOW-UP
More Than the Common Cold?

Based on the initial physical examination and Bill's medical history, it doesn't take Dr. Knapp long to make a diagnosis. He informs Bill that he has measles. Given his travel history and lack of vaccination, Bill likely became infected while he was traveling. There isn't any specific antiviral medication for measles, so Dr. Knapp tells Bill that he should stay at home, rest, and minimize his contact with other people for the next few days. Dr. Davis reports the measles case to public health authorities.

Measles is a highly contagious virus, and it is likely that Bill has infected other people. It will be important for health officials to track Bill's movements and activities since he re-entered the country.

Over the next week, Bill makes a full recovery. In the ten days since Bill was diagnosed, an additional 12 people in the town have been diagnosed with measles. Several of these people were at the street fair that Bill attended with Connie. None of the infected individuals had been vaccinated against measles.

1. The measles vaccine uses an attenuated (less virulent) virus to generate long-term protective immunity. Explain how an adaptive immune response is generated and how this response will lead to long-term protection.

2. Vaccination of a high proportion of the population can provide a protective effect known as "herd immunity"; this may explain why only 12 additional cases of measles were identified after Bill's diagnosis. How does a high level of acquired immunity within a population offer protection against contagious diseases such as measles?

Check your answers to Micro in the Clinic Follow-Up questions in the **Mastering Microbiology** Study Area.

CHAPTER SUMMARY

 Go to **Mastering Microbiology** for Interactive Microbiology, Dr. Bauman's Video Tutors, Micro Matters, MicroFlix, Micro-Boosters, practice quizzes, and more.

 Play a MicroFlix 3D animation on immunology @ **Mastering Microbiology**

Overview of Adaptive Immunity (pp. 473–474)

1. **Adaptive immunity** is the body's ability to recognize and defend against distinct species or strains of invaders. Adaptive immunity, which is present only in vertebrates, is characterized by specificity, inducibility, clonality, unresponsiveness to self, and memory. Adaptive immunity involves the actions of **lymphocytes**.

2. **B lymphocytes (B cells)** attack extracellular pathogens in **antibody immune responses** (also called humoral immune responses), involving soluble proteins called antibodies. **T lymphocytes (T cells)** carry out **cell-mediated immune responses** against intracellular pathogens.

Play *Host Defenses: The Big Picture; Cell-Mediated Immunity: Overview; Humoral Immunity: Overview* @ **Mastering Microbiology**

Elements of Adaptive Immunity (pp. 474–489)

1. The **lymphatic system** is composed of **lymphatic vessels**, which conduct the flow of **lymph**, and lymphoid tissues and organs that are directly involved in specific immunity. The latter include **lymph nodes**, the thymus, the spleen, the tonsils, and mucosa-associated lymphoid tissue (MALT). Lymphocytes originate in the red bone marrow. They mature in the marrow or in the thymus. Mature lymphocytes express characteristic membrane proteins. They migrate to and persist in various lymphoid organs, where they are available to encounter foreign invaders in the blood and lymph.

2. **Antigens** are substances that trigger specific immune responses. Effective antigen molecules are large, usually complex, stable, degradable, and foreign to their host. An **epitope** (or antigenic determinant) is the three-dimensional shape of a region of an antigen that is recognized by the immune system.

3. **Exogenous antigens** are found on microorganisms that multiply outside the cells of the body; **endogenous antigens** are produced by pathogens multiplying inside the body's cells.

4. Ideally, the body does not attack antigens on the surface of its normal cells, called **autoantigens**; this phenomenon is called self-tolerance.

5. Nucleated cells display epitopes of their own proteins and epitopes from intracellular pathogens, such as viruses, on **major histocompatibility complex (MHC)** class I proteins.

6. In the initial step in mounting an immune response, antigens are captured, ingested, and degraded into epitopes by **antigen-presenting cells (APCs)**, such as B cells, macrophages, and **dendritic cells**. Epitopes are inserted into major histocompatibility complex (MHC) class II proteins displayed by APCs.

 Play *Antigen Processing and Presentation: Overview, Steps, MHC* @ **Mastering Microbiology**

7. T cells have **T cell receptors (TCRs)**, which have an antigen-binding site for a specific epitope. T cells mature under the influence of signals from the thymus, and they attack cells that harbor endogenous pathogens during cell-mediated immune responses.

8. In cell-mediated immunity, **cytotoxic T cells (Tc cells)**, also called CD8 positive ($CD8^+$) cells because they have membrane protein CD8, act against infected or abnormal body cells, including virus-infected cells, bacteria-infected cells, some fungus- or protozoan-infected cells, some cancer cells, and foreign cells that enter the body as a result of organ transplantation.

9. Two types of **helper T cells (Th cells)**—Th1 and Th2—are characterized by a membrane protein, CD4, so they are $CD4^+$. They direct cell-mediated and antibody immune responses, respectively.

 Play *Cell-Mediated Immunity: Helper T Cells* @ **Mastering Microbiology**

10. T cells that do not recognize MHC I protein and most T cells that recognize MHC I protein in conjunction with autoantigens are removed by **apoptosis**. This is **clonal deletion**. A few self-recognizing T cells are retained and become **regulatory T cells (Tr cells)**. T cells that recognize MHC I protein but not autoantigens become the repertoire of immature T cells.

11. B lymphocytes (B cells), which mature in the red bone marrow, make immunoglobulins (Ig) of two types—**B cell receptors (BCRs)** and **antibodies**. Immunoglobulins are complementary to epitopes and consist of two light chains and two heavy chains joined via disulfide bonds to form Y-shaped molecules. BCRs are inserted into the cytoplasmic membranes of B cells via transmembrane polypeptides, whereas antibodies are secreted.

12. Together the variable regions of an immunoglobulin arm's heavy chain and light chain form an **antigen-binding site** that is called an F_{ab} region. Each B cell randomly selects (once in its life) genes for its F_{ab} regions; therefore, the F_{ab} regions are called variable regions because they differ from cell to cell. Each basic immunoglobulin molecule has two antigen-binding sites and can potentially bind two epitopes.

13. Antibodies function in complement activation, inflammation, **neutralization** (blocking the action of a toxin or attachment of a pathogen), as **opsonins** for **opsonization** (enhanced phagocytosis), **agglutination**, and **antibody-dependent cellular cytotoxicity (ADCC)**.

 Play *Humoral Immunity: Antibody Function* @ **Mastering Microbiology**

14. Antibodies are of five basic classes based upon their stems (F_c regions), which differ in their type of heavy chain.

15. **Immunoglobulin M (IgM)**, a pentamer with 10 antigen-binding sites, is the predominant class of antibody produced first during a primary antibody response. **Immunoglobulin G (IgG)** is the predominant antibody found in the bloodstream and is largely responsible for defense against invading bacteria. IgG can cross a placenta to protect the fetus. Two molecules of **immunoglobulin**

A (IgA) are attached via J chains and a polypeptide secretory component to produce **secretory IgA**, which is found in milk, tears, and mucous membrane secretions. **Immunoglobulin E (IgE)** triggers inflammation and allergic reactions. It also functions during helminth infections. **Immunoglobulin D (IgD)** is found in cytoplasmic membranes of some animals.

16. Through a process called **class switching**, antibody-producing cells change the class of antibody they secrete, beginning with IgM and then producing IgG or switching again to produce IgA or IgE.

17. B cells with B cell receptors that respond to autoantigens are selectively killed via apoptosis—further clonal deletion. Only B cells that respond to foreign antigens survive to defend the body.

18. **Cytokines** are soluble regulatory proteins that act as intercellular signals to direct activities in immune responses. Cytokines include **interleukins (ILs)**, **interferons (IFNs)**, **growth factors**, **tumor necrosis factors (TNFs)**, and **chemokines**.

Cell-Mediated Immune Responses (pp. 489–492)

1. Once activated by dendritic cells, cytotoxic T cells (Tc cells), recognize abnormal molecules presented by MHC I protein on the surface of infected, cancerous, or foreign cells. Sometimes cytotoxic T cells require cytokines from Th1 cells.

2. Activated Tc cells reproduce to form memory T cells and more Tc progeny in a process called **clonal expansion**.

3. Cytotoxic T cells destroy their target cells via two pathways: the perforin-granzyme pathway, which kills the affected cells by secreting **perforins** and **granzymes**; or the **CD95 pathway**, in which CD95L binds to CD95 on the target cell, triggering target cell apoptosis. Cytotoxic T cells may also form **memory T cells**, which function in **memory responses**.

 Play *Cell-Mediated Immunity: Cytotoxic T Cells* @ Mastering Microbiology

Antibody Immune Responses (pp. 492–495)

1. In **T-dependent antibody immunity**, an APC's MHC II protein–epitope complex activates a helper T cell (Th cell) bearing a complementary TCR. CD4 stabilizes the connection between the cells, which is an example of an immunological synapse. Interleukin 4 (IL-4) then induces the Th cell to become a type 2 helper T cell (Th2).

2. In **clonal selection**, an immunological synapse forms between the Th2 cell and a B cell bearing a complementary MHC II protein–epitope complex. The Th2 cell secretes IL-4, which induces the B cell to divide. Its offspring, collectively called a clone, become **plasma cells** or **memory B cells**.

 Play *Humoral Immunity: Clonal Selection and Expansion* @ Mastering Microbiology

3. Plasma cells live for only a short time but secrete large amounts of antibodies, beginning with IgM and class switching as they get older. Memory B cells migrate to lymphoid tissues to await a subsequent encounter with the same antigen.

4. The **primary response** to an antigen is slow to develop and of limited effectiveness. When that antigen is encountered a second time, the activation of memory cells ensures that the immune response is rapid and strong. This is a **secondary immune response**. Such enhanced antibody immune responses are memory responses.

 Play *Humoral Immunity: Primary Immune Response, Secondary Immune Response* @ Mastering Microbiology

Types of Acquired Immunity (pp. 495–497)

1. When the body mounts a specific immune response against an infectious agent, the result is called **naturally acquired active immunity**.

2. The passing of maternal IgG to the fetus and the transmission of secretory IgA in milk to a baby are examples of **naturally acquired passive immunity**.

3. **Artificially acquired active immunity** is achieved by deliberately injecting someone with antigens in vaccines to provoke an active response, as in the process of immunization.

4. **Artificially acquired passive immunotherapy** involves the administration of preformed antibodies in antitoxins or antisera to a patient.

QUESTIONS FOR REVIEW

Answers to the Questions for Review (except Short Answer questions) begin on p. A-1.

Multiple Choice

1. Antibodies function to _____.
   a. directly destroy foreign organ grafts
   b. mark invading organisms for destruction
   c. kill intracellular viruses
   d. directly promote cytokine synthesis
   e. stimulate T cell growth

2. MHC class II molecules bind to _____ and trigger _____.
   a. endogenous antigens; cytotoxic T cells
   b. exogenous antigens; cytotoxic T cells
   c. antibodies; B cells
   d. endogenous antigens; helper T cells
   e. exogenous antigens; helper T cells

3. Rejection of a foreign skin graft is an example of _____.
   a. destruction of virus-infected cells
   b. tolerance
   c. antibody-mediated immunity
   d. a secondary immune response
   e. a cell-mediated immune response

4. An autoantigen is _____.
   a. an antigen from normal microbiota
   b. a normal body component
   c. an artificial antigen
   d. any carbohydrate antigen
   e. nucleic acid

5. Among the key molecules that control cell-mediated cytotoxicity are _____.
   a. perforin
   b. immunoglobulins
   c. complement
   d. cytokines
   e. interferons

6. Which of the following lymphocytes predominates in blood?
   a. T cells
   b. B cells
   c. plasma cells
   d. memory cells
   e. All are about equally prevalent

7. The major class of immunoglobulin found on the surfaces of the walls of the intestines and airways is secretory _____.
   a. IgG
   b. IgM
   c. IgA
   d. IgE
   e. IgD

8. Which cells express MHC class I molecules in a patient?
   a. red blood cells
   b. antigen-presenting cells only
   c. neutrophils only
   d. all nucleated cells
   e. dendritic cells only

9. In which of the following sites in the body can B cells be found?
   a. lymph nodes
   b. spleen
   c. red bone marrow
   d. intestinal wall
   e. all of the above

10. Tc cells recognize epitopes only when the latter are held by _____.
    a. MHC proteins
    b. B cells
    c. interleukin 2
    d. granzyme

Modified True/False

Mark each statement as either true or false. Rewrite false statements to make them true by changing the underlined words.

1. _____ MHC class II molecules are found on <u>T cells</u>.
2. _____ <u>Apoptosis</u> is the term used to describe cellular suicide.
3. _____ Lymphocytes with CD8 glycoprotein are <u>helper</u> T cells.
4. _____ <u>Cytotoxic T cells</u> secrete immunoglobulin.
5. _____ Secretion of antibodies by activated B cells is a form of <u>cell-mediated</u> immunity.

Matching

Match each cell in the numbered list with its associated protein from the lettered list.

1. ___ Plasma cell
2. ___ Cytotoxic T cell
3. ___ Th2 cell
4. ___ Dendritic cell

A. MHC II molecule
B. Interleukin 4
C. Perforin and granzyme
D. Immunoglobulin

Match each type of immunity in the numbered list with its associated example from the lettered list.

1. ___ Artificially acquired passive immunotherapy
2. ___ Naturally acquired active immunity
3. ___ Naturally acquired passive immunity
4. ___ Artificially acquired active immunity

A. Production of IgE in response to pollen
B. Acquisition of maternal antibodies in breast milk
C. Administration of tetanus toxoid
D. Administration of antitoxin

VISUALIZE IT!

1. Label the parts of the immunoglobulin pictured here.

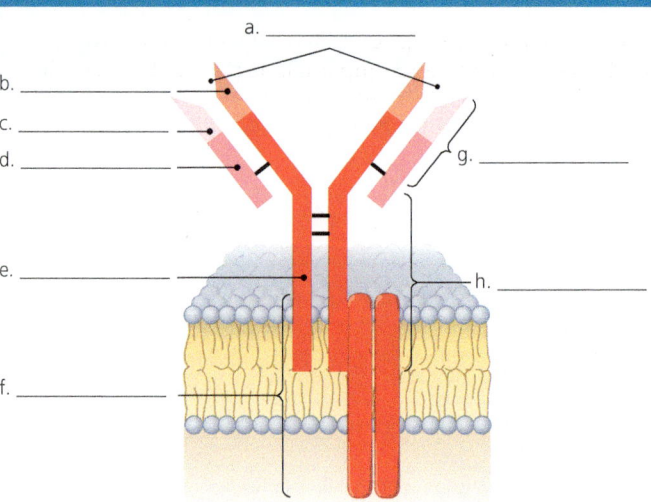

2. The nearby image is a transmission electron micrograph of a dendritic cell. Indicate where a scientist could find molecules of MHC I and MHC II. Label a pseudopod and a vesicle.

Short Answer

1. When is antigen processing an essential prerequisite for an immune response?

2. Why does the body have both antibody and cell-mediated immune responses?

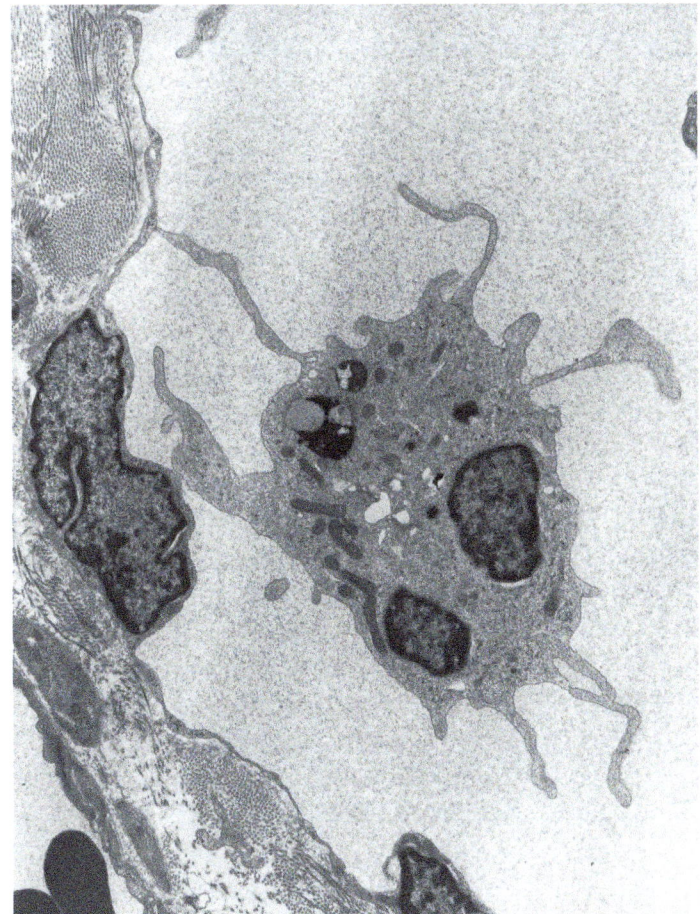

CRITICAL THINKING

1. Why is it advantageous for the lymphatic system to lack a pump?

2. Contrast innate defenses with adaptive immunity.

3. How does requiring the immune system to process antigen benefit the body?

4. Scientists can develop genetically deficient strains of mice. Describe the immunological impairments that would result in mice deficient in each of the following: class I MHC, class II MHC, TCR, BCR, IL-2 receptor, and IFN-γ.

5. Human immunodeficiency virus (HIV) preferentially destroys $CD4^+$ cells. Specifically, what effect does this have on antibody and cell-mediated immunity?

6. What would happen to a person who failed to make MHC molecules?

7. Why does the body make five different classes of immunoglobulins?

8. Some materials, such as metal bone pins and plastic heart valves, can be implanted into the body without fear of rejection by the patient's immune system. Why is this? What are the ideal properties of any material that is to be implanted?

9. What nonmembranous organelle is prevalent in plasma cells? What membranous organelle is prevalent?

10. The cross-sectional area of the afferent lymphatic vessels arriving at a lymph node is greater than the cross-sectional area of the efferent lymphatics exiting the lymph node. The result is that lymph moves slowly through a lymph node. Why is this advantageous?

11. Two students are studying for an exam on the body's defense systems. One of them insists that complement is part of the nonspecific second line of defense, but the partner insists that complement is part of an antibody immune response in the third line of defense. How would you explain to them that they are both correct?

12. In general, what sorts of pathogens might be able to more successfully attack a patient with an inability to synthesize B lymphocytes?

13. What sorts of pathogens could successfully attack a patient who is unable to produce T lymphocytes?

14. As part of the treatment for some cancers, physicians kill the cancer patients' dividing cells, including the stem cells that produce leukocytes, and then give the patients a bone marrow transplant from a healthy donor. Which cell is the most important cell in such transplanted marrow?

CONCEPT MAPPING

Using the terms provided, fill in the concept map that describes antibodies. You can also complete this and other concept maps online by going to the **Mastering Microbiology** Study Area.

Agglutination
Antibody-dependent cell-mediated cytotoxicity (ADCC)
Antigens
Antigen-stimulated B cells
Complement activation
IgA
IgD
IgE
IgG
IgM
Inflammation
Neutralization
Phagocytosis
Plasma cells
Secreted immunoglobulins

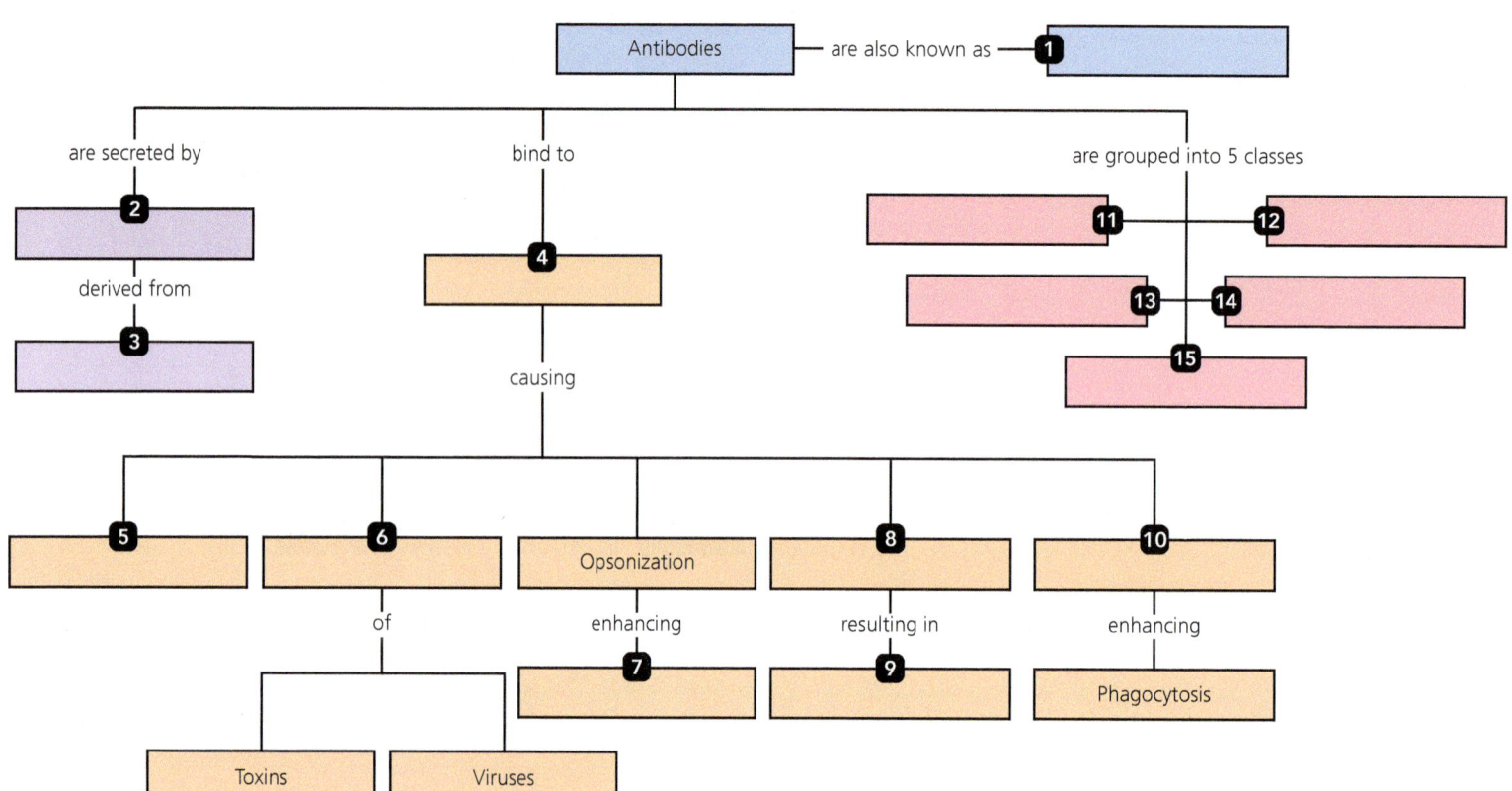

17 Immunization and Immunoassays

Before You Begin

1. Activation of which type of lymphocytes results in the production of memory cells?
2. Which type of antibody is produced first and in greater amounts upon subsequent exposures to an antigen (weeks, months, years, or decades after the initial exposure)?
3. If someone is injected with tetanus vaccine, what type of immunity results? (artificial/natural, active/passive).
4. Which part of an antibody is responsible for recognizing a specific antigen?

MICRO IN THE CLINIC

Complications of Premature Birth?

THE PAST COUPLE OF MONTHS have been difficult for Doug and Stacy. Their first baby, Ryan, arrived five weeks early. Ryan stayed in the hospital for three weeks after he was born, because there were some issues with his lungs. Nevertheless, the doctors have assured Doug and Stacy that he will be fine. The first month at home has been exciting but stressful, and Doug and Stacy are just beginning to settle into their roles as new parents.

For the past couple of days, baby Ryan has been irritable, and Stacy notices that he isn't eating as much as usual. A day or so later, Ryan has a low fever and a slight cough, and he has been sneezing. Doug and Stacy notice a "raspy" sound when Ryan breathes. Worried that there might be problems related to Ryan's earlier lung issues, they call Ryan's pediatrician immediately.

1. Are Ryan's symptoms the result of his premature birth?
2. Could there be another explanation for his symptoms?

Turn to the end of the chapter (p. 520) to find out.

Immunizations activate an adaptive immune response to protect the body from diseases.

SOLVE THE PROBLEM

Rights versus Responsibilities

In the early 20th century, almost all children contracted measles, with 4 million new infections per year, resulting in approximately 50,000 hospitalizations and 6,000 deaths. But in 2000, the Centers for Disease Control and Prevention (CDC) declared measles eliminated (defined by lack of disease transmission for a year) in the United States. This medical advance capped a campaign of immunization that began in 1963. However, multiple measles outbreaks, beginning in 2014, resulted in the greatest number of cases of measles in the United States since 1995. How did measles reemerge? Several factors contributed.

Measles is still common in other parts of the world, and travelers can bring the measles virus here. A measles outbreak occurs when there are significant numbers of unimmunized people in a community. When the immunization rate in a community is above 95 percent, then even unimmunized individuals are protected because susceptible people are too rarely in contact with infected individuals to allow disease transmission. This effect is called herd immunity. The currently recommended vaccine against measles is MMR, which protects against measles, mumps, and rubella. One dose is to be given at age 12–15 months, followed by a booster at 4–6 years, and a second booster between 19 and 40.

There is no controversy in the scientific community on MMR efficacy: the vaccine is highly effective. However, there is an ethical dilemma between the individual **rights** of people to refuse immunization as compared to their **responsibility** toward public health, since lack of immunization can allow the virus to spread, resulting in outbreaks.

If you were on a panel advising the CDC, how would you answer these questions?

1. Should we force people to be immunized?
2. What procedures should be used to ensure the highest immunization rates?
3. What other facts would help make decisions related to immunization?

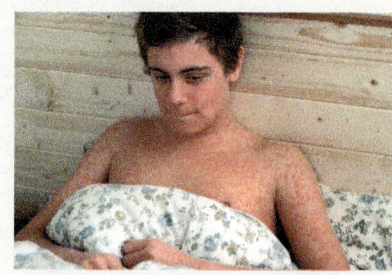

DISEASE CAN SPREAD RAPIDLY THROUGH A POPULATION... WHEN ONLY SOME PEOPLE GET VACCINATED

BEFORE | AFTER

• HEALTHY, VACCINATED • HEALTHY, NOT VACCINATED • SICK, NOT VACCINATED

http://www.cdc.gov/measles/cases-outbreaks

In this chapter, we will discuss three applications of immunology: active immunization (vaccination), passive immunotherapy using immunoglobulins (antibodies), and immune testing. Immunization has proven the most efficient and cost-effective method of controlling infectious diseases. Without the use of effective vaccines, millions more people worldwide would suffer each year from potentially fatal infectious diseases, including measles, mumps, and polio. The administration of immunoglobulins has further reduced morbidity and mortality from certain immunodeficiency diseases. Medical personnel also make practical use of immune responses as a diagnostic procedure. For example, the detection of antibodies to HIV in a person's blood indicates that the individual has been exposed to that virus and may develop AIDS. The remarkable specificity of antibodies also enables the detection of drugs in urine, recognition of pregnancy at early stages, and the identification or characterization of other biological material. The many tests developed for these purposes are the focus of the discipline of serology and are discussed in the second half of this chapter.

Immunization

An individual may be made immune to an infectious disease by two artificial methods: *active immunization*, which involves administering antigens to a patient so that the patient actively mounts an adaptive immune response, and *passive immunotherapy*, in which a patient acquires temporary immunity through the transfer of antibodies formed by other individuals or animals (see Table 16.5 on p. 497).

In the following sections, we will review the history of immunization before examining immunization and immunotherapy in more detail.

Brief History of Immunization

LEARNING OUTCOME

17.1 Discuss the history of immunization from the 12th century through the present.

As early as the 12th century, the Chinese noticed that children who recovered from smallpox never contracted the disease a second time. They therefore adopted a policy of deliberately infecting young children with particles of ground smallpox scabs from other children who had survived mild cases. By doing so, they succeeded in significantly reducing the population's overall morbidity and mortality from the disease. News of this procedure, called *variation* (var'ē-ō-lā'shŭn), spread westward through central Asia, and the technique was widely adopted.

Lady Mary Montagu (1689–1762), the wife of the English ambassador to the Ottoman Empire, learned of the procedure, had it performed on her own children, and told others about it upon her return to England in 1721. As a result, variolation came into use in England and in the American colonies. Although effective and usually successful, variolation caused death from smallpox in 1% to 2% of recipients and in people exposed to recipients, so in time variolation was outlawed.

The English physician Edward Jenner demonstrated in 1796 that protection against smallpox could also be conferred by inoculation with crusts from a person infected with cowpox—a related but very mild disease. His new technique became widely adopted. Because cowpox was also called *vaccinia*[1] (vak-sin'ē-ă), Jenner called the new technique **vaccination** (vak'si-nā'shŭn), and the protective inoculum a **vaccine** (vak-sēn'). Today, we use the term **immunization** to refer to the administration of any antigenic inoculum, which are all called vaccines. For many years thereafter, vaccination against smallpox was widely practiced, even though no one understood how it worked or whether similar techniques could protect against other diseases.

In 1879, Louis Pasteur conducted experiments on the bacterium *Pasteurella multocida* (pas-ter-el'ă mul-tŏ'si-da) and demonstrated that he could make an effective vaccine against this organism (which causes a disease in birds called fowl cholera). Once he understood the basic principle of vaccine manufacture, vaccines against anthrax and rabies rapidly followed. In 1890, researchers in Germany discovered that vaccines provide protection through the actions of antibodies. They developed the technique of transferring protective antibodies to susceptible individuals—that is, *passive immunotherapy* (im'yū-nō-thār'ă-pē).

By the late 1900s, immunologists and health care providers had formulated vaccines that significantly reduced the number of cases of many infectious diseases (**FIGURE 17.1**). We also have successful vaccines against some types of cancer. Health care providers, governments, and international organizations working together have rid the world of naturally occurring smallpox, and we hope for the worldwide eradication of polio, measles, mumps, and rubella.

Even though immunologists have produced vaccines that protect people against many deadly diseases, a variety of political, social, economic, and scientific problems prevent vaccines from reaching all those who need them. In developing nations worldwide, over 1.5 million children still die each year from vaccine-preventable infectious diseases, primarily because of political obstacles. Additionally, some pathogens, such as the protozoa of malaria and the virus of AIDS, still frustrate attempts to develop effective vaccines against them. Furthermore, the existence of vaccine-associated risks—both medical risks (the low

[1] From Latin *vacca*, meaning "cow."

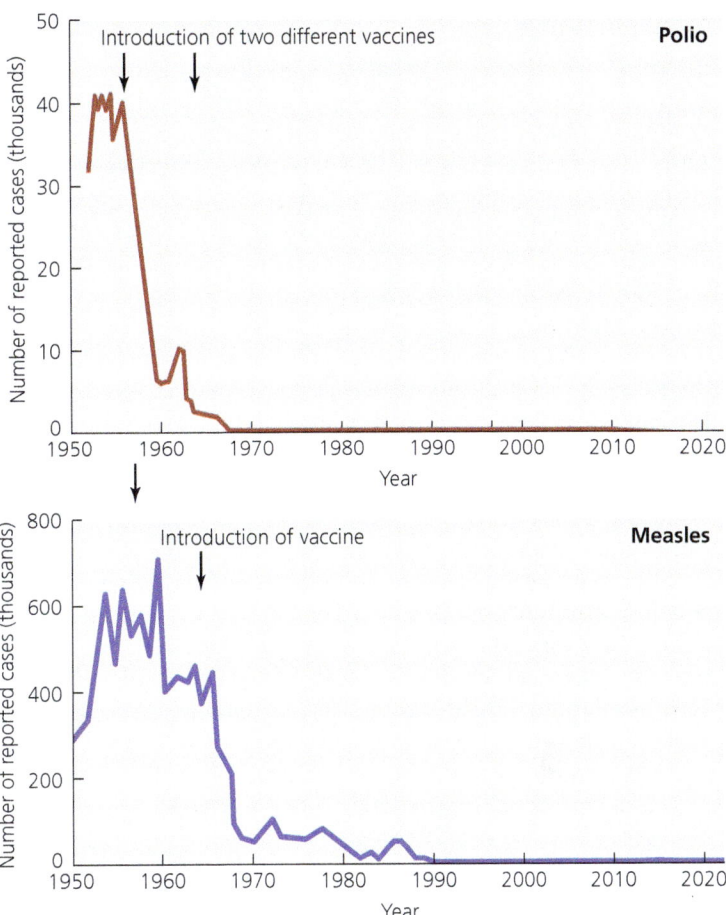

▲ **FIGURE 17.1 The effect of immunization in reducing the prevalence of two infectious diseases in the United States.** Polio is no longer endemic in the United States. Measles is nearly eradicated.

but persistent incidence of vaccine-caused diseases) and financial risks (the high costs of developing and producing vaccines and the risk of lawsuits by vaccine recipients who have adverse reactions)—has in recent years discouraged investment in new vaccines. Thus, although the history of immunization is marked by stunning advancements in public health, the future of immunization poses challenges.

Next, we take a closer look at active immunization, commonly known as vaccination.

Active Immunization

LEARNING OUTCOMES

17.2 Describe the advantages and disadvantages of five types of vaccines.

17.3 Describe three methods by which recombinant genetic techniques can be used to develop improved vaccines.

17.4 Delineate the risks and benefits of routine immunization in healthy populations, mentioning contact immunity and herd immunity.

In the following subsections, we examine types of vaccines, the roles of technology in producing modern vaccines, and issues concerning vaccine safety.

Vaccine Types

Scientists are constantly striving to develop vaccines of maximal efficacy and safety. In each case, a pathogen is altered or inactivated so that it is less likely to cause illness; however, not all types of vaccines are equally safe or effective. Effectiveness can be checked by measuring the antibody level—called the **titer** (tī'ter)—in the blood. When a titer is low, antibody production can be bolstered by administration of more antigen—a *booster immunization*.

The general types of vaccines, each of which has its own combination of strengths and weaknesses, are *attenuated (modified live) vaccines*, *inactivated (or killed) vaccines*, *toxoid vaccines*, *combination vaccines*, and *recombinant gene vaccines*. Each of these is named for the type of antigen used in the inoculum.

Attenuated (Modified Live) Vaccines Virulent microbes are not used in vaccines because they cause disease. Immunologists can reduce virulence so that, although still active, a pathogen is less likely to cause disease. The process of reducing virulence is called **attenuation** (ă-ten-yū-ā'shŭn). A common method for attenuating viruses involves raising them for numerous generations in cells in which they normally don't replicate well. As the viruses adapt to this inhospitable environment, they can lose the ability to produce disease in humans. Scientists can reduce virulence in bacteria by culturing them under unusual conditions or by using genetic manipulation.

Attenuated vaccines—those containing attenuated microbes—are also called *modified live vaccines*. Because they contain active but generally avirulent microbes, these vaccines cause very mild infections but no serious disease under normal conditions. Attenuated viruses in such a vaccine infect host cells and replicate; the infected cells then process endogenous viral antigens. Because modified live vaccines contain active microbes, a large number of antigen molecules are available. These tend to elicit vigorous antibody and cell-mediated immune responses. Further, newly immunized individuals can infect those around them, providing **contact immunity**—that is, immunity beyond the individual receiving the vaccine.

Although usually very effective, attenuated vaccines can be hazardous because modified microbes may retain enough residual virulence to cause disease in immunosuppressed people. Pregnant women should not receive live vaccines because of the danger that the attenuated pathogen might cross the placenta and harm the developing baby. Occasionally, attenuated viruses actually revert to wild type or mutate to a form that causes persistent infection or disease. For example, in 2000, a polio epidemic in the Dominican Republic and Haiti resulted from the reversion of an attenuated virus in oral polio vaccine to a virulent poliovirus. For this reason, we no longer use oral polio vaccine to immunize children in the United States.

Inactivated (Killed) Vaccines For some diseases, live vaccines have been replaced by **inactivated vaccines**, which are of two types:

- *Whole agent vaccines* are produced with inactive but whole microbes, such as rabies vaccine and hepatitis A vaccine.
- *Subunit vaccines* are produced with antigenic fragments of microbes.

Because neither whole agent nor subunit vaccines can replicate, revert, mutate, or retain residual virulence, they are safer than live vaccines. However, because they cannot replicate, booster doses must be administered to achieve full immunity, and immunized individuals do not stimulate immunity in their contacts. Also, with whole agent vaccines, nonantigenic portions of the microbe occasionally stimulate a painful inflammatory response in some individuals. As a result, whole agent pertussis vaccine has been replaced with a subunit pertussis vaccine called *acellular pertussis vaccine*, which is a mixture of four proteins from the pertussis bacterium.

When microbes are killed for use in vaccines, it is important that their antigens remain as similar to those of living organisms as possible. If chemicals are used for killing, they must not alter the antigens responsible for stimulating protective immunity. A commonly used inactivating agent is *formaldehyde*, which denatures proteins and nucleic acids. The amount of these chemical inactivating agents remaining in the final vaccine is so small that it poses no risks or health dangers.

Because the microbes of inactivated vaccines cannot reproduce, they do not present as many antigenic molecules to the body as do live vaccines; therefore, inactivated vaccines are antigenically weak. They are administered in high doses or in multiple doses, or they are incorporated with materials called **adjuvants**[2] (ad'joo-văntz), substances that increase the effective antigenicity of the vaccine by stimulating immune cell receptors and their actions. Unfortunately, high individual doses and multiple dosing increase the risk of producing allergies, and the use of adjuvants to increase antigenicity may stimulate local inflammation.

Because all types of killed vaccines are recognized by the immune system as exogenous antigens, they stimulate an antibody immune response.

Toxoid Vaccines For some bacterial diseases, notably tetanus and diphtheria, it is more efficient to induce an immune response against toxins than against cellular antigens. **Toxoid vaccines** (tok'soyd) are chemically or thermally modified toxins that are used in vaccines to stimulate active immunity. As with killed vaccines, toxoids stimulate antibody-mediated immunity. Because toxoids have few antigenic determinants, effective immunization requires multiple childhood doses as well as reinoculations every 10 years for life.

Combination Vaccines The Centers for Disease Control and Prevention (CDC) has approved several **combination vaccines** for routine use. These vaccines combine antigens from several attenuated and inactivated pathogens and toxoids that are administered simultaneously. Examples include MMR—a vaccine against measles, mumps, and rubella—and Pentacel, which is a vaccine against diphtheria, tetanus, pertussis (whooping cough), polio, and diseases of *Haemophilus influenzae* (hē-mof'i-lŭs in-flu-en'zī).

Vaccines Using Recombinant Gene Technology Although live attenuated, inactivated, and toxoid vaccines have been highly successful in controlling infectious diseases, researchers continue to seek ways to make vaccines more effective, cheaper, and safer

---
[2]From Latin *adjuvo*, meaning "to help."

and to make new vaccines against pathogens that have been difficult to protect against. For example, scientists have developed a recombinant DNA vaccine against *Blastomyces* (blas-tō-mī'sēz) infection in dogs—the first vaccine against a fungal pathogen. Scientists can also use a variety of genetic recombinant techniques to improve vaccines. For example, they might selectively delete virulence genes from a pathogen, producing an irreversibly attenuated microbe, one that cannot revert to a virulent pathogen (FIGURE 17.2a).

Scientists also use recombinant techniques to produce large quantities of very pure antigens for use in vaccines. In this process, scientists isolate the gene that codes for an antigen and insert it into a bacterium, yeast, or other cell, which then expresses and releases the antigen (FIGURE 17.2b). Vaccine manufacturers produce a hepatitis B subunit vaccine in this manner using recombinant yeast cells.

Alternatively, a genetically altered microbial cell or virus may express the antigen and act as a live vaccine (FIGURE 17.2c). Experimental recombinant vaccines of this type have used genetically modified adenoviruses, herpesviruses, poxviruses, or bacteria such as *Salmonella* (sal'mŏ-nel'ă). Vaccinia virus (cowpox virus) is often used because it is easy to administer orally or by dermal

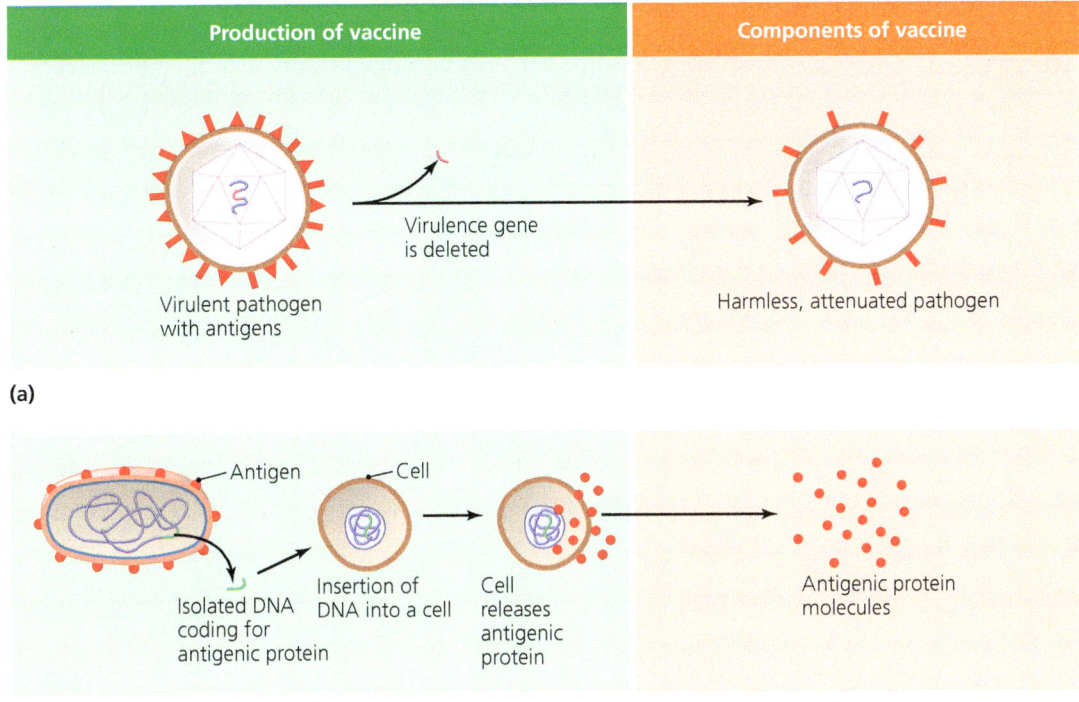
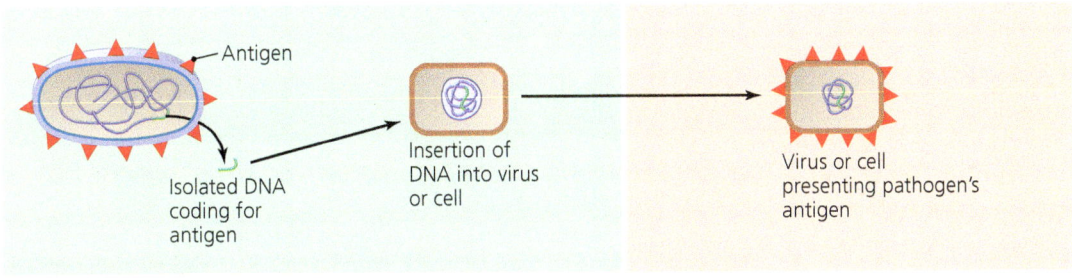
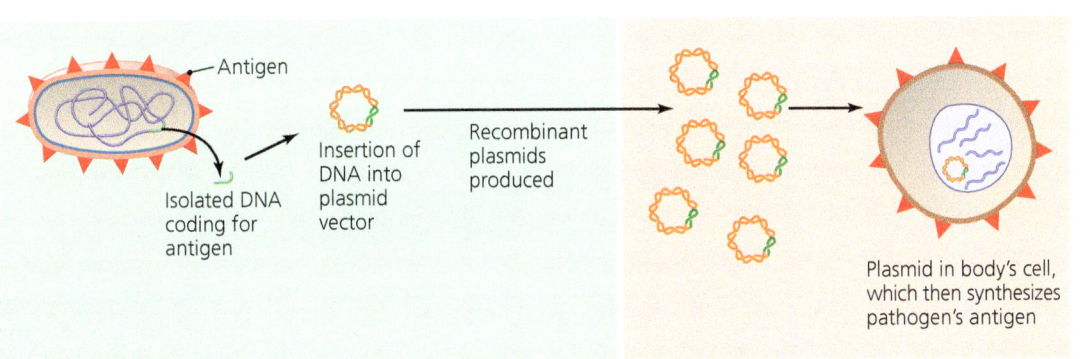

◀ FIGURE 17.2 **Some uses of recombinant DNA technology for making improved vaccines.** (a) Deletion of virulence gene(s) to create an attenuated pathogen for use in a vaccine. (b) Insertion of a gene that codes for a selected antigenic protein into a cell, which then produces large quantities of the antigen for use in a vaccine. (c) Insertion of a gene that codes for a selected antigenic protein into a cell or virus, which displays the antigen. The entire recombinant is used in a vaccine. (d) Injection of DNA containing a selected gene (in this case, as part of a plasmid) into an individual. Once some of this DNA is incorporated into the genome of a patient's cells, those cells synthesize and process the antigen, which stimulates an immune response.

scratching. Further, its large genome makes inserting a new gene into it relatively easy. Another method results in the body's own cells expressing the antigen. In this type of vaccine, DNA coding for a pathogen's antigen is inserted into the body in a vector, perhaps inside a plasmid (FIGURE 17.2d). The body's cells take up the DNA and then transcribe and translate the gene to produce antigen molecules, which trigger an immune response. No recombinant DNA vaccine of this type has yet been approved for use in humans.

Play *Vaccines: Function, Types*
@ Mastering Microbiology

Vaccine Manufacture

Manufacturers mass-produce many vaccines by growing bacteria in laboratory culture vessels, but because viruses require a host cell to reproduce, they are cultured inside chicken eggs. Availability of sterile eggs is thus critical for manufacturing viral vaccines such as flu vaccines. Because the vaccines are produced in eggs, physicians must withhold such immunizations from patients with egg allergies. Research on gene-based vaccines and development of vaccines in genetically modified plants may result in safer vaccines. Non-egg-based vaccines for viral diseases like influenza have been on the market for a few years now. Ex.: Flucelvax and Flublok.

Recommended Immunizations

The CDC and medical associations publish recommended immunization schedules for children, adults, and special populations, such as health care workers and HIV-positive individuals. The recommendations are frequently modified to reflect changes in the relationships between pathogens and the human population. FIGURE 17.3 highlights the general 2018 immunization schedules

CDC-Recommended Immunization Schedule—United States, 2018

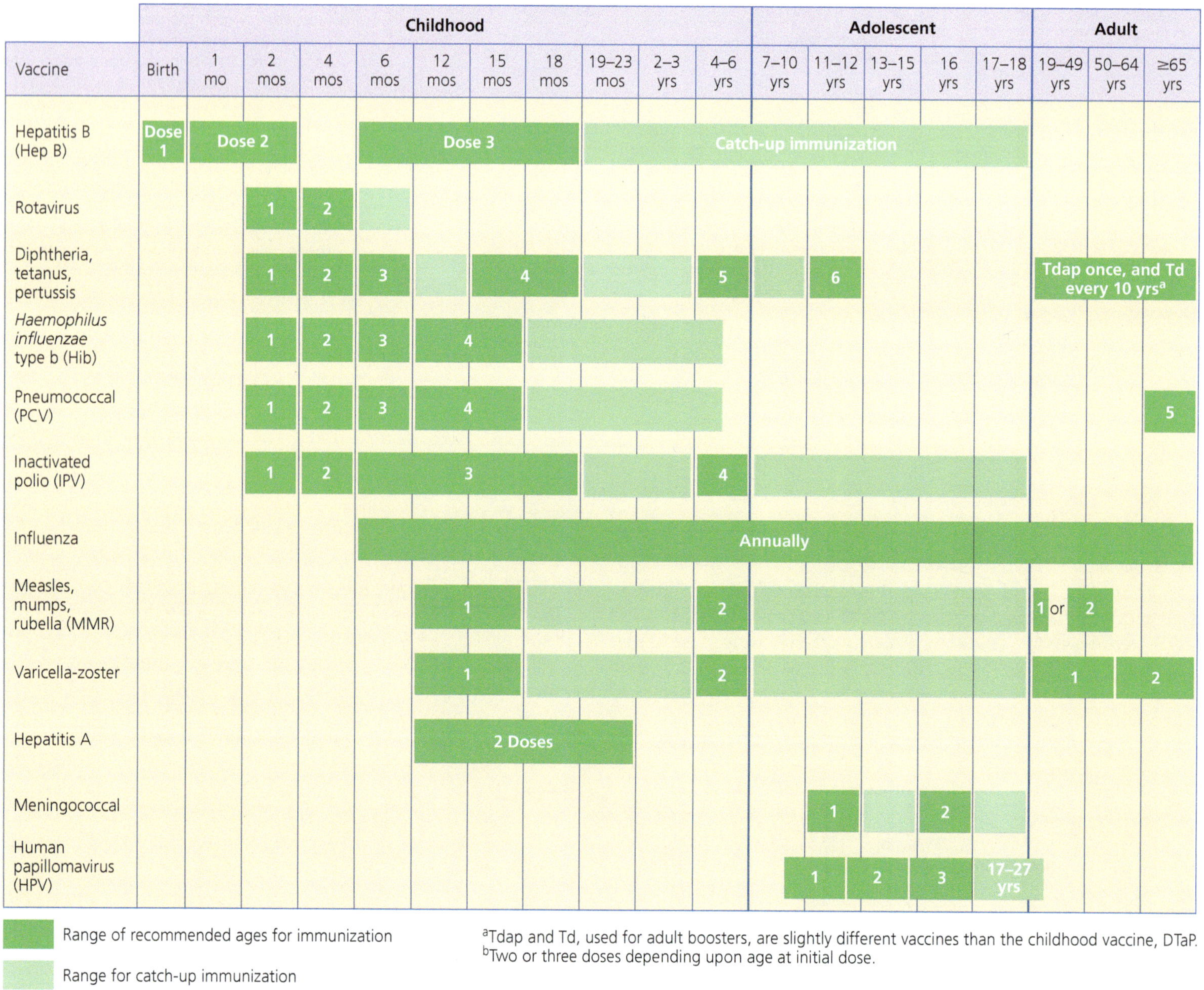

Vaccine	Birth	1 mo	2 mos	4 mos	6 mos	12 mos	15 mos	18 mos	19–23 mos	2–3 yrs	4–6 yrs	7–10 yrs	11–12 yrs	13–15 yrs	16 yrs	17–18 yrs	19–49 yrs	50–64 yrs	≥65 yrs
										Childhood			Adolescent				Adult		
Hepatitis B (Hep B)	Dose 1	Dose 2			Dose 3				Catch-up immunization										
Rotavirus			1	2															
Diphtheria, tetanus, pertussis			1	2	3		4				5		6				Tdap once, and Td every 10 yrs[a]		
Haemophilus influenzae type b (Hib)			1	2	3	4													
Pneumococcal (PCV)			1	2	3	4													5
Inactivated polio (IPV)			1	2		3					4								
Influenza									Annually										
Measles, mumps, rubella (MMR)						1					2						1 or 2		
Varicella-zoster						1					2						1		2
Hepatitis A						2 Doses													
Meningococcal													1		2				
Human papillomavirus (HPV)													1	2	3	17–27 yrs			

■ Range of recommended ages for immunization
■ Range for catch-up immunization

[a]Tdap and Td, used for adult boosters, are slightly different vaccines than the childhood vaccine, DTaP.
[b]Two or three doses depending upon age at initial dose.

▲ **FIGURE 17.3 The CDC's recommended immunization schedule for the general population.** Adult meningococcal vaccine is recommended for college students who live in dormitories.

TABLE 17.1 Principal Vaccines to Prevent Human Diseases

Vaccine	Disease Agent	Disease	Vaccine Type	Method of Administration
Recommended by the CDC				
Hepatitis B	Hepatitis B virus	Hepatitis B	Inactive subunit from recombinant yeast	Intramuscular
Rotavirus	Rotavirus	Gastroenteritis	Attenuated, recombinant	Oral
Diphtheria/tetanus/acellular pertussis (DTaP)	Diphtheria toxin	Diphtheria	Toxoid	Intramuscular
	Tetanus toxin	Tetanus	Toxoid	
	Bordetella pertussis	Whooping cough	Inactivated subunit (inactivated whole also available)	
Haemophilus influenzae type b (Hib)	Haemophilus influenzae	Meningitis, pneumonia, epiglottitis	Inactivated subunit	Intramuscular
Pneumococcal (PCV)	Streptococcus pneumoniae	Pneumonia	Inactivated subunit	Intramuscular
Polio	Poliovirus	Poliomyelitis	Inactivated (attenuated also available)	Subcutaneous or intramuscular (attenuated: oral)
Influenza	Influenzaviruses	Flu	Inactivated subunit	Intramuscular or oral
Measles/mumps/rubella (MMR)	Measles virus	Measles	Attenuated	Subcutaneous
	Mumps virus	Mumps	Attenuated	
	Rubella virus	Rubella (German measles)	Attenuated	
Varicella-zoster	Chickenpox virus	Chickenpox, shingles	Attenuated	Subcutaneous
Hepatitis A	Hepatitis A virus	Hepatitis A	Inactivated whole	Intramuscular
Human papillomavirus (HPV)	Human papillomaviruses	Genital warts, cervical cancer	Inactive recombinant	Intramuscular
Meningococcal	Neisseria meningitidis	Meningitis	Inactive	Subcutaneous or intramuscular
Available But Not Recommended for General Population in the United States				
Anthrax	Bacillus anthracis	Anthrax	Inactivated whole	Subcutaneous
BCG (bacillus of Calmette and Guérin)	Mycobacterium tuberculosis, M. leprae	Tuberculosis, leprosy	Attenuated	Intradermal
Japanese encephalitis vaccine	Japanese encephalitis virus	Encephalitis	Inactive	Subcutaneous
Rabies	Rabies virus	Rabies	Inactivated whole	Intramuscular or intradermal
Typhoid fever vaccine	Salmonella enterica	Typhoid fever	Attenuated (inactive also available)	Oral (inactive: subcutaneous or intramuscular)
Vaccinia (cowpox)	Smallpox virus, monkeypox virus	Smallpox, monkey pox	Attenuated	Subcutaneous
Yellow fever	Yellow fever virus	Yellow fever	Attenuated	Subcutaneous

recommended by the CDC. **TABLE 17.1** lists information about the vaccines available for each of the diseases in the immunization schedule, as well as some other available vaccines. Vaccines against anthrax, cholera, plague, tuberculosis, and other diseases are available, but the CDC does not recommend them for the general U.S. population.

It is important that patients follow the recommended immunization schedule not only to protect themselves but also to provide society with **herd immunity**. Herd immunity is the protection provided all individuals in a population due to the inability of a pathogen to effectively spread when a large proportion of individuals (typically more than 75%) are resistant. When immunization compliance in a population has fallen, local epidemics have resulted.

Vaccine Safety

Health care providers must carefully weigh the risks associated with all medical procedures, including vaccines, against their benefits. A common vaccine-associated problem is mild toxicity.

Some vaccines—especially whole agent vaccines that contain adjuvant—may cause pain at the injection site for several hours or days after injection. In rare cases, toxicity may result in general malaise and possibly a fever high enough to induce seizures. Although not usually life threatening, the potential for these symptoms may be sufficient to discourage people from being immunized or having their infants immunized.

A much more severe problem associated with immunization is the risk of *anaphylactic shock*, an allergic reaction that may develop to some component of the vaccine, such as egg proteins, adjuvants, or preservatives. Because people are rarely aware of such allergies ahead of time, vaccine recipients should remain for several minutes in the health care setting, where epinephrine is readily available to counter any signs of an allergic reaction.

A third major problem associated with immunization is that of residual virulence of microbes in attenuated vaccines, which we previously discussed. Attenuated viruses occasionally cause disease not only in fetuses and immunosuppressed patients but also in healthy children and adults. An example is the emergence of polio from the attenuated oral poliovirus vaccine (OPV). Though a very effective vaccine, OPV causes clinical poliomyelitis in one of every 2 million recipients or their close contacts. Medical personnel in the United States eliminated this problem by switching to inactivated polio vaccine (IPV).

Since the 1990s, lawsuits in the United States and Europe have alleged that certain vaccines against childhood diseases cause or trigger disorders such as autism, diabetes, and asthma. Extensive research has failed to substantiate these allegations. Vaccine manufacturing methods have improved tremendously in recent years, ensuring that modern vaccines are safer than those in use even in the early 2010s. The U.S. Food and Drug Administration (FDA) has established a Vaccine Adverse Event Reporting System for monitoring vaccine safety.

The CDC and FDA have determined that the problems associated with immunization are far less serious than the suffering and death that would result if we stopped immunizing people. **Beneficial Microbes: Cowpox: To Vaccinate or Not to Vaccinate?** discusses some issues surrounding the administration of smallpox vaccinations to the general public.

Passive Immunotherapy

LEARNING OUTCOMES

17.5 Identify two sources of antibodies for use in passive immunotherapy.

17.6 Compare the relative advantages and disadvantages of active immunization and passive immunotherapy.

Passive immunotherapy (sometimes called *passive immunization*) involves the administration of antibodies to a patient. Physicians use passive immunotherapy when protection against a recent infection or an ongoing disease is needed quickly. Rapid protection is achieved because passive immunotherapy does not require the body to mount a response; instead, preformed antibodies are immediately available to bind to antigen, enabling neutralization and opsonization to proceed without delay. For example, in a case of botulism poisoning (caused by the toxin of *Clostridium botulinum* (klos-trid′ē-ŭm bo-tū-lī′-num), passive immunotherapy with preformed antibodies against the toxin can prevent death.

Antibodies directed against toxins are also called *antitoxins* (an-tē-tok′sinz); *antivenom (antivenin)* used to treat venomous bites or stings is an example of an antitoxin. In some cases, infections with certain viruses—hepatitis A and B, measles, rabies, Ebola, chickenpox, and shingles—are treated with antibodies directed against the causative viruses.

To acquire antibodies for passive immunotherapy, clinicians remove blood cells and clotting factors from the blood of donors; the result is *serum*, which contains a variety of antibodies—particularly immunoglobulins in class G (IgG, or gamma globulins). When used for passive immunotherapy, such serum is called **antiserum** (an-tē-sē′rŭm) or sometimes *immune serum*. Antisera are typically collected from human blood plasma donors or from large animals intentionally exposed to a pathogen of interest, because the large blood volume of the animals contains more antibodies than can be obtained from smaller animals. Pooled antisera from a group of human donors can be administered intravenously (*intravenous immunoglobulins, IVIg*) to treat immunodeficiencies and some autoimmune and inflammatory diseases.

Passive immunotherapy has the following limitations:

- Repeated injections of animal-derived antisera can trigger an allergic response called *serum sickness*, in which the recipient mounts an immune response against animal antigens found in the antisera.

- The patient may degrade the antibodies relatively quickly; therefore, protection is not long lasting.

BENEFICIAL MICROBES

Cowpox: To Vaccinate or Not to Vaccinate?

Vaccinia necrosum

Dr. Edward Jenner developed an early use for a beneficial microbe in medicine—administering cowpox virus as the smallpox vaccine. Medical personnel in the United States followed Jenner's example by regularly vaccinating the general public until 1971, at which time the risk of contracting smallpox was deemed too low to justify mandatory vaccinations. Indeed, in 1980, the World Health Assembly declared smallpox successfully eradicated from the natural world, though the virus still exists in research laboratories. Concerns about the potential use of smallpox virus as an agent of bioterrorism have sparked debate about whether citizens should once again be vaccinated with cowpox to protect against smallpox.

Although safe and effective for most healthy adults, for others the attenuated cowpox virus can result in serious side effects—even death. Individuals with compromised immune systems (such as AIDS patients or cancer patients undergoing chemotherapy) are considered to be at particularly high risk for developing adverse reactions. Pregnant women, infants, and individuals with a history of the skin condition eczema are also considered poor candidates for the vaccine. Though rare, adverse reactions to the vaccine may also develop in certain otherwise healthy individuals. The more serious side effects include *vaccinia necrosum* (characterized by progressive cell death in the area of vaccination) and encephalitis (inflammation of the brain). Approximately one in every 1 million individuals receiving cowpox virus as a vaccine for the first time develops a fatal reaction to it.

Is the risk of a bioterrorist smallpox attack great enough to warrant the exposure to the known risks of administering cowpox virus as a vaccine to the general population? If you were a public health official, what would you decide?

- The body does not produce memory B cells in response to passive immunotherapy, so the patient is not protected against subsequent infections.

Scientists have overcome some of the limitations of antisera by developing **hybridomas** (hī-brid-ō'măz), which are tumor cells created by fusing antibody-secreting plasma cells with cancerous plasma cells called *myelomas* (mī-ĕ-lō'măz) (**FIGURE 17.4**). Each hybridoma divides continuously (because of the cancerous plasma cell component) to produce clones of itself, and each clone secretes large amounts of a single antibody molecule. These identical antibodies are called **monoclonal antibodies** (mon-ō-klō'năl) because all of them are secreted by clones originating from a single plasma cell. Once scientists have identified the hybridoma that secretes antibodies complementary to the

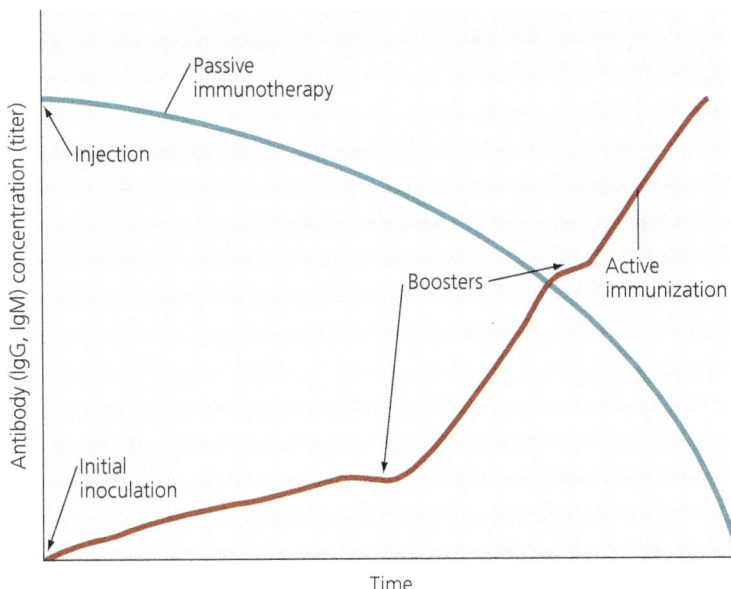

▲ **FIGURE 17.5 The characteristics of immunity produced by passive immunotherapy (blue-green) and active immunization (red).** Passive immunotherapy provides strong and immediate protection, but it disappears relatively quickly. Active immunity takes some time and may require additional booster inoculations to reach protective levels, but it is long lasting and capable of restimulation.

antigen of interest, they maintain it in tissue culture to produce the antibodies needed for passive immunotherapy. For example, physicians use such a monoclonal antibody to treat newborns infected with respiratory syncytial virus.

The FDA has approved over 70 monoclonal antibodies for medical usage in America. Repeated injection of antibodies made in lab animals, such as rabbits or mice, into humans results in immune rejection of those foreign proteins, so scientists have generated transgenic mice in which the mouse immunoglobulin genes have been replaced with human immunoglobulin genes. Hybridomas made with cells from these mice secrete antibodies identical to those of humans, eliminating rejection.

Active immunization and passive immunotherapy are used in different circumstances because the types of protection they provide have different characteristics (**FIGURE 17.5**). As just noted, passive immunotherapy with preformed antibodies is used whenever immediate protection is required. However, because preformed antibodies are removed rapidly from the blood and no memory B cells are produced, protection is temporary, and the recipient becomes susceptible again. Active immunization provides long-term protection that is capable of restimulation, but active immunity is generally slower.

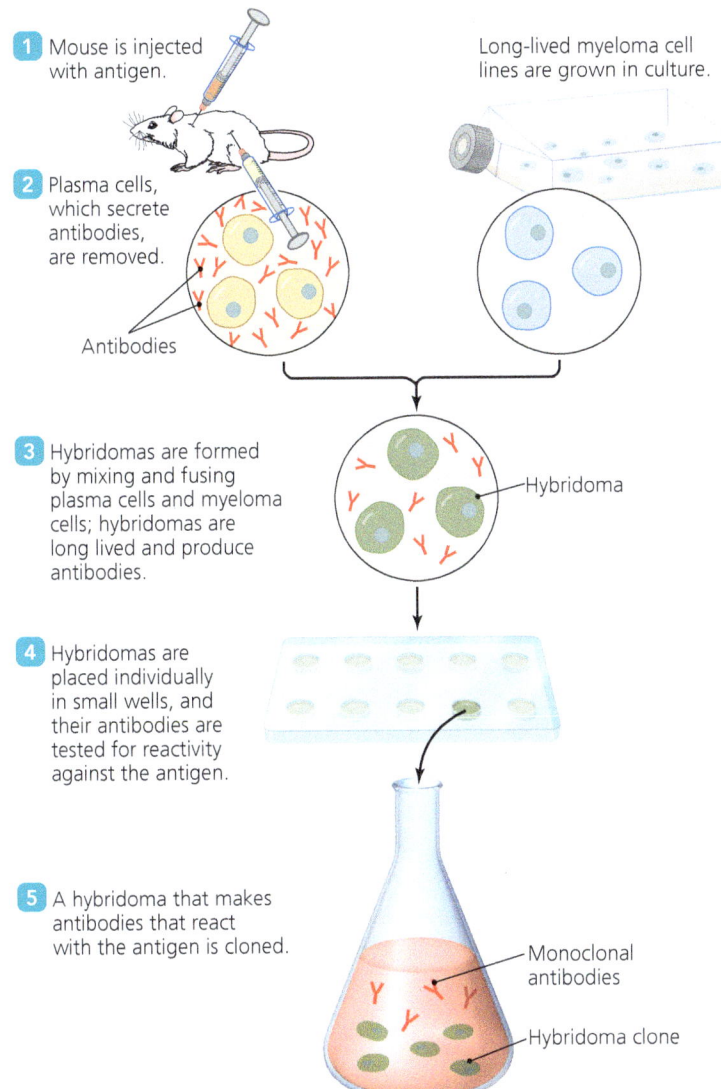

▲ **FIGURE 17.4 The production of hybridomas.** After a laboratory animal is injected with the antigen of interest **1**, plasma cells are removed from the animal and isolated **2**. When these plasma cells are fused with cultured cancer cells called *myelomas*, hybridomas result **3**. Once the hybridomas are cultured individually and the hybridoma that produces antibodies against the antigen of interest is identified **4**, it is cloned to produce a large number of hybridomas, all of which secrete identical antibodies called monoclonal antibodies **5**.

TELL ME WHY

Vaccines have drastically reduced the number of cases of many diseases, such as measles and whooping cough. Why should parents have their children immunized, given that there are so few cases?

MICRO CHECK

1. Who was the first person to develop a successful immunization against rabies?
2. Which type of vaccine most closely mimics a real infection with a pathogen?
3. What is the greatest risk of a flu shot?
4. Which of the following is an example of a live attenuated oral vaccine recommended by the CDC for the general population: rotavirus, measles, hepatitis A, varicella-zoster?
5. If an unimmunized person has possibly been exposed to the bacterium that causes tetanus, the patient is given both active and passive immune treatment against tetanus. Why are they both necessary?

Serological Tests That Use Antigens and Corresponding Antibodies

LEARNING OUTCOMES

17.7 Define *serology*.
17.8 Describe several uses of serological tests.
17.9 In general terms, compare and contrast precipitation, agglutination, neutralization, complement fixation, and labeled immunoassay methods.

The analysis of blood to determine the presence of particular antigens or specific antibodies is called **serology** (sĕ-rol'ō-jē). Scientists have developed a variety of serological tests to identify antigens or antibodies in serum. Serological methods range from simple manual procedures to complex and automated ones.

Serological tests have many uses. Epidemiologists review serological test results to monitor the spread of infection through a population. Physicians order relevant tests, which are conducted by medical laboratory scientists or lab technologists, to establish diagnoses. For example, when a physician suspects a patient might be infected by both HIV and hepatitis B virus, the doctor orders both an anti-HIV test and a hepatitis B virus surface antigen test. The first determines the presence of antibodies against HIV in the serum—strong evidence that the patient is infected with HIV. A surface antigen test indicates infection with hepatitis B virus.

In the following sections, we examine various types of serological methods: precipitations, turbidimetry, nephelometry, agglutination, neutralization, and labeled immunoassays. Some of these procedures are presented for historical reasons—more accurate and faster modern tests have replaced them. For example, one modern method—*polymerase chain reaction (PCR)*—amplifies DNA sequences so that enough copies are available to test for the presence of microbial genetic material rather than the presence of antibodies against the microbes. This allows detection of infection before the body produces antibodies.

Precipitation Tests

LEARNING OUTCOMES

17.10 Describe the general principles of precipitation testing.
17.11 Describe the technique of immunodiffusion.

One of the simplest of serological tests relies on the fact that when antigens and antibody are mixed in proper proportions, they form large, insoluble, lattice-like complexes called precipitates. For example, when a solution of a soluble antigen, such as that of the fungus *Coccidioides immitis* (kok-sid-ē-oy'dēz im'mi-tis), is mixed with an antiserum containing antibodies against the fungal antigen, the mixture quickly becomes cloudy because of the formation of a precipitate consisting of antigen-antibody complexes, also called **immune complexes**.

When a given amount of antibody is added to each of a series of test tubes containing increasing amounts of antigen, the amount of precipitate increases gradually until it reaches a maximum (**FIGURE 17.6a**). In test tubes containing still more antigen

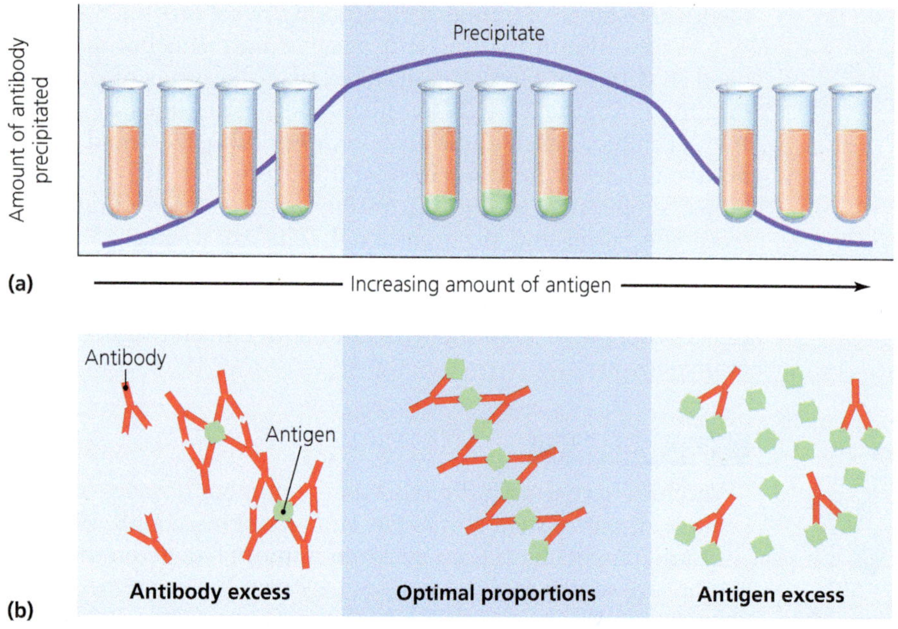

◀ **FIGURE 17.6 Characteristics of precipitation reactions. (a)** When increasing amounts of antigen are placed in tubes containing a constant amount of antibody, precipitate forms only in tubes having moderate amounts of antigen. The overlay graph shows the amount of precipitate formed versus the amount of antigen present. **(b)** In the presence of excess antigen or antibody, immune complexes are small and soluble; only when antigen and antibody are in optimal proportions do large complexes form and precipitate.

molecules, the amount of precipitate declines; in fact, in test tubes containing antigen in great excess over antibody, no precipitate at all develops. Thus, a graph of the amount of precipitate versus the amount of antigen has a maximum in the middle values.

The reasons behind this pattern of precipitation reactions are simple. Complex antigens are generally multivalent—each possesses many epitopes—and antibodies have pairs of active sites and therefore can simultaneously cross-link the same epitope on two antigen molecules (review the agglutination function of antibodies in Figure 16.14c). When there is excess antibody, each antigen molecule is covered with many antibody molecules, preventing extensive cross-linkage and thus precipitation (FIGURE 17.6b). Since there is no precipitation, an observer might conclude that there is no antigen in the solution—a negative test result. This is untrue; antigen is present. Such a *false negative* interpretation is called a *prozone phenomenon*.

When the reactants are in optimal proportions, the ratio of antigen to antibody is such that cross-linking and lattice formation are extensive. As this immune complex grows, it precipitates.

In mixtures in which antigen is in excess, little or no precipitation occurs because there are few cross-linkages. Single antibodies bound to antigen are small and soluble, so no precipitation occurs, which would result in a false negative interpretation.

Because precipitation requires the mixing of antigen and antibody in optimal proportions, it is not possible to perform a precipitation test by combining just any two solutions containing these reagents. To ensure that the optimal concentrations of antibody and antigen come together, scientists historically used a technique involving movement of the molecules through an agar gel: immunodiffusion.

Immunodiffusion

In the precipitation technique called **immunodiffusion** (im′yū-nō-di-fyū′zhŭn), a researcher cuts cylindrical holes called *wells* in an agar plate. If the researcher is examining a serum for the presence of antibodies, one well is filled with a solution of antigen and the other with the serum. Antigen and antibody molecules (if present) diffuse in all directions out of the wells and into the surrounding agar, and where they meet in optimal proportions, a line of precipitation appears (FIGURE 17.7a). If the solutions contain many different antigens and antibodies, each complementary pair of reactants reaches optimal proportions at different positions, and numerous lines of precipitation are produced—one for each interacting antigen-antibody pair (FIGURE 17.7b). Such an immunodiffusion test has been used to indicate exposure to complex mixtures of antigens from pathogens such as fungi. Only patients exposed to the fungi have serum antibodies against the fungal antigens, so only these patients' sera show precipitation. Physicians can monitor and treat such patients.

Turbidimetric and Nephelometric Tests

Turbidimetry and *nephelometry* are automated methods that measure the cloudiness of a solution, as occurs when antibodies and antigens are mixed together. As noted previously, when the

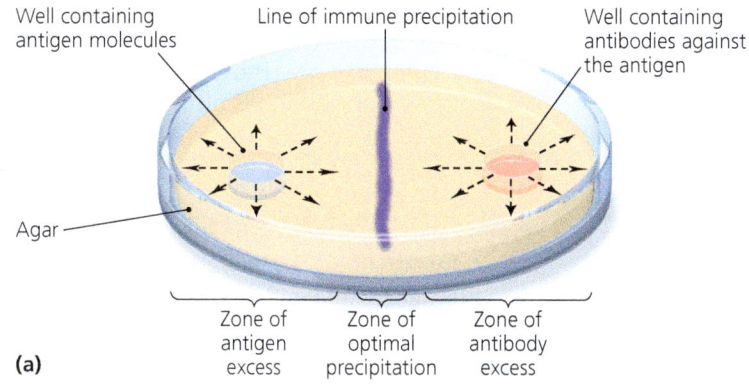
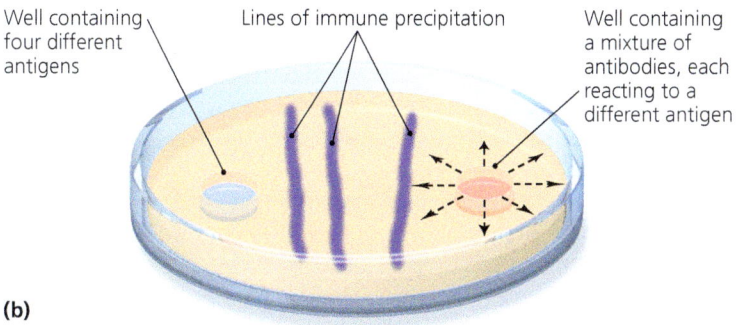

▲ FIGURE 17.7 Immunodiffusion, a type of precipitation reaction. (a) Antigen and antibody placed in wells diffuse out and through the agar; where they meet in optimal proportions, a line of precipitation forms. The artist has used color for the antigen, antibodies, and line of precipitation to clarify what happens, but in reality antigens and antibodies are colorless, and lines of precipitation are off-white. (b) When multiple antigens and antibodies are placed in the wells, multiple lines of precipitation mark the sites where different antigen-antibody combinations occurred in optimal proportions. *Why did only three lines of precipitation occur when the antigen well contained four antigens?*

Figure 17.7 None of the antibodies used in the test was complementary to the fourth antigen.

concentrations of antibodies and antigens are optimal, the initial cloudiness is followed by precipitation.

In turbidimetry, a light detector measures the amount of light passing through a solution, whereas in nephelometry, a machine measures the amount of light reflected from antigen-antibody complexes within the solution. Medical laboratory scientists use these methods to quantify the amounts of proteins, such as antibodies and complement, in serum.

Agglutination Tests

LEARNING OUTCOMES

17.12 Contrast agglutination and precipitation tests.
17.13 Describe how agglutination is used in immunological testing, including titration.

Not all antigens are soluble proteins that can be precipitated by antibody. Because of their multiple antigen-binding sites, antibodies can also cross-link particles, such as whole bacteria or antigen-coated latex beads, causing **agglutination** (ă-gloo-ti-nā'shŭn; clumping). The difference between agglutination and precipitation is that agglutination involves the clumping of insoluble particles, whereas precipitation involves the clumping of soluble molecules. Agglutination reactions are sometimes easier to see and interpret with the unaided eye.

When the particles agglutinated are red blood cells, the reaction is called *hemagglutination* (hē-mă-gloo'ti-nā'shŭn). One use of hemagglutination is to determine blood type in humans. Blood is considered type A if the red blood cells possess surface antigens called A antigen, type B if they possess B antigens, type AB if they possess both antigens, and type O if they have neither antigen. In a hemagglutination reaction to determine blood type (**FIGURE 17.8a**), two portions of a given blood sample are placed on a slide. Anti-A antibodies are added to one portion and anti-B antibodies to the other; the antibodies agglutinate those blood cells that possess corresponding antigens. Another portion of the sample may be tested with antibodies against Rh protein to determine if the blood is Rh positive (Rh+) or Rh negative (Rh−) (**FIGURE 17.8b**).

Another use of agglutination is in a type of test that determines the concentration of antibodies in a clinical sample. Although the simple *detection* of antibodies is sufficient for many purposes, it is often more desirable to measure the *amount*

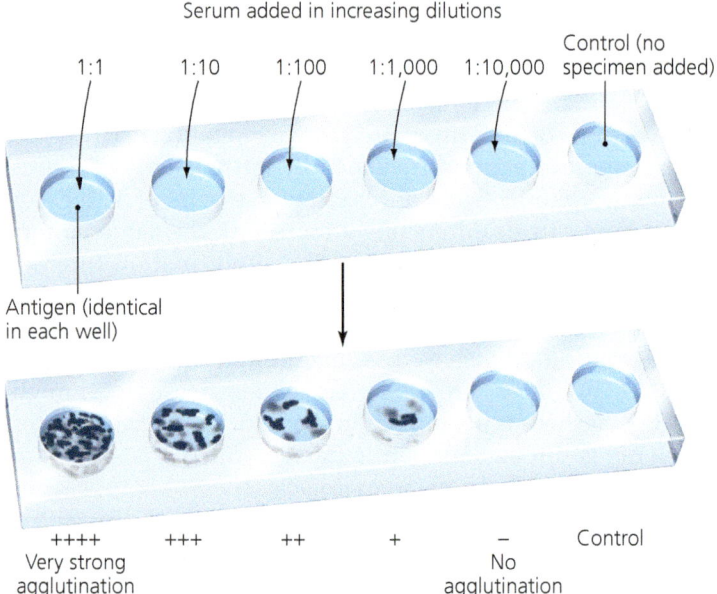

▲ **FIGURE 17.9 Titration, the use of agglutination to quantify the amount of antibody in a serum sample.** Serial dilutions of serum are added to wells containing a constant amount of antigen. At lower serum dilutions (higher concentrations of antibody), agglutination occurs; at higher serum dilutions, antibody concentration is too low to produce agglutination. The serum's titer is the highest dilution at which agglutination can be detected—in this case, 1:1000.

of antibodies in serum. By doing so, clinicians can determine whether a patient's antibody levels are rising, as occurs in response to the presence of active infectious disease, or falling, as occurs during the successful conclusion of a fight against an infection. One way of measuring antibody levels in blood sera is by **titration** (tī-trā'shŭn). In titration, the serum being tested undergoes a regular series of dilutions, and each dilution is then tested for agglutinating activity (**FIGURE 17.9**). Eventually, the antibodies in the serum become so dilute that they can no longer cause agglutination. The highest dilution of serum giving a positive reaction is a titer, which is expressed as a ratio reflecting the dilution. Thus, a serum that must be greatly diluted before agglutination ceases (for example, has been diluted a thousandfold; that is, has a titer of 1:1000) contains more antibodies than a serum that no longer agglutinates after minimal dilution (has a titer of 1:10).

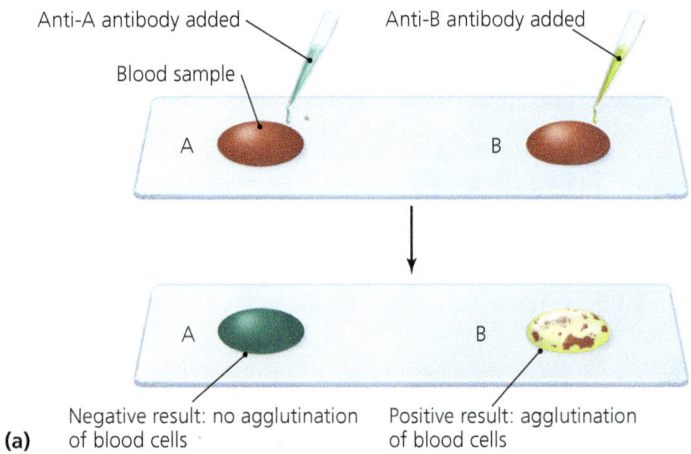

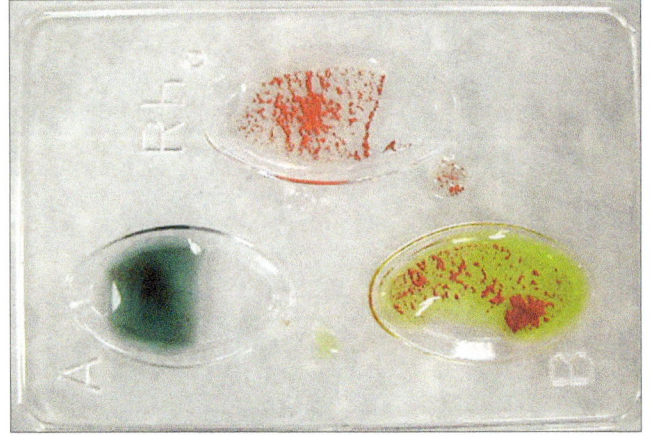

◀ **FIGURE 17.8 The use of hemagglutination to determine blood types in humans. (a)** Antibodies with active sites that bind to either of two surface antigens (antigen A or antigen B) of red blood cells are added to portions of a given blood sample. Where the antibodies react with the surface antigens, the blood cells can be seen to agglutinate, or clump together. **(b)** Photo of actual test, which also includes a test for Rh antigen, to be denoted positive (+) or negative (−). *What is the blood type of the person whose blood was used in this hemagglutination reaction?*

Figure 17.8 *The individual who donated the blood sample has type B+ blood.*

Neutralization Tests

LEARNING OUTCOMES

17.14 Explain the purpose of neutralization tests.
17.15 Contrast a viral hemagglutination inhibition test with a hemagglutination test.

Neutralization tests work because antibodies can *neutralize* the biological activity of many pathogens and their toxins. For example, combining antibodies against tetanus toxin with a sample of tetanus toxin renders the sample harmless to mice because the antibodies have reacted with and neutralized the toxin. Next we briefly consider two neutralization tests that, although not simple to perform, effectively reveal the biological activity of antibodies.

Viral Neutralization

One neutralization test is **viral neutralization**, which is based on the fact that many viruses introduced into appropriate cell cultures will invade and kill the cells, a phenomenon called a *cytopathic effect* (seen in plaque formation; see Chapter 13, Figure 13.17). However, if the viruses are first mixed with specific antibodies against them, their ability to kill cultured cells is neutralized. In a viral neutralization test, a mixture containing serum and a known pathogenic virus is introduced into a cell culture; if there are no cytopathic effects, antibodies against that virus are present in the serum. For example, if a mixture containing an individual's serum and a sample of hantavirus produces no cytopathic effect in a culture of susceptible cells, then it can be concluded that the individual's serum contains antibodies to hantavirus, and these antibodies neutralized the virus. Viral neutralization tests are sufficiently sensitive and specific to ascertain whether an individual has been exposed to a particular virus or viral strain, which may lead a physician to a diagnosis or treatment or to recommendations to prevent future infection or disease.

Viral Hemagglutination Inhibition Test

Because not all viruses are cytopathic—they do not kill their host cell—a neutralization test cannot be used to identify all viruses. However, many viruses (including influenzaviruses) have surface proteins that naturally clump red blood cells. (This natural process, called *viral hemagglutination*, must not be confused with the hemagglutination test we discussed previously—viral hemagglutination is not an antibody-antigen reaction.) Antibodies against influenzavirus inhibit viral hemagglutination; therefore, if serum from an individual stops viral hemagglutination, we know that the individual's serum contains antibodies to that particular strain of influenzavirus. Such **viral hemagglutination inhibition tests** can be used to detect antibodies against influenza, measles, mumps, and other viruses that naturally agglutinate red blood cells, thus revealing that a patient has been infected or immunized with the virus in question.

The Complement Fixation Test

LEARNING OUTCOME

17.16 Briefly explain the phenomenon that is the basis for a complement fixation test.

Activation of the classical complement system by antibodies leads to the generation of membrane attack complexes (MACs) that disrupt cytoplasmic membranes (see Figure 15.9). This phenomenon is the basis for the **complement fixation test** (kom′plĕ-ment fik-sā′shŭn), which is a complex assay used to detect the presence of specific antibodies in an individual's serum. The test can detect the presence of small amounts of antibody—amounts too small to detect by agglutination—though complement fixation tests have been replaced by other serological methods such as ELISA (discussed shortly) or genetic analysis using polymerase chain reaction (PCR) (see Figure 8.6).

Labeled Immunoassays

LEARNING OUTCOMES

17.17 List three tests that use labeled antibodies to detect either antigens or antibodies.
17.18 Compare and contrast direct and indirect fluorescent immunoassays, and identify at least three uses for these tests.
17.19 Compare and contrast the methods, purposes, and advantages of ELISA and immunoblotting tests.

A different form of serological testing involves *labeled* (or *tagged*) immunoassays, so named because these tests use antibody molecules that are linked to some molecular "label" that enables them to be detected easily. Labels include radioactive chemicals, fluorescent dyes, and enzymes. Automated machines can detect and quantify labels. For example, gamma radiation detectors can count radioactive chemicals, and fluorescence microscopes can measure fluorescent labels. Labeled immunoassays using radioactive or fluorescent labels can be used to detect either antigens or antibodies. Currently, radioactivity is rarely used, given the regulatory challenges of using and disposing of it. In the following sections, we will consider fluorescent immunoassays, ELISA, and immunoblotting tests.

Fluorescent Immunoassays

Fluorescent dyes are used as labels in several serological tests. Some fluorescent dyes can be chemically linked to an antibody without affecting the antibody's ability to bind antigen. When exposed to a specific wavelength of light (as in a fluorescence microscope), the fluorescent dye glows. Fluorescently labeled antibodies are used in direct and indirect fluorescent immunoassays.

Direct fluorescent immunoassays identify the *presence of antigens* in a tissue. The test is straightforward: A scientist floods a tissue sample suspected of containing the antigen with labeled

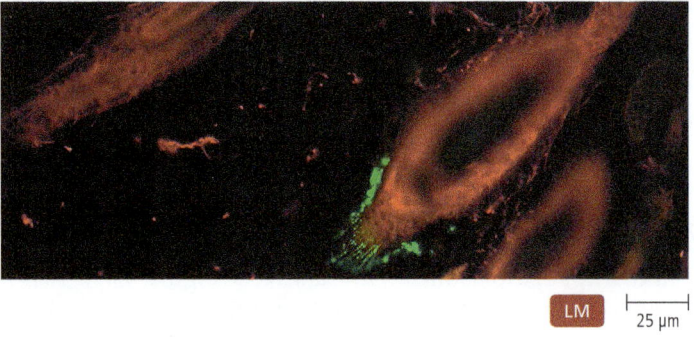

▲ FIGURE 17.10 **A direct fluorescent immunoassay.** Here rabies proteins (green) are revealed in tissue.

antibody, waits a short time to allow the antibody to bind to the antigen, washes the preparation to remove any unbound antibody, and examines it with a fluorescence microscope. If the suspected antigen is present, labeled antibody will adhere to it, and the scientist will see fluorescence. This is not a quantitative test—the amount of fluorescence observed is not directly related to the amount of antigen present.

Scientists use direct fluorescent immunoassays to identify small numbers of bacteria in patient tissues. This technique has been used to detect *Mycobacterium tuberculosis* (mī'kō-bak-tēr'ē-ŭm too-ber-kyū-lō'sis) in sputum and to detect rabies viruses infecting a brain (FIGURE 17.10).

Indirect fluorescent immunoassays are used to detect the *presence of specific antibodies* in an individual's serum via a three-step process (FIGURE 17.11a):

1. An antigen of interest is fixed to a microscope slide.
2. The individual's serum is added and remains long enough to allow serum antibodies, if present, to bind to the antigen. The serum is then washed off, leaving the antibodies bound to the antigen (but not yet visible).
3. Fluorescently labeled antibodies against human antibodies (anti–human antibody antibodies) are added to the slide and bind to the antibodies already bound to the antigen. After the slide is washed to remove unbound anti-antibodies, it is examined with a fluorescence microscope.

The presence of fluorescence indicates the presence of the labeled anti-antibodies, which are bound to serum antibodies bound to the fixed antigen; thus, fluorescence indicates that the individual has serum antibodies against the antigen of interest.

Scientists can use indirect fluorescent immunoassays to detect antibodies against many viral, protozoan, or bacterial pathogens, including *Neisseria gonorrhoeae* (nī-se'rē-ă go-nor-rē'ī), the causative agent of gonorrhea (FIGURE 17.11b). The presence of antibodies indicates that the patients have been exposed to the pathogen and may need treatment or counseling on steps to take to lower their risk of future infection.

Scientists routinely identify and separate types of white blood cells, such as lymphocytes, by using specific monoclonal antibodies produced against each cell type. The researchers can attach differently colored fluorescent dyes to the antibodies, allowing them to differentiate types of lymphocytes by the color of the dye carried by each type of antibody. Such identification tests can quantify the numbers and ratios of lymphocyte subsets,

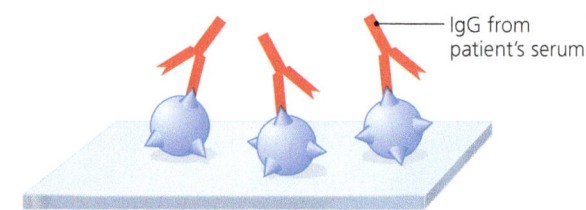
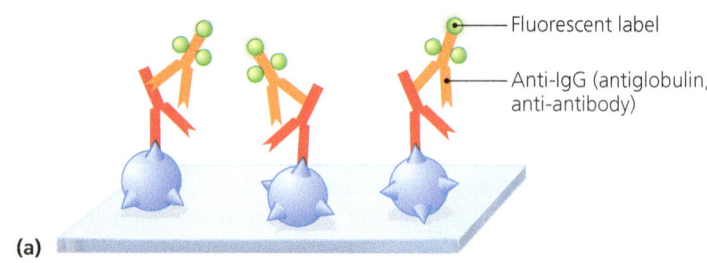

(a)

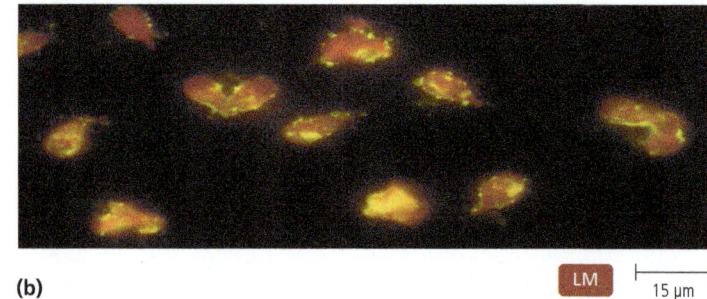

(b)

▲ FIGURE 17.11 **The indirect fluorescent immunoassay.** This test detects the presence of a specific antibody in a patient's serum. **(a)** The test procedure. 1 Antigen is attached to the slide. 2 Slide is flooded with an individual's serum to allow specific antibodies in the serum to bind to the antigen. 3 After anti-antibodies labeled with fluorescent chemical are added and then unbound anti-antibodies are washed off, the slide is examined with a fluorescence microscope. **(b)** A positive indirect fluorescent immunoassay, in which fluorescence indicates the presence of antibodies in an individual's serum against a particular antigen (here, *Giardia intestinalis*).

information critical in diagnosing and monitoring disease progression and effectiveness of treatment in patients with AIDS, other immunodeficiency diseases, leukemias, and lymphomas.

ELISAs (EIAs)

In another type of labeled immunoassay, called an **enzyme-linked immunosorbent assay (ELISA)** (im'yū-nō-sōr'bent as'sā), or simply an **enzyme immunoassay (EIA)**, the label is not a fluorescent dye but instead an enzyme that reacts with its substrate to produce a colored product that indicates a positive test. One form of ELISA is used to detect the presence and quantify the abundance of antibodies in serum—an example of indirect testing—as might be used in diagnosis of Lyme disease, legionellosis (a type of pneumonia), or cat scratch disease. An ELISA, which can take place in wells in commercially produced plates, has five basic steps with washes to remove excess chemicals between steps (FIGURE 17.12):

1. Each of the wells in the plate is coated with antigen molecules in solution.
2. Excess antigen molecules are washed off, and another protein (such as gelatin) is added to the wells to completely coat any of the surface not coated with antigen.
3. One well is assigned to each serum being tested, and a sample of the serum is added to that well. Whenever a serum sample contains antibodies against the antigen, the antibodies bind to the antigen affixed to the plate.
4. Anti-antibodies labeled with an enzyme are added to each well.
5. The enzyme's substrate is added to each well. The enzyme and substrate are chosen because their reaction results in products that cause a visible color change.

A positive reaction in a well, indicated by the development of color, can occur only if the labeled anti-antibody has bound to antibodies attached to the antigen of interest. The intensity of the color, which can be estimated visually or measured accurately using a spectrophotometer, is proportional to the amount of antibody present in the serum.

ELISA has become a test of choice for many diagnostic procedures, such as determination of HIV infection, because of its many advantages:

- ELISAs are sensitive, able to detect very small amounts of antibody (or antigen).
- Unlike some diffusion and fluorescent tests, ELISA can quantify amounts of antigen or antibody. Knowing the amount of antigen or antibody in a patient's serum can provide information concerning the course of an infection or the effectiveness of a treatment.
- ELISAs are easy to perform.
- ELISAs are relatively inexpensive.
- ELISAs can simultaneously test many samples quickly at once.
- ELISAs can be read easily, either by direct observation or by machine, lending themselves to efficient automation.
- Plates coated with antigen and gelatin can be stored for testing whenever they are needed.

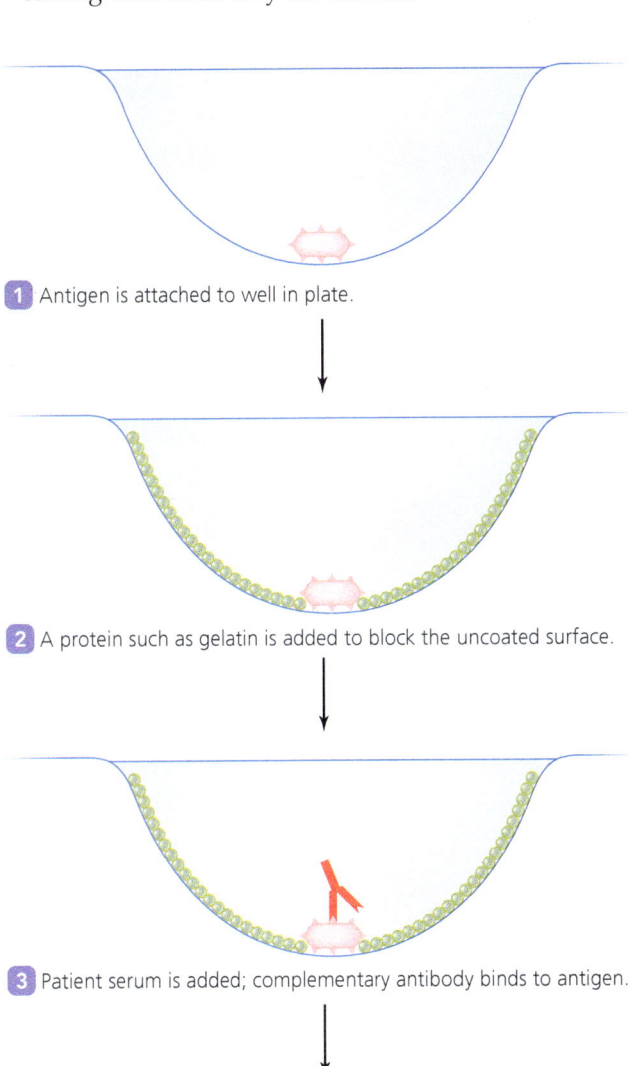

1. Antigen is attached to well in plate.

2. A protein such as gelatin is added to block the uncoated surface.

3. Patient serum is added; complementary antibody binds to antigen.

4. Enzyme-linked anti-antibody is added and binds to bound antibody.

5. Enzyme's substrate is added, and reaction produces a visible color change.

▶ FIGURE 17.12 Enzyme-linked immunosorbent assay (ELISA), also known as enzyme immunoassay (EIA). Shown is one well in a plate. The well is washed between steps. 1 Antigen added to the well attaches to it irreversibly. 2 After excess antigen is removed by washing, gelatin is added to cover any portion of the well not covered by antigen. 3 Test serum is added to the well; any specific antibodies in it bind to the antigen. 4 Enzyme-labeled anti-antibodies are added to the well and bind to any bound antibody. 5 The enzyme's substrate is added to the well, and the enzyme converts the substrate into a colored product; the amount of color, which can be measured via spectrophotometry, is directly proportional to the amount of antibody bound to the antigen.

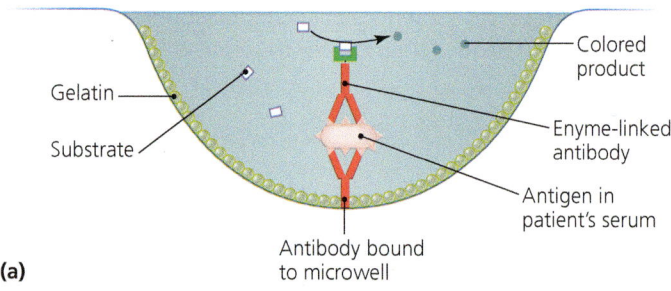

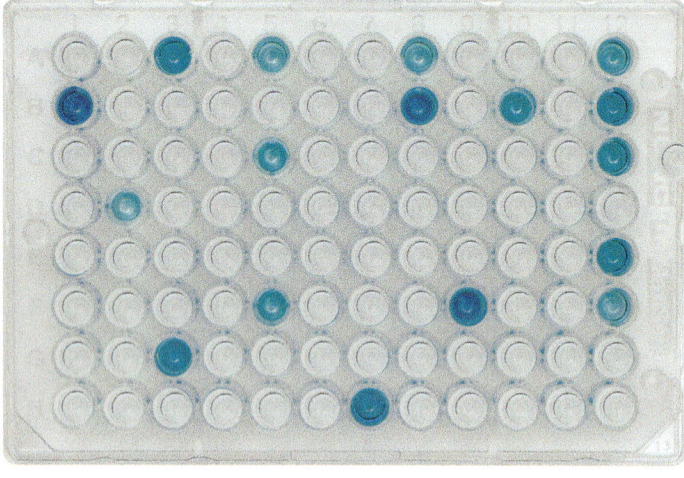

▲ FIGURE 17.13 **An antibody sandwich ELISA.** Because this variation of ELISA is used to test for the presence of antigen, antibody is attached to the well in the initial step. A second antibody sandwiches the antigen. **(a)** Artist's rendition. **(b)** Actual results. *How many wells are positive?*

Figure 17.13 Seventeen wells are positive.

A modification of the ELISA technique, called an *antibody sandwich ELISA*, can be used to detect antigen (FIGURE 17.13)—an example of direct testing. In testing for the presence of a particular virus in blood serum, for example, the plates are first coated with antibody rather than antigen. Then the sera from individuals being tested are added to the wells, and any virus in the sera will bind to the antibody attached to the well. Finally, each well is flooded with enzyme-labeled antibodies specific to the antigens of the virus. The name "antibody sandwich ELISA" refers to the fact that the antigen being tested for is "sandwiched" between two antibody molecules. Such tests can also be used to quantify the amount of an antigen in a given sample, and can be used to test for illegal drugs.

Immunoblots

An **immunoblot** (commonly called a *western blot*[3]) is a technique used to detect a specific protein, such as an antibody, in a complex mixture.

Immunoblotting tests are used to confirm the presence of proteins, including antibodies against pathogens. Physicians use immunoblots to verify the presence of HIV proteins or antibodies against the bacterium of Lyme disease in the blood serum of patients. Immunoblotting involves three steps (FIGURE 17.14a):

1. **Electrophoresis.** Antigens in a solution (in this example, HIV proteins) are placed into wells and separated by gel electrophoresis. Each of the proteins in the solution is resolved into a single band, producing invisible protein bands. (In the figure, the artist has colored the invisible bands to show you where they are.)

2. **Blotting.** The protein bands are transferred to an overlying nitrocellulose membrane. This can be done by absorbing the solution into absorbent paper—a process called *blotting*. The nitrocellulose membrane is then cut into strips.

3. **Immunoassay.** Each nitrocellulose strip is incubated with a test solution— in this example, samples from each of six individuals who are being tested for antibodies against HIV. After the strips are washed, an enzyme-labeled anti-antibody solution is added for a time; then the strips are washed again and exposed to the enzyme's substrate. Color develops wherever antibodies against the HIV proteins in the test solutions have bound to their substrates, as shown in the positive control. In this example, the individual tested in strip 3 is positive for antibodies against HIV, whereas the other five individuals are negative for antibodies against HIV (FIGURE 17.14b). Immunoblots are sensitive and can detect many types of proteins simultaneously. Colored bands common to all patients are normal serum proteins.

Point-of-Care Testing

LEARNING OUTCOMES

17.20 Discuss the benefits of using immunofiltration assays rather than an ELISA.

17.21 Contrast immunofiltration and immunochromatographic assays.

Recent years have seen the development of simple immunoassays that give clinicians useful results within minutes. These assays allow *point-of-care testing*; that is, health care providers do not have to send specimens to a laboratory for testing but can perform the test at the patient's bedside or in a doctor's office. Common point-of-care tests include *immunofiltration* and *immunochromatography assays*. These tests are not quantitative, but they do rapidly give a positive or negative result, making them very useful in arriving at a quick diagnosis.

Immunofiltration assays (im'yū-nō-fil-trā'shŭn) are rapid ELISAs based on the use of antibodies bound to a membrane filter rather than to plates. Because of the large surface area of a membrane filter, reactions proceed faster, and assay times are significantly shorter than a traditional ELISA.

Immunochromatographic assays (im'yū-nō-krō'mat-ō-graf'ik) are also fast and easy-to-read immunoassays. In these systems, an antigen solution (such as diluted blood or sputum) flowing through a porous material encounters antibody labeled with either pink colloidal[4] gold or blue colloidal selenium. Where

---

[3] A word play on Southern blot (named for biologist Edwin Southern), which is a similar procedure for detecting specific DNA sequences in samples.

[4] *Colloidal* refers to small particles suspended in a liquid or gas.

Serological Tests That Use Antigens and Corresponding Antibodies

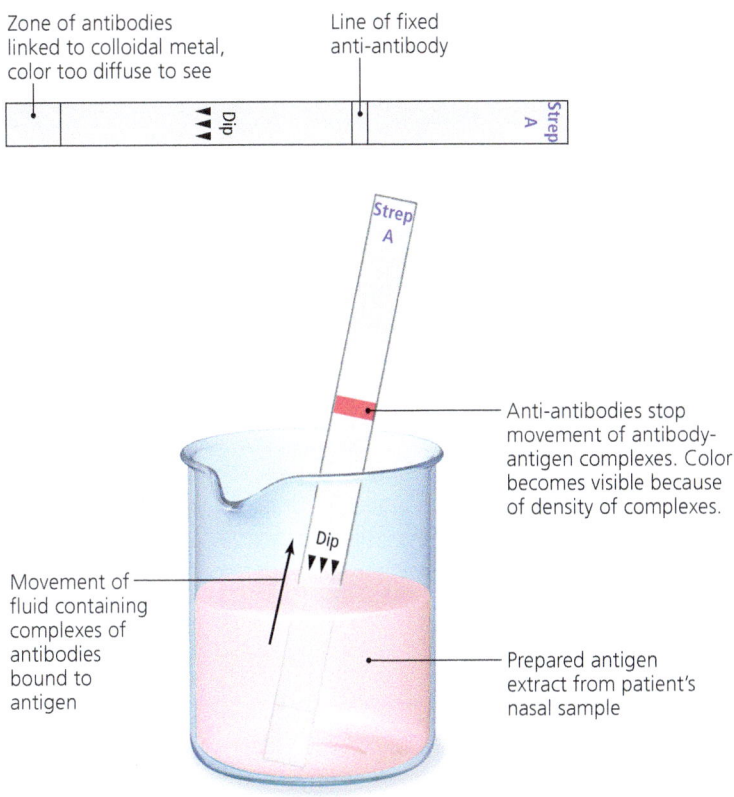

▲ FIGURE 17.15 **Immunochromatographic dipstick.** The dipstick is impregnated with colloidal metal particles linked to movable antibodies against particular antigens at one end and anti-antibodies fixed in a line closer to the other end of the membrane.

antigen and antibody bind, colored immune complexes form in the fluid, which then flows through a region where the complexes encounter antibody against them, resulting in a clearly visible pink or blue line, depending on the label used. These assays are used for pregnancy testing, which tests for *human chorionic gonadotropin*—a hormone produced only by an embryo or fetus—and for rapid identification of infectious agents such as HIV, *Escherichia coli* (esh-ĕ-rik'ē-ă kō'lī) O157:H7, group A *Streptococcus*, respiratory syncytial virus (RSV), and influenzaviruses.

In one adaptation, the antibodies are coated on membrane strips, which serve as dipsticks. At one end, anti-antibodies are fixed in a line so that they cannot move in the membrane. The lower portion of the membrane is coated with antibodies against the antigen in question. These antibodies are linked to a color indicator in the form of a colloidal metal and are free to move up the membrane by capillary action until their progress is arrested by the line of anti-antibodies. **FIGURE 17.15** illustrates the procedure as used to test for the presence of group A *Streptococcus* in the nasal secretion of a patient. A laboratory scientist prepares a solution with a nasal swab from the patient so as to

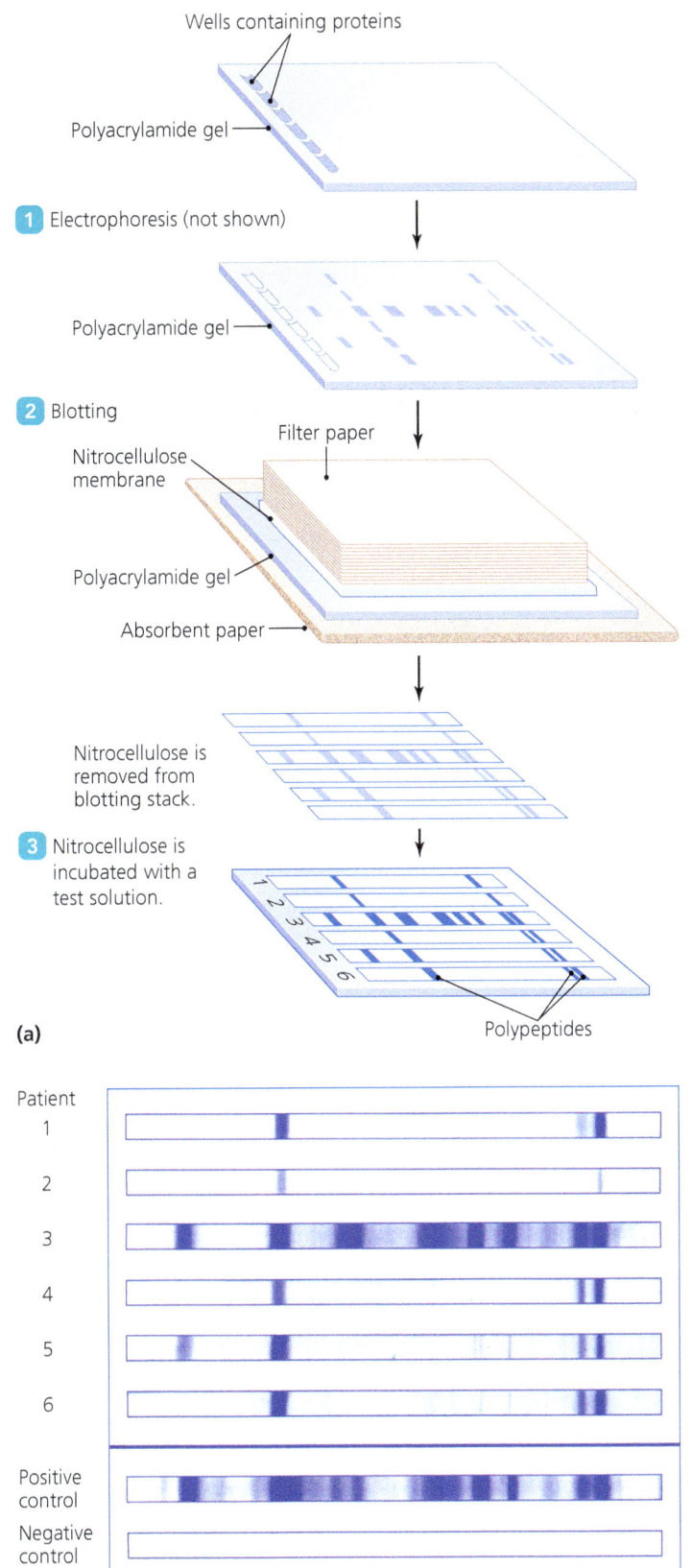

▲ FIGURE 17.14 **Immunoblotting (western blotting).** This technique detects specific proteins in a complex mixture, taking advantage of the high affinity and specificity of antibodies against the proteins. **(a)** Steps in immunoblotting test. Proteins are separated by gel electrophoresis. The artist has shown the positions of the proteins, but in reality, they would be invisible ❶. Separated proteins are transferred to a nitrocellulose membrane. This is the "blotting" ❷. Test solutions, enzyme-labeled anti-antibody, and the enzyme's substrate are added; color changes are detected wherever antibody in the test solutions has bound to proteins ❸. **(b)** Real results of an immunoblot. Patient 3 has the same proteins as the positive control.

TABLE 17.2 Antibody-Antigen Immunological Tests and Some of Their Uses

Test	Use
Immunodiffusion (precipitation)	Diagnosis of syphilis, pneumococcal pneumonia
Agglutination	Blood typing; pregnancy testing; diagnosis of salmonellosis, brucellosis, gonorrhea, rickettsial infection, mycoplasma infection, yeast infection, typhoid fever, meningitis caused by *Haemophilus*
Viral neutralization	Diagnosis of infections by specific strains of viruses
Viral hemagglutination inhibition	Diagnosis of viral infections including influenza, measles, mumps, rubella, mononucleosis
Complement fixation	In the past, diagnosis of measles, influenza A, syphilis, rubella, rickettsial infections, scarlet fever, rheumatic fever, infections of respiratory syncytial virus and *Coxiella*
Direct fluorescent antibody	Diagnosis of rabies, infections of group A *Streptococcus*, identification of lymphocyte subsets
Indirect fluorescent antibody	Diagnosis of syphilis, mononucleosis
ELISA	Pregnancy testing; presence of drugs in urine; diagnosis of hepatitis A, hepatitis B, rubella; initial diagnosis of HIV infection
Immunoblot (western blot)	Confirmation of infection with HIV; diagnosis of Lyme disease

release *Streptococcus* antigens if they are present. The scientist then dips the membrane into the solution. The membrane's antibodies bind to streptococcal antigens, forming complexes. The complexes move up the membrane by capillary action until they reach the line of anti-antibodies, where they bind and must stop because the anti-antibodies are chemically bound to the strip. Previously, the complexes were invisible because they were dilute; now they are concentrated at the line of anti-antibodies and become visible, indicating that this patient has group A *Streptococcus* in the nose. Knowing that the infection is bacterial and not viral, the physician can prescribe antibacterial drugs. The procedure from antigen preparation to diagnosis takes less than 10 minutes.

TABLE 17.2 lists some antibody-antigen immune tests that can be used to diagnose selected bacterial and viral diseases.

MICRO CHECK

6. What is the difference between an agglutination test and a precipitation test?
7. Which technique involves performing electrophoresis of proteins followed by transfer to a membrane and binding with antibody?
8. What type of assay is used to identify syphilis and mononucleosis?

TELL ME WHY

A diagnostician used an ELISA to show that a newborn had antibodies against HIV in her blood. However, six months later, the same test was negative. How can this be?

MICRO IN THE CLINIC FOLLOW-UP
Complications of Premature Birth?

Dr. Connors examines Ryan and collects a nasal wash (saline squirted into and immediately withdrawn from the nostril) to send for lab analysis. Noting Ryan's labored breath sounds and his previous lung issues, Dr. Connors admits Ryan to the hospital so he can be monitored until the lab results return.

The analysis of the nasal wash reveals that Ryan has respiratory syncytial virus (RSV). Doug and Stacy are scared by the diagnosis—Ryan's been through so much already. Dr. Connors tells them that premature babies like Ryan are at risk for acquiring RSV infection; it is actually very common. The Centers for Disease Control and Prevention (CDC) estimates that almost all children become infected with RSV before their second birthday. There is no specific treatment for RSV, but because of Ryan's medical history, Dr. Connors keeps Ryan in the hospital. Over the next couple of days, Ryan develops bronchiolitis (inflammation of the bronchioles). He is treated with supplemental oxygen and suctioning of mucus to help him breathe more easily. During the next 10 days, Ryan's condition progressively improves, and he returns home, fully recovered.

1. **The laboratory test used to identify RSV in Ryan's nasal wash is based on detecting RSV antigen in the sample. Describe one approach to identifying antigens in a sample from a patient.**

2. **Even though RSV infection in infants is common, a vaccine does not currently exist. Imagine you are designing a recombinant vaccine for RSV—what viral components would you use in your vaccine? Justify your choice.**

Check your answers to Micro in the Clinic Follow-Up questions in the **Mastering Microbiology** Study Area.

CHAPTER SUMMARY

Go to Mastering Microbiology for Interactive Microbiology, Dr. Bauman's Video Tutors, Micro Matters, MicroFlix, Micro-Boosters, practice quizzes, and more.

Immunization (pp. 504–512)

1. The first **vaccine** was developed by Edward Jenner against smallpox. He called the technique **vaccination**. **Immunization** is a more general term referring to the use of vaccines against rabies, anthrax, measles, mumps, rubella, polio, and other diseases.

2. Individuals can be protected against many infections by either active immunization or passive immunotherapy.

3. Active immunization involves giving antigen in the form of either **attenuated vaccines**, **inactivated vaccines** (killed vaccines), **toxoid vaccines**, or recombinant gene vaccines. Antibody **titer** refers to the amount of antibody produced.

4. Immunologists can reduce the virulence of microbes used in vaccines so that they are less likely to cause disease, a practice known as **attenuation**. Pathogens in attenuated vaccines are weakened so that they no longer cause disease, though they are still alive or active and can provide **contact immunity** in unimmunized individuals who associate with immunized people.

5. Inactivated vaccines are either whole agent or subunit vaccines and often contain **adjuvants**, which are chemicals added to increase their ability to stimulate active immunity.

6. Toxoid vaccines use modified toxins to stimulate antibody-mediated immunity.

7. A **combination vaccine** is composed of antigens from several pathogens so they can be administered to a patient at once.

8. Vaccine developers can use recombinant gene technology to remove virulence genes from microbes to create attenuated vaccines, to produce large quantities of antigens for use in immunizations, or to alter cells, viruses, or plasmids by introducing genes for antigens into them.

9. Having a large proportion of immunized individuals (>75%) in a population interrupts disease transmission, providing protection to unimmunized individuals. Such protection is called **herd immunity**.

10. Vaccine developers and administrators monitor vaccine safety by checking for adverse reactions to vaccines, including mild irritation at the site of inoculation, anaphylactic shock, and the possibility of an attenuated vaccine reverting to virulence.

11. **Passive immunotherapy** (a type of passive immunization) involves administration of an **antiserum** containing preformed antibodies. Serum sickness results when the patient makes antibodies against the antiserum.

12. The fusion of myelomas (cancerous plasma cells) with plasma cells results in **hybridomas**, the source of **monoclonal antibodies**, which can be used in passive immunization.

13. Health providers weigh the advantages and disadvantages of active versus passive immunization. Active immunization, in which patients synthesize their own antibodies and develop immunological memory, is longer lasting but relatively slow to develop. Passive immunity, in which preformed antibodies are administered, is fast acting but does not produce long-term effects.

Serological Tests That Use Antigens and Corresponding Antibodies (pp. 512–520)

1. **Serology** is the study and use of immunological assays on blood serum to diagnose disease or identify antibodies or antigens. Scientists use antibodies to find an antigen in a specimen and use antigen to find antibodies.

2. The simplest of the serological tests is a precipitation test, in which antigen and antibody meet in optimal proportions to form **immune complexes**, which are often insoluble. Often, this test is performed in clear gels, where it is called **immunodiffusion**.

3. Automated light detectors can measure the cloudiness of a solution—an indication of the quantity of protein in the solution. Turbidimetry measures the passage of light through the solution, whereas nephelometry measures the amount of light reflected by protein in the solution.

4. **Agglutination** tests involve the clumping of antigenic particles by antibodies. The amount of these antibodies, called a titer, is measured by diluting the serum in a process called **titration**.

5. Antibodies to viruses or toxins can be measured using a **neutralization test**, such as a **viral neutralization** test. Infection by viruses that naturally agglutinate red blood cells can be demonstrated using a **viral hemagglutination inhibition test**.

6. The **complement fixation test** is a complex assay used to determine the presence of specific antibodies.

7. Fluorescently labeled antibodies—those chemically linked to a fluorescent dye—can be used in a variety of **direct fluorescent immunoassays** and **indirect fluorescent immunoassays**. The presence of labeled antibodies is visible through a fluorescence microscope.

8. **Enzyme-linked immunosorbent assays (ELISAs)**, or **enzyme immunoassays (EIAs)**, are a family of simple tests that can be readily automated and read by machine. These tests are among the more common serological tests used. A variation of the ELISA is an **immunoblot** (western blot), which is used to detect antibodies against multiple antigens in a mixture.

9. **Immunofiltration assays** and **immunochromatographic assays** are modifications of ELISA tests that can give much more rapid diagnostic results.

QUESTIONS FOR REVIEW

Answers to the Questions for Review (except Short Answer questions) begin on p. A-1.

Multiple Choice

1. To obtain immediate immunity against tetanus, a patient should receive _____.
   a. an attenuated vaccine of *Clostridium tetani*
   b. a modified live vaccine of *C. tetani*
   c. tetanus toxoid
   d. immunoglobulin against tetanus toxin (antitoxin)

2. Which of the following vaccine types is commonly given with an adjuvant?
   a. attenuated vaccine
   b. modified live vaccine
   c. chemically killed vaccine
   d. immunoglobulin

3. Which of the following viruses was widely used in living vaccines?
   a. coronavirus
   b. poliovirus
   c. influenzavirus
   d. retrovirus

4. When antigen and antibodies combine, maximal precipitation occurs when _____.
   a. antigen is in excess
   b. antibody is in excess
   c. antigen and antibody are at equivalent concentrations
   d. antigen is added to the antibody

5. An anti-antibody is used when _____.
   a. an antigen is not precipitating
   b. an antibody is not agglutinating
   c. an antibody does not activate complement
   d. the antigen is an antibody

6. The many different proteins in serum can be analyzed by a(n) _____.
   a. anti-antibody test
   b. complement fixation test
   c. precipitation test
   d. immunodiffusion test

7. A direct fluorescent immunoassay requires which of the following?
   a. heat-inactivated serum
   b. fluorescent serum
   c. immune complexes
   d. antibodies against the antigen

8. An ELISA uses which of the following reagents?
   a. enzyme-labeled anti-antibody
   b. radioactive anti-antibody
   c. source of complement
   d. enzyme-labeled antigen

9. A direct fluorescent immunoassay can be used to detect the presence of _____.
   a. hemagglutination
   b. specific antigens
   c. antibodies
   d. complement

10. Which of the following is a good test to detect rabies virus in the brain of a dog?
    a. agglutination
    b. hemagglutination inhibition
    c. virus neutralization
    d. direct fluorescent antibody

11. Attenuation is _____.
    a. the process of reducing virulence
    b. a necessary step in vaccine manufacture
    c. a form of variolation
    d. similar to an adjuvant

12. An antiserum is _____.
    a. an anti-antibody
    b. an inactivated vaccine
    c. formed of monoclonal antibodies
    d. the liquid portion of blood used for immunization

13. Monoclonal antibodies _____.
    a. are produced by hybridomas
    b. are secreted by clone cells
    c. can be used for passive immunization
    d. all of the above

14. The study of antibody-antigen interaction in the blood is _____.
    a. attenuation
    b. agglutination
    c. precipitation
    d. serology

15. Anti–human antibody antibodies are _____.
    a. found in immunocompromised individuals
    b. used in direct fluorescent immunoassays
    c. formed by animals reacting to human immunoglobulins
    d. an alternative method in ELISA

True/False

1. _____ Passive immunotherapy provides more prolonged immunity than active immunization.
2. _____ It is standard to attenuate killed virus vaccines.
3. _____ One single serological test is inadequate for an accurate diagnosis of HIV infection.
4. _____ ELISA is very easily automated.
5. _____ ELISA has basically replaced immunoblotting.

Matching

In the blank before each description, write the letter of the matching term. Some choices may be used more than once.

1. _____ Induces rapid onset of immunity
2. _____ Induces mainly an antibody response
3. _____ Induces good cell-mediated immunity
4. _____ Increases antigenicity
5. _____ Uses antigen fragments
6. _____ Uses attenuated microbes

A. Attenuated viral vaccine
B. Adjuvant
C. Subunit vaccine
D. Immunoglobulin
E. Residual virulence

VISUALIZE IT!

1. Identify the chemicals represented by this artist's conception of an antibody sandwich ELISA.

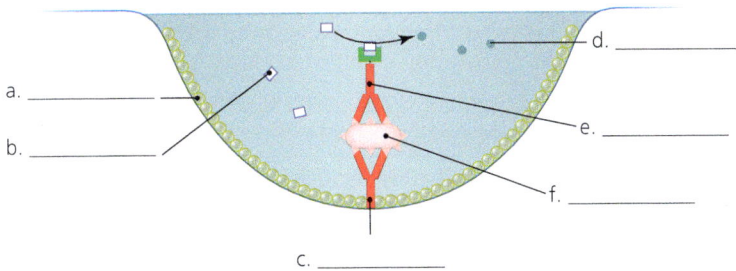

2. The two columns on the left show negative and positive immunoblot results for a particular pathogen. The numbered columns are blots of samples from 11 patients. Which patients are most likely *un*infected?

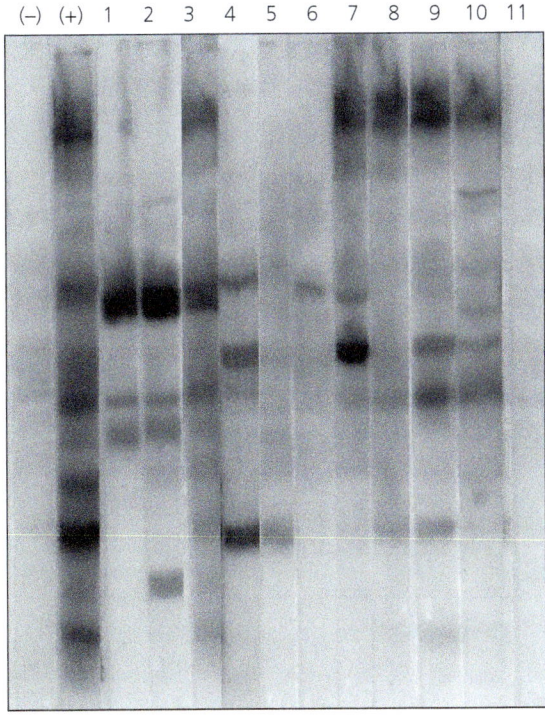

Short Answer

1. Compare and contrast the Chinese practice of variolation with Jenner's vaccination procedure.
2. What are the advantages and disadvantages of attenuated vaccines?
3. Compare the advantages and disadvantages of passive immunotherapy and active immunization.
4. How does precipitation differ from agglutination?
5. Explain how a pregnancy test works at the molecular level.
6. Compare and contrast herd immunity and contact immunity.
7. How does nephelometry differ from turbidimetry?

CRITICAL THINKING

1. Is it ethical to approve the use of a vaccine that causes significant illness in 1% of patients if it protects immunized survivors against a serious disease?

2. Which is worse: to use a diagnostic test for HIV that may falsely indicate that a patient is *not* infected (false negative) or to use one that sometimes falsely indicates that a patient *is* infected (false positive)? Defend your choice.

3. Discuss the importance of costs and technical skill in selecting a practical serological test. Under what circumstances does automation become important?

4. What bodily fluids, in addition to blood serum, might be usable for immune testing?

5. Why might a serological test give a false positive result?

6. Some researchers want to distinguish B cells from T cells in a mixture of lymphocytes. How could they do this without killing the cells?

7. Describe three ways by which genetic recombinant techniques could be used to develop safer, more effective vaccines.

8. How does a toxoid vaccine differ from an attenuated vaccine?

9. Explain why many health organizations promote breast-feeding of newborns. What risks are involved in such nursing?

10. Contrast a hemagglutination test with a viral hemagglutination inhibition test.

11. Sixty years ago, parents would have done almost anything to get a protective vaccine against polio for their children. Now parents fear the vaccine, not the disease. Why?

12. Draw a picture showing, at *both the molecular and the cellular level*, IgM agglutinating red blood cells.

CONCEPT MAPPING

Using the terms provided, fill in the concept map that describes vaccines. You can also complete this and other concept maps online by going to the Mastering Microbiology Study Area.

Active immunity
Adjuvants
Attenuated vaccines
Hepatitis B vaccine
Immunizations

Inactivated vaccines
Measles vaccine
Microorganisms
Modified toxins
Pertussis vaccine

Polio
Subunit vaccines
Tetanus toxoid
Varicella vaccine

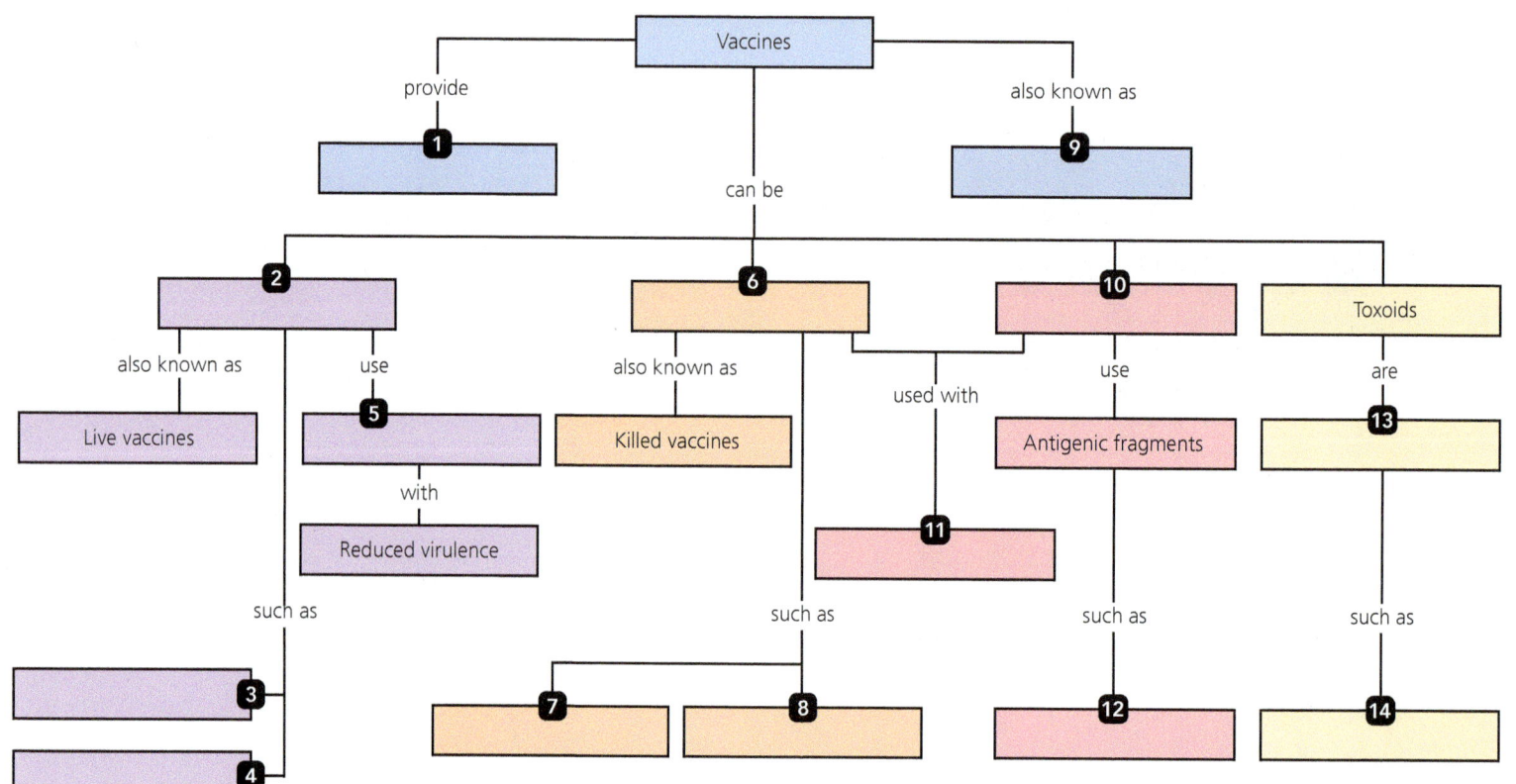

18 Immune Disorders

Before You Begin

1. Where do the stems of IgE antibodies bind?
2. What is the effect of histamine on the body?
3. What is a characteristic of an effective antigen?

MICRO IN THE CLINIC
Danger in the Blackberry Patch?

Cassie and Tim recently purchased their first home. It's been an exciting but labor-intensive process. They spent a lot of time renovating the interior of the house and have turned their attention to the yard. They had a lot of fun putting in their garden and building a swing set for their kids. But Cassie and Tim have been avoiding an unpleasant task: clearing a large patch of blackberry brambles occupying a back corner of their lot.

On a bright Saturday, Cassie and Tim decide to thin out the blackberries. It's a beautiful day outside, and the work is hard but goes more smoothly than either of them expected. Cassie is unloading a wheelbarrow full of branches into the back of the truck when she hears Tim yelling. Cassie runs to the backyard to find Tim on the ground; he tells her that when he started shoveling to dig out the roots, a swarm of hornets flew out and stung him. About fifteen minutes after the stings, Tim says that his skin feels itchy. Cassie notices that red welts are forming on Tim's face, trunk, and arms; when he says that his throat feels itchy and tight, Cassie rushes Tim to an emergency clinic, which is only two minutes away.

1. What kind of reaction is Tim having?

Turn to the end of the chapter (p. 541) to find out.

Anaphylaxis is a severe hypersensitivity to an antigen, such as hornet venom.

The immune system is an absolutely essential component of the body's defenses. However, if the immune system functions abnormally—either by overreacting or by underreacting—the malfunction may cause significant disease, even death. If the immune system functions excessively, the body develops any of a variety of *immune hypersensitivities*, such as allergies. Further, if the immune system attacks the body's own tissues, *autoimmune diseases* develop. If the immune system fails, an *immunodeficiency disease* (im′ū-nō-dē-fish′en-sē) results. AIDS is a prime example. In this chapter, we discuss each of these three categories of derangements of the immune system.

Hypersensitivities

LEARNING OUTCOME

18.1 Compare and contrast the four types of hypersensitivity.

Hypersensitivity (hī′per-sen-si-tiv′i-tē) may be defined as any immune response against a foreign antigen that is exaggerated beyond the norm. For example, most people can wear wool, smell perfume, or dust furniture without experiencing itching, wheezing, runny nose, or watery eyes. When these symptoms do occur, the person is said to be experiencing a hypersensitivity response. In the following sections, we examine each of the four main types of hypersensitivity response, designated as type I through type IV.

Type I (Immediate) Hypersensitivity

LEARNING OUTCOMES

18.2 Describe the two-part mechanism by which type I hypersensitivity occurs.
18.3 Explain the roles of three inflammatory chemicals (mediators) released from mast cell granules.
18.4 Describe three disease conditions resulting from type I hypersensitivity mechanisms.

Type I hypersensitivities are localized or systemic (whole body) reactions that result from the release of inflammatory molecules (such as histamine) in response to an antigen. These reactions are termed *immediate hypersensitivity* because they develop within seconds or minutes following contact with antigens. They are also commonly called **allergies**,[1] and the antigens that stimulate them are called **allergens** (al′er-jenz).

In the next two subsections, we will examine the two-part mechanism of a type I hypersensitivity reaction: sensitization upon initial contact with an allergen and the degranulation of sensitized cells.

Sensitization upon Initial Exposure to an Allergen

All of us are exposed to antigens in the environment, against which we typically mount immune responses that result in the production of antibodies of the gamma class (IgG). But when regulatory proteins (cytokines, especially interleukin 4) from type 2 helper T (Th2) cells stimulate B cells in allergic individuals, the B cells become plasma cells that produce class epsilon antibodies (IgE; FIGURE 18.1a). Typically, IgE is directed against parasitic worms, which rarely infect most Americans, so IgE is found at low levels in the blood serum. However, in allergic individuals, plasma cells produce a significant amount of IgE in response to allergens.

The precise reason why only some people produce high levels of IgE (and thereby suffer from allergies) is a matter of intense research. Many researchers report evidence for the *hygiene hypothesis*, which holds that children exposed to environmental antigens—such as dust mites, molds, parasitic worms, and pet hair—are less likely to develop allergies than children who have been sheltered from common environmental antigens. Conversely, other researchers have discovered evidence that environmental factors sensitize people, making them hypersensitive and more likely to develop allergies.

In any case, following initial exposure to allergens, the plasma cells of allergic individuals secrete IgE, which binds very strongly with its stem to two types of defense cells—*mast cells* and *basophils*—sensitizing these cells to respond to future exposures to the allergen.

Degranulation of Sensitized Cells

When the same allergen reenters the body, it binds to the active sites of IgE molecules on the surfaces of sensitized cells (described shortly). This binding triggers a cascade of internal biochemical reactions that causes the sensitized cells to release the inflammatory chemicals from their granules into the surrounding space—an event called *degranulation* (dē-gran-ū-lā′shŭn; FIGURE 18.1b). It is these inflammatory mediators that generate the characteristic symptoms of type I hypersensitivity reactions: respiratory distress, rhinitis (inflammation of the nasal mucous membranes, commonly called "runny nose"), watery eyes, inflammation, and reddening of the skin.

The Roles of Degranulating Cells in an Allergic Reaction

Mast cells are specialized relatives of white blood cells, deriving from other stem cells in the bone marrow. They are distributed throughout the body in connective tissues other than blood. These large, round cells are most often found in sites close to body surfaces, including the skin and the walls of the intestines and airways. Their characteristic feature is cytoplasm packed with large granules that are loaded with a mixture of potent inflammatory chemicals.

One significant chemical released from mast cell granules is **histamine** (his′tă-mēn), a small molecule related to the amino acid histidine. Histamine stimulates strong contractions in the smooth muscles of the bronchi, gastrointestinal tract, uterus, and bladder, and it also makes small blood vessels dilate (expand) and become leaky. As a result, tissues in which mast cells degranulate become red and swollen. Histamine also stimulates nerve endings, causing itching and pain. Finally, histamine is an

---

[1]From Greek *allos*, meaning "other," and *ergon*, meaning "work."

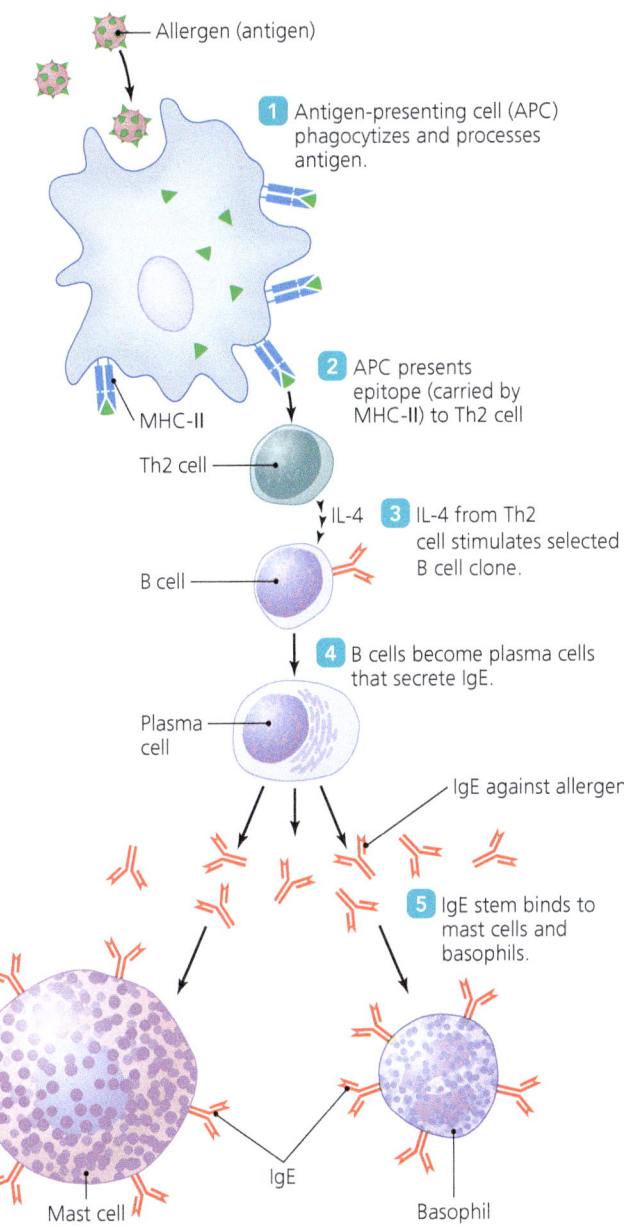

▲ **FIGURE 18.1 The mechanisms of a type I hypersensitivity reaction. (a)** Sensitization. After normal processing of an antigen (allergen) by an antigen-presenting cell 1, Th2 cells are stimulated to secrete interleukin 4 (IL-4) 2. This cytokine stimulates a B cell 3, which becomes a plasma cell that secretes IgE 4, which then binds to and sensitizes mast cells and basophils 5. **(b)** Degranulation 6. When the same allergen is subsequently encountered, its binding to the IgE molecules on the surfaces of sensitized cells triggers rapid degranulation and the release of inflammatory chemicals from the cells.

effective stimulator of bronchial mucus secretion, tear formation, and salivation.

Other triggers of allergies released by degranulating mast cells include **kinins** (kī'ninz), which are powerful inflammatory chemicals, and **proteases** (prō'tē-ās-ez)—enzymes that destroy nearby cells, activating the complement system, which in turn results in the release of more chemicals that cause inflammation. Proteases account for more than half the proteins in a mast cell granule. In addition, the binding of allergens to IgE on mast cells activates other enzymes that trigger the production of **leukotrienes** (loo-kō-trī' ēnz) and **prostaglandins** (pros-tă-glan'dinz), lipid molecules that are powerful inflammatory agents. TABLE 18.1 summarizes some inflammatory molecules released from mast cells.

Basophils, the least numerous type of leukocyte in blood, contain cytoplasmic granules that stain intensely with basic dyes (see Chapter 15, Figure 15.5a) and are filled with inflammatory chemicals similar to those found in mast cells. Sensitized basophils bind IgE and degranulate in the same way as mast cells when they encounter allergens.

The blood and tissues of allergic individuals also accumulate many **eosinophils**, leukocytes that contain numerous cytoplasmic granules that stain intensely with a red dye called *eosin* and that function primarily to destroy parasitic worms. The process during type I hypersensitivity reactions that results in the accumulation of eosinophils in the blood—a condition termed *eosinophilia*—begins with mast cell degranulation, which releases peptides that stimulate the release of eosinophils from the bone marrow. Once in the bloodstream, eosinophils are attracted to the site of mast cell degranulation, where

TABLE 18.1 Some Inflammatory Molecules Released from Mast Cells

Molecules	Role in Hypersensitivity Reactions
Released During Degranulation	
Histamine	Causes smooth muscle contraction, increased vascular permeability, and irritation
Kinins	Cause smooth muscle contraction, inflammation, and irritation
Proteases	Damage tissues and activate complement
Synthesized in Response to Allergen–IgE Binding	
Leukotrienes	Cause slow, prolonged smooth muscle contraction, inflammation, and increased vascular permeability
Prostaglandins	Some contract smooth muscle; others relax it

▲ FIGURE 18.2 **Some common allergens. (a)** Spores of the fungus *Aspergillus* on stalks. **(b)** Pollen of *Ambrosia trifida* (ragweed), a more common cause of hay fever in the United States. **(c)** Two house dust mites on a carpet. Mites' fecal pellets and dead bodies may become airborne and trigger allergic responses.

they themselves degranulate. Eosinophil granules contain unique inflammatory mediators and produce large amounts of leukotrienes, which increase movement of white blood cells from blood vessels and stimulate smooth muscle contraction, thereby contributing greatly to the severity of a hypersensitivity response.

Clinical Signs of Localized Allergic Reactions

Type I hypersensitivity reactions are usually mild and localized. The site of the reaction depends on the portal of entry of the antigens. For example, inhaled allergens may provoke a response in the upper respiratory tract commonly known as **hay fever**—a local allergic reaction marked by a runny nose, sneezing, itchy throat and eyes, and excessive tear production. Among the more common allergens are fungal (mold) spores; pollens from grasses, flowering plants, and some trees; and feces and dead bodies of house dust mites **(FIGURE 18.2)**.

If inhaled allergen particles are sufficiently small, they may reach the lungs. A type I hypersensitivity in the lungs can cause an episode of severe difficulty in breathing known as **asthma**, characterized by wheezing; coughing; excessive production of a thick, sticky mucus; and constriction of the smooth muscles of the bronchi. Asthma can be life threatening: without medical intervention, the increased mucus and bronchial constriction can quickly cause suffocation.

Other allergens—including latex; wool; certain metals; and the venom or saliva of wasps, bees, fire ants, deer flies, fleas, and other stinging or biting insects—may cause a localized inflammation of the skin. As a result of the release of histamine and other mediators, plus the ensuing leakage of serum from local blood vessels, the individual suffers raised, red areas called *hives*, or **urticaria**[2] (er'ti-kar'i-ă; FIGURE 18.3). These lesions are very itchy because histamine irritates local nerve endings.

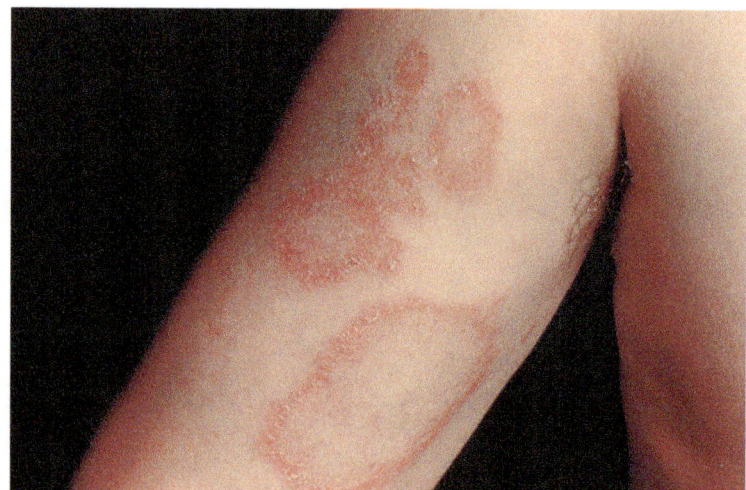

▲ FIGURE 18.3 **Urticaria.** These red, itchy patches on the skin are prompted by the release of histamine in response to an allergen. *What causes fluid to accumulate in the patches seen in urticaria?*

Figure 18.3 *The fluid accumulates because histamine from degranulated mast cells causes blood capillaries to become more permeable.*

Clinical Signs of Systemic Allergic Reactions

Following a sensitized individual's contact with an allergen, many mast cells may degranulate simultaneously, releasing massive amounts of histamine and other inflammatory mediators into the bloodstream. The release of chemicals may exceed the body's ability to adjust, resulting in a condition called **acute anaphylaxis**[3] (an'ă-fī-lak'sis), or **anaphylactic shock**.

The clinical signs of acute anaphylaxis are those of rapid suffocation. Bronchial smooth muscle, which is highly sensitive to histamine, contracts violently. In addition, increased leakage of

---

[2]From Latin *urtica*, meaning "nettle."

[3]From Greek *ana*, meaning "away from," and *phylaxis*, meaning "protection."

fluid from blood vessels causes swelling of the larynx and other tissues. The patient also experiences contraction of the smooth muscle of the intestines and bladder. Without immediate administration of *epinephrine* (ep'i-nef'rin; discussed shortly), an individual in anaphylactic shock may suffocate, collapse, and die within minutes.

A common cause of acute anaphylaxis is a bee sting in an allergic and sensitized individual. The first sting produces sensitization and the formation of IgE antibodies, and successive stings—which may occur years later—produce an anaphylactic reaction. Other allergens commonly implicated in acute anaphylaxis are certain foods (notoriously peanuts), vaccines containing egg proteins, antibiotics such as penicillin, iodine dyes, local anesthetics, blood products, and certain narcotics such as morphine.

Diagnosis of Type I Hypersensitivity

Clinicians diagnose type I hypersensitivity with a test variously called *ImmunoCAP Specific IgE blood test*, *CAP RAST*, or *Pharmacia CAP*, in which suspected allergens are mixed with samples of the patient's blood. The specifics of the test are beyond the scope of this chapter, but basically the test detects the amount of IgE directed against each allergen. High levels of a specific IgE indicate a hypersensitivity against that allergen.

Alternatively, physicians diagnose type I hypersensitivity by injecting into the skin a very small quantity of a dilute solution of the allergens being tested. In most cases, the individual is screened for more than a dozen potential allergens simultaneously, using the forearms or back as injection sites. When the individual tested is sensitive to an allergen, having been previously exposed to that allergen, local histamine release causes redness and swelling at the injection site within a few minutes **(FIGURE 18.4)**.

Prevention of Type I Hypersensitivity

Prevention of type I hypersensitivity begins with *identifying* the allergens responsible and *avoiding* them. Filtering the air and avoiding rural areas during pollen season can reduce upper respiratory allergies provoked by some pollens. Encasing bedclothes in mite-proof covers, frequent vacuuming, and avoiding home furnishings that trap dust (such as wall-to-wall carpeting and heavy drapes) can reduce the severity of other household allergies. A dehumidifier can reduce mold in a home. The best way for individuals to avoid allergic reactions to pets is to find out whether they are allergic before they get a pet. If an allergy develops afterward, they may have to find a new home for the animal.

Food allergens can be identified and then avoided by using a medically supervised elimination diet in which foods are removed one at a time from the diet to see when signs of allergy cease. Peanuts and shellfish are among the foods more commonly implicated in anaphylactic shock; allergic individuals must avoid consuming even minute amounts. For example, individuals sensitized against shellfish can suffer anaphylactic shock from consuming meat sliced on an unwashed cutting board previously used to prepare shellfish. Some vaccines contain small amounts of egg proteins and should not be given to individuals allergic to eggs.

Health care workers should assess a patient's medical history carefully before administering penicillin, iodine dyes, and other potential allergens. Therapies such as antidotes and respiratory stimulants should be immediately available if a patient's sensitivity status is unknown.

In addition to avoidance of allergens, type I hypersensitivity reactions can be prevented by *immunotherapy* (im'yū-nō-thār'ă-pē), commonly called "allergy shots," which involves the administration of a series of injections of dilute allergen, usually once a week for many months. It is unclear just how allergy shots work, but they may change the helper T cell balance from Th2 cells to Th1 cells, reducing the production of antibodies, or they may stimulate the production of IgG, which binds antigen before the antigen can react with IgE on mast cells and basophils. Immunotherapy reduces the severity of allergy symptoms by roughly 50% in about two-thirds of patients with upper respiratory allergies; however, the series of injections must be repeated every two to three years.

Treatment of Type I Hypersensitivity

One way to treat type I hypersensitivity is to administer drugs that specifically counteract the inflammatory mediators released when cells degranulate. Thus, **antihistamines** are administered to counteract histamine. However, because histamine is but one of many mediators released in type I hypersensitivity, antihistamines do not completely eliminate all clinical signs in diseases such as asthma. For asthmatics, physicians typically prescribe an inhalant containing *glucocorticoid* (gloo-kō-kōr'ti-koyd) and a *bronchodilator* (brong-kō-dī'lā-ter), which counteract the effects of many inflammatory mediators.

The hormone *epinephrine* quickly neutralizes many of the lethal mechanisms of anaphylaxis by relaxing smooth muscle tissue in the lungs, contracting smooth muscle of blood vessels, and reducing vascular permeability. Epinephrine is thus the drug of choice for the emergency treatment of both severe asthma and anaphylactic shock. Patients who suffer from severe type I hypersensitivities may carry a prescription epinephrine

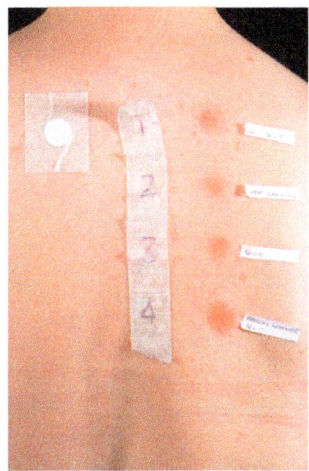

▲ **FIGURE 18.4 Skin tests for diagnosing type I hypersensitivity.** The presence of redness and swelling at an injection site indicates sensitivity to the allergen injected at that site.

kit so that they can inject themselves before their allergic reactions become life threatening.

Type II (Cytotoxic) Hypersensitivity

LEARNING OUTCOMES

18.5 Discuss the mechanisms underlying transfusion reactions.
18.6 Construct a table comparing the key features of the four blood types in the ABO blood group system.
18.7 Describe the mechanisms and treatment of hemolytic disease of the newborn.

The second major form of hypersensitivity results when cells are destroyed by an immune response—typically by the combined activities of complement and antibodies. This *cytotoxic hypersensitivity* is part of many autoimmune diseases, which we will discuss later in this chapter, but the most significant examples of type II hypersensitivity are the destruction of donor red blood cells following an incompatible blood transfusion and the destruction of fetal red blood cells. We begin our discussion by focusing on these two manifestations of type II hypersensitivity.

The ABO System and Transfusion Reactions

Red blood cells have many different glycoprotein and glycolipid molecules on their surface. Some surface molecules of red blood cells, called **blood group antigens**, have various functions, including transportation of glucose and ions across the cytoplasmic membrane.

There are several sets of blood group antigens that vary in complexity. The ABO group system is most famous and consists of just two antigens arbitrarily given the names A antigen and B antigen. Each person's red blood cells have either A antigen, B antigen, both A and B antigens, or neither antigen. Individuals with neither antigen are said to have blood type O.

As you probably know, blood can be transfused from one person to another; however, if blood is transfused to an individual with a different blood type, then the donor's blood group antigens may stimulate the production of antibodies in the recipient. These bind to and eventually destroy the transfused cells. The result can be a potentially life-threatening *transfusion reaction*. Note that it is a blood recipient's own immune system that causes problems; the donated cells merely trigger the response.

Transfusion reactions, the most problematic of which involve the ABO group, develop as follows:

- If the recipient has preexisting antibodies to foreign antigens, then the donated blood cells will be destroyed immediately—either the antibody-bound cells will be phagocytized by macrophages and neutrophils, or the antibodies will agglutinate the cells and complement will rupture them, a process called *hemolysis* (**FIGURE 18.5**). Hemolysis releases hemoglobin into the bloodstream, which may cause severe kidney damage. At the same time, the membranes of the ruptured blood cells trigger blood

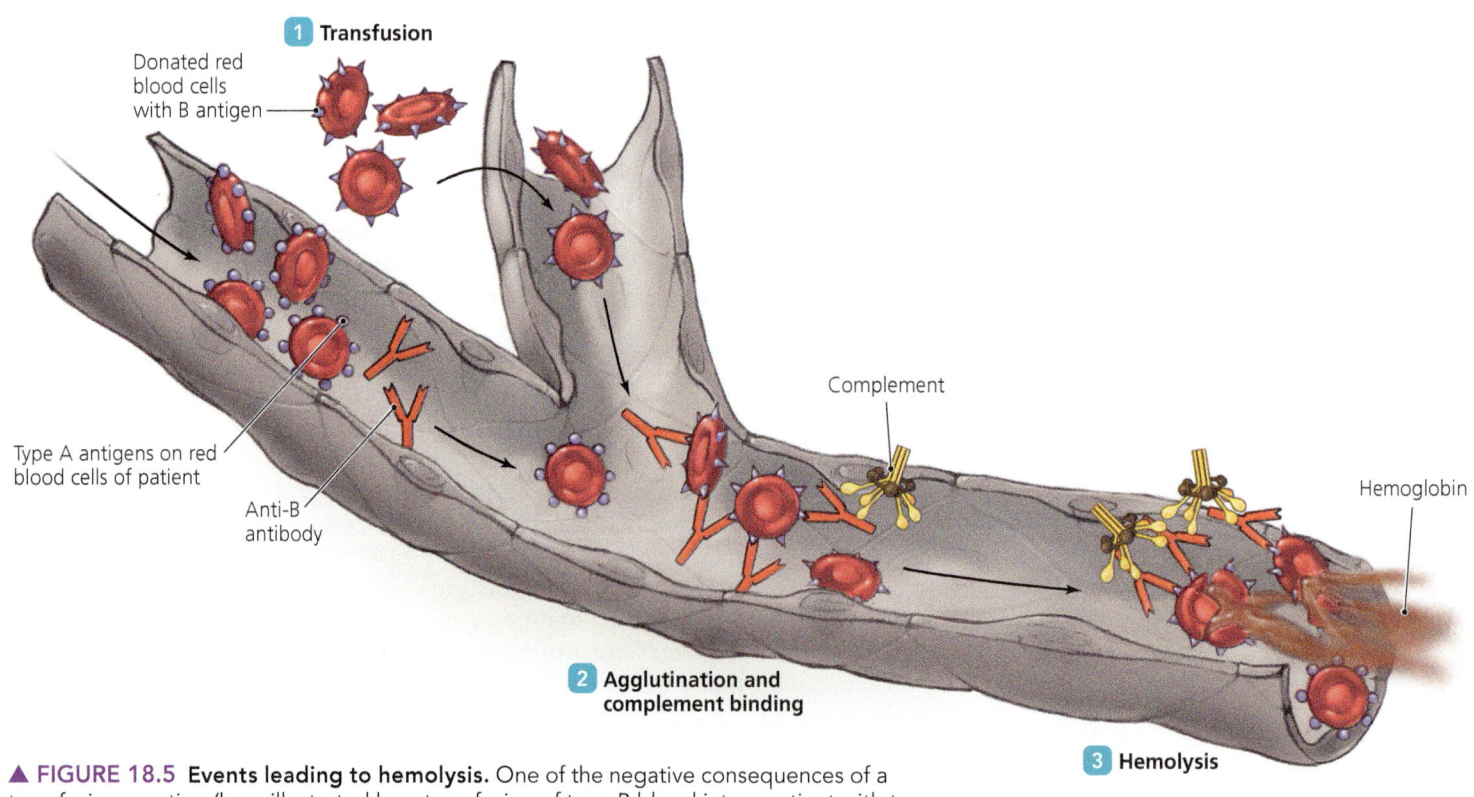

▲ **FIGURE 18.5 Events leading to hemolysis.** One of the negative consequences of a transfusion reaction (here illustrated by a transfusion of type B blood into a patient with type A blood): When antibodies against a foreign ABO antigen combine with the antigen on transfused red blood cells ❶, the cells are agglutinated, and the complexes bind complement ❷. The resulting hemolysis releases large amounts of hemoglobin into the bloodstream ❸ and produces additional negative consequences throughout the body.

clotting within blood vessels, blocking them and causing circulatory failure, fever, difficulty in breathing, coughing, nausea, vomiting, and diarrhea. If the patient survives, recovery follows the elimination of all the foreign red blood cells.

- In most people, production of antibodies against foreign ABO antigens is stimulated not by exposure to foreign blood cells but instead by exposure to antigenically similar molecules found on a wide range of plants and bacteria. Individuals encounter and make antibodies against these antigens on a daily basis, but only upon receipt of a mismatched blood transfusion do the antigens cause problems.

- If the recipient has no preexisting antibodies to the foreign blood group antigens, then the transfused cells circulate and function normally, but only for a while—that is, until the recipient's immune system mounts a primary response against the foreign antigens and produces enough antibody to destroy the foreign cells. This happens gradually over a long enough time that the severe symptoms and signs mentioned above do not occur.

To prevent transfusion reactions, laboratory personnel must cross-match ABO blood types between donors and recipients. Before a recipient receives a blood transfusion, the red cells and serum of the donor can be mixed with the serum and red cells of the recipient. If any signs of clumping are seen, then that blood is not used, and an alternative donor is sought. TABLE 18.2 presents the characteristics of the ABO blood group and the compatible donor-recipient matches that can be made.

The Rh System and Hemolytic Disease of the Newborn

Many decades ago, researchers discovered the existence of an antigen common to the red blood cells of humans and rhesus monkeys. They called this antigen, which transports anions and glucose across the cytoplasmic membrane, *rhesus* (rē'sŭs) antigen, or **Rh antigen**. Laboratory analysis of human blood samples eventually showed that the Rh antigen is present on the red blood cells of about 85% of humans; that is, about 85% of the population is *Rh positive* (Rh+), and about 15% of the population is *Rh negative* (Rh−).

In contrast to the situation with ABO antigens, preexisting antibodies against Rh antigen do not occur. Although a transfusion reaction can occur in an Rh-negative patient who receives more than one transfusion of Rh-positive blood, such a reaction is usually minor because Rh antigen molecules are less abundant than A or B antigens. Instead, the primary problem posed by incompatible Rh antigen is the risk of **hemolytic[4] disease of the newborn** (hē-mō-lit'ik; FIGURE 18.6).

This hypersensitivity reaction develops when an Rh− mother is pregnant with an Rh+ baby (who inherited an Rh gene from the father). Normally, the placenta keeps fetal red blood cells separate from the mother's blood, so fetal cells do not enter the mother's bloodstream. However, in 20–50% of pregnancies—especially during the later weeks of pregnancy, during a clinical or spontaneous abortion, or during childbirth—fetal red blood cells escape into the mother's blood. The Rh− mother's immune system recognizes these Rh+ fetal cells as foreign and initiates an antibody immune response by developing antibodies against the Rh antigen. Initially, only IgM antibodies are produced, and because IgM is a very large molecule that cannot cross the placenta, no problems arise during this first pregnancy.

However, IgG antibodies can cross the placenta, so if during a subsequent pregnancy the fetus is Rh+, anti-Rh IgG molecules produced by the mother cross the placenta and destroy the fetus's Rh+ red blood cells. Such destruction may be limited, or it may be severe enough—especially in a third or subsequent pregnancy—to cause grave problems. One classic feature of hemolytic disease of the newborn is severe *jaundice* from excessive bilirubin (bil-i-roo'bin), which is a yellowish blood pigment released during the degradation of hemoglobin from lysed red blood cells. The liver is normally responsible for removing bilirubin, but because of the immaturity of the fetal liver and overload from the hemolytic process, the bilirubin may instead be deposited in the brain, causing severe neurological damage or death during the last weeks of pregnancy or shortly after the baby's birth.

In the past, when prevention of hemolytic disease of the newborn was impossible, this terrible disease occurred in about 1 of every 300 births. Today, however, physicians can routinely and drastically reduce the number of cases of the disease by administering anti-Rh immunoglobulin (IgG), called *RhoGAM*, to Rh-negative women at 28 weeks into their pregnancy and also within 72 hours following abortion, miscarriage, or childbirth. Any fetal red cells that may have entered the mother's body are destroyed by RhoGAM before the fetal cells can trigger an immune response. As a result, sensitization of the mother does not occur, and future pregnancies are safer.

TABLE 18.2 ABO Blood Group Characteristics and Donor-Recipient Matches

ABO Blood Group	ABO Antigen(s) Present	Antibodies Present	Can Donate To	Can Receive From
A	A	Anti-B	A or AB	A or O
B	B	Anti-A	B or AB	B or O
AB	A and B	None	AB	A, B, AB, or O (universal recipient)
O	None	Both anti-A and anti-B	A, B, AB, or O (universal donor)	O

---

[4]From Greek *haima*, meaning "blood," and *lysis*, meaning "destruction."

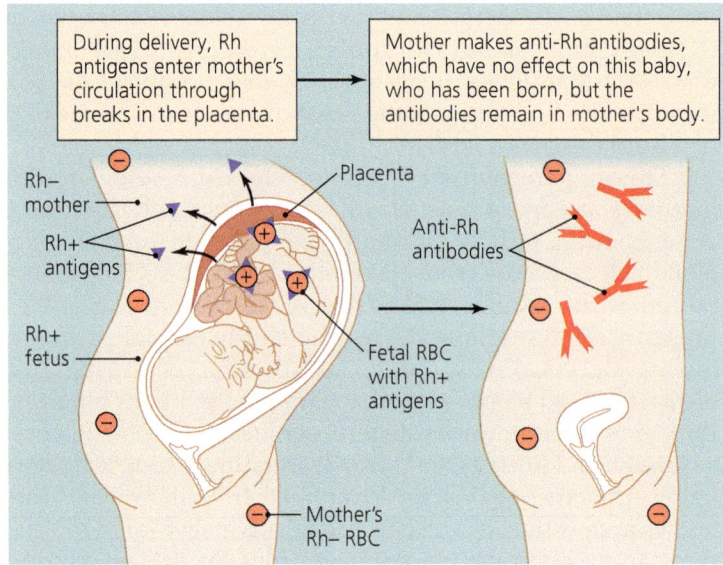

(a) First pregnancy of Rh– mother with Rh+ baby

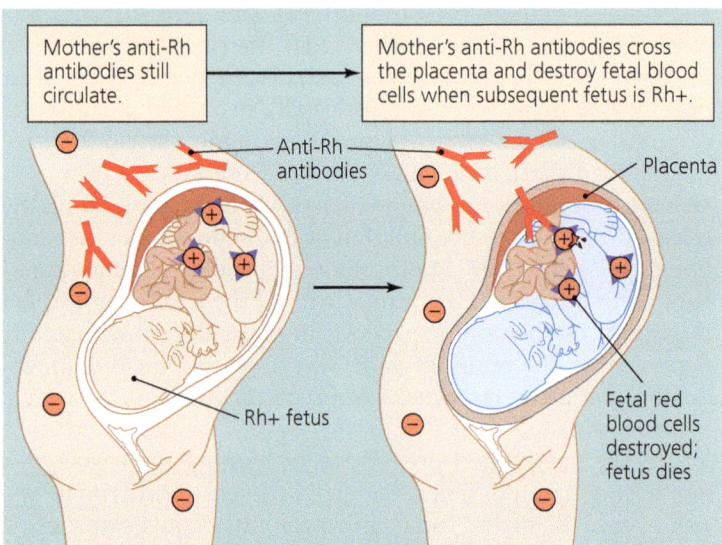

(b) Subsequent pregnancy of Rh– mother with Rh+ baby

▲ FIGURE 18.6 **Events in the development of hemolytic disease of the newborn.** (a) During an initial pregnancy, Rh red blood cells from the fetus enter the Rh– mother's circulation, often during childbirth. As a result, the mother produces anti-Rh antibodies. (Antigens and antibodies are not shown to scale.) (b) During a subsequent pregnancy with another Rh child, anti-Rh IgG molecules cross the placenta and trigger destruction of the fetus's red blood cells. *How is it possible that the child of an Rh– mother is Rh+?*

Figure 18.6 *The father of the child is Rh-positive.*

MICRO CHECK

1. Is it possible to have an allergic reaction to pollen upon the first exposure? Why or why not?
2. Which type of antibody is involved in a type I hypersensitivity response?
3. Which chemical released during type I hypersensitivity responses causes contractions of smooth muscle, such as in the bronchi?
4. What type of ABO blood group antibodies does a person with blood type B typically have?

Type III (Immune Complex–Mediated) Hypersensitivity

LEARNING OUTCOMES

18.8 Outline the basic mechanism of type III hypersensitivity.
18.9 Describe hypersensitivity pneumonitis and glomerulonephritis.
18.10 List the signs and symptoms of rheumatoid arthritis.
18.11 Discuss the cause and signs of systemic lupus erythematosus.

The formation of complexes of antigen bound to antibody, called **immune complexes**, initiates several molecular processes, including complement activation. Normally, immune complexes are removed from the body via phagocytosis. However, in *type III (immune complex–mediated) hypersensitivity* reactions, small immune complexes escape phagocytosis and circulate in the bloodstream until they become trapped in organs, joints, and tissues (such as the walls of blood vessels). In these sites, they react with complement, resulting in inflammation, and cause neutrophils to release enzymes that damage nearby tissues. FIGURE 18.7 illustrates the mechanism of type III reactions.

Type III hypersensitivities may be localized or may affect a number of body systems simultaneously. There is no cure for these diseases, though steroids that suppress the immune system may provide some relief. Two localized conditions resulting from immune complex–mediated hypersensitivity are hypersensitivity pneumonitis and glomerulonephritis; two systemic disorders are rheumatoid arthritis and systemic lupus erythematosus.

Hypersensitivity Pneumonitis

Type III hypersensitivities can affect the lungs, causing a form of pneumonia called **hypersensitivity pneumonitis** (noo-mō-nī'tus). Individuals become sensitized when minute mold spores or other antigens are inhaled deep into the lungs, stimulating the production of antibodies. A hypersensitivity reaction occurs when the subsequent inhalation of the same antigen stimulates the formation of immune complexes that then activate complement.

One form of hypersensitivity pneumonitis, called *farmer's lung*, occurs in farmers chronically exposed to spores from moldy hay. Many other syndromes in humans develop via a similar mechanism and are usually named after the source of the inhaled allergen. Thus, *pigeon breeder's lung* arises following exposure to the dust from pigeon feces, *mushroom grower's lung* is a response to the soil or fungal spores encountered in the

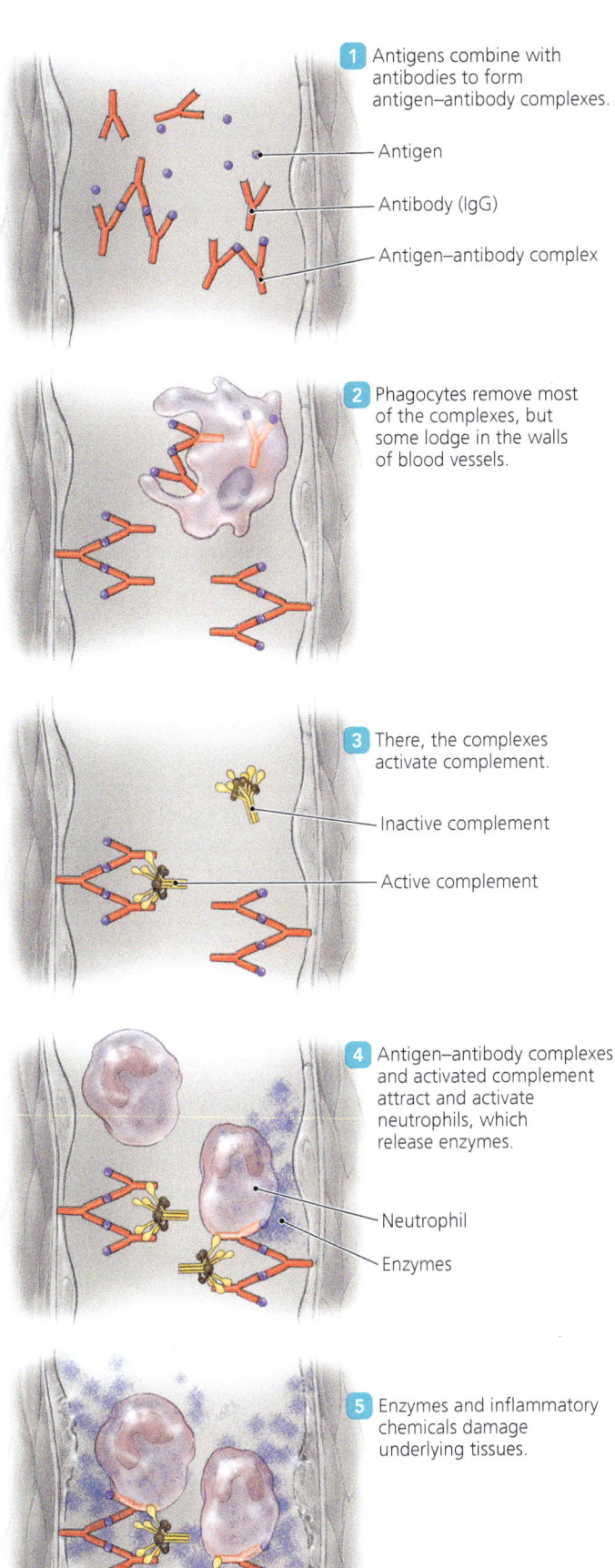

◀ **FIGURE 18.7 The mechanism of type III (immune complex-mediated) hypersensitivity.** After immune complexes form 1, some complexes are not phagocytized and lodge in tissues (in this case, the wall of a blood vessel) 2. These complexes then activate complement 3, which leads to inflammation and attracts neutrophils 4. The release of enzymes from these neutrophils leads to the destruction of tissue 5.

growing of mushrooms, and *librarian's lung* results from inhaling dust from old books.

Glomerulonephritis

Glomerulonephritis (glō-mār'ū-lō-nef-rī'tis) occurs when immune complexes circulating in the bloodstream are deposited in the walls of *glomeruli*, which are networks of minute blood vessels in the kidneys. Immune complexes damage the glomerular cells, leading to enhanced local production of cytokines that trigger nearby cells to produce more of the proteins that underlie the cells, impeding blood filtration. Sometimes, the immune complexes are deposited in the center of the glomeruli, where they stimulate local cells to divide and compress nearby blood vessels, again interfering with kidney function. The net result is kidney failure; the glomeruli lose their ability to filter wastes from the blood, ultimately resulting in death.

Rheumatoid Arthritis (RA)

Rheumatoid arthritis (RA) (roo'mă-toyd ar-thrī'tis) is a type III hypersensitivity that is also an autoimmune disease—the immune system attacks the body's own antigens. RA commences when B cells secrete IgM that binds to certain IgG molecules. IgM–IgG complexes are deposited in the joints, where they activate complement and mast cells, which release inflammatory chemicals. The resulting inflammation causes the tissues to swell, thicken, and proliferate into the joint, resulting in severe pain. As the altered tissue extends into the joint, the inflammation further erodes and destroys joint cartilage and the neighboring bony structure until the joints begin to break down and fuse; as a result, affected joints become distorted and lose their range of motion **(FIGURE 18.8)**. The course of rheumatoid arthritis is often intermittent; however, with each recurrence, the lesions and damage get progressively more severe.

The trigger of RA is not well understood. The fact that there are no animal models (because the disease appears to affect humans only) significantly hinders research on its cause. Many cases demonstrate that RA commonly follows an infectious disease in a genetically susceptible individual. Possession of certain immunity (MHC) genes appears to increase susceptibility.

Physicians treat rheumatoid arthritis by administering anti-inflammatory drugs such as ibuprofen to prevent additional joint damage and immunosuppressive drugs to inhibit the antibody immune response.

Systemic Lupus Erythematosus (SLE)

Another example of type III hypersensitivity that is also an autoimmune disorder is **systemic lupus erythematosus (SLE)**

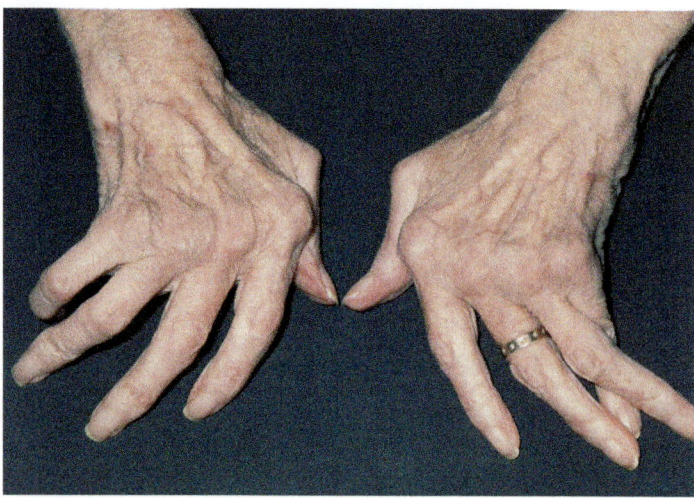

▲ FIGURE 18.8 The crippling distortion of joints characteristic of rheumatoid arthritis.

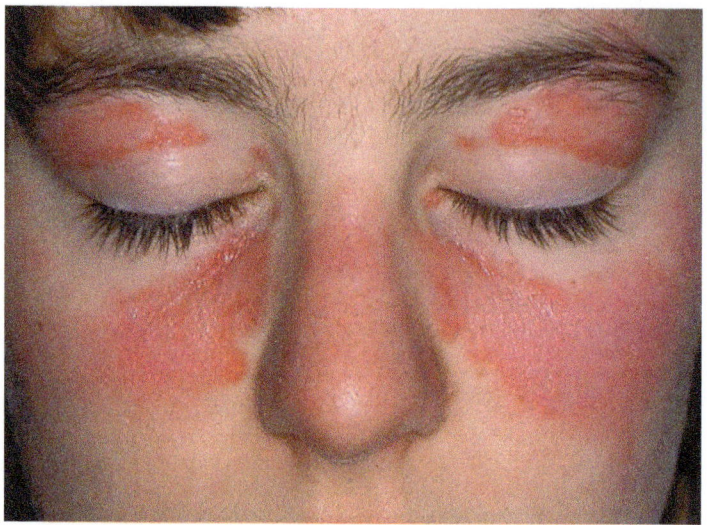

▲ FIGURE 18.9 The characteristic facial rash of systemic lupus erythematosus. Its shape corresponds to areas most exposed to sunlight, which worsens the condition.

(loo′pŭs er-ĭ-thē′mă-tō-sus), often shortened to *lupus*. Patients with lupus produce antibodies against numerous self-antigens found in normal organs and tissues. As a result, the immune system attacks multiple organs, giving rise to many different pathological lesions and clinical manifestations. One consistent feature of SLE is the development of such self-reactive antibodies, called *autoantibodies*, against nucleic acids, especially DNA. These autoantibodies combine with free DNA released from dead cells to form immune complexes that are deposited in glomeruli, causing glomerulonephritis and kidney failure. Thus, SLE is an immune complex-mediated hypersensitivity reaction. Immune complexes may also be deposited in joints, where they give rise to arthritis.

The disease's curious name—systemic lupus erythematosus—stems from two features of the disease: *Systemic* simply reflects the fact that it affects different organs throughout the body, *lupus* is Latin for "wolf," and *erythematosus* refers to a redness of the skin, so these last two words describe the characteristic red, butterfly-shaped rash that develops on the face of many patients, giving them what is sometimes described as a wolflike appearance **(FIGURE 18.9)**. This rash is caused by deposition of nucleic acid–antibody complexes in the skin and is worse in skin areas exposed to sunlight.

Although autoantibodies to nucleic acids are characteristic of SLE, many other autoantibodies are also produced. Autoantibodies to red blood cells cause hemolytic anemia, autoantibodies to platelets give rise to bleeding disorders, antilymphocyte antibodies alter immune reactivity, and autoantibodies against muscle cells cause muscle inflammation and, in some cases, damage to the heart. Because of its variety of symptoms, SLE can be misdiagnosed.

The trigger of lupus is unknown, though some drugs can induce a lupus-like disease. It is likely that SLE has many causes. Physicians treat lupus with immunosuppressive drugs that reduce autoantibody formation and with glucocorticoids that reduce the inflammation associated with the deposition of immune complexes. Scientists are currently testing a battery of novel drugs for treating lupus.

Type IV (Delayed or Cell-Mediated) Hypersensitivity

LEARNING OUTCOMES

18.12 Outline the mechanism of type IV hypersensitivity.
18.13 Describe the significance of the tuberculin test.
18.14 Identify four types of grafts.
18.15 Compare four types of drugs commonly used to prevent graft rejection.

When certain antigens contact the skin of sensitized individuals, they provoke inflammation that begins to develop at the site only after 12–24 hours. Such **delayed hypersensitivity reactions** result not from the action of antibodies but rather from interactions among antigen, antigen-presenting cells, and T cells; thus, a type IV reaction is also called a **cell-mediated hypersensitivity reaction**. The delay in this cell-mediated response reflects the time it takes for macrophages and T cells to migrate to and divide at the site of the antigen. We begin our discussion of type IV reactions by considering two common examples: the tuberculin response and allergic contact dermatitis. Then we will consider two type IV hypersensitivity reactions involving the interactions between the body and tissues grafted to (transplanted into) it—graft rejection and graft-versus-host disease—before considering donor–recipient matching and tissue typing.

The Tuberculin Response

The **tuberculin response** (too-ber′kyū-lin) is an important example of a delayed hypersensitivity reaction in which the skin of an individual exposed to tuberculosis (TB) or tuberculosis vaccine reacts to a shallow injection of *tuberculin*—a protein solution obtained from *Mycobacterium tuberculosis* (mī′kō-bak-tēr′ ē-ŭm too-ber-kyū-lō′sis). Health care providers use the tuberculin test, also called a *Mantoux test* (mahn-too′) after the French physician who perfected it, to diagnose contact with antigens of *M. tuberculosis*.

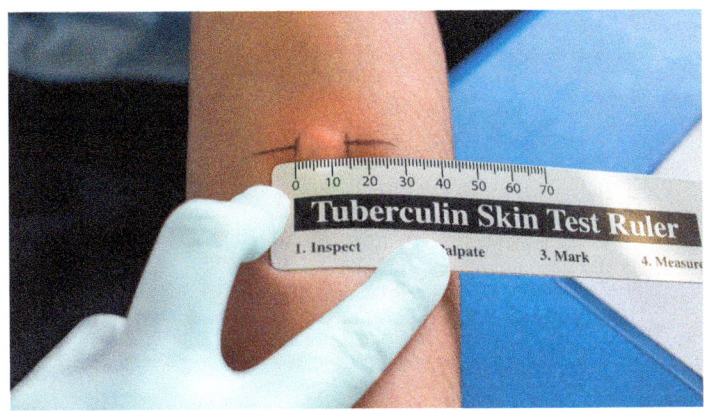

▲ FIGURE 18.10 **A positive tuberculin test, a type IV hypersensitivity response.** The hard, red swelling 10 mm or greater in diameter that is characteristic of the tuberculin response indicates that the individual has been immunized against *Mycobacterium tuberculosis* or is now or has previously been infected with the bacterium.

When tuberculin is injected into the skin of a healthy individual who has never been infected with TB nor immunized against it, no response occurs. In contrast, when tuberculin is injected into someone currently or previously infected with *M. tuberculosis* or an individual previously immunized with tuberculosis vaccine, a red, hard swelling (10 mm or greater in diameter) indicating a positive tuberculin test develops at the site (FIGURE 18.10). It isn't possible to determine whether a positive response is due to current infection, previous infection, or immunization. Inflammation in response to tuberculin reaches its greatest intensity within 24–72 hours and may persist for several weeks before fading. Microscopic examination of the lesion reveals that it is infiltrated with lymphocytes and macrophages.

A tuberculin response is mediated by memory T cells. When an individual is first infected by or immunized against *M. tuberculosis*, the resulting cell-mediated immune response generates memory T cells that persist in the body. When a sensitized individual is later injected with tuberculin, phagocytic cells migrate to the site and attract memory T cells, which secrete a mixture of cytokines that attract still more T cells and macrophages, giving rise to a slowly developing inflammation. The macrophages ingest and destroy the injected tuberculin, allowing the tissues to eventually return to normal.

Allergic Contact Dermatitis

Urushiol (ŭ-rū′shē-ŏl), the oil of poison ivy (*Toxicodendron radicans*, toks′si-kō-den′dron rā′dē-kanz) and related plants, is a small molecule that becomes antigenic when it binds to almost any protein it contacts—including proteins in the skin of most people who rub against the plant. The body regards these chemically modified skin proteins as foreign, triggering a cell-mediated immune response and resulting in an intensely irritating skin rash called **allergic contact dermatitis** (der-mă-tī′tis). In severe cases, cytotoxic T lymphocytes (Tc cells) destroy so many skin cells that acellular, fluid-filled blisters develop (FIGURE 18.11).

Other small molecules that don't trigger an immune response by themselves combine chemically with skin proteins and induce allergic contact dermatitis. Examples include formaldehyde; some cosmetics, dyes, drugs, and metal ions; and chemicals used in the production of latex for hospital gloves and tubing.

Because T cells mediate allergic contact dermatitis, epinephrine and other drugs used to treat immediate hypersensitivity reactions are ineffective. T cell activities and inflammation can, however, be suppressed by corticosteroid treatment. Good strategies for dealing with exposure to poison ivy include washing the area thoroughly and immediately with a strong soap and washing all exposed clothes as soon as possible.

CLINICAL CASE STUDY

The First Time's Not the Problem

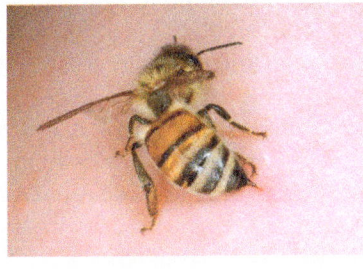

Steven, an eight-year-old boy, is brought to your office Monday morning by his father to have his upper arm checked for a possible infection. Dad is worried because the area of a bee sting on the boy's arm is getting redder, itchier, and more tender. The father gave him some children's acetaminophen yesterday, which relieved the discomfort somewhat.

There is no history of medical problems or allergies, and the child takes no medication regularly. He is otherwise feeling well, and his father tells you he is playing and eating normally. There is no previous history of bee stings, and Steven proudly tells you he "hardly even cried" when he got stung.

There is a half-dollar-sized area on his left upper arm that is puffy and red, but there is no streaking or drainage, and the area does not appear to be infected. Steven's temperature is normal, and his lungs are clear.

1. What type of hypersensitivity reaction is Steven manifesting?
2. What other over-the-counter medication might relieve the itching?
3. What mechanism is causing the signs and symptoms you are seeing?
4. Since the area is not infected, what future health risk for his son should the father be made aware of?
5. What can be done to determine future risk from a bee sting?
6. How would you recognize a severe allergic reaction?
7. What precautions can the family take to protect the boy from future reactions?

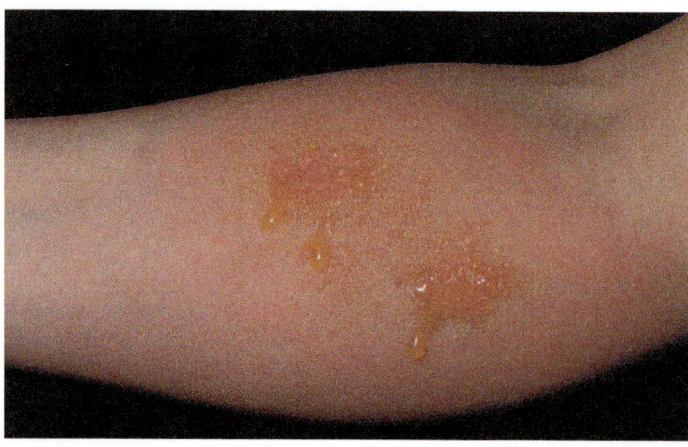

▲ FIGURE 18.11 **Allergic contact dermatitis, a type IV hypersensitivity response.** The response in this case is to poison ivy. Note the large, acellular, fluid-filled blisters that result from the destruction of skin cells.

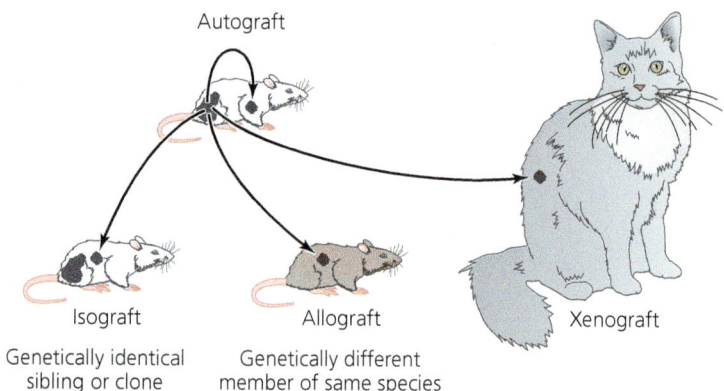

▲ FIGURE 18.12 **Types of grafts.** The names are based on the degree of relatedness between donor and recipient. Autografts are grafts moved from one location to another within a single individual. In isografts, the donor and recipient are either genetically identical siblings or clones. In allografts, the donor and recipient are genetically distinct individuals of the same species. In xenografts, the donor and recipient are of different species.

Graft Rejection

A special case of type IV hypersensitivity is the rejection of **grafts** (or transplants), which are tissues or organs, such as livers, kidneys, or hearts, that have been transplanted, whether between sites within an individual or between a donor and an unrelated recipient. Even though advances in surgical technique enable surgeons to move grafts freely from site to site, grafts recognized as foreign by a recipient's immune system may undergo **graft rejection**, a highly destructive phenomenon that can severely limit the success of organ and tissue transplantation. Graft rejection is a normal immune response against foreign major histocompatibility complex (MHC) proteins on the surface of graft cells. The likelihood of graft rejection depends on the degree to which the graft is foreign to the recipient, which in turn is related to the type of graft.

Scientists name graft types according to the degree of relatedness between the donor and the recipient (FIGURE 18.12). A graft is called an **autograft** (aw'tō-graft) when tissues are moved to a different location within the same individual. Autografts do not trigger immune responses because they do not express foreign antigens. Examples of autografts include the grafting of skin from one area of the body to another to cover a burn area or the use of a leg vein to bypass blocked coronary arteries.

Isografts (ī'sō-graftz) are grafts transplanted between two genetically identical individuals, that is, between identical siblings or clones. Because these individuals have identical MHC proteins, the immune system of the recipient cannot differentiate between the grafted cells and its own normal body cells. As a result, isografts are not rejected.

Allografts (al'ō-graftz) are grafts transplanted between genetically distinct members of the same species. Most grafts performed in humans are allografts. Because the MHC proteins of the allograft are different from those of the recipient, allografts typically induce a strong type IV hypersensitivity, resulting in graft rejection. Rejection must be stopped with immunosuppressive drugs if the graft is to survive.

Xenografts (zēn'ō-graftz) are grafts transplanted between individuals of different species. Thus, the transplant of a baboon's heart into a human is a xenograft. Because xenografts are usually very different from the tissues of the recipient, both biochemically and immunologically, they usually provoke a rapid, intense rejection that is very difficult to suppress. Therefore, xenografts from mature animals are not commonly used therapeutically.

Graft-Versus-Host Disease

Physicians often use bone marrow allografts as a component of the treatment for leukemias and lymphomas (cancers of leukocytes). In this procedure, physicians use total body irradiation combined with cytotoxic drugs to kill tumor cells in the patient's bone marrow. In the process, the patient's existing leukocytes are also destroyed, which completely eliminates the body's ability to mount any kind of immune response. Physicians then inject the patient with donated bone marrow, which produces a new set of leukocytes. Ideally, within a few months, a fully functioning bone marrow is restored.

Unfortunately, when they are transplanted, donated bone marrow T cells may regard the patient's cells as foreign, mounting an immune response against them and giving rise to a condition called **graft-versus-host disease**. If the donor and recipient mainly differ in MHC class I molecules, the grafted T cells attack all of the recipient's tissues, producing especially destructive lesions in the skin and intestine. If the donor and recipient differ mainly in MHC class II molecules, then the grafted T cells attack the antigen-presenting cells of the host, leading to immunosuppression and leaving the recipient vulnerable to infections. The same immunosuppressive drugs used to prevent graft rejection (discussed shortly) can limit graft-versus-host disease.

Donor–Recipient Matching and Tissue Typing

Although it is usually not difficult to ensure that donor and recipient have identical blood groups, MHC compatibility is much harder to achieve. The reason is the very high degree of

MHC variability, which ensures that unrelated individuals differ widely in their MHCs. In general, the more closely donor and recipient are related, the smaller their MHC difference. Given that in most cases an identical sibling is not available, it is usually preferable that grafts be donated by a parent or sibling possessing MHC antigens similar to those of the recipient.

In practice, of course, closely related donors are not always available. And even though a paired organ (such as a kidney) or a portion of an organ (such as liver tissue) may be available from a living donor, unpaired organs (such as the heart) almost always come from an unrelated cadaver. In such cases, attempts are made to match donor and recipient as closely as possible by means of *tissue typing*. Physicians examine the white cells of potential graft recipients to determine what MHC proteins they have. When a donor organ becomes available, it too is typed. Then the potential recipient whose MHC proteins most closely match those of the donor is chosen to receive the graft. Though a perfect match is rarely achieved, the closer the match, the less intense the rejection process and the greater the chance of successful grafting. A match of 50% or more of the MHC proteins is usually acceptable for most organs, but near absolute matches are required for successful bone marrow transplants.

TABLE 18.3 summarizes the major characteristics of the four types of hypersensitivity.

Next, we turn our attention to the actions of various types of immunosuppressive drugs used in situations in which the immune system is overactive.

The Actions of Immunosuppressive Drugs

The development of potent immunosuppressive drugs has played a large role in the dramatic success of modern transplantation procedures. These drugs can also be effective in combating certain autoimmune diseases (discussed shortly). In the following discussion, we consider four classes of immunosuppressive drugs: glucocorticoids, cytotoxic drugs, cylosporine, and lymphocyte-depleting therapies.

Glucocorticoids, sometimes called *corticosteroids* and commonly referred to as *steroids*, have been used as immunosuppressive agents for many years. Glucocorticoids such as *prednisone* (pred'ni-sōn) and *methylprednisolone* suppress the response of T cells to antigen and inhibit such mechanisms as T cell cytotoxicity and cytokine production. These drugs have a much smaller effect on B cell function.

Cytotoxic drugs inhibit mitosis and cytokinesis (cell division). Given that lymphocyte proliferation is a key feature of specific immunity, blocking cellular reproduction is a powerful, although very nonspecific, method of immunosuppression. Among the cytotoxic drugs that have been used, the following are noteworthy:

- *Cyclophosphamide* (sī-klō-fos'fă-mīd) cross-links daughter DNA molecules in mitotic cells, preventing their separation and blocking mitosis. It impairs both B cell and T cell responses.
- *Azathioprine* (ā-ză-thī'ō-prēn) is a purine analog; it competes with purines during the synthesis of nucleic acids, thus blocking DNA replication and suppressing both primary and secondary antibody responses.

Three other cytotoxic drugs used for immunosuppression include *mycophenolate mofetil*, which inhibits purine synthesis, and *brequinar sodium* and *leflunomide*, each of which inhibits pyrimidine synthesis and thereby inhibits cellular replication.

Drugs such as **cyclosporine** (sī-klō-spōr'ēn), a polypeptide derived from fungi, prevent production of interleukins and interferons by T cells, thereby blocking Th1 responses. Because cyclosporine acts only on activated T cells and has no effect on resting T cells, it is far less toxic than the nonspecific drugs previously described. When it is given to prevent allograft rejection, only activated T cells attacking the graft are suppressed. Because steroids have a similar effect, the combination of glucocorticoids and cyclosporine is especially potent and can enhance survival of allografts.

Scientists have developed techniques involving relatively specific *lymphocyte-depleting therapies* in an attempt to reduce the many adverse side effects associated with the use of less specific immunosuppressive drugs. One technique involves administering an antiserum called *antilymphocyte globulin*, which is specific for lymphocytes. Another, more specific, antilymphocyte technique uses monoclonal antibodies against CD3, which is found only on T cells. An even more specific monoclonal antibody is directed against the interleukin 2 receptor (IL-2R), which is expressed mainly on activated T cells. Such immunosuppressive

TABLE 18.3 The Characteristics of the Four Types of Hypersensitivity Reactions

Descriptive	Name	Cause	Time Course	Characteristic Cells Involved
Type I	Immediate hypersensitivity	Antibody (such as IgE) on sensitized cells' membranes binds antigen, causing degranulation	After initial sensitization, seconds to minutes	Previously sensitized mast cells or basophils
Type II	Cytotoxic hypersensitivity	Antibodies and complement lyse target cells	Minutes to hours	Plasma cells secrete antibodies that act against other body cells
Type III	Immune complex-mediated hypersensitivity	Nonphagocytized complexes of antibodies and antigens trigger complement activation, leading to inflammation, and cause neutrophils to release damaging enzymes	Several hours	Neutrophils
Type IV	Delayed (cell-mediated) hypersensitivity	T cells attack the body's cells	Several days	Activated T cells

TABLE 18.4 The Four Classes of Immunosuppressive Drugs

Class	Examples	Action
Glucocorticoids	Prednisone, methylprednisolone	Anti-inflammatory; kill T cells
Cytotoxic drugs	Cyclophosphamide, azathioprine, mycophenolate mofetil, brequinar sodium, leflunomide	Block cell division nonspecifically
Cyclosporine	Cyclosporine	Blocks T cell responses
Lymphocyte-depleting therapies	Antilymphocyte globulin, monoclonal antibodies	Kill T cells nonspecifically, or kill activated T cells, or inhibit IL-2 reception

therapies are effective in reversing graft rejection. Because they target a narrow range of cells, they produce fewer undesirable side effects than less specific drugs.

TABLE 18.4 lists some drugs and their actions for each of the four classes of the immunosuppressive drugs.

MICRO CHECK

5. What causes cell damage in a type III hypersensitivity?
6. What are the differences between hypersensitivity pneumonitis, rheumatoid arthritis, glomerulonephritis, and systemic lupus erythematosus?
7. What is the reason for the delayed response in a type IV hypersensitivity?
8. If a woman receives a kidney from her brother, what type of transplantation has occurred?

TELL ME WHY

During the war in Afghanistan in 2012, an army corporal with type AB blood received a life-saving blood transfusion from his sergeant, who had type O blood. Later, the sergeant was involved in a traumatic accident and needed blood desperately. The corporal wanted to help but was told his blood was incompatible. Explain why the corporal could receive blood from but could not give blood to the sergeant.

Autoimmune Diseases

Just as today's military must control its arsenal of sophisticated weapons to avert losses of its own soldiers from "friendly fire," the immune system must be carefully regulated so that it does not damage the body's own tissues. However, an immune system does occasionally produce antibodies and cytotoxic T cells that target normal body cells—a phenomenon called *autoimmunity*. Although such responses are not always damaging, they can give rise to **autoimmune diseases**, and some of these hypersensitivities are life threatening.

Causes of Autoimmune Diseases

LEARNING OUTCOME

18.16 Briefly discuss eight hypotheses concerning the causes of autoimmunity and autoimmune diseases.

Most autoimmune diseases appear to develop spontaneously and at random. Nevertheless, scientists have noted some common features of autoimmune disease. For example, they occur more often in older individuals, and they are also much more common in women than in men, although the reasons for this gender difference are unclear.

Hypotheses to explain the etiology of autoimmunity abound. They include the following:

- Estrogen may stimulate the destruction of tissues by cytotoxic T cells.
- During pregnancy, some maternal cells may cross the placenta and colonize the fetus. These cells are more likely to survive in a daughter than in a son and might trigger an autoimmune disease later in the daughter's life.
- Conversely, fetal cells may also cross the placenta and trigger autoimmunity in the mother.
- Environmental factors may contribute to the development of autoimmune disorders. Some autoimmune diseases—type 1 diabetes mellitus (dī-ă-bē′tēz mĕ-lī′tĕs) and rheumatoid arthritis, for example—develop in a few patients following their recovery from viral infections, though other individuals who develop these diseases have no history of such viral infections.
- Genetic factors may play a role in autoimmune diseases. MHC genes that in some way promote autoimmunity are found in individuals with autoimmune diseases, whereas MHC genes that somehow protect against autoimmunity may dominate in other individuals. MHC genes may also trigger autoimmune disease by preventing the elimination of some self-reactive T cells in the thymus.
- Some autoimmune diseases develop when T cells encounter self-antigens that are normally "hidden" in sites where T cells rarely go. For example, because sperm develop within the testes during puberty, long after the body has selected its T cell population, men may have T cells that recognize their own sperm as foreign. This is normally of little consequence because sperm are sequestered from the blood. But if the testes are injured, T cells may enter the site of damage and mount an autoimmune response against the sperm, resulting in infertility.
- Infections with a variety of microorganisms may trigger autoimmunity as a result of **molecular mimicry**, which occurs when an infectious agent has an epitope that is very similar or identical to a self-antigen. In responding to the invader, the body produces antibodies that are

autoantibodies (antibodies against self-antigens), which damage body tissues. For example, children infected with some strains of *Streptococcus pyogenes* (strep-tō-kok'ŭs pī-oj'en-ēz) may produce antibodies to heart muscle and so develop heart disease. Other strains of streptococci trigger the production of antibodies that cross-react with glomerular basement membranes and so cause kidney disease. It is possible that some virally triggered autoimmune diseases also result from molecular mimicry.

- Other autoimmune responses may result from failure of the normal control mechanisms of the immune system. For example, harmful, self-reactive T lymphocytes are normally destroyed via apoptosis triggered by a cell-surface receptor; however, if defects in the receptor are present, they can permit abnormal T cells to survive and cause disease.

Examples of Autoimmune Diseases

LEARNING OUTCOME

18.17 Describe a serious autoimmune disease associated with each of the following: blood cells, endocrine glands, nervous tissue, and connective tissue.

Regardless of the specific mechanism that causes an autoimmune disease, immunologists categorize them into two major groups: systemic autoimmune diseases such as lupus (discussed previously) and single-organ autoimmune diseases, which affect a single organ or tissue. Among the many single-organ autoimmune diseases recognized, common examples affect blood cells, endocrine glands, nervous tissue, or connective tissue.

Autoimmunity Affecting Blood Cells

Individuals with **autoimmune hemolytic anemia** (a type II hypersensitivity) produce antibodies against their own red blood cells. These autoantibodies speed up the destruction of the red blood cells, and the patient becomes severely anemic. Different hemolytic anemia patients make antibodies of different classes. Some patients make IgM autoantibodies, which bind to red blood cells and activate the classical complement pathway; the red blood cells are lysed, and degradation products from hemoglobin are released into the bloodstream. Other hemolytic anemia patients make IgG autoantibodies, which serve as opsonins that promote phagocytosis. In this case, red blood cells are removed by macrophages in the liver, spleen, and bone marrow. Even though these latter patients have no hemoglobin in their urine, they are still severely anemic.

The precise causes of all cases of autoimmune hemolytic anemia are unknown, but some cases follow infections with viruses or treatment with certain drugs, both of which alter the surface of red blood cells such that they are recognized as foreign and trigger an immune response.

Autoimmunity Affecting Endocrine Organs

Other common targets of autoimmune attack are the endocrine (hormone-producing) organs. For example, patients can develop autoantibodies or produce T cells against cells of the islets of Langerhans within the pancreas or against cells of the thyroid gland. In most cases, the ensuing autoimmune reaction results in damage to or destruction of the gland and in hormone deficiencies as endocrine cells are killed.

Immunological attack on the islets of Langerhans results in a loss of the ability to produce the hormone *insulin*, which leads to the development of **type 1 diabetes mellitus** (also known as *juvenile-onset diabetes*). As with other autoimmune diseases, the exact trigger of type 1 diabetes is unknown, but many patients endured a severe viral infection some months before the onset of diabetes, which may be the trigger for this type IV hypersensitivity. Additionally, some patients are known to have a genetic predisposition to developing type 1 diabetes that is associated with the possession of certain class I MHC molecules. Some physicians have been successful in delaying the onset of type 1 diabetes by treating at-risk patients with immunosuppressive drugs before damage to the islets of Langerhans becomes apparent.

An autoimmune response can lead to stimulation rather than to inhibition or destruction of glandular tissue. An example of this is **Graves' disease**, which involves the thyroid gland. This major endocrine gland located in the neck secretes iodine-containing hormones, which help regulate metabolic rate in the body. Like other autoimmune diseases, Graves' disease may be triggered by a viral infection in individuals with certain genetic backgrounds. Affected patients (usually women) make autoantibodies that bind to and stimulate receptors on the cytoplasmic membranes of thyroid cells, which elicits excessive production of thyroid hormone and growth of the thyroid gland. Such patients develop an enlarged thyroid gland, called a goiter, and protruding eyes, rapid heartbeat, fatigue, and weight loss despite increased appetite. Physicians treat Graves' disease with antithyroid medicines or radioactive iodine. Patients whose condition doesn't improve with those treatments may require surgical removal of most of the thyroid tissue.

Autoimmunity Affecting Nervous Tissue

Of the group of autoimmune diseases affecting nervous tissue, the most frequent is **multiple sclerosis (MS)** (sklĕ-rō'sis), which is another type IV hypersensitivity. The exact cause of MS is unknown, but it appears that a cell-mediated immune response against a bacterium or virus generates cytotoxic T cells that mistakenly attack and destroy the myelin sheaths that normally insulate brain and spinal cord neurons and increase the speed of nerve impulses along the length of the neurons. Consequently, MS patients experience deficits in vision, speech, and neuromuscular function that may be quite mild and intermittent or may be progressive, paralytic, and perhaps fatal.

Autoimmunity Affecting Connective Tissue

Rheumatoid arthritis (discussed previously) is another crippling autoimmune disease resulting from a type III hypersensitivity. Autoantibodies are formed against connective tissue in joints.

TELL ME WHY

Why can't scientists use the postulates of Robert Koch to determine the specific cause of Graves' disease?

Immunodeficiency Diseases

LEARNING OUTCOME

18.18 Differentiate primary from acquired immunodeficiencies, and cite one disease caused by each form of immunodeficiency.

You may have noticed that during periods of increased emotional or physical stress you are more likely to have a cold. You are not imagining this phenomenon; stress has long been known to decrease the efficiency of the immune system. Similarly, chronic defects in the immune system typically first become apparent when affected individuals become sick more often from infections of opportunistic pathogens. Such opportunistic infections are the hallmarks of *immunodeficiency diseases*, which are conditions resulting from defective immune mechanisms.

Researchers have characterized a large number of immunodeficiency diseases in humans, which are of two general types:

- **Primary immunodeficiency diseases**, which are detectable near birth and develop in infants and young children, result from some genetic or developmental defect.
- **Acquired (secondary) immunodeficiency diseases** develop in later life as a direct consequence of some other recognized cause, such as malnutrition, severe stress, or infectious disease.

Next, we consider each general type of immunodeficiency disease in turn.

Primary Immunodeficiency Diseases

Many different inherited defects have been identified in all of the body's lines of defense, affecting first and second lines of defense as well as antibody and cell-mediated immune responses.

One of the more important inherited defects in the second line of defense is **chronic granulomatous disease**, in which children have recurrent infections characterized by the inability of their phagocytes to destroy bacteria. The reason is that these children have an inherited inability to make reactive forms of oxygen, which are necessary to destroy phagocytized bacteria.

Most primary immunodeficiencies are associated with defects in the components of the third line of defense—adaptive immunity. For example, some children fail to develop any lymphoid stem cells whatsoever, and as a result they produce neither B cells nor T cells and cannot mount immune responses. The resulting defects in the immune system cause **severe combined immunodeficiency disease (SCID)**.

Other children suffer from T cell deficiencies alone. For example, **DiGeorge syndrome** results from a failure of the thymus to develop. Consequently, there are no T cells. Individuals with DiGeorge syndrome generally die of viral infections while remaining resistant to most bacteria—a phenomenon that underscores the importance of T cells in protecting against viruses. Physicians can treat DiGeorge syndrome with a thymic stem cell transplant.

B cell deficiencies also occur in children. The most severe of the B cell deficiencies, called **Bruton-type agammaglobulinemia**[5] (ā-gam′ă-glob′ū-li-nē′mē-ă), is an inherited disease in which affected babies, usually boys, cannot make immunoglobulins. These children experience recurrent bacterial infections but are usually resistant to viral, fungal, and protozoan infections.

Inherited deficiencies of a single immunoglobulin class are more common than a deficiency of all classes. Among these, IgA deficiency is the most common. Because affected children cannot produce secretory IgA, they experience recurrent infections in the respiratory and gastrointestinal tracts.

TABLE 18.5 summarizes some major primary immunodeficiency diseases.

TABLE 18.5 Some Primary Immunodeficiency Diseases

Disease	Defect	Manifestation
Chronic granulomatous disease	Ineffective phagocytes	Uncontrolled infections
Severe combined immunodeficiency disease (SCID)	Lack of T cells and B cells	No resistance to any type of infection, leading to rapid death
DiGeorge syndrome	Lack of T cells and thus no cell-mediated immunity	Overwhelming viral infections
Bruton-type agammaglobulinemia	Lack of B cells and thus lack of immunoglobulins	Overwhelming bacterial infections

MICRO CHECK

9. Which is an example of an autoimmune disease affecting the nervous system?
10. DiGeorge syndrome is an example of what type of immune disorder?

Acquired Immunodeficiency Diseases

LEARNING OUTCOMES

18.19 Describe five acquired conditions that suppress immunity.
18.20 Define *AIDS* and differentiate a disease from a syndrome.

Unlike inherited primary immunodeficiency diseases, acquired immunodeficiency diseases affect individuals who had a previously healthy immune system.

Acquired immunodeficiencies result from a number of causes. In all humans the immune system (but especially T cell production) deteriorates with increasing age; as a result, older individuals normally have less effective immunity, especially cell-mediated immunity, than younger individuals, leading to an increased incidence of both viral diseases and certain types of cancer. Severe stress can also lead to immunodeficiencies by prompting the secretion of increased quantities of corticosteroids,

---

[5] Agammaglobulinemia means "an absence of gamma globulin (IgG)."

which are toxic to T cells and thus suppress cell-mediated immunity. This is why, for example, cold sores may "break out" on the faces of students during final exams: The causative herpes simplex viruses, which are controlled by a fully functioning cell-mediated immune response, escape immune control in stressed individuals. Malnutrition and certain environmental toxins can also cause acquired immunodeficiency diseases by inhibiting the normal production of B cells and T cells.

Of course, the most significant example of the result of an acquired immunodeficiency is **acquired immunodeficiency syndrome (AIDS)**. From the time of its discovery in 1981 among homosexual males in the United States to its emergence as a worldwide pandemic, no affliction has affected modern life as much.

AIDS is not a single disease but a **syndrome**, that is, a group of signs, symptoms, and diseases associated with a common pathology. Currently, epidemiologists define this syndrome as the presence of several opportunistic or rare infections along with infection by human immunodeficiency virus (HIV) or as a severe decrease in the number of CD4 cells (< 200/μL of blood)[6] and a positive test showing the presence of antibodies against HIV. The infections include diseases of the skin, such as shingles and disseminated (widespread) herpes; diseases of the nervous system, including meningitis, toxoplasmosis, and *Cytomegalovirus* (sī-tō-meg'ă-lō-vī'rŭs) disease; diseases of the respiratory system, such as tuberculosis, *Pneumocystis* (nū-mō-sis'tis) pneumonia, histoplasmosis, and coccidioidomycosis (kok-sid-ē-oy'dō-mī-kō-sis); and diseases of the digestive system, including chronic diarrhea, thrush, and oral hairy leukoplakia. A rare cancer of blood vessels called Kaposi's sarcoma is also commonly seen in AIDS patients. AIDS often results in dementia during the final stages. (Chapter 25 examines HIV and AIDS in more detail.)

TELL ME WHY

Two boys have autoimmune diseases: One has Bruton-type agammaglobulinemia, and the other has DiGeorge syndrome. On a camping trip, each boy is stung by a bee, and each falls into poison ivy. What hypersensitivity reactions might each boy experience as a result of his camping mishaps?

---

[6]The normal value for CD4 cells is 500–1200 cells/μL.

MICRO IN THE CLINIC FOLLOW-UP
Danger in the Blackberry Patch?

By the time Cassie and Tim arrive at the emergency clinic, Tim is having a hard time catching his breath. The doctor quickly assesses Tim's condition and administers epinephrine. Within a minute, Tim is breathing much more easily. The doctor keeps Tim in the clinic for four hours to monitor his condition and make sure that no additional respiratory complications occur. The doctor tells Tim and Cassie that Tim had a type I hypersensitivity reaction to the venom that the hornet stings injected.

After a few hours with no further complications, the doctor discharges Tim from the clinic. He instructs Tim to use an antihistamine (diphenhydramine hydrochloride) to prevent further effects of histamine. This should clear up the rash and relieve his symptoms, which should fully resolve in a couple of days. The doctor also prescribes Tim a single-dose epinephrine autoinjector so that if he should have a similar reaction in the future, he can self-administer epinephrine immediately, preventing his trachea from swelling to the point that he cannot breathe.

1. **Explain how the hornet venom caused Tim's hives and his difficulty in breathing. In other words, what physiological responses are causing his symptoms?**

2. **Why will an antihistamine help treat Tim's hives?**

Check your answers to Micro in the Clinic Follow-Up questions in the Mastering Microbiology Study Area.

Make the connection among Chapters 14 through 18. Watch MICRO MATTERS videos in the Mastering Microbiology Study Area.

CHAPTER SUMMARY

1. Immunological responses may give rise to inflammatory reactions called **hypersensitivities**. An immunological attack on normal tissues gives rise to autoimmune disorders. A failure of the immune system to function normally may give rise to immunodeficiency diseases.

Hypersensitivities (pp. 526–538)

1. Type I hypersensitivity gives rise to **allergies**. Antigens that trigger this response are called **allergens**.

2. Allergies result from a two-step process: (1) allergens trigger the production of IgE antibodies that bind to **mast cells**, **basophils**, which are thereby sensitized to the allergen; and (2) the same kinds of allergens later bind to the antibodies, causing the sensitized cells to degranulate and release **histamine**, **kinins**, **proteases**, **leukotrienes**, and **prostaglandins**.

3. Depending on the amount of these molecules and the site at which they are released, the result can produce various diseases, including **hay fever**, **asthma**, **urticaria** (hives), or various

other allergies. When the inflammatory mediators exceed the body's coping mechanisms, **acute anaphylaxis**, or **anaphylactic shock**, may occur. The specific treatment for anaphylaxis is epinephrine.

4. Type I hypersensitivity can be diagnosed by skin testing and can be partially prevented by avoidance of allergens and by immunotherapy. Some type I hypersensitivities are treated with **antihistamines**.

5. Type II cytotoxic hypersensitivities, such as incompatible blood transfusions and hemolytic disease of the newborn, result when cells are destroyed by an immune response.

6. Red blood cells have **blood group antigens** on their surface. If incompatible blood is transfused into a recipient, a severe transfusion reaction can result.

7. The most important of the blood group antigens is the ABO group, which is largely responsible for transfusion reactions.

8. Approximately 85% of the human population carries **Rh antigen**, which is also found in rhesus monkeys. If an Rh-negative pregnant woman is carrying an Rh-positive fetus, the fetus may be at risk of **hemolytic disease of the newborn**, in which antibodies made by the mother against the Rh antigen may cross the placenta and destroy the fetus's red blood cells. RhoGAM administered to pregnant Rh-negative women may prevent this disease.

9. In type III hypersensitivity, excessive amounts of small **immune complexes**, which are complexes of antigen bound to antibodies, are deposited in tissues, where they cause significant tissue damage.

10. Immune complexes deposited in the lung cause a **hypersensitivity pneumonitis**, of which the most common example is farmer's lung. If large amounts of immune complexes form in the bloodstream, they may be filtered out by the glomeruli of the kidney, causing **glomerulonephritis**, which can result in kidney failure.

11. **Rheumatoid arthritis (RA)** is an autoimmune, type III hypersensitivity in which immune complexes result in the growth of inflammatory tissue within joints.

12. **Systemic lupus erythematosus (SLE)**, or lupus, is a systemic, autoimmune, type III hypersensitivity in which autoantibodies bind to many autoantigens, especially the patient's DNA.

13. Type IV hypersensitivity, also known as **delayed hypersensitivity reaction** or **cell-mediated hypersensitivity reaction** is a T cell-mediated inflammatory reaction that takes 24–72 hours to reach maximal intensity.

14. A good example of a delayed hypersensitivity reaction is the **tuberculin response**, generated when tuberculin, a protein extract of *Mycobacterium tuberculosis*, is injected into the skin of an individual who has been infected with or immunized against *M. tuberculosis*.

15. Another example of a type IV hypersensitivity reaction is **allergic contact dermatitis**, which is T cell-mediated damage to chemically modified skin cells. The best-known example is a reaction to poison ivy.

16. An organ or tissue **graft** can be made between different sites within a single individual (an **autograft**), between genetically identical individuals (an **isograft**), between genetically dissimilar individuals (an **allograft**), or between individuals of different species (a **xenograft**). Most surgical organ grafting involves allografts, which, if not treated with immunosuppressive drugs, lead to **graft rejection**.

17. In **graft-versus-host disease**, an organ donor's cells attack the recipient's body.

18. Commonly used immunosuppressive drugs include **glucocorticoids**, **cytotoxic drugs**, **cyclosporine**, and lymphocyte-depleting therapies, which involve treatment with antibodies against T cells or their receptors.

Autoimmune Diseases (pp. 538–539)

1. **Autoimmune diseases** may result when an individual begins to make autoantibodies or cytotoxic T cells against normal body components.

2. There are at least eight hypotheses concerning the cause of autoimmune disease. One involves **molecular mimicry**, in which microorganisms with epitopes similar to self-antigens trigger autoimmune tissue damage. Others implicate estrogen, cell transfer from mother to fetus or from fetus to mother during pregnancy, environmental factors such as viral infection, genetic predisposition, movement of T cells into sequestered areas such as testes, or failure of normal immunological control mechanisms.

3. One group of autoimmune diseases involves only a single organ or cell type. Examples of such diseases include **autoimmune hemolytic anemia**, **type 1 diabetes mellitus**, **Graves' disease**, **multiple sclerosis (MS)**, and rheumatoid arthritis.

4. A second group of autoimmune diseases, such as systemic lupus erythematosus, involves multiple organs or body systems.

Immunodeficiency Diseases (pp. 540–541)

1. Immunodeficiency diseases may be classified as **primary immunodeficiency diseases**, which result from mutations or developmental anomalies and occur in young children, or **acquired (secondary) immunodeficiency diseases**, which result from other known causes such as viral infections.

2. Examples of primary immunodeficiency diseases include **chronic granulomatous disease**, in which a child's neutrophils are incapable of killing ingested bacteria. Inability to produce both T cells and B cells is called **severe combined immunodeficiency disease**. In **DiGeorge syndrome**, the thymus fails to develop. In **Bruton-type agammaglobulinemia**, B cells fail to function.

3. **Acquired immunodeficiency syndrome (AIDS)** is a condition defined by the presence of antibodies against **human immunodeficiency virus (HIV)** in conjunction with certain opportunistic infections, or by HIV and a CD4+ count below 200 cells/μl of blood.

4. A **syndrome** is a complex of signs and symptoms with a common cause.

QUESTIONS FOR REVIEW

Answers to the Questions for Review (except Short Answer questions) begin on p. A-1.

Multiple Choice

1. The immunoglobulin class that mediates type I hypersensitivity is _____.
   a. IgA
   b. IgM
   c. IgG
   d. IgD
   e. IgE

2. The major inflammatory mediator released by degranulating mast cells in type I hypersensitivity is _____.
   a. immunoglobulin
   b. complement
   c. histamine
   d. interleukin
   e. prostaglandin

3. Hemolytic disease of the newborn is caused by antibodies against which major blood group antigen?
   a. MHC protein
   b. MN antigen
   c. ABO antigen
   d. rhesus antigen
   e. type II protein

4. Farmer's lung is a hypersensitivity pneumonitis resulting from _____.
   a. a type I hypersensitivity reaction to grass pollen
   b. a type II hypersensitivity to red cells in the lung
   c. a type III hypersensitivity to mold spores
   d. a type IV hypersensitivity to bacterial antigens

5. A positive tuberculin skin test indicates that a patient *not* immunized against tuberculosis _____.
   a. is free of tuberculosis
   b. is shedding *Mycobacterium*
   c. has been exposed to tuberculosis antigens
   d. is susceptible to tuberculosis
   e. is resistant to tuberculosis

6. Which of the following is an autoimmune disease?
   a. a heart attack
   b. acute anaphylaxis
   c. farmer's lung
   d. graft-versus-host disease
   e. systemic lupus erythematosus

7. When a surgeon conducts a cardiac bypass operation by transplanting a piece of vein from a patient's leg to the same patient's heart, this is a(n) _____.
   a. rejected graft
   b. autograft
   c. allograft
   d. type IV hypersensitivity
   e. cardiograft

8. A deficiency of both B cells and T cells is most likely a(n) _____.
   a. secondary immunodeficiency
   b. complex immunodeficiency
   c. acquired immunodeficiency
   d. primary immunodeficiency
   e. induced immunodeficiency

9. Infection with HIV causes _____.
   a. primary immunodeficiency disease
   b. acquired hypersensitivity syndrome
   c. acquired immunodeficiency syndrome
   d. anaphylactic immunodeficiency diseases
   e. combined immunodeficiency diseases

10. What do medical personnel administer to counteract various type I hypersensitivities?
    a. antihistamine
    b. bronchodilator
    c. corticosteroid
    d. epinephrine
    e. all of the above

Modified True/False

Indicate whether each statement is true or false. If the statement is false, change the underlined word or phrase to make the statement true.

1. _____ <u>Cyclosporine</u> is released by degranulating mast cells.

2. _____ <u>Type III</u> hypersensitivity reactions may lead to the development of glomerulonephritis.

3. _____ ABO blood group antigens are found on <u>nucleated</u> cells.

4. _____ The tuberculin reaction is a <u>type I</u> hypersensitivity.

5. _____ Graft-versus-host disease can follow a bone marrow <u>isograft</u>.

Matching

Put the number of the type of immune system hypersensitivity in the blank next to each manifestation. Each of the four types may be used more than once or not at all. If the manifestation is not an immune hypersensitivity, put zero in the blank.

1. ____ Acute anaphylaxis
2. ____ Allergic contact dermatitis
3. ____ Systemic lupus erythematosus
4. ____ Allograft rejection
5. ____ AIDS
6. ____ Graft-versus-host disease
7. ____ Milk allergy
8. ____ Rheumatoid arthritis (RA)
9. ____ Asthma
10. ____ Hay fever

I. Type I hypersensitivity
II. Type II hypersensitivity
III. Type III hypersensitivity
IV. Type IV hypersensitivity
0. Not a hypersensitivity

Short Answer

1. Why is AIDS more accurately termed a syndrome rather than a mere disease?
2. Why is a child born to an Rh+ mother not susceptible to Rh-related hemolytic disease of the newborn?
3. Why is a person who produces a large amount of IgE more likely to experience anaphylactic shock than a person who instead produces a large amount of IgG?
4. Contrast autografts, isografts, allografts, and xenografts.
5. Compare and contrast the functions of four classes of immunosuppressive drugs.

VISUALIZE IT!

1. Label the four types of grafts on the accompanying figure.

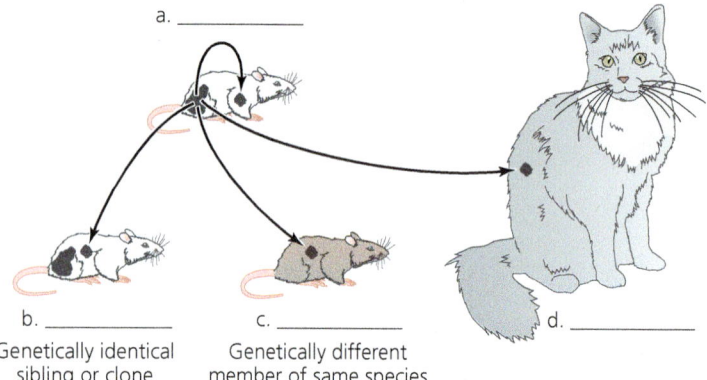

Genetically identical sibling or clone

Genetically different member of same species

2. Identify the type of hypersensitivity reaction in each photo.

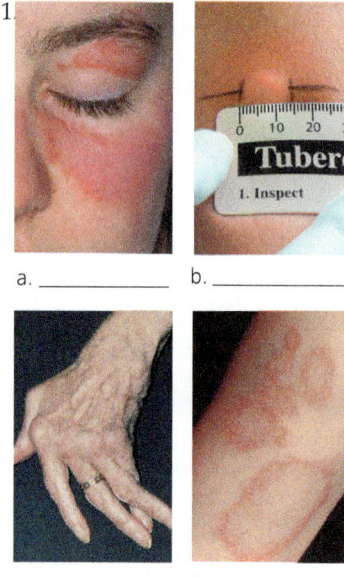

CRITICAL THINKING

1. What possible advantages might an individual gain from making IgE?
2. Why can't physicians use skin tests similar to the tuberculin reaction to diagnose other bacterial diseases?
3. In both Graves' disease and type 1 diabetes mellitus, autoantibodies are directed against cytoplasmic membrane receptors. Speculate on the clinical consequences of an autoimmune response to estrogen receptors.
4. In general, people with B cell defects acquire numerous bacterial infections, whereas those with T cell defects get viral diseases. Explain why this is so.
5. What types of illnesses cause death in patients with combined immunodeficiencies or AIDS?
6. Because of the severe shortage of organ donors for transplants, many scientists are examining the possibility of using organs from nonhuman species such as pigs. What special clinical problems might be encountered when these xenografts are used?
7. Why do the blisters of positive tuberculin reactions resemble the blisters of poison ivy?
8. A patient arrives at the doctor's office with a rash covering her legs. How could you determine whether the rash is a type I or a type IV hypersensitivity?
9. A 43-year-old woman has been diagnosed with rheumatoid arthritis. Unable to find relief from her symptoms, she seeks treatment from a doctor in another country, who injects antibodies and complement into her afflicted joints. Do you expect the treatment to improve her condition? Explain your reasoning.

CONCEPT MAPPING

Using the terms provided, fill in this concept map that describes immediate hypersensitivity. You can also complete this and other concept maps online by going to the **Mastering Microbiology** Study Area.

Allergens
Allergies
Anaphylactic shock
Antihistamine
Asthma
Basophils
Bee venom
Dust mites
Epinephrine
Hay fever
IgE
Increased vascular permeability
Mast cells
Peanuts
Pollen
Smooth muscle contraction
Type I hypersensitivity
Urticaria
Vasodilation

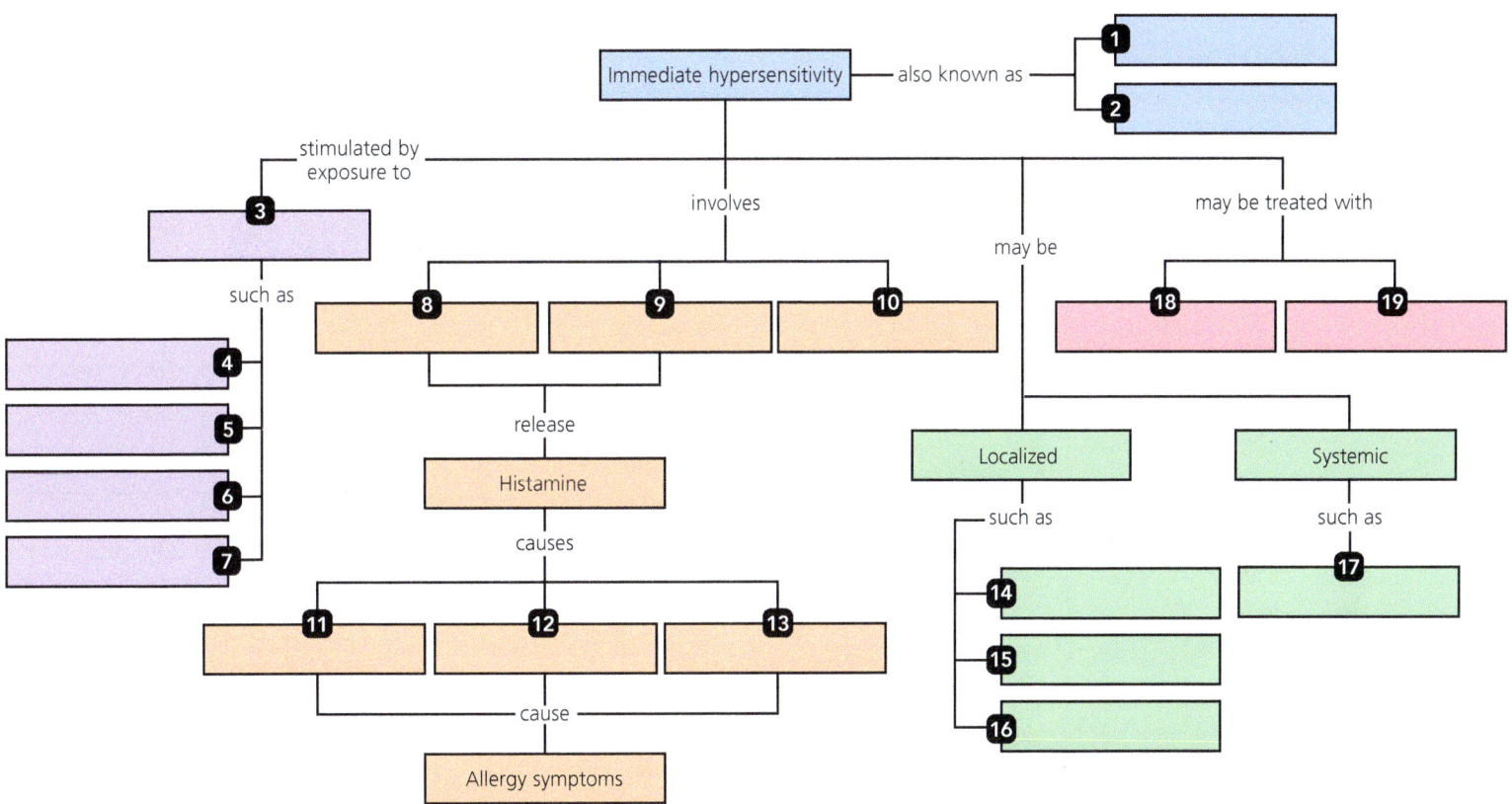

19 Pathogenic Gram-Positive Bacteria

Before You Begin

1. What is the difference between the cell walls of Gram-positive bacteria and those of Gram-negative bacteria?
2. What is the difference between Gram-positive bacteria and mycoplasmas?
3. What is the function of endospores?
4. How does coagulase facilitate bacterial virulence?

MICRO IN THE CLINIC

A Case of Dehydration?

ETHAN WORKS AS A CIVIL ENGINEER IN ALASKA where he was born and raised and has lived his entire life. He just finished the biggest project of his career today and is really looking forward to some rest and relaxation. After work on Friday night, Ethan and some of his coworkers go out for a celebratory dinner at a local seafood restaurant. They have a great time, share an amazing meal, and enjoy Ethan's success.

Saturday morning, Ethan and his wife enjoy a long bike ride and stop at a café for brunch. Late Saturday afternoon, Ethan notices that his mouth is exceptionally dry and he is beginning to have a sore throat. He had sweated a lot on the bike ride, so he assumes that he is just a bit dehydrated and needs to drink more water. When Ethan wakes up on Sunday morning, he still has a dry mouth and is having a hard time swallowing because of his sore throat. He begins having abdominal cramping with episodes of diarrhea. A few hours later, he tells his wife that he is having a difficult time breathing. She suggests they go to the emergency room.

1. Are Ethan's symptoms merely the result of dehydration?
2. Could there be another explanation for his condition?

Turn to the end of the chapter (p. 581) to find out.

Several Gram-positive bacteria, such as the endospore-forming *Clostridium botulinum* shown here, cause human diseases.

In this chapter, we will learn about medically important Gram-positive pathogens, including *Staphylococcus, Streptococcus, Bacillus,* and *Clostridioides, Clostridium.* A general characteristic of Gram-positive pathogens is that the disease symptoms they cause often result from the toxins they secrete.

In this chapter, we explore the major Gram-positive bacterial pathogens in more detail. Gram-positive bacteria stain purple when Gram stained and generally fall within the phylum Firmicutes. We will discuss the major pathogens of this group.

Taxonomists currently group Gram-positive bacteria into two major groups: low G + C Gram-positive bacteria and high G + C Gram-positive bacteria. As the names indicate, these classifications are based on the content of guanine-cytosine nucleotide base pairs versus adenine-thymine nucleotide base pairs in these organisms' DNA. Low G + Cw Gram-positive bacteria include four genera of pathogenic spherical cells (cocci): *Staphylococcus, Streptococcus,* and *Enterococcus;* four genera of pathogenic rod-shaped cells (bacilli): *Bacillus, Clostridioides, Clostridium,* and *Listeria;* and the *mycoplasmas,* a group of bacteria that lack cell walls. Mycoplasmas have historically been classified as Gram-negative bacteria because they stain pink when Gram stained. However, studies of their nucleotide sequences have revealed that they are genetically more similar to low G + C Gram-positive bacteria.

The high G + C Gram-positive pathogens include the rod-shaped genera *Corynebacterium, Mycobacterium,* and *Propionibacterium* and the filamentous, fungus-like *Nocardia* and *Actinomyces.*

We begin our discussion of Gram-positive bacteria with *Staphylococcus.*

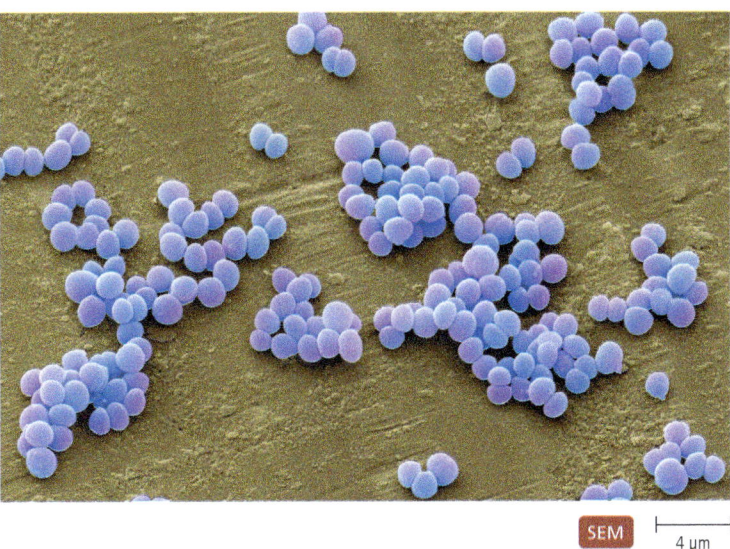

▲ FIGURE 19.1 *Staphylococcus.* Gram-positive cells appear in grapelike clusters.

Staphylococcus

Bacteria in the genus *Staphylococcus* (staf′i-lō-kok′ŭs) are living and reproducing on almost every square inch of your skin right now. They are normal members of every human's microbiota, which usually go unnoticed. However, they can be opportunistic pathogens, causing minor to life-threatening diseases.

Structure and Physiology

LEARNING OUTCOME

19.1 Contrast the virulence of *S. aureus* with that of *S. epidermidis* in humans.

Staphylococcus[1] is a genus of Gram-positive, facultatively anaerobic prokaryotes whose spherical cells are typically clustered in grapelike arrangements (FIGURE 19.1). This arrangement results from two characteristics of cell division: cell divisions occur in successively different planes, and daughter cells remain attached to one another. Staphylococcal cells are 0.5–1.0 μm in diameter and nonmotile. They are salt tolerant, capable of growing in media saturated with NaCl—about 28% salt—which explains how they tolerate the salt deposited on human skin by sweat glands. Additionally, they are tolerant of desiccation, radiation, and heat (up to 60°C for 30 minutes), allowing them to survive on environmental surfaces in addition to skin. Further, *Staphylococcus* synthesizes catalase—a characteristic that distinguishes this genus from other low G + C Gram-positive cocci.

Two species are commonly associated with staphylococcal diseases in humans:

- *Staphylococcus aureus*[2] (o′rē-ŭs) is the more virulent, producing a variety of disease conditions and symptoms depending on the site of infection.
- *Staphylococcus epidermidis* (e-pi-der-mid′is), as its name suggests, is a part of the normal microbiota of human skin, but it is an opportunistic pathogen in immunocompromised patients or when introduced into the body via intravenous catheters or on prosthetic devices, such as artificial heart valves.

Pathogenicity

LEARNING OUTCOME

19.2 Discuss the structural features, enzymes, and toxins of *Staphylococcus* that enable it to be pathogenic.

What are commonly called "staph" infections result when staphylococci breach the body's physical barriers (skin or mucous membranes); entry of only a few hundred bacteria can ultimately result in disease. The pathogenicity of *Staphylococcus* results from at least three features: structures that enable it to evade phagocytosis, the production of enzymes, and the production of toxins.

---

[1]From Greek *staphle,* meaning "small bunch of grapes," and *kokkos,* meaning "a berry."

[2]Latin for "golden," for the golden color of its colonies growing on the surface of a solid medium.

Structural Defenses Against Phagocytosis

The cells of *S. aureus* are uniformly coated with a protein, called *protein A*, that interferes with antibody immune responses by binding to the stems (F_c portion) of class G antibodies (IgG). Antibodies are opsonins—they enhance phagocytosis—precisely because phagocytic cells have receptors for antibody stems; therefore, protein A, by binding the stems, effectively inhibits opsonization. Protein A also inhibits the complement cascade, which is triggered by antibody molecules bound to antigen (see Chapter 15, Figure 15.9).

The outer surfaces of most pathogenic strains of *S. aureus* also contain *bound coagulase*, an enzyme that converts the soluble blood protein fibrinogen into long, insoluble fibrin molecules, which are threads that form blood clots around the bacteria. Fibrin clots around the bacteria in effect hide *Staphylococcus* from phagocytic cells.

Both *S. aureus* and *S. epidermidis* also evade the body's defenses by synthesizing loosely organized polysaccharide slime layers (sometimes called capsules) that inhibit chemotaxis of and endocytosis by leukocytes, particularly neutrophils. The slime layer also facilitates attachment of *Staphylococcus* to artificial surfaces, such as catheters, shunts, artificial heart valves, and synthetic joints.

In summary, protein A, bound coagulase, and a slime layer allow *S. aureus* to evade the body's defenses, whereas *S. epidermidis* relies almost exclusively on its slime layer.

Enzymes

Staphylococci produce a number of enzymes that contribute to their survival and pathogenicity:

- *Cell-free coagulase*, like bound coagulase, triggers blood clotting. Cell-free coagulase does not act on fibrin directly but instead combines with a blood protein before becoming enzymatic and converting fibrinogen to fibrin threads. Only *S. aureus* synthesizes coagulase; *S. epidermidis* and other species of *Staphylococcus* are coagulase negative.
- *Hyaluronidase* breaks down hyaluronic acid, which is a major component of the matrix between cells. Hyaluronidase, found in 90% of *S. aureus* strains, enables the bacteria to spread between cells throughout the body.
- *Staphylokinase* (produced by *S. aureus*) dissolves fibrin threads in blood clots, allowing *S. aureus* to free itself from clots. Thus, *Staphylococcus* can escape the immune system by enclosing itself in a fibrin clot (via coagulase), and then, when space and nutrients become limiting, it can digest its way out of the clot with staphylokinase and spread to new locations.
- *Lipases* digest lipids, allowing staphylococci to grow on the surface of the skin and in cutaneous oil glands. All staphylococci produce lipases.
- *β-lactamase (penicillinase)*, now present in over 90% of *S. aureus* strains, breaks down penicillin. Though β-lactamase plays no role in inhibiting the natural defenses of the body, it does allow the bacteria to survive treatment with beta-lactam antimicrobial drugs, such as penicillin and cephalosporin.

Toxins

Various strains of *Staphylococcus aureus* also possess several toxins that contribute to their pathogenicity, including the following:

- **Cytolytic toxins.** Alpha, beta, gamma, and delta toxins are proteins, coded by chromosomal genes, that disrupt the cytoplasmic membranes of a variety of cells, including leukocytes (white blood cells). Leukocidin is a fifth cytolytic toxin that lyses leukocytes specifically, providing *Staphylococcus* with some protection against phagocytosis.
- **Exfoliative toxins.** Each of two distinct proteins causes the dissolution of epidermal *desmosomes* (intercellular bridge proteins that hold adjoining cytoplasmic membranes together), causing the patient's skin cells to separate from each other and slough off the body.
- **Toxic shock syndrome (TSS) toxin.** This protein causes toxic shock syndrome (discussed shortly).
- **Enterotoxins.** These five proteins (designated A through E) stimulate the intestinal muscle contractions, nausea, and intense vomiting associated with staphylococcal food poisoning. Enterotoxins are heat stable, remaining active at 100°C for up to 30 minutes.

TABLE 19.1 compares and contrasts the virulence factors of *S. aureus* and *S. epidermidis*.

Epidemiology

Staphylococcus epidermidis is ubiquitous on human skin, whereas *S. aureus* is commonly found only on moist skin folds. Both species also grow in the upper respiratory, gastrointestinal, and urogenital tracts of humans. Both bacteria are transmitted through direct contact between individuals as well as via fomites such as contaminated clothing, bedsheets, and medical instruments; as a result, proper hand washing and aseptic techniques are essential in preventing their transfer in health care settings.

TABLE 19.1 A Comparison of the Virulence Factors of Staphylococcal Species

	S. aureus	S. epidermidis
Protein A	+	−
Coagulase	+	−
Slime layer	+	+
Catalase	+	+
Hyaluronidase	+	−
Staphylokinase	+	−
Lipase	+	+
β-lactamase (penicillinase)	+	−
Toxins (cytolytic, exfoliative, toxic-shock, and entero-)	+	−

Staphylococcal Diseases

LEARNING OUTCOMES

19.3 Describe the symptoms and prevention of staphylococcal food poisoning.

19.4 List and describe six pyogenic lesions caused by *Staphylococcus aureus*.

19.5 Discuss five systemic and potentially fatal diseases caused by *Staphylococcus*.

Staphylococcus causes a variety of medical problems, depending on the site of infection, the immune state of its host, and the toxins and enzymes a particular species or strain secretes. Staphylococcal syndromes and diseases can be categorized as noninvasive, cutaneous, and systemic diseases.

Noninvasive Disease

Staphylococcus aureus is one of the more common causes of food poisoning (more specifically, this is food intoxication because disease is caused by enterotoxin left in food from the growth of bacteria rather than by a bacterial infection). Commonly affected foods include processed meats, custard pastries, potato salad, and ice cream that have been contaminated with bacteria from human skin. (In *S. aureus* food poisoning, unlike many other forms of food poisoning, animals are not involved.) The food must remain at room temperature or warmer for several hours for the bacteria to grow, reproduce, and secrete toxin. Warming or reheating inoculated food does not inactivate enterotoxins, which are heat stable, although heating does kill the bacteria. Food contaminated with staphylococci does not appear or taste unusual.

Symptoms, which include nausea, severe vomiting, diarrhea, headache, sweating, and abdominal pain, usually appear within four hours following ingestion. Consumed staphylococci do not continue to produce toxins, so the course of the disease is rapid, usually lasting 24 hours or less.

Cutaneous Diseases

Staphylococcus aureus causes localized *pyogenic*[3] (pī-ō-jen'ik) lesions. **Staphylococcal scalded skin syndrome** is a reddening of the skin that typically begins near the mouth, spreads over the entire body, and is followed by large blisters that contain clear fluid lacking bacteria or white blood cells. These are lacking because the syndrome is caused by a toxin released by bacteria growing on the skin rather than in the body. Within two days, the affected outer layer of skin (epidermis) peels off in sheets, as if it had been dipped into boiling water (**FIGURE 19.2**). The seriousness of scalded skin syndrome results from secondary bacterial infections in denuded areas.

Small, flattened, red patches on the face and limbs, particularly of children whose immune systems are not fully developed, characterize **impetigo** (im-pe-tī'gō; **FIGURE 19.3**). The patches develop into pus-filled vesicles that eventually crust over. The pus is filled with bacteria and white blood cells, which distinguishes impetigo from scalded skin syndrome. *S. aureus* acting alone causes about 80% of impetigo cases; about 20% of cases also involve streptococci.

[3]From Greek *pyon*, meaning "pus," and *genein*, meaning "to produce."

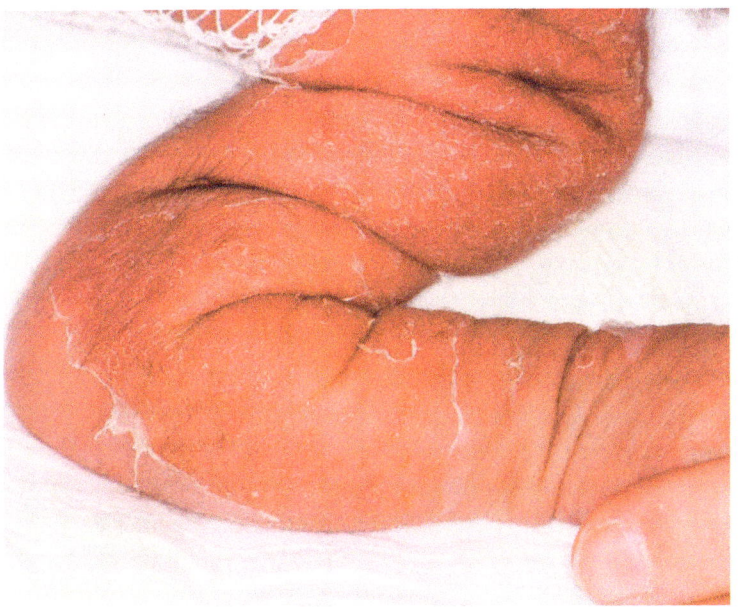

▲ **FIGURE 19.2 Staphylococcal scalded skin syndrome.** Exfoliative toxin, produced by some strains of *Staphylococcus aureus*, causes reddened patches of the epidermis to slough off.

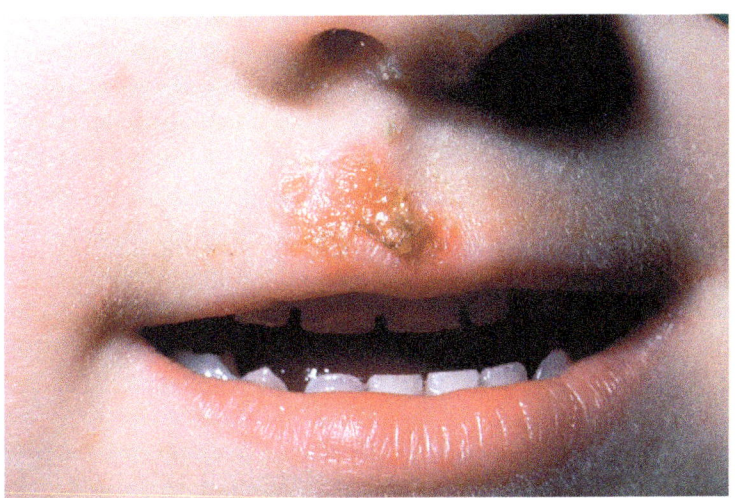

▲ **FIGURE 19.3 Impetigo.** Reddened patches of skin become pus-filled vesicles that eventually crust over.

Folliculitis (fo-lik'ū-lī'tis) is an infection of a hair follicle in which the base of the follicle becomes red, swollen, and pus filled. When this condition occurs at the base of an eyelid, it is called a **sty**. A *furuncle* (fū'rŭng-kl) or boil is a large, painful, raised nodular extension of folliculitis into surrounding tissue. When several furuncles coalesce, they form a *carbuncle* (kar'bŭng-kl), which extends deeper into the tissues, triggering the fever and chills that are characteristic of innate immunity. As staphylococci spread into underlying tissues, multiple organs and systems can become involved.

Systemic Diseases

Staphylococcus aureus and, to a lesser extent, *S. epidermidis* cause a wide variety of potentially fatal systemic infections when they are introduced into deeper tissues of the body, including the blood, heart, lungs, and bones.

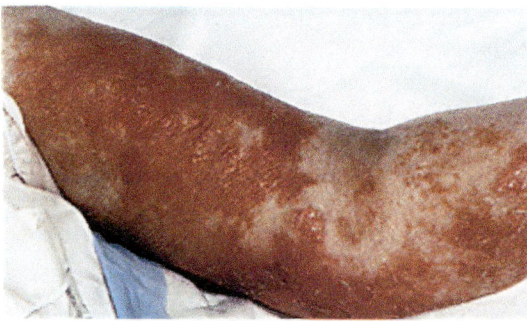

▲ FIGURE 19.4 **Toxic shock syndrome (TSS).** Fatal infections involve not only red rash (as shown here) but shock and death of internal organs as well.

Toxic Shock Syndrome (Non-Streptococcal) When strains of *Staphylococcus* that produce TSS toxin grow in a wound or in an abraded vagina, the toxin can be absorbed into the blood and cause **toxic shock syndrome, non-streptococcal (TSS)**, characterized by fever, vomiting, red rash, extremely low blood pressure, and loss of sheets of skin (FIGURE 19.4). TSS is fatal to 5% of patients when their blood pressure falls so low that the brain, heart, and other vital organs have an inadequate supply of oxygen—a condition known as *shock*.

Toxic shock syndrome occurs in both males and females, but in 1980, epidemiologists noted an epidemic of TSS among menstruating women. Researchers subsequently discovered that *S. aureus* grows exceedingly well in superabsorbent tampons, especially when the tampon remains in place for a prolonged period. As a result of the withdrawal of this type of tampon from the market, plus government-mandated reduction in the absorbency of all tampons and mandatory inclusion of educational information concerning the risks of TSS on every package of tampons, the number of cases of TSS declined rapidly (FIGURE 19.5).

Bacteremia *S. aureus* is a common cause of **bacteremia** (bak-tēr-ē'mē-ă), the presence of bacteria in the blood. After staphylococci enter the blood from a site of infection, they travel to other organs of the body, which may become infected. Furuncles, vaginal infections, infected surgical wounds, and contaminated medical devices such as intravascular catheters have all been implicated in cases of bacteremia. Healthcare-associated infections (HAIs, previously known as nosocomial infections) account for about half of all cases of staphylococcal bacteremia. Physicians fail to identify the initial site of infection in about a third of patients, but it is presumed to be an innocuous break in the skin.

Endocarditis *S. aureus* may attack the lining of the heart (including its valves), producing a condition called **endocarditis**[4] (en'dō-kar-dī'tis). Typically, patients with endocarditis have nonspecific, flulike symptoms, but their condition quickly deteriorates as the amount of blood pumped from the heart drops precipitously. About 50% of patients with endocarditis do not survive.

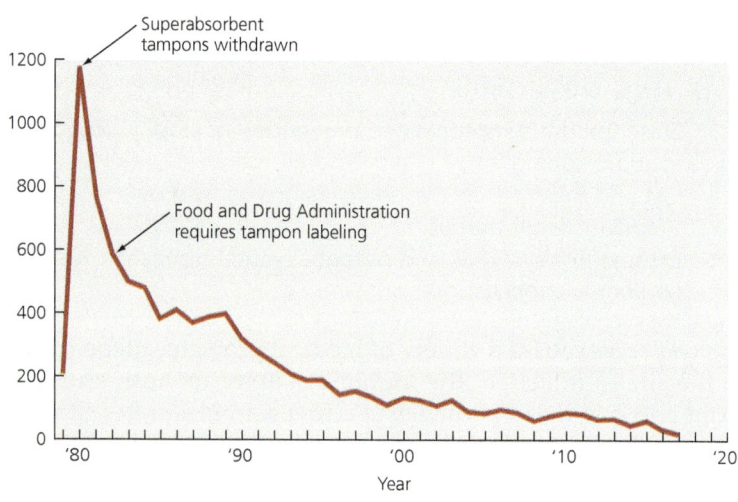

▲ FIGURE 19.5 **The incidence of toxic shock syndrome in the United States, 1979–2017.**

Pneumonia and Empyema *Staphylococcus* in the blood can invade the lungs, causing **pneumonia** (noo-mō'nē-ă)—an inflammation of the lungs in which the alveoli (air sacs) and bronchioles (smallest airways) become filled with fluid. In 10% of patients with staphylococcal pneumonia, infection spreads to the space between a lung and the chest wall where pus builds up—a condition known as **empyema**[5] (em-pī-ē'mă).

Osteomyelitis When *Staphylococcus* invades a bone, either through a traumatic wound or via the blood during bacteremia, it causes **osteomyelitis**[6] (os'tē-ō-mī-e-lī'tis)—inflammation of the bone marrow and the surrounding bone. Osteomyelitis is characterized by pain in the infected bone accompanied by high fever. In children, the disease typically occurs in the growing regions of long bones, which are areas with well-developed blood supplies. In adults, osteomyelitis is more commonly seen in vertebrae.

Diagnosis, Treatment, and Prevention

LEARNING OUTCOMES

19.6 Describe how staphylococcal species are distinguished from one another during diagnosis.

19.7 Discuss briefly the history of staphylococcal resistance to antimicrobial drugs.

Physicians diagnose staphylococcal infection by detecting grape-like arrangements of Gram-positive bacteria isolated from pus, blood, or other fluids. If staphylococci isolated from an infection are able to clot blood, then they are coagulase-positive *S. aureus*. Coagulase-negative staphylococci are usually *S. epidermidis*, which is a normal part of the microbiota of the skin; their presence in a clinical sample is usually not indicative of a staphylococcal infection.

---

[4]From Greek *endon*, meaning "within;" *kardia*, meaning "heart;" and *itis*, meaning "inflammation."

[5]From Greek *en*, meaning "in," and *pyon*, meaning "pus."

[6]From Greek *osteon*, meaning "bone"; *myelos*, meaning "marrow"; and *itis*, meaning "inflammation."

During the latter half of the 20th century, genes for β-lactamase, which convey resistance to natural penicillin, spread among *Staphylococcus* species. Thus, 90% of staphylococci were susceptible to penicillin in 1945, but only 5% are susceptible today. For this reason, the semisynthetic form of penicillin—methicillin, which is not inactivated by β-lactamase—became the drug of choice for staphylococcal infections. Unfortunately, **methicillin-resistant *Staphylococcus aureus* (MRSA)** has emerged as a major problem, initially in health care settings, where it's known as *healthcare-associated MRSA (HA-MRSA)*, but now increasingly in day care centers, high school locker rooms, and prisons, where it's called *community-associated MRSA (CA-MRSA, i.e., not acquired in a healthcare setting)*. In fact, more people die from MRSA than from HIV in the United States.

MRSA is also resistant to many other common antimicrobial drugs, including penicillin, macrolides, aminoglycosides, and cephalosporin; as a result, *vancomycin* has been used to treat MRSA infections. Physicians are particularly concerned about the increasing prevalence of **vancomycin-resistant *Staphylococcus aureus* (VRSA)** because vancomycin has become the drug of choice for treating MRSA infections.

Clinical practice has shown how crucial it is that abscesses be drained of pus and cleansed to remove the majority of bacteria so that subsequent antibiotic therapy can be effective. Systemic infections such as endocarditis and osteomyelitis require long-term therapy with antimicrobial drugs.

Because strains of *Staphylococcus* resistant to antimicrobial agents have become more common, especially in hospitals, it is imperative that health care workers take precautions against introducing the bacteria into patients. Of course, given that *Staphylococcus* is ubiquitous on human skin, staphylococcal infection cannot be eliminated. Fortunately, because a large inoculum is required to establish an infection, proper cleansing of wounds and surgical openings, attention to aseptic use of catheters and indwelling needles, and the appropriate use of antiseptics will prevent infections in most healthy patients. Health care workers with staphylococcal infections may be barred from delivery rooms, nurseries, and operating rooms. The most important measure for protecting against healthcare-associated infection is frequent hand washing.

Scientists are currently testing the efficacy and safety of a vaccine that has proven effective in protecting dialysis patients from *S. aureus* infections. Such immunization, if it proves safe and effective in preventing other infections, may have a significant effect on healthcare-associated disease. It would be especially good news for health care providers who are battling resistant strains of *Staphylococcus*.

MICRO CHECK

1. Which virulence factor allows *Staphylococcus aureus* to escape fibrin clots to spread to new locations in the body?
2. How is impetigo distinguished from staphylococcal scalded skin syndrome?
3. Which systemic disease caused by *Staphylococcus aureus* involves the buildup of pus between the lungs and the chest wall?
4. How are infections that are caused by *Staphylococcus aureus* distinguished from other staphylococcal infections?

TELL ME WHY

Staphylococcus aureus can cause disease in most every body system: integumentary, skeletal, muscular, nervous, lymphatic, respiratory, urinary, and digestive. Why is this microbe so versatile regarding the areas it can attack?

CLINICAL CASE STUDY

A Fatal Case of Methicillin-Resistant *Staphylococcus aureus* (MRSA)

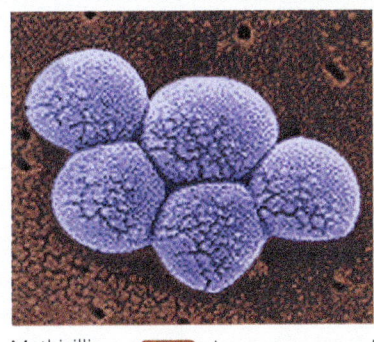

Methicillin-resistant *S. aureus* — SEM — 1 μm

An 83-year-old woman is admitted to a clinic with a wound on her left shin, a wound that looks bad, though she had just noticed it that morning. She is very concerned and is feeling quite ill. Her temperature is 102°F (38.8°C), and her heart rate is much faster than normal, though her blood pressure is within normal limits. She reports that she is diabetic.

The nurse practitioner examines the wound and diagnoses her with cellulitis—a bacterial infection of the skin and underlying tissues. He treats her prophylactically with a third-generation cephalosporin and orders a blood culture. Seventy-two hours later, the blood culture reveals clusters of Gram-positive cocci resistant to methicillin—MRSA! The NP immediately switches from cephalosporin to another antimicrobial; however, the woman's condition worsens, her blood pressure drops, and she dies five days later.

1. How might the woman have developed such a severe infection?
2. How might her shin have become infected?
3. What is the series of diseases she suffered?
4. What was likely the second antibiotic she received?

Streptococcus

LEARNING OUTCOME

19.8 Describe the classification of streptococcal strains.

The genus *Streptococcus* (strep-tō-kok'ŭs) is a diverse assemblage of Gram-positive cocci 0.5–1.2 μm in diameter and arranged in pairs or chains. They are catalase negative (unlike *Staphylococcus*), although they do synthesize peroxidase and thus are facultatively anaerobic.

Researchers differentiate species of *Streptococcus* using several different, overlapping schemes, including serological classification based on the reactions of antibodies to specific bacterial antigens, type of hemolysis (alpha, beta, or gamma; see Figure 6.13), cell arrangement, physiological properties as revealed by biochemical tests, and genetic analysis. In this chapter, we will largely use a serological classification scheme, developed by Rebecca Lancefield (1895–1981), that divides streptococci into serotype groups based on the bacteria's antigens (known, appropriately, as Lancefield antigens). The serotypes in this scheme include Lancefield groups A through H and K through V. Whereas the more significant streptococcal pathogens of humans are in groups A and B, two other significant streptococcal pathogens of humans lack Lancefield antigens. We begin our survey of the streptococci with group A *Streptococcus*.

Group A *Streptococcus: Streptococcus pyogenes*

LEARNING OUTCOMES

19.9 Describe two structures in *Streptococcus pyogenes* that enable this organism to survive the body's defenses.

19.10 Identify four kinds of enzymes that facilitate the spread of *S. pyogenes* in the body and the action of pyrogenic toxins in causing disease symptoms.

19.11 Describe seven diseases caused by *S. pyogenes* and the treatments available.

19.12 Identify the conditions under which group A *Streptococcus* causes disease.

Group A *Streptococcus* (GAS), which is synonymously known as *S. pyogenes* (pī-oj'en-ēz), is a coccus that forms white colonies 1–2 mm in diameter surrounded by a large zone of beta-hemolysis after 24 hours on blood agar plates. Pathogenic strains of this species often form capsules. The following sections discuss the pathogenesis and epidemiology of this species, as well as the diagnosis, treatment, and prevention of the diseases it causes.

Pathogenicity

Strains of *Streptococcus pyogenes* have a number of structures, enzymes, and toxins that enable them to survive as pathogens in the body.

Two main structural features enable cells of *S. pyogenes* to evade phagocytosis:

- *M protein.* A membrane protein called *M protein* destabilizes complement, thereby interfering with opsonization and lysis.

- *Hyaluronic acid capsule.* Because hyaluronic acid is normally found in the body, white blood cells may ignore bacteria "camouflaged" by this type of capsule.

Researchers have identified two *streptokinases* that break down blood clots, presumably enabling group A *Streptococcus* to rapidly spread through infected and damaged tissues. Similarly, four distinct *deoxyribonucleases* depolymerize DNA that has been released from dead cells in abscesses, reducing the firmness of the pus surrounding the bacteria and facilitating bacterial spread. *C5a peptidase* breaks down the complement protein C5a, which acts as a chemotactic factor (see Figure 15.9). Thus, *S. pyogenes* decreases the movement of white blood cells into a site of infection. Finally, *hyaluronidase* facilitates the spread of streptococci through tissues by breaking down hyaluronic acid.

GAS secretes three distinct *pyrogenic toxins*[7] (pī-rō-jen'ik) that stimulate macrophages and helper T lymphocytes to release cytokines that in turn stimulate fever, a widespread rash, and shock. Because these toxins cause blood capillaries near the surface to dilate, producing a red rash, some scientists call them *erythrogenic*[8] *toxins* (e-rith-rō-jen'ik). The genes for these toxins are carried on temperate bacteriophages, so only lysogenized bacteria—bacteria in which a virus has become part of the bacterial chromosome—secrete the toxins. Fever-stimulating *pyrogenic* toxins should not be confused with pus-producing *pyogenic* toxins.

S. pyogenes also produces two different, membrane-bound proteins, called *streptolysins*, which lyse red blood cells, white blood cells, and platelets; thus, these proteins interfere with the oxygen-carrying capacity of the blood, immunity, and blood clotting. After GAS has been phagocytized, it releases streptolysins into the cytoplasm of the phagocyte, causing lysosomes to release their contents, lysing the phagocyte and releasing the streptococci.

Epidemiology

GAS frequently infects the pharynx or skin, but the resulting abscesses are usually temporary, lasting only until adaptive immune responses against bacterial antigens (particularly M protein and streptolysins) clear the pathogens. Typically, *Streptococcus pyogenes* causes disease only when normal competing microbiota are depleted, when a large inoculum enables the streptococci to gain a rapid foothold before antibodies are formed against them, or when adaptive immunity is impaired. Following colonization of the skin or a mucous membrane, *S. pyogenes* can invade deeper tissues and organs through a break in such barriers. People spread *S. pyogenes* among themselves via respiratory droplets, especially under crowded conditions, such as those in classrooms and day care centers.

GAS was a major cause of human disease before the discovery of antimicrobial drugs, claiming the lives of millions. Because group A streptococci are sensitive to penicillin and other drugs, their significance as pathogens has declined; still, GAS sickens thousands of Americans annually.

---

[7]From Greek *pyr*, meaning "fire," and *genein*, meaning "to produce."
[8]From Greek *erythros*, meaning "red," and *genein*, meaning "to produce."

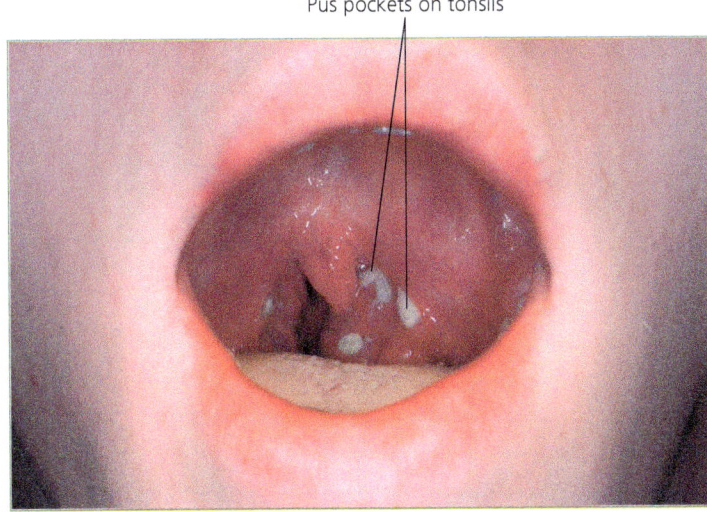

▲ FIGURE 19.6 Pharyngitis.

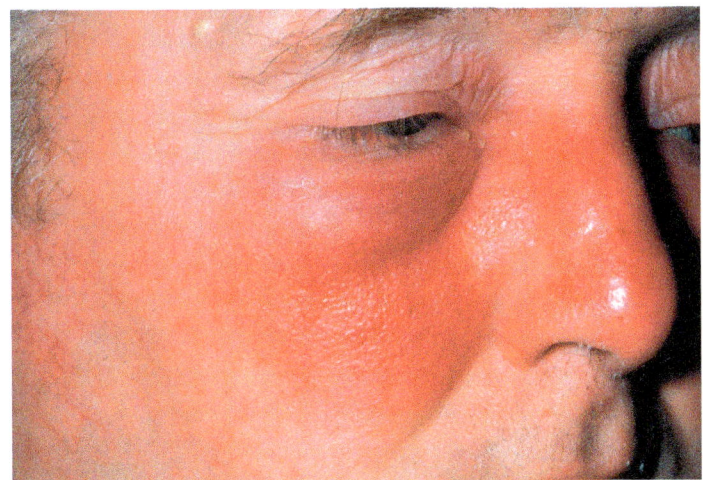

▲ FIGURE 19.7 Erysipelas. The localized, pus-filled lesions are caused by group A streptococci (Streptococcus pyogenes). What is the meaning of the word pyogenes?

Figure 19.7 Pyogenes means pus producing.

Group A Streptococcal Diseases

Group A *Streptococcus* causes a number of diseases, depending on the site of infection, the strain of bacteria, and the immune responses of the patient.

Pharyngitis A sore throat caused by streptococci, commonly known as "strep throat," is a kind of **pharyngitis** (far-in-jī'tis)—inflammation of the pharynx—that is accompanied by fever, malaise,[9] and headache. The back of the pharynx typically appears red, with swollen lymph nodes and *purulent* (pus-containing) abscesses covering the tonsils (FIGURE 19.6). Some microbiologists estimate that only 50% of patients diagnosed with strep throat actually have it; the rest have viral pharyngitis. Given that the symptoms and signs for the two diseases are identical, a sure diagnosis requires bacteriological or serological tests. Correct diagnosis is essential because bacterial pharyngitis is treatable with antibacterial drugs, which of course have no effect on viral pharyngitis.

Scarlet Fever The disease known as **scarlet fever** or as *scarlatina* often accompanies streptococcal pharyngitis when the infection involves a lysogenized strain of *S. pyogenes*. After one to two days of pharyngitis, pyrogenic toxins released by the streptococci trigger a diffuse rash that typically begins on the chest and spreads across the body. The tongue usually becomes strawberry red. The rash disappears after about a week and is followed by sloughing of the skin.

Rheumatic Fever A potential complication of *S. pyogenes* pharyngitis or of scarlet fever is **rheumatic fever** (rū-mat'ik), in which inflammation leads to damage of heart valves and muscle. Rheumatic fever is most common in five- to fifteen-year-old children. The exact relationship between streptococcal pharyngitis and rheumatic fever is not totally clear; however, failure to eradicate the organism from the pharynx is a significant risk factor for rheumatic fever. It appears that rheumatic fever is not caused directly by *Streptococcus* but instead is an autoimmune response in which antibodies directed against streptococcal antigens cross-react with heart antigens, though the exact cause of heart damage is unknown. Damage to the heart valves may be so extensive that they must be replaced when the patient reaches middle age. Rheumatic fever was much more prevalent before the advent of antimicrobial drugs.

Pyoderma, Erysipelas, and Cellulitis A **pyoderma** (pī-ō-der'ma) is a confined, pus-producing lesion that usually occurs on the exposed skin of the face, arms, or legs. One cause is group A streptococcal infection following direct contact with an infected person or contaminated fomites. This condition is also known as *impetigo* because its appearance is similar to that of the staphylococcal disease of the same name. After a pus-filled lesion breaks open, it forms a yellowish crust. This stage is highly contagious, and scratching may convey bacteria to the surrounding skin, spreading the lesions.

When a streptococcal infection also involves surrounding superficial lymphatic tissue and triggers pain and inflammation, the condition is called **erysipelas**[10] (er-i-sip'ĕ-las; FIGURE 19.7). Erysipelas occurs most commonly on the faces of children and the elderly.

Bacteria, often GAS, entering breaks in the skin may invade deeper layers of the skin to cause red, painful, localized inflammation—a condition called *cellulitis*. In many cases, the breaks in the skin are microscopic and go unnoticed by patients. Cellulitis is more common in the lower extremities, particularly in the immunocompromised, the obese, and diabetics who do not manage their condition well. Cellulitis differs from similar conditions in that cellulitis does not involve lymphatic tissue and more commonly has indistinct and spreading margins.

---

[9]French, meaning "discomfort."

[10]From Greek *erythros*, meaning "red," and *pella*, meaning "skin."

DISEASE IN DEPTH

NECROTIZING FASCIITIS

Group A *Streptococcus* and other bacteria

SIGNS AND SYMPTOMS

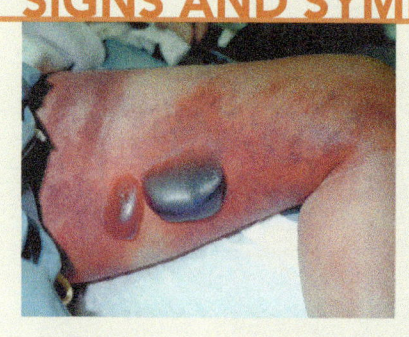

Necrotizing fasciitis is usually characterized redness, intense pain, and swelling at the of infection. Initially, the pain does not see. proportionate to the appearance of the infected area. As the bacterium digests mu: fascia—the connective tissue surrounding muscles—and fat tissue, the overlying skin becomes distended and discolored. Patients develop fever, nausea, and malaise, and the may become mentally confused as their blo pressure drops severely.

Multiple bacteria cause necrotizing fasciitis most commonly including *Streptococcus pyogenes* (also known as group A *Streptococcus* [GAS]), sensationalized in the news as "flesh-eating strep."

PATHOGENESIS

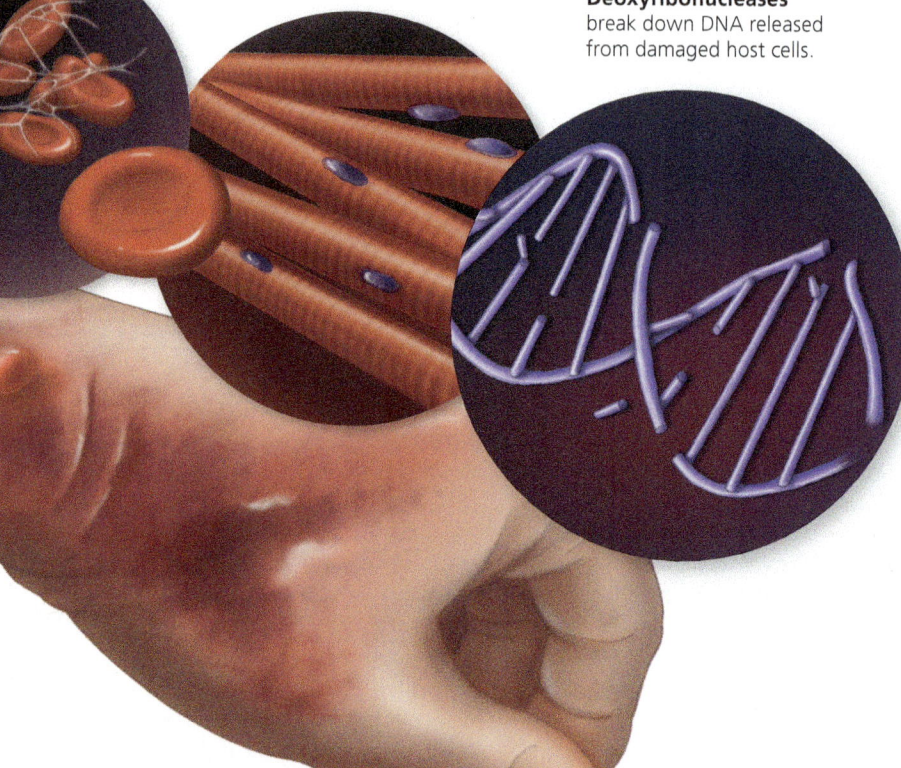

2 *S. pyogenes* secretes enyzmes that allow the bacterium to invade body tissues.

Streptokinases dissolve blood clots.

Hyaluronidase breaks down hyaluronic acid between cells.

Deoxyribonucleases break down DNA released from damaged host cells.

1 *S. pyogenes* is passed from person to person and enters the body through breaks in the skin. It spreads rapidly along muscle fascia.

 Watch Dr. Bauman's Video Tutor explore necrotizing fascitis, or go to the **Mastering Microbiology** Study Area for more information.

EPIDEMIOLOGY

Though there has been much recent press coverage, necrotizing fasciitis is not a new disease; it is possible that Hippocrates described the disease, and U.S. doctors in China definitively reported cases in 1871. Researchers estimate that there are about 500,000 cases of necrotizing fasciitis worldwide each year. About 20% of necrotizing fasciitis patients die.

Greek physician Hippocrates (460–370 B.C.) may have described necrotizing fasciitis as early as the 5th century B.C. in *Of the Epidemics*. "…And there were great fallings off (sloughing) of the flesh, sinews, and bones…" (Book II, section III, part 4).

PATHOGEN AND VIRULENCE FACTORS

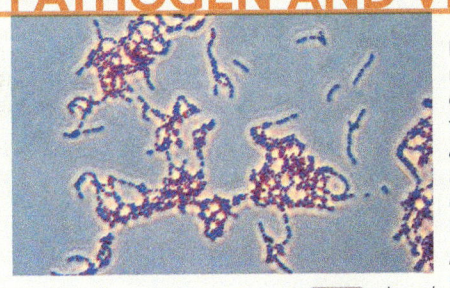

Researchers now know that necrotizing fasciitis is usually caused by multiple bacteria acting together, including *Staphylococcus aureus*, *Clostridium perfringens*, and *Bacterioides fragilis*, but most commonly involved is a low G + C, Gram-positive bacterium, *Streptococcus pyogenes* (group A *Streptococcus*), shown Gram-stained at left.

LM 15 μm

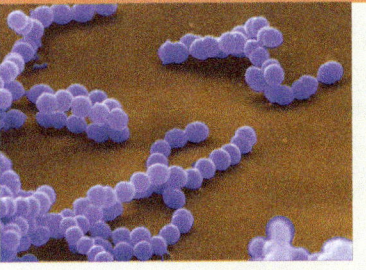

Strains of *S. pyogenes* that cause the condition have enzymes, such as deoxyribonucleases, hyaluronidase, and streptokinases, that allow the bacterium to invade body tissues. Streptococcal M protein allows the bacterium to attach to nasopharyngeal cells and to resist phagocytosis. The toxins streptolysin S and exotoxin A damage cells and tissue.

SEM 4 μm

3 Other virulence factors include M protein.

M protein on the surface of streptococcal cells helps the bacterium attach to nose and throat cells, and after entry into the body, M protein allows *S. pyogenes* to survive phagocytosis.

M proteins

4 *S. pyogenes* also secretes toxins that damage tissue.

Streptolysin S can kill many types of human cells, including neutrophils and erythrocytes.

Exotoxin A triggers an overactive immune response that further damages healthy tissue.

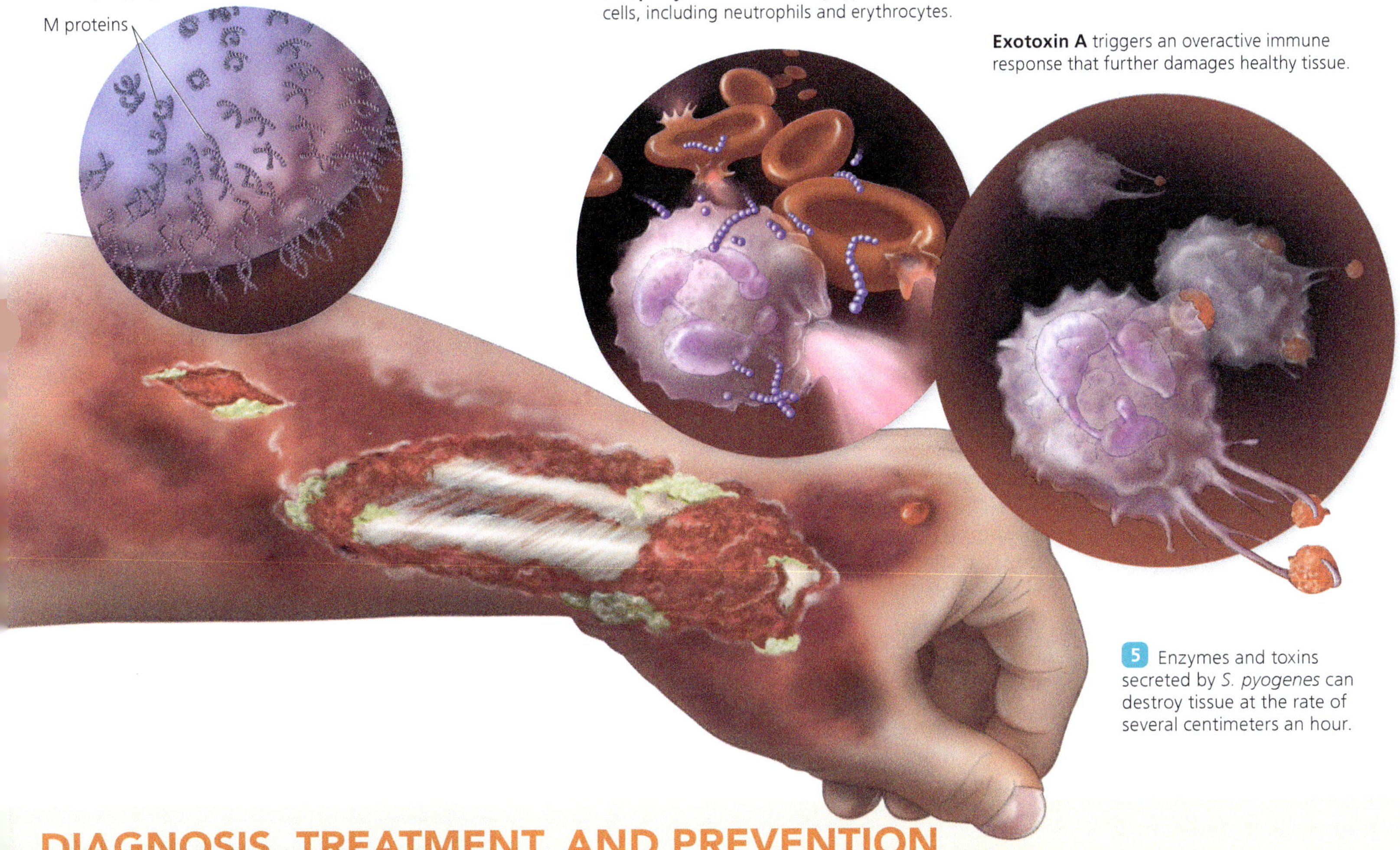

5 Enzymes and toxins secreted by *S. pyogenes* can destroy tissue at the rate of several centimeters an hour.

DIAGNOSIS, TREATMENT, AND PREVENTION

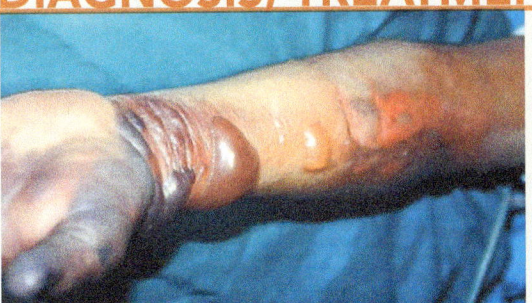

Diagnosis is difficult because early symptoms are general and flulike. Extreme pain that seems out of proportion to an injury should be treated as a possible case of necrotizing fasciitis or infection with *S. pyogenes*. Affected tissue must be removed completely to prevent the spread of bacteria. Intravenous broad-spectrum antimicrobial drugs help curb the severity of the disease. It is difficult to prevent necrotizing fasciitis because *S. pyogenes* is commonly found on humans. Patients with diabetes, cancer, or chickenpox are at a greater risk for necrotizing fasciitis. Any opening in the skin, even a minor cut, is a potential site of invasion by the bacterium; therefore, such openings should be washed thoroughly.

Streptococcal Toxic Shock Syndrome Group A streptococci can spread, albeit rarely, from an initial site of infection, particularly in patients infected with HIV or suffering with cancer, heart disease, pulmonary disease, or diabetes mellitus. Such spread leads to bacteremia and severe multisystem infections producing **streptococcal toxic shock syndrome (STSS).** Patients experience inflammation at sites of infection as well as pain, fever, chills, malaise, nausea, vomiting, and diarrhea. These signs and symptoms are followed by increased pain, organ failure, shock—and over 40% of patients die.

Necrotizing Fasciitis Another serious disease sometimes caused by group A *Streptococcus* is **necrotizing**[11] **fasciitis** (ne′kro-tī-zing fas-ē-ī′tis), sensationalized by the news media as "flesh-eating bacterial disease." **Disease in Depth: Necrotizing Fasciitis** (pp. 554–555) presents this condition in more detail.

Glomerulonephritis For an undetermined reason, antibodies bound to the antigens of some strains of group A *Streptococcus* are not removed from circulation but instead accumulate in the *glomeruli* (small blood vessels) of the kidneys' *nephrons* (filtering units). The result is **glomerulonephritis** (glō-mar′ū-lō-nef-rī′tis)—inflammation of the glomeruli and nephrons—which obstructs blood flow through the kidneys and leads to hypertension (high blood pressure) and low urine output. Blood and proteins are often secreted in the urine. Young patients usually recover fully from glomerulonephritis, but progressive and irreversible kidney damage may occur in adults.

Diagnosis, Treatment, and Prevention

Because *Streptococcus* is not a normal member of the microbiota of the skin, the observation of Gram-positive bacteria in short chains or pairs in cutaneous specimens can provide a rapid preliminary diagnosis of pyoderma, erysipelas, and necrotizing fasciitis. In contrast, streptococci are normally in the pharynx, so their presence in a respiratory sample is of little diagnostic value; instead, physicians use an immunological test called a *rapid strep test* that identifies the presence of group A streptococcal antigens.

Penicillin is usually effective against *S. pyogenes*. Erythromycin or cephalosporin is used to treat penicillin-sensitive patients. *S. pyogenes* is also susceptible to the topical antimicrobial bacitracin—a characteristic that distinguishes it from group B *Streptococcus* (discussed shortly). Necrotizing fasciitis must be treated with aggressive surgical removal of nonviable tissue and infecting bacteria. Because rheumatic fever and glomerulonephritis are the result of an immune response against group A streptococci, they cannot be treated directly; instead, the underlying infection must be arrested.

Antibodies against M protein provide long-term protection against *S. pyogenes*. However, antibodies directed against the M protein of one strain provide no protection against other strains; this explains why a person can have strep throat more than once.

Group B *Streptococcus*: *Streptococcus agalactiae*

LEARNING OUTCOMES

19.13 Contrast group B *Streptococcus* with group A *Streptococcus* in terms of colony characteristics when cultured.
19.14 Discuss the epidemiology, diagnosis, treatment, and prevention of infections with *Streptococcus agalactiae*.

Group B *Streptococcus*, or *S. agalactiae* (a-ga-lak′tē-ī), is a Gram-positive coccus, 0.6–1.2 μm in diameter, that divides to form chains. Like group A *Streptococcus*, *S. agalactiae* is beta-hemolytic, but it can be distinguished from the former by three qualities: it has group-specific, polysaccharide cell wall antigens; it forms buttery colonies that are 2–3 mm in diameter and have a small zone of beta-hemolysis after 24 hours of growth on blood agar; and it is bacitracin resistant.

Pathogenicity

Even though *S. agalactiae* forms capsules, antibodies target its capsular antigens, so the capsules don't provide the bacterium protection from the immune system. For this reason, *S. agalactiae* is most pathogenic in newborns who have not yet formed type-specific antibodies and whose mothers are uninfected (and so do not provide passive immunity across the placenta or in milk).

Group B streptococci produce enzymes—proteases (that catabolize protein), hemolysins (that lyse red blood cells), deoxyribonuclease, and hyaluronidase—that probably play a role in causing disease, though such a role has not been proven.

Epidemiology

Group B streptococci normally colonize the lower gastrointestinal (GI), genital, and urinary tracts. Diseases in adults primarily follow wound infections and childbirth, though group B *Streptococcus* is emerging as a significant pathogen in the elderly.

Sixty percent of newborns are inoculated with group B streptococcal strains either during passage through the birth canal or by health care personnel. Such infections do not cause disease when maternal antibodies have crossed the placenta, but mortality rates can exceed 50% in children of mothers who don't have antibodies. Infections in newborns less than one week old result in *early-onset disease;* infants infected at one week to three months of age have *late-onset disease.*

Diseases

Even though microbiologists initially described *Streptococcus agalactiae* as the cause of *puerperal*[12] *fever* (pyu-er′per-ăl; also called *childbirth fever*) in women, today the bacterium is most often associated with neonatal bacteremia, meningitis, and pneumonia, at least one of which occurs in approximately 3 of every 1000 newborns in the United States. Mortality has been reduced to about 5% as a result of rapid diagnosis and supportive care, but about 25% of infants surviving group B streptococcal meningitis have

---

[11]From Greek *nekros*, meaning "corpse."

[12]From Latin *puer*, meaning "child," and *pario*, meaning "to bring forth."

permanent neurological damage, including blindness, deafness, or severe intellectual disability. Immunocompromised older patients are also at risk from group B streptococcal infections, and about 25% of them die from streptococcal diseases.

Diagnosis, Treatment, and Prevention

Medical laboratory technologists identify group B streptococcal infections by means of ELISA tests utilizing antibodies directed against the bacteria's distinctive cell wall polysaccharides. Samples of clinical specimens can also be incubated in blood media containing the antimicrobial drug bacitracin, which inhibits the growth of other beta-hemolytic bacteria.

Penicillin or ampicillin work against group B *Streptococcus*, though some strains tolerate concentrations of the drugs more than 10 times greater than that needed to inhibit group A *Streptococcus*. For this reason, physicians may prescribe vancomycin instead of a penicillin.

The U.S. Centers for Disease Control and Prevention (CDC) recommends prophylactic administration of penicillin at birth to children whose mothers' urinary tracts are colonized with group B streptococci. Implementation of this guideline in 1996 reduced early-onset disease morbidity and mortality by 70% by 2001.

Other Beta-Hemolytic Streptococci

LEARNING OUTCOME

19.15 Contrast *Streptococcus equisimilis* and *S. anginosus* in terms of polysaccharide composition, diseases caused, and treatment.

Streptococcus equisimilis (ek-wi-si′mil-is) and *S. anginosus* (an-ji-nō′sŭs) are the only other pathogenic beta-hemolytic streptococci. Although members of both species typically have group C polysaccharides, some strains of *S. anginosus* have group F or group G antigens instead—an example of the confusing status of the classification of the streptococci.

S. equisimilis causes pharyngitis (and occasionally glomerulonephritis), but, unlike group A streptococci, these cases of pharyngitis never lead to rheumatic fever. *S. anginosus* causes pharyngitis. Penicillin is effective against both species.

Alpha-Hemolytic Streptococci: The Viridans Group

LEARNING OUTCOME

19.16 Identify the normal sites of viridans streptococci in the human body and list three serious diseases they cause.

Many alpha-hemolytic streptococci lack group-specific carbohydrates, and thus they are not part of any Lancefield group. Instead, microbiologists classify them as the **viridans**[13] **streptococci** (vir′i-danz strep′tō-kok′sī) because many of them produce a green pigment when grown on blood media. The taxonomic relationships among these microorganisms are poorly understood;

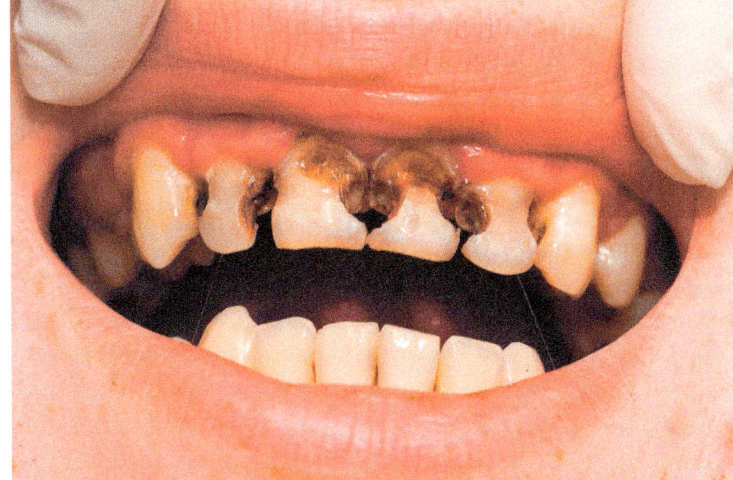

▲ **FIGURE 19.8 Dental caries.** Viridans streptococci and other bacteria in dental biofilms secrete acid that dissolves tooth enamel.

European and American microbiologists do not always agree about the names assigned to species, and some microbiologists place these microbes in a separate genus: *Abiotrophia* (a′bī-ō-trō′fē-ă). Some of the names given by Americans to viridans streptococci are *S. mitis* (mī′tis), *S. mutans* (mū′tanz), and *S. sanguis* (sang′wis). Members of the group are alpha-hemolytic and susceptible to penicillin.

Viridans streptococci normally inhabit the mouth, pharynx, GI tract, genital tract, and urinary tract of humans. They are opportunists that produce pus-filled abdominal lesions, and they are one cause of dental **caries**[14] (kar′ēz; cavities). The bacteria stick to dental surfaces via an insoluble polysaccharide called *dextran*, which they produce from glucose. Large quantities of dextran allow viridans streptococci to colonize the enamel of teeth and, with other bacteria, form a biofilm known as *dental plaque* (see Figure 14.6). Bacteria in the biofilm, particularly *Streptococcus mutans*, produce acids that dissolve tooth enamel (**FIGURE 19.8**). Viridans streptococci are not highly invasive, entering the blood only through surgical wounds and lacerations of the gums, including undetectable cuts produced by chewing hard candy, brushing the teeth, or dental procedures. Once in the blood, they can cause meningitis and endocarditis.

Streptococcus pneumoniae

LEARNING OUTCOMES

19.17 Describe how the structure of *Streptococcus pneumoniae* affects its pathogenicity.

19.18 Describe the route of *Streptococcus pneumoniae* through the body, explaining what chemical and physical properties allow it to cause pneumonia.

19.19 Discuss the diagnosis, treatment, and prevention of pneumococcal diseases.

Louis Pasteur discovered *Streptococcus pneumoniae* (nū-mō′nē-ī) in pneumonia patients in 1881. The bacterium is a Gram-positive

---

[13]From Latin *viridis*, meaning "green."

[14]Latin, meaning "dry rot."

coccus, 0.5–1.2 µm in diameter, that forms short chains or, more commonly, pairs (FIGURE 19.9). In fact, it was once classified in its own genus, "Diplococcus." Ninety-two different strains of *S. pneumoniae*, collectively called *pneumococci*, are known to infect humans.

Colonies of pneumococcus grown for 24 hours are 1–3 mm in diameter, round, mucoid, unpigmented, and dimpled in the middle because of the death of older cells. Colonies are alpha-hemolytic on blood agar when grown aerobically and beta-hemolytic when grown anaerobically. This bacterium lacks Lancefield antigens but does incorporate a species-specific teichoic acid into its cell wall.

Pathogenesis and Epidemiology

Streptococcus pneumoniae is a normal member of the pharyngeal microbiota that can colonize the lungs, sinuses, and middle ear. Even though microbiologists have studied the pneumococci extensively—the entire genomes of more than 10 strains have been sequenced—they still do not fully understand their pathogenicity; nevertheless, certain structural and chemical properties are known to be required.

The cells of virulent strains of *S. pneumoniae* are surrounded by a polysaccharide capsule, which protects them from digestion after endocytosis. A capsule is required for virulence; unencapsulated strains are avirulent. Microbiologists distinguish 90 unique serotypes based on differences in the antigenic properties of the capsules among various strains.

In addition, cells of *S. pneumoniae* insert into their cell walls a chemical called *phosphorylcholine*. When this chemical binds to receptors on cells in the lungs, in the meninges, or in blood vessel walls, it stimulates the cells to engulf the bacteria, though the cells don't digest the bacterium because of its capsule. Together, the polysaccharide capsule and phosphorylcholine enable pneumococci to "hide" inside body cells. *S. pneumoniae* can then pass across infected cells and enter the blood and brain.

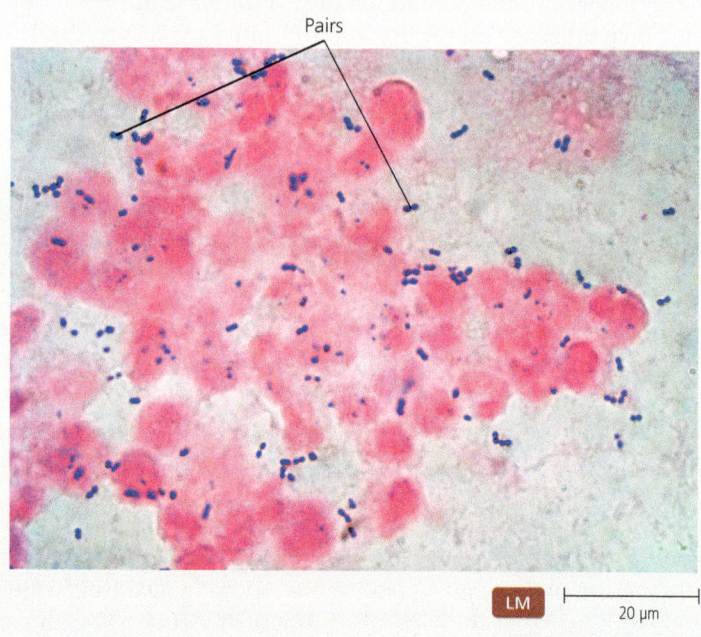

MICROBE AT A GLANCE 19.1

Streptococcus pneumoniae

Taxonomy: Domain Bacteria, phylum Firmicutes, class "Bacilli," order Bacillales, family Streptococcaceae

Other names: Pneumococcus, formerly "Diplococcus pneumoniae"

Cell morphology and arrangement: Cocci in pairs

Gram reaction: Positive

Virulence factors: Polysaccharide capsule, phosphorylcholine, protein adhesin, pneumolysin

Diseases caused: Pneumococcal pneumonia, community-acquired pneumonia (CAP), sinusitis, otitis media, bacteremia, endocarditis, pneumococcal meningitis

Treatment for diseases: Penicillin, cephalosporin, macrolide, fluoroquinolone

Prevention of disease: Two vaccines, one against 13 common strains and another against 23 strains, each lasts at least five years in adults with unimpaired immunity

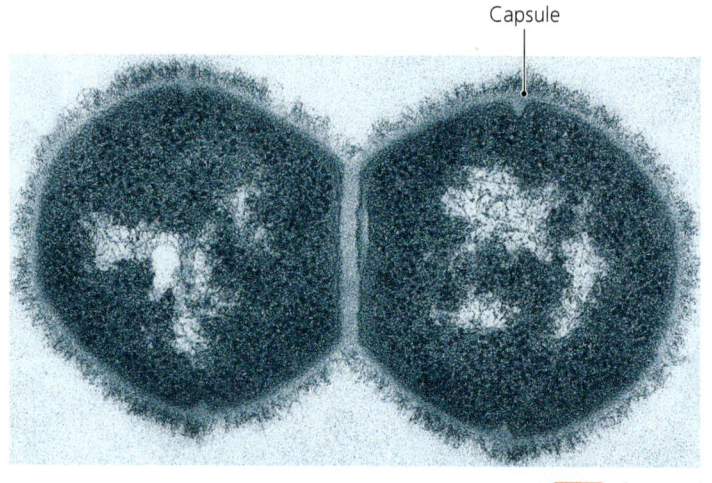

▲ **FIGURE 19.9** *Streptococcus pneumoniae.* Cells of this most common cause of pneumonia are paired and covered with a capsule. *What is pneumonia?*

Figure 19.9 Pneumonia is inflammation of the lungs, resulting in fluid buildup.

Pathogenic pneumococci secrete *adhesin*, a little-understood protein that mediates binding of the cells to epithelial cells of the pharynx. From there, the bacteria enter the lungs.

The body limits migration of bacteria into the lungs by binding the microbes with the active sites of secretory IgA. The rest of the antibody molecule then binds to mucus, trapping the bacteria, where they can be swept from the airways by the action of ciliated epithelium. The bacterium counteracts this defense by secreting *secretory IgA protease*, which destroys IgA, and *pneumolysin*, which binds to cholesterol in the cytoplasmic membranes of ciliated epithelial cells, producing transmembrane pores that result in the lysis of the cells. Pneumolysin also suppresses the digestion of endocytized bacteria by interfering with the action of lysosomes.

Streptococcus pneumoniae grows in the mouths and pharynges of 75% of humans, without causing harm; however, when pneumococci travel to the lungs, they cause disease. Typically, the incidence of pneumococcal disease is highest in children and the elderly, groups whose immune responses are not fully active.

Pneumococcal Diseases

Streptococcus pneumoniae causes a variety of diseases, which we explore next.

Pneumococcal Pneumonia The most prevalent disease caused by *S. pneumoniae* is **pneumococcal pneumonia** (nū-mō'nē-ă), which constitutes about 85% of all cases of pneumonia. The disease results when pneumococci are inhaled from the pharynx into lungs damaged either by a previous viral disease, such as influenza or measles, or by other conditions, such as alcoholism, congestive heart failure, or diabetes mellitus. As the bacteria multiply in the alveoli (air sacs), they damage the alveolar lining, allowing fluid, red blood cells, and leukocytes to enter the lungs. The leukocytes attack *Streptococcus*, in the process secreting inflammatory and pyrogenic chemicals. The onset of clinical symptoms is abrupt and includes a fever of 39–41°C and severe shaking chills. Most patients have a productive cough, slightly bloody sputum, and chest pain.

Sinusitis and Otitis Media Following viral infections of the upper respiratory tract, *S. pneumoniae* can also invade the sinuses and middle ear, where it causes **sinusitis** (sī-nŭ-sī'tis; inflammation of the nasal sinuses) and **otitis media** (ō-tī'tis mē'dē-ă; inflammation of the middle ear). Pus production and inflammation in these cavities create pressure and pain. Sinusitis occurs in patients of all ages. Otitis media is more prevalent in children because their narrow auditory tubes connecting the pharynx with the middle ears are nearly horizontal, which facilitates the flow of infected fluid from the pharynx into the middle ears. The tubes become wider and more vertical when the shape of the head changes as children grow, making infection less likely in adults.

Bacteremia and Endocarditis *Streptococcus pneumoniae* can enter the blood either through lacerations (as might result from vigorous tooth brushing, chewing hard foods, or dental procedures) or as a result of tissue damage in the lungs during pneumonia. Bacteria generally do not enter the blood during sinusitis or otitis media. As with *Staphylococcus*, *S. pneumoniae* can colonize the lining of the heart, causing endocarditis. The heart valves, once involved, are typically destroyed.

Pneumococcal Meningitis Pneumococci can spread to the meninges via bacteremia, during sinusitis or otitis media, or following head or neck surgery or trauma that opens a passage between the pharynx and the subarachnoid space of the meninges. The mortality rate of pneumococcal meningitis, which is primarily a disease of children, is up to 20 times that of meningitis caused by other microorganisms.

Diagnosis, Treatment, and Prevention

Medical laboratory technologists can quickly identify pneumococci in Gram stains of sputum smears and confirm their presence with the *Quellung*[15] *reaction*, in which anticapsular antibodies cause the capsule to swell. Antibodies against particular strains trigger Quellung reactions only against those strains. Antibody agglutination tests can also be used to identify specific strains. Culture of sputum samples is often difficult because pneumococci have fastidious nutritional requirements and cultures are often overgrown by normal oral microbiota. *S. pneumoniae* is sensitive to most antimicrobial drugs; therefore, samples for culture must be obtained before antibacterial therapy has begun. Laboratory technologists can distinguish pneumococcal colonies from other colonies of alpha-hemolytic strains by adding a drop of bile to a colony. Bile triggers chemicals present in pneumococci to lyse the cells, dissolving the colony in just a few minutes.

Penicillin has long been the drug of choice against *S. pneumoniae*, though penicillin-resistant strains have emerged since the late 1980s; about one-third of pneumococcal strains are now resistant. Cephalosporin, erythromycin, and chloramphenicol are also effective treatments.

Prevention of pneumococcal diseases is focused on a vaccine made from purified capsular material from the 23 most common pathogenic strains. The vaccine is immunogenic and long lasting in adults with healthy immune systems, but unfortunately, it is not as efficacious in patients at the greatest risk, such as the elderly, young children, and AIDS patients.

MICRO CHECK

5. How does streptokinase facilitate virulence of *Streptococcus pyogenes*?
6. Which diseases associated with *Streptococcus pyogenes* result from an adaptive immune response to bacterial antigens?
7. Which streptococcus usually causes dental caries?
8. What is the best way to prevent infection by *Streptococcus pneumoniae*?

TELL ME WHY

In 1928, Frederick Griffith discovered genetic transformation in which *Streptococcus pneumoniae* picked up DNA from the environment. Why did Griffith call the bacterium "Diplococcus pneumoniae"?

Enterococcus

LEARNING OUTCOME

19.20 Identify two species of *Enterococcus* and describe their pathogenicity and the diagnosis, treatment, and prevention of their diseases.

We have discussed two Gram-positive cocci that are pathogenic in humans: catalase-positive *Staphylococcus* and catalase-negative *Streptococcus*. Now we turn our attention to another genus of

---

[15]German, meaning "swelling."

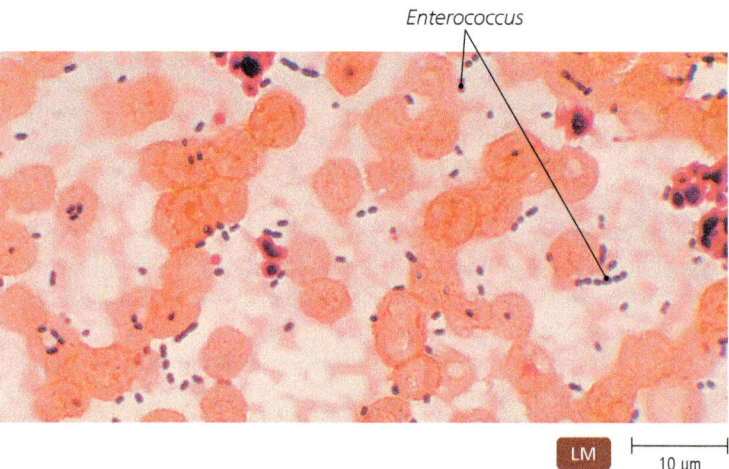

▲ FIGURE 19.10 **Enterococcus faecalis in lung tissue.** The Gram-positive cells are arranged in pairs and short chains but (unlike *S. pneumoniae*) lack capsules. *Why are these bacteria called "enterococci"?*

Figure 19.10 They are called "enterococci" because they are spherical bacteria that live in the intestinal tracts of animals.

Gram-positive, catalase-negative cocci—*Enterococcus* (en'ter-ō-kok'ŭs), so named because all enterococci are spherical and live in the intestinal tracts of animals. Lancefield classified enterococci with group D streptococci, but they differ significantly from other members of group D in that *Enterococcus* is unencapsulated, produces gas during fermentation of sugars, and is typically nonhemolytic (also called gamma-hemolytic). Because of these and other differences, microbiologists now classify *Enterococcus* as a separate genus.

Structure and Physiology

Enterococci form short chains and pairs; they do not form capsules (**FIGURE 19.10**). Enterococci grow at temperatures up to 45°C, at pH as high as 9.6, and in 6.5% NaCl or 40% bile salt broths—all conditions that severely inhibit the growth of *Streptococcus*. The two species that are significant pathogens of humans are *E. faecalis*[16] (fē-kă'lis) and *E. faecium*[17] (fē-sē'ŭm).

Pathogenesis, Epidemiology, and Diseases

E. faecalis is ubiquitous in the human colon. *E. faecium* is found less often. Because both species lack structural and chemical elements that make them virulent, they are rarely pathogenic in the intestinal tract. Nevertheless, they do have the ability to adhere to human epithelial cells, and they secrete *bacteriocins*, which are chemicals that inhibit the growth of other bacteria.

E. faecalis and *E. faecium* can cause serious disease if they are introduced into other parts of the body, such as the lungs, urinary tract, or bloodstream, via poor personal hygiene or intestinal laceration. Enterococci account for about 10% of health-care-associated infections and cause bacteremia, endocarditis, and wound infections.

Diagnosis, Treatment, and Prevention

Even though a Gram stain of *Enterococcus* looks similar to one of *S. pneumoniae*, the two genera can be distinguished readily in that *Enterococcus* is not sensitive to bile.

Enterococcal infections may be difficult to treat because strains resistant to frequently used antimicrobials—lactams, aminoglycosides, and vancomycin—are relatively common. This is particularly troublesome in that the genes for antimicrobial resistance occur on plasmids that can be transmitted to other bacteria.

It is difficult to prevent enterococcal infections, particularly in health care settings, where patients' immune systems are weakened. Health care workers should always use good hygiene and aseptic techniques to minimize the transmission of these microorganisms.

TABLE 19.2 summarizes the features of the major pathogenic, Gram-positive streptococci—*Streptococcus* and *Enterococcus*.

> **TELL ME WHY**
> Why is it unsurprising that *Enterococcus* can tolerate bile salts?

Now we turn our attention to the Gram-positive pathogenic bacilli, beginning with the low G + C, endospore-forming genera *Bacillus* and *Clostridium*.

Bacillus

Scientists divide Gram-positive bacilli (rod-shaped cells) into endospore-forming and non-endospore-forming genera. The endospore-forming genera are *Bacillus* and *Clostridium*. In this section, we examine *Bacillus anthracis*—one of the 51 species of *Bacillus* that is usually a pathogen of animals and humans. The following section discusses *Clostridium*.

Structure, Physiology, and Pathogenicity

LEARNING OUTCOME

19.21 Identify the structural features of *Bacillus* that contribute to its pathogenicity.

Bacillus anthracis[18] (ba-sil'ŭs an-thrā'sis) is a large (1 μm × 3–5 μm), rod-shaped, facultatively anaerobic, endospore-forming bacterium that normally dwells in soil. Its cells are arranged singly, in pairs, or in chains (**FIGURE 19.11**). The tough external coat and the internal chemicals of endospores make these structures resistant

---

[16]Latin, meaning "pertaining to feces."
[17]Latin, meaning "of feces."
[18]Greek, meaning "charcoal."

TABLE 19.2 Characteristics of Some Pathogenic Streptococci

Lancefield Group	Scientific Name	Hemolytic Pattern	Significant Characteristics	Characteristic Diseases
A	S. pyogenes	Large zone of beta-hemolysis	1- to 2-mm white colonies on blood agar; bacitracin sensitive	Pharyngitis, rheumatic fever, scarlet fever, pyoderma, erysipelas, cellulitis, streptococcal toxic shock syndrome, necrotizing fasciitis, glomerulonephritis
B	S. agalactiae	Small zone of beta-hemolysis	2- to 3-mm buttery colonies on blood agar; bacitracin resistant	Puerperal fever, neonatal bacteremia, meningitis, pneumonia
C	S. equisimilis	Large zone of beta-hemolysis	1- to 2-mm white colonies on blood agar	Pharyngitis, glomerulonephritis
C, F, or G	S. anginosus	Small zone of beta-hemolysis	1- to 2-mm white colonies on blood agar	Pharyngitis
—	S. mutans	Alpha-hemolysis	Viridans group (produce green pigment when grown on blood agar)	Dental caries; rarely bacteremia, meningitis, endocarditis
—	S. pneumoniae	Alpha-hemolysis (aerobic); beta-hemolysis (anaerobic)	Diplococci; capsule required for pathogenicity; bile sensitive	Pneumonia, sinusitis, otitis media, bacteremia, endocarditis, meningitis
D	Enterococcus faecalis, E. faecium, S. bovis [a]	None (gamma-hemolysis)	Diplococci or short chains; no capsule; bile insensitive	Urinary tract infections, bacteremia, endocarditis, wound infections

[a]Lancefield grouped Enterococcus with Streptococccus, but it is now considered a separate genus.

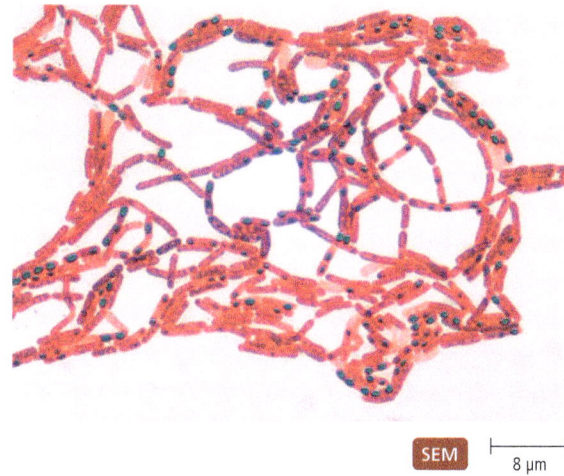

FIGURE 19.11 Bacillus anthracis. In this acid-fast stained specimen, endospores appear green inside red cells.

to harsh environmental conditions, enabling *Bacillus* to survive in the environment for centuries or perhaps even longer. A vegetative (non-endospore) cell of *Bacillus* can survive in the body because it has multiple copies of a plasmid (nonchromosomal DNA) coding for a capsule, which is composed solely of glutamic acid. This capsule inhibits effective phagocytosis by white blood cells.

Pathogenic strains of *B. anthracis* cause disease because they contain multiple copies of another plasmid coding for *anthrax toxins*—three distinct polypeptides that work together in a lethal combination. The toxin genes are "turned on" by bicarbonate, a molecule found in blood. Seven copies of one toxin combine to form a protein pore that breaches the cytoplasmic membrane of a host cell. The other two toxins enter the cell through the pore. Once in the cell, the toxins interfere with intracellular signaling, severely affect cellular metabolism, and ultimately cause the cell to undergo apoptosis (programmed cell suicide). Scientists are investigating the precise mechanisms by which the three components of anthrax toxin work in hopes of developing techniques for neutralizing them. Anthrax can be deadly even after treatment, because antimicrobial drugs do not inactivate accumulated anthrax toxin.

Epidemiology

LEARNING OUTCOME

19.22 List three methods of transmission of anthrax.

Anthrax is primarily a disease of herbivores; humans contract the disease from infected animals. Anthrax is not normally transmitted from human to human. It can invade via one of three routes: inhalation of endospores, inoculation of endospores into the body through a break in the skin, or ingestion of endospores. Ingestion anthrax is the normal means of transmission among animals but is rare in humans. In the 25 years between January 1976 and September 2001, only 15 human cases of anthrax were reported in the United States; thus, epidemiologists suspected bioterrorism when over a dozen cases of anthrax were reported in New York, Florida, and Washington, D.C. in the fall of 2001.

Disease

LEARNING OUTCOME

19.23 List and describe three clinical manifestations of *Bacillus anthracis* infections.

Bacillus anthracis causes only one disease—anthrax—but it can have three clinical manifestations. The first, *gastrointestinal anthrax*, is very rare in humans but is common in animals; it results in intestinal hemorrhaging and eventually death. *Inhalation anthrax* is also rare in humans, as it requires inhalation

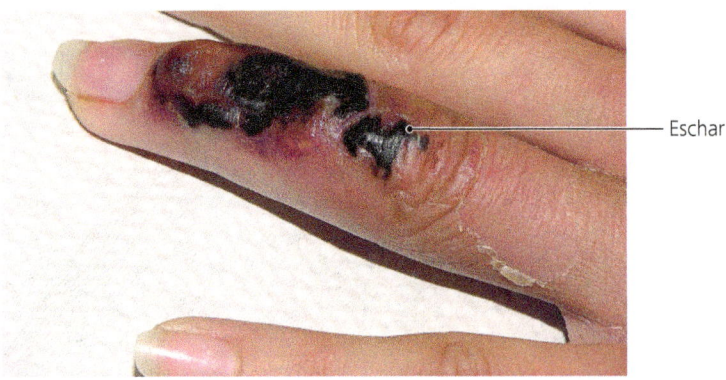

▲ FIGURE 19.12 **Cutaneous anthrax.** Black eschars are characteristic of cutaneous anthrax.

inhalation of airborne endospores. After endospores germinate in the lungs, they secrete toxins that are absorbed into the bloodstream, producing toxemia. Early signs and symptoms include fatigue, malaise, fever, aches, and cough—all of which are common to many pulmonary diseases. In a later phase, victims of inhalation anthrax have a high fever and labored breathing due to localized swelling, and they go into shock. Historically, mortality rates have been high—approaching 100% even when treated—perhaps in part because the disease is often not suspected until it is irreversible and because antimicrobial drugs do not neutralize toxins released during the course of the disease. During the bioterrorism attack of 2001, physicians learned that early and aggressive treatment of inhalation anthrax with antimicrobial drugs accompanied by persistent drainage of fluid from around the lungs increased the survival rate to greater than 50%.

Cutaneous anthrax is the most common in humans. It begins when a painless, solid, raised nodule forms on the skin at the site of infection. Cells in the affected area die, and the nodule spreads to form a painless, swollen, black, crusty ulcer called an *eschar* (es'kar; **FIGURE 19.12**). It is from the black color of eschars that anthrax, which means "charcoal" in Greek, gets its name. *B. anthracis* growing in the eschar releases anthrax toxin into the blood, producing toxemia. Untreated cutaneous anthrax is fatal for 20% of patients.

Diagnosis, Treatment, and Prevention

LEARNING OUTCOME

19.24 Identify the diagnosis, treatment, and prevention of anthrax.

Large, nonmotile, Gram-positive bacilli in clinical samples from the lungs or skin are diagnostic. As mentioned previously, endospores are not typically seen in clinical samples but are produced after a few days in culture.

Penicillin or doxycycline are preferred to treat nonbioterrorist cutaneous or inhalational anthrax; quinolones or doxycycline are used for bioterrorist anthrax. Physicians additionally prescribe monoclonal antibodies (IgG) against *B. anthracis* for prevention and treatment of inhalational anthrax following a bioterrorist event.

Prevention of naturally occurring disease in humans requires control of the disease in animals. Farmers in areas where anthrax is endemic must vaccinate their stock and deeply bury or burn the carcasses of infected animals. Anthrax vaccine has proven effective and safe for humans but requires five doses over 18 months plus annual boosters. Researchers are developing an alternative vaccine.

MICRO CHECK

9. Why is treatment of inhalational anthrax with antimicrobial drugs often not successful unless treatment starts very early?
10. Why can cutaneous anthrax be lethal in as much as 20% of cases if left untreated?

TELL ME WHY

Why is it necessary to burn or deeply bury the carcasses of animals that have died from anthrax?

Clostridia: *Clostridioides* and *Clostridium*

LEARNING OUTCOME

19.25 Characterize the major pathogenic species of *Clostridioides*. and the three major pathogenic species of *Clostridium*.

Clostridia are anaerobic, Gram-positive, endospore-forming bacilli that are ubiquitous in soil, water, sewage, and the gastrointestinal tracts of animals and humans. Several species in two genera—*Clostridioides* (klos-trid′ē-oy′dēz) and *Clostridium* (klos-trid′ē-ŭm) are significant human pathogens. Pathogenicity is due in great part to the ability of endospores to survive harsh conditions and to the secretion by vegetative cells of potent *histolytic toxins* (tissue-destroying toxins), *enterotoxins* (toxins affecting the GI tract), and *neurotoxins* (toxins affecting the nervous system). The most common pathogenic clostridia are *Clostridioides difficile* (formerly known as *Clostridium difficile*) and three species in the genus *Clostridium*: *C. botulinum*, *C. tetani*, and *C. perfringens*.

Clostridioides difficile

LEARNING OUTCOMES

19.26 Discuss the role of antimicrobial drugs in the development of gastrointestinal diseases caused by *Clostridioides difficile*.
19.27 Discuss the diagnosis, treatment, and prevention of *C. difficile* infections.

Clostridioides difficile (di′fe-sēl) is a motile, anaerobic intestinal bacterium with cells about in width and in length that form oval, subterminal endospores. The bacterium produces two toxins (called toxins A and B) and the enzyme hyaluronidase. Health care workers typically call the bacterium and its disease *C. diff.*

Pathogenesis, Epidemiology, and Disease

Although *C. difficile* is a common member of the intestinal microbiota, it can be an opportunistic pathogen in patients treated with broad-spectrum antimicrobial drugs, such as penicillin and cephalosporin. In such patients, the normal proportions of different bacteria in the colon can be significantly altered. In many cases the hardy endospores of *C. difficile* germinate, enabling it to become the predominant intestinal bacterium and produce enough toxins and enzymes to cause hemorrhagic death of the intestinal wall. In minor infections, these lesions result in a recurrent, persistent, explosive, watery diarrhea; however, in more serious cases, *C. difficile* produces life-threatening **pseudomembranous colitis,** in which large sections of the colon wall slough off, potentially perforating the colon and leading to massive internal infection by fecal bacteria and eventual death. *C. diff.* is a major cause of death of elderly patients resulting from dehydration, loss of electrolytes, and the complications of pseudomembranous colitis.

Clostridioides difficile can spread from asymptomatic carriers via airborne endospores that can survive five months or longer. In some hospitals, about 30% of healthcare-associated *C. diff.* infections are acquired from people with no symptoms.

Diagnosis, Treatment, and Prevention

Diarrhea in patients undergoing antimicrobial therapy is suggestive of *C. difficile* infection, and laboratory microbiologists confirm the diagnosis either by isolating the organism from feces using selective media or by demonstrating the presence of the toxins via immunoassays.

Discontinuation of the implicated antimicrobial drug, which allows normal microbiota to return, usually resolves minor infections with *C. difficile*. More serious cases are treated with either oral vancomycin or metronidazole, though endospores survive such therapy in about a third of patients, causing a relapse. Further treatment of relapses with either drug is often successful.

C. difficile is frequently found in hospitals, and hospital personnel can easily transmit it between patients. Proper hygiene—particularly frequent hand washing—is critical for limiting healthcare-associated infections. Its endospores survive ordinary floor cleaners, but bleach is effective in killing them.

Clostridium botulinum

LEARNING OUTCOMES

19.28 Contrast the three manifestations of botulism.
19.29 Describe the use of mice in the diagnosis of botulism.
19.30 Describe three treatments of botulism and explain how to prevent it.

Clostridium botulinum[19] (bo-tū-lī'num) is an anaerobic, endospore-forming, Gram-positive bacillus that is common in soil and water worldwide. Its endospores survive improper canning of food, germinating to produce vegetative cells that grow and release into the jar or can a powerful neurotoxin that causes **botulism** (bot'ū-lizm).

Pathogenesis

Strains of *C. botulinum* produce one of seven antigenically distinct *botulism toxins* (A through G). Scientists consider botulism toxins the deadliest toxins known—approximately one kilogram of pure toxin would be enough to kill every person in the world. Botulism toxins are extremely potent; even a small taste of contaminated food can cause full-blown illness or death.

Each of the seven toxins is a quaternary protein composed of a single neurologically active polypeptide associated with one or more nontoxic polypeptides that prevent the inactivation of the toxin by stomach acid. To understand the action of botulism toxins, we must consider the way the nervous system controls muscle contractions.

Each of the many ends of a motor neuron (nerve cell that stimulates the body) forms an intimate connection with a muscle cell at a *neuromuscular junction*; however, the two cells do not actually touch—a small gap, called a *synaptic cleft*, remains between them (**FIGURE 19.13a**). The neuron stores a chemical, called *acetylcholine*, in vesicles near its terminal cytoplasmic membrane. Acetylcholine is one of a family of chemicals called neurotransmitters that mediate communication among neurons and between neurons and other cells.

When a signal arrives at the neuromuscular junction of a motor neuron **1**, the vesicles containing acetylcholine fuse with the neuron's cytoplasmic membrane, releasing acetylcholine into the synaptic cleft **2**. Molecules of acetylcholine then diffuse across the cleft and bind to receptors on the cytoplasmic membrane of the muscle cell. The binding of acetylcholine to muscle cell receptors triggers a series of events inside the muscle cell that results in muscle contraction (not shown).

Botulism toxins act by binding irreversibly to neuronal cytoplasmic membranes **3**, thereby preventing the fusion of vesicles and secretion of acetylcholine into the synaptic cleft (**FIGURE 19.13b**). Thus, these neurotoxins prevent muscular contraction, resulting in a *flaccid paralysis*.

Epidemiology and Diseases

Clinicians recognize three manifestations of botulism: foodborne botulism, pediatric (infant) botulism, and wound botulism. Fortunately, all three are rare.

Foodborne botulism is not an infection but instead an *intoxication* (poisoning) caused by botulism toxin. About 25 cases of foodborne botulism occur in the United States each year, usually within one to two days following the consumption of toxin in home-canned foods or preserved fish. Contaminated food may not appear or smell spoiled. Patients are initially weak and dizzy and have blurred vision, dry mouth, dilated pupils, constipation, and abdominal pain, followed by a progressive paralysis that eventually affects the diaphragm. The patient remains mentally alert throughout the ordeal. In fatal cases, death results from the inability of muscles of respiration to effect inhalation; victims asphyxiate because they cannot inhale. Survivors recover very

---

[19]From Latin *botulus*, meaning "sausage."

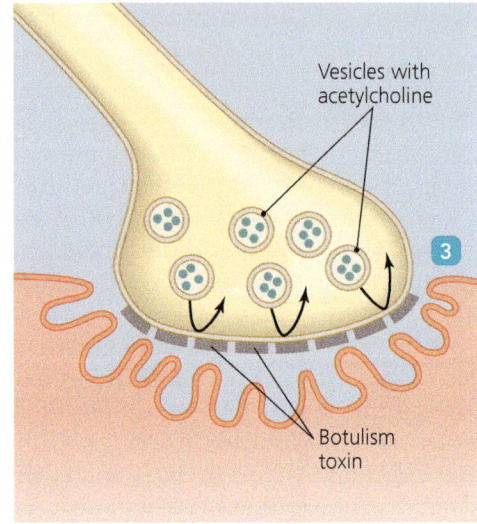

▲ FIGURE 19.13 **How botulism toxin acts at a neuromuscular junction.** (a) Normal function at a neuromuscular junction. ① A nerve impulse from the central nervous system causes vesicles filled with acetylcholine to fuse with the neuron's cytoplasmic membrane, ② releasing acetylcholine into the synaptic cleft. The binding of acetylcholine to receptors on the muscle cell's cytoplasmic membrane stimulates a series of events that result in contraction of the muscle cell (not shown). (b) Botulism toxin blocks the fusion of the vesicles with the neuron's cytoplasmic membrane, thereby preventing secretion of the neurotransmitter into the synaptic cleft ③; as a result, the muscle cell does not contract.

slowly as their nerve cells grow new endings over the course of months or years, replacing the debilitated termini.

In contrast to foodborne botulism, pediatric botulism results from the ingestion of endospores, which then germinate and colonize the infant's gastrointestinal (GI) tract. Infants are susceptible to colonization because their GI tracts do not have a sufficient number of benign microbiota to compete with *C. botulinum* for nutrients and space. Botulism toxin is absorbed into the blood from an infected infant's GI tract, causing nonspecific symptoms: crying, constipation, and "failure to thrive." Fortunately, paralysis and death are rare. About 100 cases of pediatric botulism are reported in the United States each year, though some cases reported as sudden infant death syndrome may be pediatric botulism instead.

Wound botulism usually begins four or more days following the contamination of a wound by endospores. With the exception that the gastrointestinal system is not typically involved, the symptoms are the same as those of foodborne botulism.

Diagnosis, Treatment, and Prevention

The symptoms of botulism are diagnostic; culturing the organism from contaminated food, from feces, or from the patient's wounds confirms the diagnosis. Further, toxin activity can be detected by using a mouse bioassay. In this laboratory procedure, specimens of food, feces, and/or serum are divided into two portions. Botulism antitoxin is mixed with one of the portions, and the portions are then inoculated into two sets of mice. If the mice receiving the antitoxin survive while the other mice die from botulism, the diagnosis is further confirmed.

Treatment of botulism depends on the type of disease:

- Foodborne botulism—induction of vomiting or repeated washing of the intestinal tract to remove *Clostridium*. Antitoxin (antibodies against all seven known *C. botulinum* toxins) can neutralize toxins before they bind to neurons. Antimicrobial drugs are not effective in treating foodborne botulism, which results from the ingestion of toxin, not of the bacteria themselves.
- Pediatric botulism—administration of antitoxin and supportive respiratory care while administering gentamicin or tobramycin.
- Wound botulism—administration of penicillin or chloramphenicol to kill clostridia.

Note that damage resulting from prior binding of toxin to any particular nerve ending is irreversible, so signs and symptoms of botulism may persist for years despite aggressive medical care.

Foodborne botulism is prevented by destroying all endospores in contaminated food by proper canning techniques, by preventing endospores from germinating by using refrigeration or establishing an acidic environment (pH < 4.5), or by destroying the toxin by heating to at least 80°C for 20 minutes or more. Because pediatric botulism is often associated with the consumption of honey, parents are advised not to feed honey to infants under age one because their intestinal microbiota are not sufficiently developed to inhibit the germination of *C. botulinum* endospores and their growth as toxin-producing vegetative cells.

MICROBE AT A GLANCE 19.2

Clostridium botulinum

Taxonomy: Domain Bacteria, phylum Firmicutes, class "Clostridia," order Clostridiales, family Clostridiaceae

Cell morphology and arrangement: Endospore-forming bacillus

Gram reaction: Positive

Virulence factors: Endospore, botulism toxins A–G

Diseases caused: Foodborne botulism, infant botulism, and wound botulism

Treatment for diseases: Repeated washing of intestinal tract to remove *Clostridium*, administration of antitoxin (antibodies against the toxins), antimicrobial drugs used for cases of infant botulism and wound botulism; damage to nerve endings irreversible

Prevention of disease: Proper canning to kill endospores; refrigeration to prevent endospores from germinating; heating to 80°C for 20 minutes to destroy toxin; refrain from feeding honey to infants

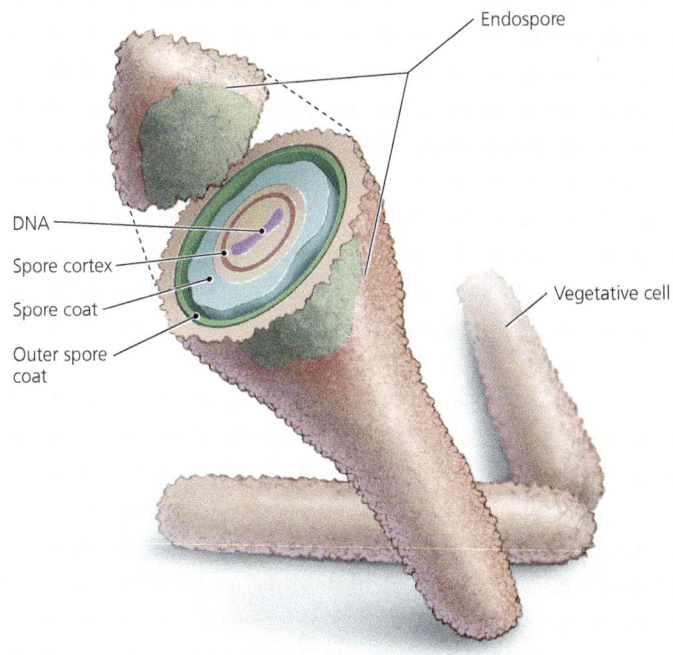

Clostridium tetani

LEARNING OUTCOMES

19.31 Describe the epidemiology of tetanus.
19.32 List treatments for and preventive measures against *Clostridium tetani* infections.

Clostridium tetani[20] (te′tan-ē) is a small, motile, obligate anaerobe that produces a terminal endospore, giving the cell a distinctive lollipop appearance (FIGURE 19.14). *C. tetani* is ubiquitous in soil, dust, and the GI tracts of animals and humans. Its vegetative cells are extremely sensitive to oxygen and live only in anaerobic environments, but its endospores survive for years. Its toxin causes the disease **tetanus.**

Pathogenesis

Contrary to popular belief, deep puncture wounds by rusty nails are not the only (or even primary) source of tetanus. Even a tiny break in the skin or mucous membranes can allow endospores of *C. tetani* access to an anaerobic environment in which they germinate, grow, and produce a fatal disease caused by **tetanospasmin** (tetanus toxin)—a potent neurotoxin released by *C. tetani* cells when they die. To understand the action of tetanospasmin, we must further consider the control of muscles by the central nervous system.

As we have seen, the secretion of acetylcholine by motor neurons at neuromuscular junctions stimulates muscles to contract. Nerves can only stimulate muscles; there is no inhibitory neurotransmitter released at the junction that could relax a muscular contraction. Instead, muscle cells naturally return to a relaxed state when the muscle remains unstimulated by the neurons. In contrast, a motor neuron can be inhibited by other neurons that release inhibitory neurotransmitters in the brain or spinal cord. In other words, inhibitory neurons of the central nervous system can inhibit motor neurons that then do not stimulate the muscle, which remains relaxed (FIGURE 19.15a).

Tetanospasmin released from *C. tetani* is composed of two polypeptides held together by a disulfide bond. The heavier of the two polypeptides binds to a receptor on a neuron's cytoplasmic membrane. The neuron then endocytizes the toxin, removes the lighter of the two polypeptides, and transports the lighter portion to the central nervous system. There the small polypeptide enters an inhibitory neuron and blocks the release of inhibitory neurotransmitter. With inhibition blocked, excitatory activity is unregulated, and muscles are signaled to contract

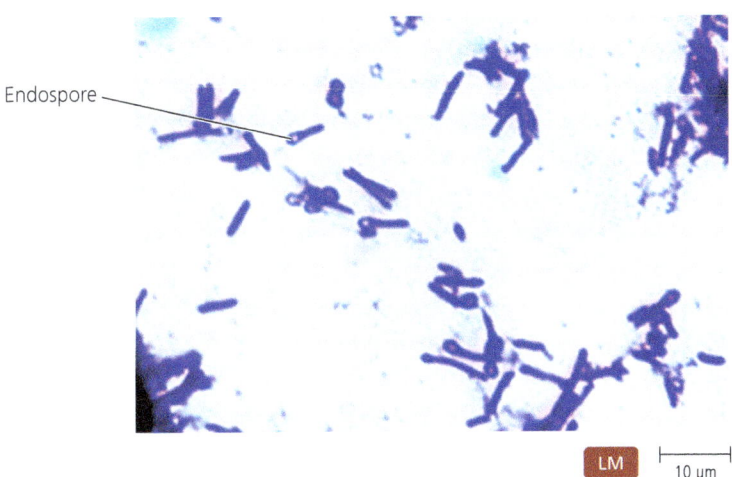

▲ **FIGURE 19.14 Cells of Clostridium tetani, with terminal endospores.** The cells have a distinctive lollipop shape. *How does the location of* C. tetani *endospores distinguish this bacterium from* Bacillus anthracis?

Figure 19.14 The endospores of *C. tetani* are terminal, whereas those of Bacillus anthracis are located centrally.

---
[20]From Greek *tetanos*, meaning "to stretch."

simultaneously (FIGURE 19.15b). The result is that muscles on both sides of joints contract and do not relax. Opposing contractions can be so severe that they break bones.

Epidemiology and Disease

Over one million cases of tetanus occur annually worldwide, mostly in less developed countries where vaccination is unavailable or medical care is inadequate. The effect of tetanospasmin is irreversible at any particular synapse, so recovery depends on the growth of new neuronal terminals to replace those affected. The mortality rate of tetanus is about 50% among all patients, though the mortality of neonatal tetanus, resulting most commonly from infection of the umbilical stump, exceeds 90%.

The incubation period of tetanus ranges from a few days to a week depending on the distance of the site of infection from the central nervous system. Typically the initial and diagnostic sign of tetanus is tightening of the jaw and neck muscles, which is why tetanus is also called *lockjaw*. Other early symptoms include sweating, drooling, grouchiness, and constant back spasms. If the toxin spreads to autonomic neurons, then heartbeat irregularities, fluctuations in blood pressure, and extensive sweating result. Spasms and contractions may spread to other muscles, becoming so severe that the arms and fists curl tightly, the feet curl down,

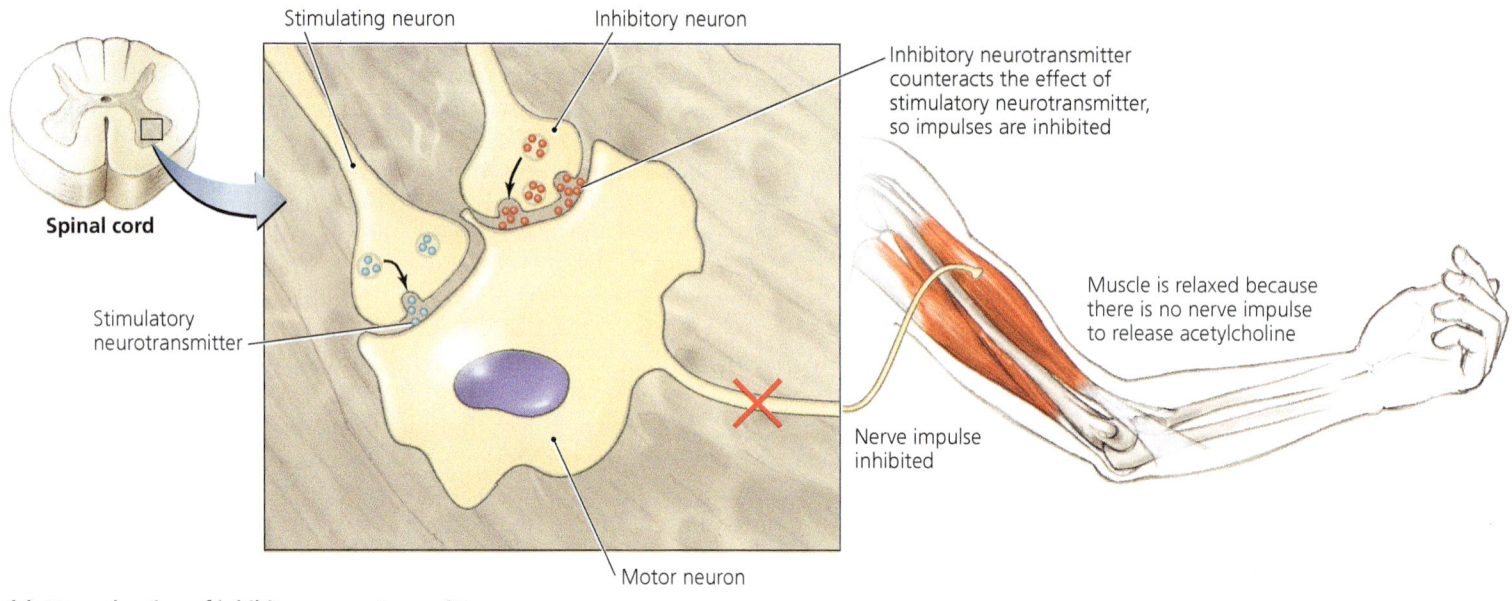

(a) Normal action of inhibitory neurotransmitter

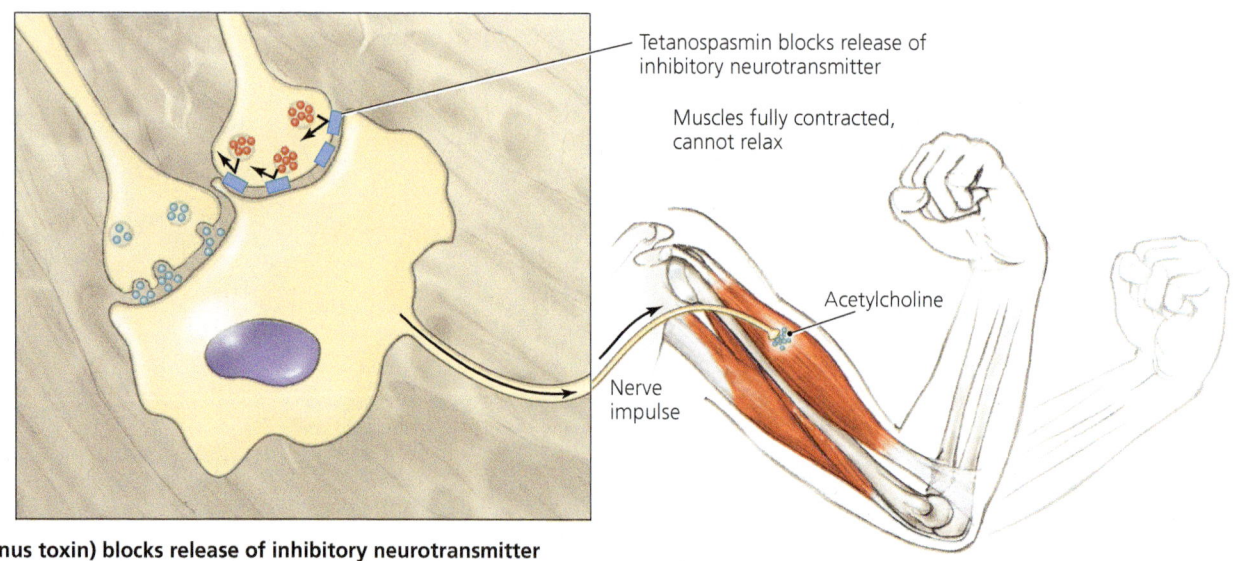

(b) Tetanospasmin (tetanus toxin) blocks release of inhibitory neurotransmitter

▲ FIGURE 19.15 **The action of tetanospasmin (tetanus toxin) on a pair of antagonistic muscles.** (a) Normally, when the muscle on one side of a joint is stimulated to contract, neurons controlling the antagonistic muscle are inhibited, so the antagonistic muscle relaxes. (b) Tetanospasmin blocks the inhibitory neurotransmitter. As a result, motor neurons controlling the antagonistic muscle are not inhibited but instead can generate nerve impulses that stimulate muscle contraction.

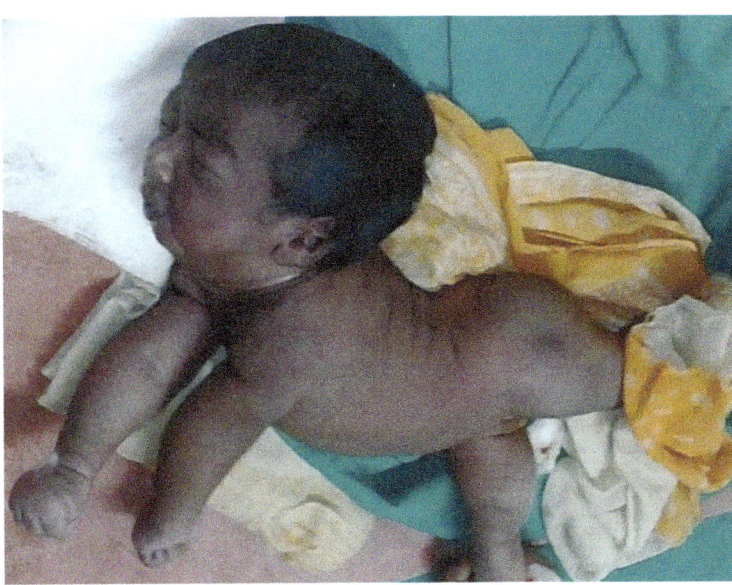

▲ FIGURE 19.16 **A patient with tetanus.** Prolonged muscular contractions result from the action of tetanus toxin.

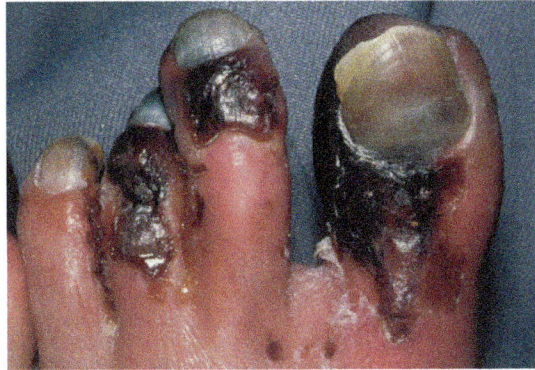

▲ FIGURE 19.17 **Gas gangrene, a life-threatening disease caused by *Clostridium perfringens*.** The blackening is the result of the death of muscle tissue (myonecrosis), whereas the "bubbling" appearance results from the production of gaseous waste products by the bacteria.

and the body assumes a stiff backward arch as the heels and back of the head bend toward one another (FIGURE 19.16). Complete, unrelenting contraction of the diaphragm results in a final inhalation; patients die because they cannot exhale.

Diagnosis, Treatment, and Prevention

The diagnostic feature of tetanus is the characteristic muscular contraction, which is often noted too late to save the patient. The bacterium itself is rarely isolated from clinical samples because it grows slowly in culture and is extremely sensitive to oxygen.

Treatment involves thorough cleaning of wounds to remove all endospores, immediate passive immunization with immunoglobulin directed against the toxin, the administration of antimicrobials such as metronidazole, and active immunization with tetanus toxoid. Cleansing and antimicrobials eliminate the bacteria, whereas the immunoglobulin binds to and neutralizes tetanospasmin before it can attach to neurons. Active immunization by one of four vaccines (DTaP, DT, Tdap, Td) stimulates the formation of antibodies that continue to neutralize toxin for a longer time. Once the toxin binds to a neuron, treatment is limited to supportive care.

The number of cases of tetanus in the United States has steadily declined as a result of effective immunization with tetanus toxoid. The CDC currently recommends six doses beginning at two months of age, followed by a booster every 10 years for life.

Clostridium perfringens

LEARNING OUTCOMES

19.33 Identify the mechanisms accounting for the pathogenesis of *Clostridium perfringens* infections.
19.34 Describe the diagnosis, treatment, and prevention of *Clostridium perfringens* infections.

Clostridium perfringens (per-frin'jens), the clostridium most frequently isolated from clinical specimens, is a large, almost rectangular, Gram-positive bacillus. Although it is nonmotile, its rapid growth enables it to proliferate across the surface of laboratory media, resembling the spread of motile bacteria. Endospores are rarely observed either in clinical samples or in culture. *C. perfringens* type A, known by its specific antigens, is the most virulent serotype.

Pathogenesis, Epidemiology, and Disease

C. perfringens produces 11 toxins that lyse erythrocytes and leukocytes, increase vascular permeability, reduce blood pressure, and kill muscle cells, resulting in irreversible damage. Because *C. perfringens* commonly grows in the digestive tracts of animals and humans, it is nearly ubiquitous in fecally contaminated soil and water.

The severity of diseases caused by *C. perfringens* ranges from mild food poisoning to life-threatening illness. Clostridial food poisoning is a relatively benign disease characterized by abdominal cramps and watery diarrhea but not fever, nausea, or vomiting. It lasts for less than 24 hours. Such food poisoning typically results from the ingestion of large numbers (10^8 or more) of *C. perfringens* type A in contaminated meat.

C. perfringens is not invasive, but when some traumatic event (such as a surgical incision, a puncture, a gunshot wound, crushing trauma, or a compound fracture) introduces endospores into the body, they can germinate in the anaerobic environment of deep tissues. The immediate result is intense pain at the initial site of infection as clostridial toxins induce swelling and tissue death. The rapidly reproducing bacteria can then spread into the surrounding tissue, causing the death of muscle and connective tissue that is typically accompanied by the production of abundant, foul-smelling, gaseous, bacterial waste products—hence the common name for the disease: **gas gangrene**[21] (FIGURE 19.17). Shock, kidney failure, and death can follow, often within a week of infection.

Diagnosis, Treatment, and Prevention

Medical laboratory scientists show that *Clostridium* is involved in food poisoning by demonstrating more than 10^5 bacteria in a

---
[21]From Greek *gangraina*, meaning "an eating sore."

gram of food or 10^6 cells per gram of feces. The appearance of gas gangrene is usually diagnostic by itself, though the detection of large Gram-positive bacilli is confirmatory.

Clostridial food poisoning is typically self-limiting—the pathogens and their toxins are eliminated in the resulting watery stool. In contrast, physicians must quickly and aggressively intervene to stop the spread of necrosis in gas gangrene in muscle by surgically removing dead tissue and administering large doses of antitoxin and penicillin. Oxygen applied under pressure may also be effective. Despite all therapeutic care, mortality of gas gangrene exceeds 40%.

It is difficult to prevent infections of *C. perfringens* because the organism is so common, though refrigeration of food prevents toxin formation and reduces the chance of clostridial food poisoning. Further, reheating contaminated food destroys any toxin that has formed. Given that gas gangrene occurs when endospores are introduced deep into tissues, proper cleaning of wounds can prevent many cases.

TELL ME WHY

Why does use of broad-spectrum antibacterial drugs increase the likelihood of a patient developing an opportunistic infection with *C. diff.*?

In the previous two sections, we examined two pathogenic, Gram-positive bacilli that form endospores: *Bacillus*, which is facultatively anaerobic, and *Clostridium*, which is anaerobic. Next, we consider the low G + C, rod-shaped *Listeria* and the unusual mycoplasmas, which are low G + C Gram-positive microbes that Gram stain pink instead of purple.

MICRO CHECK

11. What is the most significant risk factor associated with pseudomembranous colitis?
12. Which of the *Clostridium* species causes a disease that is vaccine-preventable?
13. What clostridial disease can be treated with oxygen applied under pressure?

Listeria

LEARNING OUTCOMES

19.35 Describe the characteristics of *Listeria monocytogenes* that account for its pathogenicity.
19.36 Discuss the signs and symptoms, diagnosis, treatment, and prevention of listeriosis.

Listeria monocytogenes (lis-tēr′ē-ă mo-nō-sī-tah′je-nēz) is a Gram-positive bacterium that enters the body in contaminated food or drink and causes listeriosis. *Listeria* is rarely pathogenic in healthy adults; infection in pregnant women, fetuses, newborns, the elderly, and immunocompromised patients can result in meningitis. *Listeria* is tolerant of cold temperatures, so it can grow in refrigerated food, unlike most other foodborne bacteria. Prevention of infection is difficult because the bacterium is ubiquitous, though the U.S. Food and Drug Administration (FDA) has approved treatment of cheeses and meats with active viruses specific against *Listeria*. Bacteriophage-treated food is safe for human consumption. **Disease in Depth: Listeriosis** (pp. 570–571) illustrates aspects of this disease.

TELL ME WHY

Why do doctors caution patients at risk for listeriosis to avoid deli meats, cold cuts, and soft cheeses?

Mycoplasmas

LEARNING OUTCOMES

19.37 List at least four characteristics of all mycoplasmas.
19.38 Explain why mycoplasmas have been classified with both Gram-negative and Gram-positive organisms.
19.39 Compare and contrast mycoplasmas and viruses.

Mycoplasmas[22] (mī′kō-plaz-măz) are unique bacteria named for their most common representative, the genus *Mycoplasma*. Bacteria in the genus *Ureaplasma* (yū-rē′a-plaz-mă) are also part of this group. Mycoplasmas lack cytochromes (which are present in many other organisms' electron transport chains), the enzymes of the citric acid cycle, and cell walls—indeed, they are incapable of synthesizing peptidoglycan and its precursors. Moreover, most mycoplasmas have *sterols* in their cytoplasmic membranes, a feature lacking in other prokaryotes. They are distinct from all other organisms in that the codon UGA neither serves as a stop codon nor codes for selenocysteine but instead codes for the amino acid tryptophan.

Before analysis of the nucleic acid sequences of mycoplasmal rRNA revealed that they are genetically Gram-positive organisms, mycoplasmas were classified in a separate phylum of Gram-negative bacteria—phylum Mollicutes. Modern taxonomists and the second edition of *Bergey's Manual of Systematic Bacteriology* now categorize mycoplasmas as low G + C, Gram-positive bacteria in the phylum Firmicutes and class Mollicutes. Despite being Gram-positive bacteria, they appear pink when stained with the Gram stain (because they lack cell walls).

Because they lack cell walls, mycoplasmas are pleomorphic, taking on a variety of shapes, including cocci and thin, unicellular filaments up to 150 μm long that resemble the hyphae of fungi. Mycoplasmas avoid osmotic stress by colonizing osmotically protected habitats, such as animal and human bodies, and the sterols in their membranes provide sufficient strength and rigidity.

Mycoplasmas have diameters ranging from 0.1 μm to 0.8 μm, making them the smallest *free-living* microbes—that is, those that can grow and reproduce independently of other cells. Originally, many mycoplasmas were thought to be viruses because their small, flexible cells enabled them to squeeze through the 0.45-μm pores of filters that were at one time used to remove bacteria from solutions; however, mycoplasmas contain both functional RNA and DNA, and they divide by binary fission—traits that viruses lack.

---

[22]From Greek *mycos*, meaning "fungus," and *plasma*, meaning "anything formed or molded."

Mycoplasmas require organic growth factors, including cholesterol, fatty acids, vitamins, amino acids, and nucleotides; these factors must be either acquired from a host or supplied in laboratory media. With exception of the strictly aerobic *Mycoplasma pneumoniae* (nū-mō'nē-ī), mycoplasmas are facultative anaerobes.

On solid media, most mycoplasmas form distinctive colonies that are so small they must often be viewed through a microscope. Colonies of most species resemble fried eggs, because the cells at the center of a colony tend to grow into the agar, whereas those at the periphery remain on the surface (FIGURE 19.18). However, two important pathogenic mycoplasmas, *M. pneumoniae* and *Ureaplasma*, are exceptions in that their colonies do not resemble fried eggs.

Mycoplasmas can colonize mucous membranes of the respiratory and urinary tracts and are associated with pneumonia and urinary tract infections. Of the two genera that cause diseases in humans, *Mycoplasma* is unable to utilize urea (i.e., it is urease negative), whereas *Ureaplasma* does hydrolyze urea to form ammonia.

Over 100 species of *Mycoplasma* have been identified, but only a few cause significant diseases in humans. We begin our examination of these species with *M. pneumoniae*.

Mycoplasma pneumoniae

LEARNING OUTCOME

19.40 Describe the damage done to respiratory epithelial cells by *Mycoplasma pneumoniae*.

Pathogenesis, Epidemiology, and Disease

Mycoplasma pneumoniae is a pleomorphic bacterium (FIGURE 19.19). It has an adhesive protein that attaches specifically to receptors located at the bases of cilia on epithelial cells lining the respiratory tracts of humans. Attachment causes the cilia to stop beating, and mycoplasmal colonization eventually kills the epithelial cells. This interrupts the normal removal of mucus from the respiratory tract, allowing colonization by other bacteria and causing a buildup of mucus that irritates the upper respiratory tract. Early symptoms of *M. pneumoniae* infections—including fever, malaise, headache, and sore throat—are not typical of other types of pneumonia; thus, the disease is called **primary atypical pneumonia**, or *walking pneumonia*. The body subsequently responds with a persistent, unproductive cough in an attempt to clear the lungs.

Primary atypical pneumonia may last for several weeks, but it is usually not severe enough to require hospitalization or to cause death. Symptoms can be mild, which is the reason for the name *walking pneumonia*.

Nasal secretions spread *M. pneumoniae* among people in close contact, such as classmates and family members. The disease appears to be uncommon in children under age 5 or in adults older than 20, though it is probably the most common form of pneumonia seen in children 5 to 15 years old. However, because primary atypical pneumonia is not a reportable disease and is difficult to diagnose, the actual incidence of infection is unknown.

Primary atypical pneumonia occurs throughout the year. This lack of seasonality is in contrast to pneumococcal pneumonia (caused by *Streptococcus pneumoniae*), which is more commonly seen in the fall and winter.

Diagnosis, Treatment, and Prevention

Diagnosis is difficult because mycoplasmas are small and difficult to detect in clinical specimens or tissue samples. Further, mycoplasmas grow slowly in culture, requiring two to six weeks before colonies can be seen. As previously noted, the colonies of *M. pneumoniae* differ from the colonies of other mycoplasmas in lacking a fried egg appearance; instead, the colonies have a uniform granular appearance. Hemagglutination, indirect fluorescent antibody tests, or enzyme-linked immunosorbent assays (ELISAs) are sometimes used to confirm a diagnosis.

Physicians treat primary atypical pneumonia with macrolides (e.g., erythromycin, azithromycin, or clarithromycin), doxycycline, or fluoroquinolones (e.g., levofloxacin or moxifloxacin). Prevention is difficult because patients are often infectious for long periods of time without signs or symptoms, and they

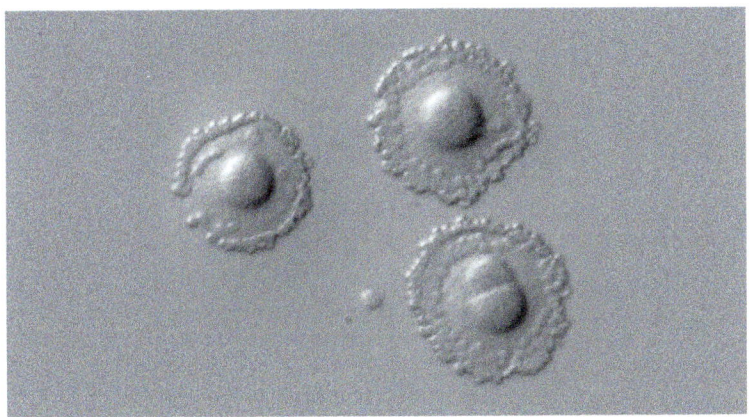

▲ FIGURE 19.18 Colonies of *Mycoplasma*. The small colonies resemble fried eggs, because central cells grow into the medium, while cells on the edges grow only on the surface.

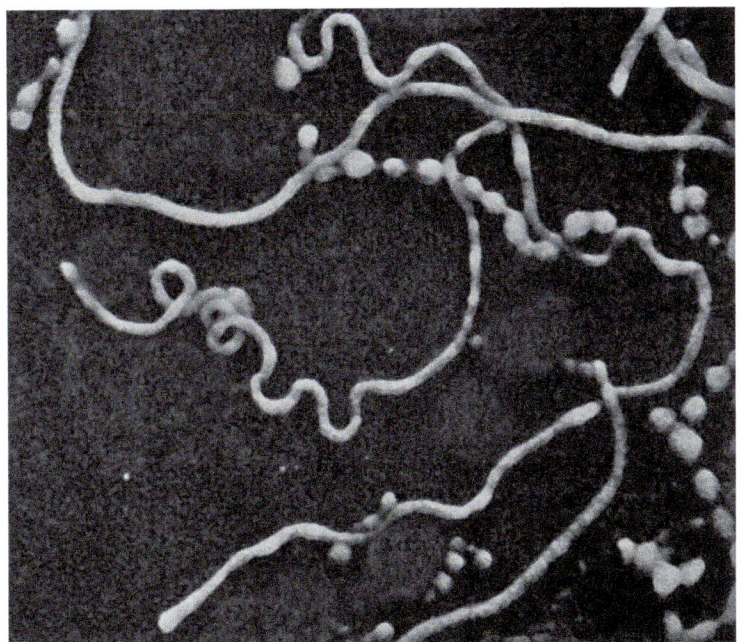

▲ FIGURE 19.19 *Mycoplasma pneumoniae*. This bacterium is pleomorphic; its cells can assume many shapes because they lack cell walls.

DISEASE IN DEPTH

LISTERIOSIS

Listeria monocytogenes with actin "tails"

SIGNS AND SYMPTOMS

Listeria usually produces nonspecific flulike symptoms in healthy, nonpregnant adults. In contrast, infection in pregnant women, fetuses, newborns, the elderly, and immunocompromised patients can be severe, causing bacterial meningitis and possibly death. Human-to-human transmission is limited to the transfer of *Listeria* from pregnant women to fetuses, leading to premature delivery, miscarriage, stillbirth, or meningitis in the newborn.

Researchers have identified more than 50 species of bacteria that can cause bacterial meningitis. One species that is responsible for many cases is *Listeria monocytogenes*. *L. monocytogenes* has virulence factors that allow it to resist phagocytosis and cause disease.

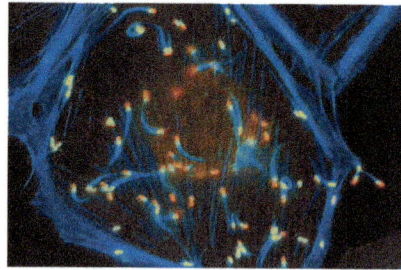

Photomicrograph of *Listeria* showing blue-stained actin "tails" LM 10 μm

1. *Listeria* enters the body, usually through contaminated drink or food.

2. *Listeria* triggers its own phagocytosis by a host cell, typically in the gallbladder.

3. The intracellular bacterium escapes the phagosome.

4. The bacterium reproduces within the phagocyte. Some of the mRNA in a *Listeria* cell activate at 37°C, so the bacterium reproduces more rapidly inside human cells.

5. The bacterium polymerizes the host cell's actin filaments into a "tail."

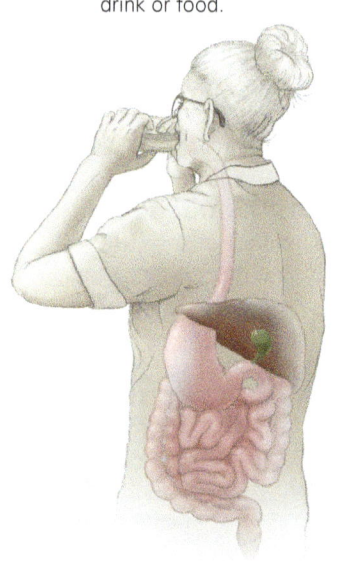

INFECTED HOST CELL

Listeria

Actin "tail"

Watch Dr. Bauman's Video Tutor explore listeriosis, or go to the **Mastering Microbiology** Study Area to investigate further.

EPIDEMIOLOGY

According to the CDC, in 2011, *Listeria*-contaminated cantaloupes "caused the deadliest foodborne disease outbreak in the United States in nearly 90 years," resulting in 30 deaths, including a miscarriage. The death toll could have been higher had health officials not quickly located the source of the infection (fruit grown on a single farm in Colorado) and issued a national warning within days of the first reports. Other outbreaks occurred after patients ate contaminated ice cream in 2015 and soft cheese in 2017.

PATHOGEN AND PATHOGENESIS

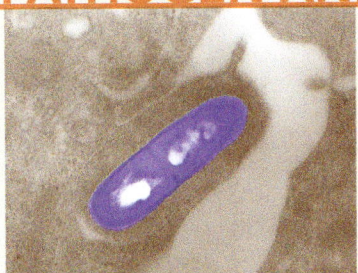
TEM 1 μm

Listeria is a low G + C, Gram-positive, non-endospore-forming, rod-shaped bacterium with end-over-end, tumbling motility outside the body. It is found in soil, water, and animals, including insects. It enters the body through consumption of contaminated drink or food, typically deli meats, hot dogs, soft cheeses, and yogurt.

Once ingested, *Listeria* binds to the surface of a cell, triggering its own endocytosis.

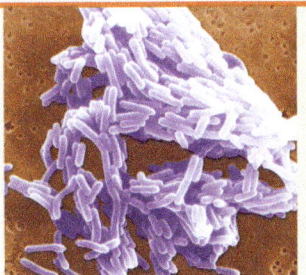
SEM 5 μm

Inside the cell's phagosome, *Listeria* synthesizes listeriolysin O, an enzyme that breaks open the phagosome before a lysosome can fuse with it; thus, *Listeria* avoids digestion by the cell. It also produces listeriolysin S, which overwhelms resident microbiota, reducing microbial antagonism against *Listeria*. The bacterium grows and reproduces in a cell's cytosol, sheltered from antibody immune responses.

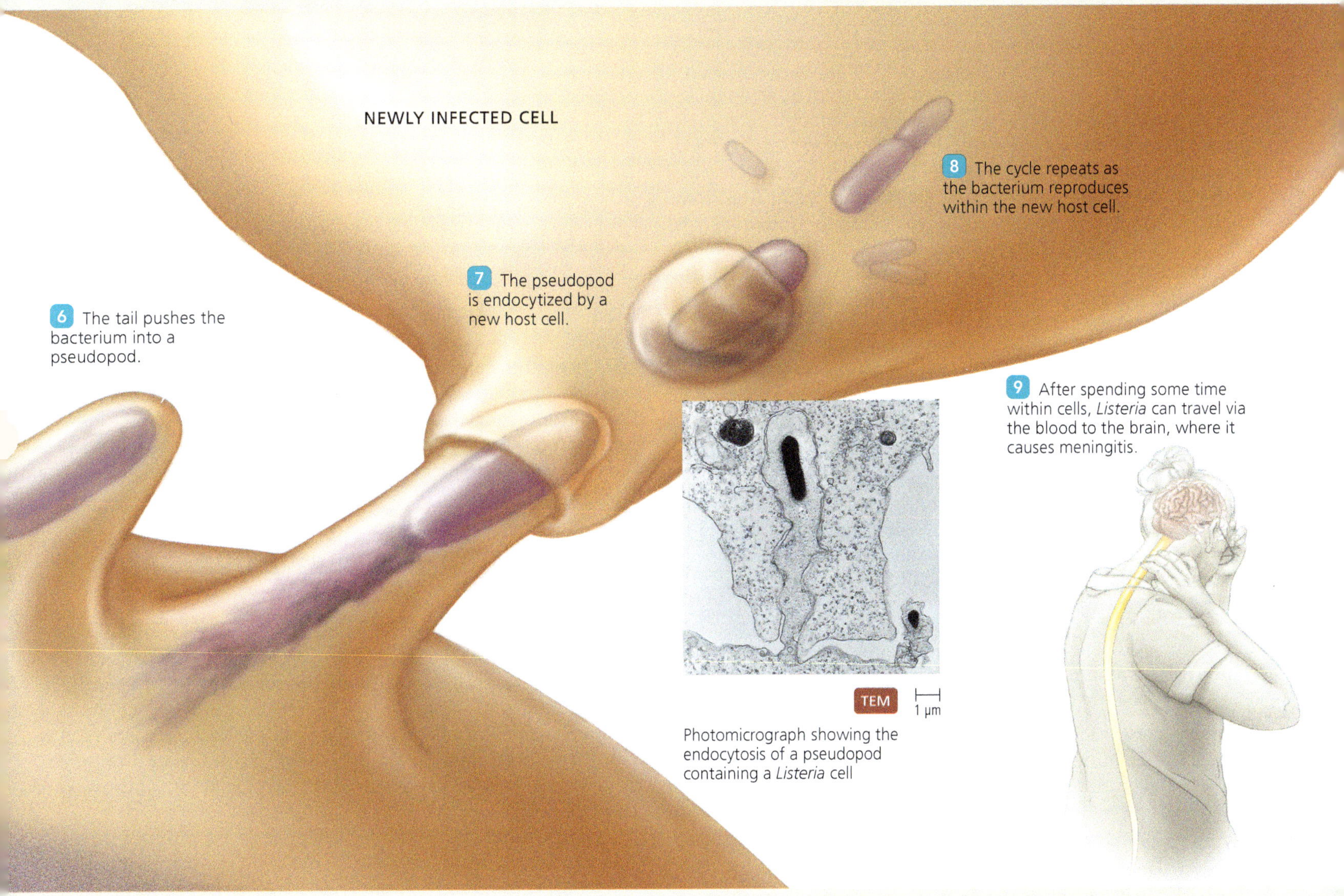

NEWLY INFECTED CELL

6 The tail pushes the bacterium into a pseudopod.

7 The pseudopod is endocytized by a new host cell.

8 The cycle repeats as the bacterium reproduces within the new host cell.

9 After spending some time within cells, *Listeria* can travel via the blood to the brain, where it causes meningitis.

TEM 1 μm
Photomicrograph showing the endocytosis of a pseudopod containing a *Listeria* cell

DIAGNOSIS

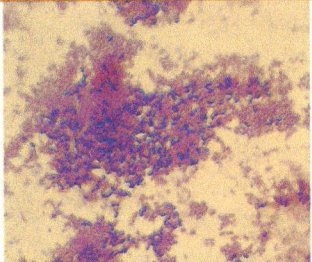
LM 15 μm

Diagnosis can be made by the discovery of *Listeria* in the cerebrospinal fluid of people with symptoms of meningitis. Unfortunately, only a few *Listeria* cells are required to produce disease, so the bacterium is seen in Gram-stained preparations of phagocytes only about a third of the time. Cultured bacteria exhibit a distinctive end-over-end "tumbling" motility that can be seen at room temperature but not at 37°C.

TREATMENT AND PREVENTION

Many antimicrobial drugs, primarily ampicillin, inhibit *Listeria*. Physicians prescribe gentamicin or trimethoprim-sulfamethoxazole for patients allergic to penicillins. Immunocompromised patients (e.g., AIDS patients) should avoid undercooked meats, packaged salads, frozen vegetables, unpasteurized dairy products, and all soft cheeses.

remain infectious even while undergoing antimicrobial treatment. Nevertheless, frequent hand antisepsis, avoidance of contaminated fomites, and reducing aerosol dispersion can limit the spread of the pathogen and the number of cases of disease. No vaccine against *M. pneumoniae* is available.

Other Mycoplasmas

LEARNING OUTCOMES

19.41 List three mycoplasmas associated with urinary and genital tract infections.
19.42 Describe pelvic inflammatory disease.

Three other mycoplasmas are associated with diseases of humans: *M. hominis* (ho'mi-nis), *M. genitalium* (jen-ē-tal'ē-ŭm), and *Ureaplasma urealyticum* (ū-rē'a-li'ti-kŭm) often colonize the urinary and genital tracts of newborn girls, though such infections rarely persist through childhood. However, the incidence of genital reinfections increases after puberty as a result of sexual activity; approximately 15% of sexually active adult Americans (males and females) are currently infected with *M. hominis*, and 70% are infected with *Ureaplasma*.

M. genitalium and *U. urealyticum* cause *nongonococcal urethritis*—that is, inflammation of the urethra not caused by *Neisseria gonorrhoeae* (nī-se'rē-a go-nor-rē'ī). By contrast, *M. hominis* can cause **pelvic inflammatory disease (PID)** in women. Pelvic inflammatory disease is characterized by inflammation of the organs in the pelvic cavity, fever, and abdominal pain.

Infections of either *M. genitalium* or *U. urealyticum* are treated with a macrolide (e.g., erythromycin or azithromycin) or a tetracycline, whereas clindamycin is used to treat *M. hominis*, which is often resistant to macrolides and tetracyclines. Abstinence, mutually faithful monogamy, and proper use of condoms prevent the spread of these sexually transmitted organisms.

TELL ME WHY

Why do pediatricians refrain from using tetracycline to treat mycoplasmal infections in children?

We have examined low G + C, pathogenic, Gram-positive cocci; endospore-forming bacilli; non-endospore-forming, rod-shaped *Listeria*; and wall-less mycoplasmas. Now we turn our attention to the high G + C, Gram-positive pathogens, beginning with *Corynebacterium*.

Corynebacterium

LEARNING OUTCOMES

19.43 Characterize the arrangements of *Corynebacterium* cells.
19.44 Describe the transmission of *Corynebacterium diphtheriae* and the effect of diphtheria toxin.
19.45 Discuss the diagnosis, treatment, and prevention of diphtheria.

Corynebacterium (kŏ-rī'nē-bak-tēr'ē-ŭm) is a genus of high G + C, pleomorphic, non-endospore-forming bacteria that are

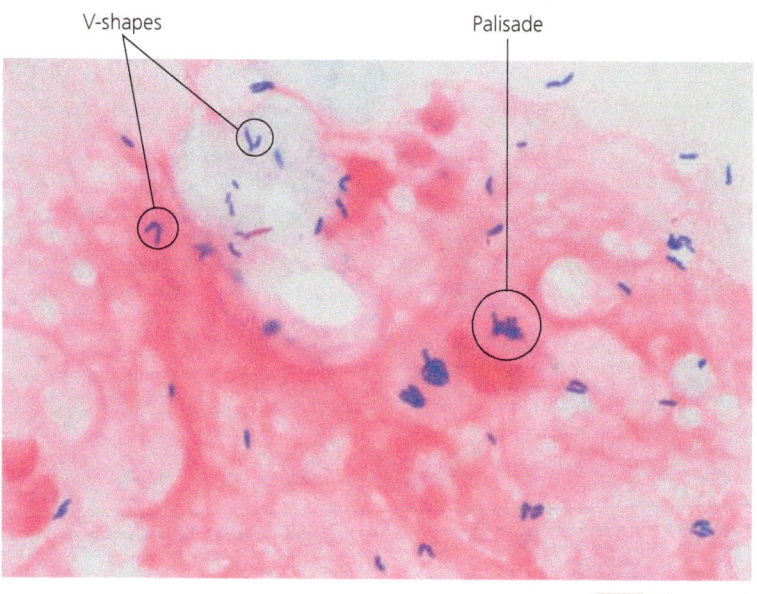

▲ **FIGURE 19.20 Gram-stained *Corynebacterium diphtheriae*.** The characteristic arrangement of the cells results from the type of binary fission called snapping division.

ubiquitous on plants and in animals and humans, where they colonize the skin and the respiratory, gastrointestinal, urinary, and genital tracts. The bacteria divide via a type of binary fission called *snapping division* (see Figure 11.4), in which daughter cells remain attached to form characteristic V-shapes and side-by-side *palisade* arrangements (**FIGURE 19.20**). Although all species of corynebacteria can be pathogenic, the agent of diphtheria is most widely known.

Pathogenesis, Epidemiology, and Disease

Corynebacterium diphtheriae[23] (dif-thi'rē-ī) is transmitted from person to person via respiratory droplets or skin contact. Diphtheria is endemic in poorer parts of the world that lack adequate immunization. The bacterium normally contains a lysogenic bacteriophage that codes for *diphtheria toxin*, which is directly responsible for the signs and symptoms of diphtheria. Cells lacking the phage do not produce toxin and are not pathogenic.

Diphtheria toxin inhibits synthesis of polypeptides in eukaryotes. Because the action of the toxin is enzymatic, a single molecule of toxin can completely block all polypeptide synthesis, resulting in cell death. Diphtheria toxin is thus one of the more potent toxins known.

Respiratory infections are most severe, resulting in the sudden and rapid signs and symptoms of **diphtheria** (dif-thē'rē-ă), including fever, pharyngitis, and the oozing of a fluid composed of intracellular fluid, blood-clotting factors, leukocytes, bacteria, and the remains of dead cells of the throat. The fluid thickens into a *pseudomembrane* (**FIGURE 19.21**) that can adhere tightly to

---

[23]From Greek *koryne*, meaning "club"; *bakterion*, meaning "small rod"; and *diphthera*, meaning "leather membrane."

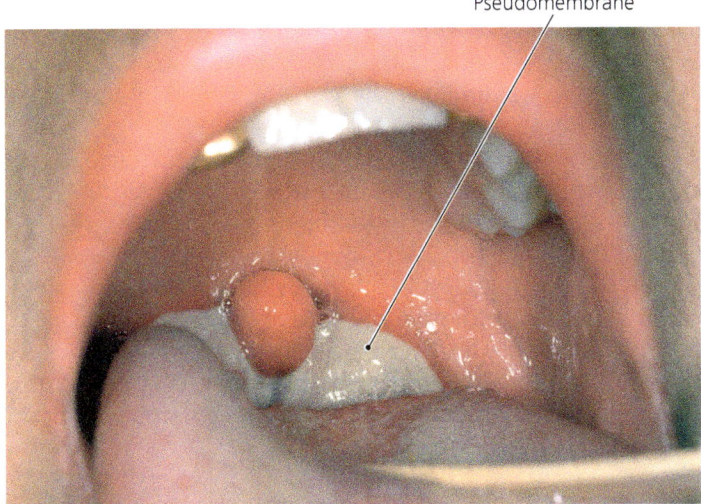

▲ FIGURE 19.21 **A pseudomembrane.** This feature is characteristic of diphtheria.

the underlying tissues, completely occluding the respiratory passages and resulting in death by suffocation.

Diagnosis, Treatment, and Prevention

Initial diagnosis is based on the presence of a pseudomembrane. Laboratory examination of the membrane or of tissue collected from the site of infection does not always reveal bacterial cells because the effects are due largely to the action of diphtheria toxin and not the cells directly. Culture of specimens on *Loffler's medium,* which was developed especially for the culture of *C. diphtheriae,* produces distinct colonial morphologies. Even though observations of these distinctive colonies are useful, absolute certainty of diagnosis results from an immunodiffusion assay called an **Elek test,** in which antibodies against the diphtheria toxin react with toxin in a sample of fluid from the patient.

The most important aspect of treatment is the administration of antitoxin (immunoglobulins against the toxin) to neutralize toxin before it binds to cells; once the toxin binds to a cell, it enters via endocytosis and kills the cell. Penicillin or erythromycin kills the bacterium, preventing the synthesis of more toxin. In severe cases, a blocked airway must be opened surgically or bypassed with a tracheostomy tube.

Because humans are the only known host for *C. diphtheriae,* the most effective way to prevent diphtheria is immunization. Before immunization, hundreds of thousands of cases occurred in the United States each year; in contrast, only 14 cases total were reported in the 21 years from 1997 to 2017. Toxoid (deactivated toxin) is administered in five injections as part of the DTaP vaccine, which combines diphtheria and tetanus toxoids with antigens of the pertussis bacterium, at 2, 4, 6, 18, and about 60 months of age, followed by booster immunizations with a slightly different vaccine (Tdap) at age 11 or 12, and every 10 years thereafter with a vaccine against tetanus and diphtheria only (Td).

MICRO CHECK

14. How does the motility of *Listeria monocytogenes* in human cells differ from its motility in the environment?
15. Based on genetic analysis, *Mycoplasma* spp. are classified as Gram-positive bacteria. Why were they classified as Gram-negative bacteria in the past?
16. Which sign is characteristic of diphtheria?

TELL ME WHY

Why is it necessary to get a diphtheria immunization "booster" every 10 years?

Mycobacterium

LEARNING OUTCOME

19.46 Characterize mycobacteria in terms of endospore formation, cell wall composition, growth rate, and resistance to antimicrobial drugs.

Another devastating high G + C, non-endospore-forming pathogen is *Mycobacterium* (mī′kō-bak-tēr′ē-ŭm). Species in this genus have cell walls containing an abundance of waxlike lipids, called **mycolic acids** (mī-kol′ik), that are composed of chains of 60 to 90 carbon atoms. This unusual cell wall is directly responsible for the unique characteristics of mycobacteria. Specifically, these bacteria:

- Grow slowly (because of the time required to synthesize numerous molecules of mycolic acid). The generation time varies from hours to several days.
- Are protected from lysis once they are phagocytized.
- Are capable of intracellular growth.
- Are protected from desiccation. They can remain viable in dried aerosol droplets for eight months.
- Are resistant to Gram staining, detergents, and many common antimicrobial drugs. Because mycobacteria stain only weakly with the Gram procedure (if at all), the acid-fast staining procedure was developed to differentially stain mycobacteria (see Figure 4.16).

Even though almost 75 species of mycobacteria are known, most mycobacterial diseases in humans are caused by two species: *M. tuberculosis* and *M. leprae,* which cause tuberculosis and leprosy, respectively. *M. avium-intracellulare* and *M. ulcerans* cause emerging mycobacterial diseases.

Tuberculosis

LEARNING OUTCOMES

19.47 Identify two effects of cord factor of *Mycobacterium tuberculosis*.
19.48 Describe the transmission of *M. tuberculosis* and its subsequent action within the human body.

19.49 Discuss the diagnosis, treatment, and prevention of tuberculosis.

19.50 Define and contrast multi-drug-resistant and extensively drug-resistant tuberculosis.

Tuberculosis (TB), formerly called *consumption*, is fundamentally a respiratory disease caused by *Mycobacterium tuberculosis* (too-ber-kyū-lō´sis). This bacterium forms dull-yellow raised colonies after growth for weeks on a special differential medium called Lowenstein–Jensen agar. The pathogen is not particularly virulent, as only about 5% of people infected with the bacterium develop disease; however, it kills about half of untreated diseased patients. **Disease in Depth: Tuberculosis** (pp. 576–577) examines the pathogenesis, symptoms, epidemiology, diagnosis, treatment and prevention of tuberculosis.

Common antimicrobials such as penicillin and erythromycin have little effect on *Mycobacterium tuberculosis* because it grows so slowly that the drugs are cleared from the body by the liver and kidneys before they have any significant effect. Further, *M. tuberculosis* can live and multiply within macrophages, where many antimicrobial drugs have little effect. The currently recommended treatment is a combination of four antimicrobials—isoniazid (INH), rifampin, ethambutol or streptomycin, and pyrazinamide for eight weeks, followed by isoniazid and rifampin for an additional eighteen weeks.

Epidemiologists and physicians are concerned about strains of *M. tuberculosis* that are resistant to at least isoniazid and rifampin. Such *multi-drug-resistant (MDR)* strains have developed in many countries—an estimated 153,000 new cases in 2016. MDR-TB must be treated with more expensive antimicrobials, such as fluoroquinolones in combination with an injectable drug (kanamycin, amikacin, or capreomycin) for as long as two years. This greatly increases the cost of treatment over that for drug-sensitive TB.

Even more serious *extensively drug-resistant TB (XDR-TB)* strains have emerged, especially in populations with a high incidence of infection with human immunodeficiency virus (HIV). XDR-TB resists treatment with isoniazid, rifampin, fluoroquinolones, and at least one of the injectable anti-TB drugs.

And now, some researchers report the occurrence of totally drug-resistant TB (TDR-TB). Scientists are urgently seeking new antimicrobials to treat tuberculosis lest the progress made toward TB control and eradication be reversed.

The prediction that tuberculosis would be eliminated in the United States by 2000 was not realized because of a resurgence of tuberculosis among AIDS patients and its subsequent spread to drug abusers and the homeless. Nevertheless, renewed efforts to detect and stop infection have reduced the number of reported U.S. cases.

Leprosy

LEARNING OUTCOMES

19.51 Compare and contrast tuberculoid leprosy with lepromatous leprosy.

19.52 Discuss the diagnosis, treatment, and prevention of leprosy.

Mycobacterium leprae (lep´rī) causes **leprosy** (lep´rō-sē)—which is also called by the less dreaded name *Hansen's disease*, after Gerhard Hansen (1841–1912), a Norwegian bacteriologist who discovered its cause in 1873. *M. leprae* is a high G + C, Gram-positive bacillus. Because of the abundance of mycolic acid in the cell wall, these bacilli do not Gram stain purple and must instead be stained with an acid-fast stain. *M. leprae* grows best at 30°C, showing a preference for cooler regions of the human body, particularly peripheral nerve endings and skin cells in the

CLINICAL CASE STUDY

This Cough Can Kill

A jumbo jet finally takes off from the airport in Johannesburg, South Africa, and Lance relaxes in his business-class seat. The next stop, 16 hours later, will be Atlanta, Georgia. Lance has just spent six weeks interning with a South African law firm that specializes in human rights issues. His work took him to clients living in crowded conditions in several poor townships. Now Lance's thoughts turn to the upcoming fall term, when he will practically live in a study carrel, law books his constant companions.

The deep coughing of a woman sitting beside him interrupts his reverie. He glances at her; she is thin and huddles against the armrest, looking pale and tired. She often rubs her chest as if it hurts.

When the flight attendant offers the woman something to eat and drink, she turns those down but requests something for her low-grade fever. Her loud, incessant coughing continues for the entire flight, interrupting Lance's sleep.

Four weeks later, Lance opens a letter from the state department of health and learns that he may have breathed in a potentially deadly bacterium on that flight from South Africa. The letter advises Lance to have a physician administer a skin test to determine if he has been infected, but it also tells him that he must wait eight more weeks before being tested.

1. What bacterium was Lance exposed to on the flight?
2. Why does he need to wait so long before being tested?
3. What is the name of the skin test?
4. If the skin test is positive, what will be the physician's recommendation to confirm the test?

fingers, toes, lips, and earlobes. The bacterium does not grow in cell-free laboratory culture, a fact that has hindered research and diagnostic studies. Armadillos and European red squirrels are its only other known hosts. Armadillos have proven valuable in studies on leprosy and on the efficacy of leprosy treatments.

Pathogenesis, Epidemiology, and Disease

Leprosy has two different manifestations depending on the immune response of the patient: Patients with a strong cell-mediated immune response are able to kill cells infected with the bacterium, resulting in a nonprogressive form of the disease called *tuberculoid leprosy*. Regions of the skin that have lost sensation as a result of nerve damage are characteristic of this form of leprosy.

By contrast, patients with a weak cell-mediated immune response develop *lepromatous leprosy* (lep-rō'mă-tŭs; FIGURE 19.22), in which bacteria multiply in skin and nerve cells, gradually destroying tissue and leading to the progressive loss of facial features, digits (fingers and toes), and other body structures. Because of severe disfigurements and the fear of contagion, throughout history lepers were often shunned, banished, or quarantined.

Development of signs and symptoms is very slow; incubation may take years before the disease is evident. Death from leprosy is rare and usually results from the infection of leprous lesions by other pathogens.

Lepromatous leprosy is the more virulent form of the disease, but fortunately it is becoming relatively rare: In the 1990s, about 12 million cases were diagnosed annually worldwide, but by the first quarter of 2014 this number had decreased to 180,464 cases. Leprosy is transmitted via person-to-person contact. Given that the nasal secretions of patients with lepromatous leprosy are loaded with mycobacteria, infection presumably occurs via inhalation of respiratory droplets. Leprosy is not particularly virulent; individuals are typically infected only after years of intimate social contact with a victim. Patients with leprosy are no longer quarantined because the disease is rarely transmitted and is fully treatable.

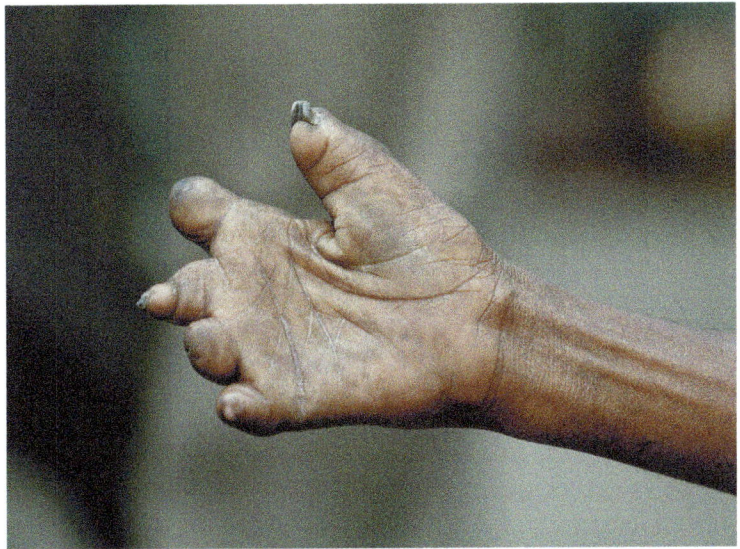

▲ FIGURE 19.22 Lepromatous leprosy can result in severe deformities.

Diagnosis, Treatment, and Prevention

Diagnosis of leprosy is based on signs and symptoms of disease—a loss of sensation in skin lesions in the case of tuberculoid leprosy and disfigurement in the case of lepromatous leprosy. Diagnosis is confirmed by a positive skin test with leprosy antigen (similar to the tuberculin skin test) or through direct observation of *acid-fast bacilli (AFBs)* in tissue samples or nasal secretions (in the case of lepromatous leprosy).

As with *M. tuberculosis*, *M. leprae* quickly develops resistance to single antimicrobial agents, so therapy consists of administering multiple drugs, such as clofazimine, rifampin, or dapsone, for 12 months, though treatment can be lifelong for some patients. Clofazimine is no longer commercially available in the United States but is still prescribed worldwide.

BCG[24] vaccine and another vaccine introduced in India in 2018 provide some protection against leprosy, but prevention is achieved primarily by limiting exposure to the pathogen and by the prophylactic use of antimicrobial agents when exposure occurs. The World Health Organization (WHO) has set a goal to reduce the prevalence of leprosy below one case per 10,000 population in every locality where the disease is endemic. WHO epidemiologists predict that such a low prevalence will permanently interfere with the spread of the bacterium among humans, and leprosy will be eliminated.

Other Mycobacterial Infections

LEARNING OUTCOME

19.53 Explain why the incidence of infections with *Mycobacterium avium-intracellulare*, long thought to be harmless to humans, has been increasing.

Mycobacterium avium-intracellulare (ā'vē-ŭm in'tra-sel-yu-la'rē), which is commonly found in soil, water, and food, was long thought to be a harmless occasional member of the respiratory microbiota. However, the advent of the AIDS epidemic has unmasked it as an important opportunistic pathogen and the cause of an emerging disease that is the most common mycobacterial infection among AIDS patients in the United States. (*M. tuberculosis* is more common in countries where tuberculosis is epidemic.) Infections are believed to result from the ingestion of contaminated food or water; direct person-to-person transmission does not occur. The bacterium spreads throughout the body via the lymph.

In contrast to infections with other mycobacteria, *M. avium-intracellulare* simultaneously affects almost every organ of the body. In some tissues, every cell is packed with mycobacteria, and during the terminal stages of AIDS the blood is often filled with thousands of bacteria per milliliter. The disease remains asymptomatic until organ failure occurs on a massive scale.

Treatment consists of trial-and-error administration of antimicrobial agents; the disseminated nature of the infection often makes treatment ineffectual.

Emerging Disease Case Study: Buruli Ulcer examines another disease of mycobacteria.

---

[24]For "bacillus of Calmette and Guérin," named after the two French developers of the vaccine.

DISEASE IN DEPTH

TUBERCULOSIS

Mycobacterium tuberculosis

SIGNS AND SYMPTOMS

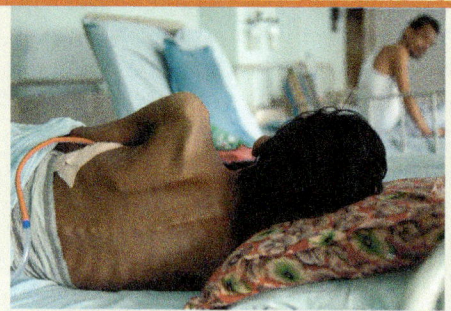

Signs and symptoms of TB are not always apparent, often limited to a minor cough and mild fever. Breathing difficulty, fatigue, malaise, weight loss, chest pain, wheezing, and coughing up blood characterize the disease as it progresses.

Many people think that tuberculosis (TB) is a disease of the past, one that has little importance to people living in industrialized countries. In part, this attitude results from the success health care workers have had in reducing the number of cases. Nevertheless, epidemiologists warn that complacency can allow this terrible killer to reemerge.

PATHOGENESIS

Primary tuberculosis

1 *Mycobacterium* typically infects the respiratory tract via inhalation of respiratory droplets from infected individuals.

2 Macrophages in alveoli phagocytize mycobacteria but are unable to digest them, in part because the bacterium inhibits fusion of lysosomes to endocytic vesicles.

3 Instead, bacteria replicate freely within macrophages, gradually killing the phagocytes. Bacteria released from dead macrophages are phagocytized by other macrophages, beginning the cycle anew.

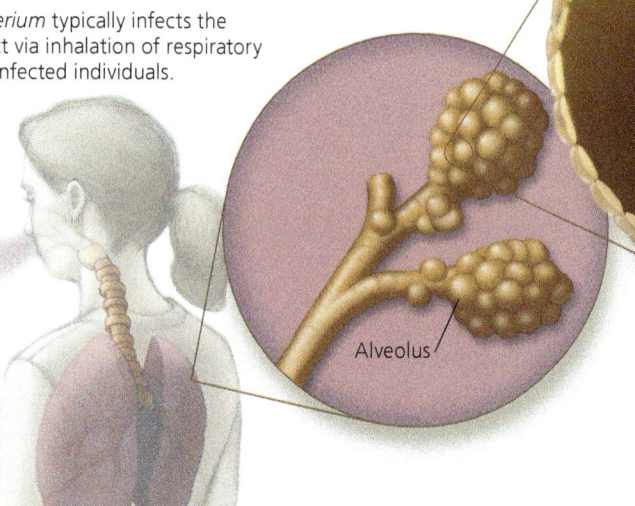

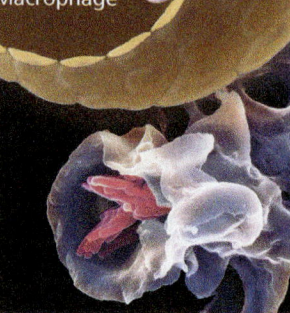

Macrophage engulfing *Mycobacterium*. SEM 5 µm

 Watch Dr. Bauman's Video Tutor to explore tuberculosis, or go to the **Mastering Microbiology** Study Area for more information.

EPIDEMIOLOGY

Someone in the world is infected with tuberculosis every 3 seconds and about three people die every minute on average, mostly in Asia and Africa. TB is on the decline in the U.S., though there were almost 9100 new cases in 2017. One-third of the world's population is infected with *M. tuberculosis*, and over 10 million new cases are seen worldwide each year.

Incidence per 100,000 population 2016
- 0–24
- 25–99
- 100–199
- 200–299
- ≥300
- No data

PATHOGEN AND VIRULENCE FACTORS

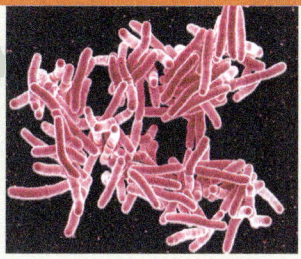

Mycobacterium tuberculosis is a high G + C, aerobic, Gram-positive bacillus (rod). Virulent strains produce cord factor, a cell wall component that produces strands of daughter cells that remain attached to one another in parallel alignments. Cord factor also inhibits migration of neutrophils and is toxic to mammalian cells. Multi-drug- resistant (MDR-TB) and extensively drug-resistant (XDR-TB) strains of *Mycobacterium* make it more difficult to rid the world of TB.

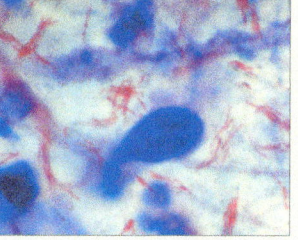

Acid-fast bacilli (pink)

Cell walls contain mycolic acid, a waxy lipid that is responsible for unique characteristics of this pathogen, including slow growth (due in part to the time required to synthesize mycolic acid), protection from lysis when cells are phagocytized, intracellular growth within host cells, and resistance to drying out, detergents, many common antimicrobial drugs, and stains. Because mycobacteria do not Gram stain well, scientists developed *acid-fast staining*, which stains mycobacteria pink and most other cells blue. Such stained cells are called *acid-fast bacilli (AFBs)*.

4 Infected macrophages present antigen to T lymphocytes, which produce lymphokines that attract and activate more macrophages and trigger inflammation. Tightly packed macrophages surround the site of infection, forming a tubercle over a two- to three-month period.

5 Other cells deposit collagen fibers, enclosing infected macrophages and lung cells within the tubercle. Infected cells in the center die, releasing *M. tuberculosis* and producing caseous necrosis—the death of tissue that takes on a cheese-like consistency due to protein and fat released from dying cells. A stalemate between the bacterium and the body's defenses develops.

Secondary/reactivated tuberculosis results when *M. tuberculosis* breaks the stalemate, ruptures the tubercle, and reestablishes an active infection. Reactivation occurs in about 10% of patients; patients whose immune systems are weakened by disease, poor nutrition, drug or alcohol abuse, or by other factors.

Disseminated tuberculosis results when macrophages carry the pathogen via blood and lymph nodes to other sites, including bone marrow, spleen, kidneys, spinal cord, and brain.

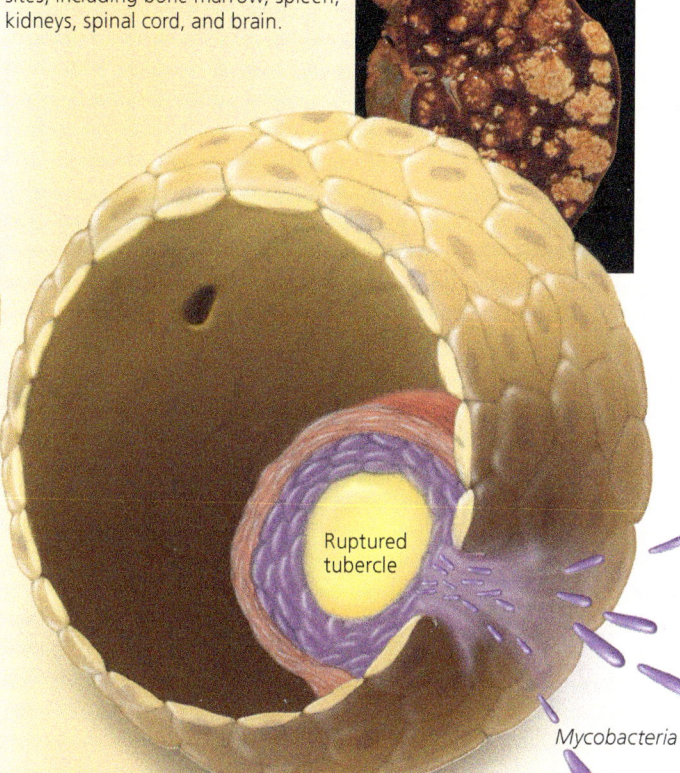

Tuberculosis lesions in spleen.

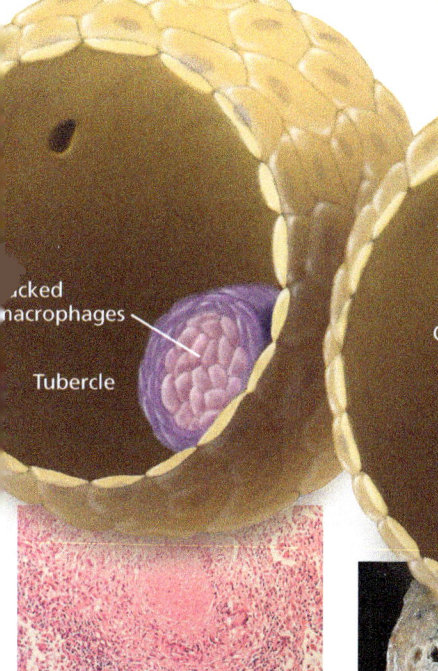

Tubercle in lung tissue.

Lung lesions caused by TB.

DIAGNOSIS

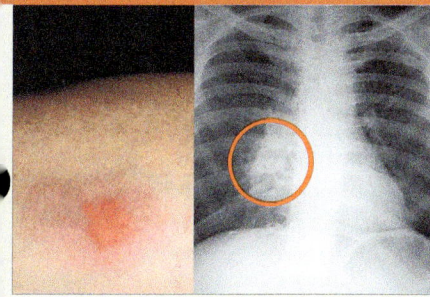

Clinicians use a tuberculin skin test to screen patients for TB exposure. A positive reaction is an enlarged, reddened, hardened, swollen, and raised lesion at the inoculation site. Chest X-ray films can reveal the presence of tubercles in the lungs. Primary TB usually occurs in the lower and central areas of the lung; secondary TB commonly appears higher.

TREATMENT AND PREVENTION

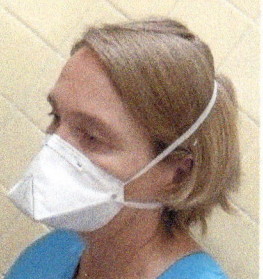

Treatment combines isoniazid, rifampin, pyrazinamide, and either ethambutol or streptomycin for eight weeks, followed by isoniazid and rifampin for an additional eighteen weeks. MDR-TB and XDR-TB are treated with fluoroquinolones or aminoglycosides. In countries where tuberculosis is common, health care workers immunize patients with BCG vaccine, which is not recommended for the immunocompromised, because it can cause TB. Patients are on airborne precautions to protect workers from inhaling respiratory droplets from TB patients.

EMERGING DISEASE CASE STUDY

Buruli Ulcer

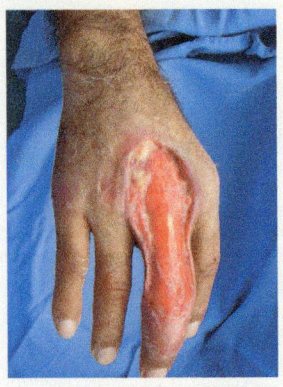

Jacques liked living in the Republic of Côte d'Ivoire (Ivory Coast) in Africa—opportunities abounded for exploration, adventure, and wildlife photography, the countryside was beautiful, and he found the people generally friendly. It was on a photographic excursion to the east that Jacques met a not-so-friendly resident of the country, an emerging mycobacterial pathogen.

The photographer thought little of the small scratch he received while documenting wildlife in the swamps of southern Côte d'Ivoire but he should have been concerned; *Mycobacterium ulcerans* had found a new home in his hand. Jacques would pay a grievous price for his lack of care.

He continued to ignore the infection when it produced a small, painless nodule. He even ignored it when his finger swelled to twice its normal size; it was painless, and he could still meet his busy schedule. But the bacterium was producing a potent toxin known as mycolactone, which destroys cells below the skin, as deep as muscles, bones, and joints. Though his hand continued to swell, making it difficult to work, and he had a slight fever, there still was no pain. His condition worsened over the course of two weeks, until the swollen finger ruptured, and a foul-smelling fluid saturated his camera. It was time to see a doctor.

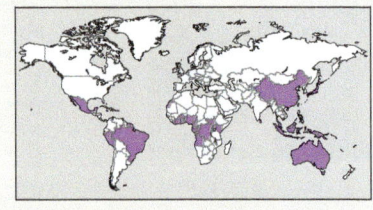

The physician diagnosed Buruli ulcer, an emerging disease that affects more people each year as a result of human encroachment into the swamps where *Mycobacterium ulcerans* lives in stagnant water. After two surgeries to remove dead tissue and the focus of the infection, several skin grafts, and two months of treatment with the intravenous antimicrobial drugs rifampicin and clindamycin, Jacques was released with scars that forever remind him of his adventure with *M. ulcerans*.

1. What might be a reason why Buruli ulcer is initially painless?
2. Why was it necessary to administer antibacterial drugs for two months rather than two weeks?
3. What environmental similarities exist in endemic countries?

MICRO CHECK

17. *Mycobacterium* spp. have cell walls with mycolic acids. Which characteristics does the mycolic acid confer?
18. In people who develop tuberculosis, tubercles form in the lungs. What does the inside of a tubercle consist of?
19. MDR-TB strains are resistant to which drugs at a minimum?
20. What is the most common way to prevent leprosy in the U.S.?

TELL ME WHY

Why is it possible for a patient to have a positive tuberculin test but not be infected with *Mycobacterium tuberculosis*?

Propionibacterium

LEARNING OUTCOMES

19.54 Identify the species of *Propionibacterium* most commonly involved in infections of humans.
19.55 Explain the role of *P. acnes* in the formation of acne.

Propionibacteria are small, Gram-positive, anaerobic rods commonly found growing on the skin. They are so named because they produce propionic acid as a by-product of the fermentation of carbohydrates. The species most commonly involved in infections of humans is *Propionibacterium acnes* (prō-pē-on-i-bak-tēr′ē-ŭm ak′nēz), which is the most common cause of **acne**, causing about 85% of acne cases in adolescents and young adults. The bacterium can also be an opportunistic pathogen in patients with intrusive medical devices such as catheters, artificial heart valves, artificial joints, and cerebrospinal fluid shunts.

In its role in the development of acne (**FIGURE 19.23**), *Propionibacterium* typically grows in the oil glands of the skin ①. Excessive production of oil—called sebum—is triggered by the hormones of adolescence, particularly testosterone in males. The excess oil stimulates the growth and reproduction of the bacterium, which secretes chemicals that attract leukocytes. The leukocytes phagocytize the bacteria and release chemicals that stimulate local inflammation. The combination of dead bacteria and dead and living leukocytes makes up the white pus associated with the pimples of acne ②. A blackhead is formed when a plug of dead and dying bacteria blocks the gland's pore ③. In *cystic acne*, a particularly severe form of the disease, bacteria form larger, inflamed pustules (cysts) ④ that rupture, triggering the formation of scar tissue.

There are many common misconceptions about acne. Cleaning the surface of the skin is not particularly effective in preventing the development of acne because the bacteria live deep in the sebaceous glands, though frequent cleansing may dry the skin, which helps loosen plugs from hair follicles. Scientists have not shown any definitive connection between acne and diet, including chocolate or oily foods.

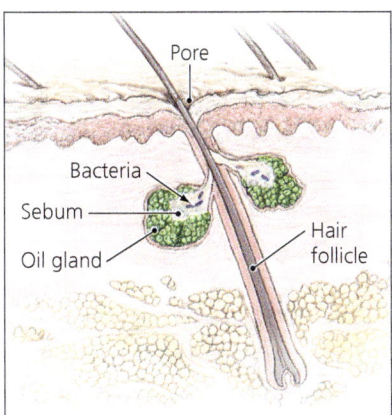

1 Normal skin
Oily sebum produced by glands reaches the hair follicle and is discharged onto the skin surface via the pore.

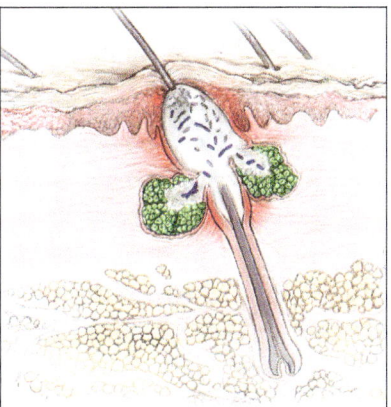

2 Whitehead
Inflamed skin swells over the pore when bacteria infect the hair follicle, leading to accumulation of sebum, colonizing bacteria, leukocytes, and pus.

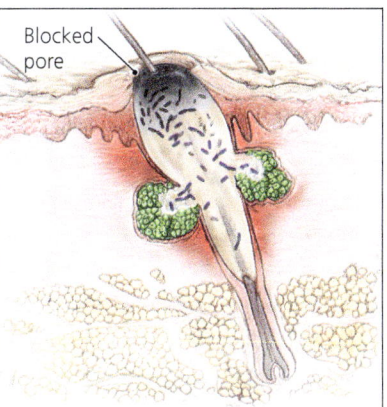

3 Blackhead
Dead and dying bacteria and sebum form a blockage of the pore.

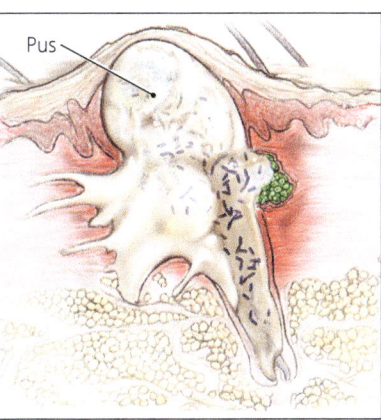

4 Pustule formation
Severe inflammation of the hair follicle causes pustule formation and rupture, producing cystic acne, which is often resolved by scar tissue formation.

◀ **FIGURE 19.23 The development of acne.** Normally, oily sebum produced by glands reaches the hair follicle and is discharged from the pore and onto the skin surface ❶. When inflammation resulting from bacteria infecting the hair follicle causes the skin to swell over the pore, sebum and the colonizing bacteria accumulate to form a whitehead ❷. The blockage of the pore by a plug of dead and dying bacteria and sebum produces a blackhead ❸. When the inflammation of the follicle becomes severe enough, pustules form ❹ and rupture, producing cystic acne, which is often resolved by the formation of scar tissue. *Which bacterial pathogen is a common cause of cystic acne?*

Figure 19.23 *Propionibacterium acnes* is a common cause of acne.

In most cases, the immune system is able to control *Propionibacterium*, and no treatment is required. Dermatologists may prescribe antimicrobial drugs, such as erythromycin and clindamycin, which have proven effective in controlling the bacterium. However, long-term antibiotic use can destroy susceptible normal bacterial microbiota, making the individual more vulnerable to drug-resistant opportunistic fungal and bacterial infections. *Retinoic acid (Accutane)*, a derivative of vitamin A, inhibits the formation of body oil and thereby deprives *P. acnes* of food. Because this drug can cause intestinal bleeding, it is prescribed only for severe cases of acne. Further, it causes birth defects and thus should not be used by pregnant women.

TELL ME WHY
Why do teenagers typically have more acne than adults?

In the previous few sections, we have discussed pathogenic, non-endospore-forming, high G + C rods in the genera *Corynebacterium*, *Mycobacterium*, and *Propionibacterium*. Now we will consider filamentous, high G + C pathogens in the genera *Nocardia* and *Actinomyces*.

Nocardia and Actinomyces

LEARNING OUTCOME

19.56 Compare and contrast *Nocardia* and *Actinomyces* species in terms of appearance, cell wall composition, Gram staining reaction, and role in producing disease.

Both *Nocardia* and *Actinomyces* have elongated, filamentous cells that resemble fungal hyphae. The cell walls of *Nocardia* contain mycolic acid, so the cells are difficult to Gram stain, but they are acid fast. *Actinomyces* stains purple in a Gram stain.

Nocardia asteroides

Nocardia asteroides (nō-kar′dē-ă as-ter-oy′dēz), a common inhabitant of soils rich in organic matter, is the most common pathogen in this genus, accounting for about 90% of infections.

Pathogenesis, Epidemiology, and Disease

N. asteroides is an opportunistic bacterial pathogen that infects numerous sites, including the lungs, skin, and central nervous system. Pulmonary infections develop following inhalation, cutaneous infections result from introduction of the bacteria into wounds, and infections of the central nervous system and other internal organs follow the spread of the bacterium in the blood.

In the lungs, *Nocardia* causes pneumonia accompanied by signs and symptoms typical of pulmonary infections—cough, shortness of breath, and fever. Cutaneous infections may produce **mycetoma**, a painless, long-lasting infection characterized by swelling, pus production, and draining sores (FIGURE 19.24).

Diagnosis, Treatment, and Prevention

Microscopic examination of samples of skin, sputum, pus, or cerebrospinal fluid is usually sufficient to suspect infection with *Nocardia*; the presence of long, acid-fast, hypha-like cells is diagnostic. Treatment usually involves six weeks of an appropriate antimicrobial drug; sulfonamides are the drugs of choice. Although such treatment is usually sufficient for localized infections in patients with normal immune systems, the prognosis for immunocompromised patients with widespread infections is poor. Prevention of nocardial diseases involves avoiding exposure to the bacterium in soil, especially by immunocompromised patients.

Actinomyces

Actinomyces (ak′ti-nō-mī′sēz) is another genus characterized by hypha-like cells (FIGURE 19.25a), as reflected in the name of the genus, which means "ray fungus" in Greek. Despite the name, *Actinomyces* is not a fungus but a bacterium. Unlike *Nocardia*, *Actinomyces* is not acid fast and stains purple when Gram stained.

Colonies of *Actinomyces* form visible concretions resembling grains of sand. Even though the yellow of these multicellular structures has prompted the name "sulfur granules," they are in fact held together by calcium phosphate.

Pathogenesis, Epidemiology, and Disease

Actinomyces is a normal member of the surface microbiota of human mucous membranes. It can become an opportunistic pathogen of the respiratory, gastrointestinal, urinary, and female genital tracts and the skin, and it sometimes causes dental caries (cavities). It also causes a disease called **actinomycosis** when it enters breaks in the mucous membranes resulting from trauma, surgery, or infection by other pathogens. Actinomycosis is characterized by the formation of multiple abscesses connected by channels in the skin or mucous membranes (FIGURE 19.25b).

Diagnosis, Treatment, and Prevention

Diagnosis of actinomycosis is difficult because other organisms cause similar symptoms, and it is difficult to show that the presence of *Actinomyces* is not merely contamination from its normal site on the mucous membranes. Often a health care worker collects and crushes a "sulfur granule" to reveal the presence of filamentous cells.

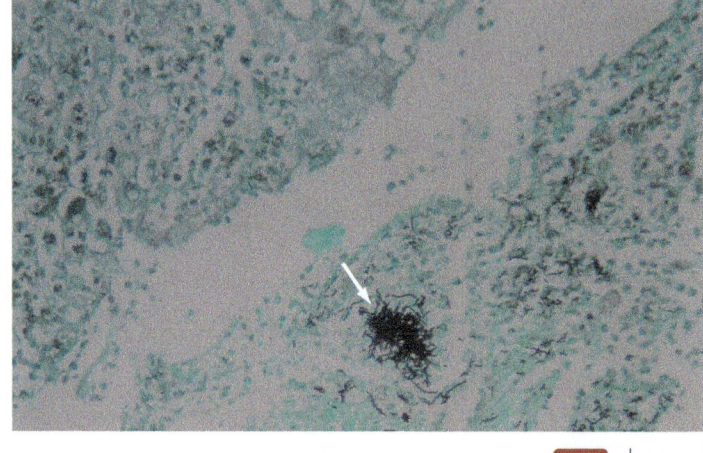

(a)

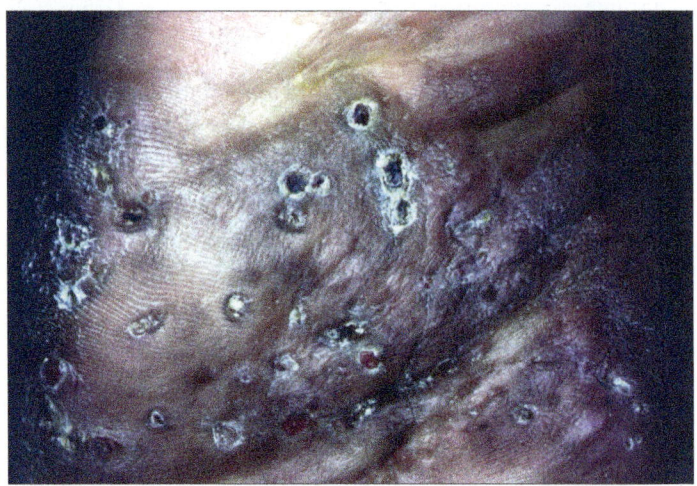

▲ FIGURE 19.24 Lesions of *Nocardia* on the sole of a left foot.

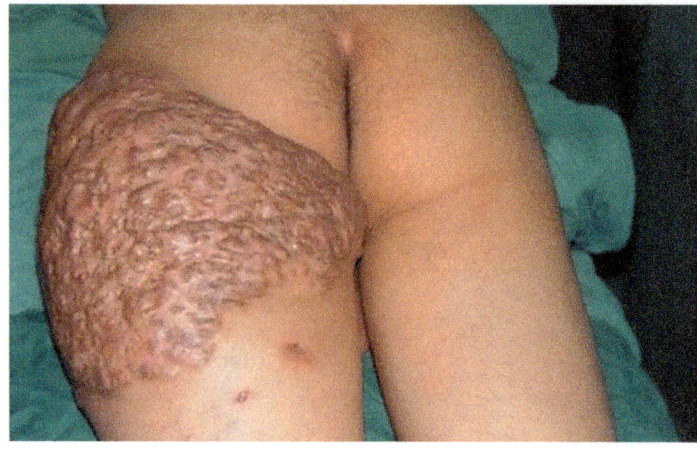

(b)

▲ FIGURE 19.25 *Actinomyces.* (a) A Gomori methenamine silver stain of pus containing the bacteria, which resemble fungal hyphae (compare with Figure 12.14) and may form visible yellow aggregations that look like grains of sand. (b) Lesions of actinomycosis.

Treatment involves surgical removal of the infected tissue and the administration of penicillin for 4 to 12 months. Prevention of *Actinomyces* infections involves good oral hygiene and the prolonged use of prophylactic antimicrobials if trauma or surgery broaches the mucous membranes.

MICRO IN THE CLINIC FOLLOW UP

A Case of Dehydration?

The initial examination at the emergency room doesn't immediately reveal a cause for Ethan's symptoms. The blood tests come back normal. In the meantime, Ethan's breathing has become more labored, and he starts having blurred vision, fixed, dilated pupils, and weakness in his facial muscles. A few hours pass and Ethan's facial paralysis seems to be descending into his torso and arms. His condition continues to deteriorate, and he is put on a mechanical ventilator to assist his breathing. Dr. Mitchell suspects a neurological disease and orders a computed tomography (CT) of Ethan's head as well as an electromyography (EMG) exam to review the electrical activity produced by Ethan's muscles. The head CT results are negative, but the EMG exam indicates Ethan's condition may be botulism. Dr. Mitchell orders an antibody test for botulinum toxins, produced by *Clostridium botulinum*.

The results of the toxin test come back positive. Ethan is diagnosed with *C. botulinum* type E. Since Ethan's condition is critical, the Centers for Disease Control and Prevention (CDC) sends antitoxin (antibodies against botulism toxin) to Ethan's hospital, and Dr. Mitchell begins treating Ethan immediately. In addition, local health authorities work with the CDC to identify the source of Ethan's intoxication.

Ethan requires mechanical ventilation for two weeks. After he can breathe independently, he is discharged from the hospital. When Ethan returns home, he finds that he still tires easily, but ultimately he makes a complete recovery. The public health inspection reveals the source of infection to be aged smoked salmon from the restaurant where Ethan enjoyed the celebratory dinner. Three other people were infected with *C. botulinum*; all were successfully treated with antitoxin.

1. Why was Ethan's botulism treated with antitoxin rather than antibiotics?
2. Propose one hypothesis for how the salmon became contaminated with *Clostridium botulinum*.
3. What type of bacterium is *C. botulinum*?

Check your answers to Micro in the Clinic Follow-Up questions in the **Mastering Microbiology** Study Area.

TELL ME WHY

Why does *Actinomyces* appear purple with a Gram stain but the equally Gram-positive *Nocardia* does not?

Go to **Mastering** Microbiology for Interactive Microbiology, Dr. Bauman's Video Tutors, Micro Matters, MicroFlix, Micro-Boosters, practice quizzes, and more.

CHAPTER SUMMARY

Staphylococcus (pp. 547–551)

1. *Staphylococcus aureus* and *Staphylococcus epidermidis* are found on the skin and in the upper respiratory, gastrointestinal, and urogenital tracts. *S. aureus* is the more virulent, partly because of its production of various enzymes (including coagulase and staphylokinase) and several toxins.

2. Food poisoning is a noninvasive disease caused by *Staphylococcus*. Cutaneous diseases caused by *Staphylococcus* include **staphylococcal scalded skin syndrome, impetigo, folliculitis, sties**, furuncles, and carbuncles.

3. Potentially fatal systemic infections caused by *Staphylococcus* include **toxic shock syndrome, non-streptococcal (TSS); bacteremia; endocarditis; pneumonia; empyema**; and **osteomyelitis**.

4. **Methicillin-resistant *Staphylococcus aureus* (MRSA)**, which is resistant to many antimicrobials including methicillin, is an emerging healthcare-associated (nosocomial) infection and community-acquired threat. **Vancomycin-resistant *S. aureus* (VRSA)** is also emerging as a threat as vancomycin is used against MRSA.

Streptococcus (pp. 552–559)

1. Scientists classify strains of *Streptococcus* based on their reactions to biochemical tests and on the presence of certain bacterial antigens known as Lancefield antigens.

2. Cells of **group A *Streptococcus*** (**GAS**, *Streptococcus pyogenes*), with Lancefield antigen A, produce M protein and a hyaluronic acid capsule, each of which contributes to the virulence of the species. Enzymes and toxins produced by *S. pyogenes* dissolve blood clots, stimulate fever, and lyse blood cells.

3. Group A *Streptococcus* produces enzymes (such as streptokinases, deoxyribonucleases, C5a peptidase, and hyaluronidase) that facilitate the spread of the bacterium through tissues, as well as pyrogenic toxins that stimulate fever, rash, and shock.

4. The diseases caused by group A streptococci include **pharyngitis** ("strep throat"), **rheumatic fever, scarlet fever** (scarlatina), **pyoderma** (impetigo), **erysipelas, streptococcal toxic shock syndrome (STSS), necrotizing fasciitis** ("flesh-eating bacteria"), and **glomerulonephritis**.

5. Cells of **group B *Streptococcus*** (*S. agalactiae*), with Lancefield antigen B, occur normally in the lower GI, genital, and urinary tracts. They are associated with neonatal diseases and are treated with penicillin.

6. The **viridans streptococci**, a group in the alpha-hemolytic streptococci, include members that inhabit the mouth, pharynx, GI tract, and genitourinary tracts; cause dental **caries** (cavities); and can enter the blood to cause bacteremia, meningitis, and endocarditis.

7. Virulent strains of *Streptococcus pneumoniae* have adhesin, which attaches the bacterium to the cells of the pharynx, and phosphorylcholine, which binds to receptors in the lungs, meninges, and blood vessels, triggering endocytosis of *Streptococcus*. Inside a human cell, a polysaccharide capsule protects the bacterium from digestion, enabling it to cross the cell and enter the blood and brain.

8. Virulent *S. pneumoniae* also secretes the proteins secretory IgA protease and pneumolysin. The latter lyses cells in the lungs.

9. **Pneumococcal pneumonia** is the most prevalent disease caused by *Streptococcus pneumoniae* infection. Other diseases include **sinusitis** (inflammation of the nasal sinuses), **otitis media** (inflammation of the middle ear), bacteremia, endocarditis, and meningitis.

Enterococcus (pp. 559–560)

1. *Enterococcus faecalis* and *E. faecium* are normal members of the intestinal microbiota that can cause opportunistic healthcare-associated bacteremia, endocarditis, and wound infections.

Bacillus (pp. 560–562)

1. *Bacillus anthracis* secretes anthrax toxin, which can cause three forms of **anthrax:** gastrointestinal anthrax (rare and fatal, with intestinal hemorrhaging), cutaneous anthrax (can be fatal if untreated), and inhalation anthrax (often fatal).

2. Clinicians diagnose anthrax by finding the distinctive large, Gram-positive, nonmotile cells. They treat the disease with antimicrobial drugs and monoclonal antibodies. Control of the disease in animals is the best way to prevent human anthrax.

Clostridia: Clostridoides and Clostridium (pp. 562–568)

1. Antimicrobial drugs can reduce the number of normal microbiota in the intestines, giving endospores of *Clostridioides difficile* the opportunity to germinate and grow. *C. diff.* causes a self-limiting explosive diarrhea or a life-threatening **pseudomembranous colitis**.

2. *Clostridium botulinum* is an anaerobic, endospore-forming bacterium that can release an extremely poisonous toxin in improperly canned food, causing **botulism**. Pediatric botulism occurs when the pathogen grows in the gastrointestinal tract of an infant. Wound botulism results when the bacterial endospores germinate in surgical or traumatic wounds.

3. Treatment of botulism involves purging the digestive tract, administration of antitoxin, respiratory support, and, in the case of pediatric or wound botulism, antimicrobial drugs.

4. *Clostridium tetani* is a ubiquitous bacterium that enters the body via a break in the skin and causes **tetanus** due to the action of **tetanospasmin**. Immunization is effective in preventing this disease.

5. *Clostridium perfringens* produces 11 toxins that lyse blood cells and cause diseases such as food poisoning and **gas gangrene**.

Listeria (pp. 568–571)

1. *Listeria monocytogenes* is rarely pathogenic but can cause fatal meningitis in pregnant women, newborns, the elderly, and immunocompromised patients. The bacterium triggers its own endocytosis by human cells, escapes from phagosomes into the cytoplasm, and moves within and between human cells by polymerizing actin fibers.

2. *Listeria* tolerates cold and can grow on deli meats, hot dogs, cheeses, and other foods in refrigerators.

Mycoplasmas (pp. 568–572)

1. **Mycoplasmas**, the smallest free-living microbes, have DNA sequences typical of Gram-positive bacteria but lack cell walls, so they stain pink like Gram-negative cells. They have both DNA and RNA and can metabolize, making them different from viruses despite their small size. The colonies of most mycoplasmas have a "fried egg" appearance.

2. *Mycoplasma pneumoniae* is a human pathogen that affects the epithelial cells of the respiratory tract and causes **primary atypical pneumonia** (walking pneumonia). It does not form "fried-egg" colonies on agar.

3. *Mycoplasma hominis* can cause **pelvic inflammatory disease**—inflammation of organs in the pelvic cavity—in women.

4. *Mycoplasma genitalium* and *Ureaplasma urealyticum* cause nongonococcal urethritis.

5. Venereal mycoplasmal diseases can be prevented by abstinence, mutually faithful monogamy, and the proper use of condoms.

Corynebacterium (pp. 572–573)

1. Corynebacteria undergo a form of binary fission called "snapping division," which results in V-shaped and palisade arrangements of cells.

2. *Corynebacterium diphtheriae*, transmitted via respiratory droplets, contains a bacteriophage that codes for diphtheria toxin, which causes the symptoms of the potentially fatal disease **diphtheria**. Diagnosis results from an immunodiffusion assay called an **Elek test**.

Mycobacterium (pp. 573–578)

1. Cell walls of *Mycobacterium* contain **mycolic acid**, so scientists stain them using an acid-fast stain rather than a Gram stain.

2. *Mycobacterium tuberculosis* with a cell wall component called cord factor causes **tuberculosis**, a disease in which the bacteria replicate within macrophages in the lungs and trigger the formation of small, hard nodules called tubercles. The tuberculin skin test screens for possible TB exposure. BCG vaccine may be used to prevent tuberculosis.

3. Multi-drug-resistant tuberculosis (MDR-TB) and extensively drug-resistant tuberculosis (XDR-TB) greatly concern clinicians and epidemiologists.

4. *Mycobacterium leprae* causes **leprosy**. Tuberculoid leprosy is a nonprogressive form of the disease, whereas lepromatous leprosy results in progressive destruction of body structures. Clinicians treat leprosy with antimicrobial drugs and can prevent the disease with BCG vaccine.

5. *Mycobacterium avium-intracellulare* is an opportunistic pathogen that can become pathogenic in AIDS patients. It adversely affects almost every organ of the body.

Propionibacterium (pp. 578–579)

1. *Propionibacterium acnes* living in the sebaceous glands of the skin causes adolescent **acne**. The bacterium may also infect patients who have intrusive medical devices. Antimicrobial drugs are more effective than topical products in the treatment of acne.

Nocardia and Actinomyces (pp. 579–581)

1. The opportunistic pathogen *Nocardia asteroides* has elongated cells resembling fungal hyphae; it is acid fast because of the presence of mycolic acid in its cell wall. Cutaneous infections may produce **mycetoma**.

2. *Actinomyces* is another opportunistic pathogen with elongated cells, but it is not acid fast. It causes **actinomycosis**.

QUESTIONS FOR REVIEW

Answers to the Questions for Review (except Short Answer questions) begin on p. A-1.

Multiple Choice

1. Which of the following bacteria causes a common type of food poisoning?
   a. *Streptococcus sanguis*
   b. *Clostridium tetani*
   c. *Staphylococcus aureus*
   d. *Streptococcus pyogenes*

2. How does *Staphylococcus aureus* affect the matrix between cells in the human body?
   a. *S. aureus* triggers blood clotting, which coats the matrix and inhibits cellular communication.
   b. *S. aureus* produces an enzyme that dissolves hyaluronic acid and thus enables it to pass between the cells.
   c. *S. aureus* possesses a hyaluronic acid capsule that causes leukocytes to ignore the bacterium as if it were camouflaged.
   d. *S. aureus* does not affect the matrix but instead produces a necrotizing agent that dissolves body cells.

3. Which of the following conditions is a systemic disease caused by *Staphylococcus*?
   a. impetigo
   b. folliculitis
   c. carbuncle
   d. toxic shock syndrome

4. A bacterium associated with bacteremia, meningitis, and pneumonia in newborns is _____.
   a. *Staphylococcus aureus*
   b. *Staphylococcus epidermidis*
   c. *Streptococcus pyogenes*
   d. *Streptococcus agalactiae*

5. Which type of anthrax is more common in animals and in humans?
   a. cutaneous anthrax
   b. inhalation anthrax
   c. gastrointestinal anthrax
   d. mucoid anthrax

6. Of the following genera, which can survive the harshest conditions?
   a. *Staphylococcus*
   b. *Clostridium*
   c. *Mycobacterium*
   d. *Actinomyces*

7. Pathogenic strains that have become resistant to antimicrobial drugs are found in which of the following genera?
   a. *Staphylococcus*
   b. *Mycobacterium*
   c. *Enterococcus*
   d. all of the above

8. The bacterium causing pseudomembranous colitis is _____.
   a. *Clostridioides difficile*
   b. *Streptococcus pyogenes*
   c. *Mycobacterium avium-intracellulare*
   d. *Corynebacterium diphtheriae*

9. Mycoplasmas _____.
   a. lack cell walls
   b. are pleomorphic
   c. have sterol in their membranes
   d. all of the above

10. In which of the following diseases would a patient experience a pseudomembrane covering the tonsils, pharynx, and larynx?
    a. tuberculoid leprosy
    b. diphtheria
    c. arrhythmia
    d. tetanus

11. Which of the following is *not* characteristic of mycoplasmas?
    a. cytochromes
    b. sterols in cytoplasmic membranes
    c. use of UGA codon for tryptophan
    d. rRNA nucleotide sequences similar to those of Gram-positive bacteria

Matching

For each of the following diseases or conditions, indicate the genus (or genera) of bacterium that causes it.

1. ___ Scalded skin syndrome
2. ___ Osteomyelitis
3. ___ Pharyngitis
4. ___ Scarlet fever
5. ___ Pyoderma
6. ___ Rheumatic fever
7. ___ Glomerulonephritis
8. ___ Sinusitis
9. ___ Otitis media
10. ___ Anthrax
11. ___ Myonecrosis
12. ___ Diphtheria
13. ___ Leprosy
14. ___ Dental caries
15. ___ Acne

A. *Staphylococcus*
B. *Streptococcus*
C. *Mycobacterium*
D. *Listeria*
E. *Propionibacterium*
F. *Corynebacterium*
G. *Bacillus*
H. *Clostridium*
I. *Actinomyces*

Short Answer

1. Why are mycoplasmas able to survive a relatively wide range of osmotic conditions, even though these bacteria lack cell walls?

2. *Mycobacterium avium-intracellulare* was considered relatively harmless until the late 20th century, when it became common in certain infections. Explain how this bacterium's pathogenicity changed.

3. Contrast tuberculoid leprosy with lepromatous leprosy in terms of pathogenesis. How does the cellular immune response of a patient affect the form of the disease?

4. Explain how mice are used in the diagnosis of botulism poisoning.

5. Why do pediatricians recommend that children under one year never be fed honey?
6. Explain why Gram-positive mycoplasmas appear pink in a Gram-stained smear.
7. Explain the different actions of pyogenic and pyrogenic toxins.
8. Explain why *Staphylococcus epidermidis* is rarely pathogenic while the similar *S. aureus* is more commonly virulent.
9. Why did epidemiologists immediately suspect terrorism in the cases of anthrax in the fall of 2001?
10. Explain the action of the toxin of *Clostridium tetani*.
11. Why is mycolic acid a virulence factor for mycobacteria?
12. Compare and contrast mycoplasmas and viruses.

VISUALIZE IT!

1. Match the genera of pathogens to their appearance in stained smears: *Actinomyces, Bacillus, Clostridium, Mycobacterium, Staphylococcus, Streptococcus*.

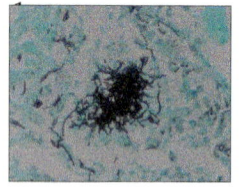
(a) Methenamine silver

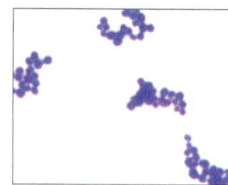
(b) Gram

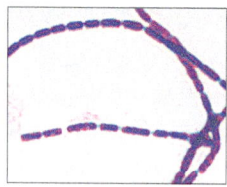
(c) Gram

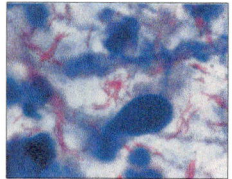
(d) Acid fast

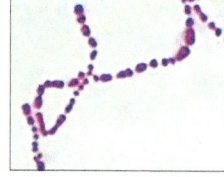
(e) Gram

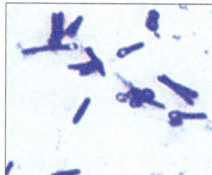
(f) Gram

2. Label acetylcholine. Color the sites of action of botulism toxin on a nerve cell.

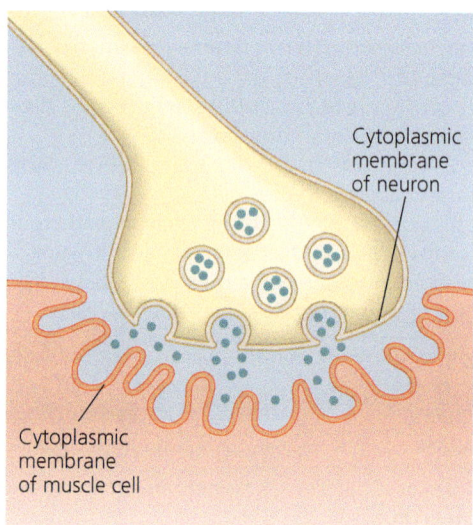

CRITICAL THINKING

1. A few days after the death of a child hospitalized with a MRSA infection, another child who had been admitted to the hospital with viral pneumonia worsened and died. An autopsy revealed that the second child also died from complications of MRSA. By what route was the second child likely infected? What should hospital personnel do to limit the transfer of MRSA and other bacteria among patients?

2. An elderly man is admitted to the hospital with severe pneumonia, from which he eventually dies. What species is the most likely cause of his infection? What antimicrobial drug is effective against this species? How could the man have been protected from infection? Is the hospital staff at significant risk of infection from the man? Which groups of patients would be at risk if the man had visited their rooms before he died?

3. Botulism toxin can be used as an antidote for tetanus. Can tetanus toxin be used as an antidote for botulism? Why or why not?

4. A blood bank refused to accept blood from a potential donor who had just had his teeth cleaned by a dental hygienist. Why did they refuse the blood?

5. *Mycoplasma pneumoniae*, like many pathogenic bacteria, is resistant to penicillin; however, unlike most resistant species, *Mycoplasma* does not synthesize β-lactamase. Explain why *Mycoplasma* is resistant to penicillin despite its inability to make β-lactamase.

6. Why don't physicians try to prevent the spread of TB by simply administering prophylactic antimicrobial drugs to everyone living in endemic areas?

CONCEPT MAPPING

Using the terms provided, fill in the concept map that describes tuberculosis. You can also complete this and other concept maps online by going to the Mastering Microbiology Study Area.

Acid-fast stain of sputum
Antibiotic resistance
Blood-tinged sputum
Chest X ray
Cough
Culture on special media
Disseminated tuberculosis
Dormant infection
Ethambutol
Extensive lung damage
For 26 weeks
Isoniazid (INH)
MDR-TB
Mycobacterium tuberculosis
Primary infection
Pyrazinamide
Rifampin
Streptomycin
Tuberculin skin test
XDR-TB

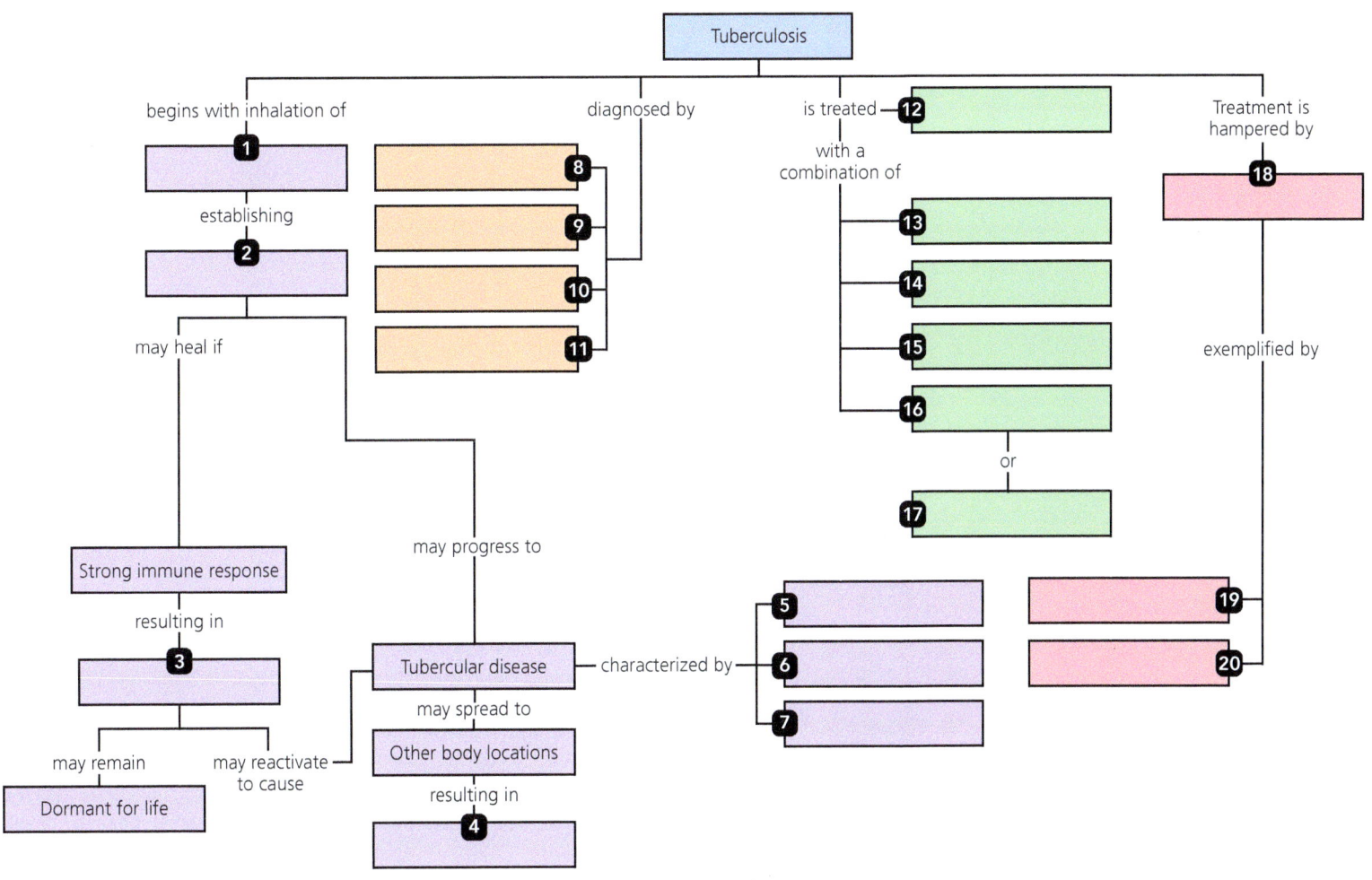

20 Pathogenic Gram-Negative Cocci and Bacilli

Before You Begin

1. Which component of a Gram-negative cell wall can cause fever, vasodilation, inflammation, shock, and blood clotting?
2. In the case of gastrointestinal disease, how can an intoxication be distinguished from an infection?
3. Most microbes that reside in the intestines are facultative or obligate anaerobes. Describe the growth of facultative anaerobes in the presence of and also in the absence of molecular oxygen.

Bordetella pertussis is a Gram-negative, rod-shaped bacterium, shown here in a colorized transmission electron micrograph.

MICRO IN THE CLINIC
Another Day Care Cold?

CLAIRE IS A YOUNG WOMAN TRYING to get used to life as a single parent. Since her husband died, she's had to find a new place to live and a new day care for her three-year-old daughter, Hailey. It's been a hard transition, but Claire is starting to feel like things are back on track.

Early one afternoon, Claire receives a call from the day care—Hailey has a fever and a severe cough that almost seems like she cannot breathe. Claire rushes to the day care to pick her up as soon as possible. When she arrives, one of the caregivers tells Claire that Hailey had been a bit fussy and cranky in the morning before she worsened. She also said that a couple of other kids had been sick earlier in the week with a similar cough. Claire takes Hailey home.

After a few days, Claire is more concerned—Hailey has a fever, has started coughing more frequently, and seems to be worse at night, whenever she is active, and when yawning. Hailey is taking a nap when Claire notices that Hailey sounds like she is wheezing or gasping for air after each cough. Claire immediately takes Hailey to her pediatrician.

1. **Does Hailey have a cold, or are her symptoms a sign of a more serious infection?**

Turn to the end of the chapter (p. 615) to find out.

Numerous Gram-negative cocci, bacilli, and coccobacilli are bacterial pathogens of humans. In fact, Gram-negative bacteria constitute the largest group of human bacterial pathogens, in part because the outer membrane of a Gram-negative bacterial cell wall contains lipid A, also known as endotoxin (see Chapter 3, Figure 3.16). Lipid A triggers fever, vasodilation, inflammation, shock, and **disseminated intravascular coagulation (DIC)**—the formation of blood clots within blood vessels throughout the body (FIGURE 20.1). Almost every Gram-negative bacterium that can breach the skin or mucous membranes, grow at 37°C, and evade the immune system can cause disease and death in humans; however, most of the Gram-negative bacteria that cause disease fall into fewer than 30 genera.

We have seen that it is possible to group these organisms according to a variety of criteria. Gram-negative bacteria are placed in *Bergey's Manual of Systematic Bacteriology* based largely on genomic similarities; here we primarily consider them according to a different organizational scheme, one that traditionally has been used for many clinical purposes—according to their shapes, oxygen requirements, and other biochemical properties.

First, we survey the primary genus of Gram-negative cocci that are pathogenic to humans, then we consider 15 genera of pathogenic, facultatively anaerobic bacilli in the families Enterobacteriaceae and Pasteurellaceae, which account for almost half of all Gram-negative pathogens. Finally, we examine a heterogeneous collection of 10 genera of pathogenic, Gram-negative aerobic bacilli before concluding by considering two genera of strictly anaerobic, pathogenic Gram-negative bacilli.

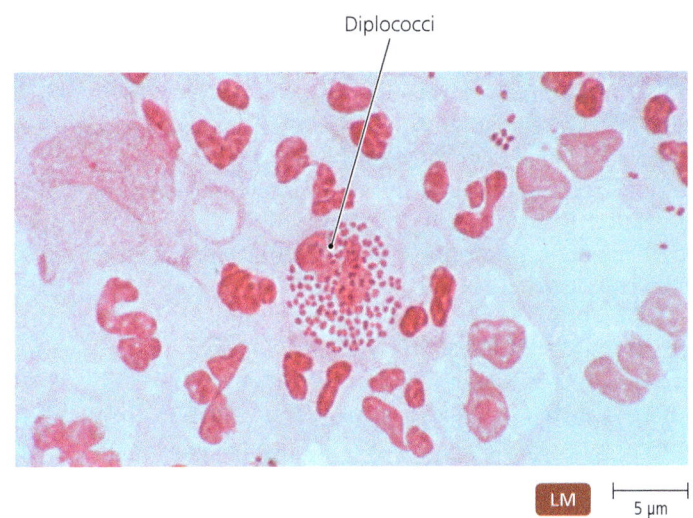

▲ **FIGURE 20.2 Diplococci of *Neisseria gonorrhoeae*.** What function of gonococcal fimbriae contributes to the bacterium's pathogenicity?

Figure 20.2 The fimbriae act as adhesion factors, enabling gonococci to adhere to cells lining the cervix and urethra and to sperm, which allows Neisseria to travel far into the female reproductive tract.

Pathogenic Gram-Negative Cocci: *Neisseria*

Neisseria (nī-se′rē-ă) is the only genus of Gram-negative cocci that regularly causes disease in humans. *Neisseria* is placed in the class Betaproteobacteria of the phylum Proteobacteria. Morphologically similar pathogenic bacteria in the genera *Moraxella* (mōr′ak-sel′ă) and *Acinetobacter* (as-i-nē′tō-bak′ter), considered coccobacilli, are now known to be more closely related to aerobic bacilli in class Gammaproteobacteria and thus are discussed in a later section.

Structure and Physiology of *Neisseria*

LEARNING OUTCOME

20.1 List three structural features of *Neisseria* that contribute to its pathogenicity.

The cells of all strains of *Neisseria* are nonmotile and typically arranged as diplococci (pairs) with their common sides flattened in a manner reminiscent of coffee beans (FIGURE 20.2). In addition to their shape, these aerobic bacteria are distinguished from many other Gram-negative pathogens by being *oxidase positive* (see Figure 20.5); that is, their electron transport chains contain the enzyme *cytochrome oxidase*. Pathogenic strains of *Neisseria* also have fimbriae and polysaccharide capsules, as well as a major cell wall antigen called *lipooligosaccharide* (lip′ō-ol-i-gō-sak′a-rīd; *LOS*), composed of lipid A (endotoxin) and sugar molecules—all of which enable the bacteria to attach to and invade human cells.

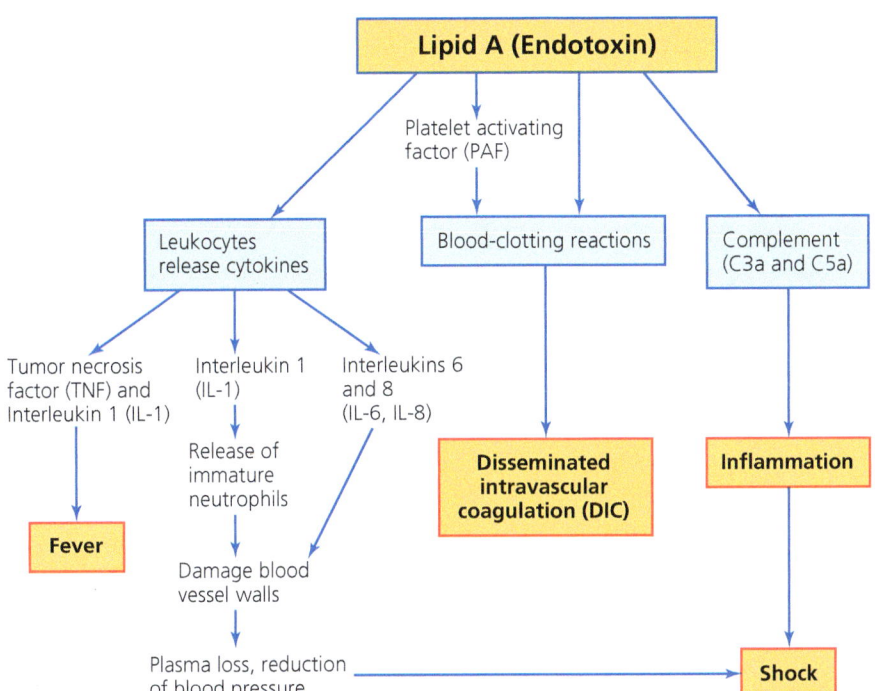

▲ **FIGURE 20.1 Potential effects of lipid A (endotoxin).** What type of bacteria produce endotoxin?

Figure 20.1 Gram-negative bacteria produce endotoxin (lipid A) as part of the outer layer of their outer membrane.

Cells of *Neisseria* that lack any one of these three structural features are typically avirulent.

In the laboratory, *Neisseria* is fastidious in its growth requirements. Microbiologists culture these bacteria on either autoclaved blood agar (called *chocolate agar* because of its appearance) or a selective medium called *modified Thayer-Martin medium*. *Neisseria* is particularly susceptible to drying and extremes of temperature. A moist culture atmosphere containing 5% carbon dioxide enhances its growth. Laboratory personnel may use sugar fermentation tests to distinguish among strains of *Neisseria*; pathogenic strains, for example, are unable to ferment maltose, sucrose, or lactose.

Two species of *Neisseria* are pathogenic to humans: the so-called gonococcus, *N. gonorrhoeae* (go-nor-rē′ī), which causes gonorrhea, and the meningococcus, *N. meningitidis* (me-nin-ji′ti-dis), which causes a type of meningitis.

The Gonococcus: *Neisseria gonorrhoeae*

LEARNING OUTCOMES

20.2 Compare and contrast the symptoms of gonorrhea in men and women.

20.3 Discuss the difficulties researchers face in developing an effective vaccine against *Neisseria gonorrhoeae*.

Pathogenesis, Epidemiology, and Disease

Neisseria gonorrhoeae causes **gonorrhea** (gon-ō-rē′ă), which has been known as a sexually transmitted disease for centuries, although the disease was confused with syphilis until the 19th century. In the 2nd century A.D., the Roman physician Claudius Galen named gonorrhea for what he thought was its cause—an excess of semen (gonorrhea means "flow of seed" in Greek). The disease is sometimes called "clap" from an archaic French word *clapoir*, meaning brothel.

Gonorrhea occurs in humans only. The World Health Organization (WHO) estimates that there are about 78 million new cases worldwide every year, and it is one of the more common sexually transmitted diseases in the United States, though the number of cases has been declining over the past few decades (**FIGURE 20.3**). Most U.S. cases of gonorrhea occur in adolescents. The U.S. Centers for Disease Control and Prevention (CDC) has a goal to reduce the incidence of gonorrhea to below 19 cases per 100,000 population. This incidence will theoretically eliminate the disease in the United States.

The disease is more common among females than males. Whereas women have a 50% chance of becoming infected during a single sexual encounter with an infected man, men have only a 20% chance of infection during a single encounter with an infected woman. An individual's risk of infection increases with increasing numbers of sexual encounters.

Gonococci adhere, via their fimbriae and capsules, to epithelial cells of the mucous membranes lining the genital, urinary, and digestive tracts of humans. As few as 100 pairs of cells can cause disease. The cocci protect themselves from the immune system by

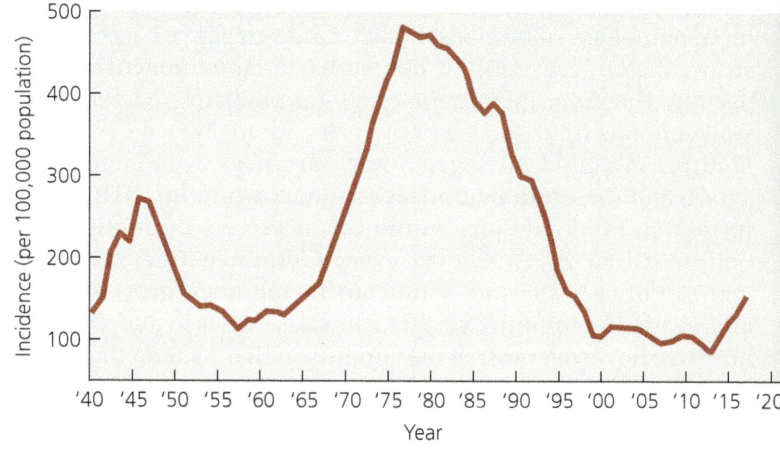

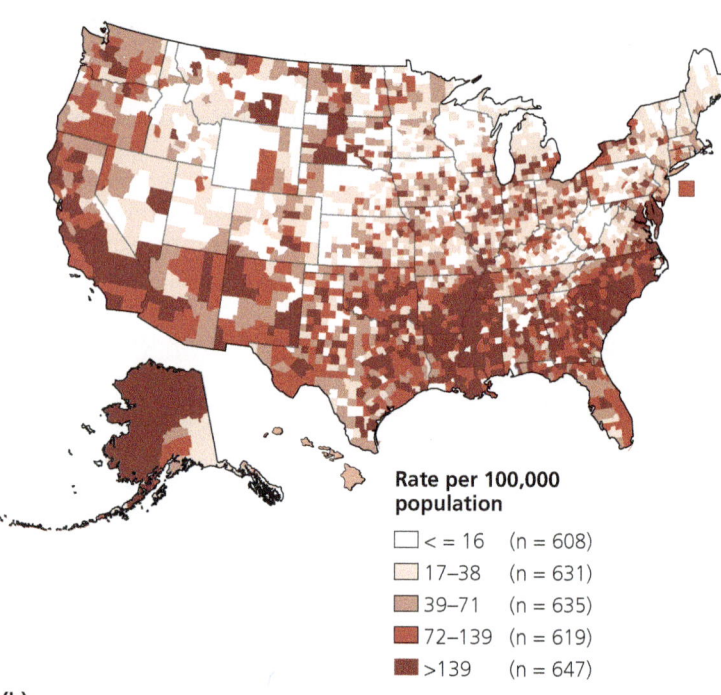

▲ **FIGURE 20.3 Incidence of gonorrhea in the United States.** (a) The number of new cases annually (incidence), 1977–2017. (b) The geographic distribution of cases reported in 2016 by county per 100,000 population. *What organism causes gonorrhea?*

Figure 20.3 *Gonorrhea is caused by Neisseria gonorrhoeae.*

secreting a protease enzyme that cleaves secretory IgA in mucus. Further, endocytized bacteria survive and multiply within neutrophils. As the bacteria multiply, they invade deeper connective tissues, often while being transported intracellularly by phagocytes.

In men, gonorrhea is usually insufferably symptomatic—acute inflammation typically occurs two to five days after infection in the urethra of the penis, causing extremely painful urination and a purulent (pus-filled) discharge. Rarely, gonococci invade the prostate or epididymis, where the formation of scar tissue can render the man infertile. In contrast, recently acquired gonorrhea in women is often asymptomatic; about 50% of infected women have no symptoms or obvious signs of infection. Even when women do have symptoms, they often mistake them for a bladder infection or vaginal yeast infection.

Neisseria cannot attach to cells lining the vagina; instead, the bacteria most commonly infect the cervix of the uterus. Gonococci can infect deep in the uterus and even the uterine (Fallopian) tubes by "hitchhiking" to these locations on sperm to which they attach by means of fimbriae or LOS. In the uterine tubes they can trigger inflammation, fever, and abdominal pain—a condition known as **pelvic inflammatory disease (PID)**. Chronic infections can lead to scarring of the tubes, resulting in *ectopic*[1] *pregnancies* or sterility.

Gonococcal infection of organs outside the reproductive tracts can also occur. Because a woman's urethral opening is close to her vaginal opening, the urethra can become infected with gonococci during sexual intercourse. Anal intercourse can lead to *proctitis* (inflammation of the rectum), and oral sexual intercourse can infect the pharynx or gums, resulting in *pharyngitis* and *gingivitis*, respectively. Oral and anal infections are most commonly seen in men who have sex with men. In very rare cases, gonococci enter the blood and travel to the joints, meninges, or heart, causing arthritis, meningitis, and endocarditis, respectively.

Gonococcal infection of a child during birth is one cause of inflammation of the conjunctiva (outer part of the eye) called *ophthalmia neonatorum*.[2] Gonococci can also infect the respiratory tracts of newborns during their passage through the birth canal. Infection of older children with *N. gonorrhoeae* is strong evidence of sexual abuse by an infected adult.

Diagnosis, Treatment, and Prevention

The presence of Gram-negative diplococci in pus from an inflamed penis is sufficient for a diagnosis of symptomatic gonorrhea in men. For diagnosis of asymptomatic cases of gonorrhea in men and women, commercially available genetic probes provide direct, accurate, rapid detection of *N. gonorrhoeae* in clinical specimens. Alternatively, the bacterium can be cultured on selective media, such as Thayer–Martin agar.

The treatment of gonorrhea has been complicated in recent years by the worldwide spread of gonococcal strains resistant to penicillin, tetracycline, erythromycin, and aminoglycosides. Currently, the CDC recommends broad-spectrum intramuscular cephalosporins (e.g., ceftriaxone) plus azithromycin or doxycycline to treat gonorrhea.

Long-term specific immunity against *N. gonorrhoeae* does not develop, and thus promiscuous people can be infected multiple times. This lack of immunity is explained in part by the highly variable surface antigens of this bacterium; immunity against one strain often provides no protection against other strains. Further, the existence of many different strains has prevented the development of an effective vaccine.

Except for the routine administration of antimicrobial agents to newborns' eyes, which successfully prevents ophthalmic disease, chemical prophylaxis is ineffective in preventing disease. In fact, the use of antimicrobials to prevent genital disease may

---

[1] From Greek *ektopos*, meaning "out of place."
[2] From Greek *ophthalmikos*, meaning "of the eye," *neo*, meaning "new," and *natalis*, meaning "birth."

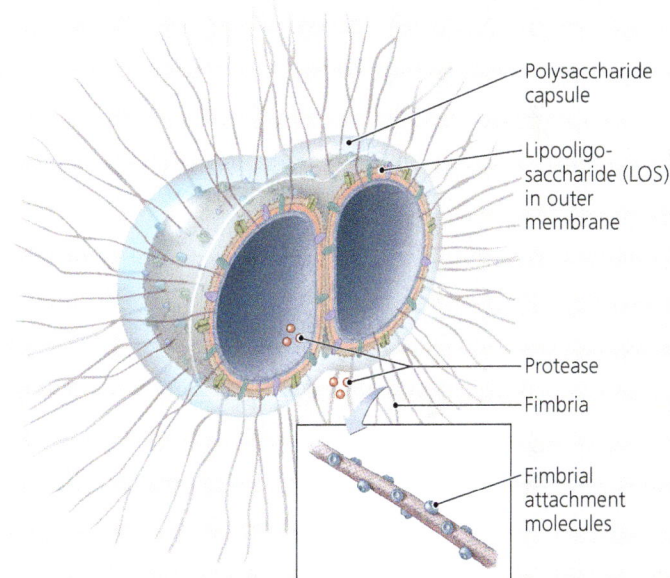

MICROBE AT A GLANCE 20.1
Neisseria gonorrhoeae

Taxonomy: Domain Bacteria, phylum Proteobacteria, class Betaproteobacteria, order Neisseriales, family Neisseriaceae

Other names: Gonococcus

Cell morphology and arrangement: Cocci in pairs

Gram reaction: Negative

Virulence factors: Capsule, fimbriae, and lipooligosaccharide adhere to human cells; IgA protease; can survive inside phagocytic cells; variable surface antigens

Diseases caused: Gonorrhea, pelvic inflammatory disease, ophthalmia neonatorum, proctitis, pharyngitis

Treatment for diseases: Cephalosporins (e.g., ceftriaxone) plus azithromycin, or doxycycline if patient is allergic or intolerant to azithromycin

Prevention of disease: Sexual abstinence, mutually faithful monogamy, consistent and proper use of condoms

Labels: Polysaccharide capsule; Lipooligosaccharide (LOS) in outer membrane; Protease; Fimbria; Fimbrial attachment molecules

---

select for hardier resistant strains, worsening the situation. The most effective methods of prevention, in order, are sexual abstinence; monogamy with an uninfected, faithful partner; and consistent and proper use of condoms. Efforts to stem the spread of gonorrhea focus on education, aggressive detection, and the screening of all sexual contacts of carriers.

The Meningococcus: *Neisseria Meningitidis*

LEARNING OUTCOMES
20.4 Describe how meningococci survive and thrive in humans.
20.5 Discuss the epidemiology of meningococcal diseases.

Neisseria meningitidis causes life-threatening diseases when it invades the blood or cerebrospinal fluid.

Pathogenesis, Epidemiology, and Disease

Of the 13 known antigenic strains of meningococci, strains known as A, B, C, and W135 cause most cases of disease in humans. The polysaccharide capsule of *N. meningitidis* resists lytic enzymes of the body's phagocytes, allowing phagocytized meningococci to survive, reproduce, and be carried throughout the body within neutrophils and macrophages. In the United States, the bacterium causes most cases of meningitis in children and adults under 20. Much of the damage caused by *N. meningitidis* results from *blebbing*—a process in which the bacterium sheds extrusions of outer membrane. The lipid A component of LOS thereby released into extracellular spaces triggers fever, vasodilation, inflammation, shock, and DIC.

N. meningitidis is a common member of the normal microbiota in the upper respiratory tracts of up to 40% of healthy people. As a result of crowded living conditions, meningococci are more prevalent in children and young adults from lower economic groups.

Respiratory droplets transmit the bacteria among people living in close contact, especially families, soldiers living in barracks, prisoners, and college students living in dormitories. In fact, meningococcal disease can be 23 times more prevalent in students living in dormitories than in the general population. Individuals whose lungs have been irritated by dust are more susceptible to airborne *Neisseria*. For example, annual outbreaks of meningococcal meningitis can kill tens of thousands in sub-Saharan Africa when dry winter winds create irritating dust storms. Even though meningococcal diseases occur worldwide, cases are relatively rare in countries with advanced health care systems and access to antimicrobial drugs.

The incidence of meningococcal disease in the United States is about 0.3 cases per 100,000 population. Most of these cases are meningitis, usually involving abrupt sore throat, fever, headache, stiff neck, vomiting, and convulsions. Very young children may have only fever and vomiting. Arthritis and partial loss of hearing occasionally result. Meningococcal meningitis can progress so rapidly that death results within six hours of the initial symptoms.

Meningococcal septicemia (blood poisoning) can also be life threatening. LOS may trigger shock, devastating blood coagulation in many organs, and *petechiae* (pe-tē'kē-ē)—minute hemorrhagic skin lesions—on the trunk and lower extremities (FIGURE 20.4a). Petechiae may coalesce and combine with regions of cell death to form large black lesions (FIGURE 20.4b). In some patients, however, septicemia produces only mild fever, arthritis, and petechiae.

Diagnosis, Treatment, and Prevention

Because meningococcal meningitis constitutes a medical emergency, rapid diagnosis is critical. The demonstration of Gram-negative diplococci within cerebrospinal phagocytes is diagnostic for meningococcal disease. Physicians obtain a sample of cerebrospinal fluid using a needle inserted into the lower spinal canal—a procedure called a **spinal tap**. *N. meningitidis* can oxidize maltose in laboratory culture, which distinguishes it from *N. gonorrhoeae*. Further, serological tests can demonstrate the presence of antibodies against *N. meningitidis*, though one strain (strain B) is relatively nonimmunogenic and therefore does not react well in such tests.

Despite the fact that meningococcal meningitis proceeds rapidly and is ordinarily 100% fatal when left untreated, immediate treatment with antimicrobial drugs has reduced mortality to less than 10%. Treatment is with a third-generation, semisynthetic cephalosporin such as ceftriaxone or cefotaxime.

Because healthy, asymptomatic carriers are common, eradication of meningococcal disease is unlikely. Prevention involves prophylactic treatment—administering the same drugs to anyone exposed to individuals with meningococcal disease, including family, classmates, day care center contacts, roommates, and health care workers. The CDC recommends immunization with meningococcal vaccine against strains A, C, and Y, though this vaccine cannot be used with children under two years old. As previously noted, strain B is weakly immunogenic, so immunity

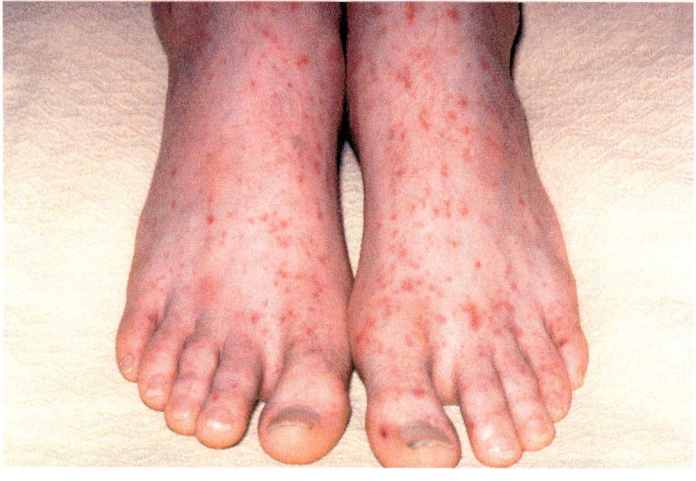

(a)

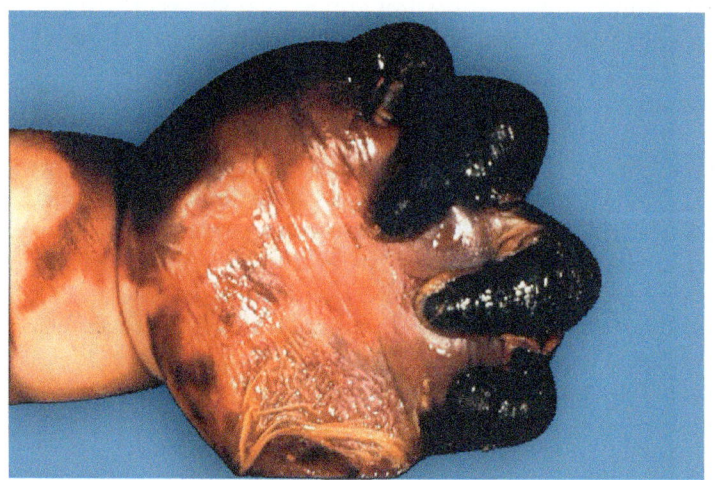

(b)

▲ FIGURE 20.4 **Petechiae in meningococcal septicemia.** These lesions can be diffuse (a) or may coalesce in areas containing dead cells (b).

to it cannot be induced artificially and must instead develop following a natural infection.

> ### MICRO CHECK
>
> 1. How do the structural characteristics of pathogenic *Neisseria* species contribute to their ability to cause disease?
> 2. How is gonorrhea experienced differently in men and women?
> 3. How is *Neisseria meningitidis* transmitted?

> **TELL ME WHY**
>
> *Neisseria gonorrhoeae* can be deposited in the vagina during sex but can wind up infecting the uterine tubes and the pelvic cavity. Why is it able to do this, despite being nonmotile?

Pathogenic, Gram-Negative, Facultatively Anaerobic Bacilli

LEARNING OUTCOME

20.6 Describe how members of the family Enterobacteriaceae are distinguished from members of the family Pasteurellaceae.

Two families of facultatively anaerobic bacilli and coccobacilli in the class Gammaproteobacteria—Enterobacteriaceae and Pasteurellaceae—contain most of the Gram-negative pathogens of humans. Scientists distinguish between members of the two families by performing an oxidase test: members of the Enterobacteriaceae are oxidase negative, whereas members of the Pasteurellaceae are oxidase positive (**FIGURE 20.5**).

We begin our survey of these microorganisms by examining the family Enterobacteriaceae, which contains about 150 species in 41 genera, including some of the more important causative agents of healthcare-associated infection (HAIs; formerly called *nosocomial infections*) of humans (**FIGURE 20.6**).

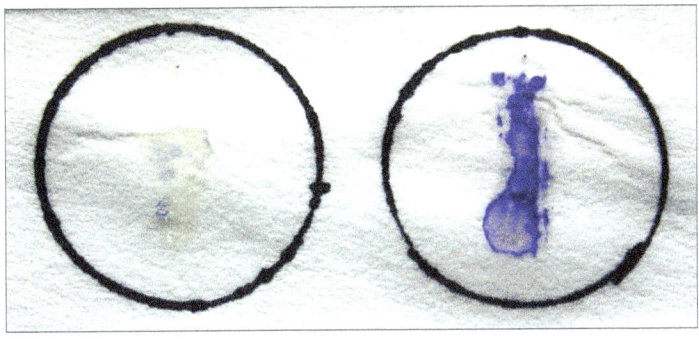

▲ **FIGURE 20.5 The oxidase test.** The test distinguishes members of the family Enterobacteriaceae (left), which are oxidase negative, from members of the family Pasteurellaceae, which are oxidase positive, as indicated by a purple color.

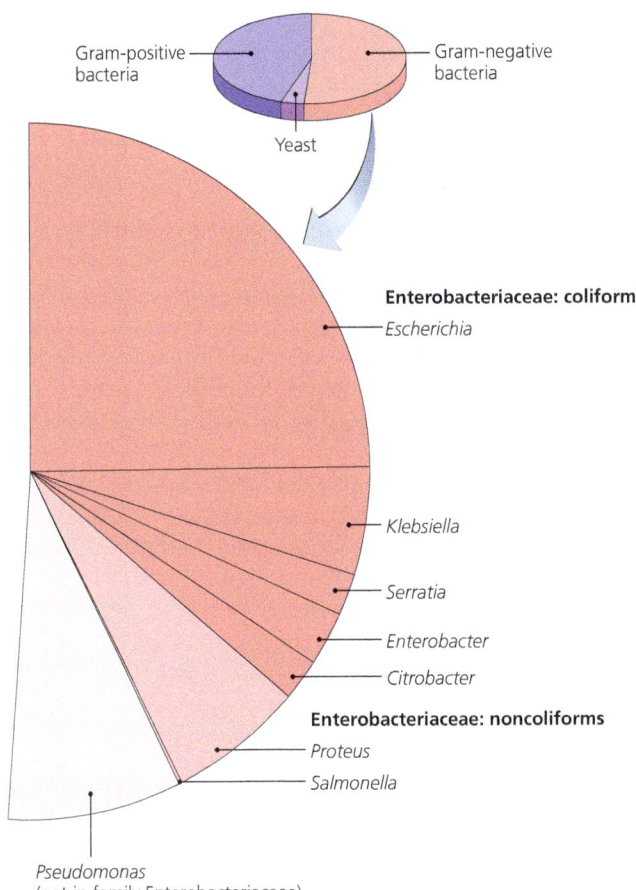

▲ **FIGURE 20.6 Relative causes of healthcare-associated infections in the United States.** Gram-negative bacteria in the family Enterobacteriaceae account for most healthcare-associated infections. *What is a nosocomial infection?*

Figure 20.6 *An infection acquired in a health care setting; the newer term is healthcare-associated infection.*

The Enterobacteriaceae: An Overview

LEARNING OUTCOMES

20.7 Discuss how to distinguish between members of the family Enterobacteriaceae in the laboratory.

20.8 List six virulence factors found in members of the family Enterobacteriaceae.

20.9 Describe the diagnosis, treatment, and prevention of diseases of enteric bacteria.

As their name suggests, prokaryotes in the family **Enterobacteriaceae** (en′ter-ō-bak-tēr-ē-ā′sē-ē)—also called **enteric bacteria**—are members of the intestinal microbiota of most animals and humans. They are also ubiquitous in water, soil, and decaying vegetation. Some species are always pathogenic in humans, whereas others are opportunists normally found in the intestinal microbiota that become pathogenic when introduced into other body sites.

As a group, enteric bacteria are the most common Gram-negative pathogens of humans. Though most enteric bacteria can be pathogenic in humans, over 95% of the medically important species are in the 13 genera we discuss shortly.

Structure and Physiology

Enteric bacteria are coccobacilli or bacilli that measure about 0.5 μm × 1.21 − 3 μm (**FIGURE 20.7**). Those members that are motile have peritrichous flagella. Some have prominent capsules, whereas others have only a loose slime layer. All members of the family reduce nitrate to nitrite and ferment glucose anaerobically, though most of them grow better in aerobic environments. As previously noted, all of them are also oxidase negative.

Given that all enteric bacteria are similar in microscopic appearance and staining properties, scientists traditionally distinguish among them by using biochemical tests, motility, and colonial characteristics on a variety of selective and nonselective media (e.g., MacConkey agar and blood agar). **FIGURE 20.8** presents one dichotomous key, based on motility and biochemical tests, for distinguishing among the 13 genera of enteric bacteria discussed in this chapter.

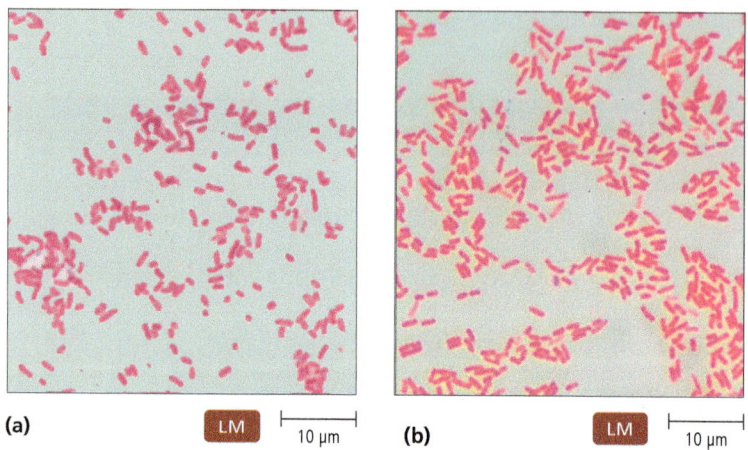

▲ **FIGURE 20.7 Gram stains of bacteria in the family Enterobacteriaceae.** (a) *Enterobacter aerogenes*, a coccobacillus. (b) *Escherichia coli*, a bacillus. Note the similarity in staining properties.

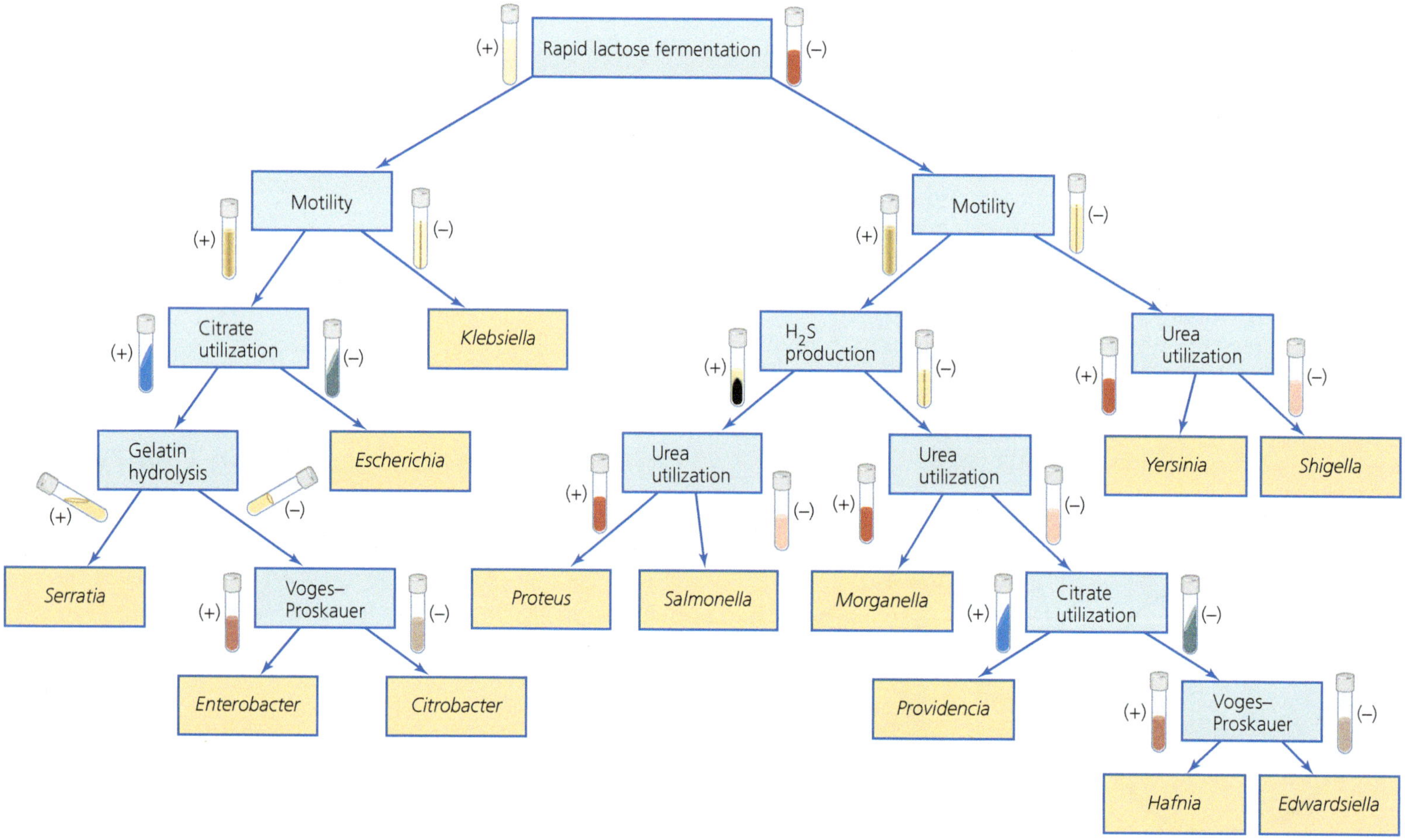

▲ **FIGURE 20.8 A dichotomous key for distinguishing among enteric bacteria.** This key, based on biochemical tests and motility, is but one of many possible keys. *What characteristics do all enteric bacteria share?*

Figure 20.8 All enteric bacteria are Gram-negative, oxidase negative, and able to metabolize nitrate to nitrite.

Pathogenesis

Members of the Enterobacteriaceaea have an outer membrane that contains lipopolysaccharide composed of three antigenic components:

- A core polysaccharide, shared by all of the enteric bacteria and called *common antigen*.
- So-called *O polysaccharide*, which has various antigenic varieties among strains and species.
- Lipid A, which, when released in a patient's blood, can trigger fever, vasodilation, inflammation, shock, and DIC (see Figure 20.1). Because of the presence of lipid A, most of the enteric bacteria can cause serious disease and death.

Other antigens of certain enteric bacteria include protein and polysaccharide capsular antigens called K antigens (but called Vi antigens in *Salmonella*) and flagellar proteins called H antigens. By controlling the genetic expression of K and H antigens, alternately producing and not producing the antigens, the bacteria survive by evading their host's immune system. Scientists use serological identification of antigens to distinguish among strains and species of enteric bacteria. For example, *Escherichia coli* O157:H7, a potentially deadly bacterium, is so designated because of its specific O and H antigenic components. The specificity of serological tests can also be used in diagnosis and in epidemiological studies to identify the source of a particular infecting agent.

Among the variety of virulence factors possessed by pathogenic enteric bacteria, some (for example, lipid A) are common to all genera, whereas others are unique to certain strains. Other virulence factors seen in the Enterobacteriaceae include the following:

- *Capsules* that protect the bacteria from phagocytosis and antibodies, and provide a poorly immunogenic surface.
- *Fimbriae* and proteins called *adhesins*, which enable the bacteria to attach tightly to human cells.
- *Exotoxins* that cause a variety of symptoms such as diarrhea. The genes for exotoxins, fimbriae, and adhesins are frequently located on plasmids, which increases the likelihood that they will be transferred among bacteria.
- *Iron-binding compounds* called *siderophores* that capture iron and make it available to the bacteria.
- *Hemolysins*, which release nutrients such as iron by lysing red blood cells.
- A so-called *type III secretion system*, a complex structure composed of 20 different polypeptides that is synthesized by several pathogenic enteric species. Once assembled, the system spans the two membranes and peptidoglycan of the bacterial cell and inserts through a host cell's cytoplasmic membrane like a hypodermic needle. The bacterium then introduces proteins into the host cell. Because bacteria synthesize type III systems only after they have contacted a host cell, the proteins of a type III system are not exposed to immune surveillance.
- Plasmid-encoded enzymes that degrade antimicrobial drugs, conveying drug resistance. One example is carbapenemase, produced by *carbapenem-resistant enterobacteriaceae* (CREs)—bacteria that are resistant to many if not most antimicrobial drugs, including carbapenems. Some physicians consider carbapenems to be drugs of last resort for infections in patients with compromised immunity or invasive devices such as endoscopic tubes.

FIGURE 20.9 illustrates the locations of the major types of antigens and virulence factors of enteric bacteria.

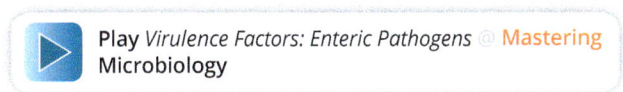

Play *Virulence Factors: Enteric Pathogens* @ Mastering Microbiology

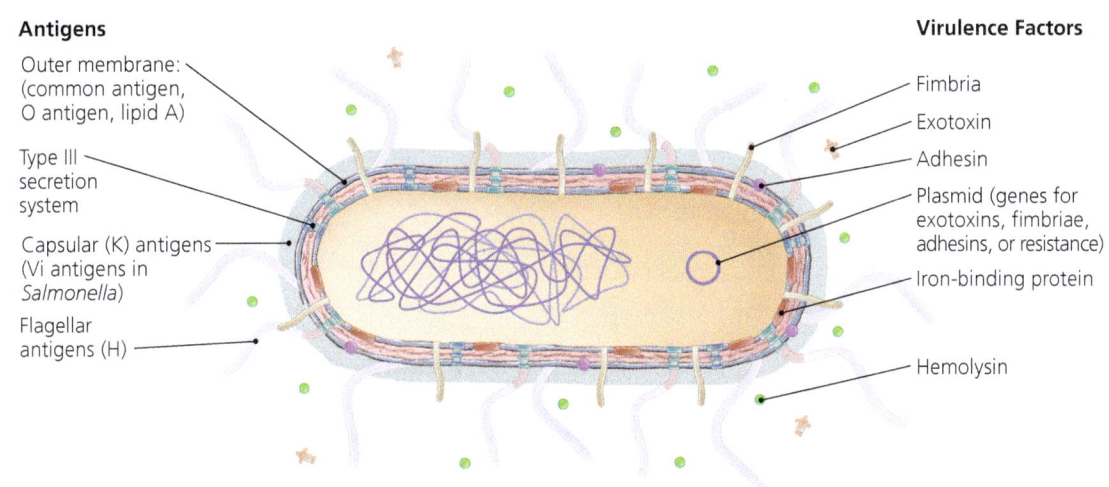

▲ **FIGURE 20.9 Antigens and virulence factors of typical enteric bacteria.**
What is the function of a type III system?

Figure 20.9 A type III secretion system is a virulence factor that enables enteric bacteria to insert chemicals directly into the cytosol of a eukaryotic cell.

Diagnosis, Treatment, and Prevention

The presence of enteric bacteria in urine, blood, or cerebrospinal fluid is always diagnostic of infection or disease. To culture members of the Enterobacteriaceae from clinical specimens, clinicians use selective and differential media such as eosin methylene blue (EMB) agar and MacConkey agar. In their role as selective media, these media inhibit the growth of many members of the normal microbiota, particularly Gram-positive bacteria, and allow the growth of enteric bacteria (FIGURE 20.10a); in their role as differential media, they enable laboratory technicians to distinguish between enteric bacteria that can and cannot ferment lactose (FIGURE 20.10b). Commercially available systems

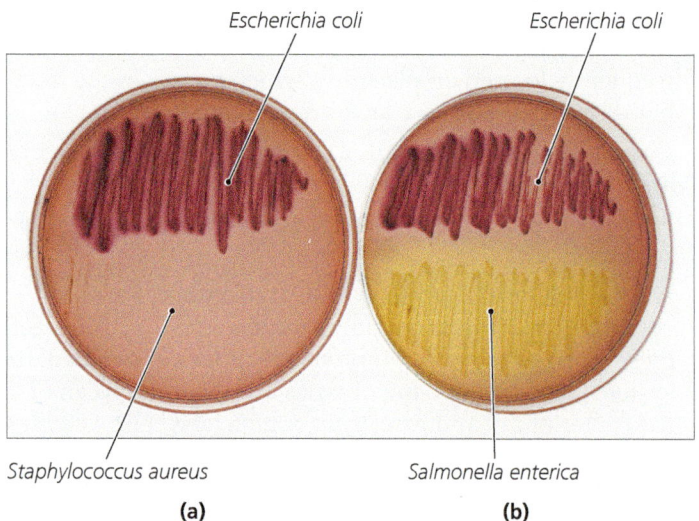

▲ FIGURE 20.10 **The use of MacConkey agar. (a)** This selective and differential medium allows growth of Gram-negative bacteria (here, *E. coli*) and inhibits Gram-positive bacteria (here, *S. aureus*). **(b)** Additionally, colonies of lactose-fermenters (here, *E. coli*) are pink, differentiating them from other enteric bacteria that do not ferment lactose (here, *S. enterica*).

using sophisticated biochemical tests can accurately distinguish among enteric bacteria in 4 to 24 hours.

Treatment of diarrhea involves treating the symptoms with fluid and electrolyte replacement. Often, antimicrobial drugs are not used to treat diarrhea because diarrhea is typically self-limiting—expulsion of the organisms from the body is often a more effective therapy than the action of antimicrobials. Also, antimicrobial therapy can worsen the prognosis by killing many bacteria at once, which releases a large amount of toxins, including lipid A.

Treatment of internal infections involves selection of an appropriate antimicrobial drug, which often necessitates a susceptibility test (Kirby–Bauer test; see Figure 10.9). Susceptible bacteria may develop resistance to antimicrobial drugs, presumably because enteric bacteria exchange plasmids readily; for example, many enteric bacteria can exchange DNA via conjugation. Antibiotics administered to control infections in the blood also kill the normal enteric bacteria, opening a niche for opportunistic resistant bacteria, such as the endospore-forming, Gram-positive bacterium *Clostridium difficile* (klos-trid'ē-ŭm di'fe-sēl).

Given that enteric bacteria are a major component of the normal microbiota, preventing infections with them is almost impossible. Health care practitioners give prophylactic antimicrobial drugs to patients whenever mucosal barriers are breached by trauma or surgery, but they avoid prolonged unrestricted use of antimicrobials, a practice that selects for resistant strains of bacteria. Good personal hygiene (particularly hand washing) and proper sewage control are important in limiting the risk of infection. Specific methods of control and prevention associated with particular genera of enteric bacteria are discussed in the following sections.

Scientists, researchers, and health care practitioners find it useful to categorize the pathogenic members of the Enterobacteriaceae into three groups:

- Coliforms, which rapidly ferment lactose, are part of the normal microbiota and may be opportunistic pathogens.
- Noncoliform members of the normal microbiota do not ferment lactose but can be opportunistic pathogens.
- True pathogens.

In the next three sections, we will examine the important members of the Enterobacteriaceae according to this scheme, beginning with opportunistic coliforms.

MICRO CHECK

4. How can members of the Enterobacteriaceae and Pasteurellaceae be distinguished from each other?
5. Members of the Enterobacteriaceae produce siderophores. What function do siderophores perform?
6. What is the best way to prevent healthcare-associated infections (HAIs) by members of the Enterobacteriaceae?

CLINICAL CASE STUDY

A Painful Problem

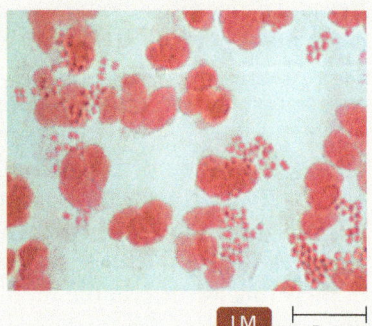

A 20-year-old male reports to his physician that he is experiencing extremely painful urination, as if he were urinating molten lava. He has also noticed a pus-like discharge from his penis. A stained smear of the discharge is shown in the photo.

The patient reports having been sexually active with two or three women in the previous six months. He last had intercourse approximately one week ago. Because his partners were "absolutely sure" that they wouldn't get pregnant, and they had no sexually transmitted diseases, he hadn't used condoms. The doctor orders analysis of DNA in the first catch of his urine.

1. What disease does this patient most likely have? What are the medical and common names for this disease?
2. How did the patient acquire the disease?
3. What could explain the lack of any history of sexually transmitted disease in his sexual partners?
4. What is the likely treatment?
5. Is the patient immune to future infections with this bacterium?

CLINICAL CASE STUDY

A Heart-Rending Experience

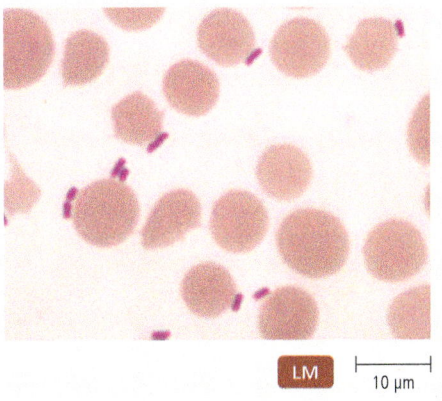

A construction worker in his mid-30s visited his physician complaining of shortness of breath and a persistent cough that had lasted more than three weeks. Upon examination, the doctor noted that the client had a history of asthma and intravenous substance abuse. The patient stated he had not used IV drugs in over a month. Believing the patient, the physician ordered a cough suppressant and told him to return if symptoms did not alleviate.

Two weeks later, the construction worker's condition was worse; the cough suppressant wasn't helping, so he sought help at an emergency room (ER). When he arrived, he had a fever, increased heart rate, was sweaty, had chills, and complained of chest pain that was worse when he inhaled. The ER physician ordered an ultrasound to visualize the heart valves, drew blood for labs and bacterial culture, obtained a chest X ray, administered an electrocardiogram (ECG), and administered intravenous normal saline. Shortly after the infusion, the man went into respiratory arrest and required mechanical ventilation for airway support.

The lab reported that the blood contained *Escherichia coli* (see photo). The physician was surprised at the blood culture results. *E. coli* rarely infects the heart.

1. How could the patient have gotten *E. coli* in his blood?
2. What had *E. coli* done to the inner structures of his heart?
3. What factors in the medical treatment may have exacerbated the disease? Why?

Coliform Opportunistic Enterobacteriaceae

LEARNING OUTCOMES

20.10 Compare and contrast *Escherichia*, *Klebsiella*, *Serratia*, *Enterobacter*, *Hafnia*, and *Citrobacter*.

20.11 Describe the pathogenesis and diseases of *Escherichia coli* O157:H7.

Coliforms (kol′i-formz) are defined as aerobic or facultatively anaerobic, Gram-negative, rod-shaped bacteria that ferment lactose to form gas within 48 hours of being placed in a lactose broth at 35°C. Coliforms are found in the intestinal tracts of animals and humans, in soil, and on plants and decaying vegetation. The presence of fecal coliforms in water is indicative of impure water and of poor sewage treatment. The most common of the opportunistic coliform pathogens are bacteria in the genera *Escherichia*, *Klebsiella*, *Serratia*, *Enterobacter*, *Hafnia*, and *Citrobacter*.

Escherichia coli

Escherichia coli (esh-ĕ-rik′ē-ă kō′lī)—the most common and important of the coliforms—is well known as a laboratory research organism.

Scientists have described numerous O, H, and K antigens of *E. coli*, which epidemiologists use to identify particular strains. Some antigens, such as O157, O111, H8, and H7, are associated with virulence. Virulent strains also have genes (located on virulence plasmids) for fimbriae, adhesins, and a variety of exotoxins, enabling these strains to colonize human tissue and cause disease. Many bacteria share such plasmids among themselves.

Pathogenic strains of *E. coli* are named for the conditions they cause, such as *enterohemorrhagic E. coli* (EHEC, life-threatening bloody diarrhea), *uropathogenic E. coli* (UPEC, urinary tract infections), and *enteroinvasive E. coli* (EIEC, severe diarrhea with high fever).

E. coli is responsible for a number of diseases, including septicemia, neonatal meningitis, pneumonia, gastroenteritis, and *urinary tract infections (UTIs)*. **Disease in Depth: Bacterial Urinary Tract Infections** (pp. 597–598) illustrates the role *Escherichia* and other bacteria play in infections of the urinary system. *E. coli* in the blood can colonize the lining of the heart, triggering inflammation and the destruction of heart valves (endocarditis).

Gastroenteritis, the most common disease associated with *E. coli*, is often mediated by toxins released from the bacterium. The toxins bind to proteins on cells lining the intestinal tract and so are called *enterotoxins*. A portion of the toxin then enters the cell and triggers a series of chemical reactions that cause the loss of sodium, chloride, potassium, bicarbonate, and water from the cells, producing watery diarrhea, cramps, nausea, and vomiting. Enterotoxin-producing strains are common in developing countries and are important causes of pediatric diarrhea, accounting for up to a third of the cases of this life-threatening disease in some countries.

Epidemiologists first described *E. coli* O157:H1 (one type of EHEC) in 1982. This type of *E. coli* is now one of the more prevalent strains of *E. coli* in developed countries. The ingestion of as few as 10 organisms can result in disease, which ranges from bloody diarrhea, to severe or fatal hemorrhagic colitis, to a severe kidney disorder called *hemolytic uremic syndrome* in which renal function ceases.

Urinary tract and blood infections with this organism can be fatal, especially in children and immunocompromised adults. Most epidemics of *E. coli* O157:H7 have been associated with the consumption of undercooked ground beef, unpasteurized milk, or fruit juice contaminated with feces. Despite its notoriety, fewer than 1 in 10 million deaths in the United States are caused by strain O157:H7.

E. coli O157:H7 produces a type III secretion system through which it introduces two types of proteins into intestinal cells. Proteins of one type disrupt the cell's metabolism; proteins of the other type become lodged in the cell's cytoplasmic membrane, where they act as receptors for the attachment of additional *E. coli* O157:H7 bacteria. Such attachment apparently enables this strain of *E. coli* to displace normal, harmless strains.

E. coli O157:H7 also produces *Shiga-like toxin*, which inhibits protein synthesis in host cells. Shiga-like toxin, which was first identified in *Shigella* (discussed shortly), attaches to the surfaces of neutrophils and is spread by them throughout the body, causing widespread death of host cells and tissues. Antimicrobial drugs that prevent protein synthesis induce *E. coli* O157:H7 to increase its production of Shiga-like toxin, worsening the illness. Generally, physicians recommend letting diarrhea run its course, because it tends to eliminate its cause.

Klebsiella

Species of the genus *Klebsiella* (kleb-sē-el'ă) grow in the digestive and respiratory systems of humans and animals. They are nonmotile, and their prominent capsules (FIGURE 20.11) give *Klebsiella* colonies a mucoid appearance and protect the bacteria from phagocytosis, enabling them to become opportunistic pathogens.

The most commonly isolated pathogenic species is *K. pneumoniae* (nu-mō'nē-ī), which causes pneumonia and may be involved in bacteremia, meningitis, wound infections, and UTIs. Pneumonia caused by *K. pneumoniae* often involves destruction of alveoli and the production of bloody sputum. Alcoholics and patients with compromised immunity are at greater risk of pulmonary disease because of their poor ability to clear aspirated oral secretions from their lower respiratory tracts.

Serratia

The motile coliform *Serratia marcescens* (se-rat'ē-a mar-ses'enz) produces a red pigment when grown at room temperature (FIGURE 20.12). For many years, scientists considered *Serratia* to be totally benign; in fact, researchers intentionally introduced this bacterium into various environments to track the movement of bacteria in general. Using this method, scientists have followed the movement of *Serratia* from the mouth into the blood of dental patients and throughout hospitals via air ducts. In a series of tests performed between 1949 and 1968, the U.S. government released large quantities of *S. marcescens* into the air over San Francisco to mimic the movement of a discharged biological weapon. About a dozen people are known to have fallen ill and at least one man died, though there is some question about the source of the victims' infections.

Now, we know that *Serratia* can grow on catheters, in saline solutions, and on other hospital supplies and that it is a potentially life-threatening opportunistic pathogen in the urinary and respiratory tracts of immunocompromised patients. The bacterium is frequently resistant to antimicrobial drugs.

Enterobacter, *Hafnia*, and *Citrobacter*

Other motile coliforms that can be opportunistic pathogens are *Enterobacter* (en'ter-ō-bak'ter), *Hafnia* (haf'nē-ă), and *Citrobacter* (sit'rō-bak-ter). These bacteria, like other coliforms, ferment lactose and reside in the digestive tracts of animals and humans as well as in soil, water, decaying vegetation, and sewage. *Enterobacter* can be a contaminant of dairy products. All three genera are involved in healthcare-associated infections of the blood, wounds, surgical incisions, and urinary tracts of immunocompromised patients. Treatment of such infections can be difficult because these prokaryotes, especially *Enterobacter*, are often resistant to most antibacterial drugs.

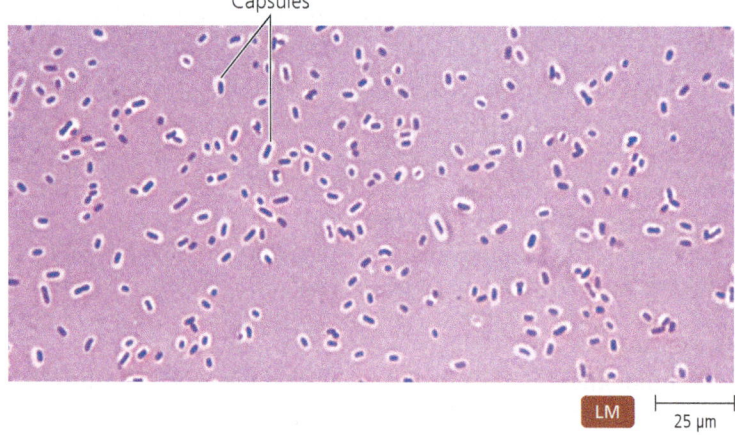

▲ FIGURE 20.11 **The prominent capsule of *Klebsiella pneumoniae*.** How does the capsule function as a virulence factor?

Figure 20.11 Capsules inhibit phagocytosis and intracellular digestion by phagocytic cells.

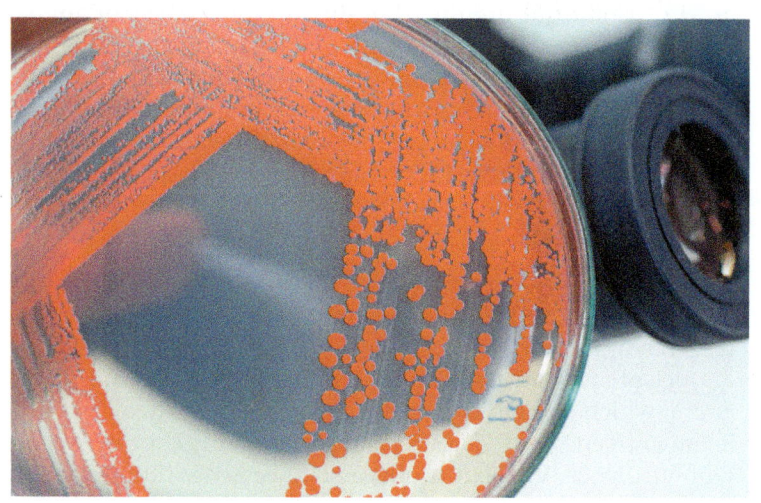

▲ FIGURE 20.12 **Red colonies of *Serratia marcescens*.** Color develops in colonies grown at room temperature.

MICRO CHECK

7. What characteristic do all coliform bacteria share?
8. What is the function of the type III secretion system in *E. coli* O157:H7?
9. Why are women more likely to contract urinary tract infections than men?
10. What part of the body is infected in patients with pyelonephritis?
11. Which of the coliform bacteria may produce a red pigment at room temperature?

Noncoliform Opportunistic Enterobacteriaceae

LEARNING OUTCOMES

20.12 Differentiate between coliform and noncoliform opportunists.
20.13 Describe the diseases caused by noncoliform opportunistic enteric bacteria.

The family Enterobacteriaceae also contains genera that cannot ferment lactose; several of these genera are opportunistic pathogens. These noncoliform opportunists include *Proteus*, *Morganella*, *Providencia*, and *Edwardsiella*.

Proteus

Like the mythical Greek god Proteus, who could change shape, the Gram-negative, facultative anaerobe *Proteus* (prō'tē-ŭs) is a typical rod-shaped bacterium with a few polar flagella when cultured in broth but differentiates into an elongated cell that swarms by means of numerous peritrichous flagella when cultured on agar. Swarming cells produce concentric wavelike patterns on agar surfaces (FIGURE 20.13). The roles of the two morphologies in human infections are not clear.

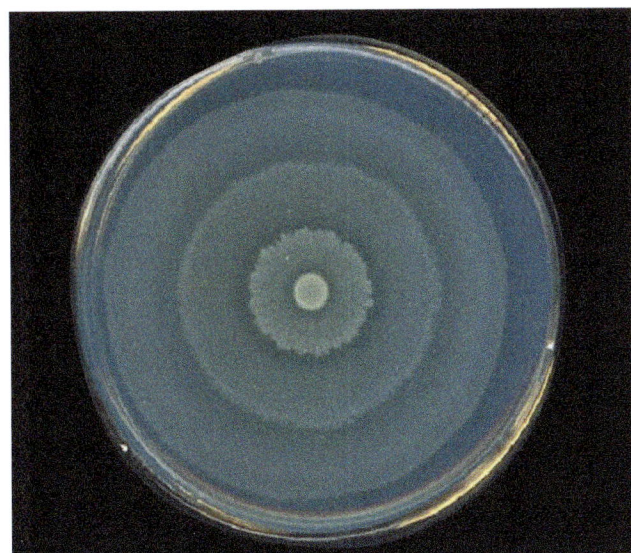

▲ FIGURE 20.13 **The wavelike concentric rings of the swarming *Proteus mirabilis*.** Swarming by the bacterium may create beauty, but it makes isolation of *Proteus* more difficult.

Proteus mirabilis (mi-ra'bi-lis) is the most common species of *Proteus* associated with disease in humans, particularly with UTIs in patients with long-term urinary catheters. One study showed that 44% of catheterized patients had *P. mirabilis* growing in their urine.

In the presence of urea, as is found in the bladder, this microorganism releases a large amount of the enzyme *urease*, which breaks down urea into carbon dioxide and ammonia. Ammonia raises the pH of urine in the bladder and kidneys such that ions that are normally soluble at the acidic pH of urine (around pH 6.0) precipitate, often around bacterial cells, to form infection-induced kidney stones composed of magnesium ammonium phosphate or calcium phosphate. Kidney stones produced in this manner can contain hundreds of *P. mirabilis* cells.

Because *Proteus* is resistant to many antimicrobial drugs, health care workers must determine proper treatment following a susceptibility test. Researchers have been able to immunize mice against UTIs with *Proteus*. This success is encouraging to those seeking protection for patients who must have long-term urinary catheters.

Morganella, *Providencia*, and *Edwardsiella*

Healthcare-associated infections of immunocompromised patients by *Morganella* (mōr'gan-el'ă), *Providencia* (prov'i-den'sē-ă), and *Edwardsiella* (ed'ward-sē-el'lă) are becoming more frequent. These organisms are primarily involved in UTIs. *Providencia* contains a plasmid that codes for urease and thus may trigger the formation of kidney stones.

Truly Pathogenic Enterobacteriaceae

LEARNING OUTCOMES

20.14 Describe the diseases caused by truly pathogenic enteric bacteria.
20.15 Contrast salmonellosis and shigellosis.
20.16 Describe the life cycle of *Yersinia pestis* and contrast bubonic and pneumonic plague.

Three important non-lactose-fermenting bacteria in the family Enterobacteriaceae are *Salmonella*, *Shigella*, and *Yersinia*. These genera are not considered members of the normal microbiota of humans because they are almost always pathogenic due to their numerous virulence factors. All three genera synthesize type III secretion systems through which they introduce proteins that inhibit phagocytosis, rearrange the cytoskeletons of eukaryotic cells, or induce apoptosis. We begin our discussion with *Salmonella*, which is the most common of the three pathogens.

Salmonella

Pathogenesis, Epidemiology, and Disease *Salmonella* (sal'mŏ-nel'ă) is a genus of motile, Gram-negative, peritrichous bacilli that live in the intestines of virtually all birds, reptiles, and mammals and are eliminated in their feces. These bacteria do not ferment lactose, but they do ferment glucose, usually with gas production. They are urease negative and oxidase negative, and most produce hydrogen sulfide (H_2S). Scientists have identified

DISEASE IN DEPTH

BACTERIAL URINARY TRACT INFECTIONS

Escherichia coli and other bacteria

SIGNS AND SYMPTOMS

Patients with mild urethritis, cystitis, or prostatitis may have only a slight fever or no symptoms, but most of these diseases involve frequent, urgent, and painful urination—a condition called dysuria. The urine may be cloudy and bloody and have a strong, foul odor. Mental confusion, which results when the bacteria spread from the urinary system to the blood (bacteremia), is typically seen in the elderly.

Millions of people in the United States, mostly girls and women, suffer bacterial urinary tract infections (UTIs) each year. This includes about 600,000 patients who annually acquire healthcare-associated UTIs. Urinary tract infections are rare in boys and men. Bacteria may trigger inflammation and pain in any or all of the urinary tract, including the urethra, urinary bladder, or kidneys—conditions called urethritis, cystitis, and pyelonephritis.

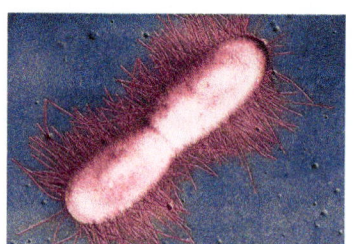

E. coli in early stage of binary fission TEM 1 μm

PATHOGENESIS
[Illustrated: uropathogenic *E. coli* (UPEC)]
In the Urethra (Urethritis)

1 *E. coli* is introduced into the urethra, often in fecal contamination.

2 *E. coli* adhere to epithelial cells in the urethra via fimbriae.

3 *E. coli* colonize epithelial cells and begins to multiply via binary fission.

4 *E. coli* swim up the urethra.

 Watch Dr. Bauman's Video Tutor explore bacterial urinary tract infections, or go to the **Mastering Microbiology** Study Area for more information.

EPIDEMIOLOGY

UTIs are more common in females than males because female urethras are shorter and closer to the anus. Diabetics, nursing home patients, elderly men who have trouble emptying their bladders because of prostate enlargement, patients with urinary catheters, women who use diaphragms for birth control, and people who do not drink adequate fluids are also at risk for UTIs.

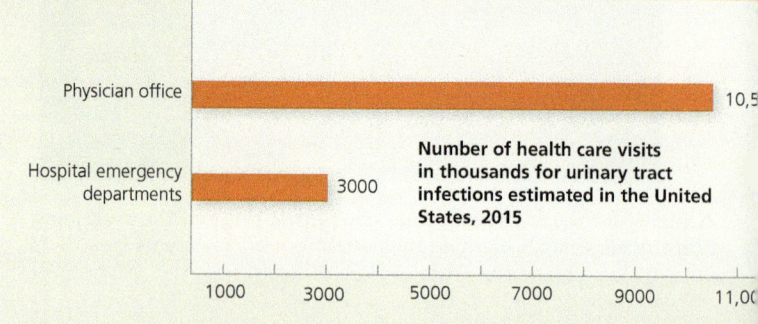

Number of health care visits in thousands for urinary tract infections estimated in the United States, 2015

Physician office: 10,5
Hospital emergency departments: 3000

PATHOGEN

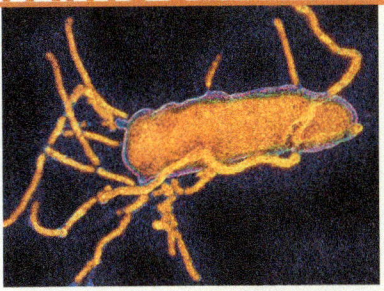
SEM 0.5 µm

Enteric bacteria—Gram-negative bacteria that are part of the intestinal microbiota—are the most common cause of urinary tract infections. *Escherichia coli* causes about 70% of cases; other bacteria from the intestinal tract, such as *Proteus* and *Klebsiella*, cause about 10% of UTIs. Nonintestinal bacteria, such as *Pseudomonas* and *Staphylococcus*, occasionally cause UTIs.

VIRULENCE FACTORS

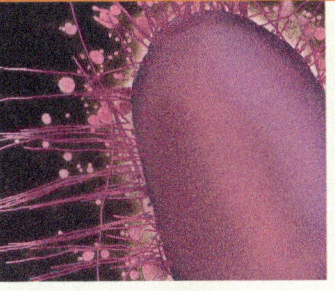
TK   TK µm

E. coli, *Proteus*, and *Pseudomonas* have flagella that can propel the bacteria up the urinary tract. Strains of *E. coli* that infect the bladder have attachment fimbriae that bind specifically to epithelial cells lining the urinary bladder. Using fimbriae, these strains of *E. coli* move into bladder cells by an unknown mechanism and form biofilm-like aggregations within bladder cells' cytosol and avoid the body's defenses.

In the Blood (Bacteremia)

6 Cells of *E. coli* are released into the bloodstream (bacteremia).

In the Bladder (Cystitis)

5 *E. coli* invades bladder cells and multiplies within the cytosol, escaping many of the body's defenses.

In the Kidney (Pyelonephritis)

7 *E. coli* ascend to the kidneys, where they trigger inflammation.

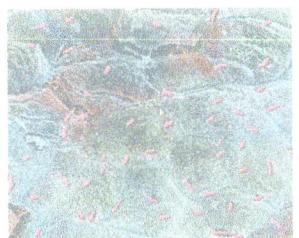
E. coli infection of bladder cells
SEM 10 µm

DIAGNOSIS

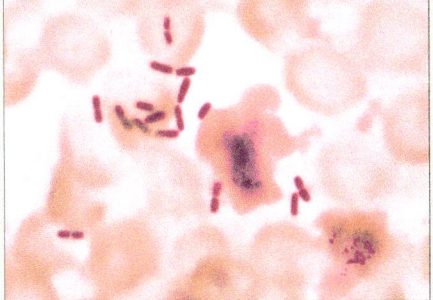
LM 4 µm

Urinalysis of patients with the signs and symptoms of urinary tract infections reveals leukocytes, erythrocytes, and the causative bacterium, usually *E. coli*.

TREATMENT AND PREVENTION

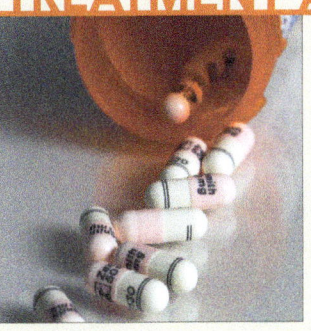

Mild UTIs resolve on their own without treatment, though antimicrobial drugs, such as nitrofurantoin or trimethoprim-sulfamethoxazole, can prevent the spread of infection to the kidneys and blood. Wiping from front to back after defecation to avoid dragging bacteria from the anus into the urethra is the most important step females can take to prevent UTIs.

more than 2500 unique serotypes (strains) of *Salmonella*. Analysis of the DNA sequences of their genes indicates that all of them properly belong to a single species—*S. enterica* (en-ter′i-kă). This taxonomically correct approach has not been well received by researchers or medical professionals, so we will consider the various strains of this single species with modern terminology as well as according to their traditional specific epithets, such as *typhi* (tī′fē) and *paratyphi* (pa′ra-tī′fē).

Most infections of humans with salmonellae result from the consumption of food contaminated with animal feces, often from pet reptiles. The pathogen is also common in foods containing poultry or eggs—about one-third of chicken eggs are contaminated. *Salmonella* released during the cracking of an egg on a kitchen counter and inoculated into other foods can reproduce into millions of cells in just a few hours. Some strains, such as *S. enterica* serotype Dublin, are acquired through the consumption of inadequately pasteurized contaminated milk. Infections with fewer than 1 million cells of most strains of *Salmonella* are usually asymptomatic in people with healthy immune systems.

Larger infective doses can result in **salmonellosis** (sal′-mo-nel-o′sis), which is characterized by nonbloody diarrhea, nausea, and vomiting. Fever, myalgia (muscle pain), headache, and abdominal cramps are also common. The events in the course of salmonellosis are depicted in FIGURE 20.14. After the salmonellae pass through the stomach, they attach to the cells lining the small intestine 1. The bacteria then use type III secretion systems to insert proteins into the host cells, inducing the normally nonphagocytic cells to endocytize the bacteria 2. The salmonellae then reproduce within endocytic vesicles 3, eventually killing the host cells 4 and inducing the signs and symptoms of salmonellosis. Cells of some strains can subsequently enter the blood, causing bacteremia 5 and localized infections throughout the body, including in the lining of the heart, the bones, and the joints.

Humans are the sole hosts of *Salmonella enterica* serotype Typhi (previously known as *S. typhi*), which causes **typhoid fever**. (A mild form of typhoid fever is produced by serotype Paratyphi.) Infection occurs via the ingestion of food or water contaminated with sewage containing bacteria from carriers, who are often asymptomatic. An infective dose of *S. enterica* Typhi is only about 1000 to 10,000 cells. The bacteria pass through the intestinal cells into the bloodstream; there they are phagocytized but not killed by phagocytic cells, which carry the bacteria to the liver, spleen, bone marrow, and gallbladder. Patients typically experience gradually increasing fever, headache, muscle pains, malaise, and loss of appetite that may persist for a week or more. Without treatment, about 15% of patients die.

Bacteria frequently reproduce in the gallbladder and can be released to reinfect the intestines, producing gastroenteritis, abdominal pain, and recurrent bacteremia. In some patients the bacteria ulcerate and perforate the intestinal wall, allowing bacteria from the intestinal tract to enter the abdominal cavity and cause *peritonitis*. FIGURE 20.15 contrasts the incidences of salmonellosis and typhoid fever in the United States since 1935.

Diagnosis, Prevention, and Treatment Clinicians diagnose salmonellosis by finding the bacterium in a clinical specimen of a patient's stool (evacuated feces) or blood. Prevention of infection

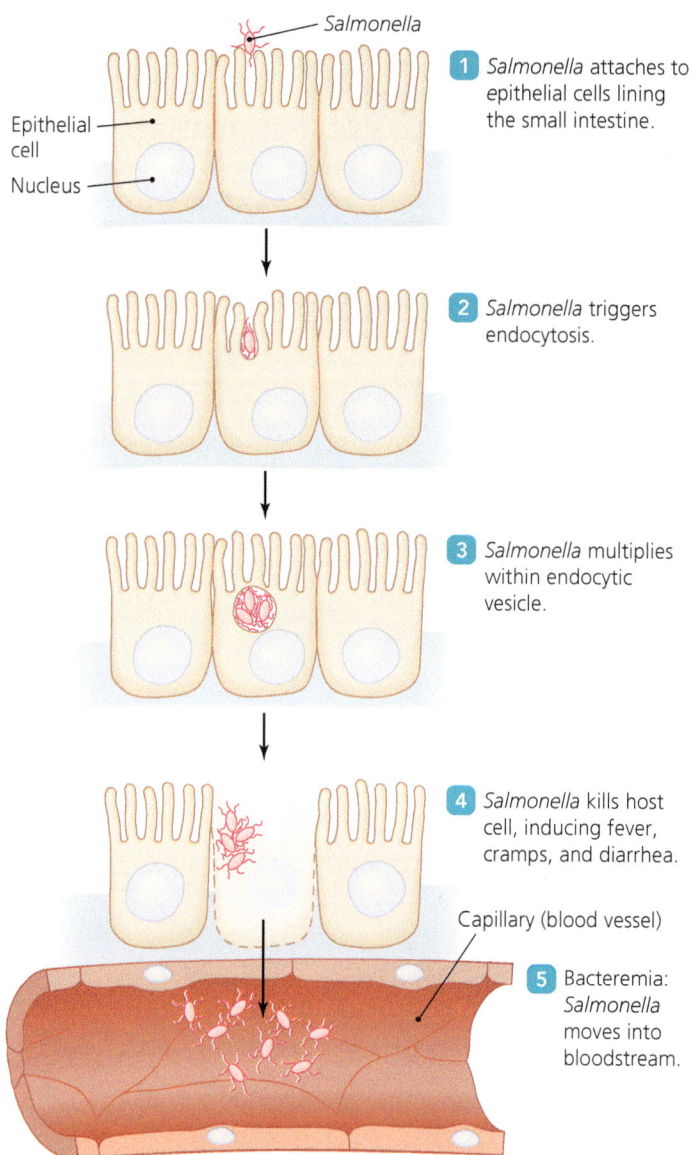

▲ FIGURE 20.14 The events in salmonellosis.

involves limiting fecal contamination of food and water supplies with good sewage and water treatment.

Food processors can add bacteriophages against *Salmonella* to food, which reduces the number of salmonellae in food. Such phage usage is currently approved in Europe but not in the United States. Researchers have developed vaccines that provide temporary protection for travelers to areas where typhoid fever is endemic.

Treatment for salmonellosis involves fluid and electrolyte replacement. Typhoid fever is treated with antimicrobial drugs such as ampicillin or ciprofloxacin, though an extensively drug resistant strain (XDR *Salmonella typhi*) emerged in Pakistan in 2018. Physicians sometimes remove the gallbladder of a carrier to reduce the chance that others will become infected.

Shigella

Pathogenesis, Epidemiology, and Disease The genus *Shigella* (shē-gel′lă) contains Gram-negative, oxidase-negative, nonmotile pathogens of the family Enterobacteriaceae that are primarily

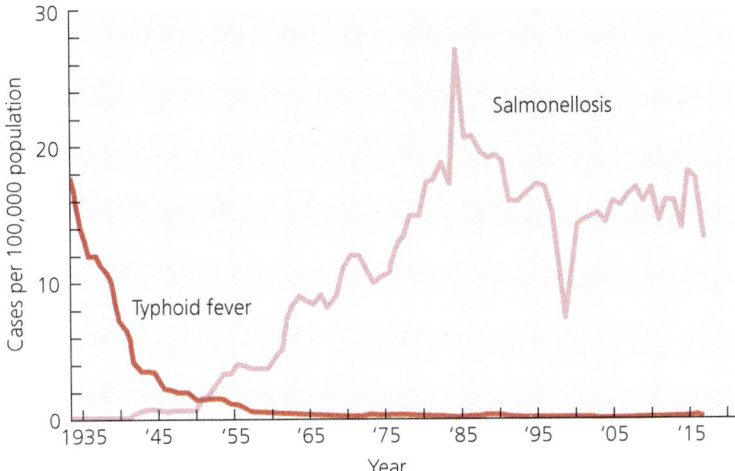

▲ FIGURE 20.15 **The incidences of diseases caused by *Salmonella* in the United States.** Whereas the incidence of typhoid fever has declined because of improved sewage treatment and personal hygiene, the incidence of reported cases of salmonellosis has increased. *What factors might explain the increase in the incidence of salmonellosis?*

Figure 20.15 *Salmonellosis* and typhoid fever are caused by the same species (*Salmonella enterica*) and transmitted via fecal contamination; however, the two diseases are caused by different strains of the bacterium. The strain causing typhoid (Typhi) is transmitted in human waste, and human waste is better managed than animal waste, which is the predominant source of strains causing salmonellosis. Further, the historical number of cases of salmonellosis is not known. The observed increase may be due to better diagnosis and reporting rather than to an actual increase in the number of cases.

parasites of the digestive tracts of humans. These bacteria produce a diarrhea-inducing enterotoxin, are urease negative, and do not produce hydrogen sulfide gas. Some scientists suggest that *Shigella* may actually be a strain of *E. coli* that has become nonmotile and oxidase negative, while other researchers go so far as to consider *Shigella* an invasive, toxin-producing *E. coli* that is cloaked in *Shigella* antigens. Still other scientists contend the converse—that *E. coli* is really a disguised species of *Shigella* that has acquired genes for flagella.

In any case, taxonomists have historically identified four well-defined species of *Shigella*: *S. dysenteriae* (dis-en-te′rē-ī), *S. flexneri* (fleks′ner-ē), *S. boydii* (boy′dē-ē), and *S. sonnei* (sōn′ne-ē). *Shigella sonnei* is most commonly isolated in industrialized nations, whereas *S. flexneri* is the predominant species in developing countries. All four species cause a severe form of dysentery called **shigellosis** (shig-ĕ-lō′sis), which is characterized by abdominal cramps, fever, diarrhea, and pus-containing, bloody stools.

Shigellosis is primarily associated with poor personal hygiene and ineffective sewage treatment. People become infected primarily by ingesting bacteria on their own contaminated hands and secondarily by consuming contaminated food. *Shigella* is little affected by stomach acid, so an infective dose may be as few as 200 cells; therefore, person-to-person spread is also possible, particularly among children and among homosexual men.

Initially, *Shigella* colonizes cells of the small intestine, causing an enterotoxin-mediated diarrhea; the main events in shigellosis, however, begin once bacteria invade cells of the large intestine (**FIGURE 20.16**). When *Shigella* attaches to an epithelial cell of the large intestine ❶, the cell is stimulated to endocytize the bacterium ❷, which then multiplies within the cell's cytosol ❸. (Note that this differs from salmonellosis, in which bacteria multiply within endocytic vesicles.) *Shigella* then polymerizes the host's actin fibers, which push the bacteria out of the host cell and into

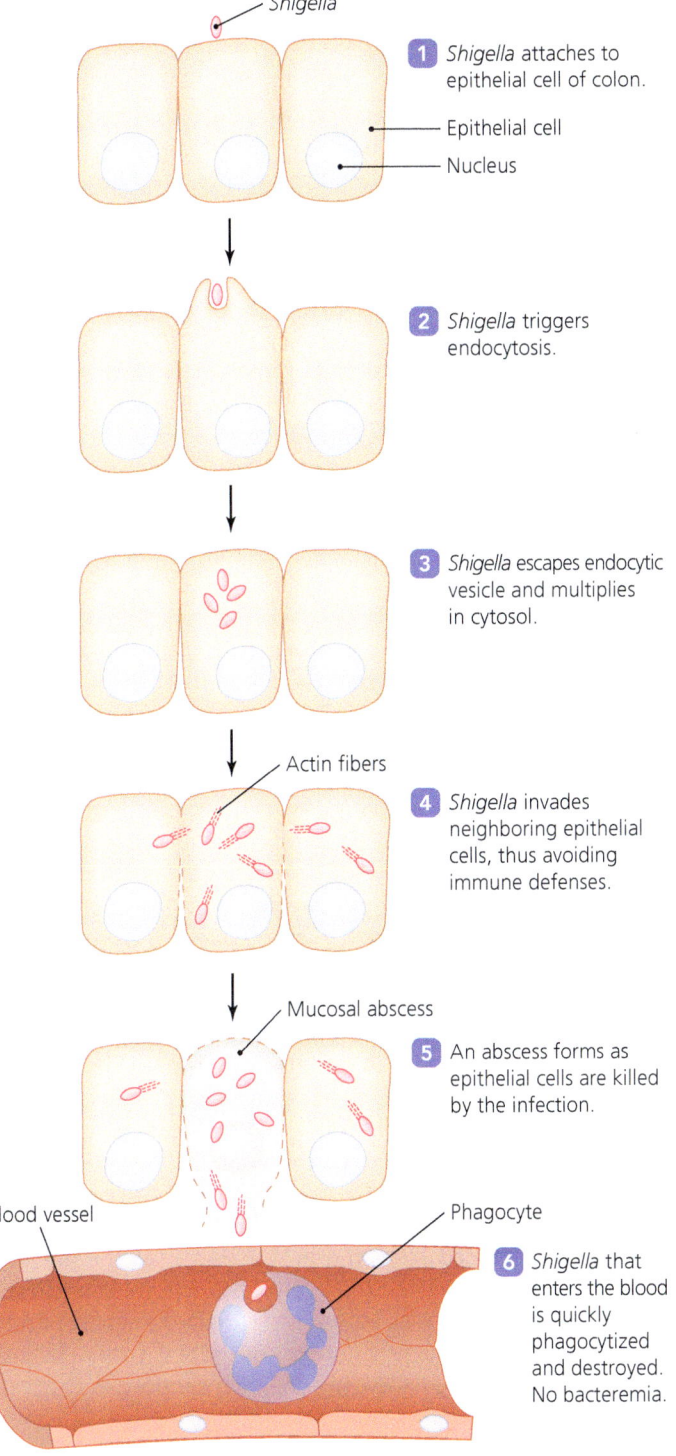

▲ FIGURE 20.16 **The events in shigellosis.**

adjacent cells ④, in the process evading the host's immune system. As the bacteria kill host cells, abscesses form in the mucosa ⑤; any bacteria that enter the blood from a ruptured abscess are quickly phagocytized and destroyed ⑥, so bacteremia is rarely a part of shigellosis. About 125,000,000 cases of shigellosis occur annually worldwide, but because mild cases are rarely reported, this figure may represent only a small percentage of the actual total.

Shigella dysenteriae secretes an exotoxin called **Shiga toxin**, which stops protein synthesis in its host's cells. Shiga toxin is similar to the Shiga-like toxin of *Escherichia coli* O157:H7. Shigellosis caused by *S. dysenteriae* is more serious than that caused by other species of *Shigella*, with a mortality rate as high as 20%. Shiga toxin kills more than 1 million people every year.

Diagnosis, Prevention, and Treatment Though physicians may suspect shigellosis on the basis of its signs and symptoms, diagnosis is dependent on finding the bacterium in the stool.

A live, attenuated vaccine against *S. flexneri* has been successful in preventing dysentery caused by this species, although the participants in the study experienced mild diarrhea and fever as a result of immunization. Researchers are working to perfect a vaccine that will not cause symptoms.

Treatment of shigellosis involves replacement of fluids and electrolytes. The disease is usually self-limiting as the bacteria are voided in feces, but oral antimicrobial drugs, such as ciprofloxacin, sulfonamides, penicillin, or cephalosporin, can reduce the spread of *Shigella* to close contacts of the patient.

Yersinia

Pathogenesis, Epidemiology, and Disease The genus *Yersinia* (yer-sin'ē-ă) contains three notable species—*Y. enterocolitica* (en'ter-ō-kō-lit'ĭ-kă), *Y. pseudotuberculosis* (soo-dō-too-ber-kyu-lō'sis), and *Y. pestis* (pes'tis)—that are normally pathogens of animals. All three species contain virulence plasmids that code for adhesins and type III secretion systems. The adhesins allow *Yersinia* to attach to human cells, after which the type III system is used to inject proteins that trigger apoptosis in macrophages and neutrophils.

Yersinia enterocolitica and *Y. pseudotuberculosis* are enteric pathogens acquired via the consumption of food or water contaminated with animal feces. *Y. enterocolitica* is a common cause of painful inflammation of the intestinal tract; such inflammation is accompanied by diarrhea and fever that can last for weeks or months. Involvement of the terminal portion of the small intestine can result in painful inflammation of the mesenteric lymph nodes, which mimics appendicitis. *Y. pseudotuberculosis* produces a less severe inflammation of the intestines.

Yersinia pestis is an extremely virulent, nonenteric pathogen that has two clinical manifestations: **bubonic plague** (boo-bon'ik) and **pneumonic plague** (noo-mō'nik). A major pandemic of plague that lasted from the mid-500s A.D. to the late 700s is estimated to have claimed the lives of more than 40 million people. This devastation was surpassed during a second pandemic that killed 30% to 60% of the population of Europe (25 million people) in just five years during the 14th century. A third major pandemic spread from China across Asia, Africa, Europe, and the Americas in the 1860s and claimed over 10 million lives. So great was the devastation of these three great pandemics that the word "plague" can still provoke a sense of dread today.

Rats, mice, and voles are the hosts for the natural endemic cycle of *Yersinia pestis* (FIGURE 20.17a); they harbor the bacteria but do not develop disease. Among rodents, fleas are the vectors for the spread of the bacteria. As the bacteria multiply within a flea, they block the esophagus such that the flea can no longer ingest blood from a host. The starving flea jumps from host to host seeking a blood meal and infecting each new host with every bite. When other animals—including prairie dogs, rabbits, deer, camels, dogs, and cats—become infected via flea bites, they act as *amplifying hosts*; that is, they support increases in the numbers of bacteria and infected fleas, even though the amplifying hosts die from plague. Humans usually become infected when bitten by infected fleas that have left their normal animal hosts (FIGURE 20.17b). Though direct contact with infected animals may transmit *Yersinia*, bubonic plague does not spread from person to person.

Bubonic plague is characterized by high fever and swollen, painful lymph nodes called *buboes*[3] (FIGURE 20.18), which develop within a week of infection. Untreated cases progress to bacteremia, which results in disseminated intravascular coagulation, subcutaneous hemorrhaging, and death of tissues, which may become infected with *Clostridium* and develop gangrene. Because of the extensive darkening of dead skin, plague has been called the "Black Death." Untreated bubonic plague is fatal in 50% of cases. Even with treatment, up to 15% of patients die. In fatal infections, death usually occurs within a week of the onset of symptoms.

Pneumonic plague occurs when *Yersinia* in the bloodstream infects the lungs. Disease develops very rapidly—patients develop fever, malaise, and pulmonary distress within a day of infection. Pneumonic plague can spread from person to person through aerosols and sputum (FIGURE 20.17c) and if left untreated is fatal in nearly 100% of cases.

Diagnosis, Prevention, and Treatment Because plague is so deadly and can progress so rapidly, diagnosis and treatment must also be rapid. The characteristic symptoms, especially in patients who have traveled in areas where plague is endemic, are usually sufficient for diagnosis. Rodent control and better personal hygiene have almost eliminated plague in industrialized countries, although wild animals remain as reservoirs. A successful vaccine has been tested in humans but is not commercially available. Many antibacterial drugs, including streptomycin, gentamicin, doxycycline, ciprofloxacin, and levofloxacin, are effective against *Yersinia*.

In summary, FIGURE 20.19 illustrates the general sites of infections by the more common members of the Enterobacteriaceae.

> ### MICRO CHECK
> 12. Which of the noncoliform opportunistic Enterobacteriaceae display swarming motility on agar surfaces?
> 13. Which of the diseases caused by truly pathogenic noncoliform Enterobacteriaceae is most commonly associated with pet reptiles and contaminated eggs?
> 14. How do *Salmonella* and *Shigella* infections differ?
> 15. What is the natural host for *Yersinia pestis*?

---

[3]From Greek *boubon*, meaning "groin," referring to a swelling in the groin.

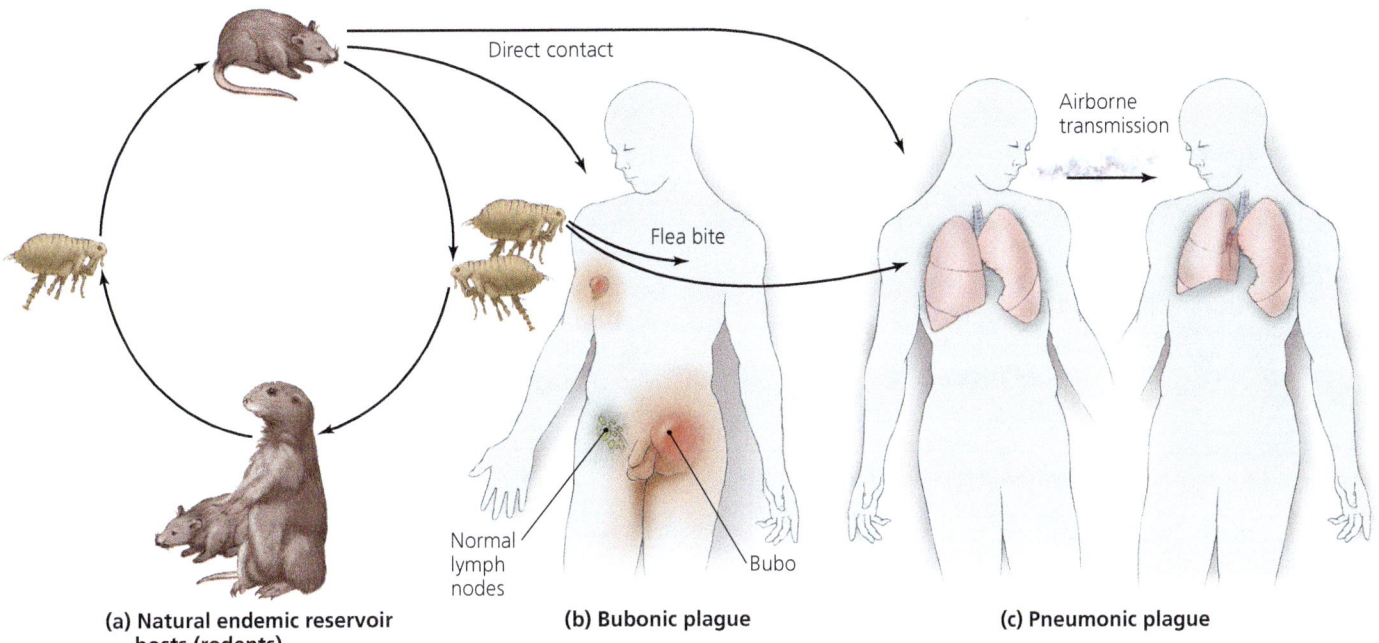

▲ **FIGURE 20.17 The natural history and transmission of *Yersinia pestis*.** This bacterium causes plague. **(a)** The natural endemic cycle of *Yersinia* among rodents. **(b, c)** Humans develop one of two forms of plague as the result of the bite of infected fleas or direct contact with infected hosts. *Yersinia* can move from the bloodstream into the lungs to cause pneumonic plague, which can be spread between humans via the airborne transmission of the bacteria in aerosols. *Why is the plague so much less devastating today than it was in the Middle Ages?*

Figure 20.17 Urban living has reduced contact between people and flea vectors living on wildlife and farm animals; improved hygiene and the use of insecticides have reduced contact with flea vectors living on pets; and antimicrobials are effective against *Y. pestis*.

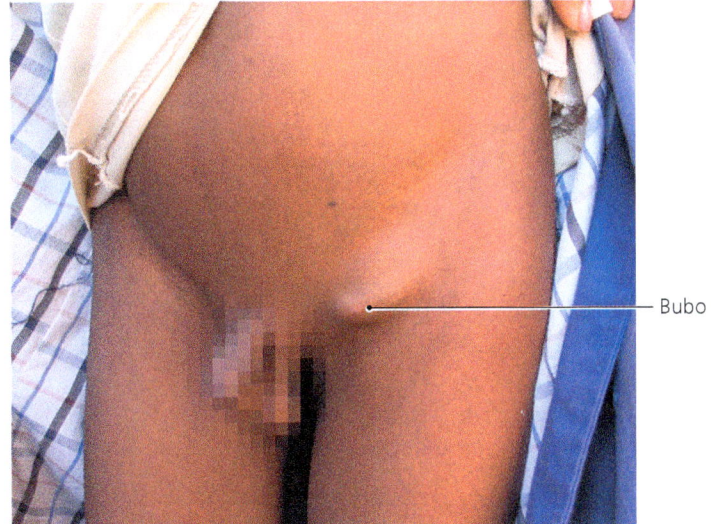

▲ **FIGURE 20.18 Bubo in an eight-year-old patient.** Such lymph nodes swollen by infection are a characteristic feature of bubonic plague.

The Pasteurellaceae

LEARNING OUTCOMES

20.17 Describe two pathogenic genera in the family Pasteurellaceae.

20.18 Identify and describe three diseases caused by species of *Haemophilus*.

In the previous sections, we examined opportunistic and pathogenic Gram-negative, facultative anaerobes of the family Enterobacteriaceae, all of which are oxidase negative. Now we consider another family of gammaproteobacteria—the **Pasteurellaceae** (pas-ter-el-ă′sē-ē), which contains species that are oxidase positive (see Figure 20.5). These microorganisms are mostly facultative anaerobes that are small, nonmotile, and fastidious in their growth requirements—they require heme or cytochromes. Two genera, *Pasteurella* and *Haemophilus*, contain most of this family's pathogens of humans.

Pasteurella

Pasteurella (pas-ter-el′ă) is part of the normal microbiota in the oral and nasopharyngeal cavities of animals, including

CLINICAL CASE STUDY

A Sick Camper

An otherwise healthy 24-year-old woman goes to her doctor complaining of a sudden onset of high fever, chills, uneasiness, and a severe headache. She also shows the doctor a painful swelling she is experiencing in her groin area. The doctor asks her about recent travel. She reports that she returned two days prior from a weeklong camping and hiking trip in Texas.

1. How did the woman most likely contract the disease?
2. What are the potential problems associated with diagnosing this disease, and how crucial is prompt diagnosis?
3. The doctor asks you, as a nursing student rotating through his clinic, your opinion on the disease diagnosis and causative agent. What is your response?
4. How should the patient be treated?
5. Who should be notified once the diagnosis is confirmed? Why?

---

domestic cats and dogs. Humans are typically infected via animal bites and scratches or via inhalation of aerosols from animals. Most cases in humans are caused by *P. multocida* (mul-tŏ'si-da), which produces localized inflammation and swelling of lymph nodes at the site of infection. Patients with suppressed immunity are at risk of widespread infection and bacteremia.

Diagnosis of infection with *Pasteurella* depends on identification of the bacterium in cultures of specimens collected from patients. As noted, *Pasteurella* is fastidious in its growth requirements, so it must be cultured on blood or chocolate agar (autoclaved blood agar). A wide variety of oral antibacterial drugs are effective against *Pasteurella*, including penicillins and fluoroquinolones.

Haemophilus

Haemophilus[4] (hē-mof'i-lŭs) species are generally small, sometimes pleomorphic bacilli (**FIGURE 20.20**) that require heme and NAD^+ for growth; as a result, they are obligate parasites, colonizing mucous membranes of humans and some animals. *Haemophilus influenzae* (in-flu-en'zī) is the most notable pathogen in the genus, though *H. ducreyi* (doo-krā'ē) is an agent of a sexually transmitted disease. Other species in the genus primarily cause opportunistic infections.

Pioneering microbiologist Margaret Pittman (1901–1995) discovered six strains of *Haemophilus*. She distinguished them by noting differences in the K antigens of their polysaccharide capsules. Pittman's work led directly to the development of the Hib vaccine against *H. influenzae* type b, which is the strain that causes about 95% of *Haemophilus* diseases, including life-threatening childhood meningitis.

Belying its name, *H. influenzae* does not cause the flu; in fact, it is rarely found in the upper respiratory tract. Instead, *H. influenzae* b was the most common cause of meningitis in children 3 to 18 months of age before immunization brought it under control. The pathogen also causes inflammation of subcutaneous tissue, infantile arthritis, and life-threatening epiglottitis in children under four years old. In the latter disease, swelling of the epiglottis and surrounding tissues can completely block the pharynx.

▲ **FIGURE 20.19 Sites of infection by some common members of the Enterobacteriaceae.** For each site, genera are listed in their relative order of prevalence.

Central nervous system
Escherichia

Lower respiratory tract
Klebsiella
Enterobacter
Escherichia
Yersinia

Bloodstream
Escherichia
Klebsiella
Enterobacter

Gastrointestinal tract
Salmonella
Shigella
Escherichia
Yersinia

Urinary tract
Escherichia
Proteus
Klebsiella
Morganella

---

[4]From Greek *haima*, meaning "blood," and *philos*, meaning "love."

Pathogenic, Gram-Negative, Aerobic Bacilli

Thus far, we have examined pathogenic Gram-negative cocci and the more commonly pathogenic, Gram-negative, facultatively anaerobic bacilli. Now we turn our attention to the pathogenic, Gram-negative, aerobic bacilli.

Pathogenic, Gram-Negative, Aerobic Bacilli

Pathogenic, Gram-negative, aerobic bacilli are a diverse group of bacteria in several classes of the phylum Proteobacteria. The following sections examine some of the more common genera in the order they are presented in *Bergey's Manual*: *Bartonella* and *Brucella* in the class Alphaproteobacteria, *Bordetella* and *Burkholderia* in the class Betaproteobacteria, and several members of the class Gammaproteobacteria, including the pseudomonads (*Pseudomonas*, *Moraxella*, and *Acinetobacter*), *Francisella*, *Legionella*, and *Coxiella*.

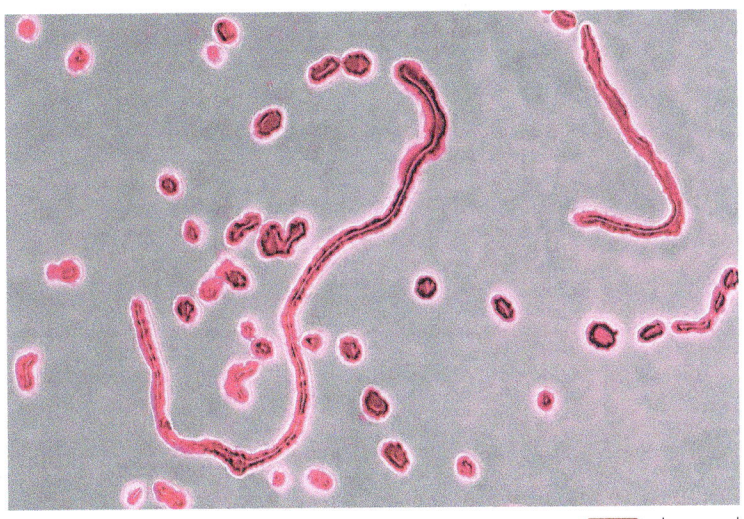

▲ **FIGURE 20.20** *Haemophilus influenzae* is pleomorphic, that is it can assume many different shapes. *What simple lab test can be used to distinguish* Haemophilus *from* Escherichia?

Figure 20.20 An oxidase test can distinguish Haemophilus, which is oxidase positive, from Escherichia, which is oxidase negative.

Pediatricians diagnose *Haemophilus* meningitis by noting lethargy, altered crying, nausea, fever, vomiting, photophobia (intolerance of bright light), stiff neck, and possible seizures. Finding pleomorphic, Gram-negative cells in cerebrospinal fluid obtained by a spinal tap confirms a diagnosis.

Prompt treatment with intravenous cephalosporin is recommended for *Haemophilus* meningitis and epiglottitis. Over the past decade, the use of an effective vaccine—Hib, which is composed of capsular antigens conjugated to protein molecules—has virtually eliminated all disease caused by *H. influenzae* in the United States. Cuban scientists have produced a synthetic polysaccharide vaccine that is cheaper and purer and that may help poorer countries eliminate the disease as well.

The incidence of infections with strains other than type b has remained fairly constant in the United States. These strains of *H. influenzae* cause a variety of diseases, including conjunctivitis, sinusitis, dental abscesses, middle-ear infections, meningitis, bronchitis, and pneumonia. Strain *aegyptius* causes Brazilian purpuric fever, an extremely rapid pediatric disease characterized by conjunctivitis that within a few days is followed by fever, vomiting, abdominal pain, shock, and death. The pathogenesis of this disease is not understood, but fortunately it is rare in the United States.

TELL ME WHY

Why is it that resistance to antimicrobial agents is more commonly seen in hospital-acquired infections with enteric bacteria than in community-based infections with the same species?

Bartonella

LEARNING OUTCOME

20.19 Distinguish among bartonellosis, trench fever, and cat scratch disease.

Members of *Bartonella* (bar-tō-nel'ă), a genus of aerobic bacilli in the class Alphaproteobacteria, are found in animals, but they are known to cause disease only in humans. They have fastidious growth requirements and must be cultured on special media such as blood agar. Many require prolonged incubation in a humid, 37°C atmosphere supplemented with carbon dioxide before they form visible colonies. Three species (formerly classified in the genus *Rochalimaea*) are pathogenic in humans.

Bartonella bacilliformis (ba-sil'li-for'mis) invades and weakens erythrocytes, causing **bartonellosis**—an often fatal disease characterized by fever, severe anemia, headache, muscle and joint pain, and chronic skin infections. Very small bloodsucking sand flies of the genus *Phlebotomus* (fle-bot'ō-mŭs) transmit the bacterium, so the disease is endemic only in Peru, Ecuador, and Colombia, where such flies live.

Bartonella quintana (kwin'ta-nă) causes **trench fever** *(five-day fever)*, which was prevalent among soldiers during World War I. Human body lice transmit the bacterium from person to person. Many infections are asymptomatic, although in other patients severe headaches, fever, and pain in the long bones characterize the disease. The fever can be recurrent, returning every five days, giving the disease its alternative name. *B. quintana* also causes two newly described diseases in immunocompromised patients: *Bacillary angiomatosis* is characterized by fever, inflamed nodular skin lesions, and proliferation of blood vessels. *Bacillary peliosis hepatis* is a disease in which patients develop blood-filled cavities in their livers.

Bartonella henselae (hen'sel-ī) causes **cat scratch disease** when the bacterium is introduced into humans through cat scratches and bites. Fleas may also transmit the bacterium from cats to people. Cat scratch disease has emerged as a relatively common and occasionally serious infection of children, affecting an estimated 32,000 children annually in the United States, particularly in warm and

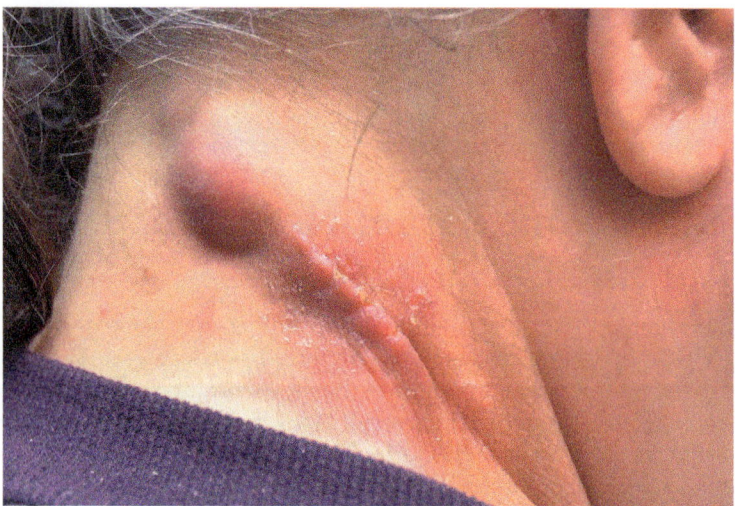

▲ FIGURE 20.21 **Cat scratch disease.** The localized swelling at the site of cat scratches or bites is characteristic.

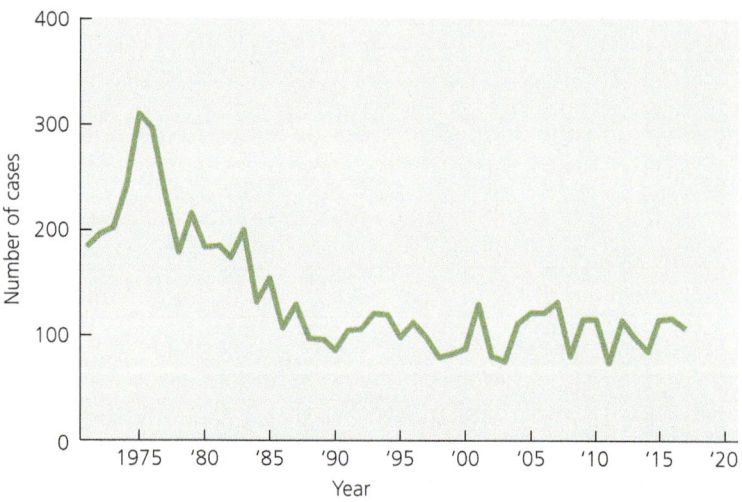

▲ FIGURE 20.22 **The incidence of brucellosis in humans in the United States, 1971–2017.** The decline in cases is largely the result of improvements in livestock management.

humid states. This disease involves prolonged fever and malaise, plus localized swelling at the site of infection (FIGURE 20.21) and of local lymph nodes for several months. Serological testing of individuals who exhibit these characteristic signs and symptoms following exposure to cats confirms a diagnosis of cat scratch disease.

Macrolides, tetracyclines, or aminoglycosides are used to treat most *Bartonella* infections, though there may be a high rate of relapse.

Brucella

LEARNING OUTCOME

20.20 Describe the cause, signs, symptoms, and treatment of brucellosis.

Brucella (broo-sel′lă) is a genus of small, nonmotile, aerobic coccobacilli that lack capsules but survive phagocytosis by preventing lysosomes from fusing with phagosomes containing the bacterium. Analysis of rRNA nucleotide base sequences reveals that *Brucella* is closely related to *Bartonella* in the class Alphaproteobacteria. Although *Brucella*-caused diseases in humans have been ascribed to four species, rRNA analysis has revealed the existence of only a single true species: *Brucella melitensis* (me-li-ten′sis); the other variants are strains of this one species. Because the classical specific epithets are so well known, they are mentioned here.

In animal hosts, the bacterium lives as an intracellular parasite in organs such as the uterus, placenta, and epididymis, but these organs are not infected in humans. Typically, infections in animals are either asymptomatic or cause a mild disease—**brucellosis** (broo-sel-ō′sis)—though they can cause sterility or abortion. Historically named *Brucella melitensis* infects goats and sheep; *B. abortus* (a-bort′us), cattle; *B. suis* (soo′is), swine; and *B. canis* (kā′nis), dogs, foxes, and coyotes.

Humans become infected either by consuming unpasteurized contaminated dairy products or through contact with animal blood, urine, or placentas in workplaces such as slaughterhouses, veterinary clinics, and feedlots. The bacterium enters the body through breaks in mucous membranes of the digestive and respiratory tracts.

Brucellosis in humans is characterized by a fluctuating fever—which gives the disease one of its common names: *undulant fever*—and chills, sweating, headache, myalgia, and weight loss. The disease in humans has been given a variety of other names, including *Bang's disease*, after microbiologist Bernhard Bang (1848–1932), who investigated the disease, and *Malta fever, rock fever of Gibraltar*, and *fever of Crete*, after localized epidemics in those locales.

Physicians treat brucellosis with a variety of antimicrobial drugs, including doxycycline in combination with gentamicin, rifampin, or streptomycin. An attenuated vaccine for animals exists but is not used in humans because the vaccine can cause disease. Through the immunization of uninfected domesticated animals and the slaughter of infected ones, the threat of brucellosis has been reduced for U.S. residents (FIGURE 20.22).

Bordetella

LEARNING OUTCOMES

20.21 Describe five virulence factors of *Bordetella pertussis*.
20.22 Identify the four phases of pertussis.

Bordetella pertussis (bōr-dĕ-tel′ă per-tus′is) is a small, aerobic, nonmotile, Gram-negative coccobacillus in the class Betaproteobacteria that is responsible for the disease **pertussis**,[5] commonly called *whooping cough*. *B. parapertussis* causes a milder form of pertussis.

Pathogenesis, Epidemiology, and Disease

Bordetella pertussis causes disease by interfering with the action of ciliated epithelial cells of the trachea. Various adhesins and toxins mediate the disease.

The bacterium attaches to certain lipids in the cytoplasmic membranes of tracheal cells via two adhesins: *filamentous hemagglutinin* and *pertussis toxin*. Filamentous hemagglutinin

---
[5]From Latin *per*, meaning "intensive," and *tussis*, meaning "cough."

also binds to certain glycoproteins on the cytoplasmic membranes of neutrophils, initiating phagocytosis of the bacteria. *B. pertussis* survives within phagocytes, in the process evading the immune system. Pertussis toxin causes infected cells to produce more receptors for filamentous hemagglutinin, leading to further bacterial attachment and phagocytosis.

Four *B. pertussis* toxins are the following:

- *Pertussis toxin*, a portion of which interferes with the ciliated epithelial cell's metabolism, resulting in increased mucus production. (Note that pertussis toxin is both an adhesin and a toxin.)
- *Adenylate cyclase toxin*, which triggers increased mucus production and inhibits leukocyte movement, endocytosis, and killing. This function may provide protection for the bacterium early in an infection.
- *Dermonecrotic toxin*, which causes localized constriction and hemorrhage of blood vessels, resulting in cell death and tissue destruction.
- *Tracheal cytotoxin*, which at low concentrations inhibits the movement of cilia on ciliated respiratory cells, and at high concentrations causes the expulsion of the cells from the lining of the trachea.

Pertussis is considered a pediatric disease, as most cases are reported in children younger than five years old. More than 60 million children worldwide suffer from pertussis each year. There were over 15,000 reported cases in the United States in 2017 (FIGURE 20.23). However, these figures may considerably underestimate the actual number of cases because patients with chronic coughs are not routinely tested for infection with *Bordetella* and because the disease in older children and adults is typically less severe and is frequently misdiagnosed as a cold or influenza.

Pertussis begins when bacteria, inhaled in aerosols, attach to and multiply in ciliated epithelial cells (FIGURE 20.24). As shown in FIGURE 20.25, pertussis then progresses through four stages:

1. During 7 to 10 days of **incubation** the bacteria multiply, but no symptoms are apparent.

2. The **catarrhal**[6] phase (kă-tă'răl) is characterized by signs and symptoms that resemble a common cold. During this phase, which lasts one to two weeks, the bacteria are most abundant and the patient is most infectious.

3. The **paroxysmal**[7] phase (par-ok-siz'mal) begins as the ciliary action of the tracheal cells is impaired, even as copious mucus is secreted. The condition worsens as ciliated cells are expelled. To clear the accumulating mucus, the body initiates a series of deep coughs, each of which is followed by a characteristic "whoop" caused by the intake of air through the congested trachea. Each day, a patient may experience 40 to 50 coughing spells that often end with vomiting and exhaustion. During this phase, which can last two to four weeks, coughing may be so severe that oxygen exchange is limited such that the patient may turn blue or even die. Severe coughing breaks ribs of some patients.

[6]From Greek *katarrheo*, meaning "to flow down."
[7]From Greek *paroxysmos*, meaning "to irritate."

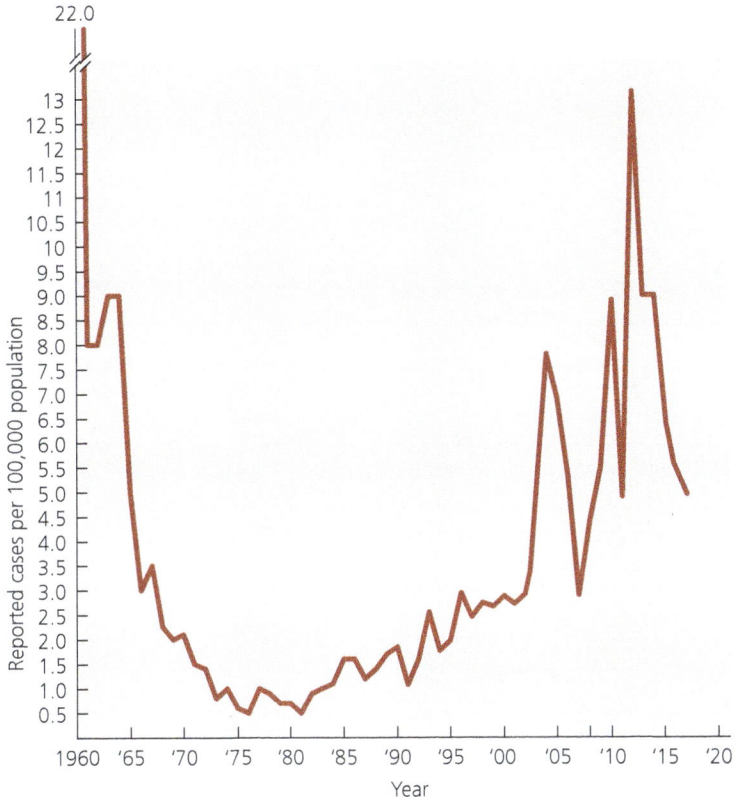

▲ FIGURE 20.23 Reported cases of pertussis in the United States, 1960–2017.

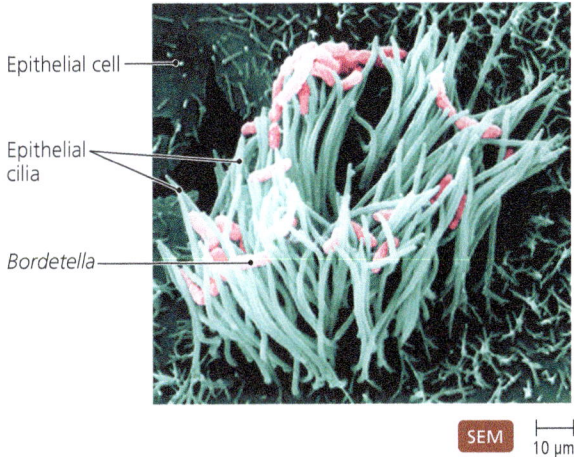

▲ FIGURE 20.24 *Bordetella*. The bacteria attach to and infect ciliated epithelial cells, such as these of the trachea, eventually causing the death of the cells.

4. During the **convalescent phase**, the number of bacteria present becomes quite small, the ciliated lining of the trachea grows back, and the severity and frequency of coughing diminish. During this phase, which typically lasts three to four weeks or longer, secondary bacterial infections (such as with *Staphylococcus* or *Streptococcus*) in the damaged epithelium may lead to bacteremia, pneumonia, seizures, and encephalopathy.

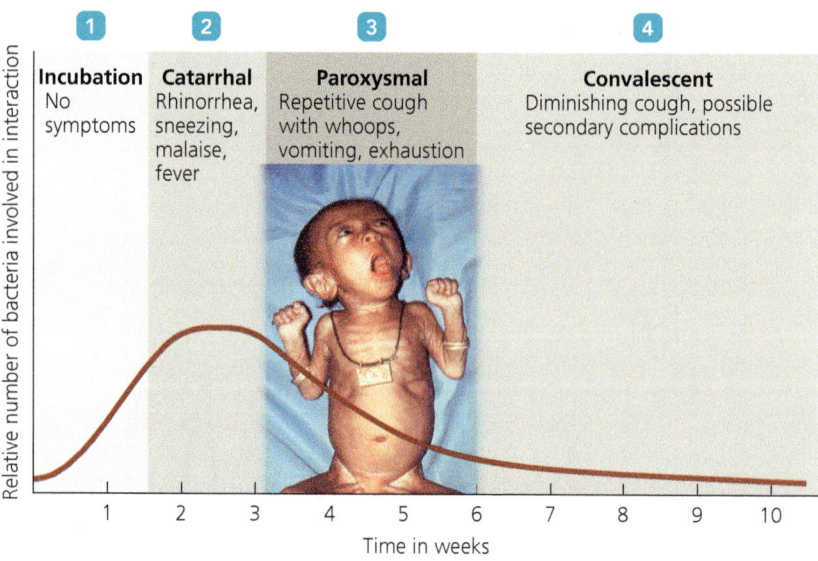

▲ FIGURE 20.25 The approximate time course for the progression of pertussis.

Diagnosis, Treatment, and Prevention

The symptoms of pertussis are usually diagnostic, particularly when a patient is known to have been exposed to *Bordetella*. Even though health care workers can isolate *B. pertussis* from respiratory specimens, the bacterium is extremely sensitive to desiccation, so specimens must be inoculated at the patient's bedside onto *Bordet–Gengou* medium, which is specially designed to support the growth of this bacterium. If this is not practical, clinicians must use special transport media to get a specimen to a laboratory.

Treatment for pertussis is primarily supportive. By the time the disease is recognized by the distinctive cough, the immune system has often already "won the battle." Recovery mostly depends on regeneration of the tracheal epithelium rather than reduction of the number of bacteria; therefore, although antibacterial drugs reduce the number of bacteria and the patient's infectivity to others, they have little effect on the course of the disease. The American Academy of Pediatrics recommends erythromycin for everyone in close contact with a whooping cough patient.

Given that *B. pertussis* has no animal or environmental reservoir and that effective vaccines (the P of DTaP and of Tdap) are available, whooping cough could be eradicated. Despite this possibility, over 10,000 cases occur each year in the United States. This is partly due to the refusal of parents to immunize their children and partly due to the fact that immunity, whether acquired artificially or naturally, lasts only about 10 years. The CDC urges parents to immunize their children and now recommends that all adults receive one dose of acellular pertussis vaccine. This is especially recommended for adults who are around infants.

Burkholderia

LEARNING OUTCOME

20.23 Describe the advantageous metabolic features of *Burkholderia* as well as its pathogenicity.

Burkholderia cepacia (burk-hol-der′ē-ă se-pā′se-ă) is a soil-dwelling, aerobic, flagellated betaproteobacterium. *Burkholderia* is noteworthy for its ability to decompose a broad range of organic molecules, making it a likely bacterium to assist in the cleanup of contaminated environmental sites. For example, *Burkholderia* is capable of digesting polychlorinated biphenyls (PCBs) and Agent Orange—an herbicide used extensively during the Vietnam War—both of which can persist for long periods in the environment. Further, farmers can use *Burkholderia* to reduce the number of fungal infections of many plant crops, including peas, beans, alfalfa, canola, and cucumbers.

Unfortunately, *Burkholderia* can also grow in health care settings, metabolizing a variety of organic chemicals and resisting many antimicrobial drugs. The bacterium is one of the more important opportunistic pathogens of the lungs of cystic fibrosis patients, in whom it metabolizes the copious mucous secretions in the lungs. Evidence for patient-to-patient spread is clear, making it imperative that infected cystic fibrosis patients avoid contact with uninfected patients. *Burkholderia* is often resistant to many antimicrobial drugs; therefore, physicians must decide which drug to use on a case-by-case basis.

Another soil-dwelling species, *Burkholderia pseudomallei* (soo-dō-mal′e-ē), causes *melioidosis* (mel′ē-oy-dō′sis), which is an Asian and Australian tropical disease that is emerging as a threat in other locales (see **Emerging Disease Case Study: Melioidosis** on p. 610). The CDC considers *B. pseudomallei* a potential agent of biological terrorism.

MICRO CHECK

16. Which species of *Bartonella* causes cat scratch disease?
17. Which toxin of *Bordetella pertussis* inhibits cells that line the trachea?
18. What is the characteristic sign of pertussis?
19. What is the best way to prevent pertussis?

Pseudomonads

LEARNING OUTCOMES

20.24 Reconcile the apparent discrepancy between the ubiquitous distribution of pseudomonads with numerous virulence factors and the fact they cause so few diseases.
20.25 Identify three genera of opportunistic pathogenic pseudomonads.
20.26 Describe *Pseudomonas aeruginosa* as an opportunistic pathogen of burn victims and cystic fibrosis patients.

Pseudomonads (soo-dō-mō′nadz) are Gram-negative, aerobic bacilli in the class Gammaproteobacteria. Unlike the fastidious members of *Bartonella*, *Brucella*, and *Bordetella*, pseudomonads are not particular about their growth requirements. They are ubiquitous in soil, decaying organic matter, and almost every moist environment, including swimming pools, hot tubs, washcloths, and contact lens solutions. In hospitals they are frequently found growing in moist food, vases of cut flowers, sinks, sponges, toilets, floor mops, dialysis machines, respirators, humidifiers, and in disinfectant solutions. Some species can even grow using trace

CLINICAL CASE STUDY

When "Health Food" Isn't

In a single day, two 19-year-old women and one 20-year-old man sought treatment at a university health clinic, complaining of acute diarrhea, nausea, and vomiting. No blood was found in their stools. One of the women was found to have a urinary tract infection. All three had eaten lunch at a nearby health food store the previous day. The man had a sandwich with tomato, avocado, sprouts, pickles, and sunflower seeds. One woman had a pocket sandwich with turkey, sprouts, and mandarin oranges; the other woman had the lunch special, described in the menu as a "delightful garden salad of fresh organic lettuces, sprouts, tomatoes, and cucumbers with zesty raspberry vinaigrette dressing." All had bottled water to drink.

1. Which of the foods is the most likely source of the infections?
2. What media would you use to culture and isolate enteric contaminants in the food?
3. Which enteric bacteria could cause these symptoms?
4. How did the woman likely acquire the urinary tract infection?
5. What is the likely treatment?
6. What steps can the food store's manager and the students take to reduce the chance of subsequent infections?

---

nutrients in distilled water. Pseudomonads utilize a wide range of organic carbon and nitrogen sources. The Entner–Doudoroff pathway is their major means of glucose catabolism rather than the more common Embden–Meyerhof–Parnas pathway of glycolysis.

Here, we examine three genera that are commonly isolated opportunistic pathogens—*Pseudomonas, Moraxella,* and *Acinetobacter*—beginning with the species of greatest medical importance: *Pseudomonas aeruginosa*.

Pseudomonas aeruginosa

Pathogenesis, Epidemiology, and Disease *Pseudomonas aeruginosa* (soo-dō-mō′nas ā-roo-ji-nō′sā) is somewhat of a medical puzzle in that even though it expresses a wide range of virulence factors, it rarely causes disease. This is fortunate because its ubiquity and inherent resistance to a wide range of antimicrobial agents would render a more virulent *Pseudomonas* a nearly insurmountable challenge to health care professionals. That *P. aeruginosa* is *only* an opportunistic pathogen is testimony to the vital importance of the body's protective tissues, cells, and chemical defenses.

P. aeruginosa has fimbriae and other adhesins that enable its attachment to host cells. The fimbriae are similar in structure to those of *N. gonorrhoeae*. The bacterial enzyme *neuraminidase* modifies the fimbriae receptors on a host cell in such a way that attachment of fimbriae is enhanced. A mucoid polysaccharide capsule also plays a role in attachment, particularly in the respiratory system of cystic fibrosis patients (as discussed shortly). The capsule also shields the bacterium from phagocytosis.

P. aeruginosa synthesizes other virulence factors, including toxins and enzymes. Lipid A (endotoxin) is prevalent in the cell wall of *Pseudomonas* and triggers fever, vasodilation, inflammation, shock, and other symptoms. Two toxins—*exotoxin A* and *exoenzyme S*—inhibit protein synthesis in eukaryotic cells, contributing to cell death and interfering with phagocytic killing.

The enzyme *elastase* breaks down elastic fibers, degrades complement components, and cleaves IgA and IgG. The bacterium also produces a blue-green pigment, called *pyocyanin*, that triggers the formation of superoxide radical (O_2^-) and peroxide anion (O_2^{2-}), two reactive forms of oxygen that contribute to tissue damage in *Pseudomonas* infections.

Although a natural inhabitant of bodies of water and moist soil, *P. aeruginosa* is rarely part of the normal human microbiota. Nevertheless, because of its ubiquity and its virulence factors, this opportunistic pathogen colonizes immunocompromised patients and is involved in about 10% of healthcare-associated infections. Once it breaches the skin or mucous membranes, *P. aeruginosa* can successfully colonize almost any organ or system. It can be involved in urinary, ear, eye, central nervous system, gastrointestinal, muscular, skeletal, and cardiovascular infections. Infections in burn victims and cystic fibrosis patients are so common they deserve special mention.

Pseudomonas infections of severe burns are pervasive (**FIGURE 20.26**). The surface of a burned area provides a warm, moist environment that is quickly colonized by this ubiquitous opportunist; almost two-thirds of burn victims develop environmental or healthcare-associated *Pseudomonas* infections.

P. aeruginosa also typically infects the lungs of cystic fibrosis patients, forming a biofilm that protects the bacteria from phagocytes. Such infections exacerbate the decline in pulmonary function in these patients by causing certain lung cells to synthesize large amounts of mucus. As the bacteria feed on the mucus, they signal host cells to secrete more of it, creating a positive feedback loop; the result of such *P. aeruginosa* infections is that cystic fibrosis patients are more likely to require hospitalization and more likely to die.

Diagnosis and Treatment Diagnosis of *Pseudomonas* infection is not always easy because its presence in a culture may represent contamination acquired during collection, transport, or inoculation. Certainly, pyocyanin discoloration of tissues is indicative of massive infection.

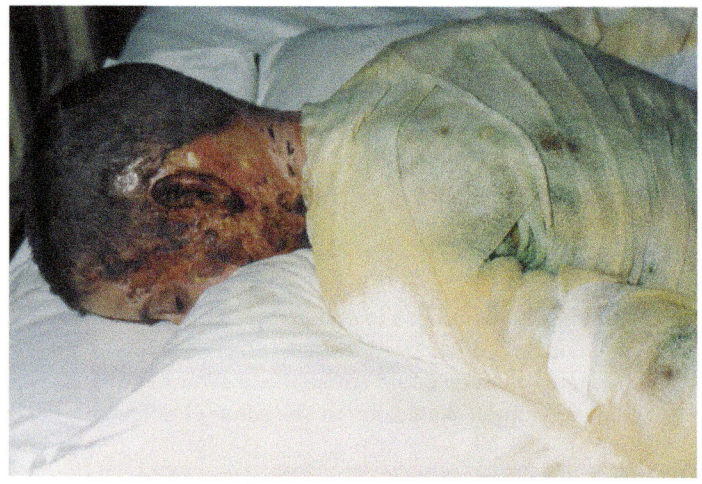

▲ FIGURE 20.26 **A *Pseudomonas aeruginosa* infection.** Bacteria growing under the bandages of this burn victim produce the green color. *What chemical is responsible for the blue-green appearance of this infection?*

Figure 20.26 Pyocyanin, a blue-green pigment secreted by *P. aeruginosa*, produces the color.

Treatment of *P. aeruginosa* can also be frustrating. The bacterium is notoriously resistant to a wide range of antibacterial agents, including many antimicrobial drugs, soaps, antibacterial dyes, and quaternary ammonium disinfectants. In fact, *Pseudomonas* has been reported to live in solutions of some antibacterial drugs and disinfectants.

Resistance of the bacterium is due to the ability of *Pseudomonas* to metabolize many drugs, to the presence of nonspecific proton/drug antiports that pump some types of drugs out of the bacterium, and to the ability of *Pseudomonas* to form biofilms, which resist the penetration of antibacterial drugs and detergents. Physicians treat infections with combinations of aminoglycoside, beta-lactam, and fluoroquinolone antimicrobials that have first proven efficacious against a particular isolate in a susceptibility test.

In summary, though *Pseudomonas aeruginosa* is ubiquitous and possesses a number of virulence factors, the bacterium rarely causes disease in healthy individuals because it cannot normally penetrate the skin and mucous membranes or ultimately evade the body's other defenses. Only in debilitated patients does this opportunist thrive.

Moraxella and *Acinetobacter*

Moraxella (mōr'ak-sel'ă) and *Acinetobacter* (as-i-nē'tō-bak'ter) are aerobic, short, plump bacilli that formerly were classified in the same family as *Neisseria*. Analysis of rRNA has shown that they are more properly classified as pseudomonads.

Moraxella catarrhalis (kă-tah'răl-is, formerly *Branhamella* [bran-hă-mel'ă] *catarrhalis*) is rarely pathogenic but can cause opportunistic infections of the sinuses, bronchi, ears, and lungs. The bacterium is susceptible to fluoroquinolones, erythromycin, tetracycline, and most other antibacterial drugs (with the exception of beta-lactams).

EMERGING DISEASE CASE STUDY

Melioidosis

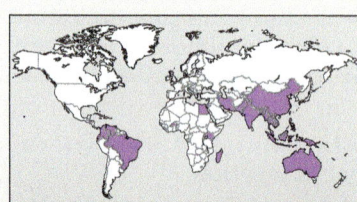

Isabella felt lucky. How many community college students had the opportunity to help a professor do research in northern Australia's Kakadu National Park for a month, get college credit for the experience, and not have to pay for the trip? Though she was the oldest person on the trip and often felt her 45 years as they hiked the trails, she didn't complain. Of course things would have been better if it didn't seem to rain all the time in Australia and if the thorns didn't tear at her legs quite so regularly; still, the trip was an adventure.

Three weeks after her return to Houston, Texas, Isabella developed a high fever (39°C/102°F) with general weakness and shortness of breath. All other signs, including the results of blood work, appeared normal. The doctors, suspecting flu, told her to get plenty of rest and fluids and discharged her home. Two days later, Isabella was intermittently drowsy and confused, her breathing was very labored, and a small cut on her leg was inflamed and had a pus-filled lesion. She was admitted to the hospital with pneumonia. Two days later, Isabella went into septic shock and died.

She died of melioidosis, an emerging disease caused by a Gram-negative bacterium, *Burkholderia pseudomallei*, which produces a toxin that inhibits protein synthesis by infected human cells. Melioidosis is endemic to the tropics of Southeast Asia, northern Australia, India, and China. Isabella had been infected either via inhalation or through the cut on her leg. Even with treatment, nearly 90% of melioidosis patients die when the bacterium spreads throughout their blood, including Isabella. (For more information about melioidosis, see p. 608.)

1. If a sample of Isabella's cerebral spinal fluid (CSF) were Gram-stained, what color would the bacterial cells appear?
2. Why did the doctor suspect flu rather than melioidosis?
3. Why is it unlikely that there will be an epidemic of melioidosis in Sweden or Norway?

Acinetobacter grows in soil, water, and sewage and is only rarely associated with disease in humans, though it is often isolated from clinical specimens. It is an opportunistic pathogen that causes infections of the respiratory, urinary, and central nervous systems. Endocarditis and septicemia have also been reported in infections with *Acinetobacter*. It is often resistant to most antimicrobial drugs, so susceptibility tests must guide the choice of an effective treatment.

Francisella

LEARNING OUTCOMES

20.27 Describe the modes of transmission of *Francisella tularensis*.

20.28 List practical measures that can be taken to prevent infection by *F. tularensis*.

Pathogenesis, Epidemiology, and Disease *Francisella tularensis*[8] (fran'si-sel'lă too-lă-ren'sis) is a very small (0.2 μm × 0.2–0.7 μm), nonmotile, strictly aerobic, Gram-negative coccobacillus of the class Gammaproteobacteria that causes a zoonotic disease called **tularemia** (too-lă-rē'mē-ă). *F. tularensis* has a capsule that discourages (by unknown mechanisms) phagocytosis and intracellular digestion.

F. tularensis is found in temperate regions of the Northern Hemisphere living in water and as an intracellular parasite of animals and amoebas. Scientists do not know why it is not found below the equator. *Francisella* has an amazingly diverse assortment of hosts. It lives in mammals, birds, fish, bloodsucking ticks and insects, and amoebas. Indeed, one is hard pressed to find an animal that cannot be its host. The most common reservoirs in the United States are rabbits, muskrats, and ticks, which give the disease two of its common names—*rabbit fever* and *tick fever*.

Francisella is also incredibly varied in the ways it can be spread among hosts. Tularemia in humans is most often acquired either through the bite of an infected tick or via contact with an infected animal. The bacterium, by virtue of its small size, can pass through apparently unbroken skin or mucous membranes. Bloodsucking flies, mosquitoes, mites, and ticks also transmit *Francisella*, and humans can be infected by consuming infected meat, drinking contaminated water, or inhaling bacteria in aerosols produced during slaughter or in a laboratory. Fortunately, human-to-human spread does not occur.

Francisella is one of the more infectious of all bacteria: Infection requires as few as 10 organisms when transmitted by a biting arthropod or through unbroken skin or mucous membranes. For example, only about 10 cells must be inhaled to cause disease, although consumption of 10^8 cells in food or drink is necessary to contract the disease via the digestive tract.

Infections from bites, scratches, or through breaks in the skin cause lesions at the site of infection as well as swollen regional lymph nodes (buboes). Inhalation may produce buboes in the chest that put pressure on lungs and induce dry cough, pain during breathing, and death.

Only 213 cases of tularemia were reported in the United States in 2017, but the actual number of infections was probably much higher considering its virulence, prevalence in animals, and multiple modes of transmission. Tularemia frequently remains unsuspected because its symptoms—fever, chills, headache, sore throat, muscle aches, and nausea—are not notably different from those of many other bacterial and viral diseases and because it is difficult to confirm tularemia by using laboratory tests. Though other forms of tularemia are typically innocuous, respiratory tularemia is fatal to more than 30% of untreated patients. Tularemia was removed from the list of nationally notifiable diseases in 1994, but concern about the possible use of *Francisella* by bioterrorists led officials in 2000 to relist it.

Diagnosis, Prevention, and Treatment Physicians diagnose tularemia with an enzyme-linked immunosorbent assay (ELISA) to demonstrate antibodies against *Francisella* in the patient's blood. To prevent infection, people should avoid the major reservoirs of *Francisella* (rabbits, muskrats, and ticks), especially in endemic areas. They can protect themselves from tick bites by wearing long clothing with tight-fitting sleeves, cuffs, and collars and by using repellent chemicals. Because *Francisella* is not present in tick saliva but only in its feces, prompt removal of ticks can mitigate infection. Hikers and hunters should never handle ill-appearing wild animals or their carcasses and should wear gloves and masks when field dressing game.

The bacterium produces beta-lactamase, so penicillins and cephalosporins are ineffective, but other antimicrobial drugs have been used successfully. Currently, intramuscular gentamicin is recommended for use against *Francisella*.

MICRO CHECK

20. What is the role of the capsule of *Pseudomonas aeruginosa* in patients with cystic fibrosis?
21. Considering the fact that *Pseudomonas aeruginosa* has so many virulence factors, why is it not a more common source of infection in the general public, rather than only an opportunistic pathogen?
22. Bacteria from which genus cause tularemia?
23. What is the best way to prevent tularemia?

Legionella

LEARNING OUTCOMES

20.29 Explain why *Legionella pneumophila* was unknown before 1976.

20.30 Describe the symptoms and treatment of Legionnaires' disease.

---

[8] Named for Tulare County, California.

CLINICAL CASE STUDY

Nightmare on The Island

Peggy loves her time on The Island each year. Her parents had taken her every year to the resort destination as a girl, and now she was doing the same with her children. Their seaside home near Cape Cod, Massachusetts, was modest in comparison to some of the neighbors', but it had been in Peggy's family for over a hundred years, and she had fond remembrances from every stage of her life. One of Peggy's fondest memories is playing tag with her friends on the lawns of The Island's homes. Now, she smiles as she watches her eight-year-old son, Jacob, help the older son of one of her childhood girlfriends mow the grassy expanse. "Building memories—that's what it's about," she thought. Little does she know that some memories can build nightmares.

Three days later, Jacob wakes up complaining of a scratchy throat, headache, and "soreness all over." Peggy is concerned about his minimally productive cough and 103°F fever. "A summer cold?" She keeps Jacob in bed, which isn't difficult because his breathing becomes more labored and painful. Two days later, he begins coughing up blood and Peggy recognizes that this isn't an ordinary summertime cold.

Peggy becomes very concerned and rushes Jacob to the emergency room. The doctor immediately orders blood cultures, arterial blood gas, and obtains a chest X ray. Jacob's respiratory status worsens and requires mechanical ventilation for airway support. The doctors begin intravenous broad-spectrum antibiotics and admit him to the pediatric ICU for close monitoring. The chest X ray shows nodular infiltrates with unilateral pleural effusion. The physician tells Peggy that Jacob has a very serious infection in his blood and also has severe pneumonia. He questions her about Jacob's activities on the island: Has the boy touched any animals? Done any outdoor activities? Been bitten by a tick? "No, no, no." Then she recalls that Jacob helped mow the grass earlier in the week. The blood cultures come back confirming a Gram-negative coccobacillus; the doctor orders intramuscular gentamicin every 12 hours for a week.

Within days, Jacob feels better and can answer questions. He tells the doctor that the lawnmower had run over the dried body of a small dead rabbit. The physician suspects the mower had spewed bacteria into the air; Jacob had inhaled a near-fatal dose.

The grassy lawn will no longer bring to mind the fond memories of Peggy's childhood; instead, she will remember men in biohazard suits taking samples, documenting the nightmarish time she almost lost her son.

1. What bacterium infected Jacob?
2. What is the common name of the disease afflicting Jacob?
3. What do the laboratory scientists at the hospital determine about the Gram reaction of the bacterium?
4. Why didn't the physician use penicillin instead of gentamicin?

In 1976, celebration of the 200th anniversary of the Declaration of Independence was curtailed in Philadelphia when hundreds of American Legion members attending a convention were stricken with severe pneumonia; 29 died. After extensive epidemiological research, this disease—dubbed **Legionnaires' disease** or *legionellosis*—was found to be caused by a previously unknown pathogen, which was subsequently named *Legionella* (lē-jŭ-nel′lă).

Pathogenesis, Epidemiology, and Disease

To date, scientists have identified over 40 species of *Legionella*. These aerobic, slender, pleomorphic bacteria in the class Gammaproteobacteria are extremely fastidious in their nutrient requirements, and laboratory media must be enhanced with iron salts and the amino acid cysteine. FIGURE 20.27 shows colonies growing on one commonly used medium—buffered charcoal yeast extract agar. *Legionella* species are almost universal inhabitants of water, but they had not been isolated previously because they stain poorly and cannot grow on common laboratory media. Nineteen species are known to cause disease in humans, but 85% of all infections in humans are caused by *L. pneumophila*[9] (noo-mō′fi-lă).

Legionella pneumophila presented a conundrum for early investigators: How can such a fastidious bacterium be nearly ubiquitous in moist environmental samples? In the original epidemic, for example, *Legionella* was cultured from condensation in hotel air conditioning ducts, an environment that seems unsuitable for a microorganism with such demanding nutritional requirements. Investigations revealed that *Legionella* living in the environment invade freshwater protozoa, typically amoebas, and reproduce inside phagocytic vesicles. Thus, the bacteria survive in the environment much as they survive in humans—as intracellular parasites.

Protozoa release bacteria-filled vesicles into the environment; alternatively, *Legionella* forms exit pores through a host cell's vesicular membrane and then through its cytoplasmic membrane. Humans acquire the disease by inhaling *Legionella* in aerosols produced by showers, vaporizers, spa whirlpools, hot tubs, air

---

[9]From Greek *pneuma*, meaning "breath," and *philos*, meaning "love."

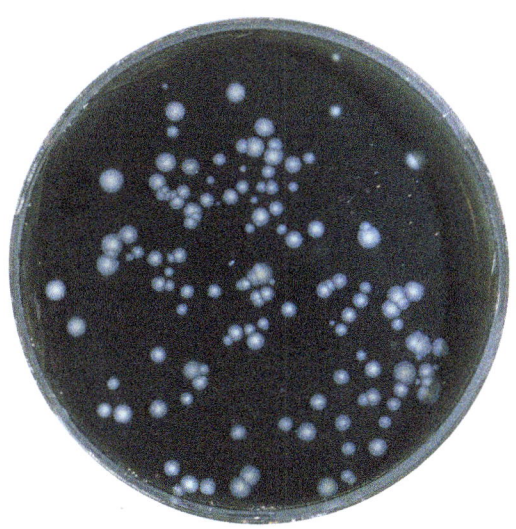

▲ FIGURE 20.27 *Legionella pneumophila* growing on buffered charcoal yeast extract agar. This bacterium cannot be grown in culture without such special media.

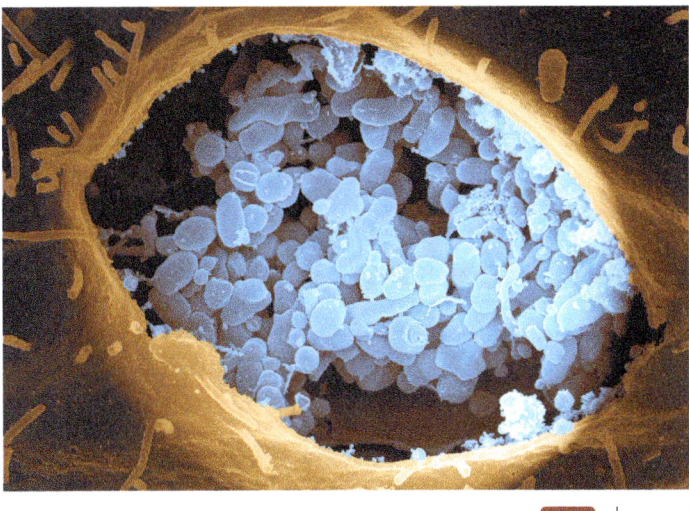

▲ FIGURE 20.28 *Coxiella burnetii* inside a host cell broken open to show the bacteria This intracellular pathogen grows and reproduces within phagolysosomes. An infective body of *Coxiella* can persist in the environment for years.

conditioning systems, cooling towers, and grocery store misters. *Legionella* was not a notable pathogen until such devices provided a suitable means of transmitting the bacterium to humans.

Legionnaires' disease is characterized by fever, chills, a dry nonproductive cough, headache, and pneumonia. Complications involving the gastrointestinal tract, central nervous system, liver, and kidneys are common. If not promptly treated, pulmonary function rapidly decreases, resulting in death in 20% of patients with normal immunity. Mortality is much higher in immunocompromised individuals, particularly kidney and heart transplant recipients.

L. pneumophila also causes a flulike illness called *Pontiac fever* (after the Michigan city where it was first described). This disease has symptoms similar to those of Legionnaires' disease, but it does not involve pneumonia and is not fatal.

Diagnosis, Treatment, and Prevention

Diagnosis is made by fluorescent antibody staining or other serological tests that reveal the presence of *Legionella* in clinical samples. The bacterium can also be cultured on suitable commercially available selective media such as buffered charcoal yeast extract.

Physicians use intravenous fluoroquinolone in adults and azithromycin in children to treat Legionnaires' disease. Pontiac fever is self-limiting and requires no treatment.

Completely eliminating *Legionella* from water supplies is not feasible, as chlorination and heating are only moderately successful. However, the bacterium is not highly virulent, so reducing its numbers is typically a successful control measure.

Coxiella

LEARNING OUTCOMES

20.31 Explain how *Coxiella burnetii* survives desiccation.
20.32 Describe the mode of transmission of Q fever.

Coxiella burnetii (kok-sē-el'ă ber-ne'tē-ē) is an extremely small, aerobic, obligate intracellular parasite (FIGURE 20.28) that grows and reproduces in the acidic environment within phagolysosomes. Its small size and dependence on the cytoplasm of eukaryotes for growth led early investigators to think it was a virus. However, its bacterial nature is unquestionable: It has a typical Gram-negative cell wall (albeit with minimal peptidoglycan), RNA and DNA (viruses typically have one or the other), functional ribosomes, and citric acid cycle enzymes. *Coxiella* was originally classified with other obligate intracellular bacteria called *rickettsias* (discussed in Chapter 21), but rRNA analysis reveals that *Coxiella* is more properly classified with *Legionella* in the class Gammaproteobacteria.

Coxiella forms an internal, stable, resistant *infective body* (sometimes called a spore) that is similar in structure and function to the endospores of some Gram-positive species. The infective body enables the bacterium to survive harsh environmental conditions (such as desiccation and heat) for years.

C. burnetii infects a wide range of mammalian and avian hosts and is transmitted among them by feeding ticks. Farm animals and pets are the reservoirs most often associated with disease in humans. Transmission to humans occurs infrequently via feeding ticks; instead, humans are most frequently infected by inhaling infective bodies that become airborne when released from dried tick feces or from the dried urine, feces, or placentas of host animals. Humans can also become infected by consuming contaminated unpasteurized milk and rarely from having sex with an infected person.

Most *Coxiella* infections are asymptomatic, though this bacterium can cause **Q fever**, so named because its cause was *questionable* (unknown) for many years. Q fever occurs worldwide, particularly among ranchers, veterinarians, and food handlers, and it may be either an acute or a chronic condition. Acute Q fever follows an incubation period of 20 days or more; besides a high fever, it involves severe headache, chills, muscle pain, and mild pneumonia. In chronic Q fever, months to years may pass from the time of infection until life-threatening endocarditis develops. Inflammation of the lungs and liver may occur simultaneously.

Q fever is diagnosed via serological testing. Physicians use long-term antimicrobial therapy to treat chronic Q fever. Researchers have most studied the use of doxycycline with another antimicrobial, such as fluoroquinolone. The biggest problem in treating Q fever is getting antimicrobial drugs to work in the acidic environment of phagolysosomes, where the bacteria live. Researchers have developed an effective vaccine for Q fever, but it is not available in the United States. Prevention involves avoiding the inhalation of dust contaminated with barnyard and pet wastes.

> **MICRO CHECK**
>
> 24. How is legionellosis transmitted?

> **TELL ME WHY**
>
> Why do most U.S. cases of tularemia occur in the late spring and summer months, while few cases occur in January?

Pathogenic, Gram-Negative, Anaerobic Bacilli

LEARNING OUTCOME

20.33 Compare and contrast *Bacteroides* and *Prevotella* with one another and with other Gram-negative opportunists.

Over 50 species of strictly anaerobic Gram-negative bacteria are known to colonize the human body. Indeed, such anaerobic bacteria are the predominant microbiota of most people's gastrointestinal, urinary, reproductive, and lower respiratory tracts. They typically outnumber aerobic and facultatively anaerobic bacteria by a factor of 100 or more. The abundance of anaerobic bacteria in these locations is important for human health: they inhibit the growth of most pathogens, and in the intestinal tract, they synthesize vitamins and vitamin precursors and assist in the digestion of food.

Relatively few of these anaerobic normal microbiota cause disease and then only when they are introduced into other parts of the body by trauma or surgery. Almost all of the opportunistic species are aerotolerant anaerobes; they produce catalase or superoxide dismutase, which are enzymes that inactivate hydrogen peroxide and superoxide radicals. The more important anaerobic opportunists are *Bacteroides* and *Prevotella*.

Bacteroides

Bacteria in the genus *Bacteroides* (bak-ter-oy′dēz) are anaerobes that live as part of the normal microbiota of the intestinal tract and the upper respiratory tract. Being Gram-negative, they have an outer wall membrane containing lipopolysaccharide. However, in contrast to the lipid A of most other Gram-negative bacteria discussed in this chapter, the lipid A formed by members of this genus has little endotoxin activity.

The most important pathogen in this genus is *Bacteroides fragilis* (fra′ji-lis). It is associated with about 85% of gastrointestinal diseases, even though it accounts for only about 1% of the bacteria in many people's colons. *B. fragilis* is a pleomorphic bacillus that produces a number of virulence factors. It attaches to host cells via fimbriae and a polysaccharide capsule, the latter of which also inhibits phagocytosis. Should the bacteria become phagocytized, short-chain fatty acids produced during anaerobic metabolism inhibit the activity of lysosomes, enabling the bacteria to survive within phagocytes.

B. fragilis is frequently involved in a variety of conditions, including abdominal infections (e.g., those following a ruptured appendix), genital infections in females (e.g., pelvic abscesses), and wound infections of the skin (which can result in life-threatening necrosis of muscle tissue). It is also involved in about 5% of all cases of bacteremia.

If anaerobic bacteria are to be isolated from patients, clinicians must maintain anaerobic conditions while collecting specimens, transporting them to the laboratory, and culturing them. Selective media are often used to reveal the presence of these bacteria. For example, because *B. fragilis* grows well in the presence of bile, it can be cultured on bile-esculin agar in an anaerobic environment—conditions that inhibit aerobic and most anaerobic bacteria (**FIGURE 20.29**). Metronidazole is the antimicrobial of choice for treating *Bacteroides* infections.

Prevotella

Members of the genus *Prevotella* (prev′ō-tel′ă) are also anaerobic, Gram-negative bacilli and were previously classified in *Bacteroides*. They differ from *Bacteroides* in that they are sensitive to bile; as a result, they do not grow in the intestinal tract but are instead found in the urinary, genital, and respiratory tracts as part of the normal microbiota. Their virulence factors, including adhesive and antiphagocytic capsules and proteases that destroy antibodies, make them potential pathogens.

Prevotella is involved with other Gram-negative bacteria in about half of all sinus and ear infections and in almost all periodontal

▲ FIGURE 20.29 *Bacteroides fragilis*. The bacterium produces a black color on bile-esculin agar. Bile suppresses the growth of most aerobes and facultative anaerobes.

infections. Additionally, these bacteria cause gynecological infections such as pelvic inflammatory disease, pelvic abscesses, and endometriosis; brain abscesses; and abdominal infections.

Treatment of *Prevotella* infections involves surgical removal of infected tissue and the use of intravenous carbapenem.

TELL ME WHY

Why are *Prevotella* and *Bacteroides* unlikely to opportunistically infect the skin?

MICRO IN THE CLINIC FOLLOW-UP
Another Day Care Cold?

Dr. Fischer completes a preliminary examination of Hailey. She learns that Claire and her husband had not felt childhood immunizations were necessary for Hailey. Based on Hailey's symptoms, in particular her acute cough that has been occurring for about a week with the distinctive coughing sound upon inspiration, she believes that Hailey has pertussis, otherwise known as whooping cough. Dr. Fischer collects a nasopharyngeal swab to send for laboratory confirmation.

Dr. Fischer tells Claire that Hailey most likely became infected at day care. It is a highly infectious disease that is transmitted via droplets from infected persons. Because Hailey isn't immunized against the bacterium that causes the disease, *Bordetella pertussis*, she was susceptible to infection.

The lab analysis confirms Dr. Fischer's diagnosis. Because Hailey's pertussis was diagnosed relatively early, Dr. Fischer prescribes a course of azithromycin: 500 mg the first day, then 250 mg for four days. After completing five days of antibiotics, Hailey should be able to safely return to day care without spreading the disease further.

Pertussis is a mandatory reportable disease to the state and local health departments, so Dr. Fischer reports the case to her local health department. The other sick kids at the day care also had whooping cough—none of them had been immunized against pertussis.

1. Explain how *B. pertussis* affects the trachea and how this damage leads to the characteristic cough associated with pertussis.

2. Hailey's infection could have been prevented with the DTaP vaccine. Hailey's parents' argument against the vaccine was that immunization was a personal choice that affected only Hailey. As a future health care professional, do you agree with Claire's argument? Why or why not?

Check your answers to Micro in the Clinic Follow-Up questions in the Mastering Microbiology Study Area.

 Go to Mastering Microbiology for Interactive Microbiology, Dr. Bauman's Video Tutors, Micro Matters, MicroFlix, MicroBoosters, practice quizzes, and more.

CHAPTER SUMMARY

1. Lipid A in the outer membranes of Gram-negative bacteria can stimulate symptoms of fever, vasodilation, inflammation, and shock as well as blood clots throughout the body, a condition known as **disseminated intravascular coagulation (DIC)**.

Pathogenic Gram-Negative Cocci: *Neisseria* (pp. 587–591)

1. *Neisseria* is a pathogenic, Gram-negative, oxidase-positive coccus; its virulence results from the presence of fimbriae, polysaccharide capsules, and lipooligosaccharide (containing lipid A) in its cell walls.

2. *Neisseria gonorrhoeae* causes **gonorrhea**, a sexually transmitted disease of humans. In men it results in acute inflammation of the urethra, whereas in women it is generally asymptomatic. It can infect the uterus and uterine tubes to cause **pelvic inflammatory disease (PID)**.

3. *Neisseria meningitidis* causes a type of meningitis; the bacterium is transmitted on respiratory droplets and is life threatening when it enters the bloodstream or central nervous system. Blebbing by the bacterium can release a dangerous level of lipid A.

Pathogenic, Gram-Negative, Facultatively Anaerobic Bacilli (pp. 591–605)

1. Members of the **Enterobacteriaceae**, the **enteric bacteria**, can be pathogenic, are oxidase negative, reduce nitrate to nitrite, and ferment glucose anaerobically. Their outer membranes contain lipopolysaccharide with lipid A, contributing to their virulence. Many species have virulence factors such as capsules, fimbriae, adhesins, exotoxins, iron-binding siderophores, hemolysins, or type III secretion systems.

 Play *Virulence Factors: Enteric Pathogens* @ Mastering Microbiology

2. Pathogenic enteric bacteria are grouped as coliform opportunists, noncoliform opportunists, and true pathogens. **Coliforms** are found in the intestinal tracts of animals and humans.

3. Coliforms can cause diarrhea and other gastrointestinal disease. Physicians treat the symptoms by providing fluid and electrolytes. For internal infections, which are rarer, clinicians prescribe an appropriate antimicrobial drug as determined by a Kirby–Bauer test.

4. *Escherichia coli* is the most common and most widely studied coliform. It causes gastroenteritis, non–healthcare-associated urinary tract infections, fatal hemorrhagic colitis, and hemolytic uremic syndrome.

5. *Klebsiella, Serratia, Enterobacter, Hafnia,* and *Citrobacter* are genera of coliform bacteria. *K. pneumoniae* causes a type of pneumonia. All are involved in healthcare-associated infections.

6. Noncoliform opportunistic pathogens of the family Enterobacteriaceae include *Proteus* (urinary tract infections) and *Morganella, Providencia,* and *Edwardsiella,* which can cause healthcare-associated infections in immunocompromised patients.

7. Truly pathogenic enteric bacteria include *Salmonella enterica,* which causes **salmonellosis**, a serious form of diarrhea. *S. enterica* serotypes Typhi and Paratyphi cause **typhoid fever**.

8. Members of the genus *Shigella* cause **shigellosis**, a severe form of diarrhea. **Shiga toxin**, secreted by some *Shigella*, arrests protein synthesis in host cells.

9. Two enteric species, *Yersinia enterocolitica* and *Y. pseudotuberculosis,* cause different degrees of intestinal distress.

10. The virulent, nonenteric bacterium *Y. pestis* causes **bubonic plague** and **pneumonic plague** and has had a major historical impact. Infected fleas spread the bacterium among rodents and into humans—the most common way for plague to enter people. *Y. pestis* moving into the lungs from the blood causes pneumonic plague, which can spread among humans through respiratory aerosols and sputum.

11. Two genera in the family **Pasteurellaceae**—*Pasteurella* and *Haemophilus*—are significant human pathogens. They differ from enteric bacteria (family Enterobacteriaceae) by being oxidase positive.

12. *Haemophilus influenzae* causes infantile meningitis, epiglottitis, arthritis, and inflammation of the skin. Widespread immunization with Hib vaccine has almost eliminated disease caused by this pathogen in the United States. *H. ducreyi* is responsible for a sexually transmitted disease.

Pathogenic, Gram-Negative, Aerobic Bacilli
(pp. 605–614)

1. *Bartonella* includes pathogenic aerobic bacilli. *B. bacilliformis,* transmitted by sand flies, causes **bartonellosis**. *B. quintana,* transmitted by lice, causes five-day fever, or **trench fever**. *B. henselae,* transmitted through cat scratches, bites, and fleas, causes **cat scratch disease**.

2. *Brucella,* transmitted in unpasteurized contaminated milk, causes **brucellosis**, also known as Bang's disease, undulant fever, Malta fever, and other names.

3. *Bordetella pertussis* is responsible for **pertussis** (whooping cough), a pediatric disease in which the ciliated epithelial cells of the trachea are damaged. After **incubation**, the disease progresses through three additional phases: the **catarrhal phase** resembles a cold; in the **paroxysmal stage** the patient coughs deeply to expel copious mucus from the trachea; and in the **convalescent phase** the disease subsides, but secondary bacterial infections may ensue.

4. Virulence factors of *B. pertussis* include filamentous hemagglutinin (an adhesin), and four toxins: pertussis, adenylate cyclase, dermonecrotic, and tracheal.

5. *Burkholderia* can decompose numerous environmental pollutants and can assist farmers by inhibiting fungal pathogens of plants; however, the bacterium also grows in the lungs of cystic fibrosis patients.

6. **Pseudomonads**, including *Pseudomonas, Moraxella,* and *Acinetobacter,* are opportunistic pathogens. The latter two rarely cause disease except in immunocompromised patients.

7. *Pseudomonas aeruginosa* is involved in some healthcare-associated infections and is common in burn victims and cystic fibrosis patients. The skin and mucous membranes protect the body from this ubiquitous pathogen, which can use many organic compounds, including human tissues, as a carbon source.

8. **Tularemia** (rabbit fever or tick fever) is a zoonotic disease caused by extremely virulent *Francisella tularensis,* which is so small that it can pass through apparently unbroken skin. *F. tularensis* can infect most animals from protozoa to insects to mammals and birds and humans. Humans have been infected by touching or being bitten or scratched by an infected animal, and by drinking, eating, or inhaling the bacterium.

9. People can reduce their chance of infection by avoiding tick bites and ill-appearing animals and by wearing gloves and masks while field dressing animals.

10. **Legionnaires' disease** (legionellosis)—a type of pneumonia—and flulike Pontiac fever are caused by *Legionella,* a bacterium transmitted in aerosols such as those produced by air conditioning systems. *Legionella* can live inside amoebas, which delayed the discovery of this ubiquitous yet fastidious bacterium. Physicians treat Legionnaires' disease with fluoroquinolones. Pontiac fever is a milder disease and generally requires no treatment.

11. The small, almost viral-sized bacterium *Coxiella burnetii* lives in phagolysosomes of mammal and bird cells; its sporelike infective bodies allow it to survive outside its host. It is transmitted via aerosolized dried feces and urine and causes **Q fever**.

Pathogenic, Gram-Negative, Anaerobic Bacilli
(pp. 614–615)

1. The Gram-negative anaerobes *Bacteroides* and *Prevotella* may be opportunistic pathogens that produce virulence factors, such as capsules, fimbriae, and proteases that degrade antibodies, particularly when the bacteria are introduced into novel sites in the body.

2. *Bacteroides* can grow in the presence of bile and so inhabit the colon where they are part of the normal microbiota and produce precursors of some vitamins.

3. *Prevotella* is sensitive to bile, so it doesn't inhabit the colon but instead is part of many people's normal microbiota in the urinary, genital, and respiratory tracts.

QUESTIONS FOR REVIEW

Answers to the Questions for Review (except Short Answer questions) begin on p. A-1.

Multiple Choice

1. The presence of lipid A in the outer membranes of Gram-negative bacteria _____.
   a. affects the formation of blood clots in the host
   b. causes these bacteria to be oxidase positive
   c. triggers the secretion of a protease enzyme to cleave IgA in mucus
   d. enables enteric bacteria to ferment glucose anaerobically

2. The only genus of Gram-negative cocci that causes significant disease in humans is _____.
   a. *Pasteurella*
   b. *Salmonella*
   c. *Klebsiella*
   d. *Neisseria*

3. Which of the following bacterial cells is most likely to be virulent?
   a. a cell with fimbriae and LOS
   b. a cell with a polysaccharide capsule and lipooligosaccharide
   c. a cell with fimbriae, lipooligosaccharide, and a polysaccharide capsule
   d. a cell with fimbriae but no capsule

4. Which of the following statements is true?
   a. PID is a severe type of diarrhea in which infection spreads from the intestines to the bloodstream.
   b. PID can result from *Neisseria* infection.
   c. PID is more common in men than women.
   d. Members of the family Enterobacteriaceae usually cause PID.

5. A coliform bacterium that likely contaminates dairy products is _____.
   a. *Bartonella*
   b. *Serratia*
   c. *Enterobacter*
   d. *Proteus*

6. Capsules of pathogenic enteric bacteria are virulence factors because they _____.
   a. capture iron from hemoglobin and store it in the bacteria
   b. release hemolysins that destroy red blood cells
   c. produce fimbriae that enable the bacteria to attach to human cells
   d. protect the bacteria from phagocytosis and from some antibodies

7. Which of the following bacteria might be responsible for the formation of petechiae in a host?
   a. *Neisseria meningitidis*
   b. *Escherichia coli* O157:H7
   c. *Klebsiella*
   d. *Proteus mirabilis*

8. The pathogen *Haemophilus influenzae* b causes _____.
   a. meningitis in children
   b. upper respiratory flu
   c. endocarditis
   d. genital chancroid

9. Which of the following diseases is typically mild?
   a. Brazilian purpuric fever
   b. bartonellosis
   c. pediatric meningitis
   d. cat scratch disease

10. Which bacterium causes infections in many burn victims?
    a. *Moraxella catarrhalis*
    b. *Pseudomonas aeruginosa*
    c. *Escherichia coli*
    d. *Bartonella bacilliformis*

11. Which of the following statements is true of Q fever?
    a. For many years its cause was questionable.
    b. It was first described in 1976 during an outbreak in Quincy, Massachusetts.
    c. Researchers found it could be effectively treated with quinine.
    d. The sharp spikes of fever on patients' temperature charts resemble porcupine quills.

12. Which of the following is a bile-tolerant anaerobe?
    a. *Bacteroides*
    b. *Escherichia*
    c. *Shigella*
    d. *Prevotella*

Matching

Match each bacterium with the disease or manifestations it causes.

1. ___ *Escherichia coli*
2. ___ *Klebsiella pneumoniae*
3. ___ *Proteus mirabilis*
4. ___ *Salmonella enterica* serotype Typhi
5. ___ *Shigella flexneri*
6. ___ *Yersinia pestis*

A. Bubonic plague
B. Typhoid fever
C. Gastroenteritis
D. Kidney stones
E. Pus-filled, bloody stools; cramps; fever; and diarrhea
F. Pneumonia

VISUALIZE IT!

1. Label the following drawing using these words: adhesin, exotoxin, H antigens, hemolysin, iron-binding protein, K antigens, outer membrane (with common antigen, lipid A, and O antigen), fimbria, plasmid virulence genes, type III secretion system.

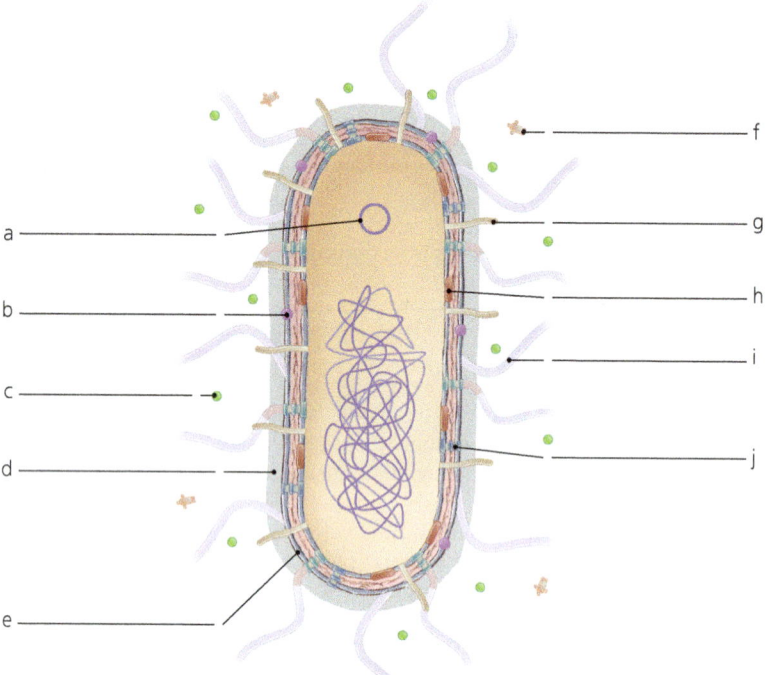

2. Shown is a MacConkey agar plate. Describe the Gram reaction and lactose fermentation of the two different bacteria shown.

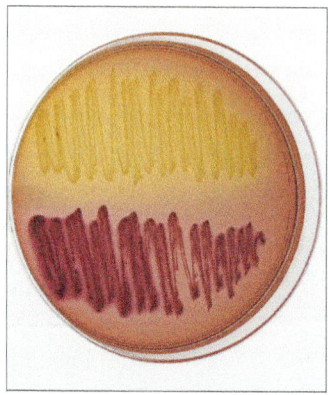

Short Answer

1. A physician prescribes fluid replacement to treat a patient with diarrhea. Although tests showed that a pathogenic enteric bacterium was the cause of the intestinal distress, an antimicrobial drug was not prescribed; why not?

2. Distinguish among the pathogenicity of coliforms, noncoliforms, and truly pathogenic enteric bacteria.

3. Why do nurses place antimicrobial agents in babies' eyes at birth?

4. Statistics show that meningococcal diseases are more frequent in college dormitories and military barracks than in the population at large. Suggest an explanation of this observation.

5. Why might an alcoholic be susceptible to pulmonary disease?

6. Name six factors that facilitate the production of disease by *Bordetella pertussis*.

7. Given that pseudomonads are present in almost every moist environment, why do they cause less disease than other, less prevalent Gram-negative bacteria?

8. Although an effective vaccine is available to eradicate pertussis in the United States, why has the number of reported cases increased since the 1970s?

9. What two illnesses can be caused by *Legionella pneumophila*?

10. What attribute of *Coxiella burnetii* enables it to survive desiccation and heat for an extended time?

11. What virulence factors allow Gram-negative anaerobes to cause disease?

12. What single biochemical test result distinguishes gammaproteobacteria in the family Enterobacteriaceae from gammaproteobacteria in the family Pasteurellaceae?

13. Describe transovarian transmission of a pathogen.

CRITICAL THINKING

1. A three-year-old boy complains to his day care provider that his head hurts. The worker calls the child's mother, who arrives 30 minutes later to pick up her son. She is concerned that he is now listless and unresponsive, so she drives straight to the hospital emergency room. Though the medical staff immediately treats the boy with penicillin, he dies—only four hours after initially complaining of a headache. What caused the boy's death? Were the day care workers or the hospital staff to blame for his death? What steps should be taken to protect other children and the staff at the day care facility?

2. An epidemiologist notices a statistical difference in the fatality rates between cases of Gram-positive bacteremia treated with antimicrobial drugs and treated cases of Gram-negative bacteremia—that is, patients with Gram-positive bacteremia are much more likely to respond to treatment and survive. Explain one reason why this might be so.

3. In one summer month, local physicians reported 11 cases of severe diarrhea in children less than three years of age. All the children lived in the same neighborhood and played in the same park, which has a wading pool. Cultures of stool specimens from the children revealed a Gram-negative bacillus that gave negative results for the following biochemical tests: oxidase, lactose, urease, and hydrogen sulfide production. The organism was also

nonmotile. What is the organism? How was it transmitted among the children? What could officials do to limit infections of other children?

4. Several years ago, epidemiologists noted that the number of cases of salmonellosis increased dramatically in the summer and was lower the rest of the year. In contrast, the number of cases of typhoid fever, caused by the same bacterium, was constant throughout the year. Explain these observations.

5. Most physicians maintain that the so-called "stomach flu" is really a bacterial infection. Which genera of bacteria discussed in this chapter are good candidates for causing this disease?

6. A hunter reports to his physician that he has been suffering with fever, chills, malaise, and fatigue. What Gram-negative bacteria may be causing his symptoms? How can a laboratory scientist distinguish among these species?

7. Ear piercings resulted in a rash of infections in teenagers, all of whom reported having their ears pierced at a kiosk in a local mall. What bacterial species might be responsible? Investigators traced the infections to a bacterium living in a bottle of disinfectant and in a sink used by employees. What bacterium is the likely agent? Why was successful treatment difficult?

8. A 21-month-old child was admitted to the hospital with fever, severe abdominal cramps, and bloody diarrhea. The family revealed they had purchased an iguana one month previously and the child had helped clean the cage. What bacterium was the likely cause of diarrhea? What was the possible treatment? How could the family prevent a recurrence?

9. Penicillin-resistant strains of *N. gonorrhoeae* produce a plasmid-coded β-lactamase, which degrades penicillin. What changes in structure or metabolism could explain resistance to other antimicrobial drugs in this bacterium?

10. Investigations of *Mycobacterium tuberculosis* have clearly demonstrated that this pathogen can be transmitted among airline passengers; however, as of 2008, no cases of in-flight transmission of the similarly transmitted *N. meningitidis* had been reported to the CDC. What are some possible explanations for the lack of documented cases of in-flight transmission of meningococci?

11. *Haemophilus influenzae* was so named because researchers isolated the organism from flu patients. Specifically, how could a proper application of Koch's postulates have prevented misnaming this bacterium?

12. Why do physicians substitute trimethoprim and sulfanilamide for tetracycline when treating children and pregnant women infected with *Brucella*?

CONCEPT MAPPING

Using the terms provided, fill in this concept map that describes bacterial meningitis. You can also complete this and other concept maps online by going to the Mastering **Microbiology** Study Area.

Antimicrobials
Cerebrospinal fluid
Gram stain
Haemophilus influenzae

Healthy carriers
Hib vaccine
Infants and young children
Neisseria meningitidis

Penicillin G
Respiratory route
Serological test
Vaccines

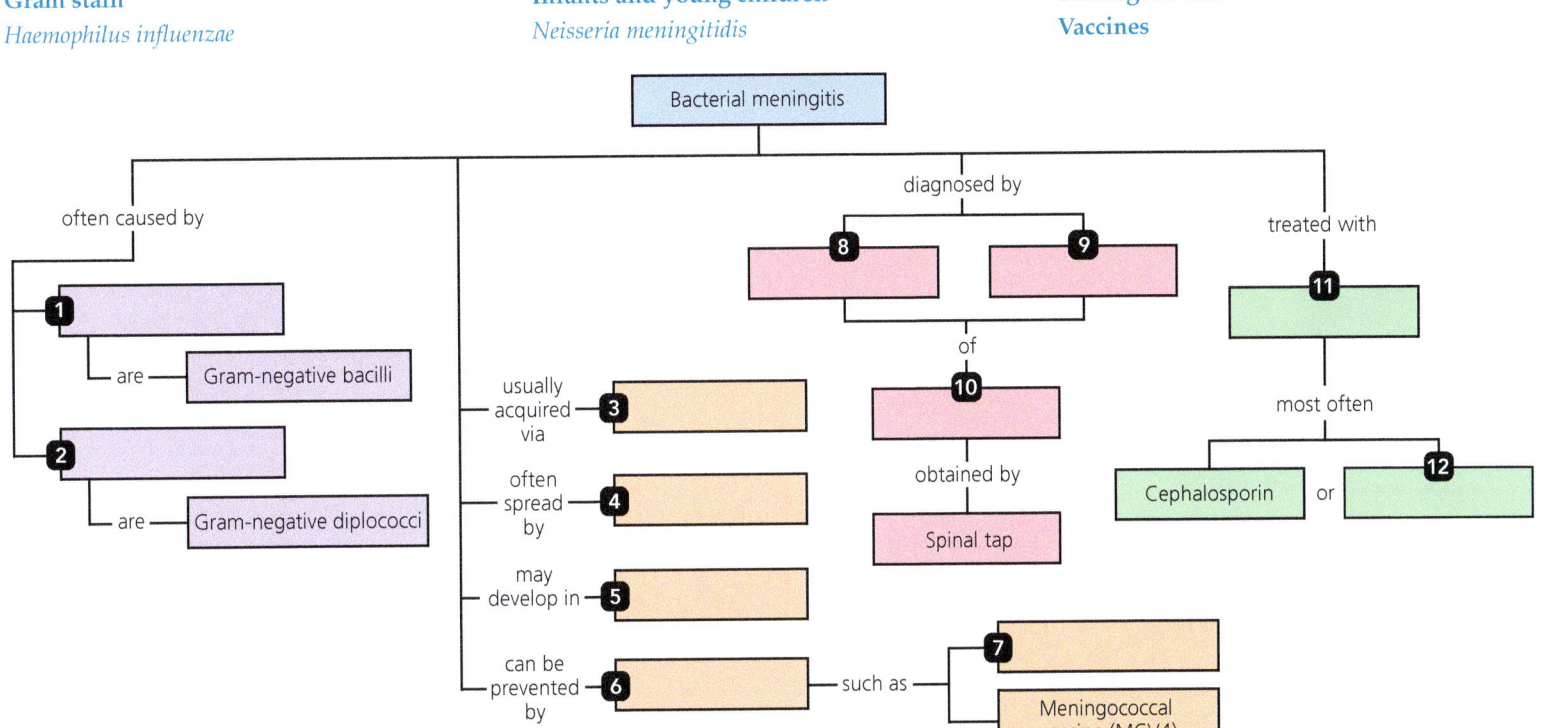

21 Rickettsias, Chlamydias, Spirochetes, and Vibrios

Before You Begin

1. What is the term for flexible, spiral-shaped bacteria?
2. What type of flagella do spirochetes have?
3. What is the difference between a mechanical and a biological vector of a disease?

Vibrio cholerae, which has a vibrio—curved rod—shape, causes cholera

MICRO IN THE CLINIC
Too Much Spice?

TREVOR AND HIS COLLEGE ROOMMATE, CHANDU, are taking a trip to Chandu's native India for a family reunion. They've been traveling around the country for three weeks—while they've visited some of India's tourist destinations, the best parts of the trip for Trevor were all of the spots that only a native like Chandu knows. They spend the last days of their trip in Delhi; Trevor is flying home, and Chandu is going to stay a couple of extra days to visit with family. On their last night together in India, Trevor and Chandu spend their time wandering, stopping whenever something catches their interest. They also enjoy some of Delhi's street food, including kebabs, aloo chaat (spicy potatoes), and kulfi (Indian ice cream). It is a great end to a great trip.

Trevor departs Delhi early the next morning. He's thankful for his nonstop flight back to New York—he woke up that morning with a slight headache and abdominal cramping. He and Chandu did drink quite a bit of alcohol, and some of the street food was pretty spicy. Perhaps his body isn't used to that much spice! Trevor falls asleep almost immediately after takeoff and is sleeping quite well until he is jolted awake by the need to vomit. Over the next several hours, Trevor continues to vomit, and he also begins having severe diarrhea. He feels terrible—there's no way this can be due to just his night on the town with Chandu.

1. **Do you think that Trevor's symptoms are the result of spicy food and too much drinking?**

   Turn to the end of the chapter (p. 642) to find out.

A number of bacteria that stain pink in a Gram stain differ from typical Gram-negative organisms in morphology, growth habit, or reproductive strategy. One such group, the mycoplasmas, are genetically low G + C, Gram-positive bacteria that stain pink because they lack cell walls. Other unusual pathogens include the obligate intracellular rickettsias and chlamydias, the spiral-shaped spirochetes, and the slightly curved vibrios. Because of their unique features, these bacteria have traditionally been discussed separately—a tradition we continue here, beginning with the rickettsias.

Rickettsias

Rickettsias (ri-ket'sē-ăz) are tiny, Gram-negative, obligate intracellular parasites that synthesize only a small amount of peptidoglycan and thus appear almost wall-less. The group as a whole is named after the most common genus of them, *Rickettsia*, named for Howard Ricketts (1871–1910), who first identified rickettsias and described the transmission of one species via its tick vector. Rickettsias are extremely small (0.3 μm × 1.0 μm); in fact, rickettsias were originally considered viruses because of their small size, but closer examination has revealed that they contain both DNA and RNA, functional ribosomes, and citric acid cycle enzymes and that they reproduce via binary fission—all characteristics of cells, not viruses.

Researchers have proposed several hypotheses to explain why rickettsias are obligate parasites, even though they have functional genes for protein synthesis, ATP production, and reproduction. Primary among these hypotheses is that rickettsias have very "leaky" cytoplasmic membranes and lose small cofactors (such as NAD^+) unless they are in an environment that contains an equivalent amount of these cofactors—such as the cytosol of a host cell.

Scientists classify rickettsias in the class Alphaproteobacteria of the bacterial phylum Proteobacteria based on the sequence of nucleotides in their rRNA molecules. Four genera—*Rickettsia, Orientia, Ehrlichia,* and *Anaplasma*—are the main rickettsias that cause diseases in humans. We begin our discussion with *Rickettsia*.

Rickettsia

LEARNING OUTCOMES

21.1 List three species of *Rickettsia* that are responsible for human infections and identify their vectors.

21.2 Describe the rash and petechiae of spotted fever rickettsiosis (Rocky Mountain spotted fever, RMSF).

21.3 Explain the relationship between epidemic typhus and Brill-Zinsser disease.

Rickettsia is a genus of nonmotile, aerobic, intracellular parasites that live in the cytosol of their host cells. They possess an outer membrane of lipopolysaccharide with endotoxin activity. A loosely organized slime layer surrounds each cell.

Outside of its host's cells, a rickettsia is unstable and dies quickly; therefore, it requires a vector for transmission from host to host. Each rickettsial species is transmitted by a different arthropod vector, which also acts as host and reservoir. Once inside a host cell, a rickettsia secretes an enzyme that digests the membrane of the endocytic vesicle, releasing the bacterium into the host cell's cytosol.

Physicians categorize rickettsial human infections into two main categories: the spotted fever group, which causes **spotted fever rickettsioses**, and the typhus group, which causes different forms of typhus. **Disease in Depth: Rocky Mountain Spotted Fever** (pp. 624–625) examines the most common of the spotted fever rickettsias, *Rickettsia rickettsii* (ri-ket'sē-ē), which causes **Rocky Mountain spotted fever (RMSF). Emerging Disease Case Study: A New Cause of Spots** considers a newly recognized spotted fever rickettsia.

Now, we examine three rickettsias that cause forms of typhus[1]—*Rickettsia prowazekii, R. typhi,* and *Orientia tsutsugamushi*.

Rickettsia prowazekii

Rickettsia prowazekii (prō-wă-ze'kē-ē) causes **epidemic typhus**, which is also called *louse-borne typhus* because it is vectored by the human body louse, *Pediculis humanus* (pĕ-dik'yu-lŭs hū-man'us; see Chapter 12, Figure 12.29d). The bacterium lives in the gut of the louse and is shed when the louse defecates as it feeds. Human victims scratch the bacteria-laden feces into the wound. Rickettsias parasitize cells of blood vessels and can travel in the blood, producing a bacteremia. *R. prowazekii* differs from other rickettsias in at least two ways: it has humans as its primary host, and the bacterium fills a host cell until the host cell breaks open, releasing the rickettsias.

Epidemic typhus occurs in crowded, unsanitary living conditions that favor the spread of body lice; currently it is most prevalent in Central America, South America, Africa, and northern China. Inhabitants of or visitors to areas with large homeless populations, impoverished areas, or refugee camps are most at risk. Outbreaks are more common in winter when it is often more difficult to wash clothes and bedding to rid them of lice.

Epidemic typhus in humans manifests with high fever, mental and physical depression, and a rash that lasts for about two weeks. In severe cases, particularly among the elderly, blood flow to the extremities may be severely restricted, leading to loss of digits or limbs. Death results in about 60% of untreated elderly patients and about 20% of younger healthy patients. Physicians diagnose epidemic typhus by observing the signs and symptoms following exposure to infected lice.

Epidemic typhus can recur many years (even decades) following an initial episode. The recurrent disease, called *Brill-Zinsser[2] disease,* is more mild and brief than the initial occurrence. Brill-Zinsser resembles murine typhus (discussed shortly).

Epidemic typhus is treated with doxycycline or chloramphenicol. Prevention involves controlling lice populations and maintaining good personal hygiene. An attenuated vaccine against epidemic typhus is available for use in high-risk populations.

---

[1]From Greek *typhos*, meaning "stupor."
[2]After physician Nathan Brill and bacteriologist Hans Zinsser, who studied the condition.

EMERGING DISEASE CASE STUDY

A New Cause of Spots

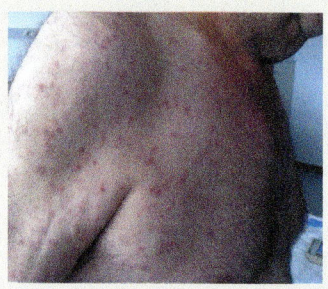

Fifty-two-year-old David has a good life. After 30 years serving the country as an army officer, he has retired to the coast of North Carolina—a region of beautiful trees, moderate winter weather, and great outdoor spaces. It's a great place to retire and enjoy hiking through the woods and meadows photographing wildlife, especially in the early fall. It would be nearly perfect if some of the wildlife didn't bite. Pervasive ants, pesky mosquitoes, and bloodsucking ticks seem to always be on the prowl.

It's a tick that brings him to his doctor today. After a hike he realized he had had a tick latched onto his shoulder for an unknown period of time. He very gently and carefully removed the tick with a pair of tweezers. Within 48 hours, David developed a low-grade fever, severe headache, and generalized malaise. The following day, he began to notice a rash running down his arm and toward his trunk. The rash had very small dark spots that seemed to resemble blood clots. He decided to make an appointment with his doctor.

The physician was able to squeeze David onto the schedule the same day. As soon as the physician assesses and inspects David's worsening rash, he suspects Rocky Mountain spotted fever (RMSF), since it is common in the Carolinas, and orders a laboratory test using anti–*Rickettsia rickettsii* antibodies. The test comes back negative; David is not infected with *R. rickettsii*. He does not have RMSF. The doctor takes a skin sample from the infected area and prescribes 100 mg of doxycycline twice daily for seven days, since David's infection is not too severe.

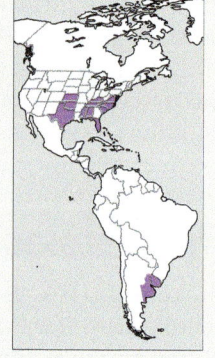

The rash resolves in a week with help from the antibiotic. Polymerase chain reaction (PCR) testing on bacteria found in the sample of skin reveals *Rickettsia parkeri*, another bacterium that causes spotted fever, though it was long thought to be harmless to humans. David is one of the first of several dozen patients to tangle with this emerging threat in the southeast United States where its vector, the Gulf Coast tick, *Amblyomma maculatum*, commonly lives.

1. Why didn't the doctor prescribe amoxicillin?
2. Why didn't the antibody test show infection?
3. What is PCR testing?
4. In a Gram-stained sample of David's skin, what color would the rickettsias be?

Rickettsia typhi

Rickettsia typhi (tī'fē) causes **murine[3] typhus** (mū'rēn), which is so named because the major reservoir for the bacterium is rodents. The disease is also known as *endemic typhus*. The rat flea *Xenopsylla cheopis* (zen-op-sil'ă chē-op'is; see Figure 12.29c) and the cat flea *Ctenocephalides felis* (tē-nō-se-fal'i-dez fē'lis; which also feeds on opossums, raccoons, and skunks) transmit the bacteria among the animal hosts and to humans. About 12 days following the bite of an infected flea, an abrupt fever, severe headache, chills, muscle pain, and nausea occur. A rash typically restricted to the chest and abdomen occurs in less than 50% of cases. The disease usually lasts about three weeks if left untreated and is rarely fatal. Patients who have recovered from murine typhus have long-lasting immunity to reinfection.

Murine typhus is endemic in every continent except Antarctica; in the United States, it is most often seen in the southern states from Florida to California. It is estimated that 50 cases occur annually in the United States, though the extent and incidence of the disease is not known because national notification requirements for murine typhus were discontinued in 1994.

Diagnosis is initially based on signs and symptoms following exposure to fleas. A fluorescent immunoassay of a blood smear provides specific confirmation. Treatment is with doxycycline. Prevention, as with other rickettsial diseases, involves avoiding bites by fleas, wearing protective clothing, and using chemical repellents. No vaccine is available for murine typhus.

Orientia tsutsugamushi

LEARNING OUTCOME

21.4 Identify the causative agent, vector, and reservoir of scrub typhus.

The rickettsial organism *Orientia tsutsugamushi* (ōr-ē-en'tē-ă tsoo-tsoo-gă-mū'shē) was formerly classified in the genus *Rickettsia*, but taxonomists now assign it to the new genus *Orientia*. It differs from *Rickettsia* by having significantly different rRNA nucleotide sequences, a thicker cell wall, and a minimal slime layer. Mites of the genus *Leptotrombidium* (lep'tō-trom-bid'ē-ŭm), also known as red mites or chiggers, transmit *Orientia* among rodents and humans. Because these arachnids feed on a host only once in their lives, an individual mite cannot transmit rickettsia directly from a rodent to a human. Instead, infected female mites transmit the bacterium to their offspring via transovarian transmission, and the offspring transmit it to a rodent or human while feeding. In addition to being the vector, mites are the only known reservoir of *Orientia*.

---

[3]From Latin *murinus*, meaning "relating to mice."

O. tsutsugamushi causes **scrub typhus**, a disease endemic to eastern Asia, Australia, and India. It occurs in the United States among immigrants who arrived from endemic areas. Scrub typhus is characterized by fever, headache, and enlarged and tender lymph nodes, all of which develop about 11 days after a mite bite. Less than half of patients with scrub typhus also develop a spreading rash on their trunks and appendages. In less than 1% of cases, death results from failure of the heart and inflammation in the central nervous system.

Physicians treat scrub typhus in nonpregnant adults with doxycycline, chloramphenicol, or macrolides; they treat children and pregnant women with azithromycin. No vaccine for scrub typhus is available. Prevention involves avoiding exposure to mites by wearing appropriate clothing and using repellent chemicals.

Ehrlichia and Anaplasma

LEARNING OUTCOMES

21.5 Identify the diseases caused by *Ehrlichia* and *Anaplasma*, respectively.

21.6 Identify the three developmental stages of *Ehrlichia* and *Anaplasma*.

21.7 Discuss the difficulties in diagnosing ehrlichiosis and anaplasmosis.

Ehrlichia chaffeensis (er-lik′ē-ă chaf-ē-en′sis) and *Anaplasma phagocytophilum* (an-ă-plaz′mă fag-ō-sī-to′fil-ŭm; previously called *Ehrlichia equi*, ē′kwē) cause **ehrlichiosis** (also called *human monocytic ehrlichiosis, HME*) and **anaplasmosis** (previously *human granulocytic ehrlichiosis, HGE*), respectively. Both diseases are considered *emerging diseases* in the United States because they were unknown before 1987, and the number of reported cases has increased from a few per year in the 1980s to nearly 6000 cases in 2017.

Ticks, including the Lone Star tick (*Amblyomma*, am-blē-ō′ma), the deer tick (*Ixodes*, ik-sō′dēz), and the dog tick (*Dermacentor*), transmit *Ehrlichia* and *Anaplasma* to humans. Once in the blood, each bacterium triggers its own phagocytosis by a monocyte in ehrlichiosis or by a neutrophil in anaplasmosis. Unlike other rickettsias, *Ehrlichia* and *Anaplasma* grow and reproduce within a host cell's phagosomes. Because the bacteria would be killed if the phagosome fused with a lysosome, the bacteria must somehow prevent fusion, but the mechanism is unknown. The bacteria grow and reproduce through three developmental stages: an infective *elementary body* and, inside a leukocyte, an *initial body* and a *morula* (FIGURE 21.1). The release of the bacteria from the morula and into the blood makes them available to feeding ticks.

Ehrlichiosis and anaplasmosis resemble spotted fever rickettsiosis but without the rash, which only rarely occurs in ehrlichiosis or anaplasmosis. *Leukopenia* (loo-kō-pē′nē-ă), which is an abnormally low leukocyte count, is typically seen. Death from untreated ehrlichiosis is about 5%, and that from untreated anaplasmosis approaches 10%, especially in elderly patients.

Diagnosis of ehrlichiosis or anaplasmosis is difficult because the symptoms resemble those of other diseases. Physicians consider ehrlichiosis or anaplasmosis in any case of otherwise unexplained acute fever in patients exposed to ticks in endemic regions (FIGURE 21.2). Fluorescent immunoassays for *Ehrlichia* or for *Anaplasma* can demonstrate the bacterium within blood cells, confirming a diagnosis.

Doxycycline is the preferred drug against both *Ehrlichia* and *Anaplasma*, even in children. Normally, pediatricians don't administer doxycycline to children, because of possible damage to developing teeth; however, the American Academy of Pediatrics recommends the drug for ehrlichiosis and anaplasmosis because the benefits outweigh the risks. Treatment should start immediately, even before the diagnosis is confirmed by serological testing, because the complications of infection and mortality rates increase when treatment is delayed. Prevention

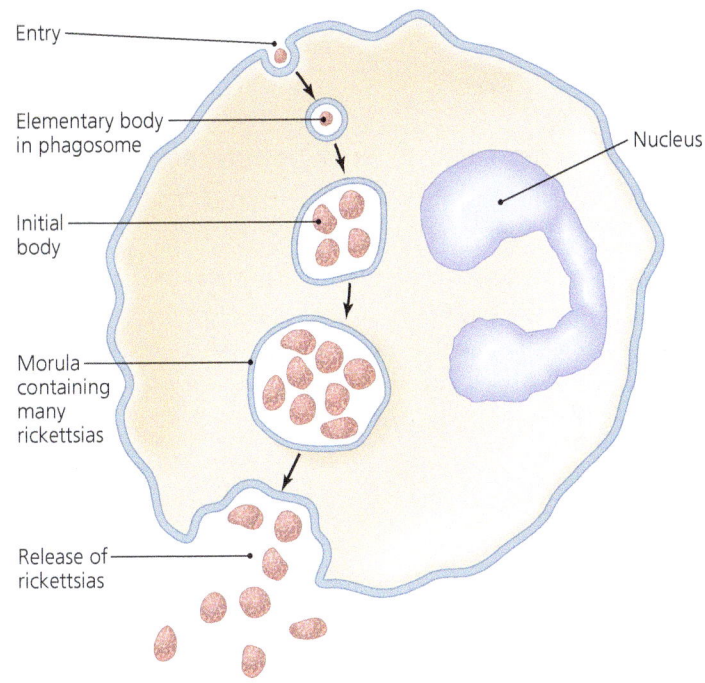

▲ FIGURE 21.1 The growth and reproduction cycle of *Ehrlichia* and *Anaplasma* in an infected leukocyte. *If the rickettsia shown here is* Ehrlichia chaffeensis, *then what kind of leukocyte is it infecting?*

Figure 21.1 *Ehrlichia chaffeensis infects human monocytes.*

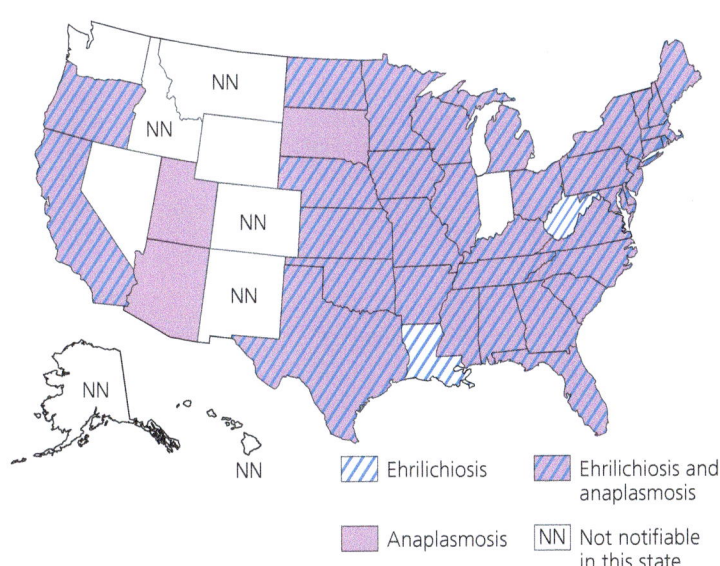

▲ FIGURE 21.2 The geographical distribution of ehrlichiosis and anaplasmosis in the United States.

DISEASE IN DEPTH

ROCKY MOUNTAIN SPOTTED FEVER

Rickettsia

SIGNS AND SYMPTOMS

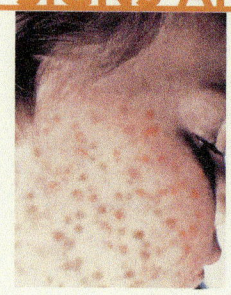

About 1 week after infection, patients experience fever, headache, chills, muscle pain, nausea, and vomiting. In most cases (90%), a spotted, non-itchy rash develops on the trunk and appendages, including palms and soles, sites not involved in rashes caused by the chickenpox or measles viruses. In about 50% of patients, the rash develops into subcutaneous hemorrhages called petechiae. In severe cases, the respiratory, central nervous, gastrointestinal, and renal systems fail. With treatment, only about 0.5% of patients die.

Rickettsia rickettsii causes Rocky Mountain spotted fever (RMSF), the most severe and most reported spotted fever rickettsiosis. Hard ticks in the genus *Dermacentor* transmit *R. rickettsii* among humans and rodents. *R. rickettsii* is typically dormant in the salivary glands of its tick vectors; only when the arachnids feed for several hours is the bacterium infective.

PATHOGENESIS

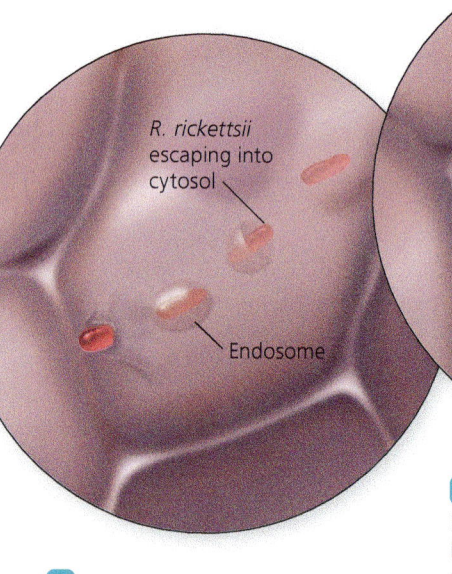

INSIDE BLOOD VESSEL

Endothelial cells lining small blood vessel

R. rickettsii escaping into cytosol

Endosome

Engorged *Dermacentor*

1 Infected tick introduces *R. rickettsii* in its saliva. This occurs only after the tick has fed for at least six hours. Active bacteria are released into the mammalian host's circulatory system.

2 *R. rickettsii* triggers endocytosis by cells lining blood vessels (endothelium); then it lyses the endosome's membrane, escaping into the cytosol.

3 Rickettsias divide every 8–12 hours in the host cell's cytosol. Daughter rickettsias escape from long cytoplasmic extensions of the host cell and infect other endothelial cells.

▶ Watch Dr. Bauman's Video Tutor explore Rocky Mountain spotted fever, or go to the **Mastering Microbiology** Study Area for more information.

EPIDEMIOLOGY

Though the earliest documented cases of Rocky Mountain spotted fever were in the Rocky Mountains, the disease is actually more prevalent in the Appalachian Mountains.

Cases of Rocky Mountain spotted fever in the United States, 2014.
- NN
- 0–1.0
- 1.0–2.3
- 2.3–6.6
- 6.6+

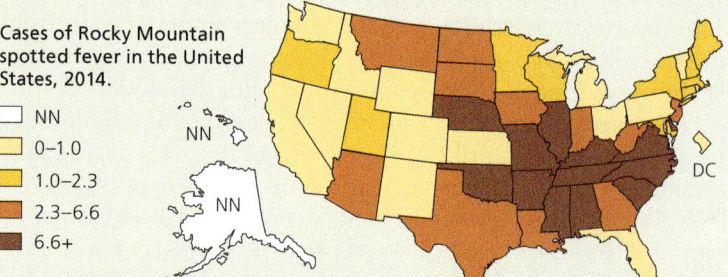

PATHOGEN

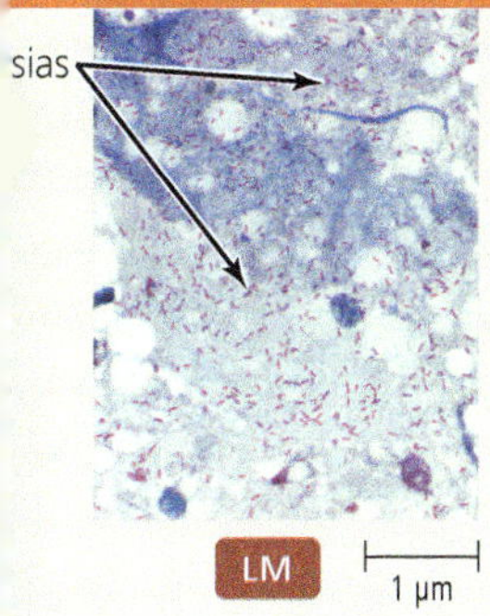

Rickettsia rickettsii is a small (0.3–1 μm), nonmotile, aerobic, Gram-negative, intracellular parasite that has a cell wall of peptidoglycan and an outer membrane of lipopolysaccharide surrounded by an organized slime layer. Rickettsias do not Gram stain well, so scientists use Gimenez-stained yolk sac smear (shown here).

Rickettsias cannot use glucose as a nutrient; instead, they oxidize amino acids and citric acid cycle intermediates, such as glutamic acid and succinic acid. For this reason, rickettsias are obliged to live inside other cells, where these nutrients are provided.

VECTOR

Rickettsias require a vector for transmission between hosts. For *R. rickettsii*, this vector is a hard tick of the genus *Dermacentor*. Male ticks infect female ticks during mating. Female ticks transmit bacteria to eggs forming in their ovaries—a process called *transovarian transmission*. *Dermacentor* can survive without feeding for more than four years, making tick elimination in the wild problematic.

4 *R. rickettsii* secretes no toxins, and disease is not the product of immune response. Apparently, damage to the endothelial cells leads to leakage of blood into the tissues, which results in low blood pressure and insufficient nutrient and oxygen delivery to the body's organs.

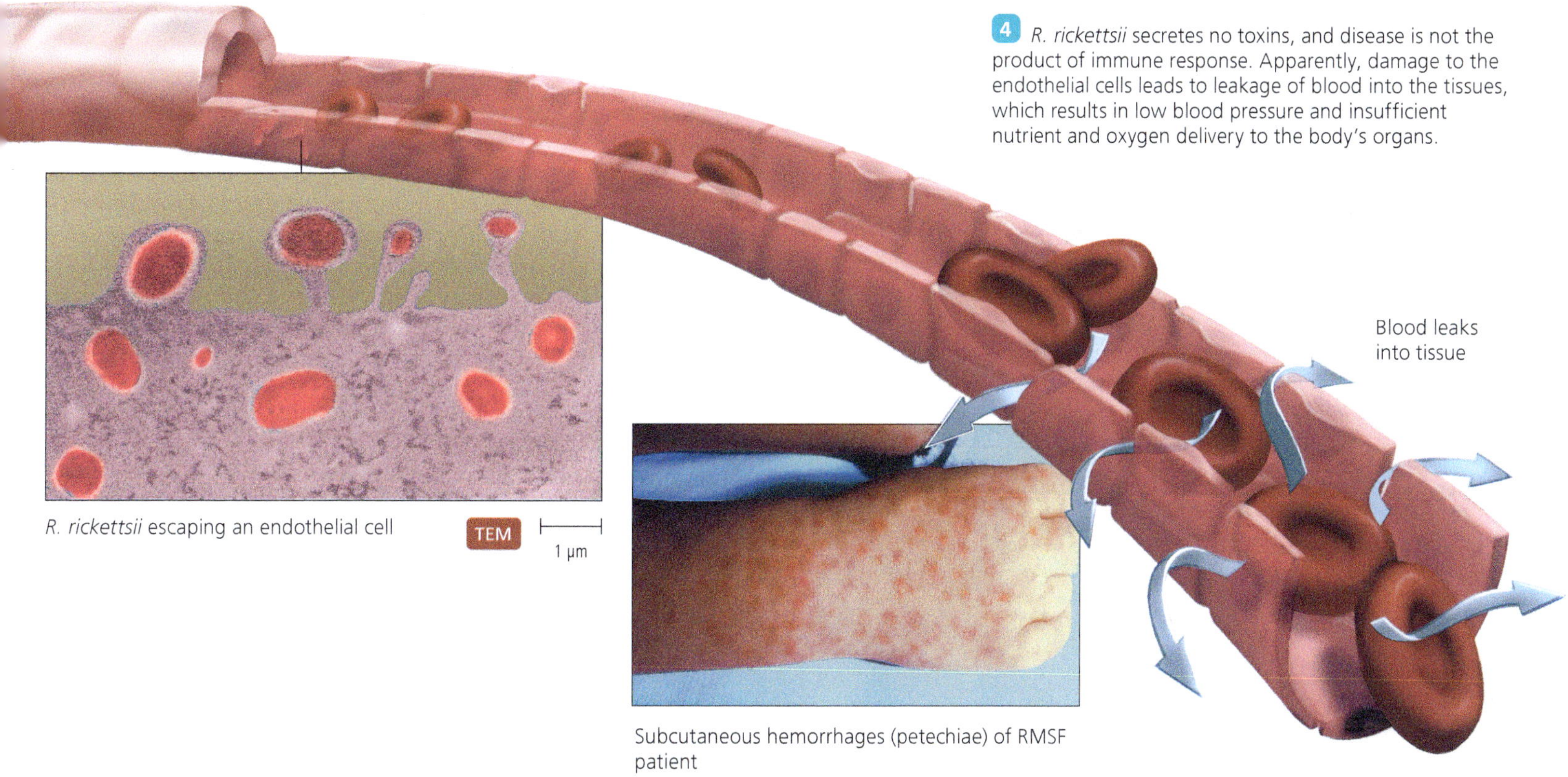

Blood leaks into tissue

R. rickettsii escaping an endothelial cell

Subcutaneous hemorrhages (petechiae) of RMSF patient

DIAGNOSIS

Negative / Positive
Latex agglutination test

Serological tests such as latex agglutination or fluorescent immunoassays are used to confirm an initial diagnosis based on sudden fever and headache following exposure to hard ticks, plus a rash on the soles or palms. Nucleic acid probes of specimens from rash lesions provide specific and accurate diagnosis. Early diagnosis is crucial because prompt treatment often makes the difference between recovery and death.

TREATMENT AND PREVENTION

Physicians treat RMSF by removing the tick and prescribing doxycycline for most adults or chloramphenicol for children and pregnant women. An effective vaccine is not available.

Wearing tight-fitting clothing, using tick repellents, promptly removing ticks, and avoiding tick-infested areas, especially in spring and summer when ticks are most voracious, help prevent infection.

TABLE 21.1 Characteristics of Some Rickettsias

Organism	Primary Vectors	Reservoirs	Diseases
R. rickettsii	Hard ticks: wood tick (Dermacentor andersoni) and dog tick (D. variabilis)	Ticks, rodents	Spotted fever rickettsiosis (Rocky Mountain spotted fever)
R. prowazekii	Human body louse (Pediculus humanus)	Humans, squirrels, squirrel fleas	Epidemic (louse-borne) typhus
R. typhi	Rat flea (Xenopsylla cheopis) and cat flea (Ctenocephalides felis)	Rodents	Murine (endemic) typhus
Orientia tsutsugamushi	Mite (chigger; Leptotrombidium spp.[a])	Mites	Scrub typhus
Ehrlichia chaffeensis	Hard tick: Lone Star tick (Amblyomma americanum)	Ticks	Ehrlichiosis (human monocytic ehrlichiosis, HME)
Anaplasma phagocytophilum	Hard tick: Ixodes spp.	Ticks	Anaplasmosis (formerly human granulocytic ehrlichiosis, HGE)

[a] spp. is an abbreviation used to indicate more than one species of a genus.

involves avoiding tick-infested areas, promptly removing ticks, wearing tight-fitting clothing, and using repellent chemicals. Vaccines against the two species are not available.

TABLE 21.1 summarizes the main rickettsial parasites of humans and their vectors, reservoirs, and diseases.

MICRO CHECK

1. Which bacterium causes Rocky Mountain spotted fever?
2. Why can't scrub typhus be transmitted directly from one person to another by an arthropod vector?
3. Which developmental stage of *Ehrlichia* and of *Anaplasma* (which cause ehrlichiosis and anaplasmosis respectively) is infective? (Which stage enters the body?)

TELL ME WHY

Why do most cases of rickettsial diseases occur from May to August? Why is epidemic typhus an exception?

Chlamydias

LEARNING OUTCOMES

21.8 Describe the life cycle of chlamydias, including the two developmental forms.
21.9 Discuss the term *energy parasite* as it relates to chlamydias.

Chlamydias are a group of bacteria that vie with rickettsias for the title "smallest bacterium." Like rickettsias, chlamydias are nonmotile and grow and multiply only within vesicles in host cells. Scientists once considered chlamydias to be viruses because of their small size, obligate intracellular lifestyle, and ability to pass through 0.45-μm pores in filters, which were thought to trap all cells. However, chlamydias are cellular and possess DNA, RNA, and functional 70S ribosomes. Each chlamydial cell is surrounded by two membranes, similar to a typical Gram-negative bacterium but without peptidoglycan between the membranes—chlamydias lack cell walls. In contrast to rickettsias, chlamydias do not have arthropods as vectors or hosts. Because of their unusual features and unique rRNA nucleotide sequences, taxonomists now classify chlamydias in their own phylum: Chlamydiae.

Chlamydias have a unique developmental cycle involving two forms, both of which can occur within endocytic vesicles of a host cell (**FIGURE 21.3**): tiny (0.2–0.4 μm) cocci called **elementary bodies (EBs)** and larger (0.6–1.5 μm) pleomorphic **reticulate bodies (RBs)**. Elementary bodies are relatively dormant, are resistant to environmental extremes, can survive outside cells, and are the infective forms. Reticulate bodies are noninfective, obligate intracellular forms that replicate via binary fission within phagosomes, where they survive by inhibiting the fusion of a lysosome with the phagosome. Chlamydias lack the metabolic enzymes needed to synthesize ATP, so they must depend on their host cells for the high-energy phosphate compounds they require; thus, chlamydias have been called *energy parasites*.

In the life cycle of chlamydias (**FIGURE 21.3a**), once an EB attaches to a host cell via adhesion factors complementary to receptors on the host cell ❶, it enters by triggering its own endocytosis ❷. Once inside the endosome, the EB converts into an RB ❸, which then divides rapidly into multiple RBs ❹. Once an infected vesicle becomes filled with RBs, it is called an **inclusion body**. About 21 hours after infection, RBs within an inclusion body begin converting back to EBs ❺, and about 19 hours after that, the EBs are released from the host cell via exocytosis ❻, becoming available to infect new cells and completing the life cycle.

Three chlamydias cause disease in humans. In order of the prevalence with which they infect humans, they are *Chlamydia trachomatis*, *Chlamydophila pneumoniae*, and *Chlamydophila psittaci*. We begin our discussion with the most common species—*C. trachomatis*.

Chlamydia trachomatis

LEARNING OUTCOMES

21.10 List the types of cells in the human body that are most often infected by *Chlamydia*.
21.11 Explain how sexually inactive children may become infected with *C. trachomatis*.
21.12 Describe the cause and symptoms of lymphogranuloma venereum.
21.13 Discuss the prevention of chlamydial infections.

Chlamydia trachomatis (kla-mid′ē-a tra-kō′ma-tis) has a very limited host range—humans. (One exceptional strain, which may eventually be classified as a separate species, causes pneumonia in mice, but all other strains are pathogens of humans.)

Pathogenesis and Epidemiology

C. trachomatis enters a human body through abrasions and lacerations and infects a limited array of cells—those that have receptors for elementary bodies, including cells of the conjunctiva and cells lining the mucous membranes of the trachea, bronchi, urethra, uterus, uterine (Fallopian) tubes, anus, and rectum. The clinical manifestations of chlamydial infection result from the destruction of infected cells at the site of infection and from the inflammatory response this destruction stimulates. Reinfection in the same site by the same or a similar strain triggers a vigorous hypersensitive immune response that can result in blindness, sterility, or sexual dysfunction.

Infection with *C. trachomatis* is the most commonly reported sexually transmitted disease in the United States; 1,571,492 cases were reported in 2017, but epidemiologists estimate that about 3.5 million asymptomatic cases go unreported annually. Sexually transmitted chlamydial infections are prevalent among women under the age of 20 because they are physiologically more susceptible to chlamydial infection.

The pathogen can be transmitted from eye to eye via droplets, hands, contaminated fomites, or flies. Infected children also harbor the bacterium in their digestive and respiratory tracts, so the bacterium can be transmitted to other children or adults via fecal contamination or respiratory droplets. Infants may also be infected during birth as they pass through an infected birth canal. Chlamydial eye infections are endemic in crowded, poor communities where people have inadequate personal hygiene, inadequate sanitation, or inferior medical care, particularly in the Middle East, North Africa, and India.

Diseases

The diseases caused by the various strains of *Chlamydia trachomatis* are of two main types: sexually transmitted diseases and an ocular disease called trachoma.

Sexually Transmitted Diseases So-called *LGV* strains of *C. trachomatis* cause **lymphogranuloma venereum**[4] (lim′fō-gran-ū-lō′mă ve-ne′rē-ŭm). The initial lesion of lymphogranuloma venereum occurs at the site of infection on the penis, urethra, scrotum, vulva, vagina, cervix, or rectum. This lesion is often overlooked because it is small and painless and heals rapidly. Headache, muscle pain, and fever may also occur at this stage of the disease.

The second stage of the disease involves the development of buboes (swollen lymph nodes) associated with lymphatic vessels draining the site of infection (**FIGURE 21.4**). The buboes, which are accompanied by fever, chills, anorexia (any medical condition involving loss of appetite for food), and muscle pain, may enlarge to the point that they rupture, producing draining sores.

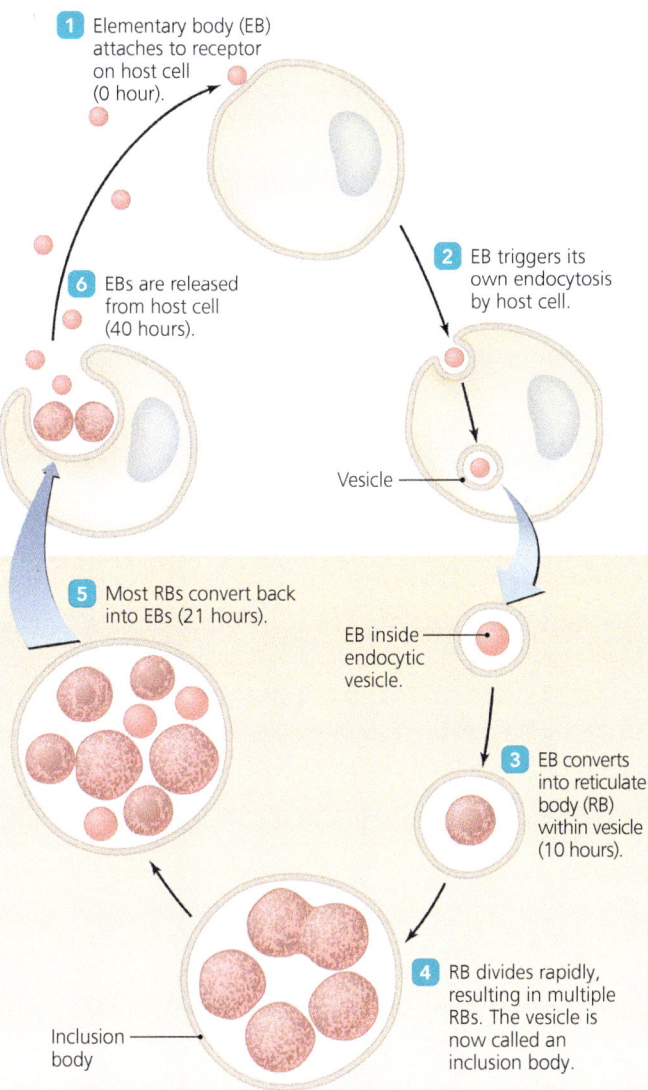

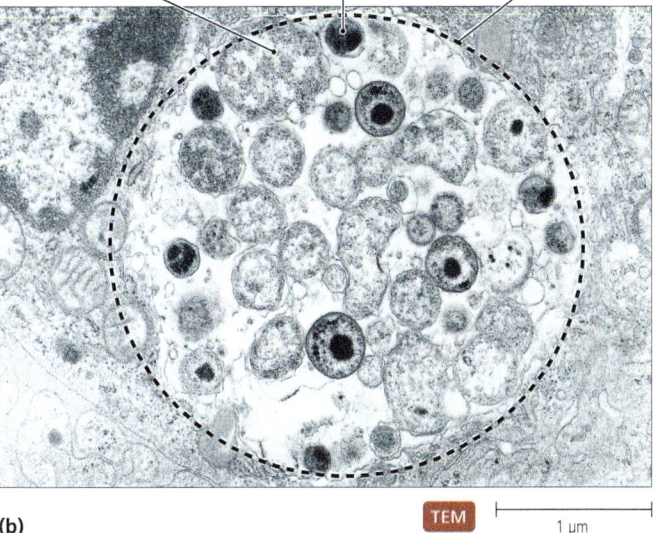

▲ **FIGURE 21.3 Development of chlamydias. (a)** The life cycle of chlamydias. Times in parentheses refer to hours since infection. **(b)** An inclusion body containing elementary bodies and reticulate bodies, here of *Chlamydia* inside a human lung cell.

---

[4]From Latin *lympha*, meaning "clear water"; Latin *granulum*, meaning "a small grain"; Greek *oma*, meaning "tumor" (swelling); and Latin *Venus*, the goddess of sexual love.

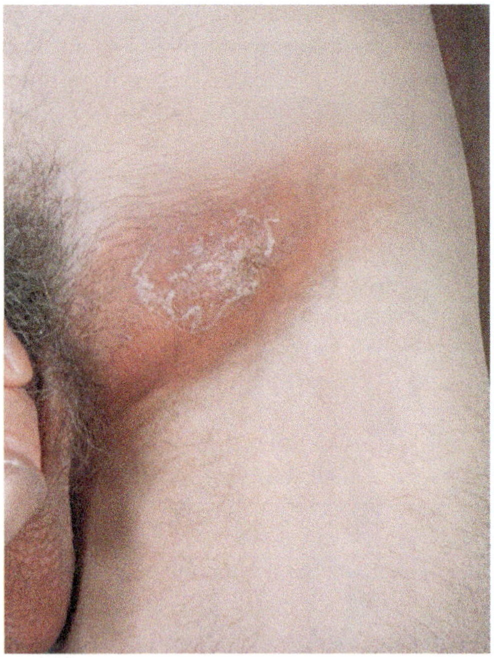

through contact with contaminated bedding, hands, clothing, or eye-seeking flies. However, LGV strains of *C. trachomatis* may produce blindness by a similar process in adults when bacteria from the genitalia are introduced into the eyes via fomites or fingers.

Diagnosis, Treatment, and Prevention

Diagnosis of chlamydial infection involves demonstration of bacteria inside cells from the site of infection. Giemsa-stained specimens may reveal bacteria or inclusion bodies within cells, but the most specific method of diagnosis involves amplifying the number of chlamydia by inoculating the specimen into a culture of susceptible cells. Laboratory technicians then demonstrate the presence of *Chlamydia* in the cell culture by means of specific fluorescent antibodies (FIGURE 21.6) or nucleic acid probes.

Physicians prescribe doxycycline or erythromycin for 21 days to eliminate genital infections of LGV strains of *C. trachomatis* in most adults and their sexual partners.

▲ FIGURE 21.4 **An advanced case of lymphogranuloma venereum in a man.** The enlarged lymph node (bubo) is a common sign of this disease. *Which microorganism causes lymphogranuloma venereum?*

Figure 21.4 Chlamydia trachomatis.

Although about 85% of genital tract infections with *C. trachomatis* in women are asymptomatic, more than 75% of infections in men have symptoms. Infected men also develop urethral inflammation (*urethritis*), which cannot be distinguished from gonorrhea based on symptoms alone. Chlamydial urethritis accounts for about 50% of cases of *nongonococcal urethritis*.

Proctitis[5] may occur in men and women as a result of lymphatic spread of the bacterium from the vagina, vulva, cervix, or urethra to the rectum. About 15% of the cases of proctitis in homosexual men result from the spread of *C. trachomatis* via anal intercourse.

An immune response against reinfections of *C. trachomatis* in women can have serious consequences, causing pelvic inflammatory disease (PID). PID involves chronic pelvic pain; irreversible damage to the uterine tubes, uterus, and ovaries; and sterility.

Trachoma So-called *trachoma* strains of *C. trachomatis* cause a disease of the eye called **trachoma** (tră-kō′mă; FIGURE 21.5), which is the leading cause of nontraumatic blindness in humans. The pathogen multiplies in cells of the conjunctiva and kills them, triggering a copious, purulent[6] (pus-filled) discharge that causes the conjunctiva to become scarred. Such scarring in turn causes the patient's eyelids to turn inward such that the eyelashes abrade, irritate, and scar the cornea, triggering an invasion of blood vessels into this normally clear surface of the eye. A scarred cornea filled with blood vessels is no longer transparent, eventually resulting in blindness.

Trachoma is typically a disease of children, who can be infected during birth, or of people of any age who acquire the pathogen

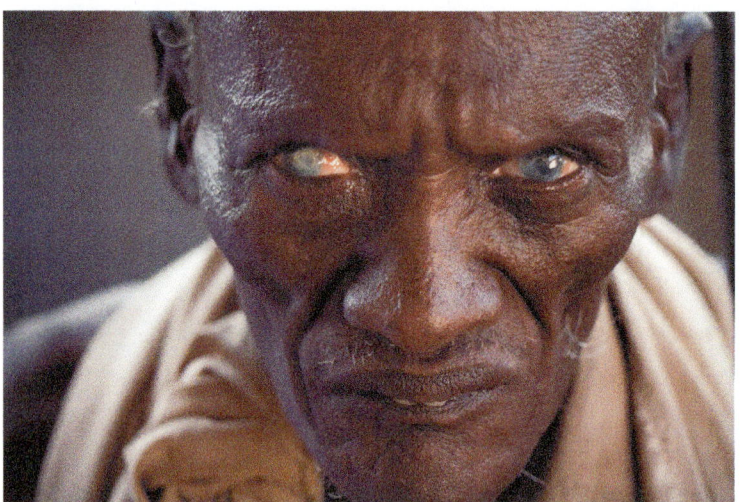

▲ FIGURE 21.5 **A man suffering from the blinding effects of trachoma.**

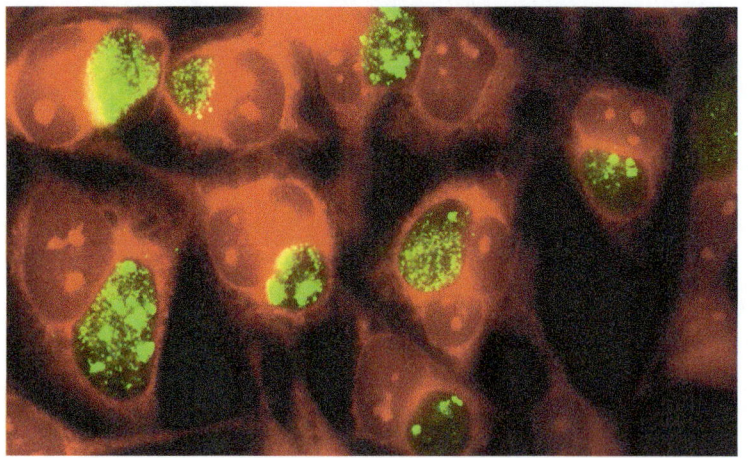

▲ FIGURE 21.6 **A direct fluorescent immunoassay for *Chlamydia trachomatis*.** The bright green color reveals the presence of *Chlamydia* within cells; this result is diagnostic for a *C. trachomatis* infection. Red indicates human cells.

---

[5]From Greek *proktos*, meaning "rectum," and *itis*, meaning "inflammation."
[6]From Latin *pur*, meaning "pus."

Erythromycin is recommended for treatment of pregnant women. Pediatricians treat trachoma in newborns with azithromycin cream for 10 to 14 days. Surgical correction of eyelid deformities may prevent the abrasion, scarring, and blindness that typically result from ocular infections.

A vaccine against *C. trachomatis* has showed promise in mice. Until scientists develop a vaccine for humans, prevention of sexually transmitted chlamydial infections is best achieved by abstinence or faithful mutual monogamy. Condoms provide some protection, though researchers warn that condoms always used correctly and consistently still fail about 10% of the time to prevent transmission of *Chlamydia*. Blindness from trachoma can be prevented only by prompt treatment with antibacterial agents and prevention of reinfection. Unfortunately, genital chlamydial infections are often asymptomatic and frequently occur among populations that have limited access to medical care.

Chlamydophila pneumoniae

LEARNING OUTCOME

21.14 Describe three diseases associated with *Chlamydophila pneumoniae*.

Chlamydophila pneumoniae (kla-mē-dof' ĭ-lă nū-mō'nē-ī; formerly called *Chlamydia pneumoniae*) causes about 10% of U.S. cases of community-acquired pneumonia and 5% of the cases of bronchitis and sinusitis. Most infections with *Chlamydophila pneumoniae* are mild, producing only malaise and a chronic cough, and do not require hospitalization. Some cases, however, are characterized by the development of a severe pneumonia that cannot be distinguished from walking pneumonia (atypical pneumonia), which is caused by *Mycoplasma pneumoniae* (mī'kō-plaz-mă nū-mō'nē-ī). Based on epidemiological data, some researchers implicated *Chlamydophila pneumoniae* as a cause of some cases of atherosclerosis—"hardening of the arteries"; however, analysis of all the studies to date do not show a definitive role for *C. pneumoniae* causing atherosclerosis.

Fluorescent antibodies demonstrate the intercellular presence of *C. pneumoniae*, which is diagnostic. Azithromycin or doxycycline for 14 days is used to treat infections of *C. pneumoniae*, but the drugs are not always effective, and infections may persist. Prevention of infection is difficult because the bacterium is ubiquitous and spreads via respiratory droplets.

Chlamydophila psittaci

LEARNING OUTCOME

21.15 Describe the diseases caused by *Chlamydophila psittaci*.

Chlamydophila psittaci (sit'ă-sē; formerly called *Chlamydia psittaci*) causes **ornithosis**[7] (ōr-ni-thō'sis), a disease of birds that can be transmitted to humans, in whom it typically causes flulike symptoms. The disease is sometimes called *parrot fever* because it was first identified in parrots. In some cases, severe pneumonia occurs. Rarely, nonrespiratory conditions such as endocarditis, hepatitis, arthritis, conjunctivitis, and encephalitis are observed.

Ornithosis is endemic to eastern Asia, Australia, and India. It occurs in the United States among immigrants who arrived from endemic areas. Fewer than 10 cases of ornithosis are reported in the United States annually, often in adults, though the disease is underreported because symptoms are often mild and diagnosis is difficult. Zookeepers, veterinarians, poultry farmers, pet shop workers, and owners of pet birds are at greatest risk of infection.

Elementary bodies of *Chlamydophila psittaci* may be inhaled in aerosolized bird feces or respiratory secretions or ingested from fingers or fomites that have contacted infected birds. Pet birds may transmit the disease to humans via beak-to-mouth contact. Because very brief exposure to birds can be sufficient for the transmission of elementary bodies, some patients cannot recall having any contact with birds. Symptoms usually occur within 10 days of exposure to the chlamydias. Without treatment, the mortality rate of ornithosis is about 20%, but with treatment death is rare.

Because the symptoms of ornithosis are the same as those of many respiratory infections, diagnosis requires demonstration of *C. psittaci* by serological testing. Tetracycline or azithromycin for one week is the preferred treatment. Prevention involves wearing protective clothing when handling infected birds, 30-day quarantine of and tetracycline treatment for all imported birds, and preventive husbandry, in which bird cages are regularly cleaned and are positioned so as to prevent the transfer of feces, feathers, food, and other materials from cage to cage. No vaccine for either birds or humans is available.

TABLE 21.2 on p. 630 compares and contrasts the features of the smallest microbes—rickettsias, chlamydias, mycoplasmas, and viruses.

Next, we turn our attention to the spirochetes.

MICRO CHECK

4. What are the only types of cells that can be infected by *Chlamydia trachomatis*?
5. Which disease is caused by an inflammatory response to a reinfection by *Chlamydia trachomatis* and can result in infertility?

TELL ME WHY

Why are penicillins and cephalosporins useless against *Chlamydia*?

Spirochetes

LEARNING OUTCOME

21.16 Describe the morphology and locomotion of spirochetes.

Spirochetes (spī'rō-kētz), which means *coiled hairs* in Greek, are thin (0.1 – 0.5 μm in diameter), tightly coiled, helically shaped, Gram-negative bacteria that share certain unique features—most notably axial filaments. Axial filaments are composed of endoflagella located in the periplasmic space between

---
[7]From Greek *ornith*, meaning "bird."

TABLE 21.2 Characteristics of the Smallest Microbes

Feature	Rickettsias	Chlamydias	Mycoplasmas[a]	Viruses[b]
Cellular Structure	Small cells with little peptidoglycan in cell walls	Small, wall-less cells with two membranes	Small, wall-less, pleomorphic cells with sterol in cytoplasmic membranes	Acellular
Diameter	0.3 μm	EBs: 0.2 – 0.4 μm RBs: 0.6 – 1.5 μm	0.1 – 0.8 μm	0.01 – 0.3 μm
Lifestyle	Obligate intracellular parasites within cytosol	Obligate intracellular parasites within vesicles	Free-living	Obligate intracellular parasites within cytosol or nuclei
Replication	Binary fission	Binary fission of reticulate bodies	Binary fission	Chemical assembly
Nucleic Acid(s)	Both DNA and RNA	Both DNA and RNA	Both DNA and RNA	DNA or RNA
Functional Ribosomes	Present	Present	Present	Absent
Metabolism	Present	Present	Limited—lack citric acid cycle, electron transport chains, and enzymes for cell wall synthesis	Completely dependent on metabolic enzymes of host cell
ATP-Generating System	Present	Absent	Present	Absent
Phylum	Proteobacteria	Chlamydiae	Firmicutes	None officially recognized

[a]Discussed in Chapter 11.
[b]Discussed in Chapter 13.

the cytoplasmic membrane and the outer (wall) membrane (see Figure 3.8b). As its axial filament rotates, a spirochete corkscrews through its environment—a type of locomotion thought to enable pathogenic spirochetes to burrow through their hosts' tissues. Mutants lacking endoflagella are rod shaped rather than helical, indicating that the axial filaments play a role in maintaining cell shape.

Taxonomists place spirochetes in their own phylum—Spirochaetes. Three genera, *Treponema*, *Borrelia*, and *Leptospira*, cause diseases in humans; we begin by discussing *Treponema*.

Treponema

LEARNING OUTCOMES

21.17 Describe the venereal disease caused by *Treponema pallidum*.
21.18 Describe the four phases of untreated syphilis and the treatment for each.
21.19 Describe the diseases caused by nonvenereal strains of *Treponema*.

Treponema (trep-ō-nē′mă) is a pathogen of humans only. Four types cause diseases, the most widespread of which is *Treponema pallidum pallidum* (i.e., subspecies, strain, or serotype *pallidum*). We will first examine this sexually transmitted pathogen and the venereal disease it causes, and then we will briefly consider the nonvenereal diseases caused by the other three strains—*T. pallidum endemicum*, *T. pallidum pertenue*, and *T. carateum*.

Treponema pallidum pallidum

Treponema pallidum pallidum (pal′li-dŭm), which is usually simply called *T. pallidum*, causes **syphilis**. Its flattened helical cells are so narrow (0.1 μm) that they are difficult to see by regular light microscopy in Gram-stained specimens. Therefore, scientists use phase-contrast or dark-field microscopy, or they intensify the contrast of the cells with dye linked to anti-treponemal antibodies or with special stains using silver atoms (**FIGURE 21.7**).

The bacterium lives naturally in humans only. It is destroyed by exposure to heat, disinfectants, soaps, drying, the concentration of oxygen in air, and pH changes, so it cannot survive in the environment. Scientists have not successfully cultured *T. pallidum* in cell-free media, though they have coaxed it to multiply in rabbits, monkeys, and rabbit epithelial cell cultures. In animals, it multiplies slowly (binary fission occurs once every 30 hours) and only for a few generations.

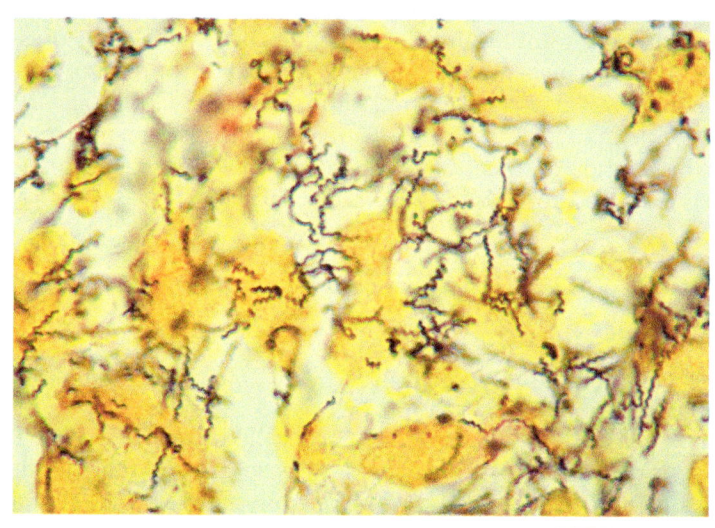

▲ **FIGURE 21.7 Spirochetes of *Treponema pallidum pallidum*.** Here, stained with a silver stain.

MICROBE AT A GLANCE 21.1

Treponema pallidum

Taxonomy: Domain Bacteria, phylum Spirochaetes, class "Spirochaetes," order Spirochaetales, family Spirochaetaceae

Cell morphology and arrangement: Flattened spirochete

Gram reaction: Negative

Virulence factors: Adhesion proteins, corkscrew motility, glycocalyx, hyaluronidase

Diseases caused: Syphilis: primary syphilis is lesion at site of infection, secondary syphilis is body rash, tertiary syphilis can have widespread and varied signs and symptoms including gummas; strain *endemicum* causes bejel; strain *pertenue* causes yaws; related species *T. carateum* causes pinta

Treatment for diseases: Penicillin for primary and secondary syphilis; tertiary syphilis not treatable; penicillin, tetracycline, or chloramphenicol to treat bejel, yaws, and pinta

Prevention of disease: Abstinence, mutual monogamy, condom usage for syphilis; for bejel, yaws, and pinta, avoid contact with diseased patients

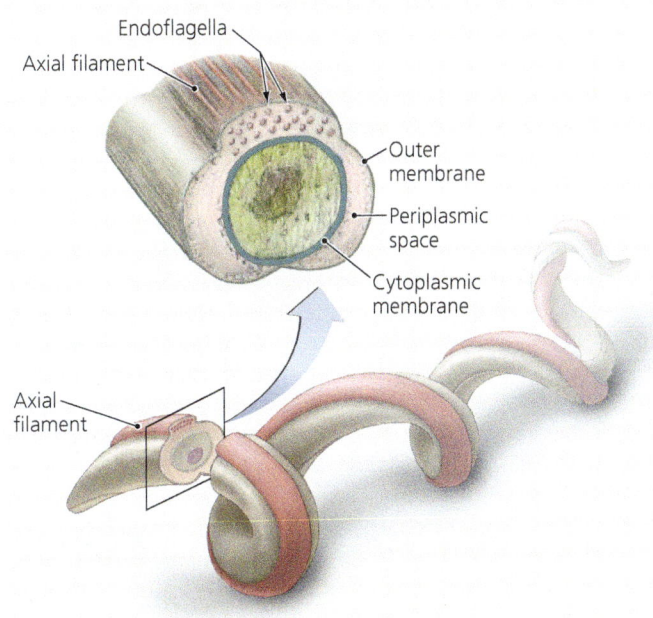

Pathogenicity and Epidemiology Scientists have had difficulty identifying the virulence factors of *T. pallidum* because the pathogen hasn't been cultured outside of cells. Researchers have used recombinant DNA techniques to insert genes from *Treponema* into *Escherichia coli* (esh-ĕ-rik′ē-ă kō′lī) and have then isolated the proteins the genes coded. Apparently, some of these proteins enable *Treponema* to adhere to human cells. Virulent strains also produce hyaluronidase, which may enable *Treponema* to infiltrate intercellular spaces. The bacterium has a glycocalyx, which may protect it from phagocytosis by leukocytes.

Syphilis occurs worldwide. Europeans first recognized the disease in 1495, leading some epidemiologists to hypothesize that syphilis was brought to Europe from the Western Hemisphere by returning Spanish explorers. Another hypothesis is that *T. pallidum* evolved from a less pathogenic strain of *Treponema* endemic to North Africa and that its spread throughout Europe was coincidental to Europeans' explorations of the Americas. In any case, the discovery and development of antimicrobial drugs over four centuries later has greatly reduced the number of syphilis cases (**FIGURE 21.8a**). The disease remains prevalent among sex workers, men who have sex with men, and users of illegal drugs. It occurs throughout the United States (**FIGURE 21.8b**).

Because of its fastidiousness and sensitivity, *T. pallidum* is an obligate parasite of humans and is transmitted almost solely via sexual contact, usually during the early stage of infection, when the spirochetes are most numerous. The risk of infection from a single, unprotected sexual contact with an infected partner is 10% to 30%. *T. pallidum* can rarely spread through blood

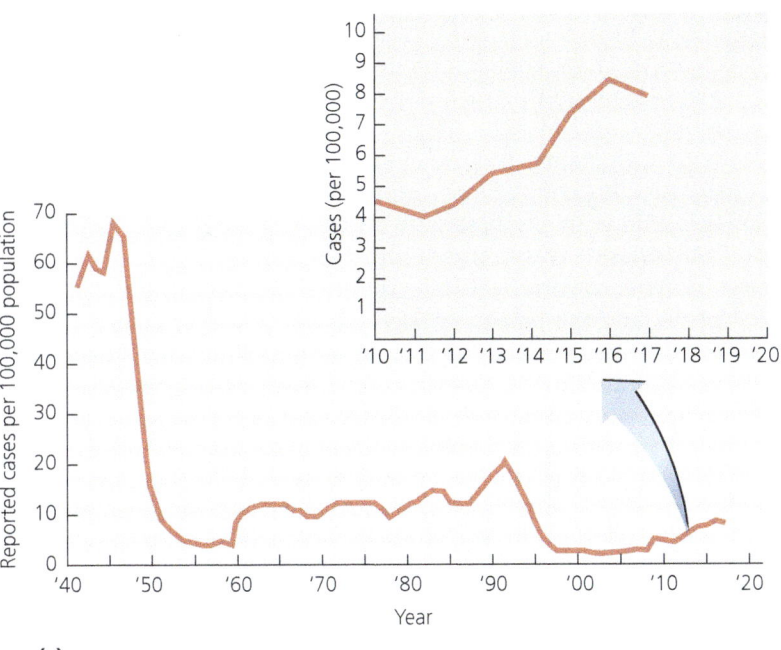

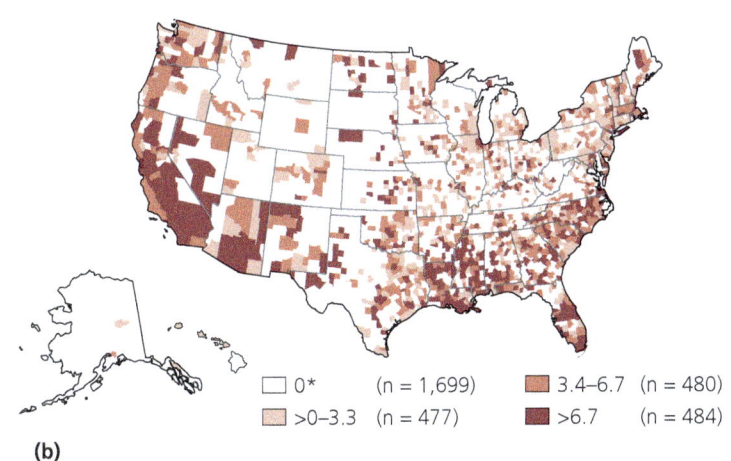

▲ **FIGURE 21.8 The incidence of adult syphilis in the United States.** (a) Nationwide incidence of syphilis, 1941–2014. (b) Reported incidence of adult syphilis by county per 100,000 population, 2016.

transfusion; it cannot spread by fomites such as toilet seats, eating utensils, or clothing.

Disease Untreated syphilis has four phases: primary, secondary, latent, and tertiary syphilis. In *primary syphilis*, a small, painless, reddened lesion called a **chancre** (shan'ker) forms at the site of infection 10 to 21 days following exposure (FIGURE 21.9a). Although chancres typically form on the external genitalia, about 20% form in the mouth, around the anus, or on the fingers, lips, or nipples. Chancres are often unobserved, especially in women, in whom these lesions frequently form on the cervix. The center of a chancre fills with serum that is extremely infectious because of the presence of millions of spirochetes. Chancres remain for three to six weeks and then disappear without scarring.

In about a third of cases, the disappearance of the chancre is the end of the disease. However, in most infections *Treponema* has invaded the bloodstream and spreads throughout the body to cause the symptoms and signs of *secondary syphilis:* sore throat, headache, mild fever, malaise, myalgia (muscle pain), lymphadenopathy (abnormal lymph nodes), and a widespread rash (FIGURE 21.9b) that can include the palms and the soles of the feet. Although this rash does not itch or hurt, it can persist for months, and like the primary chancre, rash lesions are filled with spirochetes and are extremely contagious. People, including health care workers, can become infected when fluid from the lesions enters breaks in the skin, though such nonsexual transmission of syphilis is rare.

After several weeks or months, the rash gradually disappears, and the patient enters a *latent (clinically inactive) phase* of the disease. The majority of cases do not advance beyond this point, especially in developed countries where physicians have treated the disease with antimicrobial drugs.

In untreated patients, latency may last 30 or more years, after which perhaps a third of the originally infected patients proceed to *tertiary syphilis*. This phase is associated not with the direct effects of *Treponema* but rather with severe complications resulting from inflammation and a hyperimmune response against the pathogen. Tertiary syphilis may affect virtually any tissue or organ and can cause dementia, blindness, paralysis, heart failure, and syphilitic lesions called **gummas** (gŭm'ăz), which are rubbery, painfully swollen lesions that can occur in bones, in nervous tissue, or on the skin (FIGURE 21.9c).

Congenital syphilis results when *Treponema* crosses the placenta from an infected mother to her fetus. Transmission to the fetus from a mother experiencing primary or secondary syphilis often results in the death of the fetus. If transmission occurs while the mother is in the latent phase of the disease, the result can be a latent infection in the fetus that causes intellectual disability and malformation of many fetal organs. After birth, newborns with latent infections usually exhibit a widespread rash at some time during their first two years of life.

Diagnosis, Treatment, and Prevention The diagnosis of primary, secondary, and congenital syphilis is relatively easy and rapid using specific antibody tests against antigens of *Treponema pallidum pallidum*. Spirochetes can be observed in fresh discharge from lesions but only when microscopic observations of clinical samples are made immediately—*Treponema* usually

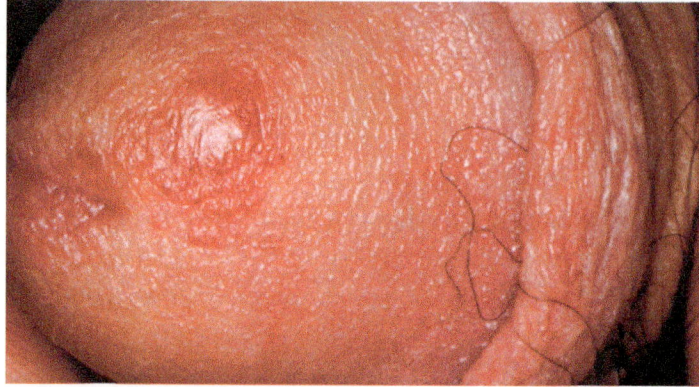
(a)

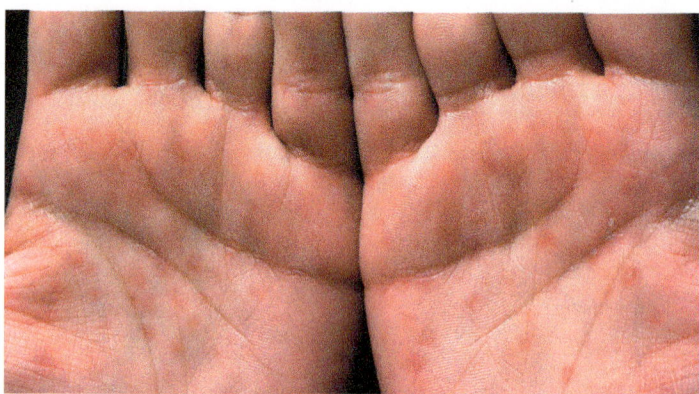
(b)

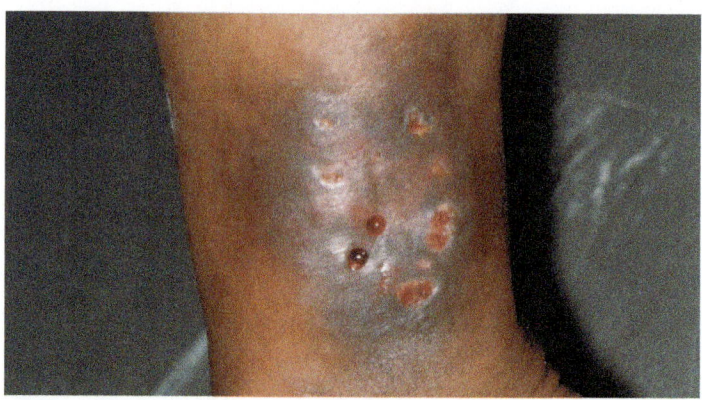
(c)

▲ FIGURE 21.9 The lesions of syphilis. (a) A chancre, a hardened and painless lesion of primary syphilis that forms at the site of the infection, here the head of a penis. (b) A widespread rash characteristic of secondary syphilis. (c) A gumma, a painful, rubbery lesion that often occurs on the skin or bones during tertiary syphilis.

does not survive transport to a laboratory. Nonpathogenic spirochetes, which are a normal part of the oral microbiota, can yield false-positive results, so clinical specimens from the mouth cannot be tested for syphilis. Tertiary syphilis is extremely difficult to diagnose because it mimics many other diseases, because few (if any) spirochetes are present, and because the signs and symptoms may occur years apart and seem unrelated to one another.

Penicillin is the drug of choice for treating primary, secondary, latent, and congenital syphilis, but it is not efficacious for tertiary syphilis because this phase is caused by a hyperimmune response, not an active infection. Physicians try to desensitize patients who have penicillin allergy and afterward treat with penicillin.

The Centers for Disease Control and Prevention has established a national goal for control of syphilis: maintaining an incidence of fewer than 0.2 cases of syphilis per 100,000 population in every county in the United States. This measure should reduce the incidence of syphilis to the point that there will not be enough cases to sustain the spread of *T. pallidum* and the disease will be eliminated in the United States. A vaccine against syphilis is not available, so abstinence, faithful mutual monogamy, and consistent and proper condom usage are the primary ways to avoid contracting syphilis. All sexual partners of syphilis patients must be treated with prophylactic penicillin to prevent spread of the disease.

Nonvenereal Treponemal Diseases

Other members of *Treponema* cause three nonsexually transmitted diseases in humans: bejel, yaws, and pinta. These diseases are seen primarily in impoverished children in Africa, Asia, and South America who have lived in unsanitary conditions. The spirochetes that cause these diseases look like *Treponema pallidum pallidum.*

T. pallidum endemicum (en-de′mi-kŭm) causes **bejel** (be′jel), a disease seen in children in Africa, Asia, and Australia. In bejel, the spirochetes are spread by contaminated eating utensils, so it is not surprising that the initial lesion is an oral lesion, which typically is so small that it is rarely observed. As the disease progresses, larger and more numerous secondary lesions form around the lips and inside the mouth. In the later stages of the disease, gummas form on the skin, bones, or nasopharyngeal mucous membranes.

T. pallidum pertenue[8] (per-ten′ū-ē) causes **yaws**, a disease of tropical South America, central Africa, and Southeast Asia that is characterized initially by oozing lesions that, although unsightly, are painless. Over time, the lesions can erode skin, cartilage, and bones (**FIGURE 21.10**). The disease is spread via contact with spirochetes in fluid draining from the lesions.

In 2012, scientists discovered that a single oral dose of azithromycin is effective in curing yaws. Many epidemiologists hope that mass treatment of endemic populations might lead to worldwide eradication.

T. carateum[9] (kar-a′tē-ŭm) causes **pinta** (pēn′tă), a skin disease seen in children in Central and South America. The spirochetes are spread among the children by skin-to-skin contact. After one to three weeks of incubation, hard, pus-filled lesions called papules form at the site of infection; the papules enlarge and persist for months or years, resulting in scarring and disfigurement.

Physicians diagnose bejel, yaws, and pinta by their distinctive appearance in children from endemic areas. Clinicians cannot detect spirochetes in specimens from the lesions of bejel, but they are present in patients with pinta and yaws.

Penicillin is the most favored antimicrobial agent for treating these three diseases. Prevention involves limiting the spread of the bacteria by preventing contact with the lesions.

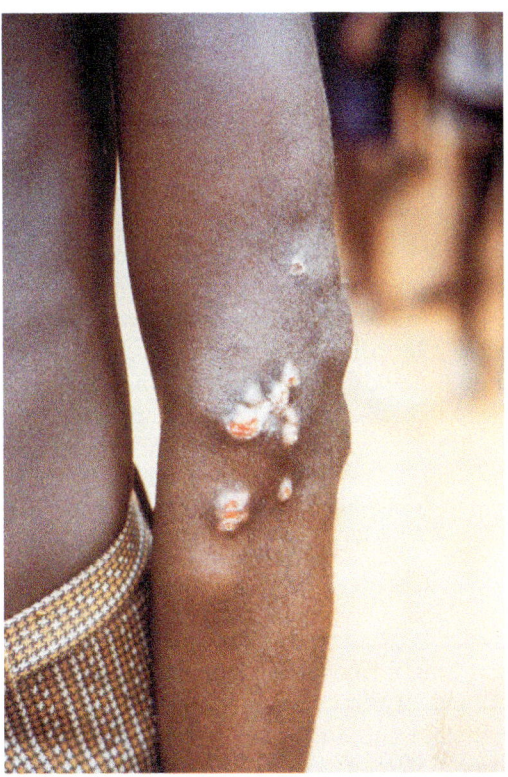

▲ FIGURE 21.10 **Yaws.** The draining lesions are characteristic of the later stages of the disease.

MICRO CHECK

6. Which strain of *Treponema* causes syphilis?
7. Which feature allows *Treponema pallidum pallidum* to enter host cells?
8. Why can't syphilis be contracted from fomites (inanimate objects)?
9. In syphilis, which stage of the disease involves chancres?
10. The drug of choice to treat syphilis is penicillin. What happens if a patient who has syphilis is allergic to penicillin?

Borrelia

LEARNING OUTCOMES

21.20 Describe Lyme disease, its vector, and its causative agent.
21.21 Discuss the life cycle of the *Ixodes* tick as it relates to Lyme disease.
21.22 Compare and contrast the two types of relapsing fever, including their causes and vectors.

Members of the genus *Borrelia* (bō-rē′lē-ă) are lightly staining, Gram-negative spirochetes that are larger than species of *Treponema*. Unusual for a Gram-negative bacterium, they lack a lipopolysaccharide outer layer. Instead they have a unique *Borrelia*

---

[8]Some taxonomists consider this organism to be a separate species: *T. pertenue*.
[9]Some taxonomists consider this organism to be a subspecies: *T. pallidum carateum*.

glycolipid with surface proteins that act in an unknown manner as virulence factors. Species of *Borrelia* cause two main diseases in humans: Lyme disease and relapsing fevers.

Lyme Disease

In 1975, epidemiologists noted that the incidence of childhood rheumatoid arthritis in Lyme, Connecticut, was over 100 times higher than expected. Upon investigation, they discovered that ticks transmitted the spirochete *Borrelia burgdorferi*[10] (burg-dōr′fer-ē; FIGURE 21.11) to human hosts to cause what became known as Lyme disease. **Lyme disease** is characterized by dermatological, cardiac, and neurological abnormalities in addition to the observed arthritis. Infected children may have paralysis of one side of their face (Bell's palsy). Even though its discovery was relatively recent, retrospective epidemiological studies have shown that Lyme disease was present in the United States for decades before its discovery.

B. burgdorferi is an unusual bacterium in that it lacks iron-containing enzymes and iron-containing proteins in its electron transport chains. By utilizing manganese rather than iron, the spirochete circumvents one of the body's natural defense mechanisms: the lack of free iron in human tissues and fluids.

Hard ticks of the genus *Ixodes*—*I. scapularis* (scap-ū-lar′is) in the northeastern and central United States, *I. pacificus* (pas-i′fi-kŭs) on the Pacific coast, *I. ricinus* (ri-ki′nŭs) in Europe, and *I. persulcatus* (per-sool-ka′tŭs) in eastern Europe and Asia—are the vectors of Lyme disease. An understanding of the life cycle of these ticks is essential to an understanding of Lyme disease.

An *Ixodes* tick lives for two years, during which it passes through three stages of development: a six-legged larva, an eight-legged nymph, and an eight-legged adult. During each stage it

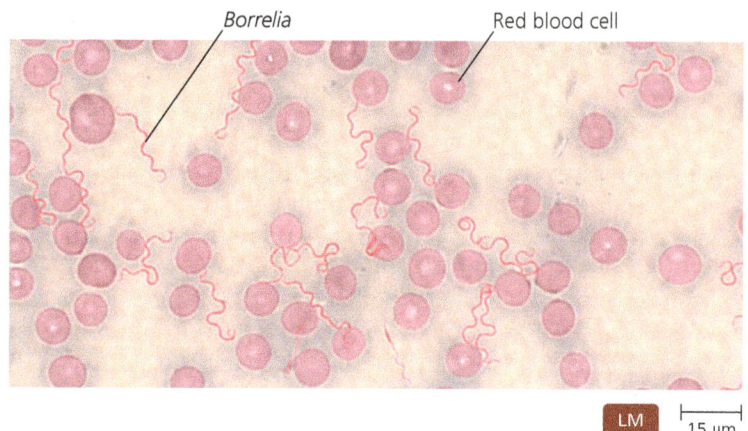

▲ FIGURE 21.11 *Borrelia burgdorferi*. This Gram-negative spiral-shaped spirochete lives in the blood and causes Lyme disease.

attaches to an animal host for a single blood meal usually lasting three to four days. After each of its three feedings, the tick drops off its host and lives in leaf litter or on brush.

The different stages of each species of *Ixodes* typically feed on different hosts. For example, the larvae and nymphs of *I. scapularis* tend to feed on deer mice and may feed on birds, whereas the adults most frequently feed on deer. In contrast, *I. pacificus* adults feed frequently on lizards, a host that does not support the growth of *Borrelia*. All stages of all species of *Ixodes* may feed on humans.

Transovarian transmission of *Borrelia* is rare, so ticks that hatch in the spring are uninfected (FIGURE 21.12 ①). Larvae become infected during their first blood meal ②. Over the

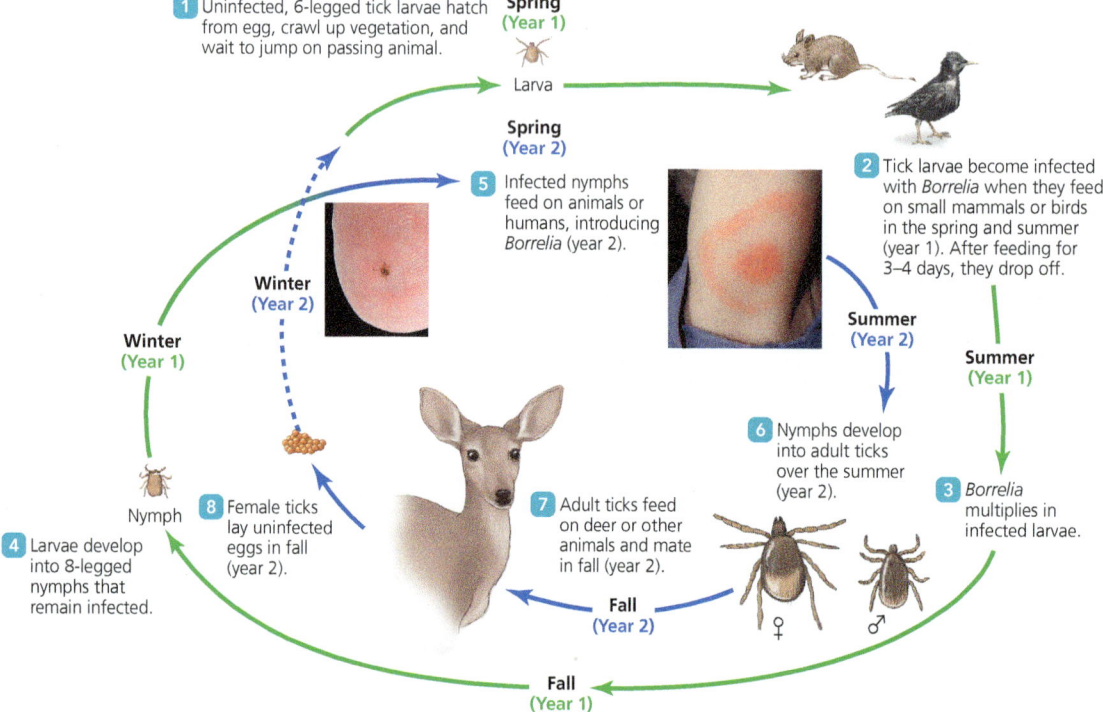

▲ FIGURE 21.12 **The life cycle of the deer tick *Ixodes* and its role as the vector of Lyme disease.** At what other stage besides the nymph stage can a tick infect a human with *B. burgdorferi*?

Figure 21.12 Adult ticks infected with B. burgdorferi can also infect humans during a blood meal.

---

[10]Named for the French bacteriologist Amedee Borrel and a specialist in tick-borne diseases, Willy Burgdorfer.

remainder of the year, larvae digest their blood meals while *Borrelia* replicates in the ticks' guts ③.

In the spring of their second year, the ticks molt into nymphs ④ and feed a second time, infecting the new hosts with *Borrelia* via saliva ⑤. Uninfected nymphs can become infected at this time if they feed on an infected host. Laboratory studies have shown that infected ticks must remain on a host for 48 hours in order to transmit enough spirochetes to establish a *Borrelia* infection in that host.

Nymphs drop off and undergo further development into adults ⑥. In the fall, adult ticks feed a final time ⑦, mate, lay eggs ⑧, and die. Adults infected with *Borrelia* infect their hosts as they feed. Adult ticks are much larger than nymphs, so humans usually see and remove adults before they can transmit *Borrelia*; thus, nymphs most often infect humans.

Lyme disease mimics many other diseases, and its range of signs and symptoms is vast. The disease typically has three phases in untreated patients:

1. An expanding red rash, which often resembles a bull's-eye, occurs at the site of infection within 30 days. About 75% of patients have such a rash, which lasts for several weeks. Other early signs and symptoms include malaise, headaches, dizziness, stiff neck, severe fatigue, fever, chills, muscle and joint pain, and lymphadenopathy.

2. Neurological symptoms (e.g., meningitis, encephalitis, and peripheral nerve neuropathy) and cardiac dysfunction typify the second phase, which is seen in only 10% of patients.

3. The final phase is characterized by severe arthritis that can last for years.

The pathological conditions of the latter phases of Lyme disease are due in large part to the body's immunological response; rarely is *Borrelia* seen in the involved tissue or isolated in cultures of specimens from these sites.

Two major events have contributed to the increase in reported cases of Lyme disease in the United States over the past decades (FIGURE 21.13a): The human population has encroached on woodland areas, and the deer population has been protected and even encouraged to feed in suburban yards. Thus, humans have been brought into closer association with deer ticks infected with *Borrelia*.

A diagnosis of Lyme disease, which is typically indicated by observations of its usual signs and symptoms, is rarely confirmed by detecting *Borrelia* in blood smears; instead, the diagnosis is confirmed through the use of serological tests.

Treatment with doxycycline (for adults) or penicillin (for children) for two weeks effectively cures most cases of Lyme disease in the first phase. Treatment of later phases is more difficult because later symptoms result primarily from immune responses rather than the presence of the spirochetes.

People hiking, picnicking, and working outdoors in areas where Lyme disease is prevalent (FIGURE 21.13b) should take precautions to reduce the chances of infection, particularly during spring and early summer, when nymphs are feeding. People who must be in the woods should wear long-sleeved shirts and long, tight-fitting pants and should tuck the cuffs of their pants into their socks to deny ticks access to skin. Repellents containing *DEET* (*N,N*-diethyl-*m*-toluamide), which is noxious to ticks,

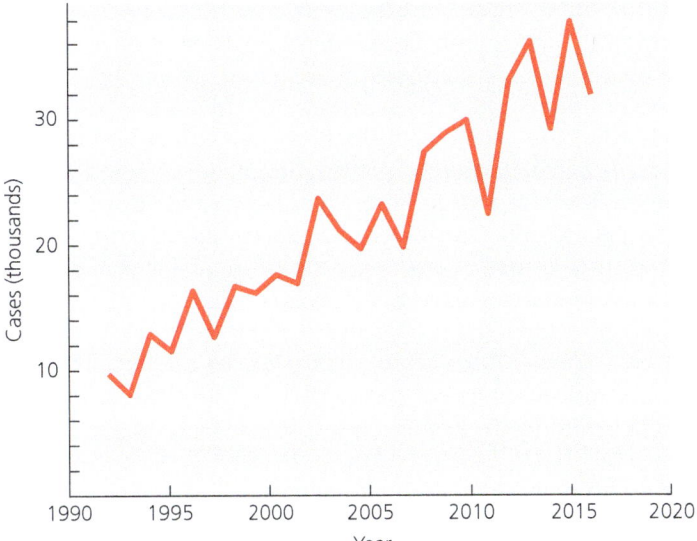

(a)

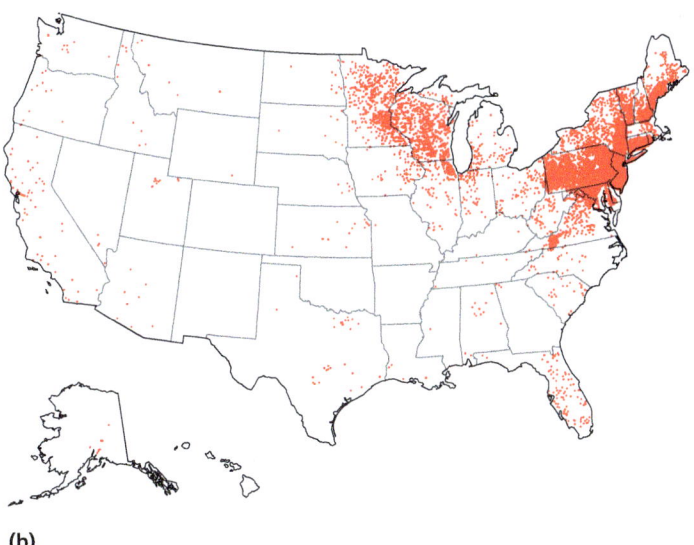

(b)

▲ FIGURE 21.13 **The occurrence of Lyme disease in the United States.** (a) Incidence of cases, 1992–2016. (b) The geographic distribution of the disease, 2016. Each dot represents a reported case of Lyme disease. The dots are placed randomly within the county where the case was reported, but do not indicate whether the disease was acquired in that county.

should be used. As soon as possible after leaving a tick-infected area, people should thoroughly examine their bodies for ticks or their bites. The manufacturer of a vaccine against one strain of *B. burgdorferi* withdrew the vaccine because some patients claimed it caused arthritis. Researchers began testing a new vaccine against six strains of the bacteria in 2018.

Relapsing Fever

Other species of spirochetes in the genus *Borrelia* can cause two types of relapsing fever. A disease called **louse-borne relapsing fever** results when *Borrelia recurrentis* (re-kur-ren′tis) is transmitted between humans by the human body louse *Pediculus humanus*. A disease known as **tick-borne relapsing fever** occurs when any of several species of *Borrelia* are transmitted between humans by soft ticks in the genus *Ornithodoros* (or-ni-thod'ō-rŭs).

Lice become infected with *B. recurrentis* when they feed on infected humans, the only reservoir for this spirochete. When infected lice are crushed, spirochetes can infect bite wounds. Because lice live only a few months, maintenance of louse-borne relapsing fever in a population requires crowded, unsanitary conditions such as occur in poor neighborhoods and during wars and natural disasters. Louse-borne relapsing fever is endemic in central and eastern Africa and in South America.

In contrast to hard ticks, soft ticks of the genus *Ornithodoros* have a brief (fewer than 30 minutes), painless, nocturnal bite that can transmit tick-borne relapsing fever. The ticks can live for years between feedings and pass spirochetes to their offspring via transovarian transmission. Soft ticks can also transmit *Borrelia* spp. (species) to animals, particularly mice and rats. Both ticks and rodents are reservoirs of infection. Tick-borne relapsing fever has a worldwide distribution; in the United States it is found primarily in the West.

Both types of relapsing fever are characterized by recurrent episodes of septicemia and fever separated by symptom-free intervals—a pattern that results from the body's repeated efforts to remove the spirochetes, which continually change their antigenic surface components (**FIGURE 21.14**). Within the first week of infection, the spirochetes induce a fever ❶, which is associated with chills, myalgia, and headache. In response to infection, the body produces antibodies that are directed against the spirochetes' surface molecules; once the antibodies have targeted the spirochetes, phagocytes clear most pathogens from the blood, and the body's temperature returns to normal ❷. However, some *Borrelia* evade the immune response by changing their antigenic surface molecules; when these spirochetes multiply, they induce a relapse of fever ❸. The body mounts a second immune response, this time against the new antigens, eventually again clearing the blood of most spirochetes and again returning body temperature to normal ❹. But a few spirochetes yet again change their antigenic components, leading to still another recurrence of fever, continuing the pattern indefinitely ❺.

Because of their relatively large size and abundance in the blood, observation of spirochetes in blood smears taken during periods of recurrent fever is the primary method of diagnosis.

Physicians successfully treat both types of relapsing fever with doxycycline or with erythromycin in pregnant women and children. Prevention involves avoidance of ticks and lice, good personal hygiene, and the use of arthropod-repelling chemicals. Effective rodent control is crucial to the control of relapsing fever. Vaccines are not available.

Leptospira

LEARNING OUTCOME

21.23 Describe *Leptospira interrogans* and zoonotic leptospirosis.

The last of the three groups of spirochetes that infect humans is *Leptospira interrogans* (lep'tō-spī'ră in-ter'ră-ganz). The specific epithet *interrogans* alludes to the fact that the end of the spirochete is hooked in a manner reminiscent of a question mark (**FIGURE 21.15**). This thin, pathogenic spirochete is an obligate aerobe that is highly motile by means of two axial filaments, each of which is anchored at one end. Clinicians and researchers grow it on special media enriched with bovine serum albumin or rabbit serum.

L. interrogans normally occurs in many wild and domestic animals—in particular, rats, raccoons, foxes, dogs, horses, cattle, and pigs—in which it grows asymptomatically in the kidney tubules. Humans contract the zoonotic disease **leptospirosis** (lep'tō-spī-rō'sis) through direct contact with the urine of infected animals or indirectly via contact with the spirochetes in contaminated streams, lakes, or moist soil, environments in which the organisms can remain viable for six weeks or more. Person-to-person spread has not been observed.

After *Leptospira* gains initial access to the body through invisible cuts and abrasions in the skin or mucous membranes, it corkscrews its way through these tissues. It then travels via the bloodstream throughout the body, including the central

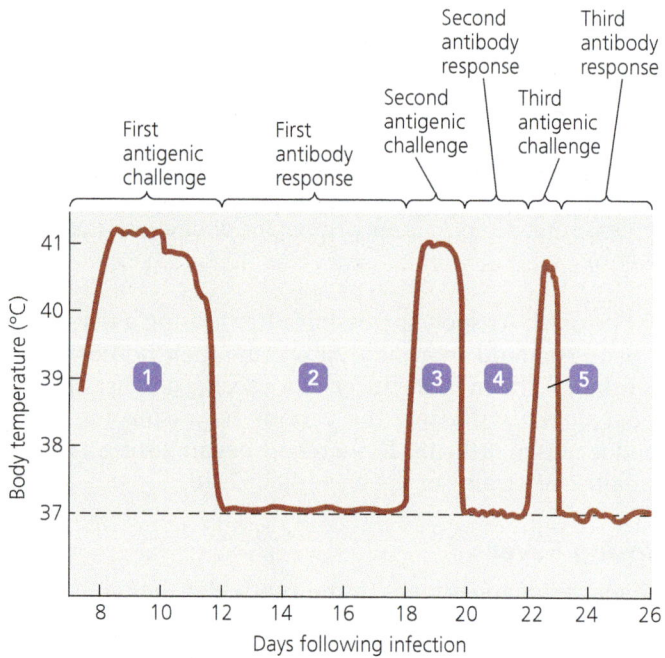

▲ **FIGURE 21.14** The time course of recurring episodes of fever in relapsing fevers.

▲ **FIGURE 21.15** *Leptospira interrogans*. The end of this spirochete is hooked like a question mark.

nervous system, damaging cells lining the small blood vessels and triggering fever and intense pain. Infection may lead to hemorrhaging and to liver and kidney dysfunction. Eventually, the bacteremia resolves, and the spirochetes are found only in the kidneys. As the disease progresses, spirochetes are excreted in urine. Leptospirosis is usually not fatal.

Leptospirosis occurs throughout the world. National reporting in the United States has ceased in part because the disease is rare; a total of only 89 cases were reported in the last two years (1993–1994) that leptospirosis was listed as a nationally reportable disease.

Because *Leptospira* is very thin and does not stain well with Gram, silver, or Giemsa stain, specific antibody tests revealing the presence of the spirochete in clinical specimens are the preferred method of diagnosis. Intravenous penicillin is used to treat infections. Rodent control is the most effective way to limit the spread of *Leptospira*, but eradication is impractical because of the spirochete's many animal reservoirs. An effective vaccine is available for livestock and pets.

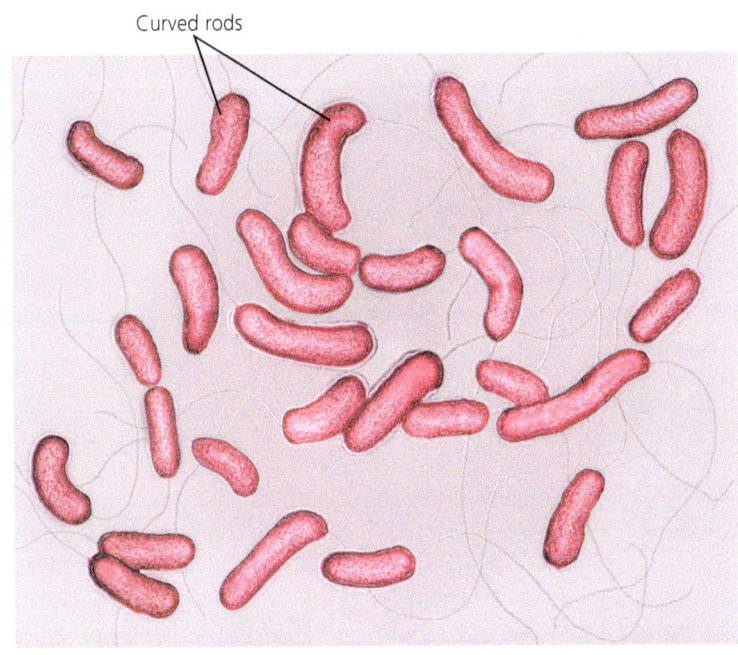

▲ FIGURE 21.16 *Vibrio cholerae.* This bacterium causes cholera.

MICRO CHECK

11. Why are tick nymphs more likely to transmit Lyme disease than adult ticks?
12. What is a first distinctive sign of Lyme disease?
13. How is leptospirosis transmitted?

TELL ME WHY

Why do health departments recommend that outdoor enthusiasts wear light-colored pants while hiking in the northeastern United States?

Pathogenic Gram-Negative Vibrios

To this point, we have discussed a number of unrelated Gram-negative bacteria: rickettsias, chlamydias, and the spirochetes. We will now turn our attention to a final group—the slightly curved bacteria called *vibrios*. The more important pathogenic vibrios of humans are in the genera *Vibrio, Campylobacter,* and *Helicobacter*.

Vibrio

LEARNING OUTCOMES

21.24 Contrast *Vibrio* with enteric bacteria in terms of their flagella and biochemical properties.
21.25 Describe the action of cholera toxin in causing cholera.
21.26 Name three species of *Vibrio* and describe the resulting diseases.

Vibrio (vib′rē-ō) is a genus of Gram-negative, slightly curved bacilli in class Gammaproteobacteria and phylum Proteobacteria (**FIGURE 21.16**). *Vibrio* shares many characteristics with enteric bacteria, such as *Escherichia* and *Salmonella,* including O polysaccharide antigens, which enable epidemiologists to distinguish among strains. *Vibrio* is oxidase positive and has a polar flagellum, unlike enteric bacteria, which are oxidase negative and have peritrichous flagella. The bacterium appears to vibrate when moving—giving rise to the genus name.

Vibrio lives naturally in estuarine and marine environments worldwide. It prefers warm, salty, and alkaline water, and most pathogenic species can multiply inside shellfish. Eleven of the 34 species of *Vibrio* cause diseases in humans; of these, only *V. cholerae* (kol′er-ī) can survive in freshwater, making it the most likely species to infect humans (via contaminated drinking water). In the following sections, we examine *V. cholerae* and the serious disease it causes.

Pathogenesis, Epidemiology, and Disease

Vibrio cholerae, particularly strain O1 El Tor, causes **cholera** (kol′er-ă; sometimes called **epidemic cholera**), one of the world's more pernicious diseases. The science of epidemiology began with Dr. John Snow's efforts to understand and limit the 1854 cholera epidemic in London (see Chapter 14).

Ships have introduced the pathogen into the harbors of every continent through their practice of taking on ballast water in one port and dumping it in another; two world wars only exacerbated the situation. In the seven major pandemics that have occurred since 1800, millions of people have been sickened and thousands have died, causing significant socioeconomic upheaval.

The seventh pandemic of O1 El Tor began in 1961 in Asia, spread to Africa and Europe in the 1970s and 1980s, and reached Peru, which had not seen cholera for a century, in January 1991. **FIGURE 21.17** illustrates the progression of this cholera pandemic throughout Latin America. Over 1 million people reported symptoms, and researchers documented over 2300 deaths during the 1990s. By 2002, the pandemic had subsided in South America,

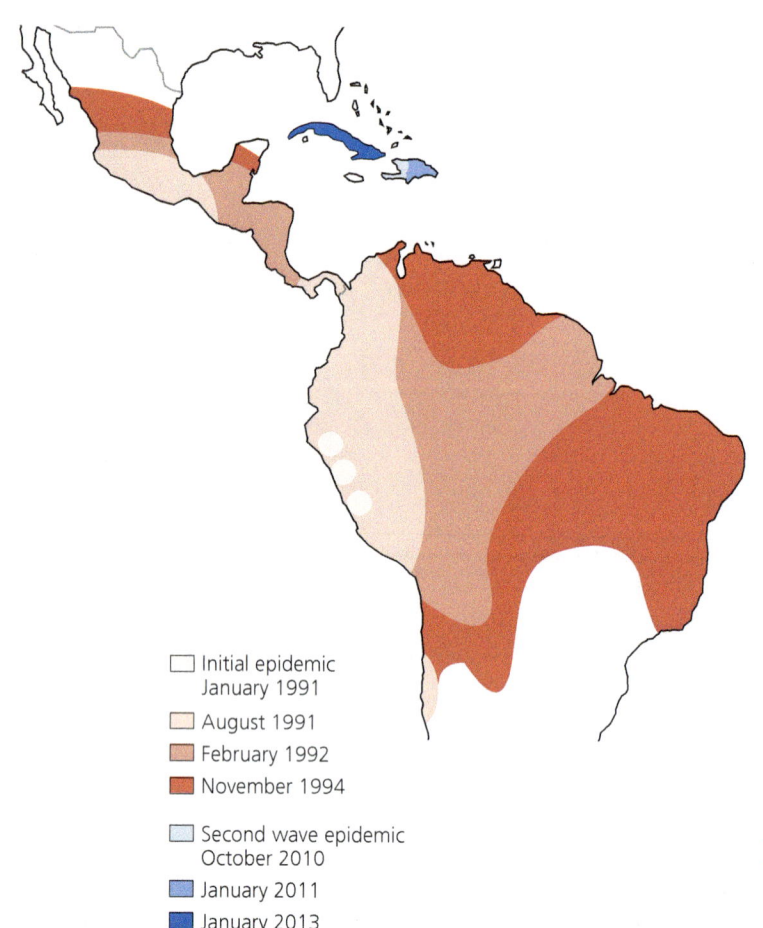

▲ FIGURE 21.17 **The spread of cholera in the Americas during the most recent pandemic.** Cholera caused by *Vibrio cholerae* O1 El Tor, which started in Indonesia, spread throughout Latin America in the first half of the 1990s, and was reintroduced into Haiti in 2010.

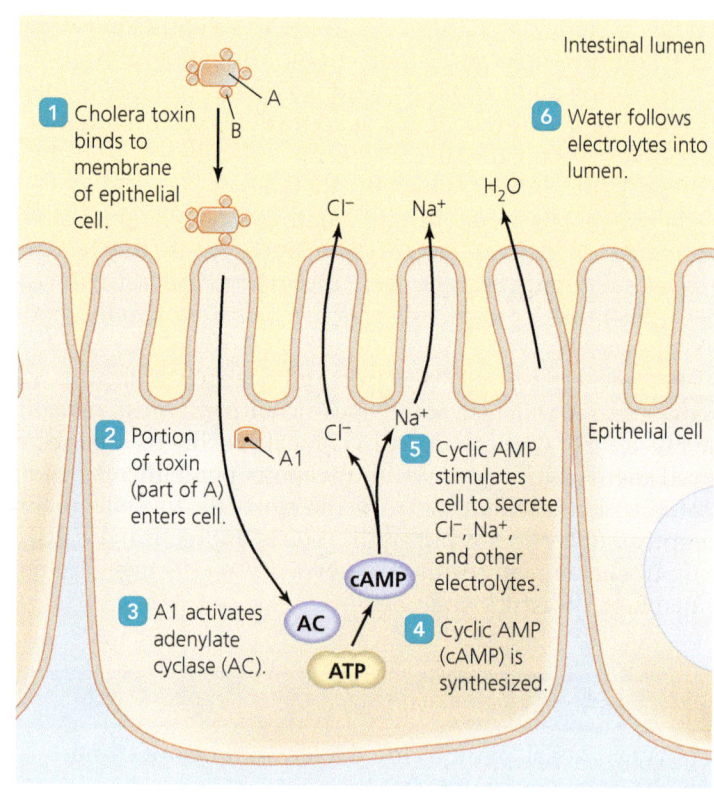

▲ FIGURE 21.18 The action of cholera toxin in intestinal epithelial cells.

with only 23 cases reported all year. O1 El Tor was reintroduced by aid workers from Asia in 2010 into Haiti. From there it spread to the Dominican Republic in 2011 and to Cuba in 2013.

A new pandemic strain, *V. cholerae* O139 Bengal, arose in India in 1992 and is spreading across Asia; this strain is the first non-O1 strain capable of causing epidemic disease. Strains of *V. cholerae* other than O1 El Tor and O139 do not produce epidemic cholera, only milder gastroenteritis.

Humans become infected with *Vibrio cholerae* by ingesting contaminated food and water. Cholera is most frequent in communities with poor sewage and water treatment. After entering the digestive system, *Vibrio* is confronted with the inhospitable acidic environment of the stomach. Most cells die, which is why a high inoculum—at least 10^8 cells—is typically required for the disease to develop. Infection can result from fewer cells in people who use antacids.

The warmer temperature of a human body compared to seawater activates *Vibrio* virulence genes. Thus, *Vibrio* shed in feces is more virulent than its counterparts in the environment. Scientists hypothesize that such gene activation explains the rapid, almost explosive, nature of cholera epidemics.

Although cholera infections may be asymptomatic or cause mild diarrhea, some result in rapid, severe, and fatal fluid and electrolyte loss. Symptoms usually begin two to three days following infection, with explosive watery diarrhea and vomiting. As the disease progresses, the colon is emptied, and the stool becomes increasingly watery, colorless, and odorless. The stool, which is typically flecked with mucus, is called *rice-water stool* because it resembles water poured off a pan of boiled rice. Some patients lose 1 liter of fluid an hour, resulting in a 50% loss of body weight over the course of infection.

The most important virulence factor of *V. cholerae* is a potent exotoxin called **cholera toxin**, which is composed of five identical B subunits and a single A subunit. FIGURE 21.18 illustrates the action of cholera toxin in producing the severe diarrhea that characterizes cholera. The process proceeds as follows:

1. One of the B subunits binds to a glycolipid receptor in the cytoplasmic membrane of an intestinal epithelial cell.
2. The A subunit is cleaved, and a portion (called A1) enters the cell's cytosol.
3. A1 acts as an enzyme that activates adenylate cyclase (AC).
4. Activated AC enzymatically converts ATP into cyclic AMP (cAMP).
5. cAMP stimulates the active secretion of excess amounts of electrolytes (sodium, chlorine, potassium, and bicarbonate ions) from the cell.
6. Water follows the movement of electrolytes from the cell and into the intestinal lumen via osmosis.

Severe fluid and electrolyte losses result in dehydration, metabolic acidosis (decreased pH of body fluids) due to loss of bicarbonate ions, hypokalemia,[11] and hypovolemic shock caused by reduced blood volume in the body. These conditions can produce muscle cramping, irregularities in heartbeat, kidney failure, and coma. Death may occur within hours of onset.

Cholera mortality is 60% in untreated patients; death can occur less than 48 hours after infection. In nonfatal cases, cholera ends spontaneously after a few days as the pathogens and toxins are flushed from the system by the severe diarrhea.

Diagnosis, Treatment, and Prevention

Rarely, an experienced observer may find the quick darting cells of *Vibrio* in the watery stool of a patient, but diagnosis is usually based on the characteristic diarrhea. *Vibrio* can be cultured on many laboratory media designed for stool cultures, but clinical specimens must be collected early in the disease (before the volume of stool dilutes the number of cells) and inoculated promptly because *Vibrio* is extremely sensitive to drying. Strains are distinguished by means of immunoassays specific for each antigenic variant.

Health care providers must promptly treat cholera patients with fluid and electrolyte replacement before hypovolemic shock ensues. Antimicrobial drugs can reduce the production of exotoxin and reduce the volume of diarrhea. A single dose of doxycycline is the recommended treatment for cholera in most adults and children. While doxycycline is typically not used in children, because of its adverse effects on developing teeth, the risk from a single dose is small.

Because *Vibrio cholerae* can grow in estuarine and marine water and because up to 75% of patients still asymptomatically harbor the pathogen, it is unlikely that cholera will be eradicated worldwide. Nevertheless, adequate sewage and water treatment can limit the spread of *Vibrio* and prevent epidemics.

The standard oral vaccine developed against the O1 strains of *V. cholerae* provides protection for at least two years. A new oral vaccine against both O1 and O139 provides longer-term protection in children. Antibiotic prophylaxis of those who travel to endemic areas has not proven effective. The infective dose for *V. cholerae* is high, so proper hygiene is the best protection.

Other Diseases of *Vibrio*

Vibrio parahaemolyticus (pa-ră-hē-mō-li'ti-kŭs) causes cholera-like gastroenteritis following ingestion of shellfish harvested from contaminated estuaries; fortunately, only rarely is it severe enough to be fatal. Typically, the disease is characterized by a self-limiting, explosive diarrhea accompanied by headache, nausea, vomiting, and cramping for 72 hours.

V. vulnificus (vul-nif'i-kŭs) is responsible for septicemia (blood poisoning) following consumption of contaminated shellfish and for infections resulting from the washing of wounds with contaminated seawater. Wound infections are characterized by swelling and reddening at the site of infection and are accompanied by fever and chills. Infections with *V. vulnificus* are fatal for 50% of untreated patients; prompt treatment with doxycycline is effective.

MICRO CHECK

14. How can *Vibrio* species be distinguished from enteric bacteria?
15. Which *Vibrio* sp. causes fatal diarrhea?

Campylobacter jejuni

LEARNING OUTCOMES

21.27 List several possible reservoirs of *Campylobacter jejuni*.
21.28 Describe gastroenteritis caused by *Campylobacter jejuni*.

Campylobacter jejuni[12] (kam'pi-lō-bak'ter jē-jū'nē) is likely the most common cause of bacterial gastroenteritis in the United States. Like *Vibrio*, it is Gram negative, slightly curved, oxidase positive, and motile by means of polar flagella; however, genetic analysis has shown that this comma-shaped pathogen is properly classified in the class Epsilonproteobacteria (and not in Gammaproteobacteria). *Campylobacter* also differs from *Vibrio* in being microaerophilic and capneic.

Campylobacter infections are zoonotic—many domesticated animals, including poultry, dogs, cats, rabbits, pigs, cattle, and minks, serve as reservoirs. Humans acquire the bacterium by consuming food, milk, or water contaminated with infected animal feces. The most common source of infection is contaminated poultry.

The pathogenicity of *C. jejuni* is not well understood, although the bacterium possesses adhesins, cytotoxins, and endotoxins that appear to enable colonization and invasion of the jejunum, ileum, and colon, producing bleeding lesions and triggering inflammation. Interestingly, nonmotile mutants are avirulent. *C. jejuni* infections commonly produce malaise, fever, abdominal pain, and bloody and frequent diarrhea—10 or more bowel movements per day are not uncommon. The disease is self-limiting; as bacteria are expelled from the intestinal tract, the symptoms abate, although a typical infection may last 7–10 days. Physicians may treat patients with azithromycin.

Scientists estimate that over 2.4 million cases of *Campylobacter* gastroenteritis occur each year in the United States—more than those caused by *Salmonella* and *Shigella* combined. However, *Campylobacter* gastroenteritis is not a reportable disease, so its actual incidence is unknown.

The number of infections can be reduced by proper food preparation, including thoroughly washing poultry carcasses, hands, and utensils that have contacted carcasses; cooking food sufficiently to kill bacteria; and pasteurizing cow's milk. Steps must also be taken to prevent contamination of the water supply with feces from stockyards, feedlots, and slaughterhouses.

---

[11]From Greek *hypo*, meaning "under"; Latin *kalium*, meaning "potassium"; and Greek *haima*, meaning "blood."

[12]From Greek *kampylos*, meaning "curved," and *jejunum*, the middle portion of the small intestine.

Helicobacter pylori

LEARNING OUTCOMES

21.29 Discuss the major change in medical opinion concerning the cause of peptic ulcers.

21.30 Describe the effect of *Helicobacter pylori* on the lining of the human stomach.

Helicobacter pylori (hel'ĭ-kō-bak'ter pī'lō-rē) is a slightly helical, highly motile bacterium that colonizes the stomachs of its hosts (**FIGURE 21.19**). Based on rRNA nucleotide sequences, *Helicobacter* is classified in the class Epsilonproteobacteria. It was originally placed in the genus *Campylobacter*; however, unlike *Campylobacter*, *Helicobacter* cannot reduce nitrate, has flagella, and is urease positive, a characteristic that is essential for colonizing the stomach.

Today, it is known that *H. pylori* causes *gastritis*[13] and most **peptic ulcers**, which are erosions of the mucous membrane of the stomach or of the initial portion of the small intestine (duodenum). Only a few decades ago, this hypothesis was very controversial. What was once the prevailing view—that stress, alcohol consumption, spicy food, or excess stomach acid production caused ulcers—was discredited beginning in 1982, when Australian gastroenterologists Robin Warren (1937–) and Barry Marshall (1951–) detected *Helicobacter* colonizing the majority of their patients' stomachs. Following treatment with antibiotics, the bacteria and the ulcers disappeared. Dr. Marshall provided final proof of *H. pylori*'s involvement by fulfilling Koch's postulates himself—he drank one of his cultures of

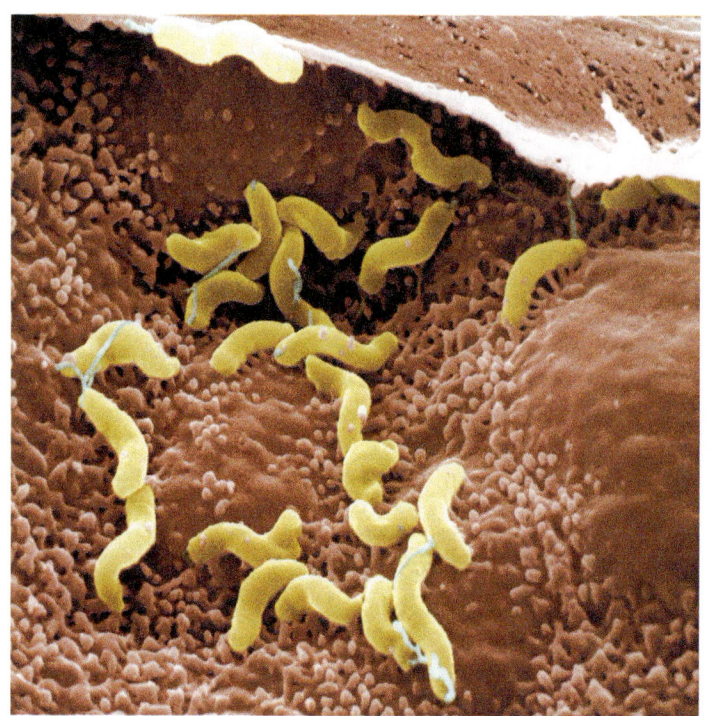

▲ **FIGURE 21.19** *Helicobacter pylori* **in a stomach.** The bacterium causes peptic ulcers.

[13]From Greek *gaster*, meaning "belly," and Greek *itis*, meaning "inflammation."

MICROBE AT A GLANCE 21.2

Helicobacter pylori

Taxonomy: Domain Bacteria, phylum Proteobacteria, class Epsilonproteobacteria, order Campylobacterales, family Campylobacteraceae

Cell morphology and arrangement: Slightly helical bacillus

Gram reaction: Negative

Virulence factors: Protein that inhibits acid production, urease, flagella, adhesins, antiphagocytic enzymes

Diseases caused: Peptic ulcer, gastritis

Treatment for diseases: Antibacterial drugs in conjunction with drugs that inhibit acid production in the stomach

Prevention of disease: Good personal hygiene, adequate sewage treatment, water purification, proper food handling

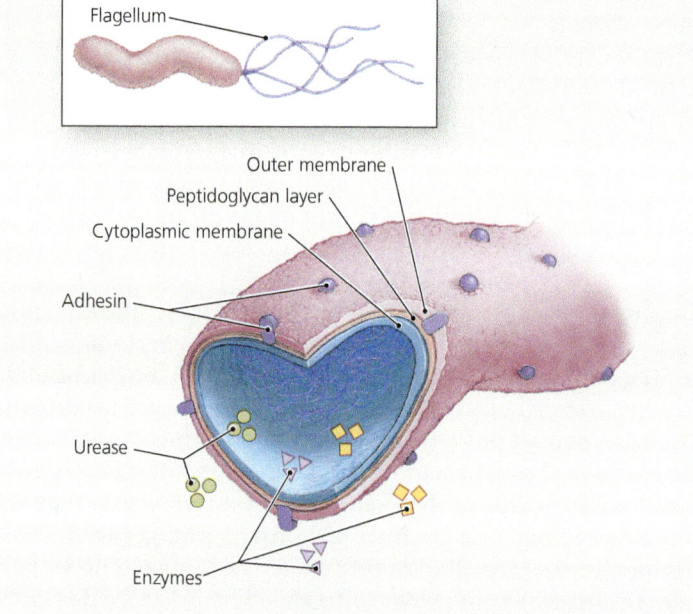

Helicobacter! As he expected, he developed painful gastritis and was able to isolate *H. pylori* from his diseased stomach. His sacrifice was worth it: Marshall and Warren received the 2005 Nobel Prize in Physiology or Medicine for their discovery. Since then, scientists have shown that long-term infection with *Helicobacter* is a significant risk factor for stomach cancer.

H. pylori possesses numerous virulence factors that enable it to colonize the human stomach: a protein that inhibits acid production by stomach cells, flagella that enable the pathogen to burrow through mucus lining the stomach, adhesins that facilitate binding to gastric cells, enzymes that inhibit phagocytic killing, and urease, an enzyme that degrades urea, which is present in gastric juice, to produce highly alkaline ammonia, which neutralizes stomach acid.

The portal of entry for *H. pylori* is the mouth. Studies have shown that *H. pylori* in feces on the hands, in well water, or on fomites may infect humans. In addition, cat feces may be a source of infection.

The formation of a peptic ulcer occurs as follows (**FIGURE 21.20**): The process begins when *H. pylori* (protected

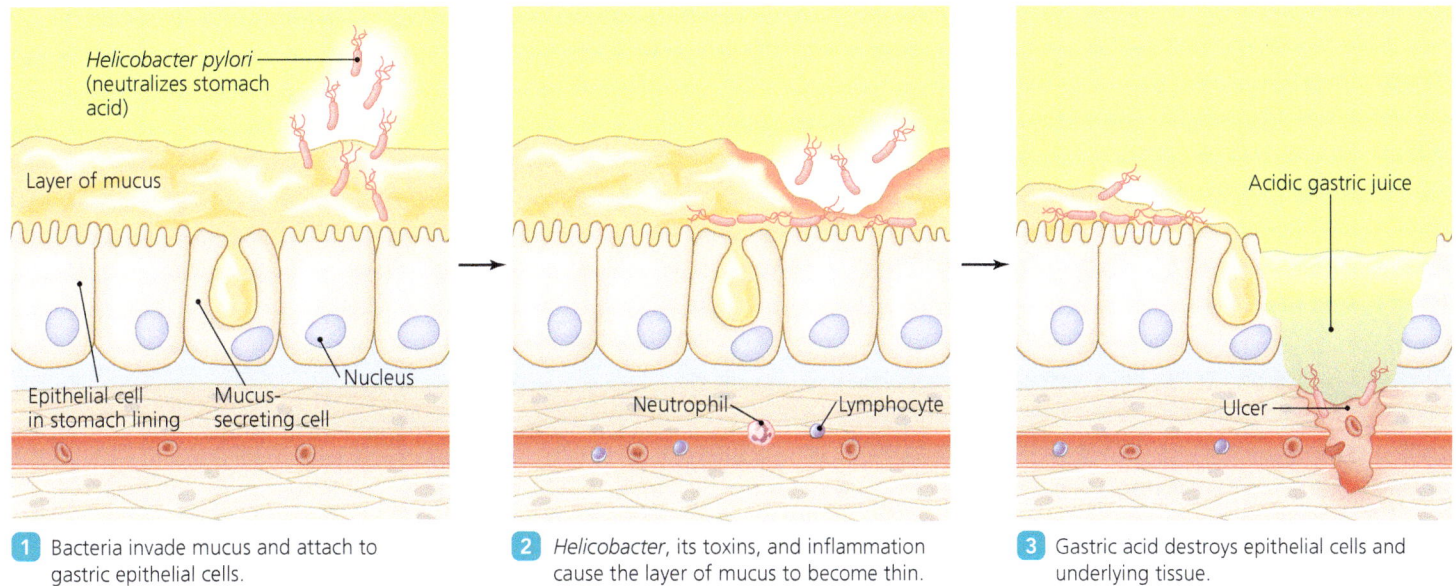

▲ FIGURE 21.20 The role of *Helicobacter pylori* in the formation of peptic ulcers.

by urease) burrows through the stomach's protective layer of mucus to reach the underlying epithelial cells ①, where the bacteria attach to the cells' cytoplasmic membranes and multiply. A variety of factors—the triggering of inflammation by bacterial exotoxin and perhaps the destruction of mucus-producing cells by the bacteria—causes the layer of mucus to become thin ②, allowing acidic gastric juice to digest the stomach lining. Once the epithelial layer has been ulcerated by gastric juice, *H. pylori* gains access to the underlying muscle tissue and blood vessels ③. Those bacteria that are phagocytized survive in part through the actions of catalase and superoxide dismutase, enzymes that neutralize part of the phagocytes' killing mechanism.

The presence of *H. pylori* in specimens from the stomach can be demonstrated by a positive urease test within one to two hours of culturing; the bacterium can also be seen in Gram-stained specimens. Definitive identification is based on a series of biochemical tests.

Physicians treat ulcers with three antimicrobial drug combinations in succession. First, metronidazole with tetracycline is given in combination with bismuth subsalicylate, which is a drug that inhibits diarrhea and acid production, allowing the stomach lining to regenerate. These drugs are followed with amoxicillin and clarithromycin in combination with other drugs that inhibit acid production by the stomach. Prevention of infection involves good personal hygiene, adequate sewage treatment, water purification, and proper food handling.

Interestingly, research indicates that treating patients with antimicrobial drugs to remove their *H. pylori* may leave the patients more susceptible to esophageal cancer, asthma, and allergies.

TELL ME WHY

Why is a cholera pandemic unlikely to become established in the United States?

CLINICAL CASE STUDY

The Case of the Lactovegetarians

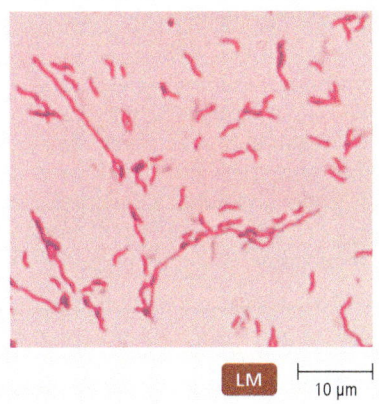

Two patients—a woman and her husband, ages 23 and 22, respectively—arrive at the health clinic one morning. They report having had severe abdominal cramps, grossly bloody diarrhea, nausea, and fever for 48 hours. Cultures of stool samples grown under microaerophilic, capneic conditions contain comma-shaped, Gram-negative bacilli. Both the patients are lactovegetarians and report being part of a "cow leasing" program at a local dairy in which patrons lease part of a cow's milk production so that they can drink natural, whole, raw milk. The couple devised the program so that they and several neighbors could circumvent state regulations prohibiting the sale of unpasteurized milk. Investigators obtained and cultured a milk sample from the dairy's bulk milk tank: the cultures contained the bacterium pictured.

1. What is the pathogen?
2. How did the couple become infected?
3. Are the couple's colleagues at work at risk of acquiring an infection from the couple?
4. What other foods that are common sources of this bacterium can be ruled out in this case?

MICRO IN THE CLINIC FOLLOW-UP

Too Much Spice?

The rest of Trevor's flight home is miserable—he spends most of it in the lavatory vomiting and has at least ten episodes of diarrhea. Since he's stuck on a plane, he figures the best he can do is drink as much water as possible until he gets home.

Trevor feels so terrible by the time he gets off the plane that he calls his doctor's office while he is waiting for his baggage at the airport. He goes straight to Dr. Getty's office. The initial examination reveals that Trevor shows signs of dehydration, low blood pressure (86/52 with his baseline at 115/64), and an elevated heart rate (112 beats per minute). Dr. Getty suspects that Trevor may have cholera based on his symptoms—vomiting and profuse diarrhea, which has now turned from stool-colored to a watery whitish-grey color with a distinctive fish odor. They will perform a rapid immunoassay with a dipstick that detects the presence of antigen O1 or O139 in stool samples. The standard treatment for cholera is aggressive intravascular or oral volume replacement of fluids (depending on the severity of dehydration or shock) and electrolyte replacement (i.e. sodium, potassium, and magnesium). Trevor starts to feel better within a day and completely recovers a couple of days later.

Analysis of Trevor's fecal samples indicates that he was infected with *Vibrio cholerae* O139. Dr. Getty believes he became infected through consumption of contaminated food (perhaps rice, vegetables, fruit, or ice cream) or water while in India. Public health officials are notified. Given that Trevor was sick on the plane and *V. cholerae* is spread via the fecal-oral route, there is a high possibility that other passengers may have been infected.

1. **Explain how cholera toxin leads to the profuse, watery diarrhea associated with *V. cholerae* infection.**

2. **The World Health Organization (WHO) recommends that antibiotics be given only to cholera patients with severe dehydration; in cases like Trevor's, oral rehydration therapy is sufficient. Can you think of reasons for limiting antibiotic use to the most severe cases?**

Check your answers to Micro in the Clinic Follow-Up questions in the Mastering Microbiology Study Area.

CHAPTER SUMMARY

> Go to Mastering Microbiology for Interactive Microbiology, Dr. Bauman's Video Tutors, Micro Matters, MicroFlix, MicroBoosters, practice quizzes, and more.

Rickettsias (pp. 621–626)

1. **Rickettsias** are extremely small, Gram-negative, obligate intracellular parasites.

2. *Rickettsia rickettsii* causes **spotted fever rickettsiosis (Rocky Mountain spotted fever, RMSF)**, a serious illness transmitted by ticks. Rash, petechiae, and death (occasionally even in treated cases) characterize the disease.

3. *Rickettsia prowazekii* causes louse-borne or **epidemic typhus**, characterized by fever, depression, and rash. The disease may recur as Brill-Zinsser disease.

4. *Rickettsia typhi* causes endemic typhus, also called **murine typhus** because the major reservoir is rodents. Various fleas act as vectors to transmit the bacterium to humans.

5. Mites (chiggers) transmit *Orientia tsutsugamushi* to cause **scrub typhus**, which is characterized by fever, headache, enlarged and tender lymph nodes, and a spreading rash.

6. *Ehrlichia chaffeensis* causes **ehrlichiosis** (human monocytic ehrlichiosis, HME), and *Anaplasma phagocytophilum* causes **anaplasmosis** (previously human granulocytic ehrlichiosis, HGE). These bacteria are transmitted to humans via the Lone Star tick, the deer tick, or the dog tick. The bacteria enter leukocytes as elementary bodies and go through two other developmental stages: an initial body and a morula.

7. Diagnosis of ehrlichiosis and anaplasmosis can be difficult because the signs and symptoms resemble those of many other diseases.

Chlamydias (pp. 626–629)

1. **Chlamydias** are small, nonmotile, obligate intracellular parasites that are energy parasites of their host cell; that is, they depend on their hosts for high-energy phosphate compounds for metabolism.

2. The developmental cycle of chlamydias includes infectious **elementary bodies (EBs)** and noninfectious **reticulate bodies (RBs)**. An endocytic vesicle full of RBs is called an **inclusion body**.

3. *Chlamydia trachomatis* enters the body through abrasions in mucous membranes of the genitalia, eyes, or respiratory tract. *Chlamydia* infection is the most-reported sexually transmitted infection in the United States. Children may be infected at birth as they pass through their infected mother's birth canal or by transmission from eye to eye.

4. *Chlamydia trachomatis* causes **lymphogranuloma venereum**. The majority of infections in women are asymptomatic, whereas the majority of infections in men result in buboes, fever, chills, anorexia, and muscle pain. Other symptoms and signs may include proctitis (in men and women) and pelvic inflammatory disease and sterility (in women). The disease can be prevented by abstinence or faithful mutual monogamy.

5. **Trachoma** is a serious eye disease that is caused by *C. trachomatis* and may result in blindness.

6. *Chlamydophila pneumoniae* causes bronchitis, pneumonia, and sinusitis.

7. *Chlamydophila psittaci* causes **ornithosis** (also called parrot fever), a respiratory disease of birds that can be transmitted to humans.

Spirochetes (pp. 629–637)

1. **Spirochetes** are helical bacteria with axial filaments that allow the organism to corkscrew, enabling it to burrow into a host's tissues.
2. **Syphilis** is caused by *Treponema pallidum pallidum*, a sexually transmitted obligate parasite of humans.
3. Primary syphilis is characterized by a **chancre**, a red lesion at the infection site. If *Treponema* moves from the chancre to the bloodstream, secondary syphilis results, causing rash, aches, and pains. After a period of latency, untreated cases may progress to tertiary syphilis, characterized by swollen **gummas**, dementia, blindness, paralysis, and heart failure.
4. **Congenital syphilis** results when an infected mother infects her fetus.
5. Penicillin is used to treat all except tertiary syphilis; no vaccine is available.
6. *Treponema pallidum endemicum* causes **bejel,** an oral disease observed in children in Africa, Asia, and Australia.
7. *Treponema pallidum pertenue* causes **yaws,** a skin disease observed in South America, central Africa, and Southeast Asia.
8. *Treponema carateum* (*T. pallidum carateum*) causes **pinta,** a disfiguring skin disease observed in children in Central and South America.
9. *Borrelia burgdorferi* causes **Lyme disease,** a disease transmitted via the bite of *Ixodes* ticks. These ticks feed only a few times during their two-year life cycle. They become infected with *Borrelia* when they feed on an infected animal and pass the bacterium on to other hosts, including humans, when they next feed.
10. Lyme disease is characterized by a bull's-eye rash, neurological and cardiac dysfunction, and severe arthritis.
11. Human body lice transmit *Borrelia recurrentis*, which causes **louse-borne relapsing fever** in Africa and South America. In the western United States, soft ticks transmit several different species of *Borrelia* causing **tick-borne relapsing fever.**
12. *Leptospira interrogans* causes **leptospirosis,** a zoonotic disease in humans transmitted via animal urine and characterized by pain, headache, and liver and kidney dysfunction.

Pathogenic Gram-Negative Vibrios (pp. 637–641)

1. *Vibrio* is a genus of Gram-negative curved bacteria with polar flagella that naturally live in marine environments.
2. *Vibrio cholerae* causes **cholera,** a disease contracted via the ingestion of contaminated food and water. Cholera has been pandemic through the centuries.
3. Via a series of biochemical steps, **cholera toxin**—an exotoxin produced by *Vibrio cholerae*—causes the movement of water out of the intestinal epithelium, resulting in potentially fatal diarrhea.
4. *Vibrio parahaemolyticus* enters the body via ingestion of shellfish from contaminated waters; it causes a milder form of cholera-like gastroenteritis.
5. *Vibrio vulnificus*, contracted either by ingestion of contaminated shellfish or by contamination of wounds with seawater, causes a potentially fatal blood poisoning.
6. *Campylobacter jejuni*, which is found in domestic animal reservoirs, commonly causes gastroenteritis when ingested in contaminated food, water, or milk.
7. Once *Helicobacter pylori* reduces the amount of mucus produced in the stomach, acidic gastric juice eats away the stomach lining. *H. pylori* is the major cause of **peptic ulcers,** a discovery that was not readily embraced when first proposed.

QUESTIONS FOR REVIEW

Answers to the Questions for Review (except Short Answer questions) begin on p. A-1.

Multiple Choice

1. Most human infections caused by species of *Rickettsia* _____.
   a. are acquired from fomites
   b. could be prevented by hand washing
   c. are transmitted via vectors
   d. are sexually transmitted

2. The bacterium that causes spotted fever rickettsiosis (RMSF) is more likely to infect a human _____.
   a. if an infected tick feeds for several hours
   b. when an infected tick initially penetrates the skin
   c. when contaminated tick feces dry and become airborne
   d. if the human is exposed to rodent feces containing the bacterium

3. The most severe rickettsial illness is caused by _____.
   a. *Rickettsia typhi*
   b. *Rickettsia rickettsii*
   c. *Orientia tsutsugamushi*
   d. *Ehrlichia chaffeensis*

4. The smallest cellular microbes are _____.
   a. rickettsias
   b. mycoplasmas
   c. chlamydias
   d. both a and c

5. The most commonly reported sexually transmitted disease in the United States is caused by the bacterium _____.
   a. *Mycoplasma genitalium*
   b. *Chlamydia trachomatis*
   c. *Chlamydophila proctitis*
   d. *Ureaplasma urealyticum*

6. Which of the following diseases would be *least* likely in rural areas of the United States?
   a. epidemic typhus
   b. Rocky Mountain spotted fever
   c. murine typhus
   d. lymphogranuloma venereum

7. Treatment of chlamydial infections involves _____.
   a. erythromycin cream
   b. doxycycline creams
   c. surgical correction of eyelid deformities
   d. all of the above

8. Which of the following organisms is transmitted via sexual contact?
   a. *Treponema pallidum endemicum*
   b. *Treponema pallidum pertenue*
   c. *Treponema pallidum pallidum*
   d. *Treponema carateum*

9. Which of the following is *not* true of cholera?
   a. The causative agent lives naturally in marine water.
   b. There is an effective vaccine for cholera.
   c. Strain O1 El Tor has been responsible for several pandemics.
   d. Rice-water stool is a symptom.

10. During which stage of syphilis is penicillin ineffective?
    a. primary syphilis
    b. secondary syphilis
    c. tertiary syphilis
    d. all of the above

11. Two weeks after a backpacking trip in Tennessee, a hiker experienced flulike symptoms and noticed a red rash on his thigh. What is the likely cause of his illness?
    a. *Treponema pallidum pertenue*
    b. *Borrelia burgdorferi*
    c. *Borrelia recurrentis*
    d. *Leptospira interrogans*

12. The most common cause of bacterial gastroenteritis in the United States is _____.
    a. *Vibrio parahaemolyticus*   c. *Helicobacter pylori*
    b. *Campylobacter jejuni*   d. *Vibrio cholerae*

13. Historical journals have described gummas on patients. What disease most likely caused these lesions?
    a. ornithosis   c. trachoma
    b. syphilis   d. pneumonia

VISUALIZE IT!

1. Label the following stages and structures of the chlamydia life cycle: *elementary body, endocytosis, vesicle, host cell, inclusion body, reticulate body.* For B–D, indicate how many hours have typically transpired since infection.

2. Label inclusion, elementary, and reticulate bodies of *Chlamydia*:

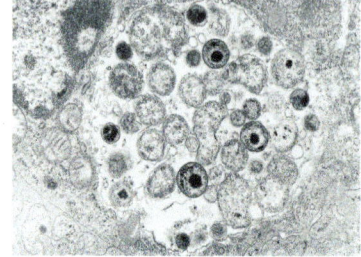

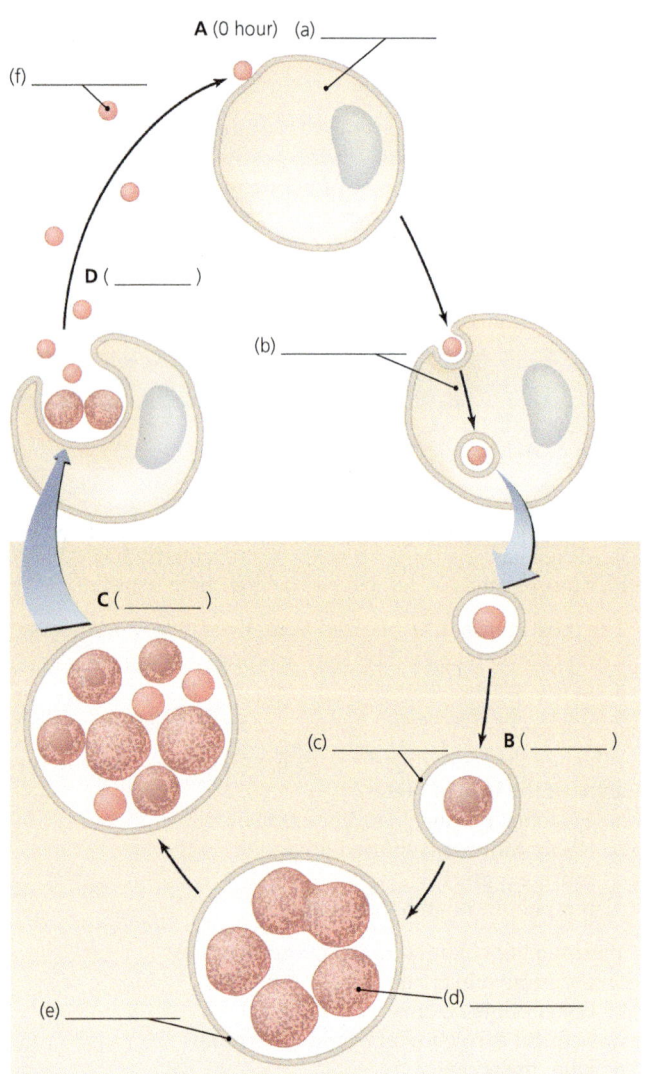

Short Answer

1. Suggest a hypothesis to explain why rickettsias are obligate parasites.
2. Describe the three developmental stages of the bacteria *Ehrlichia* and *Anaplasma*.
3. Why have scientists had problems identifying the virulence factors of *Treponema pallidum pallidum*?
4. Describe the phases of untreated syphilis.
5. Discuss the prospects for the eradication of leptospirosis.
6. Beginning with the ingestion of water contaminated with *V. cholerae* O1 El Tor, describe the course of the disease it causes.

Matching

1. Match the disease with the causative pathogen in the list.

    ___ Spotted fever rickettsiosis
    ___ Murine typhus
    ___ Epidemic typhus
    ___ Scrub typhus
    ___ Ehrlichiosis

    A. *Rickettsia typhi*
    B. *Rickettsia prowazekii*
    C. *Rickettsia rickettsii*
    D. *Orientia tsutsugamushi*
    E. *Ehrlichia chaffeensis*

2. Match the pathogen with the listed vector responsible for transmitting it to humans.

    ___ *Rickettsia typhi*
    ___ *Rickettsia prowazekii*
    ___ *Rickettsia rickettsii*
    ___ *Orientia tsutsugamushi*
    ___ *Ehrlichia chaffeensis*
    ___ *Borrelia burgdorferi*
    ___ *Borrelia recurrentis*
    ___ *Anaplasma phagocytophilum*

    A. Rat flea
    B. Body louse
    C. Hard tick
    D. Mite
    E. Soft tick

3. Match the pathogen with the listed disease(s) it causes.

    ___ *Chlamydophila psittaci*
    ___ *Chlamydophila pneumoniae*
    ___ *Chlamydia trachomatis*
    ___ *Treponema pallidum pallidum*
    ___ *Treponema pallidum pertenue*
    ___ *Treponema pallidum endemicum*
    ___ *Treponema carateum*
    ___ *Borrelia burgdorferi*

    A. Syphilis
    B. Trachoma
    C. Sinusitis
    D. Lymphogranuloma venereum
    E. Proctitis
    F. Pelvic inflammatory disease
    G. Ornithosis
    H. Yaws
    I. Bejel
    J. Pinta
    K. Lyme disease

4. Match the following diseases with the causative pathogen(s).

    ___ Peptic ulcers
    ___ Gastroenteritis (various forms)
    ___ Blood poisoning
    ___ Cholera

    A. *Vibrio cholerae*
    B. *Vibrio parahaemolyticus*
    C. *Vibrio vulnificus*
    D. *Campylobacter jejuni*
    E. *Helicobacter pylori*

CRITICAL THINKING

1. Why is it more difficult to rid a community of a disease transmitted by arthropods (e.g., Lyme disease or spotted fever rickettsiosis) than a disease transmitted via contaminated drinking water (e.g., cholera)?
2. Some scientists think that syphilis is a New World disease brought to Europe by returning Spanish explorers; others think that syphilis traveled the opposite way. Design an experiment to test the two hypotheses.
3. A patient arrives at a medical emergency room in Austin, Texas, complaining of 48 hours of severe abdominal cramping and persistent diarrhea. While waiting, he falls into a coma. He is admitted into the hospital but never regains consciousness; a week later he dies. No one else in his family is ill. They report that the man has not been out of the country but that they all dined at a seafood restaurant the night before the man became sick. He alone ate sea urchins. Culture of the man's stool reveals the presence of curved bacilli. What bacterium might account for the man's death?
4. Scientists discover a mutant strain of *Helicobacter pylori* that is urease negative. The strain is found to cause ulcers only in patients that either consume large quantities of antacids and/or take drugs to block acid production. Explain why only these patients develop ulcers when infected by the mutant strain.
5. Thirty-nine members of an extended family sought medical treatment for headaches, myalgia, and recurring fevers, each lasting about three days. The outbreak occurred shortly after a one-day family gathering in a remote, seldom-used mountain cabin in New Mexico. What disease did they contract? What causes this disease? What is the appropriate treatment? How could this family have protected itself?

CONCEPT MAPPING

Using the terms provided, fill in the concept map that describes syphilis. You can also complete this and other concept maps online by going to the **Mastering Microbiology** Study Area.

Body-wide rash
Cardiovascular syphilis
Chancre
Dark-field microscopy
Gummas
Infect fetus in pregnant women
Latent stage
Neurosyphilis
Never progress
Penicillin
Primary syphilis
Secondary syphilis
Serological tests
Spirochete
Tertiary syphilis
Treponema pallidum

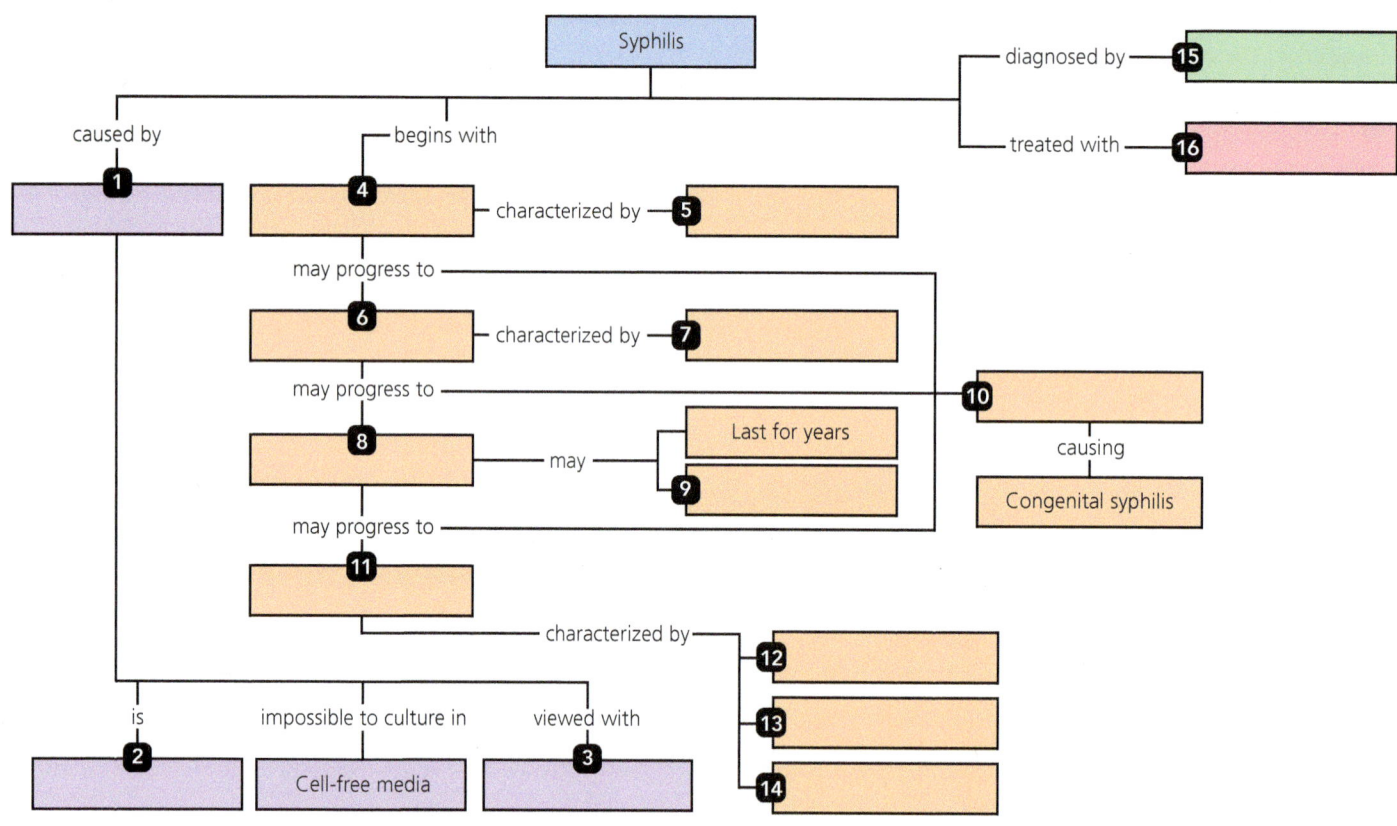

22 Pathogenic Fungi

Before You Begin

1. How are fungal and animal cytoplasmic membranes different from each other?
2. What is the difference between fungal spores and bacterial endospores?
3. When viewed through a microscope, some fungi with branched hyphae look somewhat like plants. Why are fungi not plants?
4. What are dimorphic fungi?

MICRO IN THE CLINIC

A Vacation Cold or Something Else?

VALERIE LOPEZ WAS HAPPY TO BE HOME from her road trip to Tucson, Arizona. She had truly enjoyed her trip to visit the aunt and uncle she hadn't seen since her aunt had come to her high school graduation seven years earlier. Valerie had taken advantage of a few weeks' break before starting her new job in Richmond, Virginia, driving more than 2000 miles to visit her favorite relatives even though she seemed to be recovering from a cold. It had been like old times: they had shared stories, gone camping, devoured her aunt's cooking, and played board games into the wee hours of the night. Although she had been sad to leave, she was also excited by the prospect of returning home and beginning her new job.

The three-day trip to the East Coast seemed much longer than the trip to Arizona, and Valerie felt unusually tired. She even started to cough. Could she be coming down with yet another cold? Exhausted, Valerie unpacked only the essentials from her car, took a shower, and went straight to bed. "Nothing that a good night's sleep can't cure," she thought. After a few hours, however, she woke up with a fever and a more productive cough. She downed some cold medicine and managed a little restless sleep. Valerie experienced chest tightness and difficulty breathing the next morning, so she headed to the local urgent care clinic.

1. Why is it unlikely that Valerie has just a cold?
2. Is going to the doctor a wise decision?
3. Should Valerie mention her recent travel to her doctor? Why or why not?

Turn to the end of the chapter (p. 669) to find out.

An ascomycete fungus, *Microsporum*, is one cause of athlete's foot, among other fungal skin diseases. Shown here are its asexual spores.

Most fungi exist as saprobes (absorbing nutrients from dead organisms) and function as the major decomposers of organic matter in the environment. More than 100,000 species of fungi have been classified in three divisions—Zygomycota, Ascomycota, and Basidiomycota—and of these, fewer than 200 have been demonstrated to cause diseases in humans.

This chapter is not arranged according to any taxonomic scheme (unlike other chapters in this text), because the classification of fungi is difficult. For example, most fungi have both a sexual stage, known as a *teleomorph*, and an asexual stage, known as an *anamorph*. The rules of taxonomy for fungi allow scientists to give each stage its own unique name. For example, the anamorph *Aspergillus* (as-per-jil'ŭs) is the same fungus as the teleomorph *Neosartorya* (nē-ō-sar-tōr'ya).

To minimize confusion, this chapter considers fungi using only a single name (usually the anamorph) and examines mycoses according to the site of disease. This convention is imperfect, as some fungi can invade tissues where they are not typically found. We also consider mycoses based on whether the fungus is a true pathogen (able to cause disease in people with normal immunity) or an opportunist. First, however, we will briefly consider some of the basic topics in medical mycology.

An Overview of Medical Mycology

LEARNING OUTCOME

22.1 Summarize some of the complexities of identifying and treating fungal infections.

Human mycoses are among the more difficult diseases to diagnose and treat properly. **Medical mycology** is the field of medicine that is concerned with the diagnosis, management, and prevention of fungal diseases, or **mycoses** (mī-kō'sēz). Signs of mycoses are often missed or misinterpreted, and once identified, fungi often prove to be remarkably resistant to antimicrobial agents. Before we begin our discussion of fungal diseases, therefore, it will be helpful to consider some of the characteristics that set fungal infections apart from other microbial infections.

The Epidemiology of Mycoses

LEARNING OUTCOMES

22.2 State the more significant modes of transmission for mycoses.
22.3 Explain why the actual prevalence of fungal infections is unknown.

Fungi and the spores they produce are almost everywhere in the environment—in soil, in water, and on or in most multicellular organisms, including plants and animals. They coat the surfaces of almost every object, whether made of wood, glass, metal, or plastic; thus, it is not surprising that most of us will experience a mycosis at some time.

Mycoses are typically acquired via inhalation, trauma, or ingestion; only very infrequently are fungi spread from person to person. Therefore, most mycoses are *not* contagious. Epidemics of mycoses can and do occur but not as a result of person-to-person contact. Instead, they result from mass exposure to some environmental source of fungi. For example, the cleanup of bird droppings near a building's air conditioning intake vents stirs up fungal spores in the droppings. The airborne spores could then be drawn inside the building and distributed throughout its duct system, potentially infecting numerous people.

One group of fungi that *are* contagious are *dermatophytes* (der'mă-tō-fītz), which live on the dead layers of skin and which may be transmitted between people via *fomites* (fōm'i-tēz, inanimate objects). Species of the genera *Candida* (kan'did-ă) and *Pneumocystis* (nū-mō-sis'tis) also appear to be transmitted at least some of the time by contact among humans.

Because most mycoses are not contagious, they are not typically reportable; that is, neither local public health agencies nor the U.S. Centers for Disease Control and Prevention (CDC) must be notified when they are diagnosed. Again, there are a few exceptions: Pathogenic fungi are usually reported in geographic areas where they are endemic, and certain opportunistic fungal pathogens of AIDS patients are also tracked.

The fact that most mycoses are not reportable creates a problem for epidemiologists. Without reliable data, it is impossible to track the effects of mycoses on the population, to ascertain their risks, and to identify actions to prevent them.

Categories of Fungal Agents: True Fungal Pathogens and Opportunistic Fungi

LEARNING OUTCOMES

22.4 Compare and contrast true fungal pathogens with opportunistic fungi.
22.5 Identify factors that predispose people to opportunistic fungal infections.

Of all the fungi known to cause disease in humans, four—*Blastomyces dermatitidis*, *Coccidioides immitis*, *Histoplasma capsulatum*, and *Paracoccidioides brasiliensis*—are considered true pathogens; that is, they can cause disease in otherwise healthy individuals. Other fungi, such as the common yeast *Candida albicans* (al'bi-kanz), are *opportunistic fungi*, which lack genes for proteins that aid in colonizing body tissues, though they can take advantage of some weakness in a host's defenses to cause disease. Thus, whereas true fungal pathogens can infect anyone, regardless of immune status, opportunists infect only immunologically weakened individuals.

Four main factors increase an individual's risk of experiencing opportunistic mycoses: invasive medical procedures, medical therapies, certain disease conditions, and specific lifestyle factors (TABLE 22.1). Surgical insertion of devices such as heart-valve implants can introduce fungal spores and provide a site for fungal colonization, while a variety of medical therapies leave patients with a weakened or dysfunctional immune system. AIDS, diabetes, other serious illnesses, and malnutrition also lessen immunity. Poor hygiene results in reduced skin defenses, and IV drug use can directly introduce fungi into the blood. Note that many of the factors that contribute to opportunistic fungal disease are associated with health care; thus,

An Overview of Medical Mycology

TABLE 22.1 Factors That Predispose Individuals to Opportunistic Mycoses

Factors	Examples
Medical procedures	Surgery; insertion of medical implants (heart valves, artificial joints); catheterization
Medical therapies	Immunosuppressive therapies accompanying transplantation; radiation and other cancer therapies; steroid treatments; long-term use of antibacterial agents
Disease conditions	Inherited immune defects; leukemia and lymphomas; AIDS; diabetes and other metabolic disorders; severe burns; preexisting chronic illnesses
Lifestyle factors	Malnutrition; poor hygiene; IV drug abuse

correct antisepsis and medical procedures by health care providers play a crucial role in reducing the incidence of mycoses of opportunistic fungi.

In addition to differences in pathogenesis and matters related to host susceptibility, fungal pathogens and opportunists differ with respect to geographical distribution. Whereas the four pathogenic fungi are endemic to certain regions, primarily in the Americas, opportunistic fungi are distributed throughout the world.

Dermatophytes—fungi that normally live on the skin, nails, and hair—are the only fungi that do not fall comfortably into either the pathogenic or the opportunistic grouping. They are considered by some researchers to be "emerging" pathogens. Dermatophytes can infect all individuals, not just the immunocompromised, which makes them similar to the pathogens. They are not, however, intrinsically invasive, being limited to body surfaces. They also have a tendency to occur in people with the same predisposing factors that allow access by opportunistic fungi. For these reasons, dermatophytes will be discussed as opportunists rather than as pathogens.

Clinical Manifestations of Fungal Diseases

LEARNING OUTCOME

22.6 Describe three primary clinical manifestations of mycoses.

Fungal diseases are grouped into the following three categories of clinical manifestations:

- *Fungal infections*, which are the most common mycoses, are caused by the presence in the body of either true pathogenic or opportunistic fungi. As previously noted, this chapter discusses mycoses according to general location within the body. Systemic mycoses are discussed first, followed by superficial, cutaneous, and subcutaneous mycoses.
- *Fungal toxicoses* (poisonings) are acquired through ingestion, as occurs when poisonous mushrooms are eaten. Although relatively rare, fungal toxicoses are discussed briefly near the end of the chapter.
- *Allergies* (hypersensitivity reactions) most commonly result from the inhalation of fungal spores and are the subject of the chapter's final section.

The Diagnosis of Fungal Infections

LEARNING OUTCOME

22.7 Discuss why the diagnosis of opportunistic fungal infections can be difficult.

Most patients reporting to a hospital emergency room with a severe respiratory illness are routinely tested for influenza, bacterial pneumonia, and perhaps tuberculosis; without some concrete reason to suspect a fungal infection, physicians may not look for them. However, if an emergency room patient with respiratory distress reveals, for example, that she breeds exotic birds, she is much more likely to also be tested for mycoses. Thus, a patient's history—including occupation, hobbies, travel history, and the presence of preexisting medical conditions—is critical for diagnosis of most mycoses. Even when fungal infections are relatively obvious, as when distinctive fungal growth is observed, definitive diagnosis requires isolation, laboratory culture, and morphological analysis of the fungus involved.

Microbiologists culture fungi collected from patients on *Sabouraud dextrose agar*, a medium that favors fungal growth over bacterial growth (see Chapter 6, Figure 6.12). The appearance of colonies and the microscopic appearance of yeast cells, mycelia, or mold spores are usually diagnostic.

Several techniques are commonly used in identifying fungi. *Potassium hydroxide (KOH) preparations* dissolve keratin in skin cells, leaving only fungal cells for examination. *Gomori methenamine silver (GMS) stain* is used on tissue sections to stain fungal cells black (other cells remain unstained; **FIGURE 22.1a**). *Direct immunofluorescence stain* (also called direct fluorescent antibody test) can be used to detect fungal cells in tissues (**FIGURE 22.1b**); however, immunological tests, though very useful for other microbial infectious agents, are not always useful for fungi. Because fungi are so prevalent in the environment and many are part of the normal microbiome, it is often impossible to distinguish between actual infection and simple exposure.

Diagnosis of opportunistic fungal infections is especially challenging. When a fungal opportunist infects tissues in which it is normally not found, it may display abnormal morphology that complicates identification. In addition, the fungi that

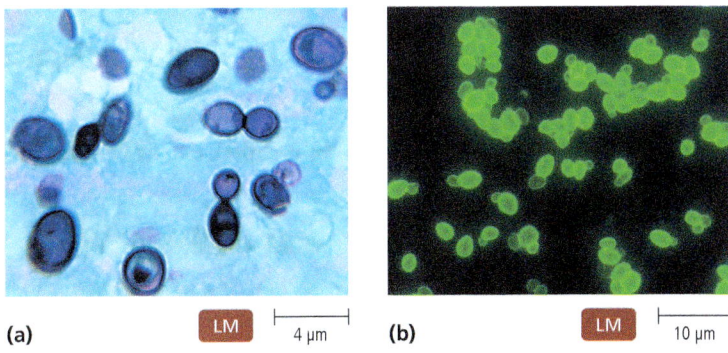

▲ **FIGURE 22.1 Fungal stains.** (a) GMS (Gomori methenamine silver) stain. *Histoplasma capsulatum*, shown here, causes histoplasmosis. (b) Direct immunofluorescence stain of *Candida albicans*.

produce pulmonary infections—the true pathogens and a few opportunists—produce symptoms and imaging profiles (X-ray studies, CT scans) that strongly resemble those of tuberculosis. Fungal masses may also resemble tumors.

Antifungal Therapies

LEARNING OUTCOME

22.8 Discuss the advantages and disadvantages of fungicidal and fungistatic medications.

Mycoses are difficult to heal for at least two reasons:

- Fungi often possess the biochemical ability to resist T cells during cell-mediated immune responses.
- Fungi are biochemically similar to human cells, which means that most fungicides are toxic to human tissues.

Many antifungal agents exploit one of the few differences between human and fungal cells—instead of cholesterol, the membranes of fungal cells contain a related molecule, *ergosterol*. Antifungal drugs target either ergosterol synthesis or its insertion into fungal membranes. However, cholesterol and ergosterol are not sufficiently different to prevent some damage to human tissues by such antifungal agents. Serious side effects associated with long-term use of almost all antifungal agents include anemia, headache, rashes, gastrointestinal upset, and serious liver and kidney damage.

The "gold standard" of antifungal agents is the anti-ergosterol fungicidal drug *amphotericin B*, considered the best drug for treating systemic mycoses and other fungal infections that do not respond to other drugs. Unfortunately, among the antifungal agents it is one of the more toxic to humans. Some major anti-ergosterol alternatives are the azole drugs—*ketoconazole*, *itraconazole*, and *fluconazole*—which are fungistatic (inhibitory) and less toxic to humans.

Three antifungal drugs that do not target ergosterol are griseofulvin, 5-fluorocytosine, and echinocandins. *Griseofulvin* interferes with microtubule formation and chromosomal separation in mitosis. Griseofulvin accumulates in the outer epidermal layers of the skin, preventing fungal penetration and growth. Since these skin cells are scheduled to die as they move toward the surface, griseofulvin's toxicity does not permanently damage humans. Patients generally tolerate the drug for the duration of treatment. For example, griseofulvin may be administered orally for up to one year to clear fungal infections of the nails without harming the patient.

5-Fluorocytosine is a nucleoside analog that inhibits RNA and DNA synthesis. Physicians give 5-fluorocytosine in conjunction with amphotericin B or an azole drug to treat infections of *Candida* and *Cryptococcus*.

Echinocandins inhibit the synthesis of 1,3-D-glucan, which is a sugar that makes up part of the cell wall of a fungus. This sugar does not occur in mammals; therefore, echinocandins are generally safe for adults. Researchers are still assessing the safety of the use of echinocandins in infants and pregnant patients.

Treatment of opportunistic fungal infections in immunocompromised patients involves two steps: a high-dose treatment to eliminate or reduce the number of fungal pathogens, followed by long-term (usually lifelong) maintenance therapy involving the administration of antifungal agents to control ongoing infections and prevent new infections.

With just a few antifungal drugs being used for long periods and treating more and more patients, scientists predict that the drugs should select drug-resistant strains from the fungal population. Fortunately, this is rarely the case; naturally occurring resistance, especially against amphotericin B, is extremely rare, though researchers cannot explain why resistance does not develop more frequently, as it has in bacterial populations under similar conditions of long-term use.

Antifungal Vaccines

LEARNING OUTCOME

22.9 Explain why there are few antifungal vaccines.

Prevention of fungal infections generally entails avoiding endemic areas and keeping one's immune system healthy. Scientists have not developed vaccines against fungi for several reasons, including:

- Most fungal infections are opportunistic and occur in patients with suppressed immunity. Vaccines may not work in such patients, because there is so little capacity for immune responses.
- In the past, there have been relatively few patients with severe fungal infections, so vaccine development was not thought to be the best use of scarce laboratory resources and personnel.
- It requires millions of American dollars to develop vaccines in compliance with government standards, and vaccine developers have not been sure of a return on their stockholders' investment, because there have been so few good candidates for vaccination.

Scientists have developed attenuated (weakened) vaccines against *Coccidioides*, but there is concern that the weakened microbe could revert to a virulent form and cause disease in people receiving the vaccine. Recently, researchers have created vaccines that protect mice against candidiasis, aspergillosis, and cryptococcosis. They do not yet know if the vaccines will work safely in humans.

> **MICRO CHECK**
>
> 1. Which type of fungal disease is most likely to be caused by a true fungal pathogen?
> 2. Why is it often hard to diagnose opportunistic fungal infections?
> 3. The "gold standard" of antifungal drugs is amphotericin B. What is its target?
> 4. What is the main advantage of fungistatic drugs over fungicidal drugs?

> **TELL ME WHY**
>
> Why would an antifungal drug do little to reduce the effects of a fungal toxicosis?

Now that we have considered some of the basics of medical mycology, we turn our attention to the first category of important mycoses: systemic mycoses caused by the four truly pathogenic fungi.

Systemic Mycoses Caused by Pathogenic Fungi

LEARNING OUTCOMES

22.10 Compare and contrast the endemic areas for the four genera of pathogenic fungi that cause systemic mycoses.

22.11 Compare the clinical appearances of the diseases resulting from each of the four pathogenic fungi and describe the treatment of their diseases.

22.12 Identify the laboratory techniques used to distinguish among the four pathogenic fungi.

Systemic mycoses—those fungal infections that spread throughout the body—result from infections by one of the four pathogenic fungi: *Histoplasma*, *Blastomyces*, *Coccidioides*, or *Paracoccidioides*. All are in the fungal division Ascomycota. These pathogenic fungi are uniformly acquired through inhalation, and all begin as a generalized pulmonary infection that then spreads via the blood to the rest of the body.

All four fungi are **dimorphic**[1] (dī-mōr′fik); that is, they have two growth forms. In the environment, where the temperature is typically below 30°C, they appear as mycelial bodies composed of hyphae, whereas within the body (37°C) they grow as spherical yeasts (**FIGURE 22.2**). The two forms differ not only structurally but also physiologically. Yeast forms are invasive because they express a variety of enzymes and other virulence factors that aid their growth and reproduction in the body. For example, they are tolerant of higher temperatures and are relatively resistant to phagocytosis by immune cells.

Dimorphic fungi are extremely hazardous to laboratory personnel working with the fungi. They must take specific precautions to avoid exposure to spores, particularly when culturing the organisms, by using biological safety cabinets with HEPA filters (see Figure 9.11), protective clothing, and masks.

In the following sections, we consider each of the systemic conditions caused by true fungal pathogens, beginning with histoplasmosis.

Histoplasmosis

Histoplasma capsulatum (his-tō-plaz′mă kap-soo-lā′tŭm), the causative agent of **histoplasmosis** (his′tō-plaz-mō′sis), is an ascomycete and the most common fungal pathogen affecting humans. *H. capsulatum* is particularly prevalent in the eastern United States along the Ohio River valley, but endemic areas also exist in Africa and South America (**FIGURE 22.3** on the next page). *H. capsulatum*

[1] Form Greek *di*, meaning "two," and *morphe*, meaning "shape."

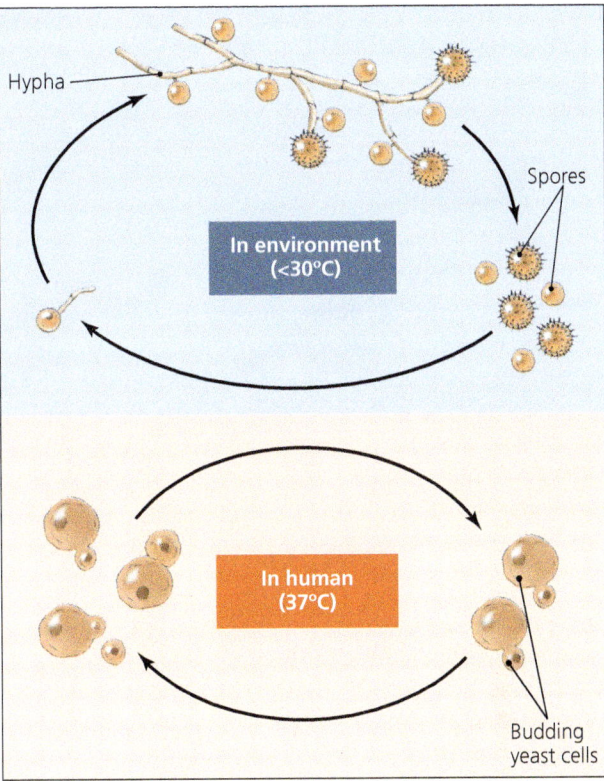

▲ **FIGURE 22.2 The dimorphic nature of true fungal pathogens.** The mycelial form grows in the environment, and the yeast form grows within a human host. Shown here is *Histoplasma capsulatum*. *What are the other three genera of pathogenic, dimorphic fungi?*

Figure 22.2 *In addition to Histoplasma, members of the genera Blastomyces, Coccidioides, and Paracoccidioides are pathogenic fungi.*

is found in moist soils containing high levels of nitrogen, such as from the droppings of bats and birds. Spores may become airborne and inhaled when soil containing the fungus is disturbed by wind or by human activities. Cutaneous inoculation can also lead to disease, but such infections are extremely rare.

H. capsulatum is an intracellular parasite that survives inhalation and subsequent phagocytosis by macrophages in air sacs of the lungs. These macrophages then disperse the fungus beyond the lungs via the blood and lymph. Cell-mediated immunity eventually develops, clearing the organism from healthy patients.

About 95% of individuals infected with pulmonary *Histoplasma* are asymptomatic, and their subclinical cases resolve without damage. Some individuals may experience mild or nonspecific respiratory symptoms (e.g., pains while breathing or mild coughing) that also disappear on their own. Slightly more pronounced symptoms—fever, night sweats, and weight loss—may occur in a few individuals. About 5% of patients develop clinical histoplasmosis, which manifests as one of four diseases:

- *Chronic pulmonary histoplasmosis* is characterized by severe coughing, blood-tinged sputum, night sweats, loss of appetite, and weight loss. It is often seen in individuals with pre-existing lung disease. It can be mistaken for tuberculosis.
- *Chronic cutaneous histoplasmosis*, characterized by ulcerative skin lesions, can follow the spread of infection from the lungs.

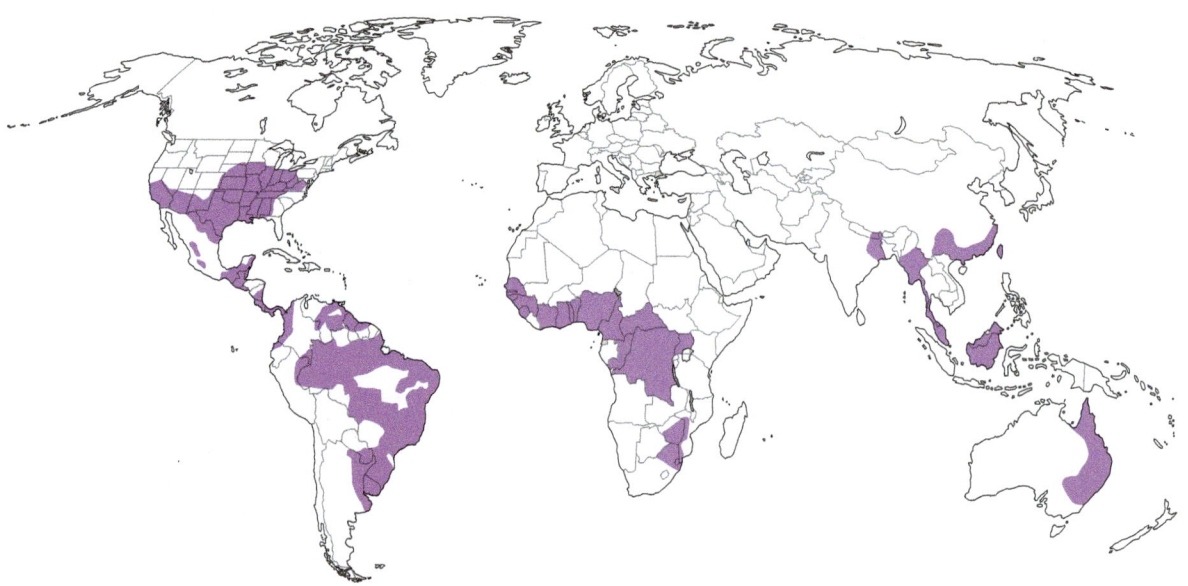

◀ FIGURE 22.3 Endemic areas for histoplasmosis.

- *Systemic histoplasmosis* can also follow if infection spreads from the lungs, but it is usually seen only in AIDS patients. This syndrome, characterized by enlargement of the spleen and liver, can be rapid, severe, and fatal.
- *Ocular histoplasmosis* is a type I hypersensitivity reaction against *Histoplasma* in the eye; it is characterized by inflammation and redness.

Diagnosis of histoplasmosis is based on the identification of the distinctive budding yeast in KOH- or GMS-prepared samples of skin scrapings, sputum, cerebrospinal fluid, or various tissues. The diagnosis is confirmed by the observation of dimorphism in cultures grown from such samples. Cultured *H. capsulatum* produces distinctively spiny spores that are also diagnostic (FIGURE 22.4). Immunoassays are not useful indicators of *Histoplasma* infection because many people have been exposed without contracting disease. In the endemic regions of the United States, close to 90% of the population tests positive for *H. capsulatum* exposure.

Infections in immunocompetent individuals typically resolve without treatment. When symptoms do not resolve, amphotericin B is prescribed. Ketoconazole can be used to treat mild infections. Maintenance therapy for AIDS patients is recommended.

CLINICAL CASE STUDY

What's Ailing the Bird Enthusiast?

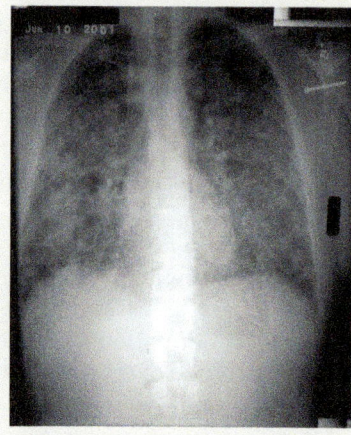

A 72-year-old man arrives at a hospital emergency room with signs and symptoms of serious pulmonary infection. He complains of a deep cough, blood-tinged sputum, night sweats, and weight loss. His liver and spleen feel enlarged upon physical examination. A chest X ray suggests tuberculosis, but there are no acid-fast bacilli (AFBs) in his sputum, which would be present with TB. The doctor suspects community-acquired pneumonia and prescribes seven days of intravenous antibiotics. After a week, the patient's symptoms are only getting worse.

The man has recently traveled to Africa and Asia. In addition, he is an avid bird enthusiast who likes to feed pigeons. He also keeps many bird feeders on his balcony, which overlooks the Ohio River and attracts numerous birds. He spends an hour a day cleaning up bird droppings.

The ICU doctors decide to test for other possible diseases. Skin tests for coccidioidomycosis and blastomycosis are inconclusive. They do a bronchoalveolar lavage in which fluid is washed through part of the patient's lung via a flexible tube inserted through the mouth. The fluid is collected and analyzed with an enzyme immunoassay. It detects fungal antigens, and his physicians diagnose pulmonary histoplasmosis.

1. What is the most likely infecting agent? How do you suppose this individual acquired the disease?
2. In people who work with birds, is this disease unusual, and are those who do have it normally symptomatic?
3. Given that some tests were inconclusive, what other tests or lab work would aid in arriving at a specific diagnosis?
4. What treatment would most likely be prescribed?

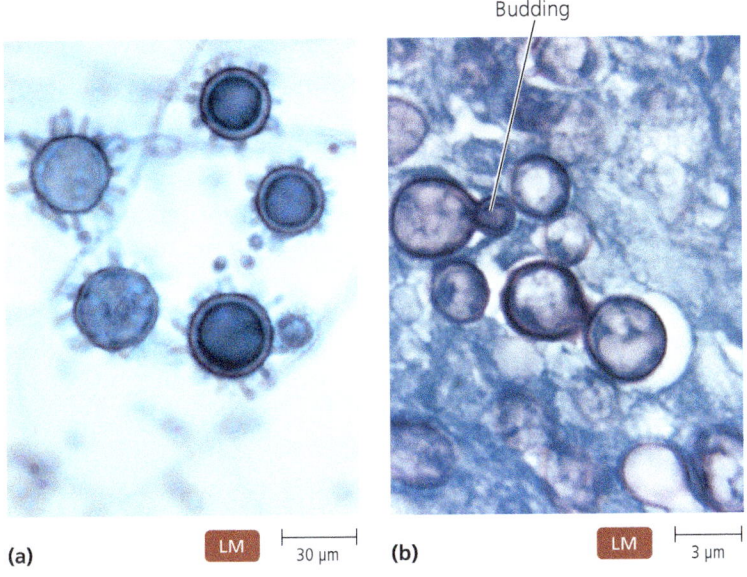

▲ FIGURE 22.4 Characteristics of *Histoplasma capsulatum* (a) Spiny spores of mycelial form. (b) Budding yeast cells in the lungs of a patient.

Blastomycosis

Blastomycoses (blas'tō-mī-kō'sēz) are caused by another ascomycete, *Blastomyces dermatitidis* (blas-tō-mī'sēz der-mă-tit'i-dis), which is endemic across the southeastern United States north to Canada (FIGURE 22.5). Outbreaks have also been reported in Latin America, Africa, Asia, and Europe. *B. dermatitidis* normally grows and sporulates in cool, damp soil that is rich in organic material, such as decaying vegetation and animal wastes. In humans, both recreational and occupational exposure occurs when fungal spores in soil become airborne and are inhaled. A relatively small inoculum can produce disease. The incidence of human infection is increasing in part because of increases in the number of immunocompromised individuals in the population.

Pulmonary blastomycosis is the most common manifestation of *Blastomyces* infection. After spores enter the lungs, they convert to yeast forms and multiply. Initial pulmonary lesions are asymptomatic in most individuals. If symptoms do develop, they are vague and include muscle aches, cough, fever, chills, malaise, and weight loss. Purulent (pus-filled) lesions develop in the lungs and expand as the yeasts multiply, resulting in death of tissues and

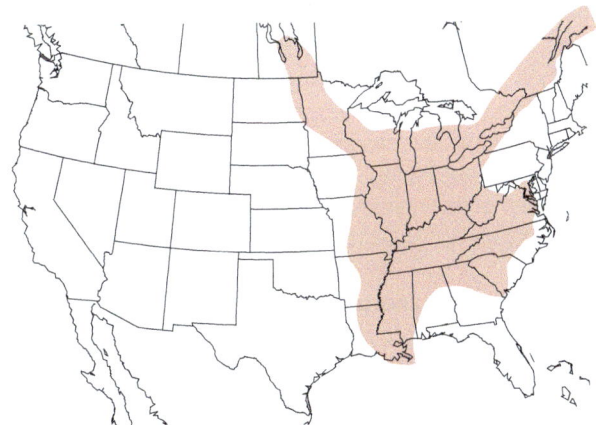

▲ FIGURE 22.5 Geographic distribution of *Blastomyces*.

MICROBE AT A GLANCE 22.1

Histoplasma capsulatum

Taxonomy: Domain Eukarya, kingdom Fungi, phylum Ascomycota, class Ascomycetes, order Onygenales, family Onygenaceae

Morphology: Thermally dimorphic: below 30°C, it forms septate hyphae, whereas at 37°C, it forms single-celled yeasts

Virulence factors: Survives phagocytosis to live inside macrophages, which can deliver the fungus throughout the body

Diseases caused: Chronic pulmonary histoplasmosis, symptomatic pulmonary histoplasmosis, acute diffuse pulmonary histoplasmosis, mediastinal granuloma

Treatment for diseases: Itraconazole for mild infections; amphotericin B for more severe cases

Prevention of disease: Ninety percent of people living in endemic areas (Ohio River valley in North America, equatorial West Africa, and isolated areas of South America) have been infected. A healthy immune system limits course of disease.

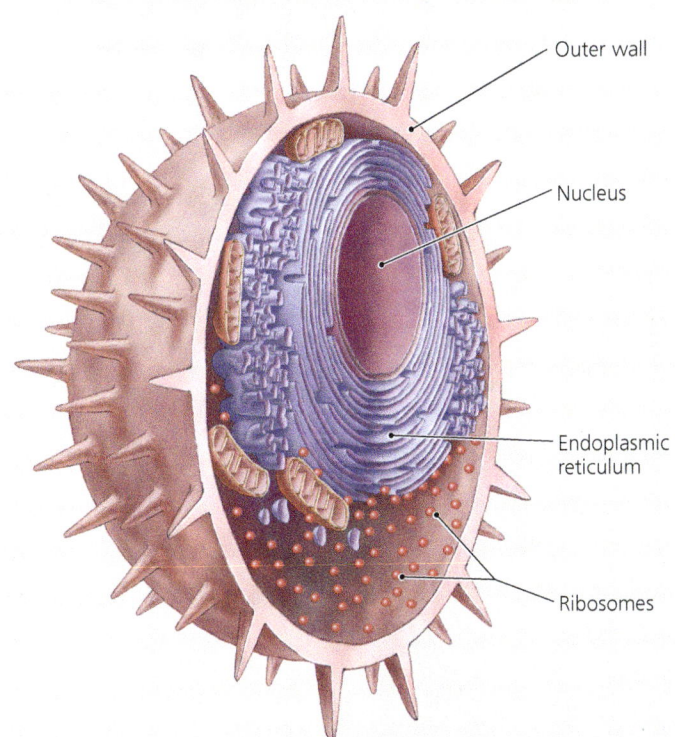

Histoplasma capsulatum spore

cavity formation. Occasionally, a deep productive cough and chest pain accompany the generalized pulmonary symptoms. In otherwise healthy people, pulmonary blastomycosis typically resolves successfully, although it may become chronic. Respiratory failure and death occur at a high frequency among AIDS patients.

The fungus can spread beyond the lungs. *Cutaneous blastomycosis* occurs in 60% to 70% of cases and consists of generally painless lesions on the face and upper body (FIGURE 22.6 on the next page). The lesions can be raised and wartlike, or they may be craterlike if tissue death occurs. In roughly 30% of cases, the fungus spreads to the spine, pelvis, cranium, ribs, long bones, or subcutaneous tissues

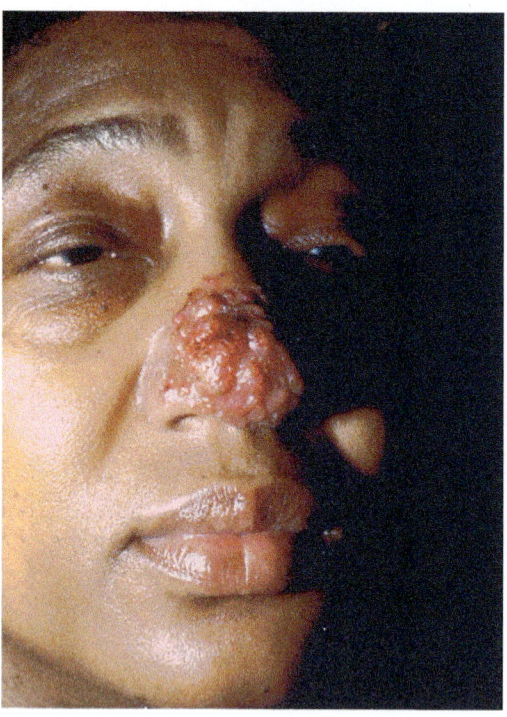

▲ FIGURE 22.6 **Cutaneous blastomycosis.** This condition typically results from the spread of *Blastomyces dermatitidis* from the lungs to the skin.

surrounding joints, a condition called *osteoarticular*[2] *blastomycosis* (os'tē-ō-ar-tik'ū-lar). About 40% of AIDS patients experience meningitis resulting from dissemination of *Blastomyces* to the central nervous system. Abscesses may form in other tissue systems as well.

Diagnosis relies on identification of *B. dermatitidis* following culture or direct examination of various samples, such as sputum, bronchoalveolar lavage (bronchial washings), biopsies, cerebrospinal fluid, or skin scrapings. Observation of dimorphism in laboratory cultures coupled with microscopic examination is diagnostic.

Physicians treat mild to moderate blastomycosis with oral itraconazole, and severe blastomycosis with amphotericin B for a few days before switching to itraconazole. Relapse is common in AIDS patients, and suppressive maintenance therapy with itraconazole is recommended.

Coccidioidomycosis

Coccidioidomycosis (kok-sid-ē-oy'dō-mī-kō'sis), caused by ascomycete *Coccidioides immitis* (kok-sid-ē-oy'dēz im'mi-tis), is found almost exclusively in the southwestern United States (Arizona, California, Nevada, New Mexico, and Texas) and northern

---

[2]From Greek *osteon*, meaning "bone," and *arthron*, meaning "joint."

EMERGING DISEASE CASE STUDY

Pulmonary Blastomycosis

Phil was glad he had spent the summer fishing, hiking, and camping in western Ontario. Canada is beautiful, and he had needed time off from his two jobs and full schedule of nursing classes. Now he was back in college, and remembrances of Canada were helping him cope with the stress.

However, Phil didn't feel well. He thought he was coming down with the flu—productive cough, fever, shortness of breath, night sweats and chills, muscle aches, tiredness, a general feeling of yuckiness, though no runny nose; yes, it must be the flu. Or was it? He knew from microbiology class that many diseases of bacteria, protozoa, and viruses have flulike symptoms. Phil had forgotten fungi.

Phil tried every over-the-counter remedy, with no benefit. He tried the health clinic on campus, but the antibiotics he received seemed to make things worse, not better. He was losing weight, and pus-filled, raised sores had appeared on his face, neck, and legs.

Phil went to the ER and was admitted immediately. The pulmonologist performed a bronchoscopy with bronchoalveolar lavage in which some fluid was washed through part of his lung and collected for laboratory testing. Two days later, the culture from the bronchial washing revealed *Blastomyces*, an emerging, dimorphic fungus. Phil had evidently become infected by inhaling conidia (spores from fungal hyphae) growing on wet leaves. The yeast phase of the fungus was multiplying and spreading throughout his body, producing the skin lesions that lasted for months, finally resolving into raised, wartlike scars.

Researchers don't know why reports of blastomycosis are becoming more prevalent. Perhaps it has to do with better diagnosis and reporting, perhaps it has to do with a growing number of AIDS patients who are susceptible to infection, or perhaps it has to do with more people like Phil adventuring into the wilderness.

The good news is that with proper diagnosis, Phil got the treatment he needed—200 mg of itraconazole orally three times a day for three days, then twice a day for six months. He graduated nursing school, and now he is a registered nurse who knows that flulike symptoms can sometimes indicate serious fungal infection. (For more about pulmonary blastomycosis, see p. 653.)

1. *Blastomyces* is dimorphic; what does this mean?
2. Why does AIDS increase the incidence of blastomycosis?
3. Given that *Blastomyces* is a mold that grows on dead leaves, how does it cause disease in people?

Mexico. Small, focally endemic areas also exist in semiarid parts of Central and South America (FIGURE 22.7). Fungi can be recovered from desert soil, rodent burrows, archeological remains, and mines. Dust that coats materials from endemic areas, including Native American pots and blankets sold to tourists, can serve as a vehicle of infection. Infection rates in endemic areas have risen in recent years as a result of population expansion and increased recreational activities, such as the use of off-road vehicles in the desert. More ominous threats to human health, however, are windstorms and earthquakes, which can disturb large tracts of contaminated soil, spreading the fungal elements for miles downwind and exposing thousands of people.

In the warm and dry summer and fall months, particularly in drought cycles, *C. immitis* grows as a mycelium and produces sturdy chains of asexual spores called *arthroconidia*. When mature, arthroconidia germinate into new mycelia in the environment, but if inhaled, arthroconidia germinate in the lungs to produce a parasitic form called a *spherule* (sfer'ool; FIGURE 22.8). As each spherule matures, it enlarges and generates a large number of spores via multiple cleavages until it ruptures and releases the spores into the surrounding tissue. Each spore then forms a new spherule to continue the cycle of division and release. This type of growth accounts for the seriousness of *Coccidioides* infection.

The major manifestation of coccidioidomycosis is pulmonary. About 60% of patients experience either no symptoms or mild, unremarkable symptoms that go unnoticed and typically resolve on their own. Other patients develop more severe infections characterized by fever, cough, chest pain, difficulty breathing, coughing up or spitting blood, headache, night sweats, weight loss, and pneumonia; in some individuals a diffuse rash may appear on the trunk. Occasionally, the mild forms of the pulmonary disease become more chronic, in which cases the continued multiplication of spherules results in large, permanent cavities in the lungs similar to those seen in tuberculosis patients.

In a very small percentage of cases, generally in those who are severely immunocompromised, *C. immitis* spreads from the lungs to various other sites. Invasion of the central nervous system (CNS) may result in meningitis, headache, nausea, and emotional disturbance. Infection can also spread to the bones, joints, and painless subcutaneous tissues; subcutaneous lesions are inflamed masses of granular material (FIGURE 22.9).

The diagnosis of coccidioidomycosis can be based on the identification of spherules in KOH- or GMS-treated samples collected from patients. Health care workers administer a *coccidioidin skin test* to screen patients for contact with *Coccidioides*. In the test, antigens of *Coccidioides* are injected under the skin. If the body has

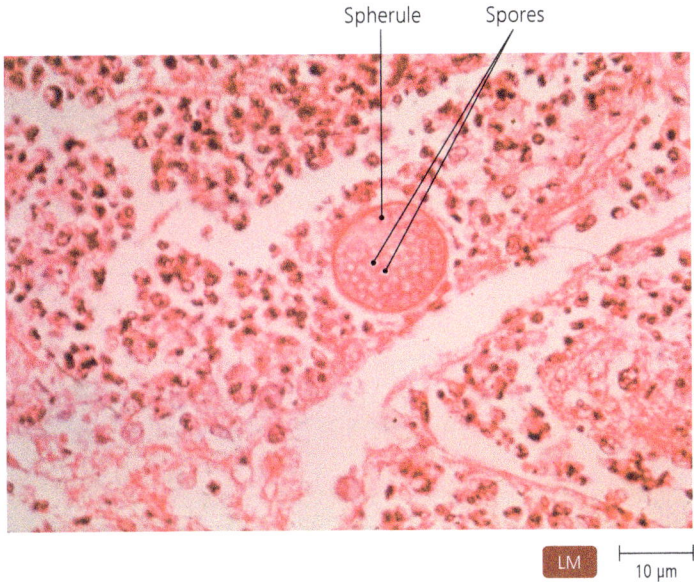

▲ FIGURE 22.8 **Spherules of *Coccidioides immitis*.** Note the numerous spores within a spherule.

▲ FIGURE 22.7 **Endemic areas of *Coccidioides*.**

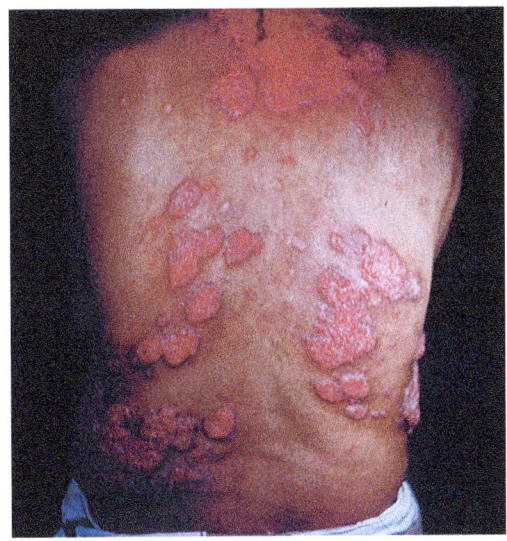

▲ FIGURE 22.9 **Coccidioidomycosis lesions in subcutaneous tissue.** Painless lesions result from the spread of *Coccidioides immitis* from the lungs.

antibodies against the fungus—that is, the patient has been or is infected—then the site will become inflamed. Diagnosis is confirmed by the isolation of the mycelial form of *C. immitis* in laboratory culture or isolation of encapsulated yeasts in the patient's tissues.

Infections in otherwise healthy patients generally resolve on their own without damage. Amphotericin B is the drug of choice for more severe or long-lasting cases. CNS involvement is fatal without treatment. In AIDS patients, maintenance therapy with itraconazole or fluconazole is recommended to prevent relapse or reinfection. The wearing of protective masks in endemic areas can prevent exposure to spores, but constant, daily wearing of masks may be impractical for all but those whose occupations put them at clear risk of infection.

Paracoccidioidomycosis

Another ascomycete, *Paracoccidioides brasiliensis* (par'ă-kok-sid-ē-oy'dēz bră-sil-ē-en'sis), causes **paracoccidioidomycosis** (par'ă-kok-sid-ē-oy'dō-mī-kō'sis), a chronic fungal disease similar to blastomycosis and coccidioidomycosis. *P. brasiliensis* is found in cool, damp soil from southern Mexico to regions of South America, particularly in Brazil. Because this fungus is far more geographically limited than the other true fungal pathogens, paracoccidioidomycosis is not a common disease. Those most at risk include farm workers in endemic areas.

Infections range from asymptomatic to systemic, and disease first becomes apparent as a pulmonary form that is slow to develop but manifests as chronic cough, fever, night sweats, malaise, and weight loss. The fungus can spread and create lesions, which are often on the face and neck.

KOH or GMS preparations of tissue samples reveal yeast cells with multiple buds in a "steering wheel" formation that is diagnostic for this organism (FIGURE 22.10). Laboratory culture at 25°C and at 37°C demonstrating dimorphism is confirmatory. Serological identification of antibodies also is useful. Treatment is with itraconazole, ketoconazole, or amphotericin B.

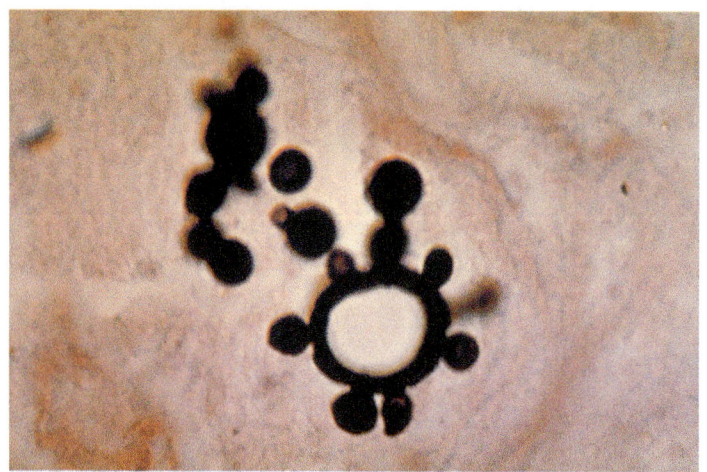

▲ FIGURE 22.10 The "steering wheel" formation of buds characteristic of *Paracoccidioides brasiliensis*.

MICRO CHECK

5. In what part of the United States is blastomycosis endemic?
6. Which systemic fungus is associated with inhaling fungal spores from bird or bat droppings?
7. How can paracoccidioidomycosis be distinguished from blastomycosis, coccidioidomycosis, and histoplasmosis in the laboratory?

TELL ME WHY

Why is it that 90% of people in areas endemic for histoplasmosis test positive for exposure to *Histoplasma* yet do not have histoplasmosis?

Systemic Mycoses Caused by Opportunistic Fungi

LEARNING OUTCOMES

22.13 Discuss the conditions that allow opportunistic fungi to cause disease.

22.14 Discuss the difficulties in diagnosing opportunistic fungal infections.

Opportunistic mycoses do not typically affect healthy humans because the fungi involved lack genes for virulence factors that make them actively invasive. Instead, opportunistic mycoses are generally limited to people with poor immunity and patients whose normal microbiome has been disrupted, as occurs following long-term use of antibacterial drugs. Because of the growing number of AIDS patients, opportunistic fungal infections have become one of the more significant causes of human disease and death. Even though any fungus can become an opportunist, five genera of fungi are routinely encountered: *Pneumocystis*, *Candida*, *Aspergillus*, *Cryptococcus*, and *Mucor*.

Opportunistic infections present a formidable challenge to clinicians. Because they appear only when their hosts are weakened, they often display "odd" clinical signatures; that is, their symptoms are often atypical, or they occur in individuals not residing in endemic areas for a particular fungus. Increased severity of symptoms is often a reflection of the poor immune status of the patient and frequently results in higher fatality rates, even if the fungus is normally nonthreatening.

In the following sections, we consider some of the mycoses caused by the "classical" fungal opportunists, beginning with *Pneumocystis* pneumonia.

Pneumocystis Pneumonia

LEARNING OUTCOMES

22.15 List the characteristics of *Pneumocystis* that distinguish it from other fungal opportunists.

22.16 Describe the course of *Pneumocytsis* pneumonia and its diagnosis and treatment.

Pneumocystis jirovecii (nū-mō-sis'tis jē-rō-vět'zē-ē), an ascomycete of the normal respiratory microbiome formerly known as *P. carinii*, is one of the more common opportunistic fungal infections seen in AIDS patients. Prior to the AIDS epidemic, disease caused by *Pneumocystis* was extremely rare.

Even though the organism was first identified in 1909, little is known about it. Originally considered a protozoan, it has been reclassified as a fungus based on rRNA nucleotide sequences and biochemistry. Its morphological and developmental characteristics, however, resemble those of protozoa more than those of fungi. Because *P. jirovecii* is an obligate parasite and cannot survive on its own in the environment, transmission most likely occurs through inhalation.

P. jirovecii is distributed worldwide in humans; based on serological confirmation of antibodies, the majority of healthy children (75%) have been exposed to the fungus by the age of five. In immunocompetent people, infection is asymptomatic, and generally clearance of the fungus from the body is followed by lasting immunity. However, some individuals may remain infected indefinitely; in such carriers, the organism remains in the alveoli (air sacs of lungs) and can be passed in respiratory droplets.

Before the AIDS epidemic, the disease ***Pneumocystis* pneumonia** was observed only in malnourished, premature infants and debilitated elderly patients. Now, the disease is almost diagnostic for AIDS. *Pneumocystis* pneumonia is abbreviated **PCP**, which originally stood for *Pneumocystis carinii* pneumonia but now can be considered to stand for *Pneumocystis* pneumonia.

Once the fungus enters the lungs of an AIDS patient, it multiplies rapidly, extensively colonizing the lungs because the patient's defenses are impaired. Widespread inflammation, fever, difficulty in breathing, and a nonproductive cough are characteristic. If left untreated, PCP involves more and more lung tissue until death occurs.

Diagnosis relies on clinical and microscopic findings. Chest X-ray studies usually reveal abnormal lung features. Stained smears of fluid from the lungs or from biopsies can reveal distinctive morphological forms of the fungus (**FIGURE 22.11**). The use of fluorescent antibody on samples taken from patients is more sensitive and provides a more specific diagnosis.

Though *Pneumocystis* is a fungus, antifungal drugs are not effective against it. Because this fungus has many of the characteristics of protozoa, both primary treatment and maintenance therapy are with the antiprotozoan drugs trimethoprim and sulfamethoxazole. Pentamidine, another antiprotozoan drug, is an alternative. The drugs may be aerosolized to allow for direct inhalation into the lungs.

Candidiasis

LEARNING OUTCOMES

22.17 Explain how candidiasis can develop from a localized infection to a systemic mycosis.

22.18 Describe the diagnosis and treatment of candidiasis.

Candidiasis (kan-di-dī'ă-sis) is any opportunistic fungal infection or disease caused by various species of the genus *Candida*—most commonly *Candida albicans* (kan'did-ă al'bi-kanz). *Candida* is one of a very few fungi that can be transmitted between individuals. **TABLE 22.2** summarizes the wide variety of clinical manifestations of candidiasis. **Disease in Depth: Candidiasis** (pp. 660–661) examines this disease in detail.

Aspergillosis

LEARNING OUTCOME

22.19 Describe the clinical manifestations, diagnosis, and treatment of aspergillosis.

Aspergillosis (as'per-ji-lō-sis) is not a single disease but instead a term for several diseases resulting from the inhalation of spores of fungi in the genus *Aspergillus* (as-per-jil'ŭs) in the division Ascomycota. *Aspergillus* is found in soil, food, compost, agricultural buildings, and air vents of homes and offices worldwide. Although exposure to *Aspergillus* most commonly causes only allergies, more serious diseases can occur, and aspergillosis is a growing problem for AIDS patients. Because the fungi are so common in the environment, little can be done to prevent exposure to the spores.

Even though *Aspergillus* species can be opportunistic pathogens of almost all body tissues, these fungi are chiefly responsible for causing three clinical pulmonary diseases:

- *Hypersensitivity aspergillosis* manifests as asthma or other allergic symptoms and results most commonly from inhalation of *Aspergillus* spores. Symptoms may be mild and result in no damage, or they may become chronic, with recurrent episodes leading to permanent damage.

- *Noninvasive aspergillomas*—ball-like masses of fungal hyphae—can form in the cavities left by a previous case of pulmonary tuberculosis. Most cases are asymptomatic, though coughing of blood-tinged sputum may occur.

- *Acute invasive pulmonary aspergillosis* is more serious in that lung tissue dies, which can lead to significant respiratory impairment. Signs and symptoms, which include fever, cough, and pain, may present as pneumonia.

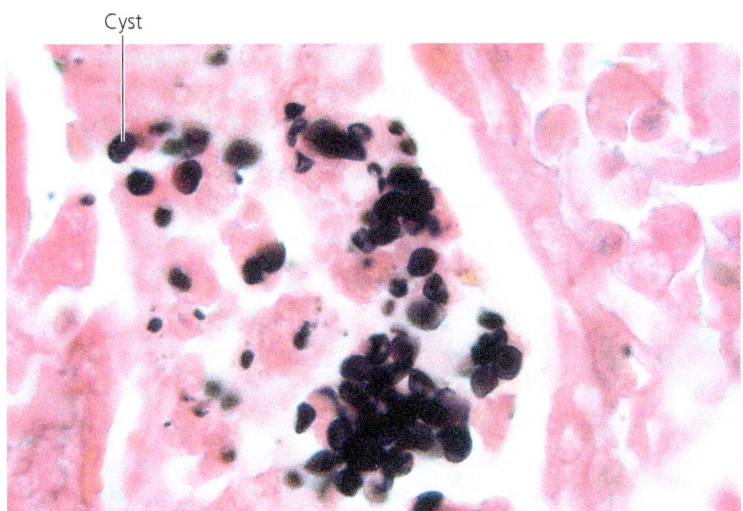

▲ **FIGURE 22.11 Cysts of *Pneumocystis jirovecii* in lung tissue.** Such microscopic findings are diagnostic for *Pneumocystis* pneumonia.

TABLE 22.2 The Clinical Manifestations of Candidiasis

Type	Clinical Signs and Symptoms	Predisposing Factors
Oropharyngeal (thrush)	White plaques on the mucosa of the mouth, tongue, gums, palate, and/or pharynx	Diabetes, AIDS, cancers, various chemotherapeutic drug regimens
Cutaneous	Moist, macular red rash between skin folds	Moisture, heat, and friction of skin in skin folds and between digits, particularly in the obese
	Diaper rash: raised red rash, pustules in the gluteal region	Infrequently changed, soiled diapers
	Onychomycosis: painful red swelling around the nails, destruction of nail tissue, and loss of the nail	Immunocompromised individuals
Vulvovaginal	Creamy white, curdlike discharge; burning; redness; painful intercourse	Use of broad-spectrum antibiotics, pregnancy, diabetes, changes to the vaginal microbiome
Chronic mucocutaneous	Lesions associated with the skin, nails, and mucous membranes; severe forms may occur	Various metabolic problems related to impaired cell-mediated immunity (e.g., diabetes) in children
Neonatal and congenital	Meningitis; renal disorders; or generalized, body-wide, red, vesicular rash	Young age, low birth weight, use of antibiotics by the mother
Esophageal	White plaques along the esophagus, burning pain, nausea, and vomiting	AIDS, immunocompromised status
Gastrointestinal	Ulceration of the stomach and intestinal mucosa	Hematological cancers
Pulmonary	Generalized, nonspecific symptoms; often remains undiagnosed until autopsy	Infection spread from other types of candidiasis
Peritoneal	Fever, pain, tenderness	Indwelling catheters for dialysis, gastrointestinal perforation
Urinary tract	Urinary tract infection: painful urination, possible discharge	Chemotherapeutic drug regimens, catheterization, diabetes, preexisting bladder problems
	Diaper rash	
	Renal: fever, pain, rigors, fungus ball	
Meningeal	Swelling of the meninges, fever, headache, stiffness of the neck	Spread of *Candida* during systemic infection
Hepatic and splenic	Fever, swelling of the liver and spleen, liver dysfunction, lesions and abscesses	Leukemia
Endocardial, myocardial, pericardial	Fever, heart murmur, congestive heart failure, anemia	Preexisting heart valve disease, catheterization plus use of antibiotics, IV drug abuse, valve prosthetics
Ocular	Cloudy vision, lesions within the eye	Spread of candidiasis, indwelling catheters, IV drug abuse, trauma
Osteoarticular	Pain when weight is placed on joint	Spread of candidiasis, prosthetic implants
Candidemia	Antibiotic-resistant fevers, tachycardia, hypotension, skin lesions, small abscesses in various organ systems	IV or urinary catheters, use of antibacterial drugs, surgery, severe burns, antibacterial therapy

Aspergillus also causes nonpulmonary disease when aspergillomas form in paranasal sinuses, ear canals, eyelids, the conjunctivas, brain, or eye sockets (FIGURE 22.12). Rarely, *cutaneous aspergillosis* results when the fungus is either introduced into the skin by trauma or spreads from the lungs in AIDS patients. Lesions begin as raised, red papules that progressively die. *Systemic aspergillosis*, which involves invasion of the major organ systems, occurs in AIDS patients and IV drug abusers. The fungus produces abscesses in the brain, kidneys, heart, bones, and gastrointestinal tract. Systemic aspergillosis is often fatal, especially when the brain is involved.

Clinical history and radiographs demonstrating abnormal lung structure are suggestive of aspergillosis, but diagnosis must be confirmed via laboratory techniques. The presence of septate hyphae and distinctive conidia in KOH- and GMS-prepared samples taken from a patient are diagnostic. Immunological tests to detect antigens in the blood are confirmatory.

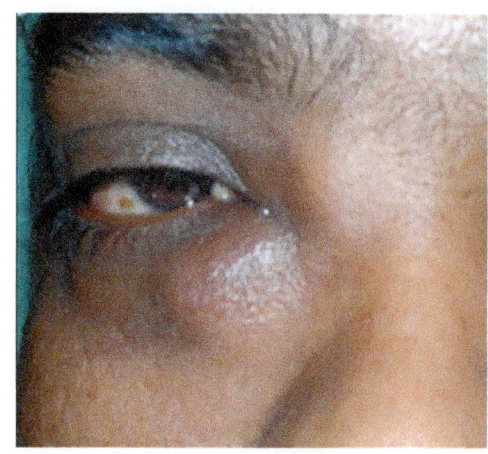

▲ FIGURE 22.12 **An invasive aspergilloma near the eye.** *In what group of individuals is aspergillosis an emerging disease?*

Figure 22.12 Aspergillosis is an emerging disease of AIDS patients.

Treatment of hypersensitivity involves either the use of various allergy medications or desensitization to the allergen. Invasive disease is treated by surgical removal of aspergillomas and surrounding tissues plus high-dose, intravenous administration of voriconazole. Maintenance therapy with itraconazole is suggested for preventing relapse in AIDS patients.

Cryptococcosis

LEARNING OUTCOMES

22.20 Describe the development of cryptococcosis and its diagnosis and treatment.

22.21 Discuss the characteristics of *Cryptococcus* that contribute to the severity of cryptococcoses in immunocompromised patients.

A basidiomycete, *Cryptococcus neoformans* (krip-tō-kok'ŭs nē-ō-for'manz), is the primary species causing **cryptococcoses** (krip'tō-kok-ō'sēz). Two varieties are known. *C. neoformans gattii* (găt'ē-ē) is found in Australia, Papua New Guinea, parts of Africa, the Mediterranean, India, Southeast Asia, parts of Central and South America, and Southern California. It infects primarily immunocompetent individuals. *C. neoformans neoformans* is found worldwide and infects mainly AIDS patients. Approximately 50% of all cryptococcal infections reported each year are due to strain *neoformans*.

Human infections follow the inhalation of spores or dried yeast in aerosols from the droppings of birds. People who work around buildings where birds roost are at increased risk of infection. Hospitals, convalescent homes, and other such facilities often place devices that deter the roosting of birds near air-intake vents in an effort to prevent *Cryptococcus*-contaminated air from entering a building.

The pathogenesis of *Cryptococcus* is enhanced by several characteristics of this fungus: the presence of a phagocytosis-resistant capsule surrounding the yeast form; the ability of the yeast to produce melanin, which further inhibits phagocytosis; and the organism's predilection for the central nervous system (CNS), which is isolated from the immune system by the blood-brain barrier. Scientists are actively investigating the processes by which *Cryptococcus* crosses the barrier. Cryptococcal infections also tend to appear in terminal AIDS patients when little immune function remains.

Primary pulmonary cryptococcosis is asymptomatic in most individuals, although some individuals experience a low-grade fever, cough, and mild chest pain. In a few cases, *invasive pulmonary cryptococcosis* occurs, resulting in chronic pneumonia in which cough and pulmonary lesions progressively worsen over a period of years.

Cryptococcal meningitis, the most common clinical form of cryptococcal infection, follows dissemination of the fungus to the CNS. Symptoms develop slowly and include headache, dizziness, drowsiness, irritability, confusion, nausea, vomiting, and neck stiffness. In late stages of the disease, loss of vision and coma occur. Acute onset of rapidly fatal cryptococcal meningitis occurs in individuals with widespread infection.

Cryptococcoma (krip'tō-kok-ō'mă) is a very rare condition in which solid fungal masses form in the cerebral hemispheres, cerebellum, or (rarely) in the spinal cord. The symptoms of this condition, which can be mistaken for cerebral tumors, are similar to those of cryptococcal meningitis but also include motor and neurological impairment.

Cutaneous cryptococcosis may also develop. Primary infections manifest as ulcerated skin lesions or as inflammation of subcutaneous tissues. Infection may resolve on its own, but patients should be monitored for infection spreading to the CNS. Secondary cutaneous lesions occur following spread of *Cryptococcus* to other areas of the body. In AIDS patients, cutaneous cryptococcosis is the second most common manifestation of *Cryptococcus* infection (after meningitis). Lesions are common on the head and neck.

MICROBE AT A GLANCE 22.2

Aspergillus

Taxonomy: Domain Eukarya, kingdom Fungi, phylum Ascomycota, class Ascomycetes, order Eurotiales, family Trichocomaceae; principally three species cause disease in humans: *A. fumigatus*, *A. niger*, and *A. flavus*

Morphology: Septate hyphae

Virulence factors: Allergenic surface molecules; opportunistic pathogen with little virulence except in immunocompromised patients via respiratory invasion

Diseases caused: Hypersensitivity aspergillosis (allergy), acute invasive pulmonary aspergillosis, cutaneous aspergillosis, systemic aspergillosis, endocarditis, gastrointestinal aspergillosis, disseminated aspergillosis (brain, eye, liver, kidneys), *Aspergillus* tracheobronchitis, chronic necrotizing aspergillosis

Treatment for diseases: Allergy: anti-inflammatory drugs and desensitization to allergens; other disease: primarily, voriconazole with or without echinocandins; secondarily, amphotericin B or isavuconazole; debridement of necrotic tissues and surgical removal of aspergillomas

Prevention of disease: Maintain a healthy immune system.

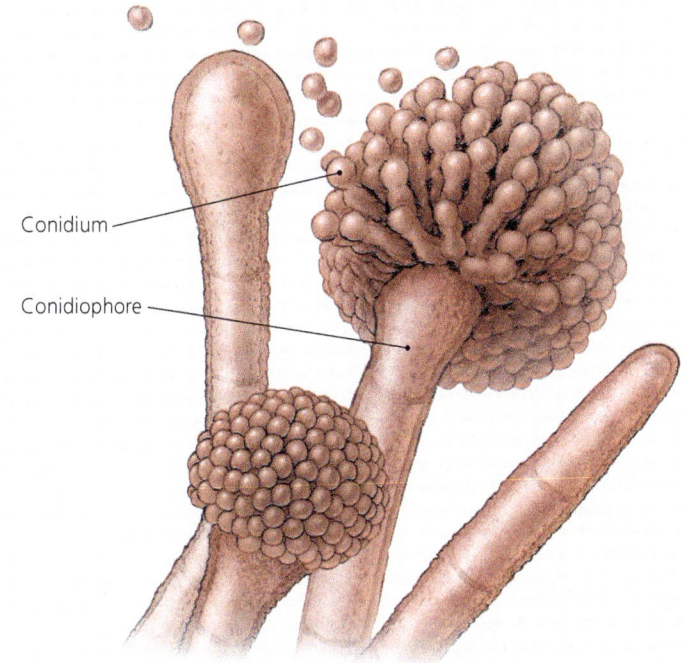

DISEASE IN DEPTH

CANDIDIASIS

Pseudohyphae stage of *Candida albicans*

SIGNS & SYMPTOMS

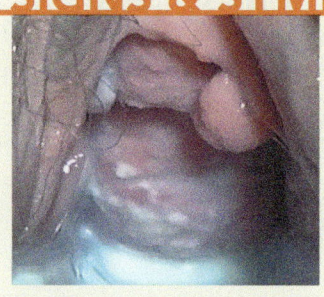

Candidiasis typically manifests as white plaques on the affected tissue. Vaginal candidiasis, shown on the left, manifests as white mucoid colonies growing on the mucous membrane of the vagina and the skin covering the vaginal labia. The yeast produces severe vaginal itching and burning, which urination intensifies. Sexual intercourse can be very painful. The vaginal discharge is often curdlike.

Candidiasis is any opportunistic fungal infection caused by various species of the genus *Candida*—most commonly *Candida albicans*. *Candida* species can infect tissues in essentially every body system, producing a wide range of disease manifestation in humans. In all cases, the fungus is opportunistic. Systemic disease is seen almost universally in immunocompromised individuals.

Cutaneous
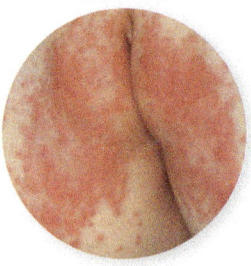

Onychomycosis (nail infection)
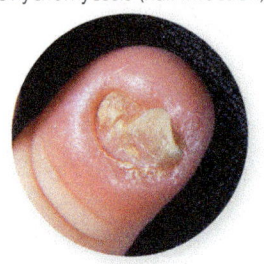

Body system manifestations (labeled on figure):
- Meningeal
- Ocular
- Oropharyngeal (thrush)
- Candidemia (in blood)
- Esophageal
- Pulmonary
- Endocardial, myocardial, pericardial
- Gastrointestinal
- Hepatic and splenic
- Neonatal and congenital (in females)
- Peritoneal
- Urinary tract
- Cutaneous
- Vulvovaginal (in females)
- Onychomycosis (nail infection)
- Osteoarticular

Oropharyngeal (thrush)
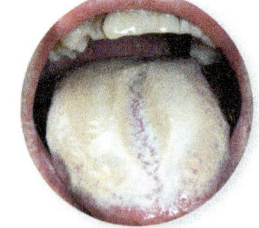

Esophageal
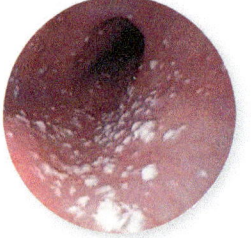

Hepatic candidiasis

 Watch Dr. Bauman's Video Tutor explore candidiasis, or go to the **Mastering Microbiology** Study Area for more information.

EPIDEMIOLOGY

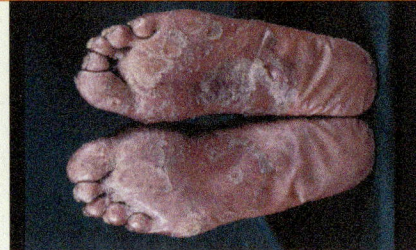

Cutaneous candidiasis in AIDS patient

There are 60,000–70,000 cases of candidiasis in the U.S. each year. Different segments of the population experience a given manifestation of candidiasis at different rates. For example, oral candidiasis is rare in healthy individuals, but it affects 10% of the elderly and nearly all those infected with HIV once they develop clinical AIDS. Predisposing factors for the development of candidiasis include cancer, invasive hospital procedures, antibacterial treatments (which inhibit normal bacteria that compete with *Candida*), diabetes, severe burns, intravenous drug abuse, and AIDS. Even with treatment, the death rate from systemic *Candida* infections can be as high as 40%.

PATHOGEN

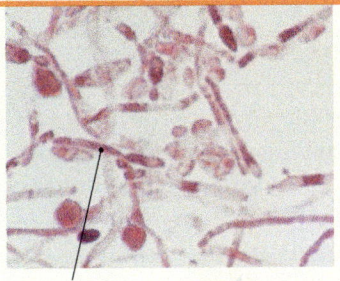

Pseudohypha — LM — 10 μm

Candida albicans, the most common species causing candidiasis, is an opportunistic ascomycete yeast. It forms long cellular extensions called **pseudohyphae** because they appear similar to the true hyphae of filamentous fungi. *Candida* spp. are common members of the microbiome of the skin and mucous membranes; for example, the digestive and reproductive tracts of 40–80% of all healthy individuals harbor the yeast.

DIAGNOSIS

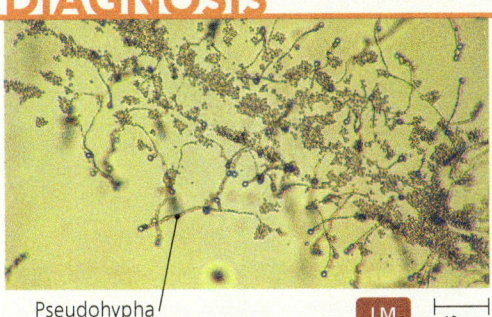

Pseudohypha — LM — 40 μm

Physicians diagnose candidiasis on the basis of signs and demonstration of clusters of budding yeasts and pseudohyphae, which are buds remaining attached to the parent cell and appearing filamentous, like hyphae.

PATHOGENESIS (Vaginal Candidiasis)

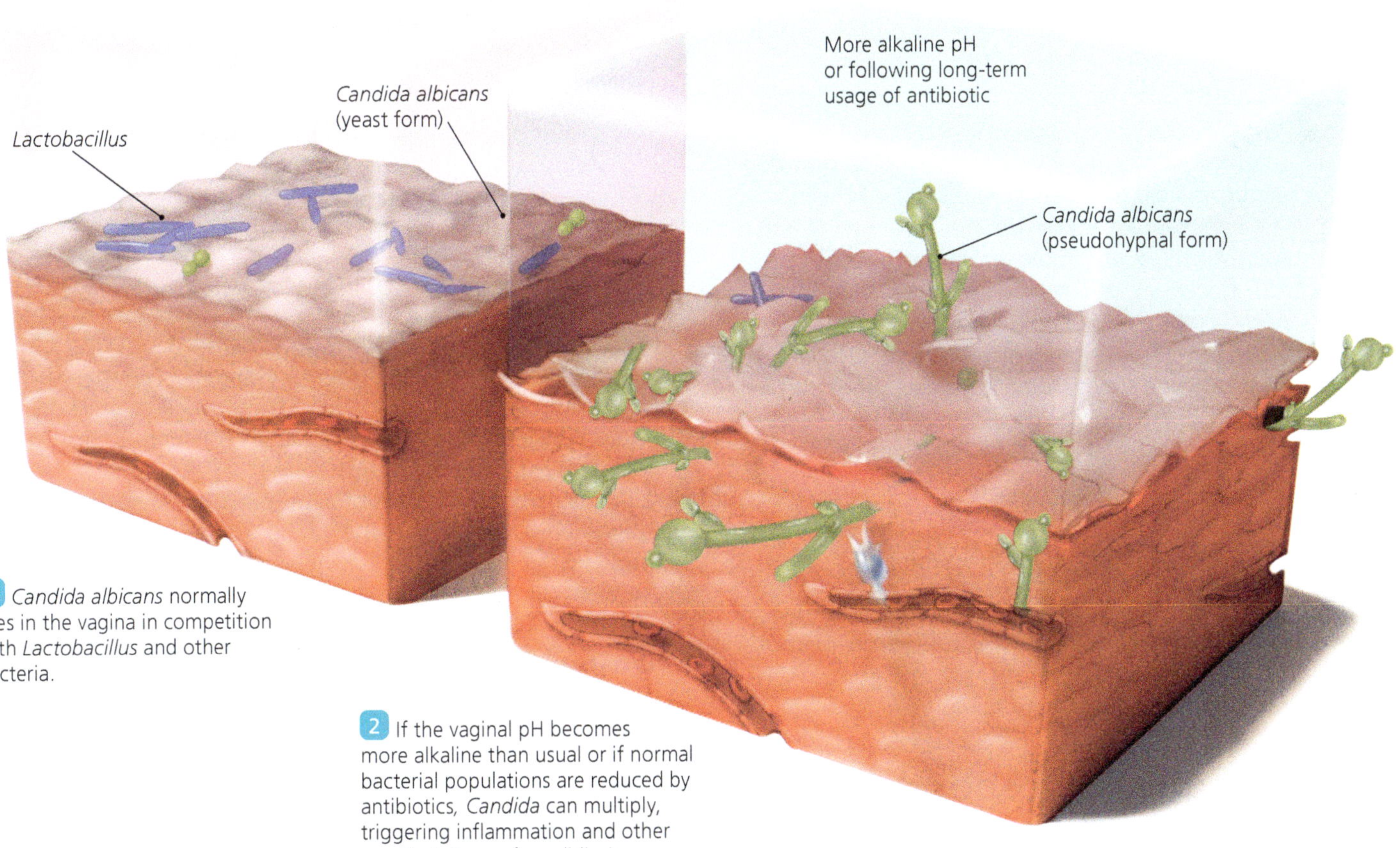

Normal pH

Lactobacillus

Candida albicans (yeast form)

More alkaline pH or following long-term usage of antibiotic

Candida albicans (pseudohyphal form)

1 *Candida albicans* normally lives in the vagina in competition with *Lactobacillus* and other bacteria.

2 If the vaginal pH becomes more alkaline than usual or if normal bacterial populations are reduced by antibiotics, *Candida* can multiply, triggering inflammation and other manifestations of candidiasis.

TREATMENT & PREVENTION

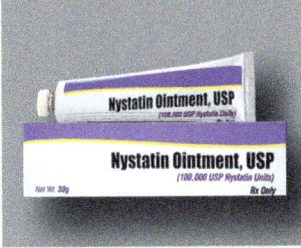

In immunocompetent patients, for whom excessive friction and body moisture in skin folds or preexisting diseases are the reasons for extensive, external fungal growth, resolution involves treating the underlying problems in addition to administering topical antifungal agents. Oral candidiasis in infants and children is treated with nystatin. Azole suppositories, creams, and/or oral administration of fluconazole are used for vaginal candidiasis. Avoid persistent moisture in genital area; for example, do not wear wet bathing suits for prolonged periods.

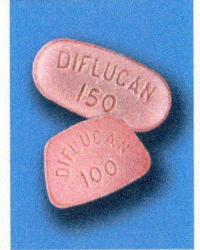

The same infections in AIDS patients are much more difficult to treat because most superficial or mucous membrane infections in these patients do not respond well to topical antifungal treatments. Orally administered fluconazole is used for primary and maintenance therapy for candidiasis in AIDS patients. Invasive candidiasis requires treatment with amphotericin B and often with 5-fluorocytosine. Prophylactic treatment in AIDS patients may include oral fluconazole.

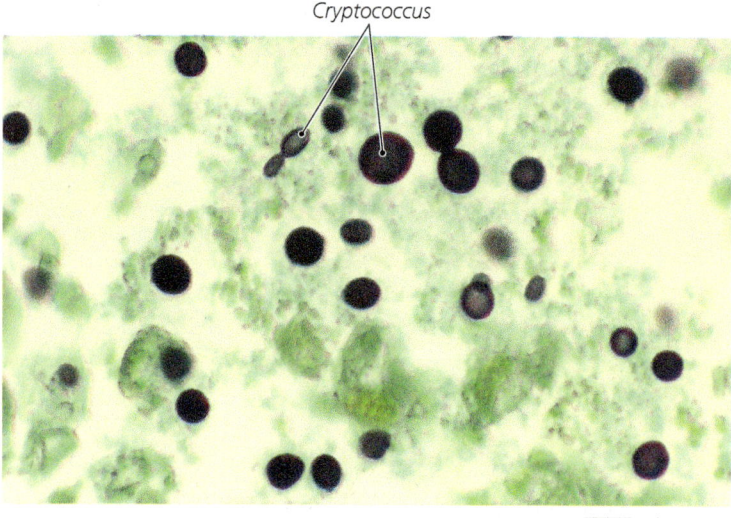

▲ FIGURE 22.13 **GMS stain of *Cryptococcus*.** Yeasts of this dimorphic fungus multiply in the lungs, as shown, and may spread to other parts of the body.

Diagnosis involves collecting specimen samples that correlate with the symptoms. GMS stains revealing the presence of encapsulated yeast in cerebrospinal fluid are highly suggestive of cryptococcal meningitis, even if no overt symptoms are present (FIGURE 22.13). Recovery of *Cryptococcus* from respiratory secretions is generally not diagnostic because of general environmental exposure, but any patient testing positive for *C. neoformans* should be monitored for systemic disease. The preferred method of confirming cryptococcal meningitis is detection of fungal antigens in cerebrospinal fluid. In AIDS patients, antigens can be detected in serum as well.

Treatment for cryptococcoses is with amphotericin B and 5-fluorocytosine administered together for 6–10 weeks. The synergistic action of the two drugs allows lower doses of amphotericin B to be used, but toxicity is not completely eliminated. Although AIDS patients may appear well following primary treatment, the fungus typically remains and must be actively suppressed by lifelong use of oral fluconazole.

Zygomycoses and Microsporidiosis

LEARNING OUTCOMES

22.22 Describe the clinical forms of zygomycoses.
22.23 Describe microsporidia and their unique method of infection.

Zygomycete fungi and a genetically similar group of very small intracellular parasites called *microsporidia* can cause disease opportunistically.

Zygomycoses

Zygomycoses (zī′gō-mī-kō′sez) are opportunistic fungal infections caused by various genera of fungi classified in the division Zygomycota, especially *Mucor* (mū′kōr; or sometimes *Rhizopus* [rī-zō′pŭs] or *Absidia* [ab-sid′ē-ă]). All three have a worldwide distribution and are extremely common in soil, on decaying organic matter, or as contaminants that cause food spoilage (*Rhizopus* is the classical bread mold).

Zygomycoses are commonly seen in patients with uncontrolled diabetes, in people who inject illegal drugs, in some cancer patients, and in some patients receiving antimicrobial agents. Infections generally develop in the face and head area but can develop elsewhere and may spread throughout the bodies of severely immunocompromised individuals, resulting in the following conditions:

- *Rhinocerebral zygomycosis* begins with infection of the nasal sinuses following inhalation of spores. The fungus spreads to the mouth and nose, producing macroscopic cottonlike growths. *Mucor* can subsequently invade blood vessels, where it produces fibrous clots, causes tissue death, and subsequently invades the brain, which is fatal within days, even with treatment.
- *Pulmonary zygomycosis* follows inhalation of spores (as from moldy foods). The fungus kills lung tissue, resulting in the formation of cavities.
- *Gastrointestinal zygomycosis* involves ulcers in the intestinal tract.
- *Cutaneous zygomycosis* results from the introduction of fungi through the skin after trauma (such as burns or needle punctures). Lesions range from pustules and ulcers to abscesses and dead patches of skin.

Diagnosis usually involves microscopic findings of fungus. Samples may be obtained from scrapings of lesions, lung aspirates, or biopsies. KOH- or GMS-stained tissue sections reveal hyphae with irregular branching and few septate divisions.

Treatment, which should begin as soon as infection is suspected, involves physical removal of infected tissues and management of predisposing factors. Amphotericin B administered intravenously for 8–10 weeks is the drug of choice for zygomycoses.

Microsporidiosis

Microsporidia (mī′krō-spor-i′dē-ă; singular: microsporidium) are obligate, intracellular, fungal pathogens of invertebrates (primarily insects) and vertebrates, including humans. These unique microbes were long classified as protozoa, but genetic analysis has revealed them to be more closely related to zygomycetes, though the exact nature of that relationship is unknown.

Microsporidia, such as members of the genera *Microsporidium*, *Encephalitozoon* (en-sef-a-līt′o-zō′an), and *Nosema* (nō-sē′mă), have the smallest genomes of any eukaryotes. They are so small that they have no room for all of the genes coding for metabolic proteins; thus, microsporidia must invade other cells in order to acquire enzymes for metabolism. Microsporidia do this in a unique way (FIGURE 22.14): Their infective spores have an internal, hollow, coiled tubule ❶. The spore unwinds its tubule and penetrates a host cell as if the tubule were a hypodermic needle ❷. The microsporidium injects all of its own cytoplasm into the host ❸. Once ensconced, it uses the host's metabolic chemicals to grow and divide, eventually consuming the host

CLINICAL CASE STUDY

Disease from a Cave

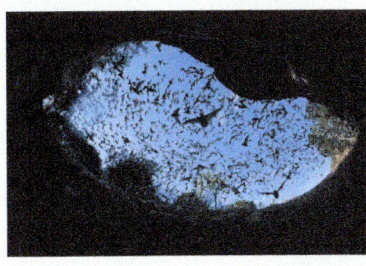

Nate lives an exciting life. His part of Kentucky is beautiful, and he has opportunities for adventure with his friends in Boy Scout Troop 138. They have been hiking, climbing, backpacking, and mountaineering; and just 10 days ago, they had spent all night in a cave deep below the town. After watching what seemed like millions of bats rush and swoop out of the cave, the troop had entered the cavern system through another entrance—a "squeeze entrance." Nate wasn't sure what a squeeze entrance was until he had to inch forward for over a hundred feet with his arms stretched in front of him, pulling with his fingertips as his back scraped the ceiling. That had been the worst part. But after clearing the entrance, the teenagers had explored, scrambled, and crawled the whole night. They had even eaten underground. It was extreme, and a little spooky, with bizarre geological formations, pools of ice-cold water, and total darkness. The guys were still talking about the time they stayed up all night inside the Earth.

Today, though, Nate doesn't feel so excited. His nonproductive coughing hurts; he feels cold and likely has a fever; and he has red bumps on the skin of his legs. His dad takes him to the doctor, who prescribes a medicine that neither Nate nor his parents had heard of—keto-something.

1. What disease is likely troubling Nate?
2. What is the name of the drug that Nate is taking?
3. How did Nate contract the disease?
4. Is Nate contagious to any of his friends in Troop 138?
5. How likely is it that other teenagers from the troop are infected?
6. How likely is it that other troop members are diseased?

---

cell from within ④. The microsporidia become new spores ⑤ that are released when the host cell dies ⑥.

Diseases of microsporidia, called microsporidioses, occur primarily in immunocompromised patients, such as AIDS patients. The most common clinical manifestation is severe diarrhea, but respiratory, muscular, and nervous system infections may manifest with coughing, myalgia (muscle pain), and headache and seizures, respectively.

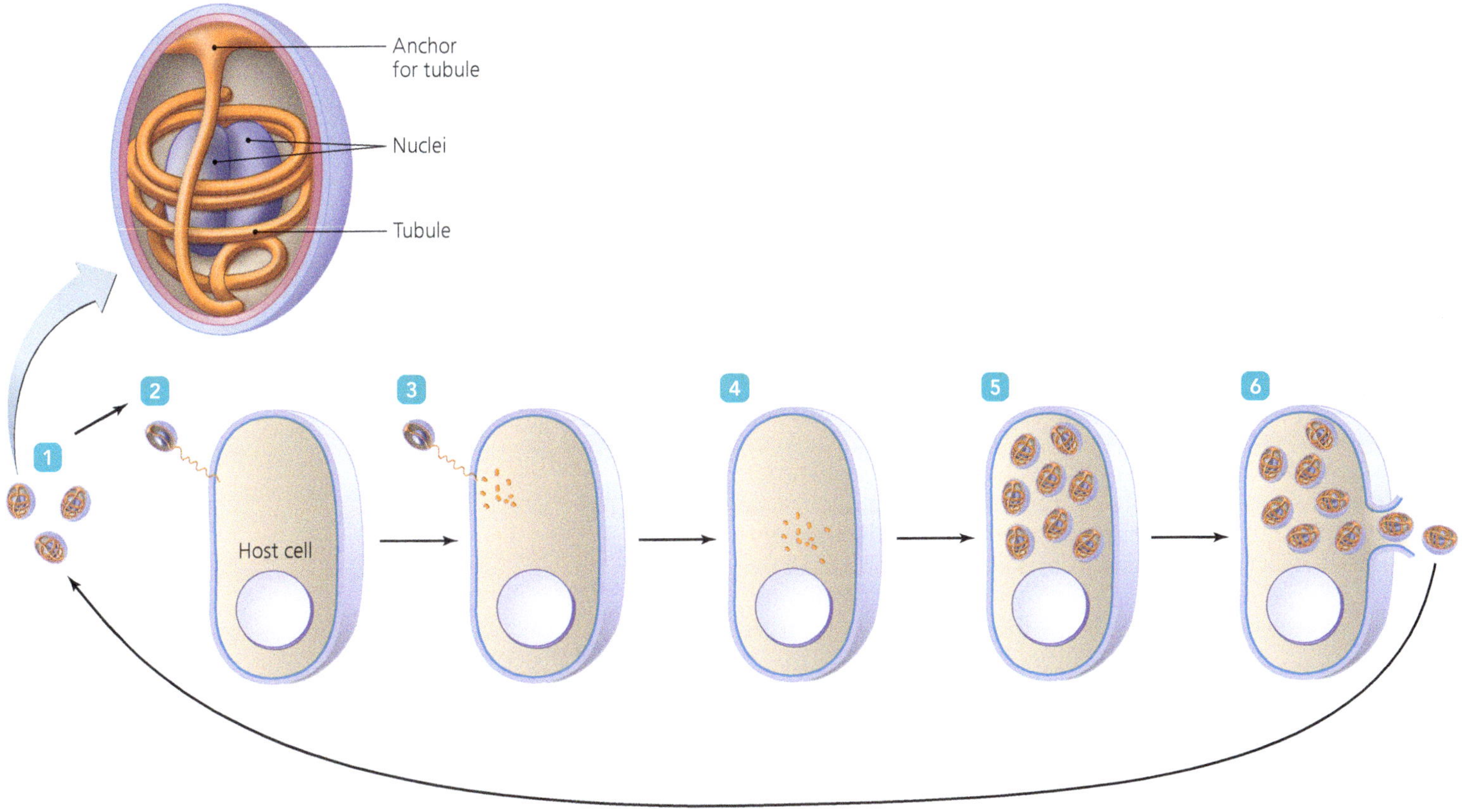

▲ FIGURE 22.14 Life cycle of a typical microsporidium.

Health care practitioners treat microsporidiosis by replacing fluids and electrolytes lost in diarrhea. Albendazole is the drug of choice.

The Emergence of Fungal Opportunists in AIDS Patients

LEARNING OUTCOMES

22.24 Identify several emerging fungal opportunists seen among AIDS patients.
22.25 Identify the factors that contribute to the emergence of new fungal opportunists.

Immunosuppression is not unique to AIDS patients, but because HIV systematically destroys functional immunity, AIDS patients are extremely vulnerable to opportunistic fungal infections. In otherwise healthy people, immunosuppressive episodes are transitory and result from surgery, chemotherapy, or an acute illness. With AIDS, by contrast, immune dysfunction is permanent. For this reason, once an opportunistic infection is established in an AIDS patient, it is not likely ever to be cured fully. In fact, mycoses account for most deaths associated with AIDS. *Pneumocystis jirovecii*, *Candida albicans*, *Aspergillus fumigatus* (fū-mĭ-gā'tus), and *Cryptococcus neoformans* are so common in HIV-positive individuals that their mycoses are part of what defines end-stage AIDS.

The emergence of new fungal opportunists results from a combination of factors. First, the number of immunocompromised individuals is increasing as a result of the spread of AIDS. Then the widespread use of antifungal drugs in this enlarging group of immunocompromised individuals selects fungi resistant to the drugs. Thus, as the immunocompromised population grows, fungi can be expected to increasingly cause illness and death. An AIDS patient's lack of immunity allows a normally superficial fungus to gain access to internal systems, with significant, even fatal, results. Since the 1980s, several new fungal opportunists have been identified, including *Fusarium* spp. and *Talaromyces marneffei*, which are ascomycetes, and *Trichosporon beigelii*, which is a basidiomycete.

Fusarium (fū-zā'rē-ŭm) species cause respiratory distress, disseminated infections, and *fungemia* (fungi in the bloodstream). These species also produce toxins that can accumulate to dangerous levels when ingested in food. *Fusarium* spp. are resistant to most antifungal agents.

Talaromyces marneffei (tal-a-rō-mī'sēz mar-nef-ē'ī) is a dimorphic, invasive fungus that causes pulmonary disease upon inhalation. It is the third most frequent illness seen in AIDS patients in Southeast Asia (behind tuberculosis and cryptococcosis).

When *Trichosporon beigelii* (trik-ō-spōr'on bā-gĕl-ē'ē) enters an AIDS patient through the lungs, the gastrointestinal tract, or catheters, it causes a drug-resistant systemic disease that is typically fatal.

To combat the threat posed by these opportunists, health care providers maintain rigorous hygiene during medical procedures. Researchers are developing new therapies that limit immunosuppression in all patients, including those with AIDS.

MICRO CHECK

8. How is *Pneumocystis jirovecii* different from other opportunistic fungal pathogens?
9. Which opportunistic fungus causes thrush?
10. What is the most common form of aspergillosis?
11. How does melanin production increase the severity of *Cryptococcus neoformans* infections in immunocompromised patients?
12. Which form of zygomycosis results in the formation of macroscopic cottonlike fungal growths in the mouth and nose?

TELL ME WHY

Why is candidiasis the most common opportunistic fungal pathogen?

Superficial, Cutaneous, and Subcutaneous Mycoses

LEARNING OUTCOME

22.26 Describe the general manifestations of superficial, cutaneous, and subcutaneous mycoses.

All the mycoses we discuss in the following sections are localized at sites at or near the surface of the body. They are the most commonly reported fungal diseases. All are opportunistic infections, but unlike those we have just discussed, they can be acquired both through environmental exposure and more frequently via person-to-person contact. Most of these fungi are not life threatening, but they often cause chronic, recurring infections and diseases.

Superficial Mycoses

LEARNING OUTCOME

22.27 Compare and contrast the clinical and diagnostic features of dermatophytosis and *Malassezia* infections.

Superficial mycoses are the most common fungal infections. They are confined to the outer, dead layers of the skin, nails, or hair, all of which are composed of dead cells filled with a protein called *keratin*—the primary food of these fungi. In AIDS patients, superficial mycoses can spread to cover significant areas of skin or become systemic.

Dermatophytoses

Dermatophytoses (der'mă-tō-fī-tō'sēz) are infections caused by *dermatophytes*, which are fungi that grow on skin, nails, and hair. The most common dermatophytosis is *athlete's foot* (FIGURE 22.15). In the past, such infections were often called *ringworms* because dermatophytes can produce circular, scaly patches that resemble a worm lying just below the surface of

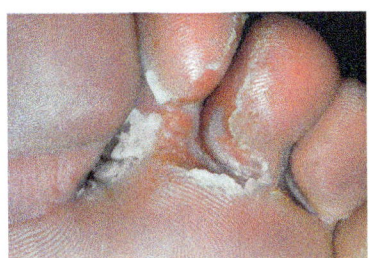

▲ FIGURE 22.15 **Athlete's foot.** Dermatophytosis of skin on the foot.

the skin. Even though we know worms are not involved, the medical names for many dermatophytoses use the term *tinea* (tin'ē-a), which is Latin for "worm." Dermatophytes use keratin as a nutrient source and thus colonize only dead tissue. The fungi may provoke cell-mediated immune responses, which can damage living tissues. Dermatophytes are among the few contagious fungi, that is, fungi that spread from person to person. Spores and bits of hyphae are constantly shed from infected individuals, making recurrent infections common.

Three genera of ascomycetes are responsible for most dermatophytoses, including athlete's foot: (1) *Trichophyton* (trik-ō-fī'ton) species, (2) *Epidermophyton floccosum* (ep'i-der-mof'i-ton flŏk'ō-sŭm), and (3) *Microsporum* (mī-kros'po-rŭm) species. All three cause skin and nail infections, but *Trichophyton* species also infect hair. Rare subcutaneous infections occur in AIDS patients.

Dermatophytoses can have a variety of clinical manifestations, some of which are summarized in TABLE 22.3. Most such diseases are clinically distinctive and so common that they are readily recognized. KOH preparations of skin or nail scrapings or hair samples reveal hyphae and arthroconidia, which confirm a diagnosis. Determination of a specific dermatophyte requires laboratory culture, but it is usually unnecessary because treatment is the same for all.

Limited infections are treated with topical antifungal agents, but more widespread infections of the scalp or skin, as well as nail infections, must be treated with oral agents. Terbinafine, administered for 6–12 weeks, is effective in most cases. Chronic or stubborn cases are treated with griseofulvin until cured.

Malassezia Infections

Malassezia furfur (mal-ă-sē'zē-ă fur'fur) is a dimorphic basidiomycete that is a normal member of the microbiome of the skin of humans worldwide. It feeds on the skin's oil and causes common, chronic superficial infections.

Infections with *M. furfur* result in **pityriasis** (pit-i-rī'ă-sis), characterized by depigmented or hyperpigmented patches of scaly skin resulting from fungal interference with melanin production (FIGURE 22.16 on the next page). This condition typically occurs on the trunk, shoulders, and arms, and rarely on the face and neck. KOH preparations of clinical specimens reveal masses of budding yeast and short hyphal forms that are diagnostic for *M. furfur*.

Superficial *Malassezia* infections are treated topically with solutions of antifungal imidazoles, such as ketoconazole shampoos. Alternatively, topical applications of selenium sulfide lotions can be used. Oral therapy with ketoconazole may be required to treat extensive infections or infections that do not respond to topical treatments. Relapses of *Malassezia* infections are common, and prophylactic topical treatment may be necessary. The skin takes months to regain its normal pigmentation following successful treatment of pityriasis.

TABLE 22.3 Common Dermatophytoses

Disease	Agents	Common Signs	Source
Tinea pedis ("athlete's foot")	*Trichophyton rubrum*; *T. mentagrophytes* var. *interdigitale*; *Epidermophyton floccosum*	Red, raised lesions on and around the toes and soles of the feet; webbing between the toes is heavily infected	Human reservoirs in toe webbing; carpeting holding infected skin cells
Tinea cruris ("jock itch")	*T. rubrum*; *T. mentagrophytes* var. *interdigitale*; *E. floccosum*	Red, raised lesions on and around the groin and buttocks	Usually spreads from the feet
Tinea unguium (onychomycosis)	*T. rubrum*; *T. mentagrophytes* var. *interdigitale*	*Superficial white onychomycosis:* patches or pits on the nail surface	Humans
		Invasive onychomycosis: yellowing and thickening of the distal nail plate, often leading to loss of the nail	
Tinea corporis	*T. rubrum*; *Microsporum gypseum*; *M. canis*	Red, raised, ringlike lesions occurring on various skin surfaces (tinea corporis on the trunk, tinea capitis on the scalp, tinea barbae of the beard)	Can spread from other body sites; can be acquired following contact with contaminated soil or animals
Tinea capitis	*M. canis*; *M. gypseum*; *T. equinum*; *T. verrucosum*; *T. tonsurans*; *T. violaceum*; *T. schoenleinii*	*Ectothrix invasion:* fungus develops arthroconidia on the outside of the hair shafts, destroying the cuticle	Humans; can be acquired following contact with contaminated soil or animals
		Endothrix invasion: fungus develops arthroconidia inside the hair shaft without destruction	
		Favus: crusts form on the scalp, with associated hair loss	

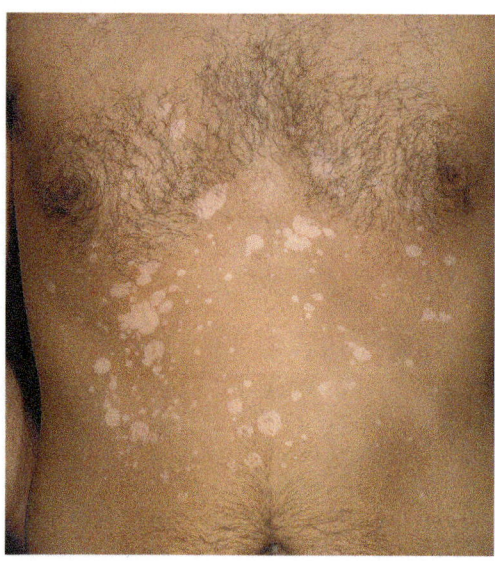

▲ FIGURE 22.16 **Pityriasis.** The variably pigmented skin patches are caused by *Malassezia furfur*.

Cutaneous and Subcutaneous Mycoses

LEARNING OUTCOMES

22.28 Explain why cutaneous and subcutaneous mycoses are not as common as superficial mycoses.
22.29 State the specific difference between chromoblastomycosis and phaeohyphomycosis.
22.30 Describe a mycetoma.
22.31 Describe how the lesions of lymphocutaneous sporotrichosis correlate with its spread through the lymphatic system.

The fungi involved in cutaneous and subcutaneous mycoses are common soil saprobes (organisms that live on dead organisms), but the diseases they produce are not as common as superficial mycoses because infection requires traumatic introduction of fungi through the dead outer layers of skin into the deeper, living tissue. Most lesions remain localized just below the skin, though infections may rarely become systemic.

Chromoblastomycosis and Phaeohyphomycosis

Chromoblastomycosis (krō′mō-blas′tō-mī-kō′sis) and **phaeohyphomycosis** (fē′ō-hī′fō-mī-kō′sis) are similar-appearing cutaneous and subcutaneous mycoses caused by dark-pigmented ascomycetes. TABLE 22.4 lists some ascomycetes that commonly cause these mycoses.

Initially, chromoblastomycosis uniformly presents as small, scaly, itchy, but painless lesions on the skin surface resulting from fungal growth in subcutaneous tissues near the site of inoculation. Over the course of years and decades, the lesions progressively worsen, becoming large, flat to thick, tough, and wartlike. They become tumorlike and extensive if not treated (FIGURE 22.17). Inflammation, fibrosis, and abscess formation occur in surrounding tissues. The fungus can spread throughout the body.

TABLE 22.4 Some Ascomycete Genera of Cutaneous and Subcutaneous Mycoses

| **Chromoblastomycosis** |
| *Fonsecaea* |
| *Phialophora* |
| *Cladophialophora* |
| **Phaeohyphomycosis** |
| *Alternaria* |
| *Exophiala* |
| *Wangiella* |
| *Cladophialophora* |
| **Mycetoma** |
| *Madurella* |
| *Pseudallescheria* |
| *Exophiala* |
| *Acremonium* |
| **Sporotrichosis** |
| *Sporothrix* |

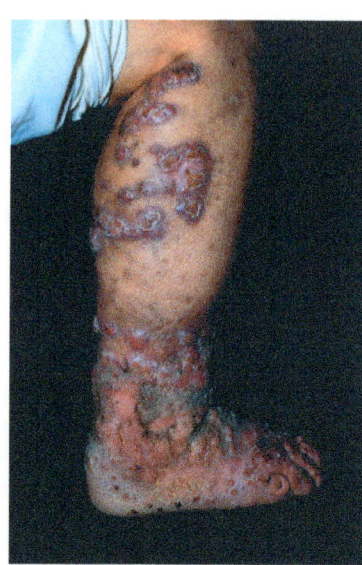

▲ FIGURE 22.17 **A leg with extensive lesions of chromoblastomycosis.** In this case they are caused by *Fonsecaea pedrosoi*.

Phaeohyphomycoses are more variable in presentation, spreading from the skin to colonize the nasal passages and sinuses in allergy sufferers and AIDS patients or the brains of AIDS patients. Fortunately, brain infection is the rarest form of phaeohyphomycosis and occurs only in the severely immunocompromised.

KOH preparations and GMS staining of skin scrapings, biopsy material, or cerebrospinal fluid reveal the key distinguishing feature between the two diseases: the microscopic morphology of the fungal cells within tissues. Tissue sections from chromoblastomycosis cases contain golden brown *sclerotic bodies* that are distinctive and distinguishable from budding yeast forms (FIGURE 22.18a), whereas tissues from phaeohyphomycosis cases contain brown-pigmented hyphae (FIGURE 22.18b). Determining

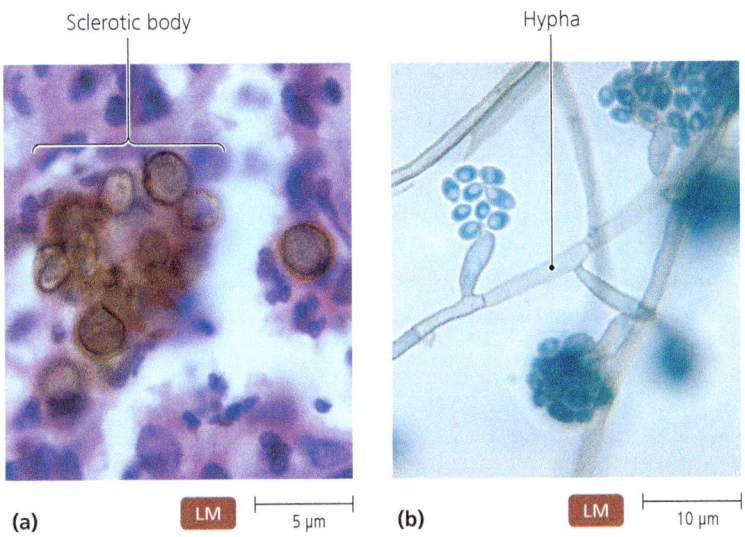

▲ FIGURE 22.18 **Microscopic differences between chromoblastomycosis and phaeohyphomycosis.** (a) Sclerotic bodies seen in chromoblastomycoses. (b) Hyphal forms found in tissues during phaeohyphomycosis, here caused by *Exophiala*.

the species of fungus causing these two diseases requires laboratory culture, macroscopic examination of colonies on agar, and microscopic examination of spores; however, determination of the specific cause does not affect the treatment options.

Both diseases are difficult to treat, especially in advanced cases. Some cases can be treated with itraconazole, but the diseases can be permanently destructive to tissues. They require surgical removal of infected and surrounding tissues followed by antifungal therapy. Extensive lesions may require amputation. The earlier treatment begins, the more likely it will be successful.

Despite the worldwide occurrence of the relevant fungi, the overall incidence of infection is relatively low. People who work daily in the soil with bare feet are at risk if they incur foot wounds. The simple act of wearing shoes would greatly reduce the number of infections.

Mycetomas

Fungal **mycetomas** (mī-sē-tō′măs) are tumorlike infections caused by fungi of several genera in the division Ascomycota (see Table 22.4 on p. 666). These fungi are distributed worldwide, but infection is most prevalent in countries near the equator. The cases that occur in the United States are almost always caused by *Pseudallescheria* (sood′al-es-kē-rē-ă) or *Exophiala* (ek-so-fī′ă-lă).

Mycetoma-producing fungi live in the soil and are introduced into humans via wounds caused by twigs, thorns, or leaves contaminated with fungi. As with chromoblastomycosis and phaeohyphomycosis, those who work barefoot in soil are most at risk, and wearing protective shoes or clothing can greatly reduce incidence.

Infection begins near the site of inoculation with the formation of small, hard, subsurface nodules that slowly worsen and spread as time passes. Local swelling occurs, and ulcerated lesions begin to produce pus. Infected areas release an oily fluid containing fungal "granules" (spores and fungal elements). The fungi spread to more tissues, destroying bone and causing permanent deformity (FIGURE 22.19).

A combination of the symptoms and microscopic demonstration of fungi in samples from the infected area is diagnostic for mycetomas. Laboratory culture of specimens produces macroscopic colonies for which the species can be identified.

Treatment involves surgical removal of the mycetoma, which may, in severe cases, involve amputation of an infected limb. Surgery is followed by one to three years of antifungal therapy with a combination of four drugs—trimethoprim, sulfamethoxazole, dapsone, and streptomycin. Even combinations of surgery and antifungal agents are not always effective, and the disease is chronic, progressively more destructive, and can be fatal.

Sporotrichosis

Sporothrix schenckii (spōr′ō-thriks shen′kē-ē) is a dimorphic ascomycete that causes **sporotrichosis** (spōr′ō-tri-kō′sis), a subcutaneous infection usually limited to the arms or legs. *S. schenckii* resides in the soil and is most commonly introduced by thorn pricks or wood splinters. Avid gardeners, farmers, and artisans who work with natural plant materials have the highest incidence of sporotrichosis. Though distributed throughout the tropics and subtropics, most cases occur in Latin America, Mexico, and Africa. The disease is also common in warm, moist areas of the United States.

Sporotrichosis initially appears as painless, nodular lesions that form around the site of inoculation. With time, these lesions produce a pus-filled discharge, but they remain localized and do not spread. If the fungus enters the lymphatic system from a primary lesion, it gives rise to secondary lesions on the skin surface along the course of lymphatic vessels (FIGURE 22.20 on the next page). The fungus remains restricted to subcutaneous tissues and does not enter the blood.

Microscopic observation of pus or biopsy tissue stained with GMS can reveal budding yeast forms in severe infections, but often the fungus is present at a low density, making direct examination of clinical samples a difficult method of diagnosis. The patient's history and clinical signs plus the observation of

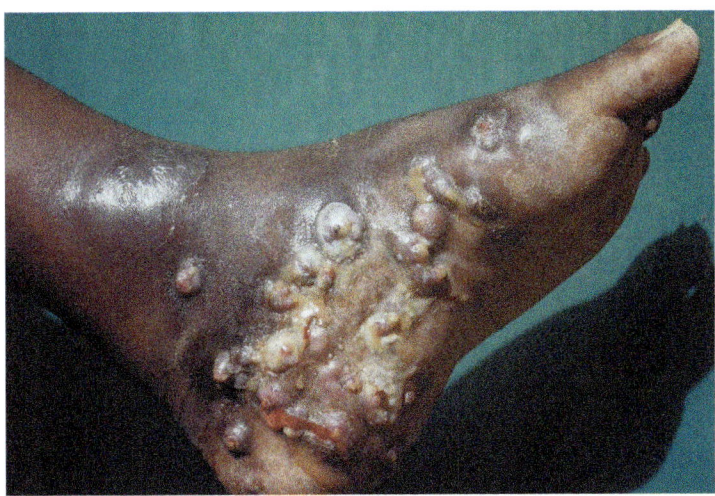

▲ FIGURE 22.19 **A mycetoma of the ankle.** Here, it results from the invasion and destruction of bone and other tissues by the fungus *Madurella mycetomatis*.

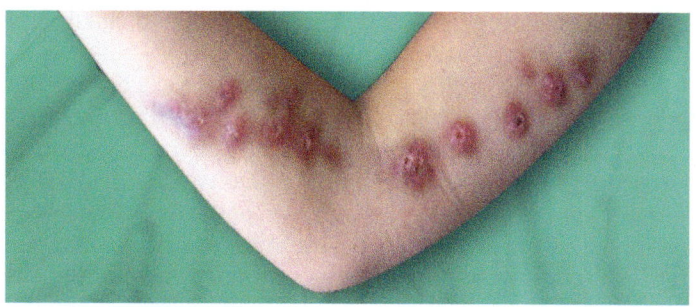

▲ FIGURE 22.20 **Sporotrichosis on the arm.** The locations of these secondary, subcutaneous lesions correspond to the course of lymphatic vessels leading from sites of primary, surface lesions. *How is sporotrichosis contracted?*

Figure 22.20 *Sporotrichosis is contracted when the soilborne fungus is inoculated into the skin by thorn pricks and scratches.*

the dimorphic nature of *S. schenckii* in laboratory culture are considered diagnostic for sporotrichosis.

Cutaneous lesions can be treated successfully with topical applications of saturated potassium iodide for several months, though some patients poorly tolerate this salt. Itraconazole or fluconazole are also useful treatments. Prevention requires the wearing of gloves, long clothing, and shoes to prevent inoculation.

> **TELL ME WHY**
>
> Why are many superficial mycoses, such as athlete's foot, chronic, recurring problems?

Fungal Intoxications and Allergies

LEARNING OUTCOMES

22.32 Compare and contrast mycotoxicosis, mycetismus, and fungal allergies.
22.33 Define *mycotoxin*.

Some fungi produce toxins or cause allergies. Fungal toxins, called **mycotoxins** (mī′kō-tok-sinz), are low-molecular-weight metabolites that can harm humans and animals that ingest them, causing *toxicosis* (poisoning). **Mycotoxicosis** (mī′kō-tok-si-kō-sis) is caused by eating mycotoxins; the fungus itself is not present. **Mycetismus** (mī′sē-tiz′mŭs) is mushroom poisoning resulting from eating mycotoxins still inside the fungus. Fungal *allergens* are usually proteins or glycoproteins that elicit hypersensitivity reactions in sensitive people who contact them.

Mycotoxicoses

LEARNING OUTCOME

22.34 Identify the most common group of mycotoxins.

Fungi produce mycotoxins during their normal metabolic activities. People most commonly consume mycotoxins in grains or vegetables that have become contaminated with fungi. Some mycotoxins can also be ingested in milk from a cow that has ingested toxin-contaminated feed. Up to 25% of the world's food supply is contaminated with mycotoxins, but only 20 of the 300 or so known toxins are ever present at dangerous levels. Long-term ingestion of mycotoxins can cause liver and kidney damage, gastrointestinal or gynecological disturbances, or cancers; each mycotoxin produces a specific clinical manifestation.

Aflatoxins (af′lă-tok′sinz) produced by the ascomycete *Aspergillus* are the best-known mycotoxins. Aflatoxins are fatal to many vertebrates and are carcinogenic at low levels when consumed continually. Aflatoxins cause liver damage and liver cancer throughout the world, but *aflatoxicosis* is most prevalent in the tropics, where mycotoxins are more common because of subsistence farming, poor food-storage conditions, and warm, moist conditions that foster the growth of *Aspergillus* in harvested foods.

Some mycotoxins are considered useful. Among them are ergot alkaloids, produced by some strains of another ascomycete, *Claviceps purpurea* (klav′i-seps poor-poo′rē-ă). For example, *ergometrine* is used to stimulate labor contractions and is used to constrict the mother's blood vessels after birth (when she is at risk of bleeding excessively). Another mycotoxin, *ergotamine*, is used to treat migraine headaches.

Mushroom Poisoning (Mycetismus)

LEARNING OUTCOME

22.35 Describe at least two types of mushroom poisoning.

Most mushrooms (the spore-bearing structures of certain basidiomycetes) are not toxic, but some produce extremely dangerous mycotoxins capable of causing neurological dysfunction or hallucinations, organ damage, or even death. *Mushroom poisoning (mycetismus)* typically occurs when untrained individuals pick and eat wild mushrooms. Poisonous mushrooms are commonly called *toadstools*.

The deadliest mushroom toxin is produced by the "death cap" mushroom, *Amanita phalloides* (am-ă-nī′tă fal-ōy′dēz; FIGURE 22.21). The death cap contains two related polypeptide toxins: *phalloidin*, which irreversibly binds actin within cells, disrupting cell structure, and *alpha-amanitin*, which inhibits mRNA synthesis. Both toxins cause liver damage.

Other deadly mushrooms include the false morel, *Gyromitra esculenta* (gī-rō-mē′tră es-kū-len′tă), which causes bloody diarrhea, convulsions, and death within two days after ingestion, and *Cortinarius gentilis* (kōr′ti-nar-ē-us jen′til-is), which causes excessive thirst, nausea, and kidney failure between three days and three weeks after ingestion.

Psilocybe cubensis (sil-ō-sī′bē kū-ben′sis) produces hallucinogenic *psilocybin*, and *Amanita muscaria* (mus-ka′rē-ă) produces two hallucinogenic toxins—*ibotenic acid* and *muscimol*. These toxins may also cause convulsions in children.

Treatment involves inducing vomiting followed by oral administration of activated charcoal to absorb toxins. Severe mushroom poisoning may necessitate a liver transplant.

▲ FIGURE 22.21 *Amanita phalloides*, the "death cap" mushroom.

Allergies to Fungi

LEARNING OUTCOME

22.36 Identify the types of hypersensitivity reactions that fungal allergens produce.

Fungal allergens are common and can be found both indoors and out. Over 80 genera of fungi have been demonstrated to trigger fungal allergies. Epidemiologists estimate that 3% to 10% of humans worldwide are affected. However, determining the specific cause of allergies is often difficult—the presence of spores in an environment does not necessarily mean that they are responsible for allergies. Because skin sensitivity testing (discussed in Chapter 18) is not performed on most people, the true impact of fungal allergens remains undetermined, even though they continue to be blamed for a variety of illnesses.

Fungal spores dispersed in the air are common allergens. The concentrations of such spores may peak in autumn or spring, although they are present year-round. Fungal allergens typically cause type I hypersensitivities in which immunoglobulin E binds the allergen, triggering responses such as asthma, eczema, hay fever, and watery eyes and nose.

Less frequently, type III hypersensitivities result from chronic inhalation of particular fungal allergens. In these cases, allergens that have penetrated deep into the lungs encounter complementary antibodies and form immune complexes in the alveoli that lead to inflammation, fibrosis, and in some cases, death. Type III fungal hypersensitivities are associated with certain occupations, such as farming, in which workers are constantly exposed to fungal spores in moldy vegetation.

MICRO CHECK

13. Why do dermatophytes only colonize dead tissue, such as the outer layers of skin, nails, and hair?
14. How are chronic or stubborn cases of dermatophytes treated?
15. Which superficial mycosis is characterized by pityriasis resulting from interference with melanin production, which presents as depigmented (lighter colored) or hyperpigmented (darker colored) patches of scaly skin?
16. What type of mycosis is caused by *Sporothrix schenckii*?
17. What is the difference between mycotoxicosis and mycetismus?

TELL ME WHY

Why might young children be at greater risk of mushroom poisoning than adults?

MICRO IN THE CLINIC FOLLOW-UP

A Vacation Cold or Something Else?

Valerie told Dr. Gupta about her recent travel and that she had a cold just before her trip. Dr. Gupta asked her about the car ride and her activities during the vacation. He listened to her chest and ordered chest X rays, as well as blood and sputum cultures. The X ray revealed that Valerie had pneumonia. Based on the results of the X rays and Valerie's recent travel history, the doctor thought that she might have coccidioidomycosis, a fungal pneumonia endemic to the southwestern United States that is transmitted by inhaling fungal spores from the dry desert soil. Because coccidioidomycosis (also known as valley fever) does not occur in the eastern United States, the doctor might not have suspected valley fever had Valerie not mentioned her trip to Arizona.

Coccidioidomycosis is a reportable disease, so Dr. Gupta informed the local department of health about Valerie's case. He also reported Valerie's trip so that the health department in Pima County, Arizona, could be notified. The sputum culture indicated that *Coccidioides immitis* was present. An inflammatory reaction after an injection of *Coccidioides* antigen beneath Valerie's skin confirmed the diagnosis.

Valley fever usually resolves on its own, but in Valerie's case, Dr. Gupta decided to treat it with intravenous (IV) fluconazole and keep her in the hospital during treatment.

1. **How could Valerie have contracted valley fever?**
2. **Explain how Valerie could have gotten the disease, while her relatives did not.**
3. **Why does Valerie need to stay in the hospital for her treatment?**

Check your answers to Micro in the Clinic Follow-Up questions in the **Mastering Microbiology** Study Area.

CHAPTER SUMMARY

 Go to **Mastering Microbiology** for Interactive Microbiology, Dr. Bauman's Video Tutors, Micro Matters, MicroFlix, MicroBoosters, practice quizzes, and more.

An Overview of Medical Mycology (pp. 648–651)

1. **Medical mycology** is the study of the diagnosis, management, and prevention of **mycoses** (fungal diseases). Few fungi cause serious diseases in humans. The actual incidence of most mycoses is unknown because they are not reportable.

2. Most fungi infect humans after fungal spores are inhaled, ingested, or introduced by trauma. Fungi are not considered contagious agents because they rarely spread from person to person. A few exceptions include the **dermatophytes** and species of *Candida* and *Pneumocystis*.

3. True fungal pathogens are capable of actively invading the tissues of normal, healthy people. Opportunists infect only weakened patients.

4. Clinically, mycoses include infections, toxicoses, and hypersensitivity reactions. Infections can be superficial, cutaneous, subcutaneous, or systemic mycoses.

5. Diagnosis of mycoses usually correlates signs and symptoms with microscopic examination of tissues or laboratory cultures for identifying unique morphological features. Biochemical and immunological tests exist for some fungi.

6. Amphotericin B kills most fungi and can be used to treat most fungal infections, but because it is somewhat toxic to humans, long-term treatment with fungistatic drugs is often used instead.

7. Few antifungal vaccines exist, in part because there have been relatively few victims of serious fungal infections and because patients who are infected are often severely immunocompromised and so would not be candidates for vaccination.

Systemic Mycoses Caused by Pathogenic Fungi (pp. 651–656)

1. **Systemic mycoses**—fungal infections that spread throughout the body—are caused by one of four pathogenic, **dimorphic** fungi: *Histoplasma*, *Blastomyces*, *Coccidioides*, or *Paracoccidioides*. Each is geographically limited. They all cause similar pulmonary diseases and can spread beyond the lungs.

2. *Histoplasma capsulatum*, which causes **histoplasmosis**, is associated primarily with bat and bird droppings in soil in the Ohio River valley. *H. capsulatum* can be carried from the lungs inside macrophages.

3. **Blastomycosis** is found in the eastern United States and is caused by *Blastomyces dermatitidis*, which normally lives in soil rich in organic material. *B. dermatitidis* often spreads beyond the lungs to produce cutaneous lesions.

4. *Coccidioides immitis*, which is limited to deserts, causes **coccidioidomycosis**, which is common in AIDS patients. Contaminated dust is a major source of transmission.

5. *Paracoccidioides brasiliensis* is found in Brazil and some other regions of South and Central America. It causes **paracoccidioidomycosis**. Following a pulmonary phase, the fungus can spread and create lesions, which are often on the face and neck.

6. Pathogenic fungi are diagnosed by morphological analysis and by demonstration of dimorphism. Serological tests are also available. If treated promptly, they can be controlled.

Systemic Mycoses Caused by Opportunistic Fungi (pp. 656–664)

1. Opportunistic infections are difficult to diagnose and treat successfully. AIDS patients require both primary treatment to curb infections and maintenance therapy to keep fungal pathogens and opportunists under control.

2. *Pneumocystis jirovecii* causes **Pneumocystis pneumonia (PCP)**, a leading cause of death in AIDS patients in the United States. The organism shows a blend of characteristics similar to those of both protozoa and fungi. PCP is debilitating because the fungus multiplies rapidly.

3. **Candidiasis**, caused by various species of *Candida*, is one of the most important pathogens of AIDS patients, but it can also cause infections in relatively healthy individuals. Disease ranges from superficial infections to systemic candidiasis.

4. **Aspergillosis** is a group of diseases caused by *Aspergillus* species, including noninvasive fungal balls in the lungs and invasive diseases that can be fatal. *Aspergillus* is very common in the environment, from which it is inhaled.

5. *Cryptococcus neoformans*, common in bird droppings and soil, causes **cryptococcoses**, which most frequently manifest as cryptococcal meningitis in AIDS patients. Other clinical manifestations are possible.

6. **Zygomycoses**, caused by *Mucor* and various other genera of zygomycetes, can involve the brain, resulting in death. Less severe infections also occur.

7. Microsporidioses, caused by obligate, intracellular microsporidia, usually manifest as diarrhea.

8. Diseases caused by *Pneumocystis*, *Candida*, *Aspergillus* spp., and *Cryptococcus* are defining illnesses of AIDS.

9. *Fusarium*, *Penicillium*, and *Trichosporon* are emerging opportunists. Factors contributing to their emergence include AIDS and the selective pressure of antifungal agents, which select for resistant strains of fungi.

Superficial, Cutaneous, and Subcutaneous Mycoses (pp. 664–668)

1. Superficial, cutaneous, and subcutaneous mycoses are caused by opportunistic fungi.

2. **Dermatophytoses** (tineas, ringworms) encompass superficial skin, nail, and hair infections caused by a variety of fungi transmitted from individual to individual. Three genera predominate: *Trichophyton*, *Microsporum*, and *Epidermophyton*.

3. *Malassezia furfur* is a fungus that infects the skin. Clinical manifestations include **pityriasis**, in which fungal growth disrupts melanin production to produce discolored patches.

4. Most superficial infections are diagnosed by simple clinical observation of symptoms; all can be treated successfully. Rarely are such infections severe or invasive, but they have a tendency to recur.

5. Cutaneous and subcutaneous mycoses require traumatic introduction of fungi into deeper, living tissue and so are less common than superficial mycoses, which affect the outer layers of dead cells.

6. **Chromoblastomycosis** and **phaeohyphomycosis** are similar diseases resulting from infection with dark-pigmented fungi. Both are acquired by traumatic introduction of fungi into the skin. Lesions may be extensive and can spread internally. The two diseases are distinguished by differences in fungal morphology in tissue sections.

7. **Mycetomas** are invasive and destructive infections following introduction of soil fungi through scrapes or pricks from vegetation. Cutaneous lesions can spread to adjacent bone and can be permanently damaging. Surgery or amputation is required to remove infected tissues.

8. **Sporotrichosis** also is acquired when soil fungi are inoculated into the body by thorn pricks. Dispersal of the fungi follows the movement of lymph draining from the site of infection, which can manifest as multiple lesions along the course of lymphatic vessels.

Fungal Intoxications and Allergies (pp. 668–669)

1. Fungal metabolism may produce **mycotoxins**, which if eaten can cause neurological and physiological damage and lead to death. **Mycotoxicosis** is fungal intoxication caused by eating mycotoxin. **Aflatoxins**, produced by *Aspergillus*, are well-known mycotoxins.

2. **Mycetismus** (mushroom poisoning) results from eating mushrooms (sometimes called toadstools) containing mycotoxins.

3. Fungal allergens (spores or other fungal elements) cause type I hypersensitivities or, more rarely, type III hypersensitivities.

QUESTIONS FOR REVIEW

Answers to the Questions for Review (except for Short Answer questions) begin on p. A-1.

Multiple Choice

1. A fungus that can infect both healthy and immunocompromised patients is called a(n)_____.
   a. true pathogen
   b. opportunistic pathogen
   c. commensal organism
   d. symbiotic organism

2. Of the following fungi, which is usually transmitted from person to person?
   a. *Blastomyces dermatitidis*
   b. *Coccidioides immitis*
   c. *Tricophyton rubrum*
   d. *Aspergillus fumigatus*

3. Which of the following is not used to identify true fungal pathogens?
   a. growth at 25°C and 37°C to show dimorphism
   b. GMS staining of infected tissues
   c. serological testing
   d. clinical symptoms alone

4. Because amphotericin B is toxic to humans, most clinicians prescribe it only for_____.
   a. dermatophyte infections
   b. *Malassezia* infections
   c. systemic infections
   d. mushroom poisoning

5. Ringworm is caused by a_____.
   a. helminth
   b. dermatophyte
   c. dimorphic fungus
   d. commensal fungus

6. Which of the following is considered a classical opportunistic fungus?
   a. *Blastomyces*
   b. *Histoplasma*
   c. *Fonsecaea*
   d. *Aspergillus*

7. Subcutaneous infections tend to be acquired through_____.
   a. inhalation and remain localized
   b. inhalation and become systemic
   c. trauma and remain localized
   d. trauma and become systemic

8. The term *dermatophyte* refers to_____.
   a. pathogenicity
   b. where a fungus grows
   c. method of spread
   d. pigmentation

9. Which of the following subcutaneous mycoses may exhibit respiratory and cerebral forms?
   a. chromoblastomycosis
   b. mycetoma
   c. phaeohyphomycosis
   d. sporotrichosis

10. Which of the following systemic mycoses is endemic to the deserts of the southwestern United States?
    a. blastomycosis
    b. coccidioidomycosis
    c. histoplasmosis
    d. paracoccidioidomycosis

11. A spherule stage is seen in humans infected with what organism?
    a. *Blastomyces dermatitidis*
    b. *Coccidioides immitis*
    c. *Histoplasma capsulatum*
    d. *Paracoccidioides brasiliensis*

12. Of the following fungal diseases, which is found in almost all terminal AIDS patients?
    a. chromoblastomycosis
    b. blastomycosis
    c. candidiasis
    d. mycetoma

13. The number of mycoses and cases of microsporidiosis worldwide are rising, in part because_____.
    a. the number of fungi in the environment is rising
    b. the number of immunocompromised individuals in the population is rising
    c. fungi have become more pathogenic
    d. fungi are developing a new tendency to spread between people

14. Fungal allergens generally stimulate what type of reaction?
    a. type I hypersensitivity
    b. type II hypersensitivity
    c. type III hypersensitivity
    d. type IV hypersensitivity

15. A pathogenic feature of *Cryptococcus neoformans* is _____.
    a. production of destructive enzymes
    b. production of a capsule
    c. infection of immune cells
    d. variation of surface antigens to avoid immune system recognition

16. The most common manifestation of *Cryptococcus* infection in AIDS patients is _____.
    a. blindness
    b. cutaneous infection
    c. meningitis
    d. pneumonia

17. Bread mold can cause which disease?
    a. aspergillosis
    b. dermatophytosis
    c. mycetoma
    d. zygomycosis

18. Mycetismus is caused by _____.
    a. inhalation of fungal allergens
    b. ingestion of mushrooms
    c. traumatic inoculation of fungi beneath the skin
    d. close contact with infected individuals

19. One of the more poisonous mycotoxins is produced by _____.
    a. *Amanita phalloides*
    b. *Amanita muscaria*
    c. *Psilocybe cubensis*
    d. *Claviceps purpurea*

20. Which of the following predisposing factors would leave a patient with the greatest long-term risk of acquiring a fungal infection?
    a. invasive medical procedures
    b. AIDS
    c. chronic illness such as diabetes
    d. short-term treatment with antibacterial agents

Modified True/False

Indicate whether each statement is true or false. If the statement is false, change the underlined word or phrase to make the statement true.

1. _____ Fungi are generally <u>not transmitted</u> from person to person.
2. _____ On the whole, fungal infections are relatively <u>easy</u> to treat.
3. _____ <u>Dermatophytoses</u> are always contracted from the environment.
4. _____ Chromoblastomycosis and phaeohyphomycosis are both caused by <u>dark-pigmented</u> ascomycetes.
5. _____ <u>Sporotrichosis</u> is often caused by introduction of fungi beneath the skin by a thorn prick.
6. _____ Coccidioidomycosis does not occur normally outside the <u>Western Hemisphere</u>.
7. _____ Treatment of individuals with broad-spectrum antibacterial agents is a predisposing factor for <u>opportunistic</u> fungal infections.
8. _____ *Candida albicans* generally causes localized opportunistic infections but can become systemic, particularly in the <u>immunocompetent</u>.
9. _____ Relapse of fungal diseases is common in <u>AIDS patients</u>.
10. _____ <u>Almost everyone has</u> allergies to fungal elements.

Fill in the Blanks

1. Dimorphic fungi exist as _____ forms in the environment and as _____ forms in their hosts.
2. The true fungal pathogens are _____ _____, _____ _____, _____ _____, and _____ _____ (give genus and specific epithet).
3. Many antifungal agents target the compound _____ in fungal cytoplasmic membranes.
4. _____ are tumorlike fungal infections.
5. Sporotrichosis is caused by the traumatic introduction of _____ _____ into the skin (give genus and species).
6. Which pathogenic fungus is associated with bird droppings? _____ _____ (give genus and species).
7. The five more common agents of opportunistic fungal infections are _____, _____, _____, _____, and _____ (genus names).
8. Thrush is caused by _____ (genus name).
9. *Pneumocystis* was once classified as a _____, but now it is classified as a _____.
10. Ergot alkaloids are produced by some strains of the genus _____.

Matching

Match each of the following diseases with the manner(s) in which fungi enter the body. Answers can be used more than once.

1. _____ Aspergillosis
2. _____ Candidiasis
3. _____ Chromoblastomycosis
4. _____ Coccidioidomycosis
5. _____ Cryptococcosis
6. _____ Dermatophytosis
7. _____ Histoplasmosis
8. _____ Hypersensitivity reactions
9. _____ Mushroom poisoning
10. _____ Mycetoma
11. _____ Sporotrichosis

A. Inhalation
B. Contact
C. Trauma
D. Ingestion

Short Answer

1. Amphotericin B is considered the "gold standard" of antifungal agents. Technically, its mode of action works against most fungal infections. Why, then, isn't it prescribed for most fungal infections?

2. Discuss why it is difficult in many cases to determine the source of superficial fungal infections (i.e., from other humans, animals, or the environment).

3. Given that superficial fungal infections are only on the surface, why is it necessary to even try to identify the source of infection?

4. AIDS patients usually die of bacterial, fungal, or microsporidial infections. Why do so many fungal infections appear in these individuals, and why are mycoses severe while fungi, for the most part, are benign residents of the environment?

5. How does mycotoxicosis differ from mycetismus?

VISUALIZE IT!

1. Color each map below to show the general area where each disease is endemic.

Blastomycosis

Coccidioidomycosis

Histoplasmosis

2. Identify these fungal genera.

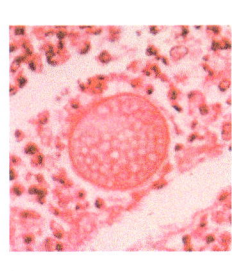 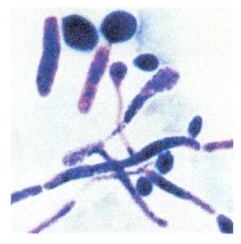 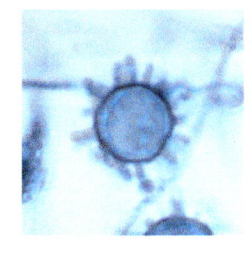

(a)   (b)   (c)

CRITICAL THINKING

1. Correlate the observation that the majority of fungal infections are caused by opportunists with the fact that most infections are acquired from the environment. How does this make the control and diagnosis of fungal infections difficult?

2. The four pathogenic fungi (*Blastomyces, Coccidioides, Histoplasma,* and *Paracoccidioides*) are all dimorphic. Antifungal agents, like all antimicrobials, are generally designed to target something specific about the organism to avoid harming human tissues. Propose how you could use the idea of dimorphism to produce new antifungal agents.

3. What actions could be taken to limit the contamination of foods with fungal toxins? Would these actions differ depending on where in the world the food is grown? Explain.

4. Onychomycoses are nail infections that can be caused by several species of fungi. Explain why these mycoses are so difficult to treat and why it is generally necessary to treat patients with oral antifungal agents for long periods of time.

5. What factors contribute to the pathogenicity of *Cryptococcus* infections?

6. Clinically, all fungi that cause subcutaneous mycoses produce lesions on the skin around the site of inoculation. What are some of the things you would look for when attempting to distinguish among them?

7. Given the various predisposing factors that make humans susceptible to opportunistic infections, how can health care providers curtail the rising incidence of such infections?

CONCEPT MAPPING

Using the terms provided, fill in this concept map that describes systemic mycoses. You can also complete this and other concept maps online by going to the Mastering Microbiology Study Area.

5-Fluorocytosine
Amphotericin B (2)
Blastomyces dermatitidis
Candida albicans
Coccidioides immitis

Cryptococcus neoformans
Histoplasma capsulatum
Inhalation
Opportunistic fungi
Other organs

Paracoccidioides brasiliensis
Pneumocystis jirovecii
Trimethoprim/Sulfamethoxazole
Voriconazole

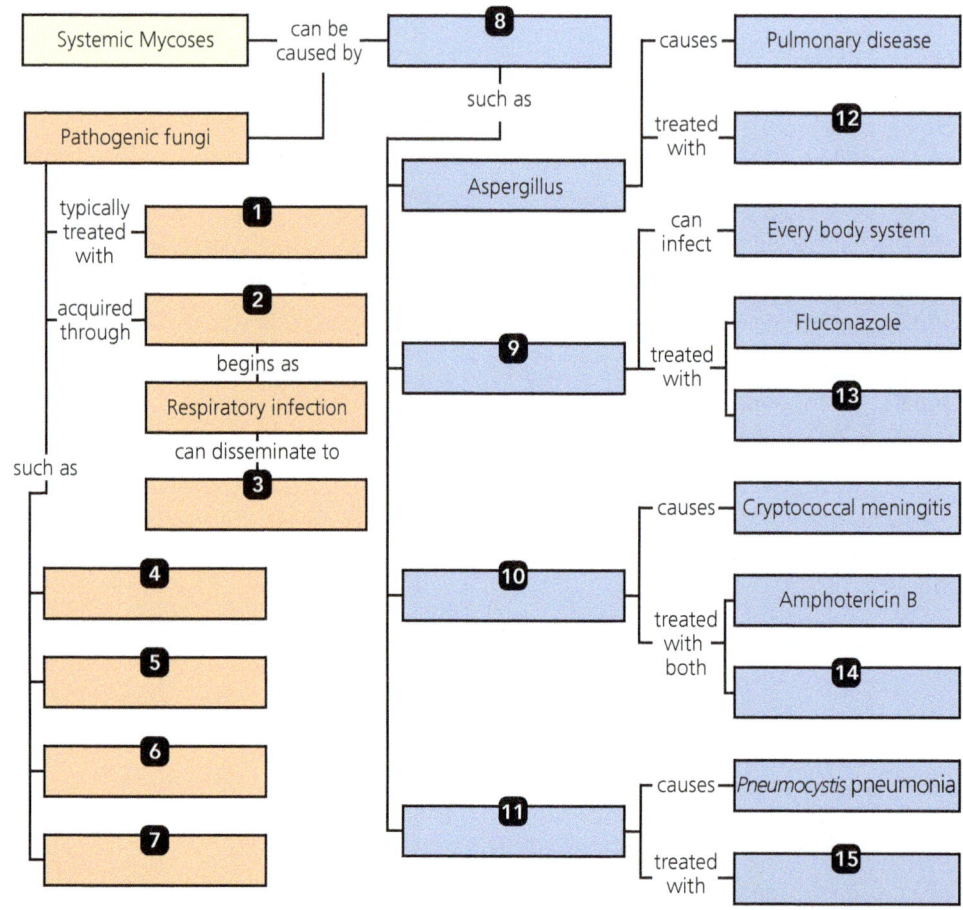

23 Parasitic Protozoa, Helminths, and Arthropod Vectors

Before You Begin

1. What is the difference between cilia and flagella in eukaryotic cells?
2. What term describes diseases that develop gradually and cause signs and symptoms for years after the initial infection, either recurring or continuous?
3. What are the major characteristics of protozoa?

MICRO IN THE CLINIC

More Than Stress?

SAM AND HIS SON GREG HAVE spent the last three weeks backpacking and exploring Mayan ruins on the Yucatán Peninsula, Mexico. It's been an adventure and a great way for Sam and Greg to spend time together before Greg starts college in the fall. About a month after they return from the trip, Sam starts experiencing a fairly consistent headache and muscle pain. Work has been hectic and Greg is getting ready to leave for school—Sam figures that his symptoms are stress; once things calm down, he'll feel better. A couple of weeks later, Sam's wife notices that the lymph nodes under his chin are swollen and big enough to feel. Since his other symptoms still haven't gone away, she insists that Sam make an appointment to see their family physician.

1. Are Sam's swollen lymph nodes related to his other symptoms?

Turn to the end of the chapter (p. 703) to find out.

Human tapeworms (*genus Taenia*) can grow to great lengths in the intestine.

Parasitology

LEARNING OUTCOMES

23.1 Define *parasite*.
23.2 Contrast *definitive host* with *intermediate host*.

Parasitologists are scientists who study **parasites**—organisms that live, feed, and grow in or on another organism at the expense of their host's metabolism. Sometimes no obvious damage is done to the host, but in other cases parasitism can lead to a host's death. Many protozoan and helminthic parasites exist worldwide, especially in the tropics and subtropics, and particularly among people living in rural, undeveloped, or overcrowded places. Parasitic diseases are also emerging as serious threats among developed nations in nontropical regions. This rise is due to many factors but can most generally be attributed to human migration, inadequate inspection and disinfection of food and water supplies, and a changing world climate that alters parasites' transmission patterns, particularly those transmitted by vectors. As a whole, parasites infect billions of humans each year, making parasitic infections medically, socially, and economically important.

Parasitic infections often involve several hosts—a **definitive host** in which mature (often sexual) forms of the parasite are present and usually reproducing and, with many parasites, one or more **intermediate hosts** in which immature parasites undergo various stages of maturation. In general, parasites infect human hosts in one of three major ways: by being ingested, through vector-borne transmission, or via direct contact and penetration of the skin or mucous membranes **(FIGURE 23.1)**.

We begin our consideration of parasitic infections by briefly discussing the role arthropod vectors play in human disease.

TELL ME WHY

Why do some parasitologists refer to definitive hosts as *primary hosts* and to intermediate hosts as *secondary hosts*?

Arthropod Vectors

LEARNING OUTCOMES

23.3 Compare and contrast biological and mechanical vectors.
23.4 List the principal arthropod vectors of human pathogens.

Vectors are animals—usually *arthropods* (ar'thrō-podz)—that carry microbial pathogens. An **arthropod** is an animal with a

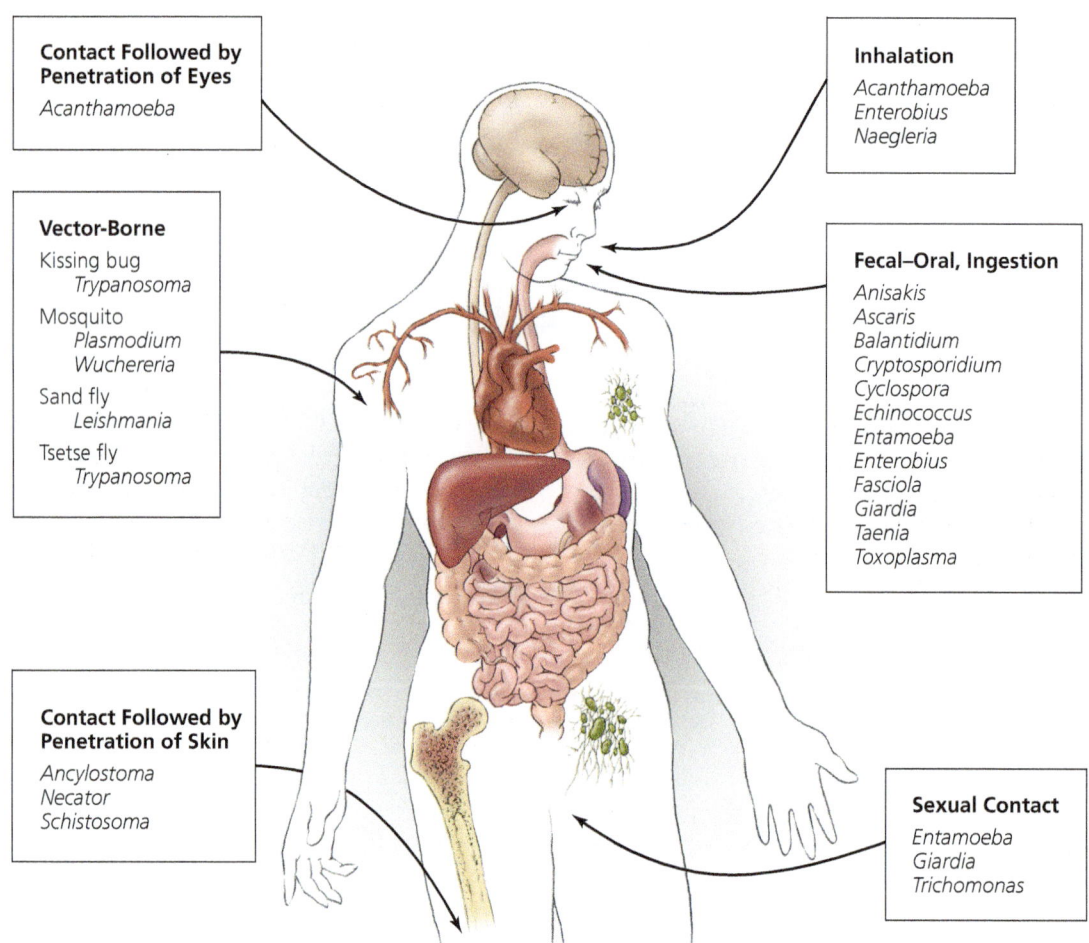

▲ **FIGURE 23.1 The major routes by which humans acquire parasitic infections.** They are the fecal–oral route, the vector-borne route, and direct contact. Occasionally parasites are inhaled. Listed genera are discussed in the chapter.

segmented body, a hard exoskeleton (external skeleton), and jointed legs. Arthropods are extremely diverse and abundant—insects alone account for more than half of all the known species on Earth.

Some arthropods are **biological vectors,** meaning that they also serve as hosts for the pathogens they transmit. **Mechanical vectors,** in contrast, carry pathogens, but the pathogens do not use the vector as a host.

Given that arthropods are usually small organisms (so small that we don't notice them until they bite us) and given that they produce large numbers of offspring, controlling arthropod vectors to eliminate their role in the transmission of important human diseases is an almost insurmountable task.

Disease vectors belong to two classes of arthropods: *Arachnida* (ticks and mites), which have eight legs as adults, and the more common *Insecta* (fleas, lice, flies, and true bugs), which have six legs as adults (see Chapter 12, Figure 12.29). Mosquitoes—a type of fly—are the most important vector of human diseases. Most arthropod vectors are found on a host only when they are actively feeding. Lice are the only vectors that may spend their entire lives in association with a single individual. (Chapter 12 discusses vectors in more detail.)

TELL ME WHY

Why are insect-vectored diseases more likely to spread over a larger geographical area than are arachnid-borne diseases?

Protozoan Parasites of Humans

LEARNING OUTCOME

23.5 Describe the morphology and epidemiology of the more important protozoan parasites.

Protozoa are unicellular eukaryotes found in a wide range of habitats. Although the majority of protozoa are free-living and do not adversely affect humans or animals, some are parasites that can cause debilitating and deadly diseases.

Among the many protozoa that enter the body via ingestion, most have two morphological forms: A feeding and reproducing stage called a *trophozoite*, which lives within the host, and a dormant *cyst* stage, which can survive in the environment and is infective to new hosts. Once ingested by a host, cysts undergo **excystment** and develop into new trophozoites, which resume feeding and reproducing. In most cases, trophozoites undergo **encystment** before leaving the host in the feces, becoming available to infect other hosts.

This chapter presents the protozoa in four groups, an approach that is different from a taxonomic presentation (see Chapter 12). The scheme presented in this chapter reflects the largely obsolete but clinically useful classical groupings for these parasites, which were based primarily on mode of locomotion: the ciliates, the amoebas, the flagellates, and the typically nonmotile apicomplexans.

MICRO CHECK

1. Malaria is a parasitic disease transmitted to humans by mosquito bites. The sexual stage of the parasite occurs in the mosquito. What type of host and what type of vector are mosquitoes?
2. Which are the most important vectors of parasitic diseases in humans?
3. Which is the infective stage of an intestinal protozoan?
4. Which group of protozoa is nonmotile?

Ciliates

LEARNING OUTCOME

23.6 Describe the characteristics of *Balantidium coli*.

Ciliates (sil'ē-āts) are protozoa that in their trophozoite stages use cilia for locomotion, for acquiring food, or both. *Balantidium coli* (bal-an-tid'ē-ŭm kō'lī), a relatively large (50 μm × 100 μm) ciliate commonly found in animal intestinal tracts, is the only ciliate known to cause disease in humans. Pigs are its most common host, but it is also found in rodents and nonhuman primates. Humans become infected by consuming food or water contaminated with feces containing cysts.

Following ingestion, excystment occurs in the small intestine, releasing trophozoites that use their cilia to attach to (and then burrow through) the mucosal epithelium lining the intestine. Eventually, some trophozoites undergo encystment, and both cysts and trophozoites are shed in feces. Trophozoites die outside the body, but cysts are hardy and infective.

In healthy adults, *B. coli* infection is generally asymptomatic. For those in poor health, however, **balantidiasis** (bal'an-ti-dī'ă-sis) occurs. Persistent diarrhea, abdominal pain, and weight loss characterize the disease. Severe infections produce *dysentery* (frequent and painful diarrhea, often containing blood and mucus) and possibly ulceration and bleeding from the intestinal mucosa. *Balantidium* infection is rarely fatal.

Paradoxically, cysts are few and so rare that they usually cannot be recovered from stool (fecal) samples, although they are the infective stage. Noninfective trophozoites, by contrast, can be detected, and their presence is diagnostic for the disease. Fresh stool samples must be used for diagnostic purposes because the trophozoites do not survive long outside the intestinal tract. The treatment of choice for balantidiasis is the antibacterial drug tetracycline, which does not kill *B. coli* directly but instead modifies the normal microbiota of the digestive tract such that the small intestine is unsuitable for infection by the ciliate.

Prevention of balantidiasis relies on good personal hygiene, especially for those who live around or work with pigs. Additionally, efficient water sanitation is necessary to kill cysts or remove them from drinking water.

Amoebas

LEARNING OUTCOME

23.7 Compare and contrast three amoebas that cause disease in humans.

Amoebas (ă-mē′băz) are protozoa that have no truly defined shape and that move and acquire food through the use of pseudopods. Although amoebas are abundant throughout the world in freshwater, seawater, and moist soil, few cause disease. The most important amoebic pathogen is *Entamoeba*.

Entamoeba

Entamoeba histolytica (ent-ă-mē′bă his-tō-li′ti-kă) is carried asymptomatically in the digestive tracts of roughly 10% of the world's human population. When disease develops, it can be fatal, causing a worldwide annual mortality of over 100,000 people. Carriers predominate in less developed countries, especially in rural areas, where human feces are used to fertilize food crops and where water sanitation is deficient. Travelers, immigrants, and institutionalized populations are at greatest risk within industrialized nations. No animal reservoirs exist, but human carriers are sufficiently numerous to ensure continued transmission of the parasite.

Infection occurs most commonly through the drinking of water contaminated with feces that contain cysts. The parasite can also be ingested following fecal contamination of hands or food or during oral-anal intercourse.

Excystment in the small intestine releases trophozoites that migrate to the large intestine and multiply. The organism uses pseudopods to attach to the intestinal mucosa, where it feeds and reproduces via binary fission. Both trophozoites and cysts are shed into the environment in feces, but the trophozoites die quickly, leaving infective cysts.

Depending on the health of the host and the virulence of the particular infecting strain, three types of **amebiasis** (ă-mē-bī′ă-sis) can result. The least severe form, *luminal amebiasis*, occurs in otherwise healthy individuals. Infections are asymptomatic; trophozoites remain in the lumen of the intestine, where they do little tissue damage. Invasive *amebic dysentery* is a more serious and more common form of infection characterized by severe diarrhea, colitis (inflammation of the colon), appendicitis, and ulceration of the intestinal mucosa. Bloody, mucus-containing stools and pain are characteristic of amebic dysentery, which affects about 500 million people worldwide. In the most serious disease, *invasive extraintestinal amebiasis*, trophozoites invade the peritoneal cavity and the bloodstream, which carries them throughout the body. Lesions of dead cells formed by the trophozoites occur most commonly in the liver but can also be found in the lungs, spleen, kidneys, and brain. Amebic dysentery and invasive extraintestinal amebiasis can be fatal, especially without adequate treatment.

Diagnosis is based on the identification of microscopic cysts or trophozoites (FIGURE 23.2) recovered from either fresh stool specimens or intestinal biopsies. Microscopic analysis is necessary to distinguish amebic dysentery from dysentery caused by bacteria. Serological identification using antibodies to detect antigens can aid in distinguishing *E. histolytica* from nonpathogenic amoebas.

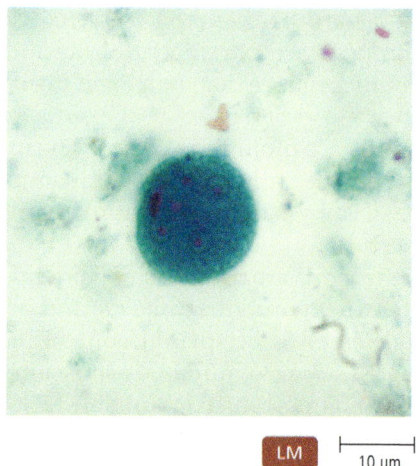

▲ FIGURE 23.2 Cyst of *Entamoeba histolytica* in fecal smear.

For asymptomatic infections, an antibacterial aminoglycoside called paromomycin is effective. Physicians prescribe iodoquinol for symptomatic amebiasis. They may also prescribe other antibacterial agents at the same time to prevent secondary bacterial infections. The coupling of oral rehydration therapy with drug treatment is recommended and is vital in severe cases.

Several preventive measures interrupt the transmission of *Entamoeba*. Discontinuing the use of human wastes as fertilizer reduces the transmission of amebiasis. Normal methods of treating wastewater and drinking water are helpful but not completely effective because the infectious cysts are hardy. Effective processing of water requires extra chemical treatment, filtration, or extensive boiling to eliminate all cysts. Good personal hygiene can eliminate transmission via intimate contact.

Acanthamoeba and *Naegleria*

Two other amoebas—*Acanthamoeba* (ă-kan-thă-mē′bă) and *Naegleria* (nā-glē′rē-ă)—cause rare and usually fatal infections of the brain. These amoebas are common free-living inhabitants of warm lakes, ponds, puddles, ditches, mud, and moist soil. They are also found in artificial water systems such as swimming pools, air conditioning units, humidifiers, and dialysis units. Contact lens wearers who use tap water (as opposed to sterile saline solution) to wash and store their lenses create an additional focal point for infection. *Naegleria* may assume a flagellated form in addition to its amoeboid and cyst stages.

Acanthamoeba usually enters a host through cuts or scrapes on the skin, through the conjunctiva via abrasions from contact lenses or trauma, or through inhalation of contaminated water while swimming. *Acanthamoeba* trophozoites when inoculated into the eye can invade and perforate the eye, resulting in *keratitis*. Damage can be extensive enough to require corneal replacement. The more common disease caused by infection with *Acanthamoeba* is **amebic encephalitis** (inflammation of the brain), characterized by headache, altered mental state, and neurological deficit. Symptoms progressively worsen over a period of weeks until the patient dies.

Infection with *Naegleria* occurs when swimmers inhale water contaminated with trophozoites. Since trophozoites are motile, they can invade the nasal mucosa and replicate. The trophozoites migrate to the brain, where they cause an **amebic meningoencephalitis** (inflammation of the brain and its membranes). Severe headache, fever, vomiting, and neurological tissue destruction lead to hemorrhage, coma, and usually death within three to seven days after the onset of symptoms.

Physicians diagnose *Acanthamoeba* and *Naegleria* infections by detecting trophozoites in corneal scrapings, cerebrospinal fluid, or biopsy material. Keratitis can be treated topically with anti-inflammatory drugs. Physicians treat *Acanthamoeba* encephalitis with ketoconazole or pentamidine and *Naegleria* infection with amphotericin B, but by the time they have diagnosed the disease, it is often too late to begin effective treatment.

As both these amoebas are environmentally hardy, control and prevention of infection can be difficult. Swimmers and bathers should avoid waterways in which the organisms are known to be endemic. Contact lens users should never use nonsterile solutions to clean or store their lenses. People who use neti pots for nasal passage irrigation must ensure that the water used and the pot itself are sterile before use. Swimming pools should be properly chlorinated and tested periodically to ensure their safety. Air conditioning systems, dialysis units, and other devices that routinely use water should also be cleaned thoroughly and regularly to prevent amoebas from becoming resident.

MICRO CHECK

5. How does tetracycline work to treat *Balantidium coli* infections?
6. Of *Balantidium*, *Entamoeba*, *Acanthamoeba*, and *Naegleria* species, which more commonly causes a deadly diarrhea?
7. What is dysentery?
8. What is the best way to prevent amebiasis?

Flagellates

LEARNING OUTCOMES

23.8 Contrast the life cycles of *Trypanosoma cruzi* and *Trypanosoma brucei*.
23.9 Describe the life cycle of *Leishmania* and the clinical forms of leishmaniasis.
23.10 Compare *Giardia* infections to those caused by *Entamoeba* and *Balantidium*.
23.11 Identify the risk factors and preventive measures for *Trichomonas vaginalis* infection.

In 1832, Charles Darwin (1809–1882) explored South America as one of the naturalists sailing aboard HMS *Beagle*. His explorations were to affect him profoundly—intellectually and medically. During the voyage, he began to formulate his ideas concerning the evolution of species, and he was bitten by a bloodsucking bug that may have infected him with *Trypanosoma cruzi*, a flagellated protozoan that afflicted him for the rest of his life. The following sections explore this and other parasitic flagellates that infect humans.

Flagellates (flaj′e-lātz) are protozoa that possess at least one long flagellum, which is used for movement. The number and arrangement of the flagella are important features for determining the species of flagellate present within a host.

Here, we will examine some parasitic kinetoplastids (*Trypanosoma* and *Leishmania*), *Giardia* (a diplomonad), and *Trichomonas* (a parabasalid). We begin our discussion of the more important flagellate parasites with *Trypanosoma cruzi*.

Trypanosoma cruzi

Trypanosoma cruzi (tri-pan′ō-sō-mǎ kroo′zē) causes **Chagas' disease.** This disease, named for Brazilian doctor Carlos Chagas (1879–1934), is endemic throughout Central and South America. Localized outbreaks have also occurred in California and Texas. Opossums and armadillos are the primary reservoirs for *T. cruzi*, but most mammals, including humans, can harbor the organism. There are an estimated 8 million infected people worldwide. Transmission occurs through the bite of true bugs—a type of insect—in the genus *Triatoma* (trī-ă-tō′mǎ). These bloodsucking bugs feed preferentially from blood vessels in the lips, which gives the bugs their common name—kissing bugs.

FIGURE 23.3 illustrates the basic life cycle of *T. cruzi*. Within the hindgut of the kissing bug, *T. cruzi* becomes infective 1 and is shed in the bug's feces while the bug feeds on a mammalian host 2, usually at night. When the host scratches the itchy wounds created by the bug bites, infective trypanosomes deposited in the feces enter the nearby wounds 3. Alternatively, the trypanosomes can enter the body through an open cut, the eyes, or mouth, or into breast-feeding infants in milk from their mothers. The bloodstream carries the trypanosomes throughout the body and they penetrate cells, especially macrophages and heart muscle cells, where they transform into small, nonflagellated forms 4. Nonflagellated trypanosomes multiply by binary fission 5, and each then develops a flagellum attached to the trypanosome down its length with an undulating membrane 6. These flagellated forms do not reproduce. The host cell bursts, releasing flagellated trypanosomes that either infect other cells 7 or circulate in the bloodstream. Circulating flagellated trypanosomes are subsequently ingested by a kissing bug when it takes a blood meal 8. Within the midgut of the kissing bug, binary fission of the flagellated cells produce another reproductive form that multiplies by binary fission 9.

Chagas' disease progresses over the course of several months through the following four stages:

1. An acute stage characterized by *chagomas*, which are swellings at the sites of the bites
2. A generalized stage characterized by fever, swollen lymph nodes, myocarditis (inflammation of the heart muscle), and enlargement of the spleen, esophagus, and colon
3. A chronic stage, which is asymptomatic and can last for years
4. A symptomatic stage characterized primarily by congestive heart failure following the formation of clusters of nonflagellated trypanosomes in heart muscle tissue

Trypanosoma-induced heart disease is one of the leading causes of death in Latin America.

Diagnosis involves detection of *Trypanosoma cruzi* DNA sequences in blood samples by PCR. Diagnosis can be confirmed

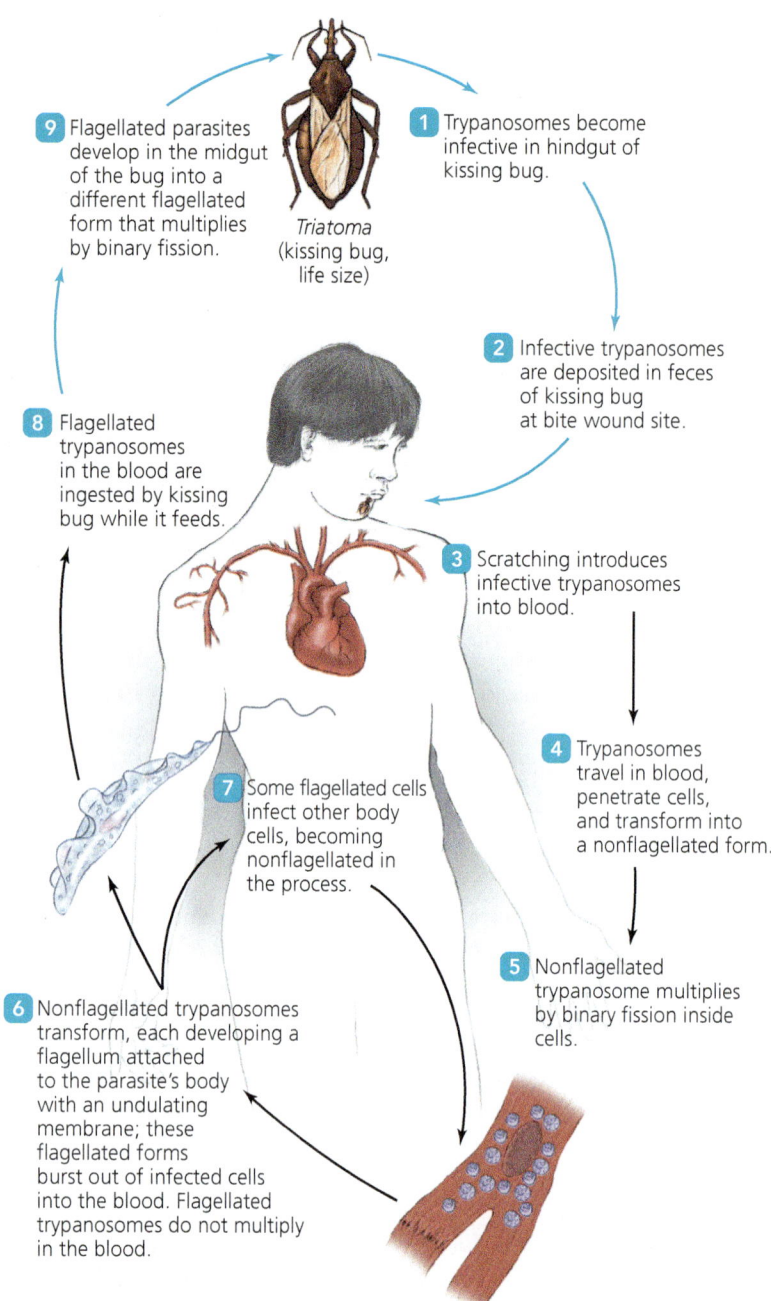

▲ FIGURE 23.3 The life cycle of *Trypanosoma cruzi*. *Triatoma*, the vector, is shown lifesize.

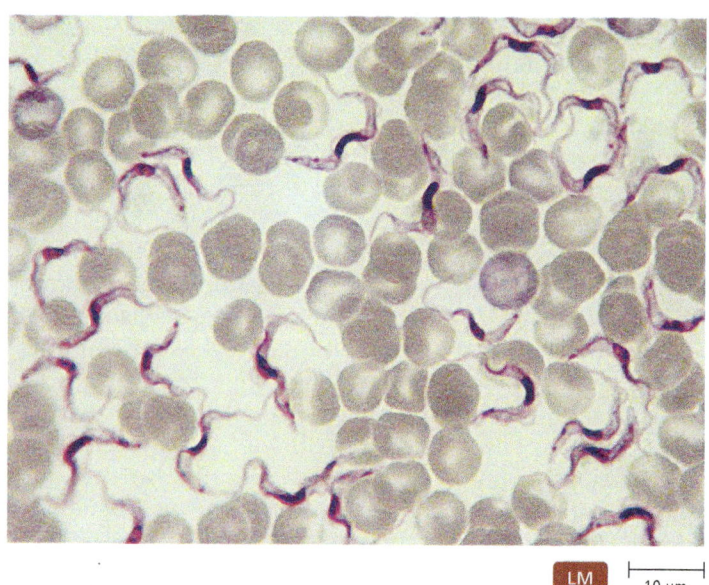

▲ FIGURE 23.4 Mature, flagellated trypanosomes of *T. cruzi* among erythrocytes.

by microscopic identification of flagellated trypanosomes or detection of their antigens in either blood, lymph, spinal fluid, or a tissue biopsy (FIGURE 23.4). A historical and simple diagnostic method, which is especially practical for clinicians without access to modern diagnostic tools, is *xenodiagnosis*. In this procedure, uninfected kissing bugs are allowed to feed on a person suspected of having a *T. cruzi* infection. Four weeks later, the bugs are dissected. The presence of parasites within the hindgut of the bug indicates the person is infected.

In its generalized and early chronic stages, Chagas' disease can be treated with benznidazole or nifurtimox; the late chronic and symptomatic stages cannot be treated. Prevention of Chagas' disease involves replacing thatch and mud building materials, which provide homes for the bugs, with concrete and brick. The use of insecticides, both personally and in the home, and sleeping under insecticide-impregnated netting can prevent insect feeding. No vaccines exist for Chagas' disease.

Trypanosoma brucei

Trypanosoma brucei (brūs′ē) causes **African trypanosomiasis**, also called **African sleeping sickness**, which afflicts more than 10,000 people annually in equatorial and subequatorial savanna, agricultural, and riverine areas of Africa. Its geographical range depends on the range of its insect vector, the tsetse (tset′sē) fly (*Glossina*, glo-sī′nă). There are two variants of *Trypanosoma brucei—T. brucei gambiense* (gam′bē-en′sē), which occurs primarily in western Africa, and *T. brucei rhodesiense* (rō-dē′zē-en′sē) in eastern and southern Africa.

Whereas cattle and sheep are the reservoirs for *T. brucei gambiense*, several species of wild animals apparently serve as reservoirs for *T. brucei rhodesiense*. Although humans usually become infected when bitten by tsetse flies previously infected while feeding on infected animals, flies may transmit both variants of *T. brucei* between humans.

The life cycle of *T. brucei* proceeds as follows (FIGURE 23.5): Within the salivary gland of a male or female tsetse fly, a form of the trypanosome parasite matures into an infective form 1. During the course of taking a blood meal from an animal or human, tsetse flies inject infective trypanosomes into the wounds 2. Blood and lymph circulation carry trypanosomes throughout the body 3. They reproduce in bodily fluids by binary fission 4. Eventually, some trypanosomes enter the central nervous system 5a, while others continue to circulate in the blood, where they can be picked up by feeding tsetse flies 5b. In the midgut of the fly, trypanosomes multiply by binary fission, producing immature, noninfective forms that migrate to the salivary glands 6, where they mature and become infective forms and begin the cycle anew.

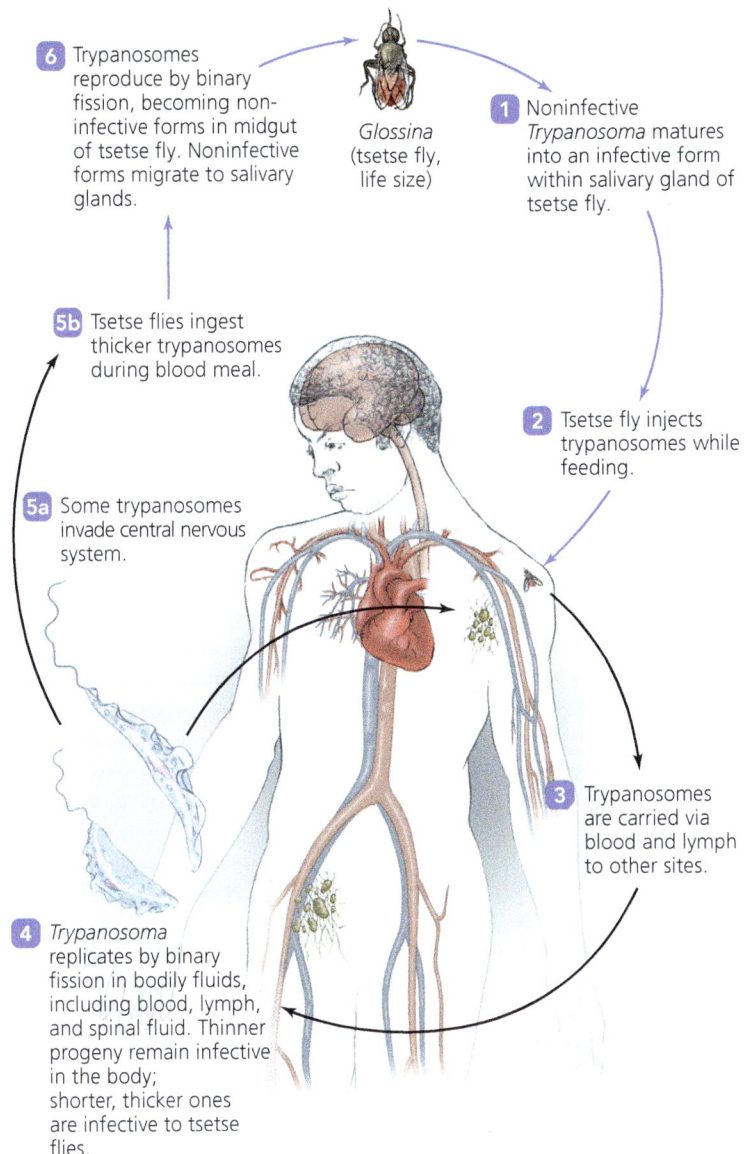

▲ FIGURE 23.5 **The life cycle of *Trypanosoma brucei*.** *Glossina* is shown life size. *By which mode of transmission do trypanosomes infect humans?*

Figure 23.5 *Trypanosomes are transmitted to humans by arthropod vectors.*

The life cycle of *T. brucei* differs from that of *T. cruzi* in several ways:

- *T. brucei* matures in the salivary gland of a tsetse fly, whereas *T. cruzi* matures in the hindgut of a kissing bug.
- Tsetse flies directly inject *T. brucei*. In contrast, a host rubs *T. cruzi* found in the kissing bug's feces into a wound.
- *T. brucei* remains outside its hosts' cells, whereas trypanosomes of *T. cruzi* live inside host cells.

Untreated African trypanosomiasis progresses through three clinical stages. First, the wound created at the site of each fly bite becomes a lesion containing dead tissue and rapidly dividing parasites. Next, the presence of parasites in the blood triggers fever, swelling of lymph nodes, and headaches. Finally, invasion of the central nervous system results in meningoencephalitis, characterized by headache, extreme drowsiness, abnormal neurological function, and coma. Untreated, a patient will die perhaps within six months of onset of disease. Symptoms may take years to develop with *T. brucei rhodesiense* but begin within three to six months with *T. brucei gambiense*.

All *T. brucei* infections are characterized by cyclical waves of *parasitemia* (parasites in the blood) that occur roughly every 7 to 10 days. Although the presence of parasites in the blood is in itself serious, these cycles are particularly dangerous because with each wave of replication, *T. brucei* changes its surface glycoproteins and thus its surface antigens. The result is that by the time the host's immune system has produced antibodies against a given set of glycoproteins, the parasite has already produced a new set, continually leaving the host's immune system one step behind the parasite. Once infected, a patient is incapable of clearing the infection and never becomes immune.

Microscopic observation of trypanosomes in blood, lymph, spinal fluid, or a tissue biopsy is diagnostic for both strains of *T. brucei*. These diagnostic forms of *T. brucei* are long and thin with a single long flagellum running along the cell and extending past the posterior end. The bodies of *T. brucei* are less curled than those of *T. cruzi*.

African sleeping sickness is one of the few human diseases that is 100% fatal if not treated. Treatment must begin as soon as possible after infection. However, immediate treatment is rare because poor health care infrastructure in endemic areas often prevents identification of the disease at its earliest stages. Pentamidine or suramin is used to treat the early stages. Melarsoprol is used when the disease progresses to the central nervous system because this drug crosses the blood-brain barrier. Eflornithine, a newer and more expensive drug, is remarkably effective against *T. brucei gambiense* (but not against *T. brucei rhodesiense*). Private pharmaceutical companies guarantee availability of this drug through the World Health Organization (WHO) to those who need it.

Clearing of tsetse fly habitats and broad application of insecticides have reduced the occurrence of African sleeping sickness in some localities. However, large-scale spraying of insecticides is impractical and expensive and may have disastrous long-term environmental consequences. Another tactic some countries have tried is to release up to a million sterile male tsetse flies a week. With the sterile files outnumbering wild flies 10 to 1, most female flies mate with sterile males and produce no offspring—a kind of birth control for tsetse flies. Some regions of the world have eradicated tsetse flies using this method. Personal protection includes use of insecticide, use of insecticide-impregnated netting, and wearing long, loose-fitting clothing, which can prevent insect feeding. No vaccine currently exists for African sleeping sickness.

Leishmania

Leishmania (lēsh-man'ē-ă) is a genus of kinetoplastid protozoa commonly hosted by wild and domestic dogs and small rodents. *Leishmania* is endemic in parts of the tropics and subtropics, including Central and South America, central and southern Asia, Africa, Europe, and the Middle East. **Leishmaniasis** (lēsh'mă-nī'ă-sis) is a **zoonosis**—a disease of animals transmitted to humans. Twenty-one of the 30 known species of *Leishmania* can infect humans.

CLINICAL CASE STUDY

A Protozoan Mystery

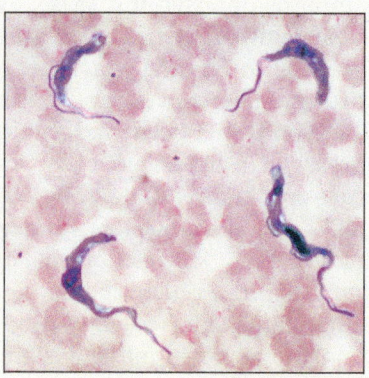

A 20-year-old student was admitted to his college's student health center with headaches, generalized muscle aches, and fever shortly after beginning the fall semester. He had spent his summer working with an international aid organization in the Democratic Republic of Congo and had returned to the United States only one week earlier. Gross examination revealed a diffuse rash and numerous insect bites. The young man had spent most of his summer outdoors in rural areas, had taken no prophylactic antimalarial medication, and had spent some time on African game reserves working with the families of local guides. The patient could not specifically remember receiving any of the bite wounds on his body, and he did not always use insect repellent in the field. The young man was admitted to the local hospital, where intermittent fever, nausea, and headache continued, and he became mentally confused and started having seizures. Initial blood smears proved negative for malaria.

1. What are some possible protozoan diseases the patient could have contracted in Africa?
2. Can this disease be identified from the symptoms alone?
3. Based on the pictured blood smear, what would you conclude about the cause of the disease?
4. What would the treatment be if the patient had tested positive for malaria?
5. What treatment would you now recommend?
6. What prevention would you have suggested to this individual?
7. Is there a local threat of anyone else contracting this disease from the young man?

Leishmania has two developmental stages: *amastigotes*, which lack flagella and multiply within a mammalian host's macrophages and monocytes (types of white blood cells), and *promastigotes*, each of which has a single anterior flagellum and develops extracellularly within a vector's gut. In the life cycle of *Leishmania* (FIGURE 23.6), a sand fly of either of the genera *Phlebotomus* (fle-bot'ō-mŭs) or *Lutzomyia* (lūts-ŏm'yē-a) ingests white blood cells containing amastigotes during a blood meal 1. The amastigotes are released from the phagocytes in the fly's midgut, where they transform into promastigotes 2. Rapidly dividing promastigotes fill the fly's digestive tract and migrate to the fly's long, flexible mouth tube 3 so that each time the fly feeds, it injects promastigotes into a new host 4. Macrophages near the bite site phagocytize the promastigotes 5, which then transform into amastigotes inside the macrophage 6. The amastigotes reproduce via binary fission 7 until the macrophage ruptures, releasing amastigotes that circulate in the bloodstream and/or infect other macrophages 8, within which they can be ingested by another sand fly.

The clinical manifestations of leishmaniasis depend on the species of *Leishmania* and the immune response of the infected host. Upon initial infection, macrophages become activated but not sufficiently to kill the intracellular parasites. The macrophages stimulate inflammatory responses that continue to be propagated by the infection of new macrophages. Eventually, infected macrophages die due to the reproduction of trypanosomes within them; depletion of macrophage numbers decreases the efficiency of an immune response. The severity of the resulting immune dysfunction depends on the overall number of macrophages infected. Of the more than 1.3 million cases of leishmaniasis estimated each year, over 50,000 are fatal.

CLINICAL CASE STUDY

A Sick Soldier

A 20-year-old soldier who served in the Middle East reports to his local Veteran's Administration hospital and complains of more than four large, painless, pink sores on his head and face. Some of the ulcers had healed, leaving scars. A sample taken from a lesion and examined microscopically shows the presence of flagellated protozoa.

1. Can a diagnosis be made from this information? If so, what is the diagnosis?
2. What treatment should the medic order?
3. How did the soldier become infected?
4. The soldier asks how he can prevent this disease in the future. How should the medic respond?
5. The soldier is a strong, healthy 20-year-old with a fully functioning immune system. Why was his body not able to sufficiently fight off the parasite?

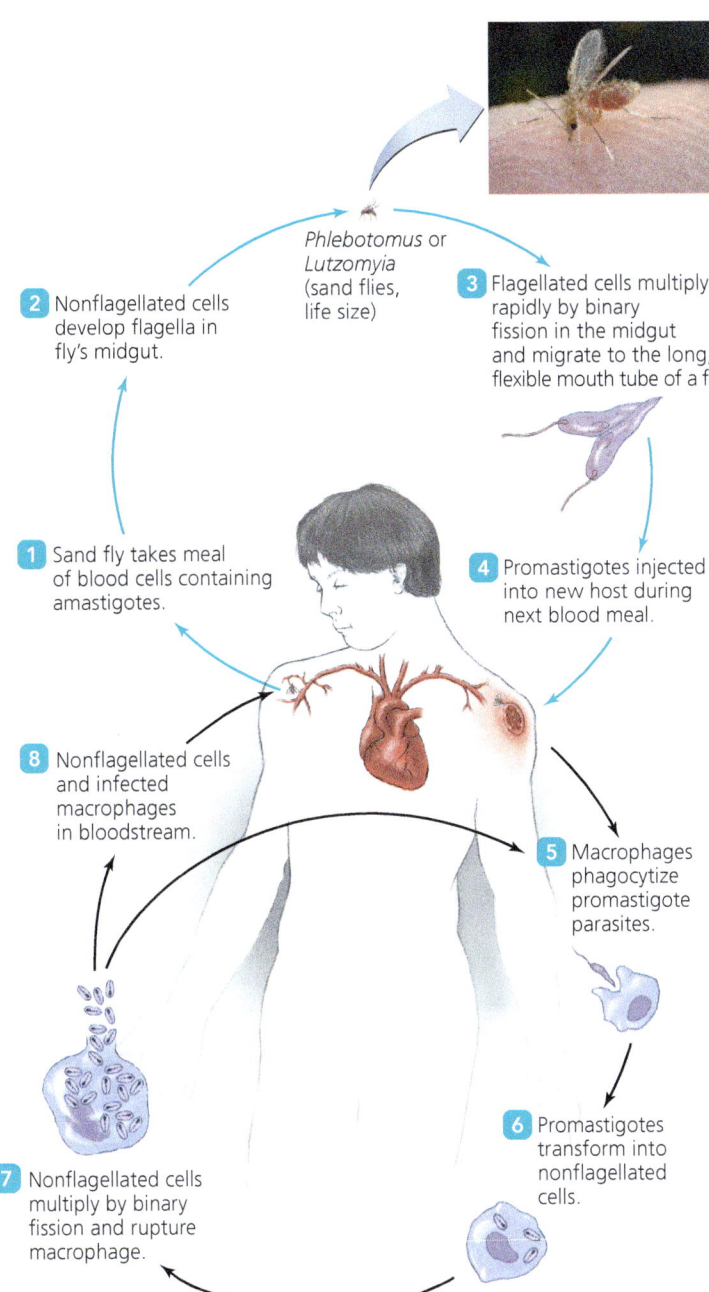

▲ FIGURE 23.6 The life cycle of *Leishmania*. Sand fly is shown life size in the art and magnified in the photo. *What is the effect of amastigote replication on the host's immune system?*

Figure 23.6 *The loss of macrophages severely inhibits the host's immune response.*

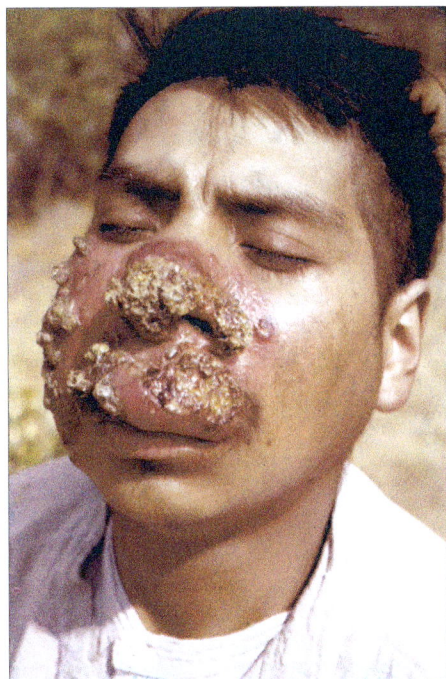

▲ FIGURE 23.7 Mucocutaneous leishmaniasis. The large skin lesions are permanently disfiguring.

Three clinical forms of leishmaniasis are commonly observed. *Cutaneous leishmaniasis* involves large, painless skin ulcers that form around the bite wounds. Such lesions often become secondarily infected with bacteria. Scars remain when the lesions heal. *Mucocutaneous leishmaniasis* results when skin lesions enlarge to encompass the mucous membranes of the mouth, nose, or soft palate. Damage can be severe and permanently disfiguring (FIGURE 23.7). Neither of these forms of leishmaniasis is fatal. However, *visceral leishmaniasis* (also known as *kala-azar*) is fatal in 95% of untreated cases. In this disease, macrophages spread the parasite to the liver, spleen, bone marrow, and lymph nodes. Inflammation, fever, weight loss, and anemia increase in severity as the disease progresses. Visceral leishmaniasis is becoming increasingly problematic as an opportunistic infection among AIDS patients.

Microscopic identification of trypanosomes in samples from cutaneous lesions, the spleen, or bone marrow is diagnostic of *Leishmania* infection. Immunoassays using antibodies to detect antigen can be used to confirm the diagnosis and to identify the strain. Molecular techniques such as PCR are needed to determine the species.

Most cases of cutaneous or mucocutaneous leishmaniasis heal without treatment (though scars may remain) and confer immunity upon the recovered patient. At one time, lesions were purposefully encouraged on the buttocks of small children to induce immunity and prevent lesions from leaving scars in more visible places. Treatment, which is required for more serious infections and for visceral leishmaniasis, generally involves administering paromomycin (not available in the United States), sodium stibogluconate, or amphotericin B. Pentamidine has been used with some success to treat resistant strains of *Leishmania*.

Prevention is essentially limited to reducing exposure by controlling reservoir host and sand fly populations. For example, rodent nesting sites and burrows can be destroyed around human habitations to reduce contact with potentially infected populations. In some areas, infected dogs are destroyed. Spraying insecticide around homes can reduce the number of sand flies. Personal use of insect repellents, protective clothing, and netting further limits exposure. Scientists are testing a possible vaccine for leishmaniasis.

DISEASE IN DEPTH

GIARDIASIS

SIGNS AND SYMPTOMS

Giardiasis is often asymptomatic, but after an incubation period of 7–14 days it can cause significant gastrointestinal distress, including severe, greasy, frothy, fatty diarrhea; abdominal pain; flatus; nausea; vomiting; loss of appetite; ineffective absorption of nutrients; and low-grade fever. Patients' stools often have a "rotten egg" smell of hydrogen sulfide. Acute disease generally lasts one to four weeks following an incubation period of about two weeks. Giardiasis can be a chronic condition, waxing and waning for months, particularly in animals.

Giardiasis is one of the more common waterborne gastrointestinal diseases in the United States. The life cycle of Giardia is shown below:

GIARDIA LIFE CYCLE

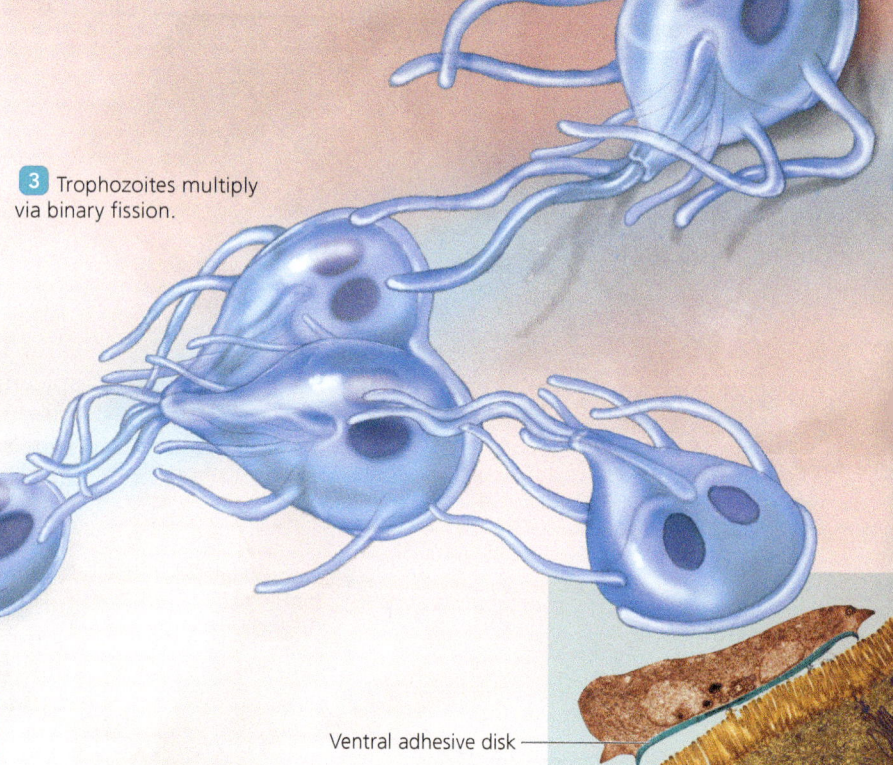

1 A host, who can be an animal or person, ingests a cyst from contaminated food, water, or hands. An infective dose can be as few as ten cysts.

2 An ingested cyst survives passage through the acidic stomach and excysts to release a trophozoite in the small intestine.

3 Trophozoites multiply via binary fission.

4 Trophozoites attach to the intestinal lining via a ventral adhesive disk or remain free.

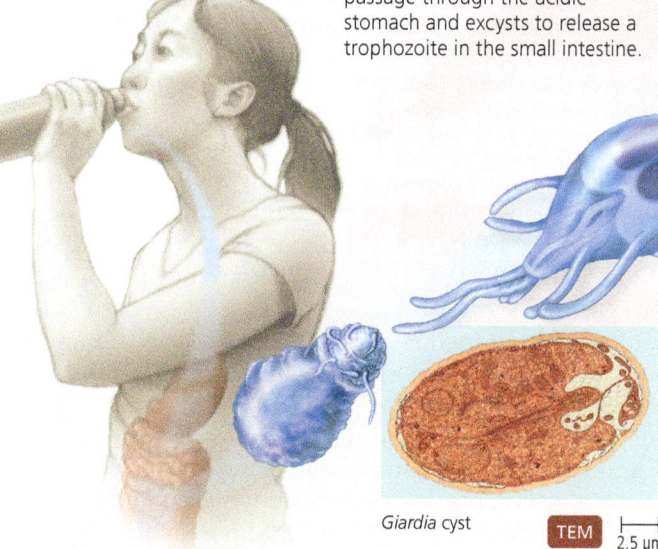

Giardia cyst — TEM 2.5 µm

Ventral adhesive disk

Side view of Giardia attached to intestinal wall. The flagella were cut off during preparation of the protozoan for microscopy. LM 2.5 µ

 *Watch Dr. Bauman's Video Tutor to explore giardiasis, or go to the* **Mastering Microbiology** *Study Area for more information.*

EPIDEMIOLOGY

Giardiasis occurs in both developed and developing countries; recent trends have shown increases in the number of cases worldwide. In the United States, backpacking, camping, swimming, and contact with certain animals may increase the risk for giardiasis.

U.S. 2012 confirmed/probable giardiasis cases per 100,000

- Not reportable
- ≤5.0
- ≥5.1–7.5
- ≥7.6–10.0
- ≥10.1–12.5
- ≥12.6

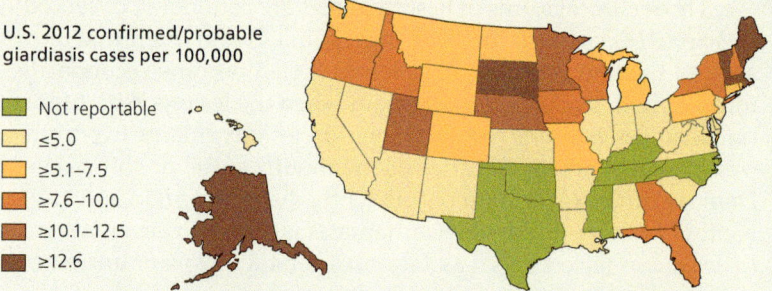

Protozoan Parasites of Humans

PATHOGEN

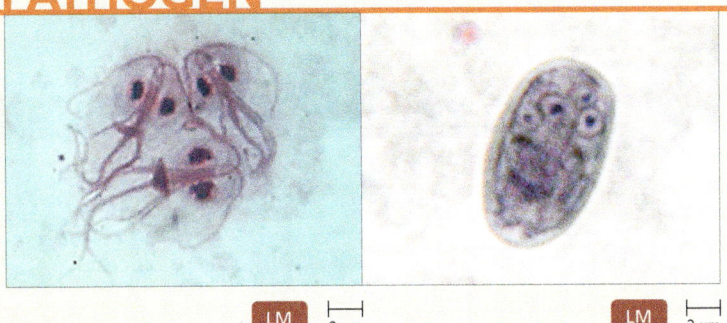

A flagellated diplomonad (a protozoan with two identical nuclei) named *Giardia intestinalis* (previously called *G. lamblia*) causes giardiasis. The protozoan has two forms—a motile feeding trophozoite shown on the left and a dormant cyst shown on the right, which has a tough shell composed of the polysaccharide chitin. The cyst is resistant to chlorine, heat, drying, and stomach acid.

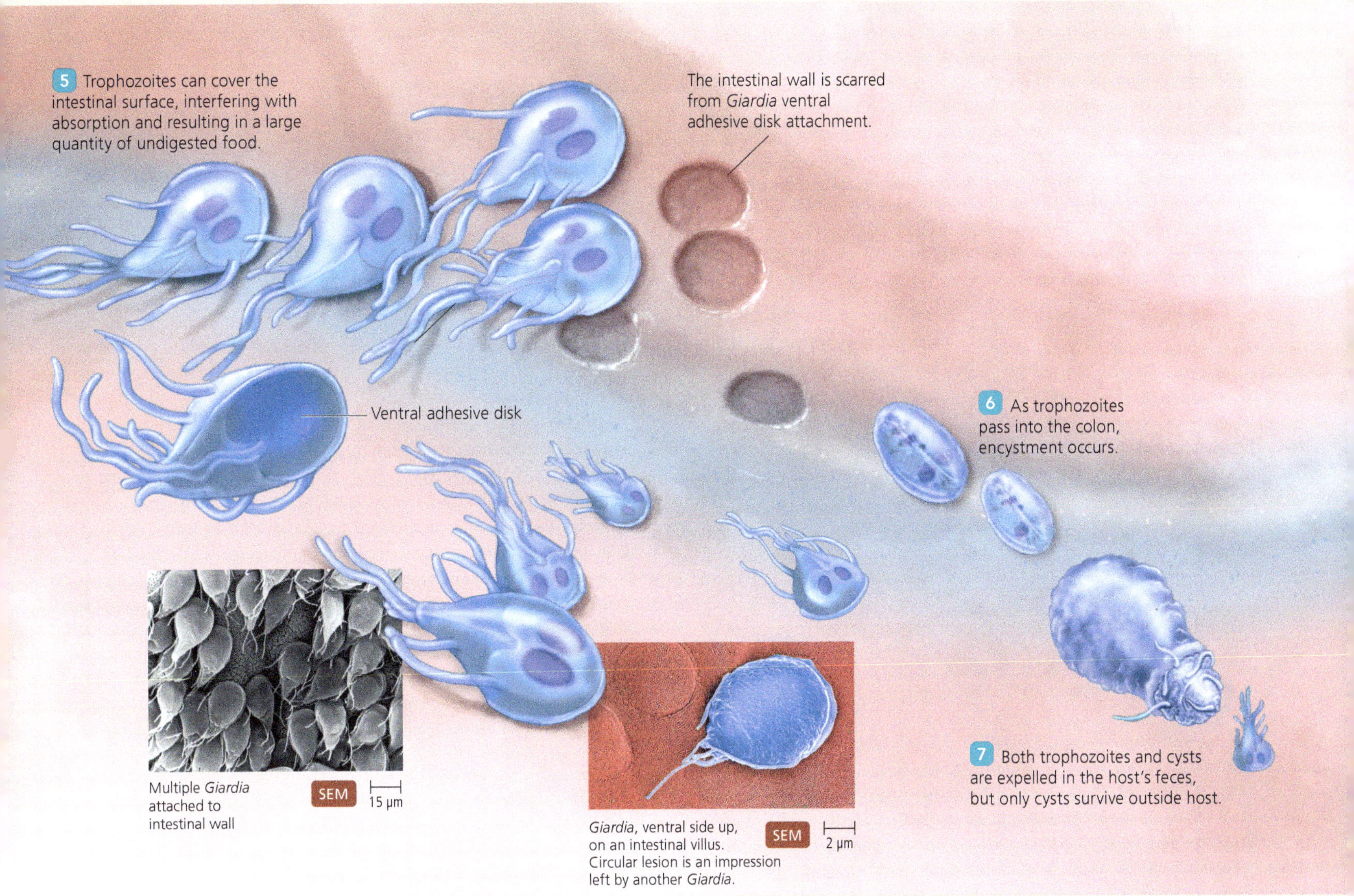

5. Trophozoites can cover the intestinal surface, interfering with absorption and resulting in a large quantity of undigested food.

The intestinal wall is scarred from *Giardia* ventral adhesive disk attachment.

Ventral adhesive disk

6. As trophozoites pass into the colon, encystment occurs.

Multiple *Giardia* attached to intestinal wall

Giardia, ventral side up, on an intestinal villus. Circular lesion is an impression left by another *Giardia*.

7. Both trophozoites and cysts are expelled in the host's feces, but only cysts survive outside host.

DIAGNOSIS AND TREATMENT

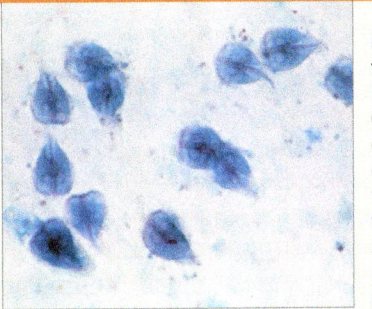

Diagnosis of giardiasis relies on immunoassays for *Giardia* antigens, identification of *Giardia* DNA using a gastrointestinal pathogen panel (GPP) test, or less reliably on microscopic examination of stool specimens to reveal oval cysts. Some infections resolve spontaneously without treatment. Drugs of choice are tinidazole or metronidazole. For patients who are intolerant to nitroimidazoles, nitazoxanide is recommended. Oral rehydration therapy is generally required for infected children regardless of severity and for adults in severe cases.

PREVENTION

In areas where *Giardia* is endemic, water should be filtered to prevent infection. When hiking, humans and their pets shouldn't drink unfiltered stream or river water. In child day care settings, excellent hand hygiene practices and the separation of feeding and diaper-changing areas are essential to avoid transmission.

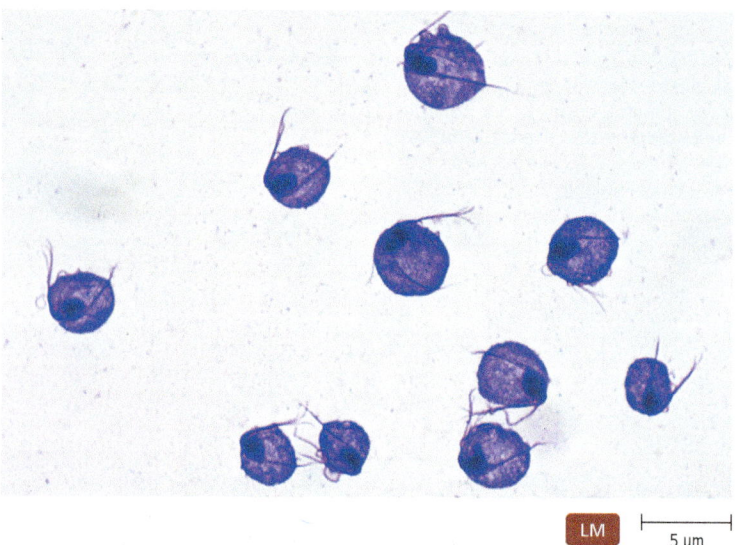

▲ **FIGURE 23.8 Trophozoites of *Trichomonas vaginalis*.** Five anterior flagella, one associated with an undulating membrane, characterize this parabasalid.

Giardia

Giardia intestinalis (jē-ar'dē-ă in-tes'ti-năl'is) causes **giardiasis** (jē-ar-dī'ă-sis), one of the more common waterborne gastrointestinal diseases in the United States. **Disease in Depth: Giardiasis** (pp. 684–685) examines *Giardia* and its disease.

Trichomonas

Trichomonas vaginalis (trik-ō-mō'nas va-jin-al'is) is globally distributed, and its disease is the most common protozoan disease in people of industrialized nations. The parasite lives on the vulvas and in the vaginas of women and in the urethras and prostates of men. This obligate parasite, which is incapable of surviving long outside a human host, is transmitted almost exclusively via sex. *T. vaginalis* occurs most frequently in people with a preexisting sexually transmitted disease, such as chlamydial infection, and in people with multiple sex partners. Further, *trichomoniasis*, as the disease is called, increases the risk of HIV transmission in both women and men.

In women, infection results in **vaginosis** (vaj-i-nō'sis), which is accompanied by a purulent (pus-filled), odorous discharge, vaginal and cervical lesions, abdominal pain, painful urination, and painful intercourse. Since inflammation is not typically involved, it is "vaginosis" rather than "vaginitis." Trophozoites feed on vaginal tissue, leading to erosion of the epithelium. *T. vaginalis* infection may cause inflammation of the urethra or bladder of men, but more typically men are asymptomatic.

Microscopic observation of actively motile trophozoites in vaginal and urethral secretions is diagnostic. The trophozoites are flat and possess five flagella and a sail-like undulating membrane **(FIGURE 23.8)**. An immunofluorescent assay can be performed when infection is suspected and microscopy is too insensitive. Patients and all their sexual partners must be treated to prevent reinfection. The nitroimidazole drugs (e.g., metronidazole) are effective against *Trichomonas*, though some resistant strains exist. Prevention of infection involves abstinence, mutual monogamy, or consistent and correct condom usage. Treating an infected mother before she gives birth can prevent genital and respiratory infection of the baby during birth.

MICRO CHECK

9. Are tsetse flies and kissing bugs biological or mechanical vectors of *Trypanosoma brucei* and *T. cruzi*, respectively?
10. What type of disease does *Trypanosoma cruzi* cause?
11. Which form of leishmaniasis is most likely to be fatal?
12. *Trichomonas vaginalis* causes vaginosis. What is a difference between vaginosis and vaginitis?

Apicomplexans

LEARNING OUTCOMES

23.12 Describe the life cycle of *Plasmodium* and relate malarial symptoms to stages in the life cycle.
23.13 Describe the clinical manifestations of *Toxoplasma gondii* infections.
23.14 Compare and contrast the intestinal diseases caused by *Cryptosporidium* and *Cyclospora*.

Apicomplexans (ap-i-kom-plek'sănz) are alveolate protozoa with infective forms characterized by a complex of organelles at their apical ends, which gives the group its name. They have been called *sporozoa* because they assume nonmotile, sporelike shapes, but this term is not accurate because they are not spores, as in fungi. Apicomplexans are parasites of animals and have complicated life cycles involving at least two hosts. Apicomplexans undergo *schizogony*—a form of asexual reproduction in which multinucleate *schizonts* form before the cells divide (see Figure 12.3).

Plasmodium, *Toxoplasma*, *Cryptosporidium*, and *Cyclospora* are four important apicomplexan parasites. We begin our discussion with *Plasmodium*.

Plasmodium

Four species of *Plasmodium* (plaz-mō'dē-ŭm) typically cause **malaria** in humans: *P. falciparum* (fal-sip'ar-ŭm), *P. vivax* (vī'vaks), *P. ovale* (ō-vă'lē), and *P. malariae* (mă-lār'ē-ī). A fifth species, *P. knowlesi* (nō-les'ē), is an emerging human pathogen.

The life cycle of *Plasmodium* is complex and has three prominent phases as the parasite goes through many different stages in humans and mosquitoes. **Disease in Depth: Malaria** (pp. 688–689) examines the three phases and aspects of the disease.

Virulence factors of *Plasmodium* include the following:

- Part of the reproductive cycle occurs within erythrocytes, which hide the parasite from immune surveillance since red blood cells do not present antigen in conjunction with major histocompatibility protein.
- A special protein assemblage called the *malaria secretome* injects toxins and enzymes into host cells.
- Adhesins enable infected red blood cells to adhere to certain body tissues such as the linings of blood vessels, which allows the parasites to avoid clearance in the spleen.

- Merozoites, one stage of *Plasmodium*, form within vesicles of liver cells that are directly secreted into blood vessels, thereby avoiding immune cells of the liver.
- Some cells of *P. vivax* and *P. ovale* can become dormant inside liver cells in a stage called a *hypnozoite*. Hypnozoites can reactivate to cause malaria several weeks to many years after an initial infection.
- Infective forms of at least one species (*P. falciparum*) trigger changes in human body chemistry, presumably breath or body odor, such that mosquitos are more attracted; thus, *Plasmodium* induces a "bite me" chemical signal in humans carrying the stage ready to be picked up by a mosquito.

People living in endemic areas for malaria and their descendants throughout the world have evolved to have one or more of the following genetic traits that increase their resistance to malaria:

- *Sickle-cell trait.* Individuals with this gene produce an abnormal type of hemoglobin called hemoglobin S (hemoglobin A is normal). Hemoglobin S causes erythrocytes to become sickle shaped, and somehow it also makes erythrocytes resist penetration by *Plasmodium*.
- *Hemoglobin C.* Humans with two genes for hemoglobin C are invulnerable to malaria. The mechanism by which this mutation provides protection is unknown.
- Genetic deficiency for the enzyme *glucose-6-phosphate dehydrogenase*. Trophozoites must acquire this enzyme from a human host before the trophozoite can synthesize DNA; thus, humans without the enzyme are spared malaria.
- Lack of so-called *Duffy*[1] antigens on erythrocytes. Because *P. vivax* requires Duffy antigens in order to attach to and infect erythrocytes, Duffy-negative individuals are resistant to this species.

The general symptoms of malaria are associated with the synchronous and nearly simultaneous reproduction of *Plasmodium* inside red blood cells—a phase called the **erythrocytic cycle.** Signs and symptoms include fever, chills, diarrhea, headache, and (occasionally) pulmonary or cardiac dysfunction. Loss of erythrocytes leads to anemia, weakness, and fatigue. The inability of the liver to process the inordinate amount of hemoglobin released from dying erythrocytes results in *jaundice*.

P. falciparum causes a form of malaria called *blackwater fever*, which is characterized by extreme fever, large-scale erythrocyte lysis, renal failure, and dark urine discolored by excreted hemoglobin. Protozoan proteins inserted on the surfaces of infected erythrocytes cause erythrocytes to become rigid and inelastic such that they cannot squeeze through capillaries, blocking blood flow and causing small hemorrhages in various tissues (and ultimately tissue death). *Cerebral malaria* results when tissue death occurs in the brain. Falciparum malaria can be fatal within 24 hours of the onset of symptoms.

If a victim survives the acute stages of malaria, immunity gradually develops. Periodic episodes become less severe over time—unless the victim becomes immunocompromised, in which case episodes can return to original levels of severity.

[1]Named for the patient in which the antigen was discovered.

Toxoplasma

Toxoplasma gondii (tok-sō-plaz′mă gon′dē-ē) is one of the world's most widely distributed protozoan parasites—30% of the world's human population is infected. Wild and domestic mammals and birds are major reservoirs for *Toxoplasma*, and cats are the definitive host, in which the protozoan reproduces sexually.

Humans typically become infected by ingesting undercooked meat containing the parasite. People at greatest risk include butchers, hunters, and anyone who tastes food while preparing it. Ingestion or inhalation with contaminated soil can also be a source of infection. The protozoan can also cross a placenta to infect the fetus. Historically, contact with infected cats and their feces was proposed as a major risk for infection, but recent studies have shown that cats are not the major source of infection for humans because cats shed *Toxoplasma* only briefly.

In the life cycle of *Toxoplasma* (FIGURE 23.9), male and female *Toxoplasma* gametes in a cat's digestive tract fuse to form *zygotes*,

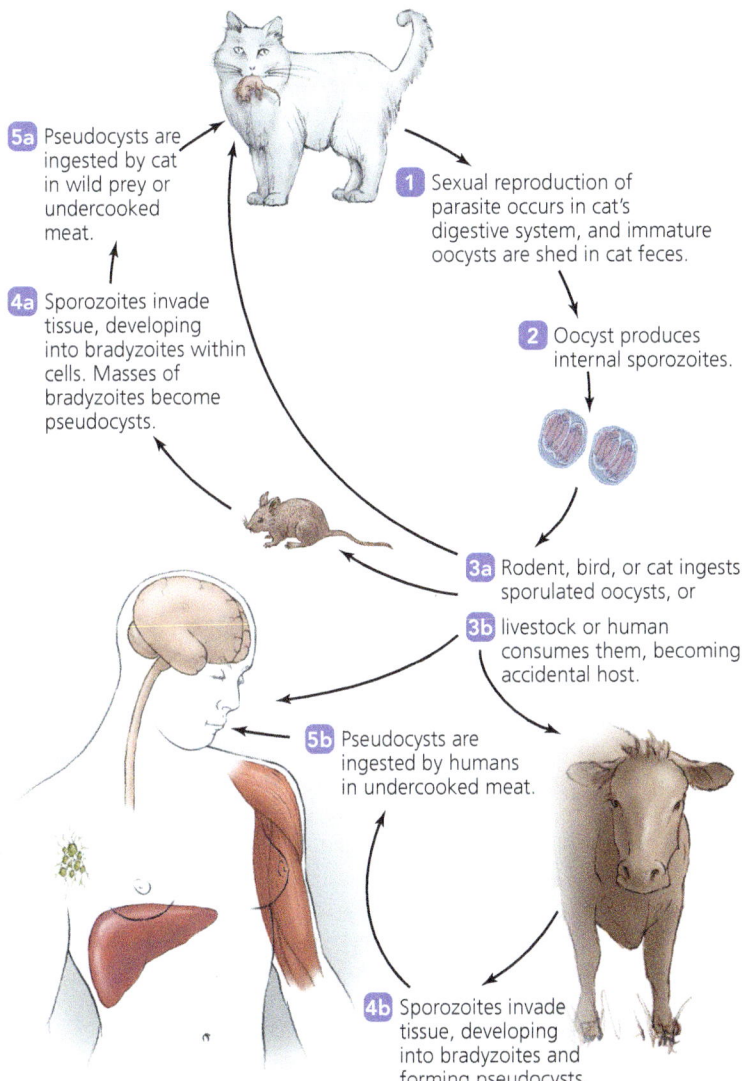

▲ FIGURE 23.9 The life cycle of *Toxoplasma gondii*. Which two groups of humans are most at risk from toxoplasmosis?

Figure 23.9 AIDS patients and first-trimester fetuses are at greatest risk from toxoplasmosis.

DISEASE IN DEPTH

MALARIA

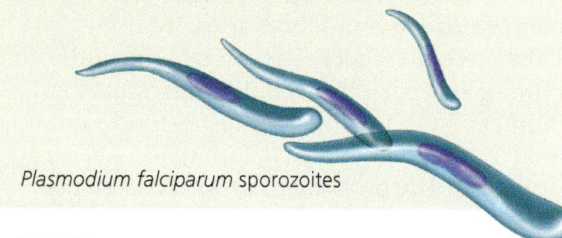
Plasmodium falciparum sporozoites

SIGNS AND SYMPTOMS

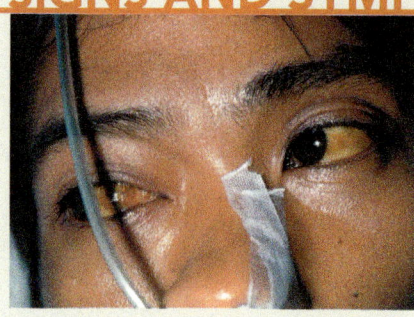

Many of malaria's symptoms are associated with immune response against the parasites, cellular debris, and toxins following the synchronous cycles of the parasite's life within erythrocytes. Two weeks after the cycle begins, enough parasites exist to cause fever, chills, diarrhea, headache, and (occasionally) pulmonary or cardiac dysfunction. Anemia, weakness, and fatigue gradually set in, and patients may become jaundiced, as seen in this malaria patient from Vietnam.

Malaria is the most prevalent and infamous of the protozoan infections. The life cycle of *Plasmodium*, the causative agent of malaria, has three prominent stages: the **liver phase**, the **erythrocytic cycle**, and the **sporogonic phase**.

PLASMODIUM LIFE CYCLE

Liver phase (in human) 1–3

1 A female *Anopheles* mosquito injects sporozoites into the human during a blood meal.

2 Sporozoites travel through the bloodstream, invade liver cells, and undergo schizogony.

3 Generally, two weeks later, the liver cells rupture and release 30,000 to 40,000 merozoites into the blood. The liver is damaged.

Erythrocytic cycle (in human) 4–7

4 Free merozoites penetrate erythrocytes.

5 A merozoite becomes a trophozoite, shown in its ring state below.

6 Trophozoites undergo schizogony to produce merozoites, which are released when the erythrocytes break open. This occurs simultaneously and cyclically every 48 to 72 hours depending on the species of *Plasmodium*. The damaged liver cannot effectively process the amount of hemoglobin released, leading to jaundice.

7 Released merozoites adhere to and then infect new erythrocytes.

8 Some merozoites develop into male and female gametocytes within erythrocytes.

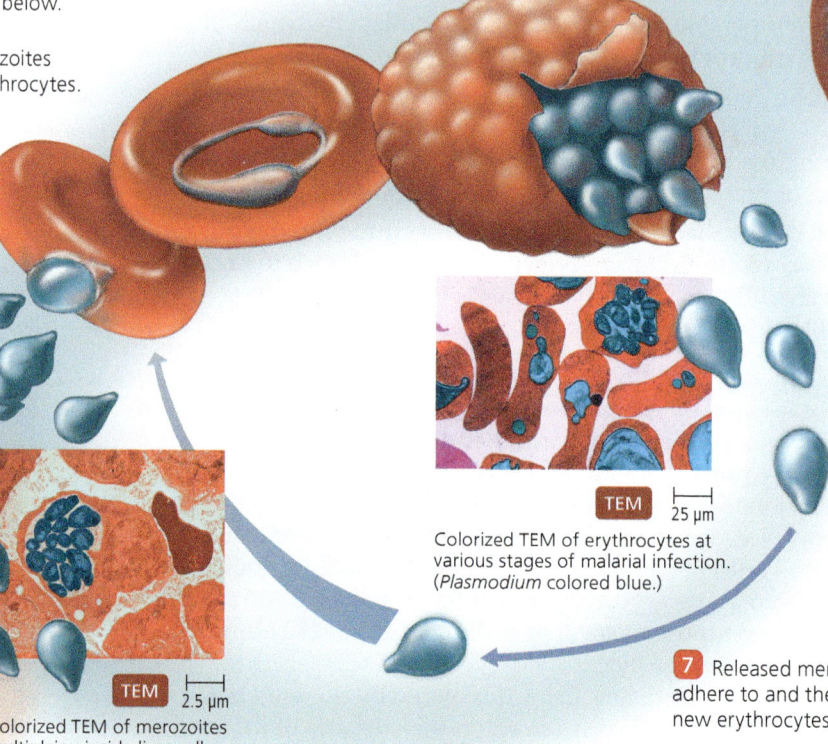

Colorized TEM of merozoites multiplying inside liver cell. 2.5 μm

Colorized TEM of erythrocytes at various stages of malarial infection. (*Plasmodium* colored blue.) 25 μm

Watch Dr. Bauman's Video Tutor to explore malaria, or go to the **Mastering Microbiology** Study Area for more information.

EPIDEMIOLOGY

Malaria is endemic in over 90 countries in the tropics and subtropics, where the parasite's mosquito vector, *Anopheles*, breeds. The World Health Organization estimates that 216 million people are infected with *Plasmodium*, and an estimated 445,000 (usually children) die annually. Malaria was prevalent in the United States until mosquito eradication programs eliminated the disease decades ago. About 1500 cases seen each year in the United States involve immigrants or travelers from endemic areas.

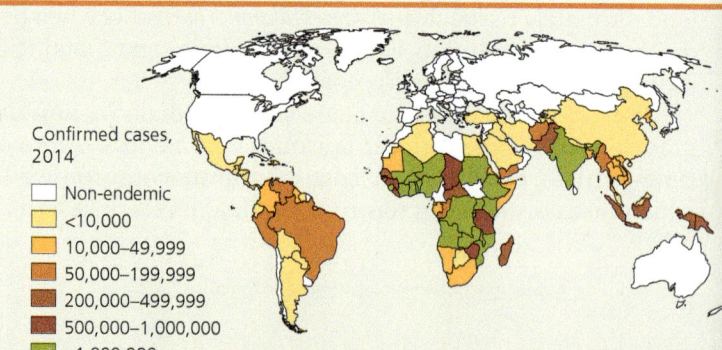

Confirmed cases, 2014
- Non-endemic
- <10,000
- 10,000–49,999
- 50,000–199,999
- 200,000–499,999
- 500,000–1,000,000
- >1,000,000

PATHOGEN AND PATHOGENESIS

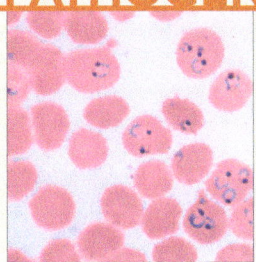
LM 7 μm

At least four species of *Plasmodium* cause malaria in humans: *P. falciparum*, *P. vivax*, *P. ovale*, and *P. malariae* are most significant. Disease severity depends upon the species: *P. ovale* causes mild disease, *P. vivax* results in chronic malaria, and *P. malariae* and *P. falciparum* (left) cause more serious malaria, known as *blackwater fever* because hemoglobin from damaged red blood cells is excreted in the urine.

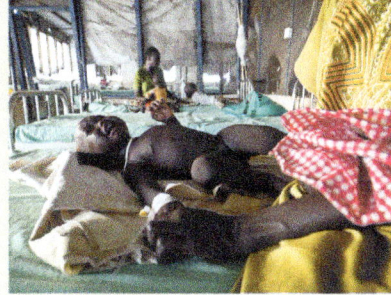

P. falciparum causes fever, erythrocyte lysis, renal failure, and dark urine. Protozoan proteins on the surfaces of erythrocytes cause erythrocytes to adhere to capillary lining, blocking blood flow and leading to small hemorrhages and ultimately tissue death. Children, such as this infant with malaria in South Sudan, are among the most vulnerable.

Sporogonic phase (in mosquito) 9–13

9 The *Anopheles* mosquito ingests erythrocytes containing gametocytes during a blood meal.

10 Gametocytes leave the erythrocyte and turn into gametes. These fuse to form a zygote.

11 The zygote differentiates into an ookinete and attaches to the gut wall.

12 The ookinete becomes an immobile, stationary oocyst in the gut wall. Sporozoites form within the oocyst.

13 Sporozoites exit the oocyst and migrate to the mosquito's salivary glands.

Gamete Zygote Ookinete Oocyst emerging from mosquito gut

1 Sporozoites are injected into a human during a blood meal. The life cycle begins again.

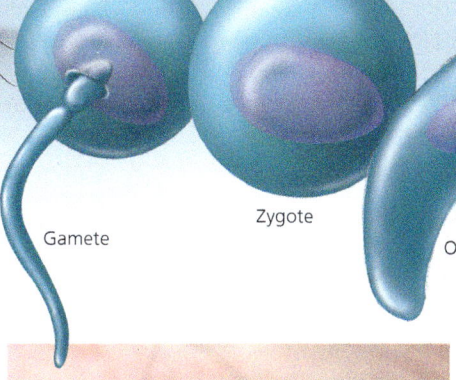
Malarial vector, female *Anopheles* mosquito, during blood meal

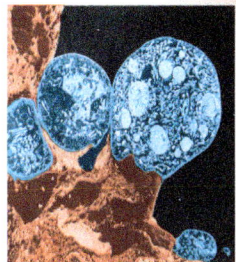
TEM 25 μm
Colorized TEM of oocysts filled with sporozoites

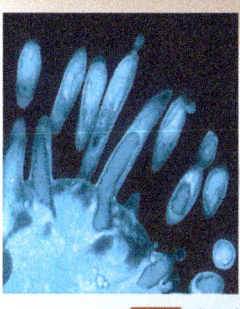
TEM 8 μm
Colorized TEM of sporozoites emerging from an oocyst

DIAGNOSIS AND TREATMENT

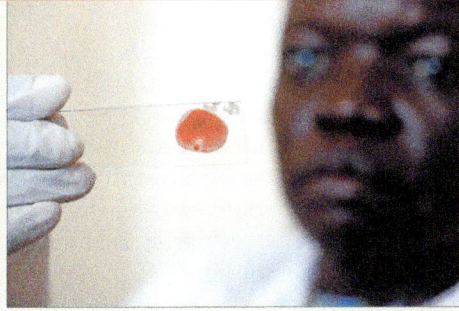

Diagnosis of malaria is commonly based on identification of trophozoites and other stages of *Plasmodium* in blood smears. Standard antimalarial drugs include chloroquine and, in areas where *Plasmodium* is resistant, pyrimethamine with artesunate.

PREVENTION

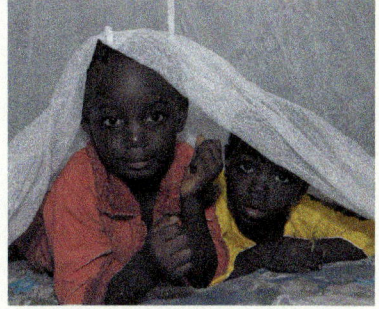

Control of malaria involves limiting contact with mosquitoes carrying *Plasmodium*. Widespread use of insecticides, drainage of wetlands, and removal of standing water can reduce mosquito breeding rates. WHO recommends DDT-impregnated netting, personal use of insect repellents, and protective clothing, such as long-sleeved shirts and long pants, to reduce mosquito bites. Travelers to endemic areas can take atovaquone and proguanil several days before and for one week after travel. Several malaria vaccines are currently under development.

which develop into immature *oocysts* that are shed for one to three weeks in the cat's feces 1. During this time, the cat excretes up to 10 million oocysts daily. An infected cat is not harmed and shows no other signs of infection. Oocysts survive and mature in moist soil for several months, contaminating vegetation. As each oocyst matures, it produces *sporozoites* internally 2. Intermediate hosts—warm-blooded animals, commonly rodents or birds—ingest sporulated oocysts on vegetation 3a. Digestion releases the sporozoites, which invade the animal's muscles, lymph nodes, digestive organs, or brain 4a, where they proliferate and develop into numerous *bradyzoites* surrounded by a thickened wall called a *pseudocyst*. Pseudocysts can remain dormant for the life of the host, particularly within neurons in the brain, until the tissue is consumed by another host.

Toxoplasma must return to a cat to reproduce, so pseudocysts preferentially form on and inhibit those parts of a rodent's brain that process cat odors and those that induce fear. With its fear of the smell of cats gone, the rodent is easily caught and eaten 5a. Bradyzoites released from digested pseudocysts infect the cat's intestinal cells, where the parasites become gametes, completing the life cycle.

Nonrodent animals (and occasionally people) become accidental hosts when they ingest oocysts containing sporozoites 3b. These develop into pseudocysts containing bradyzoites 4b. Humans most often are infected by eating pseudocysts in undercooked meat 5b.

Although the majority of people infected with *T. gondii* have no symptoms, a small percentage develop **toxoplasmosis** (tok'sō-plaz-mō-sis), a fever-producing illness with headache, muscle pain, sore throat, and enlarged lymph nodes in the head and neck. Toxoplasmosis generally results in no permanent damage and is self-limiting, resolving spontaneously within a few months to a year. However, toxoplasmosis is more severe in two populations: AIDS patients and fetuses. In AIDS patients, symptoms are thought to result when tissue pseudocysts reactivate as the immune system fails. Spastic paralysis, blindness, myocarditis, encephalitis, and death result.

Transplacental transfer of *Toxoplasma* from mother to fetus is most dangerous in the first trimester of pregnancy; it can result in spontaneous abortion, stillbirth, or epilepsy, mental retardation, an abnormally small head (microcephaly), inflammation of the retina, blindness, anemia, jaundice, and other conditions. Ocular infections may remain dormant for years, at which time blindness develops.

Physicians diagnose toxoplasmosis via microscopic identification of parasites in tissue biopsies **(FIGURE 23.10)** or via molecular identification of *T. gondii* genetic material or products in specimens using PCR, Southern blot, or DNA probes. Serology is the most common diagnostic method.

Asymptomatic patients do not need treatment for toxoplasmosis. For those with signs and symptoms, physicians in the United States prescribe pyrimethamine plus sulfadiazine or clindamycin. Treatment of infected pregnant women, regardless of their symptoms or lack thereof, prevents most transplacental infections. More aggressive treatment may be needed for AIDS patients, including the addition of steroids to reduce tissue inflammation.

Controlling the incidence of *T. gondii* infection is difficult because so many hosts can harbor the parasites. A vaccine for cats is currently under development to reduce the chance of

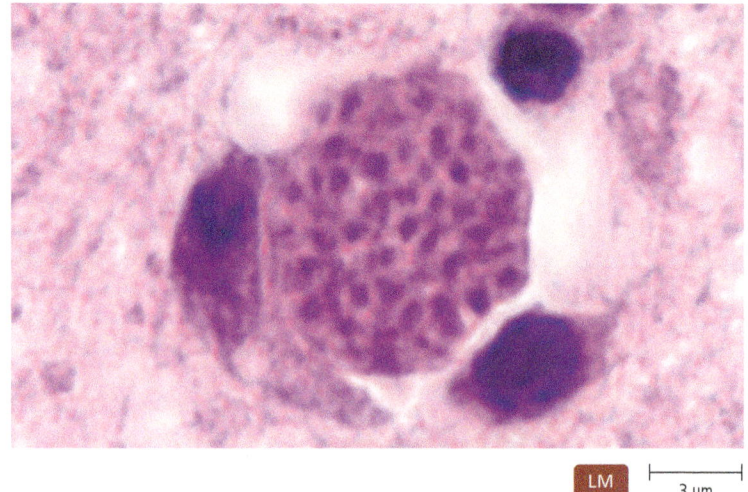

▲ **FIGURE 23.10 Pseudocysts of *Toxoplasma gondii*.** Pseudocysts contain bradyzoites and are one infective stage of this protozoan.

pet-to-owner transmission. The best prevention is to thoroughly cook or deep-freeze meats and to avoid contact with contaminated soil. Physicians advise pregnant women to avoid cat feces and cat litter boxes.

Cryptosporidium

***Cryptosporidium* enteritis** (krip'tō-spō-rid'ē-ŭm en-ter-ī'tis), also known as *cryptosporidiosis* (krip'tō-spō-rid-ē-ō'sis), is a zoonosis; that is, it is a disease of animals that is transmitted to people. *Cryptosporidium parvum* (par'vŭm), an apicomplexan, causes the disease. Once thought to infect only livestock and poultry, *Cryptosporidium* is carried asymptomatically by about 30% of people living in developing nations. It is estimated that most natural waterways in the United States are contaminated with the oocysts of *Cryptosporidium* from livestock wastes.

Infection most commonly results from drinking water contaminated with oocysts, but direct fecal-oral transmission resulting from poor hygienic practices also occurs, particularly in day care facilities. In the intestine, oocysts release sporozoites, which invade the intestinal mucosa and become intracellular parasites. Eventually, new oocysts are produced, released into the lumen of the intestine, and shed in the feces.

Signs and symptoms of *Cryptosporidium* enteritis include severe diarrhea that lasts from one to two weeks accompanied by headache, muscular pain, cramping, and severe fluid and weight loss. In HIV-positive individuals, chronic *Cryptosporidium* enteritis is life threatening and is one of the indicator diseases revealing that a person has AIDS.

Most diagnostic techniques are not very sensitive for *Cryptosporidium*, but microscopic examination of concentrated fecal samples or biopsy material can reveal oocysts **(FIGURE 23.11)**. A fluorescent immunoassay reveals low concentrations of oocysts in appropriate specimens.

Treatment consists of oral rehydration therapy and nitazoxanide, paromomycin, or azithromycin. Drinking from rivers and streams should be avoided in areas where *Cryptosporidium* is found; filtration is required to remove oocysts from drinking water because they cannot be killed by chlorination. Good personal hygiene can eliminate fecal-oral transmission of the parasite.

EMERGING DISEASE CASE STUDY

Babesiosis

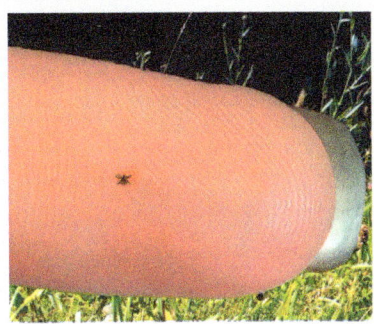

Leslie hated being sick, but here she was on a fine summer day with a headache, muscle pains, and sore joints. It was just as well she didn't know that the worst was yet to come.

A small deer tick *(Ixodes scapularis)* had taken a blood meal from Leslie's leg three weeks before. The tick was so small and the bite so painless that Leslie was unaware of the incident. She was equally unaware that the tick had infected her with an apicomplexan parasite, *Babesia microti*, normally found in mice. The protozoan was eating her red blood cells from the inside out.

Leslie became anemic and fatigued, she was losing weight, and she experienced rapid, alarming changes in mood and emotional temperament. As she became weaker, she had trouble catching her breath, and within the week, her kidneys began to fail. Intermittent fever, shaking chills, drenching sweat, nausea, and anorexia came next. Leslie was utterly miserable, but with intensive medical care, including the use of the intravenous antimicrobials clindamycin and quinine, she pulled through.

In the past decade, the number of reported cases of babesiosis has quadrupled. Perhaps this emerging disease is spreading because its tick vector has an expanding geographical range. Perhaps physicians are more aware of a disease whose incidence has been constant. Whatever the reason for our increased awareness of babesiosis, Leslie was grateful that she was diagnosed and treated effectively so that her nightmare could finally end.

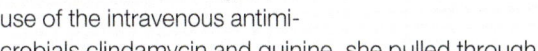

1. What aspect of the infection caused Leslie to suffer anemia and fatigue?
2. Why doesn't penicillin work for treating babesiosis?
3. How many legs does the adult vector of babesiosis have?

▲ FIGURE 23.11 **Oocysts of *Cryptosporidium parvum*.** Here they are in a stained fecal smear.

Cyclospora

Cyclospora cayetanensis (sī-klō-spōr'ă kī'ē-tan-en'sis) is a waterborne apicomplexan responsible for an emerging disease, *cyclosporiasis*, that has affected hundreds in the United States since the early 1980s. The disease is not transmitted between individuals; rather, it is acquired by eating or drinking sporulated oocysts in contaminated food or water (FIGURE 23.12). Outbreaks have been linked in particular to raspberries imported from Central and South America. The environmental reservoir for *Cyclospora* remains unknown.

Once *Cyclospora* enters the intestine, it invades the mucosal layer, causing symptoms, which develop after about a week. Manifestations include cramps, watery diarrhea, myalgia (muscle pain), and fever. Symptoms are more severe in AIDS patients, leading to severe dehydration and weight loss. The disease usually resolves in days or weeks in immunocompetent patients.

Microscopic examination of stool samples can sometimes reveal the presence of oocysts in extensive infections, but fluorescent DNA probes more reliably detect the presence of *Cyclospora*. *Cyclospora* infections are treated with trimethoprim and sulfamethoxazole given in combination for seven days.

Reliable food testing does not currently exist but would be an important tool in preventing contaminated food from reaching U.S. consumers. The only reliable methods for reducing risk of infection in the United States are thoroughly washing fruits and vegetables prior to eating them raw. Cooking or freezing also kills many oocysts.

TABLE 23.1 (p. 692) lists the general characteristics of the protozoan parasites discussed in this chapter.

MICRO CHECK

13. Which stage of *Plasmodium* is formed within vesicles inside the liver and is released into the bloodstream?
14. What is the most common mode of transmission of *Toxoplasma gondii*?
15. What signs and symptoms does toxoplasmosis cause in AIDS patients?
16. For which protozoan are humans the definitive host: *Cyclospora cayetanensis*, *Toxoplasma gondii*, *Plasmodium falciparum*, or *Trypanosoma brucei*?

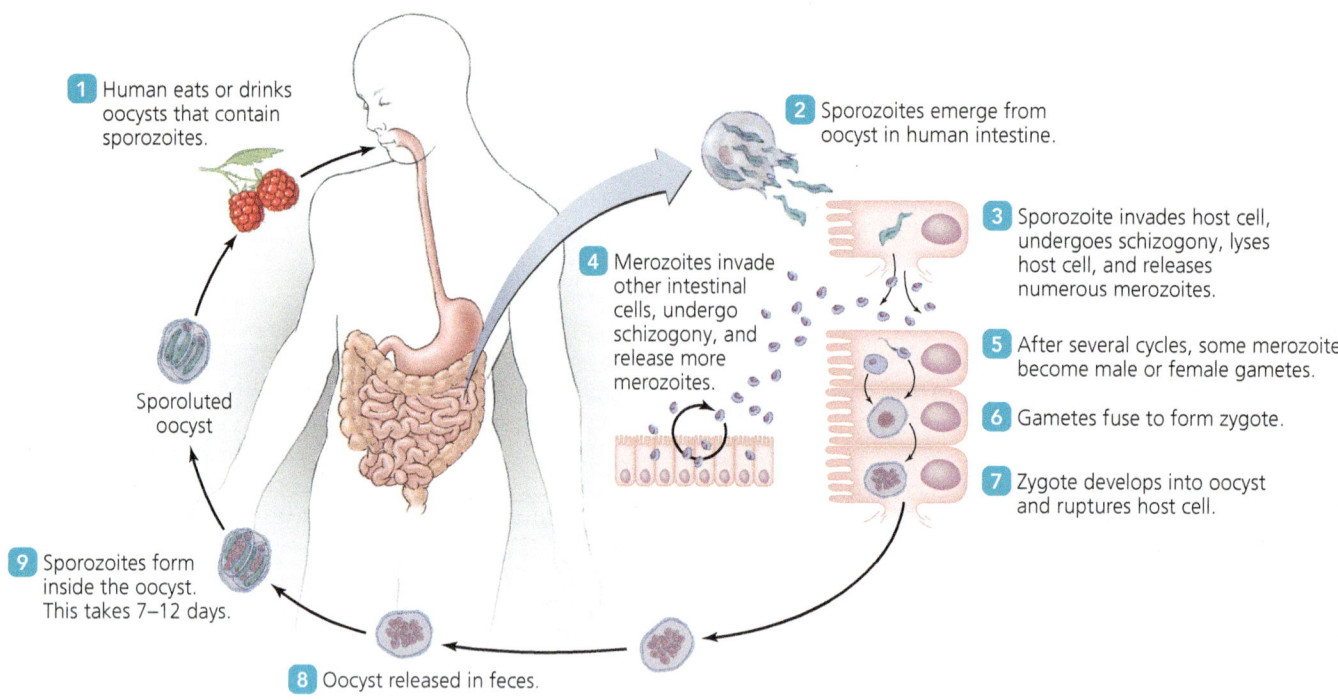

▲ FIGURE 23.12 Life cycle of *Cyclospora*

TABLE 23.1 Key Features of Protozoan Parasites of Humans

Organism	Primary Diseases	Geographical Distribution	Mode of Transmission	Host Organisms
Ciliates				
Balantidium coli	Balantidiasis, dysentery	Worldwide	Fecal-oral	Pigs, rodents, primates, humans
Amoebas				
Entamoeba histolytica	Luminal amebiasis, amebic dysentery, invasive extraintestinal amebiasis	Worldwide	Fecal-oral	Humans
Acanthamoeba spp.	Ulcerative keratitis, amebic encephalitis	Worldwide	Contact	Humans
Naegleria	Amebic meningoencephalitis	Worldwide	Inhalation	Humans
Flagellates				
Trypanosoma cruzi	Chagas' disease	Central and South America	Kissing bug (*Triatoma*)	Opossums, armadillos, humans
Trypanosoma brucei	African trypanosomiasis (African sleeping sickness)	Africa	Tsetse fly (*Glossina*)	Wild game, cattle, sheep, humans
Leishmania spp.	Cutaneous, mucocutaneous, or visceral leishmaniasis	Tropics, subtropics	Sand flies (*Phlebotomus, Lutzomyia*)	Canines, rodents, humans
Giardia intestinalis (lamblia)	Giardiasis	Worldwide	Fecal-oral	Humans, wild animals
Trichomonas vaginalis	Vaginosis	Developed nations	Sexual contact	Humans
Apicomplexans				
Plasmodium spp.	Malaria	Tropics, subtropics	Mosquitoes (*Anopheles*)	Humans
Toxoplasma gondii	Toxoplasmosis	Worldwide	Fecal-oral	Cats, livestock, humans
Cryptosporidium parvum	*Cryptosporidium* enteritis (cryptosporidiosis)	Worldwide	Fecal-oral	Livestock, poultry, humans
Cyclospora cayetanensis	Cyclosporiasis, gastrointestinal disorders	North, Central, and South America	Fecal-oral	Humans

> **TELL ME WHY**
>
> Why are diseases of *Entamoeba, Balantidium,* and *Giardia* generally more deadly in infants than in adults?

Helminthic Parasites of Humans

LEARNING OUTCOME

23.15 Describe the general morphological and physiological features of parasitic worms.

Helminths are macroscopic, multicellular, eukaryotic worms found throughout the natural world, some in parasitic associations with other animals. Helminthologists generally use the term *infestation* to indicate presence of parasitic worms in people or animals. Though helminths are not microorganisms, microbiologists often study them in part because the diagnostic signs of infestation—eggs or larvae—are microscopic.

Taxonomists divide parasitic helminths into three groups: *cestodes* (ses′tōdz), *trematodes* (trem′a-tōdz), and *nematodes* (nēm′a-tōdz).

The life cycles of parasitic helminths are complex. For many, intermediate hosts are required to support larval (immature) stages needed for the worms to reach maturity. Adult worms are either *dioecious*,[2] meaning that the male and female sex organs are in separate worms, or *monoecious*,[3] meaning that each worm has both sex organs. Monoecious organisms either fertilize each other during copulation or are capable of self-fertilization, depending on the species. Because most parasitic helminths release large numbers of fertilized eggs into the environment, transmission to new hosts is likely to occur.

Cestodes

LEARNING OUTCOMES

23.16 Describe the common features of the life cycles of tapeworms that infest humans.
23.17 List the predominant modes of infestation for *Taenia* and *Echinococcus* and suggest measures to prevent infestation.

All **cestodes**, commonly called *tapeworms*, are flat, segmented, intestinal parasites that completely lack digestive systems. Though they differ in size when mature, all possess the same general body plan **(FIGURE 23.13)**. The **scolex** (skō′leks) is a small attachment organ that possesses suckers and/or hooks used to attach the worm to host tissues (see Figure 23.13b). Anchorage is the only role of a scolex; there is no mouth. Cestodes acquire nutrients by absorption through the worm's *cuticle* (outer "skin").

Behind the scolex is the neck region. Body segments, called **proglottids** (prō-glot′idz), grow from the neck continuously as long as the worm remains attached to its host. New proglottids displace older ones, moving the older ones farther from the neck. Proglottids mature, producing both male and female reproductive organs. Thus, a chain of proglottids, called a *strobila* (strō′bi-lă;

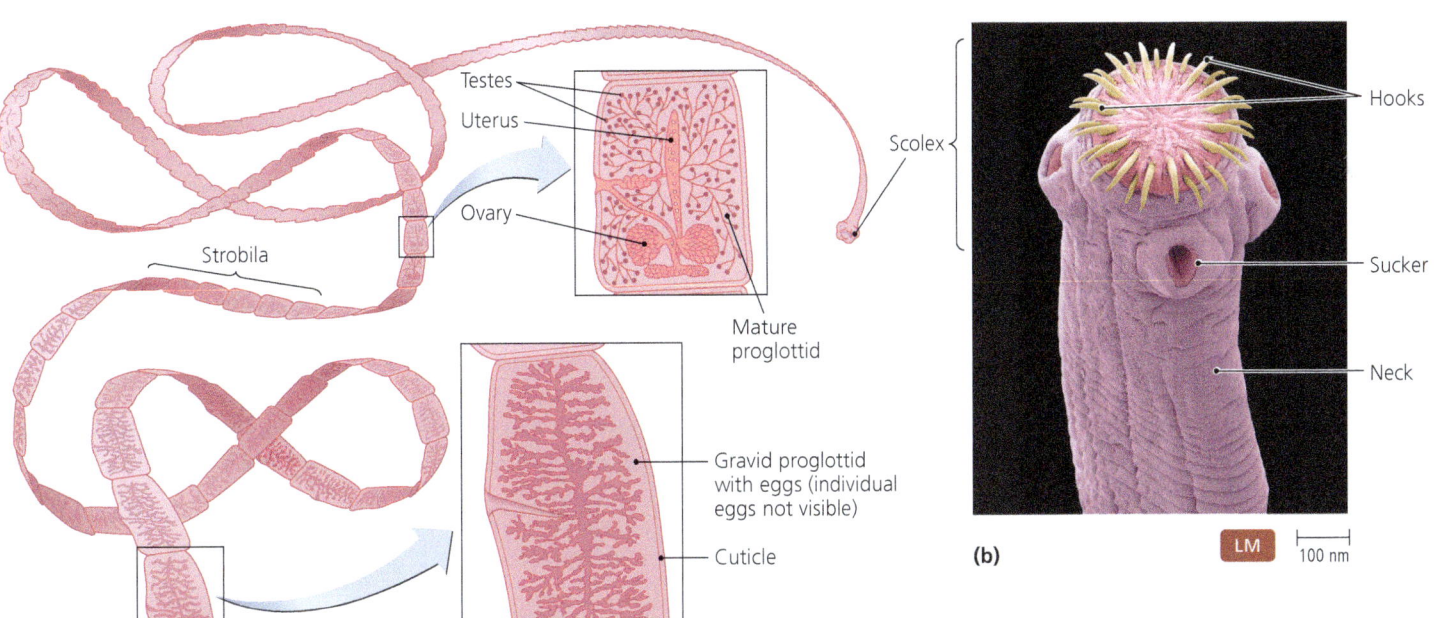

▲ **FIGURE 23.13 Features of tapeworm morphology.** Each tapeworm consists of an organ of attachment called a scolex, a neck, and a long chain of segments called proglottids. **(a)** Artist's rendition. **(b)** Micrograph. *How does a tapeworm acquire food?*

Figure 23.13 Tapeworms absorb food through their cuticles.

---

[2]From Greek *di*, meaning "two," and *oikos*, meaning "house."
[3]From Greek *mono*, meaning "one," and *oikos*, meaning "house."

plural: *strobilae*, strō'bi-lī), reflects a sequence of development: Proglottids near the neck are immature, those in the middle are mature, and those near the end are *gravid*[4]—full of fertilized eggs. Each proglottid is monoecious and may fertilize other proglottids of the same or different tapeworm; a proglottid is not capable of fertilizing itself, although the tapeworm is self-fertile.

After fertilization occurs and the proglottids fill with eggs, gravid proglottids break off the strobila and pass out of the intestine along with feces. In a few cases, the proglottids rupture within the intestine, releasing eggs directly into the feces. Some tapeworms produce proglottids large enough to be obviously visible in stools.

Generalized Tapeworm Life Cycle

The generalized life cycle of most tapeworms of humans is depicted in **FIGURE 23.14**. Gravid proglottids and/or eggs enter the environment in feces from infested humans, who are the definitive (primary) hosts 1. Intermediate (secondary) hosts, which vary with the species of tapeworm, become infected by ingesting vegetation contaminated with gravid proglottids or eggs 2. Eggs hatch into larvae within the intermediate host's intestine, and the larvae penetrate the intestinal wall and migrate to other tissues, often muscle 3, where the larvae develop into immature forms called **cysticerci** (sis-ti-ser'sī) 4. Cysticerci are generally harmless to intermediate hosts. Humans are infested by consuming undercooked meat containing cysticerci 5. A cysticercus excysts in the human's intestine to become a scolex that attaches to the intestinal wall and matures 6, developing into a new adult tapeworm 7, which eventually sheds gravid proglottids, completing the cycle.

Taenia

Taenia saginata (tē'nē-a sa-ji-na'ta), the beef tapeworm, and *Taenia solium* (sō'lī-um), the pork tapeworm, got their common names because cattle and swine, respectively, serve as intermediate hosts. Both tapeworm species are distributed worldwide in areas where beef and pork are eaten. Poor, rural areas with inadequate sewage treatment and where humans and livestock live in close proximity have the highest incidence of human infestation. Overall, *Taenia* infestations in humans are rare in the United States.

Rarely, humans become accidental intermediate hosts of *T. solium* when they ingest eggs or gravid proglottids. Larvae released from the eggs can encyst in the person's muscles, a condition called *cysticercosis* (**FIGURE 23.15**). For the parasite, this is a dead end (except following cannibalism). Humans are not intermediate hosts for *T. saginata*.

Tapeworms attached to the intestinal epithelium can grow quite large. At maturity, *T. saginata* worms have 1000 to 2000 proglottids, each of which can contain 100,000 eggs. *T. solium* adults average 1000 proglottids, each of which contains about 50,000 eggs. An infested human passes strobilae of about six proglottids per day.

Most people actively shed strobilae without experiencing symptoms, though in some cases, nausea, abdominal pain, weight loss, and diarrhea accompany infestation. If the tapeworm is particularly large, blockage of the intestine is possible.

---
[4]From Latin *gravidus*, meaning "heavy"—that is, pregnant.

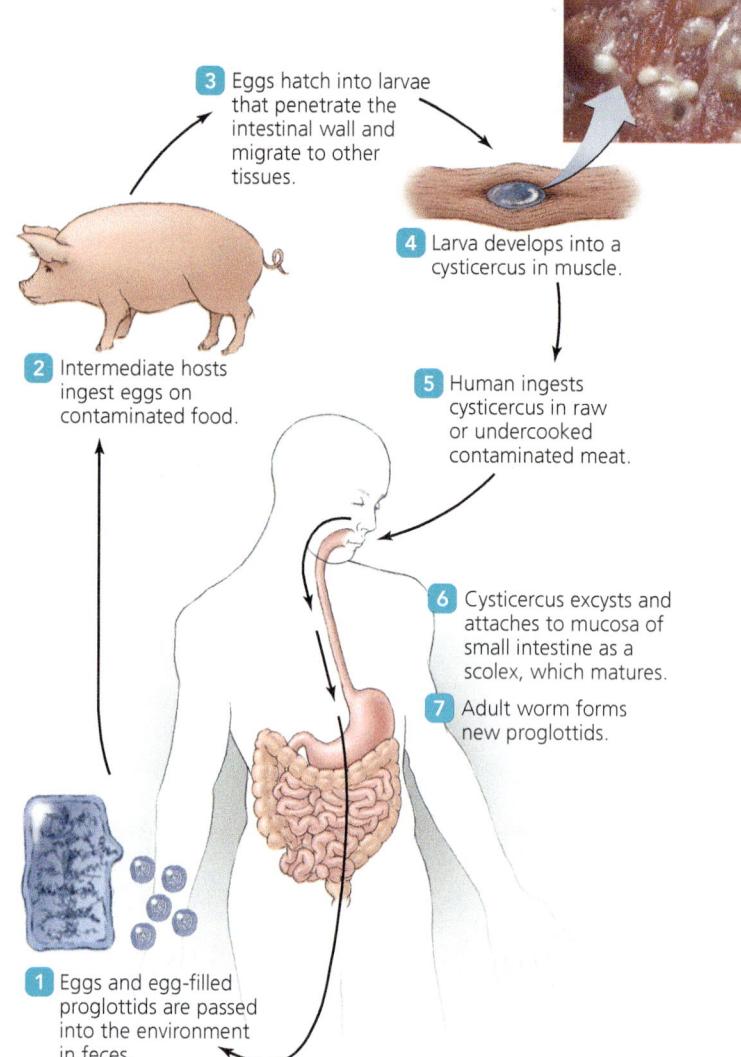

▲ **FIGURE 23.14** A generalized life cycle of some tapeworms of humans. *Which species of* Taenia *is shown in this representation?*

Figure 23.14 *Taenia solium* is illustrated, as indicated by the intermediate host being a pig.

Diagnosis is achieved by microscopic identification of proglottids in fecal samples at least three months after initial infestation. Examination of the scolex is required to differentiate between *T. saginata* and *T. solium*.

Normally, treatment is with praziquantel. Thoroughly cooking or freezing meat is the easiest method of prevention. Because cysticerci are readily visible in meat, giving it a "mealy" look, inspecting meat can reduce the chances of eating infested meat.

Echinococcus

Echinococcus granulosus (ĕ-kī'nō-kok'ŭs gra-nū-lō'sŭs) is an unusual tapeworm of canines in that its body consists of only three proglottids—one immature, one sexually mature, and one gravid. A gravid proglottid is released each time the neck forms a new immature proglottid. Canines are infected by eating cysticerci in various herbivorous hosts, such as cattle, sheep, and

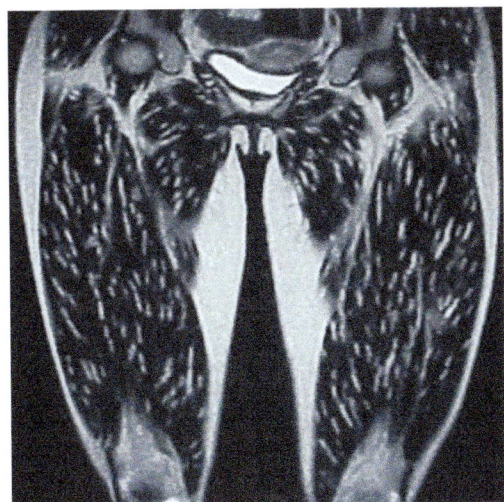

▲ FIGURE 23.15 **Cysticercosis.** Thousands of cysts of *Taenia solium* are visible as white dots in a CT Scan.

deer. Incidence is highest in areas wherever livestock are kept, including the western United States.

Humans become accidental intermediate hosts by consuming food or water contaminated with *Echinococcus* eggs shed in dogs' feces. The eggs release larvae into the intestine; the larvae invade the circulatory system and are carried throughout the body, where they form **hydatid**[5] (hī′da-tid) cysts and cause **hydatid disease.** Hydatid cysts form in any organ but occur primarily in the liver.

Over several years, hydatid cysts can calcify and enlarge to the size of grapefruits (20 cm in diameter), filling with fluid and several million granular *protoscoleces*. Each protoscolex is the potential scolex of a new worm, but since dogs do not normally eat humans, the protoscoleces of hydatid cysts in humans are a dead end for *Echinococcus*.

Symptoms of hydatid disease follow the enlargement of cysts in infested tissue and result from tissue dysfunction. For example, hydatid cysts in the liver produce abdominal pain and obstruction of the bile ducts. If a hydatid cyst ruptures, protoscoleces spread throughout the body via the bloodstream, and each protoscolex grows into a new hydatid cyst. Hydatid cysts in the brain, lungs, bones, heart, or kidneys can be fatal.

Diagnosis involves visualization of cyst masses using ultrasound, CT scan, or radiography **(FIGURE 23.16)**, followed by serological confirmation via immunoassays. Biopsies should be avoided, as rupture of a cyst can lead to the spread of protoscoleces.

Treatment involves surgery to remove cysts, followed by a benzimidazole anthelmintic drug, such as albendazole or mebendazole. Human infestation is prevented by good hygiene practices to avoid fecal-oral transmission of eggs from infested dogs. It is also a good idea to avoid drinking untreated water from streams or other waterways in areas where foxes, coyotes, or wolves roam.

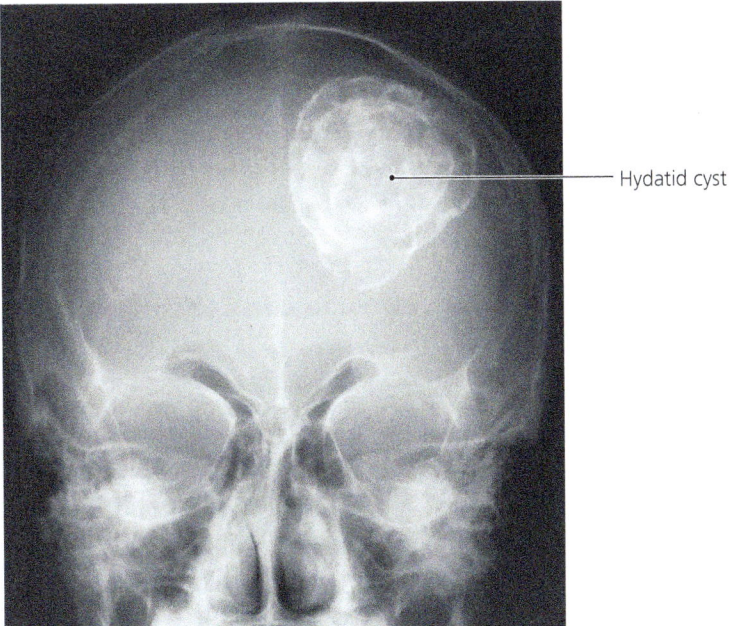

▲ FIGURE 23.16 **X ray of hydatid cyst of *Echinococcus*.** Here, the large, calcified cyst is in the cerebrum of a 20-year-old man.

Trematodes

LEARNING OUTCOMES

23.18 Explain how the life cycles of blood flukes differ from those of other flukes.

23.19 Discuss the life cycles and disease manifestations of *Schistosoma* and *Fasciola*.

Trematodes, commonly known as *flukes* (flūks), are flat, leaf-shaped worms **(FIGURE 23.17)**. A fluke has no anus and so is said to have an *incomplete digestive tract*. A ventral sucker enables

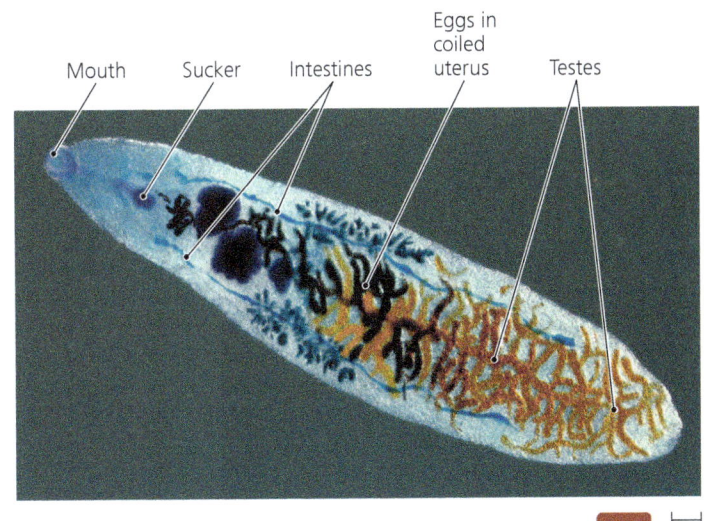

▲ FIGURE 23.17 **Some features of adult fluke morphology.** Note the sucker, which keeps the leaf-shaped flatworm, here a liver fluke *Opisthorchis sinensis*, attached to its host's tissues.

---

[5]From Greek *hydatis*, meaning "a drop of water," presumably in reference to the fluid that accumulates within hydatid cysts.

a parasite to attach to host tissues, from which position it can obtain nutrients. The geographical distribution of flukes is extremely limited by the geographical distribution of the specific species of snails they require as intermediate hosts.

Researchers group flukes together according to the sites in the body they parasitize. The following sections examine some representative flukes, but first we consider common features of fluke life cycles.

Common Features of the Fluke Life Cycle

Flukes that cause disease in humans share similar complex life cycles (FIGURE 23.18): Fluke eggs pass from the body in feces or urine, depending on the site of infestation 1. Eggs deposited in freshwater hatch to release free-swimming larvae called *miracidia* (mir-a-sid′ē-a) 2. Within 24 hours, miracidia actively seek out and burrow into a snail of a specific species 3. Two or three cycles of asexual reproduction within the snail's muscles and other organs produce larvae called *cercariae* (ser-kār′ē-ī) 4, which leave the snail as free-swimming forms. The next steps differ depending on whether the parasite is a blood, liver, lung, or intestinal fluke. In blood flukes, the cercariae aggressively seek out and penetrate the skin of humans 5a; in some intestinal and liver flukes, cercariae encyst to become *metacercariae* on vegetation 5b, while in lung flukes and other intestinal and liver flukes, cercariae encyst to become metacercariae within a second intermediate host 5c. For liver, lung, and intestinal flukes, humans become infected by ingesting metacercariae 6. Once inside a human, the parasites migrate to appropriate sites and develop into adults 7, which then produce perhaps 25,000 eggs per day that pass into the environment in feces or urine, depending on the site of infestation of each species of trematode.

With the exception of one dioecious genus (*Schistosoma*), all flukes are monoecious. They are not self-fertile.

Now, we consider examples of specific species of flukes, beginning with *Fasciola*, a liver fluke.

Representative Liver Fluke: *Fasciola*

Two species of liver fluke—*Fasciola hepatica* (fa-sē′ō-lă he-pa′ti-kă) and *F. gigantica* (ji-gan′ti-kă)—infest sheep and cattle worldwide. *Fasciola* also can infest humans, who become accidental definitive hosts when they ingest metacercariae encysted on aquatic vegetation such as watercress. Following excystment in the intestine, the larvae burrow through the intestinal wall, the peritoneal cavity, or the liver to reach the bile ducts, where they mature within three to four months.

Acute disease characterized by tissue death, abdominal pain, fever, nausea, vomiting, and diarrhea accompanies the migration of the parasite from the intestine to the liver; chronic infestation begins when flukes take up residence in the bile ducts. Symptoms coincide with episodes of bile duct obstruction and inflammation, and heavy infestations cause liver failure. Lesions can occur in other tissues as well.

Diagnosis involves the correlation of symptoms and patient history with microscopic identification of eggs in fecal matter, although *Fasciola* eggs are not always readily distinguishable from those of other intestinal and liver flukes, and symptoms alone mimic other liver disorders. However, liver failure following the consumption of watercress or other raw vegetables grown in endemic areas is suggestive.

Treatment is dependent on accurate diagnosis, as *Fasciola* does not respond to praziquantel as other flukes do. Oral bithionol or triclabendazole are effective. The latter is a veterinary drug not approved for use in humans in the United States.

Blood Flukes: *Schistosoma*

Blood flukes in the genus *Schistosoma* (shis-tō-sō-mă) are dioecious and cause **schistosomiasis** (shis′tō-sō-mī-ă-sis)—a potentially fatal disease and one of the major public health problems in the world. A common name for the disease is "snail fever," which refers to the important role that snails play in the spread of the disease. Over 850 million people worldwide are at risk of infestation with *Schistosoma* and more than 207 million require treatment every year. Of these, about 200,000 die. Most cases of schistosomiasis occur in sub-Saharan Africa, but *Schistosoma* is endemic in countries in Asia and South America as well (see **Emerging Disease Case Study: Snail Fever Reemerges in China** on p. 699). Cases occur in the United States in immigrants from endemic countries and in tourists who contact untreated freshwater in those countries. However, human schistosomiasis does not spread in the United States because of good sewage treatment and because the appropriate snails are not resident.

Three geographically limited species of *Schistosoma* infest humans:

- *S. mansoni* (man-sō′nē) is common to the Caribbean, Venezuela, Brazil, Arabia, and large areas of Africa.
- *S. haematobium* (hē′mă-tō′bē-ŭm) is found only in Africa and India.
- *S. japonicum* (jă-pon′i-kŭm) occurs in China, Taiwan, the Philippines, and Japan, although infestations in Japan are relatively rare.

Humans are the principal definitive host for most species of *Schistosoma*. Cercariae burrow through the skin of humans who contact contaminated water while washing clothes and utensils, bathing, or swimming. The larvae enter the circulatory system, where they mature and mate. Females lay eggs, which have distinctive spines (FIGURE 23.19), in the walls of the blood vessels. Muscle contractions in the blood vessels and organs of the human host move eggs to the lumen of the intestine (*S. mansoni* and *S. japonicum*) or lumens of the urinary bladder and ureters (*S. haematobium*) to be eliminated into the environment.

A mild, temporary dermatitis called *swimmer's itch* may occur in the skin where cercariae burrow.

S. mansoni and *S. japonicum* infestations become chronic when eggs lodge in the liver, lungs, brain, or other organs. Trapped eggs die and calcify in the liver, leading to tissue damage that is generally fatal. In *S. haematobium* infestations, movement of eggs into the bladder and ureters results in blood in the urine, blockage due to long-term fibrosis and calcification, and, in some geographical regions, fatal bladder cancer.

CLINICAL CASE STUDY

A Fluke Disease?

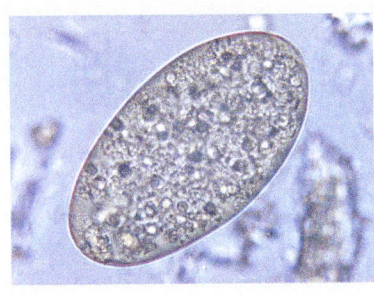

A 40-year-old female of Cambodian descent was admitted to a Michigan hospital with fatigue, constipation, weight loss, abdominal pain, and abdominal swelling. Before the onset of symptoms, the woman was healthy, as were the majority of her close relatives. Laboratory testing indicated signs of acute liver failure and a high eosinophil count, prompting a CT scan. Multiple hepatic lesions and inflammation of the bile ducts were evident. Stool samples examined over several successive days revealed the presence of numerous eggs. The woman had visited Cambodia recently, but her symptoms had begun before she traveled there.

1. What type of patient history would you take? Why is this necessary and important?
2. What would you do to ascertain the parasite causing the woman's illness?
3. What is the treatment for this condition?
4. What type of preventive measures could have kept this woman disease free?
5. Are this woman's relatives at risk of contracting the disease from her?

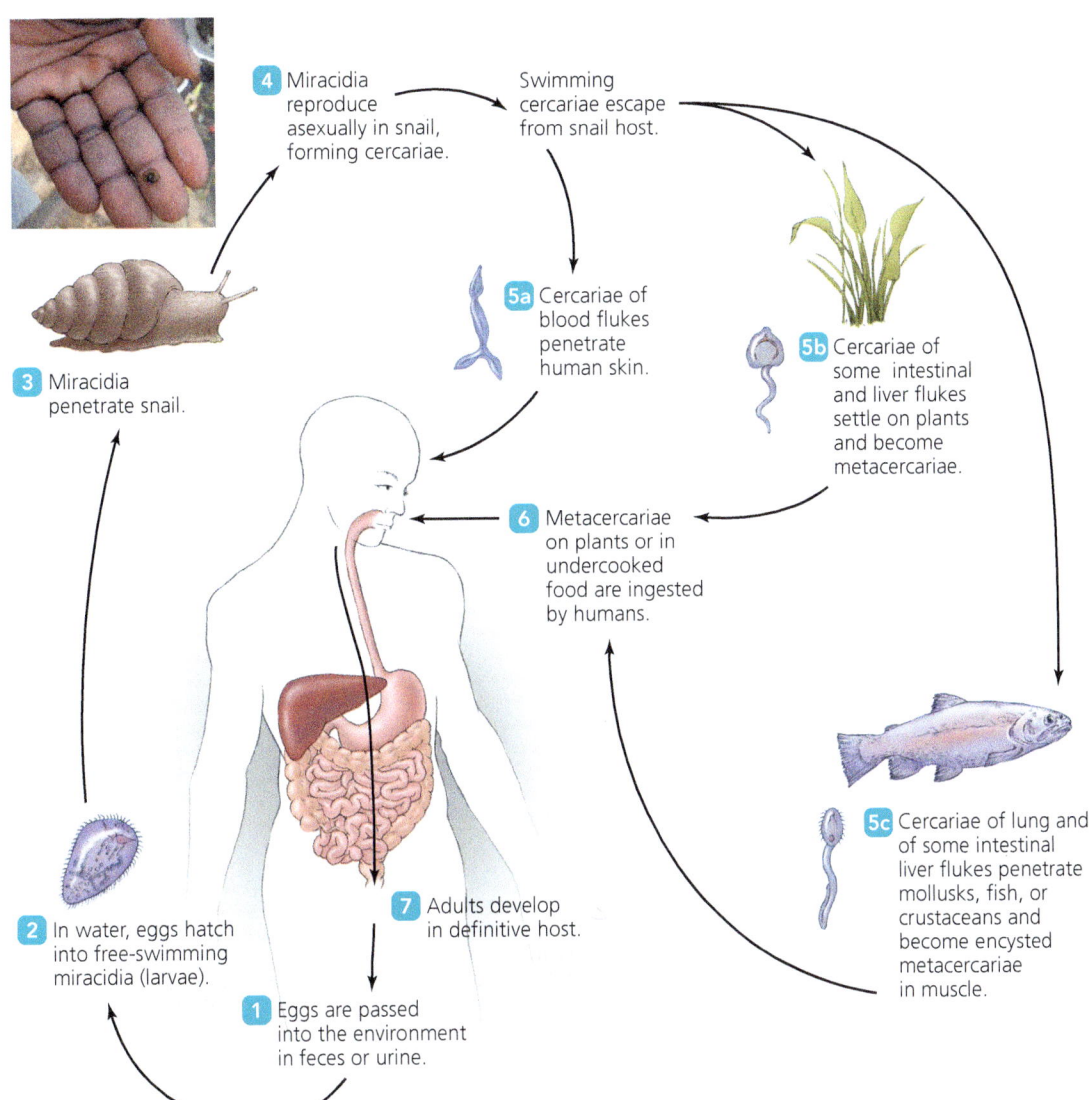

▲ FIGURE 23.18 Representations of three types of human flukes' life cycles. Depending on the species of schistosome, larval cercariae can infect humans directly or become metacercariae, which humans can consume. All trematodes spend part of their lives in freshwater snails. Whereas most flukes infect humans who ingest them, blood flukes penetrate the skin. *Can a human become infested with flukes by ingesting fluke eggs?*

Figure 23.18 No, fluke eggs are not infective to humans. Flukes develop into an infective stage only after development in a specific type of snail.

Diagnosis is most effectively made by microscopic identification of spiny eggs in either stool or urine samples. The species of blood fluke causing a given infestation can be ascertained from the shape of the egg and the location of its spine. If eggs are not seen but infestation is suspected, immunoassays can be used to identify antigen.

The drug of choice for treatment of schistosomiasis is praziquantel, which is taken five weeks after infection because the drug only works on mature worms. Prevention of infection depends on improved sanitation, particularly sewage treatment, and avoiding contact with contaminated water. Researchers are currently conducting clinical trials of a recombinant vaccine for *S. mansoni*.

> **MICRO CHECK**
> 17. How does a cestode attach to host tissues?
> 18. What is cysticercosis?
> 19. What is the role that snails play in the transmission of flukes?
> 20. Which drug is used in the United States to treat liver fluke infestations (*Fasciola* spp.)?
> 21. Which species of *Schistosoma* can cause bladder cancer, and where is it endemic?

Nematodes

LEARNING OUTCOMES

23.20 Describe the common characteristics of nematodes.
23.21 Describe the life cycle of *Ascaris lumbricoides* and state its diagnosis and treatment.
23.22 Compare hookworm infestations, their diagnosis, and their treatment with those for ascariasis.
23.23 Describe the life cycle of pinworms and explain why prevention of reinfection is difficult.
23.24 Describe the life cycle of *Anisakis* and explain how humans can become infested.
23.25 Discuss the life cycle of filarial nematodes and contrast it with the life cycles of intestinal nematodes.

Nematodes, or roundworms, are long, cylindrical worms that taper at each end, possess complete digestive tracts, and have a protective outer layer called a *cuticle*.

Features of the Life Cycles of Roundworms

Nematodes are highly successful parasites of almost all vertebrates, including humans. They have a number of reproductive strategies:

- Most intestinal nematodes shed eggs into the lumen of the intestine, where they are eliminated with the feces. The eggs

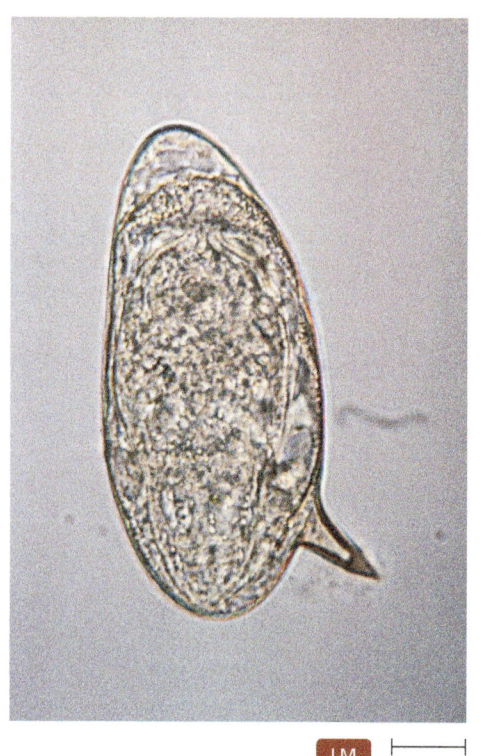
(a) *S. mansoni*

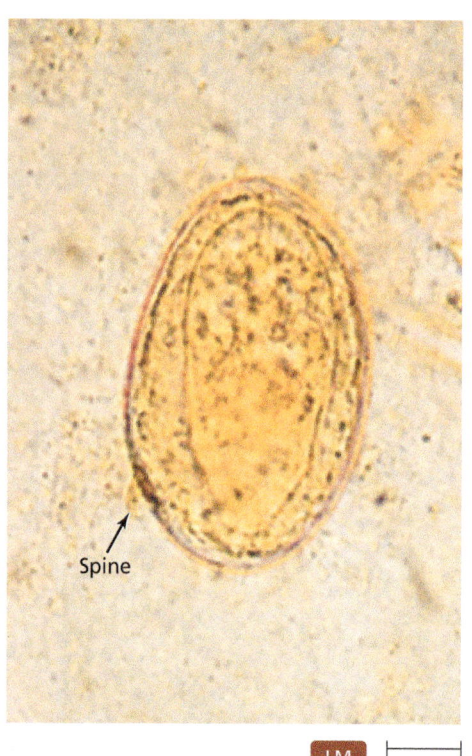
(b) *S. japonicum*

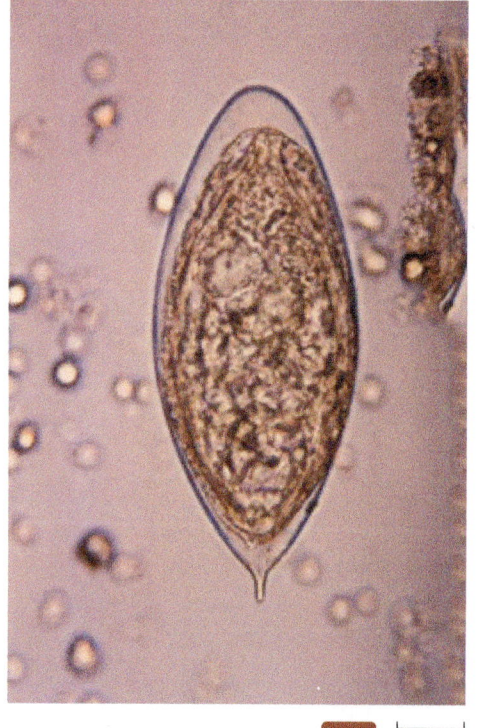
(c) *S. haematobium*

▲ **FIGURE 23.19 Eggs of *Schistosoma* species.**

are then consumed by the definitive host either in contaminated food (particularly raw fruits and vegetables that grow close to the soil) or in contaminated drinking water.

- For a few intestinal nematodes, larvae hatch in the soil and actively penetrate the skin of new hosts. Once in the body, they travel a circuitous route (discussed in a later section) to the intestine.
- Other nematodes encyst in muscle tissue and are consumed in raw or undercooked meat.
- Mosquitoes transmit a few species of nematodes without eggs, called *filarial nematodes*, among hosts.

All nematodes are dioecious and develop through larval stages either within eggs, in intermediate hosts, or in the environment before becoming adults. Adult, sexually mature stages are found only in definitive hosts. With the exception of filarial worms, copulation of male and female worms leads to the production of fertilized eggs and perpetuation of the cycle.

We examine representative intestinal and filarial nematodes in the following sections. We begin with the largest nematode parasite of humans, *Ascaris lumbricoides*.

Ascaris

Ascaris lumbricoides (as′kă-ris lŭm′bri-koy′dēz), the causative agent of **ascariasis** (as-kă-rī′ă-sis), is the most common nematode infestation of humans worldwide, infesting approximately 1 billion people, primarily in tropical and subtropical regions. *Ascaris* is also the largest nematode to infest humans, growing as large as 35 cm in length. The nematode is endemic in rural areas of the southeastern United States.

Adult worms grow and reproduce in the small intestine, where females produce about 200,000 eggs every day. Eggs passed with feces can remain viable for years in moist, warm soil, where an embryo develops within each egg. After eggs are ingested in water or on vegetables, the larvae invade the intestinal wall to enter the circulatory or lymphatic system. From there, the larvae reach the lungs, where they develop through two more larval stages in approximately two weeks. The worms are then coughed up into the pharynx and swallowed. The parasites subsequently develop into adults in the intestine.

Most infestations are asymptomatic, though if the worm burden (number of worms) is high, intestinal symptoms and signs can include abdominal pain, nausea, vomiting, and complete intestinal obstruction, which may be fatal. Transitory fever and pulmonary symptoms, including dry cough, difficulty in breathing, and bloody sputum, may occur during larval migration.

Diagnosis is made by the microscopic identification of eggs in the stool, larvae in sputum, or (rarely) by identification

EMERGING DISEASE CASE STUDY

Snail Fever Reemerges in China

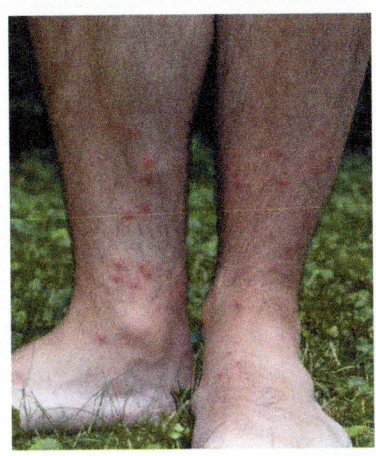

Chen is concerned about his health. First, there was the itchy rash (urticaria) covering his legs several weeks ago, which he attributed to an allergy against something in the water. Now his abdomen is definitely getting bigger. His wife calls him "Big Belly." Though some poor people in China might consider fat a good thing, Chen's increased size is actually accompanied by weight loss. Additionally, he has headaches, muscle pain, fever, vomiting, diarrhea, and—most alarming of all—blood in his urine. He suspects he has "snail fever" and so heads for the hospital.

The physicians inform him that "snail fever" is really a disease called schistosomiasis, which is caused by a parasitic helminth (worm)—the fluke *Schistosoma*. Actually, they explain, snails don't cause the disease, though immature flukes live in snails. Immature parasites had left their snail hosts and burrowed into Chen's legs while he was wading. His body's defensive reaction to this attack had produced the urticaria. Now, *Schistosoma* is attacking his intestinal tract, lungs, liver, and spleen.

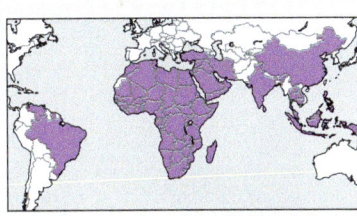

Treatment with prednisone for the itchiness and with anthelmintic praziquantel brings relief to Chen. Not every victim of schistosomiasis is so blessed; some people die a lingering and painful death.

Chen's grandparents would have been more familiar with the disease, which was nearly eradicated in the 1950s when the government put people to work clearing snails from lakes and streams. The disease is reemerging because economic stability allows the introduction of improved irrigation systems and the creation of more water reservoirs that provide a habitat for the snails that play host to *Schistosoma*.

1. Why is schistosomiasis called "snail fever"?
2. Why does good sewage treatment help control schistosomiasis?
3. Why is the initial symptom itching skin?

of adult worms passed in the stool or exiting via the nose or mouth—a distinctly unpleasant experience with larger worms (FIGURE 23.20). The treatment of *Ascaris* with albendazole or mebendazole over one to three days is 90% effective. Surgery may be required to alleviate intestinal obstruction. Proper sanitation and hygiene, including treatment of sewage and drinking water, are important for prevention. Good personal hygiene and cooking practices are also important methods of preventing infection with *Ascaris*.

Ancylostoma and *Necator*

Hookworms, so called because adults resemble shepherd's hooks, are the second most common nematode infesting humans, responsible for more than 600 million infestations worldwide, sucking the equivalent of all the blood contained in 1.5 million people per day. Two hookworms infest humans. *Ancylostoma duodenale* (an-si-los′tō-mă doo′ō-de-nā-lē) is distributed throughout Africa, Asia, the Americas, the Middle East, North Africa, and southern Europe. *Necator americanus* (nē-kā′tor ă-mer-i-ka′nŭs) predominates in the Americas and Australia but can be found in Asia and Africa as well.

Eggs passed in the stools of infested humans are deposited in soil, where they hatch in a couple of days if the soil is warm and moist. The initial larvae develop over a span of 5 to 10 days into infective larvae that can survive up to five weeks in soil. The larvae burrow through human skin and are carried by the circulatory system to the heart and the lungs. In the lungs, larvae burrow into the air cavity and migrate up the trachea to the esophagus, where they are swallowed. Larvae attach to the small intestine, mature into adults, and mate. Most adults survive in the intestine for one to two years or longer.

Adult worms have ghastly looking mouths (FIGURE 23.21) and suck the blood of their hosts, consuming as much as 150 μL of blood per day per worm. This results in chronic anemia, iron deficiency, and protein deficiency.

Itching, rash, and inflammation that can persist for several weeks characterize *ground itch*, which occurs at the site of skin penetration by larvae. Ground itch is generally worse with successive infestations. Pulmonary symptoms can also occur.

Microscopic identification of eggs in the stool along with blood in the feces and anemia is diagnostic.

Treatment is with albendazole, mebendazole, or pyrantel pamoate for several days. *N. americanus* is more difficult to treat effectively. Proper sanitation (sewage treatment) is essential for prevention. In areas where hookworm is endemic, going barefoot should be avoided to prevent exposure.

Enterobius

Enterobius vermicularis (en-ter-ō′bī-ŭs ver-mi-kū-lar′is) infests about 500 million people worldwide, particularly in temperate climates, in school-age children, and in conditions of overcrowding. *Enterobius* is the most common parasitic worm found in the United States, affecting 40 million Americans. It is commonly known as the pinworm, after the shape of the female worm's tail. Humans are the only host for *Enterobius*.

After mating in the colon, female pinworms migrate at night to the anus, where they deposit eggs perianally. Scratching dislodges eggs onto clothes or bedding, where they dry out, become aerosolized, and settle in water or on food that is then ingested. Alternatively, scratching deposits eggs on the skin and under the fingernails such that infested individuals can continually reinfect themselves by ingesting the eggs on their hands or in food.

▲ FIGURE 23.20 *Ascaris lumbricoides.* This mass of worms passed from a child in Kenya, Africa. Infestation also occurs in the southeastern United States.

▲ FIGURE 23.21 Mouth of a hookworm. The teeth allow the parasite to attach to the intestinal wall and feed on blood.

Adult worms mature in several weeks and live in the intestinal tract for approximately two months.

One-third of all *Enterobius* infestations are asymptomatic. When symptoms do occur, intense perianal itching is the chief complaint. Scratching can lead to secondary bacterial infections. Rarely, worms enter a female's genital tract, where they can cause inflammation of the vulva.

Microscopy is used for diagnosis. In the morning, before bathing or defecation, transparent sticky tape is applied to the perianal area to collect the readily identifiable microscopic eggs (FIGURE 23.22). Adult worms, if recovered, are also diagnostic.

Treatment is with an initial dose of pyrantel pamoate or mebendazole followed by a second treatment two weeks later to kill any newly acquired worms. Prevention of reinfection and spread to family members requires thorough laundering of all clothes and bedding of infested individuals. Further, infested individuals should not handle food to be consumed by others.

Anisakis

Several parasitic nematodes, most notably *Anisakis simplex* (an-a-se′kis sim′pleks), cause **anisakiasis** (an-a-se-kī′a-sis). *Anisakis* is a parasite of marine animals, but humans can become accidental hosts by eating infested raw or undercooked fish.

Anisakis has a complex life cycle with several larval stages (FIGURE 23.23). Infested marine mammals excrete the nematode's eggs into the ocean ❶. Two larval stages develop one after the other inside the egg ❷, which subsequently hatches in seawater. The emergent second-stage larvae swim in the ocean ❸, and small shrimplike crustaceans called krill eat them ❹. Inside krill, the larvae develop into third-stage larvae, which are infective to fish that eat krill ❺. Third-stage larvae inside a fish's stomach invade other tissues in the fish. Other fish eat the infested fish, and third-stage larvae infect each predator fish in turn ❻. In this manner, *Anisakis* third-stage larvae are maintained in the fish population. Eventually, a marine mammal, such as a porpoise

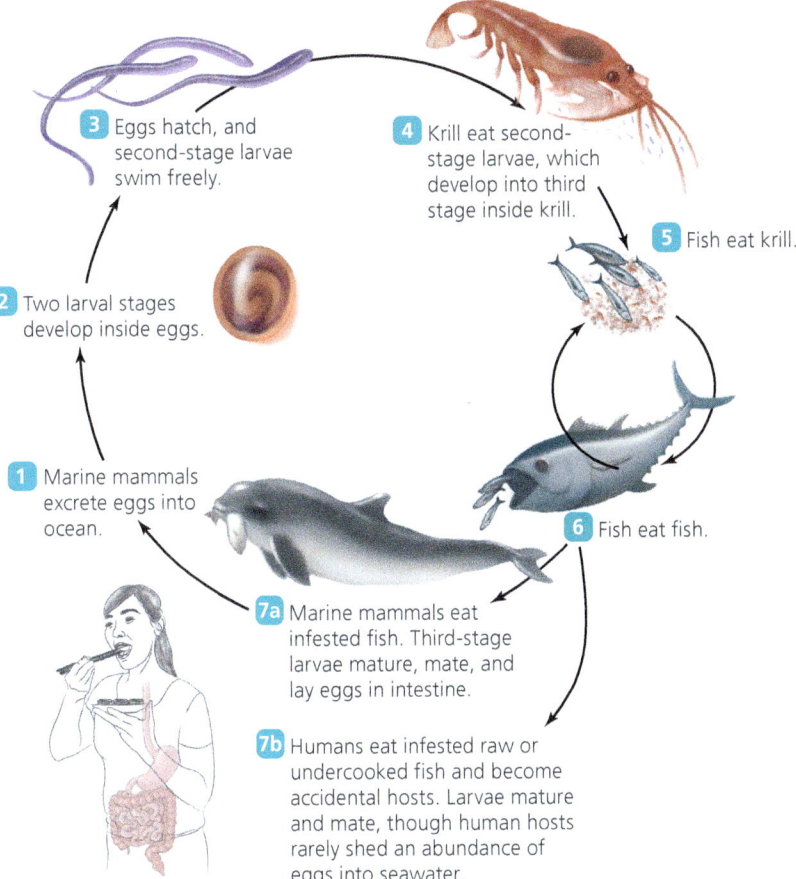

▲ FIGURE 23.23 Life cycle of *Anisakis*

or whale, eats the infested fish. In mammals, third-stage larvae mature into adult worms and mate. The female worm lays eggs that are excreted in fish feces ❼a.

Humans who eat infested fish become accidental hosts to *Anisakis*, which matures in a human and attaches to the esophagus, stomach, or intestine, leading to anisakiasis ❼b and possible hypersensitivity reactions if there are future infestations. Humans are dead-end hosts because they don't shed an abundance of eggs into seawater. Infested patients are typically asymptomatic, but some people feel the larval stage of the worm moving around in their mouths or throats. Patients can also experience sudden and violent abdominal pain, nausea, vomiting, fever, and sometimes intestinal hemorrhaging. Following recovery from an initial infestation, a few patients become sensitized and have acute, antibody-mediated allergic reactions to subsequent infestations or even exposure to dead worms in food. Allergic patients may develop a widespread rash, and some cough up adult worms—a distressing sign indeed. A very few allergic patients develop life-threatening anaphylactic shock.

Physicians report about 20,000 cases of anisakiasis worldwide annually among people who eat raw or undercooked fish. Diagnosis is generally made by endoscopy in which a fiber-optic camera is inserted rectally into the voided intestinal tract. The worms can be seen crawling in the lumen or burrowing into the intestinal wall. Treatment involves removing the worms from the intestine with small forceps attached to a cable. Prevention of anisakiasis involves avoiding raw and undercooked marine fish.

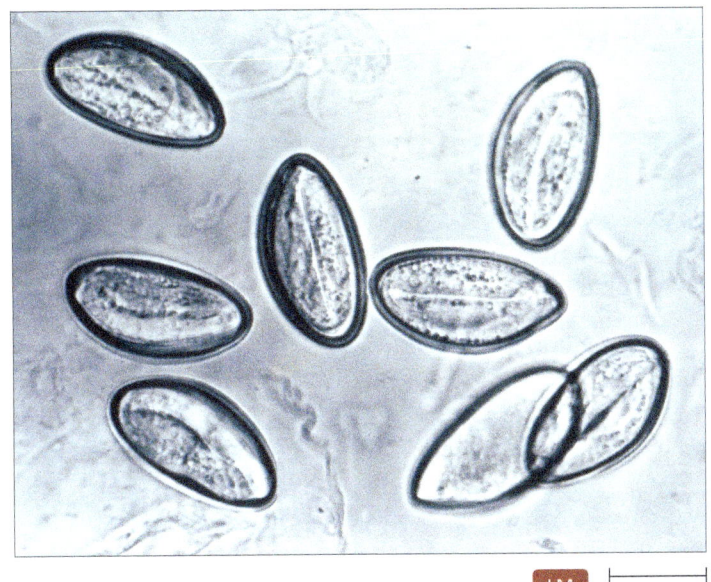

▲ FIGURE 23.22 Eggs of *Enterobius vermicularis*. Note the characteristic flattening of one side of each egg.

Wuchereria

Wuchereria bancrofti (voo-ker-e′rē-ă ban-krof′tē) is a *filarial nematode*—a type of nematode that infests not the intestinal tract of vertebrate hosts but rather the lymphatic system, causing **filariasis** (fil-ă-rī′ă-sis). *W. bancrofti* infests the lymphatic system of about 120 million people throughout the tropics of Africa, India, Southeast Asia, Indonesia, the Pacific Islands, South America, and the Caribbean. Female mosquitoes of the genera *Culex* (kyu′leks), *Aedes* (ā-ē′dēz), and *Anopheles* transmit the worm.

Mosquitoes ingest circulating larvae called *microfilariae* (mī′krō-fi-lar′ē-ī) while feeding on humans (FIGURE 23.24). The microfilariae develop as larvae in the mosquito and eventually migrate to the salivary glands. When the mosquito next feeds, parasites are injected into a new person.

Larvae migrate via the circulatory system to deeper tissues, where they mature. Adults live and reproduce for up to 17 years in lymphatic vessels. During the day, microfilariae stay in capillaries of internal organs; they swim freely in the bloodstream only at night, coinciding with the feeding habits of most mosquitoes.

Filariasis remains asymptomatic for years. As the disease progresses, lymphatic damage occurs—subcutaneous tissues swell grotesquely because blocked lymphatic vessels cannot drain properly. The end result is **elephantiasis** (el-ě-fan-tī′ă-sis), generally in the lower extremities, in which tissues enlarge and harden in areas where lymph has accumulated (FIGURE 23.25). Elephantiasis can be further associated with secondary bacterial infections in affected portions of the body.

Diagnosis is made by the microscopic identification of microfilariae in the blood. Blood samples are collected at night, when microfilariae circulate. In addition to microscopy, immunoassays using antibodies to detect *W. bancrofti* antigens in the blood can be used for diagnosis.

Treatment with diethylcarbamazine effectively kills microfilariae and some adult worms.

Prevention relies on using insect repellents or mosquito netting; wearing loose, long, light-colored clothing; or remaining indoors when mosquitoes are most active. In endemic areas, particularly urban areas, eliminating standing water (a requirement of mosquitoes' life cycle) around human dwellings decreases mosquito numbers. Widespread spraying with insecticides to kill mosquitoes has reduced the prevalence of filariasis.

TABLE 23.2 (p. 703) summarizes the key features of some helminthic parasites of humans.

Protozoan and helminthic diseases are found worldwide, and much of the world's population is at risk of infection. Many of the drugs used to treat parasitic diseases have been around for decades, and, whereas most remain effective, resistant parasites are occurring with greater frequency. No effective vaccines have been produced for any of the diseases discussed in this chapter. The only way to protect oneself, therefore, is to avoid exposure.

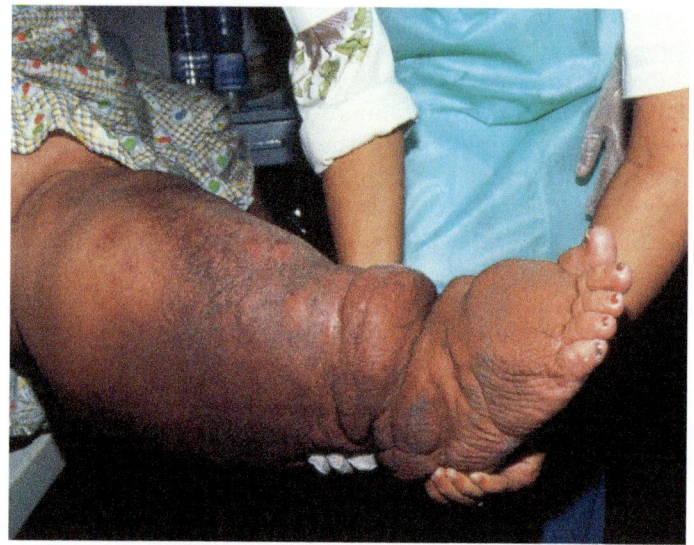

▲ **FIGURE 23.25 Elephantiasis in a leg.** This condition follows years of infestation with *Wuchereria bancrofti*. Adult *Wuchereria* are found in what human body system?

Figure 23.25 Adults of *Wuchereria* live in the lymphatic system.

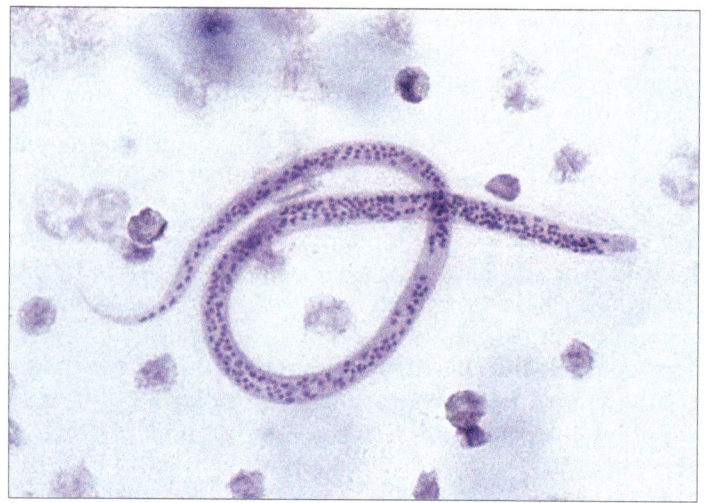

▲ **FIGURE 23.24 A microfilaria of *Wuchereria bancrofti* in blood.**

MICRO CHECK

22. What is the most common way to diagnose ascariasis?
23. Which is the most common parasitic worm in the United States?
24. What is the most common symptom of pinworm infestation?
25. How is filariasis transmitted?

TELL ME WHY

Thoroughly cooking raw vegetables prior to consumption would kill the metacercariae of *Fasciola* and prevent infection. Why is this method of prevention not practical?

TABLE 23.2 Key Features of Representative Helminthic Parasites of Humans

Organism	Primary Infection or Disease	Geographical Distribution	Mode of Transmission	Length of Adult Worms
Cestodes (Tapeworms)				
Taenia saginata; Taenia solium	Beef tapeworm infestation; pork tapeworm infestation	Worldwide with local endemic areas	Consumption of undercooked meat	*T. saginata*: 5–25 m; *T. solium*: 2–7 m
Echinococcus granulosus	Hydatid disease	Worldwide with local endemic areas	Consumption of eggs shed in dog feces	3–6 mm
Trematodes (Flukes)				
Fasciola hepatica; F. gigantica	Fascioliasis	*F. hepatica:* Europe, Middle East, Asia; *F. gigantica:* Asia, Africa, Hawaii	Consumption of watercress or lettuce	Up to 30 mm; up to 75 mm
Schistosoma spp.	Schistosomiasis	*S. mansoni:* Caribbean, South America, Arabia, Africa; *S. haematobium:* Africa, India; *S. japonicum:* eastern Asia	Direct penetration of the skin	7–20 mm
Nematodes (Roundworms)				
Ascaris lumbricoides	Ascariasis	Tropics and subtropics worldwide	Fecal-oral	Females: 20–35 cm; males: 15–30 cm
Ancylostoma duodenale, Necator americanus	Hookworm disease	*Ancylostoma:* Africa, Asia, the Americas, Middle East, North Africa, southern Europe; *Necator:* the Americas, Australia, Asia, Africa	Direct penetration of the skin	*Ancylostoma* females: 10–13 mm, males: 8–11 mm; *Necator* females: 9–11 mm, males: 7–9 mm
Enterobius vermicularis	Pinworm	Worldwide	Anal-oral and fecal-oral, inhalation	Females: 8–13 mm; males: 2–5 mm
Anisakis	Anisakiasis	Wherever people consume raw marine fish, e.g., Japan	Consumption of infested raw or undercooked marine fish	*Anisakis simplex:* 15 mm
Wuchereria bancrofti	Filariasis, elephantiasis	Worldwide, tropics	Mosquitoes	Females: 80–100 mm; males: 40 mm

MICRO IN THE CLINIC FOLLOW-UP

At his initial examination, Dr. Stephens doesn't note anything remarkable about Sam's condition. His lymph nodes are swollen, but his vital signs are stable and within normal limits. Dr. Stephens draws a blood sample for further analysis. The results should be back the next day. Dr. Stephens tells Sam to get as much rest as possible and to make a follow-up appointment in two days.

At the follow-up appointment, Dr. Stephens reports that the blood tests are normal. Sam is surprised—even though he doesn't feel extremely sick, he knows that there is something wrong with him. Dr. Stephens agrees—he has been Sam's primary care physician for years and knows that if Sam says something is wrong with him, something probably is. After talking more with Sam about his recent activities, including his trip to Mexico, Dr. Stephens orders a microscopic examination of Sam's blood serum.

Microscopy reveals the presence of motile trypanosomes, *Trypanosoma cruzi*, in Sam's blood.

Trypanosomiasis is rare in the United States, and the drugs to treat the acute stage of the disease are available only from the Centers for Disease Control and Prevention (CDC). The treatment protocol uses two antiparasitic drugs: benznidazole, which must be taken for 60 days, combined with nifurtimox, which must be taken for 90 days. The downside of treatment is that side effects are common. Over the course of treatment, Sam experiences weight loss, headaches, neuropathy (damage or dysfunction of nervous tissue), vertigo (dizziness), and nausea; however, at the end, there are no detectable parasites in Sam's blood. The side effects disappear after treatment ends. Sam is asked to follow up in three to six months to ensure that *Trypanosoma* hasn't damaged his heart.

1. Describe a virulence mechanism that *T. cruzi* uses to cause disease.
2. Although Sam had to suffer through the unpleasant side effects of treatment, he could have developed cardiac disease if he had not been diagnosed so quickly. How does infection with *T. cruzi* lead to heart disease?

Check your answers to Micro in the Clinic Follow-Up questions in the Mastering Microbiology Study Area.

CHAPTER SUMMARY

 Go to **Mastering Microbiology** for Interactive Microbiology, Dr. Bauman's Video Tutors, Micro Matters, MicroFlix, MicroBoosters, practice quizzes, and more.

Parasitology (p. 676)

1. Parasitology is the study of **parasites,** which live in or on other organisms, deriving benefit at their hosts' expense. All parasites have a **definitive host** in which adult or sexual stages reproduce, and many parasites also have one or more **intermediate hosts,** in which immature stages live.

Arthropod Vectors (pp. 676–677)

1. **Arthropods** are organisms with segmented bodies, jointed legs, and a hard exoskeleton. They are found worldwide in nearly all habitats. Some may serve as **biological vectors,** in which a parasite lives and grows; some others are **mechanical vectors,** which carry a parasite but don't serve as the parasite's host.
2. Arthropod vectors are arachnids (adults having eight legs) or insects (adults having six legs).
3. Arachnid vectors include mites and ticks.
4. Insect vectors include fleas, lice, flies, and true bugs. Mosquitoes, which are a type of fly, are the most important arthropod vectors of human diseases.

Protozoan Parasites of Humans (pp. 677–693)

1. Many protozoa have two life stages: a trophozoite (feeding) stage and a dormant cyst stage. **Encystment** is the process of cyst formation; **excystment** refers to the process by which cysts become trophozoites.
2. *Balantidium coli* is the only **ciliate** to cause disease in humans. It is responsible for **balantidiasis,** usually a mild gastrointestinal illness.
3. **Amoebas** are protozoa that move and feed using pseudopods. *Entamoeba histolytica* is a relatively common protozoan parasite that causes **amebiasis,** which can be mild luminal amebiasis, more severe amebic dysentery, or very severe invasive extraintestinal amebiasis, in which the parasite causes lesions in the liver, lungs, brain, and other organs.
4. *Acanthamoeba* and *Naegleria* are free-living amoebas that rarely infect humans. Both are acquired from water. *Acanthamoeba* causes keratitis and **amebic encephalitis,** which progresses slowly and is usually fatal. *Naegleria* causes a more rapid **amebic meningoencephalitis,** which is also fatal.
5. Parasitic **flagellates** include *Trypanosoma, Leishmania, Giardia,* and *Trichomonas.*
6. *Trypanosoma cruzi,* transmitted by the kissing bug (*Triatoma*), causes **Chagas' disease** in the Americas, where it is a frequent cause of heart disease. *T. brucei,* transmitted by the tsetse fly (*Glossina*), causes **African trypanosomiasis,** also called **African sleeping sickness.**
7. Both trypanosomes have similar life cycles. Stages of the parasites in an insect vector's gut develop into infective flagellated trypanosomes introduced to the human host by the vector. *T. cruzi* develops intracellularly in macrophages and heart muscle as nonflagellated cells. *T. brucei* retains its flagella and remains in the blood, is antigenically variable, can spread to the CNS, and can be fatal.
8. Many species of *Leishmania* infect humans. Each parasite has two developmental stages: an intracellular amastigote in mammals and a promastigote form in sand flies, which transmit the disease. Three clinical forms of **leishmaniasis** are seen, each with increasing severity: cutaneous, mucocutaneous, and visceral. Leishmaniasis is a **zoonosis**—a disease of animals transmitted to humans.
9. *Giardia intestinalis* is a frequent cause of gastrointestinal disease (**giardiasis**) and is endemic worldwide. Giardiasis is rarely fatal.
10. *Trichomonas vaginalis* is the most common protozoan disease of people in industrialized nations. This sexually transmitted protozoan causes **vaginosis.** Prevention relies on sexual abstinence, monogamy, and condom usage.
11. *Plasmodium* species, *Toxoplasma, Cryptosporidium,* and *Cyclospora* are all **apicomplexans,** in which sexual reproduction produces oocysts that undergo schizogony to produce various forms.
12. Five species of *Plasmodium* cause **malaria** and are transmitted by female *Anopheles* mosquitoes. The complex life cycle of *Plasmodium* has three stages: Sporozoites travel to the liver to initiate the **liver phase.** Cells of *P. vivax* and *P. ovale* can remain dormant inside liver cells as hypnozoites.
13. Rupture of liver cells releases merozoites from intracellular schizonts. Merozoites infect erythrocytes to initiate the **erythrocytic cycle.**
14. The **sporogonic phase** in the mosquito has an oocyst that divides to form sporozoites, which are injected into humans when the mosquito feeds. Several host traits influence whether an individual is susceptible to malaria.
15. *Toxoplasma gondii* causes **toxoplasmosis,** a disease of cats that can be acquired by other animals and humans, primarily via the consumption of meat containing the parasites. Toxoplasmosis is generally mild in humans, but *Toxoplasma* can cross the placenta to cause serious harm to fetuses.
16. Both *Cryptosporidium parvum* and *Cyclospora cayetanensis* cause waterborne gastrointestinal disease in the United States. *Cryptosporidium,* which causes **Cryptosporidium enteritis** (cryptosporidiosis), is resistant to standard water treatment methods. Cyclosporiasis is an emerging disease in the United States.

Helminthic Parasites of Humans (pp. 693–703)

1. **Helminths** are multicellular eukaryotic worms, some of which are parasitic.
2. **Cestodes** (tapeworms) all share similar morphologies and general life cycles. A **scolex** attaches the tapeworm to the intestine of its definitive host. From this extends a neck region from which **proglottids** grow. When fertilized eggs (sometimes within gravid proglottids) are eaten by an intermediate host, they hatch to release larval stages that invade tissues of the intermediate host and form **cysticerci.** When the cysticerci are eaten by a definitive host, they undergo development into adult tapeworms.
3. Humans are the definitive hosts of *Taenia saginata,* the beef tapeworm, and *Taenia solium,* the pork tapeworm. Humans become infected by eating cysticerci in undercooked beef or pork, respectively.

4. *Echinococcus granulosus* forms **hydatid** cysts to cause **hydatid disease** in humans who are accidental intermediate hosts when they ingest the helminth's eggs. The definitive host is a canine.

5. **Trematodes,** or flukes, are often divided into four groups: blood, intestinal, liver, and lung flukes. Eggs deposited in water hatch to release miracidia, which burrow into freshwater snails. Following asexual reproduction, cercariae are released into the water. In intestinal, liver, and lung flukes, cercariae encyst on plants or in a second intermediate host as metacercariae, which are ingested by human hosts. Cercariae of blood flukes directly penetrate the skin of humans. After maturation occurs in the human, flukes release eggs back into the environment via the feces or urine.

6. *Fasciola* is a liver fluke. Humans acquire *Fasciola* by ingesting metacercariae on watercress or other vegetables.

7. Species of *Schistosoma* (blood flukes) affect millions of people worldwide, causing **schistosomiasis.** Three species infest humans: *S. mansoni, S. japonicum,* and *S. haematobium,* each of which is geographically limited. Extensive infestations significantly damage the liver.

8. **Nematodes** are round, unsegmented worms with pointed ends, complete digestive systems, and a protective outer cuticle.

9. *Ascaris lumbricoides,* the largest nematode to infect humans, causes **ascariasis,** which is the most common nematode disease of humans in the world. Eggs containing infective larvae are consumed in water or with contaminated vegetables. The eggs hatch in the intestine, and larvae enter the circulatory or lymphatic system. Worm migration through the body may be associated with various symptoms, but infections are usually not fatal.

10. *Ancylostoma duodenale* and *Necator americanus* are both hookworms, the second most common nematodes to infect humans worldwide. Eggs deposited in soil mature to form infective larvae that leave the egg and actively burrow through human skin to initiate infection. The circulatory system carries the worms through the lungs and heart. Eventually worms migrate up the respiratory tract to the pharynx, where they are swallowed. They become adults in the intestine, where they attach to the intestinal mucosa and feed on blood. Hookworm infection can lead to anemia and other blood-related dysfunction.

11. The pinworm *Enterobius vermicularis* is the most common parasitic worm infestation of humans in the United States and the third most common in the world. Eggs deposited in the perianal area cause itching; scratching moves eggs where they can be introduced into the mouth, reinfecting the human host. Transmission among family members is also common.

12. People become accidental hosts of *Anisakis simplex,* a nematode that causes **anisakiasis,** if they eat infested fish. This gastrointestinal disease manifests with abdominal pain, vomiting, nausea, fever, and sometimes bleeding.

13. *Wuchereria bancrofti* is a filarial (threadlike) nematode of humans, causing **filariasis.** It lives in the lymphatic system, where adult worms block lymph flow, causing **elephantiasis,** in which lymph accumulates, especially in the lower extremities. Bacterial infections may ensue, and subcutaneous tissues harden, producing permanent disfigurement. The parasite is transmitted when microfilariae, which circulate in the blood, are picked up by feeding mosquitoes.

QUESTIONS FOR REVIEW

Answers to the Questions for Review (except for Short Answer questions) begin on p. A-1.

Multiple Choice

1. Parasitology is the study of parasitic _____.
   a. a.viruses
   b. prokaryotes
   c. fungi
   d. eukaryotes

2. The only ciliate to cause disease in humans is _____.
   a. *Naegleria*
   b. *Balantidium*
   c. *Fasciola*
   d. *Trypanosoma*

3. Which of the following organisms is regularly transmitted sexually? _____
   a. *Trichomonas*
   b. *Entamoeba*
   c. *Trypanosoma*
   d. *Enterobius*

4. *Leishmania* species are transmitted by _____.
   a. sand flies
   b. tsetse flies
   c. kissing bugs
   d. mosquitoes

5. Which of the following is the name of the intracellular infection stage of *Leishmania*? _____
   a. miracidia
   b. metacercaria
   c. bradyzoite
   d. amastigote

6. In malaria, which stage occurs in red blood cells? _____
   a. liver phase
   b. erythrocytic cycle
   c. sporogonic phase
   d. amastigote cycle

7. The definitive host for *Toxoplasma gondii* is _____.
   a. humans
   b. cats
   c. birds
   d. mosquitoes

8. Tapeworms are generally transmitted via .
   a. consumption of an intermediate host
   b. consumption of the definitive host
   c. vectors such as mosquitoes
   d. consumption of adult tapeworms

9. *Cryptosporidium* cannot be killed by routine boiling. Another parasite resistant to such boiling is _____.
   a. *Giardia*
   b. *Trypanosoma*
   c. *Toxoplasma*
   d. *Plasmodium*

10. The immature fluke stages that infect snail intermediate hosts are called _____.
    a. metacercariae
    b. cercariae
    c. cysticerci
    d. miracidia

11. The beef tapeworm is known by what scientific name? _____
    a. *Taenia solium*
    b. *Taenia saginata*
    c. *Ancylostoma duodenale*
    d. *Echinococcus granulosus*

12. *Enterobius vermicularis* is commonly called _____.
    a. hookworm
    b. pinworm
    c. whipworm
    d. tapeworm

13. The infective larvae of *Necator americanus* must pass through which human organ to mature? _____
    a. bladder
    b. brain
    c. lung
    d. liver

14. *Plasmodium* species as well as *Wuchereria bancrofti* can be carried by mosquitoes in the genus _____.
    a. Aedes
    b. Anopheles
    c. Culex
    d. Ctenocephalides

15. Which of the following arthropods is responsible for transmitting the most parasitic diseases? _____
    a. fleas
    b. ticks
    c. mosquitoes
    d. true bugs

16. The majority of cestodes are transmitted via _____.
    a. ingestion
    b. vectors
    c. direct contact
    d. inhalation

17. Which of the following is most effective in preventing infection by *Giardia*? _____
    a. sexual abstinence
    b. drinking only bottled water
    c. use of insect repellent
    d. cooking all food

18. The sporogonic phase of *Plasmodium* occurs in _____.
    a. red blood cells
    b. liver cells
    c. schizonts
    d. *Anopheles* mosquitoes

19. The tapeworm attachment organ is a _____.
    a. scolex
    b. proglottid
    c. strobila
    d. cuticle

20. Trophozoite-cyst conversion is vital to the life of _____.
    a. *Balantidium*
    b. *Entamoeba*
    c. *Giardia*
    d. all of the above

Modified True/False

Mark each statement as true or false. Rewrite the false statements to make them true by changing the underlined word(s).

1. _____ <u>Sexual contact</u> is the most common method of transmission of parasites.
2. _____ Examination of <u>stool samples</u> can reveal the presence of *Naegleria* parasites.
3. _____ <u>Trichomonas vaginalis</u> is the most common parasitic protozoan of humans in the industrialized world.
4. _____ *Trypanosoma brucei* is transmitted by <u>tsetse flies</u>, and *Trypanosoma cruzi* is transmitted by <u>kissing bugs</u>.
5. _____ *Plasmodium falciparum* causes the <u>most serious</u> form of malaria.
6. _____ Toxoplasmosis can be transmitted across a <u>placenta</u>.
7. _____ Humans can become intermediate hosts for *Taenia <u>saginata</u>*.
8. _____ *Fasciola hepatica* can be acquired by eating infected <u>sheep</u>.
9. _____ The number of cases of schistosomiasis has <u>increased</u> worldwide because of improved technology and economic stability in endemic areas.
10. _____ *Wuchereria bancrofti* is a filarial nematode that infects the <u>lymphatic system</u>.

Fill in the Blanks

1. *Balantidium coli* can be distinguished from *Entamoeba histolytica* microscopically because *B. coli* has _____.
2. _____ may be transmitted to humans from cat litter boxes.
3. African sleeping sickness is caused by *Trypanosoma* _____ but not by *Trypanosoma* _____.
4. The parasitic amoeba _____ can be acquired by ingestion.
5. Both *Plasmodium vivax* and *P. ovale* can form dormant _____.
6. Of the parasitic helminths discussed in this chapter, the only one transmitted by mosquitoes is _____.
7. The following helminths can directly penetrate the skin of humans to establish infection: , _____, _____, and _____ (give genera only).
8. A trematode that can be acquired by eating raw or undercooked vegetables is _____.
9. Hookworm disease is caused by _____ in the Middle East.
10. Both of the following parasites demonstrate nocturnal movement, which is important during diagnosis: _____ and _____.

Matching

Match the numbered organism with the correct lettered term. Answers may be used more than once, and an organism may have more than one answer.

1. ____ *Balantidium coli*
2. ____ *Echinococcus granulosus*
3. ____ *Fasciola* spp.
4. ____ *Leishmania* spp.
5. ____ *Plasmodium falciparum*
6. ____ *Plasmodium vivax*
7. ____ *Taenia* spp.
8. ____ *Toxoplasma gondii*
9. ____ *Trypanosoma*
10. ____ *Wuchereria bancrofti*

A. Miracidia
B. Bradyzoites
C. Schizogony
D. Microfilaria
E. Hydatid cyst
F. Cysticerci
G. Trophozoites
H. Amastigotes
I. Hypnozoites
J. Flagellated trypanosomes

Questions for Review

VISUALIZE IT!

1. Label the three stages of the *Plasmodium* life cycle, and label the forms of the parasite where indicated.

2. Identify the genera of the parasites in these clinical specimens.

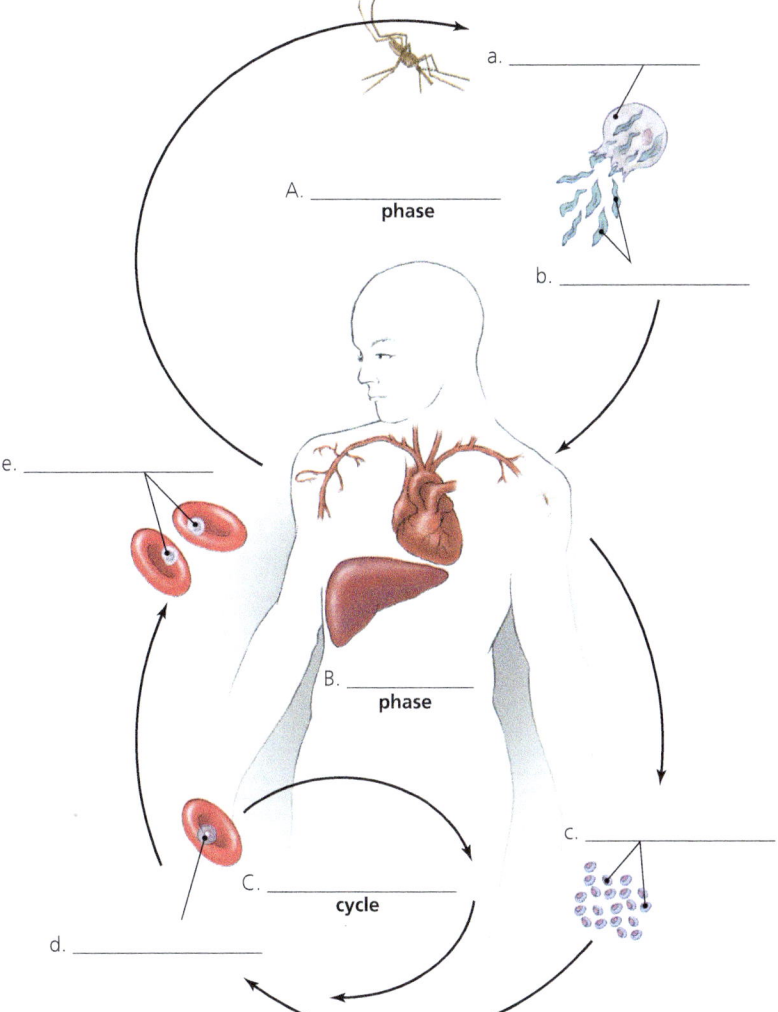

A. _____ phase

B. _____ phase

C. _____ cycle

a. _____
b. _____
c. _____
d. _____
e. _____

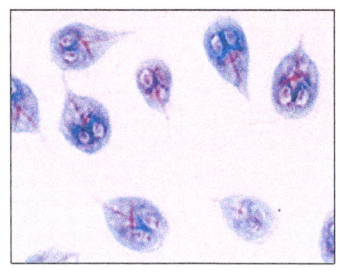

fecal
a. _____

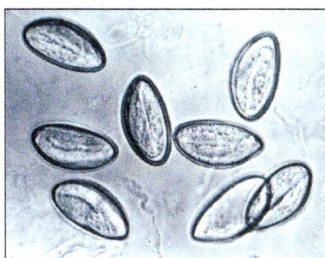

anal
b. _____

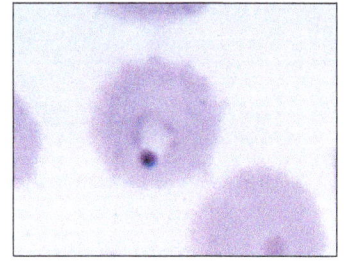

blood
c. _____

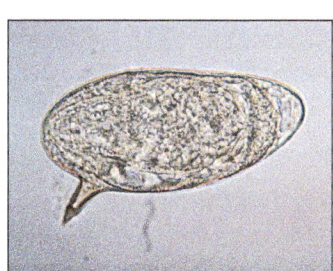

fecal
d. _____

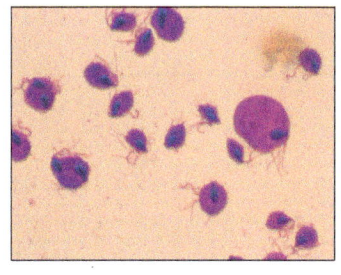

vaginal
e. _____

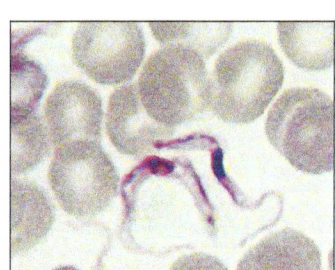

blood
f. _____

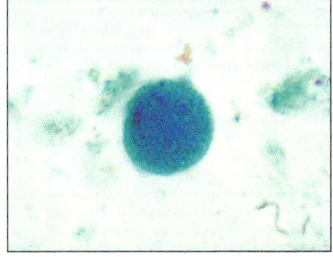

fecal
g. _____

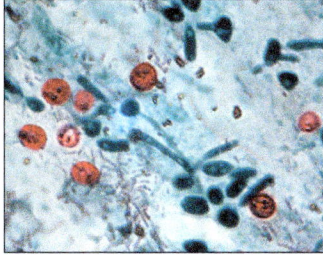

fecal
h. _____

Short Answer

1. Why do insect vectors and animal reservoirs increase the difficulty of preventing and controlling parasitic infections in humans?

2. How is the transmission of the amoeba *Entamoeba* different from the transmission of the amoebas *Acanthamoeba* and *Naegleria*?

CRITICAL THINKING

1. Compare *Entamoeba*, *Acanthamoeba*, and *Naegleria*. Based on incidence, which of these parasites is a threat to more people? Clinically, which is the most serious? Explain why your answer is not the same for both questions.

2. Explain how each of the following could lead to the reemergence of malaria in the United States: (a) global warming, (b) increased travel of individuals from endemic regions to the United States, (c) increased immigration of individuals from endemic regions to

the United States, (d) regulations against the use of insecticides, and (e) laws protecting wetlands.

3. People who suffer from AIDS are severely immunocompromised because HIV destroys immune cells; thus, all the diseases in this chapter could be a threat to an AIDS patient. Which diseases, however, would specifically exacerbate the immune dysfunction of AIDS? Why?

4. Discuss why sickle-cell trait is advantageous to people living in areas where malaria is endemic but not advantageous in malaria-free areas.

5. Present a logical argument to explain the differences between the clinical manifestations of *Trypanosoma brucei* and *T. cruzi*. Relate your argument to the respective life cycles.

6. What feature of the life cycle of *T. brucei* makes it difficult to create a successful vaccine?

7. On July 10, a man in Palm Beach, Florida, was admitted to the hospital with a month-long history of fever, chills, headache, and vomiting. A parasite was observed in a blood smear. What parasite causes these signs and symptoms? What is the treatment? Is this a rare disease in Florida?

8. A Peace Corps volunteer was evacuated from Malawi in Africa because of a two-week history of headaches, vision loss, and an episode of unconsciousness. Doctors discovered a granular swelling in his brain and eggs with peripheral spines in his urine. What is the diagnosis? What species is involved? What is the treatment?

9. Humans can accidentally become intermediate hosts of *Taenia solium*. Why is this a dead end for the parasite?

CONCEPT MAPPING

Using the terms provided, fill in this concept map that describes intestinal protozoan parasites. You can also complete this and other concept maps online by going to the Mastering Microbiology Study Area.

Apicomplexans
Ciliates
Cryptosporidium parvum
Cyclospora cayetanensis
Diarrhea
Dormant stage

Entamoeba histolytica
Environment
Fecal-oral route
Flagellates
New host

People with AIDS
Reproducing/feeding stage
Schizonts
Trophozoites
Water

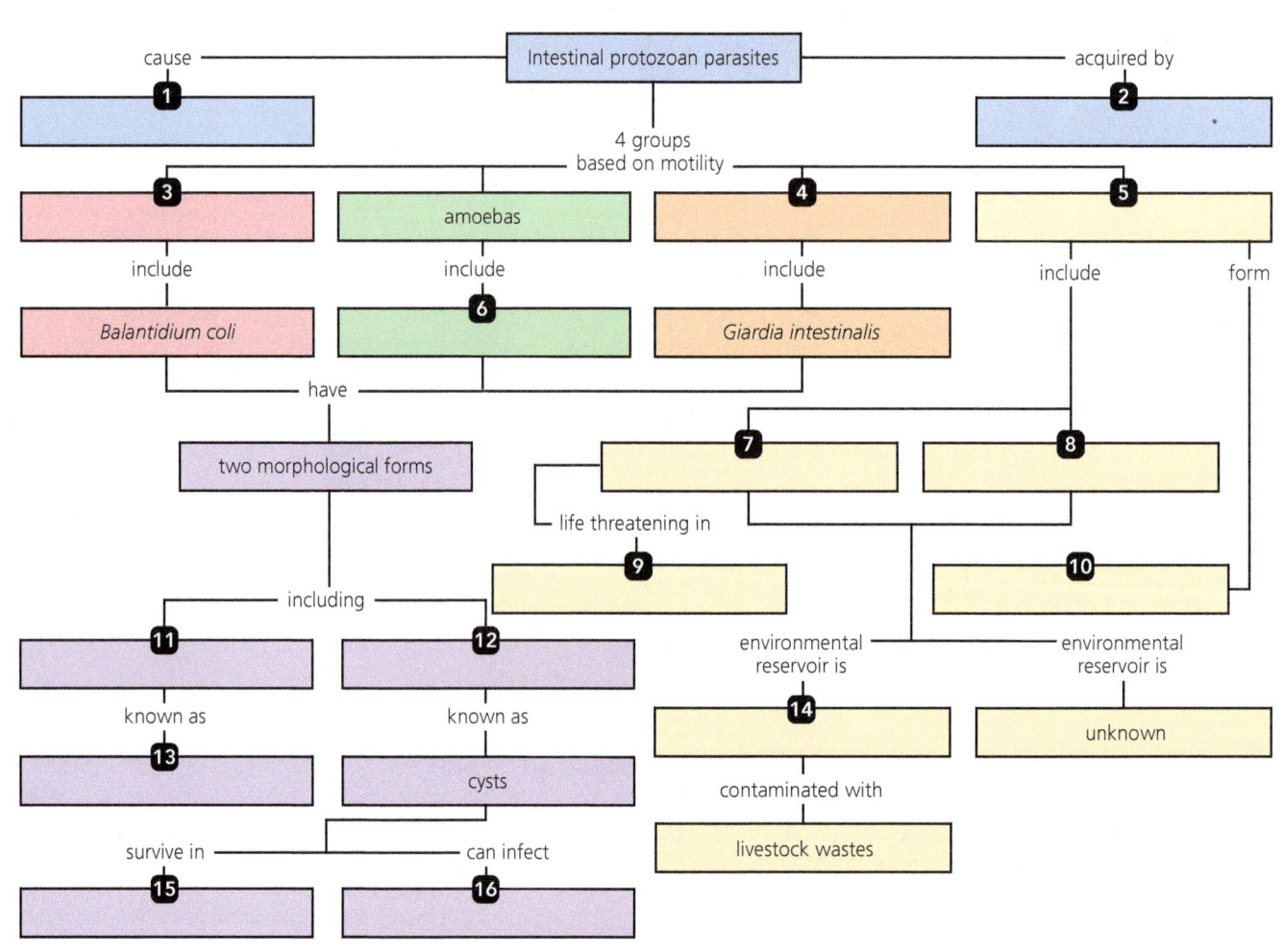

24 Pathogenic DNA Viruses

Before You Begin

1. How can viruses be distinguished from intracellular bacteria?
2. How do viral envelopes differ from cellular cytoplasmic membranes?
3. How can animal viruses be cultured in the laboratory?
4. How can enveloped viruses enter animal cells?

DNA viruses such as these papilloma (wart) viruses, use DNA for genetic material.

MICRO IN THE CLINIC

Are Cold Sores Caused by Cold Viruses?

ONE MORNING, DAKOTA CAME DOWN from her room and told her mother that she did not feel well enough to go to school. Initially, her mother was skeptical. After all, it was the third time this week that Dakota wanted to stay home. She measured Dakota's temperature and noted a slight fever, so she sent Dakota back to bed. Back in her room, Dakota called her friend Emily. In a whisper, Dakota told her friend that she had a weird, painful sore on the inside of her lip. Emily advised Dakota to tell her mother, even if she was a bit embarrassed. Dakota texted her new boyfriend to let him know that she would not be in school today. He texted her back to say he was not feeling well and was also staying home, but Dakota was fast asleep by the time he replied.

When she woke up in the early afternoon, Dakota told her mother about the sore on the inside of her lip. Dakota's mother examined her lip and saw a few closely clustered fluid-filled vesicles. Dakota's mother said that it was probably just a cold sore, but that it was a good idea to go to the clinic anyway. Dakota texted Emily an update. Emily reminded her that cold sores are not the result of a cold, but are caused by a herpesvirus. Dakota was shocked. She always thought that herpes was a sexually transmitted disease, and she had not had any sexual contact.

1. Was it necessary for Dakota to visit a doctor?
2. Are all herpesviruses transmitted through sexual contact?

Turn to the end of the chapter (p. 729) to find out.

DNA viruses that cause human diseases are grouped into seven families based on several factors, including the type of DNA they contain, the presence or absence of an envelope, size, and the host cells they attack. DNA viruses have either double-stranded DNA (dsDNA) or single-stranded DNA (ssDNA) for their genomes, and none of them contains functional RNA (see Chapter 13). Virologists classify viruses in families and genera with Latinized names, but English names are used for viral species, in contrast to species of prokaryotes and eukaryotes. Often, there is a single species in a viral genus, so virologists refer to a species solely by its genus or by its English name.

Double-stranded DNA viruses belong to the families *Poxviridae, Herpesviridae, Papillomaviridae, Polyomaviridae,* and *Adenoviridae.* The family *Parvoviridae* contains viruses with ssDNA genomes. Viruses in the family *Hepadnaviridae* are unusual because they have a genome that is partly dsDNA and partly ssDNA. In the following sections, we examine the infections and diseases caused by the viruses in each of these families.

Poxviridae

LEARNING OUTCOMES

24.1 Name two diseases caused by poxviruses and discuss their signs and symptoms.
24.2 Describe the progression of disease in poxvirus infections.
24.3 Discuss the historical importance of poxviruses in immunization and their use in disease eradication.

Poxviruses are double-stranded DNA viruses with complex capsids and envelopes **(FIGURE 24.1)**. They are the largest of human pathogenic viruses, being up to 300 nm in diameter, which is just visible with high-quality, optical, ultraviolet light microscopes. Their large size allows them to be potential vectors for the introduction of genetic material in immunizations and gene therapy.

Specific poxviruses infect many mammals, including mice, camels, elephants, cattle, and monkeys. (Note that chickenpox, discussed later in this chapter, is not caused by a poxvirus, nor does it occur in chickens; instead, it is caused by a herpesvirus that affects people.) Most animal poxviruses are species specific; they are unable to infect humans because they cannot attach to human cells, though a few poxviruses of monkeys, sheep, goats, and cattle can infect some humans because these people have surface proteins similar to proteins of those animals.

All poxviruses produce lesions that progress through a series of stages **(FIGURE 24.2)**. The lesions begin as flat, reddened **macules**[1] (mak′ūlz), which then become raised sores called **papules**[2] (pap′ūlz); when the lesions fill with clear fluid, they are called **vesicles,** which then progress to pus-filled **pustules**[3] (pŭs′choolz), which are also known as **pocks** or **pox**. These terms for the stages of poxvirus lesions are also used to describe the lesions of other skin infections. Poxvirus pustules dry up to form a crust, and because these lesions penetrate the dermis, they result in characteristic scars.

Infection with poxviruses occurs primarily through inhalation of viruses in droplets or particles of dried crusts. Because the envelopes of poxviruses are relatively unstable outside a host's body, close contact is necessary for infection. The two main poxvirus diseases of humans are *smallpox* and *molluscum contagiosum.* Three diseases of animals—*orf* (sheep and goat pox), *cowpox,* and *monkeypox*—may be transmitted to humans as well.

Smallpox

Smallpox has played important roles in the history of microbiology, immunization, epidemiology, and medicine. During the Middle Ages, 80% of the European population contracted smallpox. Later, European colonists introduced smallpox into susceptible Native Americans, resulting in the death of as many as 3.5 million people. Edward Jenner demonstrated immunization using the relatively mild cowpox virus to protect against smallpox. He was successful because the antigens of cowpox virus are chemically similar to those of smallpox virus so that exposure to cowpox results in immunological memory and subsequent resistance against both cowpox and smallpox.

The official name for smallpox virus is *Orthopoxvirus variola,* but it is commonly known as **variola** (vă-rī′ō-lă). The virus exists in two forms, or strains: **Variola major** causes severe disease with a mortality rate of 20% or higher, depending upon the age and general health of the host. **Variola minor** causes a less severe disease with a mortality rate of less than 1%. Both types of variola infect internal organs and produce fever—42°C (107°F) has been recorded—malaise, delirium, and prostration

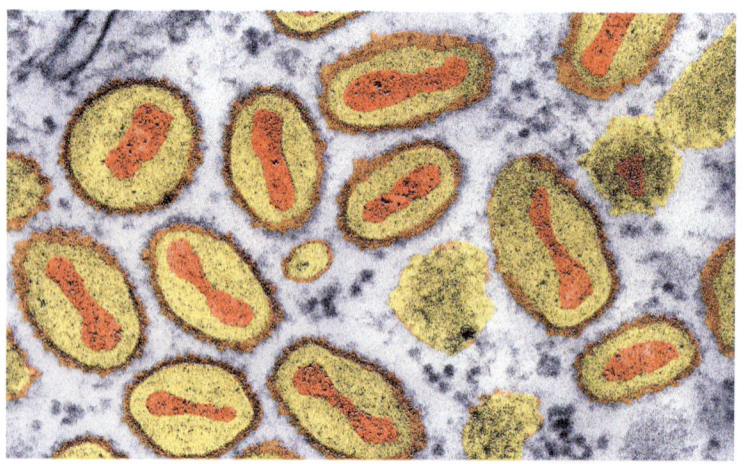

▲ **FIGURE 24.1 Poxviruses.** Vaccinia, the agent of cowpox (shown here), and variola, the causative agent of smallpox, are in the genus *Orthopoxvirus.*

---
[1]From Latin *maculatus,* meaning "spotted."
[2]From Latin *papula,* meaning "pimple."
[3]From Latin *pustula,* meaning "blister."

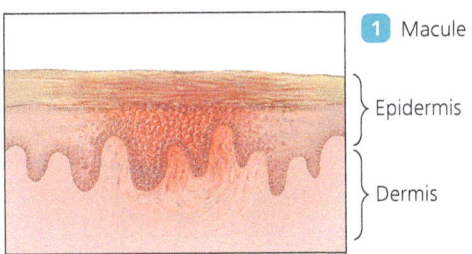
1 Macule

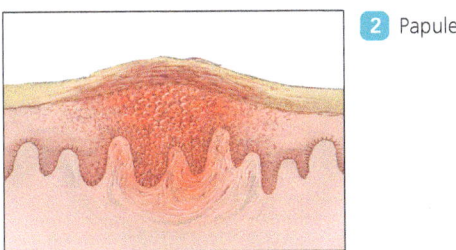
2 Papule

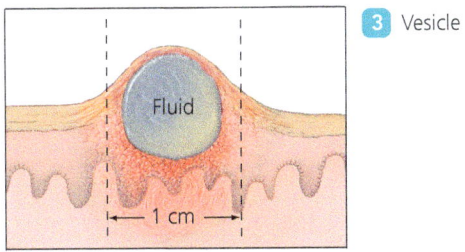
3 Vesicle

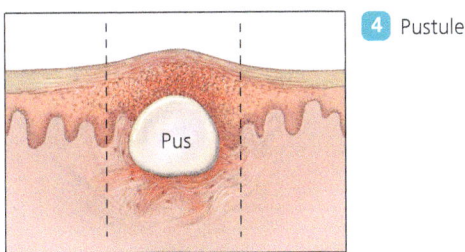
4 Pustule

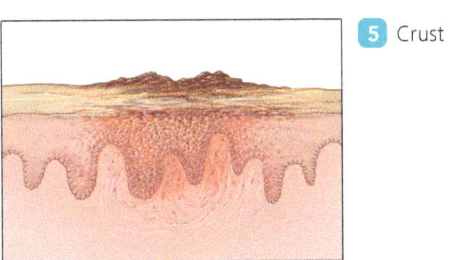
5 Crust

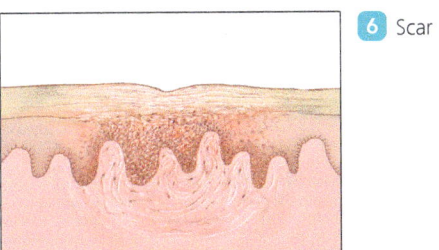
6 Scar

◀ **FIGURE 24.2 The stages of the lesions in poxvirus infections.** Red macules become hard papules, then fluid-filled vesicles that become pus-filled pustules (pox), which then crust over and leave characteristic scars. *Are the dried viruses in the crusts of smallpox and cowpox lesions infective?*

Figure 24.2 *Yes, the dried crusts of poxvirus lesions may contain active viruses. The Chinese used the viruses in crusts for variolation, an early attempt at vaccination.*

before moving via the bloodstream to the skin, where they produce the recognizable pox **(FIGURE 24.3)**. Both strains also leave severe scars on the skin of survivors, especially on the face.

Smallpox is the first human disease to be eradicated globally in nature. In 1967, the World Health Organization (WHO) began an extensive campaign of identification and isolation of smallpox victims, as well as vaccination of their contacts, with the goal of completely eliminating the disease. The efforts of WHO were successful, and in 1980, smallpox was declared to be eradicated—one of the great accomplishments of 20th-century medicine. The factors that made it possible for smallpox to be eliminated included the following:

- An inexpensive, stable, and effective vaccine (called *vaccinia*, which may have been derived from cowpox virus) is available.
- Smallpox is specific to humans; there are no animal reservoirs.
- The severe, obvious signs and symptoms of smallpox enable quick and accurate diagnosis and quarantine.
- There are no asymptomatic cases.
- The virus is spread only via close contact.

Stocks of the virus are still maintained in laboratories in the United States and Russia as research tools for investigations

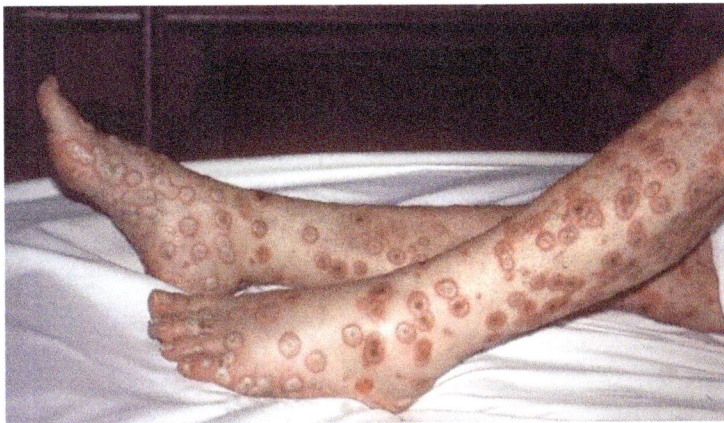

▲ **FIGURE 24.3 Smallpox lesions.** Smallpox pustules typically begin in the mouth, spread to the face, and then downward to cover the body as shown here on Janet Parker who contracted the disease as a result of an accident in a research facility in Great Britain in 1978. She was the world's last victim of this dreaded disease.

concerning virulence, pathogenicity, protection against bioterrorism, vaccination, and recombinant DNA technology.

Routine vaccination against smallpox has ceased in the United States, though some health care workers and military personnel were vaccinated following the 2001 bioterrorism attack. Studies have shown that older citizens who were vaccinated as children still retain some immunity. One-tenth the regular vaccine dose is effective for revaccination of these adults.

Now that smallpox vaccination has been discontinued in many countries, experts are concerned that the world's population is once again susceptible to smallpox epidemics if the virus were accidentally released from storage or used as a biological weapon. For this reason, many scientists advocate the destruction of all smallpox virus stocks; other scientists insist that we must maintain the virus in secure laboratories so that it is available for the development of more effective vaccines and treatments if the virus were disseminated in a bioterrorism attack. In 2018, the United States Food and Drug Administration (FDA) approved the first drug to treat smallpox—tecovirimat.

Molluscum Contagiosum

Another poxvirus, *Molluscipoxvirus*, causes **molluscum contagiosum** (mo-lŭs′kŭm kon-taj-ē-ō′sum), a skin disease characterized by smooth, waxy papules that frequently occur in lines where the patient has spread the virus by scratching (FIGURE 24.4). Lesions may occur anywhere but typically appear on the face, trunk, and external genitalia.

Molluscum contagiosum occurs worldwide and is typically spread by contact among infected children, sexually active individuals (particularly adolescents), and AIDS patients. Treatment involves removing the infected nodules by chemicals or freezing, though people with normal immune systems heal within months without treatment. Sexual abstinence prevents the genital form of the disease. Condoms do not afford protection, as the virus can spread from and to areas not covered by condoms.

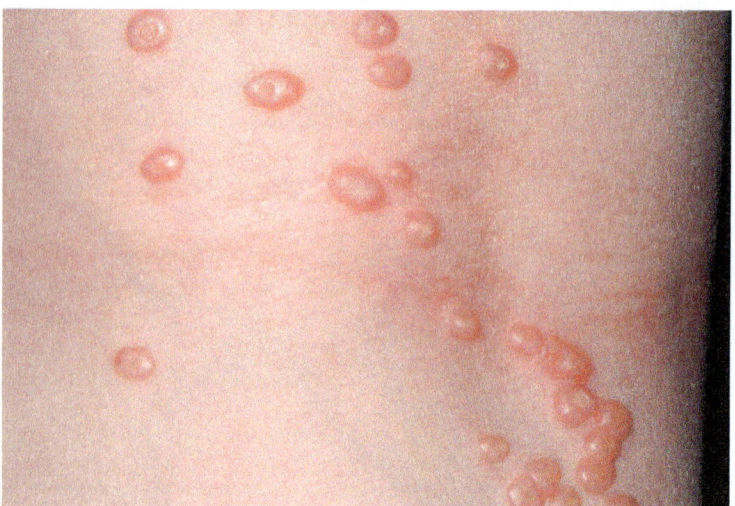

▲ FIGURE 24.4 **Lesions of molluscum contagiosum.** *Molluscipoxvirus*, a poxvirus, causes the pearly white to light pink, tumorlike lesions.

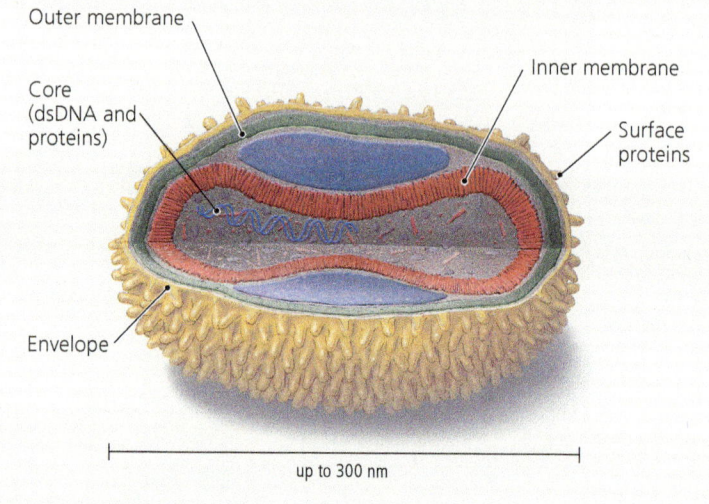

MICROBE AT A GLANCE 24.1

Orthopoxvirus variola (Smallpox Virus)

Taxonomy: Family *Poxviridae*

Genome and morphology: Single molecule of double-stranded DNA; enveloped, brick-shaped or pleomorphic, complex capsid, 250 nm × 250–300 nm × 150–300 nm

Host: Humans

Virulence factors: Remains very stable in aerosol form; highly contagious between people; can remain viable for as long as 24 hours

Disease caused: Smallpox

Treatment for disease: Tecovirimat; anti-cowpox (anti-vaccinia) immunoglobulin; cidofovir; an experimental drug called brincidofovir may be effective as well

Prevention of disease: Smallpox has been eliminated in nature and presumably could reappear only if released from secure storage facilities; if it were released, vaccination of exposed individuals and quarantine of patients would need to be implemented immediately; administration of vaccine within four days of infection prevents disease in most patients

Other Poxvirus Infections

Poxvirus infections also occur in animals. Infections of humans with these animal viruses are usually relatively mild, resulting in pox and scars but little other damage. In the case of cowpox, the advent of milking machines has reduced the number of human cases to near zero in the industrialized world. Other poxvirus infections, including those in sheep, goats, and monkeys, are less easily contracted by humans.

Transmission of these poxviruses to humans requires close contact with infected animals, similar to the contact Edward Jenner observed for milkmaids and cowpox. Still, an upsurge in the number of monkeypox cases in humans has been seen since the early 2000s. (See **Emerging Disease Case Study: Monkeypox** on p. 713.) This may be the result of human encroachment into monkey habitats or changes in viral antigens. Some researchers speculate that the cessation of smallpox vaccination may also be involved because evidence suggests that smallpox vaccination

EMERGING DISEASE CASE STUDY

Monkeypox

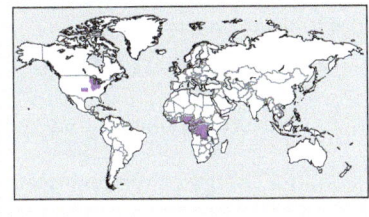

Jacob was excited about the pet he got for his birthday; he was the only boy at school who had a prairie dog! Common dogs were boring; his prairie dog could climb up the drapes, run through plastic tube mazes, and sit on Jacob's shoulder to eat. Doggie was the best. In fact, everything would be just about perfect if only Jacob weren't sick in bed.

Two weeks after his birthday, Jacob was exhausted and felt terrible. He had a fever accompanied by chills headache, and muscle aches, so his mother had kept him home from school. Then the lymph nodes in his neck and armpits had swollen, and he had gotten frightening, large, raised lesions all over his face and hands. He was really sick.

For the next three weeks, the lesions had become crusty and then fallen off, only to be replaced by new lesions. Jacob felt awful and missed playing with Doggie, who had been confined to his cage, while Jacob stayed in bed. Jacob's doctor had been unable to diagnose the disease, but an infectious disease specialist had finally confirmed monkeypox—a disease named for monkeys who can get the disease. In reality, monkeypox is a disease of rodents and is endemic to Africa, but can be transmitted to prairie dogs in America. Final investigation revealed that the pet wholesaler who had supplied Doggie to Jacob's pet store in Illinois had also supplied African monkeys to stores in Texas. At least one of those monkeys had been infected with monkeypox virus—a virus closely related to smallpox virus. The monkey had infected Doggie who had infected Jacob, most likely through a scratch.

Jacob was sick for four weeks, but monkeypox is not as virulent as smallpox, so Jacob recovered with only some scars as a reminder that not all unusual pets are a blessing.

1. Why did Jacob's lymph nodes swell during the course of a skin disease?
2. Why didn't the doctor prescribe penicillin, erythromycin, or ciprofloxacin?
3. How can a virus of rodents possibly cause a disease in monkeys and humans?

also protects against monkeypox. Thus, the elimination of smallpox and the cessation of smallpox vaccination may have led to increased human susceptibility to monkeypox. Some scientists have recommended that smallpox vaccination be reinstated in areas where monkeypox is endemic.

TELL ME WHY

Why does inoculation with vaccinia produce immunological memory and protection against smallpox virus—variola?

Herpesviridae

The family *Herpesviridae* contains a large group of viruses with enveloped icosahedral capsids containing linear dsDNA. A herpesvirus attaches to a host cell's receptor and enters the cell through the fusion of its envelope with the cytoplasmic membrane. After the viral genome is replicated and assembled in the cell's nucleus, the virion acquires its envelope from the nuclear membrane and exits the cell via exocytosis or cell lysis.

Many types of herpesviruses infect humans, and their high infection rates make them the most prevalent of DNA viruses. Indeed, three types of herpesvirus—herpes simplex, varicella-zoster, and Epstein-Barr (discussed shortly)—infect 90% of the adult population in the United States, and several other herpesviruses have infection rates over 50%.

Herpesviruses are often *latent*; that is, they may remain inactive inside infected cells, often for years. Such latent viruses may reactivate as a result of aging, chemotherapy, immunosuppression, or physical and emotional stress, causing a recurrence of the manifestations of the diseases. Additionally, some latent herpesviruses insert into a host's chromosomes, where they may cause genetic changes and potentially induce cancer.

Besides causing the diseases for which the group is named (oral herpes and genital herpes), herpesviruses cause a variety of other diseases, including chickenpox, shingles, mononucleosis, and some cancers. **Human herpesviruses** have been assigned species names combining "HHV" (for "human herpesvirus") with numbers corresponding to the order in which they were discovered. Thus, the genus *Simplexvirus* contains two species (as discussed in the next section) known as HHV-1 and HHV-2. *Varicellovirus HHV-3* causes chickenpox and shingles; *Lymphocryptovirus HHV-4*, also known as Epstein-Barr virus, causes several human diseases; human herpesvirus 5 is in the genus *Cytomegalovirus*; *Roseolovirus* includes both HHV-6 and HHV-7; and *Rhadinovirus HHV-8* is associated with a particular type of cancer in AIDS patients. All these viruses share similar capsids and genes but can be distinguished by genetic variations among them.

HHV-7 is an interesting virus in that scientists identified it as HHV-7 on the basis of its genome, envelope, capsid, and antigens, but at this time, no disease has been attributed to this herpesvirus. It is an *orphan virus*—a virus without a known disease.

Infections of *Human Herpesvirus 1* and *2*

LEARNING OUTCOMES

24.4 Describe the diseases caused by *human herpesvirus 1* and *2*, including their signs, symptoms, treatment, and prevention.

24.5 List conditions that may reactivate latent herpesviruses.

The name "herpes," which is derived from a Greek word meaning "to creep," is descriptive of the slowly spreading skin lesions that are seen with *human herpesvirus 1* (HHV-1) and *human herpesvirus 2* (HHV-2), which are in the genus *Simplexvirus*. They were known formerly as herpes simplex viruses (HSV-1 and HSV-2).

Latency and Recurrent Lesions

One of the more distressing aspects of herpes infections is the recurrence of lesions. About two-thirds of patients with HHV-1 or HHV-2 will experience recurrences during their lives as a result of the activation of latent viruses. After a primary infection, these herpesviruses often enter sensory nerve cells and are carried by cytoplasmic flow to the base of the nerve, where they remain latent in the trigeminal, brachial, sacral, or other ganglia[4] (FIGURE 24.5). Such latent viruses may reactivate later in life as a result of immune suppression caused by stress, fever, trauma, sunlight, menstruation, or disease and travel down the nerve to produce recurrent lesions as often as every two weeks. Fortunately, the lesions of recurrent infections are rarely as severe as those of initial infections because of immunological memory.

Types of HHV-1 and HHV-2 Infections

Historically, HHV-1 has also been known as "above the waist herpes," whereas HHV-2 has been known as "below the waist herpes." In the following subsections, we consider the variety of infections caused by these two species.

Oral Herpes *Human herpesvirus 1* was the first human herpesvirus to be discovered. About two weeks after infection, the virions of HHV-1 produce painful, itchy skin lesions on the lips, called **fever blisters** or **cold sores** (FIGURE 24.6), which last 7 to 10 days; initial infections may also be accompanied by flulike signs and symptoms, including malaise, fever, and muscle pain. Severe infections in which the lesions extend into the oral cavity, called *herpetic gingivostomatitis*,[5] are most often seen in young patients and in patients with lowered immune function due to disease, chemotherapy, or radiation treatment. Young adults with sore throats resulting from other viral infections may also develop *herpetic pharyngitis*.

Genital Herpes HHV-1 can infect the genitalia. The virus causes about 15% of genital herpes lesions. The remainder are caused by *human herpesvirus 2*. HHV-2 differs slightly from HHV-1 in its surface antigens, primary location of infection, and mode of

[4]A ganglion is a collection of nerve cell bodies containing the nuclei of the cells.
[5]From Latin *gingival*, meaning "gum"; Greek *stoma*, meaning "mouth"; and Greek *itis*, meaning "inflammation."

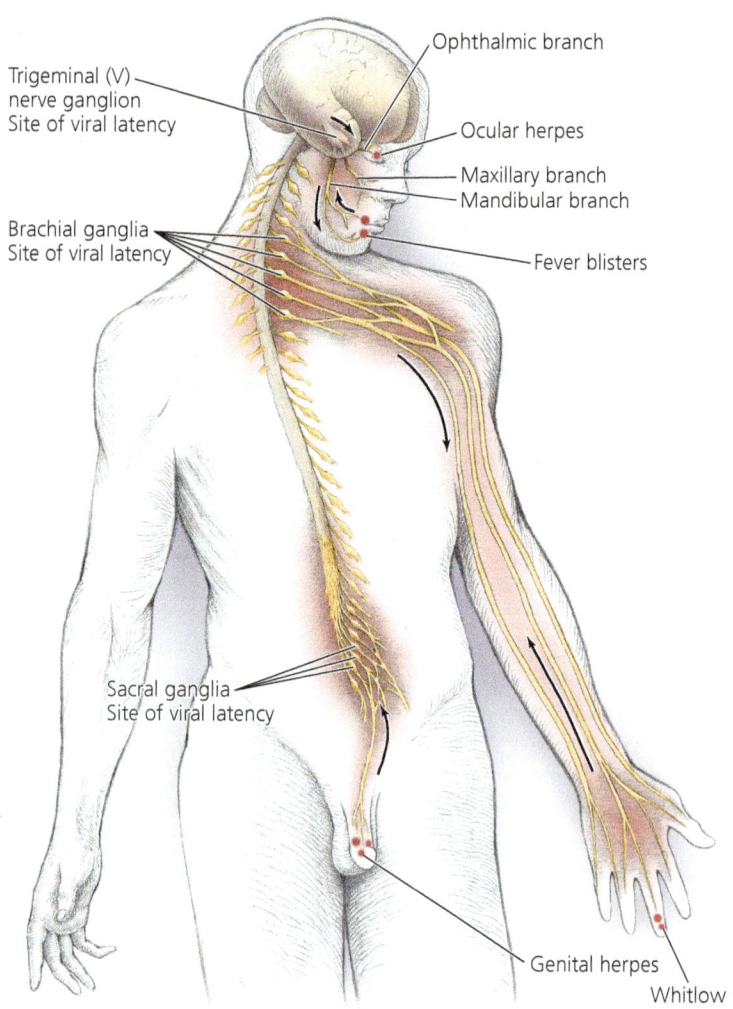

▲ **FIGURE 24.5 Sites of events in herpesvirus infections.** Infections typically occur when viruses invade the mucous membranes of either the lips (for the fever blisters of oral herpes) or the genitalia (for genital herpes) or through broken skin of a finger (for a whitlow). Viruses may remain latent for years in the trigeminal, brachial, or sacral ganglia before traveling down nerve cells to cause recurrent symptoms in the lips, genitalia, fingers, or eyes. *What factors may trigger the reactivation of latent herpesviruses and the recurrence of symptoms?*

Figure 24.5 Stress, fever, trauma, sunlight, menstruation, or diseases such as AIDS may cause recurrent symptoms as a result of immune suppression.

transmission. Type 2 herpes is usually associated with painful lesions on the genitalia (FIGURE 24.7) because the virions are typically transmitted sexually. However, HHV-2 can cause oral lesions when it infects the mouth. Thus, about 10% of oral cold sores result from HHV-2. The presence of HHV-2 in the oral region and of HHV-1 in the genital area is presumed to be the result of oral sex.

Ocular Herpes Ocular herpes (ok'yū-lăr) or *ophthalmic herpes* is an infection caused by latent herpesviruses (FIGURE 24.8a). In ocular herpes, latent HHV-1 in the trigeminal ganglion travels down the ophthalmic branch of the trigeminal nerve instead of the mandibular or maxillary branches (see Figure 24.5). Symptoms and signs usually occur in only one eye and include

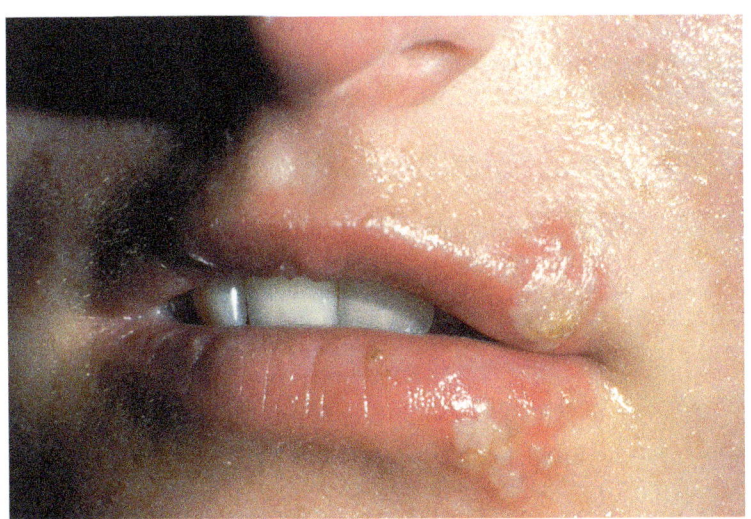

▲ FIGURE 24.6 **Oral herpes lesions.** *Which is the more likely cause of these lesions, HHV-1 or HHV-2?*

Figure 24.6 Approximately 90% of oral herpes lesions are caused by human herpesvirus 1.

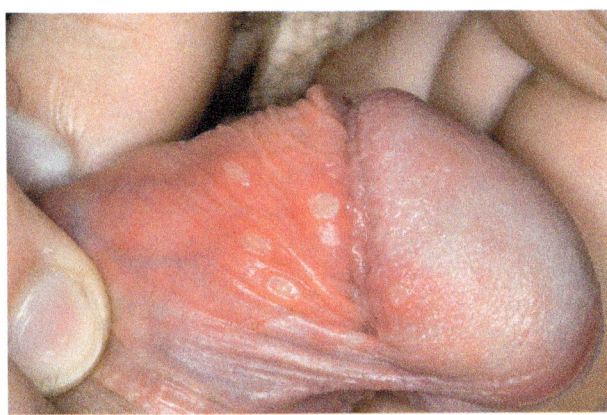

▲ FIGURE 24.7 **Genital herpes lesions.** *Are these lesions usually caused by human herpesvirus type 1, or type 2?*

Figure 24.7 Either type of herpes can cause genital lesions, but "below the waist" lesions are usually caused by HHV-2.

a gritty feeling, conjunctivitis, pain, and sensitivity to light. The viruses may also cause corneal lesions. Recurrent ocular infections can lead to blindness as a result of corneal damage.

Whitlow An inflamed blister called a **whitlow**[6] (hwit′lō) may result if either HHV-1 or HHV-2 enters a cut or break in the skin of a finger **(FIGURE 24.8b)**. A whitlow usually occurs on only one finger. Whitlow is a hazard for children who suck their thumbs and for health care workers in the fields of obstetrics, respiratory care, gynecology, and dentistry who come into contact with lesions. Such workers should always wear gloves when treating patients with herpes lesions. Since a whitlow is also a potential source of infection, symptomatic personnel should not work with patients until the worker's whitlow has healed. The viruses involved in whitlow may become latent in brachial ganglia and cause recurrent lesions.

Neonatal Herpes Although most herpes simplex infections in adults are painful and unpleasant, they are not life threatening. This is not the case for herpes infections in newborns. HHV-2 is usually the species involved in neonatal herpes. Neonatal herpes infections can be severe, with a mortality rate of 30% if the infection is cutaneous or oral and 80% if the central nervous system is infected.

A fetus can be infected *in utero* if the virus crosses the placental barrier, but it is more likely for a baby to be infected at birth through contact with lesions in the mother's reproductive tract. Also, mothers with oral lesions can infect their babies if they kiss them on the mouth. Pregnant women with herpes infections, even if asymptomatic, should inform their doctors. To protect the baby, delivery should be by cesarean section if genital lesions are present at the time of birth.

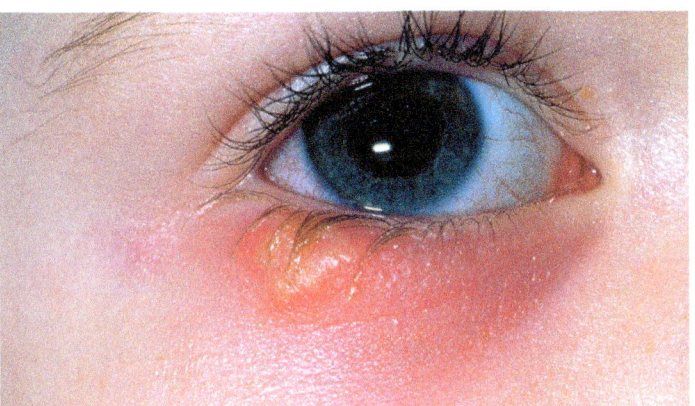

(a)

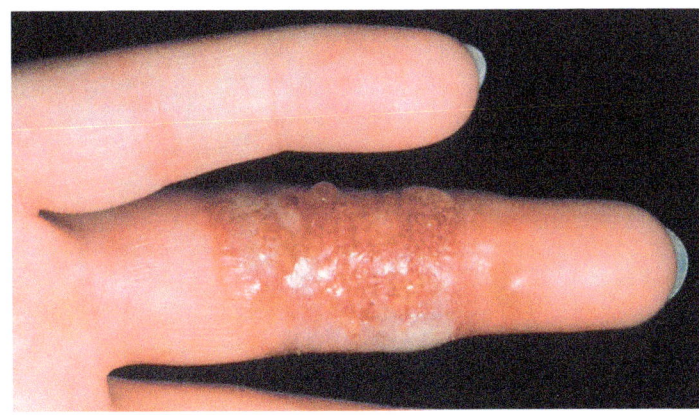

(b)

▲ FIGURE 24.8 **Two manifestations of herpesvirus infections.** (a) Ophthalmic or ocular herpes, the result of the reactivation of latent viruses in the trigeminal ganglion. (b) A whitlow on a health provider's finger, resulting from the entry of herpesviruses through a break in the skin of the finger. *How can health care workers protect themselves from a whitlow?*

Figure 24.8 Health providers should always wear gloves when treating patients who have herpes lesions on any part of their bodies.

---
[6]From Scandinavian *whick*, meaning "nail," and *flaw*, meaning "crack."

TABLE 24.1 Comparative Epidemiology and Pathology of *Human Herpesvirus 1* and *2* Infections

	HHV-1 (HSV-1)	HHV-2 (HSV-2)
Usual Diseases	90% of cold sores/fever blisters; whitlow	85% of genital herpes cases
Mode of Transmission	Close contact	Sexual intercourse
Site of Latency	Trigeminal and brachial ganglia	Sacral ganglia
Locations of Lesions	Face, mouth, and rarely trunk	External genitalia and less commonly thighs, buttocks, and anus
Other Complications	15% of genital herpes cases; pharyngitis; gingivostomatitis; ocular/ophthalmic herpes; herpes gladiatorum; 30% of neonatal herpes cases	10% of oral herpes cases; 70% of neonatal herpes cases

Other Herpes Simplex Infections Athletes may develop herpes lesions almost anywhere on their skin as a result of contact with herpes lesions during wrestling. *Herpes gladiatorum,* as this condition is called, is most often caused by HHV-1.

Human herpesviruses 1 and *2* may also cause other diseases, including encephalitis, meningitis, and pneumonia. Patients with severe immune suppression, such as those with AIDS, are most likely to develop these rare disorders.

TABLE 24.1 compares infections caused by herpesviruses 1 and 2.

Epidemiology and Pathogenesis

HHV-1 and HHV-2 occur worldwide and are transmitted through close body contact. Active lesions are the usual source of infection, though asymptomatic carriers shed HHV-2 genitally. After entering the body through cracks or cuts in the skin or mucous membranes, viruses reproduce in epithelial cells near the site of infection and produce inflammation and cell death, resulting in painful, localized lesions on the skin. By causing infected cells to fuse with uninfected neighboring cells to form a structure called a *syncytium,* herpes virions spread from cell to cell, in the process avoiding the host's antibodies.

Initial infections with herpes simplex viruses are usually age specific. Primary HHV-1 infections typically occur via casual contact during childhood and usually produce no signs or symptoms; in fact, WHO estimates that 67% of the world's children have been asymptomatically infected with HHV-1. Most HHV-2 infections, by contrast, are acquired between the ages of 15 and 29 as a result of sexual activity, making genital herpes one of the more common sexually transmitted infections in the United States.

Approximately 60% of adult Americans have been exposed to HHV-1. HHV-2 infects about 17% of Americans over age 12.

Diagnosis, Treatment, and Prevention

Physicians often diagnose herpes infections by the presence of their characteristic, recurring lesions, especially in the genital region and on the lips. Microscopic examination of infected tissue reveals syncytia. Positive diagnosis is achieved by serological testing that demonstrates the presence of viral antigen.

Infections with HHV-1 or HHV-2 are among the few viral diseases that can be controlled with chemotherapeutic agents, notably valaciclovir for both cold sores and genital herpes, and other nucleoside analogs (see Chapter 10, Figure 10.7), such as iododeoxyuridine or trifluridine, for ocular herpes. Topical applications of the drugs limit the duration of the lesions and reduce viral shedding, though none of these drugs cures the diseases or frees nerve cells of latent viral infections. Evidence suggests that a daily oral dose of acyclovir over a period of 6 to 12 months can be effective in preventing recurrent genital herpes and in reducing the spread of herpes infections to sexual partners. Many obstetricians prescribe acyclovir during the final weeks of pregnancy for women with a history of genital herpes. (Chapter 10 examines the actions of these drugs.)

Health care workers can reduce their exposure to infection with herpes simplex by wearing latex gloves. Health providers with whitlows should not have contact with patients, especially newborns.

Sexual abstinence or mutual monogamy between uninfected partners is the only sure way of preventing genital infections. People with active lesions should abstain from sexual activity until the crusts have disappeared, and either male or female condoms should be used even in the absence of lesions because asymptomatic individuals still shed viruses. Herpes lesions in women are usually on the external genitalia, so use of a condom during sexual intercourse provides little protection for their partners. Some studies indicate that circumcision reduces by 25% a man's chance of getting genital herpes.

Human Herpesvirus 3 (Varicella-Zoster Virus) Infections

LEARNING OUTCOME

24.6 Compare and contrast chickenpox in children with shingles in adults.

Human herpesvirus 3, commonly called **varicella-zoster virus (VZV),** takes its name from the two diseases it causes: *varicella,* which is typically a disease of children, and *herpes zoster,* which usually occurs in adults. Taxonomically, VZV is in the genus *Varicellovirus*—note that the viral name is spelled differently than the disease.

Epidemiology and Pathogenesis

Varicella (var-i-sel'ă), commonly known as **chickenpox,** is a highly infectious disease most often seen in children. Viruses enter the body through the respiratory tract or eyes, replicate in cells at the site of infection, and then travel via the blood throughout the body, triggering fever, malaise, and skin lesions.

Viruses are shed before and during the symptoms. Usually, chickenpox is mild, though the disease can be fatal, especially if associated with secondary bacterial infections. Chickenpox is typically more severe in adults than in children, presumably because much of the tissue damage results from the patient's immune response, and adults have a more developed immune system than children do.

About two to three weeks after infection, characteristic skin lesions first appear on the back and trunk and then spread to the face, neck, and limbs (FIGURE 24.9). In severe cases, the rash may spread into the mouth, pharynx, and vagina. Chickenpox lesions begin as macules, progress in one to two days to papules, and finally become thin-walled, fluid-filled vesicles on red bases, which have been called "dewdrops on rose petals." The vesicles turn cloudy, dry up, and crust over in a few days. Successive crops of lesions appear over a period of three to five days so that at any given time all stages can be seen. Viruses are shed through respiratory droplets and the fluid in lesions; dry crusts are not infective.

Like many herpesviruses, VZV can become latent within sensory nerves and may remain dormant for years. In about 15% of individuals who have had chickenpox, conditions such as stress, aging, or immune suppression cause the viruses to reactivate, travel down the nerve they inhabit, and produce an extremely painful skin rash near the distal end of the nerve (FIGURE 24.10). This rash is known as **shingles** or **herpes zoster** (FIGURE 24.11).

Zoster, a Greek word that means "belt," refers to the characteristic localization of shingle lesions along a band of skin called a *dermatome,* which is innervated by a single sensory nerve.

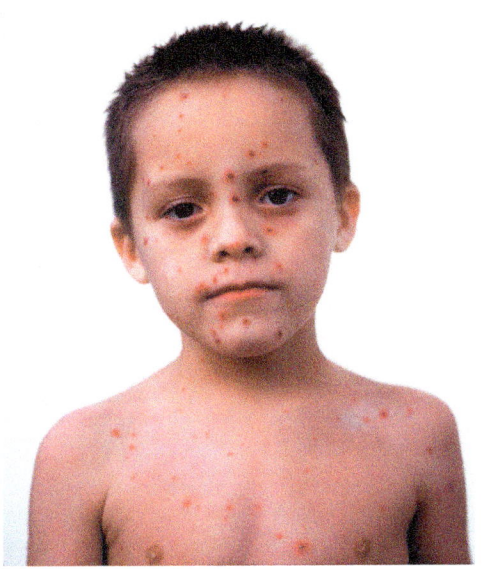

▲ **FIGURE 24.9 Characteristic chickenpox lesions.** *Varicellovirus* causes these thin-walled, fluid-filled vesicles on red bases.

Shingles lesions form only in the dermatome associated with an infected nerve, and thus the lesions are localized and on the same side of the body as the infected nerve cell. Occasionally, lesions from latent infections of cranial nerves form in the eye, ear, or other part of the head. In some patients, the pain associated with shingles remains for months or years after the lesions have healed.

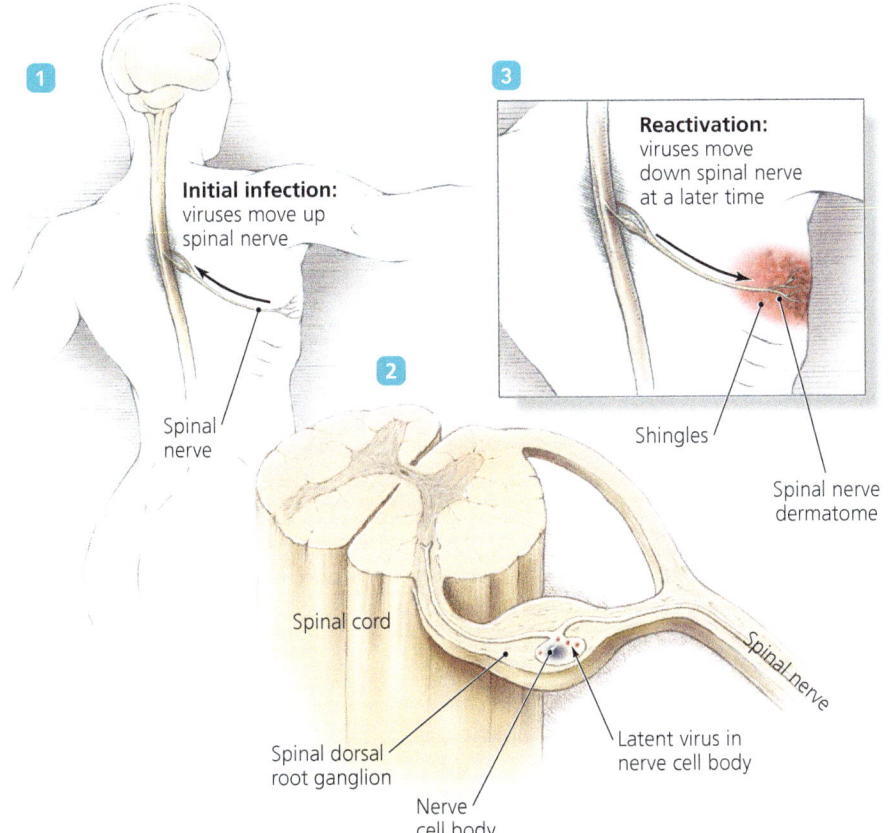

◀ **FIGURE 24.10 Latency and reactivation of varicella-zoster virus.** After initial infection, VZV can move up the spinal nerve and become latent in nerve cell bodies in spinal root ganglia. After subsequent reactivation (often years later), the viruses move down the spinal nerve to the skin, causing a rash known as shingles.

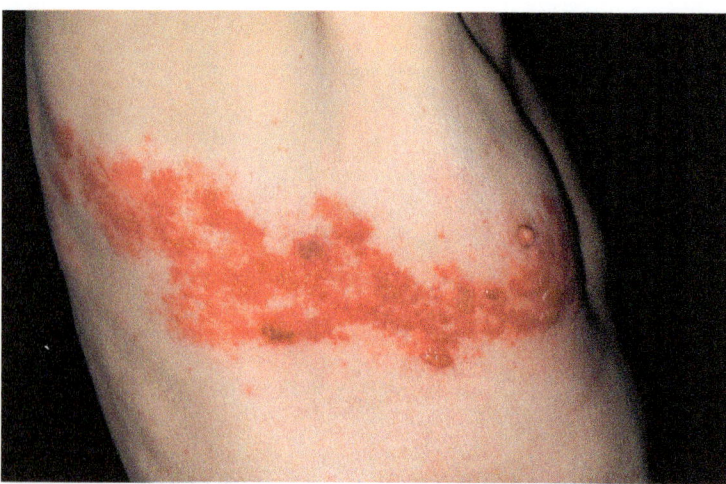

▲ **FIGURE 24.11 Shingles, a rash caused by *Varicellovirus*.** These painful lesions occur on the dermatome innervated by an infected nerve.

of immunity to chickenpox. A single dose of shingles vaccine, which contains the same antigens as the chickenpox vaccine, but more than 14 times as much, is recommended for all adults age 60 and older.

MICRO CHECK

1. Which two serious human diseases are caused by poxviruses?
2. Which of the herpesviruses is the most likely cause of genital herpes?
3. Whitlow is caused by a herpesvirus and is characterized by inflamed blisters on which body part?
4. What is the best explanation for the fact that chickenpox is more severe in adults than in children?
5. What is the most common location of the rash associated with shingles (herpes zoster)?

Unlike the multiple recurrences of herpes simplex lesions, shingles only occurs once or twice in an individual's life, usually after age 45. Before vaccination became routine, the majority of children developed chickenpox, but only a small percentage of adults developed shingles, presumably because most individuals develop complete immunity to VZV following chickenpox and therefore never acquire shingles. The observation that VZV from the lesions of shingles can cause chickenpox in individuals who had not had the disease revealed that VZV causes both shingles and chickenpox.

Diagnosis, Treatment, and Prevention

Chickenpox is usually diagnosed from the characteristic appearance of the lesions. In contrast, it is sometimes difficult to distinguish shingles from other herpesvirus lesions, though the localization within a dermatome is characteristic. Antibody tests are available to verify diagnosis of VZV infection.

Uncomplicated chickenpox is typically self-limiting and requires no treatment other than relief of the symptoms with acetaminophen and antihistamines. Aspirin should not be given to children or adolescents with symptoms of chickenpox because of the risk of contracting Reye's syndrome, a condition associated with several viral diseases and aspirin usage.

Treatment of shingles involves management of the symptoms and bed rest. Nonadherent dressings and loose-fitting clothing may help prevent irritation of the lesions. Acyclovir or nucleotide analogs derived from it provide relief from the painful rash for some patients, but they are not a cure.

It is difficult to prevent exposure to VZV because viruses are shed from patients before obvious signs appear. The U.S. Food and Drug Administration (FDA) has approved an attenuated vaccine, though the vaccine has caused some cases of shingles in Japan, and immunocompromised individuals are at risk of developing disease from the vaccine. Nevertheless, the benefits of vaccination appear to outweigh the risks, and the U.S. Centers for Disease Control and Prevention (CDC) recommends two doses of chickenpox vaccine for all children who are least one year old and for adults up to age 50 who do not have evidence

Human Herpesvirus 4 (Epstein-Barr Virus) Infections

LEARNING OUTCOME

24.7 Describe the diseases associated with Epstein-Barr virus.

A fourth human herpesvirus is *Epstein-Barr virus (EBV)*, which is in the genus *Lymphocryptovirus*. EBV is named after its codiscoverers. The official name for the species is *human herpesvirus 4*, or HHV-4.

Diseases Associated with HHV-4 (EBV) Infection

Like other herpesviruses, HHV-4 (EBV) is associated with a variety of diseases. First we consider two diseases EBV is known to cause—Burkitt's lymphoma and infectious mononucleosis—before discussing several other diseases for which EBV is implicated as a cause. As you read about the diseases with which EBV is associated, bear in mind that they are in part the result of the patient's immune status, which itself can be related to the patient's age.

In the 1950s, Denis Burkitt (1911–1993) described a malignant neoplasm of the jaw, primarily in African boys **(FIGURE 24.12)**, that appeared to be infectious. This cancer is now known as **Burkitt's lymphoma.** Intrigued by the idea of an infectious cancer, Michael Epstein (1921–) and Yvonne Barr (1932–2016) cultured samples of the tumors and isolated Epstein-Barr virus, the first virus shown to be responsible for a human cancer. (A way in which a latent virus such as EBV can trigger cancer was discussed in Chapter 13.)

Epstein-Barr virus also causes **infectious mononucleosis** (mon'ō-noo-klē-ō'sis; **mono**). Discovery of the connection between EBV and mono was accidental. A laboratory technician contracted mononucleosis while on vacation. Upon her return to work, her serum was found to contain antibodies against EBV antigens. Serological studies of large numbers of college students have provided statistical verification that EBV causes mono.

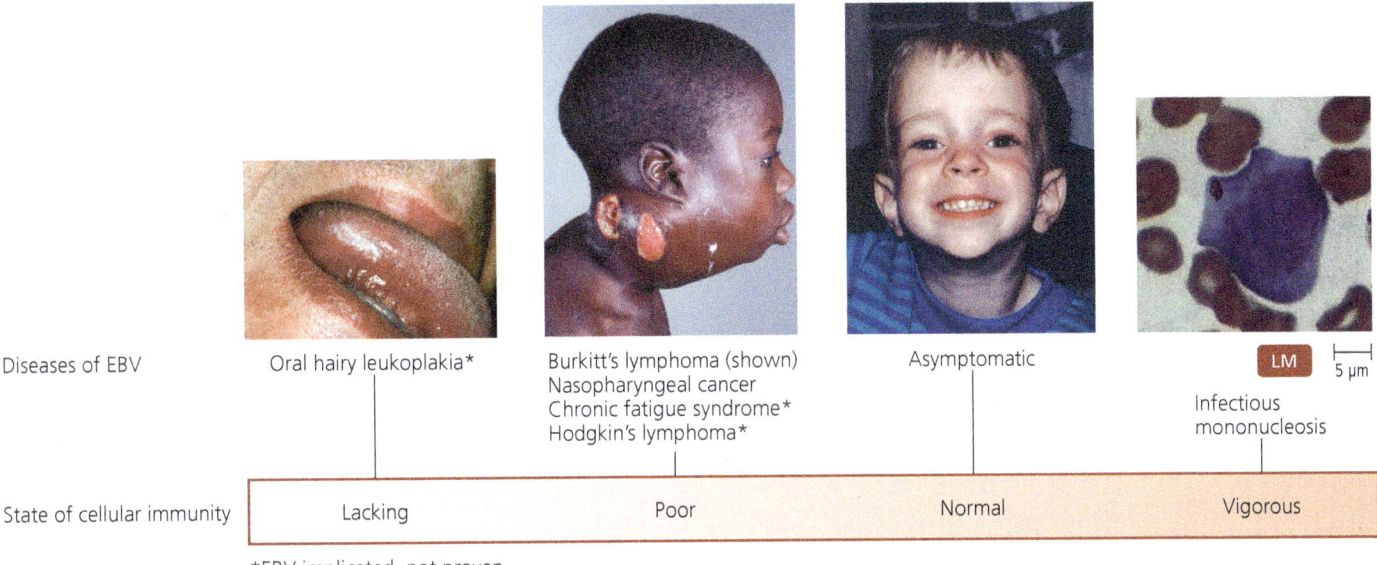

Diseases of EBV	Oral hairy leukoplakia*	Burkitt's lymphoma (shown) Nasopharyngeal cancer Chronic fatigue syndrome* Hodgkin's lymphoma*	Asymptomatic	Infectious mononucleosis
State of cellular immunity	Lacking	Poor	Normal	Vigorous

*EBV implicated, not proven

▲ FIGURE 24.12 **Diseases associated with Epstein-Barr virus.** Which disease results from infection with EBV appears to depend on the relative vigor of the host's cellular immune response, which itself can be related to age. Oral hairy leukoplakia, a precancerous change in the mucous membrane of the tongue, occurs in EBV-infected hosts with severely depressed cellular immunity (e.g., AIDS patients). Burkitt's lymphoma, a cancer of the jaw, occurs primarily in African boys whose immune systems have been suppressed by the malaria parasite. Nasopharyngeal cancer is most common in China. EBV is implicated in Hodgkin's lymphoma and chronic fatigue syndrome, which also are associated with impaired immune function. Asymptomatic EBV infections are typical for children exposed at a young age, before the immune response has become vigorous. Infectious mononucleosis, which is characterized by enlarged B lymphocytes with lobed nuclei and a high number of cytotoxic T (CD8+) cells, results from infection with EBV during adulthood, when the cellular immune response is vigorous.

EBV is an etiological agent of *nasopharyngeal cancer* in patients from southern China and is implicated as a cause of *chronic fatigue syndrome, oral hairy leukoplakia* in AIDS patients, and *Hodgkin's lymphoma.* Chronic fatigue syndrome is associated with debilitating fatigue, memory loss, and enlarged lymph nodes. Hairy leukoplakia is a precancerous change in the mucous membrane of the tongue. Hodgkin's lymphoma is a treatable cancer of B lymphocytes that spreads from one lymph node to another.

CLINICAL CASE STUDY

Grandfather's Shingles

The Wilsons were excited about their newborn twin boys and couldn't wait to take them to see Mr. Wilson's father. Grandfather Wilson was excited to see his first grandsons as well and thought their visit might help take his mind off the pain of his shingles, which had suddenly appeared only days before and had not begun to dry and crust over.

1. What virus is responsible for Grandfather Wilson's shingles?
2. What likely triggered his case of shingles?
3. Are the new parents at risk of catching shingles from Grandfather Wilson?
4. Are the newborns at risk of contracting shingles from their grandfather? If Mrs. Wilson asked your advice about visiting her father-in-law with the twins, what would you recommend?
5. Could the twins be vaccinated against shingles before the visit?

Epidemiology and Pathogenesis

Transmission of Epstein-Barr viruses usually occurs via saliva, often during the sharing of drinking glasses or while kissing. ("Kissing disease" is a common name for infectious mononucleosis.) After initially infecting the epithelial cells of the pharynx and parotid salivary glands, EBV virions enter the blood. In general, having viruses in the blood is called **viremia** (vī-rē′mē-ă). EBV invades B lymphocytes,[7] becomes latent, and immortalizes the B cells by suppressing *apoptosis* (programmed cell death). Infected B cells are one source of lymphomas, including Burkitt's and Hodgkin's lymphomas.

Infectious mononucleosis results from a "civil war" between the antibody and cellular branches of specific immunity: Cytotoxic T lymphocytes of the cell-mediated branch kill infected B lymphocytes. This "civil war" is responsible for the symptoms and signs of mono, including sore throat, fever, enlargement of the spleen and liver, and fatigue.

Seventy percent of Americans over the age of 30 have antibodies against EBV. As with other herpesviruses, the age of the host at the time of infection is a determining factor in the seriousness of the disease, in this case because the competence of cytotoxic T cells is in part related to age (see Figure 24.12). Infection during childhood, which is more likely to occur in countries with poor sanitation and inadequate standards of hygiene, is usually asymptomatic because a child's cellular immune system is immature and cannot cause severe tissue damage. Where living standards are higher, childhood infection is less likely, and the postponement of infection until adolescence or later results in a more vigorous cellular immune response that produces the signs and symptoms of mononucleosis.

Cofactors appear to play a role in the development of cancers by EBV. For example, Burkitt's lymphoma is almost exclusively limited to young males exposed to malaria parasites, and the restriction of EBV-induced nasopharyngeal cancer to certain geographical regions is suggestive of cofactors in food, environment, or the genes of regional populations.

Extreme disease, such as oral hairy leukoplakia, arises in individuals with a T cell deficiency, as occurs in malnourished children, the elderly, AIDS patients, and transplant recipients. These individuals are more susceptible to EBV because infected cells are not removed by cytotoxic T lymphocytes and therefore remain a site of virus proliferation.

Diagnosis, Treatment, and Prevention

Large, lobed, B lymphocytes with atypical nuclei and a lower than normal number of neutrophils in the blood (neutropenia) are characteristic of EBV infection. Some diseases associated with EBV, such as hairy leukoplakia and Burkitt's lymphoma, are easily diagnosed by their characteristic signs. Other EBV infections, such as infectious mononucleosis, have symptoms common to many pathogens. Fluorescent antibodies directed against anti-EBV immunoglobulins or ELISA tests provide specific diagnosis.

Burkitt's lymphoma responds well to chemotherapy, and tumors can be removed from affected jaws of patients in geographic areas where surgery is available. Care of mono patients involves relief of the symptoms. Radiation or chemotherapy often successfully treats Hodgkin's lymphoma. There is no effective treatment for other EBV-induced conditions.

Prevention of EBV infection is almost impossible because the virus is widespread, latent, and transmitted readily in saliva. However, only a small proportion of EBV infections result in disease, often because the immune system is too immature to cause cellular damage (in children).

Human Herpesvirus 5 (Cytomegalovirus) Infections

LEARNING OUTCOME

24.8 Describe the epidemiology of *Cytomegalovirus* infection.

A fifth herpesvirus that affects humans is *Cytomegalovirus* (sī-tō-meg′ă-lō-vī′rŭs; CMV), so called because cells infected with this virus become enlarged. *Cytomegalovirus* infection is one of the more common infections of humans. Studies have shown that 50% of the adult population of the United States is infected with CMV; in some other countries, 100% of the population tests positive for antibodies against CMV. Like other herpesviruses we have studied, HHV-5 becomes latent in various cells, and infection by CMV lasts for life.

Epidemiology and Pathogenesis

CMV is transmitted in bodily secretions, including saliva, mucus, milk, urine, feces, semen, and cervical secretions. Individual viruses are not highly contagious, so transmission requires intimate contact involving a large exchange of secretion. Transmission usually occurs via sexual intercourse but can result from *in utero* exposure, vaginal birth, blood transfusions, and organ transplants. CMV infects 7.5% of all neonates, making it the most prevalent viral infection in this age-group.

Most people infected with CMV are asymptomatic, but fetuses, newborns, and immunodeficient patients are susceptible to severe complications of CMV infection. About 10% of congenitally infected newborns develop signs of infection, including enlarged liver and spleen, jaundice, microcephaly, and anemia. CMV may also be **teratogenic**[8] (ter′ă-tō-jen′ik)—that is, it may cause birth defects—when the virus infects *stem cells*[9] in an embryo or fetus. In the worst cases, intellectual disability, hearing and visual damage, or death may result.

AIDS patients and other immunosuppressed adults may develop pneumonia, blindness (if the virus targets the retina), or *cytomegalovirus mononucleosis*, which is similar to infectious mononucleosis caused by Epstein-Barr virus. CMV diseases result from initial infections or from latent viruses.

Diagnosis, Treatment, and Prevention

Diagnosis of CMV-induced diseases is dependent on laboratory procedures that reveal the presence of abnormally enlarged cells

---

[7]These lymphocytes are also known as mononuclear leukocytes, from which the name mononucleosis is derived.

[8]From Greek *teratos*, meaning "monster."

[9]Formative cells, the daughter cells of which develop into several different types of mature cells.

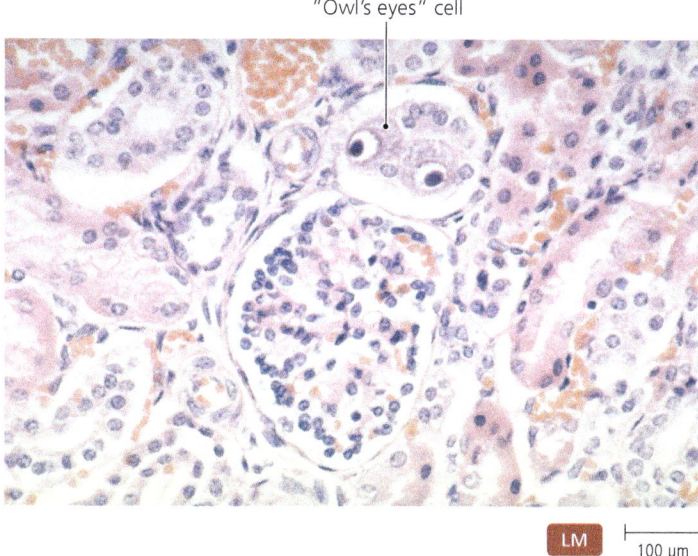

▲ FIGURE 24.13 **An owl's eyes cell.** The cell is diagnostic for *Cytomegalovirus* infection. Such abnormally enlarged cells have enlarged nuclei and contain inclusion bodies that are sites of viral assembly.

and inclusions within the nuclei of infected cells (FIGURE 24.13). Viruses and antibodies against them can be detected by ELISA tests and DNA probes. Treatment for fetuses and newborns with complications of CMV infection is difficult because in most cases damage is done before an infection is discovered.

Treatment of adults can also be frustrating. Interferon and gamma globulin treatment can slow the release of CMV from adults but does not affect the course of disease.

Abstinence, mutual monogamy, and use of condoms can reduce the chances of infection. Organs harvested for transplantation can be made safer by treatment with monoclonal antibodies against CMV, which passively reduces the CMV load before the organs are transplanted.

Other Herpesvirus Infections

LEARNING OUTCOME

24.9 Describe the diseases that are associated with HHV-6 and HHV-8.

Human herpesvirus 6 (HHV-6), in the genus *Roseolovirus*, causes **roseola** (rō-zē'ō-lă). Roseola is a nearly universal endemic illness of children characterized by an abrupt fever, sore throat, enlarged lymph nodes, and a faint pink rash on the face, neck, trunk, and thighs (FIGURE 24.14). The name of the disease is derived from the rose-colored rash.

Some researchers have linked HHV-6 to multiple sclerosis (MS). These investigators point to the fact that over 75% of MS patients have antibodies against the virus, that patients with higher antibody levels are more likely to develop MS later in life, and that the brain lesions of MS patients typically contain HHV-6. Further, MS patients respond well to treatment with interferons, which are natural antiviral compounds. This evidence is not proof of causation, but it suggests tantalizing possibilities for further research.

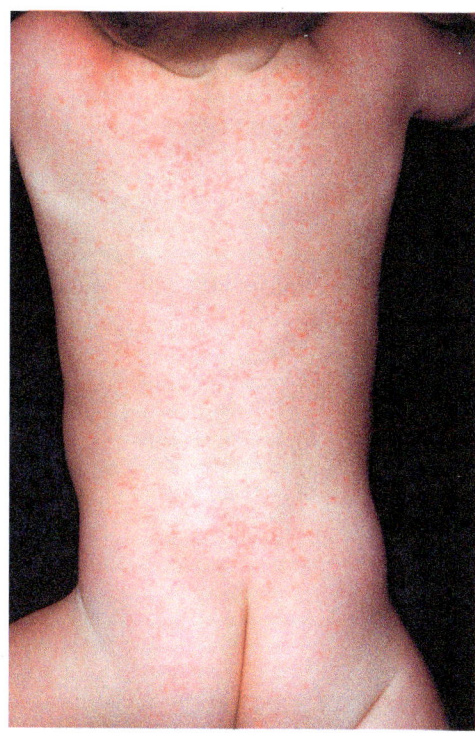

▲ FIGURE 24.14 **Roseola.** This rose-colored rash on the body results from infection with *human herpesvirus 6*.

HHV-6 may also cause mononucleosis-like symptoms and signs, including enlargement of lymph nodes. There is some evidence that *human herpesvirus 6* may also be a cofactor in the pathogenesis of AIDS; that is, infection with HHV-6 may make individuals more susceptible to AIDS.

Human herpesvirus 8 (HHV-8, *Rhadinovirus*) causes Kaposi's sarcoma, a cancer often seen in AIDS patients (FIGURE 24.15). The virus is not found in cancer-free patients or in normal tissues of victims and has been shown to make cells lining blood vessels cancerous.

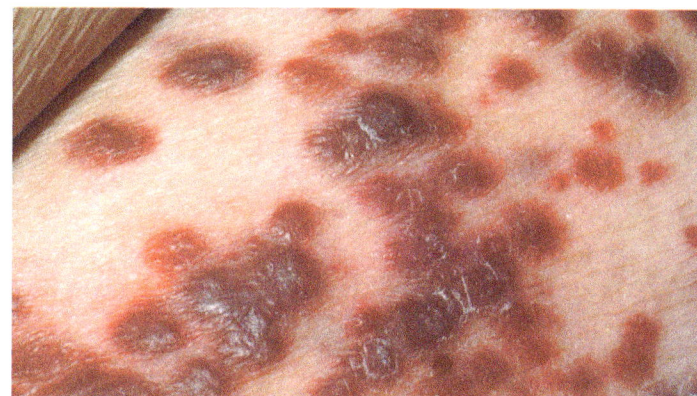

▲ FIGURE 24.15 **Kaposi's sarcoma.** A rare and malignant neoplasia of blood and blood vessels, this cancer is most often seen in AIDS patients. *What observations support the hypothesis that Kaposi's sarcoma is caused by HHV-8?*

Figure 24.15 *HHV-8 is not found in healthy patients, but it is found in Kaposi tumor cells. The virus also causes some cells to become cancerous.*

> **MICRO CHECK**
>
> 6. What disease does EBV (Epstein-Barr virus, or *human herpesvirus 4*) cause in patients with a vigorous cellular immune response?
> 7. Infection with Epstein-Barr virus causes neutropenia. What is neutropenia?
> 8. For which disease are "owl's eyes" diagnostic, when observed through microscopic examination of infected tissue?

> **TELL ME WHY**
>
> Whereas many doctors are convinced that Epstein-Barr virus causes chronic fatigue syndrome, others deny the association between EBV and the syndrome. Why is the etiology of chronic fatigue syndrome debated even though Epstein-Barr virus is present?

Papillomaviridae and *Polyomaviridae*

Papillomaviruses and polyomaviruses each have a single molecule of double-stranded DNA contained in a small, naked, icosahedral capsid. They were previously classified in a single family, "Papovaviridae," but have recently been separated into two families, *Papillomaviridae* and *Polyomaviridae*, based in part on the different diseases they cause.

Papillomavirus Infections

LEARNING OUTCOMES

24.10 Describe four kinds of warts associated with papillomavirus infections.
24.11 Describe the pathogenesis, treatment, and prevention of genital warts.

Papillomaviruses cause **papillomas** (pap-i-lō'măz), which are usually benign growths of the epithelium of the skin or mucous membranes commonly known as **warts**. Disease in Depth: **Papillomas** (p. 723) examines papillomaviruses and their diseases in detail. **Clinical Case Study: A Child with Warts** (p. 724) concerns a young patient with papillomas.

Polyomavirus Infections

LEARNING OUTCOME

24.12 Describe two diseases caused by polyomavirus infection in humans.

The search for cancer-causing viruses led scientists to the discovery of a group of viruses capable of causing several different tumors in animals and humans. Researchers named these viruses **polyomaviruses** (genus *Polyomavirus*) by combining *poly* (meaning "many") and *oma* (meaning "tumor").

Two human polyomaviruses, the *BK* and *JC* viruses, are endemic worldwide. The initials are derived from the names of the patients from which the viruses were first isolated. Most everyone appears to be infected with BK and JC by age 15, probably via the respiratory route. The viruses initially infect lymphocytes, and what happens subsequently depends at least in part on the state of an individual's immune system. If the immune response is normal, latent infections are typically prevented; if the immune system is compromised, however, then latent infections can become established in the kidneys, where reactivation of the viruses leads to the release of new virions. In the case of BK virus, the new virions can cause potentially severe urinary tract infections.

In individuals with latent infections of JC virus, reactivated virions are shed from kidney cells into the blood, spreading the virions throughout the body. Virions that reach the brain of an immunosuppressed individual cause a rare and fatal disease called **progressive multifocal leukoencephalopathy (PML)**, in which the viruses infect and kill cells called oligodendrocytes in the white matter of the central nervous system. Oligodendrocytes produce myelin, a white lipid that surrounds and insulates some nerve cells. As myelin in multiple sites is lost, PML patients progressively lose various brain functions, as the name of the disease implies, because the conduction of nerve impulses becomes increasingly impaired. Vision, speech, coordination, and cognitive skills are affected. Eventually paralysis and death result. Physicians diagnose PML based on the observation of symptoms.

Beta interferon treatments can prevent kidney damage by BK. The receptor for JC virus is a serotonin receptor, so researchers hope that serotonin receptor inhibitors will block JC virus. Unfortunately, by the time a diagnosis of JC infection can be made, damage to the brain is often severe and irreversible.

> **TELL ME WHY**
>
> Why are double-stranded DNA viruses, such as papillomaviruses, more likely to be associated with cancers than are RNA viruses?

Adenoviridae

LEARNING OUTCOME

24.13 Discuss the epidemiology and pathogenicity of diseases caused by adenoviruses.

The last family of double-stranded DNA viruses we discuss is *Adenoviridae*. Adenoviruses have a single, linear dsDNA genome contained in a naked polyhedral capsid with spikes **(FIGURE 24.16)**. Adenoviruses are so named because they were first discovered infecting cultures of human adenoid[10] cells.

Adenoviruses are but one of the many causative agents of "common colds." Even though colds typically have a common set of signs and symptoms, their causes are quite varied. At least

---
[10]Adenoids are swollen pharyngeal tonsils, which are masses of lymphatic tissue located in the pharynx, posterior to the nose.

DISEASE IN DEPTH

PAPILLOMAS

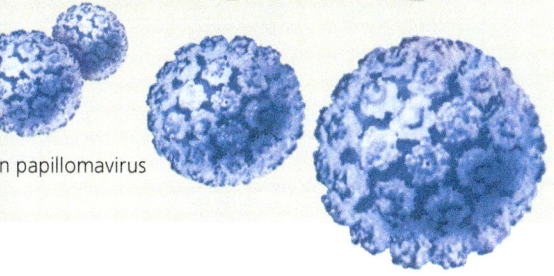

Human papillomavirus

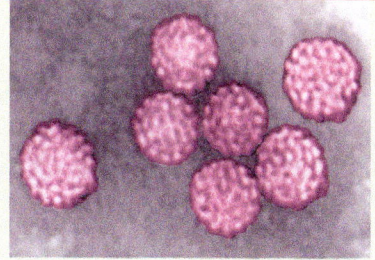

Papillomaviruses shed from a wart
TEM   60 nm

There are more than 170 strains of human papillomaviruses (HPV, family *Papillomaviridae*), each having a single molecule of double-stranded DNA contained in a naked icosahedral capsid. Half of HPVs can permanently integrate into human chromosomes, where they can trigger genital, anal, and oropharyngeal cancers, particularly when they infect a patient who is also infected with herpesvirus.

Papillomaviruses cause papillomas, usually benign growths of the epithelium of the skin or mucous membranes commonly known as warts. Papillomas are most often found on fingers or toes *(seed warts)*; on soles of feet *(plantar warts)*; on the trunk, face, elbows, or knees *(flat warts)*; or on the anus or external genitalia, most extremely as condylomata acuminata. Certain strains of papillomaviruses can also trigger cancers, such as those of genitalia.

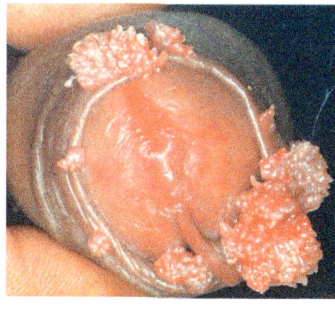

Condylomata acuminata

PATHOGENESIS

1 Papillomavirus is transmitted via direct contact or from contaminated fomites.

2 Virus multiplies within cells.

3 Virus invades adjacent cells.

4 Rapid growth of virus-infected cells creates wart.

5 Virus is shed when dead skin cells are sloughed off.

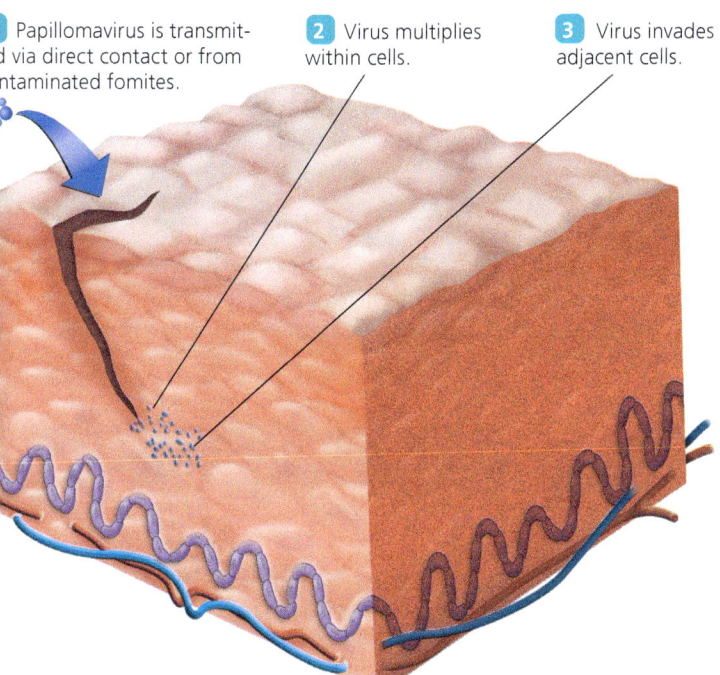

Papillomaviruses shedding from wart
TEM   600 nm

DIAGNOSIS, TREATMENT, & PREVENTION

Watch Dr. Bauman's Video Tutor explore papillomas, or go to the **Mastering Microbiology** Study Area for more information.

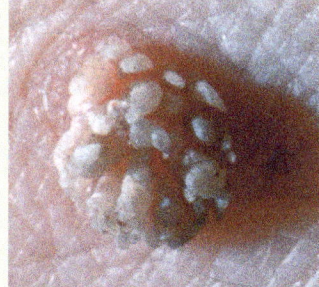

Diagnosis of warts is usually a simple matter of observation, though only DNA probes can elucidate the exact strain of *Papillomavirus* involved. Warts usually regress over time as a cell-mediated immune response recognizes and attacks virally infected cells. Cosmetic concerns and pain associated with some warts may necessitate removing infected tissue via surgery, freezing, cauterization (burning), laser, or the use of caustic chemicals. Despite treatment, viruses may remain latent in neighboring tissue and produce new warts at a later time. Prevention is difficult, though genital warts may be prevented by abstinence, mutual monogamy, or a vaccine against the more common papillomaviral strains associated with genital cancers. Use of condoms can reduce the sexual transmission of papillomaviruses.

▲ FIGURE 24.16 **Adenovirus.** Notice the distinctive spikes.

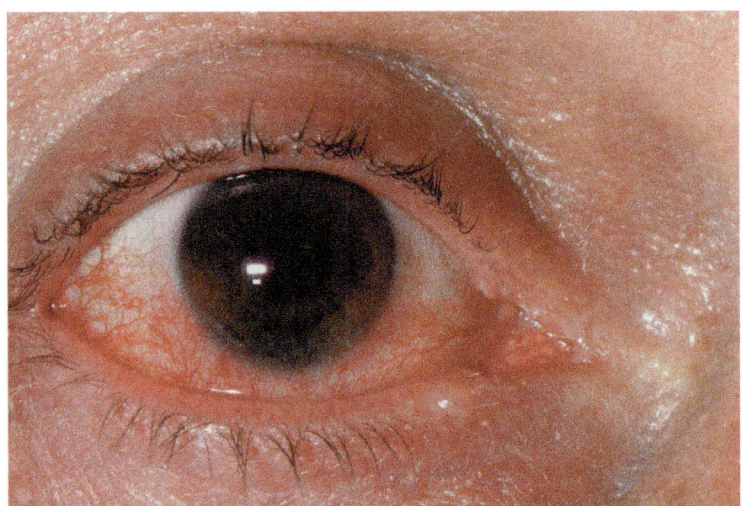

▲ FIGURE 24.17 **Adenoviral conjunctivitis (pinkeye).** *What signs and symptoms can result from infections with various adenoviruses?*

Figure 24.17 Respiratory adenoviruses cause sore throat, cough, fever, headache, and malaise; intestinal adenoviruses cause mild diarrhea in children; still other adenoviruses cause conjunctivitis (pinkeye).

30 different respiratory adenoviruses (DNA viruses) and over 150 RNA viruses cause colds.

Adenoviruses, which are spread via respiratory droplets and often enter the body through the eyes, are very stable outside the body and can survive on fomites and in improperly chlorinated drinking water. Respiratory adenovirus infections become established when adenoviruses are taken into cells lining the respiratory tract via endocytosis. The resulting infections have a variety of manifestations, including sneezing, sore throat, cough, headache, and malaise. (Table 25.1 on p. 736 compares the signs and symptoms of colds with those of other respiratory infections.) For some reason, epidemics of respiratory adenoviruses occur on military bases but rarely occur under the similar conditions found in college dormitories.

Adenoviruses can also cause infections in other systems of the body. For example, after entering cells lining the intestinal tract, intestinal adenoviruses can cause diarrhea in children. Still other adenoviruses infect the conjunctiva, resulting in conjunctivitis, which is also called *pinkeye* (FIGURE 24.17). Researchers have shown that human adenovirus 36 (Adv36) began infecting humans around 1980—the same year that the World Health Organization recognizes the beginning of a worldwide pandemic of obesity in both developed and developing countries. Is

CLINICAL CASE STUDY

A Child with Warts

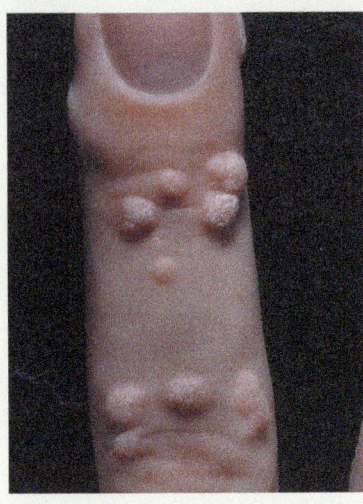

Ten-year-old Rudy has several large warts on the fingers of his right hand. They do not hurt, but their unsightly appearance causes him to shy away from people. He is afraid to shake hands, or to play with other children out of fear that he may transfer the warts to them. Initially, his mother tells him not to worry about them, but Rudy cannot help feeling self-conscious. Furthermore, Rudy fears that the warts may somehow spread on his own body. After consulting with a physician, Rudy and his mother decide to have the warts surgically removed.

1. Is it possible that Rudy's warts will spread to other parts of his body?
2. Is it possible for someone else to "catch" Rudy's warts by shaking his hand?
3. If Rudy's warts disappear without treatment, are they likely to return to the same sites on his hand?
4. Are Rudy's warts likely to become cancerous?
5. Following surgical removal of the warts, is there a chance that Rudy will develop warts again?

Adv36 a cause of human obesity? Some research suggests that it may be.

Adv36 directly affects adipocytes (fat cells) in both animals and humans: Infected monkeys gain four times as much weight and 60% more body fat than uninfected monkeys on the same diet and exercise regimen. In identical human twins where one sibling is infected with Adv36 and the other is not, the infected twin is heavier than the uninfected co-twin. At the cellular level, human stem cells infected with Adv36 more frequently become adipocytes than do uninfected stem cells, and infected adipocytes make and store more fat than uninfected cells. Perhaps someday there will be a vaccine against this potential cause of obesity.

Adenoviruses often produce semi-crystallized viral masses within the nuclei of infected cells (see **Microbe at a Glance: Adenovirus**), a feature that is diagnostic for adenoviral infections. Although this feature is similar in appearance to the nuclear inclusions seen in *Cytomegalovirus* infections, adenoviruses do not cause infected cells to enlarge.

Adenovirus infection can be treated with cidofovir, and a live, attenuated vaccine is available. Currently in the United States, the vaccine is available only to select groups, such as members of the military.

TELL ME WHY

Why isn't there a vaccine for the "common cold?"

Hepadnaviridae

LEARNING OUTCOME

24.14 Describe the unique feature of hepadnaviruses.

Viruses of the family *Hepadnaviridae*[11] are enveloped DNA viruses with icosahedral capsids that invade and replicate in liver cells. From these characteristics, human **hepatitis B virus (HBV)** acquires the name of its genus, *Orthohepadnavirus*. Its scientific species name, *hepatitis B virus*, is also its common name.

The genome of a hepadnavirus is unique in that it is partly double-stranded DNA and partly single-stranded DNA **(FIGURE 24.18)**. The amount of ssDNA in a hepadnaviral genome varies. Some virions are almost entirely dsDNA, whereas others have a significant amount of ssDNA. How does this variation in genomic structure occur?

HBV replicates through an RNA intermediary, a phenomenon that is nearly unique among DNA viruses. During replication, one strand of the DNA genome of hepatitis B virus is transcribed into RNA. This RNA molecule then serves as a template for a new DNA strand. Since such RNA to DNA synthesis is the opposite of regular transcription (which is DNA to RNA), the responsible enzyme is called **reverse transcriptase**. The result of this first step in reverse transcription is an RNA–DNA hybrid molecule.

---

[11]From Greek *hepar*, meaning "liver," and "DNA."

MICROBE AT A GLANCE 24.2

Adenovirus

Taxonomy: Family *Adenoviridae*

Genome and morphology: Single molecule of double-stranded DNA; naked icosahedral capsid, 60–90 nm with spikes

Host: Various species infect mammals, birds, frogs, or fungi

Virulence factors: Infection through aerosol, ingestion, hand-to-eye transfer, or sexual contact

Diseases caused: Immunocompetent patients: Respiratory illness, such as common cold, mild diarrhea, conjunctivitis, keratoconjunctivitis
Immunocompromised: Pneumonia, gastroenteritis, hepatitis, hemorrhagic cystitis, interstitial nephritis, meningoencephalitis

Treatment for disease: Cidofovir

Prevention of disease: Get attenuated vaccine against Adv4 and 7; avoid contact with infected people or contaminated food, beverages, or environmental surfaces, such as towels

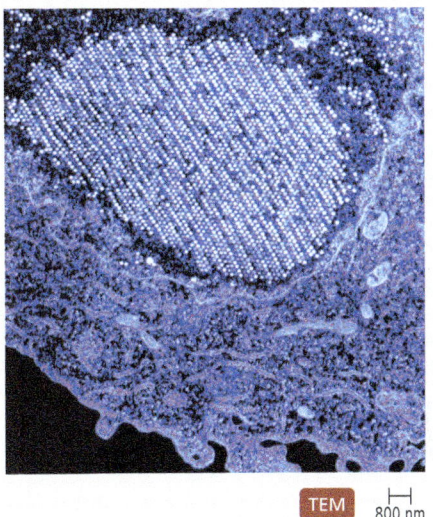

TEM 800 nm

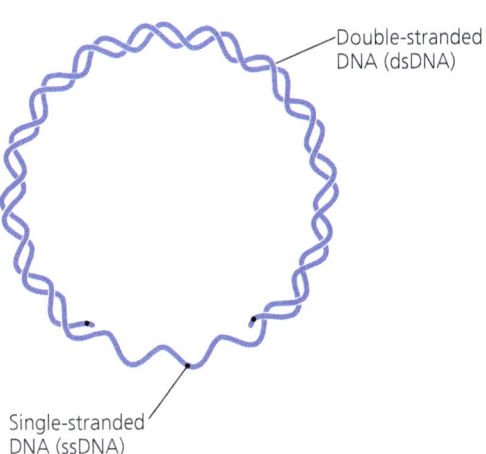

▲ **FIGURE 24.18 The genome of a hepadnavirus.** This unusual genome is composed of partly single-stranded and partly double-stranded DNA resulting from incomplete DNA replication by reverse transcriptase.

Reverse transcriptase then continues with the removal of the RNA strand from the hybrid, followed by synthesis of a DNA strand complementary to the first, thereby forming dsDNA. (Figure 25.18 in Chapter 25 illustrates the process of reverse transcription as it occurs in HIV replication.) A partly double-stranded DNA and partly single-stranded DNA molecule, which is a distinctive feature of hepatitis B viruses, results when virions are assembled before the complementary DNA strand is completely copied. Thus, the earlier that assembly occurs, the more ssDNA will remain in the viral genome; the later that assembly occurs, the more complete DNA replication will be, and the more dsDNA will be in the hepatitis B viral genome.

Hepatitis B Infections

LEARNING OUTCOME

24.15 Describe the epidemiology, treatment, and prevention of hepatitis B infections.

Hepatitis (hep-ă-tī′tis) is an inflammatory condition of the liver. The liver has many functions, including making blood-clotting factors, storing sugar and other nutrients, assisting in the digestion of lipids, and removing wastes from the blood. When a patient's liver is severely damaged by viral infection, all these functions are disturbed. **Jaundice** (jawn′dis), a yellowing of the skin and eyes, occurs when a greenish-yellow waste product called *bilirubin* accumulates in the blood **(FIGURE 24.19)**. Hepatitis is also characterized by enlargement of the liver, abdominal distress, and bleeding into the skin and internal organs. The patient becomes stuporous and eventually goes into a coma as a result of the accumulation of nitrogenous wastes in the blood.

Hepatitis B virus is the only DNA virus that causes hepatitis; all other cases of viral hepatitis result from RNA viruses (see Table 25.3 on p. 739). Hepatitis B is also called *serum hepatitis*. HBV infection results in serious liver damage in less than 10% of cases, but **coinfection** (simultaneous infection) with hepatitis D virus (an RNA virus also called *delta agent*) increases the risk of permanent liver damage.

Epidemiology and Pathogenesis

Hepatitis B viruses replicate in liver cells and are released by exocytosis rather than cell lysis. Infected liver cells thus serve as a source for the continual release of virions into the blood, resulting in billions of virions per milliliter of blood. Virions in the blood are shed into saliva, semen, and vaginal secretions.

HBV is transmitted when infected bodily fluids, particularly blood, come into contact with breaks in the skin or mucous membranes. The viral load is typically so high and the infective dose so low that infection can result from sharing a razor or toothbrush, making it difficult to determine the source of infection. Improperly sterilized needles used in acupuncture, ear piercing, tattooing, intravenous drug usage, and blood transfusion have spread the virions, which can remain infective for at least one week outside the body. They may also be spread by sexual intercourse, especially anal intercourse, which can cause tearing of the rectal lining. Infected mothers can pass HBV to their babies during childbirth.

Many infected individuals are asymptomatic, and most manifest only mild jaundice, low-grade fever, and malaise. Severe liver damage is most likely to result from such asymptomatic, chronic hepatitis B infections after 20 to 40 years. The CDC estimates that approximately 350 million people worldwide have chronic hepatitis B infections. The carrier state is age related: newborns are much more likely to remain chronically infected than are individuals infected as adults.

Today, fewer cases of hepatitis B are seen in the United States as a result of immunization programs and safer sexual practices **(FIGURE 24.20)**. Still, approximately 3000 people die each year from hepatitis B infection.

Diagnosis, Treatment, and Prevention

Laboratory diagnosis of hepatitis B infection involves the use of labeled antibodies to detect the presence of viral antigens released from HBV-infected cells. The bodily fluids of infected individuals contain copious amounts of viral protein of three forms: *Dane particles, spherical particles,* and *filamentous particles* **(FIGURE 24.21)**. Dane[12] particles are complete and infective virions, whereas spherical and filamentous particles are merely

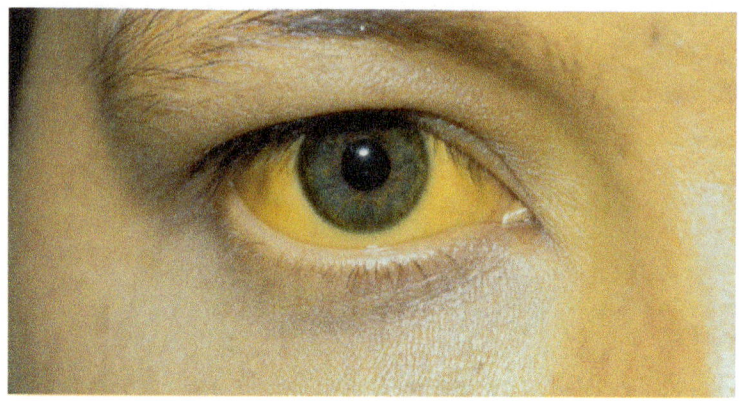

▲ FIGURE 24.19 **Jaundice.** This condition is characterized by yellow skin and eyes. One cause of jaundice is liver damage by hepatitis B virus.

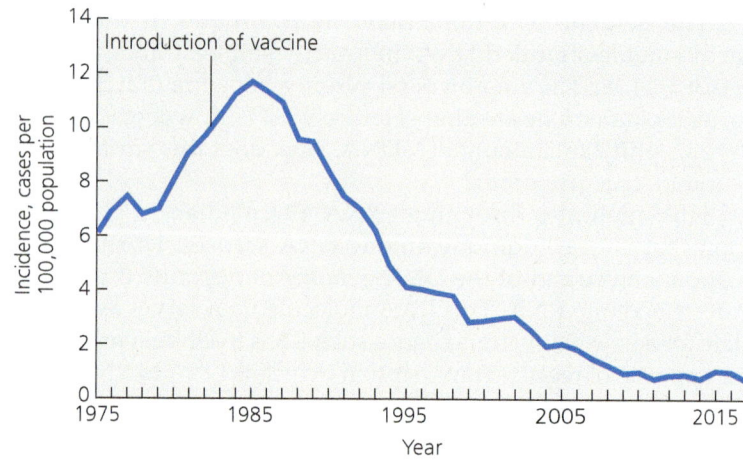

▲ FIGURE 24.20 Estimated incidence of acute hepatitis B in the United States.

---

[12]Named for D. S. Dane, who discovered HBV using electron microscopy.

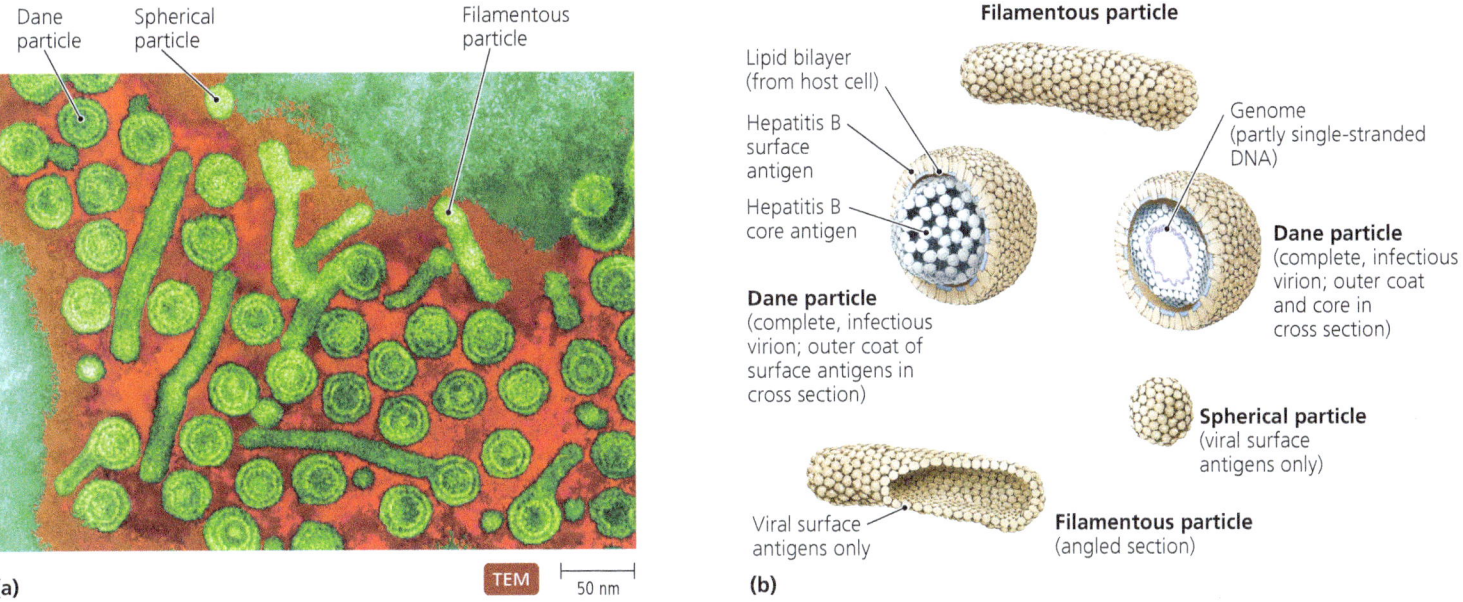

▲ **FIGURE 24.21 The three types of viral particles produced by hepatitis B viruses.** Dane bodies are complete virions, whereas filamentous particles and spherical particles are capsomeres that have assembled without genomes. (a) Micrograph. (b) Artist's rendition.

capsids composed of surface antigens without genomes. Within infected individuals, the release of so much excess viral surface antigen in the form of spherical and filamentous particles in effect serves as a decoy: The binding of antibody to empty capsids reduces the effective antibody response against infective Dane particles. The large amount of viral antigen released provides one benefit to a patient: It ensures a plentiful substrate for binding labeled antibodies in diagnostic tests.

There is no universally effective treatment for hepatitis B, though alpha interferon, entecavir, or tenofovir help in 40% of cases. Liver transplant is the only treatment for end-stage chronic hepatitis B.

Prevention is possible because an effective vaccine is available. Three doses of HBV vaccine result in protection against the virus in 95% of individuals. Several studies indicate that immunity against HBV lasts for at least 20 years and probably for life. The CDC recommends HBV immunization for all children because children are at a greater risk of chronic infection, life-threatening liver damage, and (as we will see shortly) liver cancer. Immunization is also recommended for high-risk groups,

CLINICAL CASE STUDY

The Eyes Have It

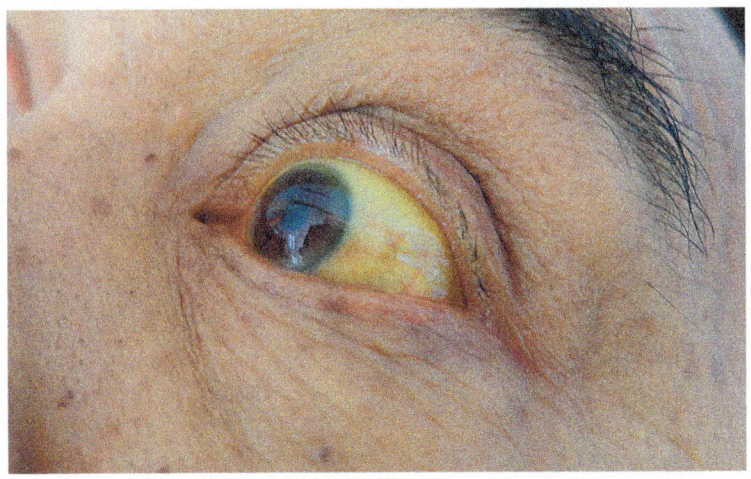

Duyen presents herself to her campus health clinic at a liberal arts college in Oregon with a two-week history of nausea, low-grade fever, fatigue, and mild pain in her right upper abdominal quadrant. As a resident assistant in her dorm, she is aware that students with similar signs and symptoms have been diagnosed with flu for the past month. She greets the health care staff with a sheepish grin and tells them her self-diagnosis. The doctor takes one look at Duyen's yellow-tinged eyes and skin, and realizes that this is not another case of influenza; it's something far worse.

1. What is the name of the condition that alarms the doctor?
2. What viral disease might Duyen have?
3. How might Duyen have been infected?
4. Which DNA virus causes this disease?

including health care workers, people who share needless to inject drugs, people receiving blood or blood products, and people who are sexually active with multiple partners, especially those who engage in anal intercourse.

To prevent the spread of HBV, special care should be taken with needles and other sharp instruments in health care settings. Unlike most enveloped viruses, HBV is resistant to detergents, ether, freezing, acid, and moderate heat, but contaminated materials can be disinfected with 10% bleach solutions. Hepatitis B virus can transfer from mother to child, but immunization of the baby within 12 hours of birth usually prevents disease in the child. Abstinence or mutual monogamy is the only sure way to prevent sexually transmitted infection.

The Role of Hepatitis B Virus in Hepatic Cancer

LEARNING OUTCOMES

24.16 Describe evidence that hepatitis B virus causes hepatic cancer.

HBV has been shown to be associated with hepatic cancer based on the following evidence:

- Epidemiological studies reveal large numbers of hepatic cancer cases in geographic areas that also have a high prevalence of HBV.
- The HBV genome has been found integrated into hepatic cancer cells.
- Hepatic cancer cells typically express HBV antigen.
- Chronic carriers of HBV are 200 times more likely to develop hepatic cancer than are noncarriers.

How does HBV produce cancer? Perhaps the gap in the double strand of DNA allows HBV to easily integrate into the DNA of a liver cell, in the process possibly activating oncogenes or suppressing oncogene repressor genes. Another theory is that repair and cell growth in response to liver damage proceeds out of control, resulting in cancer. In any case, because of its association with HBV, hepatic cancer may become the first cancer eliminated as a result of immunization.

TELL ME WHY

The hepatitis B vaccine was introduced in 1982. Why didn't the incidence of hepatitis B begin to immediately decline?

Parvoviridae

LEARNING OUTCOME

24.17 Describe the pathogenesis of erythema infectiosum.

Parvoviruses are the only pathogens of humans with a single-stranded DNA (ssDNA) genome. They are also the smallest of

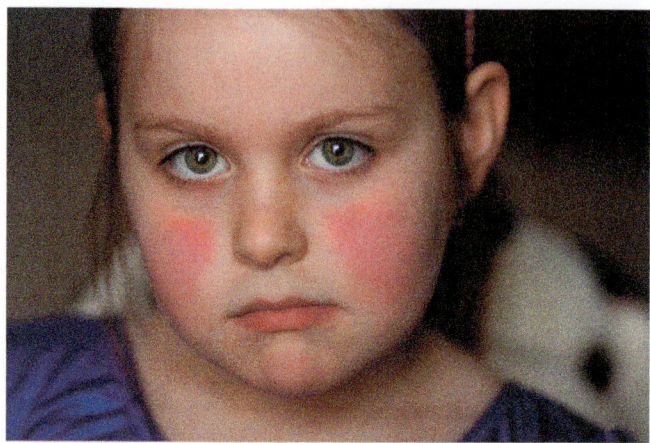

▲ FIGURE 24.22 **A case of erythema infectiosum (fifth disease).** Distinct reddening of facial skin that resembles the result of a slap is characteristic of this parvovirus-caused disease.

the DNA viruses and have an icosahedral capsid. They cause a number of endemic diseases of animals, including a potentially fatal disease in puppies.

The primary parvovirus of humans is *B19 virus* in the genus *Erythrovirus*, which is the cause of **erythema infectiosum** (er-ĭ-thē′mă in-fek-shē-ō′sŭm), also known as **fifth disease** because it was the fifth childhood rash[13] on a 1905 list. Erythema infectiosum, as the name implies, is an infectious reddening of the skin, beginning on the cheeks **(FIGURE 24.22)** and appearing later on the arms, thighs, buttocks, and trunk. The harmless rash is distinctive in appearance, typically looking like the child has been slapped. The rash is usually sufficient for diagnosis. As the rash progresses, earlier lesions fade, though they can reappear if the patient is exposed to sunlight. Sunlight aggravates the eruptions at all stages. No treatment is available.

TABLE 24.2 summarizes features of DNA viruses of humans.

MICRO CHECK

9. Where on the body are flat warts located?
10. What is the genome structure of hepadnavirus?
11. Which hepadnavirus particles are infectious?
12. What is the best way to guard against hepatitis B infection?

TELL ME WHY

Why are children with fifth disease often reported to child protective services?

---

[13]The other four rashes are those of scarlet fever, rubella (German measles), roseola, and measles.

TABLE 24.2 Taxonomy and Characteristics of DNA Viruses of Humans

Family	Strand Type	Enveloped or Naked	Capsid Symmetry	Size (diameter, nm)	Representative Genera (disease)
Poxviridae	Double	Enveloped	Complex	200–300	*Orthopoxvirus* (smallpox, cowpox), *Molluscipoxvirus* (molluscum contagiosum)
Herpesviridae	Double	Enveloped	Icosahedral	150–200	*Simplexvirus*—HHV-1 (fever blisters, respiratory infections, encephalitis), HHV-2 (genital infections), *Varicellovirus* (chickenpox), *Lymphocryptovirus* Epstein-Barr virus (infectious mononucleosis, Burkitt's lymphoma), *Cytomegalovirus* (birth defects), *Roseolovirus* (roseola)
Papillomaviridae	Double	Naked	Icosahedral	45–55	*Papillomavirus* (benign tumors, warts, cervical, penile, throat, and oral cancers)
Polyomaviridae	Double	Naked	Icosahedral	45–55	*Polyomavirus* (progressive multifocal leukoencephalopathy)
Adenoviridae	Double	Naked	Icosahedral	60–90	*Mastadenovirus* (conjunctivitis, respiratory infections)
Hepadnaviridae	Partial single and partial double	Enveloped	Icosahedral	42	*Orthohepadnavirus* (hepatitis B)
Parvoviridae	Single	Naked	Icosahedral	18–26	*Erythrovirus* (fifth disease)

MICRO IN THE CLINIC

Are Cold Sores Caused by Cold Viruses?

Dakota asked Dr. Chang if the painful sore on the inside of her lip was herpes. She was worried that her boyfriend had herpes and that she contracted it from him. Dr. Chang confirmed that Dakota indeed did have oral herpes and quickly added that there are different kinds of herpesviruses. Oral herpes, or fever blisters, which is what Dakota had, is just one kind. It was possible that she contracted this from her boyfriend, but she could have gotten it by sharing a drink with someone, or this could be an outbreak from a virus she contracted a long time ago. Oral herpes is transmitted through contact with mucous membranes. Dr. Chang explained that the virus can cause repeated outbreaks weeks, months, or even years apart, because *human herpesvirus 1* enters a latent phase in nerve cells, reactivating in response to conditions such as stress, fever, sunlight, and hormone fluctuations. Genital herpes is usually caused by *human herpesvirus 2* and is transmitted through sexual contact.

Dr. Chang explained that there are no drugs that cure herpes, but he prescribed acyclovir, to shorten the appearance of the sores and reduce the chance of spreading the virus. Dakota could take acetaminophen for the pain, but not aspirin, which can cause serious complications.

While her mother drove her home, Dakota texted Emily and explained all she had learned about the different herpesviruses.

1. **Was Dakota's boyfriend likely to have the disease?**
2. **What was Dakota's prognosis?**
3. **Should Dakota have stayed home from school?**

Check your answers to Micro in the Clinic Follow-Up questions in the Mastering Microbiology Study Area.

Go to Mastering Microbiology for Interactive Microbiology, Dr. Bauman's Video Tutors, Micro Matters, MicroFlix, Micro-Boosters, practice quizzes, and more.

CHAPTER SUMMARY

Poxviridae (pp. 710–713)

1. Poxviruses, among the largest of viruses, cause skin lesions that begin as flat, red **macules**; enlarge into **papules**; fill with clear fluid to become **vesicles**; and then become pus-filled **pustules**, also called **pocks** or **pox**. These crust over and may scar.
2. **Smallpox**, eradicated in nature by 1980, was caused by a poxvirus (**variola**). The smallpox virus strain commonly known as **variola minor** caused less severe cases of smallpox than **variola major**. Governments maintain smallpox viruses in secure laboratories as research tools for studies in pathogenicity, for vaccination and recombinant DNA technology, and as material for developing protection against bioterrorism attacks.
3. The poxvirus *Molluscipoxvirus* causes **molluscum contagiosum**, which results in tumorlike growths on the skin.
4. Although monkeypox and cowpox can infect humans, such infections are rare. A form of cowpox is used to vaccinate against smallpox and monkeypox.

Herpesviridae (pp. 713–722)

1. Herpesviruses include viruses that cause fever blisters, genital herpes, chickenpox, shingles, mononucleosis, and cancer. Individual species are named **human herpesviruses 1** through **8** (HHV-1 through HHV-8).

2. *Human herpesvirus 1* (HHV-1), which was known previously as herpes simplex type 1 (HSV-1), and *human herpesvirus 2* (HHV-2) (previously HSV-2) typically produce painful skin lesions on lips and genitalia, respectively. Oral herpes lesions are called **fever blisters** or **cold sores.**

3. HHV-1 and HHV-2 enter sensory nerve cells, where they may remain latent for years. Lesions recur in about two-thirds of patients when the latent viruses are allowed to reactivate as a result of immune suppression caused by emotional stress, sunlight, or other factors.

4. **Ocular herpes** usually affects one eye to various degrees, from recurring discomfort and light sensitivity to blindness.

5. A **whitlow** is an inflamed blister resulting from infection of HHV-1 or HHV-2 via a cut or break in the skin.

6. Infections of HHV-1 and HHV-2 are usually treatable and are often preventable.

7. **Varicella (chickenpox)** in children and **herpes zoster (shingles)** in adults are both caused by *human herpesvirus 3,* which is also known as **varicella-zoster virus (VZV)**, in the genus *Varicellovirus.* Skin lesions progress from macules to papules to vesicles to crusts. Latent viruses in the sensory nerves can be activated in adults by stress, age, or immune suppression.

8. **Burkitt's lymphoma** is an infectious cancer caused by *human herpesvirus 4,* which is also known as Epstein-Barr virus (EBV). This herpesvirus also causes **infectious mononucleosis (mono)** after entering the blood; having viruses in the blood is called **viremia.** EBV also causes nasopharyngeal cancer and is implicated in chronic fatigue syndrome, Hodgkin's lymphoma, and hairy leukoplakia seen in AIDS patients.

9. The *human herpesvirus 5, Cytomegalovirus* (CMV), causes infected cells to enlarge. This virus is widespread, transmitted by bodily secretions, and often asymptomatic, though infections in immunosuppressed adults and congenitally infected newborns can be severe and may be fatal. CMV in fetuses may be **teratogenic.**

10. *Human herpesvirus 6* (HHV-6) causes **roseola**, a rose-colored rash of children, particularly on their faces.

11. *HHV-7* is an orphan virus; that is, it causes no known disease.

12. *Human herpesvirus 8* (HHV-8) causes Kaposi's sarcoma, a cancer often seen in AIDS patients.

Papillomaviridae and Polyomaviridae (pp. 722, 723)

1. A **wart,** also known as a **papilloma,** is a benign growth of epithelium or mucous membrane caused by a papillomavirus. Seed warts are most often found on fingers or toes; plantar warts on the soles of the feet; flat warts on the trunk, face, elbow, or knees; and genital warts on the external genitalia. Condylomata acuminata are large, cauliflower-like genital warts.

2. Physicians treat warts by removing infected tissue. A vaccine is recommended to protect against genital warts.

3. **Polyomaviruses,** such as BK and JC viruses, infect lymphocytes of most people at some point but cause disease only in immunosuppressed patients. In these cases, BK viruses compromise kidney function, and JC viruses cause **progressive multifocal leukoencephalopathy (PML),** which results in severe, irreversible brain damage.

Adenoviridae (pp. 722, 724–725)

1. Adenoviruses are one cause of the "common cold" and a cause of one form of conjunctivitis (pinkeye).

Hepadnaviridae (pp. 725–728)

1. Hepadnaviruses are unique in being partially dsDNA and partially ssDNA—a condition resulting from incomplete DNA replication involving an RNA intermediary.

2. **Hepatitis B virus (HBV)** infects the liver, is transmitted in blood and other bodily fluids, and is the only DNA virus that causes **hepatitis,** the signs of which include **jaundice.** A vaccine against HBV is available, but no type of treatment is effective in all cases.

3. HBV is implicated as a cause of liver cancer based on epidemiological studies and the fact that hepatitis B virus is latent in cancerous liver cells, which express viral antigens.

Parvoviridae (pp. 728–729)

1. **Parvoviruses,** the only genus of human pathogens with ssDNA genomes, include a virus that causes a fatal disease in puppies and *B19 virus,* which causes **erythema infectiosum (fifth disease),** a harmless red rash occurring in children.

QUESTIONS FOR REVIEW

Answers to the Questions for Review (except for Short Answer questions) begin on p. A-1.

Multiple Choice

1. If the envelope of a particular virus were unstable outside the host's body, which of the following statements would you expect to be true concerning this virus?
   a. It would be a dsRNA virus.
   b. It would be transmitted by intimate contact.
   c. Touching a doorknob would easily transmit it.
   d. That virus would eventually cease to be a threat to the population.

2. For which of the following reasons are most animal poxviruses unable to infect humans?
   a. Affected animals are not in frequent contact with humans.
   b. The human immune system makes it impossible for the foreign viral particles to reproduce effectively.
   c. Attachment to human cells is unlikely.
   d. Human cells lack the necessary enzymes for infection.

3. The initial flat, red skin lesions of poxviruses are called _____.
   a. macules
   b. papules
   c. pustules
   d. pocks

4. Which of the following statements is true concerning variola major?
   a. It carries a mortality rate of less than 1%.
   b. It affects internal organs before appearing on the skin.
   c. The causative virus has been totally eradicated from Earth.
   d. The skin lesions it causes are smooth, waxy, tumorlike nodules on the face.

5. Which of the following herpesvirus infections would be potentially most serious?
   a. a whitlow
   b. ocular herpes
   c. shingles
   d. *Cytomegalovirus* in a fetus

6. A man experienced a laboratory accident in which he was infected with adenovirus. What signs or symptoms might he exhibit?
   a. sore throat
   b. headache
   c. pinkeye
   d. all of the above

7. Which of the following is an inflammatory condition of the liver?
   a. fifth disease
   b. PML
   c. seed warts
   d. hepatitis

8. Which of the following statements is *false*?
   a. B19 virus is the primary parvovirus of humans.
   b. Erythema infectiosum is caused by a parvovirus.
   c. In children, parvovirus infections are accompanied by a high mortality rate.
   d. Parvovirus infection in humans results in infectious reddening of the skin.

9. A distinguishing feature of poxvirus is _____.
   a. its large size
   b. a polyhedral capsid
   c. the type of RNA it contains
   d. the production of several types of warts

10. Monkeypox has been diagnosed in several humans in the Democratic Republic of the Congo. What might be recommended to prevent further risk of infection?
    a. Catch the monkeys for inoculation with monkeypox vaccine.
    b. Remove infected tissues from humans with chemicals, by surgery, or by freezing.
    c. Reinstate smallpox vaccinations for the country's population.
    d. The diagnosis must have been incorrect because humans are unaffected by monkeypox.

11. Which of the following viral families is most likely to contain viruses that exist in a latent state in humans?
    a. *Herpesviridae*
    b. *Poxviridae*
    c. *Adenoviridae*
    d. *Parvoviridae*

12. DNA viruses in which of the following families are relatively large and thus potentially well suited for the introduction of genetic material in gene therapy?
    a. *Herpesviridae*
    b. *Poxviridae*
    c. *Papillomaviridae*
    d. *Hepadnaviridae*

13. Being habitually careful not to touch or rub your eyes with unwashed hands would reduce your risk of contracting _____.
    a. chickenpox
    b. infectious mononucleosis
    c. seed warts
    d. a cold

14. *Human herpesvirus 2* _____.
    a. can cause genital herpes
    b. may infect a baby at birth
    c. causes about 10% of cold sores
    d. all of the above

15. Epstein-Barr virus _____.
    a. can be asymptomatic
    b. causes mononucleosis
    c. can cause cancer
    d. all of the above

VISUALIZE IT!

1. Label the successive stages of skin lesions as exemplified by smallpox.

a. _____

b. _____

c. _____

d. _____

e. _____

f. _____

CHAPTER 24 Pathogenic DNA Viruses

2. Name the disease shown in each photo.

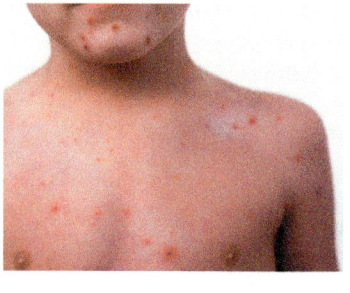
a. _____

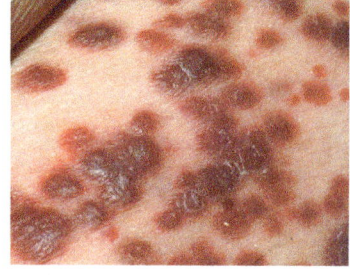
b. _____

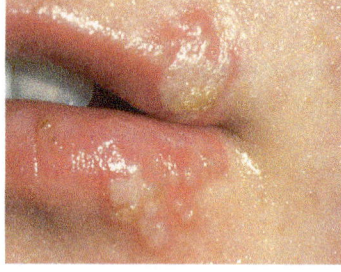
c. _____

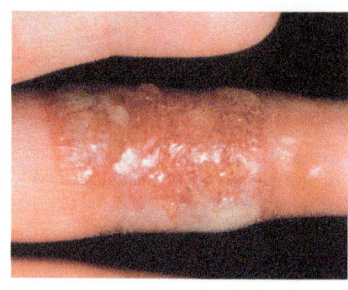
d. _____

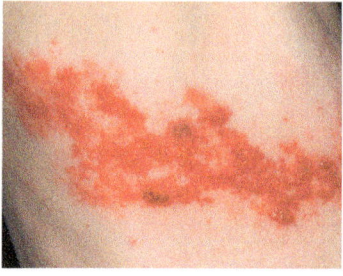
e. _____

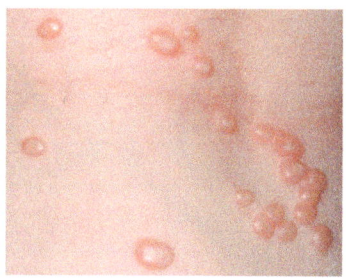
f. _____

Short Answer

1. The Smith family seems to get fever blisters regularly. Suggest an explanation for this observation.
2. You have been given a large grant to do postgraduate research on live smallpox viruses. Where in the world would you find samples?
3. What was the difference in the effects of variola major and variola minor?
4. What observation led scientists to understand the relationship between shingles and chickenpox?
5. Most of the world's population in less developed and developing countries have been infected with EBV *human herpesvirus 4* by age one and show no ill effects, even where medical care is poor. In contrast, individuals in industrialized countries are ordinarily infected after puberty, and these older patients tend to have more severe reactions to infection despite better overall health and access to medical care. Explain this apparent paradox.

Matching

Identify the viral family causing or implicated in each of the following diseases. A viral family name may be used more than once.

1. _____ Chickenpox
2. _____ Smallpox
3. _____ Cowpox
4. _____ Molluscum contagiosum
5. _____ HHV-1 infection
6. _____ Whitlow
7. _____ Shingles
8. _____ Burkitt's lymphoma
9. _____ Infectious mononucleosis
10. _____ Chronic fatigue syndrome
11. _____ *Cytomegalovirus* infection
12. _____ Genital warts
13. _____ Roseola
14. _____ Plantar warts
15. _____ Progressive multifocal leukoencephalopathy
16. _____ Common cold
17. _____ Hepatitis B
18. _____ Fifth disease

A. *Poxviridae*
B. *Herpesviridae*
C. *Papillomaviridae*
D. *Adenoviridae*
E. *Hepadnaviridae*
F. *Parvoviridae*
G. *Polyomaviridae*

CRITICAL THINKING

1. Most DNA viruses replicate within nuclei of host cells, using host enzymes to replicate their DNA. In contrast, poxviruses replicate in the cytoplasm. What problem does this create for poxvirus replication? How could the virus overcome this problem?
2. Mrs. Rathbone called the pediatrician concerning her young daughter Renee, who had a rosy facial rash and coldlike sniffles for two weeks. What is the most likely cause of Renee's problem?
3. Certain features of smallpox viruses allowed them to be eradicated in nature. Which other DNA viruses are suitable candidates for eradication, and what features of their biology make them suitable candidates?
4. A 10-year-old girl at summer camp complains of fever, runny nose, cough, sore throat, and tiredness. Within several hours, 36 other girls report to the infirmary with the same signs and symptoms. Although the girls were assigned to several different cabins and ate in two different dining halls, all of them had participated in outdoor archery, had gone horseback riding, and had been swimming in the camp pool. Infection with what DNA virus could account for their symptoms? The facts point to a common source of infection; what is it? What could the camp management do to limit such an outbreak in the future?

5. One week after spending their vacation rafting down the Colorado River, all five members of the Chen family developed cold sores on their lips. Their doctor told them that the lesions were caused by a herpesvirus. Mr. and Mrs. Chen were stunned: Isn't herpes a sexually transmitted disease? How could it have affected their young children?

6. The World Health Organization (WHO), the United States, and Russia have repeatedly discussed the destruction of the two countries' stocks of the smallpox virus, but deadlines for destruction have been postponed numerous times. In the meantime, the entire genome of variola major has been sequenced. What reasons can governments cite for maintaining smallpox viruses? Should all laboratory stores of smallpox viruses be destroyed? Given that the genome of the virus has been sequenced and that DNA can be synthesized, would elimination of all laboratory stocks really be the extinction of the smallpox virus?

7. After a patient complains that his eyes are extremely sensitive to light and feel gritty, his doctor informs him that he has ocular herpes. What causes ocular herpes? Which human herpesvirus, type 1 or type 2, is more likely to cause ocular herpes? Why?

CONCEPT MAPPING

Using the terms provided, fill in this concept map that describes herpes simplex virus. You can also complete this and other concept maps online by going to the Mastering Microbiology Study Area.

Acyclovir
Envelope
Fever blisters
Genital herpes
Gingivostomatitis

Herpetic whitlow
Human herpesvirus 2
Latency
Life threatening
Ocular herpes

Oral herpes
Icosahedral capsid
Recurrences
Simplexvirus

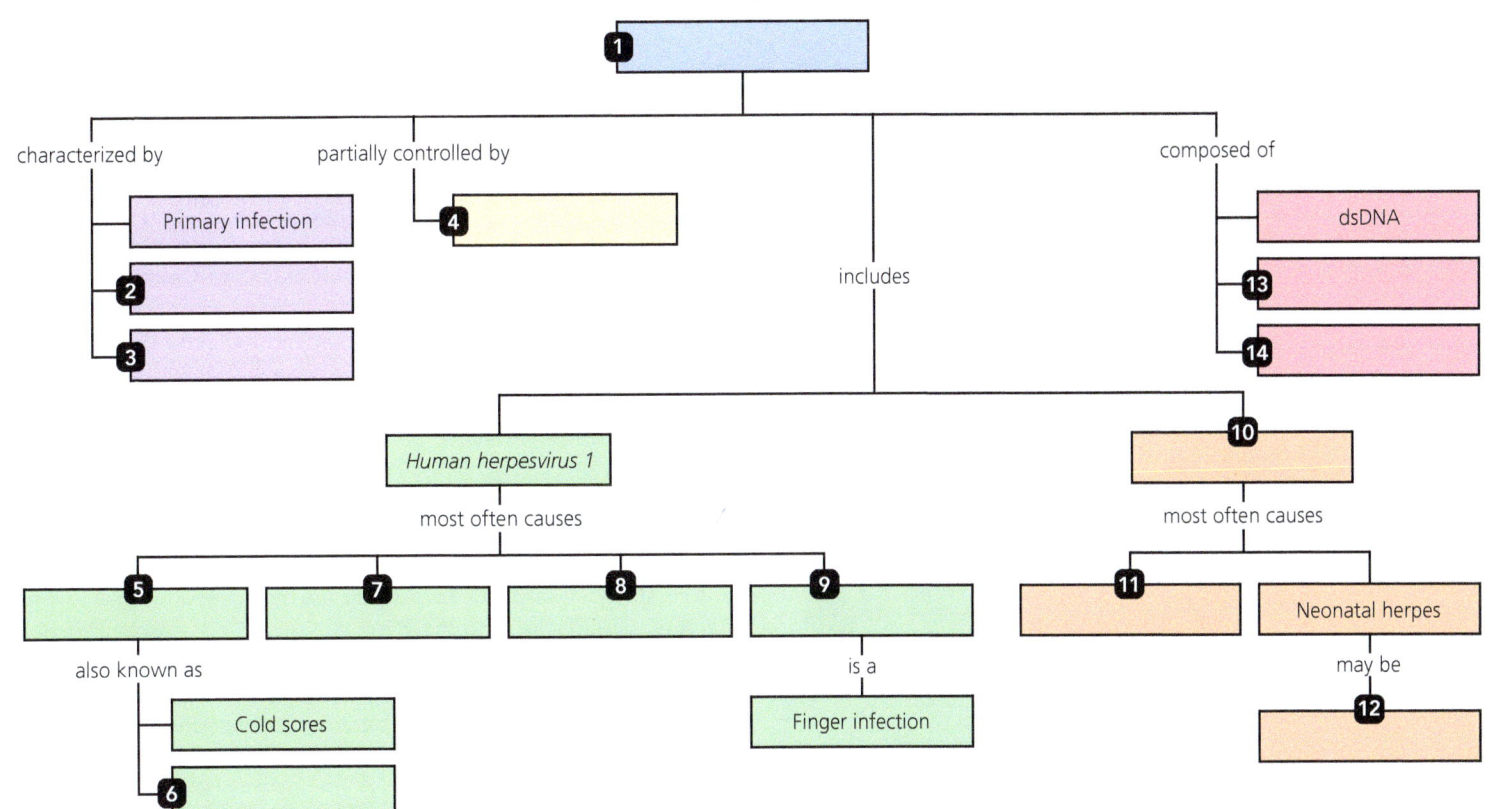

25 Pathogenic RNA Viruses

Before You Begin

1. Inside eukaryotic cells, where do most DNA and RNA viruses replicate?
2. Which group of viruses uses the enzyme reverse transcriptase during its replication cycle?
3. Why do RNA viruses generally accumulate mutations faster?

Artists rendition of Zika virus, one of many RNA viruses that cause human diseases.

MICRO IN THE CLINIC

Brain Dysfunction After a Hard Hike?

MERCEDES IS A 68-YEAR-OLD GRANDMOTHER of six. She has remained very active since she retired from her job as a park ranger. During the spring and summer, one of her favorite activities is leading a weekly hike for retirees in the Cascade Mountains in Washington State. On the most recent hike, they were able to see six different species of birds, and the wildflowers were in full bloom. The only downside was that it was unseasonably warm, and there were a lot of bugs (mosquitoes, bees, and flies).

A couple of days after the hike, Mercedes wakes up with aching joints and weakness in her legs. The hike that week had been intense, although she thought it was well within her ability. She thinks that perhaps she pushed herself too hard. She takes some ibuprofen for her aches and tries to rest. When Mercedes wakes up the next day, she has a low-grade fever and a headache, and the weakness in her legs has gotten worse.

When Mercedes's daughter, Paula, stops in for a visit, she becomes worried because she finds her mom lying on the couch in pain. As Paula talks with Mercedes, she becomes more concerned—her mom seems disoriented and has a tremor in her right hand. Seeing the tremor brings back memories of her father's battle with Parkinson's disease. Paula calls her mother's physician and makes an appointment for the next morning.

1. Does Mercedes have a neurodegenerative disease like Parkinson's?
2. Could there be another explanation for her symptoms?

Turn to the end of the chapter (p. 772) to find out.

All cells and all DNA viruses use DNA as their genetic material. In contrast, RNA viruses and viroids are the only agents that use RNA molecules to store their genetic information.

There are four basic types of RNA viruses (see Chapter 13):

- Positive single-stranded RNA (+ssRNA) viruses
- Retroviruses (which are +ssRNA viruses that convert their genome to DNA once they are inside a cell)
- Negative single-stranded RNA (−ssRNA) viruses
- Double-stranded RNA (dsRNA) viruses

Positive RNA is RNA that can be used by a ribosome to translate protein; it is essentially messenger RNA (mRNA). Negative RNA cannot be processed by a ribosome; it must first be transcribed into mRNA. (Chapter 13 examined the replication strategies of these four types of RNA viruses, and Table 13.3 on p. 399 summarizes the information.) In addition to grouping RNA viruses by the kind of RNA they contain, scientists also note that many RNA viruses have more than one molecule of RNA—in other words, their genomes are *segmented*.

RNA viruses of humans are categorized into 15 families on the basis of the following:

- Their genomic structure
- The presence of an envelope
- The size and shape of their capsid

RNA viral genera have Latinized names; however, as with DNA viruses, English names are used for the specific epithets of species. Researchers and clinicians often refer to a virus by its English name or, less frequently, solely by its genus. Virologists classify positive single-stranded RNA viruses of humans into eight families: *Picornaviridae, Caliciviridae, Astroviridae, Hepeviridae, Togaviridae, Flaviviridae, Coronaviridae,* and *Retroviridae*. Negative single-stranded RNA viruses are in six families: *Paramyxoviridae, Rhabdoviridae, Filoviridae, Orthomyxoviridae, Bunyaviridae,* and *Arenaviridae*. The only family containing double-stranded RNA viruses is *Reoviridae*. We begin our discussion of pathogenic RNA viruses by considering four families of positive ssRNA viruses that lack envelopes.

Naked, Positive ssRNA Viruses: *Picornaviridae, Caliciviridae, Astroviridae,* and *Hepeviridae*

The families *Picornaviridae, Caliciviridae, Astroviridae,* and *Hepeviridae* contain positive single-stranded RNA viruses with naked polyhedral capsids; that is, they lack membranous envelopes. The family *Picornaviridae* is a large viral family containing many human pathogens, including viruses that cause common colds, poliomyelitis, and hepatitis A. Picornaviruses range in size from 22 to 30 nm in diameter, making them the smallest of animal viruses. Five hundred million picornaviruses could sit side by side on the head of a pin. Their small size and RNA genome are reflected in the name **picornavirus** (pi-kōr-nă-vī′rus), which means "small RNA virus." Representative picornaviruses that cause human diseases are in the genera *Rhinovirus, Enterovirus,* and *Hepatovirus*.

Caliciviruses, astroviruses, and hepeviruses are generally larger than picornaviruses (27–40 nm in diameter) and have a six-pointed star shape; they cause gastrointestinal diseases.

In the following sections, we will consider in turn three diseases caused by picornaviruses—common cold, polio, and hepatitis A—before examining diseases caused by caliciviruses, astroviruses, and hepeviruses.

Common Colds Caused by Rhinoviruses

LEARNING OUTCOME

25.1 Discuss treatment of the common cold.

Many Americans sniffle and sneeze their way through at least two colds each year. Though the symptoms of all colds are similar—sneezing, *rhinorrhea* (runny nose), congestion, mild sore throat, headache, malaise, and cough for 7 to 10 days—there is no single common cause of the "common cold." Rather, many different viruses, including picornaviruses, adenoviruses, coronaviruses, reoviruses, and paramyxoviruses, cause colds. However, the over 100 *serotypes* (varieties, strains) of picornaviruses in the genus *Rhinovirus,* commonly known as **rhinoviruses**[1] (rī′nō-vī′rŭs-ĕz), cause most colds (FIGURE 25.1). TABLE 25.1 compares the symptoms of colds and other respiratory infections.

[1] From Greek *rhinos,* meaning "nose."

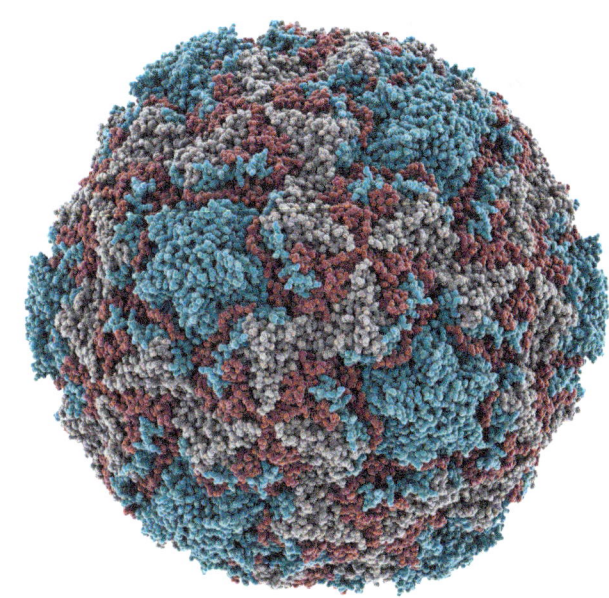

▲ FIGURE 25.1 Artist's rendition of rhinovirus 14. *What other types of viruses cause common colds?*

Figure 25.1 Besides rhinoviruses (in the family Picornaviridae), some adenoviruses, coronaviruses, reoviruses, and paramyxoviruses cause colds.

TABLE 25.1 Manifestations of Respiratory Infections

Ailment	Manifestations
Common Cold (viral)	Sneezing, rhinorrhea, congestion, sore throat, headache, malaise, cough
Influenza (viral)	Fever, rhinorrhea, headache, body aches, fatigue, dry cough, pharyngitis, congestion
"Strep" Throat (bacterial)	Fever, red and sore throat, swollen lymph nodes in neck
Viral Pneumonia	Fever, chills, mucus-producing cough, headache, body aches, fatigue
Bacterial Pneumonia	Fever, chills, congestion, cough, chest pain, rapid breathing, and possible nausea and vomiting
Bronchitis (viral or bacterial)	Mucus-producing cough, wheezing
Inhalation Anthrax (bacterial)	Fever, malaise, cough, chest discomfort, vomiting
Severe Acute Respiratory Syndrome (SARS)	High fever (>38°C), chills, shaking, headache, malaise, myalgia
Middle East Respiratory Syndrome (MERS)	Fever, cough, shortness of breath

Epidemiology

Rhinoviruses are limited to infecting the upper respiratory tract. They replicate best at a temperature of 33°C, which is the temperature of the nasal cavity.

Rhinoviruses are extremely infective—entry of a single virus into a person is sufficient to cause a cold in 50% of individuals. Symptomatic or not, an infected person can spread viruses by releasing them into the surrounding environment, where they are transmitted in aerosols produced by coughing or sneezing, via fomites (nonliving carriers of pathogens), or via hand-to-hand contact. Direct person-to-person contact is the most common means of transmitting these infections.

When cold symptoms are most severe, over 100,000 virions may be present in each milliliter of nasal mucus. They remain viable for hours outside the body. Infection often results from inoculation by hand into the mucous membranes of the eyes, where the viruses are washed by tears into the nasal cavity and readily infect nasal cells.

Although people of all ages are susceptible to rhinoviruses, people acquire some immunity against serotypes that have infected them in the past. For this reason, children typically have six to eight colds per year, younger adults have two to four, and adults over age 60 have one or fewer. An isolated population may acquire a certain amount of *herd immunity* to specific varieties of rhinoviruses; that is, when more than about 75% of the population is resistant because of previous infection, then the viruses are less likely to spread for lack of hosts (see Chapter 17). However, new serotypes introduced into a local population by outsiders or by mutations in the RNA of endemic viruses ensure that no group is free of all colds.

Diagnosis, Treatment, and Prevention

The manifestations of rhinoviruses are usually diagnostic (see Table 25.1), and laboratory tests are required only if the serotype is to be identified.

While many home remedies and over-the-counter medicines exist to treat the signs and symptoms of common colds, none prevents colds or provides a cure. Antihistamines, decongestants, and pain relievers relieve the symptoms but do not reduce the duration of the disease. Rest and fluids allow the body to mount a more effective immune response.

Studies of two common "cures"—zinc in lozenge or syrup form or large amounts of vitamin C—have yielded conflicting results: The majority of studies indicate no benefit for most people, whereas a few studies suggest that colds are shorter but not less frequent in patients who take either zinc or vitamin C before cold symptoms begin. One drug, pleconaril, taken at the onset of symptoms, reduces the seriousness and duration of colds caused by rhinoviruses.

Because rhinoviruses do not all share any antigens accessible to the immune system, an effective vaccine for the common cold would have to immunize against hundreds of strains, which is impractical. Antisepsis is probably the most important preventive measure, especially if you have touched the hands of an infected person. Disinfection of fomites is also effective in limiting the spread of colds.

Diseases of Enteroviruses

LEARNING OUTCOMES

25.2 Describe the effects of polioviruses on humans.
25.3 Describe the contributions of Jonas Salk and Albert Sabin toward eliminating polio.
25.4 Compare and contrast the diseases caused by coxsackieviruses and echoviruses.

A second genus of picornaviruses is *Enterovirus*. Contrary to their name, **enteroviruses** do not usually cause diseases of the digestive system. They infect the pharynx and intestine, where they multiply in the mucosa and lymphatic tissue without causing symptoms. Instead, enteroviruses are so named because they are transmitted via the fecal-oral route. Enteroviruses spread from their initial sites of infection through the blood (viremia) and infect different target cells, depending on the particular virus. Enteroviruses are cytolytic,[2] killing their host cells. The clinical manifestations of enteroviruses depend on the viral strain, the size of the infecting dose, the target organ, and the patient's health, gender, and age. There are three main types of enteroviruses: polioviruses, coxsackieviruses, and echoviruses.

Poliomyelitis

Older Americans still remember the dreaded **poliomyelitis** (pō′lē-ō-mī′ĕ-lī′tis), or **polio,** epidemics of the early 20th century, when hospitals were filled with crippled patients. Because polioviruses are stable for prolonged periods in swimming pools and lakes and can be acquired by swallowing contaminated water, parents in the 1930s and 1940s often feared to let their children swim. Thankfully, those days appear to be nearly over around the world. The last case of wild-type poliomyelitis in the Americas occurred in 1979. The World Health Organization (WHO) has worked for years to eradicate polio in Africa and Asia. Progress

---
[2]From Greek *cytos*, meaning "cell," and *lysis*, meaning "dissolution."

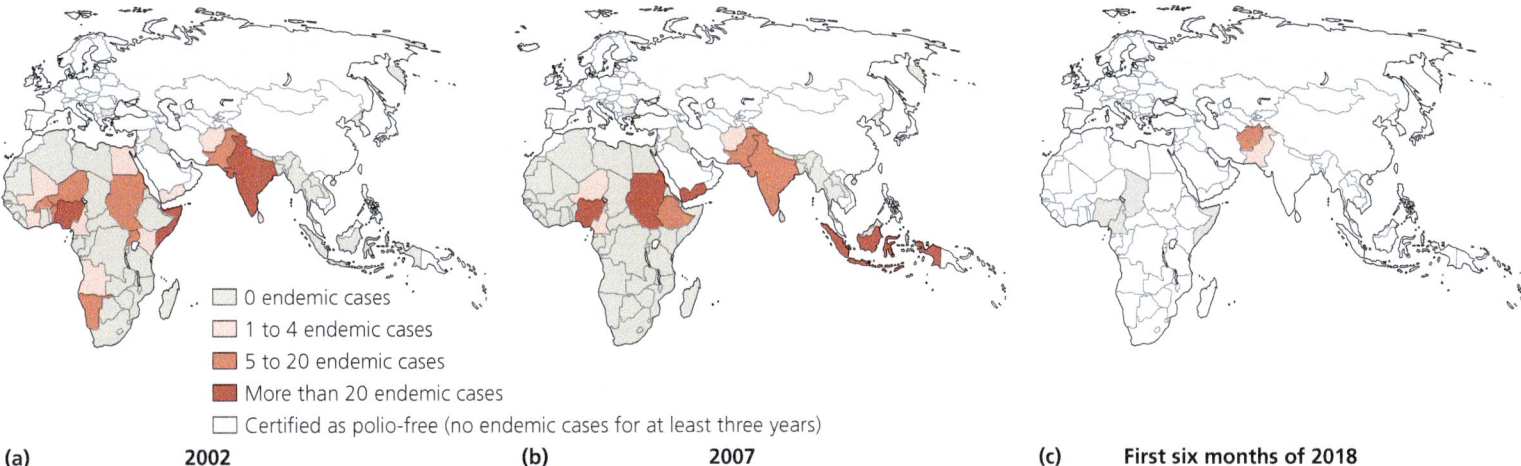

▲ FIGURE 25.2 Reports of naturally occurring polio. (a) Cases in 2002. (b) Cases in 2007. (c) Cases in first half of 2018. The goal of global eradication of polio is within reach. No cases occurred in the Americas, Australia, or New Zealand.

has been sporadic but hopeful, especially since there had been no African cases of wild polio since 2016, as of September 2018 (FIGURE 25.2).

There are three types of poliovirus. Health officials have eradicated type 2 and type 3 from humans worldwide. In the first half of 2018, type 1 poliovirus was endemic only in Pakistan (3 cases) and Afghanistan (8 cases). After being ingested and infecting pharyngeal and intestinal cells, any of these strains could cause one of the following four conditions:

- *Asymptomatic* infections. Most infections (almost 90%) are asymptomatic.
- *Minor polio*, which includes nonspecific symptoms such as temporary fever, headache, malaise, and sore throat. Approximately 5% of cases are minor polio.
- *Nonparalytic polio*, resulting from polioviruses invading the meninges and central nervous system, producing muscle spasms and back pain in addition to the general symptoms of minor polio. Nonparalytic polio occurs in about 2% of cases.
- *Paralytic polio*, in which the viruses invade cells of the spinal cord and motor cortex of the brain, producing paralysis by limiting nerve impulse conduction. The degree of paralysis varies with the type of poliovirus involved, the infective dose, and the health and age of the patient. In **bulbar poliomyelitis,** the brain stem and medulla are infected, resulting in paralysis of muscles in the limbs or of respiratory muscles. In most paralytic cases, complete recovery results after 6 to 24 months, but in some cases paralysis is lifelong (FIGURE 25.3). Paralytic polio occurs in less than 2% of infections. Famous victims of paralytic polio include novelist Sir Walter Scott, President Franklin Delano Roosevelt, and violinist Itzhak Perlman (FIGURE 25.4).

Postpolio syndrome is a crippling deterioration in the function of polio-affected muscles that occurs in up to 80% of recovered polio patients some 30 to 40 years after their original bout with poliomyelitis. This condition is not caused by a reemergence of polioviruses, as viruses are not present. Instead, the effects

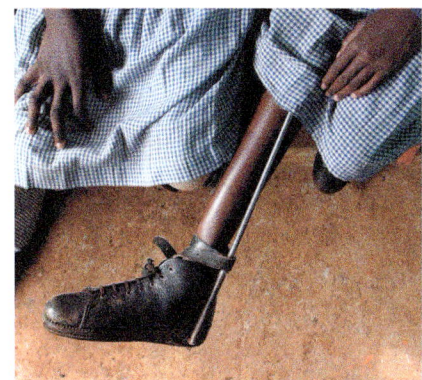

◀ FIGURE 25.3 Polio can result in severe paralysis.

◀ FIGURE 25.4 Violinist Itzhak Perlman. This world-famous violinist did not let paralytic polio stop him from achieving in his field of endeavor.

appear to stem from an aging-related aggravation of nerve damage that occurred during the original infection.

The near elimination of polio stands as one of the great achievements of 20th-century medicine. It was made possible by the development of two effective vaccines. Jonas Salk (1914–1995) developed an **inactivated polio vaccine (IPV)** in 1955. Six years

TABLE 25.2 Comparison of Polio Vaccines

	Advantages	Disadvantages
Salk Vaccine: Inactivated Polio Vaccine (IPV)	Is effective and inexpensive; is stable during transport and storage; poses no risk of vaccine-related disease	Requires booster to achieve lifelong immunity; is injected and can be painful; requires higher community immunization rate than does OPV
Sabin Vaccine: Oral Polio Vaccine (OPV)	Induces secretory antibody response similar to natural infection; is easy to administer; can result in herd immunity	Requires boosters to achieve immunity; is more expensive than IPV; is less stable than IPV; can mutate to disease-causing form; poses risk of polio developing in immunocompromised patients

later, it was replaced in the United States by a live, attenuated, **oral polio vaccine (OPV)** developed by Albert Sabin (1906–1993). Both vaccines are effective in providing immunity against all three strains of poliovirus, though OPV occasionally mutates into a virulent form and causes polio. For this reason, the United States uses IPV only. TABLE 25.2 compares the advantages and disadvantages of the two vaccines.

Other Diseases of Enteroviruses

Other species in the genus *Enterovirus* that cause human disease are **coxsackieviruses**[3] and **echoviruses**. As with all enteroviruses, infection is by the fecal-oral route. Most infections are subclinical or result in mild fever, muscle aches, and malaise. There are two types of coxsackieviruses (type A and type B), each with several strains, as well as numerous serotypes of echoviruses, which cause a variety of human diseases.

Coxsackieviruses Coxsackie A viruses are associated with lesions and fever that last for a few days to weeks and are self-limiting. Several strains cause lesions of the mouth and pharynx called *herpangina* because of their resemblance to herpes lesions. Sore throat, pain in swallowing, and vomiting accompany herpangina.

Another strain of coxsackie A virus causes *hand-foot-and-mouth disease*, mostly in children under age 10. The disease is aptly named because it involves lesions on the extremities and in the mouth (FIGURE 25.5). Yet another coxsackie A virus causes an extremely contagious *acute hemorrhagic conjunctivitis*. Other coxsackie A viruses cause some colds.

Coxsackie B viruses are associated with *myocarditis* (inflammation of the heart muscle) and *pericardial*[4] *infections*. The symptoms in young adults may resemble myocardial infarction (heart attack), though coxsackie infections involve fever, which is not seen in heart attacks. Newborns are particularly susceptible to coxsackie B myocarditis resulting in a rapid onset of heart failure with a high mortality rate.

One type of coxsackie B virus causes *pleurodynia* (ploor-ō-din'ē-ă), also known as *devil's grip*. This disease involves the sudden onset of fever and unilateral, severe, low thoracic pain, which may be excruciating. Coxsackie B virus can be transmitted across a placenta to produce severe and sometimes fatal disseminated disease of the brain, pancreas, and liver in the fetus. Infection of the pancreas by coxsackie B is suspected to be a cause of type 1 diabetes because the viruses destroy cells in the islets of Langerhans that produce insulin.

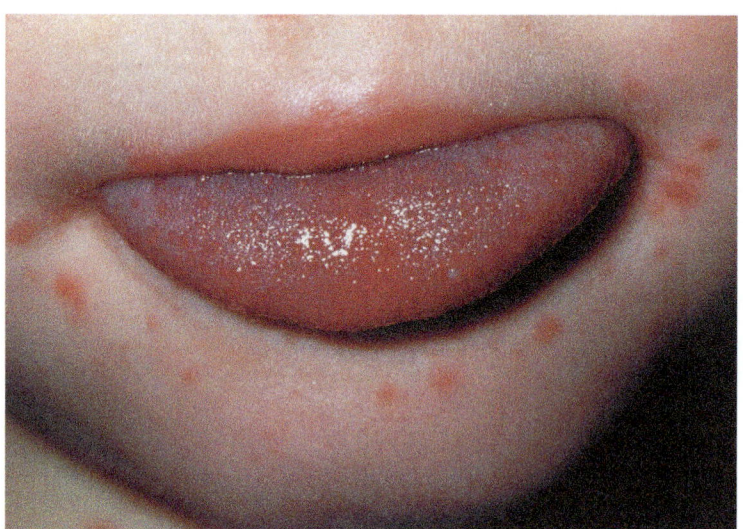

▲ **FIGURE 25.5 Lesions characteristic of hand-foot-and-mouth disease.** The malady is caused by a coxsackie A virus. *Why are coxsackieviruses considered enteroviruses when they do not cause infections of the gastrointestinal tract?*

Figure 25.5 *All enteroviruses, including coxsackievirus, are infective through the gastrointestinal tract, though they may affect other organ systems.*

Both coxsackie A and B viruses can cause *viral meningitis*, an acute disease accompanied by headache and, with type A infections, a skin rash. Coxsackievirus meningitis is usually self-limiting and uneventful unless the patient is less than one year old.

Echoviruses The name *echovirus* is derived from *enteric cytopathic human orphan* virus because these viruses are acquired intestinally and were not initially associated with any disease—they were *orphan viruses*. It is now known that echoviruses cause viral meningitis and some colds.

Epidemiology

Enteroviruses have a worldwide distribution and occur particularly in areas with inadequate sewage treatment. Transmission is by the fecal-oral route involving ingestion of contaminated food or water or oral contact with infected hands or fomites. For some reason, enterovirus diseases are more common in summer.

As we have seen, enteroviruses pose the greatest risk to fetuses and newborns, with the exception of poliovirus, which results in more serious symptoms in older children and young adults.

---

[3] From Coxsackie, New York, where the virus was first isolated.
[4] The pericardium is the membrane that surrounds and protects the heart.

TABLE 25.3 Comparison of Hepatitis Viruses

Feature	Hepatitis A	Hepatitis B	Hepatitis C	Hepatitis D	Hepatitis E
Common Names of Disease	Infectious hepatitis	Serum hepatitis	Non-A, non-B hepatitis; chronic hepatitis	Delta agent hepatitis	Hepatitis E, enteric hepatitis
Virus Family (genus)	*Picornaviridae* (*Hepatovirus*)	*Hepadnaviridae* (*Orthohepadnavirus*)	*Flaviviridae* (*Hepacivirus*)	*Arenaviridae* (*Deltavirus*)	*Hepeviridae* (*Orthohepevirus*)
Genome	+ssRNA	dsDNA	+ssRNA	−ssRNA	+ssRNA
Envelope?	Naked	Enveloped	Enveloped	Enveloped	Naked
Transmission	Fecal-oral	Needles; sex	Needles; sex	Needles; sex	Fecal-oral
Severity (mortality rate)	Mild (< 0.5%)	Occasionally severe (1–2%)	Usually subclinical (0.5–4%)	Requires coinfection with hepatitis B virus; may be severe (high)	Mild (1–2%) except in pregnant women (20%)
Chronic Carrier State?	No	Yes	Yes	No	No
Other Disease Associations	—	Hepatic cancer	Hepatic cancer	Liver damage; hepatic cancer (with hepatitis B virus)	—

Diagnosis, Treatment, and Prevention

Enterovirus infections are usually benign and have mild symptoms, so they are not often diagnosed except in severe cases such as paralytic polio and viral meningitis. The cerebrospinal fluid (CSF) of viral meningitis patients has a normal glucose level, in contrast to that seen with bacterial meningitis. Viruses are rarely found in the CSF, so their presence there is diagnostic for viral meningitis; serological testing can confirm enterovirus infection.

No antiviral therapy is effective against enterovirus infection. Treatment involves support and limitation of pain and fever. Good hygiene and adequate sewage treatment can prevent infection with enteroviruses. No vaccines are available for coxsackievirus and echovirus infections, though human clinical trials of a vaccine against strains of coxsackie B virus suspected of causing diabetes were begun in 2018. Two effective vaccines exist for polio. (Figure 17.3 summarizes current immunization recommendations.)

Hepatitis A

LEARNING OUTCOMES

25.5 Describe the signs and symptoms of hepatitis.
25.6 Compare and contrast the five viruses that cause hepatitis.

Like enteroviruses, **hepatitis A virus** (genus *Hepatovirus*) is transmitted through the fecal-oral route, but unlike enteroviruses it is not cytolytic. Hepatitis A virus can survive on surfaces such as countertops and cutting boards for days and resists common household disinfectants such as chlorine bleach.

Hepatitis A has an incubation period of about one month before fever, fatigue, nausea, anorexia, and jaundice abruptly start. As with hepatitis B (caused by a DNA virus), the patient's own cellular immune system kills infected liver cells, resulting in the signs and symptoms. Children are less likely than adults to develop symptoms because children's cell-mediated immune responses are not fully developed. Hepatitis A virus does not cause chronic liver disease, and complete recovery occurs 99% of the time. Patients release virions in their feces and are infective even without developing symptoms. To prevent hepatitis A, two doses of hepatitis A vaccine are recommended for all children and adults.

TABLE 25.3 compares five viruses known to cause viral hepatitis. Four of them (hepatitis A, C, D, and E) are RNA viruses discussed in this chapter. The fifth is hepatitis B virus (discussed in Chapter 24).

Acute Gastroenteritis

LEARNING OUTCOME

25.7 Compare and contrast caliciviruses and astroviruses.

Two families of small (about 30–40 nm in diameter), round viruses that cause acute gastroenteritis are **caliciviruses** (kă-lis′ĭ-vī′rŭs-ĕz) and **astroviruses** (as′trō-vī′rŭs-ĕz), which are slightly larger than picornaviruses and have naked, star-shaped, polyhedral capsids (FIGURE 25.6). Like enteroviruses, caliciviruses and astroviruses enter the body through the digestive system, but in contrast to enteroviruses they cause gastrointestinal disease.

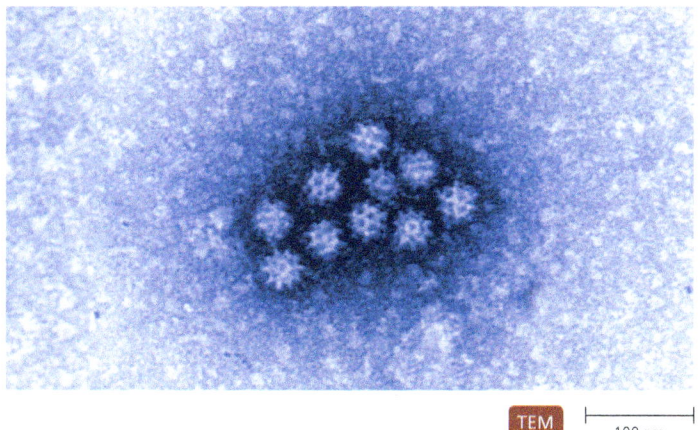

▲ FIGURE 25.6 Viruses of the families *Caliciviridae* and *Astroviridae* have naked, star-shaped capsids.

The capsids of caliciviruses have indentations, whereas those of astroviruses do not. Antigens of viruses in the two groups can be distinguished through serological tests. The incubation period for these two types of viruses is about 24 hours. Both caliciviruses and astroviruses have caused outbreaks of gastroenteritis in day care centers, schools, hospitals, nursing homes, restaurants, and on cruise ships. The symptoms resolve within 12 to 60 hours.

Caliciviruses cause diarrhea, nausea, and vomiting, though the symptoms of any given strain vary from patient to patient. The best studied of the caliciviruses is *Norovirus*. Investigators first discovered **noroviruses** in the stools of victims during an epidemic of diarrhea in Norwalk, Ohio. Noroviruses cause 90% of viral cases of gastroenteritis. Some cases can be quite severe, as demonstrated in **Emerging Disease Case Study:** *Norovirus* **in the Dorm.**

Astroviruses also cause diarrhea but no vomiting, and they are less likely to infect adults. In the United States, about three-fourths of children have antibodies against astroviruses by age seven.

There is no specific treatment for caliciviral or astroviral gastroenteritis except support and replacement of lost fluid and electrolytes. Prevention of infection involves adequate sewage treatment, purification of water supplies, frequent hand washing, and disinfection of contaminated surfaces and fomites.

Hepatitis E

LEARNING OUTCOME

25.8 Describe the prevention of hepatitis E viral infection.

Hepatitis E virus (genus *Hepevirus*) was formerly considered a calicivirus based on structure and size (27–34 nm), but now it is placed in its own family, *Hepeviridae*. Hepatitis E virus, acquired in food, drink, or from a blood transfusion, infects the liver and causes **hepatitis E,** also known as *enteric hepatitis*. Hepatitis E is fatal to 4% of patients, though that rises to 20% for infected pregnant women. Doctors do not prescribe any specific medicine to treat hepatitis E, usually merely recommending rest, plenty of fluids, and good nutrition. The disease is usually self-limiting; that is, patients typically recover on their own. Ribavarin may help the immune system clear the virus from the body. Prevention involves interrupting the fecal-oral route of infection with good personal hygiene, water purification, and sewage treatment. China has approved a vaccine against hepatitis E.

MICRO CHECK

1. Why are no vaccines available against rhinoviruses?
2. Which type of polio results in muscle spasms and back pain?
3. Which vaccine can occasionally result in polio and why?
4. How is hepatitis A transmitted and what is the best way to prevent it?

TELL ME WHY

Why might enteroviruses, which infect the body through the fecal-oral route, be more common in the summer than in the winter?

EMERGING DISEASE CASE STUDY

Norovirus in the Dorm

Zac, Tran, Roy, and Justin were not happy roommates; in fact, things could not be much worse for the college friends. Each of the men had stomach cramps, fever, chills, muscle aches, extreme tiredness, nausea, and, most distressing, horrible diarrhea and persistent vomiting. They had been fighting over the toilet, and whoever wasn't in the bathroom often had his head in a trashcan. None of the four had left their suite for two days, being unable to venture more than a dozen feet. The friends had never experienced such an attack of gastroenteritis. *Norovirus* had arrived in the dorm.

Norovirus gastroenteritis afflicts people living in close quarters: prisoners, nursing home residents, vacationers on cruise ships, and students in college dormitories. People spread the hardy virus on

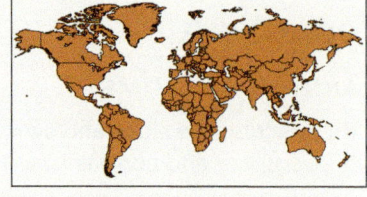

contaminated hands and fomites due to poor personal hygiene. Noroviruses are often transmitted in contaminated water and can spread in undercooked contaminated food.

The four roommates recovered as their bodies eliminated viruses from their digestive tracts. They also learned the value of hand washing, disinfecting bathrooms, and keeping the dorm room disinfected.

1. Why is vigorously rubbing the hands with hot water and soap for at least 30 seconds necessary to limit the spread of noroviruses?
2. Alcohol disrupts lipids. Why isn't hand antiseptic effective against noroviruses?
3. How can noroviruses spread via laundromats?

So far, we have considered naked, positive ssRNA viruses in the families *Picornaviridae, Caliciviridae, Astroviridae,* and *Hepeviridae.* Now we turn our attention to enveloped, positive ssRNA viruses.

Enveloped, Positive ssRNA Viruses: *Togaviridae, Flaviviridae,* and *Coronaviridae*

Members of the *Togaviridae* and *Flaviviridae* are enveloped, icosahedral, positive, single-stranded RNA viruses. Flaviviruses are smaller and have a protein matrix between the capsid and envelope. Both flaviviruses and togaviruses differ from the viruses we examined previously in being enveloped (*toga* is Latin for "cloak"); however, the envelopes of togaviruses and flaviviruses are often tightly appressed to the capsids (like shrink wrapping), which allows individual capsomeres to be visualized **(FIGURE 25.7)**. Because most togaviruses and flaviviruses are transmitted by arthropods (including mosquitoes, ticks, flies, mites, and lice), these viruses are designated **arboviruses**[5] (ar′bō-vī′rŭs-ĕz), and this shared characteristic is the reason taxonomists originally included the flaviviruses in family *Togaviridae*. However, differences in antigens, replication strategy, and RNA sequence indicate that flaviviruses belong in their own family. Thus, the term *arbovirus* has no taxonomic significance; some arboviruses are +ssRNA, some are −ssRNA, and some are dsRNA.

Viruses of the family *Coronaviridae* are also enveloped, positive, single-stranded RNA viruses. However, in contrast to togaviruses and flaviviruses, coronaviruses have helical capsids **(FIGURE 25.8)**, and none are arthropod borne.

In this section, we consider diseases of the arboviruses from the families *Togaviridae* and *Flaviviridae,* discuss the togaviruses and flaviviruses that are not transmitted via arthropods, and finally discuss the coronaviruses.

---
[5]From *arthropod borne.*

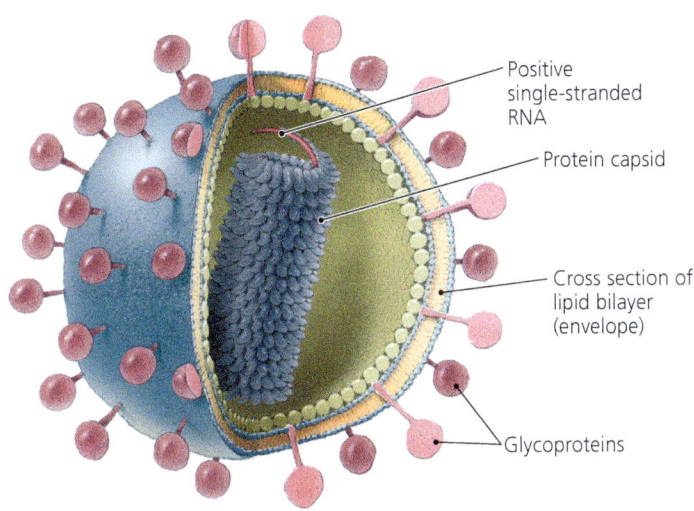

▲ **FIGURE 25.8 Enveloped +ssRNA coronavirus.** Viruses in the family *Coronaviridae* have helical capsids, as shown here, whereas viruses in the families *Togaviridae* and *Flaviviridae* have icosahedral capsids.

Diseases of + RNA Arboviruses

LEARNING OUTCOMES

25.9 Define *zoonosis* and *arbovirus.*
25.10 Compare and contrast EEE, WEE, and VEE as to their endemic ranges and natural hosts.
25.11 Describe the transmission and spread of West Nile virus encephalitis.
25.12 Contrast the two types of dengue fever.
25.13 Discuss the signs, symptoms, and prevention of yellow fever.
25.14 Describe Zika virus and its diseases: Zika fever and Zika congenital syndrome.

Specific genera of mosquitoes and ticks transmit arboviruses among animal hosts, usually small mammals or birds, and into humans. Animal diseases that spread naturally to humans are called **zoonoses,**[6] so the diseases considered in this section are both arboviral and zoonotic. Arthropod vectors remain infected with arboviruses and are a continual source of new infections when they bite their hosts. In animals, virions are released from infected cells into the blood (viremia), but persistent viremia does not occur as readily in humans. Thus, humans are dead-end hosts for these zoonotic viruses because there generally aren't enough viruses in the blood to infect a vector.

Arboviruses enter target cells through endocytosis and replicate within them. Most cause mild, flulike symptoms in humans within three to seven days of infection. Arboviral disease does not usually proceed beyond this initial manifestation. Occasionally, however, arboviruses in the blood infect the brain, liver, skin, or blood vessels. Diseases associated with such second-stage infections include several kinds of encephalitis, dengue fever, and yellow fever.

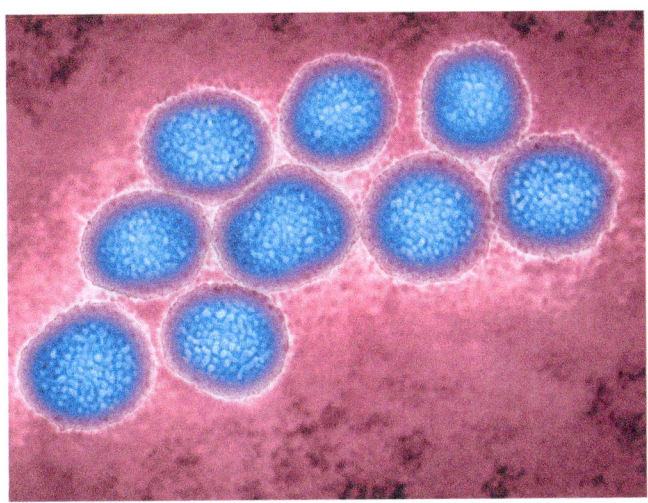

▲ **FIGURE 25.7 Togaviruses.** Each virus has a closely appressed envelope around its capsid.

---
[6]Plural of Greek *zoos,* meaning "animal," and *nosis,* meaning "disease."

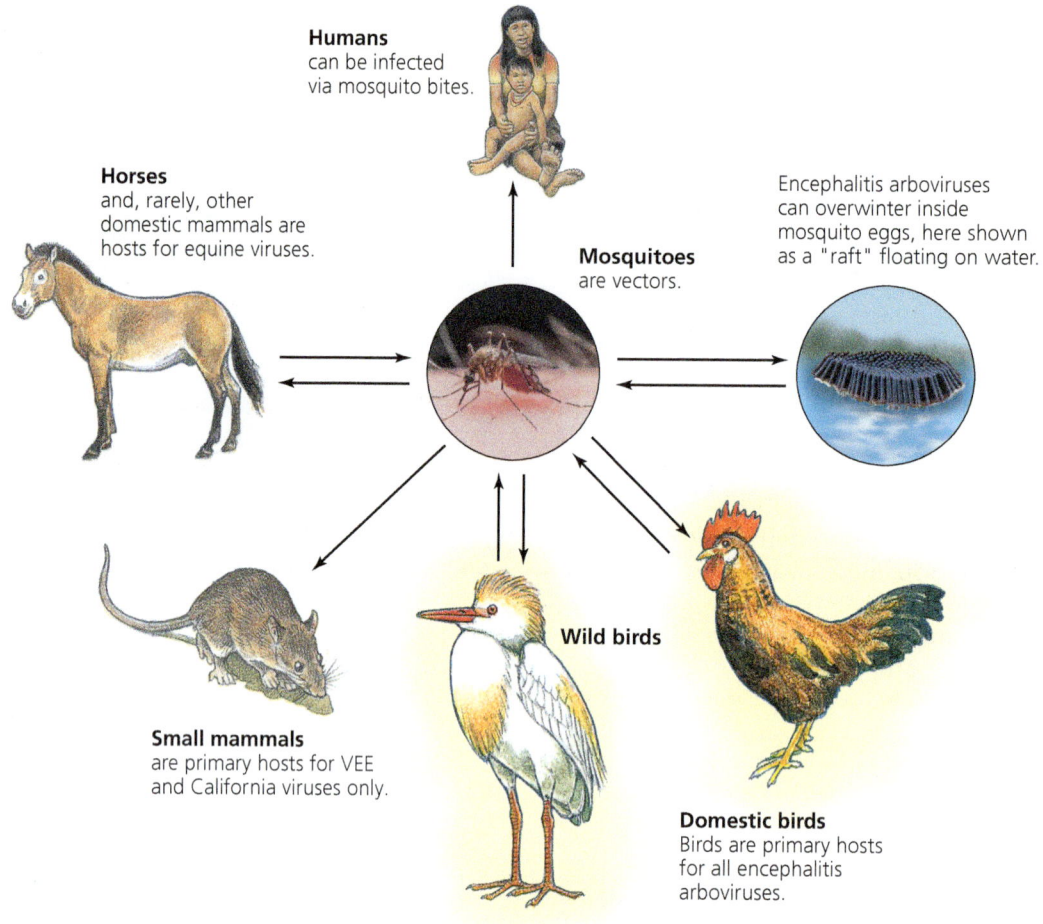

▲ **FIGURE 25.9 Hosts and transmission of viruses of equine encephalitis and West Nile encephalitis.** Mosquitoes carry Western equine encephalitis (WEE), Eastern equine encephalitis (EEE), Venezuelan equine encephalitis (VEE), and West Nile viruses from wild birds (the primary reservoir) to domesticated birds, horses, other mammals, and humans. Except for rodents carrying VEE, mammals do not normally develop viremia with these viruses, making them dead-end hosts for the virus.

Encephalitis

A number of different viruses cause encephalitis in humans. RNA togaviruses cause **Eastern equine encephalitis (EEE)**, **Western equine encephalitis (WEE)**, and **Venezuelan equine encephalitis (VEE)**. As the names indicate, viral replication can occur in the brains of horses as well as in humans, though the normal host for these viruses is either a bird (EEE, WEE, VEE) or a rodent (VEE) **(FIGURE 25.9)**. Of the three, EEE causes the most severe disease in humans, though any of them may produce fatal brain infections.

In 1999, hundreds of birds—mostly crows—suddenly began dying in the New York City area. At the same time, there were reports of people feeling ill with flulike symptoms and encephalitis after being bitten by mosquitoes. As it turned out, all these events were related. The culprit was West Nile virus (WNV). An arbovirus in the family *Flaviviridae* and endemic to Africa and Israel, West Nile virus had never before been seen in the Americas. After the initial outbreak, infected migratory birds carried the virus across the lower 48 United States. Mosquitoes further spread West Nile virus among bird populations, between birds and horses, and between birds and humans. Animals feeding on dead birds also become infected.

Eighty percent of humans infected with WNV have no symptoms; 20% have fever, headache, fatigue, and body aches. In the most severe cases—about 1 of 750 human infections—West Nile virus invades the nervous system, causing **West Nile encephalitis,** which is fatal in about 13% of acute cases. **FIGURE 25.10** illustrates the number of human cases and deaths due to neuroinvasive West Nile virus.

St. Louis encephalitis and *Japanese encephalitis* are caused by two other mosquito-borne arboviruses. These flaviviruses usually cause benign viral encephalitis. **Emerging Disease Case Study: Tick-Borne Encephalitis** examines a third flavivirus arbovirus that is not carried by a mosquito.

EMERGING DISEASE CASE STUDY

Tick-Borne Encephalitis

Analiesa's head hurt; rather, her head felt as if it were being pounded from within by gorillas with sledgehammers. This, in addition to last week's fever, nausea, and vomiting, indicated she was more than worn down by her climb two weeks ago to Krimml waterfalls—Austria's highest. There must be something else causing her muscle aches, back pain, and inability to move her shoulders properly.

Indeed, Analiesa is a victim of an emerging disease in Europe and Asia—tick-borne encephalitis (TBE). Several viruses of rodents, transmitted to humans by *Ixodes* ticks, cause TBE. Genetic analysis reveals that the viruses are related to one another and are in the family *Flaviviridae*. Doctors diagnose about 10,000 cases of TBE worldwide each year, but TBE is not a reportable disease, so many more cases likely occur. If global warming allows ticks to survive in higher latitudes and at higher elevations, and as more and more people trek into wilderness areas, scientists expect that tick-borne encephalitis will become more common.

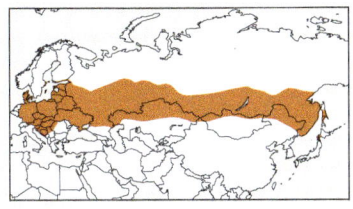

Analiesa's physician explained that there is no specific drug therapy for TBE other than painkiller for the muscle and headaches and corticosteroids for the inflammation. Her body's immune response would have to clear the virus from her system, which would take several weeks. Fortunately, it is unlikely that the virus would cause long-term complications. TBE is not considered a fatal disease, though it is classified as a category C (third highest priority) bioterrorist threat.

And should Analiesa forgo treks in the wilderness? No, but she should take advantage of the TBE vaccine, use chemical tick repellent, and avoid consuming raw milk from goats or cows—infected animals pass the virus in their milk.

1. How does encephalitis differ from meningitis?
2. Why does a brain infection manifest as muscle and joint pain?
3. Why don't standard antibiotics provide relief to Analiesa?

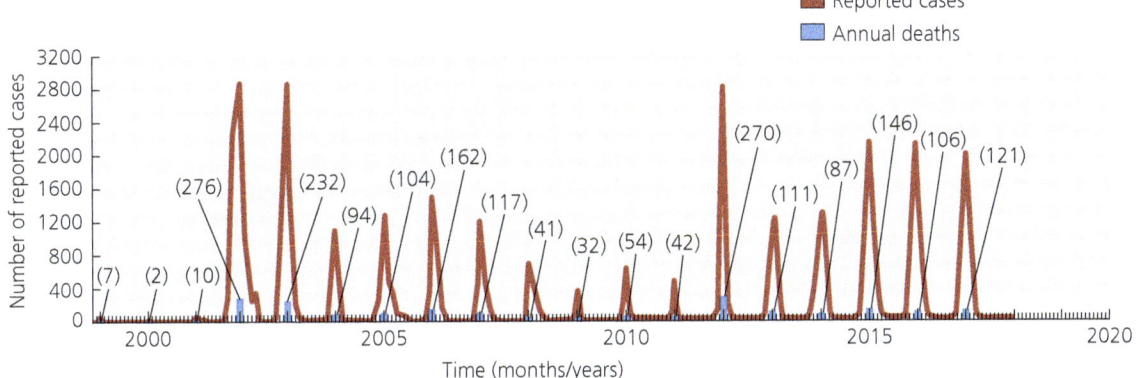

▲ **FIGURE 25.10** Human West Nile neuroinvasive infections and deaths in the United State. Note the seasonal nature of the disease.

Dengue Fever (Dengue)

Aedes mosquitoes transmit four strains of flavivirus that cause **dengue fever** (also simply called *dengue*, den'gā). Dengue viruses afflict approximately 390 million people in tropical and subtropical areas of the world. Forty percent of the world's population lives in the range of *Aedes* and is thereby susceptible to infection with dengue virus **(FIGURE 25.11)**. Dengue occurs in one or two phases separated by 24 hours of remission. First, the patient suffers for about a week from high fever, weakness, edema (swelling) of the extremities, and severe pain in the head, back, and muscles. The severity of the pain is indicated by the common name for the disease, *breakbone fever*. Some patients recover completely; others suffer the second phase, which involves a return of the fever and a bright red rash for about three days. Dengue fever is self-limiting, and a patient is usually immune to reinfection with the same strain of virus.

Dengue hemorrhagic fever is a more serious disease caused by subsequent infection with a different dengue virus strain;

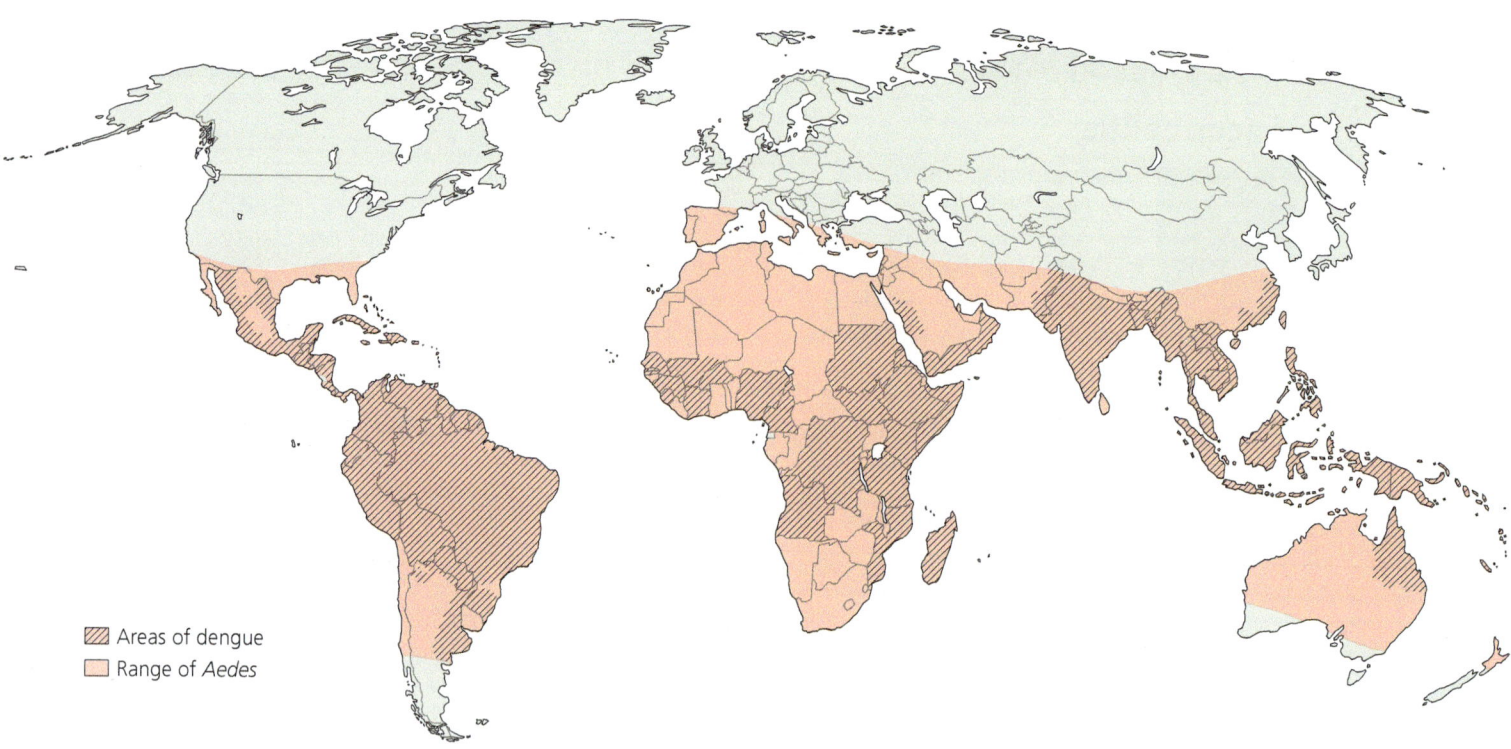

▲ FIGURE 25.11 **Dengue fever.** Global distribution of dengue fever in 2017. The U.S. population is susceptible to this disease, which afflicts millions of people worldwide because of the presence of the disease's vectors, mosquitoes in the genus *Aedes*.

it involves a hyperimmune response against the new virus (FIGURE 25.12). Inflammatory cytokines released by activated memory T cells cause rupture of blood vessels, internal bleeding, shock, and possibly death. The reason for a hyperimmune response rather than a regular response is not known.

Dengue and dengue hemorrhagic fever are epidemic in Asia, South America, and Mexico. Given that *Aedes* mosquitoes already live in the southern United States, dengue could potentially become established there if immigrants or returning travelers introduced the virus into the United States.

There is no treatment for dengue; scientists have developed an experimental vaccine, but it is limited in its efficacy and has produced adverse effects so severe that the Philippines halted its use. Mosquito control can limit the spread of the disease. **Beneficial Microbes: Eliminating Aedes-Borne Diseases** (p. 746) examines a novel approach to mosquito control.

Yellow Fever

Another virus—*Flavivirus yellow fever virus*—causes **yellow fever,** a disease involving degeneration of the liver, kidneys, and heart as well as massive hemorrhaging. Hemorrhaging in the intestines may result in "black vomit." Liver damage causes jaundice, from which the disease acquires its name and its nickname, "yellow plague." The mortality rate of yellow fever can approach 20%. Before the 20th century, yellow fever epidemics ravaged the Americas, killing thousands. For example, in 1793, it killed over 4000 people in Philadelphia (then the capital of the United States), and during the Spanish-American War of 1898, yellow fever killed more American soldiers than bullets did. With mosquito control and the development of a vaccine in the 1960s, however, yellow fever has since been eliminated from the United States and held in check in Latin America.

Yellow fever remains a significant disease, with 200,000 estimated cases and 30,000 deaths annually worldwide, 90% of which occur in tropical Africa. A single dose of yellow fever vaccine provides protection to 95% of individuals, probably for their lifetimes.

Zika Fever and Microcephaly

Flavivirus Zika virus is another enveloped, unsegmented, +ssRNA virus with an icosahedral capsid. *Aedes* mosquitoes transmit this arbovirus. The World Health Organization has verified that Zika virus can also spread from pregnant women across the placenta to infect unborn children and can be sexually transmitted vaginally, orally, or anally.

Scientists discovered the virus in 1947 in the Zika forest in Uganda, Africa, but the first cases of Zika fever in people were reported in the Federated States of Micronesia in 2007. The virus spread across the Pacific in infected people and was first seen in the Western Hemisphere in Brazil in 2015. By 2016, the virus and its disease had spread throughout South and Central American nations, to the Caribbean, and into North America.

Many people infected with Zika virus have no signs or symptoms of disease. Others have relatively mild fever, rash,

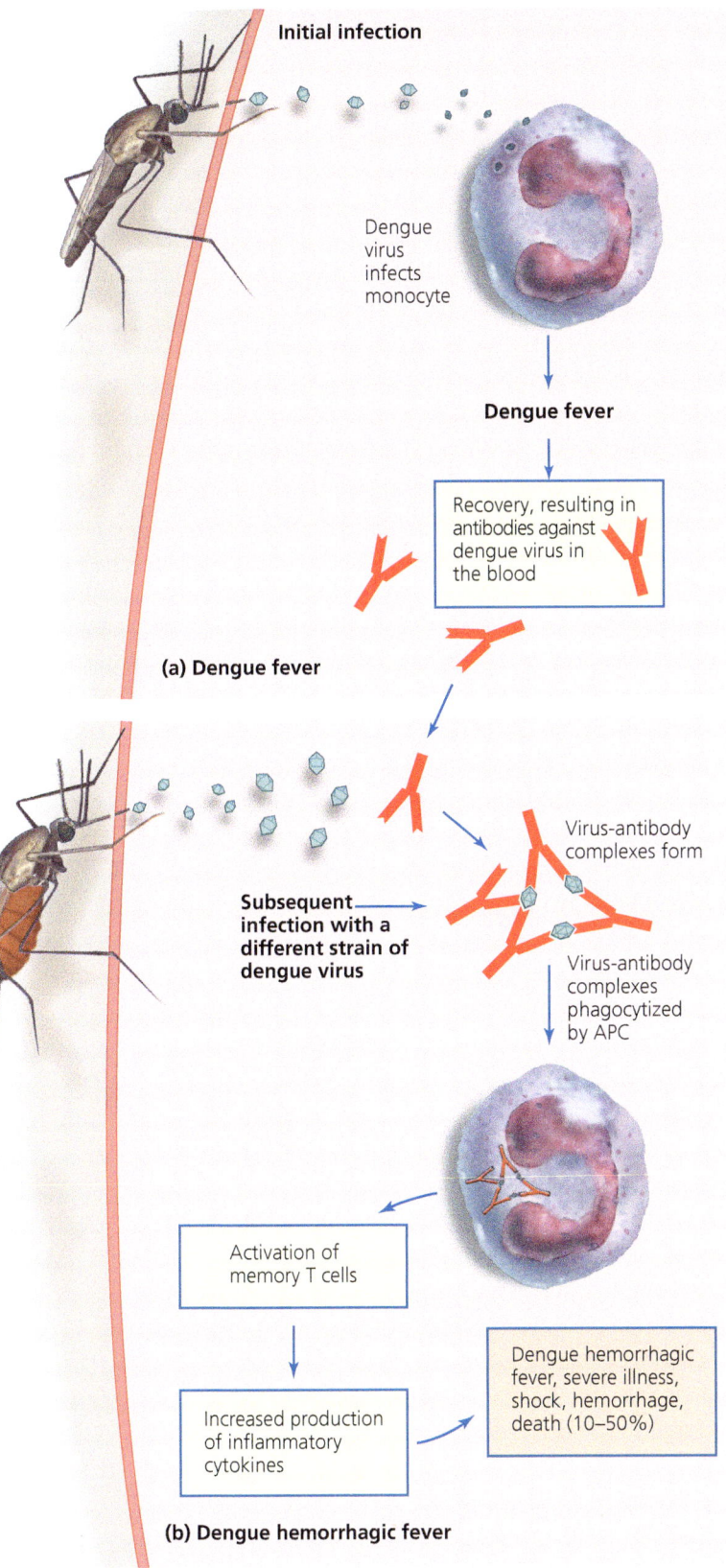

◀ **FIGURE 25.12 Dengue and dengue hemorrhagic fevers. (a)** Initial infection with any one of the four dengue viruses causes dengue fever. Recovered patients have antibodies against dengue virus circulating in their blood and, in general, are immune to reinfection with that strain. **(b)** Subsequent infection with another strain of dengue virus results in virus-antibody complexes that trigger hyperimmunity, resulting in hemorrhaging and shock.

include developmental and learning defects; problems with sight, hearing, or limb movements; *microcephaly*,[7] in which the child is born with a small head; or death (see **Emerging Disease Case Study: A Deadly Mosquito Bite?** on p. 746). Zika virus may also adversely affect normal development of the heart, digestive system, and genitalia. Many manifestations of Zika virus occur in infants who appear normal at birth.

Physicians can use real-time PCR to detect Zika RNA in patients, or they can detect anti-Zika antibodies in the serum of patients with an enzyme-linked immunosorbent assay (ELISA, or simply EIA). There is no standard treatment for Zika virus diseases; health professionals treat the symptoms and encourage patients to drink fluids to prevent dehydration, get plenty of rest, and take acetaminophen to reduce fever and pain. A Zika vaccine is being developed; in the meantime, prevention involves avoiding regions where Zika is endemic, especially when pregnant; mosquito control; and prevention of mosquito bites by using repellent, sleeping under mosquito nets, and covering the skin with clothing. Condoms can provide some protection against sexual transmission of the virus.

TABLE 25.4 p. 747 summarizes the arboviruses and their diseases, including additional viruses in the families *Bunyaviridae* and *Reoviridae* discussed below.

Diagnosis, Treatment, and Prevention of Arbovirus Infections

Serological tests, such as ELISA and agglutination of latex beads to which arbovirus antigens are affixed, are used for the diagnosis of arboviral infections. The only treatment for arboviral diseases is to provide supportive care. Only acetaminophen is recommended for managing pain and fever associated with dengue infections, as the anticoagulant properties of aspirin could aggravate the hemorrhagic properties of dengue virus infections.

Prevention of infection involves control of the vectors, which has resulted in elimination of many arboviral diseases in certain geographic areas. Insect repellents and netting reduce infections with arboviruses.

Immunization is recommended for people traveling to areas where arboviral diseases are prevalent. Vaccines for humans are available against yellow fever, Japanese encephalitis, and Russian spring-summer encephalitis viruses. Animal vaccines against the viruses that cause VEE, EEE, and West Nile encephalitis are also available. No vaccine is available for dengue.

conjunctivitis, headache, and muscle or joint pain for several days to a week. Alarmingly, Zika virus in pregnant women causes about 6% of these babies to develop *Zika congenital syndrome*. The syndrome manifests as brain abnormalities. These

---

[7]From Greek *mikro*, meaning "small," and *kephalos*, meaning "head."

BENEFICIAL MICROBES

Eliminating *Aedes*-Borne Diseases?

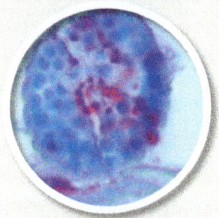

Wolbachia (stained red) inside insect cells

Aedes aegypti has distinctive stripes and a fierce bite. Unlike many mosquitoes, these aggressive bloodsuckers bite during the day. They prefer to live in urban areas, resting in the shade of houses and laying their eggs in modern containers holding a small amount of water, such as tires, cans, and water gutters. Almost 3 billion people share their neighborhoods with *Aedes*. Worst of all, these mosquitoes carry viruses that cause human diseases such as dengue.

There is no vaccine and no treatment for dengue or other viral diseases carried by *Aedes*, such as chikungunya and Zika, so prevention involves controlling the mosquitoes, which has had limited success, especially in developing countries. Enter Australian scientist Scott O'Neill and a novel beneficial bacterium—a special strain of *Wolbachia pipientis*.

Wolbachia is a Gram-negative, intracellular parasitic bacterium that infects about 70% of all insect species, including *Aedes* mosquitoes. *Wolbachia* often changes the biology of its insect hosts, and O'Neill's *Wolbachia*—strain *w*Mel—is not an exception. When male mosquitoes infected with *w*Mel mate with infected females, the females lay eggs normally, and all the offspring are infected with *Wolbachia*, which they get from their mother. However, when an infected male mates with an uninfected female, all her eggs are sterile. Thus, only Wolbachia-infected offspring are produced; *Wolbachia* ensures its own reproductive success.

It gets better! For some reason, it turns out that viruses do not replicate well in mosquitoes infected with the strain *w*Mel. So as the *w*Mel spreads through the *Aedes* population generation by generation, the success of *Wolbachia* is the demise of the viruses.

O'Neill and his collaborators have released thousands of infected mosquitoes in northern Australia and have successfully altered the mosquito population in the area so that they are unable to transmit dengue. The team plans to release millions of infected mosquitoes into neighborhoods in tropical developing nations with the goal of spreading *Wolbachia pipientis* strain *w*Mel throughout the world. They hope that this will eliminate the scourge of dengue and other *Aedes*-borne diseases forever.

EMERGING DISEASE CASE STUDY

A Deadly Mosquito Bite?

Maria is not feeling well; she woke up this morning with fever, muscle aches, and joint pain in her hands and feet. She is especially concerned because she is eight months pregnant with her first child and she's been hearing about a new disease called Zika that moves from expectant mothers across the placenta to affect their unborn children.

Maria feels nauseated by worry and fear. Could her baby be infected? Would he be malformed? Maybe she doesn't have it. Maybe she shouldn't worry. Still, she calls her obstetrician and is able to get an appointment for that afternoon.

Later, Dr. Stephens explains that Zika viral infection is rare in adults and also rarely passed to a fetus. Studies suggest that most babies born to Zika-infected mothers are normal. In any case,

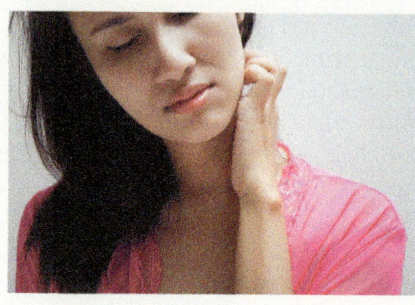

the disease in adults seldom lasts more than a week, and babies are most at risk when the pregnancy is in the first trimester; so even if Maria currently has Zika virus, her third-trimester baby is most likely safe.

Nevertheless, the doctor orders a diagnostic test for Zika—reverse transcription polymerase chain reaction (RT-PCR)—that can detect viral RNA in a patient's blood plasma. Maria quickly agrees to the test, and a nurse draws

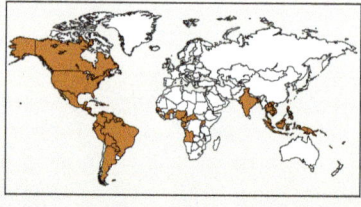

a blood sample. The doctor sends Maria home with acetaminophen for the pain and orders her to drink plenty of fluids and get lots of rest. It will take a week to get the results of the RT-PCR test from the laboratory.

A week later, Maria gets the good news—she doesn't have Zika virus. Her baby is safe! For the next six weeks she will be extra careful to avoid mosquitoes by staying inside at dawn and dusk, wearing long-sleeved blouses and long pants, and using mosquito repellent. She has had enough worry about Zika virus, mosquitoes, deformed babies, and death.

1. How does microcephaly differ from both meningitis and encephalitis?
2. Why don't standard antibiotics such as penicillin and ciprofloxacin help cure a Zika infection?
3. How would you describe PCR (polymerase chain reaction) to a patient like Maria?

TABLE 25.4 Some Diseases Caused by Arboviruses, by Viral Family

Disease	Vector	Natural Host(s)	Distribution	Symptoms
Togaviridae (enveloped, icosahedral, +ssRNA)				
Eastern Equine Encephalitis (EEE)	*Aedes, Culex,* and *Culiseta* mosquitoes	Birds	Americas	Flulike symptoms and encephalitis
Western Equine Encephalitis (WEE)	*Aedes* and *Culex* mosquitoes	Birds	Americas	Flulike symptoms and encephalitis
Venezuelan Equine Encephalitis (VEE)	*Aedes* and *Culex* mosquitoes	Birds, rodents	Americas	Flulike symptoms and encephalitis
Flaviviridae (enveloped, icosahedral, +ssRNA)				
Japanese Encephalitis	*Culex* mosquitoes	Birds, pigs	Asia	Flulike symptoms and encephalitis
West Nile Encephalitis	*Aedes, Anopheles,* and *Culex* mosquitoes	Birds	Africa, Europe, Asia, North America	Flulike symptoms and potentially fatal encephalitis
St. Louis Encephalitis	*Culex* mosquitoes	Birds	North America	Flulike symptoms and encephalitis
Russian Spring-Summer Encephalitis	*Ixodes* and *Dermacentor* ticks	Birds	Russia	Flulike symptoms and encephalitis
Dengue and Dengue Hemorrhagic Fever	*Aedes* mosquitoes	Monkeys, humans	Worldwide, especially tropics	Severe pain, hemorrhaging, hepatitis, shock
Yellow Fever	*Aedes* mosquitoes	Monkeys, humans	Africa, South America	Hepatitis, hemorrhagic fever, shock
Chikungunya	*Aedes* mosquitoes	Monkeys	Africa, Asia	Rash, nausea, joint pain
Zika Fever; Zika Congenital Syndrome	*Aedes* mosquitoes	Monkeys, apes	Americas, Africa, Southeast Asia, Pacific Islands	Fever, rash, conjunctivitis, headache, and muscle or joint pain
Bunyaviridae (enveloped, filamentous segmented, −ssRNA)				
La Crosse Encephalitis	*Aedes* mosquitoes	Rodents, small mammals, birds	North America	Fever, rash, encephalitis
Rift Valley Fever	*Aedes* mosquitoes	Sheep, goats, cattle	Africa, Asia	Hemorrhagic fever, encephalitis
Sand Fly Fever	*Phlebotomus* flies	Sheep, cattle	Africa	Hemorrhagic fever, encephalitis
Crimean-Congo Hemorrhagic Fever	*Hyalomma* ticks	Horses, cattle, goats, seabirds	Africa, Crimea	Hemorrhagic fever
Reoviridae (naked, dsRNA)				
Colorado Tick Fever	*Dermacentor* ticks	Small mammals, deer	Western North America	Fever, chills, headache, photophobia, rash, and, in children, hemorrhage

Other Diseases of Enveloped +ssRNA Viruses

LEARNING OUTCOMES

25.15 Describe the signs, effects, and prevention of rubella.
25.16 Compare and contrast hepatitis C with the other types of viral hepatitis.
25.17 Describe the structure of coronaviruses and the diseases they cause.

Rubella

Rubella (rū-bel′ă) virus (taxonomically *Rubivirus rubella virus*) shares the structure and reproductive strategy of other togaviruses, but unlike them it is transmitted by the respiratory route and not by arthropod vectors. **Rubella** has been considered one of five childhood viral diseases that produce skin lesions. (The other four are measles, caused by a −ssRNA virus discussed later in this chapter, and chickenpox, roseola, and fifth disease, which are caused by DNA viruses.) Rubella, first distinguished as a separate disease by German physicians, is commonly known as "German measles" or "three-day measles."

Rubella infects only humans, entering the respiratory system and infecting cells of the upper respiratory tract. It spreads from there to lymph nodes and into the blood and then throughout the body. Afterward, the characteristic rash of flat, pink to red spots (macules) develops (FIGURE 25.13) and lasts about three days. Rubella in children is usually not serious, but adults may develop arthritis or encephalitis. Patients shed virions in respiratory droplets for approximately two weeks before and two weeks after the rash.

Rubella was not considered a serious disease until 1941, when Norman Gregg (1892–1966), an Australian ophthalmologist, recognized that rubella infections of pregnant women resulted in severe congenital defects in their babies. These effects include cardiac abnormalities, deafness, blindness, intellectual disabilities, microcephaly, and growth retardation. Death of the fetus is also common. A mother is able to transmit the virus across the placenta even if she is asymptomatic.

Diagnosis of rubella is usually made by observation and by serological testing for IgM against rubella. No treatment is

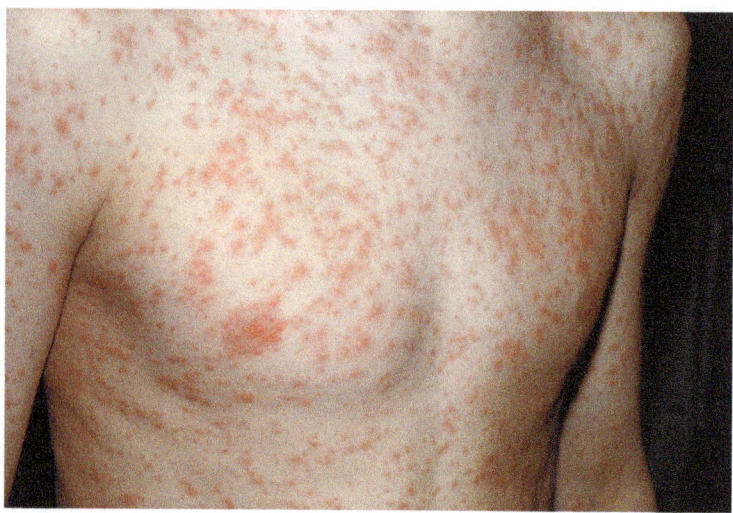

▲ FIGURE 25.13 **The rash of rubella.** The disease, also known as German measles or three-day measles, is characterized by flattened red spots (macules).

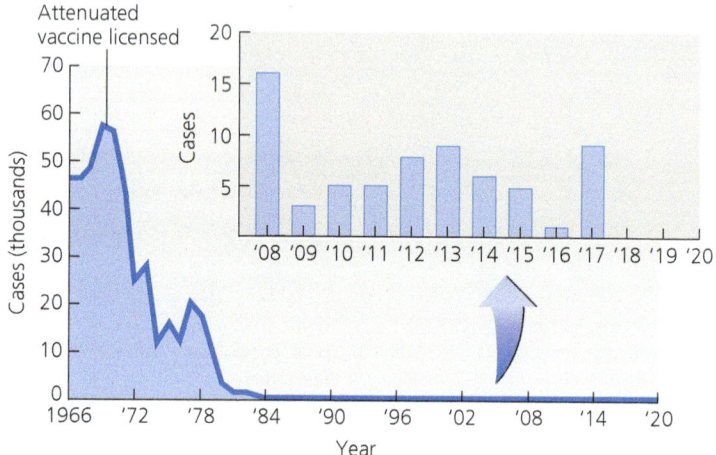

▲ FIGURE 25.14 **The efficacy of immunization against rubella.** The use of a live, attenuated virus vaccine has eliminated rubella in the United States except for a few cases acquired by travelers returning from outside the Americas.

available, but immunization has proven effective at reducing the incidence of rubella in industrialized countries (**FIGURE 25.14**). The rubella vaccine is made from a live, attenuated virus and therefore should not be given to pregnant women or immunocompromised patients. Immunization is aimed at reducing the number of rubella cases that might serve to introduce rubella into regions where home-grown rubella has been eradicated, particularly both of the Americas.

Hepatitis C

A flavivirus named *Hepacivirus hepatitis C virus,* or more commonly called **hepatitis C virus (HCV),** causes about 20% of the cases of hepatitis in the United States. About 71 million people worldwide are chronically infected by HCV, which is bloodborne and spread via needles, organ transplants, and sexual activity, but not by arthropod vectors. In the past, blood transfusions accounted for many cases of hepatitis C, but testing blood for HCV has considerably reduced the risk of infection by this means.

Hepatitis C is a chronic disease often lasting for more than 20 years, with few if any symptoms. Severe liver damage (20% of cases) and liver failure (5% of cases) occur over time, making hepatitis C the most deadly infectious disease in the U.S. Hepatic cancer sometimes results from HCV infection. **FIGURE 25.15** illustrates the prevalence of antibodies against hepatitis C across the United States.

Health care practitioners prescribe sofosbuvir, a nucleotide analog that inhibits RNA replication by HCV, curing 90% of patients with hepatitis C, as shown by inability to detect the virus on standard blood tests 12 weeks after treatment stops. There is no vaccine for hepatitis C.

Diseases of Coronaviruses

Coronaviruses (kō-rō′nă-vī′rŭs-ĕz) are enveloped, positive, single-stranded RNA viruses with helical capsids. Their envelopes form corona-like halos around the capsids, giving these viruses their name (**FIGURE 25.16**). Two coronaviruses cause colds and are the second most common cause of colds after rhinoviruses. Coronaviruses are transmitted from epithelial cells of the upper respiratory tract in large droplets (sneezes and coughs) and, like rhinoviruses, replicate best at 33°C, which is the temperature of the nasal cavity. Scientists have found MERS virus (discussed next) in bats and camels. Humans are most often infected through contact with camels or other infected people.

The most severe diseases of coronaviruses are **coronavirus respiratory syndromes**. In the spring of 2003, *severe acute respiratory syndrome (SARS)* literally changed the face of China as people donned masks to prevent spread of the virus (**FIGURE 25.17a**). Similarly, another coronavirus disease, *Middle East respiratory syndrome (MERS),* is changing the face of the Middle East and Asia (**FIGURE 25.17b**).

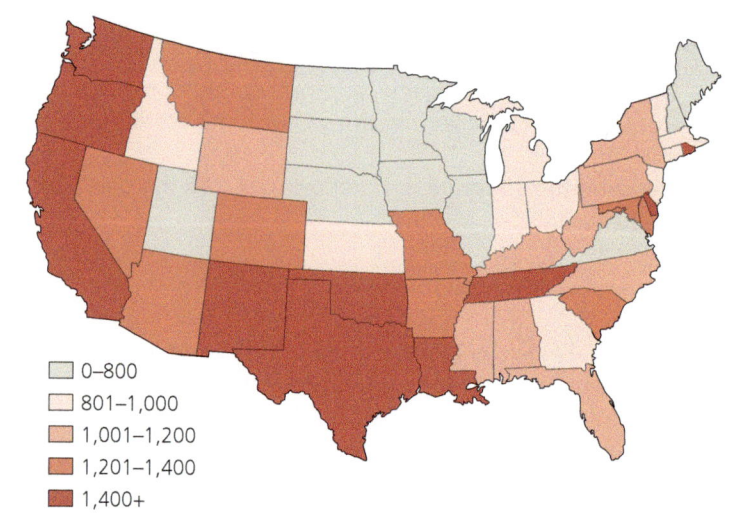

▲ FIGURE 25.15 **Estimated prevalence of hepatitis C.** Number of cases per 100,000 population, 2010.

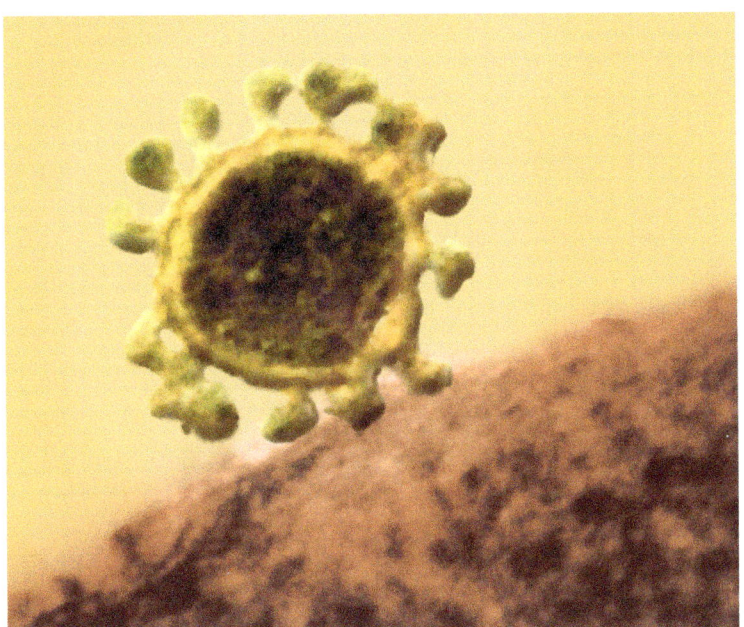

▲ FIGURE 25.16 **Coronaviruses.** These +ssRNA viruses have distinctive envelopes with glycoprotein spikes resembling flares from the sun.

Coronavirus respiratory syndromes manifest with a high fever (greater than 38°C) accompanied by shortness of breath, difficulty in breathing, malaise, and body aches. About 10–20% of patients have diarrhea. After about a week, patients develop a dry cough and pneumonia, and about 10% of SARS patients die. More than 30% of MERS patients die.

SARS initially taxed health care systems in several Asian countries, but unprecedented worldwide cooperation allowed epidemiologists to identify the virus, track its spread, sequence its genome, investigate and implement effective quarantine procedures, and bring a halt to the pandemic—all within six months—an unprecedented genomic and epidemiological accomplishment.

Progress against MERS has not been as rapid or effective, partly because there have been fewer cases, so people, including health care workers, have not been as motivated to cooperate with quarantine and infection control procedures. The spread of MERS from the Arabian Peninsula to Europe and into Korea in 2015 has alarmed epidemiologists and spurred scientists to find effective treatments and a vaccine.

Symptoms of colds can be alleviated with antihistamines and analgesics. Some physicians prescribe antiviral drugs, such as ribavirin or interferon, for SARS and MERS, but there is no consistent proof that these treatments are efficacious. Scientists have developed recombinant DNA vaccines against the SARS and MERS viruses and are testing the vaccines for safety and effectiveness. The MERS vaccine will likely be used to immunize young camels—a stratagem that may be the best way to protect people.

Primarily, physicians provide supportive care for patients with coronavirus respiratory syndromes. Epidemiologists and health care providers stopped the SARS pandemic in 2003 by quarantining patients and protecting the uninfected with face masks, and similar steps appear to work against MERS also. Scientists have developed recombinant DNA vaccines against SARS virus and are testing the vaccines for safety and effectiveness in humans. No vaccine against MERS exists.

MICRO CHECK

5. Which is the primary reservoir for West Nile virus?
6. Which arbovirus causes breakbone fever?
7. How is hepatitis C different from hepatitis A and B?
8. Which togavirus can cause severe birth defects?

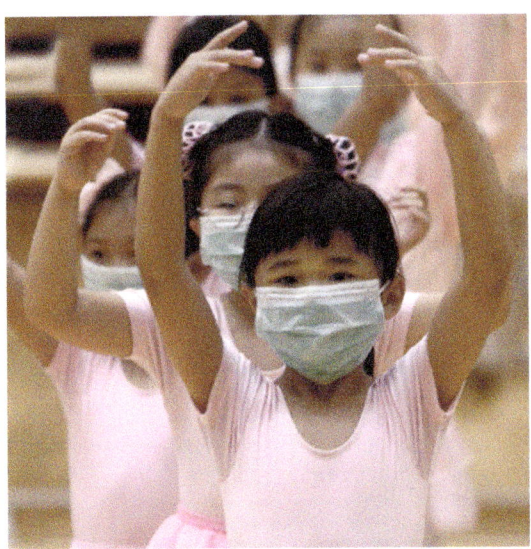
(a)

(b)

▲ FIGURE 25.17 **Prevention of coronavirus respiratory syndromes.** (a) Ballet students in Hong Kong in 2003 donned face masks to slow the spread of SARS virus. (b) MERS has struck the Middle East and Korea.

> **TELL ME WHY**
>
> Why would a vaccine that stimulated the production of memory T cells against dengue be a bad idea?

Now we examine a unique type of +ssRNA—retroviruses.

Enveloped, Positive ssRNA Viruses with Reverse Transcriptase: *Retroviridae*

LEARNING OUTCOMES

25.18 Explain how retroviruses do not conform to the central dogma of molecular biology.
25.19 Describe the steps of reverse transcription.
25.20 List two genera of retroviruses and generally describe their diseases.

Scientists have studied **retroviruses** (re′trō-vī′rŭs-ĕz) more than any other group of viruses because of their unique features and the diseases they cause. These viruses have polyhedral capsids with spiked envelopes 80 to 146 nm in diameter and genomes composed of two molecules of positive, single-stranded RNA. Each virion also contains two tRNA molecules and multiple copies of the enzymes *reverse transcriptase, protease,* and *integrase,* whose functions will be described shortly.

The unique characteristic of retroviruses is that they do not conform to what is considered the *central dogma of molecular biology.* The central dogma is a principle that genetic information flows from DNA to RNA before being translated to proteins. Retroviruses[8] transcribe DNA from RNA, which reverses the standard flow of information.

Retroviruses accomplish this amazing feat by means of a complex enzyme, **reverse transcriptase,** which transcribes double-stranded DNA from single-stranded RNA. Transcription occurs as the enzyme moves along an RNA molecule and can

---
[8]From Latin *retro,* meaning "reverse."

best be understood as occurring in three steps, though in reality the events occur simultaneously (FIGURE 25.18):

1 An RNA–DNA hybrid is made from the +RNA genome using tRNA carried by the virion as a primer for DNA synthesis. The DNA portion of the hybrid molecule is negative, single-stranded DNA (−ssDNA); that is, it is a DNA nucleotide sequence that is opposite a genetic sequence.

2 The RNA portion of the hybrid molecule is degraded by reverse transcriptase, leaving the −ssDNA.

3 The reverse transcriptase transcribes a complementary +ssDNA strand to form double-stranded DNA (dsDNA), and the tRNA primer is removed.

The discovery of reverse transcriptase was a momentous event in biology, and it made possible an ongoing revolution in recombinant DNA technology and biotechnology. For example, it is much easier to isolate mRNA for a particular protein (e.g., human insulin) and then use reverse transcriptase to make a gene than it is to find the gene itself amid the DNA of 23 pairs of human chromosomes.

In the following sections, we consider two types of retroviruses, both of which infect humans: those that are primarily oncogenic (genus *Deltaretrovirus*) and those that are primarily immunosuppressive (genus *Lentivirus*).

Oncogenic Retroviruses (*Deltaretrovirus*)

LEARNING OUTCOME

25.21 Describe one way oncogenic retroviruses may induce cancer.

Retroviruses have been linked with cancer ever since they were first isolated from tumors in chickens. In 1981, Robert Gallo (1937–) and his associates isolated for the first time a human retrovirus, **human T-lymphotropic virus 1 (HTLV-1),** from a patient with *adult acute T-cell lymphocytic leukemia,* a cancer of humans. Since then, two other human retroviruses of this type have been discovered: HTLV-2, which causes a rare cancer called *hairy cell leukemia* because of distinctive extensions of the cytoplasmic

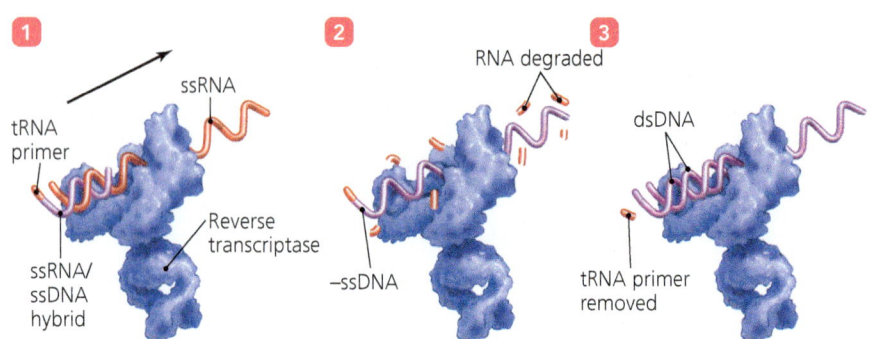

◀ **FIGURE 25.18 Action of reverse transcriptase, depicted here as three distinct steps.** **1** The enzyme transcribes a complementary −ssDNA molecule to form a DNA-RNA hybrid. It uses tRNA brought from a previous host cell as a primer for transcription. **2** The RNA portion of the hybrid is degraded, leaving −ssDNA. **3** The enzyme synthesizes a complementary +ssDNA strand, forming dsDNA.

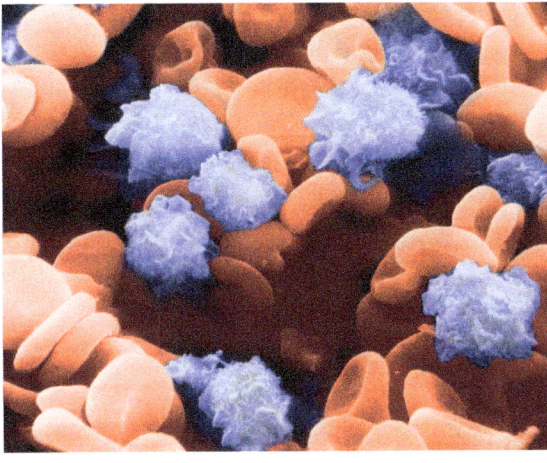

▲ FIGURE 25.19 **The characteristic extensions of the cytoplasmic membrane in hairy cell leukemia.** The infected white blood cells (colored purple) appear ruffled; red blood cells are also shown.

TABLE 25.5 Opportunistic Infections and Tumors of AIDS Patients

Type	Manifestations
Bacterial Infections (19–21)[a]	Tuberculosis, especially extrapulmonary (*Mycobacterium*)
	Rectal gonorrhea (*Neisseria*)
	Recurrent fever and septicemia due to *Salmonella*, *Haemophilus*, or *Streptococcus*
Fungal Infections (22)	*Pneumocystis* pneumonia
	Thrush, disseminated in trachea, lungs, esophagus (*Candida*)
	Histoplasmosis (*Histoplasma*)
Protozoal Infections (23)	Toxoplasmosis (*Toxoplasma*)
	Chronic diarrhea (*Cryptosporidium*, *Isospora*)
Viral-Induced Tumors (24)	Kaposi's sarcoma (especially when associated with HHV-8)
	Lymphoma (induced by Epstein-Barr virus)
Viral Infections (24)	Cytomegalovirus disseminated in brain, lungs, retina, etc.
	Human herpesvirus 1 and *2* disseminated in lungs, GI tract, etc.
	Progressive multifocal leukoencephalopathy (JC virus)
	Oral hairy leukoplakia (Epstein-Barr virus)
Others	Wasting disease; called *slim* in Africa (cause unknown)
	Dementia

[a]Numbers in parentheses refer to chapters where relevant material is discussed.

membrane of infected cells (**FIGURE 25.19**), and HTLV-5, which has not been linked to cancer or any other disease. Oncogenic retroviruses cause cancer in various ways. For example, HTLV-1 codes for a protein called *Tax* that activates genes for cell growth and division in helper T lymphocytes.

HTLV-1 and HTLV-2 infect lymphocytes and are transmitted within these cells to other people via sexual intercourse, blood transfusion, and contaminated needles. HTLV-1 is found primarily in populations in Japan and the Caribbean and among African Americans. HTLV-2 is not as well studied as HTLV-1. An enzyme-linked immunosorbent assay (ELISA) can reveal infection with an HTLV. As with many viral diseases, there is no specific antiviral treatment. Infections are chronic, and the long-term prognosis of patients is poor. Prevention involves the same changes in behavior needed to prevent HIV infection (see p. 757).

Immunosuppressive Retroviruses (*Lentivirus*) and Acquired Immunodeficiency Syndrome

LEARNING OUTCOMES

25.22 Differentiate between a disease and a syndrome, using AIDS as an example.
25.23 Describe the virion that causes AIDS.
25.24 Explain the roles of gp41 and gp120 in HIV infection.
25.25 Describe the relationship of helper T cells to the course of AIDS.
25.26 List four measures that can reduce the spread of AIDS.

From the time of its discovery in 1981 among homosexual males in the United States to its emergence as a worldwide pandemic, no affliction has affected modern life as much as **acquired immunodeficiency syndrome (AIDS)**. AIDS is not a single disease but a **syndrome**, that is, a complex of signs, symptoms, and diseases associated with a common cause. Currently, epidemiologists define AIDS as the presence of several opportunistic or rare infections along with infection by **human immunodeficiency virus (HIV)**, or as a severe decrease in the number of lymphocytes called helper T cells (<200/μL of blood) and a positive test showing the presence of HIV. Helper T cells are also known as CD4 cells because the protein CD4 is found in their membranes.

The defining infections and diseases of AIDS include those of several bacteria, fungi, protozoa, and viruses as well as a rare cancer of blood vessels (Kaposi's sarcoma) and dementia (**TABLE 25.5**). It is important to note that HIV infection is not AIDS, though infection with HIV almost always leads to AIDS in untreated people.

There are two major types of HIV. Luc Montagnier (1932–) and his colleagues at the Pasteur Institute in Paris discovered **HIV-1** in 1983. This virus is more prevalent in the United States and in Europe. **HIV-2**, the prevalent strain in West Africa, shares about 50% of its nucleic acid sequence with HIV-1. Note that the acronym HIV indicates that it is a virus. It is repetitious and incorrect to call this pathogen "HIV virus"; a more appropriate name is "AIDS virus" or simply HIV. Taxonomists classify HIV in the genus *Lentivirus*, family *Retroviridae*. Researchers have studied HIV-1 more thoroughly, so the following sections focus on this strain.

TABLE 25.6 Characteristics of HIV That Challenge the Immune System

Characteristic	Effect(s)
Retrovirus with a genome that consists of two copies of +ssRNA	Reassortment of viral genes possible; reverse transcription produces much mutation and thus genetic variation; genome integrates into host's chromosome
Targets helper T cells especially but also macrophages, dendritic cells, and muscle cells and possibly liver, nerve, and epithelial cells	Permanently infects key cells of host's immune system
Antigenic variability	Numerous antigenic variations due to mutations helps virus evade host's immune response
Induces formation of syncytia (fusion of infected cells with neighboring cells)	Increases routes of infection; intracellular site helps virus evade immune detection

Structure of HIV

HIV-1 is a typical retrovirus in shape, components, and size (see **Microbe at a Glance: *Lentivirus human immunodeficiency virus* [HIV]**, p. 754). Two antigenic glycoproteins characterize its envelope. The larger glycoprotein, named **gp120**,[9] is the primary attachment molecule of HIV. Its antigenicity changes during the course of prolonged infection, making it difficult for the body to make an effective antibody response against it. A smaller glycoprotein, **gp41,** promotes fusion of the viral envelope with the cytoplasmic membrane of a CD4 cell. The effects of these structural characteristics—antigenic variability and the ability to fuse with host cells—interfere with clearance of HIV from a patient.

TABLE 25.6 summarizes some of the characteristics that make HIV particularly difficult for the immune system to combat.

Origin of HIV

Evidence suggests that HIV arose from mutation of a similar virus—*simian immunodeficiency virus (SIV)*—found in monkeys, chimpanzees, and other simians; the nucleotide sequence of SIV is similar to those of the two strains of HIV. Based on mutation rates and the rate of antigenic change in HIV, researchers have determined that HIV likely emerged in the human population in the 1920s in Kinshasa, a city in the Democratic Republic of Congo (DRC), Africa. Scientists have identified antibodies against HIV in human blood stored since 1959, though they did not document the first case of AIDS until 1981.

The historical relationships between the two types of HIV and SIV are not clear. HIV-1 and HIV-2 may be derived from different strains of SIV, or one type of HIV may be derived from the other. We may never know the exact evolutionary relationships among the immunodeficiency viruses.

Replication of HIV

Human immunodeficiency virus replicates as a typical retrovirus. We can consider replication as occurring in eight steps **(FIGURE 25.20)**. We first look at the eight-step replication process and then examine details of attachment and entry.

1. **Attachment.** HIV primarily attaches to four kinds of cells: helper T cells; cells of the macrophage lineage, including monocytes, macrophages, and microglia (special phagocytic cells of the central nervous system); smooth muscle cells, such as those in arterial walls; and dendritic cells. HIV can also rarely infect nerve cells, liver cells, and some epithelial cells.

    Additionally, B lymphocytes attach to HIV that has been covered with complement proteins. Though HIV does not infect these B cells, the B cells deliver HIV to helper T cells, which then become infected. Similarly, infected macrophages can pass HIV to helper T cells.

2. **Entry.** HIV triggers the cell to endocytize the virus; that is, the cell's cytoplasmic membrane forms a pocket and folds in to surround the virus, forming a vesicle holding the virus.

3. **Uncoating.** The viral envelope fuses with the vesicle's membrane, and the intact capsid of HIV enters the cytosol. The virus then uncoats the capsid and releases its two ssRNA molecules from the capsid into the cell's cytoplasm.

4. **Synthesis of DNA.** Reverse transcriptase, which has been released from the capsid, synthesizes double-stranded DNA (dsDNA) using viral ssRNA as a template (see Figure 25.18).

5. **Integration.** The dsDNA made by reverse transcriptase enters the nucleus and becomes part of a human DNA molecule. There it remains as a part of the cell for life—a condition known as **latency**.

6. **Synthesis of RNA and polypeptides.** An infected cell transcribes integrated HIV genes to produce messenger RNA and multiple copies of viral ssRNA that will act as genomes for new viruses. Ribosomes within the infected cell translate mRNA to make viral-encoded polypeptides. These include attachment proteins, integrase, and a large polypeptide composed of inactive reverse transcriptase and capsomeres. The attachment proteins are inserted in the host's cytoplasmic membrane.

7. **Release.** Two molecules of genomic RNA, molecules of tRNA, and several viral polypeptides bud from the host's cytoplasmic membrane to form an immature virion.

8. **Assembly and maturation.** HIV that buds from a cell is nonvirulent because its capsid is not fully functional, and reverse transcriptase is inactive. **Protease,** a viral enzyme packaged in the virion, cleaves the large polypeptide to release reverse transcriptase and capsomeres. This action of protease, which occurs only after the virus has budded from the cell, allows final maturation of the viral capsid. HIV is now active.

Now, let us examine HIV replication in more detail.

---
[9]gp120 is short for glycoprotein with a molecular weight of 120,000 daltons.

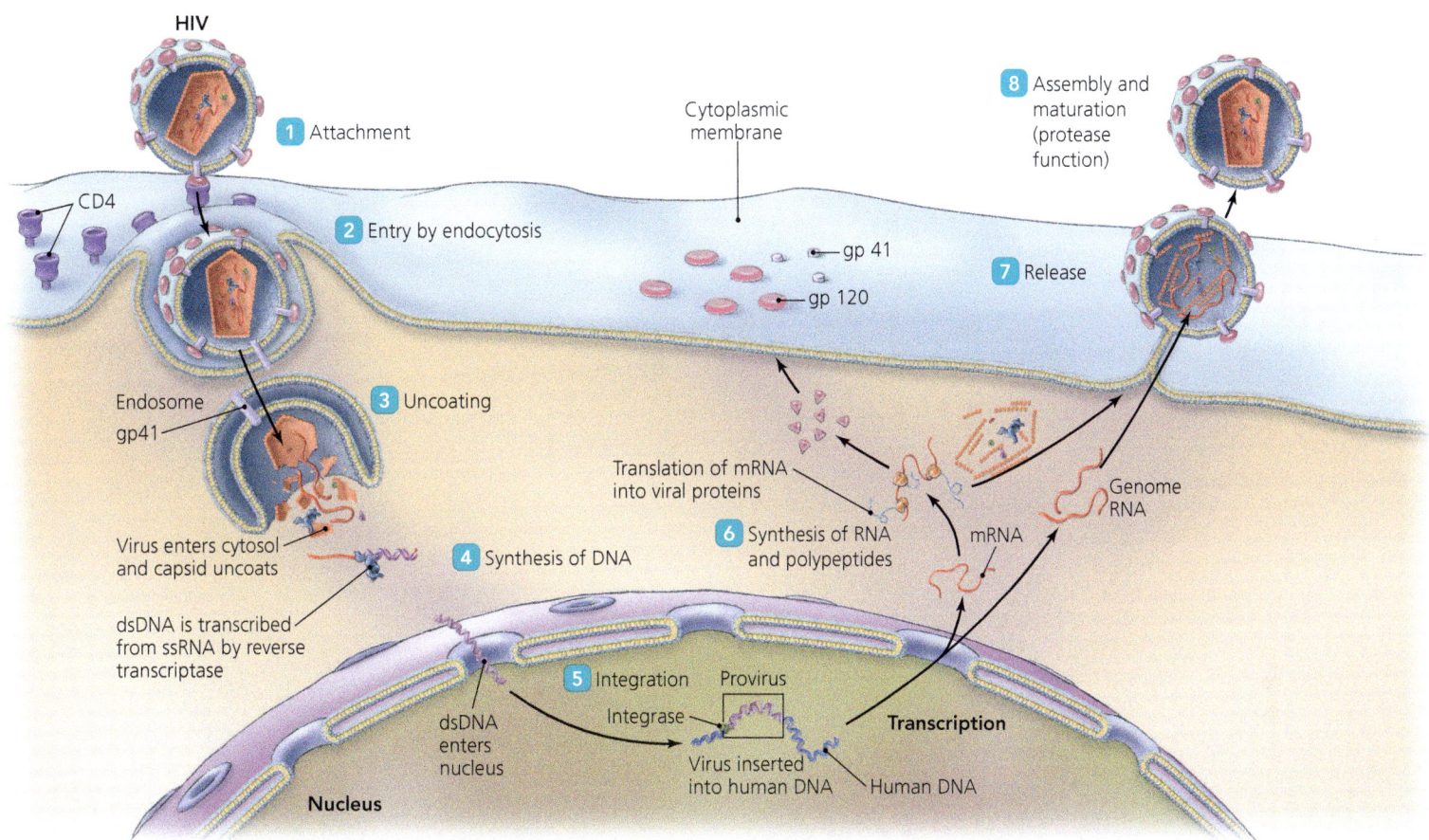

▲ FIGURE 25.20 **The replication cycle of HIV.** The artist's rendition depicts eight steps involved in the replication of the virus within a helper T cell. *How does the replication of HIV differ from the replication of bacteriophage T4?*

Figure 25.20 Bacteriophage T4 is a DNA virus; therefore, it does not have reverse transcription. Further, T4 does not integrate into a bacterium's chromosome; it assembles completely before being released from the cell; it has no envelope; and it does not carry enzymes.

Details of Attachment, Entry, and Uncoating

HIV uses gp120 and gp41 to attach to and enter target cells (FIGURE 25.21). CD4[10] on a target cell's cytoplasmic membrane is the receptor for gp120; that is, gp120 attaches HIV to CD4 ❶. The CD4-gp120 complex binds to another membrane receptor, called *fusin* (also known as *CXCR4*), which triggers the cell's membrane to move out and surround the virus—a process called *endocytosis* ❷. Endocytosis forms a bubble of membrane with the virus inside. This structure is called an *endosome* ❸.

HIV remains intact within an endosome for about 30 minutes. Glycoprotein 41 on the viral envelope evidently then facilitates fusion of the envelope with the endosome membrane. The two fuse ❹, and the viral capsid is introduced intact into the cell's cytosol ❺. The viral capsid uncoats, releasing viral RNA and proteins ❻.

Details of Synthesis and Latency

Reverse transcriptase, which was carried inside the capsid, becomes active in the cytosol. It uses tRNA as a primer to transcribe dsDNA from the ssRNA genome of the virus (see Figure 25.18). Reverse transcriptase is very error prone, making about five errors per genome. This generates multiple antigenic variations of HIV. Billions of variants may develop in a single patient over the course of the syndrome.

HIV is a latent virus. The dsDNA made by reverse transcriptase is known as a *provirus,* and it enters the nucleus. A viral enzyme known as **integrase** inserts the dsDNA provirus into a human chromosome. Once integrated, the provirus permanently remains part of the cellular DNA. It may remain dormant for years or be activated immediately, depending on its location in the human genome and the availability of promoter DNA sequences (see Figure 7.9). Infected macrophages and monocytes are major reservoirs of integrated HIV, and they serve as a means of distribution of the virus throughout the body.

An infected cell replicates integrated DNA every time cellular DNA is replicated. In this way, HIV ends up infecting all

---
[10]Scientists commonly name membrane proteins with letter-number combinations; CD4 stands for the fourth-discovered *cluster of differentiation.*

MICROBE AT A GLANCE 25.1

Lentivirus human immunodeficiency virus (HIV)

Taxonomy: Family *Retroviridae*

Genome and morphology: Two molecules of positive single-stranded RNA; polyhedral capsid with spiked envelope, about 110 to 145 nm in diameter

Host: Human

Virulence factors: Attaches to CD4 of helper T lymphocytes, vigorous mutation rate due to transcription errors of reverse transcriptase, latency

Conditions caused: HIV infection (HIV), acquired immunodeficiency syndrome (AIDS)

Treatment for disease: Antiretroviral therapy (ART), which consists of two to three antiviral drugs, such as nucleoside reverse transcriptase inhibitors (NRTIs), protease inhibitors, integrase inhibitors, and drugs called boosters that increase the half-life of the other drugs

Prevention of disease: Sexual abstinence; monogamy; refrain from sharing intravenous needles; correct, consistent condom usage can reduce but not eliminate risk of infection; male circumcision reduces risk of infection

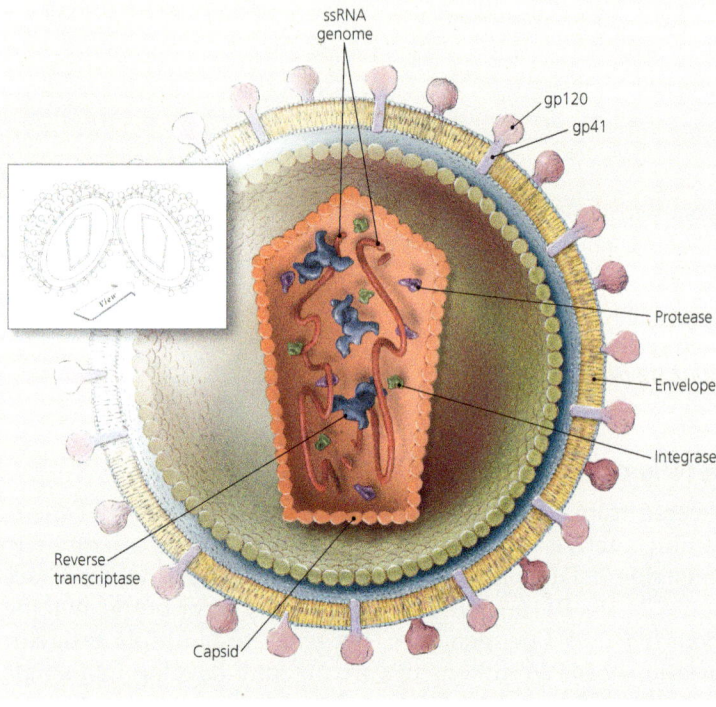

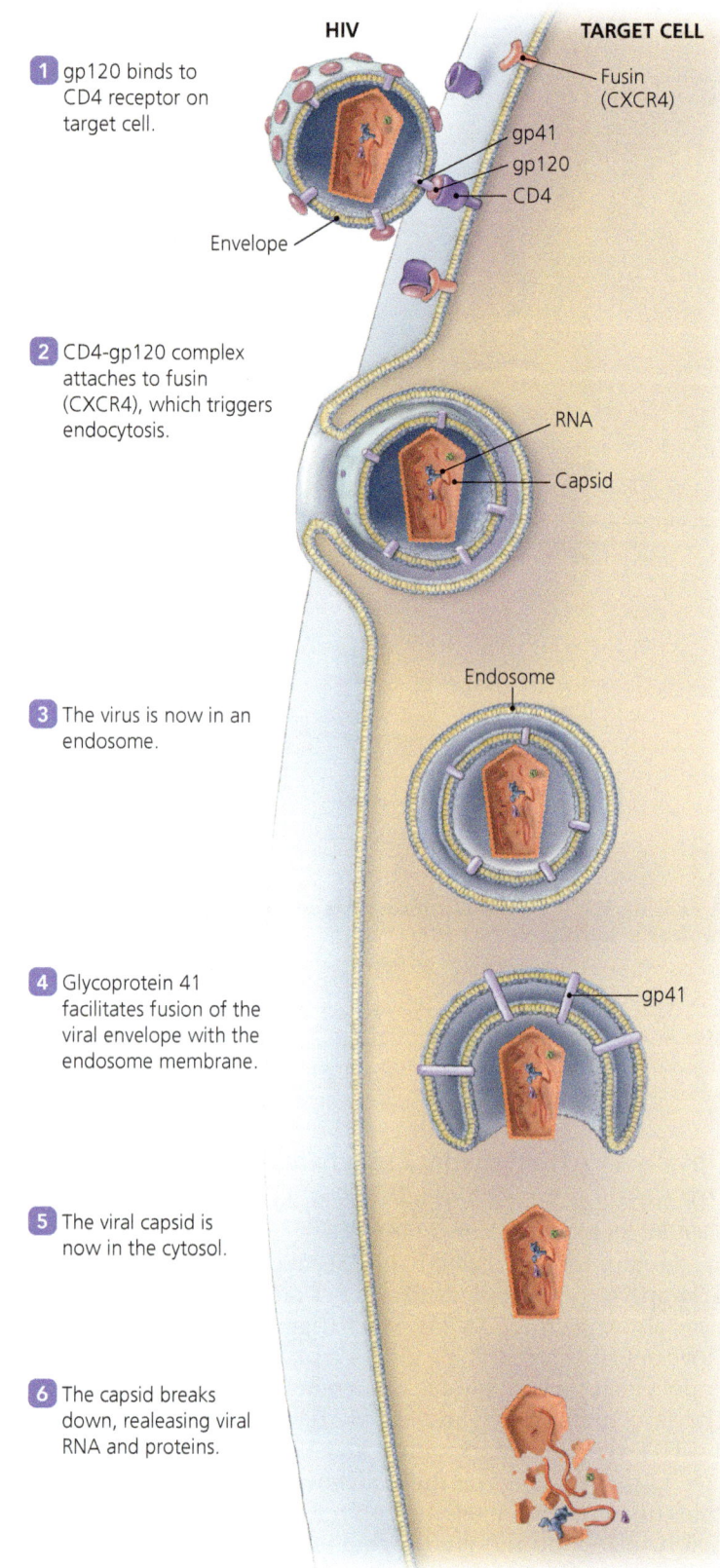

▲ FIGURE 25.21 The process by which HIV attaches to and enters a host cell.

progeny of infected cells. An infected cell may also transcribe integrated HIV to make viral messenger RNA molecules as well as entire RNA copies of the whole HIV genome.

Details of Release, Assembly, and Maturation

HIV actively participates in its release from the cell. A viral protein selects a *lipid raft*—a region of regularly packed lipids—in the cytoplasmic membrane as the point of exit. As the virus blebs from the cell, components of the raft become the envelope of the virion. Once the virus is outside the cell, capsomeres organize to form an immature capsid, and viral *protease* cleaves a

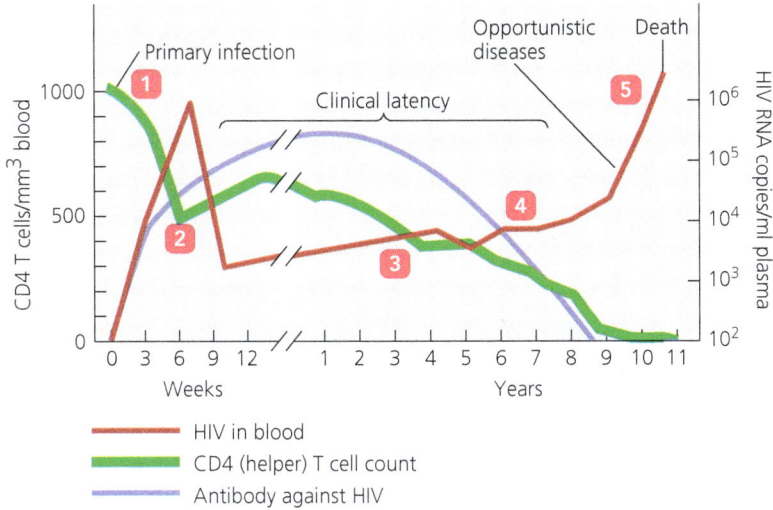

▲ FIGURE 25.22 The course of AIDS follows the course of helper T cell destruction.

polypeptide within the capsid to release functional proteins. The proteins cause the virus to mature and become infective. *Protease inhibitors*—drugs that interfere with the function of protease—have become a standard therapeutic agent in the treatment of HIV infection and have significantly lengthened the life expectancy of patients.

Pathogenesis of AIDS

Only human cells replicate HIV effectively, and, as its name indicates, the virus causes destruction of a human's immune system. **FIGURE 25.22** illustrates the observation that the destruction of helper T (CD4) cells directly relates to the course of AIDS. Initially, there is a burst of virion production and release from infected cells ❶. Fever, fatigue, weight loss, diarrhea, and body aches accompany this primary infection.

The immune system responds by producing antibodies, and the number of free virions (red line) plummets ❷. The body and HIV are waging an invisible war with few signs or symptoms. During this period, the body destroys almost a billion virions each day, but cytotoxic T cells kill about 100 million virally infected CD4 cells in the same time period. No specific symptoms accompany this stage, and the patient is often unaware of the infection.

Integrated viruses continue to replicate and virions are released into the blood to such an extent that the body cannot make enough helper T cells (green line) ❸. Over the course of 5 to 10 years, the number of helper T cells declines to a level that severely impairs the immune response. The rate of antibody formation (purple line) falls precipitously as helper T cell function is lost.

HIV production climbs (red line) ❹, and eventually the patient dies ❺. Many of the diseases associated with the loss of immune function in AIDS (see Table 25.5 on p. 751) are nonlethal infections in other patients, but AIDS patients cannot effectively resist them. Diseases such as Kaposi's sarcoma, disseminated herpes, toxoplasmosis, and *Pneumocystis* pneumonia occur rarely except in AIDS patients (**FIGURE 25.23**).

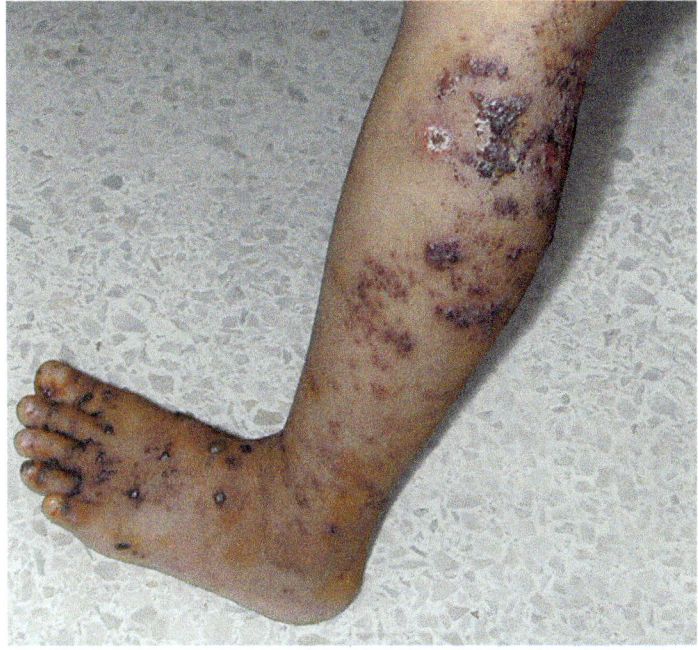

(a)

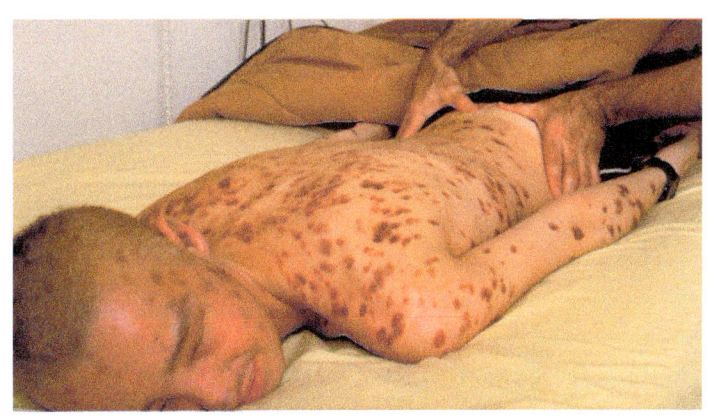

(b)

▲ FIGURE 25.23 Some diseases associated with AIDS. (a) Disseminated herpes. (b) Kaposi's sarcoma. *Which viruses cause herpes?*

Figure 25.23 Human herpesvirus 1 and human herpesvirus 2 cause herpes.

Epidemiology of AIDS

Epidemiologists first identified AIDS in young male homosexuals in the United States, but now AIDS is worldwide. WHO estimated as of December 2016 AIDS had killed more than 35 million people and that another approximately 37 million people are infected with HIV worldwide. Approximately 240 new infections occur each hour worldwide, including over 100 new cases in the United States daily. About one-third of those infected have already developed AIDS; most of the rest will eventually develop the syndrome if they don't get treatment. The spread of AIDS in sub-Saharan Africa is particularly horrific (**FIGURE 25.24**).

All body secretions of AIDS patients contain HIV; however, viruses typically exist in sufficient concentration to cause infection only in blood, semen, vaginal secretions, and breast milk.

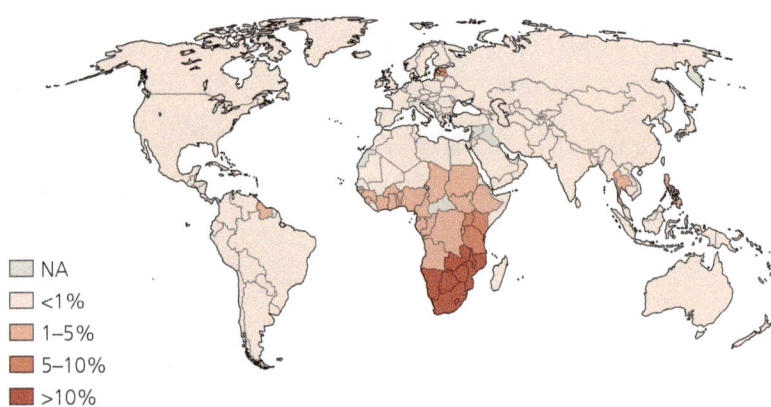

▲ FIGURE 25.24 **The global distribution of HIV/AIDS.** Figures indicate the estimated prevalence rate for adults (15 and older) living with HIV/AIDS as of the end of 2017. Note the alarmingly high number of people infected in sub-Saharan Africa.

Virions may be free or inside infected leukocytes. Infected blood contains 1000 to 100,000 virions per milliliter, while semen has 10 to 50 virions per milliliter. Other secretions have lower concentrations and are less infective than blood or semen.

Infected fluid must be injected into the body or encounter a tear or lesion in the skin or mucous membranes. Because HIV is only about 90 nm in diameter, a break in the protective membranes of the body allowing entry of HIV may be too small to bleed or to see or feel. Sufficient numbers of virions must be transmitted to target cells to establish an infection. Though scientists do not know the exact number of virions required, the number probably varies with the strain of HIV and the overall health of the patient's immune system.

HIV is transmitted primarily via sexual contact (including vaginal, anal, and oral sex, either homosexual or heterosexual) and needle sharing (FIGURE 25.25). Blood transfusions, organ transplants, tattooing, and accidental medical needlesticks have also transmitted HIV, though rarely. The risk of infection from a needlestick injury involving an HIV-infected patient is less than 1%. Mothers also transmit HIV to their babies across the placenta and in breast milk; HIV infects approximately one-third of babies born to HIV-positive women.

Factors that increase the risk of infection include the following:

- Anal intercourse, especially receptive anal intercourse
- Sex with more than one partner
- Needle sharing
- Sexual intercourse with anyone who has practiced any behavior named in the previous three categories

The mode of infection is unknown for a small number of AIDS patients, who are often unable or unwilling to answer questions about their sexual and drug abuse history. The U.S. Centers for Disease Control and Prevention (CDC) has documented a few cases of casual spread of HIV, including infections from sharing razors and toothbrushes and from mouth-to-mouth kissing. In all cases of casual spread, researchers suspect that small amounts of the donor's blood may have entered abrasions in the recipient's skin or mouth.

Diagnosis, Treatment, and Prevention

Physicians diagnose AIDS on the basis of unexplained weight loss, fatigue, fever, and fewer than 200 CD4 lymphocytes per microliter[11] of blood combined with other signs and symptoms, which vary according to the diseases involved, and the demonstration of antibodies against HIV. Recall that by one definition AIDS is the presence of one or more rare diseases and anti-HIV antibodies.

HIV itself is difficult to locate in a patient's secretions and blood because it becomes a provirus inserted into chromosomes and because for years it can be kept in check by the immune system. Therefore, diagnosis involves detecting antibodies against HIV in the blood using enzyme-linked immunosorbent assay (ELISA) and immunoblot testing. Most individuals develop antibodies within six months of infection, though some remain without detectable antibodies for up to three years. A positive test for antibodies does not mean that HIV is currently present or that the patient has or will develop AIDS; it merely indicates that the patient has been exposed to HIV. Doctors use a PCR test for HIV RNA to make a definitive diagnosis.

A small percentage of infected individuals, called *long-term nonprogressors,* do not develop AIDS even years or decades after infection. It appears that either these individuals are infected with defective virions, they have a mutated fusin receptor that does not bind effectively to HIV, or they have unusually well-developed immune systems.

Discovering new and more effective treatments for AIDS is an area of intense research and development. Currently physicians prescribe **antiretroviral therapy (ART),** previously known as *highly active antiretroviral therapy (HAART).* ART is a "cocktail" of three or more antiviral drugs to reduce viral replication, including nucleotide analogs, integrase inhibitors, protease inhibitors (e.g., darunavir), and reverse transcriptase inhibitors. ART is expensive and generally must be taken on a strict

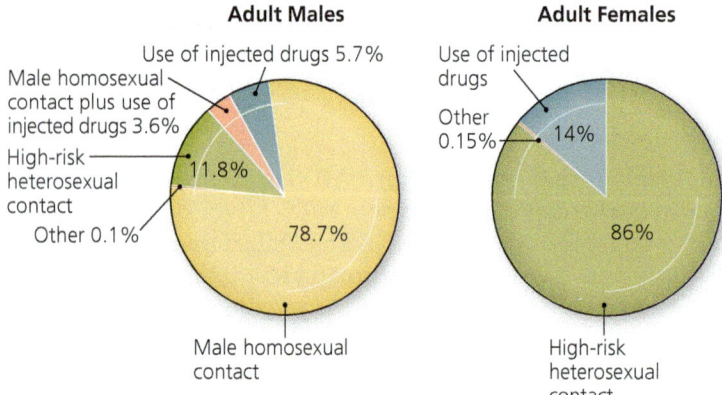

▲ FIGURE 25.25 **Modes of HIV transmission in people over 12 years of age in the United States.** Percentages represent the estimated proportions of HIV infections that result from each type of transmission. (Estimates are rounded to the nearest 1% and thus do not add up to 100%.)

---

[11]The normal value for CD4 cells is 500 to 1500 cells/μL.

schedule. Studies indicate that the therapy stops the replication of HIV because strains of the virus are unlikely to develop resistance to all of the drugs simultaneously. As long as treatment continues, a patient can live a relatively normal life; however, treatment is not a cure because the infection remains. Some scientists estimate that it would take 60 years on ART for all HIV-infected cells to die. In addition to ART, physicians manage and treat individual diseases associated with AIDS on a case-by-case basis. Doctors can use a PCR test for the presence of HIV RNA to measure the effectiveness of ART.

Researchers are working to develop methods that will prevent attachment of HIV to target cells, block the entry of HIV into cells, and halt HIV synthesis and release. Pre-exposure prophylaxis (PrEP), which involves a daily dose of tenofovir and emtricitabine, prevents replication of the virus in patients. Other studies show that treatment should begin immediately after diagnosis of infection.

Progress in developing a vaccine against HIV has been disappointing. Among the problems that must be overcome in developing an effective vaccine are the following:

- A vaccine must generate both antibody immunity, especially secretory antibody (IgA), and cell-mediated immunity, so that IgA in mucus would reduce sexual transmission of the virus and cytotoxic T lymphocytes could eliminate infected cells.
- Induction of synthesis of gamma class antibodies (IgG), a necessary function of a vaccine, can actually be detrimental to a patient. Because IgG-HIV complexes bind to B cells and also remain infective inside phagocytic cells, a vaccine must stimulate cell-mediated immune response more than antibody immune response so as to kill infected cells.
- HIV is highly mutable, generating antigenic variants that enable it to evade the immune response. Scientists estimate the diversity of HIV sequences found in one person at any one time is greater than the diversity of flu viruses in everyone worldwide during a full year.
- HIV can spread directly from cell to cell because gp41 remains on an infected cell's cytoplasmic membrane after HIV has entered. gp41 can fuse the cell to neighboring cells, allowing HIV access to the neighbor while evading some immune surveillance.
- HIV infects and inactivates macrophages, dendritic cells, and helper T cells—cells that combat infections.
- Testing a vaccine presents ethical and medical problems because HIV is a pathogen of humans only. For example, researchers stopped a study in 1994 when HIV infected five volunteers despite their having received an experimental vaccine.

Scientists have developed a vaccine that protects monkeys from SIV disease but have not been able to translate this success into an effective vaccine for humans. In the meantime, individuals can slow the AIDS epidemic with personal decisions:

- Abstinence and mutual monogamy between uninfected individuals are the only truly safe sexual behaviors. Studies have shown that whereas condoms reduce a heterosexual individual's risk of acquiring HIV by about 69%, the benefit of condom usage to a population can be undone by an overall increase in sexual activity that may result from a false sense of security that condoms make sex safe.
- Use of new, clean needles and syringes for all injections, as well as caution in dealing with sharp, potentially contaminated objects, can reduce HIV infection rates. If clean supplies are not available, 10% household bleach deactivates HIV on surfaces uncontaminated by a large amount of organic material (blood, mucus).
- Antiviral drugs given to pregnant women have reduced transfer of HIV across the placenta and in breast milk. Generally, HIV-infected mothers should not breast-feed their infants, though feeding with formula made with contaminated water is often more dangerous, triggering life-threatening diarrhea. WHO recommends formula only when it is "acceptable, feasible, affordable, sustainable, and safe."
- Screening blood, blood products, and organ transplants for HIV and anti-HIV antibodies has virtually eliminated the risk of HIV infection from these sources.
- Proper use of gloves, protective eyewear, and masks can prevent contact with infected blood.
- Researchers have determined that men who are circumcised reduce their risk of infection through sexual activity by more than 60%. Male circumcision lowers a female partner's risk by 30%.
- Women who apply 1% tenofovir gel vaginally before and after sex significantly reduce the chance of infection from HIV-positive men.

MICRO CHECK

9. What is the first step in the synthesis of dsDNA from +ssRNA by reverse transcriptase?
10. What types of disease do deltaretroviruses cause, and how are they transmitted?
11. How is the +ssRNA genome of HIV used differently than genomes of other +ssRNA viruses?
12. What are the functions of the HIV glycoproteins gp120 and gp41?
13. Reverse transcriptase needs a primer to start synthesis of DNA from an RNA template. Where does the primer come from?

TELL ME WHY

Why is it unlikely that a medical treatment will be able to rid the body's cells of HIV?

We have considered positive single-stranded RNA viruses, including retroviruses. Now we turn our attention to negative single-stranded RNA viruses.

Enveloped, Unsegmented, Negative ssRNA Viruses: *Paramyxoviridae*, *Rhabdoviridae*, and *Filoviridae*

Viruses of the families *Paramyxoviridae, Rhabdoviridae,* and *Filoviridae* are enveloped, helical, negative, single-stranded RNA (−ssRNA) viruses. Recall that a ribosome cannot directly translate the information in a negative RNA genome into a protein because the information is nonsensical to a ribosome. However, the nucleotides of negative single-stranded RNA are the complements of the nucleotides of +ssRNA, which are readable. Therefore, a −ssRNA virus must make a positive copy of its genome. This +ssRNA copy serves both as mRNA for protein translation by ribosomes and also as a template for transcription of more −ssRNA to be incorporated into new virions.

Paramyxoviruses (family *Paramyxoviridae*) have similar morphologies and antigens. They also share the ability to cause infected cells to fuse with their neighbors, forming giant, multinucleate **syncytia,** which enable paramyxoviruses to pass from an infected cell into neighboring cells and to evade immune surveillance and antibodies. The family *Paramyxoviridae* contains four genera that infect humans: *Morbillivirus* (measles virus), *Respirovirus* (two species of parainfluenza viruses), *Rubulavirus* (mumps virus and two other species of parainfluenza viruses), and *Pneumovirus* (respiratory syncytial virus).

Rhabdoviruses *(Rhabdoviridae)* have bullet-shaped envelopes and include a variety of plant and animal pathogens. Rabies is the most significant pathogen among the rhabdoviruses.

Filoviruses *(Filoviridae)* are particularly significant because they are pathogens that cause a number of emerging and frightening diseases, including Ebola and Marburg hemorrhagic fevers.

Now, we examine some diseases caused by viruses in these families, starting with the paramyxovirus that causes measles.

Measles

LEARNING OUTCOME

25.27 Describe the signs and symptoms of rubeola and SSPE and some ways of preventing them.

Morbillivirus measles virus causes **measles,** also known in the United States as *rubeola*[12] or *red measles*. Measles virus is the most contagious human virus and results in serious childhood diseases. Measles should not be confused with the generally milder rubella (German measles). **TABLE 25.7** compares and contrasts measles and rubella.

[12]Unfortunately, rubeola is the name given to rubella (German measles) in some other countries.

Pathogenesis and Epidemiology

Coughing, sneezing, and talking spread measles virus in the air via respiratory droplets. The virus infects cells of the respiratory tract before spreading via lymph and blood throughout the body. In addition to the respiratory tract, the conjunctiva, urinary tract, small blood vessels, lymphatics, and central nervous system become infected.

Humans are the only host for measles virus, and a large, dense population of susceptible individuals must be present for the virus to spread. In the United States before immunization was instituted, measles outbreaks occurred in one- to three-year epidemic cycles as the critical number of susceptible individuals increased in the population. More than 80% of susceptible patients exposed to the virus developed symptoms 7 to 13 days after exposure.

Signs and symptoms include high fever, sore throat, headache, dry cough, and conjunctivitis. After two days of illness, lesions called **Koplik's spots** appear on the mucous membrane of the mouth **(FIGURE 25.26a)**. These lesions, which look like grains of salt surrounded by a red halo, last one to two days and provide a definitive diagnosis of measles. Red, raised (maculopapular) lesions then appear on the head and spread over the body **(FIGURE 25.26b)**. These lesions are extensive and often fuse to form red patches, which gradually turn brown as the disease progresses. Measles killed 145,000 people, mostly children, in Africa alone in 2013.

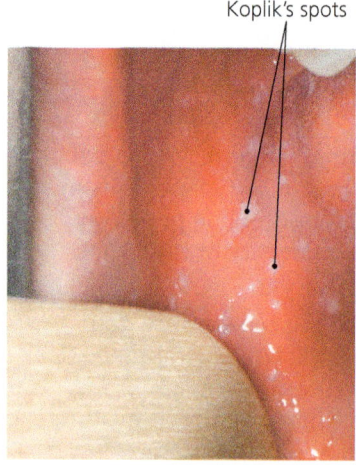
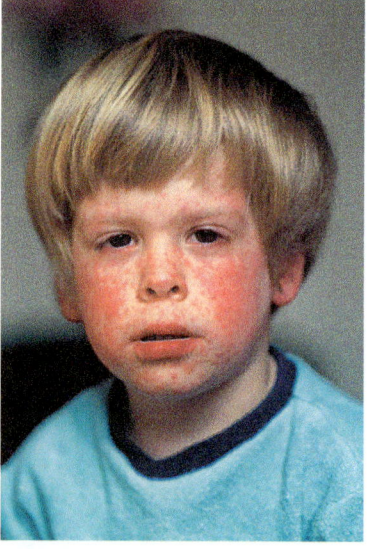

▲ **FIGURE 25.26 Signs of measles.** (a) White Koplik's spots on the oral mucous membrane. (b) Raised lesions begin on the face and spread across the body. They may grow and fuse into large reddish patches.

TABLE 25.7 A Comparison of Measles and Rubella

Disease	Causative Agent	Primary Patient(s)	Complications	Skin Rash	Koplik's Spots
Measles (also known as rubeola and red measles)	Paramyxoviridae: *Morbillivirus measles virus*	Child	Pneumonia, encephalitis, subacute sclerosing panencephalitis	Extensive	Present
Rubella (also known as German measles and three-day measles)	Togaviridae: *Rubivirus rubella virus*	Child, fetus	Birth defects	Mild	Absent

Rare complications of measles include pneumonia, encephalitis, and a fatal disease of the central nervous system called **subacute sclerosing panencephalitis (SSPE)**. SSPE involves personality changes, loss of memory, muscle spasms, and blindness. The disease begins 1 to 10 years after infection with measles virus and lasts a few years before resulting in death. A defective measles virus, which cannot make a capsid, causes SSPE. The virus replicates and moves from brain cell to brain cell via syncytia formation, limiting the functioning of infected cells and resulting in the symptoms. SSPE afflicts fewer than 7 measles patients in 1 million and is becoming much rarer as a result of immunization against measles

An effective, live, attenuated vaccine for measles introduced in 1963 had eliminated measles as an endemic disease in the United States by 2000; however, laxity in maintaining childhood immunization rates and introduction of the measles virus by travelers from endemic countries has resulted in reemergence of measles cases in the U.S. **(FIGURE 25.27)**. Measles remains a cause of death in less developed countries as well as in countries where fears about immunization have resulted in a large population of unimmunized individuals.

Diagnosis, Treatment, and Prevention

The signs of measles, particularly Koplik's spots, are sufficient for diagnosis, but serological testing can confirm the presence of measles antigen in respiratory and blood specimens.

Administration of anti-measles immunoglobulin to patients within six days of exposure to the virus can prevent disease in most. Vitamin A treatment reduces the seriousness of measles in children.

Measles vaccine, in combination with vaccines against mumps and rubella (MMR vaccine), is given to children in the United States who are younger than 15 months old.

MICROBE AT A GLANCE 25.2

Morbillivirus measles virus

Taxonomy: Family *Paramyxoviridae*

Genome and morphology: Nonsegmented, single-stranded, negative RNA (−ssRNA) in an enveloped, helical capsid

Host: Humans

Virulence factors: Virus spreads via respiratory droplets; attaches with fusion glycoproteins to human cells; some viral proteins antagonize interferon production by infected cells

Disease caused: Measles, also known in the United States as rubeola or red measles; subacute sclerosing panencephalitis (SSPE)

Treatment for disease: Primarily supportive, as no antiviral therapy is approved in the U.S; some physicians use antimeasles immunoglobulin within six days of exposure to the virus; vitamin A reduces severity of measles in children

Prevention of disease: Immunization (MMR) at 12–15 months with two boosters: one at age 4–6 years of age and a second at 19–49 years. A third booster may be given at 40–65 years of age

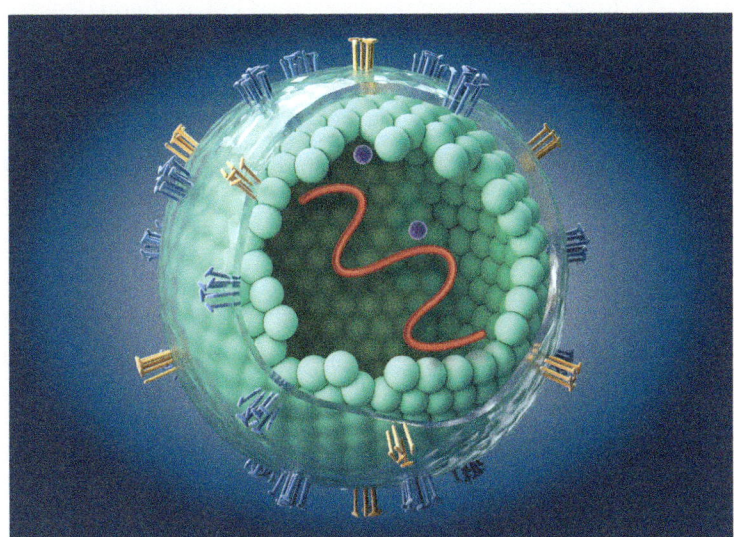

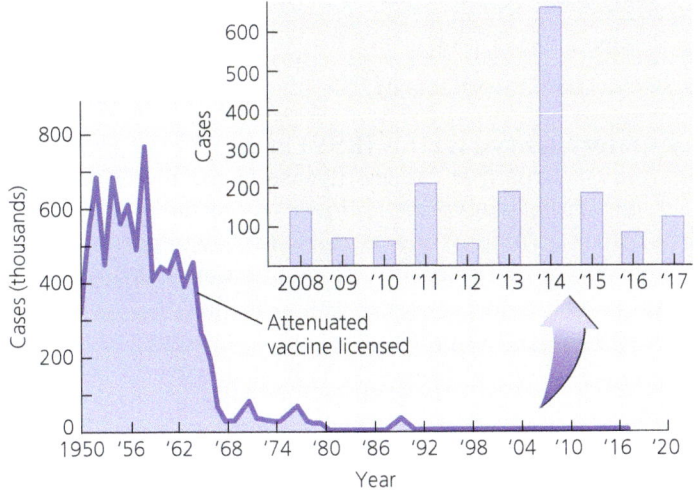

▲ **FIGURE 25.27 Measles cases in the United States since 1950.** The number of cases has declined dramatically since immunization began in 1963. *Why does the measles vaccine pose risks for immunocompromised contacts of immunized children?*

Figure 25.27 *Measles vaccine is a live attenuated vaccine and can cause disease in the immunocompromised.*

The measles portion of the vaccine is effective 95% of the time but is boosted before grade school to ensure protection against measles introduced from outside the country. Eradication of measles worldwide is possible, but there is no consensus among politicians and scientists that eradication would be worth the cost.

Diseases of Parainfluenza Virus

LEARNING OUTCOMES

25.28 Identify the specific cause of croup.
25.29 Describe the cause, signs, and prevention of mumps.

Four strains of paramyxoviruses called **parainfluenza viruses** (*Respirovirus:* human parainfluenza viruses 1 and 3 [HPIV-1 and HPIV-3]; *Rubulavirus:* HPIV-2 and HPIV-4) cause respiratory tract disease, particularly in young children. Note that scientists named the viruses in the order they were discovered, but genetic analysis has shown that HPIV-1 and HPIV-3 are more closely related to one another than they are to HPIV-2 and -4, which are more similar to one another. HPIV-1 and -2 cause colds and **croup** (kroop), a severe childhood disease characterized by inflammation and swelling of the larynx, trachea, and bronchi and a "seal bark" cough. HPIV-3 causes lower respiratory infections and one form of pneumonia, whereas HPIV-4 is limited to mild, upper respiratory tract infections. Respiratory droplets and person-to-person contact transmit the parainfluenzaviruses.

Most patients recover from parainfluenza infections within two days. There is no specific antiviral treatment beyond support and careful monitoring to check that airways do not become completely occluded. If they do, *intubation,* insertion of a tube into the airways, is necessary. Immunization with deactivated virus has not proven effective, possibly because it fails to produce significant secretory antibody to inhibit the virus outside the cells. Researchers have not developed a live attenuated vaccine against parainfluenza.

Mumps

The **mumps** virus (*Rubulavirus*) is another −ssRNA paramyxovirus. Mumps virions are spread in respiratory secretions and infect cells of the upper respiratory system. As virions are released from these areas, viremia develops, and a number of other organs are subsequently infected, resulting in fever and pain in swallowing. The parotid salivary glands, located just in front of the ears under the skin covering the lower jaw, are particularly susceptible to the virus and become painfully enlarged, a condition known as *parotitis* (FIGURE 25.28). In some patients, the mumps virus causes inflammation of the testes (orchitis), meningitis, pancreatitis, or permanent deafness in one ear. In many other patients, infection is asymptomatic. Recovery from mumps is typically complete, and the patient has effective lifelong immunity.

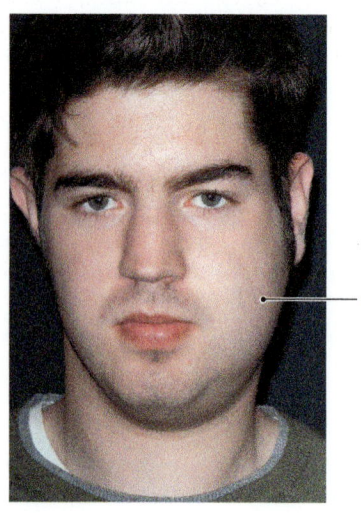

▲ FIGURE 25.28 **Parotitis.** This inflammation of the parotid salivary glands is a common sign of mumps.

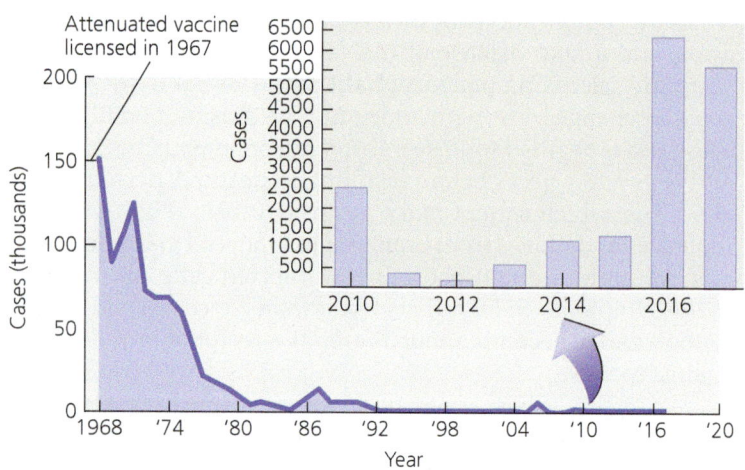

▲ FIGURE 25.29 **Incidence of mumps in the United States.** Mumps cases have declined since immunization began in 1967. *With which other vaccines is mumps vaccine administered?*

Figure 25.29 *The mumps vaccine is administered with the measles and rubella vaccines.*

There is no specific treatment for mumps. However, because an effective attenuated vaccine (MMR) is available and humans are the only natural host for the virus, mumps has almost been eradicated in the industrialized world (FIGURE 25.29). Epidemics of mumps in the late winter and early spring still occur because even with two doses of the vaccine, 12% of people are still susceptible. People living in densely populated areas are most likely to contract mumps, especially in countries without immunization programs.

Disease of Respiratory Syncytial Virus

LEARNING OUTCOME
25.30 Compare the effects of RSV infection in infants and adults.

Respiratory syncytial virus (RSV) in the genus *Pneumovirus* is a paramyxovirus that causes disease of the lower respiratory tract, particularly in young children. As its name indicates, the virus causes formation of syncytia in the lower part of the lungs (FIGURE 25.30). Additionally, plugs of mucus, fibrin, and dead cells fill the smaller air passages, resulting in *dyspnea* (dispnē'ă)—difficulty in breathing—and sometimes croup. RSV is the leading cause of fatal respiratory disease in infants and young children worldwide, especially those who are premature, immune impaired, or otherwise weakened. The disease in older children and adults is manifested as a cold or is asymptomatic. Most otherwise healthy people fully recover in one to two weeks. Reinfection can occur at all ages, but recurrent disease is milder.

Damage to the lungs is increased by the action of cytotoxic T cells and other specific immune responses to infection. Attempts at developing a vaccine with deactivated RSV have proven difficult because the vaccine enhances the severity of the cellular immune response and lung damage.

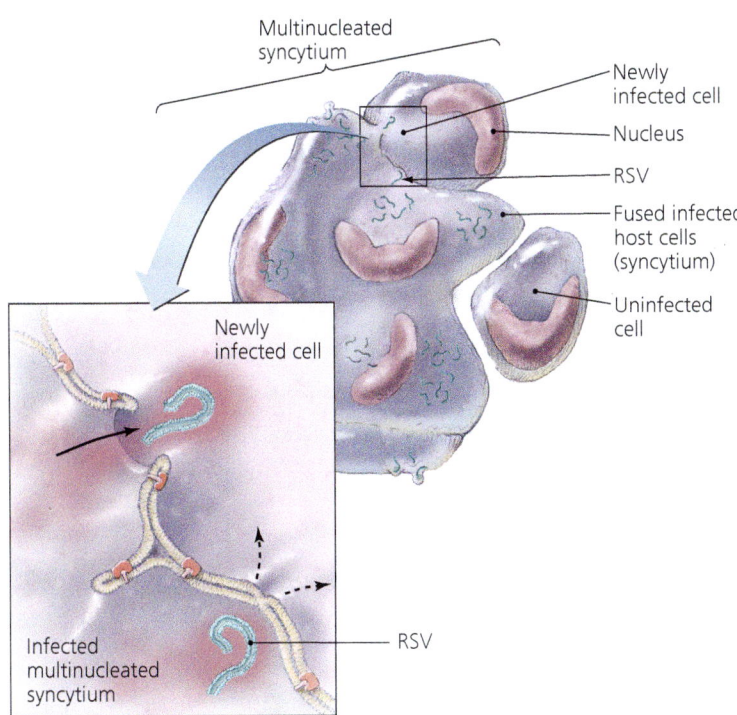

▲ FIGURE 25.30 **A syncytium forms when RSV triggers infected cells to fuse with uninfected cells.** Virions moving through these large multinucleated cells can infect new cells, all the while evading the host's immune system.

Epidemiology

RSV is prevalent in the United States: epidemiological studies reveal that 4–5 million children under age four are infected every year, approximately 125,000 require hospitalization, and about 50 die. The virus is transmitted on fomites, on hands, and less frequently via respiratory droplets. Because the virus is also extremely contagious, its introduction into a nursery can be devastating; therefore, good aseptic technique by health care workers and good hygiene by day care center employees are essential.

Diagnosis, Treatment, and Prevention

Prompt diagnosis is essential if infected infants are to get the care they need. The signs of respiratory distress provide some diagnostic clues, but verification of RSV infection is made by serological testing. Specimens of respiratory fluid may be tested by immunofluorescence, ELISA, or complementary DNA probes.

Supportive treatment involves administration of oxygen. The nucleoside analog ribavirin inhibits viral replication. Antisense RNA has shown promise in reducing the spread of RSV infection.

Control of RSV is limited to attempts to prevent the spread of RSV by frequent hand washing and the use of gowns, goggles, masks, and gloves to reduce healthcare-associated infections.

Rabies

LEARNING OUTCOME

25.31 Describe the effects, treatment, and prevention of rabies.

Although the cry "Mad dog! Mad dog!" may not make you apprehensive, in the 19th century—before Pasteur developed

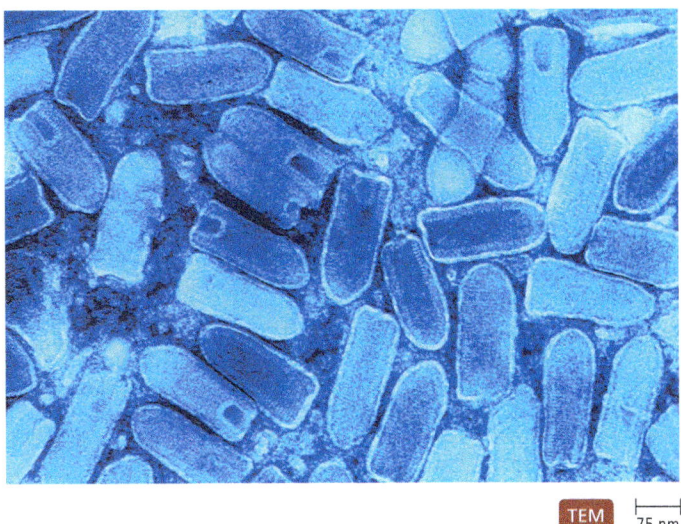

▲ FIGURE 25.31 **Vesicular stomatitis virus.** Note the bullet-shaped envelope.

a rabies vaccine and treatment—a bite from a rabid dog was a death sentence, and the cry "Mad dog!" often spawned panic. Few people survived rabies before Pasteur's pivotal work.

Rabies[13] virus is a negative, single-stranded RNA (−ssRNA) virus in the genus *Lyssavirus*, family *Rhabdoviridae*. Like viruses in the family *Paramyxoviridae*, all of the rhabdoviruses have helical capsids, but those of rhabdoviruses are supercoiled into cylinders, which gives them a striated appearance, and are surrounded by bullet-shaped envelopes (FIGURE 25.31).

Glycoprotein spikes on the surface of the envelope serve as attachment proteins. Attachment to nerve cells triggers endocytosis, and rabies viruses then replicate in the cytoplasm of these cells. The virus is carried to the central nervous system via cytoplasmic flow within the neurons. The spinal cord and brain degenerate as a result of infection, though infected cells show little damage when examined microscopically. Viruses travel back to the periphery, including the salivary glands, through nerve cells. Viruses secreted in the saliva of infected mammals are typically the infective agents.

Initial signs and symptoms of rabies include pain or itching at the site of infection, fever, headache, malaise, and anorexia. Once viruses reach the central nervous system, neurological manifestations characteristic of rabies develop: **hydrophobia**[14] (triggered by the pain involved in attempts to swallow water), seizures, disorientation, hallucinations, and paralysis. Death results from respiratory paralysis and other neurological complications.

Epidemiology

Rabies is a classical zoonosis of mammals, though not all mammals are reservoirs. Rodents, for instance, rarely get rabies. The distribution of rabies in humans follows the distribution in animals. The primary animals involved differ from locale to locale and change over time as a result of dynamics of animal populations and interactions among animals and humans. The main reservoir of rabies in urban areas is the dog. In the wild, rabies can

---

[13]From Latin *rabere*, meaning "rage" or "madness."
[14]From Greek *hydro*, meaning "water," and *phobos*, meaning "fear."

CLINICAL CASE STUDY

A Threat from the Wild

A very agitated, almost combative, man arrived at an emergency room with musculoskeletal and neurological symptoms: uncontrolled neck and back twitching, confusion, altered mentation (mental activity), and new-onset feelings of fear of water. He also had a sore throat and difficulty swallowing. He remained alive for several days but became increasingly confused. Eventually paralytic symptoms arose prior to his going into a coma. He lived with the use of mechanical ventilation, but ultimately went into respiratory arrest and died a few days later.

An autopsy revealed dark-staining bodies in the cells of his brain and a raised antibody titer in his blood. There were no obvious bites or scratches on his skin, though interviews with friends and family indicated that a bat had landed on the man's face about a month before he was admitted to the hospital.

1. What did the man suffer from?
2. What were the antibodies against?
3. What were the dark-staining bodies in his brain cells?
4. What preventive measures might health care providers have used to save the man's life?

Reference: Adapted from *MMWR* 52:47–48, 2003.

---

be found in foxes, badgers, raccoons, skunks, cats, bats, coyotes, and feral dogs (FIGURE 25.32). Bats are the source of most human cases of rabies in the United States, causing about 75% of cases.

Transmission of rabies viruses in the saliva of infected animals usually occurs via a bite. Infection is also possible via the introduction of viruses into breaks in the skin or mucous membranes or rarely through inhalation.

Diagnosis, Treatment, and Prevention

The neurological symptoms of rabies are unique and generally sufficient for diagnosis. Serological tests for antibodies can confirm a diagnosis. Unfortunately, by the time symptoms and antibody production occur, it is too late to intervene, and the disease will follow its natural course. Postmortem laboratory tests are often conducted to determine whether a suspected animal in fact carries rabies virus. These tests include antigen detection by immunofluorescence and the identification of aggregates of virions (called **Negri bodies**) in the brain (FIGURE 25.33).

Treatment of rabies begins with treatment of the site of infection. The wound should be thoroughly cleansed with water and soap or another substance that deactivates viruses. WHO recommends anointing the wound with antirabies serum, which is composed of gamma globulin purified from the blood of people who have been immunized against rabies. Initial treatment also involves injection of *human rabies immune globulin (HRIG)*. Subsequent treatment involves five vaccine injections. Rabies is one of the few infections that can be treated with active immunization because the progress

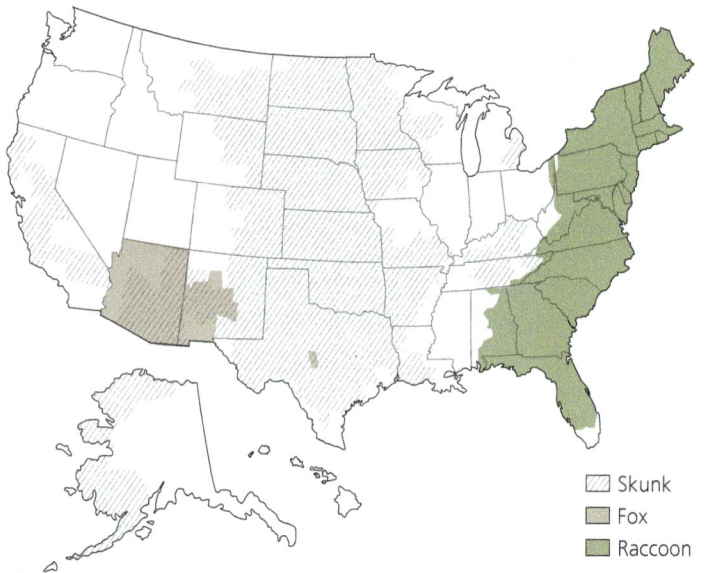

▲ FIGURE 25.32 Portions of the United States in which skunks, foxes, or raccoons are the predominant wildlife reservoirs for rabies. Absence of a color does not indicate the absence of rabies in that region, only that no wildlife reservoir is predominant. Bats, for instance, harbor rabies throughout the country but are rarely the predominant reservoir, though they are the major source of human rabies in the United States.

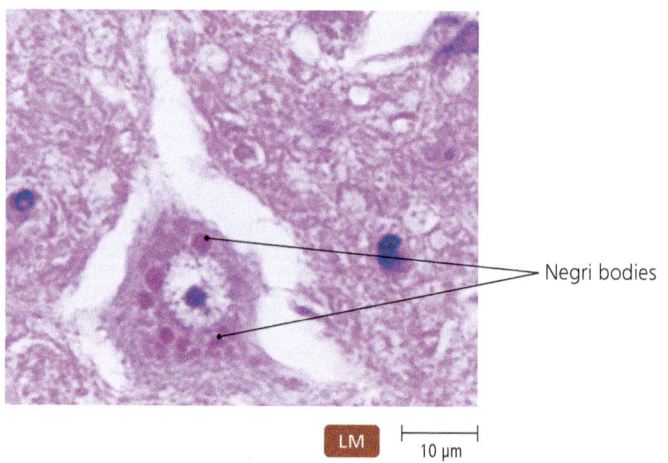

▲ FIGURE 25.33 Negri bodies, a characteristic of rabies infection. Here the viruses are in human brain cells. *What are Negri bodies?*

Figure 25.33 Negri bodies are stained aggregates of rabies virions.

of viral replication and movement to the brain is slow enough to allow effective immunity to develop before disease develops.

The rabies vaccine—*human diploid cell vaccine (HDCV)*—is prepared from deactivated rabies viruses cultured in a laboratory culture of human diploid cells. It is administered intramuscularly on days 0, 3, 7, and 14 after exposure. The vaccine can also be administered prophylactically, in three doses, to workers who regularly come into contact with animals (for example, veterinarians, zookeepers, and animal control workers) and to people traveling to areas of the world where rabies is prevalent.

Ultimately, control of rabies involves immunization of domestic dogs and cats and the removal of unwanted strays from urban areas. Little can be done to completely eliminate rabies in wild animals because rabies virus can infect so many species, though in the late 1990s, an epidemic of rabies was stopped in southern Texas by the successful immunization of the wild coyote population. This was accomplished by lacing meat with an oral vaccine and dropping it from airplanes into the coyotes' range.

Filoviruses (Hemorrhagic Fever Viruses)

LEARNING OUTCOME

25.32 Describe the effects of filoviruses on the human body.

Filamentous viruses that cause **hemorrhagic fever** were originally placed in the *Rhabdoviridae* but have now been assigned to their own family, *Filoviridae*,[15] primarily on the basis of the symptoms they cause. These enveloped viruses form long filaments, sometimes up to 14,000 nm long, that sometimes curve back upon themselves.

There are two main hemorrhagic fever filoviruses: Marburg virus and Ebola virus. **Marburg virus** was first isolated from laboratory workers in Marburg, Germany, and it has also been found in Zimbabwe, Kenya, Uganda, Angola, and the Democratic Republic of the Congo (DRC). Outbreaks of **Ebola virus** hemorrhagic fever were first recorded in the DRC in 1976 near the Ebola River. **Disease in Depth: Ebola** (pp. 764–765) examines Ebola hemorrhagic fever in more detail.

The natural reservoirs of filoviruses appear to be fruit bats, but the initial method of transmission to humans is unknown. The virions attack many cells of the body, especially macrophages and liver cells. One predominant manifestation is uncontrolled bleeding. Up to 90% of human victims of Ebola die, and about 25% of Marburg victims die. Epidemics sometimes cease only when there are not enough living humans to sustain viral transmission.

Marburg and Ebola viruses spread from person to person via contaminated bodily fluids, primarily blood. Even a small amount of infected blood contacting a mucous membrane or conjunctiva can induce disease. Some strains of Ebola virus can spread among monkeys through the air, but aerosol transmission from monkeys to humans or among humans has not been observed.

Researchers have created an experimental anti-Ebola virus antibody called ZMapp using genetic recombination technology to combine genes from mice and humans and insert them into hamster and plant cells. These cells produced a small amount of ZMapp, which clinicians used to treat health care workers during the Ebola epidemics in 2014 and 2018. At the time of this writing, no clinical trials have been conducted using ZMapp, so the Federal Drug Administration of the United States (FDA) had not verified the efficacy and safety of this treatment.

Scientists have developed a vaccine that protects monkeys who are exposed to Ebola in as few as 28 days after immunization. It is not known whether the vaccine fully protects humans.

MICRO CHECK

14. Which sign is diagnostic in the early stages of measles?
15. What is the best way to prevent mumps?
16. Which is the most common cause of a fatal respiratory disease in infants and young children?
17. Which disease causes hydrophobia (fear of water)?

TELL ME WHY

Ebola hemorrhagic fever belongs to a group of diseases called *emerging diseases*—diseases that were previously unidentified or had never been identified in human populations. Why do emerging diseases first appear in developing countries more often than in industrialized countries?

Now, we turn our attention to other groups of enveloped—ssRNA viruses, though these have segmented genomes.

Enveloped, Segmented, Negative ssRNA Viruses: *Orthomyxoviridae*, *Bunyaviridae*, and *Arenaviridae*

A second group of negative, enveloped, single-stranded RNA viruses differ from those in the previous section in having a **segmented genome**; that is, their capsids contain more than one molecule of nucleic acid. These viruses include the flu viruses of the family *Orthomyxoviridae* and hundreds of viruses in the families *Bunyaviridae* and *Arenaviridae* that normally infect animals but can be transmitted to humans.

Influenza

LEARNING OUTCOMES

25.33 Describe the roles of hemagglutinin and neuraminidase in the replication cycle of influenzaviruses.
25.34 Explain the difference between *antigenic drift* and *antigenic shift*.

Imagine being the only remaining elementary school child of your sex in your midsize town because all of your peers died six winters ago. Or imagine returning to college after a break, only to learn that half of your dorm friends had died during the previous month. These vignettes are not fictional; they happened to relatives of this author who lived through the great flu pandemic in the winter of 1918–1919. During that winter, 40% of the world's 1.9 billion people were infected

[15]From Latin *filo*, meaning "thread," and *virus*, meaning "poison."

DISEASE IN DEPTH

EBOLA VIRAL DISEASE

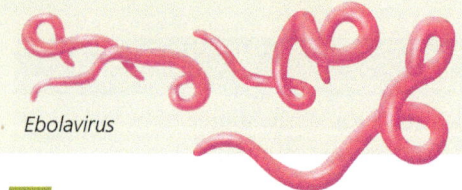

Ebolavirus

PATHOGENESIS

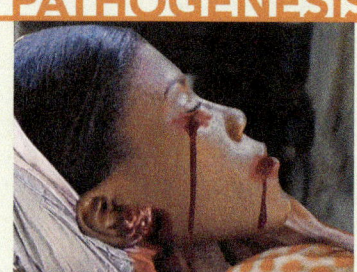

Ebolavirus spreads via direct contact with the bodily fluids, including semen, and wastes of its victims. Initial infections likely occur from contact with an animal reservoir, which may be bats, though this is not proven. Infected macrophages trigger localized blood clotting, which depletes the serum of clotting proteins to such an extent that there is massive uncontrolled bleeding. Ebola is fatal for 25–90% of patients.

Ebolavirus, named for the Ebola River in Africa where it was discovered, is of concern to health care workers, epidemiologists, and governments.

CONTAINMENT

While working with Ebola patients, health care workers must be sure that none of their skin is exposed. The U.S. Centers for Disease Control and Prevention (CDC) guidelines emphasize the importance of:

- **Training:** Prior to working with Ebola patients, workers must have received repeated training.
- **Practice:** Workers must practice putting on and removing personal protective equipment (PPE).
- **Competency:** Workers must demonstrate competency in performing all Ebola-related infection control practices and procedures, particularly PPE.
- **Observation:** A trained observer must supervise each step by every worker to ensure proper completion of established PPE protocols.

Guidelines include these measures:
- Isolation of Ebola patients in rooms with private bathrooms.
- To limit the number of people exposed to patients, doctors and nurses should disinfect patient care areas, equipment, and their own PPE immediately after use.
- Trained observers should monitor workers from a separate location to ensure proper procedures are being followed.
- Researchers studying *Ebolavirus* should follow Biosafety Level 4 (BSL-4) protocols, which include a disinfecting shower and decontamination of all materials prior to leaving the facility.
- Workers must wear an air-supplied, full-body "space" suit that maintains positive air pressure within the suit. This ensures that contaminating Ebola viruses are forcibly expelled by escaping air if there is a break in the suit.

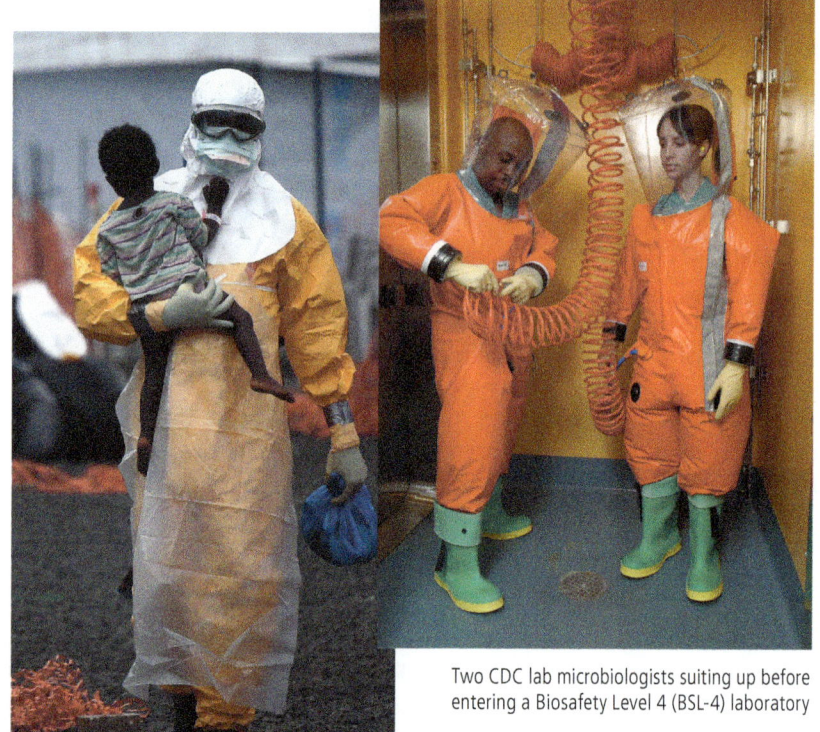

Medical worker in protective gear during Ebola outbreak in West Africa

Two CDC lab microbiologists suiting up before entering a Biosafety Level 4 (BSL-4) laboratory

EPIDEMIOLOGY

Watch Dr. Bauman's Video Tutor explore Ebola hemorrhagic fever, or go to the **Mastering Microbiology** Study Area for more information.

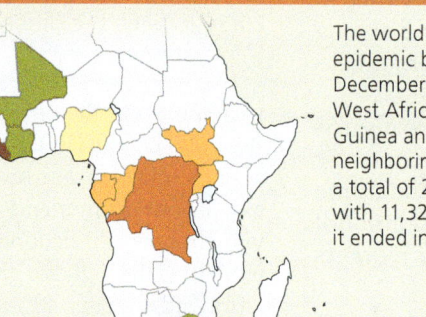

Total indigenous Ebola cases in Africa, 1976–May 2018

- 0
- 1–10
- 11–100
- 101–1000
- 1001–10,000
- >10,000

Guinea

The world's worst Ebola epidemic began in December 2013 in the West African nation of Guinea and spread into neighboring countries for a total of 28,616 cases with 11,325 deaths when it ended in January 2016.

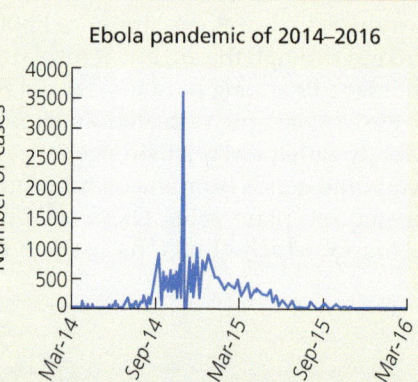

Ebola pandemic of 2014–2016

PATHOGEN & VIRULENCE FACTORS

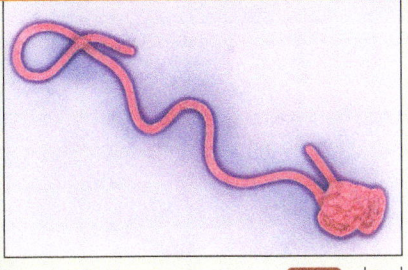
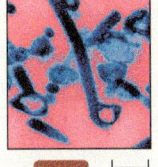

There are at least five strains of ebolaviruses. These enveloped, filamentous viruses in the family *Filoviridae* are 80 nm in diameter and vary in length up to 14,000 nm, often curving back on themselves in spherical, curved, hooked, or pleomorphic shapes. They have a surface glycoprotein that binds to neutrophils, inhibiting early and effective immune responses. Another surface protein binds to cells lining blood vessels, allowing the virus to enter and destroy the cells, which results in severe hemorrhaging. *Ebolavirus* spreads in bodily fluids, including semen. Scientists have detected *Ebolavirus* in semen from some surviving men 18 months after infection. In at least one case, *Ebolavirus* has been sexually transmitted.

SIGNS & SYMPTOMS

The disease progression can vary. The illustration below shows a typical progression of symptoms.

EARLY SYMPTOMS
Signs and symptoms initially appear 2–21 days after infection with *Ebolavirus*. Ebola first presents with headache, fever, fatigue, dizziness, sore throat, and muscle pain, followed by minor red and purple blood spots under the skin (petechiae).

MIDDLE STAGE
High fever, massive bloody vomiting, and diarrhea.

LATE SYMPTOMS
Damage to blood vessels causes severe internal hemorrhaging as well as bleeding from other body orifices, including mouth, eyes, and ears, and organ damage. Many survivors have long-lasting, severe joint pain and eye complications ranging from vision changes to inflammation leading to blindness.

SURVIVORS' SYMPTOMS
Many survivors experience long-term joint and muscle pain, hearing loss, abdominal pain, malaise, and eye damage, including blindness.

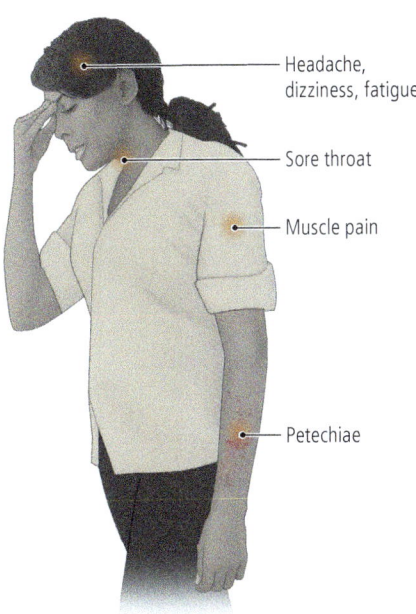
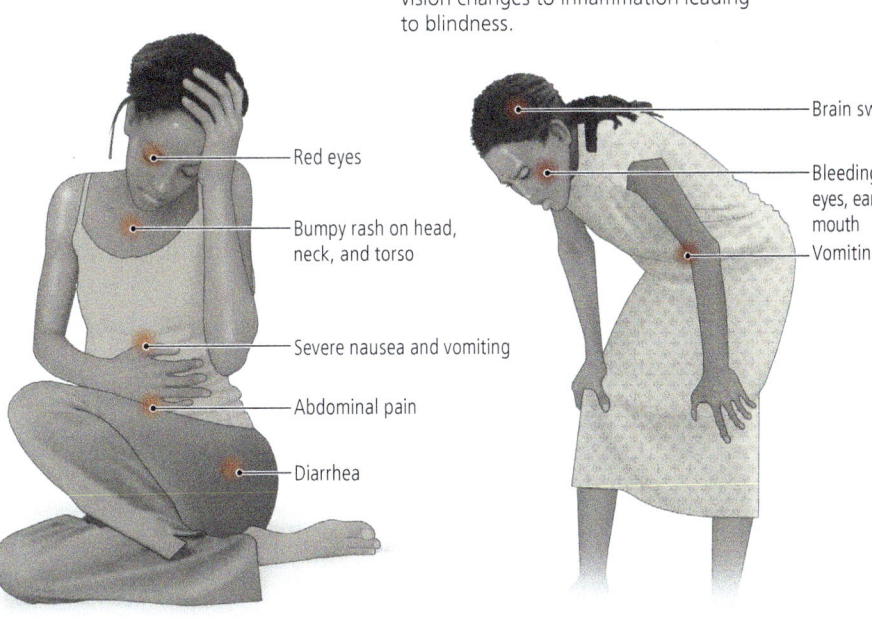
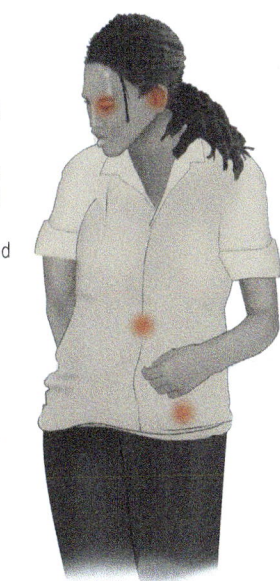

DIAGNOSIS, TREATMENT, & PREVENTION

Left: The latest in the care of contagious patients: rooms, made of clear, flexible plastic with sleeves, gloves and bodysuits built into the walls that allow nurses to safely perform about 80 percent of their care.

Right: Screening for Ebola with an infrared thermometer

Physicians diagnose Ebola based on symptoms and demonstration of filoviruses in the blood via ELISA or PCR. Treatment includes supportive care involving fluid and electrolyte replacement, and maintenance of normal oxygenation and blood pressure. In 2018, physicians tested three potential antiviral treatments: ZMapp, which is antibodies against *Ebolavirus*; favipiravir, which inhibits viral RNA polymerase; and remdesivir, which is a nucleotide analog. Prevention involves screening for people who have been exposed to Ebola, followed by their isolation and observation for three weeks to be sure they do not have the disease. Two vaccines tested against *Ebolavirus* during the West African epidemic (2014–2016) and during an outbreak in the DRC in 2018 have proven effective.

CLINICAL CASE STUDY

A Case of AIDS

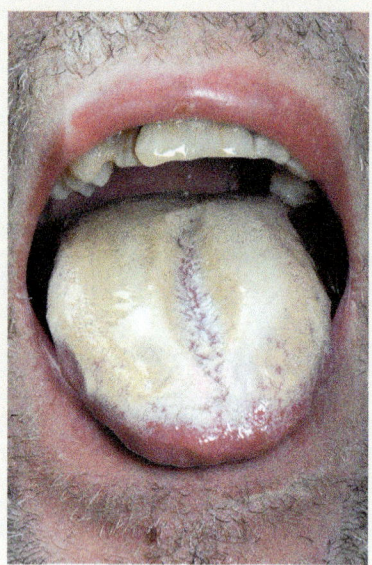

A 25-year-old male was admitted to the hospital with thrush, diarrhea, unexplained weight loss, and difficulty in breathing. Cultures of pulmonary fluid revealed the presence of *Pneumocystis jirovecii*. The man admitted to being a heroin addict and to sharing needles in a "shooting gallery."

1. What laboratory tests could confirm a diagnosis of AIDS?
2. How did the man most likely acquire an HIV infection?
3. What changes to the man's immune system allowed the opportunistic infections of *Candida* (thrush) to arise?
4. What precautions should be taken with the patient's blood?

---

with a new, extremely virulent strain of influenzavirus, and an estimated 40 million died during that single flu season. In some U.S. cities, 10,000 people died each week for several months. Could it happen again? In this section, we will examine the characteristics of influenzaviruses that enable flu—a common respiratory disease, second only to common colds—to produce such devastating epidemics and some ways to protect ourselves from the flu.

The causes of **influenza**[16] (in-flŭ-en'ză), or **flu,** are two strains of orthomyxoviruses, designated A and B. The genomes of flu viruses are extremely variable, especially with respect to the genes that code for attachment molecules. **Disease in Depth: Influenza** (pp. 768–769) examines changes in the attachment molecules, the viruses' effects on the body, viral spread in populations, and diagnosis, treatment, and prevention in more detail. Table 25.1 compares and contrasts the symptoms of influenza to those of other respiratory infections.

Note that true flu is a respiratory infection; influenzaviruses rarely attack cells outside the lungs. Cases of so-called stomach flu are probably caused by caliciviruses or astroviruses.

Diseases of Bunyaviruses

LEARNING OUTCOME

25.35 Contrast hantaviruses with other bunyaviruses.

Most **bunyaviruses** (bun'ya-vī'rŭs-ĕz) are zoonotic pathogens that are usually transmitted to humans by biting arthropods (mosquitoes, ticks, and flies); thus, they are arboviruses. Bunyaviruses have a segmented genome of three −ssRNA molecules (FIGURE 25.34), which distinguishes them from the unsegmented

[16]Influenza, which is Italian for "influence," derives from the mistaken idea that the alignment of celestial objects caused, or "influenced," the disease.

+ssRNA arboviruses we have already discussed in the families *Togaviridae* and *Flaviviridae*.

Epidemiology

Four genera of *Bunyaviridae* contain human pathogens. Some of these viruses are arboviruses; that is, they are transferred via the

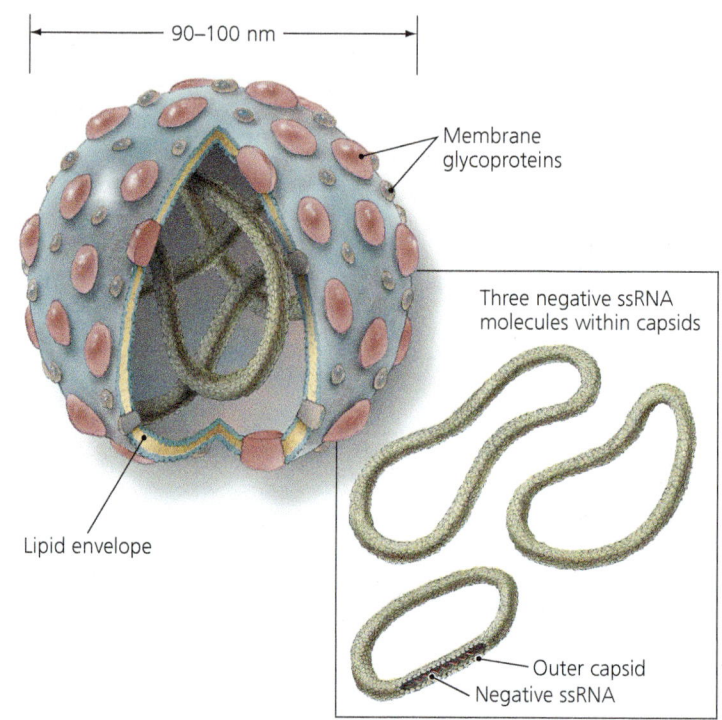

▲ FIGURE 25.34 **A bunyavirus is an enveloped helical virion.** The genome consists of three −ssRNA molecules.

bite of an arthropod. The viruses of *Rift Valley fever, California encephalitis,* and *hemorrhagic fevers* are examples. All these diseases start as an initial viremia. The blood carries viruses to target organs, such as the brain, kidney, liver, or blood vessels. Each virus has a specific target, and different bunyaviruses cause encephalitis, fever, or rash. Most patients have mild symptoms, but symptomatic cases can be severe and fatal.

Hantaviruses are an exceptional type of bunyaviruses: They are not arboviruses but are transmitted to humans via inhalation of virions in dried mouse urine or feces. Two American strains[17] of hantaviruses infect the lungs only, causing a rapid, severe, and often fatal pneumonia called ***Hantavirus* pulmonary syndrome (HPS)** (han'tă-vī-rŭs). This syndrome was first recognized during an epidemic in the Four Corners[18] area of the United States in May 1993.

Bunyaviruses cause human diseases around the world. Each virus is typically found only in certain geographic areas. However, with modern travel patterns, viruses can be introduced into new areas relatively easily.

Diagnosis, Treatment, and Prevention

The signs and symptoms caused by bunyavirus infections are indistinguishable from illnesses caused by several other viruses, so laboratory tests must be employed for diagnosis. ELISAs are used to detect viral antigens associated with Rift Valley fever, the La Crosse strain of California encephalitis, and hemorrhagic fevers. Hantaviruses in the United States were first identified in patients' tissue using the *reverse transcriptase polymerase chain reaction:* Viral RNA was transcribed to DNA using retroviral reverse transcriptase, multiple copies of the DNA were generated through polymerase chain reaction, and the resulting strands were compared to known hantavirus sequences.

No specific treatment exists for most bunyavirus infections. Human disease caused by arboviruses is prevented by limiting contact with the vectors (the use of netting and insect repellents) and by reducing arthropod abundance (elimination of breeding sites and the use of insecticides). Rodent control is necessary to prevent *Hantavirus* pulmonary syndrome. A vaccine has been developed to protect humans and animals against Rift Valley fever, and ribavirin has proved effective in treating Crimean-Congo hemorrhagic fever.

Diseases of Arenaviruses

LEARNING OUTCOMES

25.36 Identify the distinguishing characteristics of arenaviruses.
25.37 Describe the unique feature of satellite viruses such as hepatitis D virus.

Arenaviruses[19] (ă-rē'nă-vī'rŭs-ĕz) are the final group of segmented, −ssRNA viruses we will consider. Their segmented genomes are each composed of two RNA molecules. Most

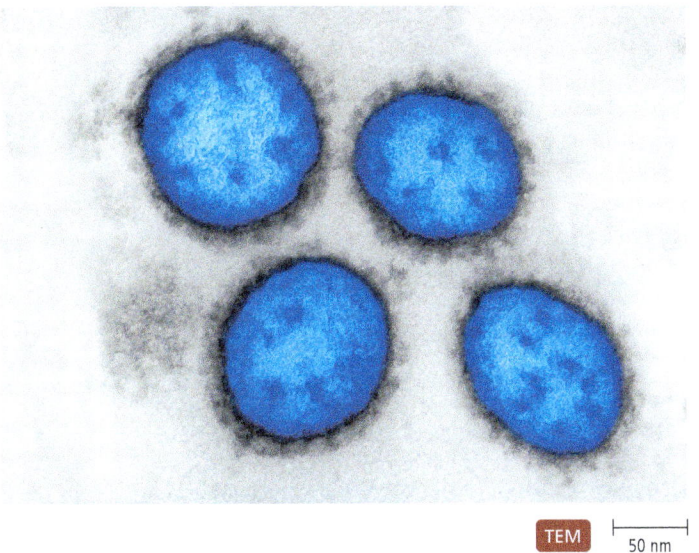

▲ **FIGURE 25.35 The sandy appearance of arenaviruses.** The graininess results from the presence of numerous ribosomes within the capsid. Although the ribosomes are functional, the virions apparently do not use them.

arenaviruses contain ribosomes, which gives them a sandy appearance in electron micrographs **(FIGURE 25.35)**. The ribosomes are functional *in vitro,*[20] but there is no evidence that the viruses utilize them in nature.

Zoonoses

Arenaviruses cause zoonoses that include **lymphocytic choriomeningitis (LCM)** (lim-fō-sit'ik kō-rē-ō-men-in-jī'tis) and hemorrhagic fevers (HF) named for locales where they occur, for example, **Lassa HF** was first reported near Lassa, Nigeria, in Africa; **Junin HF** from Junin, Argentina; **Sabiá HF** from Brazil; and **Machupo HF** from Bolivia. Hemorrhagic fever viruses in the family *Arenaviridae* are endemic in rodent populations in Africa and South America. These viruses cause severe bleeding under the skin and into internal organs. LCM` virus has been found infecting pet hamsters in the United States. Contrary to its name, LCM is rarely associated with meningitis but more typically manifests with flulike symptoms.

Most arenaviruses infect humans by being inhaled in aerosols or being consumed in contaminated food. Lassa fever virus can spread from human to human in bodily fluids.

Diagnosis of arenaviral diseases is based on a combination of symptoms and immunoassay. Fluid samples from hemorrhagic patients are too dangerous for normal laboratories and instead must be processed in maximum biocontainment facilities. Supportive care is usually all that can be done for patients with arenavirus zoonoses, though ribavirin has been beneficial in treating Lassa fever. Prevention of arenavirus diseases involves rodent control and limiting contact with rodents, their droppings, and their dried urine.

---

[17]Sin Nombre and Convict Creek strains.
[18]Four Corners is the geographic area where Arizona, Utah, New Mexico, and Colorado meet.
[19]From Greek *arenosa,* meaning "sandy."

[20]From Latin *in,* meaning "within," and *vitreus,* meaning "glassware"; in other words, "in the laboratory."

DISEASE IN DEPTH

INFLUENZA

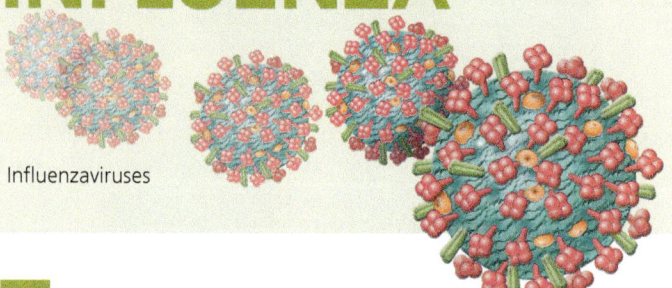
Influenzaviruses

SIGNS & SYMPTOMS

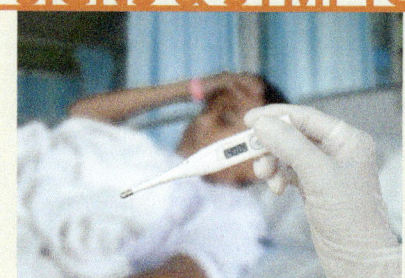

Following infection, influenza has an incubation period of about one day. The signs and symptoms of influenza usually include sudden fever between 39°C and 41°C (102–106°F), pharyngitis, congestion, dry cough, malaise, headache, and myalgia. Fever distinguishes a flu from a common cold. Most people recover in one to two weeks.

Two strains of orthomyxoviruses, designated types A and B, cause influenza. The flu is a common respiratory disease, second in prevalence only to common colds, yet it has characteristics that can produce devastating epidemics.

VIRULENCE FACTORS

Each flu virion is segmented, having eight different −ssRNA molecules, and is surrounded by an envelope studded with prominent glycoprotein spikes composed of either **hemagglutinin (HA)** or **neuraminidase (NA)**. Both HA and NA play roles in attachment: NA spikes provide the virus access to cell surfaces by hydrolyzing mucus in the lungs, whereas HA spikes bind to pulmonary epithelial cells and trigger endocytosis.

Scientists name influenza A viruses for the type of glycoproteins they have. For example, influenzavirus H5N1 has hemagglutinin subtype five and neuraminidase subtype one.

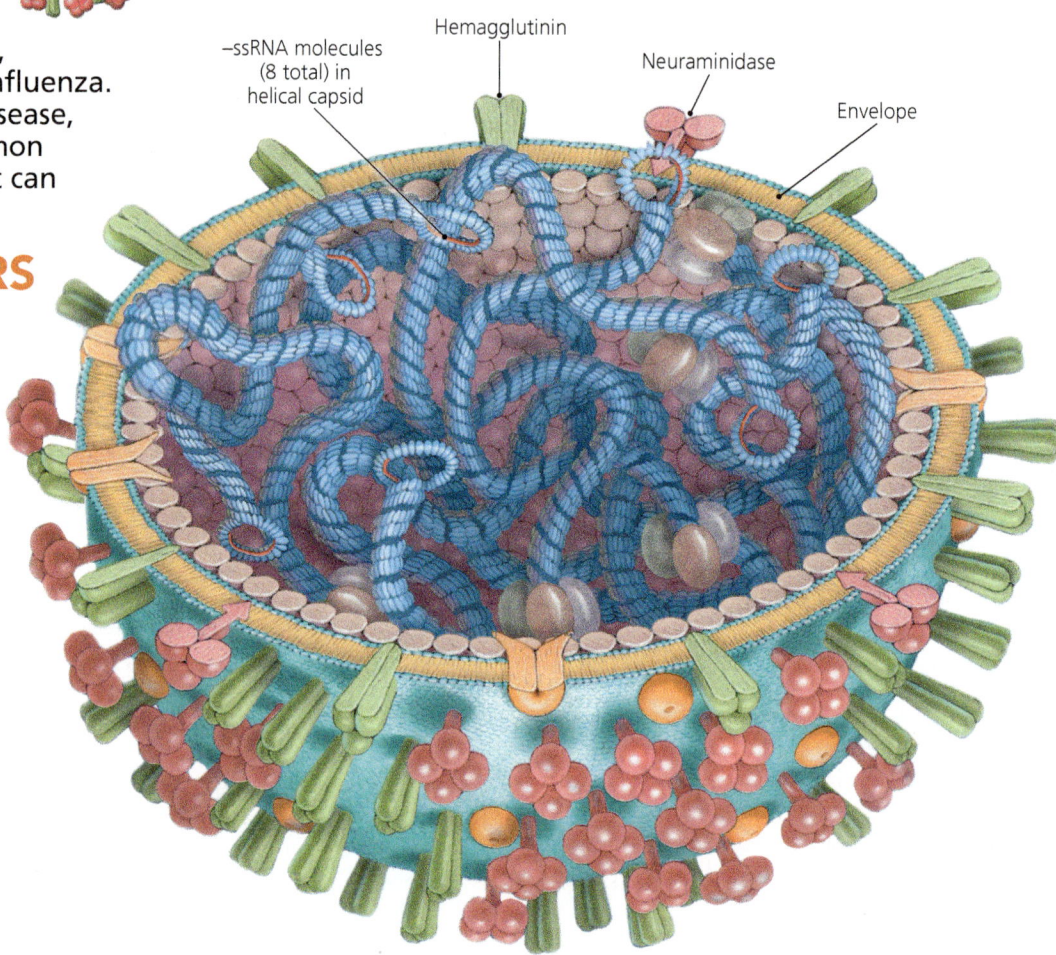

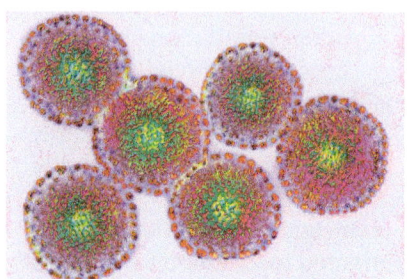
Avian influenzaviruses 50 nm

 Watch Dr. Bauman's Video Tutor explore influenza, or go to the **Mastering Microbiology** Study Area for more information.

EPIDEMIOLOGY

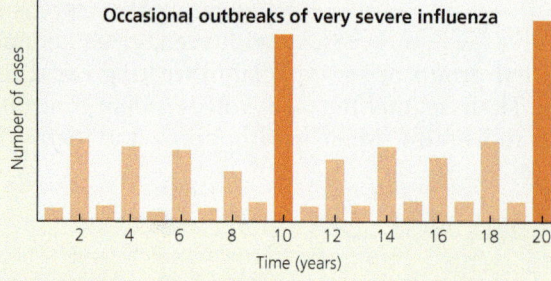
Occasional outbreaks of very severe influenza
Number of cases
Time (years)

People who get the flu one year are immunologically protected the next year from similar strains resulting from antigenic drift; thus, the total number of cases is typically lower every other year. However, antigenic shift results in major changes in antigens about every 10 years, so epidemics occur about once a decade. Epidemiologists are concerned that a deadly pandemic of influenza A virus could reoccur if genes for antigens similar to those of the 1918 flu virus should combine in a new virus by antigenic shift. A virus of concern is H5N1, which kills more than 60% of people who contract the virus from birds. Another virus, H7N9, is resistant to all anti-flu drugs approved by the U.S. Food and Drug Administration.

PATHOGENESIS

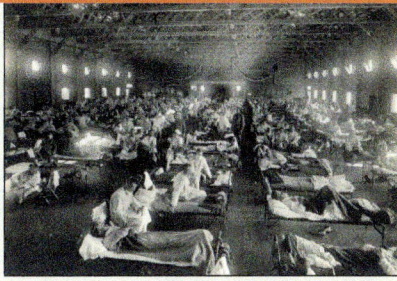

Scene from the 1918–19 flu pandemic

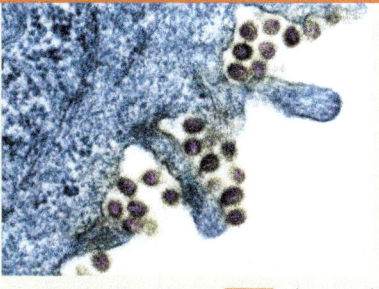

Influenzaviruses enter the body via the respiratory route. Epithelial cells lining the lungs endocytize the viruses; viral envelopes then fuse with the membranes of endocytic vesicles. Flu viruses multiply using +ssRNA both for translation and as a template for transcription of new −ssRNA viral genomes. Death of lung epithelial cells reduces the lungs' first line of defense; as a result, flu patients are more susceptible to bacterial infections, such as with *Haemophilus influenzae*, which was mistakenly named as the cause of flu.

ANTIGENIC DRIFT

Mutations and recombinations in the genes coding for HA and NA spikes are responsible for the production of new strains of influenza A and B viruses via processes known as antigenic drift and antigenic shift.

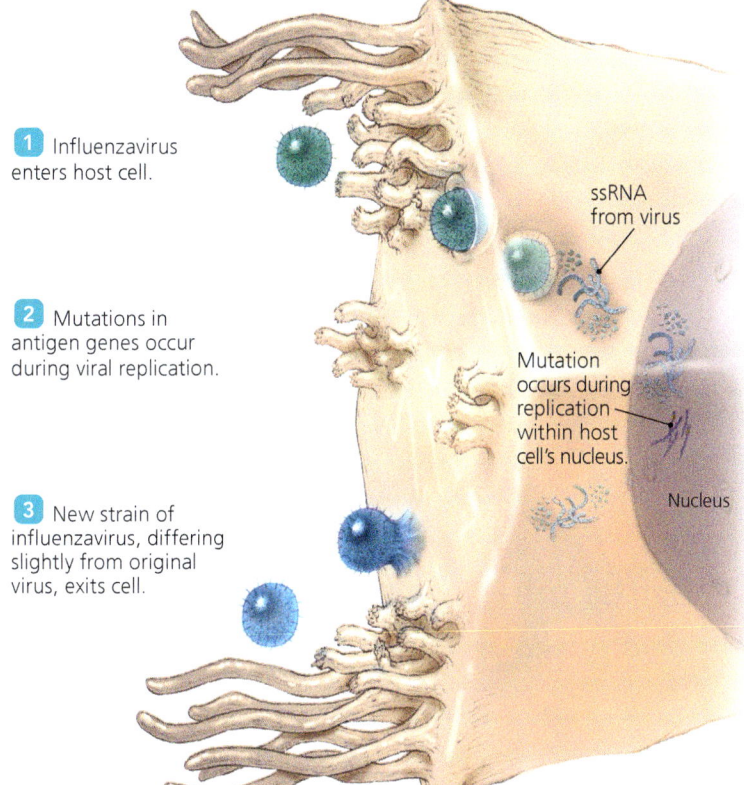

1 Influenzavirus enters host cell.

2 Mutations in antigen genes occur during viral replication.

3 New strain of influenzavirus, differing slightly from original virus, exits cell.

ANTIGENIC SHIFT

Antigenic shift by influenza A virus occurs about once a decade. Influenza B virus does not undergo antigenic shift.

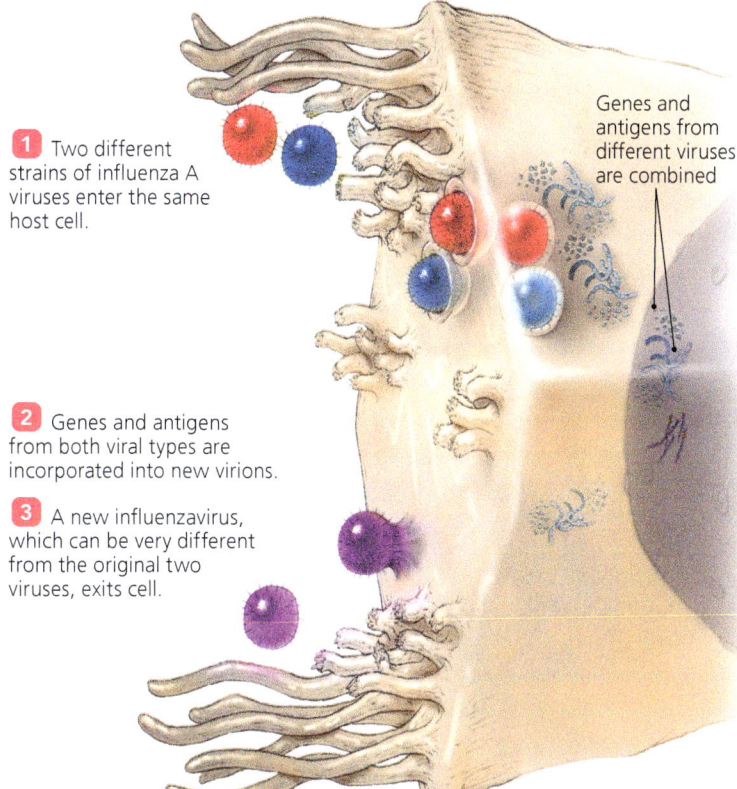

1 Two different strains of influenza A viruses enter the same host cell.

2 Genes and antigens from both viral types are incorporated into new virions.

3 A new influenzavirus, which can be very different from the original two viruses, exits cell.

DIAGNOSIS, TREATMENT, & PREVENTION

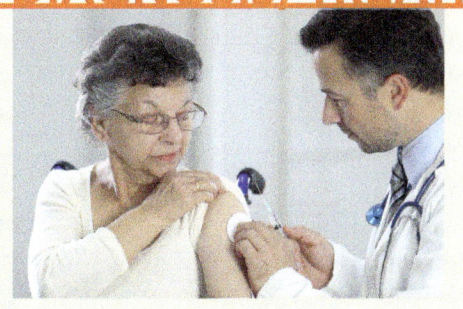

The signs and symptoms of flu during a community-wide outbreak are sufficient for an initial diagnosis of influenza. Lab tests such as immunofluorescence, ELISA, polymerase chain reaction, and rapid antigen testing can distinguish strains of the flu virus. Early and accurate diagnosis is important because antiviral therapy must begin within 48 hours of infection to be effective.

The Centers for Disease Control and Prevention (CDC) recommends the use of either of two drugs to treat flu: oseltamivir pills or inhaled zanamivir mist inhibit both type A and type B neuraminidase. The CDC discourages use of two older drugs, amantadine and rimantadine, because influenza A viruses have grown resistant to both.

Prevention is by immunization with multivalent vaccines, which are at least 70% effective. Scientists track changes in the HA and NA antigens of emerging viruses and use them to create new flu vaccines for each fall's flu season. Good personal hygiene, such as hand washing and the use of hand antisepsis, can reduce the spread of flu.

Hepatitis D

Hepatitis D virus (genus *Deltavirus*), which is also known as *delta agent*, is an arenavirus that causes damage to the liver (called hepatitis D) and plays a role along with hepatitis B virus in triggering liver cancer. Delta agent is unusual in that it does not possess genes for the glycoproteins it requires to attach to liver cells. To become infective, it "steals" the necessary glycoprotein molecules from hepatitis B viruses that are simultaneously infecting the same liver cell. For this reason, hepatitis D virus is called a **satellite virus** because its replication cycle "revolves around" a helper virus. Interestingly, in this case, the satellite virus is −ssRNA and the helper virus is dsDNA.

Like hepatitis B virus, hepatitis D virus is spread in bodily fluids via sexual activity and the use of contaminated needles. Prevention involves the same precautions used to prevent infection with hepatitis B virus—abstinence, mutual monogamy, condom usage, the use of sterile needles, and avoidance of accidental needlesticks. Since hepatitis D virus requires hepatitis B virus to become virulent, hepatitis B vaccine limits the spread of hepatitis D viruses.

> **TELL ME WHY**
>
> Why are zoonotic viruses more difficult to eradicate than viruses that only infect humans?

Naked, Segmented dsRNA Viruses: *Reoviridae*

LEARNING OUTCOME
25.38 Discuss the meaning of the name "reovirus."

Reoviruses (rē′ō-vī′rŭs-ĕz) are respiratory and enteric viruses with naked icosahedral capsids. They are unique in being the only microbes, viral or cellular, with genomes composed of double-stranded RNA (dsRNA). Each virus contains 10 to 12 segments of dsRNA.

Reoviruses were originally orphan viruses—not initially associated with any diseases—though now many have been linked to specific conditions. The acronym *reo* is derived from *respiratory, enteric, orphan*. Reoviruses include the genera *Rotavirus* and *Coltivirus*.

Rotavirus Infections

LEARNING OUTCOME
25.39 Describe the effects of rotavirus infections.

Rotavirus[21] (rō′tă-vī′rŭs) is an almost spherical reovirus with prominent glycoprotein spikes **(FIGURE 25.36)**. During replication, rotaviruses acquire and then lose an envelope. Even without envelopes, they function like enveloped viruses because glycoprotein spikes on their capsids act as attachment molecules that trigger endocytosis, a process common to enveloped viruses.

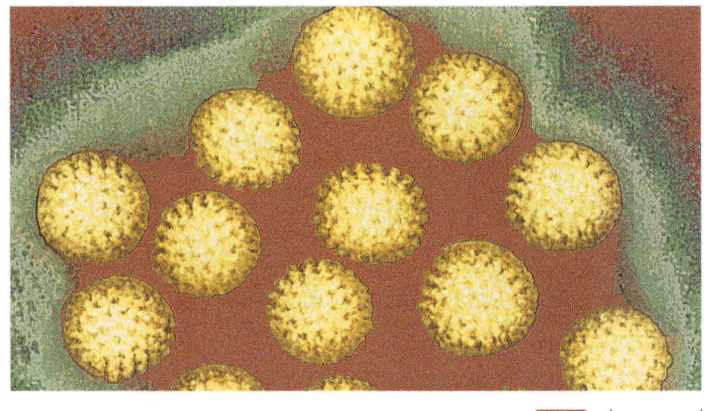

▲ **FIGURE 25.36 Rotaviruses.** The wheel-like appearance of rotaviruses, from which they get their name.

Rotaviruses infect almost all children everywhere in the first few years of life via the fecal-oral route.

Before a vaccine was available, rotaviruses were the most common cause of infantile gastroenteritis and account annually in the United States for about 63,000 cases of diarrhea in children requiring hospitalization as a result of fluid and electrolyte loss. Worldwide, rotaviruses account for an estimated 500,000 deaths annually from uncontrolled diarrhea **(FIGURE 25.37)**. Infected children may pass as many as 100 trillion virions per gram of stool.

Rotavirus infection is self-limiting, and treatment involves supportive care, such as replacement of lost water and electrolytes. Prevention of infection involves good personal hygiene, proper sewage treatment, and immunization. The vaccine approved for use in the United States prevents 75% of rotavirus cases and reduces hospitalizations and deaths by more than 96%.

Coltivirus Infections

LEARNING OUTCOME
25.40 Discuss the most common disease caused by a coltivirus.

Coltivirus (kol′tē-vī′rŭs) is an arbovirus that causes a zoonosis, **Colorado tick fever** (see Table 25.4, p. 747), for which these viruses are named. Colorado tick fever is usually a mild disease involving fever and chills, though severe cases can manifest as headache, photophobia, myalgia, conjunctivitis, rash, and, in children, hemorrhaging. Colorado tick fever should not be confused with Rocky Mountain spotted fever, a tick-borne bacterial disease (discussed in Chapter 21). Colorado tick fever is self-limiting, but red blood cells in all stages of development retain the virus for some time. Patients should not donate blood for several months after recovery.

Health care providers diagnose Colorado tick fever by detecting antigens in the blood using ELISA or other immunoassays. As with many viral diseases, there is no specific treatment. Prevention involves limiting contact with infected ticks.

---
[21]From Latin *rota*, meaning "wheel."

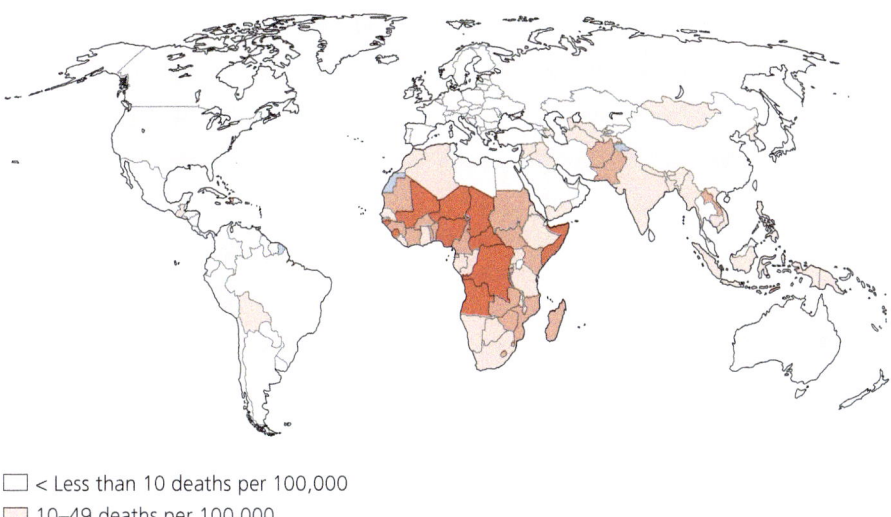

◀ **FIGURE 25.37 Deaths of children under five from rotaviral diarrhea (2013).** Mortality is most common in developing countries.

☐ < Less than 10 deaths per 100,000
☐ 10–49 deaths per 100,000
☐ 50–99 deaths per 100,000
☐ >100 deaths per 100,000
☐ Not available

RNA viruses are a heterogeneous group of virions capable of causing diseases of varying severity, including several life-threatening ones. **TABLE 25.8** summarizes the physical characteristics of most RNA viruses of humans as well as the diseases they cause. As we have seen, RNA viruses pose significant problems for health care personnel with respect to diagnosis, treatment, and prevention, and some of these viruses also pose threats to health care workers who must handle them. Emerging RNA

TABLE 25.8 Taxonomy and Characteristics of Human RNA Viruses

Family	Strand Type	Enveloped or Naked	Capsid Symmetry	Size (nm)	Representative Genera (Diseases)
Picornaviridae	Single, positive	Naked	Icosahedral	22–30	*Enterovirus* (polio) *Rhinovirus* (common cold) *Hepatovirus* (hepatitis A)
Caliciviridae	Single, positive	Naked	Icosahedral	35–40	*Norovirus* (acute gastroenteritis)
Astroviridae	Single, positive	Naked	Icosahedral	30	*Astrovirus* (gastroenteritis)
Hepeviridae	Single, positive	Naked	Icosahedral	27–34	*Orthohepevirus* (hepatitis E)
Togaviridae	Single, positive	Enveloped	Icosahedral	40–75	*Alphavirus* (encephalitis) *Rubivirus* (rubella)
Flaviviridae	Single, positive	Enveloped	Icosahedral	37–50	*Flavivirus* (yellow fever) *Hepacivirus* (hepatitis C)
Coronaviridae	Single, positive	Enveloped	Helical	80–160	*Coronavirus* (common cold, severe acute respiratory syndrome)
Retroviridae	Single, positive, segmented	Enveloped	Icosahedral	80–146	*Deltaretrovirus* (leukemia) *Lentivirus* (AIDS)
Paramyxoviridae	Single, negative	Enveloped	Helical	125–250	*Respirovirus* (colds, respiratory infections) *Pneumovirus* (respiratory syncytial disease, rarely croup or common cold) *Morbillivirus* (measles) *Rubulavirus* (mumps)
Rhabdoviridae	Single, negative	Enveloped	Helical	75 × 130 – 240	*Lyssavirus* (rabies)
Filoviridae	Single, negative	Enveloped	Helical	Up to 14,000; 80 in diameter	*Ebolavirus* (Ebola hemorrhagic fever) *Marburgvirus* (Marburg hemorrhagic fever)
Orthomyxoviridae	Single, negative, segmented	Enveloped	Helical	80–120	*Influenzavirus* (flu, rarely croup)
Bunyaviridae	Single, negative, segmented	Enveloped	Helical	90–100	*Orthobunyavirus* (encephalitis) *Hantavirus* (pneumonia)
Arenaviridae	Single, negative, segmented	Enveloped	Helical	50–300	*Lassavirus* (hemorrhagic fever) *Deltavirus* (hepatitis D)
Reoviridae	Double, segmented	Naked	Icosahedral	78–80	*Rotavirus* (diarrhea) *Coltivirus* (Colorado tick fever)

viruses are of particular concern as people move deeper into rain forests and come into contact with new viral agents, but such viruses have also provided the impetus for expanded research into the fields of immunization and treatment of viral diseases.

TELL ME WHY

Why are rotaviral deaths more rare in developed nations than in developing countries?

MICRO CHECK

18. Why is there an influenza epidemic about every 10 years?
19. How can the signs and symptoms of influenza be distinguished from those of a cold?
20. What is the best way to prevent diarrhea caused by rotavirus?

MICRO IN THE CLINIC FOLLOW-UP

Brain Dysfunction After a Hard Hike?

The initial assessment at Dr. Miller's office reveals that Mercedes has a fever (101°F), a tremor in her right hand, weakness in both legs, and she is disoriented. A sample of cerebrospinal fluid (CSF) is collected; her white blood cell count is elevated, glucose is normal, and protein is elevated. Bacterial and viral analysis of the CSF is negative.

Upon his initial assessment, Dr. Miller admits Mercedes to the hospital for monitoring. Her condition has continued to worsen. An electroencephalogram (a test of brain activity) reveals mild encephalopathy (disease or disorder of the brain).

Eight days after her admission to the hospital, serological analysis of CSF from Mercedes indicates a positive result for West Nile virus (WNV)–specific IgM. There is no specific treatment for WNV. Mercedes remains in the hospital, where she receives supportive care and careful monitoring for intracranial pressure and seizures. Three weeks later, Mercedes is released from the hospital with some residual weakness in her right leg and no remaining mental alterations.

1. Why was Mercedes's initial CSF sample negative for WNV-specific IgM?
2. Why must Mercedes be monitored for changes in intracranial pressure and seizure?

Check your answers to the Micro in the Clinic Follow-Up questions in the **Mastering** Microbiology Study Area.

> Go to Mastering Microbiology for Interactive Microbiology, Dr. Bauman's Video Tutors, Micro Matters, MicroFlix, MicroBoosters, practice quizzes, and more.

CHAPTER SUMMARY

Naked, Positive ssRNA Viruses: *Picornaviridae, Caliciviridae, Astroviridae,* and *Hepeviridae*
(pp. 735–741)

1. RNA viruses and viroids are the only infective agents that use RNA molecules as their genetic material.
2. **Picornaviruses** are the smallest of animal viruses and include the **rhinoviruses** and **enteroviruses**. Rhinoviruses are extremely infective in the upper respiratory tract and cause most "common" colds.
3. Antihistamines, decongestants, and pain relievers treat the signs and symptoms of colds, but do not cure them. Vitamin C and zinc, though popular home remedies, have not been proven efficacious.
4. Enteroviruses, which are mostly cytolytic picornaviruses that infect via the fecal-oral route, include poliovirus, coxsackievirus, and echovirus.
5. Three types of polioviruses cause **poliomyelitis**, or **polio**, which is a disease of varied degrees of severity from asymptomatic infections to **bulbar poliomyelitis**, to postpolio syndrome occurring 30 to 40 years after infection. Eradication of polio, due to effective immunization, is imminent.
6. Jonas Salk developed **inactivated polio vaccine (IPV)** in 1955. It was replaced by live **oral polio vaccine (OPV),** which was developed by Albert Sabin in 1961. Now, health professionals in many countries have switched back to IPV because OPV, which is a live, attenuated vaccine, can revert to a disease-causing form.
7. Coxsackie A virus causes lesions in the mouth and throat, respiratory infections, hand-foot-and-mouth disease, viral meningitis, and acute hemorrhagic conjunctivitis. Coxsackie B virus causes myocarditis and pericardial infections, which may be fatal in newborns. In addition, coxsackie B viruses cause pleurodynia and viral meningitis and may be a cause of type 1 diabetes.
8. So-called orphan viruses are initially unassociated with a specific disease.
9. **Echoviruses** (enteric cytopathic human orphan viruses) cause viral meningitis and colds.
10. **Hepatitis A virus** is a noncytolytic picornavirus that infects liver cells but is rarely fatal.
11. **Caliciviruses** and **astroviruses** are small viruses that enter through the digestive system, multiply in the cells of the intestinal tract, and cause gastroenteritis and diarrhea. **Noroviruses** are representative caliciviruses.
12. **Hepatitis E virus** causes **hepatitis E,** which is fatal to about 20% of infected, pregnant women. Prevention of hepatitis E involves interrupting the fecal-oral route of infection.

Enveloped, Positive ssRNA Viruses: *Togaviridae, Flaviviridae,* and *Coronaviridae* (pp. 741–750)

1. Many flaviviruses and togaviruses are **arboviruses** (viruses transmitted by arthropods) and are enveloped, icosahedral, positive, ssRNA viruses that cause flulike symptoms, encephalitis, dengue, and yellow fever.

2. Various viruses, most transmitted via mosquito vectors, cause **zoonoses**—diseases of animals that are spread to humans. Those caused by togaviruses are **Eastern equine encephalitis (EEE), Western equine encephalitis (WEE),** and **Venezuelan equine encephalitis (VEE).** A flavivirus, West Nile virus, causes **West Nile encephalitis.** All these may be fatal in humans.

3. **Dengue fever** and **dengue hemorrhagic fever** are endemic diseases in South America caused by a flavivirus and carried by the *Aedes* mosquito.

4. **Yellow fever,** which is contracted through the bite of a mosquito carrying a flavivirus, results in internal hemorrhaging and damage to liver, kidneys, and heart.

5. Zika virus, carried by *Aedes* mosquitoes and also transmitted sexually, causes Zika congenital syndrome.

6. *Rubivirus,* a togavirus that is not transmitted by an arthropod vector, causes **rubella,** commonly known as "German measles" or "three-day measles." It can cause serious birth defects if the virus crosses the placenta into a fetus.

7. **Hepatitis C virus (HCV),** a flavivirus transmitted via needles, organ transplants, and sexual activity, causes a disease that often results in serious liver damage. The RNA replication inhibitor sofosbuvir can cure hepatitis C.

8. **Coronaviruses** have a corona-like envelope and cause colds. Emerging coronaviruses cause two severe conditions called **coronavirus respiratory syndromes:** severe acute respiratory syndrome (SARS) and Middle East respiratory syndrome (MERS).

Enveloped, Positive ssRNA Viruses with Reverse Transcriptase: *Retroviridae* (pp. 750–757)

1. **Retroviruses** have **reverse transcriptase** and **integrase** enzymes. The first allows retroviruses to make dsDNA from RNA templates; the latter allows them to integrate into a host cell's chromosomes. Two types of retroviruses cause human disease: deltaretroviruses cause cancer, and lentiviruses, such as HIV, are immunosuppressive.

2. **Human T-lymphotropic viruses** are oncogenic retroviruses associated with cancer of lymphocytes. HTLV-1 causes adult acute T-cell lymphocytic leukemia; HTLV-2 causes hairy cell leukemia. Sexual intercourse, blood transfusion, and contaminated needles transmit them.

3. **Acquired immunodeficiency syndrome (AIDS)** is a condition defined by the presence of antibodies against **HIV** in conjunction with certain opportunistic infections or by HIV and a CD4 count below 200/μL of blood. A **syndrome** is a complex of signs, symptoms, and diseases with a common cause.

4. **Human immunodeficiency viruses (HIV-1** or **HIV-2)** destroy the immune system. HIV is characterized by glycoproteins such as **gp120** and **gp41,** which enable attachment and have significant antigenic variability. The destruction of the immune system results in vulnerability to any infection.

5. HIV converts its genome into double-stranded DNA, which is integrated permanently into the host DNA—a condition called **latency.**

6. HIV infects helper T cells, macrophages, and dendritic cells. Using gp120, HIV fuses to a cell, uncoats, enters the cell, and transcribes dsDNA to become a provirus, which inserts into a cellular chromosome. After HIV is replicated and released, an internal viral enzyme, **protease,** cleaves a polypeptide, allowing HIV to assemble and become virulent. HIV is spread primarily through contact with an infected individual's secretions or blood.

7. **Antiretroviral therapy (ART)** is a combination of antiviral drugs for the treatment of AIDS.

8. Abstinence, mutual monogamy between uninfected partners, and abstention from use of contaminated needles are the only ways to prevent the spread of AIDS. Correct condom usage, circumcision, bleaching of needles, prophylactic use of anti-HIV drugs, screening blood and blood products, and proper use of gloves and eye guards are each effective in reducing the transmission of HIV.

Enveloped, Unsegmented, Negative ssRNA Viruses: *Paramyxoviridae, Rhabdoviridae,* and *Filoviridae* (pp. 758–763)

1. Paramyxoviruses can cause infected cells to fuse together into giant, multinucleate **syncytia.** One virus causes **measles** ("rubeola" or "red measles"), recognized by **Koplik's spots** in the mouth followed by lesions on the head and trunk. Complications of measles include **subacute sclerosing panencephalitis (SSPE),** a progressive fatal disease of the central nervous system. Immunization is effective and eradication is possible.

2. **Parainfluenza viruses** usually cause mild coldlike diseases, including **croup.** There is no effective vaccine.

3. **Mumps** virus is a paramyxovirus that causes painfully enlarged parotid glands. Mumps vaccine is very effective, and eradication is possible.

4. **Respiratory syncytial virus (RSV)** is a paramyxovirus that affects the lungs and can be fatal in infants and young children, while in older children and adults the symptoms are similar to those of a cold.

5. **Rabies** is caused by a rhabdovirus—a bullet-shaped virus with glycoprotein spikes that attaches to nerve cells, is replicated there, moves into the central nervous system, and eventually causes brain degeneration. Infection usually results from bites by infected mammals. Symptoms include pain in swallowing (which results in **hydrophobia**), seizures, paralysis, and death. Rabies may be identified by aggregates of virions called **Negri bodies** in the brain of an affected mammal. Human and animal vaccines are effective.

6. **Marburg virus** and **Ebola virus** are filamentous viruses that cause **hemorrhagic fever.** There is no known treatment or prevention for hemorrhagic fever, though an experimental vaccine has been effective in preventing disease in monkeys.

Enveloped, Segmented, Negative ssRNA Viruses: *Orthomyxoviridae, Bunyaviridae,* and *Arenaviridae* (pp. 763–770)

1. A **segmented genome** consists of more than one molecule of nucleic acid. The term usually refers to viruses, as the genomes of most of the known viral families are not segmented.

2. **Influenza (flu)** is caused by type A or B orthomyxoviruses, which enter lung cells by means of **hemagglutinin (HA)** and **neuraminidase (NA)** glycoprotein spikes. Gene mutations for HA and NA account for continual formation of new variants of flu viruses.

3. **Antigenic drift** occurs every two or three years when a single strain mutates within a local population. **Antigenic shift** of influenza A virus occurs about every 10 years and results in reassortment of genomes from different strains of animal and human viruses infecting a common cell. Vaccines for the flu in any given flu season are prepared against newly identified variants.

4. **Bunyaviruses**, with three −ssRNA molecules as a genome, are zoonotic pathogens. They are usually arboviruses. Each bunyavirus has a specific target and may cause encephalitis, fever, rash, or death.

5. **Hantaviruses** are bunyaviruses that are transmitted to humans via inhalation of virions in dried mouse urine. They cause **Hantavirus pulmonary syndrome**.

6. **Arenaviruses** are segmented −ssRNA viruses with apparently nonfunctional ribosomes. They cause **Lassa, Junin, Sabiá, and Machupo hemorrhagic fevers; lymphocytic choriomeningitis (LCM);** and hepatitis D.

7. **Hepatitis D virus,** also known as delta agent, is a **satellite virus;** that is, it requires proteins coded by another virus (in this case, hepatitis B virus) to complete its replication cycle.

Naked, Segmented, dsRNA Viruses: *Reoviridae* (pp. 770–772)

1. The **reoviruses** are unique in having double-stranded RNA as a genome. They include **rotavirus,** which has glycoprotein spikes for attachment and causes potentially fatal infantile gastroenteritis, and *Coltivirus*, which causes **Colorado tick fever.**

QUESTIONS FOR REVIEW

Answers to the Questions for Review (except for Short Answer questions) begin on p. A-1.

Multiple Choice

1. A segmented genome is one that has _____.
   a. more than one strand of nucleic acid
   b. double-stranded RNA
   c. both RNA and DNA strands
   d. both +ssRNA and −ssRNA molecules

2. What do viruses in the families *Picornaviridae, Caliciviridae, Astroviridae, Coronaviridae, Togaviridae, Flaviviridae,* and *Retroviridae* have in common?
   a. They are arboviruses.
   b. They are nonpathogenic.
   c. They have positive single-stranded RNA genomes.
   d. They have negative single-stranded RNA genomes.

3. The smallest animal viruses are in the family _____.
   a. *Caliciviridae*
   b. *Astroviridae*
   c. *Togaviridae*
   d. *Picornaviridae*

4. Which of the following viruses cause most colds?
   a. rhinoviruses
   b. parainfluenza viruses
   c. pneumoviruses
   d. bunyaviruses

5. Arboviruses are _____.
   a. zoonotic pathogens
   b. deactivated viruses used in vaccines
   c. viruses that are transmitted to humans via the bite of an arthropod
   d. found in arbors

6. Negri bodies are associated with which of the following?
   a. Marburg virus
   b. hantavirus
   c. coltivirus
   d. rabies virus

7. If mosquitoes were eradicated from an area, which of the following diseases would be most affected?
   a. mumps
   b. *Hantavirus* pulmonary syndrome
   c. hepatitis E
   d. breakbone fever

8. Koplik's spots are oral lesions associated with _____.
   a. mumps
   b. measles
   c. flu
   d. colds

9. Which of the following is an accurate statement concerning zoonoses?
   a. They are animal diseases that spread to humans.
   b. They are diseases specifically transmitted by mosquitoes and ticks.
   c. They are mucus-borne viruses, which are transmitted in the droplets of moisture in a sneeze or cough.
   d. They are diseases that can be transmitted from humans to an animal population.

10. A horror movie portrays victims of biological warfare with uncontrolled bleeding from the eyes, mouth, nose, ears, and anus. What actual virus causes these symptoms?
    a. Ebola virus
    b. bunyavirus
    c. hantavirus
    d. human immunodeficiency virus

11. Reoviruses, such as rotaviruses and coltiviruses, are unique in _____.
    a. being naked
    b. having double-stranded RNA
    c. being both arboviruses and zoonotic
    d. having protein spikes

Matching

What viral family causes each of these diseases? (A family may be used more than once.)

1. _____ Myocarditis
2. _____ Colorado tick fever
3. _____ Rabies
4. _____ Influenza
5. _____ Dengue fever
6. _____ German measles
7. _____ Acute gastroenteritis
8. _____ Ebola virus
9. _____ RSV
10. _____ Western equine encephalitis
11. _____ No known disease

A. *Rhabdoviridae*
B. *Paramyxoviridae*
C. *Reoviridae*
D. *Coronaviridae*
E. *Togaviridae*
F. *Flaviviridae*
G. *Orthomyxoviridae*
H. Orphan virus
I. *Caliciviridae*
J. *Filoviridae*
K. *Picornaviridae*

True/False

1. _____ A single virion is sufficient to cause a cold.
2. _____ All infections of polio are crippling.
3. _____ Postpolio syndrome is due to latent polioviruses that become active 30 to 40 years after the initial infection.
4. _____ Because the oral polio vaccine contains live attenuated viruses, mutations of these viruses can cause polio.
5. _____ A typical host for a togavirus is a horse.

VISUALIZE IT!

1. Label the steps in retroviral replication shown for HIV.

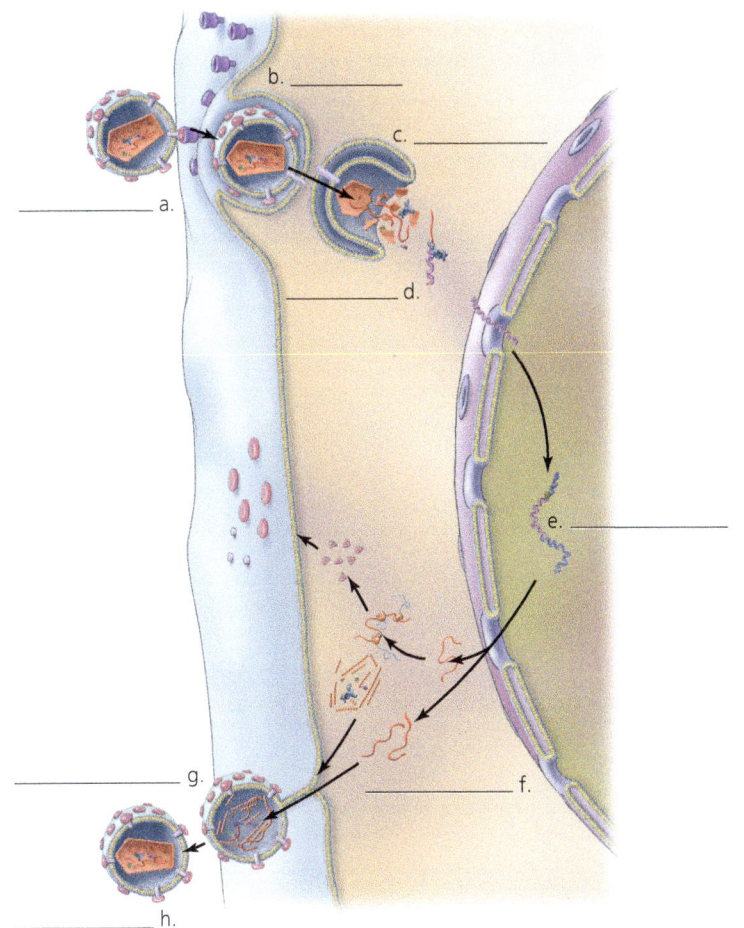

2. Label the flu epidemics. How can you best explain the biennial fluctuation in the number of cases? How can you explain the epidemics?

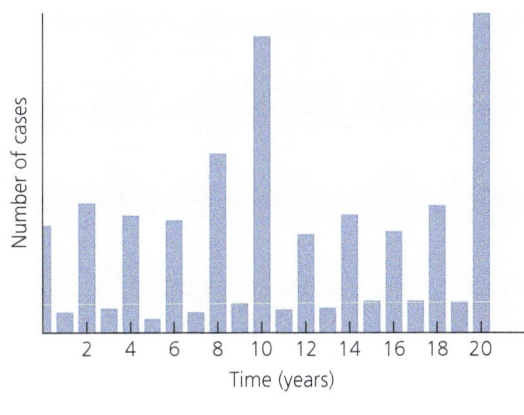

Short Answer

1. Why are humans considered "dead-end" hosts for many arboviruses?

2. Young Luis has skin lesions. His mother knows from microbiology class that five childhood diseases can produce spots. Name those five diseases and the viruses that cause each. List some questions to ask to determine which of these viruses Luis has.

3. The patient in room 519 exhibits yellowing skin and eyes, and it is suspected among the nursing staff that the diagnosis will be some kind of viral hepatitis. Make a chart of five kinds of hepatitis mentioned in this chapter, the infecting pathogen, how the patient might have become infected, and the relative degree of seriousness.

4. Why is AIDS more accurately termed a "syndrome" instead of a "disease"?

5. Consider the viruses you studied in this chapter. Which three would you rank as the deadliest?

6. Support or refute the following statement: "Rubeola is common and of little concern as a childhood disease."

7. Translate the following identification label on a vial of influenzavirus: B/Kuwait (H1N3).

8. Polio and smallpox have been eliminated as natural threats to human health in the United States. (Some risk from bioterrorism remains.) You have considered the features of these diseases that allowed them to be eliminated. From your studies of other viruses, what other viral diseases are candidates for elimination? Why hasn't AIDS been eliminated?

9. Several laboratory tests are used to identify viruses. From your study of this chapter alone, which tests would you surmise are the most common?

10. Why are there more cases of West Nile virus encephalitis in summer than in winter of every year?

11. Compare influenzavirus A 2009 (H1N1) to the 1918–1919 pandemic influenzavirus.

CRITICAL THINKING

1. A dichotomous key uses a series of questions, each with only two possible answers, to guide the user to the identification of an item (see p. 119). Design a key for the RNA viruses discussed in this chapter.

2. A 20-year-old man is brought to a South Carolina hospital's emergency room suffering from seizures, disorientation, hallucinations, and an inordinate fear of water. His family members report that he had suffered with fever and headache for several days. The patient had been bitten by a dog approximately 12 days before admission. What disease does the patient have? Will a vaccine be an effective treatment? If the dog is found to be disease free, what wild animals are the likely source of infection in South Carolina? What treatment should the patient's family, coworkers, friends, and caregivers receive?

3. How did coltiviruses get their name?

4. Retroviruses such as HIV use RNA as a primer for DNA synthesis. Why is a primer necessary?

5. Alligator farms have reported significant loss of animals to West Nile virus. Since alligator hide is generally too thick for mosquitoes to feed, how can you explain the alligators' infections?

6. Reverse transcriptase is notoriously sloppy in making DNA copies from ssRNA templates. How do retroviruses survive when their genomes are subject to such shoddy replication?

CONCEPT MAPPING

Using the terms provided, fill in this concept map that describes viral hepatitis. You can also complete this and other concept maps online by going to the Mastering Microbiology Study Area.

Acute
Blood test for antibodies (IgM)
Carrier state
Chronic
Chronic disease
Hepatic cancer

Hepatitis A
Hepatitis A vaccine
Hepatitis A virus (HAV)
Hepatitis B
Hepatitis B vaccine

Hepatitis B virus (HBV)
Hepatitis C
Hepatitis C virus (HCV)
Mild
PCR or blood test for antibodies

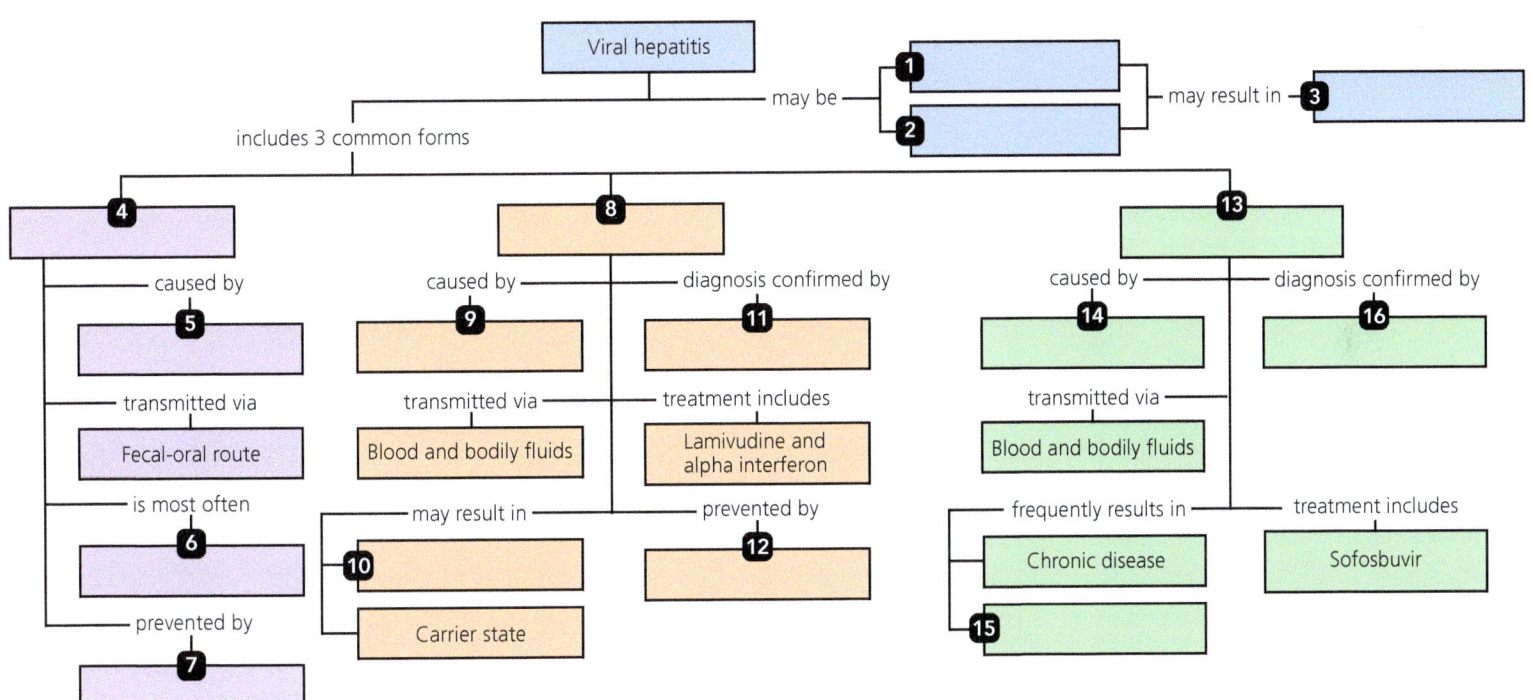

26 Applied and Industrial Microbiology

Before You Begin

1. What is the difference between fermentation and respiration?
2. At which stage of a microbial growth curve does a population first slow down active growth?
3. What is biotechnology?

Cheeses are just one product of microbial action. The holes in Swiss cheese result from gases released by metabolism.

MICRO IN THE CLINIC

A Case of Acne or Something More?

KATE IS FINISHING UP HER FIRST SEMESTER as a nursing student. She has enjoyed her classes, but she is also really looking forward to winter break. After spending the holiday with her family, Kate and her boyfriend, Mark, are planning a week of skiing at a resort in Colorado. It will be a great way to relax and recharge before spring semester.

Kate and Mark have a great trip—there's fresh snow every day, and the skiing is top-notch. The resort has a sauna and hot tub, and Kate and Mark have used both as a way to unwind and soothe stressed muscles after each day of skiing. The trip has been everything they wanted, and they return home ready for the new semester.

The day after they return, Kate notices that she has some redness on her legs, stomach, and back. Kate's condition worsens over the next 24 hours and she develops red lesions or pustules that look similar to acne on her legs, stomach, and back. She complains of a low-grade fever and the pustules are very itchy. Kate calls Mark to see how he is feeling—he says that he has some itchy red "acne" spots around his waist and on his upper thighs and buttocks, but nothing like Kate's symptoms. Kate decides to make an appointment with her doctor just to make sure it's nothing too serious.

1. What is causing Kate's symptoms?
2. Even though they've been together for the past week, Mark's symptoms are different from Kate's; what do you think might be the reason?

Turn to the end of the chapter (p. 796) to find out.

The metabolic activities of microorganisms shape much of our environment, and microbial reactions are essential to life on Earth. Bacteria and fungi in particular are capable of many metabolic processes that are useful to humans. In this chapter, we examine how the vast metabolic capability of microbes is harnessed to serve human needs. (Chapter 8 discusses some aspects of recombinant DNA technology that give organisms enhanced or new functions to make them more valuable in industry, agriculture, or medicine.)

Applied microbiology, the commercial use of microorganisms, encompasses two distinct fields: **food microbiology**, which includes the use of microorganisms in food production and the prevention of food spoilage and food-related illnesses, and **industrial microbiology**, which involves both the application of microbes to industrial manufacturing processes and solutions to environmental, health, and agricultural problems. We begin by considering food microbiology.

Food Microbiology

Microorganisms are involved in producing many of our favorite foods and beverages, from bread to wine to yogurt. Indeed, fermented foods are among the oldest foods known and are culturally very diverse. The characteristic flavors, aromas, and consistencies of such foods result from the presence of acids or sugars made by microbes during fermentation. Besides conferring taste and aroma, microbial metabolism also acts as a preservative, destroys many pathogenic microbes and toxins, and, in some cases, adds nutritional value in the form of vitamins or other nutrients. In the following sections, we explore how our knowledge of microbial growth and metabolism enables us to use microbes in food production and to control microbial activity that results in food spoilage.

The Roles of Microorganisms in Food Production

LEARNING OUTCOME

26.1 Describe how microbial metabolism can be manipulated for food production.

Biochemists use the word *fermentation* to refer to the partial oxidation of sugars to release energy using organic molecules as electron acceptors (see Chapter 5). In food microbiology, however, **fermentation** may refer to any desirable changes that occur to a food or beverage as a result of microbial growth. In contrast, **spoilage** denotes unwanted change to a food that occurs from undesirable metabolic reactions, the growth of pathogens, or the presence of unwanted microorganisms.

Fermentative microbes naturally occur on grains, fruits, and vegetables. In antiquity, people relied on these naturally occurring microbes to produce fermented foods and drinks. However, the same microbes are not always present on a food from harvest to harvest, yielding varying results. Most modern commercial food and beverage production relies on **starter cultures** composed of known microorganisms that perform specific fermentations consistently. In addition to an initial starter culture, *secondary cultures* may be added to further modify the flavor or aroma of foods. For example, the same starter culture is used to initiate formation of both Swiss cheese and blue cheese; the two cheeses differ because different secondary cultures are used in their production.

Bread

Saccharomyces cerevisiae (sak-ă-rō-mī′sēz se-ri-vis′ē-ī) metabolizes sugars to leaven[1] bread. Bakers add the yeast to flour, salt, and other ingredients to make dough, which is kneaded to introduce oxygen. The dough rises when metabolic reactions of yeast release CO_2, producing expanding pockets within the dough. Ethanol produced by fermentation evaporates during baking. Sourdough bread is made using starter cultures consisting of yeast and lactic acid bacteria. Lactic acid produced by the bacteria gives sourdough its characteristic taste.

Fermented Vegetables

People around the world ferment many types of vegetables. Most of these vegetable products are the result of the actions of lactic acid bacteria, such as *Streptococcus* (strep-tō-kok′ŭs), *Leuconostoc* (loo′kō-nos-tŏk), *Lactobacillus* (lak′tō-bă-sil′ŭs), or *Lactococcus* (lak-tō-kok′ŭs), which specifically produce lactic acid during fermentation. Lactic acid acidifies the food and produces a "sour" flavor, as in sourdough bread.

Food products derived from the fermentation of cabbage include Korean kimchi (kim-chē′) and German sauerkraut. Soy sauce is made by the fermentation of soybeans and wheat by lactobacilli, yeast *(S. cerevisiae)*, and the fungus *Aspergillus oryzae* (as-per-jil′ŭs o′ri-zī). Chocolate is derived from the fermentation of cacao seeds. Coffee production relies on natural fermentation to release the coffee bean from the outer layers of the coffee berry so that the bean can then be dried and roasted.

Pickles are another common fermented food. People often equate "pickles" with cucumbers, but other foods, such as beets and eggs, can also be pickled. *Pickling* refers to the process of preserving foods with brine (saturated salt water) or acid (vinegar). Microbial fermentation can be the source of the acid—as it is with traditional dill pickles. Various spices can be added to the pickling solution to enhance taste. Because pickled foods are acidic, few pathogenic microorganisms survive, making pickling an excellent preservation method.

Fermented foods are not made for human consumption alone. *Silage*, a product used as animal feed on many farms, is made by the natural fermentation of potatoes, corn, grass, grain stalks, or other types of green foliage. The vegetation is cut up and stored in silos that are closed to air. Under such moist, anaerobic conditions, the vegetation in the silos ferments, producing many organic compounds that make the silage aromatic and tasty to livestock and more easily digested by them. Some farmers produce silage from baled hay and grasses wrapped in plastic.

---

[1]From Latin *levare*, meaning "to raise."

Fermented Meat Products

Unless dried, cured, smoked, or fermented in some way, meat has a tendency to spoil rapidly. The combination of fermentation and drying or smoking has been used for centuries to preserve meats, particularly pork and beef. Dry sausages, such as salami and pepperoni, are made by grinding meat, mixing in various spices and starter cultures, allowing the mixture to ferment in the cold, and then stuffing it into casings. Fermented fish—common in Asian countries—are prepared by grinding fish in brine. Naturally occurring microbes ferment the mixture. The solids are removed and pressed to form a fish paste, while the liquid portion is drained off and mixed with flavoring agents to form fish sauce.

Fermented Dairy Products

Milk fermentation relies, for the most part, on the activities of lactic acid bacteria. Milk in an udder is sterile, but because milking introduces microorganisms, which may cause spoilage or diseases, we pasteurize "raw" milk. The metabolic products of starter cultures give fermented milk products, such as buttermilk, yogurt, and cheeses, their characteristic textures and aromas.

Buttermilk is made from fat-free milk (skim milk) by adding a starter culture of *Lactococcus lactis* (lak'tis) subspecies *cremoris* (kre-mōr'is) and *Leuconostoc citrovorum* (sit-rō-vō'rum). Yogurt production utilizes starter cultures of *Streptococcus thermophilus* (ther-mo'fil-us) and *Lactobacillus bulgaricus* (bul-gă'ri-kŭs). Yogurt manufacturers mix pasteurized milk, milk solids, sweeteners, and other ingredients to a uniform consistency and then add the starter culture. They ferment the mix at 43°C and then cool it to stop microbial activity. They may add flavors prior to packaging. The microorganisms are left alive and may have beneficial effects; they are so-called **probiotics**; that is, live microorganisms administered to improve health and prevent disease.

Cheeses can be hard or soft and mild or sharp. Regardless of the end product, the cheese-making process typically begins with pasteurized milk (FIGURE 26.1). One pound of cheese requires about five gallons of milk. Cultures of *Lactococcus*, *Lactobacillus*, or *Streptococcus* coagulate protein in milk to form *curds* (solids) and *whey* (liquids). For cheeses sold as soon as they are made (called *unripened cheeses*, such as cottage cheese), the curd is removed, cut into small pieces, and packaged. Alternatively, the curdling process can be accelerated by the addition of the enzyme *rennin*, a type of protease (protein-digesting enzyme). Rennin was originally obtained from the stomach of cud-chewing animals, but now most of it is produced by the genetically modified fungus *Aspergillus*.

Other cheeses (so-called *ripened cheeses*) are aged until they have the desired texture or taste. Hard cheeses, such as Parmesan and cheddar, are pressed until little water remains, while soft cheeses retain enough moisture to make them spreadable. Once curds are pressed, a cheese is aged for months to years while continued microbial activity imparts characteristic smells and tastes. Cheese producers utilize numerous different microorganisms and their enzymes along with various ripening regimens and culture conditions to make numerous, distinct types of cheeses. The average American ate 35 pounds of cheese in 2016, a clear sign of the size and importance of this microbe-dependent industry.

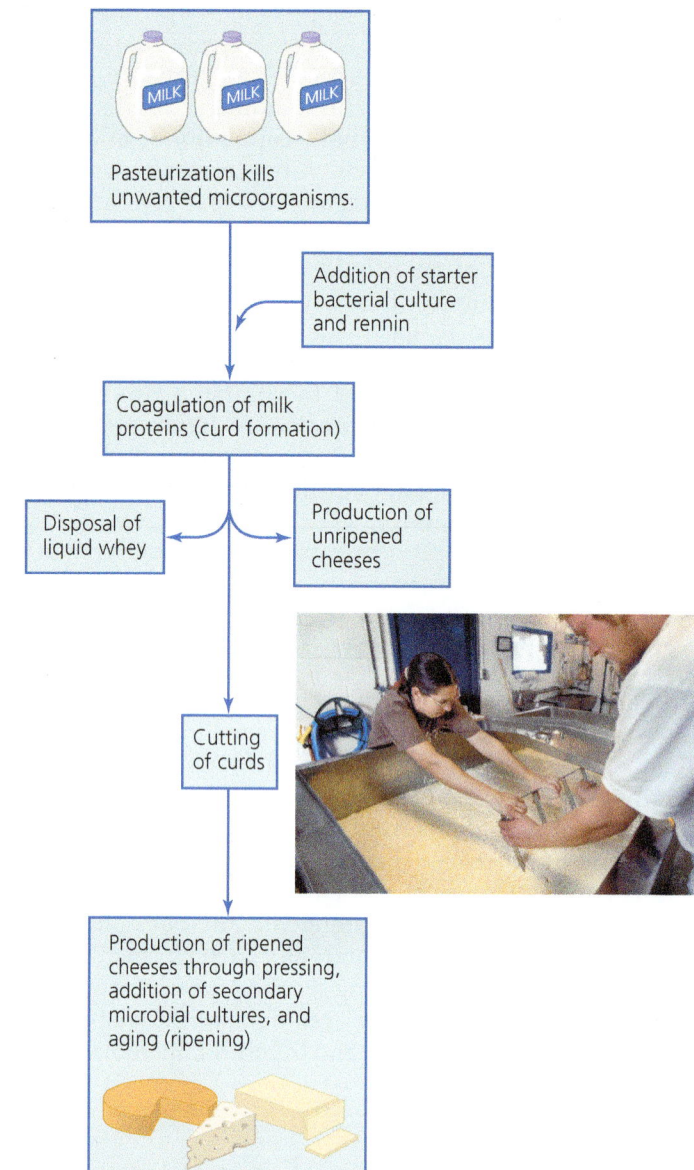

▲ FIGURE 26.1 **The cheese-making process.** The commercial production of all cheeses begins with the pasteurization of milk; the wide variety of cheeses available on the market results from the nature of subsequent processing steps. *Why do cheeses that are aged longer have more acidic, "sharper" tastes?*

Figure 26.1 Fermentation during aging results in continued production of acid; the longer the cheese is aged, the more acid produced and the "sharper" the taste.

Products of Alcoholic Fermentation

Alcoholic fermentation is the process by which various microorganisms convert simple sugars such as glucose into ethanol (drinking alcohol) and carbon dioxide (CO_2). These two products of fermentation take on different importance for different foods and beverages. As is the case for the production of fermented dairy products, manufacturers use specific starter cultures in the large-scale commercial applications of alcohol fermentation. In the next sections we consider the role of fermentation in the production of wine, spirits, beer, sake, and vinegar.

Wine and Spirits Generally, wine production proceeds as follows (FIGURE 26.2):

1. *Preparation of must.* Winemakers crush the fruit and remove the stems to form *must* (fruit solids and juices). They make red wines from the entire must of dark grapes, whereas they make white wines using only the juice of either dark or light grapes. Weather conditions and soil composition affect the sugar content of fruit, which in turn greatly affects the quality of the wine.

2. *Fermentation.* A thin film of bacteria and yeast naturally covers grapes and other fruit, and in some wineries these natural microbes produce the wine. However, in most commercial operations, vintners add sulfur dioxide (SO_2) to retard growth of naturally occurring bacteria and then add starter cultures of yeast and bacteria to ensure that wine of a consistent quality is generated from year to year. *Saccharomyces* ferments sugar in fruit juice to alcohol to make wine. *Leuconostoc* removes acids naturally present in grapes.

3. *Clarification.* Filtration or settling removes solids—a process called clarification.

4. *Aging.* Wine is aged in wooden barrels. Continued yeast metabolism and the leaching of compounds from the wood add taste and aroma.

5. *Bottling.* The wine is then bottled and distributed.

Dry wines are those in which all the sugar has fermented, whereas sweet or dessert wines retain some sugar. Most table wines have a final alcohol content of 10% to 12%, but fortified wines have an alcohol content of about 20% because distilled spirits (discussed next) are added. Sparkling wine is produced by the addition of sugar and a second fermentation that produces CO_2 inside the bottle.

Distilled spirits are made in a process similar to wine making in that *Saccharomyces* ferments fruits, grains, or vegetables. The difference is that during the aging process, the alcohol is concentrated by distillation. In this process, the liquid is heated to drive off the alcohol, which is then condensed and added back. The net result is an increase in the alcohol content (100-proof spirits are 50% alcohol). Brandies are made from fruit juices, whiskeys are made from cereal grains, and vodka is often made from potatoes.

Beer and Sake Beer, the most commonly consumed alcoholic beverage in the world, is made from barley in the following steps (FIGURE 26.3):

1. *Malting.* Barley is moistened and germinated, a process that produces catabolic enzymes that convert starch into sugars, primarily maltose. Drying halts germination, and the barley is crushed to produce *malt*.

2. *Mashing.* Malt and additional sources of carbohydrate called adjuncts (such as starch, sugar, rice, sorghum, or corn) are combined with warm water, allowing enzymes released during malting to generate more sugars. The sugary liquid is called *wort*.

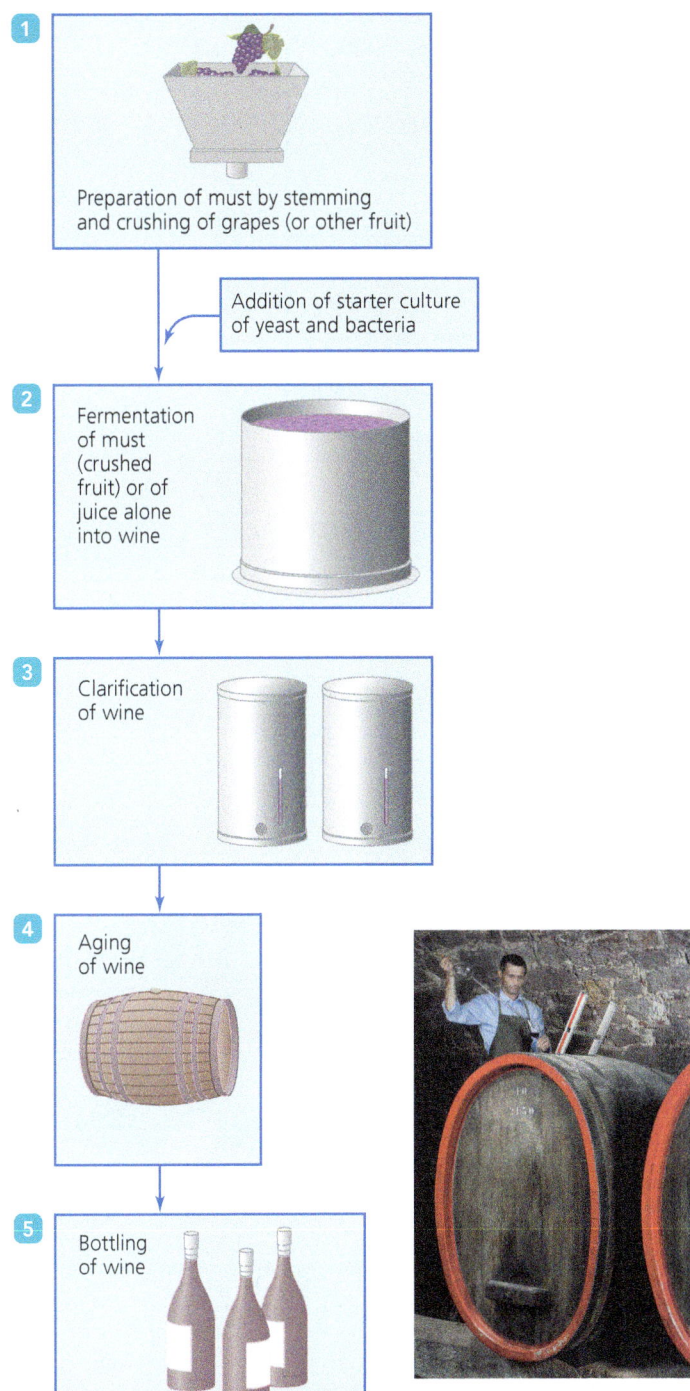

▲ FIGURE 26.2 **The wine-making process.** Red wine is produced by fermenting must (fruit solids and juice) from dark grape varieties, whereas white wine results from the fermentation of juice alone from either dark or light grape varieties.

3. *Preparation for fermentation.* The brewer removes the spent grain from the wort and adds *hops*—the dried flower-bearing parts of the vinelike hops plant. The wort mixture is boiled to halt enzymatic activity, extract flavors from the hops, kill most microorganisms, and concentrate the mixture. The remains of the hops are removed after boiling.

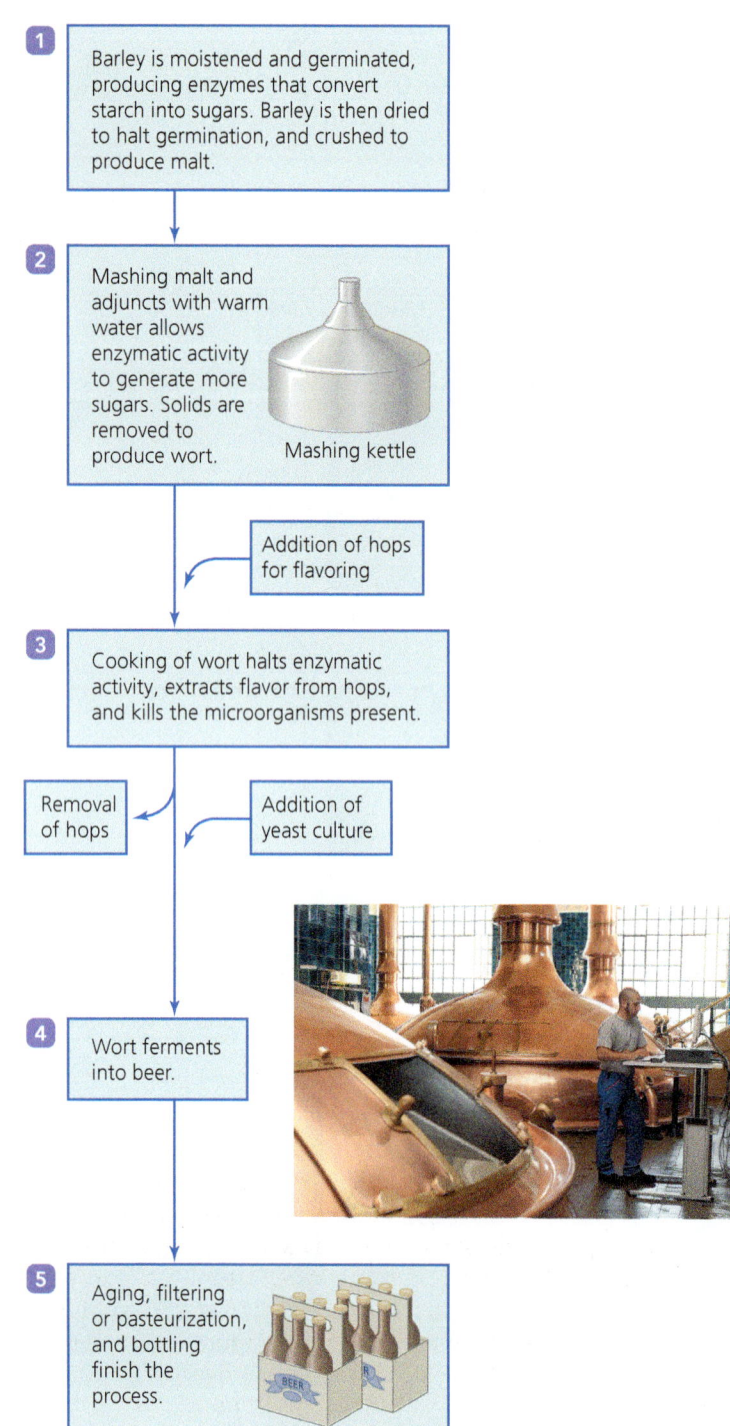

▲ FIGURE 26.3 **The beer-brewing process.** The type of beer produced (ale, lager) is determined largely by the type of yeast used (top fermenting or bottom fermenting, respectively).

4. *Fermentation.* A starter culture of *Saccharomyces* is added to the wort. Most beers—called lagers—are made with the bottom-fermenting yeast *S. carlsbergensis* (karlz-bur-jen'sis). Other beers—called ales—are produced with the top-fermenting yeast *S. cerevisiae* (so called because the yeast floats in the vat).

5. *Aging.* The beer is aged, pasteurized or filtered, and bottled or canned. Beer has a considerably lower alcohol content (about 4%) than either wine or distilled spirits.

Sake (rice beer) is made from cooked rice in which the starch is first converted to sugar (by the fungus *Aspergillus oryzae*) and the sugar then is fermented by *Saccharomyces*. Sake has an alcohol content of about 14%.

Vinegar Vinegar[2] is produced when ethanol resulting from the fermentation of a fruit, grain, or vegetable is oxidized to acetic acid. *Acetic acid bacteria* such as *Acetobacter* (a-sē'tō-bak-ter) or *Gluconobacter* (gloo-kon'ō-bak-ter) perform this secondary metabolic conversion, generating about 4% acetic acid. Different flavors of vinegar are derived from different starting materials. Cider vinegar, for example, is made from apples.

TABLE 26.1 summarizes types of fermented foods along with the organisms involved in their production.

MICRO CHECK

1. Which microbe is primarily used to ferment bread dough?
2. How does pickling preserve food?
3. Why is milk pasteurized before cheese production starts?

The Causes and Prevention of Food Spoilage

LEARNING OUTCOMES

26.2 Describe how food characteristics and the presence of microorganisms in food can lead to food spoilage.
26.3 List several methods for preventing food spoilage.

Whereas many chemical reactions involving food are desirable because they enhance taste, aroma, or preservation, other reactions are not. Spoilage of food involves adverse changes in nutritive value, taste, or appearance. Spoilage leads not only to economic loss but also potentially to illness and even death. In this section, we examine the causes of food spoilage before considering ways to prevent it.

Causes of Food Spoilage

Foods may spoil because of inherent properties of the food itself. Such *intrinsic factors* include nutrient content, water activity, pH, and physical structure of the food as well as the competitive activities of microbial populations. Alternatively, food spoilage results from *extrinsic factors* that have nothing to do with the food itself but instead with the way it is processed or handled.

---
[2]From French *vinaigre,* meaning "sour wine."

TABLE 26.1 Some Fermented Foods and the Microorganisms Used in Their Production

Food	Starting Material	Representative Culture Microorganisms
Fermented Vegetables		
Sauerkraut/kimchi	Cabbage	Various lactic acid bacteria
Pickle	Cucumbers, peppers, beets	Various lactic acid bacteria
Soy sauce	Soybeans and wheat	*Aspergillus oryzae* and *Lactobacillus* spp.[a]
Miso	Rice and soybeans or rice and other grains	*A. oryzae*, *Lactobacillus* spp., and *Torulopsis etchellsii*
Fermented Meat Products		
Dry salami	Pork, beef, chicken	Various lactic acid bacteria
Fish sauce/paste	Ground fish	Various naturally occurring bacteria
Fermented Dairy Products		
Milks		
Buttermilk	Pasteurized skim milk	*Lactococcus lactis* subspecies *cremoris* and *Leuconostoc citrovorum*
Yogurt	Pasteurized skim milk	*Streptococcus thermophilus* and *Lactobacillus bulgaricus*
Cheeses		
Cottage cheese	Pasteurized milk	*L. lactis*, including subspecies *cremoris*
Hard cheese (e.g., cheddar)	Milk curd	Starter culture as in cottage cheese without further additions
Soft cheese (e.g., Camembert)[b]	Milk curd	Starter culture as in cottage cheese plus *Penicillium camemberti*
Mold-ripened (e.g., Roquefort)	Milk curd	Starter culture as in cottage cheese plus *Penicillium roqueforti*
Animal Feed		
Silage	Corn, grains, vegetation	Various naturally occurring bacteria
Alcoholic Fermentations		
Wine	Grapes	*Saccharomyces cerevisiae*
Distilled spirits	Fruits, vegetables, grains	*S. cerevisiae*
Beer	Barley	*S. cerevisiae* or *S. carlsbergensis*
Sake	Cooked rice	*A. oryzae* and *Saccharomyces* spp.
Vinegar	Fruits, vegetables, grains	*S. cerevisiae* and *Acetobacter* or *Gluconobacter*
Bread	Flour, salt, etc.	*S. cerevisiae*

[a] spp. indicates more than one species in a genus.
[b] In addition to being a soft cheese, Camembert is also a mold-ripened cheese.

Intrinsic Factors in Food Spoilage The nutritional composition of food determines both the types of microbes present and whether they will grow. Some foods contain natural antimicrobial agents, such as the benzoic acid in cranberries, and thus are not prone to spoilage. In contrast, *fortified foods*—those enriched in vitamins and minerals to improve the health of humans—may inadvertently facilitate growth of microorganisms by providing more nutrients.

Water activity refers to water that is not bound physically by solutes or to surfaces and is thus available to microbes. The water activity of pure water is set at 1.0, and most microbes require environments with a water activity of at least 0.90. Moist foods with water activities near 0.90, such as fresh meat, support microbial growth. Dry foods (e.g., uncooked dry pasta) with water activities near zero do not support microbial growth. Food processors reduce water activity by drying foods or by adding salts or sugars. Thus, even though jam is moist, its sugar content is very high, and its water activity is too low to support much microbial growth.

Acidity of food can either be an intrinsic chemical property of the food, as in the case of citrus fruit, or result from fermentation, as with dill pickles. In either case, a pH below 5.0 typically reduces microbial growth except by a few molds or lactic acid bacteria. A pH closer to neutral (7.0) supports a wider range of microbial growth. Since the pH of most foods is close to neutral, pH is generally not a deterrent to spoilage microorganisms.

Physical structure is a visible intrinsic factor. Rinds or thick skins usually protect fruits and vegetables, and eggs have shells. These coverings are dry and nutritionally poor, and thus they support little microbial growth. If the outer covering is broken or cut, however, microbes can reach the moist interior of the food and cause it to rot.

Ground meat has more surface area and more oxygen and may have bacteria mixed within it during the grinding process. Thus, ground meat supports microbial growth and spoils faster than whole cuts of meat. In uncut meat, the largest volume of the meat is anaerobic and not exposed to microorganisms.

To some extent, microbial competition can also be an intrinsic factor in food spoilage, especially in fermented foods. Fermented foods are populated with large numbers of fermentative bacteria but few pathogens because the latter do not readily grow in the environment produced by the fermenters. Furthermore, pathogens that require the rich nutrient levels found in the human body are not typically capable of growing on "lower-nutrient" foods.

TABLE 26.2 Factors Affecting Food Spoilage

	Foods at Greatest Risk of Spoilage	Foods at Least Risk of Spoilage
Intrinsic Factors		
Nutritional Composition	Chemically rich or fortified foods (steak, bread, milk)	Chemically limited foods (flour, cereals, grains)
Water Activity	Moist foods (meat, milk)	Dry foods or those with low water activity (dried pasta, jam)
pH	Foods with neutral pH (bread)	Foods with low pH (orange juice, pickles)
Physical Structure	Foods without rinds, skins, or shells; ground meat	Foods with rinds, skins, or shells; intact foods
Microbial Competition	Foods that lack natural microbe populations (ground beef)	Foods with resident microbial populations (pickles)
Extrinsic Factors		
Degree of Processing	Unprocessed foods (raw milk, fruit)	Processed foods (pasteurized foods)
Amount of Preservatives	Foods without preservatives (meats, natural foods)	Foods with either naturally occurring or added preservatives (garlic, spices, sulfur dioxide)
Storage Temperature	Foods left in warm conditions	Foods kept cold
Storage Packaging	Foods stored exposed	Foods wrapped or sealed

TABLE 26.2 summarizes the effects of intrinsic factors on food spoilage.

Extrinsic Factors in Food Spoilage Extrinsic factors, including how food is processed, handled, and stored, also govern food spoilage. Microorganisms can enter food in a variety of ways; some are introduced accidentally during harvesting from soil or contaminated water, and commercial food processing introduces others. Mechanical vectors, such as flies, also deposit microbes onto food. However, most microbial contaminants leading to spoilage can be traced to improper handling by consumers. Methods to prevent the introduction of microbes into food or to limit their growth will be presented shortly.

Classifying Foods in Terms of Potential for Spoilage

Based on the previous considerations, foods can be grouped into three broad categories, depending on their likelihood of spoilage. *Perishable* foods, such as milk, tend to be nutrient rich, moist, and unprotected by coverings. They need to be kept cold, or they spoil relatively quickly (within days). *Semiperishable* foods, such as tomato sauce, can be stored in sealed containers for months without spoiling as long as they are not opened. Once opened, however, they may spoil within several weeks. *Nonperishable* foods are usually dry foods, such as pasta, or canned goods that can be stored almost indefinitely without spoiling. Nonperishable foods are typically either nutritionally poor, dried, fermented, or preserved (discussed shortly).

Prevention of Food Spoilage

Food spoilage can begin during production, processing, packaging, or handling. Spoilage results in significant economic losses not only to producers—in the form of lost productivity, food recalls, and even the loss of jobs resulting from the shutdown of unsafe food processing facilities—but also to consumers in the form of medical expenses and time lost from work. In the next sections, we examine some of the many techniques used to prevent food spoilage: food processing methods, the use of preservatives, and attention to temperature and other storage conditions.

Food Processing Methods *Industrial canning* is a major food packaging methodology for preserving foods (FIGURE 26.4). After food is prepared by washing, sorting, and processing, it is packaged into cans or jars and subjected to high heat (115°C, 239°F) under pressure for a given amount of time, depending on the food. This temperature is not that of a sterilizing autoclave in a microbiology laboratory, so the food must be heated longer. (Autoclave temperature [121°C, 250°F] would severely change the food's taste, texture, and nutritional content.) The heating is followed by rapid cooling. The heat kills vegetative mesophilic bacteria and destroys endospores formed by *Bacillus* (ba-sil′ŭs) and *Clostridium* (klos-trid′ē-ŭm). Although canning results in the elimination of most contaminating microorganisms, it does not

▲ FIGURE 26.4 **Industrial canning.** Food is processed, put into cans or jars, and heated under pressure for a length of time and a temperature sufficient to kill most microbes, including endospore-forming bacteria. *Why isn't the worker wearing a mask and gloves while handling open cans of food?*

Figure 26.4 *The food is about to be steam treated, so any microbes that would spoil the food or cause disease will be killed.*

sterilize food in the strictest sense; hyperthermophilic microbes remain, but because they cannot grow at room temperature, they do not pose a threat.

Spoilage can occur if the food is underprocessed or if mesophilic microbes contaminate the can during cooling or thereafter. The two most frequent contaminants found in canned goods are *Clostridium* spp. and coliforms, organisms that grow well anaerobically in low-acid foods. Contamination with the resistant endospores of *Clostridium* is especially problematic because if the cans or jars are not heated sufficiently, the endospores germinate, grow, and release toxins inside the sealed container, which can result in botulism.

Pasteurization is less rigorous than canning and is used primarily with beer, wine, and dairy products because it does not degrade the flavor of these more delicate foods. For example, high-temperature, short-time (HTST) pasteurization heats milk to 72°C for only 15 seconds. By briefly heating foods only enough to kill mesophilic, non-endospore-forming microbes (including most pathogens), pasteurization lowers the overall number of microbes, but because some microbes survive, pasteurized foods will spoil without refrigeration.

Moisture is essential for microbial growth; therefore, desiccation (drying) is an excellent food preservation technique. Drying greatly reduces or eliminates microbial growth, but it does not kill bacterial endospores. Although some fruits are still dried by being left in the sun, today most commercially dried foods are processed in ovens or heated drums that evaporate the water from the foods.

Lyophilization (lī-of'i-li-zā'shŭn), or *freeze-drying*, involves freezing foods and then using a vacuum to draw off the ice crystals. Freeze-dried foods such as soups and sauces can be reconstituted by mixing with water.

Gamma radiation, produced primarily by the isotope cobalt-60, penetrates fruits, vegetables, and meats, including fish, poultry, and beef, to cause irreparable and fatal damage to the DNA of microbes. Such ionizing *irradiation* is controversial because some consumers believe erroneously that irradiated foods become radioactive, are less nutritious, or contain toxins. Nevertheless, because irradiation can achieve complete sterilization, it is used routinely to preserve some foods, such as spices.

Ultraviolet (UV) light is not ionizing and does not penetrate very far. It is used to treat packing and cooling water used in industrial canning and to treat work surfaces and utensils in industrial meat processing plants.

In *aseptic packaging*, paper or plastic containers that cannot withstand the rigors of canning or pasteurization are sterilized with hot peroxide solutions, UV light, or superheated steam. Then a conventionally processed food is added to the aseptic package, which is sealed without the need for additional processing.

The Use of Preservatives Humans have preserved foods with salt or sugar throughout history. Both chemicals draw water by osmosis out of foods and microbes alike, killing microbes on the food and retarding the growth of any subsequent microbial contaminants. Bacon is an example of a high-salt food; jellies are examples of high-sugar foods.

Whereas salt and sugar act by removing water, some natural preservatives actively inhibit microbial enzymes or disrupt cytoplasmic membranes. For example, garlic contains *allicin*, which inhibits enzyme function. Benzoic acid, produced naturally by cranberries, also interferes with enzymatic function. Cloves, cinnamon, oregano, and thyme (and, to some degree, sage and rosemary) produce oils that interfere with the functions of membranes of microorganisms.

As we have seen, fermentation preserves some foods by producing an acidic environment that is inhospitable to most microbes. For other foods such as meats, the use of wood smoke during the drying process introduces growth inhibitors that help preserve the food. Other naturally occurring and synthetic chemicals can be purposely added to foods as preservatives. Acceptable preservatives are harmless and do not alter the taste or appearance of the food to any great extent. Organic acids, such as benzoic acid, sorbic acid, and propionic acid, are commonly used in beverages, dressings, baked goods, and a variety of other foods. Gases, such as sulfur dioxide and ethylene oxide, are used to preserve dried fruit, spices, and nuts. All such chemical preservatives inhibit some aspect of microbial metabolism, but many do not actually kill microbes; in other words, they are *germistatic* rather than *germicidal*. Some chemicals work better against bacteria than against fungi and vice versa. Benzoic acid, for example, has a largely antifungal function and does not affect the growth of many bacteria.

Attention to Temperature During Processing and Storage In general, higher temperatures are desirable during food processing and preparation to prevent food spoilage, whereas lower temperatures are desirable for food storage. High temperatures, such as those of pasteurization, canning, or cooking, kill potential pathogens because proteins and enzymes become irreversibly denatured. However, heat does not inactivate many toxins; for example, botulism toxin may remain in cooked foods even when the bacteria that produced it are dead.

Unlike heat, cold rarely kills microbes but instead merely retards their growth by slowing metabolism. Even freezing fails to kill all microorganisms and may only lower the level of microbial contamination enough to reduce the likelihood of food poisoning after frozen foods have been thawed.

To prevent food spoilage, foods should be prepared at sufficiently high temperatures and then stored under conditions that do not facilitate microbial growth. Foods should be properly stored in appropriate containers, cold foods should be kept cold, foods should be cooked thoroughly, and leftovers should be refrigerated or frozen to reduce spoilage. Wrapping leftovers or putting them in containers enables foods to be stored away from exposure to air or contact that could result in contamination.

One microbe for which cold storage does not suppress growth is *Listeria monocytogenes* (lis-tēr'ē-ă mo-nō-sī-tah'je-nēz), the causative agent of listeriosis and a common environmental bacterium. This bacterium is prevalent in certain dairy products, such as soft cheeses, and grows quite well under refrigeration; therefore, it is best to prevent its entry into foods.

Foodborne Illnesses

LEARNING OUTCOME

26.4 Discuss the basic types of illnesses caused by food spoilage or food contamination, and describe how they can be avoided.

Spoiled food, if consumed, can result in illness from toxins produced by the microbes or from the microbes themselves, but not all foodborne illnesses result from actual food spoilage. Foodborne illnesses may also result from the consumption of harmful microbes or their products in food.

Foodborne illnesses *(food poisoning)* can be divided into two types: **food infections**, caused by the consumption of living microorganisms, and **food intoxications**, caused by the consumption of microbial toxins instead of the microbes themselves. Typical signs and symptoms of food poisoning are generally the same regardless of the cause and include nausea, vomiting, diarrhea, fever, fatigue, and muscle cramps. Symptoms occur within 2–48 hours after ingestion, and the effects of the illness can linger for days. Most outbreaks of food poisoning are *common-source epidemics*, meaning that a single food source is responsible for many individual cases of illness.

The Centers for Disease Control and Prevention (CDC) estimates that about 48 million cases of food poisoning occur in the United States each year. Of these, about 128,000 people require hospitalization and 3000 die. Researchers identify the microbes involved in only about 14 million of the total cases. The U.S. Department of Agriculture estimates the economic cost of food poisoning—due to loss of productivity, medical expenses, and death—at roughly $5 billion to $10 billion per year.

More than 250 different foodborne diseases have been described. **TABLE 26.3** lists 10 common causes of foodborne illness and their sources. Except for the protozoan *Toxoplasma gondii* (tok-sō-plaz′mă gon′dē-ē), all are bacterial agents. A few fungi (*Aspergillus* and *Penicillium* [pen-i-sil′ē-ŭm]) and viruses (e.g., hepatitis A, noroviruses) also cause food poisoning.

MICRO CHECK

4. Are nutritional content, food processing, and food handling intrinsic or extrinsic factors that affect spoilage?
5. Why is ground meat more likely to spoil than larger cuts of meat?
6. What is lyophilization?
7. Why is the incubation period for food infections usually longer than that for food intoxications?

TELL ME WHY

Even though microbes are naturally present in raw milk, raw milk is not generally used for the production of cheeses. Why not?

Industrial Microbiology

The potential uses of microorganisms for producing valuable compounds, as environmental sensors, and in the genetic modification of plants and animals makes industrial microbiology one of the more important fields of study within the microbiological sciences. In the sections that follow, we will examine the use of microbes in industrial fermentation, in the production of several industrial products, and in the treatment of water and wastewater.

The Roles of Microbes in Industrial Fermentations

LEARNING OUTCOME

26.5 Describe the role of genetically manipulated microorganisms in industrial and agricultural processes and the basics of industrial-scale fermentation.

TABLE 26.3 Most Common Bacterial and Protozoan Agents of Foodborne Illnesses

Organism	Affected Food Products	Comments
Campylobacter jejuni	Raw and undercooked meats; raw milk; untreated water	Most common cause of diarrhea of all foodborne agents
Clostridium botulinum	Home-prepared foods	Produces a neurotoxin
Escherichia coli O157:H7	Meat; raw milk	Produces an enterotoxin
Listeria monocytogenes	Dairy products; raw and undercooked meats; seafood; produce	Common in soils and water, so contamination from these sources occurs easily; grows at refrigerator temperature
Salmonella spp.	Raw and undercooked eggs; meat; dairy products; fruits and vegetables	Second most common cause of foodborne illness in the United States
Shigella spp.	Salads; milk and other dairy products; water	Third most common cause of foodborne illness in the United States
Staphylococcus aureus	Cooked high-protein foods	Produces a potent toxin that is not destroyed by cooking
Toxoplasma gondii	Meat (pork in particular)	Parasitic protozoan
Vibrio vulnificus	Raw and undercooked seafood	Causes primary septicemia (bacteria in the blood)
Yersinia enterocolitica	Pork; dairy products; produce	Causes generalized diarrhea and severe cramping that mimics appendicitis; grows at refrigerator temperature

In industry, the word *fermentation* is used differently from its use in food microbiology or in the study of metabolism. *Industrial fermentations* involve the large-scale growth of particular microbes for producing beneficial compounds, such as amino acids and vitamins. Temperature, aeration, and pH are all regulated to maintain optimal microbial growth conditions. Generally, industrial fermentations start with the cheapest growth medium available, often the waste product from another process (such as whey from cheese production). Because of its scale, industrial fermentation is performed in huge vats that can be readily filled, emptied, and sterilized (FIGURE 26.5). Vats are typically made from stainless steel so that they can be cleaned and sterilized more easily.

There are two types of industrial fermentation. In *batch production*, organisms ferment their substrate until it is exhausted, and then the end product is harvested all at once. In *continuous flow production*, the vat, also called a *bioreactor*, is continuously fed new medium while wastes and product are continuously removed. For this setup to work, the organism must secrete its product into the surrounding medium.

Industrial products are produced as either primary or secondary metabolites of the microorganisms. *Primary metabolites*, such as ethanol, are produced during active growth and metabolism because they are either required for reproduction or by-products of active metabolism. *Secondary metabolites*, such as penicillin, are produced after the culture has moved from the log phase of growth and entered the stationary phase (see Chapter 6, Figure 6.20), during which time the substances produced are not immediately needed for growth.

Recombinant DNA techniques are used for the production of *recombinant microorganisms*. (Chapter 8 examines these techniques and some of their uses.) Most genetically modified organisms used in industry have been specifically designed to produce a stable, high-yield output of desirable chemicals or to perform certain novel functions.

Industrial Products of Microorganisms

LEARNING OUTCOME

26.6 List some of the various commercial products produced by microorganisms.

Microorganisms, particularly bacteria, metabolically produce many industrially useful chemicals, including enzymes, food additives and supplements, dyes, plastics, and fuels. Recombinant organisms add to this diversity by producing pharmaceuticals, such as human insulin, which are not normally manufactured by microbial fermentations. In the sections that follow we consider a variety of industrial products produced by microbes, including enzymes, alternative fuels, pharmaceuticals, pesticides, agricultural products, biosensors, and bioreporters.

Enzymes and Other Industrial Products

Enzymes are among the more important products made by microbes. Most are naturally occurring enzymes for which industrial uses have been devised. For example, amylase, produced by *Aspergillus oryzae*, is used as a spot remover. Pectinase, obtained from species of *Clostridium*, enzymatically releases cellulose fibers from flax, which are then made into linen. Proteases from a variety of microbes are used in meat tenderizers, spot removers, and cheese production. Streptokinases and hyaluronidase are used in medicine to dissolve blood clots and enhance the absorption of injected fluids, respectively. Microbes also make the enzymatic tools of recombinant DNA technology, including restriction enzymes, ligases, and polymerases.

Some products made naturally by microorganisms are useful to humans as food additives and food supplements. *Food additives* generally enhance a food in some way, such as by improving color or taste, whereas *food supplements* make up for nutritional deficiencies. Amino acids and vitamins are two important microbial products used as supplements. Vitamins are added directly to foods or sold as vitamin tablets. Amino acids are either sold in tablet or powder form or combined to make new compounds, such as the sweetener aspartame (made from the amino acids phenylalanine and aspartic acid). Other organic acids, such as citric acid, gluconic acid, and acetic acid (vinegar), are also microbially produced to be used in food manufacturing. Citric acid is used as an antioxidant in foods, and gluconic acid is used medically to facilitate calcium uptake.

Other industrial products made by microbes include dyes and cellulose fibers used in woven fabrics. Microbially produced biodegradable plastics can replace nonbiodegradable, petroleum-based plastics. This is possible because many bacteria produce carbon-based storage molecules such as PHB that have a structure similar to petroleum-based plastics (see **Beneficial Microbes: Plastics Made Perfect?** on p. 75).

▲ FIGURE 26.5 **Fermentation vats.** Such large containers are used for growing microorganisms in the vast quantities needed for the large-scale production of many industrial, agricultural, and medical products. *Why are industrial fermentation vats made of stainless steel instead of wood?*

Figure 26.5 *Stainless steel can be cleaned more easily to avoid contamination of the product being produced.*

Alternative Fuels

Photosynthetic microorganisms use the energy in sunlight to convert CO_2 into compounds that can be used as fuels—so-called *biofuels*. Other microorganisms convert biomass (organic materials such as plants or animal wastes) into renewable biofuels, including ethanol, methane, and hydrogen.

Ethanol, which is made during alcoholic fermentation and is the simplest alternative biofuel to synthesize, can be mixed with gasoline to make *gasohol*, which is used in existing cars. Currently, in the United States, food crops such as corn are fermented to ethanol, but production of alternative fuel from nonfood crops or crop wastes—for example, inedible switchgrass, corn stalks, wood chips, or seaweed—may use less water and fertilizer, and won't impact food supplies. Some research currently is focused on developing microbes to generate a more complex alcohol—butanol, which contains more energy per mass than ethanol, does not absorb water from the atmosphere like ethanol, and can be burned in most engines designed for gasoline without modification.

All microorganisms synthesize hydrocarbons as part of their normal metabolism, but only some microbes produce hydrocarbons that could be useful fuels. The colonial alga *Botryococcus braunii* (bot′rē-ō-kok′ŭs brow′nē-ē), for example, produces hydrocarbons that account for 30% of its dry weight. One day, this alga may be an important source of fuel, but the technology to harvest such hydrocarbons does not yet exist. One exception is the harvesting of methane gas, which can be collected and piped through natural gas lines (natural gas is mostly methane) to be used for cooking and heating. The largest potential source of methane is from landfills, where methanogens anaerobically convert wastes into methane via anaerobic respiration. Some communities already use methane from landfills to produce energy. Methane can also be used as a fuel in properly equipped cars.

Although a variety of microbes release hydrogen as part of their normal metabolism, the hydrogen-producing metabolic pathway does not cost-effectively produce fuel. An economically efficient method for the production of hydrogen fuel using sunlight and photosynthetic microbes would have widespread applications and enormous appeal, but much research is needed to develop technologies to biologically produce hydrogen as a viable energy source. Biofuels may play an important role in the future of the energy industry, possibly serving as a renewable replacement for fossil fuels.

Pharmaceuticals

Foremost among the pharmaceutical substances that microorganisms produce are antimicrobial drugs. About 6000 antimicrobial substances have been described since penicillin was first produced during World War II. Of these, 100 or so have current medical applications. The bacteria *Streptomyces* (strep-tō-mī′sēz) and *Bacillus* and the fungus *Penicillium* synthesize the majority of useful antimicrobials, though recombinant DNA techniques enable the modification of other microbes to produce new versions of old drugs.

In addition to antimicrobials, genetically modified microbes are producing hormones and other cell regulators. TABLE 26.4 lists some products of recombinant DNA technology used in medicine.

Pesticides and Agricultural Products

Farmers use microbes and their products in a variety of agricultural applications, particularly with regard to crop management. *Bacillus thuringiensis* (thur-in-jē-en′sis) is one of the more widely used organisms because during sporulation it produces *Bt toxin*, which, when digested by the caterpillars of such pests as gypsy moths, diamondback moths, and tomato hornworms, destroys the lining of the insect's gut wall, eventually killing it. Bt toxin isolated from the bacterium can be spread as a dust on plants, but with recombinant DNA technology the Bt gene can be added to a plant's genome. In this way plants such as corn, soybeans, and cotton protect themselves by manufacturing their own Bt toxin. There is little evidence of insect resistance to this Bt toxin despite the fact that large tracts of farmland are devoted to raising Bt-toxin-expressing plants. Some people are leery of the prospect

TABLE 26.4 Some Products of Recombinant DNA Technology Used in Medicine

Product	Modified Cell	Uses of Product
Interferons	*Escherichia coli, Saccharomyces cerevisiae*	To treat cancer, multiple sclerosis, chronic granulomatous disease, hepatitis, and warts
Interleukins	*E. coli*	To enhance immunity
Tumor necrosis factor	*E. coli*	In cancer therapy
Erythropoietin	Mammalian cell culture	To stimulate red blood cell formation; to treat anemia
Tissue plasminogen activator	Mammalian cell culture	To dissolve blood clots
Human insulin	*E. coli*	For diabetes therapy
Taxol	*E. coli*	In ovarian cancer therapy
Factor VIII	Mammalian cell culture	In hemophilia therapy
Macrophage colony stimulating factor	*E. coli, S. cerevisiae*	To stimulate bone marrow to produce more white blood cells; to counteract side effects of cancer treatment
Relaxin	*E. coli*	To ease childbirth
Human growth hormone	*E. coli*	To correct childhood deficiency of growth hormone
Hepatitis B vaccine	Carried on a plasmid of *S. cerevisiae*	To stimulate immunity against hepatitis B virus

of transgenic foods derived from these plants, whereas others contend that the genetic manipulation of crops is crucial if we are to continue to feed the world's population. In any case, one thing is clear: The genetic manipulation of plants is big business.

Pseudomonas syringae (soo-dō-mō'nas sēr'in-jī) is another example of a bacterium with agricultural applications. This bacterium produces a protein that serves in the formation of ice crystals, but the protein is not essential to the survival of the organism, and scientists can remove its gene. When sprayed on crops such as strawberries, the strain lacking the gene, known as *P. syringae* ice⁻ ("ice-minus"), inhibits the formation of ice, protecting the plants from freeze damage.

TABLE 26.5 lists selected industrial products produced by microorganisms and some of their uses.

Biosensors and Bioreporters

Among the relatively new applications of microorganisms to solve environmental problems are biosensors and bioreporters. **Biosensors** are devices that combine bacteria or microbial products (such as enzymes) with electronic measuring devices to detect other bacteria, bacterial products, or chemical compounds in the environment. For example, one such biosensor measures blood glucose levels in diabetic patients by using the enzyme glucose oxidase to convert glucose into an electrical signal. **Bioreporters** are somewhat simpler sensors that are composed of microbes (again usually bacteria) with innate signaling capabilities, such as the ability to glow in the presence of biological or chemical compounds.

Currently, biosensors and bioreporters are used to detect the presence of environmental pollutants (e.g., petroleum) and to monitor efforts to remove harmful substances; they may also be useful in detecting bioterrorist attacks. Because bacteria are very sensitive to their environments, they can detect compounds in very small amounts. Biosensors and bioreporters could serve as early warning systems to give officials more time to respond by quickly detecting the metabolic waste products of weaponized biological agents.

MICRO CHECK

8. What are secondary metabolites?
9. What is *batch production* in food production?
10. What is a main advantage of biofuels compared to fossil fuels?

Water Treatment

LEARNING OUTCOMES
26.7 Contrast water contamination and water pollution.
26.8 Describe two waterborne illnesses.
26.9 Explain how water for drinking and wastewater are treated to make them safe and usable.

Water contamination refers to the presence of anything in the water. Contaminants can be natural or artificial; they can be physical, as in the case of silt; chemical, as in the case of dissolved minerals; or biological, as in the case of amoebas. *Water pollution*, in contrast, generally refers to something added to water by human activity beyond the capacity of natural processes to get rid of it. So, all water pollution is contaminating, but not all contaminants are pollutants. Water treatment facilities process water to remove harmful contaminants and pollution.

Water Pollution

Many pollutants are not readily visible. Although physical and chemical pollutants are important, of greater concern is biological pollution of water with human pathogens such as might enter water in sewage. Biological pollutants can cause significant human diseases, which can be prevented only through water treatment.

Waterborne Illnesses

Consuming contaminated water, either as drinking water or in water added to foods, can result in a variety of bacterial, viral, or protozoan diseases. No known fungal water contaminants cause diseases.

Each year, contaminated drinking water results in roughly 3 billion to 5 billion episodes of diarrheal disease worldwide, including over 3 million deaths among children ages five years or younger. Intoxication can also occur from the presence of microbial toxins in the water.

Water treatment removes most waterborne pathogens, so waterborne illnesses are rare in the United States as compared to countries with inadequate water treatment facilities. Outbreaks that do occur in the United States are usually *point-source infections*, in which a single source of contaminated water leads to illness in individuals that consume the water.

TABLE 26.5 Selected Industrial Products Produced by Microorganisms

Products Made	Use of Product
Enzymes	
Amylase, proteases	Spot removers
Streptokinase	Breakdown of blood clots
Restriction enzymes, ligases, polymerases	Molecular biology, recombinant DNA technology
Food Additives/Supplements	
Amino acids, vitamins	Health supplements
Citric acid	Antioxidant
Sorbic acid, lysozyme	Food preservatives
Other Industrial Products	
Indigo	Dye used in manufacturing clothes
Plastics	Biodegradable substitutes for petroleum-based plastics
Alternative Fuels	
Ethanol	Used in gasohol
Methane	Burned to generate heat and electricity
Hydrogen, hydrocarbons	Potential fuels
Pesticides and Agricultural Products	
Bt toxin	Insecticide

In some marine environments, massive overgrowth of dinoflagellates, such as *Gonyaulax* (gon-ē-aw'laks), chemically pollutes water with their toxins. These overgrowths are sometimes called *red tides* because of the color that might be imparted to the water by the huge number of dinoflagellates. Their toxins can be absorbed and concentrated by shellfish, and human intoxication results from consumption of these mollusks rather than from water consumption, so such intoxication is not a true waterborne illness.

TABLE 26.6 lists some common human waterborne pathogens and toxins. **Emerging Disease Case Study: Attack in the Lake** considers a pathogen that has recently acquired more attention.

Clean water is vital for people and their activities. In the following sections, we consider the treatment of drinking water and the treatment of wastewater (sewage). In both cases, treatment is designed to remove microorganisms, chemicals, and other pollutants to prevent human illness.

Treatment of Drinking Water

Potable (pō′tăbl) water is water that is considered safe to drink, but the term *potable* does not imply that the water is devoid of all microorganisms and chemicals. Rather, it implies that the levels of microorganisms or chemicals in the water are low enough

TABLE 26.6 Selected Waterborne Agents and the Diseases They Cause

Organism	Disease
Bacteria	
Campylobacter jejuni	Acute gastroenteritis
Escherichia coli	Acute gastroenteritis
Salmonella spp.	Salmonellosis, typhoid fever
Shigella spp.	Shigellosis (bacterial dysentery)
Vibrio cholerae	Cholera
Viruses	
Hepatitis A virus	Infectious hepatitis
Norovirus	Acute gastroenteritis
Poliovirus	Poliomyelitis
Eukaryotic Parasites	
Cryptosporidium parvum	Cryptosporidiosis
Entamoeba histolytica	Amebic dysentery
Giardia intestinalis	Giardiasis
Schistosoma spp.	Schistosomiasis
Toxin Producers	
Gonyaulax (saxitoxin)	Paralytic shellfish poisoning

EMERGING DISEASE CASE STUDY

Attack in the Lake

Callie was more than worried—she was terrified. Her 10-year-old son wasn't acting normally. He was incoherent, and his eyes were rolling. It was as if he looked right through her, eyes unfocused and staring.

How life can change. Just last week the family had been enjoying their summer vacation at the lake. Everyone had been so happy—splashing, swimming, and diving, even though the lake had been warmer than normal. "Almost like a bathtub," her husband had said. Still, it had been a grand week, the opposite of this, the second week since they returned. This week seemed to have stretched to an eternity.

First, Caleb had complained of a severe headache, then he developed a high fever, and on the second day nausea and vomiting had started—he threw up 20 times by Callie's estimation. However, she could handle these things; after all, the flu or a "stomach bug" was not that uncommon to a mother of five older children. But now, Caleb was unable to move his head. He'd had a seizure and had been hallucinating. What was happening? Would it never end?

The doctors at the emergency room suspected bacterial meningitis and started an aggressive course of broad-spectrum antibiotics; however, Caleb was not going to get better. Callie's worry was about to turn to anguish. Her son's brain was being consumed by *Naegleria fowleri*.

Naegleria is an amoeba that lives in warm freshwater, where it feeds on bacteria, grows, and divides. It normally doesn't bother people. But when people jump or dive into water where *Naegleria* makes its home, the amoeba can enter the nose, penetrate the nasal mucous membrane, and crawl up the olfactory nerves to the brain, where it causes primary amebic meningoencephalitis (PAM). PAM results in part from the body's efforts to ward off the infection with inflammation; the brain swells, producing headaches and other neurological symptoms. No antimicrobial drugs are known to kill *Naegleria* in the brain. Finally, the patient lapses into coma and usually dies within one to two weeks of infection; only three survivors have been verified in hundreds of cases. Scientists have not been able to determine why some people are infected while millions of others swim and play in warm lakes without consequence.

Epidemiologists have been unable to develop water safety standards to protect against *Naegleria*: they haven't developed a method to accurately measure the number of amoebas in the water, determine the concentration of *Naegleria* required to initiate an infection, or design preventive measures other than avoiding recreation in warm lakes and wearing nose clips while swimming. Fortunately, PAM is rare; only about 133 cases have occurred in the United States since 1962, though more cases are being reported now than in the past.

For Callie and her family, this emerging disease is not rare enough.

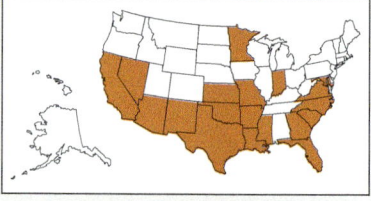

that they are not a health concern. Water that is not potable is **polluted**; that is, it contains organisms or chemicals in excess of acceptable values.

The permissible levels of microbes and chemicals in potable water vary from state to state. Nationally, the U.S. Environmental Protection Agency (EPA) requires that drinking water have a count of zero coliform bacteria per 100 ml of water and that recreational waters have no more than 200 coliforms per 100 ml of water. Recall that coliforms are intestinal bacteria such as *Escherichia coli* (esh-ĕ-rik′ē-ă kō′lī). The presence of coliforms in water indicates fecal contamination and thus an increased likelihood that disease-causing microbes are present.

Typical drinking water treatment can be summarized in four steps (FIGURE 26.6):

1 **Coagulation and flocculation.** The operator pumps water into holding tanks, where particulate materials (sand, silt, and organic material) settle. The operator then adds a chemical with a positive charge such as alum (aluminum potassium sulfate) that causes negatively charged, suspended particles and microorganisms to coagulate in large aggregates, called flocs.

2 **Sedimentation.** During sedimentation, the flocs settle to the bottom of the tank because they are denser than water (see Figure 11.24). The treatment facility then pumps the water above the coagulated flocs into a different tank for filtration.

3 **Filtration.** In this stage, the number of microbes is reduced by about 90% in one of several ways. One method uses sand and other materials to which microbes adhere and form biofilms that trap and remove other microbes. *Slow sand filters* are composed of a 1-meter layer of fine sand or diatomaceous earth and are used in smaller cities or towns to process 3 million gallons per acre of filter per day. Large cities use *rapid sand filters* that contain larger particles and gravel and can process 200 million gallons per acre per

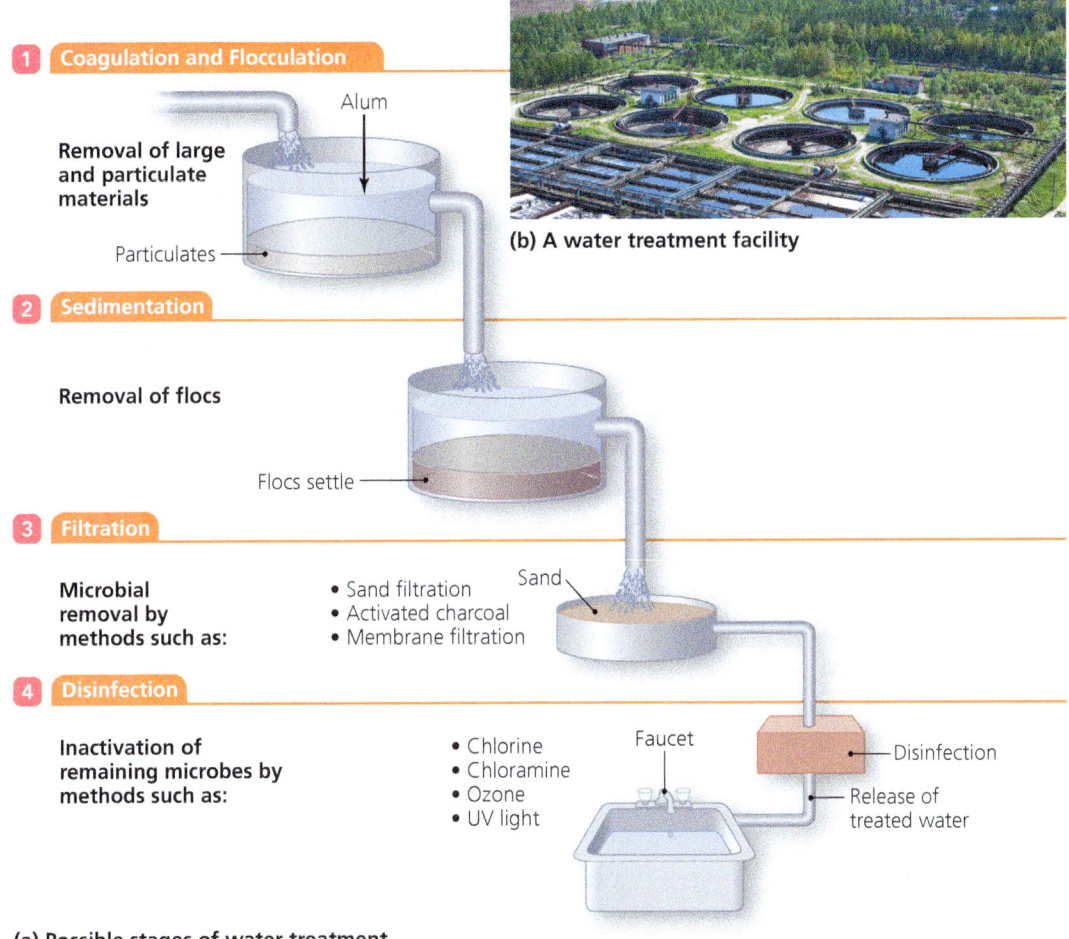

(a) Possible stages of water treatment

▲ FIGURE 26.6 **The treatment of drinking water.** (a) The stages of water treatment: coagulation and flocculation, sedimentation, filtration, and disinfection; (b) A water treatment facility. *Why is it that chemical treatment cannot destroy most viruses?*

Figure 26.6 Most chemicals are designed either to inhibit some aspect of active metabolism or to damage cellular structures; therefore, they do not damage acellular and nonmetabolizing viruses.

day. Both types of filters are cleaned by back-flushing with water. Two other filtration methods are *membrane filtration*, which uses a filter with a pore size of 0.2 μm, and filtration with *activated charcoal*, which provides the added benefit of removing some organic chemicals from the water.

4. **Disinfection.** In this stage, disinfectants, such as ozone, UV light, chlorine, or chloramine, are used to kill most microorganisms before the water is released for public consumption. Many European communities use ultraviolet light. Communities in the United States commonly use chlorine, partly because it is least expensive. Chlorine gas, an oxidizing agent, is thought to kill bacteria, algae, fungi, and protozoa by denaturing their proteins within approximately 30 minutes of treatment. Chlorine levels must be constantly adjusted to reflect estimates of *microbial load*, the number of microbes in a unit of water—a higher load requires more chlorination. Chlorination does not kill all microbes: Most viruses are not inactivated by chlorine, and bacterial endospores and protozoan cysts are generally unaffected by this chemical treatment. Only filtration can completely remove all viruses, endospores, and cysts.

Water Quality Testing

Water quality testing is a technique that uses the presence of certain **indicator organisms** to indicate the possible presence of pathogens in water. Because the majority of waterborne illnesses are caused by fecal contamination, the presence of *E. coli* and other coliforms in water indicates a probability that pathogens are present as well. *E. coli* is consistently prevalent in human waste as long as (if not longer than) most pathogens and is easily detected.

Several testing methods can be used to assess water quality. One testing method is the membrane filtration method (FIGURE 26.7a), which is simple to perform: A 100-ml water sample is poured through a fine membrane, which is then placed on selective media agar plates and incubated. Coliform colonies exhibit unique characteristics. The colonies are then counted and reported as number of colonies per 100 ml. This method gives a total coliform count.

In another test, water samples are added to small bottles containing both *ONPG* (o-nitrophenyl-β-D-galactopyranoside) and *MUG* (4-methylumbelliferyl-β-D-glucuronide) as sole nutrients. Most coliforms produce β-galactosidase, an enzyme that reacts with ONPG to produce a yellow color, but the fecal coliform *E. coli* produces an additional enzyme, β-glucuronidase, which reacts with MUG to form a compound that fluoresces blue when exposed to long-wave UV light (FIGURE 26.7b). This test allows for the rapid detection of coliforms but gives an estimate for the number of bacteria rather than an actual count. Scientists now often use PCR-based tests instead because they are more specific and more rapid.

The presence of viruses and particular bacterial strains cannot be determined with these tests; their presence must be confirmed by genetic "fingerprinting" techniques, in which water samples are collected and enriched to cultivate the organisms present. The DNA content of the enriched sample is then genetically screened for the identification of potential pathogens.

Governments are currently reconsidering the use of coliform tests to indicate fecal contamination because some coliforms grow naturally on plants even when there is little fecal contamination, giving a false-positive result for fecal pollution. Regulators are considering replacing coliform tests with genetic assays that would specifically indicate the presence of *E. coli*.

Treatment of Wastewater

Sewage, or **wastewater**, is typically defined as any water that leaves homes or businesses after being used for washing or flushed from toilets. (Some municipalities also include industrial water and rainwater as wastewater.) Wastewater contains a variety of contaminants, including suspended solids, biodegradable and nonbiodegradable organic and inorganic compounds, toxic

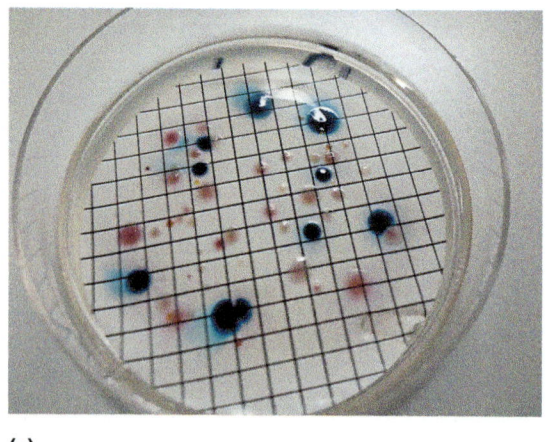

(a)

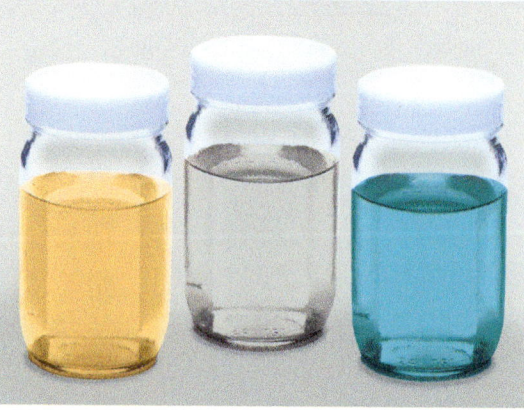

(b)

▲ **FIGURE 26.7 Two water quality tests. (a)** Membrane filtration. The grid on the membrane makes it easier to count the colonies of fecal coliforms, which are green on this selective medium. **(b)** The ONPG and MUG tests. The yellow color of the ONPG bottle indicates the presence of coliforms, whereas the blue fluorescence in the MUG bottle indicates the presence of the coliform *E. coli*; the clear bottle is the negative control.

metals, and pathogens. The objective of wastewater treatment is to remove or reduce these contaminants to an acceptable level for releasing the water into a natural environment, using it for irrigation, or sending it to a water treatment facility to be repurposed as drinking water.

At one time, "raw" (unprocessed) sewage was simply dumped into the nearest river or ocean; the idea was that wastes would be diluted to a point at which they would be harmless. With today's increasing population and greater use of water in homes for flushing, washing, and cleaning, we need to give nature a helping hand.

A key concept in the processing of wastewater is reducing **biochemical oxygen demand (BOD)**, which is a measure of the amount of oxygen required by aerobic bacteria to fully metabolize organic wastes in water. This amount is proportional to the amount of waste in the water; the higher the concentration of degradable chemicals, the more oxygen is required to catabolize them, and the higher the BOD. Effective wastewater treatment reduces the BOD to levels too low to support microbial growth, thus reducing the likelihood that pathogens will survive.

The following sections consider wastewater treatments of various types: the traditional sewage treatment used in municipal systems, treatments used in rural areas, a treatment used for agricultural wastes, and the use of artificial wetlands.

Municipal Wastewater Treatment Today, people living in larger towns and cities are usually connected to municipal sewer systems—pipes that collect wastewater and deliver it to sewage treatment plants for processing. Traditional sewage treatment consists of four phases (FIGURE 26.8):

1. *Primary treatment.* Wastewater is pumped into settling tanks, where lightweight solids, grease, and floating particles are skimmed off and heavier materials settle onto the bottom as **sludge** (raw primary biosolids). After the sludge is removed, the partially clarified water is further treated. Primary treatment removes 25% to 35% of the BOD in the water.

2. *Secondary treatment.* The water is aerated to facilitate the growth of aerobic microbes that oxidize dissolved organic chemicals to CO_2 and H_2O. In an *activated*

▼ FIGURE 26.8 **Traditional sewage treatment.** (a) A municipal wastewater treatment facility. (b) The traditional sewage treatment process. Microbial digestion during secondary treatment removes most of the biochemical oxygen demand (BOD) before the water is chemically treated and released; dried sludge is recycled as landfill.

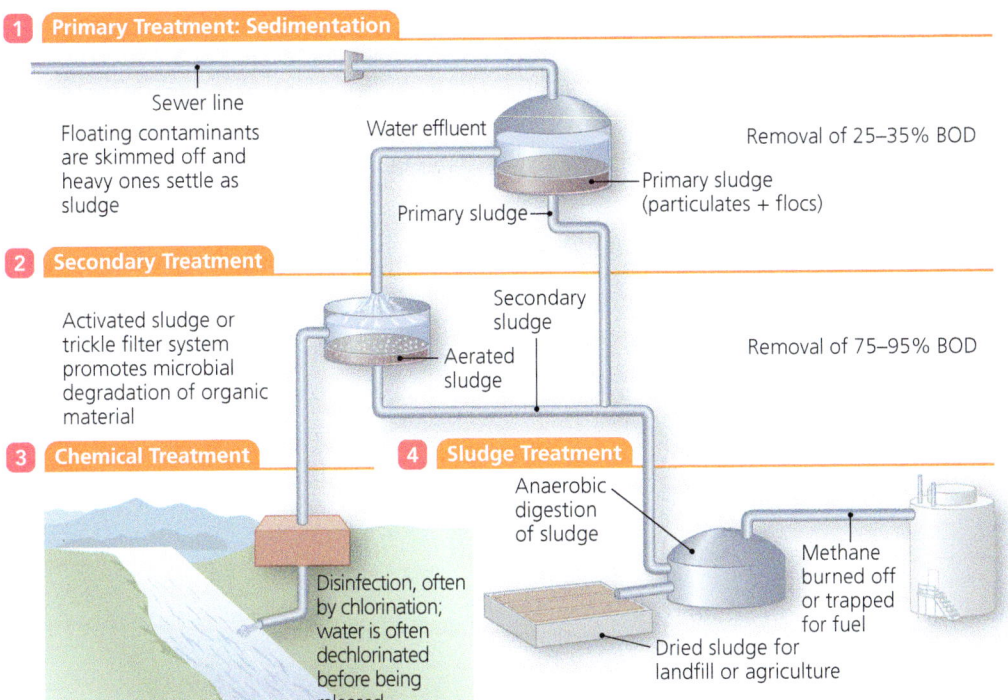

(a)

(b)

sludge system, aerated water is seeded with primary sludge, which contains millions of metabolizing bacteria; flocculation also occurs during this step. Any remaining solid material settles and is added to the sludge from primary treatment. The combined sludge is pumped into anaerobic holding tanks. Some smaller communities accomplish secondary treatment using a *trickle filter system*, which is similar to the slow sand filters used in treating drinking water but less effective in removing BOD than activated sludge systems. The biological activity in this phase reduces the BOD to 5% to 25% of the original. Most pathogenic microorganisms are also removed.

3. *Chemical treatment.* Water from secondary treatment is disinfected, usually by chlorination, which is often followed by dechlorination, after which the water is either released into rivers or the ocean or, in some states, used to irrigate crops and highway vegetation. Some communities remove nitrates, phosphates, and any remaining BOD or microorganisms from the water by passing it over fine sand filters and/or activated charcoal filters. Nitrate is converted to ammonia and discharged into the air (removes roughly 50% of the nitrogen content), whereas phosphorus is precipitated using lime or alum (removes 70% of the phosphorus content). Such tertiary treatment is generally used in environmentally sensitive areas or in areas where the only outlet for the water is a closed-lake system.

4. *Sludge treatment.* Sludge is digested anaerobically in three steps: First, anaerobic microbes ferment organic materials to produce CO_2 and organic acids. Second, other microbes metabolize these organic acids to H_2, more CO_2, and simpler organic acids, such as acetic acid. Finally, the simpler organic acids, H_2, and CO_2 are converted to methane gas, which may be used as an energy source. Any leftover sludge is then dried for use as landfill or fertilizer.

Nonmunicipal Wastewater Treatment Houses in rural areas, which typically are not connected to sewer lines, often use **septic tanks**, essentially the home equivalent of primary treatment (FIGURE 26.9). Sewage from the house enters a sealed concrete holding tank. Solids settle to the bottom, and the liquid flows from the top of the tank into an underground *leach field* that acts as a filter. Sludge in the tank and organic chemicals in the water are digested by microorganisms. However, because the tank is sealed, it must occasionally be pumped out to remove sludge buildup. **Cesspools** are similar to septic tanks except that they are not sealed. As wastes enter a system of porous concrete rings buried underground, the water is released into the surrounding soil; solid wastes accumulate at the bottom and are digested by anaerobic microbes.

Treatment of Agricultural Wastes Farmers and ranchers often use **oxidation lagoons** to treat animal waste from livestock raised in feedlots—a penned area where animals are fattened for market. Oxidation lagoons accomplish the equivalent of primary and secondary sewage treatment. Wastes are pumped into deep lagoons and left to sit for up to three months, sludge settles to the bottom of the lagoon, and anaerobic microorganisms break down the sediment. The remaining liquid is pumped into shallow, secondary lagoons, where wave action aerates the water. Aerobic microorganisms, particularly algae, break down organic chemicals suspended in the water. Eventually, the microbes die, and the clarified water is released into rivers or streams. One problem with oxidation lagoons is that they are open, which can be dangerous if floodwaters inundate the lagoons and spread largely untreated animal wastes over a wide area.

Artificial Wetlands Since the 1970s, small planned communities and some factories have constructed **artificial wetlands** to treat wastewater. Wetlands use natural processes to break down wastes and to remove microorganisms and chemicals from water before its final release. Individual septic tanks are not needed; instead, wastewater flows into successive ponds where microbial digestion occurs (FIGURE 26.10). The first pond in the series is aerated to allow aerobic digestion of wastes; anaerobic digestion occurs in the sludge at the bottom. The water then flows through marshland, where soil microbes further digest organic chemicals. A second pond, which is still and contains algae, removes additional organic material, and the water then passes through open meadowland, where grasses and plants trap pollutants. By the time the water reaches a final pond, most of the BOD

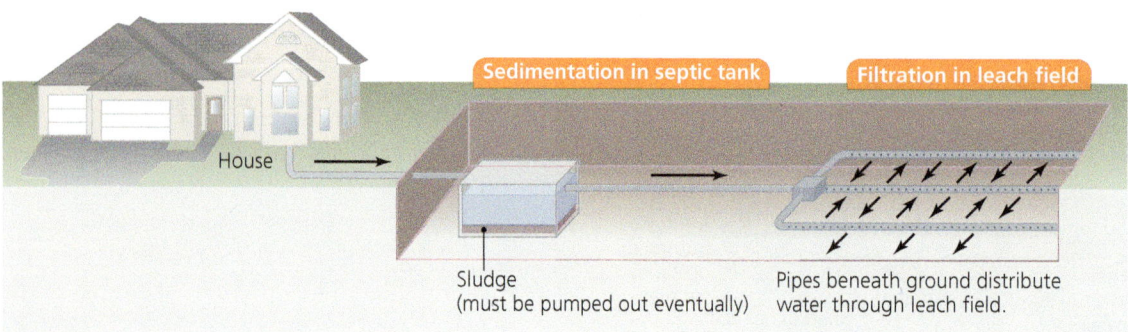

▲ FIGURE 26.9 **A home septic system.** After wastewater from the house enters the septic tank, solids settle out as sludge, and the effluent liquid is filtered by the soil in the leach fields.

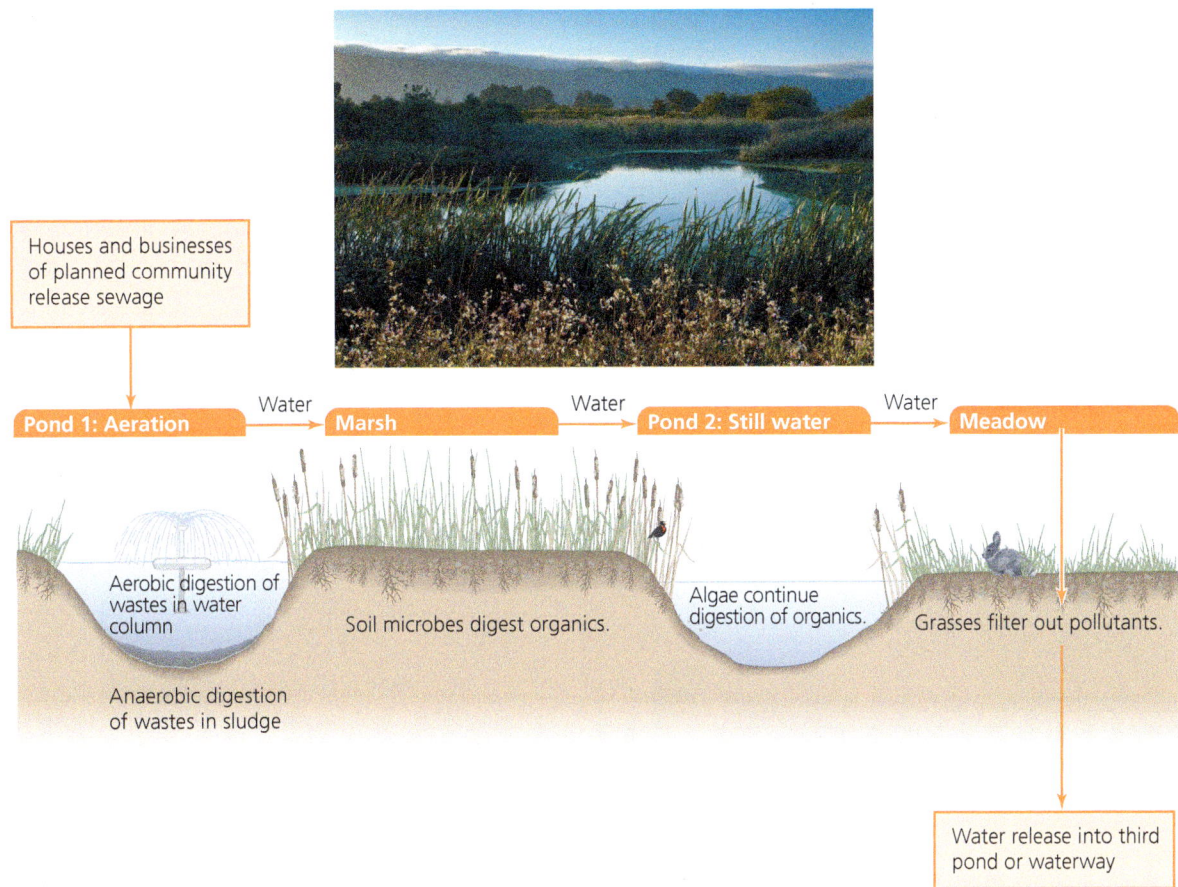

▲ **FIGURE 26.10 Wastewater treatment in an artificial wetland.** The majority of the BOD is removed by microbial action in the first pond; natural filtration by plants and the soil removes pollutants and the rest of the BOD. *Why aren't artificial wetlands feasible for major metropolitan areas?*

Figure 26.10 *Major metropolitan areas produce too much sewage and have too little available space for artificial wetlands.*

and microorganisms have been removed, and the water can be released for recreational purposes or irrigation. One drawback of an artificial wetland is that it requires considerable space—an artificial wetland to serve a small community can cover 50 acres of land or more.

MICRO CHECK

11. What is potable water?
12. What is the purpose of flocculation during the treatment of drinking water?
13. Why are coliform bacteria used to indicate water's potability?
14. What should be done to wastewater before it is released into the environment?

TELL ME WHY

In a polluted lake, the microbes reproduced prolifically, died, and then sank to the bottom, "feeding" anaerobes in the sediment. Even though the surface water looks clear, why is it still unsafe to drink the lake water?

MICRO IN THE CLINIC FOLLOW-UP

A Case of Acne or Something More?

After examining Kate and hearing about her trip to Colorado, Dr. Moore tells Kate that she most likely got a bacterial infection from the hot tub at the resort. The doctor takes a culture from Kate's pustules to send for lab analysis. The pustules are the result of bacterial folliculitis caused by Gram-negative *Pseudomonas aeruginosa*; while Kate is currently feeling some discomfort, the condition should clear up on its own. Dr. Moore recommends the use of an over-the-counter anti-itch cream containing hydrocortisone to help relieve some of the discomfort. He then prescribes an antibiotic to treat the *P. aeruginosa* infection; 250 mg of ciprofloxacin orally, twice a day, for 7 days. He also recommends that Kate contact the resort and inform them of her condition; other guests may also have been infected.

When Kate contacts the resort, she finds out that four other guests have already reported similar symptoms after using the hot tub. The resort manager informs Kate that the local health department tested the hot tub and found *Pseudomonas aeruginosa in the water*. Kate follows up on her findings with her physician. She then reports that her symptoms have improved with the help of the medications after one week.

1. What are two virulence factors that *P. aeruginosa* uses to cause infection? Explain how each of these contributes to pathogenicity.

2. Describe the steps in the development of Kate's folliculitis. Start with Kate's soaking in the hot tub through the formation of her lesions.

Check your answers to Micro in the Clinic Follow-Up questions in the Mastering Microbiology Study Area.

CHAPTER SUMMARY

Go to Mastering Microbiology for Interactive Microbiology, Dr. Bauman's Video Tutors, Micro Matters, MicroFlix, Micro-Boosters, practice quizzes, and more.

Food Microbiology (pp. 779–786)

1. The commercial use of microorganisms is referred to as **applied microbiology** and includes two distinct fields: food microbiology and industrial microbiology.

2. **Food microbiology** involves the use of microorganisms in food production and the prevention of foodborne illnesses. In this context, **fermentations** involve desirable changes to food; **spoilage** involves undesirable changes to food.

3. Food fermentations involve the use of **starter cultures**—known organisms that carry out specific and reproducible fermentation reactions. For most of the great variety of fermented vegetables, meats, and dairy products, starter cultures of lactic acid bacteria are used. The acid produced results in "sour" flavors. Some of the live organisms remaining in the fermented foods are **probiotics**—they can have health benefits.

4. Alcoholic fermentations, usually performed by yeasts, convert sugars to ethanol and carbon dioxide. Alcoholic fermentation is used in the production of wine, distilled spirits, beer, vinegar, and bread.

5. Intrinsic factors of food spoilage are properties of the food itself, such as moisture content and physical structure, that determine a food's susceptibility to spoilage. Extrinsic factors of food spoilage include ways in which the food is handled.

6. Industrial processes preserve food via canning, pasteurization, drying, freeze-drying (**lyophilization**), irradiation, and aseptic packaging techniques. Natural and artificial preservatives are added to some foods to inhibit microbial growth. In stores and at home, foods should be properly stored in appropriate containers, cold foods should be kept cold, foods should be cooked thoroughly, and leftovers should be refrigerated to reduce spoilage.

7. Food poisoning is a general term that describes instances of **food infections** (illnesses due to the consumption of living microbes) or **food intoxications** (illnesses due to the consumption of microbial toxins). Food poisoning frequently follows poor food handling.

Industrial Microbiology (pp. 786–795)

1. **Industrial microbiology** is concerned with the use of microorganisms for the production of commercially valuable materials. Such industrial fermentations synthesize desired products and can use genetically modified microbes.

2. Batch production is the growth of organisms followed by harvesting of the entire culture and its products. Continuous flow production involves the constant addition of nutrients to a culture and the removal of the products formed. Products may be either primary metabolites (produced during active growth) or secondary metabolites (produced during the stationary phase).

3. Microorganisms produce a variety of useful products, including enzymes, dyes, alternative fuels, plastics, pharmaceuticals, pesticides, biosensors, and bioreporters. Alternative fuels (biofuels) can be produced from products of photosynthesis or by fermentation of biomass into fuel. **Biosensors** combine microbes and electronics to detect microbial activity in the environment. **Bioreporters** use microbes alone as sensors.

4. Contaminated water has something in it. Contamination can be physical, chemical, or biological. Water pollution refers to human contamination of water.

5. **Potable** drinking water is derived via water treatment, which involves the removal of microbes and of organic and inorganic contaminants from water. **Polluted** water contains organisms or chemicals at unsafe levels. Some diseases, such as diarrhea or shellfish intoxication, can result from consumption of polluted water or of food harvested from polluted water.

6. Drinking water treatment involves four steps: **coagulation** and **flocculation**, **sedimentation**, **filtration**, and **disinfection**. In coagulation and flocculation, alum combines with suspended materials to make them precipitate. Sedimentation removes large materials. Filtration, using either slow or rapid sand filters, activated charcoal filters, or membrane filters, removes microorganisms and chemicals. Disinfection, usually chlorination, kills most microbes that remain after filtration. Water is potable following treatment if it has zero detected coliforms (**indicator organisms**) in 100 ml of water, as determined by one of several testing methods (membrane filtration, ONPG/MUG test).

7. **Wastewater** (sewage) refers to water used for washing or flushed from toilets. Wastewater treatment involves the removal of solids, organic chemicals, and microorganisms. **Biochemical oxygen demand (BOD)** is a measure of the amount of oxygen required to fully metabolize organic wastes.

8. Municipal wastewater treatment involves four phases. Primary treatment entails sedimentation of large materials (primary **sludge**, or biosolids). Secondary treatment involves sedimentation of secondary sludge, as well as the removal of microorganisms and organic material using activated sludge systems or trickle filters. In the third phase, effluent water is chemically treated (chlorinated and then dechlorinated) and released. In the fourth phase, primary and secondary sludge is digested and dried to produce landfill. Methane gas can also be recovered during the processes.

9. **Septic tanks** and **cesspools** are home equivalents of municipal wastewater treatment. After wastewater leaves the home, it is deposited in underground tanks. Sludge settles, and the water is released into the soil, where natural processes remove organic chemicals and microorganisms.

10. **Oxidation lagoons** are used by farmers and ranchers to process animal wastes. Waste is pumped into successive lagoons, where microorganisms digest wastes before the water is released into natural water systems.

11. In **artificial wetlands**—found in some planned communities and industrial sites—ponds, marshes, and meadowland remove organic compounds, chemicals, and microorganisms from sewage as the water moves through them

QUESTIONS FOR REVIEW

Answers to the Questions for Review (except Short Answer questions) begin on p. A-1.

Multiple Choice

1. Food fermentations do all of the following *except* _____.
   a. give foods a characteristic taste
   b. lower the risk of food spoilage
   c. sterilize foods
   d. increase the shelf life of the food

2. Commercially produced beers and wines are usually fermented with the aid of _____.
   a. naturally occurring bacteria
   b. naturally occurring yeast
   c. specific cultured bacteria
   d. specific cultured yeast

3. Which of the following lists foods in order, from perishable to nonperishable?
   a. dried pasta, cheese, fruit, uncooked ground beef
   b. dried pasta, fruit, uncooked ground beef, cheese
   c. uncooked ground beef, fruit, cheese, dried pasta
   d. uncooked ground beef, fruit, dried pasta, cheese

4. Which of the following would be the best growth medium to use for industrial fermentations?
   a. corn
   b. synthetic medium made by hand
   c. whey from cheese production
   d. brewing mash

5. Biodegradable plastics can be made from which of the following microbial metabolites?
   a. sludge
   b. PHB
   c. BOD
   d. alum

6. Strains of the bacterium *Pseudomonas syringae* have been identified as being capable of _____.
   a. producing plastics
   b. producing alternative fuels
   c. fermenting foods
   d. preventing ice formation

7. Which of the following is added during water or sewage treatment to promote flocculation?
   a. sludge
   b. PHA
   c. BOD
   d. alum

8. During chemical treatment of drinking water and wastewater, which of the following microbes is *least* likely to be inactivated or killed?
   a. algae
   b. viruses
   c. fungal spores
   d. bacteria

9. In which step is most of the organic content of sewage removed?
   a. primary treatment
   b. secondary treatment
   c. tertiary treatment
   d. sludge treatment

10. Industrial fermentation _____.
    a. always involves alcohol production
    b. involves the large-scale production of any beneficial compound
    c. refers to the oxidation of sugars using organic electron acceptors
    d. is any desirable change to food by microbial metabolism

11. Lyophilization in food preservation is by _____.
    a. cell lysis
    b. gamma radiation
    c. rapid heating
    d. freeze-drying

Modified True/False

Indicate whether each of the following statements is true or false. Rewrite the underlined phrase to make a false statement true.

1. _____ The fermentation of dairy products relies on <u>mixed acid</u> fermentation.
2. _____ Sauerkraut production involves the <u>alcoholic</u> fermentation of cabbage.
3. _____ Pasteurization kills <u>mesophilic</u> microorganisms except endospore formers.
4. _____ Methane is a gas produced by microbial metabolism that can be used directly as a <u>fuel source</u>.
5. _____ The treatment of drinking water and the treatment of sewage involve <u>similar</u> processes.

Fill in the Blanks

1. Intrinsic factors affecting food spoilage are properties of _____ rather than _____.
2. Leaving foods out at room temperature _____ the likelihood of food spoilage.
3. The two types of industrial fermentation equipment are designed for _____ production or _____ production.
4. Potable water is allowed to have _____ coliforms per 100 ml of water tested.
5. A primary goal in wastewater treatment is to reduce the _____.
6. The presence of _____ in a water sample usually means pathogens have contaminated the source.
7. _____ and _____ are the two types of food poisoning.
8. Compounds that can be used for energy and are made by microbes are called _____.
9. A _____ is a device composed of microbes and electronics used to detect other microbes or their products.
10. _____ is the amount of oxygen required by aerobic organisms to fully metabolize organic waste in water.

Matching

Match each term with its correct definition.

1. _____ Organisms whose presence in water indicates contamination from feces
2. _____ Refers to water that is fit to drink
3. _____ Used in the processing of animal wastes; mimics primary and secondary wastewater treatment
4. _____ Water that is not bound by solutes
5. _____ Undesirable fermentation reactions in food, leading to poor taste, smell, or appearance
6. _____ Brief heating of foods during processing
7. _____ Descriptor of the level of organic material present in wastewater
8. _____ Fermentative products produced by microorganisms during stationary phase

A. Spoilage
B. Water activity
C. Coliforms
D. Pasteurization
E. Secondary metabolites
F. Potable
G. BOD
H. Oxidation lagoon

VISUALIZE IT!

1. Label the steps in the cheese-making process.

2. Label the steps in the wine-making process.

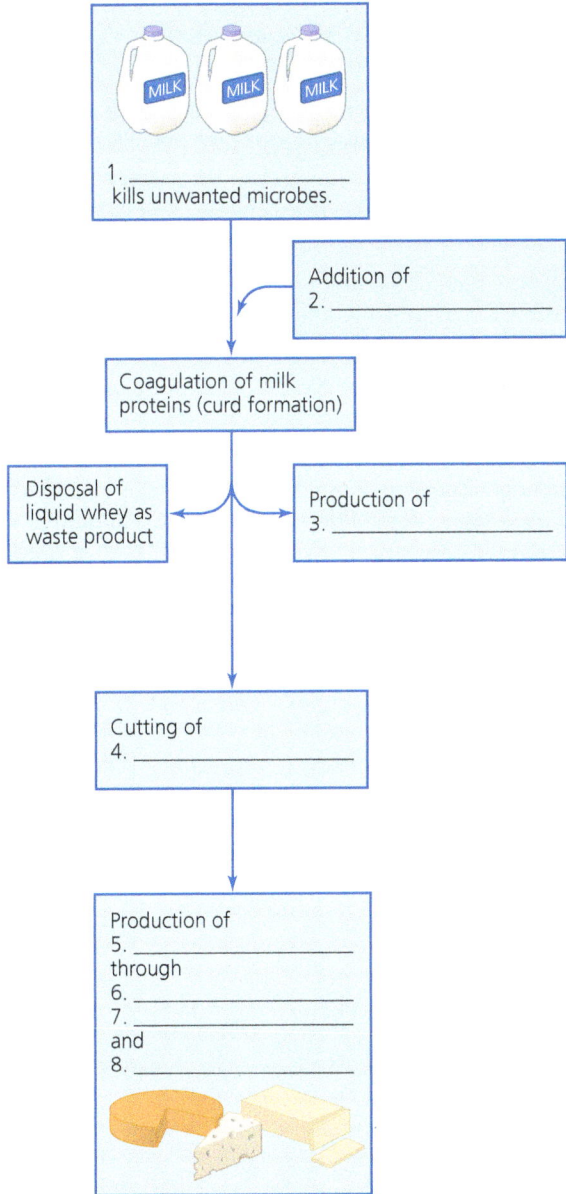

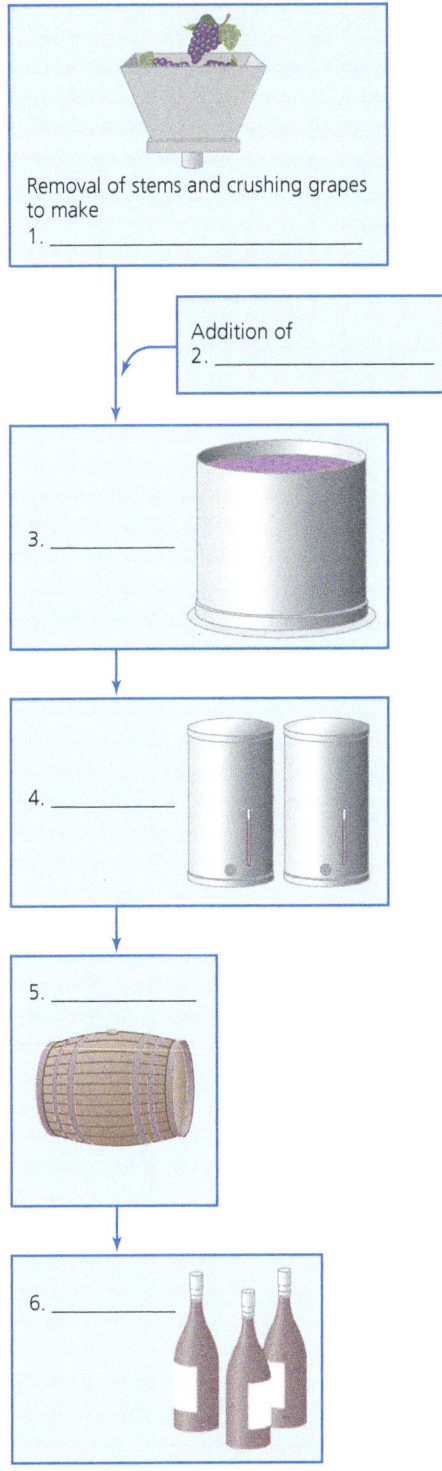

CRITICAL THINKING

1. Why does the application of recombinant DNA technology to food production enhance not only food quality but also food output? Given that it has the potential to feed more people, why are some people opposed to genetic modification of foods?

2. Given what you know about microbial nutrition and metabolism, explain why it is technically more difficult to achieve high yields of a secondary metabolite than of a primary metabolite.

3. Compare the types of alternative fuels that could be produced by microbes. Based on starting materials, which would provide the most renewable energy?

4. One way that farmers are attempting to prevent the development of widespread insect resistance to Bt toxin is by planting non-Bt-producing crops in fields adjacent to Bt-containing crops. Insects can infest both fields, but the insects eating the non-Bt plants should survive at a higher rate than those in the Bt-producing field. Why would this prevent the dissemination of resistance to Bt toxin among insects?

5. Even though water and wastewater undergo essentially the same forms of treatment, treated wastewater usually is not put into the water system from which drinking water is derived. Why not?

6. Take a critical look at the garbage in all of your wastebaskets. How much of the material present could possibly be degraded by microbes? How much could be recycled either at a recycling center or in compost? What would be left if the degradable or recyclable materials were removed?

7. Explain why sake—sometimes called rice wine—would be more accurately described as "rice beer."

8. Inexpensive bulk wines and wines that have been left exposed to air for too long often acquire a "vinegary" smell or taste. Explain why this is so.

9. Describe the intrinsic and extrinsic factors that would contribute to the spoilage of butter, an apple, and a steak.

10. Suggest why neither the original food involved nor the organism responsible can ever be identified in more than 80% of food poisoning cases.

11. Which process would you expect to yield more product from the same-sized vats over the same amount of time: batch production or continuous flow production? Why?

12. Explain how biosensors might be used for water quality testing. Would biosensors identify the microorganisms present in water more accurately than an ONPG or MUG test? Is such specificity truly necessary?

CONCEPT MAPPING

Using the terms provided, fill in this concept map that describes microbial roles in food production. You can also complete this and other concept maps online by going to the Mastering Microbiology Study Area.

Acetic acid in vinegar
Alcohol in wine
Bread to rise
Fermentation

Gluconobacter
Lactobacillus bulgaricus
Maltose from grains

Saccharomyces cerevisiae (yeast)
Soybeans and wheat
Sugars in milk

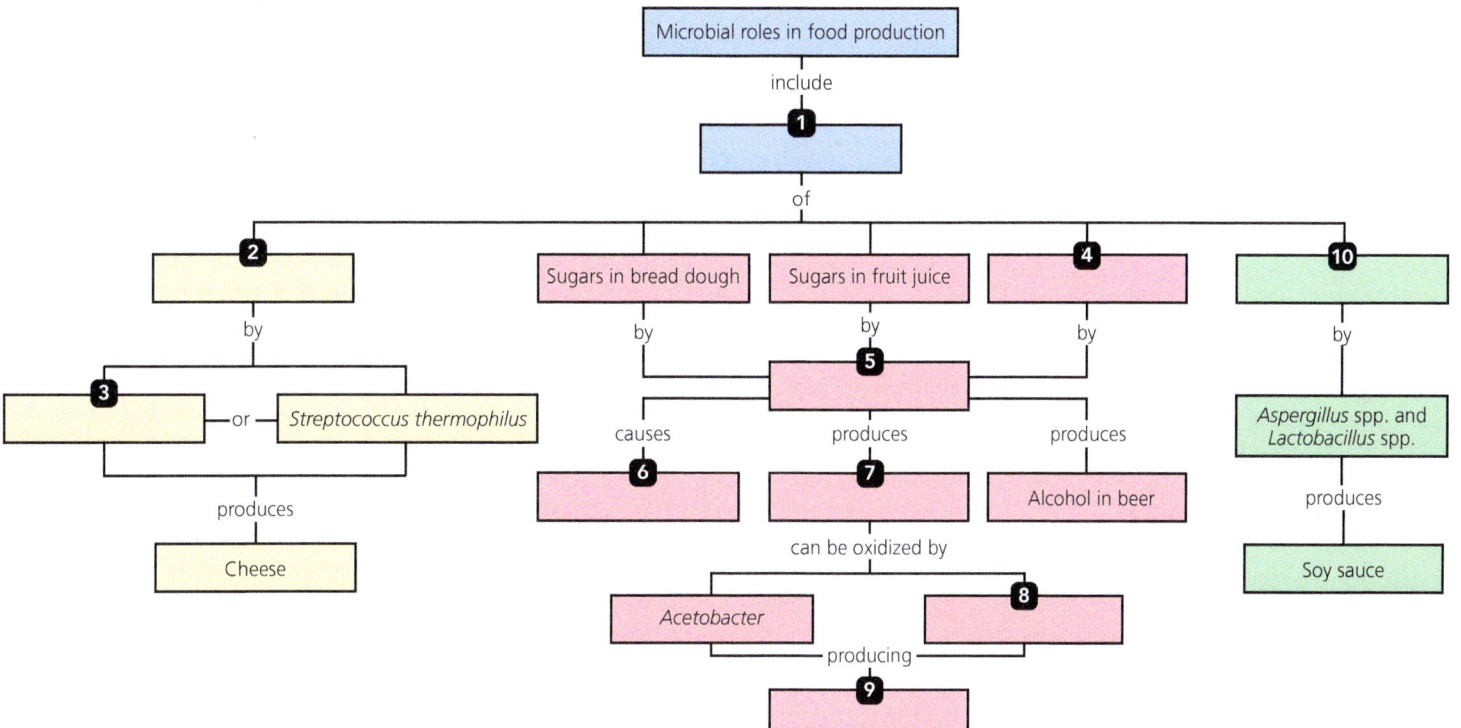

27 Microbial Ecology and Microbiomes

Before You Begin

1. Which organisms acquire carbon for organic molecules from other organisms and use chemical compounds as a source of energy?
2. By which mechanism do microbes communicate with each other in biofilm formation?
3. Which unique structure do archaea use to attach to surfaces?

A microbiome consists of all the microorganisms living in a given area, such as these bacteria living as a mat in Mammoth Hot Springs in Yellowstone National Park U.S.A.

MICRO IN THE CLINIC

A New Treatment?

PENELOPE WAS DIAGNOSED WITH CYSTIC FIBROSIS (CF) when she was five years old. As with many CF patients, the diagnosis followed an earlier diagnosis of "failure to thrive" (insufficient weight gain for age). She has had many health issues throughout the 20 years since her diagnosis, and five years ago, Penelope was diagnosed with cystic fibrosis–related diabetes (CFRD).

Penelope was recently admitted to the hospital with a lower respiratory tract infection of *Pseudomonas aeruginosa* (her third such infection in the past eight months). She is treated with a combination of three antibiotics, and after about 10 days, her symptoms begin to improve. Her physician, Dr. Kasper, also orders chest physiotherapy (vibrations used to clear the airway) and treatment with mucolytic drugs that help her cough up more mucus.

On the day that she is being discharged from the hospital, Penelope expresses to Dr. Kasper her frustration over being sick so often. Dr. Kasper expresses his sympathy to Penelope and further adds that there is another serious factor to consider. The chronic nature of Penelope's infections is damaging her lung tissue; if future infections cannot be controlled with antibiotics, there is a possibility that Penelope will require a lung transplant. There is another possibility—Dr. Kasper tells Penelope about a clinical trial looking at a new way to treat lung infections in CF patients. If Penelope is interested, Dr. Kasper can set up an appointment for them to meet with the study coordinators.

1. Should Penelope consider enrolling in the clinical study investigating a new treatment approach?

Turn to the end of the chapter (p. 816) to find out.

SOLVE THE PROBLEM

Fecal Microbiome Transfer: Medicine or Magic?

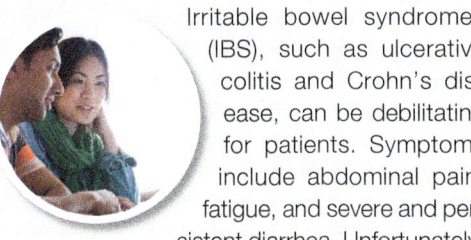

Irritable bowel syndromes (IBS), such as ulcerative colitis and Crohn's disease, can be debilitating for patients. Symptoms include abdominal pain, fatigue, and severe and persistent diarrhea. Unfortunately, there is no medical cure for these conditions, and existing treatments to alleviate symptoms are expensive and are able to help only certain patients. Researchers are considering an alternative treatment based on the idea of "rebooting" the gut microbiome.

A microbiome is the community of all microorganisms in a location, in this case, the human colon. The treatment, called *fecal microbiome transfer (FMT)*, involves transferring a portion of the microbiome from a healthy person into the IBS patient—basically a transfer of feces. The procedure involves emptying the colon (as is done for a colonoscopy) and then introducing fecal microbiota via a tube about the size of a colonoscopy tube. FMT has helped some patients

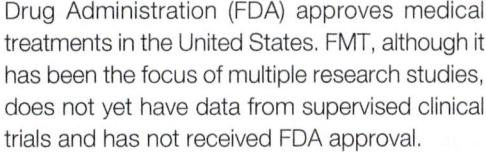

with a related disease caused by *Clostridioides difficile* (formerly known as *Clostridium difficile*), but evidence for efficacy of FMT in treating irritable bowel syndrome is mixed; that is, some studies suggest that the procedure might be beneficial, whereas other studies show no positive effect. One FMT recipient experienced significant weight gain, and there is concern about transfer of pathogens, especially enteroviruses.

A foundational principle in medicine and science is to base practices and procedures on evidence. Such evidence-based treatments and cures are determined by laboratory research, studies on animals, and numerous human clinical trials. Based on the results of research and human clinical trials, the United States Food and Drug Administration (FDA) approves medical treatments in the United States. FMT, although it has been the focus of multiple research studies, does not yet have data from supervised clinical trials and has not received FDA approval.

Imagine you and your team have to make decisions concerning FMT. How would you initially answer these questions?

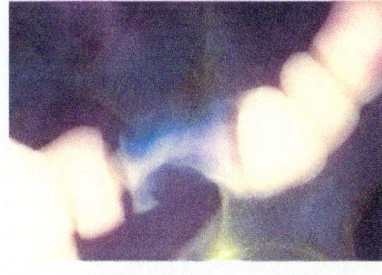

X ray showing a narrowed section of a colon, resulting from inflammation

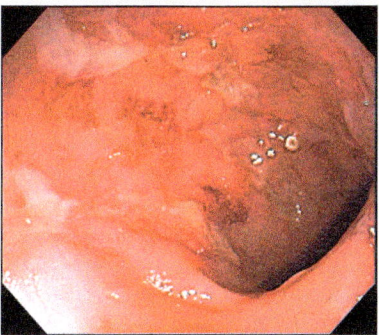

Endoscopic view of severe inflammation in a colon

1. Should FMT be used routinely to relieve the suffering of IBS patients, even though the procedure has not completed clinical trials and lacks FDA approval?
2. Should all procedures and drugs be held to the same standard of evidence?
3. What other information do you need to make an evidence-based decision?

Environmental Microbiology

Environmental microbiology is the study of microorganisms as they occur in their natural **habitats**—the physical localities in which organisms are found. Because of their vast metabolic capacities, microbes flourish in every habitat on Earth, from Antarctic ice to boiling hot springs to bedrock.

In the following sections, we will explore the roles microbes play in cycling chemical elements in soil and aquatic habitats. We will also consider the community of microbes in and on humans in hopes of developing new ways to diagnose, treat, or prevent disease. First, however, we turn our attention to the relationships among microbes in communities and between microbes and their habitats.

Microbial Ecology

LEARNING OUTCOMES

27.1 Define the terms used to describe microbial relationships within the environment.
27.2 Explain the influences of competition, antagonism, and cooperation on microbial survival.

Microorganisms use a variety of energy sources and grow under a variety of conditions. They adapt to changing conditions, compete with other organisms for scarce resources, and change their habitat in many ways. In some cases, their effects on the environment are undesirable from a human perspective, but in most cases they are beneficial and essential. The study of the interrelationships among microorganisms and the environment is called **microbial ecology**.

Ecologists use the term **biodiversity** to refer to the number of species living within a given ecosystem, whereas the term **biomass** refers to the mass of all organisms in an ecosystem. By either measure, microorganisms are the most abundant of all living things: The number of species of microbes exceeds the number of larger organisms, and the biomass of microorganisms living throughout the biosphere—including places that are uninhabitable by plants, animals, or humans—is enormous.

The first aspects of microbial ecology we will examine are the levels of microbial associations in the environment.

Levels of Microbial Associations in the Environment

A variety of terms are used to describe levels of microbial associations in the environment:

- All the members of a single species in a location is termed a *population*.
- Populations of microorganisms performing metabolically related processes make up groups called *guilds*. For example, anaerobic fermenters in a habitat make up a guild.
- A **microbiome**, or community, includes all of the microbes in one location, which includes sets of guilds. Microbiomes are typically quite heterogeneous. Only rarely—and usually only in extreme environments—does a microbiome consist of a single population.
- Populations and guilds within a community typically reside in their own distinct *microhabitats*—specific small spaces where conditions are optimal for survival.
- Groups of microhabitats form habitats in which microorganisms interact with larger organisms and the environment. Together, the organisms, the environment, and the relationships between the two constitute an **ecosystem**.
- All of the ecosystems taken together constitute the *biosphere*, that region of Earth inhabited by living organisms.

To illustrate these concepts, consider two examples, a plot of soil and the human body (FIGURE 27.1). Soil is composed of tiny particles of rock and organic material, each of which forms a microhabitat for populations of microorganisms. Populations are distributed among the soil particles according to conditions necessary for their growth, including moisture, nutrients, and chemical properties such as pH. Populations of photosynthetic microbes, for example, live near the surface and form a photosynthetic guild, which provides nutrients to other guilds living deeper within the soil. A soil community is composed of all of these populations and guilds, whose activities are a major factor affecting soil quality.

Your body is also an ecosystem with many habitats, such as the mouth or the large intestine. An example of a guild in the mouth is sucrose fermenters, which live within biofilms on the surface of teeth.

Scientists can now routinely sample an ecosystem, such as soil or the human mouth, and determine the entire composition of all bacteria present in that sample by using next-generation DNA sequencing (NGS; see Chapter 8, Figure 8.11). Such modern methods have dramatically advanced our understanding of microbial ecology. The human intestine is the habitat with the greatest density and diversity of microbes, typically over 1000 species. Microbial guilds in the intestines have more metabolic functions than those of your own cells, contributing to digestion of food and producing nutrients such as vitamins. The microbiome of humans controls our health by affecting nutrition, digestion, immune system activation and inflammation, and possibly diabetes and obesity.

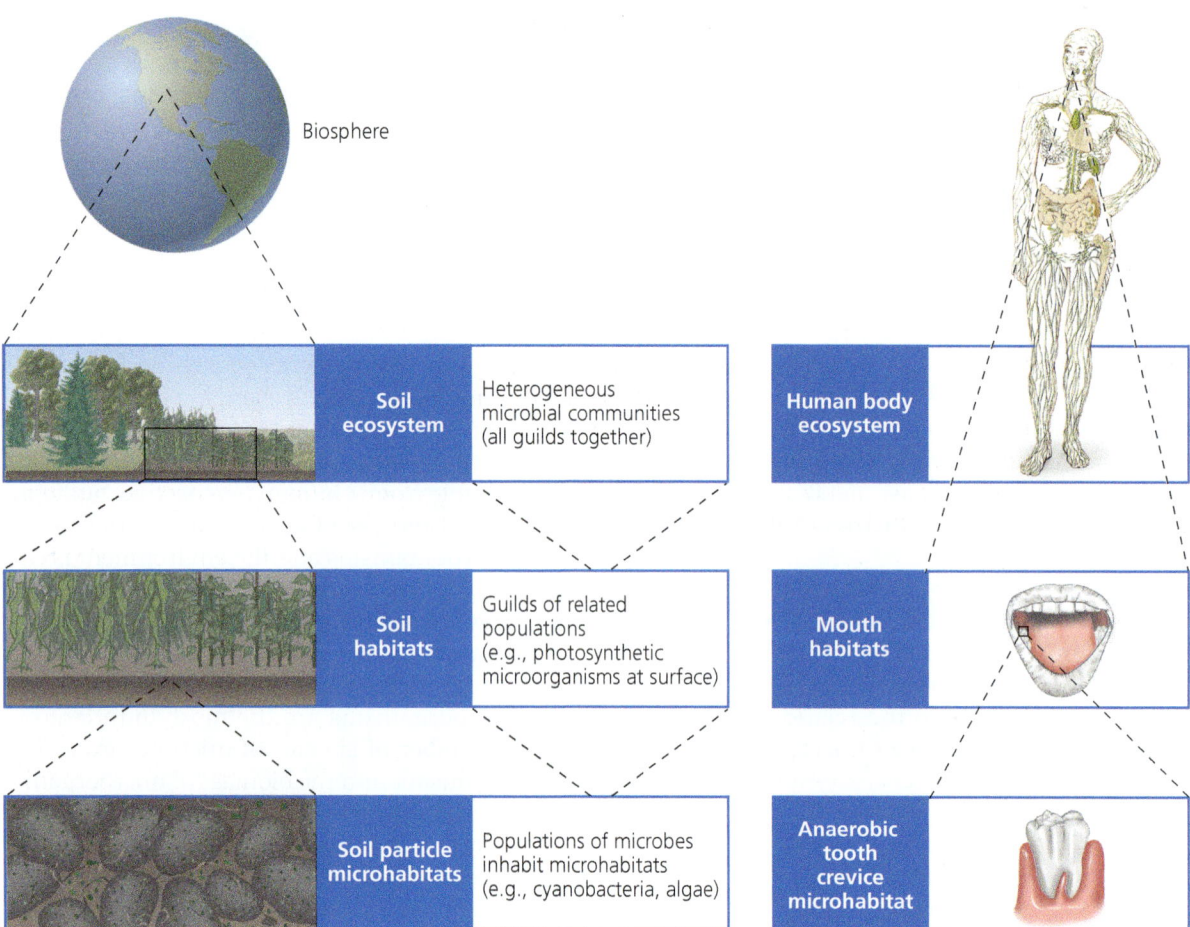

▲ **FIGURE 27.1 Terms to describe levels of microbial association in the environment.** Two environments are considered, soil and the human body.

The Role of Adaptation in Microbial Survival

What we might consider a lush ecosystem supports great biodiversity, but many microorganisms also live in what we consider harsh environments. The harsher the environment, the more specially adapted a microbe must be to survive. Some environments cycle between periods of excess and depletion, and the microbes that live in them must be capable of adapting to such constantly varying conditions. *Extremophiles*—microbes adapted to harsh temperatures, pH levels, or salt concentrations—are so adapted to extreme conditions that they cannot survive anywhere else.

Microbial growth can be held in check by *competition*. The best-adapted microorganisms have traits that provide them advantages in nutrient uptake, reproduction, response to environmental changes, or some other factor. *Antagonism*, in which a microbe makes some product that actively inhibits the growth of another, may also occur.

Although competition may limit microbial growth, rarely does a microorganism outcompete all other rivals; diversity is the norm, not the exception. Microbes often use the waste products of other microbes for their own metabolism, resulting in networks of *cooperation*. Moreover, one microbe's metabolic activities sometimes make the environment more favorable for other microbes. Biofilms are examples of microbial cooperation.

The relationships among microorganisms and their environment constantly change; each microbe adapts to subtle changes in its environment. Most habitats support successions of microbial populations within a community. By studying how these successions occur, we learn much about the microbes' effects on the environment.

Bioremediation

LEARNING OUTCOME
27.3 Describe the process of bioremediation.

Each year, Americans produce more than 150 million tons of solid wastes, accumulated from household, industrial, medical, and agricultural sources; most of it ends up in landfills. A landfill is essentially a large, open pit into which wastes are dumped, compacted, and buried. Soil microbes anaerobically break down biodegradable wastes; methanogens, which are all *Archaea*, degrade organic molecules to methane. When a landfill is full, it is often covered with soil and revegetated.

To prevent leaching of potentially hazardous materials into the soil and groundwater, a landfill pit is lined with clay or plastic, and sand and drainage pipes lining the bottom filter out small particulates and some microorganisms. Unfortunately, chemicals and hazardous compounds might still leak from landfills. Some of these substances are potentially carcinogenic (benzene), others are toxic (gasoline, lead), and still others are *recalcitrant* (resistant to decay, degradation, or reclamation by natural means).

Bioremediation is the use of organisms, particularly microorganisms, to clean up toxic, hazardous, or recalcitrant compounds by degrading them to harmless compounds. Although microorganisms eventually degrade most naturally occurring organic compounds, synthetic compounds are not always easily removed from the environment. Petroleum, pesticides, munitions, herbicides, and industrial chemicals that accumulate in soil and water are degraded only slowly by naturally existing microbes.

The reason that one synthetic molecule is biodegradable while another is recalcitrant relates to chemical structure; sometimes a subtle variation—perhaps in a single atom or bond—can be enough to make a difference. Most recalcitrant molecules resist microbial degradation because the microbes lack enzymes capable of degrading them; after all, until very recently, synthetic compounds didn't even exist. A widely known example of bioremediation is the microbial degradation of complex hydrocarbons released in an oil spill. Cleanup crews can enhance bioremediation in some cases by applying nitrogen and phosphorus fertilizers to stimulate microbial growth. Further, scientists can use recombinant DNA technology to create microbes that degrade some recalcitrant pollutants.

The Problem of Acid Mine Drainage

LEARNING OUTCOME
27.4 Discuss the problem of acid mine drainage.

Acid mine drainage is a serious environmental problem resulting from the exposure of certain metal ores to oxygen and microbial action (**FIGURE 27.2**). Coal deposits are often found associated with reduced metal compounds such as pyrite (FeS_2). Strip-mining for coal exposes pyrite to oxygen in the air, which oxidizes the iron, and bacteria such as *Thiobacillus* (thī-ō-bă-sil'ŭs) oxidize the sulfur. Rainwater then leaches the oxidized compounds from the soil to form sulfuric acid (H_2SO_4), which is carried into streams and rivers, reducing the pH enough (pH 2.5–4.5) to kill fish, plants, and other organisms. Such acidic water is also unfit for human consumption or recreational use. The EPA requires that strip mines be reburied as soon as possible to halt the processes.

Underground mining operations pose similar (but somewhat less severe) problems resulting from runoff from mine tailings,

▲ **FIGURE 27.2 The effects of acid mine drainage.** Upon exposure to air, iron and sulfur in water leached from mine tailings are oxidized to Fe^{3+}. The activity of iron–sulfur bacteria can reduce the pH of the water to a level that kills plants and animals.

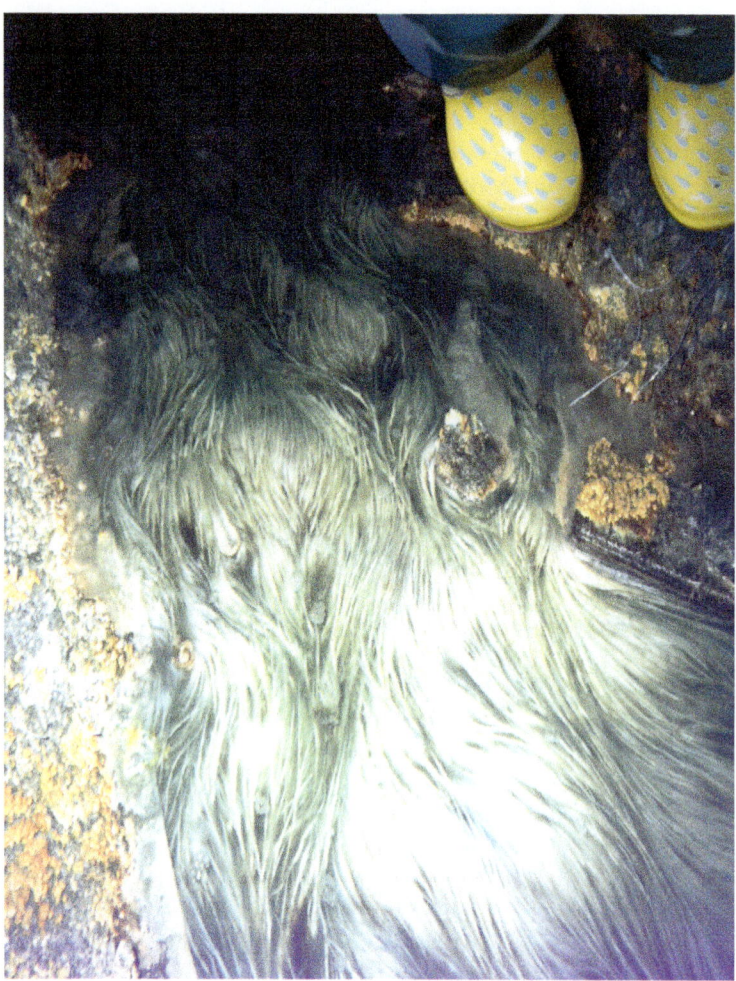

▲ FIGURE 27.3 **An acid-loving microbe.** The filamentous archaeon *Ferroplasma acidarmanus* growing as long filaments in acid mine runoff (pH 3.6) in California. The ore from this mine is rich in iron.

which are the low-grade ores remaining after the extraction of higher-grade ores. Typically, subsurface mines are backfilled as richer veins of minerals are depleted, thus limiting the exposure of iron sulfides to oxygen.

Although acid mine drainage is generally devastating to the environment, some microbes—mostly archaea—actually flourish in acidic conditions. One unique archaeal species, found in mine drainage in California, is *Ferroplasma acidarmanus* (fe′rō-plaz-ma a-sid′ar-mă-nŭs; FIGURE 27.3), which lives at a pH near zero and obtains its energy from oxidizing pyrite in mine sediments. Such an organism is just one example of the incredible diversity of microorganisms that colonize every habitat on Earth.

MICRO CHECK

1. What is the term used for groups of microbes that have similar metabolic requirements and therefore live in similar environmental niches?
2. What do we call the use of microbes to break down toxic or harmful compounds in the environment?
3. Why are open coal mines (strip mines) more likely to result in acid mine drainage than are underground mines?

The Roles of Microorganisms in Biogeochemical Cycles

LEARNING OUTCOMES

27.5 Contrast the processes by which microorganisms cycle carbon, nitrogen, sulfur, phosphorus, and trace metals.
27.6 Explain the work of microorganisms in the carbon cycle.
27.7 Contrast the actions of microbes involved in nitrogen fixation, nitrification, ammonification, denitrification, and anammox reactions.
27.8 Describe the reduction and oxidation of sulfur by microbes.
27.9 Define *eutrophication*, and explain in general how it arises.
27.10 Explain the use of microbes in biomining.

Six chemical elements make up most macromolecules—the building blocks of cells. These are hydrogen, oxygen, carbon, nitrogen, sulfur, and phosphorus.

Most elements are tied up in chemical and physical forms unavailable to organisms. The release of many elements from rock, for example, requires years of degradation by rain, wind, and microbes, so these processes contribute little to the day-to-day availability of nutrients for living things. As a consequence, the actions of organisms in recycling elements are the major components of **biogeochemical cycles**—the processes by which organisms convert elements from one form to another, typically between oxidized and reduced forms. Microbes are the primary biological components of the biogeochemical cycles.

The following sections examine the biogeochemical cycles for carbon, nitrogen, sulfur, phosphorus, and some trace metals.

The Carbon Cycle

Carbon is the fundamental element of all organic chemicals. The continual cycling of carbon in the form of organic molecules constitutes the majority of the **carbon cycle** (FIGURE 27.4). Carbon in rocks and sediments has a very low *turnover rate* (rate of conversion to other forms)—this form of carbon is incorporated into organic chemicals only slowly over long periods of time.

The start of the carbon cycle is autotrophy. Photoautotrophic *primary producers* (cyanobacteria, green and purple sulfur bacteria, green and purple nonsulfur bacteria, algae, photosynthetic protozoa, and plants) convert CO_2 to organic molecules—a process known as *carbon fixation*. Primary producers use the Calvin-Benson cycle to fix carbon (see Figure 5.26). The enzyme that fixes carbon dioxide, rubisco, is likely the most abundant enzyme on Earth. Photoautotrophs are restricted to the surfaces of soil and water systems because phototrophs require light. Chemoautotrophs can also fix carbon, but they acquire energy from H_2S or other inorganic molecules; therefore, chemoautotrophs are found in a greater variety of habitats. However, they do not fix as much carbon as photoautotrophs and are not as important in the carbon cycle.

Heterotrophs catabolize some organic molecules for energy in respiration, resulting in the release of CO_2. Other organic molecules made by autotrophs are subsequently incorporated into the tissues of heterotrophs. There they remain until the organism dies, when *decomposers* catabolize the organic materials,

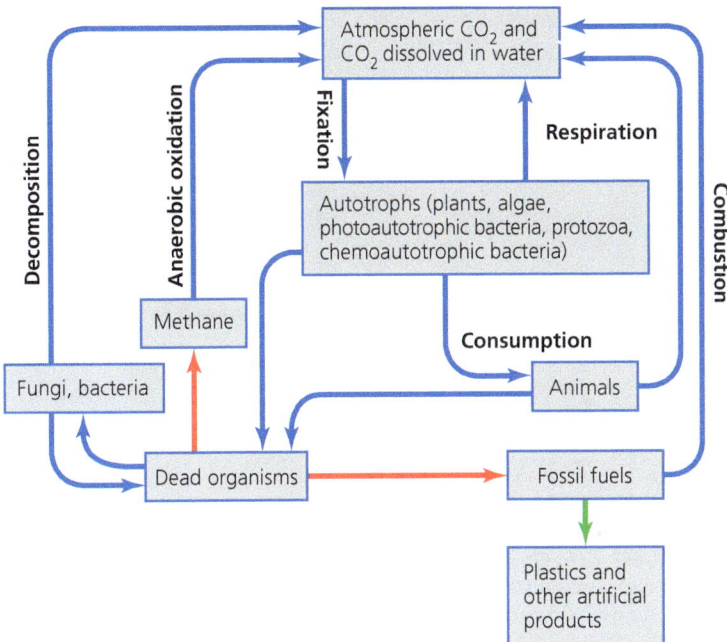

▲ FIGURE 27.4 Simplified carbon cycle. The most mobile form of carbon in the cycle is CO_2; this inorganic molecule is fixed by autotrophs, which incorporate its carbon into organic molecules. Processes that are performed solely by prokaryotes are represented by red lines—only prokaryotes produce methane and generate fossil fuels. Blue lines indicate processes carried out by both prokaryotes and eukaryotes. Only humans synthesize plastics and other artificial products from fossil fuels (green arrow).

releasing CO_2—the reverse of autotrophy. Human and animal waste is also broken down by decomposers.

The release of CO_2 starts the cycle over again as primary producers fix CO_2 once more into organic material. A rough balance exists between CO_2 fixation and CO_2 release.

Many scientists are concerned by a growing imbalance in the carbon cycle due to an overabundance of CO_2 in the atmosphere. The burning of fossil fuels and wood sends tons of CO_2 into the atmosphere each year. Furthermore, in waterlogged soils, sewage treatment plants, landfills, and the digestive systems of ruminants, methanogens actively release methane gas (CH_4), which can be photochemically transformed into CO (carbon monoxide) and CO_2. Methane-oxidizing bacteria, working in conjunction with methanogens, can also directly convert methane to CO_2. Worldwide, the rate of CO_2 production exceeds the rate at which it is being incorporated into organic material.

Carbon dioxide and methane are called "greenhouse gases" because their presence in the atmosphere prevents the escape of some infrared radiation into space, redirecting heat back to Earth, much as the glass panes of a greenhouse trap heat. Although this "greenhouse effect" is essential to life on Earth, it may result in climate change.

The Nitrogen Cycle

Nitrogen is an important nutritional element required by organisms as a component of proteins, nucleic acids, and other compounds. Most nitrogen in the environment is in the atmosphere as dinitrogen gas (N_2), which is unusable by most organisms. The majority of organisms acquire nitrogen as part of organic molecules or from soluble inorganic nitrogen compounds found in limited quantities in soil and water. These include nitrate, nitrite, and ammonia. Microbes cycle nitrogen atoms from dead organic materials and animal wastes to these soluble forms of nitrogen. They also cycle nitrogen between the biosphere and the atmosphere. FIGURE 27.5 summarizes the basic processes involved in the **nitrogen cycle**—nitrogen fixation, ammonification, nitrification, denitrification, and anammox reactions.

Nitrogen fixation, a process whereby gaseous nitrogen (N_2) is reduced to ammonia (NH_3), is an energy-expensive process in which the extremely stable nitrogen-nitrogen triple bond is broken in a reaction catalyzed by the enzyme *nitrogenase*. A limited number of prokaryotes—but no eukaryotes—fix nitrogen.

Nitrogenase functions only in the complete absence of oxygen, a condition that presents no problem for anaerobes. Aerobic nitrogen fixers, in contrast, must protect nitrogenase from oxygen. They can do this in a number of ways. For example, some aerobes use oxygen at such a high rate that it does not diffuse into the interior of the cell, where nitrogenase is sequestered. Some cyanobacteria form thick-walled, nonphotosynthetic cells called *heterocysts* to protect nitrogenase from oxygen in the environment as well as from the oxygen generated during photosynthesis (see Figure 11.14a). Other cyanobacteria fix nitrogen only at night, when oxygen-producing photosynthesis does not occur, thereby separating nitrogen fixation from photosynthesis in time rather than in space.

Nitrogen fixers may be free-living or symbiotic. Among the free-living nitrogen fixers are aerobic species of *Azotobacter*

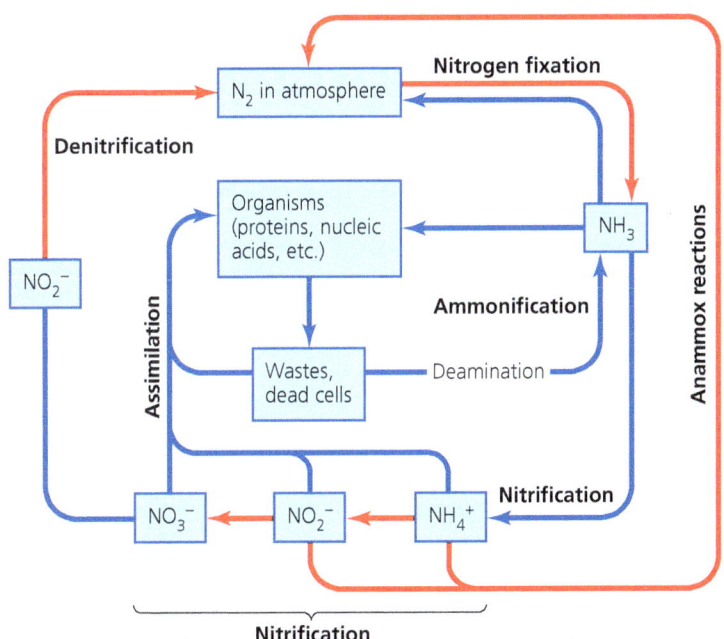

▲ FIGURE 27.5 Simplified nitrogen cycle. Even though nitrogen gas (N_2) is the most common form of nitrogen in the environment, most organisms cannot use it. To become available for their use, nitrogen from the atmosphere must be converted into ammonia (NH_3), a process that is carried out by the very few species of nitrogen-fixing prokaryotes. Processes that are performed solely by prokaryotes are represented by red lines. Blue lines indicate processes carried out by both prokaryotes and eukaryotes. Nitrification is the multistep conversion of ammonia or ammonium to nitrate.

(ā-zō-tō-bak'ter), anaerobic species of *Bacillus* and *Clostridium*, and deep-sea communities of archaea and bacteria. When they die, these microbes release fixed nitrogen into the soil or water, where it becomes available to other organisms, notably plants. None of the free-living genera are directly associated with the plants they fertilize.

Symbiotic nitrogen fixers, in contrast, live in direct association with plants, forming *root nodules* on legumes (e.g., peas and peanuts; see Figure 11.21). The predominant genus of nitrogen-fixing, symbiotic bacteria is *Rhizobium* (rī-zō'bē-ŭm). In the symbiosis, the plants provide nutrients and an anaerobic environment in the nodules, while the bacteria provide usable nitrogen to the plant. Farmers do not have to apply nitrogen fertilizers to legume crops because nitrogen fixers provide enough.

Bacteria and fungi in the soil decompose wastes and dead organisms, disassembling proteins into their constituent amino acids, which then undergo *deamination* (removal of their amino groups). Amino groups are converted to ammonia (NH_3)—a process called **ammonification**. In dry or alkaline soils, NH_3 escapes as a gas into the atmosphere, but in moist soils, NH_3 is converted to ammonium ion (NH_4^+), which organisms absorb, or ammonium is oxidized.

In **nitrification**, ammonium is oxidized to nitrate (NO_3^-) via a two-step process requiring autotrophic archaea and bacteria. In the best-studied nitrification pathway, species of the bacteria *Nitrosomonas* (nī-trō-sō-mō'nas) convert NH_4^+ to nitrite (NO_2^-), which is toxic to plants. Fortunately, *Nitrosomonas* spp. are usually found in association with species of the bacterium *Nitrobacter* (nī-trō-bak'ter) that rapidly convert nitrite to nitrate (NO_3^-), which is soluble and can be used by plants. The soluble nature of nitrate, however, means that it is leached from the soil by water and accumulates in groundwater, lakes, and rivers. Certain microorganisms in waterlogged soils perform **denitrification**, in which NO_3^- is oxidized to N_2 by anaerobic respiration. N_2 gas escapes into the atmosphere. Many organisms can incorporate inorganic nitrogen compounds into organic compounds.

An important aspect of the nitrogen cycle is *anaerobic ammonium oxidation*, or **anammox**. Anammox prokaryotes oxidize 30% to 50% of the world's ammonium into nitrogen gas using nitrite as an electron acceptor.

The Sulfur Cycle

The **sulfur cycle** involves moving sulfur between several oxidation states (FIGURE 27.6). Bacteria decompose dead organisms, releasing sulfur-containing amino acids into the environment. Sulfur released from amino acids is converted to its most reduced form, hydrogen sulfide (H_2S), by microorganisms via a process called sulfur *dissimilation*. H_2S is oxidized to elemental sulfur (S^0) and then to sulfate (SO_4^{2-}) under various conditions and by various organisms, including nonphotosynthetic autotrophs such as *Thiobacillus* and *Beggiatoa* (bej'jē-a-tō'a), photoautotrophic green and purple sulfur bacteria. Sulfate is the most readily usable form of sulfur for plants and algae, which animals then eat. Anaerobic respiration by the bacterium *Desulfovibrio* (dē'sul-fō-vib're-ō) reduces SO_4^{2-} back to H_2S. Thus, the two major inorganic constituents of the sulfur cycle are H_2S and SO_4^{2-}.

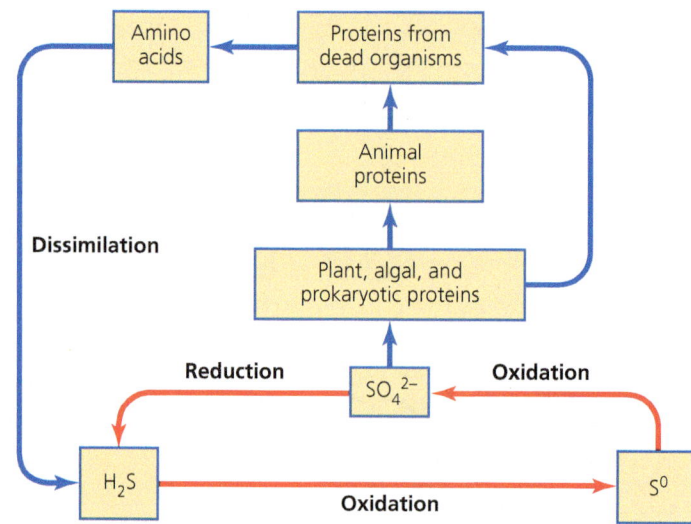

▲ **FIGURE 27.6 Simplified sulfur cycle.** The two main constituents of this cycle are hydrogen sulfide (H_2S) and sulfate (SO_4^{2-}), the fully reduced and oxidized forms of sulfur, respectively. Processes that are performed solely by prokaryotes are represented by red lines. Blue lines indicate processes carried out by both prokaryotes and eukaryotes.

The Phosphorus Cycle

Unlike nitrogen and sulfur, phosphorus undergoes little change in oxidation state in the environment. Phosphorus usually exists in the environment, though often in a metabolically limited amount. Organisms typically use phosphorus in phosphate ion (PO_4^{3-}). The **phosphorus cycle** involves the movement of phosphorus from insoluble to soluble forms available for uptake by organisms and the conversion of phosphorus from organic to inorganic forms by pH-dependent processes. No gaseous form exists to be lost to the atmosphere, but dissolved phosphates do accumulate in water, particularly the oceans, and organic forms of phosphorus are deposited in surface soils following the decomposition of dead animals and plants.

Too much phosphorus can be a problem in a habitat; for example, agricultural fertilizers rich in phosphate are easily leached from fields by rain. The resulting runoff into rivers and lakes can result in **eutrophication** (yū-trō'fi-kā'shŭn)—the overgrowth of microorganisms (particularly algae and cyanobacteria) in nutrient-rich waters (FIGURE 27.7). Such overgrowth, called a *bloom*, depletes oxygen from the water, killing aerobic organisms, such as fish. Anaerobic organisms then take over the water system, leading to an increased production of H_2S and the release of foul odors. When excess phosphate (and nitrogen) are removed, such a water system recovers over time.

The Cycling of Metals

Metal ions, including Fe^{2+}, Zn^{2+}, Cu^{2+}, Cd^{2+}, and Mg^{2+}, are important microbial nutrients. Though they are needed only in trace amounts, they can nonetheless be limiting factors in the growth of organisms. Many metals—including the most important trace metal, iron—are present in the natural environment in the form of poorly soluble molecules in rocks, soils, and sediments and are generally unavailable for uptake by organisms. The cycling of metal ions involves primarily a transition from an

▲ FIGURE 27.7 **Eutrophication.** Such overgrowth of cyanobacteria and algae can occur when there is an excess of nutrients in water, as seen here in Dianchi Lake in Kunming, Yunnan province, People's Republic of China.

insoluble to a soluble form, allowing them to be used by organisms and to move through the environment. In the ecosystem of the human body, iron is sequestered so as to not be available for invading pathogens to use; this constitutes part of innate immunity (see Chapter 15).

Generally, oxidized metal ions (those having fewer electrons) dissolve more readily than reduced ions of the same metal. Miners have successfully used **biomining**—a process in which microbes (typically archaea) oxidize copper, gold, uranium, or other metals so that the oxidized metallic ions dissolve in water. The miners can then extract the mineral-laden water and reduce the ions, which causes the metals to come out of solution where the miner can collect them.

Biogeochemical cycles are sustained by microorganisms, most of which live in soil. Next we examine aspects of soil microbes and their habitats.

MICRO CHECK

4. What is carbon fixation?
5. What does the enzyme nitrogenase do?
6. An excess of which element is most often the cause of eutrophication?

Soil Microbiology

LEARNING OUTCOMES

27.11 Identify five factors affecting microbial abundance in soils.
27.12 Describe three human and plant diseases caused by soil microbes.

Soil microbiology examines the roles played by organisms living in soil. They rarely cause human disease, though plant pathogens are prevalent in soils and are agriculturally and economically important.

The Nature of Soils

Soil arises both from the weathering of rocks and through the actions of microorganisms, which produce wastes and organic materials needed to support more complex life forms, such as plants. Soil is composed of two major layers (**FIGURE 27.8**): *topsoil*, which is rich in *humus* (organic chemicals); and *subsoil*, which is composed primarily of inorganic materials. Soil overlies bedrock, which contains little organic material. Most microorganisms are found in topsoil, where the richness of the organic deposits sustain a large biomass. Topsoil itself, however, is highly heterogeneous, and therefore different kinds and amounts of microbes are found in various soils around the world.

Factors Affecting Microbial Abundance in Soils

Several environmental factors influence the density and the composition of the microbiome within a soil, including the amount of water, availability of oxygen and other electron acceptors, acidity, temperature, and the availability of nutrients.

Moisture and oxygen content are closely linked in soils. Moisture is essential for microbial survival; microbes exhibit lower metabolic activity, are present in lower numbers, and are less diverse in dry soils than in moist soils. Because oxygen dissolves poorly in water, moist soils have a lower oxygen content than drier soils. When soil is waterlogged, anaerobes predominate, even at the surface. Weather patterns also affect oxygen content, as the presence or absence of rainwater determines moisture and thus dissolved oxygen.

The pH of a soil determines in part whether it favors bacteria or fungi. Highly acidic and highly basic soils favor fungi over bacteria, though fungi typically prefer acidic conditions. Bacteria dominate when soil pH is closer to 7.

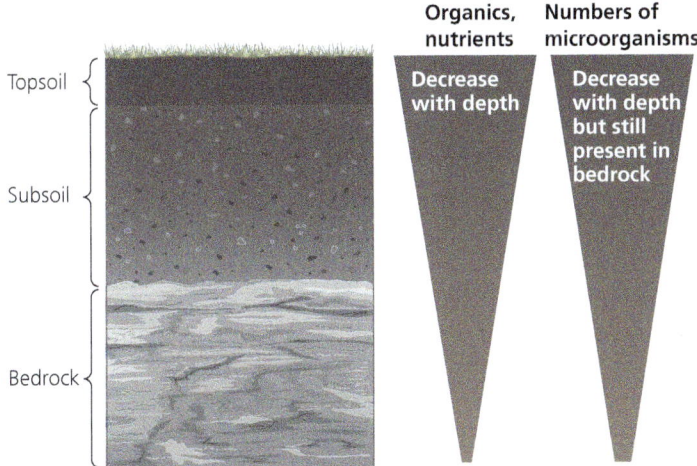

▲ FIGURE 27.8 **The soil layers and the distributions of nutrients and microorganisms within them.** Although topsoils in general are richer in nutrients and microbes than are subsoils, the nutrient and microbial content of topsoils is highly variable.

Most soil organisms are mesophiles and prefer temperatures between 20°C and 50°C. Thus, most soil microbes live quite well in areas where winters and summers are not too extreme. Psychrophiles grow only in consistently cold environments and cannot survive in soils that experience spring thawing; the opposite is true for thermophiles, which cannot survive where winters are harsh.

Nutrient availability also affects microbial diversity in soil habitats. Most soil microbes are heterotrophic, utilizing organic matter in the soil. The biomass of a microbial community is determined more by the amount of organic material than by the kind of organic material: any soil that has a relatively constant input of organic material can support a wider array of microorganisms than soil that is more barren.

Microbial Populations in Soils

Because of the variety of soils, microbial populations differ tremendously from soil to soil and even within the same soil over the course of a season. Bacteria are numerous, and diverse inhabitants of soil and are found in all soil layers. Archaea are present in soils, but the inability to culture many of them has limited our ability to study them. The fungi are also a nearly ubiquitous group of soil microorganisms. Free-living and symbiotic fungi are found only in topsoil, where they can form gigantic mycelia that cover acres. Many viruses, mostly bacteriophages, are found within soils.

Some algae and protozoa also live in soils. Soil algae live on or near the surface because, as photoautotrophs, they require light. Protozoa are mobile and move through the soil, grazing on other microbes. For the most part, protozoa require oxygen and remain in the topsoil. Neither algae nor protozoa can withstand dramatic environmental changes or the introduction of pollutants.

Wherever present, microbes perform a variety of necessary functions. They cycle nitrogen, sulfur, phosphorus, and other elements, converting them into usable forms. Microbes degrade dead organisms and their wastes, and some can clean up industrial pollutants. Further, microbes produce an incredible variety of compounds that have potential human uses. Researchers search natural populations of microbes to discover species that produce valuable chemicals or have useful biological functions.

The capability of modern scientists to directly extract DNA from soil, skipping the need to culture soil microorganisms, allows dramatic advances in our understanding of the structure and function of soil microbiomes.

Soilborne Diseases of Humans and Plants

Although the majority of soil microorganisms are harmless, there are exceptions. Soilborne infections of humans generally result from either direct contact with, ingestion of, or inhalation of microorganisms deposited in soil in animal or human feces or urine. In some cases, the microbes live and replicate in the soil, but in most cases soil is simply a vehicle for moving the pathogen from one host to another. The majority of soilborne disease agents are fungal or bacterial. Few soil protozoa or viruses cause disease.

Soil pathogens include the bacterium *Bacillus anthracis* (an-thrā'sis), the causative agent of anthrax, which produces endospores shed from the skins of infected livestock. Endospores may remain dormant in soil for decades or centuries. Disturbing the soil can lead to infection if endospores enter cuts or abrasions on the skin (cutaneous anthrax) or are inhaled into the lungs (inhalation anthrax).

Histoplasma capsulatum (his-tō-plaz'mă kap-soo-lā'tŭm) is a fungus that causes histoplasmosis—a serious respiratory tract infection. *Histoplasma* grows in soil and is also deposited there as spores in the droppings of infected birds and bats. The spores can be inhaled by humans when contaminated soil is disturbed.

Hantavirus (han'ta-vī-rŭs) pulmonary syndrome is a life-threatening viral respiratory disease acquired via the inhalation of soil contaminated by mouse droppings and urine containing *Hantavirus*. *Hantavirus* has been found throughout North America.

Soil contains many more plant pathogens than human pathogens. Microbial plant infections are generally characterized by one or more of the following signs: necrosis (rot), cankers/lesions, wilt (droopiness), blight (loss of foliage), galls (tumors), growth aberrations (too much or too little), or bleaching (loss of chlorophyll). Bacteria, fungi, and viruses all cause diseases in plants and spread either as airborne spores, through roots or wounds, or by insects.

TABLE 27.1 lists selected bacterial, fungal, and viral soilborne diseases of humans and plants.

TABLE 27.1 Selected Soilborne Diseases of Humans and Plants

Microorganism	Host	Disease
Bacteria		
Bacillus anthracis	Humans	Anthrax
Clostridium tetani	Humans	Tetanus
Agrobacterium tumefaciens	Plants	Crown gall disease
Ralstonia solanacearum	Plants	Potato wilt
Streptomyces scabies	Plants	Potato scab
Fungi		
Histoplasma capsulatum	Humans	Histoplasmosis
Blastomyces dermatitidis	Humans	Blastomycosis
Coccidioides immitis	Humans	Coccidioidomycosis
Polymyxa spp.	Plants	Root rot in cereals
Fusarium oxysporum	Plants	Root rot in many plants
Phytophthora cinnamomi	Plants	Potato blight; root rot in many plants
Viruses		
Hantavirus	Humans	*Hantavirus* pulmonary syndrome
Tobamovirus tobacco mosaic virus	Plants	Necrotic spots in various plants
Furovirus soil-borne wheat mosaic virus	Plants	Mosaic disease in winter wheat and barley

Aquatic Microbiology

LEARNING OUTCOME

27.13 Compare the characteristics and microbial populations of freshwater and marine ecosystems.

Aquatic microbiologists study microorganisms living in freshwater and marine environments. Compared to soil habitats, water ecosystems support fewer microbes overall because nutrients are diluted. Many organisms that live in aquatic systems exist in biofilms attached to surfaces. Biofilms allow aquatic organisms to concentrate enough nutrients to sustain growth; without forming biofilms, they likely would starve.

Types of Aquatic Habitats

Aquatic habitats are divided primarily into freshwater and marine systems. *Freshwater* systems, which are characterized by low salt content (about 0.05%), include groundwater, water from deep wells and springs, and surface water in the form of lakes, streams, rivers, shallow wells, and springs. *Marine* environments, characterized by a salt content of about 3.5%, encompass the open ocean and coastal waters, such as bays, estuaries, and lagoons.

Natural aquatic systems can be greatly affected by the release of so-called *domestic water*, which is water resulting from the treatment of sewage and industrial waste. Domestic water released into the environment affects water chemistry and the microorganisms living in the water. Changing levels of chemicals cause increases or decreases in microbial numbers. Furthermore, faulty treatment of sewage leads to contamination of natural water systems with pathogenic microorganisms.

Freshwater Ecosystems Microorganisms become distributed vertically within lake systems according to oxygen availability, light intensity, and temperature. Surface waters are high in oxygen, well lighted, and warmer than deeper waters. In large lakes, wave action continually mixes nutrients, oxygen, and organisms, which allows efficient utilization of resources. In stagnant waters, oxygen is readily depleted, resulting in more anaerobic metabolism and poorer water quality.

Scientists observe four zones in deep lakes (**FIGURE 27.9a**): The **littoral zone** (li'ter-al) is the area along the shoreline where nutrients enter the lake. The littoral zone is shallow, and light penetrates it; most microbes live here. The **limnetic zone** (lim-net'ik) is the upper layer of water away from the

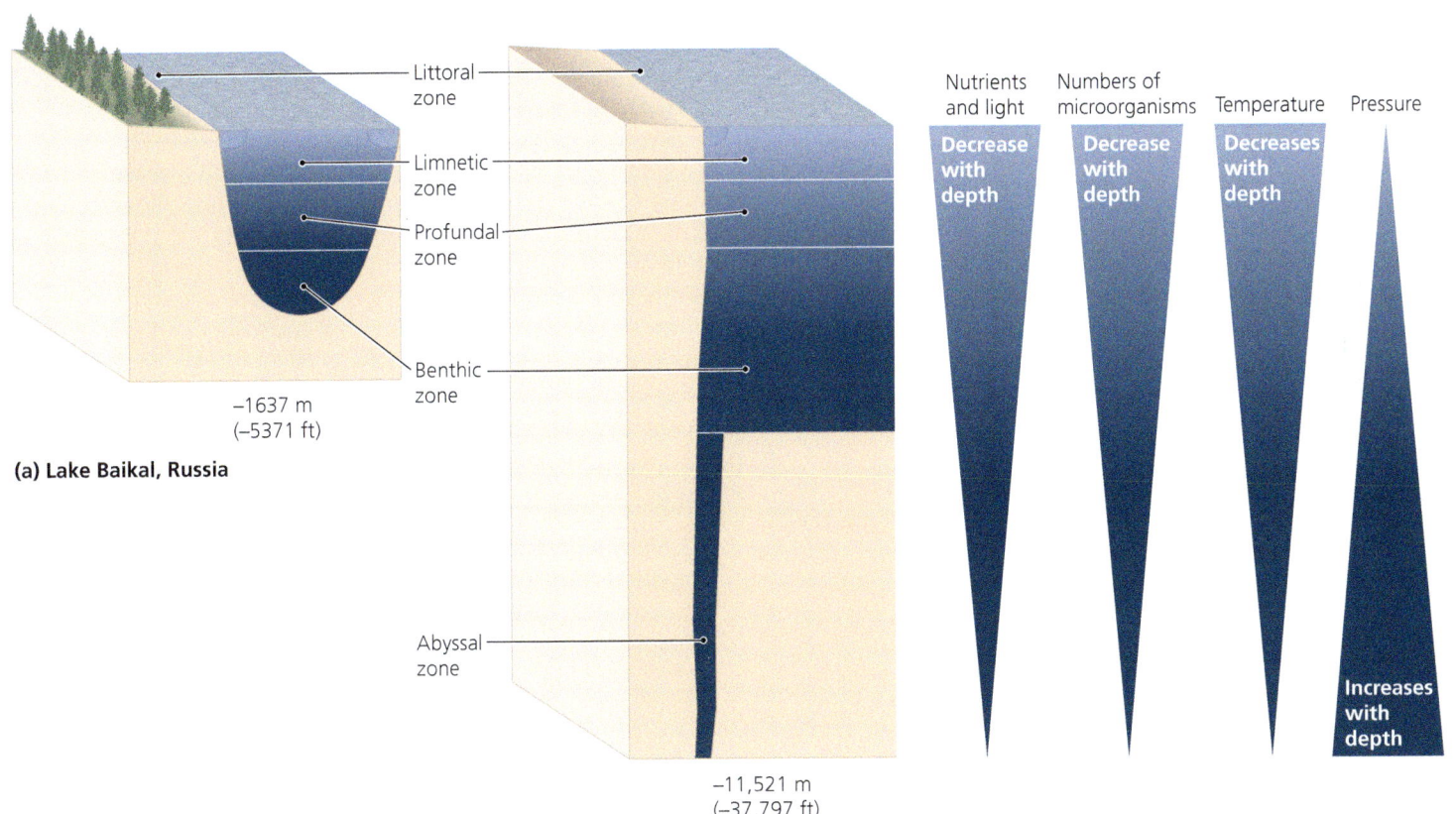

▲ **FIGURE 27.9 Vertical zonation in deep bodies of water.** (a) Lake Baikal, Russia, the world's deepest freshwater lake, and (b) the Mariana Trench in the Pacific Ocean, the deepest place in the ocean. Both deep lakes and oceans can be divided into zones that vary with respect to light penetration, concentration of nutrients, temperature, and pressure—and thus the types and abundance of microorganisms. *Why would a bacterium from the bottom of Lake Baikal die in the Mariana Trench?*

Figure 27.9 *A bacterium from the bottom of Lake Baikal could not survive the even greater pressure and the salinity of the seawater deep in the Mariana Trench.*

shore. Photoautotrophs reside here and in the littoral zone. The **profundal zone** (prō-fun'dal) is the deeper water beneath the limnetic zone. It has a lesser oxygen content and more diffuse light than the previous two zones. Some photosynthetic organisms, such as purple and green sulfur bacteria, perform anaerobic photosynthesis here. Beneath this is the **benthic zone** (ben'thik), which encompasses deeper lake water and sediments. Anaerobic bacteria in the sediments produce H_2S, which is used by organisms nearer the surface.

In contrast to lakes, streams and rivers are more uniform because organisms and nutrients are swept along and mixed. Biofilms are particularly important in moving waterways, and the majority of organisms live toward the edges, where surfaces are available, currents are less severe, and organic materials enter the water.

Marine Ecosystems Marine ecosystems are typically nutrient poor, dark, cold, and subject to great pressure. Photoautotrophic prokaryotes, diatoms, dinoflagellates, and algae are found near the surface. Most marine waters are extreme environments inhabited only by highly specialized microorganisms. All microbes in marine systems must be salt tolerant and possess highly efficient nutrient-uptake mechanisms to compensate for the scarcity of nutrients.

As with freshwater lake systems, scientists delineate zones in the oceans (**FIGURE 27.9b**). The majority of marine microorganisms are found in the littoral zone, where nutrient levels are high and light is available for photosynthesis. The benthic zone makes up the majority of the marine environment. Oceans have a fifth zone—the **abyssal zone** (a-bis'sal), which encompasses deep ocean trenches. Even though the benthic and abyssal zones have sparse nutrients, they still support microbial growth, particularly around **hydrothermal vents** located in the abyssal zone. Such vents spew superheated, nutrient-rich water, providing nutrients and an energy source for thermophilic chemoautotrophic anaerobes, which in turn support a variety of invertebrate and vertebrate animals.

Specialized Novel Aquatic Ecosystems In addition to the two broad categories of water systems just described, many distinctive aquatic ecosystems also exist, including salt lakes, iron springs, and sulfur springs. Each of these systems is inhabited by specialized microorganisms that are highly adapted to the conditions. The Great Salt Lake in Utah, for example, has a salt concentration of 5% to 7%, depending on its level, and contains the extreme halophile *Halobacterium salinarum* (hā'lō-bak-tēr'ē-ŭm sal-ē-nar'ŭm), an archaeon that thrives in highly saline water.

> ## MICRO CHECK
> 7. Topsoil contains a lot of organic material and light does not penetrate through soil, so which group of microbes is more likely abundant in soil?
> 8. What are the signs of the plant disease called blight?
> 9. In which deep lake zone do anaerobic photosynthetic bacteria live?

> **TELL ME WHY**
> A blogger stated that germs (microorganisms) are dangerous and should be avoided in all cases. Why is this idea wrong?

Biological Warfare and Bioterrorism

The properties of microorganisms that allow scientists to manipulate organisms to make advancements in medicine, food production, and industry also enable humans to fashion microbes into *biological weapons*, which can be directed at people, livestock, or crops. **Bioterrorism**, the use of microbes or their toxins to terrorize human populations, is a topic of major concern in today's world. A topic of growing concern is **agroterrorism**—the bioterrorist use of microbes to destroy a food supply. International treaties and laws of the United States, Great Britain, and other countries prohibit the use of biological weapons.

Assessing Microorganisms as Potential Agents of Warfare or Terror

LEARNING OUTCOME
27.14 Identify the criteria used to assess microorganisms for potential use as biological weapons or agents of bioterrorism.

Many microorganisms cause disease, but not all disease-causing organisms have potential as biological weapons. Governments establish criteria for evaluating the potential of microorganisms to be "weaponized." Establishing such criteria helps focus research and defense efforts where they are needed most and facilitates efforts to develop better response and deterrence capabilities.

Criteria for Assessing Biological Threats to Humans

In the United States, the assessment of a potential biological threat to humans is based on the following four criteria:

- *Public health impact.* This criterion relates to the ability of hospitals and clinics to deal effectively with numerous casualties. The more casualties, the more difficult it is for hospitals and clinics to effectively respond to the needs of all patients. If an agent causes numerous serious cases, emergency response systems could be overwhelmed and even cease to function. If an agent is highly lethal, proper disposal of bodies might become difficult, which would contribute to the spread of disease.

- *Delivery potential.* This criterion evaluates how easily an agent can be introduced into a population. The more people that can be infected at one time, the more devastating a primary attack. If an introduced agent spreads on its own through a population, secondary and tertiary waves of illness will augment the number of casualties. Also assessed as part of delivery potential are ease of mass production (the easier an agent is to produce in quantity, the greater its potential threat), availability (the more prevalent it is in the environment, the easier it is to obtain and weaponize), and environmental stability (the longer it remains infective once released, the greater the threat).

- *Public perception.* This criterion evaluates the effect of public fear on the ability of response personnel to control a disease outbreak following an attack. Agents with high mortality, few treatment options, and no vaccine instill greater fear into a populace, making quarantine (isolating infected and sick patients from the rest of the population) difficult to enforce. The resulting chaos could dramatically decrease the ability of response personnel to control disease transmission and treat patients.
- *Public health preparedness.* This criterion assesses existing response measures and attempts to identify improvements needed in the health care infrastructure to prepare for a biological attack. Diagnosis and recognition involve proper surveillance and training medical personnel to ascertain whether an attack has occurred. Once an attack has been confirmed, predetermined responses are required to reduce confusion and allow rapid control of the situation. Public health preparedness also involves funding for research and development of new vaccines, treatments, and diagnostic capabilities.

When assessing a threat, government authorities assign each potential bioterrorist agent a score for each of the criteria. Agents with the highest total scores are considered the most serious threats.

Criteria for Assessing Biological Threats to Livestock and Poultry

The criteria used to evaluate biological threats to livestock and poultry include agricultural impact, delivery potential, and analysis to determine if the outbreak is the result of terrorism or is a naturally occurring event.

Infectious agents prove the most devastating for agricultural livestock kept in large herds or flocks. Many animal pathogens exist naturally in soil or are already endemic in livestock and poultry, so they can be easily obtained and readily grown in large quantities. The highly infectious nature of some agents means that by the time a disease is recognized, much of a herd is infected already, so all must be destroyed to control an outbreak. Further, some pathogens can persist in the soil even after the animals are destroyed. Many animal diseases are spread either by contact or by inhalation, thus making attack easier for a terrorist. Though highly contagious among animals, most are not infectious to humans, making them "safe" for terrorists to handle.

Criteria for Assessing Biological Threats to Agricultural Crops

Plant diseases are generally not as contagious as animal or human diseases. Threats to crops are evaluated on predicted extent of crop loss, delivery and dissemination potential, and containment potential.

Plant pathogens that either cause severe crop loss or produce toxins are considered the greatest threats. Such agents already exist in the environment and can be readily obtained; however, plant pathogens are not as easily mass-produced as animal agents. Plant pathogens that can be spread systematically through fields by natural means, such as through contaminated soil or by insects, could remain in the environment even after destruction of the target crop. Successive plantings in contaminated fields could result in continued crop loss for as long as the agent persists. Because the causes of many plant diseases are not easily diagnosed, such a pathogen could be widely disseminated before response measures are instituted. Economic losses would be staggering, particularly given that embargos on affected crops would likely remain in place for years after an attack.

Known Microbial Threats

In the following sections, we briefly discuss some of the microorganisms currently considered threats as agents of bioterrorism. Note that as threat assessments become more refined and technology advances, the list of known bioterrorist agents is likely to change.

Human Pathogens

The National Institute of Allergy and Infectious Diseases (NIAID) categorizes biological agents that could be used against humans into three categories (TABLE 27.2 on p. 814). *Category A agents* are those with the greatest potentials as weapons. *Category B agents* have some potential as weapons but for various reasons are not as dangerous as category A agents (most lack the potential to cause mass casualties). *Category C agents* are possible threats that likely have low risk, though not enough is currently known about them to determine their true potential as weapons. Agents in all three categories are known as *biological select agents* or more simply as *select agents*. Biomedical research with select agents is carefully regulated. Research that could be misapplied to bioterrorism is called *dual use research of concern (DURC)*. Government agencies have additional special oversight over DURC.

Smallpox currently tops the list of bioterrorist threats. Fortunately, it is difficult for would-be terrorists to acquire viral samples for propagation; it also takes a high degree of skill, in addition to specific containment facilities, to work with smallpox virus. An effective vaccine is available, and the vaccine is effective when administered soon after infection.

Animal Pathogens

Biological agents against animals are also divided into categories, with category A agents being the most dangerous. Whereas many potential agents are spread via inhalation, others are spread by insect vectors, making them less likely to be used as weapons. Some agents infect wild animal populations in addition to livestock, potentially amplifying any outbreak that might occur.

Foot-and-mouth disease virus, the most dangerous of potential agroterrorism agents, affects all wild and domestic cloven-hoofed animals. The virus is spread by aerosols and by direct or indirect contact. Humans can transport it from herd to herd on their person, on farm equipment, or through the

TABLE 27.2 National Institute of Allergy and Infectious Diseases (NIAID) Microbial Bioterrorist Threats to Humans

Disease	Agent	Natural Source
Category A Threats: Highest Priority		
Smallpox	Variola (*Orthopoxvirus*)	None
Anthrax	*Bacillus anthracis* (bacterium)	Soil
Plague	*Yersinia pestis* (bacterium)	Small rodents
Botulism	*Clostridium botulinum* toxin (bacterial)	Soil
Tularemia	*Francisella tularensis* (bacterium)	Wild animals
Viral hemorrhagic fevers	Filoviruses and arenaviruses	Unknown in most cases
Category B Threats: Moderate		
Q fever (fever + flulike syndrome)	*Coxiella burnetii* (bacterium)	Sheep, goats, cattle
Brucellosis (severe flulike syndrome)	*Brucella* spp. (bacteria)	Livestock
Glanders (pulmonary syndrome)	*Burkholderia mallei* (bacterium)	Horses
Melioidosis (severe pulmonary syndrome)	*Burkholderia pseudomallei* (bacterium)	Horses, livestock, rodents, soil
Viral encephalitis	Alphaviruses	Rodents, birds
Epidemic typhus	*Rickettsia prowazekii* (bacterium)	Humans
Toxins (especially of *Clostridium perfringens* and *Staphylococcus*)	Various bacteria	Various
Psittacosis (pneumonia-like syndrome)	*Chlamydophila psittaci* (bacterium)	Birds
Food safety threats (including *Salmonella* spp., *Escherichia coli* O157:H7, *Shigella*)	Various bacteria and viruses	Soil or animals
Water safety threats (e.g., *Vibrio cholerae*, *Cryptosporidium parvum*)	Various bacteria and viruses	Water
Category C Threats: Potential Threats Likely to Have Low Risk		
Nipah and *Hendra* viruses (encephalitis)	*Henipavirus Nipah* virus; *H. Hendra* virus	Bats
Hantavirus pulmonary syndrome	Hantavirus	Rodents, soil
Tick-borne encephalitis viruses	*Flavivirus* species	Ticks
Yellow fever	*Flavivirus yellow fever* virus	Monkeys, *Aedes* mosquito
Tuberculosis	*Mycobacterium tuberculosis*	Humans
Influenza	*Influenzavirus*	Pigs, birds, humans
Coronavirus respiratory syndromes	Severe acute respiratory syndrome (SARS) virus; Middle East respiratory syndrome (MERS) virus	Bats, badgers, civets; Camels, bats
Rabies	*Lyssavirus rabies* virus	Nonrodent mammals
Chikungunya	*Alphavirus chikungunya* virus	*Aedes* mosquito
Variant Creutzfeldt-Jacob disease	prion PrP	Humans, bovines
Coccidioidomycosis	*Coccidioides* species	Soil

movement of animals between auctions and farms. Any appearance of foot-and-mouth disease on a farm requires destruction of entire herds, complete disinfection of all areas occupied by the herd, and disposal of all animals by burning or burial. A vaccine exists but not in sufficient quantities to protect all animals.

Plant Pathogens

Many plant pathogens exist, but the categorization of plant pathogens as terrorist agents lags behind similar efforts for humans and animals. Most potential agents are fungi whose dissemination could easily result in contamination of soils, which would be difficult to neutralize. Attacks against grains, corn, rice, and potatoes are considered the most dangerous because they would have significant negative effects on national economies and food supplies. All of the agents are naturally present, so detecting the difference between a natural outbreak and an intentional attack would be difficult.

Defense Against Bioterrorism

No defense can completely prevent a carefully planned biological attack against any group or nation. However, much can be done to limit the effect of an attack. The key is coupling *surveillance*—the active diagnosis and tracking of human, animal, and plant diseases—with *effective response protocols*.

Because category A human biological agents are not common, the appearance of more than a few scattered cases is highly suggestive that an attack of some kind has occurred. This is one reason why the diseases of category A are *reportable*; that is, they must be reported to state departments of health whenever they occur. Active monitoring of reportable diseases allows epidemiologists to quickly determine when unusual outbreaks are occurring. Once an unexpected pattern is seen, diagnostic confirmation can be obtained and, if an attack is deemed to have occurred, appropriate responses implemented (FIGURE 27.10). Such responses may entail forced quarantine, distribution of antimicrobial drugs, or mass vaccination.

Agroterrorism has become more of a concern with the realization that very little security protects the nation's agricultural enterprises. Livestock and poultry are routinely moved around the country without being tested for disease and without being quarantined prior to introduction into new herds or flocks. Infected animals could therefore spread disease as they pass from facility to facility. Compounding the problem is the fact that farms, ranches, auction houses, livestock shows, and irrigation facilities are all open to the public and impose few security measures to prevent purposeful infection of animals. It has been suggested that a step in defending against agroterrorism would be to restrict public access to such facilities. Additionally, effective screening of imported animals and plants would help prevent the introduction of foreign pathogens into the United States. Further, better diagnostic techniques, vaccines, and treatments need to be developed for animal pathogens.

Scientific research is critical to our efforts to be prepared for bioterrorist attacks. Since the Eisenhower administration, only defensive research has been legal in the United States. Scientists who research pathogens, including select agents, work at four biosafety levels: Investigations with microbes that do not commonly cause disease in humans occurs in BSL-1 labs, which include many university student labs. BSL-2 is suitable for organisms that pose moderate hazards, including HIV, MRSA, and *Shigella*. For agents that spread easily, often via inhalation, and cause serious disease, labs are at BSL-3. Pathogens requiring BSL-3 include *Mycobacterium tuberculosis*, West Nile virus, and *Francisella tularensis*. BSL-3 rooms have negative airflow, which draws air into the room, and all exiting air is sterilized by HEPA filtration. The maximum containment level, BSL-4, is for agents that are often fatal and untreatable, such as Ebola and Nipah viruses. Workers wear sealed suits and do not breathe the air in the lab (see Disease in Depth: Ebola Viral Disease on pp. 764–765).

Roles of Recombinant Genetic Technology in Bioterrorism

Recombinant genetic technology could be used to create new biological threats or modify existing ones so that, for example, vaccines against them no longer work. Traits of various agents could be combined to create novel agents for which no immunity exists in the population.

In addition to allowing the manipulation of existing threat agents, the techniques of recombinant DNA technology enable agents to be synthesized from scratch. Researchers were able to generate infectious poliovirus particles in the lab, based on genetic sequences that were published on the Internet. However, poliovirus is a relatively simple virus. There is no guarantee that such a process would work for more complex agents such as smallpox, but there is concern that terrorists could make deadly microbes.

The techniques of recombinant genetic technology may also be used to thwart bioterrorism. Scientists can identify unique genetic sequences—"fingerprints," or signatures—of recombinants, which may aid in tracking biological agents and determining their source. Genetic techniques may also help in developing vaccines and treatments, and recombinant DNA technology could be used to create pathogen-resistant crops.

▲ **FIGURE 27.10 One aspect of the response to a bioterrorist attack.** Shown here are biohazard-suited personnel conducting an investigation at the time of the delivery of anthrax-containing mail in Washington, D.C., in October 2001.

MICRO CHECK

10. If a biological agent used in a bioterrorist attack spreads easily on its own through a population, which criteria would this agent meet for being listed as a potential bioweapon?
11. How are threats posed by agricultural pathogens different from those of human pathogens?
12. Which biosafety level would be recommended for microbiological work on MRSA (methicillin-resistant *Staphylococcus aureus*)?

TELL ME WHY

Compare human, animal, and plant pathogens that could be used as biological agents in terms of environmental survivability. Why are animal and plant pathogens more common in environmental reservoirs than are human pathogens?

CLINICAL CASE STUDY

Bioterrorism in the Mail?

As a 56-year-old man, John had experienced plenty of coughs, fever, and nausea, but nothing like this. He was drenched in sweat as well as coughing nonproductively, all while being sick to his stomach. His symptoms were so bad that he had gone to the hospital, thinking he had the flu. After the doctors ruled out the flu, John was placed into airborne isolation immediately. All of his health care workers were dressed with personal protective equipment, as if he had plague from the Middle Ages. He didn't, but he did have a disease nearly as severe—John had inhalational anthrax, a very rare disease. He wasn't the only anthrax patient: several others had been admitted with equally severe conditions in recent days. Bioterrorists had struck in the United States.

Eventually, in October 2001, the Centers for Disease Control and Prevention (CDC) recorded nine other cases of inhalational anthrax. All the cases resulted from a bioterrorist who had mailed endospores of *Bacillus anthracis* in letters and packages.

Anthrax in the lungs progresses rapidly into respiratory distress or shock, and death often occurs within two weeks of the initial onset of symptoms. Diagnosis is difficult because the disease is rare and because the initial symptoms of anthrax mimic influenza or pneumonia.

However, when laboratory scientists discovered large Gram-positive rods in the patients' sputum, they suspected anthrax. The diagnosis was confirmed by PCR specific for *Bacillus anthracis* and antibody detection of the *Bacillus* capsule. Health care workers drained fluid from the lining of the patients' lungs several times over the subsequent week, and administered multiple intravenous antibacterial drugs, including ciprofloxacin and either clindamycin or linezolid), as well as immediate IV infusion of anthrax antitoxin. Thanks to their rapid actions, six of the patients survived—a vast improvement over the historical mortality rate of over 90%. Why was survival so improved? Quick diagnosis and treatment with a combination of antibiotics and antitoxin may be reasons. Biodefense requires constant vigilance as well as innovation.

1. In what bioterrorism threat category does the U.S. government list anthrax?
2. What are the other two types of anthrax?
3. What are advantages of multiple antibiotic therapy, specifically bacteriocidal agents?
4. Why don't doctors treat inhalation anthrax with penicillin or cephalosporins?

MICRO IN THE CLINIC FOLLOW-UP

A New Treatment?

The next week, Penelope and Dr. Kasper meet with the coordinators of the clinical study. Their approach is to examine Penelope's lung microbiome and compare it to microbiomes of healthy patients. The proposed treatment is to try to reestablish a healthy lung microbiome in an effort to decrease the frequency and intensity of Penelope's lung infections. The coordinators point out that the treatment won't cure Penelope of cystic fibrosis, but it might reduce the infections that are leading to tissue damage. If successful, the approach will decrease the likelihood that Penelope will need a lung transplant.

After further discussion with Dr. Kasper and consideration, Penelope decides to enroll in the clinical trial. Although it will take time to determine whether the new approach is effective, Penelope is hopeful that the microbiome-based treatment will lead to fewer and less severe infections and an improved quality of life.

1. **What do you think about the possibility of microbiome-based treatment of diseases like cystic fibrosis? Does this type of approach offer hope for patients? Why or why not?**
2. **What do you envision as some of the challenges to microbiome-based treatment of diseases?**

Check your answers to Micro in the Clinic Follow-Up questions in the Mastering Microbiology Study Area.

> Go to Mastering Microbiology for Interactive Microbiology, Dr. Bauman's Video Tutors, Micro Matters, MicroFlix, MicroBoosters, practice quizzes, and more.

CHAPTER SUMMARY

Environmental Microbiology (pp. 803–812)

1. Microorganisms live in microhabitats within larger **habitats**, or physical localities, in the environment. Organisms and habitats together form **ecosystems**. Single cells give rise to populations, populations performing similar functions form guilds, and many guilds together form a **microbiome**, or community of organisms living in a habitat. The study of the interactions of microorganisms among themselves and with their environment is **microbial ecology**, which is a part of **environmental microbiology** along with studies of microbial habitats.

2. All the ecosystems on Earth form the biosphere. **Biodiversity** describes the number of species living in a given ecosystem, whereas **biomass** refers to the mass of all organisms in an ecosystem.

3. Microbes compete for the scarce resources that characterize the majority of habitats. Some microbes actively oppose the growth

of other microbes (antagonism), but many microbes cooperate, forming complex biofilms.

4. **Bioremediation** is the use of microorganisms to metabolize toxins in the environment to reclaim soils and waterways. Industrial products are either biodegradable or recalcitrant (resistant to degradation by natural means). Recombinant DNA technology enables scientists to create microbes to degrade some recalcitrant chemicals.

5. Acid mine drainage is an environmental problem in places where ores contain iron. Microbial action on leached sulfur or iron in water from mines results in the production of acid and ferric iron deposits that acidify water, which is destructive to most plant and animal life.

6. **Biogeochemical cycling** involves the movement of elements and nutrients from unusable forms to usable forms by the activities of microorganisms. These processes involve production of new biomass, consumption of existing biomass, and decomposition of dead biomass for reuse in the cycle. The four major biogeochemical cycles are the carbon, nitrogen, sulfur, and phosphorus cycles. The cycling of trace metals is also important.

7. In the **carbon cycle**, CO_2 is fixed by photoautotrophs and chemoautotrophs into organic molecules, which are used by other organisms. Organic carbon is converted back to CO_2 in aerobic respiration, by decomposition, and by combustion.

8. In the **nitrogen cycle**, nitrogen gas in the atmosphere is converted to ammonia via a process called **nitrogen fixation**. Ammonia may be converted to nitrate via a two-step process called **nitrification**. Organisms also use ammonia and nitrate to make nitrogenous compounds. Such compounds in wastes and dead cells are converted back to ammonia via **ammonification**. Nitrate can be converted to nitrogen gas by **denitrification**. **Anammox** prokaryotes oxidize ammonium anaerobically into nitrogen.

9. In the **sulfur cycle**, sulfur moves between several inorganic oxidation states (primarily H_2S, SO_4^{2-}, and S^0) and proteins.

10. The **phosphorus cycle** involves the conversion of PO_4^{3-} among organic and inorganic forms.

11. **Eutrophication**—the overgrowth of microorganisms in aquatic systems—can result from the presence of excess nitrogen and phosphorus, which act as fertilizers. The overgrowth of microbes depletes the oxygen in the water, resulting in the death of fish and other animals.

12. **Biomining** is a process that uses microorganisms (usually archaea) to oxidize metals in rocks to make a soluble ion. Mineral-laden water is collected and subjected to a reducing regimen, and the metal ions solidify.

13. Soil microbiology is the study of the roles of microbes in the ground. Soils are fairly diverse and differ greatly in nutrients, water content, pH, oxygen content, and temperature. Microbes inhabit topsoil in high numbers and are less abundant in deeper rock and sediments.

14. Pathogenic microorganisms found in soil can be acquired through contact, but often disease follows the consumption of contaminated soil in food or water.

15. Some human diseases caused by soil microbes are anthrax (bacterial), histoplasmosis (fungal), and *Hantavirus* pulmonary syndrome. Soil diseases of plants include rots, blights, and galls.

16. Aquatic habitats include freshwater and marine water systems. Scientists recognize four zones in freshwater based on temperature, light, and nutrient levels: the nutrient-rich **littoral zone** along the shore, the sunlit **limnetic zone** at and near the surface, the **profundal zone** just below the limnetic zone, and the **benthic zone** on the bottom, which is generally devoid of light and nutrients. Marine environments have the same four zones plus an **abyssal zone** (below the benthic zone), which is virtually devoid of life except around **hydrothermal vents**.

17. Microorganisms living in aquatic environments typically form biofilms to better accumulate nutrients that are limiting in most nonpolluted water systems.

Biological Warfare and Bioterrorism (pp. 812–816)

1. Of the relatively few microorganisms that can cause disease in humans, animals, and plants, some might be used to purposely infect individuals and are thus potential agents of biological warfare and **bioterrorism**. It is illegal to deploy such weapons. **Agroterrorism** is the deliberate infection of livestock or crops.

2. The degree to which an organism is considered a biological threat depends on several criteria concerning public health impact, dissemination or delivery potential, public perception, and public health preparedness.

3. The National Institute of Allergy and Infectious Diseases (NIAID) categorizes human pathogens by threat level. Category A agents have the greatest potential to be used for bioterrorism, and category C refers to agents whose threat potential needs further study.

4. Defense against bioterrorism begins with surveillance—the reporting and monitoring required for effective response to biological attacks. To limit the impact of any attack that may occur, diagnoses must be reliable, and efficient control measures must be in place.

5. Recombinant genetic technology could potentially lead to the development of novel agents or the modification of existing agents to make them more difficult to control in the event of an attack. Such technology could also lead to potential vaccines, cures, or pathogen-resistant crops.

QUESTIONS FOR REVIEW

Answers to the Questions for Review (except Short Answer questions) begin on p. A-1.

Multiple Choice

1. A microbiome is composed of _____.
   a. single, pure populations
   b. all organisms in a locale
   c. all microbes in a location
   d. only the bacteria are considered

2. When examining the microbiome of the human intestine, you would expect to find how many bacterial species present?
   a. one
   b. three
   c. hundreds
   d. thousands

3. In the environment, nutrients are generally _____.
   a. limiting
   b. present in excess
   c. stable
   d. artificially induced

4. Most chemical elements exist in the environment as _____.
   a. usable forms in soil and rock
   b. usable forms in water
   c. unusable forms in soil and rock
   d. unusable forms in water

5. In the carbon cycle, microbes convert _____.
   a. CO_2 into organic material for consumption
   b. CO_2 into inorganic material for storage
   c. fossil fuels into usable organic compounds
   d. oxygen into water as a by-product of photosynthesis

6. Nitrification converts _____.
   a. organic nitrogen to NH_3
   b. NH_3 to NH_4^+
   c. NH_4^+ or NH_3 to NO_3^-
   d. NO_3^- to N_2

7. In aquatic environments, most microbial life is found in the _____.
   a. littoral zone
   b. limnetic zone
   c. profundal zone
   d. benthic zone

8. Which of the following diseases is *not* caused by NIAID category A biological weapons agents?
   a. smallpox
   b. plague
   c. Q fever
   d. tularemia

9. Of the following characteristics, which would contribute most to making a microorganism an effective biological warfare agent?
   a. is readily available in the environment
   b. can be spread by contact after original dissemination
   c. cannot be treated well outside a hospital
   d. is easily identified by symptoms

10. Anammox reactions are _____.
    a. anaerobic and part of nitrogen cycling
    b. anaerobic and part of carbon cycling
    c. aerobic and part of sulfur cycling
    d. aerobic and part of metal ion oxidation

Modified True/False

Indicate whether each of the following statements is true or false. Rewrite the underlined word to make a false statement true.

1. _____ Biofilms of microorganisms form in <u>aquatic</u> environments only.

2. _____ <u>Cooperation</u> is common among microorganisms living in microhabitats.

3. _____ Aquatic microorganisms are <u>more</u> prevalent near the surface than at the bottom of waterways.

4. _____ <u>Abyssal</u> organisms are found near shores of oceans.

5. _____ Plants and bacteria start the carbon cycle with <u>consumption</u>.

Fill in the Blanks

1. Leaching of compounds from mine tailings often results in the oxidation of two elements: _____ and _____.

2. Biogeochemical cycling involves three primary steps: _____, _____, and _____.

3. Nitrogen exists primarily as _____ in the environment.

4. Organisms typically make use of phosphorus in the form of _____.

5. *Carbon fixation* is the conversion of _____ into organic compounds.

VISUALIZE IT!

1. Label the general phases in the carbon cycle.

2. Label the processes of the nitrogen cycle.

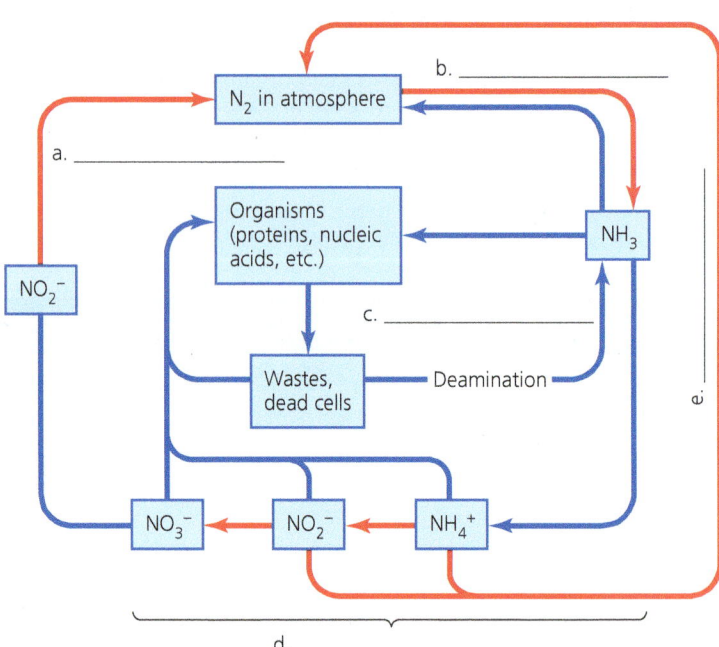

CRITICAL THINKING

1. Explain why influenzaviruses could be potentially devastating biological weapons.

2. Microbes are found mostly in topsoil, but some are found miles deep in bedrock. Nutritionally, how do deeply buried microbes survive?

3. What effect might explain the fact that the intestinal microbiome of humans is typically stable over time, despite constant intake of new microbes in food?

4. How has the technology of next-generation DNA sequencing advanced the field of microbial ecology so significantly?

5. Given the fact that microbes are present in almost every niche on Earth, why is it not surprising that bacterial genomes contain a vast diversity of genes (as you learned in Chapter 7)?

CONCEPT MAPPING

Using the terms provided, fill in this concept map that describes microbial ecology and biogeochemical cycles. You can also complete this and other concept maps online by going to the Mastering Microbiology Study Area.

Autotrophs
Carbon cycle
Guilds

Metabolic activity
Nitrogen cycle
Phosphorus cycle

Pollution
Sulfur cycle

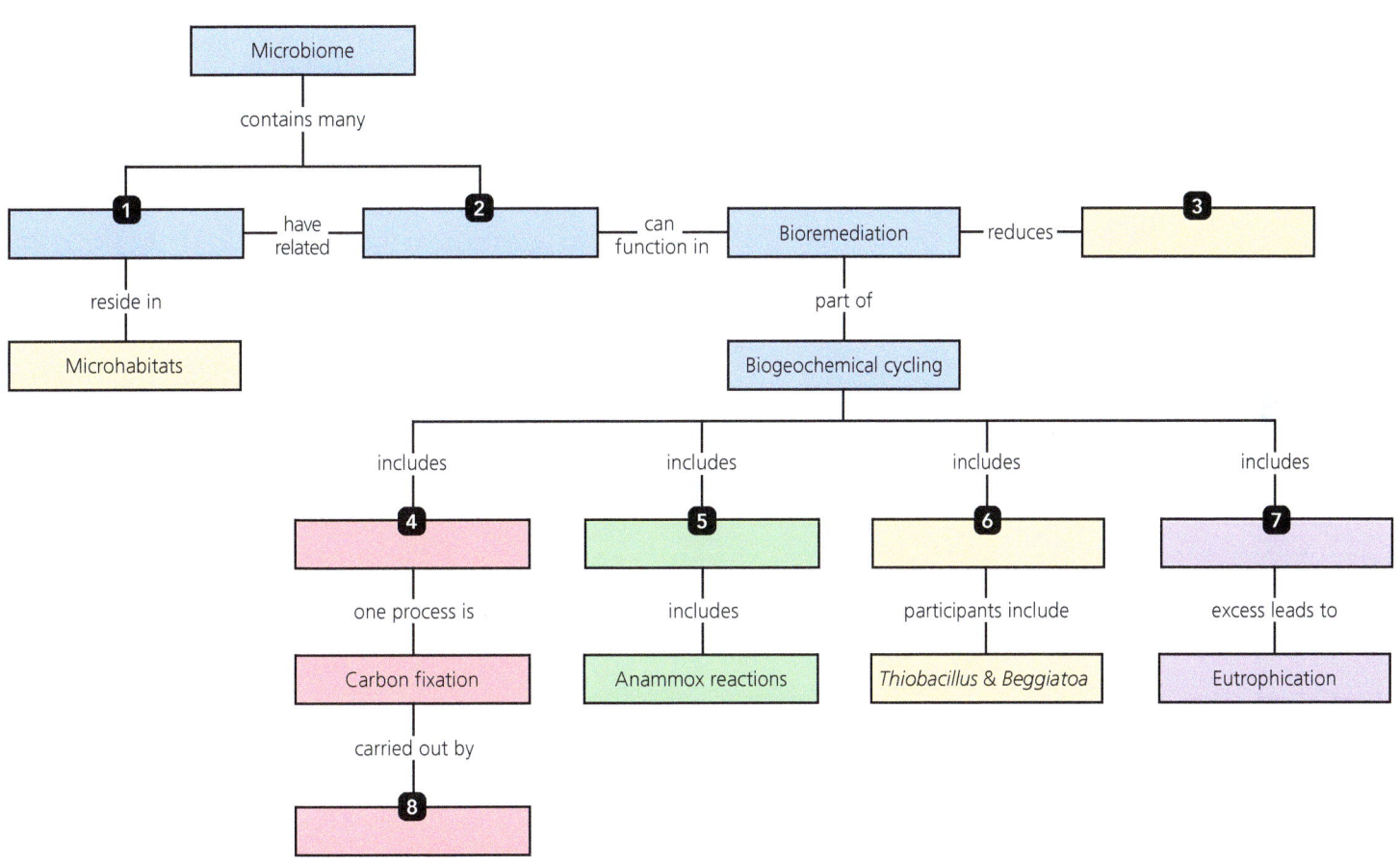

ANSWERS TO QUESTIONS FOR REVIEW

Answers to Multiple Choice, Fill in the Blanks, Matching, True/False, Visualize It!, and Concept Mapping questions are listed here. Answers to Short Answer and Critical Thinking questions are available for instructors only in the Instructor's Manual that accompanies this text.

CHAPTER 1
Multiple Choice
1. a; 2. c; 3. d; 4. a; 5. c; 6. d; 7. a; 8. b; 9. d; 10. d

Fill in the Blanks
1. Martinus Beijerinck and Sergei Winogradsky; 2. Louis Pasteur and Eduard Buchner; 3. Paul Ehrlich; 4. Edward Jenner; 5. John Snow; 6. Robert Koch; 7. John Snow; 8. Louis Pasteur; 9. Louis Pasteur

Visualize It!
1. 1. cilium; 2. flagellum; 3. pseudopod; 4. nucleus; 2. Microbes were only on dust in neck's bend.

Matching
1. J; 2. H; 3. M; 4. C, H, K; 5. B; 6. A; 7. C; 8. E; 9. D; 10. D; 11. I; 12. L

Concept Mapping
1. Eukaryotes; 2. Molds; 3. Yeasts; 4. Protozoa; 5. Unicellular; 6. Algae; 7. Unicellular; 8. Multicellular; 9–10. Bacteria, Archaea; 11–12. Acellular, Obligate intracellular parasites

CHAPTER 2
Multiple Choice
1. c; 2. d; 3. c; 4. c; 5. b; 6. c; 7. a; 8. a; 9. a; 10. c

Fill in the Blanks
1. valence; 2. nonpolar covalent; 3. ATP; 4. fat, wax, amylose (starch), glycogen; 5. functional groups; 6. hydrolysis; 7. exothermic; 8. products; 9. pH; 10. ribose

Visualize It!
1. Primary structure: light blue strands. Secondary structure: green α-helices, gray β-pleated sheets. The entire molecule represents tertiary structure; 2. See Figure 2.21; every angle formed by lines without a lettered designation represents a carbon atom. Purple box = amino group, orange box = carboxyl group, blue box = side group

Concept Mapping
1. RNA; 2–4. messenger RNA, ribosomal RNA, transfer RNA; 5. Ribonucleotides; 6. Ribose; 7. Nitrogenous bases; 8. Guanine; 9. DNA; 10. Double-stranded; 11. Deoxyribonucleotides; 12. Deoxyribose or Phosphate; 13. Nitrogenous bases; 14. Deoxyribose or Phosphate; 15–16. Adenine, Cytosine

CHAPTER 3
Multiple Choice
1. b; 2. b; 3. c; 4. c; 5. c; 6. d; 7. a; 8. a; 9. a; 10. b; 11. c; 12. d; 13. d; 14. b; 15. c

Matching
1. D Glycocalyx; B, H, I Flagella; F Axial filaments; H Cilia; A Fimbriae; C, G Pili; E Hami; 2. A Ribosome; D Cytoskeleton; F Centriole; E Nucleus; I Mitochondrion; G Chloroplast; C Endoplasmic reticulum; H Golgi body; B Peroxisome

Visualize It!
1. a. cytoplasm—contains metabolic chemicals; b. nucleoid—site of DNA (genes); c. glycocalyx—adhesion; d. cell wall—protects against osmotic forces; e. inclusions—stored chemicals; f. flagellum—motility; g. cytoplasmic membrane—controls import and export; h. nucleolus—site of RNA synthesis; i. cilium—motility; j. 80S ribosomes—make proteins; k. nuclear envelope—bounds DNA (genes); l. mitochondrion—makes ATP (energy source); m. centriole—plays a role in cell division; n. Golgi body—packages secretions; o. rough endoplasmic reticulum (RER)—transports proteins; p. smooth endoplasmic reticulum (SER)—lipid synthesis; q. cytoskeleton—helps maintain cell shape; 2. a. axial filament (endoflagella); b. polar (tuft); c. peritrichous; d. polar (single); 3. Chemical A enters the cell via facilitated diffusion through a protein channel, which is likely specific for chemical A. Chemical B does not have such a route. At some concentration of chemical A, all the protein channels are filled, such that the rate of diffusion cannot increase. The cell could increase the rate of diffusion of chemical A by inserting more of the channel protein into the membrane. The rate of diffusion of chemical B could be increased by removing chemical B from the cytoplasm at an increased rate. This would have the effect of increasing its concentration gradient.

Concept Mapping
1. Gram-positive cell wall; 2. Teichoic acids; 3. Gram-negative cell wall; 4. Periplasm; 5. Peptidoglycan; 6. Glycan chains; 7. N-acetylglucosamine; 8. Lipopolysaccharide (LPS); 9. Porins; 10. Lipid A

CHAPTER 4
Multiple Choice
1. c; 2. d; 3. c; 4. d; 5. d; 6. d; 7. b; 8. a; 9. a; 10. d

Fill in the Blanks
1. 600×; 2. heat fixation; 3. increases, increases, more; 4. Contrast; 5. negatively

Visualize It!
1. a. scanning electron; b. bright-field light; c. phase-contrast light; d. fluorescence light; e. transmission electron; f. differential interference contrast (Nomarski); 2. See Figure 4.4.

Concept Mapping
1. Iodine; 2. Ethanol and acetone; 3. Safranin; 4. Primary stain; 5. Decolorizer; 6. Gram-positive bacteria; 7. Purple; 8. Gram-negative bacteria; 9. Thin peptidoglycan layer

CHAPTER 5
Multiple Choice
1. c; 2. a; 3. c; 4. a; 5. a; 6. c; 7. b; 8. d; 9. a; 10. c; 11. d; 12. d; 13. a; 14. a; 15. a; 16. c; 17. a; 18. c; 19. a; 20. c

Matching
1. C; 2. B; 3. E; 4. A

Fill in the Blanks
1. the original reaction center chlorophyll; 2. 2; 3. pentose phosphate, Entner-Doudoroff; 4. The citric acid cycle; 5. Oxygen or $\frac{1}{2}O_2$; 6. NO_3^-, SO_4^{2-}; CO_3^{2-}; 7. inorganic; 8.

Category of Enzyme	Description
Hydrolase	Catabolizes substrate by adding water
Isomerase	Rearranges atoms
Ligase/polymerase	Joins two molecules together
Transferase	Moves functional groups
Oxidoreductase	Adds or removes electrons
Lyase	Splits large molecules

9. chemiosmosis; 10. NAD^+, FAD

Visualize It!
1. a. Glycolysis (cytosol); b. Electron transport chains (cristae—inner mitochondrial membranes); c. citric acid cycle (matrix); 2. See Figure 5.13; net ATP production: glycolysis, 2; citric acid cycle, 2; electron transport chain, 34; 3. Erythrose 4-phosphate is from the pentose phosphate pathway. PEP is from glycolysis.

Concept Mapping
1. Glycolysis; 2. Synthesis of acetyl-CoA; 3. Citric acid cycle; 4. Electron transport chain; 5. Chemiosmosis; 6. 2 ATP; 7. 2 ATP; 8. 34 ATP; 9. Substrate-level phosphorylation; 10. Substrate-level phosphorylation; 11. Oxidative phosphorylation

CHAPTER 6
Multiple Choice
1. b; 2. c; 3. b; 4. a; 5. b; 6. b; 7. d; 8. a; 9. c; 10. a; 11. b; 12. d; 13. c; 14. a; 15. a

Fill in the Blanks
1. electrons; 2. singlet; 3. nitrogen; 4. Growth factors; 5. minimum growth temperature; 6. osmotic; 7. halophiles; 8. Carotenoid; 9. fixation; 10. streak plate; 11. inorganic

Visualize It!
1. See Figure 6.2; 2. Beta-hemolysis

Concept Mapping
1. General purpose; 2. Trypticase soy agar; 3. Enriched; 4. Fastidious microorganisms; 5. Differential; 6. Fermentation broths (phenol red); 7. Selective; 8. Unwanted microorganisms; 9. Selective and differential

CHAPTER 7
Multiple Choice
1. a; 2. c; 3. d; 4. c; 5. d; 6. a; 7. a; 8. d; 9. a; 10. c; 11. c; 12. d; 13. a; 14. d; 15. c; 16. b; 17. d; 18. c; 19. b; 20. c; 21. b; 22. b; 23. a; 24. d; 25. b

AN-1

ANSWERS

Fill in the Blanks
1. initiation of transcription, elongation of the RNA transcript, termination of transcription; 2. codon; 3. silence, missense, nonsense; 4. frameshift; 5. promoter, operator, a series of genes; 6. inducible; 7. semiconservative; 8. transformation, transduction, bacterial conjugation; 9. Transposons; 10. Crossing over; 11. Transfer; 12. small interfering, micro

Visualize It!
1. a. replication fork; b. stabilizing proteins; c. nucleotide (triphosphate); d. leading strand; e. helicase; f. primase; g. DNA polymerase III; h. RNA primer; i. Okazaki fragments; j. DNA polymerase I; k. lagging strand; l. DNA ligase; 2. GC base pairing is more stable than AT pairing because GC base pairs are held together by three hydrogen bonds. Therefore, the portion of the pictured molecule that appears as a single, thick structure has more GC base pairs than does the portion that has separated into two strands, forming a loop; 3. These drugs are analogs of the nucleotide bases in DNA. When incorporated into a strand of DNA, the drugs prevent the replication of new DNA strands because they lack an —OH on the 3′ carbon of the sugar needed for continued elongation.

Concept Mapping
1. One or a few nucleotide base pairs; 2. Change in DNA sequence; 3. Inserted; 4. Deleted; 5. Substituted; 6. Insertion mutation; 7. Substitution mutation; 8. Effect on amino acid sequence; 9. Silent mutation; 10. Nonsense mutation; 11. Incorrect amino acid substituted

CHAPTER 8

Multiple Choice
1. d; 2. b; 3. c; 4. d; 5. a; 6. c; 7. c; 8. b; 9. a; 10. d

Modified True/False
1. cut; 2. True; 3. Electrophoresis; 4. True; 5. Southern blotting

Visualize It!
1. Step 1, denaturation: 94°C; step 2, priming: DNA primer, deoxyribonucleotide triphosphates, DNA polymerase, cool to 65°C; step 3, extension: same reagents as step 2, 72°C; step 4, repeat.

Concept Mapping
1. Mutagens; 2. Reverse transcriptase; 3. Restriction enzymes; 4. Blunt ends; 5–6. Transposons, Plasmids; 7. Southern blot; 8. DNA probes; 9–10. Radioactive, Fluorescent; 11–13. Electroporation, Protoplast fusion, Injection

CHAPTER 9

Multiple Choice
1. a; 2. b; 3. b; 4. d; 5. d; 6. a; 7. d; 8. d; 9. c; 10. a; 11. d; 12. d; 13. a; 14. c; 15. c; 16. b; 17. a; 18. d; 19. d; 20. a

Visualize It!
1. a. 1 min; b. 2.5 minutes; 2. Ions of copper and zinc from the brass have antimicrobial effects.

Concept Mapping
1. Boiling water; 2. Thermoduric microorganisms; 3–6. Vegetative bacterial cells, Fungi, Most viruses, Protozoan trophozoites; 7. Autoclave; 8. Sterilization technique; 9. Pasteurization; 10. Thermoduric microorganisms; 11. 63°C, 30 min; 12. Flash pasteurization; 13. Ultra-high-temperature pasteurization; 14. Equivalent treatments; 15. Ultra-high-temperature sterilization; 16. 140°C, 1–3 sec

CHAPTER 10

Multiple Choice
1. d; 2. a; 3. a; 4. c; 5. d; 6. c; 7. a; 8. d; 9. a; 10. d

Visualize It!
1. See Figure 10.4; 2. Etest; antimicrobial sensitivity and minimum inhibitory concentration; increasing the drug concentration would result in a zone of inhibition wider and longer than shown. Your oval should extend to at least 0.75.

Concept Mapping
1. Mutation; 2. Cell division; 3–5. Conjugation, Transduction, Transformation; 6. Pathogen's enzymes; 7. Beta-lactamase; 8. Penicillin; 9. Entry of antimicrobials into cell; 10. Altered targets; 11. Efflux pumps

CHAPTER 11

Modified True/False
1. asexually; 2. vibrio; 3. True; 4. lack; 5. rRNA; 6. True; 7. True; 8. True; 9. *Epulopiscium*; 10. True

Multiple Choice
1. c; 2. a; 3. c; 4. d; 5. a; 6. d; 7. a; 8. b; 9. c; 10. d

Matching
1. E; 2. B; 3. C; 4. D; 5. L; 6. K; 7. H; 8. Q; 9. I; 10. N; 11. P; 12. O; 13. M; 14. F; 15. A

Visualize It!
1. See Figure 11.1; 2. (a) central (b) subterminal

Concept Mapping
1. Prokaryotes; 2. Peptidoglycan; 3. Disease; 4. Extremophiles; 5. Thermophiles; 6. >80°C; 7. Hydrothermal vents; 8. Acidophiles; 9. Halophiles; 10. Great Salt Lake; 11. Methane; 12. Animal colons; 13. Sewage treatment

CHAPTER 12

Multiple Choice
1. a; 2. d; 3. c; 4. b; 5. a; 6. d; 7. b; 8. c; 9. a; 10. c; 11. c; 12. a; 13. d; 14. a; 15. d

Matching
First section: 1. E; 2. B; 3. D; 4. C; 5. A
Second section: 1. A; 2. F; 3. E; 4. C; 5. D; 6. B
Third section: 1. C; 2. E; 3. B; 4. D; 5. A

Visualize It!
1. a. ascospore, sexual; b. basidiospore, sexual; c. conidia, asexual; d. sporangiospore, asexual; 2. See Figure 12.17.

Fill in the Blanks
1. protozoology; 2. mycology; 3. phycology; 4. mycoses; 5. radiolarians

Concept Mapping
1. Protozoa; 2. Unicellular; 3–5. *Cryptosporidium*, *Giardia*, *Plasmodium*; 6. Fungi; 7. Yeasts; 8. Molds; 9. Chitin; 10. Algae; 11–12. Colonial, Unicellular; 13. Cell walls; 14–15. Alginic acid, Carrageenan

CHAPTER 13

Multiple Choice
1. c; 2. c; 3. a; 4. a; 5. b; 6. c; 7. a; 8. d; 9. d; 10. b

Matching
1. H; 2. G; 3. C; 4. B; 5. D; 6. E; 7. F; 8. A; 9. J; 10. I

Visualize It!
1. See Figure 13.8; 2. See Figure 13.5.

Concept Mapping
1. Attachment; 2. Uncoating; 3. Synthesis; 4. Release; 5. Host cell; 6–7. Endocytosis, Fusion; 8. Viral nucleic acid; 9–14. dsDNA, +ssRNA, +ssRNA (retrovirus), −ssRNA, dsRNA; 15. Viral proteins; 16. New virions; 17–18. Budding, Exocytosis

CHAPTER 14

Multiple Choice
1. a; 2. b; 3. b; 4. a; 5. d; 6. a; 7. d; 8. c; 9. a; 10. d; 11. d; 12. b; 13. c; 14. d; 15. b

Fill in the Blanks
1. pathogen; 2. asymptomatic or subclinical; 3. etiology; 4. epidemiology; 5. zoonoses; 6. fomites; 7. Healthcare-associated; 8. prevalence; 9. biological; 10. lipopolysaccharide (LPS) or lipid A

Visualize It!
1. a. Endemic (or sporadic); b. Epidemic; c. Pandemic; 2. The red epidemic affected more people during the first 3 days. The red epidemic has a short incubation time and is much more contagious; as a result, the susceptible population was infected in a short time. The blue epidemic has a longer incubation time than the red, resulting in a longer time until cases show up and a longer time over which new infections are occurring.

Concept Mapping
1. Contact transmission; 2. Body; 3. Indirect contact; 4. Droplet transmission; 5. Vehicle transmission; 6–7. Waterborne, Foodborne; 8. Vector transmission; 9. Arthropod; 10. Biological; 11. Mosquito

CHAPTER 15

Multiple Choice
1. d; 2. d; 3. b; 4. a; 5. d; 6. c; 7. d; 8. b; 9. a; 10. d

Modified True/False
1. dead; 2. True; 3. True; 4. True; 5. monocytes; 6. True; 7. pathogen; 8. ingestion; 9. phagolysosomes; 10. pathogen's cytoplasmic membrane; 11. inflammation; 12. True; 13. True; 14. antimicrobial peptides; 15. True

Matching
First section: 1. B; 2. B; 3. B; 4. B; 5. A; 6. B; 7. A; 8. B; 9. A; 10. A; 11. A; 12. B; 13. C; 14. C; 15. A
Second section: 1. J; 2. E; 3. G; 4. D; 5. C; 6. H; 7. A; 8. B; 9. F; 10. I

Visualize It!
1. See Figure 15.6; 2. See Figure 15.5.

Concept Mapping
1–4. Dendritic cells, Eosinophils, Macrophages, Neutrophils; 5. Chemotaxis; 6. Adhesion; 7. Ingestion; 8. Killing; 9. Elimination; 10. Chemotactic factors; 11. Opsonins; 12. Phagosome; 13. Lysosome; 14. Phagolysosome; 15. Exocytosis

CHAPTER 16

Multiple Choice
1. b; 2. e; 3. e; 4. b; 5. d; 6. a; 7. c; 8. d; 9. e; 10. a

Modified True/False
1. antigen-presenting cells; 2. True; 3. cytotoxic; 4. Plasma cells; 5. antibody

Matching
First section: 1. D; 2. C; 3. B; 4. A
Second section: 1. D; 2. A; 3. B; 4. C

Visualize It!
1. See Figures 16.10 and 16.13; 2. MHC I and MHC II are found on the cytoplasmic membrane; pseudopods are the thin, finger-like extensions of the cell; vesicles are the white structures within the cell.

Concept Mapping
1. Secreted immunoglobulins; 2. Plasma cells; 3. Antigen-stimulated B cells; 4. Antigens; 5. Antibody-dependent cell-mediated cytotoxicity (ADCC) 6. Neutralization; 7. Phagocytosis; 8. Complement activation; 9. Inflammation; 10. Agglutination; 11–14. IgA, IgD, IgE, IgG, IgM

CHAPTER 17
Multiple Choice
1. d; 2. c; 3. b; 4. c; 5. d; 6. d; 7. d; 8. a; 9. b; 10. d; 11. a; 12. d; 13. d; 14. d; 15. c

True/False
1. False; 2. False; 3. True; 4. True; 5. False

Matching
1. D; 2. C; 3. A; 4. B; 5. C; 6. A

Visualize It!
1. See Figure 17.13; 2. Patients 5, 6, and 11 are most likely uninfected; patients 1 and 7 may be uninfected. The other patients have likely been infected.

Concept Mapping
1. Active immunity; 2. Attenuated vaccines; 3–4. Measles vaccine, Varicella vaccine; 5. Microorganisms; 6. Inactivated vaccines; 7–8. Pertussis, Polio; 9. Immunizations; 10. Subunit vaccines; 11. Adjuvants; 12. Hepatitis B vaccine; 13. Modified toxins; 14. Tetanus toxoid

CHAPTER 18
Multiple Choice
1. e; 2. c; 3. d; 4. c; 5. c; 6. e; 7. b; 8. d; 9. c; 10. e

Modified True/False
1. Histamine, kinins, and/or proteases are; 2. True; 3. red blood; 4. type IV; 5. allograft or xenograft

Matching
1. I; 2. IV; 3. III; 4. IV; 5. 0; 6. IV; 7. I; 8. III; 9. I; 10. I

Visualize It!
1. See Figure 18.12; 2. a. systemic lupus erythematosus (SLE), b. positive tuberculin test, c. rheumatoid arthritis, d. urticaria (hives)

Concept Mapping
1–2. Type I hypersensitivity, Allergies; 3. Allergens; 4–7. Bee venom, Dust mites, Peanuts, Pollen; 8–9. Basophils, Mast cells; 10. IgE; 11–13. Increased vascular permeability, Smooth muscle contraction, Vasodilation; 14–16. Asthma, Hay fever, Urticaria; 17. Anaphylactic shock; 18–19. Antihistamine, Epinephrine

CHAPTER 19
Multiple Choice
1. c; 2. b; 3. d; 4. d; 5. c; 6. b; 7. d; 8. a; 9. d; 10. b; 11. a

Matching
1. A; 2. A; 3. B, F; 4. B; 5. B; 6. B; 7. B; 8. B; 9. B; 10. G; 11. H; 12. F; 13. C; 14. B, I; 15. E

Visualize It!
1. a. *Actinomyces*; b. *Staphylococcus*; c. *Bacillus*; d. *Mycobacterium*; e. *Streptococcus*; f. *Clostridium*; 2. See Figure 19.13.

Concept Mapping
1. *Mycobacterium tuberculosis*; 2. Primary infection; 3. Dormant infection; 4. Disseminated tuberculosis; 5–7. Cough, Blood-tinged sputum, Extensive lung damage; 8–11. Acid-fast stain of sputum, Chest X ray, Culture on special media, Tuberculin skin test; 12. For 26 weeks; 13–15. Isoniazid (INH), Rifampin, Pyrazinamide; 16–17. Ethambutol, Streptomycin; 18. Antibiotic resistance; 19–20. MDR-TB, XDR-TB

CHAPTER 20
Multiple Choice
1. a; 2. d; 3. c; 4. b; 5. c; 6. d; 7. a; 8. a; 9. d; 10. b; 11. a; 12. a

Matching
1. C; 2. F; 3. D; 4. B; 5. E; 6. A

Visualize It!
1. See Figure 20.9; 2. Top: Gram-negative, lactose negative; Bottom: Gram-negative, lactose positive

Concept Mapping
1. *Haemophilus influenzae*; 2. *Neisseria meningitidis*; 3. Respiratory route; 4. Healthy carriers; 5. Infants and young children; 6. Vaccines; 7. Hib vaccine; 8–9. Gram stain, Serological tests; 10. Cerebrospinal fluid; 11. Antimicrobials; 12. Penicillin G

CHAPTER 21
Multiple Choice
1. c; 2. a; 3. b; 4. d; 5. d; 6. a; 7. d; 8. c; 9. b; 10. c; 11. b; 12. b; 13. b

Visualize It!
1. See Figure 21.3a; 2. See Figure 21.3b.

Matching
1. C spotted fever rickettsiosis; A Murine typhus; B Epidemic typhus; D Scrub typhus; E Ehrlichiosis; 2. A *Rickettsia typhi*; B *Rickettsia prowazekii*; C *Rickettsia rickettsii*; D *Orientia tsutsugamushi*; C *Ehrlichia chaffeensis*; C *Borrelia burgdorferi*; B *Borrelia recurrentis*; C *Anaplasma phagocytophilum*; 3. G *Chlamydophila psittaci*; C *Chlamydophila pneumoniae*; B, D, E, F *Chlamydia trachomatis*; A *Treponema pallidum pallidum*; H *Treponema pallidum pertenue*; I *Treponema pallidum endemicum*; J *Treponema carateum*; K *Borrelia burgdorferi*; 4. E Peptic ulcers; A, B, D Gastroenteritis; C Blood poisoning; A Cholera

Concept Mapping
1. *Treponema pallidum*; 2. Spirochete; 3. Dark-field microscopy; 4. Primary syphilis; 5. Chancre; 6. Secondary syphilis; 7. Body-wide rash; 8. Latent stage; 9. Never progress; 10. Infect fetus in pregnant women; 11. Tertiary syphilis; 12–14. Cardiovascular syphilis, Gummas, Neurosyphilis; 15. Serological tests; 16. Penicillin

CHAPTER 22
Multiple Choice
1. a; 2. c; 3. d; 4. c; 5. b; 6. d; 7. c; 8. b; 9. c; 10. b; 11. b; 12. c; 13. b; 14. a; 15. b; 16. c; 17. d; 18. b; 19. a; 20. b

Modified True/False
1. True; 2. difficult; 3. Systemic mycoses; 4. True; 5. True; 6. True; 7. True; 8. immunocompromised; 9. True; 10. 3–10% of individuals have

Fill in the Blanks
1. mycelial, yeast; 2. *Blastomyces dermatitidis, Coccidioides immitis, Histoplasma capsulatum, Paracoccidioides brasiliensis*; 3. ergosterol; 4. Mycetomas; 5. *Sporothrix schenckii*; 6. *Cryptococcus neoformans*; 7. *Aspergillus, Candida, Cryptococcus, Pneumocystis, Mucor*; 8. *Candida*; 9. protozoan, fungus; 10. *Aspergillus*

Matching
1. A, C; 2. B; 3. C; 4. A; 5. A; 6. B; 7. A, C; 8. A; 9. D; 10. C; 11. C

Visualize It!
1. See Figures 22.3, 22.5, and 22.7; 2. a. *Coccidioides*, b. *Candida*, c. *Histoplasma*

Concept Mapping
1. Amphotericin B; 2. Inhalation; 3. Other organs; 4–7. *Blastomyces dermatitidis, Coccidioides immitis, Histoplasma capsulatum, Paracoccidioides brasiliensis*; 8. Opportunistic fungi; 9. *Candida albicans*; 10. *Cryptococccus neoformans*; 11. *Pneumocystis jirovecii*; 12. Voriconazole; 13. Amphotericin B; 14. 5-fluorocytosine; 15. Trimethoprim/Sulfamethoxazole

CHAPTER 23
Multiple Choice
1. d; 2. b; 3. a; 4. a; 5. d; 6. b; 7. b; 8. a; 9. a; 10. d; 11. b; 12. b; 13. c; 14. b; 15. c; 16. a; 17. b; 18. d; 19. a; 20. d

Modified True/False
1 Ingestion; 2. corneal scrapings, cerebrospinal fluid, or biopsy material; 3. True; 4. True; 5. True; 6. True; 7. *solium*; 8. water plants; 9. True; 10. True

Fill in the Blanks
1. cilia; 2. *Toxoplasma gondii*; 3. *brucei, cruzi*; 4. *Entamoeba*; 5. hypnozoites; 6. *Wuchereria*; 7. *Ancylostoma, Necator, Schistosoma*; 8. *Fasciola*; 9. *Ancylostoma*; 10. *Enterobius vermicularis* and *Wuchereria bancrofti*

Matching
1. G; 2. E; 3. A; 4. H; 5. C, G; 6. C, G, I; 7. F; 8. B; 9. J; 10. D

Visualize It!
1. See Disease in Depth: Malaria on pp. 688–689. 2. a. *Giardia* b. *Enterobius* c. *Plasmodium* d. *Schistosoma* e. *Trichomonas* f. *Trypanosoma* g. *Entamoeba* h. *Cryptosporidium*

Concept Mapping
1. Diarrhea; 2. Fecal-oral route; 3. Ciliates; 4. Flagellates; 5. Apicomplexans; 6. *Entamoeba histolytica*; 7. *Crytosporidium parvum*; 8. *Cyclospora cayetanensis*; 9. People with AIDS; 10. Schizonts; 11. Reproducing/feeding stage; 12. Dormant stage; 13. Trophozoites; 14. Water; 15. Environment; 16. New host

CHAPTER 24
Multiple Choice
1. b; 2. c; 3. a; 4. b; 5. d; 6. d; 7. d; 8. c; 9. a; 10. c; 11. a; 12. b; 13. d; 14. d; 15. d

Visualize It!
1. See Figure 24.2; 2. a. chickenpox b. Kaposi's sarcoma c. oral herpes d. whitlow e. shingles f. molluscum contagiosum

Matching
1. B; 2. A; 3. A; 4. A; 5. B; 6. B; 7. B; 8. B; 9. B; 10. B; 11. B; 12. C; 13. B; 14. C; 15. G; 16. D; 17. E; 18. F

Concept Mapping
1. *Simplexvirus*; 2. Latency; 3. Recurrences; 4. Acyclovir; 5. Oral herpes; 6. Fever blisters; 7–8. Gingivostomatitis, Ocular herpes; 9. Herpetic whitlow; 10. *Human herpesvirus 2*; 11. Genital herpes; 12. Life threatening; 13–14. Icosahedral capsid, Envelope

CHAPTER 25
Multiple Choice
1. a; 2. c; 3. d; 4. a; 5. c; 6. d; 7. d; 8. b; 9. a; 10. a; 11. b

Matching
1. K; 2. C; 3. A; 4. G; 5. F; 6. D; 7. I; 8. J; 9. B; 10. E; 11. H

True/False
1. True; 2. False; 3. False; 4. True; 5. True

Visualize It!
1. See Figure 25.20; 2. Flu epidemics occurred in years 10 and 20. The biennial fluctuation is due to antigenic drift; epidemics result from antigenic shift.

Concept Mapping
1. Acute; 2. Chronic; 3. Hepatic cancer; 4. Hepatitis A; 5. Hepatitis A virus (HAV); 6. Mild; 7. Hepatitis A vaccine; 8. Hepatitis B; 9. Hepatitis B virus (HBV); 10. Chronic disease; 11. Blood test for antibodies (IgM); 12. Hepatitis B vaccine; 13. Hepatitis C; 14. Hepatitis C virus (HCV); 15. Carrier state; 16. PCR or blood test for antibodies

CHAPTER 26
Multiple Choice
1. c; 2. d; 3. c; 4. c; 5. b; 6. d; 7. d; 8. b; 9. b; 10. b; 11. d

Modified True/False
1. lactic acid; 2. lactic acid; 3. True; 4. True; 5. True

Fill in the Blanks
1. the food, processing or handling; 2. increases; 3. batch, continuous flow; 4. zero; 5. BOD (biochemical oxygen demand); 6. indicator organisms; 7. food infection, food intoxication; 8. biofuels; 9. biosensor; 10. BOD (biochemical oxygen demand)

Matching
1. C; 2. F; 3. H; 4. B; 5. A; 6. D; 7. G; 8. E

Visualize It!
1. See Figure 26.1; 2. See Figure 26.2.

Concept Mapping
1. Fermentation; 2. Sugars in milk; 3. *Lactobacillus bulgaricus*; 4. Maltose from grains; 5. *Saccharomyces cerevisiae* (yeast); 6. Bread to rise; 7. Alcohol in wine; 8. *Gluconobacter*; 9. Acetic acid in vinegar; 10. Soybeans and wheat

CHAPTER 27
Multiple Choice
1. c; 2. d; 3. a; 4. c; 5. a; 6. c; 7. a; 8. c; 9. b; 10. a

Modified True/False
1. True; 2. Competition; 3. True; 4. Littoral; 5. autotrophy

Fill in the Blanks
1. iron, sulfur; 2. production, consumption, decomposition; 3. dinitrogen gas (N_2); 4. phosphate ion (PO_4^{3-}); 5. inorganic carbon (carbon dioxide, CO_2)

Visualize It!
1. See Figure 27.4; 2. See Figure 27.5

Concept Mapping
1. Guilds; 2. Metabolic activity; 3. Pollution; 4. Carbon cycle; 5. Nitrogen cycle; 6. Sulfur cycle; 7. Phosphorus cycle; 8. Autotrophs

APPENDIX A — ABRIDGED Microbiology in Nursing and Allied Health (MINAH) Undergraduate Curriculum Guidelines

These are abridged from the full MINAH guidelines, which can be found using the QR code located on the inside back cover of this textbook.

Impact of Microorganisms in Health and Disease

1. **The microbiome includes diverse cellular and acellular microbes that impact human health; a microbiome dysbiosis (imbalance) that changes the level, location, or diversity of our microbiota may lead to disease.**
   ASM Recommended Curriculum Guideline: 23
   NCLEX-RN Alignment: Basic Care & Comfort; Physiological Adaptation

2. **Microorganisms are everywhere and live in diverse and dynamic ecosystems, including the human body.**
   Microbes are classified into different taxonomic groups; understanding these groups supports infection management and informs patient care plans. Knowing what parts of the body microbes colonize is important in infection control and for diagnosing infectious diseases.
   ASM Recommended Curriculum Guideline: 20
   NCLEX-RN Alignment: Pharmacological & Parenteral Therapies; Safety & Infection Control

3. **Most bacteria in nature live in biofilm communities.**
   Biofilm production presents unique challenges to health care, such as providing a continuously available pathogen source for renewed infections and conferring resistance to antimicrobial agents. Implanted medical devices are a common location for biofilm formation.
   ASM Recommended Curriculum Guideline: 21
   NCLEX-RN Alignment: Pharmacological & Parenteral Therapies; Reduction of Risk Potential

4. **Microbes interact with human hosts in beneficial, neutral, or detrimental ways.**
   Nurses and allied health workers should understand the microbiological and epidemiological features of pathogens. Host factors impact infectious disease development.
   ASM Recommended Curriculum Guideline: 23
   NCLEX-RN Alignment: Pharmacological & Parenteral Therapies; Safety & Infection Control; Health Promotion & Maintenance; Basic Care & Comfort; Physiological Adaptation

5. **Humans use microorganisms and their products to make pharmaceuticals.**
   ASM Recommended Curriculum Guideline: 26
   NCLEX-RN Alignment: Health Promotion & Maintenance

Microbial Pathogenicity

6. **Pathogens have diverse virulence factors that influence their pathogenesis and impact treatment options and clinical management.**
   Understanding virulence factors and pathogenesis mechanisms allows healthcare workers to identify, properly treat, and reduce infectious disease transmission.
   ASM Recommended Curriculum Guidelines: 8, 9, 10, 23
   NCLEX-RN Alignment: Health Promotion & Maintenance; Safety & Infection Control

7. **Pathogens are continuously evolving and virulence is not a static property. Understanding mechanisms that impact pathogen evolution (e.g., vertical and horizontal genetic variation, mutations, recombination, etc.) is central to limiting pathogen evolution.**
   ASM Recommended Curriculum Guidelines: 2, 3, 15
   NCLEX-RN Alignment: Safety & Infection Control

Identifying and Managing Infectious Diseases

8. **Koch's postulates are used to identify the etiological agent of certain infectious diseases.**
   ASM Recommended Curriculum Guideline: 23
   NCLEX-RN Alignment: Management of Care

9. **A variety of methods are used to identify infectious agents.**
   Serology, diverse molecular methods, biochemical tests, and staining methods are used to definitively diagnose infections.
   ASM Recommended Curriculum Guideline: 34
   NCLEX-RN Alignment: Physiological Adaptation

10. **Vaccines are safe and effective tools for preventing disease.**
    Vaccines come in different formulations, and have different recommended administration schedules; they train the immune system and promote herd immunity against a

particular pathogen. Nurses and other allied health workers must understand how vaccines work and be able to speak intelligibly about vaccines to all stakeholders.
ASM Recommended Curriculum Guideline: 31
NCLEX-RN Alignment: Reduction of Risk Potential; Communication & Documentation; Physiological Adaptation

Healthcare-Acquired Infections and Epidemiology

11. **Healthcare-acquired infections (HAIs, nosocomial infections) are costly and often have a poorer prognosis than community-acquired infections.**
    Standard/universal precautions, transmission precautions, surgical asepsis, and biosafety level precautions limit HAIs and allow healthcare workers to safely manage patients and safely collect/analyze patient samples.
    ASM Recommended Curriculum Guidelines: 23, 37
    NCLEX-RN Alignment: Safety & Infection Control; Health Promotion & Maintenance; Therapeutic Environment

12. **Tracking and reducing the incidence of healthcare-acquired infections is a collaborative effort that saves lives.**
    Epidemiologists use diverse surveillance techniques to monitor certain infectious diseases. Being familiar with emerging and re-emerging pathogens is important in managing potential outbreaks. An appreciation of nationally notifiable diseases is essential for compliance with reporting protocols.
    ASM Recommended Curriculum Guidelines: 23, 37
    NCLEX-RN Alignment: Safety & Infection Control; Health Promotion & Maintenance; Therapeutic Environment

13. **There are many strategies (such as quarantine, vector control, and patient education) to break the epidemiological triangle and prevent disease transmission.**
    ASM Recommended Curriculum Guideline: 23
    NCLEX-RN Alignment: Safety & Infection Control; Behavioral Intervention; Therapeutic Communication

Controlling Microbial Growth to Limit Disease

14. **A microbe's survival and growth in a given environment depends on its metabolic characteristics.**
    Understanding a pathogen's metabolic features helps in recognizing where they can thrive and their potential for introduction into humans.
    ASM Recommended Curriculum Guidelines: 11, 13
    NCLEX-RN Alignment: Physiological Adaptation; Basic Care & Comfort

15. **Microbial growth is controlled using physical, chemical, mechanical, and biological means.**
    Controlling microbial growth lowers healthcare-acquired infection (HAI) incidence. Specific and nonspecific immune defenses are biological controls against pathogens. An understanding of microbial control helps healthcare workers understand how critical, semicritical, and noncritical equipment should be managed, as well as how to properly prepare patient body sites for medical procedures like injections and surgery.
    ASM Recommended Curriculum Guidelines: 7, 8, 14
    NCLEX-RN Alignment: Pharmacological & Parenteral Therapies; Health Promotion & Maintenance; Physiological Adaptation; Basic Care & Comfort

16. **Antimicrobial compounds combat bacteria, fungi, helminths, protozoans, and viruses.**
    Understanding structural and functional microbial features allows us to develop new antimicrobial drugs and assess drug specificity mechanisms to limit adverse drug effects. The type of antimicrobial drug used to treat a particular pathogen depends on patient and microbe features.
    ASM Recommended Curriculum Guidelines: 14, 15
    NCLEX-RN Alignment: Safety & Infection Control

17. **Proper stewardship of antimicrobial drugs is essential to limit antimicrobial resistance.**
    Antimicrobial drug stewardship includes testing and tracking antimicrobial resistance; prescribing antimicrobial drugs only when truly needed; and promoting patient compliance with drug dosing regimens.
    ASM Recommended Curriculum Guidelines: 14, 15
    NCLEX-RN Alignment: Pharmacological & Parenteral Therapies; Safety & Infection Control; Psychosocial Integrity

Scientific Thinking and Critical Thinking Skills

18. **Applying the process of science is relevant to nursing.**
    Understanding the process of science is central to science literacy and fundamental to nursing practices. Analyzing and interpreting results from a variety of microbiological tests and applying analytical reasoning to solve problems are central to nursing practices.
    ASM Recommended Curriculum Guideline: 28
    NCLEX-RN Alignment: Nursing Process

19. **Using quantitative reasoning ties into nursing practice.**
    Healthcare workers should: (1) be competent in drawing conclusions from charts and graphs related to patient medical history; (2) understand the metric system and scientific notation, as this terminology is used in patient medical history and is used in calculating medication dosages; (3) appreciate that microbe levels impact disease development and prognosis.
    ASM Recommended Curriculum Guideline: 29
    NCLEX-RN Alignment: Pharmaceutical & Parenteral Therapies; Reduction of Risk Potential

20. **The ability to communicate and collaborate with other disciplines is important for a cross-disciplinary healthcare team.**
    Healthcare workers should be able to effectively communicate about microbiology-related topics in written and oral formats and effectively work as individuals and in groups.

ASM Recommended Curriculum Guideline: 30
NCLEX-RN Alignment: Communication & Documentation; Management of Care

21. **Understanding the relationship between science and society improves clinical practice and promotes the human aspect of medicine.**
    Nurses should be able to identify and discuss ethical issues in microbiology, especially with regard to vaccines.
    ASM Recommended Curriculum Guideline: 31
    NCLEX-RN Alignment: Management of Care; Communication & Documentation; Psychosocial Integrity

Microbiology Laboratory Skills

22. **Aseptic technique is central to collecting clinical samples and to protecting healthcare providers and patients.**
    ASM Recommended Curriculum Guidelines: 32, 34
    NCLEX-RN Alignment: Management of Care; Safety & Infection Control; Health Promotion & Maintenance; Psychosocial Integrity; Basic Care & Comfort

23. **Microbiological and molecular lab techniques are key to identifying pathogens and implementing effective treatment options.**
    ASM Recommended Curriculum Guidelines: 33, 34, 35, 36
    NCLEX-RN Alignment: Management of Care

24. **All healthcare providers must understand protective procedures for handling infectious materials to prevent the spread of disease.**
    Understanding biosafety levels and emergency procedures is central to safe nursing. Understanding proper biomedical waste management is important to reduce risk of pathogen exposure and to limit infections.
    ASM Recommended Curriculum Guideline: 37
    NCLEX-RN Alignment: Safety & Infection Control; Physiological Adaptation; Reduction of Risk Potential; Basic Care & Comfort

25. **The ability to document and report on experimental protocols, results, and conclusions is key to patient treatment.**
    Nurses must accurately label specimens and keep records.
    ASM Recommended Curriculum Guideline: 38
    NCLEX-RN Alignment: Communication & Documentation

APPENDIX B PRONUNCIATIONS OF SELECTED ORGANISMS AND VIRUSES

Absidia (ab-sid'ē-ă)
Abiotrophia (ā'bī-ō-trō'fē-ă)
Acanthamoeba (ă-kan-thă-mē'bă)
Acetobacter (a-sē'tō-bak-ter)
Acinetobacter (as-i-nē'tō-bak'ter)
Acremonium (ak'rê-mō'nē-ŭm)
Actinomyces israelii (ak'ti-nō-mī'sēz is-rā'el-ē-ē)
Agaricus (a-gār'i-kus)
Agrobacterium tumefaciens (ag'rō-bak-tēr'ē-um tū'me-fāsh-enz)
Alternaria (al-ter-nā'rē-ă)
Amanita muscaria (am-ă-nī'tă mus-ka'rē-ă)
Amanita phalloides (am-ă-nī'tă fal-ōy'dēz)
Amoeba (am-ē'bă)
Amycolatopsis orientalis (am-ē-kō'la-top-sis o-rē-en-tal'is)
Anaplasma phagocytophilum (an-ă-plaz'mă fag-ō-sī-to'fil-ŭm)
Ancylostoma duodenale (an-si-los'tō-mă doo'ō-de-nā-lē)
Anisakis simplex (an-a-se'kis sim'pleks)
Aquifex (ăk'wē-feks)
Ascaris lumbricoides (as'kă-ris lŭm'bri-koy'dēz)
Aspergillus fumigatus (as-per-jil'ŭs fū-mē-gā'tus)
Aspergillus oryzae (as-per-jil'ŭs o'ri-zī)
Azomonas (ā-zō-mō'nas)
Azospirillum (ā-zō-spī'ril-ŭm)
Azotobacter (ā-zō-tō-bak'ter)

Bacillus anthracis (ba-sil'ŭs an-thrā'sis)
Bacillus cereus (ba-sil'ŭs se'rē-ŭs)
Bacillus licheniformis (ba-sil'ŭs lī-ken-i-for'mis)
Bacillus polymyxa (ba-sil'ŭs po-lē-miks'ă)
Bacillus stearothermophilus (ba-sil'ŭs ste-rō-ther-ma'fil-ŭs)
Bacillus subtilis (ba-sil'ŭs sŭt'i-lis)
Bacillus thuringiensis (ba-sil'ŭs thur-in-jē-en'sis)
Bacteroides fragilis (bak-ter-oy'dēz fra'ji-lis)
Balantidium coli (bal-an-tid'ē-ŭm kō'lē)
Bartonella bacilliformis (bar-tō-nel'ă ba-sil'li-for'mis)
Bartonella henselae (bar-tō-nel'ă hen'sel-ī)
Bartonella quintana (bar-tō-nel'ă kwin'ta-nă)
Bdellovibrio (del-lō-vib'rē-ō)
Beggiatoa (bej'jē-a-tō'ă)
Blastomyces dermatitidis (blas-tō-mī'sēz der-mă-tit'i-dis)

Bordetella pertussis (bōr-dê-tel'ă per-tus'is)
Borrelia burgdorferi (bō-rē'lē-ă burg-dōr'fer-ē)
Borrelia recurrentis (bō-rē'lē-ă re-kur-ren'tis)
Botryococcus braunii (bot'rē-ō-kok'ŭs brow'nē-ē)
Brucella abortus (broo-sel'lă a-bort'us)
Brucella canis (broo-sel'lă kā'nis)
Brucella melitensis (broo-sel'lă me-li-ten'sis)
Brucella suis (broo-sel'lă soo'is)
Burkholderia cepacia (burk-hol-der'ē-ă se-pā'se-ă)
Burkholderia pseudomallei (burk-hol-der'ē-ă soo-dō-mal'e-ē)

Campylobacter jejuni (kam'pi-lō-bak'ter jē-jū'nē)
Candida albicans (kan'did-ă al'bi-kanz)
Caulobacter (kaw'lō-bak-ter)
Chlamydia trachomatis (kla-mid'ē-ă tra-kō'ma-tis)
Chlamydophila pneumoniae (kla-mē-dof'î-lă noo-mō'nē-ī)
Chlamydophila psittaci (kla-mē-dof'î-lă sit'ă-sē)
Chondrus crispus (kon'drŭs krisp'ŭs)
Citrobacter (sit'rō-bak-ter)
Cladophialophora (klă-dŏf'ē-ă-lof'ŏ-ră)
Claviceps purpurea (klav'i-seps poor-poo'rē'ă)
Clostridium botulinum (klos-trid'ē-ŭm bo-tū-lī'num)
Clostridium difficile (klos-trid'ē-ŭm di'fe-sēl)
Clostridium perfringens (klos-trid'ē-ŭm per-frin'jens)
Clostridium tetani (klos-trid'ē-ŭm te'tan-ē)
Coccidioides immitis (kok-sid-ē-oy'dēz im'mi-tis)
Codium (kō'dē-ŭm)
Coltivirus (kol'tē-vī'rŭs)
Cortinarius gentilis (kōr'ti-nar-ē-us jen'til-is)
Corynebacterium diphtheriae (kŏ-rī'nē-bak-tēr'ē-ŭm dif-thi'rē-ī)
Coxiella burnetii (kok-sē-el'ă ber-ne'tē-ē)
Cryptococcus neoformans (krip-tō-kok'ŭs nē-ō-for'manz)
Cryptosporidium parvum (krip-tō-spō-rid'ē-ŭm par'vŭm)
Cyclospora cayetanensis (sī-klō-spōr'ă kī'ē-tan-en'sis)
Cytomegalovirus (sī-tō-meg'ă-lō-vī'rŭs)
Cytophaga (sī-tof'ă-gă)

Deinococcus radiodurans (dī-nō-kok'ŭs rā-dē-ō-dur'anz)
Desulfovibrio (dē'sul-fō-vib'rē-ō)
Dictyostelium (dik-tē-ō-stē'lē-um)

Didinium (dī-di′nē-ŭm)
Diplococcus pneumoniae (dip′lō-kok′ŭs nū-mō′nē-ī)

Echinococcus granulosus (ê-kī′nō-kok′ŭs gra-nū-lō′sŭs)
Edwardsiella (ed′ward-sē-el′ă)
Ehrlichia chaffeensis (er-lik′ē-ă chaf-ē-en′sis)
Encephalatizoon intestinalis (en-sef-a-lat-e′zō-an in-tes′ti-năl′is)
Entamoeba histolytica (ent-ă-mē′bă his-tō-li′ti-kă)
Enterobacter (en′ter-ō-bak′ter)
Enterobius vermicularis (en-ter-ō′bī-ŭs ver-mi-kū-lar′is)
Enterococcus faecalis (en′ter-ō-kok′ŭs fē-kă′lis)
Enterococcus faecium (en′ter-ō-kok′ŭs fē-sē′ŭm)
Epidermophyton floccosum (ep′i-der-mof′i-ton flŏk′ō-sŭm)
Epulopiscium (ep′yoo-lō-pis′sē-ŭm)
Escherichia coli (esh-ê-rik′ē-ă kō′lī)
Euglena granulata (yū-glēn′ă gran-yū-lă′tă)
Exophiala (ek-sō-fī′ă-lă)

Fasciola gigantica (fa-sē′ō-lă ji-gan′ti-kă)
Fasciola hepatica (fa-sē′ō-lă he-pa′ti-kă)
Ferroplasma acidarmanus (fe′rō-plaz-ma a-sid′ar-mă-nŭs)
Fonsecaea (fon-sē-sē′ă)
Francisella tularensis (fran′si-sel′ă too-lă-ren′sis)
Fusarium (fū-zā′rē-ŭm)

Gelidium (jel-li′dē-ŭm)
Geogemma barossii (jē′ō-jem-ă ba-rōs′ē-ē)
Giardia intestinalis (jē-ar′dē-ă in-tes′ti-năl′is)
Gluconobacter (gloo-kon′ō-bak-ter)
Gonyaulax (gon-ē-aw′laks)
Gymnodinium (jim-nō-din′ē-ŭm)
Gyromitra esculenta (gī-rō-mē′tră es-kū-len′tă)

Haemophilus ducreyi (hē-mof′i-lŭs doo-krā′ē)
Haemophilus influenzae (hē-mof′i-lŭs in-flu-en′zī)
Hafnia (haf′nē-ă)
Halobacterium salinarum (hā′lō-bak-tēr′ē-ŭm sal-ē-nar′ŭm)
Hantavirus (han′tā-vī-rŭs)
Helicobacter pylori (hel′î-kō-bak′ter pī′lō-rē)
Histoplasma capsulatum (his-tō-plaz′mă kap-soo-lā′tŭm)

Izziella abbottiae (iz-ē-el′lă ab′ot-ē-ī)

Klebsiella pneumoniae (kleb-sē-el′ă nū-mō′nē-ī)

Lactobacillus bulgaricus (lak′tō-bă-sil′ŭs bul-gā′ri-kŭs)
Lactococcus lactis (lak-tō-kok′ŭs lak′tis)
Legionella pneumophila (lē-jŭ-nel′lă noo-mō′fi-lă)
Leishmania (lēsh-man′ē-ă)
Leptospira interrogans (lep′tō-spī′ră in-ter′ră-ganz)

Leuconostoc citrovorum (loo′kō-nos-tŏk sit-rō-vō′rum)
Listeria monocytogenes (lis-tēr′ē-ă mo-nō-sī-tah′je-nēz)
Lyssavirus (lis′ă-vī-rŭs)

Madurella (mad′ū-rel′ă)
Malassezia furfur (mal-ă-sē′zē-ă fur′fur)
Mariprofundus ferrooxydans (mĕ-rē-prō-fun′dus fĕ-rō-oks′ē-danz)
Methanopyrus (meth′a-nō-pī′rŭs)
Microsporidium (mī-krō-spor-i′dē-ŭm)
Microsporum (mī-kros′po-rŭm)
Moraxella catarrhalis (mōr′ak-sel′ă kă-tah′răl-is)
Morganella (mōr′gan-el′ă)
Mucor (mū′kōr)
Mycobacterium avium-intracellulare (mī′kō-bak-tēr′ē-ŭm ā′vē-ŭm in′tra-sel-yu-la′rē)
Mycobacterium bovis (mī′kō-bak-tēr′ē-ŭm bō′vis)
Mycobacterium leprae (mī′kō-bak-tēr′ē-ŭm lep′rī)
Mycobacterium tuberculosis (mī′kō-bak-tēr′ē-ŭm too-ber-kyū-lō′sis)
Mycoplasma genitalium (mī′kō-plaz-mă jen-ē-tal′ē-ŭm)
Mycoplasma hominis (mī′kō-plaz-mă ho′mi-nis)
Mycoplasma pneumoniae (mī′kō-plaz-mă nū-mō′nē-ī)

Naegleria (nā-glē′rē-ă)
Nasuia deltocephalinicola (nă-soo′ē-ă del-to-sef′ă-lĭn-ĭ-kō-lă)
Necator americanus (nē-kā′tor ă-mer-i-ka′nus)
Neisseria gonorrhoeae (nī-se′rē-ă go-nor-rē′ī)
Neisseria meningitidis (nī-se′rē-ă me-nin-ji′ti-dis)
Neosartorya (nē-ō-sar-tōr′ya)
Neurospora crassa (noo-ros′pōr-ă kras′ă)
Nitrobacter (nī-trō-bak′ter)
Nitrosomonas (nī-trō-sō-mō′nas)
Nocardia asteroides (nō-kar′dē-ă as-ter-oy′dēz)
Nosema (nō-sē′mă)

Orientia tsutsugamushi (ōr-ē-en′tē-ă tsoo-tsoo-gă-mū′shē)
Orthopoxvirus variola (ōr-thō-poks′vī-rŭs vă-rī′ō-lă)

Paracoccidioides brasiliensis (par′ă-kok-sid-ē-oy′dēz bră-sil-ē-en′sis)
Paramecium (par-ă-mē′sē-ŭm)
Pasteurella haemolytica (pas-ter-el′ă hē-mō-lit′i-kă)
Pasteurella multocida (pas-ter-el′ă mul-tŏ′si-da)
Penicillium chrysogenum (pen-i-sil′ē-ŭm krī-so′jên-ŭm)
Penicillium marneffei (pen-i-sil′ē-ŭm mar-nef-ē′ī)
Pfiesteria (fes-tēr′ē-ă)
Phialophora (fī-ă-lof′ŏ-ră)
Physarum (fī-sar′um)

Phytophthora infestans (fī-tof′tho-ră in-fes′tanz)
Planctomyces (plank-tō-mī′sēz)
Plasmodium falciparum (plaz-mō′dē-ŭm fal-sip′ar-ŭm)
Plasmodium knowlesi (plaz-mō′dē-ŭm nō-les′ē)
Plasmodium malariae (plaz-mō′dē-ŭm mă-lār′ē-ī)
Plasmodium ovale (plaz-mō′dē-ŭm ō-vă′lē)
Plasmodium vivax (plaz-mō′dē-ŭm vī′vaks)
Pneumocystis jirovecii (nū-mō-sis′tis jē-rō-vêt′zē-ē)
Prevotella (prev′ō-tel′ă)
Propionibacterium acnes (prō-pē-on-i-bak-tēr′ē-ŭm ak′nēz)
Proteus mirabilis (prō′tē-ŭs mi-ra′bi-lis)
Prototheca (prō-tō-thē′kă)
Providencia (prov′i-den′sē-ă)
Pseudallescheria (sood′al-es-kē-rē-ă)
Pseudomonas aeruginosa (soo-dō-mō′nas ā-roo-ji-nō′să)
Pseudomonas putida (soo-dō-mō′nas pyoo′ti-dă)
Pseudomonas syringae (soo-dō-mō′nas sēr′in-jī)
Psilocybe cubensis (sil-lō-sī′bē kū-ben′sis)
Pyrodictium (pī-rō-dik′tē-ŭm)

Rhizobium (rī-zō′bē-ŭm)
Rhizopus nigricans (rī-zō′pŭs ni′gri-kanz)
Rhodopseudomonas palustris (rō-dō′soo-dō-mō′nas pal-us′tris)
Rickettsia prowazekii (ri-ket′sē-ă prō-wă-ze′kē-ē)
Rickettsia rickettsii (ri-ket′sē-ă ri-ket′sē-ē)
Rickettsia typhi (ri-ket′sē-ă tī′fē)
Rotavirus (rō′tă-vī′rŭs)

Saccharomyces carlsbergensis (sak-ă-rō-mī′sēz karlz-bur-jen′sis)
Saccharomyces cerevisiae (sak-ă-rō-mī′sēz se-ri-vis′ē-ī)
Saccharopolyspora erythraea (sak-ă-rō-pol-ē-spō′ra e-rith′rē-a)
Salmonella enterica (sal′mŏ-nel′ă en-ter′i-kă)
    serotype Choleraesuis (kol-er-a-su′is)
    serotype Paratyphi (pa′ra-tī′fē)
    serotype Typhi (tī′fē)
    serotype Typhimurium (tī′fē-mur-ē-ŭm)
Sarcina (sar′si-nă)
Schistosoma haematobium (shis-tō-sō′mă hē′mă-tō′bē-ŭm)
Schistosoma japonicum (shis-tō-sō′mă jă-pon′i-kŭm)
Schistosoma mansoni (shis-tō-sō′mă man-sō′nē)
Selenomonas (sê-lē′nō-mō′nas)
Serratia marcescens (ser-rat′ē-a mar-ses′enz)
Shigella boydii (shē-gel′lă boy′dē-ē)
Shigella dysenteriae (shē-gel′lă dis-en-te′rē-ī)
Shigella flexneri (shē-gel′lă fleks′ner-ē)
Shigella sonnei (shē-gel′lă sōn′ne-ē)
Sphaerotilus (sfēr-ō′til-us)
Spiroplasma (spī′rō-plaz-mă)

Sporothrix schenckii (spōr′ō-thriks shen′kē-ē)
Staphylococcus aureus (staf′i-lō-kok′ŭs o′rē-ŭs)
Staphylococcus epidermidis (staf′i-lō-kok′ŭs ep-i-der-mid′is)
Streptococcus agalactiae (strep-tō-kok′ŭs a-ga-lak′tē-ī)
Streptococcus anginosus (strep-tō-kok′ŭs an-ji-nō′sŭs)
Streptococcus equisimilis (strep-tō-kok′ŭs ek-wi-si′mil-is)
Streptococcus mitis (strep-tō-kok′ŭs mī′tis)
Streptococcus mutans (strep-tō-kok′ŭs mū′tanz)
Streptococcus pneumoniae (strep-tō-kok′ŭs nū-mō′nē-ī)
Streptococcus pyogenes (strep-tō-kok′ŭs pī-oj′en-ēz)
Streptococcus sanguis (strep-tō-kok′ŭs sang′wis)
Streptococcus thermophilus (strep-tō-kok′ŭs ther-mo′fil-us)
Streptomyces (strep-tō-mī′sēz)

Taenia saginata (tē′nē-a sa-ji-na′ta)
Taenia solium (tē′nē-a sō′lī-um)
Thermococcus (ther-mō-kok′us)
Thermus aquaticus (ther′mŭs a-kwa′ti-kŭs)
Thiobacillus (thī-ō-bă-sil′ŭs)
Toxoplasma gondii (tok-sō-plaz′mă gon′dē-ē)
Trebouxia (tre-book′sē-a)
Treponema carateum (trep-ō-nē′mă kar-a′tē-ŭm)
Treponema pallidum (trep-ō-nē′mă pal′li-dŭm)
Trichomonas vaginalis (trik-ō-mō′nas va-jin-al′is)
Trichonympha (trik-ō-nimf′ă)
Trichophyton (trik-ō-fī′ton)
Trichosporon beigelii (trik-ō-spor′on bā-gêl′ē-ē)
Trypanosoma brucei gambiense
    (tri-pan′ō-sō-mă brūs′ē gam′bē-en′sē)
Trypanosoma brucei rhodesiense
    (tri-pan′ō-sō-mă brūs′ē rō-dē′zē-en′sē)
Trypanosoma cruzi (tri-pan′ō-sō-mă kroo′zē)

Ureaplasma urealyticum (yū-rē′ă-plaz-mă ū-rē′ă-li′ti-kŭm)

Veillonella (vī-lō-nel′ă)
Vibrio cholerae (vib′rē-ō kol′er-ī)
Vibrio parahaemolyticus (vib′rē-ō pa-ră-hē-mō-li′ti-kŭs)
Vibrio vulnificus (vib′rē-ō vul-nif′i-kŭs)
Vorticella (vōr-ti-sel′ă)

Wangiella (wang-gē-el′ă)
Wuchereria bancrofti (voo-ker-e′rē-ă ban-krof′tē)

Yersinia enterocolitica (yer-sin′ē-ă en′ter-ō-kō-lit′ĭ-kă)
Yersinia pestis (yer-sin′ē-ă pes′tis)
Yersinia pseudotuberculosis (yer-sin′ē-ă soo-dō-too-ber-kyū-lō′sis)

Zoogloea (zō′ō-glē-ă)

GLOSSARY

A site In a ribosome, a binding site that accommodates tRNA delivering an amino acid.

Abscess An isolate site of infection, such as a pimple, boil, or pustule.

Abyssal zone In marine habitats, the zone of water beneath the benthic zone, virtually devoid of life except around hydrothermal vents.

Acellular Noncellular.

Acetyl-CoA Combination of two-carbon acetate and coenzyme A.

Acid Compound that dissociates into one or more hydrogen ions and one or more anions.

Acid-fast bacilli (AFB) (*acid-fast rod, AFR*) Bacilli that retain stain during decolorization by acid-alcohol, particularly species of *Mycobacterium*.

Acid-fast rod (AFR) Bacilli that retain stain during decolorization by acid-alcohol, particularly species of *Mycobacterium*.

Acid-fast stain In microscopy, a differential stain used to penetrate waxy cell walls.

Acidic dye In microscopy, an anionic chromophore used to stain alkaline structures. Works most effectively in acidic environments.

Acidophile Microorganism requiring acidic pH.

Acne Skin disorder characterized by presence of whiteheads, blackheads, and, in severe cases, cysts; typically caused by infection with *Propionibacterium acnes*.

Acquired immunodeficiency syndrome (AIDS) The presence of several opportunistic or rare infections along with infection by human immunodeficiency virus (HIV) or a severe decrease in the number of CD4+ cells (200/μL) along with a positive test showing the presence of HIV.

Acquired (secondary) immunodeficiency diseases Any of a group of immunodeficiency diseases that develop in older children, adults, and the elderly as a direct consequence of some other recognized cause, such as infectious disease.

Actinomycetes High G+C Gram-positive bacteria that form branching filaments and produce spores, thus resembling fungi.

Actinomycosis Disease caused by *Actinomyces*; characterized by formation of multiple, interconnected abscesses in skin or mucous membrane.

Activation energy The amount of energy needed to trigger a chemical reaction.

Active site Functional site of an enzyme, the shape of which is complementary to the shape of the substrate.

Active transport The movement of a substance against its electrochemical gradient via carrier proteins and requiring cell energy from ATP.

Acute anaphylaxis Condition in which the release of inflammatory mediators overwhelms the body's coping mechanisms.

Acute disease Any disease that develops rapidly but lasts only a short time, whether it resolves in convalescence or death.

Acute inflammation Type of inflammation that develops quickly, is short lived, and is usually beneficial.

Adaptive immunity Resistance against pathogens that acts more effectively upon subsequent infections with the same pathogen.

Adenine Ring-shaped nitrogenous base found in nucleotides of DNA and RNA.

Adenosine triphosphate (ATP) The primary short-term, recyclable energy molecule fueling cellular reactions.

Adherence Process by which phagocytes attach to microorganisms through the binding of complementary chemicals on the cytoplasmic membranes.

Adhesins Molecules that attach pathogens to their target cells.

Adhesion The attachment of microorganisms to host cells.

Adhesion factors A variety of structures or attachment proteins by which microorganisms attach to host cells.

Adjuvant Chemical added to a vaccine to increase its ability to stimulate active immunity.

Aerobe An organism that uses oxygen as a final electron acceptor.

Aerobic respiration Type of cellular respiration requiring oxygen atoms as final electron acceptors.

Aerosol A cloud of water droplets that travels more than 1 meter in airborne transmission and less than 1 meter in droplet transmission.

Aerotolerant anaerobe Microorganism that prefers anaerobic conditions but can tolerate exposure to low levels of oxygen.

Aflatoxin Carcinogenic mycotoxin produced by *Aspergillus*.

African trypanosomiasis (African sleeping sickness) Potentially fatal disease caused by a bite from a tsetse fly carrying *Trypanosoma brucei* and characterized by formation of a lesion at the site of the bite, followed by parasitemia and central nervous system invasion.

Agar Gel-like polysaccharide isolated from red algae and used as thickening agent.

Agglutination Aggregation (clumping) caused when antibodies bind to two antigens, perhaps hindering the activity of pathogenic microorganisms and increasing the chance that they will be phagocytized.

Agglutination test In serology, a procedure in which antiserum is mixed with a sample that potentially contains its target antigen.

Agranulocyte Type of leukocyte having a uniform cytoplasm lacking large granules.

Agroterrorism The use of microbes to terrorize humans by destroying the livestock and crops that constitute their food supply.

Airborne transmission Spread of pathogens to the respiratory mucous membranes of a new host via the air or in droplets carried more than 1 meter.

Alcohol Intermediate-level disinfectant that denatures proteins and disrupts cell membranes.

Aldehyde Compound containing terminal groups; used as a high-level disinfectant because it cross-links organic functional groups in proteins and nucleic acids.

Algae Eukaryotic unicellular or multicellular photosynthetic organisms with simple reproductive structures.

Alginic acid Cell wall polysaccharide of brown algae.

Alkalinophile Microorganism requiring alkaline pH environments.

Allergen An antigen that stimulates an allergic response.

Allergic contact dermatitis Type of delayed hypersensitivity reaction in which chemically modified skin proteins trigger a cell-mediated immune response.

Allergy An immediate hypersensitivity response against an antigen.

Allograft Type of graft in which tissues are transplanted from a donor to a genetically dissimilar recipient of the same species.

Allylamines Class of antifungal drugs that disrupt cytoplasmic membranes.

Alpha interferons (IFN-α) Interferons secreted by virally infected monocytes, macrophages, and some lymphocytes within hours after infection.

Alphaproteobacteria Class of aerobic Gram-negative bacteria in the phylum Proteobacteria capable of growing at very low nutrient levels.

Alternation of generations In algae, method of sexual reproduction in which diploid bodies alternate with haploid bodies.

Alveolar macrophage Fixed macrophage of the lungs.

Alveolates Protozoa with small membrane-bound cavities called alveoli beneath their cell surfaces.

Amebiasis A mild to severe dysentery that, if invasive, can cause the formation of lesions in the liver, lungs, brain, and other organs; caused by infection with *Entamoeba histolytica*.

Amensalism Symbiotic relationship in which one member harms another without receiving any benefit or harm itself.

American trypanosomiasis (*Chagas' disease*) Potentially fatal disease caused by a bite from a kissing bug carrying *Trypanosoma cruzi* and characterized by the formation of swelling at the site of the bite, followed by fever, swollen lymph nodes, myocarditis, organ enlargement, and eventually congestive heart failure.

Ames test Method for screening mutagens that is commonly used to identify potential carcinogens.

Amination Reaction involving the addition of an amine group to a metabolite to make an amino acid.

Amino acid A monomer of polypeptides.

Aminoglycoside Antimicrobial agent that inhibits protein synthesis by changing the shape of the 30S ribosomal subunit.

Ammonification Process by which microorganisms disassemble proteins in soil wastes into amino acids, which are then converted to ammonia.

Amoebas (*amoebae*) Protozoa that move and feed by pseudopods.

Amphibolic reaction A reversible metabolic reaction; that is, a reaction that can be catabolic or anabolic.

Anabolism All of the synthesis reactions in an organism taken together.

Anaerobe An organism that cannot tolerate oxygen.

Anaerobic respiration Type of cellular respiration not requiring oxygen atoms as final electron acceptors.

Analog Chemicals that have similar molecular shapes.

Analytical epidemiology Detailed investigation of a disease, including analysis of data to determine the probable cause, mode of transmission, and possible means of prevention.

Anammox Anaerobic ammonium oxidation, which is one aspect of the nitrogen cycle.

Anaphase Third stage of mitosis, during which sister chromatids separate and move to opposite poles of the spindle to form chromosomes. Also used for the comparable stage of meiosis.

Anaphylactic shock Condition in which the release of inflammatory mediators overwhelms the body's coping mechanisms, causing suffocation, edema, smooth muscle contraction, and often death.

Anaplasmosis (*human granulocytic anaplasmosis, HGA*) Tick-borne disease caused by a rickettsia, *Anaplasma phagocytophilum*, manifesting with flu-like signs and symptoms. Formerly called human granulocytic ehrlichiosis.

Anion A negatively charged ion.

Anisakiasis Gastrointestinal disease caused by fish nematode parasite *Anisakis*, which may also manifest as an allergic reaction to the worm.

Anthrax Gastrointestinal, cutaneous, or pulmonary disease that is can be fatal without aggressive treatment; caused by ingestion, inoculation, or inhalation of spores of *Bacillus anthracis*.

Antibiotic Antimicrobial agent that is produced naturally by an organism.

Antibody (*immunoglobulin*) Proteinaceous antigen-binding molecule secreted by plasma cells.

Antibody-dependent cellular cytotoxicity (ADCC) Process whereby natural killer lymphocytes (NK cells) lyse cells covered with antibodies.

Antibody immune response (humoral immune response) The immune response centered around B lymphocytes and immunoglobulins.

Anticodon Portion of tRNA molecule that is complementary to a codon on mRNA.

Antigen Molecule that triggers a specific immune response.

Antigen-binding site Site formed by the variable regions of a heavy and light chain of an antibody.

Antigen-presenting cell (APC) Dendritic cells, macrophages, and B cells, which process antigens and activate cells of the immune system.

Antigenic determinant (*epitope*) The three-dimensional shape of a region of an antigen that is recognized by the immune system.

Antigenic drift Phenomenon that occurs every two to three years when a single strain of influenzavirus mutates within a local population.

Antigenic shift Major antigenic change that occurs on average every 10 years and results from the reassortment of genomes from different influenzavirus strains within host cells.

Antihistamines Drugs that specifically neutralize histamine.

Antimicrobial (*antimicrobial drug*) Any compound used to treat infectious disease; may also function as intermediate-level disinfectant.

Antimicrobial-associated diarrhea (*C. diff.-associated diarrhea*) Severe diarrhea caused by opportunistic growth of *Clostridium difficile* following inhibition of other bacteria by antimicrobial drugs.

Antimicrobial drug (*antimicrobial*) Chemotherapeutic agent used to treat microbial infection.

Antimicrobial enzyme Enzyme that acts against microbes.

Antimicrobial peptide (*defensin*) Chain of about 20 to 50 amino acids that acts against microorganisms.

Antiretroviral therapy (ART) A cocktail of antiviral drugs including nucleotide analogs, integrase inhibitors, protease inhibitors, and reverse transcriptase inhibitors.

Antisense nucleic acid RNA or single-stranded DNA with a nucleotide sequence complementary to a molecule of mRNA; used to control translation of polypeptide.

Antisense RNA RNA with a nucleotide sequence complementary to a molecule of mRNA; used to control translation of polypeptide.

Antisepsis The inhibition or killing of microorganisms on skin or tissue by the use of a chemical antiseptic.

Antiseptic Chemical used to inhibit or kill microorganisms on skin or tissue.

Antiserum In serology, blood fluid containing antibodies that bind to the antigens that triggered their production.

Antitoxin Antibodies formed by the host that bind to and protect against toxins.

Antivenin (antivenom) Antitoxin used to treat snake bites.

Antiviral proteins (AVPs) Proteins triggered by alpha and beta interferons that prevent viral replication.

Apicomplexans In protozoan taxonomy, group of pathogenic alveolate protozoa characterized by the complex of special intracellular organelles located at the apices of the infective stages of these microbes.

Apoenzyme The protein portion of protein enzymes that is inactive unless bound to one or more cofactors.

Apoptosis Programmed cell suicide.

Applied microbiology Branch of microbiology studying the commercial use of microorganisms in industry and foods.

Arachnid Group of arthropods distinguished by the presence of eight legs, such as spiders, ticks, and mites.

Arboviral encephalitis Inflammation of the brain and/or meninges caused by viruses transmitted by bloodsucking arthropods.

Arboviruses Viruses that are transmitted by arthropods; they include members of several viral families.

Archaea (*archaeon*, sing.) In Woese's taxonomy, domain that includes all prokaryotic cells having archaeal rRNA sequences.

Archaezoa Protozoa that lack mitochondria, Golgi bodies, chloroplasts, and peroxisomes.

Arenaviruses Group of segmented, negative ssRNA viruses that cause zoonotic diseases.

ART (antiretroviral therapy) A cocktail of antiviral drugs including nucleotide analogs, integrase inhibitors, protease inhibitors, and reverse transcriptase inhibitors.

Arthropod Animal with a segmented body, hard exoskeleton, and jointed legs, including arachnids and insects.

Arthropod vector Animals with segmented bodies, external skeletons, and jointed legs that carry pathogens; vectors include ticks, mites, fleas, flies, and true bugs.

Artificial wetlands Use of successive ponds, marshes, and meadowlands to remove wastes from sewage as the water moves through the wetlands.

Artificially acquired active immunity Type of immunity that occurs when the body receives antigens by injection, as with vaccinations, and mounts a specific immune response.

Artificially acquired passive immunotherapy Treatment in which patient receives via injection preformed antibodies in antitoxins or antisera, which can destroy fast-acting and potentially fatal antigens, such as rattlesnake venom.

Ascariasis Symptomatic but not typically fatal disease caused by infection with the nematode *Ascaris lumbricoides*.

Ascomycota Division of fungi characterized by the formation of haploid ascospores within sacs called asci.

Ascospore Haploid germinating structure of ascomycetes.

Ascus Sac in which haploid ascospores are formed and from which they are released in fungi of the division Ascomycota.

Aseptate Lacking cross walls.

Aseptic Characteristic of an environment or procedure that is free of contamination by pathogens.

Aspergillosis Term for several localized and invasive diseases caused by infection with *Aspergillus* species.

Assembly In virology, fourth stage of the lytic replication cycle, in which new virions are assembled in the host cell.

Asthma Hypersensitivity reaction affecting the lungs and characterized by bronchial constriction and excessive mucus production.

Astroviruses Group of small, round enteric viruses that cause diarrhea, typically in children.

Asymptomatic (*subclinical*) Characteristic of disease that may go unnoticed because of absence of symptoms, even though clinical tests may reveal signs of disease.

Atom The smallest chemical unit of matter.

Atomic force microscope (AFM) Type of probe microscope that uses a pointed probe to traverse the surface of a specimen. A laser beam detects vertical movements of the probe, which a computer translates to reveal the atomic topography of the specimen.

Atomic mass (*atomic weight*) The sum of the masses of the protons, neutrons, and electrons in an atom.

Atomic number The number of protons in the nucleus of an atom.

ATP synthase (ATPase) Enzyme that phosphorylates ATP in oxidative phosphorylation and photophosphorylation.

Attachment In virology, first stage of the lytic replication cycle, in which the virion attaches to the host cell.

Attenuated vaccine Inoculum in which pathogens are weakened so that, theoretically, they no longer cause disease; residual virulence can be a problem.

Attenuation The process of reducing vaccine virulence.

Autoantigens Antigens on the surface of normal body cells.

Autoclave Device that uses steam heat under pressure to sterilize chemicals and objects that can tolerate moist heat.

Autograft Type of graft in which tissues are moved to a different location within the same patient.

Autoimmune disease Any of a group of diseases that result when an individual begins to make autoantibodies or cytotoxic T cells against normal body components.

Autoimmune hemolytic anemia Disease resulting when an individual produces antibodies against his or her own red blood cells.

Avirulent Harmless.

Axenic Having only one organism present.

Axial filament In cell morphology, structure composed of rotating endoflagella that allows a spirochete to "corkscrew" through its medium.

Azoles Class of antifungal drugs that disrupt cytoplasmic membranes.

B cell B lymphocyte.

B cell receptor (BCR) Antibody integral to the cytoplasmic membrane and expressed by B lymphocytes.

B lymphocyte (*B cell*) Lymphocyte that arises and matures in the red bone marrow in adults and is found primarily in the spleen, lymph nodes, red bone marrow, and Peyer's patches of the intestines and that secretes antibodies.

Bacillus Rod-shaped prokaryotic cell.

Bacitracin Antimicrobial that blocks NAG and NAM secretion from the cytoplasm, prompting cell lysis.

Bacteremia The presence of bacteria in the blood; often caused by infection with *Staphylococcus aureus* or *Streptococcus pneumoniae*.

Bacteria Prokaryotic microorganisms typically having cell walls composed of peptidoglycan. In Woese's taxonomy, domain that includes all prokaryotic cells having bacterial rRNA sequences.

Bacterial gastroenteritis Inflammation of the mucous membrane of the stomach and intestines, caused by a bacterial pathogen.

Bacterial intoxication (*toxification*) Food poisoning caused by bacterial toxin.

Bacteriophage (*phage*) Virus that infects and usually destroys bacterial cells.

Bacteriorhodopsin Purple protein synthesized by *Halobacterium* that absorbs light energy to synthesize ATP.

Bacteroid Diverse group of Gram-negative microbes that have similar rRNA nucleotide sequences, such as *Bacteroides*.

Balantidiasis A mild gastrointestinal illness caused by *Balantidium coli*.

Barophile Microorganism requiring the extreme hydrostatic pressure found at great depth below the surface of water.

Base Molecule that binds with hydrogen ions when dissolved in water.

Base pair (bp) A complementary arrangement of nucleotides in a strand of DNA or RNA. For example, in both DNA and RNA, guanine and cytosine pair.

Base-excision repair Mechanism by which enzymes excise a section of a DNA strand containing an error, and then DNA polymerase fills in the gap.

Basic dye In microscopy, a cationic chromophore used to stain acidic structures. Works most effectively in alkaline environments.

Basidiocarp Fruiting body of basidiomycetes; includes mushrooms, puffballs, stinkhorns, jelly fungi, bird's nest fungi, and bracket fungi.

Basidiomycota Division of fungi characterized by production of basidiospores and basidiocarps.

Basophil Type of granulocyte that stains blue with the basic dye methylene blue.

Bejel Childhood disease caused by the spirochete *Treponema pallidum endemicum*; characterized by rubbery oral lesions.

Benign tumor Mass of neoplastic cells that remains in one place and is not generally harmful.

Benthic zone Bottom zone of freshwater or marine water, devoid of light and with scarce nutrients.

Beta interferons (IFN-β) Interferons secreted by virally infected fibroblasts within hours after infection.

Beta-lactam Antimicrobial whose functional portion is composed of beta-lactam rings, which inhibit peptidoglycan formation by irreversibly binding to the enzymes that cross-link NAM subunits.

Beta (β)-lactamase Bacterial enzyme that breaks the beta-lactam rings of penicillin and similar molecules, rendering them inactive.

Beta-oxidation A catabolic process in which enzymes split pairs of hydrogenated carbon atoms from a fatty acid and join them to coenzyme A to form acetyl-CoA.

Betaproteobacteria Class of diverse Gram-negative bacteria in the phylum Proteobacteria capable of growing at very low nutrient levels.

Binary fission The most common method of asexual reproduction of prokaryotes, in which the parental cell disappears with the formation of progeny.

Binomial nomenclature The classification method used in the Linnaean system of taxonomy, which assigns each species both a genus name and a specific epithet.

Biochemical oxygen demand (BOD) A measure of the amount of oxygen aerobic bacteria require to metabolize organic wastes in water.

Biochemistry Branch of chemistry that studies the chemical reactions of living things.

Biodiversity The number of species living in a given ecosystem.

Biofilm A slimy community of microbes growing on a surface.

Biogeochemical cycling The movement of elements and nutrients from unusable forms to usable forms because of the activities of microorganisms.

Biological vector Biting arthropod or other animal that transmits pathogens and serves as host for the multiplication of the pathogen during some stage of the pathogen's life cycle.

Biomass The quantity of all organisms in a given ecosystem.

Biomining The use of microbes to convert metals to soluble forms that can be more easily extracted.

Bioremediation The use of microorganisms to metabolize toxins in the environment to reclaim soils and waterways.

Bioreporter Type of biosensor composed of microbes with innate signaling capabilities.

Biosensor Device that combines bacteria or microbial products such as enzymes with electronic measuring devices to detect other bacteria, bacterial products, or chemical compounds in the environment.

Biosphere The region of Earth inhabited by living organisms.

Biotechnology Branch of microbiology in which microbes are manipulated to manufacture useful products.

Bioterrorism The use of microbes or their toxins to terrorize human populations.

Blastomycosis Pulmonary disease found in the southeastern United States, caused by infection with *Blastomyces dermatitidis*.

Blood group antigens The surface molecules of red blood cells.

Bodily fluid transmission Spread of pathogenic microorganisms via blood, urine, saliva, or other bodily fluids.

Botulism Potentially fatal intoxication with botulism toxin; three types include foodborne botulism, infant botulism, and wound botulism.

Bradykinin Peptide chain of nine amino acids that is a potent mediator of inflammation.

Broad-spectrum drug Antimicrobial that works against many different kinds of pathogens.

Bronchitis Inflammation of the bronchi.

Broth A liquid, nutrient-rich medium used for cultivating microorganisms.

Broth dilution test Test for determining the minimum inhibitory concentration in which a standardized amount of bacteria is added to serial dilutions of antimicrobial agents in tubes or wells containing broth.

Brucellosis Disease caused by *Brucella*; usually asymptomatic or mild, though it can result in sterility or abortion in animals.

Bruton-type agammaglobulinemia An inherited disease in which affected babies cannot make immunoglobulins and experience recurrent bacterial infections.

Bt toxin (Bt) Insecticidal poison produced by *Bacillus thuringiensis* bacteria.

Bubo Swollen inflamed lymph node.

Bubonic plague Severe systemic disease, fatal if untreated in 50% of patients, and characterized by fever, tissue necrosis, and the presence of buboes; caused by infection with *Yersinia pestis*.

Budding In prokaryotes and yeasts, reproductive process in which an outgrowth of the parent cell receives a copy of the genetic material, enlarges, and detaches. In virology, extrusion of enveloped virions through the host's cell membrane.

Buffer A substance, such as a protein, that prevents drastic changes in pH.

Bulbar poliomyelitis Infection of the brain stem and medulla resulting in paralysis of muscles in the limbs or respiratory system; caused by infection with poliovirus.

Bunyaviruses Group of zoonotic pathogens that have a segmented genome of three $-$ssRNA molecules and are transmitted to humans via arthropods.

Burkitt's lymphoma Infectious cancer of the jaw caused by infection with Epstein-Barr virus.

Caliciviruses Group of small, round enteric viruses that cause diarrhea, nausea, and vomiting.

Calor Heat.

Calvin-Benson cycle Stage of photosynthesis in which atmospheric carbon dioxide is fixed and reduced to produce glucose.

Cancer Disease characterized by the presence of one or more malignant tumors.

Candidiasis Term for several opportunistic diseases caused by infection with *Candida* species.

Candin Antifungal drug that inhibits cell wall synthesis.

Capnophile Microorganism that grows best with high levels of carbon dioxide in addition to low levels of oxygen.

Capsid A protein coat surrounding the nucleic acid core of a virion.

Capsomere A proteinaceous subunit of a capsid.

Capsule Glycocalyx composed of repeating units of organic chemicals firmly attached to the cell surface.

Capsule stain (*negative stain*) In microscopy, a staining technique used primarily to reveal bacterial capsules and involving application of an acidic dye that leaves the specimen colorless and the background stained.

Carbohydrate Organic macromolecule consisting of atoms of carbon, hydrogen, and oxygen.

Carbon cycle Biogeochemical cycle in which carbon is cycled in the form of organic molecules.

Carbon fixation The attachment of atmospheric carbon dioxide to ribulose 1,5-bisphosphate (RuBP).

Carbuncle The coalescence of several furuncles extending deep into underlying tissues; caused by infection with *Staphylococcus aureus*.

Carcinogen Chemical capable of causing cancer.

Caries (*cavities*) Tooth decay; caused by viridans streptococci and other bacteria.

Carotenoid Plant pigment that acts as an antioxidant.

Carrageenan Gel-like polysaccharide isolated from red algae and used as thickening agent.

Carrier In human pathology, continuous asymptomatic human source of infection.

Catabolism All of the decomposition reactions in an organism taken together.

Catarrhal phase In pertussis, initial phase lasting 1 to 2 weeks and characterized by signs and symptoms resembling those of a common cold.

Cation A positively charged ion.

Cat scratch disease Common and occasionally serious infection in children; characterized by fever and malaise plus localized swelling; caused by infection with *Bartonella henselae*.

CD4 Distinguishing cytoplasmic membrane protein of helper T cells, which is the initial binding site of HIV.

CD4+ cell (*helper T cell, Th cell*) In cell-mediated immune response, a type of cell characterized by CD4 cell-surface glycoprotein; regulates the activity of B cells and cytotoxic T cells.

CD8 Distinguishing cytoplasmic membrane protein of cytotoxic T cells.

CD8+ cell (*cytotoxic T cell, Tc cell*) In cell-mediated immune response, type of cell characterized by CD8 cell-surface glycoprotein; secretes perforins and granzymes that destroy infected or abnormal body cells.

CD95 pathway In cell-mediated cytotoxicity, pathway involving CD95 protein that triggers apoptosis of infected cells.

C. diff. (Clostridium difficile) diarrhea (*antimicrobial-associated diarrhea*) Watery stools resulting from elimination of normal intestinal microbiota by antimicrobial drug therapy, which allows *Clostridium difficile* endospores to germinate.

Cell culture Cells isolated from an organism and grown on the surface of a medium or in broth. Viruses can be grown in a cell culture.

Cell wall In most cells, structural boundary composed of polysaccharide or protein chains that provides shape and support against osmotic pressure.

Cell-mediated hypersensitivity reaction (*type IV hypersensitivity, delayed hypersensitivity*) T cell–mediated inflammatory reaction that takes 24 to 72 hours to reach maximal intensity.

Cell-mediated immune response Immune response used by T cells to fight intracellular pathogens and abnormal body cells.

Cellular respiration Metabolic process that involves the complete oxidation of substrate molecules and production of ATP via a series of redox reactions.

Cellular slime mold Individual haploid myxamoeba that phagocytizes bacteria, yeasts, dung, and decaying vegetation.

Central dogma In genetics, fundamental description of protein synthesis that states that genetic information is transferred from DNA to RNA to polypeptides, which function alone or in conjunction as proteins.

Centrioles Nonmembranous organelles in animal cells that appear to function in the formation of flagella and cilia and in cell division.

Centrosome Region of a cell containing centrioles.

Cesspool Home equivalent of primary wastewater treatment in which wastes enter a series of porous concrete rings buried underground and are digested by microbes.

Cestodes (*tapeworms*) Group of helminths that are long, flat, and segmented and lack digestive systems.

Chagas' disease (*American trypanosomiasis*) Potentially fatal disease caused by a bite from a kissing bug carrying *Trypanosoma cruzi* and characterized by the formation of swellings at the site of the bite, followed by fever, swollen lymph nodes, myocarditis, organ enlargement, and eventually congestive heart failure.

Chancre Painless red lesion that appears at the site of infection with *Treponema pallidum*, the agent of syphilis.

Chancroid Soft, painful, venereal ulcer at site of infection by *Haemophilus ducreyi*.

Chemical bond An interaction between atoms in which electrons are either shared or transferred in such a way as to fill their valence shells.

Chemical fixation In microscopy, a technique that uses methyl alcohol or formalin to attach a smear to a slide.

Chemical reaction The making or breaking of a chemical bond.

Chemiosmosis Use of ion gradients to generate ATP.

Chemoautotroph Microorganism that uses carbon dioxide as a carbon source and catabolizes organic molecules for energy.

Chemoheterotroph Microorganism that uses organic compounds for both energy and carbon.

Chemokine An immune system cytokine that signals leukocytes to rush to the site of inflammation or infection and activate other leukocytes.

Chemostat A continuous culture device controlled by adding fresh medium at the same rate old medium is removed.

Chemotactic factors Chemicals, such as peptides derived from complement and cytokines, that attract cells.

Chemotaxis Cell movement that occurs in response to chemical stimulus.

Chemotherapeutic agent Chemical used to treat disease.

Chemotherapy A branch of medical microbiology in which chemicals are studied for their potential to destroy pathogenic microorganisms.

Chickenpox (*varicella*) Highly infectious disease characterized by fever, malaise, and skin lesions, caused by infection with varicella-zoster virus.

Chitin Strong, flexible nitrogenous polysaccharide found in fungal cell walls and in the exoskeletons of insects and other arthropods.

Chlamydia Any of the small Gram-negative pathogenic cocci that grow and reproduce within the cells of mammals, birds, and a few invertebrates and that spread as elementary bodies.

Chloramphenicol Antimicrobial drug that blocks the enzymatic site of the 50S ribosomal subunit, inhibiting polypeptide synthesis.

Chlorophyll Pigment molecule that captures light energy for use in photosynthesis.

Chlorophyta Green-pigmented division of algae that have chlorophylls *a* and *b*, store sugar and starch as food reserves, and have rRNA sequences similar to plants. Considered the progenitors of plants.

Chloroplast Light-harvesting organelle found in photosynthetic eukaryotes.

Cholera Disease contracted through the ingestion of food and water contaminated with *Vibrio cholerae* and characterized by vomiting and watery diarrhea.

Cholera toxin Exotoxin produced by *Vibrio cholerae* that causes the movement of water out of the intestinal epithelium.

Chromatin Threadlike mass of DNA and associated histone proteins that becomes visible during mitosis as chromosomes.

Chromatin fiber An association of nucleosomes and proteins found within the chromosomes of eukaryotic cells.

Chromoblastomycosis Cutaneous and subcutaneous disease characterized by lesions that can spread internally; caused by traumatic introduction of ascomycete fungi into the skin.

Chromosome A molecule of DNA associated with protein. In prokaryotes, typically circular and localized in a region of the cytosol called the nucleoid. In eukaryotes, chromosomes are threadlike and are most visible during mitosis and meiosis.

Chronic disease Any disease that develops slowly, usually with less severe symptoms, and is continual or recurrent.

Chronic granulomatous disease Primary immunodeficiency disease in which children have recurrent infections characterized by the development of large masses of inflammatory cells in lymph nodes, lungs, bones, and skin.

Chronic inflammation Type of inflammation that develops slowly, lasts a long time, and can cause damage (even death) to tissues, resulting in disease.

Chrysophyta Division of algae including the golden algae, yellow-green algae, and diatoms.

Cilia Short, hairlike, rhythmically motile projections of some eukaryotic cells.

Ciliate In protozoan taxonomy, group of alveolate protozoa characterized by the presence of cilia in their trophozoite stages.

Citric acid cycle (*Krebs cycle*) Series of eight enzymatically catalyzed reactions that transfer stored energy from acetyl-CoA to coenzymes NAD^+ and FAD.

Class Taxonomical grouping of similar orders of organisms.

Class switching The process in which a plasma cell changes the type of antibody F_C region (stem) that it synthesizes and secretes.

Clinical laboratory scientist An expert in health care–related microbiological laboratory procedures who has at least a bachelor degree.

Clinical specimen Sample of human material, such as feces or blood, that is examined or tested for the presence of microorganisms.

Coinfection Simultaneous infection with more than one pathogen.

Clonal deletion Process by which cells with receptors that respond to autoantigens are selectively killed via apoptosis.

Clonal expansion In immunology, the reproduction of activated lymphocytes.

Clonal selection In antibody immunity, recognition and activation only of B lymphocytes with BCRs complementary to a specific antigenic determinant.

Clustered, regularly interspaced short palindromic repeats (CRISPR) A genetic location coding for a mechanism that protects prokaryotes from infections by bacteriophages by cleaving DNA sequences unique to the phage.

Coccidioidomycosis Pulmonary disease found in the southwestern United States, caused by infection with *Coccidioides immitis*.

Coccobacillus A prokaryotic cell intermediate in shape between a sphere and a rod, such as an elongated coccus.

Coccus Spherical prokaryotic cell.

Codon Triplet of mRNA nucleotides that codes for specific amino acids. For example, AAA is a codon for lysine.

Coenocyte Multinucleate cell resulting from repeated mitosis but postponed or absent cytokinesis.

Coenzyme Organic cofactor.

Cofactor Inorganic ions or organic molecules that are essential for enzyme action.

Coinfection Simultaneous infection with more than one pathogen.

Cold enrichment Incubation of a specimen in a refrigerator to enhance the growth of cold-tolerant species.

Cold sore Common name for oral lesion of a herpesvirus.

Coliforms Enteric Gram-negative bacteria that ferment lactose to gas and are found in the intestinal tracts of animals and humans.

Colony Visible population of microorganisms living in one place; an aggregation of cells arising from a single parent cell.

Colony-forming unit (CFU) A single cell or group of related cells that produce a colony.

Colorado tick fever Zoonosis caused by *Coltivirus* that is typically characterized by mild fever and chills.

Combination vaccine Inoculum composed of antigens from several pathogens that are administered simultaneously.

Commensalism Symbiotic relationship in which one member benefits without significantly affecting the other.

Communicable disease Any infectious disease that comes either directly or indirectly from another host.

Competence Ability of a cell to take up DNA from the environment.

Competitive inhibition (in immunity) Activities of members of the microbiome of the body that make it less likely that a pathogen can produce disease.

Competitive inhibitor (metabolism) Inhibitory substance that blocks enzyme activity by blocking active sites.

Complement fixation test A complex assay used to determine the presence of specific antibodies in serum.

Complement system (complement) Set of blood plasma proteins that act as chemotactic attractants, trigger inflammation and fever, and ultimately effect the destruction of foreign cells.

Complementary DNA (cDNA) DNA synthesized from an mRNA template using reverse transcriptase.

Complex medium Culturing medium that contains nutrients released by the partial digestion of yeast, beef, soy, or other proteins; thus, the exact chemical composition is unknown.

Complex transposon Transposon containing genes not connected with transposition.

Compound A molecule containing atoms of more than one element.

Compound microscope Microscope using a series of lenses for magnification.

Concentration gradient The difference in concentration of a chemical on the two sides of a membrane. Also called a *chemical gradient*.

Condenser lens In a compound microscope, a lens that directs light through the specimen as well as one or more mirrors that deflect the light's path.

Condylomata acuminata Large, cauliflower-like genital warts caused by infection with a papillomavirus.

Confocal microscope Type of light microscope that uses ultraviolet lasers to illuminate fluorescent chemicals in a single plane of the specimen.

Congenital syphilis Disease characterized by mental retardation, organ malformation, and, in some cases, death of the fetus of a woman infected with *Treponema pallidum*, the agent of syphilis.

Conjugation In genetics: method of horizontal gene transfer in which a bacterium containing a fertility plasmid forms a conjugation pilus that attaches and transfers plasmid genes to a recipient; in reproduction of ciliates: coupling of mating cells.

Conjugation pilus Proteinaceous, rodlike structure extending from the surface of a cell; mediates conjugation.

Conjunctivitis (*pinkeye*) Inflammation of the lining of an eyelid.

Consumption Tuberculosis; refers to wasting away of a body affected with TB at several sites.

Contact immunity Immunity conferred to an unvaccinated individual following contact with an individual vaccinated with an attenuated vaccine.

Contagious disease A communicable disease that is easily transmitted from a reservoir or patient.

Contamination The presence of microorganisms in or on the body or other site.

Continuous cell culture Type of cell culture created from tumor cells.

Contrast The difference in visual intensity between two objects, or between an object and its background.

Convalescence In the infectious disease process, final stage during which the patient recovers from the illness, and tissues and systems are repaired and return to normal.

Convalescent phase In pertussis, final phase lasting three to four weeks, during which the ciliated lining of the trachea grows back and frequency of coughing spells diminishes, but secondary bacterial infections may ensue.

Cord factor A cell wall component of pathogenic *Mycobacterium tuberculosis* that produces strands of daughter cells, inhibits migration of neutrophils, and is toxic to body cells.

Coronaviruses Group of enveloped ssRNA viruses that cause colds as well as severe acute respiratory syndrome.

Coronavirus respiratory syndrome Severe respiratory disease caused by coronaviruses SARS virus and MERS virus.

Corticosteroids Another name for immunosuppressive agents that suppress the action of T cells.

Counterstain In a Gram stain, red stain that provides contrasting color to the primary stain, causing Gram-negative cells to appear pink.

Covalent bond The sharing of a pair of electrons by two atoms.

Coxsackieviruses Group of enteroviruses that cause a variety of diseases in humans, ranging from mild fever and colds to myocarditis and heart failure.

Crenation Shriveling of a cell caused by osmosis in a hypertonic environment.

CRISPR (clustered, regularly interspaced short palindromic repeats) A genetic location coding for a mechanism that protects prokaryotes from infections by bacteriophages by cleaving DNA sequences unique to the phage.

Cristae Folds within the inner membrane of a mitochondrion that increase its surface area.

Cross resistance Phenomenon in which resistance to one antimicrobial drug confers resistance to similar drugs.

Crossing over Process in which portions of homologous chromosomes are recombined during the formation of gametes.

Croup Inflammation and swelling of the larynx, trachea, and bronchi and a "seal bark" cough, often caused by infection with a parainfluenza or rarely by other respiratory viruses.

Cryptococcosis Disease caused by the dimorphic fungus *Cryptococcus*; typically manifests as meningitis.

Cryptosporidiosis (*Cryptosporidium enteritis*) A gastrointestinal disease caused by infection with *Cryptosporidium parvum*; in humans, characterized by diarrhea, fluid loss, and weight loss; may be fatal in HIV-positive patients.

Culture Act of cultivating microorganisms or the microorganisms that are cultivated.

Cutaneous anthrax Disease of the skin caused by inoculation with spores of *Bacillus anthracis*.

Cuticle Outer protective "skin" of a nematode.

Cyanobacteria Gram-negative photosystem bacteria that vary greatly in shape, size, and method of reproduction.

Cyclic photophosphorylation Return of electrons to the original reaction center of a photosystem after passing down an electron transport chain.

Cycloserine Semisynthetic antimicrobial used to treat infections with Gram-positive bacteria.

Cyclosporine Immunosuppressive drug that inhibits action of activated T cells.

Cyst In protozoan morphology, the hardy resting stage characterized by a thick capsule and a low metabolic rate.

Cysticercus Immature tapeworm, usually in muscle of intermediate host.

Cytokines Proteins secreted by many types of cells that regulate adaptive immune responses.

Cytokinesis Division of a cell's cytoplasm.

Cytoplasm General term used to describe the semiliquid, gelatinous material inside a cell.

Cytoplasmic membrane Membrane surrounding all cells, and composed of a fluid mosaic of phospholipids and proteins.

Cytosine Ring-shaped nitrogenous base found in nucleotides of DNA and RNA.

Cytoskeleton Internal network of fibers contributing to the basic shape of eukaryotic and rod-shaped prokaryotic cells.

Cytosol The liquid portion of the cytoplasm.

Cytotoxic drugs Group of drugs that inhibit cells.

Cytotoxic T cell (*Tc cell, CD8+ cell*) In cell-mediated immune response, type of cell characterized by CD8 cell-surface glycoprotein; secretes perforins and granzymes that destroy infected or abnormal body cells.

Dark-field microscope Microscope used for studying pale or small specimens; deflects light rays so that they miss the objective lens.

Dark repair (*nucleotide excision repair*) Mechanism by which enzymes cut a section of a single strand of damaged DNA from a molecule, creating a gap that is repaired by DNA polymerase and DNA ligase.

Deamination Process in which amine groups are split from amino acids.

Death phase Phase in a growth curve in which the organisms are dying more quickly than they are being replaced by new organisms.

Decimal reduction time (D) The time required to destroy 90% of the microbes in a sample.

Decline In the infectious disease process, period in which the body gradually returns to normal as the patient's immune response and any medical treatments vanquish the pathogens.

Decolorizing agent In a stain, a solution that washes the primary stain away.

Decomposition reaction A chemical reaction in which the bonds of larger molecules are broken to form smaller atoms, ions, and molecules.

Deep-freezing Long-term storage of cultures at temperatures ranging from −50° C to −95° C.

Deeply branching bacteria Prokaryotic autotrophs with rRNA sequences and growth characteristics thought to be similar to those of earliest bacteria.

Defensins (*antimicrobial peptides*) Small peptide chains that act against a broad range of pathogens.

Defined medium (*synthetic medium*) Culturing medium of which the exact chemical composition is known.

Definitive host In the life cycle of parasites, host in which mature and sometimes sexual forms of the parasite are present and usually reproducing.

Degerming The removal of microbes from a surface by scrubbing.

Dehydration synthesis Type of synthesis reaction in which two smaller molecules are joined together by a covalent bond, and a water molecule is formed.

Delayed hypersensitivity reaction (*type IV hypersensitivity, cell-mediated hypersensitivity*) T cell-mediated inflammatory reaction that takes 24 to 72 hours to reach maximal intensity.

Deletion Type of mutation in which a nucleotide base pair is deleted.

Deltaproteobacteria Group of Proteobacteria that includes *Desulfovibrio, Bdellovibrio,* and myxobacteria.

Denaturation Process by which a protein's three-dimensional structure is altered, eliminating function.

Dendritic cells Cells of the epidermis and mucous membranes that devour pathogens.

Dengue fever Self-limiting but extremely painful disease caused by a flavivirus transmitted by *Aedes* mosquitoes.

Dengue hemorrhagic fever (DHF) Potentially fatal disease involving a hyperimmune response to reinfection with dengue virus and causing ruptured blood vessels, internal bleeding, and shock.

Denitrification The conversion of nitrate into nitrogen gas by anaerobic respiration.

Deoxyribonucleic acid (DNA) Nucleic acid consisting of nucleotides made up of phosphate, a deoxyribose pentose sugar, and an arrangement of the bases adenine, guanine, cytosine, and thymine.

Dermatophyte Fungus that normally lives on skin, nails, or hair.

Dermatophytoses Any of a variety of superficial skin, nail, and hair infections caused by dermatophytes.

Dermis The layer of the skin deep to the epidermis and containing hair follicles, glands, and nerve endings.

Descriptive epidemiology The careful recording of data concerning a disease.

Desiccation Inhibition of microbial growth by drying.

Detergent Positively charged organic surfactant.

Deuteromycetes Informal grouping of fungi having no known sexual stage.

Diapedesis (*emigration*) Process whereby leukocytes leave intact blood vessels by squeezing between lining cells.

Diatom Type of alga in the division Chrysophyta; has cell walls made of silica arranged in nesting halves called frustules.

Dichotomous key Method of identifying organisms in which information is arranged in paired statements, only one of which applies to any particular organism.

Differential interference contrast microscope Type of phase microscope that uses prisms to split light beams, giving images a three-dimensional appearance.

Differential medium Culturing medium formulated such that either the presence of visible changes in the medium or differences in the appearances of colonies help microbiologists differentiate among kinds of bacteria growing on the medium.

Differential stain In microscopy, a stain using more than one dye so that different structures can be distinguished. The Gram stain is the most commonly used.

Differential white blood cell count Lab technique that indicates the relative numbers of leukocytes.

Diffusion The net movement of a chemical down its concentration gradient.

Diffusion susceptibility test (*Kirby-Bauer test*) Simple, inexpensive test widely used to reveal which drug is most effective against a particular pathogen. Procedure involves inoculating a Petri plate uniformly with a standardized amount of the pathogen in question and arranging on the plate disks soaked in the drugs to be tested.

DiGeorge syndrome Failure of the thymus to develop, and thus, absence of T cells.

Dimorphic Having two forms; for example, dimorphic fungi have both yeastlike and moldlike bodies.

Dinoflagellate In protozoan taxonomy, group of unicellular, flagellated, alveolate protozoa characterized by photosynthetic pigments.

Dioecious Male and female sex organs are in separate individuals.

Diphtheria Mild to potentially fatal respiratory disease caused by diphtheria toxin following infection with *Corynebacterium diphtheriae*.

Diphtheroids Generally nonpathogenic pleomorphic bacilli named for the similarity of their appearance to *Corynebacterium diphtheriae*.

Diplococcus A pair of cocci.

Diploid A nucleus with two copies of each chromosome.

Diploid cell culture Type of cell culture created from embryonic animal, plant, or human cells that have been isolated and provided appropriate growth conditions.

Dipstick immunochromatographic assay Rapid modification of ELISA test in which an antigen solution flows through a porous strip, encountering labeled antibody; used for pregnancy testing and for rapid identification of infectious agents.

Direct antibody test Immune test allowing direct observation of the presence of antigen.

Direct contact transmission Spread of pathogens from one host to another involving body contact between the hosts.

Direct fluorescent antibody test Immune test allowing direct observation of the presence of antigen in a tissue sample flooded with labeled antibody.

Disaccharide Carbohydrate consisting of two monosaccharide molecules joined together.

Disease Any adverse internal condition severe enough to interfere with normal body functioning.

Disease process Definite sequence of events following contamination and infection.

Disinfectant Physical or chemical agent used to inhibit or destroy microorganisms on inanimate objects.

Disinfection The use of physical or chemical agents to inhibit or destroy microorganisms on inanimate objects. In water treatment, ozone, UV light, or chlorination kill most microorganisms.

Disseminated intravascular coagulation (DIC) The formation of blood clots within blood vessels throughout the body; triggered by lipid A.

DNA fingerprinting (*genetic fingerprinting*) Technique that identifies unique sequences of DNA to determine paternity; connect blood, semen, or skin cells to suspects in criminal investigations; or identify pathogens.

DNA microarray Numerous distinct ssDNA molecules bound to a substrate and used to probe for complementary sequences.
Dolor Pain.
Domain Any of three basic types of cell groupings distinguished by Carl Woese, containing the Linnaean taxon of kingdoms.
Donor cell In horizontal gene transfer, a cell that contributes part of its genome to a recipient.
Droplet transmission Spread of pathogens from one host to another via aerosols, which exit the body during exhaling, coughing, and sneezing and travel less than 1 meter.
Dysentery Disease characterized by severe diarrhea, often with stools containing blood and mucus.
Dyspnea Difficulty in breathing.
Dysuria Painful urination.

E site In translation, site at which tRNA exits from the ribosome.
Eastern equine encephalitis (EEE) Potentially fatal infection of the brain caused by a togavirus.
Ebola hemorrhagic fever Often fatal disease beginning with flu-like symptoms and progressing to severe internal hemorrhaging. Caused by *Ebolavirus*.
Echinocandin Antifungal drug that inhibits cell wall synthesis.
Echoviruses (*enteric cytopathic human orphan viruses*) Group of enteroviruses that cause viral meningitis and colds.
Ecosystem All of the organisms living in a particular habitat and the relationships between the two.
Efflux pump Transmembrane pump that removes antimicrobial drugs from a cell or from the periplasm.
Ehrlichiosis (*human monocytic ehrlichiosis, HME*) Tick-borne disease caused by a rickettsia, *Ehrlichia chaffeensis*, manifesting with flulike signs and symptoms. Also previously used to refer to human granulocytic ehrlichiosis, now called anaplasmosis.
Electrical gradient Voltage across a membrane created by the electrical charges of the chemicals on either side.
Electrochemical gradient The chemical and electrical gradients across a cell membrane.
Electrolyte Any hydrated cation or anion; can conduct electricity through a solution.
Electron A negatively charged subatomic particle.
Electron transport chain Series of redox reactions that pass electrons from one membrane-bound carrier to another and then to a final electron acceptor.
Electronegativity The attraction of an atom for electrons.
Elek test Immunodiffusion assay used to detect the presence of diphtheria toxin in a fluid sample.
Element Matter that is composed of a single type of atom.
Elephantiasis Enlargement and hardening of tissues, especially in the lower extremities, where lymph has accumulated following infection with *Wuchereria bancrofti*.
Empyema In patients with staphylococcal pneumonia, the presence of pus in the alveoli of the lungs.
Encephalitis Inflammation of the brain.
Encystment In the life cycle of protozoa, stage in which cysts form in host tissues.

Endemic In epidemiology, a disease that occurs at a relatively stable frequency within a given area or population.
Endemic typhus (*murine typhus*) Disease transmitted by fleas and characterized by high fever, headache, chills, muscle pain, and nausea; caused by infection with *Rickettsia typhi*.
Endocarditis Potentially fatal inflammation of the endocardium; typically caused by infection with *Staphylococcus aureus* or *Streptococcus pneumoniae*.
Endocytosis Active transport process, used by some eukaryotic cells, in which pseudopods surround a substance and move it into the cell.
Endoflagellum A special flagellum of spirochetes that spirals tightly around a cell rather than protruding from it.
Endogenous antigen Antigen produced by microbes that multiply inside the cells of the body.
Endogenous healthcare-associated infection An infection arising within a patient from opportunistic pathogens.
Endoplasmic reticulum (ER) Netlike arrangement of hollow tubules continuous with the outer membrane of the nuclear envelope and functioning as a transport system.
Endosome A sac formed during endocytosis containing the endocytized substance.
Endospore Environmentally resistant structure produced by the transformation of a vegetative cell of the Gram-positive genera *Bacillus* or *Clostridium*.
Endosymbiotic theory Proposal that eukaryotes were formed from the phagocytosis of small prokaryotes by larger prokaryotes, forming organelles.
Endothermic reaction Any chemical reaction that requires energy.
Endotoxins Lipopolysaccharides released from the outer membrane of the cell wall of dead and dying Gram-negative bacteria whose lipid portion (lipid A) is toxic.
Enrichment culture Technique used to enhance the growth of less abundant microorganisms by using a selective medium.
Enterobacteriaceae (*enteric bacteria*) Family of oxidase-negative Gram-negative bacteria, which can be pathogenic.
Enteroviruses Group of picornaviruses that are transmitted via the fecal-oral route but cause disease in any of a variety of target organs.
Entner-Doudoroff pathway Series of reactions that catabolize glucose to pyruvic acid using different enzymes from those used in either glycolysis or the pentose phosphate pathway.
Entry In virology, second stage of the lytic replication cycle, in which the virion or its genome enters the host cell.
Envelope In virology, membrane surrounding the viral capsid.
Environmental microbiology Branch of microbiology studying the role of microorganisms in soils, water, and other habitats.
Environmental specimen Sample of material taken from such sources as ponds, soil, or air and tested for the presence of microorganisms.
Enzyme An organic catalyst.
Enzyme immunoassay (EIA), enzyme-linked immunosorbent assay (ELISA) A family of simple immune tests that use enzymatic products as a label and that can be readily automated and read by machine.

Eosinophil Type of granulocyte that stains red to orange with the acidic dye eosin.
Eosinophilia An abnormal blood condition in which the number of eosinophils is greater than normal.
Epidemic In epidemiology, a disease that occurs at a greater than normal frequency for a given area or population.
Epidemiology Study of the occurrence, distribution, and spread of disease in humans.
Epidermis The outermost layer of the skin.
Epididymitis Inflammation of the epididymis.
Epitope (*antigenic determinant*) The three-dimensional shape of a region of an antigen that is recognized by the immune system.
Epsilonproteobacteria Group of Gram-negative rods, vibrios, and spiraled bacteria in the phylum Proteobacteria.
Erysipelas Impetigo spreading to lymph nodes, accompanied by pain and inflammation and caused by infection with group A *Streptococcus*.
Erythema infectiosum (*fifth disease*) Harmless red rash occurring in children and caused by infection with B19 virus.
Erythrocyte Red blood cell.
Erythrocytic cycle In the life cycle of *Plasmodium*, stage during which merozoites infect and cause lysis of erythrocytes.
Eschar Black, swollen, crusty, painless skin ulcer of anthrax.
Etest Test for determining minimum inhibitory concentration; a plastic strip containing a gradient of the antimicrobial agent being tested is placed on a plate inoculated with the pathogen of interest.
Ethambutol Antimicrobial drug that disrupts formation of arabinogalactan-mycolic acid by mycobacteria.
Etiology The study of the causation of disease.
Euglenids Protozoa that store food as paramylon, lack cell walls, and have eyespots used in positive phototaxis.
Eukarya In Woese's taxonomy, domain that includes all eukaryotic cells.
Eukaryote Any organism made up of cells containing a nucleus composed of genetic material surrounded by a distinct membrane. Classification includes animals, plants, algae, fungi, and protozoa.
Eutrophication The overgrowth of microorganisms in aquatic systems.
Evolution Changes in the genetic makeup of a population leading to the production of new varieties.
Exchange reaction Type of chemical reaction in which atoms are moved from one molecule to another by means of the breaking and forming of covalent bonds.
Excystment In the life cycle of protozoa, stage following ingestion by the host, in which cysts become trophozoites.
Exfoliative toxins Toxins of certain strains of *Staphylococcus aureus* that break down desmosomes in the skin, causing the outer layers of skin to slough off.
Exocytosis Active transport process, used by some eukaryotic cells, in which vesicles fuse with the cytoplasmic membrane and export their substances from the cell.
Exoerythrocytic phase (*liver phase*) In the life cycle of *Plasmodium* (cause of malaria), stage during which sporozoites, injected into the blood by an *Anopheles* mosquito, infect the liver.

Exogenous antigen Antigen produced by microorganisms that multiply outside the cells of the body.

Exogenous healthcare-associated infection An infection caused by pathogens acquired from the health care environment.

Exon Coding sequence of mRNA. Exons are connected to produce a functional mRNA molecule.

Exothermic reaction Any chemical reaction that releases energy.

Exotoxin Toxin secreted by a pathogenic microorganism into its environment.

Experimental epidemiology The testing of hypotheses resulting from analytical epidemiology concerning the cause of a disease.

Exponential (logarithmic) growth Increase in size of a microbial population in which the number of cells doubles in a fixed interval of time.

Extensively drug-resistant (XDR) tuberculosis Tuberculosis caused by extensively drug-resistant *Mycobacterium*.

Extremophile Microbe that requires extreme conditions of temperature, pH, and/or salinity to survive.

F (fertility) plasmid (*F factor*) Small, circular, extrachromosomal molecule of DNA coding for conjugation pili. Bacterial cells that contain an F plasmid are called F^+ cells and serve as donors during conjugation.

F_C region The stem region of an antibody.

Facilitated diffusion Movement of substances across a cell membrane via protein channels.

Facultative anaerobe Microorganism that can live with or without oxygen.

Family Taxonomical grouping of similar genera of organisms.

Fats Compounds composed of three fatty acid molecules linked to a molecule of glycerol.

Fecal-oral infection Spread of pathogenic microorganisms in feces to the mouth, such as results from drinking sewage-contaminated water.

Fecal transplant Injection of feces into a patient, typically via an enema tube; used to restore normal microbiota of the colon especially in recurrent cases with *Clostridium difficile* infection.

Feedback inhibition (*negative feedback*) Method of controlling the action of enzymes in which the end product of a series of reactions inhibits an enzyme in an earlier part of the pathway.

Fermentation In metabolism, the partial oxidation of sugar to release energy using an endogenous organic molecule rather than an electron transport chain as the final electron acceptor. In food microbiology, any desirable change to food or beverage induced by microbes.

Fever Body temperature above 37°C.

Fever blisters (*cold sores*) Painful, itchy lesions on the lips; characteristic of infection with *human herpesvirus 1*.

Filariasis Infection of the lymphatic system caused by a filarial nematode.

Filtration The passage of air or liquid through a material that traps and removes microbes. In water treatment, a process in which microbial biofilms on sand particles trap and remove other microbes.

Fimbriae Sticky, proteinaceous extensions of some bacterial cells that function to adhere cells to one another and to environmental surfaces.

Firmicutes Phylum of bacteria that includes clostridia, mycoplasmas, and low G+C Gram-positive bacilli and cocci.

Flagellates Group of protozoa that possess at least one long flagellum, generally used for movement.

Flagellum A long, whiplike structure protruding from a cell.

Flavin adenine dinucleotide (FAD) Important vitamin-derived electron carrier molecule.

Flea Vertically flattened, bloodsucking, wingless insect; vector of some pathogens.

Flocculation Process in water treatment in which alum (aluminum ammonium sulfate) added to the water forms sediments with particles and microorganisms.

Fluid mosaic model Model describing the arrangement and motion of the proteins within the cytoplasmic membrane.

Fluorescence microscope Type of light microscope that uses an ultraviolet light source to fluoresce objects.

Fluoroquinolone A type of quinolone antimicrobial that inhibits gyrase, the enzyme that corrects supercoiling by replicating bacterial DNA.

Fly Insect with transparent wings that are not hidden or covered, including mosquitoes; vectors for many pathogens.

Folliculitis Infection of a hair follicle by *Staphylococcus aureus*.

Fomites (*fomes*, sing.) Objects inadvertently used to transfer pathogens to new hosts, such as a glass or towel.

Food infection Type of food poisoning in which living organisms are consumed.

Food intoxication Type of food poisoning resulting from consumption of microbial toxin.

Food microbiology The use of microorganisms in food production and the prevention of foodborne illnesses.

Food vesicle Sac formed during endocytosis of a solid; also called an endosome or phagosome.

Foodborne transmission Spread of pathogenic microorganisms in or on foods that are poorly processed, undercooked, or improperly refrigerated.

Foraminifera Type of armored marine amoeba.

Formed elements Cells and cell fragments suspended in blood plasma.

Frameshift mutation Type of mutation in which nucleotide triplets subsequent to an insertion or deletion are displaced, creating new sequences of codons that result in vastly altered polypeptide sequences.

Functional group An arrangement of atoms common to all members of a class of organic molecules, such as the amine group found in all amino acids.

Fungi Eukaryotic organisms that have cell walls and obtain food from other organisms.

Furuncle A large, painful, nodular extension of folliculitis into surrounding tissue; may be caused by infection with *Staphylococcus aureus*.

Gametocyte In sexual reproduction of protozoa, cell that can fuse with another gametocyte to form a diploid zygote.

Gamma interferons (IFN-γ) Interferon produced by T lymphocytes and NK lymphocytes; activates macrophages and neutrophils days after an infection.

Gammaproteobacteria Largest and most diverse class of Proteobacteria, including purple sulfur bacteria, methane oxidizers, pseudomonads, and others.

Gas gangrene Death of muscle and connective tissues accompanied by gaseous waste, caused by *Clostridium perfringens*.

Gaseous agent High-level disinfecting gas used to sterilize heat-sensitive equipment and large objects.

Gastroenteritis Inflammation of the mucous membrane of the stomach and intestines.

Gastrointestinal anthrax Disease that can be fatal without aggressive treatment; caused by ingestion of spores of *Bacillus anthracis*.

GC content The percentage of a cell's DNA bases that are guanine and cytosine.

Gel electrophoresis Technique used in recombinant DNA technology to separate molecules by size, shape, and electrical charge.

Gene A specific sequence of nucleotides that codes for a polypeptide or an RNA molecule.

Gene library Collection of bacterial or phage clones, each of which carries a fragment of an organism's genome.

Gene therapy The use of recombinant DNA technology to insert a missing gene or repair a defective gene in human cells.

Genera Plural of genus.

Generation time Time required for a cell to grow and divide.

Genetic engineering The manipulation of genes via recombinant DNA technology for practical applications.

Genetic fingerprinting (*DNA fingerprinting*) Technique that identifies unique sequences of DNA to determine paternity; connect blood, semen, or skin cells to suspects in criminal investigations; or identify pathogens.

Genetic mapping Application of recombinant DNA technology in which genes are located on a nucleic acid molecule.

Genetic recombination The exchange of segments, typically genes, between two DNA molecules.

Genetic screening Procedure by which laboratory tests are used to screen patient and fetal DNA for mutant genes.

Genetics The study of inheritance and heritable traits as expressed in an organism's genetic material.

Genital herpes Painful, itchy lesions in or around the genitalia or anus caused by herpesviruses 1 or 2.

Genome The sum of all the genetic material in a cell or virus.

Genomics The sequencing, analysis, and comparison of genomes.

Genotype Actual set of genes in an organism's genome.

Genus Taxonomical grouping of similar species of organisms.

Germ theory of disease Hypothesis formulated by Pasteur in 1857 that microorganisms are responsible for disease.

German measles (*rubella*) Disease caused by infection with *Rubivirus* resulting in characteristic rash lasting about 3 days; mild in children but potentially teratogenic to fetuses of infected women.

Giardiasis A mild to severe gastrointestinal illness caused by ingestion of cysts of *Giardia intestinalis*.

Gingivitis Inflammation of the gums.

Glomerulonephritis Deposition of immune complexes in the walls of the glomeruli—networks of minute blood vessels in the kidneys—that may result in kidney failure; typically caused by infection with group A *Streptococcus*.

Glucocorticoids (*corticosteroids*) Immunosuppressive agents, including prednisone and methylprednisolone, that suppress the response of T cells.

Glycocalyx (*glycocalyces*, pl.) Sticky external sheath of prokaryotic and eukaryotic cells.

Glycolysis (*Embden-Meyerhof-Parnas pathway*) First step in the catabolism of glucose via respiration and fermentation.

Goblet cells Mucus-secreting cells in the epithelium of mucous membranes.

Golgi body In eukaryotic cells, a series of flattened, hollow sacs surrounded by phospholipid bilayers and functioning to package large molecules for export in secretory vesicles.

Gonorrhea A sexually transmitted disease caused by infection with *Neisseria gonorrhoeae*.

gp41 Antigenic HIV glycoprotein that promotes fusion of the viral envelope with a target cell.

gp120 Antigenic glycoprotein that is the primary attachment molecule of HIV.

Graft Tissue or organ transplanted to a new site.

Graft rejection Rejection of donated tissue or organs by a transplant recipient.

Graft-versus-host disease Disease resulting when donated bone marrow cells mount an immune response against the recipient's cells.

Gram-negative cell Generally, a prokaryotic cell having a wall composed of a thin layer of wall material, an external membrane, and a periplasmic space between; appears pink after the Gram-staining procedures.

Gram-positive cell Prokaryotic cell having a thick wall; in bacteria, composed of a thick layer of peptidoglycan containing teichoic acids; Gram-positive cells retain the crystal violet dye used in the Gram-staining procedure, appearing purple.

Gram stain Technique for staining microbial samples by applying a series of dyes that leave some microbes purple and others pink. Developed by Christian Gram in 1884.

Granulocyte Type of leukocyte having large granules in the cytoplasm.

Granzyme Protein molecule in the cytoplasm of cytotoxic T cells that causes an infected cell to undergo apoptosis.

Graves' disease Production of autoantibodies that stimulate excessive production of thyroid hormone and growth of the thyroid gland.

Gross mutation Major change in the nucleotide sequence of DNA resulting from inversions, duplications, transpositions, or large insertions or deletions of nucleotides.

Group A Streptococcus (*GAS*, *S. pyogenes*) A Gram-positive coccus that produces protein M and a hyaluronic acid capsule, both of which contribute to the pathogenicity of the species.

Group B Streptococcus (*S. agalactiae*) A Gram-positive coccus that normally resides in the lower GI, genital, and urinary tracts but can cause disease in newborns.

Group translocation Active process, occurring in some prokaryotes, by which a substance being actively transported across a cell membrane is chemically changed during transport.

Growth An increase in size; in bacteriology, an increase in population.

Growth curve Graph that plots the number in a population over time.

Growth factor Organic chemical, such as a vitamin, required in very small amounts for metabolism. In immunology, an immune system cytokine that stimulates stem cells to divide, ensuring that the body is supplied with sufficient leukocytes of all types.

Guanine Ring-shaped nitrogenous base found in nucleotides of DNA and RNA.

Gumma Lesion that occurs in bones, in nervous tissue, or on skin in patients with tertiary syphilis.

Gyrase Bacterial enzyme that relieves DNA supercoils that form when a bacterium replicates circular DNA.

HAART (highly active antiretroviral therapy) A cocktail of antiviral drugs including nucleoside analogs, integrase inhibitors, protease inhibitors, and reverse transcriptase inhibitors.

Habitat The physical localities in which organisms are found.

Halogen One of the four very reactive, nonmetallic chemical elements: iodine, chlorine, bromine, and fluorine. Used in disinfectants and antiseptics.

Halophile Microorganism requiring a saline environment (greater than 9% NaCl).

Hamus Proteinaceous, filamentous, helical extension of some archaeal cells that functions to attach the cells to one another and environmental surfaces.

Hansen's disease (*leprosy*) Disease caused by infection with *Mycobacterium leprae* that produces either a nonprogressive tuberculoid form or a progressive lepromatous form that destroys tissues, including facial features, digits, and other structures.

Hantavirus pulmonary syndrome (HPS) Rapid, severe, and often fatal pneumonia caused by infection with *Hantavirus*.

Hantaviruses Group of bunyaviruses that are transmitted to humans via inhalation of virions in dried deer-mouse excreta and that cause *Hantavirus* pulmonary syndrome.

Haploid A nucleus with a single copy of each chromosome.

Haustoria Modified hyphae that penetrate the tissue of the host to withdraw nutrients.

Hay fever Allergic reaction localized to the upper respiratory tract and characterized by nasal discharge, sneezing, itchy throat and eyes, and excessive tear production.

Healthcare-associated disease (*nosocomial disease*) A disease acquired in a health care facility.

Healthcare-associated infection (HAI) (*nosocomial infection*) An infection acquired in a health care facility.

Heat fixation In microscopy, a technique that uses the heat from a flame to attach a smear to a slide.

Heat shock Forcing a bacterial cell to become competent by placing it in a warm environment followed by a cold treatment such that the cell takes up DNA from its environment.

Heavy-metal ions The ions of relatively high density metals, such as arsenic and mercury, that are toxic at low concentrations. They can be used as antimicrobial agents, though they are seldom used to treat patients, because they are toxic to humans.

Helminths Multicellular eukaryotic worms, some of which are parasitic.

Helper T cell (*Th cell, CD4+ cell*) In cell-mediated immune response, a type of cell characterized by CD4 cell-surface glycoprotein; regulates the activity of B cells and cytotoxic T cells.

Hemagglutinin (HA) Component of glycoprotein spikes in the lipid envelope of influenzaviruses that help them attach to pulmonary epithelial cells.

Hemolytic disease of the newborn Disease that results when antibodies made by an Rh-negative woman cross the placenta and destroy the red blood cells of an Rh-positive fetus.

Hemorrhagic fever Viral syndrome characterized by fever, bleeding in the skin and mucous membranes, low blood pressure, and shock.

HEPA (high-efficiency particulate air) filter Filters built into biological safety cabinets; they prevent exposure to microbes by maintaining a barrier of moving filtered air across the cabinet's openings.

Hepatitis Inflammation of the liver.

Hepatitis A Inflammation of the liver resulting from infection with hepatitis A virus.

Hepatitis B Inflammatory condition of the liver caused by infection with hepatitis B virus and characterized by jaundice.

Hepatitis C Chronic inflammation of the liver that can cause permanent liver damage or hepatic cancer; caused by infection with hepatitis C virus.

Hepatitis D Inflammation of liver that can cause severe liver damage or hepatic cancer; caused by infection with hepatitis D virus.

Hepatitis E (*enteric hepatitis*) Inflammation of the liver, which is fatal in 20% of infected pregnant women; caused by hepatitis E virus.

Herd immunity Protection against illness provided to a population when a pathogen cannot spread because the majority of the group are resistant to the pathogen.

Herpangina Lesions of the mouth and pharynx caused by coxsackie A virus; resemble those of herpesvirus.

Herpes Painful, itchy skin lesions caused by herpesviruses.

Herpes zoster (*shingles*) Extremely painful skin rash caused by reactivation of latent varicella-zoster virus.

Heterocyst Thick-walled nonphotosynthetic cell of cyanobacteria; reduces nitrogen.

Hfr (high frequency of recombination) cell Cell containing an F plasmid that is integrated into the prokaryotic chromosome. Hfr cells form pili and transfer cellular genes more frequently than normal F cells.

Highly active antiretroviral therapy (HAART) A cocktail of antiviral drugs including nucleotide analogs, integrase inhibitors, protease inhibitors, and reverse transcriptase inhibitors.

Histamine Inflammatory chemical released from damaged cells that causes vasodilation of capillaries.

Histone Globular protein found in eukaryotic and archaeal chromosomes.

Histoplasmosis Pulmonary, cutaneous, ocular, or systemic disease found in the Ohio River valley and caused by infection with *Histoplasma capsulatum*.

Holoenzyme The combination of an apoenzyme and its cofactors.

Horizontal (lateral) gene transfer Process in which a donor cell contributes part of its genome to a recipient cell, which may be a different species or genus from the donor.

Host In symbiosis, member of a parasitic relationship that supports the parasite.

Human herpesviruses Group of viruses of humans that cause skin lesions, which are often creeping; diseases include herpes, chickenpox, mononucleosis, and roseola.

Human immunodeficiency viruses Retroviruses that destroy the immune system.

Human T-lymphotropic viruses Group of oncogenic retroviruses associated with cancer of lymphocytes.

Humoral immune response (antibody immune response) The immune response centered around B lymphocytes and immunoglobulins.

Hybridomas Tumor cells created by fusing antibody-secreting plasma cells with cancerous plasma cells called myelomas.

Hydatid Fluid-filled.

Hydatid disease Potentially fatal disease caused by infection with the canine tapeworm *Echinococcus granulosus* and characterized by the presence of fluid-filled cysts in the liver or other tissues.

Hydrogen bond The electrical attraction between a partially charged hydrogen atom and a full or partial negative charge on a different region of the same molecule or another molecule. Hydrogen bonds confer unique properties to water molecules.

Hydrolysis A decomposition reaction in which a covalent bond is broken and the ionic components of water are added to the products.

Hydrophilic Attracted to water.

Hydrophobia Literally, a fear of water; symptom caused by painful swallowing characteristic of rabies infection.

Hydrophobic Insoluble in water.

Hydrothermal vent Vent in marine abyssal zone that spews superheated, nutrient-rich water.

Hydroxyl radical Most reactive of the toxic forms of oxygen.

Hypersensitivity Any immune response against a foreign antigen that is exaggerated beyond the norm.

Hypersensitivity pneumonitis A form of pneumonia.

Hyperthermophile Microorganism requiring temperatures above 80°C.

Hypertonic Characteristic of a solution having a higher concentration of solutes than another.

Hyphae Long, branched, tubular filaments in the bodies of molds.

Hypotonic Characteristic of a solution having a lower concentration of solutes than another.

Iatrogenic infections A subset of healthcare-associated (nosocomial) infections that are the direct result of a medical procedure or treatment, such as surgery.

Illness In the infectious disease process, the most severe stage, in which signs and symptoms are most evident.

Immune complexes Antigen-antibody complexes.

Immunization Administration of an antigenic inoculum to stimulate an adaptive immune response and immunological memory.

Immunoblot (*western blot*) Variation of an ELISA test that can detect the presence of proteins, such as antibodies against multiple antigens.

Immunochromatographic assay Immune test in which antigen molecules form visible immune complexes with antibodies labeled with a colored substance.

Immunodiffusion An immune test in which antibodies and antigens diffuse from separate wells in agar to form a line of precipitate.

Immunofiltration assay Rapid modification of ELISA test using membrane filters rather than plates.

Immunoglobulin (Ig) (*antibody*) Proteinaceous antigen-binding molecule secreted by plasma cells.

Immunoglobulin A (IgA) The antibody class most commonly associated with various body secretions, including tears and milk. IgA pairs with a secretory component to form secretory IgA.

Immunoglobulin D (IgD) A membrane-bound antibody molecule found in some animals as a B cell receptor.

Immunoglobulin E (IgE) Signal antibody molecule that triggers the inflammatory response, particularly in allergic reactions and infections by parasitic worms.

Immunoglobulin G (IgG) The predominant antibody class found in the bloodstream and the primary defender against invading bacteria.

Immunoglobulin M (IgM) The second most common antibody class and the predominant antibody produced first during a primary humoral immune response.

Immunological synapse Interface between cells of the immune system that involves cell-to-cell signaling.

Immunology Study of the body's specific defenses against pathogens.

Immunophilins Immunosuppressive drugs, such as cyclosporine, that inhibit T cell function.

Immunotherapy Administration of antibodies (passive immunization) or dilute antigen so as to provide immunological protection against antigens.

Impetigo Presence of red, pus-filled vesicles on the face and limbs of children; caused by infection with *Staphylococcus aureus* or *Streptococcus pyogenes*.

Inactivated polio vaccine (IPV) Inoculum developed by Jonas Salk in 1955 for vaccination against poliovirus.

Inactivated vaccine Inoculum containing either whole agents or subunits and often adjuvants.

Incidence In epidemiology, the number of new cases of a disease in a given area or population during a given period of time.

Inclusion Deposited substance such as a lipid, gas vesicle, or magnetite stored within the cytosol of a cell.

Inclusion bodies In the life cycle of chlamydias, eukaryotic phagosomes full of chlamydial reticulate bodies.

Incubation period Stage in infectious disease process between infection and occurrence of the first symptoms or signs of disease. In a laboratory culture, the period between adding a sample to a plate and the development of colonies.

Index case In epidemiology, the first instance of the disease in a given area or population.

Indicator organisms Fecal microbe found in the environment that reveals potential contamination by feces.

Indirect contact transmission Spread of pathogens from one host to another via inanimate objects called fomites.

Indirect fluorescent antibody test Immune test allowing observation through a fluorescence microscope of the presence of antigen in a tissue sample flooded with labeled antibody.

Indirect selection (*negative selection*) Process by which auxotrophic mutants are isolated and cultured.

Induced-fit model Description of way in which an enzyme changes its shape slightly after binding to its substrate so as to bind it more tightly.

Inducible operon Type of operon that is not normally transcribed and must be activated by inducers.

Induction In virology, excision of a prophage from the host chromosome, at which point the prophage reenters the lytic phase.

Industrial microbiology Branch of microbiology in which microbes are manipulated to manufacture useful products.

Infection Successful invasion of the body by a pathogenic microorganism.

Infection control Branch of microbiology studying the prevention and control of infectious disease.

Infectious mononucleosis (*mono*) Disease characterized by sore throat, fever, fatigue, and enlargement of the spleen and liver; caused by infection with Epstein-Barr virus.

Influenza (flu) Infectious disease caused by two species of orthomyxoviruses and characterized by fever, malaise, headache, and myalgia; certain strains can be fatal.

Innate immunity Resistance to pathogens conferred by barriers, chemicals, cells, and processes that remain unchanged upon subsequent infections with the same pathogens.

Inoculum Sample of microorganisms.

Inorganic chemical Molecule lacking carbon.

Insect Arthropod with three distinct body divisions: head, thorax, abdomen; vectors for some helminthic, protozoan, bacterial, and viral pathogens.

Insertion Type of mutation in which a base pair is inserted into a genetic sequence.

Insertion sequence (IS) A simple transposon consisting of no more than two inverted repeats and a gene that encodes the enzyme transposase.

Integrase Enzyme carried by the virions of HIV that allows integration into a human chromosome.

Interferons (IFNs) Protein molecules that inhibit the spread of viral infections.

Interleukins (ILs) Immune system cytokines that signal among leukocytes.

Intermediate host In the life cycle of parasites, host in which immature forms of the parasite are present and undergoing various stages of maturation.

Intoxication (*bacterial*) Food poisoning caused by bacterial toxin.

Intron Noncoding sequence of mRNA that is removed to make functional mRNA.

In-use test Method of evaluating the effectiveness of a disinfectant or antiseptic that tests efficacy under specific, real-life conditions.

Inverted repeat (IR) Palindromic sequence found at each end of a transposon.

Ion An atom or group of atoms that has either a full negative charge or a full positive charge.

Ionic bond A type of bond formed from the attraction of opposite electrical charges. Electrons are not shared.

Ionizing radiation Form of radiation with wavelengths shorter than 1 nm that are energetic enough to create ions by ejecting electrons from atoms.

Ischemia Local anemia due to interruption of blood supply by mechanical blockage.

Isograft Type of graft in which tissues are moved between genetically identical individuals (identical twins).

Isoniazid (INH) Antimicrobial drug that disrupts formation of arabinogalactan-mycolic acid by mycobacteria.
Isotonic Characteristic of a solution having the same concentration of solutes and water as another.
Isotopes Atoms of a given element that differ only in the number of neutrons they contain.

Jaundice Yellowing of skin and eyes due to accumulation of bilirubin in the blood.

Kelsey-Sykes capacity test Standard assessment approved by the European Union to determine the ability of a given chemical to inhibit microbial growth.
Keratitis Inflammation of the cornea.
Kinetoplastid Euglenozoan protozoan with a single large mitochondrion that contains an apical region of mitochondrial DNA called a kinetoplast.
Kingdom Taxonomical grouping of similar phyla of organisms.
Kinins Powerful inflammatory chemicals released by mast cells.
Kissing bug Blood-eating insect of family Reduviidae that seemingly prefers oral blood vessels.
Koch's postulates A series of steps, elucidated by Robert Koch, that must be taken to prove the cause of any infectious disease.
Koplik's spots Mouth lesions characteristic of measles.
Korarchaeota Phylum of archaea; known only from environmental RNA samples.
Krebs cycle (*citric acid cycle*) Series of eight enzymatically catalyzed reactions that transfer stored energy from acetyl-CoA to coenzymes NAD^+ and FAD.

Lag phase Phase in a growth curve in which the organisms are adjusting to their environment.
Lagging strand Daughter strand of DNA synthesized in short segments that are later joined. Synthesis of the lagging strand always moves away from the replication fork and lags behind synthesis of the leading strand.
Laryngitis Inflammation of the larynx.
Latency In virology, process by which an animal virus, sometimes not incorporated into the chromosomes of the cell, remains inactive in the cell, possibly for years.
Latent disease Any disease in which a pathogen remains inactive for a long period of time before becoming active.
Latent syphilis Clinically inactive phase of syphilis.
Latent virus (*provirus*) An animal virus that remains inactive in a host cell.
Lateral (horizontal) gene transfer Process in which a donor cell contributes part of its genome to a recipient cell, which may be a different species or genus from the donor.
Leading strand Daughter strand of DNA synthesized continuously toward the replication fork as a single long chain of nucleotides.
Legionnaires' disease (*legionellosis*) Severe pneumonia caused by infection with a *Legionella* species, usually *L. pneumophila*.
Leishmaniasis Any of three clinical syndromes caused by a bite from a sand fly carrying *Leishmania* and ranging from painless skin ulcers to disfiguring lesions to visceral leishmaniasis, which is systemic and fatal in 95% of untreated cases.

Lepromin test Assay utilizing antigens of *Mycobacterium leprae* used in diagnosis of leprosy.
Leprosy (*Hansen's disease*) Disease caused by infection with *Mycobacterium leprae* that produces either a nonprogressive tuberculoid form or a progressive lepromatous form that destroys tissues, including facial features, digits, and other structures.
Leptospirosis Zoonotic disease contracted by humans upon exposure to infected animals; characterized by pain, headache, and liver and kidney disease; caused by infection with *Leptospira interrogans*.
Leukocyte White blood cell.
Leukopenia Decrease in the number of white blood cells in the blood.
Leukotrienes Inflammatory chemicals released from damaged cells that increase vascular permeability.
Lichen Organism composed of a fungus living in partnership with photosynthetic microbes, either green algae or cyanobacteria.
Light-dependent reaction Reaction of photosynthesis requiring light.
Light-independent reaction Reaction of photosynthesis not requiring light and synthesizing glucose from carbon dioxide and water.
Light repair Mechanism by which prokaryotic DNA photolyase breaks the bonds between adjoining pyrimidine nucleotides, restoring the original DNA sequence.
Limnetic zone Sunlit, upper layer of freshwater or marine water away from the shore.
Lincosamides Antimicrobial drugs that bind to the 50S subunit of bacterial ribosomes, preventing ribosomal movement.
Lipid Any of a diverse group of organic macromolecules not composed of monomers and insoluble in water.
Lipid A The lipid component of lipopolysaccharide, which is released from dead Gram-negative bacterial cells and can trigger shock and other symptoms in human hosts.
Lipoglycopeptide Antimicrobial drugs consisting of side chains of carbohydrates (glycans) and of lipids attached to relatively short chains of amino acids (peptide).
Lipopolysaccharide (LPS) (*endotoxin*) Toxic molecule composed of lipid A, which is the toxic portion, and polysaccharide found in the external membrane of Gram-negative cell walls.
Listeriosis Disease caused by *Listeria monocytogenes* and usually manifesting as meningitis and bacteremia.
Lithotroph Microorganism that acquires electrons from inorganic sources.
Littoral zone Shoreline zone of freshwater or marine water.
Liver phase (*exoerythrocytic phase*) In the life cycle of *Plasmodium* (cause of malaria), stage during which sporozoites, injected into the blood by an *Anopheles* mosquito, infect the liver.
Log phase Phase in a growth curve in which the population is most actively growing.
Logarithmic (exponential) growth Increase in size of a microbial population in which the number of cells doubles in a fixed interval of time.
Louse (*lice*, pl.) Sucking or biting parasitic insects that vector some bacterial pathogens.
Louse-borne relapsing fever Disease caused by *Borrelia recurrentis* transmitted between humans by the body louse *Pediculus humanus*.

Lumbar puncture (spinal tap) Collection of cerebrospinal fluid from the lower vertebral column for diagnostic purposes.
Lyme disease Disease carried by ticks infected with *Borrelia burgdorferi* and characterized by a "bull's-eye" rash, neurologic and cardiac dysfunction, and severe arthritis.
Lymph Fluid found in lymphatic vessels that is similar in composition to blood serum and intercellular fluid.
Lymph nodes Organs that monitor the composition of lymph.
Lymphangitis Condition in which inflamed lymphatic vessels become visible as red streaks under the skin.
Lymphatic system Body system composed of lymphatic vessels and lymphoid tissues and organs.
Lymphatic vessels Tubes that conduct lymph.
Lymphocyte Type of small agranulocyte that originates in the red bone marrow and has nuclei that nearly fill the cell.
Lymphocytic choriomeningitis (LCM) Zoonosis caused by an arenavirus; characterized by flulike symptoms and rarely by meningitis.
Lymphogranuloma venereum Sexually transmitted disease caused by infection with *Chlamydia trachomatis* and leading in some cases to proctitis or, in women, pelvic inflammatory disease.
Lyophilization Removal of water from a frozen culture or other substance by means of vacuum pressure. Used for the long-term preservation of cells and foods.
Lysogenic conversion Change in phenotype due to insertion of a lysogenic bacteriophage into a bacterial chromosome.
Lysogenic phage Bacteriophage that does not immediately kill its host cell.
Lysogenic replication cycle (*lysogeny*) Process of viral replication in which a bacteriophage enters a bacterial cell, inserts into the DNA of the host, and remains inactive. The phage is then replicated every time the host cell replicates its chromosome. Later, the phage may leave the chromosome.
Lysogeny (*lysogenic replication cycle*) Process of viral replication in which a bacteriophage enters a bacterial cell, inserts into the DNA of the host, and remains inactive. The phage is then replicated every time the host cell replicates its chromosome. Later, the phage may leave the chromosome.
Lysosome Vesicle in animal cells that contains digestive enzymes.
Lysozyme Antibacterial protein secreted in sweat.
Lytic replication cycle Process of viral replication consisting of five stages ending with lysis of and release of new virions from the host cell.

Macrolide Antimicrobial agent that inhibits protein synthesis by inhibiting the ribosomal 50S subunits.
Macrophage Mature form of monocyte, which is a phagocyte of bacteria, fungi, spores, and dust, as well as dead cells.
Macule Any flat, reddened skin lesion; characteristic of early infection with a poxvirus.
Magnification The apparent increase in size of an object viewed via microscopy.
Major histocompatibility complex (MHC) A cluster of genes, located on each copy of chromosome 6 in humans, that codes for membrane-bound glycoproteins called major histocompatibility antigens.

Malaise Feeling of general discomfort.

Malaria A mild to potentially fatal disease caused by a bite from an *Anopheles* mosquito carrying any of four species of *Plasmodium*; characterized by fever, chills, hemorrhage, and potential destruction of brain tissue.

MALDI-TOF mass spectrometry (MALDI-TOF) Matrix-assisted laser desorption/ionization time-of-flight mass spectrometry is a technique that uses a matrix to stabilize large organic molecules, such as proteins, that have been ionized with laser energy so that the ions retain structure when sorted based on mass-to-charge ratio by mass spectrometry.

Malignant tumor Mass of neoplastic cells that can invade neighboring tissues and may metastasize to cause tumors in distant organs or tissues.

Marburg hemorrhagic fever Often fatal disease beginning with flu-like symptoms and progressing to severe internal hemorrhaging. Caused by *Marburgvirus*.

Margination Process by which leukocytes stick to the walls of blood vessels at the site of infection.

Mast cells Specialized cells located in connective tissue that release histamine when they are exposed to complement.

Matter Anything that takes up space and has mass.

MDR TB (multi-drug-resistant TB) Tuberculosis caused by *Mycobacterium* resistant to at least isoniazid and rifampin.

Measles (*rubeola*) Contagious disease characterized by fever, sore throat, headache, dry cough, conjunctivitis, and lesions called Koplik's spots; caused by infection with *Morbillivirus*.

Mechanical vector Housefly, cockroach, or other animal that passively carries pathogens to new hosts on its feet or other body parts and is not infected by the pathogens it carries.

Medical technologist An expert in health care-related microbiological laboratory procedures.

Medium A collection of nutrients used for cultivating microorganisms.

Meiosis Nuclear division of diploid eukaryotic cells resulting in four haploid nuclei.

Membrane attack complexes (MACs) The end products of the complement cascade, which form circular holes in a pathogen's membrane.

Membrane filters Thin circles of nitrocellulose or plastic containing specific pore sizes, some small enough to trap viruses.

Membrane filtration Direct method of estimating population size in which a large sample is poured through a filter small enough to trap cells.

Membrane raft In a eukaryotic membrane, a distinct assemblage of lipids and proteins that remains together as a functional group.

Memory B cell B lymphocyte that migrates to lymphoid tissues to await a subsequent encounter with antigen previously encountered.

Memory response The rapid and enhanced immune response to a subsequent encounter with a familiar antigen.

Memory T cell Type of T cell that persists in lymphoid tissues for months or years awaiting subsequent contact with an antigenic determinant matching its TCR, at which point it produces cytotoxic T cells.

Meningitis Inflammation of the meninges, which can be caused by bacteria, viruses, fungi, or protozoa.

Meningoencephalitis Inflammation of the brain and of its meninges.

Mesophile Microorganism requiring temperatures ranging from 20°C to about 40°C.

Messenger RNA (mRNA) Form of ribonucleic acid that carries genetic information from DNA to a ribosome.

Metabolism The sum of all chemical reactions, both anabolic and catabolic, within an organism.

Metachromatic granules Inclusions of *Corynebacteria* that store phosphate and stain differently from the rest of the cytoplasm.

Metagenomic sequencing Sequencing all DNA from an environmental site to discover all microbes present.

Metaphase Second stage of mitosis, during which chromosomes line up and attach to microtubules of the spindle. Also used for the comparable stage of meiosis.

Metastasis The spreading of malignant cancer cells to nonadjacent organs and tissues, where they produce new tumors.

Methane oxidizer Any Gram-negative bacterium that utilizes methane both as a carbon and as an energy source.

Methanogen Obligate anaerobe that produces methane gas.

Methicillin-resistant Staphylococcus aureus (MRSA) Strain of *S. aureus* that is resistant to many common antimicrobial drugs and has emerged as a major nosocomial problem.

Methylation Process in which a cell adds a methyl group to one or two bases that are part of specific nucleotide sequences.

Microaerophile Microorganism that requires low levels of oxygen.

Microbe An organism or virus too small to be seen without a microscope.

Microbial antagonism (microbial competition) Normal condition in which established microbiota use up available nutrients and space, reducing the ability of arriving pathogens to colonize.

Microbial death Permanent loss of reproductive capacity of a microorganism.

Microbial death rate A measurement of the efficacy of an antimicrobial agent.

Microbial ecology The study of the interactions of microorganisms among themselves and their environment.

Microbiome All the microbes in a particular environment.

Microbiota The group of microbes that normally inhabit the body, usually without causing disease.

Microglia Fixed macrophages of the nervous system.

Micrograph A photograph of a microscopic image.

Microorganism An organism too small to be seen without a microscope.

MicroRNA (miRNA) Short (about 21-nucleotide) RNA molecule that binds to complementary segment of messenger RNA (mRNA), preventing translation.

Microscopy The use of light or electrons to magnify objects.

Microsporidia Unicellular, intracellular, parasitic fungi previously classified as protozoa.

Middle East respiratory syndrome (MERS) Manifestation of infection by a coronavirus called MERS virus.

Minimum bactericidal concentration (MBC) test An extension of the MIC test in which samples taken from clear MIC tubes are transferred to plates containing a drug-free growth medium and monitored for bacterial replication.

Minimum inhibitory concentration (MIC) The smallest amount of a drug that will inhibit a pathogen.

Mismatch repair Mechanism by which enzymes scan newly synthesized, nonmethylated DNA for mismatched bases, remove them, and replace them.

Missense mutation A substitution in a nucleotide sequence resulting in a codon that specifies a different amino acid: What is transcribed makes sense but not the right sense.

Mite Minute arachnid, which vectors *Orientia*, the agent of scrub typhus.

Mitochondria Spherical to elongated structures found in most eukaryotic cells that produce most of the ATP in the cell.

Mitosis Nuclear division of a eukaryotic cell resulting in two nuclei with the same ploidy as the original.

Mold A typically multicellular fungus that grows as long filaments called *hyphae* and reproduces by means of spores.

Molecular biology Branch of biology combining aspects of biochemistry, cell biology, and genetics to explain cell function at the molecular level, particularly via the use of genome sequencing.

Molecular mimicry Process in which microorganisms with epitopes similar to self-antigens trigger autoimmune tissue damage.

Molecule Two or more atoms held together by chemical bonds.

Molluscum contagiosum Skin disease caused by *Molluscipoxvirus*; characterized by smooth, waxy papules.

Monoclonal antibodies Identical antibodies secreted by a cell line originating from a single plasma cell.

Monocyte Type of agranulocyte that has slightly lobed nuclei.

Monoecious One individual contains both male and female organs.

Monomer A subunit of a macromolecule, such as a protein.

Monosaccharide (*simple sugar*) A monomer of carbohydrate, such as a molecule of glucose.

Morbidity Any change from a state of health.

Mordant In microscopy, a substance that binds to a dye and makes it less soluble.

Mosquito Type of fly with bloodsucking females; vector for many pathogens.

Most probable number (MPN) method Statistical estimation of the size of a microbial population based upon the dilution of a sample required to eliminate microbial growth.

Multi-drug-resistant (MDR) tuberculosis Tuberculosis caused by *Mycobacterium* resistant to at least isoniazid and rifampin.

Multiple drug-resistant pathogens Disease-causing microbes insensitive to three or more different types of antimicrobial agents; colloquially called superbugs.

Multiple sclerosis (MS) Autoimmune disease in which cytotoxic T cells attack and destroy the myelin sheath that insulates neurons.

Mumps Disease caused by infection with the mumps virus and characterized by fever, parotitis, pain in swallowing, and, in some cases, meningitis or deafness.

Mupirocin Antimicrobial drug that selectively binds to isoleucyl-tRNA synthetase, preventing this tRNA from functioning.

Mutagen Physical or chemical agent that introduces a mutation.

Mutant A cell with an unrepaired genetic mutation, or any of its descendants.

Mutation In genetics, a permanent change in the nucleotide base sequence of a genome.

Mutualism Symbiotic relationship in which both members benefit from their interaction.

Myalgia Muscle pain.

Mycelium Tangled mass of hyphae.

Mycetismus Mushroom poisoning.

Mycetoma Destructive, tumorlike infection of the skin, fascia, and/or bones of the hands or feet caused by mycelial fungi of several genera in the division Ascomycota.

Mycolic acid Long carbon-chain waxy lipid found in the walls of cells in the genus *Mycobacterium* that makes them resistant to desiccation and staining with water-based dyes.

Mycology The scientific study of fungi.

Mycoplasmas Class of low G + C bacteria that lack cytochromes, enzymes of the Krebs cycle, and cell walls and are pleomorphic.

Mycosis Fungal disease.

Mycotoxicosis Poisoning caused by eating food contaminated with fungal toxins.

Mycotoxins Secondary metabolites produced by fungi and toxic to humans.

Myxobacteria Gram-negative, aerobic, soil-dwelling bacteria with a unique life cycle including a stage of differentiation into fruiting bodies containing resistant myxospores.

Nanoarchaeum Small archaeon genus that possibly represents a fourth phylum of archaea; known only from environmental RNA samples.

Narrow-spectrum drug Antimicrobial that works against only a few kinds of pathogens.

Natural killer (NK) lymphocyte (NK cell) Type of defensive leukocyte of innate immunity that secretes toxins onto the surfaces of virally infected cells and neoplasms.

Naturally acquired active immunity Type of immunity that occurs when the body responds to exposure to antigens by mounting specific immune responses.

Naturally acquired passive immunity Type of immunity that occurs when a fetus, newborn, or child receives antibodies across the placenta or within breast milk.

Necrosis Death of a tissue or organ.

Necrotizing fasciitis Potentially fatal condition marked by toxemia, organ failure, and destruction of muscle and fat tissue following infection with group A *Streptococcus*.

Negative feedback (*feedback inhibition*) Method of controlling the action of enzymes in which the end-product of a series of reactions inhibits an enzyme in an earlier part of the pathway.

Negative selection (*indirect selection*) Process by which auxotrophic mutants are isolated and cultured.

Negative stain (*capsule stain*) In microscopy, a staining technique used primarily to reveal bacterial capsules and involving application of an acidic dye that leaves the specimen colorless and the background stained.

Negative-sense single-stranded RNA (negative single-stranded RNA, −ssRNA) Viral single-stranded RNA that is not recognized by ribosomes.

Negri bodies Aggregates of virions in the brains of rabies patients.

Nematodes Group of round, unsegmented helminths with pointed ends that have a complete digestive tract.

Neoplasia Uncontrolled cell division in a multicellular animal.

Nephelometry Automated method that measures the cloudiness of a solution by quantifying the amount of light it scatters.

Neuraminidase (NA) Component of glycoprotein spikes in the lipid envelope of influenzaviruses that provides access to cell surfaces by hydrolyzing mucus in the lungs.

Neutralization Antibody function in which the action of a toxin or attachment of a pathogen is blocked.

Neutralization test Immune test that measures the ability of antibodies to neutralize the biological activity of pathogens and toxins.

Neutron An uncharged subatomic particle.

Neutrophil Type of granulocyte that stains lilac with a mixture of acidic and basic dyes.

Neutrophile Microorganism requiring neutral pH.

Next-generation sequencing (NGS) Automated techniques for determining large numbers of nucleotide sequences simultaneously. NGS largely replaces first-generation, Sanger sequencing methods.

Nicotinamide adenine dinucleotide (NAD$^+$) Important vitamin-derived electron carrier molecule.

Nicotinamide adenine dinucleotide phosphate (NADP$^+$) Important vitamin-derived electron carrier molecule.

Nitrification The process by which bacteria convert reduced nitrogen compounds such as ammonia into nitrate, which is more available to plants.

Nitrifying bacteria Chemoautotrophic bacteria that derive electrons from the oxidation of nitrogenous compounds.

Nitrogen cycle Biogeochemical cycle involving nitrogen fixation, ammonification, nitrification, denitrification, and anammox reactions.

Nitrogen fixation The conversion of atmospheric nitrogen to ammonia.

NOD protein In innate immunity, intracellular receptor for microbial component.

Noncommunicable disease An infectious disease that arises from outside of hosts or from normal microbiota.

Noncompetitive inhibitor Inhibitory substance that blocks enzyme activity by binding to an allosteric site on the enzyme other than the active site.

Noncyclic photophosphorylation The production of ATP by noncyclic electron flow.

Nonionizing radiation Electromagnetic radiation with a wavelength greater than 1 nm.

Nonliving reservoir of infection Soil, water, food, or inanimate object that is a continuous source of infection.

Nonpolar covalent bond Type of chemical bond in which there is equal sharing of electrons between atoms with similar electronegativities.

Nonsense mutation A substitution in a nucleotide sequence that causes an amino acid codon to be replaced by a stop codon.

Normal microbiota Microorganisms that colonize the surfaces of the human body without normally causing disease. They may be resident or transient.

Noroviruses Group of caliciviruses that cause diarrhea.

Nosocomial disease (*healthcare-associated disease*) A disease acquired in a health care facility.

Nosocomial infection (*healthcare-associated infection*) An infection acquired in a health care facility.

Nuclear envelope Double membrane composed of phospholipid bilayers surrounding a cell nucleus.

Nuclear pores Spaces in the nuclear envelope that function to control the transport of substances through it.

Nucleoid Region of the prokaryotic cytosol containing the cell's chromosome(s).

Nucleolus Specialized region in a cell nucleus where RNA is synthesized.

Nucleoplasm The semiliquid matrix of a cell nucleus.

Nucleoside Component of a nucleotide consisting of a nitrogenous base and a five-carbon sugar.

Nucleoside analog Chemical with a structure similar to a natural nucleoside.

Nucleosome Bead of DNA bound to histone in a eukaryotic chromosome.

Nucleotide Monomer of a nucleic acid, which is composed of a nucleoside and a phosphate.

Nucleotide analog Compound structurally similar to a normal nucleotide that can be incorporated into DNA; may result in mismatched base pairing.

Nucleotide excision repair (*dark repair*) Mechanism by which enzymes cut a section of a single strand of damaged DNA from a molecule, creating a gap that is repaired by DNA polymerase and ligase.

Nucleus Spherical to ovoid membranous organelle containing a eukaryotic cell's primary genetic material.

Numerical aperture Measure of the ability of a lens to gather light.

Nutrient Any chemical, such as carbon, hydrogen, and so on, required for growth of microbial populations.

Objective lens In microscopy, the lens immediately above the object being magnified.

Obligate aerobe Microorganism that requires oxygen as the final electron acceptor of the electron transport chain.

Obligate anaerobe Microorganism that cannot tolerate oxygen and uses a final electron acceptor other than oxygen.

Obligate halophile Microorganism requiring high osmotic pressure.

Occult septicemia The condition of an unidentified bacterial pathogen being present in the blood and causing signs of illness.

Ocular herpes (*ophthalmic herpes*) Disorder characterized by conjunctivitis, a gritty feeling in the eye, and pain and characteristic of latent *human herpesvirus 1*.

Ocular lens In microscopy, the lens closest to the eyes. May be single (monocular) or paired (binocular).

Operator Regulatory element in an operon where repressor protein binds to stop transcription.

Operon A series of genes, a promoter, and often an operator sequence controlled by one regulatory gene. The operon model explains gene regulation in prokaryotes.

Opportunistic pathogens Microorganisms that cause disease when the immune system is suppressed, when microbial antagonism is reduced, or when introduced into an abnormal area of the body.

Opsonin Antimicrobial protein that enhances phagocytosis.

Opsonization The coating of pathogens by proteins called opsonins, making them more vulnerable to phagocytes.

Optimum growth temperature Temperature at which a microorganism's metabolic activities produce the highest growth rate.

Oral herpes Painful, itchy skin lesions around the mouth and lips, caused by herpesvirus 1 or 2.

Oral polio vaccine (OPV) Inoculum developed by Albert Sabin in 1961 for vaccination against poliovirus.

Orchitis Inflammation of a testis.

Order Taxonomical grouping of similar families of organisms.

Organelle Cellular structure that acts as a tiny organ to carry out one or more cell functions.

Organic compounds Molecules that contain both carbon and hydrogen atoms.

Organotroph Microorganism that acquires electrons from organic sources.

Ornithosis (*psittacosis, parrot fever*) A respiratory disease of birds that can be transmitted to humans and is caused by infection with *Chlamydophila psittaci*.

Orphan virus A virus that has not been specifically linked to any particular disease.

Osmosis The diffusion of water molecules across a semipermeable membrane.

Osmotic pressure The pressure exerted across a selectively permeable membrane by the solutes in a solution on one side of the membrane. The osmotic pressure exerted by high-salt or high-sugar solutions can be used to inhibit microbial growth in certain foods.

Osteomyelitis Inflammation of the bone marrow and surrounding bone; often caused by infection with *Staphylococcus*.

Otitis media Inflammation of the middle ear, often caused by *Streptococcus pneumoniae*.

Oxazolidinone Antibacterial drug that inhibits initiation of polypeptide synthesis in Gram-positive bacteria.

Oxidase test Chemical test for presence of cytochrome oxidase in a cell.

Oxidation lagoons Successive wastewaster treatment areas (lagoons) used by farmers and ranchers to treat animal wastes and from which water is released into natural water systems.

Oxidation-reduction reaction (*redox reaction*) Any metabolic reaction involving the transfer of electrons from an electron donor to an electron acceptor. Reactions in which electrons are accepted are called reduction reactions, whereas reactions in which electrons are donated are oxidation reactions.

Oxidative phosphorylation The use of energy from redox reactions to attach inorganic phosphate to ADP.

Oxidizing agent Antimicrobial agent that releases oxygen radicals.

P site In a ribosome, a binding site that holds a tRNA and the growing polypeptide.

Palisade In cell morphology, a folded arrangement of bacilli.

Pandemic In epidemiology, the occurrence of an epidemic on more than one continent simultaneously.

Papilloma (*wart*) Benign growth of the epithelium of the skin or mucous membranes.

Papule Any raised, reddened skin lesion that progresses from a macule; characteristic of infection with a poxvirus.

Parabasalid Group of single-celled, animal-like microorganisms that contain a Golgi-like parabasal body.

Paracoccidioidomycosis Pulmonary disease found from southern Mexico to South America caused by infection with *Paracoccidioides brasiliensis*.

Parainfluenzaviruses Group of enveloped, negative ssRNA viruses that cause respiratory disease, particularly in children.

Parasite A microbe that derives benefit from its host while harming it or even killing it.

Parasitism Symbiotic relationship in which one organism derives benefit while harming or even killing its host.

Parasitology The study of parasites.

Parenteral route A means by which pathogenic microorganisms can be deposited directly into deep tissues of the body, as in puncture wounds and hypodermic injections.

Paroxysmal phase In pertussis, second phase lasting two to four weeks and characterized by exhausting coughing spells.

Parrot fever (psittacosis, ornithosis) A respiratory disease of birds that can be transmitted to humans and is caused by infection with *Chlamydophila psittaci*.

Parvoviruses Group of extremely small, pathogenic ssDNA viruses.

Passive immunotherapy (*passive immunization*) Delivery of preformed antibodies against pathogens to patients.

Pasteurellaceae Family of gammaproteobacteria, two genera of which—*Pasteurella* and *Haemophilus*—are pathogenic.

Pasteurization The use of heat to kill pathogens and reduce the number of spoilage microorganisms in food and beverages.

Pathogen A microorganism capable of causing disease.

Pathogen-associated molecular patterns (PAMPs) Molecules that are shared by a variety of microbes, are absent in humans, and trigger immune responses.

Pathogenicity A microorganism's ability to cause disease.

Pelvic inflammatory disease (PID) Infection of the uterus and uterine tubes; may be caused by infection with any of several bacteria.

Pentose phosphate pathway Enzymatic formation of phosphorylated pentose sugars from glucose 6-phosphate.

Peptic ulcer Erosion of the mucous membrane of the stomach or duodenum, usually caused by infection with *Helicobacter pylori*.

Peptide bond A covalent bond between amino acids in proteins.

Peptidoglycan Large, interconnected polysaccharide composed of chains of two alternating sugars and crossbridges of amino acids. Main component of bacterial cell walls.

Perforin Protein molecule in the cytoplasm of cytotoxic T cells that forms channels (perforations) in an infected cell's membrane.

Periodontal disease Inflammation and infection of the tissues surrounding and supporting the teeth.

Periplasmic space In Gram-negative cells, the space between the cell membrane and the outer membrane containing peptidoglycan and periplasm.

Peritrichous Term used to describe a cell having flagella covering the cell surface.

Persister cell Bacterial cell that resists the action of antimicrobial drugs even though it is genetically identical to susceptible cells.

Peroxide anion Toxic form of oxygen which is detoxified by catalase or peroxidase.

Peroxisome Vesicle found in all eukaryotic cells that degrades poisonous metabolic wastes.

Pertussis (*whooping cough*) Pediatric disease characterized by development of copious mucus, loss of tracheal cilia, and deep "whooping" cough; caused by infection with *Bordetella pertussis*.

Petechiae Subcutaneous hemorrhages.

Petri plate Dish filled with solid medium used in culturing microorganisms.

pH scale A logarithmic scale used for measuring the concentration of hydrogen ions in a solution.

Phaeohyphomycosis Cutaneous and subcutaneous disease characterized by lesions that can spread internally; caused by traumatic introduction of ascomycetes into the skin.

Phaeophyta Brown-pigmented division of algae having cell walls composed of cellulose and alginic acid, a thickening agent.

Phage (*bacteriophage*) Virus that infects and usually destroys bacterial cells.

Phage typing Method of classifying microorganisms in which unknown bacteria are identified by observing plaques.

Phagocytes Cells, often leukocytes, that are capable of phagocytosis.

Phagocytosis Type of endocytosis in which solids are moved into the cell.

Phagolysosome Digestive vesicle formed by the fusing of a lysosome with a phagosome.

Phagosome A sac formed by a phagocyte's pseudopods; an intracellular food vesicle.

Pharyngitis (*strep throat*) Inflammation of the throat, often caused by infection with group A *Streptococcus*.

Phase microscope Type of microscope used to examine living microorganisms or fragile specimens.

Phase-contrast microscope Type of phase microscope that produces sharply defined images in which fine structures can be seen in living cells.

Phenol coefficient Method of evaluating the effectiveness of a disinfectant or antiseptic that compares the agent's efficacy to that of phenol.

Phenolic Compound derived from phenol molecules that have been chemically modified to denature proteins and disrupt cell membranes in a wide variety of pathogens.

Phenotype The physical features and functional traits of an organism expressed by genes in the genotype.

Phospholipid Phosphate-containing lipid made up of molecules with two fatty acid chains.

Phospholipid bilayer Two-layered structure of a cell's membranes.
Phosphorus cycle Biogeochemical cycle in which phosphorus is cycled between oxidation states.
Photoautotroph Microorganism that requires light energy and uses carbon dioxide as a carbon source.
Photoheterotroph Microorganism that requires light energy and gains nutrients via catabolism of organic compounds.
Photophosphorylation The use of energy from light to attach inorganic phosphate to ADP.
Photosynthesis Process in which light energy is captured by chlorophylls and transferred to ATP and metabolites.
Photosystem Network of light-absorbing chlorophyll molecules and other pigments held within a protein matrix on thylakoids.
Phototaxis Cell movement that occurs in response to light stimulus.
Phycoerythrin Red accessory pigment of photosynthesis in red algae.
Phycology Study of algae.
Phylum Taxonomical grouping of similar classes of organisms.
Picornaviruses Family of viruses that contain positive single-stranded RNA with naked polyhedral capsids; many are human pathogens.
Piedra Firm, irregular nodules on hair shafts caused by aggregates of fungal hyphae and spores.
Pilus (*conjugation pilus*) A tubule involved in bacterial conjugation.
Pinocytosis Type of endocytosis in which liquids are moved into the cell.
Pinta Childhood disease caused by the spirochete *Treponema carateum*; characterized by hard, pus-filled lesions.
Pinworm Common name of *Enterobius vermicularis*, whose adult female has a tail like a straight pin.
Pityriasis versicolor Condition characterized by depigmented or hyperpigmented patches of scaly skin resulting from infection with *Malassezia furfur*.
Plague Disease caused by *Yersinia pestis*, often manifesting with enlarged lymph nodes (bubonic plague) or with severe pulmonary distress (pneumonic plague).
Plaque In phage typing, the clear region within the bacterial lawn where growth is inhibited by bacteriophages.
Plaque assay Technique for estimating phage numbers in which each plaque corresponds to a single phage in the original bacterium/virus mixture.
Plasma The liquid portion of blood.
Plasma cells B cells that are actively fighting against exogenous antigens and secreting antibodies.
Plasmid A small, circular molecule of DNA that replicates independently of the chromosome. Each carries genes for its own replication and often for one or more nonessential functions such as resistance to antimicrobial drugs.
Plasmodial slime mold (*acellular slime mold*) Streaming, coenocytic, colorful filaments of cytoplasm that phagocytize organic debris and bacteria.
Platelet Cell fragments involved in blood clotting.
Platelet activating factor (PAF) Cytokine that is a potent trigger for blood coagulation.
Pleomorphic In cell morphology, term used to describe a variably shaped prokaryotic cell.

Pneumococcal pneumonia Inflammation of the lungs caused by *Streptococcus pneumoniae*—the pneumococcus.
Pneumococcus Common name of *Streptococcus pneumoniae*.
Pneumocystis pneumonia (PJP) Debilitating fungal pneumonia that is a leading cause of death in AIDS patients and is caused by opportunistic infection with *Pneumocystis jirovecii*.
Pneumonia Inflammation of the lungs; typically caused by infection with *Streptococcus pneumoniae*.
Pneumonic plague Fever and severe respiratory distress caused by infection of the lungs with *Yersinia pestis*; fatal if untreated in nearly 100% of cases.
Point mutation A genetic mutation affecting only one or a few base pairs in a genome. Point mutations include substitutions, insertions, and deletions.
Polar In cell morphology, pertaining to either end of a cell, such as polar flagella.
Polar covalent bond Type of bond in which there is unequal sharing of electrons between atoms with opposite electrical charges.
Poliomyelitis (*polio*) Infection of varied degrees of severity from asymptomatic to crippling and caused by infection with poliovirus.
Polluted Containing microorganisms or chemicals in excess of acceptable values.
Polyenes Group of antimicrobial drugs such as amphotericin B that disrupt the cytoplasmic membrane of targeted cells by becoming incorporated into the membrane and damaging its integrity.
Polymer Repeating chains of covalently linked monomers found in macromolecules.
Polymerase chain reaction (PCR) Technique of recombinant DNA technology that allows researchers to produce a large number of identical DNA molecules *in vitro*.
Polyomavirus A cancer-causing virus.
Polysaccharide Carbohydrate polymer composed of several to thousands of covalently linked monosaccharides.
Polyunsaturated fatty acid A long-chain organic acid with several double bonds between adjacent carbon atoms.
Portal of entry Entrance site of pathogenic microorganisms, including the skin, mucous membranes, and placenta.
Portal of exit Exit site of pathogenic microorganisms, including the nose, mouth, and urethra.
Positive selection Process by which mutants are selected by eliminating wild-type phenotypes.
Positive-sense single-stranded RNA (positive-strand RNA, +ssRNA) Viral single-stranded RNA that can act directly as mRNA.
Postpolio syndrome Crippling deterioration of muscle function, likely due to aging-related aggravation of nerve damage by poliovirus.
Potable Fit to drink.
Pour-plate Method of culturing microorganisms in which colony-forming units are separated from one another using a series of dilutions.
Pox (*pocks; pustule*) Any raised, pus-filled skin lesion; characteristic of infection with a poxvirus.
Precursor metabolite Any of 12 molecules typically generated by a catabolic pathway and essential to the synthesis of organic macromolecules in cells.
Prevalence In epidemiology, the total number of cases of a disease in a given area or population during a given period of time.

Primary amebic meningoencephalopathy Often fatal inflammation of the brain characterized by headache, vomiting, fever, and destruction of neurological tissue; caused by infection with *Naegleria* or *Acanthamoeba*.
Primary atypical pneumonia So-called walking pneumonia, characterized by mild respiratory symptoms that last for several weeks; caused by infection with *Mycoplasma pneumoniae*.
Primary immunodeficiency diseases Any of a group of diseases detectable near birth and resulting from a genetic or developmental defect.
Primary response The slow and limited immune response to a first encounter with an unfamiliar antigen.
Primary stain In staining, the initial dye, which colors all cells.
Primary syphilis Initial phase of the sexually-transmitted disease caused by *Treponema pallidum* characterized by a small, painless, red, hard lesion called a chancre.
Prion Proteinaceous infectious particle that lacks nucleic acids and replicates by converting similar normal proteins into new prions.
Probe Nucleic acid molecule with a specific nucleotide sequence that has been labeled with a radioactive or fluorescent chemical so that its location can be detected.
Probiotic Live microorganisms administered to improve health and prevent disease.
Prodromal period In the infectious disease process, the short stage of generalized, mild symptoms that precedes illness.
Products The atoms, ions, or molecules that remain after a chemical reaction is complete.
Profundal zone Zone of freshwater or marine water beneath the limnetic zone and above the benthic zone.
Proglottids Body segments of a tapeworm, produced continuously as long as the worm remains attached to its host.
Progressive multifocal leukoencephalopathy (PML) Progressive, fatal disease in which JC virus (a polyomavirus) kills cells of the central nervous system.
Prokaryote Any unicellular microorganism that lacks a nucleus. Classification includes bacteria and archaea.
Promoter Region of DNA where transcription begins.
Prophage An inactive bacteriophage, which is inserted into a host's chromosome.
Prophase First stage of mitosis, during which DNA condenses into chromatids and the spindle apparatus forms. Also used for the comparable stage of meiosis.
Prostaglandins Inflammatory chemicals released from damaged cells that increase vascular permeability.
Protease Enzyme secreted by microorganisms that digests proteins into amino acids outside a microbe's cell wall; in inflammatory reactions, chemicals released by mast cells that activate the complement system; in virology, an internal viral enzyme that makes HIV virulent.
Protein A complex macromolecule consisting of carbon, hydrogen, oxygen, nitrogen, and sulfur that is important to many cell functions.
Proteobacteria Phylum of prokaryotes that includes six classes (designated alpha, beta, gamma,

delta, epsilon, and zeta) of Gram-negative bacteria sharing common 16S rRNA nucleotide sequences.
Proton A positively charged subatomic particle, which is also the nucleus of a hydrogen atom.
Proton gradient Electrochemical gradient of hydrogen ions across a membrane.
Protozoa Single-celled eukaryotes that lack a cell wall and are similar to animals in their nutritional needs and structure.
Provirus (*latent virus*) Inactive virus in an animal cell.
Pseudohyphae Long cellular extension of the yeast *Candida* that look like the filamentous hyphae of molds.
Pseudomembranous colitis Inflammation of the colon, which is covered by a membrane consisting of connective tissue and dead and dying cells; condition is last state of *Clostridium difficile*–associated diarrhea (*C. diff.*–associated diarrhea).
Pseudomonad Any Gram-negative, aerobic, rod-shaped bacterium in the class Gammaproteobacteria that catabolizes carbohydrates by the Entner-Doudoroff and pentose phosphate pathways.
Pseudopods Movable extensions of the cytoplasm and membrane of some eukaryotic cells.
Psittacosis (*ornithosis, parrot fever*) A respiratory disease of birds that can be transmitted to humans and is caused by infection with *Chlamydophila psittaci*.
Psychrophile Microorganism requiring cold temperatures (below 20°C).
Psychrotolerant Ability of organisms to live in cold temperatures, though they do not grow best at these temperatures.
Pulmonary anthrax Disease of the respiratory system that is usually fatal without aggressive treatment; caused by inhalation of spores of *Bacillus anthracis*.
Pure culture (*axenic culture*) Culture containing cells of only one species.
Purple sulfur bacteria Group of gammaproteobacteria, which are obligate anaerobes and oxidize hydrogen sulfide to sulfur.
Pustule (*pox*) Any raised, pus-filled skin lesion; characteristic of infection with a poxvirus.
Pyoderma Any confined, pus-producing lesion on the exposed skin of the face, arms, or legs; often caused by infection with group A *Streptococcus*.
Pyrimidine dimer Mutation in which adjacent pyrimidine bases covalently bond to one another; caused by nonionizing radiation in the form of ultraviolet light.
Pyrogen Chemical that triggers the hypothalamic "thermostat" to reset at a higher temperature, inducing fever.

Q fever Fever caused by the bacterium *Coxiella burnetii*; cause was questionable (thus "Q") for many years.
Quaternary ammonium compound (quat) Detergent antimicrobial that is harmless to humans.
Quinolone An antimicrobial that inhibits gyrase, the enzyme that corrects supercoiling of replicating bacterial DNA.
Quorum sensing Process by which bacteria measure their density in an environment by utilizing signal and receptor molecules.

Rabies Neuromuscular disease characterized by hydrophobia, seizures, hallucinations, and paralysis; fatal if untreated. Caused by infection with the rabies virus.
Radial immunodiffusion Type of immunodiffusion test in which an antigen solution is allowed to diffuse into agar containing specific concentrations of antibodies, causing a ring of precipitate to form.
Radiation The release of high-speed subatomic particles or waves of electromagnetic energy from atoms.
Radiolaria Amoebas with threadlike pseudopods and silica shells.
Reactants The atoms, ions, or molecules that exist at the beginning of a chemical reaction.
Reaction center chlorophyll In a photosystem, a chlorophyll molecule in which electrons excited by light energy are passed to an acceptor molecule that is the initial carrier of an electron transport chain.
Real-time PCR Technique of recombinant DNA technology that allows researchers to determine how many copies of DNA are being made as a polymerase chain reaction (PCR) proceeds.
Recipient cell In horizontal gene transfer, a cell that receives part of the genome of a donor cell.
Recombinant Cell or DNA molecule resulting from genetic recombination between donated and recipient nucleotide sequences.
Recombinant DNA technology Type of biotechnology in which scientists change the genotypes and phenotypes of organisms.
Recombinant vaccine Vaccine produced using recombinant genetic technology.
Red measles (*rubeola, measles*) Contagious disease characterized by fever, sore throat, headache, dry cough, conjunctivitis, and lesions called Koplik's spots; caused by infection with *Morbillivirus*.
Red tide Abundance of red-pigmented dinoflagellates in marine water.
Redox reaction (*oxidation-reduction reaction*) Any metabolic reaction involving the transfer of electrons from an electron donor to an electron acceptor. Reactions in which electrons are accepted are called reduction reactions, whereas reactions in which electrons are donated are oxidation reactions.
Reducing medium Special culturing medium containing compounds that combine with free oxygen and remove it from the medium.
Regulatory RNA Form of ribonucleic acid used to control gene expression.
Regulatory T cell (*Tr cell, suppressor T cell*) Thymus-matured lymphocyte that serves to repress adaptive immune responses and prevent autoimmune diseases.
Release In virology, final stage of the lytic replication cycle, in which the new virions are released from the host cell, which lyses.
Reoviruses Group of naked, segmented, dsRNA viruses that cause respiratory and gastrointestinal disease.
Repressible operon Type of operon that is continually transcribed until deactivated by repressors.
Reproduction An increase in number.
Reservoir of infection Living or nonliving continuous source of infectious disease.
Resolution The ability to distinguish between objects that are close together.
Respiratory syncytial virus (RSV) infection Disease caused by a virus in genus *Pneumovirus* that causes fusion of cells in the lungs and difficulty in breathing.
Responsiveness An ability to respond to environmental stimuli.
Restriction enzyme Enzyme that cuts DNA at specific nucleotide sequences and is used to produce recombinant DNA molecules.
Reticulate bodies Noninfectious stage in the life cycle of chlamydias.
Retrovirus Any +ssRNA virus that uses the enzyme reverse transcriptase carried within its capsid to transcribe DNA from its RNA.
Reverse transcriptase Complex enzyme that allows retroviruses to make dsDNA from RNA templates; used in recombinant DNA technology to make cDNA.
Revolving nosepiece Portion of a compound microscope on which several objective lenses are mounted.
Rh antigens Cytoplasmic membrane proteins common to the red blood cells of 85% of humans as well as rhesus monkeys.
Rheumatic fever A complication of untreated group A streptococcal pharyngitis in which inflammation leads to damage of the heart valves and muscle.
Rheumatoid arthritis (RA) A crippling, systemic autoimmune disease resulting from a type III hypersensitivity reaction in which antibody complexes are deposited in the joints, causing inflammation.
Rhinosinusitis Inflammation of lining of nose and sinuses.
Rhinoviruses Group of picornaviruses that cause upper respiratory tract infection and the "common cold."
Rhodophyta Red algae, generally containing the pigment phycoerythrin, the storage molecule floridean starch, and cell walls of agar or carrageenan.
Ribonucleic acid (RNA) Nucleic acid consisting of nucleotides made up of phosphate, a ribose pentose sugar, and an arrangement of the bases adenine, guanine, cytosine, and uracil.
Ribosomal RNA (rRNA) Form of ribonucleic acid that, together with polypeptides, makes up the structure of ribosomes.
Ribosome Nonmembranous organelle found in prokaryotes and eukaryotes that is composed of protein and ribosomal RNA and functions to make polypeptides.
Riboswitch RNA molecule that changes shape in response to shifts in environmental conditions, which results in genetic regulation.
Ribozyme RNA molecule functioning as an enzyme.
Rickettsias Group of extremely small, Gram-negative, obligate intracellular parasites that appear almost wall-less.
RNA polymerase Enzyme that synthesizes RNA by linking RNA nucleotides that are complementary to genetic sequences in DNA.
RNA primer RNA molecule used by DNA polymerase or reverse transcriptase as a starting point for DNA synthesis.
Rocky Mountain spotted fever (RMSF) Serious illness caused by infection with *Rickettsia rickettsii* transmitted by ticks and characterized by rash, malaise, petechiae, encephalitis, and death in 5% of cases.
Roseola Endemic illness of children characterized by an abrupt fever, sore throat, enlarged lymph nodes, and faint pink rash; caused by infection with *human herpesvirus 6*.

Rotaviruses Group of reoviruses that cause a potentially fatal infantile gastroenteritis.
Rough endoplasmic reticulum (RER) Type of endoplasmic reticulum that has ribosomes adhering to its outer surface; these produce proteins for transport throughout the cell.
R plasmid Extrachromosomal piece of DNA containing genes for resistance to antimicrobial drugs.
Rubella (*German measles*) Disease caused by infection with *Rubivirus* resulting in characteristic rash lasting about three days; mild in children but potentially teratogenic to fetuses of infected women.
Rubeola (*measles, red measles*) Contagious disease characterized by fever, sore throat, headache, dry cough, conjunctivitis, and lesions called Koplik's spots; caused by infection with *Morbillivirus*.
Rubor Redness.

Salmonellosis A serious diarrheal disease resulting from consumption of food contaminated with the enteric bacterium *Salmonella*.
Salt A crystalline compound formed by ionic bonding of metallic with nonmetallic elements.
Sanitization The process of disinfecting surfaces and utensils used by the public.
Saprobe Fungus that absorbs nutrients from dead organisms.
Sarcina A cuboidal packet of cocci.
Satellite virus A virus, such as hepatitis D virus, that requires glycoproteins coded by another virus to complete its replication cycle.
Saturated fatty acid A long-chain, organic acid in which all but the terminal carbon atoms are covalently linked to two hydrogen atoms.
Scabies Skin disease caused by a burrowing mite.
Scalded skin syndrome Reddening and blistering of the skin caused by infection with *Staphylococcus aureus*.
Scanning electron microscope (SEM) Type of electron microscope that uses magnetic fields within a vacuum tube to scan a beam of electrons across a specimen's metal-coated surface.
Scanning tunneling microscope (STM) Type of probe microscope in which a metallic probe passes slightly above the surface of the specimen, revealing surface details at the atomic level.
Scarlet fever Diffuse rash and sloughing of skin caused by infection with group A *Streptococcus*.
Schaeffer-Fulton endospore stain In microscopy, staining technique that uses heat to drive a malachite green primary stain into an endospore.
Schistosomiasis A potentially fatal disease caused by infection with a blood fluke in the genus *Schistosoma*; may cause tissue damage in the liver, lungs, brain, or other organs.
Schizogony Special type of asexual reproduction in which the protozoan *Plasmodium* undergoes multiple mitoses to form a multinucleate schizont.
Schizont Multinucleate body that undergoes cytokinesis to release several cells.
Scientific method Process by which scientists attempt to prove or disprove hypotheses through observations of the outcomes of carefully controlled experiments.
Scolex Small attachment organ that possesses suckers and/or hooks used to attach a tapeworm to host tissues.
Sebum Oily substance secreted by the sebaceous glands of the skin that lowers pH.

Secondary immune response Enhanced immune response following a second contact with an antigen.
Secondary syphilis Second phase of the sexually-transmitted disease caused by *Treponema pallidum* characterized by sore throat, headache, mild fever, malaise, diseased lymph nodes, and a widespread rash.
Secretory IgA The combination of IgA and a secretory component, found in tears, mucous membrane secretions, and breast milk, where it agglutinates and neutralizes antigens.
Secretory vesicle In eukaryotic cells, vesicles containing secretions packaged by the Golgi body that fuse to the cytoplasmic membrane and then release their contents outside the cell via exocytosis.
Sedimentation Settling of particulate matter; the first step in treating water for drinking.
Segmented genome Genetic material consisting of more than one molecule of nucleic acid, used particularly for viruses.
Selective medium Culturing medium containing substances that either favor the growth of particular microorganisms or inhibit the growth of unwanted ones.
Selective toxicity Principle by which an effective antimicrobial agent must be more toxic to a pathogen than to the pathogen's host.
Selectively permeable In cell physiology, characteristic of a membrane that allows some substances to cross while preventing the crossing of others.
Semisynthetic antimicrobial Antimicrobial that has been chemically altered.
Septate Characterized by the presence of cross walls.
Septic shock Extremely low blood pressure resulting from dilation of blood vessels triggered by bacteria or bacterial toxins.
Septic tank The home equivalent of primary wastewater treatment, consisting of a sealed concrete holding tank in which solids settle to the bottom and the effluent flows into a leach field that acts as a filter.
Septicemia (*sepsis*) The condition of pathogens being present in the blood and causing signs of illness.
Serial dilution A stepwise dilution of a liquid culture in which the dilution factor at each step is constant.
Serology The study and use of immunological tests to diagnose and treat disease or identify antibodies or antigens.
Serum Blood plasma with clotting factors removed.
Serum sickness Type III hypersensitivity resulting from antibodies directed against antisera.
Severe acute respiratory syndrome (SARS) Manifestation of infection by a coronavirus called SARS virus.
Severe combined immunodeficiency disease (SCID) Primary immunodeficiency disease in children that affects both T cells and B cells and causes recurrent infections.
Sexually transmitted disease (STD) Disease resulting from a sexually transmitted infection.
Sexually transmitted infection (STI) Invasion of a pathogen into the body resulting from sexual activity.
Shiga toxin Exotoxin secreted by *Shigella dysenteriae* that stops protein synthesis in host cells.

Shigellosis A severe form of dysentery caused by any of four species of *Shigella*.
Shine-Dalgarno sequence The sequence of nucleotides in a molecule of mRNA where the smaller ribosomal subunit initiates translation. The sequence is named for its discoverers.
Shingles (*herpes zoster*) Extremely painful skin rash caused by reactivation of latent varicella-zoster virus.
Shock Severe disturbance of blood circulation resulting in insufficient delivery of oxygen to vital organs.
Siderophore An iron-binding molecule released by some bacteria and fungi.
Signs In pathology, objective manifestations of a disease that can be observed or measured by others.
Silent mutation Mutation produced by base-pair substitution that does not change the amino acid sequence, because of the redundancy of the genetic code.
Simple microscope Microscope containing a single magnifying lens.
Simple stain In microscopy, a stain composed of a single dye such as crystal violet.
Singlet oxygen Toxic form of oxygen, neutralized by pigments called carotenoids.
Sinusitis Inflammation of the nasal sinuses; typically caused by *Streptococcus pneumoniae*.
Slant tube (slant) Test tube containing agar media that solidified while the tube was resting at an angle.
Slime layer Loose, water-soluble glycocalyx.
Slime mold Eukaryotic microbe resembling a filamentous fungus but lacking a cell wall and phagocytizing rather than absorbing nutrients.
Sludge After primary treatment of wastewater, the heavy material remaining at the bottom of settling tanks.
Small interfering RNA (siRNA) RNA molecule complementary to a portion of a molecule of mRNA, tRNA, or a gene, rendering the target ineffective.
Smallpox Infectious disease eradicated in nature by 1980 and characterized by high fever, malaise, delirium, pustules, and death in about 20% of untreated cases.
Smear In microscopy, the thin film of organisms on the slide.
Smooth endoplasmic reticulum (SER) Type of endoplasmic reticulum that lacks ribosomes and plays a role in lipid synthesis and transport.
Snapping division A variation of binary fission in Gram-positive prokaryotes in which the parent cell's outer wall tears apart with a snapping movement to create the daughter cells.
SOS response Mechanism by which prokaryotic cells with extensive DNA damage use a variety of processes to induce DNA polymerase to copy the damaged DNA.
Southern blot Technique used in recombinant DNA technology that allows researchers to stabilize specific DNA sequences from an electrophoresis gel and then localize them using DNA dyes or probes.
Species Taxonomic category of organisms that can successfully interbreed.
Species resistance Property that protects a type of organism from infection by pathogens of other, very different organisms.
Specific epithet In taxonomy, latter portion of the descriptive name of a species.

Specific immunity The ability of a vertebrate to recognize and defend against distinct species or strains of invaders.

Spectrum of action The number of different kinds of pathogens a drug acts against.

Spinal tap (*lumbar puncture*) Collection of cerebrospinal fluid from the lower vertebral column for diagnostic purposes.

Spiral In cell morphology, a spiral-shaped prokaryotic cell.

Spirillus A stiff spiral-shaped prokaryotic cell.

Spirochetes Group of helical, Gram-negative bacteria with axial filaments that cause the organism to corkscrew, enabling it to burrow into a host's tissues.

Spliceosome Protein-RNA complex that removes introns from eukaryotic RNA.

Spoilage Any unwanted change to a food.

Spontaneous generation The theory that living organisms can arise from nonliving matter.

Sporadic In epidemiology, a disease that occurs in only a few scattered cases within a given area or population during a given period of time.

Spore Reproductive cell of actinomycetes and fungi.

Sporogonic phase In the life cycle of *Plasmodium*, stage during which sporozoites are produced in the mosquito's digestive tract and migrate into the mosquito's salivary glands.

Sporotrichosis Subcutaneous infection usually limited to the arms and legs; lesions form around the site of infection with *Sporothrix schenckii*.

Spotted fever rickettsiosis Serious illness caused by infection with a rickettsial bacterium transmitted by ticks and characterized by rash, malaise, petechiae, and encephalitis, and death in 5% of cases.

Spp. Abbreviation used to indicate several species of a genus.

Staining Coloring microscopy specimens with stains called dyes.

Staphylococcal scalded skin syndrome (SSSS) Disease caused by exfoliative toxin of *Staphylococcus aureus* in which epidermis peels off.

Staphylococcal toxic shock syndrome Potentially fatal syndrome characterized by fever, vomiting, red rash, low blood pressure, and loss of sheets of skin, usually caused by systemic infection with strains of *Staphylococcus* that produce toxic shock syndrome toxins.

Staphylococcus A cluster of cocci.

Starter culture Group of known microorganisms that carry out specific and reproducible fermentation reactions.

Stationary phase Phase in a growth curve in which new organisms are being produced at the same rate at which older organisms are dying.

Stem cells Generative cells capable of dividing to form daughter cells of a variety of types.

Sterile Free of microbial contamination.

Sterilization The eradication of all organisms, including bacterial endospores and viruses, although not prions, in or on an object.

Steroid Lipids consisting of four fused carbon rings attached to various side chains and functional groups.

Streak-plate Method of culturing microorganisms in which a sterile inoculating loop is used to spread an inoculum across the surface of a solid medium in Petri dishes.

Streptococcal pharyngitis (*strep throat*) Inflammation of the throat, often caused by group A *Streptococcus*.

Streptococcal toxic shock syndrome (TSS) Shock produced by toxins of *Streptococcus* with manifestations similar to toxic shock syndrome of *Staphylococcus*.

Streptococcus A chain of cocci.

Streptogramins Antimicrobial drugs that bind to the 50S ribosomal subunit and prevent ribosome movement along messenger RNA.

Structural analog Chemical that competes with a structurally similar molecule.

Sty Inflamed bacterial infection of the base of an eyelid.

Subacute disease Any disease that has a duration and severity that lies somewhere between acute and chronic.

Subacute sclerosing panencephalitis (SSPE) Slow, progressive disease of the central nervous system that results in memory loss, muscle spasms, and death several years after infection with a defective measles virus.

Subclinical (*asymptomatic*) Characteristic of disease that may go unnoticed because of absence of symptoms, even though clinical tests may reveal signs of disease.

Substitution Type of mutation in which a nucleotide base pair is replaced.

Substrate The molecule upon which an enzyme acts.

Substrate-level phosphorylation The transfer of phosphate to ADP from another phosphorylated organic compound.

Subunit vaccine Type of vaccine developed using recombinant DNA technology that exposes the recipient's immune system to a pathogen's antigens but not the pathogen itself.

Sulfonamide Antimetabolic drug that is a structural analog of para-aminobenzoic acid (PABA).

Sulfur cycle Biogeochemical cycle in which sulfur is cycled between oxidation states.

Superficial mycoses Fungal infections of the surface of the skin.

Superinfection A new infection in addition to one already present, particularly following the use of antimicrobials.

Superoxide radical Toxic form of oxygen that is detoxified by superoxide dismutase.

Surfactant Chemical that acts to reduce the surface tension of solvents such as water by decreasing the attraction among solvent molecules.

Symbiont A member of a symbiotic relationship.

Symbiosis A continuum of close associations between two or more organisms that ranges from mutually beneficial to associations in which one member damages the other member.

Symptoms Subjective characteristics of a disease that can be felt by the patient alone.

Synapse In immunology, the interface between cells of the immune system that involves cell-to-cell signaling.

Syncytium Giant, multinucleated cytoplasmic mass formed by fusion of a virally infected cell to its neighbors.

Syndrome A group of symptoms, signs, and diseases that collectively characterizes a particular abnormal condition.

Synergism Interplay between drugs that results in efficacy that exceeds the efficacy of either drug alone.

Synthesis In virology, the production of new viral proteins and nucleic acids using the metabolic machinery of the host cell; third stage of lytic replication cycle.

Synthesis reaction A chemical reaction involving the formation of larger, more complex molecules.

Synthetic drug Antimicrobial that has been completely synthesized in a laboratory.

Synthetic medium (*defined medium*) Culturing medium of which the exact chemical composition is known.

Syphilis Sexually transmitted disease caused by infection with *Treponema pallidum*. Untreated, it can have four phases: primary syphilis, which manifests as a chancre; secondary syphilis, which typically manifests as a widespread rash; latent syphilis, which is clinically inactive; and tertiary syphilis, which involves severe hyperimmune complications.

Systemic diseases Diseases caused by microbes spread via the blood and lymph that affect other body systems.

Systemic lupus erythematosus (*lupus*) A systemic autoimmune disease in which the individual produces autoantibodies against numerous antigens, including nucleic acids.

T cell T lymphocyte.

Tc cell (*cytotoxic T cell, CD8+ cell*) In cell-mediated immune response, type of cell characterized by CD8 cell-surface glycoprotein; secretes perforins and granzymes that destroy infected or abnormal body cells.

Th cell (*helper T cell, CD4+ cell*) In cell-mediated immune response, a type of cell characterized by CD4 cell-surface glycoprotein; regulates the activity of B cells and cytotoxic T cells.

Tr cell (*regulatory T cell, suppressor T cell*) Thymus-matured lymphocyte that serves to repress adaptive immune responses and prevent autoimmune diseases.

T cell receptor (TCR) Antigen receptor generated in the cytoplasmic membrane of T lymphocytes.

T lymphocyte (*T cell*) Lymphocyte that matures in the thymus and acts primarily against endogenous antigens in cell-mediated immune responses.

T-dependent antibody immunity Adaptive immune response resulting in immunoglobulin production that requires the action of a specific helper T cell (Th2).

T-dependent antigens Molecules that stimulate an immune response only with the involvement of a helper T cell.

T-independent antibody immunity Adaptive immune response resulting in immunoglobulin production following cross-linking of BCRs on numerous B cells and lacking involvement of helper T cells.

T-independent antigens Large molecules with repeating subunits that trigger an antibody immune response without the activation of T cells.

Tapeworms (*cestodes*) Group of long, flat, and segmented helminths.

Taxa Nonoverlapping groups of organisms sorted on the basis of mutual similarities.

Taxis Cell movement that occurs as a positive or negative response to light or chemicals.

Taxonomic system A system for naming and grouping similar organisms together.

Taxonomy The science of classifying and naming organisms.
Telophase Final stage of mitosis, during which nuclear envelopes form around the daughter nuclei. Also used for the comparable stage of meiosis.
Temperate phage (*lysogenic phage*) Bacteriophage that does not immediately kill its host cell.
Templating The process by which prions guide normal proteins with the same amino acid sequence as the prion to misfold into more prions.
Teratogenic Characterized by an ability to cause birth defects.
Terminator Region of DNA where transcription ends.
Tertiary syphilis Final phase of the sexually-transmitted disease caused by *Treponema pallidum* occurring years after the secondary phase and characterized by severe hyperimmune complications resulting in dementia, blindness, paralysis, heart failure, or rubbery, swollen lesions called gummas.
Tetanospasmin Neurotoxin of *Clostridium tetani* that blocks the release of inhibitory neurotransmitters in the central nervous system.
Tetanus Potentially fatal infection with *Clostridium tetani*, which produces tetanospasmin, a potent neurotoxin.
Tetracycline Antimicrobial agent that inhibits protein synthesis by blocking the tRNA docking site.
Tetrad In genetics: two chromosomes, which are each made up of two DNA molecules, physically associated together during prophase I and metaphase I of meiosis; in cellular arrangements, four cocci remaining attached following cell division.
Therapeutic index (TI) Ratio of the largest dose of a drug that is not toxic to the drug's smallest effective dose.
Therapeutic window Range of drug concentrations that are effective without being toxic to a patient.
Thermal death point The lowest temperature that kills all cells in a broth in 10 minutes.
Thermal death time The time it takes to completely sterilize a particular volume of liquid at a set temperature.
Thermophile Microorganism requiring temperatures above 45°C.
Thrombocytopenia Decrease in the number of platelets in the blood.
Thylakoid In photosynthetic cells, portion of cellular membrane containing light-absorbing photosystems.
Thymine Ring-shaped nitrogenous base found in nucleotides of DNA.
Tick Bloodsucking arachnid, which vectors a number of bacterial and viral pathogens.
Tick-borne relapsing fever Disease caused by *Borrelia* spp. transmitted between humans by soft ticks.
Tincture Solution of antimicrobial chemical in alcohol.
Titer In serology, a measure of the level of antibody in blood serum, determined by titration and expressed as a ratio reflecting the dilution.
Titration Serial dilution of blood serum to test for agglutination activity.
Toll-like receptors (TLRs) Integral membrane proteins that bind to specific microbial chemicals.
Total magnification A multiple of the magnification achieved by the objective and ocular lenses of a compound microscope.

Toxemia Presence in the blood of poisons called toxins.
Toxic shock syndrome (nonstreptococcal; TSS) Potentially fatal condition characterized by fever, vomiting, red rash, low blood pressure, and loss of sheets of skin, caused by systemic infection with strains of *Staphylococcus*.
Toxin Chemical that either harms tissues or triggers host immune responses that cause damage.
Toxoid vaccine Inoculum using modified toxins to stimulate antibody-mediated immunity.
Toxoplasmosis A disease affecting animals and caused by infection with *Toxoplasma gondii*. In humans, characterized by mild, febrile symptoms but may be fatal in AIDS patients; transplacental transmission may result in miscarriage, stillbirth, or severe birth defects.
Trace element Element required in very small amounts for microbial metabolism.
Trachoma Serious eye disease caused by *Chlamydia trachomatis*.
Transamination Reaction involving transfer of an amine group from one amino acid to another.
Transcription Process in which the genetic code from DNA is copied as RNA nucleotide sequences.
Transcriptome All of the mRNA produced by a cell under a given set of circumstances.
Transcriptomics Study of transcriptomes and their functions.
Transducing phage Virus that transfers bacterial DNA from one bacterium to another.
Transduction Method of horizontal gene transfer in which DNA is transferred from one cell to another via a replicating virus.
Transfer RNA (tRNA) Form of ribonucleic acid that carries amino acids to the ribosome.
Transformation Method of horizontal gene transfer in which a recipient cell takes up DNA from the environment.
Transgenic organism Plant or animal that has been genetically altered by the inclusion of genes from other organisms.
Translation Process in which the sequence of genetic information carried by mRNA is used by ribosomes to construct polypeptides with specific amino acid sequences.
Transmission electron microscope (TEM) Type of electron microscope that generates a beam of electrons that passes through the specimen and produces an image on a fluorescent screen.
Transport medium A special type of medium used to move clinical specimens from one location to another while preserving the relative abundance of organisms and preventing contamination of the specimen or environment.
Transposition Mutation in which a genetic segment is transferred to a new position through the action of a DNA segment called a transposon.
Transposon Segment of DNA found in most prokaryotes, eukaryotes, and viruses that codes for the enzyme transposase and can move from one location in a DNA molecule to another location in the same or a different molecule.
Trematodes (*flukes*) Group of helminths that are flat, leaf-shaped, have incomplete digestive systems, and have oral and ventral suckers.
Trench fever A disease common among World War I soldiers; caused by the bacterium *Bartonella quintana*.

Trichomoniasis Inflammation of the genitalia caused by *Trichomonas vaginalis*.
Trophozoite The motile feeding stage of a protozoa.
Tubercle Hard pulmonary nodule resulting from infection with mycobacteria.
Tuberculin response Type of delayed hypersensitivity reaction in which the skin of an individual exposed to tuberculosis or tuberculosis vaccine reacts to a subcutaneous injection of tuberculin.
Tuberculin skin test Test for a delayed hypersensitivity reaction to a subcutaneous injection of tuberculin.
Tuberculosis (TB) A respiratory disease caused by infection with *Mycobacterium tuberculosis*; its disseminated form can result in wasting away of the body and death.
Tularemia Zoonotic disease causing fever, chills, malaise, and fatigue, and caused by infection with *Francisella tularensis*.
Tumor In the pathology of cancer, a mass of neoplastic cells. In inflammation, a symptom of swelling (edema).
Tumor necrosis factor (TNF) An immune system cytokine secreted by macrophages and T cells to kill tumor cells and to regulate immune responses and inflammation.
Turbidimetry Automated method that measures the cloudiness of a solution by passing light through it.
Type 1 diabetes mellitus Immunological attack on the islets of Langerhans cells in the pancreas resulting in the inability to produce the hormone insulin.
Type III secretion systems Complex proteinaceous structure that inserts into target cells, forming a channel for the secretion of bacterial toxin or enzymes.
Typhoid fever Fever, headache, and malaise produced by infection with *Salmonella enterica* serotypes Typhi and Paratyphi; severe infections may cause peritonitis.
Typhus (*epidemic typhus, murine typhus, scrub typhus*) A group of diseases caused by rickettsias transmitted by arthropod vectors.

Uncoating In animal viruses, the removal of a viral capsid within a host cell.
Unsaturated fatty acid A long-chain, organic acid with at least one double bond between adjacent carbon atoms, and thus at least one carbon atom bound to only a single hydrogen atom.
Uracil Ring-shaped nitrogenous base found in nucleotides of RNA.
Urinary tract infection (UTI) Invasion and multiplication of microorganisms, usually bacteria, in the kidneys, ureters, bladder, or urethra.
Urticaria Hives.
Use-dilution test Method of evaluating the effectiveness of a disinfectant or antiseptic against specific microbes in which the most effective agent is the one that entirely prevents microbial growth at the highest dilution.

Vaccination Active immunization; specifically against smallpox.
Vaccine The inoculum used in active immunization.
Vacuole General term for membranous sac that stores or carries a substance in a cell.
Vaginosis Noninflammatory infection of the vagina.

Valence The combining capacity of an atom.

Vancomycin Antimicrobial drug that disrupts formation of Gram-positive bacterial cell walls by interfering with alanine-alanine crossbridges linking N-acetylglucosamine subunits.

Vancomycin-resistant *Staphylococcus aureus* (VRSA) Strain of *S. aureus* that is resistant to vancomycin and usually resistant to many common antimicrobial drugs as well.

Variant Creutzfeldt-Jakob disease (vCJD) Dementia caused by a prion that destroys brain tissue such that the brain appears spongelike—full of holes.

Varicella (*chickenpox*) Highly infectious disease characterized by fever, malaise, and skin lesions, and caused by infection with varicella-zoster virus.

Varicella-zoster virus (VZV) Virus that causes chickenpox (varicella) and shingles (herpes zoster).

Variola Common name for the smallpox virus.

Variola major Variant of smallpox virus, which causes severe disease with a mortality rate of 20% or higher.

Variola minor Variant of smallpox virus, which causes less severe disease and mortality rate of less than 1%.

Vector In genetics and recombinant DNA technology, nucleic acid molecule such as a viral genome, transposon, or plasmid that is used to deliver a gene into a cell. In epidemiology, an animal (typically an arthropod) that transmits disease from one host to another.

Vegetations Bulky masses of platelets and clotting proteins that surround and bury the bacteria involved in endocarditis.

Vehicle transmission Spread of pathogens via air, drinking water, and food, as well as bodily fluids being handled outside the body.

Venezuelan equine encephalitis (VEE) Potentially fatal infection of the brain caused by a togavirus.

Vesicle General term for membranous sac that stores or carries a substance in a cell; in human pathology, any raised skin lesion filled with clear fluid.

Viable plate count Estimation of the size of a microbial population based upon the number of colonies formed when diluted samples are plated onto agar media.

Vibrio A slightly curved rod-shaped prokaryotic cell.

Viral gastroenteritis Inflammation of the mucous membrane of the stomach and intestines, caused by a viral pathogen.

Viral hemagglutination inhibition test Immune test commonly used to detect antibodies against influenza, measles, and other viruses that naturally agglutinate red blood cells.

Viral neutralization Test of serum for the presence of antibodies against a particular virus in which test serum is mixed with the virus, and then the mixture is added to a cell culture. Survival of the cells indicates antibodies in the serum neutralized the viruses.

Viremia Viral infection of the blood.

Viridans streptococci Group of alpha-hemolytic streptococci, which produce a green pigment when grown on blood media and normally inhabit the mouth and throat, as well as the GI, genital, and urinary tracts.

Virion A virus outside of a cell, consisting of a proteinaceous capsid surrounding a nucleic acid core.

Viroid Extremely small, circular piece of RNA that is infectious and pathogenic in plants.

Virulence A measure of pathogenicity.

Virulence factors Enzymes, toxins, and other factors that affect the relative ability of a pathogen to infect and cause disease.

Virus Tiny infectious acellular agent with nucleic acid surrounded by proteinaceous capsomeres that form a covering called a capsid.

Viviparity Process by which live offspring are produced in the body of a mother.

Wandering macrophage Type of macrophage that leaves the blood via diapedesis to travel to distant sites of infection.

Warts (*papillomas*) Benign epithelial growths caused by papillomaviruses.

Wastewater (*sewage*) Any water that leaves homes or businesses after being used for washing or flushed from toilets.

Water mold Eukaryotic microbe resembling a filamentous fungus but having tubular cristae in their mitochondria, cell walls of cellulose, two flagella, and true diploid bodies.

Waterborne transmission Spread of pathogenic microorganisms via water.

Wavelength The distance between two corresponding points of a wave.

Wax Alcohol-containing lipid made up of molecules with one fatty acid chain.

West Nile encephalitis Potentially fatal infection of the brain caused by an arbovirus in the family *Flaviviridae*.

Western blot test (*immunoblot*) Variation of an ELISA test that can detect the presence of antibodies against multiple antigens; used to verify the presence of antibodies against HIV in the serum of individuals who have tested positive by ELISA.

Western equine encephalitis (WEE) Potentially fatal infection of the brain caused by a togavirus.

Whitlow Inflamed blister that may result from infection with *human herpesvirus 1* or *human herpesvirus 2* via a cut or break in the skin.

Whooping cough (*pertussis*) Pediatric disease characterized by the development of copious mucus, loss of tracheal cilia, and deep "whooping" cough; caused by infection with *Bordetella pertussis*.

Wild-type cell A cell normally found in nature (in the wild); a nonmutant.

Wound Trauma to body's tissue.

XDR-TB Tuberculosis caused by extensively drug-resistant *Mycobacterium*.

Xenodiagnosis Method of diagnosing Chagas' disease in which an uninfected *Triatoma* vector is allowed to feed on a patient. Subsequent presence of trypanosomes in the bug's gut indicates the patient is infected.

Xenograft Type of graft in which tissues are transplanted between individuals of different species.

Xenotransplant Technique involving recombinant DNA technology in which human genes are inserted into animals to produce cells, tissues, or organs that are then introduced into the human body.

Yaws Large, destructive, pain-free lesions of the skin, bones, and lymph nodes caused by *Treponema pallidum pertenue*.

Yeast A unicellular, typically oval or round fungus that usually reproduces asexually by budding.

Yellow fever Often fatal hemorrhagic disease contracted through a mosquito bite carrying a flavivirus.

Zetaproteobacteria Class of usually autotrophic, Gram-negative bacteria in the phylum Proteobacteria discovered through metagenomic sequencing.

Zone of inhibition In a diffusion susceptibility test, a clear area surrounding the drug-soaked disk where the microbe does not grow.

Zoonoses Diseases that are naturally spread from their usual animal host to humans.

Zygomycoses Opportunistic fungal infections caused by various genera of fungi classified in the division Zygomycota.

Zygomycota Division of fungi including coenocytic molds called zygomycetes. Most are saprobes.

Zygosporangium Thick, black, rough-walled sexual structure of zygomycetes that can withstand desiccation and other harsh environmental conditions.

Zygospores Haploid spores formed from the surviving nuclei within zygosporangia.

Zygote In sexual reproduction, diploid cell formed by the union of gametes.

CREDITS

Photo Credits

Cover: Kateryna Kon/Science Photo Library/Getty Images

Solve The Problem: Hero Images Inc./Alamy Stock Photo

CHAPTER 1 Opener: Vladumir Kolosov/Alamy Stock Photo, **inset:** Barry Diomede/Alamy Stock Photo; **p.** 2 **left to right:** James Gathany/CDC, Jean Roy/CDC, Eye of Science/Science Source; **1.1:** Robert Thom/Alamy Stock Photo; **1.2:** Biophoto Associates/Science Source; **1.3:** Laguna Design/Science Source; **1.4:** L. Brent Selinger, Pearson Education; **1.5a:** Dr Jeremy Burgess/Science Source; **1.5b:** Steve Gschmeissner/Science Source; **1.6a:** NHPA/Superstock; **1.6b:** blickwinkel/Guenther/Alamy Stock Photo; **1.6c:** Bruce J. Russell/Biomedia Associates/Science Source; **1.7a:** M. I. Walker/Science Source; **1.7b:** Jan Hinsch/Science Source; **1.8:** Sinclair Stammers/Science Source; **1.9:** Lee D. Simon/Science Source; **1.11:** Pictorial Press Ltd/Alamy Stock Photo; **p.** 10: HappyAlex/Fotolia; **1.15:** Wilhelm Fechner, U.S. National Library of Medicine; **1.16:** U.S. National Library of Medicine; **1.17:** Satirus/Fotolia; **1.18:** ASM/Science Source; **p.** 16, **1.19:** U.S. National Library of Medicine; **1.20:** Science History Images/Photo Researchers/Alamy Stock Photo; **1.22:** Chronicle/Alamy Stock Photo; **p.** 21 **top:** SGO/BSIP SA/Alamy Stock Photo, **bottom:** Keith Weller, Agricultural Research Service, USDA.

CHAPTER 2 Opener: Eye of Science/Science Source, **inset:** Ariel Skelley/Getty Images; **2.12b:** Felix Büscher/Age Fotostock; **p.** 39: Fernando Fernández/Pixtal/age fotostock; **p.** 40: Artwell/Fotolia.

CHAPTER 3 Opener: Science History Images/Alamy Stock Photo, **inset:** Jose Luis Pelaez, Blend Images/Getty Images, Kim M Smith/Shutterstock, Denys Kovtun/Alamy Stock Photo; **3.1a:** Dr. Gopal Murti/Science Source; **3.1b:** SPL/Science Source; **3.3:** Photo Researchers, Inc./Science Source; **3.5a:** I. Rantala/Science Source; **3.5b:** Dr. Kari Lounatmaa/Science Source; **3.7a:** Biophoto Associates/Science Source; **3.7b:** Eye of Science/Science Source; **3.7c:** HEATHER DAVIES/Science Photo Library/Alamy Stock Photo; **3.8a:** Juergen Berger/Science Source; **3.10:** Thomas Deerinck/Science Source; **3.11:** Public Library of Science; **3.12:** David Scharf/Science Source; **p.** 70: P. MARAZZI/Science Source; **3.24:** From: Using Microbes to Improve Polymer Beads PolyBatics Produces Novel Biobeads with Attached Proteins, Antibodies, and Ligands. Carol Potera. Genetic Engineering & Biotechnology News, Column, January 15, 2013 (Vol. 33, No. 2). Fig. 1; **p.** 75: Josh Reynolds/AP Images; **3.25b:** Science History Images/Photo Researchers/Alamy Stock Photo; **3.26:** Rut Carballido-Lopez; **3.27:** From: "The unique structure of archaeal 'hami', highly complex cell appendages with nano-grappling hooks." C. Moissl, R. Rachel, A. Briegel, H. Engelhardt, R. Huber, *Mol Microbiol.* 2005 Apr; 56(2):361-70; Fig. 1 and 2; **3.28a:** Springer Publishing; **3.28b:** Scanning electron micrograph by William Hixon; **3.28c:** Mike Dyall-Smith; **3.29:** Robert Bauman; **3.30:** Don W. Fawcett/Science Source; **3.31:** Donald L. Ferry; **3.32a:** SPL/Science Source; **3.32b:** Steve Gschmeissner/Science Source; **3.32c:** Omnikron/Science Library/Science Source; **3.34b:** Photo Researchers, Inc./Science Source; **3.35a:** Conly L. Rieder; **3.36:** Photo Researchers, Inc./Science Source; **3.37:** Don W. Fawcett/Science Source; **3.38:** Biophoto Associates/Science Source; **3.39:** Photo Researchers, Inc./Science Source; **3.41:** Don W. Fawcett/Science Source; **3.42:** Electron micrograph by Wm. P. Wergin, courtesy of E. H. Newcomb, University of Wisconsin.

CHAPTER 4 Opener: STEVE GSCHMEISSNER/Science Source, **inset:** Johner Images/Alamy Stock Photo; **4.4a:** Charles D. Winters/Science Source; **4.8b:** CDC; **4.9:** From: Application of confocal and multi-photon imaging to geomicrobiology. T. Kawaguchi and A. W. Decho. Bio-Rad Application Note 30, 1-6, (2000). Courtesy of Bio-Rad Laboratories, Inc., © 2000; **4.10b:** Seelevel.com; **4.10c:** Dr. Tony Brain/Science Source; **4.11a:** Steve Gschmeissner/Science Source; **4.11b, d:** Eye of Science/Science Source; **4.11c:** Andrew Syred/Science Source; **4.12a:** Science Source; **4.12b:** Torunn Berge/Science Source; **Table 4.2 top to bottom:** Elisabeth Pierson/Pearson Education, Inc., From: Application of confocal and multi-photon imaging to geomicrobiology. T. Kawaguchi and A. W. Decho. Bio-Rad Application Note 30, 1-6, (2000). Courtesy of Bio-Rad Laboratories, Inc., © 2000, Dr. Tony Brain/Science Source, Steve Gschmeissner/Science Source, Photo Researchers, Inc./Science Source; **4.18:** ASM/Science Source; **Table 4.3:** ASM/Science Source; **p.** 113: Yasunori Tanji, Institute of Technology, Department of Biotechnology, Tokyo; **4.22:** Microscan panel. © Beckman Coulter, Inc. Used with permission; **4.24:** borzywoj/shutterstock; **p.** 118: Bill Olmsted/The Janesville Gazette/AP Images; **p.** 122(a): Eye of Science/Science Source, **(b):** John Durham/Science Source, **(d):** CDC, **(e):** M. Wurtz/Biozentrum/Science Source, **(f):** M. I. Walker/Science Source.

CHAPTER 5 Opener: Eye of Science/Science Source, **inset:** Siri Stafford/Getty Images; **5.23b:** Biophoto Associates/Science Source; **p.** 157: Don W. Fawcett/Science Source.

CHAPTER 6 Opener: Don W. Fawcett/Science Source, **inset:** Image Point Fr/Shutterstock; **p.** 162 **left:** Hemis/Alamy Stock Photo, **top right:** Scimat/Science Source, **bottom right:** DuneBug/Stockimo/Alamy Stock Photo; **6.3b:** L. Brent Selinger/Pearson Education, Inc.; **6.5a:** Doug Allen/Nature Picture Library; **6.5b:** Wayne P. Armstrong; **p.** 169: Savannah River National Laboratory/U.S. Department of Energy; **p.** 170: Yuri Arcurs/Shutterstock; **p.** 171: Ted Croll/Science Source; **6.8b, 6.9b:** L. Brent Selinger/Pearson Education, Inc.; **6.10b:** Khamkhlai Thanet/Shutterstock; **6.11-6.15:** L. Brent Selinger/Pearson Education, Inc.; **6.17b:** Lee D. Simon/Science Source; **p.** 182: CDC; **6.24b:** BSIP SA/Alamy Stock Photo, Janice Carr/CDC; **6.24c:** CDC; **6.26a:** Richard Megna/Fundamental Photographs, NYC; **6.26b:** Science Stock Photography/Science Source; **p.** 190: HansN.

CHAPTER 7 Opener: KTSDESIGN/Science Photo Library/Alamy Stock Photo, **inset:** Peter Dazeley/Photographer's Choice/Getty Images; **7.2a:** Dr. Klaus Boller/Science Source; **7.2b:** Huntington Potter, Byrd Alzheimers Institute and University of South Florida and David Dressler, Oxford University and Balliol College; **7.3a:** Victoria E. Foe; **7.3b:** Barbara Hamkalo; **7.3c, d:** G F Bahr/Armed Forces Institute of Pathology; **7.19:** E V Kiseleva; **7.36a:** Charles C Brinton and Louis Baron, "The Properties of Sex pili, The Viral Nature of "Conjugal" Genetic Transfer Systems, and Some Possible Approaches to the Control of Bacterial Drug Resistance," *CRC Critical Reviews in Microbiology*, Vol. 1, Issues 1, 1971; **p.** 230: Mike Miller/CDC; **p.** 235: Ross Inman/University of Wisconsin.

CHAPTER 8 Opener: Illustration created by Stephen D. Dixon and Feng Zhang, **inset:** Blend Images/Alamy Stock Photo; **p.** 246: National Parks Service; **8.7b:** Photo Researchers, Inc./Science Source; **8.8b:** RGB Ventures/SuperStock/Alamy Stock Photo; **8.9d:** University of California, Irvine Transgenic Mouse Facility collection; **8.10:** AdvanDx; **p.** 253: National Institutes of Health (NIH); **8.12:** Dennis Gonsalves.

CHAPTER 9 Opener: Gary Gaugler/Science Source, **inset:** auremar/Fotolia; **p.** 263 **left:** pegbes/Fotolia, **top right:** Steve Gschmeissner/Science Source, **center right:** Cultura RM/Alamy Stock Photo, **bottom left:** Maxine Bolton/EyeEm/Getty Images; **p.** 267 **top:** From: An update on Acanthamoeba keratitis: diagnosis, pathogenesis and treatment. Jacob Lorenzo-Morales et al. Parasite 22: 10. (2015). DOI:10.1051/parasite/2015010. PMID 25687209. ISSN 1776.1042, **bottom:** P. Garg & G. N. Rao/International Centre for Eye Health (ICEH); **9.4:** Xinhua/Alamy Stock Photo; **9.7a:** Photofusion Picture Library/Alamy Stock Photo; **9.9:** John Mock/Lonely Planet Images/Getty Images; **9.10b:** Photo Researchers, Inc./Science Source; **9.11b:** Tyler Olson/Shutterstock; **9.12a:** U.S. Department of Agriculture; **9.12b, 9.14:** Richard Megna/Fundamental Photographs, NYC; **p.** 277: Scimat/Science Source; **9.16:** Robert Bauman; **p.** 285: Robert Reed.

CHAPTER 10 Opener: Fahroni/Shutterstock, **inset:** Jose Luis Pelaez Inc/Blend Images/Getty Images; **10.1:** Rich Robison/Pearson Education, Inc.; **p.** 299: Dr P. Marazzi/Science Source; **10.10:** Microrao, Dept. of Microbiology, JJMMC, Davangere; **10.11:** Tim Pietzcker, Universitatsklinikum Ulm, Germany; **p.** 301: P. Marazzi/Science Source; **10.14a:** Com4; **10.14b:** Ted Croll/Science Source; **p.** 304: Scimat/Science Source; **p.** 305: CDC; **10.18:** Eddy Vercauteren/Department of Microbiology/University Hospital/Antwerp/Belgium; **p.** 307: Ariel Skelley/DigitalVision/Getty Images; **p.** 321: Tim Pietzcker, Universitatsklinikum Ulm, Germany.

CHAPTER 11 Opener: Eye of Science/Science Source, **inset:** Westend61/Getty Images; **11.1a:** SPL/Science Source; **11.1b:** David M. Phillips/Science Source; **11.1c:** Juergen Berger/Science Source; **11.1d:** Ami image/Science Source; **11.1e:** Andrew Syred/Science Source; **11.1f:** Janice Haney Carr/CDC; **11.1g:** Hans-Peter Klenk; **11.1h:** Bergey's Manual Trust; **11.2:** Rich Robison/Pearson Education, Inc.; **11.4b:** From: Ultrastructural Explanation for Snapping Postfission Movements in Arthrobacter crystallopoietes, Terry A. Krulwich, Jack L. Pate, *Journal of Bacteriology* Jan 1971, 105 (1) 408-412; **11.5:** David Scharf/Science Source; **11.7:** Esther R. Angert, Cornell University; **11.8a:** Mike Dyall-Smith/Dept. of Microbiology and Immunology/University of Melbourne/Australia; **11.8b:** SIRIKWAN DOKUTA/Shutterstock; **11.8c-e:** L. Brent Selinger/Pearson Education, Inc.; **11.9:** William A. Clark/CDC; **11.11a:** Angels Tapias, Microphotograph of the Archea Thermococcus Gammatolerans, 2005; **11.11b:** Reinhard Rachel/University of Regensburg, Germany; **11.12:** Sascha Burkard/Shutterstock; **11.13:** Nnehring/E+/Getty Images; **11.14a:** MCC-NIES Microbial Culture Collection National Institute for Environmental Studies; **11.14b:** M. I. Walker/Science Source; **11.14c:** Robin A. Matthews, Western Washington University, Bellingham WA; **11.15:** David J. Patterson; **11.16:** F. Thiaucourt/CIRAD/Montpellier/France; **11.17:** SCIMAT/Science Source; **11.18:** Cecil H. Fox/Science Source; **11.19:** David Berd/CDC; **11.20:** Chao Jiang; **11.21:** Seelevel.com; **11.22:** From: Phylogeny by a polyphasic approach of the order Caulobacterales, proposal of Caulobacter mirabilis sp. nov., Phenylobacterium haematophilum sp. nov. and Phenylobacterium conjunctum sp. nov., and emendation of the genus Phenylobacterium. W R Abraham et al. *Int J Syst Evol Microbiol.* 2008 Aug;58 (Pt 8):1939-49. 1; **11.23:** Custom Life Science Images/Alamy Stock Photo; **p.** 341: Yves Brun; **11.24:** Ecologix Environmental Systems; **11.25a:** David J. Patterson; **11.25b:** SEM Image by W. Urbancik/University of Vienna; **11.26:** Tony Brain/Science Source; **p.** 345 **top:** James Cavallini/Science Source, **bottom:** Koji Niino/SOURCENEXT/Alamy Stock Photo; **11.27b:** Alfred Pasieka/Photolibrary/Getty Images; **11.28b:** Dr. Heinrich Lünsdorf of HZI Braunschweig; **p.** 351: Rich Robison/Pearson Education, Inc.

CHAPTER 12 Opener: Kent Wood/Science Source, **inset:** Greg Vaughn/VWPics/age fotostock; **12.3a:** Dartmouth College Electron Microscope Facility; **12.3b:** Michael V. Danilchik/Program in Molecular and Cellular Bioscences/Oregon Health Sciences University; **12.3c:** Scimat/Science Source; **12.6:** Walter Dawn/Science Source; **12.7:** Ed Reschke/Photolibrary/Getty Images; **12.8b:** Biophoto Associates/Science Source; **12.9:** Guy Brugerolle/Universitad Clearmont Ferrand; **12.10:** Meckes/Ottowas/Eye on Science/Science Source; **12.11:** Biophoto Associates/Science Source; **12.12a:** Eye of Science/Science Source; **12.12b:** Claude Carre/Science Source; **12.13a:** Biophoto Associates/Science Source; **12.13b:** Blickwinkel/Alamy Stock Photo; **12.14a, b:** L. Brent Selinger/Pearson Education, Inc.; **12.14c:** Biophoto Associates/Science Source; **12.14d:** Christine Case/Skyline College;

C-2 CREDITS

12.15: Claude Carre/Science Source; **12.16:** George L. Barron; **12.17a:** Lucile K George/CDC; **12.17b:** London Scientific Films/Oxford Scientific/Getty Images; **12.17c:** David Scharf/Science Source; **12.19:** Ed Reschke/Photolibrary/Getty Images; **12.20a:** Tomasztc/Fotolia; **12.20b:** Ed Reschke/Photolibrary/Getty Images; **p.** 370 **top:** From "Pulmonary Aspergilloma: A Significant Cause of Life-Threatening Massive Hemoptysis." Z. A. Ali, A. Mahmoud, F. R. Krueger, S. Kapre, C. Mead. Resident & Staff Physician. 2008; 54(1):27-32; Fig. 2, **bottom:** CDC; **p.** 371: Pixelmania/Fotolia; **12.21a:** Arve Bettum/Shutterstock, **inset:** Biophoto Associates/Science Source; **12.21b:** Paul J Fusco/Science Source; **12.22b:** Eye of Science/Science Source; **12.23:** Fred M. Rhoades; **12.25:** D. P. Wilson/Science Source; **12.26:** John van den Hoff, Australian Antarctic Division; **12.27:** Ethan Daniels/Shutterstock; **12.28:** Steve Gschmeissner/Science Source; **12.29:** Fred Rhoades; **12.30a:** Kent Wood/Science Source; **12.30b:** © Stekolnikov et al.; licensee BioMed Central Ltd. 2014; **12.30c:** CDC; **12.30d:** *PLoS Biology*. 2(11); e340. Photo by Vincent S. Smith; **12.30e:** Scott Camazine/Alamy Stock Photo; **12.30f:** Fullerene/Getty Images; **12.30g:** The Natural History Museum/Alamy Stock Photo.

CHAPTER 13 Opener: Carol and Mike Werner/Science Source, **inset:** Whit Richardson/Alamy Stock Photo; **p.** 385 **left:** Merijn Salverda and Richard Stouthamer/Science Source, **top right:** LM Otero/AP Images, **bottom right:** James Gathany/CDC; **13.1b:** David M. Phillips/Science Source; **13.2:** Huntington Potter, Byrd Alzheimers Institute and University of South Florida and David Dressler, Oxford University and Balliol College; **13.3a:** Nigel Cattlin/Science Source; **13.3b:** Ami Images/Science Source; **13.3c:** Eye of Science/Science Source; **13.5a:** Science History Images/Photo Researchers/Alamy Stock Photo; **13.5b:** Biophoto Associates/Science Source; **13.5c:** Dr. Fred. A. Murphy/CDC; **13.5d:** Abergel Chantal/Information Genomique et Structurale UMR7256 CNRS, Aix-Marseille Université; **13.6a:** Department of Microbiology, Biozentrum, University of Basel/Science Source; **p.** 390: NASA; **13.7b:** NIAID/Science Source; **p.** 395: Biophoto Associates/Science Source; **13.10:** M. Wurtz/Science Source; **p.** 401 James Gathany/CDC; **13.17:** L Brent Selinger/Pearson Education, Inc.; **13.19:** Phanie/Science Source; **13.20:** Theodor O. Diener, Agricultural Research Service, USDA; **13.21 left:** Barba Mauro/Fotolia, **right:** Nigel Cattlin/Alamy Stock Photo; **13.24a:** Michael Abbey/Science Source; **13.24b:** Sherif Zaki and Wun-Ju Shieh/CDC; **p.** 407: Yuri Arcurs/Alamy Stock Photo.

CHAPTER 14 Opener: Omikron/Science Source, **inset:** Wavebreak Media Ltd/123rf, Anna Henley/Getty Images; **p.** 414 **left:** Richard Wayman/Alamy Stock Photo, **right:** BSIP SA/Alamy Stock Photo; **14.1:** jeridu/E+/Getty Images; **p.** 415: Ralph E. Berry/Oregon State University Libraries; **14.2:** Dr. Tony Brain/Science Source; **p.** 419: Bettmann/Getty Images; **14.5b:** J. William Costerton/Montana State University; **14.6:** David Scharf/Science Source; **14.10:** Images from the History of Medicine, NLM/Courtesy of the National Library of Medicine; **14.12:** Custom Medical Stock Photo/Alamy Stock Photo; **14.13:** Akintunde Akinleye/Reuters; **p.** 432: Glowimages RF/Stock Photos/Glow Images; **14.18:** CDC Morbidity and Mortality Weekly Report (MMWR).

CHAPTER 15 Opener: Science Picture Co/Alamy Stock Photo, **inset:** Chris Rout/Alamy Stock Photo; **15.1:** Eye of Science/Science Source; **15.5a top:** Gwen V. Childs, **bottom:** Science Source; **15.5b top:** J. O. Ballard, M.D., Professor of Medicine & Pathology, Penn State University College of Medicine, **bottom:** chirawan Somsanuk/123RF; **15.6:** Alison K. Criss/University of Virginia Health Sciences Center; **p.** 460: Science Photo Library - CNRI/SPL/Getty Images; **15.10:** From: Immune lysis of normal human and paroxysmal nocturnal hemoglobinuria (PNH) red blood cells. III. The membrane defects caused by complement lysis. W. F. Rosse et al. J Exp Med. 1966 Jun 1;123(6):969-84. Fig. 7; **p.** 470a: chirawan Somsanuk/123RF, **b:** Michael Ross/Science Source, **c:** Gwen V. Childs, **d:** J. O. Ballard, M.D., Professor of Medicine & Pathology, Penn State University College of Medicine, **e:** Gwen V Childs.

CHAPTER 16 Opener: Kateryna Kon/shutterstock, **inset:** Blue Jean Images/Alamy Stock Photo; **16.1:** Biophoto Associates/Science Source; **16.5, 16.11:** Steve Gschmeissner /Science Source; **16.13b:** Tim Evans/Science Source; **p.** 496 **top:** Eye of Science/Science Source, **bottom:** Fullerene/Getty Images; **Table 16.5 top left:** PBWPIX/Alamy Stock Photo, **top right:** Diane Macdonald/Stockbyte/Getty Images, **bottom left:** Supri/Reuters, **bottom right:** Barbara Rice/CDC; **p.** 501: David M. Phillips/Science Source.

CHAPTER 17 Opener: Numstocker/Shutterstock, **inset:** Chris Rout/Alamy Stock Photo; **p.** 504 **left:** Sherry Yates Young/Stock Photo/123RF, **top right:** Sally and Richard Greenhill/Alamy Stock Photo, **bottom right:** CDC; **p.** 510: CDC; **17.8b:** Karen E. Petersen; **17.10:** CDC; **17.11b:** CDC/BSIP SA/Alamy Stock Photo; **17.13b:** L. Brent Selinger/Pearson Education, Inc.

CHAPTER 18 Opener: J. Fieber Agency/Arco Images GmbH/Alamy Stock Photo, **inset:** Jupiterimages/Stockbyte/Getty Images; **18.2a:** Microfield Scientific Ltd./Science Source; **18.2b:** David Scharf/Science Source; **18.2c:** Andrew Syred/Science Source; **18.3:** BSIP/Science Source; **18.4:** Mediscan/Alamy Stock Photo; **18.8:** P. Marazzi/Science Source; **18.9:** Dr. Riccardo Rondinone; **18.10:** Adam Murphy/Alamy Stock Photo; **p.** 535: Scott Camazine/Science Source; **18.11:** Seelevel.com.

CHAPTER 19 Opener: Eye of Science/Science Source, **inset:** ndoeljindoel/Shutterstock, RGBCODE/Shutterstock; **19.1:** David McCarthy/Science Source; **19.2:** J.R. Mekkes Dermatology AMC; **19.3:** BSIP/Science Source; **19.4:** American Family Physician; **p.** 551: Janice Haney Carr/CDC; **19.6:** Scott Camazine/Alamy Stock Photo; **19.7:** P. Marazzi/Science Source; **p.** 554 **top:** Lawrence B. Stack, **bottom:** National Library of Medicine; **p.** 555 **top:** Eye of Science/Science Source, **bottom:** Courtesy of Naiel Bisharat; **19.8:** Lighthunter/Shutterstock; **19.9:** Vincent A. Fischetti, Ph.D.; **p.** 558: Eye of Science/Science Source; **19.10:** Mike Miller/CDC; **19.11:** Larry Stauffer, Oregon State Public Health Laboratory, CDC; **19.12:** Cutaneous Anthrax, Belgian Traveler/CDC; **19.14:** Rich Robison/Pearson Education, Inc.; **19.16:** CDC; **19.17:** SPL/Science Source; **19.18:** From: P110 and P140 cytadherence-related proteins are negative effectors of terminal organelle duplication in Mycoplasma genitalium. O Q Pich et al. *PLoS One*. 2009 Oct 14;4(10) Fig. 3:e7452. doi: 10.1371/journal.pone.0007452; **19.19:** Alan W. Rodwell, Animal Health Research Laboratory, C.S.I.R.O.; **p.** 570 **top to bottom:** Igor Mojzes/Fotolia, Julie A. Theriot; Timothy J. Mitchison, Lorry Eason/ Photodisc/Getty Images; **p.** 571 **top left:** Balasubr Swaminathan/Peggy Hayes/CDC, **top right:** Pr. Courtieu/Science Source, **center:** Lewis G. Tilney, **bottom left:** Biophoto Associates/Science Source, **bottom right:** Maren Caruso/Getty Images; **19.20:** Hiroshi Eguchi; **19.21:** Mediscan/Alamy Stock Photo; **p.** 574: Blend Images/Shutterstock; **19.22:** Rupak De Chowdhur/Reuters; **p.** 576 **top:** imageBROKER/Alamy, **bottom:** SPL/Science Source; **p.** 577 **top, left to right:** National Institute of Allergy and Infectious Diseases, Biophoto Associates/Science Source, **center, left to right:** CNRI/Science Source, Biophoto Associates/Science Source, CNRI/Science Source, **bottom, left to right:** From: Images in Clinical Infectious Diseases: Purified Protein Derivative - PPD. A. Q. Sousa. *Braz J Infect Dis* vol.10 no. 6 Salvador Dec. 2006. Fig. A, Santibhavank P/Shutterstock, Seelevel.com; **p.** 578: Humberto Guerra, Juan Carlos Palomino, Eduardo Falconí, Francisco Bravo, Ninoska Donaires, Eric Van Marck, and Françoise Portaels; **19.24:** DermNet NZ; **19.25a:** Dr. Haitham Alfalah; **19.25b:** Mohammad S Fazeli/Hamed Bateni/Regents of the University of California; **p.** 581: ndoeljindoel/Shutterstock; **p.** 581: RGBCODE/Shutterstock; **p.** 584: L. Brent Selinger/Pearson Education, Inc.; **p.** 584: CDC; **p.** 584: Biophoto Associates/Science Source; **p.** 584: Image Courtesy of Eastman Kodak Company; **p.** 584: Rich Robison/Pearson Education, Inc.; **p.** 584: Dr. Haitham Alfalah.

CHAPTER 20 Opener: A. Barry Dowsett/Science Source, **inset:** LWA/Getty Images; **20.2:** BSIP SA/Alamy Stock Photo; **20.4a:** Mediscan/Alamy Stock Photo; **20.4b:** Gust/CDC; **20.5:** L. Brent Selinger/Pearson Education, Inc.; **20.7a:** Mike Peres RBP SPAS/Newscom; **20.7b:** toeytoey/Shutterstock; **20.10b:** L. Brent Selinger/Pearson Education, Inc.; **p.** 594: Norman Jacobs/CDC; **20.10a, p.** 595: L. Brent Selinger/Pearson Education, Inc.; **20.11:** M I (Spike) Walker/Alamy Stock Photo; **20.12:** Rattiya Thongdumhyu/shutterstock; **20.13:** From: Anaerobic respiration using a complete oxidative TCA cycle drives multicellular swarming in Proteus mirabilis. C. J. Alteri. *MBio*. 2012 Oct 30;3(6). Fig. 1; **p.** 598 **top:** Alliance/Shutterstock, **bottom:** CNRI/Science Source; **p.** 599 **top left:** James Cavallini/Science Source, **top right:** SERCOMI/Science Source, **center:** Pietro M. Motta/Science Source, **bottom left:** Michael Abbey/Science Source, **bottom right:** Science History Images/Alamy Stock Photo; **20.18:** Brook Yockey/CDC; **p.** 604: Mike Schirf/Getty Images; **20.20:** James Cavallini/Science Source; **20.21:** Seelevel.com; **20.24:** From: Infect Immun. 2005 Jun;73(6): Cover related to Bordetella bronchiseptica adherence to cilia is mediated by multiple adhesin factors and blocked by surfactant protein A.J A Edwards et al. © 2005, American Society for Microbiology. All Rights Reserved. (Special assistance on scanning electron microscopy by M. K. McCuskey); **20.25:** CDC; **p.** 609: ene/Shutterstock; **20.26:** Leonard Morse; **p.** 610: THPStock/Fotolia; **p.** 612: Russell E. Enscore/CDC; **20.28:** Photo Researchers Agency/Science History Images/Alamy Stock Photo; **20.27:** Jim Feeley/CDC; **20.29:** Used with permission from Thermo Fisher Scientific.

CHAPTER 21 Opener: Kiyoshi Takahase Segundo/Alamy Stock Photo, **inset:** Blend Images/Alamy Stock Photo, hakat/Fotolia; **p.** 622, **p.** 624, **p.** 625: CDC; **bottom right:** Denis Brien/YAY Micro/Age Fotostock; **21.3b:** Moredun Animal Health/Science Source; **21.4:** Dr. Milton Reisch/Corbis Documentary/Getty Images; **21.5:** DMA/Alamy Stock Photo; **21.6:** M. Singla and B. Bal. Infectivity Assays For Chlamydia Trachomatis. The Internet Journal of Microbiology. 2006; 2(2). Fig. 2; **21.7:** Jr. Edwin P. Ewing/CDC; **21.9b:** Sidea Revuz/BSIP/Science Source; **21.9a:** BSIP SA/Alamy Stock Photo; **21.9c:** Martin M. Rotker/Science Source; **21.10:** Peter Perine/CDC; **21.12a:** Steve Heap/Zoonar GmbH/Alamy Stock Photo; **21.11:** From: Cutting Edge: The Spirochetemia of Murine Relapsing Fever Is Cleared by Complement-Independent Bactericidal Antibodies, Sean E. Connolly and Jorge L. Benach. J *Immunol* September 15, 2001, 167 (6) 3029-3032; DOI: https://doi.org/10.4049/jimmunol.167.6.3029; **21.12b:** James Gathany/CDC; **21.15:** Ben Adler/Australian Research Council Centre of Excellence in Structural and Functional Microbial Genomics; **21.16:** James Cavallini/Science Source; **21.19:** Eye of Science/Science Source; **p.** 641: Robert Weaver/CDC; **p.** 642: **p.** 644: Moredun Animal Health/Science Source.

CHAPTER 22 Opener: Pr. Bouree/Science Source, **inset:** Anton Foltin/Shutterstock; **22.1a:** Libero Ajello/CDC; **22.1b:** Maxine Jalbert/CDC; **p.** 652: From: Severe histoplasmosis in travelers to Nicaragua. M. Weinberg. Emerg Infect Dis. 2003 Oct;9(10):1322-5. Figure 1; **22.4a:** Libero Ajello/CDC; **22.4b:** CDC; **22.6:** John W. Rippon/University of Chicago Medical Center; **p.** 654: Robert Bauman; **22.8:** CDC/Science Source; **22.9:** Dermnet.com; **22.10:** Lucille K. Georg/CDC; **22.11:** PR BOUREE/BSIP SA/Alamy Stock Photo; **22.12:** From: Invasive orbital Aspergillosis in an immunocompetent individual after functional endoscopic sinus surgery. M. Ashok Kumar et al. *Annals of Tropical Medicine and Public Health*. 2013; 6 (2): 254-255. Fig. 1; **p.** 660 **top:** Mikael Häggström, **left (top):** CDC, **left (bottom):** Gust/CDC, **right (top):** SPL/Science Source, **right (center):** Gastrolab/Science Source, **right (bottom):** Dr. M.A. Ansary/Science Source, **bottom:** Martin M. Rotker/Science Source; **p.** 661 **top left:** Richard J. Green/Science Source, **top right:** Martin M. Rotker, **bottom left:** Seelevel.com, **bottom right:** NIH; **22.13:** Jr. Edwin P. Ewing/CDC; **p.** 663: Philip Dalton/Nature Picture Library; **22.15:** SPL/Science Source; **22.16:** Dermnet.com; **22.17:** Robert A Schwartz; **22.18:** Ajello Libero/CDC; **22.19:** Dr. Michel Develoux, Societe de Pathologie Exotique; **22.20:** From: Cryosurgery for the treatment of cutaneous sporotrichosis in four pregnant women. V. Fichman et al. *PLoS Negl Trop Dis*. 2018 Apr 23;12(4):e0006434. doi: 10.1371/journal.pntd.0006434. eCollection 2018 Apr. Fig 1; **22.21:** Angelo Gandolfi/Nature Picture Library; **p.** 673(a): CDC/Science Source, **(b):** John Durham/Science Source, **(c):** CDC.

CHAPTER 23 Opener: Science Photo Library/Alamy Stock Photo, **inset:** grafikplusfoto/Fotolia, Barna Tanko/Fotolia; **23.2:** Mae Melvin, CDC; **23.4:** Nicholas Despo/Biology and Life Sciences Department, Thiel College, Greenville, PA; **p.** 682 **top:** CDC, **bottom:** LESLIE E. KOSSOFF/AP Images; **23.6:** James Gathany, CDC; **23.7:** Dr Zuño Burstein, Prof. Emeritus The National University of San Marcos; **23.8:** J. L. Carson/Newscom; **p.** 684 **top:** Cristina Pedrazzini/Science Source, **bottom left:** Courtesy of ATLAS OF HUMAN INTESTINAL PROTOZOA (www.atlas-protozoa.com),

CREDITS C-3

bottom right: CDC; **p. 685 top left:** SPL/Science Source, **top right, center:** CDC, **bottom left:** Michael Abbey/Science Source, **bottom right:** David P. Lewis/Shutterstock; **p. 688 top:** Sue Ford/Science Source, **bottom left:** Omikron/Science Source, **bottom right:** Dr. Gopal Murti/Science Source; **p. 689 top left:** CDC, **top right:** AFP/Stringer/Getty Images, **center, left to right:** Kletr/Fotolia, London School of Hygiene & Tropical Medicine/Science Source, **bottom left:** Mauro Fermariello/Science Source, **bottom right:** Irene Abdou Stock Connection Worldwide/Newscom; **23.10:** Dimitri P. Agamanolis/neuropathology-web.org; **p. 691:** California Department of Public Health; **23.11:** CDC; **23.13b:** Steve Gschmeissner/Science Source; **23.15:** From: Taenia solium cysticercosis. H. H. García et al. Lancet. 2003 Aug 16;362(9383):547-56, Fig. 4bc; **23.16:** Dr. William Thomas, McLean, Virginia; **23.17:** Eye of Science/Science Source; **23.18:** From: PLOS. "Stop the snails: Control of snail hosts critical to schistosomiasis elimination." *Science Daily*, 21 July 2016. www.sciencedaily.com/releases/2016/07/160721151049.htm. Photo: Susanne Sokolow; **p. 697:** CDC; **23.19a:** CDC; **23.19b:** Moore/CDC; **23.19c:** D.S. Martin/CDC; **p. 699:** John Cornellier; **23.20:** CDC; **23.21:** Thierry Berrod, Mona Lisa Production/Science Source; **23.22:** CDC; **23.24:** Steve Upton; **23.25:** Andy Crump, TDR, World Health Organization/Science Source.

CHAPTER 24 Opener: Linda M. Stannard, University of Cape Town/Science Source, **inset:** Sergio Azenha/Alamy Stock Photo; **24.1:** Dr. Klaus Boller/Science Source; **24.3:** From: The History of Smallpox. A.Geddes. Clinics in Dermatology, Volume 24, Issue 3, Pages 152-157 (May 2006), Fig. 3. © 2006 Elsevier Inc.; **24.4:** Dr P. Marazzi/Science Source; **p. 713:** Henk Bentlage/Shutterstock; **24.6-24.8:** Dr. Zara/Science Source; **24.9:** Newscom; **24.11:** Dr P. Marazzi/Science Source; **24.12 left to right:** SPL/Science Source, Dr M.A. Ansary/Science Source, Robert Bauman, Jatin M. Vyas/US National Library of Medicine; **p. 719:** Peter Sherrard/Photographer's Choice/Getty Images; **24.13:** New England Journal of Medicine; **24.15:** National Cancer Institute; **24.14:** Scott Camazine/Science Source; **p. 723 top to bottom:** BSIP/UIG/Getty Images, Clinical Photography, Central Manchester University Hospitals NHS Foundation Trust, UK/Science Source, Dr. Klaus Boller/Science Source, Biophoto Associates/Science Source; **24.16:** M. Wurtz/Biozentrum, University of Basel/Science Source; **24.17:** Dr P. Marazzi/Science Source; **p. 724:** Dermnet.com; **p. 725:** CDC; **24.19:** Garry Watson/Science Source; **24.21a:** Science Source; **p. 727:** Casa nayafana/Shutterstock; **24.22:** Michelle Vilasineekul of Veronicavphotography; **p. 732:** Newscom.

CHAPTER 25 Opener: Maurizio De Angelis/Science Source, **inset:** Buburuzaproductions/E+/Getty Images; **25.1:** Laguna Design/Science Source; **25.3:** Agencja Fotograficzna Caro/Alamy Stock Photo; **25.4:** Kathy Willens/AP Images; **25.5:** Dr P. Marazzi/Science Source; **25.6:** © CVUA Stuttgart; **p. 740:** Piotr Marcinski/Fotolia; **25.7:** James Cavallini/Science Source; **25.9 left:** Steven Ellingson/123rf, **right:** Steven Doggett; **p. 743:** erik_karits/Fotolia; **p. 746 top to bottom:** © 2011 Fischer et al., Associated Press, Tharakorn Arunothai/123rf; **25.13:** Science Source; **25.16:** Science History Images/Photo Researchers/Alamy Stock Photo; **25.17a:** Associated Press; **25.17b:** Fahad Shadeed/Reuters; **25.19:** Professor Aaron Polliack/Science Source; **25.23a:** Virat Sirisanthana M.D, Professor of Pediatrics, Chiang Mai University; **25.23b:** A. Ramey/PhotoEdit; **25.26a:** Dr P. Marazzi/Science Source; **25.26b:** mediacolor's/Alamy Stock Photo; **p. 759:** Springer Medizin/Science Source; **25.28:** Dr P. Marazzi/Science Source; **25.31:** Dr. Fred. A. Murphy/CDC; **p. 762:** U.S. Geological Survey; **25.33:** Dr. Daniel P. Perl/CDC; **p. 764 top:** Jack Hollingsworth/Blend Images/Getty Images, **left:** John Moore/Getty Images News/Getty Images, **right:** Dr. Scott Smith/CDC; **p. 765 top left:** Elena Ryabchikova/Voisin/Science Source, **top right:** CAMR/A. Barry Dowsett/Science Source, **bottom left:** Caroline Thirion/ALIMA, **bottom right:** Samuel Mambo/REUTERS; **p. 766:** SPL/Science Source; **25.35:** BSIP SA/Alamy Stock Photo; **p. 768 top:** saiyood/Fotolia, **bottom:** James Cavallini/Science Source; **p. 769 top left:** Otis Historical Archives, NMHM, **top right:** Thomas Deerinck, NCMIR/Science Source, **bottom:** goodluz/Fotolia; **25.36:** Dr. Linda M. Stannard, University of Cape Town/Science Source.

CHAPTER 26 Opener: Valentyn Volkov/Alamy Stock Photo, **inset:** MBI/Alamy Stock Photo; **26.1:** Associated Press; **26.2:** Paul Hudson/fStop/Age Fotostock; **26.3:** Arnd Wiegmann/Reuters; **26.4:** EPA Photos; **26.5:** Novo Nordisk; **p. 790 top:** CDC, **bottom:** Alex Skelly/Moment/Getty Images; **26.6:** Kekyalyaynen/Shutterstock; **26.7a:** Bastian Schnabel/Shutterstock; **26.7b:** Jade ThaiCatwalk/Shutterstock; **26.8:** mariusz szczygiel/Fotolia; **26.10:** Gary Crabbe/Enlightened Images/Alamy Stock Photo.

CHAPTER 27 Opener: Stephen Street/Alamy Stock Photo, **inset:** Michael Tatman/Alamy Stock Photo; **p. 803 left:** 7activestudio/Fotolia, **inset:** Scimat/Science Source, **top right:** CNRI/Science Source, **bottom right:** David Musher/Science Source; **27.2:** Anna H.Kaksonen; **27.3:** Radhoose/Shutterstock; **27.7:** CPRESS PHOTO LIMITED/Alamy Stock Photo; **27.10:** Associated Press.

SUBJECT INDEX

NOTE: A *t* following a page number indicates tabular material, an *f* following a page number indicates a figure, and a *b* following a page number indicates a boxed feature.

A. *See* Adenine
AAD. *See* Antimicrobial-associated (*C. difficile*) diarrhea
A antigen, 514, 514*f*, 530, 530*f*, 531*t*
Abbott, Isabella, 113
Abdominal infections, *Bacteroides fragilis* causing, 614
Abiogenesis (spontaneous generation), 7–10, 8*f*, 9*f*, 10*b*
Abiotrophia genus/spp., 557
ABO system, 514, 514*f*, 530–531, 530*f*, 531*t*
Abscess, 464
  in group A streptococcal pharyngitis, 553, 553*f*
Absidia genus/spp., 662
Absorbance, spectrophotometer measuring percentage of, 186*f*, 187
Abyssal zone, 811*f*, 812
Acanthamoeba genus/spp., 364, 365*t*, 678–679, 692*t*
  encephalitis caused by, 678, 692*t*
  keratitis caused by, 267*b*, 678, 679, 692*t*
Acceptor stem, transfer RNA, 209, 209*f*
Accessory photosynthetic pigments, 373, 374
Accutane. *See* Retinoic acid
Acellular (plasmodial) slime molds, 364, 365*t*
Acellular agents, 385, 391*t*. *See also* Prion(s); Viroid(s); Viruses
Acellular obligatory parasites, viruses as, 6, 7*f*
Acellular pertussis vaccine (diphtheria/tetanus/acellular pertussis [DTaP]), 506, 508*f*, 509*t*, 567, 573, 608
Acetaldehyde, fermentation producing, 141*f*
Acetic acid (vinegar), microbial production of, 12*t*, 782, 783*t*, 787
Acetic acid bacteria, 782, 783*t*
Acetobacter genus/spp., 340
  in vinegar production, 340, 782, 783*t*
Acetone
  in Gram staining, 109, 109*f*
  microbial fermentation producing, 142*f*
Acetylcholine
  botulism toxins affecting, 563, 564*f*
  tetanus toxin affecting, 565, 566*f*
Acetyl-CoA (acetyl-coenzyme A), 52, 149*t*, 153*f*
  in carbohydrate biosynthesis, 150*f*, 153*f*
  in citric acid/Krebs cycle, 133*f*, 136, 137*f*, 149*t*, 153*f*
  in glucose catabolism, 133*f*, 153*f*
  in lipid biosynthesis, 150, 151*f*, 153*f*
  in lipid catabolism/beta-oxidation, 143, 143*f*, 153*f*
  synthesis of, 133*f*, 135–136, 135*f*
Acetyl-CoA (acetyl-coenzyme A) synthetase, 128*f*
N-Acetylglucosamine (NAG), 45*f*, 66–67, 66*f*, 67*f*, 290, 292*f*
  Caulobacter glue and, 341*b*
N-Acetylmuramic acid (NAM), 66–67, 66*f*, 67*f*, 290, 292*f*
  antimicrobials affecting, 290–291, 292*f*
ACh. *See* Acetylcholine
Acid(s), 38–39, 38*f*
  pH of, 38, 38*f*
Acid-base balance, 38. *See also* pH
Acid-fast bacilli/rods, 110, 337, 337*f*
  in leprosy diagnosis, 575
  in tuberculosis diagnosis, 432*b*, 577*b*
Acid-fast bacteria, 67
Acid-fast stain, 67, 109–110, 110*f*, 111*t*, 337, 337*f*, 573, 574
Acidic dyes, 108
  negative staining and, 110, 111*f*
Acidity. *See* pH
Acid mine drainage, 805–806, 805*f*, 806*f*
Acidophiles, 39, 167
Acid rain, 39*b*
Acid-tolerant microbes, 39, 40*b*, 167
Acinetobacter genus/spp., 611
Acne, 39, 578–579, 579*f*

Acquired immunity, 473, 473*t*, 495–496, 497*t*. *See also* Adaptive immunity
  artificial, 495, 496, 497*t*. *See also* Immunization; Vaccines
    active/active immunization, 495, 496, 497*t*, 505–510, 507*f*, 508*f*, 509*f*, 511*f*. *See also* Immunization; Vaccines
    passive immunotherapy, 496, 497*t*, 504, 505, 510–511, 511*f*
  natural, 495, 497*t*
    active, 495, 497*t*
    passive, 487, 495, 497*t*
Acquired immunodeficiency diseases, 540–541. *See also* HIV infection/AIDS
Acquired immunodeficiency syndrome (AIDS), 541, 751–757, 751*t*, 752*t*, 753*f*, 754*b*, 754*t*, 755*f*, 765*f*. *See also* HIV infection/AIDS
Acremonium genus/spp.
  antimicrobials produced by, 290*t*, 309*t*
  mycetoma caused by, 666*t*
Acridine, as frameshift mutagen, 219, 219*f*
Actin
  in cytokinesis, 357
  in eukaryotic cytoskeleton, 83, 84*f*
Actinobacteria (phylum), 330*f*, 336, 337*t*
  biofuel production and, 162*b*
Actinomyces genus/spp./*Actinomyces israelii*, 337*t*, 338, 580–581, 580*f*
Actinomycetes, 337–338, 338*f*
  reproduction in, 326, 327*f*, 338
Actinomycin, mechanism of action of, 296
Actinomycosis, 580–581, 580*f*
Activated charcoal filters
  in drinking water treatment, 791*f*, 792
  in wastewater treatment, 794
Activated sludge system, wastewater treatment and, 793–794, 793*f*
Activation, enzyme, 131, 131*f*
Activation energy, enzymes affecting, 128–129, 129*f*
Active immunity
  artificially acquired/active immunization, 495, 496, 497*t*, 505–510, 507*f*, 508*f*, 509*f*, 511*f*. *See also* Immunization; Vaccines
  naturally acquired, 495, 497*t*
Active site
  chlorophyll, 144, 145*f*
  enzyme, 128*f*, 129, 129*f*
    activation/inhibition and, 131–132, 131*f*, 132*f*
Active transport, 73, 73*f*, 74*t*
Active transport processes, 70, 73–74, 73*f*, 74*t*
  active transport, 73, 73*f*, 74*t*
  endocytosis/exocytosis, 81, 81*f*, 81*t*
  group translocation, 73, 73–74, 73*f*, 74*t*
Acute anaphylaxis. *See* Anaphylaxis/anaphylactic shock
Acute disease, 433, 434*t*
Acute hemorrhagic conjunctivitis, 738
Acute inflammation, 462
ACV. *See* Acyclovir
Acyclovir, 296, 297*f*, 314*t*
  spectrum of action of, 299*f*, 314*t*
Adaptation, microbial survival and, 805
Adaptive immunity, 447, 472–502, 473*t*
  acquired, 473, 473*t*. *See also* Immunization; Vaccines
  antibody immune response in, 474
  cell-mediated immune response in, 474, 480–482, 482*f*, 483*f*, 489–492, 490*f*, 491*f*, 495
  defects in, immunodeficiency diseases and, 540, 540*t*
  elements of, 474–489. *See also specific element*
  antigens, 476–478, 477*f*
    cells presenting, 478–479, 478*f*
    processing of, 479–480, 479*f*, 480*f*

B lymphocytes (B cells) and antibodies, 473, 474, 483–488, 483*f*, 484*f*, 485*f*, 486*f*, 487*f*, 488*t*. *See also* Antibody immune response; B lymphocytes
cytokines, 481, 488, 488–489, 489*t*
lymphatic organs/tissues, 474, 474–476, 475*f*
preparation and, 478–482, 478*f*, 479*f*, 480*f*, 481*f*, 482*f*
T lymphocytes (T cells), 473, 474, 474*f*, 482*t*. *See also* Cell-mediated immune response; T lymphocytes
  major histocompatibility complex in, 478–479, 478*f*
  types of, 495–496, 497*t*
ADCC. *See* Antibody-dependent cellular cytotoxicity
Adefovir, 297*f*, 314*t*
Adenine (A), 50, 50*f*, 51, 51*f*, 52, 52*t*, 194, 195*f*
Adenine deoxyribonucleoside, 50
Adenosine, 52*f*, 297*f*
Adenosine arabinoside, 297*f*, 314*t*
Adenosine diphosphate. *See* ADP
Adenosine monophosphate. *See* AMP
Adenosine triphosphate. *See* ATP
Adenosine triphosphate deoxyribonucleotide (dATP). *See also* Triphosphate deoxyribonucleotides
  in DNA replication, 198
Adenoviridae (adenoviruses), 392*t*, 722–725, 724*f*, 725*b*, 729*t*
  common cold/pharyngitis caused by, 722–724, 725*b*
  as genetic vectors, 242
  immunization against, 725, 725*b*
Adenovirus 36, obesity and, 724
Adenylate cyclase toxin
  Bordetella pertussis, 607
  Vibrio cholerae, 638, 638*f*
Adhesins, 422
  Bordetella, 422, 606
  Enterobacteriaceae/*E. coli*, 593, 593*f*, 595
  Helicobacter pylori, 640, 640*b*
  Neisseria gonorrhoeae, 422
  Plasmodium, 422, 686
  Pseudomonas aeruginosa, 609
  Streptococcus pneumoniae, 558, 558*b*
  Yersinia pestis, 602
Adhesion, 421–422, 422*f*. *See also* Attachment
  in phagocytosis, 455, 455*f*
Adhesion disks, 422
Adhesion factors, 422, 422*f*
Adhesive disk, in *Giardia* life cycle, 684*b*, 685*b*
Adipocytes, adenoviruses affecting, 725
Adjuvants, for inactivated vaccines, 506
Administration route, antimicrobial drug, 300–302, 301*f*
ADP, 52, 52*f*
  in active transport, 73, 73*f*
  phosphorylation of, 127. *See also* Phosphorylation
  in Calvin-Benson cycle, 148*f*, 153*f*
  in carbohydrate biosynthesis/gluconeogenesis, 150*f*
  in chemiosmosis, 139, 140
  in citric acid/Krebs cycle, 136, 137*f*
  in electron transport chain, 133*f*, 138*f*, 153*f*
  in glycolysis, 133*f*, 134*f*, 135, 135*f*
  in lipid catabolism, 143*f*
  photophosphorylation (photosynthesis), 146, 147*f*
Adult acute T-cell lymphocytic leukemia, HTLV-1 causing, 750
Adv36. *See* Adenovirus 36
Aedes mosquitoes
  as disease vectors, 433*t*, 747*t*
  for arboviral encephalitis, 747*t*
  for chikungunya, 401*b*, 747*t*

for dengue fever, 743, 744, 744*f*, 745*f*, 747*t*
  mosquito control and, 385*b*, 746*b*
  Wolbachia in elimination of, 385*b*, 746*b*
for yellow fever, 16*b*, 747*t*
for Zika fever, 744, 747*t*
range of, 744, 744*f*
Aerobes/aerobic bacteria, 139, 164
  Gram-negative, 605–614
  obligate, 163, 165*f*
Aerobic respiration, 139, 140*t*, 142*t*
Aerosols, airborne transmission of disease and, 431, 432*b*, 433*t*
Aerotolerant anaerobes, 164, 165*f*
  culturing, 178
AFBs. *See* Acid-fast bacilli
Afferent lymphatic vessels, 475*f*, 476
Aflatoxicosis, 668
Aflatoxins, 668
  mutagenic effects of, 219, 668
AFM. *See* Atomic force microscopy/microscopes
African trypanosomiasis (African sleeping sickness), 15*t*, 680–681, 681*f*, 682*b*, 692*t*
AFRs (acid-fast rods). *See* Acid-fast bacilli
Agammaglobulinemia, Bruton-type, 540, 540*t*
Agar, 6, 13, 13*f*, 174–175, 175*f*
  in algal cell wall, 6, 80, 80*f*, 374, 376*t*
  for cultures, 174–175, 175*f*
  nutrient, 174, 178*f*
Agaricus genus/spp., 370, 373*t*
Agarose, for gel electrophoresis, 246, 247*f*
Age/aging, immunity and, 540
Agglutination, antibody, 117, 117*f*, 486, 486*f*, 514
  in blood typing/transfusion reactions, 514, 514*f*, 530, 530*f*
  by IgA, 487
  in serology, 117, 117*f*, 513–514, 514*f*, 520*t*
Agglutination tests, 117, 117*f*, 513–514, 514*f*, 520*t*
Aging
  in beer production, 782, 782*f*
  in wine and spirits production, 781, 781*f*
Agranulocytes, 452*f*, 453, 453*f*
Agricultural crops. *See also* Agricultural microbiology
  in alternative fuel production, 788
  biological threats to, 812, 813, 814, 815. *See also* Agroterrorism
  Burkholderia in protection of, 608
  genetically modified, 254–256, 255*f*, 788–789, 789*t*
    ethical/safety issues and, 256, 788–789
Agricultural microbiology, 19*t*, 254–256, 255*f*, 788–789, 789*t*
  crop threats and, 812, 813, 814, 815. *See also* Agroterrorism
  livestock/poultry threats and, 813
  recombinant DNA technology in, 254–256, 255*f*, 788–789, 789*t*
  Bt toxin and, 255, 335–336, 336*f*, 788–789, 789*t*
    ethical/safety issues and, 256, 788–789
  waste treatment and, 794
Agrobacterium genus/spp./*Agrobacterium tumefaciens*, 340, 341*f*, 344*t*
  chromosomes of, 195
  conjugation in, 227
  glyphosate tolerance and, 254–255
  soilborne transmission of, 810*t*
Agroterrorism, 812. *See also* Bioterrorism
  animal pathogens and, 813, 813–814
  assessing threat potential and, 813
  defense against, 815
  plant pathogens and, 813, 814
AIDS (acquired immunodeficiency syndrome), 541, 751–757, 751*t*, 752*t*, 753*f*, 754*b*, 754*t*, 755*f*, 765*f*. *See also* HIV infection/AIDS
AIDS virus. *See* HIV

I-1

SUBJECT INDEX

Airborne transmission of disease, 431, 432b, 433t
Air-drying, in specimen preparation, 106, 108f
Akinetes, of cyanobacteria, 333, 333f
Alanine, 47f, 48f
Albendazole, 316t, 317t, 496b
Alcaligenes faecalis, culture of, 177f
Alcohol(s). *See also* Alcoholic beverages; Alcoholic fermentation; Ethanol
  antimicrobial actions of, 276–277, 281t
  functional groups in, 40, 41t
Alcohol–butanol, as biofuel, 788
Alcoholic beverages, microbial production of, 5, 10–12, 10b, 11f, 12t, 18f, 141, 141f, 142f
Alcoholic fermentation, 10–12, 10b, 11f, 12t, 141, 141f, 142f, 780–782, 781f, 782f, 783f
Aldehydes, antimicrobial action of, 279, 281t
Algae, 6, 6f, 373–376, 376t
  beneficial/industrial uses of, 6
  blue-green. *See* Cyanobacteria
  brown, 375, 375f, 376t
  cell walls of, 6, 80, 80f
  classification of, 357, 357–358, 374–376, 376t
  cytokinesis in, 357
  distribution of, 373
  golden, 375, 376t
  green, 374, 376t
  hydrocarbons produced by, 788
  in lichens, 371, 372f, 374
  morphology of, 373
  photosynthesis by, 6, 6f
  red, 80f, 373, 375f, 376t
  reproduction of, 373–374, 374f
  in soil, 810
  viruses killing, 390b
  yellow-green, 375, 376t
Algal blooms, 808
  intoxications caused by, 362–363, 790
  viruses affecting, 390b
Alginic acid/alginate/algin, in algal cell wall, 80, 375, 376t
Alkalinophiles, 167
Alkalis. *See* Base(s) (chemical)
Allergens, 477, 526, 527f, 528, 528f. *See also* Allergic reactions/allergies
  avoidance of, 529
  fungal, 528, 668, 669
  inhaled, 528
  initial exposure to, 263b, 526, 527f
  testing for, 529, 529f
Allergic contact dermatitis, 535, 536f
Allergic reactions/allergies, 526–530, 527f, 527t, 528f, 529f, 537t
  anisakiasis and, 701
  antigens and, 476
  to antimicrobial drugs/penicillin, 302
  aspergillus causing, 657, 659, 659b
  bee/insect stings causing, 525b, 528, 535b, 541b
  clinical signs of
    in localized reaction, 528, 528f
    in systemic reactions, 528–529
  degranulation/degranulating cells in, 526, 527f, 527t
  diagnosis of, 529, 529f
  to food, 529
  to fungi, 528, 649, 668, 669
  hygiene hypothesis and, 263b, 526
  in immunotherapy (serum sickness), 510
  immunotherapy for, 529
  initial exposure/sensitization and, 263b, 526, 527f
  poison ivy as, 535, 536f
  prevention of, 529
  transgenic organisms and, 256
  treatment of, 529, 541b
  vaccines causing, 509
"Allergy shots," 529
Allicin, in garlic, food preservation and, 785
Allografts, 536, 536f
Allolactose, in operons, 214, 214f
Allosteric activation, 131, 131f
Allosteric inhibition, 131–132, 131f, 132f
  feedback, 132, 132f
  noncompetitive, 132, 132f
Allosteric site, 131, 131f
Allylamines, 315t
  mechanism of action of, 294, 315t

Alpha-amanitin, 668
Alpha (α)-carbon, amino acid, 47, 47f
Alpha F$_c$ region/heavy chains, 485, 488t
Alpha- (α-) glucose, 44, 45f
Alpha (α)-helices, 48, 49f
  cellular PrP and, 48, 405, 406f
Alpha-hemolysis/alpha-hemolytic streptococci
  Streptococcus pneumoniae, 557–559, 558b, 558f, 561t
  viridans group, 557, 557f, 561t
Alpha (α)-interferon, 458, 458f, 459t
Alpha (α)-ketoglutaric acid, 149f
  in citric acid/Krebs cycle, 137f, 149t
  in transamination, 151f
Alphaproteobacteria, 330f, 334t, 338–340, 339f, 340f, 344t
  pathogenic, 339–340, 605, 606, 621
Alphavirus genus/spp./alphaviruses, 392f, 771f
  as biological weapons, 814t
Alternaria genus/spp., phaeohyphomycosis caused by, 666f
Alternation of generations, 373, 374f
Alternative fuels, 162b, 788, 789t
Alternative pathway of complement activation, 459f, 460
Alternative splicing, 206
Alum
  in drinking water treatment, 791, 791f
  in wastewater treatment, 794
Aluminum, 30f
Alveolar macrophages, 454
Alveolata (kingdom), 357, 359f
Alveolates, 362–363, 362f, 363f, 365t
Alveoli, of protozoa, 362, 362f
Amanita genus/spp. (*Amanita muscaria/phalloides*), 371f, 373t, 668, 669f
Amanitin, alpha, 668
Amantadine, 313t, 769b
  mechanism of action of, 295, 313t
Amastigotes, in *Leishmania* life cycle, 682, 683f
Amblyomma (Lone Star) ticks, as disease vectors
  for ehrlichiosis and anaplasmosis, 623, 626t
  for *Rickettsia parkeri* infection, 622b
Amblyomma maculatum (Gulf Coast tick), in spotted fever rickettsiosis, 622b
Ambrosia trifida (ragweed), 528f
Amebae. *See* Amoebas (amoebae)
Amebiasis/amebic dysentery, 364, 678, 692t
  waterborne transmission of, 678, 790b
Amebic encephalitis, *Acanthamoeba* causing, 678, 692t
Amebic meningoencephalitis, *Naegleria* causing, 679, 692t, 790b
Amensalism, 415, 416t
American trypanosomiasis (Chagas' disease), 675b, 679–680, 680f, 692t, 703b
Ames, Bruce, 222
Ames test, 222–223, 223f
Amikacin, 310t
Amination, 150, 151f
Amino acid(s), 38, 47, 47f
  amination of, 150, 151f
  biosynthesis of, 150, 151f, 153f
  in carbohydrate biosynthesis, 150f
  deamination of, 143, 144f, 807f, 808
  essential, 150
  genetic code determining, 207–208, 208f
  for microbial growth, 165t
  microbial production of, 787, 789t
  in nitrogen cycle, 808
  peptide bonds linking, 47, 48f
  in protein catabolism, 143, 144f, 153f
  sequence of (primary structure of protein), 48, 49f
  in sulfur cycle, 808, 808f
  transamination of, 150, 151f
  in translation, 207, 208, 209, 209f
Amino acid substitutions, 216–217, 217f
Aminoacyl-tRNA synthetases, 209, 293
Amino functional group, 40, 41f
  in amino acids, 47, 47f, 48f
Aminoglycosides, 310t
  mechanism of action of, 291f, 293, 294f, 310t
  toxicity of, 302, 310t

Ammonia, 38
  in amino acid biosynthesis, 150, 153f
  with chlorine (chloramine)
    antimicrobial action of, 278
    in drinking water treatment, 791f, 792
    phenol coefficient of, 281
  nitrogen cycle/fixation and, 165, 339, 340, 807, 807f, 808
  wastewater treatment producing, 794
Ammonification, in nitrogen cycle, 807f, 808
Ammonium/ammonium ion, 38, 278f
  in detergents, 278–279, 278f
  in nitrogen cycle, 807f, 808
Ammonium oxidation, anaerobic (anammox), 807f, 808
Amoeba proteus, 5f
Amoebas (amoebae), 360–361, 678–679, 678f, 692t
  in Amoebozoa, 363, 364, 365t
  free-living, 364, 365t, 677, 678–679
  isolation of, 174
  in Rhizaria, 363–364, 363f, 364f, 365t
Amoeboid action, 81
Amoebozoa (kingdom), 357, 359f, 364, 365t
Amoxicillin, 309t
  with clavulanic acid, 307f
AMP, 52, 52f, 127
  cyclic (cAMP), in lactose (*lac*) operon regulation, 214, 214f
Amphibolic reactions, 149
Amphotericin B, 315t, 650
  mechanism of action of, 293, 294f, 315t, 650
  microbial production of, 290t, 315t
Ampicillin, 309t, 557
Amplifying hosts, for *Yersinia pestis*, 602
Amycolatopsis orientalis/rifamycinia, antimicrobials produced by, 290t, 313t
Amylase, microbial production of, 787, 789t
Amylose, 45, 46f
Anabaena genus/spp., 333f
Anabolism/anabolic pathways, 36, 125, 125f, 126, 127, 144–151, 149t, 153f
  amino acid biosynthesis, 150, 151f, 153f
  carbohydrate/glucose biosynthesis, 149, 150f, 153f
  integration and regulation of, 151–152, 153f
  lipid biosynthesis, 150, 151f, 153f
  nucleotide biosynthesis, 151, 152f, 153f
  photosynthesis, 144–148, 145f, 147f, 148f, 153f
Anaerobes/anaerobic bacteria, 12, 164
  aerotolerant, 164, 165f
  culturing, 178
  culturing, 177, 178f
  facultative, 164, 165f
    glycolytic, 342, 343t, 344t
    Gram-negative bacilli, 591–605
    yeasts as, 11
  in fermentation, 11–12, 11f
  obligate, 163, 165f
    Gram-negative bacilli, 614, 614f
Anaerobic ammonium oxidation (anammox), 807f, 808
Anaerobic glove box, 177
Anaerobic media, 177, 178f
Anaerobic respiration, 139, 142t
Analytical epidemiology, 437
Anammox (anaerobic ammonium oxidation), 807f, 808
Anamorph, 648
Anaphase
  meiosis, 356, 356f, 357t
  mitosis, 355, 355f, 357t
Anaphylaxis/anaphylactic shock, 528–529
  antimicrobial drugs causing, 302
  bee/insect stings causing, 525b, 529, 541b
  food allergies causing, 529
  vaccines causing, 509, 529
Anaplasma genus/spp./*Anaplasma phagocytophilum*, 623–626, 623f, 626t
Anaplasmosis, 623–626, 623f, 626t
Ancylostoma duodenale, 700, 703t
Anemia, autoimmune hemolytic, 539
Angiomatosis, bacillary, 605
Animal bites, *Pasteurella* infecting, 604
Animal cultures, 178
  viral, 178, 403
Animal feed (silage), fermentation in production of, 779, 783t

Animalia (kingdom), 113, 114f, 359f
Animal models, in biomedical research, recombinant DNA technology and, 254
Animal pathogens, as biological weapons, 813–814
  threat to livestock/poultry and, 813
Animal reservoirs, 418, 419t
Animal viruses, 387
  culturing, 178, 403
  latency of, 400, 401t
  replication of, 396–400, 397f, 399f, 399t, 400f, 401t
    assembly and release in, 398–400, 400f, 401t
    attachment in, 396–397, 401t
      antimicrobials affecting, 290, 291f, 298, 313t
    entry and uncoating in, 397, 397f, 401t
      antiviral drug action and, 295, 313t, 313–314t
    persistent infection and, 400, 400f, 401t
    synthesis in, 397–398, 398f, 399f, 399t, 401t
Animal wastes, oxidation lagoons in treatment of, 794
Anion(s), 33, 38, 38f
  of salts, 39, 106
Anionic chromophores/dyes, 108
Anisakiasis, 701, 703t
Anisakids, raw fish in transmission of, 701, 701f, 703t
Anisakis genus/spp./*Anisakis simplex*, 701, 701f, 703t
Anopheles mosquitoes, as disease vectors, 433t
  for arboviral encephalitis, 747t
  for filariasis (*Wuchereria bancrofti*), 702
  for malaria (*Plasmodium* species), 688b, 689b, 692t
Anoxygenic organisms
  green and purple bacteria as, 333, 334t
  noncyclic photophosphorylation in, 146
Antacids, bacteria in stomach affected by, 39, 40b, 53b
Antagonism (microbial), 168
  adaptation/survival and, 805
  antimicrobial drug combinations and, 307
  food spoilage and, 783
  in host defense, 298, 450
  lactobacilli and, 336
  opportunistic infections/superinfections and, 298, 299b, 302, 418, 439
Anthelminthic drugs, 299f, 316t
Anthrax, 13, 335, 419t, 561–562, 562f, 736t. *See also Bacillus anthracis*
  bioterrorism and, 256, 336, 562, 814t
  cutaneous, 324b, 336, 348b, 562, 562f
  discovery of cause of, 13, 424
  gastrointestinal, 561
  immunization/vaccine against, 509t, 562
  inhalation, 336, 561–562, 736t
  safe handling of bacteria causing, 268
  soilborne transmission of, 810, 810f
Anthrax toxins, 336, 561
Anthrax vaccine, 509t, 562
Anti-A antibodies, blood typing/transfusion reactions and, 514, 514f, 530, 531t
Anti-antibodies
  in ELISAs, 517, 517f
  in immunochromatographic assays, 520
  in indirect fluorescent immunoassays, 516, 516f
Antibacterial drugs, 290t, 309–313t. *See also* Antibiotics
Anti-B antibodies, blood typing/transfusion reactions and, 514, 514f, 530, 531t
Antibiotics, 280, 281t, 289, 290t. *See also specific drug and* Antimicrobial drugs
  adverse/side effects of, 302, 302f, 303f
  bacterial cell wall synthesis affected by, 290–292, 291f, 292f, 309–310t
  diarrhea associated with use of, 288b, 302, 319b, 439, 563
  fecal microbiome transfer for, 803b
  discovery of, 18, 21, 21f, 289
  mechanisms of action of, 264–266, 289–298, 291f, 292f, 294f, 296f, 297f, 309–318t

microbial production of, 12t, 289
microbiome/normal microbiota/
opportunistic infection and,
298, 299b, 302, 304b, 418
resistance to. See Resistance
spectrum of action of, 298, 299f
Antibodies, 47, 428, 452, 455, 474, 483–488,
483f, 484f, 485f, 486f, 487f,
488t. See also Antibody
immune response;
Immunoglobulin(s)
antitoxins as, 428
binding sites of, 483, 483f, 488t
classes of, 486–487, 488t
in complement activation/cascade,
459–460, 459f, 461f
deficient, 540, 540t
diagnostic uses and. See Serology
fluorescent dye linked to, 103, 103f
for confocal microscopy, 103
for fluorescence microscopy, 103, 103f
for serological testing, 515–516, 516f
function of, 485–486, 486f
in immune response. See Antibody
immune response
labeled, in serological testing, 515–518,
516f, 517f, 518f, 519f
monoclonal, 511
in fluorescent immunoassays, 516
immunosuppressive actions of, 537,
538t
in passive immunotherapy, 511, 511f
in newborn, naturally acquired passive
immunity and, 487, 495, 497t
in passive immunotherapy, 496, 497t, 504,
505, 510, 511, 511f
phagocytosis and, 455
plasma cell secretion of, 493f, 494
quantifying, titration for, 514, 514f
in serology, 117, 512–520, 520t
specificity of, 485, 485f
structure of, 485, 485f
Antibody-antigen immune tests. See
Serology/serological tests
Antibody-dependent cellular cytotoxicity
(ADCC), 486, 486f
Antibody immune response, 474, 492–494,
493f, 495, 495f
memory cells/immunological memory
and, 493f, 494–495, 495, 495f,
496, 497t
T-dependent, 492–494, 493f
T-independent, 492
Antibody sandwich ELISA, 518, 518f
Antibody titers, 506, 514, 514f
booster immunizations and, 506, 511f
Antibody wells, in immunodiffusion, 513,
513f
Anticodons, 209, 209f
Anti-Ebola virus antibody (ZMapp), 763,
765b
Antifungal drugs, 290t, 299f, 315t, 650
cell walls affected by, 291f, 292
cytoplasmic membranes affected by, 291f,
293, 315t
spectrum of action of, 299f
Antigen(s), 476–478, 477f
B cell receptor specificity and, 473
blood group, 514, 514f, 530, 530f
in complement activation/cascade, 459f,
460, 461f
dye-tagged antibodies binding to
in fluorescence microscopy, 103, 103f
in fluorescent immunoassays, 515–516,
516f
of Enterobacteriaceae, 593, 593f
presentation of. See also Antigen-
presenting cells
in antibody immune response, 492,
493f
in cell-mediated immune response,
490, 490f
processing of, 479–480, 479f, 480f
self (autoantigens), 477, 477f
autoimmunity and, 481–482, 482f, 487,
487f, 538, 538–539
clonal deletion and, 481–482, 482f,
487, 487f
graft rejection and, 478, 536
T cell regulation and, 482, 482f
tolerance to, 473, 481–482, 482f, 487,
487f. See also Self tolerance

serological tests and, 117, 117f, 512–520,
515–516, 516f, 520t
in type I (immediate) hypersensitivity,
526, 527f
in vaccines, recombinant DNA
technology in introduction
of, 253, 507, 507f
Antigen-antibody (immune) complexes, 512,
513, 532, 533f
in glomerulonephritis, 533, 556
in hypersensitivity pneumonitis, 532
in immunochromatographic assays, 519,
519f
in rheumatoid arthritis, 533
in systemic lupus erythematosus, 534
Antigen-antibody reactions
in serology, 117, 117f. See also Serology/
serological tests
in taxonomy/microbial classification,
117, 117f
Antigen-binding site/groove
in cell-mediated immune response,
490–491, 491f
immunoglobulin/B cell receptor, 483,
483f, 485, 485f, 488t
MHC molecule, 478, 478–479, 478f
Antigenic determinants. See Epitope(s)
Antigenic drift, influenzavirus, 768b, 769b
Antigenic shift, influenzavirus, 768b, 769b
Antigen-presenting cells (APCs), 478–479, 478f
in antibody immune response, 492, 493f
antigen processing and, 480, 480f
in cell-mediated immune response,
490, 490f
MHC proteins and, 478–479, 479f, 490, 490f
in type I (immediate) hypersensitivity, 527t
Antigen wells
in enzyme-linked immunosorbent assay
(ELISA), 517, 517f, 518f
in immunodiffusion, 513, 513f
Anti-Hantavirus antibodies, 408b
Antihistamines
for inflammation, 464
for type I (immediate) hypersensitivity,
529, 541b
Anti-human antibody antibodies. See also
Anti-antibodies
in indirect fluorescent immunoassays,
516, 516f
Anti-Listeria phage, 277b, 568
Antilymphocyte globulin, 537, 538t
Antimetabolic agents/antimetabolites, 290,
291f, 295, 296f, 312t, 316t
anthelminthic, 316t
antiprotozoan, 317–318t
antiviral, 295
resistance to, 306
Antimicrobial-associated (C. difficile)
diarrhea/pseudomembranous
colitis, 288b, 302, 319f, 426f,
439, 563
fecal microbiome transfer for, 803b
Antimicrobial chemoprophylaxis
for meningococcal infection, 590
for ophthalmia neonatorum, 589
Antimicrobial drugs/antimicrobials, 280,
281t, 288–323, 290t, 309–318t
allergies to, 302
attachment affected by, 290, 291f, 298,
313t
biofilms and, 170
candidiasis and, 298, 299b, 302, 660b, 661b
cell wall affected by, 66, 67, 265, 290,
290–292, 291f, 292f, 309–310t
combination therapy with, resistance
and, 307, 307f
cytoplasmic membranes affected by, 265,
269, 290, 291f, 293–295, 294f,
311–312t, 315t
discovery/history of, 18, 21, 21f, 289
distribution of, 302
effectiveness of, 298–300, 300f, 301b, 301f
exposure to, development of resistance
and, 303–304, 304f
indiscriminate use of, 307–308
mechanisms of action of, 264–266,
289–298, 291f, 292f, 294f, 296f,
297f, 309–318t
metabolic pathways affected by, 290, 291f,
295, 296f, 312t, 316t, 317t
microbial production/sources of, 289,
290f, 788, 788t

microbiome/normal microbiota affected
by, 298, 299b, 302
probiotics and, 304b
new drug development and, 308
nucleic acids affected by, 265–266, 290,
291f, 295–297, 296f, 297f,
312–313t, 314t, 318t
prescribing considerations and, 298–303,
299b, 299f, 300f, 301b, 301f,
302f
proteins synthesis affected by, 265–266,
290, 291f, 293, 294f, 310–311t,
314t, 317t
resistance to. See Resistance
ribosomes affected by, 76–77, 210,
293, 294f
routes of administration of, 300–302, 301f
safety/side effects/toxicity of, 302, 302f, 303f
selective toxicity and, 290
semisynthetic, 280, 281t, 289, 290t, 308,
309t, 310t, 313t, 315t, 316t,
318t
spectrum of action of, 298, 299f, 309–318t
synthetic, 280, 281t, 289, 311t, 312t, 313t,
314t, 315t, 316t, 317t, 318t
Antimicrobial enzymes, 280, 281t
Antimicrobial methods. See also specific type
and Antimicrobial drugs
action of, 264–266
chemical, 276–282, 281t
cytoplasmic membranes affected by,
265, 269
desiccation, 272, 272f, 275t, 785
efficacy of
evaluation of, 280–282
factors affecting, 266–268, 266f, 267f
environmental conditions and, 267–268,
267f
filtration, 272–273, 273f, 273t, 274f, 275t
in drinking water treatment, 791–792,
791f
in wastewater treatment, 793f, 794
heat-related, 268–271, 269f, 270f, 271f,
271t
lyophilization, 272, 275t, 785
microbial death rates and, 264, 265f
microbial susceptibility and, 266–267,
266f
osmotic pressure, 168, 273, 275t
physical, 268–275, 275t
principles of use of, 263–266, 265f, 265t
radiation, 273–275, 274f, 275t, 785
refrigeration/freezing, 271–272, 275t
resistance and, 282
selection of, 266–268, 266f, 267f
terminology used in, 263–264, 265t
treatment site and, 266
Antimicrobial peptides (defensins), 448,
449f, 451, 455
Antimicrobial therapy. See Antimicrobial
drugs
Antimony, antimicrobial action of, 295
Antioxidants, 164
Antiparallel strands, DNA, 51–52, 194, 195f
Antiparasitic drugs, cytoplasmic membrane
affected by, 295
Antiphagocytic chemicals/factors, 427f,
428–429
Bacillus anthracis and, 561
Pseudomonas aeruginosa and, 609
staphylococci and, 548
streptococci and, 429
Yersinia pestis and, 602
Antiports, 73, 73f
Antiprotozoan drugs, 293, 299f, 317–318t
Antirabies serum, 762
Anti-Rh antibodies, in hemolytic disease of
newborn, 531, 532f
Anti-Rh immunoglobulin, hemolytic disease
of newborn prevention
and, 531
Antisense nucleic acids
as antimicrobials, 291f, 293, 294f, 314t
synthetic nucleic acids in synthesis
of, 240
yield/nutritional value of food and, 255
Antisepsis/antiseptics, 16, 18f, 264, 265t. See
also Hygiene
chemical agents used for, 276–280, 276f,
277f, 278f, 279f, 281t

evaluation of efficacy of, 280–282
resistance and, 282
Antiserum (antisera), 117, 510
microbial classification and, 117
passive immunotherapy and, 496, 497t, 510
serological tests and, 117
Antithamnion genus/spp., 376t
Antitoxins, 428, 510
botulism, 510, 564, 565b, 581b
diphtheria, 573
passive immunotherapy and, 496, 497t
Antivenom (antivenin), 510
Antiviral drugs, 265, 299f, 313–314t
antiretroviral therapy (ART), 754b,
756–757
attachment antagonists, 298, 313t
entry inhibitors, 295, 313t
nucleic acid synthesis affected by,
296–297, 297f, 314t
nucleotide/nucleoside analogs as, 297,
314t
protein synthesis affected by, 314t
spectrum of action of, 299f
uncoating inhibited by, 295, 313–314t
viral metabolism affected by, 295
viral proteins inhibited by, 314t
Antiviral proteins (AVPs), 314t,
458–459, 458f
Anus
as portal of entry, 420f
as portal of exit, 430f
APCs. See Antigen-presenting cells
Apicomplexans, 362, 365t, 686–693, 692t
Apoenzymes, 127, 128f
Apoptosis
in ADCC, 486, 486f
NOD proteins in, 457
in self-tolerance/clonal deletion, 481–482,
482f, 487, 487f
suppression of, HHV-4 (Epstein-Barr
virus) and, 720
TLRs/PAMPs in, 457
Appendix, 475f
Applied microbiology, 19t, 779, 779–786
food microbiology, 10b, 12t, 18f, 19t,
779–786, 783t, 786t. See also
Food microbiology
industrial microbiology/biotechnology,
12, 12b, 18f, 19t, 238, 239f, 779,
786–795. See also Recombinant
DNA technology
Aquarius remigis, 37f
Aquatic microbiology, 811–812, 811f
aquatic habitats and, 811–812, 811f
waterborne illnesses and, 419b, 431, 433t,
789–790, 790b, 790t
water pollution and, 789, 791
Aquifex genus/spp., 332
Aquificae (phylum), 330f, 332
Ar. See Argon
Arachnida/arachnids/arachnid vectors, 377,
377–378, 378f, 677
taxonomic classification of, 114f
Arachnoidiscus, 105f
Arboviruses, 385b, 741, 747t
bunyaviruses, 392t, 735, 747t, 763,
766–767, 766f, 771t
coltiviruses, 392t, 770, 771t
diagnosis/treatment/prevention of
infection caused by, 745
encephalitis caused by, 742, 742f, 743b,
743f. See also specific type
immunization against, 745
reoviruses, 392t, 735, 747t, 770–771, 771t
togaviruses and flaviviruses, 392t, 735,
741–746, 741f, 747t, 771t
Archaea, 79t, 89t, 329–332, 330f, 331f. See also
specific type
acid mine drainage and, 806, 806f
in biomining, 809
in bioremediation, 805
cell walls of, 4, 78, 79f, 79t, 89t
chromosomes of, 89t, 195, 198t
classification of, 4, 114f, 115, 330f
nucleotide sequencing and, 118–119,
330f
cytoplasmic membranes of, 78, 79t, 89t
cytoplasm of, 79, 79t
as domain, 60, 114f, 115, 329
external structure of, 77–78, 78f, 79f, 79t
genome of, 194–195, 196f, 198t
glycocalyces of, 77, 79t, 89t

Archaea (Continued)
life processes in, 59t
plasmids in, 196, 198t
reproduction in, 329
shapes of, 78, 79f
size of, 4
in soil, 810
temperature range of. See Hyperthermophiles; Thermophiles
translation in, 212
Arenaviridae (arenaviruses), 392t, 735, 739t, 763, 767–770, 767f, 771t
as biological weapons, 814t
Argon, 30f
Arildone, 313t
mechanism of action/viral uncoating and, 291f, 298, 313t
spectrum of action of, 299f, 313t
Aristotle, spontaneous generation theory of, 8
Arithmetic graph, 181, 181f
Arithmetic growth, 180, 180f
Arm(s), immunoglobulin/B cell receptor, 483, 483f, 485, 485f
Armadillos, in study of leprosy, 575
Armillaria genus/spp., 366
Arsenic, antimicrobial action of, 279, 281t, 289, 317t
ART (antiretroviral therapy), 754b, 756–757
during pregnancy, 757
Artemisinins, 317t
Artesunate, 317t
Arthritis
in Lyme disease, 635
rheumatoid, 533, 534f, 538, 539
Arthrobacter genus/spp., 330f
Arthroconidia, of *Coccidioides immitis*, 655
Arthropod, 676–677
Arthropod-borne viruses. See Arboviruses
Arthropod vectors, 377–379, 378f, 432, 433t, 676–677, 676f
arbovirus diseases transmitted by, 741, 741–746, 747t
bunyavirus diseases transmitted by, 766–767
public health measures in control of, 440
rickettsial diseases transmitted by, 340, 622b, 624f, 625b
Artificial immunity/artificially acquired active immunity/active immunization, 495, 496, 497t, 505–510, 507f, 508f, 509f, 511f. See also Immunization; Vaccines
Artificially acquired passive immunotherapy, 496, 497t, 504, 505, 510–511, 511f
Artificial sweetener, microbial production of, 12t, 787
Artificial wetlands, in wastewater treatment, 794–795, 795f
Ascariasis, 699–700, 703t
Ascaris lumbricoides, 699–700, 700f, 703t
Asci (ascus), 369, 369f
Ascocarps, 369, 369f, 371b, 372f
Ascomycota (division)/ascomycetes, 368, 369–370, 369f, 370f, 371, 373t, 648, 657
chromoblastomycosis caused by, 666, 666t
dermatophytoses caused by, 665
in lichens, 371, 372, 372f
mycetomas caused by, 666t, 667
phaeohyphomycosis caused by, 666, 666t
reproduction in, 369, 369f, 373t
sporotrichosis caused by, 666t, 667
Ascospores, 369, 369f, 373t
Ascus (asci), 369, 369f
-ase (suffix), for enzyme, 127
Aseptate hyphae, 365, 365f
Aseptic/aseptic technique, 173, 264, 265t
Aseptic packaging, in food preservation, 785
Asexual reproduction, 4, 59
in algae, 373
in ascomycetes, 369
in eukaryotes, 354
in fungi, 4, 367, 368f
zygomycetes, 369, 369f
in prokaryotes, 326, 326f, 327f
in protozoa, 5–6, 360
Asexual spores, 4, 326, 327f, 367, 368f. See also Spores
A site, ribosomal, 210, 210f, 211, 211f
antimicrobials affecting, 293, 294f

Aspartic acid
in amination and transamination, 150, 151f
in nucleotide biosynthesis, 151, 152f
Aspergilloma, 370b, 657, 658f, 659b
Aspergillosis, 370b, 657–659, 658f, 659b
Aspergillus genus/spp. (*Aspergillus flavus/fumigatus/niger/oryzae*), 105f, 369, 370b, 648, 657–659, 658f, 659b, 664
aflatoxins produced by, 668
mutagenic effects of, 219, 668
in amylase production, 787
echinocandins produced by, 315t
fermentation products of, 142f
sake, 782, 783f
vegetables, 779, 783t
foodborne illnesses caused by, 786
in rennin production, 780
Aspirin, Reye's syndrome and, 718
Assembly, in viral replication, 392
animal virus, 398, 399f, 400f, 401t
bacteriophage, 393f, 394, 396, 396f, 401t
HIV, 752, 753f, 754–755
Assimilation, in nitrogen cycle, 807f
Associations (microbial). See Microbial associations
Asthma, 528
aspergillosis causing, 657
Astroviridae (astroviruses), 392t, 735, 739–740, 739f, 771t
Asymptomatic infection/disease, 423, 434t
poliomyelitis, 737. See also Poliomyelitis
Atherosclerosis, *Chlamydophila pneumoniae* and, 437, 629
Athlete(s), herpes infection in, 716, 716t
Athlete's foot (tinea pedis), 365, 664, 665f, 665t
Atom(s), 28–31, 28f, 29f, 29t, 30f. See also under Atomic
chemical bonds joining, 31–35, 32f, 33f, 33t, 34f, 35f, 35t
electronegativity of, 31, 33f
Atomic force microscopy/microscopes, 106, 106f, 107t
resolving power of, 99f
Atomic mass/atomic weight, 28, 29t
Atomic mass unit (dalton), 28, 29t
Atomic number, 28
Atomic structure, 28, 28f
Atopobium genus/spp., 330f
Atovaquone, 315t, 317t
mechanism of action of, 295, 315t, 317t
Atovaquone-proguanil, 441b
ATP, 52, 52f, 59
in active transport, 70, 73, 73f
anabolic pathways requiring, 149, 153f
in Calvin-Benson cycle, 146, 148f, 153f
in carbohydrate biosynthesis/gluconeogenesis, 150f
catabolic pathways producing, 126, 153f
glucose catabolism, 133, 133f, 134f, 135, 140t, 142f
lipid catabolism, 143, 143f
in cellular respiration, 133f, 135, 140t, 142f
in chemiosmosis, 139, 140
in citric acid/Krebs cycle, 133, 133f, 136, 137f, 140t
in electron transport chain, 133, 133f, 138f, 139, 140t, 153f
energy storage/release and, 125, 125–126, 125f, 126–127
in Entner-Doudoroff pathway, 140
in fermentation, 133f, 142f
in glycolysis, 133, 133f, 134, 134f, 135, 135f, 140t
mitochondrial production of, 87
in nucleotide biosynthesis, 151
in pentose phosphate pathway, 140
in photophosphorylation/photosynthesis, 145, 146, 147f
ATP synthases (ATPases), 140
in active transport, 73
in chemiosmosis, 140
in electron transport chain, 138f
in photophosphorylation/photosynthesis, 146, 147f
Attachment, 422, 422f. See also specific organism
antimicrobials preventing, 290, 291f, 298, 313t
in phagocytosis, 455, 455f
in viral replication, 392, 401t

animal virus, 396–397, 401t
bacteriophage, 393f, 394, 395, 396f, 401t
HIV, 752, 753, 753f, 754f
Attachment antagonists, 290, 291f, 298, 313t
mechanism of action of, 298, 313t
Attachment fimbriae. See Fimbriae
Attachment proteins, 394, 422
HIV, 752, 754f
rabies virus, 761
Attenuated (modified live) vaccines, 506
recombinant, 507–508, 507f
residual virulence/safety and, 506, 509–510
Attenuation, in vaccine preparation, 506
Atypical (mycoplasmal) pneumonia, 460b, 569
Auramine O dye, 103, 103f
Aureobasidium genus/spp., asexual spores of, 368f
Autism, vaccination and, 510
Autoantibodies, 539
to blood cells, 539
in blood disorders, 539
in connective tissue disease, 539
endocrine disorders and, 539
molecular mimicry and, 538–539
in nervous system disorders, 539
in rheumatoid arthritis, 533, 539
in systemic lupus erythematosus, 534
Autoantigens (self-antigens), 477, 477f
autoimmunity and, 481–482, 482f, 487, 487f, 538, 538–539
clonal deletion and, 481–482, 482f, 487, 487f
graft rejection and, 478, 536
T cell regulation and, 482, 482f
tolerance to, 473, 481–482, 482f, 487, 487f. See also Self-tolerance
Autoclave/autoclaving, in microbial control, 269–270, 270f, 271f, 275t
Autograft, 536, 536f
Autoimmune diseases, 481, 526, 538–539. See also specific type
Autoimmune hemolytic anemia, 539
Autoimmunity, 477, 481, 538. See also Autoimmune diseases
T cell regulation in prevention of, 482, 482f, 487, 487f, 539
Autotrophs, 163, 163f
in carbon cycle, 806, 807f
in nitrogen cycle, 808
in sulfur cycle, 808
Auxotroph, identification of, 222, 222f
Avery, Oswald, 19, 224
Avian influenza, 768b
Avirulent organisms, 422, 426
attenuation for vaccines and, 506
Avoidance therapy, in type I (immediate) hypersensitivity/allergies, 529
AVPs. See Antiviral proteins
Axenic culture, 172–173. See also Pure culture
Axenic environment, 416
defense mechanisms affected by, 450
Axial filament, 64, 64f
Azathioprine, 537, 538f
Azidothymidine (AZT), 296–297, 297f, 314t
Azithromycin, 288b, 311t, 460b, 615b
Azoles, 315t, 317t, 650
mechanism of action of, 294, 315t, 317t, 650
spectrum of action of, 299f, 315t, 317t
Azomonas genus/spp., 343, 344t
Azospirillum genus/spp., 339, 344t
Azotobacter genus/spp./*Azotobacter chroococcum*, 343, 344t
nitrogen fixation by, 343, 807–808
PHB granules in, 75f
AZT. See Azidothymidine
Aztreonam, 309t

B. See Boron
B19 virus (*Erythrovirus* genus/spp.), 392t, 728, 728f
placenta crossed by, 421t
Babesia microti, 691b
Babesiosis, 691b
Bacillariophyceae (diatoms), 6, 6f, 354, 375, 375–376, 376f, 376t
Bacillary angiomatosis, 605
Bacillary peliosis hepatis, 605
Bacilli (bacillus), 116, 325, 325f, 330f, 337t
arrangements of, 66, 66f, 327–328, 328f, 560, 561f

Gram-negative, 330f. See also specific agent
aerobic, 605–614
anaerobic, 614, 614f
facultatively anaerobic, 591–605
Gram-positive, 330f, 335–336, 336f, 337t
Bacillus genus/spp., 335–336, 336f, 337t, 560–562, 561f, 562f
antimicrobials produced by, 290t, 294, 309t, 311t, 312t, 335, 788
beneficial, 335–336, 336f
endospores of, 75, 76, 76f, 325, 326f, 335, 560, 562
staining, 110, 110f
as sterilization indicator, 270, 271f
susceptibility and, 266–267
food contaminated by, canning and, 784
in nitrogen cycle, 139, 808
shape of organisms in, 328, 328f
Bacillus amyloliquefaciens, restriction enzymes produced by, 240f
Bacillus anthracis, 335, 348b, 419t, 424, 560–562, 561f, 562f. See also Anthrax
bioterrorism and, 256, 336, 562, 814t
endospores of, 336, 560, 562
Schaeffer-Fulton endospore stain of, 110f
fluorescein isothiocyanate dye staining of, 103
fluorescent phages in identification of, 113b
immunization against, 509t, 562
safe handling of, 268
soilborne transmission of, 810, 810t
Bacillus brevis, antimicrobials produced by, 311t
Bacillus of Calmette and Guérin (BCG) vaccine, 509t
for leprosy, 509t, 575
for tuberculosis, 509t, 577b
tuberculin skin test and, 534, 535
Bacillus cereus, 328f
Bacillus licheniformis
antimicrobials produced by, 290t, 309t, 335
binary fission in, 179f
Bacillus polymyxa, antimicrobials produced by, 290t, 294, 312t, 335
Bacillus subtilis, 105f
cytoskeleton of, 77f
Bacillus thuringiensis, Bt toxin produced by, 255, 335–336, 336f, 788–789, 789t
Bacitracin, 309t
mechanism of action of, 291, 291f, 309t
microbial production of, 290t, 309t, 335
Bacteremia/septicemia
Bacteroides fragilis causing, 614
in CA-MRSA, 305b
E. coli causing, 89b, 595, 595b, 599b, 604f
Enterobacteriaceae causing, 604f
in leptospirosis, 636–637
Listeria causing, 336
meningococcal, 590, 590f
neonatal, *Streptococcus agalactiae* causing, 556
pneumococcal, 558b, 559
Salmonella causing, 600, 600f
staphylococcal, 550
UTIs and, 89b, 599b
Vibrio vulnificus causing, 40b, 639, 786t
Bacteria, 4, 4f, 79t, 89t, 330f, 332–347. See also specific type
acid-fast, 67
antimicrobials produced by, 290t
beneficial/industrial uses of, 4, 12, 12b, 18f, 19t. See also Applied microbiology; Beneficial microbes; Industrial microbiology
cell walls of, 4, 66–69, 66f, 67f, 68f, 79t, 89t
absence of in some bacteria, 4, 69
antimicrobials and, 66, 67, 265, 290–292, 292f, 309–310t
peptidoglycan in, 4, 45, 66–67, 66f, 67f, 68f, 290
chromosomes of, 74, 89t, 195–196, 196t, 198t
classification/identification of, 4, 4f, 114f, 115, 115–119, 116f, 117f, 118f. See also Taxonomy/microbial classification and identification
biochemical tests in, 116–117, 116f, 117f

culture in, 116, 171, 172f, 173–174, 173f, 174f, 176f
MALDI-TOF mass spectrometry in, 118
nucleotide sequencing and, 115, 118–119
phage typing in, 117, 118f
physical characteristics and, 116
serological tests in, 117, 117f
colony formation by, 171, 172f
conjugation in, 226–227, 227f, 228f, 228t
culturing viruses in, 402–403, 403f. *See also* Bacteriophage(s)
cytoplasmic membranes of, 69–74, 69f, 71f, 72f, 73f, 74f, 79t, 89t
antimicrobials affecting, 265, 294, 311t–312t
cytoplasm of, 74–77, 75f, 76f, 77f, 79t
deeply branching, 330f, 332
DNA in, 74, 195–196, 196f, 198t
replication and, 199–201, 200f, 202f
as domain, 60, 114f, 115, 329
endospores of. *See* Endospores
enteric, 342, 343t, 591–602, 592f, 593f, 594f, 595b. *See also* Enterobacteriaceae
external structure of, 62–66, 62f, 63f, 64f, 65f, 66f, 79t
in fermentation, 11–12, 11f
fimbriae/pili of, 65–66, 65f, 66f, 79t, 89t, 226, 227f, 228f
flagella of, 62–65, 63f, 64f, 79t, 89t
staining, 110, 111f, 111t
genome of, 194–195, 196f, 198t
glycocalyces of, 62, 62f, 79t, 89t
Gram-negative, 15, 15f, 67, 330f, 338–347, 344t. *See also* Gram-negative bacteria
Gram-positive, 15, 15f, 67, 330f, 337t, 546–585. *See also* Gram-positive bacteria
horizontal (lateral) gene transfer among, 223–227, 225f, 226f, 227f, 228f, 228t, 230b
S. aureus and, 336
life processes in, 59t
motility of, 62–65, 63f, 64f, 65, 89t
nitrifying, 330f, 339
opportunistic infections caused by. *See also* Opportunistic pathogens
in HIV infection/AIDS, 751t
organelles in, 76–77, 77f
pathogenic. *See* Pathogen(s)
photosynthetic. *See specific type and* Photosynthesis
phototrophic, 163, 163f, 332–334, 333f, 334f, 334t
placenta crossed by, 421t
plasmids in, 196, 198t
resistant strains of. *See* Resistance
shapes/arrangements of, 66, 66f
size of, 4, 4f, 61f
soilborne disease caused by, 810, 810t
transcription in, 203–204, 205f, 206f
transduction in, 225, 226f, 228t
transformation in, 224–225, 225f, 228t
translation in, 203, 203f, 207–212, 208f, 209f, 210f, 211f, 212f
antimicrobials affecting, 290, 293, 294f
urinary system infection caused by, 598–599b
viruses infecting. *See* Bacteriophage(s)
viruses/viroids/prions compared with, 408t
waterborne illnesses caused by, 789, 790t
zoonoses and, 419t
Bacterial food poisoning. *See* Food poisoning
Bacterial gastroenteritis. *See specific cause and* Foodborne illnesses; Gastroenteritis
Bacterial pneumonia(s), 736t. *See also* Pneumonia
Bacterial viruses. *See* Bacteriophage(s)
Bactericides/bactericidal drug, 264, 300. *See also* Antibiotics
Bacteriochlorophyll(s), 144
green and purple bacteria using, 144, 333, 334t, 339
Bacteriocin(s), 196, 308, 560
Bacteriocin plasmids, 196
Bacteriology, 18f, 19t. *See also* Bacteria
Bacteriophage(s) (phages), 113b, 117, 277b, 387, 387f, 395b

beneficial effects of, 113b, 277b, 390b, 395b
CRISPR-Cas systems and, 242, 243f
culturing, 402–403, 403f
fluorescent, 113b
in identification/classification of bacteria, 113b, 117, 118f
lysogenic, 395, 395f
in microbial control, 277b
replication of, 225, 226f, 401t
lysogenic, 395–396, 395f, 396f
lytic, 392–394, 393f, 394f, 396, 396f
shapes of, 389, 390f
sizes of, 388f
temperate, 395, 395f
therapeutic use of, 395b
in transduction, 225, 226f, 394
Bacteriophage (phage) attachment, 393f, 394, 395, 396f
Bacteriophage (phage) clones, in gene libraries, 243, 244f
Bacteriophage (phage) therapy, 395b
Bacteriophage (phage) typing, 113b, 117, 118f
Bacteriorhodopsins, 331
Bacteriostatic drug, 300
Bacteroid(s), 344t, 347
Bacteroides genus/spp./*Bacteroides fragilis*, 344t, 347, 614, 614f
necrotizing fasciitis caused by, 555b
Bacteroidetes (phylum), 330f
Bacteroidia, 344t
Baker's yeast. *See Saccharomyces* genus/spp.
Baking soda (sodium bicarbonate), 38
Balantidiasis, 677, 692t
Balantidium genus/spp./*Balantidium coli*, 362, 365t, 677, 692t
*Bam*HI restriction enzyme, 240t
Banfield, Jill, 20
Bang, Bernhard, 606
Bang's disease (brucellosis), 15t, 340, 344t, 606, 606f
bioterrorism and, 814t
B antigen, 514, 514f, 530, 530f, 531t
Barophiles, 168
Barriers, in innate immunity, 420, 421f, 447–448, 448f, 449t
Barr, Yvonne, 718
Bartonella genus/spp. (*Bartonella bacilliformis*/*henselae*/*quintana*), 605–606, 606f
Bartonellosis, 605
Basal body
of archaea flagellum, 77
of bacterial flagellum, 62, 63, 63f
of eukaryotic flagellum, 82, 83f
Base(s) (chemical), 38–39, 38f
pH of, 38, 38f
Base(s) (nucleic acid), 50, 50f, 51, 51f, 52
Base-excision repair, 220, 220f
Base pairs (bp), 194, 195f
point mutations affecting, 216, 217, 217f, 218f, 218t
in taxonomy/microbial classification, G + C content and, 118–119, 330f, 547
Base-pair substitutions, 216, 217, 217f, 218f, 218t. *See also* Mutations
Basic dyes, 108
Basidia, 370, 371f
Basidiocarps, 370, 371f. *See also* Mushroom(s)
Basidiomycota (division)/basidiomycetes, 368, 370, 373f, 648
in lichens, 371, 372f
Basidiospores, 370, 371f, 373f
Basophil(s), 452f, 453, 453f, 527
in inflammation, 452f, 453, 463, 463f
in type I (immediate) hypersensitivity degranulation of, 527, 527f
sensitization of, 526, 527f
Bassler, Bonnie, 20
Bat(s)
as filovirus (Ebola and Marburg) reservoir, 763, 764b, 765b
histoplasmosis and, 651, 663b
as rabies virus reservoir, 418, 762, 762b
Batch method, for pasteurization, 271, 271t
Batch production, in industrial fermentation, 787
B cell receptors, 483, 483f, 494
diversity of, 483–485, 484f
on memory B cells, 494
specificity of, 483, 483f

B cells. *See* B lymphocytes
BCG (bacillus of Calmette and Guérin) vaccine, 509t
for leprosy, 509t, 575
for tuberculosis, 509t, 577b
tuberculin skin test and, 534, 535
BCR. *See* B cell receptors
Bdellovibrio genus/spp., 343, 344t, 345f
Be. *See* Beryllium
Beadle, George, 19, 365
Bedaquiline, 312t
Bedrock, 809, 809f
Beef. *See* Meat/pork
Beef tapeworm (*Taenia saginata*), 694, 703t
Bee/insect stings, allergic reaction/anaphylaxis caused by, 525b, 528, 529, 535b, 541b
Beer, microbes/fermentation in production of, 10b, 142f, 781–782, 782f, 783t
Beggiatoa genus/spp., in sulfur cycle, 808
von Behring, Emil, 18f, 21
Beijerinck, Martinus, 14, 15t, 18f, 20, 176
Bejel, 631b, 633
Beneficial microbes, 4
algae, 6
antibacterial phages, 277b, 395b
architecture-preserving, 39b
ascomycetes, 370
Bacillus, 335–336, 336f
bacteria, 4
bacteriophages, 113b, 277b, 390b, 395b
Bdellovibrio bacteria/Gram-negative infection and, 343, 344t, 345f
in biofuel production, 162b, 788, 789t
bioremediation and, 19t, 20, 169b, 805
bread/wine/beer production and, 5, 10b, 12t, 779, 781–782, 781f, 782f, 783t
calcite formation and, 39b
cowpox, 2b
dengue/mosquito control and, 746b
fluorescent phages, 113b
fungi as, 169b, 365
glue production and, 341b
lactobacilli, 304b, 336
lichens, 372–373
methanogens, 332, 788
microbiome and, 253b
mutualistic symbiosis and, 415b
mycotoxins, 668
Nocardia, 338
nuclear waste clean-up and, 169b
paper/cardboard decomposition and, 162b
plastics production and, 74, 75b, 787, 789t
polymerase chain reaction and, 246b
probiotics and, 10b, 304b, 780
smallpox vaccination and, 510b
Streptomyces, 289, 290f, 309t, 310t, 311t, 316t, 338
truffles and, 371b
viruses, 113b, 390b, 395b
wastewater treatment and, 793–794, 793f, 794, 795f
water molds, 376, 377b
Wolbachia, 385b, 746b
yeasts, 10b
Benign tumor, 401
Benthic zone, 811f, 812
Benzalkonium chloride, 278f, 279
Benzimidazoles, 316t
Benznidazole, 313t, 318t, 703b
Benzoic acid, in food preservation, 785
Benzopyrene, as frameshift mutagen, 219
Bergey's Manual of Determinative Bacteriology, 116
Bergey's Manual of Systematic Bacteriology, 116, 329
Beryllium (Be), 30f
Beta-carotene gene, insertion of into rice, 255
Betadine, antimicrobial action of, 277, 277f
Beta (β)-glucose, 44, 45f
Beta-hemolysis/beta-hemolytic streptococci
Streptococcus agalactiae, 556–557, 561t
Streptococcus equisimilis and *Streptococcus anginosus*, 557, 561t
Streptococcus pyogenes, 552–556, 553f, 554–555b, 561t
Beta (β)-interferon, 458, 458f, 459t
Beta-lactam antibiotics, 290–291, 292f, 309t
mechanism of action of, 290–291, 292f, 309t
resistance to, 306, 306f, 309t

Beta (β)-lactamase
antimicrobial resistance and, 306, 306f, 309t
staphylococcal pathogenicity/virulence and, 548, 548t, 551
Beta (β)-lactam rings, 290–291, 292f
Beta-oxidation, of fatty acids, 143, 143f
Beta (β)-pleated sheets, 48, 49f
prion PrP and, 48, 405, 406f
Beta-propiolactone, for gas sterilization, 280, 281t
Betaproteobacteria, 330f, 334t, 340–341, 342f, 344t
pathogenic, 340–341, 587, 605
Beverages/alcoholic beverages, microbial production of, 5, 10–12, 10b, 11f, 12t, 18f, 19t, 141, 141f, 142f
BGH. *See* Bovine growth hormone
Bicarbonate, 38
Helicobacter pylori producing, 167
Bidirectionality, of DNA replication, 201, 202f
Bilayers, phospholipid. *See* Phospholipid bilayers
Bile, in host defense, 451t
Bile-esculin agar, for *Bacteroides* culture, 614, 614f
Bilirubin, accumulation of
in hemolytic disease of newborn, 531
in hepatitis, 726, 726f
Binary fission, 179–187, 179f, 326, 326f. *See also* Population (microbe), growth of
in amoebas, 363
in myxobacteria, 346, 346f
in protozoa, 360
Binocular microscope, 100
Binomials/binomial nomenclature, 113
Biochemical oxygen demand (BOD), reduction of in wastewater treatment, 793, 793f, 794
Biochemistry, 12, 19, 35. *See also* Chemical reactions
in taxonomy/microbial classification, 116–117, 116f, 117f
Biodiversity, 803
smallpox virus destruction and, 2b
soil nutrients and, 810
Biofilm(s), 62, 65–66, 162, 168–170, 169f, 170b, 171b, 171f, 422, 422f
adaptation/survival and, 805
in aquatic ecosystems, 811
archaea glycocalyces in, 77
confocal microscopes for examination of, 103, 104f
dental plaque as, 62, 66, 168, 170, 171b, 422, 422f, 557
Selenomonas in, 335
viridans streptococci in, 557
drinking water treatment and, 791
fimbriae in, 65–66, 65f
in-use test of organisms in, 282
microbial growth as, 162, 168–170, 169f, 170b, 171b, 171f
Pseudomonas forming, 610
resistance and, 306
Biofuels, 162b, 788, 789t
Biogeochemical cycles, 20, 806–809, 807f, 808f, 809f
Biological select agents, as biological weapons, 813
Biological vectors, 377, 432, 433t, 677. *See also* Arthropod vectors
Biological warfare/weapons, 812–815, 814t, 815f. *See also* Bioterrorism
assessing threat potential and, 812–813
defense against, 814–815, 815f
known threats and, 813–814, 814t
phage typing and, 113b
recombinant DNA technology and, 256, 812, 815
Bioluminescence, 362
Biomass, 803, 810
in biofuel production, 788
Biomedical animal models, recombinant DNA technology and, 254
Biomining, 809
Bioreactor, for industrial fermentation, 787
Bioremediation, 19t, 20, 169b, 805
Burkholderia for, 608
Kineococcus radiotolerans for, 169b
Bioreporters, microbes as, 789

Biosafety levels, 268, 268f, 815
Biosensors, microbes as, 789
Biosphere, 804, 804f
Biotechnology/industrial microbiology, 12, 12b, 18f, 19t, 238, 239f, 779, 786–795. *See also* Recombinant DNA technology
   agricultural applications/products of. *See* Agricultural microbiology
   biofuel production and, 162b, 788, 789t
   bioremediation and, 19t, 20, 169b, 805
   biosensors and bioreporters and, 789
   enzyme production and, 12t, 787
   fermentations and, 786–787, 787f
   pesticide production and, 12t, 788–789, 789t
   pharmaceutical applications/products of, 12t, 18f, 19t, 252–254, 788, 788t
   products of, 12, 12t, 787–789, 789t
   recombinant DNA technology in, 238, 239f
   water treatment and, 789–795, 790b, 790t, 791f, 792f, 793f, 794f, 795f
Bioterrorism, 812–815, 814t, 815f. *See also specific disease or agent*
   animal pathogens and, 813, 813–814
   assessing microorganisms' threat potential and, 812–813
   biosensors and bioreporters and, 789
   defense against, 814–815, 815f
   food supply and, 812, 813, 813–814, 814, 814t, 815. *See also* Agroterrorism
   human pathogens and, 812–813, 813, 814t
   known microbial threats and, 813–814, 814t
   mutualistic symbiosis and, 415b
   plant pathogens and, 813, 814
      assessing threat potential and, 813
      defense against, 815
   recombinant genetic technology and, 256, 812, 815
Bird(s)
   arboviral encephalitis and, 742, 742f, 747t
   chlamydial pneumonia and, 629
   cryptococcosis and, 659
   fungal infections and, 652b
   histoplasmosis and, 651, 652b
   influenza and, 768b
   West Nile virus/encephalitis and, 734b, 742, 742f, 747t
Bisphenolics, 276, 276f, 281t
1,3-Bisphosphoglyceric acid
   in Calvin-Benson cycle, 148f
   in glycolysis, 134f
Bites
   Pasteurella infecting, 604
   rabies transmitted by, 762
Bithionol, 316t
BK virus, 722
"Black Death," 602. *See also* Plague
Black-eyed ticks. *See Ixodes* ticks; Tick(s)
"Black hairy tongue," metronidazole causing, 302, 302f
Blackhead, 578, 579f. *See also* Acne
"Black vomit," in yellow fever, 744
Blackwater fever, 689b. *See also* Malaria
Bladder (urinary)
   infections of, 15t, 598b, 599b
   microbiome/normal microbiota of, 417t
Blades, 373
Blastomyces genus/spp./*Blastomyces dermatitidis*, 648, 653–654, 653f, 654b, 654f
   recombinant DNA in vaccine development for, 507
   soilborne transmission of, 653, 810t
Blastomycosis, 653–654, 653f, 654b, 654f
   soilborne transmission of, 653, 810t
   vaccine against, 507
Blatella species, as disease vectors, 433t
Bleach. *See* Sodium hypochlorite
Blebbing, in meningococcal infection, 590
Blepharisma americana, cilia of, 5f, 83f
Blindness, trachoma causing, 628, 629
Blood. *See also specific component*
   HIV infection of, 756
   in host defense, 451–454, 451t, 452f, 453f, 454b
   microbial infection of. *See* Bacteremia/septicemia
   movement of, in host defense, 451t
   as portal of exit/disease transmission and, 430, 430f
   screening
      for hepatitis C, 748
      for HIV, 757
   types/typing, 514, 514f, 530, 530f, 531, 531t
Blood agar, 173f, 175, 176, 176f, 177t
Blood cells, 452–454, 452f, 453f, 454b. *See also specific type*
   autoimmunity affecting, 539
   defensive, 452–454, 454b
   development of (hematopoiesis), 452–453, 452f
Blood clots. *See* Coagulation
Blood flukes (*Schistosoma* species), 696–698, 697f, 698f, 699b, 703t
   waterborne transmission of, 431, 697f, 699b, 790t
Blood group antigens, 514, 514f, 530, 530f. *See also* ABO system
Blood poisoning. *See* Bacteremia/septicemia; Toxemia
Blood specimen, collection of, 173t
Blood stem cells, 452, 452f, 453, 473. *See also* Stem cells
Bloodsucking bugs. *See* Bugs (true/bloodsucking bugs)
Bloodsucking flies. *See* Flies
Blood transfusion. *See* Transfusion
Blood types/typing, 514, 514f, 530, 530f, 531, 531t
   hemagglutination tests in, 514, 514f
Blood vessel dilation/permeability, in inflammation, 462–464, 462f, 463f, 465f, 465t
Bloom(s), 808
   intoxications caused by, 362–363, 790
   viruses affecting, 390b
Blotting, in immunoblot/western blot, 518, 518f
Blue dye (indigo), microbial production of, 789f
Blue-green bacteria/blue-green algae. *See* Cyanobacteria
Blunt ends, of restriction enzymes, 240, 241f
B lymphocytes/B cells, 473, 474, 483–488, 483f, 484f, 485f, 486f, 487f, 488t
   activation of, 493f, 494
   in antibody immune response, 474, 492
   clonal deletion of, 487–488, 487f
   clonal selection of, 492–494, 493f
   deficient, 540, 540t
   Epstein-Barr virus infection of, 719, 719f, 720
   in HIV replication, 752
   memory, 493f, 494–495
   production/maturation of, 473
   proliferation and differentiation of, 493f, 494
   receptors on, 483, 483f, 494. *See also* B cell receptors
   in type I (immediate) hypersensitivity, 526, 527t
Boceprevir, 314t
BOD. *See* Biochemical oxygen demand
Body defenses. *See* Host defenses
Body fluids
   in host defense, 451t
   as portals of exit/disease transmission and, 430, 430f, 432
      Ebola/Marburg hemorrhagic fevers and, 763, 764b, 765b
      HIV infection/AIDS and, 755–756, 757
Body temperature
   elevated. *See* Fever
   hypothalamus in regulation of, 464, 465f
Bohr, Niels, 28
Bohr models/diagrams, 28, 28f, 30f
Boil (furuncle), 182b, 464, 549
Boiling, in microbial control, 269, 275t
Bond(s). *See* Chemical bonds
Bond angles, 33, 33f
Bone infection (osteomyelitis), staphylococcal, 550
Bone marrow, 452, 452f, 475f
   clonal deletion of B cells in, 487, 487f
   lymphocyte formation/maturation and, 473
   red, 473
Bone marrow suppression, chemotherapy causing, 454b
Bone marrow transplants, graft-versus-host disease and, 536
Booster immunization, 506, 508f, 511f. *See also specific disorder*
Bordetella genus/spp. (*Bordetella parapertussis/pertussis*), 340, 344t, 345b, 426f, 606–608, 607f, 608f
   adhesins produced by, 422, 606
   immunization against, 345b, 508f, 509t, 608
   pertussis (whooping cough) caused by, 345b, 508f, 509t, 606–608, 607f, 608f, 615b
   virulence of, 426f, 606–607, 607, 607f
Bordet-Gengou medium, for *Bordetella pertussis* culture, 608
Boron, 29t, 30f
Borrelia genus/spp.(*Borrelia burgdorferi/recurrentis*), 120b, 344t, 419t, 633–636, 634f, 635f, 636f. *See also* Lyme disease
   motility of, 64, 64f
   serological tests for, 117, 120b, 517, 518, 635
Botryococcus braunii, hydrocarbons produced by, 788
Bottling, in wine and spirits production, 781, 781f
Botulism, 546b, 563–564, 564f, 565b, 581b, 785
   bioterrorism and, 814t
Botulism antitoxin, 510, 564, 565b, 581b
Botulism (botulinum) toxins, 563, 564f, 581b, 785, 786t
   as biological weapons, 814t
   type E, 581b
Bovine growth hormone (BGH), 255
Bovine spongiform encephalitis (BSE/mad cow disease/variant Creutzfeldt-Jakob disease), 21b, 48, 406f, 407, 407b
   bioterrorism and, 814t
   enzyme for elimination of prion causing, 280
bp. *See* Base pairs
Brachial ganglion, latent herpetic infection of, 714, 714f, 716t
Bradykinin, 463
Bradyzoites, in *Toxoplasma* life cycle, 687f, 690
Brain, infection involving. *See* Encephalitis; Meningitis
Branhamella (Moraxella) catarrhalis, 610
   restriction enzymes produced by, 240t
Brazilian purpuric fever, 605
Bread, microbes in production of, 5, 10b, 12t, 779, 783t
Breakbone (dengue) fever, 237b, 257b, 734b, 743–744, 744f, 745f, 747t
   vaccine development and, 744
Breast milk
   immunoglobulins in, passively acquired natural immunity and, 487, 495, 497f
   as portal of exit, 430, 430f
   for HIV, 756
Brequinar sodium, 537, 538t
Brewer's yeast. *See Saccharomyces* genus/spp.
Bright-field microscopy/microscopes, 100, 100–101, 101f, 102f, 107t
   compound, 100–101, 101f, 102f
   resolving power of, 99f
   simple, 100
Brill-Zinsser disease, 621
Brincidofovir, 314t
Broad-spectrum drugs, 298
Bromine, antimicrobial action of, 278, 281t
5´Bromouracil, point mutations and, 218, 218f
Bronchi (bronchus), 449f
   microbiome/normal microbiota of, 417t
Bronchitis, 288f, 520b, 736t
   Chlamydophila (Chlamydia) pneumoniae causing, 629
Bronchodilators, for asthma, 529
Broth(s), 171
   nutrient, 174, 175
   turbid, microbial growth estimation and, 186–187, 186f
Broth dilution test, 300, 300f
Broth tubes, carbohydrate utilization in, 116f, 176, 177f
Brown algae, 375, 375f, 376t

Bruce, David, 15t
Brucella genus/spp. (*Brucella abortus/canis/suis/melitensis*), 340, 344t, 606, 606f
   as biological weapon, 814t
   pasteurization in control of, 270, 606
Brucellosis (undulant fever), 15t, 340, 344t, 606, 606f
   bioterrorism and, 814t
Bruton-type agammaglobulinemia, 540, 540t
BSE. *See* Bovine spongiform encephalitis
BSL-1 to 4 (biosafety levels), 268, 268f, 815
Bt toxin, 255, 335–336, 336f, 788–789, 789t
   ethical/safety issues and, 256, 788–789
Bubo/buboes
   in lymphogranuloma venereum, 627, 628f
   in plague, 602, 603f, 604b
   in tularemia, 611
Bubonic plague, 15t, 419t, 602, 603f, 604b. *See also* Plague
Buchner, Eduard, experiments in fermentation by, 12, 18f
Budding
   fungus/yeast reproduction by, 5, 5f, 357, 358f, 365, 366f, 367
   prokaryote reproduction by, 326, 327f
   virus release by, 400, 400f
   HIV, 752
Buffers, 38–39
Bugs (true/bloodsucking bugs), 378f
   as disease vectors, 433t, 677. *See also specific type*
   in Chagas' disease (*Trypanosoma cruzi*), 379, 679, 680f, 692f
Bulbar poliomyelitis, 737. *See also* Poliomyelitis
"Bull's-eye" rash, in Lyme disease, 120b, 635
Bunyaviridae (bunyaviruses), 392t, 735, 747t, 763, 766–767, 766f, 771t
   encephalitis caused by, 392t, 747t, 767, 771t
Burkholderia genus/spp. (*Burkholderia cepacia/mallei/pseudomallei*), 340–341, 344t, 608, 610b
   as biological weapon, 608, 814t
   bioremediation and, 608
Burkitt, Denis, 718
Burkitt's lymphoma, HHV-4 (Epstein-Barr virus) causing, 392t, 402, 718, 719f, 720
Burn injury/wound, *Pseudomonas aeruginosa* infection of, 426f, 609, 610f
Burst size, 394, 394f
Burst time, 394, 394f
Buruli ulcer, 578b
Butanol (ethyl alcohol), biofuel production and, 162b, 788
Buttermilk, fermentation in production of, 780, 783t

C. *See* Carbon; Complement; Cytosine
^{12}C (carbon-12), 28, 29f
^{13}C (carbon-13), 28, 29f
^{14}C (carbon-14), 29, 29f
C1, 459f, 461f
C2, 459f, 461f
C2a, 461f
C2b, 461f
C3, 459f, 460, 461f
C3a, 459f, 460, 461f, 462
   in inflammation, 462, 463, 463f
C3b, 459f, 460, 461f, 462
C4, 459f, 461f
C4a, 461f
C4b, 461f
C5, 461f
C5a, 461f, 462
   in inflammation, 462, 463, 463f
   phagocyte migration and, 464
C5a peptidase, group A streptococcal pathogenicity/virulence and, 552
C5b, 461f
C6, 461f
C7, 461f
C8, 461f
C9, 461f
Ca. *See* Calcium
Cadmium, cycling of, 808–809
Calcite, *Myxococcus xanthus* forming, 39f
Calcium, 29t, 30f
Calcium carbonate

in algal cell wall, 374, 376t
in foraminifera shells, 363, 363f
Calcium hypochlorite, antimicrobial action of, 277
Caliciviridae (caliciviruses), 392t, 735, 739–740, 739f, 740b, 771t
California (LaCrosse) encephalitis, 392t, 747t, 767
Calor, in inflammation, 462
Calvin-Benson cycle, 146, 148f, 153f
　carbohydrate biosynthesis and, 146, 148f, 150f, 153f
　in carbon cycle, 146, 148f, 806
　lipid biosynthesis and, 150, 151f
cAMP. *See* Cyclic adenosine monophosphate
Campylobacter genus/spp./*Campylobacter jejuni*, 344t, 346, 639
　diarrhea caused by, 282b, 639, 786t
　food/waterborne transmission of, 282b, 639, 790t
Campylobacteria, 330t, 344t, 346
CA-MRSA. *See* Community-associated methicillin-resistant *Staphylococcus aureus*
Cancer, 401–402
　mutagens causing, 222–223, 223f
　viruses in, 401–402, 402f
　　Epstein-Barr virus, 392t, 402, 718, 719, 719f, 720
　　hepatitis and, 728
　　herpesviruses, 713
　　HIV infection/AIDS and, 751t
　　papillomaviruses, 392t, 723b
　　polyomaviruses, 722
　　retroviruses, 750–751, 751f
Candida genus/spp./*Candida albicans*, 367, 426f, 648, 657, 658t, 660–661b, 664
　antimicrobial drug use and, 298, 299b, 302, 660b, 661b
　budding in, 367, 661b
　in HIV infection/AIDS, 541, 660b, 661b, 664, 766b
　opportunistic infection/superinfection caused by, 298, 299b, 302, 418, 657, 660–661b
　oropharyngeal infection caused by (thrush), 299b, 302, 426f, 658t, 660b
　　antimicrobial drug use and, 299b, 302
　　in HIV infection/AIDS, 541, 660b, 766b
　transmission of, 648, 657
　vaginal, 298, 302, 658t, 660b, 661b
　　antimicrobial therapy and, 298, 302
Candidemia, 658t, 660b
Candidiasis, 658t, 660–661b
　antimicrobial therapy and, 298, 299b, 302, 660b, 661b
　in HIV infection/AIDS, 541, 660b, 661b, 664, 766b
　oropharyngeal, 299b, 302, 426f, 658t, 660b
　vaginal, 298, 302, 426f, 658t, 660b, 661b
Candle jars, 178
Canning, industrial, in food preservation, 784–785, 784f
CAP. *See* Catabolite activator protein
Capillaries
　blood, in inflammation, 462–463, 462f
　lymphatic, 475f, 476
Capnophiles, culturing, 178
Capping, in transcription, 206
CAP RAST, 529
Capreomycin, 310t
Capsid(s) (viral), 386, 386f, 389, 389f, 391f, 391t
　replication and assembly, 393f, 394
　　entry/uncoating, 394, 397, 397f, 401t
Capsomeres/capsomers, 389, 389f
　in viral assembly, 394
Capsular antigens, 593, 593f
Capsule (lymph node), 475f
Capsule (polysaccharide), 62, 62f. *See also specific organism and* Glycocalyx
　as antiphagocytic factor, 427f
　Bacillus anthracis and, 561
　cryptococci and, 659
　Enterobacteriaceae and, 593, 593f
　Francisella tularensis and, 611
　gonococci and, 588, 589b
　Haemophilus influenzae and, 604

Klebsiella and, 596, 596f
meningococci and, 590
Neisseria and, 587, 588, 589b, 590
Pseudomonas aeruginosa and, 609
staphylococci and, 548, 548t
streptococci and, 62f
　group A streptococci, 429
　group B streptococci, 556
　Streptococcus pneumoniae (pneumococcus), 62, 558, 558b
Capsule (negative) stains, 110, 111f, 111t
Carbapenemase, 593
Carbapenems, 309t
　mechanism of action of, 290–291, 291f, 292f, 309t
Carbepenem-resistant Enterobacteriaceae (CRE), 230b, 593
Carbohydrate(s), 44–46, 45f, 46f
　biosynthesis of, 149, 150f, 153f
　catabolism of, 133–140, 133f, 134f, 135f, 137f, 139f, 140t, 142t, 153f. *See also* Glucose, catabolism of
Carbohydrate utilization broth tubes, 116f, 176, 177f
Carbohydrate utilization test, 116, 116f
Carbolfuchsin
　in acid-fast staining, 109
　as flagellar stain, 110
Carbolic acid. *See* Phenol
Carbon, 29t, 30f, 806. *See also* Carbon dioxide
　alpha (α), amino acid, 47, 47f
　atomic mass of, 29t
　atomic number of, 28, 29t
　isotopes of, 28–29, 29f
　for microbial growth, 163, 163f
　in nonpolar covalent bonds, 32, 32–33, 32f
　recycling of, 20, 806–807, 807f
Carbon-12, 28, 29f
Carbon-13, 28, 29f
Carbon-14, 29, 29f
Carbon chains, 32–33
Carbon cycle, 20, 806–807, 807f
Carbon dioxide (CO_2). *See also* Carbon
　in carbon cycle, 806, 807, 807f
　fixation of. *See* Carbon fixation
　as greenhouse gas, 332, 807
　microbial fermentation producing, 142f
　in microbial growth, 163, 163f
Carbon dioxide incubators, 178
Carbon fixation, 146, 806
　in carbon cycle, 806
　in light-independent photosynthesis/Calvin-Benson cycle, 146, 148f, 806
Carbonyl group, 41t
Carboxyl group, 41t
　in amino acids, 47, 47f, 48f
　in fatty acids, 41, 41t
Carbuncle, 549
carcino- (prefix), 423t
Carcinogens, 222
　identification of, 222–223, 223f
Cardiac valves
　endocarditis and, 550
　rheumatic fever and, 553
Cardiovascular system, in host defense, 451t
Caries (dental/cavities), 62, 170, 171b
　viridans streptococci causing, 557, 557f
Carotene
　in algae, 376t
　in brown algae, 375, 376t
　in chrysophytes, 375, 376t
　in dinoflagellates, 362
　in euglenids, 361
　in green algae, 376t
β-Carotene gene, insertion of into rice, 255
Carotenoids, 150
　singlet oxygen affected by, 164
Carrageenan, in algal cell wall, 80, 374, 376t
Carrier molecules/proteins
　in active transport, 73, 73f
　antigenic response to molecules and, 476
　electron, 126
　in electron transport chains, 136, 139
　in facilitated diffusion, 71, 71f
Carriers (disease), 418, 419b. *See also specific disease or organism*
　of cryptosporidiosis, 690
　of *Entamoeba histolytica*, 678
　of hepatitis, 726, 739t
　of HHV-1/HHV-2, 716

during incubation and convalescence, 430
of meningococcal infection, 590
of *Pneumocystis jirovecii*, 657
of typhoid/*Salmonella*, 419b, 600
Cas. *See* CRISPR-associated enzymes
Cascade, complement, 460, 460–462, 461f
　staphylococcal protein A affecting, 548
Caseous necrosis, in tuberculosis, 577b
Caspofungin, 292, 315t
Castor bean ticks. *See Ixodes* ticks; Tick(s)
Cat(s)/cat feces, *Toxoplasma* transmission and, 687, 687f, 690
Catabolism/catabolic pathways, 36, 125, 125–126, 125f, 126–127, 133–144, 153f
　carbohydrate/glucose, 133–140, 133f, 134f, 135f, 137f, 139f, 140t, 142t, 153f
　integration and regulation of, 151–152, 153f
　lipid, 143, 143f, 153f
　protein, 143, 144f, 153f
Catabolite activator protein (CAP), in lactose (lac) operon regulation, 213–214, 214f
Catalase, 164
　hydrogen peroxide affected by, 164, 278
　in peroxisomes, 87
　staphylococcal pathogenicity/virulence and, 547, 548t
Catalysts (enzyme), 47, 127. *See also* Enzyme(s)
Catarrhal phase, of pertussis, 607, 608f
Categories A/B/C agents, bioterrorism and, 813, 814t
Cat fleas. *See* Fleas
Cathode ray machines, 274
Cation(s), 33, 38, 38f
　of salts, 39, 106
Cationic chromophores/dyes, 108
Cat scratch disease, 20, 605, 606f
Caulobacter genus/spp./*Caulobacter crescentus*, 340, 340f, 344t
　glue secreted by, 341f
Cavities (dental caries), 62, 170, 171b
　viridans streptococci causing, 557, 557f
CBC (complete blood count), 454b
Cd. *See* Cadmium
CD (cluster of differentiation) designations, 481
CD3, monoclonal antibodies against, 537
CD4. *See also* Helper T lymphocytes
　in HIV infection/AIDS, 541, 751, 753, 753f, 754b
　regulatory T cells and, 481, 482t
　T helper cells and, 481, 482t, 492, 493f
CD4 positive ($CD4^+$) T cells, 481, 482t. *See also* Helper T lymphocytes
CD8, 481, 482t
CD8 positive ($CD8^+$) T cells, 481, 482t. *See also* Cytotoxic T cells
CD25, 481, 482t
CD40, in B cell activation, 493f, 494
CD40L, in B cell activation, 493f, 494
CD95, 491, 491f
CD95 cytotoxic pathway, 491, 491f
CD95L, 482t, 491, 491f
CDC. *See* Centers for Disease Control and Prevention
C. diff. (antimicrobial associated) diarrhea/pseudomembranous colitis, 288b, 302, 319b, 426f, 439, 563
　fecal microbiome transfer for, 803b
cDNA. *See* Complementary DNA
Cefotaxime, 309t
Ceftriaxone, 301b, 309t
Cefuroxime, 309t
Cell(s), 59–60, 60f, 61f. *See also* Eukaryotes/eukaryotic cells; Prokaryotes/prokaryotic cells
　life processes and, 59, 59t, 60
　size of, 60, 61f
　solution tonicity affecting, 72–73, 72f
　structure/function of, 58–95
　　cell walls. *See* Cell walls
　　characteristics of life and, 59, 59t
　　cytoplasm
　　　of eukaryotes, 82–88, 83f, 84f, 85t, 86f, 87f, 88f, 89t
　　　of prokaryotes, 60f, 79t
　　　archaea, 79, 79t
　　　bacteria, 74–77, 75f, 76f, 77f, 79t

cytoplasmic membranes. *See also* Cytoplasmic (cell/plasma) membranes
　eukaryotic, 61f, 80–82, 80f, 81f, 81t, 89t
　prokaryotic, 60f, 69f, 79t
　　archaeal, 78, 79t, 89t
　　bacterial, 69–74, 69f, 71f, 72f, 73f, 74t, 79t, 89t
endosymbiotic theory and, 88
external structure
　of archaea, 77–78, 78f, 79t
　of bacteria, 62–66, 62f, 63f, 64f, 65f, 66f, 79t
　of eukaryotes, 80
Saccharomyces in study of, 10b
types of, 60, 60f
viruses compared with, 391t
Cell counter, for microbial growth estimation, 183, 183f
Cell (tissue) cultures, 178
　pharmaceuticals produced in, 788t
　viral, 178, 403–404, 404f
Cell death, programmed. *See* Apoptosis
Cell division (cytokinesis), 83, 354, 356f, 357, 358f
　cancer and, 401, 402
　centrosomes in, 84
Cell lysis
　in ADCC, 486, 486f
　in virus release, 394, 396, 396f. *See also* Lytic replication
Cell-mediated (type IV) hypersensitivity, 534–535, 535b, 535f, 536f, 537t
　allergic contact dermatitis as, 535, 536f
　donor-recipient matching/tissue typing and, 536–537
　graft rejection as, 536
　graft-versus-host disease and, 536
　tuberculin response as, 534–535, 535f
Cell-mediated immune response, 474, 480–482, 482f, 483f, 489–492, 490f, 491f, 495. *See also* T lymphocytes/T cells
Cell membranes. *See* Cytoplasmic (cell/plasma) membranes
Cell nucleus, 60, 85t, 196
Cell plate, 357, 358f
Cell suicide, programmed. *See* Apoptosis
Cellular cytotoxicity, antibody-dependent (ADCC), 486, 486f
Cellular inclusions
　in cytomegalovirus disease, 721, 721f
　in eukaryotic cells, 89t
　in prokaryotic cells, 60f, 89t
　　archaea, 89t
　　bacteria, 74, 75f, 89t
Cellular PrP (c-PrP), 48, 405, 406f
　diseases caused by changes in, 48, 405–407, 406f, 407b
　templating and, 405, 406f
Cellular respiration, 133, 133f, 135–140, 135f, 137f, 138f, 142t
　acetyl-CoA synthesis and, 133f, 135–136, 135f
　aerobic, 139, 140t, 142t
　anaerobic, 139, 142t
　chemiosmosis and, 136, 139–140
　citric acid/Krebs cycle in, 133, 133f, 136, 137f
　electron transport in, 133f
　fermentation compared with, 142t
Cellular slime molds, 364, 365t
Cellulases, in biofuel production, 162b
Cellulitis, 553
Cellulose, 45, 46f
　in cell walls, 45, 80
　　algal, 376t
　　of water molds, 376
　conversion of to biofuel, 162b
　microbial production of, 787
Cell walls, 89t
　absence of in some bacteria, 4, 69
　antimicrobials and, 66, 67, 265, 290, 290–292, 291f, 292f, 309–310t
　of eukaryotic cells, 80, 80f, 89t
　　algae/diatoms, 6, 6f, 80, 80f
　　fungi, 4, 292
　hyper- and hypotonic solutions and, 72f, 73, 168
　of prokaryotic cells, 60f, 66, 79t, 89t
　　archaea, 4, 78, 79t, 89t
　　bacteria, 4, 66–69, 66f, 67f, 68f, 79t, 89t

SUBJECT INDEX

Cell walls (Continued)
　Gram-negative, 67, 67–68, 68f
　Gram-positive, 67, 68f
　peptidoglycan in, 4, 45, 66–67, 66f, 67f, 68f
Centers for Disease Control and Prevention (CDC)
　biosafety guidelines of, 268, 268f, 815
　data sharing and, 435, 437t, 438f, 440
　immunizations recommended by, 508–509, 508f, 509t
　Morbidity and Mortality Weekly Report published by, 435, 437t
　notifiable/reportable diseases listed by/MMWR, 435, 437t, 438f
　bioterrorism defense and, 815, 815f
Centimeter (cm), 97, 97t
Central dogma, 203, 203f
　retroviruses and, 750
Central nervous system. *See* Nervous system/nervous tissue
Central vacuole, 87f
Centrioles, 61f, 83–85, 84f
Centromere, 355, 355f
Centrosome, 83–85, 84f, 85t
Cepacol. *See* Cetylpyridinium
Cephalexin, 309t
Cephalosporins, 309t
　mechanism of action of, 290–291, 291f, 292f, 309t
　staphylococcal resistance to, 551, 551b
Cephalosporium (Acremonium) genus/spp.
　antimicrobials produced by, 290t, 309t
　mycetoma caused by, 666t
Cephalothin, 309t
　mechanism of action of, 292f, 309t
　microbial production of, 290t
Cephamycin, 309t
Cercariae, in fluke life cycle, 696, 697f
　Schistosoma, 696, 697f
Cerebral malaria, 687
Cerebrospinal fluid analysis
　in cryptococcal meningitis, 662
　in meningococcal meningitis, 590
　specimen collection/spinal tap and, 173t
Cervical cancer, HPV infection and, 402
　vaccination and, 508t, 509t
Cesspools, 794
Cestodes (tapeworms), 6, 378, 419t, 693, 693–695, 693f, 694f, 703t
　beef (*Taenia saginata*), 694, 703t
　dog (*Echinococcus granulosus*), 694–695, 695f, 703t
　life cycle of, 694, 694f
　pork (*Taenia solium*), 694, 694f, 695f, 703t
Cetylpyridinium, 278f, 279
CF. *See* Cystic fibrosis
CFU. *See* Colony-forming units
CGD. *See* Chronic granulomatous disease
Chagas, Carlos, 679
Chagas' disease (American trypanosomiasis), 675b, 679–680, 680f, 692t, 703b
Chagoma, 679
Chain, binary fission forming, 179
Chancre, in syphilis, 632, 632f
Channel proteins, 47
　in active transport, 73, 73f
　in facilitated diffusion, 71, 71f, 80–81
Charcoal, activated
　in drinking water treatment, 791f, 792
　in wastewater treatment, 794
Cheese
　microbes/fermentation in production of, 12t, 142f, 780, 780f, 783t
　raw/unpasteurized, disease caused by, 262b, 282b
Chemical behavior, of atoms, electron configurations and, 29–31, 30f
Chemical bonds, 31–35, 32f, 33f, 33t, 34f, 35f, 35t
　angles of, 33, 33f
　electronegativity and, 31, 33f
　hydrogen, 34–35, 34f, 35f, 35t
　ionic, 33–34, 34f, 35t
　nonpolar covalent, 31–33, 32f, 33f, 35t
　peptide, 47, 48f
　polar covalent, 33, 33f, 35t
Chemical chemoprophylaxis. *See* Antimicrobial chemoprophylaxis
Chemical defense, in innate immunity, 457–462, 457f, 458f, 459f, 459t, 460f, 461f, 462f. *See also* Complement; Interferon(s)

Chemical elements, 28, 29t, 30f
　for microbial growth, 162–165, 163f, 165f, 165t
　periodic table of, 30f
　trace, 165
Chemical fixation, 106
Chemical (concentration) gradient, 70
　diffusion and, 71, 71f, 72, 72f
Chemical mutagens, 218–219, 218f, 219f
Chemical reactions, 19, 35–37, 36f. *See also specific type and* Chemistry
　enzymes in, 127–132
　　control of (activation/inhibition), 131–132, 131f, 132f
　　enzyme concentration and, 131
　　enzyme-substrate specificity and, 129, 129f
　　pH and, 130f, 131
　　substrate concentration and, 130f, 131
　　temperature and, 129–130, 130f, 131f
　in metabolism, 19, 35–37, 36f
Chemical requirements, for microbial growth, 162–165, 163f, 165f, 165t
Chemiosmosis, 136, 139–140
　in photophosphorylation/photosynthesis, 139, 146, 147f
Chemistry, 27–57. *See also under* Chemical *and* Biochemistry
　atoms and, 28–31, 28f, 29f, 29t, 30f
　chemical bonds and, 31–35, 32f, 33f, 33t, 34f, 35f, 35t
　chemical reactions and, 19, 35–37, 36f
　inorganic substances (water/acids/bases/salts) and, 37–39, 37f, 38f
　organic substances and, 40–53
Chemoautotrophs, 163, 163f
　in carbon cycle, 806, 807f
Chemoheterotrophs, 163, 163f
Chemokine(s), 455, 489. *See also* Cytokine(s)
Chemoprophylaxis. *See* Antimicrobial chemoprophylaxis
Chemostat, 182, 182f
Chemotactic factors, 455, 464, 465t
Chemotaxis, 65, 65f, 82
　in phagocytosis, 455, 455f
Chemotherapy/chemotherapeutic agents, 289. *See also specific type and* Antimicrobial drugs
　bone marrow suppression caused by, 454b
　discovery/history of, 18, 18f, 19t
Chemotrophs, 163, 163f
Chestnut blight, 369
Chest X-rays, in tuberculosis diagnosis, 577b
Chicken eggs, 61f
　embryonated, culturing viruses in, 403, 403f
　　vaccine manufacture and, 403, 508
　Salmonella contamination of, 600, 786t
Chickenpox (varicella), 392t, 716–717, 717f
　immunization against, 508f, 509t, 718
Chiggers (mites), as disease vectors, 377, 378f, 433t, 677
　for scrub typhus (*Orientia tsutsugamushi*), 622, 626t
Chikungunya/chikungunya virus, 401b, 747t
　bioterrorism and, 814t
Childbirth (puerperal) fever, streptococcal, 16, 556
Chill(s), fever and, 464–465
Chitin, in fungal cell walls, 80, 364
Chlamydia(s), 344t, 347, 630. *See also under Chlamydia; Chlamydophila*
　animal/cell cultures for growth of, 178
　antimicrobials effective against, 299f
　developmental forms/life cycle of, 626, 627f
　molecular methods for growth estimation of, 187
Chlamydiae (phylum), 330f, 344t, 626, 630
Chlamydia trachomatis, 626–629, 628f
　LGV strain of, 627
　pelvic inflammatory disease caused by, 628
　sexual transmission of, 347, 627–628, 628f
　trachoma/trachoma strain and, 628, 628f
Chlamydomonas nivalis, 167
Chlamydophila (Chlamydia) pneumoniae, 629
　atherosclerosis and, 437, 629
Chlamydophila (Chlamydia) psittaci, 629
　as biological weapon, 814t
Chlamydospores, 367, 368f

Chloramine(s)
　antimicrobial action of, 278
　drinking water treatment and, 791f, 792
　phenol coefficient of, 281
Chloramphenicol, 310t
　mechanism of action of, 291f, 293, 294f, 310t
　microbial production of, 290t, 310t
Chlorine, 29t, 30f. *See also* Chloramine(s)
　antimicrobial action of, 277–278, 281f
　in drinking water treatment, 277, 791f, 792
　in sodium chloride formation, 33, 34f
　in wastewater treatment, 277, 793f, 794
Chlorine dioxide, antimicrobial action of, 277–278
Chlorobi (phylum), 330f, 333, 334, 334t
Chloroflexi (phylum), 330f, 333, 334, 334t
Chlorophyll(s), 144–145, 145f, 376t
　algae using, 373
　　brown algae, 375, 376t
　　chrysophytes, 375, 376t
　　green algae, 374, 376t
　　red algae, 373, 374, 376t
　cyanobacteria using, 144, 333, 334t
　dinoflagellates using, 362
　euglenids using, 361
　fluorescent, 103
　reaction center, 145–146, 145f
Chlorophyta (division), 274, 376t
Chloroplasts, 85t, 87–88, 88f, 145, 147f
　DNA/chromosomes in, 87, 88, 197
　in euglenids, 357, 361, 361f
　evolution/endosymbiotic theory and, 87, 88, 333
Chloroquine, 318t
Chloroquine-resistant *Plasmodium*, 441b
Chocolate, fermentation in production of, 779
Chocolate agar, for *Neisseria* culture, 588
Cholera, 12, 17, 620b, 637–639, 638f, 641b
　bioterrorism and, 814t
　discovery of cause/spread of, 17, 437, 439f
　immunization against, 639
　incubation period for, 429t
　waterborne transmission of, 17, 437, 439f, 637, 790t
Cholera toxin, 638, 638f
Cholesterol, 44, 44f
　in eukaryotic membranes, 44, 44f, 81
　for microbial growth, 165t
Chondrus genus/spp./*Chondrus crispus*, 113, 374, 376t
Choriomeningitis, lymphocytic, 767
Chromatids, 355, 355f, 356, 356f
Chromatin/chromatin fibers, 85, 86f, 196–197, 197f, 354
Chromoblastomycosis, 666–667, 666f, 666t, 667f
Chromophore, 108
Chromosomes, 85, 89t
　in binary fission, 179, 179f
　in chloroplasts, 87, 88, 197
　eukaryotic/nuclear, 85, 89t, 196–197, 197f, 198t, 354
　extranuclear, 197–198
　homologous, 356, 357t
　in nuclear division (mitosis/meiosis), 85, 197, 197f, 354, 355, 355f, 356, 356f, 357t
　prokaryotic, 89t, 195–196, 196f, 198t
　　archaeal, 89t, 195, 198t
　　bacterial, 74, 89t, 195–196, 196f, 198t
　in transposition, 229, 229f
Chronic disease, 433, 434t
Chronic fatigue syndrome, HHV-4 (Epstein-Barr virus) and, 719
Chronic granulomatous disease, 540, 540t
Chronic hepatitis, 726, 748. *See also* Hepatitis
Chronic infection. *See* Persistent infection
Chronic inflammation, 462
Chronic obstructive pulmonary disease (COPD), 449
Chronic wasting disease, 407
Chrysolaminarin, 375
Chrysophyta, 375–376, 376f, 376t
Ciclopirox, 315t
-cide/-cidal (suffix), 264, 265t
Cidifovir, 314t
Cilia, 61f, 82, 83f, 84f, 89t
　mucous membrane
　　Bordetella pertussis affecting, 606, 607, 607f
　　in host defense, 449, 449t
　protozoan, 5, 5f, 362, 362f, 677

Ciliary escalator, 449
　Bordetella pertussis affecting, 606, 607, 607f
Ciliated columnar cells, in host defense, 449
Ciliates, 362, 362f, 365t, 677, 692t
　reproduction in, 360
Ciliophora, 360, 362
Ciprofloxacin, 313t, 348b, 796b
Circumcision, HIV transmission and, 754b, 757
Citric acid, microbial production of, 787, 789t
Citric acid (Krebs) cycle, 133, 133f, 136, 137f, 149t, 153f
Citrobacter genus/spp., 592f, 596
　healthcare-associated (nosocomial) infection caused by, 591f, 596
CJD. *See* Creutzfeldt-Jakob/variant Creutzfeldt-Jakob disease
Cl. *See* Chlorine
Cladophialophora genus/spp.
　chromoblastomycosis caused by, 666t
　phaeohyphomycosis caused by, 666t
Cladosporium genus/spp., radioactive waste remediation and, 169b
Clap. *See* Gonorrhea
Clarification, in wine and spirits production, 781, 781f
Clarithromycin, 311t
Class(es), 113, 114f
Class I major histocompatibility complex proteins, 478, 478f
　antigen processing and, 479, 479f, 480
　in cell-mediated immune response, 490, 490f
Class II major histocompatibility complex proteins, 478, 478f
　in antibody immune response, 492, 493f
　antigen processing and, 480, 480f
　in cell-mediated immune response, 490, 490f
Classical pathway of complement activation/cascade, 459–460, 459f, 461f
Classification (microbe), 112. *See also* Taxonomy/microbial classification and identification
Class switching (immunoglobulin), 487, 494
Claviceps genus/spp./*Claviceps purpurea*, 369, 373t, 668
Clavulanic acid, synergistic effects of, 307
Climate change
　carbon dioxide and methane and, 332, 342, 807
　viruses affecting, 390b
Clindamycin, 310t, 317t, 348b
Clinical specimens, for microbial culture, 171, 172, 173t
　transport media for, 172, 177
ClO₂. *See* Chlorine dioxide
Clofazimine, 312t
　mechanism of action of, 291f, 297, 312t
Clonal deletion
　of B cells, 487–488, 487f
　of T cells, 481–482, 482f
Clonal expansion, in cell-mediated immune response, 490, 490f
Clonality, in adaptive immunity, 473
Clonal selection, 246
　antibody immune response and, 492–494, 493f
　T-dependent antibody immunity and, 492–494, 493f
Clones. *See also under* Clonal
　in adaptive immunity, 473
　　antibody immune response, 492–494, 493f
　　cell-mediated immune response, 489–491, 490f, 491f
　in gene libraries, 243, 244f
Closed system, for culture, 182
Clostridium genus/spp./clostridia, 330f, 335, 337t, 562–568. *See also specific organism*
　as biological weapons, 814t
　endospores of, 75, 76, 76f, 269, 325, 326f, 335, 562, 563, 565
　staining, 110
　susceptibility and, 266–267
　fermentation products of, 141, 142f
　food contaminated with, 546b, 563–564, 567, 568, 581b, 784, 785. *See also* Botulism
　nitrogen fixation by, 808

in pectinase production, 787
in soil, 419, 810t
toxins produced by, 335, 428
Clostridium botulinum, 335, 563–564, 564f, 565b, 581b, 786t. *See also* Botulism
　as biological weapon, 814t
　D value for, 269
　endospores of, 269, 326f, 563, 565b
　toxins produced by, 563, 564f, 565b, 581b, 785, 786t
　　as biological weapon, 814t
Clostridium difficile, 335, 426f, 562–563
　antimicrobial-associated diarrhea/ pseudomembranous colitis caused by, 288b, 302, 319b, 439, 563
　fecal microbiome transfer for, 803b
　pathogenicity/virulence of, 426f, 562
　resistant strains of, 305
Clostridium perfringens, 15t, 335, 567–568
　fermentation products of, 141
　gas gangrene caused by, 15t, 567, 568
　necrotizing fasciitis caused by, 555b
　toxins produced by, 567
　　as biological weapon, 814t
Clostridium tetani, 15t, 335, 565–567, 565f, 566f, 567f. *See also* Tetanus
　endospores of, 565, 565f
　soilborne transmission of, 810t
Clots (blood). *See* Coagulation
Clotting factor, microbial production of, 253
Clouds (electron shells), 30, 30f
Cloxacillin, 309t
Cluster, binary fission forming, 179f, 180
Cluster of differentiation (CD) designations, 481
cm (centimeter), 97, 97t
CMV. *See Cytomegalovirus* genus/spp.; *Cytomegalovirus* disease;
Co. *See* Cobalt
Coagulase, as virulence factor, 426, 427f
　staphylococcal, 548, 548t
Coagulase-negative staphylococci, 548, 548t, 550
Coagulase-positive staphylococci, 548, 548t, 550
Coagulation
　disseminated intravascular (DIC), 587, 587f
　in host defense/inflammation, 451t, 452, 464, 465f
Coagulation (particulate), in drinking water treatment, 791, 791f
Coal mining, acid mine drainage and, 805–806, 805f
Cobalt, 29t
　radioactive, in food preservation, 785
Cocci (coccus), 116, 325, 325f, 330f
　archaeal, 79t
　arrangements of, 66, 66f, 327, 328f
　Gram-negative/Gram-positive. *See also specific agent*
Coccidioides immitis, 366, 648, 654–656, 655f, 669b
　as biological weapon, 814t
　immunization against, 650
　precipitation test for, 512
　transmission of, 431, 655, 656, 810t
Coccidioidin skin test, 655–656
Coccidioidomycosis, 366, 647b, 654–656, 655f, 669b
　bioterrorism and, 814t
　in HIV infection/AIDS, 541, 656
　immunization against, 650
　transmission of, 431, 655, 810t
Coccobacillus, 325, 325f
Cockroaches, as disease vectors, 433t
Codium genus/spp., 374, 376t
Codons, 207–208, 208t
　in translation, 208–212, 208f, 209f, 210f, 211f, 212f
Coenocytes, 357
Coenzyme(s), 52, 127, 128, 128f, 128t
　ATP in formation of, 52
　in citric acid/Krebs cycle, 133f, 136, 137f
Coenzyme A, 52, 128t
　in acetyl-CoA synthesis, 135, 135f
　in citric acid/Krebs cycle, 136, 137f
　in lipid catabolism/beta-oxidation, 143, 143f
Coenzyme Q, in electron transport chain, 138f, 139
Coevolution. *See also* Evolution
　parasitism and, 416

Cofactors, 127, 128, 128f, 128t
　organic. *See* Coenzyme(s)
Coffee, fermentation in production of, 779
Cohesiveness, of water, 37, 37f
Coinfection, hepatitis B and hepatitis D, 726, 770
col-/colo- (prefix), 423t
Cold. *See* Freezing; Refrigeration
Cold (common), 735–736, 735f, 736t
　adenoviruses causing, 722–724, 725b
　coronaviruses causing, 392t, 748, 749, 771t
　immunization and, 736
　parainfluenza viruses causing, 760
　paramyxoviruses causing, 392t, 760, 771t
　rhinoviruses causing, 735–736, 735f, 736t, 771t
Cold enrichment, 176
Cold sore (oral herpes), 392t, 709b, 714, 714f, 715f, 716t, 729b
Cold vaccine, impracticality of, 736
Coliforms, 594, 595–596, 596f, 598–599b. *See also Escherichia coli*
　canned food contaminated by, 785
　fecal, 595
　healthcare-associated (nosocomial) infection caused by, 591f
　water contaminated by, 595, 791, 792, 792f
Colitis, pseudomembranous, *Clostridium difficile* causing, 302, 439, 563
　fecal microbiome transfer for, 803b
Collagen, 448
Collagenase, as virulence factor, 426, 427f
Colon, microbiome/normal microbiota of, 416, 417t
　transfer of, 803b
Colony, microbial, 162, 171, 172f
Colony-forming units, 172
　microbial growth estimation affected by
　　membrane filtration and, 185
　　viable plate counts and, 184
　pour plate isolation and, 173, 174f
　streak plate isolation and, 173
Colorado tick fever, 392t, 747t, 770, 771t
Coltivirus genus/spp./coltiviruses, 392t, 770, 771t
Columnar cells, ciliated, in host defense, 449
Combination vaccines, 506
Combustion, in carbon cycle, 807f
Commensalism, 415, 416t
Commercial sterilization, 264
　heat-related methods for, 269
Common antigen, 593, 593f
Common cold, 735–736, 735f, 736t
　adenoviruses causing, 722–724, 725b
　coronaviruses causing, 392t, 748, 749, 771t
　immunization and, 736
　parainfluenza viruses causing, 760
　paramyxoviruses causing, 392t, 760, 771t
　rhinoviruses causing, 735–736, 735f, 736t, 771t
Common-source epidemics, food poisoning and, 786
Communicable disease, 433, 434t
Communities (microbial), 20, 252. *See also* Microbiome
Community-associated methicillin-resistant *Staphylococcus aureus* (CA-MRSA), 305, 305b, 551
Competent cells/competence, 224–225
　heat shock causing, 248
Competition (microbial), 298
　adaptation/survival and, 805
　antimicrobial drug combinations and, 307
　food spoilage and, 783
　in host defense, 298, 450
　lactobacilli and, 336
　opportunistic pathogens/superinfections and, 298, 299b, 302, 418, 439
Competitive inhibition/inhibitors, 131–132, 131f
　antimicrobial action and, 132
　host defense and, 450
Complement/complement system, 47, 455, 459–462, 459f, 461f, 462f
　activation of, 459–460, 459f, 486
　antibodies in, 459–460, 459f, 461f, 486
　staphylococcal protein A affecting, 548
　in host defense, 452
　in immune complex–mediated (type III) hypersensitivity, 532, 533f
　in inflammation, 463, 463f

phagocyte migration and, 464
　in transfusion reactions, 530, 530f
Complementary DNA (cDNA), 51, 239
　replication and, 198, 198f
Complement cascade, 460, 460–462, 461f
　staphylococcal protein A affecting, 548
Complement fixation test, 515, 520t
Complement proteins. *See* Complement/ complement system
Complete blood count (CBC), 454b
Complex media, 175, 177t
Complex transposons, 229f, 230
Complex viruses, 389, 389f
Compound, chemical bonds in formation of, 31
Compound microscopes, 100–101, 101f, 102f
　resolving power of, 99f
Compromised host. *See* Immunocompromised host; Opportunistic pathogens
Concentration (chemical) gradient, 70
　diffusion and, 71, 71f, 72, 72f
Condenser lenses, 100, 101f
　for dark-field microscopes, 101
Condom, in STI/STD prevention, 440, 589, 589b
　HIV infection/AIDS, 754b, 757
Condylomata acuminata (genital warts), 723b
Confocal microscopy/microscopes, 103, 104f, 107t
Congenital diseases, 423, 424t
　candidiasis, 658t, 660b
　cytomegalovirus, 720
　rubella, 747
　syphilis, 632
　toxoplasmosis, 690
　Zika, 745, 747t
Conidiophores, 367, 368f
Conidiospores/conidia, 367, 368f
　in ascomycete reproduction, 369
Conjugation, 66, 226–227, 227f, 228f, 228t
　antimicrobial resistance and, 226, 303
　bacterial, 66, 226–227, 227f, 228f, 228t
　fertility (F) plasmids and, 196, 226, 227f, 228f
　in protozoa, 360
Conjugation (sex) pili, 66, 66f, 79t, 89t, 226, 227f, 228f
Conjunctiva
　as portal of entry, 420, 420f, 421
　trachoma affecting, 628, 628f
Conjunctivitis, 392t, 446b, 466b
　adenoviruses causing (pinkeye), 724, 724f, 725b
　coxsackie A virus causing (acute hemorrhagic conjunctivitis), 738
　gonococcal, in newborn, 279, 589
Connecting chains, short, 67, 67f
Connective tissue, 448
　autoimmunity affecting, 539
Constant region
　immunoglobulin, 485f
　T cell receptor, 481f
Constitutive genes, 212
Consumption, in carbon cycle, 807f
Consumption (tuberculosis), 574. *See also* Tuberculosis
Contact dermatitis, allergic, 535, 536f
Contact immunity, 506
Contact lens use, *Acanthamoeba* keratitis and, 267t, 678, 679
Contact transmission of disease, 431, 431f, 433t
Contagious disease, 433, 434t
Contamination, microbial, 420. *See also specific type*
　pure cultures and, 172, 173
Continuous cultures
　cell, for viruses, 404
　in chemostat, 182, 182f
Continuous flow production, in industrial fermentation, 787
Contractile vacuoles, in protozoa, 359, 360f
Contrast, 99, 100
　phase light rays and, 100
　staining and, 100. *See also* Staining
Contrast microscopy/microscopes
　differential interference (Nomarski), 101, 102f, 107t
　phase, 101, 102f, 107t
Control groups, in scientific method, 9f, 10

Convalescence, 430
　in pertussis, 607, 608f
Cooperation, microbial, adaptation/survival and, 805
COPD (chronic obstructive pulmonary disease), 449
Copper, 29t
　antimicrobial action of, 279, 281t
　cycling of, 808–809
　in electron transport trains, 139
Cord factor, *Mycobacterium tuberculosis* virulence and, 577b
Corepressor, in tryptophan (*trp*) operon, 215, 215f
Core temperature
　elevated. *See* Fever
　hypothalamus in regulation of, 464, 465f
Corn
　ethanol production for biofuels and, 788
　recombinant DNA technology improving nutritional value of, 256
Cornea, trachoma affecting, 628
Coronaviridae (coronaviruses), 392t, 735, 741, 741f, 748–749, 749f, 771t
　as biological weapons, 814t
　common cold caused by, 392t, 748, 749, 771t
　Middle East respiratory syndrome (MERS) caused by, 736t, 748, 749, 749f
　severe acute respiratory syndrome (SARS) caused by, 392t, 736t, 748, 749, 749f, 771t
Coronavirus respiratory syndromes, 392t, 736t, 748–749, 749f
　bioterrorism and, 814t
Cortex, lymph node, 475f
Corticosteroids
　for allergic contact dermatitis, 535
　for asthma, 529
　immunosuppressive actions of, 537, 538t, 540–541
Cortinarius gentilis, 668
Corynebacterium genus/spp./*Corynebacterium diphtheriae*, 15t, 289, 330f, 337, 337t, 572–573, 572f, 573f. *See also* Diphtheria
　immunization against, 506, 508f, 509t, 573
　palisade/V-shape arrangements in, 328, 328f, 337
Coulter counter, 184
Counterstain
　in acid-fast staining, 109
　in Gram staining, 109, 109f
　in Schaeffer-Fulton endospore staining, 110, 110f
Coupled transport, 73, 73f
Covalent bonds, 31, 32f, 35t
　in dehydration synthesis, 36, 36f
　in exchange/transfer reactions, 36–37
　in hydrolysis, 36, 36f
　nonpolar, 31–33, 32f, 33f, 35t
　nucleic acid, 50–51, 51f, 52
　peptide, 47, 48, 48f
　polar, 33, 33f, 35t
Cowpox, 710, 710f
　smallpox prevention and, 2b, 17, 505, 509t, 510b, 710, 711, 712b
Coxiella genus/spp./*Coxiella burnetii*, 342, 344t, 613–614, 613f
　as biological weapon, 814t
Coxsackieviruses (coxsackie A and B viruses), 738, 738f
c-PRP. *See* Cellular PrP
CRE. *See* Carbepenem-resistant Enterobacteriaceae
Crenarchaeota (phylum), 330, 330f
Crenation, 72, 72f, 73, 168
Creutzfeldt-Jakob/variant Creutzfeldt-Jakob disease, 21b, 48, 406f, 407, 407f
　bioterrorism and, 814t
　enzyme for elimination of prion causing, 280
Crimean-Congo hemorrhagic fever, 747t, 767
Crisis of fever, 465
CRISPR, 242–243, 243f
CRISPR-associated enzymes (Cas), 242, 243f
CRISPR-Cas9, 242–243, 243f
　ethical/safety issues and, 256
CRISPR-RNA (crRNA), 242, 243f
Cristae, 87, 88f, 359, 362
Crohn's disease, fecal microbiome transfer for, 803b

SUBJECT INDEX

Crop(s). *See also* Agricultural microbiology
  in alternative fuel production, 788
  biological threats to, 812, 813, 814, 815.
    See also Agroterrorism
  Burkholderia in protection of, 608
  genetically modified, 254–256, 255f
    ethical/safety issues and, 256, 788–789
Crop management/yields, industrial
    microbiology/recombinant
    DNA technology and, 254–256,
    255f, 788–789, 789t
  Bt toxin and, 255, 335–336, 336f, 788–789, 789t
  ethical/safety issues and, 256, 788–789
Crossbridges (peptide), 66–67, 67f, 291, 292f
  antimicrobials affecting, 291, 292f
Crossing over, 223, 356, 356f, 357f
Cross-matching, for blood transfusions, 531, 531t
Cross resistance, 307
Cross walls (septa), in hyphae, 365, 366f
Croup, 760, 771t
Crown gall disease. *See also* Galls
  soilborne transmission of, 810t
crRNA (CRISPR-RNA), 242, 243f
Crust, in poxvirus infection, 710, 711f
Crustose lichens, 372, 372f
Cryptococcal (fungal) meningitis, 370, 659, 662
Cryptococcoma, 659
Cryptococcosis, 659–662, 662f
Cryptococcus neoformans/Cryptococcus neoformans gatti/neoformans, 370, 373f, 659–662, 662f, 664
Cryptosporidiosis, 362, 690, 692t
  bioterrorism and, 814t
  waterborne transmission of, 690, 790t
Cryptosporidium genus/spp./*Cryptosporidium parvum*, 362, 365t, 690, 691f, 692t
  bioterrorism and, 814t
  cyst stage of, 266f, 690, 691f
  waterborne transmission of, 690, 790t
Crystal violet, 108, 108f, 111t
  in Gram staining, 109, 109f
CSF. *See* Cerebrospinal fluid
Ctenocephalides felis, as disease vector, in murine (endemic) typhus (*Rickettsia typhi*), 622, 626t
Cu. *See* Copper
Culex mosquitoes, as disease vectors, 433t, 747t
  for arboviral encephalitis, 747t
  for filariasis (*Wuchereria bancrofti*), 702
Culicidae family, 385b. *See also* Mosquitoes
Culiseta mosquitoes, as disease vectors, 433t, 747t
Culture (microbial), 171–179, 172f, 173t
  animal, 178
    for viruses, 178, 403
  for biochemical testing, 116
  cell, 178
    for viruses, 178, 403–404, 404f
  clinical sampling for, 171, 172, 173t
  in commercial food/beverage production, 779
  continuous
    cell, for viruses, 404
    in chemostat, 182, 182f
  enrichment, 176
  environmental sampling for, 171
  low-oxygen, 178
  media for, 171, 174–178, 175f. *See also* Culture media
  preserving, 178–179
  pure, 172–174, 173f, 174f
  special techniques in, 177–178
  viral, 402–404, 403f, 404f
Culture media, 171, 174–178, 175f
  anaerobic, 177, 178f
  complex, 175, 177f
  defined (synthetic), 175, 175t
  differential, 176–177, 177f, 177t, 178f
  enriched, 175
  reducing, 177
  selective, 18f, 175–176, 176f
  selective and differential, 176–177, 178f
  transport, 172, 177
Cuprum. *See* Copper
Curds, in cheese production, 780, 780f
Cutaneous anthrax, 324b, 336, 348b, 562, 562f. *See also* Anthrax
Cutaneous aspergillosis, 658, 659b
Cutaneous blastomycosis, 653–654, 654f

Cutaneous candidiasis, 658t, 660b
Cutaneous cryptococcosis, 659
Cutaneous histoplasmosis, 651, 663b
Cutaneous leishmaniasis, 682b, 683, 692t
Cutaneous mycoses, 651, 666–668, 666f, 666t, 667f, 668f
Cutaneous sporotrichosis, 666, 667, 668, 668f
Cutaneous zygomycosis, 662
Cuticle
  nematode, 698
  tapeworm, 693, 693f
CWD. *See* Chronic wasting disease
CXCR4 (fusin), as HIV receptor, 753, 754f
  in long-term nonprogressors, 756
Cyanobacteria (phylum), 330f, 333, 334t
Cyanobacteria (blue-green bacteria), 333, 333f, 334t
  bacteriophage of, oxygen production and, 390b
  evolution/endosymbiotic theory and, 88, 333
  gas vesicles in, 74
  in lichens, 370, 371, 372
  nitrogen fixation by, 165, 333, 807
  photosynthesis by, 333
  pH range tolerated by, 39
Cyclic adenosine monophosphate (cAMP), in lactose (*lac*) operon regulation, 214, 214f
Cyclic photophosphorylation, 146, 147f
Cyclophosphamide, 537, 538t
Cycloserine, 291, 309t
Cyclospora genus/spp./*Cyclospora cayetanensis*, 691, 692f, 692t
Cyclosporiasis, 691, 692f, 692t
Cyclosporine, 365, 537, 538t
Cysteine, 47
Cystic acne, 578, 579f. *See also* Acne
Cysticerci, tapeworm, 694, 694f
Cysticercosis, 694, 694f
Cystic fibrosis, 802b, 816b
  Burkholderia infections and, 608
  Pseudomonas infections and, 609, 802b, 816b
Cystitis (bladder infection), 15t, 598b, 599b
Cyst stage
  of cestodes
    Echinococcus (hydatid cysts), 695, 695f
    Taenia, 694, 695f
  of protozoa, 359, 677
    Entamoeba histolytica, 678, 678f
    Giardia, 684b, 685b
    Naegleria, 678
  susceptibility and, 266f, 267
Cytidine, 297f
Cytidine triphosphate deoxyribonucleotide (dCTP). *See also* Triphosphate deoxyribonucleotides
  in DNA replication, 198
Cytochrome(s)
  in electron transport chains, 138f, 139
  in photosynthesis, 147f
Cytochrome oxidase, 139
  in *Neisseria*, 139, 587
Cytokine(s), 481, 488, 488–489, 489t. *See also specific type*
  in type I (immediate) hypersensitivity, 526, 527f
Cytokine network, 488–489
Cytokinesis (cell/cytoplasmic division), 83, 354, 356f, 357, 358f
  cancer and, 401, 402
  centrosomes in, 84
Cytolytic toxins, staphylococcal, 548, 548t
Cytomegalovirus genus/spp., 392t, 713, 720–721, 721f, 729t
  placenta crossed by, 421t, 720
Cytomegalovirus disease, 392t, 720–721, 721f
  in HIV infection/AIDS, 541, 720
  mononucleosis, 720
  transplacental transmission and, 421t, 720
Cytometry, flow, for microbial growth estimation, 184
Cytopathic effect, viral neutralization tests and, 515
Cytophaga genus/spp., 344t, 347
Cytoplasm, 74. *See also* Cytosol; Inclusion(s); Organelles
  division of (cytokinesis), 83, 354, 356f, 357, 358f
  cancer and, 401, 402
  centrosomes in, 84

  in eukaryotic cells, 82–88, 83f, 84f, 85t, 86f, 87f, 88f, 89t
  in prokaryotic cells, 60
    archaea, 79, 79t
    bacteria, 74–77, 75f, 76f, 77f, 79t
Cytoplasmic (cell/plasma) membranes, 89t
  antimicrobials affecting, 265, 269, 290, 291f, 293–295, 294f, 311–312t
  in binary fission, 179, 179f, 326, 326f
  eukaryotic, 61f, 80–82, 80f, 81f, 89t
  function of/movement across, 70–74, 71f, 72f, 73f, 74f
    active processes in, 70, 73–74, 73f, 74t
    in bacterial cells, 70–74, 71f, 72f, 73f, 74f
    in eukaryotic cells, 81, 81f, 81t
    passive processes in, 70, 71–73, 71f, 72f, 74t
  permeability of, 70
  phospholipid bilayers forming, 42, 43f, 67, 68f, 69, 69f, 81
  cholesterol in, 44, 44f, 81
  prokaryotic, 60f, 69, 89t
  archaeal, 78, 79t, 89t
  bacterial, 69–74, 69f, 71f, 72f, 73f, 74f, 79t, 89t
    Gram-positive/negative cell walls and, 67, 68f
  electron transport chains in, 136, 138f
  TLRs in, 457, 457f
Cytoplasmic streaming, fungal nutrition and, 366
Cytosine (C), 50, 50f, 51, 51f, 52, 52f, 194, 195f. *See also* G + C content/percentage/ratio
Cytoskeleton, 77, 85t
  of bacteria/prokaryotic cells, 77, 77f, 85t
  of eukaryotic cells, 61f, 83, 84f, 85t
Cytosol
  in archaea, 89t
  in bacteria, 74, 89t
  in eukaryotic cells, 82, 89t
Cytotoxic drugs, immunosuppressive actions of, 537, 538t
Cytotoxic (type II) hypersensitivity, 530–531, 530f, 531t, 532f, 537t
  ABO system/transfusion reactions and, 530–531, 530f, 531t
  Rh system/hemolytic disease of newborn and, 531, 532f
Cytotoxicity, antibody-dependent cellular (ADCC), 486, 486f
Cytotoxic T cells (Tc cells), 481, 482t
  activation/function of, 489–491, 490f, 491f
  CD95 pathway used by, 491, 491f
  perforin-granzyme pathway used by, 491, 491f
Cytotoxin(s), 427f, 428
  tracheal, 607

D. *See* Decimal reduction time
d4T. *See* Stavudine
Dairy products. *See also* Cheese; Milk
  Brucella contamination of, 270, 606
  fermented, 780, 780f, 783t
  pasteurization/safety and, 12, 270, 271, 271t, 440, 785
  raw/unpasteurized, disease caused by, 262b, 282b
  ultra-high-pasteurization/sterilization of, 271, 271t
D-alanine, 48f
Dalbavancin, 310t
Dalfopristin, 311t
Dalton (atomic mass unit), 28, 29t
Dalton, John, 28
Dane particles, 726, 726–727, 727f
Dapsone, 312t
  mechanism of action of, 291f, 312t
Dark-field microscopy/microscopes, 100, 101, 102f, 107t
Dark reactions. *See* Light-independent reactions
Dark repair, 220–221, 220f
Darunavir, 314t
Darwin, Charles, 115, 679. *See also* Evolution
Data reporting, in epidemiology, 434–435, 435f
Data sharing, by public health agencies, 435, 437t, 438f, 440
dATP. *See* Adenosine triphosphate deoxyribonucleotide
Daughter cells
  in binary fission, 179, 179f, 326, 326f

  in cytokinesis, 357, 358f
  in fungal reproduction, 367
  in schizogony, 357, 359f
  in snapping division, 326, 327f
Daughter nuclei, 355, 355f, 356, 356f
dCTP. *See* Cytidine triphosphate deoxyribonucleotide
ddC. *See* Dideoxycytidine
ddI. *See* Dideoxyinosine
Deamination, 143, 144f
  in nitrogen cycle, 807f, 808
Death, microbial, 264
"Death-cap mushroom," 668, 669f
Death (decline) phase, of microbial growth, 181–182, 181f
Decarboxylation
  in acetyl-CoA synthesis, 135, 135f
  in citric acid/Krebs cycle, 136
Decimal reduction time (D), 269, 269f
Decimeter (dm), 97, 97t
Decline phase of disease, 430
Decline (death) phase of microbial growth, 181–182, 181f
Decolorizing agent
  in acid-fast staining, 109
  in Gram staining, 109, 109f
  in Schaeffer-Fulton endospore staining, 110
Decomposition/decomposers
  in carbon cycle, 806–807, 807f
  fungi as, 365, 366, 370
  water molds as, 376, 377f
Decomposition reactions, 36, 36f
Deep-freezing, for culture preservation, 179
Deeply branching bacteria, 330f, 332
Deer mouse, in *Hantavirus* transmission, 384f, 408b, 767, 810
Deer ticks. *See Ixodes* ticks; Tick(s)
DEET (*N,N*-diethyl-*m*-toluamide), in Lyme disease prevention, 120b, 635
Defecation, in host defense, 451t
Defensins (antimicrobial peptides), 448, 449t, 451, 455
Defined (synthetic) media, 175, 175t
Definitive host, 676
Degenerative diseases, 424, 424t
Degerming, 264, 265t, 277f
Degranulation/degranulating cells, in type I (immediate) hypersensitivity, 526, 527f, 527t
Dehydration synthesis, 36, 36f
  of fats, 41, 42f
  in peptide bond formation, 47, 48f
  of polysaccharides, 45
  of sucrose (disaccharide formation), 45, 45f
Dehydrogenation reactions, 126. *See also* Oxidation
Deinococcus genus/spp./*Deinococcus radiodurans*, 250–251, 332
  radiation resistance of, 250–251, 332
Deinococcus-Thermus (phylum), 330f, 332
Delayed (type IV) hypersensitivity, 534–537, 535b, 535f, 536f, 537t
  allergic contact dermatitis as, 535, 536f
  donor-recipient matching/tissue typing and, 536–537
  graft rejection as, 536
  graft-versus-host disease and, 536
  tuberculin response as, 534–535, 535f
Deletion, clonal. *See* Clonal deletion
Deletion (frameshift mutation), 216, 217f, 218t, 219, 219f
Delivery potential, biological threat of microorganism and, 812–813
Delta agent (hepatitis D virus), 739t, 770, 771t
  hepatitis B virus coinfection and, 726, 770
Delta F_c region/heavy chains, 485, 487, 488t
Deltaproteobacteria, 330f, 343–346, 344t
Deltaretrovirus genus/spp., 392t, 750–751, 751f, 771t
Deltavirus genus/spp. (delta agent), 739t, 770, 771t
  hepatitis B virus coinfection and, 726, 770
Dementia
  in Creutzfeldt-Jakob/variant Creutzfeldt-Jakob disease, 21b, 407b
  in HIV infection/AIDS, 541, 751
Denaturation
  DNA, in polymerase chain reaction, 244, 245f, 246b
  protein/enzyme, 49, 130, 131f
    antimicrobial action and, 265, 269, 271

Dendrites, on dendritic cells, 478, 479f
Dendritic cells, 447–448, 449, 454, 478, 479f.
  See also Antigen-presenting cells
  in antibody immune response, 492, 493f
  antigen presentation/processing and, 478, 479f, 480, 490, 490f
  in cell-mediated immune response, 478, 490, 490f
  HIV affecting, 752t
  in host defense, 447–448, 449, 454, 478
Dengue fever/dengue hemorrhagic fever/dengue virus, 237b, 257b, 734f, 743–744, 744f, 745f, 747t
  vaccine development and, 744
Denitrification, in nitrogen cycle, 807f, 808
Dental caries (cavities), 62, 170, 171b
  viridans streptococci causing, 557, 557f
Dental plaque (biofilm), 62, 66, 168, 170, 171b, 422, 422f, 557
  Selenomonas in, 335
  viridans streptococci in, 557
Deoxyribonucleases
  group A streptococci (*Streptococcus pyogenes*) producing, 552
    necrotizing fasciitis and, 554b, 555b
  group B streptococci (*Streptococcus agalactiae*) producing, 556
Deoxyribonucleic acid. *See* DNA
Deoxyribonucleoside, 50
Deoxyribonucleotides, 194. *See also* Triphosphate deoxyribonucleotides triphosphate
  in DNA replication, 198, 199f, 201
  polymerase chain reactions and, 244, 245f
Deoxyribose, 44, 50, 50f, 52t, 151, 194
Dermacentor ticks
  as disease vectors, 433t
    for arboviral encephalitis, 747t
    for ehrlichiosis and anaplasmosis, 623, 626t
    for Rocky Mountain spotted fever (*Rickettsia rickettsii*), 624b, 625b, 626t
  taxonomic classification of, 114f
Dermatitis, allergic contact, 535, 536f
dermato- (prefix), 423t
Dermatomes, shingles lesions following, 717, 717f, 718f
Dermatophytes, 419t, 648, 649, 664–665, 665f, 665t
  transmission of, 648, 665
Dermatophytoses/ringworm, 419t, 649, 664–665, 665f, 665t
Dermcidins, 448
Dermis, 421f, 447
  in host defense, 448
Dermonecrotic toxin, 607
Descriptive epidemiology, 436–437, 439f
Desiccation
  in microbial control/food preservation, 272, 272f, 275t, 783, 785
  in specimen preparation, 106
Desmosomes, staphylococcal exfoliative toxin affecting, 548
Desulfovibrio genus/spp., 343, 344t
  in sulfur cycle, 139, 343, 808
Detergents, 278–279, 278f, 281t
Deuteromycota (division)/deuteromycetes, 368, 370–371. *See also* Ascomycota
Devil's grip (pleurodynia), coxsackie B virus causing, 738
Dextran, in tooth decay, 557
D forms, 47, 48f
D (diversity) gene segment, 484, 484f
dGTP. *See* Guanosine triphosphate deoxyribonucleotide
D$_H$ (diversity) heavy chain gene segment, 484
DHAP. *See* Dihydroxyacetone phosphate
d'Herelle, Felix, 395b
DHF. *See* Dengue fever/dengue hemorrhagic fever/dengue virus
Diabetes mellitus
  autoimmunity in, 538, 539
  coxsackie B virus infection and, 738
  cystic fibrosis-related, 802b
Diapedesis, 453, 464, 465f
Diaper rash, candidal, 658t

Diarrhea. *See also specific cause and* Dysentery; Foodborne illnesses; Gastroenteritis
  adenoviruses causing, 724, 725b
  in amebiasis, 678
  antimicrobial-associated (*C. difficile*/pseudomembranous colitis), 288b, 302, 319b, 426f, 439, 563
    fecal microbiome transfer for, 803b
  astroviruses causing, 392t, 739–740, 739f, 771t
  caliciviruses causing, 739–740, 739f, 740b
  Campylobacter, 282b, 639, 786t
  in cholera, 12, 620b, 638, 641b
  clostridial, 546b, 581b
    fecal microbiome transfer for, 803b
  contaminated drinking water causing, 789
  in cryptosporidiosis, 690
  E. coli causing, 15t, 428, 595, 596, 604f, 786t
  Enterobacteriaceae causing, 594
  in giardiasis, 684b
  in HIV infection/AIDS, 541, 766b
  microsporidial, 496b, 662
  Norovirus causing, 1b, 22b, 392t, 740, 740b, 771t, 790t
  rotaviruses causing, 392t, 770, 771f, 771t
  Salmonella causing, 600, 600f, 604f, 786t
  Shigella causing (shigellosis), 15t, 601–602, 601f, 604f, 786t
  in staphylococcal intoxication (food poisoning), 549, 786t
  traveler's (*E. coli*), 15t
  vibrios causing, 638, 639, 786t
Diatom(s), 6f, 354, 355, 375, 375–376, 376f
Diatomaceous earth, 12t, 375–376
DIC. *See* Disseminated intravascular coagulation
Dichotomous keys, 119, 119f
  for Enterobacteriaceae, 592, 592f
Dicloxacillin, 309t
Dictyosome, 86
Dictyostelium genus/spp., 365t
Dideoxycytidine (ddC), 297f
Dideoxyinosine (ddI), 297f
Didinium genus/spp., 362, 362f, 365t
Diethylcarbamazine, 316t
N,N-Diethyl-m-toluamide. *See* DEET
Differential interference contrast (Nomarski) microscopy/microscopes, 101, 102f, 107t
Differential media, 176–177, 177f, 177t, 178f
Differential stains, 108–110, 109f, 110f, 111t. *See also specific type*
Differential white blood cell count, 454, 454b
Diffusion, 71, 71f, 74t
  chemiosmosis and, 139
  facilitated, 71, 71f, 74t
Diffusion susceptibility (Kirby-Bauer) tests, 299, 300f, 301b
  for Enterobacteriaceae, 594
DiGeorge syndrome, 540, 540t
Digestive system. *See* Gastrointestinal (digestive) tract
Digestive vesicles (phagolysosomes), 87f, 455f, 456
Dihydrofolic acid, antimicrobials affecting, 295, 296f
Dihydroxyacetone-P, 130f
Dihydroxyacetone phosphate (DHAP)
  in carbohydrate biosynthesis/gluconeogenesis, 150f
  in glycolysis, 134, 134f
  in lipid biosynthesis, 151f
  in lipid catabolism, 143, 143f
Dikaryon, 367, 368f
  in ascomycete reproduction, 369
  in basidiomycetes, 370
  in zygomycete reproduction, 369
Dilation, blood vessel, in inflammation, 462–464, 462f, 463f, 465f, 465t
Dilutions
  in MPN method, 185, 185f
  pour-plate technique and, 173, 174f
  in titration, 514, 514f
  viable plate counts and, 184, 184f
Dilution tests
  broth, 300, 300f
  use, 281–282
Dimers, ultraviolet light causing formation of, 218, 218f, 275
  repair of, 220, 220f

Dimorphic fungi, 365–366, 366f, 651, 651f
Dinitrogen gas, 807, 807f, 808. *See also* Nitrogen
Dinoflagellates, 354, 355, 359, 362–363, 363f, 365t
  intoxications caused by, 362–363, 790, 790t
Dioecious helminths, 693
Dipeptide, 47, 48f
Diphenhydramine hydrochloride, for type I (immediate) hypersensitivity, 541b
Diphosphate (pyrophosphate), in DNA replication, 199, 199f
Diphtheria, 15t, 289, 572–573, 572f, 573f
  immunization against, 508t, 509t, 573
Diphtheria antitoxin, 573
Diphtheria/tetanus/acellular pertussis vaccine (DTaP), 506, 508f, 509t, 567, 573, 608
Diphtheria toxin, 572, 573
Diphtheria toxoid, 506, 509t, 573
Diphtheroids, virulence of, 426f
Diplobacilli (arrangement), 328f
Diplococci (arrangement), 327, 328f
Diplococcus genus/spp./diplococci, 113. *See also Neisseria* genus/spp.; *Streptococcus* genus/spp.
Diplococcus pneumoniae. *See Streptococcus pneumoniae*
Diploid cell(s), 196, 354
Diploid cell cultures, for viruses, 404
Diploid cell vaccine, human (HDCV), 763. *See also* Rabies vaccine
Diploid generation, in algal reproduction, 373, 374f
Diploid nucleus, 354, 355, 355f, 356, 356f
  in fungal reproduction, 367, 368f
Diplomonadida (kingdom)/diplomonads, 357, 359f, 361, 365t
Dipylidium genus/spp./*Dipylidium caninum*, 378, 419t
Direct contact transmission of disease, 431, 433t
Direct fluorescent immunoassays, 515–516, 516f, 520f
  for chlamydia, 628, 628f
  for fungi, 649, 649f
Direct immunofluorescence stain, for fungi, 649, 649f
Direct penetration, in animal virus replication, 397, 397f
Disaccharides, 45, 45f
Disease, 422. *See also* Infection; Infectious disease
  categories of, 423–424, 424t
  causes of (etiology), 12–15, 13f, 14f, 15f, 15t, 18f, 19t, 423–426, 424t, 425f
  defenses against, 20–22, 21f, 447. *See also* Host defenses; Immune response
    nonspecific (innate immunity), 446–471, 446t, 466t, 473, 473t. *See also* Innate immunity
    specific (adaptive immunity), 447, 472–502, 473t. *See also* Adaptive immunity
  diagnosis of
    clinical sampling and, 172, 173t
    recombinant DNA techniques in, 254
    DNA microarrays, 248, 253
  epidemiology of, 17, 18f, 19t, 434–441. *See also* Epidemiology
  frequency of, epidemiology and, 434–435, 434f, 435f, 436f, 437f, 438f
  germ theory of, 13, 20–21, 424
  manifestations of, 423
  prevention of, 15–18, 16b, 17f
    probiotics in, 304b, 780
  reportable, 435, 437t, 438f
    bioterrorism defense and, 815, 815f
  terminology of, 423t
Disease process, 429–430, 429t
Disease surveillance, bioterrorism defense and, 815, 815f
Disease vectors. *See* Vectors, in disease transmission
Disinfection/disinfectants, 264, 265t
  chemical agents for, 264, 276–280, 276f, 277f, 278f, 279f, 281t
  in drinking water treatment, 791f, 792
  evaluation of efficacy of, 280–282
  in wastewater treatment, 793f, 794

Disseminated intravascular coagulation (DIC), 587, 587f
Dissimilation, sulfur, 808, 808f
Dissociation (ionization), 34, 34f
Distilled spirits, fermentation in production of, 781, 783t
Distribution, antimicrobial drug, 302
Disulfide bridges, 48
Diversity
  human microbiome and, 253b
  lymphocyte receptor, 483–485, 484f
  metabolic, 140
  microbial, 803
    smallpox virus destruction and, 2b
  soil nutrients and, 810
Diversity (D) gene segment, 484, 484f
Divisions, 357
  classification of algae into, 374
  classification of fungi into, 368
dm (decimeter), 97, 97t
DNA, 19, 50–54, 51f, 52t, 85. *See also* Nucleic acid(s)
  antimicrobial activity of, 456
  in archaea, 195, 198t
  autoantibodies against, in systemic lupus erythematosus, 534
  in bacteria, 74, 195–196, 196f, 198t
  in chloroplasts, 87, 88
  comparison of with RNA, 52t
  complementary (cDNA), 51, 239
  replication and, 198, 198f
  in conjugation, 226, 227f, 228f
  double helix structure of, 51f, 52, 194, 195f
    hydrogen bonds and, 35, 51f, 52, 194, 195f
  double-stranded, 51, 52t, 194, 195f, 397–398. *See also* Double-stranded DNA
    in viruses, 51, 52t, 386, 391t, 392t, 397–398, 399t, 710, 729t. *See also* Double-stranded DNA viruses
  eukaryotic, 196–197, 197f, 198t, 354
  function of, 52, 52t
  in hepatitis B replication, 725–726
  insertion of into cells, 248
    gene therapy and, 20, 254
    vaccine development and, 253, 506–508, 507, 507f
  linker, 197f
  in lysogeny, 395, 396f
  mitochondrial, 87, 88, 197
    antimicrobial activity/killing by eosinophils and, 456
  in nuclear division (mitosis/meiosis), 196–197, 197f, 354, 355, 355f, 356, 356f, 357f
  prokaryotic, 194–196, 196f, 198t
  recombinant, 223, 224f. *See also* Recombinant DNA technology
  in recombinant vaccines, 253, 506–508, 507, 507f
  repair of, 219–221, 220f
    direct, 220, 220f
    error-prone, 221
    methylation in, 201
    single-strand, 220–221, 220f
  replication/synthesis of, 151, 152f, 198–202, 198f, 199f, 200f, 202f, 213f
    antimicrobials affecting, 291f, 295, 295–297, 296f, 318t
    in binary fission, 179–181, 179f, 326, 326f
    eukaryotic DNA differences and, 202
    methylation in, 201
    polymerase chain reaction and, 244–246, 245f, 246f, 252t
    viral DNA and, 393f, 394, 395, 396, 396f
  single-stranded, 51, 398. *See also* Single-stranded DNA
    in viruses, 51, 386, 391t, 392t, 398, 399t, 710, 729t. *See also* Single-stranded DNA viruses
  structure of, 51f, 52, 194, 195f
  synthetic, 239–240
  taxonomy/microbial classification and, 115, 118–119, 329, 330f, 357, 359f, 360
  in transduction, 225, 226f
  in transformation, 224–225, 225f
  in transposons/transposition of, 228–230, 229f
  viral, 51, 52t, 385, 386, 391t

SUBJECT INDEX

DNA (Continued)
    synthesis of, 392
        in bacteriophage replication, 393f, 394, 395, 396, 396f, 401t
        in dsDNA viruses, 393f, 394
        in HIV replication, 752, 753–754, 753f
        in ssDNA viruses, 398, 399t
DNA fingerprints/signatures, 252
    bioterrorism defense and, 815
    prokaryote classification and, 115
    in water quality testing, 792
DNA gyrase. See Gyrase
DNA helicases, 199, 200f, 201
DNA ligase
    in DNA repair, 220f, 221
    in genetic recombination, 223, 224f
    in lagging strand synthesis, 200f, 201
DNA microarrays, 246–248, 247f, 252t
    for genetic screening, 253
DNA polymerases
    antimicrobials affecting, 297
    antiviral drugs and, 297
    in DNA repair, 220f, 221
    in DNA replication, 199, 200f, 201, 202, 213t
    in lagging strand synthesis, 200f, 201, 202
    in leading strand synthesis, 200f, 201, 202
    in polymerase chain reaction, 244, 245f, 246b
    in viral synthesis, 398
DNA primer, in polymerase chain reaction, 244, 245f
    synthetic nucleic acids in synthesis of, 240
DNA probes. See also Probe(s)
    in clone selection, 246
    synthetic nucleic acids in synthesis of, 240
DNA sequencing. See Genomics
DNA triplets, 207
DNA vaccines, 253, 506–508, 507, 507f
DNA viruses, 51, 52t, 385, 386, 391t, 392t, 709–733, 729t. See also specific type
    adenoviruses (Adenoviridae), 392t, 722–725, 724f, 725b, 729t
    assembly of, 393f, 394, 396, 396f
    cancer and, 402
    hepadnaviruses (Hepadnaviridae), 391, 392t, 725–728, 725f, 726f, 727b, 727f, 729t, 739f
    herpesviruses (Herpesviridae), 391, 392t, 713–722, 714f, 729t
    papillomaviruses (Papillomaviridae), 392t, 722, 723b, 724b, 729t
    parvoviruses (Parvoviridae), 392t, 728, 729t
    pathogenic, 709–733, 729t
    placenta crossed by, 421f
    polyomaviruses (Polyomaviridae), 392t, 722, 729t
    poxviruses (Poxviridae), 392t, 710–713, 710f, 711f, 713b, 729t
    synthesis of, 397–398, 399t
Dog fleas. See Fleas
Dog tapeworm
    Dipylidium genus/spp., 378, 419t
    Echinococcus granulosus, 694–695, 695f, 703t
Dog ticks. See Dermacentor ticks; Tick(s)
Dolor (pain), in inflammation, 462
Domagk, Gerhard, 21, 289
Domains, 60, 114f, 115, 329
Domestic water, aquatic systems affected by, 811
Donor cell/DNA, in horizontal (lateral) gene transfer, 224
    conjugation, 226, 227f, 228f
    transduction, 225, 226f
    transformation, 224–225
Donor organs, rejection of, 536
    donor-recipient matching/tissue typing and, 536–537
    immunosuppressive drugs and, 537–538, 538f
    MHC antigens and, 478, 536, 536–537
Donor-recipient matching
    blood transfusions and, 531, 531t
    prevention of graft rejection and, 536–537
Double covalent bond, 31, 32f
Double helix, of DNA, 51f, 52, 194, 195f
    hydrogen bonds and, 35, 51f, 52, 194, 195f

Double-stranded DNA, 51, 52t, 194, 195f, 397–398
Double-stranded DNA viruses, 51, 52t, 386, 391t, 392t, 397–398, 399t, 710, 729t
    synthesis in, 397–398, 399t
Double-stranded RNA, 386, 398, 399f, 399t
Double-stranded RNA viruses, 386, 391t, 392t, 398, 399f, 399t, 735
    interferons affecting, 458, 458f
    naked segmented, 770–771, 771t
    synthesis in, 398, 399f, 399t
    toll-like receptors affecting, 457
Doxycycline, 311f
Drain opener, microbial production of, 12t
Drechslerella genus/spp., 367f
Drinking (potable) water, 440
    contamination of, 431, 437, 439f, 440, 789, 790. See also Waterborne illnesses
    bioterrorism and, 814t
    public health and, 440
    testing quality of, 792, 792f
    treatment of, 790–792, 791f
Droplet nuclei, 431
Droplet transmission of disease, 431, 431f, 432b, 433t
Drug(s), 289. See also specific type and Antimicrobial drugs
Drug abuse
    aspergillosis and, 658
    candidiasis and, 660b
    hepatitis and, 726, 728
    HIV infection/AIDS and, 754b
Drug administration routes, 300–302, 301f
Drug allergies, 302
Drug distribution, 302
Drug resistance. See Resistance
Dry heat, in microbial control, 271, 275t
Drying (desiccation)
    in microbial control/food preservation, 272, 272f, 275t, 783, 785
    in specimen preparation, 106
Dry weight method, for microbial growth estimation, 187
dsDNA. See Double-stranded DNA
dsDNA viruses. See Double-stranded DNA viruses
dsRNA. See Double-stranded RNA
dsRNA viruses. See Double-stranded RNA viruses
DTaP (acellular pertussis) vaccine, 506, 508f, 509f, 567, 573, 608
dTTP. See Thymidine triphosphate deoxyribonucleotide
DT vaccine, 567. See also DTaP (acellular pertussis) vaccine
Dual use research of concern, 813
Duffy antigens, absence of, malaria resistance and, 687
Duplication (mutation), 216
DURC. See Dual use research of concern
Durham tube, 177f
Dust mites, 528, 528f, 529
Dutch elm disease, 365, 369
Dutton, Joseph, 15t
Dye(s)
    microbial production of, 787
    for specimen preparation. See Staining
Dysentery. See also specific causative agent and Diarrhea
    amebic, 364, 678, 692t
    waterborne transmission of, 678, 790t
    in balantidiasis, 677, 692t
    Shigella, 601–602, 601f, 786t
Dyspnea, in respiratory syncytial virus infection, 760
Dysuria, in UTIs, 599b

Ear(s)
    as portal of entry, 420f
    as portal of exit, 430, 430f
Ear infections, Prevotella causing, 614
Eastern equine encephalitis (EEE), 742, 742f, 747t
Ebolavirus/Ebola virus/Ebola hemorrhagic fever, 392t, 763, 764–765b, 771t
    polymerase chain reaction in identification of, 244, 765b
    safe handling of virus and, 268, 764b
    vaccine development and, 763

EBs. See Elementary bodies
EBV. See Epstein-Barr virus
Echinocandins, 315t, 650
    mechanism of action of, 291f, 292, 315t, 650
Echinococcus granulosus, 694–695, 695f, 703t
Echoviruses, 738
E. coli. See Escherichia coli
Ecological microbiology, 18f, 803–805, 804f
    adaptation/survival and, 805
    associations and
        levels of, 803–804, 804f
        microbial growth and, 168–170, 169f, 170b, 171f
EcoRI restriction enzyme, 240, 240t
EcoRII restriction enzyme, 240
Ecosystems, 804, 804f
    aquatic
        freshwater, 811, 811–812, 811f
        marine, 811, 811f, 812
        specialized novel, 812
    human, 804, 804f
    soil, 804, 804f
Ectopic pregnancy, pelvic inflammatory disease and, 589
Ectothrix, in tinea capitis, 665t
Edema, in inflammation, 462
ED pathway. See Entner-Doudoroff pathway
Edwardsiella genus/spp., 592f, 597
EEE. See Eastern equine encephalitis
Efferent lymphatic vessels, 475f, 476
Efficacy
    of antimicrobial drugs, 298–300, 300f, 301b, 301f
    of antimicrobial methods
        antisepsis and disinfection, 280–282
        evaluation of, 280–282
        factors affecting, 266–268, 266f, 267f
        microbial death rate and, 264, 265f
Efflux pumps, 306
Eflornithine, 318t, 681
Eggs
    culturing viruses in, 403, 403f
        vaccine manufacture and, 403, 508
    Salmonella contamination of, 600, 786t
EHEC (enterohemorrhagic E. coli), 595
Ehrlich, Paul, 18, 18f, 21, 289
Ehrlichia genus/spp./Ehrlichia chaffeensis, 623–626, 623f, 626t
Ehrlichia equi. See Anaplasma genus/spp.
Ehrlichiosis, 623–626, 623f, 626t
EIA (enzyme immunoassay). See Enzyme-linked immunosorbent assay/enzyme immunoassay
EIEC (enteroinvasive E. coli), 595
80S ribosomes. See Eukaryotic ribosomes
Elastase, Pseudomonas aeruginosa producing, 609
Electrical gradient, 70, 71f. See also Electrochemical gradient
Electrochemical gradient, 139
    in active processes, 73, 73f, 74t
    in chemiosmosis, 139
    in passive processes/diffusion, 71, 71f, 72, 72f
Electrolytes, 34, 39
Electromagnetic radiation, 273. See also Radiation; Ultraviolet light
    in microbial control/food preservation, 273, 274, 274f, 275t, 785
    microscopy and, 98, 98f
Electromagnetic spectrum, microscopy and, 98, 98f
Electron(s), 28, 28f, 98
    atomic mass of, 28
    chemical bonds and, 31–35, 32f, 34f, 35f, 35t
    configurations of, behavior of atom and, 29–31, 30f
    in microbial growth, 163
    valence, 30–31, 30f
    wavelengths of, 98, 104
Electron acceptor, 126, 126f
    chlorophyll reaction center, 145, 145f
    final. See Final electron acceptor
Electron beams, 98, 274
    in microbial control, 274, 275t
    microscopy and. See also Electron microscopy/microscopes
        scanning electron microscopy, 104
        scanning tunneling microscopy, 105
        transmission electron microscopy, 104
Electron carrier molecules, 126, 128, 128t

Electron configurations, 29–31, 30f
Electron-dense stains, 104, 112
Electron donor, 126, 126f
Electronegativity, 31, 33f
    in ionic bonds, 33
    in nonpolar covalent bonds, 31, 33f
    in polar covalent bonds, 33
Electronic counters, for microbial growth estimation, 184
Electron microscopy/microscopes, 104–105, 105f, 107f
    preparation of specimens for, 104, 106
    resolving power of, 99f, 104
    staining specimens for, 111–112
        preparation and, 106
Electron shells/clouds, 30, 30f
Electron transport/electron transport chain, 133, 133f, 136–139, 138f, 153f
    fermentation pathways and, 139–140
    in photosynthesis, 145–146, 146
Electrophoresis, 246, 247f, 252t
    in immunoblot/western blot, 518, 518f
    in Southern blot, 246
Electroporation, 248, 249f, 252t
Eleftheria terrae, 308
Elek test, in diphtheria, 573
Element(s) (chemical), 28, 29t, 30f
    for microbial growth, 162–165, 163f, 165f, 165t
    periodic table of, 30f
    trace, 165
Elementary bodies
    in chlamydia growth/reproduction, 626, 627f, 629
    in Ehrlichia and Anaplasma growth/reproduction, 623, 623f
Elephantiasis, in filariasis, 702, 702f, 703t
Elimination, in phagocytosis, 455f, 456
Elion, Gertrude, 296
ELISA (enzyme-linked immunosorbent assay), 516–518, 517f, 518f
    antibody sandwich, 518, 518f
    in HIV diagnosis, 517, 756
    rapid (immunofiltration assays), 518
Elongation
    RNA transcript, 204
    in translation, 211–212, 211f, 212f
Elongation factors, 206, 211
Embden-Meyerhof pathway (glycolysis), 133–135, 133f, 134f, 135f, 153f
Embryonated chicken eggs, culturing viruses in, 403, 403f
    vaccine manufacture and, 403, 508
Emerging/reemerging diseases, 21–22, 21b. See also specific disease
    Acanthamoeba keratitis, 267b, 678, 679, 692t
    antimicrobial resistance and, 305b
    aspergillosis, 370b, 657–659, 658f, 659b
    babesiosis, 691b
    blastomycosis, 653–654, 653f, 654b, 654f
    Buruli ulcer, 578b
    Chikungunya, 401t, 747t
    community-associated MRSA, 305, 305b
    coronavirus respiratory syndromes, 392t, 736b, 748–749, 749f
    cyclosporiasis, 691, 692f, 692t
    dengue, 237b, 257b, 734f, 743–744, 744f, 745f, 747t
    dermatophytoses causing, 649, 664–665, 665f, 665t
    ehrlichiosis and anaplasmosis, 623–626, 623f, 626t
    filovirus infections, 392t, 763, 771t
    fungal opportunists in HIV infection/AIDS and, 664, 751t
    hemorrhagic fevers (Ebola/Marburg), 392t, 763, 764–765b, 771t
    malaria (P. knowlesi), 686
    measles, 472b, 497b, 504b, 758, 758–759, 758f, 758t, 759b, 759f
    melioidosis, 608, 610b
    microsporidiosis, 369, 496b, 663–664, 663f
    Middle East respiratory syndrome (MERS), 736b, 748, 749, 749f
    monkeypox, 712–713, 713b
    mycobacterial (nontubercular), 573, 575, 578b
    Naegleria amebic meningoencephalitis, 679, 692t, 790b
    necrotizing fasciitis, 118b, 554–555b, 556
    Norovirus gastroenteritis, 1b, 22b, 392t, 740, 740b, 771t, 790t

pertussis (whooping cough), 344t, 345b, 586b, 606–608, 607f, 608f, 615b
Rickettsia parkeri infection, 622b
schistosomiasis/snail fever, 696–698, 697f, 698f, 699b, 703t
severe acute respiratory syndrome (SARS), 391f, 392t, 736t, 748, 749, 749f
spotted fever rickettsiosis, 621, 622b, 626t
tick-borne encephalitis, 743b, 814t
tuberculosis, 573–574, 574b, 576–577b
variant Creutzfeldt-Jakob disease, 21b, 48, 406f, 407, 407b
Vibrio vulnificus infection, 40b, 170b, 639, 786t
West Nile virus disease, 387, 435f, 734b, 742, 742f, 743f, 747t, 772b
Zika fever/virus, 744–745, 746b, 747t
-emia (suffix), 423t
EMP. *See* Embden-Meyerhof pathway
Empty magnification, 99
Empyema, staphylococcal, 550
Emtricitabine, 297, 314t
  HIV infection prevention and, 757
Encephalitis, 392t
  amebic
    Acanthamoeba causing, 678, 692t
    Naegleria causing, 679, 692t, 790b
  arboviral, 742, 742f, 743b, 747t. *See also specific type*
    bunyaviruses causing, 392t, 747t, 767, 771t
    flaviviruses/togaviruses causing, 392t, 742, 742f, 743b, 747t, 771t
  bioterrorism and, 743b, 814t
  Japanese, 392t, 742, 747t
    immunization against, 509t
  measles, 759
  Nipah virus, bioterrorism and, 814t
  smallpox vaccine and, 510b
  in *Trypanosoma brucei* infection, 681
  West Nile, 435f, 734b, 742, 742f, 743f, 747t, 772b
Encephalitozoon genus/spp./*Encephalitozoon intestinalis*, 369, 496b, 662
Encephalopathies, spongiform, 405–407, 406f, 407b
  bovine (BSE/mad cow disease/variant Creutzfeldt-Jakob disease), 21b, 48, 406f, 407, 407b
  bioterrorism and, 814t
  enzyme for elimination of prion causing, 280
Encystment, 677
Endemic disease, 435, 436f
Endemic (murine) typhus, 622, 626t
Endergonic pathways, 126
endo- (prefix), 423t
Endocarditis
  candidal, 658t, 660b
  E. coli causing, 595b
  pneumococcal, 558b, 559
  staphylococcal, 550
  viridans streptococci causing, 557
Endocrine (hormonal) diseases, 424, 424t
  autoimmunity and, 539
Endocytosis, 81, 81f, 81t, 87f
  in animal virus replication, 397, 397f
  in HIV replication, 752, 753, 753f, 754f
Endoflagella, 64, 64f
Endogenous antigens, 477
  processing of, 479–480, 479f
Endogenous healthcare-associated infections, 439
Endonucleases, restriction. *See* Restriction enzymes/endonucleases
Endoplasmic reticulum, 61f, 85f, 86, 86f
  rough, 61f, 86, 86f
  smooth, 61f, 86, 86f, 87f
Endosome, in HIV replication, 753, 753f, 754f
Endospores, 74, 75–76, 76f, 89t, 325, 326f. *See also specific organisms*
  in anthrax, 336, 560
  antimicrobial susceptibility and, 266–267, 266f
  Bacillus, 75, 76, 76f, 325, 326f, 335, 560, 562
  clostridial, 75, 76, 76f, 269, 325, 326f, 335, 562
    botulism and, 269, 326f, 563, 565b, 785
    gas gangrene and, 567, 568
    tetanus and, 565, 565f

death phase of microbial growth and, 182
food contamination and, 76, 325, 784, 785
industrial canning and, 784
in soilborne diseases, 810
staining, 110, 110f, 111t
as sterilization indicator, 270, 271f
Endosymbiotic theory, 88, 333
Endothermic reactions, 36
Endothrix, in tinea capitis, 665t
Endotoxin (lipid A), 67, 68f, 427f, 428t, 587, 587f
  Enterobacteriaceae and, 593, 593f
  gonococci and, 589, 589b
  meningococci and, 590
  Pseudomonas aeruginosa and, 609
  Shiga-like, 596
End-product inhibition (feedback inhibition/negative feedback), 132, 132f
Energy. *See also* ATP
  activation, enzymes affecting, 128–129, 129f
  in anabolic and catabolic pathways, 125, 125–126, 125f, 126–127
  decomposition reactions producing, 36
  in glycolysis, 133, 134f, 135, 135f, 234–235
  for microbial growth, 163, 163f
  storage/release of
    ATP and, 125, 125–126, 125f, 126–127
    fats and, 42
  synthesis reactions requiring, 36
Energy parasites, chlamydias as, 626
Enfuvirtide, 313t
  mechanism of action of, 291f, 313t
Enhancer, in *lac* operon regulation, 214, 214f
Enriched media, 175
Enrichment culture, 176
Entamoeba genus/spp./*Entamoeba histolytica*, 364, 365t, 678, 678t, 692t
  dysentery/amebiasis caused by, 364, 678, 692t
  virulence factors of, 426–428
  waterborne transmission of, 678, 790t
Entecavir, 314t
Enteric bacteria, 342, 343t, 591, 591–602, 592f, 593f, 594f, 595b. *See also* Enterobacteriaceae
Enteric cytopathic human orphan virus (echovirus), 738
Enteric hepatitis (hepatitis E), 739t, 740, 771t
Enteritis, *Cryptosporidium*, 690, 692t
Enterobacter genus/spp./*Enterobacter aerogenes*, 343t, 344t, 592f, 596, 604f
  fermentation in identification of, 141
  healthcare-associated (nosocomial) infection caused by, 591f, 596
  sites of infection caused by, 591f
Enterobacteriaceae, 342, 343t, 591, 591–602, 592f, 593f, 594f, 595b. *See also specific types*
  coliforms, 594, 595–596, 596f, 598–599b
  healthcare-associated (nosocomial) infection caused by, 591, 591f
  dichotomous key for identification of, 592, 592f
  noncoliforms, 594, 597, 597f
  healthcare-associated (nosocomial) infection caused by, 591f
  opportunistic, 595–597, 596f, 598–599b
  oxidase test and, 591, 591f
  pathogenicity of/virulence and, 593, 593f, 597–605
  resistant strains of, 230b, 593
Enterobius vermicularis (pinworm), 700–701, 701f, 703t
Enterococcus genus/spp. (*Enterococcus faecalis/faecium*) (enterococci), 113, 336, 559–560, 560f, 561t
  Entner-Doudoroff pathway used by, 140
  resistant/multiple-drug-resistant strains of, 230b, 273t, 306, 336, 560
  horizontal gene transfer and, 230b
Enterohemorrhagic *E. coli*, 595
Enteroinvasive *E. coli*, 595
Enterotoxin(s), 428
  clostridial, 428, 562
  in *E. coli* diarrhea/gastroenteritis, 428, 595, 596
  in *Shigella* diarrhea, 601, 602
  in staphylococcal intoxication (food poisoning), 548, 548t, 549
Enterovirus genus/spp./enteroviruses, 392t, 735, 736–741, 771t. *See also* Coxsackieviruses; Echoviruses; Poliovirus

meningitis caused by, 738, 739
polio caused by, 392t, 736–738, 737f, 738t, 771t
transmission of, 431
Entner-Doudoroff pathway, 140, 609
Entry
  in infection. *See* Portals of entry
  virus
    antiviral drugs affecting, 295, 313t
    in replication, 392
      animal virus, 397, 397f, 400f, 401t
      bacteriophage, 393f, 394, 401t
      HIV, 752, 753, 753f, 754f
Entry inhibitors, 295, 313t
Envelope
  nuclear, 60, 61f, 85, 86f, 196, 355, 355f
  viral/enveloped viruses/virions, 385, 386, 389–390, 391f
    antimicrobial action/susceptibility and, 265, 266, 266f
    budding in release and, 400, 400f
    entry of host cell and, 397, 397f
    negative ssRNA viruses
      segmented, 763–770, 771t
      unsegmented, 758–763, 771t
    persistent infections and, 400, 400f
    positive ssRNA viruses, 741–750, 741f, 747t, 771t
    with reverse transcriptase (retroviruses), 239, 391t, 392t, 398, 399t, 735, 750–757, 750f, 771t
Environment
  antimicrobial efficacy affected by, 267–268, 267f
  autoimmunity and, 538
  microbial population of soil and, 804, 804f, 810
  response to, as characteristic of life, 59, 59t
  role of microbes in, 18f, 19t, 20. *See also* Environmental microbiology
Environmental antigens, in type I (immediate) hypersensitivity, 526
Environmental microbiology, 18f, 19t, 20, 803–812
  acid mine drainage and, 805–806, 805f, 806f
  applied, 19t, 779, 779–786
  aquatic microbiology and, 811–812, 811f
  biogeochemical cycles and, 20, 806–809, 807f, 808f, 809f
  bioremediation and, 19t, 20, 169b, 805
  biosensors and bioreporters in, 789
  DNA microarrays in, 248
  microbial ecology and, 18f, 803–805, 804f
  microbiomes and, 803–812, 804f. *See also* Microbiome
  soil microbiology and, 804, 804f, 809–810, 809f, 810t
Environmental Protection Agency (EPA), water standards set by, 791
Environmental specimens, for microbial culture, 171
Enzymatic catalyst, 47
Enzyme(s), 12, 47, 127–132
  activation of, 131, 131f
  active site of, 128f
    activation/inhibition and, 131–132, 131f, 132f
  activity of, 128–132, 129f, 130f, 131f, 132f
    control of (activation/inhibition), 131–132, 131f, 132f
    enzyme concentration and, 131
    enzyme-substrate specificity and, 129, 129f
    pH affecting, 130f, 131
    substrate concentration and, 130f, 131
    temperature affecting, 129–130, 130f, 131f
  antimicrobial, 280, 281t
  concentration of, enzymatic activity affected by, 131
  denaturation of, 130, 131f
  extracellular, as virulence factors, 426–428, 427f. *See also specific enzyme and specific organism*
  in fermentation, Buchner's demonstration of, 12

inhibition of, 131–132, 131f, 132f
makeup of, 127, 128f
in metabolism, 12, 125, 127–132
microbial production of, 12t, 787, 789t
naming/classification of, 127–128, 128t
restriction. *See* Restriction enzymes/endonucleases
in serological testing, 516–518, 517f, 518f
Enzyme cofactors, 127, 128f, 128t
Enzyme immunoassay (EIA). *See* Enzyme-linked immunosorbent assay/enzyme immunoassay
Enzyme-linked immunosorbent assay/enzyme immunoassay (ELISA/EIA), 516–518, 517f, 518f
  antibody sandwich, 518, 518f
  in HIV diagnosis, 517, 756
  rapid (immunofiltration assays), 518
Enzyme-substrate complex, 129, 130f
Enzyme-substrate specificity, 129, 129f
Eosin, 108
  as histological stain, 110
  as negative stain, 110
Eosinophil(s), 452f, 453, 453t
  in innate immunity/host defense, 452f, 453, 456
  in type I (immediate) hypersensitivity, 527–528, 527f
Eosinophilia, 456, 527
EPA (Environmental Protection Agency), water standards set by, 791
Epidemic cholera, 12, 17, 637–639, 638f. *See also* Cholera
Epidemic disease, 435, 436f
  common-source, food poisoning and, 786
Epidemic (louse-born) typhus, 621, 626t
  bioterrorism and, 814t
Epidemiology, 17, 18f, 19t, 434–441
  analytical, 437
  bioterrorism defense and, 815
  causation theories and, 426
  data reporting and, 434–435, 435f
  descriptive, 436–437, 439f
  disease frequency and, 434–435, 434f, 435f, 436f, 437t, 438f
  experimental, 437
  hospital, 437–440, 439f. *See also* Healthcare-associated (nosocomial) infections/diseases
  Koch's experiments/postulates and, 13–14, 14f, 437
  public health and, 435, 437f, 438f, 440–441
  research studies used in, 436–437, 439f
Epidermis, 421f, 447, 448f
  exfoliative toxin/staphylococcal scalded skin syndrome and, 548, 549
  in host defense, 447–448, 448f
Epidermophyton floccosum, 419t, 665, 665t
Epiglottitis, *Haemophilus influenzae* causing, 604
Epinephrine, for anaphylactic shock/type I (immediate) hypersensitivity, 529, 529–530, 541b
Epithelium, in host defense, 448–449
Epithet, specific, 113
  for viruses, 391, 391t
Epitope(s) (antigenic determinants), 476, 477f
  antibody binding/immune response and, 485, 492, 493f
  antigen processing and, 479, 479f, 480, 480f
  autoimmunity and, 538–539
  B cell receptor specificity and, 483, 483f
  in cell-mediated immune response, 490, 490f, 491, 491f
  clonal deletion of T cells and, 482, 482f
  MHC proteins and, 478–479, 490, 490f
  T cell receptor specificity and, 481, 481f
Epitope-binding site
  B cell receptor, 483f, 484
  T cell receptor, 481, 481f
Epsilon antibodies. *See* Immunoglobulin E
Epsilon F_c region/heavy chains, 485, 487, 488f
Epsilonproteobacteria, 330f, 344t, 346, 639, 640
EPS matrix. *See* Extracellular polymeric substance (EPS) matrix

Epstein-Barr virus (EBV/human herpesvirus-4), 392t, 713, 718–720, 719f, 729t
Epstein, Michael, 718
Epulopiscium genus/spp./*Epulopiscium fishelsoni*, 60f, 326, 335, 337t
   genome of, 195
   reproduction in, 326, 327f
Equine encephalitis, Eastern (EEE)/Western (WEE)/Venezuelan (VEE), 742, 742f, 747t
ER. *See* Endoplasmic reticulum
Ergometrine, 668
Ergosterol, in cytoplasmic membrane, drugs affecting, 293, 294, 294f, 650
Ergot/ergot alkaloids, 369, 668
Ergotamine, 668
Error-prone repair, 221
Erysipelas, 553, 553f
Erythema infectiosum (fifth disease), 392t, 728, 728f
   transplacental transmission of, 421t
Erythrocyte(s) (red blood cells), 452, 452f, 453
   blood group antigens on, 514, 514f, 530, 530f
   in malaria/*Plasmodium* life cycle, 687, 688f, 689b
   size of, 388f
Erythrocytic cycle, in *Plasmodium* life cycle, 687, 688f
Erythrogenic toxins. *See* Pyrogens/pyrogenic (erythrogenic) toxins
Erythromycin, 311t
   mechanism of action of, 210, 293, 311t
   microbial production of, 290t, 311t
   resistance to, 305, 311t
   spectrum of action of, 299f, 311t
Erythropoietin, recombinant DNA in production of, 788t
Erythrose 4-phosphate, 149t
   in pentose phosphate pathway, 149t
Erythrovirus genus/spp., 392t, 728
   placenta crossed by, 421t
Eschar, in cutaneous anthrax, 562, 562f
Escherich, Theodore, 15t, 113
Escherichia genus/spp., 113, 343t, 344t, 592f. *See also Escherichia coli*
   fermentation in identification of, 141
   healthcare-associated (nosocomial) infection caused by, 591f
   sites of infection caused by, 604f
Escherichia coli, 15f, 60f, 108f, 164, 342, 592f, 595–596, 598–599b, 786f
   adhesins produced by, 422f, 595
   arrangement of, 328f
   clotting factor produced by, 20
   conjugation/conjugation pili in, 66f, 226–227, 227f, 228f
   culture of, 173f, 174, 175f, 177f, 178f, 594f
   diarrhea/foodborne illness caused by, 15t, 428, 595, 596, 604f, 786t
   DNA microarray of, 246
   endocarditis caused by, 595b
   gene expression regulation in, 213
   generation time of, 180, 181f
   genes/genome of, 194
   as indicator organism, 792, 792f
   isoleucine inhibition and, 132
   kidney infection caused by, 89b
   knockout gene of, 251–252
   lambda phage of, 395, 395f
      replication of, 395–396, 396f
   metabolism in, 126, 140, 149, 150
   O157:H7, 304b, 473, 593, 595–596, 786t
      bioterrorism and, 113b, 814t
      fluorescent phages in identification of, 113b
   operons in, 213–215, 214f, 215f
   pasteurization in control of, 270
   pharmaceuticals produced by, 238, 252, 788t
   probiotics and, 304b
   in recombinant DNA technology/genetic engineering, 251–252, 342, 788t
   resistant strains of, 196
   restriction enzymes produced by, 240, 240t
   safe handling of, 268
   septicemia/bacteremia caused by, 89b, 595, 595b
   size of, 60f, 388f

T4 (type 4) bacteriophage of, 388f, 390f, 392
   replication of, 392–394, 393f, 394f
   temperature affecting growth of, 166f
   urinary tract infection (UTI) caused by, 15t, 89b, 504f, 598–599b
   virulence factors/strains and, 196, 428, 595, 599b
   water contamination/waterborne transmission and, 595, 790t, 791, 792, 792f
Escherichia coli Keio Knockout collection, 251–252
E site, ribosomal, 210, 210f
Esophageal candidiasis, 658t, 660b
Essential amino acids, 150
Ester group, 41t
Estrogen, autoimmunity and, 538
Etest, 300, 300f
Ethambutol, 310t
   mechanism of action of, 291–292, 291f, 310t
Ethanol (ethyl alcohol)
   in acid-fast staining, 109
   antimicrobial action of, 276–277
   fermentation producing, 10–12, 10b, 11f, 141, 141f, 142f, 780–782, 781f, 782f, 783t. *See also* Alcoholic fermentation
      as alternative fuel source, 162b, 788, 789t
      beverage production and, 10–12, 10b, 11f, 141, 141f, 142f, 780–782, 781f, 782f, 783t
   in Gram staining, 109, 109f
Ether group, 41t
Ether linkages, in archaeal cytoplasmic membrane, 78
Ethidium bromide, as frameshift mutagen, 219
Ethyl alcohol. *See* Ethanol
Ethylene oxide
   in food preservation, 785
   for gas sterilization, 280, 281t
Etiology, 12–15, 13f, 14f, 15f, 15t, 18f, 19t, 423–426, 424t, 425f
   Koch's experiments/postulates and, 13–14, 14f, 15f, 18f, 424–426, 425f
   experimental epidemiology and, 437
Euchromatin, 197, 197f
Euglena genus/spp./*Euglena granulata*, 357, 361, 361f, 365f
   flagellum of, 83f, 361f
   isolation of, 174
   for vitamin B₁₂ identification, 175
Euglenids, 361, 361f, 365t
Euglenoid movement, 361
Euglenozoa (kingdom), 357, 359f, 361, 361f, 365t
Eukarya (domain), 60, 114f, 115, 329
Eukaryotes/eukaryotic cells, 4, 59–60, 61f, 89t, 353–383. *See also specific type*
   antimicrobials effective against, 293
   cell walls of, 80, 80f, 89t
   characteristics of, 354–357, 355f, 356f, 357t, 358f, 359f
   chromosomes of, 85, 89t, 196–197, 197f, 198t, 354
   classification of, 114f, 115, 357–358, 358, 359f
      nucleotide sequencing and, 115, 357–358
   cytoplasmic membrane of, 61f, 80–82, 80f, 81f, 81t, 89t. *See also* Cytoplasmic (cell/plasma) membranes
   cytoplasm of, 82–88, 83f, 84f, 85t, 86f, 87f, 88f, 89t
   DNA in, 196–197, 197f, 198t, 354
      replication and, 202
   evolution/endosymbiotic theory of formation of, 88
   external structure of, 80
   genomes of, 196–198, 197f, 198t
   glycocalyces of, 80, 89t
   life processes in, 59t
   mRNA of, 206, 207f, 209
   organelles of, 83–88, 84f, 85t, 86f, 87f, 88f
   plasmids in, 197–198, 198t
   reproduction of, 354–357, 355f, 356f, 357f, 358f, 359f
   size of, 60, 61f

   transcription in, 206, 207f
   translation in, 209, 212
Eukaryotic ribosomes, 61f, 76, 82–83, 85t, 89t, 209f, 210, 293
   antimicrobials affecting, 210, 293
   in translation, 209f, 210, 212
Euryarchaeota (phylum), 330, 330f, 331, 332
Eutrophication, phosphorus and, 808, 809f
Evolution, 115
   archaeal classification and, 115, 330f
   bacterial classification and, 115, 330f
   bacteroid classification and, 347
   Brucella classification and, 606
   chlamydia classification and, 347, 626
   of chloroplasts, 87, 88
   chrysophyte classification and, 375
   Coxiella classification and, 613
   crop resistance to pests and, 255
   cyanobacteria and, 88, 333
   Darwin's theory of, 115, 679
   of diplomonads, 361
   endosymbiotic theory and, 88
   eukaryotes/eukaryotic classification and, 88, 115, 357–358, 359f
   flavivirus classification and, 741
   genomics/metagenomic sequencing and, 20, 118–119, 347
   Gram-negative bacteria and, 330f
   Gram-positive bacteria and, 330f
   green algae/plants and, 374
   Helicobacter classification and, 640
   of HIV, 752
   influenzaviruses and, 768b, 769b
   malaria resistance and, 687
   microbial ecology and, 803, 805, 812
   microbiome members and, 253b
   microsporidia classification and, 662
   of mitochondria, 87, 88
   mutations and, 216, 219
      cold virus/cold vaccine and, 736
      emerging/reemerging diseases and, 21b
   mycoplasma classification and, 330f, 547, 568
   nucleotide sequencing and, 20, 115, 118–119, 329, 330f, 357, 359f, 360
   Orientia classification and, 622
   osmotic pressure tolerance and, 168
   paramyxovirus classification and, 760
   parasitism and, 416
   Pneumocystis jirovecii classification and, 657
   prokaryotic classification and, 115, 329, 330f
   protozoan classification and, 360
   resistant microbes and, 219
   ribosomal RNA relationships and, 115, 329, 330f
   rickettsia classification and, 621
   Salmonella classification and, 600
   Shigella classification and, 601
   taxonomy/microbial classification and, 113–115, 118–119, 329, 330f, 357, 359f, 360
   Treponema (syphilis) and, 631
   viral classification and, 391
   water mold classification and, 376
Exchange (transfer) reactions, 36–37
Excystment, 677, 678, 684b, 694, 694f, 696
Exergonic pathways, 125–126
Exfoliative toxins, 548, 548t
Exocytosis, 81, 81t, 87f
   in phagocytosis, 455f, 456
   in virus release, 400
Exoenzyme S, *Pseudomonas aeruginosa* producing, 609
Exogenous antigens, 477, 477f
   processing of, 480, 480f
Exogenous healthcare-associated infections, 439
Exons, 206, 207f
Exophiala genus/spp.
   mycetoma caused by, 666t, 667
   phaeohyphomycosis caused by, 666t, 667f
Exothermic reactions, 36
Exotoxin(s), 427f, 428, 428t. *See also specific type*
   Enterobacteriaceae producing, 593, 593f
Exotoxin A
   in necrotizing fasciitis, 555b
   Pseudomonas aeruginosa producing, 609
Experiment(s), in scientific method, 9, 9f
Experimental epidemiology, 437
Exponential (logarithmic) growth, 180, 180f, 181, 181f

Exponential (log) phase, of microbial growth, 181, 181f
Extension, in polymerase chain reaction, 244, 245f
Extensively drug-resistant (XDR) tuberculosis, 574, 577b
Extracellular enzymes, as virulence factors, 426–428, 427f. *See also specific type*
Extracellular pathogens. *See also* Bacteria; Fungi
   antibodies/B cells attacking, 474
Extracellular polymeric substance (EPS)/matrix, in biofilms, 168–169
Extraintestinal amebiasis, invasive, 678, 692t
Extranuclear chromosomes, 197–198
Extreme thermophiles. *See* Hyperthermophiles
Extremophiles, 169b, 329, 329–332, 331f, 805. *See also specific type*
   radioactivity and, 169b
Eye(s)
   lacrimal apparatus of, in host defense, 449, 450f
   microbial diseases of, 446b, 466b
      Acanthamoeba keratitis, 267b, 678, 679, 692t
      candidal, 658t, 660b
      Chlamydia trachomatis causing/trachoma, 627, 628, 628f
      herpetic, 714–715, 714f, 715f, 716f
      histoplasmosis, 652
      Neisseria gonorrhoeae, 279, 589
   microbiome/normal microbiota of, 417t
   as portal of entry, 420, 420f, 421
      for parasites, 676f
   as portal of exit, 430, 430f
"Eyespot," in euglenids, 361, 361f

F. *See* Fluorine
F_{ab} region, 485, 485f
F_c region, 485, 485f
Facilitated diffusion, 71, 71f, 74t
Factor(s). *See* Plasmids
Factor VIII, recombinant DNA in production of, 788t
Factors B/D/P, in complement activation, 459f, 460
Facultative anaerobes, 164, 165f
   glycolytic, 342, 343t, 344t
   Gram-negative bacilli, 591–605
   yeasts as, 11
Facultative halophiles, 168
FAD (flavin adenine dinucleotide), 52, 126, 128, 128t, 153f
   in citric acid/Krebs cycle, 136, 137f, 153f
   in electron transport chain, 136, 138f, 153f
$FADH_2$, 126, 153f
   in beta-oxidation, 143, 143f
   in chemiosmosis, 140
   in citric acid/Krebs cycle, 133f, 136, 137f, 140f, 153f
   in electron transport chain, 136, 138f, 139, 140t, 153f
False morel, 668
False negative, 513
Families, 113, 114f, 115
   for viruses, 391
Farmer's lung, 532
Fas. *See* CD95
Fasciitis, necrotizing ("flesh-eating disease"), 118b, 336, 554–555b, 556
Fasciola genus/spp. (*Fasciola gigantica/hepatica*) (liver flukes), 419t, 696, 697f, 703t
Fascioliasis, 697f, 703t
Fastidious organisms, 175
Fat(s), 41–42, 42f
   biosynthesis of, 150, 151f
   catabolism of, 143, 143f
Fat cells, adenoviruses affecting, 725
Fatigue/chronic fatigue syndrome, HHV-4 (Epstein-Barr virus) and, 719
Fatty acids, 41, 42, 42f, 43f
   biosynthesis of, 150, 153f
   in fats, 41
   lipid biosynthesis and, 41, 42f, 150, 151f, 153f
   lipid catabolism/beta-oxidation and, 143, 143f, 153f
   in phospholipids, 42
   in waxes, 42

SUBJECT INDEX I-15

Favus, in tinea capitis, 665*t*
F⁺ cells, 226, 227*f*, 228*f*
F⁻ cells, 226, 227*f*, 228*f*
F_c region, 485, 485*f*
Fe. *See* Iron
Fecal coliforms
  as indicator organisms, 792, 792*f*
  water contaminated by, 595, 791, 792, 792*f*
Fecal microbiome transfer, 803*b*
Fecal-oral infection
  foodborne transmission and, 432
  parasitic infection and, 676*f*
  waterborne transmission and, 431
Feces
  for fecal microbiome transfer, 803*b*
  as portal of exit, 430, 430*f*
Feedback inhibition (negative feedback/end-product inhibition), 132, 132*f*
Female reproductive system. *See also* Reproductive system
  microbiome/normal microbiota of, 417*t*
Fermentation/fermentation pathways, 10–12, 10*b*, 11*f*, 140–141, 141*f*, 142*f*, 142*t*
  aerobic/anaerobic respiration compared with, 142*t*
  alcoholic, 10–12, 10*b*, 11*f*, 141, 141*f*, 142*f*, 780–782, 781*f*, 782*f*, 783*t*
  bacteria in, 11–12, 11*f*
  biofuel production and, 162*b*, 788
  in carbohydrate/glucose catabolism, 133, 133*f*, 140–141, 141*f*, 142*f*, 142*t*
  cellular respiration compared with, 142*t*
  in food microbiology, 10*b*, 779–782, 780*f*, 781*f*, 782*f*, 783*t*, 785
    alcoholic beverage production and, 10–12, 10*b*, 11*f*, 141, 141*f*, 142*f*, 780–782, 781*f*, 782*f*, 783*t*
    bread production and, 10*b*, 12*t*, 779, 783*t*
    dairy products and, 780, 780*f*, 783*t*
    food preservation/spoilage and, 779, 783, 785
    meat product production/preservation and, 780, 783*t*
    silage and, 779, 783*t*
    vegetables and, 779, 783*t*
  industrial, 786–787, 787*f*
  microbial identification and, 141
  products of, 10*b*, 141, 142*f*, 779–782, 780*f*, 781*f*, 782*f*, 783*t*
  yeasts in, 10*b*, 11, 11*f*
Fermentation tests, 141
Fermentation vats, 787, 787*f*
Ferritin, 452
Ferroplasma acidarmanus, 806, 806*f*
Ferrum. *See* Iron
Fertility (F) plasmids, 196, 226, 227*f*, 228*f*
Fetus. *See also* Newborn; Pregnancy
  autoimmunity and, 538
  coxsackie B virus infection of, 738
  cytomegalovirus infection of, 720, 721
  herpesvirus infection of, 715
  HIV infection of, 421*t*, 756
  ART during pregnancy and, 757
  listeriosis and, 336, 421*t*, 568, 570*b*
  maternal antimicrobial use affecting, 302
  Rh status of, hemolytic disease of newborn and, 531, 532*f*
  rubella infection of, 747
  syphilis and, 421*t*, 632
  toxoplasmosis and, 421*t*, 687, 690
  Zika fever/virus infection of, 744, 745, 746*b*, 747*t*
Fever, 464–466, 465*f*, 587*f*
  pyrogens/pyrogenic toxins and, 464, 552
  in relapsing fever, 636, 636*f*
Fever blister (oral herpes), 392*t*, 709*b*, 714, 714*f*, 715*f*, 716*t*, 729*b*
Fever of Crete (brucellosis), 15*t*, 340, 344*t*, 606, 606*f*
  bioterrorism and, 814*t*
Fibrinogen, in inflammation, 464
Fibrobacteres (phylum), 330*f*
Fibroblasts, in tissue repair, 464
Fifth disease (erythema infectiosum), 392*t*, 728, 728*f*
  placental transmission of, 421*t*
Filament
  of archaeal flagellum, 77
  of bacterial flagellum, 62, 63*f*
    axial, 64, 64*f*
  of eukaryotic flagellum, 82

Filamentous hemagglutinin, *Bordetella pertussis* producing, 606–607
Filamentous particles, hepatitis B virus, 726, 727*f*
Filarial nematodes, 699
  Wuchereria, 702, 702*f*, 703*t*
Filariasis, 702, 702*f*, 703*t*
Filoviridae (filoviruses), 392*t*, 735, 758, 763, 765*b*, 771*t*
  as biological weapons, 814*t*
Filterable viruses, 14, 272. *See also* Viruses
Filtration, 272
  in microbial control, 272–273, 273*f*, 274*f*, 275*t*
  drinking water treatment and, 791–792, 791*f*
  wastewater treatment and, 793*f*, 794
  in microbial growth estimation, 184–185, 185*f*
Fimbriae, 89*t*
  archaeal, 78, 79*t*, 89*t*
  bacterial, 65–66, 65*f*, 79*t*, 89*t*
  of Enterobacteriaceae, 593, 593*f*, 599*b*
  of *Neisseria*, 65, 587, 588, 589, 589*b*
  of *Pseudomonas aeruginosa*, 609
Final electron acceptor, 136, 138*f*, 139, 142*t*
  in cyclic photophosphorylation, 146
  fermentation pathways and, 141, 142*t*
  oxygen as, 138*f*, 139, 163, 164. *See also* Aerobes/aerobic bacteria
Fingerprints. *See* DNA fingerprints/signatures; Ribosomal fingerprints
Firmicutes (phylum), 330*f*, 335, 337*t*, 568, 630*t*
  biofuel production and, 162*b*
First line of defense, 447, 447–451, 466*t*
FISH. *See* Fluorescent *in situ* hybridization
Fish
  anisakids transmitted in, 701, 701*f*, 703*t*
  fermented, 780, 783*t*
  genetically modified, for biomedical animal models, 254
Fission, binary, 179–180, 179*f*, 326, 326*f*. *See also* Population (microbe), growth of
5-day fever, 605
Five-kingdom taxonomic scheme, 113–115
Five prime (5′) end, of nucleic acid, 51, 51*f*, 194, 195*f*
Fixation
  carbon, 806
    in carbon cycle, 806
    in light-independent photosynthesis/Calvin-Benson cycle/carbon cycle, 146, 148*f*, 806
  chemical, 106
  heat, 106
  nitrogen, 165, 333, 807, 807*f*
    by alphaproteobacteria, 330*f*, 339, 339*f*
    by cyanobacteria, 165, 333, 807
  of specimen, 106, 108*f*
Fixed macrophages, 454
Fixed nitrogen. *See* Nitrogen fixation
Flaccid paralysis, botulism toxins causing, 563, 581*b*
Flagella, 89*t*
  of brown algae, 375, 375*f*
  of dinoflagellates, 362, 363*f*
  of diplomonads, 361
  of euglenids, 357, 361, 361*f*
  of eukaryotic cells, 82, 83*f*, 84*f*, 89*t*
  of prokaryotic cells, 60*f*, 89*t*
    archaea, 77–78, 79*t*, 89*t*
    bacteria, 62–65, 63*f*, 64*f*, 79*t*, 89*t*
  of protozoa, 5*f*, 361, 361*f*, 679
  of *Pseudomonas*, 343*f*, 599*b*
  staining of, 110, 111*f*, 111*t*
  of *Trichomonas vaginalis*, 686, 686*f*
  of *Trypanosoma cruzi*, 679, 680*f*
  of water molds, 376
Flagellar (H) antigens, 593, 593*f*
  of *E. coli*, 595
Flagellar stain, 110, 111*f*, 111*t*
Flagellates, 679–686, 692*t*
Flagellin, 62
Flagyl. *See* Metronidazole
Flash pasteurization, 271, 271*t*
Flat warts, 723*b*
Flavin adenine dinucleotide (FAD), 52, 126, 128, 128*f*, 153*t*
  in citric acid/Krebs cycle, 136, 137*f*, 153*t*
  in electron transport chain, 136, 138*f*, 153*t*

Flavin mononucleotide (FMN), in electron transport chain, 136, 138*f*
Flavirus genus/spp., 392*t*, 419*t*
Flaviviridae (flaviviruses), 392*t*, 735, 739*t*, 741, 747*t*, 771*t*
  encephalitis caused by, 392*t*, 742, 742*f*, 743*b*, 747*t*
  as biological weapons, 743*b*, 814*t*
  yellow fever caused by (*Flavivirus yellow fever virus*), 15*t*, 392*t*, 744, 747*t*, 771*t*
  as biological weapon, 814*t*
  Zika fever caused by (*Flavivirus Zika virus*), 744–745, 746*b*, 747*t*
Flavoproteins, in electron transport chains, 136
Fleas, as disease vectors, 378, 378*f*, 433*t*, 677
  for cat scratch disease (*Bartonella henselae*), 605
  for murine (endemic) typhus (*Rickettsia typhi*), 622, 626*f*
  for plague (*Yersinia pestis*), 602, 603*f*
Fleming, Alexander, 18*f*, 21, 238, 289
"Flesh-eating disease" (necrotizing fasciitis), 118*b*, 336, 554–555*b*, 556
Flies, as disease vectors, 378–379, 378*f*, 433*t*, 677
  for African trypanosomiasis/sleeping sickness (*Trypanosoma brucei*), 379, 680, 681, 681*f*, 692*t*
  for bartonellosis, 605
  for leishmaniasis, 378, 682, 683*f*, 692*t*
Flocs/flocculation
  in drinking water treatment, 791, 791*f*
  in wastewater treatment, 341, 342*f*, 793*f*, 794
Flora, normal. *See* Microbiome
Flow cytometry, for microbial growth estimation, 184
Flu. *See* Influenza
Fluconazole, 315*t*, 650
  mechanism of action of, 294, 315*t*, 650
Fluid mosaic model, of membrane structure, 69, 80
Flukes (trematodes), 693, 695–698, 695*f*, 697*b*, 697*f*, 703*t*
  blood (*Schistosoma* species), 696–698, 697*f*, 698*f*, 699*b*, 703*t*
  life cycle of, 696, 697*f*
  liver (*Fasciola* species), 419*t*, 696, 697*f*, 703*t*
Fluorescein isothiocyanate, 103
Fluorescence, 100
Fluorescence microscopy/microscopes, 100, 102–103, 103*f*, 107*t*
Fluorescent antibodies/antigens, 103, 103*f*
  for confocal microscopy, 103
  for fluorescence microscopy, 103, 103*f*
  for serology, 515–516, 516*f*. *See also* Fluorescent immunoassays
Fluorescent dyes/stains, 84*f*, 103, 103*f*, 515
  for bacteriophages, 113*b*
  for confocal microscopy, 103
  for fluorescence microscopy, 103, 103*f*
Fluorescent immunoassays, 515–516, 516*f*, 520*t*
  in chlamydial infection, 628, 628*f*
  in fungal infection, 649, 649*f*
Fluorescent *in situ* hybridization (FISH), 249–250, 250*f*
Fluorescent phages, 113*b*
Fluorescent wavelengths, 102
Fluoride, antimicrobial action of/tooth decay prevention and, 278
Fluorine, 30*f*
  antimicrobial action of, 278, 281*t*
5-Fluorocytosine, 315*t*, 650
Fluoroquinolones, 313*t*, 318*t*
  diarrhea associated with use of, 319*b*
  mechanism of action of, 291*f*, 296, 313*t*, 318*t*
  resistance to, 306
Flu shot/vaccine (influenza immunization), 508*t*, 509*t*, 769*b*
FMD. *See* Foot-and-mouth disease
FMN (flavin mononucleotide), in electron transport chain, 136, 138*f*
FMT. *See* Fecal microbiome transfer
Focal infection, 434*t*
Focal point, light refraction/magnification and, 98, 98*f*
Folic acid
  antimicrobial drug action and, 132, 296*f*
  in nucleotide biosynthesis, 151, 152*f*, 296*f*

Foliose lichens, 372, 372*f*
Folliculitis
  pseudomonal, 778*b*, 796*b*
  staphylococcal, 549
Fomites (fomes), 431
  adenoviruses transmitted by, 724
  dermatophytes transmitted by, 648
  noroviruses transmitted by, 740*b*
  papillomaviruses transmitted by, 723*b*
  respiratory syncytial virus transmitted by, 761
  rhinoviruses/common cold transmitted by, 736
  staphylococci transmitted by, 548
  streptococci transmitted by, 553
Fomivirsen, 314*t*
  mechanism of action of, 293, 314*t*
Fonsecaea genus/spp./*Fonsecaea pedrosoi*, chromoblastomycosis caused by, 666*f*, 666*t*
Food. *See also* Foodborne illnesses; Food microbiology
  allergic reactions/anaphylactic shock and, 529
  genetically modified, 254–256, 255*f*
    ethical/safety issues and, 256
  microbes in production of, 10*b*, 12*t*, 18*f*, 19*t*, 779–782, 780*f*, 781*f*, 782*f*, 783*t*
  as nonliving reservoir, 419
  nutritional value of, recombinant DNA technology affecting, 255–256
    ethical/safety issues and, 256
  processing of
    endospore contamination and, 76, 325
    spoilage prevention and, 784, 784–785, 784*f*
  safety of
    bioterrorism and (agroterrorism), 812, 813, 813–814, 814, 814*t*, 815. *See also* Agroterrorism
    genetic modification/recombinant DNA technology and, 256, 788–789
  spoilage of. *See* Food spoilage
  storage of
    refrigeration/freezing and, 271–272, 275*t*
    spoilage prevention and, 784, 785
  for vaccine delivery, 253
Food additives, microbial production of, 787, 789*t*
Foodborne illnesses, 262*b*, 282*b*, 353*b*, 379*b*, 431–432, 431*f*, 433*t*, 440, 786, 786*t*. *See also specific type* and Food microbiology
  amebiasis, 678
  bioterrorism and, 814*t*
  Brucella causing, 606
  Campylobacter causing, 282*b*, 639, 786*t*
  clostridial/botulism, 546*b*, 563–564, 567, 568, 581*b*, 785
  Cyclospora causing, 691, 692*f*
  Escherichia coli causing, 15*t*, 428, 595, 596, 604*f*, 786*t*
  Fasciola (liver flukes) causing, 696, 703*t*
  infections, 786
  intoxications, 353*b*, 379*b*, 785, 786
  Listeria causing, 272, 336, 568, 570–571*b*, 785, 786*t*
  mycotoxins causing, 668
  nematodes (roundworms) causing, 699
  Norovirus causing, 1*b*, 22*b*, 740*b*
  pneumonia and, 414*b*
  prions causing, 21*b*, 407
  public health and, 440
  raw/unpasteurized dairy products causing, 262*b*, 282*b*
  Salmonella causing (salmonellosis/typhoid fever), 15*t*, 600, 600*f*, 786*t*
  Shigella causing (shigellosis), 601–602, 601*f*, 604*f*, 786*t*
  staphylococcal, 549, 786*t*
    incubation period for, 429*t*
  tapeworm (cestode) infestation, 694, 694*f*, 695*f*, 703*t*
  Toxoplasma causing (toxoplasmosis), 687, 687*f*, 690, 786, 786*t*
  traveler's diarrhea, 15*t*
  Vibrio causing/cholera, 40*b*, 620*b*, 638, 641*b*, 786*t*
  Yersinia causing, 602, 786*t*

SUBJECT INDEX

Foodborne transmission of disease, 431–432, 431f, 433t. *See also* Foodborne illnesses
   public health and, 440
Food infections, 786. *See also* Foodborne illnesses
Food intoxications, 353b, 379b, 785, 786. *See also* Foodborne illnesses
Food microbiology, 10b, 12t, 18f, 19t, 779–786, 783t, 786t
   decimal reduction time and, 269, 269f
   endospore formation and, 76, 325
   foodborne illnesses and, 431–432, 431f, 433t, 786, 786t
   food production and, 10b, 12t, 18f, 19t, 779–782, 780f, 781f, 782f, 783t
   food spoilage and, 779, 782–785, 784f, 784t. *See also* Food spoilage
   pH tolerance and, 783
   public health and, 440
Food poisoning (infections/intoxications), 262b, 282b, 353b, 379b, 431–432, 431f, 433t, 440, 785, 786, 786t. *See also* Foodborne illnesses
Food preservatives, 785
Food spoilage, 779, 782–785, 784f, 784t
   causes of, 782–784, 784t
   extrinsic factors in, 782, 784
   illnesses associated with, 431–432, 431f, 433t, 440, 786, 786t. *See also* Foodborne illnesses
   intrinsic factors in, 782, 783
   potential for, food classified by, 784
   prevention of, 784–785, 784f. *See also specific method*
Food supplements, microbial production of, 787, 789t
Food supply, agroterrorism and, 812, 813, 813–814, 814, 814t, 815. *See also* Agroterrorism
Food vesicle (phagosome), 81, 81f, 87f, 455f, 456
   maturation of/microbial killing and, 455f, 456
   TLRs in membranes of, 457, 457t
Foot-and-mouth disease virus, as biological weapon, 813–814
Foraminifera, 363, 363f, 365t
Forde, Robert, 15t
Formaldehyde, 279
   antimicrobial action of, 279, 281t
   covalent bonds in formation of, 32, 32f
   in vaccine preparation, 506
Formalin, antimicrobial action of, 279
Formed elements, 452, 452f. *See also* Blood cells
N-Formylmethionine, 207, 208f, 211, 212
   in archaea, 329
Fortified foods, spoilage and, 783
Fosamprenavir, 314t
Four-color reversible termination sequencing, 250, 251f
Fox, George, 20, 115
F (fertility) plasmids, 196, 226, 227f, 228f
Fracastoro, Girolamo, 12
Fraenkel, Albert, 15t
Fragment(s)
   complement, 459f, 460–462, 461f
   in inflammation, 463, 463f
   Okazaki, 201, 202
Fragment, antigen-binding (F_{ab}) region, 485, 485f
Fragment (crystallizable) (F_c) region, 485, 485f
Fragmentation
   in algal reproduction, 373, 374f
   in prokaryotic reproduction, 326
Frameshift mutagens, 219, 219f
Frameshift mutations, 216, 217, 217f, 218f, 219, 219f
   chemicals causing, 219, 219f
   deletions, 216, 217f, 218f, 219, 219f
   insertions, 216, 217f, 218f, 219, 219f
   transpositions, 216, 228–230, 229f
Francisella genus/spp./*Francisella tularensis*, 426f, 611, 612b
   as biological weapon, 611, 814t
Free-living amoebas, 364, 365t, 677, 678–679
Free-living nitrogen fixers, 807–808
Freeze-drying (lyophilization)
   for culture preservation, 179
   in microbial control/food preservation, 272, 275t, 785

Freeze resistance, recombinant DNA technology and, 255, 789
Freezing
   in culture preservation, 179
   in food preservation/microbial control, 271–272, 275t
Freshwater ecosystems, 811, 811–812, 811f
Fructose, 44, 45f
Fructose 1,6-bisphosphate
   in carbohydrate biosynthesis/gluconeogenesis, 150f, 153f
   lysis of, 129, 130f
   in glycolysis, 133, 134, 134f, 153f
Fructose 1,6-bisphosphate aldolase, 128t, 130f
Fructose 6-phosphate, 149
   in carbohydrate biosynthesis/gluconeogenesis, 150f
   in glycolysis, 134f, 149t
Fruit bats. *See* Bat(s)
Fruiting bodies
   of ascomycetes (ascocarps), 369, 369f, 371b
   of basidiomycotes (basidiocarps), 370, 371f
   of molds, 366
   of myxobacteria, 346, 346f
Frustules, 375
Fruticose lichens, 372, 372f
Fuels, alternative (biofuels), 162b, 788, 789t
Fumaric acid, in citric acid/Krebs cycle, 137f
Functional genomics, 251–252
Functional groups, 40, 41f
Fungal allergens, 528, 668, 669
Fungal spores, 4, 367, 368f, 648
Fungal toxins, 669
Fungal viruses, 388
Fungemia, *Fusarium* causing, 664
Fungi, 4, 5f, 364–373, 373t, 647–674. *See also specific organism*
   antimicrobials effective against, 299f, 650. *See also* Antifungal drugs
   antimicrobials produced by, 290f, 365
   beneficial/industrial uses of, 169b, 365
   as biological threat, 814
   cell walls of, 4, 80, 292, 364
      antimicrobial drugs affecting, 292
   classification of, 4, 5f, 114f, 357, 359f, 368–371, 373t, 648
   culture of, 175, 176f, 649
   cytokinesis in, 357, 358f
   diseases/infections caused by, 365, 419t, 647–674. *See also* Mycoses
      categories of agents causing (pathogenic/opportunistic), 648–649, 649t
      clinical manifestations of, 649
      cutaneous, 651, 666–668, 666f, 666t, 667f, 668f
      diagnosis of, 649–650, 649f
      intoxications/allergies, 528, 649, 668, 669, 669f
      prevention of, 650
      soilborne diseases, 810, 810t
      superficial, 664–665, 665f, 665t, 666t
      systemic, 648, 651–656, 651f, 656–664
      treatment of, 650. *See also* Antifungal drugs
   drug resistance and, 650
   hypertonic environment tolerated by, 273
   imperfect, 371
   in lichens, 371, 372f
   microbial genetics research and, 365
   morphology of, 365–366, 366f
   nutrition of, 366–367, 367f
   opportunistic, 648–649, 649t, 656–664
      in HIV infection/AIDS, 664, 751t
   identification of, 649–660
   treatment of infections caused by, 650
   pathogenic, 648, 651–656, 651f. *See also* Fungi, diseases/infections caused by
   pH range tolerated by, 39
   radioactive waste remediation and, 169b
   reproduction of, 367, 368f
   significance of, 365
   in soil, 810
      disease caused by, 810, 810t
   susceptibility of, 266f
   vaccines against, 650
      recombinant DNA in development of, 507
   viroidlike agents affecting, 405
Fungi (kingdom), 113, 114f, 357, 359f, 364, 368
Fungicides, 264. *See also* Antifungal drugs

Furovirus soil-borne wheat mosaic virus, 810t
Furuncle (boil), 182b, 464, 549
Fusarium genus/spp./*Fusarium oxysporum,* 664
   soilborne transmission of, 810t
Fusin (CXCR4), as HIV receptor, 753, 754f
   in long-term nonprogressors, 756
Fusobacteria (phylum), 330f

G. *See* Guanine
G + C content/percentage/ratio, 118–119, 547
   high, 330f, 336–338, 547
   low, 330f, 334–336, 337t, 547
   taxonomy/microbial classification and, 118–119, 330f, 547
G3P. *See* Glyceraldehyde 3-phosphate
Galen, Claudius, 588
Galilei, Galileo, 100
Gallo, Robert, 750
Galls, 340, 341f
Gametes/gametocytes, 59, 354
   algal, 373, 374f
   protozoan, 360
      in *Cyclospora* life cycle, 692f
      in *Plasmodium* life cycle, 688b, 689b
      in *Toxoplasma* life cycle, 685, 690
Gamma F_c region/heavy chains, 485, 488t
Gamma globulins. *See* Immunoglobulin G
Gamma-hemolysis, by enterococci, 560, 561t
Gamma (γ)-interferon (macrophage activation factor), 458, 459, 459t, 489, 489t
Gammaproteobacteria, 330f, 334t, 341–343, 344f, 591, 605, 608, 637
Gamma radiation, 98f
   Kineococcus radiotolerans resistance to, 169b
   for microbial control/food preservation, 274, 274f, 275t, 785
   mutagenic effects of, 217
Ganciclovir, 297f, 314t
Ganglia/ganglion (neuronal)
   latent herpesviruses in, 714, 714f, 715f, 716f
   latent varicella-zoster virus in, 717f
Gangrene (gas), 15t, 567, 568
Garlic, allicin in, food preservation and, 785
Gartner, A. A., 15t
GAS. *See* Group A streptococcus
Gas(es), antimicrobial action of, 280, 281t
Gas gangrene, 15t, 567, 568
Gasohol, 788
Gas sterilization, 280, 281t
Gastric acid. *See* Stomach acid
Gastric ulcer, 167. *See also* Peptic ulcers
Gastritis, *Helicobacter pylori* infection and, 640, 640b
Gastroenteritis. *See also specific cause and* Diarrhea; Foodborne illnesses; Waterborne illnesses
   astroviruses causing, 392t, 739–740, 739f, 771t
   caliciviruses causing, 739–740, 739f, 740b
   Campylobacter diarrhea, 282b, 639, 786t
   cholera, 620b, 638, 641b
   cryptosporidiosis, 690, 692t
   Escherichia coli causing, 15t, 428, 595, 596, 604f, 786t
   giardiasis, 684b, 692t
   Norovirus causing, 1b, 22b, 392t, 740, 740b, 771t, 790f
   rotaviruses causing, 392t, 770, 771f, 771t
   Salmonella causing (salmonellosis/typhoid fever), 15t, 600, 600f, 604f, 786t
   Shigella causing (shigellosis), 601–602, 601f, 604f, 786t
   traveler's diarrhea, 15t
   vibrios causing, 638, 639, 786t
   waterborne transmission of, 740b, 790f
Gastroferritin, in host defense, 451t
Gastrointestinal anthrax, 561. *See also* Anthrax
Gastrointestinal candidiasis, 658t, 660b
Gastrointestinal pathogen panel, 685b
Gastrointestinal (digestive) tract
   in host defense, 450, 451t
   microbial diseases of. *See also* Foodborne illnesses; Gastroenteritis
      Enterobacteriaceae causing, 604f
   microbiome/normal microbiota of, 416, 417t

host defenses and, 450
transfer of, 803b
as portal of entry, 420, 420f, 421
Gastrointestinal zygomycosis, 662
Gas vesicles, 74
Gated channels/ports, in active transport, 73, 73f
GDP, in citric acid/Krebs cycle, 137f
Gel electrophoresis, 246, 247f, 252t. *See also* Electrophoresis
   in immunoblot/western blot, 518, 518f
   in Southern blot, 246
Gelidium genus/spp., 374, 376t
   cell wall of, 80f
gen-/-gen (prefix or suffix), 423t
Gene(s), 19–20. *See also* Genetics
   autoimmunity and, 538
   expression of. *See* Gene expression
   function of, 202–216
      determining (functional genomics), 251–252
   insertion of into cells, 248, 249t, 252t
      agricultural applications of, 254–256, 255f, 788–789, 789t
      gene therapy and, 20, 254
      vaccine production and, 253, 506–508, 507f, 788t
      xenotransplantation and, 254
   locating, 249–250, 250f
   mapping, 248–252, 250f, 251f, 252t
   mutations of, 216–223. *See also specific type and* Mutations
   overexpression of, 252
   regulatory, 214, 214f, 215, 215f
Gene activation, in cholera, 638
Gene expression. *See also* Transcription; Translation
   control of, 212–216, 213f, 214f, 215f, 215t
   metabolic regulation and, 152
   DNA microarrays in monitoring of, 247
   in functional genomics, 252
   methylation affecting, 201
   vectors ensuring, 241
Gene gun, for DNA insertion into cells, 248, 249f, 252t
Gene knockout, 251–252
Gene libraries, 243, 244f, 252t
Gene overexpression, in functional genomics, 252
Genera. *See* Genus
Generalized transduction, 225
Generation time, 180
Gene sequencing. *See* Genomics
Gene silencing, 216
Gene therapy, 20, 238, 254
   ethical/safety issues and, 256–257
Genetically modified food, 254–256, 255f
   safety of, 256, 788–789
Genetically modified organism (GMO), 238, 254–256, 255f
   in biofuel production, 162b, 788
   ethical/safety issues and, 162b, 256, 788–789
   industrial use and, 162b, 238
Genetic code, 207–208, 208f
   redundancy in, 208, 209, 240
   synthetic nucleic acids in elucidation of, 239
Genetic engineering/recombinant DNA technology, 18f, 19t, 20, 237–261, 239f. *See also* Recombinant DNA technology
Genetic fingerprints/signatures, 252
   bioterrorism defense and, 815
   prokaryote classification and, 115, 329
   in water quality testing, 792
Genetic information, transfer of, 203–212, 203f, 213t. *See also* Transcription; Translation
Genetic mapping, 248–252, 250f, 251f, 252t
Genetic markers, in vectors, 241
Genetic recombination/transfer, 223–230, 224f, 230f. *See also* Recombinant DNA technology
   horizontal (lateral) gene transfer and, 223–227, 225f, 226f, 227f, 228f, 228t, 230b
   mutations caused by, 217
   transposons/transposition and, 228–230, 229f
Genetics, 18f, 19–20, 19t, 194

central dogma of, 203, 203f
retroviruses and, 750
microbial, 18f, 19–20, 19t, 193–236. *See also specific aspect*
  DNA replication and, 151, 152f, 198–202, 198f, 199f, 200f, 202f
  eukaryotic genomes and, 196–198, 197f, 198t
  gene expression regulation and, 212–216, 213f, 214f, 215f, 215t
  gene function and, 202–216
    determining (functional genomics), 251–252
  genome structure and replication and, 194–202
  genotype/phenotype relationship and, 202–203
  information transfer and, 203–212, 203f, 213t. *See also* Transcription; Translation
  mutations and, 216–223
  nucleic acid structure and, 50–53, 51f, 194, 195f
  prokaryotic genomes and, 194–195, 196f, 198t
  recombination and transfer and, 223–230, 224f, 230f
Genetic screening, 253
  ethical issues and, 256–257
Gene transfer
  horizontal (lateral), 223–227, 225f, 226f, 227f, 228f, 228t, 230b
    antimicrobial resistance and, 227, 229f, 230, 230b, 303
    bacterial conjugation, 226–227, 227f, 228f, 228t
    S. aureus strains and, 336
    transduction, 225, 226f, 228t
    transformation, 224–225, 225f, 228t
  vertical, 224
Genital herpes/ulcers, 392t, 714, 714f, 715f, 716t, 729b
  incubation period for, 429t
Genital warts, 723t
Genome(s), 194–202, 198t. *See also specific organism*
  DNA replication and, 151, 152f, 198–202, 198f, 199f, 200f, 202f
  eukaryotic, 196–198, 197f, 198t
  manipulation of. *See* Recombinant DNA technology
  nucleic acid structure and, 50–53, 51f, 194, 195f
  prokaryotic (bacterial and archaeal), 194–195, 196f, 198t
  sequencing. *See* Genomics
  structure/replication of, 194–202
  viral, 386, 387f, 391t
    segmented, 735, 763
Genomics (genome/nucleic acid sequencing), 20, 250–252, 251f, 252t
  antimicrobial drug development and, 308
  ethical and safety issues and, 256–257
  functional, 251–252
  metagenomic, 20, 118–119, 252, 347
  microbial classification/evolution and, 20, 115, 118–119, 329, 330f, 357, 359f, 360. *See also* Evolution
  in microbial growth estimation, 187
  synthetic nucleic acid probes in, 240
Genotype, 202, 203
Gentamicin, 310t
  mechanism of action of, 293, 310t
  microbial production of, 290t
Genus (genera), 113, 114f
  binomial nomenclature and, 113
Geogemma genus/spp./*Geogemma barossii*, 166, 330
Germ(s)/germ theory of disease, 12, 13, 20–21, 424
German measles. *See* Rubella
Germicides, 264
  effectiveness of, 267
Germistatic chemicals, in food preservation, 785
Giardia genus/spp./*Giardia intestinalis (lamblia)*, 361, 365t, 684–685b, 692t
  indirect immunofluorescence immunoassay in detection of, 516f

size of, 61f
waterborne transmission of, 684b, 685b, 686, 790t
Giardiasis, 684–685b, 686, 692t
  waterborne transmission of, 684b, 685b, 686, 790t
Gingivitis, gonococcal, 589
Gingivostomatitis, herpetic, 714, 716t
Glanders, bioterrorism and, 814t
Glass lenses. *See* Lenses
Gliding
  cyanobacterial movement by, 333, 334t
  Cytophaga movement by, 347
Global warming
  carbon dioxide and methane and, 332, 342, 807
  viruses affecting, 390b
Globigerina genus/spp., 363f
Glomeruli. *See* Glomerulus
Glomerulonephritis
  immune complexes causing, 533, 556
  after streptococcal infection, 539, 556
Glomerulus (glomeruli)
  immune complex–mediated (type III) hypersensitivity affecting, 533
  streptococcal infection affecting, 539, 556
Glossina (tsetse) flies, as disease vectors, 378–379, 378f, 433f, 680, 681f
  for African trypanosomiasis/sleeping sickness (*Trypanosoma brucei*), 379, 680, 681, 681f, 692t
Glove box, anaerobic, 177
Glucan
  in fungal cell walls, 292
    echinocandin mechanism of action and, 292
  in tooth decay, 170
Glucocorticoids (corticosteroids)
  for allergic contact dermatitis, 535
  for asthma, 529
  immunosuppressive actions of, 537, 538f, 540–541
Glucomannan, in fungal cell walls, 80
Gluconeogenesis, 149, 150f, 153f
Gluconic acid, microbial production of, 787
Gluconobacter genus/spp., 340
  in vinegar production, 340, 782, 783t
Glucose, 44, 45f
  antimicrobials affecting, 295
  in Calvin-Benson cycle, 146, 148f
  in carbohydrate biosynthesis/gluconeogenesis, 149, 150f, 153f
  catabolism of, 133–140, 133f, 134f, 135f, 137f, 139f, 140t, 142t, 153f
    cellular respiration in, 133, 133f, 135–140, 135f, 137f, 138f, 142t. *See also* Cellular respiration
    Entner-Doudoroff pathway in, 140
    fermentation in, 133, 133f, 140–141, 141f, 142f, 142t
    glycolysis in, 133, 133–135, 133f, 134f, 135f, 153f
    pentose phosphate pathway in, 140, 153f
  conversion of to biofuel, 162b
  phosphorylation of
    as exchange reaction, 37
    group translocation and, 74
Glucose 6-phosphate, 149f, 150f, 153f
  in Calvin-Benson cycle, 148f
  in carbohydrate biosynthesis/gluconeogenesis, 150f, 153f
  in glycolysis, 134f, 149f, 153f
  group translocation and, 74
  in pentose phosphate pathway, 152f, 153f
Glucose-6-phosphate dehydrogenase deficiency, malaria resistance and, 687
Glue, *Caulobacter*, 341b
Glutamine/glutamic acid, 47f
  in nucleotide biosynthesis, 151, 152f
  in transamination, 151f
Glutaraldehyde, 279
  antimicrobial action of, 279, 281t
Glyceraldehyde 3-phosphate (G3P), 130f, 149t, 153f
  in Calvin-Benson cycle, 146, 148f
  in carbohydrate biosynthesis/gluconeogenesis, 149, 150f, 153f
  in glycolysis, 134, 134f, 149f, 153f
  in lipid biosynthesis, 150, 151f, 153f

Glycerol, 153f
  in carbohydrate biosynthesis, 150f, 153f
  in lipid biosynthesis, 41, 42f, 150, 151f, 153f
  in lipid catabolism, 143, 143f, 153f
  in phospholipids, 42, 43f
Glycine, 47, 47f
Glycocalyx (glycocalyces), 89t. *See also* Capsule
  of eukaryotic cells, 80, 89t
  of prokaryotic cells, 60f, 79f, 89t
    archaea, 77, 79f, 89t
    bacteria, 62, 62f, 79f, 89t
Glycogen, 44, 45, 46f, 149, 150f
  in algae, 374, 376t
Glycolipids, 46
  Borrelia, 634
Glycolysis, 133–135, 133f, 134f, 135f, 149t, 153f
  lipid biosynthesis and, 150, 151f
Glycolytic facultative anaerobes, 342, 343t, 344t
Glycoproteins, 49
  in adhesion (infection), 422, 422f
  phagocytosis and, 455, 455f
  in algal cell walls, 376t
  Ebolavirus, 765b
  HIV, 752, 753, 753f, 754b, 754f
  in host defense, 449
  influenzavirus, 768b
  rabies virus, 761
  rotavirus, 770, 770f
  in viral envelope, 389, 391f
Glyphosate, tolerance to, recombinant DNA technology and, 254–255
GMO. *See* Genetically modified organism
GMS stain. *See* Gomori methenamine silver (GMS) stain
Goat pox (orf), 710
Goblet cells, in host defense, 449
Goiter, 539
Gold, 28
Golden Age of Microbiology, 7–18, 15t, 18f
Golden algae, 375, 376t
Golgi body/complex/apparatus, 61f, 85t, 86–87, 86f, 87f
Golgi, Camillo, 86
Gomori methenamine silver (GMS) stain, 110, 649, 649f
  in cryptococcosis, 662, 662f
Gonococcus. *See* Gonorrhea; *Neisseria gonorrhoeae*
Gonorrhea, 344t, 588–589, 588f, 589b, 594b. *See also Neisseria gonorrhoeae*
  drug-resistant, 589
  indirect fluorescent immunoassay in diagnosis of, 516
  pelvic inflammatory disease and, 589, 589b
Gonyaulax genus/spp., 362–363, 363f, 365t
  intoxications caused by, 362–363, 790, 790t
gp41, 752, 753, 753f, 754b, 754f
gp120, 752, 753, 753f, 754b, 754f
Grafts/graft rejection, 536, 536f
  donor-recipient matching/tissue typing and, 536–537
  immunosuppressive drugs and, 537–538, 538f
  MHC antigens and, 478, 536, 536–537
Graft-versus-host disease, 536
Gram, Hans Christian, 15, 18f, 108
Gramicidin, 311t
  mechanism of action of, 293, 311t
Gram-negative archaea, cell walls of, 78
Gram-negative bacteria, 15, 15f, 67, 330f, 338–347, 344t, 586–619. *See also specific agent*
  antimicrobials effective against, 299f
  bacilli, 330f
    aerobic, 605–614
    anaerobic, 614, 614f
    facultatively anaerobic, 591–605
  Bdellovibrio/bacterial pathogens of, 343, 344t, 345f
  cell walls of, 67, 67–68, 68f
  cocci, 330f, 587–591
  endotoxins released by, 427f, 428, 428t
  flagellar structure and, 63f
  healthcare-associated (nosocomial) infection caused by, 591, 591f
  proteobacteria, 330f, 333, 334, 334t, 338–347, 344t

staining, 15, 15f, 67, 108–109, 109f, 111t, 592f
  susceptibility of, 266f
  vibrios, 637–641
Gram-positive archaea, cell walls of, 78
Gram-positive bacteria, 15, 15f, 67, 330f, 337t, 546–585. *See also specific agent*
  antimicrobials effective against, 299f
  cell walls of, 67, 68f
  flagellar structure and, 63f
  high G + C, 330f, 336–338, 337t, 547
  low G + C, 330f, 334–336, 337t, 547
  staining, 15, 15f, 67, 108–109, 109f, 111t, 547
  susceptibility of, 266f
Gram stain, 15, 15f, 67, 108–109, 109f, 111t
Grana (granum), chloroplast, 88f, 144
Granulocytes, 452f, 453, 453f
Granulomatous disease, chronic, 540, 540t
Granzyme/perforin-granzyme cytotoxic pathway, 486, 486f, 491, 491f
Graves' disease, autoimmunity and, 539
Gravid proglottids, tapeworm, 693f, 694, 694f
Green algae, 374, 376t
  in lichens, 371, 374
Greenhouse gas
  carbon dioxide as, 332, 807
  methane as, 332, 342, 807
Green phototrophic bacteria, 333–334
  nonsulfur, 330f, 333, 334, 334t
  sulfur, 330f, 333, 334, 334t
Gregg, Norman, 747
Griffith, Frederick, 224
Griseofulvin, 315t, 650
  microbial production of, 290t, 315t
Gross mutations, 216
Ground itch, 700
Ground meat, spoilage and, 783
Group A streptococcus (*Streptococcus pyogenes*), 552–556, 553f, 554–555b, 561t
  autoimmunity and, 539
  diseases caused by, 553–556, 553f, 554–555b, 561t
  glomerulonephritis and, 539, 556
  immunochromatographic assays in identification of, 519–520, 519f
  necrotizing fasciitis/"flesh-eating disease" caused by, 118b, 336, 554–555b, 556
  pharyngitis caused by, 70b, 553, 553f, 736t
  rheumatic fever and, 553
  resistant strains of, 305
  rheumatic fever caused by, 553
  scarlet fever caused by, 553
  skin infections caused by, 124b, 154b, 553, 553f
  toxic shock syndrome (STSS) caused by, 556
  virulence factors of, 429, 552, 558, 558b
Group B streptococcus (*Streptococcus agalactiae*), 556–557, 561t
Group C streptococcus, 557, 561t
Group D streptococcus, 560, 561t
Group F streptococcus, 557, 561t
Group G streptococcus, 557, 561t
Group translocation, 73, 73–74, 73f, 74t
Growth
  as characteristic of life, 59, 59t
  microbial, 59, 59t, 161–192. *See also* Microbial growth
Growth curve, 180–181, 180f, 181f
Growth factors, 165, 165t, 489
  fastidious organisms for identification of, 175
Growth hormone
  bovine (BGH), 255
  human, microbial production of, 12t, 788t
GTP, in citric acid/Krebs cycle, 136, 137f
Guanine (G), 50, 50f, 51, 51f, 52, 52f, 194, 195f. *See also* G + C content/percentage/ratio
Guanosine, 297f
Guanosine triphosphate (GTP), in citric acid/Krebs cycle, 136, 137f
Guanosine triphosphate deoxyribonucleotide (dGTP). *See also* Triphosphate deoxyribonucleotides
  in DNA replication, 198, 199f

SUBJECT INDEX

Guilds, 804, 804f
Gummas, syphilitic, 631b, 632, 632f
Gymnodinium genus/spp., 362–363, 365t
Gyrase/DNA gyrase, 201
   quinolones/fluoroquinolones affecting, 296
   resistance and, 306
Gyromitra esculenta, 668

H. *See* Hydrogen
H1N1 influenzavirus, 435
H$_2$S. *See* Hydrogen sulfide
H5N1 influenzavirus, 768b
H7N9 influenzavirus, 768b
HA. *See* Hemagglutinin
HAART (highly active antiretroviral therapy). *See* Antiretroviral therapy
Habitats, microbial, 803, 804, 804f. *See also* Environmental microbiology
   aquatic, 811–812, 811f
   human, 804, 804f
   soil, 804, 804f
*Hae*III restriction enzyme, 240t
Haemophilus genus/spp. (*Haemophilus ducreyi/parainfluenzae*), 343t, 344t, 604–605, 605f. *See also Haemophilus influenzae*
   restriction enzymes produced by, 240t
Haemophilus influenzae, 113, 604–605
   immunization against, 508f, 509t, 604, 605
   infection in flu victims and, 424, 604, 769b
   meningitis caused by, 424, 509t, 604, 605
   restriction enzymes produced by, 240, 240t
   genetic mapping and, 249
   strain *aegyptius*, 240t, 605
Haemophilus influenzae type b vaccine, 508f, 509t, 604, 605
Hafnia genus/spp., 592f, 596
HAI. *See* Healthcare-associated (nosocomial) infections/diseases
Hair, dermatophytosis/ringworm involving, 665
Hairy-cell leukemia, HTLV-2 causing, 750–751, 751f
Hairy leukoplakia
   HHV-4 (Epstein-Barr virus) and, 719, 719f, 720
   in HIV infection/AIDS, 541, 719, 720
Halobacterium genus/spp./*Halobacterium salinarium*, 331–332, 812
Halogens, antimicrobial action of, 277–278, 277f, 281t
Halophiles, 168, 330f, 331–332, 331f
Haloquadratum walsbyi, 79f
Hami (hamus), 77, 78, 78f, 79t, 89t
Hand-foot-and-mouth disease, 738, 738f
Handwashing, in infection prevention/control, 16, 440
Hansen, Gerhard, 574
Hansen's disease (leprosy), 337, 425, 574–575, 575f
   immunization against (BCG vaccine), 509t, 575
   incubation period for, 429t
Hantavirus genus/spp., 392t, 419t, 767, 771t
   as biological weapon, 814t
   reverse transcriptase polymerase chain reaction in identification of, 767
   transmission of, 384b, 408b, 431, 767, 810, 810t
Hantavirus pulmonary syndrome, 384b, 408b, 419t, 767, 771t
   bioterrorism and, 814t
   transmission of, 384b, 408b, 767, 810, 810t
H antigens, 593, 593f
   of *E. coli*, 595
Haploid cell(s), 195, 354
Haploid generation, in algal reproduction, 373, 374, 374f
Haploid nucleus, 354, 355, 356, 356f
   in fungal reproduction, 367, 368f
Haustoria, 366
   lichens and, 372
HAV. *See* Hepatitis A virus
Hay fever, 528
HBV. *See* Hepatitis B virus
HCV. *See* Hepatitis C virus
HDCV. *See* Human diploid cell vaccine
HDV. *See* Hepatitis D virus
He. *See* Helium

Healthcare-associated (nosocomial) infections/diseases (HAI), 16, 193b, 230b, 424t, 437–440, 439f, 591f
   biofilms causing, 168
   Burkholderia causing, 608
   Clostridium difficile causing, 302, 439, 563
   control of, 16, 438–440
   Enterobacteriaceae causing, 591, 591f
   enterococcal, 230b, 560
   factors affecting, 439, 439f
   horizontal gene transfer and, 230b
   Morganella/Providencia/Edwardsiella causing, 597
   MRSA, 305, 305b, 551
   Pseudomonas aeruginosa causing, 591f, 609, 610, 610f
   respiratory syncytial virus, 760–761, 761f
   staphylococcal, 550, 551
   types of, 439
Healthcare-associated MRSA, 305, 305b, 551. *See also* Methicillin-resistant *Staphylococcus aureus*
Heart disease
   candidal, 658t, 660b
   Chlamydophila pneumoniae and, 437, 629
   streptococcal infection and, 553
   Trypanosoma-induced, 679
Heart valves
   endocarditis and, 550
   rheumatic fever and, 553
Heat, localized, in inflammation, 462
Heat fixation, 106
Heat-related microbial control methods, 268–271, 269f, 270f, 271f, 271t. *See also specific type*
   in food preservation. *See also* Pasteurization
Heat shock, for DNA insertion into cells, 248
Heavy chains, immunoglobulin/B cell receptor, 483, 483f, 484, 484f, 485f
Heavy metals/heavy metal ions, antimicrobial action of, 279, 279f, 281t, 295, 317t
HeLa cell culture, 404
Helical cytoskeleton, 77, 77f
Helical viruses, 389, 389f
Helicases, DNA, 199, 200f, 201, 202
Helicobacter genus/spp./*Helicobacter pylori*, 164, 344t, 346, 426, 640–641, 640b, 640f, 641f
   peptic ulcers and, 39, 53b, 167, 640–641, 640b, 641f
   pH range/stomach acid and, 39, 53b, 167, 640, 641, 641f
Helium, 30, 30f
Helix
   α, 48, 49f
      in cellular PrP, 48, 405, 406f
   double, of DNA, 51f, 52, 194, 195f
      hydrogen bonds and, 35, 51f, 52, 194, 195f
Helminths (parasitic worms), 6, 6f, 377, 693–702, 703t. *See also specific type*
   antimicrobials effective against, 299f, 316t. *See also* Anthelminthic drugs
   diseases caused by, 419t, 693–702, 703t
   eosinophils/eosinophilia and, 456
Helper T lymphocytes (Th cells), 481, 482t
   activation and, 492, 493f
   in antibody immune response, 492, 493f, 494
   B cell activation and, 493f, 494
   in cell-mediated immune response, 490, 490f
   differentiation of, 490, 490f, 492–494, 493f
   in HIV infection/AIDS, 387, 541, 751, 752, 752f, 755, 755f, 756
      attachment and, 752, 753, 753f, 754f
   type 1 (Th1), 481, 482t, 490, 490f
   type 2 (Th2), 481, 482t, 492, 492–494, 493f, 494
   in type I (immediate) hypersensitivity, 526, 527t
Hemagglutination, 514
   viral, 515
Hemagglutination inhibition test, viral, 515, 520t

Hemagglutinin
   filamentous, *Bordetella pertussis* producing, 606–607
   influenzavirus, 768b
      mutations in, 769b
Hematopoiesis, 452, 452f
Hematopoietic (blood) stem cells, 452, 452f, 453, 473
Hematoxylin and eosin stain, 110
Heme
   cytochromes associated with, in electron transport chains, 139
   Haemophilus influenzae requiring, 604
   for microbial growth, 165t
Hemoglobin C, malaria resistance and, 687
Hemoglobin S, malaria resistance and, 687
Hemolysin, 452
   Enterobacteriaceae producing, 593, 593f
   group B streptococci (*Streptococcus agalactiae*) producing, 556
   Staphylococcus aureus producing, 452
Hemolysis
   by streptococci, 176, 176f, 552, 561t. *See also specific type*
   in transfusion reaction, 530–531, 530f
Hemolytic anemia, autoimmune, 539
Hemolytic disease of newborn, 531, 532f
Hemolytic uremic syndrome
   E. coli O157:H7 causing, 595
   as epidemic, 435
Hemorrhagic conjunctivitis, acute, 738
Hemorrhagic fevers, 763
   arenavirus, 767, 771t
   bioterrorism and, 814t
   bunyavirus (Crimean–Congo), 747t, 767
   dengue, 237b, 257b, 743–744, 745f, 747t
   filovirus (Ebola and Marburg), 392t, 763, 764–765b, 771t
Hendra virus encephalitis, bioterrorism and, 814t
Henipavirus Hendra virus, as biological weapon, 814t
Henipavirus Nipah virus, as biological weapon, 814t
HEPA (high-efficiency particulate air) filters, in microbial control, 272–273, 274f
   biosafety and, 268
Hepacivirus genus/spp./*Hepacivirus hepatitis C virus*, 392t, 739t, 748, 771t
Hepadnaviridae (hepadnaviruses), 391, 392t, 725–728, 726f, 727b, 727f, 729t, 739t
   genome of, 725, 725f
hepat- (prefix), 423t
Hepatic cancer, hepatitis B virus and, 728
Hepatic candidiasis, 658t, 660b
Hepatitis, 739t
   safe handling of viruses causing, 268
   type A, 392t, 739, 739t, 771t
      immunization against, 508f, 509t, 739
      waterborne transmission of, 790t
   type B, 392t, 725, 726–728, 726f, 727b, 727f, 739t
      hepatic cancer and, 728
      hepatitis D virus coinfection and, 726, 770
      immunization against, 508f, 509t, 726, 726f, 727–728, 728
      incubation period for, 429t
   type C (non-A/non-B), 739t, 748, 748f, 771t
   type D (delta agent), 739t, 770, 771t
      hepatitis B coinfection and, 726, 770
   type E (enteric), 739t, 740, 771t
Hepatitis A virus, 392t, 735, 739, 739t, 771t
   waterborne transmission of, 790t
Hepatitis A virus vaccine, 508f, 509t, 739
Hepatitis B virus, 392t, 725, 726–728, 726f, 727b, 727f, 729t, 739t. *See also* Hepatitis, type B
   hepatic cancer and, 728
   replication of, 398, 725–726
Hepatitis B virus vaccine, 508f, 509t, 726, 726f, 727–728, 728
   recombinant DNA in production of, 253, 507, 788t
Hepatitis C virus, 392t, 739t, 748, 771t
Hepatitis D virus (delta agent), 739t, 770, 771t
   hepatitis B virus coinfection and, 726, 770
Hepatitis E virus, 392t, 739t, 740, 771t

Hepatitis G virus, genetic mapping in identification of, 248
Hepatovirus genus/spp./hepatitis A virus, 392t, 735, 739, 739t, 771t
   vaccine for, 508f, 509t, 739
   waterborne transmission of, 790t
Hep B vaccine. *See* Hepatitis B virus vaccine
Hepeviridae (hepeviruses), 392t, 735, 739t, 740, 771t
Hepevirus genus/spp., 392t, 739t, 740
Herbicide tolerance, recombinant DNA technology and, 254–255
Herd immunity, 497b, 504b, 509
   common cold and, 736
Hereditary diseases, 423, 424t. *See also* Genetics
d'Herelle, Felix, 395b
Herpangina, 738
Herpes gladiatorum, 716, 716t
Herpes infections, 391, 392t, 713–722, 714f. *See also under* Human herpesvirus
   of eye, 714–715, 714f, 715f, 716t
   genital, 392t, 714, 714f, 715f, 716t, 729b
   incubation period for, 429t
   in HIV infection/AIDS, 716, 755, 755f
   latency/recurrence and, 713, 714, 714f, 729b
   neonatal, 715, 716t
   oral, 392t, 709b, 714, 714f, 715f, 716t, 729b
Herpes simplex viruses. *See* Human herpesvirus(es) 1 and 2
Herpesviridae (herpesviruses), 391, 392t, 713–722, 714f, 729t. *See also* Herpes infections
   latency/recurrence and, 713, 714, 714f, 729b
Herpes zoster (shingles), 717–718, 717f, 718, 718f, 719b
   immunization against, 508f, 509t, 718
Herpetic gingivostomatitis, 714, 716t
Herpetic pharyngitis, 714, 716t
Herpetic whitlow, 714f, 715, 715f, 716t
Hesse, Fanny, 13, 13f
HE stain. *See* Hematoxylin and eosin stain
Heterochromatin, 197, 197f
Heterocysts, in cyanobacteria, 333, 333f, 807
Heterotrophs, 163, 163f
   in carbon cycle, 806–807
HEV. *See* Hepatitis E virus
Hexokinase, in glycolysis, 128t
Hexose(s), 44
Hfr (high frequency of recombination) cells, 226, 228f
HGA. *See* Human granulocytic anaplasmosis
HGE (human granulocytic ehrlichiosis). *See* Anaplasmosis
HHV. *See under* Human herpesvirus
Hib vaccine, 508f, 509t
Hierarchy, phylogenetic, 115. *See also* Evolution
High-efficiency particulate air (HEPA) filters, in microbial control, 272–273, 274f
   biosafety and, 268
High-energy phosphate bonds, ATP release and, 52, 126–127
   in substrate-level phosphorylation, 127, 135, 135f
High frequency of recombination (Hfr) cells, 226, 228f
High G + C Gram-positive bacteria, 330f, 336–338, 337t, 547
High-level germicides, 267
Highly active antiretroviral therapy/HAART. *See* Antiretroviral therapy
High-power/high dry objective lens, 100
High-temperature, short-time pasteurization, 785
*Hind*II restriction enzyme, 240, 240t
*Hind*III restriction enzyme, 240, 240t
*Hinf*I restriction enzyme, 240t
Hinge region, immunoglobulin, 485f
Histamine
   in inflammation, 463, 465f, 526
   in probiotic mechanism of action, 304b
   in type I (immediate) hypersensitivity, 526, 526–527, 527f, 527t, 528, 528f

Histidine
  in Ames test, 223, 223f
  in probiotic mechanism of action, 304b
Histidine auxotrophs, in Ames test, 223, 223f
Histological stains, 110
Histolytic toxins, clostridial, 562
Histones, 85, 195, 196, 197f, 198t
Histoplasma capsulatum, 366, 648, 651–652, 652b, 652f, 653b, 653f, 663b
  transmission of, 431, 651, 810, 810t
Histoplasmosis, 366, 652b, 653b, 663b
  in HIV infection/AIDS, 541, 652
  transmission of, 431, 651, 810, 810t
HIV (human immunodeficiency virus), 387, 387f, 391t, 392t, 541, 751–757, 751t, 752t, 753f, 754b, 754f, 755f, 765f, 771f. See also HIV infection/AIDS
  antibodies to, 755, 755f
    diagnosis and, 756
  antigenic variability and, 752t, 753
  CD4/CD4 (helper) T cells and, 387, 541, 751, 752, 752t, 754b, 755, 755f
  as genetic vector, 242
  immune system challenges and, 752t
  nomenclature for, 391t
  origin of, 752
  as provirus (latent virus), 400, 752, 753, 753f
  replication of, 752–755, 753f, 754f
  resistance and, 754b
  reverse transcriptase in, 398, 752, 753, 753f, 754b
  structure of, 752, 754b
  synthesis in, 398, 753–754, 753f
  transplacental transmission of, 421t, 756
    ART during pregnancy and, 757
  type 1, 751, 752
  type 2, 751, 752
  vaccine development and, 253, 757
HIV infection/AIDS, 526, 541, 751–757, 751t, 752t, 753f, 754b, 754f, 755f, 765f, 766b, 771t. See also HIV
  aspergillosis in, 657, 658, 659, 664
  blastomycosis in, 653, 654
  candidiasis in, 541, 660b, 661b, 664, 766b
  coccidioidomycosis in, 541, 656
  course of infection and, 755, 755f
  cryptococcosis in, 659, 662, 664
  cryptosporidiosis in, 690
  Cyclospora infection in, 691
  cytomegalovirus infection in, 541, 720
  diagnosis of, 756
    ELISA in, 517, 756
    immunoblot/western blot in, 518
  drug resistance and, 754b
  epidemiology of, 755–756, 756f
  hairy leukoplakia in, 541, 719, 720
  helper T lymphocytes in, 387, 541, 751, 752, 752t, 755, 755f, 756
    attachment and, 752, 753, 753f, 754f
  HHV-1/HHV-2 infections and, 716, 755, 755f
  HHV-6 infection and, 721
  histoplasmosis in, 541, 652
  incidence/prevalence of, 434, 434f, 755–756, 756f
  incubation period for, 429t
  Kaposi's sarcoma in, 541, 721, 721f, 755, 755f
  leishmaniasis in, 683
  microsporidiosis in, 369, 496b, 663
  mycobacterial infections in, 575
  opportunistic infections/tumors in, 418, 541, 664, 751, 751t, 755, 755f, 766b. See also specific type
    fungal infections, 664, 751t
  pathogenesis of, 755, 755f
  PCR in, 245, 756
  Pneumocystis pneumonia (PCP) in, 541, 657, 664, 766b
  pregnancy and, 421t, 756, 757
  prevention of, 754b, 757
  toxoplasmosis in, 541, 690
  transmission of, 421t, 431, 437, 755–756, 756f
    transplacental, 421t, 756
      ART during pregnancy and, 757
  treatment of, 756–757
  tuberculosis in, 541
  vaccine development and, 253, 757
  virus causing. See HIV

Hives (urticaria), in type I (immediate) hypersensitivity, 525b, 528, 528f, 541b
HME. See Human monocytic ehrlichiosis
HMP. See Human microbiome/Human Microbiome Project
Hodgkin, Dorothy, 290
Hodgkin lymphoma/disease, HHV-4 (Epstein-Barr virus) and, 402, 719, 720
Holdfasts, 373
Holoenzymes, 127, 128f
  in substrate-level phosphorylation, 135, 135f
Homologous chromosomes, 356, 357t
Homologous sequences, in genetic recombination, 223, 224f
Honey, pediatric (infant) botulism and, 564
Hook
  of archaea flagellum, 77
  of bacterial flagellum, 62, 63
Hookworms (*Ancylostoma* and *Necator*), 700, 700f, 703t
Hopanoids, 69
Hops, in beer production, 781, 782f
Horizontal (lateral) gene transfer, 223–227, 225f, 226f, 227f, 228f, 228t, 230b
  antimicrobial resistance and, 227, 229f, 230, 230b, 303
  bacterial conjugation, 226–227, 227f, 228f, 228t
  S. aureus strains and, 336
  transduction, 225, 226f, 228t
  transformation, 224–225, 225f, 228t
Hormonal (endocrine) diseases, 424, 424t
  autoimmunity and, 539
Horses, arboviral encephalitis and, 742, 742f. See also Equine encephalitis
Hospital-acquired infections. See Healthcare-associated (nosocomial) infections/diseases
Hospital epidemiology, 437–440, 439f. See also Healthcare-associated (nosocomial) infections/diseases
Hospital (medical/pharmaceutical) microbiology, 18f, 19t, 252–254, 788, 788t
Host(s)
  amplifying, for *Yersinia pestis*, 602
  movement of microbes into, 420–421, 420f, 421f, 421t. See also Infection; Portals of entry
  movement of microbes out of, 430, 430f
  of parasite, 415–416, 415f, 676f
  symbiotic relationships with microbes and, 168, 414–418, 415b, 415f, 416t
  of viruses, 386–388, 387f
Host defenses, 20–22, 21f, 447. See also Immune response
  adaptive immunity, 447, 472–502, 473t. See also Adaptive immunity
  first line of, 447, 447–451, 466t
  innate immunity, 446–471, 466t. See also Innate immunity
    proteins in, 47
  second line of, 447, 451–466, 466t
  species resistance and, 447
  third line of, 447
Hot air, in microbial control, 271, 275t
Hot tub (*Pseudomonas aeruginosa*) folliculitis, 778b, 796b
Houseflies, as disease vectors, 433t
HpaI restriction enzyme, 240t
HPIV. See Human parainfluenza viruses
HPS. See Hantavirus pulmonary syndrome
HPV. See Human papillomaviruses
HPV vaccine, 508f, 509t
HRIG. See Human rabies immune globulin
HSV (herpes simplex viruses). See Human herpesvirus(es) 1 and 2
HTLV. See Human T-lymphotropic viruses
HTST. See High-temperature, short-time pasteurization
Human adenovirus 36, obesity and, 724
Human chorionic gonadotropin, immunochromatographic assays for, 519

Human diploid cell vaccine (HDCV), 763. See also Rabies vaccine
Human genome, 194, 250
  sequencing of. See Genomics
Human Genome Project, 250
Human granulocytic anaplasmosis (HGA), 623–626, 623f, 626f
Human granulocytic ehrlichiosis. See Anaplasmosis
Human growth hormone, microbial production of, 12t, 253, 788t
Human herpesvirus(es), 713. See also specific type
Human herpesvirus(es) 1 and 2 (HHV-1/HHV-2), 713, 714–716, 714f, 715f, 716t, 729t
  diagnosis/treatment/prevention of infections caused by, 716
  epidemiology and pathogenesis of infections caused by, 716, 716t
  eye infection caused by, 714–715, 714f, 715f, 716t
  genital infection and, 714, 714f, 715f, 716t, 729b
  incubation period for, 429t
  in HIV infection/AIDS, 716, 755, 755f
  latency/recurrent lesions and, 714, 714f, 729b
  neonatal infection and, 715, 716t
  ocular/ophthalmic infection and, 714–715, 714f, 715f, 716t
  oral infection and, 392t, 709b, 714, 714f, 715f, 716t, 729b
  whitlow caused by, 714f, 715, 715f, 716t
Human herpesvirus 3 (HHV-3), 713, 716–718, 717f, 718f, 719b. See also Varicella-zoster virus
Human herpesvirus 4 (HHV-4), 392t, 713, 718–720, 719f
Human herpesvirus 5 (HHV-5), 713, 720–721, 721f. See also *Cytomegalovirus* genus/spp.
Human herpesvirus 6 (HHV-6), 392t, 713, 721, 721f
Human herpesvirus 7 (HHV-7), 713
Human herpesvirus 8 (HHV-8), 713, 721, 721f
Human immunodeficiency virus. See HIV; HIV infection/AIDS
Human insulin, microbial production of, 12t, 238, 252, 788t
Human microbiome/Human Microbiome Project (HMP), 253b, 804, 804f. See also Microbiome
Human monocytic ehrlichiosis (HME), 623–626, 623f, 626f
Human papillomaviruses, 392t, 722, 723b, 724b, 729t. See also *Papillomaviridae*
  cancer and, 392t, 723b
  skin infection (warts) caused by, 392t, 722, 723b, 724b
Human papillomavirus vaccine, 508f, 509t
Human parainfluenza viruses, 758, 759–760
Human pathogens, as biological weapons, 813, 814t
  threat potential and, 812–813
Human rabies immune globulin (HRIG), 762
Human T-lymphotropic viruses (HTLV-1/HTLV-2/HTLV-5), 750–751
Humoral immune response. See also Antibody immune response
Humus, in topsoil, 809
Hyalomma ticks, as disease vectors, for arboviruses, 747t
Hyaluronidase/hyaluronic acid
  microbial production of, 787
  as virulence factor, 426, 427f
  of *Clostridium difficile*, 562
  of group A streptococci, 552, 554b, 555b
  of group B streptococci, 556
  in necrotizing fasciitis, 554b, 555b
  of staphylococci, 548, 548t
  of *Treponema*, 631, 631b
Hybridization, 187
  fluorescent *in situ* (FISH), 249–250, 250f
Hybridomas, in passive immunotherapy, 511, 511f
Hydatid cysts/hydatid disease, 695, 695f, 703t

Hydrocarbons
  bioremediation of, 805
  microbial production of, 788, 789t
Hydrochloric acid, 38
  in acid-fast staining, 109
Hydrogen, 29t, 30f
  as alternative fuel, 339, 788, 789t
  atomic number of, 28, 29t
  covalent bonds in formation of, 31–32, 32f. See also Hydrogen bonds
  electrons of, 30
  for microbial growth, 163
Hydrogen bonds, 34–35, 34f, 35f, 35t
  nucleic acid, 25, 50–51, 51f, 52, 194, 195f
  water cohesiveness and, 37, 37f
Hydrogen ions, 38, 38f. See also pH
  acid-base balance and, 38
Hydrogen peroxide, 164, 278
  antimicrobial action of, 164, 278, 281t
    peroxide anion and, 164
  in killing by neutrophils, 456
Hydrogen sulfide (H_2S)
  sulfate-reducing microbes producing, 332, 339, 343, 808, 808f
  in sulfur cycle, 808, 808f
Hydrogen sulfide (H_2S) test, 116, 116f
Hydrolases, 127, 128t
Hydrolysis, 36, 36f, 127, 128f
  of disaccharides/sucrose, 45, 45f
  in lipid catabolism, 143, 143f
  in protein catabolism, 144f
Hydrophilic molecules, in phospholipids, 42, 43f, 69, 69f
Hydrophobia, in rabies, 761, 762b
Hydrophobic molecules, in lipids/phospholipids, 40, 42, 43f, 69, 69f
Hydrostatic pressure, microbial growth affected by, 168
Hydrothermal vents, 330, 332, 812
Hydroxyl functional group, 40, 41t
  in nucleic acid, 51
Hydroxyl ions, 38, 38f
  in acid-base balance, 38, 38f
  in dehydration synthesis, 36, 36f
Hydroxyl radicals, 164
Hygiene, infection control/prevention and, 16, 440. See also Infection control
  Lister's contribution/antisepsis and, 16, 18f
  Nightingale's contribution/nursing and, 17, 17f
  Semmelweis' contribution/handwashing and, 16, 440
  Snow's contribution/epidemiology and, 17
Hygiene hypothesis, 263b, 526
Hyperimmune response, dengue hemorrhagic fever and, 744, 745f
Hypersensitivity, 526, 526–538, 537t. See also specific type
  anisakiasis and, 701
  immunosuppressive drugs and, 537–538, 538t
  type I (immediate), 526–530, 527f, 527t, 528f, 529f, 537t
    fungal allergens causing, 528, 669
  type II (cytotoxic), 530–531, 530f, 531t, 532f, 537t
  type III (immune–complex mediated), 532–534, 533f, 534f, 537t
    fungal allergens causing, 532, 669
  type IV (delayed-cell-mediated), 534–537, 535b, 535f, 536f, 537t
Hypersensitivity aspergillosis, 657, 659, 659b
Hypersensitivity pneumonitis, 532–533
Hyperthermophiles, 130, 166–167, 167f, 330–331, 331f
  commercial sterilization and, 264, 269
  in polymerase chain reaction, 245, 246b, 331
Hypertonic solutions, 72, 72f, 168
  food preservation/microbial control and, 168, 273
Hyphae, 365, 366f, 651, 651f
  in ascomycetes, 369
  in basidiomycetes, 370
  in lichens, 371, 372f
  in phaeohyphomycosis, 666, 667f
Hypnozoites, in *Plasmodium* life cycle, 687
Hypochlorite, antimicrobial action of, 277

Hypothalamus, body temperature controlled by, 464, 465f
Hypothesis, in scientific method, 9, 9f
Hypotonic solutions, 72, 72f, 168

I. See Iodine
Iatrogenic diseases/infections, 424t, 439
Ibotenic acid, 668
"Ice-minus" *Pseudomonas syringae*, 789
ICTV (International Committee on Taxonomy of Viruses), 390–391
Identification (microbe), 112, 115–119, 116f, 117f, 118f. See also Taxonomy/microbial classification and identification
idio- (prefix), 423t
Idiopathic diseases, 424t
IFNs. See Interferon(s)
Ig. See Immunoglobulin(s)
Iglewski, Barbara, 20
IL(s). See Interleukin(s)
IL-2. See Interleukin 2
IL-2R. See Interleukin 2 receptor
IL-4. See Interleukin 4
IL-12. See Interleukin 12
Illness stage of disease, 430
IM. See Intramuscular (IM) drug administration
Imipenem, 291, 292f, 309t
Immediate (type I) hypersensitivity, 526–530, 527f, 527t, 528f, 529f, 537t. See also Allergic reactions/allergies
 antigens and, 476
 to antimicrobial drugs/penicillin, 302
 bee/insect stings causing, 525b, 528, 535b, 541b
 clinical signs of
  in localized reaction, 528, 528f
  in systemic reactions, 528–529
 degranulation/degranulating cells in, 526, 526–527, 527f, 527t
 diagnosis of, 529, 529f
 food causing, 529
 fungal allergens causing, 528, 669
 hygiene hypothesis and, 263b, 526
 immunotherapy for, 529
 initial exposure/sensitization and, 263b, 526, 527f
 prevention of, 529
 transgenic organisms and, 256
 treatment of, 529–530
 vaccines causing, 509
Immersion oil, 100, 102f
Immortal B cells, HHV-4 (Epstein-Barr virus) producing, 720
Immune complexes, 512, 513, 532, 533f
 in glomerulonephritis, 533, 556
 in hypersensitivity pneumonitis, 532
 in immunochromatographic assays, 519, 519f
 in rheumatoid arthritis, 533
 in systemic lupus erythematosus, 534
Immune complex–mediated (type III) hypersensitivity, 532–534, 533f, 534f, 537t
 fungal allergens causing, 532, 669
Immune disorders, 525–545. See also specific type
 autoimmune, 481, 526, 538–539. See also Autoimmune diseases
 hygiene hypothesis and, 263b, 526
 hypersensitivities, 526, 526–538, 537t. See also Allergic reactions; Hypersensitivity
 immunodeficiency diseases, 526, 540–541, 540t. See also HIV infection/AIDS
 immunosuppressive drugs and, 537–538, 538f
Immune globulin, human rabies (HRIG), 762
Immune hypersensitivities, 526, 526–538, 537t. See also specific type and Hypersensitivity
Immune receptors, 473t. See also B cell receptors; T cell receptors
Immune response (immunity), 18
 acquired, 473, 473t. See also Acquired immunity
 adaptive, 447, 472–502, 473t. See also Adaptive immunity
 antibody, 474, 492–494, 493f, 495, 495f
 cell-mediated, 474, 480–482, 482f, 483f, 489–492, 490f, 491f, 495
 contact, 506
 diagnostic uses and, 512–520, 520t. See also Immune testing
 disorders of/defects in. See specific type and Immune disorders
 herd, 497b, 504b, 509
 common cold and, 736
 innate, 446–471, 466t, 473, 473t. See also Innate immunity
 microorganisms triggering, 117
 preparation for, 478–482, 478f, 479f, 480f, 481f, 482f
 species resistance and, 447
Immune response cytokines. See Cytokines
Immune serum. See Antiserum
Immune suppression
 healthcare-associated infections and, 439
 opportunistic pathogens/infections and, 418
Immune system cytokines. See Cytokines
Immune testing, 512–520, 520t. See also specific test and Serology
Immunity. See Immune response (immunity)
Immunization, 18, 494, 495, 496, 497t, 504–512, 505f, 507f, 508f, 509f. See also specific organism or disease and Vaccines
 active/artificially acquired immunity and, 495, 496, 497t, 505–510, 507f, 508f, 509f, 511f
 booster, 506, 508f, 511f. See also specific disorder
 ethical issues and, 504b
 history of, 2b, 504–505, 505f
 immunologic memory and, 491, 494, 495, 495f, 496, 497t
 passive (passive immunotherapy), 496, 497t, 504, 505, 510–511, 511f
 recommendations for, 508–509, 508f, 509t
 safety concerns and, 509–510
 vaccine types and, 506–508, 507f
Immunoassays, 512–520, 520t. See also specific test and Serology
 in immunoblot/western blot, 518, 518f
 labeled, 515–518, 515f, 517f, 518f, 519f
Immunoblots (Western blots), 518, 519f
 in HIV diagnosis, 518
ImmunoCAP Specific IgE blood test, 529
Immunochromatographic assays, 518–519, 519f
Immunocompromised host. See also HIV infection/AIDS; Opportunistic pathogens
 aspergillosis in, 657, 658, 659, 659b
 attenuated (live) vaccines contraindicated in, 506
 candidiasis in, 541, 660b, 661b
 coccidioidomycosis in, 541, 655, 656
 cytomegalovirus infection in, 720
 healthcare-associated infections and, 439
 Listeria infection in, 336, 568, 570b
 microsporidiosis in, 369, 496b, 663
 Morganella/Providencia/Edwardsiella infection in, 597
 opportunistic pathogens/infections and, 418, 540–541, 540t, 664. See also specific type
 Pneumocystis pneumonia in, 541, 657
 Pseudomonas infection in, 609, 610, 610f
 toxoplasmosis in, 541, 690
 Vibrio vulnificus infection and, 40b
 zygomycoses in, 662
Immunodeficiency diseases, 526, 540–541, 540t
 acquired (secondary), 540–541. See also HIV infection/AIDS
 opportunistic infections and, 418, 540–541, 541t
 primary, 540, 540t
 severe combined (SCID), 540, 540t
Immunodiffusion, 513, 513f, 520t
Immunofiltration assays, 518
Immunofluorescence, 103, 103f
Immunoglobulin(s), 452, 483, 483f, 485. See also Antibodies
 classes of, 486–487, 488t
 deficient, 540, 540t
 function of, 485–486, 486f
 intravenous, 510
 in passive immunotherapy, 496, 497t, 504, 505, 510, 511, 511f
 plasma cell secretion of, 493f, 494
 specificity of, 485, 485f
 structure of, 485, 485f
Immunoglobulin A (IgA), 485, 487, 488t
 deficiency of, 540
 in neutralization, 487
 plasma cell secretion of, 494
 secretory, 487, 488t
  naturally acquired passive immunity and, 487, 495, 497t
 protease against
  gonococcal pathogenicity/virulence and, 588, 589b
  pneumococcal pathogenicity/virulence and, 558
Immunoglobulin D (IgD), 485, 487, 488t
Immunoglobulin E (IgE), 485, 487, 488t
 plasma cell secretion of, 494
 in type I (immediate) hypersensitivity/allergies, 487, 526, 527, 527f, 527t
 testing for, 529, 529f
Immunoglobulin G (IgG), 485, 487, 488t, 495f, 510
 in hemolytic disease of newborn, 531, 532f
 in naturally acquired passive immunity, 487, 495, 497t
 passive immunotherapy and, 510
 plasma cell secretion of, 494, 495f
Immunoglobulin M (IgM), 485, 486–487, 488t, 495f
 in complement activation/inflammation, 485, 486, 487, 488t
 plasma cell secretion of, 494, 495f
Immunological diseases, 424, 424t. See also specific type
Immunological memory, 473, 474, 491–492, 493f, 494–495, 495, 495f, 496, 497t
Immunological synapse, 490
 T helper cell activation and, 492
Immunological tests. See also specific test and Serology/serological tests
Immunology/immunologists, 18, 18f, 19t, 21, 473. See also Immune response
Immunosuppressive drugs, 537–538, 538f
 in diabetes prevention, 539
Immunosuppressive retroviruses, 751–757, 751f, 752f, 753f, 754b, 754f, 755f, 765f. See also HIV
Immunotherapy
 passive, 496, 497t, 504, 505, 510–511, 511f
 for type I (immediate) hypersensitivity, 529
Imperfect fungi, 371
Impetigo (pyoderma), 553
Impetigo
 staphylococcal, 549, 549f
 streptococcal, 549
Inactivated polio vaccine (IPV/Salk vaccine), 508f, 509t, 510, 737–738, 738t
Inactivated (killed) vaccines, 506
Incidence of disease, 434, 434f
Incineration, in microbial control, 271, 275b
Inclusion(s), cellular
 in cytomegalovirus disease, 721, 721f
 in eukaryotic cells, 89t
 in prokaryotic cells, 60f, 89t
  archaea, 89t
  bacteria, 74, 75f, 89t
Inclusion bodies, in chlamydial growth/reproduction, 626, 627f
Incubation
 measurement techniques not requiring, 183–184, 183f
 measurement techniques requiring, 184–186, 184f, 185f, 186t
 in streak plate method of isolation, 173
Incubation period, 429, 429t. See also specific disease
Index case, 435–436
Indicator organisms, in water quality testing, 792, 792f
Indigenous microbiota. See Microbiome
Indigo, microbial production of, 789t
Indirect contact transmission of disease, 431, 433t
Indirect fluorescent immunoassays, 516, 516f, 520t
Indirect (negative) selection, of mutants, 222, 222f
Induced fit model, 129, 129f
Inducers, in operon, 213, 214f
Inducibility, in adaptive immunity, 473, 492
Inducible operons, 213, 215f
 lactose (*lac*) operon as, 213–214, 214f
Induction
 in lactose (lac) operon regulation, 214, 214f
 in lysogeny, 395, 396f
Industrial canning, in food preservation, 784–785, 784f
Industrial fermentations, 786–787, 787f
Industrial microbiology/biotechnology, 12, 12b, 18f, 19t, 238, 239f, 779, 786–795. See also Recombinant DNA technology
 agricultural applications/products of, 19t, 254–256, 255f, 788–789, 789t. See also Agricultural microbiology
 alternative fuel production and, 162b, 788, 789t
 bioremediation and, 19t, 20, 169b, 805
 biosensors and bioreporters and, 789
 enzyme production and, 12t, 787, 789t
 fermentations and, 786–787, 787f
 pesticide production and, 12t, 788–789, 789t
 pharmaceutical applications/products of, 12t, 18f, 19t, 252–254, 788, 788t
 products of, 12, 12t, 787–789, 789t
 recombinant DNA technology in, 238, 239f
 water treatment and, 789–795, 790b, 790t, 791f, 792f, 793f, 794f, 795f
Infant. See Newborn
Infant (pediatric) botulism, 564
Infection, 420–422, 420f, 421f, 421t, 422f. See also Infectious disease
Infection control, 17, 18f, 19t, 262–287. See also Microbial growth, control of
 handwashing in, 16, 440
 healthcare-associated (nosocomial) infections and, 16, 439–440
 Lister's contribution and, 16
 Nightingale's contribution and, 17, 17f
 public health policies/education and, 435, 437f, 438f, 440–441
 Semmelweis' contribution and, 16
 Snow's contribution/epidemiology and, 17
Infectious disease, 413–445, 423t, 424t. See also specific type or causative agent
 adhesion in, 421–422, 422f
  phagocytosis and, 455, 455f
 animal reservoirs of, 418, 419t
 autoimmunity and, 538–539
 causes (etiology) of, 12–15, 13f, 14f, 15f, 15t, 18f, 19t, 423–426, 424t, 425f
 classification of, 432–434, 434t
 contamination and, 420
 defenses against, 20–22, 21f, 447
 diagnosis of
  clinical sampling and, 172, 173t
  recombinant DNA techniques in, 254
  DNA microarrays, 248, 253
 epidemiology of, 17, 18f, 19t, 434–441. See also Epidemiology
 fever and, 464–466, 465f
 germ theory of, 13, 424
 healthcare-associated (nosocomial). See Healthcare-associated (nosocomial) infections/diseases
 human carriers of, 418, 419b
 iatrogenic, 424t, 439
 manifestations of, 423, 423t
 microbe–host symbiosis and, 414–418, 415b, 415f, 416t
 nature of, 422–430
 nonliving reservoirs of, 419
 nosocomial. See Healthcare-associated (nosocomial) infections/diseases
 notifiable/reportable, 435, 437t, 438t
  bioterrorism defense and, 815, 815f
 opportunistic, 298, 299b, 302, 417–418

persistent, enveloped viruses causing, 400, 400f, 401t
portals of entry and, 420–421, 420f, 421f, 421t
portals of exit and, 430, 430f
prevention of, 15–18, 16b, 17f
  probiotics in, 304b, 780
public health records/policies and, 435, 437t, 438f
reservoirs of, 418–420, 419b, 419t
secondary/superinfection, antimicrobial use and, 298, 299b, 302, 439
stages of, 429–430, 429t
terminology of, 423t, 433, 434t
transmission modes of, 430–432, 431f, 432b, 433t
  public health in interruption and, 440
treatment of. See Antimicrobial drugs
virulence and, 426, 426f, 428t
Infectious hepatitis. See Hepatitis, type A
Infectious mononucleosis (mono)
  cytomegalovirus causing, 720
  HHV-4 (Epstein-Barr virus) causing, 392t, 718, 719f, 720
  HHV-6 infection and, 721
Infective body, *Coxiella* forming, 613, 613f
Infestation, 693. See also Parasite(s)
Inflammation, 462–464, 463f, 465f, 465t, 587f
  antibodies in, 486
  cells in, 452f
  fever and, 464–466, 465f
  NOD proteins in, 457
  TLRs/PAMPs in, 457
Inflammatory mediators, 462f, 463–464, 463f, 465t
  in type I (immediate) hypersensitivity, 526, 527f, 527t
Influenza/influenzaviruses (influenza viruses A and B), 392t, 736t, 763–766, 768–769b, 771t
  avian, 768b
  as biological weapons, 814t
  H1N1, 435
  H5N1, 768b
  H7N9, 768b
  immunization against, 508f, 509t, 769b
  incubation period for, 429t, 768b
  mutations in, 768b, 769b
  pandemics caused by, 435
  resistant strains of, 769b
  safe handling of, 268
  transmission of, 769b
  virulence factors of, 768b
Influenza virus vaccine, 508f, 509t, 769b
Infra-red light, 98f
Ingestion, in phagocytosis, 455f, 456
INH. See Isoniazid
Inhalation. See also Respiratory system
  parasitic infection and, 676f
Inhalation anthrax, 336, 561–562, 736t. See also Anthrax
  bioterrorism and, 336, 562
Inhaled allergens, 528
Inheritance. See Genetics
Inhibition/inhibitors, enzyme, 131–132, 131f, 132f
  competitive, 131–132, 131f
    antimicrobial action and, 132
  host defense and, 450
  feedback, 132, 132f
  noncompetitive/allosteric, 132, 132f
Inhibition zone
  in diffusion susceptibility test, 299, 300f
  in Etest, 300, 300f
  synergism and, 307f
Inhibitory neurotransmitter, tetanospasmin (tetanus toxin) affecting, 565, 566f
Initial bodies, in *Ehrlichia* and *Anaplasma* growth/reproduction, 623, 623f
Initiation
  DNA replication, 199, 200f, 202, 213t
    methylation affecting, 201
  transcription, 204, 205f
  translation, 210, 210f
    antimicrobials affecting, 293, 294f
Initiation complex, 210, 210f
Injection
  for DNA insertion into cells, 248, 249f
  intramuscular (IM)/intravenous (IV), for drug administration, 301, 301f

Injection drug use. See Drug abuse
Innate immunity, 446–471, 466f, 473, 473t
  antimicrobial peptides (defensins) in, 448, 449t, 451, 455
  blood/blood cells in, 452–454, 452f, 453f, 454b. See also specific blood component
  chemicals/secretions in, 451, 451t, 457–462, 457t, 458f, 459f, 459t, 460b, 461f, 462f
  complement/complement system in, 455, 459–462, 459f, 461f, 462f
  fever in, 464–466, 465f
  first line of defense and, 447, 447–451, 466t
  inflammation in, 462–464, 463f, 465f, 465t
  interferons in, 457–459, 458f, 459f
  lacrimal apparatus in, 449, 450f
  microbiome in, 449–450
  mucous membranes/mucus in, 448–449, 449f, 449t
  NOD proteins in, 457
  nonphagocytic killing in, 456
  phagocytosis in, 452f, 453, 454–456, 455f
  second line of defense and, 447, 451–466, 466t
  skin in, 420, 421f, 447–448, 448f, 449t
  toll-like receptors (TLRs) in, 457, 457t
Inner protein rings, bacterial flagellar motion and, 63f
Inoculum/inocula, 171
Inorganic molecules/chemicals, 37–39, 37f, 38f. See also Acid(s); Base(s); Salt(s); Water
Insecta/insects/insect vectors, 377, 377–378, 378f, 677. See also specific type
  taxonomic classification of, 114f
Insect bites/stings
  allergic reaction/anaphylaxis caused by, 525b, 528, 529, 535b, 541b
  as portal of entry, 420f
Insecticides, microbial production of, 12t, 788–789, 789t
Insertion (frameshift mutation), 216, 217f, 218f, 219, 219f
  transposition as, 229
Insertion sequences (IS), 229, 229f
Insulin
  autoimmunity affecting production of, 539
  microbial production of, 12t, 238, 252, 788t
Integral protein(s)
  in bacterial cell walls, 67, 68f
  in bacterial flagella, 63f
  in cytoplasmic membrane, 69, 69f
    facilitated diffusion and, 71, 71f
  in electron transport chains, 136, 139
Integrase
  HIV, 752, 753, 754b
  in retroviruses, 750
Integration, in HIV replication, 752, 753, 753f
Interference (differential interference contrast/Nomarski) microscopy/microscopes, 101, 102f, 107t
Interfering RNA, small (siRNA), 216
Interferon(s), 457–459, 458f, 459t, 489, 489t
  in adaptive immunity, 489
  in innate immunity, 457–459, 458f, 459t
    TLRs/PAMPs and, 457
  recombinant DNA in production of, 253, 788t
  type I (alpha and beta), 458, 458f, 459t
  type II (gamma) (macrophage activation factor), 458, 459, 459t, 489
Interleukin(s), 489, 489t
  recombinant DNA in production of, 788t
Interleukin 2, 489t
  in cell-mediated immune response, 490, 490f
Interleukin 2 receptor
  in cell-mediated immune response, 490, 490f
  monoclonal antibodies against, 537
Interleukin 4, 489t
  in antibody immune response, 492–494, 493f
  in B cell activation, 493f, 494
  in helper T cell differentiation, 492–494, 493f
  in type I (immediate) hypersensitivity, 526, 527f

Interleukin 12, 489t
  in helper T cell differentiation, 490, 490f
Intermediate filaments, in eukaryotic cytoskeleton, 83, 84f
Intermediate hosts, 676
Intermediate-level germicides, 267
Internal carbonyl group, 41t
International Committee on Taxonomy of Viruses (ICTV), 390–391
Interphase, 354–355, 357f
Intestinal secretions, in host defense, 451t
Intestines. See Gastrointestinal (digestive) tract
Intoxications. See also Toxins
  contaminated food causing, 353b, 379b, 785, 786. See also Botulism; Foodborne illnesses
  contaminated water causing, 789, 790t. See also Waterborne illnesses
  fungi causing/mushroom poisoning, 370, 649, 668, 669f
Intracellular pathogens, 342. See also Viruses
  T cells/cell-mediated immune response and, 474, 480
Intramuscular (IM) drug administration, 301, 301f
Intravenous (IV) drug abuse. See Drug abuse
Intravenous (IV) drug administration, 301, 301f
Intravenous immunoglobulins (IVIg), 510
Introns, 206, 207f
Intubation, 760
In-use test, 282
Invasive extraintestinal amebiasis, 678, 692t
Invasive pulmonary cryptococcosis, 659
Inversion (mutation), 216
Inverted repeat (IR), 229, 229f
Iodine, 29t
  antimicrobial action of, 277, 277f, 281t
  in Gram staining, 109, 109f
Iododeoxyuridine, 297
Iodophors, antimicrobial action of, 277, 277f
Iodoquinol, 317t
Ion(s), 33. See also specific type
Ion gradients, in chemiosmosis, 139
Ionic bonds, 33–34, 34f, 35t
Ionization (dissociation), 34, 34f
Ionizing radiation, 273, 274–275. See also Radiation
  as energy source for fungi, 367
  in microbial control/food preservation, 274–275, 274f, 275f, 785
  mutagenic effects of, 217–218
IPV. See Inactivated polio vaccine
IR. See Inverted repeat
Iron/iron salts, 29t
  acid mine drainage and, 805, 805f, 806, 806f
  cycling of, 808–809
  in electron transport chains, 139
  in host defense, 452
Iron-binding compounds, of Enterobacteriaceae, 593, 593f
Iron-binding proteins, in host defense, 452
Iron-sulfur bacteria, in acid mine drainage, 805f
Iron-sulfur proteins, in electron transport chains, 139
Irradiation. See Radiation
Irritable bowel syndrome, fecal microbiome transfer for, 803b
IS. See Insertion sequences
Islets of Langerhans, autoimmunity affecting, 539
Isocitric acid, in citric acid/Krebs cycle, 137f
Isografts, 536, 536f
Isolation (microbe), for pure culture, 172–174, 173f, 174f
Isoleucine
  feedback inhibition and, 132
  tRNA, mupirocin affecting, 293
Isoleucyl-rRNA synthetase, mupirocin affecting, 293
Isomer, 127
Isomerases, 127, 128t
Isomerization, in citric acid/Krebs cycle, 136
Isoniazid, 310t
  mechanism of action of, 291–292, 291f, 310t
  spectrum of action of, 299f, 310t
Isonicotinic acid hydrazide. See Isoniazid
Isopropanol
  antimicrobial actions of, 276–277
  microbial fermentation producing, 142f
Isotonic solutions, 72, 72f

Isotopes, 28–29, 29f
  radioactive, 29
-itis (suffix), 423t
Itraconazole, 315t, 650
IV. See under Intravenous
Ivanovsky, Dmitri, 14, 15t, 18f, 388
Ivermectin, 316t
  mechanism of action of, 295, 316t
  microbial production of, 290t, 316t
IVIg. See Intravenous immunoglobulins
Ixodes ticks
  as disease vectors, 433t
    for arboviral encephalitis, 743b, 747t
    for babesiosis, 691b
    for ehrlichiosis and anaplasmosis, 623, 626f
    for Lyme disease (*Borrelia burgdorferi*), 634–635, 634f
  life cycle of, 634–635, 634f
  taxonomic classification of, 114f
Izziella abbottiae, 113

Japanese encephalitis, 392t, 742, 747t
Japanese encephalitis vaccine, 509t
Jaundice
  in hemolytic disease of newborn, 531
  in hepatitis, 726, 726f, 727b
  in malaria, 687
  in yellow fever, 744
J (joining) chain, 487, 488t
JC virus, 722
Jenner, Edward, 2b, 17–18, 18f, 505, 510b, 710
J_H (junction) heavy chain gene segment, 484, 484f
J_L (junction) light chain gene segment, 484
"Jock itch" (tinea cruris), 665f
Joining (J) chain, 487, 488t
"Jumping genes," 229, 229f. See also Transposons
Junction (J_H) heavy chain gene segment, 484, 484f
Junction (J_L) light chain gene segment, 484
Junction segment genes, 484, 484f
Junin hemorrhagic fever, 767
Juvenile-onset diabetes, autoimmunity in, 538, 539

K. See Potassium
Kala-azar (visceral leishmaniasis), 683, 692t
Kalium. See Potassium
Kanamycin, 310t
  transposons in resistance to, 229f
K antigens, 593, 593f
  of *E. coli*, 595
  of *Haemophilus influenzae*, 604
Kaposi's sarcoma, 402, 721, 721f, 751
  HHV-8 causing, 721
  in HIV infection/AIDS, 541, 721, 721f, 755, 755f
Kappa gene locus, BCR, 484
KCl. See Potassium chloride
Keio Knockout collection, *E. coli*, 251–252
Kelps, 6, 375, 375f
Kelsey-Sykes capacity test, 282
Keratin, superficial mycoses and, 664, 665
Keratinase, 426
Keratitis, *Acanthamoeba*, 267b, 678, 679, 692t
Ketoconazole, 315t, 317t, 650
α-Ketoglutaric acid, 149t
  in citric acid/Krebs cycle, 137f, 149t
  in transamination, 151f
Kidneys
  bacterial infection of, 58b, 89b, 598b, 599b
  in *E. coli* O157 infection (hemolytic uremic syndrome), 595
  immune complex–mediated (type III) hypersensitivity affecting, 533
  in leptospirosis, 636
  streptococcal infection affecting, 539, 556
Killing
  nonphagocytic, 456
  in phagocytosis, 455f, 456
Kimchi, fermentation in production of, 779, 783t
Kinases, as virulence factors, 426, 427f
Kineococcus radiotolerans, 169b
Kinetoplast, 361, 361f
Kinetoplastids, 361, 361f, 365t, 679
Kingdoms, 113, 114f, 115, 357–358, 359f
Kinins, in type I (immediate) hypersensitivity, 527, 527f, 527t

SUBJECT INDEX

Kirby-Bauer (diffusion susceptibility) tests, 299, 300f, 301t
  for Enterobacteriaceae, 594
Kissing bugs (*Triatoma*), as disease vectors, 379, 433t
  in Chagas' disease (*Trypanosoma cruzi*), 379, 679, 680f, 692t
Kissing disease. *See* Mononucleosis
Kitasato, Shibasaburo, 15t, 18f, 21
Klebs, Edwin, 15t
Klebsiella genus/spp./*Klebsiella pneumoniae*, 343t, 344t, 592t, 596, 596f
  capsules of, 62, 596, 596f
    staining, 111f
  healthcare-associated (nosocomial) infection caused by, 591f
  pneumonia caused by, 596
  resistant strains of, 230b
  sites of infection caused by, 604f
  UTIs caused by, 596, 599b, 604f
Kluyver, Albert, 19
Knockout gene, 251–252
Koch, Robert, 13, 13f, 18f, 106, 424
  experiments/postulates of, 13–14, 14f, 15t, 18f, 424–426, 425f, 437
Koch's postulates, 14, 15t, 18f, 424–426, 425f
  exceptions to, 425–426
  experimental epidemiology and, 437
KOH (potassium hydroxide) preparations, fungi identified by, 649
Koplik's spots, 758, 758f, 758t
Korarchaeota (phylum), 330f
Krebs (citric acid) cycle, 133, 133f, 136, 137f, 149t, 153f
Kupffer cells (stellate macrophages), 478
Kuru, 407

Labeled immunoassays, 515–518, 516f, 517f, 518f, 519f. *See also specific type*
Lacks, Henrietta, 404
lac operator, 214, 214f
lac operon, 213–214, 214f
Lacrimal apparatus, in host defense, 449, 450f
La Crosse (California) encephalitis, 392t, 747t, 767
β-Lactam antibiotics, 290–291, 292f, 309t
  mechanism of action of, 290–291, 292f, 309t
  resistance to, 306, 306f, 309t
β-Lactamase
  antimicrobial resistance and, 306, 306f, 309t
  staphylococcal pathogenicity/virulence and, 548, 548t, 551
β-Lactam rings, 290–291, 292f
Lactic acid, microbial fermentation producing, 141, 141f, 142f
  bread production and, 779
  cheese production and, 142f
  dairy products and, 780
  vegetable fermentation and, 779, 783t
Lactic acid dehydrogenase, 128t
Lactobacillus genus/spp./*Lactobacillus bulgaricus*/lactobacilli, 330f, 336, 337t, 426t
  fermentation products of, 142f, 336
    cheese, 780
    vegetables, 779, 783t
    yogurt, 336, 780, 783t
  as probiotic, 304b
  vulvovaginal candidiasis and, 661b
Lactococcus genus/spp./*Lactococcus lactis*
  in buttermilk production, 780, 783t
  in cheese production, 780, 783t
  nisin produced by, 311t
  in vegetable fermentation, 779
Lactose, 45
Lactose (*lac*) operon, 213–214, 214f
Laetiporus sulphureus, 371f
Lagging strand, in DNA replication, 199, 200f, 201
Lagoons, oxidation, in wastewater treatment, 794
Lag phase, of microbial growth, 181, 181f
L-alanine, 48f
Lambda gene locus, BCR, 484
Lambda phage/lambda phage replication, 395–396, 395f, 396f
Lamellae, photosynthetic, 332
Laminarin, 375
Lamivudine, 297f, 314t
Lancefield, Rebecca, 117, 552

Lancefield antigens/groups, 117, 552, 561t
Landfills
  bioremediation and, 805
  microbial production of methane and, 332, 788, 807
Langerhans, islets of, autoimmunity affecting, 539
Large intestine (colon), microbiome/normal microbiota of, 416, 417t
  transfer of, 803b
Larynx, 449f
Lasers
  atomic force microscopes using, 106
  confocal microscopes using, 100, 103, 104f
Lassavirus genus/spp./Lassa fever virus/Lassa hemorrhagic fever, 392t, 767, 771f
  safe handling and, 268
Latency/latent diseases, 400, 401t, 433, 434t
  animal virus, 400, 401t. *See also* Latent viruses
  herpesvirus, 713, 714f, 729b
  HIV, 400, 752, 753, 753f
  malaria, 687
  polyomavirus, 722
  varicella-zoster virus, 717–718, 717f, 719b
Latent phase, in syphilis, 632
Latent viruses (proviruses), 400. *See also* Latency
  HIV as, 400, 752, 753, 753f
Lateral (horizontal) gene transfer, 223–227, 225f, 226f, 227f, 228f, 228t, 230b
  antimicrobial resistance and, 227, 229f, 230, 230b, 303
  bacterial conjugation, 226–227, 227f, 228f, 228t
  S. aureus strains and, 336
  transduction, 225, 226f, 228t
  transformation, 224–225, 225f, 228t
Laundry enzymes, microbial production of, 12t
Laws/theories, in scientific method, 9f, 10
LCM. *See* Lymphocytic choriomeningitis
Leach fields, 794, 794f
Leading strand, in DNA replication, 199, 199–201, 200f
Lectin(s)/lectin pathway of complement activation, 459f, 460
Lederberg, Esther, 20, 226, 395
Ledipasvir, 314t
van Leeuwenhoek, Antoni, 2–4, 3f, 6–7, 18f, 97
Leflunomide, 537, 538t
Legionella genus/spp./*Legionella pneumophila*, 342, 344t, 611–613, 613f
Legionnaires' disease (legionellosis), 344t, 611–613, 613f
  serological testing for, 517, 613
Leishmania genus/spp., 361, 365f, 681–683, 682b, 683f, 692t
Leishmaniasis, 378, 681–683, 682b, 683f, 692t
Lenses
  in compound microscope, 100, 101f, 102f
  light refraction/magnification and, 98–99, 98f
  numerical aperture of, resolution affected by, 99
  immersion oil and, 100, 102f
Lentivirus genus/spp./*Lentivirus* human immunodeficiency virus, 391t, 392t, 751–757, 751f, 752f, 753f, 754b, 754t, 755f, 765f, 771t. *See also* HIV
  placenta crossed by, 421t, 756
Lepromatous leprosy, 575, 575f
Leprosy (Hansen's disease), 574–575, 575f
  immunization against (BCG vaccine), 509f, 575
  incubation period for, 429t
Leptospira genus/spp./*Leptospira interrogans*/leptospirosis, 636–637, 636f
Leptotrombidium species, as disease vectors, 433t
  for scrub typhus (*Orientia tsutsugamushi*), 622, 626t
Leuconostoc genus/spp./*Leuconostoc citrovorum*
  in buttermilk production, 780, 783t
  in vegetable fermentation, 779
  in wine and spirits production, 781
Leukemia
  adult acute T-cell lymphocytic, HTLV-1 causing, 750

  deltaretrovirus causing, 392t, 750–751, 751f, 771f
  hairy-cell, HTLV-2 causing, 750–751, 751f
Leukocidins, 429
  staphylococcal, 548
Leukocyte(s) (white blood cells), 452f, 453
  in host defense, 452–454, 452f, 453f
  in inflammation, 452f, 463f, 464
  lab analysis of, 454, 454f
  polymorphonuclear. *See* Neutrophil(s)
Leukocyte (white blood cell) count, 454, 454f
Leukoencephalopathy, progressive multifocal (PML), 392t, 722
Leukopenia, in ehrlichiosis and anaplasmosis, 623
Leukoplakia, hairy
  HHV-4 (Epstein-Barr virus) and, 719, 719f, 720
  in HIV infection/AIDS, 541, 719, 720
Leukotrienes
  in inflammation, 463, 465f
  in type I (immediate) hypersensitivity, 527, 527f, 527t
Levofloxacin, 288b, 313t
  diarrhea associated with use of, 319b
LGV strain, of *C. trachomatis*, 627
Li. *See* Lithium
Librarian's lung, 533
Libraries (gene), 243, 244f, 252t
Lice, as disease vectors, 378, 378f, 433t, 677
  for epidemic typhus (*Rickettsia prowazekii*), 621, 626t
  for relapsing fever (*Borrelia recurrentis*), 635, 636
  for trench fever (*Bartonella quintana*), 605
Lichens, 370, 371–373, 372f
Life processes/characteristics, 59, 59t
  in bacteria/archaea/eukaryotes, 59t
  chemical/metabolic, 19, 59, 59t
  in viruses, 59t, 404
Ligands, as adhesion factors, 422, 422f. *See also* Adhesins; Attachment proteins
Ligases, 127, 128t
  DNA
    in DNA repair, 220f, 221
    in genetic recombination, 223, 224f, 240, 241f
    in lagging strand synthesis, 200f, 201
  microbial production of, 787, 789t
Light
  in microbial growth, 163, 163f
  in phase, contrast and, 100
  in photosynthesis. *See* Photosynthesis
  refraction of, image magnification and, 98–99, 98f
  oil immersion lens affecting, 100, 102f
  transmission of, spectrophotometer measuring, 186–187, 186f
  wavelengths of, 98, 98f
    chlorophyll absorption and, 144
    microscopy and, 98, 98f, 104
Light chains, immunoglobulin/B cell receptor, 483, 483f, 484, 484f, 485f
Light-dependent reactions, 145, 145–146, 145f, 148f
Light-independent reactions, 145, 146–148, 148f
Light micrograph, 101
Light microscopy/microscopes, 100–104, 101f, 102f, 103f, 104f
  bright-field, 100, 100–101, 101f, 102f, 107t
  confocal, 103, 104f, 107t
  dark-field, 100, 101, 102f, 107t
  differential interference contrast (Nomarski), 101, 102f, 107t
  fluorescence, 100, 102–103, 103f, 107t
  phase, 100, 101, 102f, 107t
  resolving power of, 99f
  stains used for, 106–111, 111t
Light rays/waves, 98, 98f
Light reactions. *See* Light-dependent reactions
Light repair, 220, 220f
Lime, in wastewater treatment, 794
Limiting nutrient
  hydrogen as, 163
  nitrogen as, 165
Limnetic zone, 811–812, 811f

Lincosamides, 310t, 317t
  mechanism of action of, 293, 294f, 310t, 317t
Linezolid, 311t
Linker DNA, 197f
Linnaean system of taxonomy, 112–115, 114f, 357, 359f, 360
Linnaeus, Carolus, 4, 18f, 112, 357
Linoleic acid, 43f
Lipase(s)
  in lipid catabolism, 128t, 143, 143f
  as virulence factors of staphylococci, 548, 548t
Lipid(s), 40–44, 42f, 43f, 44f. *See also specific type*
  biosynthesis of, 150, 151f, 153f
  catabolism of, 143, 143f, 153f
Lipid A (endotoxin), 67, 68f, 427f, 428t, 587, 587f
  Enterobacteriaceae and, 593, 593f
  gonococci and, 589, 589b
  meningococci and, 590
  Pseudomonas aeruginosa and, 609
  Shiga-like, 596
Lipid rafts, 754
  PrP in, 405
Lipoglycopeptides, 291, 291f, 310t
Lipooligosaccharides (LOS), in *Neisseria* cell walls, 587
  gonococcal virulence and, 589, 589b
  meningococcal virulence and, 590
Lipopolysaccharide (LPS), 428. *See also* Endotoxin
  in Gram-negative bacteria, 67, 68f, 428
Lipoproteins, 49
  in adhesion (infection), 422
Lipoteichoic acids, in Gram-positive cell walls, 67, 68f
Lister, Joseph, 16, 18f, 276, 281
Listeria genus/spp./*Listeria monocytogenes*, 166, 336, 337t, 568, 570–571b, 786t
  anti-*Listeria* phage in control of, 277b, 568
  food/drink contaminated by, 272, 336, 568, 570–571b, 785, 786t
  growth in refrigerated products and, 272, 336, 568, 785, 786t
  meningitis caused by, 336, 568, 570b
  placenta crossed by, 336, 421t, 568, 570b
  reuterin affecting, 304b
Listeriolysin O, 571b
Listeriosis, 336, 568, 570–571b, 785
  transplacental transmission of, 336, 421t, 568, 570b
Lithium (Li), 30, 30f
Lithotroph(s), 136, 163, 165
Lithotrophic photoautotrophs, 165
Littoral zone, 811, 811f, 812
Liver
  Bartonella quintana infection of, 605
  candidal infection of, 658t, 660b
  in *Fasciola* (liver flukes) infection, 696
  hepatitis affecting, 726, 748, 770
  in mushroom poisoning, 668
  in *Plasmodium* life cycle, 688b
Liver flukes (*Fasciola* species), 419t, 696, 697f, 703t
Liver phase, in *Plasmodium* life cycle, 688b
Livestock
  assessing biological threats to, 813
  defense against biological threats to, 815
LM. *See* Light micrograph
Local drug administration, 300
Local infection, 434t
Loci (locus), receptor diversity and, 484, 484f
Lockjaw, 566. *See also* Tetanus
Loffler's medium, for *Corynebacterium diphtheriae* culture, 573
Logarithmic (exponential) growth, 180, 180f, 181, 181f
Log (exponential) phase, of microbial growth, 181, 181f
Lone Star (*Amblyomma*) tick, as disease vector
  for ehrlichiosis and anaplasmosis, 623, 626t
  for *Rickettsia parkeri* infection, 622b
Long-term nonprogressors, in HIV infection, 756
LOS. *See* Lipooligosaccharide
Louse. *See* Lice
Louse-borne (epidemic) typhus, 621, 626t
  bioterrorism and, 814t
Louse-borne relapsing fever, 635, 636

Lowenstein-Jensen agar, for *Mycobacterium tuberculosis* culture, 574
Lower respiratory system. *See* Lung(s); Respiratory system
Low G + C Gram-positive bacteria, 330f, 334–336, 337t, 547
Low-level germicides, 267
Low-oxygen cultures, 178
Low-power objective lens, 100
LPS. *See* Lipopolysaccharide
Lumefantrine, 317t
Luminal amebiasis, 678, 692t
Luminescence, protozoal, 362
Lung(s). *See under* Pulmonary and Respiratory system
Lupus (systemic lupus erythematosus), 533–534, 534f
Lutzomyia genus/spp., leishmaniasis transmitted by, 682, 683f, 692t
Lyases, 127, 128t
Lyme disease, 96b, 120b, 344t, 419t, 634–635, 634f, 635f
 serological tests for, 117, 120b, 517, 518, 635
Lymph/lymph flow, 474, 475f, 476
Lymphatic capillaries, 475f, 476
Lymphatic ducts, 475f, 476
Lymphatic nodule, 475f
Lymphatic system, 474, 474–476, 475f. *See also* Lymph nodes
 Wuchereria infection of, 702, 702f
Lymphatic tissue, mucosa-associated (MALT), 475f, 476
Lymphatic valves, 475f, 476
Lymphatic vessels, 474–476, 475f
Lymph nodes, 475f, 476
 inflamed (buboes)
  in lymphogranuloma venereum, 627, 628f
  in plague, 602, 603f, 604b
Lymphocryptovirus genus/spp., 392t, 713, 718, 729t. *See also* Epstein-Barr virus
Lymphocyte(s), 452f, 453, 453f, 473, 474f. *See also* B lymphocytes; T lymphocytes
 in adaptive immunity, 447, 452f, 453, 473–474, 474f
 natural (NK)/innate immunity and, 453
Lymphocyte-depleting therapies, 537, 538t
Lymphocytic choriomeningitis, 767
Lymphogranuloma venereum, 344t, 347, 627–628, 628f
Lymphoid organs, 475f, 476
Lymphoid tissues, 475f, 476
Lymphoma, HHV-4 (Epstein-Barr virus) and Burkitt's, 392t, 402, 718, 719f, 720
 Hodgkin, 402, 719, 720
Lyophilization (freeze-drying)
 for culture preservation, 179
 in microbial control/food preservation, 272, 275f, 785
Lysergic acid, *Claviceps purpurea* producing, 369
Lysogenic conversion, 395
Lysogenic phages, 395, 395f. *See also* Lysogeny/lysogenic replication cycle
Lysogeny/lysogenic replication cycle, 395–396, 395f, 396f
Lysosomes, 61f, 85t, 87, 87f, 455f, 456
Lysozyme, 280, 394, 448, 449, 449t
 antimicrobial action of, 280, 448, 449, 449t
 microbial production of, 789t
Lyssavirus genus/spp., 391f, 392t, 419t, 761, 771t
 as biological weapon, 814t
Lytic replication, 392, 392–394, 393f, 394f, 396, 396f

m (meter), 97, 97t
MAC. *See* Membrane attack complex
MacConkey agar, 175, 176–177, 177t, 178f
 for Enterobacteriaceae culture, 592, 593, 594f
Machupo hemorrhagic fever, 767
MacLeod, Colin, 19, 224
Macrocystis genus/spp., 375, 375f, 376t
Macrolides, 311t
 mechanism of action of, 291f, 293, 294f, 311t
Macromolecules. *See also specific type*
 organic, 40–53

polymerization of, 125, 125f, 153f. *See also* Metabolism
Macronucleus, in protozoa, 359, 362
Macrophage(s), 428–429, 453–454, 465f. *See also* Phagocytes; Phagocytosis
 alveolar, 454
 fixed, 454
 stellate (Kupffer cells), 478
 wandering, 453–454, 464
Macrophage activation factor (gamma-interferon), 458, 459, 459t, 489
Macrophage colony stimulating factor, recombinant DNA in production of, 788t
Macule(s)
 in poxvirus infection, 710, 711f
 in rubella, 747, 748f
Mad cow disease (variant Creutzfeldt-Jakob disease), 21b, 48, 406f, 407, 407b
 bioterrorism and, 814t
 enzyme for elimination of prion causing, 280
Madurella genus/spp./*Madurella mycetomatis*, mycetoma caused by, 666t, 667f
"Magic bullet," 18, 21, 289
Magnesium, 29t, 30f
 in chlorophyll, 144
 cycling of, 808–809
 as enzyme cofactor, 127, 128t
Magnesium hydroxide, 39
Magnetic fields/lenses, electron beam refraction and, 98, 104
Magnification, 98–99, 98f
 compound microscope and, 100
 electron microscope and, 104
 empty, 99
 oil immersion lens affecting, 100, 102f
 total, 100
Major histocompatibility complex, 478–479, 478f
 autoimmunity and, 538
 binding sites of, 478, 478f, 479
Major histocompatibility complex proteins, 478, 478f
 in antibody immune response, 492, 493f
 antigen presentation and, 478–479, 478f, 490, 490f
 antigen processing and, 479–480, 479f, 480f
 cell-mediated immune response and, 490, 490f
 clonal deletion of T cells and, 482, 482f
 donor-recipient matching/tissue typing and, 536–537
 graft-versus-host disease and, 536
 rheumatoid arthritis and, 533
 transplantation/rejection and, 478, 536
Malachite green, in Schaeffer-Fulton endospore staining, 110, 110f
Malaria, 362, 413b, 419t, 441b, 686–687, 688–689b, 692t
 Plasmodium drug resistance and, 306, 441b, 689b
 vaccine development and, 689b
Malaria secretome, 686
Malassezia genus/spp./*Malassezia furfur*, 665, 666f
MALDI-TOF mass spectrometry, 118
Male reproductive system. *See also* Reproductive system
 microbiome/normal microbiota of, 417t
Malic acid, in citric acid/Krebs cycle, 137f
Malignant tumor, 401–402. *See also* Cancer
MALT (mucosa-associated lymphatic tissue), 475f, 476
Malt/malting, in beer production, 781, 782f
Malta fever (brucellosis), 15t, 340, 344t, 606, 606f
 bioterrorism and, 814t
Maltose (malt sugar), 45
Manganese, 29t
 Borrelia burgdorferi virulence and, 633
Mannose, in lectin pathway, 459f, 460
Mantoux (tuberculin) test, 534–535, 535f, 577f
Marburgvirus/Marburg virus/Marburg hemorrhagic fever, 392t, 763, 771t
Margination, 464
Margulis, Lynn, 88
Marine ecosystems, 811, 811f, 812

Mariprofundus genus/spp./*Mariprofundus ferrooxydans*, 344t, 347
Marshall, Barry, 640
Mashing, in beer production, 781, 782f
Mass, atomic, 28, 29t
Mass spectrometry, MALDI-TOF, 118
Mastadenovirus genus/spp., 392t, 729t
Mast cell(s), 463, 526
 in inflammation, 463, 463f
 in type I (immediate) hypersensitivity
  degranulation of, 526–527, 527f, 527t
  sensitization of, 526, 527f
Mastigophora, 360
Matrix, of mitochondrion, 87, 88f
Matrix (extracellular polymeric substance/EPS), in biofilms, 168–169
Matrix-assisted laser desorption/ionization time-of-flight (MALDI-TOF) mass spectrometry, 118
Matrix proteins, viral, 389, 391f
Matter, 28
Maturation, in HIV replication, 752, 753f, 754–755
Maximum growth temperature, 166, 166f
MBC test. *See* Minimum bactericidal concentration (MBC) test
McCarty, Maclyn, 19, 224
McClintock, Barbara, 229
MDR-TB (multi-drug-resistant tuberculosis), 306, 307b, 574, 577b
Measles (rubeola/red measles)/measles virus, 392t, 472b, 497b, 504b, 758, 758–759, 758f, 758t, 759b, 759f, 771t
 immunization against, 497b, 504b, 505f, 508f, 509f, 759, 759b, 759f. *See also* Measles/mumps/rubella (MMR) vaccine
 transmission of, 431, 497b, 758, 759b
Measles, German. *See* Rubella
Measles/mumps/rubella (MMR) vaccine, 497b, 504b, 505f, 506, 508f, 509f, 748, 748f, 759, 759b, 759f, 760, 760f
 contraindications to during pregnancy, 748
Measurement, units of, 97, 97t
Meat/pork
 contaminated
  E. coli O157:H7 infection and, 595, 786b
  tapeworm infestation and, 694, 694f, 695f, 703f
  toxoplasmosis caused by, 687, 687f, 690
  variant Creutzfeldt-Jakob disease caused by, 21b, 407
 fermentation in production/preservation of, 780, 783f, 785
 spoilage and, 783, 785
Mebendazole, 316t, 317t
Mechanical vectors, 377, 432, 433t, 677. *See also* Arthropod vectors
Mediators, inflammatory, 462f, 463–464, 463f, 465f
 in type I (immediate) hypersensitivity, 526, 527f, 527t
Medical mycology, 19t, 648–674. *See also* Fungi; Mycoses
 antifungal therapies and. *See* Antifungal drugs
 antifungal vaccines and, 507, 650
 categories of fungal agents and (pathogenic/opportunistic), 648–649, 649t
 clinical manifestations of fungal disease and, 649
 diagnosis of fungal infections and, 649–650, 649f
 epidemiology of mycoses and, 648
Medical/pharmaceutical microbiology, 18f, 19t, 252–254, 788, 788t
Medium (culture), 171. *See also* Culture media
Medulla, lymph node, 475f, 476
Mefloquine, 318t
Megakaryocytes, 453
Megavirus genus/spp., 388, 389f
Meglumine, 317t
Meiosis I and II, 356–357, 356f, 357t
Melanin
 Cryptococcus producing, 659
 fungal use of radioactivity and, 367
 fungi affecting production of, 665
 microbial remediation of nuclear radiation and, 169b

Melarsoprol, 317t
Melioidosis, 608, 610b
 bioterrorism and, 608, 814t
Membrane(s). *See also* Cytoplasmic (cell/plasma) membranes; Mucous membranes
 in eggs (embryonic tissues), viral cultures and, 403, 403f
 phospholipid bilayers forming, 42
 cholesterol in, 44, 44f, 81
Membrane attack complex (MAC), 461f, 462, 462f
 complement fixation test and, 515
Membrane filters/filtration
 in microbial control, 272, 273f, 273t
 drinking water treatment and, 791f, 792
 in microbial growth estimation, 184–185, 185f, 272
 in water quality testing, 792, 792f
Membrane fusion, in animal virus replication, 397, 397f
Membrane proteins, 69, 69f
 in electron transport chains, 136, 139
 in facilitated diffusion, 71, 71f
Membrane rafts, 81
Membranous organelles, 76f, 85–88, 85t, 87f, 88f, 89t. *See also specific type*
Memory (immunological), 473, 474, 491–492, 493f, 494–495, 495f
Memory B cells, 493f, 494–495
Memory response, 492
Memory T cells, 491–492, 494
 tuberculin response mediated by, 535
Meningeal candidiasis, 658t, 660b
Meningitis, 15t. *See also specific causative agent*
 in blastomycosis, 654
 candidal, 658t, 660b
 coxsackieviruses causing, 738
 cryptococcal (fungal meningitis/cryptococcosis), 370, 659, 662
 Haemophilus influenzae causing, 424, 509t, 604, 605
 in HIV infection/AIDS, 541
 Listeria causing, 336, 568, 570b
 Neisseria meningitidis causing (meningococcal), 15t, 344t, 590–591
  immunization against, 508f, 509t, 590–591
 Streptococcus agalactiae (group B streptococci) causing, 556
 Streptococcus pneumoniae causing (pneumococcal), 558b, 559
Meningococcal meningitis, 15t, 344t, 590–591
 immunization against, 508f, 509t, 590–591
Meningococcal septicemia, 590, 590f
Meningococcal vaccine, 508f, 509t, 590–591
Meningococcus. *See Neisseria meningitidis*
Meningoencephalitis
 amebic, *Naegleria* causing, 679, 692t, 790b
 in *Trypanosoma brucei* infection, 681
Menstruation/menstrual flow, in host defense, 451t
Mental disease, 424, 424t
Mercury, antimicrobial action of, 279, 281t, 317t
Merismopedia genus/spp., 333f
Meropenem, 230b, 309t
Merozoites
 in *Cyclospora* life cycle, 692f
 in *Plasmodium* life cycle, 687, 688b
MERS. *See* Middle East respiratory syndrome
Mesophiles, 166, 167f
 soil organisms as, 810
Messenger RNA (mRNA), 206, 207f, 208, 208f, 209
 antimicrobials affecting, 291f, 295, 295–297, 296f
 eukaryotic, 206, 207f, 209
 prokaryotic, 208, 208f
 real-time PCR in identification of, 246
 in reverse transcription, 239
 as riboswitch, 216
 in transcription, 203, 206, 207f
 transcriptomes, 250
 in translation, 207, 208–209, 208f, 210, 210f, 213t
 in viral synthesis, 398, 399f, 399t
  HIV, 752, 753f, 754

SUBJECT INDEX

Metabolic activity, in microbial growth estimation, 187
Metabolic expression, control of, 152
Metabolism, 12, 18f, 19t, 37, 124–160, 153f, 295. *See also specific compound*
  acid-base balance/pH range and, 38–39
  amino acid biosynthesis, 150, 151f, 153f
  antimicrobials affecting, 290, 291f, 295, 296f, 312t, 316t, 317t
  ATP production/energy storage and, 125, 125–126, 125f, 126f, 126–127
  biochemistry in study of, 12, 19
  carbohydrate biosynthesis, 149, 150f, 153f
  carbohydrate catabolism, 133–140, 133f, 134f, 135f, 137f, 139f, 140t, 142t, 153f
  catabolism and anabolism and, 125–126, 125f, 126–127, 153f. *See also* Anabolism; Catabolism
  as characteristic of life, 19, 59, 59t
  chemical reactions underlying, 19, 35–37, 36f, 125–133, 125f
  diversity of, 140
  enzymes in, 12, 125, 127–132. *See also* Enzyme(s)
  human microbiome and, 253b
  integration/regulation of functions of, 151–152, 153f
  intermediate pathways in, 153f
  lipid biosynthesis, 150, 151f, 153f
  lipid catabolism, 143, 143f, 153f
  microbial growth and, 162, 187
  nucleotide biosynthesis, 151, 152f, 153f
  oxidation-reduction (redox) reactions, 125, 126, 126f
  pathways for, 125, 153f. *See also specific type*
  photosynthesis and, 144–148, 145f, 147f, 148f, 153f
  protein catabolism, 143, 144f, 153f
  protein synthesis, 153f
  in resistant cells, 306
Metabolites
  precursor, 125, 125f, 126, 149, 149f, 153f
    for amino acid conversion, 150
    in citric acid/Krebs cycle, 136
    in glycolysis, 133
    in nucleic acid biosynthesis, 151, 152f
    pentose phosphate pathway producing, 140
  primary, of industrial fermentation, 787
  secondary, of industrial fermentation, 787
Metabolomics, 250
Metacercariae, in fluke life cycle, 696, 697f
Metachromatic granules, in corynebacteria, 337
Metagenomics, 252
Metagenomic sequencing, 20, 118–119, 252, 347
Metal-containing proteins
  in electron transport chains, 139
  superoxide dismutases and, 164
Metal ions
  antimicrobial action of, 279, 279f, 281t, 295, 317t
  cycling of, 808–809
Metaphase
  meiosis, 356, 356f, 357t
  mitosis, 355, 355f, 357t
Metastasis, 401
Meter (m), 97, 97t
Methane
  as alternative fuel source, 332, 788, 789t
  covalent bonds in formation of, 32f
  as greenhouse gas, 332, 342, 807
  methanogens producing, 115, 139, 332, 788, 807
  wastewater treatment producing, 332
Methane oxidizers, 342
Methanogens, 115, 139, 330f, 332, 788, 805, 807
Methanopyrus genus/spp./*Methanopyrus kandleri*, 78, 79f, 332
Methicillin, 291, 292f, 309t, 551. *See also* Methicillin-resistant *Staphylococcus aureus*
Methicillin-resistant *Staphylococcus aureus* (MRSA), 305, 305b, 336, 551, 551b
  community-associated, 305b, 551
  safe handling of, 268

Methionine, 203f, 207, 207f, 208f, 212
  in archaea, 329
  PrP misfolding and, 405
Methyl alcohol, for specimen preparation, 106
Methylation
  in DNA replication, 201, 202
  in mismatch repair, 201
Methylene blue, 108, 111t
  in acid-fast staining, 109
Methylococcus genus/spp., 344t
Methylprednisolone, 537, 538t
4-Methylumbilliferyl-β-D-glucuronide (MUG) water quality test, 792, 792f
Metric units of measurement, 97, 97t
Metrifonate, 316t
Metronidazole, 313t, 318t, 319b
  toxicity of, 302, 302f, 318t
Mevinic acids, 365
MfpA protein, *Mycobacterium tuberculosis* resistance and, 306
Mg. *See* Magnesium
MHC. *See* Major histocompatibility complex
MHC proteins. *See* Major histocompatibility complex proteins
Mice. *See also* Rodents
  genetically modified, for biomedical animal models, 254
  in *Hantavirus* transmission, 384b, 408b, 767, 810
Microaerophiles, 164
  culturing, 178
Microarrays, DNA, 246–248, 247f, 252t
  for genetic screening, 253
Microbes/microorganisms, 3–4, 3f. *See also* Microbiology
  adaptation in survival of, 805
  antagonism/antagonistic relationships and, 168. *See also* Antagonism; Competition
  antigenicity of, 117
  in aquatic habitats, 811–812, 811f
  association and biofilms of, 168–170, 169f, 170b, 171f, 803–804, 804f. *See also* Biofilm(s); Microbial associations
  beneficial. *See specific organism* and Beneficial microbes
  in biogeochemical cycles, 2–7, 806–809, 807f, 808f, 809f
  as biological weapons, 812–815, 814t, 815f. *See also* Bioterrorism
  in biomining, 809
  in bioremediation, 19, 20, 169b, 805
  classification of, 4–7, 4f, 5f, 7f, 8f, 112, 171, 172f. *See also* Taxonomy/microbial classification and identification
    biochemical tests in, 116–117, 116f, 117f
    evolution/nucleotide sequencing and, 20, 115, 118–119, 329, 330f, 357, 359f, 360
    MALDI-TOF mass spectrometry and, 118
    phage typing in, 117, 118f
    physical characteristics and, 116
    serological tests in, 117, 117f
  culturing, 171–179, 172f, 173t. *See also* Culture
  death of, 264. *See also* Microbial death rate
  epidemiology in study of, 17, 18f, 19t, 434–441
  evolution of. *See* Evolution
  genetics of. *See* Genetics, microbial
  growth of. *See* Microbial growth
  in human body, 253b, 804, 804f. *See also* Microbiome
  hygiene hypothesis and, 263b, 526
  identification of, 112, 115–119, 116f, 117f, 118f. *See also* Microbes/microorganisms, classification of; Taxonomy/microbial classification and identification
  culture in, 171, 172f
  DNA microarrays in, 248
  evolution/nucleotide sequencing and, 113–115, 118–119, 329, 330f, 357, 359f, 360
  industrial uses of, 12, 12t, 18f, 19t. *See also* Industrial microbiology

  isolation of, 172–174, 173f, 174f
  life processes in, 59, 59t
  microscopes in discovery/observation of. *See* Microscopy
  movement of into hosts, 420–421, 420f, 421f, 421t. *See also* Infection; Portals of entry
  movement of out of hosts, 430, 430f. *See also* Portals of exit
  normal. *See* Microbiome
  pathogenic. *See specific type and* Pathogen(s)
  pH range and, 38–39
  placenta crossed by, 420f, 421, 421t, 431
  recombinant, 223, 224f, 787
  requirements of for existence, 162–171. *See also* Chemistry; Microbial growth
  resistant. *See* Resistance
  role of in environment, 18f, 19t, 20, 803–812. *See also* Environmental microbiology
  in soil, 804, 804f, 809–810, 809f, 810t
    diseases caused by, 810, 810t
  spontaneous generation (abiogenesis) of, 7–10, 8f, 9f, 10b
  susceptibility of to antimicrobial methods/agents, 266–267, 266f, 298–300, 300f, 301b, 301f. *See also* Resistance
  symbiotic relationships and, 168, 414–418, 415b, 415f, 416f
  synergistic relationships and, 168
  virulence/virulence factors of, 426–429, 427f, 428t
  in water, levels allowed, 791
    chlorine disinfection and, 277, 791f, 792
Microbial adaptation, survival and, 805
Microbial antagonism. *See* Antagonism
Microbial associations. *See also* Biofilm(s)
  growth and, 168–170, 169f, 170b, 171f
  levels of, 803–804, 804f
Microbial communities, 20, 252. *See also* Microbiome
Microbial competition. *See* Competition
Microbial control. *See* Antimicrobial drugs; Antimicrobial methods; Microbial growth, control of
Microbial cooperation, adaptation/survival and, 805
Microbial cultures. *See* Culture
Microbial death, 264
Microbial death rate, 264, 265f
  environmental conditions affecting, 267–268
  susceptibility of organisms and, 266
  thermal methods of microbial control and, 269, 269f
Microbial ecology, 18f, 803–805, 804f
  adaptation/survival and, 805
  associations and
  levels of, 803–804, 804f
  microbial growth and, 168–170, 169f, 170b, 171f
Microbial genetics. *See* Genetics, microbial
Microbial growth, 59, 59t, 161–192
  arithmetic, 180, 180f
  associations and biofilms and, 168–170, 169f, 170b, 171f. *See also* Biofilm(s)
  continuous culture in chemostat and, 182, 182f
  control of
    actions of agents/methods used in, 264–266
    bacteriophages in, 277b
    biosafety levels and, 268, 268f
    in body, 288–323. *See also* Antimicrobial drugs
    chemical methods for, 276–282, 281t
    efficacy of methods and, 266–268, 266f, 267f
    in environment, 262–287. *See also* Antimicrobial methods
    microbial death rates and, 264, 265f
    physical methods for, 268–275, 275t
    principles of, 263–266, 265f, 265t
    resistance and, 282
    selection of methods for, 266–268, 266f, 267f
    terminology of, 263–264, 265t
  in culture, 171–179, 172f, 173t. *See also* Culture

  generation time and, 180
  logarithmic (exponential), 180, 180f, 181, 181f
  mathematical considerations in, 180, 180f
  measuring, 183–187
    direct methods for, 183–186, 183f, 184f, 185f, 186f
    indirect methods for, 186–187, 186f
  phases of, 180–182, 181f
  of populations, 162, 179–187. *See also* Population (microbe)
  requirements for, 162–171
    nutrients (chemical and energy), 162–165, 163f, 165f, 165t
    physical, 165–168, 166f, 167f
Microbial growth curve, 180–181, 180f, 181f
Microbial guilds, 804, 804f
Microbial load, chlorine disinfection of water and, 277, 792
Microbial metabolism. *See* Metabolism
Microbial morphology. *See* Morphology
Microbial nutrition, 161–192. *See also* Microbial growth
Microbial resistance. *See* Resistance
Microbiology
  agricultural, 19t, 254–256, 255f, 788–789, 789t
    crop threats and, 812, 813, 814, 815. *See also* Agroterrorism
    livestock/poultry threats and, 813
    recombinant DNA technology in, 254–256, 255f, 788–789, 789t
    Bt toxin and, 255, 335–336, 336f, 788–789, 789t
    ethical/safety issues and, 256, 788–789
  applied, 19t, 779, 779–786. *See also specific type*
  aquatic, 811–812, 811f
  bioterrorism and, 812–815, 814t, 815f. *See also* Bioterrorism
  chemistry of, 27–57. *See also* Chemistry
  ecological, 18f, 803–805, 804f
  environmental, 18f, 19t, 20, 803–812
  fields of, 18–19, 19t. *See also specific type*
  food, 10b, 12t, 18f, 19t, 779–786, 783f, 786t. *See also* Food microbiology
  history of, 1–26
    early years, 2–7
    Golden Age, 7–18, 15t, 18f
    modern age, 18–22, 19t
  hospital (medical/pharmaceutical), 18f, 19t, 252–254, 788, 788f
  industrial/biotechnology, 12, 12b, 12t, 18f, 19t, 238, 239f, 779, 786–795. *See also* Industrial microbiology; Recombinant DNA technology
  mathematical considerations in, 180, 180f
  public health, 19t, 435, 437t, 438f
  soil, 804, 804f, 809–810, 809f, 810t
Microbiome/normal microbiota/flora, 172, 252, 253b, 416–418, 416f, 417t, 803–812, 804f. *See also specific structure or organism*
  acquisition of, 416–417
  anaerobic Gram-negative bacilli, 614
  antimicrobials affecting, 298, 299b, 302
  probiotics and, 304b
  disease caused by, 417–418. *See also* Opportunistic pathogens
    antimicrobial use and, 298, 299b, 302, 304b
  in disease treatment, 816b
  fecal, transfer of, 803b
  health maintenance and, 804
  human, 253b, 804, 804f
  in innate immunity, 449–450
  isolation of in pure culture, 172
  of soil, 804, 804f, 809–810, 809f, 810t
Microbiota, normal. *See* Microbiome
Microcephaly, Zika virus infection and, 745
Micrococcus luteus, arrangement of, 328f
Microfilaments
  in cytokinesis, 357
  in eukaryotic cytoskeleton, 83, 84f
Microfilariae, of *Wuchereria bancrofti*, 702, 702f
Microglia/microglial cells, 454
  antigen presentation by, 478
Micrographs, 101
  light, 101

scanning electron, 104, 105f
transmission electron, 104, 105f
Microhabitats, 804, 804f
human, 804, 804f
soil, 804, 804f
Microinjection, for DNA insertion into cells, 248, 249f, 252t
Micrometer (μm), 97, 97t
Micromonospora genus/spp./*Micromonospora purpurea*, antimicrobials produced by, 290t, 310t
Micronucleus, in protozoa, 359, 362
Microorganisms. *See* Microbes/microorganisms
Micropipettes
for DNA insertion into cells, 248, 249f, 252t
for microbial isolation, 174
MicroRNAs (miRNAs), 215
MicroScan panel, 116–117, 117f
Microscopic counts, for microbial growth estimation, 183, 183f
Microscopy/microscopes, 97–106, 107t. *See also specific type*
atomic force, 106, 106f, 107t
bright-field, 100, 100–101, 101f, 102f, 107t
confocal, 103, 104f, 107t
dark-field, 100, 101, 102f, 107t
differential interference contrast (Nomarski), 101, 102f, 107t
electron, 104–105, 105f, 107t
fluorescence, 100, 102–103, 103f, 107t
Leeuwenhoek's development of, 3–4, 3f, 6–7, 97
light, 100–104, 101f, 102f, 103f, 104f
phase, 100, 101, 102f, 107t
phase-contrast, 101, 102f, 107t
principles of, 98–100, 98f, 99f
probe, 105–106, 106f, 107t
resolving power of, 99, 99f
scanning electron, 104–105, 105f, 107t
scanning tunneling, 105, 106f, 107t
stains used for, 106–112, 111t
transmission electron, 105, 105f, 107t
Microsporidia/microsporidiosis, 369, 496b, 662–664, 663f
as emerging/reemerging infection, 369, 496b, 662–664, 663f
Microsporidium genus/spp., 369, 662
Microsporum genus/spp. (*Microsporum canis/gypseum*), 419t, 665, 665t
Microtubules
triplets of
in centrioles, 83–84, 84f
in cilia/flagella, 82, 83f
tubulin
in centrioles, 83–84, 84f
in cilia/flagella, 82, 83f
in cytoskeleton, 83, 84f
griseofulvin mechanism of action and, 650
Microwaves, 98f, 275
microbial control and, 275
MIC test. *See* Minimum inhibitory concentration (MIC) test
Middle East respiratory syndrome (MERS/coronavirus respiratory syndrome), 736t, 748, 749, 749f
bioterrorism and, 814t
Migration, phagocyte, 464, 465f
Milk
Campylobacter jejuni contamination of, 282b, 639
in cheese production, 780, 780f
Coxiella contamination of, 613
disease transmission and, 262b, 282b, 432, 440
fermentation products of, 783t
pasteurization/safety and, 12, 271, 271t, 432, 440, 785
Salmonella contamination of, 600, 786t
ultra-high-temperature pasteurization/sterilization of, 271, 271t
Milk curds, in cheese production, 780, 780f
Milk sugar (lactose), 45
Millimeter (mm), 97, 97t
Mimicry, molecular, 538–539
Mine tailings, acid mine drainage and, 167, 805–806, 805f, 807f
Minimum bactericidal concentration (MBC) test, 300, 301f
Minimum growth temperature, 166, 166f

Minimum inhibitory concentration (MIC) test, 299–300, 300f
Minocycline, 311t
Minor polio, 737. *See also* Poliomyelitis
Miracidia, in fluke life cycle, 696, 697f
miRNA-induced silencing complex (miRISC), 215–216
miRNAs (microRNAs), 215
Mismatch repair, 220f, 221
Missense mutations, 216–217, 217f, 218t
Mites
as disease vectors, 377, 378, 378f, 433t, 677
for scrub typhus (*Orientia tsutsugamushi*), 622, 626t
dust, 528, 528f, 529
taxonomic classification of, 114f
Mitochondria, 61f, 85t, 87, 88f
evolution/endosymbiotic theory and, 87, 88
in protozoa, 359
Mitochondrial DNA/chromosomes, 87, 88, 197
antimicrobial activity/killing by eosinophils and, 456
replication of, 202
Mitochondrial membranes, 87, 88f
electron transport chains in, 136
Mitochondrial ribosomes, 87, 88f
Mitosis, 202, 354–355, 355f, 357t
centrosomes in, 84
chromatin/chromatin fibers during, 85, 197
chromosomes in, 85, 197, 197f, 354, 355, 355f
Mitosomes, 361
mm (millimeter), 97, 97t
MMR. *See* Measles/mumps/rubella (MMR) vaccine
MMWR (Morbidity and Mortality Weekly Report), 435, 438f
Mn. *See* Manganese
Mo. *See* Molybdenum
Modified live vaccines (attenuated/live vaccines), 506
recombinant, 507–508, 507f
residual virulence/safety and, 506, 509–510
Modified Thayer-Martin medium, for *Neisseria* culture, 588
Moist heat, in microbial control, 269–271, 270f, 271f, 271t, 275f
Moisture, in soil, microbial survival and, 809
Mold(s), 4, 365, 366, 366f. *See also* Fungi; Water molds
allergies caused by, 528, 669
asexual/sexual spores of, 4, 367, 368f
morphology of, 365, 366, 366f
slime, 365f
classification of, 357, 364
zygomycetes, 369, 662
Molecular biology/techniques, 20. *See also* Biotechnology/industrial microbiology; Recombinant DNA technology
for microbial growth estimation, 187
vaccine development/production and, 253, 506–508, 507f, 749, 788t
Molecular formula, 32
Molecular mimicry, 538–539
Molecule
chemical bonds in formation of, 31
nonpolar covalent bonds, 31–33, 32f, 33f, 35t
polar covalent bonds, 33, 33f, 35t
decomposition reactions in formation of, 36, 36f
synthesis reactions in formation of, 35–36, 36f
Mollicutes, 337f, 568
Molluscipoxvirus, 712, 729t
Molluscum contagiosum, 712, 712f
Molybdenum, 29t
Monkeypox, 710, 712–713, 713b
immunization and, 509t, 712–713
Mono. *See* Mononucleosis
Monobactams, 309t
Monoclonal antibodies, 511
in fluorescent immunoassays, 516
immunosuppressive actions of, 537, 538f
in passive immunotherapy, 511, 511f
Monocular microscope, 100
Monocytes, 452f, 453, 453f, 464

Monoecious helminths, 693
Monomer(s), 44
Mononegavirales (order), 391t
Mononucleosis (mono)
cytomegalovirus causing, 720
HHV-4 (Epstein-Barr virus) causing, 392t, 718, 719f, 720
HHV-6 infection and, 721
Monosaccharides, 44, 45f
Monounsaturated fats/fatty acids, 41, 42f, 43f
Montagnier, Luc, 751
Montagu, Lady Mary, 17, 17f, 505
Moraxella genus/spp./*Moraxella catarrhalis*, 610
restriction enzymes produced by, 240t
Morbidity, 422. *See also* Disease
Morbidity and Mortality Weekly Report (*MMWR*), 435, 438f
Morbillivirus genus/spp./*Morbillivirus* measles virus, 392t, 497b, 758, 758–759, 758f, 758t, 759b, 759f, 771t. *See also* Measles
Morchella esculenta, 369f
Mordants, 110
in flagellar staining, 110
in Gram staining, 109, 109f
Morel, 369f
false, 668
Morganella genus/spp., 592f, 597
sites of infection caused by, 604f
Morphology, 18f
of algae, 373
of fungi, 365–366, 366f
of prokaryotic cells, 325, 325f
of protozoa, 359, 360f
in taxonomy/microbial classification, 116
Morula, in *Ehrlichia* and *Anaplasma* growth/reproduction, 623, 623f
Mosaic disease, wheat, soilborne transmission of, 810t
Mosquitoes
as disease vectors, 378f, 379, 385f, 433t, 677
for arboviruses, 741–746, 747f
for chikungunya, 401b, 747t
for dengue fever, 257b, 743, 744, 744f, 745f, 747t
mosquito control and, 385b, 746b
for encephalitis, 742, 742f
for filarial nematodes, 702, 703t
for malaria (*Plasmodium* species), 687, 688b, 689b, 692t
for West Nile virus, 734b, 742, 742f, 747t
Wolbachia in elimination of, 385b, 746b
for yellow fever, 16b, 744, 747t
for Zika virus, 744, 746b, 747t
Wolbachia-infected, 385b, 746b
Most probable number (MPN) method, for microbial growth estimation, 185–186, 185f, 186t
Motility, 89t. *See also* Cilia; Flagella
by archaea, 77–78, 89t
by bacteria, 89t
cyanobacteria, 333
fimbriae in, 65
flagellar, 62–65, 63f, 64f, 89t
by eukaryotic cells, 82, 83f, 84f, 89t
by protozoa, classification and, 360, 679
Mouth
microbiome/normal microbiota of, 417t
as portal of entry, 420f, 421
for *Helicobacter pylori*, 640
as portal of exit, 430, 430f
Moxifloxacin, 313t
MPN. *See* Most probable number
M protein, 429, 552, 555f, 556
mRNA. *See* Messenger RNA
MRSA. *See* Methicillin-resistant *Staphylococcus aureus*
MS. *See* Multiple sclerosis
MS2
genome of, 386, 387f
size of, 388f
*Msp*I restriction enzyme, 240t
Mucin(s), in host defense, 449
Mucinase, 426–428
Mucocutaneous candidiasis, 658t
Mucocutaneous leishmaniasis, 683, 683f, 692t
Mucor genus/spp./*Mucor rouxii*, 366f, 662
asexual spores of, 368f
Mucosa-associated lymphatic tissue (MALT), 475f, 476

Mucous membranes
in host defense, 448–449, 449f, 449t
as portals of entry, 420–421, 420f, 448
respiratory, 449, 449f
specimen collection from, 173t
Mucus, 448, 449
in host defense, 449, 449t
Mu F_c region/heavy chains, 485, 488t
MUG (4-methylumbelliferyl-β-D-glucuronide) water quality test, 792, 792f
Muller, Hermann, 217
Multi-drug resistance/multiple-drug-resistant pathogens, 230b, 306–307
efflux pumps and, 306
Enterococcus, 230b, 273t, 306, 336, 560
Klebsiella pneumoniae, 230b
in tuberculosis (MDR-TB), 306, 307b, 574, 577b
Multiple sclerosis (MS)
autoimmunity and, 539
HHV-6 and, 721
Mumps/mumps virus, 392t, 758, 760, 760f, 771t
immunization against, 508f, 509t, 760, 760f. *See also* Measles/mumps/rubella (MMR) vaccine
Municipal wastewater treatment/municipal sewer systems, 793–794, 793f
Mupirocin, 311t
mechanism of action of, 291f, 293, 311t
microbial production of, 290t, 311t
Murine (endemic) typhus, 622, 626t
Musca genus/spp., as disease vectors, 433t
Muscimol, 668
Muscles
tapeworm encystment in, 694, 694f
tetanus toxin affecting, 565–566, 566–567, 566f, 567f
Mushroom(s), 370. *See also* Basidiomycota; Fungi
poisonous, 370, 668, 669f
Mushroom grower's lung, 532–533
Mushroom poisoning, 370, 668, 669f
Must, in wine-making, 781, 781f
Mutagens, 217–219, 218f, 219f, 238
frameshift, 219, 219f
identification of, 222–223, 223f
radiation, 217–218
in recombinant DNA technology, 238, 252t
Mutants, 221
identification of, 221–222, 221f, 222f
Mutations, 216–223
agents causing (mutagens), 217–219, 218f, 219f
DNA repair and, 219–221, 220f
effects of, 216–217, 217f, 218t
emerging/reemerging diseases and, 21b
frameshift, 216, 217, 217f, 218t, 219, 219f. *See also* Frameshift mutations
chemicals causing, 219, 219f
frequency of, 219
gross, 216
HIV origins and, 752, 754b
identification of mutants/mutagens/carcinogens and, 221–223, 221f, 222f, 223f
influenzavirus, 768b, 769b
missense, 216–217, 217f, 218t
nonsense, 217, 217f, 218t
point, 216, 216–217, 217f, 218t
recombinant DNA technology and, 238
screening for, 253
ethical issues and, 256–257
silent, 216, 217f, 218t
spontaneous, 217
transposition as, 216, 228–230, 229f
types of, 216, 218t
Mutualism, 414–415, 415b, 415f, 416t
Mycelium/mycelia, 366, 366f
of basidiomycete, 370, 371f
of *Coccidioides immitis*, 655
Mycetismus (mushroom poisoning), 370, 668, 669f
Mycetomas
fungal, 666t, 667, 667f
Nocardia causing, 580, 580f

SUBJECT INDEX

Mycobacterium genus/spp. (*Mycobacterium avium-intracellulare/bovis/ulcerans*) (mycobacteria), 330f, 337, 337t, 573–578, 578b
  acid-fast staining of, 67, 109–110, 110f, 337, 337f, 573
  antimicrobials effective against, 291–292, 299f
  mycolic acid/waxes in cell walls of, 42, 67, 291, 337, 573, 574
  pasteurization in control of, 270
  susceptibility of, 266f, 267
Mycobacterium leprae, 337, 425, 574–575, 575f. See also Leprosy
  animal culture for growth of, 178
  clofazimine affecting, 297, 312t
  generation time of, 180
  immunization against (BCG vaccine), 509t, 575
Mycobacterium tuberculosis, 14, 337, 424, 573–574, 574b, 576–577b. See also Tuberculosis
  auramine O dye staining, 103, 103f
  as biological weapon, 814t
  clofazimine affecting, 297, 312t
  direct fluorescent immunoassay in detection of, 516
  disseminated infection caused by, 577b
  HIV infection/AIDS and, 541
  immunization against (BCG vaccine), 509t, 577b
    tuberculin skin test and, 534, 535
  mycolic acid/waxes and, 42, 67, 291, 577b
  primary infection caused by, 576–577b
  pyrazinamide affecting, 295, 312t
  resistant strains of, 306, 307b, 574, 577b
  safe handling of, 268
  secondary/reactivated infection caused by, 577b
  susceptibility of, 267
  tuberculin test in detection of, 534–535, 535f
  water requirements and, 167
Mycolactone, 578b
Mycolic acid, in *Mycobacterium* cell walls, 67, 291, 337, 573
  antimicrobial agents and, 292
  Mycobacterium leprae, 67, 291, 574
  Mycobacterium tuberculosis, 42, 67, 291, 577b
Mycology/mycologists, 18f, 19t, 364. See also Fungi; Mycoses
  medical, 648–674
    antifungal therapies and. See Antifungal drugs
    antifungal vaccines and, 507, 650
    categories of fungal agents and (pathogenic/opportunistic), 648–649, 649t
    clinical manifestations of fungal disease and, 649
    diagnosis of fungal infections and, 649–650, 649f
    epidemiology of mycoses and, 648
Mycophenolate mofetil, 537, 538t
Mycoplasma genus/spp. (*Mycoplasma genitalium/hominis/pneumoniae*) (mycoplasmas) 187b, 330f, 335, 335f, 337t, 460b, 547, 568–572, 569f, 630b
  absence of cell wall and, 69, 335, 568
  pneumonia caused by, 187b, 460b, 569
  sterol in membranes of, 44, 335, 568
Mycorrhizae, 365
Mycoses, 365, 419t, 648–674. See also specific type and Fungi
  categories of agents causing (pathogenic/opportunistic), 648–649, 649t
  clinical manifestations of, 649
  cutaneous/subcutaneous, 651, 666–668, 666f, 666t, 667f, 668f
  diagnosis of, 649–650, 649f
  epidemiology of, 648
  in HIV infection/AIDS, 664, 751t
  opportunistic, 648–649, 649t, 664. See also Opportunistic mycoses
  pathogenic, 648, 651–656, 651f
  prevention of/vaccines for, 650
  superficial, 664–665, 665f, 665t, 666t
  systemic
    opportunistic, 656–664
    pathogenic, 648, 651–656, 651f
  treatment of, 650. See also Antifungal drugs
Mycotoxicoses, 668

Mycotoxins, 668
Myelin sheaths, autoimmune destruction of, in multiple sclerosis, 539
Myelomas, in hybridoma production, 511, 511f
Myocardial candidiasis, 658t, 660b
Myocarditis, coxsackie B virus causing, 738
Myonecrosis (gas gangrene), 567, 568
Myxobacteria, 330f, 344t, 346, 346f
Myxococcus xanthus, beneficial uses of, 39b
Myxospores, 346, 346f

N. See Nitrogen
NA. See Neuraminidase
Na. See Sodium
N-acetylglucosamine (NAG), 45f, 66–67, 66f, 67f, 290, 292f
  Caulobacter glue and, 341b
N-acetylmuramic acid (NAM), 66–67, 66f, 67f, 290, 292f
  antimicrobials affecting, 290–291, 292f
NaCl. See Sodium chloride
NAD$^+$ (nicotinamide adenine dinucleotide), 126, 128, 128t, 153f
  in acetyl-CoA synthesis, 135
  in citric acid/Krebs cycle, 136, 137f, 153f
  in electron transport chain, 136, 138f, 153f
  in fermentation, 141, 141f, 142f
  in glycolysis, 134f
  Haemophilus influenzae requiring, 604
NADH, 126, 153f
  acetyl-CoA synthesis producing, 135, 135f, 140t
  in beta-oxidation, 143, 143f
  in chemiosmosis, 140
  in citric acid/Krebs cycle, 133f, 136, 137f, 140t, 153f
  in electron transport chain, 136, 138f, 139, 140t, 153f
  Entner-Doudoroff pathway producing, 140
  in fermentation, 133f, 141, 141f, 142f
  glycolysis producing, 133, 133f, 140t
  for microbial growth, 165t
NADP$^+$ (nicotinamide adenine dinucleotide phosphate), 126, 128, 128t, 153f
  in Calvin-Benson cycle, 148f, 153f
  in photosynthesis/photophosphorylation, 146, 147f, 153f
NADPH, 126, 153f
  in Calvin-Benson cycle, 146, 148f, 153f
  Entner-Doudoroff pathway producing, 140
  pentose phosphate pathway producing, 140
  in photophosphorylation/photosynthesis, 145, 146, 147f, 153f
Naegleria genus/spp., 364, 365t, 678–679, 692t, 790b
Nafcillin, 309t
NAG. See N-acetylglucosamine
Nail(s)
  candidal infection of, 658t, 660b
  dermatophytosis/ringworm involving, 419t, 649, 664–665, 665f, 665t
Naked (nonenveloped) viruses/virions, 389, 390
  antimicrobial action/susceptibility of, 265, 266, 266f
  entry/uncoating of, 397, 397f
  positive ssRNA, 735–741, 771t
  release of, 400
  segmented dsRNA, 770–771, 771t
Nalidixic acid, 313t
NAM. See N-acetylmuramic acid
Nanoarchaeota (phylum), 330f
Nanometer (nm), 97, 97t
Narrow-spectrum drugs, 298
Nasal cavity, microbiome/normal microbiota of, 417t
Nasopharyngeal cancer, HHV-4 (Epstein-Barr virus) and, 719, 719f, 720
Nasuia deltocephalinicola, 194, 195
Natamycin, 280, 311t
National Institute of Allergy and Infectious Diseases (NIAD), bioterrorist threats categorized by, 813, 814t
Native immunity, 447. See also Innate immunity
Natrium. See Sodium
Natural immunity, 446–471, 466t. See also Innate immunity
Natural killer lymphocytes (NK cells/lymphocytes), 453, 456, 486, 486f

Naturally acquired immunity
  active, 495, 497t
  passive, 487, 495, 497t
Ne. See Neon
Necator americanus, 700, 703t
Necrosis, tissue
  gas gangrene and, 567, 568
  Streptococcus pyogenes causing, 124b, 154b
Necrotizing fasciitis ("flesh-eating disease"), 118b, 336, 554–555b, 556
Needham, John T., experiments in spontaneous generation by, 8
Needles, contaminated
  hepatitis transmission and, 431, 728, 770
  HIV infection/AIDS transmission and, 431, 754b, 756, 757
Neelsen, Friedrich, 109
Negative chemotaxis/taxis, 65, 82, 455
Negative feedback (feedback/end-product inhibition), 132, 132f
Negative (indirect) selection, of mutants, 222, 222f
Negative-sense RNA, 398, 398f, 735
Negative (negative-sense) single-stranded RNA viruses, 398, 399f, 735, 771t
  segmented, 763–770, 771t
  synthesis in, 398, 399f, 399t
  unsegmented, 758–763, 771t
Negative (capsule) stains, 110, 111f, 111t
  for electron microscopy, 112
Negri bodies, 762, 762f
Neisseria genus/spp./neisserias, 330f, 340, 344t, 587–591, 587f, 588f, 589b, 590f
  cytochromes in, 139, 587
Neisseria gonorrhoeae (gonococcus), 587f, 588–589, 588f, 589b, 594b. See also Gonorrhea
  adhesins produced by, 422
  culturing, 178, 588
  eye infection caused by, 279, 589
  fimbriae on, 65, 588, 589, 589b
  indirect fluorescent immunoassay in detection of, 516
  pelvic inflammatory disease caused by, 589, 589b
  phagocytosis of, 455f
  resistant strains of, 589
Neisseria meningitidis (meningococcus), 589–591, 590f
  immunization/vaccine against, 508f, 509t, 590–591
  iron needs of, 452
  meningitis caused by, 15t, 344t, 590–591
  septicemia caused by, 590, 590f
Nematodes (roundworms), 693, 698–702, 703t
  Ancylostoma and *Necator* (hookworms), 700, 700f, 703t
  Anisakis, 701, 701f, 703t
  Ascaris, 699–700, 700f, 703t
  Enterobius (pinworms), 700–701, 701f, 703t
  filarial, 699
    Wuchereria, 702, 702f, 703t
  life cycles of, 698–699
  mutualism in, 415b
  predation by fungi and, 366, 367f
Neomycin, 310t
  microbial production of, 290t
Neon, 30f
Neonatal candidiasis, 658t, 660b
Neonatal herpes, 715, 716f
Neonatal meningitis, group B streptococcal (*Streptococcus agalactiae*), 556
Neonate. See Newborn
Neoplasia/neoplastic disease, 401, 424, 424t. See also Cancer
Neosartorya genus/spp., 648
Nephelometry, 513
Nephron(s), streptococcal infection affecting, 556
Nerve(s), botulism toxins affecting, 546b, 581b
Nervous system/nervous tissue
  autoimmunity affecting, 539
  Enterobacteriaceae infection of, 604f
  toxins affecting. See Neurotoxins
Neti pots, 679
NETs (neutrophil extracellular traps), 456
Neuraminidase
  influenzavirus, 768b
    mutations in, 769b
  Pseudomonas aeruginosa, 609

Neuraminidase inhibitors, 313t
Neuromuscular junction
  botulism toxins affecting, 563, 564f
  tetanospasmin affecting, 565, 566f
Neurospora genus/spp./*Neurospora crassa*, 370, 373t
  microbial genetics research and, 19–20, 365, 370
Neurotoxins, 362–363, 428
  Clostridium botulinum producing, 562, 563, 564f, 565b, 581b, 786t. See also Botulism (botulinum) toxins
  Clostridium tetani producing, 565–566, 566f. See also Tetanospasmin
  dinoflagellates producing, 362–363, 790, 790t
Neurotransmitters
  botulism toxins affecting, 563, 564f
  tetanospasmin (tetanus toxin) affecting, 565, 566f
Neutralization, antibody, 486, 486f, 515, 520t
  by IgA, 487
Neutralization tests, 515, 520t
Neutron(s), 28, 28f
  atomic mass of, 28
Neutropenia, in Epstein-Barr infection, 720
Neutrophil(s), 452f, 453, 453f
  in inflammation, 464, 465f
  in nonphagocytic killing, 456
  in phagocytosis, 452f, 453, 455f
Neutrophil extracellular traps (NETs), 456
Neutrophile(s), 167
Newborn
  candidiasis in, 658t, 660b
  coxsackie B virus myocarditis in, 738
  cytomegalovirus infection in, 720, 721
  gonococcal infection in, 279, 589
  hemolytic disease of, 531, 532f
  hepatitis B in, 726, 728
  herpes infection in, 715
  HIV transmission to, 421t, 756
    ART during pregnancy and, 757
  listeriosis in, 336, 568, 570b
  naturally acquired passive immunity and, 487, 495, 497t
  ophthalmia neonatorum in, 589
  respiratory syncytial virus infection in, 503b, 520b, 760–761
  rubella infection in, 747
  Streptococcus agalactiae (group B streptococcus) infection in, 556
  syphilis in, 632
  toxoplasmosis in, 687, 690
  Zika virus infection and, 745, 747t
Next-generation sequencing (NGS), 250–251, 251f
  ecosystem sampling and, 804
N-formylmethionine, 207, 208f, 211, 212
  in archaea, 329
NGS. See Next-generation sequencing
NGU. See Nongonococcal urethritis
NH$_3$. See Ammonia
NH$_4^+$. See Ammonium/ammonium ion
Niacin (nicotinic acid/vitamin B$_3$), for microbial growth, 165t
NIAD (National Institute of Allergy and Infectious Diseases), bioterrorist threats categorized by, 813, 814t
Niclosamide, 316t
  spectrum of action of, 299f, 316t
Nicotinamide adenine dinucleotide (NAD$^+$), 126, 128, 128t, 153f
  in acetyl-CoA synthesis, 135
  in citric acid/Krebs cycle, 136, 137f, 153f
  in electron transport chain, 136, 138f, 153f
  in fermentation, 141, 141f, 142f
  in glycolysis, 134f
  Haemophilus influenzae requiring, 604
Nicotinamide adenine dinucleotide phosphate (NADP$^+$), 126, 128, 128t, 153f
  in Calvin-Benson cycle, 148f, 153f
  in photosynthesis/photophosphorylation, 146, 147f, 153f
Nicotinamide nucleotide, 52
Nicotinic acid (niacin), for microbial growth, 165t
van Niel, C.B., 19
Nifurtimox, 317t, 703b

SUBJECT INDEX I-27

Nightingale, Florence, 17, 17f, 18f
Nigrosin, as negative stain, 110
"9 + 0" arrangement, in cilia/flagella, 82, 83f
"9 + 2" arrangement, in cilia/flagella, 82, 83f
Nipah virus encephalitis, bioterrorism and, 814t
Niridazole, 316t
Nisin, 280, 311t
Nitazoxanide, 312t, 317t
Nitrates
  in nitrogen cycle, 139, 339, 807f, 808
  wastewater treatment producing, 794
Nitric oxide, in killing by neutrophils, 456
Nitrification, in nitrogen cycle, 339, 807f, 808
Nitrifying bacteria, 330f, 339
Nitrites, in nitrogen cycle, 139, 807, 808
Nitrobacter genus/spp., 339, 344t, 808
Nitrogen, 29t, 30f, 807
  microbial growth and, 164–165
  recycling of, 20, 807–808, 807f
Nitrogenase, in nitrogen fixation, 807
Nitrogen cycle, 20, 807–808, 807f
Nitrogen fixation, 165, 333, 807, 807f
  by alphaproteobacteria, 330f, 339, 339f
  by cyanobacteria, 165, 333, 807
Nitrogenous wastes, 143
Nitroimidazoles, 313t, 318t
Nitrosomonas genus/spp., 340, 344t, 808
Nitrous acid, mutagenic effects of, 219
NK cells/lymphocytes (natural killer lymphocytes), 453, 456, 486, 486f
nm (nanometer), 97, 97t
N,N-diethyl-*m*-toluamide. See DEET
Nocardia genus/spp./*Nocardia asteroides*, 330f, 337t, 338, 579–580, 580f
  acid-fast staining of, 109–110, 110f
NOD proteins, 457
Nomarski (differential interference contrast) microscopy/microscopes, 101, 102f, 107t
Nomenclature, 112. See also Taxonomy/ microbial classification and identification
  binomial, 113
Non-A/non-B hepatitis. See Hepatitis, type C
Noncoliform Enterobacteriaceae, 594, 597, 597f
  healthcare-associated (nosocomial) infection caused by, 591f
Noncommunicable disease, 433, 434t
Noncompetitive inhibition/inhibitors, 132, 132f
Noncyclic photophosphorylation, 146, 147f
Nonenveloped (naked) viruses/virions, 389, 390
  antimicrobial action/susceptibility and, 265, 266, 266f
  entry/uncoating of, 397, 397f
  positive ssRNA, 735–741, 771t
  release of, 400
  segmented dsRNA, 770–771, 771t
Nongonococcal urethritis (NGU)
  Chlamydia trachomatis causing, 628
  mycoplasmas causing, 572
Noninvasive aspergillomas, 657
Nonionizing radiation, 273, 275. See also Radiation
  in microbial control/food preservation, 275, 275t, 785
  mutagenic effects of, 218
Nonliving reservoirs, 419
Nonmembranous organelles, 85t. See also specific type
  in bacteria/prokaryotic cells, 76–77, 77f, 85t
  in eukaryotic cells, 83–85, 84f, 85t
Nonmunicipal wastewater treatment, 794f
Nonparalytic polio, 737. See also Poliomyelitis
Nonperishable food, 784
Nonpolar covalent bonds, 31–33, 32f, 33f, 35t
Nonprofessional antigen-presenting cells, 478
Nonsense mutations, 217, 217f, 218f
Non-streptococcal toxic shock syndrome (TSS), 548, 548t, 550, 550f
Nonsulfur bacteria
  green, 330f, 333, 334, 334t
  purple, 330f, 333, 334, 334t, 339, 344t
Nonvenereal treponemal diseases, 633, 633f
Normal microbiota (normal flora), 172, 252, 253b, 416–418, 416f, 417t, 450. See also Microbiome

Norovirus genus/spp./noroviruses, 392t, 740, 740b, 771t
  gastroenteritis caused by, 1b, 22b, 392t, 740, 740b, 771t, 790t
  waterborne transmission of, 740b, 790t
Northern blot, 246
Nose
  microbiome/normal microbiota of, 417t
  as portal of entry, 420f, 421
  as portal of exit, 430, 430f
Nosema genus/spp., 369, 662
Nosepiece, revolving, 100
Nosocomial infections/diseases. See Healthcare-associated (nosocomial) infections/ diseases
Notifiable/reportable diseases, 435, 437t 438f
  bioterrorism defense and, 815, 815f
Nuclear chromosomes, 85, 89t, 196–197, 197f
  in nuclear division (mitosis/meiosis), 85, 197, 197f, 354, 355, 355f, 356, 356f, 357f
Nuclear division, 354–357, 355f, 356f, 357f. See also Meiosis; Mitosis
Nuclear envelope, 60, 61f, 85, 86f, 196, 355, 355f, 356, 356f
Nuclear pores, 61f, 85, 86f
Nuclear waste, microbial remediation of, 169b
Nucleic acid(s), 38, 50–53, 51f, 52t, 198t. See also DNA; Nucleotide(s); RNA
  antisense
    as antimicrobials, 291f, 293, 294f, 314t
    synthetic nucleic acids in synthesis of, 240
  autoantibodies against, in systemic lupus erythematosus, 534
  comparison of, 52t
  function of, 52, 52t
  hydrogen bonds in, 35, 50–51, 51f, 52, 194, 195f
  microbial identification/classification and, 115, 118–119, 329, 330f, 357, 359f, 360
  structure of, 50–53, 51f, 194, 195f
  synthesis of, 151, 152f, 153f, 239
    antimicrobials affecting, 265–266, 290, 291f, 295–297, 296f, 297f, 312–313t, 314t, 316t, 317t
    anthelminthic drugs, 316t
    antiprotozoan drugs, 318t
    antiviral agents, 296–297, 297f, 314t
    synthetic, in recombinant DNA technology, 239–240, 252t
  viral, 385, 386, 386f
    synthesis of, 392, 393f, 394
    antimicrobials and, 296–297, 297f, 314t
    strategies for, 399t
Nucleic acid hybridization. See Hybridization
Nucleic acid probes, 252t
  in clone selection, 246
  diagnostic applications of, 254
  synthetic nucleic acids in synthesis of, 240
Nucleic acid sequencing. See Genomics
Nucleic acid synthesis machines, 239
Nucleocapsid, 386, 386f
Nucleoid, 60f, 74, 195, 196f
Nucleoli (nucleolus), 61f, 85, 86f, 206
Nucleoplasm, 85, 86f
Nucleoproteins, 49
Nucleoside(s), 50, 194, 195f, 297f
Nucleoside analogs, as antimicrobials, 218–219, 291f, 296, 297f
Nucleosomes, 196–197, 197f
Nucleotide(s), 50–54, 50f, 194, 195f
  biosynthesis of, 151, 152f, 153f
    antimicrobials affecting, 295–297, 296f, 297f
  chemicals altering, mutagenic effects of, 219
Nucleotide analogs, 218–219, 218f, 314t
  as antimicrobials, 296, 297, 314t
  mutagenic effects of, 218–219, 218f
Nucleotide bases, 194, 195f
Nucleotide-excision repair, 220–221, 220f
Nucleotide sequencing, 20, 250–252, 251f, 252t
  antimicrobial drug development and, 308
  ethical and safety issues and, 256–257

functional, 251–252
metagenomic, 20, 118–119, 252, 347
microbial classification/evolution and, 20, 115, 118–119, 329, 330f, 357, 359f, 360. See also Evolution
  in microbial growth estimation, 187
  synthetic nucleic acid probes in, 240
Nucleus, 89t. See also under Nuclear
  atomic, 28, 28f
  cell/eukaryotic, 60, 85, 85t, 86f, 89t, 196
  division of, 354–357, 355f, 356f, 357f. See also Meiosis; Mitosis
Numerical aperture of lens, resolution affected by, 99
  immersion oil and, 100, 102f
Nursing, Florence Nightingale/infection prevention and, 17, 17f, 18f
Nutrient(s)
  cycling of, 808–809
  food spoilage and, 782, 783, 784t
  improving value of, recombinant DNA technology and, 255–256
  limiting
    hydrogen as, 163
    nitrogen as, 165
  for metabolism, 125, 125f. See also Cytoplasmic (cell/plasma) membranes, function of/ movement across
  for microbial growth, 162–165, 163f, 165f, 165t
  in soil, microbial diversity and, 810
Nutrient agar, 174, 178f
Nutrient broth, 174, 175
Nutrition, microbial, 161–192. See also Microbial growth
Nutritional diseases, 424, 424t
Nutritional value of food, recombinant DNA technology affecting, 255–256
  ethical/safety issues and, 256
Nuttall, George, 15t
Nystatin, 315t
  mechanism of action of, 293, 315t
  microbial production of, 290t

O_2. See Oxygen
O_3. See Ozone
O antigen (O polysaccharide), 593, 593f
  of *E. coli*, 595
  of *Vibrio*, 637
Obesity
  human adenovirus 36 and, 724
  Selenomonas and, 335
Objective lenses, 100, 101f
Obligate acidophiles, 167
Obligate aerobes, 163, 165f
Obligate anaerobes, 163, 165f
  culture of, 177, 178f
  Gram-negative bacilli, 614, 614f
Obligate halophiles, 168
Ocular infections
  Acanthamoeba causing, 267b, 678, 679, 692t
  candidal, 658t, 660b
  Chlamydia trachomatis causing, 627, 628, 628f
  herpetic, 714–715, 714f, 715f, 716t
  in histoplasmosis, 652
Ocular lenses, 100, 101f
Ofloxacin, 313t
OH. See Hydroxyl radicals
Oil glands, 421f, 448
  acne and, 578, 579f
  in host defense, 448
Oil immersion objective lens, 100, 102f
Oil spills, bioremediation in clean up of, 805
Okazaki, Reiji, 201
Okazaki fragments, 201, 202
Oleic acid, 43f
Oligoadenylate synthetase, 459
Oltipraz, 316t
-oma (suffix), 423t
Oncogenes, 402, 402f
Oncogenic retroviruses, 750–751, 751f
O'Neill, Scott, 746b
ONPG (o-nitrophenyl-β-ᴅ-galactopyranoside) water quality test, 792, 792f
Onychomycosis
  Candida causing, 658t, 660b
  dermatophytes causing (tinea unguium), 419t, 649, 664–665, 665f, 665t
Oocysts
  of *Cryptosporidium*, 690, 691f
  of *Cyclospora*, 692f

of *Plasmodium*, 689b
  of *Toxoplasma*, 687f, 690
Ookinetes, in *Plasmodium* life cycle, 689b
Open system, for culture, chemostat as, 182, 182f
Operator, in operon, 213, 213f, 214f, 215, 215f
Operons, 212–215, 213f, 214f, 215f, 215f
Ophthalmia neonatorum, 279, 589
Ophthalmic/ocular herpes, 714–715, 714f, 715f, 716t
Opisthorchis sinensis, 695f
O polysaccharide (O antigen), 593, 593f
  of *E. coli*, 595
  of *Vibrio*, 637
Opportunistic mycoses, 648–649, 649t, 664. See also specific type
  diagnosis of, 649–650
  in HIV infection/AIDS, 664, 751t
  systemic, 656–664
Opportunistic pathogens (opportunists), 417–418, 418t. See also specific organism or disease and Immunocompromised host
  anaerobic bacilli, 614, 614f
  antimicrobial therapy and, 298, 299b, 302, 418
  enteric
    coliform, 595–596, 596f, 598–599b
    noncoliform, 597, 597f
  fungi, 648–649, 649t
    identification of, 649–650
    treatment of infection caused by, 650
  in HIV infection/AIDS, 418, 541, 664, 751, 751t, 755, 755f
  Pseudomonas aeruginosa, 609
Opsonins, 455, 456
  antibodies as, 456, 486, 486f
  complement as, 459, 460, 461f, 462
Opsonization, 455, 486, 486f
  staphylococcal protein A affecting, 548
Optical density, light refraction/ magnification and, 98, 98f
Optimum growth temperature, 166, 166f
OPV. See Oral polio vaccine
Oral candidiasis. See Oropharyngeal candidiasis
Oral drug administration, 301, 301f
Oral hairy leukoplakia
  HHV-4 (Epstein-Barr virus) and, 719, 719f, 720
  in HIV infection/AIDS, 541, 719, 720
Oral herpes (fever blisters/cold sores), 392t, 709b, 714, 714f, 715f, 716t, 729b
Oral polio vaccine (OPV/Sabin vaccine), 738, 738t
  safety of, 506, 510, 738
Orbivirus genus/spp., 392t
Orchitis, mumps virus causing, 760
Order(s), 113, 114f
  for viruses, 391
Orf, 710
Organelles, 60, 60f, 85t, 89t. See also specific type
  membranous, 76f, 85–88, 85t, 87f, 88f, 89t
  nonmembranous, 85t
    in bacteria/prokaryotic cells, 76–77, 77f, 85t
    in eukaryotic cells, 83–85, 84f, 85t
Organic acids, 38, 39
  in food preservation, 785
  microbial production of, 787
Organic compounds/macromolecules, 33, 40–53. See also specific type
  functional groups and, 40
  for microbial growth, 165, 165f
Organic phosphate group, 41t
  in nucleotides, 50, 50f, 52
  in phospholipids, 43, 43f
Organotrophs, 163
Orientia tsutsugamushi, 622–623, 626t
Origin
  in DNA replication, 199, 200f, 201f
  of transfer (conjugation), 226, 227f, 228f
Oritavancin, 310f
Ornithodoros ticks, as disease vectors, for relapsing fever (*Borrelia* genus/spp.), 636
Ornithosis (psittacosis), 629
  bioterrorism and, 814t
Oropharyngeal candidiasis (thrush), 299b, 302, 426f, 658t, 660b
  antimicrobial drug use and, 299b, 302
  in HIV infection/AIDS, 541, 660b, 766b

Orphan viruses
    enteric cytopathic human (echovirus) classification and, 738
    HHV-7 as, 713
    respiratory enteric orphan (reoviruses), 392t, 735, 747t, 770–771, 771t
Orthobunyavirus genus/spp., 771t
Orthocresol, 276f
Orthohepadnavirus genus/spp., 392t, 725, 729t, 739t. See also Hepatitis B virus
Orthohepevirus genus/spp., 739t, 771t
Orthomyxoviridae (orthomyxoviruses), 392t, 735, 763, 766, 768–769b, 771t. See also Influenza/influenzaviruses
Orthopoxvirus genus/spp./Orthopoxvirus variola (smallpox/variola virus), 2b, 392t, 710–712, 710f, 711f, 712b, 729t
    bioterrorism and, 510b, 712, 813, 814t
    safe handling of, 268
    shape of, 389
    size of, 61f, 388f
    vaccination against. See Smallpox vaccination
Oscillatoria genus/spp., 333f
-ose (suffix), 44
Oseltamivir, 313t, 769b
-osis (suffix), 423t
Osmium tetraoxide, 112
Osmosis, 71f, 72–73, 72f, 74t, 168, 273
Osmotic pressure, 168
    microbial growth affected by, 168, 273, 275t
    cell walls and, 72f, 73, 168
OsO$_4$ (osmium tetraoxide), 112
Osteoarticular blastomycosis, 654
Osteoarticular candidiasis, 658t, 660b
Osteomyelitis, staphylococcal, 305b, 550
Otitis media, pneumococcal, 558b, 559
Outer membrane, in Gram-negative cell walls, 67, 68f
Outer protein rings, bacterial flagellar motion and, 63f
Overexpression (gene), in functional genomics, 252
Owl's eye cells, in cytomegalovirus disease, 721f
Oxacillin, 309t
Oxaloacetate/oxaloacetic acid, 149t
    in amination and transamination, 150, 151f
    in carbohydrate biosynthesis/gluconeogenesis, 150f
    in citric acid/Krebs cycle, 136, 137f, 149t
Oxamniquine, 316t
Oxazolidinones, 311t
    mechanism of action of, 293, 294f, 311t
Oxidase
    in Neisseria, 139, 587
    in pasteurellae, 591, 591f
    in peroxisomes, 87
    in Vibrio, 637
Oxidase negative/positive bacteria, 139, 591, 591f
Oxidase test, 591, 591f
Oxidation, 126, 126f. See also Oxidation-reduction (redox) reactions
    anaerobic ammonium (anammox), 807f, 808
    antimicrobial effects of, 278, 281t
    beta, of fatty acids, 143, 143f
    in carbon cycle, 807f
Oxidation lagoons, in wastewater treatment, 794
Oxidation-reduction (redox) reactions, 125, 126, 126f
    in cellular respiration, 136, 139
    in microbial growth, 163
Oxidative phosphorylation, 127, 148t
    chemiosmosis and, 139, 140
Oxidizing agents, antimicrobial action of, 278, 281t
Oxidoreductases, 127, 128t
Oxygen, 29t, 30f
    in aquatic ecosystems, 811, 812
    atomic number of, 28, 29t
    covalent bonds in formation of, 32f
    as final electron acceptor, 138f, 139, 163, 164. See also Aerobes/aerobic bacteria
    microbial cultures and
        anaerobic culture, 177, 178f
        low-oxygen culture, 178
    microbial growth and, 163–164, 165f
    singlet, 164
    in soil, microbial survival and, 809
    toxic forms of, 164
Oxygen demand, biochemical (BOD), reduction of in wastewater treatment, 793, 793f, 794
Oxygenic organisms
    cyanobacteria as, 333, 334t
    noncyclic photophosphorylation in, 146, 147f
Oxygen radicals, antimicrobial action of, 278
Oysters, contaminated, Vibrio vulnificus causing food poisoning and, 40b
Ozone, antimicrobial action of, 278, 281t
    drinking water treatment and, 791f, 792

P. See Phosphorus
PABA. See Para-aminobenzoic acid
Packaging (food), aseptic, 785
Pain, in inflammation, 462
Palindromes/palindromic sequences
    in restriction sites, 240, 241f
    in transposons, 229
Palisades, in Corynebacterium, 328, 328f, 337, 572
Palmitic acid, 43f
PAM. See Primary amebic meningoencephalitis
PAMPs. See Pathogen-associated molecular patterns
Pancreas, autoimmunity affecting, 539
Pandemic, 435, 436f
    cholera, 637–638, 638f
    influenza, 435
    plague, 602
Panencephalitis, subacute sclerosing (SSPE), 759
Papaya ringspot virus resistance, recombinant DNA technology and, 255, 255f
Papillomas (warts), 392t, 722, 723b, 724b
Papillomaviridae (papillomaviruses), 392t, 722, 723b, 724b, 729t
    cancer and, 392t, 723b
    skin infection (warts) caused by, 392t, 722, 723b, 724b
    vaccine against, 508f, 509t
Papovaviridae. See Papillomaviridae; Polyomaviridae
Papule(s), in poxvirus infection, 710, 711f
Para-aminobenzoic acid (PABA)
    for microbial growth, 165t
    structural analogs of, antimicrobial action of, 132, 295, 296f
Parabasala (kingdom)/parabasalids, 357, 359f, 360, 360f, 365f
Parabasal body, 360
Paracoccidioides brasiliensis, 648, 656, 656f
Paracoccidioidomycosis, 656
Parainfluenzaviruses, 758, 759–760
Paralia sulcata, 375f
Paralysis, botulism toxins causing, 563, 581b
Paralytic polio, 737, 737f. See also Poliomyelitis
Paralytic shellfish poisoning (PSP), 353b, 379b, 790t
Paramecium genus/spp., 60f, 105f, 360f, 362f, 365t
    phagocytosis in, 362, 362f
Paramylon, in euglenids, 361, 361f
Paramyxoviridae (paramyxoviruses), 392t, 735, 758, 759, 760, 771t
    common cold/respiratory infection caused by, 392t, 760, 771t
Pararosaniline, as flagellar stain, 110
Parasite(s), 415–416, 416t, 676–708, 676f. See also Protozoa
    acellular obligatory, viruses as, 6, 7f
    energy, chlamydias as, 626
    helminthic, 693–702, 703f. See also Helminths
    hosts of, 415–416, 676, 676f
    protozoan, 677–691, 692t. See also Protozoa
    raw fish in transmission of, 701, 701f, 703t
    sexually transmitted infections and, 676f
    size of, 61
    skin infections and, 676f
    transmission of to humans, 676, 676f
    waterborne diseases caused by, 790t

Parasitemia, in Trypanosoma brucei infection, 681
Parasitism, 415–416, 416t
Parasitology/parasitologists, 18f, 19t, 676, 676f. See also Parasite(s)
Parenteral route, pathogen portal of entry and, 420
Paromomycin, 310t, 317t
Parotid glands/parotitis, in mumps, 760, 760f
Paroxysmal phase, of pertussis, 607, 608f
Parrot fever (ornithosis), 629
    bioterrorism and, 814t
Particulate radiation, 273. See also Radiation
Parvoviridae (parvoviruses), 392t, 728, 729t
    synthesis in, 398
Parvovirus B19 (Erythrovirus genus/spp.), 392t, 728, 728f
    placenta crossed by, 421t
Passive immunity
    artificially acquired (immunotherapy), 496, 497t, 504, 505, 510–511, 511f
    naturally acquired, 487, 495, 497t
Passive immunotherapy/immunization, 496, 497t, 504, 505, 510–511, 511f
Passive transport processes, 70, 71–73, 71f, 72f, 74t
Pasteur, Louis, 8–9, 8f, 18f, 270, 424
    experiments in fermentation by, 10–12, 11f, 141
    experiments in spontaneous generation by, 8–9, 9f
    germ theory and, 13, 424
    Streptococcus pneumoniae discovered by, 557
    vaccine development and, 18, 505
Pasteurella genus/spp. (Pasteurella hemolytica/multocida), 113, 603–604
    development of vaccine for, 505
Pasteurellaceae, 343t, 591, 603–605
    oxidase test and, 591, 591f, 603, 605f
Pasteurization, 12, 18f, 264, 265t, 270–271, 271t, 275t, 432, 440, 785
    ultra-high-temperature, 271, 271t
patho-/-patho (prefix or suffix), 423t
Pathogen(s), 13, 20. See also specific type
    animal, as biological weapons, 813–814
        assessing threat to livestock/poultry and, 813
    asepsis and, 264
    as biological weapons, 812–815, 814t, 815f
        assessing threat potential and, 812–813
        defense against, 814–815, 815f
        known threats and, 813–814, 814t
        recombinant DNA technology and, 815
    biosafety levels and, 268, 268f, 815
    defenses against. See Host defenses
    epidemiology in study of, 17, 18f, 19t, 434–441
    extracellular. See also Bacteria; Fungi
        antibodies/B cells attacking, 474
    human, as biological weapons, 813, 814t
        assessing threat potential and, 812–813
    intracellular, 342. See also Viruses
        T cells/cell-mediated immune response and, 474, 480
    Koch's postulates and, 13–14, 424–426, 425f, 437
    movement of into hosts, 420–421, 420f, 421f, 421t. See also Infection; Portals of entry
    movement of out of hosts, 430, 430f
    parasitic, 415–416, 676–708
    pasteurization in elimination of, 12, 18f, 264, 265t, 270–271, 271t, 275t, 785
    plant, in bioterrorism, 813, 814, 815. See also Agroterrorism
    resistance and, 303–308, 304b, 304f, 305b, 306f, 307b, 307f
    soil, 810, 810t
    spread/transmission of, 430–432, 431f, 432b, 433t
    vaccination/immunization against, 18, 495, 496, 497t, 504–512, 505f, 507f, 508f, 509f. See also Vaccines
    virulence/virulence factors of, 426–429, 426f, 427f, 428t

Pathogen-associated molecular patterns (PAMPs), 457, 457t
Pathogenicity, 426. See also Virulence
Pathway (metabolic), 125
Pattern recognition receptors, 457
Pauling, Linus, 20
Pauling scale, 33f
PCP. See Pneumocystis pneumonia
PCR. See Polymerase chain reaction
PCV. See Pneumococcal/pneumococcal conjugate vaccine
PEAS. See Possible estuary-associated syndrome
Pectinase, microbial production of, 787
Pediatric botulism, 564
Pediculus genus/spp./Pediculus humanus (lice), as disease vectors, 378, 378f, 433t, 677
    for epidemic typhus (Rickettsia prowazekii), 621, 626t
    for relapsing fever (Borrelia recurrentis), 635, 636
    for trench fever (Bartonella quintana), 605
Peliosis hepatis, bacillary, 605
Pellicle
    of euglenids, 361
    microbial growth estimates and, 187
Pelvic infections
    Bacteroides fragilis causing, 614
    Prevotella causing, 614
Pelvic inflammatory disease
    chlamydial, 628
    gonococcal, 589, 589b
    mycoplasmal, 572
Penciclovir, 297f, 314t
Penetration
    in animal virus replication, 397, 397f, 401t
    in bacteriophage replication, 393f, 394, 395, 396f, 401t
Penicillin, 309t
    allergic reaction to, 302
    discovery of, 21, 21f, 289
    mechanism of action of/bacterial cell wall and, 290–291, 291f, 292f, 309t
    Penicillium producing, 4, 5f, 238, 289, 289f, 290t, 309t, 370, 788
    resistance to, 221, 221f, 309t
        beta-(β) lactamase and, 306, 306f, 309t, 548, 548t, 551
    spectrum of action of, 298, 299f, 309t
    for streptococcal infections, 70b, 556, 557, 559
    for syphilis, 632–633
Penicillinase (beta (β)-lactamase)
    antimicrobial resistance and, 306, 306f, 309t
    staphylococcal pathogenicity/virulence and, 548, 548t, 551
Penicillium genus/spp. (Penicillium chrysogenum/griseofulvin/marneffei), 4, 5f, 289f, 290t, 315t, 370, 373f
    in cheese production, 783
    hypertonic environment tolerated by, 273
    penicillin produced by, 4, 5f, 238, 289, 289f, 290t, 309t, 370, 788
    reproduction in, 368f
Penis, as portal of entry, 420f
Pentacel vaccine, 506
Pentamidine, 318t
    mechanism of action of, 297, 318t
Pentose(s), 44, 140
    in nucleotides, 50, 50f
Pentose phosphate pathway, 140, 149t, 153f
    in nucleotide biosynthesis, 151, 152f
PEP. See Phosphoenolpyruvic acid
Peptic ulcers, 27b, 39, 53b, 167, 640–641, 640b, 641f
Peptide bonds, 47, 48f
Peptide crossbridges, 66–67, 67f
Peptidoglycan, 45
    in bacterial cell wall, 4, 45, 66–67, 66f, 67f, 68f, 290
    antimicrobials and, 290–291, 292f
Peracetic acid, antimicrobial action of, 278, 281t
Peramivir, 313t
Peranema genus/spp., flagella of, 5f
Percentage of transmission/absorbance, spectrophotometer measuring, 186f, 187

Perforin/perforin-granzyme cytotoxic pathway, 486, 486f, 491, 491f
Pericardial infection
 in candidiasis, 658t, 660b
 coxsackie B virus causing, 738
Periodic table of elements, 30f
Periodontal infections, *Prevotella* causing, 614
Peripheral nervous system. See Nerve(s); Nervous system/nervous tissue
Peripheral protein(s), in cytoplasmic membrane, 69, 69f
Periplaneta genus/spp., as disease vectors, 433t
Periplasm, 67–68
Periplasmic space, 67, 68f
Perishable foods, 784
Peristalsis, in host defense, 451t
Peritonitis
 candidal, 658t, 660b
 Salmonella causing, 600
Peritrichous flagellum/bacterium, 63, 64f, 65f
Permeability
 blood vessel, in inflammation, 462–464, 462f, 463f, 465f
 cell membrane, selective, 70
Permeases
 in active transport, 73
 in facilitated diffusion, 71, 71f
Peromyscus maniculatus (deer mouse), in Hantavirus transmission, 384b, 408b, 767, 810
Peroxidase, hydrogen peroxide affected by, 164
Peroxide(s), antimicrobial action of, 164, 278, 281t
 food preservation and, 785
Peroxide anion, 164
Peroxisomes, 85t, 87
Persistent infections, 400, 400f, 401t
Persister cells, 303–304
Person-to-person spread of disease, 431
Pertussis (whooping cough), 344t, 345b, 426f, 586b, 606–608, 607f, 608f, 615b
 immunization against, 345b, 508f, 509t, 608
 reemergence of, 345b, 608
 virulence and, 426f, 606–607, 607f
Pertussis toxin, 606–607, 607
Pesticides, microbial production of, 12t, 788–789, 789t
Pest resistance, recombinant DNA technology and, 255, 255f
 Bt toxin and, 255, 335–336, 336f, 788–789, 789t
Pet(s)
 allergy risks and, 526
 salmonellosis and, 600
Petechiae
 in meningococcal septicemia, 590, 590f
 in Rocky Mountain spotted fever, 624b, 625b
Petri dishes, 173, 175
Petri plates, 173, 174f, 175
 in anaerobic cultures, 177, 178f
Petroff-Hauser counting chamber, 183, 183f
Peyer's patches, 475f, 476
Pfiesteria genus/spp., 363, 365t
Pfiesteria toxin, 363
pH, 38, 38f, 167
 antimicrobial efficacy affected by, 267–268
 enzymatic activity affected by, 130f, 131
 of food, spoilage/preservation and, 783
 in host defense, 450
 skin and, 448
 microbial growth affected by, 131, 167
 microbial identification by fermentation and, 141
 microorganism tolerance of ranges of, 38–39, 40b
 soil, microbial population affected by, 809
 of stomach, microbial growth and, 53b, 167
 vaginal, microbial growth and, 167, 661b
pH buffers, 38–39
pH meter, 39
pH scale, 38, 38f
Phaeohyphomycosis, 666–667, 667f
Phaeophyta, 375, 375f, 376f
Phage(s). See Bacteriophage(s)
Phage attachment, 393f, 394, 395, 396f

Phage clones, in gene libraries, 243, 244f
Phage therapy, 395b
Phage typing, 113b, 117, 118f
Phagocytes, 428–429, 454, 455f. See also Macrophage(s); Phagocytosis
 defective, in immunodeficiency diseases, 540
 migration of, 464, 465f
Phagocytosis, 81, 81t, 452f, 453, 454–456, 455f
 evasion of, virulence and, 427f. See also Antiphagocytic chemicals/factors
 in host defense, 453, 454–456, 455f
Phagolysosomes, 87f, 455f, 456
 Coxiella growing in, 613, 613f
Phagosome (food vesicle), 81, 81f, 87f, 455f, 456
 maturation of/microbial killing and, 455f, 456
 TLRs in membranes of, 457, 457t
Phalloidin, 668
Pharmaceutical microbiology, 18f, 19t, 252–254, 788, 788t
Pharmacia CAP, 529
Pharyngitis
 in diphtheria, 572
 gonococcal, 589, 589b
 herpetic, 714, 716t
 streptococcal, 70b, 553, 553f, 557
  group A streptococci causing, 70b, 553, 553f
  rheumatic fever and, 553
Pharynx, 449f, 450f
 microbiome/normal microbiota of, 417t
Phase (light), contrast affected by, 100
Phase-contrast microscopy/microscopes, 101, 102f, 107t
Phase microscopy/microscopes, 100, 101, 102f, 107t
 differential interference contrast (Nomarski), 101, 102f, 107t
 phase-contrast, 101, 102f, 107t
PHB. See Polyhydroxybutyrate
Phenol/phenolics, 16, 276, 276f, 281, 281t
Phenol coefficient, 281
Phenotype, 202–203
Phenylalanine, 47f
Phialophora genus/spp., chromoblastomycosis caused by, 666t
Phlebotomus genus/spp.
 bartonellosis transmitted by, 605
 leishmaniasis transmitted by, 378, 682, 683f, 692t
 sand fly fever transmitted by, 747t
Phosphate
 in ATP production. See Phosphorylation
 in phosphorus cycle, 808
Phosphate bonds, ATP release and, 52
 in substrate-level phosphorylation, 135, 135f
Phosphate (organic) group, 41t
 in nucleotides, 50, 50f, 52
 in phospholipids, 42, 43f
Phosphoenolpyruvic acid (PEP), 129, 149t
 in glycolysis, 134f, 135, 135f, 149t
Phosphoglucoisomerase, in glycolysis, 128t
Phosphogluconate (pentose phosphate) pathway, 140
3-Phosphoglyceric acid, 149t, 153f
 in Calvin-Benson cycle, 146, 148f
 in glycolysis, 134f, 149t, 153f
Phospholipid(s), 42, 43f, 44f, 69, 69f, 150, 151f. See also Phospholipid bilayers
Phospholipid bilayers, 42, 43f, 67, 68f, 69, 69f
 cholesterol in, 44, 44f, 81
 in nuclear envelope, 85, 86f
 permeability of, 70
 in viral envelope, 389
Phosphorus, 29t, 30f, 808
 for microbial growth, 165
 wastewater treatment producing, 794
Phosphorus cycle, 808, 809f
Phosphorylation, 126–127, 127, 148t. See also Photophosphorylation
 as exchange reaction, 37
 in glycolysis, 133, 134f
 in group translocation, 73f, 74
 oxidative, 127, 148t
  chemiosmosis and, 139, 140
 substrate-level, 127, 135, 135f, 148t
  in citric acid/Krebs cycle, 136
  in glycolysis, 134f, 135, 135f

Phosphorylcholine, in *Streptococcus pneumoniae* cell wall, 558, 558b
Photic zone, algae in, 373
Photoautotrophs, 163, 163f
 in carbon cycle, 806, 807f
 lithotrophic, 165
 in sulfur cycle, 808
Photoheterotrophs, 163, 163f
Photolyase, 220, 220f
Photophosphorylation, 127, 148t
 chemiosmosis and, 139, 146, 147f
 cyclic, 146, 147f
 noncyclic, 146, 147f
Photosynthesis, 144–148, 145f, 147f, 148f, 148t, 153f
 by algae, 6, 6f
 bacteriophages affecting, 390b
 carbon cycle and, 146, 148, 806, 807f
 chemicals and structures in, 144–145, 145f
 chloroplasts in, 87–88, 88f, 145
 by cyanobacteria, 333
 by green and purple phototrophic bacteria, 333–334
 by lichens, 371
 light-dependent reactions in, 145
 light-independent reactions in, 145
 nitrogen fixation and, 807
 pentose phosphate pathway and, 140, 152f
Photosynthetic guild, 804, 804f
Photosynthetic lamellae, 332
Photosynthetic pigments, 144. See also specific type and Chlorophyll(s)
 accessory, 373, 374
 phototrophic bacteria using, 144, 333
Photosystem(s), 145, 145f, 146, 147f
Photosystem I, 145, 145f, 146, 147f
Photosystem II, 145, 146, 147f
Phototaxis, 65
Phototrophic bacteria/phototrophs, 163, 163f, 332–334, 333f, 334f, 334t
 in lichens, 371, 372f
Phycocyanin, 376t
Phycoerythrin, 374, 376t
Phycology/phycologists, 18f, 19t, 373. See also Algae
Phyla (phylum), 113, 114f, 329, 330f
Phylogenetic hierarchy, 115. See also Evolution
Physarum genus/spp., 365t
Phytophthora cinnamomi/infestans, 255, 376
 resistance to, recombinant DNA technology and, 255
 soilborne transmission of, 810t
Phytoplankton, 375
Pickles, fermentation in production of, 779, 783t
Pickling, for food preservation/flavoring, 779
Picornaviridae (picornaviruses), 391, 392t, 735–741, 771t. See also specific type
 common cold caused by, 735–736, 735f, 736f
 hepatitis A caused by, 739, 739f, 771t
 polio caused by, 736–738, 737f, 738t, 771t
PID. See Pelvic inflammatory disease
Pigeon breeder's lung, 532
Pili (conjugation/sex pili), 66, 66f, 79, 89t, 226, 227f, 228f
Pimple, 464
 in acne, 578, 579f
Pinkeye, 446b, 466b, 724, 724f. See also Conjunctivitis
Pinocytosis, 81, 81t
Pinta, 631b, 632
Pinworm (*Enterobius vermicularis*), 700–701, 701f, 703t
Piperacillin, 309t
Piperaquine, 318t
Pittman, Margaret, 604
Pityriasis, 665, 666f
Placenta
 IgG crossing, in naturally acquired passive immunity, 487, 495, 497f
 as portal of entry, 420f, 421, 421t, 431
  for coxsackie B virus, 738
  for hepatitis B virus, 726
  for herpesviruses, 715
  for HIV, 421t, 756
   ART during pregnancy and, 757

 for listeriosis, 336, 421t, 568, 570b
 for rubella, 421t, 747
 for syphilis, 421t, 632
 for toxoplasmosis, 421t, 687, 690
 for Zika virus, 744, 745, 746b
Plague, 15t, 419t, 426f, 602, 603f, 604b
 bioterrorism and, 814t
 bubonic, 15t, 419t, 602, 603f, 604b
 pneumonic, 602, 603f
Planctomyces genus/spp., reproduction in, 326
Planctomycetes (phylum), 330f
Plankton, 358, 362
Plant(s)
 cell walls in, 80
 culturing viruses in, 403
 cytokinesis in, 357, 358f
 diseases of
  in biological warfare, 812, 813, 814, 815. See also Agroterrorism
  soilborne, 810, 810t
  viroids causing, 404, 405f
Plantae (kingdom), 113, 114f, 358, 359f, 374, 376t
Plantar warts, 723b
Plant galls, 340, 341f
Plant pathogens
 as biological weapons, 813, 814, 815. See also Agroterrorism
 in soil, 810, 810t
Plant viruses, 387–388, 387f
 culturing, 403
Plaque(s)
 in phage typing, 117, 118f
 viral, 394, 403, 403f
Plaque (dental/biofilm), 62, 66, 168, 170, 171b, 422, 422f, 557
 Selenomonas in, 335
 viridans streptococci in, 557
Plaque assay, 403
Plasma/plasma proteins, 451
 in host defense, 47, 451–452
Plasma cells, 483, 484f, 486, 493f, 494
 in hybridoma production, 511, 511f
 in type I (immediate) hypersensitivity, 526, 527f
Plasma membranes. See Cytoplasmic (cell/plasma) membranes
Plasmids, 196, 196f, 197–198, 198t
 bacteriocin, 196
 fertility (F), 196, 226, 227f, 228f
 as genetic vectors, 241, 242f
 resistance (R), 196, 227, 303
  multiple resistance and, 306
  transposons and, 230
 transposition and, 229, 229f
 in vaccine development, 253
 virulence, 196. See also Adhesins
Plasmodial (acellular) slime molds, 364, 365t
Plasmodium genus/spp. (*Plasmodium falciparum/knowlesi/malariae/ovale/vivax*), 5, 362, 365t, 686–687, 688–689b, 692t
 adhesins produced by, 422, 686
 life cycle/reproduction of, 357, 686, 688–689b
 malaria caused by, 419t, 441b, 686–687, 688–689b, 692t
 resistant/multi-drug-resistant strains of, 306, 441b, 689b
 vaccine development and, 689b
 virulence factors of, 686–687
Plastics, biodegradable, microbial production of, 74, 75b, 787, 789t
Plate (cell), 357, 358f
Platelet(s), 452f, 453
Pleconaril, 313t
 virus attachment affected by, 291f, 298, 313t
Pleomorphic forms, 325, 325f. See also specific type
 of mycoplasmas, 335, 568, 569f
Pleurodynia, coxsackie B virus causing, 738
PML. See Progressive multifocal leukoencephalopathy
PMNs (polymorphonuclear leukocytes). See Neutrophil(s)
Pneumococcal infection, 15t, 559, 561t
 immunization against, 508f, 509t, 558b, 559
 meningitis, 15t, 558b, 559
 pneumonia, 15t, 558b, 559

Pneumococcal/pneumococcal conjugate vaccine (PCV), 508f, 509t, 558b, 559
Pneumococci/pneumococcus. *See Streptococcus pneumoniae*
Pneumocystis genus/spp./*Pneumocystis jirovecii (Pneumocystis carinii),* 656–657, 657f, 664
 transmission of, 648, 657
Pneumocystis pneumonia (PCP), 656–657, 657f
 in HIV infection/AIDS, 541, 657, 664, 766b
Pneumocysts, 373
Pneumolysin, pneumococcal pathogenicity/virulence and, 558, 558b
Pneumonia, 161b, 187b, 288b, 414b
 Chlamydophila (Chlamydia) pneumoniae causing, 629
 Chlamydophila (Chlamydia) psittaci causing, 629
 coccidioidomycosis and, 647b, 669b
 cryptococcal, 659
 hantavirus causing, 384b, 392t, 408b, 419t, 771t
  transmission and, 384b, 408b, 767, 810, 810f
 in HIV infection/AIDS, 541, 657, 766b
 Klebsiella pneumoniae causing, 596, 604f
 Legionella pneumophila causing (Legionnaire's disease), 344t, 611–613, 613f
 manifestations of, 736t
 mycoplasmal (primary atypical/walking), 460b, 569
 Nocardia causing, 580
 parainfluenza viruses causing, 760
 pneumococcal, 15t, 558b, 559
  immunization against, 508f, 509t, 558b, 559
 Pneumocystis (PCP), 541, 656–657, 657f
  in HIV infection/AIDS, 541, 657, 664, 766b
 staphylococcal, 550
 viral, 392t, 736t. *See also specific causative agent*
Pneumonic plague, 602, 603f. *See also* Plague
Pneumonitis, hypersensitivity, 532–533
Pneumovirus genus/spp./respiratory syncytial virus, 392t, 503b, 520b, 758, 760–761, 761f, 771t
PO_4^{3-}. *See* Phosphate
Pock(s) (pustules), in poxvirus infection, 710, 711f
Point mutations, 216, 216–217, 217f, 218t
Point-of-care testing, 518–519, 519f
Point-source infections, waterborne illnesses and, 789
Poisonings. *See* Toxicoses
Poison ivy, 535, 536f
Polar covalent bonds, 33, 33f, 35t
Polar flagella, 63, 64f
Poliomyelitis (polio), 392t, 736–738, 737f, 738t, 771t
 immunization against, 505f, 508f, 509t, 737–738, 738t
  vaccine-induced, 506, 510
 waterborne transmission of, 736, 790t
Polio vaccine, 505f, 508f, 509t, 510, 737–738, 738t
 inactivated (IPV/Salk), 508f, 509t, 510, 737–738, 738t
 oral (OPV/Sabin), 738, 738t
 safety of, 506, 510, 738
 polio (vaccine-induced polio) and, 506, 510
Poliovirus, 392t, 398, 736–738, 737f, 738t, 771t
 bioterrorism and, 815
 immunization against, 505f, 508f, 509t, 510, 737–738, 738t. *See also* Polio vaccine
 size of, 388f
 synthesis in, 398
 waterborne transmission of, 736, 790t
Poliovirus vaccine. *See* Polio vaccine
Pollen, allergies caused by, 528
Pollution
 biosensors and bioreporters in detection of, 789
 water, 789, 791. *See also* Waterborne illnesses
  acid mine drainage and, 805–806, 805f

Polyadenylation, in transcription, 206, 207f
Polycistronic operon, 213, 215
Polyenes, 315t
 cytoplasmic membranes affected by, 291f, 293, 315t
Polyhedral viruses, 389, 389f
Polyhydroxybutyrate, 74, 75b, 75f
Polymer(s), 44
Polymerase(s), 127, 128t. *See also* DNA polymerases; RNA polymerase
 microbial production of, 787, 789t
Polymerase chain reaction (PCR), 118, 187, 244–246, 245f, 246b, 252t, 512
 diagnostic applications of, 254, 512
 C. difficile identification and, 319b
 dengue diagnosis and, 257b
 HIV detection and, 245, 756
 hyperthermophiles in, 245, 246b, 331
 primer in, synthetic nucleic acid as, 240
 reverse transcriptase (RT-PCR)
  for dengue identification, 257b
  for hantavirus detection, 767
  for Zika virus detection, 746b
Polymerization, 125, 125f
 DNA replication and, 198, 213t
 nucleic acid structure and, 51, 51f
Polymorphonuclear leukocytes. *See* Neutrophil(s)
Polymyxa genus/spp., soilborne transmission of, 810t
Polymyxin, 312t
 mechanism of action of, 291f, 294, 312t
 microbial production of, 290t, 294, 312t, 335
 spectrum of action of, 299f, 312t
 toxicity of, 302, 312t
Polyomaviridae (polyomaviruses/*Polyomavirus* genus/spp.), 392t, 722, 729t
Polypeptides, 47
 synthesis of
  DNA/RNA in, 203, 203f, 204, 206, 207, 207f, 208f, 210, 213t. *See also* Transcription; Translation
  elongation and, 211–212, 211f, 212f
  in HIV replication, 752, 753f
  regulation of, 212–216, 213f, 214f, 215f, 215t
Polyribosome, 211, 212f
Polysaccharides, 45–46, 46f, 153f. *See also* Carbohydrate(s)
 microbial virulence and. *See* Capsule
Polyunsaturated fats/fatty acids, 41, 43f
Pontiac fever, 613
Population (human), epidemiology in study of pathogens affecting, 17, 18f, 19t, 434–441. *See also* Epidemiology
Population (microbe), 804, 804f
 drug resistance in, 282, 303–308, 304b, 304f, 305b, 306f, 307b, 307f
 growth of, 162, 179–187. *See also* Microbial growth
  continuous culture in chemostat and, 182, 182f
  generation time and, 180
  mathematical considerations in, 180, 180f
  measuring, 183–187
   direct methods for, 183–186, 183f, 184f, 185f, 186f
   indirect methods for, 186–187, 186f
  phases of, 180–182, 181f
 in soil, 804, 804f, 810
Pore(s), nuclear, 61f, 85, 86f
Pore size, for filters, 272, 273t
Porins, in Gram-negative cell walls, 67, 68f
 resistance and, 306
Pork. *See* Meat/pork
Pork tapeworm (*Taenia solium*), 694, 694f, 695f, 703t
Portals of entry, 420–421, 420f, 421f, 421t
Portals of exit, 430, 430f
Ports, in active transport, 73, 73f
Positive chemotaxis/taxis, 65, 65f, 82
 in phagocytosis, 455, 455f
Positive selection, of mutants, 221–222, 221f
Positive-sense single-stranded RNA, 398, 399f, 399t, 735
Positive (positive-sense) single-stranded RNA viruses, 398, 399f, 399t, 735

 with reverse transcriptase (retroviruses), 239, 391t, 392t, 398, 399t, 735, 750–757, 750f, 771t
 naked, 735–741, 771t
 synthesis in, 398, 399f, 399t
Possible estuary-associated syndrome (PEAS), 363
Postpolio syndrome, 737
Postulates, Koch's, 14, 15t, 18f, 424–426, 425f
 exceptions to, 424–426
 experimental epidemiology and, 437
Potable (drinking) water, 440
 contamination of, 431, 437, 439f, 440, 789, 790. *See also* Waterborne illnesses
  bioterrorism and, 814t
 public health and, 440
 testing quality of, 792, 792f
 treatment of, 790–792, 791f
Potassium, 29t, 30f
Potassium chloride, ionic bonds in formation of, 33
Potassium dihydrogen phosphate, 38
Potassium hydroxide (KOH) preparations, fungi identified by, 649
Potato blight, soilborne transmission of, 810t
Potato pathogens
 resistance to, recombinant DNA technology and, 255
 soilborne transmission of, 810t
Potato scab, soilborne transmission of, 810t
Potato wilt, soilborne transmission of, 810t
Poultry
 assessing biological threats to, 813
 Campylobacter jejuni contamination of, 639
 defense against biological threats to, 815
 Salmonella contamination of, 600, 786t
Pour plates, 173–174, 174f
Pox (pocks/pustules), in poxvirus infection, 710, 711f
Poxviridae (poxviruses), 392t, 710–713, 710f, 711f, 713b, 729t
 as genetic vectors, 242
p-PRP. *See* Prion PrP
Prairie dogs, monkeypox transmitted by, 713b
Praziquantel, 316t
 mechanism of action of, 295, 316t
 spectrum of action of, 299f, 316t
Precipitation tests, 512–513, 512f, 520t
Precursor metabolites, 125, 125f, 126, 149, 149t, 153f
 for amino acid conversion, 150
 in citric acid/Krebs cycle, 136
 in glycolysis, 133
 in nucleotide biosynthesis, 151, 152f
 pentose phosphate pathway producing, 140
Prednisone, 537, 538t
Pre-exposure prophylaxis, HIV infection prevention and, 757
Pregnancy
 antimicrobial drug use during, 302
 attenuated (modified live) vaccines contraindicated in, 506, 748
 autoimmunity and, 538
 coxsackie B virus infection during, 738
 cytomegalovirus disease during, 421t, 720, 721
 disease transmission and, 421t, 431. *See also* Placenta, as portal of entry
 ectopic, pelvic inflammatory disease and, 589
 gonorrhea during, 279, 589
 ophthalmia neonatorum and, 589
 herpes infection during/neonatal herpes, 715, 716b
 HIV infection/AIDS and, 421t, 756, 757
 Listeria infection during, 336, 421t, 568, 570b
 Rh system/hemolytic disease of newborn and, 531, 532f
 rubella/rubella vaccination during, 747, 748
 Streptococcus agalactiae (group B streptococcus) infection during, 557
 syphilis during, 421t, 633
 toxoplasmosis during, 421t, 687, 690
 Zika virus infection and, 744, 745, 746b, 747t
Pregnancy tests, immunochromatographic, 519

Pre-messenger RNA (pre-mRNA), 206, 207f
PrEP. *See* Preexposure prophylaxis
Preparedness (public health), assessing biological threats and, 813
Preservatives, food, 785
Pressure cooking (autoclaving), in microbial control, 269–270, 270f, 271f, 275t
Prevalence of disease, 434, 434f
Prevotella genus/spp., 614
Primaquine, 318t
Primary amebic meningoencephalitis, *Naegleria* causing, 679, 692t, 790b
Primary atypical (mycoplasmal) pneumonia, 460b, 569
Primary follicles, lymph node, 475f, 476
Primary host. *See* Definitive host
Primary immune response, 494, 495f
Primary immunodeficiency diseases, 540, 540t
Primary infection, 434t
Primary lymphoid organs, 475f, 476
Primary metabolites, of industrial fermentation, 787
Primary producers, in carbon cycle, 806, 807
Primary pulmonary cryptococcosis, 659
Primary stain
 in acid-fast staining, 109
 in Gram staining, 109, 109f
 in Schaeffer-Fulton endospore staining, 110
Primary structure of protein, 48, 49f
Primary syphilis, 631b, 632
Primary tuberculosis, 576–577b
Primase
 in lagging strand synthesis, 200f, 201
 in leading strand synthesis, 199, 200f
Priming/primers, in polymerase chain reaction, 244, 245f
 synthetic nucleic acid and, 240
Prion(s), 385, 405–407, 406f, 407f, 408t
 bacteria/viruses/viroids compared with, 408t
 as biological weapon, 814t
 disease caused by, 48, 405–407, 406f
 enzyme for elimination of, 280
 portals of entry used by, 421
 sterilization ineffective against, 407
 susceptibility of, 266f, 267
 templating action of, 405, 406f
Prion protein. *See* PrP
Prion PrP (p-PRP), 405, 406f, 407, 407b
 as biological weapon, 814t
 spongiform encephalopathies and, 48, 405–407, 406f, 407b
 in templating, 405, 406f
Prionzyme, 280
Probe(s), 252t
 in clone selection, 246
 diagnostic applications of, 254
 electronic, in microscopy, 105, 106, 107t
 for fluorescent *in situ* hybridization (FISH), 249–250, 250f
 synthetic nucleic acids in synthesis of, 240
Probe microscopy/microscopes, 105–106, 106f, 107t
Probiotics, 10b, 304b, 450, 780
Proctitis
 chlamydial, 628
 gonococcal, 589, 589b
Prodromal period, 430
Product(s), 35
 in industrial microbiology, 12, 12t, 787–789, 789t
Production/producers, in carbon cycle, 806, 807
Professional antigen-presenting cells, 478, 478f
Profundal zone, 811f, 812
Proglottids, tapeworm, 693–694, 693f, 694f
Programmed cell death/suicide. *See* Apoptosis
Progressive multifocal leukoencephalopathy (PML), 392t, 722
Proguanil, 318t
 with atovaquone, 441b
Prokaryotae (kingdom), 113
Prokaryotes/prokaryotic cells, 59–60, 60f, 89f, 324–352. *See also* Archaea; Bacteria
 arrangements of, 327–328, 328f
 cell walls of, 60f, 66, 79t

characteristics of, 79t, 325–328, 325f, 326f, 327f, 328f
chromosomes of, 89t, 195–196, 196f, 198t
classification of, 4, 115, 329, 330f
    biochemical tests in, 116–117, 116f, 117f
    nucleic acid analysis/nucleotide sequencing and, 115, 118–119
CRISPR regions in, 242–243, 243f
cytoplasmic membrane of, 60f, 69f, 79t. *See also* Cytoplasmic (cell/plasma) membranes
cytoplasm of, 60f, 79t. *See also* Cytoplasm
evolution of eukaryotic cells and, 88
genomes of, 194–195, 196f, 198t
horizontal (lateral) gene transfer among, 223–227, 225f, 226f, 227f, 228f, 228t, 230b
life processes in, 59t
morphology of, 325, 325f
mRNA of, 208, 208f
organelles in, 76–77, 77f, 85t
plasmids in, 196, 198t
reproduction of, 326, 326f, 327f
size of, 4, 4f
transcription in, 203–205, 205f, 206f
translation in, 203, 203f, 207–212, 208f, 209f, 210f, 211f, 212f
Prokaryotic operons, 212–215, 213f, 214f, 215f, 215t
Prokaryotic ribosomes, 60f, 85t, 87, 88, 89t, 209–210, 209f, 210f, 293
    antimicrobials affecting, 76–77, 210, 293, 294f
    archaeal, 78, 85t, 89t
    bacterial, 76–77, 85t, 89t
    endosymbiotic theory/evolution and, 87, 88
    in translation, 209–210, 209f, 210f
Promastigotes, in *Leishmania* life cycle, 682, 683f
Promoter
    in operon, 213, 213f, 214f, 215, 215f
    in transcription, 204, 205f, 208f, 213f
Proofreading exonuclease function, 201, 213t
Propamidine isethionate, mechanism of action of, 297
Prophage, 395, 396f
    induction of, 395, 396f
Prophase
    meiosis, 356, 356f, 357t
    mitosis, 355, 355f, 357t
Propionibacterium genus/spp./*Propionibacterium acnes*, 578–579, 579f
    acne caused by, 39, 578–579, 579f
    fermentation products of, 142f, 578
    pH range tolerated by, 39
Propionic acid
    food preservation and, 785
    microbial fermentation producing, 142f, 578
Propylene oxide, for gas sterilization, 280, 281t
Prostaglandins
    in fever, 464
    in inflammation, 463, 464, 465f
    in type I (immediate) hypersensitivity, 527, 527f, 527t
Prostate gland secretions, in host defense, 451t
Prosthecae, 339, 339f
    of *Caulobacter*, 340, 340f, 341b
Protease(s), 527, 752
    group B streptococci (*Streptococcus agalactiae*) producing, 556
    HIV, 752, 753f, 754–755, 754b
    microbial production of, 787, 789t
    in protein catabolism, 143, 144f
    in retroviruses, 750
    secretory IgA
        gonococcal pathogenicity/virulence and, 588, 589b
        pneumococcal pathogenicity/virulence and, 558
    in type I (immediate) hypersensitivity, 527, 527f, 527t
Protease inhibitors, 314t
    for HIV infection/AIDS, 755
    mechanism of action of, 295, 314t
Protein(s), 46–49, 47f, 48f, 49f. *See also specific type and* Amino acid(s)
    antiviral (AVPs), 314t
    in bacterial cell walls, 67, 68f

    biosynthesis of. *See* Protein synthesis
    catabolism of, 143, 144f, 153f
    complement. *See* Complement/complement system
    denaturation of, 49, 130, 131f
        antimicrobial action and, 265, 269, 271
    enzymes. *See* Enzyme(s)
    genes for, synthetic nucleic acids in creation of, 240
    in host defense/offense, 47
    matrix, 389, 391f
    membrane, 69, 69f
        in electron transport chains, 136, 139
        in facilitated diffusion, 71, 71f
    regulatory, 47
    structural, 47
    structure of, 48–49, 49f
        antimicrobial action and, 265
        function and, 48, 49
    synthesis of. *See* Protein synthesis
    transport, 47
Protein A, staphylococcal pathogenicity/virulence and, 548, 548t
Protein adhesin. *See* Adhesins
Protein kinase R, 459
Protein rings, bacterial flagellar motion and, 63, 63f
Protein synthesis, 153f
    amino acid biosynthesis and, 150, 151f, 153f. *See also* Amino acid(s), biosynthesis of
    antimicrobials affecting, 265–266, 290, 291f, 293, 294f, 310–311f, 314t, 317t
        antiprotozoan drugs, 317t
        antiviral drugs, 314t
    protein kinase R affecting, 459
    recombinant DNA technology in, 252–253
    in ribosomes, 76
    in viral replication, 392
        animal virus, 397–398, 399f, 399t, 401t
        bacteriophage, 393f, 394, 401t
Proteobacteria (phylum), 330f, 333, 334, 334f, 338–347, 344t, 630t. *See also* Gram-negative bacteria
    α, 330f, 334f, 338–340, 339f, 340f, 344t, 605, 606, 621
    β, 330f, 334f, 340–341, 342f, 344t, 587, 605
    δ, 330f, 343–346, 344t
    ε, 330f, 344t, 346, 639, 640
    γ, 330f, 334f, 341–343, 344t, 591, 605, 608, 637
    ζ, 330f, 344t, 347
Proteus genus/spp. (*Proteus mirabilis/vulgaris*), 343t, 344t, 592f, 597, 597f
    fermentation in identification of, 141
    fimbriae and flagella of, 65f, 597
    flagellar stain of, 111f
    healthcare-associated (nosocomial) infection caused by, 591f, 597
    sites of infection caused by, 604f
    UTIs caused by, 597, 599b, 604f
Protista (kingdom), 113, 357, 359f, 360
Proton(s), 28, 28f
    atomic mass of, 28
    in chemiosmosis, 139, 140, 146
    in electron transport chain, 136, 139
    in photosynthesis, 146
Proton gradient, 136, 139, 140
    halobacteria establishing, 331–332
Proton motive force, 140
    in photosynthesis, 146
Protooncogenes, 402, 402f
Protoplasts
    in electroporation, 248
    fusion of, 248, 249f, 252f
Protoscoleces, of hydatid cysts, 695
Prototheca genus/spp., 374, 376t
Protozoa, 5–6, 5f, 358–364, 365f, 677–691, 692t. *See also* Parasite(s)
    antimicrobials effective against, 293, 299f, 317–318t
    classification of, 357, 360–364
    cysts of, 359. *See also* Cyst stage, of protozoa
    cytokinesis in, 357, 358f
    diseases caused by/pathogenic, 419t
    distribution of, 358
    locomotive structures of, 5, 5f
        classification and, 360
    morphology of, 359, 360f
    nutrition of, 359

    opportunistic infections caused by, in HIV infection/AIDS, 751t
    placenta crossed by, 421t
    reproduction of, 360
    sizes of, 61f
    in soil, 810
    susceptibility of, 266f, 267
    in termites (mutualism), 415, 415f
Protozoology/protozoologists, 18f, 19t, 358. *See also* Protozoa
Providencia genus/spp., 592f, 597
Proviruses (latent viruses), 400. *See also* Latency
    HIV as, 400, 752, 753, 753f
Prozone phenomenon, 513
PrP, 405–407, 406f. *See also* Prion(s)
    cellular (c-PRP), 48, 405, 406f
    prion (p-PRP), 48, 405, 406f, 407, 407b
    as biological weapon, 814t
    spongiform encephalopathies and, 48, 405–407, 406f, 407b
    in templating, 405, 406f
Prusiner, Stanley, 405
PS. *See* Photosystem(s)
Pseudallescheria genus/spp., mycetoma caused by, 666f, 667
Pseudocysts, of *Toxoplasma*, 687f, 690, 690f
Pseudohyphae (candidal), 367, 660b, 661b
Pseudomembrane, in diphtheria, 289, 572–573, 573f, 573f
Pseudomembranous colitis, *Clostridium difficile* causing, 302, 439, 563
    fecal microbiome transfer for, 803b
Pseudomonads, 342, 343, 343f, 608–611
    Entner-Doudoroff pathway used by, 140, 342, 343, 609
    healthcare-associated (nosocomial) infection caused by, 591f
Pseudomonas genus/spp. (*Pseudomonas aeruginosa/fluorescens/syringae*), 343, 343f, 344t, 426f, 609–610, 610f
    agricultural applications for, 789
    antimicrobials produced by, 290f, 311t
    in biofilm, 610
    burn wound/skin infection caused by, 609, 610f
    crop freeze resistance and, 255, 789
    cytochromes in, 139
    Entner-Doudoroff pathway used by, 140
    fluorescent, 103
    folliculitis caused by, 796b
    gene expression regulation in, 213
    healthcare-associated (nosocomial) infection caused by, 591f
    in Kelsey-Sykes capacity test, 282
    in nitrogen cycle, 139
    quats inactive against, 279, 610
    resistant/multiple-drug-resistant strains of, 196, 306, 609, 610
    respiratory infection caused by, in cystic fibrosis, 609, 802b, 816b
    in use-dilution test, 281
    UTIs caused by, 599b
    virulence of, 426f, 599b, 609
Pseudopods
    dendritic cell, 449
    in endocytosis/phagocytosis, 81, 81f, 455, 455f
    in motility, 5, 5f, 81
    of protozoa, 5, 5f
        amoebas, 363, 363f, 364, 364f, 678
Psilocybe cubensis, 370, 668
Psilocybin, 370, 668
P site, ribosomal, 210, 210f, 211, 211f
Psittacosis (ornithosis), 629
    bioterrorism and, 814t
Psychrobacter arcticus, 166
Psychrophiles, 166, 167f
    growth of in refrigerated food, 272
    soil organisms as, 810
    study of genetic basis of, 251
Psychrotolerant organisms (psychrotrophs), 166, 167f
Public health
    assessing biological threats and, 812, 813
    epidemiology and, 435, 437t, 438f, 440–441
    vaccine requirements and, 504b
Public health education, 440–441
Public health microbiology, 19t, 435, 437t, 438f
Public health preparedness, assessing biological threats and, 813
Public Health Service, 440

Public perception, assessing biological threats and, 813
Puerperal (childbirth) fever, streptococcal, 16, 556
Pulmonary anthrax, 336, 561–562, 736t. *See also* Anthrax
    bioterrorism and, 336, 562
Pulmonary aspergillosis, 370b, 657, 658, 659b
Pulmonary blastomycosis, 653, 654b
Pulmonary candidiasis, 658t, 660b
Pulmonary coccidioidomycosis, 647b, 655, 669b
Pulmonary cryptococcosis, 659
Pulmonary histoplasmosis, 651, 652b, 653b, 663b
Pulmonary paracoccidioidomycosis, 656
Pulmonary tuberculosis. *See* Tuberculosis
Pulmonary zygomycosis, 662
Pure culture, 172–174, 173f, 174f
Purification (water), 790–792, 791f
Purines, 50, 50f, 52t
    biosynthesis of, 151, 152f
    antimicrobials affecting, 295, 296f
Purple phototrophic bacteria, 333–334, 334f
    nonsulfur, 330f, 333, 334, 334f, 339, 344t
    sulfur, 333, 334, 334f, 334t, 342, 342f, 344t
Purpuric fever, Brazilian, 605
Purulent abscesses, in group A streptococcal pharyngitis, 553, 553f
Pus, 464, 465f
Pustule(s), 464
    in acne, 578, 579f
    in poxvirus infection, 710, 711f
Pyelonephritis, 598b, 599b
Pyocyanin, in *Pseudomonas aeruginosa* infection, 609
Pyoderma (streptococcal impetigo), 553
Pyogenic lesions, staphylococcal, 549, 549f
Pyrantel pamoate, 316t
Pyrazinamide, 312t
    mechanism of action of, 295, 312t
Pyridoxal phosphate, 128t
    in transamination, 150, 151f
Pyrimethamine, 318t
Pyrimidine(s), 50, 50f, 52t
    biosynthesis of, 151, 152f
    antimicrobials affecting, 295, 296f
Pyrimidine dimers, ultraviolet light causing formation of, 218, 218f, 275
    repair of, 220, 220f
Pyrite, acid mine drainage and, 805, 806
Pyrococcus furiosus, 78f, 79f
Pyrodictium genus/spp., 330, 331f
Pyrogens/pyrogenic (erythrogenic) toxins, 464
    group A streptococcal pathogenicity/virulence and, 552
    in scarlet fever, 553
Pyrophosphate, in DNA replication, 199, 199f
Pyruvate dehydrogenase, cellular respiration/acetyl-CoA synthesis, 135, 135f
Pyruvic acid/pyruvate, 149t, 153f
    in acetyl-CoA synthesis, 135, 135f
    in carbohydrate biosynthesis/gluconeogenesis, 150f, 153f
    in fermentation, 141, 141f, 142f
    in glycolysis, 133, 133f, 134, 134f, 135f, 149t, 153f

Q fever, 344t, 613–614
    bioterrorism and, 814t
Quantitative polymerase chain reaction (qPCR), 245–246
Quaternary ammonium compounds (quats), 278–279, 278f
Quaternary structure of protein, 48, 49f
Quellung reaction, 559
Quinine, 318t
Quinolones, 313t, 318t
    diarrhea associated with use of, 319b
    mechanism of action of, 291f, 296, 313t, 318t
    resistance to, 306
Quinupristin, 311t
Quorum sensing, 20, 169, 171f, 213
    amplification of, biofilm prevention/disruption and, 170
Quorum-sensing molecules, 169, 171f
    Vibrio vulnificus secreting, 170b

R (residue), 40
RA. *See* Rheumatoid arthritis
Rabbit fever (tularemia), 426f, 611, 612b
    bioterrorism and, 611, 814t

Rabies, 392t, 418, 419t, 761–763, 761f, 762b, 762f, 771t
  bioterrorism and, 814t
  immunization against, 509t, 762–763
  reservoirs for, 418, 761–762, 762b
Rabies immune globulin, human (HRIG), 762
Rabies vaccine, 509t, 762–763
Rabies virus, 391t, 392t, 761, 761f, 771t
  bioterrorism and, 814t
  direct fluorescent immunoassay in detection of, 516, 516f, 762
  shape of, 389f
Radiation
  Deinococcus resistance to, 250–251, 332
  as energy source for fungi, 367
  in microbial control/food preservation, 273–275, 274f, 275t, 785
  mutagenic effects of, 217–218
    DNA repair and, 220, 220f
  wavelengths of, 98, 98f, 273
    chlorophyll absorption and, 144
    microscopy and, 98, 98f, 104
Radioactive decay, 29
  microbial remediation of waste caused by, 169b
Radioactive isotopes, 29
Radioactive waste, microbial resistance/remediation and, 169b
Radiolaria/radiolarians, 364, 364f, 365t
RAG (recombination activating gene) protein, 484, 484f
Ragweed (*Ambrosia trifida*), 528f
Ralstonia solanacearum, soilborne transmission of, 810t
Rapid identification tests, 116–117, 118f
  streptococcal antigen, 556
Rapid sand filters, in drinking water treatment, 791–792
Rapid streptococcal antigen test (rapid strep test), 556
Rashes. *See also* Skin, diseases/infections of
  in chickenpox, 717, 717f, 718
  in erythema infectiosum (fifth disease), 728, 728f
  in Lyme disease, 96b, 120b, 635
  in measles, 472t, 497f, 758, 758f, 758t
  in *Rickettsia parkeri* infection, 622b
  in *Rickettsia rickettsii* infection (Rocky Mountain spotted fever), 624b, 625b
  in roseola, 721, 721f
  in rubella, 747, 748f, 758t
  in shingles, 717, 718f, 719b
  in syphilis, 631b, 632, 632f
  in systemic lupus erythematosus, 534, 534f
Rat fleas. *See* Fleas; Rodents
Raw fish, anisakids transmitted in, 701, 701f, 703t
Rays, light, 98, 98f
RBs. *See* Reticulate bodies
Reactant(s), 35
Reaction center/reaction center chlorophyll, 145–146, 145f, 147f
Real-time polymerase chain reaction (PCR), 245–246
Recalcitrant substances, bioremediation and, 805
Receptor blocking drugs, biofilm formation and, 170
Receptor diversity, 483–485, 484f
Recipient cell, in horizontal (lateral) gene transfer, 224
Recombinant(s)/recombinant cells, 223, 224f, 787
  in horizontal gene transfer, 224
    conjugation and, 226, 228f
  restriction enzymes producing, 240, 241f, 787, 789t
  selecting clone of, 246
Recombinant DNA, 223, 224f. *See also* Recombinant DNA technology
Recombinant DNA technology/genetic engineering, 18f, 19t, 20, 237–261, 239f. *See also* Molecular biology/techniques
  in agriculture, 254–256, 255f, 788–789, 789t
    Bt toxin and, 255, 335–336, 336f, 788–789, 789t
    ethical/safety issues and, 256, 788–789
  applications of, 238, 248–256
  biofuel production and, 162b, 788, 789t
  biomedical animal models and, 254
  in biotechnology, 238, 239f
  bioterrorism and, 256, 812, 815
  competent cells and, 224–225, 248
  CRISPR in, 242–243, 243f
  diagnostic applications of, 254
  for DNA fingerprinting, 252
    in water quality testing, 792
  DNA insertion into cells and, 248, 249f
  in environmental studies, 248
  ethics of, 162b, 256–257
  gene libraries and, 243, 244f, 252t
  gene therapy and, 20, 238, 254
  genetic mapping and, 248–252, 250f, 251f, 252t
  for genetic screening, 253
  hyperthermophiles in, 245, 246b, 331
  industrial uses of, 162b, 238
  microbial community studies and, 252
  mutagens in, 238, 252t
  for pharmaceutical/therapeutic uses, 238, 252–254, 788, 788t
  for protein synthesis, 252–253
  restriction enzymes in, 240, 240t, 241f, 252t
  reverse transcriptase/transcription in, 239, 252t
  safety of, 256
  SARS/MERS vaccine development and, 749
  in soil microbiology, 810
  synthetic nucleic acids in, 239–240, 252t
  techniques of, 243–248, 245f, 246b, 247f, 249f, 252t
  tools of, 238–243, 240t, 241f, 242f, 243f, 252t
    microbial production of, 787
  in vaccine development/production, 253, 506–508, 507f, 749, 788t
  vectors in, 241–242, 242f, 252t
  for xenotransplants, 254
Recombinant microorganisms, 223, 224f, 787. *See also* Recombinant DNA technology
Recombination (genetic), 223–230, 224f, 230f. *See also* Recombinant DNA technology
  horizontal (lateral) gene transfer and, 223–227, 225f, 226f, 227f, 228f, 228t, 230b
  mutations caused by, 217
  transposons/transposition and, 228–230, 229f
Recombination activating gene (RAG) protein, 484, 484f
Red algae, 80f, 373, 374, 375f, 376f
Red blood cells. *See* Erythrocyte(s)
Red bone marrow, 473, 475f. *See also* Bone marrow
  lymphocyte formation/maturation and, 473
Redi, Francesco, experiments in spontaneous generation by, 8, 8f
Red measles. *See* Measles
Redness, in inflammation, 462
Redox (oxidation-reduction) reactions, 125, 126, 126f
  in cellular respiration, 136, 139
  in microbial growth, 163
Red tide, 362, 790
Reducing media, 177
Reduction, 126, 126f. *See also* Redox (oxidation-reduction) reactions
  in Calvin-Benson cycle, 146, 148f
Redundancy, genetic code, 208, 209, 240
Reed, Walter, 14, 15f
Reemerging diseases. *See also* Emerging/reemerging diseases
Refraction, image magnification and, 98–99, 98f
  oil immersion lens affecting, 100, 102f
Refrigeration
  for culture preservation, 179
  in food preservation/microbial control, 271, 275t
    Listeria resistance to, 272, 336, 568, 785, 786t
Regulatory genes, 214, 214f, 215, 215f
Regulatory proteins, 47

Regulatory RNA
  in transcription, 204
  in translation, 215–216
Regulatory (suppressor) T cells (Tr cells), 481, 482t, 492
  clonal deletion and, 482, 482f
Rejection, transplant (graft), 536
  donor-recipient matching/tissue typing and, 536–537
  immunosuppressive drugs and, 537–538, 538f
  MHC antigens and, 478, 536, 536–537
Relapsing fever, 635–636, 636f
Relaxin, recombinant DNA in production of, 788f
Release, in viral replication, 392
  animal virus, 400, 400f, 401t
  bacteriophage, 393f, 394, 396, 396f, 401t
  HIV, 752, 753f, 754–755
Release factors, in translation, 212, 212f
Rennin, in cheese production, 780, 780f
Reoviridae (reoviruses), 392t, 735, 747t, 770–771, 771t
Repair
  DNA, 219–221, 220f
    direct, 220, 220f
    error-prone, 221
    methylation in, 201
    single-strand, 220–221, 220f
  tissue, 464
Repeat(s), in CRISPR, 242, 243f
Replica plating, in negative (indirect) selection, 222, 222f
Replicating transposons, 229f
Replication
  DNA, 151, 152f, 198–202, 198f, 199f, 200f, 202f. *See also* DNA, replication/synthesis of
  viral, 391f, 392–400, 401t. *See also* Viral replication
Replication fork, 199, 200f, 201, 202, 202f
Reportable/notifiable diseases, 435, 437t, 438f
  bioterrorism defense and, 815, 815f
Repressible operons, 213, 215t
  tryptophan (*trp*) operon as, 215, 215f
Repression, in lactose (lac) operon regulation, 214, 214f
Repressor protein/repressors
  in lysogeny, 395
  in operons, 213, 214, 214f, 215, 215f
Reproduction. *See also specific organism*
  in algae, 373–374, 374f
  archaeal, 329
  by binary fission, 179–187, 179f, 326, 326f
  as characteristic of life, 59, 59t, 354
  eukaryotic, 354–357, 355f, 356f, 357t, 358f, 359f
  fungal, 367, 368f
  prokaryotic, 326, 326f, 327f
  protozoal, 360
Reproductive spores, 326, 327f
  asexual, 4, 326, 327f, 367, 368f
  sexual, 4, 367, 368f
Reproductive system
  in host defense, 451t
  microbiome/normal microbiota of, 417t
  sexually transmitted (venereal) diseases and. *See* Sexually transmitted infections/diseases
Reptiles (pet), salmonellosis and, 600
RER. *See* Rough endoplasmic reticulum
Reservoirs of infection, 418–420, 419b, 419t
Resident microbiota, 416, 417t. *See also* Microbiome
Residual body, 455f, 456
Residual virulence, vaccine safety and, 506, 509–510
Resistance (host). *See* Host defenses
Resistance (microbial), 303–308, 304b, 304f, 305b, 306f, 307b, 307f, 309–318t. *See also specific organism*
  antimicrobial drug exposure and, 303–304, 304f
  biofilms and, 306
  choice of control method and, 266, 266f
  complex transposons and, 229f, 230
  conjugation and, 226
  cross, 307
  development of, 282, 303–304, 304f
  diffusion susceptibility tests in determination of, 299, 301b
  emerging/reemerging diseases and, 305b

endospores and, 266–267, 266f
evolution/mutations and, 219
healthcare-associated (nosocomial) infections and, 230b
horizontal gene transfer and, 227, 229f, 230, 230b, 303
mechanisms of, 305–306, 306f
multiple drug, 230b, 306–307. *See also* Multi-drug resistance
phage therapy and, 395b
plasmids and, 196, 227, 303
  multiple resistance and, 306
  transposons and, 230
probiotics and, 304b
retarding, 307–308, 307f
susceptibility and, 266–267, 266f
transposons and, 229f, 230
Resistance (disease), species, 447
Resistance (R) plasmids/factors, 196, 227, 303
  multiple resistance and, 306
  transposons and, 230
Resistance pumps, 306
Resolution/resolving power, 99, 99f. *See also specific type of microscope*
  of electron microscope, 99f, 104
  immersion oil affecting, 100, 102f
Respiration
  in carbon cycle, 806, 807f
  cellular. *See* Cellular respiration
Respiratory syncytial virus (RSV/RSV infection/*Pneumovirus*), 392t, 503b, 520b, 758, 760–761, 761f, 771t
Respiratory system. *See also under* Pulmonary *and* Pneumonia
  adenovirus infection of, 722–724, 725b
  anthrax affecting, 336, 561–562, 736t. *See also* Anthrax
    bioterrorism and, 336, 562
  aspergillosis affecting, 370b, 657, 658, 659b
  blastomycosis affecting, 653, 654b
  candidal infection of, 658t, 660b
  coccidioidomycosis affecting, 647t, 655, 669b
  cryptococcal infection of, 659
  cystic fibrosis and, 609, 802b, 816b
  Enterobacteriaceae infection of, 604f
  histoplasmosis affecting, 651, 652b, 653b, 663b
  in host defense, 448, 449, 449f
  in immune complex–mediated (type III) hypersensitivity, 532–533
  manifestations of infection of, 736t
  microbiome/normal microbiota of, 417t
    cystic fibrosis treatment and, 816b
  nocardial infection of, 580
  paracoccidioidomycosis affecting, 656
  parainfluenza virus infection of, 760
  as portal of entry, 420f, 421
    for parasites, 676f
  as portal of exit, 430, 430f
  respiratory syncytial virus infection of, 503b, 520b, 758, 760–761, 761f, 771t
  rhinovirus infection of, 735–736, 735f, 736t, 771t
  smoking affecting, 449, 460b
  specimen collection from, 173t
  tuberculosis affecting, 573–574, 574b, 576–577b
  Yersinia infection of (pneumonic plague), 602, 603f, 604f
  zygomycosis affecting, 662
Respirovirus genus/spp., 758, 759, 771t
Responsiveness, as characteristic of life, 59, 59t
Restriction enzymes/endonucleases, 240, 240t, 241f, 252t
  gene library production and, 243, 244f
  in genetic mapping, 249
  microbial production of, 787, 789t
  vector production and, 241, 242f
Restriction fragmentation, for genetic mapping, 249
Restriction sites, 240, 241f
Reticulate (initial) bodies, in *Ehrlichia* and *Anaplasma* growth/reproduction, 623, 623f
Reticulate bodies, in chlamydial growth/reproduction, 626, 627f
Reticulitermes genus/spp., protozoa (*Trichonympha*) in, 415, 415f

Retinoic acid, for acne, 579
Retrospective studies, analytical studies as, 437
Retroviridae (retroviruses), 239, 391*t*, 392*t*, 398, 399*t*, 735, 750–757, 750*f*, 771*f*
 immunosuppressive, 751–757, 751*t*, 752*t*, 753*f*, 754*b*, 754*t*, 755*f*, 765*f*. *See also* HIV
 oncogenic, 750–751, 751*f*
 synthesis in, 398, 399*t*
Reuterin, 304*b*
Reverse transcriptase, 239, 252*t*, 398, 400, 725, 725*f*, 750, 750*f*. *See also Retroviridae*
 in DNA/cDNA synthesis, 239, 750, 750*f*
 in hepatitis B virus replication, 398, 725–726
 HIV, 398, 752, 753, 753*f*, 754*b*
 inhibition of, 297
 in recombinant DNA technology, 239, 252*t*
 in retroviruses, 239, 398, 750–757, 750*f*
 RNA viruses with, 398. *See also Retroviridae*
Reverse transcriptase inhibitors, mechanism of action of, 297
Reverse transcriptase polymerase chain reaction (RT-PCR)
 for dengue identification, 257*b*
 for hantavirus identification, 767
 for Zika virus detection, 746*b*
Reverse transcription, 239, 252*t*. *See also* Reverse transcriptase
Revertant cells, in Ames test, 223, 223*f*
Revolving nosepiece, 100
Reye's syndrome, aspirin use and, 718
R factors. *See* R (resistance) plasmids/factors
R group, 40, 41*f*
 in amino acids, 47, 47*f*, 48
Rhabdoviridae (rhabdoviruses), 391*t*, 392*t*, 735, 758, 761, 771*t*
Rhadinovirus genus/spp., 713, 721
Rh (Rhesus) antigen, 514, 514*f*, 523*f*, 531
Rheumatic fever, 553
Rheumatoid arthritis, 533, 534*f*, 538, 539
Rhinocerebral zygomycosis, 662
Rhinorrhea, in common cold, 735
Rhinovirus genus/spp./rhinoviruses, 735, 735–736, 735*f*, 736*t*, 771*t*
 common cold caused by, 735–736, 735*f*, 736*t*, 771*t*
Rhizaria (kingdom), 357, 359*f*, 363–364, 363*f*, 364*f*, 365*t*
Rhizobium genus/spp., 339, 339*f*, 344*t*
 nitrogen fixation by, 165, 339, 339*f*, 808
Rhizopus genus/spp., 369, 373*t*, 662
Rho-dependent termination, of transcription, 204, 205*f*
Rhodocyclus genus/spp., 339
Rhodophyta (kingdom), 358, 359*f*, 374, 375*f*, 376*f*
Rhodopseudomonas palustris, 339
RhoGAM, 531
Rho protein, 205*f*
Rh system, 514, 514*f*, 523*f*, 531
Ribavirin, 297*f*, 314*t*
 spectrum of action of, 299*f*, 314*t*
Ribonucleic acid. *See* RNA
Ribonucleotide, 50
Ribose, 50, 50*f*, 52*t*, 151, 194
Ribose 5-phosphate, 149*t*
 in nucleotide synthesis, 151, 152*f*
 in pentose phosphate pathway, 149*t*
Ribosomal enzymes. *See* Ribozymes
Ribosomal fingerprints, prokaryotic classification and, 115, 118–119, 329, 330*f*
Ribosomal RNA (rRNA), 76
 taxonomy/microbial classification and, 115, 118–119, 329, 330*f*. *See also specific organism*
 in transcription, 203
 in translation, 209–210, 209*f*, 210*f*
Ribosome(s), 76–77, 85*t*, 89*t*, 209, 209*f*
 antimicrobial drugs and, 76–77, 210, 290, 293, 294*f*, 310*t*
 binding sites for, 210, 210*f*
 in chloroplasts, 87, 88
 endosymbiotic theory/evolution and, 87, 88
 eukaryotic, 61*f*, 82–83, 85*t*, 89*t*, 209*f*, 210
 antimicrobials affecting, 210, 293

microbial taxonomy and, 115
mitochondrial, 87, 88*f*
prokaryotic, 60*f*, 85*t*, 89*t*, 209–210, 209*f*, 210*f*
 antimicrobials affecting, 76–77, 210, 293, 294*f*, 310*t*
 archaeal, 79, 85*t*, 89*t*
 bacterial, 76–77, 85*t*, 89*t*
 in translation, 209–210, 209*f*, 210*f*
 protein synthesis in, 76
 on rough endoplasmic reticulum, 86, 86*f*
 translation/polypeptide synthesis and, 203*f*, 207, 208*f*, 209–210, 209*f*, 210*f*, 211, 211*f*, 213*t*. *See also* Translation
Ribosome-binding site (Shine-Dalgarno sequence), 210
Riboswitch, 216
Ribozymes, 128
 antimicrobial drugs affecting, 266
 in transcription, 204
 in translation, 211, 211*f*
Ribulose 1,5-bisphosphate (RuBP), in Calvin-Benson cycle, 146, 148*f*
Rice, recombinant DNA technology improving nutritional value of, 255, 256
Rice beer (sake), fermentation in production of, 782, 783*t*
"Rice-water stool," in cholera, 12, 638
Ricketts, Howard, 621
Rickettsia(s), 330*f*, 621–626, 622*b*, 623*f*, 624–625*b*, 626*t*, 630*t*
 animal/cell cultures for growth of, 178
 antimicrobials effective against, 299*t*
Rickettsia genus/spp. (*Rickettsia parkeri/prowazekii/rickettsii/typhi*), 340, 344*t*, 621–622, 622*b*, 624–625*b*, 626*t*
 as biological weapon, 814*t*
Rickettsia tsutsugamushi. *See Orientia tsutsugamushi*
Rickettsiosis, spotted fever, 621, 622*b*, 626*t*
 Rocky Mountain spotted fever, 344*t*, 621, 624–625*b*, 626*t*
Rifampin, 313*t*
 mechanism of action of, 291*f*, 297, 313*t*
 microbial production of, 290*t*, 313*t*
 resistance to, 219, 313*t*
Rifamycin, 313*t*
Rifaximin, 313*t*
Rift Valley fever, 747*t*, 767
Rimantadine, 314*t*, 769*b*
 mechanism of action of, 295, 314*t*
Ringspot virus resistance, recombinant DNA technology and, 255, 255*f*
Ringworm (dermatophytoses), 419*t*, 649, 664–665, 665*f*, 665*t*
Ripened cheese, 780
RMSF. *See* Rocky Mountain spotted fever
RNA, 50–54, 50*f*, 52*t*. *See also* Nucleic acid(s)
 antimicrobials affecting, 291*f*, 293, 294*f*, 295, 295–297, 296*f*
 antisense, yield/nutritional value of food and, 255
 comparison of with DNA, 52*t*
 CRISPR (crRNA), 242, 243*f*
 double-stranded, 386, 398, 399*f*, 399*t*
 in viruses, 386, 391*t*, 392*t*, 398, 399*f*, 399*t*, 735. *See also* Double-stranded RNA viruses
 function of, 52, 52*t*
 HIV, 752, 753, 753*f*, 754*b*, 754*f*
 messenger (mRNA), 206, 207*f*, 208, 208*f*, 209. *See also* Messenger RNA
 micro (miRNA), 215
 microbial taxonomy/microbial classification and, 115, 118–119, 329, 330*f*, 357, 359*f*, 360
 negative-sense single-stranded, 398, 399*f*, 735
 positive-sense single-stranded, 398, 399*f*, 399*t*, 735
 pre-messenger (pre-mRNA), 206, 207*f*
 regulatory
 in transcription, 204
 in translation, 215–216
 ribosomal (rRNA), 76
 taxonomy/microbial classification and, 115, 118–119, 329, 330*f*. *See also specific organism*
 in transcription, 203
 in translation, 209–210, 209*f*, 210*f*

single-stranded, 52*t*, 398, 399*f*. *See also* Single-stranded RNA
 in viruses, 52*t*, 386, 391*t*, 392*t*, 398, 399*f*, 399*t*. *See also* Single-stranded RNA viruses
 small interfering (siRNA), 216
 structure of, 50, 50*f*, 51, 52*t*, 194, 195*f*
 synthesis of, 151, 152*f*
 antimicrobials affecting, 291*f*, 295, 295–297, 296*f*
 in hepatitis B virus replication, 398, 725–726
 in HIV replication, 752, 753, 753*f*, 754*f*
 in nucleoli, 85
 viral, 398, 399*f*, 399*t*
 synthetic, 239–240
 transcription of, 203, 203–206, 205*f*, 206*f*, 207*f*. *See also* Transcription
 reverse, 239
 transfer (tRNA), 209, 209*f*. *See also* Transfer RNA
 in translation, 203, 207–212, 208*f*, 209*f*, 210*f*, 211*f*, 212*f*
 viral, 385, 386, 391*t*
 interferons affecting, 458, 458*f*
 synthesis of, 398, 399*f*, 399*t*
 Toll-like receptors affecting, 457
 in viroids, 404, 405*f*
RNA-dependent RNA transcriptase, 398
RNA enzymes. *See* Ribozymes
RNA polymerase, 204, 205*f*, 206, 213*t*
 antimicrobials affecting, 297
 operon regulation and, 213, 214, 214*f*, 215, 215*f*
 in transcription, 204, 205*f*, 213*t*
 in viral synthesis, 398, 399*f*
RNA primer
 in lagging strand synthesis, 200*f*, 201
 in leading strand synthesis, 199, 200*f*
 transcription of, 203
RNA probes. *See also* Probe(s)
 in clone selection, 246
 synthetic nucleic acids in synthesis of, 240
RNA transcript, elongation of, 204
RNA transcriptase, RNA-dependent, 398
RNA viruses, 385, 386, 391*t*, 392*t*, 734–777, 771*t*. *See also specific type*
 cancer and, 402
 double-stranded, 386, 391*t*, 392*t*, 398, 399*f*, 399*t*, 735, 771*t*
 naked segmented, 770–771, 771*t*
 pathogenic, 734–777, 771*t*
 picornaviruses (*Picornaviridae*), 391, 392*t*, 735–741, 771*t*
 placenta crossed by, 421*t*
 single-stranded, 52*t*, 386, 391*t*, 392*t*, 398, 399*f*, 399*t*, 771*t*
 negative-sense, 398, 399*f*, 735, 771*t*
 enveloped
 segmented, 763–770, 771*t*
 unsegmented, 758–763, 771*t*
 positive-sense, 398, 399*f*, 399*t*, 735, 771*t*
 enveloped, 741–750, 741*f*, 747*t*, 771*t*
 with reverse transcriptase (retroviruses), 239, 391*t*, 392*t*, 398, 399*t*, 735, 750–757, 750*f*, 771*t*
 naked, 735–741, 771*t*
 synthesis of, 398, 399*f*, 399*t*
Rochalimaea genus/spp., 605
Rock fever of Gibraltar (brucellosis), 15*t*, 340, 344*t*, 606, 606*f*
 bioterrorism and, 814*t*
Rocky Mountain spotted fever, 344*t*, 621, 624–625*b*, 626*t*
 safe handling of microbes causing, 268
Rod, of bacterial flagellum, 63, 63*f*
Rodents. *See also* Fleas
 in arboviral encephalitis transmission, 742*f*
 in *Babesia microti* transmission, 691*b*
 genetically modified, for biomedical animal models, 254
 in *Hantavirus* transmission, 384*b*, 408*b*, 767, 810
 in *Rickettsia typhi* transmission, 622, 626*t*
 in *Toxoplasma* transmission, 687*f*, 690
 in *Yersinia pestis* transmission, 602, 603*f*
Root nodules, in nitrogen fixation, 339, 339*f*, 808
Root rot, soilborne transmission of, 810*t*
Rose-gardener's disease (sporotrichosis), 666*t*, 667, 668, 668*f*

Roseola, 392*t*, 721, 721*f*
Roseolovirus genus/spp. (human herpesviruses 6 and 7/HHV-6 and 7), 392*t*, 713, 721, 721*f*, 729*t*
Rotavirus genus/spp./rotaviruses, 392*t*, 770, 770*f*, 771*f*, 771*t*
Rotavirus vaccine, 508*f*, 509*t*, 770
Rough endoplasmic reticulum, 61*f*, 86, 86*f*
Roundup. *See* Glyphosate
Roundworms (nematodes), 693, 698–702, 703*t*
Ancylostoma and *Necator* (hookworms), 700, 700*f*, 703*t*
Anisakis, 701, 701*f*, 703*t*
Ascaris, 699–700, 700*f*, 703*t*
Enterobius (pinworms), 700–701, 701*f*, 703*t*
 filarial, 699
Wuchereria, 702, 702*f*, 703*t*
 life cycles of, 698–699
 mutualism in, 415*b*
 predation by fungi and, 366, 367*f*
Rous, F. Peyton, 402
R (resistance) plasmids/factors, 196, 227, 303
 multiple resistance and, 306
 transposons and, 230
rRNA. *See* Ribosomal RNA
RSV. *See* Respiratory syncytial virus
Rubbing alcohol. *See* Isopropanol
Rubella (German/three-day measles), 392*t*, 747–748, 748*f*, 758*t*, 771*t*
 immunization against, 508*f*, 509*t*, 748, 748*f*. *See also* Measles/mumps/rubella (MMR) vaccine
 transplacental transmission of, 421*t*, 747
Rubella virus (*Rubivirus rubella virus*), 392*t*, 747–748, 758*t*, 771*t*
 immunization against, 508*f*, 509*t*, 748, 748*f*. *See also* Measles/mumps/rubella (MMR) vaccine
 placenta crossed by, 421*t*, 747
Rubeola (measles/red measles), 392*t*, 472*b*, 497*b*, 504*b*, 758, 758–759, 758*f*, 758*t*, 759*b*, 759*f*
 immunization against, 497*b*, 504*b*, 505*f*, 508*f*, 509*t*, 759, 759*f*. *See also* Measles/mumps/rubella (MMR) vaccine
 transmission of, 431, 497*b*, 758
Rubisco, in carbon fixation, 146, 806
Rubivirus genus/spp./*Rubivirus rubella virus* (rubella virus), 392*t*, 747–748, 758*t*, 771*t*. *See also* Rubella
 immunization against, 508*f*, 509*t*, 748, 748*f*. *See also* Measles/mumps/rubella (MMR) vaccine
 placenta crossed by, 421*t*, 747
Rubor, in inflammation, 462
RuBP. *See* Ribulose 1,5-bisphosphate
Rubulavirus genus/spp./mumps virus, 392*t*, 758, 760, 771*t*
 immunization against, 508*f*, 509*t*, 760, 760*f*. *See also* Measles/mumps/rubella (MMR) vaccine
Runs, in bacterial motility, 64–65, 65*f*
Russian spring–summer encephalitis, 747*t*
Rusts, 370

S. *See* Sulfur; Svedbergs
Sabiá hemorrhagic fever, 767
Sabin, Albert, 738
Sabin vaccine (oral polio vaccine/OPV), 738, 738*t*
 safety of, 506, 510, 738
Sabouraud dextrose agar, 175, 176*f*, 649
Saccharomyces genus/spp. (*Saccharomyces carlsbergensis/cerevisiae*), 5, 5*f*, 365, 366*f*, 367, 370, 373*t*
 cellular/genetic study and, 10*b*, 365
 chromosomes in, 354
 fermentation products of, 10*b*, 142*f*, 367, 370
 alcoholic beverages, 10*b*, 142*f*, 781, 782, 783*t*
 bread, 10*b*, 779, 783*t*
 vegetables, 779
 pharmaceuticals produced by, 238, 788*t*
 plasmids of, 197–198
 as probiotic, 10*b*

SUBJECT INDEX

Saccharopolyspora erythraea, antimicrobials produced by, 290t, 311t
Sacral nerve ganglion, latent herpetic infection of, 714, 714f, 716t
Safranin, 108
  in Gram staining, 109, 109f
  in Schaeffer-Fulton endospore staining, 110, 110f
Sake, fermentation in production of, 782, 783t
Salami, fermentation in production of, 780, 783t
Saliva
  in host defense, 451t
  microorganisms in, 417t
  as portal of exit/disease transmission and, 430f
Salivary glands, mumps affecting, 760, 760f
Salk, Jonas, 737
Salk vaccine (inactivated polio vaccine/IPV), 508f, 509t, 510, 737–738, 738t
Salmonella genus/spp./*Salmonella enterica*, 15t, 343t, 344t, 419t, 592f, 597–600, 600f, 786t
  in Ames test, 223, 223f
  bioterrorism and, 814t
  carriers of, 419b, 600
  fluorescent phages in identification of, 113b
  foodborne transmission of, 15t, 600, 786t
  healthcare-associated (nosocomial) infection caused by, 591f
  immunization against, 509t, 600
  phage typing in identification of, 113b, 597–600, 600f
  salmonellosis/typhoid fever caused by, 15t, 600, 600f, 601f, 604f, 786t
  serotype Choleraesuis (*Salmonella choleraesuis*)
    culture of, 178f, 594f
    in use-dilution test, 281
  serotype Dublin, 600
  serotype Paratyphi (*Salmonella paratyphi*), 600
  serotype Typhi (*Salmonella typhi*), 419b, 509t, 600
  sites of infection caused by, 604f
  toxins produced by, 428
  Vi antigens produced by, 593, 593f
  waterborne transmission of, 419b, 600, 790t
Salmonellosis, 15t, 419t, 597–600, 600f, 601f
  bioterrorism and, 814t
  waterborne transmission of, 600, 790t
Salt. *See* Sodium; Sodium chloride
Salt(s), 33, 39
  dyes as, 106
  ionic bonds in formation of, 33, 34f
Salt tolerance, in plants, recombinant DNA technology and, 255
Salvarsan, 317
Sand filters
  in drinking water treatment, 791–792, 791f
  in wastewater treatment, 794
Sand flies, as disease vectors
  for bartonellosis, 605
  for leishmaniasis, 378, 682, 683f, 692t
  for sand fly fever, 747t
Sand fly fever, 747t
Sanger, Frederick, 250
Sanger method, for nucleotide sequencing, 250
Sanitation. *See* Wastewater (sewage), treatment of; Water treatment
Sanitization, 264, 265t
Saprobes, fungi as, 366, 367f, 666
Sarcina genus/spp., shape of organisms in, 66, 66f, 328, 328f
Sarcinae (arrangement), 66, 66f, 327, 328f
Sarcodina, 360
Sarcoma, Kaposi's, 402, 721, 721f
Sarcomastigophora, 360
SARS. *See* Severe acute respiratory syndrome
Satellite virus, hepatitis D virus as, 770
Saturated fats/fatty acids, 41, 41–42, 42f, 43f
Saturation point, enzymatic activity and, 130f, 131
Sauerkraut, fermentation in production of, 779, 783t

Saxitoxin. *See also Gonyaulax* genus/spp.
  waterborne illness caused by, 379b, 790t
Scalded skin syndrome, staphylococcal, 549, 549f
Scanning electron micrographs, 104, 105f
Scanning electron microscopy/microscopes, 104–105, 105f, 107t
  resolving power of, 99f, 104
Scanning objective lens, 100
Scanning tunneling microscopy/microscopes, 105, 106f, 107t
  resolving power of, 99f
Scar(s)
  in poxvirus infection, 710, 711, 711f
  tissue repair and, 464
Scarlet fever (scarlatina), 553
Schaeffer-Fulton endospore stain, 110, 110f, 111f
Schistosoma genus/spp. (*Schistosoma haematobium/japonicum/mansoni*) (blood flukes), 696–698, 697f, 698f, 699b, 703t
  waterborne transmission of, 431, 697f, 699b, 790t
Schistosomiasis, 696–698, 698f, 699b, 703t
  waterborne transmission of, 431, 697f, 699b, 790t
Schizogony, 354, 357, 359f, 686
  in *Cyclospora* life cycle, 692f
  in *Plasmodium* life cycle, 357, 688b
Schizont(s), 357, 359f, 686
Schleiden, Matthias, 59
Schwann, Theodor, 59
SCID. *See* Severe combined immunodeficiency disease
Science, definition of, 2
Scientific method, 9–10, 9f
  study of fermentation and, 11–12, 11f
Scientific notation, 180
Sclerotic bodies, in chromoblastomycosis, 666, 667f
Scolex, tapeworm, 693, 693f
Scrapie, 48, 407
Scrub typhus, 623, 626t
Seaweeds, 6, 373
Sebaceous glands, 421f, 448
  in host defense, 448
Sebum, 448
  acne and, 578, 579f
  in host defense, 448, 449t
Secondary cultures, in commercial food/beverage production, 779
Secondary hosts. *See* Intermediate hosts
Secondary immune response, 494, 495f
Secondary infection, 434t. *See also* Opportunistic pathogens
  antimicrobial use and, 298, 299b, 302
Secondary lymphoid organs, 475f, 476
Secondary metabolites, of industrial fermentation, 787
Secondary structure of protein, 48, 49f
Secondary syphilis, 631b, 632
Second-generation drugs, 308
Second line of defense, 447, 451–466, 466t
Secretions
  in host defense, 451, 451t
  as portals of exit, 430, 430f
Secretion systems, type III. *See* Type III secretion systems
Secretome, malaria, 686
Secretory component, 487, 488t
Secretory IgA, 487, 488t
  naturally acquired passive immunity and, 487, 495, 497t
Secretory IgA protease
  gonococcal pathogenicity/virulence and, 588, 589b
  pneumococcal pathogenicity/virulence and, 558
Secretory vesicles, 61f, 86f, 87, 87f
Sections, 113
Sediment, microbial growth estimates and, 187
Sedimentation
  in drinking water treatment, 791, 791f
  in wastewater treatment, 793f
Sedimentation rate, 76
Seed warts, 723b
Segmented genome, RNA virus, 735, 763
Select agents, as biological weapons, 813

Selective and differential media, 176–177, 178f
Selective media, 175–176, 176f, 178f
Selective permeability, cell membrane, 70
Selective toxicity, 290
Selenium, for microbial growth, 165
Selenocysteine, 47f
Selenomonas genus/spp./*Selenomonas noxia*, 330f, 335, 337t
  obesity and, 335
Self-antigens (autoantigens), 477, 477f
  autoimmunity and, 481–482, 482f, 487, 487f, 538, 538–539
  clonal deletion and, 481–482, 482f, 487, 487f
  graft rejection and, 478, 536
  T cell regulation and, 482, 482f
  tolerance to, 473, 481–482, 482f, 487, 487f. *See also* Self-tolerance
Self-stimulation, in cell-mediated immune response, 490, 490f
Self-termination, of transcription, 204, 205f
Self-tolerance, 473, 477
  autoimmunity and, 481–482, 482f, 487, 487f, 538, 538–539
  clonal deletion and, 481–482, 482f, 487, 487f
  graft rejection and, 478, 536
  Toll-like receptors and, 457
SEM. *See* Scanning electron microscopy/microscopes
Semen, as portal of exit, 430, 430f
  for *Ebolavirus*, 764b, 765b
  for HIV, 756
Semiconservative DNA replication, 198, 198f, 326
Semilogarithmic graph, 181, 181f
Seminal vesicles, as portal of exit, 430f
Semiperishable food, 784
Semisynthetic antimicrobials, 280, 281t, 289, 290t, 308, 309t, 310t, 313t, 315t, 316t, 318t. *See also* Antibiotics; Antimicrobial drugs
  new drug development and, 308
Semmelweis, Ignaz, 16, 18f
Sensitization, in type I (immediate) hypersensitivity, 263b, 526, 527f
Sepsis. *See* Bacteremia/septicemia
Septa (cross walls), in hyphae, 365, 366f
Septate hyphae, 365, 366f
septi- (prefix), 423t
Septicemia. *See* Bacteremia/septicemia
Septic shock, 587f. *See also* Bacteremia/septicemia
Septic tanks/system, 794, 794f
SER. *See* Smooth endoplasmic reticulum
Serial dilution, 184, 184f
  pour-plate technique and, 173, 174f
  in titration, 514, 514f
  viable plate counts and, 184, 184f
Serology/serological tests, 18f, 19t, 21, 117, 117f, 512–520, 520t
  agglutination tests, 117, 117f, 513–514, 514f, 520t
  complement fixation tests, 515, 520t
  labeled immunoassays, 515–518, 516f, 517f, 518f, 519f
  nephelometric tests, 513
  neutralization tests, 515, 520t
  point-of-care testing, 518–519, 519f
  precipitation tests, 512–513, 512f, 520t
  in taxonomy/microbial classification, 117, 117f
  turbidimetric tests, 513
Serotypes, 112, 113
  rhinovirus, 735, 736
  Salmonella, 600
  for streptococci, 117, 552, 558
Serovars, 63
Serratia genus/spp./*Serratia marcescens*, 172f, 343t, 344t, 592f, 596, 596f
  healthcare-associated (nosocomial) infection caused by, 591f, 596
  restriction enzymes produced by, 240, 240t
Serum, 117, 452, 510
  for passive immunotherapy. *See* Antiserum
  study of. *See* Serology
Serum hepatitis. *See* Hepatitis, type B
Serum sickness, 510
70S ribosomes. *See* Prokaryotic ribosomes

Severe acute respiratory syndrome (SARS/coronavirus respiratory syndrome)/SARS virus, 391f, 392t, 736t, 748, 749, 749f, 771f
  bioterrorism and, 814t
Severe combined immunodeficiency disease (SCID), 540, 540t
Sewage (wastewater)
  drinking water contamination and, 431, 437, 439f, 440, 789, 790
  treatment of, 792–795, 793f, 794f, 795f
  domestic water affecting aquatic systems and, 811
  microbes in, 793–794, 793f, 794f, 795f
Sewer systems, municipal, 793–794, 793f
Sex (conjugation) pili, 66, 66f, 79t, 89t, 226, 227f, 228f
Sexual abuse, gonorrhea in child and, 589
Sexually transmitted infections/diseases (STIs/STDs). *See also specific disease*
  chlamydial, 347, 627–628, 628f
  cytomegalovirus infection, 720
  education in prevention of, 440, 441
  genital herpes, 714, 714f, 715f, 716, 716t, 729b
  genital warts, 723b
  gonorrhea, 344t, 588–589, 588f, 589b, 594b
  Haemophilus ducreyi causing, 604
  hepatitis, 726, 770
  HIV infection/AIDS, 754b, 756, 756f, 757
  mycoplasma causing, 572
  parasitic, 676f
  protozoan/trichomoniasis, 360, 686, 692t
  syphilis, 630–633, 631b, 631f
  Zika fever, 744, 745
Sexual reproduction, 4, 59
  in algae, 373, 373–374, 374f
  in ascomycetes, 369, 369f, 373t
  in eukaryotes, 354
  in fungi, 4, 367, 368f
  zygomycetes, 369, 369f, 373t
  in protozoa, 360
Sexual spores, 4, 367, 368f. *See also* Spores
Sheath, of cyanobacteria, 333, 333f
Sheep pox (orf), 710
Shellfish
  toxin contamination of, 353b, 363, 379b, 790
  Vibrio contamination of, 639
Shellfish poisoning
  dinoflagellates causing, 362–363, 790, 790t
  paralytic (PSP), 353b, 379b, 790t
Shells, electron/valence, 30–31, 30f, 31
Shiga, Kiyoshi, 15t
Shiga-like toxin, *E. coli* O157:H7 producing, 596
Shiga toxin, 602
Shigella genus/spp. (*Shigella boydii/dysenteriae/flexneri/sonnei*), 15t, 343t, 344t, 592f, 600–602, 601f, 786t
  bioterrorism and, 814t
  foodborne/waterborne transmission of, 601–602, 601f, 604f, 786t, 790t
  immunization/vaccine development and, 602
  sites of infection caused by, 604f
  toxins produced by, 428, 602
Shigellosis, 15t, 600–602, 601f
  bioterrorism and, 814t
  foodborne/waterborne transmission of, 601
Shine-Dalgarno sequence (ribosome-binding site), 210
Shingles (herpes zoster), 717–718, 717f, 718, 718f, 719b
  immunization against, 508f, 509t, 718
Shock, 550
  anaphylactic. *See* Anaphylaxis/anaphylactic shock
  in non-streptococcal toxic shock syndrome, 550
  septic, 587f
  in streptococcal toxic shock syndrome, 556
Short connecting chains, 67, 67f
Si. *See* Silicon
Sickle-cell trait, malaria resistance and, 687
Side group, in amino acids, 47, 47f
Siderophores, 452, 593

Sigma factor, 204, 205f
Signs of disease, 423, 423t
Silage, fermentation in production of, 779, 783t
Silent mutations, 216, 217f, 218t
Silica
　in diatom cell walls, 375, 376t
　radiolarian shells composed of, 364, 364f
Silicates, in algal cell wall, 80
Silicon, 29t, 30f
Silver, antimicrobial action of, 279, 281t
Silver nitrate, 279
Simian immunodeficiency virus (SIV), 752
　vaccine against, 757
Simple microscope, 100
Simple stains, 108, 108f, 111t
Simple sugars (monosaccharides), 44, 45f
Simplexvirus genus/spp., 392t, 713, 714–716, 714f, 715f, 716t, 729t. *See also* Human herpesvirus(es) 1 and 2
Simulium flies, as disease vectors, 433t
Single-stranded DNA, 51, 398
　in microarrays, 246, 247, 247f
　in next-generation sequencing, 250, 251f
Single-stranded DNA viruses, 51, 386, 391t, 392t, 398, 399t, 710, 729t
　synthesis in, 398, 399t
Single-stranded RNA, 52t
　negative-sense, 398, 399f, 399t
　positive-sense, 398, 399f, 399t
Single-stranded RNA viruses, 52t, 386, 391t, 392t, 398, 399t, 735
　negative-sense, 398, 399f, 399t, 735, 771t
　　enveloped
　　　segmented, 763–770, 771t
　　　unsegmented, 758–763, 771t
　positive-sense, 398, 399f, 399t, 735
　　enveloped, 741–750, 741f, 747t, 771t
　　　with reverse transcriptase (retroviruses), 239, 391t, 392t, 398, 399t, 735, 750–757, 750f, 771t
　　naked, 735–741, 771t
　synthesis in, 398, 399f, 399t
Single-strand repair, 220–221, 220f
Singlet oxygen, 164
Sinusitis
　Chlamydophila (Chlamydia) pneumoniae causing, 629
　pneumococcal, 558b, 559
　Prevotella causing, 614
siRNA (small interfering RNA), 216
SIV (simian immunodeficiency virus), 752
　vaccine against, 757
Skin
　diseases/infections of, 182b. *See also specific type and under Cutaneous*
　　acne, 578–579, 579f
　　actinomycosis, 580–581, 580f
　　anthrax, 562, 562f
　　Bacteroides fragilis causing, 614
　　dermatophytosis/ringworm, 419t, 649, 664–665, 665t
　　fungal (mycoses), 651, 666–668, 666f, 666t, 667f, 668f
　　　superficial, 664–665, 665t, 666t
　　mycetoma (fungal), 666t, 667, 667f
　　mycetoma (*Nocardia*), 580, 580f
　　parasitic, 676f
　　staphylococcal, 549, 549f
　in host defense, 420, 421f, 447–448, 448f, 449f
　microbiome/normal microbiota of, 416, 417t
　　staphylococci, 336, 547, 548
　as portal of entry, 420, 420f
　　for parasites, 676f
　as portal of exit, 430, 430f
　specimen collection from, 173t, 182b
　structure of, 420, 421f
Skin tests
　for leprosy, 575
　for tuberculosis (tuberculin response), 534–535, 535f, 577b
　for type I (immediate) hypersensitivity, 529, 529f
Slant tubes/slants, 175, 175f
SLE. *See* Systemic lupus erythematosus

Sleeping sickness, African (African trypanosomiasis), 15t, 680–681, 681f, 682b, 692t
Slime layer, 62, 62f. *See also* Capsule
　of *Rickettsia*, 621
　of staphylococci, 548, 548t
Slime molds, 364, 365t
　classification of, 357, 364
Slow sand filters, in drinking water treatment, 791
Slow viruses. *See* Prion(s)
Sludge, wastewater treatment and, 332, 793, 793f, 794, 794f
SmaI restriction enzyme, 240, 240t
Small interfering RNA (siRNA), 216
Small intestine, microbiome/normal microbiota of, 417t
Smallpox, 2b, 710–712, 711f, 712b
　bioterrorism and, 510b, 712, 813, 814t
Smallpox vaccination, 509t, 510b, 710, 711, 712, 712b
　history/discovery of, 2b, 505, 710
　monkeypox prevention and, 509t, 712–713
Smallpox virus (variola virus), 2b, 710–712, 711f, 712b, 729t
　bioterrorism and, 510b, 712, 813, 814t
　safe handling of, 268
　shape of, 389
　size of, 61f, 388f
　vaccination against. *See* Smallpox vaccination
Smears, preparation of, 106, 108f
Smoking, compromised immunity and, 449, 460b
Smooth endoplasmic reticulum, 61f, 86, 86f, 87f
Smuts, 370
Snail(s)
　in fluke life cycle, 696, 697f
　Schistosoma, 696, 697f, 699b
Snail fever (schistosomiasis), 696–698, 698f, 699b, 703t
　waterborne transmission of, 431, 697f, 699b, 790t
Snapping division, 326, 327f
　by *Corynebacterium*, 337, 572, 572f
Snow, John, 17, 18f, 437, 439f, 637
SO_4^{2-}. *See* Sulfate
Soap, 278, 281t
Sodium (salt), 29t, 30f
　in aquatic habitats, 811, 812
　in food preservation, 168, 273, 783, 785
　microbial growth and, 168, 273
　　halophiles and, 168, 331, 331f
　plant tolerance for, recombinant DNA technology and, 255
　on skin, in host defense, 448
　in sodium chloride formation, 33, 34f
Sodium bicarbonate, 38
Sodium chloride
　dissociation of, 34, 34f
　ionic bonds in formation of, 33, 34f
　in sweat, in host defense, 448, 547
Sodium hypochlorite, antimicrobial action of, 277
Sodium stibogluconate, 317t
Sofosbuvir, 314t
Soil
　microorganisms living in, 804, 804f, 809–810, 809f, 810t
　　associations and, 804, 804f
　　diseases caused by, 810, 810t
　　factors affecting abundance of, 809–810
　　nature of, 809, 809f
　as nonliving disease reservoir, 419
Soilborne diseases, 810, 810t
Soilborne wheat mosaic virus, 810t
Soil microbiology, 804, 804f, 809–810, 809f, 810t
Solute(s), in passive transport processes, 72, 72f
Solutions, isotonic/hypertonic/hypotonic, 72, 72f, 168
Solvent(s)
　in passive transport processes, 72
　water as, 37
Sorbic acid
　in food preservation, 785
　microbial production of, 789t
Soredia, 372, 372f

Sore throat. *See* Pharyngitis
SOS response, in DNA repair, 221
Sour cream, microbes in production of, 12t
Southern, Ed, 246
Southern blot, 246, 252t
Soy sauce, fermentation in production of, 12t, 142f, 779, 783t
Spacer(s), in CRISPR, 242, 243f
Spallanzani, Lazzaro, experiments in spontaneous generation by, 8
Specialized transduction, 225
Species, 112, 114f
Species Plantarum (book), 112
Species resistance, 447
Specific epithet, 113
　for viruses, 391, 391t
Specificity
　in adaptive immunity, 473
　　antibody structure and, 485
　　B cell receptor, 483, 483f
　　T cell receptor, 481, 481f
　enzyme-substrate, 129, 129f
Specimen collection, for culture
　clinical sampling, 171, 172, 173t
　environmental sampling, 171
Specimen preparation
　for staining, 106
　for transmission electron microscopy, 104, 106
Specimen transport, 172, 177
Spectrophotometer, in microbial growth estimation, 186–187, 186f
Spectrum of action, of antimicrobial drug, 298, 299f
Spermatia, 374
Sphaerotilus genus/spp., 341, 342f, 344t
Spherical particles, hepatitis B virus, 726, 727f
Spherule, of *Coccidioides immitis*, 655, 655f
Sphingobacteria, 344t
Spinal tap, 590
Spindle, 355, 355f, 356
Spiral-shaped prokaryotes, 325, 325f
Spiramycin, 310t
Spirilla (spirillum), 325, 325f
Spirits (distilled), fermentation in production of, 781, 783t
Spirochaetes (phylum), 330f, 344t
Spirochete(s), 325, 325f, 344t, 347, 629–637
　endoflagella/axial filaments of, 63–64, 64f
Spirogyra genus/spp., 6f, 376t
Spiroplasma genus/spp., cytoskeleton of, 77
Spleen, 475f, 476
Splenic candidiasis, 658t, 660b
Spliceosome, 206, 207f
Splicing, in transcription, 206, 207f
Spoilage. *See* Food spoilage
Spongiform encephalopathies, 405–407, 406f, 407b
　bovine (BSE/mad cow disease/variant Creutzfeldt-Jakob disease), 21b, 48, 406f, 407, 407b
　bioterrorism and, 814t
　enzyme for elimination of prion causing, 280
Spontaneous generation (abiogenesis), 7–10, 8f, 9f, 10b
Spontaneous mutations, 217
Sporadic disease, 435, 436f
Sporangiophores, 367, 368f
Sporangiospores, 367, 368f
Sporangium/sporangia, 367, 368f
　of myxobacteria, 346, 346f
　of zygomycete, 369, 369f
Spores. *See also specific organism and* Endospores
　fungal, 4, 367, 368f, 648
　reproductive, 326, 327f
　　asexual, 4, 326, 327f, 367, 368f
　　sexual, 4, 367, 368f
Sporogonic phase, in *Plasmodium* life cycle, 689b
Sporothrix genus/spp./*Sporothrix schenckii*, sporotrichosis caused by, 666t, 667–668, 668f
Sporotrichosis, 666t, 667–668, 668f
Sporozoa, 360. *See also* Apicomplexans
Sporozoites
　in *Cryptosporidium parvum* life cycle, 690
　in *Cyclospora* life cycle, 692f
　in *Plasmodium* life cycle, 688b, 689b
　in *Toxoplasma* life cycle, 687f, 690
Sporulation, 75, 76f, 325

Spotted fever rickettsioses, 621, 622b
Spring–summer encephalitis, Russian, 747t
Sputum, as portal of exit, 430, 430f
ssDNA. *See* Single-stranded DNA
ssDNA viruses. *See* Single-stranded DNA viruses
SSPE. *See* Subacute sclerosing panencephalitis
ssRNA. *See* Single-stranded RNA
ssRNA viruses. *See* Single-stranded RNA viruses
Stab culture, 177
Staining, 106–112, 111t
　acid-fast stain and, 67, 109–110, 110f, 111t, 337, 337f
　contrast and, 100
　differential stains and, 108–110, 109f, 110f, 111t
　for electron microscopy, 111–112
　endospore stain and, 110, 110f, 111t
　flagellar stain and, 110, 111f, 111t
　fluorescent dyes for, 84f, 103, 103f, 515
　Gram stain and, 15, 15f, 67
　histological, 110
　for light microscopy, 106–111, 111t
　negative (capsule) stain and, 110, 111f, 111t
　principles of, 106–108
　simple stains and, 108, 108f, 111t
　special stains and, 110–111, 111f, 111t
　specimen preparation for, 106, 108f
Standard precautions, 172
Stanley, Wendell, 388
Staphylococcal intoxication (food poisoning), 549, 786t
　bioterrorism and, 814t
　incubation period for, 429t
Staphylococcal scalded skin syndrome, 549, 549f
Staphylococcal toxic shock syndrome (TSS), 548, 548t, 550, 550f
Staphylococcal toxins, 336, 548, 548t
　as biological weapons, 814t
Staphylococci (arrangement), 66, 66f, 327, 328f, 547, 547f
Staphylococcus genus/spp./staphylococci, 336, 337t, 547–551, 547f
　acid-fast staining of, 110f
　capsule/slime layer of, 548, 548t
　diseases caused by, 549–551, 549f, 550f, 551b. *See also specific disease*
　endocarditis caused by, 550
　enzymes produced by, 548, 548t
　epidemiology of, 548
　FISH in identification of, 250f
　food intoxication/poisoning caused by, 549, 786t
　incubation period for, 429t
　in microbiome/normal microbiota, 336, 547
　on skin, 336, 547, 548
　pathogenicity/virulence factors of, 547–548, 548t
　pneumonia caused by, 550
　resistant/multiple-drug-resistant strains of, 305, 305b, 306, 336, 551, 551b
　shape of organisms in, 66, 66f, 327, 328f
　size of, 61f
　skin infection caused by, 549, 549f
　toxic shock syndrome caused by, 548, 548t, 550, 550f
　toxins of, 336, 548, 548t
　　as biological weapons, 814t
　transmission of, 431, 548
　UTIs caused by, 599b
Staphylococcus aureus, 15f, 108f, 336, 547
　arrangement of, 328f
　culture of, 175, 178f
　diseases caused by, 547–550, 549f, 550, 550f, 551b
　epidemiology of, 548
　FISH in identification of, 250f
　food intoxication/poisoning caused by, 549, 786t
　generation time of, 180
　iron needs of, 452
　in Kelsey-Sykes capacity test, 282
　methicillin-resistant, 305, 305b, 336, 551, 551b
　safe handling of, 268
　in microbiome/normal microbiota, 336

Staphylococcus aureus (Continued)
  necrotizing fasciitis caused by, 555b
  osmotic pressure affecting, 168
  pathogenicity/virulence factors of, 548, 548t
  resistant strains of, 305, 305b, 336, 551, 551b
  siderophores secreted by, 452
  skin infection caused by, 549, 549f
  toxins of, 336, 428, 548t
    as biological weapons, 814t
  transmission of, 431, 548
  in use-dilution test, 281
  vaccine development and, 551
  vancomycin resistance and, 551
Staphylococcus epidermidis, 448, 547
  as commensal/normal flora, 448, 547
  diseases caused by, 547, 549, 550
  epidemiology of, 548
  folliculitis caused by, 549
  pathogenicity/virulence factors of, 548, 548t
  transmission of, 548
Staphylokinase, as virulence factor, 548, 548t
Starch, plant (amylose), 45, 46f
Star-shaped prokaryotes, 325, 325f
Start codon, 208, 208f, 211, 213t
Starter cultures, in commercial food/beverage production, 779, 780, 780f, 781, 781f, 782
-stasis/-static (suffix), 264, 265t
Stationary phase, of microbial growth, 181, 181f
Stavudine (d4T), 297f
STDs. *See* Sexually transmitted infections/diseases
Steam
  in autoclaving, 269–270, 270f
  in ultra-high temperature sterilization, 271
Stearic acid, 43f
Steinernema genus/spp., mutualism in, 415b
Stellate macrophages (Kupffer cells), 478
Stem, immunoglobulin, 485, 485f
Stem cells, 449
  blood, 452, 452f, 453, 473
  cytomegalovirus infection of, teratogenic effects of, 720
  in hematopoiesis/hematopoietic, 452, 452f, 453
  in host defense/epithelial cell replacement, 449
  tissue repair and, 464, 465f
Stephanodiscus genus/spp., 376f
Stereoisomers, 47, 48f
Sterilization, 173, 263–264, 265t
  with autoclave, 269–270, 270f, 271f, 275t
  commercial, 264, 269
  gas, 280, 281f
  heat-related methods for, 269–270, 270f, 271f, 275t
  indicators of, 270, 271f
  ultra-high-temperature, 271, 271f, 275t
Steroids, 44, 44f
  for allergic contact dermatitis, 535
  for asthma, 529
  biosynthesis of, 150
  immunosuppressive actions of, 537, 538f, 540–541
Sterols, 44
  in eukaryotic membranes, 44, 81
  in mycoplasma membranes, 44, 335, 568
Stibogluconate, 317f
Sticky ends, of restriction enzymes, 240, 241f
Stipes, 373
STIs. *See* Sexually transmitted infections/diseases
St. Louis encephalitis, 742, 747t
STM. *See* Scanning tunneling microscopy/microscopes
Stomach
  pH of, microbial growth and, 53b, 167
  specimen collection from, 173t
Stomach acid
  bacteria affected by, 53b, 167
  in host defense, 53b, 167, 451, 451t
  antacids and, 39, 40b, 53b
Stomach flu, 766. *See also* Gastroenteritis
Stop codon, 208, 208f, 212, 212f, 213t
Strachan, David, 263b
Strain(s), 112, 113. *See also* Serotypes
Strain R/strain S bacteria, transformation and, 224, 225f

Stramenopila (kingdom), 357, 358, 359f, 375, 376, 376f
Streak plates, 173, 173f
Streamers, in biofilm, 170
Streaming, cytoplasmic, fungal nutrition and, 366
"Strep throat" (streptococcal pharyngitis), 70b, 553, 553f, 736t
  rheumatic fever and, 553
Streptobacilli (arrangement), 328f
Streptococcal toxic shock syndrome (STSS), 556
Streptococci (arrangement), 66, 66f, 327, 328f, 552
Streptococcus genus/spp./streptococci (*Streptococcus agalactiae/anginosus/bovis/eqiuisimilis/mitis/mutans/sanguis/thermophilus*), 4f, 105f, 336, 337t, 552–559, 561t. *See also Streptococcus pneumoniae; Streptococcus pyogenes*
  alpha-hemolytic, 557–559, 558b, 558f, 561t
  beta-hemolytic, 552–557, 553f, 554–555b, 561t
  fermentation products of, 142f
    cheese, 780
    vegetables, 779
    yogurt, 780, 783t
  glomerulonephritis after infection with, 539, 556
  group A. *See Streptococcus pyogenes*
  group B, 556–557, 561t
  hemolytic pattern of, 176, 176f, 552, 561t
  Lancefield groups in classification of, 117, 552, 561t. *See also* specific type under Group
  multi-drug-resistant strains of, 306, 336
  pharyngitis caused by, 70b, 553, 553f, 736t
    rheumatic fever and, 553
  puerperal fever caused by, 16
  shape of organisms in, 66, 66f, 327, 328f
  transmission of, 431
  viridans, 557, 557f
  virulence factors of, 429
  in yogurt production, 142f, 780, 783t
Streptococcus pneumoniae (pneumococcus), 15t, 113, 557–559, 558b, 558f, 561t
  arrangement of, 328f, 557–558, 558b, 558f
  capsules of, 62, 558, 558b
  culture of, 559
  diseases caused by, 558b, 559, 561t
  immunization against, 508f, 509t, 558b, 559
  microbial genetics research and, 19
  in microbiome/normal microbiota, 558, 559
  resistant strains of, 559
  transformation of, 224–225, 225f
Streptococcus pyogenes (group A streptococcus), 552–556, 553f, 554–555b, 561t
  arrangement of, 328f
  autoimmunity and, 539
  diseases caused by, 553–556, 553f, 554–555b, 561t
  glomerulonephritis and, 539, 556
  immunochromatographic assays in identification of, 519–520, 519f
  necrotizing fasciitis/"flesh-eating disease" caused by, 118b, 336, 554–555b, 556
  pharyngitis caused by, 70b, 553, 553f, 736t
    rheumatic fever and, 553
  resistant strains of, 305
  rheumatic fever caused by, 553
  scarlet fever caused by, 553
  skin infection caused by, 124b, 154b, 553, 553f
  toxic shock syndrome (STSS) caused by, 556
  virulence factors of, 429, 552, 558, 558b
Streptococcus sanguis, 557
Streptococcus thermophilus, in yogurt production, 780, 783t
Streptogramins, 311t
  mechanism of action of, 293, 311t
Streptokinase(s)
  group A streptococcal pathogenicity/virulence and, 552, 554b, 555b

microbial production of, 787, 789t
  in necrotizing fasciitis, 554b, 555b
Streptolysins/streptolysin S, 552
  in necrotizing fasciitis, 555b
Streptomyces genus/spp., 330f, 337t, 338
  antimicrobials produced by, 289, 290t, 309t, 310t, 311t, 315t, 316t, 338, 788
  reproduction of, 327f
  soilborne transmission of, 810t
Streptomycin, 310t
  mechanism of action of, 293, 294f, 310t
  microbial production of, 290t, 310t
  spectrum of action of, 299f, 310t
Stress, immunodeficiencies/opportunistic infections and, 418, 540, 540–541
Strip-mining, acid mine drainage and, 805
Strobila (strobilae), 693–694, 693f
Stroma, chloroplast, 88f, 145
Structural analogs, of para-aminobenzoic acid (PABA), antimicrobial action of, 132, 295, 296f
Structural formula, 32
Structural proteins, 47
STSS. *See* Streptococcal toxic shock syndrome
Sty, 549
Subacute disease, 433, 434t
Subacute sclerosing panencephalitis (SSPE), 759
Subclinical infection, 423
Subcutaneous coccidioidomycosis, 655, 655f
Subcutaneous mycoses, 666–668, 666f, 666t, 667f, 668f
Subfamilies, 113
Sublimation, in lyophilization, 272
Subsoil, 809, 809f
Subspecies, 113, 630. *See also* Serotypes
Substitutions (mutation), 216, 218t
Substrate, 127
  enzymatic activity and, 129, 130f, 131, 131f
  concentration and, 130f, 131
  inhibition and, 131, 131f, 132, 132f
  specificity and, 129, 129f
Substrate-level phosphorylation, 127, 135, 135f, 148t
  in citric acid/Krebs cycle, 136
  in glycolysis, 134f, 135, 135f
Subunit vaccines, 253, 506
Succinic acid, in citric acid/Krebs cycle, 137f
Succinyl-CoA, 149t
  in citric acid/Krebs cycle, 137f, 149t
Sucrose, 45, 45f
Sudan black dye, 108
Sugars. *See also* Carbohydrate(s)
  in food preservation/microbial control, 168, 273, 783, 785
  simple (monosaccharides), 44, 45f
Suicide, cell (programmed). *See* Apoptosis
Sulfadiazine, 312t, 318t
Sulfadoxine, 296f, 312t, 318t
Sulfa drugs. *See* Sulfonamides/sulfa drugs
Sulfamethoxazole, 296f, 312t
Sulfanilamide, 289, 296f, 312t, 318t
  mechanism of action of, 132, 295, 296f, 318t
Sulfate
  reduction of, 335, 808, 808f
  in sulfur cycle, 808, 808f
Sulfhydryl group, 41t
Sulfisoxazole, 296f
Sulfonamides/sulfa drugs, 295, 296f, 312t, 318t
  discovery of, 21
  mechanism of action of, 291f, 295, 296f, 312t, 318t
  spectrum of action of, 299f, 312t, 318t
Sulfur, 29t, 30f, 808
  acid mine drainage and, 167, 341, 805, 805f
  in electron transport trains, 139
  for microbial growth, 165
  recycling of, 20, 341, 343, 808, 808f
Sulfur bacteria
  in carbon cycle, 806
  colorless, 341. *See also Thiobacillus* genus/spp.
  green, 330f, 333, 334, 334f
  purple, 333, 334, 334f, 334t, 342, 342f, 344f
  in sulfur cycle, 808
Sulfur cycle, 20, 808, 808f
Sulfur dioxide, in food preservation, 785

Sulfur dissimilation, 808, 808f
Sulfuric acid, 38
  acid mine drainage and, 167, 341, 805
Superbugs, 306. *See also* Multi-drug resistance/multiple-drug-resistant pathogens; Resistance
Supercoils, in DNA replication, 201
Superficial mycoses, 664–665, 665f, 665t, 666f
Superinfections, 298, 299b, 302, 439. *See also* Opportunistic pathogens
  antimicrobial therapy and, 298, 299b, 302, 439
Superoxide dismutases, 164
Superoxide radicals, 164
  neutrophil killing and, 456
Suppressor T cells. *See* Regulatory T cells
Suramin, 318t
Surface tension, of water, 37, 37f
Surfactants, 278, 278f
  antimicrobial action of, 278–279, 281
Surveillance, bioterrorism defense and, 815
Susceptibility
  antimicrobial drug (effectiveness), 298–300, 300f, 301b, 301f
    diffusion susceptibility tests in determination of, 299, 300f, 301b
    minimum bactericidal concentration (MBC) test in determination of, 300, 301f
    minimum inhibitory concentration (MIC) test in determination of, 299–300, 300f
  antimicrobial method, 266–267, 266f
Suspected causative agent, 14, 15
Svedbergs (S), 76
"Swamp gas," 332
Swarmer cell, in *Caulobacter* reproduction, 340, 340f, 341b
Sweat/sweat glands, 421f
  in host defense, 448, 466t, 547
Sweetener, artificial, microbial production of, 12t, 787
Swelling, in inflammation, 462
Swimmer's itch, 696
Swine flu virus, 435
Symbionts, 414
Symbiosis/symbiotic relationships, 168, 414–418, 415b, 415f, 416t
  nitrogen fixers and, 808
Symports, 73, 73f
Symptoms of disease, 423, 423t
Synapse(s), immunological, 490
  T helper cell activation and, 492
Synaptic cleft, botulism toxins affecting, 563, 564f
Syncytia
  herpes infection and, 716
  HIV infection and, 752t
  paramyxovirus infection and, 758
  respiratory syncytial virus infection and, 760, 761f
Syndrome, 423
  AIDS as, 423, 541, 751
Synergistic relationship/synergism, 168
  resistance and, 307, 307f
Synthesis, nucleic acid, 151, 152f
  in viral replication, 392
    animal virus, 397–398, 399f, 399t, 400f, 401t
    bacteriophage, 393f, 394, 401t
    HIV, 752, 753–754, 753f
Synthesis reactions, 35–36, 36f. *See also* Dehydration synthesis
Synthetic antimicrobials, 280, 281t, 289, 311t, 312t, 313t, 314t, 315t, 316t, 317t, 318t. *See also* Antibiotics; Antimicrobial drugs
Synthetic (defined) media, 175, 175f
Synthetic nucleic acids, in recombinant DNA technology, 239–240, 252t
Syphilis, 344t, 630–633, 631b, 631f. *See also Treponema pallidum pallidum*
  congenital, 632
  incubation period for, 429t
  transplacental transmission of, 421t, 632
Systemic aspergillosis, 370b, 658, 659b
Systemic histoplasmosis, 652
Systemic infection, 434t
  fungal/mycotic, 648, 651–656, 651f, 656–664

opportunistic fungi causing, 656–664
pathogenic fungi causing, 648, 651–656, 651f
staphylococcal, 549–550, 550f, 551b
Systemic lupus erythematosus, 533–534, 534f

T. *See* Thymine
T4 (type 4) bacteriophage, 388f, 390f, 392
replication of, 392–394, 393f, 394f
Taenia genus/spp./*Taenia saginata* (beef tapeworm)/*Taenia solium* (pork tapeworm), 694, 694f, 695f, 703t
life cycle of, 694, 694f
Tagged (labeled) immunoassays, 515–518, 516f, 517f, 518f, 519f. *See also specific type*
Talaromyces marneffei, 664
Tampon use, toxic shock syndrome and, 550, 550f
Tapeworms (cestodes), 6, 378, 419t, 693, 693–695, 693f, 694f, 695f, 703t
beef (*Taenia saginata*), 694, 703t
dog (*Echinococcus granulosus*), 694–695, 695f, 703t
life cycle of, 694, 694f
pork (*Taenia solium*), 694, 694f, 695f, 703t
Taq DNA polymerase, 245, 246b
Tattoos, community-associated MRSA and, 305b
Tatum, Edward, 19, 365
Taxa, 60, 112, 113, 329. *See also* Taxonomy/microbial classification and identification
Taxis, 59, 65
Taxol, recombinant DNA in production of, 788t
Taxonomic keys, 119, 119f
Taxonomic system, 4. *See also* Taxonomy/microbial classification and identification
Taxonomists, 329
Taxonomy/microbial classification and identification, 4–7, 4f, 5f, 7f, 8f, 18f, 112–119, 114f, 116f, 117f, 118f, 119f, 329. *See also specific type of microbe*
biochemical tests in, 116–117, 116f, 117f
categories in, 112–115, 114f
eukaryotic classification and, 114f, 115, 357–358, 359f
evolution/nucleotide sequencing and, 115, 118–119, 329, 330f, 357, 359f, 360
historical basis of, 4, 112–115, 114f
identifying characteristics and, 115–119, 116f, 117f, 118f
infectious disease classification and, 433
Linnaean, 112–115, 114f, 357, 359f, 360
MALDI-TOF mass spectrometry in, 118
morphology/physical characteristics and, 116
phage typing in, 117, 118f
prokaryotic classification and, 114f, 115, 329
serological tests in, 117, 117f
TB. *See* Tuberculosis
TBE. *See* Tick-borne encephalitis
TCA cycle. *See* Tricarboxylic acid (TCA/Krebs/citric acid) cycle
Tc cell. *See* Cytotoxic T cells
T-cell lymphocytic leukemia, adult acute, HTLV-1 causing, 750
T cell receptors, 481, 481f
in antibody immune response, 492, 493f, 494
in cell-mediated immune response, 490, 490f, 491, 491f
specificity of, 481, 481f
T cells. *See* T lymphocytes
TCR. *See* T cell receptor
Tdap vaccine, 508f, 567, 573, 608
T-dependent antibody immunity, 492–494, 493f
TDR-TB (totally drug-resistant tuberculosis), 574
Td vaccine, 508f, 567, 573
Tears
in host defense, 449, 450f
microbiome/normal microbiota of, 417t
as portal of exit, 430, 430f
Tecovirimat, for smallpox, 712, 712b

Tedizolid, 311t, 313t
Teeth
antimicrobial agents affecting, 302, 302f
decay of, 62, 170, 171b, 557, 557f
Teichoic acids, in Gram-positive cell walls, 67, 68f
Teixobactin, 308
Telaprevir, 314t
Teleomorphs, 648
Telithromycin, 311t
Telophase
meiosis, 356, 356f, 357t
mitosis, 355, 355f, 357t
TEM. *See* Transmission electron microscopy/microscopes
TEM images. *See* Transmission electron micrographs/images
Temperate phages, 395, 395f. *See also* Lysogeny/lysogenic replication cycle
Temperature
core body
elevated. *See* Fever
hypothalamus in regulation of, 464, 465f
enzymatic activity affected by, 129–131, 130f, 131f
during food processing/storage, 271–272
spoilage prevention and, 784, 785
microbial control methods and, 267, 267f
heat, 268–271, 269f, 270f, 271f, 271t
refrigeration and freezing, 271–272, 275f
microbial growth and, 166–167, 166f, 167f
soil, microbial population affected by, 810
Temperature range, 166
Templating of prions, 405, 406f
Tenofovir, 297f, 314t
HIV infection prevention and, 757
terato- (prefix), 423t
Teratogenicity, of cytomegalovirus, 720
Terbinafine, 315t
mechanism of action of, 294, 315t
Terminal carbonyl group, 41t
Termination
DNA replication, 201, 202f, 213t
transcription, 204, 205f, 208f, 213t
translation, 212, 212f, 213t
Terminator sequence, 204, 205f, 213t
Termites, protozoa (*Trichonympha*) in, 415, 415f
Tertiary structure of protein, 48, 49f
Tertiary syphilis, 631b, 632
Tetanospasmin (tetanus toxin), 565–566, 566f
immunological memory and, 494
muscles affected by, 565–566, 566–567, 566f, 567f
Tetanus, 15t, 565–567, 565f, 566f, 567f
immunization against, 508f, 509t, 567
immunological memory and, 494, 495f
incubation period for, 429t, 566
soilborne transmission of, 810t
Tetanus toxin, 565–566, 566f
immunological memory and, 494
muscles affected by, 565–566, 566–567, 566f, 567f
Tetanus toxoid, 506
immunological memory and, 494
Tetracyclines, 311t
fetal toxicity of, 302
mechanism of action of, 291f, 293, 294f, 311t
microbial production of, 290t, 311t
spectrum of action of, 298, 299f, 311t
toxicity of, 302, 302f, 311t
Tetrads (arrangement), 327, 328f
Tetrads (chromosomal), 356, 356f, 357t
Tetrahydrofolate/tetrahydrofolic acid (THF), 128t
antimicrobials affecting, 295, 296f
Th1 cells. *See* Helper T lymphocytes (Th cells), type 1
Th2 cells. *See* Helper T lymphocytes (Th cells), type 2
Thaumarchaeota (phylum), 330f
Thayer-Martin medium, modified, for *Neisseria* culture, 588
Th cells. *See* Helper T lymphocytes
T-helper (Th) cells. *See* Helper T lymphocytes
Theories/laws, in scientific method, 9f, 10
Therapeutic index (TI), 302, 302f, 303f
Therapeutic window, 302
Thermal death point, 269

Thermal death time, 269
Thermococcus genus/spp., 330, 331f
Thermocycler, for polymerase chain reaction, 245, 331
Thermoduric organisms, 166, 270–271
Thermophiles/thermophilic organisms, 166–167, 167f, 270–271, 330–331, 330f, 331f
hyperthermophiles, 130, 166–167, 167f, 330–331, 331f
commercial sterilization and, 264, 269
in polymerase chain reaction, 245, 246b, 331
Thermoplasma acidophilum, 78, 79f
Thermus aquaticus, 245, 246b
THF. *See* Tetrahydrofolate/tetrahydrofolic acid
Thiabendazole, 316t
Thiamine pyrophosphate, 128t
Thiazolides, 312t
Thimerosal, 279
Thiobacillus genus/spp., 341, 344t
pH affected by, 39
in sulfur cycle, 341, 808
acid mine drainage and, 39, 341, 805
Third-generation drugs, 308
Third line of defense, 447. *See also* Adaptive immunity
Three-day measles. *See* Rubella
Three prime (3′) end, of nucleic acid, 51, 51f, 194, 195f
Thrush (oropharyngeal candidiasis), 299b, 302, 426f, 658f, 660b
antimicrobial drug use and, 299b, 302
in HIV infection/AIDS, 541, 660b, 766b
Thylakoids, 87–88, 88f, 145, 145f
Thylakoid space, 88f, 145, 146
Thymidine, 297f
Thymidine triphosphate deoxyribonucleotide (dTTP). *See also* Triphosphate deoxyribonucleotides
in DNA replication, 198
Thymine (T), 50, 50f, 51, 51f, 52, 52t, 194, 195f
Thymine dimer, ultraviolet light causing, 218f
repair of, 220, 220f
Thymus, 473–474, 475f
absence of, in DiGeorge syndrome, 540, 540f
clonal deletion of T cells in, 482, 482f
T lymphocyte maturation in, 473–474, 480, 482, 482t
Thyroid hormone, autoimmunity affecting production of, 539
TI. *See* Therapeutic index
Tick(s)
as disease vectors, 377, 378f, 433t, 677
for arboviruses, 741–746, 747t
for babesiosis, 691b
for *Coxiella* infection, 613
for ehrlichiosis and anaplasmosis, 623, 626t
for encephalitis (tick-borne encephalitis), 743b
for Lyme disease (*Borrelia burgdorferi*), 120b, 634–635, 634f
for relapsing fever (*Borrelia* genus/spp.), 635–636, 636
for rickettsial diseases, 340, 622b, 624b, 625f
for Rocky Mountain spotted fever (*Rickettsia rickettsii*), 624b, 625b, 626t
for tularemia, 611
taxonomic classification of, 114f
Tick-borne encephalitis, 743b
bioterrorism and, 743b, 814t
Tick fever
Colorado, 392t, 747t, 770, 771t
tularemia, 426f, 611, 612b
bioterrorism and, 611, 814t
Tigecycline, 230f, 311t
Tincture(s), 276
of iodine, 277
T-independent antibody immunity, 492
Tinea (tinea barbae/capitis/corporis/cruris/pedis/unguium) (dermatophytoses), 664–665, 665f, 665t
athlete's foot, 365, 664, 665f
"jock itch," 665t
onychomycosis, 665t

Tinidazole, 313t, 318t
Tissue, diseased, specimen collection from, 173t
Tissue cultures. *See* Cell (tissue) cultures
Tissue plasminogen activating factor, recombinant DNA in production of, 788t
Tissue repair, 464
Tissue typing, prevention of graft rejection and, 537
Titer, antibody, 506, 514, 514f
booster immunizations and, 506, 511f
Titration, 514, 514f
TLRs. *See* Toll-like receptors
T lymphocytes/T cells, 473, 474, 474f, 482t
activation/function of clones of, 489–491, 490f, 491f
in allergic contact dermatitis, 535
autoimmunity and, 538, 539
in cell-mediated immune response, 474, 480–482, 481f, 482f, 482t, 495
clonal deletion/development of, 481–482, 482f
cytotoxic, 481, 482t. *See also* Cytotoxic T cells
deficient, 540, 540t
helper, 481, 482t. *See also* Helper T lymphocytes
in HIV infection/AIDS, 387, 541, 751, 752t, 755, 755f, 756
attachment and, 752, 753, 753f, 754f
memory, 491–492, 494
tuberculin response mediated by, 535
production/maturation of, 473–474, 480–481, 482, 482f
receptors on, 481, 481f. *See also* T cell receptor
regulation of, 492
regulatory (Tr cells/suppressor T cells), 481, 482t, 492t
selective, 481, 482t
in type I (immediate) hypersensitivity, 526, 527t
TMV. *See* Tobamovirus (tobacco mosaic virus)
TNF. *See* Tumor necrosis factor
Toadstools, 370, 668, 669f
Tobacco mosaic disease, 15t, 388, 810t
Tobamovirus (tobacco mosaic virus), 15t, 387, 388
shape of, 389f
size of, 388f
soilborne transmission of, 810t
Tobramycin, 310t
Togaviridae (togaviruses), 392t, 735, 741, 741f, 747t, 771t
encephalitis caused by, 742, 742f, 747t, 771t
Tolerance, to autoantigens. *See* Self-tolerance
Toll-like receptors (TLRs), 457, 457t
interferons and, 458
Tongue, 449f
Tonsils, 475f, 476
Tooth decay (dental caries), 62, 170, 171b
viridans streptococci causing, 557, 557f
Topical drug administration, 300
Topoisomerase, 201, 202. *See also* Gyrase/DNA gyrase
Topsoil, 809, 809f
Torulopsis etchellsii, in vegetable fermentation, 783t
Totally drug-resistant (TDR) tuberculosis, 574
Total magnification, 100
tox- (prefix), 423t
Toxemia (blood poisoning), 428. *See also* Bacteremia/septicemia
Toxicity, antimicrobial drug, 302, 302f, 303f
selective, 290
Toxicodendron radicans, allergic reaction to (poison ivy), 535, 536f
Toxicoses (poisonings)
contaminated food causing, 353b, 379b, 785, 786. *See also* Botulism; Foodborne illnesses
fungal/mushroom poisoning, 370, 649, 668, 669f
water, 789, 790t. *See also* Waterborne illnesses
Toxic shock syndrome
non-streptococcal (TSS), 548, 548t, 550, 550f
streptococcal (STSS), 556
Toxic shock syndrome (TSS) toxin, 548, 548t, 550
Toxifications, food, 353b, 379b, 785, 786. *See also* Foodborne illnesses

SUBJECT INDEX

Toxin A/toxin B, clostridial, 562, 565b
Toxins. *See also specific type and* Endotoxin; Exotoxin(s)
  adenylate cyclase
    Bordetella pertussis, 607
    Vibrio cholerae, 638, 638f
  anthrax, 336, 561
  as biological weapons, 814t
  botulinum. *See* Toxins, clostridial/botulinum/botulinum
  Bt, 255, 335–336, 336f, 788–789, 789t
  cholera, 638, 638f
  clostridial/botulinum/botulinum, 335, 428, 562, 563, 564f, 565b, 567, 581b, 785, 786t
    as biological weapons, 814t
  dermonecrotic, 607
  diphtheria, 572, 573
  of Enterobacteriaceae, 593, 593f
  food poisoning caused by, 353b, 379b, 785, 786. *See also* Foodborne illnesses
  fungal, 668
  immunodeficiency and, 541
  neurologic diseases caused by. *See* Neurotoxins
  pertussis, 606–607, 607
  in pseudomembranous colitis, 562
  of *Pseudomonas aeruginosa*, 609
  pyrogenic (erythrogenic). *See* Pyrogens/pyrogenic (erythrogenic) toxins
  in shellfish
    dinoflagellate ingestion and, 362–363, 790, 790t
    paralytic shellfish poisoning and, 353b, 379b, 790t
  Shiga, 602
  Shiga-like, 596
  staphylococcal, 336, 548, 548t
    as biological weapons, 814t
  streptococcal, 552. *See also* Pyrogens/pyrogenic (erythrogenic) toxins
  tetanus. *See* Tetanospasmin
  as virulence factors, 427f, 428, 428t
  waterborne illnesses and, 789, 790t. *See also* Waterborne illnesses
Toxoid vaccines, 428, 506
  diphtheria, 506, 509t, 573
  tetanus, 506
    immunological memory and, 494
Toxoplasma genus/spp./*Toxoplasma gondii*, 362, 365t, 419t, 687–690, 687f, 690f, 692t
  foodborne transmission of, 687, 687f, 690, 786, 786t
  life cycle of, 687–690, 687f
  placenta crossed by, 421t, 687, 690
Toxoplasmosis, 362, 419t, 690, 692t
  in HIV infection/AIDS, 541, 690
  transplacental transmission of, 421t, 687, 690
Trace elements, for microbial growth, 165
Trachea, 449, 449f
  in host defense, 449
  microbiome/normal microbiota of, 417t
Tracheal cytotoxin, 607
Trachoma, 628, 628f
Transamination, 150, 151f
Transcription, 203, 203–206, 203f, 205f, 207f, 213f, 239
  antimicrobials affecting, 296, 297
  concurrent, 204, 206f
  DNA microarrays in monitoring of, 247
  eukaryote differences and, 206, 207f
  operons in regulation of, 212–215, 213f, 214f, 215f, 215t
  reverse, 239, 252t. *See also* Reverse transcriptase
  in viral replication, 239, 394, 398, 399f
Transcription factors, 206
Transcriptomics/transcriptomes, 250
Transducing phages, 225, 226f, 394
Transducing virion, 225, 226f, 394
Transduction, 225, 226f, 228t, 394
  antimicrobial resistance and, 303
  bacteriophage replication and, 225, 226f, 394
Transferases, 127, 128t
Transfer (exchange) reactions, 36–37
Transferrin, 452

Transfer RNA (tRNA), 209, 209f
  in HIV replication, 752, 753
  ribosomal-binding sites for, 210, 210f
  antimicrobials affecting, 293, 294f
  in transcription, 203
  in translation, 207, 209, 209f, 210, 210f, 211, 211f
Transfer RNA synthetases, aminoacyl, 209, 293
Transformation, 224–225, 225f, 228t
  antimicrobial resistance and, 303
Transfusion
  ABO system/transfusion reactions and, 530–531, 530f, 531t
  hepatitis B virus transmitted by, 726
  hepatitis C virus transmitted by, 748
  HIV transmitted by, 756
Transfusion reactions, 530–531, 530f, 531t
Transgenic (genetically modified) organisms, in agricultural microbiology, 254–256, 255f, 788–789, 789t
  safety of, 256, 788–789
Transient microbiota, 416. *See also* Microbiome
Translation, 203, 203f, 207–212, 208f, 209f, 210f, 211f, 212f, 213t
  antimicrobials affecting, 290, 293, 294f
  eukaryote differences and, 209, 212
  events in, 210–212, 210f, 211f, 212f
  genetic code and, 207–208, 208f
  participants in, 208–210, 209f, 210f
  RNA in control of, 215–216
  in viral replication, 394, 397, 398, 399f
Translocation, group, 73, 73–74, 73f, 74t
Transmembrane portion, immunoglobulin/B cell receptor, 483, 483f
Transmission (infectious disease), modes of, 430–432, 431f, 432b, 433t
  public health agencies in interruption of, 440
Transmission (light), spectrophotometer measuring, 186–187, 186f
Transmission electron micrographs/images, 104, 105f
Transmission electron microscopy/microscopes, 105, 105f, 107t
  resolving power of, 99f, 104
  specimen preparation for, 104
  staining for, 104, 111–112
Transovarian transmission
  in Lyme disease (*Borrelia burgdorferi*), 634–635, 634f
  in relapsing fever (*Borrelia recurrentis*), 636
  in Rocky Mountain spotted fever (*Rickettsia rickettsii*), 625b
  in scrub typhus (*Orientia*), 622
Transplacental transmission of disease. *See* Placenta, as portal of entry
Transplantation, rejection and, 536
  donor-recipient matching/tissue typing and, 536–537
  immunosuppressive drugs and, 537–538, 538f
  MHC antigens and, 478, 536, 536–537
Transportation proteins, 47
Transport media, 172, 177
Transport processes
  active, 70, 73–74, 73f, 74t
  in bacterial cells, 70–74, 71f, 72f, 73f, 74t
  in eukaryotic cells, 81, 81f, 81t
  passive, 70, 71–73, 71f, 72f, 74t
Transport vesicles, 61f, 87f
Transposase/transposase gene, 229, 229f
Transposition, 216, 228–230, 229f
Transposons, 228–230, 229f
  complex, 229f, 230
  as genetic vectors, 242
  simple (insertion sequences), 229, 229f
Traveler's diarrhea, 15t
Tr cells. *See* Regulatory T cells
Trebouxia genus/spp., 374, 376f
Trematodes (flukes), 693, 695–698, 695f, 697b, 697f, 703t
  blood (*Schistosoma* species), 696–698, 697f, 698f, 699b, 703t
  life cycle of, 696, 697f
  liver (*Fasciola* species), 419t, 696, 697f, 703t
Trench fever, 605
Treponema genus/spp. (*Treponema carateum/pallidum endemicum/pallidum*

pertenue), 344t, 347, 630–633, 631b, 633f
  nonvenereal diseases caused by, 633, 633f
Treponema pallidum pallidum, 630–633, 630f, 631b, 631f. *See also* Syphilis
  animal culture for growth of, 178, 630
  motility of, 64, 631b
  placenta crossed by, 421t, 632
Triatoma genus/spp. (kissing bugs), as disease vectors, 379, 433t
  in Chagas' disease (*Trypanosoma cruzi*), 379, 679, 680f, 692t
Tribes, 113
Tricarboxylic acid (TCA/Krebs/citric acid) cycle, 133, 133f, 136, 137f
Trichomonas genus/spp./*Trichomonas vaginalis*, 360, 360f, 365t, 686, 686f, 692t
Trichomoniasis, 360, 686, 686f, 692t
Trichonympha genus/spp., 415, 415f
Trichophyton genus/spp. (*Trichophyton equinum/mentagrophytes/rubrum/schoenleinii/tonsurans/verrucosum/violaceum*), 419t, 665, 665f
Trichosporon beigelii, 664
Trickle filter system, in wastewater treatment, 793f, 794
Triclabendazole, 316t
Triclosan, 276, 276f
Trifluridine, 297t
Trigeminal nerve ganglion, latent herpes infection of, 714, 714f, 715f, 716f
Triglycerides, 41, 42f, 150. *See also* Fat(s)
  biosynthesis of, 150, 151f
  catabolism of, 143, 143f
Trimethoprim, 312t, 318t
  mechanism of action of, 291f, 295, 312t, 318t
Triphosphate deoxyribonucleotides, in DNA replication, 198, 199f, 201
  polymerase chain reactions and, 244, 245f
Triphosphate nucleotides, in DNA replication, 199f, 201
Triphosphate ribonucleotides, in transcription, 204, 205f
Triplets
  DNA, 207
  of microtubules
    in centrioles, 83–84, 84f
    in cilia/flagella, 82, 83f
tRNA. *See* Transfer RNA
Trophozoites, 359, 677
  in *Acanthamoeba* infection, 678
  in *Balantidium coli* infection, 677
  in *Entamoeba histolytica* infection, 678, 679
  in *Giardia* infection, 684b, 685b
  in *Naegleria* infection, 678–679, 679
  in *Plasmodium* infection, 688b
  susceptibility of, 266f
  in *Trichomonas* infection, 686, 686f
trp operon, 215, 215f
True (bloodsucking) bugs, 378f
  as disease vectors, 433t, 677. *See also specific type*
  in Chagas' disease (*Trypanosoma cruzi*), 379, 679, 680f, 692t
Truffles, 370
Trypanosoma genus/spp., 361, 365t, 692t
Trypanosoma brucei, 680–681, 681f, 692t
  gambiense, 15t, 680, 681
  rhodesiense, 680, 681
Trypanosoma cruzi, 679, 679–680, 680f, 692t, 703b
Trypanosome(s)
  Leishmania, 682, 683
  Trypsanosoma brucei, 681, 681f
  Trypsanosoma cruzi, 679, 680f
Trypanosomiasis
  African (African sleeping sickness), 15t, 680–681, 681f, 692t
  American (Chagas' disease), 675b, 679–680, 680f, 692t, 703b
Trypticase soy agar, 175
Tryptophan auxotroph, identification of, 222, 222f
Tryptophan (*trp*) operon, 215, 215f
Tsetse (*Glossina*) flies, as disease vectors, 378–379, 378f, 433t, 680, 681f
  for African trypanosomiasis/sleeping sickness (*Trypanosoma brucei*), 379, 680, 681, 681f, 692t

TSS. *See* Non-streptococcal toxic shock syndrome
Tubercles, 577b
Tuberculin, 534
Tuberculin skin test/tuberculin response, 534–535, 535f, 577b
Tuberculoid leprosy, 575
Tuberculosis, 14, 307b, 573–574, 574b, 576–577b. *See also Mycobacterium tuberculosis*
  bioterrorism and, 814t
  discovery of cause of, 14, 424
  disseminated, 577b
  drug-resistant, 306, 574, 577b
  extensively drug-resistant (XDR), 574, 577b
  in HIV infection/AIDS, 541
  immunization against (BCG vaccine), 509t, 577b
  tuberculin skin test and, 534, 535
  multi-drug-resistant, 306, 307b, 574, 577b
  primary, 576–577b
  reemergence of, 574, 576–577b
  safe handling of bacteria causing, 268
  secondary/reactivated, 577b
  skin test in diagnosis of, 534–535, 535f, 577b
  transmission of, 431, 432b, 574b, 577b
Tuber genus/spp. (*Tuber magnatum/melanosporum*), 370, 371b, 373t
Tubulin microtubules
  antimicrobials affecting, 295
  in centrioles, 83–84, 84f
  in cilia/flagella, 82, 83f
  in cytoskeleton, 83, 84f
  griseofulvin mechanism of action and, 650
Tubulin polymerization, antimicrobials affecting, 295
Tularemia (rabbit fever), 426t, 611, 612b
  bioterrorism and, 611, 814t
Tumbles, in bacterial motility, 64, 65, 65f
Tumor, 401, 424, 424t. *See also* Cancer
Tumor necrosis factor, 489, 489t
  recombinant DNA in production of, 788t
Tunneling current, 105
Turbidimetry, 513
Turbidity, in microbial growth estimation, 186–187, 186f
Turnover rate, in carbon cycle, 806
2-μm circle, 198
Type 1 diabetes mellitus, autoimmunity in, 538, 539
Type 1 helper T cells (Th1 cells). *See* Helper T lymphocytes (Th cells), type 1
Type I (immediate) hypersensitivity, 526–530, 527f, 527t, 528f, 529f, 537t. *See also* Immediate (type I) hypersensitivity
Type I interferons, 458, 458f, 459t
Type 2 helper T cells (TH2 cells). *See* Helper T lymphocytes (Th cells), type 2
Type II (cytotoxic) hypersensitivity, 530–531, 530f, 531t, 532f, 537t. *See also* Cytotoxic (type II) hypersensitivity
Type II interferons, 458, 459t
Type III (immune complex–mediated) hypersensitivity, 532–534, 533f, 534f, 537t
  fungal allergens causing, 532, 669
Type III secretion systems, of Enterobacteriaceae, 593, 593f, 597
  E. coli O157:H7, 596
  Salmonella, 600
  Shigella, 600
  Yersinia, 602
Type IV (delayed/cell-mediated) hypersensitivity, 534–537, 535b, 535f, 536f, 537t. *See also* Delayed (type IV) hypersensitivity
Typhoid fever, 600, 601f
  carriers of, 419b, 600
  immunization against, 509t, 600
  waterborne transmission of, 419b, 600, 790t
Typhoid fever vaccine, 509t, 600
Typhus, 344t, 621, 626t
  bioterrorism and, 814t

SUBJECT INDEX I-39

epidemic (louse-borne), 621, 626t
  bioterrorism and, 814t
murine (endemic), 622, 626t
scrub, 623, 626t
U. *See* Uracil
Ubiquinones, in electron transport chain, 138f, 139
Ulcerative colitis, fecal microbiome transfer for, 803b
Ulcers
  in gastrointestinal zygomycosis, 662
  genital, herpesvirus. *See* Genital herpes/ulcers
  peptic, 27b, 53b, 167, 640–641, 640b, 641f
Ultra-high-temperature pasteurization, 271, 271t
Ultra-high-temperature sterilization, 271, 271t, 275t
Ultramicrotome, 104
Ultrastructure, 104
Ultraviolet light, 98f, 275. *See also* Light
  fluorescence/confocal microscopes using, 100, 102, 103
  in microbial control, 275, 275t
    drinking water treatment and, 275, 791f, 792
    food preservation and, 785
    mutagenic effects of, 218, 218f, 275
    DNA repair and, 220, 220f
Ulva genus/spp., alternation of generations in, 374f
Uncoating
  in animal virus replication, 397, 397f, 401t
  antiviral drugs affecting, 295, 313–314t
  in HIV replication, 752, 753, 753f
Undulant fever (brucellosis), 15t, 340, 344t, 606, 606f
  bioterrorism and, 814t
Uniports, 73, 73f
Units of measurement, 97, 97t
Unripened cheese, 780
Unsaturated fats/fatty acids, 41, 42, 42f, 43f
UPEC (uropathogenic *E. coli*), 595, 598–599b
Upper respiratory system, microbiome/normal microbiota of, 417t
Uracil (U), 50, 50f, 51, 52t, 194, 195f
Uracil ribonucleotide, 50
Ureaplasma genus/spp./*Ureaplasma urealyticum*, 568, 569, 572. *See also Mycoplasma* genus/spp.
Urease
  Helicobacter pylori producing, 167, 640, 640b, 641
  Proteus producing, 597
Urethra
  bacterial infection of, 598b. *See also* Urethritis
  microbiome/normal microbiota of, 417t
  as portal of entry, 420f
  as portal of exit, 430, 430f
Urethritis, 598b
  E. coli causing, 598b
  gonococcal, 589, 594b
  nongonococcal (NGU)
    Chlamydia causing, 628
    mycoplasmas causing, 572
  in trichomoniasis, 686
Urinary bladder, infections of, 15t, 598b, 599b
Urinary system. *See also* Urinary tract infections
  in host defense, 451t
  microbiome/normal microbiota of, 417t
Urinary tract infections, 58b, 89b, 598–599b
  bacterial, 15t, 598–599b, 604f
  candidal, 658t, 660b
  E. coli causing, 89b, 595, 598–599b, 604f
  Morganella/Providencia/Edwardsiella causing, 597, 604f
  Proteus causing, 597, 604f
Urine
  in host defense, 451t
  leptospirosis transmission via, 636
  as portal of exit/disease transmission and, 430, 430f
  in UTIs, 58b, 89b, 599b
Urine specimen/urinalysis
  collection and, 173t
  in UTI, 599b
Uropathogenic *E. coli*, 595, 598–599b
Urticaria, in type I (immediate) hypersensitivity, 525b, 528, 528f, 541b

Urushiol, allergic reaction to (poison ivy), 535, 536f
Use-dilution test, 281–282
US Public Health Service, 440
UTIs. *See* Urinary tract infections
UV light. *See* Ultraviolet light

Vaccination, 2b, 18, 495, 496, 497t, 505. *See also* Immunization; Vaccines
Vaccine Adverse Event Reporting System, 510
Vaccine-induced polio, 506, 510
Vaccines, 18, 495, 496, 497t, 504–512, 505f, 507f, 508f, 509f. *See also* specific type and specific organism or disease
  active/artificially acquired immunity and, 495, 496, 497t, 505–510, 507f, 508f, 509f, 511
  attenuated (modified live), 506
    recombinant, 507–508, 507f
    residual virulence/safety and, 506, 509–510
  combination, 506
  egg cultures for preparation of, 403, 508
  ethical issues and, 504b
  history/discovery of, 2b, 504–505, 505f
  inactivated (killed), 506
  manufacture of, 508
  recombinant DNA technology in production of, 253, 506–508, 507f, 788t
  recommendations for, 508–509, 508f, 509t
  safety of, 509–510
  subunit, 253, 506
  thimerosal in, 279
  toxoid, 428, 506
    diphtheria, 506, 509t, 573
    tetanus, 506
      immunological memory and, 494
  types of, 506–508, 507f
  whole agent, 506
Vaccinia necrosum, 510b
Vaccinia virus, 18, 710f
  smallpox prevention/vaccine and, 18, 505, 509t, 510b, 710, 711, 712b
Vacuoles, 85t, 87, 87f
  contractile, in protozoa, 359, 360f
Vacuum filtration, in microbial control, 272, 273f
Vagina
  microbiome/normal microbiota of, 417t
  pH of, microbial growth and, 167, 661b
  as portal of entry, 420f
  as portal of exit, 430, 430f
Vaginal candidiasis, 298, 302, 426f, 658t, 660b, 661b
  antimicrobial therapy and, 298, 302
Vaginal discharge/secretions
  in host defense, 451, 451t
  as portal of exit, 430, 430f
  in trichomoniasis, 686
Vaginosis, *Trichomonas vaginalis* causing, 686, 692t
Valacyclovir, 297t, 314t
Valence, 31
Valence electrons, 30–31, 30f, 31
Valence shell, 31
Valley fever (coccidioidomycosis), 366, 647b, 654–656, 655f, 669b
  bioterrorism and, 814t
  in HIV infection/AIDS, 541, 656
  immunization against, 650
  transmission of, 431, 655, 810t
Valves
  cardiac
    endocarditis and, 550
    rheumatic fever and, 553
  lymphatic, 475f, 476
Vancomycin, 310t, 551, 557
  enterococcal resistance (VRE) and, 273f, 560
  mechanism of action of, 291, 310t
  microbial production of, 290
  Staphylococcus aureus resistance (VRSA) and, 551
van Leeuwenhoek, Antoni, 2–4, 3f, 6–7, 18f, 97
van Niel, C.B., 19
Variable (V_H) heavy chain gene segment, 484t, 484f, 485f
Variable (V_L) light chain gene segment, 484, 485f

Variable region
  immunoglobulin/B cell receptor, 483, 483f, 484, 484f, 485f
  T cell receptor, 481, 481f
Variant Creutzfeldt-Jakob disease, 21b, 48, 406f, 407, 407b
  bioterrorism and, 814t
  enzyme for elimination of prion causing, 280
Varicella (chickenpox), 392t, 716–717, 717f
  immunization against, 508f, 509t, 718
Varicella-zoster virus (VZV/HHH-3), 713, 716–718, 717f, 718f, 719b
  immunization against, 508f, 509t, 718
  latency/recurrent infection and, 717–718, 717f, 719b
Varicellovirus genus/spp., 392t, 713, 716, 729t. *See also* Varicella-zoster virus
Varieties, 113. *See also* Serotypes
Variola (smallpox) virus (variola major/minor), 2b, 710–712, 711f, 712b, 729t
  bioterrorism and, 510b, 712, 813, 814t
  safe handling of, 268
  shape of, 389
  size of, 61f, 388f
  vaccination against. *See* Smallpox vaccination
Variolation, 17, 505
Vascular system. *See* Blood vessels
Vasodilation, in inflammation, 462–464, 462f, 463f, 465f, 465t
vCJD. *See* Variant Creutzfeldt-Jakob disease
Vectors
  in disease transmission, 377–379, 378f, 432, 433t. *See also* specific disease and specific vector
    arthropod, 377–379, 378f, 432, 433t, 676–677, 676f
    public health measures in control of, 440
  genetic (recombinant), 241–242, 242f, 252t
    vaccine development and, 507–508, 507f
VEE. *See* Venezuelan equine encephalitis
Vegetables, fermented, 779, 783t
Vegetative cells, 75, 76f, 325, 326f
  on Schaeffer-Fulton endospore stain, 110, 110f, 325
Vegetative fungi, susceptibility of, 266f
Vehicle transmission of disease, 431–432, 431f, 432b, 433t
Venereal diseases. *See* Sexually transmitted infections/diseases
Venezuelan equine encephalitis (VEE), 742, 742f, 747t
Vertical gene transfer, 224
Vesicle(s), 85t, 87, 87f
  digestive (phagolysosomes), 87f, 455f, 456
  food (phagosome), 81, 81f, 87f, 455f, 456
  maturation of/microbial killing and, 455f, 456
  TLRs in membranes of, 457, 457t
  gas, 74
  secretory, 61f, 86f, 87, 87f
  transport, 61f, 87f
Vesicle (skin lesion), in poxvirus infection, 710, 711f
Vesicular stomatitis virus, 761f
V_H (variable) heavy chain gene segment, 484, 484f
Viable plate counts, 184, 184f
Vi antigens, 593, 593f
Vibrio(s), 325, 325f, 637–641, 641b
Vibrio genus/spp. (*Vibrio parahaemolyticus*/*vulnificus*), 40b, 170b, 343t, 344t, 637–639, 786t. *See also Vibrio cholerae*
  antacids affecting, 40b
  culturing, 176, 639
  foodborne illnesses caused by, 40b, 638, 786t
Vibrio cholerae, 637–639, 638f, 641b. *See also* Cholera
  bioterrorism and, 814t
  chromosomes of, 74
  culturing, 176, 639
  immunization against, 639
  pH range tolerated by, 167
  strain O1 El Tor, 637, 638
  strain O139 Bengal, 638, 641b
  waterborne transmission of, 17, 437, 439t, 637, 638, 639, 790t
Vibrionaceae, 343t

Villa-Komaroff, Lydia, 252
Vinegar (acetic acid), microbial production of, 12t, 782, 783t, 787
Viral cultures, 402–404, 403f, 404f
  animal cultures for, 178, 403
  bacteria for, 402–403, 403f. *See also* Bacteriophage(s)
  cell/tissue cultures for, 178
  embryonated chicken eggs for, 403, 403f
    vaccine development and, 403, 508
  plants for, 403
  for vaccines, 506
Viral encephalitis, 742, 742f, 743b
Viral envelope, 385, 386, 389–390, 391f, 400, 400f. *See also* Envelope, viral
Viral gastroenteritis. *See* specific causative agent and Foodborne illnesses; Gastroenteritis
Viral genome, 386, 387f, 391t
Viral hemagglutination, 515
Viral hemagglutination inhibition test, 515, 520t
Viral hemorrhagic fevers. *See* Hemorrhagic fevers
Viral hepatitis. *See* Hepatitis
Viral meningitis, coxsackieviruses causing, 738
Viral neutralization, 486, 486f, 515, 520t
Viral pneumonia. *See also* Pneumonia
  hantavirus causing, 384b, 392t, 408b, 419t, 771t
    transmission and, 384b, 408b, 767, 810, 810t
  parainfluenza viruses causing, 760
Viral proteins, antiviral drugs affecting, 314t
Viral replication, 391t, 392–400, 401t
  of animal viruses, 396–400, 397f, 399f, 399f, 400f, 401t
  of bacteriophages, 401t
    lysogenic, 395–396, 395f, 396f
    lytic, 392–394, 393f, 394f, 396, 396f
    transduction and, 225, 226f, 394
Viremia, 720
  in arboviral infection, 741, 742f
  in bunyavirus infection, 767
  in enterovirus infection, 736
  in Epstein-Barr virus infection, 720
  in mumps, 760
Viridans streptococci, 557, 557f
Virion(s), 386, 386f. *See also* Viruses
  enveloped, 385, 386, 389–390, 391f. *See also* Envelope, viral
  nonenveloped (naked), 389, 390. *See also* Nonenveloped (naked) viruses/virions
  release of
    in bacteriophage replication, 393f, 394, 394f, 396, 396f, 401t
    in persistent infection, 400, 400f
  shapes of, 389, 389f
  sizes of, 388f, 391t
Viroid(s), 385, 404–405, 405f, 408t
  bacteria/viruses/prions compared with, 408t
Viroidlike agents, 405
Virology, 18f, 19t. *See also* Viruses
Virucides, 264. *See also* Antiviral drugs
Virulence, 426, 426f, 428t. *See also* specific organism and Virulence factors
  attenuation of, for vaccines, 506
    recombinant DNA technology and, 507, 507f
  residual, vaccine safety and, 506, 509–510
Virulence factors, 426–429, 427f, 428t. *See also* specific organism
  adhesion and, 422, 422f, 455
Virulence plasmids, 196. *See also* Adhesins
Viruses, 6–7, 7f, 14, 272, 385–404, 408t, 630t. *See also* specific type and under Viral
  activation of, drugs blocking, 295
  animal, 387. *See also* Animal viruses
  antimicrobials effective against, 265, 299f, 313–314t. *See also* Antiviral drugs
  attachment of
    antimicrobial drugs affecting, 290, 291f, 298, 313t
    in viral replication, 392, 401t
      animal virus, 396–397, 401t
      bacteriophage, 393f, 394, 395, 396f, 401t
      HIV, 752, 753, 753f, 754f

SUBJECT INDEX

Viruses (Continued)
  bacteria infected by. See Bacteriophage(s)
  in bacterial identification/classification, 113b, 117, 118f
  bacteria/viroids/prions compared with, 408t
  beneficial/industrial uses of, 113b, 277b, 390b, 395b
  as biological weapons, 813–814, 814t, 815
  in cancer, 401–402, 402f. See also Cancer
  capsids of, 386, 386f, 389, 389f, 391f, 391t
    replication and assembly, 393f, 394
    entry/uncoating, 394, 397, 397f, 401t
  cells compared with, 391t
  characteristics of, 385–390, 386f, 391t
  classification of, 116, 390–391, 391t, 392t
    shape and, 389, 389f
  culturing, 402–404, 403f, 404f. See also Viral cultures
  discovery of, 14, 272
  diseases/infections caused by, 419t
    autoimmunity and, 538, 539
    in HIV infection/AIDS, 751t, 755f
    DNA, 51, 52t, 385, 386, 391t, 392t, 709–733, 729t. See also DNA viruses
  entry of into cells
    antiviral drugs affecting, 295, 313t
    in replication
      animal virus, 397, 397f, 401t
      bacteriophage, 393f, 394, 401t
      HIV, 752, 753, 753f, 754f
  enveloped, 385, 386, 389–390, 391f, 400, 400f. See also Envelope, viral
  filterable, 14, 272
  filtration in control of, 272
  fungal, 388
  genetic material of, 386, 387f
  as genetic vectors, 242
  genomes of, 386, 387f, 391t
    segmented, 735, 763
  hosts of, 386–388, 387f
  interferons acting against, 457–459, 458f, 459t
  life processes in, 59t, 404
  methylation affecting, 201
  nonenveloped (naked), 389, 390. See also Nonenveloped (naked) viruses/virions
    antimicrobial action/susceptibility and, 265, 266, 266f
    entry/uncoating of, 397, 397f
    positive ssRNA, 735–741, 771t
    release of, 400
    segmented dsRNA, 770–771, 771t
  nucleic acids of. See also DNA viruses; RNA viruses
    antimicrobials affecting, 296–297, 297f, 314t
  plant, 387–388, 387f
  replication of, 391t, 392–400, 401t. See also Viral replication
  RNA, 385, 391t, 392t, 734–777, 771t. See also RNA viruses
  shapes of, 389, 389f
  sizes of, 61f, 388, 388f, 391t
  slow. See Prion(s)
  in soil, 810
    disease caused by, 810t
  T cells/cell-mediated immune response and, 474, 480
  in transduction, 225, 226f, 394
  uncoating of
    in animal virus replication, 397, 397f, 401t
    antiviral agents affecting, 295, 313–314t
    waterborne illness caused by, 789, 790t
Visceral leishmaniasis (kala-azar), 683, 692t
Vitamin(s)
  as coenzyme source, 127, 128t
  for microbial growth, 165
  microbial production of, 12t, 787, 789t
Vitamin B_3 (niacin), for microbial growth, 165t
Vitamin B_{12}, Euglena granulata in identification of, 175
Viviparity, 326, 327f
V_L (variable) light chain gene segment, 484
Vomiting. See also Gastroenteritis
  in host defense, 451f
von Behring, Emil, 18f, 21
Voriconazole, 315t, 370b

Vorticella genus/spp., 362
VRE (vancomycin-resistant enterococci), 273f, 560
VRSA (vancomycin-resistant Staphylococcus aureus), 551
V-shapes, in Corynebacterium, 328, 328f, 337, 572
Vulvovaginal candidiasis, 298, 302, 426f, 658t, 660b, 661b
  antimicrobial therapy and, 298, 302
VZV. See Varicella-zoster virus

Waksman, Selman, 289
Walking pneumonia (primary atypical/mycoplasmal pneumonia), 460b, 569
Wandering macrophages, 453–454, 464
Wangiella genus/spp.
  phaeohyphomycosis caused by, 666t
  radioactive waste remediation and, 169b
Warren, Robin, 640
Warts (papillomas), 392t, 722, 723b, 724b
  genital, 723b
Wastewater (sewage)
  drinking water contamination and, 431, 437, 439f, 440, 789, 790
  treatment of, 792–795, 793f, 794f, 795f
    domestic water affecting aquatic systems and, 811
    microbes in, 793–794, 793f, 794, 795f
Wasting (chronic wasting disease), 407
Water, 33, 33f, 37, 37f. See also Aquatic microbiology; Wastewater (sewage)
  bond angle for, 33, 33f
  chemical bonds in formation of, 37, 37f, 38
  in chemical reactions, 35–36, 36, 36f
  contamination of, 431, 437, 439f, 440, 789, 790. See also Waterborne illnesses
    bioterrorism and, 814t
  diffusion of. See Osmosis
  diseases transmitted by. See Waterborne illnesses
  domestic, aquatic systems affected by, 811
  fluoridation of, prevention of tooth decay and, 278
  microbial growth and, 167–168
  microorganisms living in, 811–812, 811f. See also Aquatic microbiology
  as nonliving disease reservoir, 419
  physical effects of, microbial growth and, 167–168
  pollution of, 789, 791. See also Waterborne illnesses
    acid mine drainage and, 805–806, 805f
  potable, 440
    contamination of, 431, 437, 439f, 440, 789, 790. See also Waterborne illnesses
    bioterrorism and, 814t
    public health and, 440
    treatment of, 790–792, 791f
  salt dissociation/ionization and, 34, 34f
  as solvent, 37
  passive transport and, 71f, 72–73, 72f
  testing quality of, 792, 792f
Water activity of food, spoilage and, 783
Waterborne illnesses, 419b, 431, 433t, 789–790, 790b, 790t
  Acanthamoeba causing, 267b, 678, 679
  adenoviruses causing, 724
  bioterrorism and, 814t
  Campylobacter jejuni causing, 639, 790t
  Cryptosporidium causing, 690, 790t
  Cyclospora causing, 691, 692f
  Echinococcus causing (hydatid cysts), 695
  Entamoeba causing, 678, 790t
  Giardia causing, 684b, 685b, 686, 790t
  hepatitis, 790t
  Legionnaires' disease, 612–613
  leptospirosis, 636
  Naegleria causing, 678, 679, 790b
  Norovirus gastroenteritis, 740b, 790t
  polio, 736, 790t
  public health and, 440
  roundworms (nematodes) causing, 699
  Salmonella causing (salmonellosis/typhoid fever), 419t, 600, 790t
  Schistosoma (blood flukes) causing, 431, 697, 699b, 790t
  Shigella causing (shigellosis), 601, 790t

Vibrio causing/cholera, 17, 437, 439f, 637, 638, 639, 790t
Yersinia causing, 602
Waterborne transmission of disease, 416b, 431, 433t. See also Waterborne illnesses
  public health and, 440
Water molds, 376–377, 377f
  classification of, 357, 376
  resistance to, recombinant DNA technology and, 255
Water quality testing, 792, 792f
Water treatment, 789–795, 790b, 790t, 791f, 792f, 793f, 794f, 795f
  copper in, 279
Wavelengths of radiation, 98, 98f, 273
  chlorophyll absorption and, 144
  microscopy and, 98, 98f, 104
Waves, light, 98, 98f
Waxes, 42, 67. See also Mycolic acid
Weaponization, assessing microorganisms' potential for, 812–813
WEE. See Western equine encephalitis
Weichselbaum, Anton, 15t
Weight, atomic, 28
Welch, William, 15t
Wells
  antibody/antigen, in immunodiffusion, 513, 513f
  antigen, in enzyme-linked immunosorbent assay (ELISA), 517, 517f, 518f
Western blot (immunoblot), 518, 519f
  in HIV diagnosis, 518
Western equine encephalitis (WEE), 742, 742f, 747t
West Nile virus/West Nile encephalitis, 387, 435f, 734b, 742, 742f, 743f, 747t, 772b
  incidence of, 435f
Wetlands, artificial, in wastewater treatment, 794–795, 795f
Wheat mosaic virus, soilborne, 810t
Whey, in cheese production, 780, 780f
White blood cell(s). See Leukocyte(s)
White blood cell count, 454, 454b
Whitehead, 579f. See also Acne
Whitlow, herpetic, 714f, 715, 715f, 716t
WHO (World Health Organization), 440
Whole agent vaccines, 506
Whooping cough (pertussis), 344t, 345b, 426f, 586b, 606–608, 607f, 608f, 615b
  immunization against, 345b, 508f, 509t, 608
  reemergence of, 345b, 608
  virulence and, 426f, 606–607, 607f
Wild-type cells, 221
Wine, fermentation and, 10b, 141, 142f, 781, 781f, 783t
Winogradsky, Sergei, 18f, 20
WNV. See West Nile virus
Wobble, anticodon, 209
Woese, Carl, 20, 115
Wöhler, Friedrich, 59
Wolbachia genus/spp./Wolbachia pipientis, mosquito/dengue control and, 385b, 746b
Working distance, 100
World Health Organization (WHO), 440
Worms. See Helminths
Wort, in beer production, 781, 782f
Wound(s). See also Wound care/infections
  bite, rabies and, 762
  specimen collection from, 173t
Wound botulism, 564
Wound care/infections
  Bacteroides fragilis and, 614
  rabies and, 762
  tetanus and, 567
  Vibrio vulnificus and, 170b, 639
Wuchereria bancrofti, 702, 702f, 703t

Xanthophylls, 375, 376t
XDR-TB (extensively drug-resistant tuberculosis), 574, 577b
Xenodiagnosis, in Chagas disease (Trypanosoma cruzi), 680
Xenografts/xenotransplants, 254, 536, 536f
Xenopsylla fleas, as disease vectors, 433t
  for murine (endemic) typhus (Rickettsia typhi), 622, 626t

Vibrio causing/cholera, 17, 437, 439f, 637, 638, 639, 790t

Xenorhabdus genus/spp., mutualism in, 415b
X rays, 98f, 274
  in microbial control, 275, 275t
  mutagenic effects of, 217

Yaws, 631b, 633, 633f
Yeasts, 5, 5f, 651, 651f
  beneficial uses of, 10b
  budding in, 5, 5f, 357, 358f, 365, 366f, 367
  in fermentation, 10b, 11, 11f
    alcoholic beverage production and, 10b, 11, 11f, 142f, 781, 782, 783t
    bread production and, 10b, 12t, 779, 783t
    vegetable preparation and, 779
  morphology of, 365, 366f
  opportunistic infection caused by, 298, 302, 418. See also Candida
  as probiotic, 10b
Yellow fever, 15t, 16b, 392t, 419t, 744, 747t
  bioterrorism and, 814t
  immunization against, 509t, 744
Yellow fever virus, 15t, 392t, 744, 747t
  as biological weapon, 814t
  safe handling and, 268
Yellow fever virus vaccine, 509t, 744
Yellow-green algae, 375, 376t
"Yellow Jack." See Yellow fever
Yersin, Alexandre, 15t
Yersinia genus/spp. (Yersinia enterocolitica/pseudotuberculosis), 343t, 344t, 592f, 602, 603f. See also Yersinia pestis
  foodborne illnesses/growth in refrigerated products and, 272, 602, 786t
  sites of infection caused by, 604f
Yersinia pestis, 15t, 419t, 426f, 602, 603f, 604b. See also Plague
  as biological weapon, 814t
  mRNA for virulence regulator of, 216
  pneumonia caused by, 602, 603f
Yogurt, fermentation in production of, 12t, 142f, 780, 783t

Zanamivir, 313t, 769b
Zephiran. See Benzalkonium chloride
Zetaproteobacteria, 330f, 344t, 347
Ziehl, Franz, 109
Ziehl-Neelsen acid-fast stain, 109–110, 110f, 111t, 337, 337f
Zika congenital syndrome, 745, 747t
Zika fever/virus, 744–745, 746b, 747t
Zinc, 29t
  antimicrobial action of, 279, 281t
  cycling of, 808–809
ZMapp, 763, 765b
Zn. See Zinc
Zone of inhibition
  in diffusion susceptibility test, 299, 300f
  in Etest, 300, 300f
  synergism and, 307f
Zoogloea genus/spp., 341, 342f, 344t
Zoonoses/zoonotic diseases, 418, 419t, 741. See also specific organism and type
  arbovirus infections as, 741
  arenavirus infections as, 767
  brucellosis as, 606, 606f
  bunyavirus infections as, 766
  Campylobacter infections as, 639
  cryptosporidiosis as, 690
  leishmaniasis as, 681
  leptospirosis as, 636
  rabies as, 418, 761
  tularemia as, 611, 612b
Zoster. See Herpes zoster
Zoster (shingles), 717–718, 717f, 718, 718f, 719b
  immunization against, 508f, 509t, 718
Zygomycoses, 662
Zygomycota (division)/zygomycetes, 368, 369, 369f, 373t, 648, 662
Zygosporangia, 369, 369f
Zygospores, 369, 373t
Zygote(s), 354
  algal, 374
  protozoan, 360
    in Cyclospora life cycle, 692f
    in Plasmodium life cycle, 689b
    in Toxoplasma life cycle, 687–690